RSMeans
FROM THE GORDIAN GROUP®

Light Commercial Cost Data

2016

35th annual edition

Robert J. Kuchta, Senior Editor

Engineering Director
Bob Mewis, CCP *(1, 2, 4, 13, 14, 31, 32, 33, 34, 35, 41, 44, 46)*

Contributing Editors
Christopher Babbitt
Adrian C. Charest, PE
Cheryl Elsmore
Wafaa Hamitou
Joseph Kelble
Charles Kibbee
Robert J. Kuchta *(8)*
Michael Landry
Thomas Lane (6, 7)
Genevieve Medeiros
Elisa Mello
Chris Morris *(26, 27, 28, 48)*
Melville J. Mossman, PE *(21, 22, 23)*
Marilyn Phelan, AIA *(9, 10, 11, 12)*
Stephen C. Plotner *(3, 5)*
Stephen Rosenberg
Kevin Souza
Keegan Spraker
Tim Tonello
David Yazbek

Vice President Data & Engineering
Chris Anderson

Product Manager
Andrea Sillah

Production Manager
Debbie Panarelli

Production
Sharon Larsen
Jonathan Forgit
Sheryl Rose
Mary Lou Geary

Technical Support
Gary L. Hoitt
Kathryn S. Rodriguez

Cover Design
Blaire Gaddis

Numbers in italics are the divisional responsibilities for each editor. Please contact the designated editor directly with any questions.

RSMeans
Construction Publishers & Consultants
1099 Hingham Street, Suite 201
Rockland, MA 02370
United States of America
1-888-607-2575
www.RSMeans.com

Printed in the United States of America
ISSN 0896-7601
ISBN 978-1-943215-12-6

0186

$219.99 per copy (in United States)
Price is subject to change without prior notice.

Related RSMeans Products and Services

The engineers at RSMeans suggest the following products and services as companion information resources to *RSMeans Light Commercial Cost Data:*

Construction Cost Data

Open Shop Construction Cost Data 2016
Residential Cost Data 2016

Reference Books

Residential & Light Commercial Construction Standards
Square Foot & Assemblies Estimating Methods
Designing & Building with the IBC
Estimating Building Costs
RSMeans Estimating Handbook
Green Building: Project Planning & Estimating
How to Estimate with RSMeans Data
Plan Reading & Material Takeoff
Project Scheduling and Management for Construction

Seminars and In-House Training

Unit Price Estimating
RSMeans Online® Training
Practical Project Management for Construction Professionals
Scheduling with MSProject for Construction Professionals
Mechanical & Electrical Estimating

RSMeans Online Store

Visit RSMeans at www.RSMeans.com for the most reliable and current resources available on the market. Learn more about our more than 20 data sets available in Online, Book, eBook, and CD formats. Our library of reference books is also available, along with professional development seminars aimed at improving cost estimating, project management, administration, and facilities management skills.

RSMeans Electronic Data

Receive the most up-to-date cost information with RSMeans Online. This web-based service is quick, intuitive, and easy to use, giving you instant access to RSMeans' comprehensive database. Learn more at: **www.RSMeans.com/Online**.

RSMeans Custom Solutions

Building owners, facility managers, building product manufacturers, attorneys, and even insurance firms across the public and private sectors have engaged in RSMeans' custom solutions to solve their estimating needs including:

Custom Cost Engineering Solutions

Knowing the lifetime maintenance cost of your building, how much it will cost to build a specific building type in different locations across the country and globe, and if the estimate from which you are basing a major decision is accurate are all imperative for building owners and their facility managers.

Market & Custom Analytics

Taking a construction product to market requires accurate market research in order to make critical business development decisions. Once the product goes to market, one way to differentiate your product from the competition is to make any cost saving claims you have to offer.

Third-Party Legal Resources

Natural disasters can lead to reconstruction and repairs just as new developments and expansions can lead to issues of eminent domain. In these cases where one party must pay, it is plausible for legal disputes over costs and estimates to arise.

Construction Costs for Software Applications

More than 25 unit price and assemblies cost databases are available through a number of leading estimating and facilities management software partners. For more information, see "Other RSMeans Products and Services" at the back of this publication.

RSMeans data is also available to federal, state, and local government agencies as multi-year, multi-seat licenses.

For procurement construction cost data such as Job Order Contracting and Change Order Management please refer to The Gordian Group's procurement solutions.

For information on our current partners, call 1-888-607-2575.

Table of Contents

Foreword

Who We Are

Since 1942, RSMeans has delivered construction cost estimating information and consulting throughout North America. In 2014, RSMeans was acquired by The Gordian Group, combining two industry-leading construction cost databases. Through the RSMeans line of products and services, The Gordian Group provides innovative construction cost estimating data to organizations pursuing efficient and effective construction planning, estimating, procurement, and information solutions.

Our Offerings

With RSMeans' construction cost estimating data, contractors, architects, engineers, facility owners, and managers can utilize the most up-to-date data available in many formats to meet their specific planning and estimating needs.

When you purchase information from RSMeans, you are, in effect, hiring the services of a full-time staff of construction and engineering professionals.

Our thoroughly-experienced and highly-qualified staff works daily to collect, analyze, and disseminate comprehensive cost information to meet your needs.

These staff members have years of practical construction experience and engineering training prior to joining the firm. As a result, you can count on them not only for accurate cost figures, but also for additional background reference information that will help you create a realistic estimate.

The RSMeans organization is equipped to help you solve construction problems through its variety of data solutions.

Access our comprehensive database electronically with RSMeans Online. Quick, intuitive, easy to use, and updated continuously throughout the year, RSMeans Online is the most accurate and up-to-date cost estimating data.

This up-to-date, accurate, and localized data can be leveraged for a broad range of applications, such as Custom Cost Engineering solutions, Market and Custom Analytics, and Third-Party Legal Resources.

To ensure you are getting the most from your cost estimating data, we also offer a myriad of reference guides, training, and professional seminars (learn more at **www.RSMeans.com/Learn**).

In short, RSMeans can provide you with the tools and expertise for developing accurate and dependable construction estimates and budgets in a variety of ways.

Our Commitment

Today at RSMeans, we do more than talk about the quality of our data and the usefulness of the information. We stand behind all of our data—from historical cost indexes to construction materials and techniques—to craft current costs and predict future trends.

If you have any questions about our products or services, please call us toll-free at 1-888-607-2575. You can also visit our website at: **www.RSMeans.com**.

How the Cost Data Is Built: An Overview

A Powerful Construction Tool

You now have one of the most powerful construction tools available today. A successful project is built on the foundation of an accurate and dependable estimate. This tool will enable you to construct such an estimate.

For the casual user the information is designed to be:

- quickly and easily understood so you can get right to your estimate.
- filled with valuable information so you can understand the necessary factors that go into a cost estimate.

For the professional user, the information is designed to be:

- a handy reference that can be quickly referred to for key costs.
- a comprehensive, fully reliable source of current construction costs and productivity rates so you'll be prepared to estimate any project.
- a source for preliminary project costs, product selections, and alternate materials and methods.

To meet all of these requirements, we have organized the information into the following clearly defined sections.

Estimating with RSMeans Unit Price Cost Data*

Please refer to these steps for guidance on completing an estimate using RSMeans unit price cost data.

How to Use the Information: The Details

This section contains an in-depth explanation of how the information is arranged and how you can use it to determine a reliable construction cost estimate. It includes how we develop our cost figures and how to prepare your estimate.

Unit Prices*

All cost data has been divided into 50 divisions according to the MasterFormat system of classification and numbering.

Assemblies*

The cost data in this section has been organized in an "Assemblies" format. These assemblies are the functional elements of a building and are arranged according to the 7 elements of the UNIFORMAT II classification system. For a complete explanation of a typical "Assembly", see "How RSMeans Assemblies Data Works."

*Residential Models**

Model buildings for four classes of construction—economy, average, custom, and luxury—are developed and shown with complete costs per square foot.

*Commercial/Industrial/Institutional Models**

This section contains complete costs for 77 typical model buildings expressed as costs per square foot.

*Green Commercial/Industrial/Institutional Models**

This section contains complete costs for 25 green model buildings expressed as costs per square foot.

*References**

This section includes information on Equipment Rental Costs, Crew Listings, Historical Cost Indexes, City Cost Indexes, Location Factors, Reference Tables, and Change Orders, as well as a listing of abbreviations.

- **Equipment Rental Costs:** Included are the average costs to rent and operate hundreds of pieces of construction equipment.
- **Crew Listings**: This section lists all the crews referenced in the cost data. A crew is composed of more than one trade classification and/or the addition of power equipment to any trade classification. Power equipment is included in the cost of the crew. Costs are shown both with bare labor rates and with the installing contractor's overhead and profit added. For each, the total crew cost per eight-hour day and the composite cost per labor-hour are listed.
- **Historical Cost Indexes**: These indexes provide you with data to adjust construction costs over time.
- **City Cost Indexes**: All costs in this data set are U.S. national averages. Costs vary by region. You can adjust for this by CSI Division to over 700 locations throughout the U.S. and Canada by using this data.
- **Location Factors**: You can adjust total project costs to over 900 locations throughout the U.S. and Canada by using the weighted number, which applies across all divisions.
- **Reference Tables**: At the beginning of selected major classifications in the Unit Prices are reference numbers indicators. These numbers refer you to related information in the Reference Section. In this section, you'll find reference tables, explanations, and estimating information that support how we develop the unit price data, technical data, and estimating procedures.
- **Change Orders**: This section includes information on the factors that influence the pricing of change orders.
- **Abbreviations**: A listing of abbreviations used throughout this information, along with the terms they represent, is included.

Index (printed versions only)

A comprehensive listing of all terms and subjects will help you quickly find what you need when you are not sure where it occurs in MasterFormat.

Conclusion

This information is designed to be as comprehensive and easy to use as possible.

The Construction Specifications Institute (CSI) and Construction Specifications Canada (CSC) have produced the 2014 edition of MasterFormat®, a system of titles and numbers used extensively to organize construction information.

All unit prices in the RSMeans cost data are now arranged in the 50-division MasterFormat® 2014 system.

* Not all information is available in all data sets

Note: The material prices in RSMeans cost data are "contractor's prices." They are the prices that contractors can expect to pay at the lumberyards, suppliers'/distributors' warehouses, etc. Small orders of specialty items would be higher than the costs shown, while very large orders, such as truckload lots, would be less. The variation would depend on the size, timing, and negotiating power of the contractor. The labor costs are primarily for new construction or major renovation rather than repairs or minor alterations. With reasonable exercise of judgment, the figures can be used for any building work.

Estimating with RSMeans Unit Prices

Following these steps will allow you to complete an accurate estimate using RSMeans Unit Price cost data.

1. Scope Out the Project

- Think through the project and identify the CSI Divisions needed in your estimate.
- Identify the individual work tasks that will need to be covered in your estimate.
- The Unit Price data has been divided into 50 Divisions according to the CSI MasterFormat® 2014.
- In printed versions, the Unit Price Section Table of Contents on page 1 may also be helpful when scoping out your project.
- Experienced estimators find it helpful to begin with Division 2 and continue through completion. Division 1 can be estimated after the full project scope is known.

2. Quantify

- Determine the number of units required for each work task that you identified.
- Experienced estimators include an allowance for waste in their quantities. (Waste is not included in RSMeans Unit Price line items unless otherwise stated.)

3. Price the Quantities

- Use the search tools available to locate individual Unit Price line items for your estimate.
- Reference Numbers indicated within a Unit Price section refer to additional information that you may find useful.
- The crew indicates who is performing the work for that task. Crew codes are expanded in the Crew Listings in the Reference Section to include all trades and equipment that comprise the crew.
- The Daily Output is the amount of work the crew is expected to do in one day.
- The Labor-Hours value is the amount of time it will take for the crew to install one unit of work.
- The abbreviated Unit designation indicates the unit of measure upon which the crew, productivity, and prices are based.
- Bare Costs are shown for materials, labor, and equipment needed to complete the Unit Price line item. Bare costs do not include waste, project overhead, payroll insurance, payroll taxes, main office overhead, or profit.
- The Total Incl O&P cost is the billing rate or invoice amount of the installing contractor or subcontractor who performs the work for the Unit Price line item.

4. Multiply

- Multiply the total number of units needed for your project by the Total Incl O&P cost for each Unit Price line item.
- Be careful that your take off unit of measure matches the unit of measure in the Unit column.
- The price you calculate is an estimate for a completed item of work.
- Keep scoping individual tasks, determining the number of units required for those tasks, matching each task with individual Unit Price line items, and multiplying quantities by Total Incl O&P costs.
- An estimate completed in this manner is priced as if a subcontractor, or set of subcontractors, is performing the work. The estimate does not yet include Project Overhead or Estimate Summary components such as general contractor markups on subcontracted work, general contractor office overhead and profit, contingency, and location factors.

5. Project Overhead

- Include project overhead items from Division 1-General Requirements.
- These items are needed to make the job run. They are typically, but not always, provided by the general contractor. Items include, but are not limited to, field personnel, insurance, performance bond, permits, testing, temporary utilities, field office and storage facilities, temporary scaffolding and platforms, equipment mobilization and demobilization, temporary roads and sidewalks, winter protection, temporary barricades and fencing, temporary security, temporary signs, field engineering and layout, final cleaning, and commissioning.
- Each item should be quantified and matched to individual Unit Price line items in Division 1, then priced and added to your estimate.
- An alternate method of estimating project overhead costs is to apply a percentage of the total project cost, usually 5% to 15% with an average of 10% (see General Conditions).
- Include other project related expenses in your estimate such as:
    - Rented equipment not itemized in the Crew Listings
    - Rubbish handling throughout the project (see 02 41 19.19)

6. Estimate Summary

- Include sales tax as required by laws of your state or county.
- Include the general contractor's markup on self-performed work, usually 5% to 15% with an average of 10%.
- Include the general contractor's markup on subcontracted work, usually 5% to 15% with an average of 10%.
- Include the general contractor's main office overhead and profit:
    - RSMeans gives general guidelines on the general contractor's main office overhead (see section 01 31 13.60 and Reference Number R013113-50).
    - RSMeans gives no guidance on the general contractor's profit.
    - Markups will depend on the size of the general contractor's operations, projected annual revenue, the level of risk, and the level of competition in the local area and for this project in particular.
- Include a contingency, usually 3% to 5%, if appropriate.
- Adjust your estimate to the project's location by using the City Cost Indexes or the Location Factors in the Reference Section:
    - Look at the rules in "How to Use the City Cost Indexes" to see how to apply the Indexes for your location.
    - When the proper Index or Factor has been identified for the project's location, convert it to a multiplier by dividing it by 100, then multiply that multiplier by your estimated total cost. The original estimated total cost will now be adjusted up or down from the national average to a total that is appropriate for your location.

Editors' Note:
We urge you to spend time reading and understanding the supporting material. An accurate estimate requires experience, knowledge, and careful calculation. The more you know about how we at RSMeans developed the data, the more accurate your estimate will be. In addition, it is important to take into consideration the reference material such as Equipment Listings, Crew Listings, City Cost Indexes, Location Factors, and Reference Tables.

How to Use the Cost Data: The Details

What's Behind the Numbers? The Development of Cost Data

The staff at RSMeans continually monitors developments in the construction industry in order to ensure reliable, thorough, and up-to-date cost information. While overall construction costs may vary relative to general economic conditions, price fluctuations within the industry are dependent upon many factors. Individual price variations may, in fact, be opposite to overall economic trends. Therefore, costs are constantly tracked, and complete updates are performed yearly. Also, new items are frequently added in response to changes in materials and methods.

Costs—$ (U.S.)

All costs represent U.S. national averages and are given in U.S. dollars. The RSMeans City Cost Indexes (CCI) can be used to adjust costs to a particular location. The CCI for Canada can be used to adjust U.S. national averages to local costs in Canadian dollars. No exchange rate conversion is necessary because it has already been factored in.

G The processes or products identified by the green symbol in our publications have been determined to be environmentally responsible and/or resource-efficient solely by the RSMeans engineering staff. The inclusion of the green symbol does not represent compliance with any specific industry association or standard.

Material Costs

The RSMeans staff contacts manufacturers, dealers, distributors, and contractors all across the U.S. and Canada to determine national average material costs. If you have access to current material costs for your specific location, you may wish to make adjustments to reflect differences from the national average. Included within material costs are fasteners for a normal installation. RSMeans engineers use manufacturers' recommendations, written specifications, and/or standard construction practices for size and spacing of fasteners. Adjustments to material costs may be required for your specific application or location. The manufacturer's warranty is assumed. Extended warranties are not included in the material costs. **Material costs do not include sales tax.**

Labor Costs

Labor costs are based upon a mathematical average of trade-specific wages in 30 major U.S. cities. The type of wage (union, open shop, or residential) is identified on the inside back cover of printed publications or is selected by the estimator when using the electronic products. Markups for the wages can also be found on the inside back cover of printed publications and/or under the labor references found in the electronic products.

- If wage rates in your area vary from those used, or if rate increases are expected within a given year, labor costs should be adjusted accordingly.

Labor costs reflect productivity based on actual working conditions. In addition to actual installation, these figures include time spent during a normal weekday on tasks, such as material receiving and handling, mobilization at site, site movement, breaks, and cleanup.

Productivity data is developed over an extended period so as not to be influenced by abnormal variations and reflects a typical average.

Equipment Costs

Equipment costs include not only rental, but also operating costs for equipment under normal use. The operating costs include parts and labor for routine servicing, such as repair and replacement of pumps, filters, and worn lines. Normal operating expendables, such as fuel, lubricants, tires, and electricity (where applicable), are also included. Extraordinary operating expendables with highly variable wear patterns, such as diamond bits and blades, are excluded. These costs are included under materials. Equipment rental rates are obtained from industry sources throughout North America—contractors, suppliers, dealers, manufacturers, and distributors.

Rental rates can also be treated as reimbursement costs for contractor-owned equipment. Owned equipment costs include depreciation, loan payments, interest, taxes, insurance, storage, and major repairs.

Equipment costs do not include operators' wages.

Equipment Cost/Day—The cost of equipment required for each crew is included in the Crew Listings in the Reference Section (small tools that are considered essential everyday tools are not listed out separately). The Crew Listings itemize specialized tools and heavy equipment along with labor trades. The daily cost of itemized equipment included in a crew is based on dividing the weekly bare rental rate by 5 (number of working days per week), then adding the hourly operating cost times 8 (the number of hours per day). This Equipment Cost/Day is shown in the last column of the Equipment Rental Costs in the Reference Section.

Mobilization/Demobilization—The cost to move construction equipment from an equipment yard or rental company to the job site and back again is not included in equipment costs. Mobilization (to the site) and demobilization (from the site) costs can be found in the Unit Price Section. If a piece of equipment is already at the job site, it is not appropriate to utilize mob./demob. costs again in an estimate.

Overhead and Profit

Total Cost including O&P for the installing contractor is shown in the last column of the Unit Price and/or Assemblies. This figure is the sum of the bare material cost plus 10% for profit, the bare labor cost plus total overhead and profit, and the bare equipment cost plus 10% for profit. Details for the calculation of overhead and profit on labor are shown on the inside back cover of the printed product and in the Reference Section of the electronic product.

General Conditions

Cost data in this data set is presented in two ways: Bare Costs and Total Cost including O&P (Overhead and Profit). General Conditions, or General Requirements, of the contract should also be added to the Total Cost including O&P when applicable. Costs for General Conditions are listed in Division 1 of the Unit Price Section and in the Reference Section.

General Conditions for the installing contractor may range from 0% to 10% of the Total Cost including O&P. For the general or prime contractor, costs for General Conditions may range from 5% to 15% of the Total Cost including O&P, with a figure of 10% as the most typical allowance. If applicable, the Assemblies and Models sections use costs that include the installing contractor's overhead and profit (O&P).

Factors Affecting Costs

Costs can vary depending upon a number of variables. Here's a listing of some factors that affect costs and points to consider.

Quality—The prices for materials and the workmanship upon which productivity is based represent sound construction work. They are also in line with industry standard and manufacturer specifications and are frequently used by federal, state, and local governments.

Overtime—We have made no allowance for overtime. If you anticipate premium time or work beyond normal working hours, be sure to make an appropriate adjustment to your labor costs.

Productivity—The productivity, daily output, and labor-hour figures for each line item are based on working an eight-hour day in daylight hours in moderate temperatures. For work that extends beyond normal work hours or is performed under adverse conditions, productivity may decrease.

Size of Project—The size, scope of work, and type of construction project will have a significant impact on cost. Economies of scale can reduce costs for large projects. Unit costs can often run higher for small projects.

Location—Material prices are for metropolitan areas. However, in dense urban areas, traffic and site storage limitations may increase costs. Beyond a 20-mile radius of metropolitan areas, extra trucking or transportation charges may also increase the material costs slightly. On the other hand, lower wage rates may be in effect. Be sure to consider both of these factors when preparing an estimate, particularly if the job site is located in a central city or remote rural location. In addition, highly specialized subcontract items may require travel and per-diem expenses for mechanics.

Other Factors—

- season of year
- contractor management
- weather conditions
- local union restrictions
- building code requirements
- availability of:
  - adequate energy
  - skilled labor
  - building materials
- owner's special requirements/restrictions
- safety requirements
- environmental considerations
- access

Unpredictable Factors—General business conditions influence "in-place" costs of all items. Substitute materials and construction methods may have to be employed. These may affect the installed cost and/or life cycle costs. Such factors may be difficult to evaluate and cannot necessarily be predicted on the basis of the job's location in a particular section of the country. Thus, where these factors apply, you may find significant but unavoidable cost variations for which you will have to apply a measure of judgment to your estimate.

Rounding of Costs

In printed publications only, all unit prices in excess of $5.00 have been rounded to make them easier to use and still maintain adequate precision of the results.

How Subcontracted Items Affect Costs

A considerable portion of all large construction jobs is usually subcontracted. In fact, the percentage done by subcontractors is constantly increasing and may run over 90%. Since the workers employed by these companies do nothing else but install their particular product, they soon become experts in that line. The result is, installation by these firms is accomplished so efficiently that the total in-place cost, even adding the general contractor's overhead and profit, is no more, and often less, than if the principal contractor had handled the installation. Companies that deal with construction specialties are anxious to have their product perform well and, consequently, the installation will be the best possible.

Contingencies

The allowance for contingencies generally provides for unforeseen construction difficulties. On alterations or repair jobs, 20% is not too much. If drawings are final and only field contingencies are being considered, 2% or 3% is probably sufficient, and often nothing needs to be added. Contractually, changes in plans will be covered by extras. The contractor should consider inflationary price trends and possible material shortages during the course of the job. These escalation factors are dependent upon both economic conditions and the anticipated time between the estimate and actual construction. If drawings are not complete or approved, or a budget cost is wanted, it is wise to add 5% to 10%. Contingencies, then, are a matter of judgment.

Important Estimating Considerations

The productivity, or daily output, of each craftsman or crew assumes a well-managed job where tradesmen with the proper tools and equipment, along with the appropriate construction materials, are present. Included are daily set-up and cleanup time, break time, and plan layout time. Unless otherwise indicated, time for material movement on site (for items that can be transported by hand) of up to 200' into the building and to the first or second floor is also included. If material has to be transported by other means, over greater distances, or to higher floors, an additional allowance should be considered by the estimator.

While horizontal movement is typically a sole function of distances, vertical transport introduces other variables that can significantly impact productivity. In an occupied building, the use of elevators (assuming access, size, and required protective measures are acceptable) must be understood at the time of the estimate. For new construction, hoist wait and cycle times can easily be 15 minutes and may result in scheduled access extending beyond the normal work day. Finally, all vertical transport will impose strict weight limits likely to preclude the use of any motorized material handling.

The productivity, or daily output, also assumes installation that meets manufacturer/designer/standard specifications. A time allowance for quality control checks, minor adjustments, and any task required to ensure the proper function or operation is also included. For items that require connections to services, time is included for positioning, leveling, securing the unit, and for making all the necessary connections (and start up where applicable), ensuring a complete installation. Estimating of the services themselves (electrical, plumbing, water, steam, hydraulics, dust collection, etc.) is separate.

In some cases, the estimator must consider the use of a crane and an appropriate crew for the installation of large or heavy items. For those situations where a crane is not included in the assigned crew and as part of the line item cost, equipment rental costs, mobilization and demobilization costs, and operator and support personnel costs must be considered.

Labor-Hours

The labor-hours expressed in this publication are derived by dividing the total daily labor-hours for the crew by the daily output. Based on average installation time and the assumptions listed above, the labor-hours include: direct labor, indirect labor, and nonproductive time. A typical day for a craftsman might include, but is not limited to:

- Direct Work
  - Measuring and layout
  - Preparing materials
  - Actual installation
  - Quality assurance/quality control
- Indirect Work
  - Reading plans or specifications
  - Preparing space
  - Receiving materials
  - Material movement
  - Giving or receiving instruction
  - Miscellaneous
- Non-Work
  - Chatting
  - Personal issues
  - Breaks
  - Interruptions (i.e., sickness, weather, material or equipment shortages, etc.)

If any of the items for a typical day do not apply to the particular work or project situation, the estimator should make any necessary adjustments.

Final Checklist

Estimating can be a straightforward process provided you remember the basics. Here's a checklist of some of the steps you should remember to complete before finalizing your estimate.

Did you remember to:

- factor in the City Cost Index for your locale?
- take into consideration which items have been marked up and by how much?
- mark up the entire estimate sufficiently for your purposes?
- read the background information on techniques and technical matters that could impact your project time span and cost?
- include all components of your project in the final estimate?
- double check your figures for accuracy?
- call RSMeans if you have any questions about your estimate or the data you've used? Remember, RSMeans stands behind its information. If you have any questions about your estimate, about the costs you've used from our data, or even about the technical aspects of the job that may affect your estimate, feel free to call the RSMeans editors at 1-888-607-2575.

Free Quarterly Updates

Stay up-to-date throughout the year with RSMeans' free cost data updates four times a year. Sign up online to make sure you have access to the newest data. Every quarter we provide the city cost adjustment factors for hundreds of cities and key materials.

Register at: **http://info.TheGordianGroup.com/2016Updates.html**.

Commercial/Industrial/ Institutional Section

Table of Contents

Introduction to the Commercial/ Industrial/Institutional Section

General

The Commercial/Industrial/Institutional section of this manual contains base building costs per square foot of floor area for 50 model buildings. Each model has a table of square foot costs for combinations of exterior wall and framing systems. This table is supplemented by a list of common additives and their unit costs. A breakdown of the component costs used to develop the base cost for the model is included. The total cost derived from the base cost and additives must be modified using the appropriate location factor from the Reference Section.

This section may be used directly to estimate the construction cost of most types of buildings knowing only the floor area, exterior wall construction, and framing systems. To adjust the base cost for components which are different from the model, use the tables from the Assemblies Section.

Building Identification & Model Selection

The building models in this section represent structures by use. Occupancy, however, does not necessarily identify the building, i.e., a restaurant could be a converted warehouse. In all instances, the building should be described and identified by its own physical characteristics. The model selection should also be guided by comparing specifications with the model. In the case of converted use, data from one model may be used to supplement data from another.

Green Models

Consistent with expanding green trends in the design and construction industries, 16 green building models are currently included. Although similar to our standard models in building type and structural system, the green models meet or exceed Energy Star requirements and address many of the items necessary to obtain LEED certification. Although our models do not include site-specific information, our design assumption is that they are located in climate zone 5. DOE's eQuest software was used to perform an energy analysis of the model buildings. By reducing energy use, we were able to reduce the size of the service entrances, switchgear, power feeds, and generators. Each of the following building systems was researched and analyzed: building envelope, HVAC, plumbing fixtures, lighting, and electrical service. These systems were targeted because of their impact on energy usage and green building.

Wall and roof insulation was increased to reduce heat loss and a recycled vapor barrier was added to the foundation. White roofs were specified for models with flat roofs. Interior finishes include materials with recycled content and low VOC paint. Stainless steel toilet partitions were selected because it is a recycled material and is easy to maintain. Plumbing fixtures were selected to conserve water. Water closets are low-flow and equipped with auto-sensor valves. Waterless urinals are specified throughout. Faucets use auto-sensor flush valves and are powered by a hydroelectric power unit inside the faucet. Energy efficient, low flow water coolers were specified to reduce energy consumption and water usage throughout each building model.

Lighting efficiency was achieved by specifying LED fixtures in place of standard fluorescents. Daylight on/off lighting control systems were incorporated into the models. These controls are equipped with sensors that automatically turn off the lights when sufficient daylight is available. Energy monitoring systems were also included in all green models.

Adjustments

The base cost tables represent the base cost per square foot of floor area for buildings without a basement and without unusual special features. Basement costs and other common additives are listed below the base cost table. Cost adjustments can also be made to the model by using the tables from the Assemblies Section. This table is for example only.

Dimensions

All base cost tables are developed so that measurements can be readily made during the inspection process. Areas are calculated from exterior dimensions and story heights are measured from the top surface of one floor to the top surface of the floor above. Roof areas are measured by horizontal area covered and costs related to inclines are converted with appropriate factors. The precision of measurement is a matter of the user's choice and discretion. For ease in calculation, consideration should be given to measuring in tenths of a foot, i.e., 9 ft. 6 in. = 9.5 ft., 9 ft. 4 in. = 9.3 ft.

Floor Area

The term "Floor Area" as used in this section includes the sum of the floor plate at grade level and above. This dimension is measured from the outside face of the foundation wall. Basement costs are calculated separately. The user must exercise his/her own judgment, where the lowest level floor is slightly below grade, whether to consider it at grade level or make the basement adjustment.

How to Use the Commercial/ Industrial/Institutional Section

The following is a detailed explanation of a sample entry in the Commercial/Industrial/Institutional Square Foot Cost Section. Each bold number below corresponds to the described item in the following list with the appropriate component or cost of the sample entry in parentheses.

Prices listed are costs that include overhead and profit of the installing contractor and additional markups for general conditions and architects' fees.

COMMERCIAL/INDUSTRIAL/ INSTITUTIONAL	M.010 **1**	Apartment, 1-3 Story **2**

Costs per square foot of floor area

Exterior Wall **3**	S.F. Area	8000	12000	15000	19000	22500 **4**	25000	29000	32000	36000
	L.F. Perimeter	213	280	330	350	400	433	442	480	520
Fiber Cement Siding	Wood Frame	166.25	159.10	156.20	151.25	149.80	148.90	146.30	145.70	144.90
Stone Veneer	Wood Frame	184.50	175.15	171.35	164.00	**162.10** **5**	160.90	156.90	156.15	154.95
Brick Veneer	Steel Frame	166.10	157.65	154.30	147.85	146.25	145.20	141.70	141.05	140.00
EIFS on Metal Studs	Steel Frame	150.25	144.40	142.10	138.55	137.35	136.70	134.85	134.40	133.75
Brick Veneer	Reinforced Concrete	180.90	171.05	167.10	159.15	157.20	155.95	151.60	150.85	149.55
Stucco on Concrete Block	Reinforced Concrete	170.40	161.85	158.40	151.90	150.15	149.10	145.60	144.90	143.80
Perimeter Adj., Add or Deduct **6**	Per 100 L.F.	13.05	8.75	6.95	5.50	4.65 **7**	4.25	3.60	3.25	2.90
Story Hgt. Adj., Add or Deduct	Per 1 Ft.	2.05	1.75	1.70	1.40	1.35	1.35	1.20	1.15	1.05
8 ***For Basement**, add $32.45 per square foot of basement area*										

The above costs were calculated using the basic specifications shown on the facing page. These costs should be adjusted where necessary for design alternatives and owner's requirements. Reported completed project costs, for this type of structure, range from $63.55 to $236.40 per S.F. **9**

Common additives **10**

Description	Unit	$ Cost
Appliances		
Cooking range, 30" free standing		
1 oven	Ea.	595 - 2275
2 oven	Ea.	1450 - 3150
30" built-in		
1 oven	Ea.	1075 - 2225
2 oven	Ea.	1725 - 3275
Counter top cook tops, 4 burner	Ea.	480 - 2075
Microwave oven	Ea.	275 - 850
Combination range, refrig. & sink, 30" wide	Ea.	1575 - 2100
72" wide	Ea.	2250
Combination range, refrigerator, sink, microwave oven & icemaker	Ea.	5900
Compactor, residential, 4-1 compaction	Ea.	860 - 1450
Dishwasher, built-in, 2 cycles	Ea.	510 - 855
4 cycles	Ea.	650 - 1675
Garbage disposer, sink type	Ea.	174 - 340
Hood for range, 2 speed, vented, 30" wide	Ea.	222 - 1150
42" wide	Ea.	310 - 2400
Refrigerator, no frost 10-12 C.F.	Ea.	600 - 745
18-20 C.F.	Ea.	910 - 2075

Description	Unit	$ Cost
Closed Circuit Surveillance, One station		
Camera and monitor	Ea.	1975
For additional camera stations, add	Ea.	1075
Elevators, Hydraulic passenger, 2 stops		
2000# capacity	Ea.	64,900
2500# capacity	Ea.	66,900
3500# capacity	Ea.	71,900
Additional stop, add	Ea.	6825
Emergency Lighting, 25 watt, battery operated		
Lead battery	Ea.	325
Nickel cadmium	Ea.	665
Laundry Equipment		
Dryer, gas, 16 lb. capacity	Ea.	950
30 lb. capacity	Ea.	3900
Washer, 4 cycle	Ea.	1175
Commercial	Ea.	1575
Smoke Detectors		
Ceiling type	Ea.	214
Duct type	Ea.	540

1 Model Number (M.010)

"M" distinguishes this section and stands for model. The number designation is a sequential number.

2 Type of Building (Apartment, 1-3 Story)

There are 50 different types of commercial/industrial/ institutional buildings highlighted in this section.

3 Exterior Wall Construction and Building Framing Options (Face Brick with Concrete Block Back-up and Open Web Steel Bar Joists)

Three or more commonly used exterior walls and, in most cases, two typical building framing systems are presented for each type of building. The model selected should be based on the actual characteristics of the building being estimated.

4 Total Square Foot of Floor Area and Base Perimeter Used to Compute Base Costs (22,500 Square Feet and 400 Linear Feet)

Square foot of floor area is the total gross area of all floors at grade and above and does not include a basement. The perimeter in linear feet used for the base cost is generally for a rectangular, economical building shape.

5 Cost per Square Foot of Floor Area ($162.10)

The highlighted cost is for a building of the selected exterior wall and framing system and floor area. Costs for buildings with floor areas other than those calculated may be interpolated between the costs shown.

6 Building Perimeter and Story Height Adjustments

Square-foot costs for a building with a perimeter or floor-to-floor story height significantly different from the model used to calculate the base cost may be adjusted (add or deduct) to reflect the actual building geometry.

7 Cost per Square Foot of Floor Area for the Perimeter and/or Height Adjustment ($4.65 for Perimeter Difference and $1.35 for Story Height Difference)

Add (or deduct) $4.65 to the base square-foot cost for each 100 feet of perimeter difference between the model and the actual building. Add (or deduct) $1.35 to the base square-foot cost for each 1 foot of story height difference between the model and the actual building.

8 Optional Cost per Square Foot of Basement Floor Area ($32.45)

The cost of an unfinished basement for the building being estimated is $32.45 times the gross floor area of the basement.

9 Range of Cost per Square Foot of Floor Area for Similar Buildings ($63.55 to $236.40)

Many different buildings of the same type have been built using similar materials and systems. RSMeans historical cost data of actual construction projects indicates a range of $63.55 to $236.40 for this type of building.

10 Common Additives

Common components and/or systems used in this type of building are listed. These costs should be added to the total building cost. Additional selections may be found in the Assemblies Section.

How to Use the Commercial/Industrial/Inst. Section (Continued)

The following is a detailed explanation of the specification and costs for a model building in the Commercial/Industrial/Institutional Square Foot Cost Section. Each bold number below corresponds to the described item in the following list with the appropriate component of the sample entry in parentheses.

Prices listed are costs that include overhead and profit of the installing contractor.

Model costs calculated for a 3 story building with 10′ story height and 22,500 square feet of floor area **1**

2 **Apartment, 1-3 Story**

				Unit	Unit Cost	Cost Per S.F.	% Of Sub-Total
A. SUBSTRUCTURE							
1010	Standard Foundations	Poured concrete; strip and spread footings		S.F. Ground	4.59	1.53	
1020	Special Foundations	N/A		–	–	–	
1030	Slab on Grade	4″ reinforced concrete		S.F. Slab	5.10	1.70	3.8%
2010	Basement Excavation	Site preparation for slab and trench for foundation wall and footing		S.F. Ground	.29	.10	
2020	Basement Walls	4′ Foundation wall		L.F. Wall	37	1.31	
B. SHELL							
	B10 Superstructure						
1010	Floor Construction	Wood Beam and joist on wood columns, fireproofed		S.F. Floor	24	15.83	15.0%
1020	Roof Construction	Wood roof, truss, 4/12 slope		S.F. Roof	6.93	2.31	
	B20 Exterior Enclosure						
2010	Exterior Walls	Ashlar stone veneer on wood stud back up, insulated	*80% of wall*	S.F. Wall	36	15.21	
2020	Exterior Windows	Aluminum horizontal sliding	*20% of wall*	Each	495	3.52	15.7%
2030	Exterior Doors	Steel door , hollow metal with frame		Each	1975	.35	
	B30 Roofing						
3010	Roof Coverings	Asphalt roofing, strip shingles, flashing		S.F. Roof	6.60	2.20	1.8%
3020	Roof Openings	N/A		–	–	–	
C. INTERIORS **3**							
1010	Partitions	Gypsum board on wood studs **4**	*10 S.F. Floor/L.F. Partition*	S.F. Partition **6**	6.68 **7**	5.94 **8**	
1020	Interior Doors	Single leaf solid and hollow core, fire doors	*130 S.F. Floor/Door* **5**	Each	600	4.55	
1030	Fittings	Residential wall and base wood cabinets, laminate counters		S.F. Floor	4.24	4.24	
2010	Stair Construction	Wood Stairs with wood rails		Flight	2675	.71	21.8% **9**
3010	Wall Finishes	95% paint, 5% ceramic wall tile		S.F. Surface	1.25	2.22	
3020	Floor Finishes	80% Carpet tile, 10% vinyl composition tile, 10% ceramic tile		S.F. Floor	4.94	4.94	
3030	Ceiling Finishes	Painted gypsum board ceiling on resilient channels		S.F. Ceiling	3.85	3.85	
D. SERVICES							
	D10 Conveying						
1010	Elevators & Lifts	Hydraulic passenger elevator		Each	105,300	4.68	3.9%
1020	Escalators & Moving Walks	N/A		–	–	–	
	D20 Plumbing						
2010	Plumbing Fixtures	Kitchen, bath, laundry and service fixtures, supply and drainage	*1 Fixture/285 S.F. Floor*	Each	1753	6.15	
2020	Domestic Water Distribution	Electric water heater		S.F. Floor	7.43	7.43	11.6%
2040	Rain Water Drainage	Roof drains		S.F. Roof	1.32	.44	
	D30 HVAC						
3010	Energy Supply	Oil fired hot water, baseboard radiation		S.F. Floor	7.93	7.93	
3020	Heat Generating Systems	N/A		–	–	–	
3030	Cooling Generating Systems	Chilled water, air cooled condenser system		S.F. Floor	8.39	8.39	13.5%
3050	Terminal & Package Units	N/A		–	–	–	
3090	Other HVAC Sys. & Equipment	N/A		–	–	–	
	D40 Fire Protection						
4010	Sprinklers	Wet pipe spinkler system, light hazard		S.F. Floor	3.18	3.18	2.6%
4020	Standpipes	N/A		–	–	–	
	D50 Electrical						
5010	Electrical Service/Distribution	800 ampere service, panel board and feeders		S.F. Floor	2.55	2.55	
5020	Lighting & Branch Wiring	Incandescent fixtures, receptacles, switches, A.C. and misc. power		S.F. Floor	6.93	6.93	9.2%
5030	Communications & Security	Addressable alarm systems, emergency lighting, internet and phone wiring		S.F. Floor	1.65	1.65	
5090	Other Electrical Systems	N/A		–	–	–	
E. EQUIPMENT & FURNISHINGS							
1010	Commercial Equipment	N/A		–	–	–	
1020	Institutional Equipment	N/A		–	–	–	1.1 %
1030	Vehicular Equipment	N/A		–	–	–	
1090	Other Equipment	Residential freestanding gas ranges, dishwashers		S.F. Floor	1.35	1.35	
F. SPECIAL CONSTRUCTION							
1020	Integrated Construction	N/A		–	–	–	0.0 %
1040	Special Facilities	N/A		–	–	–	
G. BUILDING SITEWORK		**N/A**					
					10 **Sub-Total**	121.19	**100%**
	CONTRACTOR FEES (General Requirements: 10%, Overhead: 5%, Profit: 10%) **11**				25%	30.31	
	ARCHITECT FEES				7%	10.60	
					12 **Total Building Cost**	**162.10**	

1 Building Description (Model costs are calculated for a 3-story apartment building with 10' story height and 22,500 square feet of floor area)
The model highlighted is described in terms of building type, number of stories, typical story height, and square footage.

2 Type of Building (Apartment, 1-3 Story)

3 Division C Interiors (C1020 Interior Doors)
System costs are presented in divisions according to the 7-division UNIFORMAT II classifications. Each of the component systems is listed.

4 Specification Highlights (Single leaf solid and hollow core, fire doors)
All systems in each subdivision are described with the material and proportions used.

5 Quality Criteria (130 S.F. Floor/Door)
The criteria used in determining quantities for the calculations are shown.

6 Unit (Each)
The unit of measure in this column is the unit of measure of the particular system shown that corresponds to the unit cost.

7 Unit Cost ($600)
The cost per unit of measure of each system subdivision.

8 Cost per Square Foot ($4.55)
The cost per square foot for each system is the unit cost of the system times the total number of units, divided by the total square feet of building area.

9 % of Sub-Total (21.8%)
The percent of sub-total is the total cost per square foot of all systems in the division divided by the sub-total cost per square foot of the building.

10 Sub-Total ($121.19)
The sub-total is the total of all the system costs per square foot.

11 Project Fees
(Contractor Fees) (25%); (Architect's Fees) (7%)
Contractor fees to cover the general requirements, overhead, and profit of the general contractor are added as a percentage of the sub-total. An architect's fee, also as a percentage of the sub-total, is also added. These values vary with the building type.

12 Total Building Cost ($162.10)
The total building cost per square foot of building area is the sum of the square foot costs of all the systems, the general contractor's general requirements, overhead and profit, and the architect's fee. The total building cost is the amount which appears shaded in the Cost per Square Foot of Floor Area table shown previously.

Costs per square foot of floor area

Exterior Wall	S.F. Area	8000	12000	15000	19000	22500	25000	29000	32000	36000
	L.F. Perimeter	213	280	330	350	400	433	442	480	520
Fiber Cement Siding	Wood Frame	166.25	159.10	156.20	151.25	149.80	148.90	146.30	145.70	144.90
Stone Veneer	Wood Frame	184.50	175.15	171.35	164.00	162.10	160.90	156.90	156.15	154.95
Brick Veneer	Steel Frame	166.10	157.65	154.30	147.85	146.25	145.20	141.70	141.05	140.00
EIFS on Metal Studs	Steel Frame	150.25	144.40	142.10	138.55	137.35	136.70	134.85	134.40	133.75
Brick Veneer	Reinforced Concrete	180.90	171.05	167.10	159.15	157.20	155.95	151.60	150.85	149.55
Stucco on Concrete Block	Reinforced Concrete	170.40	161.85	158.40	151.90	150.15	149.10	145.60	144.90	143.80
Perimeter Adj., Add or Deduct	Per 100 L.F.	13.05	8.75	6.95	5.50	4.65	4.25	3.60	3.25	2.90
Story Hgt. Adj., Add or Deduct	Per 1 Ft.	2.05	1.75	1.70	1.40	1.35	1.35	1.20	1.15	1.05
***For Basement**, add $ 32.45 per square foot of basement area*										

The above costs were calculated using the basic specifications shown on the facing page. These costs should be adjusted where necessary for design alternatives and owner's requirements. Reported completed project costs, for this type of structure, range from $ 63.55 to $ 236.40 per S.F.

Common additives

Description	Unit	$ Cost
Appliances		
Cooking range, 30" free standing		
1 oven	Ea.	595 - 2275
2 oven	Ea.	1450 - 3150
30" built-in		
1 oven	Ea.	1075 - 2225
2 oven	Ea.	1725 - 3275
Counter top cook tops, 4 burner	Ea.	480 - 2075
Microwave oven	Ea.	275 - 850
Combination range, refrig. & sink, 30" wide	Ea.	1575 - 2100
72" wide	Ea.	2250
Combination range, refrigerator, sink, microwave oven & icemaker	Ea.	5900
Compactor, residential, 4-1 compaction	Ea.	860 - 1450
Dishwasher, built-in, 2 cycles	Ea.	510 - 855
4 cycles	Ea.	650 - 1675
Garbage disposer, sink type	Ea.	174 - 340
Hood for range, 2 speed, vented, 30" wide	Ea.	222 - 1150
42" wide	Ea.	310 - 2400
Refrigerator, no frost 10-12 C.F.	Ea.	600 - 745
18-20 C.F.	Ea.	910 - 2075

Description	Unit	$ Cost
Closed Circuit Surveillance, One station		
Camera and monitor	Ea.	1975
For additional camera stations, add	Ea.	1075
Elevators, Hydraulic passenger, 2 stops		
2000# capacity	Ea.	64,900
2500# capacity	Ea.	66,900
3500# capacity	Ea.	71,900
Additional stop, add	Ea.	6825
Emergency Lighting, 25 watt, battery operated		
Lead battery	Ea.	325
Nickel cadmium	Ea.	665
Laundry Equipment		
Dryer, gas, 16 lb. capacity	Ea.	950
30 lb. capacity	Ea.	3900
Washer, 4 cycle	Ea.	1175
Commercial	Ea.	1575
Smoke Detectors		
Ceiling type	Ea.	214
Duct type	Ea.	540

Important: See the Reference Section for Location Factors.

Model costs calculated for a 3 story building with 10′ story height and 22,500 square feet of floor area

Apartment, 1-3 Story

			Unit	Unit Cost	Cost Per S.F.	% Of Sub-Total
A. SUBSTRUCTURE						
1010	Standard Foundations	Poured concrete; strip and spread footings	S.F. Ground	4.59	1.53	3.8%
1020	Special Foundations	N/A	—	—	—	
1030	Slab on Grade	4″ reinforced concrete	S.F. Slab	5.10	1.70	
2010	Basement Excavation	Site preparation for slab and trench for foundation wall and footing	S.F. Ground	.29	.10	
2020	Basement Walls	4′ Foundation wall	L.F. Wall	37	1.31	
B. SHELL						
	B10 Superstructure					
1010	Floor Construction	Wood Beam and joist on wood columns, fireproofed	S.F. Floor	24	15.83	15.0%
1020	Roof Construction	Wood roof, truss, 4/12 slope	S.F. Roof	6.93	2.31	
	B20 Exterior Enclosure					
2010	Exterior Walls	Ashlar stone veneer on wood stud back up, insulated *80% of wall*	S.F. Wall	36	15.21	15.7%
2020	Exterior Windows	Aluminum horizontal sliding *20% of wall*	Each	495	3.52	
2030	Exterior Doors	Steel door , hollow metal with frame	Each	1975	.35	
	B30 Roofing					
3010	Roof Coverings	Asphalt roofing, strip shingles, flashing	S.F. Roof	6.60	2.20	1.8%
3020	Roof Openings	N/A	—	—	—	
C. INTERIORS						
1010	Partitions	Gypsum board on wood studs *10 S.F. Floor/L.F. Partition*	S.F. Partition	6.68	5.94	21.8%
1020	Interior Doors	Single leaf solid and hollow core, fire doors *130 S.F. Floor/Door*	Each	600	4.55	
1030	Fittings	Residential wall and base wood cabinets, laminate counters	S.F. Floor	4.24	4.24	
2010	Stair Construction	Wood Stairs with wood rails	Flight	2675	.71	
3010	Wall Finishes	95% paint, 5% ceramic wall tile	S.F. Surface	1.25	2.22	
3020	Floor Finishes	80% Carpet tile, 10% vinyl composition tile, 10% ceramic tile	S.F. Floor	4.94	4.94	
3030	Ceiling Finishes	Painted gypsum board ceiling on resilient channels	S.F. Ceiling	3.85	3.85	
D. SERVICES						
	D10 Conveying					
1010	Elevators & Lifts	Hydraulic passenger elevator	Each	105,300	4.68	3.9%
1020	Escalators & Moving Walks	N/A	—	—	—	
	D20 Plumbing					
2010	Plumbing Fixtures	Kitchen, bath, laundry and service fixtures, supply and drainage *1 Fixture/285 S.F. Floor*	Each	1753	6.15	11.6%
2020	Domestic Water Distribution	Electric water heater	S.F. Floor	7.43	7.43	
2040	Rain Water Drainage	Roof drains	S.F. Roof	1.32	.44	
	D30 HVAC					
3010	Energy Supply	Oil fired hot water, baseboard radiation	S.F. Floor	7.93	7.93	13.5%
3020	Heat Generating Systems	N/A	—	—	—	
3030	Cooling Generating Systems	Chilled water, air cooled condenser system	S.F. Floor	8.39	8.39	
3050	Terminal & Package Units	N/A	—	—	—	
3090	Other HVAC Sys. & Equipment	N/A	—	—	—	
	D40 Fire Protection					
4010	Sprinklers	Wet pipe spinkler system, light hazard	S.F. Floor	3.18	3.18	2.6%
4020	Standpipes	N/A	—	—	—	
	D50 Electrical					
5010	Electrical Service/Distribution	800 ampere service, panel board and feeders	S.F. Floor	2.55	2.55	9.2%
5020	Lighting & Branch Wiring	Incandescent fixtures, receptacles, switches, A.C. and misc. power	S.F. Floor	6.93	6.93	
5030	Communications & Security	Addressable alarm systems, emergency lighting, internet and phone wiring	S.F. Floor	1.65	1.65	
5090	Other Electrical Systems	N/A	—	—	—	
E. EQUIPMENT & FURNISHINGS						
1010	Commercial Equipment	N/A	—	—	—	1.1 %
1020	Institutional Equipment	N/A	—	—	—	
1030	Vehicular Equipment	N/A	—	—	—	
1090	Other Equipment	Residential freestanding gas ranges, dishwashers	S.F. Floor	1.35	1.35	
F. SPECIAL CONSTRUCTION						
1020	Integrated Construction	N/A	—	—	—	0.0 %
1040	Special Facilities	N/A	—	—	—	
G. BUILDING SITEWORK		**N/A**				
				Sub-Total	121.19	**100%**
	CONTRACTOR FEES (General Requirements: 10%, Overhead: 5%, Profit: 10%)			*25%*	30.31	
	ARCHITECT FEES			*7%*	10.60	
				Total Building Cost	**162.10**	

Costs per square foot of floor area

Exterior Wall	S.F. Area	2000	3000	4000	5000	7500	10000	15000	25000	50000
	L.F. Perimeter	188	224	260	300	356	400	504	700	1200
Brick Veneer	Wood Frame	228.00	202.20	189.25	182.05	169.00	161.75	155.15	149.60	145.55
	Steel Frame	254.40	229.00	216.25	209.10	196.25	189.15	182.65	177.25	173.25
Brick Veneer on Block	Wood Truss	231.20	204.20	190.70	183.10	169.25	161.60	154.55	148.65	144.35
	Bearing Wall	264.60	237.55	224.00	216.45	202.55	194.90	187.85	181.90	177.65
Wood Siding	Wood Frame	215.00	191.60	179.85	173.20	161.60	155.25	149.35	144.45	140.85
Vinyl Siding	Wood Frame	219.25	195.20	183.20	176.45	164.55	158.00	152.05	146.95	143.35
Perimeter Adj., Add or Deduct	Per 100 L.F.	28.40	18.95	14.25	11.30	7.50	5.70	3.80	2.25	1.15
Story Hgt. Adj., Add or Deduct	Per 1 Ft.	3.55	2.75	2.50	2.25	1.75	1.50	1.25	1.00	0.90

***For Basement**, add $29.25 per square foot of basement area*

The above costs were calculated using the basic specifications shown on the facing page. These costs should be adjusted where necessary for design alternatives and owner's requirements. Reported completed project costs, for this type of structure, range from $95.75 to $240.05 per S.F.

Common additives

Description	Unit	$ Cost
Appliances		
Cooking range, 30" free standing		
1 oven	Ea.	595 - 2275
2 oven	Ea.	1450 - 3150
30" built-in		
1 oven	Ea.	1075 - 2225
2 oven	Ea.	1725 - 3275
Counter top cook tops, 4 burner	Ea.	480 - 2075
Microwave oven	Ea.	275 - 850
Combination range, refrig. & sink, 30" wide	Ea.	1575 - 2100
60" wide	Ea.	2100
72" wide	Ea.	2250
Combination range, refrigerator, sink, microwave oven & icemaker	Ea.	5900
Compactor, residential, 4-1 compaction	Ea.	860 - 1450
Dishwasher, built-in, 2 cycles	Ea.	510 - 855
4 cycles	Ea.	650 - 1675
Garbage disposer, sink type	Ea.	174 - 340
Hood for range, 2 speed, vented, 30" wide	Ea.	222 - 1150
42" wide	Ea.	310 - 2400

Description	Unit	$ Cost
Appliances, cont.		
Refrigerator, no frost 10-12 C.F.	Ea.	600 - 745
14-16 C.F.	Ea.	720 - 870
18-20 C.F.	Ea.	910 - 2075
Laundry Equipment		
Dryer, gas, 16 lb. capacity	Ea.	950
30 lb. capacity	Ea.	3900
Washer, 4 cycle	Ea.	1175
Commercial	Ea.	1575
Sound System		
Amplifier, 250 watts	Ea.	2125
Speaker, ceiling or wall	Ea.	206
Trumpet	Ea.	395

Model costs calculated for a 1 story building with 10' story height and 10,000 square feet of floor area

Assisted - Senior Living

				Unit	Unit Cost	Cost Per S.F.	% Of Sub-Total
A. SUBSTRUCTURE							
1010	Standard Foundations	Poured concrete strip footings		S.F. Ground	2.40	2.40	9.4%
1020	Special Foundations	N/A		—	—	—	
1030	Slab on Grade	4" reinforced concrete with vapor barrier and granular base		S.F. Slab	5.10	5.10	
2010	Basement Excavation	Site preparation for slab and trench for foundation wall and footing		S.F. Ground	.51	.51	
2020	Basement Walls	4' foundation wall		L.F. Wall	74	2.94	
B. SHELL							
	B10 Superstructure						
1010	Floor Construction	Steel column		S.F. Floor	.57	.57	6.5%
1020	Roof Construction	Plywood on wood trusses (pitched); fiberglass batt insulation		S.F. Roof	7.02	7.02	
	B20 Exterior Enclosure						
2010	Exterior Walls	Brick veneer on wood studs, insulated	*80% of wall*	S.F. Wall	22	7.15	10.0%
2020	Exterior Windows	Double hung wood	*20% of wall*	Each	603	3.02	
2030	Exterior Doors	Aluminum and glass; 18 ga. steel flush		Each	2413	1.45	
	B30 Roofing						
3010	Roof Coverings	Asphalt shingles with flashing (pitched); al. gutters & downspouts		S.F. Roof	2.73	2.73	2.3%
3020	Roof Openings	N/A		—	—	—	
C. INTERIORS							
1010	Partitions	Gypsum board on wood studs	*6 S.F. Floor/L.F. Partition*	S.F. Partition	4.93	6.57	26.3%
1020	Interior Doors	Single leaf wood	*80 S.F. Floor/Door*	Each	593	6.97	
1030	Fittings	N/A		—	—	—	
2010	Stair Construction	N/A		—	—	—	
3010	Wall Finishes	Paint on drywall		S.F. Surface	.76	2.03	
3020	Floor Finishes	80% carpet, 20% ceramic tile		S.F. Floor	6.82	6.82	
3030	Ceiling Finishes	Gypsum board on wood furring, painted		S.F. Ceiling	8.22	8.22	
D. SERVICES							
	D10 Conveying						
1010	Elevators & Lifts	N/A		—	—	—	0.0 %
1020	Escalators & Moving Walks	N/A		—	—	—	
	D20 Plumbing						
2010	Plumbing Fixtures	Kitchen, toilet and service fixtures, supply and drainage	*1 Fixture/200 S.F. Floor*	Each	2698	13.49	13.5%
2020	Domestic Water Distribution	Gas water heater		S.F. Floor	2.23	2.23	
2040	Rain Water Drainage	N/A		—	—	—	
	D30 HVAC						
3010	Energy Supply	N/A		—	—	—	3.6 %
3020	Heat Generating Systems	N/A		—	—	—	
3030	Cooling Generating Systems	N/A		—	—	—	
3050	Terminal & Package Units	Gas fired hot water boiler; terminal package heat/A/C units		S.F. Floor	4.22	4.22	
3090	Other HVAC Sys. & Equipment	N/A		—	—	—	
	D40 Fire Protection						
4010	Sprinklers	Wet pipe sprinkler system		S.F. Floor	3.28	3.28	3.6%
4020	Standpipes	Standpipes		S.F. Floor	.86	.86	
	D50 Electrical						
5010	Electrical Service/Distribution	600 ampere service, panel board, feeders and branch wiring		S.F. Floor	8.34	8.34	23.4%
5020	Lighting & Branch Wiring	Fuorescent & incandescent fixtures, receptacles, switches, A.C. and misc. power		S.F. Floor	10.76	10.76	
5030	Communications & Security	Addressable alarm, fire detection and command center, communication system		S.F. Floor	8.02	8.02	
5090	Other Electrical Systems	Emergency generator, 7.5 kW		S.F. Floor	.22	.22	
E. EQUIPMENT & FURNISHINGS							
1010	Commercial Equipment	Residential washers and dryers		S.F. Floor	.43	.43	1.4%
1020	Institutional Equipment	N/A		—	—	—	
1030	Vehicular Equipment	N/A		—	—	—	
1090	Other Equipment	Residential ranges & refrigerators		S.F. Floor	1.24	1.24	
F. SPECIAL CONSTRUCTION							
1020	Integrated Construction	N/A		—	—	—	0.0 %
1040	Special Facilities	N/A		—	—	—	
G. BUILDING SITEWORK	**N/A**						
					Sub-Total	116.59	**100%**
	CONTRACTOR FEES (General Requirements: 10%, Overhead: 5%, Profit: 10%)				*25%*	29.13	
	ARCHITECT FEES				*11%*	16.03	
					Total Building Cost	**161.75**	

Costs per square foot of floor area

Exterior Wall	S.F. Area	2000	2700	3400	4100	4800	5500	6200	6900	7600
	L.F. Perimeter	180	208	236	256	280	303	317	337	357
Face Brick with Concrete Block Back-up	Steel Frame	249.30	234.80	226.30	218.85	214.40	210.85	206.85	204.45	202.45
	R/Conc. Frame	262.60	248.10	239.55	232.20	227.75	224.20	220.20	217.75	215.75
Precast Concrete Panel	Steel Frame	257.85	242.00	232.75	224.65	219.80	215.95	211.55	208.85	206.70
	R/Conc. Frame	275.75	257.85	247.35	238.45	233.00	228.70	224.00	221.05	218.70
Limestone with Concrete Block Back-up	Steel Frame	280.25	261.30	250.10	240.35	234.45	229.80	224.40	221.20	218.60
	R/Conc. Frame	293.55	274.60	263.45	253.65	247.80	243.15	237.75	234.55	231.95
Perimeter Adj., Add or Deduct	Per 100 L.F.	44.80	33.15	26.25	21.85	18.65	16.30	14.45	12.95	11.80
Story Hgt. Adj., Add or Deduct	Per 1 Ft.	4.25	3.60	3.25	2.95	2.75	2.65	2.40	2.30	2.20
***For Basement**, add $ 30.80 per square foot of basement area*										

The above costs were calculated using the basic specifications shown on the facing page. These costs should be adjusted where necessary for design alternatives and owner's requirements. Reported completed project costs, for this type of structure, range from $ 147.10 to $ 362.05 per S.F.

Common additives

Description	Unit	$ Cost
Bulletproof Teller Window, 44" x 60"	Ea.	5475
60" x 48"	Ea.	7300
Closed Circuit Surveillance, One station		
Camera and monitor	Ea.	1975
For additional camera stations, add	Ea.	1075
Counters, Complete	Station	6000
Door & Frame, 3' x 6'-8", bullet resistant steel		
with vision panel	Ea.	7200 - 9375
Drive-up Window, Drawer & micr., not incl. glass	Ea.	8900 - 12,500
Emergency Lighting, 25 watt, battery operated		
Lead battery	Ea.	325
Nickel cadmium	Ea.	665
Night Depository	Ea.	9475 - 14,200
Package Receiver, painted	Ea.	1900
stainless steel	Ea.	2950
Partitions, Bullet resistant to 8' high	L.F.	320 - 520
Pneumatic Tube Systems, 2 station	Ea.	31,400
With TV viewer	Ea.	59,000

Description	Unit	$ Cost
Service Windows, Pass thru, steel		
24" x 36"	Ea.	3800
48" x 48"	Ea.	3950
72" x 40"	Ea.	6275
Smoke Detectors		
Ceiling type	Ea.	214
Duct type	Ea.	540
Twenty-four Hour Teller		
Automatic deposit cash & memo	Ea.	52,500
Vault Front, Door & frame		
1 hour test, 32"x 78"	Opng.	8075
2 hour test, 32" door	Opng.	9900
40" door	Opng.	11,400
4 hour test, 32" door	Opng.	10,500
40" door	Opng.	12,300
Time lock, two movement, add	Ea.	2275

Model costs calculated for a 1 story building with 14′ story height and 4,100 square feet of floor area

Bank

			Unit	Unit Cost	Cost Per S.F.	% Of Sub-Total
A. SUBSTRUCTURE						
1010	Standard Foundations	Poured concrete; strip and spread footings	S.F. Ground	3.94	3.94	8.6%
1020	Special Foundations	N/A	—	—	—	
1030	Slab on Grade	4″ reinforced concrete with vapor barrier and granular base	S.F. Slab	5.10	5.10	
2010	Basement Excavation	Site preparation for slab and trench for foundation wall and footing	S.F. Ground	.29	.29	
2020	Basement Walls	4′ Foundation wall	L.F. Wall	74	5.51	
B. SHELL						
	B10 Superstructure					
1010	Floor Construction	Cast-in-place columns	L.F. Column	108	5.27	11.5%
1020	Roof Construction	Cast-in-place concrete flat plate	S.F. Roof	14.50	14.50	
	B20 Exterior Enclosure					
2010	Exterior Walls	Face brick with concrete block backup *80% of wall*	S.F. Wall	30	21.05	16.9%
2020	Exterior Windows	Horizontal aluminum sliding *20% of wall*	Each	555	6.47	
2030	Exterior Doors	Double aluminum and glass and hollow metal	Each	3228	1.57	
	B30 Roofing					
3010	Roof Coverings	Built-up tar and gravel with flashing; perlite/EPS composite insulation	S.F. Roof	7.32	7.32	4.3%
3020	Roof Openings	N/A	—	—	—	
C. INTERIORS						
1010	Partitions	Gypsum board on metal studs *20 S.F. of Floor/L.F. Partition*	S.F. Partition	10.04	5.02	13.8%
1020	Interior Doors	Single leaf hollow metal *200 S.F. Floor/Door*	Each	1133	5.67	
1030	Fittings	N/A	—	—	—	
2010	Stair Construction	N/A	—	—	—	
3010	Wall Finishes	50% vinyl wall covering, 50% paint	S.F. Surface	1.33	1.33	
3020	Floor Finishes	50% carpet tile, 40% vinyl composition tile, 10% quarry tile	S.F. Floor	5.22	5.22	
3030	Ceiling Finishes	Mineral fiber tile on concealed zee bars	S.F. Ceiling	6.57	6.57	
D. SERVICES						
	D10 Conveying					
1010	Elevators & Lifts	N/A	—	—	—	0.0 %
1020	Escalators & Moving Walks	N/A	—	—	—	
	D20 Plumbing					
2010	Plumbing Fixtures	Toilet and service fixtures, supply and drainage *1 Fixture/580 S.F. Floor*	Each	5150	8.88	6.9%
2020	Domestic Water Distribution	Gas fired water heater	S.F. Floor	1.53	1.53	
2040	Rain Water Drainage	Roof drains	S.F. Roof	1.39	1.39	
	D30 HVAC					
3010	Energy Supply	N/A	—	—	—	7.3 %
3020	Heat Generating Systems	Included in D3050	—	—	—	
3030	Cooling Generating Systems	N/A	—	—	—	
3050	Terminal & Package Units	Single zone rooftop unit, gas heating, electric cooling	S.F. Floor	12.60	12.60	
3090	Other HVAC Sys. & Equipment	N/A	—	—	—	
	D40 Fire Protection					
4010	Sprinklers	Wet pipe sprinkler system	S.F. Floor	4.57	4.57	4.3%
4020	Standpipes	Standpipe	S.F. Floor	2.91	2.91	
	D50 Electrical					
5010	Electrical Service/Distribution	200 ampere service, panel board and feeders	S.F. Floor	2.64	2.64	13.0%
5020	Lighting & Branch Wiring	Fluorescent fixtures, receptacles, switches, A.C. and misc. power	S.F. Floor	9	9	
5030	Communications & Security	Alarm systems, internet and phone wiring, emergency lighting, and security television	S.F. Floor	10.09	10.09	
5090	Other Electrical Systems	Emergency generator, 15 kW, Uninterruptible power supply	S.F. Floor	.66	.66	
E. EQUIPMENT & FURNISHINGS						
1010	Commercial Equipment	Automatic teller, drive up window, night depository	S.F. Floor	7.81	7.81	6.2%
1020	Institutional Equipment	Closed circuit TV monitoring system	S.F. Floor	2.89	2.89	
1030	Vehicular Equipment	N/A	—	—	—	
1090	Other Equipment	N/A	—	—	—	
F. SPECIAL CONSTRUCTION						
1020	Integrated Construction	N/A	—	—	—	7.1 %
1040	Special Facilities	Security vault door	S.F. Floor	12.19	12.19	
G. BUILDING SITEWORK	**N/A**					
			Sub-Total		171.99	**100%**
	CONTRACTOR FEES (General Requirements: 10%, Overhead: 5%, Profit: 10%)			*25%*	43.01	
	ARCHITECT FEES			*8%*	17.20	
			Total Building Cost		**232.20**	

Bowling Alley

Costs per square foot of floor area

Exterior Wall	S.F. Area	12000	14000	16000	18000	20000	22000	24000	26000	28000
	L.F. Perimeter	460	491	520	540	566	593	620	645	670
Concrete Block	Steel Roof Deck	165.10	162.15	159.85	157.80	156.30	155.10	154.15	153.20	152.50
Decorative Concrete Block	Steel Roof Deck	169.35	166.10	163.50	161.15	159.45	158.15	157.00	156.00	155.15
Face Brick on Concrete Block	Steel Roof Deck	176.55	172.65	169.55	166.75	164.75	163.15	161.80	160.65	159.60
Jumbo Brick on Concrete Block	Steel Roof Deck	173.50	169.85	167.05	164.45	162.55	161.05	159.80	158.70	157.75
Stucco on Concrete Block	Steel Roof Deck	168.20	164.95	162.45	160.20	158.55	157.30	156.20	155.20	154.40
Precast Concrete Panel	Steel Roof Deck	173.80	170.10	167.20	164.60	162.70	161.20	160.00	158.85	157.90
Perimeter Adj., Add or Deduct	Per 100 L.F.	4.90	4.25	3.70	3.25	3.00	2.70	2.45	2.35	2.10
Story Hgt. Adj., Add or Deduct	Per 1 Ft.	1.10	1.00	0.95	0.90	0.80	0.75	0.70	0.75	0.65
***For Basement**, add $30.80 per square foot of basement area*										

The above costs were calculated using the basic specifications shown on the facing page. These costs should be adjusted where necessary for design alternatives and owner's requirements. Reported completed project costs, for this type of structure, range from $65.05 to $171.35 per S.F.

Common additives

Description	Unit	$ Cost
Bowling Alleys, Incl. alley, pinsetter,		
scorer, counter & misc. supplies, average	Lane	61,500
For automatic scorer, add (maximum)	Lane	11,000
Emergency Lighting, 25 watt, battery operated		
Lead battery	Ea.	325
Nickel cadmium	Ea.	665
Lockers, Steel, Single tier, 60" or 72"	Opng.	222 - 365
2 tier, 60" or 72" total	Opng.	115 - 164
5 tier, box lockers	Opng.	69.50 - 78.50
Locker bench, lam., maple top only	L.F.	33
Pedestals, steel pipe	Ea.	71
Seating		
Auditorium chair, all veneer	Ea.	305
Veneer back, padded seat	Ea.	315
Upholstered, spring seat	Ea.	310
Sound System		
Amplifier, 250 watts	Ea.	2125
Speaker, ceiling or wall	Ea.	206
Trumpet	Ea.	395

Important: See the Reference Section for Location Factors.

Model costs calculated for a 1 story building with 14' story height and 20,000 square feet of floor area

Bowling Alley

			Unit	Unit Cost	Cost Per S.F.	% Of Sub-Total
A. SUBSTRUCTURE						
1010	Standard Foundations	Poured concrete; strip and spread footings	S.F. Ground	1.16	1.16	7.5%
1020	Special Foundations	N/A	—	—	—	
1030	Slab on Grade	4" reinforced concrete with vapor barrier and granular base	S.F. Slab	5.10	5.10	
2010	Basement Excavation	Site preparation for slab and trench for foundation wall and footing	S.F. Ground	.51	.51	
2020	Basement Walls	4' Foundation wall	L.F. Wall	74	2.08	
B. SHELL						
	B10 Superstructure					
1010	Floor Construction	N/A	—	—	—	8.0 %
1020	Roof Construction	Metal deck on open web steel joists with columns and beams	S.F. Roof	9.39	9.39	
	B20 Exterior Enclosure					
2010	Exterior Walls	Concrete block *90% of wall*	S.F. Wall	12.26	4.37	7.0%
2020	Exterior Windows	Horizontal pivoted *10% of wall*	Each	1985	3.27	
2030	Exterior Doors	Double aluminum and glass and hollow metal	Each	2419	.61	
	B30 Roofing					
3010	Roof Coverings	Built-up tar and gravel with flashing; perlite/EPS composite insulation	S.F. Roof	5.61	5.61	4.9%
3020	Roof Openings	Roof hatches	S.F. Roof	.15	.15	
C. INTERIORS						
1010	Partitions	Concrete block *50 S.F. Floor/L.F. Partition*	S.F. Partition	9.80	1.96	6.2%
1020	Interior Doors	Hollow metal single leaf *1000 S.F. Floor/Door*	Each	1133	1.14	
1030	Fittings	N/A	—	—	—	
2010	Stair Construction	N/A	—	—	—	
3010	Wall Finishes	Paint and block filler	S.F. Surface	5.58	2.23	
3020	Floor Finishes	Vinyl tile *25% of floor*	S.F. Floor	2.36	.59	
3030	Ceiling Finishes	Suspended fiberglass board *25% of area*	S.F. Ceiling	5.51	1.38	
D. SERVICES						
	D10 Conveying					
1010	Elevators & Lifts	N/A	—	—	—	0.0 %
1020	Escalators & Moving Walks	N/A	—	—	—	
	D20 Plumbing					
2010	Plumbing Fixtures	Toilet and service fixtures, supply and drainage *1 Fixture/2200 S.F. Floor*	Each	3256	1.48	2.0%
2020	Domestic Water Distribution	Gas fired water heater	S.F. Floor	.28	.28	
2040	Rain Water Drainage	Roof drains	S.F. Roof	.61	.61	
	D30 HVAC					
3010	Energy Supply	N/A	—	—	—	12.0 %
3020	Heat Generating Systems	Included in D3050	—	—	—	
3030	Cooling Generating Systems	N/A	—	—	—	
3050	Terminal & Package Units	Single zone rooftop unit, gas heating, electric cooling	S.F. Floor	14.10	14.10	
3090	Other HVAC Sys. & Equipment	N/A	—	—	—	
	D40 Fire Protection					
4010	Sprinklers	Sprinklers, light hazard	S.F. Floor	3.28	3.28	3.3%
4020	Standpipes	Standpipe, wet, Class III	S.F. Floor	.60	.60	
	D50 Electrical					
5010	Electrical Service/Distribution	800 ampere service, panel board and feeders	S.F. Floor	2.19	2.19	11.9%
5020	Lighting & Branch Wiring	High efficiency fluorescent fixtures, receptacles, switches, A.C. and misc. power	S.F. Floor	10.45	10.45	
5030	Communications & Security	Addressable alarm systems and emergency lighting	S.F. Floor	1.29	1.29	
5090	Other Electrical Systems	Emergency generator, 7.5 kW	S.F. Floor	.08	.08	
E. EQUIPMENT & FURNISHINGS						
1010	Commercial Equipment	N/A	—	—	—	37.3 %
1020	Institutional Equipment	N/A	—	—	—	
1030	Vehicular Equipment	N/A	—	—	—	
1090	Other Equipment	Bowling Alley, ballrack, automatic scorers	S.F. Floor	44	44.04	
F. SPECIAL CONSTRUCTION						
1020	Integrated Construction	N/A	—	—	—	0.0 %
1040	Special Facilities	N/A	—	—	—	
G. BUILDING SITEWORK		**N/A**				
				Sub-Total	117.95	**100%**
	CONTRACTOR FEES (General Requirements: 10%, Overhead: 5%, Profit: 10%)			*25%*	29.50	
	ARCHITECT FEES			*6%*	8.85	
				Total Building Cost	**156.30**	

Costs per square foot of floor area

Exterior Wall	S.F. Area	6000	8000	10000	12000	14000	16000	18000	20000	22000
	L.F. Perimeter	320	386	453	475	505	530	570	600	630
Face Brick with Concrete Block Back-up	Bearing Walls	169.25	163.45	160.00	154.75	151.50	148.80	147.30	145.80	144.50
	Steel Frame	172.25	166.55	163.15	157.95	154.75	152.10	150.60	149.15	147.85
Decorative Concrete Block	Bearing Walls	161.55	156.50	153.45	149.05	146.30	144.00	142.75	141.45	140.35
	Steel Frame	164.55	159.55	156.55	152.25	149.50	147.30	146.05	144.80	143.70
Precast Concrete Panels	Bearing Walls	169.70	163.90	160.40	155.15	151.85	149.15	147.60	146.10	144.75
	Steel Frame	172.75	167.00	163.55	158.35	155.05	152.40	150.95	149.40	148.10
Perimeter Adj., Add or Deduct	Per 100 L.F.	13.05	9.80	7.85	6.60	5.55	4.90	4.45	3.95	3.60
Story Hgt. Adj., Add or Deduct	Per 1 Ft.	2.10	2.00	1.85	1.60	1.45	1.35	1.35	1.20	1.20
Basement—Not Applicable										

The above costs were calculated using the basic specifications shown on the facing page. These costs should be adjusted where necessary for design alternatives and owner's requirements. Reported completed project costs, for this type of structure, range from $78.60 to $189.50 per S.F.

Common additives

Description	Unit	$ Cost
Directory Boards, Plastic, glass covered		
30" x 20"	Ea.	555
36" x 48"	Ea.	1425
Aluminum, 24" x 18"	Ea.	600
36" x 24"	Ea.	715
48" x 32"	Ea.	1000
48" x 60"	Ea.	2100
Emergency Lighting, 25 watt, battery operated		
Lead battery	Ea.	325
Nickel cadmium	Ea.	665
Benches, Hardwood	L.F.	137 - 228
Ticket Printer	Ea.	8050
Turnstiles, One way		
4 arm, 46" dia., manual	Ea.	2500
Electric	Ea.	3375
High security, 3 arm		
65" dia., manual	Ea.	9450
Electric	Ea.	11,900
Gate with horizontal bars		
65" dia., 7' high transit type	Ea.	7250

Model costs calculated for a 1 story building with 14′ story height and 12,000 square feet of floor area

Bus Terminal

				Unit	Unit Cost	Cost Per S.F.	% Of Sub-Total
A. SUBSTRUCTURE							
1010	Standard Foundations	Poured concrete; strip and spread footings		S.F. Ground	2.24	2.24	11.3%
1020	Special Foundations	N/A		—	—	—	
1030	Slab on Grade	6″ reinforced concrete with vapor barrier and granular base		S.F. Slab	7.70	7.70	
2010	Basement Excavation	Site preparation for slab and trench for foundation wall and footing		S.F. Ground	.29	.29	
2020	Basement Walls	4′ Foundation wall		L.F. Wall	76	2.99	
B. SHELL							
	B10 Superstructure						
1010	Floor Construction	N/A		—	—	—	5.5 %
1020	Roof Construction	Metal deck on open web steel joists with columns and beams		S.F. Roof	6.45	6.45	
	B20 Exterior Enclosure						
2010	Exterior Walls	Face brick with concrete block backup	*70% of wall*	S.F. Wall	30	11.67	15.9%
2020	Exterior Windows	Store front	*30% of wall*	Each	30	5.05	
2030	Exterior Doors	Double aluminum and glass		Each	5525	1.84	
	B30 Roofing						
3010	Roof Coverings	Built-up tar and gravel with flashing; perlite/EPS composite insulation		S.F. Roof	6.38	6.38	5.5%
3020	Roof Openings	N/A		—	—	—	
C. INTERIORS							
1010	Partitions	Lightweight concrete block	*15 S.F. Floor/L.F. Partition*	S.F. Partition	9.78	6.52	27.5%
1020	Interior Doors	Hollow metal	*150 S.F. Floor/Door*	Each	1133	7.55	
1030	Fittings	N/A		—	—	—	
2010	Stair Construction	N/A		—	—	—	
3010	Wall Finishes	Glazed coating		S.F. Surface	1.79	2.39	
3020	Floor Finishes	Quarry tile and vinyl composition tile		S.F. Floor	9.07	9.07	
3030	Ceiling Finishes	Mineral fiber tile on concealed zee bars		S.F. Ceiling	6.57	6.57	
D. SERVICES							
	D10 Conveying						
1010	Elevators & Lifts	N/A		—	—	—	0.0 %
1020	Escalators & Moving Walks	N/A		—	—	—	
	D20 Plumbing						
2010	Plumbing Fixtures	Toilet and service fixtures, supply and drainage	*1 Fixture/850 S.F. Floor*	Each	6715	7.90	9.5%
2020	Domestic Water Distribution	Electric water heater		S.F. Floor	2.10	2.10	
2040	Rain Water Drainage	Roof drains		S.F. Roof	1.07	1.07	
	D30 HVAC						
3010	Energy Supply	N/A		—	—	—	9.2 %
3020	Heat Generating Systems	Included in D3050		—	—	—	
3030	Cooling Generating Systems	N/A		—	—	—	
3050	Terminal & Package Units	Single zone rooftop unit, gas heating, electric cooling		S.F. Floor	10.80	10.80	
3090	Other HVAC Sys. & Equipment	N/A		—	—	—	
	D40 Fire Protection						
4010	Sprinklers	Wet pipe sprinkler system		S.F. Floor	4.22	4.22	4.5%
4020	Standpipes	Standpipe		S.F. Floor	.99	.99	
	D50 Electrical						
5010	Electrical Service/Distribution	400 ampere service, panel board and feeders		S.F. Floor	1.79	1.79	11.1%
5020	Lighting & Branch Wiring	Fluorescent fixtures, receptacles, switches, A.C. and misc. power		S.F. Floor	8.81	8.81	
5030	Communications & Security	Alarm systems and emergency lighting		S.F. Floor	1.99	1.99	
5090	Other Electrical Systems	Emergency generator, 7.5 kW		S.F. Floor	.43	.43	
E. EQUIPMENT & FURNISHINGS							
1010	Commercial Equipment	N/A		—	—	—	0.0 %
1020	Institutional Equipment	N/A		—	—	—	
1030	Vehicular Equipment	N/A		—	—	—	
1090	Other Equipment	N/A		—	—	—	
F. SPECIAL CONSTRUCTION							
1020	Integrated Construction	N/A		—	—	—	0.0 %
1040	Special Facilities	N/A		—	—	—	
G. BUILDING SITEWORK		**N/A**					
					Sub-Total	116.81	**100%**
	CONTRACTOR FEES (General Requirements: 10%, Overhead: 5%, Profit: 10%)				*25%*	29.18	
	ARCHITECT FEES				*6%*	8.76	
					Total Building Cost	**154.75**	

Costs per square foot of floor area

Exterior Wall	S.F. Area	600	800	1000	1200	1600	2000	2400	3000	4000
	L.F. Perimeter	100	114	128	139	164	189	210	235	280
Brick with Concrete Block Back-up	Steel Frame	325.40	312.10	304.10	297.45	290.10	285.70	281.80	276.80	272.25
	Bearing Walls	318.70	305.40	297.40	290.75	283.40	279.00	275.10	270.10	265.50
Concrete Block	Steel Frame	285.30	277.70	273.15	269.30	265.15	262.60	260.40	257.55	254.95
	Bearing Walls	279.40	271.80	267.20	263.40	259.25	256.70	254.50	251.65	249.00
Galvanized Steel Siding	Steel Frame	289.30	281.05	276.15	272.05	267.55	264.90	262.50	259.40	256.55
Metal Sandwich Panel	Steel Frame	285.10	276.65	271.60	267.35	262.70	259.85	257.45	254.25	251.30
Perimeter Adj., Add or Deduct	Per 100 L.F.	91.65	68.80	55.00	45.80	34.35	27.50	22.95	18.35	13.70
Story Hgt. Adj., Add or Deduct	Per 1 Ft.	5.65	4.85	4.35	3.95	3.50	3.20	3.05	2.70	2.35
Basement—Not Applicable										

The above costs were calculated using the basic specifications shown on the facing page. These costs should be adjusted where necessary for design alternatives and owner's requirements. Reported completed project costs, for this type of structure, range from $ 141.05 to $ 400.70 per S.F.

Common additives

Description	Unit	$ Cost
Air Compressors, Electric		
1-1/2 H.P., standard controls	Ea.	1100
Dual controls	Ea.	1425
5 H.P., 115/230 volt, standard control	Ea.	2925
Dual controls	Ea.	4000
Emergency Lighting, 25 watt, battery operated		
Lead battery	Ea.	325
Nickel cadmium	Ea.	665
Fence, Chain link, 6' high		
9 ga. wire, galvanized	L.F.	28
6 ga. wire	L.F.	31.50
Gate	Ea.	380
Product Dispenser with vapor recovery for 6 nozzles	Ea.	27,200
Laundry Equipment		
Dryer, gas, 16 lb. capacity	Ea.	950
30 lb. capacity	Ea.	3900
Washer, 4 cycle	Ea.	1175
Commercial	Ea.	1575

Description	Unit	$ Cost
Lockers, Steel, single tier, 60" or 72"	Opng.	222 - 365
2 tier, 60" or 72" total	Opng.	115 - 164
5 tier, box lockers	Opng.	69.50 - 78.50
Locker bench, lam. maple top only	L.F.	33
Pedestals, steel pipe	Ea.	71
Paving, Bituminous		
Wearing course plus base course	S.Y.	6.60
Safe, Office type, 1 hour rating		
30" x 18" x 18"	Ea.	2500
60" x 36" x 18", double door	Ea.	9350
Sidewalks, Concrete 4" thick	S.F.	4.23
Yard Lighting, 20' aluminum pole with 400 watt high pressure sodium fixture	Ea.	3050

Important: See the Reference Section for Location Factors.

Model costs calculated for a 1 story building with 12′ story height and 800 square feet of floor area

Car Wash

Code	Item	Description		Unit	Unit Cost	Cost Per S.F.	% Of Sub-Total
A. SUBSTRUCTURE							
1010	Standard Foundations	Poured concrete; strip and spread footings		S.F. Ground	2.98	2.98	8.5%
1020	Special Foundations	N/A		—	—	—	
1030	Slab on Grade	5″ reinforced concrete with vapor barrier and granular base		S.F. Slab	6.78	6.78	
2010	Basement Excavation	Site preparation for slab and trench for foundation wall and footing		S.F. Ground	.96	.96	
2020	Basement Walls	4′ Foundation wall		L.F. Wall	65	9.19	
B. SHELL							
	B10 Superstructure						
1010	Floor Construction	N/A		—	—	—	3.7 %
1020	Roof Construction	Metal deck, open web steel joists, beams, columns		S.F. Roof	8.70	8.70	
	B20 Exterior Enclosure						
2010	Exterior Walls	Face brick with concrete block backup	*70% of wall*	S.F. Wall	30	36.03	26.6%
2020	Exterior Windows	Horizontal pivoted steel	*5% of wall*	Each	780	7.41	
2030	Exterior Doors	Steel overhead hollow metal		Each	3715	18.58	
	B30 Roofing						
3010	Roof Coverings	Built-up tar and gravel with flashing; perlite/EPS composite insulation		S.F. Roof	8.37	8.37	3.6%
3020	Roof Openings	N/A		—	—	—	
C. INTERIORS							
1010	Partitions	Concrete block	*20 S.F. Floor/S.F. Partition*	S.F. Partition	9.14	4.57	3.3%
1020	Interior Doors	Hollow metal	*600 S.F. Floor/Door*	Each	1133	1.89	
1030	Fittings	N/A		—	—	—	
2010	Stair Construction	N/A		—	—	—	
3010	Wall Finishes	Paint		S.F. Surface	1.19	1.19	
3020	Floor Finishes	N/A		—	—	—	
3030	Ceiling Finishes	N/A		—	—	—	
D. SERVICES							
	D10 Conveying						
1010	Elevators & Lifts	N/A		—	—	—	0.0 %
1020	Escalators & Moving Walks	N/A		—	—	—	
	D20 Plumbing						
2010	Plumbing Fixtures	Toilet and service fixtures, supply and drainage	*1 Fixture/160 S.F. Floor*	Each	2566	16.04	23.5%
2020	Domestic Water Distribution	Gas fired water heater		S.F. Floor	35	34.64	
2040	Rain Water Drainage	Roof drains		S.F. Roof	4.11	4.11	
	D30 HVAC						
3010	Energy Supply	N/A		—	—	—	3.5 %
3020	Heat Generating Systems	N/A		—	—	—	
3030	Cooling Generating Systems	N/A		—	—	—	
3050	Terminal & Package Units	Single zone rooftop unit, gas heating, electric cooling		S.F. Floor	8.15	8.15	
3090	Other HVAC Sys. & Equipment	N/A		—	—	—	
	D40 Fire Protection						
4010	Sprinklers	N/A		—	—	—	0.0 %
4020	Standpipes	N/A		—	—	—	
	D50 Electrical						
5010	Electrical Service/Distribution	400 ampere service, panel board and feeders		S.F. Floor	24	24.25	27.3%
5020	Lighting & Branch Wiring	Fluorescent fixtures, receptacles, switches and misc. power		S.F. Floor	38	37.62	
5030	Communications & Security	N/A		—	—	—	
5090	Other Electrical Systems	Emergency generator, 7.5 kW		S.F. Floor	1.88	1.88	
E. EQUIPMENT & FURNISHINGS							
1010	Commercial Equipment	N/A		—	—	—	0.0 %
1020	Institutional Equipment	N/A		—	—	—	
1030	Vehicular Equipment	N/A		—	—	—	
1090	Other Equipment	N/A		—	—	—	
F. SPECIAL CONSTRUCTION							
1020	Integrated Construction	N/A		—	—	—	0.0 %
1040	Special Facilities	N/A		—	—	—	
G. BUILDING SITEWORK		**N/A**					
					Sub-Total	233.34	**100%**
	CONTRACTOR FEES (General Requirements: 10%, Overhead: 5%, Profit: 10%)				*25%*	58.34	
	ARCHITECT FEES				*7%*	20.42	
					Total Building Cost	**312.10**	

Costs per square foot of floor area

Exterior Wall	S.F. Area	2000	7000	12000	17000	22000	27000	32000	37000	42000
	L.F. Perimeter	180	340	470	540	640	740	762	793	860
Decorative Concrete Block	Wood Arch	244.95	180.75	166.85	156.75	152.95	150.55	145.85	142.75	141.45
	Steel Truss	245.85	181.65	167.75	157.65	153.85	151.45	146.75	143.65	142.35
Stone with Concrete Block Back-up	Wood Arch	302.65	211.90	191.95	177.10	171.60	168.10	161.15	156.45	154.55
	Steel Truss	296.90	206.15	186.25	171.40	165.90	162.45	155.40	150.75	148.85
Face Brick with Concrete Block Back-up	Wood Arch	271.35	195.00	178.35	166.05	161.50	158.60	152.85	149.00	147.45
	Steel Truss	265.60	189.30	172.65	160.30	155.80	152.90	147.10	143.30	141.75
Perimeter Adj., Add or Deduct	Per 100 L.F.	62.55	17.80	10.45	7.40	5.70	4.65	3.90	3.35	2.95
Story Hgt. Adj., Add or Deduct	Per 1 Ft.	3.65	1.95	1.60	1.30	1.20	1.15	1.00	0.80	0.80
***For Basement**, add $ 31.60 per square foot of basement area*										

The above costs were calculated using the basic specifications shown on the facing page. These costs should be adjusted where necessary for design alternatives and owner's requirements. Reported completed project costs, for this type of structure, range from $ 90.05 to $ 368.20 per S.F.

Common additives

Description	Unit	$ Cost
Altar, Wood, custom design, plain	Ea.	3225
Deluxe	Ea.	16,300
Granite or marble, average	Ea.	16,800
Deluxe	Ea.	45,300
Ark, Prefabricated, plain	Ea.	11,600
Deluxe	Ea.	151,500
Baptistry, Fiberglass, incl. plumbing	Ea.	7350 - 12,100
Bells & Carillons, 48 bells	Ea.	1,095,500
24 bells	Ea.	515,500
Confessional, Prefabricated wood		
Single, plain	Ea.	4425
Deluxe	Ea.	10,800
Double, plain	Ea.	8275
Deluxe	Ea.	23,400
Emergency Lighting, 25 watt, battery operated		
Lead battery	Ea.	325
Nickel cadmium	Ea.	665
Lecterns, Wood, plain	Ea.	960
Deluxe	Ea.	6675

Description	Unit	$ Cost
Pews/Benches, Hardwood	L.F.	137 - 228
Pulpits, Prefabricated, hardwood	Ea.	1975 - 11,700
Railing, Hardwood	L.F.	255
Steeples, translucent fiberglass		
30" square, 15' high	Ea.	10,900
25' high	Ea.	12,300
Painted fiberglass, 24" square, 14' high	Ea.	9150
28' high	Ea.	8275
Aluminum		
20' high, 3'- 6" base	Ea.	11,000
35' high, 8'- 0" base	Ea.	40,400
60' high, 14'- 0" base	Ea.	80,000

Important: See the Reference Section for Location Factors.

Model costs calculated for a 1 story building with 24′ story height and 17,000 square feet of floor area

Church

			Unit	Unit Cost	Cost Per S.F.	% Of Sub-Total
A. SUBSTRUCTURE						
1010	Standard Foundations	Poured concrete; strip and spread footings	S.F. Ground	3.72	3.72	10.3%
1020	Special Foundations	N/A	—	—	—	
1030	Slab on Grade	4″ reinforced concrete with vapor barrier and granular base	S.F. Slab	5.10	5.10	
2010	Basement Excavation	Site preparation for slab and trench for foundation wall and footing	S.F. Ground	.16	.16	
2020	Basement Walls	4′ Foundation wall	L.F. Wall	74	3.50	
B. SHELL						
	B10 Superstructure					
1010	Floor Construction	N/A	—	—	—	16.2 %
1020	Roof Construction	Wood deck on laminated wood arches	S.F. Roof	18.35	19.58	
	B20 Exterior Enclosure					
2010	Exterior Walls	Face brick with concrete block backup *80% of wall (adjusted for end walls)*	S.F. Wall	36	21.80	22.4%
2020	Exterior Windows	Aluminum, top hinged, in-swinging and curtain wall panels *20% of wall*	S.F. Window	31	4.76	
2030	Exterior Doors	Double hollow metal swinging, single hollow metal	Each	1426	.51	
	B30 Roofing					
3010	Roof Coverings	Asphalt shingles with flashing; polystyrene insulation	S.F. Roof	6.16	6.16	5.1%
3020	Roof Openings	N/A	—	—	—	
C. INTERIORS						
1010	Partitions	Plaster on metal studs *40 S.F. Floor/L.F. Partitions*	S.F. Partition	13.73	8.24	14.3%
1020	Interior Doors	Hollow metal *400 S.F. Floor/Door*	Each	1133	2.84	
1030	Fittings	N/A	—	—	—	
2010	Stair Construction	N/A	—	—	—	
3010	Wall Finishes	Paint	S.F. Surface	1.07	1.28	
3020	Floor Finishes	Carpet	S.F. Floor	4.94	4.94	
3030	Ceiling Finishes	N/A	—	—	—	
D. SERVICES						
	D10 Conveying					
1010	Elevators & Lifts	N/A	—	—	—	0.0 %
1020	Escalators & Moving Walks	N/A	—	—	—	
	D20 Plumbing					
2010	Plumbing Fixtures	Kitchen, toilet and service fixtures, supply and drainage *1 Fixture/2430 S.F. Floor*	Each	7460	3.07	2.8%
2020	Domestic Water Distribution	Gas fired hot water heater	S.F. Floor	.37	.37	
2040	Rain Water Drainage	N/A	—	—	—	
	D30 HVAC					
3010	Energy Supply	Oil fired hot water, wall fin radiation	S.F. Floor	9.49	9.49	15.4%
3020	Heat Generating Systems	N/A	—	—	—	
3030	Cooling Generating Systems	N/A	—	—	—	
3050	Terminal & Package Units	Split systems with air cooled condensing units	S.F. Floor	9.09	9.09	
3090	Other HVAC Sys. & Equipment	N/A	—	—	—	
	D40 Fire Protection					
4010	Sprinklers	Wet pipe sprinkler system	S.F. Floor	3.28	3.28	3.7%
4020	Standpipes	Standpipe	S.F. Floor	1.20	1.20	
	D50 Electrical					
5010	Electrical Service/Distribution	400 ampere service, panel board and feeders	S.F. Floor	.77	.77	9.7%
5020	Lighting & Branch Wiring	Fluorescent fixtures, receptacles, switches, A.C. and misc. power	S.F. Floor	7.89	7.89	
5030	Communications & Security	Alarm systems, sound system and emergency lighting	S.F. Floor	2.89	2.89	
5090	Other Electrical Systems	Emergency generator, 7.5 kW	S.F. Floor	.12	.12	
E. EQUIPMENT & FURNISHINGS						
1010	Commercial Equipment	N/A	—	—	—	0.0 %
1020	Institutional Equipment	N/A	—	—	—	
1030	Vehicular Equipment	N/A	—	—	—	
1090	Other Equipment	N/A	—	—	—	
F. SPECIAL CONSTRUCTION						
1020	Integrated Construction	N/A	—	—	—	0.0 %
1040	Special Facilities	N/A	—	—	—	
G. BUILDING SITEWORK		**N/A**				
				Sub-Total	120.76	**100%**
	CONTRACTOR FEES (General Requirements: 10%, Overhead: 5%, Profit: 10%)			*25%*	30.19	
	ARCHITECT FEES			*10%*	15.10	
				Total Building Cost	**166.05**	

M.100 Club, Country

Costs per square foot of floor area

Exterior Wall	S.F. Area	2000	4000	6000	8000	12000	15000	18000	20000	22000
	L.F. Perimeter	180	280	340	386	460	535	560	600	640
Stone Ashlar Veneer On Concrete Block	Wood Truss	242.15	216.30	**202.00**	193.35	183.40	180.55	176.25	175.10	174.15
	Steel Joists	246.65	220.25	205.50	196.60	186.40	183.45	179.00	177.85	176.85
Stucco on Concrete Block	Wood Truss	215.90	195.90	185.50	179.30	172.25	170.15	167.20	166.35	165.65
	Steel Joists	220.85	200.25	189.35	182.95	175.65	173.45	170.35	169.45	168.80
Brick Veneer	Wood Frame	246.00	207.80	190.60	180.90	170.10	166.65	162.55	161.20	160.15
Wood Shingles	Wood Frame	235.55	199.60	184.00	175.25	165.65	162.55	158.90	157.70	156.75
Perimeter Adj., Add or Deduct	Per 100 L.F.	42.80	21.40	14.25	10.70	7.10	5.70	4.75	4.30	3.90
Story Hgt. Adj., Add or Deduct	Per 1 Ft.	5.00	3.95	3.10	2.70	2.10	2.00	1.75	1.65	1.65
***For Basement**, add $27.80 per square foot of basement area*										

The above costs were calculated using the basic specifications shown on the facing page. These costs should be adjusted where necessary for design alternatives and owner's requirements. Reported completed project costs, for this type of structure, range from $89.05 to $296.35 per S.F.

Common additives

Description	Unit	$ Cost
Bar, Front bar	L.F.	415
Back bar	L.F.	330
Booth, Upholstered, custom, straight	L.F.	248 - 460
"L" or "U" shaped	L.F.	256 - 435
Fireplaces, Brick not incl. chimney		
or foundation, 30" x 24" opening	Ea.	2825
Chimney, standard brick		
Single flue 16" x 20"	V.L.F.	98
20" x 20"	V.L.F.	115
2 flue, 20" x 32"	V.L.F.	178
Lockers, Steel, single tier, 60" or 72"	Opng.	222 - 365
2 tier, 60" or 72" total	Opng.	115 - 164
5 tier, box lockers	Opng.	69.50 - 78.50
Locker bench, lam. maple top only	L.F.	33
Pedestals, steel pipe	Ea.	71
Refrigerators, Prefabricated, walk-in		
7'-6" High, 6' x 6'	S.F.	200
10' x 10'	S.F.	157
12' x 14'	S.F.	140
12' x 20'	S.F.	123

Description	Unit	$ Cost
Sauna, Prefabricated, complete, 6' x 4'	Ea.	6500
6' x 9'	Ea.	10,100
8' x 8'	Ea.	11,600
10' x 12'	Ea.	15,500
Smoke Detectors		
Ceiling type	Ea.	214
Duct type	Ea.	540
Sound System		
Amplifier, 250 watts	Ea.	2125
Speaker, ceiling or wall	Ea.	206
Trumpet	Ea.	395
Steam Bath, Complete, to 140 C.F.	Ea.	2875
To 300 C.F.	Ea.	3200
To 800 C.F.	Ea.	6950
To 2500 C.F.	Ea.	9150
Swimming Pool Complete, gunite	S.F.	97 - 121
Tennis Court, Complete with fence		
Bituminous	Ea.	46,700 - 82,500
Clay	Ea.	50,000 - 84,000

Important: See the Reference Section for Location Factors.

Model costs calculated for a 1 story building with 12′ story height and 6,000 square feet of floor area

Club, Country

				Unit	Unit Cost	Cost Per S.F.	% Of Sub-Total
A. SUBSTRUCTURE							
1010	Standard Foundations	Poured concrete; strip and spread footings		S.F. Ground	2.31	2.31	8.8%
1020	Special Foundations	N/A		—	—	—	
1030	Slab on Grade	4″ reinforced concrete with vapor barrier and granular base		S.F. Slab	5.10	5.10	
2010	Basement Excavation	Site preparation for slab and trench for foundation wall and footing		S.F. Ground	.29	.29	
2020	Basement Walls	4′ Foundation wall		L.F. Wall	74	5.41	
B. SHELL							
	B10 Superstructure						
1010	Floor Construction	6 x 6 wood columns		S.F. Floor	.06	.06	4.7%
1020	Roof Construction	Wood truss with plywood sheathing		S.F. Roof	6.94	6.94	
	B20 Exterior Enclosure						
2010	Exterior Walls	Stone ashlar veneer on concrete block	*65% of wall*	S.F. Wall	41	18.21	19.3%
2020	Exterior Windows	Aluminum horizontal sliding	*35% of wall*	Each	495	7.85	
2030	Exterior Doors	Double aluminum and glass, hollow metal		Each	2778	2.77	
	B30 Roofing						
3010	Roof Coverings	Asphalt shingles		S.F. Roof	16.44	2.89	1.9%
3020	Roof Openings	N/A		—	—	—	
C. INTERIORS							
1010	Partitions	Gypsum board on metal studs, load bearing	*14 S.F. Floor/L.F. Partition*	S.F. Partition	7.69	5.49	16.8%
1020	Interior Doors	Single leaf wood	*140 S.F. Floor/Door*	Each	593	4.23	
1030	Fittings	N/A		—	—	—	
2010	Stair Construction	N/A		—	—	—	
3010	Wall Finishes	40% vinyl wall covering, 40% paint, 20% ceramic tile		S.F. Surface	2.44	3.49	
3020	Floor Finishes	50% carpet, 30% hardwood tile, 20% ceramic tile		S.F. Floor	7.73	7.73	
3030	Ceiling Finishes	Gypsum plaster on wood furring		S.F. Ceiling	4.12	4.12	
D. SERVICES							
	D10 Conveying						
1010	Elevators & Lifts	N/A		—	—	—	0.0 %
1020	Escalators & Moving Walks	N/A		—	—	—	
	D20 Plumbing						
2010	Plumbing Fixtures	Kitchen, toilet and service fixtures, supply and drainage	*1 Fixture/125 S.F. Floor*	Each	1984	15.87	16.0%
2020	Domestic Water Distribution	Gas fired hot water		S.F. Floor	8	8	
2040	Rain Water Drainage	N/A		—	—	—	
	D30 HVAC						
3010	Energy Supply	N/A		—	—	—	18.7 %
3020	Heat Generating Systems	Included in D3050		—	—	—	
3030	Cooling Generating Systems	N/A		—	—	—	
3050	Terminal & Package Units	Multizone rooftop unit, gas heating, electric cooling		S.F. Floor	28	28	
3090	Other HVAC Sys. & Equipment	N/A		—	—	—	
	D40 Fire Protection						
4010	Sprinklers	Wet pipe sprinkler system		S.F. Floor	5.47	5.47	4.8%
4020	Standpipes	Standpipe		S.F. Floor	1.71	1.71	
	D50 Electrical						
5010	Electrical Service/Distribution	400 ampere service, panel board and feeders		S.F. Floor	3.31	3.31	9.1%
5020	Lighting & Branch Wiring	High efficiency fluorescent fixtures, receptacles, switches, A.C. and misc. power		S.F. Floor	8.22	8.22	
5030	Communications & Security	Addressable alarm systems and emergency lighting		S.F. Floor	1.91	1.91	
5090	Other Electrical Systems	Emergency generator, 11.5 kW		S.F. Floor	.23	.23	
E. EQUIPMENT & FURNISHINGS							
1010	Commercial Equipment	N/A		—	—	—	0.0 %
1020	Institutional Equipment	N/A		—	—	—	
1030	Vehicular Equipment	N/A		—	—	—	
1090	Other Equipment	N/A		—	—	—	
F. SPECIAL CONSTRUCTION							
1020	Integrated Construction	N/A		—	—	—	0.0 %
1040	Special Facilities	N/A		—	—	—	
G. BUILDING SITEWORK	**N/A**						
					Sub-Total	149.61	**100%**
	CONTRACTOR FEES (General Requirements: 10%, Overhead: 5%, Profit: 10%)				*25%*	37.43	
	ARCHITECT FEES				*8%*	14.96	
					Total Building Cost	**202**	

COMMERCIAL/INDUSTRIAL/ INSTITUTIONAL | M.170

Community Center

Costs per square foot of floor area

Exterior Wall	S.F. Area	4000	6000	8000	10000	12000	14000	16000	18000	20000
	L.F. Perimeter	260	340	420	453	460	510	560	610	600
Face Brick with Concrete Block Back-up	Bearing Walls	161.45	154.25	150.55	145.05	139.80	138.30	137.10	136.20	133.35
	Steel Frame	166.00	158.60	154.75	149.10	143.70	142.15	140.90	140.00	137.00
Decorative Concrete Block	Bearing Walls	142.15	136.60	133.80	129.70	125.80	124.70	123.85	123.10	121.00
	Steel Frame	153.60	147.75	144.75	140.40	136.40	135.15	134.20	133.50	131.30
Tilt Up Concrete Wall Panels	Bearing Walls	144.60	139.55	136.95	133.30	129.85	128.85	128.05	127.45	125.55
	Steel Frame	149.15	143.85	141.20	137.35	133.75	132.65	131.80	131.20	129.20
Perimeter Adj., Add or Deduct	Per 100 L.F.	17.85	11.90	8.95	7.15	5.90	5.05	4.50	4.00	3.55
Story Hgt. Adj., Add or Deduct	Per 1 Ft.	2.65	2.25	2.15	1.85	1.60	1.45	1.45	1.40	1.25
***For Basement**, add $30.15 per square foot of basement area*										

The above costs were calculated using the basic specifications shown on the facing page. These costs should be adjusted where necessary for design alternatives and owner's requirements. Reported completed project costs, for this type of structure, range from $87.10 to $275.20 per S.F.

Common additives

Description	Unit	$ Cost
Bar, Front bar	L.F.	415
Back bar	L.F.	330
Booth, Upholstered, custom straight	L.F.	248 - 460
"L" or "U" shaped	L.F.	256 - 435
Bowling Alleys, incl. alley, pinsetter		
Scorer, counter & misc. supplies, average	Lane	61,500
For automatic scorer, add	Lane	11,000
Emergency Lighting, 25 watt, battery operated		
Lead battery	Ea.	325
Nickel cadmium	Ea.	665
Kitchen Equipment		
Broiler	Ea.	4050
Coffee urn, twin 6 gallon	Ea.	2975
Cooler, 6 ft. long	Ea.	3875
Dishwasher, 10-12 racks per hr.	Ea.	4200
Food warmer	Ea.	770
Freezer, 44 C.F., reach-in	Ea.	6100
Ice cube maker, 50 lb. per day	Ea.	1950
Range with 1 oven	Ea.	3275

Description	Unit	$ Cost
Movie Equipment		
Projector, 35mm	Ea.	12,700 - 17,400
Screen, wall or ceiling hung	S.F.	9.10 - 13.30
Partitions, Folding leaf, wood		
Acoustic type	S.F.	79.50 - 131
Seating		
Auditorium chair, all veneer	Ea.	305
Veneer back, padded seat	Ea.	315
Upholstered, spring seat	Ea.	310
Classroom, movable chair & desk	Set	81 - 171
Lecture hall, pedestal type	Ea.	267 - 610
Sound System		
Amplifier, 250 watts	Ea.	2125
Speaker, ceiling or wall	Ea.	206
Trumpet	Ea.	395
Stage Curtains, Medium weight	S.F.	10.75 - 37.50
Curtain Track, Light duty	L.F.	81.50
Swimming Pools, Complete, gunite	S.F.	97 - 121

Model costs calculated for a 1 story building with 12′ story height and 10,000 square feet of floor area

Community Center

			Unit	Unit Cost	Cost Per S.F.	% Of Sub-Total
A. SUBSTRUCTURE						
1010	Standard Foundations	Poured concrete; strip and spread footings	S.F. Ground	1.78	1.78	10.5%
1020	Special Foundations	N/A	—	—	—	
1030	Slab on Grade	4″ reinforced concrete with vapor barrier and granular base	S.F. Slab	5.10	5.10	
2010	Basement Excavation	Site preparation for slab and trench for foundation wall and footing	S.F. Ground	.29	.29	
2020	Basement Walls	4′ foundation wall	L.F. Wall	69	4.06	
B. SHELL						
	B10 Superstructure					
1010	Floor Construction	N/A	—	—	—	6.9 %
1020	Roof Construction	Metal deck on open web steel joists	S.F. Roof	7.43	7.43	
	B20 Exterior Enclosure					
2010	Exterior Walls	Face brick with concrete block backup *80% of wall*	S.F. Wall	30	13.09	15.0%
2020	Exterior Windows	Aluminum sliding *20% of wall*	Each	580	1.97	
2030	Exterior Doors	Double aluminum and glass and hollow metal	Each	2554	1.02	
	B30 Roofing					
3010	Roof Coverings	Built-up tar and gravel with flashing; perlite/EPS composite insulation	S.F. Roof	6.62	6.62	6.2%
3020	Roof Openings	Roof hatches	S.F. Roof	.09	.09	
C. INTERIORS						
1010	Partitions	Gypsum board on metal studs *14 S.F. Floor/L.F. Partition*	S.F. Partition	7.91	5.65	26.7%
1020	Interior Doors	Single leaf hollow metal *140 S.F. Floor/Door*	Each	1133	8.09	
1030	Fittings	Toilet partitions, directory board, mailboxes	S.F. Floor	1.68	1.68	
2010	Stair Construction	N/A	—	—	—	
3010	Wall Finishes	Paint	S.F. Surface	1.66	2.37	
3020	Floor Finishes	50% carpet, 50% vinyl tile	S.F. Floor	4.33	4.33	
3030	Ceiling Finishes	Mineral fiber tile on concealed zee bars	S.F. Ceiling	6.57	6.57	
D. SERVICES						
	D10 Conveying					
1010	Elevators & Lifts	N/A	—	—	—	0.0 %
1020	Escalators & Moving Walks	N/A	—	—	—	
	D20 Plumbing					
2010	Plumbing Fixtures	Kitchen, toilet and service fixtures, supply and drainage *1 Fixture/910 S.F. Floor*	Each	2930	3.22	12.4%
2020	Domestic Water Distribution	Electric water heater	S.F. Floor	9.61	9.61	
2040	Rain Water Drainage	Roof drains	S.F. Roof	.54	.54	
	D30 HVAC					
3010	Energy Supply	N/A	—	—	—	9.3 %
3020	Heat Generating Systems	Included in D3050	—	—	—	
3030	Cooling Generating Systems	N/A	—	—	—	
3050	Terminal & Package Units	Single zone rooftop unit, gas heating, electric cooling	S.F. Floor	9.95	9.95	
3090	Other HVAC Sys. & Equipment	N/A	—	—	—	
	D40 Fire Protection					
4010	Sprinklers	Wet pipe sprinkler system	S.F. Floor	3.28	3.28	3.1%
4020	Standpipes	N/A	—	—	—	
	D50 Electrical					
5010	Electrical Service/Distribution	200 ampere service, panel board and feeders	S.F. Floor	1.09	1.09	7.9%
5020	Lighting & Branch Wiring	High efficiency fluorescent fixtures, receptacles, switches, A.C. and misc. power	S.F. Floor	5.72	5.72	
5030	Communications & Security	Addressable alarm systems and emergency lighting	S.F. Floor	1.46	1.46	
5090	Other Electrical Systems	Emergency generator, 15 kW	S.F. Floor	.19	.19	
E. EQUIPMENT & FURNISHINGS						
1010	Commercial Equipment	Freezer, chest type	S.F. Floor	.94	.94	2.1%
1020	Institutional Equipment	N/A	—	—	—	
1030	Vehicular Equipment	N/A	—	—	—	
1090	Other Equipment	Kitchen equipment, directory board, mailboxes, built-in coat racks	S.F. Floor	1.31	1.31	
F. SPECIAL CONSTRUCTION						
1020	Integrated Construction	N/A	—	—	—	0.0 %
1040	Special Facilities	N/A	—	—	—	
G. BUILDING SITEWORK		**N/A**				
				Sub-Total	107.45	**100%**
	CONTRACTOR FEES (General Requirements: 10%, Overhead: 5%, Profit: 10%)			*25%*	26.86	
	ARCHITECT FEES			*8%*	10.74	
				Total Building Cost	**145.05**	

Costs per square foot of floor area

Exterior Wall	S.F. Area	10000	12500	15000	17500	20000	22500	25000	30000	40000
	L.F. Perimeter	400	450	500	550	600	625	650	700	800
Fiber Cement	Wood Frame	299.35	296.75	295.00	293.70	292.85	291.40	290.25	288.50	286.35
Stone Veneer	Wood Frame	316.55	311.85	308.70	306.40	304.75	302.15	300.15	297.05	293.20
Curtain Wall	Steel Frame	326.45	321.10	317.55	315.00	313.20	310.25	307.90	304.50	300.20
EIFS	Steel Frame	304.95	302.05	300.10	298.65	297.60	296.05	294.80	292.95	290.60
Brick Veneer	Reinforced Concrete	327.55	322.65	319.40	317.00	315.30	312.65	310.50	307.25	303.20
Metal Panel	Reinforced Concrete	324.30	319.85	316.90	314.70	313.15	310.70	308.75	305.75	302.10
Perimeter Adj., Add or Deduct	Per 100 L.F.	6.60	5.15	4.35	3.80	3.25	2.90	2.60	2.10	1.60
Story Hgt. Adj., Add or Deduct	Per 1 Ft.	0.90	0.85	0.75	0.75	0.65	0.60	0.55	0.50	0.40
***For Basement**, add $ 33.10 per square foot of basement area*										

The above costs were calculated using the basic specifications shown on the facing page. These costs should be adjusted where necessary for design alternatives and owner's requirements. Reported completed project costs, for this type of structure, range from $ 179.75 to $ 364.75 per S.F.

Common additives

Description	Unit	$ Cost
Clock System		
20 room	Ea.	18,900
50 room	Ea.	44,300
Closed Circuit Surveillance, one station		
Camera and monitor	Total	1975
For additional camera stations, add	Ea.	1075
For zoom lens - remote control, add	Ea.	3150 - 10,700
For automatic iris for low light, add	Ea.	2800
Directory Boards, plastic, glass covered		
30" x 20"	Ea.	555
36" x 48"	Ea.	1425
Aluminum, 24" x 18"	Ea.	600
36" x 24"	Ea.	715
48" x 32"	Ea.	1000
48" x 60"	Ea.	2100
Emergency Lighting, 25 watt, battery operated		
Lead battery	Ea.	325
Nickel cadmium	Ea.	665

Description	Unit	$ Cost
Smoke Detectors		
Ceiling type	Ea.	214
Duct type	Ea.	540
Sound System		
Amplifier, 250 watts	Ea.	2125
Speakers, ceiling or wall	Ea.	206
Trumpets	Ea.	395

Important: See the Reference Section for Location Factors.

Model costs calculated for a 1 story building with 16'-6" story height and 22,500 square feet of floor area

Computer Data Center

			Unit	Unit Cost	Cost Per S.F.	% Of Sub-Total
A. SUBSTRUCTURE						
1010	Standard Foundations	Poured concrete; strip and spread footings	S.F. Ground	1.49	1.49	4.0%
1020	Special Foundations	N/A	—	—	—	
1030	Slab on Grade	4" reinforced concrete	S.F. Slab	5.10	5.10	
2010	Basement Excavation	Site preparation for slab and trench for foundation wall and footing	S.F. Ground	.16	.18	
2020	Basement Walls	4' Foundation wall	L.F. Wall	74	2.04	
B. SHELL						
	B10 Superstructure					
1010	Floor Construction	Wood columns, fireproofed	S.F. Floor	2.78	2.78	4.7%
1020	Roof Construction	Wood roof truss 4:12 pitch	S.F. Roof	7.37	7.37	
	B20 Exterior Enclosure					
2010	Exterior Walls	Fiber cement siding on wood studs, insulated *95% of wall*	S.F. Wall	16.83	7.33	4.2%
2020	Exterior Windows	Aluminum horizontal sliding *5% of wall*	Each	555	.85	
2030	Exterior Doors	Aluminum and glass single and double doors	Each	3698	.99	
	B30 Roofing					
3010	Roof Coverings	Asphalt roofing, strip shingles, gutters and downspouts	S.F. Roof	2.08	2.08	1.0%
3020	Roof Openings	Roof hatch	S.F. Roof	.19	.19	
C. INTERIORS						
1010	Partitions	Sound deadening gypsum board on wood studs *15 S.F. Floor/L.F. Partition*	S.F. Partition	8.22	7.59	11.6%
1020	Interior Doors	Solid core wood doors in metal frames *370 S.F. Floor/Door*	Each	739	2	
1030	Fittings	Plastic laminate toilet partitions	S.F. Floor	.29	.29	
2010	Stair Construction	N/A	—	—	—	
3010	Wall Finishes	90% Paint, 10% ceramic wall tile	S.F. Surface	1.56	2.88	
3020	Floor Finishes	65% Carpet, 10% porcelain tile, 10% quarry tile	S.F. Floor	5.92	5.92	
3030	Ceiling Finishes	Acoustic ceiling tiles on suspended channel grid	S.F. Ceiling	6.57	6.57	
D. SERVICES						
	D10 Conveying					
1010	Elevators & Lifts	N/A	—	—	—	0.0 %
1020	Escalators & Moving Walks	N/A	—	—	—	
	D20 Plumbing					
2010	Plumbing Fixtures	Restroom and service fixtures, supply and drainage *1 Fixture/1180 S.F. Floor*	Each	1912	1.62	1.1%
2020	Domestic Water Distribution	Gas fired water heater	S.F. Floor	.78	.78	
2040	Rain Water Drainage	Roof drains	S.F. Floor			
	D30 HVAC					
3010	Energy Supply	Hot water reheat system	S.F. Floor	5.96	5.96	41.4%
3020	Heat Generating Systems	Oil fired hot water boiler and pumps	Each	50,250	9.83	
3030	Cooling Generating Systems	Cooling tower and chiller	S.F. Floor	9.08	9.08	
3050	Terminal & Package Units	N/A	—	—	—	
3090	Other HVAC Sys. & Equipment	Plate heat exchanger, ductwork, AHUs and VAV terminals	S.F. Floor	65	65.41	
	D40 Fire Protection					
4010	Sprinklers	85% light hazard wet pipe sprinkler system, 15% preaction system	S.F. Floor	3.96	3.96	2.1%
4020	Standpipes	Standpipes and hose systems	S.F. Floor	.65	.65	
	D50 Electrical					
5010	Electrical Service/Distribution	1200 ampere service, panel boards and feeders	S.F. Floor	3.55	3.55	27.7%
5020	Lighting & Branch Wiring	Fluorescent fixtures, receptacles, switches, A.C. and misc. power	S.F. Floor	15.92	15.92	
5030	Communications & Security	Telephone systems, internet wiring, and addressable alarm systems	S.F. Floor	28	28.15	
5090	Other Electrical Systems	Emergency generator, UPS system with 15 minute pack	S.F. Floor	12.82	12.82	
E. EQUIPMENT & FURNISHINGS						
1010	Commercial Equipment	N/A	—	—	—	0.0 %
1020	Institutional Equipment	N/A	—	—	—	
1030	Vehicular Equipment	N/A	—	—	—	
1090	Other Equipment	N/A	—	—	—	
F. SPECIAL CONSTRUCTION & DEMOLITION						
1020	Integrated Construction	Pedestal access floor	S.F. Floor	4.46	4.46	2.0%
1040	Special Facilities	N/A	—	—	—	
G. BUILDING SITEWORK	**N/A**					
				Sub-Total	217.84	**100%**
	CONTRACTOR FEES (General Requirements: 10%, Overhead: 5%, Profit: 10%)			*25%*	54.50	
	ARCHITECT FEES			*7%*	19.06	
				Total Building Cost	**291.40**	

M.180 Courthouse, 1 Story

Costs per square foot of floor area

Exterior Wall	S.F. Area	16000	23000	30000	37000	44000	51000	58000	65000	72000
	L.F. Perimeter	597	763	821	968	954	1066	1090	1132	1220
Limestone with Concrete Block Back-up	R/Conc. Frame	215.90	206.25	196.45	193.50	186.75	185.05	181.75	179.60	178.65
	Steel Frame	218.95	209.30	199.50	196.55	189.80	188.10	184.80	182.65	181.70
Face Brick with Concrete Block Back-up	R/Conc. Frame	198.65	190.90	183.85	181.40	176.75	175.35	173.10	171.50	170.85
	Steel Frame	201.70	193.95	186.85	184.40	179.80	178.40	176.10	174.60	173.85
Stone with Concrete Block Back-up	R/Conc. Frame	204.45	196.10	188.10	185.50	180.10	178.65	176.00	174.25	173.45
	Steel Frame	207.50	199.10	191.10	188.50	183.15	181.65	179.05	177.30	176.45
Perimeter Adj., Add or Deduct	Per 100 L.F.	8.05	5.60	4.30	3.45	2.95	2.50	2.25	1.95	1.80
Story Hgt. Adj., Add or Deduct	Per 1 Ft.	2.95	2.55	2.10	2.05	1.65	1.60	1.45	1.35	1.30
For Basement, add $28.50 per square foot of basement area										

The above costs were calculated using the basic specifications shown on the facing page. These costs should be adjusted where necessary for design alternatives and owner's requirements. Reported completed project costs, for this type of structure, range from $153.35 to $285.70 per S.F.

Common additives

Description	Unit	$ Cost
Benches, Hardwood	L.F.	137 - 228
Clock System		
20 room	Ea.	18,900
50 room	Ea.	44,300
Closed Circuit Surveillance, One station		
Camera and monitor	Ea.	1975
For additional camera stations, add	Ea.	1075
Directory Boards, Plastic, glass covered		
30" x 20"	Ea.	555
36" x 48"	Ea.	1425
Aluminum, 24" x 18"	Ea.	600
36" x 24"	Ea.	715
48" x 32"	Ea.	1000
48" x 60"	Ea.	2100
Emergency Lighting, 25 watt, battery operated		
Lead battery	Ea.	325
Nickel cadmium	Ea.	665

Description	Unit	$ Cost
Flagpoles, Complete		
Aluminum, 20' high	Ea.	1825
40' high	Ea.	4300
70' high	Ea.	11,300
Fiberglass, 23' high	Ea.	1275
39'-5" high	Ea.	3275
59' high	Ea.	6475
Intercom System, 25 station capacity		
Master station	Ea.	3100
Intercom outlets	Ea.	181
Handset	Ea.	505
Safe, Office type, 1 hour rating		
30" x 18" x 18"	Ea.	2500
60" x 36" x 18", double door	Ea.	9350
Smoke Detectors		
Ceiling type	Ea.	214
Duct type	Ea.	540

Important: See the Reference Section for Location Factors.

Model costs calculated for a 1 story building with 14′ story height and 30,000 square feet of floor area

Courthouse, 1 Story

			Unit	Unit Cost	Cost Per S.F.	% Of Sub-Total
A. SUBSTRUCTURE						
1010	Standard Foundations	Poured concrete; strip and spread footings	S.F. Ground	1.72	1.72	6.2%
1020	Special Foundations	N/A	—	—	—	
1030	Slab on Grade	4″ reinforced concrete with vapor barrier and granular base	S.F. Slab	5.10	5.10	
2010	Basement Excavation	Site preparation for slab and trench for foundation wall and footing	S.F. Ground	.16	.16	
2020	Basement Walls	4′ foundation wall	L.F. Wall	74	2.15	
B. SHELL						
	B10 Superstructure					
1010	Floor Construction	Cast-in-place columns	L.F. Column	64	1.43	13.5%
1020	Roof Construction	Cast-in-place concrete waffle slab	S.F. Roof	18.45	18.45	
	B20 Exterior Enclosure					
2010	Exterior Walls	Limestone panels with concrete block backup *75% of wall*	S.F. Wall	63	18.11	14.6%
2020	Exterior Windows	Aluminum with insulated glass *25% of wall*	Each	705	2.94	
2030	Exterior Doors	Double wood	Each	1861	.43	
	B30 Roofing					
3010	Roof Coverings	Built-up tar and gravel with flashing; perlite/EPS composite insulation	S.F. Roof	5.59	5.59	3.9%
3020	Roof Openings	Roof hatches	S.F. Roof	.09	.09	
C. INTERIORS						
1010	Partitions	Plaster on metal studs *10 S.F. Floor/L.F. Partition*	S.F. Partition	12.50	15	32.9%
1020	Interior Doors	Single leaf wood *100 S.F. Floor/Door*	Each	593	5.93	
1030	Fittings	Toilet partitions	S.F. Floor	.49	.49	
2010	Stair Construction	N/A	—	—	—	
3010	Wall Finishes	70% paint, 20% wood paneling, 10% vinyl wall covering	S.F. Surface	2	4.80	
3020	Floor Finishes	60% hardwood, 20% carpet, 20% terrazzo	S.F. Floor	12.70	12.70	
3030	Ceiling Finishes	Gypsum plaster on metal lath, suspended	S.F. Ceiling	9.47	9.47	
D. SERVICES						
	D10 Conveying					
1010	Elevators & Lifts	N/A	—	—	—	0.0 %
1020	Escalators & Moving Walks	N/A	—	—	—	
	D20 Plumbing					
2010	Plumbing Fixtures	Toilet and service fixtures, supply and drainage *1 Fixture/1110 S.F. Floor*	Each	4229	3.81	5.7%
2020	Domestic Water Distribution	Electric hot water heater	S.F. Floor	3.45	3.45	
2040	Rain Water Drainage	Roof drains	S.F. Roof	1.07	1.07	
	D30 HVAC					
3010	Energy Supply	N/A	—	—	—	12.9 %
3020	Heat Generating Systems	Included in D3050	—	—	—	
3030	Cooling Generating Systems	N/A	—	—	—	
3050	Terminal & Package Units	Multizone unit, gas heating, electric cooling	S.F. Floor	19	19	
3090	Other HVAC Sys. & Equipment	N/A	—	—	—	
	D40 Fire Protection					
4010	Sprinklers	Wet pipe sprinkler system	S.F. Floor	2.48	2.48	2.0%
4020	Standpipes	Standpipe, wet, Class III	S.F. Floor	.48	.48	
	D50 Electrical					
5010	Electrical Service/Distribution	400 ampere service, panel board and feeders	S.F. Floor	.98	.98	8.2%
5020	Lighting & Branch Wiring	High efficiency fluorescent fixtures, receptacles, switches, A.C. and misc. power	S.F. Floor	9.32	9.32	
5030	Communications & Security	Addressable alarm systems, internet wiring, and emergency lighting	S.F. Floor	1.60	1.60	
5090	Other Electrical Systems	Emergency generator, 11.5 kW	S.F. Floor	.14	.14	
E. EQUIPMENT & FURNISHINGS						
1010	Commercial Equipment	N/A	—	—	—	0.0 %
1020	Institutional Equipment	N/A	—	—	—	
1030	Vehicular Equipment	N/A	—	—	—	
1090	Other Equipment	N/A	—	—	—	
F. SPECIAL CONSTRUCTION						
1020	Integrated Construction	N/A	—	—	—	0.0 %
1040	Special Facilities	N/A	—	—	—	
G. BUILDING SITEWORK		N/A				
			Sub-Total		146.89	**100%**
	CONTRACTOR FEES (General Requirements: 10%, Overhead: 5%, Profit: 10%)			*25%*	36.71	
	ARCHITECT FEES			*7%*	12.85	
			Total Building Cost		**196.45**	

M.195 Day Care Center

Costs per square foot of floor area

Exterior Wall	S.F. Area	2000	5000	7000	10000	12000	15000	18000	21000	25000
	L.F. Perimeter	200	310	360	440	480	520	560	600	660
Tiltup Concrete Panel	Steel Joists	197.85	169.45	162.35	157.35	154.75	151.60	149.45	147.85	146.60
Decorative Concrete Block	Bearing Walls	203.10	172.00	164.10	158.55	155.75	152.15	149.85	148.05	146.60
Brick on Block	Steel Joists	233.35	191.05	180.25	172.55	168.60	163.65	160.35	157.95	155.90
Stucco on Concrete Block	Wood Truss	196.50	169.75	162.85	158.00	155.50	152.35	150.25	148.65	147.35
Brick Veneer	Wood Frame	205.65	175.55	167.75	162.25	159.40	155.75	153.35	151.55	150.10
Wood Siding	Wood Frame	189.50	165.75	159.70	155.45	153.25	150.55	148.75	147.35	146.20
Perimeter Adj., Add or Deduct	Per 100 L.F.	25.25	10.05	7.20	5.00	4.20	3.35	2.85	2.45	1.95
Story Hgt. Adj., Add or Deduct	Per 1 Ft.	2.50	1.60	1.30	1.10	1.00	0.90	0.85	0.75	0.60
***For Basement**, add $28.25 per square foot of basement area*										

The above costs were calculated using the basic specifications shown on the facing page. These costs should be adjusted where necessary for design alternatives and owner's requirements. Reported completed project costs, for this type of structure, range from $66.45 to $242.10 per S.F.

Common additives

Description	Unit	$ Cost
Emergency Lighting, 25 watt, battery operated		
Lead battery	Ea.	325
Nickel cadmium	Ea.	665
Flagpoles, Complete		
Aluminum, 20' high	Ea.	1825
40' high	Ea.	4300
70' high	Ea.	11,300
Fiberglass, 23' high	Ea.	1275
39'-5" high	Ea.	3275
59' high	Ea.	6475
Gym Floor, Incl. sleepers and finish, maple	S.F.	13.60
Intercom System, 25 Station capacity		
Master station	Ea.	3100
Intercom outlets	Ea.	181
Handset	Ea.	505

Description	Unit	$ Cost
Lockers, Steel, single tier, 60" to 72"	Opng.	222 - 365
2 tier, 60" to 72" total	Opng.	115 - 164
5 tier, box lockers	Opng.	69.50 - 78.50
Locker bench, lam. maple top only	L.F.	33
Pedestals, steel pipe	Ea.	71
Smoke Detectors		
Ceiling type	Ea.	214
Duct type	Ea.	540
Sound System		
Amplifier, 250 watts	Ea.	2125
Speaker, ceiling or wall	Ea.	206
Trumpet	Ea.	395

Important: See the Reference Section for Location Factors.

Model costs calculated for a 1 story building with 12' story height and 10,000 square feet of floor area

Day Care Center

			Unit	Unit Cost	Cost Per S.F.	% Of Sub-Total
A. SUBSTRUCTURE						
1010	Standard Foundations	Poured concrete; strip and spread footings	S.F. Ground	1.98	1.98	8.5%
1020	Special Foundations	N/A	—	—	—	
1030	Slab on Grade	4" concrete with vapor barrier and granular base	S.F. Slab	4.64	4.64	
2010	Basement Excavation	Site preparation for slab and trench for foundation wall and footing	S.F. Ground	.29	.29	
2020	Basement Walls	4' foundation wall	L.F. Wall	74	3.23	
B. SHELL						
	B10 Superstructure					
1010	Floor Construction	Wood beams on columns	S.F. Floor	.21	.21	6.4%
1020	Roof Construction	Wood trusses	S.F. Roof	7.37	7.37	
	B20 Exterior Enclosure					
2010	Exterior Walls	Brick veneer on wood studs *85% of wall*	S.F. Wall	23	10.24	15.7%
2020	Exterior Windows	Window wall *15% of wall*	Each	48	3.79	
2030	Exterior Doors	Aluminum and glass; steel	Each	2479	4.72	
	B30 Roofing					
3010	Roof Coverings	Asphalt shingles, 9" fiberglass batt insulation, gutters and downspouts	S.F. Roof	4.32	4.32	3.6%
3020	Roof Openings	N/A	—	—	—	
C. INTERIORS						
1010	Partitions	Gypsum board on wood studs *8 S.F. Floor/S.F. Partition*	S.F. Partition	7.34	4.89	15.6%
1020	Interior Doors	Single leaf hollow metal *700 S.F. Floor/Door*	Each	1133	3.02	
1030	Fittings	Toilet partitions	S.F. Floor	.45	.45	
2010	Stair Construction	N/A	—	—	—	
3010	Wall Finishes	Paint	S.F. Surface	1.43	1.91	
3020	Floor Finishes	5% quarry tile, 95% vinyl composition tile	S.F. Floor	2.83	2.83	
3030	Ceiling Finishes	Fiberglass tile on tee grid	S.F. Ceiling	5.51	5.51	
D. SERVICES						
	D10 Conveying					
1010	Elevators & Lifts	N/A	—	—	—	0.0 %
1020	Escalators & Moving Walks	N/A	—	—	—	
	D20 Plumbing					
2010	Plumbing Fixtures	Toilet and service fixtures, supply and drainage *1 Fixture/455 S.F. Floor*	Each	2729	15.33	15.2%
2020	Domestic Water Distribution	Electric water heater	S.F. Floor	2.73	2.73	
2040	Rain Water Drainage	N/A	—	—	—	
	D30 HVAC					
3010	Energy Supply	Oil fired hot water, wall fin radiation	S.F. Floor	10.44	10.44	18.7%
3020	Heat Generating Systems	N/A	—	—	—	
3030	Cooling Generating Systems	N/A	—	—	—	
3050	Terminal & Package Units	Split systems with air cooled condensing units	S.F. Floor	11.85	11.85	
3090	Other HVAC Sys. & Equipment	N/A	—	—	—	
	D40 Fire Protection					
4010	Sprinklers	Sprinkler, light hazard	S.F. Floor	3.28	3.28	3.5%
4020	Standpipes	Standpipe	S.F. Floor	.86	.86	
	D50 Electrical					
5010	Electrical Service/Distribution	200 ampere service, panel board and feeders	S.F. Floor	1.09	1.09	8.8%
5020	Lighting & Branch Wiring	High efficiency fluorescent fixtures, receptacles, switches, A.C. and misc. power	S.F. Floor	7.76	7.76	
5030	Communications & Security	Addressable alarm systems and emergency lighting	S.F. Floor	1.48	1.48	
5090	Other Electrical Systems	Emergency generator, 15 kW	S.F. Floor	.09	.09	
E. EQUIPMENT & FURNISHINGS						
1010	Commercial Equipment	N/A	—	—	—	4.0 %
1020	Institutional Equipment	Cabinets and countertop	S.F. Floor	4.77	4.77	
1030	Vehicular Equipment	N/A	—	—	—	
1090	Other Equipment	N/A	—	—	—	
F. SPECIAL CONSTRUCTION						
1020	Integrated Construction	N/A	—	—	—	0.0 %
1040	Special Facilities	N/A	—	—	—	
G. BUILDING SITEWORK	N/A					
				Sub-Total	119.08	**100%**
	CONTRACTOR FEES (General Requirements: 10%, Overhead: 5%, Profit: 10%)			*25%*	29.77	
	ARCHITECT FEES			*9%*	13.40	
				Total Building Cost	**162.25**	

Costs per square foot of floor area

Exterior Wall	S.F. Area	12000	18000	24000	30000	36000	42000	48000	54000	60000
	L.F. Perimeter	460	580	713	730	826	880	965	1006	1045
Concrete Block	Steel Frame	129.35	122.60	119.55	115.35	113.90	112.20	111.40	110.20	109.25
	Bearing Walls	127.15	120.35	117.30	113.10	111.65	109.90	109.10	107.95	107.00
Precast Concrete Panels	Steel Frame	136.05	128.20	124.75	119.60	117.90	115.85	114.90	113.45	112.30
Insulated Metal Panels	Steel Frame	131.60	124.45	121.30	116.70	115.20	113.40	112.55	111.25	110.30
Face Brick on Common Brick	Steel Frame	146.20	136.75	132.60	126.00	124.00	121.40	120.25	118.35	116.90
Tilt-up Concrete Panel	Steel Frame	130.95	123.90	120.80	116.30	114.80	113.05	112.20	110.95	109.95
Perimeter Adj., Add or Deduct	Per 100 L.F.	5.20	3.50	2.60	2.10	1.70	1.45	1.30	1.15	1.00
Story Hgt. Adj., Add or Deduct	Per 1 Ft.	0.80	0.65	0.60	0.45	0.45	0.40	0.40	0.35	0.40
***For Basement**, add $ 28.85 per square foot of basement area*										

The above costs were calculated using the basic specifications shown on the facing page. These costs should be adjusted where necessary for design alternatives and owner's requirements. Reported completed project costs, for this type of structure, range from $ 49.30 to $ 190.85 per S.F.

Common additives

Description	Unit	$ Cost
Clock System		
20 room	Ea.	18,900
50 room	Ea.	44,300
Dock Bumpers, Rubber blocks		
4-1/2" thick, 10" high, 14" long	Ea.	85.50
24" long	Ea.	128
36" long	Ea.	174
12" high, 14" long	Ea.	102
24" long	Ea.	155
36" long	Ea.	183
6" thick, 10" high, 14" long	Ea.	100
24" long	Ea.	154
36" long	Ea.	196
20" high, 11" long	Ea.	176
Dock Boards, Heavy		
60" x 60" Aluminum, 5,000# cap.	Ea.	1500
9000# cap.	Ea.	1500
15,000# cap.	Ea.	1725

Description	Unit	$ Cost
Dock Levelers, Hinged 10 ton cap.		
6' x 8'	Ea.	5900
7' x 8'	Ea.	7775
Partitions, Woven wire, 10 ga., 1-1/2" mesh		
4' wide x 7' high	Ea.	199
8' high	Ea.	224
10' High	Ea.	268
Platform Lifter, Portable, 6'x 6'		
3000# cap.	Ea.	10,200
4000# cap.	Ea.	12,500
Fixed, 6' x 8', 5000# cap.	Ea.	12,400

Important: See the Reference Section for Location Factors.

Model costs calculated for a 1 story building with 20′ story height and 30,000 square feet of floor area

Factory, 1 Story

				Unit	Unit Cost	Cost Per S.F.	% Of Sub-Total
A. SUBSTRUCTURE							
1010	Standard Foundations	Poured concrete; strip and spread footings		S.F. Ground	1.42	1.42	12.2%
1020	Special Foundations	N/A		—	—	—	
1030	Slab on Grade	4″ reinforced concrete with vapor barrier and granular base		S.F. Slab	6.78	6.78	
2010	Basement Excavation	Site preparation for slab and trench for foundation wall and footing		S.F. Ground	.16	.16	
2020	Basement Walls	4′ foundation wall		L.F. Wall	74	2.15	
B. SHELL							
	B10 Superstructure						
1010	Floor Construction	N/A		—	—	—	10.8 %
1020	Roof Construction	Metal deck, open web steel joists, beams and columns		S.F. Roof	9.30	9.30	
	B20 Exterior Enclosure						
2010	Exterior Walls	Concrete block	*75% of wall*	S.F. Wall	8.08	2.95	8.5%
2020	Exterior Windows	Industrial horizontal pivoted steel	*25% of wall*	Each	815	3.10	
2030	Exterior Doors	Double aluminum and glass, hollow metal, steel overhead		Each	2514	1.27	
	B30 Roofing						
3010	Roof Coverings	Built-up tar and gravel with flashing; perlite/EPS composite insulation		S.F. Roof	5.76	5.76	7.1%
3020	Roof Openings	Roof hatches		S.F. Roof	.35	.35	
C. INTERIORS							
1010	Partitions	Concrete block	*60 S.F. Floor/L.F. Partition*	S.F. Partition	8.15	1.63	8.3%
1020	Interior Doors	Single leaf hollow metal and fire doors	*600 S.F. Floor/Door*	Each	1133	1.89	
1030	Fittings	Toilet partitions		S.F. Floor	.99	.99	
2010	Stair Construction	N/A		—	—	—	
3010	Wall Finishes	Paint		S.F. Surface	4.35	1.74	
3020	Floor Finishes	Vinyl composition tile	*10%of floor*	S.F. Floor	2.60	.26	
3030	Ceiling Finishes	Fiberglass board on exposed grid system	*10% of area*	S.F. Ceiling	6.57	.66	
D. SERVICES							
	D10 Conveying						
1010	Elevators & Lifts	N/A		—	—	—	0.0 %
1020	Escalators & Moving Walks	N/A		—	—	—	
	D20 Plumbing						
2010	Plumbing Fixtures	Toilet and service fixtures, supply and drainage	*1 Fixture/1000 S.F. Floor*	Each	4610	4.61	7.1%
2020	Domestic Water Distribution	Gas fired water heater		S.F. Floor	.59	.59	
2040	Rain Water Drainage	Roof drains		S.F. Roof	.93	.93	
	D30 HVAC						
3010	Energy Supply	Oil fired hot water, unit heaters		S.F. Floor	9.03	9.03	23.0%
3020	Heat Generating Systems	N/A		—	—	—	
3030	Cooling Generating Systems	Chilled water, air cooled condenser system		S.F. Floor	10.78	10.78	
3050	Terminal & Package Units	N/A		—	—	—	
3090	Other HVAC Sys. & Equipment	N/A		—	—	—	
	D40 Fire Protection						
4010	Sprinklers	Sprinklers, ordinary hazard		S.F. Floor	3.66	3.66	5.0%
4020	Standpipes	Standpipe, wet, Class III		S.F. Floor	.61	.61	
	D50 Electrical						
5010	Electrical Service/Distribution	600 ampere service, panel board and feeders		S.F. Floor	1.06	1.06	17.5%
5020	Lighting & Branch Wiring	High intensity discharge fixtures, receptacles, switches, A.C. and misc. power		S.F. Floor	12.61	12.61	
5030	Communications & Security	Addressable alarm systems and emergency lighting		S.F. Floor	1.43	1.43	
5090	Other Electrical Systems	N/A		—	—	—	
E. EQUIPMENT & FURNISHINGS							
1010	Commercial Equipment	N/A		—	—	—	0.6 %
1020	Institutional Equipment	N/A		—	—	—	
1030	Vehicular Equipment	Dock shelters		Each	.51	.51	
1090	Other Equipment	N/A		—	—	—	
F. SPECIAL CONSTRUCTION							
1020	Integrated Construction	N/A		—	—	—	0.0 %
1040	Special Facilities	N/A		—	—	—	
G. BUILDING SITEWORK	**N/A**						
					Sub-Total	86.23	**100%**
	CONTRACTOR FEES (General Requirements: 10%, Overhead: 5%, Profit: 10%)				*25%*	21.57	
	ARCHITECT FEES				*7%*	7.55	
					Total Building Cost	**115.35**	

Costs per square foot of floor area

Exterior Wall	S.F. Area	4000	4500	5000	5500	6000	6500	7000	7500	8000
	L.F. Perimeter	260	280	300	310	320	336	353	370	386
Face Brick Concrete Block Back-up	Steel Joists	178.40	174.90	172.20	168.55	**165.50**	163.60	162.15	160.90	159.60
	Bearing Walls	164.50	161.05	158.30	154.65	151.60	149.70	148.25	147.00	145.75
Decorative Concrete Block	Steel Joists	165.05	162.15	159.90	157.00	154.65	153.05	151.85	150.80	149.80
	Bearing Walls	151.15	148.20	146.00	143.10	140.65	139.15	137.95	136.90	135.90
Limestone with Concrete Block Back-up	Steel Joists	199.60	195.25	191.80	186.95	183.00	180.55	178.70	177.05	175.50
	Bearing Walls	185.70	181.30	177.90	173.05	169.05	166.65	164.75	163.10	161.50
Perimeter Adj., Add or Deduct	Per 100 L.F.	19.15	17.10	15.30	13.90	12.75	11.75	10.95	10.15	9.60
Story Hgt. Adj., Add or Deduct	Per 1 Ft.	2.60	2.50	2.35	2.25	2.10	2.05	2.00	1.95	1.95
For Basement, add $35.10 per square foot of basement area										

The above costs were calculated using the basic specifications shown on the facing page. These costs should be adjusted where necessary for design alternatives and owner's requirements. Reported completed project costs, for this type of structure, range from $79.80 to $236.10 per S.F.

Common additives

Description	Unit	$ Cost
Appliances		
Cooking range, 30" free standing		
1 oven	Ea.	595 - 2275
2 oven	Ea.	1450 - 3150
30" built-in		
1 oven	Ea.	1075 - 2225
2 oven	Ea.	1725 - 3275
Counter top cook tops, 4 burner	Ea.	480 - 2075
Microwave oven	Ea.	275 - 850
Combination range, refrig. & sink, 30" wide	Ea.	1575 - 2100
60" wide	Ea.	2100
72" wide	Ea.	2250
Combination range refrigerator, sink microwave oven & icemaker	Ea.	5900
Compactor, residential, 4-1 compaction	Ea.	860 - 1450
Dishwasher, built-in, 2 cycles	Ea.	510 - 855
4 cycles	Ea.	650 - 1675
Garbage disposer, sink type	Ea.	174 - 340
Hood for range, 2 speed, vented, 30" wide	Ea.	222 - 1150
42" wide	Ea.	310 - 2400

Description	Unit	$ Cost
Appliances, cont.		
Refrigerator, no frost 10-12 C.F.	Ea.	600 - 745
14-16 C.F.	Ea.	720 - 870
18-20 C.F.	Ea.	910 - 2075
Lockers, Steel, single tier, 60" or 72"	Opng.	222 - 365
2 tier, 60" or 72" total	Opng.	115 - 164
5 tier, box lockers	Opng.	69.50 - 78.50
Locker bench, lam. maple top only	L.F.	33
Pedestals, steel pipe	Ea.	71
Sound System		
Amplifier, 250 watts	Ea.	2125
Speaker, ceiling or wall	Ea.	206
Trumpet	Ea.	395

Important: See the Reference Section for Location Factors.

Model costs calculated for a 1 story building with 14' story height and 6,000 square feet of floor area

Fire Station, 1 Story

				Unit	Unit Cost	Cost Per S.F.	% Of Sub-Total
A. SUBSTRUCTURE							
1010	Standard Foundations	Poured concrete; strip and spread footings		S.F. Ground	2.49	2.49	11.2%
1020	Special Foundations	N/A		—	—	—	
1030	Slab on Grade	6" reinforced concrete with vapor barrier and granular base		S.F. Slab	6.78	6.78	
2010	Basement Excavation	Site preparation for slab and trench for foundation wall and footing		S.F. Ground	.51	.51	
2020	Basement Walls	4' foundation wall		L.F. Wall	74	3.92	
B. SHELL							
	B10 Superstructure						
1010	Floor Construction	N/A		—	—	—	8.5 %
1020	Roof Construction	Metal deck, open web steel joists, beams on columns		S.F. Roof	10.38	10.38	
	B20 Exterior Enclosure						
2010	Exterior Walls	Face brick with concrete block backup	*75% of wall*	S.F. Wall	30	16.85	19.1%
2020	Exterior Windows	Aluminum insulated glass	*10% of wall*	Each	815	1.91	
2030	Exterior Doors	Single aluminum and glass, overhead, hollow metal	*15% of wall*	S.F. Door	41	4.61	
	B30 Roofing						
3010	Roof Coverings	Built-up tar and gravel with flashing; perlite/EPS composite insulation		S.F. Roof	6.95	6.95	5.8%
3020	Roof Openings	Skylights, roof hatches		S.F. Roof	.15	.15	
C. INTERIORS							
1010	Partitions	Concrete block	*17 S.F. Floor/L.F. Partition*	S.F. Partition	8.77	5.16	12.7%
1020	Interior Doors	Single leaf hollow metal	*500 S.F. Floor/Door*	Each	1133	2.27	
1030	Fittings	Toilet partitions		S.F. Floor	.49	.49	
2010	Stair Construction	N/A		—	—	—	
3010	Wall Finishes	Paint		S.F. Surface	1.92	2.26	
3020	Floor Finishes	50% vinyl tile, 50% paint		S.F. Floor	2.14	2.14	
3030	Ceiling Finishes	Fiberglass board on exposed grid, suspended	*50% of area*	S.F. Ceiling	6.57	3.29	
D. SERVICES							
	D10 Conveying						
1010	Elevators & Lifts	N/A		—	—	—	0.0 %
1020	Escalators & Moving Walks	N/A		—	—	—	
	D20 Plumbing						
2010	Plumbing Fixtures	Kitchen, toilet and service fixtures, supply and drainage	*1 Fixture/375 S.F. Floor*	Each	3604	9.61	11.3%
2020	Domestic Water Distribution	Gast fired water heater		S.F. Floor	2.98	2.98	
2040	Rain Water Drainage	Roof drains		S.F. Roof	1.27	1.27	
	D30 HVAC						
3010	Energy Supply	N/A		—	—	—	18.4 %
3020	Heat Generating Systems	Included in D3050		—	—	—	
3030	Cooling Generating Systems	N/A		—	—	—	
3050	Terminal & Package Units	Rooftop multizone unit system		S.F. Floor	23	22.61	
3090	Other HVAC Sys. & Equipment	N/A		—	—	—	
	D40 Fire Protection						
4010	Sprinklers	Wet pipe sprinkler system		S.F. Floor	4.57	4.57	4.9%
4020	Standpipes	Standpipe, wet, Class III		S.F. Floor	1.43	1.43	
	D50 Electrical						
5010	Electrical Service/Distribution	200 ampere service, panel board and feeders		S.F. Floor	1.59	1.59	8.1%
5020	Lighting & Branch Wiring	High efficiency fluorescent fixtures, receptacles, switches, A.C. and misc. power		S.F. Floor	6.47	6.47	
5030	Communications & Security	Addressable alarm systems		S.F. Floor	1.91	1.91	
5090	Other Electrical Systems	N/A		—	—	—	
E. EQUIPMENT & FURNISHINGS							
1010	Commercial Equipment	N/A		—	—	—	0.0 %
1020	Institutional Equipment	N/A		—	—	—	
1030	Vehicular Equipment	N/A		—	—	—	
1090	Other Equipment	N/A		—	—	—	
F. SPECIAL CONSTRUCTION							
1020	Integrated Construction	N/A		—	—	—	0.0 %
1040	Special Facilities	N/A		—	—	—	
G. BUILDING SITEWORK	**N/A**						
					Sub-Total	122.60	**100%**
	CONTRACTOR FEES (General Requirements: 10%, Overhead: 5%, Profit: 10%)				*25%*	30.64	
	ARCHITECT FEES				*8%*	12.26	
					Total Building Cost	**165.50**	

Costs per square foot of floor area

Exterior Wall	S.F. Area	4000	5000	6000	8000	10000	12000	14000	16000	18000
	L.F. Perimeter	180	205	230	260	300	340	353	386	420
Cedar Beveled Siding	Wood Frame	204.30	196.20	190.75	182.40	178.00	175.10	171.80	170.15	168.90
Aluminum Siding	Wood Frame	201.35	193.55	188.25	180.30	176.05	173.30	170.15	168.60	167.40
Board and Batten	Wood Frame	199.60	192.85	188.30	181.20	177.55	175.10	172.05	170.75	169.65
Face Brick on Block	Bearing Wall	204.75	195.15	188.70	178.25	172.90	169.40	164.90	162.90	161.35
Stucco on Block	Bearing Wall	189.20	181.00	175.45	167.00	162.50	159.50	156.15	154.45	153.15
Decorative Block	Wood Joists	208.45	200.00	194.30	185.45	180.80	177.75	174.15	172.40	171.05
Perimeter Adj., Add or Deduct	Per 100 L.F.	15.45	12.35	10.30	7.70	6.25	5.15	4.35	3.90	3.40
Story Hgt. Adj., Add or Deduct	Per 1 Ft.	2.05	1.85	1.75	1.50	1.40	1.30	1.20	1.10	1.10
***For Basement**, add $21.70 per square foot of basement area*										

The above costs were calculated using the basic specifications shown on the facing page. These costs should be adjusted where necessary for design alternatives and owner's requirements. Reported completed project costs, for this type of structure, range from $115.75 to $253.10 per S.F.

Common additives

Description	Unit	$ Cost
Appliances		
Cooking range, 30" free standing		
1 oven	Ea.	595 - 2275
2 oven	Ea.	1450 - 3150
30" built-in		
1 oven	Ea.	1075 - 2225
2 oven	Ea.	1725 - 3275
Counter top cook tops, 4 burner	Ea.	480 - 2075
Microwave oven	Ea.	275 - 850
Combination range, refrig. & sink, 30" wide	Ea.	1575 - 2100
60" wide	Ea.	2100
72" wide	Ea.	2250
Combination range, refrigerator, sink, microwave oven & icemaker	Ea.	5900
Compactor, residential, 4-1 compaction	Ea.	860 - 1450
Dishwasher, built-in, 2 cycles	Ea.	510 - 855
4 cycles	Ea.	650 - 1675
Garbage disposer, sink type	Ea.	174 - 340
Hood for range, 2 speed, vented, 30" wide	Ea.	222 - 1150
42" wide	Ea.	310 - 2400

Description	Unit	$ Cost
Appliances, cont.		
Refrigerator, no frost 10-12 C.F.	Ea.	600 - 745
14-16 C.F.	Ea.	720 - 870
18-20 C.F.	Ea.	910 - 2075
Elevators, Hydraulic passenger, 2 stops		
1500# capacity	Ea.	63,900
2500# capacity	Ea.	66,900
3500# capacity	Ea.	71,900
Laundry Equipment		
Dryer, gas, 16 lb. capacity	Ea.	950
30 lb. capacity	Ea.	3900
Washer, 4 cycle	Ea.	1175
Commercial	Ea.	1575
Sound System		
Amplifier, 250 watts	Ea.	2125
Speaker, ceiling or wall	Ea.	206
Trumpet	Ea.	395

Important: See the Reference Section for Location Factors.

Model costs calculated for a 2 story building with 10′ story height and 10,000 square feet of floor area

Fraternity/Sorority House

			Unit	Unit Cost	Cost Per S.F.	% Of Sub-Total
A. SUBSTRUCTURE						
1010	Standard Foundations	Poured concrete; strip and spread footings	S.F. Ground	2.68	1.34	
1020	Special Foundations	N/A	—	—	—	
1030	Slab on Grade	4″ reinforced concrete with vapor barrier and granular base	S.F. Slab	5.10	2.56	4.7%
2010	Basement Excavation	Site preparation for slab and trench for foundation wall and footing	S.F. Ground	.51	.26	
2020	Basement Walls	4′ foundation wall	L.F. Wall	65	1.94	
B. SHELL						
	B10 Superstructure					
1010	Floor Construction	Plywood on wood joists	S.F. Floor	41	20.28	17.0%
1020	Roof Construction	Plywood on wood rafters (pitched)	S.F. Roof	3.52	1.97	
	B20 Exterior Enclosure					
2010	Exterior Walls	Cedar bevel siding on wood studs, insulated *80% of wall*	S.F. Wall	12.75	6.12	
2020	Exterior Windows	Double hung wood *20% of wall*	Each	603	2.89	8.2%
2030	Exterior Doors	Solid core wood	Each	2337	1.65	
	B30 Roofing					
3010	Roof Coverings	Asphalt shingles with flashing (pitched); rigid fiberglass insulation	S.F. Roof	4.88	2.44	1.9%
3020	Roof Openings	N/A	—	—	—	
C. INTERIORS						
1010	Partitions	Gypsum board on wood studs *25 S.F. Floor/L.F. Partition*	S.F. Partition	7.69	2.46	
1020	Interior Doors	Single leaf wood *200 S.F. Floor/Door*	Each	593	2.97	
1030	Fittings	N/A	—	—	—	
2010	Stair Construction	Wood	Flight	2675	.81	13.8%
3010	Wall Finishes	Paint	S.F. Surface	.75	.48	
3020	Floor Finishes	10% hardwood, 70% carpet, 20% ceramic tile	S.F. Floor	7.25	7.25	
3030	Ceiling Finishes	Gypsum board on wood furring	S.F. Ceiling	4.12	4.12	
D. SERVICES						
	D10 Conveying					
1010	Elevators & Lifts	One hydraulic passenger elevator	Each	81,000	8.10	6.2%
1020	Escalators & Moving Walks	N/A	—	—	—	
	D20 Plumbing					
2010	Plumbing Fixtures	Kitchen toilet and service fixtures, supply and drainage *1 Fixture/150 S.F. Floor*	Each	2060	13.73	
2020	Domestic Water Distribution	Gas fired water heater	S.F. Floor	3.20	3.20	13.0%
2040	Rain Water Drainage	N/A	—	—	—	
	D30 HVAC					
3010	Energy Supply	Oil fired hot water, baseboard radiation	S.F. Floor	7.16	7.16	
3020	Heat Generating Systems	N/A	—	—	—	
3030	Cooling Generating Systems	N/A	—	—	—	10.8%
3050	Terminal & Package Units	Split system with air cooled condensing unit	S.F. Floor	7	7	
3090	Other HVAC Sys. & Equipment	N/A	—	—	—	
	D40 Fire Protection					
4010	Sprinklers	Wet pipe sprinkler system	S.F. Floor	3.54	3.54	3.5%
4020	Standpipes	Standpipe	S.F. Floor	1.05	1.05	
	D50 Electrical					
5010	Electrical Service/Distribution	600 ampere service, panel board and feeders	S.F. Floor	3.95	3.95	
5020	Lighting & Branch Wiring	High efficiency fluorescent fixtures, receptacles, switches, A.C. and misc. power	S.F. Floor	9.57	9.57	20.9%
5030	Communications & Security	Addressable alarm, communication system, internet and phone wiring, and generator set	S.F. Floor	13.60	13.60	
5090	Other Electrical Systems	Emergency generator, 7.5 kW	S.F. Floor	.22	.22	
E. EQUIPMENT & FURNISHINGS						
1010	Commercial Equipment	N/A	—	—	—	
1020	Institutional Equipment	N/A	—	—	—	0.0 %
1030	Vehicular Equipment	N/A	—	—	—	
1090	Other Equipment	N/A	—	—	—	
F. SPECIAL CONSTRUCTION						
1020	Integrated Construction	N/A	—	—	—	0.0 %
1040	Special Facilities	N/A	—	—	—	
G. BUILDING SITEWORK	**N/A**					
				Sub-Total	130.66	**100%**
	CONTRACTOR FEES (General Requirements: 10%, Overhead: 5%, Profit: 10%)			25%	32.64	
	ARCHITECT FEES			9%	14.70	
				Total Building Cost	**178**	

Costs per square foot of floor area

Exterior Wall	S.F. Area	4000	6000	8000	10000	12000	14000	16000	18000	20000
	L.F. Perimeter	260	320	384	424	460	484	510	540	576
Vertical Redwood Siding	Wood Frame	154.60	141.85	135.70	130.95	127.65	124.85	122.95	121.45	120.45
Brick Veneer	Wood Frame	200.50	173.90	160.85	151.70	145.40	140.35	136.65	133.95	131.90
Aluminum Siding	Wood Frame	187.50	163.25	151.30	143.20	137.70	133.45	130.30	127.90	126.15
Brick on Block	Wood Truss	207.80	180.00	166.40	156.65	149.90	144.50	140.55	137.65	135.50
Limestone on Block	Wood Truss	230.85	198.80	183.30	171.60	163.35	156.60	151.65	148.10	145.50
Stucco on Block	Wood Truss	192.30	167.25	154.95	146.50	140.80	136.25	132.90	130.50	128.60
Perimeter Adj., Add or Deduct	Per 100 L.F.	10.75	7.15	5.35	4.30	3.55	3.15	2.65	2.40	2.20
Story Hgt. Adj., Add or Deduct	Per 1 Ft.	1.45	1.25	1.05	1.00	0.90	0.80	0.70	0.75	0.70
***For Basement**, add $ 29.10 per square foot of basement area*										

The above costs were calculated using the basic specifications shown on the facing page. These costs should be adjusted where necessary for design alternatives and owner's requirements. Reported completed project costs, for this type of structure, range from $ 113.30 to $ 348.40 per S.F.

Common additives

Description	Unit	$ Cost
Autopsy Table, Standard	Ea.	11,300
Deluxe	Ea.	18,900
Directory Boards, Plastic, glass covered		
30" x 20"	Ea.	555
36" x 48"	Ea.	1425
Aluminum, 24" x 18"	Ea.	600
36" x 24"	Ea.	715
48" x 32"	Ea.	1000
48" x 60"	Ea.	2100
Emergency Lighting, 25 watt, battery operated		
Lead battery	Ea.	325
Nickel cadmium	Ea.	665
Mortuary Refrigerator, End operated		
Two capacity	Ea.	10,800
Six capacity	Ea.	16,300

Description	Unit	$ Cost
Planters, Precast concrete		
48" diam., 24" high	Ea.	695
7" diam., 36" high	Ea.	1800
Fiberglass, 36" diam., 24" high	Ea.	775
60" diam., 24" high	Ea.	1300
Smoke Detectors		
Ceiling type	Ea.	214
Duct type	Ea.	540

Model costs calculated for a 1 story building with 12' story height and 10,000 square feet of floor area

Funeral Home

				Unit	Unit Cost	Cost Per S.F.	% Of Sub-Total
A. SUBSTRUCTURE							
1010	Standard Foundations	Poured concrete; strip and spread footings		S.F. Ground	2.38	2.38	10.9%
1020	Special Foundations	N/A		–	–	–	
1030	Slab on Grade	4" reinforced concrete with vapor barrier and granular base		S.F. Slab	5.10	5.10	
2010	Basement Excavation	Site preparation for slab and trench for foundation wall and footing		S.F. Ground	.29	.29	
2020	Basement Walls	4' foundation wall		L.F. Wall	65	2.74	
B. SHELL							
B10 Superstructure							
1010	Floor Construction	Wood beams on columns		S.F. Floor	.78	.78	6.0%
1020	Roof Construction	Plywood on wood truss		S.F. Roof	4.99	4.99	
B20 Exterior Enclosure							
2010	Exterior Walls	1" x 4" vertical T & G redwood siding on wood studs	*90% of wall*	S.F. Wall	12.95	5.93	9.5%
2020	Exterior Windows	Double hung wood	*10% of wall*	Each	483	2.05	
2030	Exterior Doors	Wood swinging double doors, single leaf hollow metal		Each	1933	1.16	
B30 Roofing							
3010	Roof Coverings	Single ply membrane, fully adhered; polyisocyanurate sheets		S.F. Roof	6.95	6.95	7.2%
3020	Roof Openings	N/A		–	–	–	
C. INTERIORS							
1010	Partitions	Gypsum board on wood studs with sound deadening board	*15 S.F. Floor/L.F. Partition*	S.F. Partition	8.79	4.69	26.6%
1020	Interior Doors	Single leaf wood	*150 S.F. Floor/Door*	Each	593	3.95	
1030	Fittings	N/A		–	–	–	
2010	Stair Construction	N/A		–	–	–	
3010	Wall Finishes	50% wallpaper, 25% wood paneling, 25% paint		S.F. Surface	3.03	3.23	
3020	Floor Finishes	70% carpet, 30% ceramic tile		S.F. Floor	8.21	8.21	
3030	Ceiling Finishes	Fiberglass board on exposed grid, suspended		S.F. Ceiling	5.51	5.51	
D. SERVICES							
D10 Conveying							
1010	Elevators & Lifts	N/A		–	–	–	0.0 %
1020	Escalators & Moving Walks	N/A		–	–	–	
D20 Plumbing							
2010	Plumbing Fixtures	Toilet and service fixtures, supply and drainage	*1 Fixture/770 S.F. Floor*	Each	2626	3.41	8.7%
2020	Domestic Water Distribution	Electric water heater		S.F. Floor	4.91	4.91	
2040	Rain Water Drainage	Roof drain		S.F. Roof			
D30 HVAC							
3010	Energy Supply	N/A		–	–	–	16.2 %
3020	Heat Generating Systems	Included in D3030		–	–	–	
3030	Cooling Generating Systems	Multizone rooftop unit, gas heating, electric cooling		S.F. Floor	15.60	15.60	
3050	Terminal & Package Units	N/A		–	–	–	
3090	Other HVAC Sys. & Equipment	N/A		–	–	–	
D40 Fire Protection							
4010	Sprinklers	Wet pipe sprinkler system		S.F. Floor	3.28	3.28	4.3%
4020	Standpipes	Standpipe, wet, Class III		S.F. Floor	.86	.86	
D50 Electrical							
5010	Electrical Service/Distribution	400 ampere service, panel board and feeders		S.F. Floor	1.90	1.90	10.5%
5020	Lighting & Branch Wiring	High efficiency fluorescent fixtures, receptacles, switches, A.C. and misc. power		S.F. Floor	6.61	6.61	
5030	Communications & Security	Addressable alarm systems and emergency lighting		S.F. Floor	1.48	1.48	
5090	Other Electrical Systems	Emergency generator, 15 kW		S.F. Floor	.09	.09	
E. EQUIPMENT & FURNISHINGS							
1010	Commercial Equipment	N/A		–	–	–	0.0 %
1020	Institutional Equipment	N/A		–	–	–	
1030	Vehicular Equipment	N/A		–	–	–	
1090	Other Equipment	N/A		–	–	–	
F. SPECIAL CONSTRUCTION							
1020	Integrated Construction	N/A		–	–	–	0.0 %
1040	Special Facilities	N/A		–	–	–	
G. BUILDING SITEWORK		**N/A**					
					Sub-Total	96.10	**100%**
	CONTRACTOR FEES (General Requirements: 10%, Overhead: 5%, Profit: 10%)				25%	24.04	
	ARCHITECT FEES				9%	10.81	
					Total Building Cost	**130.95**	

Costs per square foot of floor area

Exterior Wall	S.F. Area	12000	14000	16000	19000	21000	23000	26000	28000	30000
	L.F. Perimeter	440	474	510	556	583	607	648	670	695
E.I.F.S. on Concrete Block	Steel Frame	124.45	121.85	119.95	117.60	116.25	115.05	113.75	112.85	112.15
Tilt-up Concrete Wall	Steel Frame	122.00	119.65	117.90	115.70	114.45	113.30	112.10	111.25	110.65
Face Brick with Concrete Block Back-up	Bearing Walls	123.65	120.55	118.35	115.55	113.95	112.45	110.95	109.90	109.05
	Steel Frame	130.40	127.40	125.10	122.35	120.70	119.25	117.70	116.65	115.90
Stucco on Concrete Block	Bearing Walls	115.85	113.40	111.55	109.30	108.05	106.85	105.60	104.80	104.15
	Steel Frame	122.95	120.55	118.70	116.45	115.15	114.00	112.75	111.90	111.25
Perimeter Adj., Add or Deduct	Per 100 L.F.	6.50	5.60	4.90	4.05	3.70	3.40	3.00	2.85	2.65
Story Hgt. Adj., Add or Deduct	Per 1 Ft.	1.40	1.30	1.25	1.15	1.10	1.00	0.95	0.90	0.90
***For Basement**, add $ 32.85 per square foot of basement area*										

The above costs were calculated using the basic specifications shown on the facing page. These costs should be adjusted where necessary for design alternatives and owner's requirements. Reported completed project costs, for this type of structure, range from $ 60.30 to $ 175.35 per S.F.

Common additives

Description	Unit	$ Cost
Emergency Lighting, 25 watt, battery operated		
Lead battery	Ea.	325
Nickel cadmium	Ea.	665
Smoke Detectors		
Ceiling type	Ea.	214
Duct type	Ea.	540
Sound System		
Amplifier, 250 watts	Ea.	2125
Speaker, ceiling or wall	Ea.	206
Trumpet	Ea.	395

Important: See the Reference Section for Location Factors.

Model costs calculated for a 1 story building with 14' story height and 21,000 square feet of floor area

Garage, Auto Sales

				Unit	Unit Cost	Cost Per S.F.	% Of Sub-Total
A. SUBSTRUCTURE							
1010	Standard Foundations	Poured concrete; strip and spread footings		S.F. Ground	1.35	1.35	13.1%
1020	Special Foundations	N/A		—	—	—	
1030	Slab on Grade	4" reinforced concrete with vapor barrier and granular base		S.F. Slab	7.70	7.70	
2010	Basement Excavation	Site preparation for slab and trench for foundation wall and footing		S.F. Ground	.29	.29	
2020	Basement Walls	4' foundation wall		L.F. Wall	74	2.04	
B. SHELL							
	B10 Superstructure						
1010	Floor Construction	N/A		—	—	—	13.2 %
1020	Roof Construction	Metal deck, open web steel joists, beams, columns		S.F. Roof	11.50	11.50	
	B20 Exterior Enclosure						
2010	Exterior Walls	E.I.F.S.	*70% of wall*	S.F. Wall	16.54	4.50	16.5%
2020	Exterior Windows	Window wall	*30% of wall*	Each	54	6.34	
2030	Exterior Doors	Double aluminum and glass, hollow metal, steel overhead		Each	4106	3.52	
	B30 Roofing						
3010	Roof Coverings	Built-up tar and gravel with flashing; perlite/EPS composite insulation		S.F. Roof	6.55	6.55	7.5%
3020	Roof Openings	N/A		—	—	—	
C. INTERIORS							
1010	Partitions	Gypsum board on metal studs	*28 S.F. Floor/L.F. Partition*	S.F. Partition	4.64	1.99	13.6%
1020	Interior Doors	Hollow metal	*280 S.F. Floor/Door*	Each	1133	4.05	
1030	Fittings	N/A		—	—	—	
2010	Stair Construction	N/A		—	—	—	
3010	Wall Finishes	Paint		S.F. Surface	1	.86	
3020	Floor Finishes	50% vinyl tile, 50% paint		S.F. Floor	2.14	2.14	
3030	Ceiling Finishes	Fiberglass board on exposed grid, suspended	*50% of area*	S.F. Ceiling	5.51	2.76	
D. SERVICES							
	D10 Conveying						
1010	Elevators & Lifts	N/A		—	—	—	0.0 %
1020	Escalators & Moving Walks	N/A		—	—	—	
	D20 Plumbing						
2010	Plumbing Fixtures	Toilet and service fixtures, supply and drainage	*1 Fixture/1500 S.F. Floor*	Each	2685	1.79	6.9%
2020	Domestic Water Distribution	Gas fired water heater		S.F. Floor	2	2	
2040	Rain Water Drainage	Roof drains		S.F. Roof	2.21	2.21	
	D30 HVAC						
3010	Energy Supply	N/A		—	—	—	10.8 %
3020	Heat Generating Systems	N/A		—	—	—	
3030	Cooling Generating Systems	N/A		—	—	—	
3050	Terminal & Package Units	Single zone rooftop unit, gas heating, electric cooling		S.F. Floor	8.79	8.79	
3090	Other HVAC Sys. & Equipment	Underfloor garage exhaust system		S.F. Floor	.63	.63	
	D40 Fire Protection						
4010	Sprinklers	Wet pipe sprinkler system		S.F. Floor	4.22	4.22	5.4%
4020	Standpipes	Standpipe		S.F. Floor	.44	.44	
	D50 Electrical						
5010	Electrical Service/Distribution	200 ampere service, panel board and feeders		S.F. Floor	.47	.47	11.6%
5020	Lighting & Branch Wiring	T-8 fluorescent fixtures, receptacles, switches, A.C. and misc. power		S.F. Floor	6.33	6.33	
5030	Communications & Security	Addressable alarm systems, partial internet wiring and emergency lighting		S.F. Floor	3.18	3.18	
5090	Other Electrical Systems	Emergency generator, 7.5 kW		S.F. Floor	.07	.07	
E. EQUIPMENT & FURNISHINGS							
1010	Commercial Equipment	N/A		—	—	—	1.4 %
1020	Institutional Equipment	N/A		—	—	—	
1030	Vehicular Equipment	Hoists, compressor, fuel pump		S.F. Floor	1.20	1.20	
1090	Other Equipment	N/A		—	—	—	
F. SPECIAL CONSTRUCTION							
1020	Integrated Construction	N/A		—	—	—	0.0 %
1040	Special Facilities	N/A		—	—	—	
G. BUILDING SITEWORK	**N/A**						
					Sub-Total	86.92	**100%**
	CONTRACTOR FEES (General Requirements: 10%, Overhead: 5%, Profit: 10%)				*25%*	21.72	
	ARCHITECT FEES				*7%*	7.61	
					Total Building Cost	**116.25**	

Costs per square foot of floor area

Exterior Wall	S.F. Area	2000	4000	6000	8000	10000	12000	14000	16000	18000
	L.F. Perimeter	180	260	340	420	450	500	550	575	600
Concrete Block	Wood Joists	149.10	131.00	125.00	121.95	117.75	115.75	114.35	112.45	111.05
	Steel Joists	155.25	135.00	128.30	124.90	120.30	118.00	116.50	114.35	112.85
Cast in Place Concrete	Wood Joists	157.45	137.35	130.70	127.30	122.35	120.00	118.35	116.15	114.50
	Steel Joists	165.10	142.15	134.45	130.60	125.20	122.60	120.75	118.30	116.45
Stucco	Wood Frame	154.05	133.15	126.15	122.65	117.40	114.90	113.20	110.90	109.15
Insulated Metal Panels	Steel Frame	170.00	145.45	137.30	133.20	127.40	124.65	122.70	120.10	118.15
Perimeter Adj., Add or Deduct	Per 100 L.F.	24.20	12.15	8.10	6.10	4.80	4.00	3.40	3.10	2.70
Story Hgt. Adj., Add or Deduct	Per 1 Ft.	1.95	1.45	1.25	1.15	1.00	0.90	0.85	0.80	0.75
***For Basement**, add $31.05 per square foot of basement area*										

The above costs were calculated using the basic specifications shown on the facing page. These costs should be adjusted where necessary for design alternatives and owner's requirements. Reported completed project costs, for this type of structure, range from $77.10 to $231.70 per S.F.

Common additives

Description	Unit	$ Cost
Air Compressors		
Electric 1-1/2 H.P., standard controls	Ea.	1100
Dual controls	Ea.	1425
5 H.P. 115/230 Volt, standard controls	Ea.	2925
Dual controls	Ea.	4000
Product Dispenser		
with vapor recovery for 6 nozzles	Ea.	27,200
Lifts, Single post		
8000# cap., swivel arm	Ea.	8275
Two post, adjustable frames, 12,000# cap.	Ea.	5250
24,000# cap.	Ea.	10,900
30,000# cap.	Ea.	51,000
Four post, roll on ramp, 25,000# cap.	Ea.	20,200
Lockers, Steel, single tier, 60" or 72"	Opng.	222 - 365
2 tier, 60" or 72" total	Opng.	115 - 164
5 tier, box lockers	Opng.	69.50 - 78.50
Locker bench, lam. maple top only	L.F.	33
Pedestals, steel pipe	Ea.	71
Lube Equipment		
3 reel type, with pumps, no piping	Ea.	11,500
Spray Painting Booth, 26' long, complete	Ea.	13,700

Important: See the Reference Section for Location Factors.

Model costs calculated for a 1 story building with 14′ story height and 10,000 square feet of floor area

Garage, Repair

				Unit	Unit Cost	Cost Per S.F.	% Of Sub-Total
A. SUBSTRUCTURE							
1010	Standard Foundations	Poured concrete; strip and spread footings		S.F. Ground	2.50	2.50	15.5%
1020	Special Foundations	N/A		—	—	—	
1030	Slab on Grade	6″ reinforced concrete with vapor barrier and granular base		S.F. Slab	7.70	7.70	
2010	Basement Excavation	Site preparation for slab and trench for foundation wall and footing		S.F. Ground	.29	.29	
2020	Basement Walls	4′ foundation wall		L.F. Wall	74	3.31	
B. SHELL							
B10 Superstructure							
1010	Floor Construction	N/A		—	—	—	6.1 %
1020	Roof Construction	Metal deck on open web steel joists		S.F. Roof	5.42	5.42	
B20 Exterior Enclosure							
2010	Exterior Walls	Concrete block	*80% of wall*	S.F. Wall	12.26	6.18	10.4%
2020	Exterior Windows	Hopper type commercial steel	*5% of wall*	Each	495	1.04	
2030	Exterior Doors	Steel overhead and hollow metal	*15% of wall*	S.F. Door	22	2.07	
B30 Roofing							
3010	Roof Coverings	Built-up tar and gravel; perlite/EPS composite insulation		S.F. Roof	6.41	6.41	7.2%
3020	Roof Openings	Skylight		S.F. Roof	.02	.02	
C. INTERIORS							
1010	Partitions	Concrete block	*50 S.F. Floor/L.F. Partition*	S.F. Partition	21	4.24	10.3%
1020	Interior Doors	Single leaf hollow metal	*3000 S.F. Floor/Door*	Each	1133	.38	
1030	Fittings	Toilet partitions		S.F. Floor	.15	.15	
2010	Stair Construction	N/A		—	—	—	
3010	Wall Finishes	Paint		S.F. Surface	6.68	2.67	
3020	Floor Finishes	90% metallic floor hardener, 10% vinyl composition tile		S.F. Floor	1.30	1.30	
3030	Ceiling Finishes	Gypsum board on wood joists in office and washrooms	*10% of area*	S.F. Ceiling	3.97	.40	
D. SERVICES							
D10 Conveying							
1010	Elevators & Lifts	N/A		—	—	—	0.0 %
1020	Escalators & Moving Walks	N/A		—	—	—	
D20 Plumbing							
2010	Plumbing Fixtures	Toilet and service fixtures, supply and drainage	*1 Fixture/500 S.F. Floor*	Each	1495	2.99	6.5%
2020	Domestic Water Distribution	Gas fired water heater		S.F. Floor	.62	.62	
2040	Rain Water Drainage	Roof drains		S.F. Roof	2.14	2.14	
D30 HVAC							
3010	Energy Supply	N/A		—	—	—	11.1 %
3020	Heat Generating Systems	N/A		—	—	—	
3030	Cooling Generating Systems	N/A		—	—	—	
3050	Terminal & Package Units	Single zone AC unit		S.F. Floor	9	9	
3090	Other HVAC Sys. & Equipment	Garage exhaust system		S.F. Floor	.88	.88	
D40 Fire Protection							
4010	Sprinklers	Sprinklers, ordinary hazard		S.F. Floor	4.22	4.22	5.8%
4020	Standpipes	Standpipe		S.F. Floor	.94	.94	
D50 Electrical							
5010	Electrical Service/Distribution	200 ampere service, panel board and feeders		S.F. Floor	.44	.44	13.3%
5020	Lighting & Branch Wiring	T-8 fluorescent fixtures, receptacles, switches, A.C. and misc. power		S.F. Floor	7.88	7.88	
5030	Communications & Security	Addressable alarm systems, partial internet wiring and emergency lighting		S.F. Floor	3.40	3.40	
5090	Other Electrical Systems	Emergency generator, 15 kW		S.F. Floor	.09	.09	
E. EQUIPMENT & FURNISHINGS							
1010	Commercial Equipment	N/A		—	—	—	13.9 %
1020	Institutional Equipment	N/A		—	—	—	
1030	Vehicular Equipment	Hoists		S.F. Floor	12.42	12.42	
1090	Other Equipment	N/A		—	—	—	
F. SPECIAL CONSTRUCTION							
1020	Integrated Construction	N/A		—	—	—	0.0 %
1040	Special Facilities	N/A		—	—	—	
G. BUILDING SITEWORK	**N/A**						
					Sub-Total	89.10	**100%**
	CONTRACTOR FEES (General Requirements: 10%, Overhead: 5%, Profit: 10%)				*25%*	22.29	
	ARCHITECT FEES				*8%*	8.91	
					Total Building Cost	**120.30**	

Costs per square foot of floor area

Exterior Wall	S.F. Area	600	800	1000	1200	1400	1600	1800	2000	2200
	L.F. Perimeter	100	120	126	140	153	160	170	180	190
Face Brick with Concrete Block Back-up	Wood Truss	265.30	248.50	228.85	220.35	213.75	206.30	201.60	197.90	194.75
	Steel Joists	283.85	266.15	245.40	236.45	229.50	221.60	216.60	212.65	209.40
Enameled Sandwich Panel	Steel Frame	252.25	236.00	217.45	209.25	203.00	195.95	191.45	187.95	185.00
Tile on Concrete Block	Steel Joists	301.50	282.10	258.75	248.80	241.05	232.20	226.65	222.20	218.60
Aluminum Siding	Wood Frame	227.85	214.15	199.20	192.45	187.30	181.65	178.05	175.15	172.75
Wood Siding	Wood Frame	229.25	215.40	200.25	193.40	188.20	182.50	178.80	175.90	173.50
Perimeter Adj., Add or Deduct	Per 100 L.F.	113.90	85.45	68.40	57.00	48.80	42.70	37.95	34.15	31.10
Story Hgt. Adj., Add or Deduct	Per 1 Ft.	7.45	6.75	5.70	5.20	4.90	4.45	4.20	4.00	3.85
Basement—Not Applicable										

The above costs were calculated using the basic specifications shown on the facing page. These costs should be adjusted where necessary for design alternatives and owner's requirements. Reported completed project costs, for this type of structure, range from $55.55 to $280.65 per S.F.

Common additives

Description	Unit	$ Cost
Air Compressors		
Electric 1-1/2 H.P., standard controls	Ea.	1100
Dual controls	Ea.	1425
5 H.P. 115/230 volt, standard controls	Ea.	2925
Dual controls	Ea.	4000
Product Dispenser		
with vapor recovery for 6 nozzles	Ea.	27,200
Lifts, Single post		
8000# cap. swivel arm	Ea.	8275
Two post, adjustable frames, 12,000# cap.	Ea.	5250
24,000# cap.	Ea.	10,900
30,000# cap.	Ea.	51,000
Four post, roll on ramp, 25,000# cap.	Ea.	20,200
Lockers, Steel, single tier, 60" or 72"	Opng.	222 - 365
2 tier, 60" or 72" total	Opng.	115 - 164
5 tier, box lockers	Ea.	69.50 - 78.50
Locker bench, lam. maple top only	L.F.	33
Pedestals, steel pipe	Ea.	71
Lube Equipment		
3 reel type, with pumps, no piping	Ea.	11,500

Important: See the Reference Section for Location Factors.

Model costs calculated for a 1 story building with 12′ story height and 1,400 square feet of floor area

Garage, Service Station

				Unit	Unit Cost	Cost Per S.F.	% Of Sub-Total
A. SUBSTRUCTURE							
1010	Standard Foundations	Poured concrete; strip and spread footings		S.F. Ground	4.03	4.03	12.5%
1020	Special Foundations	N/A		—	—	—	
1030	Slab on Grade	6″ reinforced concrete with vapor barrier and granular base		S.F. Slab	7.70	7.70	
2010	Basement Excavation	Site preparation for slab and trench for foundation wall and footing		S.F. Ground	.96	.96	
2020	Basement Walls	4′ foundation wall		L.F. Wall	65	7.05	
B. SHELL							
	B10 Superstructure						
1010	Floor Construction	N/A		—	—	—	4.6 %
1020	Roof Construction	Plywood on wood trusses		S.F. Roof	7.32	7.32	
	B20 Exterior Enclosure						
2010	Exterior Walls	Face brick with concrete block backup	*60% of wall*	S.F. Wall	30	23.68	29.2%
2020	Exterior Windows	Store front and metal top hinged outswinging	*20% of wall*	Each	40	10.59	
2030	Exterior Doors	Steel overhead, aluminum & glass and hollow metal	*20% of wall*	Each	45	11.90	
	B30 Roofing						
3010	Roof Coverings	Asphalt shingles with flashing; perlite/EPS composite insulation		S.F. Roof	5.02	5.02	3.2%
3020	Roof Openings	N/A		—	—	—	
C. INTERIORS							
1010	Partitions	Concrete block	*25 S.F. Floor/L.F. Partition*	S.F. Partition	7.13	2.28	7.4%
1020	Interior Doors	Single leaf hollow metal	*700 S.F. Floor/Door*	Each	1133	1.62	
1030	Fittings	Toilet partitions		S.F. Floor	2.11	2.11	
2010	Stair Construction	N/A		—	—	—	
3010	Wall Finishes	Paint		S.F. Surface	5.30	3.39	
3020	Floor Finishes	Vinyl composition tile	*35% of floor area*	S.F. Floor	2.63	.92	
3030	Ceiling Finishes	Painted gypsum board on furring in sales area & washrooms	*35% of floor area*	S.F. Ceiling	4.12	1.44	
D. SERVICES							
	D10 Conveying						
1010	Elevators & Lifts	N/A		—	—	—	0.0 %
1020	Escalators & Moving Walks	N/A		—	—	—	
	D20 Plumbing						
2010	Plumbing Fixtures	Toilet and service fixtures, supply and drainage	*1 Fixture/235 S.F. Floor*	Each	2056	8.75	8.3%
2020	Domestic Water Distribution	Gas fired water heater		S.F. Floor	4.47	4.47	
2040	Rain Water Drainage	N/A		—	—	—	
	D30 HVAC						
3010	Energy Supply	N/A		—	—	—	14.3 %
3020	Heat Generating Systems	N/A		—	—	—	
3030	Cooling Generating Systems	N/A		—	—	—	
3050	Terminal & Package Units	Single zone AC unit		S.F. Floor	17.33	17.33	
3090	Other HVAC Sys. & Equipment	Underfloor exhaust system		S.F. Floor	5.27	5.27	
	D40 Fire Protection						
4010	Sprinklers	Wet pipe sprinkler system		S.F. Floor	8.92	8.92	8.1%
4020	Standpipes	Standpipe		S.F. Floor	3.97	3.97	
	D50 Electrical						
5010	Electrical Service/Distribution	200 ampere service, panel board and feeders		S.F. Floor	4.70	4.70	12.4%
5020	Lighting & Branch Wiring	High efficiency fluorescent fixtures, receptacles, switches, A.C. and misc. power		S.F. Floor	6	6	
5030	Communications & Security	Addressable alarm systems and emergency lighting		S.F. Floor	8.28	8.28	
5090	Other Electrical Systems	Emergency generator		S.F. Floor	.65	.65	
E. EQUIPMENT & FURNISHINGS							
1010	Commercial Equipment	N/A		—	—	—	0.0 %
1020	Institutional Equipment	N/A		—	—	—	
1030	Vehicular Equipment	N/A		—	—	—	
1090	Other Equipment	N/A		—	—	—	
F. SPECIAL CONSTRUCTION							
1020	Integrated Construction	N/A		—	—	—	0.0 %
1040	Special Facilities	N/A		—	—	—	
G. BUILDING SITEWORK		**N/A**					
					Sub-Total	158.35	**100%**
	CONTRACTOR FEES (General Requirements: 10%, Overhead: 5%, Profit: 10%)				*25%*	39.56	
	ARCHITECT FEES				*8%*	15.84	
					Total Building Cost	**213.75**	

Costs per square foot of floor area

Exterior Wall	S.F. Area	12000	16000	20000	25000	30000	35000	40000	45000	50000
	L.F. Perimeter	440	520	600	700	708	780	841	910	979
Reinforced Concrete Block	Lam. Wood Arches	165.15	160.90	158.30	156.25	152.50	151.25	150.10	149.35	148.75
	Rigid Steel Frame	173.75	169.25	166.55	164.40	160.45	159.15	157.90	157.15	156.50
Face Brick with Concrete Block Back-up	Lam. Wood Arches	186.45	179.75	175.70	172.50	166.20	164.15	162.30	161.10	160.05
	Rigid Steel Frame	193.60	186.90	182.90	179.65	173.35	171.40	169.50	168.25	167.30
Metal Sandwich Panels	Lam. Wood Arches	175.25	169.75	166.55	163.90	158.95	157.35	155.85	154.90	154.10
	Rigid Steel Frame	185.75	180.05	176.55	173.70	168.35	166.65	165.00	164.00	163.10
Perimeter Adj., Add or Deduct	Per 100 L.F.	6.40	4.75	3.85	3.10	2.55	2.25	1.90	1.70	1.50
Story Hgt. Adj., Add or Deduct	Per 1 Ft.	0.95	0.80	0.75	0.70	0.60	0.60	0.55	0.50	0.45
Basement—Not Applicable										

The above costs were calculated using the basic specifications shown on the facing page. These costs should be adjusted where necessary for design alternatives and owner's requirements. Reported completed project costs, for this type of structure, range from $84.30 to $251.50 per S.F.

Common additives

Description	Unit	$ Cost
Bleachers, Telescoping, manual		
To 15 tier	Seat	133 - 189
16-20 tier	Seat	284 - 345
21-30 tier	Seat	300 - 395
For power operation, add	Seat	57 - 89
Gym Divider Curtain, Mesh top		
Manual roll-up	S.F.	12.25
Gym Mats		
2" naugahyde covered	S.F.	5.15
2" nylon	S.F.	9.10
1-1/2" wall pads	S.F.	6.85
1" wrestling mats	S.F.	5.15
Scoreboard		
Basketball, one side	Ea.	3525 - 13,500
Basketball Backstop		
Wall mtd., 6' extended, fixed	Ea.	2525 - 2850
Swing up, wall mtd.	Ea.	2625 - 4100

Description	Unit	$ Cost
Lockers, Steel, single tier, 60" or 72"	Opng.	222 - 365
2 tier, 60" or 72" total	Opng.	115 - 164
5 tier, box lockers	Opng.	69.50 - 78.50
Locker bench, lam. maple top only	L.F.	33
Pedestals, steel pipe	Ea.	71
Sound System		
Amplifier, 250 watts	Ea.	2125
Speaker, ceiling or wall	Ea.	206
Trumpet	Ea.	395
Emergency Lighting, 25 watt, battery operated		
Lead battery	Ea.	325
Nickel cadmium	Ea.	665

Important: See the Reference Section for Location Factors.

Model costs calculated for a 1 story building with 25′ story height and 20,000 square feet of floor area

Gymnasium

				Unit	Unit Cost	Cost Per S.F.	% Of Sub-Total
A. SUBSTRUCTURE							
1010	Standard Foundations	Poured concrete; strip and spread footings		S.F. Ground	1.42	1.42	7.3%
1020	Special Foundations	N/A		—	—	—	
1030	Slab on Grade	4″ reinforced concrete with vapor barrier and granular base		S.F. Slab	5.10	5.10	
2010	Basement Excavation	Site preparation for slab and trench for foundation wall and footing		S.F. Ground	.16	.16	
2020	Basement Walls	4′ foundation wall		L.F. Wall	65	1.94	
B. SHELL							
	B10 Superstructure						
1010	Floor Construction	Steel column		S.F. Floor	2.25	2.25	17.4%
1020	Roof Construction	Wood deck on laminated wood arches		S.F. Roof	18.35	18.35	
	B20 Exterior Enclosure						
2010	Exterior Walls	Reinforced concrete block (end walls included)	*90% of wall*	S.F. Wall	10.83	7.31	10.8%
2020	Exterior Windows	Metal horizontal pivoted	*10% of wall*	Each	647	4.86	
2030	Exterior Doors	Aluminum and glass, hollow metal, steel overhead		Each	2138	.65	
	B30 Roofing						
3010	Roof Coverings	EPDM, 60 mils, fully adhered; polyisocyanurate insulation		S.F. Roof	5.19	5.19	4.4%
3020	Roof Openings	N/A		—	—	—	
C. INTERIORS							
1010	Partitions	Concrete block	*50 S.F. Floor/L.F. Partition*	S.F. Partition	8.80	1.76	19.7%
1020	Interior Doors	Single leaf hollow metal	*500 S.F. Floor/Door*	Each	1133	2.27	
1030	Fittings	Toilet partitions		S.F. Floor	.30	.30	
2010	Stair Construction	N/A		—	—	—	
3010	Wall Finishes	50% paint, 50% ceramic tile		S.F. Surface	9.30	3.72	
3020	Floor Finishes	90% hardwood, 10% ceramic tile		S.F. Floor	14.32	14.32	
3030	Ceiling Finishes	Mineral fiber tile on concealed zee bars	*15% of area*	S.F. Ceiling	6.57	.98	
D. SERVICES							
	D10 Conveying						
1010	Elevators & Lifts	N/A		—	—	—	0.0 %
1020	Escalators & Moving Walks	N/A		—	—	—	
	D20 Plumbing						
2010	Plumbing Fixtures	Toilet and service fixtures, supply and drainage	*1 Fixture/515 S.F. Floor*	Each	4378	8.50	11.2%
2020	Domestic Water Distribution	Electric water heater		S.F. Floor	4.79	4.79	
2040	Rain Water Drainage	N/A		—	—	—	
	D30 HVAC						
3010	Energy Supply	N/A		—	—	—	9.1 %
3020	Heat Generating Systems	Included in D3050		—	—	—	
3030	Cooling Generating Systems	N/A		—	—	—	
3050	Terminal & Package Units	Single zone rooftop unit, gas heating, electric cooling		S.F. Floor	10.80	10.80	
3090	Other HVAC Sys. & Equipment	N/A		—	—	—	
	D40 Fire Protection						
4010	Sprinklers	Wet pipe sprinkler system		S.F. Floor	3.28	3.28	3.6%
4020	Standpipes	Standpipe		S.F. Floor	1.01	1.01	
	D50 Electrical						
5010	Electrical Service/Distribution	400 ampere service, panel board and feeders		S.F. Floor	1.08	1.08	10.9%
5020	Lighting & Branch Wiring	High efficiency fluorescent fixtures, receptacles, switches, A.C. and misc. power		S.F. Floor	8.77	8.77	
5030	Communications & Security	Addressable alarm systems, sound system and emergency lighting		S.F. Floor	2.81	2.81	
5090	Other Electrical Systems	Emergency generator, 7.5 kW		S.F. Floor	.22	.22	
E. EQUIPMENT & FURNISHINGS							
1010	Commercial Equipment	N/A		—	—	—	5.5 %
1020	Institutional Equipment	N/A		—	—	—	
1030	Vehicular Equipment	N/A		—	—	—	
1090	Other Equipment	Bleachers, sauna, weight room		S.F. Floor	6.53	6.53	
F. SPECIAL CONSTRUCTION							
1020	Integrated Construction	N/A		—	—	—	0.0 %
1040	Special Facilities	N/A		—	—	—	
G. BUILDING SITEWORK		**N/A**					

	Sub-Total	118.37	**100%**
CONTRACTOR FEES (General Requirements: 10%, Overhead: 5%, Profit: 10%)	*25%*	29.57	
ARCHITECT FEES	*7%*	10.36	
	Total Building Cost	**158.30**	

Costs per square foot of floor area

Exterior Wall	S.F. Area	1000	2000	3000	4000	5000	10000	15000	20000	25000
	L.F. Perimeter	126	179	219	253	283	400	490	568	632
Decorative Concrete Block	Steel Frame	282.45	237.25	219.75	210.20	204.10	189.90	184.15	181.00	178.75
	Bearing Walls	279.50	233.90	216.15	206.55	200.30	185.90	180.15	176.90	174.60
Face Brick with Concrete Block Back-up	Steel Frame	309.10	256.10	235.10	223.45	215.90	198.15	190.90	186.80	183.90
	Bearing Walls	306.40	252.90	231.70	220.00	212.35	194.40	187.10	182.95	179.95
Metal Sandwich Panel	Steel Frame	282.55	237.40	219.90	210.40	204.25	190.10	184.40	181.20	178.95
Precast Concrete Panel	Bearing Walls	274.60	230.40	213.30	204.05	198.10	184.35	178.85	175.80	173.65
Perimeter Adj., Add or Deduct	Per 100 L.F.	55.10	27.55	18.40	13.80	10.95	5.45	3.65	2.75	2.20
Story Hgt. Adj., Add or Deduct	Per 1 Ft.	3.60	2.60	2.10	1.85	1.60	1.10	1.00	0.85	0.75
***For Basement**, add $ 30.35 per square foot of basement area*										

The above costs were calculated using the basic specifications shown on the facing page. These costs should be adjusted where necessary for design alternatives and owner's requirements. Reported completed project costs, for this type of structure, range from $ 112.85 to $ 275.80 per S.F.

Common additives

Description	Unit	$ Cost
Closed Circuit Surveillance, One station		
Camera and monitor	Ea.	1975
For additional camera stations, add	Ea.	1075
Emergency Lighting, 25 watt, battery operated		
Lead battery	Ea.	325
Nickel cadmium	Ea.	665
Laundry Equipment		
Dryers, coin operated 30 lb.	Ea.	3900
Double stacked	Ea.	8550
50 lb.	Ea.	4375
Dry cleaner 20 lb.	Ea.	41,100
30 lb.	Ea.	62,500
Washers, coin operated	Ea.	1575
Washer/extractor 20 lb.	Ea.	7200
30 lb.	Ea.	11,400
50 lb.	Ea.	13,400
75 lb.	Ea.	25,600
Smoke Detectors		
Ceiling type	Ea.	214
Duct type	Ea.	540

Important: See the Reference Section for Location Factors.

Model costs calculated for a 1 story building with 12′ story height and 3,000 square feet of floor area

Laundromat

				Unit	Unit Cost	Cost Per S.F.	% Of Sub-Total
A. SUBSTRUCTURE							
1010	Standard Foundations	Poured concrete; strip and spread footings		S.F. Ground	3.24	3.24	8.5%
1020	Special Foundations	N/A		—	—	—	
1030	Slab on Grade	5″ reinforced concrete with vapor barrier and granular base		S.F. Slab	5.54	5.54	
2010	Basement Excavation	Site preparation for slab and trench for foundation wall and footing		S.F. Ground	.51	.51	
2020	Basement Walls	4′ Foundation wall		L.F. Wall	65	4.71	
B. SHELL							
	B10 Superstructure						
1010	Floor Construction	Steel column fireproofing		S.F. Floor	.84	.84	6.5%
1020	Roof Construction	Metal deck, open web steel joists, beams, columns		S.F. Roof	9.90	9.90	
	B20 Exterior Enclosure						
2010	Exterior Walls	Decorative concrete block	*90% of wall*	S.F. Wall	14.99	11.82	10.9%
2020	Exterior Windows	Store front	*10% of wall*	Each	43	3.73	
2030	Exterior Doors	Double aluminum and glass		Each	3438	2.29	
	B30 Roofing						
3010	Roof Coverings	Built-up tar and gravel with flashing; perlite/EPS composite insulation		S.F. Roof	8.41	8.41	5.1%
3020	Roof Openings	N/A		—	—	—	
C. INTERIORS							
1010	Partitions	Gypsum board on metal studs	*60 S.F. Floor/L.F. Partition*	S.F. Partition	25	4.15	6.9%
1020	Interior Doors	Single leaf wood	*750 S.F. Floor/Door*	Each	593	.79	
1030	Fittings	N/A		—	—	—	
2010	Stair Construction	N/A		—	—	—	
3010	Wall Finishes	Paint		S.F. Surface	.75	.25	
3020	Floor Finishes	Vinyl composition tile		S.F. Floor	2.21	2.21	
3030	Ceiling Finishes	Fiberglass board on exposed grid system		S.F. Ceiling	3.97	3.97	
D. SERVICES							
	D10 Conveying						
1010	Elevators & Lifts	N/A		—	—	—	0.0 %
1020	Escalators & Moving Walks	N/A		—	—	—	
	D20 Plumbing						
2010	Plumbing Fixtures	Toilet and service fixtures, supply and drainage	*1 Fixture/600 S.F. Floor*	Each	6318	10.53	36.7%
2020	Domestic Water Distribution	Gas fired hot water heater		S.F. Floor	48	47.94	
2040	Rain Water Drainage	Roof drains		S.F. Roof	1.82	1.82	
	D30 HVAC						
3010	Energy Supply	N/A		—	—	—	5.0 %
3020	Heat Generating Systems	Included in D3050		—	—	—	
3030	Cooling Generating Systems	N/A		—	—	—	
3050	Terminal & Package Units	Rooftop single zone unit systems		S.F. Floor	8.23	8.23	
3090	Other HVAC Sys. & Equipment	N/A		—	—	—	
	D40 Fire Protection						
4010	Sprinklers	Sprinkler, ordinary hazard		S.F. Floor	4.99	4.99	3.0%
4020	Standpipes	N/A		—	—	—	
	D50 Electrical						
5010	Electrical Service/Distribution	200 ampere service, panel board and feeders		S.F. Floor	3.98	3.98	17.3%
5020	Lighting & Branch Wiring	High efficiency fluorescent fixtures, receptacles, switches, A.C. and misc. power		S.F. Floor	21	20.90	
5030	Communications & Security	Addressable alarm systems and emergency lighting		S.F. Floor	3.03	3.03	
5090	Other Electrical Systems	Emergency generator, 7.5 kW		S.F. Floor	.51	.51	
E. EQUIPMENT & FURNISHINGS							
1010	Commercial Equipment	N/A		—	—	—	0.0 %
1020	Institutional Equipment	N/A		—	—	—	
1030	Vehicular Equipment	N/A		—	—	—	
1090	Other Equipment	N/A		—	—	—	
F. SPECIAL CONSTRUCTION							
1020	Integrated Construction	N/A		—	—	—	0.0 %
1040	Special Facilities	N/A		—	—	—	
G. BUILDING SITEWORK		**N/A**					
					Sub-Total	164.29	**100%**
	CONTRACTOR FEES (General Requirements: 10%, Overhead: 5%, Profit: 10%)				*25%*	41.08	
	ARCHITECT FEES				*7%*	14.38	
					Total Building Cost	**219.75**	

Costs per square foot of floor area

Exterior Wall	S.F. Area	7000	10000	13000	16000	19000	22000	25000	28000	31000
	L.F. Perimeter	240	300	336	386	411	435	472	510	524
Face Brick with Concrete Block Back-up	R/Conc. Frame	180.60	171.65	163.95	160.45	156.15	**152.85**	151.20	149.95	147.70
	Steel Frame	183.00	174.05	166.30	162.90	158.60	155.25	153.65	152.40	150.15
Limestone with Concrete Block	R/Conc. Frame	214.35	199.55	187.10	181.40	174.50	169.25	166.60	164.55	161.05
	Steel Frame	211.45	198.95	187.80	182.95	176.50	171.70	169.30	167.50	164.20
Precast Concrete Panels	R/Conc. Frame	206.00	192.20	180.75	175.50	169.20	164.40	162.00	160.15	156.95
	Steel Frame	208.40	194.60	183.20	177.95	171.65	166.85	164.40	162.55	159.35
Perimeter Adj., Add or Deduct	Per 100 L.F.	21.95	15.40	11.85	9.65	8.05	7.00	6.15	5.55	5.05
Story Hgt. Adj., Add or Deduct	Per 1 Ft.	3.30	2.85	2.45	2.40	2.05	1.90	1.80	1.80	1.65
***For Basement**, add $ 43.25 per square foot of basement area*										

The above costs were calculated using the basic specifications shown on the facing page. These costs should be adjusted where necessary for design alternatives and owner's requirements. Reported completed project costs, for this type of structure, range from $ 108.45 to $ 278.25 per S.F.

Common additives

Description	Unit	$ Cost
Carrels Hardwood	Ea.	925 - 2375
Closed Circuit Surveillance, One station		
Camera and monitor	Ea.	1975
For additional camera stations, add	Ea.	1075
Elevators, Hydraulic passenger, 2 stops		
1500# capacity	Ea.	63,900
2500# capacity	Ea.	66,900
3500# capacity	Ea.	71,900
Emergency Lighting, 25 watt, battery operated		
Lead battery	Ea.	325
Nickel cadmium	Ea.	665
Flagpoles, Complete		
Aluminum, 20' high	Ea.	1825
40' high	Ea.	4300
70' high	Ea.	11,300
Fiberglass, 23' high	Ea.	1275
39'-5" high	Ea.	3275
59' high	Ea.	6475

Description	Unit	$ Cost
Library Furnishings		
Bookshelf, 90" high, 10" shelf double face	L.F.	280
single face	L.F.	176
Charging desk, built-in with counter		
Plastic laminated top	L.F.	410
Reading table, laminated		
top 60" x 36"	Ea.	775

Important: See the Reference Section for Location Factors.

Model costs calculated for a 2 story building with 14′ story height and 22,000 square feet of floor area

Library

				Unit	Unit Cost	Cost Per S.F.	% Of Sub-Total
A. SUBSTRUCTURE							
1010	Standard Foundations	Poured concrete; strip and spread footings		S.F. Ground	3.64	1.82	5.3%
1020	Special Foundations	N/A		—	—	—	
1030	Slab on Grade	4″ reinforced concrete with vapor barrier and granular base		S.F. Slab	5.10	2.56	
2010	Basement Excavation	Site preparation for slab and trench for foundation wall and footing		S.F. Ground	.29	.15	
2020	Basement Walls	4′ foundation wall		L.F. Wall	74	1.45	
B. SHELL							
B10 Superstructure							
1010	Floor Construction	Concrete waffle slab		S.F. Floor	23	11.69	18.3%
1020	Roof Construction	Concrete waffle slab		S.F. Roof	18.06	9.03	
B20 Exterior Enclosure							
2010	Exterior Walls	Face brick with concrete block backup	*90% of wall*	S.F. Wall	30	14.82	16.1%
2020	Exterior Windows	Window wall	*10% of wall*	Each	52	2.87	
2030	Exterior Doors	Double aluminum and glass, single leaf hollow metal		Each	5525	.50	
B30 Roofing							
3010	Roof Coverings	Single ply membrane, EPDM, fully adhered; perlite/EPS composite insulation		S.F. Roof	5.26	2.63	2.4%
3020	Roof Openings	Roof hatches		S.F. Roof	.08	.04	
C. INTERIORS							
1010	Partitions	Gypsum board on metal studs	*30 S.F. Floor/L.F. Partition*	S.F. Partition	11.30	4.52	16.2%
1020	Interior Doors	Single leaf wood	*300 S.F. Floor/Door*	Each	593	1.98	
1030	Fittings	N/A		—	—	—	
2010	Stair Construction	Concrete filled metal pan		Flight	8950	.81	
3010	Wall Finishes	Paint		S.F. Surface	.76	.61	
3020	Floor Finishes	50% carpet, 50% vinyl tile		S.F. Floor	3.80	3.80	
3030	Ceiling Finishes	Mineral fiber on concealed zee bars		S.F. Ceiling	6.57	6.57	
D. SERVICES							
D10 Conveying							
1010	Elevators & Lifts	One hydraulic passenger elevator		Each	84,480	3.84	3.4%
1020	Escalators & Moving Walks	N/A		—	—	—	
D20 Plumbing							
2010	Plumbing Fixtures	Toilet and service fixtures, supply and drainage	*1 Fixture/1835 S.F. Floor*	Each	6973	3.80	5.0%
2020	Domestic Water Distribution	Gas fired water heater		S.F. Floor	1.26	1.26	
2040	Rain Water Drainage	Roof drains		S.F. Roof	1.14	.57	
D30 HVAC							
3010	Energy Supply	N/A		—	—	—	17.7 %
3020	Heat Generating Systems	Included in D3050		—	—	—	
3030	Cooling Generating Systems	N/A		—	—	—	
3050	Terminal & Package Units	Multizone unit, gas heating, electric cooling		S.F. Floor	20	20	
3090	Other HVAC Sys. & Equipment	N/A		—	—	—	
D40 Fire Protection							
4010	Sprinklers	Wet pipe sprinkler system		S.F. Floor	2.79	2.79	3.3%
4020	Standpipes	Standpipe		S.F. Floor	.96	.96	
D50 Electrical							
5010	Electrical Service/Distribution	400 ampere service, panel board and feeders		S.F. Floor	1.14	1.14	12.5%
5020	Lighting & Branch Wiring	High efficiency fluorescent fixtures, receptacles, switches, A.C. and misc. power		S.F. Floor	10.77	10.77	
5030	Communications & Security	Addressable alarm systems, internet wiring, and emergency lighting		S.F. Floor	1.97	1.97	
5090	Other Electrical Systems	Emergency generator, 7.5 kW, Uninterruptible power supply		S.F. Floor	.26	.26	
E. EQUIPMENT & FURNISHINGS							
1010	Commercial Equipment	N/A		—	—	—	0.0 %
1020	Institutional Equipment	N/A		—	—	—	
1030	Vehicular Equipment	N/A		—	—	—	
1090	Other Equipment	N/A		—	—	—	
F. SPECIAL CONSTRUCTION							
1020	Integrated Construction	N/A		—	—	—	0.0 %
1040	Special Facilities	N/A		—	—	—	
G. BUILDING SITEWORK		**N/A**					
					Sub-Total	113.21	**100%**
		CONTRACTOR FEES (General Requirements: 10%, Overhead: 5%, Profit: 10%)			*25%*	28.32	
		ARCHITECT FEES			*8%*	11.32	
					Total Building Cost	**152.85**	

M.400 Medical Office, 1 Story

Costs per square foot of floor area

Exterior Wall	S.F. Area	4000	5500	7000	8500	10000	11500	13000	14500	16000
	L.F. Perimeter	280	320	380	440	453	503	510	522	560
Vinyl Siding	Wood Frame	161.90	153.80	150.85	148.90	144.75	143.65	140.75	138.75	138.00
Stone Veneer	Wood Frame	177.95	167.30	163.45	161.05	155.50	154.00	150.20	147.45	146.50
Fiber Cement	Steel Frame	163.40	155.15	152.15	150.25	146.05	144.85	141.95	139.85	139.15
E.I.F.S.	Steel Frame	168.35	159.60	156.40	154.40	149.90	148.70	145.55	143.35	142.60
Precast Concrete Panel	Reinforced Concrete	196.50	183.65	179.15	176.25	169.55	167.80	163.10	159.75	158.65
Brick Veneer	Reinforced Concrete	204.60	190.55	185.65	182.55	175.15	173.25	168.10	164.45	163.20
Perimeter Adj., Add or Deduct	Per 100 L.F.	15.05	10.90	8.60	7.10	6.05	5.25	4.60	4.10	3.75
Story Hgt. Adj., Add or Deduct	Per 1 Ft.	2.05	1.65	1.55	1.50	1.40	1.25	1.15	1.05	1.00

***For Basement**, add $28.15 per square foot of basement area*

The above costs were calculated using the basic specifications shown on the facing page. These costs should be adjusted where necessary for design alternatives and owner's requirements. Reported completed project costs, for this type of structure, range from $91.35 to $235.00 per S.F.

Common additives

Description	Unit	$ Cost
Cabinets, Hospital, base		
Laminated plastic	L.F.	400
Stainless steel	L.F.	655
Counter top, laminated plastic	L.F.	77.50
Stainless steel	L.F.	195
For drop-in sink, add	Ea.	1100
Nurses station, door type		
Laminated plastic	L.F.	450
Enameled steel	L.F.	435
Stainless steel	L.F.	770
Wall cabinets, laminated plastic	L.F.	294
Enameled steel	L.F.	345
Stainless steel	L.F.	625

Description	Unit	$ Cost
Directory Boards, Plastic, glass covered		
30" x 20"	Ea.	555
36" x 48"	Ea.	1425
Aluminum, 24" x 18"	Ea.	600
36" x 24"	Ea.	715
48" x 32"	Ea.	1000
48" x 60"	Ea.	2100
Heat Therapy Unit		
Humidified, 26" x 78" x 28"	Ea.	3450
Smoke Detectors		
Ceiling type	Ea.	214
Duct type	Ea.	540
Tables, Examining, vinyl top		
with base cabinets	Ea.	1600 - 6675
Utensil Washer, Sanitizer	Ea.	9600
X-Ray, Mobile	Ea.	18,500 - 85,500

Important: See the Reference Section for Location Factors.

Model costs calculated for a 1 story building with 10′ story height and 7,000 square feet of floor area

Medical Office, 1 Story

				Unit	Unit Cost	Cost Per S.F.	% Of Sub-Total
A. SUBSTRUCTURE							
1010	Standard Foundations	Poured concrete; strip and spread footings		S.F. Ground	3.34	3.34	10.6%
1020	Special Foundations	N/A		—	—	—	
1030	Slab on Grade	4″ reinforced slab on grade		S.F. Slab	5.10	5.10	
2010	Basement Excavation	Site preparation for slab and trench for foundation wall and footing		S.F. Ground	.29	.29	
2020	Basement Walls	4′ Foundation wall		L.F. Wall	74	3.99	
B. SHELL							
	B10 Superstructure						
1010	Floor Construction	Wood columns, fireproofed		S.F. Floor	3.08	3.08	8.7%
1020	Roof Construction	Wood roof trusses, 4:12 slope		S.F. Roof	7.37	7.37	
	B20 Exterior Enclosure						
2010	Exterior Walls	Ashlar stone veneer on wood stud back up	*70% of wall*	S.F. Wall	38	14.53	17.9%
2020	Exterior Windows	Double hung, insulated wood windows	*30% of wall*	Each	603	5.77	
2030	Exterior Doors	Single leaf solid core wood doors		Each	1426	1.23	
	B30 Roofing						
3010	Roof Coverings	Asphalt shingles with flashing, gutters and downspouts		S.F. Roof	2.50	2.50	2.1%
3020	Roof Openings	N/A		—	—	—	
C. INTERIORS							
1010	Partitions	Fire rated gypsum board on wood studs	*6 S.F. Floor/L.F. Partition*	S.F. Partition	7	9.33	20.5%
1020	Interior Doors	Single leaf solid core wood doors	*240 S.F. Floor/Door*	Each	593	2.47	
1030	Fittings	N/A		—	—	—	
2010	Stair Construction	N/A		—	—	—	
3010	Wall Finishes	Paint on drywall		S.F. Surface	.87	2.32	
3020	Floor Finishes	50% carpet , 50% vinyl composition tile		S.F. Floor	4.45	4.45	
3030	Ceiling Finishes	Acoustic ceiling tiles on suspended support system		S.F. Ceiling	6.02	6.02	
D. SERVICES							
	D10 Conveying						
1010	Elevators & Lifts	N/A		—	—	—	0.0 %
1020	Escalators & Moving Walks	N/A		—	—	—	
	D20 Plumbing						
2010	Plumbing Fixtures	Restroom, exam room and service fixtures, supply and drainage	*1 Fixture/350 S.F. Floor*	Each	1813	5.18	6.0%
2020	Domestic Water Distribution	Gas fired water heater		S.F. Floor	2	2	
2040	Rain Water Drainage	Roof drains		S.F. Roof			
	D30 HVAC						
3010	Energy Supply	N/A		—	—	—	12.1 %
3020	Heat Generating Systems	Included in D3050		—	—	—	
3030	Cooling Generating Systems	N/A		—	—	—	
3050	Terminal & Package Units	Multizone unit, gas heating , electric cooling		S.F. Floor	14.55	14.55	
3090	Other HVAC Sys. & Equipment	N/A		—	—	—	
	D40 Fire Protection						
4010	Sprinklers	Wet pipe spinkler system		S.F. Floor	4.57	4.57	4.8%
4020	Standpipes	Standpipes and hose systems, with pumps		S.F. Floor	1.22	1.22	
	D50 Electrical						
5010	Electrical Service/Distribution	200 Ampere service, panel boards and feeders		S.F. Floor	1.58	1.58	13.7%
5020	Lighting & Branch Wiring	Fluorescent fixtures, receptacles, switches, A.C. and misc. power		S.F. Floor	7.41	7.41	
5030	Communications & Security	Alarm systems, emergency lighting, internet and phone wiring		S.F. Floor	7.39	7.39	
5090	Other Electrical Systems	N/A		—	—	—	
E. EQUIPMENT & FURNISHINGS							
1010	Commercial Equipment	N/A		—	—	—	3.6 %
1020	Institutional Equipment	Exam room equipment, cabinets and countertops		S.F. Floor	4.30	4.30	
1030	Vehicular Equipment	N/A		—	—	—	
1090	Other Equipment	N/A		—	—	—	
F. SPECIAL CONSTRUCTION							
1020	Integrated Construction	N/A		—	—	—	0.0 %
1040	Special Facilities	N/A		—	—	—	
G. BUILDING SITEWORK	**N/A**						
					Sub-Total	119.99	**100%**
	CONTRACTOR FEES (General Requirements: 10%, Overhead: 5%, Profit: 10%)				25%	29.96	
	ARCHITECT FEES				9%	13.50	
					Total Building Cost	**163.45**	

M.410 Medical Office, 2 Story

Costs per square foot of floor area

Exterior Wall	S.F. Area	4000	5500	7000	8500	10000	11500	13000	14500	16000
	L.F. Perimeter	180	210	240	270	286	311	336	361	386
Vinyl Siding	Wood Frame	195.95	187.00	181.95	178.65	175.10	173.15	171.70	170.55	169.60
Stone Veneer	Wood Frame	213.20	201.50	194.85	190.55	185.70	183.15	181.15	179.60	178.30
Fiber Cement Siding	Steel Frame	202.90	193.50	188.10	184.65	180.90	178.85	177.25	176.05	175.00
E.I.F.S.	Steel Frame	194.25	185.00	179.70	176.30	172.55	170.60	169.05	167.85	166.80
Precast Concrete Panel	Reinforced Concrete	237.65	223.90	216.05	211.00	205.25	202.20	199.85	198.00	196.50
Brick Veneer	Reinforced Concrete	200.55	188.60	181.85	177.45	172.50	169.90	167.90	166.30	165.00
Perimeter Adj., Add or Deduct	Per 100 L.F.	22.30	16.30	12.80	10.55	8.95	7.90	6.90	6.10	5.55
Story Hgt. Adj., Add or Deduct	Per 1 Ft.	2.45	2.15	1.95	1.80	1.55	1.60	1.50	1.40	1.35
***For Basement**, add $31.30 per square foot of basement area*										

The above costs were calculated using the basic specifications shown on the facing page. These costs should be adjusted where necessary for design alternatives and owner's requirements. Reported completed project costs, for this type of structure, range from $93.80 to $320.35 per S.F.

Common additives

Description	Unit	$ Cost
Cabinets, Hospital, base		
Laminated plastic	L.F.	400
Stainless steel	L.F.	655
Counter top, laminated plastic	L.F.	77.50
Stainless steel	L.F.	195
For drop-in sink, add	Ea.	1100
Nurses station, door type		
Laminated plastic	L.F.	450
Enameled steel	L.F.	435
Stainless steel	L.F.	770
Wall cabinets, laminated plastic	L.F.	294
Enameled steel	L.F.	345
Stainless steel	L.F.	625
Elevators, Hydraulic passenger, 2 stops		
1500# capacity	Ea.	63,900
2500# capacity	Ea.	66,900
3500# capacity	Ea.	71,900

Description	Unit	$ Cost
Directory Boards, Plastic, glass covered		
30" x 20"	Ea.	555
36" x 48"	Ea.	1425
Aluminum, 24" x 18"	Ea.	600
36" x 24"	Ea.	715
48" x 32"	Ea.	1000
48" x 60"	Ea.	2100
Emergency Lighting, 25 watt, battery operated		
Lead battery	Ea.	325
Nickel cadmium	Ea.	665
Heat Therapy Unit		
Humidified, 26" x 78" x 28"	Ea.	3450
Smoke Detectors		
Ceiling type	Ea.	214
Duct type	Ea.	540
Tables, Examining, vinyl top with base cabinets	Ea.	1600 - 6675
Utensil Washer, Sanitizer	Ea.	9600
X-Ray, Mobile	Ea.	18,500 - 85,500

Important: See the Reference Section for Location Factors.

Model costs calculated for a 2 story building with 10' story height and 7,000 square feet of floor area

Medical Office, 2 Story

				Unit	Unit Cost	Cost Per S.F.	% Of Sub-Total
A. SUBSTRUCTURE							
1010	Standard Foundations	Poured concrete; strip and spread footings		S.F. Ground	4.80	2.40	
1020	Special Foundations	N/A		—	—	—	
1030	Slab on Grade	4" reinforced slab on grade		S.F. Slab	5.10	2.56	5.6%
2010	Basement Excavation	Site preparation for slab and trench for foundation wall and footing		S.F. Ground	.51	.26	
2020	Basement Walls	4' Foundation wall		L.F. Wall	74	2.52	
B. SHELL							
	B10 Superstructure						
1010	Floor Construction	Open web bar joist, slab form, fireproofed columns		S.F. Floor	14.72	7.36	8.9%
1020	Roof Construction	Metal deck, open web steel joists, beams, columns		S.F. Roof	9.94	4.97	
	B20 Exterior Enclosure						
2010	Exterior Walls	Fiber cement siding on metal studs	*70% of wall*	S.F. Wall	14.81	7.11	
2020	Exterior Windows	Aluminum sliding windows	*30% of wall*	Each	495	6.79	12.3%
2030	Exterior Doors	Aluminum & glass entrance doors		Each	5525	3.15	
	B30 Roofing						
3010	Roof Coverings	Single ply membrane, stone ballast, rigid insulation		S.F. Roof	7.16	3.58	3.2%
3020	Roof Openings	Roof and smoke hatches		S.F. Roof	1.68	.84	
C. INTERIORS							
1010	Partitions	Fire rated gypsum board on metal studs	*6 S.F. Floor/L.F. Partition*	S.F. Partition	5.21	6.94	
1020	Interior Doors	Single leaf solid core wood doors	*240 S.F. Floor/Door*	Each	593	2.80	
1030	Fittings	N/A		—	—	—	
2010	Stair Construction	Concrete filled metal pan, with rails		Flight	14,075	10.05	23.7%
3010	Wall Finishes	Paint on drywall		S.F. Surface	.92	2.45	
3020	Floor Finishes	50% carpet , 50% vinyl composition tile		S.F. Floor	4.45	4.45	
3030	Ceiling Finishes	Acoustic ceiling tiles on suspended support system		S.F. Ceiling	6.02	6.02	
D. SERVICES							
	D10 Conveying						
1010	Elevators & Lifts	One hydraulic elevator		Each	106,050	15.15	11.0%
1020	Escalators & Moving Walks	N/A		—	—	—	
	D20 Plumbing						
2010	Plumbing Fixtures	Restroom, exam room and service fixtures, supply and drainage	*1 Fixture/350 S.F. Floor*	Each	1904	5.44	
2020	Domestic Water Distribution	Gas fired water heater		S.F. Floor	2	2	6.1%
2040	Rain Water Drainage	Roof drain		S.F. Roof	2.06	1.03	
	D30 HVAC						
3010	Energy Supply	N/A		—	—	—	
3020	Heat Generating Systems	Included in D3050		—	—	—	
3030	Cooling Generating Systems	N/A		—	—	—	10.5 %
3050	Terminal & Package Units	Multizone unit, gas heating , electric cooling		S.F. Floor	14.55	14.55	
3090	Other HVAC Sys. & Equipment	N/A		—	—	—	
	D40 Fire Protection						
4010	Sprinklers	Wet pipe spinkler system		S.F. Floor	3.54	3.54	3.9%
4020	Standpipes	Standpipes and hose systems, with pumps		S.F. Floor	1.85	1.85	
	D50 Electrical						
5010	Electrical Service/Distribution	400 Ampere service, panel boards and feeders		S.F. Floor	2.31	2.31	
5020	Lighting & Branch Wiring	Fluorescent fixtures, receptacles, switches, A.C. and misc. power		S.F. Floor	7.41	7.41	12.4%
5030	Communications & Security	Alarm systems, emergency lighting, internet and phone wiring		S.F. Floor	7.39	7.39	
5090	Other Electrical Systems	N/A		—	—	—	
E. EQUIPMENT & FURNISHINGS							
1010	Commercial Equipment	N/A		—	—	—	
1020	Institutional Equipment	Exam room equipment, cabinets and countertops		S.F. Floor	3.15	3.15	2.3 %
1030	Vehicular Equipment	N/A		—	—	—	
1090	Other Equipment	N/A		—	—	—	
F. SPECIAL CONSTRUCTION							
1020	Integrated Construction	N/A		—	—	—	0.0 %
1040	Special Facilities	N/A		—	—	—	
G. BUILDING SITEWORK	**N/A**						
					Sub-Total	138.07	**100%**
	CONTRACTOR FEES (General Requirements: 10%, Overhead: 5%, Profit: 10%)				*25%*	34.50	
	ARCHITECT FEES				*9%*	15.53	
					Total Building Cost	**188.10**	

Costs per square foot of floor area

Exterior Wall	S.F. Area	2000	3000	4000	6000	8000	10000	12000	14000	16000
	L.F. Perimeter	240	260	280	380	480	560	580	660	740
Brick Veneer	Wood Frame	182.60	160.45	149.40	143.25	140.25	137.40	133.05	132.00	131.30
Aluminum Siding	Wood Frame	166.90	149.10	140.20	134.95	132.35	130.05	126.70	125.85	125.25
Wood Siding	Wood Frame	166.50	148.85	140.00	134.80	132.20	129.90	126.60	125.70	125.05
Wood Shingles	Wood Frame	169.70	151.05	141.80	136.40	133.75	131.25	127.75	126.90	126.25
Precast Concrete Block	Wood Truss	165.60	148.15	139.45	134.35	131.70	129.45	126.15	125.35	124.70
Brick on Concrete Block	Wood Truss	191.05	166.55	154.30	147.75	144.45	141.30	136.45	135.35	134.50
Perimeter Adj., Add or Deduct	Per 100 L.F.	24.90	16.55	12.40	8.30	6.20	5.00	4.10	3.65	3.10
Story Hgt. Adj., Add or Deduct	Per 1 Ft.	4.60	3.30	2.70	2.45	2.25	2.20	1.85	1.85	1.75
***For Basement**, add $21.50 per square foot of basement area*										

The above costs were calculated using the basic specifications shown on the facing page. These costs should be adjusted where necessary for design alternatives and owner's requirements. Reported completed project costs, for this type of structure, range from $75.70 to $358.35 per S.F.

Common additives

Description	Unit	$ Cost
Closed Circuit Surveillance, One station		
Camera and monitor	Ea.	1975
For additional camera stations, add	Ea.	1075
Emergency Lighting, 25 watt, battery operated		
Lead battery	Ea.	325
Nickel cadmium	Ea.	665
Laundry Equipment		
Dryer, gas, 16 lb. capacity	Ea.	950
30 lb. capacity	Ea.	3900
Washer, 4 cycle	Ea.	1175
Commercial	Ea.	1575
Sauna, Prefabricated, complete		
6' x 4'	Ea.	6500
6' x 6'	Ea.	7750
6' x 9'	Ea.	10,100
8' x 8'	Ea.	11,600
8' x 10'	Ea.	12,900
10' x 12'	Ea.	15,500
Smoke Detectors		
Ceiling type	Ea.	214
Duct type	Ea.	540

Description	Unit	$ Cost
Swimming Pools, Complete, gunite	S.F.	97 - 121
TV Antenna, Master system, 12 outlet	Outlet	249
30 outlet	Outlet	217
100 outlet	Outlet	207

Model costs calculated for a 1 story building with 9′ story height and 8,000 square feet of floor area

Motel, 1 Story

				Unit	Unit Cost	Cost Per S.F.	% Of Sub-Total
A. SUBSTRUCTURE							
1010	Standard Foundations	Poured concrete; strip and spread footings		S.F. Ground	2.22	2.22	11.2%
1020	Special Foundations	N/A		—	—	—	
1030	Slab on Grade	4″ reinforced concrete with vapor barrier and granular base		S.F. Slab	5.10	5.10	
2010	Basement Excavation	Site preparation for slab and trench for foundation wall and footing		S.F. Ground	.29	.29	
2020	Basement Walls	4′ foundation wall		L.F. Wall	69	4.11	
B. SHELL							
	B10 Superstructure						
1010	Floor Construction	N/A		—	—	—	6.6 %
1020	Roof Construction	Plywood on wood trusses		S.F. Roof	6.94	6.94	
	B20 Exterior Enclosure						
2010	Exterior Walls	Face brick on wood studs with sheathing, insulation and paper	*80% of wall*	S.F. Wall	23	9.85	16.7%
2020	Exterior Windows	Wood double hung	*20% of wall*	Each	483	3.73	
2030	Exterior Doors	Wood solid core		Each	1426	3.92	
	B30 Roofing						
3010	Roof Coverings	Asphalt shingles with flashing (pitched); rigid fiberglass insulation		S.F. Roof	4.21	4.21	4.0%
3020	Roof Openings	N/A		—	—	—	
C. INTERIORS							
1010	Partitions	Gypsum bd. and sound deadening bd. on wood studs	*9 S.F. Floor/L.F. Partition*	S.F. Partition	9.48	8.43	22.3%
1020	Interior Doors	Single leaf hollow core wood	*300 S.F. Floor/Door*	Each	510	1.70	
1030	Fittings	N/A		—	—	—	
2010	Stair Construction	N/A		—	—	—	
3010	Wall Finishes	90% paint, 10% ceramic tile		S.F. Surface	1.38	2.46	
3020	Floor Finishes	85% carpet, 15% ceramic tile		S.F. Floor	6.71	6.71	
3030	Ceiling Finishes	Painted gypsum board on furring		S.F. Ceiling	4.12	4.12	
D. SERVICES							
	D10 Conveying						
1010	Elevators & Lifts	N/A		—	—	—	0.0 %
1020	Escalators & Moving Walks	N/A		—	—	—	
	D20 Plumbing						
2010	Plumbing Fixtures	Toilet and service fixtures, supply and drainage	*1 Fixture/90 S.F. Floor*	Each	1652	18.36	19.3%
2020	Domestic Water Distribution	Gas fired water heater		S.F. Floor	1.87	1.87	
2040	Rain Water Drainage	N/A		—	—	—	
	D30 HVAC						
3010	Energy Supply	N/A		—	—	—	3.5 %
3020	Heat Generating Systems	Included in D3050		—	—	—	
3030	Cooling Generating Systems	N/A		—	—	—	
3050	Terminal & Package Units	Through the wall electric heating and cooling units		S.F. Floor	3.72	3.72	
3090	Other HVAC Sys. & Equipment	N/A		—	—	—	
	D40 Fire Protection						
4010	Sprinklers	Wet pipe sprinkler system		S.F. Floor	4.57	4.57	5.4%
4020	Standpipes	Standpipe, wet, Class III		S.F. Floor	1.07	1.07	
	D50 Electrical						
5010	Electrical Service/Distribution	200 ampere service, panel board and feeders		S.F. Floor	1.24	1.24	10.0%
5020	Lighting & Branch Wiring	Fluorescent fixtures, receptacles, switches and misc. power		S.F. Floor	6.63	6.63	
5030	Communications & Security	Addressable alarm systems		S.F. Floor	2.63	2.63	
5090	Other Electrical Systems	N/A		—	—	—	
E. EQUIPMENT & FURNISHINGS							
1010	Commercial Equipment	Laundry equipment		S.F. Floor	.97	.97	0.9%
1020	Institutional Equipment	N/A		—	—	—	
1030	Vehicular Equipment	N/A		—	—	—	
1090	Other Equipment	N/A		—	—	—	
F. SPECIAL CONSTRUCTION							
1020	Integrated Construction	N/A		—	—	—	0.0 %
1040	Special Facilities	N/A		—	—	—	
G. BUILDING SITEWORK	**N/A**						
					Sub-Total	104.85	**100%**
	CONTRACTOR FEES (General Requirements: 10%, Overhead: 5%, Profit: 10%)				*25%*	26.23	
	ARCHITECT FEES				*7%*	9.17	
					Total Building Cost	**140.25**	

Costs per square foot of floor area

Exterior Wall	S.F. Area	25000	37000	49000	61000	73000	81000	88000	96000	104000
	L.F. Perimeter	433	593	606	720	835	911	978	1054	1074
Decorative Concrete Block	Wood Joists	150.35	147.65	143.25	142.20	141.60	141.30	141.10	140.85	140.05
	Precast Conc.	166.25	162.45	157.30	155.85	155.05	154.60	154.25	153.90	153.10
Stucco on Concrete Block	Wood Joists	155.00	151.05	145.50	144.05	143.15	142.70	142.35	142.05	141.10
	Precast Conc.	169.95	165.55	159.55	157.90	156.90	156.35	156.00	155.60	154.65
Wood Siding	Wood Frame	147.45	145.05	141.20	140.30	139.70	139.45	139.25	139.00	138.35
Brick Veneer	Wood Frame	154.85	150.95	145.45	144.00	143.05	142.65	142.30	141.95	141.00
Perimeter Adj., Add or Deduct	Per 100 L.F.	4.05	2.70	2.05	1.70	1.40	1.25	1.10	1.05	1.00
Story Hgt. Adj., Add or Deduct	Per 1 Ft.	1.60	1.50	1.15	1.10	1.10	1.05	1.00	1.00	1.00
***For Basement**, add $ 28.15 per square foot of basement area*										

The above costs were calculated using the basic specifications shown on the facing page. These costs should be adjusted where necessary for design alternatives and owner's requirements. Reported completed project costs, for this type of structure, range from $ 69.70 to $ 358.35 per S.F.

Common additives

Description	Unit	$ Cost
Closed Circuit Surveillance, One station		
Camera and monitor	Ea.	1975
For additional camera station, add	Ea.	1075
Elevators, Hydraulic passenger, 2 stops		
1500# capacity	Ea.	63,900
2500# capacity	Ea.	66,900
3500# capacity	Ea.	71,900
Additional stop, add	Ea.	6825
Emergency Lighting, 25 watt, battery operated		
Lead battery	Ea.	325
Nickel cadmium	Ea.	665
Laundry Equipment		
Dryer, gas, 16 lb. capacity	Ea.	950
30 lb. capacity	Ea.	3900
Washer, 4 cycle	Ea.	1175
Commercial	Ea.	1575

Description	Unit	$ Cost
Sauna, Prefabricated, complete		
6′ x 4′	Ea.	6500
6′ x 6′	Ea.	7750
6′ x 9′	Ea.	10,100
8′ x 8′	Ea.	11,600
8′ x 10′	Ea.	12,900
10′ x 12′	Ea.	15,500
Smoke Detectors		
Ceiling type	Ea.	214
Duct type	Ea.	540
Swimming Pools, Complete, gunite	S.F.	97 - 121
TV Antenna, Master system, 12 outlet	Outlet	249
30 outlet	Outlet	217
100 outlet	Outlet	207

Important: See the Reference Section for Location Factors.

Model costs calculated for a 3 story building with 9′ story height and 49,000 square feet of floor area

Motel, 2-3 Story

				Unit	Unit Cost	Cost Per S.F.	% Of Sub-Total
A. SUBSTRUCTURE							
1010	Standard Foundations	Poured concrete; strip and spread footings		S.F. Ground	2.40	.80	3.0%
1020	Special Foundations	N/A		—	—	—	
1030	Slab on Grade	4″ reinforced concrete with vapor barrier and granular base		S.F. Slab	5.10	1.70	
2010	Basement Excavation	Site preparation for slab and trench for foundation wall and footing		S.F. Ground	.29	.10	
2020	Basement Walls	4′ foundation wall		L.F. Wall	74	.91	
B. SHELL							
	B10 Superstructure						
1010	Floor Construction	Precast concrete plank		S.F. Floor	10.23	6.82	8.4%
1020	Roof Construction	Precast concrete plank		S.F. Roof	9.63	3.21	
	B20 Exterior Enclosure						
2010	Exterior Walls	Decorative concrete block	*85% of wall*	S.F. Wall	16.98	4.82	11.1%
2020	Exterior Windows	Aluminum sliding	*15% of wall*	Each	555	1.85	
2030	Exterior Doors	Aluminum and glass doors and entrance with transom		Each	1966	6.50	
	B30 Roofing						
3010	Roof Coverings	Built-up tar and gravel with flashing; perlite/EPS composite insulation		S.F. Roof	6.15	2.05	1.8%
3020	Roof Openings	Roof hatches		S.F. Roof	.12	.04	
C. INTERIORS							
1010	Partitions	Concrete block	*7 S.F. Floor/L.F. Partition*	S.F. Partition	18.24	20.85	38.4%
1020	Interior Doors	Wood hollow core	*70 S.F. Floor/Door*	Each	510	7.28	
1030	Fittings	N/A		—	—	—	
2010	Stair Construction	Concrete filled metal pan		Flight	14,075	3.45	
3010	Wall Finishes	90% paint, 10% ceramic tile		S.F. Surface	1.38	3.15	
3020	Floor Finishes	85% carpet, 5% vinyl composition tile, 10% ceramic tile		S.F. Floor	6.71	6.71	
3030	Ceiling Finishes	Textured finish		S.F. Ceiling	4.12	4.12	
D. SERVICES							
	D10 Conveying						
1010	Elevators & Lifts	Two hydraulic passenger elevators		Each	104,860	4.28	3.6%
1020	Escalators & Moving Walks	N/A		—	—	—	
	D20 Plumbing						
2010	Plumbing Fixtures	Toilet and service fixtures, supply and drainage	*1 Fixture/180 S.F. Floor*	Each	3805	21.14	19.2%
2020	Domestic Water Distribution	Gas fired water heater		S.F. Floor	1.21	1.21	
2040	Rain Water Drainage	Roof drains		S.F. Roof	1.38	.46	
	D30 HVAC						
3010	Energy Supply	N/A		—	—	—	2.8 %
3020	Heat Generating Systems	Included in D3050		—	—	—	
3030	Cooling Generating Systems	N/A		—	—	—	
3050	Terminal & Package Units	Through the wall electric heating and cooling units		S.F. Floor	3.35	3.35	
3090	Other HVAC Sys. & Equipment	N/A		—	—	—	
	D40 Fire Protection						
4010	Sprinklers	Sprinklers, wet, light hazard		S.F. Floor	2.60	2.60	2.4%
4020	Standpipes	Standpipe, wet, Class III		S.F. Floor	.25	.25	
	D50 Electrical						
5010	Electrical Service/Distribution	800 ampere service, panel board and feeders		S.F. Floor	1.49	1.49	9.0%
5020	Lighting & Branch Wiring	Fluorescent fixtures, receptacles, switches and misc. power		S.F. Floor	7.32	7.32	
5030	Communications & Security	Addressable alarm systems and emergency lighting		S.F. Floor	1.80	1.80	
5090	Other Electrical Systems	Emergency generator, 7.5 kW		S.F. Floor	.13	.13	
E. EQUIPMENT & FURNISHINGS							
1010	Commercial Equipment	Commercial laundry equipment		S.F. Floor	.32	.32	0.3%
1020	Institutional Equipment	N/A		—	—	—	
1030	Vehicular Equipment	N/A		—	—	—	
1090	Other Equipment	N/A		—	—	—	
F. SPECIAL CONSTRUCTION							
1020	Integrated Construction	N/A		—	—	—	0.0 %
1040	Special Facilities	N/A		—	—	—	
G. BUILDING SITEWORK		**N/A**					
					Sub-Total	118.71	**100%**
	CONTRACTOR FEES (General Requirements: 10%, Overhead: 5%, Profit: 10%)				*25%*	29.69	
	ARCHITECT FEES				*6%*	8.90	
					Total Building Cost	**157.30**	

Movie Theater

Costs per square foot of floor area

Exterior Wall	S.F. Area	9000	10000	12000	13000	14000	15000	16000	18000	20000
	L.F. Perimeter	385	410	440	460	480	500	510	547	583
Decorative Concrete Block	Steel Joists	177.00	172.35	163.85	161.00	158.45	156.30	153.75	150.55	147.90
Painted Concrete Block	Steel Joists	170.85	166.45	158.65	155.85	153.55	151.55	149.20	146.15	143.70
Face Brick on Conc. Block	Steel Joists	187.20	182.15	172.60	169.40	166.65	164.30	161.35	157.80	154.85
Precast Concrete Panels	Steel Joists	202.65	196.95	185.85	182.20	179.05	176.30	172.90	168.80	165.40
Tilt-up Panels	Steel Joists	174.55	170.00	161.80	158.95	156.50	154.40	152.00	148.85	146.20
Metal Sandwich Panels	Steel Joists	173.40	168.95	160.80	158.00	155.55	153.50	151.10	148.00	145.45
Perimeter Adj., Add or Deduct	Per 100 L.F.	10.60	9.55	7.95	7.35	6.80	6.40	6.00	5.30	4.75
Story Hgt. Adj., Add or Deduct	Per 1 Ft.	1.60	1.55	1.45	1.30	1.30	1.25	1.20	1.15	1.10
Basement—Not Applicable										

The above costs were calculated using the basic specifications shown on the facing page. These costs should be adjusted where necessary for design alternatives and owner's requirements. Reported completed project costs, for this type of structure, range from $ 89.20 to $ 233.25 per S.F.

Common additives

Description	Unit	$ Cost
Emergency Lighting, 25 watt, battery operated		
Lead battery	Ea.	325
Nickel cadmium	Ea.	665
Seating		
Auditorium chair, all veneer	Ea.	305
Veneer back, padded seat	Ea.	315
Upholstered, spring seat	Ea.	310
Classroom, movable chair & desk	Set	81 - 171
Lecture hall, pedestal type	Ea.	267 - 610
Smoke Detectors		
Ceiling type	Ea.	214
Duct type	Ea.	540
Sound System		
Amplifier, 250 watts	Ea.	2125
Speaker, ceiling or wall	Ea.	206
Trumpet	Ea.	395

Important: See the Reference Section for Location Factors.

Model costs calculated for a 1 story building with 20′ story height and 12,000 square feet of floor area

Movie Theater

			Unit	Unit Cost	Cost Per S.F.	% Of Sub-Total
A. SUBSTRUCTURE						
1010	Standard Foundations	Poured concrete; strip and spread footings	S.F. Ground	1.41	1.41	7.8%
1020	Special Foundations	N/A	—	—	—	
1030	Slab on Grade	4″ reinforced concrete with vapor barrier and granular base	S.F. Slab	5.10	5.10	
2010	Basement Excavation	Site preparation for slab and trench for foundation wall and footing	S.F. Ground	.29	.29	
2020	Basement Walls	4′ foundation wall	L.F. Wall	74	2.70	
B. SHELL						
	B10 Superstructure					
1010	Floor Construction	Open web steel joists, slab form, concrete *mezzanine 2250 S.F.*	S.F. Floor	11.07	2.07	10.0%
1020	Roof Construction	Metal deck on open web steel joists	S.F. Roof	10.22	10.22	
	B20 Exterior Enclosure					
2010	Exterior Walls	Decorative concrete block *80% of wall*	S.F. Wall	15.15	11.11	16.6%
2020	Exterior Windows	Window wall *20% of wall*	Each	54	7.98	
2030	Exterior Doors	Sliding mallfront aluminum and glass and hollow metal	Each	2488	1.25	
	B30 Roofing					
3010	Roof Coverings	Built-up tar and gravel with flashing; perlite/EPS composite insulation	S.F. Roof	6.09	6.09	5.0%
3020	Roof Openings	N/A	—	—	—	
C. INTERIORS						
1010	Partitions	Concrete block *40 S.F. Floor/L.F. Partition*	S.F. Partition	8.78	3.51	19.9%
1020	Interior Doors	Single leaf hollow metal *705 S.F. Floor/Door*	Each	1133	1.61	
1030	Fittings	Toilet partitions	S.F. Floor	.63	.63	
2010	Stair Construction	Concrete filled metal pan	Flight	16,800	2.80	
3010	Wall Finishes	Paint	S.F. Surface	3.83	3.06	
3020	Floor Finishes	Carpet 50%, ceramic tile 5% *50% of area*	S.F. Floor	12.48	6.24	
3030	Ceiling Finishes	Mineral fiber tile on concealed zee runners suspended	S.F. Ceiling	6.57	6.57	
D. SERVICES						
	D10 Conveying					
1010	Elevators & Lifts	N/A	—	—	—	0.0 %
1020	Escalators & Moving Walks	N/A	—	—	—	
	D20 Plumbing					
2010	Plumbing Fixtures	Toilet and service fixtures, supply and drainage *1 Fixture/500 S.F. Floor*	Each	4630	9.26	9.0%
2020	Domestic Water Distribution	Gas fired water heater	S.F. Floor	.52	.52	
2040	Rain Water Drainage	Roof drains	S.F. Roof	1.25	1.25	
	D30 HVAC					
3010	Energy Supply	N/A	—	—	—	8.1 %
3020	Heat Generating Systems	Included in D3050	—	—	—	
3030	Cooling Generating Systems	N/A	—	—	—	
3050	Terminal & Package Units	Single zone rooftop unit, gas heating, electric cooling	S.F. Floor	9.95	9.95	
3090	Other HVAC Sys. & Equipment	N/A	—	—	—	
	D40 Fire Protection					
4010	Sprinklers	Wet pipe sprinkler system	S.F. Floor	3.91	3.91	3.9%
4020	Standpipes	Standpipe	S.F. Floor	.87	.87	
	D50 Electrical					
5010	Electrical Service/Distribution	400 ampere service, panel board and feeders	S.F. Floor	1.65	1.65	8.6%
5020	Lighting & Branch Wiring	High efficiency fluorescent fixtures, receptacles, switches, A.C. and misc. power	S.F. Floor	6.45	6.45	
5030	Communications & Security	Addressable alarm systems, sound system and emergency lighting	S.F. Floor	2.25	2.25	
5090	Other Electrical Systems	Emergency generator, 7.5 kW	S.F. Floor	.18	.18	
E. EQUIPMENT & FURNISHINGS						
1010	Commercial Equipment	N/A	—	—	—	11.1 %
1020	Institutional Equipment	Projection equipment, screen	S.F. Floor	9.24	9.24	
1030	Vehicular Equipment	N/A	—	—	—	
2010	Fixed Furnishings	Movie theater seating	S.F. Floor	4.33	4.33	
F. SPECIAL CONSTRUCTION						
1020	Integrated Construction	N/A	—	—	—	0.0 %
1040	Special Facilities	N/A	—	—	—	
G. BUILDING SITEWORK	**N/A**					
				Sub-Total	122.50	**100%**
	CONTRACTOR FEES (General Requirements: 10%, Overhead: 5%, Profit: 10%)			25%	30.63	
	ARCHITECT FEES			7%	10.72	
				Total Building Cost	**163.85**	

Costs per square foot of floor area

Exterior Wall	S.F. Area	10000	15000	20000	25000	30000	35000	40000	45000	50000
	L.F. Perimeter	286	370	453	457	513	568	624	680	735
Precast Concrete Panels	Bearing Walls	206.85	198.30	193.90	186.40	184.15	182.45	181.20	180.20	179.40
	Steel Frame	212.50	203.95	199.65	192.20	189.90	188.20	187.05	186.05	185.25
Face Brick with Concrete Block Back-up	Bearing Walls	196.40	189.25	185.65	179.75	177.85	176.50	175.50	174.70	174.05
	Steel Joists	202.95	195.80	192.25	186.45	184.55	183.20	182.20	181.45	180.75
Stucco on Concrete Block	Bearing Walls	191.80	185.35	182.05	176.80	175.15	173.90	173.00	172.25	171.65
	Steel Joists	198.35	191.90	188.65	183.50	181.85	180.65	179.75	179.00	178.40
Perimeter Adj., Add or Deduct	Per 100 L.F.	15.30	10.15	7.65	6.15	5.10	4.35	3.80	3.45	3.10
Story Hgt. Adj., Add or Deduct	Per 1 Ft.	3.80	3.20	3.05	2.40	2.25	2.15	2.10	2.00	1.95
***For Basement**, add $ 29.25 per square foot of basement area*										

The above costs were calculated using the basic specifications shown on the facing page. These costs should be adjusted where necessary for design alternatives and owner's requirements. Reported completed project costs, for this type of structure, range from $ 95.75 to $ 240.00 per S.F.

Common additives

Description	Unit	$ Cost
Beds, Manual	Ea.	865 - 2700
Elevators, Hydraulic passenger, 2 stops		
1500# capacity	Ea.	63,900
2500# capacity	Ea.	66,900
3500# capacity	Ea.	71,900
Emergency Lighting, 25 watt, battery operated		
Lead battery	Ea.	325
Nickel cadmium	Ea.	665
Intercom System, 25 station capacity		
Master station	Ea.	3100
Intercom outlets	Ea.	181
Handset	Ea.	505
Kitchen Equipment		
Broiler	Ea.	4050
Coffee urn, twin 6 gallon	Ea.	2975
Cooler, 6 ft. long	Ea.	3875
Dishwasher, 10-12 racks per hr.	Ea.	4200
Food warmer	Ea.	770
Freezer, 44 C.F., reach-in	Ea.	6100

Description	Unit	$ Cost
Kitchen Equipment, cont.		
Ice cube maker, 50 lb. per day	Ea.	1950
Range with 1 oven	Ea.	3275
Laundry Equipment		
Dryer, gas, 16 lb. capacity	Ea.	950
30 lb. capacity	Ea.	3900
Washer, 4 cycle	Ea.	1175
Commercial	Ea.	1575
Nurses Call System		
Single bedside call station	Ea.	281
Pillow speaker	Ea.	276
Refrigerator, Prefabricated, walk-in		
7'-6" high, 6' x 6'	S.F.	200
10' x 10'	S.F.	157
12' x 14'	S.F.	140
12' x 20'	S.F.	123
TV Antenna, Master system, 12 outlet	Outlet	249
30 outlet	Outlet	217
100 outlet	Outlet	207
Whirlpool Bath, Mobile, 18" x 24" x 60"	Ea.	6350
X-Ray, Mobile	Ea.	18,500 - 85,500

 Important: See the Reference Section for Location Factors.

Model costs calculated for a 2 story building with 10′ story height and 25,000 square feet of floor area

Nursing Home

			Unit	Unit Cost	Cost Per S.F.	% Of Sub-Total
A. SUBSTRUCTURE						
1010	Standard Foundations	Poured concrete; strip and spread footings	S.F. Ground	1.68	.84	3.5%
1020	Special Foundations	N/A	—	—	—	
1030	Slab on Grade	4″ reinforced concrete with vapor barrier and granular base	S.F. Slab	5.10	2.56	
2010	Basement Excavation	Site preparation for slab and trench for foundation wall and footing	S.F. Ground	.29	.15	
2020	Basement Walls	4′ foundation wall	L.F. Wall	65	1.17	
B. SHELL						
	B10 Superstructure					
1010	Floor Construction	Pre-cast double tees with concrete topping	S.F. Floor	12.66	6.33	8.7%
1020	Roof Construction	Pre-cast double tees	S.F. Roof	10.82	5.41	
	B20 Exterior Enclosure					
2010	Exterior Walls	Precast concrete panels	S.F. Wall	45	14.06	12.3%
2020	Exterior Windows	Wood double hung *15% of wall*	Each	555	2.03	
2030	Exterior Doors	Double aluminum & glass doors, single leaf hollow metal	Each	2246	.44	
	B30 Roofing					
3010	Roof Coverings	Built-up tar and gravel with flashing; perlite/EPS composite insulation	S.F. Roof	6.12	3.06	2.3%
3020	Roof Openings	Roof hatches	S.F. Roof	.08	.04	
C. INTERIORS						
1010	Partitions	Gypsum board on metal studs *8 S.F. Floor/L.F. Partition*	S.F. Partition	8.49	8.49	20.1%
1020	Interior Doors	Single leaf wood *80 S.F. Floor/Door*	Each	593	7.41	
1030	Fittings	N/A	—	—	—	
2010	Stair Construction	Concrete stair	Flight	6300	1.01	
3010	Wall Finishes	50% vinyl wall coverings, 45% paint, 5% ceramic tile	S.F. Surface	1.63	3.25	
3020	Floor Finishes	95% vinyl tile, 5% ceramic tile	S.F. Floor	3.02	3.02	
3030	Ceiling Finishes	Painted gypsum board	S.F. Ceiling	3.85	3.85	
D. SERVICES						
	D10 Conveying					
1010	Elevators & Lifts	One hydraulic hospital elevator	Each	106,000	4.24	3.2%
1020	Escalators & Moving Walks	N/A	—	—	—	
	D20 Plumbing					
2010	Plumbing Fixtures	Kitchen, toilet and service fixtures, supply and drainage *1 Fixture/230 S.F. Floor*	Each	7914	34.41	27.0%
2020	Domestic Water Distribution	Oil fired water heater	S.F. Floor	1.28	1.28	
2040	Rain Water Drainage	Roof drains	S.F. Roof	1.10	.55	
	D30 HVAC					
3010	Energy Supply	Oil fired hot water, wall fin radiation	S.F. Floor	7.57	7.57	10.8%
3020	Heat Generating Systems	N/A	—	—	—	
3030	Cooling Generating Systems	N/A	—	—	—	
3050	Terminal & Package Units	Split systems with air cooled condensing units	S.F. Floor	6.91	6.91	
3090	Other HVAC Sys. & Equipment	N/A	—	—	—	
	D40 Fire Protection					
4010	Sprinklers	Sprinkler, light hazard	S.F. Floor	2.79	2.79	2.4%
4020	Standpipes	Standpipe	S.F. Floor	.45	.45	
	D50 Electrical					
5010	Electrical Service/Distribution	800 ampere service, panel board and feeders	S.F. Floor	1.56	1.56	9.7%
5020	Lighting & Branch Wiring	High efficiency fluorescent fixtures, receptacles, switches, A.C. and misc. power	S.F. Floor	9.56	9.56	
5030	Communications & Security	Addressable alarm systems and emergency lighting	S.F. Floor	1.36	1.36	
5090	Other Electrical Systems	Emergency generator, 15 kW	S.F. Floor	.55	.55	
E. EQUIPMENT & FURNISHINGS						
1010	Commercial Equipment	N/A	—	—	—	0.0 %
1020	Institutional Equipment	N/A	—	—	—	
1030	Vehicular Equipment	N/A	—	—	—	
1090	Other Equipment	N/A	—	—	—	
F. SPECIAL CONSTRUCTION						
1020	Integrated Construction	N/A	—	—	—	0.0 %
1040	Special Facilities	N/A	—	—	—	
G. BUILDING SITEWORK		**N/A**				
				Sub-Total	134.35	**100%**
		CONTRACTOR FEES (General Requirements: 10%, Overhead: 5%, Profit: 10%)		*25%*	33.58	
		ARCHITECT FEES		*11%*	18.47	
				Total Building Cost	**186.40**	

M.455 Office, 1 Story

Costs per square foot of floor area

Exterior Wall	S.F. Area	2000	3000	5000	7000	9000	12000	15000	20000	25000
	L.F. Perimeter	220	260	320	360	420	480	520	640	700
Vinyl Siding	Wood Frame	195.85	175.50	157.30	147.95	143.95	139.15	135.60	133.40	130.85
Stone Veneer	Wood Frame	232.30	203.80	177.85	164.50	158.85	151.85	146.65	143.55	139.75
Fiber Cement Siding	Steel Frame	160.60	138.05	117.85	107.55	103.10	97.90	93.95	91.50	88.75
E.I.F.S.	Steel Frame	209.50	187.45	167.70	157.60	153.30	148.15	144.35	141.90	139.25
Precast Concrete Panel	Reinforced Concrete	262.30	230.10	200.35	**184.70**	178.20	170.10	163.90	160.30	155.80
Brick Veneer	Reinforced Concrete	268.50	234.20	202.60	186.00	179.20	170.60	164.00	160.20	155.45
Perimeter Adj., Add or Deduct	Per 100 L.F.	25.40	17.00	10.10	7.25	5.70	4.25	3.35	2.50	2.05
Story Hgt. Adj., Add or Deduct	Per 1 Ft.	2.70	2.20	1.55	1.25	1.15	1.00	0.90	0.75	0.70
***For Basement**, add $29.70 per square foot of basement area*										

The above costs were calculated using the basic specifications shown on the facing page. These costs should be adjusted where necessary for design alternatives and owner's requirements. Reported completed project costs, for this type of structure, range from $73.10 to $282.60 per S.F.

Common additives

Description	Unit	$ Cost
Closed circuit surveillance, one station		
Camera and monitor	Ea.	1975
For additional camera stations, add	Ea.	1075
Directory boards, plastic, glass covered		
30" x 20"	Ea.	555
36" x 48"	Ea.	1425
Aluminum, 24" x 18"	Ea.	600
36" x 24"	Ea.	715
48" x 32"	Ea.	1000
48" x 60"	Ea.	2100
Electronic, wall mounted	S.F.	2900
Free standing	S.F.	4125
Pedestal access floor system w/ plastic laminate cover		
Computer room, less than 6000 SF	S.F.	20.50
Greater than 6000 SF	S.F.	20
Office, greater than 6000 S.F.	S.F.	16.50

Description	Unit	$ Cost
Security access systems		
Metal detectors, wand type	Ea.	138
Walk-through portal type, single-zone	Ea.	5150
Multi-zone	Ea.	6300
X-ray equipment		
Desk top, for mail, small packages	Ea.	3975
Conveyer type, including monitor, minimum	Ea.	17,900
Maximum	Ea.	31,100
Explosive detection equipment		
Hand held, battery operated	Ea.	28,100
Walk-through portal type	Ea.	48,900
Uninterruptible power supply, 15 kVA/12.75 kW	kW	1815

Important: See the Reference Section for Location Factors.

Model costs calculated for a 1 story building with 12′ story height and 7,000 square feet of floor area

Office, 1 Story

			Unit	Unit Cost	Cost Per S.F.	% Of Sub-Total
A. SUBSTRUCTURE						
1010	Standard Foundations	Poured concrete; strip and spread footings	S.F. Ground	3.16	3.16	8.9%
1020	Special Foundations	N/A	—	—	—	
1030	Slab on Grade	4″ reinforced concrete	S.F. Slab	5.10	5.10	
2010	Basement Excavation	Site preparation for slab and trench for foundation wall and footing	S.F. Ground	.29	.29	
2020	Basement Walls	4′ Foundation wall	L.F. Wall	74	3.78	
B. SHELL						
	B10 Superstructure					
1010	Floor Construction	Cast in place concrete columns	S.F. Floor	2.50	2.50	11.9%
1020	Roof Construction	Concrete beam and slab	S.F. Roof	13.90	13.90	
	B20 Exterior Enclosure					
2010	Exterior Walls	Precast concrete panels, insulated *80% of wall*	S.F. Wall	47	23.19	21.6%
2020	Exterior Windows	Aluminum awning windows *20% of wall*	Each	705	3.78	
2030	Exterior Doors	Aluminum and glass single and double doors, hollow metal doors	Each	3373	2.90	
	B30 Roofing					
3010	Roof Coverings	Single ply membrane, stone ballast, rigid insulation	S.F. Roof	6.53	6.53	5.0%
3020	Roof Openings	Smoke hatches	S.F. Roof	.31	.31	
C. INTERIORS						
1010	Partitions	Gypsum board on metal studs *20 S.F. Floor/L.F. Partition*	S.F. Partition	5.64	2.82	13.6%
1020	Interior Doors	70% Single leaf solid wood doors, 30% hollow metal single leaf *320 S.F. Floor/Door*	Each	755	2.37	
1030	Fittings	Plastic laminate toilet partitions	S.F. Floor	.36	.36	
2010	Stair Construction	N/A	—	—	—	
3010	Wall Finishes	90% Paint, 10% ceramic wall tile	S.F. Surface	1.13	1.13	
3020	Floor Finishes	65% Carpet, 10% porcelain tile, 10% quarry tile	S.F. Floor	5.46	5.46	
3030	Ceiling Finishes	Acoustic ceiling tiles on suspended channel grid	S.F. Ceiling	6.57	6.57	
D. SERVICES						
	D10 Conveying					
1010	Elevators & Lifts	N/A	—	—	—	0.0 %
1020	Escalators & Moving Walks	N/A	—	—	—	
	D20 Plumbing					
2010	Plumbing Fixtures	Restroom and service fixtures, supply and drainage *1 Fixture/700 S.F. Floor*	Each	2100	3	3.8%
2020	Domestic Water Distribution	Gas fired water heater	S.F. Floor	1.49	1.49	
2040	Rain Water Drainage	Roof drains	S.F. Roof	.72	.72	
	D30 HVAC					
3010	Energy Supply	N/A	—	—	—	14.9 %
3020	Heat Generating Systems	Included in D3050	—	—	—	
3030	Cooling Generating Systems	N/A	—	—	—	
3050	Terminal & Package Units	Multizone gas heating, electric cooling unit	S.F. Floor	21	20.55	
3090	Other HVAC Sys. & Equipment	N/A	—	—	—	
	D40 Fire Protection					
4010	Sprinklers	Wet pipe light hazard sprinkler system	S.F. Floor	3.28	3.28	3.4%
4020	Standpipes	Standpipes	S.F. Floor	1.46	1.46	
	D50 Electrical					
5010	Electrical Service/Distribution	400 ampere service, panel boards and feeders	S.F. Floor	4.93	4.93	17.0%
5020	Lighting & Branch Wiring	Fluorescent fixtures, receptacles, switches, A.C. and misc. power	S.F. Floor	11.77	11.77	
5030	Communications & Security	Telephone systems, internet wiring, and addressable alarm systems	S.F. Floor	6.72	6.72	
5090	Other Electrical Systems	Emergency generator, 7.5 kW	S.F. Floor			
E. EQUIPMENT & FURNISHINGS						
1010	Commercial Equipment	N/A	—	—	—	0.0 %
1020	Institutional Equipment	N/A	—	—	—	
1030	Vehicular Equipment	N/A	—	—	—	
1090	Other Equipment	N/A	—	—	—	
F. SPECIAL CONSTRUCTION						
1020	Integrated Construction	N/A	—	—	—	0.0 %
1040	Special Facilities	N/A	—	—	—	
G. BUILDING SITEWORK		**N/A**				
				Sub-Total	138.07	**100%**
	CONTRACTOR FEES (General Requirements: 10%, Overhead: 5%, Profit: 10%)			*25%*	34.55	
	ARCHITECT FEES			*7%*	12.08	
				Total Building Cost	**184.70**	

Costs per square foot of floor area

Exterior Wall	S.F. Area	5000	8000	12000	16000	20000	35000	50000	65000	80000
	L.F. Perimeter	220	260	310	330	360	440	550	600	675
Vinyl Siding	Wood Frame	193.90	172.85	160.80	152.50	148.15	139.55	136.85	134.25	132.95
Stone Veneer	Wood Frame	230.15	199.55	181.95	169.30	**162.85**	149.75	145.75	141.65	139.70
Fiber Cement Siding	Steel Frame	192.90	171.45	159.15	150.70	146.30	137.50	134.70	132.10	130.75
E.I.F.S.	Steel Frame	195.50	173.05	160.20	151.15	146.45	137.10	134.20	131.35	129.95
Precast Concrete Panel	Reinforced Concrete	256.15	219.90	199.00	183.75	176.00	160.15	155.35	150.45	148.10
Brick Veneer	Reinforced Concrete	244.50	208.65	188.05	173.05	165.35	149.70	145.00	140.15	137.85
Perimeter Adj., Add or Deduct	Per 100 L.F.	23.75	14.80	9.80	7.40	5.90	3.40	2.35	1.80	1.50
Story Hgt. Adj., Add or Deduct	Per 1 Ft.	3.30	2.40	1.85	1.50	1.35	0.95	0.80	0.70	0.65
***For Basement**, add $30.80 per square foot of basement area*										

The above costs were calculated using the basic specifications shown on the facing page. These costs should be adjusted where necessary for design alternatives and owner's requirements. Reported completed project costs, for this type of structure, range from $76.50 to $296.35 per S.F.

Common additives

Description	Unit	$ Cost
Closed circuit surveillance, one station		
Camera and monitor	Ea.	1975
For additional camera stations, add	Ea.	1075
Directory boards, plastic, glass covered		
30" x 20"	Ea.	555
36" x 48"	Ea.	1425
Aluminum, 24" x 18"	Ea.	600
36" x 24"	Ea.	715
48" x 32"	Ea.	1000
48" x 60"	Ea.	2100
Electronic, wall mounted	S.F.	2900
Free standing	S.F.	4125
Escalators, 10' rise, 32" wide, glass balustrade	Ea.	137,900
Metal balustrade	Ea.	145,900
48" wide, glass balustrade	Ea.	145,400
Metal balustrade	Ea.	154,400
Pedestal access floor system w/ plastic laminate cover		
Computer room, less than 6000 SF	S.F.	20.50
Greater than 6000 SF	S.F.	20
Office, greater than 6000 S.F.	S.F.	16.50

Description	Unit	$ Cost
Security access systems		
Metal detectors, wand type	Ea.	138
Walk-through portal type, single-zone	Ea.	5150
Multi-zone	Ea.	6300
X-ray equipment		
Desk top, for mail, small packages	Ea.	3975
Conveyer type, including monitor, minimum	Ea.	17,900
Maximum	Ea.	31,100
Explosive detection equipment		
Hand held, battery operated	Ea.	28,100
Walk-through portal type	Ea.	48,900
Uninterruptible power supply, 15 kVA/12.75 kW	kW	1815

Important: See the Reference Section for Location Factors.

Model costs calculated for a 3 story building with 12′ story height and 20,000 square feet of floor area

Office, 2-4 Story

				Unit	Unit Cost	Cost Per S.F.	% Of Sub-Total
A. SUBSTRUCTURE							
1010	Standard Foundations	Poured concrete; strip and spread footings		S.F. Ground	3.63	1.21	3.6%
1020	Special Foundations	N/A		—	—	—	
1030	Slab on Grade	4″ reinforced slab on grade		S.F. Slab	5.10	1.70	
2010	Basement Excavation	Site preparation for slab and trench for foundation wall and footing		S.F. Ground	.29	.10	
2020	Basement Walls	4′ Foundation wall		L.F. Wall	37	1.32	
B. SHELL							
	B10 Superstructure						
1010	Floor Construction	Wood Beam and joist on wood columns, fireproofed		S.F. Floor	25	16.51	15.6%
1020	Roof Construction	Wood roof, truss, 4/12 slope		S.F. Roof	7.38	2.46	
	B20 Exterior Enclosure						
2010	Exterior Walls	Ashlar stone veneer on wood stud back up, insulated	*80% of wall*	S.F. Wall	35	18.38	19.2%
2020	Exterior Windows	Aluminum awning type windows	*20% of wall*	Each	705	3.98	
2030	Exterior Doors	Aluminum and glass single and double doors, hollow metal doors		Each	3373	1.02	
	B30 Roofing						
3010	Roof Coverings	Asphalt shingles with flashing, gutters and downspouts		S.F. Roof	2.25	.75	0.6%
3020	Roof Openings	N/A		—	—	—	
C. INTERIORS							
1010	Partitions	Gypsum board on wood studs	*20 S.F. Floor/L.F. Partition*	S.F. Partition	7.23	2.89	16.0%
1020	Interior Doors	Single leaf solid core wood doors, hollow metal egress doors	*315 S.F. Floor/Door*	Each	755	2.42	
1030	Fittings	Plastic laminate toilet partitions		S.F. Floor	.21	.21	
2010	Stair Construction	Concrete filled metal pan with rails		Flight	2675	.94	
3010	Wall Finishes	Paint on drywall		S.F. Surface	1.26	1.01	
3020	Floor Finishes	60% Carpet tile, 30% vinyl composition tile, 10% ceramic tile		S.F. Floor	5.46	5.46	
3030	Ceiling Finishes	Acoustic ceiling tiles on suspended support system		S.F. Ceiling	6.57	6.57	
D. SERVICES							
	D10 Conveying						
1010	Elevators & Lifts	Two hydraulic passenger elevators		Each	113,100	11.31	9.3%
1020	Escalators & Moving Walks	N/A		—	—	—	
	D20 Plumbing						
2010	Plumbing Fixtures	Restroom and service fixtures, supply and drainage	*1 Fixture/1050 S.F. Floor*	Each	1843	1.75	1.9%
2020	Domestic Water Distribution	Gas fired water heater		S.F. Floor	.52	.52	
2040	Rain Water Drainage	Roof drains		S.F. Roof			
	D30 HVAC						
3010	Energy Supply	N/A		—	—	—	13.3 %
3020	Heat Generating Systems	Included in D3050		—	—	—	
3030	Cooling Generating Systems	N/A		—	—	—	
3050	Terminal & Package Units	Multizone unit, gas heating , electric cooling		S.F. Floor	16.25	16.25	
3090	Other HVAC Sys. & Equipment	N/A		—	—	—	
	D40 Fire Protection						
4010	Sprinklers	Wet pipe spinkler system		S.F. Floor	3.45	3.45	3.6%
4020	Standpipes	Standpipes and hose systems		S.F. Floor	.94	.94	
	D50 Electrical						
5010	Electrical Service/Distribution	1000 Ampere service, panel boards and feeders		S.F. Floor	3.49	3.49	16.9%
5020	Lighting & Branch Wiring	Fluorescent fixtures, receptacles, switches, A.C. and misc. power		S.F. Floor	11.32	11.32	
5030	Communications & Security	Addressable alarm systems, emergency lighting, internet and phone wiring		S.F. Floor	5.59	5.59	
5090	Other Electrical Systems	Uninterruptible power supply with battery pack, 12.75kW		S.F. Floor	.19	.19	
E. EQUIPMENT & FURNISHINGS							
1010	Commercial Equipment	N/A		—	—	—	0.0 %
1020	Institutional Equipment	N/A		—	—	—	
1030	Vehicular Equipment	N/A		—	—	—	
1090	Other Equipment	N/A		—	—	—	
F. SPECIAL CONSTRUCTION							
1020	Integrated Construction	N/A		—	—	—	0.0 %
1040	Special Facilities	N/A		—	—	—	
G. BUILDING SITEWORK	**N/A**						
					Sub-Total	121.74	**100%**
	CONTRACTOR FEES (General Requirements: 10%, Overhead: 5%, Profit: 10%)				25%	30.46	
	ARCHITECT FEES				7%	10.65	
					Total Building Cost	**162.85**	

Costs per square foot of floor area

Exterior Wall	S.F. Area	5000	7500	10000	12500	15000	17500	20000	22500	25000
	L.F. Perimeter	284	350	410	450	490	530	570	600	632
Brick Veneer	Wood Truss	432.30	420.15	413.55	408.15	404.55	401.95	400.00	398.05	396.60
Brick Veneer	Steel Frame	433.00	421.35	415.00	409.85	406.45	403.95	402.15	400.30	398.90
Decorative Concrete Block	Steel Frame	435.80	423.70	417.05	411.70	408.05	405.50	403.55	401.65	400.20
Decorative Concrete Block	Bearing Wall	427.60	416.95	411.15	406.45	403.35	401.10	399.40	397.75	396.50
Tiltup Concrete Panel	Steel Frame	428.55	417.80	411.90	**407.15**	403.95	401.70	400.00	398.35	397.00
EIFS on Concrete Block	Steel Frame	431.90	419.15	412.25	406.95	403.35	400.80	398.90	397.05	395.60
Perimeter Adj., Add or Deduct	Per 100 L.F.	18.45	12.30	9.20	7.30	6.10	5.25	4.60	4.10	3.70
Story Hgt. Adj., Add or Deduct	Per 1 Ft.	2.80	2.30	2.00	1.70	1.60	1.45	1.40	1.30	1.25
***For Basement**, add $ 28.15 per square foot of basement area*										

The above costs were calculated using the basic specifications shown on the facing page. These costs should be adjusted where necessary for design alternatives and owner's requirements. Reported completed project costs, for this type of structure, range from $ 189.10 to $ 980.55 per S.F.

Common additives

Description	Unit	$ Cost
Cabinets, base, door units, metal	L.F.	320
Drawer units	L.F.	635
Tall storage cabinets, open, 7′ high	L.F.	610
With glazed doors	L.F.	890
Wall cabinets, metal, 12-1/2″ deep, open	L.F.	254
With doors	L.F.	450
Counter top, laminated plastic, no backsplash	L.F.	77.50
Stainless steel counter top	L.F.	195
For drop-in stainless 43″ x 21″ sink, add	Ea.	1100
Nurse Call Station		
Single bedside call station	Ea.	281
Ceiling speaker station	Ea.	150
Emergency call station	Ea.	155
Pillow speaker	Ea.	276
Standard call button	Ea.	172
Master control station for 20 stations	Total	5575

Description	Unit	$ Cost
Directory Boards, Plastic, glass covered		
Plastic, glass covered, 30″ x 20″	Ea.	555
36″ x 48″	Ea.	1425
Building directory, alum., black felt panels, 1 door, 24″ x 18″	Ea.	600
36″ x 24″	Ea.	715
48″ x 32″	Ea.	1000
48″ x 60″	Ea.	2100
Tables, Examining, vinyl top		
with base cabinets	Ea.	1600
Utensil washer-sanitizer	Ea.	9600
X-ray, mobile, minimum	Ea.	18,500

Important: See the Reference Section for Location Factors.

Model costs calculated for a 1 story building with 15'-4" story height and 12,500 square feet of floor area

Outpatient Surgery Center

				Unit	Unit Cost	Cost Per S.F.	% Of Sub-Total
A. SUBSTRUCTURE							
1010	Standard Foundations	Poured concrete; strip and spread footings		S.F. Ground	1.70	1.70	3.7%
1020	Special Foundations	N/A		—	—	—	
1030	Slab on Grade	4" reinforced concrete with vapor barrier and granular base		S.F. Slab	6.34	6.34	
2010	Basement Excavation	Site preparation for slab and trench for foundation wall and footing		S.F. Ground	.29	.29	
2020	Basement Walls	4' Foundation wall		L.F. Wall	74	2.65	
B. SHELL							
	B10 Superstructure						
1010	Floor Construction	N/A		—	—	—	2.5 %
1020	Roof Construction	Steel Joists and girders on columns		S.F. Roof	7.43	7.43	
	B20 Exterior Enclosure						
2010	Exterior Walls	Brick veneer	*85% of wall*	S.F. Wall	17.65	8.28	6.0%
2020	Exterior Windows	Aluminum frame, fixed	*15% of wall*	Each	799	4.40	
2030	Exterior Doors	Aluminum and glass entry; steel flush doors		Each	6730	5.38	
	B30 Roofing						
3010	Roof Coverings	Single-ply membrane, loose laid and ballasted; polyisocyanurate insulation		S.F. Roof	6.34	6.34	2.2%
3020	Roof Openings	Roof hatches		S.F. Roof	.10	.10	
C. INTERIORS							
1010	Partitions	Gypsum board on metal studs	*7 S.F. Floor/L.F. Partition*	S.F. Partition	7.40	10.57	12.8%
1020	Interior Doors	Single and double leaf fire doors	*200 S.F. Floor/Door*	Each	1330	6.65	
1030	Fittings	Stainless steel toilet partitions, steel lockers		S.F. Floor	1.66	1.66	
2010	Stair Construction	N/A		—	—	—	
3010	Wall Finishes	80% Paint, 10% vinyl wall covering, 10% ceramic tile		S.F. Surface	1.87	5.33	
3020	Floor Finishes	10% Carpet, 80% vinyl sheet, 10% ceramic tile		S.F. Floor	10.01	10.01	
3030	Ceiling Finishes	Mineral fiber tile and plastic coated tile on suspension system		S.F. Ceiling	3.95	3.95	
D. SERVICES							
	D10 Conveying						
1010	Elevators & Lifts	N/A		—	—	—	0.0 %
1020	Escalators & Moving Walks	N/A		—	—	—	
	D20 Plumbing						
2010	Plumbing Fixtures	Medical, patient & specialty fixtures, supply and drainage	*1 fixture/350 S.F. Floor*	Each	3171	9.06	4.5%
2020	Domestic Water Distribution	Electric water heater		S.F. Floor	2.19	2.19	
2040	Rain Water Drainage	Roof drains		S.F. Roof	2.14	2.14	
	D30 HVAC						
3010	Energy Supply	Hot water reheat for surgery		S.F. Floor	10.50	10.50	38.0%
3020	Heat Generating Systems	Boiler		Each	59,825	6.68	
3030	Cooling Generating Systems	N/A		—	—	—	
3050	Terminal & Package Units	N/A		—	—	—	
3090	Other HVAC Sys. & Equipment	Rooftop AC, CAV, roof vents, heat exchanger, surgical air curtains		S.F. Floor	96	96.36	
	D40 Fire Protection						
4010	Sprinklers	Wet pipe sprinkler system		S.F. Floor	3.28	3.28	1.6%
4020	Standpipes	Standpipe		S.F. Floor	1.59	1.59	
	D50 Electrical						
5010	Electrical Service/Distribution	1200 ampere service, panel board and feeders		S.F. Floor	6.40	6.40	12.8%
5020	Lighting & Branch Wiring	T-8 fluorescent light fixtures, receptacles, switches, A.C. and misc. power		S.F. Floor	16.05	16.05	
5030	Communications & Security	Addressable alarm system, internet wiring, communications system, emergency lighting		S.F. Floor	9.57	9.57	
5090	Other Electrical Systems	Emergency generator, 300 kW with fuel tank		S.F. Floor	6.22	6.22	
E. EQUIPMENT & FURNISHINGS							
1010	Commercial Equipment	N/A		—	—	—	16.0 %
1020	Institutional Equipment	Medical gas system, cabinetry, scrub sink, sterilizer		S.F. Floor	47	46.68	
1030	Vehicular Equipment	N/A		—	—	—	
1090	Other Equipment	Ice maker, washing machine		S.F. Floor	1.01	1.01	
F. SPECIAL CONSTRUCTION & DEMOLITION							
1020	Integrated Construction	N/A		—	—	—	0.0 %
1040	Special Facilities	N/A		—	—	—	
G. BUILDING SITEWORK		**N/A**					
					Sub-Total	298.81	**100%**
	CONTRACTOR FEES (General Requirements: 10%, Overhead: 5%, Profit: 5%)				*20%*	74.72	
	ARCHITECT FEES				*9%*	33.62	
					Total Building Cost	**407.15**	

Costs per square foot of floor area

Exterior Wall	S.F. Area	7000	9000	11000	13000	15000	17000	19000	21000	23000
	L.F. Perimeter	240	280	303	325	354	372	397	422	447
Limestone with Concrete Block Back-up	Bearing Walls	273.50	255.80	240.75	230.10	223.55	216.85	212.50	209.05	206.10
	R/Conc. Frame	278.15	261.15	247.00	237.00	230.75	224.55	220.45	217.15	214.45
Face Brick with Concrete Block Back-up	Bearing Walls	235.45	221.10	210.05	202.15	197.10	192.35	189.05	186.50	184.30
	R/Conc. Frame	248.60	234.35	223.25	215.55	210.45	205.70	202.40	199.90	197.70
Decorative Concrete Block	Bearing Walls	225.40	212.05	201.95	194.90	190.20	185.95	182.95	180.60	178.65
	R/Conc. Frame	239.75	226.40	216.35	209.35	204.65	200.45	197.45	195.15	193.15
Perimeter Adj., Add or Deduct	Per 100 L.F.	34.80	27.05	22.15	18.80	16.25	14.35	12.85	11.60	10.60
Story Hgt. Adj., Add or Deduct	Per 1 Ft.	6.35	5.75	5.05	4.65	4.35	4.05	3.90	3.70	3.55
***For Basement**, add $ 26.20 per square foot of basement area*										

The above costs were calculated using the basic specifications shown on the facing page. These costs should be adjusted where necessary for design alternatives and owner's requirements. Reported completed project costs, for this type of structure, range from $ 124.05 to $ 323.60 per S.F.

Common additives

Description	Unit	$ Cost
Cells Prefabricated, 5'-6' wide, 7'-8' high, 7'-8' deep	Ea.	12,500
Elevators, Hydraulic passenger, 2 stops		
1500# capacity	Ea.	63,900
2500# capacity	Ea.	66,900
3500# capacity	Ea.	71,900
Emergency Lighting, 25 watt, battery operated		
Lead battery	Ea.	325
Nickel cadmium	Ea.	665
Flagpoles, Complete		
Aluminum, 20' high	Ea.	1825
40' high	Ea.	4300
70' high	Ea.	11,300
Fiberglass, 23' high	Ea.	1275
39'-5" high	Ea.	3275
59' high	Ea.	6475

Description	Unit	$ Cost
Lockers, Steel, Single tier, 60" to 72"	Opng.	222 - 365
2 tier, 60" or 72" total	Opng.	115 - 164
5 tier, box lockers	Opng.	69.50 - 78.50
Locker bench, lam. maple top only	L.F.	33
Pedestals, steel pipe	Ea.	71
Safe, Office type, 1 hour rating		
30" x 18" x 18"	Ea.	2500
60" x 36" x 18", double door	Ea.	9350
Shooting Range, Incl. bullet traps, target provisions, and contols, not incl. structural shell	Ea.	51,000
Smoke Detectors		
Ceiling type	Ea.	214
Duct type	Ea.	540
Sound System		
Amplifier, 250 watts	Ea.	2125
Speaker, ceiling or wall	Ea.	206
Trumpet	Ea.	395

Model costs calculated for a 2 story building with 12' story height and 11,000 square feet of floor area

Police Station

				Unit	Unit Cost	Cost Per S.F.	% Of Sub-Total
A. SUBSTRUCTURE							
1010	Standard Foundations	Poured concrete; strip and spread footings		S.F. Ground	2.72	1.36	3.4%
1020	Special Foundations	N/A		—	—	—	
1030	Slab on Grade	4" reinforced concrete with vapor barrier and granular base		S.F. Slab	5.10	2.56	
2010	Basement Excavation	Site preparation for slab and trench for foundation wall and footing		S.F. Ground	.29	.15	
2020	Basement Walls	4' foundation wall		L.F. Wall	74	2.02	
B. SHELL							
	B10 Superstructure						
1010	Floor Construction	Open web steel joists, slab form, concrete		S.F. Floor	11.78	5.89	4.7%
1020	Roof Construction	Metal deck on open web steel joists		S.F. Roof	4.70	2.35	
	B20 Exterior Enclosure						
2010	Exterior Walls	Limestone with concrete block backup	*80% of wall*	S.F. Wall	63	33.32	25.6%
2020	Exterior Windows	Metal horizontal sliding	*20% of wall*	Each	1123	9.89	
2030	Exterior Doors	Hollow metal		Each	2731	1.98	
	B30 Roofing						
3010	Roof Coverings	Built-up tar and gravel with flashing; perlite/EPS composite insulation		S.F. Roof	7.06	3.53	2.0%
3020	Roof Openings	N/A		—	—	—	
C. INTERIORS							
1010	Partitions	Concrete block	*20 S.F. Floor/L.F. Partition*	S.F. Partition	13.48	6.74	18.0%
1020	Interior Doors	Single leaf kalamein fire door	*200 S.F. Floor/Door*	Each	1133	5.67	
1030	Fittings	Toilet partitions		S.F. Floor	.94	.94	
2010	Stair Construction	Concrete filled metal pan		Flight	16,800	4.58	
3010	Wall Finishes	90% paint, 10% ceramic tile		S.F. Surface	3.46	3.46	
3020	Floor Finishes	70% vinyl composition tile, 20% carpet tile, 10% ceramic tile		S.F. Floor	3.91	3.91	
3030	Ceiling Finishes	Mineral fiber tile on concealed zee bars		S.F. Ceiling	6.57	6.57	
D. SERVICES							
	D10 Conveying						
1010	Elevators & Lifts	One hydraulic passenger elevator		Each	81,070	7.37	4.2%
1020	Escalators & Moving Walks	N/A		—	—	—	
	D20 Plumbing						
2010	Plumbing Fixtures	Toilet and service fixtures, supply and drainage	*1 Fixture/580 S.F. Floor*	Each	4217	7.27	7.4%
2020	Domestic Water Distribution	Oil fired water heater		S.F. Floor	4.28	4.28	
2040	Rain Water Drainage	Roof drains		S.F. Roof	3	1.50	
	D30 HVAC						
3010	Energy Supply	N/A		—	—	—	11.6 %
3020	Heat Generating Systems	N/A		—	—	—	
3030	Cooling Generating Systems	N/A		—	—	—	
3050	Terminal & Package Units	Multizone HVAC air cooled system		S.F. Floor	21	20.55	
3090	Other HVAC Sys. & Equipment	N/A		—	—	—	
	D40 Fire Protection						
4010	Sprinklers	Wet pipe sprinkler system		S.F. Floor	3.54	3.54	2.6%
4020	Standpipes	Standpipe		S.F. Floor	1.02	1.02	
	D50 Electrical						
5010	Electrical Service/Distribution	400 ampere service, panel board and feeders		S.F. Floor	1.96	1.96	11.6%
5020	Lighting & Branch Wiring	T-8 fluorescent fixtures, receptacles, switches, A.C. and misc. power		S.F. Floor	11.45	11.45	
5030	Communications & Security	Addressable alarm systems, internet wiring, intercom and emergency lighting		S.F. Floor	6.94	6.94	
5090	Other Electrical Systems	Emergency generator, 15 kW		S.F. Floor	.23	.23	
E. EQUIPMENT & FURNISHINGS							
1010	Commercial Equipment	N/A		—	—	—	8.9 %
1020	Institutional Equipment	Lockers, detention rooms, cells, gasoline dispensers		S.F. Floor	13.19	13.19	
1030	Vehicular Equipment	Gasoline dispenser system		S.F. Floor	2.47	2.47	
1090	Other Equipment	N/A		—	—	—	
F. SPECIAL CONSTRUCTION							
1020	Integrated Construction	N/A		—	—	—	0.0 %
1040	Special Facilities	N/A		—	—	—	
G. BUILDING SITEWORK	**N/A**						
					Sub-Total	176.69	**100%**
	CONTRACTOR FEES (General Requirements: 10%, Overhead: 5%, Profit: 10%)				*25%*	44.18	
	ARCHITECT FEES				*9%*	19.88	
					Total Building Cost	**240.75**	

M.495 Post Frame Barn

Costs per square foot of floor area

Exterior Wall	S.F. Area	600	750	925	1150	1450	1800	2100	2600	3200
	L.F. Perimeter	100	112	128	150	164	182	200	216	240
Steel Siding on Wood Studs	Wood Truss	70.50	65.60	62.45	60.05	55.50	52.25	50.65	47.45	45.35
Aluminum Siding	Wood Truss	69.15	64.80	61.90	59.75	55.70	52.85	51.45	48.60	46.75
Wood Board Siding	Wood Truss	73.20	68.50	65.45	63.25	58.80	55.75	54.25	51.25	49.20
Plywood Siding	Wood Truss	69.65	65.15	62.30	60.15	56.00	53.10	51.70	48.85	46.95
Perimeter Adj., Add or Deduct	Per 100 L.F.	40.75	32.65	26.45	21.30	16.75	13.55	11.60	9.45	7.70
Story Hgt. Adj., Add or Deduct	Per 1 Ft.	2.95	2.60	2.40	2.30	1.95	1.75	1.65	1.50	1.30
Basement—Not Applicable										

The above costs were calculated using the basic specifications shown on the facing page. These costs should be adjusted where necessary for design alternatives and owner's requirements. Reported completed project costs, for this type of structure, range from $ 39.65 to $ 158.30 per S.F.

Common additives

Description	Unit	$ Cost
Slab on Grade		
4" thick, non-industrial, non-reinforced	S.F.	4.64
Reinforced	S.F.	5.10
Light industrial, non-reinforced	S.F.	5.88
Reinforced	S.F.	6.34
Industrial, non-reinforced	S.F.	9.79
Reinforced	S.F.	10.24
6" thick, non-industrial, non-reinforced	S.F.	5.51
Reinforced	S.F.	6.15
Light industrial, non-reinforced	S.F.	6.77
Reinforced	S.F.	7.70
Heavy industrial, non-reinforced	S.F.	12.65
Reinforced	S.F.	13.30
Insulation		
Fiberboard, low density, 1" thick, R2.78	S.F.	1.34
Foil on reinforced scrim, single bubble air space, R8.8	C.S.F.	62.50
Vinyl faced fiberglass, 1-1/2" thick, R5	S.F.	0.81
Vinyl/scrim/foil (VSF), 1-1/2" thick, R5	S.F.	1.01

Description	Unit	$ Cost
Roof ventilation, ridge vent, aluminum	L.F.	7.40
Mushroom vent, aluminum	Ea.	87.50
Gutter, aluminum, 5" K type, enameled	L.F.	7.60
Steel, half round or box, 5", enameled	L.F.	6.90
Wood, clear treated cedar, fir or hemlock, 4" x 5"	L.F.	23.50
Downspout, aluminum, 3" x 4", enameled	L.F.	6.65
Steel, round, 4" diameter	L.F.	6.65

Important: See the Reference Section for Location Factors.

Model costs calculated for a 1 story building with 14' story height and 1,800 square feet of floor area

Post Frame Barn

			Unit	Unit Cost	Cost Per S.F.	% Of Sub-Total
A. SUBSTRUCTURE						
1010	Standard Foundations	Included in B2010 Exterior Walls	–	–	–	0.0%
1020	Special Foundations	N/A	–	–	–	
1030	Slab on Grade	N/A	–	–	–	
2010	Basement Excavation	N/A	–	–	–	
2020	Basement Walls	N/A	–	–	–	
B. SHELL						
	B10 Superstructure					
1010	Floor Construction	N/A	–	–	–	17.9 %
1020	Roof Construction	Wood truss, 4/12 pitch	S.F. Roof	6.94	6.94	
	B20 Exterior Enclosure					
2010	Exterior Walls	2 x 6 Wood framing, 6 x 6 wood posts in concrete; steel siding panels	S.F. Wall	10.60	14.25	58.2%
2020	Exterior Windows	Steel windows, fixed	Each	516	4.06	
2030	Exterior Doors	Steel overhead, hollow metal doors	Each	1898	4.22	
	B30 Roofing					
3010	Roof Coverings	Metal panel roof	S.F. Roof	3.91	3.91	10.1%
3020	Roof Openings	N/A	–	–	–	
C. INTERIORS						
1010	Partitions	N/A	–	–	–	0.0 %
1020	Interior Doors	N/A	–	–	–	
1030	Fittings	N/A	–	–	–	
2010	Stair Construction	N/A	–	–	–	
3010	Wall Finishes	N/A	–	–	–	
3020	Floor Finishes	N/A	–	–	–	
3030	Ceiling Finishes	N/A	–	–	–	
D. SERVICES						
	D10 Conveying					
1010	Elevators & Lifts	N/A	–	–	–	0.0 %
1020	Escalators & Moving Walks	N/A	–	–	–	
	D20 Plumbing					
2010	Plumbing Fixtures	N/A	–	–	–	0.0 %
2020	Domestic Water Distribution	N/A	–	–	–	
2040	Rain Water Drainage	N/A	–	–	–	
	D30 HVAC					
3010	Energy Supply	N/A	–	–	–	0.0 %
3020	Heat Generating Systems	N/A	–	–	–	
3030	Cooling Generating Systems	N/A	–	–	–	
3050	Terminal & Package Units	N/A	–	–	–	
3090	Other HVAC Sys. & Equipment	N/A	–	–	–	
	D40 Fire Protection					
4010	Sprinklers	N/A	–	–	–	0.0 %
4020	Standpipes	N/A	–	–	–	
	D50 Electrical					
5010	Electrical Service/Distribution	60 Ampere service, panel board & feeders	S.F. Floor	1.71	1.71	13.7%
5020	Lighting & Branch Wiring	High bay fixtures, receptacles & switches	S.F. Floor	3.61	3.61	
5030	Communications & Security	N/A	–	–	–	
5090	Other Electrical Systems	N/A	–	–	–	
E. EQUIPMENT & FURNISHINGS						
1010	Commercial Equipment	N/A	–	–	–	0.0 %
1020	Institutional Equipment	N/A	–	–	–	
1030	Vehicular Equipment	N/A	–	–	–	
1090	Other Equipment	N/A	–	–	–	
F. SPECIAL CONSTRUCTION & DEMOLITION						
1020	Integrated Construction	N/A	–	–	–	0.0 %
1040	Special Facilities	N/A	–	–	–	
G. BUILDING SITEWORK		**N/A**				
				Sub-Total	38.70	**100%**
	CONTRACTOR FEES (General Requirements: 10%, Overhead: 5%, Profit: 5%)			*20%*	9.68	
	ARCHITECT FEES			*8%*	3.87	
				Total Building Cost	**52.25**	

Costs per square foot of floor area

Exterior Wall	S.F. Area	5000	7000	9000	11000	13000	15000	17000	19000	21000
	L.F. Perimeter	300	380	420	468	486	513	540	580	620
Face Brick with Concrete Block Back-up	Steel Frame	161.80	152.70	144.00	139.15	133.70	130.35	127.70	126.25	125.00
	Bearing Walls	162.20	153.00	144.20	139.25	133.80	130.40	127.70	126.25	125.00
Limestone with Concrete Block Back-up	Steel Frame	191.95	179.95	167.40	160.45	152.45	147.50	143.65	141.55	139.85
	Bearing Walls	192.35	180.25	167.60	160.60	152.55	147.55	143.70	141.55	139.85
Decorative Concrete Block	Steel Frame	151.60	143.50	136.05	131.85	127.35	124.55	122.30	121.05	120.00
	Bearing Walls	152.00	143.75	136.25	132.00	127.45	124.55	122.30	121.05	119.95
Perimeter Adj., Add or Deduct	Per 100 L.F.	16.50	11.90	9.20	7.45	6.35	5.50	4.90	4.30	3.90
Story Hgt. Adj., Add or Deduct	Per 1 Ft.	2.65	2.40	2.05	1.85	1.70	1.50	1.40	1.35	1.35
***For Basement**, add $ 27.40 per square foot of basement area*										

The above costs were calculated using the basic specifications shown on the facing page. These costs should be adjusted where necessary for design alternatives and owner's requirements. Reported completed project costs, for this type of structure, range from $ 102.90 to $ 267.15 per S.F.

Common additives

Description	Unit	$ Cost
Closed Circuit Surveillance, One station		
Camera and monitor	Ea.	1975
For additional camera stations, add	Ea.	1075
Emergency Lighting, 25 watt, battery operated		
Lead battery	Ea.	325
Nickel cadmium	Ea.	665
Flagpoles, Complete		
Aluminum, 20′ high	Ea.	1825
40′ high	Ea.	4300
70′ high	Ea.	11,300
Fiberglass, 23′ high	Ea.	1275
39′-5″ high	Ea.	3275
59′ high	Ea.	6475

Description	Unit	$ Cost
Mail Boxes, Horizontal, key lock, 15″ x 6″ x 5″	Ea.	62
Double 15″ x 12″ x 5″	Ea.	86
Quadruple 15″ x 12″ x 10″	Ea.	137
Vertical, 6″ x 5″ x 15″, aluminum	Ea.	62
Bronze	Ea.	68
Steel, enameled	Ea.	64.50
Scales, Dial type, 5 ton cap.		
8′ x 6′ platform	Ea.	12,400
9′ x 7′ platform	Ea.	14,400
Smoke Detectors		
Ceiling type	Ea.	214
Duct type	Ea.	540

Important: See the Reference Section for Location Factors.

Model costs calculated for a 1 story building with 14′ story height and 13,000 square feet of floor area

Post Office

				Unit	Unit Cost	Cost Per S.F.	% Of Sub-Total
A. SUBSTRUCTURE							
1010	Standard Foundations	Poured concrete; strip and spread footings		S.F. Ground	1.53	1.53	9.9%
1020	Special Foundations	N/A		—	—	—	
1030	Slab on Grade	4″ reinforced concrete with vapor barrier and granular base		S.F. Slab	5.10	5.10	
2010	Basement Excavation	Site preparation for slab and trench for foundation wall and footing		S.F. Ground	.29	.29	
2020	Basement Walls	4′ foundation wall		L.F. Wall	74	2.75	
B. SHELL							
	B10 Superstructure						
1010	Floor Construction	Steel column fireproofing		L.F. Column	32	.24	9.3%
1020	Roof Construction	Metal deck, open web steel joists, columns		S.F. Roof	8.90	8.90	
	B20 Exterior Enclosure						
2010	Exterior Walls	Face brick with concrete block backup	*80% of wall*	S.F. Wall	30	12.60	17.3%
2020	Exterior Windows	Double strength window glass	*20% of wall*	Each	705	3.21	
2030	Exterior Doors	Double aluminum & glass, single aluminum, hollow metal, steel overhead		Each	2627	1.21	
	B30 Roofing						
3010	Roof Coverings	Built-up tar and gravel with flashing; perlite/EPS composite insulation		S.F. Roof	6.30	6.30	6.4%
3020	Roof Openings	N/A		—	—	—	
C. INTERIORS							
1010	Partitions	Concrete block	*15 S.F. Floor/L.F. Partition*	S.F. Partition	8.78	7.02	23.7%
1020	Interior Doors	Single leaf hollow metal	*150 S.F. Floor/Door*	Each	1133	7.55	
1030	Fittings	Toilet partitions, cabinets, shelving, lockers		S.F. Floor	1.54	1.54	
2010	Stair Construction	N/A		—	—	—	
3010	Wall Finishes	Paint		S.F. Surface	2.11	3.38	
3020	Floor Finishes	50% vinyl tile, 50% paint		S.F. Floor	2.14	2.14	
3030	Ceiling Finishes	Mineral fiber tile on concealed zee bars	*25% of area*	S.F. Ceiling	6.57	1.65	
D. SERVICES							
	D10 Conveying						
1010	Elevators & Lifts	N/A		—	—	—	0.0 %
1020	Escalators & Moving Walks	N/A		—	—	—	
	D20 Plumbing						
2010	Plumbing Fixtures	Toilet and service fixtures, supply and drainage	*1 Fixture/1180 S.F. Floor*	Each	2030	1.72	3.6%
2020	Domestic Water Distribution	Gas fired water heater		S.F. Floor	.48	.48	
2040	Rain Water Drainage	Roof drains		S.F. Roof	1.33	1.33	
	D30 HVAC						
3010	Energy Supply	N/A		—	—	—	8.7 %
3020	Heat Generating Systems	Included in D3050		—	—	—	
3030	Cooling Generating Systems	N/A		—	—	—	
3050	Terminal & Package Units	Single zone, gas heating, electric cooling		S.F. Floor	8.55	8.55	
3090	Other HVAC Sys. & Equipment	N/A		—	—	—	
	D40 Fire Protection						
4010	Sprinklers	Wet pipe sprinkler system		S.F. Floor	4.22	4.22	5.2%
4020	Standpipes	Standpipe		S.F. Floor	.92	.92	
	D50 Electrical						
5010	Electrical Service/Distribution	600 ampere service, panel board and feeders		S.F. Floor	2.47	2.47	15.8%
5020	Lighting & Branch Wiring	T-8 fluorescent fixtures, receptacles, switches, A.C. and misc. power		S.F. Floor	8.51	8.51	
5030	Communications & Security	Addressable alarm systems, internet wiring and emergency lighting		S.F. Floor	4.39	4.39	
5090	Other Electrical Systems	Emergency generator, 15 kW		S.F. Floor	.14	.14	
E. EQUIPMENT & FURNISHINGS							
1010	Commercial Equipment	N/A		—	—	—	0.0 %
1020	Institutional Equipment	N/A		—	—	—	
1030	Vehicular Equipment	N/A		—	—	—	
1090	Other Equipment	N/A		—	—	—	
F. SPECIAL CONSTRUCTION							
1020	Integrated Construction	N/A		—	—	—	0.0 %
1040	Special Facilities	N/A		—	—	—	
G. BUILDING SITEWORK		**N/A**					
					Sub-Total	98.14	**100%**
	CONTRACTOR FEES (General Requirements: 10%, Overhead: 5%, Profit: 10%)				*25%*	24.52	
	ARCHITECT FEES				9%	11.04	
					Total Building Cost	**133.70**	

Racquetball Court

Costs per square foot of floor area

Exterior Wall	S.F. Area	5000	10000	15000	21000	25000	30000	40000	50000	60000
	L.F. Perimeter	287	400	500	600	660	700	834	900	1000
Face Brick with Concrete Block Back-up	Steel Frame	229.30	196.35	184.20	175.90	172.20	167.25	163.10	158.70	156.60
	Bearing Walls	227.15	193.80	181.50	173.05	169.30	164.30	160.10	155.60	153.45
Concrete Block	Steel Frame	198.35	174.85	166.20	160.50	158.00	154.70	151.90	149.00	147.60
Brick Veneer	Steel Frame	215.95	187.05	176.40	169.30	166.05	161.90	158.30	154.55	152.70
Galvanized Steel Siding	Steel Frame	189.15	168.80	161.45	156.60	154.45	151.75	149.40	147.05	145.85
Metal Sandwich Panel	Steel Frame	239.95	203.85	190.40	181.25	177.10	171.65	167.00	162.05	159.65
Perimeter Adj., Add or Deduct	Per 100 L.F.	28.10	14.15	9.35	6.65	5.55	4.75	3.50	2.85	2.25
Story Hgt. Adj., Add or Deduct	Per 1 Ft.	5.70	4.05	3.35	2.90	2.65	2.35	2.15	1.85	1.65
***For Basement**, add $26.45 per square foot of basement area*										

The above costs were calculated using the basic specifications shown on the facing page. These costs should be adjusted where necessary for design alternatives and owner's requirements. Reported completed project costs, for this type of structure, range from $97.50 to $304.05 per S.F.

Common additives

Description	Unit	$ Cost
Bar, Front Bar	L.F.	415
Back Bar	L.F.	330
Booth, Upholstered, custom straight	L.F.	248 - 460
"L" or "U" shaped	L.F.	256 - 435
Bleachers, Telescoping, manual		
To 15 tier	Seat	133 - 189
21-30 tier	Seat	300 - 395
Courts		
Ceiling	Court	9225
Floor	Court	15,700
Walls	Court	26,600
Emergency Lighting, 25 watt, battery operated		
Lead battery	Ea.	325
Nickel cadmium	Ea.	665
Kitchen Equipment		
Broiler	Ea.	4050
Cooler, 6 ft. long, reach-in	Ea.	3875
Dishwasher, 10-12 racks per hr.	Ea.	4200
Food warmer, counter 1.2 KW	Ea.	770
Freezer, reach-in, 44 C.F.	Ea.	6100
Ice cube maker, 50 lb. per day	Ea.	1950

Description	Unit	$ Cost
Lockers, Steel, single tier, 60" or 72"	Opng.	222 - 365
2 tier, 60" or 72" total	Opng.	115 - 164
5 tier, box lockers	Opng.	69.50 - 78.50
Locker bench, lam. maple top only	L.F.	33
Pedestals, steel pipe	Ea.	71
Sauna, Prefabricated, complete		
6' x 4'	Ea.	6500
6' x 9'	Ea.	10,100
8' x 8'	Ea.	11,600
8' x 10'	Ea.	12,900
10' x 12'	Ea.	15,500
Sound System		
Amplifier, 250 watts	Ea.	2125
Speaker, ceiling or wall	Ea.	206
Trumpet	Ea.	395
Steam Bath, Complete, to 140 C.F.	Ea.	2875
To 300 C.F.	Ea.	3200
To 800 C.F.	Ea.	6950
To 2500 C.F.	Ea.	9150

Important: See the Reference Section for Location Factors.

Model costs calculated for a 2 story building with 12' story height and 30,000 square feet of floor area

Racquetball Court

				Unit	Unit Cost	Cost Per S.F.	% Of Sub-Total
A. SUBSTRUCTURE							
1010	Standard Foundations	Poured concrete; strip and spread footings		S.F. Ground	2.06	1.03	5.1%
1020	Special Foundations	N/A		—	—	—	
1030	Slab on Grade	5" reinforced concrete with vapor barrier and granular base		S.F. Slab	6.78	3.39	
2010	Basement Excavation	Site preparation for slab and trench for foundation wall and footing		S.F. Ground	.29	.19	
2020	Basement Walls	4' foundation wall		L.F. Wall	74	1.71	
B. SHELL							
	B10 Superstructure						
1010	Floor Construction	Open web steel joists, slab form, concrete, columns	*50% of area*	S.F. Floor	17.96	8.98	10.3%
1020	Roof Construction	Metal deck on open web steel joists, columns		S.F. Roof	7.46	3.73	
	B20 Exterior Enclosure						
2010	Exterior Walls	Face brick with concrete block backup	*95% of wall*	S.F. Wall	30	15.82	15.3%
2020	Exterior Windows	Storefront	*5% of wall*	S.F. Window	97	2.72	
2030	Exterior Doors	Aluminum and glass and hollow metal		Each	3016	.41	
	B30 Roofing						
3010	Roof Coverings	Built-up tar and gravel with flashing; perlite/EPS composite insulation		S.F. Roof	6.48	3.24	2.7%
3020	Roof Openings	Roof hatches		S.F. Roof	.14	.07	
C. INTERIORS							
1010	Partitions	Concrete block, gypsum board on metal studs	*25 S.F. Floor/L.F. Partition*	S.F. Partition	13.45	5.38	13.6%
1020	Interior Doors	Single leaf hollow metal	*810 S.F. Floor/Door*	Each	1133	1.39	
1030	Fittings	Toilet partitions		S.F. Floor	.40	.40	
2010	Stair Construction	Concrete filled metal pan		Flight	16,800	1.68	
3010	Wall Finishes	Paint		S.F. Surface	1.30	1.04	
3020	Floor Finishes	80% carpet tile, 20% ceramic tile	*50% of floor area*	S.F. Floor	6.12	3.06	
3030	Ceiling Finishes	Mineral fiber tile on concealed zee bars	*60% of area*	S.F. Ceiling	6.57	3.94	
D. SERVICES							
	D10 Conveying						
1010	Elevators & Lifts	N/A		—	—	—	0.0 %
1020	Escalators & Moving Walks	N/A		—	—	—	
	D20 Plumbing						
2010	Plumbing Fixtures	Kitchen, bathroom and service fixtures, supply and drainage	*1 Fixture/1000 S.F. Floor*	Each	3420	3.42	5.6%
2020	Domestic Water Distribution	Gas fired water heater		S.F. Floor	3.19	3.19	
2040	Rain Water Drainage	Roof drains		S.F. Roof	.68	.34	
	D30 HVAC						
3010	Energy Supply	N/A		—	—	—	22.0 %
3020	Heat Generating Systems	Included in D3050		—	—	—	
3030	Cooling Generating Systems	N/A		—	—	—	
3050	Terminal & Package Units	Multizone unit, gas heating, electric cooling		S.F. Floor	27	27.20	
3090	Other HVAC Sys. & Equipment	N/A		—	—	—	
	D40 Fire Protection						
4010	Sprinklers	Sprinklers, light hazard		S.F. Floor	3.03	3.03	2.7%
4020	Standpipes	Standpipe		S.F. Floor	.37	.37	
	D50 Electrical						
5010	Electrical Service/Distribution	400 ampere service, panel board and feeders		S.F. Floor	.86	.86	7.9%
5020	Lighting & Branch Wiring	Fluorescent and high intensity discharge fixtures, receptacles, switches, A.C. and misc. power		S.F. Floor	7.78	7.78	
5030	Communications & Security	Addressable alarm systems and emergency lighting		S.F. Floor	1.06	1.06	
5090	Other Electrical Systems	Emergency generator, 15 kW		S.F. Floor	.06	.06	
E. EQUIPMENT & FURNISHINGS							
1010	Commercial Equipment	N/A		—	—	—	14.9 %
1020	Institutional Equipment	N/A		—	—	—	
1030	Vehicular Equipment	N/A		—	—	—	
1090	Other Equipment	Courts, sauna baths		S.F. Floor	18.42	18.42	
F. SPECIAL CONSTRUCTION							
1020	Integrated Construction	N/A		—	—	—	0.0 %
1040	Special Facilities	N/A		—	—	—	
G. BUILDING SITEWORK	**N/A**						
					Sub-Total	123.91	**100%**
	CONTRACTOR FEES (General Requirements: 10%, Overhead: 5%, Profit: 10%)				*25%*	30.95	
	ARCHITECT FEES				*8%*	12.39	
					Total Building Cost	**167.25**	

M.530 Restaurant

Costs per square foot of floor area

Exterior Wall	S.F. Area	2000	2800	3500	4200	5000	5800	6500	7200	8000
	L.F. Perimeter	180	212	240	268	300	314	336	344	368
Wood Siding	Wood Frame	205.50	194.70	189.35	185.70	182.95	179.40	177.70	175.55	174.30
Brick Veneer	Wood Frame	209.70	197.85	191.95	188.00	184.95	181.00	179.10	176.65	175.25
Stone Veneer	Steel Frame	246.00	229.55	221.40	215.85	211.60	205.70	203.05	199.30	197.35
Fiber Cement Siding	Steel Frame	212.15	201.20	195.80	192.10	189.30	185.65	184.00	181.70	180.45
E.I.F.S.	Reinforced Concrete	233.10	220.55	214.30	210.10	206.80	202.55	200.55	197.85	196.40
Stucco	Reinforced Concrete	237.00	223.90	217.40	213.05	209.60	205.15	203.10	200.30	198.75
Perimeter Adj., Add or Deduct	Per 100 L.F.	22.65	16.25	13.00	10.85	9.05	7.85	7.05	6.30	5.65
Story Hgt. Adj., Add or Deduct	Per 1 Ft.	2.20	1.90	1.70	1.65	1.45	1.35	1.35	1.20	1.15
***For Basement**, add $ 32.10 per square foot of basement area*										

The above costs were calculated using the basic specifications shown on the facing page. These costs should be adjusted where necessary for design alternatives and owner's requirements. Reported completed project costs, for this type of structure, range from $ 132.00 to $ 307.60 per S.F.

Common additives

Description	Unit	$ Cost
Bar, Front Bar	L.F.	415
Back bar	L.F.	330
Booth, Upholstered, custom straight	L.F.	248 - 460
"L" or "U" shaped	L.F.	256 - 435
Cupola, Stock unit, redwood		
30" square, 37" high, aluminum roof	Ea.	750
Copper roof	Ea.	415
Fiberglass, 5'-0" base, 63" high	Ea.	3900 - 4550
6'-0" base, 63" high	Ea.	5075 - 7300
Decorative Wood Beams, Non load bearing		
Rough sawn, 4" x 6"	L.F.	7.95
4" x 8"	L.F.	9
4" x 10"	L.F.	10.75
4" x 12"	L.F.	12.30
8" x 8"	L.F.	13.10
Emergency Lighting, 25 watt, battery operated		
Lead battery	Ea.	325
Nickel cadmium	Ea.	665

Description	Unit	$ Cost
Fireplace, Brick, not incl. chimney or foundation		
30" x 29" opening	Ea.	2825
Chimney, standard brick		
Single flue, 16" x 20"	V.L.F.	98
20" x 20"	V.L.F.	115
2 Flue, 20" x 24"	V.L.F.	144
20" x 32"	V.L.F.	178
Kitchen Equipment		
Broiler	Ea.	4050
Coffee urn, twin 6 gallon	Ea.	2975
Cooler, 6 ft. long	Ea.	3875
Dishwasher, 10-12 racks per hr.	Ea.	4200
Food warmer, counter, 1.2 KW	Ea.	770
Freezer, 44 C.F., reach-in	Ea.	6100
Ice cube maker, 50 lb. per day	Ea.	1950
Range with 1 oven	Ea.	3275
Refrigerators, Prefabricated, walk-in		
7'-6" high, 6' x 6'	S.F.	200
10' x 10'	S.F.	157
12' x 14'	S.F.	140
12' x 20'	S.F.	123

Model costs calculated for a 1 story building with 12′ story height and 5,000 square feet of floor area

Restaurant

				Unit	Unit Cost	Cost Per S.F.	% Of Sub-Total
A. SUBSTRUCTURE							
1010	Standard Foundations	Poured concrete; strip and spread footings		S.F. Ground	4.06	4.06	10.3%
1020	Special Foundations	N/A		—	—	—	
1030	Slab on Grade	4″ reinforced concrete		S.F. Slab	5.10	5.10	
2010	Basement Excavation	Site preparation for slab and trench for foundation wall and footing		S.F. Ground	.51	.51	
2020	Basement Walls	4′ Foundation wall		L.F. Wall	74	4.41	
B. SHELL							
	B10 Superstructure						
1010	Floor Construction	Wood columns		S.F. Floor	.15	.38	5.4%
1020	Roof Construction	Wood roof truss, 4:12 slope		S.F. Roof	6.20	6.94	
	B20 Exterior Enclosure						
2010	Exterior Walls	Wood siding on wood studs, insulated	*70% of wall*	S.F. Wall	18.99	9.57	13.8%
2020	Exterior Windows	Curtain glazing wall system	*30% of wall*	Each	43	4.16	
2030	Exterior Doors	Aluminum & glass double doors, hollow metal doors		Each	5126	5.14	
	B30 Roofing						
3010	Roof Coverings	Asphalt strip shingles, gutters and downspouts		S.F. Roof	13.93	2.24	1.6%
3020	Roof Openings	N/A		—	—	—	
C. INTERIORS							
1010	Partitions	Fire rated gypsum board on wood studs	*25 S.F. Floor/L.F. Partition*	S.F. Partition	6.30	2.52	13.3%
1020	Interior Doors	Single leaf hollow core wood doors	*1000 S.F. Floor/Door*	Each	510	.51	
1030	Fittings	Plastic laminate toilet partitions		S.F. Floor	.83	.83	
2010	Stair Construction	N/A		—	—	—	
3010	Wall Finishes	80% Paint, 20% ceramic tile		S.F. Surface	2.49	1.99	
3020	Floor Finishes	70% Carpet tile, 30% quarry tile		S.F. Floor	8.17	8.17	
3030	Ceiling Finishes	Gypsum board ceiling on wood furring		S.F. Ceiling	4.11	4.11	
D. SERVICES							
	D10 Conveying						
1010	Elevators & Lifts	N/A		—	—	—	0.0 %
1020	Escalators & Moving Walks	N/A		—	—	—	
	D20 Plumbing						
2010	Plumbing Fixtures	Restroom, kitchen and service fixtures, supply and drainage	*1 Fixture/350 S.F. Floor*	Each	2844	8.01	9.5%
2020	Domestic Water Distribution	Gas fired water heater		S.F. Floor	4.93	4.93	
2040	Rain Water Drainage	Roof drains		S.F. Roof			
	D30 HVAC						
3010	Energy Supply	N/A		—	—	—	26.0 %
3020	Heat Generating Systems	Included in D3050		—	—	—	
3030	Cooling Generating Systems	N/A		—	—	—	
3050	Terminal & Package Units	Multizone unit gas heating, electric cooling, kitchen exhaust system		S.F. Floor	36	35.63	
3090	Other HVAC Sys. & Equipment	N/A		—	—	—	
	D40 Fire Protection						
4010	Sprinklers	Wet pipe sprinkler system		S.F. Floor	8.53	8.53	7.7%
4020	Standpipes	Standpipe		S.F. Floor	2.04	2.04	
	D50 Electrical						
5010	Electrical Service/Distribution	400 Ampere service, panelboards and feeders		S.F. Floor	4.48	4.48	12.4%
5020	Lighting & Branch Wiring	Fluorescent fixtures, receptacles, switches, a.c. and misc. power		S.F. Floor	9.07	9.07	
5030	Communications & Security	Addressable alarm system		S.F. Floor	3.46	3.46	
5090	Other Electrical Systems	Emergency generator, 15 kW		S.F. Floor			
E. EQUIPMENT & FURNISHINGS							
1010	Commercial Equipment	N/A		—	—	—	0.0 %
1020	Institutional Equipment	N/A		—	—	—	
1030	Vehicular Equipment	N/A		—	—	—	
1090	Other Equipment	N/A		—	—	—	
F. SPECIAL CONSTRUCTION							
1020	Integrated Construction	N/A		—	—	—	0.0 %
1040	Special Facilities	N/A		—	—	—	
G. BUILDING SITEWORK		**N/A**					
					Sub-Total	136.79	**100%**
	CONTRACTOR FEES (General Requirements: 10%, Overhead: 5%, Profit: 10%)				*25%*	34.19	
	ARCHITECT FEES				*7%*	11.97	
					Total Building Cost	**182.95**	

Restaurant, Fast Food

Costs per square foot of floor area

Exterior Wall	S.F. Area	2000	2800	3500	4000	5000	5800	6500	7200	8000
	L.F. Perimeter	180	212	240	260	300	314	336	344	368
Wood Siding	Wood Frame	199.50	193.00	189.75	188.15	185.85	183.20	182.00	180.30	179.45
Brick Veneer	Wood Frame	207.60	200.25	196.50	194.65	192.10	189.05	187.75	185.75	184.80
Stone Veneer	Steel Frame	236.50	225.25	219.60	216.80	212.85	208.20	206.25	203.20	201.80
Fiber Cement Siding	Steel Frame	203.05	196.75	193.55	192.00	189.75	187.20	186.05	184.35	183.55
E.I.F.S.	Reinforced Concrete	226.00	218.00	214.05	212.00	209.30	206.00	204.60	202.45	201.45
Stucco	Reinforced Concrete	226.20	218.15	214.20	212.20	209.45	206.15	204.75	202.60	201.60
Perimeter Adj., Add or Deduct	Per 100 L.F.	22.75	16.25	13.00	11.35	9.15	7.75	7.05	6.30	5.70
Story Hgt. Adj., Add or Deduct	Per 1 Ft.	2.40	2.00	1.80	1.70	1.60	1.40	1.40	1.25	1.20
Basement—Not Applicable										

The above costs were calculated using the basic specifications shown on the facing page. These costs should be adjusted where necessary for design alternatives and owner's requirements. Reported completed project costs, for this type of structure, range from $ 131.70 to $ 256.80 per S.F.

Common additives

Description	Unit	$ Cost
Bar, Front Bar	L.F.	415
Back bar	L.F.	330
Booth, Upholstered, custom straight	L.F.	248 - 460
"L" or "U" shaped	L.F.	256 - 435
Drive-up Window	Ea.	8900 - 12,500
Emergency Lighting, 25 watt, battery operated		
Lead battery	Ea.	325
Nickel cadmium	Ea.	665
Kitchen Equipment		
Broiler	Ea.	4050
Coffee urn, twin 6 gallon	Ea.	2975
Cooler, 6 ft. long	Ea.	3875
Dishwasher, 10-12 racks per hr.	Ea.	4200
Food warmer, counter, 1.2 KW	Ea.	770
Freezer, 44 C.F., reach-in	Ea.	6100
Ice cube maker, 50 lb. per day	Ea.	1950
Range with 1 oven	Ea.	3275

Description	Unit	$ Cost
Refrigerators, Prefabricated, walk-in		
7'-6" High, 6' x 6'	S.F.	200
10' x 10'	S.F.	157
12' x 14'	S.F.	140
12' x 20'	S.F.	123
Serving		
Counter top (Stainless steel)	L.F.	195
Base cabinets	L.F.	400 - 655
Sound System		
Amplifier, 250 watts	Ea.	2125
Speaker, ceiling or wall	Ea.	206
Trumpet	Ea.	395
Storage		
Shelving	S.F.	9.95
Washing		
Stainless steel counter	L.F.	195

Important: See the Reference Section for Location Factors.

Model costs calculated for a 1 story building with 10′ story height and 4,000 square feet of floor area

Restaurant, Fast Food

				Unit	Unit Cost	Cost Per S.F.	% Of Sub-Total
A. SUBSTRUCTURE							
1010	Standard Foundations	Poured concrete; strip and spread footings		S.F. Ground	3.23	3.23	9.8%
1020	Special Foundations	N/A		—	—	—	
1030	Slab on Grade	4″ reinforced concrete		S.F. Slab	5.10	5.10	
2010	Basement Excavation	Site preparation for slab and trench for foundation wall and footing		S.F. Ground	.51	.51	
2020	Basement Walls	4′ Foundation wall		L.F. Wall	74	4.78	
B. SHELL							
	B10 Superstructure						
1010	Floor Construction	Wood columns		S.F. Floor	.38	.38	3.3%
1020	Roof Construction	Flat wood rafter roof		S.F. Roof	4.20	4.20	
	B20 Exterior Enclosure						
2010	Exterior Walls	Wood siding on wood studs, insulated	*70% of wall*	S.F. Wall	16.22	7.38	14.0%
2020	Exterior Windows	Curtain glazing wall system	*30% of wall*	Each	49	4.40	
2030	Exterior Doors	Aluminum & glass double doors, hollow metal doors		Each	3840	7.70	
	B30 Roofing						
3010	Roof Coverings	Single ply membrane, stone ballast, rigid insulation		S.F. Roof	6.19	6.19	4.8%
3020	Roof Openings	Roof and smoke hatches		S.F. Roof	.50	.50	
C. INTERIORS							
1010	Partitions	Water resistant gypsum board on wood studs	*25 S.F. Floor/L.F. Partition*	S.F. Partition	6.20	2.48	20.2%
1020	Interior Doors	Single leaf hollow core wood doors	*1000 S.F. Floor/Door*	Each	510	.51	
1030	Fittings	Stainless steel toilet partitions		S.F. Floor	1.48	1.48	
2010	Stair Construction	N/A		—	—	—	
3010	Wall Finishes	90% Paint, 10% ceramic tile		S.F. Surface	1.83	1.46	
3020	Floor Finishes	Quarry tile		S.F. Floor	15.69	15.69	
3030	Ceiling Finishes	Acoustic ceiling tiles on suspended support grid		S.F. Ceiling	6.57	6.57	
D. SERVICES							
	D10 Conveying						
1010	Elevators & Lifts	N/A		—	—	—	0.0 %
1020	Escalators & Moving Walks	N/A		—	—	—	
	D20 Plumbing						
2010	Plumbing Fixtures	Restroom, kitchen and service fixtures, supply and drainage	*1 Fixture/330 S.F. Floor*	Each	2491	7.48	9.0%
2020	Domestic Water Distribution	Gas fired water heater		S.F. Floor	3.50	3.50	
2040	Rain Water Drainage	Roof drains		S.F. Roof	1.59	1.59	
	D30 HVAC						
3010	Energy Supply	N/A		—	—	—	12.4 %
3020	Heat Generating Systems	Included in D3050		—	—	—	
3030	Cooling Generating Systems	N/A		—	—	—	
3050	Terminal & Package Units	Single zone unit gas heating, electric cooling		S.F. Floor	17.26	17.26	
3090	Other HVAC Sys. & Equipment	N/A		—	—	—	
	D40 Fire Protection						
4010	Sprinklers	Wet pipe sprinkler system, ordinary hazard		S.F. Floor	8.53	8.53	7.6%
4020	Standpipes	Standpipe		S.F. Floor	2.13	2.13	
	D50 Electrical						
5010	Electrical Service/Distribution	400 Ampere service, panelboards and feeders		S.F. Floor	4.75	4.75	13.6%
5020	Lighting & Branch Wiring	Fluorescent fixtures, receptacles, switches, a.c. and misc. power		S.F. Floor	9.83	9.83	
5030	Communications & Security	Addressable alarm, emergency lighting		S.F. Floor	4.32	4.32	
5090	Other Electrical Systems	N/A		—	—	—	
E. EQUIPMENT & FURNISHINGS							
1010	Commercial Equipment	N/A		—	—	—	5.3 %
1020	Institutional Equipment	N/A		—	—	—	
1030	Vehicular Equipment	N/A		—	—	—	
1090	Other Equipment	Prefabricated walk in refrigerators		S.F. Floor	7.40	7.40	
F. SPECIAL CONSTRUCTION							
1020	Integrated Construction	N/A		—	—	—	0.0 %
1040	Special Facilities	N/A		—	—	—	
G. BUILDING SITEWORK		**N/A**					
					Sub-Total	139.35	**100%**
	CONTRACTOR FEES (General Requirements: 10%, Overhead: 5%, Profit: 10%)				*25%*	34.86	
	ARCHITECT FEES				*8%*	13.94	
					Total Building Cost	**188.15**	

For customer support on your Light Commercial Cost Data, call 888.607.2575.

M.550 Rink, Hockey/Indoor Soccer

Costs per square foot of floor area

Exterior Wall	S.F. Area	10000	15000	20000	25000	30000	35000	40000	45000	50000
	L.F. Perimeter	450	500	600	700	740	822	890	920	966
Face Brick with Concrete Block Back-up	Steel Joists	190.00	173.60	168.60	165.55	161.05	159.35	157.60	155.15	153.60
	Lam. Wood Truss	200.45	184.70	180.10	177.30	172.85	171.25	169.60	167.15	165.65
Concrete Block	Steel Frame	183.60	172.65	169.15	167.05	**164.05**	162.90	161.75	160.15	159.15
	Lam. Wood Truss	178.10	168.05	165.00	163.15	160.35	159.30	158.30	156.75	155.80
Galvanized Steel Siding	Steel Frame	182.10	171.60	168.20	166.15	163.30	162.15	161.00	159.45	158.50
Metal Sandwich Panel	Steel Frame	184.90	173.65	170.00	167.90	164.80	163.60	162.40	160.75	159.65
Perimeter Adj., Add or Deduct	Per 100 L.F.	12.65	8.50	6.35	5.10	4.20	3.60	3.15	2.80	2.55
Story Hgt. Adj., Add or Deduct	Per 1 Ft.	2.00	1.50	1.35	1.30	1.05	1.05	1.00	0.95	0.90
Basement—Not Applicable										

The above costs were calculated using the basic specifications shown on the facing page. These costs should be adjusted where necessary for design alternatives and owner's requirements. Reported completed project costs, for this type of structure, range from $77.90 to $242.45 per S.F.

Common additives

Description	Unit	$ Cost
Bar, Front Bar	L.F.	415
Back bar	L.F.	330
Booth, Upholstered, custom straight	L.F.	248 - 460
"L" or "U" shaped	L.F.	256 - 435
Bleachers, Telescoping, manual		
To 15 tier	Seat	133 - 189
16-20 tier	Seat	284 - 345
21-30 tier	Seat	300 - 395
For power operation, add	Seat	57 - 89
Emergency Lighting, 25 watt, battery operated		
Lead battery	Ea.	325
Nickel cadmium	Ea.	665
Lockers, Steel, single tier, 60" or 72"	Opng.	222 - 365
2 tier, 60" or 72" total	Opng.	115 - 164
5 tier, box lockers	Opng.	69.50 - 78.50
Locker bench, lam. maple top only	L.F.	33
Pedestals, steel pipe	Ea.	71

Description	Unit	$ Cost
Rink		
Dasher boards & top guard	Ea.	185,500
Mats, rubber	S.F.	23
Score Board	Ea.	16,100 - 27,800

Important: See the Reference Section for Location Factors.

Model costs calculated for a 1 story building with 24′ story height and 30,000 square feet of floor area

Rink, Hockey/Indoor Soccer

			Unit	Unit Cost	Cost Per S.F.	% Of Sub-Total
A. SUBSTRUCTURE						
1010	Standard Foundations	Poured concrete; strip and spread footings	S.F. Ground	1.51	1.51	8.0%
1020	Special Foundations	N/A	—	—	—	
1030	Slab on Grade	6″ reinforced concrete with vapor barrier and granular base	S.F. Slab	6.15	6.15	
2010	Basement Excavation	Site preparation for slab and trench for foundation wall and footing	S.F. Ground	.16	.16	
2020	Basement Walls	4′ foundation wall	L.F. Wall	69	1.97	
B. SHELL						
B10 Superstructure						
1010	Floor Construction	Wide flange beams and columns	S.F. Floor	37	26.10	21.3%
1020	Roof Construction	(incl. in B1010)				
B20 Exterior Enclosure						
2010	Exterior Walls	Concrete block *95% of wall*	S.F. Wall	14.15	7.96	8.3%
2020	Exterior Windows	Store front *5% of wall*	Each	47	1.40	
2030	Exterior Doors	Aluminum and glass, hollow metal, overhead	Each	3226	.86	
B30 Roofing						
3010	Roof Coverings	Elastomeric neoprene membrane with flashing; perlite/EPS composite insulation	S.F. Roof	4.81	4.81	4.1%
3020	Roof Openings	Roof hatches	S.F. Roof	.23	.23	
C. INTERIORS						
1010	Partitions	Concrete block *140 S.F. Floor/L.F. Partition*	S.F. Partition	8.75	.75	6.7%
1020	Interior Doors	Hollow metal *2500 S.F. Floor/Door*	Each	1133	.45	
1030	Fittings	N/A	—	—	—	
2010	Stair Construction	N/A	—	—	—	
3010	Wall Finishes	Paint	S.F. Surface	11.32	1.94	
3020	Floor Finishes	80% rubber mat, 20% paint *50% of floor area*	S.F. Floor	8.74	4.37	
3030	Ceiling Finishes	Mineral fiber tile on concealed zee bar *10% of area*	S.F. Ceiling	6.57	.66	
D. SERVICES						
D10 Conveying						
1010	Elevators & Lifts	N/A	—	—	—	0.0 %
1020	Escalators & Moving Walks	N/A	—	—	—	
D20 Plumbing						
2010	Plumbing Fixtures	Toilet and service fixtures, supply and drainage *1 Fixture/1070 S.F. Floor*	Each	6056	5.66	7.9%
2020	Domestic Water Distribution	Oil fired water heater	S.F. Floor	3.19	3.19	
2040	Rain Water Drainage	Roof drains	S.F. Roof	.78	.78	
D30 HVAC						
3010	Energy Supply	Oil fired hot water, unit heaters *10% of area*	S.F. Floor	.83	.83	11.0%
3020	Heat Generating Systems	N/A	—	—	—	
3030	Cooling Generating Systems	N/A	—	—	—	
3050	Terminal & Package Units	Single zone, electric cooling *90% of area*	S.F. Floor	12.69	12.69	
3090	Other HVAC Sys. & Equipment	N/A	—	—	—	
D40 Fire Protection						
4010	Sprinklers	N/A	—	—	—	0.5 %
4020	Standpipes	Standpipe	S.F. Floor	.65	.65	
D50 Electrical						
5010	Electrical Service/Distribution	600 ampere service, panel board and feeders	S.F. Floor	1.50	1.50	9.2%
5020	Lighting & Branch Wiring	High intensity discharge and fluorescent fixtures, receptacles, switches, A.C. and misc. power	S.F. Floor	7.50	7.50	
5030	Communications & Security	Addressable alarm systems, emergency lighting and public address	S.F. Floor	2.01	2.01	
5090	Other Electrical Systems	Emergency generator	S.F. Floor	.22	.22	
E. EQUIPMENT & FURNISHINGS						
1010	Commercial Equipment	N/A	—	—	—	0.0 %
1020	Institutional Equipment	N/A	—	—	—	
1030	Vehicular Equipment	N/A	—	—	—	
1090	Other Equipment	N/A	—	—	—	
F. SPECIAL CONSTRUCTION						
1020	Integrated Construction	N/A	—	—	—	23.1 %
1040	Special Facilities	Dasher boards and rink (including ice making system)	S.F. Floor	28	28.31	
G. BUILDING SITEWORK		**N/A**				
				Sub-Total	122.66	**100%**
		CONTRACTOR FEES (General Requirements: 10%, Overhead: 5%, Profit: 10%)		*25%*	30.66	
		ARCHITECT FEES		*7%*	10.73	
				Total Building Cost	**164.05**	

Costs per square foot of floor area

Exterior Wall	S.F. Area	25000	30000	35000	40000	45000	50000	55000	60000	65000
	L.F. Perimeter	900	1050	1200	1350	1510	1650	1800	1970	2100
Fiber Cement Siding	Steel Frame	152.45	151.30	150.50	149.80	149.55	149.00	148.70	148.70	148.15
Metal Panel	Steel Frame	153.75	152.55	151.65	151.00	**150.70**	150.10	149.75	149.80	149.20
E.I.F.S.	Steel Frame	154.95	153.80	152.95	152.35	152.00	151.50	151.20	151.15	150.60
Tilt-up Concrete Panel	Reinforced Concrete	189.75	187.85	186.50	185.45	185.05	184.05	183.55	183.60	182.70
Brick Veneer	Reinforced Concrete	175.45	173.90	172.85	172.00	171.65	170.90	170.55	170.50	169.80
Stone Veneer	Reinforced Concrete	211.70	209.20	207.40	206.00	205.45	204.15	203.45	203.60	202.35
Perimeter Adj., Add or Deduct	Per 100 L.F.	2.90	2.45	2.10	1.85	1.65	1.45	1.30	1.20	1.15
Story Hgt. Adj., Add or Deduct	Per 1 Ft.	1.15	1.10	1.05	1.15	1.05	1.05	1.05	1.05	1.00
***For Basement**, add $23.75 per square foot of basement area*										

The above costs were calculated using the basic specifications shown on the facing page. These costs should be adjusted where necessary for design alternatives and owner's requirements. Reported completed project costs, for this type of structure, range from $91.00 to $231.50 per S.F.

Common additives

Description	Unit	$ Cost
Bleachers, Telescoping, manual		
To 15 tier	Seat	133 - 189
16-20 tier	Seat	284 - 345
21-30 tier	Seat	300 - 395
For power operation, add	Seat	57 - 89
Carrels Hardwood	Ea.	925 - 2375
Clock System		
20 room	Ea.	18,900
50 room	Ea.	44,300
Emergency Lighting, 25 watt, battery operated		
Lead battery	Ea.	325
Nickel cadmium	Ea.	665
Flagpoles, Complete		
Aluminum, 20' high	Ea.	1825
40' high	Ea.	4300
Fiberglass, 23' high	Ea.	1275
39'-5" high	Ea.	3275
Kitchen Equipment		
Broiler	Ea.	4050
Cooler, 6 ft. long, reach-in	Ea.	3875

Description	Unit	$ Cost
Kitchen Equipment, cont.		
Dishwasher, 10-12 racks per hr.	Ea.	4200
Food warmer, counter, 1.2 KW	Ea.	770
Freezer, 44 C.F., reach-in	Ea.	6100
Ice cube maker, 50 lb. per day	Ea.	1950
Range with 1 oven	Ea.	3275
Lockers, Steel, single tier, 60" to 72"	Opng.	222 - 365
2 tier, 60" to 72" total	Opng.	115 - 164
5 tier, box lockers	Opng.	69.50 - 78.50
Locker bench, lam. maple top only	L.F.	33
Pedestals, steel pipe	Ea.	71
Seating		
Auditorium chair, all veneer	Ea.	305
Veneer back, padded seat	Ea.	315
Upholstered, spring seat	Ea.	310
Classroom, movable chair & desk	Set	81 - 171
Lecture hall, pedestal type	Ea.	267 - 610
Sound System		
Amplifier, 250 watts	Ea.	2125
Speaker, ceiling or wall	Ea.	206
Trumpet	Ea.	395

Important: See the Reference Section for Location Factors.

Model costs calculated for a 1 story building with 15′ story height and 45,000 square feet of floor area

School, Elementary

			Unit	Unit Cost	Cost Per S.F.	% Of Sub-Total
A. SUBSTRUCTURE						
1010	Standard Foundations	Poured concrete; strip and spread footings	S.F. Ground	2.82	2.82	9.5%
1020	Special Foundations	N/A	—	—	—	
1030	Slab on Grade	4″ reinforced concrete	S.F. Slab	5.10	5.10	
2010	Basement Excavation	Site preparation for slab and trench for foundation wall and footing	S.F. Ground	.16	.28	
2020	Basement Walls	4′ Foundation wall	L.F. Wall	74	2.47	
B. SHELL						
	B10 Superstructure					
1010	Floor Construction	Fireproofing for structural columns	S.F. Floor	.94	.94	9.6%
1020	Roof Construction	Metal deck, open web steel joists, beams, columns	S.F. Roof	9.90	9.90	
	B20 Exterior Enclosure					
2010	Exterior Walls	Textured metal panel on metal stud, insulated *70% of wall*	S.F. Wall	16.23	5.72	10.8%
2020	Exterior Windows	Curtain wall system, thermo-break frame, aluminum awning windows *30% of wall*	Each	759	5.83	
2030	Exterior Doors	Aluminum & glass double doors, hollow metal doors	Each	3703	.66	
	B30 Roofing					
3010	Roof Coverings	Single ply membrane, stone ballast, rigid insulation	S.F. Roof	6.18	6.18	5.6%
3020	Roof Openings	Roof and smoke hatches	S.F. Roof	1967	.18	
C. INTERIORS						
1010	Partitions	Gypsum board on metal studs, sound attentuation insulation *20 S.F. Floor/L.F. Partition*	S.F. Partition	5.16	2.58	22.7%
1020	Interior Doors	Hollow metal doors *700 S.F. Floor/Door*	Each	1133	1.62	
1030	Fittings	Chalkboards, lockers, toilet partitions	S.F. Floor	6.26	6.26	
2010	Stair Construction	N/A	—	—	—	
3010	Wall Finishes	90% Paint, 10% ceramic tile	S.F. Surface	1.64	1.64	
3020	Floor Finishes	10% Carpet, 10% terrazzo, 60% VCT, 20% wood	S.F. Floor	6.95	6.95	
3030	Ceiling Finishes	Acoustic ceiling tiles on suspended channel grid	S.F. Ceiling	6.57	6.57	
D. SERVICES						
	D10 Conveying					
1010	Elevators & Lifts	N/A	—	—	—	0.0 %
1020	Escalators & Moving Walks	N/A	—	—	—	
	D20 Plumbing					
2010	Plumbing Fixtures	Restroom, kitchen, and service fixtures, supply and drainage *1 Fixture/325 S.F. Floor*	Each	2285	7.03	7.5%
2020	Domestic Water Distribution	Gas fired water heater	S.F. Floor	.62	.62	
2040	Rain Water Drainage	Roof drains	S.F. Roof	.83	.83	
	D30 HVAC					
3010	Energy Supply	Forced hot water heating system, fin tube radiation	S.F. Floor	9.49	9.49	18.2%
3020	Heat Generating Systems	N/A	—	—	—	
3030	Cooling Generating Systems	N/A	—	—	—	
3050	Terminal & Package Units	Split system with air cooled condensing unit	S.F. Floor	11	11	
3090	Other HVAC Sys. & Equipment	N/A	—	—	—	
	D40 Fire Protection					
4010	Sprinklers	Wet pipe spinkler system	S.F. Floor	2.48	2.48	2.6%
4020	Standpipes	Standpipe	S.F. Floor	.40	.40	
	D50 Electrical					
5010	Electrical Service/Distribution	800 Ampere service, panelboards and feeders	S.F. Floor	1.06	1.06	13.2%
5020	Lighting & Branch Wiring	Fluorescent fixtures, receptacles, switches, a.c. and misc. power	S.F. Floor	10.19	10.19	
5030	Communications & Security	Addressable alarm system, emergency lighting, internet & phone wiring	S.F. Floor	3.59	3.59	
5090	Other Electrical Systems	Emergency generator, 15kW	S.F. Floor	.08	.08	
E. EQUIPMENT & FURNISHINGS						
1010	Commercial Equipment	N/A	—	—	—	0.2 %
1020	Institutional Equipment	Laboratory equipment	S.F. Floor	.19	.19	
1030	Vehicular Equipment	N/A	—	—	—	
1090	Other Equipment	N/A	—	—	—	
F. SPECIAL CONSTRUCTION						
1020	Integrated Construction	N/A	—	—	—	0.0 %
1040	Special Facilities	N/A	—	—	—	
G. BUILDING SITEWORK	**N/A**					
				Sub-Total	112.66	**100%**
	CONTRACTOR FEES (General Requirements: 10%, Overhead: 5%, Profit: 10%)			25%	28.18	
	ARCHITECT FEES			7%	9.86	
				Total Building Cost	**150.70**	

M.580 School, Jr High, 2-3 Story

Costs per square foot of floor area

Exterior Wall	S.F. Area	50000	65000	80000	95000	110000	125000	140000	155000	170000
	L.F. Perimeter	850	1060	1280	1490	1700	1920	2140	2340	2560
Fiber Cement Siding	Steel Frame	151.25	149.50	148.70	147.90	147.35	147.05	146.85	146.45	146.30
Metal Panel	Steel Frame	153.40	151.35	150.35	149.50	148.90	148.55	148.25	147.80	147.65
E.I.F.S.	Steel Frame	153.95	152.20	151.40	150.60	150.05	149.75	149.55	149.15	149.00
Tilt-up Concrete Panel	Reinforced Concrete	175.90	173.55	172.50	171.45	170.70	170.30	170.05	169.45	169.30
Brick Veneer	Reinforced Concrete	178.95	176.60	175.50	174.45	173.70	173.30	173.05	172.45	172.30
Stone Veneer	Reinforced Concrete	198.00	194.65	193.20	191.70	190.65	190.15	189.75	189.00	188.75
Perimeter Adj., Add or Deduct	Per 100 L.F.	2.95	2.25	1.80	1.60	1.40	1.20	1.05	1.00	0.90
Story Hgt. Adj., Add or Deduct	Per 1 Ft.	1.20	1.10	1.10	1.10	1.10	1.10	1.05	1.10	1.10
For Basement, add $ 32.40 per square foot of basement area										

The above costs were calculated using the basic specifications shown on the facing page. These costs should be adjusted where necessary for design alternatives and owner's requirements. Reported completed project costs, for this type of structure, range from $ 99.85 to $ 234.65 per S.F.

Common additives

Description	Unit	$ Cost
Bleachers, Telescoping, manual		
To 15 tier	Seat	133 - 189
16-20 tier	Seat	284 - 345
21-30 tier	Seat	300 - 395
For power operation, add	Seat	57 - 89
Carrels Hardwood	Ea.	925 - 2375
Clock System		
20 room	Ea.	18,900
50 room	Ea.	44,300
Elevators, Hydraulic passenger, 2 stops		
1500# capacity	Ea.	63,900
2500# capacity	Ea.	66,900
Emergency Lighting, 25 watt, battery operated		
Lead battery	Ea.	325
Nickel cadmium	Ea.	665
Flagpoles, Complete		
Aluminum, 20' high	Ea.	1825
40' high	Ea.	4300
Fiberglass, 23' high	Ea.	1275
39'-5" high	Ea.	3275

Description	Unit	$ Cost
Kitchen Equipment		
Broiler	Ea.	4050
Cooler, 6 ft. long, reach-in	Ea.	3875
Dishwasher, 10-12 racks per hr.	Ea.	4200
Food warmer, counter, 1.2 KW	Ea.	770
Freezer, 44 C.F., reach-in	Ea.	6100
Lockers, Steel, single tier, 60" to 72"	Opng.	222 - 365
2 tier, 60" to 72" total	Opng.	115 - 164
5 tier, box lockers	Opng.	69.50 - 78.50
Locker bench, lam. maple top only	L.F.	33
Pedestals, steel pipe	Ea.	71
Seating		
Auditorium chair, all veneer	Ea.	305
Veneer back, padded seat	Ea.	315
Upholstered, spring seat	Ea.	310
Classroom, movable chair & desk	Set	81 - 171
Lecture hall, pedestal type	Ea.	267 - 610
Sound System		
Amplifier, 250 watts	Ea.	2125
Speaker, ceiling or wall	Ea.	206
Trumpet	Ea.	395

Important: See the Reference Section for Location Factors.

Model costs calculated for a 2 story building with 15' story height and 110,000 square feet of floor area

School, Jr High, 2-3 Story

			Unit	Unit Cost	Cost Per S.F.	% Of Sub-Total
A. SUBSTRUCTURE						
1010	Standard Foundations	Poured concrete; strip and spread footings	S.F. Ground	1.72	.86	4.3%
1020	Special Foundations	N/A	—	—	—	
1030	Slab on Grade	4" reinforced concrete	S.F. Slab	5.10	2.56	
2010	Basement Excavation	Site preparation for slab and trench for foundation wall and footing	S.F. Ground	.16	.14	
2020	Basement Walls	4' Foundation wall	L.F. Wall	74	1.13	
B. SHELL						
	B10 Superstructure					
1010	Floor Construction	Open web steel joists, slab form, concrete, fireproofed steel columns	S.F. Floor	14.70	7.35	11.2%
1020	Roof Construction	Metal deck, open web steel joists, beams, columns	S.F. Roof	9.94	4.97	
	B20 Exterior Enclosure					
2010	Exterior Walls	Fiber cement siding on wood studs, insulated *75% of wall*	S.F. Wall	14.78	5.14	10.9%
2020	Exterior Windows	Curtain wall system, thermo-break frame *25% of wall*	Each	54	6.28	
2030	Exterior Doors	Aluminum & glass double doors, hollow metal doors, overhead doors	Each	2358	.62	
	B30 Roofing					
3010	Roof Coverings	Single ply membrane, stone ballast, rigid insulation	S.F. Roof	6.02	3.01	2.9%
3020	Roof Openings	Roof and smoke hatches	S.F. Roof	.28	.14	
C. INTERIORS						
1010	Partitions	Gypsum board on metal studs, sound attentuation insulation *20 S.F. Floor/L.F. Partition*	S.F. Partition	5.80	3.48	25.8%
1020	Interior Doors	Hollow metal doors *750 S.F. Floor/Door*	Each	1133	1.51	
1030	Fittings	Chalkboards, lockers, toilet partitions	S.F. Floor	8	8	
2010	Stair Construction	Cement filled metal pan, picket rail	Flight	11,250	.82	
3010	Wall Finishes	90% Paint, 10% ceramic tile	S.F. Surface	1.60	1.92	
3020	Floor Finishes	10% Carpet, 10% terrazzo, 70% VCT, 10% wood	S.F. Floor	6.14	6.14	
3030	Ceiling Finishes	Acoustic ceiling tiles on suspended channel grid	S.F. Ceiling	6.57	6.57	
D. SERVICES						
	D10 Conveying					
1010	Elevators & Lifts	Hydraulic passenger elevator	Each	80,300	.73	0.7%
1020	Escalators & Moving Walks	N/A	—	—	—	
	D20 Plumbing					
2010	Plumbing Fixtures	Restroom, kitchen, and service fixtures, supply and drainage *1 Fixture/350 S.F. Floor*	Each	1953	5.58	5.9%
2020	Domestic Water Distribution	Gas fired water heater	S.F. Floor	.63	.63	
2040	Rain Water Drainage	Roof drains	S.F. Roof	.68	.34	
	D30 HVAC					
3010	Energy Supply	N/A	—	—	—	17.2 %
3020	Heat Generating Systems	Included in D3050	—	—	—	
3030	Cooling Generating Systems	N/A	—	—	—	
3050	Terminal & Package Units	Single zone unit gas heating, electric cooling	S.F. Floor	19	19	
3090	Other HVAC Sys. & Equipment	N/A	—	—	—	
	D40 Fire Protection					
4010	Sprinklers	Wet pipe spinkler system	S.F. Floor	2.16	2.16	2.3%
4020	Standpipes	Standpipe	S.F. Floor	.40	.40	
	D50 Electrical					
5010	Electrical Service/Distribution	1600 Ampere service, panelboards and feeders	S.F. Floor	1	1	13.6%
5020	Lighting & Branch Wiring	Fluorescent fixtures, receptacles, switches, a.c. and misc. power	S.F. Floor	9.47	9.47	
5030	Communications & Security	Addressable alarm system, emergency lighting, internet & phone wiring	S.F. Floor	4.15	4.15	
5090	Other Electrical Systems	Emergency generator, 100kW	S.F. Floor	.33	.33	
E. EQUIPMENT & FURNISHINGS						
1010	Commercial Equipment	N/A	—	—	—	5.2 %
1020	Institutional Equipment	Laboratory equipment	S.F. Floor	1.69	1.69	
1030	Vehicular Equipment	N/A	—	—	—	
1090	Other Equipment	Gym equipment, bleachers, scoreboards	S.F. Floor	4.05	4.05	
F. SPECIAL CONSTRUCTION						
1020	Integrated Construction	N/A	—	—	—	0.0 %
1040	Special Facilities	N/A	—	—	—	
G. BUILDING SITEWORK		**N/A**				
				Sub-Total	110.17	**100%**
	CONTRACTOR FEES (General Requirements: 10%, Overhead: 5%, Profit: 10%)			*25%*	27.54	
	ARCHITECT FEES			*7%*	9.64	
				Total Building Cost	**147.35**	

Costs per square foot of floor area

Exterior Wall	S.F. Area	1000	2000	3000	4000	6000	8000	10000	12000	15000
	L.F. Perimeter	126	179	219	253	310	358	400	438	490
Wood Siding	Wood Frame	153.05	128.60	118.65	113.10	106.80	103.25	100.80	99.05	97.20
Face Brick Veneer	Wood Frame	193.40	162.65	149.95	142.70	134.40	129.65	126.35	124.00	121.50
Stucco on Concrete Block	Steel Frame	169.15	141.15	129.70	123.15	115.70	111.50	108.60	106.50	104.25
	Bearing Walls	164.35	136.35	124.85	118.35	110.90	106.70	103.75	101.65	99.45
Metal Sandwich Panel	Steel Frame	181.35	149.80	136.70	129.20	120.65	115.75	112.35	109.95	107.30
Precast Concrete	Steel Frame	219.00	176.10	157.85	147.35	135.20	128.15	123.30	119.75	115.95
Perimeter Adj., Add or Deduct	Per 100 L.F.	40.60	20.30	13.50	10.15	6.80	5.00	4.10	3.40	2.65
Story Hgt. Adj., Add or Deduct	Per 1 Ft.	3.10	2.15	1.85	1.55	1.30	1.05	0.95	0.90	0.80
***For Basement**, add $23.65 per square foot of basement area*										

The above costs were calculated using the basic specifications shown on the facing page. These costs should be adjusted where necessary for design alternatives and owner's requirements. Reported completed project costs, for this type of structure, range from $74.25 to $257.35 per S.F.

Common additives

Description	Unit	$ Cost
Check Out Counter		
Single belt	Ea.	3575
Double belt	Ea.	5100
Emergency Lighting, 25 watt, battery operated		
Lead battery	Ea.	325
Nickel cadmium	Ea.	665
Refrigerators, Prefabricated, walk-in		
7'-6" high, 6' x 6'	S.F.	200
10' x 10'	S.F.	157
12' x 14'	S.F.	140
12' x 20'	S.F.	123
Refrigerated Food Cases		
Dairy, multi deck, 12' long	Ea.	12,800
Delicatessen case, single deck, 12' long	Ea.	9525
Multi deck, 18 S.F. shelf display	Ea.	8775
Freezer, self-contained chest type, 30 C.F.	Ea.	10,300
Glass door upright, 78 C.F.	Ea.	13,500

Description	Unit	$ Cost
Refrigerated Food Cases, cont.		
Frozen food, chest type, 12' long	Ea.	9375
Glass door reach-in, 5 door	Ea.	15,900
Island case 12' long, single deck	Ea.	8350
Multi deck	Ea.	9700
Meat cases, 12' long, single deck	Ea.	8350
Multi deck	Ea.	11,700
Produce, 12' long single deck	Ea.	7875
Multi deck	Ea.	9950
Safe, Office type, 1 hour rating		
30" x 18" x 18"	Ea.	2500
60" x 36" x 18", double door	Ea.	9350
Smoke Detectors		
Ceiling type	Ea.	214
Duct type	Ea.	540
Sound System		
Amplifier, 250 watts	Ea.	2125
Speaker, ceiling or wall	Ea.	206
Trumpet	Ea.	395

Important: See the Reference Section for Location Factors.

Model costs calculated for a 1 story building with 12′ story height and 4,000 square feet of floor area

Store, Convenience

				Unit	Unit Cost	Cost Per S.F.	% Of Sub-Total
A. SUBSTRUCTURE							
1010	Standard Foundations	Poured concrete; strip and spread footings		S.F. Ground	1.05	1.05	12.7%
1020	Special Foundations	N/A		—	—	—	
1030	Slab on Grade	4″ reinforced concrete with vapor barrier and granular base		S.F. Slab	5.10	5.10	
2010	Basement Excavation	Site preparation for slab and trench for foundation wall and footing		S.F. Ground	.51	.51	
2020	Basement Walls	4′ foundation wall		L.F. Wall	65	4.08	
B. SHELL							
B10 Superstructure							
1010	Floor Construction	N/A		—	—	—	8.2 %
1020	Roof Construction	Wood truss with plywood sheathing		S.F. Roof	6.94	6.94	
B20 Exterior Enclosure							
2010	Exterior Walls	Wood siding on wood studs, insulated	*80% of wall*	S.F. Wall	9.42	5.72	17.3%
2020	Exterior Windows	Storefront	*20% of wall*	Each	44	6.75	
2030	Exterior Doors	Double aluminum and glass, solid core wood		Each	2884	2.17	
B30 Roofing							
3010	Roof Coverings	Asphalt shingles; rigid fiberglass insulation		S.F. Roof	5.48	5.48	6.5%
3020	Roof Openings	N/A		—	—	—	
C. INTERIORS							
1010	Partitions	Gypsum board on wood studs	*60 S.F. Floor/L.F. Partition*	S.F. Partition	12.36	2.06	14.2%
1020	Interior Doors	Single leaf wood, hollow metal	*1300 S.F. Floor/Door*	Each	1213	.93	
1030	Fittings	N/A		—	—	—	
2010	Stair Construction	N/A		—	—	—	
3010	Wall Finishes	Paint		S.F. Surface	1.08	.36	
3020	Floor Finishes	Vinyl composition tile		S.F. Floor	2.61	2.61	
3030	Ceiling Finishes	Mineral fiber tile on wood furring		S.F. Ceiling	6.02	6.02	
D. SERVICES							
D10 Conveying							
1010	Elevators & Lifts	N/A		—	—	—	0.0 %
1020	Escalators & Moving Walks	N/A		—	—	—	
D20 Plumbing							
2010	Plumbing Fixtures	Toilet and service fixtures, supply and drainage	*1 Fixture/1000 S.F. Floor*	Each	3570	3.57	6.1%
2020	Domestic Water Distribution	Gas fired water heater		S.F. Floor	1.56	1.56	
2040	Rain Water Drainage	N/A		—	—	—	
D30 HVAC							
3010	Energy Supply	N/A		—	—	—	8.0 %
3020	Heat Generating Systems	Included in D3050		—	—	—	
3030	Cooling Generating Systems	N/A		—	—	—	
3050	Terminal & Package Units	Single zone rooftop unit, gas heating, electric cooling		S.F. Floor	6.74	6.74	
3090	Other HVAC Sys. & Equipment	N/A		—	—	—	
D40 Fire Protection							
4010	Sprinklers	Sprinkler, ordinary hazard		S.F. Floor	4.57	4.57	7.9%
4020	Standpipes	Standpipe		S.F. floor	2.13	2.13	
D50 Electrical							
5010	Electrical Service/Distribution	200 ampere service, panel board and feeders		S.F. Floor	2.98	2.98	19.2%
5020	Lighting & Branch Wiring	High efficiency fluorescent fixtures, receptacles, switches, A.C. and misc. power		S.F. Floor	10.16	10.16	
5030	Communications & Security	Addressable alarm systems and emergency lighting		S.F. Floor	2.71	2.71	
5090	Other Electrical Systems	Emergency generator, 7.5 kW		S.F. Floor	.37	.37	
E. EQUIPMENT & FURNISHINGS							
1010	Commercial Equipment	N/A		—	—	—	0.0 %
1020	Institutional Equipment	N/A		—	—	—	
1030	Vehicular Equipment	N/A		—	—	—	
1090	Other Equipment	N/A		—	—	—	
F. SPECIAL CONSTRUCTION							
1020	Integrated Construction	N/A		—	—	—	0.0 %
1040	Special Facilities	N/A		—	—	—	
G. BUILDING SITEWORK		**N/A**					
					Sub-Total	84.57	**100%**
	CONTRACTOR FEES (General Requirements: 10%, Overhead: 5%, Profit: 10%)				*25%*	21.13	
	ARCHITECT FEES				*7%*	7.40	
					Total Building Cost	**113.10**	

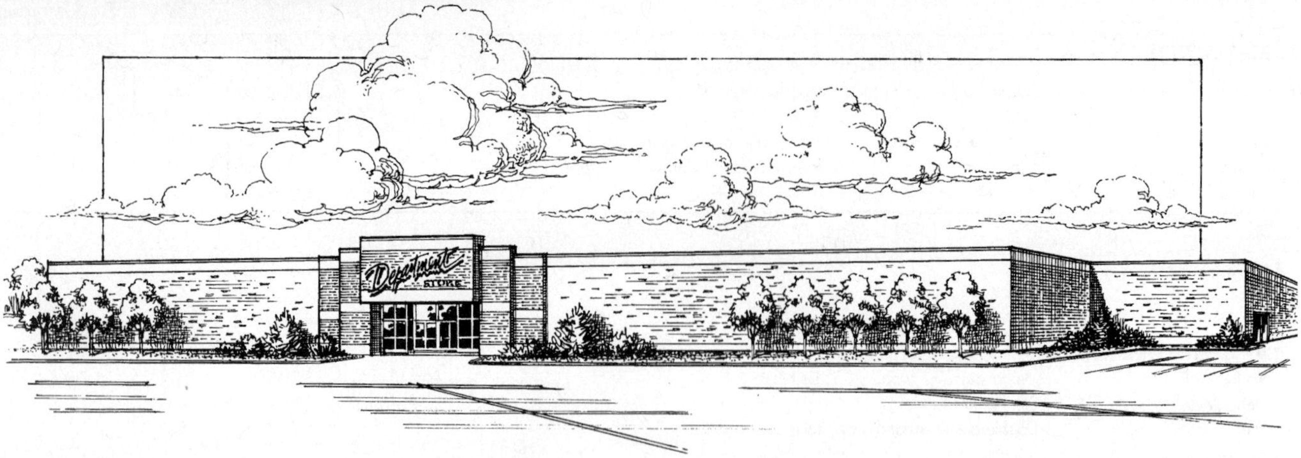

Costs per square foot of floor area

Exterior Wall	S.F. Area	50000	65000	80000	95000	110000	125000	140000	155000	170000
	L.F. Perimeter	920	1065	1167	1303	1333	1433	1533	1633	1733
Face Brick with Concrete Block Back-up	R/Conc. Frame	123.85	121.55	119.60	118.60	**117.00**	116.35	115.75	115.30	114.95
	Steel Frame	107.70	105.40	103.45	102.40	100.85	100.15	99.60	99.15	98.80
Decorative Concrete Block	R/Conc. Frame	118.95	117.15	115.70	114.90	113.80	113.30	112.85	112.50	112.20
	Steel Joists	103.80	101.95	100.35	99.55	98.30	97.75	97.25	96.95	96.60
Precast Concrete Panels	R/Conc. Frame	122.95	120.85	119.05	118.05	116.60	116.00	115.45	115.00	114.70
	Steel Joists	106.35	104.25	102.40	101.45	100.00	99.35	98.75	98.40	98.00
Perimeter Adj., Add or Deduct	Per 100 L.F.	1.75	1.35	1.10	0.95	0.80	0.65	0.70	0.60	0.50
Story Hgt. Adj., Add or Deduct	Per 1 Ft.	0.90	0.80	0.70	0.70	0.60	0.55	0.60	0.55	0.50
For Basement, add $ 23.35 per square foot of basement area										

The above costs were calculated using the basic specifications shown on the facing page. These costs should be adjusted where necessary for design alternatives and owner's requirements. Reported completed project costs, for this type of structure, range from $ 64.05 to $ 159.40 per S.F.

Common additives

Description	Unit	$ Cost
Closed Circuit Surveillance, One station		
Camera and monitor	Ea.	1975
For additional camera stations, add	Ea.	1075
Directory Boards, Plastic, glass covered		
30" x 20"	Ea.	555
36" x 48"	Ea.	1425
Aluminum, 24" x 18"	Ea.	600
36" x 24"	Ea.	715
48" x 32"	Ea.	1000
48" x 60"	Ea.	2100
Emergency Lighting, 25 watt, battery operated		
Lead battery	Ea.	325
Nickel cadmium	Ea.	665
Safe, Office type, 1 hour rating		
30" x 18" x 18"	Ea.	2500
60" x 36" x 18", double door	Ea.	9350
Sound System		
Amplifier, 250 watts	Ea.	2125
Speaker, ceiling or wall	Ea.	206
Trumpet	Ea.	395

Important: See the Reference Section for Location Factors.

Model costs calculated for a 1 story building with 14′ story height and 110,000 square feet of floor area

Store, Department, 1 Story

				Unit	Unit Cost	Cost Per S.F.	% Of Sub-Total
A. SUBSTRUCTURE							
1010	Standard Foundations	Poured concrete; strip and spread footings		S.F. Ground	.87	.87	
1020	Special Foundations	N/A		—	—	—	
1030	Slab on Grade	4″ reinforced concrete with vapor barrier and granular base		S.F. Slab	5.10	5.10	8.1%
2010	Basement Excavation	Site preparation for slab and trench for foundation wall and footing		S.F. Ground	.29	.29	
2020	Basement Walls	4′ foundation wall		L.F. Wall	74	.89	
B. SHELL							
	B10 Superstructure						
1010	Floor Construction	Cast-in-place concrete columns		L.F. Column	64	.81	26.0%
1020	Roof Construction	Precast concrete beam and plank, concrete columns		S.F. Roof	22	22.16	
	B20 Exterior Enclosure						
2010	Exterior Walls	Face brick with concrete block backup	*90% of wall*	S.F. Wall	30	4.54	
2020	Exterior Windows	Storefront	*10% of wall*	Each	66	1.11	6.9%
2030	Exterior Doors	Sliding electric operated entrance, hollow metal		Each	7662	.42	
	B30 Roofing						
3010	Roof Coverings	Built-up tar and gravel with flashing; perlite/EPS composite insulation		S.F. Roof	5.21	5.21	6.0%
3020	Roof Openings	Roof hatches		S.F. Roof	.05	.05	
C. INTERIORS							
1010	Partitions	Gypsum board on metal studs	*60 S.F. Floor/L.F. Partition*	S.F. Partition	9.90	1.65	
1020	Interior Doors	Single leaf hollow metal	*600 S.F. Floor/Door*	Each	1133	1.89	
1030	Fittings	N/A		—	—	—	
2010	Stair Construction	N/A		—	—	—	26.3%
3010	Wall Finishes	Paint		S.F. Surface	.75	.25	
3020	Floor Finishes	50% ceramic tile, 50% carpet tile		S.F. Floor	14.85	14.85	
3030	Ceiling Finishes	Mineral fiber board on exposed grid system, suspended		S.F. Ceiling	4.58	4.58	
D. SERVICES							
	D10 Conveying						
1010	Elevators & Lifts	N/A		—	—	—	0.0 %
1020	Escalators & Moving Walks	N/A		—	—	—	
	D20 Plumbing						
2010	Plumbing Fixtures	Toilet and service fixtures, supply and drainage	*1 Fixture/4075 S.F. Floor*	Each	6479	1.59	
2020	Domestic Water Distribution	Gas fired water heater		S.F. Floor	.38	.38	3.1%
2040	Rain Water Drainage	Roof drains		S.F. Roof	.78	.78	
	D30 HVAC						
3010	Energy Supply	N/A		—	—	—	
3020	Heat Generating Systems	Included in D3050		—	—	—	
3030	Cooling Generating Systems	N/A		—	—	—	8.9 %
3050	Terminal & Package Units	Single zone unit gas heating, electric cooling		S.F. Floor	7.86	7.86	
3090	Other HVAC Sys. & Equipment	N/A		—	—	—	
	D40 Fire Protection						
4010	Sprinklers	Sprinklers, light hazard		S.F. Floor	2.48	2.48	3.1%
4020	Standpipes	Standpipe		S.F. Floor	.27	.27	
	D50 Electrical						
5010	Electrical Service/Distribution	1200 ampere service, panel board and feeders		S.F. Floor	.72	.72	
5020	Lighting & Branch Wiring	High efficiency fluorescent fixtures, receptacles, switches, A.C. and misc. power		S.F. Floor	8.36	8.36	11.6%
5030	Communications & Security	Addressable alarm systems, internet wiring and emergency lighting		S.F. Floor	1.16	1.16	
5090	Other Electrical Systems	Emergency generator, 7.5 kW		S.F. Floor	.04	.04	
E. EQUIPMENT & FURNISHINGS							
1010	Commercial Equipment	N/A		—	—	—	
1020	Institutional Equipment	N/A		—	—	—	0.0 %
1030	Vehicular Equipment	N/A		—	—	—	
1090	Other Equipment	N/A		—	—	—	
F. SPECIAL CONSTRUCTION							
1020	Integrated Construction	N/A		—	—	—	0.0 %
1040	Special Facilities	N/A		—	—	—	
G. BUILDING SITEWORK	**N/A**						
					Sub-Total	88.31	**100%**
	CONTRACTOR FEES (General Requirements: 10%, Overhead: 5%, Profit: 10%)				25%	22.07	
	ARCHITECT FEES				6%	6.62	
					Total Building Cost	**117**	

Costs per square foot of floor area

Exterior Wall	S.F. Area	4000	6000	8000	10000	12000	15000	18000	20000	22000
	L.F. Perimeter	260	340	360	410	440	490	540	565	594
Vinyl Siding	Wood Frame	124.50	114.85	106.25	102.60	99.30	96.20	94.10	92.85	91.95
Fiber Cement Siding	Wood Frame	125.95	116.35	107.85	104.20	100.90	97.80	95.75	94.50	93.60
EIFS on Metal Studs	Steel Joists	134.50	124.35	115.10	111.25	107.70	104.40	102.15	100.80	99.80
Stone Veneer	Steel Frame	177.40	162.55	146.80	140.65	134.65	129.15	125.40	123.05	121.35
Brick Veneer	Reinforced Concrete	164.60	151.30	137.60	132.20	127.05	122.20	118.95	116.90	115.45
Stucco	Reinforced Concrete	172.50	158.00	142.65	136.70	130.90	125.45	121.85	119.55	117.90
Perimeter Adj., Add or Deduct	Per 100 L.F.	12.65	8.40	6.30	5.10	4.25	3.40	2.90	2.55	2.30
Story Hgt. Adj., Add or Deduct	Per 1 Ft.	1.45	1.25	1.00	0.90	0.80	0.75	0.70	0.65	0.55
***For Basement**, add $32.85 per square foot of basement area*										

The above costs were calculated using the basic specifications shown on the facing page. These costs should be adjusted where necessary for design alternatives and owner's requirements. Reported completed project costs, for this type of structure, range from $64.60 to $224.60 per S.F.

Common additives

Description	Unit	$ Cost
Emergency Lighting, 25 watt, battery operated		
Lead battery	Ea.	325
Nickel cadmium	Ea.	665
Safe, Office type, 1 hour rating		
30" x 18" x 18"	Ea.	2500
60" x 36" x 18", double door	Ea.	9350
Smoke Detectors		
Ceiling type	Ea.	214
Duct type	Ea.	540
Sound System		
Amplifier, 250 watts	Ea.	2125
Speaker, ceiling or wall	Ea.	206
Trumpet	Ea.	395

Model costs calculated for a 1 story building with 14′ story height and 8,000 square feet of floor area

Store, Retail

				Unit	Unit Cost	Cost Per S.F.	% Of Sub-Total
A. SUBSTRUCTURE							
1010	Standard Foundations	Poured concrete; strip and spread footings		S.F. Ground	1.94	1.94	13.5%
1020	Special Foundations	N/A		—	—	—	
1030	Slab on Grade	4″ reinforced concrete		S.F. Slab	5.10	5.10	
2010	Basement Excavation	Site preparation for slab and trench for foundation wall and footing		S.F. Ground	.29	.29	
2020	Basement Walls	4′ Foundation wall		L.F. Wall	74	3.31	
B. SHELL							
	B10 Superstructure						
1010	Floor Construction	Wood columns,		S.F. Floor	.38	.38	9.8%
1020	Roof Construction	Wood roof truss 4:12 pitch		S.F. Roof	7.37	7.37	
	B20 Exterior Enclosure						
2010	Exterior Walls	Vinyl siding on wood studs, insulated	*85% of wall*	S.F. Wall	13.07	7	13.0%
2020	Exterior Windows	Storefront glazing	*15% of wall*	Each	47	2.06	
2030	Exterior Doors	Aluminum and glass single doors, hollow metal doors		Each	2379	1.20	
	B30 Roofing						
3010	Roof Coverings	Asphalt roofing, strip shingles, gutters and downspouts		S.F. Roof	3.59	3.59	4.7%
3020	Roof Openings	Roof hatches		S.F. Roof	.11	.11	
C. INTERIORS							
1010	Partitions	Sound deadening gypsum board on wood studs	*60 S.F. Floor/L.F. Partition*	S.F. Partition	9.06	1.51	16.2%
1020	Interior Doors	Hollow metal doors	*1150 S.F. Floor/Door*	Each	1133	.99	
1030	Fittings	Toilet partitions		S.F. Floor	.18	.18	
2010	Stair Construction	N/A		—	—	—	
3010	Wall Finishes	90% Paint, 10% ceramic tile		S.F. Surface	2.58	.86	
3020	Floor Finishes	Vinyl composition tile		S.F. Floor	2.61	2.61	
3030	Ceiling Finishes	Acoustic ceiling tiles on suspended channel grid		S.F. Ceiling	6.57	6.57	
D. SERVICES							
	D10 Conveying						
1010	Elevators & Lifts	N/A		—	—	—	0.0 %
1020	Escalators & Moving Walks	N/A		—	—	—	
	D20 Plumbing						
2010	Plumbing Fixtures	Restroom and service fixtures, supply and drainage	*1 Fixture/1350 S.F. Floor*	Each	2439	1.83	7.8%
2020	Domestic Water Distribution	Gas fired water heater		S.F. Floor	3.08	3.08	
2040	Rain Water Drainage	Roof Drains		S.F. Roof	1.23	1.23	
	D30 HVAC						
3010	Energy Supply	N/A		—	—	—	10.0 %
3020	Heat Generating Systems	Included in D3050		—	—	—	
3030	Cooling Generating Systems	N/A		—	—	—	
3050	Terminal & Package Units	Single zone unit gas heating, electric cooling		S.F. Floor	7.86	7.86	
3090	Other HVAC Sys. & Equipment	N/A		—	—	—	
	D40 Fire Protection						
4010	Sprinklers	Wet pipe spinkler system		S.F. Floor	4.22	4.22	6.7%
4020	Standpipes	Standpipe		S.F. Floor	1.07	1.07	
	D50 Electrical						
5010	Electrical Service/Distribution	400 Ampere service, panelboards and feeders		S.F. Floor	2.70	2.70	18.2%
5020	Lighting & Branch Wiring	Flourescent fixtures, receptacles, a.c. and misc. power		S.F. Floor	10	10	
5030	Communications & Security	Addressable alarm system, emergency lighting		S.F. Floor	1.64	1.64	
5090	Other Electrical Systems	N/A		—	—	—	
E. EQUIPMENT & FURNISHINGS							
1010	Commercial Equipment	N/A		—	—	—	0.0 %
1020	Institutional Equipment	N/A		—	—	—	
1030	Vehicular Equipment	N/A		—	—	—	
1090	Other Equipment	N/A		—	—	—	
F. SPECIAL CONSTRUCTION							
1020	Integrated Construction	N/A		—	—	—	0.0 %
1040	Special Facilities	N/A		—	—	—	
G. BUILDING SITEWORK		**N/A**					
					Sub-Total	78.70	**100%**
	CONTRACTOR FEES (General Requirements: 10%, Overhead: 5%, Profit: 10%)				*25%*	19.68	
	ARCHITECT FEES				*8%*	7.87	
					Total Building Cost	**106.25**	

Costs per square foot of floor area

Exterior Wall	S.F. Area	12000	16000	20000	26000	32000	38000	44000	52000	60000
	L.F. Perimeter	450	510	570	650	716	780	840	920	1020
Stucco Face	Reinforced Concrete	161.15	147.85	140.00	132.00	126.15	122.10	118.95	115.95	114.30
Curtain Wall	Steel Frame	140.35	129.80	123.55	117.25	112.75	109.55	107.15	104.80	103.55
E.I.F.S.	Steel Frame	119.85	112.45	108.00	103.65	100.55	98.45	96.80	95.25	94.30
Metal Panel	Steel Frame	117.35	109.95	105.60	101.30	98.20	96.05	94.40	92.95	92.00
Tilt-up Concrete Panel	Reinforced Concrete	133.70	124.65	119.30	114.00	110.10	107.45	105.35	103.45	102.30
Brick Veneer	Reinforced Concrete	170.60	155.95	147.15	138.30	131.75	127.25	123.75	120.40	118.55
Perimeter Adj., Add or Deduct	Per 100 L.F.	14.95	11.25	8.95	6.85	5.60	4.70	4.10	3.45	3.00
Story Hgt. Adj., Add or Deduct	Per 1 Ft.	1.45	1.25	1.05	1.00	0.80	0.75	0.75	0.65	0.70
***For Basement**, add $23.80 per square foot of basement area*										

The above costs were calculated using the basic specifications shown on the facing page. These costs should be adjusted where necessary for design alternatives and owner's requirements. Reported completed project costs, for this type of structure, range from $75.45 to $216.10 per S.F.

Common additives

Description	Unit	$ Cost
Check Out Counter		
Single belt	Ea.	3575
Double belt	Ea.	5100
Scanner, registers, guns & memory 2 lanes	Ea.	18,300
10 lanes	Ea.	173,500
Power take away	Ea.	6000
Emergency Lighting, 25 watt, battery operated		
Lead battery	Ea.	325
Nickel cadmium	Ea.	665
Refrigerators, Prefabricated, walk-in		
7'-6" High, 6' x 6'	S.F.	200
10' x 10'	S.F.	157
12' x 14'	S.F.	140
12' x 20'	S.F.	123
Refrigerated Food Cases		
Dairy, multi deck, 12' long	Ea.	12,800
Delicatessen case, single deck, 12' long	Ea.	9525
Multi deck, 18 S.F. shelf display	Ea.	8775
Freezer, self-contained chest type, 30 C.F.	Ea.	10,300
Glass door upright, 78 C.F.	Ea.	13,500

Description	Unit	$ Cost
Refrigerated Food Cases, cont.		
Frozen food, chest type, 12' long	Ea.	9375
Glass door reach in, 5 door	Ea.	15,900
Island case 12' long, single deck	Ea.	8350
Multi deck	Ea.	9700
Meat case 12' long, single deck	Ea.	8350
Multi deck	Ea.	11,700
Produce, 12' long single deck	Ea.	7875
Multi deck	Ea.	9950
Safe, Office type, 1 hour rating		
30" x 18" x 18"	Ea.	2500
60" x 36" x 18", double door	Ea.	9350
Smoke Detectors		
Ceiling type	Ea.	214
Duct type	Ea.	540
Sound System		
Amplifier, 250 watts	Ea.	2125
Speaker, ceiling or wall	Ea.	206
Trumpet	Ea.	395

Important: See the Reference Section for Location Factors.

Model costs calculated for a 1 story building with 18′ story height and 44,000 square feet of floor area

Supermarket

				Unit	Unit Cost	Cost Per S.F.	% Of Sub-Total
A. SUBSTRUCTURE							
1010	Standard Foundations	Poured concrete; strip and spread footings		S.F. Ground	.96	.96	9.5%
1020	Special Foundations	N/A		—	—	—	
1030	Slab on Grade	4″ reinforced slab on grade		S.F. Slab	5.10	5.10	
2010	Basement Excavation	Site preparation for slab and trench for foundation wall and footing		S.F. Ground	.16	.16	
2020	Basement Walls	4′ Foundation wall		L.F. Wall	74	1.40	
B. SHELL							
	B10 Superstructure						
1010	Floor Construction	N/A		—	—	—	12.4 %
1020	Roof Construction	Metal deck on open web steel joints, columns		S.F. Roof	9.90	9.90	
	B20 Exterior Enclosure						
2010	Exterior Walls	N/A		—	—	—	20.6 %
2020	Exterior Windows	Curtain wall glazing system, thermo-break frame	*100% of wall*	Each	44	15.13	
2030	Exterior Doors	Glass sliding entrance doors, overhead doors, hollow metal doors		Each	5475	1.37	
	B30 Roofing						
3010	Roof Coverings	Single ply membrane, rigid insulation, gravel stop		S.F. Roof	4.93	4.93	6.3%
3020	Roof Openings	Roof and smoke hatches		S.F. Roof	.12	.12	
C. INTERIORS							
1010	Partitions	Concrete partitions, gypsum board on metal stud	*50 S.F. Floor/L.F. Partition*	S.F. Partition	8.33	2	15.9%
1020	Interior Doors	Hollow metal doors	*2000 S.F. Floor/Door*	Each	1133	.57	
1030	Fittings	Stainless steel toilet partitions		S.F. Floor	.10	.10	
2010	Stair Construction	N/A		—	—	—	
3010	Wall Finishes	Paint		S.F. Surface	1.79	.86	
3020	Floor Finishes	Vinyl composition tile		S.F. Floor	2.61	2.61	
3030	Ceiling Finishes	Acoustic ceiling tiles on suspended channel grid		S.F. Ceiling	6.57	6.57	
D. SERVICES							
	D10 Conveying						
1010	Elevators & Lifts	N/A		—	—	—	0.0 %
1020	Escalators & Moving Walks	N/A		—	—	—	
	D20 Plumbing						
2010	Plumbing Fixtures	Restroom, food prep and service fixtures, supply and drainage	*1 Fixture/1750 S.F. Floor*	Each	1971	1.12	2.9%
2020	Domestic Water Distribution	Gas fired water heater		S.F. Floor	.24	.24	
2040	Rain Water Drainage	Roof drains		S.F. Roof	.96	.96	
	D30 HVAC						
3010	Energy Supply	N/A		—	—	—	5.5 %
3020	Heat Generating Systems	Included in D3050		—	—	—	
3030	Cooling Generating Systems	N/A		—	—	—	
3050	Terminal & Package Units	Single zone unit gas heating, electric cooling		S.F. Floor	4.42	4.42	
3090	Other HVAC Sys. & Equipment	N/A		—	—	—	
	D40 Fire Protection						
4010	Sprinklers	Wet pipe sprinkler system		S.F. Floor	3.28	3.28	4.8%
4020	Standpipes	Standpipe		S.F. Floor	.54	.54	
	D50 Electrical						
5010	Electrical Service/Distribution	1600 ampere service, panel board and feeders		S.F. Floor	2.46	2.46	22.2%
5020	Lighting & Branch Wiring	Fluorescent fixtures, receptacles, switches, A.C. and misc. power		S.F. Floor	13.12	13.12	
5030	Communications & Security	Addressable alarm system, emergency lighting, internet & phone wiring		S.F. Floor	2.14	2.14	
5090	Other Electrical Systems	Emergency generator, 15kW		S.F. Floor	.05	.05	
E. EQUIPMENT & FURNISHINGS							
1010	Commercial Equipment	N/A		—	—	—	0.0 %
1020	Institutional Equipment	N/A		—	—	—	
1030	Vehicular Equipment	N/A		—	—	—	
1090	Other Equipment	N/A		—	—	—	
F. SPECIAL CONSTRUCTION							
1020	Integrated Construction	N/A		—	—	—	0.0 %
1040	Special Facilities	N/A		—	—	—	
G. BUILDING SITEWORK		**N/A**					
					Sub-Total	80.11	**100%**
	CONTRACTOR FEES (General Requirements: 10%, Overhead: 5%, Profit: 10%)				*25%*	20.03	
	ARCHITECT FEES				*7%*	7.01	
					Total Building Cost	**107.15**	

M.660 Telephone Exchange

Costs per square foot of floor area

Exterior Wall	S.F. Area	2000	3000	4000	5000	6000	7000	8000	9000	10000
	L.F. Perimeter	180	220	260	286	320	353	368	397	425
Face Brick with Concrete Block Back-up	Steel Frame	212.70	188.75	176.75	**167.00**	161.70	157.80	152.75	150.35	148.25
	Bearing Walls	203.15	181.75	171.05	162.05	157.30	153.70	148.95	146.75	144.85
Limestone with Concrete Block Back-up	Steel Frame	247.50	218.85	204.50	192.05	185.60	180.75	174.10	171.05	168.45
	Bearing Walls	247.20	215.90	200.30	187.10	180.15	174.95	168.00	164.75	161.95
Decorative Concrete Block	Steel Frame	199.30	177.80	167.10	158.50	153.75	150.30	145.90	143.80	141.90
	Bearing Walls	194.40	172.90	162.15	153.55	148.85	145.35	141.00	138.85	137.00
Perimeter Adj., Add or Deduct	Per 100 L.F.	46.40	30.90	23.20	18.55	15.45	13.25	11.60	10.25	9.25
Story Hgt. Adj., Add or Deduct	Per 1 Ft.	5.45	4.40	3.95	3.40	3.25	3.00	2.75	2.60	2.55

***For Basement**, add $ 34.30 per square foot of basement area*

The above costs were calculated using the basic specifications shown on the facing page. These costs should be adjusted where necessary for design alternatives and owner's requirements. Reported completed project costs, for this type of structure, range from $ 105.35 to $ 380.10 per S.F.

Common additives

Description	Unit	$ Cost
Emergency Lighting, 25 watt, battery operated		
Lead battery	Ea.	325
Nickel cadmium	Ea.	665
Smoke Detectors		
Ceiling type	Ea.	214
Duct type	Ea.	540
Emergency Generators, complete system, gas		
15 kw	Ea.	18,200
85 kw	Ea.	38,400
170 kw	Ea.	97,500
Diesel, 50 kw	Ea.	25,600
150 kw	Ea.	47,100
350 kw	Ea.	71,500

Model costs calculated for a 1 story building with 12′ story height and 5,000 square feet of floor area

Telephone Exchange

			Unit	Unit Cost	Cost Per S.F.	% Of Sub-Total
A. SUBSTRUCTURE						
1010	Standard Foundations	Poured concrete; strip and spread footings	S.F. Ground	2.37	2.37	10.2%
1020	Special Foundations	N/A	—	—	—	
1030	Slab on Grade	4″ reinforced concrete with vapor barrier and granular base	S.F. Slab	5.10	5.10	
2010	Basement Excavation	Site preparation for slab and trench for foundation wall and footing	S.F. Ground	.51	.64	
2020	Basement Walls	4′ foundation wall	L.F. Wall	74	4.20	
B. SHELL						
	B10 Superstructure					
1010	Floor Construction	Steel column fireproofing	S.F. Floor	.71	.71	8.8%
1020	Roof Construction	Metal deck, open web steel joists, beams, columns	S.F. Roof	9.90	9.90	
	B20 Exterior Enclosure					
2010	Exterior Walls	Face brick with concrete block backup *80% of wall*	S.F. Wall	30	16.53	24.7%
2020	Exterior Windows	Outward projecting steel *20% of wall*	Each	1705	11.71	
2030	Exterior Doors	Single aluminum glass with transom	Each	3790	1.52	
	B30 Roofing					
3010	Roof Coverings	Built-up tar and gravel with flashing; perlite/EPS composite insulation	S.F. Roof	6.83	6.83	5.7%
3020	Roof Openings	N/A	—	—	—	
C. INTERIORS						
1010	Partitions	Double layer gypsum board on metal studs *15 S.F. Floor/L.F. Partition*	S.F. Partition	5.85	3.90	22.2%
1020	Interior Doors	Single leaf hollow metal *150 S.F. Floor/Door*	Each	1133	7.55	
1030	Fittings	Toilet partitions	S.F. Floor	1.18	1.18	
2010	Stair Construction	N/A	—	—	—	
3010	Wall Finishes	Paint	S.F. Surface	1.98	2.64	
3020	Floor Finishes	90% carpet, 10% terrazzo	S.F. Floor	5.88	5.88	
3030	Ceiling Finishes	Fiberglass board on exposed grid system, suspended	S.F. Ceiling	5.51	5.51	
D. SERVICES						
	D10 Conveying					
1010	Elevators & Lifts	N/A	—	—	—	0.0 %
1020	Escalators & Moving Walks	N/A	—	—	—	
	D20 Plumbing					
2010	Plumbing Fixtures	Kitchen, toilet and service fixtures, supply and drainage *1 Fixture/715 S.F. Floor*	Each	3504	4.90	7.2%
2020	Domestic Water Distribution	Gas fired water heater	S.F. Floor	2.21	2.21	
2040	Rain Water Drainage	Roof drains	S.F. Roof	1.61	1.61	
	D30 HVAC					
3010	Energy Supply	N/A	—	—	—	6.3 %
3020	Heat Generating Systems	Included in D3050	—	—	—	
3030	Cooling Generating Systems	N/A	—	—	—	
3050	Terminal & Package Units	Single zone unit, gas heating, electric cooling	S.F. Floor	7.53	7.53	
3090	Other HVAC Sys. & Equipment	N/A	—	—	—	
	D40 Fire Protection					
4010	Sprinklers	Wet pipe sprinkler system	S.F. Floor	4.99	4.99	5.8%
4020	Standpipes	Standpipe	S.F. Floor	2.04	2.04	
	D50 Electrical					
5010	Electrical Service/Distribution	200 ampere service, panel board and feeders	S.F. Floor	1.90	1.90	9.0%
5020	Lighting & Branch Wiring	High efficiency fluorescent fixtures, receptacles, switches, A.C. and misc. power	S.F. Floor	5.94	5.94	
5030	Communications & Security	Addressable alarm systems and emergency lighting	S.F. Floor	2.86	2.86	
5090	Other Electrical Systems	Emergency generator, 15 kW	S.F. Floor	.19	.19	
E. EQUIPMENT & FURNISHINGS						
1010	Commercial Equipment	N/A	—	—	—	0.0 %
1020	Institutional Equipment	N/A	—	—	—	
1030	Vehicular Equipment	N/A	—	—	—	
1090	Other Equipment	N/A	—	—	—	
F. SPECIAL CONSTRUCTION						
1020	Integrated Construction	N/A	—	—	—	0.0 %
1040	Special Facilities	N/A	—	—	—	
G. BUILDING SITEWORK		**N/A**				
				Sub-Total	120.34	**100%**
	CONTRACTOR FEES (General Requirements: 10%, Overhead: 5%, Profit: 10%)			*25%*	30.11	
	ARCHITECT FEES			*11%*	16.55	
				Total Building Cost	**167**	

Costs per square foot of floor area

Exterior Wall	S.F. Area	5000	6500	8000	9500	11000	14000	17500	21000	24000
	L.F. Perimeter	300	360	386	396	435	510	550	620	680
Face Brick with Concrete Block Back-up	Steel Joists	156.15	150.15	143.25	137.20	**134.80**	131.40	126.85	124.95	123.75
	Wood Joists	154.00	148.25	141.70	136.00	133.75	130.55	126.30	124.45	123.30
Stone with Concrete Block Back-up	Steel Joists	163.75	157.20	149.40	142.55	139.90	136.05	130.85	128.70	127.30
	Wood Joists	161.60	155.30	147.85	141.35	138.85	135.15	130.30	128.20	126.85
Brick Veneer	Wood Frame	143.75	138.40	132.45	127.35	125.25	122.30	118.45	116.80	115.75
E.I.F.S.	Wood Frame	136.50	131.50	126.05	121.45	119.55	116.80	113.35	111.85	110.85
Perimeter Adj., Add or Deduct	Per 100 L.F.	15.00	11.55	9.35	7.90	6.85	5.40	4.35	3.60	3.10
Story Hgt. Adj., Add or Deduct	Per 1 Ft.	2.70	2.55	2.20	1.90	1.80	1.65	1.45	1.35	1.25
For Basement, add $28.20 per square foot of basement area										

The above costs were calculated using the basic specifications shown on the facing page. These costs should be adjusted where necessary for design alternatives and owner's requirements. Reported completed project costs, for this type of structure, range from $99.60 to $261.90 per S.F.

Common additives

Description	Unit	$ Cost
Directory Boards, Plastic, glass covered		
30" x 20"	Ea.	555
36" x 48"	Ea.	1425
Aluminum, 24" x 18"	Ea.	600
36" x 24"	Ea.	715
48" x 32"	Ea.	1000
48" x 60"	Ea.	2100
Emergency Lighting, 25 watt, battery operated		
Lead battery	Ea.	325
Nickel cadmium	Ea.	665
Flagpoles, Complete		
Aluminum, 20' high	Ea.	1825
40' high	Ea.	4300
70' high	Ea.	11,300
Fiberglass, 23' high	Ea.	1275
39'-5" high	Ea.	3275
59' high	Ea.	6475
Safe, Office type, 1 hour rating		
30" x 18" x 18"	Ea.	2500
60" x 36" x 18", double door	Ea.	9350

Description	Unit	$ Cost
Smoke Detectors		
Ceiling type	Ea.	214
Duct type	Ea.	540
Vault Front, Door & frame		
1 Hour test, 32" x 78"	Opng.	8075
2 Hour test, 32" door	Opng.	9900
40" door	Opng.	11,400
4 Hour test, 32" door	Opng.	10,500
40" door	Opng.	12,300
Time lock movement; two movement	Ea.	2275

Model costs calculated for a 1 story building with 12′ story height and 11,000 square feet of floor area

Town Hall, 1 Story

				Unit	Unit Cost	Cost Per S.F.	% Of Sub-Total
A. SUBSTRUCTURE							
1010	Standard Foundations	Poured concrete; strip and spread footings		S.F. Ground	1.60	1.60	10.0%
1020	Special Foundations	N/A		—	—	—	
1030	Slab on Grade	4″ reinforced concrete with vapor barrier and granular base		S.F. Slab	5.10	5.10	
2010	Basement Excavation	Site preparation for slab and trench for foundation wall and footing		S.F. Ground	.29	.32	
2020	Basement Walls	4′ foundation wall		L.F. Wall	74	2.91	
B. SHELL							
B10 Superstructure							
1010	Floor Construction	Steel column fireproofing		S.F. Floor	.25	.25	7.8%
1020	Roof Construction	Metal deck, open web steel joists, beams, interior columns		S.F. Roof	7.50	7.50	
B20 Exterior Enclosure							
2010	Exterior Walls	Face brick with concrete block backup	*70% of wall*	S.F. Wall	30	9.99	15.5%
2020	Exterior Windows	Metal outward projecting	*30% of wall*	Each	705	4.36	
2030	Exterior Doors	Metal and glass with transom		Each	2791	1.01	
B30 Roofing							
3010	Roof Coverings	Built-up tar and gravel with flashing; perlite/EPS composite insulation		S.F. Roof	6.38	6.38	6.4%
3020	Roof Openings						
C. INTERIORS							
1010	Partitions	Gypsum board on metal studs	*20 S.F. Floor/L.F. Partition*	S.F. Partition	8.80	4.40	23.9%
1020	Interior Doors	Wood solid core	*200 S.F. Floor/Door*	Each	593	2.97	
1030	Fittings	Toilet partitions		S.F. Floor	.46	.46	
2010	Stair Construction	N/A		—	—	—	
3010	Wall Finishes	90% paint, 10% ceramic tile		S.F. Surface	1.38	1.38	
3020	Floor Finishes	70% carpet tile, 15% terrazzo, 15% vinyl composition tile		S.F. Floor	7.82	7.82	
3030	Ceiling Finishes	Mineral fiber tile on concealed zee bars		S.F. Ceiling	6.57	6.57	
D. SERVICES							
D10 Conveying							
1010	Elevators & Lifts	N/A		—	—	—	0.0 %
1020	Escalators & Moving Walks	N/A		—	—	—	
D20 Plumbing							
2010	Plumbing Fixtures	Kitchen, toilet and service fixtures, supply and drainage	*1 Fixture/500 S.F. Floor*	Each	2290	4.58	8.5%
2020	Domestic Water Distribution	Gas fired water heater		S.F. Floor	1.67	1.67	
2040	Rain Water Drainage	Roof drains		S.F. Roof	2.15	2.15	
D30 HVAC							
3010	Energy Supply	N/A		—	—	—	8.6 %
3020	Heat Generating Systems	Included in D3050		—	—	—	
3030	Cooling Generating Systems	N/A		—	—	—	
3050	Terminal & Package Units	Multizone unit, gas heating, electric cooling		S.F. Floor	8.55	8.55	
3090	Other HVAC Sys. & Equipment	N/A		—	—	—	
D40 Fire Protection							
4010	Sprinklers	Wet pipe sprinkler system		S.F. Floor	3.28	3.28	4.1%
4020	Standpipes	Standpipe, wet, Class III		S.F. Floor	.78	.78	
D50 Electrical							
5010	Electrical Service/Distribution	400 ampere service, panel board and feeders		S.F. Floor	1.96	1.96	15.1%
5020	Lighting & Branch Wiring	High efficiency fluorescent fixtures, receptacles, switches, A.C. and misc. power		S.F. Floor	9.78	9.78	
5030	Communications & Security	Addressable alarm systems, internet wiring, and emergency lighting		S.F. Floor	3.02	3.02	
5090	Other Electrical Systems	Emergency generator, 15 kW		S.F. Floor	.16	.16	
E. EQUIPMENT & FURNISHINGS							
1010	Commercial Equipment	N/A		—	—	—	0.0 %
1020	Institutional Equipment	N/A		—	—	—	
1030	Vehicular Equipment	N/A		—	—	—	
1090	Other Equipment	N/A		—	—	—	
F. SPECIAL CONSTRUCTION							
1020	Integrated Construction	N/A		—	—	—	0.0 %
1040	Special Facilities	N/A		—	—	—	
G. BUILDING SITEWORK		**N/A**					
					Sub-Total	98.95	**100%**
	CONTRACTOR FEES (General Requirements: 10%, Overhead: 5%, Profit: 10%)				25%	24.72	
	ARCHITECT FEES				9%	11.13	
					Total Building Cost	**134.80**	

M.680 Town Hall, 2-3 Story

Costs per square foot of floor area

Exterior Wall	S.F. Area	8000	10000	12000	15000	18000	24000	28000	35000	40000
	L.F. Perimeter	206	233	260	300	320	360	393	451	493
Face Brick with Concrete Block Back-up	Steel Frame	208.45	197.50	190.10	182.75	175.75	167.00	163.70	159.70	157.75
	R/Conc. Frame	209.95	198.95	191.60	184.20	177.20	168.50	165.15	161.20	159.20
Stone with Concrete Block Back-up	Steel Frame	219.05	207.70	200.05	192.35	**184.45**	174.60	170.90	166.60	164.40
	R/Conc. Frame	219.75	207.85	199.85	191.80	183.95	174.20	170.50	166.10	163.90
Limestone with Concrete Block Back-up	Steel Frame	227.00	213.30	204.20	194.95	185.50	173.65	169.30	164.05	161.40
	R/Conc. Frame	228.50	214.80	205.70	196.45	187.00	175.15	170.80	165.50	162.90
Perimeter Adj., Add or Deduct	Per 100 L.F.	23.10	18.50	15.45	12.30	10.25	7.70	6.60	5.30	4.60
Story Hgt. Adj., Add or Deduct	Per 1 Ft.	3.50	3.15	2.95	2.70	2.40	2.05	1.90	1.75	1.65
***For Basement**, add $ 27.50 per square foot of basement area*										

The above costs were calculated using the basic specifications shown on the facing page. These costs should be adjusted where necessary for design alternatives and owner's requirements. Reported completed project costs, for this type of structure, range from $ 89.25 to $ 261.90 per S.F.

Common additives

Description	Unit	$ Cost
Directory Boards, Plastic, glass covered		
30" x 20"	Ea.	555
36" x 48"	Ea.	1425
Aluminum, 24" x 18"	Ea.	600
36" x 24"	Ea.	715
48" x 32"	Ea.	1000
48" x 60"	Ea.	2100
Elevators, Hydraulic passenger, 2 stops		
1500# capacity	Ea.	63,900
2500# capacity	Ea.	66,900
3500# capacity	Ea.	71,900
Additional stop, add	Ea.	6825
Emergency Lighting, 25 watt, battery operated		
Lead battery	Ea.	325
Nickel cadmium	Ea.	665

Description	Unit	$ Cost
Flagpoles, Complete		
Aluminum, 20' high	Ea.	1825
40' high	Ea.	4300
70' high	Ea.	11,300
Fiberglass, 23' high	Ea.	1275
39'-5" high	Ea.	3275
59' high	Ea.	6475
Safe, Office type, 1 hour rating		
30" x 18" x 18"	Ea.	2500
60" x 36" x 18", double door	Ea.	9350
Smoke Detectors		
Ceiling type	Ea.	214
Duct type	Ea.	540
Vault Front, Door & frame		
1 Hour test, 32" x 78"	Opng.	8075
2 Hour test, 32" door	Opng.	9900
40" door	Opng.	11,400
4 Hour test, 32" door	Opng.	10,500
40" door	Opng.	12,300
Time lock movement; two movement	Ea.	2275

Important: See the Reference Section for Location Factors.

Model costs calculated for a 3 story building with 12′ story height and 18,000 square feet of floor area

Town Hall, 2-3 Story

				Unit	Unit Cost	Cost Per S.F.	% Of Sub-Total
A. SUBSTRUCTURE							
1010	Standard Foundations	Poured concrete; strip and spread footings		S.F. Ground	4.08	1.36	3.4%
1020	Special Foundations	N/A		—	—	—	
1030	Slab on Grade	4″ reinforced concrete with vapor barrier and granular base		S.F. Slab	5.10	1.70	
2010	Basement Excavation	Site preparation for slab and trench for foundation wall and footing		S.F. Ground	.51	.26	
2020	Basement Walls	4′ foundation wall		L.F. Wall	37	1.31	
B. SHELL							
	B10 Superstructure						
1010	Floor Construction	Open web steel joists, slab form, concrete, wide flange steel columns		S.F. Floor	27	18.08	15.4%
1020	Roof Construction	Metal deck, open web steel joists, beams, interior columns		S.F. Roof	8.22	2.74	
	B20 Exterior Enclosure						
2010	Exterior Walls	Stone with concrete block backup	*70% of wall*	S.F. Wall	41	18.46	21.0%
2020	Exterior Windows	Metal outward projecting	*10% of wall*	Each	1100	9.18	
2030	Exterior Doors	Metal and glass with transoms		Each	2609	.73	
	B30 Roofing						
3010	Roof Coverings	Built-up tar and gravel with flashing; perlite/EPS composite insulation		S.F. Roof	6.96	2.32	1.7%
3020	Roof Openings	N/A		—	—	—	
C. INTERIORS							
1010	Partitions	Gypsum board on metal studs	*20 S.F. Floor/L.F. Partition*	S.F. Partition	9.82	4.91	21.8%
1020	Interior Doors	Wood solid core	*200 S.F. Floor/Door*	Each	593	2.97	
1030	Fittings	Toilet partitions		S.F. Floor	.27	.27	
2010	Stair Construction	Concrete filled metal pan		Flight	16,800	5.60	
3010	Wall Finishes	90% paint, 10% ceramic tile		S.F. Surface	1.38	1.38	
3020	Floor Finishes	70% carpet tile, 15% terrazzo, 15% vinyl composition tile		S.F. Floor	7.82	7.82	
3030	Ceiling Finishes	Mineral fiber tile on concealed zee bars		S.F. Ceiling	6.57	6.57	
D. SERVICES							
	D10 Conveying						
1010	Elevators & Lifts	Two hydraulic elevators		Each	113,130	12.57	9.3%
1020	Escalators & Moving Walks	N/A		—	—	—	
	D20 Plumbing						
2010	Plumbing Fixtures	Toilet and service fixtures, supply and drainage	*1 Fixture/1385 S.F. Floor*	Each	6052	4.37	4.2%
2020	Domestic Water Distribution	Gas fired water heater		S.F. Floor	.59	.59	
2040	Rain Water Drainage	Roof drains		S.F. Roof	2.31	.77	
	D30 HVAC						
3010	Energy Supply	N/A		—	—	—	6.3 %
3020	Heat Generating Systems	Included in D3050		—	—	—	
3030	Cooling Generating Systems	N/A		—	—	—	
3050	Terminal & Package Units	Multizone unit, gas heating, electric cooling		S.F. Floor	8.55	8.55	
3090	Other HVAC Sys. & Equipment	N/A		—	—	—	
	D40 Fire Protection						
4010	Sprinklers	Sprinklers, light hazard		S.F. Floor	2.95	2.95	4.6%
4020	Standpipes	Standpipe, wet, Class III		S.F. Floor	3.26	3.26	
	D50 Electrical						
5010	Electrical Service/Distribution	400 ampere service, panel board and feeders		S.F. Floor	1.64	1.64	12.3%
5020	Lighting & Branch Wiring	High efficiency fluorescent fixtures, receptacles, switches, A.C. and misc. power		S.F. Floor	11.67	11.67	
5030	Communications & Security	Addressable alarm systems, internet wiring, and emergency lighting		S.F. Floor	3.15	3.15	
5090	Other Electrical Systems	Emergency generator, 15 kW		S.F. Floor	.20	.20	
E. EQUIPMENT & FURNISHINGS							
1010	Commercial Equipment	N/A		—	—	—	0.0 %
1020	Institutional Equipment	N/A		—	—	—	
1030	Vehicular Equipment	N/A		—	—	—	
1090	Other Equipment	N/A		—	—	—	
F. SPECIAL CONSTRUCTION							
1020	Integrated Construction	N/A		—	—	—	0.0 %
1040	Special Facilities	N/A		—	—	—	
G. BUILDING SITEWORK	**N/A**						
					Sub-Total	135.38	**100%**
	CONTRACTOR FEES (General Requirements: 10%, Overhead: 5%, Profit: 10%)				25%	33.84	
	ARCHITECT FEES				9%	15.23	
					Total Building Cost	**184.45**	

M.685 Veterinary Hospital

Costs per square foot of floor area

Exterior Wall	S.F. Area	3500	5000	7500	9000	11000	12500	15000	20000	25000
	L.F. Perimeter	240	290	350	390	430	460	510	580	650
E.I.F.S.	Wood Truss	163.80	155.00	146.90	144.35	141.60	140.10	138.20	135.30	133.50
	Steel Joists	174.45	165.30	156.80	154.10	151.20	149.60	147.65	144.55	142.70
Wood Siding	Wood Truss	166.35	157.70	**149.70**	147.20	144.50	143.00	141.20	138.30	136.60
Brick Veneer	Wood Truss	175.85	165.75	156.20	153.25	149.90	148.10	145.90	142.40	140.20
Brick on Block	Wood Truss	179.05	167.90	157.25	153.95	150.25	148.20	145.70	141.70	139.25
Tilt Up Concrete	Steel Joists	166.65	158.45	150.90	148.50	146.00	144.60	142.85	140.20	138.65
Perimeter Adj., Add or Deduct	Per 100 L.F.	11.95	8.40	5.55	4.65	3.75	3.30	2.85	2.10	1.75
Story Hgt. Adj., Add or Deduct	Per 1 Ft.	1.50	1.25	1.00	0.95	0.85	0.75	0.80	0.65	0.60
***For Basement**, add $29.70 per square foot of basement area*										

The above costs were calculated using the basic specifications shown on the facing page. These costs should be adjusted where necessary for design alternatives and owner's requirements. Reported completed project costs, for this type of structure, range from $73.10 to $282.60 per S.F.

Common additives

Description	Unit	$ Cost
Closed circuit surveillance, one station		
Camera and monitor	Ea.	1975
For additional camera stations, add	Ea.	1075
Directory boards, plastic, glass covered		
30" x 20"	Ea.	555
36" x 48"	Ea.	1425
Aluminum, 24" x 18"	Ea.	600
36" x 24"	Ea.	715
48" x 32"	Ea.	1000
48" x 60"	Ea.	2100
Electronic, wall mounted	S.F.	2900
Free standing	S.F.	4125
Kennel fencing		
Kennel fencing, 1-1/2" mesh, 6' long, 3'-6" wide, 6'-2" high	Ea.	750
12' long	Ea.	975
Top covers, 1-1/2" mesh, 6' long	Ea.	200
12' long	Ea.	273

Description	Unit	$ Cost
Kennel doors		
2 way, swinging type, 13" x 19" opening	Opng.	191
17" x 29" opening	Opng.	213
9" x 9" opening, electronic with accessories	Opng.	250

Important: See the Reference Section for Location Factors.

Model costs calculated for a 1 story building with 12′ story height and 7,500 square feet of floor area

Veterinary Hospital

			Unit	Unit Cost	Cost Per S.F.	% Of Sub-Total
A. SUBSTRUCTURE						
1010	Standard Foundations	Poured concrete; strip and spread footings	S.F. Ground	1.89	1.89	10.4%
1020	Special Foundations	N/A	—	—	—	
1030	Slab on Grade	4″ and 5″ reinforced concrete with vapor barrier and granular base	S.F. Slab	5.10	5.10	
2010	Basement Excavation	Site preparation for slab and trench for foundation wall and footing	S.F. Ground	.51	.96	
2020	Basement Walls	4′ foundation wall	L.F. Wall	74	3.43	
B. SHELL						
	B10 Superstructure					
1010	Floor Construction	6 x 6 wood columns	S.F. Floor	.22	.22	6.9%
1020	Roof Construction	Wood roof truss with plywood sheathing	S.F. Roof	7.37	7.37	
	B20 Exterior Enclosure					
2010	Exterior Walls	Cedar bevel siding on wood studs, insulated *85% of wall*	S.F. Wall	12.79	6.09	10.9%
2020	Exterior Windows	Wood double hung and awning *15% of wall*	Each	1120	3.84	
2030	Exterior Doors	Aluminum and glass	Each	2607	2.08	
	B30 Roofing					
3010	Roof Coverings	Ashphalt shingles with flashing (pitched); al. gutters and downspouts	S.F. Roof	4.81	4.81	4.4%
3020	Roof Openings	N/A	—	—	—	
C. INTERIORS						
1010	Partitions	Gypsum board on wood studs *8 S.F. Floor/L.F. Partition*	S.F. Partition	4.75	5.94	20.0%
1020	Interior Doors	Solid core wood *180 S.F. Floor/Door*	Each	650	3.61	
1030	Fittings	Lockers	Each	.44	.44	
2010	Stair Construction	N/A	—	—	—	
3010	Wall Finishes	Paint on drywall	S.F. Surface	.91	2.27	
3020	Floor Finishes	90% vinyl composition tile, 5% ceramic tile, 5% sealed concrete	S.F. Floor	3.18	3.18	
3030	Ceiling Finishes	90% acoustic tile, 10% gypsum board	S.F. Ceiling	6.52	6.53	
D. SERVICES						
	D10 Conveying					
1010	Elevators & Lifts	N/A	—	—	—	0.0 %
1020	Escalators & Moving Walks	N/A	—	—	—	
	D20 Plumbing					
2010	Plumbing Fixtures	Toilet and service fixtures, supply and drainage *1 Fixture/1075 S.F. Floor*	Each	1646	2.94	4.0%
2020	Domestic Water Distribution	Electric water heater	S.F. Floor	1.49	1.49	
2040	Rain Water Drainage	N/A	—	—	—	
	D30 HVAC					
3010	Energy Supply	N/A	—	—	—	5.1 %
3020	Heat Generating Systems	N/A	—	—	—	
3030	Cooling Generating Systems	N/A	—	—	—	
3050	Terminal & Package Units	Split system with air cooled condensing units	S.F. Floor	5.55	5.55	
3090	Other HVAC Sys. & Equipment	N/A	—	—	—	
	D40 Fire Protection					
4010	Sprinklers	Wet pipe sprinkler system	S.F. Floor	3.28	3.28	3.0%
4020	Standpipes	N/A	—	—	—	
	D50 Electrical					
5010	Electrical Service/Distribution	800 ampere service, panel board and feeders	S.F. Floor	6.42	6.42	20.5%
5020	Lighting & Branch Wiring	High efficiency fluorescent fixtures, receptacles, switches, A.C. and misc. power	S.F. Floor	11.54	11.54	
5030	Communications & Security	Addressable alarm systems, internet wiring, and emergency lighting	S.F. Floor	4.52	4.52	
5090	Other Electrical Systems	Emergency generator, 100 kW	S.F. Floor	.05	.05	
E. EQUIPMENT & FURNISHINGS						
1010	Commercial Equipment	N/A	—	—	—	14.9 %
1020	Institutional Equipment	Eye wash station	S.F. Floor	14.47	14.47	
1030	Vehicular Equipment	N/A	—	—	—	
1090	Other Equipment	Countertop, plastic laminate	S.F. Floor	1.86	1.86	
F. SPECIAL CONSTRUCTION & DEMOLITION						
1020	Integrated Construction	N/A	—	—	—	0.0 %
1040	Special Facilities	N/A	—	—	—	
G. BUILDING SITEWORK		**N/A**				
				Sub-Total	109.88	**100%**
	CONTRACTOR FEES (General Requirements: 10%, Overhead: 5%, Profit: 10%)			25%	27.46	
	ARCHITECT FEES			9%	12.36	
				Total Building Cost	**149.70**	

Costs per square foot of floor area

Exterior Wall	S.F. Area	10000	15000	20000	25000	30000	35000	40000	50000	60000
	L.F. Perimeter	410	500	600	640	700	766	833	966	1000
Metal Panel	Steel Frame	102.55	93.60	89.65	85.20	82.85	81.20	80.10	78.50	76.00
Pre-Engineered Metal Building	Steel Frame	110.90	100.80	96.25	91.20	88.55	86.75	85.40	83.55	80.75
E.I.F.S.	Steel Frame	118.80	108.30	103.65	98.35	95.55	93.65	92.30	90.40	87.40
Brick Veneer	Reinforced Concrete	164.25	147.45	140.00	131.05	**126.30**	123.20	120.95	117.80	112.60
Precast Concrete Panel	Reinforced Concrete	176.05	156.90	148.40	138.05	132.65	129.10	126.55	122.90	116.85
Tilt-up Concrete Panel	R/Conc. Frame	128.70	117.20	112.05	106.15	102.95	100.85	99.35	97.20	93.85
Perimeter Adj., Add or Deduct	Per 100 L.F.	8.40	5.70	4.20	3.40	2.80	2.45	2.10	1.70	1.40
Story Hgt. Adj., Add or Deduct	Per 1 Ft.	1.05	0.90	0.75	0.65	0.55	0.60	0.55	0.45	0.45
***For Basement**, add $ 26.45 per square foot of basement area*										

The above costs were calculated using the basic specifications shown on the facing page. These costs should be adjusted where necessary for design alternatives and owner's requirements. Reported completed project costs, for this type of structure, range from $ 39.65 to $ 158.30 per S.F.

Common additives

Description	Unit	$ Cost
Dock Leveler, 10 ton cap.		
6′ x 8′	Ea.	5900
7′ x 8′	Ea.	7775
Emergency Lighting, 25 watt, battery operated		
Lead battery	Ea.	325
Nickel cadmium	Ea.	665
Fence, Chain link, 6′ high		
9 ga. wire	L.F.	28
6 ga. wire	L.F.	31.50
Gate	Ea.	380
Flagpoles, Complete		
Aluminum, 20′ high	Ea.	1825
40′ high	Ea.	4300
70′ high	Ea.	11,300
Fiberglass, 23′ high	Ea.	1275
39′-5″ high	Ea.	3275
59′ high	Ea.	6475
Paving, Bituminous		
Wearing course plus base course	S.Y.	6.60
Sidewalks, Concrete 4″ thick	S.F.	4.23

Description	Unit	$ Cost
Sound System		
Amplifier, 250 watts	Ea.	2125
Speaker, ceiling or wall	Ea.	206
Trumpet	Ea.	395
Yard Lighting, 20′ aluminum pole with 400 watt high pressure sodium fixture	Ea.	3050

Important: See the Reference Section for Location Factors.

Model costs calculated for a 1 story building with 24' story height and 30,000 square feet of floor area

Warehouse

			Unit	Unit Cost	Cost Per S.F.	% Of Sub-Total
A. SUBSTRUCTURE						
1010	Standard Foundations	Poured concrete; strip and spread footings	S.F. Ground	1.29	1.29	9.2%
1020	Special Foundations	N/A	—	—	—	
1030	Slab on Grade	5" reinforced concrete	S.F. Slab	5.54	5.54	
2010	Basement Excavation	Site preparation for slab and trench for foundation wall and footing	S.F. Ground	.16	.16	
2020	Basement Walls	4' Foundation wall	L.F. Wall	74	1.71	
B. SHELL						
	B10 Superstructure					
1010	Floor Construction	Cast in place concrete columns, beams, and slab, fireproofed	S.F. Floor	65	6.52	22.2%
1020	Roof Construction	Precast double T with 2" topping	S.F. Roof	14.42	14.42	
	B20 Exterior Enclosure					
2010	Exterior Walls	Face brick with concrete block back up *98% of wall*	S.F. Wall	35	19.37	25.7%
2020	Exterior Windows	Aluminum horizontal sliding *2% of wall*	Each	495	3.70	
2030	Exterior Doors	Aluminum and glass entry doors, hollow steel doors, overhead doors	Each	3220	1.19	
	B30 Roofing					
3010	Roof Coverings	Single ply membrane, stone ballast, rigid insulation	S.F. Roof	5.09	5.09	5.7%
3020	Roof Openings	Roof and smoke hatches	S.F. Roof	.32	.32	
C. INTERIORS						
1010	Partitions	Concrete block partitions, gypsum board on metal stud *100 S.F. Floor/L.F. Partition*	S.F. Partition	15.63	1.25	6.2%
1020	Interior Doors	Single leaf wood hollow core doors, metal doors *2500 S.F. Floor/Door*	Each	666	.26	
1030	Fittings	N/A	—	—	—	
2010	Stair Construction	Steel grate type with rails	Flight	13,850	.93	
3010	Wall Finishes	Paint	S.F. Surface	4.19	.67	
3020	Floor Finishes	90% hardener, 10% vinyl composition tile	S.F. Floor	2.05	2.05	
3030	Ceiling Finishes	Acoustic ceiling tiles on suspended channel grid	S.F. Ceiling	6.57	.66	
D. SERVICES						
	D10 Conveying					
1010	Elevators & Lifts	N/A	—	—	—	0.0 %
1020	Escalators & Moving Walks	N/A	—	—	—	
	D20 Plumbing					
2010	Plumbing Fixtures	Restroom and service fixtures, supply and drainage *1 Fixture/2500 S.F. Floor*	Each	1275	.51	1.3%
2020	Domestic Water Distribution	Gas fired water heater	S.F. Floor	.22	.22	
2040	Rain Water Drainage	Roof drains	S.F. Roof	.53	.53	
	D30 HVAC					
3010	Energy Supply	N/A	—	—	—	6.2 %
3020	Heat Generating Systems	Ventilation with heat system	Each	139,800	5.03	
3030	Cooling Generating Systems	N/A	—	—	—	
3050	Terminal & Package Units	Single zone unit gas heating, electric cooling	S.F. Floor	.79	.79	
3090	Other HVAC Sys. & Equipment	N/A	—	—	—	
	D40 Fire Protection					
4010	Sprinklers	Wet pipe sprinkler system, ordinary hazard	S.F. Floor	3.78	3.78	6.4%
4020	Standpipes	Standpipe	S.F. Floor	2.26	2.26	
	D50 Electrical					
5010	Electrical Service/Distribution	200 ampere service, panel board and feeders	S.F. Floor	.57	.57	9.5%
5020	Lighting & Branch Wiring	Fluorescent fixtures, receptacles, switches, A.C. and misc. power	S.F. Floor	5.81	5.81	
5030	Communications & Security	Addressable alarm system	S.F. Floor	2.61	2.61	
5090	Other Electrical Systems	N/A	—	—	—	
E. EQUIPMENT & FURNISHINGS						
1010	Commercial Equipment	N/A	—	—	—	1.9 %
1020	Institutional Equipment	N/A	—	—	—	
1030	Vehicular Equipment	Dock boards and levelers	S.F. Floor	1.02	1.82	
1090	Other Equipment	N/A	—	—	—	
F. SPECIAL CONSTRUCTION						
1020	Integrated Construction	Shipping and receiving air curtain	S.F. Floor	5.37	5.37	5.7%
1040	Special Facilities	N/A	—	—	—	
G. BUILDING SITEWORK		**N/A**				
				Sub-Total	94.43	**100%**
	CONTRACTOR FEES (General Requirements: 10%, Overhead: 5%, Profit: 10%)			*25%*	23.61	
	ARCHITECT FEES			*7%*	8.26	
				Total Building Cost	**126.30**	

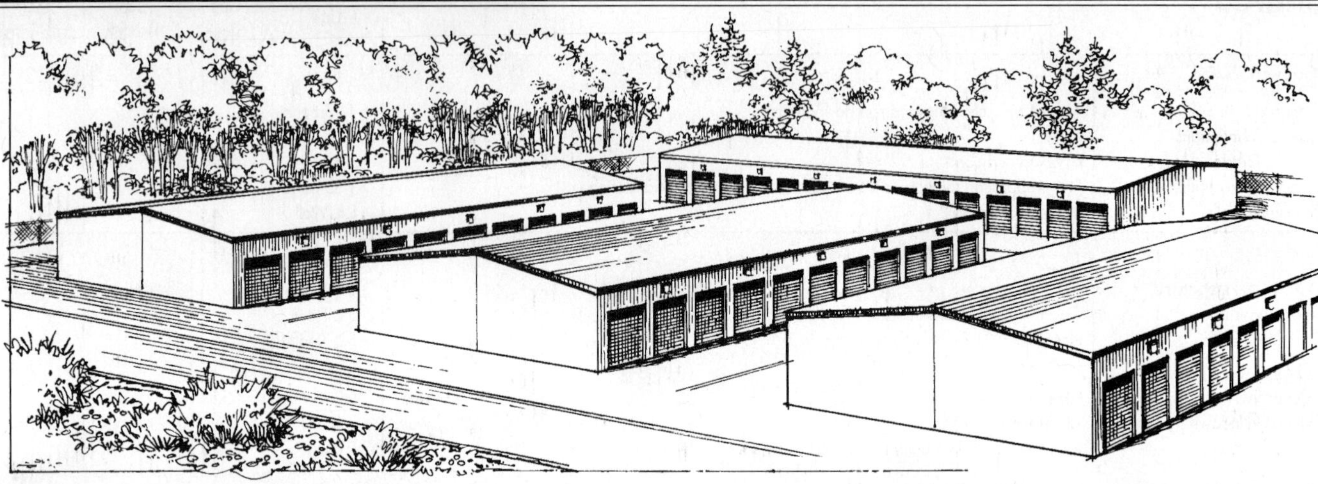

Costs per square foot of floor area

Exterior Wall	S.F. Area	2000	3000	5000	8000	12000	20000	30000	50000	100000
	L.F. Perimeter	180	220	300	420	580	900	1300	2100	4100
Concrete Block	Steel Frame	96.50	84.75	75.40	70.15	67.20	64.90	63.70	62.75	62.00
	R/Conc. Frame	96.85	88.80	82.40	78.80	76.80	75.20	74.35	73.70	73.20
Metal Sandwich Panel	Steel Frame	87.75	78.90	71.90	67.95	65.80	64.05	63.10	62.40	61.90
Tilt-up Concrete Panel	Steel and Concrete	89.35	81.30	74.95	71.35	69.40	67.80	66.95	66.30	65.80
Precast Concrete Panel	Steel Frame	108.10	95.10	84.70	78.90	75.65	73.10	71.70	70.70	69.90
	R/Conc. Frame	129.00	116.65	106.80	101.25	98.15	95.70	94.40	93.40	92.70
Perimeter Adj., Add or Deduct	Per 100 L.F.	18.85	12.60	7.55	4.75	3.15	1.90	1.25	0.75	0.40
Story Hgt. Adj., Add or Deduct	Per 1 Ft.	1.25	1.05	0.85	0.75	0.70	0.65	0.60	0.65	0.60
Basement—Not Applicable										

The above costs were calculated using the basic specifications shown on the facing page. These costs should be adjusted where necessary for design alternatives and owner's requirements. Reported completed project costs, for this type of structure, range from $38.00 to $217.80 per S.F.

Common additives

Description	Unit	$ Cost
Dock Leveler, 10 ton cap.		
6′ x 8′	Ea.	5900
7′ x 8′	Ea.	7775
Emergency Lighting, 25 watt, battery operated		
Lead battery	Ea.	325
Nickel cadmium	Ea.	665
Fence, Chain link, 6′ high		
9 ga. wire	L.F.	28
6 ga. wire	L.F.	31.50
Gate	Ea.	380
Flagpoles, Complete		
Aluminum, 20′ high	Ea.	1825
40′ high	Ea.	4300
70′ high	Ea.	11,300
Fiberglass, 23′ high	Ea.	1275
39′-5″ high	Ea.	3275
59′ high	Ea.	6475
Paving, Bituminous		
Wearing course plus base course	S.Y.	6.60
Sidewalks, Concrete 4″ thick	S.F.	4.23

Description	Unit	$ Cost
Sound System		
Amplifier, 250 watts	Ea.	2125
Speaker, ceiling or wall	Ea.	206
Trumpet	Ea.	395
Yard Lighting,		
20′ aluminum pole with		
400 watt high pressure sodium fixture	Ea.	3050

Important: See the Reference Section for Location Factors.

Model costs calculated for a 1 story building with 12′ story height and 20,000 square feet of floor area

Warehouse Self Storage

				Unit	Unit Cost	Cost Per S.F.	% Of Sub-Total
A. SUBSTRUCTURE							
1010	Standard Foundations	Poured concrete; strip and spread footings		S.F. Ground	2.11	2.11	25.4%
1020	Special Foundations	N/A		—	—	—	
1030	Slab on Grade	4″ reinforced concrete with vapor barrier and granular base		S.F. Slab	6.34	6.34	
2010	Basement Excavation	Site preparation for slab and trench for foundation wall and footing		S.F. Ground	.29	.58	
2020	Basement Walls	4′ foundation wall		L.F. Wall	74	3.31	
B. SHELL							
B10 Superstructure							
1010	Floor Construction	Steel column fireproofing		S.F. Floor	.62	.62	14.5%
1020	Roof Construction	Metal deck, open web steel joists, beams, columns		S.F. Roof	6.41	6.41	
B20 Exterior Enclosure							
2010	Exterior Walls	Concrete block	*53% of wall*	S.F. Wall	20	5.73	24.6%
2020	Exterior Windows	Aluminum projecting	*1% of wall*	Each	1100	.11	
2030	Exterior Doors	Steel overhead, hollow metal	*46% of wall*	Each	1500	6.08	
B30 Roofing							
3010	Roof Coverings	Metal panel roof; perlite/EPS composite insulation		S.F. Roof	6.54	6.54	13.5%
3020	Roof Openings	N/A		—	—	—	
C. INTERIORS							
1010	Partitions	Metal panels on metal studs	*11S.F. Floor/L.F. Partition*	S.F. Partition	1.11	1.21	2.6%
1020	Interior Doors	Single leaf hollow metal	*20,000 S.F. Floor/Door*	Each	600	.06	
1030	Fittings	N/A		—	—	—	
2010	Stair Construction	N/A		—	—	—	
3010	Wall Finishes	N/A		—	—	—	
3020	Floor Finishes	N/A		—	—	—	
3030	Ceiling Finishes	N/A		—	—	—	
D. SERVICES							
D10 Conveying							
1010	Elevators & Lifts	N/A		—	—	—	0.0 %
1020	Escalators & Moving Walks	N/A		—	—	—	
D20 Plumbing							
2010	Plumbing Fixtures	Toilet and service fixtures, supply and drainage	*1 Fixture/10,000 S.F. Floor*	Each	2600	.26	2.4%
2020	Domestic Water Distribution	Gas fired water heater		S.F. Floor	.31	.31	
2040	Rain Water Drainage	Roof drain		S.F. Roof	.59	.59	
D30 HVAC							
3010	Energy Supply	N/A		—	—	—	2.3 %
3020	Heat Generating Systems	N/A		—	—	—	
3030	Cooling Generating Systems	N/A		—	—	—	
3050	Terminal & Package Units	Single zone rooftop unit		S.F. Floor	1.13	1.13	
3090	Other HVAC Sys. & Equipment	N/A		—	—	—	
D40 Fire Protection							
4010	Sprinklers	N/A		—	—	—	0.0 %
4020	Standpipes	N/A		—	—	—	
D50 Electrical							
5010	Electrical Service/Distribution	60 ampere service, panel board and feeders		S.F. Floor	.36	.36	14.7%
5020	Lighting & Branch Wiring	High bay fixtures, receptacles, switches and misc. power		S.F. Floor	5.48	5.48	
5030	Communications & Security	Addressable alarm systems		S.F. Floor	1.29	1.29	
5090	Other Electrical Systems	N/A		—	—	—	
E. EQUIPMENT & FURNISHINGS							
1010	Commercial Equipment	N/A		—	—	—	0.0 %
1020	Institutional Equipment	N/A		—	—	—	
1030	Vehicular Equipment	N/A		—	—	—	
1090	Other Equipment	N/A		—	—	—	
F. SPECIAL CONSTRUCTION							
1020	Integrated Construction	N/A		—	—	—	0.0 %
1040	Special Facilities	N/A		—	—	—	
G. BUILDING SITEWORK	**N/A**						
					Sub-Total	48.52	**100%**
	CONTRACTOR FEES (General Requirements: 10%, Overhead: 5%, Profit: 10%)				*25%*	12.13	
	ARCHITECT FEES				*7%*	4.25	
					Total Building Cost	**64.90**	

Green Commercial/Industrial/ Institutional Section

Table of Contents

Costs per square foot of floor area

Exterior Wall	S.F. Area	2000	2700	3400	4100	4800	5500	6200	6900	7600
	L.F. Perimeter	180	208	236	256	280	303	317	337	357
Face Brick with Concrete Block Back-up	Steel Frame	277.10	259.10	248.55	239.55	234.05	229.80	224.95	222.00	219.70
	R/Conc. Frame	290.10	272.05	261.50	252.55	247.00	242.75	237.95	235.00	232.65
Precast Concrete Panel	Steel Frame	280.00	261.50	250.70	241.45	235.75	231.40	226.45	223.40	220.95
	R/Conc. Frame	296.20	276.70	265.25	255.65	249.70	245.10	239.95	236.85	234.25
Limestone with Concrete Block Back-up	Steel Frame	304.25	282.30	269.45	258.35	251.65	246.45	240.40	236.80	233.85
	R/Conc. Frame	317.95	296.00	283.15	272.05	265.35	260.10	254.10	250.45	247.50
Perimeter Adj., Add or Deduct	Per 100 L.F.	50.75	37.55	29.80	24.80	21.20	18.50	16.40	14.75	13.35
Story Hgt. Adj., Add or Deduct	Per 1 Ft.	4.95	4.20	3.75	3.40	3.20	3.00	2.80	2.70	2.55
***For Basement**, add $ 33.25 per square foot of basement area*										

The above costs were calculated using the basic specifications shown on the facing page. These costs should be adjusted where necessary for design alternatives and owner's requirements. Reported completed project costs, for this type of structure, range from $ 158.85 to $ 391.05 per S.F.

Common additives

Description	Unit	$ Cost
Bulletproof Teller Window, 44" x 60"	Ea.	5475
60" x 48"	Ea.	7300
Closed Circuit Surveillance, one station, camera & monitor	Ea.	1975
For additional camera stations, add	Ea.	1075
Counters, complete, dr & frame, 3' x 6'-8", bullet resist stl with vision panel	Ea.	7200 - 9375
Drive-up Window, drawer & micr., not incl. glass	Ea.	8900 - 12,500
Night Depository	Ea.	9475 - 14,200
Package Receiver, painted	Ea.	1900
Stainless steel	Ea.	2950
Partitions, bullet resistant to 8' high	L.F.	320 - 520
Pneumatic Tube Systems, 2 station	Ea.	31,400
With TV viewer	Ea.	59,000
Service Windows, pass thru, steel, 24" x 36"	Ea.	3800
48" x 48"	Ea.	3950
Twenty-four Hour Teller, automatic deposit cash & memo	Ea.	52,500
Vault Front, door & frame, 2 hour test, 32" door	Opng.	9900
4 hour test, 40" door	Opng.	12,300
Time lock, two movement, add	Ea.	2275

Description	Unit	$ Cost
Commissioning Fees, sustainable commercial construction	S.F.	0.24 - 3.04
Energy Modelling Fees, banks to 10,000 SF	Ea.	11,000
Green Bldg Cert Fees for comm construction project reg	Project	900
Photovoltaic Pwr Sys, grid connected, 20 kW (~2400 SF), roof	Ea.	241,700
Green Roofs, 6" soil depth, w/treated wd edging & sedum mats	S.F.	11.27
10" Soil depth, with treated wood edging & sedum mats	S.F.	12.84
Greywater Recovery Systems, prepackaged comm, 1530 gal	Ea.	53,480
Rainwater Harvest Sys, prepckged comm, 10,000 gal, sys contrller	Ea.	31,175
20,000 gal. w/system controller	Ea.	50,100
30,000 gal. w/system controller	Ea.	80,225
Solar Domestic HW, closed loop, add-on sys, ext heat exchanger	Ea.	9750
Draindown, hot water system, 120 gal tank	Ea.	12,700

Important: See the Reference Section for Location Factors.

Model costs calculated for a 1 story building with 14′ story height and 4,100 square feet of floor area

G Bank

				Unit	Unit Cost	Cost Per S.F.	% Of Sub-Total
A. SUBSTRUCTURE							
1010	Standard Foundations	Poured concrete; strip and spread footings		S.F. Ground	3.94	3.94	8.4%
1020	Special Foundations	N/A		—	—	—	
1030	Slab on Grade	4″ reinforced concrete with recycled vapor barrier and granular base		S.F. Slab	5.11	5.11	
2010	Basement Excavation	Site preparation for slab and trench for foundation wall and footing		S.F. Ground	.29	.29	
2020	Basement Walls	4′ Foundation wall		L.F. Wall	81	6.03	
B. SHELL							
	B10 Superstructure						
1010	Floor Construction	Cast-in-place columns		L.F. Column	5.27	5.27	10.9%
1020	Roof Construction	Cast-in-place concrete flat plate		S.F. Roof	14.50	14.50	
	B20 Exterior Enclosure						
2010	Exterior Walls	Face brick with concrete block backup	*80% of wall*	S.F. Wall	35	24.68	18.0%
2020	Exterior Windows	Horizontal aluminum sliding	*20% of wall*	Each	555	6.47	
2030	Exterior Doors	Double aluminum and glass and hollow metal, low VOC paint		Each	3205	1.57	
	B30 Roofing						
3010	Roof Coverings	Single-ply TPO membrane, 60 mils, heat welded seams w/4″ thk. R20 insul.		S.F. Roof	6.85	6.85	3.8%
3020	Roof Openings	N/A		—	—	—	
C. INTERIORS							
1010	Partitions	Gypsum board on metal studs w/sound attenuation	*20 SF of Flr./LF Part.*	S.F. Partition	11.36	5.68	13.6%
1020	Interior Doors	Single leaf hollow metal, low VOC paint	*200 S.F. Floor/Door*	Each	1133	5.67	
1030	Fittings	N/A		—	—	—	
2010	Stair Construction	N/A		—	—	—	
3010	Wall Finishes	50% vinyl wall covering, 50% paint, low VOC		S.F. Surface	1.39	1.39	
3020	Floor Finishes	50% carpet tile, 40% vinyl composition tile, recycled, 10% quarry tile		S.F. Floor	5.51	5.51	
3030	Ceiling Finishes	Mineral fiber tile on concealed zee bars		S.F. Ceiling	6.57	6.57	
D. SERVICES							
	D10 Conveying						
1010	Elevators & Lifts	N/A		—	—	—	0.0 %
1020	Escalators & Moving Walks	N/A		—	—	—	
	D20 Plumbing						
2010	Plumbing Fixtures	Toilet low flow, auto sensor and service fixt., supply and drainage	*1 Fixt./580 SF Flr.*	Each	5498	9.48	6.5%
2020	Domestic Water Distribution	Tankless, on demand water heaters, natural gas/propane		S.F. Floor	.89	.89	
2040	Rain Water Drainage	Roof drains		S.F. Roof	1.39	1.39	
	D30 HVAC						
3010	Energy Supply	N/A		—	—	—	6.0 %
3020	Heat Generating Systems	Included in D3050		—	—	—	
3040	Distribution Systems	Enthalpy heat recovery packages		Each	14,475	3.53	
3050	Terminal & Package Units	Single zone rooftop air conditioner		S.F. Floor	7.35	7.35	
3090	Other HVAC Sys. & Equipment	N/A		—	—	—	
	D40 Fire Protection						
4010	Sprinklers	Wet pipe sprinkler system		S.F. Floor	4.57	4.57	4.1%
4020	Standpipes	Standpipe		S.F. Floor	2.91	2.91	
	D50 Electrical						
5010	Electrical Service/Distribution	200 ampere service, panel board and feeders		S.F. Floor	1.44	1.44	15.4%
5020	Lighting & Branch Wiring	LED fixtures, daylt. dimming & ltg on/off control, receptacles, switches, A.C.		S.F. Floor	12.20	12.20	
5030	Communications & Security	Alarm systems, internet/phone wiring, and security television		S.F. Floor	10.09	10.09	
5090	Other Electrical Systems	Emergency generator, 15 kW, UPS, energy monitoring systems		S.F. Floor	4.36	4.36	
E. EQUIPMENT & FURNISHINGS							
1010	Commercial Equipment	Automatic teller, drive up window, night depository		S.F. Floor	7.81	7.81	6.6%
1020	Institutional Equipment	Closed circuit TV monitoring system		S.F. Floor	2.89	2.89	
1090	Other Equipment	Waste handling recycling tilt truck		Each	1.35	1.35	
2020	Moveable Furnishings	No smoking signage		Each	.02	.02	
F. SPECIAL CONSTRUCTION							
1020	Integrated Construction	N/A		—	—	—	6.7 %
1040	Special Facilities	Security vault door		S.F. Floor	12.19	12.19	
G. BUILDING SITEWORK	**N/A**						
					Sub-Total	182	**100%**
	CONTRACTOR FEES (General Requirements: 10%, Overhead: 5%, Profit: 10%)				*25%*	45.52	
	ARCHITECT FEES				*11%*	25.03	
					Total Building Cost	**252.55**	

Costs per square foot of floor area

Exterior Wall	S.F. Area	4000	4500	5000	5500	6000	6500	7000	7500	8000
	L.F. Perimeter	260	280	300	310	320	336	353	370	386
Face Brick Concrete Block Back-up	Steel Joists	198.40	194.40	191.25	187.00	**183.55**	181.40	179.70	178.25	176.75
	Bearing Walls	197.70	193.70	190.55	186.30	182.85	180.65	178.95	177.50	176.05
Decorative Concrete Block	Steel Joists	189.85	186.45	183.80	180.45	177.65	175.90	174.50	173.25	172.05
	Bearing Walls	182.85	179.50	176.85	173.45	170.65	168.90	167.50	166.25	165.05
Limestone with Concrete Block Back-up	Steel Joists	222.65	217.85	214.10	208.90	204.60	201.95	199.95	198.15	196.40
	Bearing Walls	215.65	210.90	207.15	201.90	197.60	194.95	192.95	191.15	189.45
Perimeter Adj., Add or Deduct	Per 100 L.F.	21.90	19.50	17.55	15.95	14.65	13.45	12.50	11.70	10.95
Story Hgt. Adj., Add or Deduct	Per 1 Ft.	2.90	2.80	2.70	2.60	2.40	2.30	2.25	2.20	2.20
***For Basement**, add $37.90 per square foot of basement area*										

The above costs were calculated using the basic specifications shown on the facing page. These costs should be adjusted where necessary for design alternatives and owner's requirements. Reported completed project costs, for this type of structure, range from $86.20 to $255.00 per S.F.

Common additives

Description	Unit	$ Cost
Appliances, cooking range, 30" free standing		
1 oven	Ea.	595 - 2275
2 oven	Ea.	1450 - 3150
Microwave oven	Ea.	275 - 850
Compactor, residential, 4-1 compaction	Ea.	860 - 1450
Dishwasher, built-in, 2 cycles	Ea.	510 - 855
4 cycles	Ea.	650 - 1675
Garbage disposer, sink type	Ea.	174 - 340
Hood for range, 2 speed, vented, 30" wide	Ea.	222 - 1150
Refrigerator, no frost 10-12 C.F.	Ea.	600 - 745
14-16 C.F.	Ea.	720 - 870
18-20 C.F.	Ea.	910 - 2075
Lockers, Steel, single tier, 60" or 72"	Opng.	222 - 365
Locker bench, lam. maple top only	L.F.	33
Pedestals, steel pipe	Ea.	71
Sound System, amplifier, 250 watts	Ea.	2125
Speaker, ceiling or wall	Ea.	206
Trumpet	Ea.	395

Description	Unit	$ Cost
Commissioning Fees, sustainable commercial construction	S.F.	0.24 - 3.04
Energy Modelling Fees, Fire Stations to 10,000 SF	Ea.	11,000
Green Bldg Cert Fees for comm construction project reg	Project	900
Photovoltaic Pwr Sys, grid connected, 20 kW (~2400 SF), roof	Ea.	241,700
Green Roofs, 6" soil depth, w/treated wd edging & sedum mats	S.F.	11.27
10" Soil depth, with treated wood edging & sedum mats	S.F.	12.84
Greywater Recovery Systems, prepackaged comm, 3060 gal.	Ea.	58,085
4590 gal.	Ea.	72,815
Rainwater Harvest Sys, prepckged comm, 10,000 gal, sys contrller	Ea.	31,175
20,000 gal. w/system controller	Ea.	50,100
30,000 gal. w/system controller	Ea.	80,225
Solar Domestic HW, closed loop, add-on sys, ext heat exchanger	Ea.	9750
Drainback, hot water system, 120 gal tank	Ea.	12,425

Model costs calculated for a 1 story building with 14′ story height and 6,000 square feet of floor area

				Unit	Unit Cost	Cost Per S.F.	% Of Sub-Total
A. SUBSTRUCTURE							
1010	Standard Foundations	Poured concrete; strip and spread footings		S.F. Ground	3.15	3.15	10.8%
1020	Special Foundations	N/A		—	—	—	
1030	Slab on Grade	6″ reinforced concrete with recycled vapor barrier and granular base		S.F. Slab	6.79	6.79	
2010	Basement Excavation	Site preparation for slab and trench for foundation wall and footing		S.F. Ground	.51	.51	
2020	Basement Walls	4′ foundation wall		L.F. Wall	81	4.29	
B. SHELL							
	B10 Superstructure						
1010	Floor Construction	N/A		—	—	—	7.3 %
1020	Roof Construction	Metal deck, open web steel joists, beams on columns		S.F. Roof	9.97	9.97	
	B20 Exterior Enclosure						
2010	Exterior Walls	Face brick with concrete block backup	*75% of wall*	S.F. Wall	35	19.77	19.3%
2020	Exterior Windows	Aluminum insulated glass	*10% of wall*	Each	815	1.91	
2030	Exterior Doors	Single aluminum and glass, overhead, hollow metal, low VOC paint	*15% of wall*	S.F. Door	2564	4.59	
	B30 Roofing						
3010	Roof Coverings	Single-ply TPO membrane, 60 mil, with flashing, R-20 insulation and roof edges		S.F. Roof	6.48	6.48	4.9%
3020	Roof Openings	Skylights, roof hatches		S.F. Roof	.15	.15	
C. INTERIORS							
1010	Partitions	Concrete block w/foamed-in insulation	*17 S.F. Floor/LF Part.*	S.F. Partition	10.22	6.01	12.7%
1020	Interior Doors	Single leaf hollow metal, low VOC paint	*500 S.F. Floor/Door*	Each	1133	2.27	
1030	Fittings	Toilet partitions		S.F. Floor	.49	.49	
2010	Stair Construction	N/A		—	—	—	
3010	Wall Finishes	Paint, low VOC		S.F. Surface	2.19	2.58	
3020	Floor Finishes	50% vinyl tile with recycled content, 50% paint, low VOC		S.F. Floor	2.58	2.58	
3030	Ceiling Finishes	Mineral board acoustic ceiling tiles, concealed grid, suspended	*50% of area*	S.F. Ceiling	3.29	3.29	
D. SERVICES							
	D10 Conveying						
1010	Elevators & Lifts	N/A		—	—	—	0.0 %
1020	Escalators & Moving Walks	N/A		—	—	—	
	D20 Plumbing						
2010	Plumbing Fixtures	Kitchen, toilet, low flow, auto sensor & service fixt., supply & drain.	*1 Fixt./375 SF Flr.*	Each	3893	10.38	10.0%
2020	Domestic Water Distribution	Water heater, tankless, on-demand, natural gas/propane		S.F. Floor	1.94	1.94	
2040	Rain Water Drainage	Roof drains		S.F. Roof	1.27	1.27	
	D30 HVAC						
3010	Energy Supply	N/A		—	—	—	17.4 %
3020	Heat Generating Systems	Included in D3050		—	—	—	
3040	Distribution Systems	Enthalpy heat recovery packages		Each	17,150	2.86	
3050	Terminal & Package Units	Multizone rooftop air conditioner, SEER 14		S.F. Floor	21	20.80	
3090	Other HVAC Sys. & Equipment	N/A		—	—	—	
	D40 Fire Protection						
4010	Sprinklers	Wet pipe sprinkler system		S.F. Floor	4.57	4.57	4.4%
4020	Standpipes	Standpipe, wet, Class III		S.F. Floor	1.43	1.43	
	D50 Electrical						
5010	Electrical Service/Distribution	200 ampere service, panel board and feeders		S.F. Floor	1.35	1.35	12.4%
5020	Lighting & Branch Wiring	LED fixt., daylt. dimming & ltg. on/off control, receptacles, switches, A.C. & misc. pwr.		S.F. Floor	9.51	9.51	
5030	Communications & Security	Addressable alarm systems		S.F. Floor	1.91	1.91	
5090	Other Electrical Systems	Energy monitoring systems		S.F. Floor	4.14	4.14	
E. EQUIPMENT & FURNISHINGS							
1010	Commercial Equipment	N/A		—	—	—	0.7 %
1020	Institutional Equipment	N/A		—	—	—	
1090	Other Equipment	Waste handling recycling tilt truck		Each	.92	.92	
2020	Moveable Furnishings	No smoking signage		Each	.04	.04	
F. SPECIAL CONSTRUCTION							
1020	Integrated Construction	N/A		—	—	—	0.0 %
1040	Special Facilities	N/A		—	—	—	
G. BUILDING SITEWORK	**N/A**						
					Sub-Total	135.95	**100%**
	CONTRACTOR FEES (General Requirements: 10%, Overhead: 5%, Profit: 10%)				*25%*	34	
	ARCHITECT FEES				*8%*	13.60	
					Total Building Cost	**183.55**	

Costs per square foot of floor area

Exterior Wall	S.F. Area	4000	6000	8000	10000	12000	14000	16000	18000	20000
	L.F. Perimeter	260	320	384	424	460	484	510	540	576
Vertical Redwood Siding	Wood Frame	154.80	144.10	138.95	134.75	131.75	129.20	127.30	126.00	125.00
Brick Veneer	Wood Frame	167.40	155.10	149.25	144.25	140.70	137.60	135.30	133.80	132.65
Aluminum Siding	Wood Frame	154.85	144.80	139.95	136.05	133.25	130.90	129.15	128.00	127.10
Brick on Block	Wood Truss	178.85	164.75	158.05	152.25	148.05	144.35	141.70	139.85	138.55
Limestone on Block	Wood Truss	200.15	182.15	173.75	165.95	160.45	155.55	151.95	149.50	147.75
Stucco on Block	Wood Truss	185.95	169.90	162.35	155.55	150.70	146.40	143.30	141.15	139.65
Perimeter Adj., Add or Deduct	Per 100 L.F.	12.50	8.30	6.25	5.00	4.20	3.55	3.10	2.75	2.55
Story Hgt. Adj., Add or Deduct	Per 1 Ft.	1.60	1.30	1.20	1.00	0.95	0.80	0.80	0.75	0.70
***For Basement**, add $ 31.40 per square foot of basement area*										

The above costs were calculated using the basic specifications shown on the facing page. These costs should be adjusted where necessary for design alternatives and owner's requirements. Reported completed project costs, for this type of structure, range from $ 122.40 to $ 376.25 per S.F.

Common additives

Description	Unit	$ Cost
Autopsy Table, standard	Ea.	11,300
Deluxe	Ea.	18,900
Directory Boards, plastic, glass covered, 30" x 20"	Ea.	555
36" x 48"	Ea.	1425
Aluminum, 24" x 18"	Ea.	600
36" x 24"	Ea.	715
48" x 32"	Ea.	1000
Mortuary Refrigerator, end operated		
Two capacity	Ea.	10,800
Six capacity	Ea.	16,300
Planters, precast concrete, 48" diam., 24" high	Ea.	695
7" diam., 36" high	Ea.	1800
Fiberglass, 36" diam., 24" high	Ea.	775
60" diam., 24" high	Ea.	1300

Description	Unit	$ Cost
Commissioning Fees, sustainable commercial construction	S.F.	0.24 - 3.04
Energy Modelling Fees, commercial buildings to 10,000 SF	Ea.	11,000
Greater than 10,000 SF add	S.F.	0.04
Green Bldg Cert Fees for comm construction project reg	Project	900
Photovoltaic Pwr Sys, grid connected, 20 kW (~2400 SF), roof	Ea.	241,700
Green Roofs, 6" soil depth, w/treated wd edging & sedum mats	S.F.	11.27
10" Soil depth, with treated wood edging & sedum mats	S.F.	12.84
Greywater Recovery Systems, prepackaged comm, 3060 gal.	Ea.	58,085
4590 gal.	Ea.	72,815
Rainwater Harvest Sys, prepckged comm, 10,000 gal, sys contrller	Ea.	31,175
20,000 gal. w/system controller	Ea.	50,100
30,000 gal. w/system controller	Ea.	80,225
Solar Domestic HW, closed loop, add-on sys, ext heat exchanger	Ea.	9750
Drainback, hot water system, 120 gal tank	Ea.	12,425
Draindown, hot water system, 120 gal tank	Ea.	12,700

Model costs calculated for a 1 story building with 12′ story height and 10,000 square feet of floor area

G Funeral Home

				Unit	Unit Cost	Cost Per S.F.	% Of Sub-Total
A. SUBSTRUCTURE							
1010	Standard Foundations	Poured concrete; strip and spread footings		S.F. Ground	1.72	1.72	10.6%
1020	Special Foundations	N/A		—	—	—	
1030	Slab on Grade	4″ reinforced concrete with recycled plastic vapor barrier and granular base		S.F. Slab	5.11	5.11	
2010	Basement Excavation	Site preparation for slab and trench for foundation wall and footing		S.F. Ground	.29	.29	
2020	Basement Walls	4′ foundation wall		L.F. Wall	72	3.36	
B. SHELL							
	B10 Superstructure						
1010	Floor Construction	N/A		—	—	—	5.0 %
1020	Roof Construction	Plywood on wood truss		S.F. Roof	4.99	4.99	
	B20 Exterior Enclosure						
2010	Exterior Walls	1″ x 4″ vertical T & G redwood siding on 2x6 wood studs with insul.	*90% of wall*	S.F. Wall	13.54	6.20	9.5%
2020	Exterior Windows	Double hung wood	*10% of wall*	Each	483	2.05	
2030	Exterior Doors	Wood swinging double doors, single leaf hollow metal, low VOC paint		Each	1933	1.16	
	B30 Roofing						
3010	Roof Coverings	Single ply membrane, TPO, 45 mil, fully adhered; polyisocyanurate sheets		S.F. Roof	5.96	5.96	6.0%
3020	Roof Openings	N/A		—	—	—	
C. INTERIORS							
1010	Partitions	Gypsum board on wood studs with sound deadening board	*15 SF Flr./LF Part.*	S.F. Partition	9.09	4.85	24.7%
1020	Interior Doors	Single leaf wood, low VOC paint	*150 S.F. Floor/Door*	Each	593	3.95	
1030	Fittings	N/A		—	—	—	
2010	Stair Construction	N/A		—	—	—	
3010	Wall Finishes	50% wallpaper, 25% wood paneling, 25% paint (low VOC)		S.F. Surface	3.06	3.26	
3020	Floor Finishes	70% carpet tile, 30% ceramic tile		S.F. Floor	6.87	6.87	
3030	Ceiling Finishes	Fiberglass board on exposed grid, suspended		S.F. Ceiling	5.51	5.51	
D. SERVICES							
	D10 Conveying						
1010	Elevators & Lifts	N/A		—	—	—	0.0 %
1020	Escalators & Moving Walks	N/A		—	—	—	
	D20 Plumbing						
2010	Plumbing Fixtures	Toilets, low flow, auto sensor, urinals & service fixt., supply & drain.	*1 Fixt./770 SF Flr.*	Each	3126	4.06	5.5%
2020	Domestic Water Distribution	Electric water heater, point-of-use		S.F. Floor	.43	.43	
2040	Rain Water Drainage	Roof drain		S.F. Roof	.90	.90	
	D30 HVAC						
3010	Energy Supply	N/A		—	—	—	19.1 %
3020	Heat Generating Systems	Included in D3050		—	—	—	
3040	Distribution Systems	Enthalpy heat recovery packages		Each	36,450	3.65	
3050	Terminal & Package Units	Multizone rooftop air conditioner, SEER 14		S.F. Floor	15.24	15.24	
3090	Other HVAC Sys. & Equipment	N/A		—	—	—	
	D40 Fire Protection						
4010	Sprinklers	Wet pipe sprinkler system		S.F. Floor	3.28	3.28	4.2%
4020	Standpipes	Standpipe, wet, Class III		S.F. Floor	.86	.86	
	D50 Electrical						
5010	Electrical Service/Distribution	400 ampere service, panel board and feeders		S.F. Floor	1.54	1.54	14.8%
5020	Lighting & Branch Wiring	LED light fixtures, daylt. dimming control & ltg. on/off sys., recept., switches, & A.C. power		S.F. Floor	9.65	9.65	
5030	Communications & Security	Addressable alarm systems		S.F. Floor	1.48	1.48	
5090	Other Electrical Systems	Emergency generator, 15 kW and energy monitoring systems		S.F. Floor	1.92	1.92	
E. EQUIPMENT & FURNISHINGS							
1010	Commercial Equipment	N/A		—	—	—	0.6 %
1020	Institutional Equipment	N/A		—	—	—	
1090	Other Equipment	Waste handling recycling tilt truck		Each	.56	.56	
2020	Moveable Furnishings	No smoking signage		Each	.04	.04	
F. SPECIAL CONSTRUCTION							
1020	Integrated Construction	N/A		—	—	—	0.0 %
1040	Special Facilities	N/A		—	—	—	
G. BUILDING SITEWORK	**N/A**						
					Sub-Total	98.89	**100%**
	CONTRACTOR FEES (General Requirements: 10%, Overhead: 5%, Profit: 10%)				25%	24.74	
	ARCHITECT FEES				9%	11.12	
					Total Building Cost	**134.75**	

Costs per square foot of floor area

Exterior Wall	S.F. Area	7000	10000	13000	16000	19000	22000	25000	28000	31000
	L.F. Perimeter	240	300	336	386	411	435	472	510	524
Face Brick with Concrete Block Back-up	R/Conc. Frame	194.35	184.25	175.65	171.85	167.05	163.40	161.50	160.15	157.70
	Steel Frame	197.80	187.70	179.10	175.30	170.50	166.80	164.90	163.55	161.10
Limestone with Concrete Block	R/Conc. Frame	223.25	209.10	196.95	191.55	184.65	179.50	176.80	174.85	171.35
	Steel Frame	224.15	210.75	198.95	193.85	187.10	182.05	179.45	177.55	174.05
Precast Concrete Panels	R/Conc. Frame	212.80	199.95	189.05	184.20	178.05	173.40	171.05	169.30	166.20
	Steel Frame	215.35	202.55	191.60	186.75	180.60	176.00	173.60	171.85	168.70
Perimeter Adj., Add or Deduct	Per 100 L.F.	24.00	16.80	12.95	10.45	8.80	7.65	6.70	5.95	5.40
Story Hgt. Adj., Add or Deduct	Per 1 Ft.	3.70	3.20	2.80	2.55	2.25	2.15	2.00	1.90	1.75

***For Basement**, add $ 46.75 per square foot of basement area*

The above costs were calculated using the basic specifications shown on the facing page. These costs should be adjusted where necessary for design alternatives and owner's requirements. Reported completed project costs, for this type of structure, range from $ 117.10 to $ 300.50 per S.F.

Common additives

Description	Unit	$ Cost
Carrels Hardwood	Ea.	925 - 2375
Closed Circuit Surveillance, one station, camera and monitor	Ea.	1975
For additional camera stations, add	Ea.	1075
Elevators, hydraulic passenger, 2 stops, 1500# capacity	Ea.	63,900
2500# capacity	Ea.	66,900
3500# capacity	Ea.	71,900
Library Furnishings, bookshelf, 90" high, 10" shelf double face	L.F.	280
Single face	L.F.	176
Charging desk, built-in with counter		
Plastic laminated top	L.F.	410
Reading table, laminated		
Top 60" x 36"	Ea.	775

Description	Unit	$ Cost
Commissioning Fees, sustainable institutional construction	S.F.	0.58 - 2.47
Energy Modelling Fees, commercial buildings to 10,000 SF	Ea.	11,000
Greater than 10,000 SF add	S.F.	0.04
Green Bldg Cert Fees for comm construction project reg	Project	900
Photovoltaic Pwr Sys, grid connected, 20 kW (~2400 SF), roof	Ea.	241,700
Green Roofs, 6" soil depth, w/treated wd edging & sedum mats	S.F.	11.27
10" Soil depth, with treated wood edging & sedum mats	S.F.	12.84
Greywater Recovery Systems, prepackaged comm, 3060 gal.	Ea.	58,085
4590 gal.	Ea.	72,815
Rainwater Harvest Sys, prepckged comm, 10,000 gal, sys contrller	Ea.	31,175
20,000 gal. w/system controller	Ea.	50,100
30,000 gal. w/system controller	Ea.	80,225
Solar Domestic HW, closed loop, add-on sys, ext heat exchanger	Ea.	9750
Drainback, hot water system, 120 gal tank	Ea.	12,425
Draindown, hot water system, 120 gal tank	Ea.	12,700

Model costs calculated for a 2 story building with 14′ story height and 22,000 square feet of floor area

G Library

			Unit	Unit Cost	Cost Per S.F.	% Of Sub-Total
A. SUBSTRUCTURE						
1010	Standard Foundations	Poured concrete; strip and spread footings	S.F. Ground	3.64	1.82	5.1%
1020	Special Foundations	N/A	—	—		
1030	Slab on Grade	4″ reinforced concrete with recycled vapor barrier and granular base	S.F. Slab	5.11	2.56	
2010	Basement Excavation	Site preparation for slab and trench for foundation wall and footing	S.F. Ground	.29	.15	
2020	Basement Walls	4′ foundation wall	L.F. Wall	81	1.59	
B. SHELL						
	B10 Superstructure					
1010	Floor Construction	Concrete waffle slab	S.F. Floor	25	12.38	17.7%
1020	Roof Construction	Concrete waffle slab	S.F. Roof	18.06	9.03	
	B20 Exterior Enclosure					
2010	Exterior Walls	Face brick with concrete block backup *90% of wall*	S.F. Wall	33	16.61	16.5%
2020	Exterior Windows	Window wall *10% of wall*	Each	52	2.87	
2030	Exterior Doors	Double aluminum and glass, single leaf hollow metal, low VOC paint	Each	5525	.50	
	B30 Roofing					
3010	Roof Coverings	Single-ply TPO membrane, 60 mil, heat welded w/R-20 insulation	S.F. Roof	5.70	2.85	2.4%
3020	Roof Openings	Roof hatches	S.F. Roof	.08	.04	
C. INTERIORS						
1010	Partitions	Gypsum board on metal studs w/sound attenuation insul. *30 SF Flr./LF Part.*	S.F. Partition	12.60	5.04	16.5%
1020	Interior Doors	Single leaf wood, low VOC paint *300 S.F. Floor/Door*	Each	593	1.98	
1030	Fittings	N/A	—	—	—	
2010	Stair Construction	Concrete filled metal pan	Flight	8950	.81	
3010	Wall Finishes	Paint, low VOC	S.F. Surface	1.81	1.45	
3020	Floor Finishes	50% carpet tile, 50% vinyl composition tile, recycled content	S.F. Floor	4.16	4.16	
3030	Ceiling Finishes	Mineral fiber on concealed zee bars	S.F. Ceiling	6.57	6.57	
D. SERVICES						
	D10 Conveying					
1010	Elevators & Lifts	One hydraulic passenger elevator	Each	84,480	3.84	3.2%
1020	Escalators & Moving Walks	N/A	—	—	—	
	D20 Plumbing					
2010	Plumbing Fixtures	Toilet, low flow, auto sensor & service fixt., supply & drain. *1 Fixt./1835 SF.Flr.*	Each	7579	4.13	4.3%
2020	Domestic Water Distribution	Tankless, on demand water heater, gas/propane	S.F. Floor	.52	.52	
2040	Rain Water Drainage	Roof drains	S.F. Roof	1.14	.57	
	D30 HVAC					
3010	Energy Supply	N/A	—	—	—	17.8 %
3020	Heat Generating Systems	Included in D3050	—	—	—	
3040	Distribution Systems	Enthalpy heat recovery packages	Each	29,975	2.73	
3050	Terminal & Package Units	Multizone rooftop air conditioner, SEER 14	S.F. Floor	18.80	18.80	
3090	Other HVAC Sys. & Equipment	N/A	—	—	—	
	D40 Fire Protection					
4010	Sprinklers	Wet pipe sprinkler system	S.F. Floor	2.79	2.79	3.1%
4020	Standpipes	Standpipe	S.F. Floor	.96	.96	
	D50 Electrical					
5010	Electrical Service/Distribution	400 ampere service, panel board and feeders	S.F. Floor	.53	.53	13.2%
5020	Lighting & Branch Wiring	LED fixtures, daylt. dim., ltg on/off, recept., switches, and A.C. power	S.F. Floor	12.17	12.17	
5030	Communications & Security	Addressable alarm systems, internet wiring, and emergency lighting	S.F. Floor	1.97	1.97	
5090	Other Electrical Systems	Emergency generator, 7.5 kW, UPS, energy monitoring systems	S.F. Floor	1.36	1.36	
E. EQUIPMENT & FURNISHINGS						
1010	Commercial Equipment	N/A	—	—	—	0.2 %
1020	Institutional Equipment	N/A	—	—	—	
1090	Other Equipment	Waste handling recycling tilt truck	Each	.26	.26	
2020	Moveable Furnishings	No smoking signage	Each			
F. SPECIAL CONSTRUCTION						
1020	Integrated Construction	N/A	—	—	—	0.0 %
1040	Special Facilities	N/A	—	—	—	
G. BUILDING SITEWORK		**N/A**				
				Sub-Total	121.04	**100%**
	CONTRACTOR FEES (General Requirements: 10%, Overhead: 5%, Profit: 10%)			*25%*	30.26	
	ARCHITECT FEES			*8%*	12.10	
				Total Building Cost	**163.40**	

Costs per square foot of floor area

Exterior Wall	S.F. Area	4000	5500	7000	8500	10000	11500	13000	14500	16000
	L.F. Perimeter	280	320	380	440	453	503	510	522	560
Face Brick with Concrete Block Back-up	Steel Joists	233.35	222.95	218.90	216.35	211.40	209.85	206.40	203.95	203.00
	Wood Truss	236.15	225.75	221.70	219.20	214.20	212.65	209.20	206.80	205.85
Stucco on Concrete Block	Steel Joists	240.85	229.20	224.75	221.95	216.25	214.55	210.60	207.80	206.80
	Wood Truss	247.80	236.05	231.70	228.85	223.10	221.45	217.50	214.75	213.75
Brick Veneer	Wood Truss	230.90	221.65	218.05	215.75	211.40	210.05	207.05	204.90	204.10
Wood Siding	Wood Frame	224.25	216.55	213.45	211.50	207.95	206.75	204.40	202.65	202.00
Perimeter Adj., Add or Deduct	Per 100 L.F.	17.20	12.50	9.90	8.15	6.85	6.05	5.30	4.80	4.35
Story Hgt. Adj., Add or Deduct	Per 1 Ft.	3.55	2.90	2.80	2.65	2.25	2.20	2.00	1.80	1.80

For Basement, add $30.40 per square foot of basement area

The above costs were calculated using the basic specifications shown on the facing page. These costs should be adjusted where necessary for design alternatives and owner's requirements. Reported completed project costs, for this type of structure, range from $98.65 to $253.80 per S.F.

Common additives

Description	Unit	$ Cost
Cabinets, hospital, base, laminated plastic	L.F.	400
Counter top, laminated plastic	L.F.	77.50
Nurses station, door type , laminated plastic	L.F.	450
Wall cabinets, laminated plastic	L.F.	294
Directory Boards, plastic, glass covered, 30″ x 20″	Ea.	555
36″ x 48″	Ea.	1425
Aluminum, 36″ x 24″	Ea.	715
48″ x 32″	Ea.	1000
Heat Therapy Unit, humidified, 26″ x 78″ x 28″	Ea.	3450
Tables, examining, vinyl top, with base cabinets	Ea.	1600 - 6675
Utensil Washer, Sanitizer	Ea.	9600
X-Ray, Mobile	Ea.	18,500 - 85,500

Description	Unit	$ Cost
Commissioning Fees, sustainable commercial construction	S.F.	0.24 - 3.04
Energy Modelling Fees, commercial buildings to 10,000 SF	Ea.	11,000
Greater than 10,000 SF add	S.F.	0.04
Green Bldg Cert Fees for comm construction project reg	Project	900
Photovoltaic Pwr Sys, grid connected, 20 kW (~2400 SF), roof	Ea.	241,700
Green Roofs, 6″ soil depth, w/treated wd edging & sedum mats	S.F.	11.27
10″ Soil depth, with treated wood edging & sedum mats	S.F.	12.84
Greywater Recovery Systems, prepackaged comm, 3060 gal.	Ea.	58,085
4590 gal.	Ea.	72,815
Rainwater Harvest Sys, prepckged comm, 10,000 gal, sys contrller	Ea.	31,175
20,000 gal. w/system controller	Ea.	50,100
30,000 gal. w/system controller	Ea.	80,225
Solar Domestic HW, closed loop, add-on sys, ext heat exchanger	Ea.	9750
Drainback, hot water system, 120 gal tank	Ea.	12,425
Draindown, hot water system, 120 gal tank	Ea.	12,700

Important: See the Reference Section for Location Factors.

Model costs calculated for a 1 story building with 10′ story height and 7,000 square feet of floor area

G Medical Office, 1 Story

				Unit	Unit Cost	Cost Per S.F.	% Of Sub-Total
A. SUBSTRUCTURE							
1010	Standard Foundations	Poured concrete; strip and spread footings		S.F. Ground	2	2	7.2%
1020	Special Foundations	N/A		—	—	—	
1030	Slab on Grade	4″ reinforced concrete with recycled vapor barrier and granular base		S.F. Slab	5.11	5.11	
2010	Basement Excavation	Site preparation for slab and trench for foundation wall and footing		S.F. Ground	.29	.29	
2020	Basement Walls	4′ foundation wall		L.F. Wall	81	4.37	
B. SHELL							
	B10 Superstructure						
1010	Floor Construction	N/A		—	—	—	4.3 %
1020	Roof Construction	Plywood on wood trusses		S.F. Roof	6.94	6.94	
	B20 Exterior Enclosure						
2010	Exterior Walls	Face brick with concrete block backup	*70% of wall*	S.F. Wall	33	12.67	12.8%
2020	Exterior Windows	Wood double hung	*30% of wall*	Each	603	5.77	
2030	Exterior Doors	Aluminum and glass doors and entrance with transoms		Each	1426	2.44	
	B30 Roofing						
3010	Roof Coverings	Asphalt shingles with flashing (Pitched); rigid fiber glass insulation, gutters		S.F. Roof	8.10	8.10	5.0%
3020	Roof Openings	N/A		—	—	—	
C. INTERIORS							
1010	Partitions	Gypsum bd. & sound deadening bd. on wood studs w/insul.	*6 SF Flr./LF Part.*	S.F. Partition	8.82	11.76	21.5%
1020	Interior Doors	Single leaf wood, low VOC paint	*60 S.F. Floor/Door*	Each	593	9.88	
1030	Fittings	N/A		—	—	—	
2010	Stair Construction	N/A		—	—	—	
3010	Wall Finishes	50% paint, low VOC, 50% vinyl wall covering		S.F. Surface	1.31	3.50	
3020	Floor Finishes	50% carpet tile, 50% vinyl composition tile, recycled content		S.F. Floor	3.85	3.85	
3030	Ceiling Finishes	Mineral fiber tile on concealed zee bars		S.F. Ceiling	6.02	6.02	
D. SERVICES							
	D10 Conveying						
1010	Elevators & Lifts	N/A		—	—	—	0.0 %
1020	Escalators & Moving Walks	N/A		—	—	—	
	D20 Plumbing						
2010	Plumbing Fixtures	Toilet, low flow, auto sensor, exam room & service fixt., supply & drain.	*1 Fixt./195 SF Flr.*	Each	4524	23.20	16.4%
2020	Domestic Water Distribution	Gas fired tankless water heater		S.F. Floor	2.54	2.54	
2040	Rain Water Drainage	Roof drains		S.F. Roof	1	1	
	D30 HVAC						
3010	Energy Supply	N/A		—	—	—	9.7 %
3020	Heat Generating Systems	Included in D3050		—	—	—	
3040	Distribution Systems	Enthalpy heat recovery packages		Each	17,150	2.45	
3050	Terminal & Package Units	Multizone rooftop air conditioner, SEER 14		S.F. Floor	13.41	13.41	
3090	Other HVAC Sys. & Equipment	N/A		—	—	—	
	D40 Fire Protection						
4010	Sprinklers	Wet pipe sprinkler system		S.F. Floor	4.57	4.57	3.6%
4020	Standpipes	Standpipe		S.F. Floor	1.22	1.22	
	D50 Electrical						
5010	Electrical Service/Distribution	200 ampere service, panel board and feeders		S.F. Floor	1.27	1.27	12.9%
5020	Lighting & Branch Wiring	LED fixtures, daylt. dim., ltg. on/off, recept., switches, and A.C. power		S.F. Floor	9.31	9.31	
5030	Communications & Security	Alarm systems, internet & phone wiring, intercom system, & emerg. ltg.		S.F. Floor	7.39	7.39	
5090	Other Electrical Systems	Emergency generator, 7.5 kW, and energy monitoring systems		S.F. Floor	3.08	3.08	
E. EQUIPMENT & FURNISHINGS							
1010	Commercial Equipment	N/A		—	—	—	6.5 %
1020	Institutional Equipment	Exam room casework and countertops		S.F. Floor	9.72	9.72	
1090	Other Equipment	Waste handling recycling tilt truck		Each	.79	.79	
2020	Moveable Furnishings	No smoking signage		Each	.09	.09	
F. SPECIAL CONSTRUCTION							
1020	Integrated Construction	N/A		—	—	—	0.0 %
1040	Special Facilities	N/A		—	—	—	
G. BUILDING SITEWORK	**N/A**						
					Sub-Total	162.74	**100%**
	CONTRACTOR FEES (General Requirements: 10%, Overhead: 5%, Profit: 10%)				25%	40.65	
	ARCHITECT FEES				9%	18.31	
					Total Building Cost	**221.70**	

Costs per square foot of floor area

Exterior Wall	S.F. Area	4000	5500	7000	8500	10000	11500	13000	14500	16000
	L.F. Perimeter	180	210	240	270	286	311	336	361	386
Face Brick with Concrete Block Back-up	Steel Joists	268.65	257.45	251.15	247.05	242.45	239.95	238.00	236.60	235.35
	Wood Joists	273.75	262.65	256.35	252.20	247.60	245.15	243.20	241.75	240.55
Stucco on Concrete Block	Steel Joists	253.10	244.25	**239.25**	235.95	232.50	230.50	229.00	227.85	226.85
	Wood Joists	263.20	253.60	248.25	244.70	240.90	238.75	237.15	235.95	234.85
Brick Veneer	Wood Frame	263.15	253.60	248.20	244.70	240.85	238.75	237.10	235.90	234.85
Wood Siding	Wood Frame	255.85	247.45	242.65	239.55	236.20	234.35	232.90	231.85	230.90
Perimeter Adj., Add or Deduct	Per 100 L.F.	31.15	22.75	17.80	14.65	12.50	10.90	9.65	8.60	7.80
Story Hgt. Adj., Add or Deduct	Per 1 Ft.	4.60	3.95	3.55	3.30	2.95	2.80	2.70	2.60	2.50
__For Basement__, add $ 33.80 per square foot of basement area										

The above costs were calculated using the basic specifications shown on the facing page. These costs should be adjusted where necessary for design alternatives and owner's requirements. Reported completed project costs, for this type of structure, range from $ 101.30 to $ 346.00 per S.F.

Common additives

Description	Unit	$ Cost
Cabinets, hospital, base, laminated plastic	L.F.	400
Counter top, laminated plastic	L.F.	77.50
Nurses station, door type , laminated plastic	L.F.	450
Wall cabinets, laminated plastic	L.F.	294
Elevators, hydraulic passenger, 2 stops, 2500# capacity	Ea.	66,900
3500# capacity	Ea.	71,900
Directory Boards, plastic, glass covered, 30" x 20"	Ea.	555
36" x 48"	Ea.	1425
Aluminum, 36" x 24"	Ea.	715
48" x 32"	Ea.	1000
Heat Therapy Unit, humidified, 26" x 7.8" x 28"	Ea.	3450
Tables, examining, vinyl top, with base cabinets	Ea.	1600 - 6675
Utensil Washer, Sanitizer	Ea.	9600
X-Ray, Mobile	Ea.	18,500 - 85,500

Description	Unit	$ Cost
Commissioning Fees, sustainable commercial construction	S.F.	0.24 - 3.04
Energy Modelling Fees, commercial buildings to 10,000 SF	Ea.	11,000
Greater than 10,000 SF add	S.F.	0.04
Green Bldg Cert Fees for comm construction project reg	Project	900
Photovoltaic Pwr Sys, grid connected, 20 kW (~2400 SF), roof	Ea.	241,700
Green Roofs, 6" soil depth, w/treated wd edging & sedum mats	S.F.	11.27
10" Soil depth, with treated wood edging & sedum mats	S.F.	12.84
Greywater Recovery Systems, prepackaged comm, 3060 gal.	Ea.	58,085
4590 gal.	Ea.	72,815
Rainwater Harvest Sys, prepckged comm, 10,000 gal, sys contrller	Ea.	31,175
20,000 gal. w/system controller	Ea.	50,100
30,000 gal. w/system controller	Ea.	80,225
Solar Domestic HW, closed loop, add-on sys, ext heat exchanger	Ea.	9750
Drainback, hot water system, 120 gal tank	Ea.	12,425
Draindown, hot water system, 120 gal tank	Ea.	12,700

Model costs calculated for a 2 story building with 10′ story height and 7,000 square feet of floor area

G Medical Office, 2 Story

				Unit	Unit Cost	Cost Per S.F.	% Of Sub-Total
A. SUBSTRUCTURE							
1010	Standard Foundations	Poured concrete; strip and spread footings		S.F. Ground	3.20	1.60	4.0%
1020	Special Foundations	N/A		—	—	—	
1030	Slab on Grade	4″ reinforced concrete with vapor barrier and granular base		S.F. Slab	5.11	2.56	
2010	Basement Excavation	Site preparation for slab and trench for foundation wall and footing		S.F. Ground	.16	.08	
2020	Basement Walls	4′ foundation wall		L.F. Wall	81	2.76	
B. SHELL							
	B10 Superstructure						
1010	Floor Construction	Open web steel joists, slab form, concrete, columns		S.F. Floor	12.96	6.48	5.4%
1020	Roof Construction	Metal deck, open web steel joists, beams, columns		S.F. Roof	5.96	2.98	
	B20 Exterior Enclosure						
2010	Exterior Walls	Stucco on concrete block		S.F. Wall	17.54	8.42	9.6%
2020	Exterior Windows	Outward projecting metal	*30% of wall*	Each	495	6.79	
2030	Exterior Doors	Aluminum and glass doors with transoms		Each	5525	1.58	
	B30 Roofing						
3010	Roof Coverings	Built-up tar and gravel with flashing; perlite/EPS composite insulation		S.F. Roof	6.80	3.40	2.0%
3020	Roof Openings	Roof hatches		S.F. Roof	.26	.13	
C. INTERIORS							
1010	Partitions	Gypsum bd. & acous. insul. on metal studs	*6 SF Floor/LF Part.*	S.F. Partition	7.34	9.78	21.1%
1020	Interior Doors	Single leaf wood	*60 S.F. Floor/Door*	Each	593	9.88	
1030	Fittings	N/A		—	—	—	
2010	Stair Construction	Concrete filled metal pan		Flight	14,075	4.03	
3010	Wall Finishes	45% paint, 50% vinyl wall coating, 5% ceramic tile		S.F. Surface	1.31	3.50	
3020	Floor Finishes	50% carpet, 50% vinyl composition tile		S.F. Floor	3.85	3.85	
3030	Ceiling Finishes	Mineral fiber tile on concealed zee bars		S.F. Ceiling	6.02	6.02	
D. SERVICES							
	D10 Conveying						
1010	Elevators & Lifts	One hydraulic hospital elevator		Each	106,050	15.15	8.6%
1020	Escalators & Moving Walks	N/A		—	—	—	
	D20 Plumbing						
2010	Plumbing Fixtures	Toilet, exam room & service fixt., supply & drain.	*1 Fixt./160 SF Flr.*	Each	4202	26.26	17.0%
2020	Domestic Water Distribution	Gas fired water heater		S.F. Floor	2.54	2.54	
2040	Rain Water Drainage	Roof drains		S.F. Roof	2.16	1.08	
	D30 HVAC						
3010	Energy Supply	N/A		—	—	—	9.0 %
3020	Heat Generating Systems	Included in D3050		—	—	—	
3040	Distribution Systems	Enthalpy heat recovery packages		Each	17,150	2.45	
3050	Terminal & Package Units	Multizone rooftop air conditioner, SEER 14		S.F. Floor	13.41	13.41	
3090	Other HVAC Sys. & Equipment	N/A		—	—	—	
	D40 Fire Protection						
4010	Sprinklers	Wet pipe sprinkler system		S.F. Floor	3.54	3.54	3.1%
4020	Standpipes	Standpipe		S.F. Floor	1.85	1.85	
	D50 Electrical						
5010	Electrical Service/Distribution	400 ampere service, panel board and feeders		S.F. Floor	1.58	1.58	14.4%
5020	Lighting & Branch Wiring	LED fixtures, receptacles, switches, A.C. and misc. power		S.F. Floor	13.11	13.11	
5030	Communications & Security	Alarm system, internet & phone wiring, intercom system, & emergency ltg.		S.F. Floor	7.39	7.39	
5090	Other Electrical Systems	Emergency generator, 7.5 kW		S.F. Floor	3.17	3.17	
E. EQUIPMENT & FURNISHINGS							
1010	Commercial Equipment	N/A		—	—	—	5.8 %
1020	Institutional Equipment	Exam room casework and contertops		S.F. Floor	9.72	9.72	
1090	Other Equipment	Waste handling recycling tilt truck		Each	.48	.48	
2020	Moveable Furnishings	No smoking signage		Each	.02	.02	
F. SPECIAL CONSTRUCTION							
1020	Integrated Construction	N/A		—	—	—	0.0 %
1040	Special Facilities	N/A		—	—	—	
G. BUILDING SITEWORK		**N/A**					
					Sub-Total	175.59	**100%**
	CONTRACTOR FEES (General Requirements: 10%, Overhead: 5%, Profit: 10%)				25%	43.91	
	ARCHITECT FEES				9%	19.75	
					Total Building Cost	**239.25**	

G.420 G Motel, 1 Story

Costs per square foot of floor area

Exterior Wall	S.F. Area	2000	3000	4000	6000	8000	10000	12000	14000	16000
	L.F. Perimeter	240	260	280	380	480	560	580	660	740
Brick Veneer	Wood Frame	209.70	184.25	171.65	164.70	**161.15**	158.00	152.95	151.85	150.95
Aluminum Siding	Wood Frame	194.25	173.05	162.55	156.40	153.30	150.65	146.60	145.65	144.85
Wood Siding	Wood Frame	193.85	172.80	162.25	156.20	153.10	150.45	146.45	145.45	144.75
Wood Shingles	Wood Frame	196.95	175.00	164.05	157.80	154.65	151.80	147.65	146.60	145.85
Precast Concrete Block	Wood Truss	196.60	174.80	163.90	157.65	154.50	151.75	147.55	146.55	145.80
Brick on Concrete Block	Wood Truss	223.75	194.40	179.75	172.00	168.10	164.40	158.50	157.15	156.25
Perimeter Adj., Add or Deduct	Per 100 L.F.	28.60	19.15	14.25	9.55	7.20	5.70	4.75	4.05	3.65
Story Hgt. Adj., Add or Deduct	Per 1 Ft.	4.55	3.35	2.60	2.40	2.30	2.10	1.85	1.80	1.80

For Basement, add $23.20 per square foot of basement area

The above costs were calculated using the basic specifications shown on the facing page. These costs should be adjusted where necessary for design alternatives and owner's requirements. Reported completed project costs, for this type of structure, range from $81.75 to $387.00 per S.F.

Common additives

Description	Unit	$ Cost
Closed Circuit Surveillance, one station, camera & monitor	Ea.	1975
For additional camera stations, add	Ea.	1075
Laundry Equipment, dryer, gas, 30 lb. capacity	Ea.	3900
Washer, commercial	Ea.	1575
Sauna, prefabricated, complete, 6′ x 9′	Ea.	10,100
8′ x 8′	Ea.	11,600
Swimming Pools, Complete, gunite	S.F.	97 - 121
TV Antenna, Master system, 12 outlet	Outlet	249
30 outlet	Outlet	217
100 outlet	Outlet	207

Description	Unit	$ Cost
Commissioning Fees, sustainable commercial construction	S.F.	0.24 - 3.04
Energy Modelling Fees, commercial buildings to 10,000 SF	Ea.	11,000
Greater than 10,000 SF add	S.F.	0.04
Green Bldg Cert Fees for comm construction project reg	Project	900
Photovoltaic Pwr Sys, grid connected, 20 kW (~2400 SF), roof	Ea.	241,700
Green Roofs, 6″ soil depth, w/treated wd edging & sedum mats	S.F.	11.27
10″ Soil depth, with treated wood edging & sedum mats	S.F.	12.84
Greywater Recovery Systems, prepackaged comm, 3060 gal.	Ea.	58,085
4590 gal.	Ea.	72,815
Rainwater Harvest Sys, prepckged comm, 10,000 gal, sys contrller	Ea.	31,175
20,000 gal. w/system controller	Ea.	50,100
30,000 gal. w/system controller	Ea.	80,225
Solar Domestic HW, closed loop, add-on sys, ext heat exchanger	Ea.	9750
Drainback, hot water system, 120 gal tank	Ea.	12,425
Draindown, hot water system, 120 gal tank	Ea.	12,700

Important: See the Reference Section for Location Factors.

Model costs calculated for a 1 story building with 9′ story height and 8,000 square feet of floor area

G Motel, 1 Story

				Unit	Unit Cost	Cost Per S.F.	% Of Sub-Total
A. SUBSTRUCTURE							
1010	Standard Foundations	Poured concrete; strip and spread footings		S.F. Ground	3.04	3.04	12.5%
1020	Special Foundations	N/A		—	—	—	
1030	Slab on Grade	4″ reinforced concrete with recycled vapor barrier and granular base		S.F. Slab	5.11	5.11	
2010	Basement Excavation	Site preparation for slab and trench for foundation wall and footing		S.F. Ground	.29	.29	
2020	Basement Walls	4′ foundation wall		L.F. Wall	81	6.61	
B. SHELL							
	B10 Superstructure						
1010	Floor Construction	N/A		—	—	—	5.8 %
1020	Roof Construction	Plywood on wood trusses		S.F. Roof	6.94	6.94	
	B20 Exterior Enclosure						
2010	Exterior Walls	Face brick on wood studs with sheathing, insulation and paper	*80% of wall*	S.F. Wall	23	9.89	14.6%
2020	Exterior Windows	Wood double hung	*20% of wall*	Each	483	3.73	
2030	Exterior Doors	Wood solid core		Each	1426	3.92	
	B30 Roofing						
3010	Roof Coverings	Asphalt strip shingles, 210-235 lbs/Sq., R-10 insul. & weather barrier-recycled		S.F. Roof	7.93	7.93	6.6%
3020	Roof Openings	N/A		—	—	—	
C. INTERIORS							
1010	Partitions	Gypsum bd. and sound deadening bd. on wood studs	*9 SF Flr./LF Part.*	S.F. Partition	9.45	8.40	18.5%
1020	Interior Doors	Single leaf hollow core wood, low VOC paint	*300 S.F. Floor/Door*	Each	510	1.70	
1030	Fittings	N/A		—	—	—	
2010	Stair Construction	N/A		—	—	—	
3010	Wall Finishes	90% paint, low VOC, 10% ceramic tile		S.F. Surface	1.50	2.66	
3020	Floor Finishes	85% carpet tile, 15% ceramic tile		S.F. Floor	5.31	5.31	
3030	Ceiling Finishes	Painted gypsum board on furring, low VOC paint		S.F. Ceiling	4.18	4.18	
D. SERVICES							
	D10 Conveying						
1010	Elevators & Lifts	N/A		—	—	—	0.0 %
1020	Escalators & Moving Walks	N/A		—	—	—	
	D20 Plumbing						
2010	Plumbing Fixtures	Toilet, low flow, auto sensor, & service fixt., supply & drainage	*1 Fixt./90 SF Flr.*	Each	1997	22.19	19.9%
2020	Domestic Water Distribution	Gas fired tankless water heater		S.F. Floor	.37	.37	
2040	Rain Water Drainage	Roof Drains		S.F. Roof	1.47	1.47	
	D30 HVAC						
3010	Energy Supply	N/A		—	—	—	3.1 %
3020	Heat Generating Systems	Included in D3050		—	—	—	
3030	Cooling Generating Systems	N/A		—	—	—	
3050	Terminal & Package Units	Through the wall electric heating and cooling units		S.F. Floor	3.72	3.72	
3090	Other HVAC Sys. & Equipment	N/A		—	—	—	
	D40 Fire Protection						
4010	Sprinklers	Wet pipe sprinkler system		S.F. Floor	4.57	4.57	4.7%
4020	Standpipes	Standpipe, wet, Class III		S.F. Floor	1.07	1.07	
	D50 Electrical						
5010	Electrical Service/Distribution	200 ampere service, panel board and feeders		S.F. Floor	1.24	1.24	12.9%
5020	Lighting & Branch Wiring	LED fixt., daylit. dim., ltg. on/off control, recept., switches and misc. pwr.		S.F. Floor	9.64	9.64	
5030	Communications & Security	Addressable alarm systems		S.F. Floor	2.63	2.63	
5090	Other Electrical Systems	Emergency generator, 7.5 kW, and energy monitoring systems		S.F. Floor	2.07	2.07	
E. EQUIPMENT & FURNISHINGS							
1010	Commercial Equipment	Laundry equipment		S.F. Floor	.97	.97	1.5%
1020	Institutional Equipment	N/A		—	—	—	
1090	Other Equipment	Waste handling recycling tilt truck		Each	.69	.69	
2020	Moveable Furnishings	No smoking signage		Each	.15	.15	
F. SPECIAL CONSTRUCTION							
1020	Integrated Construction	N/A		—	—	—	0.0 %
1040	Special Facilities	N/A		—	—	—	
G. BUILDING SITEWORK	**N/A**						
					Sub-Total	120.49	**100%**
	CONTRACTOR FEES (General Requirements: 10%, Overhead: 5%, Profit: 10%)				*25%*	30.12	
	ARCHITECT FEES				*7%*	10.54	
					Total Building Cost	**161.15**	

G Motel, 2-3 Story

Costs per square foot of floor area

Exterior Wall	S.F. Area	25000	37000	49000	61000	73000	81000	88000	96000	104000
	L.F. Perimeter	433	593	606	720	835	911	978	1054	1074
Decorative Concrete Block	Wood Joists	168.85	165.30	159.60	158.30	157.40	156.90	156.65	156.30	155.35
	Precast Conc.	180.60	176.75	170.95	169.60	168.65	168.20	167.85	167.50	166.55
Stucco on Concrete Block	Wood Joists	170.15	165.90	159.70	158.10	157.05	156.55	156.20	155.80	154.80
	Precast Conc.	182.40	178.10	171.85	170.25	169.30	168.75	168.40	168.00	166.95
Wood Siding	Wood Frame	163.60	160.35	155.65	154.45	153.65	153.20	153.00	152.75	151.90
Brick Veneer	Wood Frame	170.15	165.85	159.65	158.05	157.05	156.50	156.15	155.80	154.80
Perimeter Adj., Add or Deduct	Per 100 L.F.	5.20	3.50	2.65	2.10	1.80	1.65	1.50	1.35	1.20
Story Hgt. Adj., Add or Deduct	Per 1 Ft.	1.90	1.70	1.30	1.25	1.25	1.25	1.15	1.20	1.10

***For Basement**, add $30.40 per square foot of basement area*

The above costs were calculated using the basic specifications shown on the facing page. These costs should be adjusted where necessary for design alternatives and owner's requirements. Reported completed project costs, for this type of structure, range from $75.30 to $387.00 per S.F.

Common additives

Description	Unit	$ Cost
Closed Circuit Surveillance, one station, camera & monitor	Ea.	1975
For additional camera station, add	Ea.	1075
Elevators, hydraulic passenger, 2 stops, 2500# capacity	Ea.	66,900
3500# capacity	Ea.	71,900
Additional stop, add	Ea.	6825
Laundry Equipment, dryer, gas, 30 lb. capacity	Ea.	3900
Washer, commercial	Ea.	1575
Sauna, prefabricated, complete, 6′ x 9′	Ea.	10,100
8′ x 8′	Ea.	11,600
Swimming Pools, Complete, gunite	S.F.	97 - 121
TV Antenna, Master system, 12 outlet	Outlet	249
30 outlet	Outlet	217
100 outlet	Outlet	207

Description	Unit	$ Cost
Commissioning Fees, sustainable commercial construction	S.F.	0.24 - 3.04
Energy Modelling Fees, commercial buildings to 10,000 SF	Ea.	11,000
Greater than 10,000 SF add	S.F.	0.04
Green Bldg Cert Fees for comm construction project reg	Project	900
Photovoltaic Pwr Sys, grid connected, 20 kW (~2400 SF), roof	Ea.	241,700
Green Roofs, 6″ soil depth, w/treated wd edging & sedum mats	S.F.	11.27
10″ Soil depth, with treated wood edging & sedum mats	S.F.	12.84
Greywater Recovery Systems, prepackaged comm, 3060 gal.	Ea.	58,085
4590 gal.	Ea.	72,815
Rainwater Harvest Sys, prepckged comm, 10,000 gal, sys contrller	Ea.	31,175
20,000 gal. w/system controller	Ea.	50,100
30,000 gal. w/system controller	Ea.	80,225
Solar Domestic HW, closed loop, add-on sys, ext heat exchanger	Ea.	9750
Drainback, hot water system, 120 gal tank	Ea.	12,425
Draindown, hot water system, 120 gal tank	Ea.	12,700

Model costs calculated for a 3 story building with 9' story height and 49,000 square feet of floor area

G Motel, 2-3 Story

				Unit	Unit Cost	Cost Per S.F.	% Of Sub-Total
A. SUBSTRUCTURE							
1010	Standard Foundations	Poured concrete; strip and spread footings		S.F. Ground	1.38	.46	
1020	Special Foundations	N/A		—	—	—	
1030	Slab on Grade	4" reinforced concrete with recycled vapor barrier and granular base		S.F. Slab	5.11	1.70	3.3%
2010	Basement Excavation	Site preparation for slab and trench for foundation wall and footing		S.F. Ground	.16	.05	
2020	Basement Walls	4' foundation wall		L.F. Wall	81	1.99	
B. SHELL							
	B10 Superstructure						
1010	Floor Construction	Precast concrete plank		S.F. Floor	10.23	6.82	7.8%
1020	Roof Construction	Precast concrete plank		S.F. Roof	9.63	3.21	
	B20 Exterior Enclosure						
2010	Exterior Walls	Decorative concrete block	*85% of wall*	S.F. Wall	21	5.89	
2020	Exterior Windows	Aluminum sliding	*15% of wall*	Each	555	1.85	10.9%
2030	Exterior Doors	Aluminum and glass doors and entrance with transom		Each	1924	6.35	
	B30 Roofing						
3010	Roof Coverings	Single-ply TPO membrane, 60 mils w/R-20 insul.		S.F. Roof	6.15	2.05	1.6%
3020	Roof Openings	Roof hatches		S.F. Roof	.12	.04	
C. INTERIORS							
1010	Partitions	Concrete block w/foamed-in insul.	*7 SF Flr./LF Part.*	S.F. Partition	19.74	22.56	
1020	Interior Doors	Wood hollow core, low VOC paint	*70 S.F. Floor/Door*	Each	510	7.28	
1030	Fittings	N/A		—	—	—	
2010	Stair Construction	Concrete filled metal pan		Flight	14,075	3.45	35.6%
3010	Wall Finishes	90% paint, low VOC, 10% ceramic tile		S.F. Surface	1.50	3.42	
3020	Floor Finishes	85% carpet tile, 5% vinyl composition tile, recycled content, 10% ceramic tile		S.F. Floor	5.31	5.31	
3030	Ceiling Finishes	Textured finish		S.F. Ceiling	3.94	3.94	
D. SERVICES							
	D10 Conveying						
1010	Elevators & Lifts	Two hydraulic passenger elevators		Each	104,860	4.28	3.3%
1020	Escalators & Moving Walks	N/A		—	—	—	
	D20 Plumbing						
2010	Plumbing Fixtures	Toilet, low flow, auto sensor & service fixtures, supply & drain.	*1 Fixt./180 SF Flr.*	Each	4802	26.68	
2020	Domestic Water Distribution	Gas fired, tankless water heater		S.F. Floor	.26	.26	21.2%
2040	Rain Water Drainage	Roof drains		S.F. Roof	1.38	.46	
	D30 HVAC						
3010	Energy Supply	N/A		—	—	—	
3020	Heat Generating Systems	Included in D3050		—	—	—	
3030	Cooling Generating Systems	N/A		—	—	—	2.6 %
3050	Terminal & Package Units	Through the wall electric heating and cooling units		S.F. Floor	3.35	3.35	
3090	Other HVAC Sys. & Equipment	N/A		—	—	—	
	D40 Fire Protection						
4010	Sprinklers	Sprinklers, wet, light hazard		S.F. Floor	2.60	2.60	2.2%
4020	Standpipes	Standpipe, wet, Class III		S.F. Floor	.25	.25	
	D50 Electrical						
5010	Electrical Service/Distribution	800 ampere service, panel board and feeders		S.F. Floor	1.22	1.22	
5020	Lighting & Branch Wiring	LED fixtures, daylt. dim., ltg. on/off, recept., switches and misc. power		S.F. Floor	10.33	10.33	11.1%
5030	Communications & Security	Addressable alarm systems and emergency lighting		S.F. Floor	1.80	1.80	
5090	Other Electrical Systems	Emergency generator, 7.5 kW, energy monitoring systems		S.F. Floor	1.02	1.02	
E. EQUIPMENT & FURNISHINGS							
1010	Commercial Equipment	Commercial laundry equipment		S.F. Floor	.32	.32	
1020	Institutional Equipment	N/A		—	—	—	0.3%
1090	Other Equipment	Waste handling recycling tilt truck		Each	.07	.07	
2020	Moveable Furnishings	No smoking signage		Each	.02	.02	
F. SPECIAL CONSTRUCTION							
1020	Integrated Construction	N/A		—	—	—	0.0 %
1040	Special Facilities	N/A		—	—	—	
G. BUILDING SITEWORK		**N/A**					
					Sub-Total	129.03	**100%**
	CONTRACTOR FEES (General Requirements: 10%, Overhead: 5%, Profit: 10%)				*25%*	32.24	
	ARCHITECT FEES				*6%*	9.68	
					Total Building Cost	**170.95**	

Costs per square foot of floor area

Exterior Wall	S.F. Area	2000	3000	5000	7000	9000	12000	15000	20000	25000
	L.F. Perimeter	220	260	320	360	420	480	520	640	700
Wood Siding	Wood Truss	205.15	186.15	169.00	160.20	156.40	151.85	148.45	146.40	143.90
Brick Veneer	Wood Truss	245.60	222.70	201.95	191.25	186.65	181.10	176.95	174.50	171.40
Brick on Block	Wood Truss	196.85	171.15	147.60	135.40	130.25	123.85	119.05	116.25	112.70
	Steel Roof Deck	248.55	221.40	196.55	183.55	178.10	171.45	166.35	163.35	159.65
EIFS on Metal Studs	Steel Roof Deck	227.25	205.70	186.15	176.10	171.70	166.60	162.65	160.35	157.50
Tiltup Concrete Panel	Steel Roof Deck	233.90	210.55	189.45	178.60	173.90	168.35	164.05	161.50	158.45
Perimeter Adj., Add or Deduct	Per 100 L.F.	25.30	16.90	10.10	7.25	5.65	4.25	3.40	2.55	2.05
Story Hgt. Adj., Add or Deduct	Per 1 Ft.	3.15	2.50	1.85	1.50	1.35	1.20	1.00	0.90	0.80
***For Basement**, add $32.10 per square foot of basement area*										

The above costs were calculated using the basic specifications shown on the facing page. These costs should be adjusted where necessary for design alternatives and owner's requirements. Reported completed project costs, for this type of structure, range from $78.95 to $305.20 per S.F.

Common additives

Description	Unit	$ Cost
Closed Circuit Surveillance, one station, camera & monitor	Ea.	1975
For additional camera stations, add	Ea.	1075
Directory boards, plastic, glass covered, 30" x 20"	Ea.	555
36" x 48"	Ea.	1425
Aluminum, 36" x 24"	Ea.	715
48" x 32"	Ea.	1000
Electronic, wall mounted	S.F.	2900
Pedestal access flr sys w/ PLAM cover, comp rm,		
Less than 6000 SF	S.F.	20.50
Greater than 6000 SF	S.F.	20
Office, greater than 6000 S.F.	S.F.	16.50
Uninterruptible power supply, 15 kVA/12.75 kW	kW	1815

Description	Unit	$ Cost
Commissioning Fees, sustainable commercial construction	S.F.	0.24 - 3.04
Energy Modelling Fees, commercial buildings to 10,000 SF	Ea.	11,000
Greater than 10,000 SF add	S.F.	0.04
Green Bldg Cert Fees for comm construction project reg	Project	900
Photovoltaic Pwr Sys, grid connected, 20 kW (~2400 SF), roof	Ea.	241,700
Green Roofs, 6" soil depth, w/treated wd edging & sedum mats	S.F.	11.27
10" Soil depth, with treated wood edging & sedum mats	S.F.	12.84
Greywater Recovery Systems, prepackaged comm, 3060 gal.	Ea.	58,085
4590 gal.	Ea.	72,815
Rainwater Harvest Sys, prepckged comm, 10,000 gal, sys contrller	Ea.	31,175
20,000 gal. w/system controller	Ea.	50,100
30,000 gal. w/system controller	Ea.	80,225
Solar Domestic HW, closed loop, add-on sys, ext heat exchanger	Ea.	9750
Drainback, hot water system, 120 gal tank	Ea.	12,425
Draindown, hot water system, 120 gal tank	Ea.	12,700

Important: See the Reference Section for Location Factors.

Model costs calculated for a 1 story building with 12' story height and 7,000 square feet of floor area

G Office, 1 Story

			Unit	Unit Cost	Cost Per S.F.	% Of Sub-Total
A. SUBSTRUCTURE						
1010	Standard Foundations	Poured concrete; strip and spread footings	S.F. Ground	2.61	2.61	9.4%
1020	Special Foundations	N/A	—	—	—	
1030	Slab on Grade	4" reinforced concrete with recycled vapor barrier and granular base	S.F. Slab	5.11	5.11	
2010	Basement Excavation	Site preparation for slab and trench for foundation wall and footing	S.F. Ground	.29	.29	
2020	Basement Walls	4' foundation wall	L.F. Wall	153	4.34	
B. SHELL						
	B10 Superstructure					
1010	Floor Construction	N/A	—	—	—	7.6 %
1020	Roof Construction	Steel joists, girders & deck on columns	S.F. Roof	9.96	9.96	
	B20 Exterior Enclosure					
2010	Exterior Walls	E.I.F.S. on metal studs	S.F. Wall	16.59	8.19	11.3%
2020	Exterior Windows	Aluminum outward projecting *20% of wall*	Each	705	3.78	
2030	Exterior Doors	Aluminum and glass, hollow metal	Each	3358	2.88	
	B30 Roofing					
3010	Roof Coverings	Single-ply TPO membrane, 45 mils, heat welded seams w/R-20 insul.	S.F. Roof	9.56	9.56	7.4%
3020	Roof Openings	Roof hatch	S.F. Roof	.23	.23	
C. INTERIORS						
1010	Partitions	Gyp. bd. on mtl. studs w/sound attenuation *20 SF Flr/LF Part.*	S.F. Partition	9.50	4.75	18.2%
1020	Interior Doors	Single leaf hollow metal, low VOC paint *200 S.F. Floor/Door*	Each	1133	5.67	
1030	Fittings	Toilet partitions	S.F. Floor	.84	.84	
2010	Stair Construction	N/A	—	—	—	
3010	Wall Finishes	60% vinyl wall covering, 40% paint, low VOC	S.F. Surface	1.48	1.48	
3020	Floor Finishes	60% carpet tile, 30% vinyl composition tile, recycled content, 10% ceramic tile	S.F. Floor	4.69	4.69	
3030	Ceiling Finishes	Mineral fiber tile on concealed zee bars	S.F. Ceiling	6.57	6.57	
D. SERVICES						
	D10 Conveying					
1010	Elevators & Lifts	N/A	—	—	—	0.0 %
1020	Escalators & Moving Walks	N/A	—	—	—	
	D20 Plumbing					
2010	Plumbing Fixtures	Toilet, low flow, auto sensor, & service fixt., supply & drain. *1 Fixt./1320 SF Flr.*	Each	8778	6.65	6.3%
2020	Domestic Water Distribution	Tankless, on-demand water heater, gas/propane	S.F. Floor	.83	.83	
2040	Rain Water Drainage	Roof drains	S.F. Roof	.75	.75	
	D30 HVAC					
3010	Energy Supply	N/A	—	—	—	16.2 %
3020	Heat Generating Systems	Included in D3050	—	—	—	
3040	Distribution Systems	Enthalpy heat recovery packages	Each	17,150	2.45	
3050	Terminal & Package Units	Multizone rooftop air conditioner, SEER 14	S.F. Floor	18.90	18.90	
3090	Other HVAC Sys. & Equipment	N/A	—	—	—	
	D40 Fire Protection					
4010	Sprinklers	Sprinkler system, light hazard	S.F. Floor	3.28	3.28	3.6%
4020	Standpipes	Standpipes and hose systems	S.F. Floor	1.46	1.46	
	D50 Electrical					
5010	Electrical Service/Distribution	400 ampere service, panel board and feeders	S.F. Floor	1.94	1.94	19.4%
5020	Lighting & Branch Wiring	LED fixtures, daylt. dim., ltg. on/off, recept., switches, and A.C. power	S.F. Floor	14.49	14.49	
5030	Communications & Security	Addressable alarm systems, internet and phone wiring, and emergency lighting	S.F. Floor	6.72	6.72	
5090	Other Electrical Systems	Emergency generator, 7.5 kW, energy monitoring systems	S.F. Floor	2.39	2.39	
E. EQUIPMENT & FURNISHINGS						
1010	Commercial Equipment	N/A	—	—	—	0.6 %
1020	Institutional Equipment	N/A	—	—	—	
1090	Other Equipment	Waste handling recycling tilt truck	Each	.79	.79	
2020	Moveable Furnishings	No smoking signage	Each	.05	.05	
F. SPECIAL CONSTRUCTION						
1020	Integrated Construction	N/A	—	—	—	0.0 %
1040	Special Facilities	N/A	—	—	—	
G. BUILDING SITEWORK		**N/A**				
				Sub-Total	131.65	**100%**
		CONTRACTOR FEES (General Requirements: 10%, Overhead: 5%, Profit: 10%)		*25%*	32.93	
		ARCHITECT FEES		*7%*	11.52	
				Total Building Cost	**176.10**	

Costs per square foot of floor area

Exterior Wall	S.F. Area	5000	8000	12000	16000	20000	35000	50000	65000	80000
	L.F. Perimeter	220	260	310	330	360	440	550	600	675
Face Brick with Concrete Block Back-up	Wood Joists	253.20	221.25	203.00	190.00	183.25	169.75	165.60	161.45	159.55
	Steel Joists	253.35	221.50	203.20	190.15	**183.45**	169.95	165.80	161.60	159.70
Glass and Metal Curtain Wall	Steel Frame	304.60	261.40	236.55	218.40	209.15	190.30	184.55	178.65	175.95
	R/Conc. Frame	299.50	256.75	232.20	214.15	204.95	186.25	180.60	174.70	172.00
Wood Siding	Wood Frame	202.75	180.10	167.20	158.35	153.75	144.60	141.70	138.95	137.65
Brick Veneer	Wood Frame	217.15	190.75	175.65	165.10	159.65	148.70	145.25	142.00	140.40
Perimeter Adj., Add or Deduct	Per 100 L.F.	40.90	25.65	17.05	12.75	10.25	5.85	4.05	3.15	2.50
Story Hgt. Adj., Add or Deduct	Per 1 Ft.	6.75	5.05	3.95	3.15	2.75	1.95	1.65	1.40	1.25
***For Basement**, add $ 33.25 per square foot of basement area*										

The above costs were calculated using the basic specifications shown on the facing page. These costs should be adjusted where necessary for design alternatives and owner's requirements. Reported completed project costs, for this type of structure, range from $ 82.60 to $ 320.05 per S.F.

Common additives

Description	Unit	$ Cost
Closed Circuit Surveillance, one station, camera and monitor	Ea.	1975
For additional camera stations, add	Ea.	1075
Directory Boards, plastic, glass covered, 30" x 20"	Ea.	555
36" x 48"	Ea.	1425
Aluminum, 36" x 24"	Ea.	715
48" x 32"	Ea.	1000
Electronic, wall mounted	S.F.	2900
Escalators, 10' rise, 32" wide, glass balustrade	Ea.	137,900
48" wide, glass balustrade	Ea.	145,400
Pedestal access flr sys w/ PLAM cover, comp rm,		
Less than 6000 SF	S.F.	20.50
Greater than 6000 SF	S.F.	20
Office, greater than 6000 S.F.	S.F.	16.50
Uninterruptible power supply, 15 kVA/12.75 kW	kW	1815

Description	Unit	$ Cost
Commissioning Fees, sustainable commercial construction	S.F.	0.24 - 3.04
Energy Modelling Fees, commercial buildings to 10,000 SF	Ea.	11,000
Greater than 10,000 SF add	S.F.	0.04
Green Bldg Cert Fees for comm construction project reg	Project	900
Photovoltaic Pwr Sys, grid connected, 20 kW (~2400 SF), roof	Ea.	241,700
Green Roofs, 6" soil depth, w/treated wd edging & sedum mats	S.F.	11.27
10" Soil depth, with treated wood edging & sedum mats	S.F.	12.84
Greywater Recovery Systems, prepackaged comm, 3060 gal.	Ea.	58,085
4590 gal.	Ea.	72,815
Rainwater Harvest Sys, prepckged comm, 10,000 gal, sys contrller	Ea.	31,175
20,000 gal. w/system controller	Ea.	50,100
30,000 gal. w/system controller	Ea.	80,225
Solar Domestic HW, closed loop, add-on sys, ext heat exchanger	Ea.	9750
Drainback, hot water system, 120 gal tank	Ea.	12,425
Draindown, hot water system, 120 gal tank	Ea.	12,700

Important: See the Reference Section for Location Factors.

Model costs calculated for a 3 story building with 12′ story height and 20,000 square feet of floor area

G Office, 2-4 Story

				Unit	Unit Cost	Cost Per S.F.	% Of Sub-Total
A. SUBSTRUCTURE							
1010	Standard Foundations	Poured concrete; strip and spread footings		S.F. Ground	7.05	2.35	4.2%
1020	Special Foundations	N/A		–	–	–	
1030	Slab on Grade	4″ reinforced concrete with recycled vapor barrier and granular base		S.F. Slab	5.11	1.70	
2010	Basement Excavation	Site preparation for slab and trench for foundation wall and footing		S.F. Ground	.16	.05	
2020	Basement Walls	4′ foundation wall		L.F. Wall	153	1.70	
B. SHELL							
B10 Superstructure							
1010	Floor Construction	Open web steel joists, slab form, concrete, columns		S.F. Floor	18.17	12.11	10.5%
1020	Roof Construction	Metal deck, open web steel joists, columns		S.F. Roof	7.05	2.35	
B20 Exterior Enclosure							
2010	Exterior Walls	Face brick with concrete block backup	*80% of wall*	S.F. Wall	35	18.30	17.0%
2020	Exterior Windows	Aluminum outward projecting	*20% of wall*	Each	705	3.98	
2030	Exterior Doors	Aluminum and glass, hollow metal		Each	3358	1.02	
B30 Roofing							
3010	Roof Coverings	Single-ply, TPO membrane, 60 mils, heat welded seams w/ R-20 insul.		S.F. Roof	6.96	2.32	1.7%
3020	Roof Openings	N/A		–	–	–	
C. INTERIORS							
1010	Partitions	Gypsum board on metal studs	*20 SF Flr/LF Part.*	S.F. Partition	9.80	3.92	21.0%
1020	Interior Doors	Single leaf hollow metal, low VOC paint	*200 S.F. Floor/Door*	Each	1133	5.67	
1030	Fittings	Toilet partitions		S.F. Floor	1.77	1.77	
2010	Stair Construction	Concrete filled metal pan		Flight	14,075	4.93	
3010	Wall Finishes	60% vinyl wall covering, 40% paint, low VOC		S.F. Surface	1.49	1.19	
3020	Floor Finishes	60% carpet tile, 30% vinyl composition tile, recycled content, 10% ceramic tile		S.F. Floor	4.69	4.69	
3030	Ceiling Finishes	Mineral fiber tile on concealed zee bars		S.F. Ceiling	6.57	6.57	
D. SERVICES							
D10 Conveying							
1010	Elevators & Lifts	Two hydraulic passenger elevators		Each	113,100	11.31	8.2%
1020	Escalators & Moving Walks	N/A		–	–	–	
D20 Plumbing							
2010	Plumbing Fixtures	Toilet, low flow, auto sensor, & service fixt., supply & drain.	*1 Fixt./1320 SF Flr.*	Each	5148	3.90	3.5%
2020	Domestic Water Distribution	Gas fired, tankless water heater		S.F. Floor	.18	.18	
2040	Rain Water Drainage	Roof drains		S.F. Roof	1.98	.66	
D30 HVAC							
3010	Energy Supply	N/A		–	–	–	13.0 %
3020	Heat Generating Systems	Included in D3050		–	–	–	
3040	Distribution Systems	Enthalpy heat recovery packages		Each	17,150	2.57	
3050	Terminal & Package Units	Multizone rooftop air conditioner. SEER 14		S.F. Floor	15.23	15.23	
3090	Other HVAC Sys. & Equipment	N/A		–	–	–	
D40 Fire Protection							
4010	Sprinklers	Wet pipe sprinkler system		S.F. Floor	3.45	3.45	3.2%
4020	Standpipes	Standpipes and hose systems		S.F. Floor	.94	.94	
D50 Electrical							
5010	Electrical Service/Distribution	1000 ampere service, panel board and feeders		S.F. Floor	2.73	2.73	17.5%
5020	Lighting & Branch Wiring	LED fixtures, daylt. dim., ltg. on/off, recept., switches, and A.C. power		S.F. Floor	14.21	14.21	
5030	Communications & Security	Addressable alarm systems, internet and phone wiring, & emergency ltg.		S.F. Floor	5.59	5.59	
5090	Other Electrical Systems	Emergency generator, 7.5 kW, UPS, energy monitoring systems		S.F. Floor	1.47	1.47	
E. EQUIPMENT & FURNISHINGS							
1010	Commercial Equipment	N/A		–	–	–	0.2 %
1020	Institutional Equipment	N/A		–	–	–	
1090	Other Equipment	Waste handling recycling tilt truck		Each	.28	.28	
2020	Moveable Furnishings	No smoking signage		Each	.02	.02	
F. SPECIAL CONSTRUCTION							
1020	Integrated Construction	N/A		–	–	–	0.0 %
1040	Special Facilities	N/A		–	–	–	
G. BUILDING SITEWORK	**N/A**						
					Sub-Total	137.16	**100%**
	CONTRACTOR FEES (General Requirements: 10%, Overhead: 5%, Profit: 10%)				*25%*	34.29	
	ARCHITECT FEES				*7%*	12	
					Total Building Cost	**183.45**	

Costs per square foot of floor area

Exterior Wall	S.F. Area	7000	9000	11000	13000	15000	17000	19000	21000	23000
	L.F. Perimeter	240	280	303	325	354	372	397	422	447
Limestone with Concrete Block Back-up	Bearing Walls	275.30	259.75	246.20	236.60	230.75	224.70	220.85	217.70	215.10
	R/Conc. Frame	273.40	258.50	245.85	236.85	231.35	225.65	222.05	219.15	216.70
Face Brick with Concrete Block Back-up	Bearing Walls	233.05	220.15	209.75	202.35	197.75	193.20	190.20	187.75	185.75
	R/Conc. Frame	251.60	238.75	228.35	220.90	216.35	211.75	208.80	206.30	204.30
Decorative Concrete Block	Bearing Walls	221.15	209.35	200.20	193.70	189.55	185.60	182.90	180.75	179.00
	R/Conc. Frame	240.90	229.10	219.90	213.40	209.30	205.30	202.65	200.45	198.75
Perimeter Adj., Add or Deduct	Per 100 L.F.	33.50	26.00	21.30	18.05	15.65	13.80	12.30	11.15	10.15
Story Hgt. Adj., Add or Deduct	Per 1 Ft.	6.00	5.40	4.75	4.35	4.15	3.80	3.60	3.45	3.35
***For Basement**, add $28.30 per square foot of basement area*										

The above costs were calculated using the basic specifications shown on the facing page. These costs should be adjusted where necessary for design alternatives and owner's requirements. Reported completed project costs, for this type of structure, range from $134.00 to $349.50 per S.F.

Common additives

Description	Unit	$ Cost
Cells Prefabricated, 5'-6' x 7'-8", 7'-8' high	Ea.	12,500
Elevators, hydraulic passenger, 2 stops, 2500# capacity	Ea.	66,900
3500# capacity	Ea.	71,900
Lockers, Steel, Single tier, 60" to 72"	Opng.	222 - 365
Locker bench, lam. maple top only	L.F.	33
Pedestals, steel pipe	Ea.	71
Safe, Office type, 1 hr rating, 60"x36"x18", double door	Ea.	9350
Shooting Range, incl. bullet traps,		
Target provisions & controls, excl. struct shell	Ea.	51,000
Sound System, amplifier, 250 watts	Ea.	2125
Speaker, ceiling or wall	Ea.	206
Trumpet	Ea.	395

Description	Unit	$ Cost
Commissioning Fees, sustainable commercial construction	S.F.	0.24 - 3.04
Energy Modelling Fees, commercial buildings to 10,000 SF	Ea.	11,000
Greater than 10,000 SF add	S.F.	0.04
Green Bldg Cert Fees for comm construction project reg	Project	900
Photovoltaic Pwr Sys, grid connected, 20 kW (~2400 SF), roof	Ea.	241,700
Green Roofs, 6" soil depth, w/treated wd edging & sedum mats	S.F.	11.27
10" Soil depth, with treated wood edging & sedum mats	S.F.	12.84
Greywater Recovery Systems, prepackaged comm, 3060 gal.	Ea.	58,085
4590 gal.	Ea.	72,815
Rainwater Harvest Sys, prepckged comm, 10,000 gal, sys contrller	Ea.	31,175
20,000 gal. w/system controller	Ea.	50,100
30,000 gal. w/system controller	Ea.	80,225
Solar Domestic HW, closed loop, add-on sys, ext heat exchanger	Ea.	9750
Drainback, hot water system, 120 gal tank	Ea.	12,425
Draindown, hot water system, 120 gal tank	Ea.	12,700

Important: See the Reference Section for Location Factors.

Model costs calculated for a 2 story building with 12′ story height and 11,000 square feet of floor area

G Police Station

				Unit	Unit Cost	Cost Per S.F.	% Of Sub-Total
A. SUBSTRUCTURE							
1010	Standard Foundations	Poured concrete; strip and spread footings		S.F. Ground	2.72	1.36	3.9%
1020	Special Foundations	N/A		—	—	—	
1030	Slab on Grade	4″ reinforced concrete with recycled vapor barrier and granular base		S.F. Slab	5.11	2.56	
2010	Basement Excavation	Site preparation for slab and trench for foundation wall and footing		S.F. Ground	.29	.15	
2020	Basement Walls	4′ foundation wall		L.F. Wall	81	2.95	
B. SHELL							
	B10 Superstructure						
1010	Floor Construction	Open web steel joists, slab form, concrete		S.F. Floor	11.78	5.89	4.6%
1020	Roof Construction	Metal deck on open web steel joists		S.F. Roof	4.70	2.35	
	B20 Exterior Enclosure						
2010	Exterior Walls	Limestone with concrete block backup	*80% of wall*	S.F. Wall	60	31.52	24.0%
2020	Exterior Windows	Metal horizontal sliding	*20% of wall*	Each	1123	9.89	
2030	Exterior Doors	Hollow metal		Each	2709	1.96	
	B30 Roofing						
3010	Roof Coverings	Single-ply TPO membrane, 60 mils, heat welded w/R-20 insul.		S.F. Roof	6.58	3.29	1.8%
3020	Roof Openings	N/A		—	—	—	
C. INTERIORS							
1010	Partitions	Concrete block w/foamed-in insul.	*20 SF Flr./LF Part.*	S.F. Partition	10.22	5.11	15.6%
1020	Interior Doors	Single leaf kalamein fire door, low VOC paint	*200 S.F. Floor/Door*	Each	1133	5.67	
1030	Fittings	Toilet partitions		S.F. Floor	.94	.94	
2010	Stair Construction	Concrete filled metal pan		Flight	16,800	3.05	
3010	Wall Finishes	90% paint, low VOC, 10% ceramic tile		S.F. Surface	2.52	2.52	
3020	Floor Finishes	70% vinyl composition tile, recycled content, 20% carpet tile, 10% ceramic tile		S.F. Floor	4.41	4.41	
3030	Ceiling Finishes	Mineral fiber tile on concealed zee bars		S.F. Ceiling	6.57	6.57	
D. SERVICES							
	D10 Conveying						
1010	Elevators & Lifts	One hydraulic passenger elevator		Each	81,070	7.37	4.1%
1020	Escalators & Moving Walks	N/A		—	—	—	
	D20 Plumbing						
2010	Plumbing Fixtures	Toilet, low flow, auto sensor, & service fixt., supply & drainage	*1 Fixt./580 SF Flr.*	Each	4901	8.45	6.1%
2020	Domestic Water Distribution	Electric water heater, point-of-use, energy saver		S.F. Floor	1.12	1.12	
2040	Rain Water Drainage	Roof drains		S.F. Roof	3	1.50	
	D30 HVAC						
3010	Energy Supply	N/A		—	—	—	12.0 %
3020	Heat Generating Systems	N/A		—	—	—	
3040	Distribution Systems	Enthalpy heat recovery packages		Each	29,975	2.73	
3050	Terminal & Package Units	Multizone rooftop air conditioner, SEER 14		S.F. Floor	18.90	18.90	
3090	Other HVAC Sys. & Equipment	N/A		—	—	—	
	D40 Fire Protection						
4010	Sprinklers	Wet pipe sprinkler system		S.F. Floor	3.54	3.54	2.5%
4020	Standpipes	Standpipe		S.F. Floor	1.02	1.02	
	D50 Electrical						
5010	Electrical Service/Distribution	400 ampere service, panel board and feeders		S.F. Floor	.98	.98	16.4%
5020	Lighting & Branch Wiring	LED fixtures, daylt. dim., ltg. on/off cntrl., recept., switches, & A.C. pwr.		S.F. Floor	20	20.10	
5030	Communications & Security	Addressable alarm systems, internet wiring, intercom &emergency ltg.		S.F. Floor	6.94	6.94	
5090	Other Electrical Systems	Emergency generator, 15 kW, energy monitoring systems		S.F. Floor	1.66	1.66	
E. EQUIPMENT & FURNISHINGS							
1020	Institutional Equipment	Lockers, detention rooms, cells, gasoline dispensers		S.F. Floor	13.19	13.19	9.0%
1030	Vehicular Equipment	Gasoline dispenser system		S.F. Floor	2.47	2.47	
1090	Other Equipment	Waste handling recycling tilt truck		Each	.50	.50	
2020	Moveable Furnishings	No smoking signage		Each	.04	.04	
F. SPECIAL CONSTRUCTION							
1020	Integrated Construction	N/A		—	—	—	0.0 %
1040	Special Facilities	N/A		—	—	—	
G. BUILDING SITEWORK		**N/A**					

	Sub-Total	180.70	**100%**
CONTRACTOR FEES (General Requirements: 10%, Overhead: 5%, Profit: 10%)	*25%*	45.17	
ARCHITECT FEES	*9%*	20.33	
	Total Building Cost	**246.20**	

Costs per square foot of floor area

Exterior Wall	S.F. Area	2000	2800	3500	4200	5000	5800	6500	7200	8000
	L.F. Perimeter	180	212	240	268	300	314	336	344	368
Wood Siding	Wood Frame	262.35	248.60	241.65	237.05	233.40	228.90	226.75	223.85	222.25
Brick Veneer	Wood Frame	269.00	253.60	245.85	240.70	236.60	231.35	228.95	225.55	223.80
Face Brick with Concrete Block Back-up	Wood Joists	281.55	264.10	255.40	249.60	244.95	238.90	236.15	232.30	230.20
	Steel Joists	280.40	260.80	250.95	244.40	239.15	232.25	229.05	224.60	222.25
Stucco on Concrete Block	Wood Joists	269.10	253.60	245.90	240.75	236.60	231.40	228.95	225.60	223.85
	Steel Joists	256.10	240.60	232.90	227.75	223.60	218.40	215.95	212.60	210.85
Perimeter Adj., Add or Deduct	Per 100 L.F.	30.20	21.55	17.30	14.40	12.10	10.40	9.25	8.45	7.55
Story Hgt. Adj., Add or Deduct	Per 1 Ft.	3.20	2.70	2.45	2.35	2.20	1.90	1.85	1.75	1.65
For Basement, add $ 34.70 per square foot of basement area										

The above costs were calculated using the basic specifications shown on the facing page. These costs should be adjusted where necessary for design alternatives and owner's requirements. Reported completed project costs, for this type of structure, range from $ 142.55 to $ 332.20 per S.F.

Common additives

Description	Unit	$ Cost
Bar, Front Bar	L.F.	415
Back bar	L.F.	330
Booth, Upholstered, custom straight	L.F.	248 - 460
"L" or "U" shaped	L.F.	256 - 435
Fireplace, brick, excl. chimney or foundation, 30"x29" opening	Ea.	2825
Chimney, standard brick, single flue, 16" x 20"	V.L.F.	98
2 Flue, 20" x 24"	V.L.F.	144
Kitchen Equipment		
Broiler	Ea.	4050
Coffee urn, twin 6 gallon	Ea.	2975
Cooler, 6 ft. long	Ea.	3875
Dishwasher, 10-12 racks per hr.	Ea.	4200
Food warmer, counter, 1.2 KW	Ea.	770
Freezer, 44 C.F., reach-in	Ea.	6100
Ice cube maker, 50 lb. per day	Ea.	1950
Range with 1 oven	Ea.	3275
Refrigerators, prefabricated, walk-in, 7'-6" high, 6' x 6'	S.F.	200
10' x 10'	S.F.	157
12' x 14'	S.F.	140
12' x 20'	S.F.	123

Description	Unit	$ Cost
Commissioning Fees, sustainable commercial construction	S.F.	0.24 - 3.04
Energy Modelling Fees, commercial buildings to 10,000 SF	Ea.	11,000
Greater than 10,000 SF add	S.F.	0.04
Green Bldg Cert Fees for comm construction project reg	Project	900
Photovoltaic Pwr Sys, grid connected, 20 kW (~2400 SF), roof	Ea.	241,700
Green Roofs, 6" soil depth, w/treated wd edging & sedum mats	S.F.	11.27
10" Soil depth, with treated wood edging & sedum mats	S.F.	12.84
Greywater Recovery Systems, prepackaged comm, 3060 gal.	Ea.	58,085
4590 gal.	Ea.	72,815
Rainwater Harvest Sys, prepckged comm, 10,000 gal, sys contrller	Ea.	31,175
20,000 gal. w/system controller	Ea.	50,100
30,000 gal. w/system controller	Ea.	80,225
Solar Domestic HW, closed loop, add-on sys, ext heat exchanger	Ea.	9750
Drainback, hot water system, 120 gal tank	Ea.	12,425
Draindown, hot water system, 120 gal tank	Ea.	12,700

Important: See the Reference Section for Location Factors.

Model costs calculated for a 1 story building with 12′ story height and 5,000 square feet of floor area

G Restaurant

				Unit	Unit Cost	Cost Per S.F.	% Of Sub-Total
A. SUBSTRUCTURE							
1010	Standard Foundations	Poured concrete; strip and spread footings		S.F. Ground	2.48	2.48	
1020	Special Foundations	N/A		–	–	–	
1030	Slab on Grade	4″ reinforced concrete with recycled vapor barrier and granular base		S.F. Slab	5.11	5.11	7.4%
2010	Basement Excavation	Site preparation for slab and trench for foundation wall and footing		S.F. Ground	.51	.51	
2020	Basement Walls	4′ foundation wall		L.F. Wall	81	4.83	
B. SHELL							
	B10 Superstructure						
1010	Floor Construction	Wood columns		S.F. Floor	.78	.78	4.9%
1020	Roof Construction	Plywood on wood truss (pitched)		S.F. Roof	7.76	7.76	
	B20 Exterior Enclosure						
2010	Exterior Walls	Cedar siding on wood studs with insulation	*70% of wall*	S.F. Wall	12.50	6.30	
2020	Exterior Windows	Storefront windows	*30% of wall*	Each	53	10.81	12.7%
2030	Exterior Doors	Aluminum and glass doors and entrance with transom		Each	5117	5.13	
	B30 Roofing						
3010	Roof Coverings	Cedar shingles with flashing (pitched); rigid fiberglass insulation		S.F. Roof	11.93	11.93	6.9%
3020	Roof Openings	Skylights		S.F. Roof	.10	.10	
C. INTERIORS							
1010	Partitions	Gypsum board on wood studs	*25 SF Floor/LF Part.*	S.F. Partition	11.20	4.48	
1020	Interior Doors	Hollow core wood, low VOC paint	*250 S.F. Floor/Door*	Each	510	2.04	
1030	Fittings	Toilet partitions		S.F. Floor	1.18	1.18	
2010	Stair Construction	N/A		–	–	–	14.3%
3010	Wall Finishes	75% paint, low VOC, 25% ceramic tile		S.F. Surface	2.40	1.92	
3020	Floor Finishes	65% carpet, 35% quarry tile		S.F. Floor	8.71	8.71	
3030	Ceiling Finishes	Mineral fiber tile on concealed zee bars		S.F. Ceiling	6.57	6.57	
D. SERVICES							
	D10 Conveying						
1010	Elevators & Lifts	N/A		–	–	–	0.0 %
1020	Escalators & Moving Walks	N/A		–	–	–	
	D20 Plumbing						
2010	Plumbing Fixtures	Kitchen, toilet, low flow, auto sensor, & service fixt., supply & drain.	*1 Fixt./335 SF Flr.*	Each	3515	9.90	
2020	Domestic Water Distribution	Tankless, on-demand water heater gas/propane		S.F. Floor	2.43	2.43	8.1%
2040	Rain Water Drainage	Roof drains		S.F. Roof	1.80	1.80	
	D30 HVAC						
3010	Energy Supply	N/A		–	–	–	
3020	Heat Generating Systems	Included in D3050		–	–	–	
3040	Distribution Systems	Enthalpy heat recovery packages and kitchen exhaust/make-up air system		Each	42,225	8.45	26.7 %
3050	Terminal & Package Units	Multizone rooftop air conditioner, SEER 14		S.F. Floor	38	38.20	
3090	Other HVAC Sys. & Equipment	N/A		–	–	–	
	D40 Fire Protection						
4010	Sprinklers	Sprinklers, light hazard and ordinary hazard kitchen		S.F. Floor	8.53	8.53	6.1%
4020	Standpipes	Standpipe		S.F. Floor	2.04	2.04	
	D50 Electrical						
5010	Electrical Service/Distribution	400 ampere service, panel board and feeders		S.F. Floor	2.24	2.24	
5020	Lighting & Branch Wiring	LED fixtures, daylt. dim., ltg. on/off, receptacles, switches, and A.C. power		S.F. Floor	12.27	12.27	12.2%
5030	Communications & Security	Addressable alarm systems and emergency lighting		S.F. Floor	3.46	3.46	
5090	Other Electrical Systems	Emergency generator, 15 kW and energy monitoring systems		S.F. Floor	3.39	3.39	
E. EQUIPMENT & FURNISHINGS							
1010	Commercial Equipment	N/A		–	–	–	
1020	Institutional Equipment	N/A		–	–	–	0.7 %
1090	Other Equipment	Waste handling recycling tilt truck		Each	1.11	1.11	
2020	Moveable Furnishings	No smoking signage		Each	.06	.06	
F. SPECIAL CONSTRUCTION							
1020	Integrated Construction	N/A		–	–	–	0.0 %
1040	Special Facilities	N/A		–	–	–	
G. BUILDING SITEWORK	**N/A**						
					Sub-Total	174.52	**100%**
	CONTRACTOR FEES (General Requirements: 10%, Overhead: 5%, Profit: 10%)				*25%*	43.61	
	ARCHITECT FEES				*7%*	15.27	
					Total Building Cost	**233.40**	

Costs per square foot of floor area

Exterior Wall	S.F. Area	25000	30000	35000	40000	45000	50000	55000	60000	65000
	L.F. Perimeter	900	1050	1200	1350	1510	1650	1800	1970	2100
Face Brick with Concrete Block Back-up	Steel Frame	186.35	184.15	182.65	181.55	180.85	179.90	179.30	179.20	178.40
	Bearing Walls	178.65	176.75	175.35	174.40	173.85	173.00	172.50	172.40	171.70
Stucco on Concrete Block	Steel Frame	180.10	178.10	176.70	175.70	175.05	174.15	173.65	173.50	172.80
	Bearing Walls	173.70	171.70	170.30	169.30	168.70	167.80	167.30	167.15	166.40
Decorative Concrete Block	Steel Frame	178.30	176.35	174.90	173.90	173.25	172.40	171.90	171.75	171.05
	Bearing Walls	174.90	172.85	171.50	170.45	169.80	169.00	168.45	168.30	167.60
Perimeter Adj., Add or Deduct	Per 100 L.F.	4.55	3.90	3.25	2.80	2.55	2.30	2.10	1.90	1.75
Story Hgt. Adj., Add or Deduct	Per 1 Ft.	1.75	1.70	1.60	1.60	1.65	1.65	1.65	1.60	1.55
***For Basement**, add $ 25.65 per square foot of basement area*										

The above costs were calculated using the basic specifications shown on the facing page. These costs should be adjusted where necessary for design alternatives and owner's requirements. Reported completed project costs, for this type of structure, range from $ 98.30 to $ 250.05 per S.F.

Common additives

Description	Unit	$ Cost
Bleachers, Telescoping, manual, 16-20 tier	Seat	284 - 345
21-30 tier	Seat	300 - 395
For power operation, add	Seat	57 - 89
Carrels Hardwood	Ea.	925 - 2375
Clock System, 20 room	Ea.	18,900
50 room	Ea.	44,300
Kitchen Equipment		
Broiler	Ea.	4050
Cooler, 6 ft. long, reach-in	Ea.	3875
Dishwasher, 10-12 racks per hr.	Ea.	4200
Food warmer, counter, 1.2 KW	Ea.	770
Freezer, 44 C.F., reach-in	Ea.	6100
Ice cube maker, 50 lb. per day	Ea.	1950
Range with 1 oven	Ea.	3275
Lockers, Steel, single tier, 60" to 72"	Opng.	222 - 365
2 tier, 60" to 72" total	Opng.	115 - 164
Locker bench, lam. maple top only	L.F.	33
Pedestals, steel pipe	Ea.	71
Seating, auditorium chair, veneer back, padded seat	Ea.	315
Classroom, movable chair & desk	Set	81 - 171
Lecture hall, pedestal type	Ea.	267 - 610

Description	Unit	$ Cost
Sound System, amplifier, 250 watts	Ea.	2125
Speaker, ceiling or wall	Ea.	206
Commissioning Fees, sustainable institutional construction	S.F.	0.58 - 2.47
Energy Modelling Fees, academic buildings to 10,000 SF	Ea.	9000
Greater than 10,000 SF add	S.F.	0.20
Green Bldg Cert Fees for school construction project reg	Project	900
Photovoltaic Pwr Sys, grid connected, 20 kW (~2400 SF), roof	Ea.	241,700
Green Roofs, 6" soil depth, w/treated wd edging & sedum mats	S.F.	11.27
10" Soil depth, with treated wood edging & sedum mats	S.F.	12.84
Greywater Recovery Systems, prepackaged comm, 3060 gal.	Ea.	58,085
4590 gal.	Ea.	72,815
Rainwater Harvest Sys, prepckged comm, 10,000 gal, sys contrller	Ea.	31,175
20,000 gal. w/system controller	Ea.	50,100
30,000 gal. w/system controller	Ea.	80,225
Solar Domestic HW, closed loop, add-on sys, ext heat exchanger	Ea.	9750
Drainback, hot water system, 120 gal tank	Ea.	12,425
Draindown, hot water system, 120 gal tank	Ea.	12,700

Important: See the Reference Section for Location Factors.

Model costs calculated for a 1 story building with 15′ story height and 45,000 square feet of floor area

G School, Elementary

				Unit	Unit Cost	Cost Per S.F.	% Of Sub-Total
A. SUBSTRUCTURE							
1010	Standard Foundations	Poured concrete; strip and spread footings		S.F. Ground	4.71	4.71	11.2%
1020	Special Foundations	N/A		—	—	—	
1030	Slab on Grade	4″ reinforced concrete with recycled vapor barrier and granular base		S.F. Slab	5.11	5.11	
2010	Basement Excavation	Site preparation for slab and trench for foundation wall and footing		S.F. Ground	.16	.16	
2020	Basement Walls	4′ foundation wall		L.F. Wall	81	4.59	
B. SHELL							
	B10 Superstructure						
1010	Floor Construction	N/A		—	—	—	3.3 %
1020	Roof Construction	Metal deck on open web steel joists		S.F. Roof	4.29	4.29	
	B20 Exterior Enclosure						
2010	Exterior Walls	Face brick with concrete block backup	*70% of wall*	S.F. Wall	35	12.44	13.6%
2020	Exterior Windows	Steel outward projecting	*25% of wall*	Each	705	4.63	
2030	Exterior Doors	Metal and glass	*5% of wall*	Each	3680	.66	
	B30 Roofing						
3010	Roof Coverings	Single-ply TPO membrane, 60 mils, w/flashing; polyiso. insulation		S.F. Roof	9.86	9.86	7.6%
3020	Roof Openings	N/A		—	—	—	
C. INTERIORS							
1010	Partitions	Concrete block w/foamed-in insul.	*20 SF Flr./LF Part.*	S.F. Partition	10.62	5.31	19.8%
1020	Interior Doors	Single leaf kalamein fire doors, low VOC paint	*700 S.F. Floor/Door*	Each	1133	1.62	
1030	Fittings	Toilet partitions		S.F. Floor	1.84	1.84	
2010	Stair Construction	N/A		—	—	—	
3010	Wall Finishes	75% paint, low VOC, 15% glazed coating, 10% ceramic tile		S.F. Surface	4.45	4.45	
3020	Floor Finishes	65% vinyl composition tile, recycled content, 25% carpet tile, 10% terrazzo		S.F. Floor	5.89	5.89	
3030	Ceiling Finishes	Mineral fiber tile on concealed zee bars		S.F. Ceiling	6.57	6.57	
D. SERVICES							
	D10 Conveying						
1010	Elevators & Lifts	N/A		—	—	—	0.0 %
1020	Escalators & Moving Walks	N/A		—	—	—	
	D20 Plumbing						
2010	Plumbing Fixtures	Kitchen, toilet, low flow, auto sensor, & service fixt., supply & drain.	*1 Fixt./625 SF Flr.*	Each	8894	14.23	12.0%
2020	Domestic Water Distribution	Gas fired, tankless water heater		S.F. Floor	.14	.14	
2040	Rain Water Drainage	Roof drains		S.F. Roof	1.22	1.22	
	D30 HVAC						
3010	Energy Supply	N/A		—	—	—	15.9 %
3020	Heat Generating Systems	N/A		—	—	—	
3040	Distribution Systems	Enthalpy heat recovery packages		Each	36,450	2.43	
3050	Terminal & Package Units	Multizone rooftop air conditioner		S.F. Floor	18.30	18.30	
3090	Other HVAC Sys. & Equipment	N/A		—	—	—	
	D40 Fire Protection						
4010	Sprinklers	Sprinklers, light hazard		S.F. Floor	2.48	2.48	2.2%
4020	Standpipes	Standpipe		S.F. Floor	.40	.40	
	D50 Electrical						
5010	Electrical Service/Distribution	800 ampere service, panel board and feeders		S.F. Floor	.86	.86	14.1%
5020	Lighting & Branch Wiring	LED fixtures, daylt. dim., ltg. on/off, recept., switches, and A.C. power		S.F. Floor	12.91	12.91	
5030	Communications & Security	Addressable alarm systems, internet wiring, comm. systems & emerg. ltg.		S.F. Floor	3.59	3.59	
5090	Other Electrical Systems	Emergency generator, 15 kW, energy monitoring systems		S.F. Floor	1.01	1.01	
E. EQUIPMENT & FURNISHINGS							
1010	Commercial Equipment	N/A		—	—	—	0.2 %
1020	Institutional Equipment	Chalkboards		S.F. Floor	.19	.19	
1090	Other Equipment	Waste handling recycling tilt truck		Each	.08	.08	
2020	Moveable Furnishings	No smoking signage		Each	.01	.01	
F. SPECIAL CONSTRUCTION							
1020	Integrated Construction	N/A		—	—	—	0.0 %
1040	Special Facilities	N/A		—	—	—	
G. BUILDING SITEWORK		**N/A**					
					Sub-Total	129.98	**100%**
	CONTRACTOR FEES (General Requirements: 10%, Overhead: 5%, Profit: 10%)				*25%*	32.50	
	ARCHITECT FEES				*7%*	11.37	
					Total Building Cost	**173.85**	

Costs per square foot of floor area

Exterior Wall	S.F. Area	50000	65000	80000	95000	110000	125000	140000	155000	170000
	L.F. Perimeter	850	1060	1280	1490	1700	1920	2140	2340	2560
Face Brick with Concrete Block Back-up	Steel Frame	181.25	179.15	178.05	177.10	176.40	176.10	175.80	175.30	175.15
	Bearing Walls	172.90	170.80	169.65	168.75	168.05	167.70	167.40	166.90	166.80
Concrete Block Stucco Face	Steel Frame	186.05	183.65	182.50	181.40	180.60	180.25	179.90	179.35	179.20
	Bearing Walls	179.50	177.15	175.95	174.95	174.10	173.70	173.40	172.85	172.70
Decorative Concrete Block	Steel Frame	173.60	171.80	170.80	170.00	169.40	169.15	168.85	168.45	168.35
	Bearing Walls	165.65	163.75	162.75	162.00	161.35	161.05	160.80	160.40	160.25
Perimeter Adj., Add or Deduct	Per 100 L.F.	3.95	3.05	2.45	2.10	1.75	1.55	1.40	1.25	1.20
Story Hgt. Adj., Add or Deduct	Per 1 Ft.	2.00	1.90	1.85	1.85	1.75	1.75	1.75	1.75	1.75
***For Basement**, add $ 35.00 per square foot of basement area*										

The above costs were calculated using the basic specifications shown on the facing page. These costs should be adjusted where necessary for design alternatives and owner's requirements. Reported completed project costs, for this type of structure, range from $ 107.85 to $ 253.45 per S.F.

Common additives

Description	Unit	$ Cost
Bleachers, Telescoping, manual, 16-20 tier	Seat	284 - 345
21-30 tier	Seat	300 - 395
For power operation, add	Seat	57 - 89
Carrels Hardwood	Ea.	925 - 2375
Clock System, 20 room	Ea.	18,900
50 room	Ea.	44,300
Elevators, hydraulic passenger, 2 stops, 2500# capacity	Ea.	66,900
3500# capacity	Ea.	71,900
Kitchen Equipment		
Broiler	Ea.	4050
Cooler, 6 ft. long, reach-in	Ea.	3875
Dishwasher, 10-12 racks per hr.	Ea.	4200
Food warmer, counter, 1.2 KW	Ea.	770
Freezer, 44 C.F., reach-in	Ea.	6100
Lockers, Steel, single tier, 60" to 72"	Opng.	222 - 365
2 tier, 60" to 72" total	Opng.	115 - 164
Locker bench, lam. maple top only	L.F.	33
Pedestals, steel pipe	Ea.	71

Description	Unit	$ Cost
Seating, auditorium chair, veneer back, padded seat	Ea.	315
Upholstered, spring seat	Ea.	310
Classroom, movable chair & desk	Set	81 - 171
Lecture hall, pedestal type	Ea.	267 - 610
Sound System, amplifier, 250 watts	Ea.	2125
Speaker, ceiling or wall	Ea.	206
Commissioning Fees, sustainable institutional construction	S.F.	0.58 - 2.47
Energy Modelling Fees, academic buildings to 10,000 SF	Ea.	9000
Greater than 10,000 SF add	S.F.	0.20
Green Bldg Cert Fees for school construction project reg	Project	900
Photovoltaic Pwr Sys, grid connected, 20 kW (~2400 SF), roof	Ea.	241,700
Green Roofs, 6" soil depth, w/treated wd edging & sedum mats	S.F.	11.27
10" Soil depth, with treated wood edging & sedum mats	S.F.	12.84
Greywater Recovery Systems, prepackaged comm, 3060 gal.	Ea.	58,085
4590 gal.	Ea.	72,815
Rainwater Harvest Sys, prepckged comm, 10,000 gal, sys contrller	Ea.	31,175
20,000 gal. w/system controller	Ea.	50,100
30,000 gal. w/system controller	Ea.	80,225
Solar Domestic HW, closed loop, add-on sys, ext heat exchanger	Ea.	9750
Drainback, hot water system, 120 gal tank	Ea.	12,425
Draindown, hot water system, 120 gal tank	Ea.	12,700

Important: See the Reference Section for Location Factors.

Model costs calculated for a 2 story building with 15′ story height and 110,000 square feet of floor area

G School, Jr High, 2-3 Story

				Unit	Unit Cost	Cost Per S.F.	% Of Sub-Total
A. SUBSTRUCTURE							
1010	Standard Foundations	Poured concrete; strip and spread footings		S.F. Ground	2.32	1.16	3.8%
1020	Special Foundations	N/A		—	—	—	
1030	Slab on Grade	4″ reinforced concrete with recycled vapor barrier and granular base		S.F. Slab	5.11	2.56	
2010	Basement Excavation	Site preparation for slab and trench for foundation wall and footing		S.F. Ground	.16	.08	
2020	Basement Walls	4′ foundation wall		L.F. Wall	81	1.24	
B. SHELL							
	B10 Superstructure						
1010	Floor Construction	Open web steel joists, slab form, concrete, columns		S.F. Floor	27	13.67	14.7%
1020	Roof Construction	Metal deck, open web steel joists, columns		S.F. Roof	11.52	5.76	
	B20 Exterior Enclosure						
2010	Exterior Walls	Face brick with concrete block backup	*75% of wall*	S.F. Wall	35	12.27	14.9%
2020	Exterior Windows	Window wall	*25% of wall*	Each	61	6.71	
2030	Exterior Doors	Double aluminum & glass		Each	2321	.61	
	B30 Roofing						
3010	Roof Coverings	Single-ply TPO membrane & standing seam metal; polyiso. insulation		S.F. Roof	13.42	6.71	5.1%
3020	Roof Openings	Roof hatches		S.F. Roof	.04	.02	
C. INTERIORS							
1010	Partitions	Concrete block w/foamed-in insul.	*20 SF Flr./LF Part.*	S.F. Partition	11.27	6.76	22.3%
1020	Interior Doors	Single leaf kalamein fire doors, low VOC paint	*750 S.F. Floor/Door*	Each	1133	1.51	
1030	Fittings	Toilet partitions, chalkboards		S.F. Floor	1.40	1.40	
2010	Stair Construction	Concrete filled metal pan		Flight	14,075	.77	
3010	Wall Finishes	50% paint, low VOC, 40% glazed coatings, 10% ceramic tile		S.F. Surface	3.48	4.18	
3020	Floor Finishes	50% vinyl comp. tile, recycled content, 30% carpet tile, 20% terrrazzo		S.F. Floor	8.24	8.24	
3030	Ceiling Finishes	Mineral fiberboard on concealed zee bars		S.F. Ceiling	6.57	6.57	
D. SERVICES							
	D10 Conveying						
1010	Elevators & Lifts	One hydraulic passenger elevator		Each	80,300	.73	0.6%
1020	Escalators & Moving Walks	N/A		—	—	—	
	D20 Plumbing						
2010	Plumbing Fixtures	Kitchen, toilet, low flow, auto sensor, & service fixt., supply & drain.	*1 Fixt./1170 SF Flr.*	Each	6330	5.41	4.9%
2020	Domestic Water Distribution	Gas fired, tankless water heater		S.F. Floor	.33	.33	
2040	Rain Water Drainage	Roof drains		S.F. Roof	1.38	.69	
	D30 HVAC						
3010	Energy Supply	N/A		—	—	—	15.6 %
3020	Heat Generating Systems	Included in D3050		—	—	—	
3040	Distribution Systems	Enthalpy heat recovery packages		Each	36,450	2.32	
3050	Terminal & Package Units	Multizone rooftop air conditioner, SEER 14		S.F. Floor	18.30	18.30	
3090	Other HVAC Sys. & Equipment	N/A		—	—	—	
	D40 Fire Protection						
4010	Sprinklers	Sprinklers, light hazard	*10% of area*	S.F. Floor	2.16	2.16	1.9%
4020	Standpipes	Standpipe, wet, Class III		S.F. Floor	.40	.40	
	D50 Electrical						
5010	Electrical Service/Distribution	1600 ampere service, panel board and feeders		S.F. Floor	.72	.72	13.5%
5020	Lighting & Branch Wiring	LED fixtures, daylt. dim., ltg. on/off, recept., switches, and A.C. power		S.F. Floor	12.13	12.13	
5030	Communications & Security	Addressable alarm systems, internet wiring, comm. systems and emerg. ltg.		S.F. Floor	4.15	4.15	
5090	Other Electrical Systems	Emergency generator, 100 kW, and energy monitoring systems		S.F. Floor	.75	.75	
E. EQUIPMENT & FURNISHINGS							
1010	Commercial Equipment	N/A		—	—	—	2.7 %
1020	Institutional Equipment	Laboratory casework and counters		S.F. Floor	2.39	2.39	
1090	Other Equipment	Waste handlg. recyc. tilt truck, built-in athletic equip., bleachers & backstps.		Each	1.16	1.16	
2020	Moveable Furnishings	No smoking signage		Each	.02	.02	
F. SPECIAL CONSTRUCTION							
1020	Integrated Construction	N/A		—	—	—	0.0 %
1040	Special Facilities	N/A		—	—	—	
G. BUILDING SITEWORK	**N/A**						
					Sub-Total	131.88	**100%**
	CONTRACTOR FEES (General Requirements: 10%, Overhead: 5%, Profit: 10%)				*25%*	32.98	
	ARCHITECT FEES				*7%*	11.54	
					Total Building Cost	**176.40**	

Costs per square foot of floor area

Exterior Wall	S.F. Area	12000	16000	20000	26000	32000	38000	44000	52000	60000
	L.F. Perimeter	450	510	570	650	716	780	840	920	1020
Face Brick with Concrete Block Back-up	Steel Frame	167.80	157.45	151.25	145.05	140.65	137.50	135.20	132.80	131.65
	Bearing Walls	152.50	142.10	135.95	129.75	125.35	122.20	119.90	117.55	116.30
Stucco on Concrete Block	Steel Frame	158.35	149.45	144.10	138.75	135.05	132.35	130.40	128.45	127.40
	Bearing Walls	156.00	147.00	141.70	136.40	132.60	130.00	128.00	126.00	125.00
Precast Concrete Panels	Steel Frame	175.00	163.50	156.70	149.85	144.95	141.40	138.85	136.20	134.85
Metal Sandwich Panels	Steel Frame	168.95	158.40	152.10	145.80	141.30	138.15	135.75	133.40	132.15
Perimeter Adj., Add or Deduct	Per 100 L.F.	10.15	7.55	6.00	4.60	3.75	3.20	2.75	2.45	2.00
Story Hgt. Adj., Add or Deduct	Per 1 Ft.	2.15	1.80	1.60	1.40	1.30	1.20	1.05	1.05	0.95
***For Basement**, add $ 25.70 per square foot of basement area*										

The above costs were calculated using the basic specifications shown on the facing page. These costs should be adjusted where necessary for design alternatives and owner's requirements. Reported completed project costs, for this type of structure, range from $ 81.50 to $ 233.40 per S.F.

Common additives

Description	Unit	$ Cost
Check Out Counter, single belt	Ea.	3575
Scanner, registers, guns & memory 10 lanes	Ea.	173,500
Power take away	Ea.	6000
Refrigerators, prefabricated, walk-in, 7'-6" High, 10' x 10'	S.F.	157
12' x 14'	S.F.	140
12' x 20'	S.F.	123
Refrigerated Food Cases, dairy, multi deck, 12' long	Ea.	12,800
Delicatessen case, multi deck, 18 S.F. shelf display	Ea.	8775
Freezer, glass door upright, 78 C.F.	Ea.	13,500
Frozen food, chest type, 12' long	Ea.	9375
Glass door reach in, 5 door	Ea.	15,900
Island case 12' long, multi deck	Ea.	9700
Meat case, 12' long, multi deck	Ea.	11,700
Produce, 12' long multi deck	Ea.	9950
Safe, Office type, 1 hr rating, 60"x36"x18", double door	Ea.	9350

Description	Unit	$ Cost
Sound System, amplifier, 250 watts	Ea.	2125
Speaker, ceiling or wall	Ea.	206
Commissioning Fees, sustainable commercial construction	S.F.	0.24 - 3.04
Energy Modelling Fees, commercial buildings to 10,000 SF	Ea.	11,000
Greater than 10,000 SF add	S.F.	0.04
Green Bldg Cert Fees for comm construction project reg	Project	900
Photovoltaic Pwr Sys, grid connected, 20 kW (~2400 SF), roof	Ea.	241,700
Green Roofs, 6" soil depth, w/treated wd edging & sedum mats	S.F.	11.27
10" Soil depth, with treated wood edging & sedum mats	S.F.	12.84
Greywater Recovery Systems, prepackaged comm, 3060 gal.	Ea.	58,085
4590 gal.	Ea.	72,815
Rainwater Harvest Sys, prepckged comm, 10,000 gal, sys contrller	Ea.	31,175
20,000 gal. w/system controller	Ea.	50,100
30,000 gal. w/system controller	Ea.	80,225
Solar Domestic HW, closed loop, add-on sys, ext heat exchanger	Ea.	9750
Drainback, hot water system, 120 gal tank	Ea.	12,425
Draindown, hot water system, 120 gal tank	Ea.	12,700

Model costs calculated for a 1 story building with 18′ story height and 44,000 square feet of floor area

G Supermarket

				Unit	Unit Cost	Cost Per S.F.	% Of Sub-Total
A. SUBSTRUCTURE							
1010	Standard Foundations	Poured concrete; strip and spread footings		S.F. Ground	1.08	1.08	8.8%
1020	Special Foundations	N/A		—	—	—	
1030	Slab on Grade	4″ reinforced concrete with recycled vapor barrier and granular base		S.F. Slab	5.11	5.11	
2010	Basement Excavation	Site preparation for slab and trench for foundation wall and footing		S.F. Ground	.16	.16	
2020	Basement Walls	4′ foundation wall		L.F. Wall	81	1.54	
B. SHELL							
	B10 Superstructure						
1010	Floor Construction	N/A		—	—	—	8.7 %
1020	Roof Construction	Metal deck, open web steel joists, beams, interior columns		S.F. Roof	7.81	7.81	
	B20 Exterior Enclosure						
2010	Exterior Walls	Face brick with concrete block backup	*85% of wall*	S.F. Wall	35	10.31	16.5%
2020	Exterior Windows	Storefront windows	*15% of wall*	Each	61	3.13	
2030	Exterior Doors	Sliding entrance doors with electrical operator, hollow metal		Each	5459	1.37	
	B30 Roofing						
3010	Roof Coverings	Single-ply TPO membrane, 60 mils, heat welded seams w/R-20 insul.		S.F. Roof	4.98	4.98	5.7%
3020	Roof Openings	Roof hatches		S.F. Roof	.12	.12	
C. INTERIORS							
1010	Partitions	50% CMU w/foamed-in insul., 50% gyp. bd. on mtl. studs	*50 SF Flr./LF Part.*	S.F. Partition	8.63	2.07	16.7%
1020	Interior Doors	Single leaf hollow metal, low VOC paint	*2000 S.F. Floor/Door*	Each	1133	.57	
1030	Fittings	N/A		—	—	—	
2010	Stair Construction	N/A		—	—	—	
3010	Wall Finishes	Paint, low VOC		S.F. Surface	5.02	2.41	
3020	Floor Finishes	Vinyl composition tile, recycled content		S.F. Floor	3.33	3.33	
3030	Ceiling Finishes	Mineral fiber tile on concealed zee bars		S.F. Ceiling	6.57	6.57	
D. SERVICES							
	D10 Conveying						
1010	Elevators & Lifts	N/A		—	—	—	0.0 %
1020	Escalators & Moving Walks	N/A		—	—	—	
	D20 Plumbing						
2010	Plumbing Fixtures	Toilet, low flow, auto sensor, & service fixt., supply & drain.	*1 Fixt./1820 SF Flr.*	Each	2257	1.24	2.7%
2020	Domestic Water Distribution	Point of use water heater, electric, energy saver		S.F. Floor	.19	.19	
2040	Rain Water Drainage	Roof drains		S.F. Roof	.96	.96	
	D30 HVAC						
3010	Energy Supply	N/A		—	—	—	7.4 %
3020	Heat Generating Systems	Included in D3050		—	—	—	
3040	Distribution Systems	Enthalpy heat recovery packages		Each	29,975	2.73	
3050	Terminal & Package Units	Single zone rooftop air conditioner		S.F. Floor	3.87	3.87	
3090	Other HVAC Sys. & Equipment	N/A		—	—	—	
	D40 Fire Protection						
4010	Sprinklers	Sprinklers, light hazard		S.F. Floor	3.28	3.28	4.3%
4020	Standpipes	Standpipe		S.F. Floor	.54	.54	
	D50 Electrical						
5010	Electrical Service/Distribution	1600 ampere service, panel board and feeders		S.F. Floor	1.83	1.83	29.1%
5020	Lighting & Branch Wiring	LED fixtures, daylt. dim., ltg. on/off, receptacles, switches, and A.C.power		S.F. Floor	21	21.11	
5030	Communications & Security	Addressable alarm systems, partial internet wiring and emergency lighting		S.F. Floor	2.14	2.14	
5090	Other Electrical Systems	Emergency generator, 15 kW and energy monitoring systems		S.F. Floor	1.02	1.02	
E. EQUIPMENT & FURNISHINGS							
1010	Commercial Equipment	N/A		—	—	—	0.2 %
1020	Institutional Equipment	N/A		—	—	—	
1090	Other Equipment	Waste handling recycling tilt truck		Each	.13	.13	
2020	Moveable Furnishings	No smoking signage		Each	.02	.02	
F. SPECIAL CONSTRUCTION							
1020	Integrated Construction	N/A		—	—	—	0.0 %
1040	Special Facilities	N/A		—	—	—	
G. BUILDING SITEWORK	**N/A**						
					Sub-Total	89.62	**100%**
	CONTRACTOR FEES (General Requirements: 10%, Overhead: 5%, Profit: 10%)				*25%*	22.44	
	ARCHITECT FEES				*7%*	7.84	
					Total Building Cost	**119.90**	

G Warehouse

Costs per square foot of floor area

Exterior Wall	S.F. Area	10000	15000	20000	25000	30000	35000	40000	50000	60000
	L.F. Perimeter	410	500	600	640	700	766	833	966	1000
Tiltup Concrete Panels	Steel Frame	142.35	131.05	125.85	120.65	117.80	115.85	114.50	112.55	109.90
Brick with Block Back-up	Bearing Walls	162.35	146.50	139.35	131.60	127.40	124.65	122.65	119.75	115.55
Concrete Block	Steel Frame	148.35	138.35	133.75	128.95	126.35	124.60	123.35	121.55	119.00
	Bearing Walls	133.35	123.30	118.70	113.95	**111.35**	109.55	108.40	106.55	104.00
Galvanized Steel Siding	Steel Frame	154.35	144.05	139.40	134.45	131.75	130.00	128.70	126.85	124.20
Metal Sandwich Panels	Steel Frame	156.70	145.15	139.85	134.15	131.10	129.05	127.60	125.50	122.40
Perimeter Adj., Add or Deduct	Per 100 L.F.	8.30	5.50	4.15	3.35	2.80	2.45	2.10	1.60	1.30
Story Hgt. Adj., Add or Deduct	Per 1 Ft.	0.95	0.80	0.75	0.65	0.55	0.60	0.55	0.45	0.40
***For Basement**, add $ 28.55 per square foot of basement area*										

The above costs were calculated using the basic specifications shown on the facing page. These costs should be adjusted where necessary for design alternatives and owner's requirements. Reported completed project costs, for this type of structure, range from $ 42.80 to $ 170.95 per S.F.

Common additives

Description	Unit	$ Cost
Dock Leveler, 10 ton cap., 6' x 8'	Ea.	5900
7' x 8'	Ea.	7775
Fence, Chain link, 6' high, 9 ga. wire	L.F.	28
6 ga. wire	L.F.	31.50
Gate	Ea.	380
Paving, Bituminous, wearing course plus base course	S.Y.	6.60
Sidewalks, Concrete 4" thick	S.F.	4.23
Sound System, amplifier, 250 watts	Ea.	2125
Speaker, ceiling or wall	Ea.	206
Yard Lighting, 20' aluminum pole, 400W, HP sodium fixture	Ea.	3050

Description	Unit	$ Cost
Commissioning Fees, sustainable industrial construction	S.F.	0.20 - 1.06
Energy Modelling Fees, commercial buildings to 10,000 SF	Ea.	11,000
Greater than 10,000 SF add	S.F.	0.04
Green Bldg Cert Fees for comm construction project reg	Project	900
Photovoltaic Pwr Sys, grid connected, 20 kW (~2400 SF), roof	Ea.	241,700
Green Roofs, 6" soil depth, w/treated wd edging & sedum mats	S.F.	11.27
10" Soil depth, with treated wood edging & sedum mats	S.F.	12.84
Greywater Recovery Systems, prepackaged comm, 3060 gal.	Ea.	58,085
4590 gal.	Ea.	72,815
Rainwater Harvest Sys, prepckged comm, 10,000 gal, sys contrller	Ea.	31,175
20,000 gal. w/system controller	Ea.	50,100
30,000 gal. w/system controller	Ea.	80,225
Solar Domestic HW, closed loop, add-on sys, ext heat exchanger	Ea.	9750
Drainback, hot water system, 120 gal tank	Ea.	12,425
Draindown, hot water system, 120 gal tank	Ea.	12,700

Model costs calculated for a 1 story building with 24′ story height and 30,000 square feet of floor area

G Warehouse

				Unit	Unit Cost	Cost Per S.F.	% Of Sub-Total
A. SUBSTRUCTURE							
1010	Standard Foundations	Poured concrete; strip and spread footings		S.F. Ground	1.36	1.36	19.8%
1020	Special Foundations	N/A		—	—	—	
1030	Slab on Grade	5″ reinforced concrete with recycled vapor barrier and granular base		S.F. Slab	12.52	12.52	
2010	Basement Excavation	Site preparation for slab and trench for foundation wall and footing		S.F. Ground	.16	.16	
2020	Basement Walls	4′ foundation wall		L.F. Wall	81	2.44	
B. SHELL							
B10 Superstructure							
1010	Floor Construction	Mezzanine: open web steel joists, slab form, concrete beams, columns	*10% of area*	S.F. Floor	2.07	2.07	11.0%
1020	Roof Construction	Metal deck, open web steel joists, beams, columns		S.F. Roof	7.05	7.05	
B20 Exterior Enclosure							
2010	Exterior Walls	Concrete block	*95% of wall*	S.F. Wall	16	8.51	11.7%
2020	Exterior Windows	N/A		—	—	—	
2030	Exterior Doors	Steel overhead, hollow metal	*5% of wall*	Each	3204	1.19	
B30 Roofing							
3010	Roof Coverings	Single-ply TPO membrane, 60 mils, heat welded w/R-20 insul.		S.F. Roof	5.14	5.14	6.6%
3020	Roof Openings	Roof hatches and skylight		S.F. Roof	.32	.32	
C. INTERIORS							
1010	Partitions	Concrete block w/foamed-in insul. (office and washrooms)	*100 SF Flr./LF Part.*	S.F. Partition	10.13	.81	8.2%
1020	Interior Doors	Single leaf hollow metal, low VOC paint	*5000 S.F. Floor/Door*	Each	1133	.23	
1030	Fittings	N/A		—	—	—	
2010	Stair Construction	Steel gate with rails		Flight	13,850	.93	
3010	Wall Finishes	Paint, low VOC		S.F. Surface	13	2.08	
3020	Floor Finishes	90% hardener, 10% vinyl composition tile, recycled content		S.F. Floor	2.12	2.12	
3030	Ceiling Finishes	Suspended mineral tile on zee channels in office area	*10% of area*	S.F. Ceiling	.66	.66	
D. SERVICES							
D10 Conveying							
1010	Elevators & Lifts	N/A		—	—	—	0.0 %
1020	Escalators & Moving Walks	N/A		—	—	—	
D20 Plumbing							
2010	Plumbing Fixtures	Toilet, low flow, auto sensor & service fixt., supply & drainage	*1 Fixt./2500 SF Flr.*	Each	3975	1.59	3.8%
2020	Domestic Water Distribution	Tankless, on-demand water heater, gas/propane		S.F. Floor	.39	.39	
2040	Rain Water Drainage	Roof drains		S.F. Roof	1.22	1.22	
D30 HVAC							
3010	Energy Supply	N/A		—	—	—	6.8 %
3020	Heat Generating Systems	Ventilation with heat		Each	139,800	5.03	
3030	Cooling Generating Systems	N/A		—	—	—	
3050	Terminal & Package Units	Single zone rooftop air conditioner, SEER 14		S.F. Floor	.65	.65	
3090	Other HVAC Sys. & Equipment	N/A		—	—	—	
D40 Fire Protection							
4010	Sprinklers	Sprinklers, ordinary hazard		S.F. Floor	3.78	3.78	7.3%
4020	Standpipes	Standpipe		S.F. Floor	2.26	2.26	
D50 Electrical							
5010	Electrical Service/Distribution	200 ampere service, panel board and feeders		S.F. Floor	.29	.29	16.0%
5020	Lighting & Branch Wiring	LED fixtures, daylt. dim., ltg. on/off, receptacles, switches, and A.C. power		S.F. Floor	9.24	9.24	
5030	Communications & Security	Addressable alarm systems		S.F. Floor	2.61	2.61	
5090	Other Electrical Systems	Energy monitoring systems		S.F. Floor	1.19	1.19	
E. EQUIPMENT & FURNISHINGS							
1010	Commercial Equipment	N/A		—	—	—	2.4 %
1030	Vehicular Equipment	Dock boards, dock levelers		S.F. Floor	1.82	1.82	
1090	Other Equipment	Waste handling recycling tilt truck		Each	.19	.19	
2020	Moveable Furnishings	No smoking signage		Each	.02	.02	
F. SPECIAL CONSTRUCTION							
1020	Integrated Construction	Shipping & receiving air curtain		S.F. Floor	5.37	5.37	6.5%
1040	Special Facilities	N/A		—	—	—	
G. BUILDING SITEWORK		**N/A**					
					Sub-Total	83.24	**100%**
	CONTRACTOR FEES (General Requirements: 10%, Overhead: 5%, Profit: 10%)				*25%*	20.83	
	ARCHITECT FEES				*7%*	7.28	
					Total Building Cost	**111.35**	

Assemblies Section

Table of Contents

Table of Contents

Table of Contents

Table of Contents

How RSMeans Assemblies Data Works

Assemblies estimating provides a fast and reasonably accurate way to develop construction costs. An assembly is the grouping of individual work items, with appropriate quantities, to provide a cost for a major construction component in a convenient unit of measure.

An assemblies estimate is often used during early stages of design development to compare the cost impact of various design alternatives on total building cost.

Assemblies estimates are also used as an efficient tool to verify construction estimates.

Assemblies estimates do not require a completed design or detailed drawings. Instead, they are based on the general size of the structure and other known parameters of the project. The degree of accuracy of an assemblies estimate is generally within +/- 15%.

Most RSMeans assemblies consist of three major elements: a graphic, the system components, and the cost data itself. The **Graphic** is a visual representation showing

1 Unique 12-character Identifier

RSMeans assemblies are identified by a **unique 12-character identifier**. The assemblies are numbered using UNIFORMAT II, ASTM Standard E1557. The first 5 characters represent this system to Level 3. The last 7 characters represent further breakdown by RSMeans in order to arrange items in understandable groups of similar tasks. Line numbers are consistent across all RSMeans publications, so a line number in any RSMeans assemblies data set will always refer to the same work.

2 Narrative Descriptions

RSMeans assemblies descriptions appear in two formats: narrative and table. **Narrative descriptions** are shown in a hierarchical structure to make them readable. In order to read a complete description, read up through the indents to the top of the section. Include everything that is above and to the left that is not contradicted by information below.

Narrative Format

C20 Stairs

C2010 Stair Construction

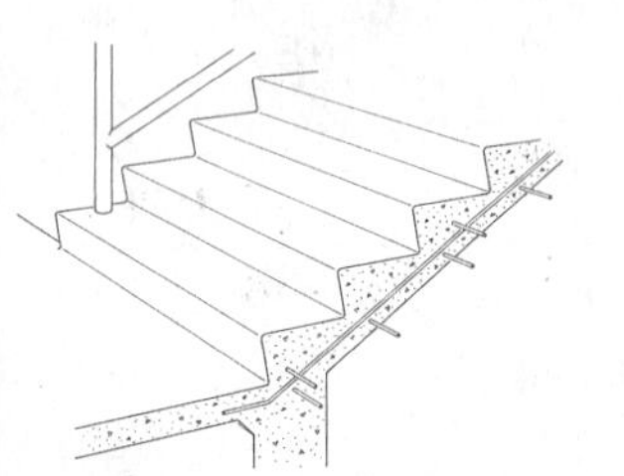

General Design: See reference section for code requirements. Maximum height between landings is 12′; usual stair angle is 20° to 50° with 30° to 35° best. Usual relation of riser to treads is:
Riser + tread = 17.5.
2x (Riser) + tread = 25.
Riser x tread = 70 or 75.
Maximum riser height is 7″ for commercial, 8-1/4″ for residential.
Usual riser height is 6-1/2″ to 7-1/4″.

Minimum tread width is 11″ for commercial and 9″ for residential.

For additional information please see reference section.

Cost Per Flight: Table below lists the cost per flight for 4′-0″ wide stairs.
Side walls are not included.
Railings are included.

C2010 110	Stairs	COST PER FLIGHT		
		MAT.	INST.	TOTAL
0470	Stairs, C.I.P. concrete, w/o landing, 12 risers, w/o nosing	1,125	2,075	3,200
0480	With nosing	2,100	2,250	4,350
0550	W/landing, 12 risers, w/o nosing	1,300	2,550	3,850
0560	With nosing	2,250	2,725	4,975
0570	16 risers, w/o nosing	1,575	3,200	4,775
0580	With nosing	2,875	3,425	6,300
0590	20 risers, w/o nosing	1,875	3,850	5,725
0600	With nosing	3,475	4,150	7,625
0610	24 risers, w/o nosing	2,150	4,500	6,650
0620	With nosing	4,100	4,850	8,950
0630	Steel, grate type w/nosing & rails, 12 risers, w/o landing	5,700	1,025	6,725
0640	With landing	7,975	1,400	9,375
0660	16 risers, with landing	9,875	1,725	11,600
0680	20 risers, with landing	11,800	2,050	13,850
0700	24 risers, with landing	13,700	2,400	16,100
0710	Concrete fill metal pan & picket rail, 12 risers, w/o landing	7,375	1,025	8,400
0720	With landing	9,725	1,525	11,250
0740	16 risers, with landing	12,200	1,875	14,075
0760	20 risers, with landing	14,600	2,200	16,800
0780	24 risers, with landing	17,100	2,525	19,625
0790	Cast iron tread & pipe rail, 12 risers, w/o landing	7,375	1,025	8,400
0800	With landing	9,725	1,525	11,250
1120	Wood, prefab box type, oak treads, wood rails 3′-6″ wide, 14 risers	2,250	425	2,675
1150	Prefab basement type, oak treads, wood rails 3′-0″ wide, 14 risers	1,100	104	1,204

For supplemental customizable square foot estimating forms, visit: **www.RSMeans.com/2016extras**

the typical appearance of the assembly in question, frequently accompanied by additional explanatory technical information describing the class of items. The **System Components** is a listing of the individual tasks that make up the assembly, including the quantity and unit of measure for each item, along with the cost of material and installation.

The **Assemblies Data** below lists prices for other similar systems with dimensional and/or size variations.

All RSMeans assemblies costs represent the cost for the installing contractor. An allowance for profit has been added to all material, labor, and equipment rental costs. A markup for labor burdens, including workers' compensation, fixed overhead, and business overhead, is included with installation costs.

The information in RSMeans cost data represents a "national average" cost. This data should be modified to the project location using the **City Cost Indexes** or **Location Factors** tables found in the Reference Section.

Table Format

A20 Basement Construction

A2020 Basement Walls

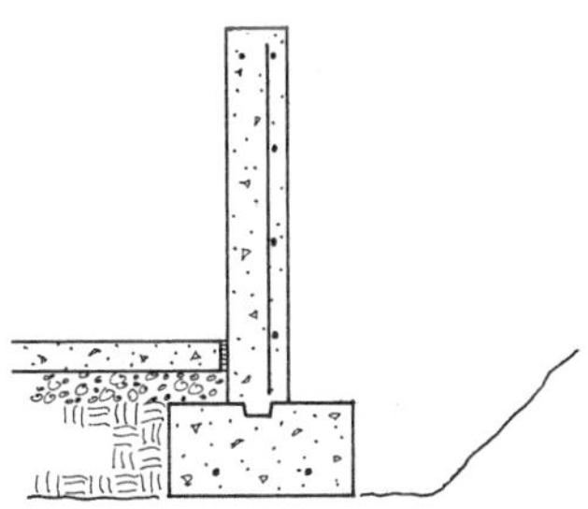

The Foundation Bearing Wall System includes: forms up to 16′ high (four uses); 3,000 p.s.i. concrete placed and vibrated; and form removal with breaking form ties and patching walls. The wall systems list walls from 6″ to 16″ thick and are designed with minimum reinforcement.

Excavation and backfill are not included.

Please see the reference section for further design and cost information.

A2020 110 — Walls, Cast in Place

	WALL HEIGHT (FT.)	PLACING METHOD	CONCRETE (C.Y. per L.F.)	REINFORCING (LBS. per L.F.)	WALL THICKNESS (IN.)	COST PER L.F. MAT.	COST PER L.F. INST.	COST PER L.F. TOTAL
1500	4′	direct chute	.074	3.3	6	17	42.50	59.50
1520			.099	4.8	8	20.50	44	64.50
1540			.123	6.0	10	24	44.50	68.50
1560			.148	7.2	12	28	45.50	73.50
1580			.173	8.1	14	31	46.50	77.50
1600			.197	9.44	16	34.50	47.50	82
3000	6′	direct chute	.111	4.95	6	25.50	64	89.50
3020			.149	7.20	8	31	65.50	96.50
3040			.184	9.00	10	36	67	103
3060			.222	10.8	12	41.50	68.50	110
5000	8′	direct chute	.148	6.6	6	34	85	119
5020			.199	9.6	8	41.50	91	132.50
5040			.250	12	10	49	89.50	138.50
5061			.296	14.39	12	55.50	92	147.50
6020	10′	direct chute	.248	12	8	52	110	162
6040			.307	14.99	10	60.50	112	172.50
6061			.370	17.99	12	69.50	114	183.50
7220	12′	pumped	.298	14.39	8	62.50	136	198.50
7240			.369	17.99	10	72.50	139	211.50
7260			.444	21.59	12	83.50	144	227.50
9220	16′	pumped	.397	19.19	8	83	182	265
9240			.492	23.99	10	96.50	185	281.50
9260			.593	28.79	12	111	191	302

3 Unit of Measure

All RSMeans assemblies data includes a typical **Unit of Measure** used for estimating that item. For instance, for continuous footings or foundation walls the unit is linear feet (L.F.). For spread footings the unit is each (Ea.). The estimator needs to take special care that the unit in the data matches the unit in the takeoff. Abbreviations and unit conversions can be found in the Reference Section.

4 Table Descriptions

Table descriptions work in a similar fashion, except that if there is a blank in the column at a particular line number, read up to the description above in the same column.

How RSMeans Assemblies Data Works (Continued)

Sample Estimate

This sample demonstrates the elements of an estimate, including a tally of the RSMeans data lines. Published assemblies costs include all markups for labor burden and profit for the installing contractor. This estimate adds a summary of the markups applied by a general contractor on the installing contractor's work. These figures represent the total cost to the owner. The RSMeans location factor is applied at the bottom of the estimate to adjust the cost of the work to a specific location.

Project Name:	**Interior Fit-out, ABC Office**			
Location:	**Anywhere, USA**	**Date: 1/1/2016**		**OPN**
Assembly Number	**Description**	**Qty.**	**Unit**	**Subtotal**
C1010 124 1200 ❶	Wood partition, 2 x 4 @ 16" OC w/5/8" FR gypsum board	560.000	S.F.	$2,464.00
C1020 114 1800	Metal door & frame, flush hollow core, 3'-0" x 7'-0"	2.000	Ea.	$2,350.00
C3010 230 0080	Painting, brushwork, primer & 2 coats	1,120.000	S.F.	$1,198.40
C3020 410 0140	Carpet, tufted, nylon, roll goods, 12' wide, 26 oz	240.000	S.F.	$787.20
C3030 210 6000	Acoustic ceilings, 24" x 48" tile, tee grid suspension	200.000	S.F.	$1,102.00
D5020 125 0560	Receptacles incl plate, box, conduit, wire, 20 A duplex	8.000	Ea.	$1,944.00
D5020 125 0720	Light switch incl plate, box, conduit, wire, 20 A single pole	2.000	Ea.	$476.00
D5020 210 0560	Fluorescent fixtures, recess mounted, 20 per 1000 SF	200.000	S.F.	$1,996.00
	Assembly Subtotal			**$12,317.60**
	Sales Tax @ ❷	**5 %**		**$ 307.94**
	General Requirements @ ❸	**7 %**		**$ 862.23**
	Subtotal A			**$13,487.77**
	GC Overhead @ ❹	**5 %**		**$ 674.39**
	Subtotal B			**$14,162.16**
	GC Profit @ ❺	**5 %**		**$ 708.11**
	Subtotal C			**$14,870.27**
	Adjusted by Location Factor ❻	**118.1**		**$ 17,561.79**
	Architects Fee @ ❼	**8 %**		**$ 1,404.94**
	Contingency @ ❽	**15 %**		**$ 2,634.27**
	Project Total Cost			**$ 21,601.00**

This estimate is based on an interactive spreadsheet.
A copy of this spreadsheet is located on the RSMeans website at **www.RSMeans.com/2016extras**.
You are free to download it and adjust it to your methodology.

1 Work Performed

The body of the estimate shows the RSMeans data selected, including line numbers, a brief description of each item, its takeoff quantity and unit, and the total installed cost, including the installing contractor's overhead and profit.

2 Sales Tax

If the work is subject to state or local sales taxes, the amount must be added to the estimate. In a conceptual estimate it can be assumed that one half of the total represents material costs. Therefore, apply the sales tax rate to 50% of the assembly subtotal.

3 General Requirements

This item covers project-wide needs provided by the general contractor. These items vary by project but may include temporary facilities and utilities, security, testing, project cleanup, etc. In assemblies estimates a percentage is used, typically between 5% and 15% of project cost.

4 General Contractor Overhead

This entry represents the general contractor's markup on all work to cover project administration costs.

5 General Contractor Profit

This entry represents the GC's profit on all work performed. The value included here can vary widely by project and is influenced by the GC's perception of the project's financial risk and market conditions.

6 Location Factor

RSMeans published data is based on national average costs. If necessary, adjust the total cost of the project using a location factor from the "Location Factor" table or the "City Cost Indexes" table found in the Reference Section. Use location factors if the work is general, covering the work of multiple trades. If the work is by a single trade (e.g., masonry) use the more specific data found in the City Cost Indexes.

To adjust costs by location factors, multiply the base cost by the factor and divide by 100.

7 Architect's Fee

If appropriate, add the design cost to the project estimate. These fees vary based on project complexity and size. Typical design and engineering fees can be found in the reference section.

8 Contingency

A factor for contingency may be added to any estimate to represent the cost of unknowns that may occur between the time that the estimate is performed and the time the project is constructed. The amount of the allowance will depend on the stage of design at which the estimate is done, and the contractor's assessment of the risk involved.

A1010 Standard Foundations

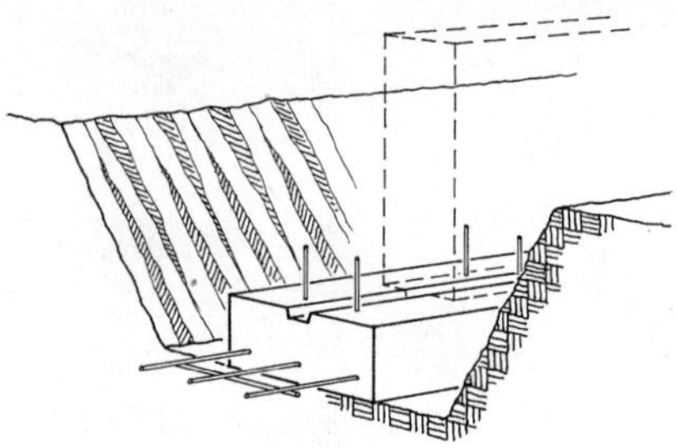

The Strip Footing System includes: excavation; hand trim; all forms needed for footing placement; forms for 2″ x 6″ keyway (four uses); dowels; and 3,000 p.s.i. concrete.

The footing size required varies for different soils. Soil bearing capacities are listed for 3 KSF and 6 KSF. Depths of the system range from 8″ and deeper. Widths range from 16″ and wider. Smaller strip footings may not require reinforcement.

Please see the reference section for further design and cost information.

A1010 110	Strip Footings	COST PER L.F.		
		MAT.	INST.	TOTAL
2100	Strip footing, load 2.6 KLF, soil capacity 3 KSF, 16″ wide x 8″ deep, plain	7.50	9.15	16.65
2300	Load 3.9 KLF, soil capacity 3 KSF, 24″ wide x 8″ deep, plain	9.50	10.35	19.85
2500	Load 5.1 KLF, soil capacity 3 KSF, 24″ wide x 12″ deep, reinf.	17.15	19.80	36.95
2700	Load 11.1 KLF, soil capacity 6 KSF, 24″ wide x 12″ deep, reinf.	17.15	19.80	36.95
2900	Load 6.8 KLF, soil capacity 3 KSF, 32″ wide x 12″ deep, reinf.	20.50	21.50	42
3100	Load 14.8 KLF, soil capacity 6 KSF, 32″ wide x 12″ deep, reinf.	20.50	21.50	42
3300	Load 9.3 KLF, soil capacity 3 KSF, 40″ wide x 12″ deep, reinf.	24	23	47
3500	Load 18.4 KLF, soil capacity 6 KSF, 40″ wide x 12″ deep, reinf.	24.50	23.50	48
4500	Load 10 KLF, soil capacity 3 KSF, 48″ wide x 16″ deep, reinf.	34.50	29.50	64
4700	Load 22 KLF, soil capacity 6 KSF, 48″ wide, 16″ deep, reinf.	35	30.50	65.50
5700	Load 15 KLF, soil capacity 3 KSF, 72″ wide x 20″ deep, reinf.	59	42	101
5900	Load 33 KLF, soil capacity 6 KSF, 72″ wide x 20″ deep, reinf.	62.50	45	107.50

A1010 Standard Foundations

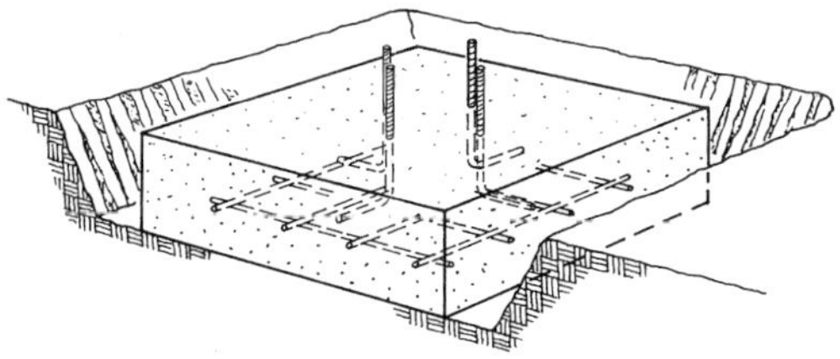

The Spread Footing System includes: excavation; backfill; forms (four uses); all reinforcement; 3,000 p.s.i. concrete (chute placed); and float finish.

Footing systems are priced per individual unit. The Expanded System Listing at the bottom shows various footing sizes. It is assumed that excavation is done by a truck mounted hydraulic excavator with an operator and oiler.

Backfill is with a dozer, and compaction by air tamp. The excavation and backfill equipment is assumed to operate at 30 C.Y. per hour.

Please see the reference section for further design and cost information.

A1010 210	Spread Footings	COST EACH		
		MAT.	INST.	TOTAL
7090	Spread footings, 3000 psi concrete, chute delivered			
7100	Load 25K, soil capacity 3 KSF, 3′-0″ sq. x 12″ deep	60.50	102	162.50
7150	Load 50K, soil capacity 3 KSF, 4′-6″ sq. x 12″ deep	129	177	306
7200	Load 50K, soil capacity 6 KSF, 3′-0″ sq. x 12″ deep	60.50	102	162.50
7250	Load 75K, soil capacity 3 KSF, 5′-6″ sq. x 13″ deep	205	249	454
7300	Load 75K, soil capacity 6 KSF, 4′-0″ sq. x 12″ deep	104	152	256
7350	Load 100K, soil capacity 3 KSF, 6′-0″ sq. x 14″ deep	259	299	558
7410	Load 100K, soil capacity 6 KSF, 4′-6″ sq. x 15″ deep	159	208	367
7450	Load 125K, soil capacity 3 KSF, 7′-0″ sq. x 17″ deep	410	430	840
7500	Load 125K, soil capacity 6 KSF, 5′-0″ sq. x 16″ deep	205	249	454
7550	Load 150K, soil capacity 3 KSF 7′-6″ sq. x 18″ deep	500	500	1,000
7610	Load 150K, soil capacity 6 KSF, 5′-6″ sq. x 18″ deep	273	315	588
7650	Load 200K, soil capacity 3 KSF, 8′-6″ sq. x 20″ deep	710	665	1,375
7700	Load 200K, soil capacity 6 KSF, 6′-0″ sq. x 20″ deep	360	385	745
7750	Load 300K, soil capacity 3 KSF, 10′-6″ sq. x 25″ deep	1,300	1,100	2,400
7810	Load 300K, soil capacity 6 KSF, 7′-6″ sq. x 25″ deep	680	655	1,335
7850	Load 400K, soil capacity 3 KSF, 12′-6″ sq. x 28″ deep	2,075	1,625	3,700
7900	Load 400K, soil capacity 6 KSF, 8′-6″ sq. x 27″ deep	945	850	1,795
8010	Load 500K, soil capacity 6 KSF, 9′-6″ sq. x 30″ deep	1,300	1,100	2,400
8100	Load 600K, soil capacity 6 KSF, 10′-6″ sq. x 33″ deep	1,750	1,425	3,175
8200	Load 700K, soil capacity 6 KSF, 11′-6″ sq. x 36″ deep	2,250	1,750	4,000
8300	Load 800K, soil capacity 6 KSF, 12′-0″ sq. x 37″ deep	2,525	1,925	4,450
8400	Load 900K, soil capacity 6 KSF, 13′-0″ sq. x 39″ deep	3,125	2,275	5,400
8500	Load 1000K, soil capacity 6 KSF, 13′-6″ sq. x 41″ deep	3,550	2,550	6,100

A1010 320	Foundation Dampproofing	COST PER L.F.		
		MAT.	INST.	TOTAL
1000	Foundation dampproofing, bituminous, 1 coat, 4′ high	1	4.01	5.01
1400	8′ high	2	8	10
1800	12′ high	3	12.40	15.40
2000	2 coats, 4′ high	1.96	4.97	6.93
2400	8′ high	3.92	9.95	13.87
2800	12′ high	5.90	15.30	21.20
3000	Asphalt with fibers, 1/16″ thick, 4′ high	1.56	4.97	6.53
3400	8′ high	3.12	9.95	13.07
3800	12′ high	4.68	15.30	19.98
4000	1/8″ thick, 4′ high	2.80	5.90	8.70
4400	8′ high	5.60	11.80	17.40
4800	12′ high	8.40	18.05	26.45

A1010 Standard Foundations

A1010 320	Foundation Dampproofing	COST PER L.F.		
		MAT.	INST.	TOTAL
5000	Asphalt coated board and mastic, 1/4" thick, 4' high	5.10	5.35	10.45
5400	8' high	10.25	10.75	21
5800	12' high	15.35	16.50	31.85
6000	1/2" thick, 4' high	7.70	7.45	15.15
6400	8' high	15.35	14.95	30.30
6800	12' high	23	23	46
7000	Cementitious coating, on walls, 1/8" thick coating, 4' high	3.24	8	11.24
7400	8' high	6.50	16	22.50
7800	12' high	9.70	24	33.70
8000	Cementitious/metallic slurry, 4 coat, 1/2"thick, 2' high	.77	7.70	8.47
8400	4' high	1.54	15.40	16.94
8800	6' high	2.31	23	25.31

A1030 Slab on Grade

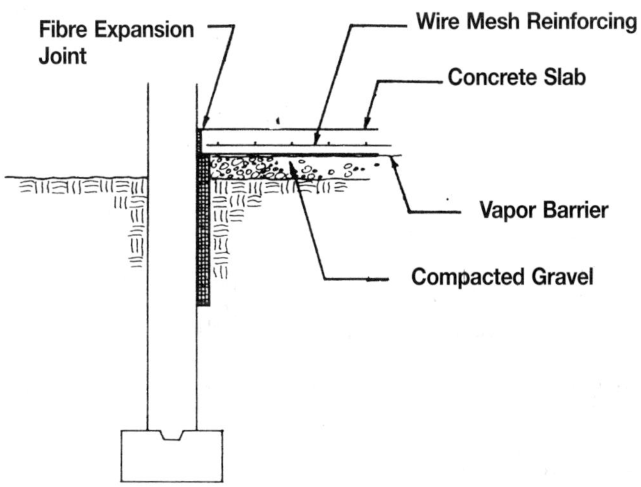

Reinforced Slab on Grade

A Slab on Grade system includes fine grading; 6″ of compacted gravel; vapor barrier; 3500 p.s.i. concrete; bituminous fiber expansion joint; all necessary edge forms and 4 uses; steel trowel finish; and sprayed on membrane curing compound. Wire mesh reinforcing used in all reinforced slabs.

Non-industrial slabs are for foot traffic only with negligible abrasion. Light industrial slabs are for pneumatic wheels and light abrasion. Industrial slabs are for solid rubber wheels and moderate abrasion. Heavy industrial slabs are for steel wheels and severe abrasion.

A1030 120	Plain & Reinforced	COST PER S.F.		
		MAT.	INST.	TOTAL
2220	Slab on grade, 4″ thick, non industrial, non reinforced	2.22	2.42	4.64
2240	Reinforced	2.37	2.73	5.10
2260	Light industrial, non reinforced	2.92	2.96	5.88
2280	Reinforced	3.07	3.27	6.34
2300	Industrial, non reinforced	3.59	6.20	9.79
2320	Reinforced	3.74	6.50	10.24
3340	5″ thick, non industrial, non reinforced	2.59	2.49	5.08
3360	Reinforced	2.74	2.80	5.54
3380	Light industrial, non reinforced	3.29	3.03	6.32
3400	Reinforced	3.44	3.34	6.78
3420	Heavy industrial, non reinforced	4.71	7.35	12.06
3440	Reinforced	4.81	7.70	12.51
4460	6″ thick, non industrial, non reinforced	3.07	2.44	5.51
4480	Reinforced	3.30	2.85	6.15
4500	Light industrial, non reinforced	3.79	2.98	6.77
4520	Reinforced	4.16	3.54	7.70
4540	Heavy industrial, non reinforced	5.20	7.45	12.65
4560	Reinforced	5.45	7.85	13.30
5580	7″ thick, non industrial, non reinforced	3.45	2.52	5.97
5600	Reinforced	3.75	2.96	6.71
5620	Light industrial, non reinforced	4.18	3.06	7.24
5640	Reinforced	4.48	3.50	7.98
5660	Heavy industrial, non reinforced	5.60	7.35	12.95
5680	Reinforced	5.80	7.80	13.60
6700	8″ thick, non industrial, non reinforced	3.82	2.56	6.38
6720	Reinforced	4.07	2.93	7
6740	Light industrial, non reinforced	4.55	3.10	7.65
6760	Reinforced	4.80	3.47	8.27
6780	Heavy industrial, non reinforced	6	7.45	13.45
6800	Reinforced	6.35	7.85	14.20

A2010 Basement Excavation

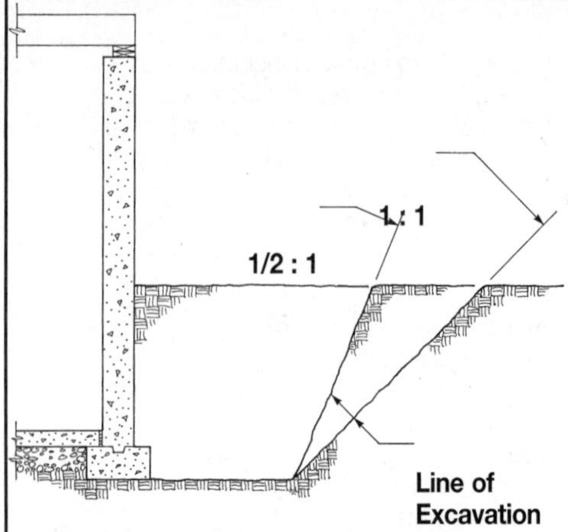

Pricing Assumptions: Two-thirds of excavation is by 2-1/2 C.Y. wheel mounted front end loader and one-third by 1-1/2 C.Y. hydraulic excavator.

Two-mile round trip haul by 12 C.Y. tandem trucks is included for excavation wasted and storage of suitable fill from excavated soil. For excavation in clay, all is wasted and the cost of suitable backfill with two-mile haul is included.

Sand and gravel assumes 15% swell and compaction; common earth assumes 25% swell and 15% compaction; clay assumes 40% swell and 15% compaction (non-clay).

In general, the following items are accounted for in the costs in the table below.

1. Excavation for building or other structure to depth and extent indicated.
2. Backfill compacted in place.
3. Haul of excavated waste.
4. Replacement of unsuitable material with bank run gravel.

Note: Additional excavation and fill beyond this line of general excavation for the building (as required for isolated spread footings, strip footings, etc.) are included in the cost of the appropriate component systems.

A2010 110	Building Excavation & Backfill	COST PER S.F.		
		MAT.	INST.	TOTAL
2220	Excav & fill, 1000 S.F., 4′ sand, gravel, or common earth, on site storage		.96	.96
2240	Off site storage		1.48	1.48
2260	Clay excavation, bank run gravel borrow for backfill	2.25	2.34	4.59
2280	8′ deep, sand, gravel, or common earth, on site storage		5.85	5.85
2300	Off site storage		13.45	13.45
2320	Clay excavation, bank run gravel borrow for backfill	9.15	10.95	20.10
2340	16′ deep, sand, gravel, or common earth, on site storage		15.90	15.90
2350	Off site storage		33	33
2360	Clay excavation, bank run gravel borrow for backfill	25.50	28	53.50
3380	4000 S.F., 4′ deep, sand, gravel, or common earth, on site storage		.51	.51
3400	Off site storage		1.05	1.05
3420	Clay excavation, bank run gravel borrow for backfill	1.15	1.18	2.33
3440	8′ deep, sand, gravel, or common earth, on site storage		4.20	4.20
3460	Off site storage		7.55	7.55
3480	Clay excavation, bank run gravel borrow for backfill	4.16	6.55	10.71
3500	16′ deep, sand, gravel, or common earth, on site storage		10	10
3520	Off site storage		20.50	20.50
3540	Clay, excavation, bank run gravel borrow for backfill	11.15	15.55	26.70
4560	10,000 S.F., 4′ deep, sand gravel, or common earth, on site storage		.29	.29
4580	Off site storage		.62	.62
4600	Clay excavation, bank run gravel borrow for backfill	.70	.73	1.43
4620	8′ deep, sand, gravel, or common earth, on site storage		3.68	3.68
4640	Off site storage		5.65	5.65
4660	Clay excavation, bank run gravel borrow for backfill	2.52	5.10	7.62
4680	16′ deep, sand, gravel, or common earth, on site storage		8.25	8.25
4700	Off site storage		14.35	14.35
4720	Clay excavation, bank run gravel borrow for backfill	6.70	11.65	18.35
5740	30,000 S.F., 4′ deep, sand, gravel, or common earth, on site storage		.16	.16
5760	Off site storage		.36	.36
5780	Clay excavation, bank run gravel borrow for backfill	.41	.42	.83
5860	16′ deep, sand, gravel, or common earth, on site storage		7.10	7.10
5880	Off site storage		10.45	10.45
5900	Clay excavation, bank run gravel borrow for backfill	3.71	9.05	12.76
6910	100,000 S.F., 4′ deep, sand, gravel, or common earth, on site storage		.10	.10
6920	Off site storage		.21	.21
6930	Clay excavation, bank run gravel borrow for backfill	.20	.22	.42
6940	8′ deep, sand, gravel, or common earth, on site storage		3.10	3.10
6950	Off site storage		3.69	3.69
6960	Clay excavation, bank run gravel borrow for backfill	.77	3.56	4.33
6970	16′ deep, sand, gravel, or common earth, on site storage		6.50	6.50
6980	Off site storage		8.30	8.30
6990	Clay excavation, bank run gravel borrow for backfill	2	7.55	9.55

A2020 Basement Walls

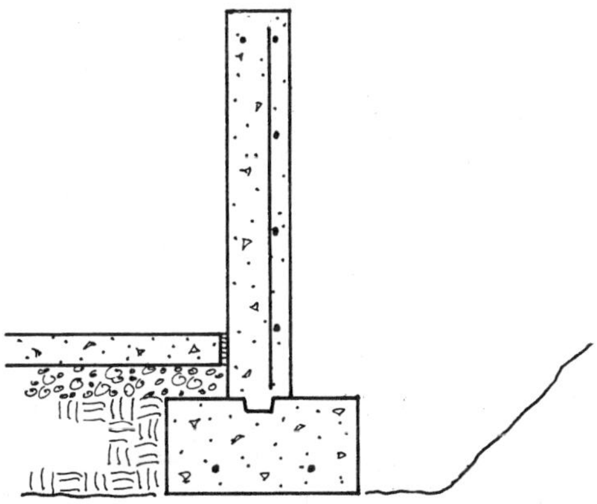

The Foundation Bearing Wall System includes: forms up to 16′ high (four uses); 3,000 p.s.i. concrete placed and vibrated; and form removal with breaking form ties and patching walls. The wall systems list walls from 6″ to 16″ thick and are designed with minimum reinforcement.

Excavation and backfill are not included.

Please see the reference section for further design and cost information.

A2020 110 Walls, Cast in Place

	WALL HEIGHT (FT.)	PLACING METHOD	CONCRETE (C.Y. per L.F.)	REINFORCING (LBS. per L.F.)	WALL THICKNESS (IN.)	COST PER L.F.		
						MAT.	INST.	TOTAL
1500	4′	direct chute	.074	3.3	6	17	42.50	59.50
1520			.099	4.8	8	20.50	44	64.50
1540			.123	6.0	10	24	44.50	68.50
1560			.148	7.2	12	28	45.50	73.50
1580			.173	8.1	14	31	46.50	77.50
1600			.197	9.44	16	34.50	47.50	82
3000	6′	direct chute	.111	4.95	6	25.50	64	89.50
3020			.149	7.20	8	31	65.50	96.50
3040			.184	9.00	10	36	67	103
3060			.222	10.8	12	41.50	68.50	110
5000	8′	direct chute	.148	6.6	6	34	85	119
5020			.199	9.6	8	41.50	91	132.50
5040			.250	12	10	49	89.50	138.50
5061			.296	14.39	12	55.50	92	147.50
6020	10′	direct chute	.248	12	8	52	110	162
6040			.307	14.99	10	60.50	112	172.50
6061			.370	17.99	12	69.50	114	183.50
7220	12′	pumped	.298	14.39	8	62.50	136	198.50
7240			.369	17.99	10	72.50	139	211.50
7260			.444	21.59	12	83.50	144	227.50
9220	16′	pumped	.397	19.19	8	83	182	265
9240			.492	23.99	10	96.50	185	281.50
9260			.593	28.79	12	111	191	302

B1010 Floor Construction

General: It is desirable for purposes of consistency and simplicity to maintain constant column sizes throughout building height. To do this, concrete strength may be varied (higher strength concrete at lower stories and lower strength concrete at upper stories), as well as varying the amount of reinforcing.

The table provides probably minimum column sizes with related costs and weight per lineal foot of story height.

B1010 203 C.I.P. Column, Square Tied

	LOAD (KIPS)	STORY HEIGHT (FT.)	COLUMN SIZE (IN.)	COLUMN WEIGHT (P.L.F.)	CONCRETE STRENGTH (PSI)	COST PER V.L.F.		
						MAT.	INST.	TOTAL
0640	100	10	10	96	4000	10.95	40.50	51.45
0680		12	10	97	4000	11.15	40.50	51.65
0700		14	12	142	4000	14.60	49	63.60
0840	200	10	12	140	4000	15.95	51	66.95
0860		12	12	142	4000	16.25	51.50	67.75
0900		14	14	196	4000	18.95	57	75.95
0920	300	10	14	192	4000	20	58.50	78.50
0960		12	14	194	4000	20.50	59	79.50
0980		14	16	253	4000	23.50	65.50	89
1020	400	10	16	248	4000	25	67	92
1060		12	16	251	4000	25.50	68	93.50
1080		14	16	253	4000	26	68.50	94.50
1200	500	10	18	315	4000	29.50	75	104.50
1250		12	20	394	4000	35.50	85	120.50
1300		14	20	397	4000	36.50	86	122.50
1350	600	10	20	388	4000	37	87	124
1400		12	20	394	4000	38	88	126
1600		14	20	397	4000	38.50	89	127.50
3400	900	10	24	560	4000	50.50	108	158.50
3800		12	24	567	4000	51.50	110	161.50
4000		14	24	571	4000	52.50	111	163.50
7300	300	10	14	192	6000	20	58	78
7500		12	14	194	6000	20.50	58.50	79
7600		14	14	196	6000	21	59	80
8000	500	10	16	248	6000	26	67	93
8050		12	16	251	6000	26.50	68	94.50
8100		14	16	253	6000	27	68.50	95.50
8200	600	10	18	315	6000	30	75.50	105.50
8300		12	18	319	6000	30.50	76.50	107
8400		14	18	321	6000	31	77	108
8800	800	10	20	388	6000	37	85.50	122.50
8900		12	20	394	6000	37.50	86	123.50
9000		14	20	397	6000	38.50	87	125.50

B1010 Floor Construction

B1010 203 C.I.P. Column, Square Tied

	LOAD (KIPS)	STORY HEIGHT (FT.)	COLUMN SIZE (IN.)	COLUMN WEIGHT (P.L.F.)	CONCRETE STRENGTH (PSI)	COST PER V.L.F.		
						MAT.	INST.	TOTAL
9100	900	10	20	388	6000	40	89	129
9300		12	20	394	6000	40.50	90	130.50
9600		14	20	397	6000	41.50	91	132.50

B1010 204 C.I.P. Column, Square Tied-Minimum Reinforcing

	LOAD (KIPS)	STORY HEIGHT (FT.)	COLUMN SIZE (IN.)	COLUMN WEIGHT (P.L.F.)	CONCRETE STRENGTH (PSI)	COST PER V.L.F.		
						MAT.	INST.	TOTAL
9913	150	10-14	12	135	4000	13.25	47.50	60.75
9918	300	10-14	16	240	4000	21.50	62	83.50
9924	500	10-14	20	375	4000	32	80.50	112.50
9930	700	10-14	24	540	4000	44	100	144
9936	1000	10-14	28	740	4000	58	122	180
9942	1400	10-14	32	965	4000	72	137	209
9948	1800	10-14	36	1220	4000	89.50	160	249.50

B1010 Floor Construction

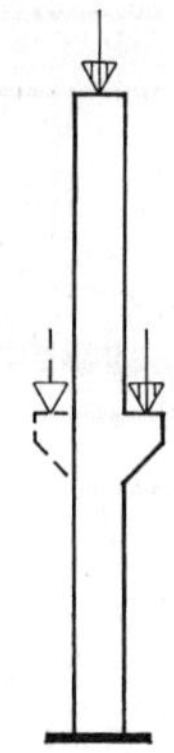

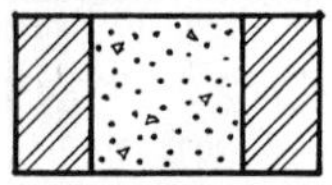

Concentric Load

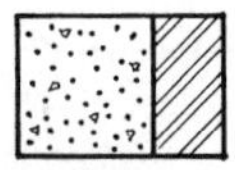

Eccentric Load

General: Data presented here is for plant produced members transported 50 miles to 100 miles to the site and erected.

Design and pricing assumptions:
Normal wt. concrete, f'c = 5 KSI

Main reinforcement, fy = 60 KSI
Ties, fy = 40 KSI

Minimum design eccentricity, 0.1t.

Concrete encased structural steel haunches are assumed where practical; otherwise galvanized rebar haunches are assumed.

Base plates are integral with columns.

Foundation anchor bolts, nuts and washers are included in price.

B1010 206 Tied, Concentric Loaded Precast Concrete Columns

	LOAD (KIPS)	STORY HEIGHT (FT.)	COLUMN SIZE (IN.)	COLUMN WEIGHT (P.L.F.)	LOAD LEVELS	COST PER V.L.F.		
						MAT.	INST.	TOTAL
0560	100	10	12x12	164	2	173	10.05	183.05
0570		12	12x12	162	2	168	8.35	176.35
0580		14	12x12	161	2	168	8.20	176.20
0590	150	10	12x12	166	3	168	8.10	176.10
0600		12	12x12	169	3	168	7.25	175.25
0610		14	12x12	162	3	168	7.15	175.15
0620	200	10	12x12	168	4	167	8.60	175.60
0630		12	12x12	170	4	168	7.75	175.75
0640		14	12x12	220	4	167	7.65	174.65

B1010 207 Tied, Eccentric Loaded Precast Concrete Columns

	LOAD (KIPS)	STORY HEIGHT (FT.)	COLUMN SIZE (IN.)	COLUMN WEIGHT (P.L.F.)	LOAD LEVELS	COST PER V.L.F.		
						MAT.	INST.	TOTAL
1130	100	10	12x12	161	2	165	10.05	175.05
1140		12	12x12	159	2	165	8.35	173.35
1150		14	12x12	159	2	165	8.20	173.20
1390	600	10	18x18	385	4	242	8.60	250.60
1400		12	18x18	380	4	243	7.75	250.75
1410		14	18x18	375	4	242	7.65	249.65
1480	800	10	20x20	490	4	220	8.60	228.60
1490		12	20x20	480	4	220	7.75	227.75
1500		14	20x20	475	4	220	7.65	227.65

B1010 Floor Construction

(A) Wide Flange

(B) Pipe

(C) Pipe, Concrete Filled

(D) Square Tube

(E) Square Tube Concrete Filled

(F) Rectangular Tube

(G) Rectangular Tube, Concrete Filled

General: The following pages provide data for seven types of steel columns: wide flange, round pipe, round pipe concrete filled, square tube, square tube concrete filled, rectangular tube and rectangular tube concrete filled.

Design Assumptions: Loads are concentric; wide flange and round pipe bearing capacity is for 36 KSI steel. Square and rectangular tubing bearing capacity is for 46 KSI steel.

The effective length factor K=1.1 is used for determining column values in the tables. K=1.1 is within a frequently used range for pinned connections with cross bracing.

How To Use Tables:

a. Steel columns usually extend through two or more stories to minimize splices. Determine floors with splices.

b. Enter Table No. below with load to column at the splice. Use the unsupported height.

c. Determine the column type desired by price or design.

Cost:

a. Multiply number of columns at the desired level by the total height of the column by the cost/VLF.

b. Repeat the above for all tiers.

Please see the reference section for further design and cost information.

B1010 208 Steel Columns

	LOAD (KIPS)	UNSUPPORTED HEIGHT (FT.)	WEIGHT (P.L.F.)	SIZE (IN.)	TYPE	COST PER V.L.F.		
						MAT.	INST.	TOTAL
1000	25	10	13	4	A	23	8.65	31.65
1020			7.58	3	B	13.40	8.65	22.05
1040			15	3-1/2	C	16.35	8.65	25
1120			20	4x3	G	18.85	8.65	27.50
1200		16	16	5	A	26	6.45	32.45
1220			10.79	4	B	17.65	6.45	24.10
1240			36	5-1/2	C	24.50	6.45	30.95
1320			64	8x6	G	38	6.45	44.45
1600	50	10	16	5	A	28.50	8.65	37.15
1620			14.62	5	B	26	8.65	34.65
1640			24	4-1/2	C	19.50	8.65	28.15
1720			28	6x3	G	25	8.65	33.65
1800		16	24	8	A	39	6.45	45.45
1840			36	5-1/2	C	24.50	6.45	30.95
1920			64	8x6	G	38	6.45	44.45
2000	50	20	28	8	A	43.50	6.45	49.95
2040			49	6-5/8	C	30.50	6.45	36.95
2120			64	8x6	G	36	6.45	42.45
2200	75	10	20	6	A	35.50	8.65	44.15
2240			36	4-1/2	C	49	8.65	57.65
2320			35	6x4	G	28	8.65	36.65
2400		16	31	8	A	50.50	6.45	56.95
2440			49	6-5/8	C	32	6.45	38.45
2520			64	8x6	G	38	6.45	44.45
2600		20	31	8	A	48	6.45	54.45
2640			81	8-5/8	C	46	6.45	52.45
2720			64	8x6	G	36	6.45	42.45
2800	100	10	24	8	A	42.50	8.65	51.15
2840			35	4-1/2	C	49	8.65	57.65
2920			46	8x4	G	34.50	8.65	43.15
3000		16	31	8	A	50.50	6.45	56.95
3040			56	6-5/8	C	47.50	6.45	53.95
3120			64	8x6	G	38	6.45	44.45

B1010 Floor Construction

B1010 208 — Steel Columns

	LOAD (KIPS)	UNSUPPORTED HEIGHT (FT.)	WEIGHT (P.L.F.)	SIZE (IN.)	TYPE	COST PER V.L.F.		
						MAT.	INST.	TOTAL
3200	100	20	40	8	A	62	6.45	68.45
3240			81	8-5/8	C	46	6.45	52.45
3320			70	8x6	G	51.50	6.45	57.95
3400	125	10	31	8	A	55	8.65	63.65
3440			81	8	C	52	8.65	60.65
3520			64	8x6	G	41	8.65	49.65
3600		16	40	8	A	65.50	6.45	71.95
3640			81	8	C	48.50	6.45	54.95
3720			64	8x6	G	38	6.45	44.45
3800		20	48	8	A	74.50	6.45	80.95
3840			81	8	C	46	6.45	52.45
3920			60	8x6	G	51.50	6.45	57.95
4000	150	10	35	8	A	62	8.65	70.65
4040			81	8-5/8	C	52	8.65	60.65
4120			64	8x6	G	41	8.65	49.65
4200		16	45	10	A	73.50	6.45	79.95
4240			81	8-5/8	C	48.50	6.45	54.95
4320			70	8x6	G	54.50	6.45	60.95
4400		20	49	10	A	76	6.45	82.45
4440			123	10-3/4	C	65.50	6.45	71.95
4520			86	10x6	G	51	6.45	57.45
4600	200	10	45	10	A	79.50	8.65	88.15
4640			81	8-5/8	C	52	8.65	60.65
4720			70	8x6	G	59	8.65	67.65
4800		16	49	10	A	80	6.45	86.45
4840			123	10-3/4	C	69	6.45	75.45
4920			85	10x6	G	63.50	6.45	69.95
5200	300	10	61	14	A	108	8.65	116.65
5240			169	12-3/4	C	91.50	8.65	100.15
5320			86	10x6	G	88	8.65	96.65
5400		16	72	12	A	118	6.45	124.45
5440			169	12-3/4	C	85	6.45	91.45
5600		20	79	12	A	122	6.45	128.45
5640			169	12-3/4	C	80.50	6.45	86.95
5800	400	10	79	12	A	140	8.65	148.65
5840			178	12-3/4	C	119	8.65	127.65
6000		16	87	12	A	142	6.45	148.45
6040			178	12-3/4	C	111	6.45	117.45
6400	500	10	99	14	A	175	8.65	183.65
6600		16	109	14	A	178	6.45	184.45
6800		20	120	12	A	186	6.45	192.45
7000	600	10	120	12	A	212	8.65	220.65
7200		16	132	14	A	216	6.45	222.45
7400		20	132	14	A	204	6.45	210.45
7600	700	10	136	12	A	240	8.65	248.65
7800		16	145	14	A	237	6.45	243.45
8000		20	145	14	A	224	6.45	230.45
8200	800	10	145	14	A	256	8.65	264.65
8300		16	159	14	A	260	6.45	266.45
8400		20	176	14	A	272	6.45	278.45
8800	900	10	159	14	A	281	8.65	289.65
8900		16	176	14	A	288	6.45	294.45
9000		20	193	14	A	299	6.45	305.45

B1010 Floor Construction

B1010 208 Steel Columns

	LOAD (KIPS)	UNSUPPORTED HEIGHT (FT.)	WEIGHT (P.L.F.)	SIZE (IN.)	TYPE	COST PER V.L.F.		
						MAT.	INST.	TOTAL
9100	1000	10	176	14	A	310	8.65	318.65
9200		16	193	14	A	315	6.45	321.45
9300		20	211	14	A	325	6.45	331.45

B1010 Floor Construction

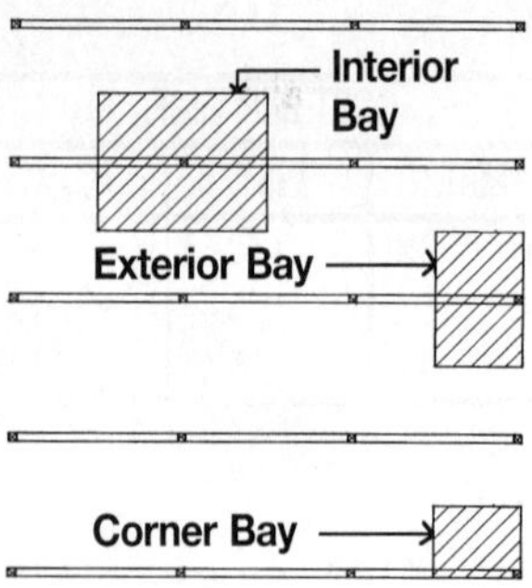

Description: Table below lists costs of columns per S.F. of bay for wood columns of various sizes and unsupported heights and the maximum allowable total load per S.F. by bay size.

Design Assumptions: Columns are concentrically loaded and are not subject to bending.

Fiber stress (f) is 1200 psi maximum.

Modulus of elasticity is 1,760,000. Use table to factor load capacity figures for modulus of elasticity other than 1,760,000.

The cost of columns per S.F. of exterior bay is proportional to the area supported. For exterior bays, multiply the costs below by two. For corner bays, multiply the cost by four.

Modulus of Elasticity	Factor
1,210,000 psi	0.69
1,320,000 psi	0.75
1,430,000 psi	0.81
1,540,000 psi	0.87
1,650,000 psi	0.94
1,760,000 psi	1.00

B1010 210 Wood Columns

	NOMINAL COLUMN SIZE (IN.)	BAY SIZE (FT.)	UNSUPPORTED HEIGHT (FT.)	MATERIAL (BF per M.S.F.)	TOTAL LOAD (P.S.F.)	COST PER S.F.		
						MAT.	INST.	TOTAL
1000	4 x 4	10 x 8	8	133	100	.18	.23	.41
1050			10	167	60	.23	.28	.51
1200		10 x 10	8	106	80	.15	.18	.33
1250			10	133	50	.18	.23	.41
1400		10 x 15	8	71	50	.10	.12	.22
1450			10	88	30	.12	.15	.27
1600		15 x 15	8	47	30	.06	.08	.14
1650			10	59	15	.08	.10	.18
2000	6 x 6	10 x 15	8	160	230	.25	.25	.50
2050			10	200	210	.32	.31	.63
2200		15 x 15	8	107	150	.17	.17	.34
2250			10	133	140	.21	.21	.42
2400		15 x 20	8	80	110	.13	.12	.25
2450			10	100	100	.16	.16	.32
2500			12	120	70	.19	.19	.38
2600		20 x 20	8	60	80	.09	.09	.18
2650			10	75	70	.12	.12	.24
2800		20 x 25	8	48	60	.08	.07	.15
2850			10	60	50	.09	.09	.18
3400	8 x 8	20 x 20	8	107	160	.15	.16	.31
3450			10	133	160	.19	.19	.38
3600		20 x 25	8	85	130	.12	.14	.26
3650			10	107	130	.15	.18	.33
3800		25 x 25	8	68	100	.11	.09	.20
3850			10	85	100	.14	.11	.25
4200	10 x 10	20 x 25	8	133	210	.22	.18	.40
4250			10	167	210	.28	.23	.51
4400		25 x 25	8	107	160	.18	.14	.32
4450			10	133	160	.22	.18	.40
4700	12 x 12	20 x 25	8	192	310	.31	.24	.55
4750			10	240	310	.38	.31	.69
4900		25 x 25	8	154	240	.25	.20	.45
4950			10	192	240	.31	.24	.55

B1010 Floor Construction

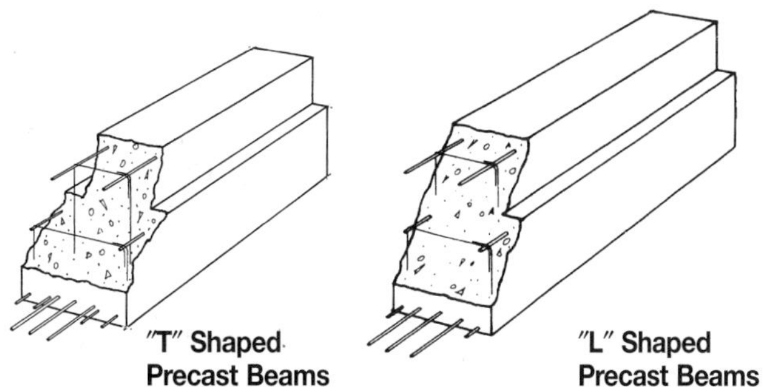

B1010 214 "T" Shaped Precast Beams

	SPAN (FT.)	SUPERIMPOSED LOAD (K.L.F.)	SIZE W X D (IN.)	BEAM WEIGHT (P.L.F.)	TOTAL LOAD (K.L.F.)	COST PER L.F.		
						MAT.	INST.	TOTAL
2300	15	2.8	12x16	260	3.06	185	14.05	199.05
2500		8.37	12x28	515	8.89	220	15.25	235.25
8900	45	3.34	12x60	1165	4.51	355	9.05	364.05
9900		6.5	24x60	1915	8.42	380	12.65	392.65

B1010 215 "L" Shaped Precast Beams

	SPAN (FT.)	SUPERIMPOSED LOAD (K.L.F.)	SIZE W X D (IN.)	BEAM WEIGHT (P.L.F.)	TOTAL LOAD (K.L.F.)	COST PER L.F.		
						MAT.	INST.	TOTAL
2250	15	2.58	12x16	230	2.81	174	14.05	188.05
2400		5.92	12x24	370	6.29	191	14.05	205.05
4000	25	2.64	12x28	435	3.08	192	9.15	201.15
4450		6.44	18x36	790	7.23	238	10.85	248.85
5300	30	2.80	12x36	565	3.37	214	8.30	222.30
6400		8.66	24x44	1245	9.90	284	11.90	295.90

B1010 Floor Construction

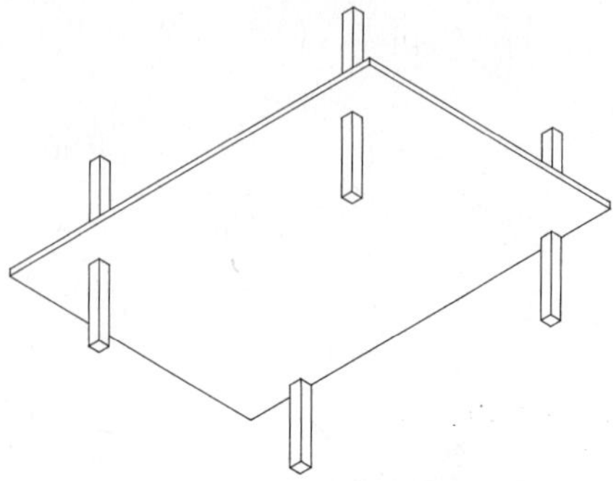

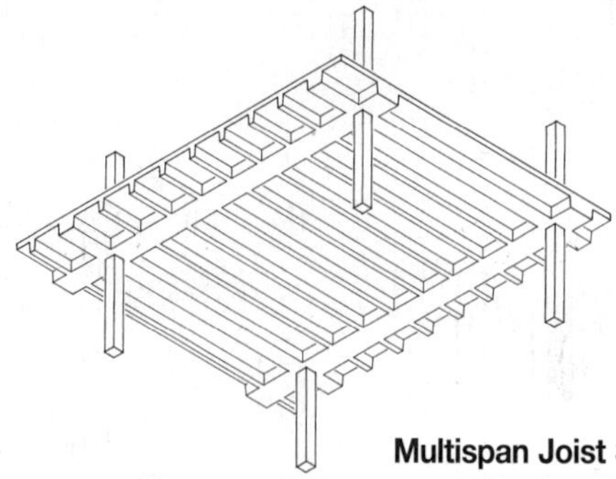

Multispan Joist Slab

General: Flat Plates: Solid uniform depth concrete two way slab without drops or interior beams. Primary design limit is shear at columns.

General: Combination of thin concrete slab and monolithic ribs at uniform spacing to reduce dead weight and increase rigidity.

B1010 223 Cast in Place Flat Plate

	BAY SIZE (FT.)	SUPERIMPOSED LOAD (P.S.F.)	MINIMUM COL. SIZE (IN.)	SLAB THICKNESS (IN.)	TOTAL LOAD (P.S.F.)	COST PER S.F.		
						MAT.	INST.	TOTAL
3000	15 x 20	40	14	7	127	5.15	7.30	12.45
3400		75	16	7-1/2	169	5.50	7.40	12.90
3600		125	22	8-1/2	231	6.05	7.65	13.70
3800		175	24	8-1/2	281	6.10	7.65	13.75
4200	20 x 20	40	16	7	127	5.15	7.25	12.40
4400		75	20	7-1/2	175	5.55	7.45	13
4600		125	24	8-1/2	231	6.05	7.60	13.65
5000		175	24	8-1/2	281	6.10	7.65	13.75
5600	20 x 25	40	18	8-1/2	146	6	7.60	13.60
6000		75	20	9	188	6.20	7.65	13.85
6400		125	26	9-1/2	244	6.75	7.95	14.70
6600		175	30	10	300	7	8.05	15.05
7000	25 x 25	40	20	9	152	6.25	7.65	13.90
7400		75	24	9-1/2	194	6.60	7.90	14.50
7600		125	30	10	250	7	8.05	15.05

B1010 226 Cast in Place Multispan Joist Slab

	BAY SIZE (FT.)	SUPERIMPOSED LOAD (P.S.F.)	MINIMUM COL. SIZE (IN.)	RIB DEPTH (IN.)	TOTAL LOAD (P.S.F.)	COST PER S.F.		
						MAT.	INST.	TOTAL
2000	15 x 15	40	12	8	115	6.85	8.85	15.70
2100		75	12	8	150	6.90	8.90	15.80
2200		125	12	8	200	7.05	9	16.05
2300		200	14	8	275	7.20	9.35	16.55
2600	15 x 20	40	12	8	115	7	8.85	15.85
2800		75	12	8	150	7.10	9.40	16.50
3000		125	14	8	200	7.35	9.50	16.85
3300		200	16	8	275	7.70	9.70	17.40
3600	20 x 20	40	12	10	120	7.15	8.75	15.90
3900		75	14	10	155	7.45	9.30	16.75
4000		125	16	10	205	7.50	9.45	16.95
4100		200	18	10	280	7.80	9.85	17.65
6200	30 x 30	40	14	14	131	7.90	9.20	17.10
6400		75	18	14	166	8.10	9.45	17.55
6600		125	20	14	216	8.55	9.95	18.50
6700		200	24	16	297	9.15	10.35	19.50

B1010 Floor Construction

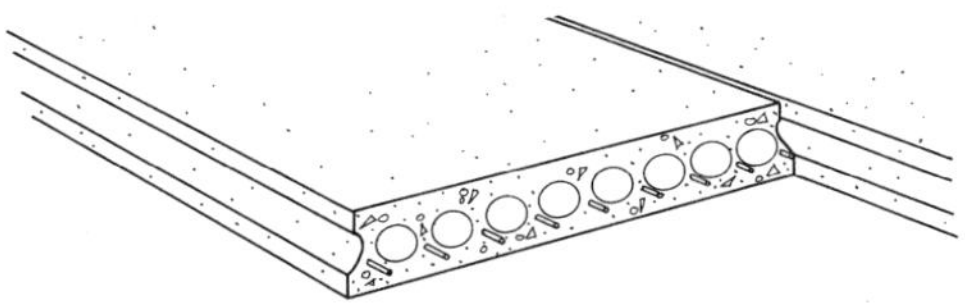

Precast Plank with No Topping

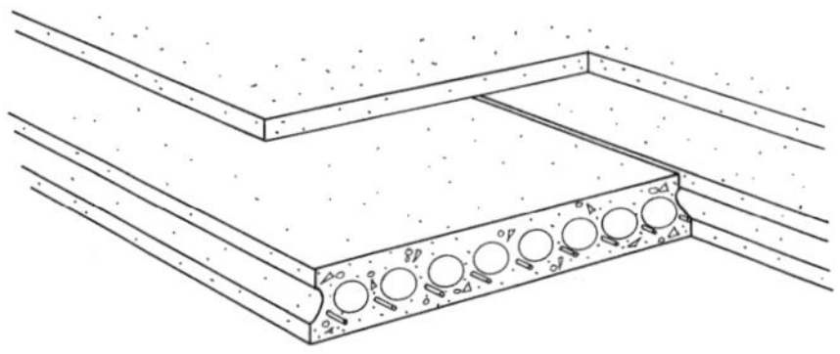

Precast Plank with 2″ Concrete Topping

B1010 229 Precast Plank with No Topping

	SPAN (FT.)	SUPERIMPOSED LOAD (P.S.F.)	TOTAL DEPTH (IN.)	DEAD LOAD (P.S.F.)	TOTAL LOAD (P.S.F.)	COST PER S.F.		
						MAT.	INST.	TOTAL
0720	10	40	4	50	90	6.75	2.38	9.13
0750		75	6	50	125	7.60	2.03	9.63
0770		100	6	50	150	7.60	2.03	9.63
0800	15	40	6	50	90	7.60	2.03	9.63
0820		75	6	50	125	7.60	2.03	9.63
0850		100	6	50	150	7.60	2.03	9.63
0950	25	40	6	50	90	7.60	2.03	9.63
0970		75	8	55	130	8.45	1.78	10.23
1000		100	8	55	155	8.45	1.78	10.23
1200	30	40	8	55	95	8.45	1.78	10.23
1300		75	8	55	130	8.45	1.78	10.23
1400		100	10	70	170	8.80	1.58	10.38
1500	40	40	10	70	110	8.80	1.58	10.38
1600		75	12	70	145	9.10	1.42	10.52
1700	45	40	12	70	110	9.10	1.42	10.52

B1010 230 Precast Plank with 2″ Concrete Topping

	SPAN (FT.)	SUPERIMPOSED LOAD (P.S.F.)	TOTAL DEPTH (IN.)	DEAD LOAD (P.S.F.)	TOTAL LOAD (P.S.F.)	COST PER S.F.		
						MAT.	INST.	TOTAL
2000	10	40	6	75	115	7.80	4.41	12.21
2100		75	8	75	150	8.65	4.06	12.71
2200		100	8	75	175	8.65	4.06	12.71
2500	15	40	8	75	115	8.65	4.06	12.71
2600		75	8	75	150	8.65	4.06	12.71
2700		100	8	75	175	8.65	4.06	12.71
3100	25	40	8	75	115	8.65	4.06	12.71
3200		75	8	75	150	8.65	4.06	12.71
3300		100	10	80	180	9.50	3.81	13.31
3400	30	40	10	80	120	9.50	3.81	13.31
3500		75	10	80	155	9.50	3.81	13.31
3600		100	10	80	180	9.50	3.81	13.31
4000	40	40	12	95	135	9.85	3.61	13.46
4500		75	14	95	170	10.15	3.45	13.60
5000	45	40	14	95	135	10.15	3.45	13.60

B1010 Floor Construction

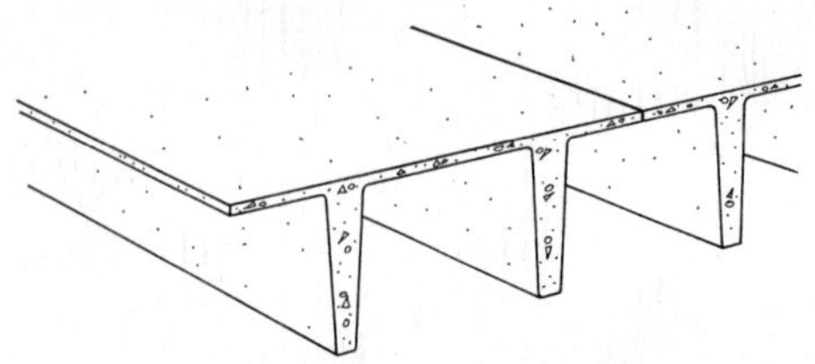

Most widely used for moderate span floors and roofs. At shorter spans, they tend to be competitive with hollow core slabs. They are also used as wall panels.

B1010 234 Precast Double "T" Beams with No Topping

	SPAN (FT.)	SUPERIMPOSED LOAD (P.S.F.)	DBL. "T" SIZE D (IN.) W (FT.)	CONCRETE "T" TYPE	TOTAL LOAD (P.S.F.)	COST PER S.F. MAT.	INST.	TOTAL
4300	50	30	20x8	Lt. Wt.	66	9.80	1.11	10.91
4400		40	20x8	Lt. Wt.	76	9.80	1.02	10.82
4500		50	20x8	Lt. Wt.	86	9.90	1.25	11.15
4600		75	20x8	Lt. Wt.	111	9.95	1.44	11.39
5600	70	30	32x10	Lt. Wt.	78	9.80	.76	10.56
5750		40	32x10	Lt. Wt.	88	9.85	.95	10.80
5900		50	32x10	Lt. Wt.	98	9.90	1.10	11
6000		75	32x10	Lt. Wt.	123	10.05	1.35	11.40
6100		100	32x10	Lt. Wt.	148	10.25	1.86	12.11
6200	80	30	32x10	Lt. Wt.	78	9.90	1.10	11
6300		40	32x10	Lt. Wt.	88	10.05	1.35	11.40
6400		50	32x10	Lt. Wt.	98	10.15	1.61	11.76

B1010 235 Precast Double "T" Beams With 2″ Topping

	SPAN (FT.)	SUPERIMPOSED LOAD (P.S.F.)	DBL. "T" SIZE D (IN.) W (FT.)	CONCRETE "T" TYPE	TOTAL LOAD (P.S.F.)	COST PER S.F. MAT.	INST.	TOTAL
7100	40	30	18x8	Reg. Wt.	120	8.60	2.52	11.12
7200		40	20x8	Reg. Wt.	130	10.05	2.40	12.45
7300		50	20x8	Reg. Wt.	140	10.15	2.62	12.77
7400		75	20x8	Reg. Wt.	165	10.20	2.73	12.93
7500		100	20x8	Reg. Wt.	190	10.35	3.05	13.40
7550	50	30	24x8	Reg. Wt.	120	10.15	2.52	12.67
7600		40	24x8	Reg. Wt.	130	10.20	2.61	12.81
7750		50	24x8	Reg. Wt.	140	10.20	2.62	12.82
7800		75	24x8	Reg. Wt.	165	10.30	2.94	13.24
7900		100	32x10	Reg. Wt.	189	10.20	2.46	12.66

B1010 Floor Construction

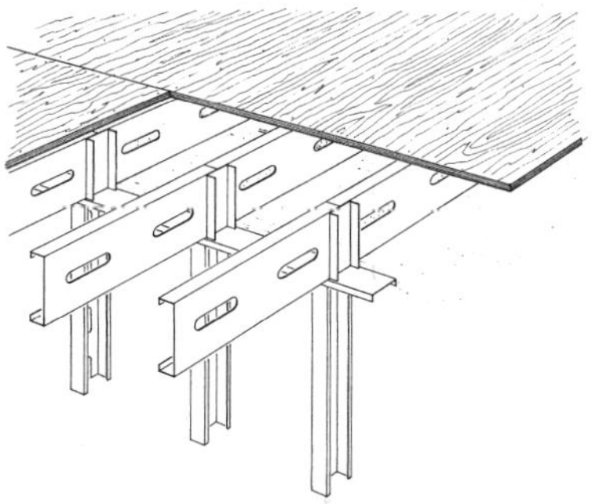

Description: Table below lists costs for light gauge CEE or PUNCHED DOUBLE joists to suit the span and loading with the minimum thickness subfloor required by the joist spacing.

Design Assumptions:

Maximum live load deflection is 1/360 of the clear span.

Maximum total load deflection is 1/240 of the clear span.

Bending strength is 20,000 psi.

8% allowance has been added to framing quantities for overlaps, double joists at openings under partitions, etc.; 5% added to glued & nailed subfloor for waste.

Maximum span is in feet and is the unsupported clear span.

B1010 244 Light Gauge Steel Floor Systems

	SPAN (FT.)	SUPERIMPOSED LOAD (P.S.F.)	FRAMING DEPTH (IN.)	FRAMING SPAC. (IN.)	TOTAL LOAD (P.S.F.)	COST PER S.F.		
						MAT.	INST.	TOTAL
1500	15	40	8	16	54	2.86	1.59	4.45
1550			8	24	54	3	1.64	4.64
1600		65	10	16	80	3.73	1.89	5.62
1650			10	24	80	3.69	1.87	5.56
1700		75	10	16	90	3.73	1.89	5.62
1750			10	24	90	3.69	1.87	5.56
1800		100	10	16	116	4.86	2.33	7.19
1850			10	24	116	3.69	1.87	5.56
1900		125	10	16	141	4.86	2.33	7.19
1950			10	24	141	3.69	1.87	5.56
2500	20	40	8	16	55	3.29	1.76	5.05
2550			8	24	55	3	1.64	4.64
2600		65	8	16	80	3.81	1.98	5.79
2650			10	24	80	3.70	1.88	5.58
2700		75	10	16	90	3.52	1.81	5.33
2750			12	24	90	4.12	1.67	5.79
2800		100	10	16	115	3.52	1.81	5.33
2850			10	24	116	4.75	2.06	6.81
2900		125	12	16	142	5.50	2.01	7.51
2950			12	24	141	5.10	2.19	7.29
3500	25	40	10	16	55	3.73	1.89	5.62
3550			10	24	55	3.69	1.87	5.56
3600		65	10	16	81	4.86	2.33	7.19
3650			12	24	81	4.57	1.78	6.35
3700		75	12	16	92	5.50	2.01	7.51
3750			12	24	91	5.10	2.19	7.29
3800		100	12	16	117	6.20	2.19	8.39
3850		125	12	16	143	7.05	2.82	9.87
4500	30	40	12	16	57	5.50	2.01	7.51
4550			12	24	56	4.55	1.78	6.33
4600		65	12	16	82	6.15	2.18	8.33
4650		75	12	16	92	6.15	2.18	8.33

B1010 Floor Construction

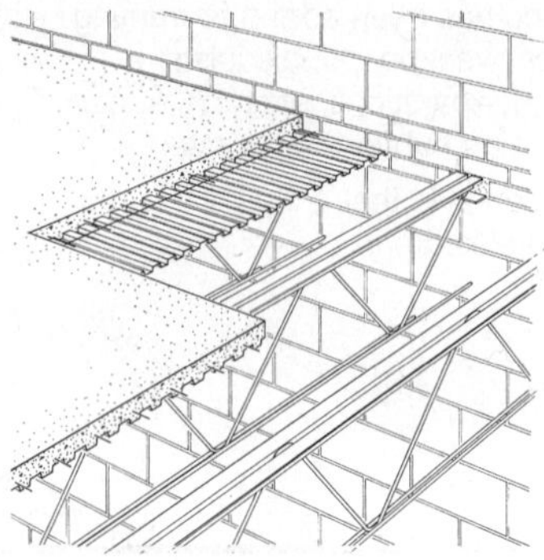

Description: Table below lists cost per S.F. for a floor system on bearing walls using open web steel joists, galvanized steel slab form and 2-1/2″ concrete slab reinforced with welded wire fabric.

Design and Pricing Assumptions:
Concrete f'c = 3 KSI placed by pump.
WWF 6 x 6 – W1.4 x W1.4 (10 x 10)
Joists are spaced as shown.
Slab form is 28 gauge galvanized.
Joists costs include appropriate bridging. Deflection is limited to 1/360 of the span. Screeds and steel trowel finish.

Design Loads	Min.	Max.
Joists	3.0 PSF	7.6 PSF
Slab Form	1.0	1.0
2-1/2″ Concrete	27.0	27.0
Ceiling	3.0	3.0
Misc.	9.0	9.4
	43.0 PSF	48.0 PSF

B1010 246 Deck & Joists on Bearing Walls

	SPAN (FT.)	SUPERIMPOSED LOAD (P.S.F.)	JOIST SPACING FT. - IN.	DEPTH (IN.)	TOTAL LOAD (P.S.F.)	COST PER S.F.		
						MAT.	INST.	TOTAL
1050	20	40	2-0	14-1/2	83	6.05	3.28	9.33
1070		65	2-0	16-1/2	109	6.50	3.47	9.97
1100		75	2-0	16-1/2	119	6.50	3.47	9.97
1120		100	2-0	18-1/2	145	6.55	3.48	10.03
1150		125	1-9	18-1/2	170	7.15	3.67	10.82
1170	25	40	2-0	18-1/2	84	6.55	3.48	10.03
1200		65	2-0	20-1/2	109	6.75	3.55	10.30
1220		75	2-0	20-1/2	119	7.20	3.70	10.90
1250		100	2-0	22-1/2	145	7.30	3.73	11.03
1270		125	1-9	22-1/2	170	8.30	4.06	12.36
1300	30	40	2-0	22-1/2	84	7.30	3.37	10.67
1320		65	2-0	24-1/2	110	7.90	3.51	11.41
1350		75	2-0	26-1/2	121	8.20	3.58	11.78
1370		100	2-0	26-1/2	146	8.70	3.70	12.40
1400		125	2-0	24-1/2	172	10.20	4.05	14.25
1420	35	40	2-0	26-1/2	85	9.65	3.92	13.57
1450		65	2-0	28-1/2	111	9.95	4	13.95
1470		75	2-0	28-1/2	121	9.95	4	13.95
1500		100	1-11	28-1/2	147	10.85	4.19	15.04
1520		125	1-8	28-1/2	172	11.90	4.45	16.35
1550	5/8″ gyp. fireproof.							
1560	On metal furring, add					.86	2.81	3.67

B1010 Floor Construction

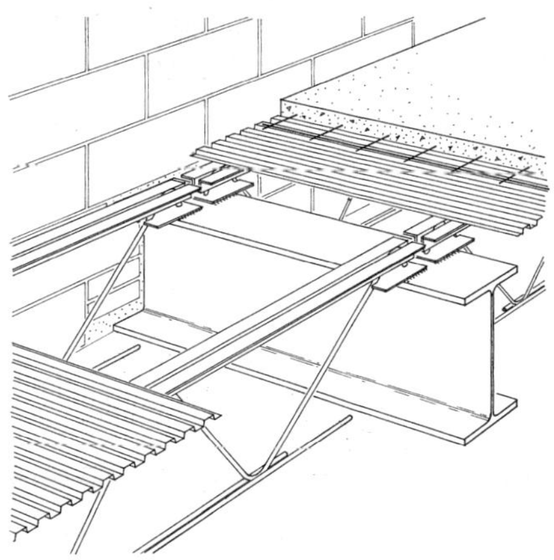

Table below lists costs for a floor system on exterior bearing walls and interior columns and beams using open web steel joists, galvanized steel slab form, 2-1/2″ concrete slab reinforced with welded wire fabric.

Design and Pricing Assumptions:
- Structural Steel is A36.
- Concrete f'c = 3 KSI placed by pump.
- WWF 6 x 6 – W1.4 x W1.4 (10 x 10)
- Columns are 12′ high.
- Building is 4 bays long by 4 bays wide.
- Joists are 2′ O.C. ± and span the long direction of the bay.

Joists at columns have bottom chords extended and are connected to columns.

Slab form is 28 gauge galvanized. Column costs in table are for columns to support 1 floor plus roof loading in a 2-story building; however, column costs are from ground floor to 2nd floor only. Joist costs include appropriate bridging. Deflection is limited to 1/360 of the span. Screeds and steel trowel finish.

Design Loads	Min.	Max.
S.S & Joists	4.4 PSF	11.5 PSF
Slab Form	1.0	1.0
2-1/2″ Concrete	27.0	27.0
Ceiling	3.0	3.0
Misc.	7.6	5.5
	43.0 PSF	48.0 PSF

B1010 248 Steel Joists on Beam & Wall

	BAY SIZE (FT.)	SUPERIMPOSED LOAD (P.S.F.)	DEPTH (IN.)	TOTAL LOAD (P.S.F.)	COLUMN ADD	COST PER S.F.		
						MAT.	INST.	TOTAL
1720	25x25	40	23	84		8.20	3.73	11.93
1730					columns	.42	.10	.52
1750	25x25	65	29	110		8.60	3.84	12.44
1760					columns	.42	.10	.52
1770	25x25	75	26	120		9.55	4.12	13.67
1780					columns	.48	.12	.60
1800	25x25	100	29	145		10.75	4.49	15.24
1810					columns	.48	.12	.60
1820	25x25	125	29	170		11.30	4.64	15.94
1830					columns	.54	.13	.67
2020	30x30	40	29	84		9.35	3.72	13.07
2030					columns	.37	.09	.46
2050	30x30	65	29	110		10.60	4.02	14.62
2060					columns	.37	.09	.46
2070	30x30	75	32	120		10.90	4.09	14.99
2080					columns	.43	.11	.54
2100	30x30	100	35	145		12.05	4.36	16.41
2110					columns	.50	.12	.62
2120	30x30	125	35	172		13.90	4.81	18.71
2130					columns	.60	.15	.75
2170	30x35	40	29	85		10.55	4.01	14.56
2180					columns	.36	.08	.44
2200	30x35	65	29	111		12.55	4.49	17.04
2210					columns	.41	.10	.51
2220	30x35	75	32	121		12.55	4.49	17.04
2230					columns	.42	.10	.52
2250	30x35	100	35	148		12.75	4.53	17.28
2260					columns	.52	.12	.64
2270	30x35	125	38	173		14.30	4.89	19.19
2280					columns	.53	.12	.65
2320	35x35	40	32	85		10.85	4.08	14.93
2330					columns	.37	.09	.46
2350	35x35	65	35	111		13.25	4.65	17.90
2360					columns	.44	.11	.55
2370	35x35	75	35	121		13.55	4.72	18.27
2380					columns	.44	.11	.55

B1010 Floor Construction

B1010 248 — Steel Joists on Beam & Wall

	BAY SIZE (FT.)	SUPERIMPOSED LOAD (P.S.F.)	DEPTH (IN.)	TOTAL LOAD (P.S.F.)	COLUMN ADD	COST PER S.F.		
						MAT.	INST.	TOTAL
2400	35x35	100	38	148		14.10	4.85	18.95
2410					columns	.55	.13	.68
2460	5/8 gyp. fireproof.							
2475	On metal furring, add					.86	2.81	3.67

B1010 Floor Construction

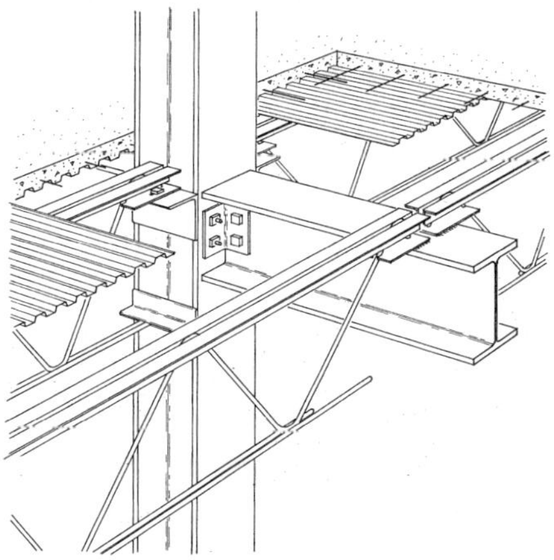

Table below lists costs for a floor system on steel columns and beams using open web steel joists, galvanized steel slab form, and 2-1/2" concrete slab reinforced with welded wire fabric.

Design and Pricing Assumptions:

- Structural Steel is A36.
- Concrete f'c = 3 KSI placed by pump.
- WWF 6 x 6 – W1.4 x W1.4 (10 x 10)
- Columns are 12' high.
- Building is 4 bays long by 4 bays wide.
- Joists are 2' O.C. ± and span the long direction of the bay.

Joists at columns have bottom chords extended and are connected to columns.

Slab form is 28 gauge galvanized. Column costs in table are for columns to support 1 floor plus roof loading in a 2-story building; however, column costs are from ground floor to 2nd floor only. Joist costs include appropriate bridging. Deflection is limited to 1/360 of the span. Screeds and steel trowel finish.

Design Loads	Min.	Max.
S.S. & Joists	6.3 PSF	15.3 PSF
Slab Form	1.0	1.0
2-1/2" Concrete	27.0	27.0
Ceiling	3.0	3.0
Misc.	5.7	1.7
	43.0 PSF	48.0 PSF

B1010 250 Steel Joists, Beams & Slab on Columns

	BAY SIZE (FT.)	SUPERIMPOSED LOAD (P.S.F.)	DEPTH (IN.)	TOTAL LOAD (P.S.F.)	COLUMN ADD	COST PER S.F.		
						MAT.	INST.	TOTAL
2350	15x20	40	17	83		8.50	3.76	12.26
2400					column	1.38	.34	1.72
2450	15x20	65	19	108		9.45	3.99	13.44
2500					column	1.38	.34	1.72
2550	15x20	75	19	119		9.80	4.11	13.91
2600					column	1.51	.36	1.87
2650	15x20	100	19	144		10.40	4.28	14.68
2700					column	1.51	.36	1.87
2750	15x20	125	19	170		10.95	4.40	15.35
2800					column	2.01	.48	2.49
2850	20x20	40	19	83		9.25	3.92	13.17
2900					column	1.13	.27	1.40
2950	20x20	65	23	109		10.20	4.19	14.39
3000					column	1.51	.36	1.87
3100	20x20	75	26	119		10.75	4.33	15.08
3200					column	1.51	.36	1.87
3400	20x20	100	23	144		11.20	4.44	15.64
3450					column	1.51	.36	1.87
3500	20x20	125	23	170		12.40	4.78	17.18
3600					column	1.81	.43	2.24
3700	20x25	40	44	83		9.90	4.12	14.02
3800					column	1.21	.28	1.49
3900	20x25	65	26	110		10.85	4.36	15.21
4000					column	1.21	.28	1.49
4100	20x25	75	26	120		11.30	4.51	15.81
4200					column	1.45	.35	1.80
4300	20x25	100	26	145		12	4.69	16.69
4400					column	1.45	.35	1.80
4500	20x25	125	29	170		13.30	5.10	18.40
4600					column	1.69	.40	2.09
4700	25x25	40	23	84		10.75	4.32	15.07
4800					column	1.16	.28	1.44
4900	25x25	65	29	110		11.35	4.47	15.82
5000					column	1.16	.28	1.44
5100	25x25	75	26	120		12.55	4.81	17.36
5200					column	1.35	.32	1.67

B1010 Floor Construction

B1010 250 Steel Joists, Beams & Slab on Columns

	BAY SIZE (FT.)	SUPERIMPOSED LOAD (P.S.F.)	DEPTH (IN.)	TOTAL LOAD (P.S.F.)	COLUMN ADD	COST PER S.F.		
						MAT.	INST.	TOTAL
5300	25x25	100	29	145		13.90	5.25	19.15
5400					column	1.35	.32	1.67
5500	25x25	125	32	170		14.70	5.45	20.15
5600					column	1.50	.36	1.86
5700	25x30	40	29	84		12.25	4.39	16.64
5800					column	1.12	.27	1.39
5900	25x30	65	29	110		12.75	4.51	17.26
6000					column	1.12	.27	1.39
6050	25x30	75	29	120		13.10	4.61	17.71
6100					column	1.24	.30	1.54
6150	25x30	100	29	145		14.20	4.86	19.06
6200					column	1.24	.30	1.54
6250	25x30	125	32	170		16	5.25	21.25
6300					column	1.43	.35	1.78
6350	30x30	40	29	84		12.25	4.39	16.64
6400					column	1.04	.24	1.28
6500	30x30	65	29	110		13.95	4.78	18.73
6600					column	1.04	.24	1.28
6700	30x30	75	32	120		14.20	4.85	19.05
6800					column	1.19	.28	1.47
6900	30x30	100	35	145		15.75	5.20	20.95
7000					column	1.39	.34	1.73
7100	30x30	125	35	172		18	5.75	23.75
7200					column	1.55	.38	1.93
7300	30x35	40	29	85		13.80	4.76	18.56
7400					column	.89	.21	1.10
7500	30x35	65	29	111		16.05	5.30	21.35
7600					column	1.15	.27	1.42
7700	30x35	75	32	121		16.05	5.30	21.35
7800					column	1.17	.28	1.45
7900	30x35	100	35	148		16.55	5.40	21.95
8000					column	1.43	.35	1.78
8100	30x35	125	38	173		18.35	5.85	24.20
8200					column	1.46	.35	1.81
8300	35x35	40	32	85		14.20	4.84	19.04
8400					column	1.02	.24	1.26
8500	35x35	65	35	111		16.90	5.50	22.40
8600					column	1.23	.30	1.53
9300	35x35	75	38	121		17.30	5.60	22.90
9400					column	1.23	.30	1.53
9500	35x35	100	38	148		18.65	5.90	24.55
9600					column	1.52	.36	1.88
9750	35x35	125	41	173		19.25	6.05	25.30
9800					column	1.54	.38	1.92
9810	5/8 gyp. fireproof.							
9815	On metal furring, add					.86	2.81	3.67

B1010 Floor Construction

Description: Table B1010 258 lists S.F. costs for steel deck and concrete slabs for various spans.

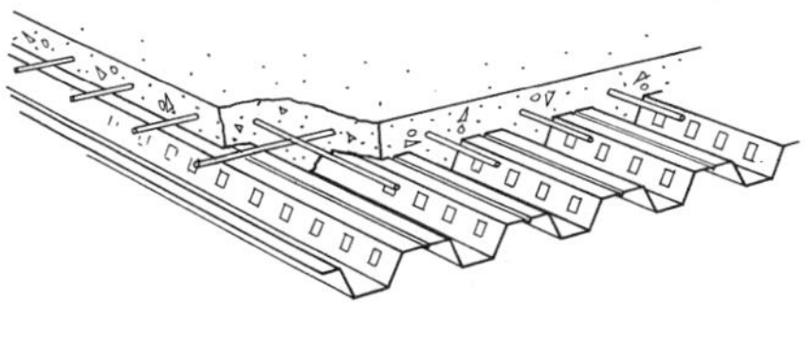

Metal Deck/Concrete Fill

Description: Table B1010 261 lists the S.F. costs for wood joists and a minimum thickness plywood subfloor.

Wood Joist

Description: Table B1010 264 lists the S.F. costs, total load, and member sizes, for various bay sizes and loading conditions.

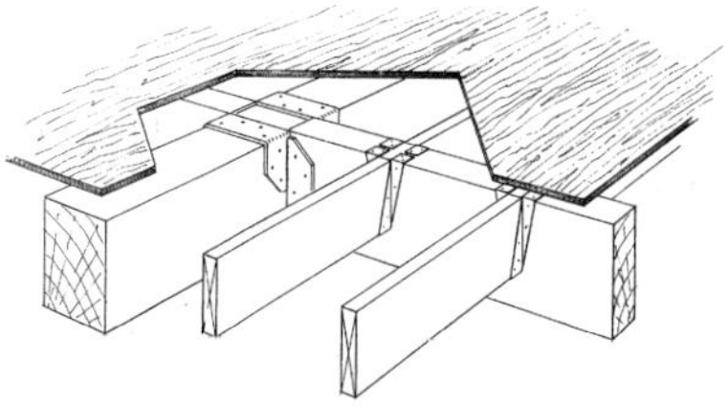

Wood Beam and Joist

B1010 258 Metal Deck/Concrete Fill

	SUPERIMPOSED LOAD (P.S.F.)	DECK SPAN (FT.)	DECK GAGE DEPTH	SLAB THICKNESS (IN.)	TOTAL LOAD (P.S.F.)	COST PER S.F.		
						MAT.	INST.	TOTAL
0900	125	6	22 1-1/2	4	164	3.08	2.16	5.24
0920		7	20 1-1/2	4	164	3.39	2.30	5.69
0950		8	20 1-1/2	4	165	3.39	2.30	5.69
0970		9	18 1-1/2	4	165	3.96	2.30	6.26
1000		10	18 2	4	165	3.86	2.42	6.28
1020		11	18 3	5	169	4.75	2.61	7.36

B1010 261 Wood Joist

		COST PER S.F.		
		MAT.	INST.	TOTAL
2900	2"x8", 12" O.C.	1.81	1.72	3.53
2950	16" O.C.	1.56	1.47	3.03
3000	24" O.C.	1.57	1.36	2.93
3300	2"x10", 12" O.C.	2.41	1.96	4.37
3350	16" O.C.	2	1.65	3.65
3400	24" O.C.	1.85	1.47	3.32
3700	2"x12", 12" O.C.	2.85	1.99	4.84
3750	16" O.C.	2.33	1.67	4
3800	24" O.C.	2.08	1.49	3.57
4100	2"x14", 12" O.C.	3.41	2.16	5.57
4150	16" O.C.	2.77	1.80	4.57
4200	24" O.C.	2.37	1.58	3.95
7101	Note: Subfloor cost is included in these prices.			

B1010 264 Wood Beam & Joist

	BAY SIZE (FT.)	SUPERIMPOSED LOAD (P.S.F.)	GIRDER BEAM (IN.)	JOISTS (IN.)	TOTAL LOAD (P.S.F.)	COST PER S.F.		
						MAT.	INST.	TOTAL
2000	15x15	40	8 x 12 4 x 12	2 x 6 @ 16	53	9.05	3.98	13.03
2050		75	8 x 16 4 x 16	2 x 8 @ 16	90	12.60	4.36	16.96
2100		125	12 x 16 6 x 16	2 x 8 @ 12	144	19.50	5.60	25.10
2150		200	14 x 22 12 x 16	2 x 10 @ 12	227	39	8.55	47.55
3000	20x20	40	10 x 14 10 x 12	2 x 8 @ 16	63	11.90	4.02	15.92
3050		75	12 x 16 8 x 16	2 x 10 @ 16	102	16.85	4.49	21.34
3100		125	14 x 22 12 x 16	2 x 10 @ 12	163	33	7.15	40.15

B1010 Floor Construction

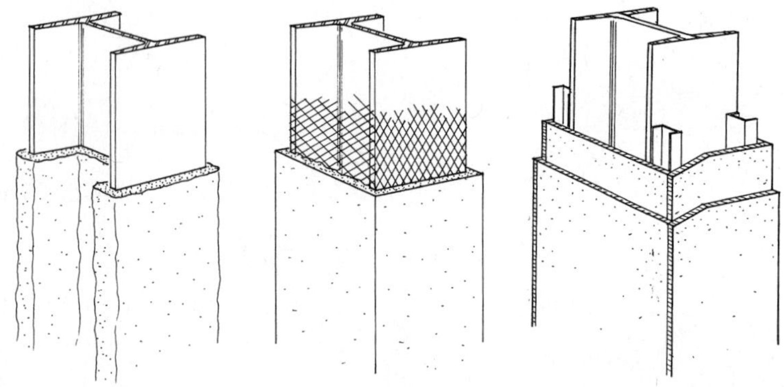

Listed below are costs per V.L.F. for fireproofing by material, column size, thickness and fire rating. Weights listed are for the fireproofing material only.

B1010 720 Steel Column Fireproofing

	ENCASEMENT SYSTEM	COLUMN SIZE (IN.)	THICKNESS (IN.)	FIRE RATING (HRS.)	WEIGHT (P.L.F.)	COST PER V.L.F.		
						MAT.	INST.	TOTAL
3000	Concrete	8	1	1	110	6.90	30	36.90
3300		14	1	1	258	11.40	43.50	54.90
3400			2	3	325	13.65	49.50	63.15
3450	Gypsum board	8	1/2	2	8	3.66	19.65	23.31
3550	1 layer	14	1/2	2	18	3.96	21	24.96
3600	Gypsum board	8	1	3	14	5.10	25	30.10
3650	1/2" fire rated	10	1	3	17	5.45	26.50	31.95
3700	2 layers	14	1	3	22	5.65	27.50	33.15
3750	Gypsum board	8	1-1/2	3	23	6.80	31.50	38.30
3800	1/2" fire rated	10	1-1/2	3	27	7.65	35	42.65
3850	3 layers	14	1-1/2	3	35	8.45	38.50	46.95
3900	Sprayed fiber	8	1-1/2	2	6.3	4	6.25	10.25
3950	Direct application		2	3	8.3	5.50	8.60	14.10
4050		10	1-1/2	2	7.9	4.83	7.50	12.33
4200	Sprayed fiber	14	1-1/2	2	10.8	6	9.35	15.35

B1020 Roof Construction

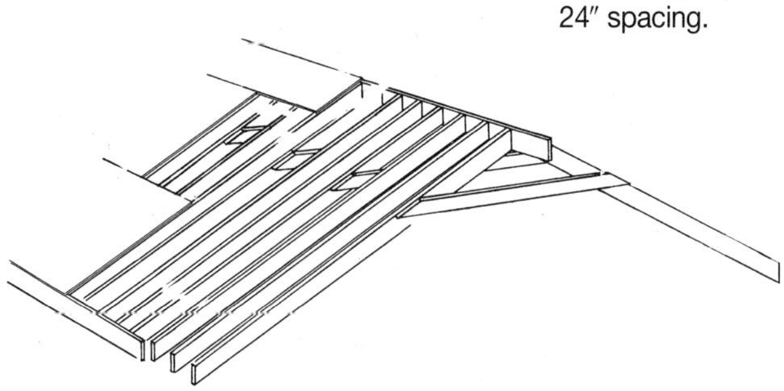

The table below lists prices per S.F. for roof rafters and sheathing by nominal size and spacing. Sheathing is 5/16″ CDX for 12″ and 16″ spacing and 3/8″ CDX for 24″ spacing.

Factors for Converting Inclined to Horizontal

Roof Slope	Approx. Angle	Factor	Roof Slope	Approx. Angle	Factor
Flat	0°	1.000	12 in 12	45.0°	1.414
1 in 12	4.8°	1.003	13 in 12	47.3°	1.474
2 in 12	9.5°	1.014	14 in 12	49.4°	1.537
3 in 12	14.0°	1.031	15 in 12	51.3°	1.601
4 in 12	18.4°	1.054	16 in 12	53.1°	1.667
5 in 12	22.6°	1.083	17 in 12	54.8°	1.734
6 in 12	26.6°	1.118	18 in 12	56.3°	1.803
7 in 12	30.3°	1.158	19 in 12	57.7°	1.873
8 in 12	33.7°	1.202	20 in 12	59.0°	1.943
9 in 12	36.9°	1.250	21 in 12	60.3°	2.015
10 in 12	39.8°	1.302	22 in 12	61.4°	2.088
11 in 12	42.5°	1.357	23 in 12	62.4°	2.162

B1020 102	Wood/Flat or Pitched	COST PER S.F.		
		MAT.	INST.	TOTAL
2500	Flat rafter, 2″x4″, 12″ O.C.	1.20	1.55	2.75
2550	16″ O.C.	1.08	1.33	2.41
2600	24″ O.C.	.95	1.14	2.09
2900	2″x6″, 12″ O.C.	1.46	1.55	3.01
2950	16″ O.C.	1.28	1.33	2.61
3000	24″ O.C.	1.08	1.14	2.22
3300	2″x8″, 12″ O.C.	1.74	1.67	3.41
3350	16″ O.C.	1.49	1.42	2.91
3400	24″ O.C.	1.23	1.20	2.43
3700	2″x10″, 12″ O.C.	2.34	1.91	4.25
3750	16″ O.C.	1.93	1.60	3.53
3800	24″ O.C.	1.51	1.31	2.82
4100	2″x12″, 12″ O.C.	2.78	1.94	4.72
4150	16″ O.C.	2.42	1.78	4.20
4200	24″ O.C.	1.74	1.33	3.07
4500	2″x14″, 12″ O.C.	3.34	2.82	6.16
4550	16″ O.C.	2.70	2.29	4.99
4600	24″ O.C.	2.03	1.78	3.81
4900	3″x6″, 12″ O.C.	2.99	1.64	4.63
4950	16″ O.C.	2.43	1.40	3.83
5000	24″ O.C.	1.85	1.18	3.03
5300	3″x8″, 12″ O.C.	3.97	1.84	5.81
5350	16″ O.C.	3.15	1.54	4.69
5400	24″ O.C.	2.33	1.28	3.61
5700	3″x10″, 12″ O.C.	4.85	2.09	6.94
5750	16″ O.C.	3.83	1.74	5.57
5800	24″ O.C.	2.78	1.41	4.19
6100	3″x12″, 12″ O.C.	5.65	2.52	8.17
6150	16″ O.C.	4.44	2.06	6.50
6200	24″ O.C.	3.19	1.62	4.81
7001	Wood truss, 4 in 12 slope, 24″ O.C., 24′ to 29′ span	3.79	2.27	6.06
7100	30′ to 43′ span	4.44	2.50	6.94
7200	44′ to 60′ span	4.39	2.27	6.66

B1020 Roof Construction

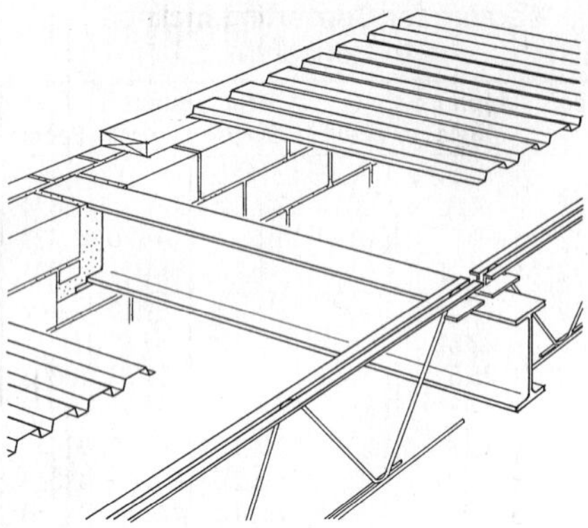

Table below lists the cost per S.F. for a roof system with steel columns, beams, and deck using open web steel joists and 1-1/2" galvanized metal deck. Perimeter of system is supported on bearing walls.

Design and Pricing Assumptions:
Columns are 18' high.
Joists are 5'-0" O.C. and span the long direction of the bay.

Joists at columns have bottom chords extended and are connected to columns. Column costs are not included but are listed separately per S.F. of floor.

Roof deck is 1-1/2", 22 gauge galvanized steel. Joist cost includes appropriate bridging. Deflection is limited to 1/240 of the span. Fireproofing is not included.

Costs/S.F. are based on a building 4 bays long and 4 bays wide.

Design Loads	Min.	Max.
Joists & Beams	3 PSF	5 PSF
Deck	2	2
Insulation	3	3
Roofing	6	6
Misc.	6	6
Total Dead Load	20 PSF	22 PSF

B1020 108 Steel Joists, Beams & Deck on Columns & Walls

	BAY SIZE (FT.)	SUPERIMPOSED LOAD (P.S.F.)	DEPTH (IN.)	TOTAL LOAD (P.S.F.)	COLUMN ADD	COST PER S.F.		
						MAT.	INST.	TOTAL
3000	25x25	20	18	40		4.31	1.16	5.47
3100					columns	.39	.07	.46
3200		30	22	50		4.62	1.22	5.84
3300					columns	.52	.09	.61
3400		40	20	60		5.10	1.34	6.44
3500					columns	.52	.09	.61
3600	25x30	20	22	40		4.54	1.11	5.65
3700					columns	.44	.08	.52
3800		30	20	50		5.15	1.24	6.39
3900					columns	.44	.08	.52
4000		40	25	60		5.35	1.26	6.61
4100					columns	.52	.09	.61
4200	30x30	20	25	42		5	1.18	6.18
4300					columns	.36	.07	.43
4400		30	22	52		5.45	1.29	6.74
4500					columns	.44	.08	.52
4600		40	28	62		5.65	1.33	6.98
4700					columns	.44	.08	.52
4800	30x35	20	22	42		5.20	1.24	6.44
4900					columns	.37	.07	.44
5000		30	28	52		5.45	1.29	6.74
5100					columns	.37	.07	.44
5200		40	25	62		5.90	1.39	7.29
5300					columns	.44	.08	.52
5400	35x35	20	28	42		5.20	1.24	6.44
5500					columns	.32	.06	.38
5600		30	25	52		6.30	1.47	7.77
5700					columns	.37	.07	.44
5800		40	28	62		6.40	1.49	7.89
5900					columns	.41	.08	.49

B1020 Roof Construction

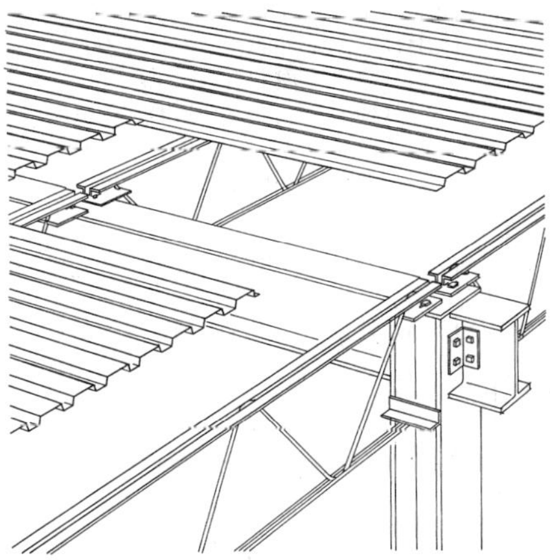

Description: Table below lists the cost per S.F. for a roof system with steel columns, beams, and deck, using open web steel joists and 1-1/2" galvanized metal deck.

Design and Pricing Assumptions:
Columns are 18' high.
Building is 4 bays long by 4 bays wide.
Joists are 5'-0" O.C. and span the long direction of the bay.
Joists at columns have bottom chords extended and are connected to columns. Column costs are not included but are listed separately per S.F. of floor.

Roof deck is 1-1/2", 22 gauge galvanized steel. Joist cost includes appropriate bridging. Deflection is limited to 1/240 of the span. Fireproofing is not included.

Design Loads	Min.	Max.
Joists & Beams	3 PSF	5 PSF
Deck	2	2
Insulation	3	3
Roofing	6	6
Misc.	6	6
Total Dead Load	20 PSF	22 PSF

B1020 112 Steel Joists, Beams, & Deck on Columns

	BAY SIZE (FT.)	SUPERIMPOSED LOAD (P.S.F.)	DEPTH (IN.)	TOTAL LOAD (P.S.F.)	COLUMN ADD	COST PER S.F.		
						MAT.	INST.	TOTAL
1100	15x20	20	16	40		4.26	1.15	5.41
1200					columns	2.26	.42	2.68
1500		40	18	60		4.75	1.27	6.02
1600					columns	2.26	.42	2.68
2300	20x25	20	18	40		4.84	1.28	6.12
2400					columns	1.35	.25	1.60
2700		40	20	60		5.45	1.43	6.88
2800					columns	1.81	.34	2.15
2900	25x25	20	18	40		5.70	1.44	7.14
3000					columns	1.08	.20	1.28
3100		30	22	50		5.90	1.48	7.38
3200					columns	1.45	.27	1.72
3300		40	20	60		6.55	1.63	8.18
3400					columns	1.45	.27	1.72
3500	25x30	20	22	40		5.55	1.30	6.85
3600					columns	1.21	.23	1.44
3900		40	25	60		6.70	1.54	8.24
4000					columns	1.45	.27	1.72
4100	30x30	20	25	42		6.30	1.44	7.74
4200					columns	1	.19	1.19
4300		30	22	52		6.95	1.58	8.53
4400					columns	1.21	.23	1.44
4500		40	28	62		7.30	1.64	8.94
4600					columns	1.21	.23	1.44
5300	35x35	20	28	42		6.85	1.56	8.41
5400					columns	.88	.16	1.04
5500		30	25	52		7.70	1.72	9.42
5600					columns	1.03	.19	1.22
5700		40	28	62		8.30	1.85	10.15
5800					columns	1.14	.21	1.35

B1020 Roof Construction

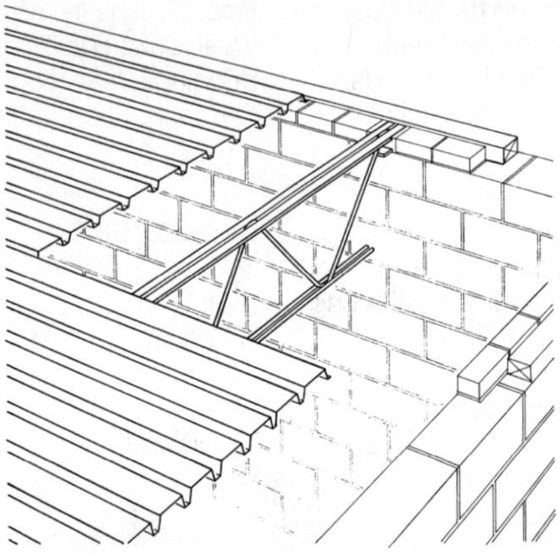

Description: Table below lists cost per S.F. for a roof system using open web steel joists and 1-1/2" galvanized metal deck. The system is assumed supported on bearing walls or other suitable support. Costs for the supports are not included.

Design and Pricing Assumptions:
Joists are 5'-0" O.C.
Roof deck is 1-1/2", 22 gauge galvanized.

B1020 116 Steel Joists & Deck on Bearing Walls

	BAY SIZE (FT.)	SUPERIMPOSED LOAD (P.S.F.)	DEPTH (IN.)	TOTAL LOAD (P.S.F.)		COST PER S.F.		
						MAT.	INST.	TOTAL
1100	20	20	13-1/2	40		2.73	.85	3.58
1200		30	15-1/2	50		2.79	.86	3.65
1300		40	15-1/2	60		2.99	.93	3.92
1400	25	20	17-1/2	40		2.99	.93	3.92
1500		30	17-1/2	50		3.22	1	4.22
1600		40	19-1/2	60		3.27	1.02	4.29
1700	30	20	19-1/2	40		3.26	1.02	4.28
1800		30	21-1/2	50		3.33	1.04	4.37
1900		40	23-1/2	60		3.58	1.12	4.70
2000	35	20	23-1/2	40		3.54	.94	4.48
2100		30	25-1/2	50		3.64	.96	4.60
2200		40	25-1/2	60		3.86	1.01	4.87
2300	40	20	25-1/2	41		3.92	1.03	4.95
2400		30	25-1/2	51		4.16	1.08	5.24
2500		40	25-1/2	61		4.30	1.12	5.42
2600	45	20	27-1/2	41		4.71	1.21	5.92
2700		30	31-1/2	51		4.98	1.28	6.26
2800		40	31-1/2	61		5.25	1.35	6.60
2900	50	20	29-1/2	42		5.25	1.35	6.60
3000		30	31-1/2	52		5.75	1.46	7.21
3100		40	31-1/2	62		6.10	1.55	7.65
3200	60	20	37-1/2	42		6.25	1.76	8.01
3300		30	37-1/2	52		6.90	1.95	8.85
3400		40	37-1/2	62		6.90	1.95	8.85
3500	70	20	41-1/2	42		6.90	1.95	8.85
3600		30	41-1/2	52		7.35	2.08	9.43
3700		40	41-1/2	64		8.95	2.51	11.46
3800	80	20	45-1/2	44		8.70	2.44	11.14
3900		30	45-1/2	54		8.70	2.44	11.14
4000		40	45-1/2	64		9.60	2.69	12.29
4400	100	20	57-1/2	44		8.55	2.32	10.87
4500		30	57-1/2	54		10.20	2.75	12.95
4600		40	57-1/2	65		11.25	3.03	14.28
4700	125	20	69-1/2	44		10.20	2.75	12.95
4800		30	69-1/2	56		11.85	3.18	15.03
4900		40	69-1/2	67		13.45	3.62	17.07

B1020 Roof Construction

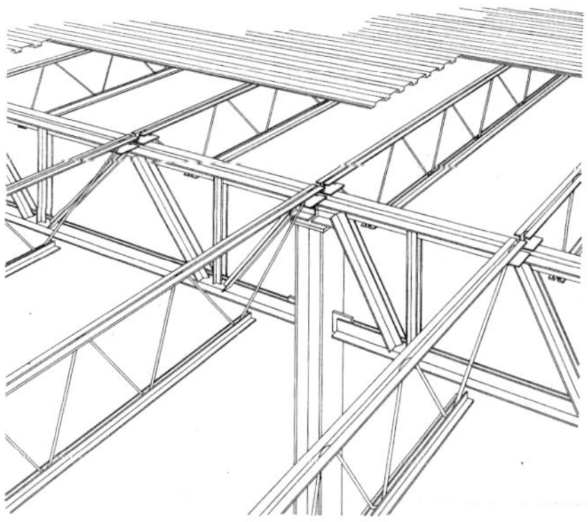

Description: Table below lists costs for a roof system supported on exterior bearing walls and interior columns. Costs include bracing, joist girders, open web steel joists and 1-1/2″ galvanized metal deck.

Design and Pricing Assumptions:
Columns are 18′ high.
Joists are 5′-0″ O.C.
Joist girders and joists have bottom chords connected to columns. Roof deck is 1-1/2″, 22 gauge galvanized steel. Costs include bridging and bracing. Deflection is limited to 1/240 of the span.

Fireproofing is not included.
Costs/S.F. are based on a building 4 bays long and 4 bays wide.
Costs for bearing walls are not included.

Column costs are not included but are listed separately per S.F. of floor.

B1020 120 Steel Joists, Joist Girders & Deck on Columns & Walls

	BAY SIZE (FT.) GIRD X JOISTS	SUPERIMPOSED LOAD (P.S.F.)	DEPTH (IN.)	TOTAL LOAD (P.S.F.)	COLUMN ADD	COST PER S.F.		
						MAT.	INST.	TOTAL
2350	30x35	20	32-1/2	40		3.98	1.17	5.15
2400					columns	.37	.07	.44
2550		40	36-1/2	60		4.46	1.29	5.75
2600					columns	.43	.08	.51
3000	35x35	20	36-1/2	40		4.34	1.46	5.80
3050					columns	.32	.06	.38
3200		40	36-1/2	60		4.86	1.59	6.45
3250					columns	.41	.08	.49
3300	35x40	20	36-1/2	40		4.40	1.32	5.72
3350					columns	.33	.07	.40
3500		40	36-1/2	60		4.95	1.47	6.42
3550					columns	.36	.07	.43
3900	40x40	20	40-1/2	41		5.10	1.48	6.58
3950					columns	.32	.06	.38
4100		40	40-1/2	61		5.70	1.62	7.32
4150					columns	.32	.06	.38
5100	45x50	20	52-1/2	41		5.75	1.69	7.44
5150					columns	.23	.04	.27
5300		40	52-1/2	61		7	1.99	8.99
5350					columns	.33	.07	.40
5400	50x45	20	56-1/2	41		5.55	1.64	7.19
5450					columns	.23	.04	.27
5600		40	56-1/2	61		7.15	2.04	9.19
5650					columns	.31	.05	.36
5700	50x50	20	56-1/2	42		6.25	1.82	8.07
5750					columns	.23	.04	.27
5900		40	59	64		7.20	2.22	9.42
5950					columns	.32	.06	.38
6300	60x50	20	62-1/2	43		6.30	1.86	8.16
6350					columns	.36	.07	.43
6500		40	71	65		7.35	2.30	9.65
6550					columns	.48	.09	.57

B1020 Roof Construction

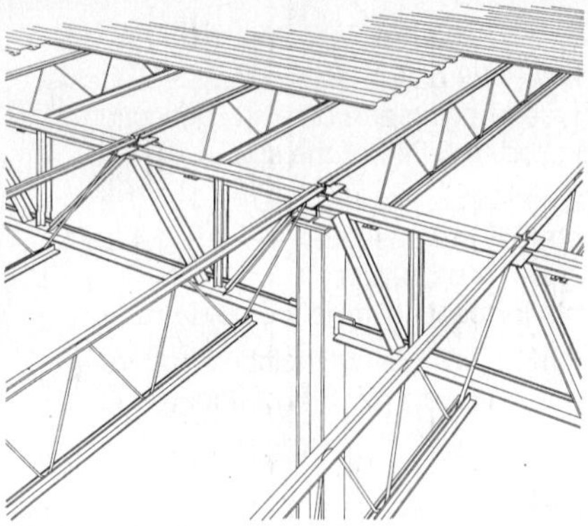

Description: Table below lists the cost per S.F. for a roof system supported on columns. Costs include joist girders, open web steel joists and 1-1/2″ galvanized metal deck.

Design and Pricing Assumptions:
Columns are 18′ high.
Joists are 5′-0″ O.C.
Joist girders and joists have bottom chords connected to columns. Roof deck is 1-1/2″, 22 gauge galvanized steel. Costs include bridging and bracing. Deflection is limited to 1/240 of the span. Fireproofing is not included.

Costs/S.F. are based on a building 4 bays long and 4 bays wide.

Costs for columns are not included, but are listed separately per S.F. of area.

B1020 124 Steel Joists & Joist Girders on Columns

	BAY SIZE (FT.) GIRD X JOISTS	SUPERIMPOSED LOAD (P.S.F.)	DEPTH (IN.)	TOTAL LOAD (P.S.F.)	COLUMN ADD	COST PER S.F. MAT.	INST.	TOTAL
2000	30x30	20	17-1/2	40		3.99	1.40	5.39
2050					columns	1.21	.23	1.44
2100		30	17-1/2	50		4.32	1.51	5.83
2150					columns	1.21	.23	1.44
2200		40	21-1/2	60		4.44	1.55	5.99
2250					columns	1.21	.23	1.44
2300	30x35	20	32-1/2	40		4.27	1.31	5.58
2350					columns	1.03	.19	1.22
2500		40	36-1/2	60		4.80	1.46	6.26
2550					columns	1.21	.23	1.44
3200	35x40	20	36-1/2	40		5.55	1.78	7.33
3250					columns	.90	.17	1.07
3400		40	36-1/2	60		6.25	1.96	8.21
3450					columns	1	.19	1.19
3800	40x40	20	40-1/2	41		5.60	1.77	7.37
3850					columns	.88	.16	1.04
4000		40	40-1/2	61		6.30	1.96	8.26
4050					columns	.88	.16	1.04
4100	40x45	20	40-1/2	41		6.05	2.04	8.09
4150					columns	.78	.15	.93
4200		30	40-1/2	51		6.55	2.16	8.71
4250					columns	.78	.15	.93
4300		40	40-1/2	61		7.30	2.34	9.64
4350					columns	.88	.16	1.04
5000	45x50	20	52-1/2	41		6.35	2.04	8.39
5050					columns	.62	.12	.74
5200		40	52-1/2	61		7.80	2.41	10.21
5250					columns	.80	.15	.95
5300	50x45	20	56-1/2	41		6.15	2.01	8.16
5350					columns	.62	.12	.74
5500		40	56-1/2	61		8	2.46	10.46
5550					columns	.80	.15	.95
5600	50x50	20	56-1/2	42		6.90	2.19	9.09
5650					columns	.63	.12	.75
5800		40	59	64		7.90	2.61	10.51
5850					columns	.89	.16	1.05

B1020 310 Canopies

		COST PER S.F. MAT.	INST.	TOTAL
0100	Canopies, wall hung, prefinished aluminum, 8′ x 10′	32.50	18.90	51.40

B2010 Exterior Walls

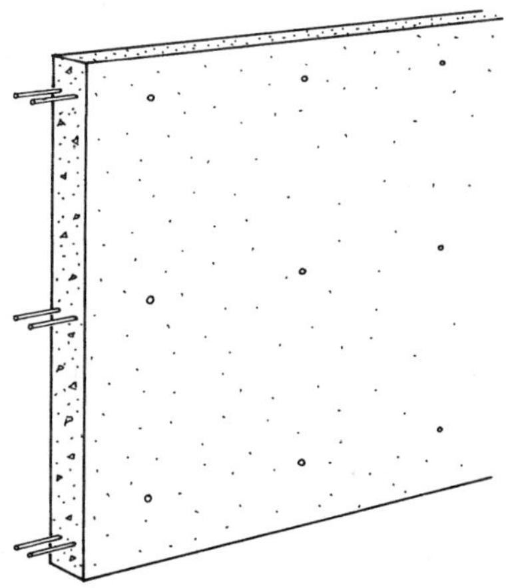

The table below describes a concrete wall system for exterior closure. There are several types of wall finishes priced from plain finish to a finish with 3/4″ rustication strip.

Design Assumptions:
Conc. f'c = 3000 to 5000 psi
Reinf. fy = 60,000 psi

B2010 101	Cast In Place Concrete	COST PER S.F.		
		MAT.	INST.	TOTAL
2100	Conc wall reinforced, 8′ high, 6″ thick, plain finish, 3000 PSI	4.83	13.45	18.28
2700	Aged wood liner, 3000 PSI	6.05	15.40	21.45
3000	Sand blast light 1 side, 3000 PSI	5.40	15.40	20.80
3700	3/4″ bevel rustication strip, 3000 PSI	4.95	14.30	19.25
4000	8″ thick, plain finish, 3000 PSI	5.70	13.80	19.50
4100	4000 PSI	5.85	13.80	19.65
4300	Rub concrete 1 side, 3000 PSI	5.70	16.45	22.15
4550	8″ thick, aged wood liner, 3000 PSI	6.95	15.80	22.75
4750	Sand blast light 1 side, 3000 PSI	6.30	15.75	22.05
5300	3/4″ bevel rustication strip, 3000 PSI	5.85	14.70	20.55
5600	10″ thick, plain finish, 3000 PSI	6.55	14.15	20.70
5900	Rub concrete 1 side, 3000 PSI	6.55	16.80	23.35
6200	Aged wood liner, 3000 PSI	7.80	16.15	23.95
6500	Sand blast light 1 side, 3000 PSI	7.15	16.10	23.25
7100	3/4″ bevel rustication strip, 3000 PSI	6.70	15.05	21.75
7400	12″ thick, plain finish, 3000 PSI	7.60	14.55	22.15
7700	Rub concrete 1 side, 3000 PSI	7.60	17.20	24.80
8000	Aged wood liner, 3000 PSI	8.80	16.55	25.35
8300	Sand blast light 1 side, 3000 PSI	8.15	16.50	24.65
8900	3/4″ bevel rustication strip, 3000 PSI	7.70	15.45	23.15

B2010 Exterior Walls

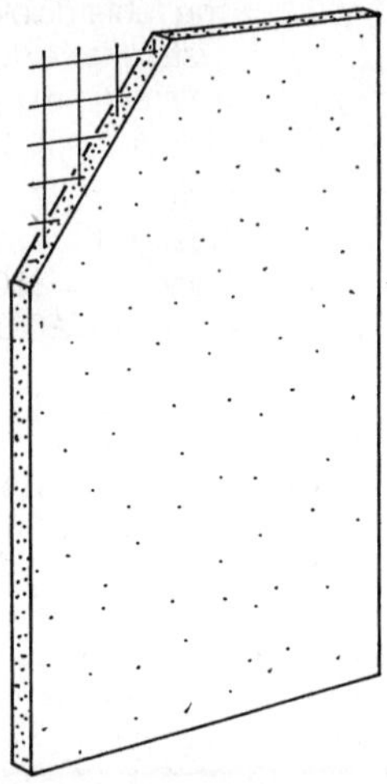

Precast concrete wall panels are either solid or insulated with plain, colored or textured finishes. Transportation is an important cost factor. Prices below are based on delivery within fifty miles of a plant. Engineering data is available from fabricators to assist with construction details. Usual minimum job size for economical use of panels is about 5000 S.F. Small jobs can double the prices below. For large, highly repetitive jobs, deduct up to 15% from the prices below.

B2010 102 Flat Precast Concrete

	THICKNESS (IN.)	PANEL SIZE (FT.)	FINISHES	RIGID INSULATION (IN)	TYPE	COST PER S.F.		
						MAT.	INST.	TOTAL
3000	4	5x18	smooth gray	none	low rise	12.65	5.45	18.10
3050		6x18				10.60	4.56	15.16
3100		8x20				21	2.20	23.20
3150		12x20				19.90	2.08	21.98
3200	6	5x18	smooth gray	2	low rise	13.70	6	19.70
3250		6x18				11.65	5.10	16.75
3300		8x20				22.50	2.79	25.29
3350		12x20				20.50	2.57	23.07
3400	8	5x18	smooth gray	2	low rise	29	3.49	32.49
3450		6x18				27.50	3.32	30.82
3500		8x20				25	3.07	28.07
3550		12x20				23	2.84	25.84
3600	4	4x8	white face	none	low rise	48.50	3.16	51.66
3650		8x8				36.50	3.76	40.26
3700		10x10				32	2.08	34.08
3750		20x10				29	1.88	30.88
3800	5	4x8	white face	none	low rise	49.50	3.22	52.72
3850		8x8				37.50	2.44	39.94
3900		10x10				33.50	2.16	35.66
3950		20x20				30.50	1.97	32.47
4000	6	4x8	white face	none	low rise	51.50	3.33	54.83
4050		8x8				39	2.53	41.53
4100		10x10				34.50	2.23	36.73
4150		20x10				31.50	2.05	33.55
4200	6	4x8	white face	2	low rise	52.50	3.90	56.40
4250		8x8				40.50	3.10	43.60
4300		10x10				35.50	2.80	38.30
4350		20x10				31.50	2.05	33.55
4400	7	4x8	white face	none	low rise	53	3.42	56.42
4450		8x8				40.50	2.63	43.13
4500		10x10				36.50	2.36	38.86
4550		20x10				33.50	2.16	35.66

B2010 Exterior Walls

B2010 102 Flat Precast Concrete

	THICKNESS (IN.)	PANEL SIZE (FT.)	FINISHES	RIGID INSULATION (IN)	TYPE	COST PER S.F.		
						MAT.	INST.	TOTAL
4600	7	4x8	white face	2	low rise	54	3.99	57.99
4650		8x8				41.50	3.20	44.70
4700		10x10				37.50	2.93	40.43
4750		20x10				34.50	2.73	37.23
4800	8	4x8	white face	none	low rise	54	3.50	57.50
4850		8x8				41.50	2.69	44.19
4900		10x10				37.50	2.42	39.92
4950		20x10				34.50	2.23	36.73
5000		4x8	white face	2	low rise	55	4.07	59.07
5050		8x8				42.50	3.26	45.76
5100		10x10				43.50	3.47	46.97
5150		20x10				43.50	3.47	46.97

B2010 103 Fluted Window or Mullion Precast Concrete

	THICKNESS (IN.)	PANEL SIZE (FT.)	FINISHES	RIGID INSULATION (IN)	TYPE	COST PER S.F.		
						MAT.	INST.	TOTAL
5200	4	4x8	smooth gray	none	high rise	29	12.50	41.50
5250		8x8				20.50	8.90	29.40
5300		10x10				40	4.21	44.21
5350		20x10				35.50	3.69	39.19
5400	5	4x8	smooth gray	none	high rise	29.50	12.70	42.20
5450		8x8				21.50	9.25	30.75
5500		10x10				42	4.39	46.39
5550		20x10				37	3.88	40.88
5600	6	4x8	smooth gray	none	high rise	30.50	13.05	43.55
5650		8x8				22	9.55	31.55
5700		10x10				43	4.52	47.52
5750		20x10				38.50	4	42.50
5800		4x8	smooth gray	2	high rise	31.50	13.60	45.10
5850		8x8				23.50	10.15	33.65
5900		10x10				44.50	5.10	49.60
5950		20x10				39.50	4.57	44.07
6000	7	4x8	smooth gray	none	high rise	31	13.35	44.35
6050		8x8				23	9.85	32.85
6100		10x10				45	4.71	49.71
6150		20x10				39.50	4.13	43.63
6200	7	4x8	smooth gray	2	high rise	32	13.90	45.90
6250		8x8				24	10.40	34.40
6300		10x10				46	5.30	51.30
6350		20x10				40.50	4.70	45.20
6400	8	4x8	smooth gray	none	high rise	31.50	13.50	45
6450		8x8				23.50	10.05	33.55
6500		10x10				46.50	4.86	51.36
6550		20x10				41.50	4.33	45.83
6600		4x8	smooth gray	2	high rise	32.50	14.05	46.55
6650		8x8				24.50	10.60	35.10
6700		10x10				47.50	5.45	52.95
6750		20x10				42.50	4.90	47.40

B2010 Exterior Walls

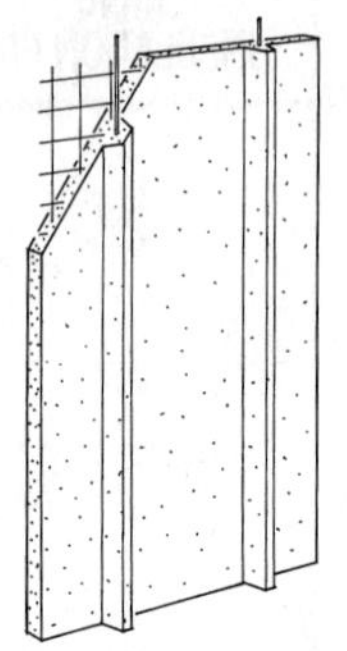

Ribbed Precast Panel

B2010 104 Ribbed Precast Concrete

	THICKNESS (IN.)	PANEL SIZE (FT.)	FINISHES	RIGID INSULATION(IN.)	TYPE	COST PER S.F.		
						MAT.	INST.	TOTAL
6800	4	4x8	aggregate	none	high rise	34.50	12.50	47
6850		8x8				25.50	9.15	34.65
6900		10x10				42.50	3.30	45.80
6950		20x10				37.50	2.92	40.42
7000	5	4x8	aggregate	none	high rise	35	12.70	47.70
7050		8x8				26	9.40	35.40
7100		10x10				44.50	3.42	47.92
7150		20x10				39.50	3.06	42.56
7200	6	4x8	aggregate	none	high rise	36	12.95	48.95
7250		8x8				26.50	9.60	36.10
7300		10x10				45.50	3.53	49.03
7350		20x10				41	3.17	44.17
7400		4x8	aggregate	2	high rise	37	13.50	50.50
7450		8x8				28	10.20	38.20
7500		10x10				47	4.10	51.10
7550		20x10				42	3.74	45.74
7600	7	4x8	aggregate	none	high rise	36.50	13.25	49.75
7650		8x8				27.50	9.85	37.35
7700		10x10				47	3.64	50.64
7750		20x10				42	3.26	45.26
7800		4x8	aggregate	2	high rise	38	13.80	51.80
7850		8x8				28.50	10.45	38.95
7900		10x10				48.50	4.21	52.71
7950		20x10				43.50	3.83	47.33
8000	8	4x8	aggregate	none	high rise	37.50	13.50	51
8050		8x8				28	10.15	38.15
8100		10x10				49	3.78	52.78
8150		20x10				44	3.41	47.41
8200		4x8	aggregate	2	high rise	38.50	14.05	52.55
8250		8x8				29.50	10.75	40.25
8300		10x10				50	5.75	55.75
8350		20x10				45	3.98	48.98

B2010 Exterior Walls

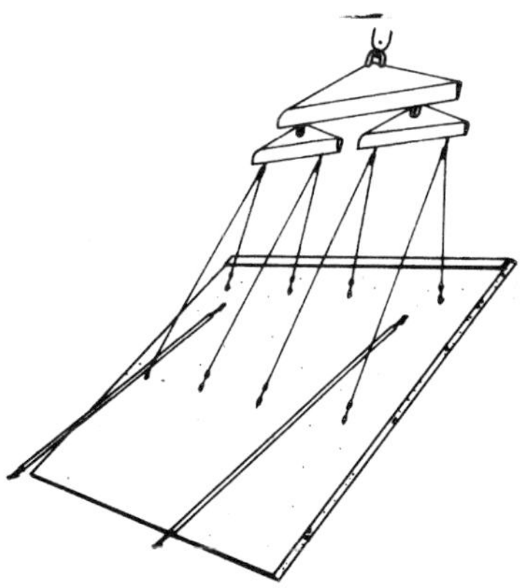

The advantage of tilt up construction is in the low cost of forms and placing of concrete and reinforcing. Tilt up has been used for several types of buildings, including warehouses, stores, offices, and schools. The panels are cast in forms on the ground, or floor slab. Most jobs use 5-1/2″ thick solid reinforced concrete panels.

Design Assumptions:
Conc. f'c = 3000 psi
Reinf. fy = 60,000

B2010 106	Tilt-Up Concrete Panel	COST PER S.F.		
		MAT.	INST.	TOTAL
3200	Tilt-up conc panels, broom finish, 5-1/2″ thick, 3000 PSI	4.37	5.15	9.52
3250	5000 PSI	4.48	5.05	9.53
3300	6″ thick, 3000 PSI	4.79	5.30	10.09
3350	5000 PSI	4.94	5.20	10.14
3400	7-1/2″ thick, 3000 PSI	6.15	5.50	11.65
3450	5000 PSI	6.30	5.40	11.70
3500	8″ thick, 3000 PSI	6.55	5.65	12.20
3550	5000 PSI	6.80	5.55	12.35
3700	Steel trowel finish, 5-1/2″ thick, 3000 PSI	4.37	5.25	9.62
3750	5000 PSI	4.48	5.15	9.63
3800	6″ thick, 3000 PSI	4.79	5.35	10.14
3850	5000 PSI	4.94	5.30	10.24
3900	7-1/2″ thick, 3000 PSI	6.15	5.55	11.70
3950	5000 PSI	6.30	5.45	11.75
4000	8″ thick, 3000 PSI	6.55	5.75	12.30
4050	5000 PSI	6.80	5.65	12.45
4200	Exp. aggregate finish, 5-1/2″ thick, 3000 PSI	4.69	5.30	9.99
4250	5000 PSI	4.81	5.20	10.01
4300	6″ thick, 3000 PSI	5.10	5.40	10.50
4350	5000 PSI	5.25	5.30	10.55
4400	7-1/2″ thick, 3000 PSI	6.45	5.60	12.05
4450	5000 PSI	6.65	5.50	12.15
4500	8″ thick, 3000 PSI	6.90	5.75	12.65
4550	5000 PSI	7.10	5.65	12.75
4600	Exposed aggregate & vert. rustication 5-1/2″ thick, 3000 PSI	7.05	6.60	13.65
4650	5000 PSI	7.15	6.50	13.65
4700	6″ thick, 3000 PSI	7.50	6.75	14.25
4750	5000 PSI	7.60	6.65	14.25
4800	7-1/2″ thick, 3000 PSI	8.80	6.95	15.75
4850	5000 PSI	9	6.85	15.85
4900	8″ thick, 3000 PSI	9.25	7.05	16.30
4950	5000 PSI	9.45	7	16.45
5000	Vertical rib & light sandblast, 5-1/2″ thick, 3000 PSI	6.85	8.70	15.55
5100	6″ thick, 3000 PSI	7.25	8.85	16.10
5200	7-1/2″ thick, 3000 PSI	8.60	9.05	17.65
5300	8″ thick, 3000 PSI	9.05	9.20	18.25
6000	Broom finish w/2″ polystyrene insulation, 6″ thick, 3000 PSI	4.21	6.40	10.61
6100	Broom finish 2″ fiberplank insulation, 6″ thick, 3000 PSI	4.75	6.35	11.10
6200	Exposed aggregate w/2″ polystyrene insulation, 6″ thick, 3000 PSI	4.43	6.40	10.83
6300	Exposed aggregate 2″ fiberplank insulation, 6″ thick, 3000 PSI	4.97	6.35	11.32

B2010 Exterior Walls

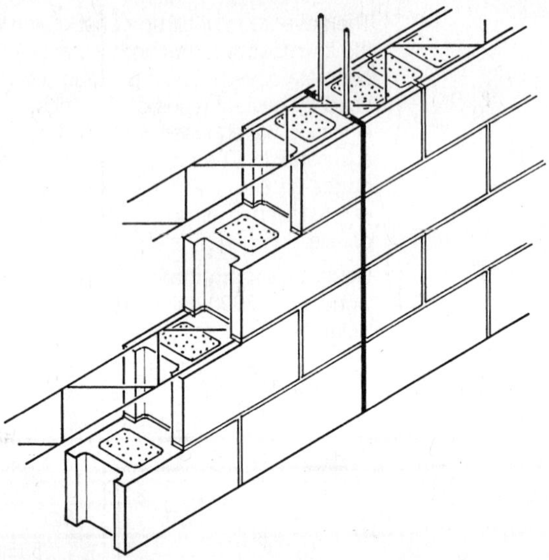

Exterior concrete block walls are defined in the following terms; structural reinforcement, weight, percent solid, size, strength and insulation. Within each of these categories, two to four variations are shown. No costs are included for brick shelf or relieving angles.

B2010 109 Concrete Block Wall - Regular Weight

	TYPE	SIZE (IN.)	STRENGTH (P.S.I.)	CORE FILL		COST PER S.F.		
						MAT.	INST.	TOTAL
1200	Hollow	4x8x16	2,000	none		2.18	5.40	7.58
1250			4,500	none		2.66	5.40	8.06
1400		8x8x16	2,000	perlite		4.46	6.55	11.01
1410				styrofoam		4.46	6.15	10.61
1440				none		2.97	6.15	9.12
1450			4,500	perlite		5.45	6.55	12
1460				styrofoam		5.45	6.15	11.60
1490				none		3.95	6.15	10.10
2000	75% solid	4x8x16	2,000	none		2.62	5.45	8.07
2100		6x8x16	2,000	perlite		3.86	6	9.86
2500	Solid	4x8x16	2,000	none		2.45	5.60	8.05
2700		8x8x16	2,000	none		4.13	6.40	10.53

B2010 110 Concrete Block Wall - Lightweight

	TYPE	SIZE (IN.)	WEIGHT (P.C.F.)	CORE FILL		COST PER S.F.		
						MAT.	INST.	TOTAL
3100	Hollow	8x4x16	105	perlite		4.40	6.20	10.60
3110				styrofoam		4.40	5.80	10.20
3200		4x8x16	105	none		2.30	5.30	7.60
3250			85	none		2.92	5.15	8.07
3300		6x8x16	105	perlite		4.14	5.95	10.09
3310				styrofoam		4.45	5.65	10.10
3340				none		3.13	5.65	8.78
3400		8x8x16	105	perlite		5.25	6.40	11.65
3410				styrofoam		5.25	6	11.25
3440				none		3.78	6	9.78
3450			85	perlite		5.45	6.25	11.70
4000	75% solid	4x8x16	105	none		3.31	5.35	8.66
4050			85	none		3.59	5.20	8.79
4100		6x8x16	105	perlite		5.05	5.85	10.90
4500	Solid	4x8x16	105	none		2.08	5.50	7.58
4700		8x8x16	105	none		4.12	6.35	10.47

B2010 Exterior Walls

B2010 111 Reinforced Concrete Block Wall - Regular Weight

	TYPE	SIZE (IN.)	STRENGTH (P.S.I.)	VERT. REINF & GROUT SPACING		COST PER S.F. MAT.	INST.	TOTAL
5200	Hollow	4x8x16	2,000	#4 @ 48"		2.34	6	8.34
5300		6x8x16	2,000	#4 @ 48"		3.03	6.45	9.48
5330				#5 @ 32"		3.25	6.75	10
5340				#5 @ 16"		3.71	7.65	11.36
5350			4,500	#4 @ 28"		3.51	6.45	9.96
5390				#5 @ 16"		4.19	7.65	11.84
5400		8x8x16	2,000	#4 @ 48"		3.36	6.90	10.26
5430				#5 @ 32"		3.53	7.30	10.83
5440				#5 @ 16"		4.08	8.40	12.48
5450		8x8x16	4,500	#4 @ 48"		4.27	6.95	11.22
5490				#5 @ 16"		5.05	8.40	13.45
5500		12x8x16	2,000	#4 @ 48"		5.20	8.85	14.05
5540				#5 @ 16"		6.30	10.35	16.65
6100	75% solid	6x8x16	2,000	#4 @ 48"		3.49	6.35	9.84
6140				#5 @ 16"		3.93	7.25	11.18
6150			4,500	#4 @ 48"		4.13	6.35	10.48
6190				#5 @ 16"		4.57	7.25	11.82
6200		8x8x16	2,000	#4 @ 48"		3.72	6.85	10.57
6230				#5 @ 32"		3.88	7.10	10.98
6240				#5 @ 16"		4.18	7.90	12.08
6250			4,500	#4 @ 48"		5	6.85	11.85
6280				#5 @ 32"		5.15	7.10	12.25
6290				#5 @ 16"		5.45	7.90	13.35
6500	Solid-double Wythe	2-4x8x16	2,000	#4 @ 48" E.W.		5.75	12.80	18.55
6530				#5 @ 16" E.W.		6.45	13.60	20.05
6550			4,500	#4 @ 48" E.W.		8.15	12.60	20.75
6580				#5 @ 16" E.W.		8.85	13.40	22.25

B2010 112 Reinforced Concrete Block Wall - Lightweight

	TYPE	SIZE (IN.)	WEIGHT (P.C.F.)	VERT REINF. & GROUT SPACING		COST PER S.F. MAT.	INST.	TOTAL
7100	Hollow	8x4x16	105	#4 @ 48"		3.23	6.60	9.83
7140				#5 @ 16"		4.02	8.05	12.07
7150			85	#4 @ 48"		4.77	6.45	11.22
7190				#5 @ 16"		5.55	7.90	13.45
7400		8x8x16	105	#4 @ 48"		4.10	6.80	10.90
7440				#5 @ 16"		4.89	8.25	13.14
7450		8x8x16	85	#4 @ 48"		4.27	6.65	10.92
7490				#5 @ 16"		5.05	8.10	13.15
7800		8x8x24	105	#4 @ 48"		3.16	7.35	10.51
7840				#5 @ 16"		3.95	8.80	12.75
7850			85	#4 @ 48"		7.15	6.25	13.40
7890				#5 @ 16"		7.90	7.70	15.60
8100	75% solid	6x8x16	105	#4 @ 48"		4.70	6.20	10.90
8130				#5 @ 32"		4.85	6.40	11.25
8150			85	#4 @ 48"		4.76	6.10	10.86
8180				#5 @ 32"		4.91	6.30	11.21
8200		8x8x16	105	#4 @ 48"		5.65	6.70	12.35
8230				#5 @ 32"		5.85	6.95	12.80
8250			85	#4 @ 48"		5	6.55	11.55
8280				#5 @ 32"		5.15	6.80	11.95

B2010 Exterior Walls

B2010 112 Reinforced Concrete Block Wall - Lightweight

	TYPE	SIZE (IN.)	WEIGHT (P.C.F.)	VERT REINF. & GROUT SPACING		COST PER S.F.		
						MAT.	INST.	TOTAL
8500	Solid-double	2-4x8x16	105	#4 @ 48"		5	12.60	17.60
8530	Wythe		105	#5 @ 16"		5.70	13.40	19.10
8600		2-6x8x16	105	#4 @ 48"		8.15	13.55	21.70
8630				#5 @ 16"		8.85	14.30	23.15
8650			85	#4 @ 48"		11.50	12.95	24.45

B2010 Exterior Walls

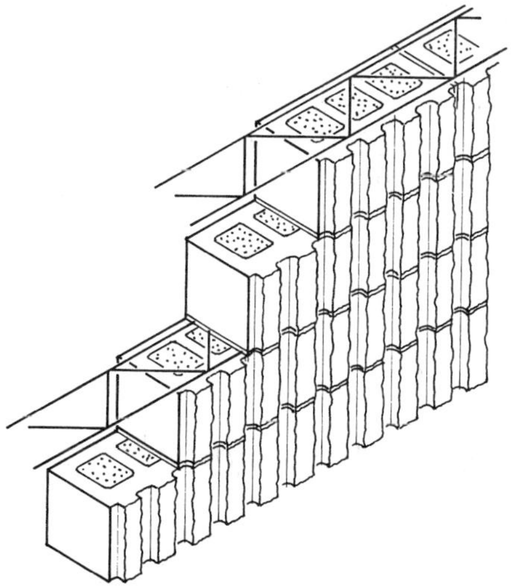

Exterior split ribbed block walls are defined in the following terms; structural reinforcement, weight, percent solid, size, number of ribs and insulation. Within each of these categories two to four variations are shown. No costs are included for brick shelf or relieving angles. Costs include control joints every 20′ and horizontal reinforcing.

B2010 113 Split Ribbed Block Wall - Regular Weight

	TYPE	SIZE (IN.)	RIBS	CORE FILL		COST PER S.F. MAT.	INST.	TOTAL
1220	Hollow	4x8x16	4	none		4.41	6.70	11.11
1250			8	none		4.78	6.70	11.48
1280			16	none		5.15	6.80	11.95
1430		8x8x16	8	perlite		7.60	7.95	15.55
1440				styrofoam		7.45	7.55	15
1450				none		6.10	7.55	13.65
1530		12x8x16	8	perlite		9.60	10.55	20.15
1540				styrofoam		8.90	9.80	18.70
1550				none		7.20	9.80	17
2120	75% solid	4x8x16	4	none		5.50	6.80	12.30
2150			8	none		6	6.80	12.80
2180			16	none		6.50	6.85	13.35
2520	Solid	4x8x16	4	none		6.25	6.85	13.10
2550			8	none		7.40	6.85	14.25
2580			16	none		7.40	7	14.40

B2010 115 Reinforced Split Ribbed Block Wall - Regular Weight

	TYPE	SIZE (IN.)	RIBS	VERT. REINF. & GROUT SPACING		COST PER S.F. MAT.	INST.	TOTAL
5200	Hollow	4x8x16	4	#4 @ 48″		4.57	7.30	11.87
5230			8	#4 @ 48″		4.94	7.30	12.24
5260			16	#4 @ 48″		5.30	7.40	12.70
5430		8x8x16	8	#4 @ 48″		6.45	8.35	14.80
5440				#5 @ 32″		6.65	8.70	15.35
5450				#5 @ 16″		7.20	9.80	17
5530		12x8x16	8	#4 @ 48″		7.65	10.65	18.30
5540				#5 @ 32″		7.95	11	18.95
5550				#5 @ 16″		8.70	12.15	20.85
6230	75% solid	6x8x16	8	#4 @ 48″		7	7.70	14.70
6240				#5 @ 32″		7.15	7.90	15.05
6250				#5 @ 16″		7.40	8.60	16
6330		8x8x16	8	#4 @ 48″		7.80	8.25	16.05
6340				#5 @ 32″		8	8.50	16.50
6350				#5 @ 16″		8.30	9.30	17.60

B2010 Exterior Walls

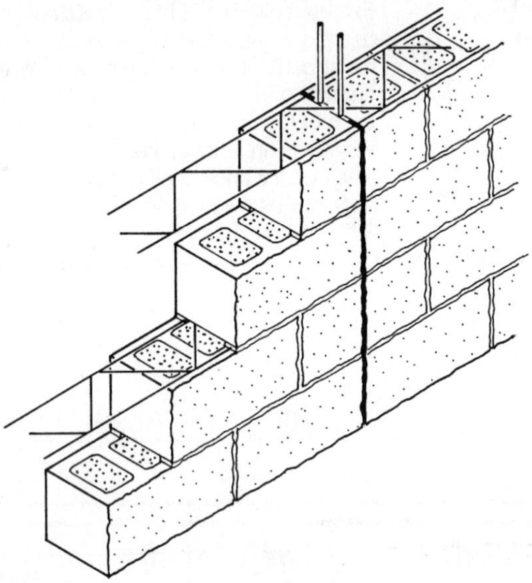

Exterior split face block walls are defined in the following terms; structural reinforcement, weight, percent solid, size, scores and insulation. Within each of these categories two to four variations are shown. No costs are included for brick shelf or relieving angles. Costs include control joints every 20′ and horizontal reinforcing.

B2010 117 Split Face Block Wall - Regular Weight

	TYPE	SIZE (IN.)	SCORES	CORE FILL		COST PER S.F.		
						MAT.	INST.	TOTAL
1200	Hollow	8x4x16	0	perlite		7.35	8.75	16.10
1210				styrofoam		7.35	8.35	15.70
1240				none		5.85	8.35	14.20
1250			1	perlite		7.75	8.75	16.50
1260				styrofoam		7.75	8.35	16.10
1290				none		6.25	8.35	14.60
1300		12x4x16	0	perlite		9.65	9.95	19.60
1310				styrofoam		8.95	9.20	18.15
1340				none		7.25	9.20	16.45
1350			1	perlite		10.10	9.95	20.05
1360				styrofoam		9.40	9.20	18.60
1390				none		7.70	9.20	16.90
1400		4x8x16	0	none		4.03	6.60	10.63
1450			1	none		4.36	6.70	11.06
1500		6x8x16	0	perlite		5.60	7.60	13.20
1510				styrofoam		5.90	7.30	13.20
1540				none		4.57	7.30	11.87
1550			1	perlite		5.95	7.70	13.65
1560				styrofoam		6.25	7.40	13.65
1590				none		4.94	7.40	12.34
1600		8x8x16	0	perlite		6.65	8.20	14.85
1610				styrofoam		6.65	7.80	14.45
1640				none		5.15	7.80	12.95
1650	Hollow	8x8x16	1	perlite		7.05	8.30	15.35
1660				styrofoam		7.05	7.90	14.95
1690				none		5.55	7.90	13.45
1700		12x8x16	0	perlite		8.60	10.75	19.35
1705			0	perlite		10.20	13	23.20
1710				styrofoam		7.90	10	17.90
1740				none		6.20	10	16.20

B2010 Exterior Walls

B2010 117 Split Face Block Wall - Regular Weight

	TYPE	SIZE (IN.)	SCORES	CORE FILL		COST PER S.F.		
						MAT.	INST.	TOTAL
1750	Hollow	8x4x16	1	perlite		9.10	10.95	20.05
1760				styrofoam		8.40	10.20	18.60
1790				none		6.70	10.20	16.90
1800	75% solid	8x4x16	0	perlite		7.85	8.70	16.55
1840				none		7.10	8.50	15.60
1850			1	perlite		8.40	8.70	17.10
1890				none		7.65	8.50	16.15
2000		4x8x16	0	none		5.05	6.70	11.75
2050			1	none		5.45	6.80	12.25
2400	Solid	8x4x16	0	none		7.95	8.65	16.60
2450			1	none		8.55	8.65	17.20
2900		12x8x16	0	none		8.65	10.35	19
2950			1	none		9.35	10.55	19.90

B2010 119 Reinforced Split Face Block Wall - Regular Weight

	TYPE	SIZE (IN.)	SCORES	VERT. REINF. & GROUT SPACING		COST PER S.F.		
						MAT.	INST.	TOTAL
5200	Hollow	8x4x16	0	#4 @ 48"		6.20	9.15	15.35
5210				#5 @ 32"		6.40	9.50	15.90
5240				#5 @ 16"		6.95	10.60	17.55
5250			1	#4 @ 48"		6.60	9.15	15.75
5260				#5 @ 32"		6.80	9.50	16.30
5290				#5 @ 16"		7.35	10.60	17.95
5700		12x8x16	0	#4 @ 48"		6.65	10.85	17.50
5710				#5 @ 32"		6.95	11.20	18.15
5740				#5 @ 16"		7.70	12.35	20.05
5750			1	#4 @ 48"		7.15	11.05	18.20
5760				#5 @ 32"		7.45	11.40	18.85
5790				#5 @ 16"		8.20	12.55	20.75
6000	75% solid	8x4x16	0	#4 @ 48"		7.25	9.10	16.35
6010				#5 @ 32"		7.45	9.35	16.80
6040				#5 @ 16"		7.75	10.15	17.90
6050			1	#4 @ 48"		7.80	9.10	16.90
6060				#5 @ 32"		8	9.35	17.35
6090				#5 @ 16"		8.30	10.15	18.45
6100		12x4x16	0	#4 @ 48"		8.95	10.10	19.05
6110				#5 @ 32"		9.15	10.35	19.50
6140				#5 @ 16"		9.55	11.05	20.60
6150			1	#4 @ 48"		9.60	10.10	19.70
6160				#5 @ 32"		9.80	10.35	20.15
6190				#5 @ 16"		10.25	11.25	21.50
6700	Solid-double Wythe	2-4x8x16	0	#4 @ 48" E.W.		12.25	15.20	27.45
6710				#5 @ 32" E.W.		12.50	15.35	27.85
6740				#5 @ 16" E.W.		12.95	16	28.95
6750			1	#4 @ 48" E.W.		13.25	15.30	28.55
6760				#5 @ 32" E.W.		13.50	15.45	28.95
6790				#5 @ 16" E.W.		13.95	16.10	30.05
6800		2-6x8x16	0	#4 @ 48" E.W.		13.70	16.75	30.45
6810				#5 @ 32" E.W.		14	16.90	30.90
6840				#5 @ 16" E.W.		14.45	17.50	31.95
6850			1	#4 @ 48" E.W.		14.80	16.95	31.75
6860				#5 @ 32" E.W.		15.10	17.10	32.20
6890				#5 @ 16" E.W.		15.55	17.70	33.25

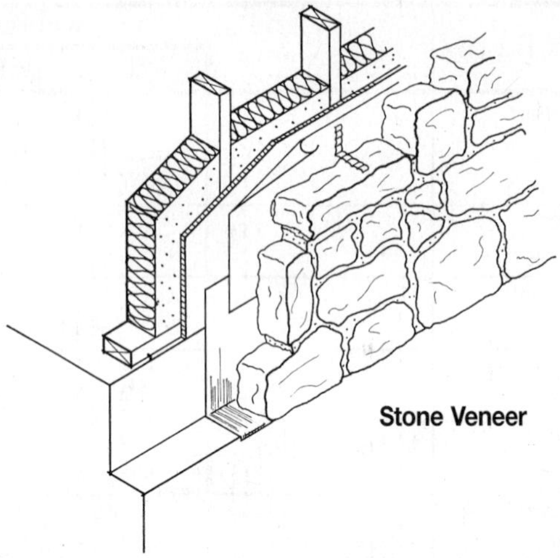

Stone Veneer

The table below lists costs per S.F. for stone veneer walls on various backup using different stone.

B2010 128	Stone Veneer	COST PER S.F.		
		MAT.	INST.	TOTAL
2000	Ashlar veneer, 4", 2" x 4" stud backup, 16" O.C., 8' high, low priced stone	15.40	19.15	34.55
2050	2" x 6" stud backup, 16" O.C.	17.15	22.50	39.65
2100	Metal stud backup, 8' high, 16" O.C.	16.25	19.40	35.65
2150	24" O.C.	15.90	18.95	34.85
2200	Conc. block backup, 4" thick	16.90	22.50	39.40
2300	6" thick	17.55	23	40.55
2350	8" thick	17.70	23.50	41.20
2400	10" thick	18.25	24.50	42.75
2500	12" thick	19.50	26.50	46
3100	High priced stone, wood stud backup, 10' high, 16" O.C.	22	22	44
3200	Metal stud backup, 10' high, 16" O.C.	23	22	45
3250	24" O.C.	22.50	21.50	44
3300	Conc. block backup, 10' high, 4" thick	23.50	25.50	49
3350	6" thick	24	25.50	49.50
3400	8" thick	25	27.50	52.50
3450	10" thick	25	27.50	52.50
3500	12" thick	26	29	55
4000	Indiana limestone 2" thk., sawn finish, wood stud backup, 10' high, 16" O.C	34	12.05	46.05
4100	Metal stud backup, 10' high, 16" O.C.	36.50	15.10	51.60
4150	24" O.C.	36.50	15.10	51.60
4200	Conc. block backup, 4" thick	36	15.65	51.65
4250	6" thick	36.50	15.85	52.35
4300	8" thick	36.50	16.35	52.85
4350	10" thick	37	17.55	54.55
4400	12" thick	38.50	19.25	57.75
4450	2" thick, smooth finish, wood stud backup, 8' high, 16" O.C.	34	12.05	46.05
4550	Metal stud backup, 8' high, 16" O.C.	35	12.15	47.15
4600	24" O.C.	34.50	11.70	46.20
4650	Conc. block backup, 4" thick	36	15.45	51.45
4700	6" thick	36.50	15.70	52.20
4750	8" thick	36.50	16.20	52.70
4800	10" thick	37	17.55	54.55
4850	12" thick	38.50	19.25	57.75
5350	4" thick, smooth finish, wood stud backup, 8' high, 16" O.C.	36.50	12.05	48.55
5450	Metal stud backup, 8' high, 16" O.C.	37.50	12.30	49.80
5500	24" O.C.	37	11.85	48.85
5550	Conc. block backup, 4" thick	38.50	15.65	54.15
5600	6" thick	39	15.85	54.85

B2010 Exterior Walls

B2010 128	Stone Veneer	COST PER S.F.		
		MAT.	INST.	TOTAL
5650	8" thick	43	20	63
5700	10" thick	39.50	17.55	57.05
5750	12" thick	41	19.25	60.25
6000	Granite, gray or pink, 2" thick, wood stud backup, 8' high, 16" O.C.	33.50	22	55.50
6100	Metal studs, 8' high, 16" O.C.	34.50	22.50	57
6150	24" O.C.	34	22	56
6200	Conc. block backup, 4" thick	35.50	25.50	61
6250	6" thick	36	26	62
6300	8" thick	36	26.50	62.50
6350	10" thick	36.50	27.50	64
6400	12" thick	38	29	67
6900	4" thick, wood stud backup, 8' high, 16" O.C.	44.50	25.50	70
7000	Metal studs, 8' high, 16" O.C.	45.50	26	71.50
7050	24" O.C.	45	25.50	70.50
7100	Conc. block backup, 4" thick	46.50	29	75.50
7150	6" thick	47	29.50	76.50
7200	8" thick	47	30	77
7250	10" thick	47.50	31	78.50
7300	12" thick	49	33	82

B2010 Exterior Walls

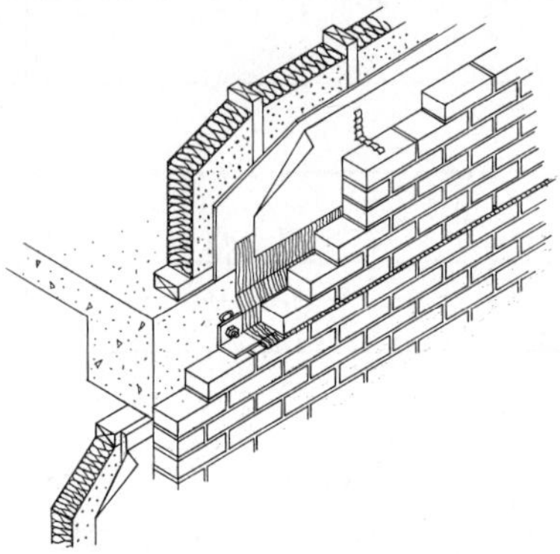

Exterior brick veneer/stud backup walls are defined in the following terms: type of brick and studs, stud spacing and bond. All systems include a back-up wall, a control joint every 20′, a brick shelf every 12′ of height, ties to the backup and the necessary dampproofing, flashing and insulation.

B2010 129 Brick Veneer/Wood Stud Backup

	FACE BRICK	STUD BACKUP	STUD SPACING (IN.)	BOND	FACE	COST PER S.F.		
						MAT.	INST.	TOTAL
1100	Standard	2x4-wood	16	running		7.40	14.95	22.35
1120				common		8.45	16.90	25.35
1140				Flemish		9.20	19.70	28.90
1160				English		10.10	20.50	30.60
1400		2x6-wood	16	running		7.90	14.90	22.80
1420				common		8.70	17	25.70
1440				Flemish		9.45	19.80	29.25
1460				English		10.35	21	31.35
1700	Glazed	2x4-wood	16	running		23	15.45	38.45
1720				common		27	17.60	44.60
1740				Flemish		29.50	20.50	50
1760				English		33	22	55
2300	Engineer	2x4-wood	16	running		7.40	13.40	20.80
2320				common		8.40	14.95	23.35
2340				Flemish		9.10	17.60	26.70
2360				English		10	18.35	28.35
2900	Roman	2x4-wood	16	running		10.75	13.75	24.50
2920				common		12.50	15.45	27.95
2940				Flemish		13.65	17.95	31.60
2960				English		15.20	19.25	34.45
4100	Norwegian	2x4-wood	16	running		7.95	10.75	18.70
4120				common		9.10	11.90	21
4140				Flemish		9.85	13.75	23.60
4160				English		10.80	14.30	25.10

B2010 130 Brick Veneer/Metal Stud Backup

	FACE BRICK	STUD BACKUP	STUD SPACING (IN.)	BOND	FACE	COST PER S.F.		
						MAT.	INST.	TOTAL
5050	Standard	16 ga x 6″LB	16	running		9	15.70	24.70
5100		25ga.x6″NLB	24	running		6.90	14.60	21.50
5120				common		7.90	16.55	24.45
5140				Flemish		8.30	18.70	27
5160				English		9.55	20.50	30.05
5200		20ga.x3-5/8″NLB	16	running		7	15.20	22.20
5220				common		8.05	17.15	25.20
5240				Flemish		8.80	19.95	28.75
5260				English		9.70	21	30.70

B2010 Exterior Walls

B2010 130 Brick Veneer/Metal Stud Backup

	FACE BRICK	STUD BACKUP	STUD SPACING (IN.)	BOND	FACE	COST PER S.F.		
						MAT.	INST.	TOTAL
5400	Standard	16ga.x3-5/8"LB	16	running		7.85	15.45	23.30
5420				common		8.85	17.40	26.25
5440				Flemish		9.60	20	29.60
5460				English		10.50	21	31.50
5500			24	running		7.50	15	22.50
5520				common		8.50	16.95	25.45
5540				Flemish		9.25	19.75	29
5560				English		10.15	21	31.15
5700	Glazed	25ga.x6"NLB	24	running		22.50	15.10	37.60
5720				common		26.50	17.25	43.75
5740				Flemish		29	20.50	49.50
5760				English		32.50	21.50	54
5800		20ga.x3-5/8"NLB	24	running		22.50	15.25	37.75
5820				common		26.50	17.40	43.90
5840				Flemish		29	20.50	49.50
5860				English		32.50	22	54.50
6000		16ga.x3-5/8"LB	16	running		23.50	15.95	39.45
6020				common		27.50	18.10	45.60
6040				Flemish		30	21	51
6060				English		33.50	22.50	56
6300	Engineer	25ga.x6"NLB	24	running		6.90	13.05	19.95
6320				common		7.85	14.60	22.45
6340				Flemish		8.55	17.25	25.80
6360				English		9.45	18	27.45
6400		20ga.x3-5/8"NLB	16	running		7	13.65	20.65
6420				common		8	15.20	23.20
6440				Flemish		8.70	17.85	26.55
6460				English		9.60	18.60	28.20
6900	Roman	25ga.x6"NLB	24	running		10.20	13.40	23.60
6920				common		11.95	15.05	27
6940				Flemish		13.10	17.60	30.70
6960				English		14.65	18.90	33.55
7000		20ga.x3-5/8"NLB	16	running		10.35	14	24.35
7020				common		12.10	15.70	27.80
7040				Flemish		13.25	18.20	31.45
7060				English		14.80	19.50	34.30
7500	Norman	25ga.x6"NLB	24	running		8.20	11.45	19.65
7520				common		9.45	12.75	22.20
7540				Flemish		21	14.85	35.85
7560				English		11.45	15.65	27.10
7600		20ga.x3-5/8"NLB	24	running		8.20	11.60	19.80
7620				common		9.45	12.90	22.35
7640				Flemish		21	15	36
7660				English		11.45	15.80	27.25
8100	Norwegian	25ga.x6"NLB	24	running		7.45	10.40	17.85
8120				common		8.55	11.55	20.10
8140				Flemish		9.30	13.40	22.70
8160				English		10.25	13.95	24.20
8400		16ga.x3-5/8"LB	16	running		8.40	11.25	19.65
8420				common		9.50	12.40	21.90
8440				Flemish		10.25	14.25	24.50
8460				English		11.20	14.80	26

B2010 Exterior Walls

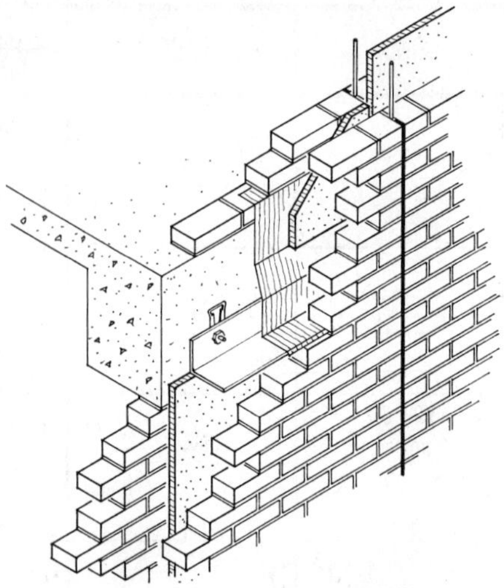

Exterior brick face cavity walls are defined in the following terms: cavity treatment, type of face brick, backup masonry, total thickness and insulation. Seven types of face brick are shown with various types of backup. All systems include a brick shelf, ties to the backups and necessary dampproofing, flashing, and control joints every 20′.

B2010 134 Brick Face Cavity Wall

	FACE BRICK	BACKUP MASONRY	TOTAL THICKNESS (IN.)	CAVITY INSULATION		COST PER S.F.		
						MAT.	INST.	TOTAL
1000	Standard	4″ common brick	10	polystyrene		10.50	23	33.50
1020				none		10.20	22.50	32.70
1040		6″ SCR brick	12	polystyrene		13	20.50	33.50
1060				none		12.70	19.95	32.65
1080		4″ conc. block	10	polystyrene		8	18.55	26.55
1100				none		7.70	17.95	25.65
1120		6″ conc. block	12	polystyrene		8.65	18.95	27.60
1140				none		8.35	18.35	26.70
1160		4″ L.W. block	10	polystyrene		8.10	18.45	26.55
1180				none		7.85	17.85	25.70
1200		6″ L.W. block	12	polystyrene		9	18.80	27.80
1220				none		8.70	18.20	26.90
1240		4″ glazed block	10	polystyrene		17.25	19.85	37.10
1260				none		16.95	19.25	36.20
1280		6″ glazed block	12	polystyrene		17.25	19.85	37.10
1300				none		16.95	19.25	36.20
1320		4″ clay tile	10	polystyrene		12.90	17.85	30.75
1340				none		12.60	17.20	29.80
1360		4″ glazed tile	10	polystyrene		18.90	23.50	42.40
1380				none		18.60	23	41.60
1500	Glazed	4″ common brick	10	polystyrene		26	23.50	49.50
1520				none		25.50	23	48.50
1580		4″ conc. block	10	polystyrene		23.50	19.05	42.55
1600				none		23	18.45	41.45
1660		4″ L.W. block	10	polystyrene		23.50	18.95	42.45
1680				none		23.50	18.35	41.85
1740		4″ glazed block	10	polystyrene		32.50	20.50	53
1760				none		32.50	19.75	52.25
1820		4″ clay tile	10	polystyrene		28.50	18.35	46.85
1840				none		28	17.70	45.70
1860		4″ glazed tile	10	polystyrene		30.50	18.85	49.35
1880				none		30.50	18.20	48.70
2000	Engineer	4″ common brick	10	polystyrene		10.50	21.50	32
2020				none		10.20	21	31.20

B2010 Exterior Walls

B2010 134 Brick Face Cavity Wall

	FACE BRICK	BACKUP MASONRY	TOTAL THICKNESS (IN.)	CAVITY INSULATION		COST PER S.F.		
						MAT.	INST.	TOTAL
2080	Engineer	4″ conc. block	10	polystyrene		8	17	25
2100				none		7.70	16.40	24.10
2160		4″ L.W. block	10	polystyrene		8.10	16.90	25
2180				none		7.80	16.30	24.10
2240		4″ glazed block	10	polystyrene		17.20	18.30	35.50
2260				none		16.95	17.70	34.65
2320		4″ clay tile	10	polystyrene		12.85	16.30	29.15
2340				none		12.60	15.65	28.25
2360		4″ glazed tile	10	polystyrene		18.85	22	40.85
2380				none		18.60	21.50	40.10
2500	Roman	4″ common brick	10	polystyrene		13.80	22	35.80
2520				none		13.50	21	34.50
2580		4″ conc. block	10	polystyrene		11.30	17.35	28.65
2600				none		11	16.75	27.75
2660		4″ L.W. block	10	polystyrene		11.45	17.25	28.70
2680				none		11.15	16.65	27.80
2740		4″ glazed block	10	polystyrene		20.50	18.65	39.15
2760				none		20.50	18.05	38.55
2820		4″ clay tile	10	polystyrene		16.20	16.65	32.85
2840				none		15.90	16	31.90
2860		4″ glazed tile	10	polystyrene		22	22.50	44.50
2880				none		22	21.50	43.50
3000	Norman	4″ common brick	10	polystyrene		11.80	19.90	31.70
3020				none		11.50	19.30	30.80
3080		4″ conc. block	10	polystyrene		9.30	15.40	24.70
3100				none		9	14.80	23.80
3160		4″ L.W. block	10	polystyrene		9.45	15.30	24.75
3180				none		9.15	14.70	23.85
3240		4″ glazed block	10	polystyrene		18.55	16.70	35.25
3260				none		18.25	16.10	34.35
3320		4″ clay tile	10	polystyrene		14.20	14.70	28.90
3340				none		13.90	14.05	27.95
3360		4″ glazed tile	10	polystyrene		20	20.50	40.50
3380				none		19.90	19.75	39.65
3500	Norwegian	4″ common brick	10	polystyrene		11.05	18.85	29.90
3520				none		10.75	18.25	29
3580		4″ conc. block	10	polystyrene		8.55	14.35	22.90
3600				none		8.25	13.75	22
3660		4″ L.W. block	10	polystyrene		8.65	14.25	22.90
3680				none		8.35	13.65	22
3740		4″ glazed block	10	polystyrene		17.75	15.65	33.40
3760				none		17.50	15.05	32.55
3820		4″ clay tile	10	polystyrene		13.40	13.65	27.05
3840				none		13.15	13	26.15
3860		4″ glazed tile	10	polystyrene		19.40	19.30	38.70
3880				none		19.15	18.70	37.85
4000	Utility	4″ common brick	10	polystyrene		11.40	17.90	29.30
4020				none		11.10	17.25	28.35
4080		4″ conc. block	10	polystyrene		8.90	13.40	22.30
4100				none		8.60	12.75	21.35
4160		4″ L.W. block	10	polystyrene		9.05	13.30	22.35
4180				none		8.75	12.65	21.40

B2010 Exterior Walls

B2010 134 Brick Face Cavity Wall

	FACE BRICK	BACKUP MASONRY	TOTAL THICKNESS (IN.)	CAVITY INSULATION		COST PER S.F.		
						MAT.	INST.	TOTAL
4240	Utility	4″ glazed block	10	polystyrene		18.15	14.70	32.85
4260				none		17.85	14.05	31.90
4320		4″ clay tile	10	polystyrene		13.80	12.65	26.45
4340				none		13.50	12.05	25.55
4360		4″ glazed tile	10	polystyrene		19.80	18.35	38.15
4380				none		19.50	17.70	37.20

B2010 135 Brick Face Cavity Wall - Insulated Backup

	FACE BRICK	BACKUP MASONRY	TOTAL THICKNESS (IN.)	BACKUP CORE FILL		COST PER S.F.		
						MAT.	INST.	TOTAL
5100	Standard	6″ conc. block	10	perlite		9.35	18.65	28
5120				styrofoam		9.65	18.35	28
5180		6″ L.W. block	10	perlite		9.70	18.50	28.20
5200				styrofoam		10	18.20	28.20
5260		6″ glazed block	10	perlite		18.85	19.85	38.70
5280				styrofoam		19.15	19.55	38.70
5340		6″ clay tile	10	none		14.95	17.70	32.65
5360		8″ clay tile	12	none		17.20	18.35	35.55
5600	Glazed	6″ conc. block	10	perlite		25	19.15	44.15
5620				styrofoam		25	18.85	43.85
5680		6″ L.W. block	10	perlite		25	19	44
5700				styrofoam		25.50	18.70	44.20
5760		6″ glazed block	10	perlite		34.50	20.50	55
5780				styrofoam		34.50	20	54.50
5840		6″ clay tile	10	none		30.50	18.20	48.70
5860		8″ clay tile	8	none		32.50	18.85	51.35
6100	Engineer	6″ conc. block	10	perlite		9.30	17.10	26.40
6120				styrofoam		9.65	16.80	26.45
6180		6″ L.W. block	10	perlite		9.70	16.95	26.65
6200				styrofoam		10	16.65	26.65
6260		6″ glazed block	10	perlite		18.85	18.30	37.15
6280				styrofoam		19.15	18	37.15
6340		6″ clay tile	10	none		14.95	16.15	31.10
6360		8″ clay tile	12	none		17.20	16.80	34
6600	Roman	6″ conc. block	10	perlite		12.65	17.45	30.10
6620				styrofoam		12.95	17.15	30.10
6680		6″ L.W. block	10	perlite		13	17.30	30.30
6700				styrofoam		13.30	17	30.30
6760		6″ glazed block	10	perlite		22	18.65	40.65
6780				styrofoam		22.50	18.35	40.85
6840		6″ clay tile	10	none		18.25	16.50	34.75
6860		8″ clay tile	12	none		20.50	17.15	37.65
7100	Norman	6″ conc. block	10	perlite		10.65	15.50	26.15
7120				styrofoam		10.95	15.20	26.15
7180		6″ L.W. block	10	perlite		11	15.35	26.35
7200				styrofoam		11.30	15.05	26.35
7260		6″ glazed block	10	perlite		20	16.70	36.70
7280				styrofoam		20.50	16.40	36.90
7340		6″ clay tile	10	none		16.25	14.55	30.80
7360		8″ clay tile	12	none		18.55	15.20	33.75

B2010 Exterior Walls

B2010 135 Brick Face Cavity Wall - Insulated Backup

	FACE BRICK	BACKUP MASONRY	TOTAL THICKNESS (IN.)	BACKUP CORE FILL		COST PER S.F.		
						MAT.	INST.	TOTAL
7600	Norwegian	6″ conc. block	10	perlite		9.85	14.45	24.30
7620				styrofoam		10.20	14.15	24.35
7680		6″ L.W. block	10	perlite		10.25	14.30	24.55
7700				styrofoam		10.55	14	24.55
7760		6″ glazed block	10	perlite		19.40	15.65	35.05
7780				styrofoam		19.70	15.35	35.05
7840		6″ clay tile	10	none		15.50	13.50	29
7860		8″ clay tile	12	none		17.75	14.15	31.90
8100	Utility	6″ conc. block	10	perlite		10.25	13.45	23.70
8120				styrofoam		10.55	13.15	23.70
8180		6″ L.W. block	10	perlite		10.60	13.30	23.90
8200				styrofoam		10.90	13	23.90
8220		8″ L.W. block	12	perlite		11.75	13.75	25.50
8240				styrofoam		11.75	13.35	25.10
8260		6″ glazed block	10	perlite		19.75	14.65	34.40
8280				styrofoam		20	14.35	34.35
8300		8″ glazed block	12	perlite		21.50	15.15	36.65
8340		6″ clay tile	10	none		15.85	12.55	28.40
8360		8″ clay tile	12	none		18.15	13.15	31.30

B2010 Exterior Walls

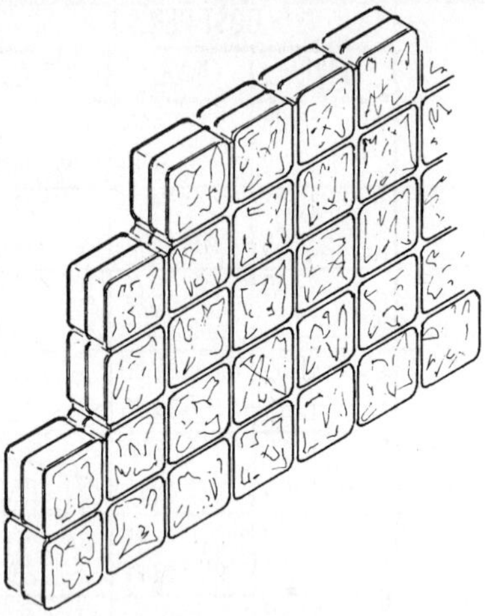

The table below lists costs per S.F. for glass block walls. Included in the costs are the following special accessories required for glass block walls.

Glass block accessories required for proper installation.

Wall ties: Galvanized double steel mesh full length of joint.

Fiberglass expansion joint at sides and top.

Silicone caulking: One gallon does 95 L.F.

Oakum: One lb. does 30 L.F.

Asphalt emulsion: One gallon does 600 L.F.

If block are not set in wall chase, use 2′-0″ long wall anchors at 2′-0″ O.C.

B2010 140	Glass Block	COST PER S.F.		
		MAT.	INST.	TOTAL
2300	Glass block 4″ thick, 6″x6″ plain, under 1,000 S.F.	30.50	21.50	52
2400	1,000 to 5,000 S.F.	30	18.80	48.80
2500	Over 5,000 S.F.	29	17.65	46.65
2600	Solar reflective, under 1,000 S.F.	43	29.50	72.50
2700	1,000 to 5,000 S.F.	42	25.50	67.50
2800	Over 5,000 S.F.	41	24	65
3500	8″x8″ plain, under 1,000 S.F.	17.75	16.25	34
3600	1,000 to 5,000 S.F.	17.40	14	31.40
3700	Over 5,000 S.F.	16.90	12.65	29.55
3800	Solar reflective, under 1,000 S.F.	25	22	47
3900	1,000 to 5,000 S.F.	24.50	18.70	43.20
4000	Over 5,000 S.F.	23.50	16.80	40.30
5000	12″x12″ plain, under 1,000 S.F.	26.50	15.05	41.55
5100	1,000 to 5,000 S.F.	26	12.65	38.65
5200	Over 5,000 S.F.	25	11.55	36.55
5300	Solar reflective, under 1,000 S.F.	37.50	20	57.50
5400	1,000 to 5,000 S.F.	36.50	16.80	53.30
5600	Over 5,000 S.F.	35	15.25	50.25
5800	3″ thinline, 6″x6″ plain, under 1,000 S.F.	24.50	21.50	46
5900	Over 5,000 S.F.	23.50	17.65	41.15
6000	Solar reflective, under 1,000 S.F.	34.50	29.50	64
6100	Over 5,000 S.F.	33	24	57
6200	8″x8″ plain, under 1,000 S.F.	14	16.25	30.25
6300	Over 5,000 S.F.	13.40	12.65	26.05
6400	Solar reflective, under 1,000 S.F.	19.50	22	41.50
6500	Over 5,000 S.F.	18.65	16.80	35.45

B2010 Exterior Walls

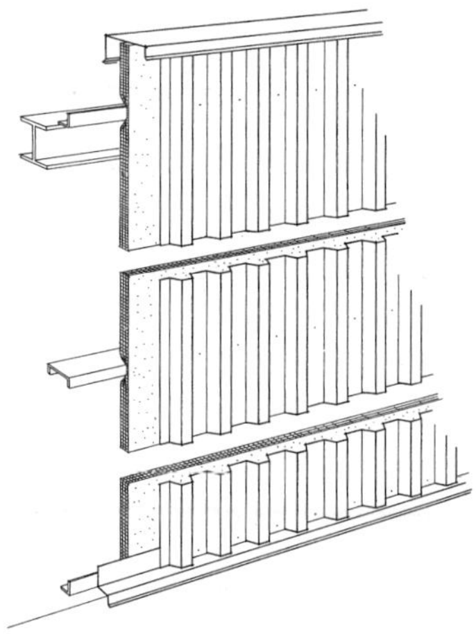

The table below lists costs for metal siding of various descriptions, not including the steel frame, or the structural steel, of a building. Costs are per S.F. including all accessories and insulation.

For steel frame support see System B2010 154.

B2010 146	Metal Siding Panel	COST PER S.F.		
		MAT.	INST.	TOTAL
1400	Metal siding aluminum panel, corrugated, .024" thick, natural	3.26	3.54	6.80
1450	Painted	3.45	3.54	6.99
1500	.032" thick, natural	3.48	3.54	7.02
1550	Painted	4.03	3.54	7.57
1600	Ribbed 4" pitch, .032" thick, natural	3.55	3.54	7.09
1650	Painted	4.09	3.54	7.63
1700	.040" thick, natural	4.07	3.54	7.61
1750	Painted	4.64	3.54	8.18
1800	.050" thick, natural	4.59	3.54	8.13
1850	Painted	5.20	3.54	8.74
1900	Ribbed 8" pitch, .032" thick, natural	3.39	3.40	6.79
1950	Painted	3.92	3.41	7.33
2000	.040" thick, natural	3.91	3.42	7.33
2050	Painted	4.46	3.45	7.91
2100	.050" thick, natural	4.42	3.43	7.85
2150	Painted	5.05	3.46	8.51
3000	Steel, corrugated or ribbed, 29 Ga., .0135" thick, galvanized	2.14	3.22	5.36
3050	Colored	3.07	3.25	6.32
3100	26 Ga., .0179" thick, galvanized	2.25	3.24	5.49
3150	Colored	3.25	3.27	6.52
3200	24 Ga., .0239" thick, galvanized	2.80	3.25	6.05
3250	Colored	3.54	3.28	6.82
3300	22 Ga., .0299" thick, galvanized	3.02	3.27	6.29
3350	Colored	4.24	3.30	7.54
3400	20 Ga., .0359" thick, galvanized	3.02	3.27	6.29
3450	Colored	4.50	3.49	7.99
4100	Sandwich panels, factory fab., 1" polystyrene, steel core, 26 Ga., galv.	5.90	5.05	10.95
4200	Colored, 1 side	7.25	5.05	12.30
4300	2 sides	9.10	5.05	14.15
4400	2" polystyrene core, 26 Ga., galvanized	10.05	5.05	15.10
4500	Colored, 1 side	8.35	5.05	13.40
4600	2 sides	10.20	5.05	15.25
4700	22 Ga., baked enamel exterior	13.35	5.30	18.65
4800	Polyvinyl chloride exterior	14.10	5.30	19.40
5100	Textured aluminum, 4' x 8' x 5/16" plywood backing, single face	4.50	3.01	7.51
5200	Double face	5.85	3.01	8.86

B20 Exterior Enclosure

B2010 Exterior Walls

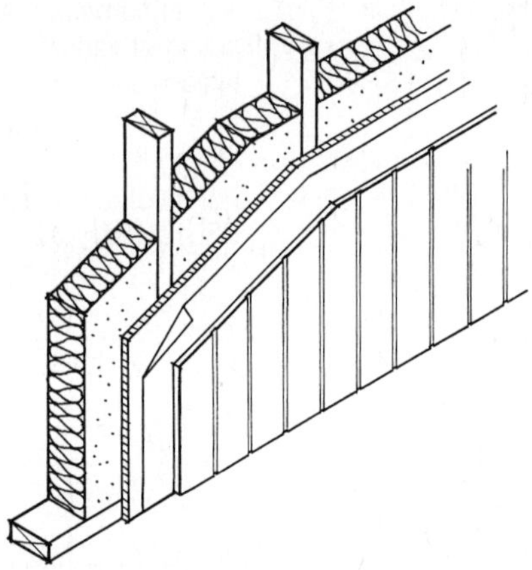

The table below lists costs per S.F. for exterior walls with wood siding. A variety of systems are presented using both wood and metal studs at 16″ and 24″ O.C.

B2010 148	Panel, Shingle & Lap Siding	COST PER S.F.		
		MAT.	INST.	TOTAL
1400	Wood siding w/2″ x 4″ studs, 16″ O.C., insul. wall, 5/8″ text 1-11 fir ply.	3.68	5.20	8.88
1450	5/8″ text 1-11 cedar plywood	5.05	5	10.05
1500	1″ x 4″ vert T.&G. redwood	5.65	6	11.65
1600	1″ x 8″ vert T.&G. redwood	7.75	5.20	12.95
1650	1″ x 5″ rabbetted cedar bev. siding	6.75	5.40	12.15
1700	1″ x 6″ cedar drop siding	6.85	5.45	12.30
1750	1″ x 12″ rough sawn cedar	3.49	5.25	8.74
1800	1″ x 12″ sawn cedar, 1″ x 4″ battens	7	4.89	11.89
1850	1″ x 10″ redwood shiplap siding	7	5.05	12.05
1900	18″ no. 1 red cedar shingles, 5-1/2″ exposed	4.88	6.85	11.73
1950	6″ exposed	4.68	6.60	11.28
2000	6-1/2″ exposed	4.48	6.40	10.88
2100	7″ exposed	4.28	6.15	10.43
2150	7-1/2″ exposed	4.08	5.95	10.03
3000	8″ wide aluminum siding	4.54	4.66	9.20
3150	8″ plain vinyl siding	3.31	4.74	8.05
3250	8″ insulated vinyl siding	3.67	5.25	8.92
3300				
3400	2″ x 6″ studs, 16″ O.C., insul. wall, w/ 5/8″ text 1-11 fir plywood	4.11	5.30	9.41
3500	5/8″ text 1-11 cedar plywood	5.45	5.30	10.75
3600	1″ x 4″ vert T.&G. redwood	6.10	6.10	12.20
3700	1″ x 8″ vert T.&G. redwood	8.20	5.35	13.55
3800	1″ x 5″ rabbetted cedar bev siding	7.20	5.55	12.75
3900	1″ x 6″ cedar drop siding	7.25	5.55	12.80
4000	1″ x 12″ rough sawn cedar	3.92	5.35	9.27
4200	1″ x 12″ sawn cedar, 1″ x 4″ battens	7.45	5.05	12.50
4500	1″ x 10″ redwood shiplap siding	7.45	5.15	12.60
4550	18″ no. 1 red cedar shingles, 5-1/2″ exposed	5.30	6.95	12.25
4600	6″ exposed	5.10	6.75	11.85
4650	6-1/2″ exposed	4.91	6.50	11.41
4700	7″ exposed	4.71	6.30	11.01
4750	7-1/2″ exposed	4.51	6.05	10.56
4800	8″ wide aluminum siding	4.97	4.80	9.77
4850	8″ plain vinyl siding	4.82	8.25	13.07
4900	8″ insulated vinyl siding	5.20	8.75	13.95
4910				
4950	8″ fiber cement siding	4.70	8.90	13.60
5000	2″ x 6″ studs, 24″ O.C., insul. wall, 5/8″ text 1-11, fir plywood	3.94	5	8.94
5050	5/8″ text 1-11 cedar plywood	5.30	5	10.30
5100	1″ x 4″ vert T.&G. redwood	7	9.20	16.20

B2010 Exterior Walls

B2010 148	Panel, Shingle & Lap Siding	COST PER S.F.		
		MAT.	INST.	TOTAL
5150	1" x 8" vert T.&G. redwood	8.05	5.05	13.10
5200	1" x 5" rabbetted cedar bev siding	7	5.25	12.25
5250	1" x 6" cedar drop siding	7.10	5.25	12.35
5300	1" x 12" rough sawn cedar	3.75	5.05	8.80
5400	1" x 12" sawn cedar, 1" x 4" battens	7.25	4.72	11.97
5450	1" x 10" redwood shiplap siding	7.30	4.86	12.16
5500	18" no. 1 red cedar shingles, 5-1/2" exposed	5.15	6.65	11.80
5550	6" exposed	4.94	6.45	11.39
5650	7" exposed	4.54	6	10.54
5700	7-1/2" exposed	4.34	5.75	10.09
5750	8" wide aluminum siding	4.80	4.49	9.29
5800	8" plain vinyl siding	3.57	4.57	8.14
5850	8" insulated vinyl siding	3.93	5.10	9.03
5875	8" fiber cement siding	4.53	8.60	13.13
5900	3-5/8" metal studs, 16 Ga. 16" O.C. insul. wall, 5/8" text 1-11 fir plywood	4.54	5.45	9.99
5950	5/8" text 1-11 cedar plywood	5.90	5.45	11.35
6000	1" x 4" vert T.&G. redwood	6.50	6.25	12.75
6050	1" x 8" vert T.&G. redwood	8.65	5.50	14.15
6100	1" x 5" rabbetted cedar bev siding	7.60	5.65	13.25
6150	1" x 6" cedar drop siding	7.70	5.70	13.40
6200	1" x 12" rough sawn cedar	4.35	5.50	9.85
6250	1" x 12" sawn cedar, 1" x 4" battens	7.85	5.15	13
6300	1" x 10" redwood shiplap siding	7.90	5.30	13.20
6350	18" no. 1 red cedar shingles, 5-1/2" exposed	5.75	7.10	12.85
6500	6" exposed	5.55	6.90	12.45
6550	6-1/2" exposed	5.35	6.65	12
6600	7" exposed	5.15	6.45	11.60
6650	7-1/2" exposed	5.15	6.25	11.40
6700	8" wide aluminum siding	5.40	4.93	10.33
6750	8" plain vinyl siding	4.17	4.83	9
6800	8" insulated vinyl siding	4.53	5.35	9.88
7000	3-5/8" metal studs, 16 Ga. 24" O.C. insul wall, 5/8" text 1-11 fir plywood	4.17	4.83	9
7050	5/8" text 1-11 cedar plywood	5.55	4.83	10.38
7100	1" x 4" vert T.&G. redwood	6.15	5.80	11.95
7150	1" x 8" vert T.&G. redwood	8.25	5.05	13.30
7200	1" x 5" rabbetted cedar bev siding	7.25	5.25	12.50
7250	1" x 6" cedar drop siding	7.35	5.25	12.60
7300	1" x 12" rough sawn cedar	3.98	5.05	9.03
7350	1" x 12" sawn cedar 1" x 4" battens	7.50	4.72	12.22
7400	1" x 10" redwood shiplap siding	7.50	4.86	12.36
7450	18" no. 1 red cedar shingles, 5-1/2" exposed	5.60	6.90	12.50
7500	6" exposed	5.35	6.65	12
7550	6-1/2" exposed	5.15	6.45	11.60
7600	7" exposed	4.97	6.20	11.17
7650	7-1/2" exposed	4.57	5.75	10.32
7700	8" wide aluminum siding	5.05	4.49	9.54
7750	8" plain vinyl siding	3.80	4.57	8.37
7800	8" insul. vinyl siding	4.16	5.10	9.26

B2010 Exterior Walls

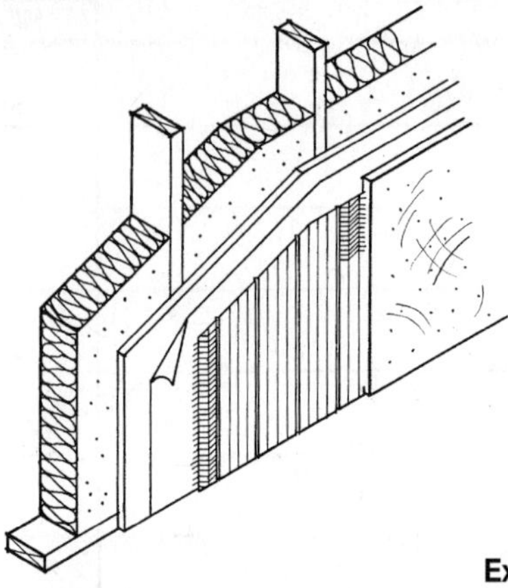

Exterior Stucco Wall

The table below lists costs for some typical stucco walls including all the components as demonstrated in the component block below. Prices are presented for backup walls using wood studs, metal studs and CMU.

B2010 151	Stucco Wall	COST PER S.F.		
		MAT.	INST.	TOTAL
2100	Cement stucco, 7/8" th., plywood sheathing, stud wall, 2" x 4", 16" O.C.	2.93	8.45	11.38
2200	24" O.C.	2.82	8.20	11.02
2300	2" x 6", 16" O.C.	3.36	8.60	11.96
2400	24" O.C.	3.19	8.30	11.49
2500	No sheathing, metal lath on stud wall, 2" x 4", 16" O.C.	2.03	7.35	9.38
2600	24" O.C.	1.92	7.10	9.02
2700	2" x 6", 16" O.C.	2.46	7.50	9.96
2800	24" O.C.	2.29	7.20	9.49
2900	1/2" gypsum sheathing, 3-5/8" metal studs, 16" O.C.	3.16	7.70	10.86
2950	24" O.C.	2.79	7.25	10.04
3000	Cement stucco, 5/8" th., 2 coats on std. CMU block, 8"x 16", 8" thick	3.43	10.10	13.53
3100	10" thick	5	10.35	15.35
3200	12" thick	5.15	11.90	17.05
3300	Std. light Wt. block 8" x 16", 8" thick	4.19	9.95	14.14
3400	10" thick	4.95	10.20	15.15
3500	12" thick	5.20	11.70	16.90
3600	3 coat stucco, self furring metal lath 3.4 Lb/SY, on 8" x 16", 8" thick	4.30	12.05	16.35
3700	10" thick	5.05	10.95	16
3800	12" thick	8.25	14.75	23
3900	Lt. Wt. block, 8" thick	4.24	10.55	14.79
4000	10" thick	5	10.80	15.80
4100	12" thick	5.25	12.30	17.55

B2010 152	E.I.F.S.	COST PER S.F.		
		MAT.	INST.	TOTAL
5100	E.I.F.S., plywood sheathing, stud wall, 2" x 4", 16" O.C., 1" EPS	3.97	8.50	12.47
5110	2" EPS	4.26	8.50	12.76
5120	3" EPS	4.54	8.50	13.04
5130	4" EPS	4.83	8.50	13.33
5140	2" x 6", 16" O.C., 1" EPS	4.40	8.65	13.05
5150	2" EPS	4.87	8.80	13.67
5160	3" EPS	4.97	8.65	13.62
5170	4" EPS	5.25	8.65	13.90
5180	Cement board sheathing, 3-5/8" metal studs, 16" O.C., 1" EPS	4.85	10.50	15.35
5190	2" EPS	5.15	10.50	15.65
5200	3" EPS	5.45	10.50	15.95
5210	4" EPS	6.60	10.50	17.10
5220	6" metal studs, 16" O.C., 1" EPS	5.45	10.55	16
5230	2" EPS	5.90	10.70	16.60

B2010 Exterior Walls

B2010 152	E.I.F.S.	COST PER S.F.		
		MAT.	INST.	TOTAL
5240	3" EPS	6	10.55	16.55
5250	4" EPS	7.15	10.55	17.70
5260	CMU block, 8" x 8" x 16", 1" EPS	5.05	11.95	17
5270	2" EPS	6.80	15.70	22.50
5280	3" EPS	5.60	11.95	17.55
5290	4" EPS	6.75	13.30	20.05
5300	8" x 10" x 16", 1" EPS	6.65	12.20	18.85
5310	2" EPS	6.90	12.20	19.10
5320	3" EPS	7.20	12.20	19.40
5330	4" EPS	7.50	12.20	19.70
5340	8" x 12" x 16", 1" EPS	6.80	13.75	20.55
5350	2" EPS	7.10	13.75	20.85
5360	3" EPS	7.35	13.75	21.10
5370	4" EPS	7.65	13.75	21.40

B2010 Exterior Walls

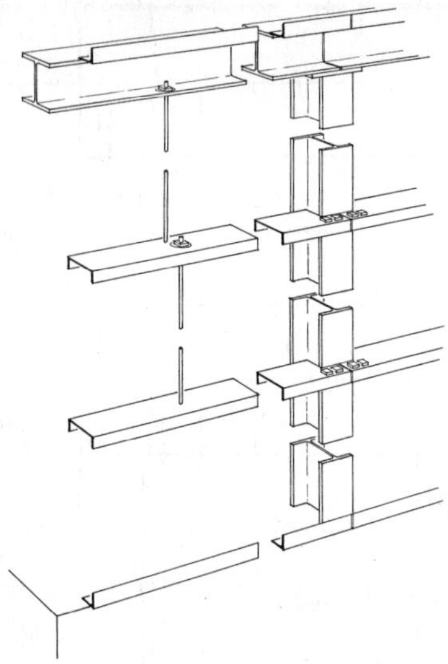

Description: The table below lists costs, $/S.F., for channel girts with sag rods and connector angles top and bottom for various column spacings, building heights and wind loads. Additive costs are shown for wind columns.

How to Use this Table: Add the cost of girts, sag rods, and angles to the framing costs of steel buildings clad in metal or composition siding. If the column spacing is in excess of the column spacing shown, use intermediate wind columns. Additive costs are shown under "Wind Columns".

B2010 154 Metal Siding Support

	BLDG. HEIGHT (FT.)	WIND LOAD (P.S.F.)	COL. SPACING (FT.)		INTERMEDIATE COLUMNS	COST PER S.F. MAT.	INST.	TOTAL
3000	18	20	20			2.42	3.32	5.74
3100					wind cols.	1.25	.23	1.48
3200		20	25			2.65	3.36	6.01
3300					wind cols.	1	.19	1.19
3400		20	30			2.92	3.41	6.33
3500					wind cols.	.83	.15	.98
3600		20	35			3.21	3.47	6.68
3700					wind cols.	.71	.13	.84
3800		30	20			2.68	3.37	6.05
3900					wind cols.	1.25	.23	1.48
4000		30	25			2.92	3.41	6.33
4100					wind cols.	1	.19	1.19
4200		30	30			3.23	3.47	6.70
4300					wind cols.	1.13	.20	1.33
4600	30	20	20			2.35	2.33	4.68
4700					wind cols.	1.77	.32	2.09
4800		20	25			2.64	2.39	5.03
4900					wind cols.	1.41	.26	1.67
5000		20	30			2.97	2.45	5.42
5100					wind cols.	1.39	.26	1.65
5200		20	35			3.32	2.51	5.83
5300					wind cols.	1.38	.26	1.64
5400		30	20			2.67	2.39	5.06
5500					wind cols.	2.09	.39	2.48
5600		30	25			2.96	2.44	5.40
5700					wind cols.	1.99	.37	2.36
5800		30	30			3.34	2.52	5.86
5900					wind cols.	1.87	.35	2.22

B2010 Exterior Walls

B2010 160	Pole Barn Exterior Wall	COST PER S.F.		
		MAT.	INST.	TOTAL
2000	Pole barn exterior wall, pressure treated pole in concrete, 8′ O.C.			
3000	Steel siding, 8′ eave	4.42	8.45	12.87
3050	10′ eave	4.23	7.75	11.98
3100	12′ eave	4.01	7.05	11.06
3150	14′ eave	3.90	6.70	10.60
3200	16′ eave	3.82	6.35	10.17
4000	Aluminum siding, 8′ eave	4.21	8.45	12.66
4050	10′ eave	3.96	7.60	11.56
4100	12′ eave	3.80	7.05	10.85
4150	14′ eave	3.69	6.70	10.39
4200	16′ eave	3.61	6.35	9.96
5000	Wood siding, 8′ eave	4.68	7.50	12.18
5050	10′ eave	4.43	6.70	11.13
5100	12′ eave	4.27	6.15	10.42
5150	14′ eave	4.16	5.75	9.91
5200	16′ eave	4.08	5.45	9.53
6000	Plywood siding, 8′ eave	3.88	7.45	11.33
6050	10′ eave	3.63	6.65	10.28
6100	12′ eave	3.47	6.10	9.57
6150	14′ eave	3.36	5.75	9.11
6200	16′ eave	3.28	5.40	8.68

B2020 Exterior Windows

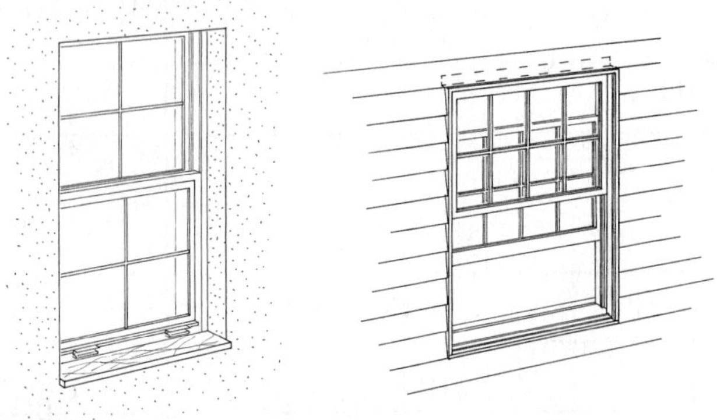

The table below lists window systems by material, type and size. Prices between sizes listed can be interpolated with reasonable accuracy. Prices include frame, hardware, and casing as illustrated in the component block below.

B2020 102 Wood Windows

	MATERIAL	TYPE	GLAZING	SIZE	DETAIL	COST PER UNIT		
						MAT.	INST.	TOTAL
3000	Wood	double hung	std. glass	2′-8″ x 4′-6″		247	220	467
3050				3′-0″ x 5′-6″		325	253	578
3100			insul. glass	2′-8″ x 4′-6″		263	220	483
3150				3′-0″ x 5′-6″		350	253	603
3200		sliding	std. glass	3′-4″ x 2′-7″		330	183	513
3250				4′-4″ x 3′-3″		345	201	546
3300				5′-4″ x 6′-0″		435	241	676
3350			insul. glass	3′-4″ x 2′-7″		405	214	619
3400				4′-4″ x 3′-3″		415	233	648
3450				5′-4″ x 6′-0″		530	275	805
3500		awning	std. glass	2′-10″ x 1′-9″		230	110	340
3600				4′-4″ x 2′-8″		375	106	481
3700			insul. glass	2′-10″ x 1′-9″		281	127	408
3800				4′-4″ x 2′-8″		460	116	576
3900		casement	std. glass	1′-10″ x 3′-2″	1 lite	340	145	485
3950				4′-2″ x 4′-2″	2 lite	600	193	793
4000				5′-11″ x 5′-2″	3 lite	900	248	1,148
4050				7′-11″ x 6′-3″	4 lite	1,275	297	1,572
4100				9′-11″ x 6′-3″	5 lite	1,675	345	2,020
4150			insul. glass	1′-10″ x 3′-2″	1 lite	340	145	485
4200				4′-2″ x 4′-2″	2 lite	615	193	808
4250				5′-11″ x 5′-2″	3 lite	950	248	1,198
4300				7′-11″ x 6′-3″	4 lite	1,350	297	1,647
4350				9′-11″ x 6′-3″	5 lite	1,675	345	2,020
4400		picture	std. glass	4′-6″ x 4′-6″		475	248	723
4450				5′-8″ x 4′-6″		535	276	811
4500		picture	insul. glass	4′-6″ x 4′-6″		580	289	869
4550				5′-8″ x 4′-6″		650	320	970
4600		fixed bay	std. glass	8′ x 5′		1,650	460	2,110
4650				9′-9″ x 5′-4″		1,175	640	1,815
4700			insul. glass	8′ x 5′		2,175	460	2,635
4750				9′-9″ x 5′-4″		1,275	640	1,915
4800		casement bay	std. glass	8′ x 5′		1,725	525	2,250
4850			insul. glass	8′ x 5′		1,875	525	2,400
4900		vert. bay	std. glass	8′ x 5′		1,875	525	2,400
4950			insul. glass	8′ x 5′		1,950	525	2,475

B2020 Exterior Windows

B2020 104 Steel Windows

	MATERIAL	TYPE	GLAZING	SIZE	DETAIL	COST PER UNIT MAT.	INST.	TOTAL
5000	Steel	double hung	1/4" tempered	2'-8" x 4'-6"		910	168	1,078
5050				3'-4" x 5'-6"		1,400	256	1,656
5100			insul. glass	2'-8" x 4'-6"		930	194	1,124
5150				3'-4" x 5'-6"		1,425	296	1,721
5200		horz. pivoted	std. glass	2' x 2'		276	56	332
5250				3' x 3'		620	126	746
5300				4' x 4'		1,100	224	1,324
5350				6' x 4'		1,650	335	1,985
5400			insul. glass	2' x 2'		281	64.50	345.50
5450				3' x 3'		635	145	780
5500				4' x 4'		1,125	258	1,383
5550				6' x 4'		1,675	390	2,065
5600		picture window	std. glass	3' x 3'		390	126	516
5650				6' x 4'		1,050	335	1,385
5700			insul. glass	3' x 3'		405	145	550
5750				6' x 4'		1,075	390	1,465
5800		industrial security	std. glass	2'-9" x 4'-1"		830	157	987
5850				4'-1" x 5'-5"		1,625	310	1,935
5900			insul. glass	2'-9" x 4'-1"		845	181	1,026
5950				4'-1" x 5'-5"		1,675	355	2,030
6000		comm. projected	std. glass	3'-9" x 5'-5"		1,350	285	1,635
6050				6'-9" x 4'-1"		1,825	385	2,210
6100			insul. glass	3'-9" x 5'-5"		1,375	330	1,705
6150				6'-9" x 4'-1"		1,875	445	2,320
6200		casement	std. glass	4'-2" x 4'-2"	2 lite	1,225	243	1,468
6250			insul. glass	4'-2" x 4'-2"		1,250	280	1,530
6300			std. glass	5'-11" x 5'-2"	3 lite	2,550	430	2,980
6350			insul. glass	5'-11" x 5'-2"		2,600	495	3,095

B2020 106 Aluminum Windows

	MATERIAL	TYPE	GLAZING	SIZE	DETAIL	COST PER UNIT MAT.	INST.	TOTAL
6400	Aluminum	double hung	insul. glass	3'-0" x 4'-0"		530	117	647
6450				4'-5" x 5'-3"		440	146	586
6500			insul. glass	3'-1" x 3'-2"		470	140	610
6550				4'-5" x 5'-3"		530	175	705
6600		sliding	std. glass	3' x 2'		239	117	356
6650				5' x 3'		365	130	495
6700				8' x 4'		385	195	580
6750				9' x 5'		585	293	878
6800			insul. glass	3' x 2'		255	117	372
6850				5' x 3'		425	130	555
6900				8' x 4'		620	195	815
6950				9' x 5'		935	293	1,228
7000		single hung	std. glass	2' x 3'		229	117	346
7050				2'-8" x 6'-8"		400	146	546
7100				3'-4" x 5'-0"		330	130	460
7150			insul. glass	2' x 3'		277	117	394
7200				2'-8" x 6'-8"		520	146	666
7250				3'-4" x 5'		365	130	495
7300		double hung	std. glass	2' x 3'		320	84	404
7350				2'-8" x 6'-8"		950	249	1,199

B2020 Exterior Windows

B2020 106 Aluminum Windows

	MATERIAL	TYPE	GLAZING	SIZE	DETAIL	COST PER UNIT		
						MAT.	INST.	TOTAL
7400	Aluminum	Double hung		3'-4" x 5'		890	233	1,123
7450			insul. glass	2' x 3'		330	97	427
7500				2'-8" x 6'-8"		975	287	1,262
7550				3'-4" x 5'-0"		910	269	1,179
7600		casement	std. glass	3'-1" x 3'-2"		283	137	420
7650				4'-5" x 5'-3"		685	330	1,015
7700			insul. glass	3'-1" x 3'-2"		296	157	453
7750				4'-5" x 5'-3"		720	380	1,100
7800		hinged swing	std. glass	3' x 4'		585	168	753
7850				4' x 5'		980	280	1,260
7900			insul. glass	3' x 4'		605	194	799
7950				4' x 5'		1,000	325	1,325
8200		picture unit	std. glass	2'-0" x 3'-0"		182	84	266
8250				2'-8" x 6'-8"		540	249	789
8300				3'-4" x 5'-0"		505	233	738
8350			insul. glass	2'-0" x 3'-0"		191	97	288
8400				2'-8" x 6'-8"		565	287	852
8450				3'-4" x 5'-0"		530	269	799
8500		awning type	std. glass	3'-0" x 3'-0"	2 lite	470	83.50	553.50
8550				3'-0" x 4'-0"	3 lite	540	117	657
8600				3'-0" x 5'-4"	4 lite	650	117	767
8650				4'-0" x 5'-4"	4 lite	715	130	845
8700			insul. glass	3'-0" x 3'-0"	2 lite	495	83.50	578.50
8750				3'-0" x 4'-0"	3 lite	620	117	737
8800				3'-0" x 5'-4"	4 lite	765	117	882
8850				4'-0" x 5'-4"	4 lite	855	130	985

B2020 108 Vinyl Clad Windows

	MATERIAL	TYPE	GLAZING	SIZE	DETAIL	COST PER UNIT		
						MAT.	INST.	TOTAL
1000	Vinyl clad	casement	insul. glass	1'-4" x 4'-0"		425	117	542
1010				2'-0" x 3'-0"		355	117	472
1020				2'-0" x 4'-0"		410	123	533
1030				2'-0" x 5'-0"		460	130	590
1040				2'-0" x 6'-0"		480	130	610
1050				2'-4" x 4'-0"		460	130	590
1060				2'-6" x 5'-0"		565	126	691
1070				3'-0" x 5'-0"		790	130	920
1080				4'-0" x 3'-0"		865	130	995
1090				4'-0" x 4'-0"		745	130	875
1100				4'-8" x 4'-0"		820	130	950
1110				4'-8" x 5'-0"		925	151	1,076
1120				4'-8" x 6'-0"		1,025	151	1,176
1130				6'-0" x 4'-0"		945	151	1,096
1140				6'-0" x 5'-0"		1,050	151	1,201
3000		double-hung	insul. glass	2'-0" x 4'-0"		440	112	552
3050				2'-0" x 5'-0"		500	112	612
3100				2'-4" x 4'-0"		455	117	572
3150				2'-4" x 4'-8"		455	117	572
3200				2'-4" x 6'-0"		515	117	632
3250				2'-6" x 4'-0"		455	117	572
3300				2'-8" x 4'-0"		680	151	831

B2020 Exterior Windows

B2020 108 — Vinyl Clad Windows

	MATERIAL	TYPE	GLAZING	SIZE	DETAIL	COST PER UNIT		
						MAT.	INST.	TOTAL
3350	Vinyl clad	double hung	insul. glass	2'-8" x 5'-0"		535	130	665
3375				2'-8" x 6'-0"		545	130	675
3400				3'-0 x 3'-6"		430	117	547
3450				3'-0 x 4'-0"		495	123	618
3500				3'-0 x 4'-8"		530	123	653
3550				3'-0 x 5'-0"		545	130	675
3600				3'-0 x 6'-0"		490	130	620
3700				4'-0 x 5'-0"		675	139	814
3800				4'-0 x 6'-0"		825	139	964

B2020 210 — Tubular Aluminum Framing

		COST/S.F. OPNG.		
		MAT.	INST.	TOTAL
1100	Alum flush tube frame, for 1/4" glass, 1-3/4"x4", 5'x6'opng, no inter horiz	14.95	10.20	25.15
1150	One intermediate horizontal	20.50	12.15	32.65
1200	Two intermediate horizontals	26	14.10	40.10
1250	5' x 20' opening, three intermediate horizontals	13.75	8.70	22.45
1400	1-3/4" x 4-1/2", 5' x 6' opening, no intermediate horizontals	17.20	10.20	27.40
1450	One intermediate horizontal	23	12.15	35.15
1500	Two intermediate horizontals	28.50	14.10	42.60
1550	5' x 20' opening, three intermediate horizontals	15.45	8.70	24.15
1700	For insulating glass, 2"x4-1/2", 5'x6' opening, no intermediate horizontals	17.50	10.80	28.30
1750	One intermediate horizontal	22.50	12.90	35.40
1800	Two intermediate horizontals	27	14.95	41.95
1850	5' x 20' opening, three intermediate horizontals	14.95	9.25	24.20
2000	Thermal break frame, 2-1/4"x4-1/2", 5'x6'opng, no intermediate horizontals	18.30	10.95	29.25
2050	One intermediate horizontal	24.50	13.45	37.95
2100	Two intermediate horizontals	30.50	15.90	46.40
2150	5' x 20' opening, three intermediate horizontals	16.55	9.70	26.25

B2020 Exterior Windows

The table below lists costs of curtain wall and spandrel panels per S.F. Costs do not include structural framing used to hang the panels.

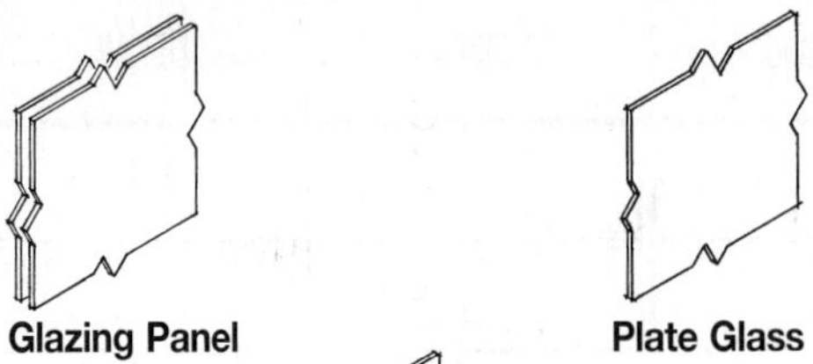

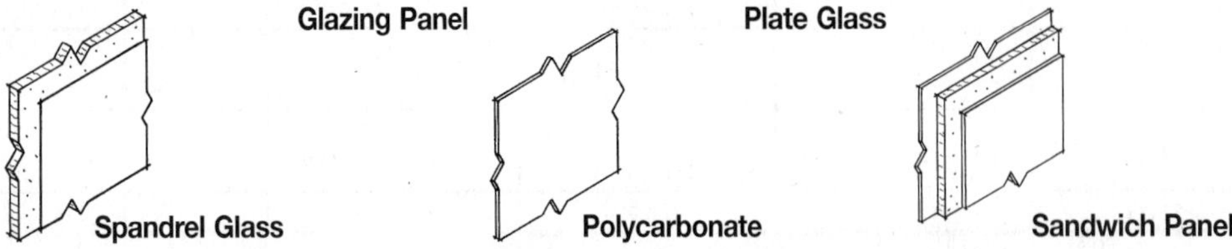

B2020 220	Curtain Wall Panels	COST PER S.F.		
		MAT.	INST.	TOTAL
1000	Glazing panel, insulating, 1/2″ thick, 2 lites 1/8″ float, clear	11.30	10.30	21.60
1100	Tinted	15.65	10.30	25.95
1200	5/8″ thick units, 2 lites 3/16″ float, clear	15.05	10.90	25.95
1300	Tinted	15.15	10.90	26.05
1400	1″ thick units, 2 lites, 1/4″ float, clear	18.85	13.10	31.95
1500	Tinted	26	13.10	39.10
1600	Heat reflective film inside	30.50	11.55	42.05
1700	Light and heat reflective glass, tinted	34.50	11.55	46.05
2000	Plate glass, 1/4″ thick, clear	7.05	8.15	15.20
2050	Tempered	9.90	8.15	18.05
2100	Tinted	10	8.15	18.15
2200	3/8″ thick, clear	11.45	13.10	24.55
2250	Tempered	18.45	13.10	31.55
2300	Tinted	17.25	13.10	30.35
2400	1/2″ thick, clear	19.40	17.85	37.25
2450	Tempered	27.50	17.85	45.35
2500	Tinted	30	17.85	47.85
2600	3/4″ thick, clear	39	28	67
2650	Tempered	45.50	28	73.50
3000	Spandrel glass, panels, 1/4″ plate glass insul w/fiberglass, 1″ thick	18.55	8.15	26.70
3100	2″ thick	22.50	8.15	30.65
3200	Galvanized steel backing, add	6.45		6.45
3300	3/8″ plate glass, 1″ thick	31.50	8.15	39.65
3400	2″ thick	35.50	8.15	43.65
4000	Polycarbonate, masked, clear or colored, 1/8″ thick	15.75	5.75	21.50
4100	3/16″ thick	19.10	5.95	25.05
4200	1/4″ thick	19.10	6.35	25.45
4300	3/8″ thick	29.50	6.55	36.05
5000	Facing panel, textured alum., 4′ x 8′ x 5/16″ plywood backing, sgl face	4.50	3.01	7.51
5100	Double face	5.85	3.01	8.86
5200	4′ x 10′ x 5/16″ plywood backing, single face	4.73	3.01	7.74
5300	Double face	6.40	3.01	9.41
5400	4′ x 12′ x 5/16″ plywood backing, single face	4.73	3.01	7.74
5500	Sandwich panel, 22 Ga. galv., both sides 2″ insulation, enamel exterior	13.35	5.30	18.65
5600	Polyvinylidene fluoride exterior finish	14.10	5.30	19.40
5700	26 Ga., galv. both sides, 1″ insulation, colored 1 side	7.25	5.05	12.30
5800	Colored 2 sides	9.10	5.05	14.15

B2030 Exterior Doors

B2030 110 Glazed Doors, Steel or Aluminum

	MATERIAL	TYPE	DOORS	SPECIFICATION	OPENING	COST PER OPNG.		
						MAT.	INST.	TOTAL
5600	St. Stl. & glass	revolving	stock unit	manual oper.	6'-0" x 7'-0"	52,000	7,750	59,750
5650				auto Cntrls.	6'-10" x 7'-0"	68,500	8,325	76,825
5700	Bronze	revolving	stock unit	manual oper.	6'-10" x 7'-0"	53,000	15,600	68,600
5750				auto Cntrls.	6'-10" x 7'-0"	69,500	16,200	85,700
5800	St. Stl. & glass	balanced	standard	economy	3'-0" x 7'-0"	10,300	1,300	11,600
5850				premium	3'-0" x 7'-0"	17,500	1,725	19,225
6300	Alum. & glass	w/o transom	narrow stile	w/panic Hrdwre.	3'-0" x 7'-0"	1,900	885	2,785
6350				dbl. door, Hrdwre.	6'-0" x 7'-0"	3,100	1,475	4,575
6400			wide stile	hdwre.	3'-0" x 7'-0"	2,100	870	2,970
6450				dbl. door, Hdwre.	6'-0" x 7'-0"	4,075	1,725	5,800
6500			full vision	hdwre.	3'-0" x 7'-0"	2,800	1,400	4,200
6550				dbl. door, Hdwre.	6'-0" x 7'-0"	3,750	1,950	5,700
6600			non-standard	hdwre.	3'-0" x 7'-0"	2,300	870	3,170
6650				dbl. door, Hdwre.	6'-0" x 7'-0"	4,600	1,725	6,325
6700			bronze fin.	hdwre.	3'-0" x 7'-0"	1,675	870	2,545
6750				dbl. door, Hrdwre.	6'-0" x 7'-0"	3,350	1,725	5,075
6800			black fin.	hdwre.	3'-0" x 7'-0"	2,350	870	3,220
6850				dbl. door, Hdwre.	6'-0" x 7'-0"	4,700	1,725	6,425
6900		w/transom	narrow stile	hdwre.	3'-0" x 10'-0"	2,400	1,000	3,400
6950				dbl. door, Hdwre.	6'-0" x 10'-0"	3,750	1,775	5,525
7000			wide stile	hdwre.	3'-0" x 10'-0"	2,725	1,200	3,925
7050				dbl. door, Hdwre.	6'-0" x 10'-0"	4,150	2,100	6,250
7100			full vision	hdwre.	3'-0" x 10'-0"	3,025	1,325	4,350
7150				dbl. door, Hdwre.	6'-0" x 10'-0"	4,525	2,325	6,850
7200			non-standard	hdwre.	3'-0" x 10'-0"	2,400	940	3,340
7250				dbl. door, Hdwre.	6'-0" x 10'-0"	4,775	1,875	6,650
7300			bronze fin.	hdwre.	3'-0" x 10'-0"	1,775	940	2,715
7350				dbl. door, Hdwre.	6'-0" x 10'-0"	3,525	1,875	5,400
7400			black fin.	hdwre.	3'-0" x 10'-0"	2,450	940	3,390
7450				dbl. door, Hdwre.	6'-0" x 10'-0"	4,875	1,875	6,750
7500		revolving	stock design	minimum	6'-10" x 7'-0"	24,900	3,125	28,025
7550				average	6'-0" x 7'-0"	29,300	3,900	33,200
7600				maximum	6'-10" x 7'-0"	34,600	5,200	39,800
7650				min., automatic	6'-10" x 7'-0"	41,500	3,700	45,200
7700				avg., automatic	6'-10" x 7'-0"	45,900	4,475	50,375
7750				max., automatic	6'-10" x 7'-0"	51,000	5,775	56,775
7800		balanced	standard	economy	3'-0" x 7'-0"	7,375	1,300	8,675
7850				premium	3'-0" x 7'-0"	8,825	1,675	10,500
7900		mall front	sliding panels	alum. fin.	16'-0" x 9'-0"	3,900	755	4,655
7950					24'-0" x 9'-0"	5,525	1,400	6,925
8000				bronze fin.	16'-0" x 9'-0"	4,550	880	5,430
8050					24'-0" x 9'-0"	6,425	1,625	8,050
8100			fixed panels	alum. fin.	48'-0" x 9'-0"	10,200	1,100	11,300
8150				bronze fin.	48'-0" x 9'-0"	11,900	1,275	13,175
8200		sliding entrance	5' x 7' door	electric oper.	12'-0" x 7'-6"	9,575	1,400	10,975
8250		sliding patio	temp. glass	economy	6'-0" x 7'-0"	1,825	253	2,078
8300			temp. glass	economy	12'-0" x 7'-0"	4,275	340	4,615
8350				premium	6'-0" x 7'-0"	2,750	380	3,130
8400					12'-0" x 7'-0"	6,425	510	6,935

B2030 Exterior Doors

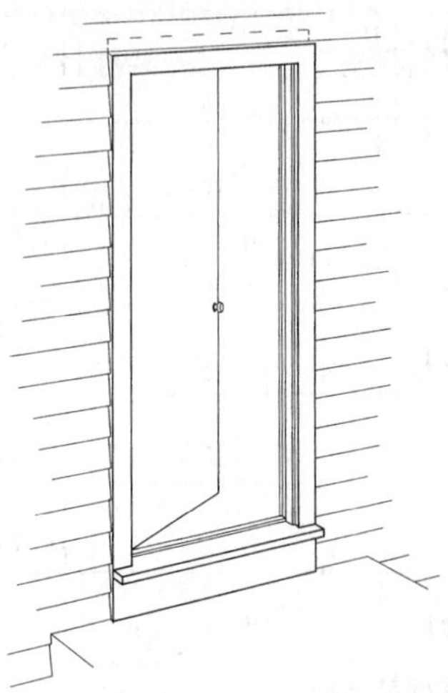

Costs are listed for exterior door systems by material, type and size. Prices between sizes listed can be interpolated with reasonable accuracy. Prices are per opening for a complete door system including frame as illustrated in the component block.

B2030 210 Wood Doors

	MATERIAL	TYPE	DOORS	SPECIFICATION	OPENING	COST PER OPNG. MAT.	INST.	TOTAL
2350	Birch	solid core	single door	hinged	2′-6″ x 6′-8″	1,150	276	1,426
2400					2′-6″ x 7′-0″	1,150	276	1,426
2450					2′-8″ x 7′-0″	1,150	276	1,426
2500					3′-0″ x 7′-0″	1,150	276	1,426
2550			double door		2′-6″ x 6′-8″	2,150	505	2,655
2600					2′-6″ x 7′-0″	2,150	505	2,655
2650					2′-8″ x 7′-0″	2,125	505	2,630
2700					3′-0″ x 7′-0″	2,000	440	2,440
2750	Wood	combination	storm & screen		3′-0″ x 6′-8″	335	67.50	402.50
2800					3′-0″ x 7′-0″	370	74.50	444.50
2850		overhead	panels, H.D.	manual oper.	8′-0″ x 8′-0″	1,250	505	1,755
2900					10′-0″ x 10′-0″	1,775	565	2,340
2950					12′-0″ x 12′-0″	2,450	675	3,125
3000					14′-0″ x 14′-0″	3,700	780	4,480
3050					20′-0″ x 16′-0″	6,200	1,550	7,750
3100				electric oper.	8′-0″ x 8′-0″	2,425	760	3,185
3150					10′-0″ x 10′-0″	2,950	820	3,770
3200					12′-0″ x 12′-0″	3,625	930	4,555
3250					14′-0″ x 14′-0″	4,875	1,025	5,900
3300					20′-0″ x 16′-0″	7,450	2,050	9,500

B2030 220 Steel Doors

	MATERIAL	TYPE	DOORS	SPECIFICATION	OPENING	COST PER OPNG. MAT.	INST.	TOTAL
3350	Steel 18 Ga.	hollow metal	1 door w/frame	no label	2′-6″ x 7′-0″	1,550	285	1,835
3400					2′-8″ x 7′-0″	1,550	285	1,835
3450					3′-0″ x 7′-0″	1,575	305	1,880
3500					3′-6″ x 7′-0″	1,675	300	1,975
3550		hollow metal	1 door w/frame	no label	4′-0″ x 8′-0″	1,975	300	2,275
3600			2 doors w/frame	no label	5′-0″ x 7′-0″	3,000	530	3,530
3650					5′-4″ x 7′-0″	3,000	530	3,530
3700					6′-0″ x 7′-0″	3,000	530	3,530
3750					7′-0″ x 7′-0″	3,250	560	3,810
3800					8′-0″ x 8′-0″	3,825	560	4,385

B2030 Exterior Doors

B2030 220 Steel Doors

	MATERIAL	TYPE	DOORS	SPECIFICATION	OPENING	COST PER OPNG. MAT.	INST.	TOTAL
3850	Steel 18 Ga.	Hollow metal	1 door w/frame	"A" label	2'-6" x 7'-0"	1,950	345	2,295
3900					2'-8" x 7'-0"	1,950	350	2,300
3950					3'-0" x 7'-0"	1,950	350	2,300
4000					3'-6" x 7'-0"	2,100	355	2,455
4050					4'-0" x 8'-0"	2,325	375	2,700
4100			2 doors w/frame	"A" label	5'-0" x 7'-0"	3,750	635	4,385
4150					5'-4" x 7'-0"	3,750	645	4,395
4200					6'-0" x 7'-0"	3,750	645	4,395
4250					7'-0" x 7'-0"	4,075	660	4,735
4300					8'-0" x 8'-0"	4,075	660	4,735
4350	Steel 24 Ga.	overhead	sectional	manual oper.	8'-0" x 8'-0"	990	505	1,495
4400					10'-0" x 10'-0"	1,350	565	1,915
4450					12'-0" x 12'-0"	1,625	675	2,300
4500					20'-0" x 14'-0"	3,775	1,450	5,225
4550				electric oper.	8'-0" x 8'-0"	2,175	760	2,935
4600					10'-0" x 10'-0"	2,525	820	3,345
4650					12'-0" x 12'-0"	2,800	930	3,730
4700					20'-0" x 14'-0"	5,025	1,950	6,975
4750	Steel	overhead	rolling	manual oper.	8'-0" x 8'-0"	1,300	730	2,030
4800					10'-0" x 10'-0"	2,150	835	2,985
4850					12'-0" x 12' 0"	2,175	975	3,150
4900					14'-0" x 14'-0"	3,475	1,475	4,950
4950					20'-0" x 12'-0"	2,425	1,300	3,725
5000					20'-0" x 16'-0"	4,175	1,950	6,125
5050				electric oper.	8'-0" x 8'-0"	2,600	965	3,565
5100					10'-0" x 10'-0"	3,450	1,075	4,525
5150					12'-0" x 12'-0"	3,475	1,200	4,675
5200					14'-0" x 14'-0"	4,775	1,700	6,475
5250					20'-0" x 12'-0"	3,625	1,525	5,150
5300					20'-0" x 16'-0"	5,375	2,175	7,550
5350				fire rated	10'-0" x 10'-0"	2,450	1,075	3,525
5400			rolling grille	manual oper.	10'-0" x 10'-0"	2,850	1,175	4,025
5450					15'-0" x 8'-0"	3,200	1,475	4,675
5500		vertical lift	1 door w/frame	motor operator	16"-0" x 16"-0"	23,700	4,925	28,625
5550					32'-0" x 24'-0"	53,500	3,275	56,775

B2030 230 Aluminum Doors

	MATERIAL	TYPE	DOORS	SPECIFICATION	OPENING	COST PER OPNG. MAT.	INST.	TOTAL
6000	Aluminum	combination	storm & screen	hinged	3'-0" x 6'-8"	340	72.50	412.50
6050					3' 0" x 7'-0"	375	80	455
6100		overhead	rolling grille	manual oper.	12'-0" x 12'-0"	4,525	2,050	6,575
6150				motor oper.	12'-0" x 12'-0"	5,900	2,300	8,200
6200	Alum. & Fbrgls.	overhead	heavy duty	manual oper.	12'-0" x 12'-0"	3,350	675	4,025
6250				electric oper.	12'-0" x 12'-0"	4,525	930	5,455

B3010 Roof Coverings

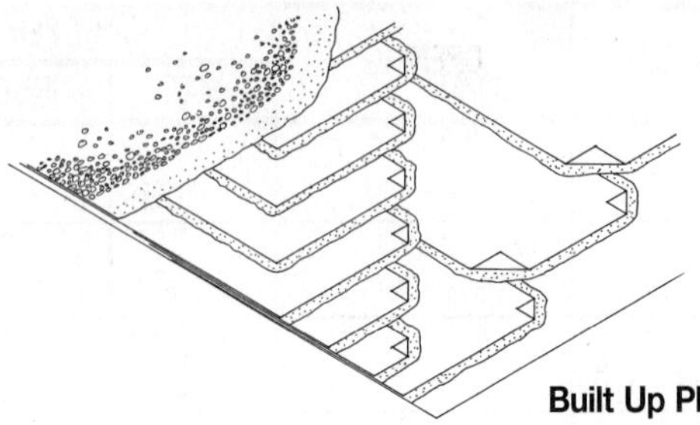

Multiple ply roofing is the most popular covering for minimum pitch roofs.

B3010 105	Built-Up	COST PER S.F.		
		MAT.	INST.	TOTAL
1200	Asphalt flood coat w/gravel; not incl. insul, flash., nailers			
1300				
1400	Asphalt base sheets & 3 plies #15 asphalt felt, mopped	1.09	1.70	2.79
1500	On nailable deck	1.13	1.79	2.92
1600	4 plies #15 asphalt felt, mopped	1.50	1.87	3.37
1700	On nailable deck	1.33	1.98	3.31
1800	Coated glass base sheet, 2 plies glass (type IV), mopped	1.21	1.70	2.91
1900	For 3 plies	1.47	1.87	3.34
2000	On nailable deck	1.39	1.98	3.37
2300	4 plies glass fiber felt (type IV), mopped	1.82	1.87	3.69
2400	On nailable deck	1.65	1.98	3.63
2500	Organic base sheet & 3 plies #15 organic felt, mopped	1.16	1.86	3.02
2600	On nailable deck	1.14	1.98	3.12
2700	4 plies #15 organic felt, mopped	1.43	1.70	3.13
2750				
2800	Asphalt flood coat, smooth surface, not incl. insul, flash., nailers			
2900	Asphalt base sheet & 3 plies #15 asphalt felt, mopped	1.16	1.57	2.73
3000	On nailable deck	1.08	1.63	2.71
3100	Coated glass fiber base sheet & 2 plies glass fiber felt, mopped	1.16	1.50	2.66
3200	On nailable deck	1.10	1.57	2.67
3300	For 3 plies, mopped	1.42	1.63	3.05
3400	On nailable deck	1.34	1.70	3.04
3700	4 plies glass fiber felt (type IV), mopped	1.68	1.63	3.31
3800	On nailable deck	1.60	1.70	3.30
3900	Organic base sheet & 3 plies #15 organic felt, mopped	1.18	1.57	2.75
4000	On nailable decks	1.09	1.63	2.72
4100	4 plies #15 organic felt, mopped	1.38	1.70	3.08
4200	Coal tar pitch with gravel surfacing			
4300	4 plies #15 tarred felt, mopped	2.20	1.79	3.99
4400	3 plies glass fiber felt (type IV), mopped	1.81	1.98	3.79
4500	Coated glass fiber base sheets 2 plies glass fiber felt, mopped	1.86	1.98	3.84
4600	On nailable decks	1.62	2.09	3.71
4800	3 plies glass fiber felt (type IV), mopped	2.52	1.79	4.31
4900	On nailable decks	2.28	1.87	4.15

B3010 Roof Coverings

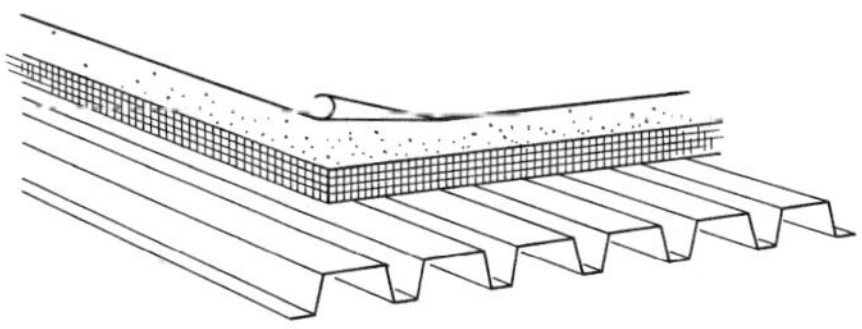

Fully Adhered

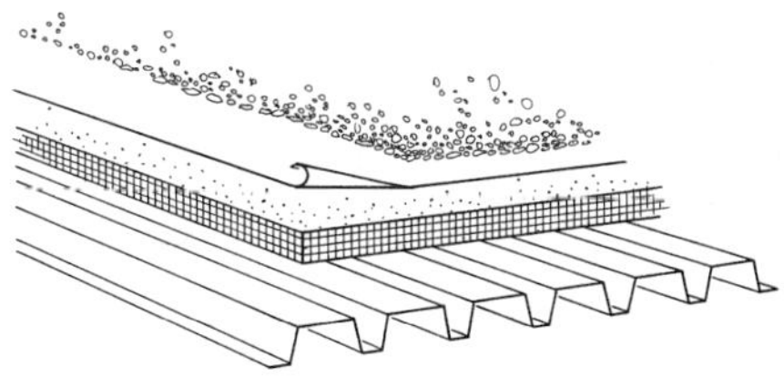

Ballasted

The systems listed below reflect only the cost for the single ply membrane.

B3010 120	Single Ply Membrane	COST PER S.F.		
		MAT.	INST.	TOTAL
1000	CSPE (Chlorosulfonated polyethylene), 45 mils, plate attached	2.76	.68	3.44
2000	EPDM (Ethylene propylene diene monomer), 45 mils, fully adhered	1.22	.91	2.13
4000	Modified bit., SBS modified, granule surface cap sheet, mopped, 150 mils	1.40	1.87	3.27
4500	APP modified, granule surface cap sheet, torched, 180 mils	1.02	1.19	2.21
6000	Reinforced PVC, 48 mils, loose laid and ballasted with stone	1.29	.47	1.76
6200	Fully adhered with adhesive	1.70	.91	2.61

B3010 130	Preformed Metal Roofing	COST PER S.F.		
		MAT.	INST.	TOTAL
0200	Corrugated roofing, aluminum, mill finish, .0175" thick, .272 P.S.F.	1.10	1.59	2.69
0250	.0215" thick, .334 P.S.F.	1.38	1.59	2.97

B3010 135	Formed Metal	COST PER S.F.		
		MAT.	INST.	TOTAL
1000	Batten seam, formed copper roofing, 3" min slope, 16 oz., 1.2 P.S.F.	13.85	5.25	19.10
1100	18 oz., 1.35 P.S.F.	15.35	5.75	21.10
2000	Zinc copper alloy, 3" min slope, .020" thick, .88 P.S.F.	14.10	4.78	18.88
3000	Flat seam, copper, 1/4" min. slope, 16 oz., 1.2 P.S.F.	10.10	4.78	14.88
3100	18 oz., 1.35 P.S.F.	11.35	4.98	16.33
5000	Standing seam, copper, 2-1/2" min. slope, 16 oz., 1.25 P.S.F.	10.85	4.43	15.28
5100	18 oz., 1.40 P.S.F.	12.10	4.78	16.88
6000	Zinc copper alloy, 2-1/2" min. slope, .020" thick, .87 P.S.F.	13.60	4.78	18.38
6100	.032" thick, 1.39 P.S.F.	19.60	5.25	24.85

B3010 Roof Coverings

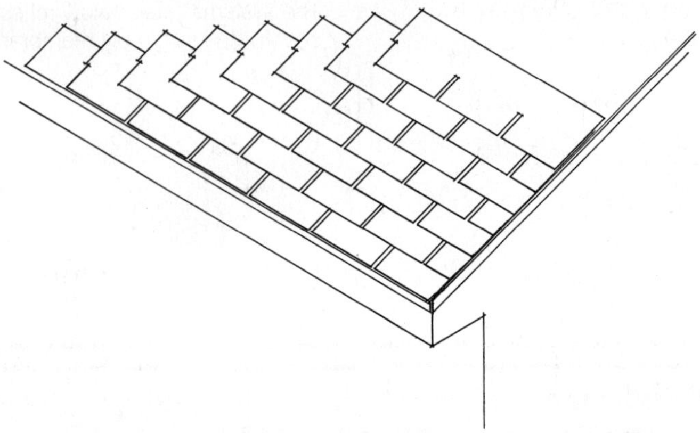

Shingles and tiles are practical in applications where the roof slope is more than 3-1/2" per foot of rise. Table below lists the various materials and the weight per S.F.

B3010 140	Shingle & Tile	COST PER S.F.		
		MAT.	INST.	TOTAL
1095	Asphalt roofing			
1100	Strip shingles, 4" slope, inorganic class A 210-235 lb./sq.	.95	1	1.95
1150	Organic, class C, 235-240 lb./sq.	1.12	1.09	2.21
1200	Premium laminated multi-layered, class A, 260-300 lb./sq.	1.80	1.49	3.29
1545	Metal roofing			
1550	Alum., shingles, colors, 3" min slope, .019" thick, 0.4 PSF	2.77	1.09	3.86
1850	Steel, colors, 3" min slope, 26 gauge, 1.0 PSF	4.53	2.24	6.77
2795	Slate roofing			
2800	Slate roofing, 4" min. slope, shingles, 3/16" thick, 8.0 PSF	6.05	2.79	8.84
3495	Wood roofing			
3500	4" min slope, cedar shingles, 16" x 5", 5" exposure 1.6 PSF	3.22	2.19	5.41
4000	Shakes, 18", 8-1/2" exposure, 2.8 PSF	2.09	2.69	4.78
5095	Tile roofing			
5100	Aluminum, mission, 3" min slope, .019" thick, 0.65 PSF	9.20	2.11	11.31
6000	Clay tile, flat shingle, interlocking, 15", 166 pcs/sq, fireflashed blend	5.20	2.45	7.65
6002				

B3010 Roof Coverings

B3010 320	Roof Deck Rigid Insulation	COST PER S.F.		
		MAT.	INST.	TOTAL
0100	Fiberboard low density			
0150	1" thick R2.78	.73	.61	1.34
0300	1 1/2" thick R4.17	1.04	.61	1.65
0350	2" thick R5.56	1.33	.61	1.94
0370	Fiberboard high density, 1/2" thick R1.3	.45	.51	.96
0380	1" thick R2.5	.74	.61	1.35
0390	1 1/2" thick R3.8	1.07	.61	1.68
0410	Fiberglass, 3/4" thick R2.78	.79	.51	1.30
0450	15/16" thick R3.70	1.01	.51	1.52
0550	1-5/16" thick R5.26	1.63	.51	2.14
0650	2 7/16" thick R10	1.97	.61	2.58
1510	Polyisocyanurate 2#/CF density, 1" thick	.64	.41	1.05
1550	1 1/2" thick	.82	.44	1.26
1600	2" thick	1.02	.48	1.50
1650	2 1/2" thick	1.33	.50	1.83
1700	3" thick	1.46	.51	1.97
1750	3 1/2" thick	2.21	.51	2.72
1800	Tapered for drainage	.77	.41	1.18
1810	Expanded polystyrene, 1#/CF density, 3/4" thick R2.89	.33	.39	.72
1820	2" thick R7.69	.69	.44	1.13
1830	Extruded polystyrene, 15 PSI compressive strength, 1" thick, R5	.78	.39	1.17
1835	2" thick R10	.98	.44	1.42
1840	3" thick R15	1.84	.51	2.35
2550	40 PSI compressive strength, 1" thick R5	.70	.39	1.09
2600	2" thick R10	1.23	.44	1.67
2650	3" thick R15	1.72	.51	2.23
2700	4" thick R20	2.22	.51	2.73
2750	Tapered for drainage	.96	.41	1.37
2810	60 PSI compressive strength, 1" thick R5	.93	.40	1.33
2850	2" thick R10	1.67	.45	2.12
2900	Tapered for drainage	1.18	.41	1.59

B3010 410	Base Flashing							
	TYPE	DESCRIPTION	SPECIFICATION	REGLET	COUNTER FLASHING	COST PER L.F.		
						MAT.	INST.	TOTAL
1000	Aluminum	mill finish	.019" thick	alum. .025	alum. .032	13.35	14.40	27.75
1100			.050" thick	.025	.032	14.90	14.40	29.30
1200		fabric, 2 sides	.004" thick	.025	.032	13.75	12.55	26.30
1300			.016" thick	.025	.032	14	12.55	26.55
1400		mastic, 2 sides	.004" thick	.025	.032	13.75	12.55	26.30
1500			.016" thick	.025	.032	14.25	12.55	26.80
1700	Copper	sheets, plain	16 oz.	copper 16 oz.	copper 16 oz.	30	15.25	45.25
1800			20 oz.	16 oz.	16 oz.	33	15.45	48.45
1900			24 oz.	16 oz.	16 oz.	37.50	15.65	53.15
2000			32 oz.	16 oz.	16 oz.	42	15.85	57.85
2200	Sheets, Metal	galvanized	20 Ga.	galv. 24 Ga.	galv. 24 Ga.	12.35	32.50	44.85
2300			24 Ga.	24 Ga.	24 Ga.	11.75	26	37.75
2400			30 Ga.	24 Ga.	24 Ga.	11.10	19.60	30.70
2600	Rubber	butyl	1/32" thick	galv. 24 Ga.	galv. 24 Ga.	12.45	12.80	25.25
2700			1/16" thick	24 Ga.	24 Ga.	13.25	12.80	26.05
2800		neoprene	1/16" thick	24 Ga.	24 Ga.	13.10	12.80	25.90
2900			1/8" thick	24 Ga.	24 Ga.	16.95	12.80	29.75

B30 Roofing

B3010 Roof Coverings

B3010 420 Roof Edges

	EDGE TYPE	DESCRIPTION	SPECIFICATION	FACE HEIGHT		COST PER L.F.		
						MAT.	INST.	TOTAL
1000	Aluminum	mill finish	.050" thick	4"		14.45	9	23.45
1100				6"		14.65	9.30	23.95
1300		duranodic	.050" thick	4"		15.50	9	24.50
1400				6"		15.95	9.30	25.25
1600		painted	.050" thick	4"		15.60	9	24.60
2000	Copper	plain	16 oz.	4"		34	9	43
2100				6"		44	9.30	53.30
2300			20 oz.	4"		42.50	10.30	52.80
2400				6"		49.50	9.95	59.45
2700	Sheet Metal	galvanized	20 Ga.	4"		17.25	10.70	27.95
2800				6"		17.65	10.70	28.35
3000			24 Ga.	4"		14.30	9	23.30
3100				6"		14.55	9	23.55

B3010 430 Flashing

	MATERIAL	BACKING	SIDES	SPECIFICATION	QUANTITY	COST PER S.F.		
						MAT.	INST.	TOTAL
0040	Aluminum	none		.019"		1.38	3.27	4.65
0050				.032"		1.49	3.27	4.76
0300		fabric	2	.004"		1.76	1.44	3.20
0400		mastic		.004"		1.76	1.44	3.20
0700	Copper	none		16 oz.	<500 lbs.	8.65	4.12	12.77
0800				24 oz.	<500 lbs.	16.05	4.51	20.56
3500	PVC black	none		.010"		.23	1.66	1.89
3700				.030"		.37	1.66	2.03
4200	Neoprene			1/16"		2.81	1.66	4.47
4500	Stainless steel	none		.015"	<500 lbs.	7.05	4.12	11.17
4600	Copper clad				>2000 lbs.	7.35	3.06	10.41
5000	Plain			32 ga.		3.58	3.06	6.64
5009								

B3010 Roof Coverings

B3010 610 Gutters

	SECTION	MATERIAL	THICKNESS	SIZE	FINISH	COST PER L.F.		
						MAT.	INST.	TOTAL
0050	Box	aluminum	.027"	5"	enameled	3.17	4.67	7.84
0100					mill	3.19	4.67	7.86
0200			.032"	5"	enameled	3.95	4.52	8.47
0500		copper	16 Oz.	5"	lead coated	20	4.52	24.52
0600					mill	8.85	4.52	13.37
1000		steel galv.	28 Ga.	5"	enameled	2.40	4.52	6.92
1200			26 Ga.	5"	mill	2.64	4.52	7.16
1800		vinyl		4"	colors	1.43	4.40	5.83
1900				5"	colors	1.76	4.40	6.16
2300		hemlock or fir		4"x5"	treated	18.45	5.05	23.50
3000	Half round	copper	16 Oz.	4"	lead coated	16.15	4.52	20.67
3100					mill	10.45	4.52	14.97
3600		steel galv.	28 Ga.	5"	enameled	2.40	4.52	6.92
4102		stainless steel		5"	mill	10.40	4.52	14.92
5000		vinyl		4"	white	1.49	4.40	5.89
5002								

B3010 620 Downspouts

	MATERIALS	SECTION	SIZE	FINISH	THICKNESS	COST PER V.L.F.		
						MAT.	INST.	TOTAL
0100	Aluminum	rectangular	2"x3"	embossed mill	.020"	1.16	2.98	4.14
0150				enameled	.020"	1.53	2.98	4.51
0250			3"x4"	enameled	.024"	2.61	4.04	6.65
0300		round corrugated	3"	enameled	.020"	2.20	2.98	5.18
0350			4"	enameled	.025"	3.25	4.04	7.29
0500	Copper	rectangular corr.	2"x3"	mill	16 Oz.	9.60	2.98	12.58
0600		smooth		mill	16 Oz.	12.15	2.98	15.13
0700		rectangular corr.	3"x4"	mill	16 Oz.	11.45	3.90	15.35
1300	Steel	rectangular corr.	2"x3"	galvanized	28 Ga.	2.13	2.98	5.11
1350				epoxy coated	24 Ga.	2.56	2.98	5.54
1400		smooth		galvanized	28 Ga.	4.02	2.98	7
1450		rectangular corr.	3"x4"	galvanized	28 Ga.	2	3.90	5.90
1500				epoxy coated	24 Ga.	3.11	3.90	7.01
1550		smooth		galvanized	28 Ga.	4.47	3.90	8.37
1600		round corrugated	2"	galvanized	28 Ga.	2.31	2.98	5.29
1650			3"	galvanized	28 Ga.	2.31	2.98	5.29
1700			4"	galvanized	28 Ga.	2.77	3.90	6.67
1750			5"	galvanized	28 Ga.	4.10	4.35	8.45
2552	S.S. tubing sch.5	rectangular	3"x4"	mill		54	3.90	57.90
2702								

B3010 630 Gravel Stop

	MATERIALS	SECTION	SIZE	FINISH	THICKNESS	COST PER L.F.		
						MAT.	INST.	TOTAL
5100	Aluminum	extruded	4"	mill	.050"	7.05	3.90	10.95
5200			4"	duranodic	.050"	8.10	3.90	12
5300			8"	mill	.050"	8.30	4.52	12.82
5400			8"	duranodic	.050"	10	4.52	14.52
6000			12"-2 pc.	duranodic	.050"	11.90	5.65	17.55
6100	Stainless	formed	6"	mill	24 Ga.	16.85	4.19	21.04

B3020 Roof Openings

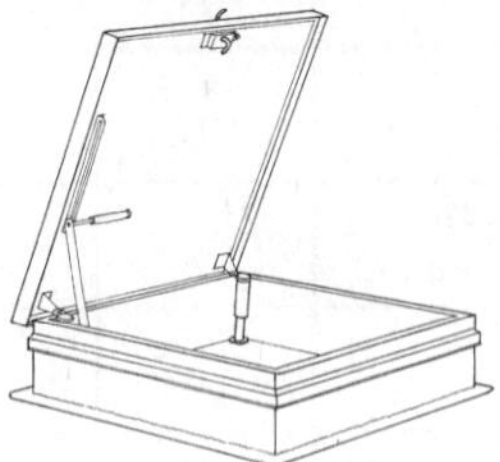

Roof Hatch

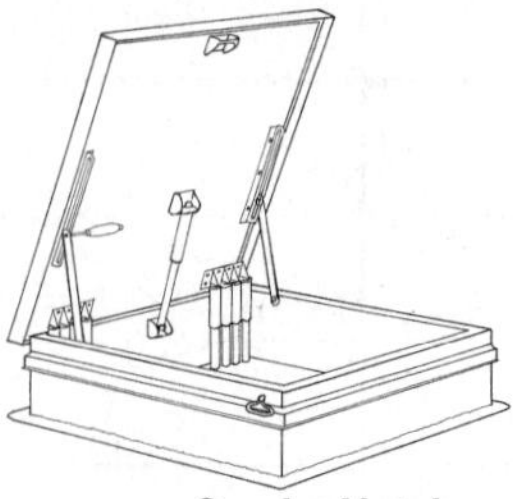

Smoke Hatch

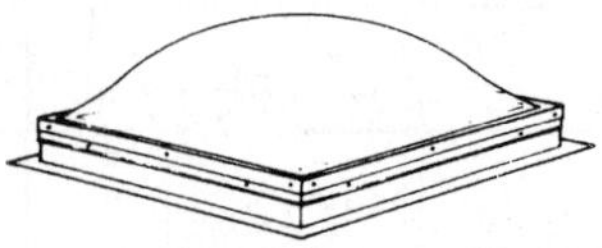

Skylight

B3020 110	Skylights	COST PER S.F.		
		MAT.	INST.	TOTAL
5100	Skylights, plastic domes, insul curbs, nom. size to 10 S.F., single glaze	32	11.95	43.95
5200	Double glazing	36.50	14.70	51.20
5300	10 S.F. to 20 S.F., single glazing	32.50	4.84	37.34
5400	Double glazing	31	6.05	37.05
5500	20 S.F. to 30 S.F., single glazing	24.50	4.11	28.61
5600	Double glazing	29.50	4.84	34.34
5700	30 S.F. to 65 S.F., single glazing	24	3.14	27.14
5800	Double glazing	30	4.11	34.11
6000	Sandwich panels fiberglass, 1-9/16" thick, 2 S.F. to 10 S.F.	20.50	9.55	30.05
6100	10 S.F. to 18 S.F.	18.25	7.20	25.45
6200	2-3/4" thick, 25 S.F. to 40 S.F.	29.50	6.50	36
6300	40 S.F. to 70 S.F.	24	5.80	29.80

B3020 210	Hatches	COST PER OPNG.		
		MAT.	INST.	TOTAL
0200	Roof hatches with curb and 1" fiberglass insulation, 2'-6"x3'-0", aluminum	1,075	191	1,266
0300	Galvanized steel 165 lbs.	700	191	891
0400	Primed steel 164 lbs.	775	191	966
0500	2'-6"x4'-6" aluminum curb and cover, 150 lbs.	1,400	213	1,613
0600	Galvanized steel 220 lbs.	975	213	1,188
0650	Primed steel 218 lbs.	1,050	213	1,263
0800	2'x6"x8'-0" aluminum curb and cover, 260 lbs.	2,075	290	2,365
0900	Galvanized steel, 360 lbs.	1,975	290	2,265
0950	Primed steel 358 lbs.	1,575	290	1,865
1200	For plexiglass panels, add to the above	495		495
2100	Smoke hatches, unlabeled not incl. hand winch operator, 2'-6"x3', galv	845	231	1,076
2200	Plain steel, 160 lbs.	940	231	1,171
2400	2'-6"x8'-0",galvanized steel, 360 lbs.	2,150	315	2,465
2500	Plain steel, 350 lbs.	1,725	315	2,040
3000	4'-0"x8'-0", double leaf low profile, aluminum cover, 359 lb.	2,925	239	3,164
3100	Galvanized steel 475 lbs.	2,425	239	2,664
3200	High profile, aluminum cover, galvanized curb, 361 lbs.	2,850	239	3,089

C Interiors

C1010 Partitions

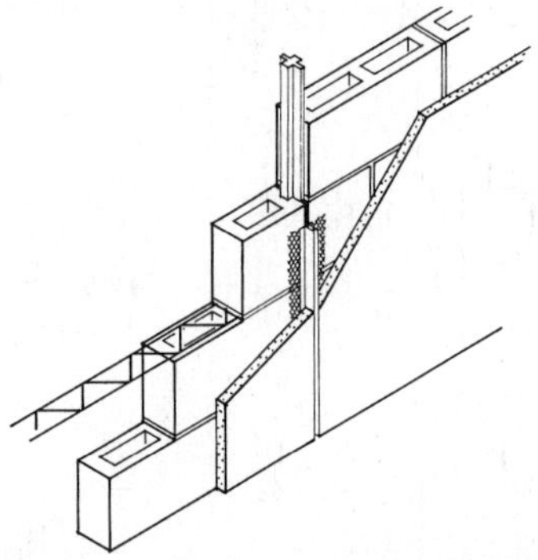

The Concrete Block Partition Systems are defined by weight and type of block, thickness, type of finish and number of sides finished. System components include joint reinforcing on alternate courses and vertical control joints.

C1010 102 Concrete Block Partitions - Regular Weight

	TYPE	THICKNESS (IN.)	TYPE FINISH	SIDES FINISHED		COST PER S.F.		
						MAT.	INST.	TOTAL
1000	Hollow	4	none	0		2.18	5.40	7.58
1010			gyp. plaster 2 coat	1		2.57	7.95	10.52
1020				2		2.95	10.50	13.45
1200			portland - 3 coat	1		2.45	8.30	10.75
1400			5/8" drywall	1		2.82	7.25	10.07
1500		6	none	0		2.77	5.80	8.57
1510			gyp. plaster 2 coat	1		3.16	8.35	11.51
1520				2		3.54	10.90	14.44
1700			portland - 3 coat	1		3.04	8.70	11.74
1900			5/8" drywall	1		3.41	7.65	11.06
1910				2		4.05	9.55	13.60
2000		8	none	0		2.97	6.15	9.12
2010			gyp. plaster 2 coat	1		3.36	8.70	12.06
2020			gyp. plaster 2 coat	2		3.74	11.30	15.04
2200			portland - 3 coat	1		3.24	9.10	12.34
2400			5/8" drywall	1		3.61	8.05	11.66
2410				2		4.25	9.90	14.15
2500		10	none	0		3.51	6.45	9.96
2510			gyp. plaster 2 coat	1		3.90	9	12.90
2520				2		4.28	11.55	15.83
2700			portland - 3 coat	1		3.78	9.35	13.13
2900			5/8" drywall	1		4.15	8.30	12.45
2910				2		4.79	10.20	14.99
3000	Solid	2	none	0		1.88	5.35	7.23
3010			gyp. plaster	1		2.27	7.90	10.17
3020				2		2.65	10.45	13.10
3200			portland - 3 coat	1		2.15	8.25	10.40
3400			5/8" drywall	1		2.52	7.20	9.72
3410				2		3.16	9.10	12.26
3500		4	none	0		2.45	5.60	8.05
3510			gyp. plaster	1		2.95	8.15	11.10
3520				2		3.22	10.70	13.92
3700			portland - 3 coat	1		2.72	8.50	11.22
3900			5/8" drywall	1		3.09	7.45	10.54
3910				2		3.73	9.35	13.08

C1010 Partitions

C1010 102 Concrete Block Partitions - Regular Weight

	TYPE	THICKNESS (IN.)	TYPE FINISH	SIDES FINISHED		COST PER S.F.		
						MAT.	INST.	TOTAL
4010	Solid	6	gyp. plaster	1		3.22	8.55	11.77
4020				2		3.60	11.10	14.70
4200			portland - 3 coat	1		3.10	8.90	12
4400			5/8" drywall	1		3.47	7.85	11.32
4410				2		4.11	9.75	13.86

C1010 104 Concrete Block Partitions - Lightweight

	TYPE	THICKNESS (IN.)	TYPE FINISH	SIDES FINISHED		COST PER S.F.		
						MAT.	INST.	TOTAL
5000	Hollow	4	none	0		2.30	5.30	7.60
5010			gyp. plaster	1		2.69	7.85	10.54
5020				2		3.07	10.40	13.47
5200			portland - 3 coat	1		2.57	8.20	10.77
5400			5/8" drywall	1		2.94	7.15	10.09
5410				2		3.58	9.05	12.63
5500		6	none	0		3.13	5.65	8.78
5520			gyp. plaster	2		3.90	10.75	14.65
5700			portland - 3 coat	1		3.40	8.55	11.95
5900			5/8" drywall	1		3.77	7.50	11.27
5910				2		4.41	9.40	13.81
6000		8	none	0		3.78	6	9.78
6010			gyp. plaster	1		4.17	8.55	12.72
6020				2		4.55	11.15	15.70
6200			portland - 3 coat	1		4.05	8.95	13
6400			5/8" drywall	1		4.42	7.90	12.32
6410				2		5.05	9.75	14.80
6500		10	none	0		4.55	6.30	10.85
6700			portland - 3 coat	1		4.82	9.20	14.02
6900			5/8" drywall	1		5.20	8.15	13.35
6910				2		5.85	10.05	15.90
7000	Solid	4	none	0		2.08	5.50	7.58
7010			gyp. plaster	1		2.47	8.05	10.52
7020				2		2.85	10.60	13.45
7200			portland - 3 coat	1		2.35	8.40	10.75
7400			5/8" drywall	1		2.72	7.35	10.07
7410				2		3.36	9.25	12.61
7500		6	none	0		3.65	5.95	9.60
7510			gyp. plaster	1		4.18	8.55	12.73
7520				2		4.42	11.05	15.47
7700			portland - 3 coat	1		3.92	8.85	12.77
7900			5/8" drywall	1		4.29	7.80	12.09
7910				2		4.93	9.70	14.63
8000		8	none	0		4.12	6.35	10.47
8010			gyp. plaster	1		4.51	8.90	13.41
8020				2		4.89	11.50	16.39
8200			portland - 3 coat	1		4.39	9.30	13.69
8400			5/8" drywall	1		4.76	8.25	13.01
8410				2		5.40	10.10	15.50

C1010 Partitions

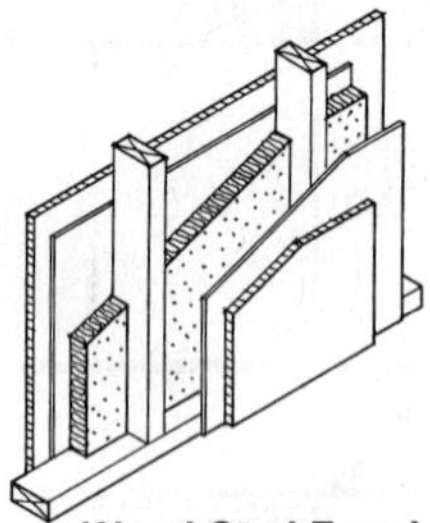

Wood Stud Framing

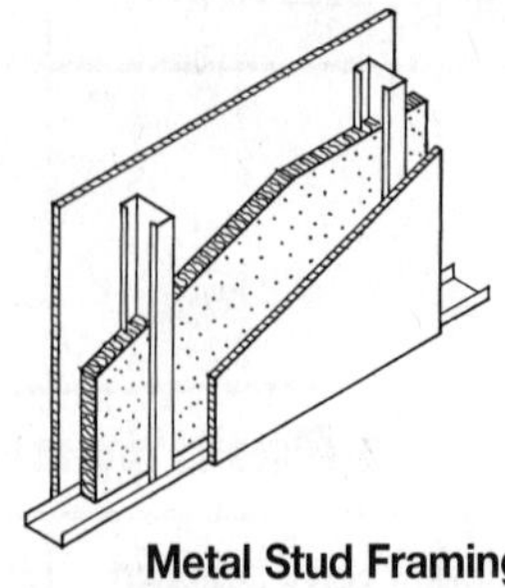

Metal Stud Framing

The Drywall Partitions/Stud Framing Systems are defined by type of drywall and number of layers, type and spacing of stud framing, and treatment on the opposite face. Components include taping and finishing.

Cost differences between regular and fire resistant drywall are negligible, and terminology is interchangeable. In some cases fiberglass insulation is included for additional sound deadening.

C1010 124 Drywall Partitions/Wood Stud Framing

	FACE LAYER	BASE LAYER	FRAMING	OPPOSITE FACE	INSULATION	COST PER S.F.		
						MAT.	INST.	TOTAL
1200	5/8" FR drywall	none	2 x 4, @ 16" O.C.	same	0	1.34	3.06	4.40
1250				5/8" reg. drywall	0	1.32	3.06	4.38
1300				nothing	0	.90	2.04	2.94
1400		1/4" SD gypsum	2 x 4 @ 16" O.C.	same	1-1/2" fiberglass	2.64	4.69	7.33
1450				5/8" FR drywall	1-1/2" fiberglass	2.20	4.13	6.33
1500				nothing	1-1/2" fiberglass	1.76	3.11	4.87
1600		resil. channels	2 x 4 @ 16", O.C.	same	1-1/2" fiberglass	2.09	5.95	8.04
1650				5/8" FR drywall	1-1/2" fiberglass	1.93	4.76	6.69
1700				nothing	1-1/2" fiberglass	1.49	3.74	5.23
1800		5/8" FR drywall	2 x 4 @ 24" O.C.	same	0	2.01	3.87	5.88
1850				5/8" FR drywall	0	1.62	3.36	4.98
1900				nothing	0	1.18	2.34	3.52
2200		5/8" FR drywall	2 rows-2 x 4	same	2" fiberglass	3.12	5.60	8.72
2250			16"O.C.	5/8" FR drywall	2" fiberglass	2.73	5.10	7.83
2300				nothing	2" fiberglass	2.29	4.07	6.36
2400	5/8" WR drywall	none	2 x 4, @ 16" O.C.	same	0	1.50	3.06	4.56
2450				5/8" FR drywall	0	1.42	3.06	4.48
2500				nothing	0	.98	2.04	3.02
2600		5/8" FR drywall	2 x 4, @ 24" O.C.	same	0	2.17	3.87	6.04
2650				5/8" FR drywall	0	1.70	3.36	5.06
2700				nothing	0	1.26	2.34	3.60
2800	5/8" VF drywall	none	2 x 4, @ 16" O.C.	same	0	2.14	3.28	5.42
2850				5/8" FR drywall	0	1.74	3.17	4.91
2900				nothing	0	1.30	2.15	3.45
3000		5/8" FR drywall	2 x 4 , 24" O.C.	same	0	2.81	4.09	6.90
3050				5/8" FR drywall	0	2.02	3.47	5.49
3100				nothing	0	1.58	2.45	4.03
3200	1/2" reg drywall	3/8" reg drywall	2 x 4, @ 16" O.C.	same	0	2.06	4.08	6.14
3250				5/8" FR drywall	0	1.70	3.57	5.27
3300				nothing	0	1.26	2.55	3.81

C1010 Partitions

C1010 126 Drywall Partitions/Metal Stud Framing

	FACE LAYER	BASE LAYER	FRAMING	OPPOSITE FACE	INSULATION	COST PER S.F.		
						MAT.	INST.	TOTAL
5200	5/8″ FR drywall	none	1-5/8″ @ 24″ O.C.	same	0	1.08	2.71	3.79
5250				5/8″ reg. drywall	0	1.06	2.71	3.77
5300				nothing	0	.64	1.69	2.33
5310			2-1/2″ @ 24″ O.C.	same	0	1.13	2.72	3.85
5320				5/8″ reg. drywall	0	1.11	2.72	3.83
5330				nothing	0	.69	1.70	2.39
5400			3-5/8″ @ 24″ O.C.	same	0	1.18	2.72	3.90
5450				5/8″ reg. drywall	0	1.16	2.72	3.88
5500				nothing	0	.74	1.70	2.44
5530		1/4″ SD gypsum	1-5/8″ @ 16″ O.C.	same	0	2.03	4.18	6.21
5535				5/8″ FR drywall	0	1.15	3.06	4.21
5540				nothing	0	1.15	2.60	3.75
5545			2-1/2″ @ 16″ O.C.	same	0	2.10	4.19	6.29
5550				5/8″ FR drywall	0	1.22	3.07	4.29
5555				nothing	0	1.22	2.61	3.83
5560			3-5/8″ @ 16″ O.C.	same	0	2.17	4.22	6.39
5565				5/8″ FR drywall	0	1.29	3.10	4.39
5570				nothing	0	1.29	2.64	3.93
5600			1-5/8″ @ 24″ O.C.	same	0	1.96	3.83	5.79
5650				5/8″ FR drywall	0	1.52	3.27	4.79
5700				nothing	0	1.08	2.25	3.33
5800			2-1/2″ @ 24″ O.C.	same	0	2.01	3.84	5.85
5850				5/8″ FR drywall	0	1.57	3.28	4.85
5900				nothing	0	1.13	2.26	3.39
5910			3-5/8″ @ 24″ O.C.	same	0	2.06	3.84	5.90
5920				5/8″ FR drywall	0	1.62	3.28	4.90
5930				nothing	0	1.18	2.26	3.44
6000		5/8″ FR drywall	2-1/2″ @ 16″ O.C.	same	0	2.10	4.36	6.46
6050				5/8″ FR drywall	0	1.71	3.85	5.56
6100				nothing	0	1.27	2.83	4.10
6110			3-5/8″ @ 16″ O.C.	same	0	2.07	4.12	6.19
6120				5/8″ FR drywall	0	1.68	3.61	5.29
6130				nothing	0	1.24	2.59	3.83
6170			2-1/2″ @ 24″ O.C.	same	0	1.91	3.74	5.65
6180				5/8″ FR drywall	0	1.52	3.23	4.75
6190				nothing	0	1.08	2.21	3.29
6200			3-5/8″ @ 24″ O.C.	same	0	1.96	3.74	5.70
6250				5/8″FR drywall	3-1/2″ fiberglass	2.09	3.61	5.70
6300				nothing	0	1.13	2.21	3.34
6310	5/8″ WR drywall	none	1-5/8″ @ 16″ O.C.	same	0	1.31	3.06	4.37
6320				5/8″ WR drywall	0	1.78	3.57	5.35
6330				nothing	0	1.26	2.55	3.81
6340			2-1/2″ @ 16″ O.C.	same	0	2.32	4.09	6.41
6350				5/8″ WR drywall	0	1.85	3.58	5.43
6360				nothing	0	1.33	2.56	3.89
6370			3-5/8″ @ 16″ O.C.	same	0	2.39	4.12	6.51
6380				5/8″ WR drywall	0	1.92	3.61	5.53
6390				nothing	0	1.40	2.59	3.99

C1010 Partitions

C1010 126 Drywall Partitions/Metal Stud Framing

	FACE LAYER	BASE LAYER	FRAMING	OPPOSITE FACE	INSULATION	COST PER S.F.		
						MAT.	INST.	TOTAL
6400	5/8" WR drywall	none	1-5/8" @ 24" O.C.	same	0	1.24	2.71	3.95
6450				5/8" WR drywall	0	1.16	2.71	3.87
6500				nothing	0	.72	1.69	2.41
6510			2-1/2" @ 24" O.C.	same	0	2.23	3.74	5.97
6520				5/8" WR drywall	0	1.21	2.72	3.93
6530				nothing	0	.77	1.70	2.47
6600			3-5/8" @ 24" O.C.	same	0	1.34	2.72	4.06
6650				5/8" WR drywall	0	1.26	2.72	3.98
6700				nothing	0	.82	1.70	2.52
6800		5/8" FR drywall	2-1/2" @ 16" O.C.	same	0	2.26	4.36	6.62
6850				5/8" FR drywall	0	1.79	3.85	5.64
6900				nothing	0	1.35	2.83	4.18
6910			3-5/8" @ 16" O.C.	same	0	2.23	4.12	6.35
6920				5/8" FR drywall	0	1.76	3.61	5.37
6930				nothing	0	1.32	2.59	3.91
6940			2-1/2" @ 24" O.C.	same	0	1.91	3.74	5.65
6950				5/8" FR drywall	0	1.60	3.23	4.83
6960				nothing	0	1.16	2.21	3.37
7000			3-5/8" @ 24" O.C.	same	0	2.12	3.74	5.86
7050				5/8"FR drywall	3-1/2" fiberglass	2.17	3.61	5.78
7100				nothing	0	1.21	2.21	3.42
7110	5/8" VF drywall	none	1-5/8" @ 16" O.C.	same	0	1.95	3.28	5.23
7120				5/8" FR drywall	0	1.62	3.57	5.19
7130				nothing	0	1.18	2.55	3.73
7140			3-5/8" @ 16" O.C.	same	0	2.23	4.12	6.35
7150				5/8" FR drywall	0	1.76	3.61	5.37
7160				nothing	0	1.32	2.59	3.91
7200		none	1-5/8" @ 24" O.C.	same	0	1.88	2.93	4.81
7250				5/8" FR drywall	0	1.48	2.82	4.30
7300				nothing	0	1.04	1.80	2.84
7400			3-5/8" @ 24" O.C.	same	0	1.98	2.94	4.92
7450				5/8" FR drywall	0	1.58	2.83	4.41
7500				nothing	0	1.14	1.81	2.95
7600		5/8" FR drywall	2-1/2" @ 16" O.C.	same	0	2.90	4.58	7.48
7650				5/8" FR drywall	0	2.11	3.96	6.07
7700				nothing	0	1.67	2.94	4.61
7710			3-5/8" @ 16" O.C.	same	0	2.87	4.34	7.21
7720				5/8" FR drywall	0	2.08	3.72	5.80
7730				nothing	0	1.64	2.70	4.34
7740			2-1/2" @ 24" O.C.	same	0	2.71	3.96	6.67
7750				5/8" FR drywall	0	1.92	3.34	5.26
7760				nothing	0	1.48	2.32	3.80
7800			3-5/8" @ 24" O.C.	same	0	2.76	3.96	6.72
7850				5/8"FR drywall	3-1/2" fiberglass	2.49	3.72	6.21
7900				nothing	0	1.53	2.32	3.85

C1010 Partitions

C1010 128	Drywall Components	COST PER S.F. MAT.	COST PER S.F. INST.	COST PER S.F. TOTAL
0060	Metal studs, 24" O.C. including track, load bearing, 20 gage, 2-1/2"	.70	.95	1.65
0080	3-5/8"	.84	.97	1.81
0100	4"	.75	.99	1.74
0120	6"	1.12	1.01	2.13
0140	Metal studs, 24" O.C. including track, load bearing, 18 gage, 2-1/2"	.70	.95	1.65
0160	3-5/8"	.84	.97	1.81
0180	4"	.75	.99	1.74
0200	6"	1.12	1.01	2.13
0220	16 gage, 2-1/2"	.82	1.08	1.90
0240	3-5/8"	.98	1.10	2.08
0260	4"	1.03	1.13	2.16
0280	6"	1.30	1.15	2.45
0300	Non load bearing, 25 gage, 1-5/8"	.20	.67	.87
0340	3-5/8"	.30	.68	.98
0360	4"	.33	.68	1.01
0380	6"	.40	.70	1.10
0400	20 gage, 2-1/2"	.33	.84	1.17
0420	3-5/8"	.37	.86	1.23
0440	4"	.44	.86	1.30
0460	6"	.52	.87	1.39
0540	Wood studs including blocking, shoe and double top plate, 2"x4", 12" O.C.	.57	1.27	1.84
0560	16" O.C.	.46	1.02	1.48
0580	24" O.C.	.35	.81	1.16
0600	2"x6", 12" O.C.	.88	1.45	2.33
0620	16" O.C.	.71	1.13	1.84
0640	24" O.C.	.54	.88	1.42
0642	Furring one side only, steel channels, 3/4", 12" O.C.	.42	2.03	2.45
0644	16" O.C.	.38	1.80	2.18
0646	24" O.C.	.25	1.36	1.61
0647	1-1/2" , 12" O.C.	.57	2.27	2.84
0648	16" O.C.	.51	1.99	2.50
0649	24" O.C.	.34	1.56	1.90
0650	Wood strips, 1" x 3", on wood, 12" O.C.	.43	.92	1.35
0651	16" O.C.	.32	.69	1.01
0652	On masonry, 12" O.C.	.46	1.02	1.48
0653	16" O.C.	.35	.77	1.12
0654	On concrete, 12" O.C.	.46	1.95	2.41
0655	16" O.C.	.35	1.46	1.81
0665	Gypsum board, one face only, exterior sheathing, 1/2"	.51	.90	1.41
0680	Interior, fire resistant, 1/2"	.40	.51	.91
0700	5/8"	.39	.51	.90
0720	Sound deadening board 1/4"	.44	.56	1
0740	Standard drywall 3/8"	.39	.51	.90
0760	1/2"	.36	.51	.87
0780	5/8"	.37	.51	.88
0800	Tongue & groove coreboard 1"	.83	2.11	2.94
0820	Water resistant, 1/2"	.45	.51	.96
0840	5/8"	.47	.51	.98
0860	Add for the following:, foil backing	.17		.17
0880	Fiberglass insulation, 3-1/2"	.52	.38	.90
0900	6"	.68	.38	1.06
0920	Rigid insulation 1"	.57	.51	1.08
0940	Resilient furring @ 16" O.C.	.22	1.59	1.81
0960	Taping and finishing	.05	.51	.56
0980	Texture spray	.04	.60	.64
1000	Thin coat plaster	.12	.64	.76
1050	Sound wall framing, 2x6 plates, 2x4 staggered studs, 12" O.C.	.69	1.24	1.93

C1010 Partitions

C1010 144	Plaster Partition Components	COST PER S.F.		
		MAT.	INST.	TOTAL
0060	Metal studs, 16" O.C., including track, non load bearing, 25 gage, 1-5/8"	.34	1.03	1.37
0080	2-1/2"	.34	1.03	1.37
0100	3-1/4"	.41	1.06	1.47
0120	3-5/8"	.41	1.06	1.47
0140	4"	.45	1.07	1.52
0160	6"	.54	1.08	1.62
0180	Load bearing, 20 gage, 2-1/2"	.95	1.32	2.27
0200	3-5/8"	1.14	1.34	2.48
0220	4"	1.18	1.37	2.55
0240	6"	1.51	1.39	2.90
0260	16 gage 2-1/2"	1.13	1.49	2.62
0280	3-5/8"	1.35	1.54	2.89
0300	4"	1.41	1.56	2.97
0320	6"	1.78	1.59	3.37
0340	Wood studs, including blocking, shoe and double plate, 2"x4" , 12" O.C.	.57	1.27	1.84
0360	16" O.C.	.46	1.02	1.48
0380	24" O.C.	.35	.81	1.16
0400	2"x6" , 12" O.C.	.88	1.45	2.33
0420	16" O.C.	.71	1.13	1.84
0440	24" O.C.	.54	.88	1.42
0460	Furring one face only, steel channels, 3/4" , 12" O.C.	.42	2.03	2.45
0480	16" O.C.	.38	1.80	2.18
0500	24" O.C.	.25	1.36	1.61
0520	1-1/2" , 12" O.C.	.57	2.27	2.84
0540	16" O.C.	.51	1.99	2.50
0560	24"O.C.	.34	1.56	1.90
0580	Wood strips 1"x3", on wood., 12" O.C.	.43	.92	1.35
0600	16"O.C.	.32	.69	1.01
0620	On masonry, 12" O.C.	.46	1.02	1.48
0640	16" O.C.	.35	.77	1.12
0660	On concrete, 12" O.C.	.46	1.95	2.41
0680	16" O.C.	.35	1.46	1.81
0700	Gypsum lath. plain or perforated, nailed to studs, 3/8" thick	.37	.62	.99
0720	1/2" thick	.30	.66	.96
0740	Clipped to studs, 3/8" thick	.37	.70	1.07
0760	1/2" thick	.30	.75	1.05
0780	Metal lath, diamond painted, nailed to wood studs, 2.5 lb.	.50	.62	1.12
0800	3.4 lb.	.53	.66	1.19
0820	Screwed to steel studs, 2.5 lb.	.50	.66	1.16
0840	3.4 lb.	.51	.70	1.21
0860	Rib painted, wired to steel, 2.75 lb	.43	.70	1.13
0880	3.4 lb	.55	.75	1.30
0900	4.0 lb	.69	.82	1.51
0910				
0920	Gypsum plaster, 2 coats	.45	2.44	2.89
0940	3 coats	.65	2.92	3.57
0960	Perlite or vermiculite plaster, 2 coats	.72	2.80	3.52
0980	3 coats	.78	3.45	4.23
1000	Stucco, 3 coats, 1" thick, on wood framing	.78	4.88	5.66
1020	On masonry	.31	3.81	4.12
1100	Metal base galvanized and painted 2-1/2" high	.72	1.99	2.71

C1010 Partitions

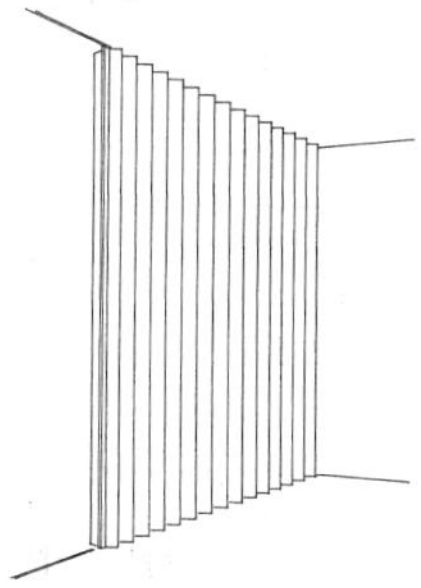

Folding Accordion

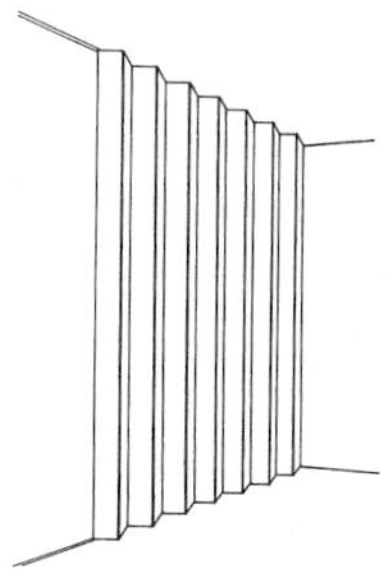

Folding Leaf

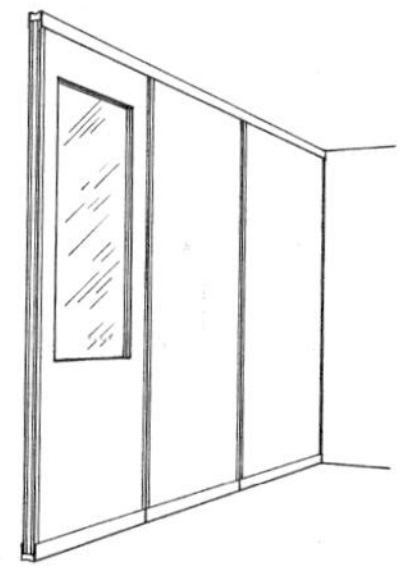

Movable and Borrow Lites

C1010 205	Partitions	COST PER S.F.		
		MAT.	INST.	TOTAL
0360	Folding accordion, vinyl covered, acoustical, 3 lb. S.F., 17 ft max. hgt	35	10.15	45.15
0380	5 lb. per S.F. 27 ft max height	48.50	10.65	59.15
0400	5.5 lb. per S.F., 17 ft. max height	56.50	11.25	67.75
0420	Commercial, 1.75 lb per S.F., 8 ft. max height	31.50	4.50	36
0440	2.0 Lb per S.F., 17 ft. max height	32.50	6.75	39.25
0460	Industrial, 4.0 lb. per S.F. 27 ft. max height	49	13.50	62.50
0480	Vinyl clad wood or steel, electric operation 6 psf	69	6.35	75.35
0500	Wood, non acoustic, birch or mahogany	37	3.38	40.38
0560	Folding leaf, alum framed acoustical 12′ high., 5.5 lb/S.F. standard trim	45	16.90	61.90
0580	Premium trim	54	34	88
0600	6.5 lb. per S.F., standard trim	47	16.90	63.90
0620	Premium trim	58.50	34	92.50
0640	Steel acoustical, 7.5 per S.F., vinyl faced, standard trim	66.50	16.90	83.40
0660	Premium trim	81.50	34	115.50
0680	Wood acoustic type, vinyl faced to 18′ high 6 psf, economy trim	62.50	16.90	79.40
0700	Standard trim	74.50	22.50	97
0720	Premium trim	96.50	34	130.50
0740	Plastic lam. or hardwood faced, standard trim	64.50	16.90	81.40
0760	Premium trim	68.50	34	102.50
0780	Wood, low acoustical type to 12 ft. high 4.5 psf	47	20.50	67.50
0840	Demountable, trackless wall, cork finish, semi acous, 1-5/8″ th, unsealed	42.50	3.12	45.62
0860	Sealed	47	5.35	52.35
0880	Acoustic, 2″ thick, unsealed	40	3.32	43.32
0900	Sealed	61.50	4.50	66
0920	In-plant modular office system, w/prehung steel door			
0940	3″ thick honeycomb core panels			
0960	12′ x 12′, 2 wall	13.60	.62	14.22
0970	4 wall	13.80	.82	14.62
0980	16′ x 16′, 2 wall	13.75	.43	14.18
0990	4 wall	9.35	.44	9.79
1000	Gypsum, demountable, 3″ to 3-3/4″ thick x 9′ high, vinyl clad	7.35	2.33	9.68
1020	Fabric clad	18.30	2.55	20.85
1040	1.75 system, vinyl clad hardboard, paper honeycomb core panel			
1060	1-3/4″ to 2-1/2″ thick x 9′ high	12.30	2.33	14.63
1080	Unitized gypsum panel system, 2″ to 2-1/2″ thick x 9′ high			
1100	Vinyl clad gypsum	15.85	2.33	18.18
1120	Fabric clad gypsum	26	2.55	28.55
1140	Movable steel walls, modular system			
1160	Unitized panels, 48″ wide x 9′ high			
1180	Baked enamel, pre-finished	17.85	1.88	19.73
1200	Fabric clad	26	2.01	28.01
1203				

C1020 Interior Doors

Sliding Panel-Mall Front

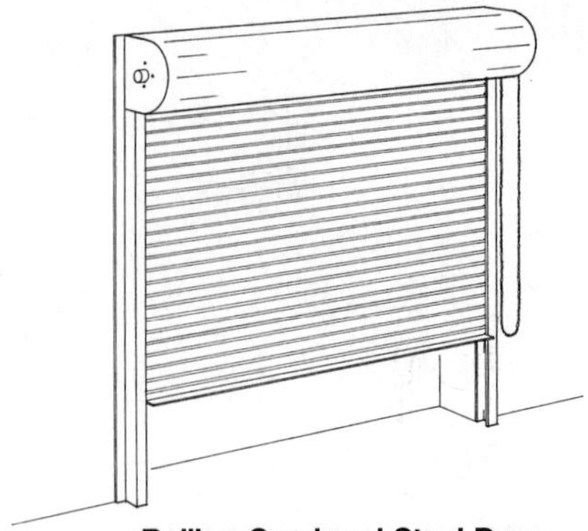

Rolling Overhead Steel Door

C1020 102	Special Doors	COST PER OPNG.		
		MAT.	INST.	TOTAL
2500	Single leaf, wood, 3′-0″x7′-0″x1 3/8″, birch, solid core	405	188	593
2510	Hollow core	330	180	510
2530	Hollow core, lauan	320	180	500
2540	Louvered pine	465	180	645
2550	Paneled pine	495	180	675
2600	Hollow metal, comm. quality, flush, 3′-0″x7′-0″x1-3/8″	935	198	1,133
2650	3′-0″x10′-0″ openings with panel	1,475	257	1,732
2700	Metal fire, comm. quality, 3′-0″x7′-0″x1-3/8″	1,175	206	1,381
3200	Double leaf, wood, hollow core, 2 - 3′-0″x7′-0″x1-3/8″	590	345	935
3300	Hollow metal, comm. quality, B label, 2′-3′-0″x7′-0″x1-3/8″	2,225	435	2,660
3400	6′-0″x10′-0″ opening, with panel	2,850	535	3,385
3500	Double swing door system, 12′-0″x7′-0″, mill finish	9,450	2,700	12,150
3700	Black finish	9,600	2,775	12,375
3800	Sliding entrance door and system mill finish	11,000	2,475	13,475
3900	Bronze finish	12,000	2,675	14,675
4000	Black finish	12,500	2,775	15,275
4100	Sliding panel mall front, 16′x9′ opening, mill finish	3,900	755	4,655
4200	Bronze finish	5,075	980	6,055
4300	Black finish	6,250	1,200	7,450
4400	24′x9′ opening mill finish	5,525	1,400	6,925
4500	Bronze finish	7,175	1,825	9,000
4600	Black finish	8,850	2,250	11,100
4700	48′x9′ opening mill finish	10,200	1,100	11,300
4800	Bronze finish	13,300	1,425	14,725
4900	Black finish	16,300	1,750	18,050
5000	Rolling overhead steel door, manual, 8′ x 8′ high	1,300	730	2,030
5100	10′ x 10′ high	2,150	835	2,985
5200	20′ x 10′ high	3,500	1,175	4,675
5300	12′ x 12′ high	2,175	975	3,150
5400	Motor operated, 8′ x 8′ high	2,600	965	3,565
5500	10′ x 10′ high	3,450	1,075	4,525
5600	20′ x 10′ high	4,800	1,400	6,200
5700	12′ x 12′ high	3,475	1,200	4,675
5800	Roll up grille, aluminum, manual, 10′ x 10′ high, mill finish	3,150	1,425	4,575
5900	Bronze anodized	4,950	1,425	6,375
6000	Motor operated, 10′ x 10′ high, mill finish	4,525	1,675	6,200
6100	Bronze anodized	6,325	1,675	8,000
6200	Steel, manual, 10′ x 10′ high	2,850	1,175	4,025
6300	15′ x 8′ high	3,200	1,475	4,675
6400	Motor operated, 10′ x 10′ high	4,225	1,400	5,625
6500	15′ x 8′ high	4,575	1,700	6,275
8970	Counter door, rolling, 6′ high, 14′ wide, aluminum	3,025	725	3,750

C1030 Fittings

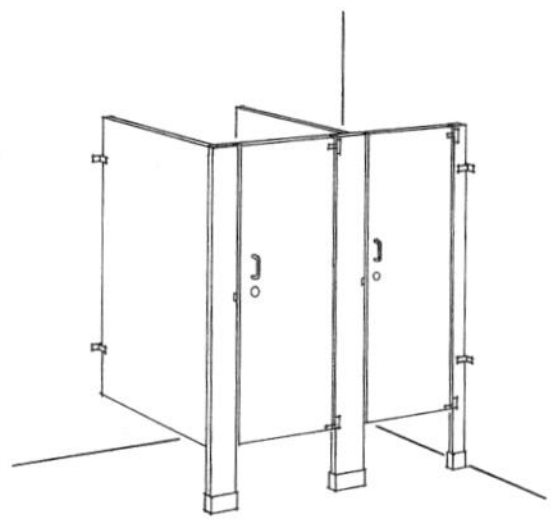

Toilet Units

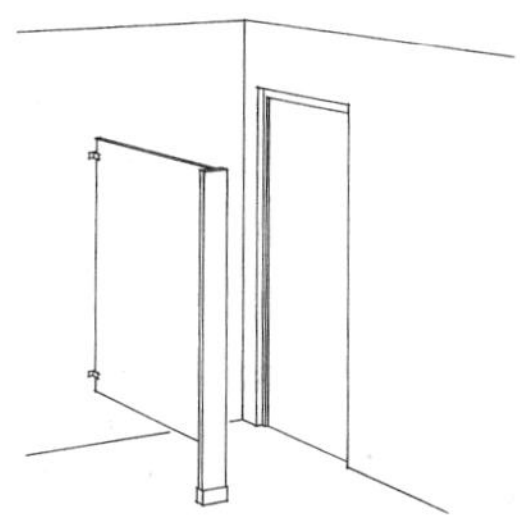

Entrance Screens

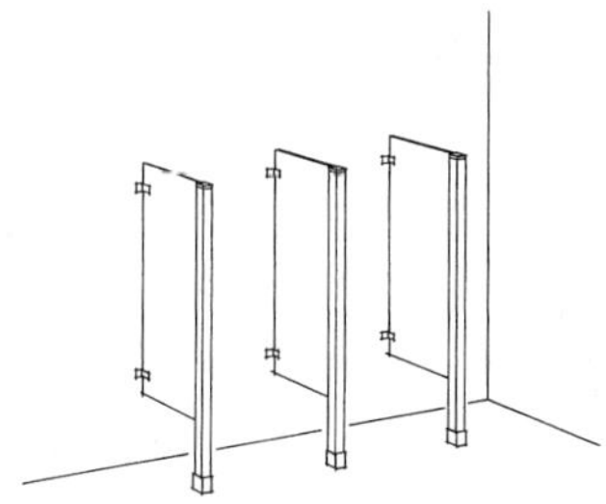

Urinal Screens

C1030 110	Toilet Partitions	COST PER UNIT		
		MAT.	INST.	TOTAL
0380	Toilet partitions, cubicles, ceiling hung, marble	1,975	480	2,455
0400	Painted metal	575	253	828
0420	Plastic laminate	580	253	833
0460	Stainless steel	1,225	253	1,478
0480	Handicap addition	495		495
0520	Floor and ceiling anchored, marble	2,175	380	2,555
0540	Painted metal	630	203	833
0560	Plastic laminate	905	203	1,108
0600	Stainless steel	1,350	203	1,553
0620	Handicap addition	405		405
0660	Floor mounted marble	1,350	320	1,670
0680	Painted metal	660	145	805
0700	Plastic laminate	585	145	730
0740	Stainless steel	1,575	145	1,720
0760	Handicap addition	370		370
0780	Juvenile deduction	45.50		45.50
0820	Floor mounted with headrail marble	1,275	320	1,595
0840	Painted metal	420	169	589
0860	Plastic laminate	845	169	1,014
0900	Stainless steel	1,125	169	1,294
0920	Handicap addition	410		410
0960	Wall hung, painted metal	700	145	845
1020	Stainless steel	1,825	145	1,970
1040	Handicap addition	410		410
1080	Entrance screens, floor mounted, 54″ high, marble	870	106	976
1100	Painted metal	266	67.50	333.50
1140	Stainless steel	1,000	67.50	1,067.50
1300	Urinal screens, floor mounted, 24″ wide, laminated plastic	224	127	351
1320	Marble	710	146	856
1340	Painted metal	253	127	380
1380	Stainless steel	640	127	767
1428	Wall mounted wedge type, painted metal	148	101	249
1460	Stainless steel	635	101	736
1500	Partitions, shower stall, single wall, painted steel, 2′-8″ x 2′-8″	1,050	226	1,276
1510	Fiberglass, 2′-8″ x 2′-8″	875	251	1,126
1520	Double wall, enameled steel, 2′-8″ x 2′-8″	1,125	226	1,351
1530	Stainless steel, 2′-8″ x 2′-8″	2,525	226	2,751
1560	Tub enclosure, sliding panels, tempered glass, aluminum frame	450	283	733
1570	Chrome/brass frame, clear glass	1,325	375	1,700

C1030 Fittings

C1030 310	Storage Specialties, EACH	COST EACH		
		MAT.	INST.	TOTAL
0200	Lockers, steel, single tier, 5′ to 6′ high, per opening, 1 wide	181	40.50	221.50
0210	3 wide	320	47	367
0600	Shelving, metal industrial, braced, 3′ wide, 1′ deep	23.50	10.05	33.55
0610	2′ deep	34	10.60	44.60

C1030 510	Identifying/Visual Aid Specialties, EACH	COST EACH		
		MAT.	INST.	TOTAL
0100	Control boards, magnetic, porcelain finish, framed, 24″ x 18″	213	127	340
0110	96″ x 48″	1,175	203	1,378
0120	Directory boards, outdoor, black plastic, 36″ x 24″	840	505	1,345
0130	36″ x 36″	970	675	1,645
0140	Indoor, economy, open faced, 18″ x 24″	183	145	328
0500	Signs, interior electric exit sign, wall mounted, 6″	102	72	174
0510	Street, reflective alum., dbl. face, 4 way, w/bracket	269	34	303
0520	Letters, cast aluminum, 1/2″ deep, 4″ high	27.50	28	55.50
0530	1″ deep, 10″ high	58	28	86
0540	Plaques, cast aluminum, 20″ x 30″	2,025	253	2,278
0550	Cast bronze, 36″ x 48″	5,125	505	5,630

C1030 520	Identifying/Visual Aid Specialties, S.F.	COST PER S.F.		
		MAT.	INST.	TOTAL
0100	Bulletin board, cork sheets, no frame, 1/4″ thick	1.52	3.49	5.01
0120	Aluminum frame, 1/4″ thick, 3′ x 5′	9.90	4.25	14.15
0200	Chalkboards, wall hung, alum, frame & chalktrough	14.90	2.25	17.15
0210	Wood frame & chalktrough	10.85	2.42	13.27
0220	Sliding board, one board with back panel	57	2.16	59.16
0230	Two boards with back panel	89	2.16	91.16
0240	Liquid chalk type, alum. frame & chalktrough	13.65	2.25	15.90
0250	Wood frame & chalktrough	33.50	2.25	35.75

C1030 710	Bath and Toilet Accessories, EACH	COST EACH		
		MAT.	INST.	TOTAL
0100	Specialties, bathroom accessories, st. steel, curtain rod, 5′ long, 1″ diam	31	39	70
0120	Dispenser, towel, surface mounted	52	31.50	83.50
0140	Grab bar, 1-1/4″ diam., 12″ long	33.50	21	54.50
0160	Mirror, framed with shelf, 18″ x 24″	218	25.50	243.50
0170	72″ x 24″	285	84.50	369.50
0180	Toilet tissue dispenser, surface mounted, single roll	21	16.90	37.90
0200	Towel bar, 18″ long	45.50	22	67.50
0300	Medicine cabinets, sliding mirror doors, 20″ x 16″ x 4-3/4″, unlighted	146	72.50	218.50
0310	24″ x 19″ x 8-1/2″, lighted	202	101	303

C1030 730	Bath and Toilet Accessories, L.F.	COST PER L.F.		
		MAT.	INST.	TOTAL
0100	Partitions, hospital curtain, ceiling hung, polyester oxford cloth	24.50	4.94	29.44
0110	Designer oxford cloth	17.20	6.25	23.45

C1030 830	Fabricated Cabinets, EACH	COST EACH		
		MAT.	INST.	TOTAL
0110	Household, base, hardwood, one top drawer & one door below x 12″ wide	298	41	339
0115	24″ wide	410	45.50	455
0120	Four drawer x 24″ wide	385	45.50	430.50
0130	Wall, hardwood, 30″ high with one door x 12″ wide	258	46	304
0140	Two doors x 48″ wide	565	55	620

C1030 Fittings

C1030 830	Fabricated Counters, L.F.	COST PER L.F.		
		MAT.	INST.	TOTAL
0150	Counter top-laminated plastic, stock, economy	19	16.90	35.90
0160	Custom-square edge, 7/8" thick	16.85	38	54.85
0170	School, counter, wood, 32" high	264	50.50	314.50
0180	Metal, 84" high	470	67.50	537.50

C1030 910	Other Fittings, EACH	COST EACH		
		MAT.	INST.	TOTAL
0500	Mail boxes, horizontal, rear loaded, aluminum, 5" x 6" x 15" deep	47	14.90	61.90
0510	Front loaded, aluminum, 10" x 12" x 15" deep	102	25.50	127.50
0520	Vertical, front loaded, aluminum, 15" x 5" x 6" deep	47	14.90	61.90
0530	Bronze, duranodic finish	53	14.90	67.90
0540	Letter slot, post office	131	63.50	194.50
0550	Mail counter, window, post office, with grille	605	253	858
0700	Turnstiles, one way, 4' arm, 46" diam., manual	2,300	203	2,503
0710	Electric	2,525	845	3,370
0720	3 arm, 5'-5" diam. & 7' high, manual	8,425	1,025	9,450
0730	Electric	10,200	1,700	11,900

C2010 Stair Construction

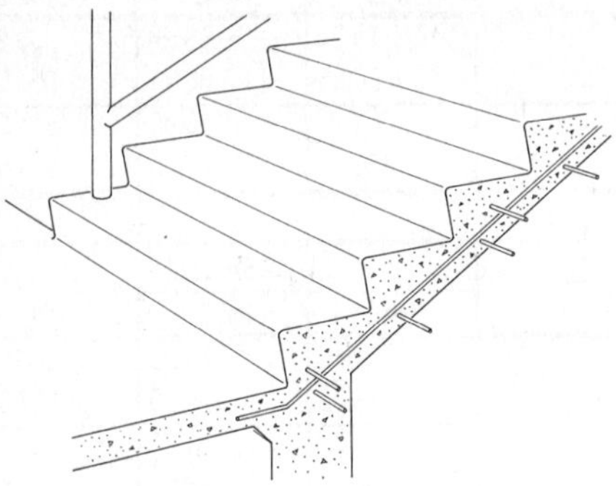

General Design: See reference section for code requirements. Maximum height between landings is 12′; usual stair angle is 20° to 50° with 30° to 35° best. Usual relation of riser to treads is:

Riser + tread = 17.5.
2x (Riser) + tread = 25.
Riser x tread = 70 or 75.

Maximum riser height is 7″ for commercial, 8-1/4″ for residential.
Usual riser height is 6-1/2″ to 7-1/4″.

Minimum tread width is 11″ for commercial and 9″ for residential.

For additional information please see reference section.

Cost Per Flight: Table below lists the cost per flight for 4′-0″ wide stairs.
Side walls are not included.
Railings are included.

C2010 110	Stairs	COST PER FLIGHT		
		MAT.	INST.	TOTAL
0470	Stairs, C.I.P. concrete, w/o landing, 12 risers, w/o nosing	1,125	2,075	3,200
0480	With nosing	2,100	2,250	4,350
0550	W/landing, 12 risers, w/o nosing	1,300	2,550	3,850
0560	With nosing	2,250	2,725	4,975
0570	16 risers, w/o nosing	1,575	3,200	4,775
0580	With nosing	2,875	3,425	6,300
0590	20 risers, w/o nosing	1,875	3,850	5,725
0600	With nosing	3,475	4,150	7,625
0610	24 risers, w/o nosing	2,150	4,500	6,650
0620	With nosing	4,100	4,850	8,950
0630	Steel, grate type w/nosing & rails, 12 risers, w/o landing	5,700	1,025	6,725
0640	With landing	7,975	1,400	9,375
0660	16 risers, with landing	9,875	1,725	11,600
0680	20 risers, with landing	11,800	2,050	13,850
0700	24 risers, with landing	13,700	2,400	16,100
0710	Concrete fill metal pan & picket rail, 12 risers, w/o landing	7,375	1,025	8,400
0720	With landing	9,725	1,525	11,250
0740	16 risers, with landing	12,200	1,875	14,075
0760	20 risers, with landing	14,600	2,200	16,800
0780	24 risers, with landing	17,100	2,525	19,625
0790	Cast iron tread & pipe rail, 12 risers, w/o landing	7,375	1,025	8,400
0800	With landing	9,725	1,525	11,250
1120	Wood, prefab box type, oak treads, wood rails 3′-6″ wide, 14 risers	2,250	425	2,675
1150	Prefab basement type, oak treads, wood rails 3′-0″ wide, 14 risers	1,100	104	1,204

C3010 Wall Finishes

C3010 230	Paint & Covering	COST PER S.F.		
		MAT.	INST.	TOTAL
0060	Painting, interior on plaster and drywall, brushwork, primer & 1 coat	.14	.65	.79
0080	Primer & 2 coats	.21	.86	1.07
0100	Primer & 3 coats	.29	1.05	1.34
0120	Walls & ceilings, roller work, primer & 1 coat	.14	.43	.57
0140	Primer & 2 coats	.21	.55	.76
0160	Woodwork incl. puttying, brushwork, primer & 1 coat	.14	.93	1.07
0180	Primer & 2 coats	.21	1.24	1.45
0200	Primer & 3 coats	.29	1.68	1.97
0260	Cabinets and casework, enamel, primer & 1 coat	.14	1.05	1.19
0280	Primer & 2 coats	.22	1.29	1.51
0300	Masonry or concrete, latex, brushwork, primer & 1 coat	.31	.88	1.19
0320	Primer & 2 coats	.40	1.25	1.65
0340	Addition for block filler	.23	1.08	1.31
0380	Fireproof paints, intumescent, 1/8" thick 3/4 hour	2.43	.86	3.29
0400	3/16" thick 1 hour	5.40	1.29	6.69
0420	7/16" thick 2 hour	7	3	10
0440	1-1/16" thick 3 hour	11.40	6	17.40
0500	Gratings, primer & 1 coat	.33	1.31	1.64
0600	Pipes over 12" diameter	.84	4.20	5.04
0700	Structural steel, brushwork, light framing 300-500 S.F./Ton	.11	1.51	1.62
0720	Heavy framing 50-100 S.F./Ton	.11	.76	.87
0740	Spraywork, light framing 300-500 S.F./Ton	.12	.34	.46
0760	Heavy framing 50-100 S.F./Ton	.12	.37	.49
0800	Varnish, interior wood trim, no sanding sealer & 1 coat	.08	1.05	1.13
0820	Hardwood floor, no sanding 2 coats	.17	.22	.39
0840	Wall coatings, acrylic glazed coatings, minimum	.34	.80	1.14
0860	Maximum	.72	1.38	2.10
0880	Epoxy coatings, solvent based	.44	.80	1.24
0900	Water based	.30	2.47	2.77
0940	Exposed epoxy aggregate, troweled on, 1/16" to 1/4" aggregate, topping mix	.68	1.79	2.47
0960	Integral mix	1.46	3.23	4.69
0980	1/2" to 5/8" aggregate, topping mix	1.32	3.23	4.55
1000	Integral mix	2.29	5.25	7.54
1020	1" aggregate, topping mix	2.33	4.67	7
1040	Integral mix	3.55	7.65	11.20
1060	Sprayed on, topping mix	.63	1.43	2.06
1080	Water based	1.16	2.90	4.06
1100	High build epoxy 50 mil, solvent based	.75	1.08	1.83
1120	Water based	1.27	4.43	5.70
1140	Laminated epoxy with fiberglass solvent based	.80	1.43	2.23
1160	Water based	1.45	2.90	4.35
1180	Sprayed perlite or vermiculite 1/16" thick, solvent based	.30	.14	.44
1200	Water based	.81	.66	1.47
1260	Wall coatings, vinyl plastic, solvent based	.36	.57	.93
1280	Water based	.90	1.75	2.65
1300	Urethane on smooth surface, 2 coats, solvent based	.34	.37	.71
1320	Water based	.65	.63	1.28
1340	3 coats, solvent based	.48	.50	.98
1360	Water based	.87	.89	1.76
1380	Ceramic-like glazed coating, cementitious, solvent based	.53	.96	1.49
1400	Water based	.89	1.22	2.11
1420	Resin base, solvent based	.42	.66	1.08
1440	Water based	.64	1.27	1.91
1460	Wall coverings, aluminum foil	1.14	1.54	2.68
1480	Copper sheets, .025" thick, phenolic backing	7.95	1.76	9.71
1500	Vinyl backing	6.10	1.76	7.86
1520	Cork tiles, 12"x12", light or dark, 3/16" thick	5.15	1.76	6.91
1540	5/16" thick	3.44	1.80	5.24

C3010 Wall Finishes

C3010 230	Paint & Covering	COST PER S.F.		
		MAT.	INST.	TOTAL
1560	Basketweave, 1/4" thick	3.85	1.76	5.61
1580	Natural, non-directional, 1/2" thick	7.20	1.76	8.96
1600	12"x36", granular, 3/16" thick	1.44	1.10	2.54
1620	1" thick	1.85	1.14	2.99
1640	12"x12", polyurethane coated, 3/16" thick	4.49	1.76	6.25
1660	5/16" thick	7.25	1.80	9.05
1661	Paneling, prefinished plywood, birch	1.32	2.41	3.73
1662	Mahogany, African	3.08	2.53	5.61
1663	Philippine (lauan)	.72	2.03	2.75
1664	Oak or cherry	2.31	2.53	4.84
1665	Rosewood	3.25	3.17	6.42
1666	Teak	3.25	2.53	5.78
1667	Chestnut	5.35	2.70	8.05
1668	Pecan	2.81	2.53	5.34
1669	Walnut	5.80	2.53	8.33
1670	Wood board, knotty pine, finished	2.49	4.18	6.67
1671	Rough sawn cedar	3.86	4.18	8.04
1672	Redwood	5.80	4.18	9.98
1673	Aromatic cedar	2.98	4.48	7.46
1680	Cork wallpaper, paper backed, natural	2.20	.88	3.08
1700	Color	3.16	.88	4.04
1720	Gypsum based, fabric backed, minimum	.85	.53	1.38
1740	Average	1.40	.59	1.99
1760	Small quantities	.77	.66	1.43
1780	Vinyl wall covering, fabric back, light weight	1.07	.66	1.73
1800	Medium weight	.99	.88	1.87
1820	Heavy weight	1.51	.97	2.48
1840	Wall paper, double roll, solid pattern, avg. workmanship	.65	.66	1.31
1860	Basic pattern, avg. workmanship	1.24	.79	2.03
1880	Basic pattern, quality workmanship	1.89	.97	2.86
1900	Grass cloths with lining paper, minimum	1.06	1.06	2.12
1920	Maximum	3.14	1.21	4.35
1940	Ceramic tile, thin set, 4-1/4" x 4-1/4"	2.74	4.22	6.96
1960	12" x 12"	5	5	10

C3010 235	Paint Trim	COST PER L.F.		
		MAT.	INST.	TOTAL
2040	Painting, wood trim, to 6" wide, enamel, primer & 1 coat	.14	.53	.67
2060	Primer & 2 coats	.22	.67	.89
2080	Misc. metal brushwork, ladders	.66	5.25	5.91
2100	Pipes, to 4" dia.	.11	1.11	1.22
2120	6" to 8" dia.	.22	2.21	2.43
2140	10" to 12" dia.	.66	3.31	3.97
2160	Railings, 2" pipe	.23	2.63	2.86
2180	Handrail, single	.18	1.05	1.23
2185	Caulking & Sealants, Polyurethane, In place, 1 or 2 component, 1/2" X 1/4"	.38	1.67	2.05

C3020 Floor Finishes

C3020 410	Tile & Covering	COST PER S.F. MAT.	COST PER S.F. INST.	COST PER S.F. TOTAL
0060	Carpet tile, nylon, fusion bonded, 18" x 18" or 24" x 24", 24 oz.	3.72	.63	4.35
0080	35 oz.	4.33	.63	4.96
0100	42 oz.	5.65	.63	6.28
0140	Carpet, tufted, nylon, roll goods, 12' wide, 26 oz.	2.61	.67	3.28
0160	36 oz.	4.27	.67	4.94
0180	Woven, wool, 36 oz.	11.10	.72	11.82
0200	42 oz.	12.20	.72	12.92
0220	Padding, add to above, 2.7 density	.73	.33	1.06
0240	13.0 density	.98	.33	1.31
0260	Composition flooring, acrylic, 1/4" thick	1.83	4.83	6.66
0280	3/8" thick	2.31	5.60	7.91
0300	Epoxy, 3/8" thick	3.09	3.72	6.81
0320	1/2" thick	4.46	5.15	9.61
0340	Epoxy terrazzo, granite chips	6.60	5.85	12.45
0360	Recycled porcelain	10.05	7.75	17.80
0380	Mastic, hot laid, 1-1/2" thick, minimum	4.57	3.64	8.21
0400	Maximum	5.85	4.83	10.68
0420	Neoprene 1/4" thick, minimum	4.50	4.62	9.12
0440	Maximum	6.15	5.85	12
0460	Polyacrylate with ground granite 1/4", granite chips	3.87	3.42	7.29
0480	Recycled porcelain	7.10	5.25	12.35
0500	Polyester with colored quartz chips 1/16", minimum	3.50	2.37	5.87
0520	Maximum	5.25	3.72	8.97
0540	Polyurethane with vinyl chips, clear	8.35	2.37	10.72
0560	Pigmented	12.15	2.92	15.07
0600	Concrete topping, granolithic concrete, 1/2" thick	.36	4.07	4.43
0620	1" thick	.71	4.17	4.88
0640	2" thick	1.43	4.80	6.23
0660	Heavy duty 3/4" thick, minimum	.48	6.30	6.78
0680	Maximum	.89	7.50	8.39
0700	For colors, add to above, minimum	.52	1.45	1.97
0720	Maximum	.86	1.60	2.46
0740	Exposed aggregate finish, minimum	.22	.74	.96
0760	Maximum	.37	.99	1.36
0780	Abrasives, .25 P.S.F. add to above, minimum	.58	.54	1.12
0800	Maximum	.87	.54	1.41
0820	Dust on coloring, add, minimum	.52	.36	.88
0840	Maximum	.86	.74	1.60
0860	Floor coloring using 0.6 psf powdered color, 1/2" integral, minimum	5.75	4.07	9.82
0880	Maximum	6.10	4.07	10.17
0900	Dustproofing, add, minimum	.21	.24	.45
0920	Maximum	.74	.36	1.10
0930	Paint	.40	1.25	1.65
0940	Hardeners, metallic add, minimum	.66	.54	1.20
0960	Maximum	1.98	.80	2.78
0980	Non-metallic, minimum	.17	.54	.71
1000	Maximum	.51	.80	1.31
1020	Integral topping and finish, 1:1:2 mix, 3/16" thick	.12	2.40	2.52
1040	1/2" thick	.32	2.53	2.85
1060	3/4" thick	.48	2.82	3.30
1080	1" thick	.64	3.20	3.84
1100	Terrazzo, minimum	3.69	15.60	19.29
1120	Maximum	6.80	19.50	26.30
1340	Cork tile, minimum	7.65	1.43	9.08
1360	Maximum	13.35	1.43	14.78
1380	Polyethylene, in rolls, minimum	4.58	1.64	6.22
1400	Maximum	7.50	1.64	9.14
1420	Polyurethane, thermoset, minimum	5.95	4.50	10.45

C3020 Floor Finishes

C3020 410	Tile & Covering	COST PER S.F. MAT.	INST.	TOTAL
1440	Maximum	7.45	9	16.45
1460	Rubber, sheet goods, minimum	7.80	3.75	11.55
1480	Maximum	13.55	5	18.55
1500	Tile, minimum	6.30	1.13	7.43
1520	Maximum	11.20	1.64	12.84
1540	Synthetic turf, minimum	5.15	2.14	7.29
1560	1/2" thick	5.55	2.37	7.92
1580	Vinyl, composition tile, minimum	1.31	.90	2.21
1600	Maximum	1.71	.90	2.61
1620	Vinyl tile, 3/32"	4.08	.90	4.98
1640	Maximum	3.58	.90	4.48
1660	Sheet goods, plain pattern/colors	4.06	1.80	5.86
1680	Intricate pattern/colors	8.25	2.25	10.50
1720	Tile, ceramic natural clay	6.35	4.38	10.73
1730	Marble, synthetic 12"x12"x5/8"	11.35	13.35	24.70
1740	Porcelain type, minimum	5.70	4.38	10.08
1760	Maximum	7.60	5.15	12.75
1800	Quarry tile, mud set, minimum	8.75	5.75	14.50
1820	Maximum	9.55	7.30	16.85
1840	Thin set, deduct		1.15	1.15
1850	Tile, natural stone, marble, in mortar bed, 12" x 12" x 3/8" thick	17.60	21	38.60
1860	Terrazzo precast, minimum	5.30	6	11.30
1880	Maximum	10.25	6	16.25
1900	Non-slip, minimum	22.50	14.60	37.10
1920	Maximum	22.50	19.50	42
2020	Wood block, end grain factory type, natural finish, 2" thick	4.94	4.05	8.99
2040	Fir, vertical grain, 1"x4", no finish, minimum	2.89	1.99	4.88
2060	Maximum	3.07	1.99	5.06
2080	Prefinished white oak, prime grade, 2-1/4" wide	5.30	2.98	8.28
2100	3-1/4" wide	5.75	2.74	8.49
2120	Maple strip, sanded and finished, minimum	5.85	4.31	10.16
2140	Maximum	6	4.31	10.31
2160	Oak strip, sanded and finished, minimum	4.65	4.31	8.96
2180	Maximum	5.50	4.31	9.81
2200	Parquetry, sanded and finished, plain pattern	6.95	4.50	11.45
2220	Intricate pattern	11.65	6.40	18.05
2260	Add for sleepers on concrete, treated, 24" O.C., 1"x2"	1.80	2.60	4.40
2280	1"x3"	1.95	2.03	3.98
2300	2"x4"	1	1.03	2.03
2340	Underlayment, plywood, 3/8" thick	1.09	.68	1.77
2350	1/2" thick	1.29	.70	1.99
2360	5/8" thick	1.42	.72	2.14
2370	3/4" thick	1.63	.78	2.41
2380	Particle board, 3/8" thick	.45	.68	1.13
2390	1/2" thick	.50	.70	1.20
2400	5/8" thick	.59	.72	1.31
2410	3/4" thick	.75	.78	1.53
2420	Hardboard, 4' x 4', .215" thick	.69	.68	1.37
9200	Vinyl, composition tile, 12" x 12" x 1/8" thick, recycled content	2.43	.90	3.33

C3030 Ceiling Finishes

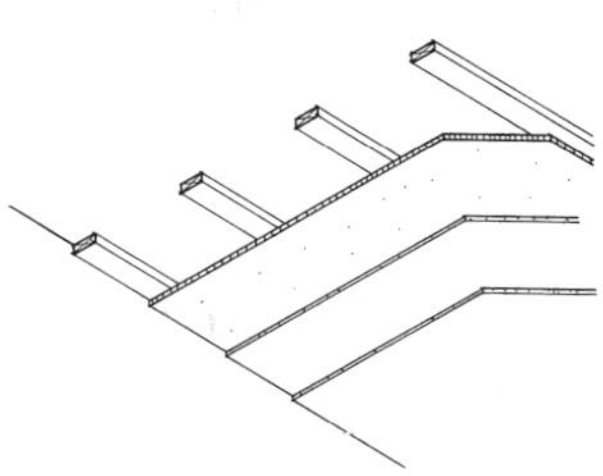

2 Coats of Plaster on Gypsum Lath on Wood Furring

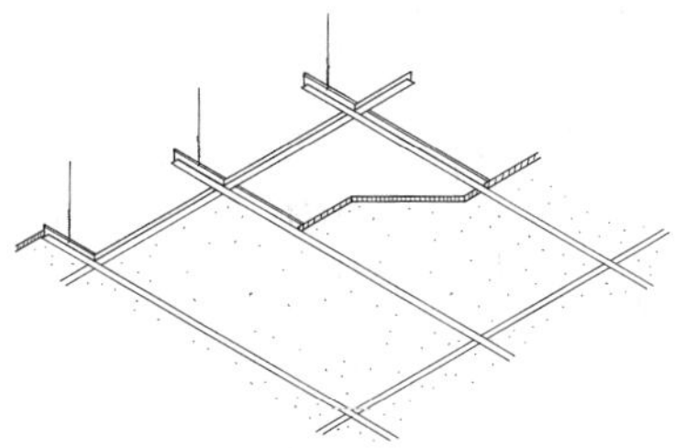

Fiberglass Board on Exposed Suspended Grid System

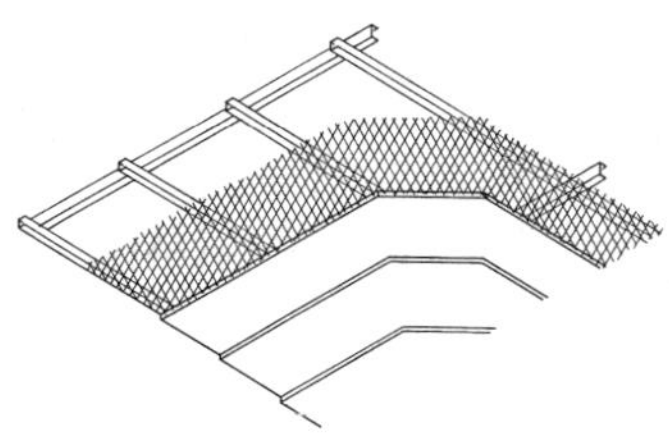

Plaster and Metal Lath on Metal Furring

C3030 105 Plaster Ceilings

	TYPE	LATH	FURRING	SUPPORT		COST PER S.F.		
						MAT.	INST.	TOTAL
2400	2 coat gypsum	3/8″ gypsum	1″x3″ wood, 16″ O.C.	wood		1.28	5.20	6.48
2500	Painted			masonry		1.31	5.30	6.61
2600				concrete		1.31	5.90	7.21
2700	3 coat gypsum	3.4# metal	1″x3″ wood, 16″ O.C.	wood		1.64	5.50	7.14
2800	Painted			masonry		1.67	5.60	7.27
2900				concrete		1.67	6.25	7.92
3000	2 coat perlite	3/8″ gypsum	1″x3″ wood, 16″ O.C.	wood		1.55	5.35	6.90
3100	Painted			masonry		1.58	5.45	7.03
3200				concrete		1.58	6.05	7.63
3300	3 coat perlite	3.4# metal	1″x3″ wood, 16″ O.C.	wood		1.71	5.45	7.16
3400	Painted			masonry		1.74	5.55	7.29
3500				concrete		1.74	6.20	7.94
3600	2 coat gypsum	3/8″ gypsum	3/4″ CRC, 12″ O.C.	1-1/2″ CRC, 48″O.C.		1.38	6.40	7.78
3700	Painted		3/4″ CRC, 16″ O.C.	1-1/2″ CRC, 48″O.C.		1.34	5.75	7.09
3800			3/4″ CRC, 24″ O.C.	1-1/2″ CRC, 48″O.C.		1.21	5.25	6.46
3900	2 coat perlite	3/8″ gypsum	3/4″ CRC, 12″ O.C.	1-1/2″ CRC, 48″O.C		1.65	6.75	8.40
4000	Painted		3/4″ CRC, 16″ O.C.	1-1/2″ CRC, 48″O.C.		1.61	6.15	7.76
4100			3/4″ CRC, 24″ O.C.	1-1/2″ CRC, 48″O.C.		1.48	5.65	7.13
4200	3 coat gypsum	3.4# metal	3/4″ CRC, 12″ O.C.	1-1/2″ CRC, 36″ O.C.		1.91	8.20	10.11
4300	Painted		3/4″ CRC, 16″ O.C.	1-1/2″ CRC, 36″ O.C.		1.87	7.60	9.47
4400			3/4″ CRC, 24″ O.C.	1-1/2″ CRC, 36″ O.C.		1.74	7.10	8.84
4500	3 coat perlite	3.4# metal	3/4″ CRC, 12″ O.C.	1-1/2″ CRC,36″ O.C.		2.04	9	11.04
4600	Painted		3/4″ CRC, 16″ O.C.	1-1/2″ CRC, 36″ O.C.		2	8.35	10.35
4700			3/4″ CRC, 24″ O.C.	1-1/2″ CRC, 36″ O.C.		1.87	7.85	9.72

C3030 110 Drywall Ceilings

	TYPE	FINISH	FURRING	SUPPORT		COST PER S.F.		
						MAT.	INST.	TOTAL
4800	1/2″ F.R. drywall	painted and textured	1″x3″ wood, 16″ O.C.	wood		.95	3.17	4.12
4900				masonry		.98	3.27	4.25
5000				concrete		.98	3.89	4.87
5100	5/8″ F.R. drywall	painted and textured	1″x3″ wood, 16″ O.C.	wood		.94	3.17	4.11
5200				masonry		.97	3.27	4.24
5300				concrete		.97	3.89	4.86

C3030 Ceiling Finishes

C3030 110 Drywall Ceilings

	TYPE	FINISH	FURRING	SUPPORT		COST PER S.F. MAT.	INST.	TOTAL
5400	1/2" F.R. drywall	painted and textured	7/8" resil. channels	24" O.C.		.77	3.08	3.85
5500			1"x2" wood	stud clips		.95	2.97	3.92
5602		painted	1-5/8" metal studs	24" O.C.		.83	2.75	3.58
5700	5/8" F.R. drywall	painted and textured	1-5/8"metal studs	24" O.C.		.82	2.75	3.57
5702								

C3030 140 Plaster Ceiling Components

		COST PER S.F. MAT.	INST.	TOTAL
0060	Plaster, gypsum incl. finish			
0080	3 coats	.65	3.27	3.92
0100	Perlite, incl. finish, 2 coats	.72	3.22	3.94
0120	3 coats	.78	4.04	4.82
0140	Thin coat on drywall	.12	.64	.76
0200	Lath, gypsum, 3/8" thick	.37	.87	1.24
0220	1/2" thick	.30	.91	1.21
0240	5/8" thick	.36	1.07	1.43
0260	Metal, diamond, 2.5 lb.	.50	.70	1.20
0280	3.4 lb.	.53	.75	1.28
0300	Flat rib, 2.75 lb.	.43	.70	1.13
0320	3.4 lb.	.55	.75	1.30
0440	Furring, steel channels, 3/4" galvanized , 12" O.C.	.42	2.27	2.69
0460	16" O.C.	.38	1.65	2.03
0480	24" O.C.	.25	1.14	1.39
0500	1-1/2" galvanized , 12" O.C.	.57	2.51	3.08
0520	16" O.C.	.51	1.84	2.35
0540	24" O.C.	.34	1.22	1.56
0560	Wood strips, 1"x3", on wood, 12" O.C.	.43	1.45	1.88
0580	16" O.C.	.32	1.09	1.41
0600	24" O.C.	.22	.73	.95
0620	On masonry, 12" O.C.	.46	1.58	2.04
0640	16" O.C.	.35	1.19	1.54
0660	24" O.C.	.23	.79	1.02
0680	On concrete, 12" O.C.	.46	2.41	2.87
0700	16" O.C.	.35	1.81	2.16
0720	24" O.C.	.23	1.21	1.44
0722				
0740				
0940	Paint on plaster or drywall, roller work, primer + 1 coat	.14	.43	.57
0960	Primer + 2 coats	.21	.55	.76

C3030 210 Acoustical Ceilings

	TYPE	TILE	GRID	SUPPORT		COST PER S.F. MAT.	INST.	TOTAL
5800	5/8" fiberglass board	24" x 48"	tee	suspended		2.45	1.52	3.97
5900		24" x 24"	tee	suspended		2.70	1.67	4.37
6000	3/4" fiberglass board	24" x 48"	tee	suspended		3.96	1.55	5.51
6100		24" x 24"	tee	suspended		4.21	1.70	5.91
6500	5/8" mineral fiber	12" x 12"	1"x3" wood, 12" O.C.	wood		2.88	3.14	6.02
6600				masonry		2.91	3.27	6.18
6700				concrete		2.91	4.10	7.01
6800	3/4" mineral fiber	12" x 12"	1"x3" wood, 12" O.C.	wood		3.65	3.14	6.79
6900				masonry		3.65	3.14	6.79
7000				concrete		3.65	3.14	6.79

C3030 Ceiling Finishes

C3030 210 Acoustical Ceilings

	TYPE	TILE	GRID	SUPPORT		COST PER S.F.		
						MAT.	INST.	TOTAL
7100 7102	3/4"mineral fiber on 5/8" F.R. drywall	12" x 12"	25 ga. channels	runners		4.10	3.93	8.03
7200 7201 7202	5/8" plastic coated Mineral fiber	12" x 12"		adhesive backed		2.89	1.69	4.58
7300 7301 7302	3/4" plastic coated Mineral fiber	12" x 12"		adhesive backed		3.66	1.69	5.35
7400 7401 7402	3/4" mineral fiber	12" x 12"	conceal 2" bar & channels	suspended		2.75	3.82	6.57

C3030 240 Acoustical Ceiling Components

		COST PER S.F.		
		MAT.	INST.	TOTAL
2480	Ceiling boards, eggcrate, acrylic, 1/2" x 1/2" x 1/2" cubes	1.97	1.01	2.98
2500	Polystyrene, 3/8" x 3/8" x 1/2" cubes	1.73	.99	2.72
2520	1/2" x 1/2" x 1/2" cubes	2	1.01	3.01
2540	Fiberglass boards, plain, 5/8" thick	1.38	.81	2.19
2560	3/4" thick	2.89	.84	3.73
2580	Grass cloth faced, 3/4" thick	2.99	1.01	4
2600	1" thick	3.64	1.04	4.68
2620	Luminous panels, prismatic, acrylic	2.90	1.27	4.17
2640	Polystyrene	1.78	1.27	3.05
2660	Flat or ribbed, acrylic	4.37	1.27	5.64
2680	Polystyrene	2.52	1.27	3.79
2700	Drop pan, white, acrylic	6.25	1.27	7.52
2720	Polystyrene	4.90	1.27	6.17
2740	Mineral fiber boards, 5/8" thick, standard	.75	.75	1.50
2760	Plastic faced	2.76	1.27	4.03
2780	2 hour rating	1.25	.75	2
2800	Perforated aluminum sheets, .024 thick, corrugated painted	2.71	1.03	3.74
2820	Plain	5.05	1.01	6.06
3080	Mineral fiber, plastic coated, 12" x 12" or 12" x 24", 5/8" thick	2.45	1.69	4.14
3100	3/4" thick	3.22	1.69	4.91
3120	Fire rated, 3/4" thick, plain faced	1.44	1.69	3.13
3140	Mylar faced	2.12	1.69	3.81
3160	Add for ceiling primer	.14		.14
3180	Add for ceiling cement	.44		.44
3240	Suspension system, furring, 1" x 3" wood 12" O.C.	.43	1.45	1.88
3260	T bar suspension system, 2' x 4' grid	.84	.63	1.47
3280	2' x 2' grid	1.09	.78	1.87
3300	Concealed Z bar suspension system 12" module	.95	.97	1.92
3320	Add to above for 1-1/2" carrier channels 4' O.C.	.13	1.08	1.21
3340	Add to above for carrier channels for recessed lighting	.24	1.10	1.34

D1010 Elevators and Lifts

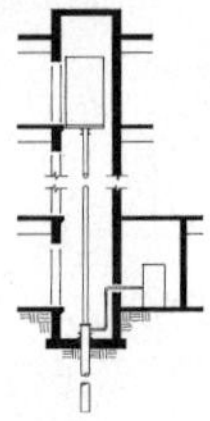

The hydraulic elevator obtains its motion from the movement of liquid under pressure in the piston connected to the car bottom. These pistons can provide travel to a maximum rise of 70′ and are sized for the intended load. As the rise reaches the upper limits the cost tends to exceed that of a geared electric unit.

D1010 110	Hydraulic	COST EACH		
		MAT.	INST.	TOTAL
1300	Pass. elev., 1500 lb., 2 Floors, 100 FPM	48,000	15,900	63,900
1400	5 Floors, 100 FPM	84,500	42,300	126,800
1600	2000 lb., 2 Floors, 100 FPM	49,000	15,900	64,900
1700	5 floors, 100 FPM	85,500	42,300	127,800
1900	2500 lb., 2 Floors, 100 FPM	51,000	15,900	66,900
2000	5 floors, 100 FPM	88,000	42,300	130,300
2200	3000 lb., 2 Floors, 100 FPM	52,500	15,900	68,400
2300	5 floors, 100 FPM	89,000	42,300	131,300
2500	3500 lb., 2 Floors, 100 FPM	56,000	15,900	71,900
2600	5 floors, 100 FPM	92,500	42,300	134,800
2800	4000 lb., 2 Floors, 100 FPM	57,500	15,900	73,400
2900	5 floors, 100 FPM	94,000	42,300	136,300
3100	4500 lb., 2 Floors, 100 FPM	60,500	15,900	76,400
3200	5 floors, 100 FPM	97,000	42,300	139,300
4000	Hospital elevators, 3500 lb., 2 Floors, 100 FPM	82,000	15,900	97,900
4100	5 floors, 100 FPM	131,500	42,300	173,800
4300	4000 lb., 2 Floors, 100 FPM	82,000	15,900	97,900
4400	5 floors, 100 FPM	131,500	42,300	173,800
4600	4500 lb., 2 Floors, 100 FPM	90,000	15,900	105,900
4800	5 floors, 100 FPM	139,500	42,300	181,800
4900	5000 lb., 2 Floors, 100 FPM	93,500	15,900	109,400
5000	5 floors, 100 FPM	143,000	42,300	185,300
6700	Freight elevators (Class "B"), 3000 lb., 2 Floors, 50 FPM	120,500	20,200	140,700
6800	5 floors, 100 FPM	193,000	53,000	246,000
7000	4000 lb., 2 Floors, 50 FPM	125,500	20,200	145,700
7100	5 floors, 100 FPM	198,000	53,000	251,000
7500	10,000 lb., 2 Floors, 50 FPM	157,000	20,200	177,200
7600	5 floors, 100 FPM	229,500	53,000	282,500
8100	20,000 lb., 2 Floors, 50 FPM	204,500	20,200	224,700
8200	5 Floors, 100 FPM	277,000	53,000	330,000

D2010 Plumbing Fixtures

Minimum Plumbing Fixture Requirements

Classification	Occupancy	Description	Water		Lavatories		Bathtubs/Showers	Drinking Fountains	Other
			Male	Female	Male	Female			
Assembly	A-1	Theaters and other buildings for the performing arts and motion pictures	1:125	1:65	1:200			1:500	1 Service Sink
	A-2	Nightclubs, bars, taverns dance halls	1:40		1:75			1:500	1 Service Sink
		Restaurants, banquet halls, food courts	1:75		1:200			1:500	1 Service Sink
	A-3	Auditorium w/o permanent seating, art galleries, exhibition halls, museums, lecture halls, libraries, arcades & gymnasiums	1:125	1:65	1:200			1:500	1 Service Sink
		Passenger terminals and transportation facilities	1:500		1:750			1:1000	1 Service Sink
		Places of worship and other religious services	1:150	1:75	1:200			1:1000	1 Service Sink
	A-4	Indoor sporting events and activities, coliseums, arenas, skating rinks, pools, and tennis courts	1:75 for the first 1500, then 1:120 for the remainder	1:40 for the first 1520, then 1:60 for the remainder	1:200	1:150		1:1000	1 Service Sink
	A-5	Outdoor sporting events and activities, stadiums, amusement parks, bleachers, grandstands	1:75 for the first 1500, then 1:120 for the remainder	1:40 for the first 1520, then 1:60 for the remainder	1:200	1:150		1:1000	1 Service Sink
Business	B	Buildings for the transaction of business, professional services, other services involving merchandise, office buildings, banks, light industrial	1:25 for the first 50, then 1:50 for the remainder		1:40 for the first 80, then 1:80 for the remainder			1:100	1 Service Sink
Educational	E	Educational Facilities	1:50		1:50			1:100	1 Service Sink
Factory and industrial	F-1 and F-2	Structures in which occupants are engaged in work fabricating, assembly or processing of products or materials	1:100		1:100		See *International Plumbing Code*	1:400	1 Service Sink
Institutional	I-1	Residential Care	1:10		1:10		1:8	1:100	1 Service Sink
	I-2	Hospitals, ambulatory nursing home care recipient	1 per room		1 per room		1:15	1:100	1 Service Sink
		Employees, other than residential care	1:25		1:35			1:100	
		Visitors, other than residential care	1:75		1:100			1:500	
	I-3	Prisons	1 per cell		1 per cell		1:15	1:100	1 Service Sink
		Reformatories, detention and correction centers	1:15		1:15		1:15	1:100	1 Service Sink
		Employees	1:25		1:35			1:100	
	I-4	Adult and child day care	1:15		1:15		1	1:100	1 Service Sink
Mercantile	M	Retail stores, service stations, shops, salesrooms, markets and shopping centers	1:500		1:750			1:1000	1 Service Sink
Residential	R-1	Hotels, Motels, boarding houses (transient)	1 per sleeping unit		1 per sleeping unit		1 per sleeping unit		1 Service Sink
	R-2	Dormitories, fraternities, sororities and boarding houses (not transient)	1:10		1:10		1:8	1:100	1 Service Sink
		Apartment House	1 per dwelling unit		1 per dwelling unit		1 per dwelling unit		1 Kitchen sink per dwelling; 1 clothes washer connection per 20 dwellings
	R-3	1 and 2 Family Dwellings	1 per dwelling unit		1:10		1 per dwelling unit		1 Kitchen sink per dwelling; 1 clothes washer connection per dwelling
	R-3	Congregate living facilities w/<16 people	1:10		1:10		1:8	1:100	1 Service Sink
	R-4	Congregate living facilities w/<16 people	1:10		1:10		1:8	1:100	1 Service Sink
Storage	S-1 and S-2	Structures for the storage of good, warehouses, storehouses and freight depots, low and moderate hazard	1:100		1:100		See *International Plumbing Code*	1;1000	1 Service Sink

Table 2902.1

Excerpted from the 2012 *International Building Code*, Copyright 2011. Washington, D.C.: International Code Council. Reproduced with permission. www.ICCSAFE.org

D2010 Plumbing Fixtures

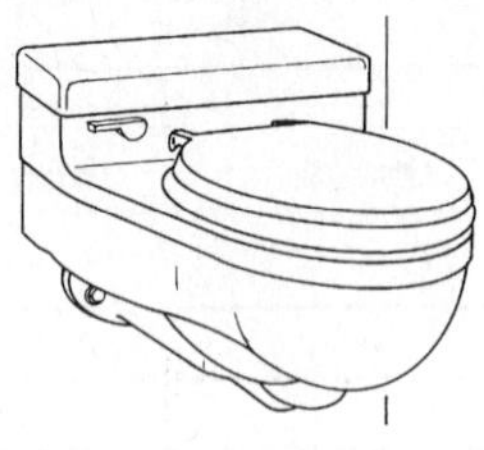

One Piece Wall Hung Water Closet

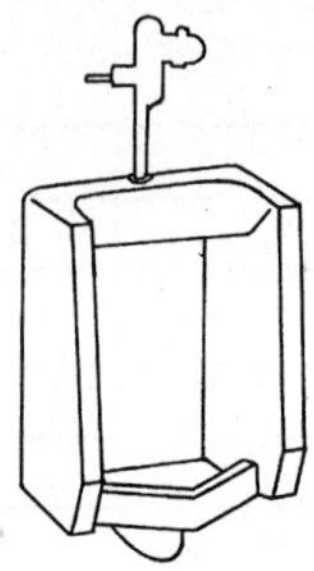

Wall Hung Urinal

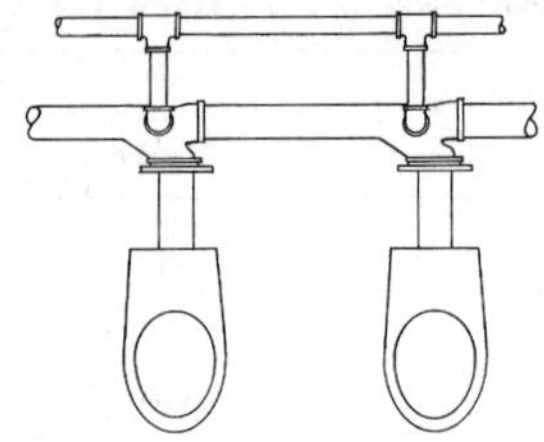

Side By Side Water Closet Group

Floor Mount Water Closet

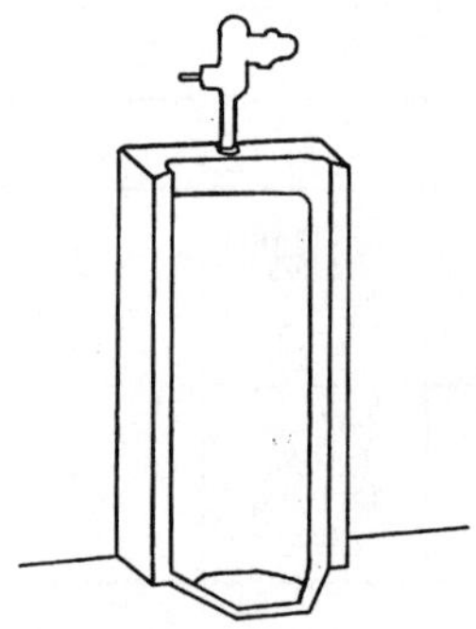

Stall Type Urinal

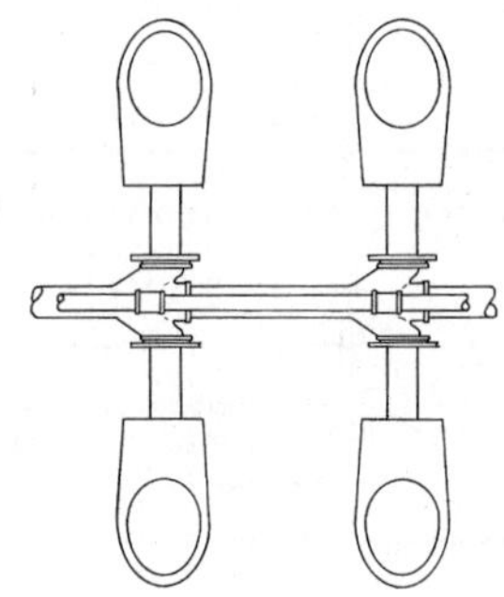

Back to Back Water Closet Group

D2010 110	Water Closet Systems	COST EACH		
		MAT.	INST.	TOTAL
1800	Water closet, vitreous china			
1840	Tank type, wall hung			
1880	Close coupled two piece	1,625	615	2,240
1920	Floor mount, one piece	1,225	655	1,880
1960	One piece low profile	1,200	655	1,855
2000	Two piece close coupled	660	655	1,315
2040	Bowl only with flush valve			
2080	Wall hung	2,225	695	2,920
2120	Floor mount	790	665	1,455
2122				
2160	Floor mount, ADA compliant with 18" high bowl	785	680	1,465

D2010 120	Water Closets, Group	COST EACH		
		MAT.	INST.	TOTAL
1760	Water closets, battery mount, wall hung, side by side, first closet	2,325	715	3,040
1800	Each additional water closet, add	2,225	675	2,900
3000	Back to back, first pair of closets	4,350	950	5,300
3100	Each additional pair of closets, back to back	4,275	935	5,210
9000	Back to back, first pair of closets, auto sensor flush valve, 1.28 gpf	4,875	1,025	5,900
9100	Ea additional pair of cls, back to back, auto sensor flush valve, 1.28 gpf	4,725	965	5,690

D2010 210	Urinal Systems	COST EACH		
		MAT.	INST.	TOTAL
2000	Urinal, vitreous china, wall hung	555	685	1,240
2040	Stall type	1,225	820	2,045

D2010 Plumbing Fixtures

Systems are complete with trim and rough-in (supply, waste and vent) to connect to supply branches and waste mains.

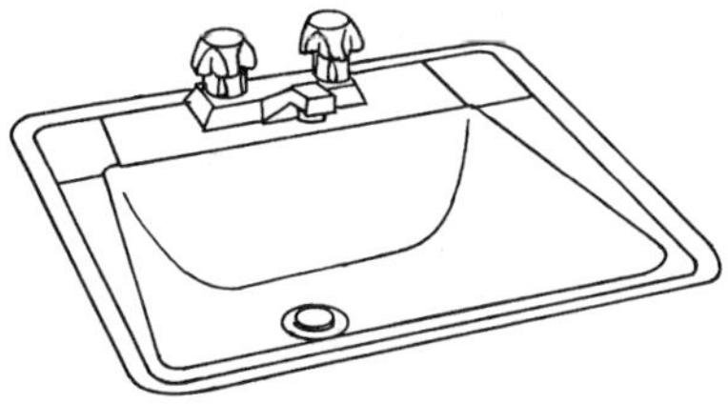

Vanity Top

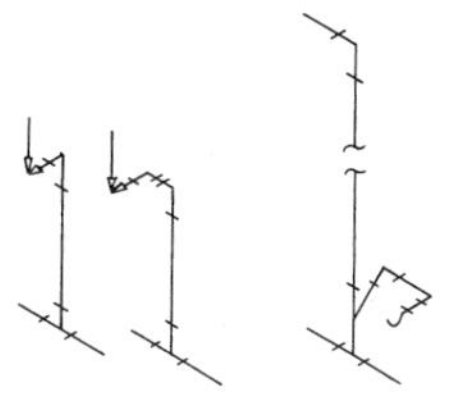

Supply Waste/Vent

Wall Hung

D2010 310	Lavatory Systems	COST EACH		
		MAT.	INST.	TOTAL
1560	Lavatory w/trim, vanity top, PE on CI, 20" x 18", Vanity top by others.	585	615	1,200
1600	19" x 16" oval	395	615	1,010
1640	18" round	700	615	1,315
1680	Cultured marble, 19" x 17"	405	615	1,020
1720	25" x 19"	435	615	1,050
1760	Stainless, self-rimming, 25" x 22"	615	615	1,230
1800	17" x 22"	605	615	1,220
1840	Steel enameled, 20" x 17"	405	630	1,035
1880	19" round	405	630	1,035
1920	Vitreous china, 20" x 16"	500	645	1,145
1960	19" x 16"	510	645	1,155
2000	22" x 13"	510	645	1,155
2040	Wall hung, PE on CI, 18" x 15"	905	680	1,585
2080	19" x 17"	935	680	1,615
2120	20" x 18"	770	680	1,450
2160	Vitreous china, 18" x 15"	685	700	1,385
2200	19" x 17"	640	700	1,340
2240	24" x 20"	770	700	1,470
2300	20" x 27", handicap	1,425	755	2,180

D2010 Plumbing Fixtures

Systems are complete with trim and rough-in (supply, waste and vent) to connect to supply branches and waste mains.

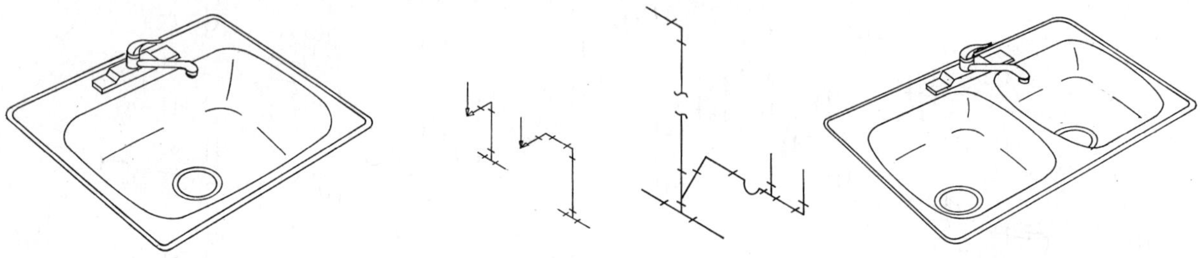

Countertop Single Bowl — Supply — Waste/Vent — Countertop Double Bowl

D2010 410	Kitchen Sink Systems	COST EACH		
		MAT.	INST.	TOTAL
1720	Kitchen sink w/trim, countertop, PE on CI, 24"x21", single bowl	600	675	1,275
1760	30" x 21" single bowl	875	675	1,550
1800	32" x 21" double bowl	705	725	1,430
1880	Stainless steel, 19" x 18" single bowl	940	675	1,615
1920	25" x 22" single bowl	1,025	675	1,700
1960	33" x 22" double bowl	1,375	725	2,100
2000	43" x 22" double bowl	1,550	740	2,290
2040	44" x 22" triple bowl	1,575	770	2,345
2080	44" x 24" corner double bowl	1,100	740	1,840
2120	Steel, enameled, 24" x 21" single bowl	855	675	1,530
2160	32" x 21" double bowl	870	725	1,595
2240	Raised deck, PE on CI, 32" x 21", dual level, double bowl	795	920	1,715
2280	42" x 21" dual level, triple bowl	1,650	1,000	2,650
2281				
2282				

D2010 Plumbing Fixtures

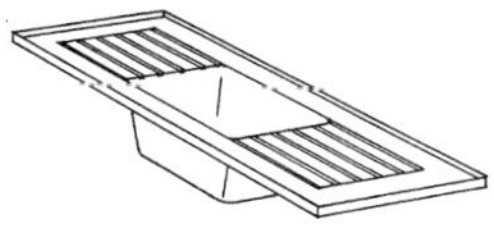

Laboratory Sink

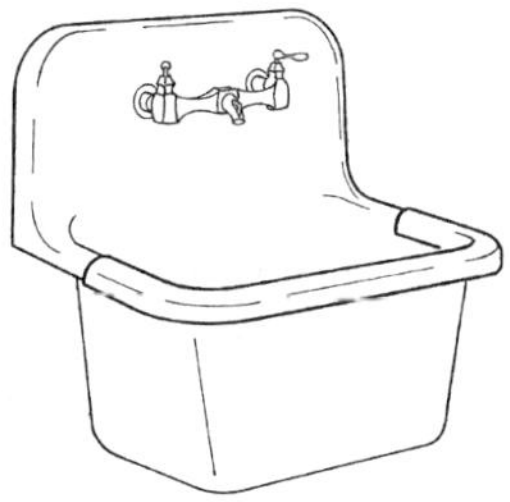

Service Sink

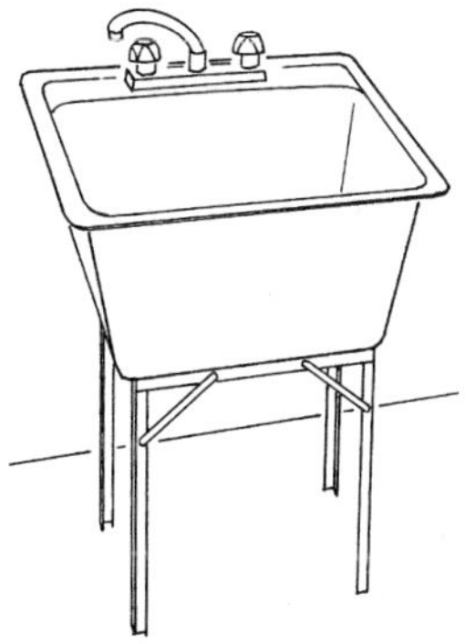

Single Compartment Sink

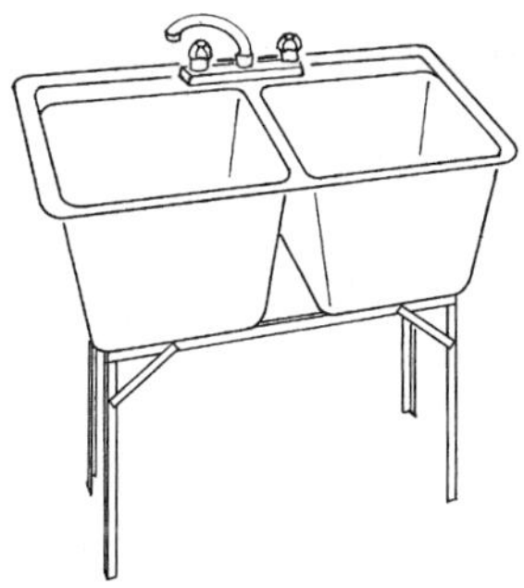

Double Compartment Sink

Systems are complete with trim and rough-in (supply, waste and vent) to connect to supply branches and waste mains.

D2010 420 Laundry Sink Systems

		COST EACH		
		MAT.	INST.	TOTAL
1740	Laundry sink w/trim, PE on CI, black iron frame			
1760	24″ x 20″, single compartment	920	660	1,580
1840	48″ x 21″ double compartment	1,175	720	1,895
1920	Molded stone, on wall, 22″ x 21″ single compartment	470	660	1,130
1960	45″x 21″ double compartment	675	720	1,395
2040	Plastic, on wall or legs, 18″ x 23″ single compartment	445	650	1,095
2080	20″ x 24″ single compartment	460	650	1,110
2120	36″ x 23″ double compartment	520	700	1,220
2160	40″ x 24″ double compartment	625	700	1,325

D2010 430 Laboratory Sink Systems

		COST EACH		
		MAT.	INST.	TOTAL
1580	Laboratory sink w/trim, stainless steel, single bowl,			
1590	Stainless steel, single bowl,			
1600	Double drainboard, 54″ x 24″ O.D.	1,450	860	2,310
1640	Single drainboard, 47″ x 24″O.D.	1,175	860	2,035
1670	Stainless steel, double bowl,			
1680	70″ x 24″ O.D.	1,575	860	2,435
1750	Polyethylene, single bowl,			
1760	Flanged, 14-1/2″ x 14-1/2″ O.D.	520	775	1,295
1800	18-1/2″ x 18-1/2″ O.D.	630	775	1,405
1840	23-1/2″ x 20-1/2″ O.D.	650	775	1,425
1920	Polypropylene, cup sink, oval, 7″ x 4″ O.D.	410	685	1,095
1960	10″ x 4-1/2″ O.D.	435	685	1,120
1961				

D2010 440 Service Sink Systems

		COST EACH		
		MAT.	INST.	TOTAL
4260	Service sink w/trim, PE on CI, corner floor, 28″ x 28″, w/rim guard	1,900	870	2,770
4300	Wall hung w/rim guard, 22″ x 18″	2,125	1,000	3,125
4340	24″ x 20″	2,200	1,000	3,200
4380	Vitreous china, wall hung 22″ x 20″	1,925	1,000	2,925

D2010 Plumbing Fixtures

Recessed Bathtub

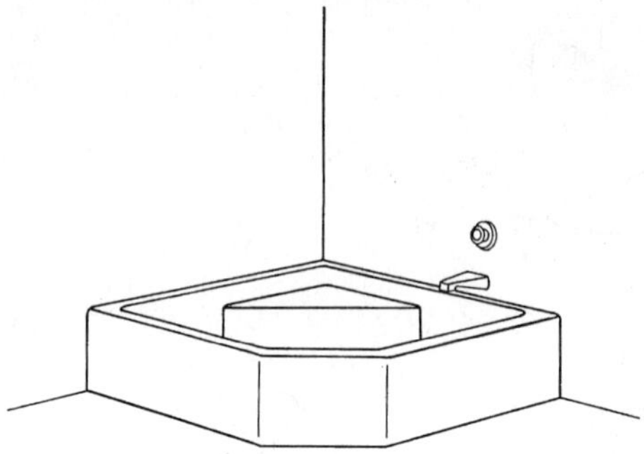

Corner Bathtub

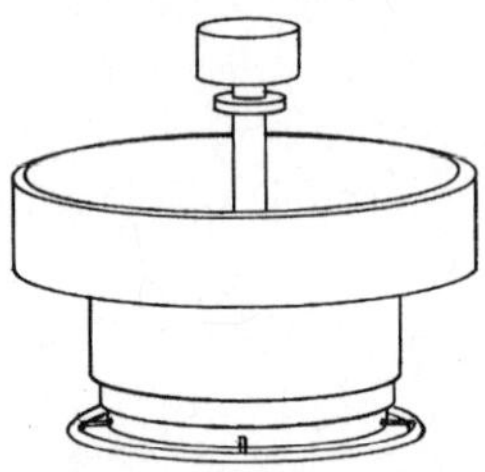

Circular Wash Fountain

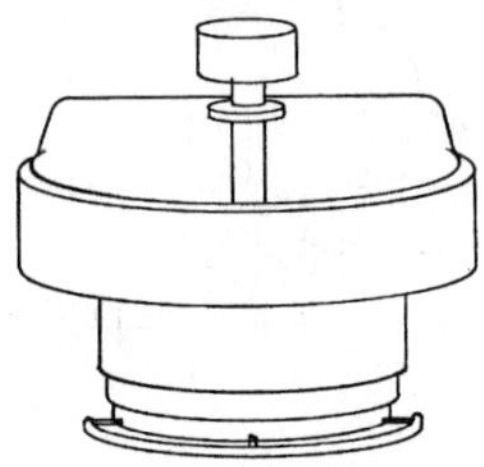

Semi-Circular Wash Fountain

Systems are complete with trim and rough-in (supply, waste and vent) to connect to supply branches and waste mains.

D2010 510	Bathtub Systems	COST EACH		
		MAT.	INST.	TOTAL
2000	Bathtub, recessed, P.E. on Cl., 48″ x 42″	3,325	765	4,090
2040	72″ x 36″	3,425	850	4,275
2080	Mat bottom, 5′ long	1,650	740	2,390
2120	5′-6″ long	2,425	765	3,190
2160	Corner, 48″ x 42″	3,325	740	4,065
2200	Formed steel, enameled, 4′-6″ long	945	680	1,625

D2010 610	Group Wash Fountain Systems	COST EACH		
		MAT.	INST.	TOTAL
1740	Group wash fountain, precast terrazzo			
1760	Circular, 36″ diameter	8,500	1,075	9,575
1800	54″ diameter	10,500	1,175	11,675
1840	Semi-circular, 36″ diameter	7,500	1,075	8,575
1880	54″ diameter	9,950	1,175	11,125
1960	Stainless steel, circular, 36″ diameter	7,700	1,000	8,700
2000	54″ diameter	9,375	1,100	10,475
2040	Semi-circular, 36″ diameter	6,025	1,000	7,025
2080	54″ diameter	8,225	1,100	9,325
2160	Thermoplastic, circular, 36″ diameter	5,375	820	6,195
2200	54″ diameter	6,175	945	7,120
2240	Semi-circular, 36″ diameter	4,975	820	5,795
2280	54″ diameter	5,975	945	6,920

D2010 Plumbing Fixtures

Square Shower Stall

Corner Angle Shower Stall

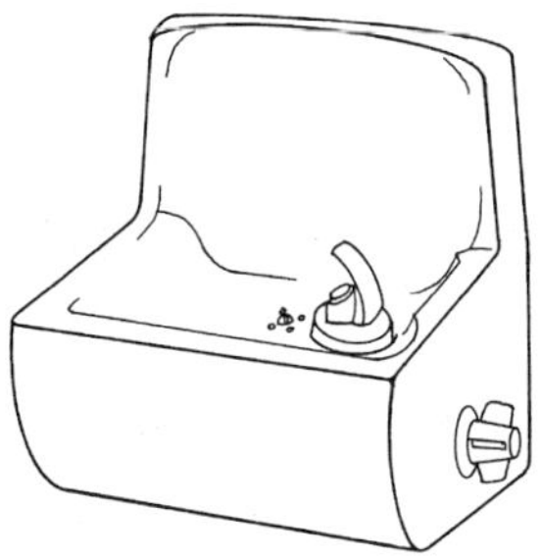

Wall Mounted, Low Back

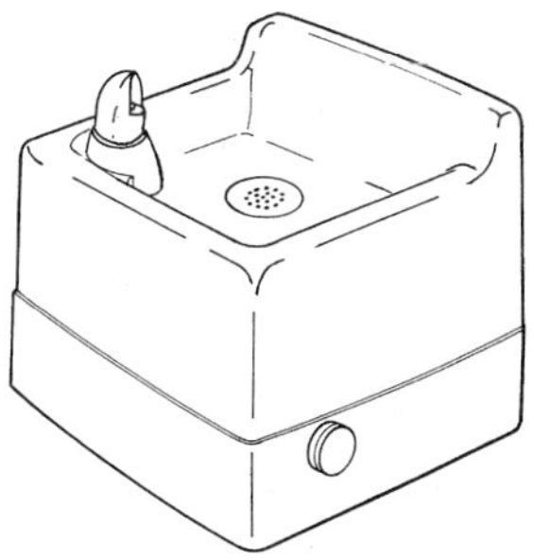

Wall Mounted, No Back

Systems are complete with trim and rough-in (supply, waste and vent) to connect to supply branches and waste mains.

D2010 710	Shower Systems	COST EACH		
		MAT.	INST.	TOTAL
1560	Shower, stall, baked enamel, molded stone receptor, 30″ square	1,575	710	2,285
1600	32″ square	1,575	715	2,290
1640	Terrazzo receptor, 32″ square	1,775	715	2,490
1680	36″ square	1,925	725	2,650
1720	36″ corner angle	2,200	725	2,925
1800	Fiberglass one piece, three walls, 32″ square	665	700	1,365
1840	36″ square	715	700	1,415
1880	Polypropylene, molded stone receptor, 30″ square	995	1,025	2,020
1920	32″ square	1,000	1,025	2,025
1960	Built-in head, arm, bypass, stops and handles	113	268	381

D2010 810	Drinking Fountain Systems	COST EACH		
		MAT.	INST.	TOTAL
1740	Drinking fountain, one bubbler, wall mounted			
1760	Non recessed			
1800	Bronze, no back	1,300	405	1,705
1840	Cast iron, enameled, low back	1,250	405	1,655
1880	Fiberglass, 12″ back	2,325	405	2,730
1920	Stainless steel, no back	1,375	405	1,780
1960	Semi-recessed, poly marble	1,300	405	1,705
2040	Stainless steel	1,675	405	2,080
2080	Vitreous china	1,225	405	1,630
2120	Full recessed, poly marble	2,025	405	2,430
2200	Stainless steel	1,875	405	2,280
2240	Floor mounted, pedestal type, aluminum	2,675	550	3,225
2320	Bronze	2,325	550	2,875
2360	Stainless steel	2,150	550	2,700

D2010 Plumbing Fixtures

D2010 820	Water Cooler Systems	COST EACH		
		MAT.	INST.	TOTAL
1840	Water cooler, electric, wall hung, 8.2 G.P.H.	1,075	520	1,595
1880	Dual height, 14.3 G.P.H.	1,375	535	1,910
1920	Wheelchair type, 7.5 G.P.H.	1,225	520	1,745
1960	Semi recessed, 8.1 G.P.H.	1,050	520	1,570
2000	Full recessed, 8 G.P.H.	2,550	560	3,110
2040	Floor mounted, 14.3 G.P.H.	1,225	455	1,680
2080	Dual height, 14.3 G.P.H.	1,575	550	2,125
2120	Refrigerated compartment type, 1.5 G.P.H.	1,875	455	2,330

D2010 Plumbing Fixtures

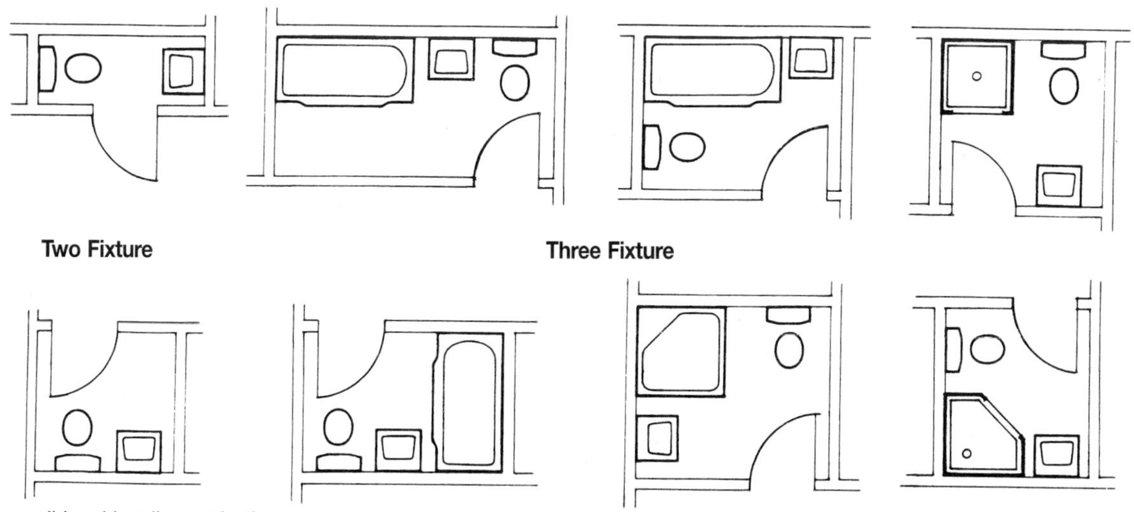

*Common wall is with adjacent bathroom.

D2010 920	Two Fixture Bathroom, Two Wall Plumbing	COST EACH		
		MAT.	INST.	TOTAL
1180	Bathroom, lavatory & water closet, 2 wall plumbing, stand alone	1,675	1,700	3,375
1200	Share common plumbing wall*	1,550	1,450	3,000

D2010 922	Two Fixture Bathroom, One Wall Plumbing	COST EACH		
		MAT.	INST.	TOTAL
2220	Bathroom, lavatory & water closet, one wall plumbing, stand alone	1,600	1,525	3,125
2240	Share common plumbing wall*	1,375	1,300	2,675

D2010 924	Three Fixture Bathroom, One Wall Plumbing	COST EACH		
		MAT.	INST.	TOTAL
1150	Bathroom, three fixture, one wall plumbing			
1160	Lavatory, water closet & bathtub			
1170	Stand alone	2,975	1,975	4,950
1180	Share common plumbing wall *	2,575	1,425	4,000

D2010 926	Three Fixture Bathroom, Two Wall Plumbing	COST EACH		
		MAT.	INST.	TOTAL
2130	Bathroom, three fixture, two wall plumbing			
2140	Lavatory, water closet & bathtub			
2160	Stand alone	3,000	2,000	5,000
2180	Long plumbing wall common *	2,700	1,600	4,300
3610	Lavatory, bathtub & water closet			
3620	Stand alone	3,200	2,275	5,475
3640	Long plumbing wall common *	3,050	2,075	5,125
4660	Water closet, corner bathtub & lavatory			
4680	Stand alone	4,700	2,025	6,725
4700	Long plumbing wall common *	4,300	1,525	5,825
6100	Water closet, stall shower & lavatory			
6120	Stand alone	3,275	2,275	5,550
6140	Long plumbing wall common *	3,150	2,100	5,250
7060	Lavatory, corner stall shower & water closet			
7080	Stand alone	3,675	2,025	5,700
7100	Short plumbing wall common *	3,175	1,350	4,525

D2020 Domestic Water Distribution

Installation includes piping and fittings within 10′ of heater. Gas and oil fired heaters require vent piping (not included with these units).

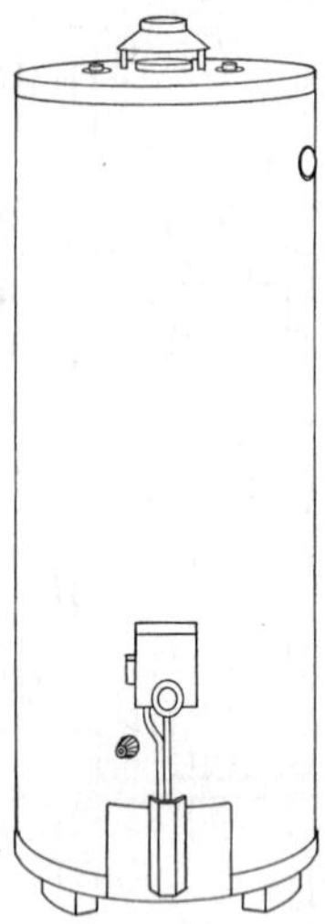

Gas Fired

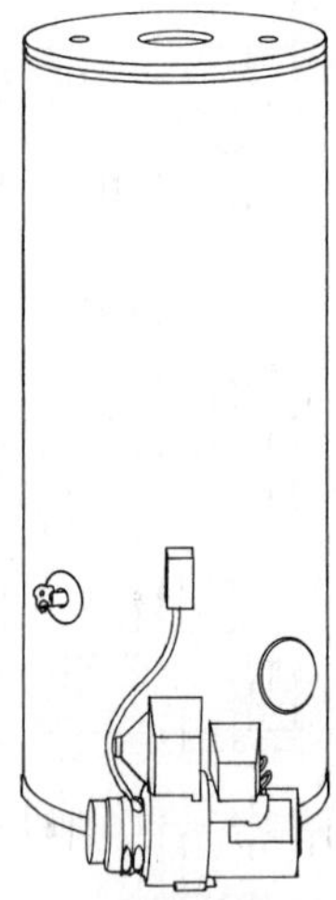

Oil Fired

D2020 220	Gas Fired Water Heaters - Residential Systems	COST EACH		
		MAT.	INST.	TOTAL
2200	Gas fired water heater, residential, 100°F rise			
2260	30 gallon tank, 32 GPH	2,475	1,200	3,675
2300	40 gallon tank, 32 GPH	2,650	1,350	4,000
2340	50 gallon tank, 63 GPH	2,775	1,350	4,125
2380	75 gallon tank, 63 GPH	3,850	1,500	5,350
2420	100 gallon tank, 63 GPH	3,975	1,575	5,550

D2020 230	Oil Fired Water Heaters - Residential Systems	COST EACH		
		MAT.	INST.	TOTAL
2200	Oil fired water heater, residential, 100°F rise			
2220	30 gallon tank, 103 GPH	2,000	1,100	3,100
2260	50 gallon tank, 145 GPH	2,400	1,250	3,650
2300	70 gallon tank, 164 GPH	3,350	1,400	4,750
2340	85 gallon tank, 181 GPH	10,800	1,450	12,250
2344				

D2020 Domestic Water Distribution

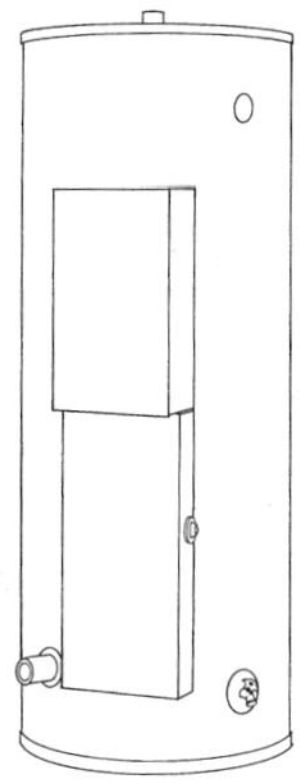

Systems below include piping and fittings within 10′ of heater. Electric water heaters do not require venting.

D2020 240	Electric Water Heaters - Commercial Systems	COST EACH		
		MAT.	INST.	TOTAL
1800	Electric water heater, commercial, 100°F rise			
1820	50 gallon tank, 9 KW 37 GPH	7,800	1,000	8,800
1860	80 gal, 12 KW 49 GPH	9,900	1,250	11,150
1900	36 KW 147 GPH	13,500	1,350	14,850
1940	120 gal, 36 KW 147 GPH	14,600	1,450	16,050
1980	150 gal, 120 KW 490 GPH	45,100	1,550	46,650
2020	200 gal, 120 KW 490 GPH	47,500	1,600	49,100
2060	250 gal, 150 KW 615 GPH	53,000	1,850	54,850
2100	300 gal, 180 KW 738 GPH	57,000	1,950	58,950
2140	350 gal, 30 KW 123 GPH	42,500	2,100	44,600
2180	180 KW 738 GPH	59,000	2,100	61,100
2220	500 gal, 30 KW 123 GPH	55,500	2,475	57,975
2260	240 KW 984 GPH	89,500	2,475	91,975
2300	700 gal, 30 KW 123 GPH	67,000	2,725	69,725
2340	300 KW 1230 GPH	99,500	2,725	102,225
2380	1000 gal, 60 KW 245 GPH	81,000	3,800	84,800
2420	480 KW 1970 GPH	131,500	3,800	135,300
2460	1500 gal, 60 KW 245 GPH	118,500	4,650	123,150
2500	480 KW 1970 GPH	162,500	4,650	167,150

D2020 Domestic Water Distribution

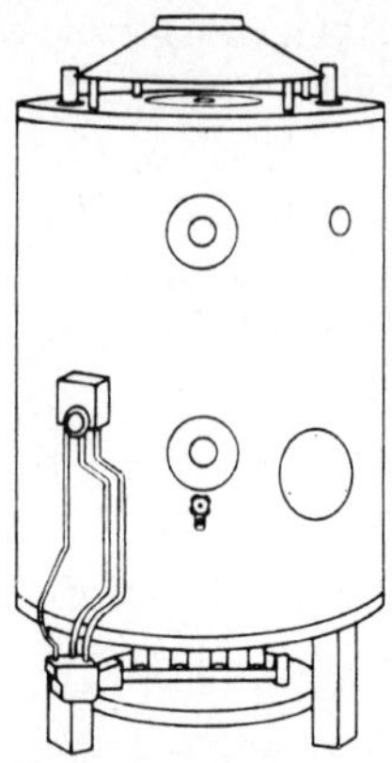

Units may be installed in multiples for increased capacity.

Included below is the heater with self-energizing gas controls, safety pilots, insulated jacket, hi-limit aquastat and pressure relief valve.

Installation includes piping and fittings within 10′ of heater. Gas heaters require vent piping (not included in these prices).

D2020 250	Gas Fired Water Heaters - Commercial Systems	COST EACH		
		MAT.	INST.	TOTAL
1760	Gas fired water heater, commercial, 100°F rise			
1780	75.5 MBH input, 63 GPH	4,950	1,550	6,500
1820	95 MBH input, 86 GPH	8,550	1,550	10,100
1860	100 MBH input, 91 GPH	8,800	1,625	10,425
1900	115 MBH input, 110 GPH	8,850	1,675	10,525
1980	155 MBH input, 150 GPH	11,100	1,875	12,975
2020	175 MBH input, 168 GPH	11,400	2,000	13,400
2060	200 MBH input, 192 GPH	11,700	2,275	13,975
2100	240 MBH input, 230 GPH	12,400	2,450	14,850
2140	300 MBH input, 278 GPH	13,500	2,800	16,300
2180	390 MBH input, 374 GPH	16,000	2,850	18,850
2220	500 MBH input, 480 GPH	21,600	3,050	24,650
2260	600 MBH input, 576 GPH	24,900	3,300	28,200

D2020 Domestic Water Distribution

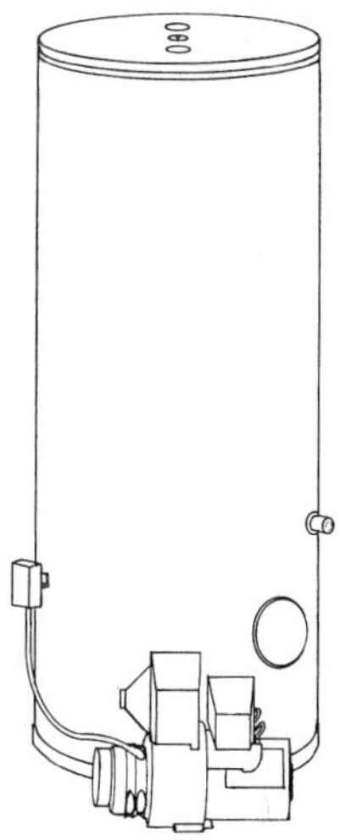

Units may be installed in multiples for increased capacity.

Included below is the heater, wired-in flame retention burners, cadmium cell primary controls, hi-limit controls, ASME pressure relief valves, draft controls, and insulated jacket.

Oil fired water heater systems include piping and fittings within 10′ of heater. Oil fired heaters require vent piping (not included in these systems).

D2020 260	Oil Fired Water Heaters - Commercial Systems	COST EACH		
		MAT.	INST.	TOTAL
1800	Oil fired water heater, commercial, 100°F rise			
1820	140 gal., 140 MBH input, 134 GPH	26,400	1,325	27,725
1900	140 gal., 255 MBH input, 247 GPH	28,400	1,650	30,050
1940	140 gal., 270 MBH input, 259 GPH	35,000	1,875	36,875
1980	140 gal., 400 MBH input, 384 GPH	36,100	2,175	38,275
2060	140 gal., 720 MBH input, 691 GPH	38,400	2,250	40,650
2100	221 gal., 300 MBH input, 288 GPH	51,000	2,475	53,475
2140	221 gal., 600 MBH input, 576 GPH	56,500	2,500	59,000
2180	221 gal., 800 MBH input, 768 GPH	57,500	2,625	60,125
2220	201 gal., 1000 MBH input, 960 GPH	58,500	2,550	61,050
2260	201 gal., 1250 MBH input, 1200 GPH	59,500	2,600	62,100
2300	201 gal., 1500 MBH input, 1441 GPH	64,500	2,675	67,175
2340	411 gal., 600 MBH input, 576 GPH	65,000	2,750	67,750
2380	411 gal., 800 MBH input, 768 GPH	68,000	2,800	70,800
2420	411 gal., 1000 MBH input, 960 GPH	71,000	3,250	74,250
2460	411 gal., 1250 MBH input, 1200 GPH	72,500	3,325	75,825
2500	397 gal., 1500 MBH input, 1441 GPH	77,000	3,425	80,425
2540	397 gal., 1750 MBH input, 1681 GPH	78,500	3,550	82,050

D2040 Rain Water Drainage

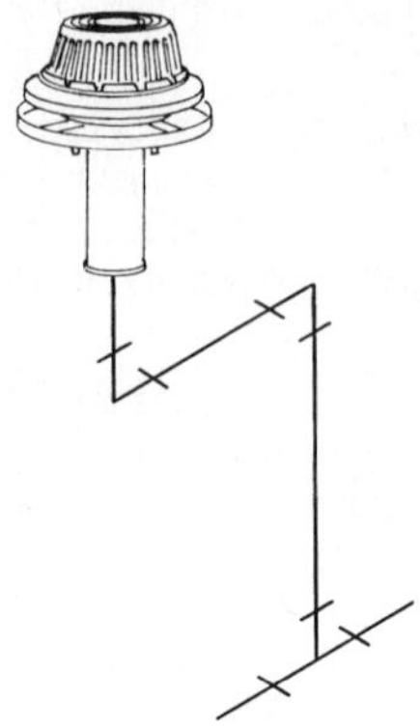

Design Assumptions: Vertical conductor size is based on a maximum rate of rainfall of 4″ per hour. To convert roof area to other rates multiply "Max. S.F. Roof Area" shown by four and divide the result by desired local rate. The answer is the local roof area that may be handled by the indicated pipe diameter.

Basic cost is for roof drain, 10′ of vertical leader and 10′ of horizontal, plus connection to the main.

Pipe Dia.	Max. S.F. Roof Area	Gallons per Min.
2″	544	23
3″	1610	67
4″	3460	144
5″	6280	261
6″	10,200	424
8″	22,000	913

D2040 210	Roof Drain Systems	COST EACH		
		MAT.	INST.	TOTAL
1880	Roof drain, DWV PVC, 2″ diam., piping, 10′ high	305	580	885
1920	For each additional foot add	6.30	17.65	23.95
1960	3″ diam., 10′ high	385	675	1,060
2000	For each additional foot add	9	19.65	28.65
2040	4″ diam., 10′ high	460	755	1,215
2080	For each additional foot add	11.25	21.50	32.75
2120	5″ diam., 10′ high	1,475	865	2,340
2160	For each additional foot add	30	24	54
2200	6″ diam., 10′ high	1,475	960	2,435
2240	For each additional foot add	18.60	26.50	45.10
2280	8″ diam., 10′ high	3,175	1,575	4,750
2320	For each additional foot add	39	31.50	70.50
3940	C.I., soil, single hub, service wt., 2″ diam. piping, 10′ high	715	635	1,350
3980	For each additional foot add	12.85	16.50	29.35
4120	3″ diam., 10′ high	920	685	1,605
4160	For each additional foot add	17	17.35	34.35
4200	4″ diam., 10′ high	1,050	750	1,800
4240	For each additional foot add	21.50	18.90	40.40
4280	5″ diam., 10′ high	1,475	775	2,250
4320	For each additional foot add	29	19.75	48.75
4360	6″ diam., 10′ high	1,850	830	2,680
4400	For each additional foot add	35.50	20.50	56
4440	8″ diam., 10′ high	3,625	1,725	5,350
4480	For each additional foot add	53	35.50	88.50
6040	Steel galv. sch 40 threaded, 2″ diam. piping, 10′ high	750	615	1,365
6080	For each additional foot add	9.30	16.25	25.55
6120	3″ diam., 10′ high	1,375	880	2,255
6160	For each additional foot add	18.55	24	42.55
6200	4″ diam., 10′ high	1,950	1,125	3,075
6240	For each additional foot add	27	29	56
6280	5″ diam., 10′ high	1,875	930	2,805
6320	For each additional foot add	30	28	58
6360	6″ diam, 10′ high	2,375	1,150	3,525
6400	For each additional foot add	37.50	36	73.50
6440	8″ diam., 10′ high	4,400	1,675	6,075
6480	For each additional foot add	56.50	40.50	97

D3010 Energy Supply

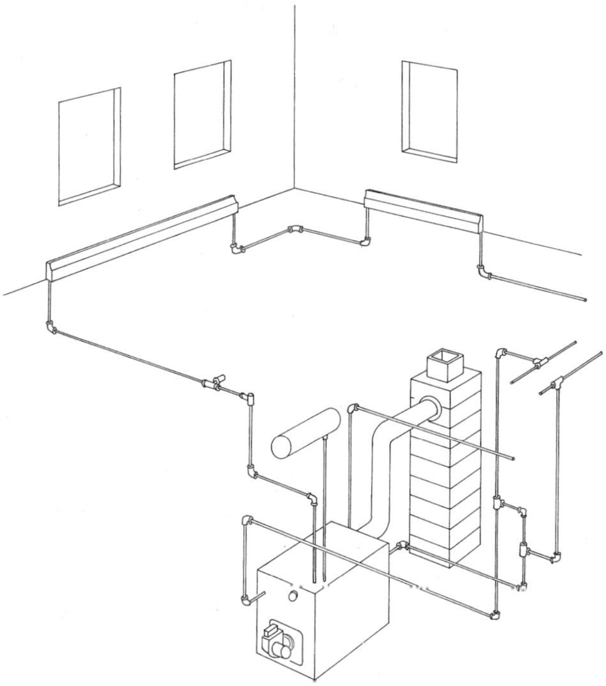

Forced Hot Water Heating System

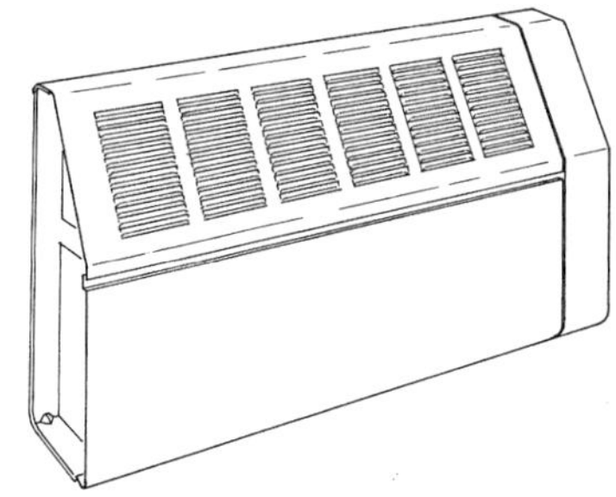

Fin Tube Radiation

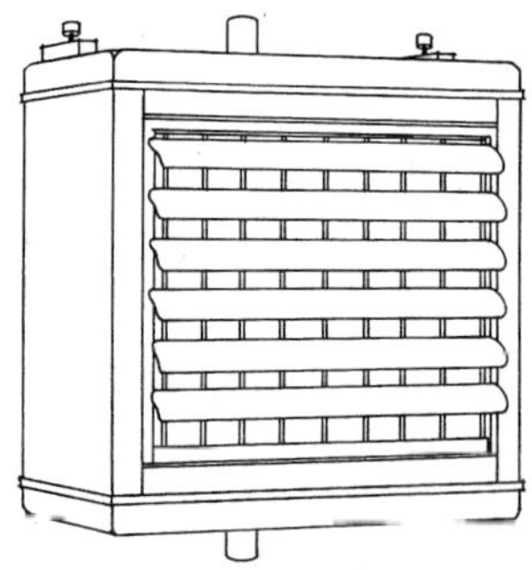

Terminal Unit Heater

D3010 510	Apartment Building Heating - Fin Tube Radiation	COST PER S.F.		
		MAT.	INST.	TOTAL
1740	Heating systems, fin tube radiation, forced hot water			
1800	10,000 S.F. area, 100,000 C.F. volume	3.14	3.37	6.51
1840	20,000 S.F. area, 200,000 C.F. volume	3.43	3.78	7.21
1880	30,000 S.F. area, 300,000 C.F. volume	3.32	3.67	6.99

D3010 520	Commercial Building Heating - Fin Tube Radiation	COST PER S.F.		
		MAT.	INST.	TOTAL
1940	Heating systems, fin tube radiation, forced hot water			
1960	1,000 S.F. bldg, one floor	18.20	12.50	30.70
2000	10,000 S.F., 100,000 C.F., total two floors	4.70	4.79	9.49
2040	100,000 S.F., 1,000,000 C.F., total three floors	1.90	2.13	4.03
2080	1,000,000 S.F., 10,000,000 C.F., total five floors	1.01	1.14	2.15

D3010 530	Commercial Bldg. Heating - Terminal Unit Heaters	COST PER S.F.		
		MAT.	INST.	TOTAL
1860	Heating systems, terminal unit heaters, forced hot water			
1880	1,000 S.F. bldg., one floor	18	11.40	29.40
1920	10,000 S.F. bldg., 100,000 C.F. total two floors	4.25	3.95	8.20
1960	100,000 S.F. bldg., 1,000,000 C.F. total three floors	1.95	1.95	3.90
2000	1,000,000 S.F. bldg., 10,000,000 C.F. total five floors	1.32	1.27	2.59

D3020 Heat Generating Systems

Small Electric Boiler Systems Considerations:

1. Terminal units are fin tube baseboard radiation rated at 720 BTU/hr with 200° water temperature or 820 BTU/hr steam.
2. Primary use being for residential or smaller supplementary areas, the floor levels are based on 7-1/2″ ceiling heights.
3. All distribution piping is copper for boilers through 205 MBH. All piping for larger systems is steel pipe.

Large Electric Boiler System Considerations:

1. Terminal units are all unit heaters of the same size. Quantities are varied to accommodate total requirements.
2. All air is circulated through the heaters a minimum of three times per hour.
3. As the capacities are adequate for commercial use, gross output rating by floor levels are based on 10′ ceiling height.
4. All distribution piping is black steel pipe.

D3020 102	Small Heating Systems, Hydronic, Electric Boilers	COST PER S.F.		
		MAT.	INST.	TOTAL
1100	Small heating systems, hydronic, electric boilers			
1120	Steam, 1 floor, 1480 S.F., 61 M.B.H.	7.50	7.30	14.80
1160	3,000 S.F., 123 M.B.H.	5.75	6.40	12.15
1200	5,000 S.F., 205 M.B.H.	4.98	5.90	10.88
1240	2 floors, 12,400 S.F., 512 M.B.H.	3.88	5.85	9.73
1280	3 floors, 24,800 S.F., 1023 M.B.H.	4.32	5.80	10.12
1320	34,750 S.F., 1,433 M.B.H.	3.97	5.60	9.57
1360	Hot water, 1 floor, 1,000 S.F., 41 M.B.H.	12.65	4.07	16.72
1400	2,500 S.F., 103 M.B.H.	8.85	7.30	16.15
1440	2 floors, 4,850 S.F., 205 M.B.H.	8.45	8.70	17.15
1480	3 floors, 9,700 S.F., 410 M.B.H.	8.70	9	17.70

D3020 104	Large Heating Systems, Hydronic, Electric Boilers	COST PER S.F.		
		MAT.	INST.	TOTAL
1230	Large heating systems, hydronic, electric boilers			
1240	9,280 S.F., 135 K.W., 461 M.B.H., 1 floor	4.13	2.70	6.83
1280	14,900 S.F., 240 K.W., 820 M.B.H., 2 floors	5.35	4.39	9.74
1320	18,600 S.F., 296 K.W., 1,010 M.B.H., 3 floors	5.25	4.75	10
1360	26,100 S.F., 420 K.W., 1,432 M.B.H., 4 floors	5.15	4.67	9.82
1400	39,100 S.F., 666 K.W., 2,273 M.B.H., 4 floors	4.43	3.90	8.33
1440	57,700 S.F., 900 K.W., 3,071 M.B.H., 5 floors	4.24	3.82	8.06
1480	111,700 S.F., 1,800 K.W., 6,148 M.B.H., 6 floors	3.77	3.29	7.06
1520	149,000 S.F., 2,400 K.W., 8,191 M.B.H., 8 floors	3.71	3.28	6.99
1560	223,300 S.F., 3,600 K.W., 12,283 M.B.H., 14 floors	3.91	3.74	7.65

D3020 Heat Generating Systems

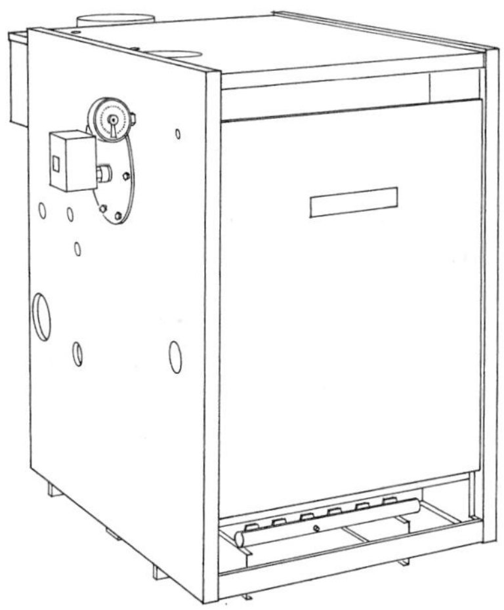

Boiler Selection: The maximum allowable working pressures are limited by ASME "Code for Heating Boilers" to 15 PSI for steam and 160 PSI for hot water heating boilers, with a maximum temperature limitation of 250°F. Hot water boilers are generally rated for a working pressure of 30 PSI. High pressure boilers are governed by the ASME "Code for Power Boilers" which is used almost universally for boilers operating over 15 PSIG. High pressure boilers used for a combination of heating/process loads are usually designed for 150 PSIG.

Boiler ratings are usually indicated as either Gross or Net Output. The Gross Load is equal to the Net Load plus a piping and pickup allowance. When this allowance cannot be determined, divide the gross output rating by 1.25 for a value equal to or greater than the next heat loss requirement of the building.

Table below lists installed cost per boiler and includes insulating jacket, standard controls, burner and safety controls. Costs do not include piping or boiler base pad. Outputs are Gross.

D3020 106	Boilers, Hot Water & Steam	COST EACH		
		MAT.	INST.	TOTAL
0600	Boiler, electric, steel, hot water, 12 K.W., 41 M.B.H.	5,450	1,275	6,725
0620	30 K.W., 103 M.B.H.	5,950	1,375	7,325
0640	60 K.W., 205 M.B.H.	6,600	1,500	8,100
0660	120 K.W., 410 M.B.H.	7,025	1,825	8,850
0680	210 K.W., 716 M.B.H.	8,375	2,750	11,125
0700	510 K.W., 1,739 M.B.H.	25,400	5,075	30,475
0720	720 K.W., 2,452 M.B.H.	31,400	5,725	37,125
0740	1,200 K.W., 4,095 M.B.H.	39,900	6,575	46,475
0760	2,100 K.W., 7,167 M.B.H.	70,000	8,275	78,275
0780	3,600 K.W., 12,283 M.B.H.	103,500	14,000	117,500
0820	Steam, 6 K.W., 20.5 M.B.H.	4,500	1,375	5,875
0840	24 K.W., 81.8 M.B.H.	5,500	1,500	7,000
0860	60 K.W., 205 M.B.H.	7,600	1,650	9,250
0880	150 K.W., 512 M.B.H.	11,000	2,525	13,525
0900	510 K.W., 1,740 M.B.H.	35,900	6,200	42,100
0920	1,080 K.W., 3,685 M.B.H.	44,700	8,950	53,650
0940	2,340 K.W., 7,984 M.B.H.	95,000	14,000	109,000
0980	Gas, cast iron, hot water, 80 M.B.H.	2,175	1,450	3,625
1000	100 M.B.H.	2,825	1,575	4,400
1020	163 M.B.H.	3,475	2,125	5,600
1040	280 M.B.H.	4,750	2,375	7,125
1060	544 M.B.H.	10,800	4,200	15,000
1080	1,088 M.B.H.	14,900	5,325	20,225
1100	2,000 M.B.H.	23,500	8,350	31,850
1120	2,856 M.B.H.	33,900	10,700	44,600
1140	4,720 M.B.H.	81,500	14,700	96,200
1160	6,970 M.B.H.	144,500	24,000	168,500
1180	For steam systems under 2,856 M.B.H., add 8%			
1520	Oil, cast iron, hot water, 109 M.B.H.	2,550	1,775	4,325
1540	173 M.B.H.	3,225	2,125	5,350
1560	236 M.B.H.	4,100	2,500	6,600
1580	1,084 M.B.H.	12,000	5,675	17,675
1600	1,600 M.B.H.	15,400	8,150	23,550
1620	2,480 M.B.H.	22,600	10,400	33,000
1640	3,550 M.B.H.	30,100	12,500	42,600
1660	Steam systems same price as hot water			

D3020 Heat Generating Systems

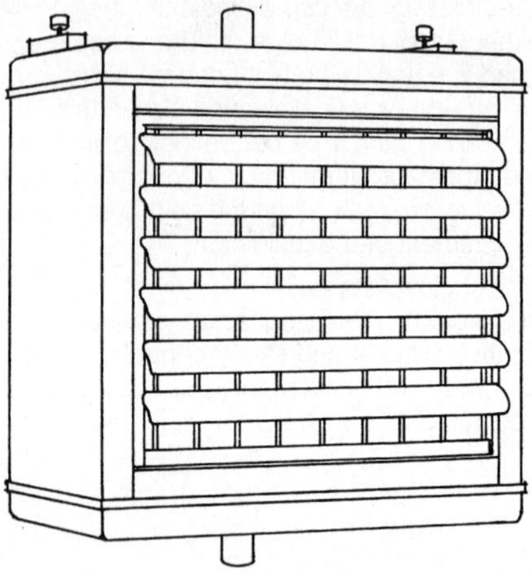

Unit Heater

Fossil Fuel Boiler System Considerations:

1. Terminal units are horizontal unit heaters. Quantities are varied to accommodate total heat loss per building.
2. Unit heater selection was determined by their capacity to circulate the building volume a minimum of three times per hour in addition to the BTU output.
3. Systems shown are forced hot water. Steam boilers cost slightly more than hot water boilers. However, this is compensated for by the smaller size or fewer terminal units required with steam.
4. Floor levels are based on 10′ story heights.
5. MBH requirements are gross boiler output.

D3020 108	Heating Systems, Unit Heaters	COST PER S.F.		
		MAT.	INST.	TOTAL
1260	Heating systems, hydronic, fossil fuel, terminal unit heaters,			
1280	Cast iron boiler, gas, 80 M.B.H., 1,070 S.F. bldg.	10.84	8.53	19.37
1320	163 M.B.H., 2,140 S.F. bldg.	7.30	5.80	13.10
1360	544 M.B.H., 7,250 S.F. bldg.	5.45	4.11	9.56
1400	1,088 M.B.H., 14,500 S.F. bldg.	4.47	3.90	8.37
1440	3,264 M.B.H., 43,500 S.F. bldg.	3.67	2.94	6.61
1480	5,032 M.B.H., 67,100 S.F. bldg.	4.21	3.02	7.23
1520	Oil, 109 M.B.H., 1,420 S.F. bldg.	12.10	7.60	19.70
1560	235 M.B.H., 3,150 S.F. bldg.	7.50	5.45	12.95
1600	940 M.B.H., 12,500 S.F. bldg.	5.35	3.54	8.89
1640	1,600 M.B.H., 21,300 S.F. bldg.	4.96	3.41	8.37
1680	2,480 M.B.H., 33,100 S.F. bldg.	4.95	3.07	8.02
1720	3,350 M.B.H., 44,500 S.F. bldg.	4.39	3.14	7.53
1760	Coal, 148 M.B.H., 1,975 S.F. bldg.	94.50	4.98	99.48
1800	300 M.B.H., 4,000 S.F. bldg.	51.50	3.89	55.39
1840	2,360 M.B.H., 31,500 S.F. bldg.	11.35	3.13	14.48

D3020 Heat Generating Systems

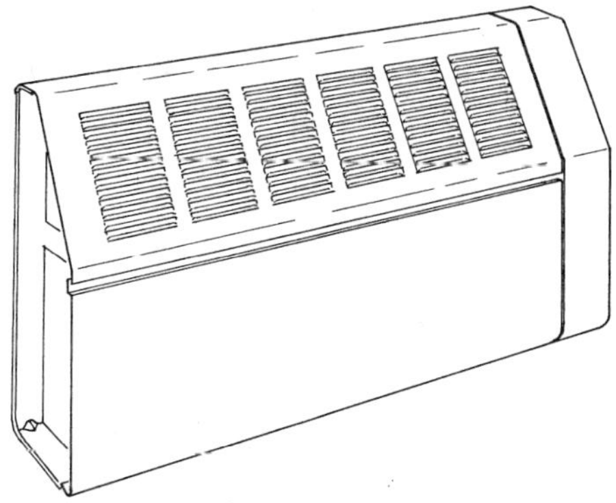

Fin Tube Radiator

Fossil Fuel Boiler System Considerations:

1. Terminal units are commercial steel fin tube radiation. Quantities are varied to accommodate total heat loss per building.
2. Systems shown are forced hot water. Steam boilers cost slightly more than hot water boilers. However, this is compensated for by the smaller size or fewer terminal units required with steam.
3. Floor levels are based on 10′ story heights.
4. MBH requirements are gross boiler output.

D3020 110	Heating System, Fin Tube Radiation	COST PER S.F.		
		MAT.	INST.	TOTAL
3230	Heating systems, hydronic, fossil fuel, fin tube radiation			
3240	Cast iron boiler, gas, 80 MBH, 1,070 S.F. bldg.	13.27	12.90	26.17
3280	169 M.B.H., 2,140 S.F. bldg.	8.40	8.15	16.55
3320	544 M.B.H., 7,250 S.F. bldg.	7.20	7	14.20
3360	1,088 M.B.H., 14,500 S.F. bldg.	6.40	6.90	13.30
3400	3,264 M.B.H., 43,500 S.F. bldg.	5.75	6.05	11.80
3440	5,032 M.B.H., 67,100 S.F. bldg.	6.30	6.15	12.45
3480	Oil, 109 M.B.H., 1,420 S.F. bldg.	16.45	14.10	30.55
3520	235 M.B.H., 3,150 S.F. bldg.	9.25	8.30	17.55
3560	940 M.B.H., 12,500 S.F. bldg.	7.30	6.55	13.85
3600	1,600 M.B.H., 21,300 S.F. bldg.	7.05	6.55	13.60
3640	2,480 M.B.H., 33,100 S.F. bldg.	7.05	6.20	13.25
3680	3,350 M.B.H., 44,500 S.F. bldg.	6.45	6.25	12.70
3720	Coal, 148 M.B.H., 1,975 S.F. bldg.	96	7.80	103.80
3760	300 M.B.H., 4,000 S.F. bldg.	53.50	6.70	60.20
3800	2,360 M.B.H., 31,500 S.F. bldg.	13.35	6.20	19.55
4080	Steel boiler, oil, 97 M.B.H., 1,300 S.F. bldg.	12.25	11.60	23.85
4120	315 M.B.H., 4,550 S.F. bldg.	6.30	5.90	12.20
4160	525 M.B.H., 7,000 S.F. bldg.	9.10	6.80	15.90
4200	1,050 M.B.H., 14,000 S.F. bldg.	8.20	6.80	15
4240	2,310 M.B.H., 30,800 S.F. bldg.	7.10	6.25	13.35
4280	3,150 M.B.H., 42,000 S.F. bldg.	6.60	6.30	12.90

D3030 Cooling Generating Systems

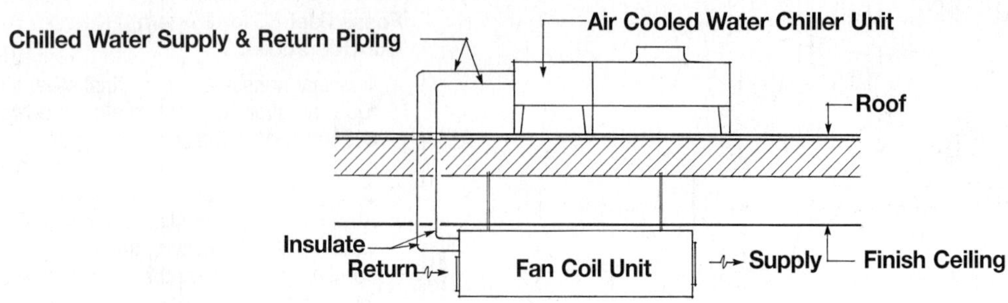

Design Assumptions: The chilled water, air cooled systems priced, utilize reciprocating hermetic compressors and propeller-type condenser fans. Piping with pumps and expansion tanks is included based on a two pipe system. No ducting is included and the fan-coil units are cooling only. Water treatment and balancing are not included. Chilled water piping is insulated. Area distribution is through the use of multiple fan coil units. Fewer but larger fan coil units with duct distribution would be approximately the same S.F. cost.

D3030 110	Chilled Water, Air Cooled Condenser Systems	COST PER S.F.		
		MAT.	INST.	TOTAL
1180	Packaged chiller, air cooled, with fan coil unit			
1200	Apartment corridors, 3,000 S.F., 5.50 ton	6.50	7.20	13.70
1240	6,000 S.F., 11.00 ton	5.40	5.70	11.10
1280	10,000 S.F., 18.33 ton	4.65	4.32	8.97
1320	20,000 S.F., 36.66 ton	3.63	3.11	6.74
1360	40,000 S.F., 73.33 ton	5.10	2.93	8.03
1440	Banks and libraries, 3,000 S.F., 12.50 ton	9.95	8.30	18.25
1480	6,000 S.F., 25.00 ton	9.10	6.75	15.85
1520	10,000 S.F., 41.66 ton	7.15	4.67	11.82
1560	20,000 S.F., 83.33 ton	8.50	3.98	12.48
1600	40,000 S.F., 167 ton*			
1680	Bars and taverns, 3,000 S.F., 33.25 ton	17.55	9.65	27.20
1720	6,000 S.F., 66.50 ton	15.75	7.35	23.10
1760	10,000 S.F., 110.83 ton	12.70	2.14	14.84
1800	20,000 S.F., 220 ton*			
1840	40,000 S.F., 440 ton*			
1920	Bowling alleys, 3,000 S.F., 17.00 ton	12.55	9.30	21.85
1960	6,000 S.F., 34.00 ton	9.70	6.50	16.20
2000	10,000 S.F., 56.66 ton	8.40	4.63	13.03
2040	20,000 S.F., 113.33 ton	9.45	4	13.45
2080	40,000 S.F., 227 ton*			
2160	Department stores, 3,000 S.F., 8.75 ton	9.15	8	17.15
2200	6,000 S.F., 17.50 ton	6.95	6.10	13.05
2240	10,000 S.F., 29.17 ton	5.70	4.38	10.08
2280	20,000 S.F., 58.33 ton	4.79	3.21	8
2320	40,000 S.F., 116.66 ton	6.20	3.02	9.22
2400	Drug stores, 3,000 S.F., 20.00 ton	14.25	9.55	23.80
2440	6,000 S.F., 40.00 ton	11.55	7.10	18.65
2480	10,000 S.F., 66.66 ton	12.90	5.70	18.60
2520	20,000 S.F., 133.33 ton	11.10	4.22	15.32
2560	40,000 S.F., 267 ton*			
2640	Factories, 2,000 S.F., 10.00 ton	8.75	8	16.75
2680	6,000 S.F., 20.00 ton	7.90	6.45	14.35
2720	10,000 S.F., 33.33 ton	6.15	4.48	10.63
2760	20,000 S.F., 66.66 ton	7.50	3.79	11.29
2800	40,000 S.F., 133.33 ton	6.70	3.10	9.80
2880	Food supermarkets, 3,000 S.F., 8.50 ton	9	7.95	16.95
2920	6,000 S.F., 17.00 ton	6.80	6.05	12.85
2960	10,000 S.F., 28.33 ton	5.40	4.27	9.67
3000	20,000 S.F., 56.66 ton	4.59	3.14	7.73

D3030 Cooling Generating Systems

D3030 110	Chilled Water, Air Cooled Condenser Systems	COST PER S.F.		
		MAT.	INST.	TOTAL
3040	40,000 S.F., 113.33 ton	6.05	3.03	9.08
3120	Medical centers, 3,000 S.F., 7.00 ton	8.10	7.75	15.85
3160	6,000 S.F., 14.00 ton	6.10	5.90	12
3200	10,000 S.F., 23.33 ton	5.45	4.46	9.91
3240	20,000 S.F., 46.66 ton	4.35	3.27	7.62
3280	40,000 S.F., 93.33 ton	5.60	2.97	8.57
3360	Offices, 3,000 S.F., 9.50 ton	8.25	7.90	16.15
3400	6,000 S.F., 19.00 ton	7.65	6.35	14
3440	10,000 S.F., 31.66 ton	6	4.44	10.44
3480	20,000 S.F., 63.33 ton	7.50	3.86	11.36
3520	40,000 S.F., 126.66 ton	6.60	3.13	9.73
3600	Restaurants, 3,000 S.F., 15.00 ton	11.15	8.60	19.75
3640	6,000 S.F., 30.00 ton	9.25	6.55	15.80
3680	10,000 S.F., 50.00 ton	8.10	4.74	12.84
3720	20,000 S.F., 100.00 ton	9.65	4.23	13.88
3760	40,000 S.F., 200 ton*			
3840	Schools and colleges, 3,000 S.F., 11.50 ton	9.45	8.20	17.65
3880	6,000 S.F., 23.00 ton	8.60	6.60	15.20
3920	10,000 S.F., 38.33 ton	6.75	4.59	11.34
3960	20,000 S.F., 76.66 ton	8.35	4.03	12.38

D3030 Cooling Generating Systems

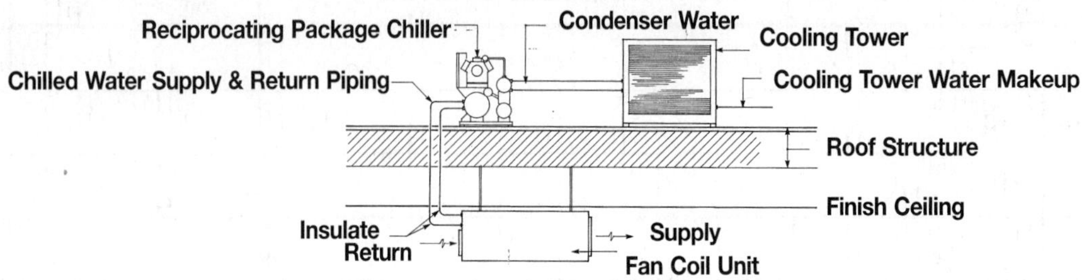

General: Water cooled chillers are available in the same sizes as air cooled units. They are also available in larger capacities.

Design Assumptions: The chilled water systems with water cooled condenser include reciprocating hermetic compressors, water cooling tower, pumps, piping and expansion tanks and are based on a two pipe system. Chilled water piping is insulated. No ducts are included and fan-coil units are cooling only. Area distribution is through use of multiple fan coil units. Fewer but larger fan coil units with duct distribution would be approximately the same S.F. cost. Water treatment and balancing are not included.

D3030 115	Chilled Water, Cooling Tower Systems	COST PER S.F.		
		MAT.	INST.	TOTAL
1300	Packaged chiller, water cooled, with fan coil unit			
1320	Apartment corridors, 4,000 S.F., 7.33 ton	6.30	7.05	13.35
1360	6,000 S.F., 11.00 ton	5.30	6	11.30
1400	10,000 S.F., 18.33 ton	4.96	4.43	9.39
1440	20,000 S.F., 26.66 ton	4	3.29	7.29
1480	40,000 S.F., 73.33 ton	4.99	3.16	8.15
1520	60,000 S.F., 110.00 ton	5.25	3.26	8.51
1600	Banks and libraries, 4,000 S.F., 16.66 ton	10.55	7.55	18.10
1640	6,000 S.F., 25.00 ton	8.95	6.65	15.60
1680	10,000 S.F., 41.66 ton	7.25	4.91	12.16
1720	20,000 S.F., 83.33 ton	8.10	4.46	12.56
1760	40,000 S.F., 166.66 ton	8.95	5.75	14.70
1800	60,000 S.F., 250.00 ton	9.20	6.10	15.30
1880	Bars and taverns, 4,000 S.F., 44.33 ton	17	9.35	26.35
1920	6,000 S.F., 66.50 ton	21	9.25	30.25
1960	10,000 S.F., 110.83 ton	17.70	7.30	25
2000	20,000 S.F., 221.66 ton	20.50	7.75	28.25
2040	40,000 S.F., 440 ton*			
2080	60,000 S.F., 660 ton*			
2160	Bowling alleys, 4,000 S.F., 22.66 ton	12.35	8.25	20.60
2200	6,000 S.F., 34.00 ton	11.05	7.25	18.30
2240	10,000 S.F., 56.66 ton	9.65	5.30	14.95
2280	20,000 S.F., 113.33 ton	10.05	4.67	14.72
2320	40,000 S.F., 226.66 ton	11.80	5.70	17.50
2360	60,000 S.F., 340 ton			
2440	Department stores, 4,000 S.F., 11.66 ton	7.20	7.50	14.70
2480	6,000 S.F., 17.50 ton	7.75	6.15	13.90
2520	10,000 S.F., 29.17 ton	6	4.59	10.59
2560	20,000 S.F., 58.33 ton	4.93	3.40	8.33
2600	40,000 S.F., 116.66 ton	6.20	3.32	9.52
2640	60,000 S.F., 175.00 ton	8.10	5.60	13.70
2720	Drug stores, 4,000 S.F., 26.66 ton	13.30	8.50	21.80
2760	6,000 S.F., 40.00 ton	11.45	7.25	18.70
2800	10,000 S.F., 66.66 ton	13.50	6.25	19.75
2840	20,000 S.F., 133.33 ton	11.50	4.87	16.37
2880	40,000 S.F., 266.67 ton	12.10	6.45	18.55
2920	60,000 S.F., 400 ton*			
3000	Factories, 4,000 S.F., 13.33 ton	9.10	7.20	16.30
3040	6,000 S.F., 20.00 ton	7.85	6.25	14.10
3080	10,000 S.F., 33.33 ton	6.60	4.73	11.33

D3030 Cooling Generating Systems

D3030 115	Chilled Water, Cooling Tower Systems	COST PER S.F.		
		MAT.	INST.	TOTAL
3120	20,000 S.F., 66.66 ton	7.65	4.02	11.67
3160	40,000 S.F., 133.33 ton	6.85	3.43	10.28
3200	60,000 S.F., 200.00 ton	8.35	5.85	14.20
3280	Food supermarkets, 4,000 S.F., 11.33 ton	7.10	7.45	14.55
3320	6,000 S.F., 17.00 ton	6.95	6.10	13.05
3360	10,000 S.F., 28.33 ton	5.90	4.56	10.46
3400	20,000 S.F., 56.66 ton	5.10	3.41	8.51
3440	40,000 S.F., 113.33 ton	6.15	3.33	9.48
3480	60,000 S.F., 170.00 ton	8.05	5.60	13.65
3560	Medical centers, 4.000 S.F., 9.33 ton	6.15	6.85	13
3600	6,000 S.F., 14.00 ton	6.75	5.95	12.70
3640	10,000 S.F., 23.33 ton	5.40	4.42	9.82
3680	20,000 S.F., 46.66 ton	4.58	3.33	7.91
3720	40,000 S.F., 93.33 ton	5.60	3.25	8.85
3760	60,000 S.F., 140.00 ton	6.95	5.65	12.60
3840	Offices, 4,000 S.F., 12.66 ton	8.80	7.15	15.95
3880	6,000 S.F., 19.00 ton	7.80	6.35	14.15
3920	10,000 S.F., 31.66 ton	6.55	4.77	11.32
3960	20,000 S.F., 63.33 ton	7.60	4.02	11.62
4000	40,000 S.F., 126.66 ton	7.50	5.55	13.05
4040	60,000 S.F., 190.00 ton	8.10	5.80	13.90
4120	Restaurants, 4,000 S.F., 20.00 ton	11.05	7.70	18.75
4160	6,000 S.F., 30.00 ton	9.95	6.85	16.80
4200	10,000 S.F., 50.00 ton	8.90	5.10	14
4240	20,000 S.F., 100.00 ton	9.45	4.67	14.12
4280	40,000 S.F., 200.00 ton	9.45	5.55	15
4320	60,000 S.F., 300.00 ton	10.15	6.30	16.45
4400	Schools and colleges, 4,000 S.F., 15.33 ton	9.95	7.40	17.35
4440	6,000 S.F., 23.00 ton	8.50	6.55	15.05
4480	10,000 S.F., 38.33 ton	6.85	4.82	11.67
4520	20,000 S.F., 76.66 ton	7.85	4.42	12.27
4560	40,000 S.F., 153.33 ton	8.45	5.65	14.10
4600	60,000 S.F., 230.00 ton	8.55	5.90	14.45
4603				

D3050 Terminal & Package Units

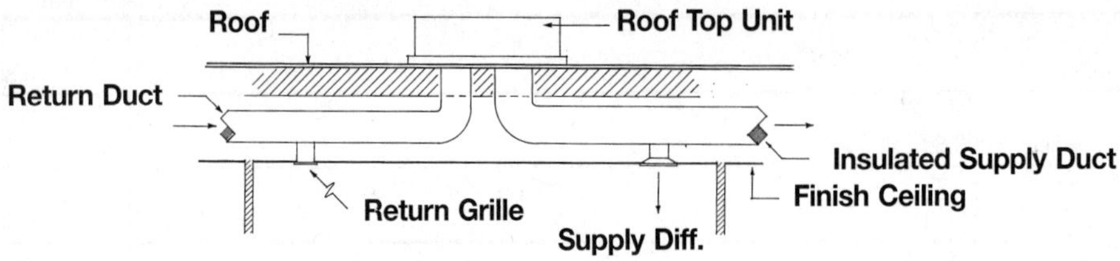

System Description: Rooftop single zone units are electric cooling and gas heat. Duct systems are low velocity, galvanized steel supply and return. Price variations between sizes are due to several factors. Jumps in the cost of the rooftop unit occur when the manufacturer shifts from the largest capacity unit on a small frame to the smallest capacity on the next larger frame, or changes from one compressor to two. As the unit capacity increases for larger areas the duct distribution grows in proportion. For most applications there is a tradeoff point where it is less expensive and more efficient to utilize smaller units with short simple distribution systems. Larger units also require larger initial supply and return ducts which can create a space problem. Supplemental heat may be desired in colder locations. The table below is based on one unit supplying the area listed. The 10,000 S.F. unit for bars and taverns is not listed because a nominal 110 ton unit would be required and this is above the normal single zone rooftop capacity.

*Size would suggest multiple units.

D3050 150	Rooftop Single Zone Unit Systems	COST PER S.F.		
		MAT.	INST.	TOTAL
1260	Rooftop, single zone, air conditioner			
1280	Apartment corridors, 500 S.F., .92 ton	3.15	3.35	6.50
1320	1,000 S.F., 1.83 ton	3.15	3.32	6.47
1360	1500 S.F., 2.75 ton	1.98	2.72	4.70
1400	3,000 S.F., 5.50 ton	2.12	2.63	4.75
1440	5,000 S.F., 9.17 ton	2.17	2.35	4.52
1480	10,000 S.F., 18.33 ton	3.32	2.21	5.53
1560	Banks or libraries, 500 S.F., 2.08 ton	7.20	7.55	14.75
1600	1,000 S.F., 4.17 ton	4.51	6.20	10.71
1640	1,500 S.F., 6.25 ton	4.82	5.95	10.77
1680	3,000 S.F., 12.50 ton	4.95	5.35	10.30
1720	5,000 S.F., 20.80 ton	7.55	5.05	12.60
1760	10,000 S.F., 41.67 ton	5.80	5	10.80
1840	Bars and taverns, 500 S.F. 5.54 ton	11.80	10.05	21.85
1880	1,000 S.F., 11.08 ton	12.10	8.35	20.45
1920	1,500 S.F., 16.62 ton	11.70	7.80	19.50
1960	3,000 S.F., 33.25 ton	15.50	7.50	23
2000	5,000 S.F., 55.42 ton	13.25	7.50	20.75
2040	10,000 S.F., 110.83 ton*			
2080	Bowling alleys, 500 S.F., 2.83 ton	6.10	8.40	14.50
2120	1,000 S.F., 5.67 ton	6.55	8.10	14.65
2160	1,500 S.F., 8.50 ton	6.75	7.25	14
2200	3,000 S.F., 17.00 ton	6.50	6.95	13.45
2240	5,000 S.F., 28.33 ton	8.45	6.80	15.25
2280	10,000 S.F., 56.67 ton	7.30	6.80	14.10
2360	Department stores, 500 S.F., 1.46 ton	5.05	5.30	10.35
2400	1,000 S.F., 2.92 ton	3.16	4.35	7.51
2480	3,000 S.F., 8.75 ton	3.47	3.73	7.20
2520	5,000 S.F., 14.58 ton	3.35	3.58	6.93
2560	10,000 S.F., 29.17 ton	4.36	3.50	7.86
2640	Drug stores, 500 S.F., 3.33 ton	7.20	9.90	17.10
2680	1,000 S.F., 6.67 ton	7.70	9.55	17.25
2720	1,500 S.F., 10.00 ton	7.90	8.55	16.45
2760	3,000 S.F., 20.00 ton	12.10	8.05	20.15
2800	5,000 S.F., 33.33 ton	9.95	8	17.95
2840	10,000 S.F., 66.67 ton	8.60	8	16.60
2920	Factories, 500 S.F., 1.67 ton	5.75	6.05	11.80
3000	1,500 S.F., 5.00 ton	3.86	4.77	8.63

D3050 Terminal & Package Units

D3050 150	Rooftop Single Zone Unit Systems	COST PER S.F.		
		MAT.	INST.	TOTAL
3040	3,000 S.F., 10.00 ton	3.96	4.27	8.23
3080	5,000 S.F., 16.67 ton	3.83	4.10	7.93
3120	10,000 S.F., 33.33 ton	4.99	4.01	9
3200	Food supermarkets, 500 S.F., 1.42 ton	4.88	5.15	10.03
3240	1,000 S.F., 2.83 ton	3.03	4.20	7.23
3280	1,500 S.F., 4.25 ton	3.28	4.06	7.34
3320	3,000 S.F., 8.50 ton	3.36	3.63	6.99
3360	5,000 S.F., 14.17 ton	3.25	3.49	6.74
3400	10,000 S.F., 28.33 ton	4.23	3.40	7.63
3480	Medical centers, 500 S.F., 1.17 ton	4.03	4.24	8.27
3520	1,000 S.F., 2.33 ton	4.03	4.23	8.26
3560	1,500 S.F., 3.50 ton	2.53	3.47	6
3640	5,000 S.F., 11.67 ton	2.77	2.99	5.76
3680	10,000 S.F., 23.33 ton	4.23	2.82	7.05
3760	Offices, 500 S.F., 1.58 ton	5.45	5.75	11.20
3800	1,000 S.F., 3.17 ton	3.43	4.72	8.15
3840	1,500 S.F., 4.75 ton	3.67	4.53	8.20
3880	3,000 S.F., 9.50 ton	3.77	4.06	7.83
3920	5,000 S.F., 15.83 ton	3.64	3.89	7.53
3960	10,000 S.F., 31.67 ton	4.74	3.81	8.55
4000	Restaurants, 500 S.F., 2.50 ton	8.65	9.05	17.70
4040	1,000 S.F., 5.00 ton	5.80	7.15	12.95
4080	1,500 S.F., 7.50 ton	5.95	6.40	12.35
4120	3,000 S.F., 15.00 ton	5.75	6.15	11.90
4160	5,000 S.F., 25.00 ton	9.05	6.05	15.10
4200	10,000 S.F., 50.00 ton	6.45	6	12.45
4240	Schools and colleges, 500 S.F., 1.92 ton	6.60	6.95	13.55
4280	1,000 S.F., 3.83 ton	4.14	5.70	9.84
4360	3,000 S.F., 11.50 ton	4.55	4.91	9.46
4400	5,000 S.F., 19.17 ton	6.95	4.64	11.59
4441	10,000 S.F., 38.33 ton	5.35	4.60	9.95

D3050 Terminal & Package Units

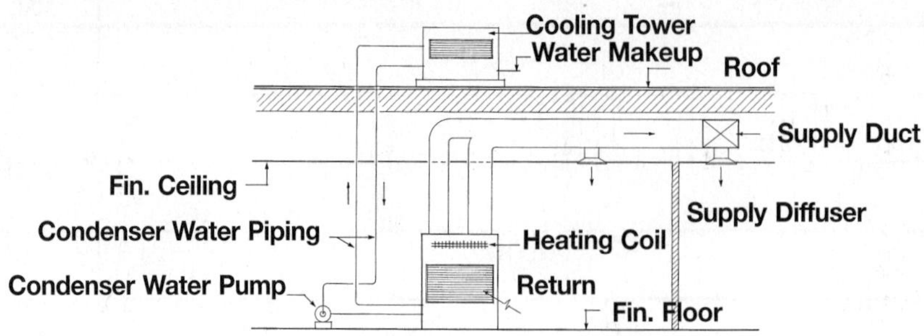

System Description: Self-contained, single package water cooled units include cooling tower, pump, piping allowance. Systems for 1000 S.F. and up include duct and diffusers to provide for even distribution of air. Smaller units distribute air through a supply air plenum, which is integral with the unit.

Returns are not ducted and supplies are not insulated.

Hot water or steam heating coils are included but piping to boiler and the boiler itself is not included.

Where local codes or conditions permit single pass cooling for the smaller units, deduct 10%.

D3050 160	Self-contained, Water Cooled Unit Systems	COST PER S.F.		
		MAT.	INST.	TOTAL
1280	Self-contained, water cooled unit	3.42	2.63	6.05
1300	Apartment corridors, 500 S.F., .92 ton	3.40	2.65	6.05
1320	1,000 S.F., 1.83 ton	3.40	2.62	6.02
1360	3,000 S.F., 5.50 ton	2.83	2.22	5.05
1400	5,000 S.F., 9.17 ton	2.73	2.06	4.79
1440	10,000 S.F., 18.33 ton	3.94	1.85	5.79
1520	Banks or libraries, 500 S.F., 2.08 ton	7.45	2.60	10.05
1560	1,000 S.F., 4.17 ton	6.45	5.05	11.50
1600	3,000 S.F., 12.50 ton	6.20	4.71	10.91
1640	5,000 S.F., 20.80 ton	8.95	4.22	13.17
1680	10,000 S.F., 41.66 ton	7.80	4.22	12.02
1760	Bars and taverns, 500 S.F., 5.54 ton	16.30	4.39	20.69
1800	1,000 S.F., 11.08 ton	16.05	8	24.05
1840	3,000 S.F., 33.25 ton	21	6.35	27.35
1880	5,000 S.F., 55.42 ton	19.85	6.70	26.55
1920	10,000 S.F., 110.00 ton	19.30	6.65	25.95
2000	Bowling alleys, 500 S.F., 2.83 ton	10.15	3.55	13.70
2040	1,000 S.F., 5.66 ton	8.75	6.85	15.60
2080	3,000 S.F., 17.00 ton	12.20	5.75	17.95
2120	5,000 S.F., 28.33 ton	11.05	5.55	16.60
2160	10,000 S.F., 56.66 ton	10.35	5.70	16.05
2200	Department stores, 500 S.F., 1.46 ton	5.25	1.83	7.08
2240	1,000 S.F., 2.92 ton	4.49	3.53	8.02
2280	3,000 S.F., 8.75 ton	4.33	3.29	7.62
2320	5,000 S.F., 14.58 ton	6.25	2.96	9.21
2360	10,000 S.F., 29.17 ton	5.70	2.86	8.56
2440	Drug stores, 500 S.F., 3.33 ton	11.95	4.17	16.12
2480	1,000 S.F., 6.66 ton	10.30	8.05	18.35
2520	3,000 S.F., 20.00 ton	14.30	6.75	21.05
2560	5,000 S.F., 33.33 ton	15.05	6.65	21.70
2600	10,000 S.F., 66.66 ton	12.15	6.75	18.90
2680	Factories, 500 S.F., 1.66 ton	5.95	2.08	8.03
2720	1,000 S.F. 3.37 ton	5.20	4.08	9.28
2760	3,000 S.F., 10.00 ton	4.96	3.77	8.73
2800	5,000 S.F., 16.66 ton	7.15	3.38	10.53
2840	10,000 S.F., 33.33 ton	6.50	3.26	9.76
2920	Food supermarkets, 500 S.F., 1.42 ton	5.05	1.77	6.82
2960	1,000 S.F., 2.83 ton	5.25	4.06	9.31

D3050 Terminal & Package Units

D3050 160	Self-contained, Water Cooled Unit Systems	COST PER S.F.		
		MAT.	INST.	TOTAL
3000	3,000 S.F., 8.50 ton	4.37	3.43	7.80
3040	5,000 S.F., 14.17 ton	4.21	3.19	7.40
3080	10,000 S.F., 28.33 ton	5.50	2.77	8.27
3160	Medical centers, 500 S.F., 1.17 ton	4.18	1.45	5.63
3200	1,000 S.F., 2.33 ton	4.35	3.35	7.70
3240	3,000 S.F., 7.00 ton	3.61	2.82	6.43
3280	5,000 S.F., 11.66 ton	3.48	2.63	6.11
3320	10,000 S.F., 23.33 ton	5	2.36	7.36
3400	Offices, 500 S.F., 1.58 ton	5.65	1.98	7.63
3440	1,000 S.F., 3.17 ton	5.90	4.54	10.44
3480	3,000 S.F., 9.50 ton	4.71	3.58	8.29
3520	5,000 S.F., 15.83 ton	6.80	3.21	10.01
3560	10,000 S.F., 31.67 ton	6.15	3.09	9.24
3640	Restaurants, 500 S.F., 2.50 ton	8.95	3.13	12.08
3680	1,000 S.F., 5.00 ton	7.70	6.05	13.75
3720	3,000 S.F., 15.00 ton	7.45	5.65	13.10
3760	5,000 S.F., 25.00 ton	10.75	5.10	15.85
3800	10,000 S.F., 50.00 ton	6.70	4.68	11.38
3880	Schools and colleges, 500 S.F., 1.92 ton	6.85	2.41	9.26
3920	1,000 S.F., 3.83 ton	5.90	4.64	10.54
3960	3,000 S.F., 11.50 ton	5.70	4.34	10.04
4000	5,000 S.F., 19.17 ton	8.25	3.89	12.14
4040	10,000 S.F., 38.33 ton	7.20	3.89	11.09

D3050 Terminal & Package Units

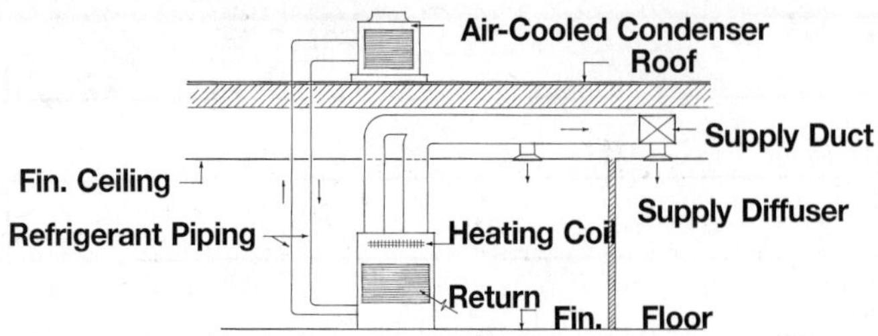

System Description: Self-contained air cooled units with remote air cooled condenser and interconnecting tubing. Systems for 1000 S.F. and up include duct and diffusers. Smaller units distribute air directly.

Returns are not ducted and supplies are not insulated.

Potential savings may be realized by using a single zone rooftop system or through-the-wall unit, especially in the smaller capacities, if the application permits.

Hot water or steam heating coils are included but piping to boiler and the boiler itself is not included.

Condenserless models are available for 15% less where remote refrigerant source is available.

D3050 165	Self-contained, Air Cooled Unit Systems	COST PER S.F.		
		MAT.	INST.	TOTAL
1300	Self-contained, air cooled unit			
1320	Apartment corridors, 500 S.F., .92 ton	5.30	3.50	8.80
1360	1,000 S.F., 1.83 ton	5.20	3.45	8.65
1400	3,000 S.F., 5.50 ton	4.85	3.18	8.03
1440	5,000 S.F., 9.17 ton	4	3.03	7.03
1480	10,000 S.F., 18.33 ton	3.34	2.79	6.13
1560	Banks or libraries, 500 S.F., 2.08 ton	11.50	4.47	15.97
1600	1,000 S.F., 4.17 ton	10.95	7.20	18.15
1640	3,000 S.F., 12.50 ton	9.10	6.90	16
1680	5,000 S.F., 20.80 ton	7.70	6.35	14.05
1720	10,000 S.F., 41.66 ton	8.50	6.20	14.70
1800	Bars and taverns, 500 S.F., 5.54 ton	26.50	9.75	36.25
1840	1,000 S.F., 11.08 ton	24	13.85	37.85
1880	3,000 S.F., 33.25 ton	22.50	12.05	34.55
1920	5,000 S.F., 55.42 ton	22.50	12.10	34.60
1960	10,000 S.F., 110.00 ton	23	12.05	35.05
2040	Bowling alleys, 500 S.F., 2.83 ton	15.80	6.10	21.90
2080	1,000 S.F., 5.66 ton	14.95	9.80	24.75
2120	3,000 S.F., 17.00 ton	10.40	8.65	19.05
2160	5,000 S.F., 28.33 ton	11.75	8.45	20.20
2200	10,000 S.F., 56.66 ton	11.80	8.45	20.25
2240	Department stores, 500 S.F., 1.46 ton	8.15	3.14	11.29
2280	1,000 S.F., 2.92 ton	7.70	5.05	12.75
2320	3,000 S.F., 8.75 ton	6.35	4.82	11.17
2360	5,000 S.F., 14.58 ton	6.35	4.82	11.17
2400	10,000 S.F., 29.17 ton	6.05	4.36	10.41
2480	Drug stores, 500 S.F., 3.33 ton	18.60	7.20	25.80
2520	1,000 S.F., 6.66 ton	17.60	11.55	29.15
2560	3,000 S.F., 20.00 ton	12.40	10.20	22.60
2600	5,000 S.F., 33.33 ton	13.85	9.95	23.80
2640	10,000 S.F., 66.66 ton	14.05	9.95	24
2720	Factories, 500 S.F., 1.66 ton	9.45	3.61	13.06
2760	1,000 S.F., 3.33 ton	8.85	5.80	14.65
2800	3,000 S.F., 10.00 ton	7.30	5.50	12.80
2840	5,000 S.F., 16.66 ton	6.05	5.10	11.15
2880	10,000 S.F., 33.33 ton	6.90	4.98	11.88

D3050 Terminal & Package Units

D3050 165	Self-contained, Air Cooled Unit Systems	COST PER S.F.		
		MAT.	INST.	TOTAL
2960	Food supermarkets, 500 S.F., 1.42 ton	7.90	3.06	10.96
3000	1,000 S.F., 2.83 ton	8.05	5.35	13.40
3040	3,000 S.F., 8.50 ton	7.50	4.91	12.41
3080	5,000 S.F., 14.17 ton	6.20	4.69	10.89
3120	10,000 S.F., 28.33 ton	5.90	4.23	10.13
3200	Medical centers, 500 S.F., 1.17 ton	6.55	2.52	9.07
3240	1,000 S.F., 2.33 ton	6.70	4.41	11.11
3280	3,000 S.F., 7.00 ton	6.20	4.04	10.24
3320	5,000 S.F., 16.66 ton	5.10	3.87	8.97
3360	10,000 S.F., 23.33 ton	4.27	3.57	7.84
3440	Offices, 500 S.F., 1.58 ton	8.85	3.41	12.26
3480	1,000 S.F., 3.16 ton	9.10	6	15.10
3520	3,000 S.F., 9.50 ton	6.95	5.25	12.20
3560	5,000 S.F., 15.83 ton	5.85	4.83	10.68
3600	10,000 S.F., 31.66 ton	6.60	4.72	11.32
3680	Restaurants, 500 S.F., 2.50 ton	13.90	5.40	19.30
3720	1,000 S.F., 5.00 ton	13.25	8.65	21.90
3760	3,000 S.F., 15.00 ton	10.95	8.30	19.25
3800	5,000 S.F., 25.00 ton	10.35	7.45	17.80
3840	10,000 S.F., 50.00 ton	10.75	7.45	18.20
3920	Schools and colleges, 500 S.F., 1.92 ton	10.70	4.13	14.83
3960	1,000 S.F., 3.83 ton	10.15	6.65	16.80
4000	3,000 S.F., 11.50 ton	8.35	6.35	14.70
4040	5,000 S.F., 19.17 ton	7	5.85	12.85
4080	10,000 S.F., 38.33 ton	7.80	5.70	13.50

D3050 Terminal & Package Units

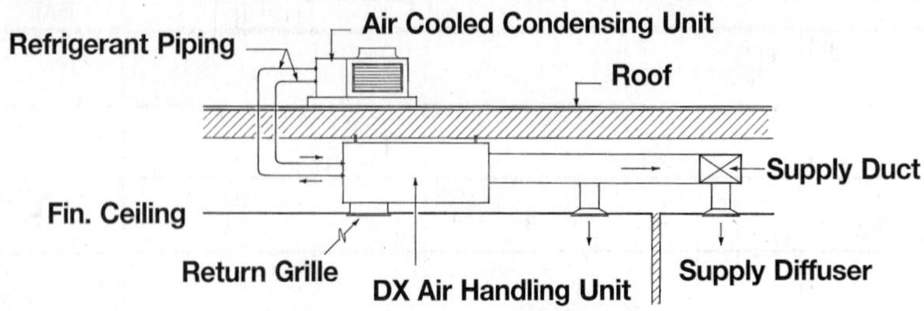

General: Split systems offer several important advantages which should be evaluated when a selection is to be made. They provide a greater degree of flexibility in component selection which permits an accurate match-up of the proper equipment size and type with the particular needs of the building. This allows for maximum use of modern energy saving concepts in heating and cooling. Outdoor installation of the air cooled condensing unit allows space savings in the building and also isolates the equipment operating sounds from building occupants.

Design Assumptions: The systems below are comprised of a direct expansion air handling unit and air cooled condensing unit with interconnecting copper tubing. Ducts and diffusers are also included for distribution of air. Systems are priced for cooling only. Heat can be added as desired either by putting hot water/steam coils into the air unit or into the duct supplying the particular area of need. Gas fired duct furnaces are also available. Refrigerant liquid line is insulated.

D3050 170	Split Systems With Air Cooled Condensing Units	COST PER S.F.		
		MAT.	INST.	TOTAL
1260	Split system, air cooled condensing unit			
1280	Apartment corridors, 1,000 S.F., 1.83 ton	2.53	2.18	4.71
1320	2,000 S.F., 3.66 ton	1.79	2.18	3.97
1360	5,000 S.F., 9.17 ton	1.63	2.73	4.36
1400	10,000 S.F., 18.33 ton	2.03	2.94	4.97
1440	20,000 S.F., 36.66 ton	2.15	2.99	5.14
1520	Banks and libraries, 1,000 S.F., 4.17 ton	4.07	4.97	9.04
1560	2,000 S.F., 8.33 ton	3.70	6.20	9.90
1600	5,000 S.F., 20.80 ton	4.63	6.70	11.33
1640	10,000 S.F., 41.66 ton	4.92	6.80	11.72
1680	20,000 S.F., 83.32 ton	5.95	6.90	12.85
1760	Bars and taverns, 1,000 S.F., 11.08 ton	8.45	9.75	18.20
1800	2,000 S.F., 22.16 ton	12.60	11.40	24
1840	5,000 S.F., 55.42 ton	10.80	10.70	21.50
1880	10,000 S.F., 110.84 ton	13.10	10.95	24.05
1920	20,000 S.F., 220 ton*			
2000	Bowling alleys, 1,000 S.F., 5.66 ton	5.60	9.30	14.90
2040	2,000 S.F., 11.33 ton	5	8.40	13.40
2080	5,000 S.F., 28.33 ton	6.30	9.10	15.40
2120	10,000 S.F., 56.66 ton	6.70	9.25	15.95
2160	20,000 S.F., 113.32 ton	8.75	9.75	18.50
2320	Department stores, 1,000 S.F., 2.92 ton	2.88	3.43	6.31
2360	2,000 S.F., 5.83 ton	2.89	4.80	7.69
2400	5,000 S.F., 14.58 ton	2.58	4.34	6.92
2440	10,000 S.F., 29.17 ton	3.24	4.69	7.93
2480	20,000 S.F., 58.33 ton	3.44	4.76	8.20
2560	Drug stores, 1,000 S.F., 6.66 ton	6.60	10.95	17.55
2600	2,000 S.F., 13.32 ton	5.90	9.90	15.80
2640	5,000 S.F., 33.33 ton	7.85	10.85	18.70
2680	10,000 S.F., 66.66 ton	8.10	11.10	19.20
2720	20,000 S.F., 133.32 ton*			
2800	Factories, 1,000 S.F., 3.33 ton	3.28	3.91	7.19
2840	2,000 S.F., 6.66 ton	3.31	5.45	8.76
2880	5,000 S.F., 16.66 ton	3.70	5.35	9.05
2920	10,000 S.F., 33.33 ton	3.93	5.45	9.38
2960	20,000 S.F., 66.66 ton	4.05	5.55	9.60
3040	Food supermarkets, 1,000 S.F., 2.83 ton	2.80	3.32	6.12
3080	2,000 S.F., 5.66 ton	2.82	4.65	7.47

D3050 Terminal & Package Units

D3050 170	Split Systems With Air Cooled Condensing Units	COST PER S.F.		
		MAT.	INST.	TOTAL
3120	5,000 S.F., 14.66 ton	2.51	4.21	6.72
3160	10,000 S.F., 28.33 ton	3.15	4.56	7.71
3200	20,000 S.F., 56.66 ton	3.34	4.63	7.97
3280	Medical centers, 1,000 S.F., 2.33 ton	2.51	2.70	5.21
3320	2,000 S.F., 4.66 ton	2.31	3.84	6.15
3360	5,000 S.F., 11.66 ton	2.07	3.48	5.55
3400	10,000 S.F., 23.33 ton	2.60	3.76	6.36
3440	20,000 S.F., 46.66 ton	2.76	3.82	6.58
3520	Offices, 1,000 S.F., 3.17 ton	3.12	3.72	6.84
3560	2,000 S.F., 6.33 ton	3.14	5.20	8.34
3600	5,000 S.F., 15.83 ton	2.81	4.71	7.52
3640	10,000 S.F., 31.66 ton	3.53	5.10	8.63
3680	20,000 S.F., 63.32 ton	3.84	5.25	9.09
3760	Restaurants, 1,000 S.F., 5.00 ton	4.97	8.20	13.17
3800	2,000 S.F., 10.00 ton	4.44	7.45	11.89
3840	5,000 S.F., 25.00 ton	5.55	8.05	13.60
3880	10,000 S.F., 50.00 ton	5.90	8.15	14.05
3920	20,000 S.F., 100.00 ton	7.75	8.60	16.35
4000	Schools and colleges, 1,000 S.F., 3.83 ton	3.74	4.59	8.33
4040	2,000 S.F., 7.66 ton	3.39	5.70	9.09
4080	5,000 S.F., 19.17 ton	4.26	6.15	10.41
4120	10,000 S.F., 38.33 ton	4.52	6.25	10.77
4160	20,000 S.F., 76.66 ton	4.65	6.35	11

D3090 Other HVAC Systems/Equip

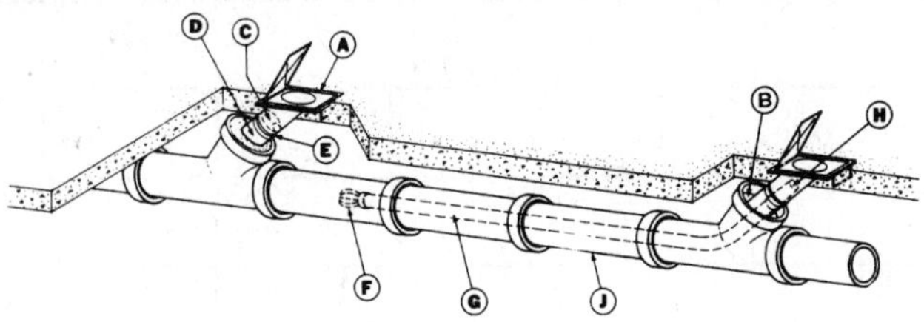

Cast Iron Garage Exhaust System

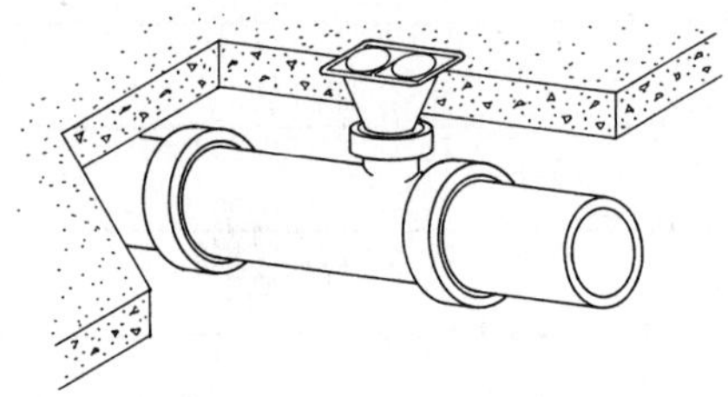

Dual Exhaust System

D3090 320	Garage Exhaust Systems	COST PER BAY		
		MAT.	INST.	TOTAL
1040	Garage, single 3" exhaust outlet, cars & light trucks, one bay	4,400	1,575	5,975
1060	Additional bays up to seven bays	1,000	405	1,405
1500	4" outlet, trucks, one bay	4,425	1,575	6,000
1520	Additional bays up to six bays	1,025	405	1,430
1600	5" outlet, diesel trucks, one bay	4,725	1,575	6,300
1650	Additional single bays up to six	1,425	480	1,905
1700	Two adjoining bays	4,725	1,575	6,300
2000	Dual exhaust, 3" outlets, pair of adjoining bays	5,475	1,950	7,425
2100	Additional pairs of adjoining bays	1,600	480	2,080

D4010 Sprinklers

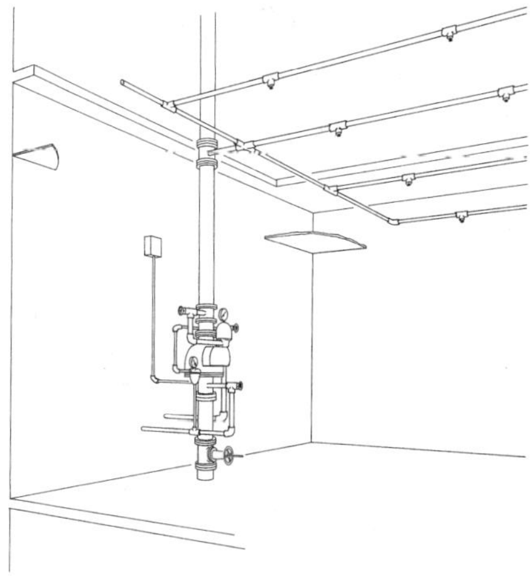

Dry Pipe System: A system employing automatic sprinklers attached to a piping system containing air under pressure, the release of which from the opening of sprinklers permits the water pressure to open a valve known as a "dry pipe valve". The water then flows into the piping system and out the opened sprinklers.

All areas are assumed to be open.

D4010 310	Dry Pipe Sprinkler Systems	COST PER S.F.		
		MAT.	INST.	TOTAL
0520	Dry pipe sprinkler systems, steel, black, sch. 40 pipe			
0530	Light hazard, one floor, 500 S.F.	10.05	5.15	15.20
0560	1000 S.F.	5.70	2.98	8.68
0580	2000 S.F.	5.50	3.04	8.54
0600	5000 S.F.	2.87	2.06	4.93
0620	10,000 S.F.	2.01	1.71	3.72
0640	50,000 S.F.	1.47	1.51	2.98
0660	Each additional floor, 500 S.F.	2.04	2.52	4.56
0680	1000 S.F.	1.85	2.04	3.89
0700	2000 S.F.	1.81	1.91	3.72
0720	5000 S.F.	1.48	1.64	3.12
0740	10,000 S.F.	1.33	1.49	2.82
0760	50,000 S.F.	1.22	1.33	2.55
1000	Ordinary hazard, one floor, 500 S.F.	10.25	5.25	15.50
1020	1000 S.F.	5.90	3.02	8.92
1040	2000 S.F.	5.70	3.17	8.87
1060	5000 S.F.	3.33	2.18	5.51
1080	10,000 S.F.	2.65	2.24	4.89
1100	50,000 S.F.	2.37	2.10	4.47
1140	Each additional floor, 500 S.F.	2.26	2.58	4.84
1160	1000 S.F.	2.14	2.30	4.44
1180	2000 S.F.	2.18	2.11	4.29
1200	5000 S.F.	2.03	1.79	3.82
1220	10,000 S.F.	1.84	1.74	3.58
1240	50,000 S.F.	1.76	1.50	3.26
1500	Extra hazard, one floor, 500 S.F.	14.40	6.45	20.85
1520	1000 S.F.	8.80	4.67	13.47
1540	2000 S.F.	6.30	4.05	10.35
1560	5000 S.F.	3.83	3.05	6.88
1580	10,000 S.F.	4.14	2.90	7.04
1600	50,000 S.F.	4.47	2.79	7.26
1660	Each additional floor, 500 S.F.	3.32	3.17	6.49
1680	1000 S.F.	3.28	3.01	6.29
1700	2000 S.F.	3	3.02	6.02
1720	5000 S.F.	2.55	2.63	5.18
1740	10,000 S.F.	3.14	2.38	5.52
1760	50,000 S.F.	3.16	2.30	5.46
2020	Grooved steel, black, sch. 40 pipe, light hazard, one floor, 2000 S.F.	5.35	2.60	7.95

D4010 Sprinklers

D4010 310	Dry Pipe Sprinkler Systems	COST PER S.F.		
		MAT.	INST.	TOTAL
2060	10,000 S.F.	2.05	1.49	3.54
2100	Each additional floor, 2000 S.F.	1.84	1.54	3.38
2150	10,000 S.F.	1.37	1.27	2.64
2200	Ordinary hazard, one floor, 2000 S.F.	5.65	2.75	8.40
2250	10,000 S.F.	2.55	1.90	4.45
2300	Each additional floor, 2000 S.F.	2.12	1.69	3.81
2350	10,000 S.F.	1.88	1.68	3.56
2400	Extra hazard, one floor, 2000 S.F.	6.35	3.47	9.82
2450	10,000 S.F.	3.64	2.46	6.10
2500	Each additional floor, 2000 S.F.	3.06	2.49	5.55
2550	10,000 S.F.	2.78	2.16	4.94
3050	Grooved steel, black, sch. 10 pipe, light hazard, one floor, 2000 S.F.	5.30	2.57	7.87
3100	10,000 S.F.	2.03	1.47	3.50
3150	Each additional floor, 2000 S.F.	1.79	1.51	3.30
3200	10,000 S.F.	1.35	1.25	2.60
3250	Ordinary hazard, one floor, 2000 S.F.	5.60	2.74	8.34
3300	10,000 S.F.	2.49	1.85	4.34
3350	Each additional floor, 2000 S.F.	2.08	1.68	3.76
3400	10,000 S.F.	1.82	1.63	3.45
3450	Extra hazard, one floor, 2000 S.F.	6.35	3.45	9.80
3500	10,000 S.F.	3.46	2.41	5.87
3550	Each additional floor, 2000 S.F.	3.03	2.47	5.50
3600	10,000 S.F.	2.72	2.12	4.84
4050	Copper tubing, type M, light hazard, one floor, 2000 S.F.	5.65	2.58	8.23
4100	10,000 S.F.	2.51	1.49	4
4150	Each additional floor, 2000 S.F.	2.18	1.55	3.73
4200	10,000 S.F.	1.84	1.28	3.12
4250	Ordinary hazard, one floor, 2000 S.F.	6.10	2.87	8.97
4300	10,000 S.F.	3.08	1.75	4.83
4350	Each additional floor, 2000 S.F.	2.84	1.77	4.61
4400	10,000 S.F.	2.37	1.50	3.87
4450	Extra hazard, one floor, 2000 S.F.	7	3.51	10.51
4500	10,000 S.F.	5.15	2.60	7.75
4550	Each additional floor, 2000 S.F.	3.69	2.53	6.22
4600	10,000 S.F.	3.83	2.28	6.11
5050	Copper tubing, type M, T-drill system, light hazard, one floor			
5060	2000 S.F.	5.70	2.41	8.11
5100	10,000 S.F.	2.40	1.25	3.65
5150	Each additional floor, 2000 S.F.	2.20	1.38	3.58
5200	10,000 S.F.	1.73	1.04	2.77
5250	Ordinary hazard, one floor, 2000 S.F.	5.90	2.47	8.37
5300	10,000 S.F.	2.99	1.58	4.57
5350	Each additional floor, 2000 S.F.	2.40	1.41	3.81
5400	10,000 S.F.	2.22	1.29	3.51
5450	Extra hazard, one floor, 2000 S.F.	6.50	2.90	9.40
5500	10,000 S.F.	4.46	1.92	6.38
5550	Each additional floor, 2000 S.F.	3.22	1.92	5.14
5600	10,000 S.F.	3.12	1.60	4.72

Reference - System Classification

System Classification

Rules for installation of sprinkler systems vary depending on the classification of occupancy falling into one of three categories as follows:

Light Hazard Occupancy

The protection area allotted per sprinkler should not exceed 225 S.F., with the maximum distance between lines and sprinklers on lines being 15′. The sprinklers do not need to be staggered. Branch lines should not exceed eight sprinklers on either side of a cross main. Each large area requiring more than 100 sprinklers and without a sub-dividing partition should be supplied by feed mains or risers sized for ordinary hazard occupancy.
Maximum system area = 52,000 S.F.

Included in this group are:

Churches
Clubs
Educational
Hospitals
Institutional
Libraries
(except large stack rooms)
Museums
Nursing Homes
Offices
Residential
Restaurants
Theaters and Auditoriums
(except stages and prosceniums)
Unused Attics

Ordinary Hazard Occupancy

The protection area allotted per sprinkler shall not exceed 130 S.F. of noncombustible ceiling and 130 S.F. of combustible ceiling. The maximum allowable distance between sprinkler lines and sprinklers on line is 15′. Sprinklers shall be staggered if the distance between heads exceeds 12′. Branch lines should not exceed eight sprinklers on either side of a cross main.
Maximum system area = 52,000 S.F.

Included in this group are:

Group 1
Automotive Parking and Showrooms
Bakeries
Beverage manufacturing
Canneries
Dairy Products Manufacturing/Processing
Electronic Plans
Glass and Glass Products Manufacturing
Laundries
Restaurant Service Areas

Group 2
Cereal Mills
Chemical Plants—Ordinary
Confectionery Products
Distilleries
Dry Cleaners
Feed Mills
Horse Stables
Leather Goods Manufacturing
Libraries—Large Stack Room Areas
Machine Shops
Metal Working
Mercantile
Paper and Pulp Mills
Paper Process Plants
Piers and Wharves
Post Offices
Printing and Publishing
Repair Garages
Stages
Textile Manufacturing
Tire Manufacturing
Tobacco Products Manufacturing
Wood Machining
Wood Product Assembly

Extra Hazard Occupancy

The protection area allotted per sprinkler shall not exceed 100 S.F. of noncombustible ceiling and 100 S.F. of combustible ceiling. The maximum allowable distance between lines and between sprinklers on lines is 12′. Sprinklers on alternate lines shall be staggered if the distance between sprinklers on lines exceeds 8′. Branch lines should not exceed six sprinklers on either side of a cross main.
Maximum system area:
Design by pipe schedule = 25,000 S.F.
Design by hydraulic calculation = 40,000 S.F.

Included in this group are:

Group 1
Aircraft hangars
Combustible Hydraulic Fluid Use Area
Die Casting
Metal Extruding
Plywood/Particle Board Manufacturing
Printing (inks with flash points < 100 degrees F
Rubber Reclaiming, Compounding, Drying, Milling, Vulcanizing
Saw Mills
Textile Picking, Opening, Blending, Garnetting, Carding, Combing of Cotton, Synthetics, Wood Shoddy, or Burlap
Upholstering with Plastic Foams

Group 2
Asphalt Saturating
Flammable Liquids Spraying
Flow Coating
Manufactured/Modular Home Building Assemblies (where finished enclosure is present and has combustible interiors)
Open Oil Quenching
Plastics Processing
Solvent Cleaning
Varnish and Paint Dipping

Reference - Sprinkler Systems (Automatic)

Sprinkler systems may be classified by type as follows:

1. **Wet Pipe System.** A system employing automatic sprinklers attached to a piping system containing water and connected to a water supply so that water discharges immediately from sprinklers opened by a fire.
2. **Dry Pipe System.** A system employing automatic sprinklers attached to a piping system containing air under pressure, the release of which as from the opening of sprinklers permits the water pressure to open a valve known as a "dry pipe valve". The water then flows into the piping system and out the opened sprinklers.
3. **Pre-Action System.** A system employing automatic sprinklers attached to a piping system containing air that may or may not be under pressure, with a supplemental heat responsive system of generally more sensitive characteristics than the automatic sprinklers themselves, installed in the same areas as the sprinklers; actuation of the heat responsive system, as from a fire, opens a valve which permits water to flow into the sprinkler piping system and to be discharged from any sprinklers which may be open.
4. **Deluge System.** A system employing open sprinklers attached to a piping system connected to a water supply through a valve which is opened by the operation of a heat responsive system installed in the same areas as the sprinklers. When this valve opens, water flows into the piping system and discharges from all sprinklers attached thereto.
5. **Combined Dry Pipe and Pre-Action Sprinkler System.** A system employing automatic sprinklers attached to a piping system containing air under pressure with a supplemental heat responsive system of generally more sensitive characteristics than the automatic sprinklers themselves, installed in the same areas as the sprinklers; operation of the heat responsive system, as from a fire, actuates tripping devices which open dry pipe valves simultaneously and without loss of air pressure in the system. Operation of the heat responsive system also opens approved air exhaust valves at the end of the feed main which facilitates the filling of the system with water which usually precedes the opening of sprinklers. The heat responsive system also serves as an automatic fire alarm system.
6. **Limited Water Supply System.** A system employing automatic sprinklers and conforming to these standards but supplied by a pressure tank of limited capacity.
7. **Chemical Systems.** Systems using halon, carbon dioxide, dry chemical or high expansion foam as selected for special requirements. Agent may extinguish flames by chemically inhibiting flame propagation, suffocate flames by excluding oxygen, interrupting chemical action of oxygen uniting with fuel or sealing and cooling the combustion center.
8. **Firecycle System.** Firecycle is a fixed fire protection sprinkler system utilizing water as its extinguishing agent. It is a time delayed, recycling, preaction type which automatically shuts the water off when heat is reduced below the detector operating temperature and turns the water back on when that temperature is exceeded. The system senses a fire condition through a closed circuit electrical detector system which controls water flow to the fire automatically. Batteries supply up to 90 hour emergency power supply for system operation. The piping system is dry (until water is required) and is monitored with pressurized air. Should any leak in the system piping occur, an alarm will sound, but water will not enter the system until heat is sensed by a firecycle detector.

Area coverage sprinkler systems may be laid out and fed from the supply in any one of several patterns as shown below. It is desirable, if possible, to utilize a central feed and achieve a shorter flow path from the riser to the furthest sprinkler. This permits use of the smallest sizes of pipe possible with resulting savings.

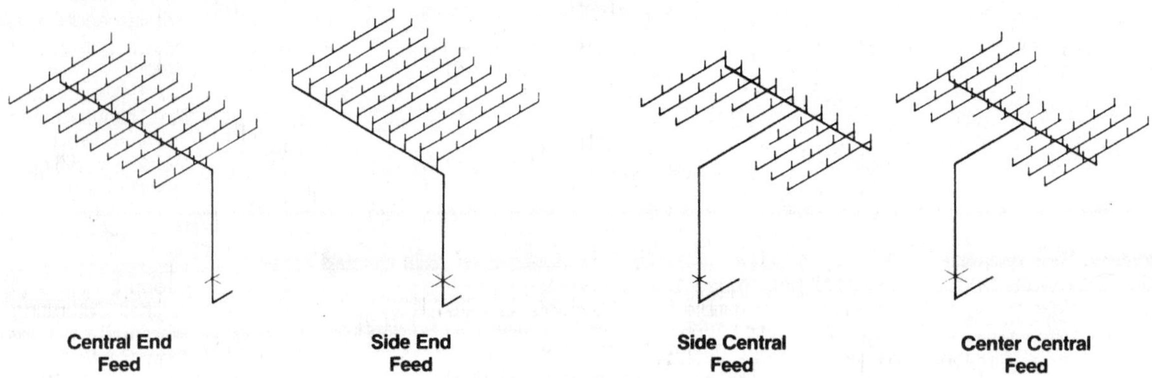

D4010 Sprinklers

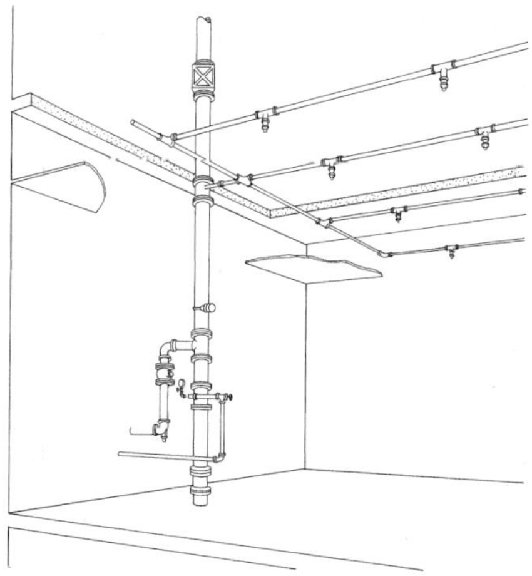

Wet Pipe System. A system employing automatic sprinklers attached to a piping system containing water and connected to a water supply so that water discharges immediately from sprinklers opened by heat from a fire.

All areas are assumed to be open.

D4010 410	Wet Pipe Sprinkler Systems	COST PER S.F. MAT.	COST PER S.F. INST.	COST PER S.F. TOTAL
0520	Wet pipe sprinkler systems, steel, black, sch. 40 pipe			
0530	Light hazard, one floor, 500 S.F.	2.96	2.66	5.62
0560	1000 S.F.	6.35	2.70	9.05
0580	2000 S.F.	5.70	2.73	8.43
0600	5000 S.F.	2.64	1.93	4.57
0620	10,000 S.F.	1.65	1.63	3.28
0640	50,000 S.F.	1.01	1.47	2.48
0660	Each additional floor, 500 S.F.	1.26	2.25	3.51
0680	1000 S.F.	1.31	2.07	3.38
0700	2000 S.F.	1.24	1.89	3.13
0720	5000 S.F.	.87	1.62	2.49
0740	10,000 S.F.	.79	1.47	2.26
0760	50,000 S.F.	.66	1.15	1.81
1000	Ordinary hazard, one floor, 500 S.F.	3.26	2.84	6.10
1020	1000 S.F.	6.30	2.66	8.96
1040	2000 S.F.	5.85	2.87	8.72
1060	5000 S.F.	2.95	2.04	4.99
1080	10,000 S.F.	2.07	2.15	4.22
1100	50,000 S.F.	1.64	2.02	3.66
1140	Each additional floor, 500 S.F.	1.50	2.55	4.05
1160	1000 S.F.	1.27	2.05	3.32
1180	2000 S.F.	1.42	2.08	3.50
1200	5000 S.F.	1.40	1.94	3.34
1220	10,000 S.F.	1.22	1.99	3.21
1240	50,000 S.F.	1.16	1.76	2.92
1500	Extra hazard, one floor, 500 S.F.	12.25	4.36	16.61
1520	1000 S.F.	7.55	3.78	11.33
1540	2000 S.F.	6.15	3.85	10
1560	5000 S.F.	3.55	3.35	6.90
1580	10,000 S.F.	3.26	3.18	6.44
1600	50,000 S.F.	3.62	3.07	6.69
1660	Each additional floor, 500 S.F.	2.18	3.13	5.31
1680	1000 S.F.	2.13	2.97	5.10
1700	2000 S.F.	1.86	2.98	4.84
1720	5000 S.F.	1.53	2.62	4.15
1740	10,000 S.F.	2.01	2.39	4.40
1760	50,000 S.F.	1.99	2.30	4.29
2020	Grooved steel, black sch. 40 pipe, light hazard, one floor, 2000 S.F.	5.70	2.31	8.01

D4010 Sprinklers

D4010 410	Wet Pipe Sprinkler Systems	COST PER S.F.		
		MAT.	INST.	TOTAL
2060	10,000 S.F.	2.19	1.47	3.66
2100	Each additional floor, 2000 S.F.	1.27	1.52	2.79
2150	10,000 S.F.	.83	1.25	2.08
2200	Ordinary hazard, one floor, 2000 S.F.	5.80	2.45	8.25
2250	10,000 S.F.	1.97	1.81	3.78
2300	Each additional floor, 2000 S.F.	1.36	1.66	3.02
2350	10,000 S.F.	1.12	1.65	2.77
2400	Extra hazard, one floor, 2000 S.F.	6.15	3.16	9.31
2450	10,000 S.F.	2.66	2.35	5.01
2500	Each additional floor, 2000 S.F.	1.92	2.45	4.37
2550	10,000 S.F.	1.62	2.11	3.73
3050	Grooved steel, black sch. 10 pipe, light hazard, one floor, 2000 S.F.	5.65	2.28	7.93
3100	10,000 S.F.	1.67	1.39	3.06
3150	Each additional floor, 2000 S.F.	1.22	1.49	2.71
3200	10,000 S.F.	.81	1.23	2.04
3250	Ordinary hazard, one floor, 2000 S.F.	5.75	2.44	8.19
3300	10,000 S.F.	1.91	1.76	3.67
3350	Each additional floor, 2000 S.F.	1.32	1.65	2.97
3400	10,000 S.F.	1.06	1.60	2.66
3450	Extra hazard, one floor, 2000 S.F.	6.10	3.14	9.24
3500	10,000 S.F.	2.48	2.30	4.78
3550	Each additional floor, 2000 S.F.	1.89	2.43	4.32
3600	10,000 S.F.	1.56	2.07	3.63
4050	Copper tubing, type M, light hazard, one floor, 2000 S.F.	6	2.29	8.29
4100	10,000 S.F.	2.15	1.41	3.56
4150	Each additional floor, 2000 S.F.	1.61	1.53	3.14
4200	10,000 S.F.	1.30	1.26	2.56
4250	Ordinary hazard, one floor, 2000 S.F.	6.25	2.57	8.82
4300	10,000 S.F.	2.50	1.66	4.16
4350	Each additional floor, 2000 S.F.	1.85	1.68	3.53
4400	10,000 S.F.	1.61	1.47	3.08
4450	Extra hazard, one floor, 2000 S.F.	6.80	3.20	10
4500	10,000 S.F.	4.15	2.49	6.64
4550	Each additional floor, 2000 S.F.	2.55	2.49	5.04
4600	10,000 S.F.	2.67	2.23	4.90
5050	Copper tubing, type M, T-drill system, light hazard, one floor			
5060	2000 S.F.	6.05	2.12	8.17
5100	10,000 S.F.	2.04	1.17	3.21
5150	Each additional floor, 2000 S.F.	1.63	1.36	2.99
5200	10,000 S.F.	1.19	1.02	2.21
5250	Ordinary hazard, one floor, 2000 S.F.	6.10	2.17	8.27
5300	10,000 S.F.	2.41	1.49	3.90
5350	Each additional floor, 2000 S.F.	1.64	1.38	3.02
5400	10,000 S.F.	1.56	1.33	2.89
5450	Extra hazard, one floor, 2000 S.F.	6.30	2.59	8.89
5500	10,000 S.F.	3.44	1.81	5.25
5550	Each additional floor, 2000 S.F.	2.18	1.91	4.09
5600	10,000 S.F.	1.96	1.55	3.51

D4020 Standpipes

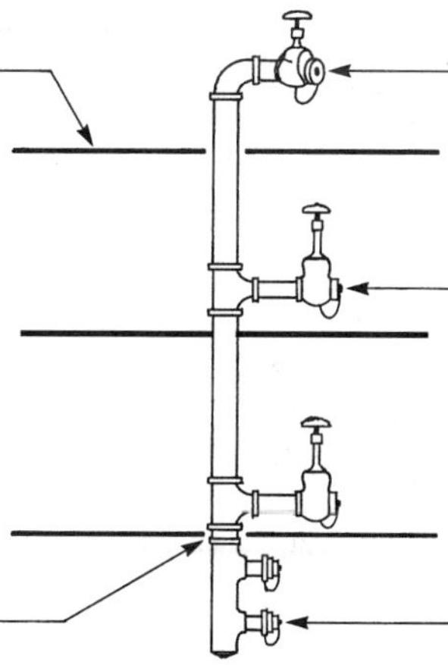

D4020 310	Wet Standpipe Risers, Class I	COST PER FLOOR		
		MAT.	INST.	TOTAL
0550	Wet standpipe risers, Class I, steel, black sch. 40, 10′ height			
0560	4″ diameter pipe, one floor	5,550	2,850	8,400
0580	Additional floors	1,325	890	2,215
0600	6″ diameter pipe, one floor	9,375	4,900	14,275
0620	Additional floors	2,375	1,375	3,750
0640	8″ diameter pipe, one floor	14,200	5,925	20,125
0660	Additional floors	3,450	1,675	5,125

D4020 310	Wet Standpipe Risers, Class II	COST PER FLOOR		
		MAT.	INST.	TOTAL
1030	Wet standpipe risers, Class II, steel, black sch. 40, 10′ height			
1040	2″ diameter pipe, one floor	2,475	1,025	3,500
1060	Additional floors	810	400	1,210
1080	2-1/2″ diameter pipe, one floor	3,425	1,525	4,950
1100	Additional floors	890	465	1,355

D4020 310	Wet Standpipe Risers, Class III	COST PER FLOOR		
		MAT.	INST.	TOTAL
1530	Wet standpipe risers, Class III, steel, black sch. 40, 10′ height			
1540	4″ diameter pipe, one floor	5,675	2,850	8,525
1560	Additional floors	1,175	745	1,920
1580	6″ diameter pipe, one floor	9,500	4,900	14,400
1600	Additional floors	2,450	1,375	3,825
1620	8″ diameter pipe, one floor	14,400	5,925	20,325
1640	Additional floors	3,525	1,675	5,200

D4020 Standpipes

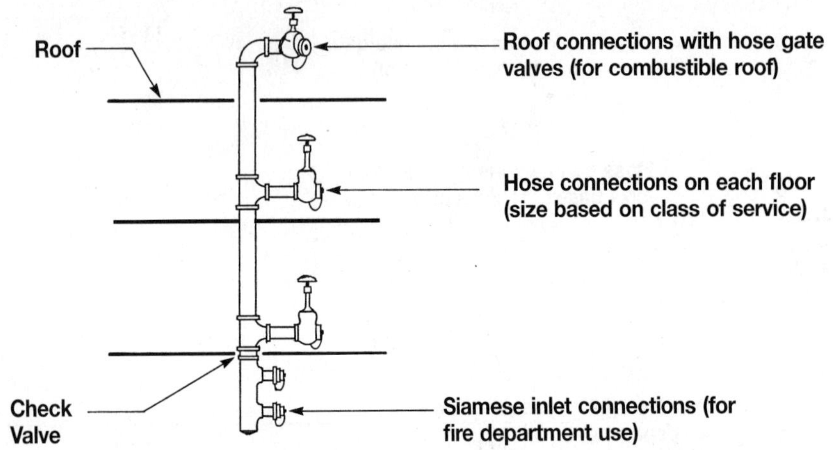

D4020 330	Dry Standpipe Risers, Class I	COST PER FLOOR		
		MAT.	INST.	TOTAL
0530	Dry standpipe riser, Class I, steel, black sch. 40, 10′ height			
0540	4″ diameter pipe, one floor	3,225	2,325	5,550
0560	Additional floors	1,225	840	2,065
0580	6″ diameter pipe, one floor	6,825	3,900	10,725
0600	Additional floors	2,275	1,350	3,625
0620	8″ diameter pipe, one floor	10,600	4,750	15,350
0640	Additional floors	3,350	1,625	4,975

D4020 330	Dry Standpipe Risers, Class II	COST PER FLOOR		
		MAT.	INST.	TOTAL
1030	Dry standpipe risers, Class II, steel, black sch. 40, 10′ height			
1040	2″ diameter pipe, one floor	2,075	1,075	3,150
1060	Additional floors	705	350	1,055
1080	2-1/2″ diameter pipe, one floor	2,750	1,250	4,000
1100	Additional floors	790	420	1,210

D4020 330	Dry Standpipe Risers, Class III	COST PER FLOOR		
		MAT.	INST.	TOTAL
1530	Dry standpipe risers, Class III, steel, black sch. 40, 10′ height			
1540	4″ diameter pipe, one floor	3,275	2,275	5,550
1560	Additional floors	1,075	760	1,835
1580	6″ diameter pipe, one floor	6,900	3,900	10,800
1600	Additional floors	2,350	1,350	3,700
1620	8″ diameter pipe, one floor	10,600	4,750	15,350
1640	Additional floors	3,425	1,625	5,050

D4020 Standpipes

D4020 410	Fire Hose Equipment	COST EACH		
		MAT.	INST.	TOTAL
0100	Adapters, reducing, 1 piece, FxM, hexagon, cast brass, 2-1/2" x 1-1/2"	64.50		64.50
0200	Pin lug, 1-1/2" x 1"	55.50		55.50
0250	3" x 2-1/2"	144		144
0300	For polished chrome, add 75% mat.			
0400	Cabinets, D.S. glass in door, recessed, steel box, not equipped			
0500	Single extinguisher, steel door & frame	134	127	261
0550	Stainless steel door & frame	237	127	364
0600	Valve, 2-1/2" angle, steel door & frame	156	84.50	240.50
0650	Aluminum door & frame	189	84.50	273.50
0700	Stainless steel door & frame	255	84.50	339.50
0750	Hose rack assy, 2-1/2" x 1-1/2" valve & 100' hose, steel door & frame	292	169	461
0800	Aluminum door & frame	430	169	599
0850	Stainless steel door & frame	575	169	744
0900	Hose rack assy & extinguisher,2-1/2"x1-1/2" valve & hose,steel door & frame	305	203	508
0950	Aluminum	550	203	753
1000	Stainless steel	605	203	808
1550	Compressor, air, dry pipe system, automatic, 200 gal., 3/4 H.P.	890	435	1,325
1600	520 gal., 1 H.P.	925	435	1,360
1650	Alarm, electric pressure switch (circuit closer)	102	21.50	123.50
2500	Couplings, hose, rocker lug, cast brass, 1-1/2"	61.50		61.50
2550	2-1/2"	81		81
3000	Escutcheon plate, for angle valves, polished brass, 1-1/2"	16.70		16.70
3050	2-1/2"	27.50		27.50
3500	Fire pump, electric, w/controller, fittings, relief valve			
3550	4" pump, 30 H.P., 500 G.P.M.	17,000	3,000	20,000
3600	5" pump, 40 H.P., 1000 G.P.M.	19,000	3,400	22,400
3650	5" pump, 100 H.P., 1000 G.P.M.	25,800	3,775	29,575
3700	For jockey pump system, add	2,975	510	3,485
5000	Hose, per linear foot, synthetic jacket, lined,			
5100	300 lb. test, 1-1/2" diameter	3.74	.39	4.13
5150	2-1/2" diameter	6.45	.46	6.91
5200	500 lb. test, 1-1/2" diameter	3.86	.39	4.25
5250	2-1/2" diameter	6.85	.46	7.31
5500	Nozzle, plain stream, polished brass, 1-1/2" x 10"	55.50		55.50
5550	2-1/2" x 15" x 13/16" or 1-1/2"	116		116
5600	Heavy duty combination adjustable fog and straight stream w/handle 1-1/2"	475		475
5650	2-1/2" direct connection	590		590
6000	Rack, for 1-1/2" diameter hose 100 ft. long, steel	71	51	122
6050	Brass	120	51	171
6500	Reel, steel, for 50 ft. long 1-1/2" diameter hose	151	72.50	223.50
6550	For 75 ft. long 2-1/2" diameter hose	245	72.50	317.50
7050	Siamese, w/plugs & chains, polished brass, sidewalk, 4" x 2-1/2" x 2-1/2"	730	405	1,135
7100	6" x 2-1/2" x 2-1/2"	945	510	1,455
7200	Wall type, flush, 4" x 2-1/2" x 2-1/2"	600	203	803
7250	6" x 2-1/2" x 2-1/2"	835	221	1,056
7300	Projecting, 4" x 2-1/2" x 2-1/2"	550	203	753
7350	6" x 2-1/2" x 2-1/2"	925	221	1,146
7400	For chrome plate, add 15% mat.			
8000	Valves, angle, wheel handle, 300 Lb., rough brass, 1-1/2"	115	47	162
8050	2-1/2"	202	80.50	282.50
8100	Combination pressure restricting, 1-1/2"	98.50	47	145.50
8150	2-1/2"	202	80.50	282.50
8200	Pressure restricting, adjustable, satin brass, 1-1/2"	350	47	397
8250	2-1/2"	410	80.50	490.50
8300	Hydrolator, vent and drain, rough brass, 1-1/2"	102	47	149
8350	2-1/2"	102	47	149
8400	Cabinet assy, incls. adapter, rack, hose, and nozzle	925	305	1,230

D4090 Other Fire Protection Systems

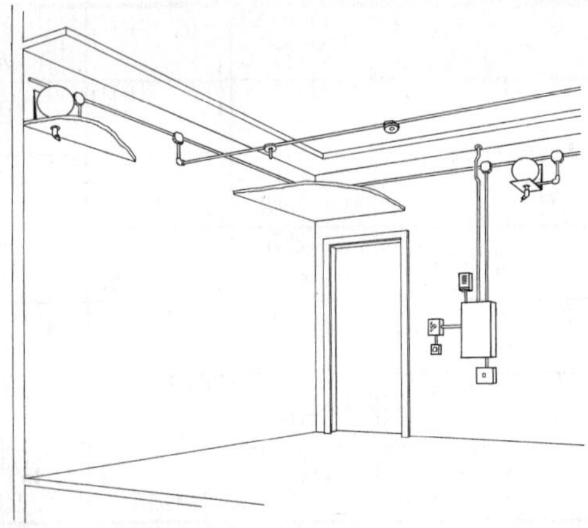

General: Automatic fire protection (suppression) systems other than water sprinklers may be desired for special environments, high risk areas, isolated locations or unusual hazards. Some typical applications would include:

Paint dip tanks
Securities vaults
Electronic data processing
Tape and data storage
Transformer rooms
Spray booths
Petroleum storage
High rack storage

Piping and wiring costs are dependent on the individual application and must be added to the component costs shown below.

All areas are assumed to be open.

D4090 910	Fire Suppression Unit Components	COST EACH		
		MAT.	INST.	TOTAL
0020	Detectors with brackets			
0040	Fixed temperature heat detector	56	72	128
0060	Rate of temperature rise detector	56	72	128
0080	Ion detector (smoke) detector	121	93	214
0200	Extinguisher agent			
0240	200 lb FM200, container	8,550	260	8,810
0280	75 lb carbon dioxide cylinder	1,400	173	1,573
0320	Dispersion nozzle			
0340	FM200 1-1/2" dispersion nozzle	73.50	41.50	115
0380	Carbon dioxide 3" x 5" dispersion nozzle	73.50	32	105.50
0420	Control station			
0440	Single zone control station with batteries	1,900	575	2,475
0470	Multizone (4) control station with batteries	3,600	1,150	4,750
0500	Electric mechanical release	184	181	365
0550	Manual pull station	66.50	96.50	163
0640	Battery standby power 10" x 10" x 17"	455	144	599
0740	Bell signalling device	106	72	178

D4090 920	FM200 Systems	COST PER C.F.		
		MAT.	INST.	TOTAL
0820	Average FM200 system, minimum			1.94
0840	Maximum			3.85

D5010 Electrical Service/Distribution

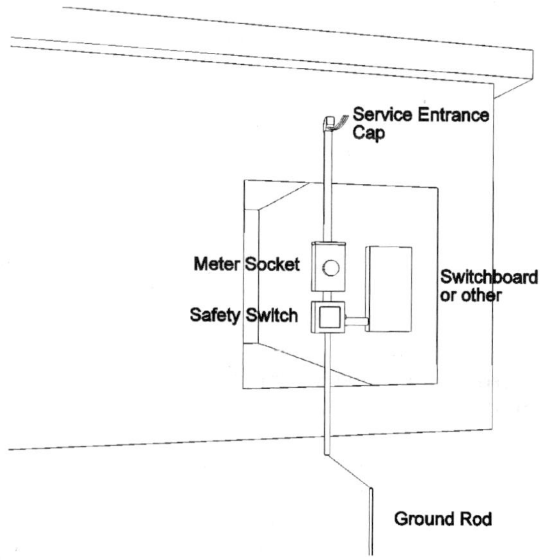

D5010 120	Overhead Electric Service, 3 Phase - 4 Wire	COST EACH		
		MAT.	INST.	TOTAL
0200	Service installation, includes breakers, metering, 20' conduit & wire			
0220	3 phase, 4 wire, 120/208 volts, 60 A	555	940	1,495
0240	100 A	725	1,075	1,800
0245	100 A w/circuit breaker	1,525	1,325	2,850
0280	200 A	1,350	1,475	2,825
0285	200 A w/circuit breaker	3,025	1,850	4,875
0320	400 A	2,700	2,925	5,625
0325	400 A w/circuit breaker	5,875	3,650	9,525
0360	600 A	4,600	4,325	8,925
0365	600 A, w/switchboard	9,175	5,475	14,650
0400	800 A	6,300	5,175	11,475
0405	800 A, w/switchboard	10,900	6,475	17,375
0440	1000 A	7,850	6,375	14,225
0445	1000 A, w/switchboard	13,300	7,825	21,125
0480	1200 A	10,400	7,275	17,675
0485	1200 A, w/groundfault switchboard	34,100	9,025	43,125
0520	1600 A	12,800	9,500	22,300
0525	1600 A, w/groundfault switchboard	38,600	11,400	50,000
0560	2000 A	17,100	11,500	28,600
0565	2000 A, w/groundfault switchboard	46,200	14,200	60,400
0610	1 phase, 3 wire, 120/240 volts, 100 A (no safety switch)	182	520	702
0615	100 A w/load center	490	1,000	1,490
0620	200 A	420	715	1,135
0625	200 A w/load center	1,250	1,525	2,775

D5010 Electrical Service/Distribution

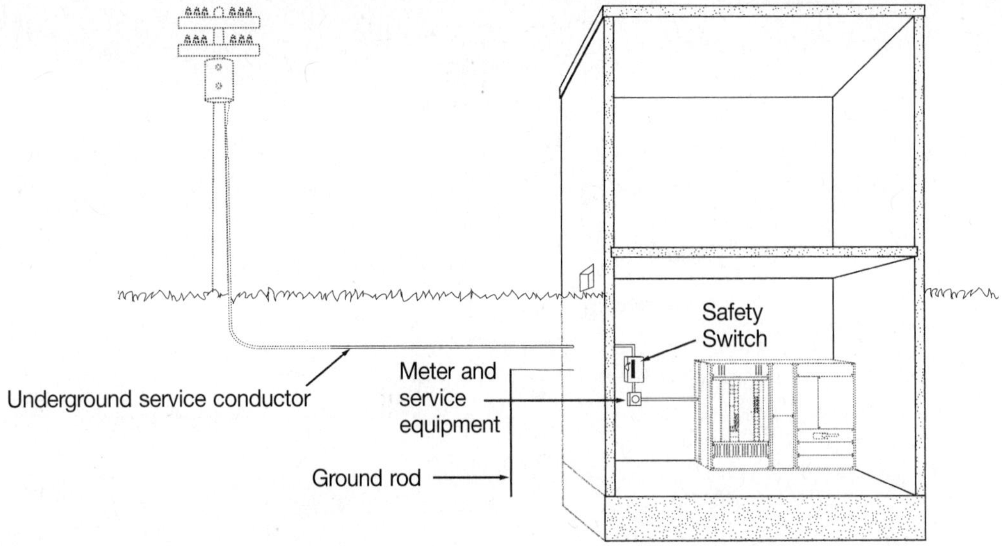

D5010 130	Underground Electric Service	COST EACH		
		MAT.	INST.	TOTAL
0950	Underground electric service including excavation, backfill and compaction			
1000	3 phase, 4 wire, 277/480 volts, 2000 A	34,400	18,400	52,800
1050	2000 A w/groundfault switch	62,000	20,500	82,500
1100	1600 A	27,300	14,800	42,100
1150	1600 A w/groundfault switchboard	53,000	16,700	69,700
1200	1200 A	20,900	11,800	32,700
1250	1200 A w/groundfault switchboard	44,600	13,500	58,100
1400	800 A	14,700	9,775	24,475
1450	800 A w/switchboard	19,300	11,100	30,400
1500	600 A	10,300	8,975	19,275
1550	600 A w/switchboard	14,900	10,100	25,000
1600	1 phase, 3 wire, 120/240 volts, 200 A	3,100	3,075	6,175
1650	200 A w/load center	3,925	3,875	7,800
1700	100 A	2,150	2,500	4,650
1750	100 A w/load center	3,000	3,300	6,300

D5010 Electrical Service/Distribution

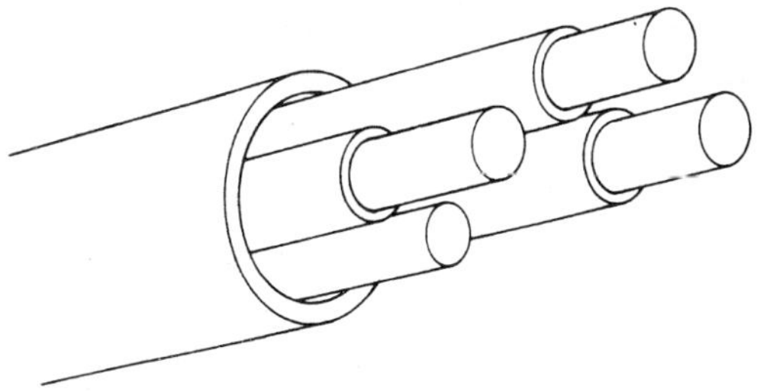

D5010 230	Feeder Installation	COST PER L.F.		
		MAT.	INST.	TOTAL
0200	Feeder installation 600 V, including RGS conduit and XHHW wire, 60 A	5.30	10.60	15.90
0240	100 A	9.50	13.75	23.25
0280	200 A	21	21	42
0320	400 A	42.50	41.50	84
0360	600 A	78.50	68	146.50
0400	800 A	102	80.50	182.50
0440	1000 A	128	105	233
0480	1200 A	157	136	293
0520	1600 A	204	161	365
0560	2000 A	256	211	467
1200	Branch installation 600 V, including EMT conduit and THW wire, 15 A	1.13	5.15	6.28
1240	20 A	1.32	5.50	6.82
1280	30 A	1.65	5.70	7.35
1320	50 A	3.27	7.90	11.17
1360	65 A	4.02	8.40	12.42
1400	85 A	6.15	9.75	15.90
1440	100 A	6.80	9.90	16.70
1480	130 A	10.20	11.60	21.80
1520	150 A	12.50	13.35	25.85
1560	200 A	17	15.20	32.20

D5010 Electrical Service/Distribution

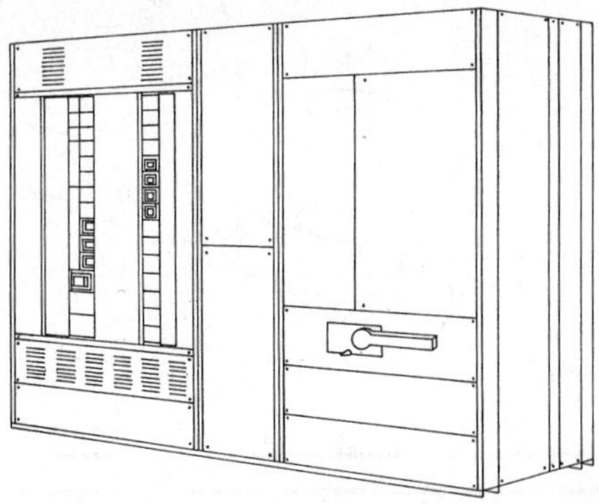

D5010 240	Switchgear	COST EACH		
		MAT.	INST.	TOTAL
0190	Switchgear installation, including switchboard, panels, & circuit breaker			
0200	120/208 V, 1 phase, 400 A	9,225	2,450	11,675
0240	600 A	13,200	2,725	15,925
0280	800 A	15,300	3,150	18,450
0300	1000 A	18,800	3,525	22,325
0320	1200 A	19,800	3,950	23,750
0360	1600 A	32,100	4,575	36,675
0400	2000 A	38,800	4,950	43,750
0500	277/480 V, 3 phase, 400 A	12,600	4,550	17,150
0520	600 A	18,400	5,200	23,600
0540	800 A	19,900	5,750	25,650
0560	1000 A	24,700	6,425	31,125
0580	1200 A	27,100	7,150	34,250
0600	1600 A	40,000	7,900	47,900
0620	2000 A	50,500	8,650	59,150

D5010 Electrical Service/Distribution

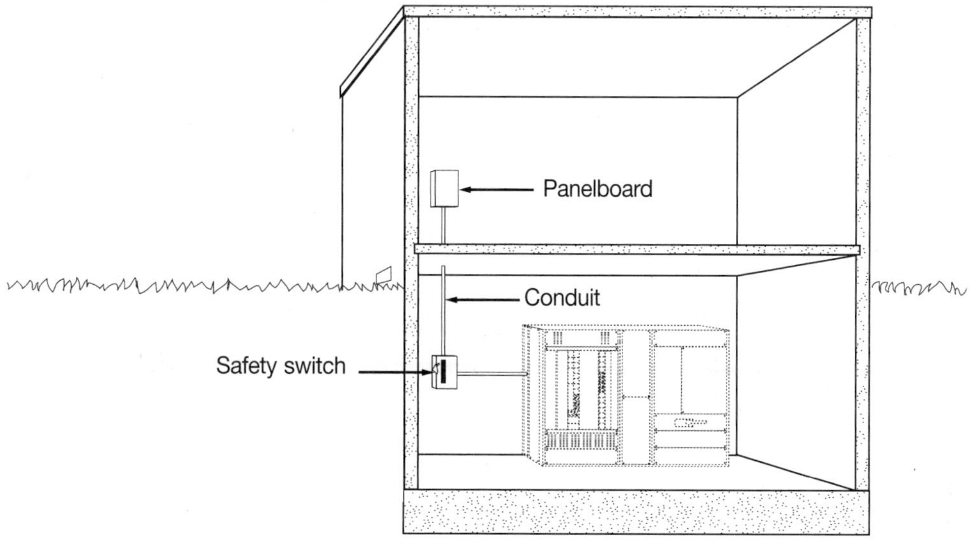

D5010 250	Panelboard	COST EACH		
		MAT.	INST.	TOTAL
0900	Panelboards, NQOD, 4 wire, 120/208 volts w/conductor & conduit			
1000	100 A, 0 stories, 0' horizontal	1,400	1,225	2,625
1020	1 stories, 25' horizontal	1,775	1,750	3,525
1040	5 stories, 50' horizontal	2,525	2,775	5,300
1060	10 stories, 75' horizontal	3,400	4,000	7,400
1080	225A, 0 stories, 0' horizontal	2,625	1,600	4,225
2000	1 stories, 25' horizontal	3,750	2,550	6,300
2020	5 stories, 50' horizontal	5,950	4,425	10,375
2040	10 stories, 75' horizontal	8,500	6,600	15,100
2060	400A, 0 stories, 0' horizontal	3,875	2,400	6,275
2080	1 stories, 25' horizontal	5,425	3,950	9,375
3000	5 stories, 50' horizontal	8,475	6,975	15,450
3020	10 stories, 75' horizontal	16,900	13,600	30,500
3040	600 A, 0 stories, 0' horizontal	5,725	2,875	8,600
3060	1 stories, 25' horizontal	8,650	5,375	14,025
3080	5 stories, 50' horizontal	14,400	10,300	24,700
4000	10 stories, 75' horizontal	30,000	20,300	50,300
4010	Panelboards, NEHB, 4 wire, 277/480 volts w/conductor, conduit, & safety switch			
4020	100 A, 0 stories, 0' horizontal, includes safety switch	3,125	1,675	4,800
4040	1 stories, 25' horizontal	3,500	2,200	5,700
4060	5 stories, 50' horizontal	4,250	3,250	7,500
4080	10 stories, 75' horizontal	5,125	4,450	9,575
5000	225 A, 0 stories, 0' horizontal	4,700	2,050	6,750
5020	1 stories, 25' horizontal	5,800	3,000	8,800
5040	5 stories, 50' horizontal	8,025	4,850	12,875
5060	10 stories, 75' horizontal	10,600	7,025	17,625
5080	400 A, 0 stories, 0' horizontal	7,450	3,150	10,600
6000	1 stories, 25' horizontal	9,000	4,675	13,675
6020	5 stories, 50' horizontal	12,100	7,725	19,825
6040	10 stories, 75' horizontal	20,400	14,400	34,800
6060	600 A, 0 stories, 0' horizontal	11,500	3,975	15,475
6080	1 stories, 25' horizontal	14,400	6,500	20,900
7000	5 stories, 50' horizontal	20,200	11,400	31,600
7020	10 stories, 75' horizontal	35,800	21,400	57,200

D5020 Lighting and Branch Wiring

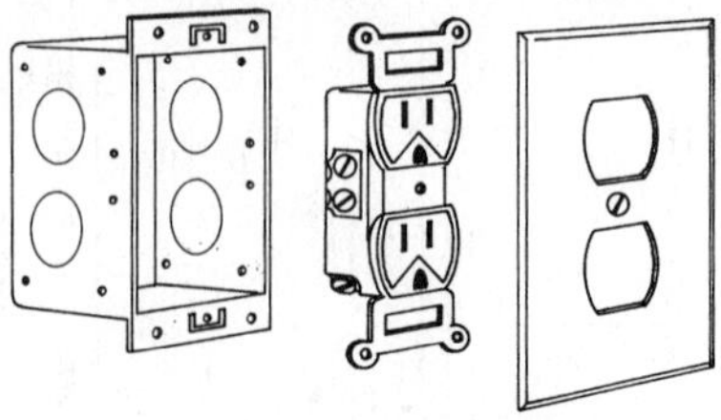

Duplex Wall Receptacle

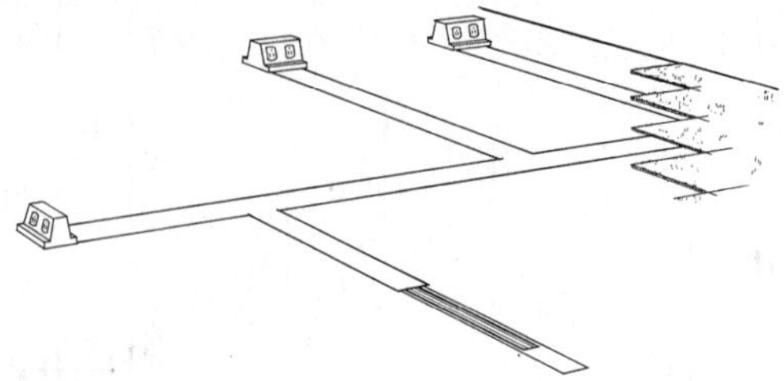

Undercarpet Receptacle System

D5020 110	Receptacle (by Wattage)	COST PER S.F.		
		MAT.	INST.	TOTAL
0190	Receptacles include plate, box, conduit, wire & transformer when required			
0200	2.5 per 1000 S.F., .3 watts per S.F.	.46	1.29	1.75
0240	With transformer	.54	1.35	1.89
0280	4 per 1000 S.F., .5 watts per S.F.	.52	1.50	2.02
0320	With transformer	.65	1.60	2.25
0360	5 per 1000 S.F., .6 watts per S.F.	.61	1.78	2.39
0400	With transformer	.78	1.91	2.69
0440	8 per 1000 S.F., .9 watts per S.F.	.64	1.96	2.60
0480	With transformer	.91	2.16	3.07
0520	10 per 1000 S.F., 1.2 watts per S.F.	.65	2.12	2.77
0560	With transformer	.98	2.38	3.36
0600	16.5 per 1000 S.F., 2.0 watts per S.F.	.78	2.66	3.44
0640	With transformer	1.33	3.08	4.41
0680	20 per 1000 S.F., 2.4 watts per S.F.	.81	2.89	3.70
0720	With transformer	1.48	3.40	4.88

D5020 115	Receptacles, Floor	COST PER S.F.		
		MAT.	INST.	TOTAL
0200	Receptacle systems, underfloor duct, 5′ on center, low density	7.45	2.84	10.29
0240	High density	8	3.65	11.65
0280	7′ on center, low density	5.90	2.44	8.34
0320	High density	6.45	3.25	9.70
0400	Poke thru fittings, low density	1.25	1.39	2.64
0440	High density	2.50	2.76	5.26
0520	Telepoles, using Romex, low density	1.19	.85	2.04
0560	High density	2.37	1.71	4.08
0600	Using EMT, low density	1.23	1.14	2.37
0640	High density	2.51	2.26	4.77
0720	Conduit system with floor boxes, low density	1.18	.98	2.16
0760	High density	2.38	1.94	4.32
0840	Undercarpet power system, 3 conductor with 5 conductor feeder, low density	1.57	.34	1.91
0880	High density	3.11	.71	3.82

D5020 Lighting and Branch Wiring

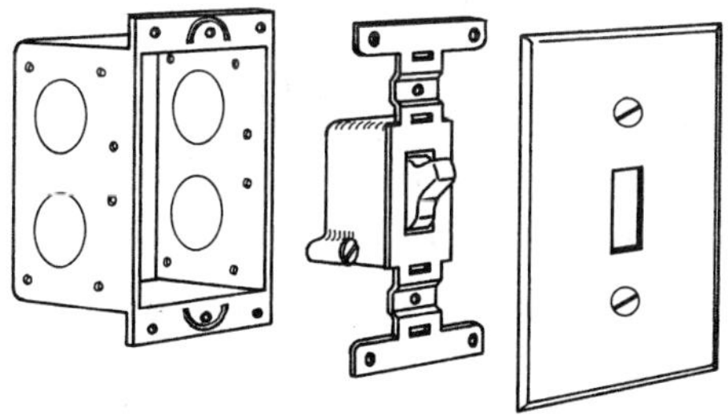

Description: Table D5020 130 includes the cost for switch, plate, box, conduit in slab or EMT exposed and copper wire. Add 20% for exposed conduit.

No power required for switches.

Federal energy guidelines recommend the maximum lighting area controlled per switch shall not exceed 1000 S.F. and that areas over 500 S.F. shall be so controlled that total illumination can be reduced by at least 50%.

D5020 130	Wall Switch by Sq. Ft.	COST PER S.F.		
		MAT.	INST.	TOTAL
0200	Wall switches, 1.0 per 1000 S.F.	.06	.21	.27
0240	1.2 per 1000 S.F.	.08	.22	.30
0280	2.0 per 1000 S.F.	.12	.34	.46
0320	2.5 per 1000 S.F.	.15	.41	.56
0360	5.0 per 1000 S.F.	.31	.91	1.22
0400	10.0 per 1000 S.F.	.62	1.80	2.42

D5020 135	Miscellaneous Power	COST PER S.F.		
		MAT.	INST.	TOTAL
0200	Miscellaneous power, to .5 watts	.04	.11	.15
0240	.8 watts	.06	.14	.20
0280	1 watt	.08	.20	.28
0320	1.2 watts	.09	.23	.32
0360	1.5 watts	.11	.27	.38
0400	1.8 watts	.12	.31	.43
0440	2 watts	.14	.38	.52
0480	2.5 watts	.18	.46	.64
0520	3 watts	.21	.53	.74

D5020 140	Central A. C. Power (by Wattage)	COST PER S.F.		
		MAT.	INST.	TOTAL
0200	Central air conditioning power, 1 watt	.08	.22	.30
0220	2 watts	.10	.26	.36
0240	3 watts	.20	.38	.58
0280	4 watts	.20	.39	.59
0320	6 watts	.35	.52	.87
0360	8 watts	.45	.56	1.01
0400	10 watts	.63	.65	1.28

D5020 Lighting and Branch Wiring

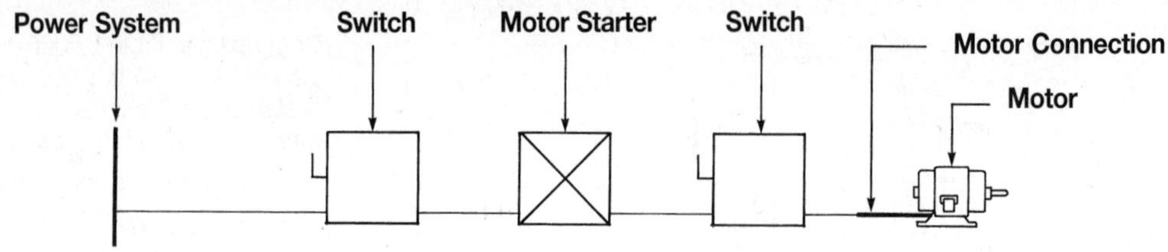

Motor Installation

D5020 145	Motor Installation	COST EACH		
		MAT.	INST.	TOTAL
0200	Motor installation, single phase, 115V, 1/3 HP motor size	530	875	1,405
0240	1 HP motor size	555	875	1,430
0280	2 HP motor size	605	930	1,535
0320	3 HP motor size	685	950	1,635
0360	230V, 1 HP motor size	535	890	1,425
0400	2 HP motor size	570	890	1,460
0440	3 HP motor size	640	955	1,595
0520	Three phase, 200V, 1-1/2 HP motor size	610	975	1,585
0560	3 HP motor size	740	1,075	1,815
0600	5 HP motor size	725	1,175	1,900
0640	7-1/2 HP motor size	755	1,200	1,955
0680	10 HP motor size	1,125	1,525	2,650
0720	15 HP motor size	1,550	1,675	3,225
0760	20 HP motor size	1,950	1,950	3,900
0800	25 HP motor size	2,000	1,950	3,950
0840	30 HP motor size	3,150	2,300	5,450
0880	40 HP motor size	3,900	2,725	6,625
0920	50 HP motor size	6,900	3,150	10,050
0960	60 HP motor size	7,200	3,350	10,550
1000	75 HP motor size	8,875	3,825	12,700
1040	100 HP motor size	26,600	4,500	31,100
1080	125 HP motor size	26,900	4,950	31,850
1120	150 HP motor size	31,300	5,825	37,125
1160	200 HP motor size	32,300	6,875	39,175
1240	230V, 1-1/2 HP motor size	585	965	1,550
1280	3 HP motor size	715	1,050	1,765
1320	5 HP motor size	700	1,175	1,875
1360	7-1/2 HP motor size	700	1,175	1,875
1400	10 HP motor size	1,075	1,425	2,500
1440	15 HP motor size	1,200	1,575	2,775
1480	20 HP motor size	1,850	1,875	3,725
1520	25 HP motor size	1,950	1,950	3,900
1560	30 HP motor size	2,000	1,950	3,950
1600	40 HP motor size	3,800	2,650	6,450
1640	50 HP motor size	4,075	2,800	6,875
1680	60 HP motor size	6,925	3,175	10,100
1720	75 HP motor size	8,225	3,600	11,825
1760	100 HP motor size	9,475	4,000	13,475
1800	125 HP motor size	27,100	4,650	31,750
1840	150 HP motor size	28,900	5,275	34,175
1880	200 HP motor size	30,300	5,875	36,175
1960	460V, 2 HP motor size	725	975	1,700
2000	5 HP motor size	855	1,050	1,905
2040	10 HP motor size	805	1,175	1,980
2080	15 HP motor size	1,075	1,350	2,425
2120	20 HP motor size	1,125	1,450	2,575

D5020 Lighting and Branch Wiring

D5020 145	Motor Installation	COST EACH		
		MAT.	INST.	TOTAL
2160	25 HP motor size	1,200	1,500	2,700
2200	30 HP motor size	1,550	1,625	3,175
2240	40 HP motor size	1,975	1,750	3,725
2280	50 HP motor size	2,200	1,950	4,150
2320	60 HP motor size	3,350	2,275	5,625
2360	75 HP motor size	3,950	2,525	6,475
2400	100 HP motor size	4,375	2,800	7,175
2440	125 HP motor size	7,225	3,175	10,400
2480	150 HP motor size	8,950	3,550	12,500
2520	200 HP motor size	10,300	4,025	14,325
2600	575V, 2 HP motor size	725	975	1,700
2640	5 HP motor size	855	1,050	1,905
2680	10 HP motor size	805	1,175	1,980
2720	20 HP motor size	1,075	1,350	2,425
2760	25 HP motor size	1,125	1,450	2,575
2800	30 HP motor size	1,550	1,625	3,175
2840	50 HP motor size	1,675	1,700	3,375
2880	60 HP motor size	3,300	2,275	5,575
2920	75 HP motor size	3,350	2,275	5,625
2960	100 HP motor size	3,950	2,525	6,475
3000	125 HP motor size	7,050	3,125	10,175
3040	150 HP motor size	7,225	3,175	10,400
3080	200 HP motor size	9,050	3,600	12,650

D5020 155	Motor Feeder	COST PER L.F.		
		MAT.	INST.	TOTAL
0200	Motor feeder systems, single phase, feed up to 115V 1HP or 230V 2 HP	2.66	6.80	9.46
0240	115V 2HP, 230V 3HP	2.79	6.90	9.69
0280	115V 3HP	3.09	7.20	10.29
0360	Three phase, feed to 200V 3HP, 230V 5HP, 460V 10HP, 575V 10HP	2.77	7.35	10.12
0440	200V 5HP, 230V 7.5HP, 460V 15HP, 575V 20HP	2.97	7.50	10.47
0520	200V 10HP, 230V 10HP, 460V 30HP, 575V 30HP	3.43	7.90	11.33
0600	200V 15HP, 230V 15HP, 460V 40HP, 575V 50HP	4.76	9.05	13.81
0680	200V 20HP, 230V 25HP, 460V 50HP, 575V 60HP	6.75	11.50	18.25
0760	200V 25HP, 230V 30HP, 460V 60HP, 575V 75HP	7.65	11.70	19.35
0840	200V 30HP	8.50	12.10	20.60
0920	230V 40HP, 460V 75HP, 575V 100HP	10.80	13.15	23.95
1000	200V 40HP	12.45	14.10	26.55
1080	230V 50HP, 460V 100HP, 575V 125HP	14.95	15.55	30.50
1160	200V 50HP, 230V 60HP, 460V 125HP, 575V 150HP	17.15	16.55	33.70
1240	200V 60HP, 460V 150HP	21	19.40	40.40
1320	230V 75HP, 575V 200HP	24	20	44
1400	200V 75HP	26.50	20.50	47
1480	230V 100HP, 460V 200HP	36.50	24	60.50
1560	200V 100HP	50.50	30	80.50
1640	230V 125HP	50.50	30	80.50
1720	200V 125HP, 230V 150HP	56.50	37.50	94
1800	200V 150HP	68	40.50	108.50
1880	200V 200HP	85	43	128
1960	230V 200HP	81	44.50	125.50

D5020 Lighting and Branch Wiring

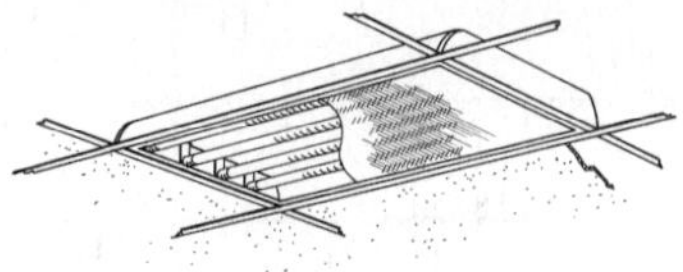

Fluorescent Fixture

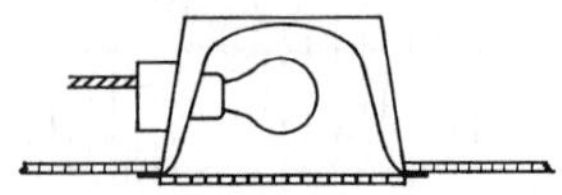

Incandescent Fixture

D5020 210	Fluorescent Fixtures (by Wattage)	COST PER S.F. MAT.	COST PER S.F. INST.	COST PER S.F. TOTAL
0190	Fluorescent fixtures recess mounted in ceiling			
0195	T12, standard 40 watt lamps			
0200	1 watt per S.F., 20 FC, 5 fixtures @40 watts per 1000 S.F.	.78	1.71	2.49
0240	2 watt per S.F., 40 FC, 10 fixtures @40 watt per 1000 S.F.	1.55	3.32	4.87
0280	3 watt per S.F., 60 FC, 15 fixtures @40 watt per 1000 S.F	2.30	5.05	7.35
0320	4 watt per S.F., 80 FC, 20 fixtures @40 watt per 1000 S.F.	3.07	6.65	9.72
0400	5 watt per S.F., 100 FC, 25 fixtures @40 watt per 1000 S.F.	3.84	8.35	12.19
0402				
0450	T8, energy saver 32 watt lamps			
0500	0.8 watt per S.F., 20 FC, 5 fixtures @32 watt per 1000 S.F.	.85	1.71	2.56
0520	1.6 watt per S.F., 40 FC, 10 fixtures @32 watt per 1000 S.F.	1.68	3.32	5
0540	2.4 watt per S.F., 60 FC, 15 fixtures @ 32 watt per 1000 S.F	2.49	5.05	7.54
0560	3.2 watt per S.F., 80 FC, 20 fixtures @32 watt per 1000 S.F.	3.33	6.65	9.98
0580	4 watt per S.F., 100 FC, 25 fixtures @32 watt per 1000 S.F.	4.17	8.35	12.52

D5020 216	Incandescent Fixture (by Wattage)	COST PER S.F. MAT.	COST PER S.F. INST.	COST PER S.F. TOTAL
0190	Incandescent fixture recess mounted, type A			
0200	1 watt per S.F., 8 FC, 6 fixtures per 1000 S.F.	1	1.36	2.36
0240	2 watt per S.F., 16 FC, 12 fixtures per 1000 S.F.	2	2.73	4.73
0280	3 watt per S.F., 24 FC, 18 fixtures, per 1000 S.F.	3	4.03	7.03
0320	4 watt per S.F., 32 FC, 24 fixtures per 1000 S.F.	3.99	5.40	9.39
0400	5 watt per S.F., 40 FC, 30 fixtures per 1000 S.F.	5	6.75	11.75

D5020 226	H.I.D. Fixture, High Bay, 16′ (by Wattage)	COST PER S.F. MAT.	COST PER S.F. INST.	COST PER S.F. TOTAL
0190	High intensity discharge fixture, 16′ above work plane			
0240	1 watt/S.F., type E, 42 FC, 1 fixture/1000 S.F.	1.12	1.42	2.54
0280	Type G, 52 FC, 1 fixture/1000 S.F.	1.12	1.42	2.54
0320	Type C, 54 FC, 2 fixture/1000 S.F.	1.28	1.58	2.86
0440	2 watt/S.F., type E, 84 FC, 2 fixture/1000 S.F.	2.24	2.89	5.13
0480	Type G, 105 FC, 2 fixture/1000 S.F.	2.24	2.89	5.13
0520	Type C, 108 FC, 4 fixture/1000 S.F.	2.56	3.18	5.74
0640	3 watt/S.F., type E, 126 FC, 3 fixture/1000 S.F.	3.34	4.31	7.65
0680	Type G, 157 FC, 3 fixture/1000 S.F.	3.34	4.31	7.65
0720	Type C, 162 FC, 6 fixture/1000 S.F.	3.84	4.76	8.60
0840	4 watt/S.F., type E, 168 FC, 4 fixture/1000 S.F.	4.49	5.80	10.29
0880	Type G, 210 FC, 4 fixture/1000 S.F.	4.49	5.80	10.29
0920	Type C, 243 FC, 9 fixture/1000 S.F.	5.55	6.65	12.20
1040	5 watt/S.F., type E, 210 FC, 5 fixture/1000 S.F.	5.60	7.20	12.80
1080	Type G, 262 FC, 5 fixture/1000 S.F.	5.60	7.20	12.80
1120	Type C, 297 FC, 11 fixture/1000 S.F.	6.85	8.20	15.05

D5020 Lighting and Branch Wiring

D5020 234	H.I.D. Fixture, High Bay, 30′ (by Wattage)	COST PER S.F.		
		MAT.	INST.	TOTAL
0190	High intensity discharge fixture, 30′ above work plane			
0240	1 watt/S.F., type E, 37 FC, 1 fixture/1000 S.F.	1.25	1.81	3.06
0280	Type G, 45 FC., 1 fixture/1000 S.F.	1.25	1.81	3.06
0320	Type F, 50 FC, 1 fixture/1000 S.F.	1.05	1.41	2.46
0440	2 watt/S.F., type E, 74 FC, 2 fixtures/1000 S.F.	2.51	3.59	6.10
0480	Type G, 92 FC, 2 fixtures/1000 S.F.	2.51	3.59	6.10
0520	Type F, 100 FC, 2 fixtures/1000 S.F.	2.13	2.87	5
0640	3 watt/S.F., type E, 110 FC, 3 fixtures/1000 S.F.	3.77	5.45	9.22
0680	Type G, 138 FC, 3 fixtures/1000 S.F.	3.77	5.45	9.22
0720	Type F, 150 FC, 3 fixtures/1000 S.F.	3.16	4.27	7.43
0840	4 watt/S.F., type E, 148 FC, 4 fixtures/1000 S.F.	5.05	7.25	12.30
0880	Type G, 185 FC, 4 fixtures/1000 S.F.	5.05	7.25	12.30
0920	Type F, 200 FC, 4 fixtures/1000 S.F.	4.24	5.75	9.99
1040	5 watt/S.F., type E, 185 FC, 5 fixtures/1000 S.F.	6.30	9.05	15.35
1080	Type G, 230 FC, 5 fixtures/1000 S.F.	6.30	9.05	15.35
1120	Type F, 250 FC, 5 fixtures/1000 S.F.	5.30	7.15	12.45

D5020 Lighting and Branch Wiring

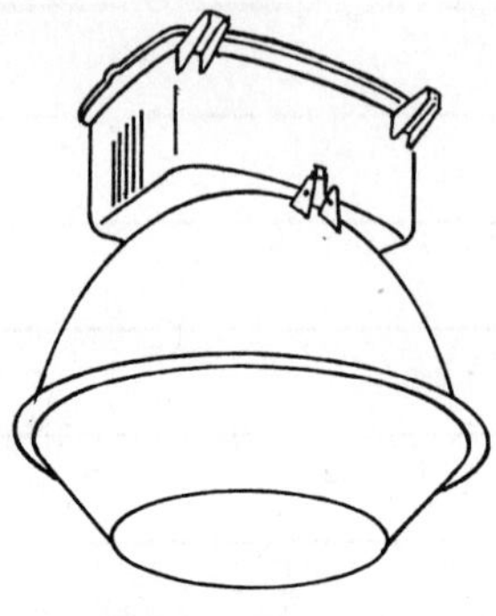

LOW BAY FIXTURES

J. Metal halide 250 watt

K. High pressure sodium 150 watt

D5020 238	**H.I.D. Fixture, Low Bay, 8'-10' (by Wattage)**	COST PER S.F.		
		MAT.	INST.	TOTAL
0190	High intensity discharge fixture, 8'-10' above work plane			
0240	1 watt/S.F., type J, 30 FC, 4 fixtures/1000 S.F.	2.51	2.94	5.45
0280	Type K, 29 FC, 5 fixtures/1000 S.F.	2.52	2.65	5.17
0400	2 watt/S.F., type J, 52 FC, 7 fixtures/1000 S.F.	4.46	5.35	9.81
0440	Type K, 63 FC, 11 fixtures/1000 S.F.	5.45	5.55	11
0560	3 watt/S.F., type J, 81 FC, 11 fixtures/1000 S.F.	6.90	8.10	15
0600	Type K, 92 FC, 16 fixtures/1000 S.F.	7.95	8.15	16.10
0720	4 watt/S.F., type J, 103 FC, 14 fixtures/1000 S.F.	8.90	10.70	19.60
0760	Type K, 127 FC, 22 fixtures/1000 S.F.	10.85	11.10	21.95
0880	5 watt/S.F., type J, 133 FC, 18 fixtures/1000 S.F.	11.35	13.50	24.85
0920	Type K, 155 FC, 27 fixtures/1000 S.F.	13.35	13.70	27.05

D5020 242	**H.I.D. Fixture, Low Bay, 16' (by Wattage)**	COST PER S.F.		
		MAT.	INST.	TOTAL
0190	High intensity discharge fixture, mounted 16' above work plane			
0240	1 watt/S.F., type J, 28 FC, 4 fixt./1000 S.F.	2.65	3.34	5.99
0280	Type K, 27 FC, 5 fixt./1000 S.F.	2.95	3.73	6.68
0400	2 watt/S.F., type J, 48 FC, 7 fixt/1000 S.F.	4.89	6.45	11.34
0440	Type K, 58 FC, 11 fixt/1000 S.F.	6.25	7.60	13.85
0560	3 watt/S.F., type J, 75 FC, 11 fixt/1000 S.F.	7.55	9.75	17.30
0600	Type K, 85 FC, 16 fixt/1000 S.F.	9.20	11.35	20.55
0720	4 watt/S.F., type J, 95 FC, 14 fixt/1000 S.F.	9.80	12.85	22.65
0760	Type K, 117 FC, 22 fixt/1000 S.F.	12.50	15.25	27.75
0880	5 watt/S.F., type J, 122 FC, 18 fixt/1000 S.F.	12.45	16.20	28.65
0920	Type K, 143 FC, 27 fixt/1000 S.F.	15.45	18.95	34.40

D5030 Communications and Security

D5030 910	Communication & Alarm Systems	COST EACH		
		MAT.	INST.	TOTAL
0200	Communication & alarm systems, includes outlets, boxes, conduit & wire			
0210	Sound system, 6 outlets	6,275	7,275	13,550
0220	12 outlets	9,075	11,600	20,675
0240	30 outlets	16,100	22,000	38,100
0280	100 outlets	49,300	73,500	122,800
0320	Fire detection systems, non-addressable, 12 detectors	3,550	6,050	9,600
0360	25 detectors	6,200	10,200	16,400
0400	50 detectors	12,200	20,300	32,500
0440	100 detectors	22,600	36,800	59,400
0450	Addressable type, 12 detectors	5,225	6,100	11,325
0452	25 detectors	9,475	10,300	19,775
0456	100 detectors	35,300	37,200	72,500
0480	Intercom systems, 6 stations	4,225	4,975	9,200
0560	25 stations	14,000	19,100	33,100
0640	100 stations	53,000	70,000	123,000
0680	Master clock systems, 6 rooms	5,575	8,125	13,700
0720	12 rooms	8,975	13,900	22,875
0760	20 rooms	12,500	19,600	32,100
0800	30 rooms	20,800	36,100	56,900
0840	50 rooms	34,000	61,000	95,000
0920	Master TV antenna systems, 6 outlets	2,775	5,150	7,925
0960	12 outlets	5,150	9,575	14,725
1000	30 outlets	12,700	22,200	34,900
1040	100 outlets	42,300	72,500	114,800

D5090 Other Electrical Systems

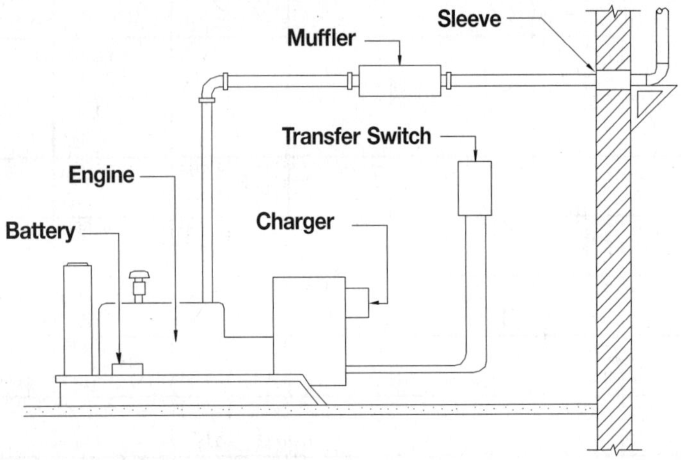

Description: System below tabulates the installed cost for generators by kW. Included in costs are battery, charger, muffler, and transfer switch.

No conduit, wire, or terminations included.

D5090 210	Generators (by kW)	COST PER kW		
		MAT.	INST.	TOTAL
0190	Generator sets, include battery, charger, muffler & transfer switch			
0200	Gas/gasoline operated, 3 phase, 4 wire, 277/480 volt, 7.5 kW	1,250	255	1,505
0240	11.5 kW	1,150	193	1,343
0280	20 kW	785	126	911
0320	35 kW	535	82.50	617.50
0360	80 kW	385	49.50	434.50
0400	100 kW	335	48	383
0440	125 kW	550	45.50	595.50
0480	185 kW	490	34.50	524.50
0560	Diesel engine with fuel tank, 30 kW	410	96	506
0600	50 kW	435	75.50	510.50
0760	150 kW	273	40.50	313.50
0840	200 kW	247	33	280
0880	250 kW	210	27.50	237.50
0960	350 kW	181	22.50	203.50
1040	500 kW	198	17.55	215.55

E1010 Commercial Equipment

E1010 110	Security/Vault, EACH	COST EACH		
		MAT.	INST.	TOTAL
0100	Bank equipment, drive up window, drawer & mike, no glazing, economy	7,725	1,175	8,900
0110	Deluxe	10,100	2,350	12,450
0120	Night depository, economy	8,300	1,175	9,475
0130	Deluxe	11,800	2,350	14,150
0140	Pneumatic tube systems, 2 station, standard	28,800	2,600	31,400
0150	Teller, automated, 24 hour, single unit	49,700	2,600	52,300
0160	Teller window, bullet proof glazing, 44″ x 60″	4,650	815	5,465
0170	Pass through, painted steel, 72″ x 40″	4,800	1,475	6,275
0300	Safe, office type, 1 hr. rating, 34″ x 20″ x 20″	2,500		2,500
0310	4 hr. rating, 62″ x 33″ x 20″	9,750		9,750
0320	Data storage, 4 hr. rating, 63″ x 44″ x 16″	14,800		14,800
0330	Jewelers, 63″ x 44″ x 16″	15,800		15,800
0340	Money, "B" label, 9″ x 14″ x 14″	595		595
0350	Tool and torch resistive, 24″ x 24″ x 20″	8,650	196	8,846
0500	Security gates-scissors type, painted steel, single, 6′ high, 5-1/2′ wide	252	293	545
0510	Double gate, 7-1/2′ high, 14′ wide	695	585	1,280

E1010 510	Mercantile Equipment, EACH	COST EACH		
		MAT.	INST.	TOTAL
0015	Barber equipment, chair, hydraulic, economy	690	21	711
0020	Deluxe	4,375	31.50	4,406.50
0100	Checkout counter, single belt	3,500	78	3,578
0110	Double belt, power take-away	5,025	87	5,112
0200	Display cases, freestanding, glass and aluminum, 3′-6″ x 3′ x 1′-0″ deep	1,425	127	1,552
0220	Wall mounted, glass and aluminum, 3′ x 4′ x 1′-4″ deep	2,375	203	2,578
0320	Frozen food, chest type, 12 ft. long	9,050	325	9,375

E1010 610	Laundry/Dry Cleaning, EACH	COST EACH		
		MAT.	INST.	TOTAL
0100	Laundry equipment, dryers, gas fired, residential, 16 lb. capacity	755	193	948
0110	Commercial, 30 lb. capacity, single	3,700	193	3,893
0120	Dry cleaners, electric, 20 lb. capacity	37,500	3,600	41,100
0130	30 lb. capacity	57,500	4,825	62,325
0140	Ironers, commercial, 120″ with canopy, 8 roll	182,500	10,300	192,800
0150	Institutional, 110″, single roll	35,900	2,875	38,775
0160	Washers, residential, 4 cycle	975	193	1,168
0170	Commercial, coin operated, deluxe	3,775	193	3,968

E1020 Institutional Equipment

E1020 110	Ecclesiastical Equipment, EACH	COST EACH MAT.	INST.	TOTAL
0090	Church equipment, altar, wood, custom, plain	2,875	360	3,235
0100	Granite, custom, deluxe	40,500	4,775	45,275
0110	Baptistry, fiberglass, economy	6,175	1,175	7,350
0120	Bells & carillons, keyboard operation	20,400	6,875	27,275
0130	Confessional, wood, single, economy	3,575	845	4,420
0140	Double, deluxe	20,900	2,525	23,425
0150	Steeples, translucent fiberglas, 30″ square, 15′ high	9,400	1,500	10,900

E1020 130	Ecclesiastical Equipment, L.F.	COST PER L.F. MAT.	INST.	TOTAL
0100	Arch. equip., church equip. pews, bench type, hardwood, economy	111	25.50	136.50
0110	Deluxe	194	34	228

E1020 210	Library Equipment, EACH	COST EACH MAT.	INST.	TOTAL
0110	Library equipment, carrels, metal, economy	289	101	390
0120	Hardwood, deluxe	2,250	127	2,377

E1020 230	Library Equipment, L.F.	COST PER L.F. MAT.	INST.	TOTAL
0100	Library equipment, book shelf, metal, single face, 90″ high x 10″ shelf	134	42	176
0110	Double face, 90″ high x 10″ shelf	490	91.50	581.50
0120	Charging desk, built-in, with counter, plastic laminate	335	72.50	407.50

E1020 310	Theater and Stage Equipment, EACH	COST EACH MAT.	INST.	TOTAL
0200	Movie equipment, changeover, economy	535		535
0210	Film transport, incl. platters and autowind, economy	5,600		5,600
0220	Lamphouses, incl. rectifiers, xenon, 1000W	7,400	288	7,688
0230	4000W	10,800	385	11,185
0240	Projector mechanisms, 35 mm, economy	12,700		12,700
0250	Deluxe	17,400		17,400
0260	Sound systems, incl. amplifier, single, economy	3,800	640	4,440
0270	Dual, Dolby/super sound	18,900	1,450	20,350
0280	Projection screens, wall hung, manual operation, 50 S.F., economy	355	102	457
0290	Electric operation, 100 S.F., deluxe	2,650	505	3,155

E1020 320	Theater and Stage Equipment, S.F.	COST PER S.F. MAT.	INST.	TOTAL
0090	Movie equipment, projection screens, rigid in wall, acrylic, 1/4″ thick	48	5.05	53.05
0100	1/2″ thick	55.50	7.55	63.05
0110	Stage equipment, curtains, velour, medium weight	9.05	1.69	10.74
0120	Silica based yarn, fireproof	16.90	20.50	37.40
0130	Stages, portable with steps, folding legs, 8″ high	37		37
0140	Telescoping platforms, aluminum, deluxe	55	26.50	81.50

E1020 330	Theater and Stage Equipment, L.F.	COST PER L.F. MAT.	INST.	TOTAL
0100	Stage equipment, curtain track, heavy duty	69	56.50	125.50
0110	Lights, border, quartz, colored	192	29	221

E10 Equipment

E1020 Institutional Equipment

E1020 610	Detention Equipment, EACH	COST EACH		
		MAT.	INST.	TOTAL
0110	Detention equipment, cell front rolling door, 7/8" bars, 5' x 7' high	5,900	1,250	7,150
0120	Cells, prefab., including front, 5' x 7' x 7' deep	10,800	1,675	12,475
0130	Doors and frames, 3' x 7', single plate	5,050	635	5,685
0140	Double plate	6,150	635	6,785
0150	Toilet apparatus, incl wash basin	3,775	780	4,555
0160	Visitor cubicle, vision panel, no intercom	3,525	1,250	4,775

E1020 710	Laboratory Equipment, EACH	COST EACH		
		MAT.	INST.	TOTAL
0110	Laboratory equipment, glassware washer, distilled water, economy	6,850	400	7,250
0120	Deluxe	15,300	720	16,020
0140	Radio isotope	18,600		18,600

E1020 720	Laboratory Equipment, S.F.	COST PER S.F.		
		MAT.	INST.	TOTAL
0100	Arch. equip., lab equip., counter tops, acid proof, economy	47	12.35	59.35
0110	Stainless steel	165	12.35	177.35

E1020 730	Laboratory Equipment, L.F.	COST PER L.F.		
		MAT.	INST.	TOTAL
0110	Laboratory equipment, cabinets, wall, open	203	50.50	253.50
0120	Base, drawer units	580	56.50	636.50
0130	Fume hoods, not incl. HVAC, economy	560	188	748
0140	Deluxe incl. fixtures	1,075	420	1,495

E1020 810	Medical Equipment, EACH	COST EACH		
		MAT.	INST.	TOTAL
0100	Dental equipment, central suction system, economy	1,750	480	2,230
0110	Compressor-air, deluxe	9,925	1,025	10,950
0120	Chair, hydraulic, economy	2,450	1,025	3,475
0130	Deluxe	8,600	2,050	10,650
0140	Drill console with accessories, economy	2,375	320	2,695
0150	Deluxe	5,375	320	5,695
0160	X-ray unit, portable	2,625	129	2,754
0170	Panoramic unit	17,600	860	18,460
0300	Medical equipment, autopsy table, standard	10,700	580	11,280
0310	Deluxe	17,900	965	18,865
0320	Incubators, economy	3,425		3,425
0330	Deluxe	13,100		13,100
0700	Station, scrub-surgical, single, economy	4,825	193	5,018
0710	Dietary, medium, with ice	18,400		18,400
0720	Sterilizers, general purpose, single door, 20" x 20" x 28"	14,200		14,200
0730	Floor loading, double door, 28" x 67" x 52"	227,000		227,000
0740	Surgery tables, standard	13,400	835	14,235
0750	Deluxe	28,900	1,175	30,075
0770	Tables, standard, with base cabinets, economy	1,250	340	1,590
0780	Deluxe	6,175	505	6,680
0790	X-ray, mobile, economy	18,500		18,500
0800	Stationary, deluxe	247,000		247,000

E1030 Vehicular Equipment

E1030 110	Vehicular Service Equipment, EACH	COST EACH		
		MAT.	INST.	TOTAL
0110	Automotive equipment, compressors, electric, 1-1/2 H.P., std. controls	515	595	1,110
0120	5 H.P., dual controls	3,100	890	3,990
0130	Hoists, single post, 4 ton capacity, swivel arms	6,050	2,225	8,275
0140	Dual post, 12 ton capacity, adjustable frame	10,400	470	10,870
0150	Lube equipment, 3 reel type, with pumps	9,675	1,775	11,450
0160	Product dispenser, 6 nozzles, w/vapor recovery, not incl. piping, installed	27,200		27,200
0800	Scales, dial type, built in floor, 5 ton capacity, 8′ x 6′ platform	9,350	3,050	12,400
0810	10 ton capacity, 9′ x 7′ platform	12,100	4,350	16,450
0820	Truck (including weigh bridge), 20 ton capacity, 24′ x 10′	13,900	5,075	18,975

E1030 210	Parking Control Equipment, EACH	COST EACH		
		MAT.	INST.	TOTAL
0110	Parking equipment, automatic gates, 8 ft. arm, one way	4,425	1,050	5,475
0120	Traffic detectors, single treadle	2,225	480	2,705
0130	Booth for attendant, economy	8,000		8,000
0140	Deluxe	29,000		29,000
0150	Ticket printer/dispenser, rate computing	9,450	825	10,275
0160	Key station on pedestal	635	281	916

E1030 310	Loading Dock Equipment, EACH	COST EACH		
		MAT.	INST.	TOTAL
0110	Dock bumpers, rubber blocks, 4-1/2″ thick, 10″ high, 14″ long	66	19.50	85.50
0120	6″ thick, 20″ high, 11″ long	137	39	176
0130	Dock boards, H.D., 5′ x 5′, aluminum, 5000 lb. capacity	1,500		1,500
0140	16,000 lb. capacity	1,725		1,725
0150	Dock levelers, hydraulic, 7′ x 8′, 10 ton capacity	6,325	1,250	7,575
0160	Dock lifters, platform, 6′ x 6′, portable, 3000 lb. capacity	10,200		10,200
0170	Dock shelters, truck, scissor arms, economy	2,050	505	2,555
0180	Deluxe	2,750	1,025	3,775

E1090 Other Equipment

E1090 210	Solid Waste Handling Equipment, EACH	COST EACH MAT.	COST EACH INST.	COST EACH TOTAL
0110	Waste handling, compactors, single bag, 250 lbs./hr., hand fed	16,900	370	17,270
0120	Heavy duty industrial, 5 C.Y. capacity	36,500	1,775	38,275
0130	Incinerator, electric, 100 lbs./hr., economy	72,500	2,800	75,300
0140	Gas, 2000 lbs./hr., deluxe	419,500	20,800	440,300
0150	Shredder, no baling, 35 tons/hr.	336,500		336,500
0160	Incl. baling, 50 tons/day	672,000		672,000

E1090 350	Food Service Equipment, EACH	COST EACH MAT.	COST EACH INST.	COST EACH TOTAL
0110	Kitchen equipment, bake oven, single deck	6,075	130	6,205
0120	Broiler, without oven	3,925	130	4,055
0130	Commercial dish washer, semiautomatic, 50 racks/hr.	8,100	800	8,900
0140	Automatic, 275 racks/hr.	34,600	3,475	38,075
0150	Cooler, beverage, reach-in, 6 ft. long	3,700	173	3,873
0160	Food warmer, counter, 1.65 kw	770		770
0170	Fryers, with submerger, single	1,375	149	1,524
0180	Double	2,650	208	2,858
0185	Ice maker, 1000 lb. per day, with bin	5,650	1,050	6,700
0190	Kettles, steam jacketed, 20 gallons	9,575	201	9,776
0200	Range, restaurant type, burners, 2 ovens and 24″ griddle	5,225	173	5,398
0210	Range hood, incl. carbon dioxide system, economy	2,075	345	2,420
0220	Deluxe	2,075	345	2,420

E1090 360	Food Service Equipment, S.F.	COST PER S.F. MAT.	COST PER S.F. INST.	COST PER S.F. TOTAL
0110	Refrigerators, prefab, walk-in, 7′-6″ high, 6′ x 6′	181	18.50	199.50
0120	12′ x 20′	114	9.25	123.25

E1090 410	Residential Equipment, EACH	COST EACH MAT.	COST EACH INST.	COST EACH TOTAL
0110	Arch. equip., appliances, range, cook top, 4 burner, economy	385	96	481
0120	Built in, single oven 30″ wide, economy	990	96	1,086
0130	Standing, single oven-21″ wide, economy	555	78	633
0135	Free standing, 30″ wide, 1 oven, average	1,000	99	1,099
0140	Double oven-30″ wide, deluxe	3,075	78	3,153
0150	Compactor, residential, economy	760	101	860
0160	Deluxe	1,275	169	1,444
0170	Dish washer, built-in, 2 cycles, economy	330	181	511
0180	4 or more cycles, deluxe	1,325	360	1,685
0190	Garbage disposer, sink type, economy	102	72	174
0200	Deluxe	267	72	339
0210	Refrigerator, no frost, 10 to 12 C.F., economy	520	78	598
0220	21 to 29 C.F., deluxe	4,225	261	4,486
0300	Washing machine, automatic	1,825	580	2,405

E1090 610	School Equipment, EACH	COST EACH MAT.	COST EACH INST.	COST EACH TOTAL
0110	School equipment, basketball backstops, wall mounted, wood, fixed	1,650	885	2,535
0120	Suspended type, electrically operated	7,250	1,700	8,950
0130	Bleachers-telescoping, manual operation, 15 tier, economy (per seat)	106	27	133
0140	Power operation, 30 tier, deluxe (per seat)	435	48.50	483.50
0150	Weight lifting gym, universal, economy	385	780	1,165
0160	Deluxe	16,100	1,575	17,675
0170	Scoreboards, basketball, 1 side, economy	2,750	765	3,515
0180	4 sides, deluxe	17,300	10,500	27,800
0800	Vocational shop equipment, benches, metal	630	203	833
0810	Wood	735	203	938

E1090 Other Equipment

E1090 610	School Equipment, EACH	COST EACH		
		MAT.	INST.	TOTAL
0820	Dust collector, not incl. ductwork, 6′ diam.	5,450	515	5,965
0830	Planer, 13″ x 6″	1,200	253	1,453

E1090 620	School Equipment, S.F.	COST PER S.F.		
		MAT.	INST.	TOTAL
0110	School equipment, gym mats, naugahyde cover, 2″ thick	5.15		5.15
0120	Wrestling, 1″ thick, heavy duty	5.15		5.15

E1090 810	Athletic, Recreational, and Therapeutic Equipment, EACH	COST EACH		
		MAT.	INST.	TOTAL
0110	Sauna, prefabricated, incl. heater and controls, 7′ high, 6′ x 4′	5,850	640	6,490
0120	10′ x 12′	14,100	1,400	15,500
0130	Heaters, wall mounted, to 200 C.F.	805		805
0140	Floor standing, to 1000 C.F., 12500 W	4,275	192	4,467
0610	Shooting range incl. bullet traps, controls, separators, ceilings, economy	40,100	6,550	46,650
0620	Deluxe	59,500	11,100	70,600
0650	Sport court, squash, regulation, in existing building, economy			40,400
0660	Deluxe			45,000
0670	Racketball, regulation, in existing building, economy	43,800	8,900	52,700
0680	Deluxe	47,400	17,800	65,200
0700	Swimming pool equipment, diving stand, stainless steel, 1 meter	10,500	375	10,875
0710	3 meter	17,200	2,525	19,725
0720	Diving boards, 16 ft. long, aluminum	4,475	375	4,850
0730	Fiberglass	3,625	375	4,000
0740	Filter system, sand, incl. pump, 6000 gal./hr.	2,225	640	2,865
0750	Lights, underwater, 12 volt with transformer, 300W	340	575	915
0760	Slides, fiberglass with aluminum handrails & ladder, 6′ high, straight	3,975	635	4,610
0780	12′ high, straight with platform	16,000	845	16,845

E1090 820	Athletic, Recreational, and Therapeutic Equipment, S.F.	COST PER S.F.		
		MAT.	INST.	TOTAL
0110	Swimming pools, residential, vinyl liner, metal sides	23.50	7.05	30.55
0120	Concrete sides	28.50	12.25	40.75
0130	Gunite shell, plaster finish, 350 S.F.	52.50	25.50	78
0140	800 S.F.	42.50	14.70	57.20
0150	Motel, gunite shell, plaster finish	65	32	97
0160	Municipal, gunite shell, tile finish, formed gutters	249	55	304

E2010 Fixed Furnishings

E2010 310	Window Treatment, EACH	COST EACH		
		MAT.	INST.	TOTAL
0110	Furnishings, blinds, exterior, aluminum, louvered, 1'-4" wide x 3'-0" long	220	50.50	270.50
0120	1'-4" wide x 6'-8" long	390	56.50	446.50
0130	Hemlock, solid raised, 1-'4" wide x 3'-0" long	79	50.50	129.50
0140	1'-4" wide x 6'-9" long	167	56.50	223.50
0150	Polystyrene, louvered, 1'-3" wide x 3'-3" long	39.50	50.50	90
0160	1'-3" wide x 6'-8" long	75	56.50	131.50
0200	Interior, wood folding panels, louvered, 7" x 20" (per pair)	90.50	30	120.50
0210	18" x 40" (per pair)	148	30	178

E2010 320	Window Treatment, S.F.	COST PER S.F.		
		MAT.	INST.	TOTAL
0110	Furnishings, blinds-interior, venetian-aluminum, stock, 2" slats	6	.86	6.86
0120	Custom, 1" slats, deluxe	5.15	.86	6.01
0130	Vertical, PVC or cloth, T&B track, economy	8.60	1.10	9.70
0140	Deluxe	13.80	1.27	15.07
0150	Draperies, unlined, economy	3.19		3.19
0160	Lightproof, deluxe	10.55		10.55
0510	Shades, mylar, wood roller, single layer, non-reflective	3.52	.74	4.26
0520	Metal roller, triple layer, heat reflective	8.95	.74	9.69
0530	Vinyl, light weight, 4 ga.	.69	.74	1.43
0540	Heavyweight, 6 ga.	2.12	.74	2.86
0550	Vinyl coated cotton, lightproof decorator shades	4.99	.74	5.73
0560	Woven aluminum, 3/8" thick, light and fireproof	6.65	1.45	8.10

E2010 420	Fixed Floor Grilles and Mats, S.F.	COST PER S.F.		
		MAT.	INST.	TOTAL
0110	Floor mats, recessed, inlaid black rubber, 3/8" thick, solid	25.50	2.52	28.02
0120	Colors, 1/2" thick, perforated	30.50	2.52	33.02
0130	Link-including nosings, steel-galvanized, 3/8" thick	27	2.52	29.52
0140	Vinyl, in colors	28.50	2.52	31.02

E2010 510	Fixed Multiple Seating, EACH	COST EACH		
		MAT.	INST.	TOTAL
0110	Seating, painted steel, upholstered, economy	144	29	173
0120	Deluxe	490	36	526
0400	Seating, lecture hall, pedestal type, economy	221	46	267
0410	Deluxe	540	70	610
0500	Auditorium chair, veneer construction	260	46	306
0510	Fully upholstered, spring seat	266	46	312

E2020 Moveable Furnishings

E2020 210	Furnishings/EACH	COST EACH		
		MAT.	INST.	TOTAL
0200	Hospital furniture, beds, manual, economy	865		865
0210	Deluxe	2,700		2,700
0220	All electric, economy	1,925		1,925
0230	Deluxe	4,175		4,175
0240	Patient wall systems, no utilities, economy, per room	1,575		1,575
0250	Deluxe, per room	2,125		2,125
0300	Hotel furnishings, standard room set, economy, per room	2,775		2,775
0310	Deluxe, per room	9,475		9,475
0500	Office furniture, standard employee set, economy, per person	705		705
0510	Deluxe, per person	2,675		2,675
0550	Posts, portable, pedestrian traffic control, economy	165		165
0560	Deluxe	465		465
0700	Restaurant furniture, booth, molded plastic, stub wall and 2 seats, economy	420	253	673
0710	Deluxe	1,550	340	1,890
0720	Upholstered seats, foursome, single-economy	860	101	961
0730	Foursome, double-deluxe	1,950	169	2,119

E2020 220	Furniture and Accessories, L.F.	COST PER L.F.		
		MAT.	INST.	TOTAL
0210	Dormitory furniture, desk top (built-in),laminated plastc, 24"deep, economy	49.50	20.50	70
0220	30" deep, deluxe	276	25.50	301.50
0230	Dressing unit, built-in, economy	204	84.50	288.50
0240	Deluxe	610	127	737
0310	Furnishings, cabinets, hospital, base, laminated plastic	300	101	401
0320	Stainless steel	555	101	656
0330	Countertop, laminated plastic, no backsplash	52	25.50	77.50
0340	Stainless steel	169	25.50	194.50
0350	Nurses station, door type, laminated plastic	350	101	451
0360	Stainless steel	670	101	771
0710	Restaurant furniture, bars, built-in, back bar	228	101	329
0720	Front bar	315	101	416
0910	Wardrobes & coatracks, standing, steel, single pedestal, 30" x 18" x 63"	120		120
0920	Double face rack, 39" x 26" x 70"	136		136
0930	Wall mounted rack, steel frame & shelves, 12" x 15" x 26"	77.50	7.20	84.70
0940	12" x 15" x 50"	42.50	3.77	46.27

F Special Construction

F1010 Special Structures

F1010 120	Air-Supported Structures, S.F.	COST PER S.F.		
		MAT.	INST.	TOTAL
0110	Air supported struc., polyester vinyl fabric, 24oz., warehouse, 5000 S.F.	29	.31	29.31
0120	50,000 S.F.	13.55	.25	13.80
0130	Tennis, 7,200 S.F.	26	.26	26.26
0140	24,000 S.F.	18.05	.26	18.31
0150	Woven polyethylene, 6 oz., shelter, 3,000 S.F.	17.60	.52	18.12
0160	24,000 S.F.	13.20	.26	13.46
0170	Teflon coated fiberglass, stadium cover, economy	63	.14	63.14
0180	Deluxe	75	.19	75.19
0190	Air supported storage tank covers, reinf. vinyl fabric, 12 oz., 400 S.F.	25.50	.44	25.94
0200	18,000 S.F.	9.30	.40	9.70

F1010 210	Pre-Engineered Structures, EACH	COST EACH		
		MAT.	INST.	TOTAL
0600	Radio towers, guyed, 40 lb. section, 50′ high, 70 MPH basic wind speed	3,075	1,175	4,250
0610	90 lb. section, 400′ high, wind load 70 MPH basic wind speed	42,400	14,100	56,500
0620	Self supporting, 60′ high, 70 MPH basic wind speed	5,000	2,475	7,475
0630	190′ high, wind load 90 MPH basic wind speed	31,400	9,875	41,275
0700	Shelters, aluminum frame, acrylic glazing, 8′ high, 3′ x 9′	3,375	1,025	4,400
0710	9′ x 12′	8,025	1,600	9,625

F1010 320	Other Special Structures, S.F.	COST PER S.F.		
		MAT.	INST.	TOTAL
0110	Swimming pool enclosure, transluscent, freestanding, economy	37	5.05	42.05
0120	Deluxe	102	14.45	116.45
0510	Tension structures, steel frame, polyester vinyl fabric, 12,000 S.F.	14.55	2.33	16.88
0520	20,800 S.F.	14.30	2.10	16.40

F1010 330	Special Structures, EACH	COST EACH		
		MAT.	INST.	TOTAL
0110	Kiosks, round, 5′ diam., 7′ high, aluminum wall, illuminated	27,600		27,600
0120	Rectangular, 5′ x 9′, 1″ insulated dbl. wall fiberglass, 7′-6″ high	28,200		28,200
0220	Silos, steel prefab, 30,000 gal., painted, economy	24,300	4,900	29,200
0230	Epoxy-lined, deluxe	50,500	9,800	60,300

F1010 340	Special Structures, S.F.	COST PER S.F.		
		MAT.	INST.	TOTAL
0110	Comfort stations, prefab, mobile on steel frame, economy	196		196
0120	Permanent on concrete slab, deluxe	211	44	255
0210	Domes, bulk storage, wood framing, wood decking, 50′ diam.	75.50	1.51	77.01
0220	116′ diam.	36.50	1.75	38.25
0230	Steel framing, metal decking, 150′ diam.	36.50	9.40	45.90
0240	400′ diam.	29.50	7.15	36.65
0250	Geodesic, wood framing, wood panels, 30′ diam.	37.50	1.57	39.07
0260	60′ diam.	24.50	1.07	25.57
0270	Aluminum framing, acrylic panels, 40′ diam.			70.50
0280	Aluminum panels, 400′ diam.			22
0310	Garden house, prefab, wood, shell only, 48 S.F.	57	5.05	62.05
0320	200 S.F.	43.50	21	64.50
0410	Greenhouse, shell-stock, residential, lean-to, 8′-6″ long x 3′-10″ wide	47.50	30	77.50
0420	Freestanding, 8′-6″ long x 13′-6″ wide	47.50	9.40	56.90
0430	Commercial-truss frame, under 2000 S.F., deluxe			14.65
0440	Over 5,000 S.F., economy			13.40
0450	Institutional-rigid frame, under 500 S.F., deluxe			30
0460	Over 2,000 S.F., economy			12.25
0510	Hangar, prefab, galv. roof and walls, bottom rolling doors, economy	13.55	3.73	17.28
0520	Electric bifolding doors, deluxe	13.40	4.85	18.25

F1020 Integrated Construction

F1020 110	Integrated Construction, EACH	COST EACH		
		MAT.	INST.	TOTAL
0110	Integrated ceilings, radiant electric, 2′ x 4′ panel, manila finish	101	23	124
0120	ABS plastic finish	114	44.50	158.50

F1020 120	Integrated Construction, S.F.	COST PER S.F.		
		MAT.	INST.	TOTAL
0110	Integrated ceilings, Luminaire, suspended, 5′ x 5′ modules, 50% lighted	4.41	7.25	11.66
0120	100% lighted	6.55	13	19.55
0130	Dimensionaire, 2′ x 4′ module tile system, no air bar	2.66	2.03	4.69
0140	With air bar, deluxe	3.90	2.03	5.93
0220	Pedestal access floor pkg., inc. stl. pnls, peds.& stringers, w/vinyl cov.	27	2.70	29.70
0230	With high pressure laminate covering	25	2.70	27.70
0240	With carpet covering	26.40	2.70	29.10
0250	Aluminum panels, no stringers, no covering	36	2.03	38.03

F1020 250	Special Purpose Room, EACH	COST EACH		
		MAT.	INST.	TOTAL
0110	Portable booth, acoustical, 27 db 1000 hz., 15 S.F. floor	4,100		4,100
0120	55 S.F. flr.	8,450		8,450

F1020 260	Special Purpose Room, S.F.	COST PER S.F.		
		MAT.	INST.	TOTAL
0110	Anechoic chambers, 7′ high, 100 cps cutoff, 25 S.F.			2,650
0120	200 cps cutoff, 100 S.F.			1,375
0130	Audiometric rooms, under 500 S.F.	59	20.50	79.50
0140	Over 500 S.F.	57.50	16.90	74.40
0300	Darkrooms, shell, not including door, 240 S.F., 8′ high	31	8.45	39.45
0310	64 S.F., 12′ high	76.50	15.85	92.35
0510	Music practice room, modular, perforated steel, under 500 S.F.	38	14.45	52.45
0520	Over 500 S.F.	32	12.65	44.65

F1020 330	Special Construction, L.F.	COST PER L.F.		
		MAT.	INST.	TOTAL
0110	Spec. const., air curtains, shipping & receiving, 8′high x 5′wide, economy	785	113	898
0120	20′ high x 8′ wide, heated, deluxe	3,675	283	3,958
0130	Customer entrance, 10′ high x 5′ wide, economy	1,175	113	1,288
0140	12′ high x 4′ wide, heated, deluxe	2,125	141	2,266

F1030 Special Construction Systems

F1030 120	Sound, Vibration, and Seismic Construction, S.F.	COST PER S.F.		
		MAT.	INST.	TOTAL
0020	Special construction, acoustical, enclosure, 4" thick, 8 psf panels	37	21	58
0030	Reverb chamber, 4" thick, parallel walls	52	25.50	77.50
0110	Sound absorbing panels, 2'-6" x 8', painted metal	13.85	7.05	20.90
0120	Vinyl faced	10.80	6.35	17.15
0130	Flexible transparent curtain, clear	8.35	7.90	16.25
0140	With absorbing foam, 75% coverage	11.65	7.90	19.55
0150	Strip entrance, 2/3 overlap	8.35	12.55	20.90
0160	Full overlap	10.55	14.75	25.30
0200	Audio masking system, plenum mounted, over 10,000 S.F.	.81	.26	1.07
0210	Ceiling mounted, under 5,000 S.F.	1.42	.48	1.90

F1030 210	Radiation Protection, EACH	COST EACH		
		MAT.	INST.	TOTAL
0110	Shielding, lead x-ray protection, radiography room, 1/16" lead, economy	11,100	3,825	14,925
0120	Deluxe	13,400	6,375	19,775
0210	Deep therapy x-ray, 1/4" lead, economy	31,100	11,900	43,000
0220	Deluxe	38,400	15,900	54,300

F1030 220	Radiation Protection, S.F.	COST PER S.F.		
		MAT.	INST.	TOTAL
0110	Shielding, lead, gypsum board, 5/8" thick, 1/16" lead	12.30	6.15	18.45
0120	1/8" lead	25.50	7	32.50
0130	Lath, 1/16" thick	11.55	7.05	18.60
0140	1/8" thick	34.50	7.95	42.45
0150	Radio frequency, galvanized steel, prefab type, economy	5.65	2.70	8.35
0160	Radio frequency, door, copper/wood laminate, 4" x 7'	9.95	7.25	17.20

F1030 910	Other Special Construction Systems, EACH	COST EACH		
		MAT.	INST.	TOTAL
0110	Disappearing stairways, folding, pine, 8'-6" ceiling	194	127	321
0220	Automatic electric, wood, 8' to 9' ceiling	9,575	1,025	10,600
0300	Fireplace prefabricated, freestanding or wall hung, painted	1,625	390	2,015
0310	Stainless steel	3,350	565	3,915
0320	Woodburning stoves, cast iron, economy, less than 1500 S.F. htg. area	1,425	780	2,205
0330	Greater than 2000 S.F. htg. area	3,075	1,275	4,350

F1040 Special Facilities

F1040 210	Ice Rinks, EACH	COAT EACH MAT.	INST.	TOTAL
0100	Ice skating rink, 85′ x 200′, 55° system, 5 mos., 100 ton			664,000
0110	90° system, 12 mos., 135 ton			751,000
0120	Dash boards, acrylic screens, polyethylene coated plywood	156,000	29,500	185,500
0130	Fiberglass and aluminum construction	179,000	29,500	208,500

F1040 510	Liquid & Gas Storage Tanks, EACH	COST EACH MAT.	INST.	TOTAL
0100	Tanks, steel, ground level, 100,000 gal.			226,500
0110	10,000,000 gal.			5,151,500
0120	Elevated water, 50,000 gal.		171,000	443,000
0130	1,000,000 gal.		795,500	1,955,500
0150	Cypress wood, ground level, 3,000 gal.	9,900	9,375	19,275
0160	Redwood, ground level, 45,000 gal.	77,500	25,500	103,000

F1040 910	Special Construction, EACH	COST EACH MAT.	INST.	TOTAL
0110	Special construction, bowling alley incl. pinsetter, scorer etc., economy	42,500	10,100	52,600
0120	Deluxe	58,500	11,300	69,800
0130	For automatic scorer, economy, add	6,400		6,400
0140	Deluxe, add	11,000		11,000
0300	Control tower, modular, 12′ x 10′, incl. instrumentation, economy			882,000
0310	Deluxe			1,386,000
0400	Garage costs, residential, prefab, wood, single car economy	6,275	1,025	7,300
0410	Two car deluxe	14,200	2,025	16,225
0500	Hangars, prefab, galv. steel, bottom rolling doors, economy (per plane)	18,000	4,125	22,125
0510	Electrical bi-folding doors, deluxe (per plane)	16,000	5,650	21,650

G1030 Site Earthwork

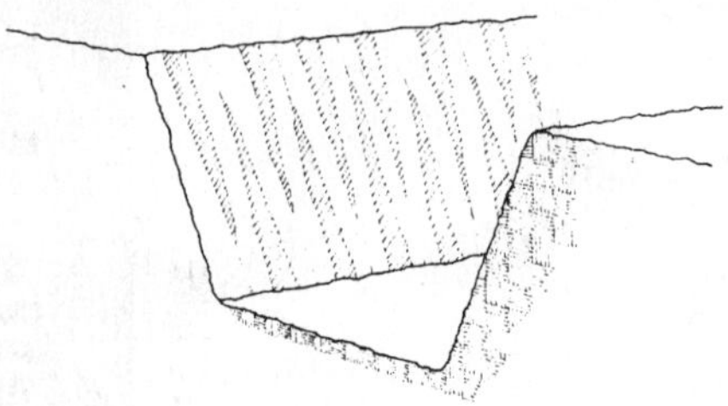

Trenching Systems are shown on a cost per linear foot basis. The systems include: excavation; backfill and removal of spoil; and compaction for various depths and trench bottom widths. The backfill has been reduced to accommodate a pipe of suitable diameter and bedding.

The slope for trench sides varies from none to 1:1.

The Expanded System Listing shows Trenching Systems that range from 2′ to 12′ in width. Depths range from 2′ to 25′.

G1030 805	Trenching Common Earth	COST PER L.F.		
		MAT.	INST.	TOTAL
1310	Trenching, common earth, no slope, 2′ wide, 2′ deep, 3/8 C.Y. bucket	.71	1.58	2.29
1330	4′ deep, 3/8 C.Y. bucket	1.29	3.15	4.44
1360	10′ deep, 1 C.Y. bucket	3.46	6.30	9.76
1400	4′ wide, 2′ deep, 3/8 C.Y. bucket	1.61	3.18	4.79
1420	4′ deep, 1/2 C.Y. bucket	2.57	5.20	7.77
1450	10′ deep, 1 C.Y. bucket	8	12.05	20.05
1480	18′ deep, 2-1/2 C.Y. bucket	13.05	18.60	31.65
1520	6′ wide, 6′ deep, 5/8 C.Y. bucket w/trench box	8.55	11.55	20.10
1540	10′ deep, 1 C.Y. bucket	11.55	14.85	26.40
1570	20′ deep, 3-1/2 C.Y. bucket	22	22	44
1640	8′ wide, 12′ deep, 1-1/2 C.Y. bucket w/trench box	17.35	20	37.35
1680	24′ deep, 3-1/2 C.Y. bucket	35.50	34.50	70
1730	10′ wide, 20′ deep, 3-1/2 C.Y. bucket w/trench box	29	32.50	61.50
1740	24′ deep, 3-1/2 C.Y. bucket	42.50	40	82.50
3500	1 to 1 slope, 2′ wide, 2′ deep, 3/8 C.Y. bucket	1.29	3.15	4.44
3540	4′ deep, 3/8 C.Y. bucket	3.69	9.45	13.14
3600	10′ deep, 1 C.Y. bucket	20.50	38	58.50
3800	4′ wide, 2′ deep, 3/8 C.Y. bucket	2.21	4.76	6.97
3840	4′ deep, 1/2 C.Y. bucket	4.66	10.55	15.21
3900	10′ deep, 1 C.Y. bucket	27.50	42.50	70
4030	6′ wide, 6′ deep, 5/8 C.Y. bucket w/trench box	16.50	22.50	39
4050	10′ deep, 1 C.Y. bucket	27	35	62
4080	20′ deep, 3-1/2 C.Y. bucket	89	96	185
4500	8′ wide, 12′ deep, 1-1/2 C.Y. bucket w/trench box	41	52	93
4650	24′ deep, 3-1/2 C.Y. bucket	139	144	283
4800	10′ wide, 20′ deep, 3-1/2 C.Y. bucket w/trench box	83.50	103	186.50
4850	24′ deep, 3-1/2 C.Y. bucket	148	152	300

G1030 Site Earthwork

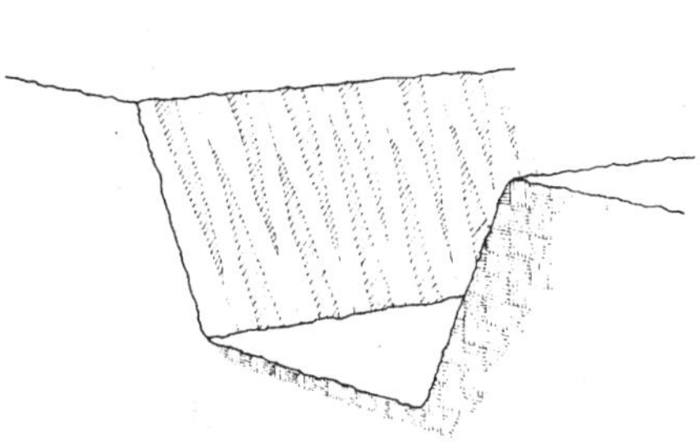

Trenching Systems are shown on a cost per linear foot basis. The systems include: excavation; backfill and removal of spoil; and compaction for various depths and trench bottom widths. The backfill has been reduced to accommodate a pipe of suitable diameter and bedding.

The slope for trench sides varies from none to 1:1.

The Expanded System Listing shows Trenching Systems that range from 2′ to 12′ in width. Depths range from 2′ to 25′.

G1030 807	Trenching Sand & Gravel	COST PER L.F.		
		MAT.	INST.	TOTAL
1310	Trenching, sand & gravel, no slope, 2′ wide, 2′ deep, 3/8 C.Y. bucket	.64	1.46	2.10
1320	3′ deep, 3/8 C.Y. bucket	1.03	2.45	3.48
1330	4′ deep, 3/8 C.Y. bucket	1.19	2.92	4.11
1340	6′ deep, 3/8 C.Y. bucket	1.71	3.42	5.15
1350	8′ deep, 1/2 C.Y. bucket	2.25	4.51	6.75
1360	10′ deep, 1 C.Y. bucket	2.50	4.69	7.20
1400	4′ wide, 2′ deep, 3/8 C.Y. bucket	1.46	2.93	4.39
1410	3′ deep, 3/8 C.Y. bucket	2.01	4.39	6.40
1420	4′ deep, 1/2 C.Y. bucket	2.34	4.86	7.20
1430	6′ deep, 1/2 C.Y. bucket	4.18	6.45	10.65
1440	8′ deep, 1/2 C.Y. bucket	6.10	9.05	15.15
1450	10′ deep, 1 C.Y. bucket	6.10	8.85	14.95
1460	12′ deep, 1 C.Y. bucket	7.70	11.10	18.80
1470	15′ deep, 1-1/2 C.Y. bucket	9.25	12.60	22
1480	18′ deep, 2-1/2 C.Y. bucket	10.85	13.40	24.50
1520	6′ wide, 6′ deep, 5/8 C.Y. bucket w/trench box	7.80	10.65	18.45
1530	8′ deep, 3/4 C.Y. bucket	10.35	13.80	24
1540	10′ deep, 1 C.Y. bucket	10	13.35	23.50
1550	12′ deep, 1-1/2 C.Y. bucket	11.45	14.60	26
1560	16′ deep, 2 C.Y. bucket	15.70	18	33.50
1570	20′ deep, 3-1/2 C.Y. bucket	19.25	20.50	40
1580	24′ deep, 3-1/2 C.Y. bucket	24	25.50	49.50
1640	8′ wide, 12′ deep, 1-1/2 C.Y. bucket w/trench box	16.10	18.90	35
1650	15′ deep, 1-1/2 C.Y. bucket	19.55	24	43.50
1660	18′ deep, 2-1/2 C.Y. bucket	23.50	26	49.50
1680	24′ deep, 3-1/2 C.Y. bucket	32.50	33	65.50
1730	10′ wide, 20′ deep, 3-1/2 C.Y. bucket w/trench box	33	33	66
1740	24′ deep, 3-1/2 C.Y. bucket	41	41	82
1780	12′ wide, 20′ deep, 3-1/2 C.Y. bucket w/trench box	39.50	39.50	79
1790	25′ deep, 3-1/2 C.Y. bucket	51.50	50.50	102
1800	1/2:1 slope, 2′ wide, 2′ deep, 3/8 C.Y. bucket	.92	2.19	3.11
1810	3′ deep, 3/8 C.Y. bucket	1.54	3.84	5.40
1820	4′ deep, 3/8 C.Y. bucket	2.29	5.85	8.15
1840	6′ deep, 3/8 C.Y. bucket	4.11	8.55	12.65
1860	8′ deep, 1/2 C.Y. bucket	6.55	13.60	20
1880	10′ deep, 1 C.Y. bucket	8.55	16.60	25
2300	4′ wide, 2′ deep, 3/8 C.Y. bucket	1.74	3.67	5.40
2310	3′ deep, 3/8 C.Y. bucket	2.63	6.05	8.70
2320	4′ deep, 1/2 C.Y. bucket	3.31	7.35	10.65
2340	6′ deep, 1/2 C.Y. bucket	7.05	11.35	18.40
2360	8′ deep, 1/2 C.Y. bucket	11.85	18.20	30
2380	10′ deep, 1 C.Y. bucket	13.45	20	33.50
2400	12′ deep, 1 C.Y. bucket	22.50	33.50	56
2430	15′ deep, 1-1/2 C.Y. bucket	26.50	36.50	63

G1030 Site Earthwork

G1030 807	Trenching Sand & Gravel	COST PER L.F. MAT.	INST.	TOTAL
2460	18' deep, 2-1/2 C.Y. bucket	43	54.50	97.50
2840	6' wide, 6' deep, 5/8 C.Y. bucket w/trench box	11.40	15.95	27.50
2860	8' deep, 3/4 C.Y. bucket	16.70	23	39.50
2880	10' deep, 1 C.Y. bucket	17.95	25	43
2900	12' deep, 1-1/2 C.Y. bucket	22.50	29.50	52
2940	16' deep, 2 C.Y. bucket	36	42.50	78.50
2980	20' deep, 3-1/2 C.Y. bucket	50.50	56	107
3020	24' deep, 3-1/2 C.Y. bucket	72	77	149
3100	8' wide, 12' deep, 1-1/4 C.Y. bucket w/trench box	27	34	61
3120	15' deep, 1-1/2 C.Y. bucket	36.50	47.50	84
3140	18' deep, 2-1/2 C.Y. bucket	49	57	106
3180	24' deep, 3-1/2 C.Y. bucket	80.50	84.50	165
3270	10' wide, 20' deep, 3-1/2 C.Y. bucket w/trench box	64.50	68.50	133
3280	24' deep, 3-1/2 C.Y. bucket	88.50	92.50	181
3370	12' wide, 20' deep, 3-1/2 C.Y. bucket w/trench box	72	76.50	149
3380	25' deep, 3-1/2 C.Y. bucket	103	107	210
3500	1:1 slope, 2' wide, 2' deep, 3/8 C.Y. bucket	2.06	3.76	5.80
3520	3' deep, 3/8 C.Y. bucket	2.16	5.50	7.65
3540	4' deep, 3/8 C.Y. bucket	3.38	8.80	12.20
3560	6' deep, 3/8 C.Y. bucket	4.11	8.55	12.65
3580	8' deep, 1/2 C.Y. bucket	10.80	22.50	33.50
3600	10' deep, 1 C.Y. bucket	14.65	28.50	43
3800	4' wide, 2' deep, 3/8 C.Y. bucket	2.01	4.40	6.40
3820	3' deep, 3/8 C.Y. bucket	3.25	7.70	10.95
3840	4' deep, 1/2 C.Y. bucket	4.26	9.85	14.10
3860	6' deep, 1/2 C.Y. bucket	9.90	16.25	26
3880	8' deep, 1/2 C.Y. bucket	17.55	27.50	45
3900	10' deep, 1 C.Y. bucket	21	31.50	52.50
3920	12' deep, 1 C.Y. bucket	30	45	75
3940	15' deep, 1-1/2 C.Y. bucket	43.50	60.50	104
3960	18' deep, 2-1/2 C.Y. bucket	59	75	134
4030	6' wide, 6' deep, 5/8 C.Y. bucket w/trench box	15	21	36
4040	8' deep, 3/4 C.Y. bucket	23	32.50	55.50
4050	10' deep, 1 C.Y. bucket	26	36.50	62.50
4060	12' deep, 1-1/2 C.Y. bucket	33.50	44.50	78
4070	16' deep, 2 C.Y. bucket	56.50	67	124
4080	20' deep, 3-1/2 C.Y. bucket	82	91.50	174
4090	24' deep, 3-1/2 C.Y. bucket	119	129	248
4500	8' wide, 12' deep, 1-1/2 C.Y. bucket w/trench box	38	49	87
4550	15' deep, 1-1/2 C.Y. bucket	53.50	71	125
4600	18' deep, 2-1/2 C.Y. bucket	74.50	88	163
4650	24' deep, 3-1/2 C.Y. bucket	128	136	264
4800	10' wide, 20' deep, 3-1/2 C.Y. bucket w/trench box	95.50	104	200
4850	24' deep, 3-1/2 C.Y. bucket	136	144	280
4950	12' wide, 20' deep, 3-1/2 C.Y. bucket w/trench box	103	110	213
4980	25' deep, 3-1/2 C.Y. bucket	155	163	320

G1030 815	Pipe Bedding	COST PER L.F. MAT.	INST.	TOTAL
1440	Pipe bedding, side slope 0 to 1, 1' wide, pipe size 6" diameter	1.88	.86	2.74
1460	2' wide, pipe size 8" diameter	4.07	1.84	5.91
1500	Pipe size 12" diameter	4.25	1.92	6.17
1600	4' wide, pipe size 20" diameter	10.15	4.61	14.76
1660	Pipe size 30" diameter	10.65	4.83	15.48
1680	6' wide, pipe size 32" diameter	18.25	8.25	26.50
1740	8' wide, pipe size 60" diameter	36	16.35	52.35
1780	12' wide, pipe size 84" diameter	66.50	30	96.50

G2010 Roadways

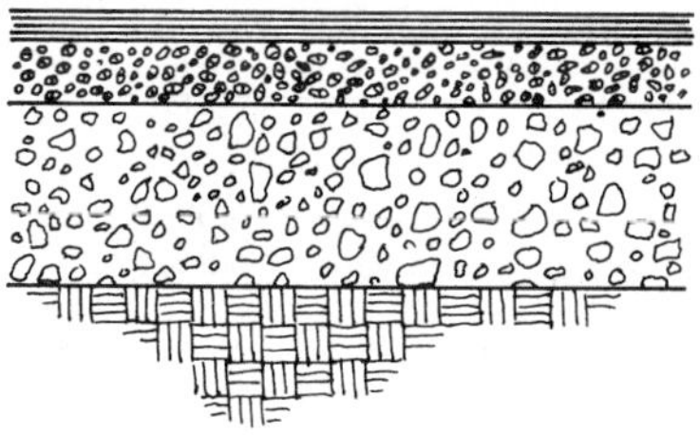

The Bituminous Roadway Systems are listed for pavement thicknesses between 3-1/2″ and 7″ and crushed stone bases from 3″ to 22″ in depth. Systems costs are expressed per linear foot for varying widths of two and multi-lane roads. Earth moving is not included. Granite curbs and line painting are added as required system components.

G2010 232	Bituminous Roadways Crushed Stone	COST PER L.F. MAT.	COST PER L.F. INST.	COST PER L.F. TOTAL
1050	Bitum. roadway, two lanes, 3-1/2″ th. pvmt., 3″ th. crushed stone,24′ wide	77.50	48	125.50
1100	28′ wide	85	48	133
1150	32′ wide	92.50	55	147.50
1300	9″ thick crushed stone, 24′ wide	91.50	50	141.50
1350	28′ wide	101	54.50	155.50
1400	32′ wide	111	58.50	169.50
1420	12″ thick crushed stone, 24′ wide	98.50	52.50	151
1550	4″ thick pavement., 4″ thick crushed stone, 24′ wide	85.50	46.50	132
1600	28′ wide	94	50	144
1650	32′ wide	103	53.50	156.50
1800	10″ thick crushed stone, 24′ wide	99.50	51.50	151
1850	28′ wide	111	55	166
1900	32′ wide	122	60.50	182.50
3050	5″ thick pavement, 6″ thick crushed stone, 24′ wide	102	50.50	152.50
3100	28′ wide	114	54.50	168.50
3150	32′ wide	125	59	184
3300	14″ thick crushed stone, 24′ wide	121	56.50	177.50
3350	28′ wide	135	61.50	196.50
3400	32′ wide	151	67	218

G2020 Parking Lots

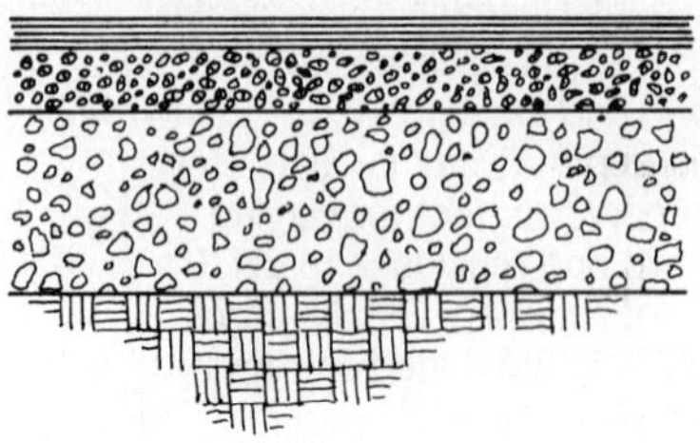

The Parking Lot System includes: compacted bank-run gravel; fine grading with a grader and roller; and bituminous concrete wearing course. All Parking Lot systems are on a cost per car basis. There are three basic types of systems: 90° angle, 60° angle, and 45° angle. The gravel base is compacted to 98%. Final stall design and lay-out of the parking lot with precast bumpers, sealcoating and white paint is also included.

The Expanded System Listing shows the three basic parking lot types with various depths of both gravel base and wearing course. The gravel base depths range from 6″ to 10″. The bituminous paving wearing course varies from a depth of 3″ to 6″.

G2020 210	Parking Lots Gravel Base	COST PER CAR		
		MAT.	INST.	TOTAL
1500	Parking lot, 90° angle parking, 3″ bituminous paving, 6″ gravel base	805	390	1,195
1540	10″ gravel base	915	460	1,375
1560	4″ bituminous paving, 6″ gravel base	1,025	415	1,440
1600	10″ gravel base	1,150	480	1,630
1620	6″ bituminous paving, 6″ gravel base	1,375	445	1,820
1660	10″ gravel base	1,500	510	2,010
1800	60° angle parking, 3″ bituminous paving, 6″ gravel base	805	390	1,195
1840	10″ gravel base	915	460	1,375
1860	4″ bituminous paving, 6″ gravel base	1,025	415	1,440
1900	10″ gravel base	1,150	480	1,630
1920	6″ bituminous paving, 6″ gravel base	1,375	445	1,820
1960	10″ gravel base	1,500	510	2,010
2200	45° angle parking, 3″ bituminous paving, 6″ gravel base	825	405	1,230
2240	10″ gravel base	935	470	1,405
2260	4″ bituminous paving, 6″ gravel base	1,050	425	1,475
2300	10″ gravel base	1,175	490	1,665
2320	6″ bituminous paving, 6″ gravel base	1,425	455	1,880
2360	10″ gravel base	1,525	525	2,050

G2040 Site Development

G2040 810	Flagpoles	COST EACH		
		MAT.	INST.	TOTAL
0110	Flagpoles, on grade, aluminum, tapered, 20' high	1,175	640	1,815
0120	70' high	9,700	1,600	11,300
0140	59' high	5,050	1,425	6,475
0150	Concrete, internal halyard, 20' high	1,500	510	2,010
0160	100' high	19,900	1,275	21,175

G2040 950	Other Site Development, EACH	COST EACH		
		MAT.	INST.	TOTAL
0110	Grandstands, permanent, closed deck, steel, economy (per seat)			24
0120	Deluxe (per seat)			27
0130	Composite design, economy (per seat)			46.50
0140	Deluxe (per seat)			82.50

G3030 Storm Sewer

G3030 210	Manholes & Catch Basins	COST PER EACH		
		MAT.	INST.	TOTAL
1920	Manhole/catch basin, brick, 4′ I.D. riser, 4′ deep	1,325	1,525	2,850
1980	10′ deep	2,975	3,625	6,600
3200	Block, 4′ I.D. riser, 4′ deep	1,175	1,250	2,425
3260	10′ deep	2,600	3,000	5,600
4620	Concrete, cast-in-place, 4′ I.D. riser, 4′ deep	1,325	2,150	3,475
4680	10′ deep	3,225	5,150	8,375
5820	Concrete, precast, 4′ I.D. riser, 4′ deep	1,550	1,100	2,650
5880	10′ deep	4,050	2,600	6,650
6200	6′ I.D. riser, 4′ deep	3,425	1,725	5,150
6260	10′ deep	7,225	4,075	11,300

G4020 Site Lighting

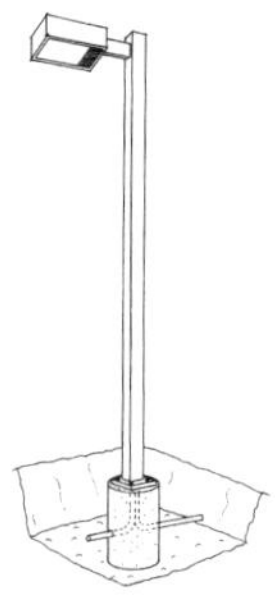

Light Pole

G4020 210	Light Pole (Installed)	COST EACH		
		MAT.	INST.	TOTAL
0200	Light pole, aluminum, 20′ high, 1 arm bracket	1,275	1,025	2,300
0240	2 arm brackets	1,400	1,025	2,425
0280	3 arm brackets	1,550	1,075	2,625
0320	4 arm brackets	1,675	1,075	2,750
0360	30′ high, 1 arm bracket	2,225	1,300	3,525
0400	2 arm brackets	2,350	1,300	3,650
0440	3 arm brackets	2,500	1,325	3,825
0480	4 arm brackets	2,625	1,325	3,950
0680	40′ high, 1 arm bracket	2,725	1,725	4,450
0720	2 arm brackets	2,850	1,725	4,575
0760	3 arm brackets	2,975	1,775	4,750
0800	4 arm brackets	3,125	1,775	4,900
0840	Steel, 20′ high, 1 arm bracket	1,425	1,100	2,525
0880	2 arm brackets	1,475	1,100	2,575
0920	3 arm brackets	1,450	1,125	2,575
0960	4 arm brackets	1,550	1,125	2,675
1000	30′ high, 1 arm bracket	1,750	1,400	3,150
1040	2 arm brackets	1,800	1,400	3,200
1080	3 arm brackets	1,800	1,425	3,225
1120	4 arm brackets	1,900	1,425	3,325
1320	40′ high, 1 arm bracket	2,300	1,875	4,175
1360	2 arm brackets	2,350	1,875	4,225
1400	3 arm brackets	2,350	1,925	4,275
1440	4 arm brackets	2,450	1,925	4,375

Unit Price Section

Table of Contents

Table of Contents (cont.)

How RSMeans Unit Price Data Works

All RSMeans unit price data is organized in the same way.

It is important to understand the structure, so that you can find information easily and use it correctly.

03 30 Cast-In-Place Concrete

03 30 53 – Miscellaneous Cast-In-Place Concrete

03 30 53.40 Concrete In Place (1)			Crew (5)	Daily Output (6)	Labor-Hours (7)	Unit (8)	2016 Bare Costs (9) Material	Labor	Equipment	Total (10)	Total Incl O&P (11)
0010	CONCRETE IN PLACE (4)	R033053-10									
0020	Including forms (4 uses), Grade 60 rebar, concrete (Portland cement	R033105-10									
0050	Type I), placement and finishing unless otherwise indicated	R033105-20									
0500	Chimney foundations (5000 psi), over 5 C.Y.	R033105-50	C-14C	32.22	3.476	C.Y.	155	124	.97	279.97	380
0510	(3500 psi), under 5 C.Y. (3)	R033105-70	"	23.71	4.724	"	182	168	1.32	351.32	480
3540	Equipment pad (3000 psi), 3' x 3' x 6" thick		C-14H	45	1.067	Ea.	47.50	39	.70	87.20	118
(2) 3550	4' x 4' x 6" thick			30	1.600		70.50	58.50	1.06	130.06	177
3560	5' x 5' x 8" thick			18	2.667		124	98	1.76	223.76	300
3570	6' x 6' x 8" thick			14	3.429		167	126	2.26	295.26	395
3580	8' x 8' x 10" thick			8	6		360	220	3.96	583.96	765
3590	10' x 10' x 12" thick			5	9.600		615	350	6.35	971.35	1,275
3800	Footings (3000 psi), spread under 1 C.Y.		C-14C	28	4	C.Y.	171	142	1.12	314.12	425
3825	1 C.Y. to 5 C.Y.			43	2.605		207	93	.73	300.73	385
3850	Over 5 C.Y.	R033105-80		75	1.493		190	53	.42	243.42	298

1 Line Numbers

RSMeans **Line Numbers** consist of 12 characters, which identify a unique location in the database for each task. The first 6 or 8 digits conform to the Construction Specifications Institute MasterFormat® 2014. The remainder of the digits are a further breakdown by RSMeans in order to arrange items in understandable groups of similar tasks. Line numbers are consistent across all RSMeans publications, so a line number in any RSMeans product will always refer to the same unit of work.

2 Descriptions

RSMeans **Descriptions** are shown in a hierarchical structure to make them readable. In order to read a complete description, read up through the indents to the top of the section. Include everything that is above and to the left that is not contradicted by information below. For instance, the complete description for line 03 30 53.40 3550 is "Concrete in place, including forms (4 uses), Grade 60 rebar, concrete (Portland cement Type 1), placement and finishing unless otherwise indicated; Equipment pad (3000 psi), 4' x 4' x 6" thick."

3 RSMeans Data

When using **RSMeans data**, it is important to read through an entire section to ensure that you use the data that most closely matches your work. Note that sometimes there is additional information shown in the section that may improve your price. There are frequently lines that further describe, add to, or adjust data for specific situations.

4 Reference Information

RSMeans engineers have created **reference** information to assist you in your estimate. **If** there is information that applies to a section, it will be indicated at the start of the section. The Reference Section is located in the back of the data set.

5 Crews

Crews include labor and/or equipment necessary to accomplish each task. In this case, Crew C-14H is used.

RSMeans selects a crew to represent the workers and equipment that are typically used for that task. In this case, Crew C-14H consists of one carpenter foreman (outside), two carpenters, one rodman, one laborer, one cement finisher, and one gas engine vibrator. Details of all crews can be found in the reference section.

Crews - Open Shop

Crew No.	Bare Costs		Incl. Subs O&P		Cost Per Labor-Hour	
Crew C-14H	Hr.	Daily	Hr.	Daily	Bare Costs	Incl. O&P
1 Carpenter Foreman (outside)	$39.80	$318.40	$66.65	$533.20	$36.66	$61.18
2 Carpenters	37.80	604.80	63.30	1012.80		
1 Rodman (reinf.)	39.75	318.00	67.15	537.20		
1 Laborer	29.20	233.60	48.90	391.20		
1 Cement Finisher	35.60	284.80	57.80	462.40		
1 Gas Engine Vibrator		31.60		34.76	0.66	0.72
48 L.H., Daily Totals		$1791.20		$2971.56	$37.32	$61.91

6 Daily Output

The **Daily Output** is the amount of work that the crew can do in a normal 8-hour workday, including mobilization, layout, movement of materials, and cleanup. In this case, crew C-14H can install thirty 4' x 4' x 6" thick concrete pads in a day. Daily output is variable and based on many factors, including the size of the job, location, and environmental conditions. RSMeans data represents work done in daylight (or adequate lighting) and temperate conditions.

7 Labor-Hours

The figure in the **Labor-Hours** column is the amount of labor required to perform one unit of work–in this case the amount of labor required to construct one 4' x 4' equipment pad. This figure is calculated by dividing the number of hours of labor in the crew by the daily output (48 labor hours divided by 30 pads = 1.6 hours of labor per pad). Multiply 1.600 times 60 to see the value in minutes: 60 x 1.6 = 96 minutes. Note: the labor-hour figure is not dependent on the crew size. A change in crew size will result in a corresponding change in daily output, but the labor-hours per unit of work will not change.

8 Unit of Measure

All RSMeans unit cost data includes the typical **Unit of Measure** used for estimating that item. For concrete-in-place the typical unit is cubic yards (C.Y.) or each (Ea.). For installing broadloom carpet it is square yard, and for gypsum board it is square foot. The estimator needs to take special care that the unit in the data matches the unit in the take-off. Unit conversions may be found in the Reference Section.

9 Bare Costs

Bare Costs are the costs of materials, labor, and equipment that the installing contractor pays. They represent the cost, in U.S. dollars, for one unit of work. They do not include any markups for profit or labor burden.

10 Bare Total

The **Total column** represents the total bare cost for the installing contractor, in U.S. dollars. In this case, the sum of $70.50 for material + $58.50 for labor + $1.06 for equipment is $130.06.

11 Total Incl O&P

The **Total Incl O&P column** is the total cost, including overhead and profit, that the installing contractor will charge the customer. This represents the cost of materials plus 10% profit, the cost of labor plus labor burden and 10% profit, and the cost of equipment plus 10% profit. It does not include the general contractor's overhead and profit. Note: See the inside back cover of the printed product or the reference section of the electronic product for details of how RSMeans calculates labor burden.

National Average

The data in RSMeans print publications represents a "national average" cost. This data should be modified to the project location using the ***City Cost Indexes*** *or* ***Location Factors*** *tables found in the Reference Section. Use the location factors to adjust estimate totals if the project covers multiple trades. Use the city cost indexes (CCI) for single trade projects or projects where a more detailed analysis is required. All figures in the two tables are derived from the same research. The last row of data in the CCI, the weighted average, is the same as the numbers reported for each location in the location factor table.*

How RSMeans Unit Price Data Works (Continued)

Project Name: Pre-Engineered Steel Building		Architect: As Shown						
Location:	**Anywhere, USA**						**01/01/16**	**OPN**
(1) **Line Number**	**Description**	**Qty**	**Unit**	**Material**	**Labor**	**Equipment**	**SubContract**	**Estimate Total**
03 30 53.40 3940	Strip footing, 12" x 24", reinforced	34	C.Y.	$4,896.00	$2,822.00	$22.10	$0.00	
03 30 53.40 3950	Strip footing, 12" x 36", reinforced	15	C.Y.	$2,070.00	$997.50	$7.80	$0.00	
03 11 13.65 3000	Concrete slab edge forms	500	L.F.	$165.00	$885.00	$0.00	$0.00	
03 22 11.10 0200	Welded wire fabric reinforcing	150	C.S.F.	$2,632.50	$3,075.00	$0.00	$0.00	
03 31 13.35 0300	Ready mix concrete, 4000 psi for slab on grade	278	C.Y.	$31,414.00	$0.00	$0.00	$0.00	
03 31 13.70 4300	Place, strike off & consolidate concrete slab	278	C.Y.	$0.00	$3,711.30	$161.24	$0.00	
03 35 13.30 0250	Machine float & trowel concrete slab	15,000	S.F.	$0.00	$7,050.00	$450.00	$0.00	
03 15 16.20 0140	Cut control joints in concrete slab	950	L.F.	$47.50	$304.00	$85.50	$0.00	
03 39 23.13 0300	Sprayed concrete curing membrane	150	C.S.F.	$1,905.00	$738.00	$0.00	$0.00	
Division 03	**Subtotal**			**$43,130.00**	**$19,582.80**	**$726.64**	**$0.00**	**$63,439.44**
08 36 13.10 2650	Manual 10' x 10' steel sectional overhead door	8	Ea.	$9,800.00	$2,680.00	$0.00	$0.00	
08 36 13.10 2860	Insulation and steel back panel for OH door	800	S.F.	$3,800.00	$0.00	$0.00	$0.00	
Division 08	**Subtotal**			**$13,600.00**	**$2,680.00**	**$0.00**	**$0.00**	**$16,280.00**
13 34 19.50 1100	Pre-Engineered Steel Building, 100' x 150' x 24'	15,000	SF Flr.	$0.00	$0.00	$0.00	$307,500.00	
13 34 19.50 6050	Framing for PESB door opening, 3' x 7'	4	Opng.	$0.00	$0.00	$0.00	$1,980.00	
13 34 19.50 6100	Framing for PESB door opening, 10' x 10'	8	Opng.	$0.00	$0.00	$0.00	$8,400.00	
13 34 19.50 6200	Framing for PESB window opening, 4' x 3'	6	Opng.	$0.00	$0.00	$0.00	$2,970.00	
13 34 19.50 5750	PESB door, 3' x 7', single leaf	4	Opng.	$2,380.00	$512.00	$0.00	$0.00	
13 34 19.50 7750	PESB sliding window, 4' x 3' with screen	6	Opng.	$2,430.00	$294.00	$67.80	$0.00	
13 34 19.50 6550	PESB gutter, eave type, 26 ga., painted	300	L.F.	$2,040.00	$600.00	$0.00	$0.00	
13 34 19.50 8650	PESB roof vent, 12" wide x 10' long	15	Ea.	$540.00	$2,400.00	$0.00	$0.00	
13 34 19.50 6900	PESB insulation, vinyl faced, 4" thick	27,400	S.F.	$12,330.00	$7,124.00	$0.00	$0.00	
Division 13	**Subtotal**			**$19,720.00**	**$10,930.00**	**$67.80**	**$320,850.00**	**$351,567.80**
		Subtotal		**$76,450.00**	**$33,192.80**	**$794.44**	**$320,850.00**	**$431,287.24**
Division 01	(2) **General Requirements @ 7%**			**5,351.50**	**2,323.50**	**55.61**	**22,459.50**	
		Estimate Subtotal		**$81,801.50**	**$35,516.30**	**$850.05**	**$343,309.50**	**$431,287.24**
	(3)	**Sales Tax @ 5%**		**4,090.08**		**42.50**	**8,582.74**	
		Subtotal A		**85,891.58**	**35,516.30**	**892.55**	**351,892.24**	
	(4)	**GC O & P**		**8,589.16**	**24,151.08**	**89.26**	**35,189.22**	
		Subtotal B		**94,480.73**	**59,667.38**	**981.81**	**387,081.46**	**$542,211.38**
	(5)	**Contingency @ 5%**						**27,110.57**
		Subtotal C						**$569,321.95**
	(6)	**Bond @ $12/1000 +10% O&P**						**7,515.05**
		Subtotal D						**$576,837.00**
	(7)	**Location Adjustment Factor**			**102.30**			**13,267.25**
		Grand Total						**$590,104.25**

This estimate is based on an interactive spreadsheet. A copy of this spreadsheet is located on the RSMeans website at **www.RSMeans.com/2016extras.** You are free to download it and adjust it to your methodology.

Sample Estimate

This sample demonstrates the elements of an estimate, including a tally of the RSMeans data lines and a summary of the markups on a contractor's work to arrive at a total cost to the owner. The RSMeans Location Factor is added at the bottom of the estimate to adjust the cost of the work to a specific location.

1 Work Performed

The body of the estimate shows the RSMeans data selected, including the line number, a brief description of each item, its take-off unit and quantity, and the bare costs of materials, labor, and equipment. This estimate also includes a column titled "SubContract." This data is taken from the RSMeans column "Total Incl O&P" and represents the total that a subcontractor would charge a general contractor for the work, including the sub's markup for overhead and profit.

2 Division 1, General Requirements

This is the first division numerically but the last division estimated. Division 1 includes project-wide needs provided by the general contractor. These requirements vary by project but may include temporary facilities and utilities, security, testing, project cleanup, etc. For small projects a percentage can be used, typically between 5% and 15% of project cost. For large projects the costs may be itemized and priced individually.

3 Sales Tax

If the work is subject to state or local sales taxes, the amount must be added to the estimate. Sales tax may be added to material costs, equipment costs, and subcontracted work. In this case, sales tax was added in all three categories. It was assumed that approximately half the subcontracted work would be material cost, so the tax was applied to 50% of the subcontract total.

4 GC O&P

This entry represents the general contractor's markup on material, labor, equipment, and subcontractor costs. RSMeans' standard markup on materials, equipment, and subcontracted work is 10%. In this estimate, the markup on the labor performed by the GC's workers uses "Skilled Workers Average" shown in Column F on the table "Installing Contractor's Overhead & Profit," which can be found on the inside-back cover of the printed product or in the Reference Section of the electronic product.

5 Contingency

A factor for contingency may be added to any estimate to represent the cost of unknowns that may occur between the time that the estimate is performed and the time the project is constructed. The amount of the allowance will depend on the stage of design at which the estimate is done and the contractor's assessment of the risk involved. Refer to section 01 21 16.50 for contingency allowances.

6 Bonds

Bond costs should be added to the estimate. The figures here represent a typical performance bond, ensuring the owner that if the general contractor does not complete the obligations in the construction contract the bonding company will pay the cost for completion of the work.

7 Location Adjustment

RSMeans published data is based on national average costs. If necessary, adjust the total cost of the project using a location factor from the "Location Factor" table or the "City Cost Index" table. Use location factors if the work is general, covering multiple trades. If the work is by a single trade (e.g.,, masonry) use the more specific data found in the "City Cost Indexes."

Division 1 General Requirements

Estimating Tips

01 20 00 Price and Payment Procedures

- Allowances that should be added to estimates to cover contingencies and job conditions that are not included in the national average material and labor costs are shown in section 01 21.
- When estimating historic preservation projects (depending on the condition of the existing structure and the owner's requirements), a 15%–20% contingency or allowance is recommended, regardless of the stage of the drawings.

01 30 00 Administrative Requirements

- Before determining a final cost estimate, it is a good practice to review all the items listed in Subdivisions 01 31 and 01 32 to make final adjustments for items that may need customizing to specific job conditions.
- Requirements for initial and periodic submittals can represent a significant cost to the General Requirements of a job. Thoroughly check the submittal specifications when estimating a project to determine any costs that should be included.

01 40 00 Quality Requirements

- All projects will require some degree of Quality Control. This cost is not included in the unit cost of construction listed in each division. Depending upon the terms of the contract, the various costs of inspection and testing can be the responsibility of either the owner or the contractor. Be sure to include the required costs in your estimate.

01 50 00 Temporary Facilities and Controls

- Barricades, access roads, safety nets, scaffolding, security, and many more requirements for the execution of a safe project are elements of direct cost. These costs can easily be overlooked when preparing an estimate. When looking through the major classifications of this subdivision, determine which items apply to each division in your estimate.
- Construction Equipment Rental Costs can be found in the Reference Section in section 01 54 33. Operators' wages are not included in equipment rental costs.
- Equipment mobilization and demobilization costs are not included in equipment rental costs and must be considered separately.
- The cost of small tools provided by the installing contractor for his workers is covered in the "Overhead" column on the "Installing Contractor's Overhead and Profit" table that lists labor trades, base rates and markups and, therefore, is included in the "Total Incl. O&P" cost of any Unit Price line item.

01 70 00 Execution and Closeout Requirements

- When preparing an estimate, thoroughly read the specifications to determine the requirements for Contract Closeout. Final cleaning, record documentation, operation and maintenance data, warranties and bonds, and spare parts and maintenance materials can all be elements of cost for the completion of a contract. Do not overlook these in your estimate.

Reference Numbers

Reference numbers are shown at the beginning of some major classifications. These numbers refer to related items in the Reference Section. The reference information may be an estimating procedure, an alternate pricing method, or technical information.

Note: Not all subdivisions listed here necessarily appear. ■

01 11 Summary of Work

01 11 31 – Professional Consultants

01 11 31.10 Architectural Fees		Crew	Daily Output	Labor-Hours	Unit	Material	Labor	Equipment	Total	Total Incl O&P
						2016 Bare Costs				
0010	**ARCHITECTURAL FEES** R011110-10									
0020	For new construction									
0060	Minimum				Project				4.90%	4.90%
0090	Maximum								16%	16%
0100	For alteration work, to $500,000, add to new construction fee								50%	50%
0150	Over $500,000, add to new construction fee								25%	25%
2000	For "Greening" of building G								3%	3%

01 11 31.20 Construction Management Fees		Crew	Daily Output	Labor-Hours	Unit	Material	Labor	Equipment	Total	Total Incl O&P
0010	**CONSTRUCTION MANAGEMENT FEES**									
0060	For work to $100,000				Project				10%	10%
0070	To $250,000								9%	9%
0090	To $1,000,000								6%	6%

01 11 31.30 Engineering Fees		Crew	Daily Output	Labor-Hours	Unit	Material	Labor	Equipment	Total	Total Incl O&P
0010	**ENGINEERING FEES** R011110-30									
1200	Structural, minimum				Project				1%	1%
1300	Maximum				"				2.50%	2.50%

01 11 31.75 Renderings		Crew	Daily Output	Labor-Hours	Unit	Material	Labor	Equipment	Total	Total Incl O&P
0010	**RENDERINGS** Color, matted, 20" x 30", eye level,									
0020	1 building, minimum				Ea.	2,050			2,050	2,275
0050	Average					2,875			2,875	3,175
0100	Maximum					4,625			4,625	5,100

01 21 Allowances

01 21 16 – Contingency Allowances

01 21 16.50 Contingencies		Crew	Daily Output	Labor-Hours	Unit	Material	Labor	Equipment	Total	Total Incl O&P
0010	**CONTINGENCIES**, Add to estimate									
0020	Conceptual stage				Project				20%	20%
0150	Final working drawing stage				"				3%	3%

01 21 63 – Taxes

01 21 63.10 Taxes		Crew	Daily Output	Labor-Hours	Unit	Material	Labor	Equipment	Total	Total Incl O&P
0010	**TAXES** R012909-80									
0020	Sales tax, State, average				%	5.08%				
0050	Maximum R012909-85					7.50%				
0200	Social Security, on first $118,500 of wages						7.65%			
0300	Unemployment, combined Federal and State, minimum						.60%			
0350	Average						9.60%			
0400	Maximum						12%			

01 31 Project Management and Coordination

01 31 13 – Project Coordination

01 31 13.30 Insurance

	01 31 13.30 Insurance	Crew	Daily Output	Labor-Hours	Unit	2016 Bare Costs Material	2016 Bare Costs Labor	2016 Bare Costs Equipment	Total	Total Incl O&P
0010	**INSURANCE** R013113-40									
0020	Builders risk, standard, minimum				Job				.24%	.24%
0050	Maximum								.64%	.64%
0200	All-risk type, minimum								.25%	.25%
0250	Maximum				↓				.62%	.62%
0400	Contractor's equipment floater, minimum				Value				.50%	.50%
0450	Maximum				"				1.50%	1.50%
0600	Public liability, average				Job				2.02%	2.02%
0800	Workers' compensation & employer's liability, average R013113-60									
0850	by trade, carpentry, general				Payroll		14.37%			
0900	Clerical						.48%			
0950	Concrete						12.56%			
1000	Electrical						5.64%			
1050	Excavation						9.34%			
1100	Glazing						13.08%			
1150	Insulation						11.56%			
1200	Lathing						8.63%			
1250	Masonry						13.73%			
1300	Painting & decorating						11.66%			
1350	Pile driving						14.63%			
1400	Plastering						10.49%			
1450	Plumbing						6.94%			
1500	Roofing						31.44%			
1550	Sheet metal work (HVAC)						8.84%			
1600	Steel erection, structural						27.45%			
1650	Tile work, interior ceramic						8.82%			
1700	Waterproofing, brush or hand caulking						6.76%			
1800	Wrecking						22.92%			
2000	Range of 35 trades in 50 states, excl. wrecking & clerical, min.						1.42%			
2100	Average						12.61%			
2200	Maximum				↓		107.12%			

01 31 13.40 Main Office Expense

	01 31 13.40 Main Office Expense	Crew	Daily Output	Labor-Hours	Unit	Material	Labor	Equipment	Total	Total Incl O&P
0010	**MAIN OFFICE EXPENSE** Average for General Contractors R013113-50									
0020	As a percentage of their annual volume									
0125	Annual volume under 1 million dollars				% Vol.				17.50%	
0145	Up to 2.5 million dollars								8%	
0150	Up to 4.0 million dollars								6.80%	
0200	Up to 7.0 million dollars				↓				5.60%	

01 31 13.60 General Contractor's Main Office Overhead

	01 31 13.60 General Contractor's Main Office Overhead	Crew	Daily Output	Labor-Hours	Unit	Material	Labor	Equipment	Total	Total Incl O&P
0010	**GENERAL CONTRACTOR'S MAIN OFFICE OVERHEAD** R013113-50									
0020	As percent of direct costs, minimum				%				5%	
0050	Average								25%	
0100	Maximum				↓				30%	

01 31 13.80 Overhead and Profit

	01 31 13.80 Overhead and Profit	Crew	Daily Output	Labor-Hours	Unit	Material	Labor	Equipment	Total	Total Incl O&P
0010	**OVERHEAD & PROFIT** Allowance to add to items in this									
0020	data set that do not include Subs O&P, average				%				25%	
0100	Allowance to add to items in this data set that									
0110	do include Subs O&P, minimum				%				5%	5%
0150	Average								10%	10%
0200	Maximum								15%	15%
0290	Typical, by size of project, under $50,000								40%	
0310	$50,000 to $100,000				↓				35%	

01 31 Project Management and Coordination

01 31 13 – Project Coordination

01 31 13.80 Overhead and Profit		Crew	Daily Output	Labor-Hours	Unit	2016 Bare Costs Material	Labor	Equipment	Total	Total Incl O&P
0320	$100,000 to $500,000				%				25%	
0330	$500,000 to $1,000,000								20%	

01 41 Regulatory Requirements

01 41 26 – Permit Requirements

01 41 26.50 Permits		Crew	Daily Output	Labor-Hours	Unit	Material	Labor	Equipment	Total	Total Incl O&P
0010	**PERMITS**									
0020	Rule of thumb, most cities, minimum				Job				.50%	.50%
0100	Maximum				"				2%	2%

01 45 Quality Control

01 45 23 – Testing and Inspecting Services

01 45 23.50 Testing		Crew	Daily Output	Labor-Hours	Unit	Material	Labor	Equipment	Total	Total Incl O&P
0010	**TESTING** and Inspecting Services									
0015	For concrete building costing $1,000,000, minimum				Project				4,725	5,200
0020	Maximum								38,000	41,800
0050	Steel building, minimum								4,725	5,200
0070	Maximum								14,800	16,300
0600	Concrete testing, aggregates, abrasion, ASTM C 131				Ea.				136	150
0650	Absorption, ASTM C 127								42	46
1800	Compressive test, cylinder, delivered to lab, ASTM C 39								12	13
1900	Picked up by lab, minimum								14	15
1950	Average								18	20
2000	Maximum								27	30
2200	Compressive strength, cores (not incl. drilling), ASTM C 42								36	40
2250	Core drilling, 4" diameter (plus technician)				Inch				23	25
2260	Technician for core drilling				Hr.				45	50
2300	Patching core holes				Ea.				22	24
9000	Thermographic testing, for bldg envelope heat loss, average 2,000 S.F. [G]				"				500	500

01 52 Construction Facilities

01 52 13 – Field Offices and Sheds

01 52 13.20 Office and Storage Space		Crew	Daily Output	Labor-Hours	Unit	Material	Labor	Equipment	Total	Total Incl O&P
0010	**OFFICE AND STORAGE SPACE**									
0020	Office trailer, furnished, no hookups, 20' x 8', buy	2 Skwk	1	16	Ea.	10,700	615		11,315	12,800
0250	Rent per month					193			193	213
0300	32' x 8', buy	2 Skwk	.70	22.857		16,000	875		16,875	19,100
0350	Rent per month					246			246	270
0700	For air conditioning, rent per month, add					48.50			48.50	53.50
1200	Storage boxes, 20' x 8', buy	2 Skwk	1.80	8.889		2,975	340		3,315	3,850
1250	Rent per month					78			78	85.50

01 52 13.40 Field Office Expense		Crew	Daily Output	Labor-Hours	Unit	Material	Labor	Equipment	Total	Total Incl O&P
0010	**FIELD OFFICE EXPENSE**									
0100	Office equipment rental average				Month	200			200	220
0120	Office supplies, average					80			80	88
0140	Telephone bill; avg. bill/month incl. long dist.					85			85	93.50
0160	Lights & HVAC					160			160	176

01 54 Construction Aids

01 54 16 – Temporary Hoists

01 54 16.50 Weekly Forklift Crew		Crew	Daily Output	Labor-Hours	Unit	Material	2016 Bare Costs Labor	Equipment	Total	Total Incl O&P
0010	**WEEKLY FORKLIFT CREW**									
0100	All-terrain forklift, 45′ lift, 35′ reach, 9000 lb. capacity	A-3P	.20	40	Week		1,525	2,675	4,200	5,450

01 54 19 – Temporary Cranes

01 54 19.50 Daily Crane Crews		Crew	Daily Output	Labor-Hours	Unit	Material	Labor	Equipment	Total	Total Incl O&P
0010	**DAILY CRANE CREWS** for small jobs, portal to portal									
0100	12-ton truck-mounted hydraulic crane	A-3H	1	8	Day		320	875	1,195	1,500
0200	25-ton	A-3I	1	8			320	1,025	1,345	1,650
0300	40-ton	A-3J	1	8			320	1,250	1,570	1,900
0400	55-ton	A-3K	1	16			605	1,650	2,255	2,800
0500	80-ton	A-3L	1	16			605	2,375	2,980	3,600
0600	100-ton	A-3M	1	16			605	2,375	2,980	3,625
0900	If crane is needed on a Saturday, Sunday or Holiday									
0910	At time-and-a-half, add				Day		50%			
0920	At double time, add				"		100%			

01 54 19.60 Monthly Tower Crane Crew		Crew	Daily Output	Labor-Hours	Unit	Material	Labor	Equipment	Total	Total Incl O&P
0010	**MONTHLY TOWER CRANE CREW**, excludes concrete footing									
0100	Static tower crane, 130′ high, 106′ jib, 6200 lb. capacity	A-3N	.05	176	Month		7,075	25,200	32,275	39,400

01 54 23 – Temporary Scaffolding and Platforms

01 54 23.60 Pump Staging		Crew	Daily Output	Labor-Hours	Unit	Material	Labor	Equipment	Total	Total Incl O&P
0010	**PUMP STAGING**, Aluminum R015423-20									
0200	24′ long pole section, buy				Ea.	420			420	460
0300	18′ long pole section, buy					325			325	355
0400	12′ long pole section, buy					219			219	241
0500	6′ long pole section, buy					115			115	127
0600	6′ long splice joint section, buy					85.50			85.50	94
0700	Pump jack, buy					174			174	192
0900	Foldable brace, buy					68.50			68.50	75.50
1000	Workbench/back safety rail support, buy					92			92	101
1100	Scaffolding planks/workbench, 14″ wide x 24′ long, buy					765			765	840
1200	Plank end safety rail, buy					350			350	385
1250	Safety net, 22′ long, buy					420			420	460
1300	System in place, 50′ working height, per use based on 50 uses	2 Carp	84.80	.189	C.S.F.	6.95	7.15		14.10	19.60
1400	100 uses		84.80	.189		3.49	7.15		10.64	15.80
1500	150 uses		84.80	.189		2.34	7.15		9.49	14.50

01 54 23.70 Scaffolding		Crew	Daily Output	Labor-Hours	Unit	Material	Labor	Equipment	Total	Total Incl O&P
0010	**SCAFFOLDING** R015423-10									
0015	Steel tube, regular, no plank, labor only to erect & dismantle									
0091	Building exterior, wall face, 1 to 5 stories, 6′-4″ x 5′ frames	3 Clab	8	3	C.S.F.		87.50		87.50	147
0201	6 to 12 stories	4 Clab	8	4			117		117	196
0310	13 to 20 stories	5 Carp	8	5			189		189	315
0461	Building interior, walls face area, up to 16′ high	3 Clab	12	2			58.50		58.50	98
0561	16′ to 40′ high		10	2.400			70		70	117
0801	Building interior floor area, up to 30′ high		150	.160	C.C.F.		4.67		4.67	7.80
0901	Over 30′ high	4 Clab	160	.200	"		5.85		5.85	9.80
0906	Complete system for face of walls, no plank, material only rent/mo				C.S.F.	35.50			35.50	39
0908	Interior spaces, no plank, material only rent/mo				C.C.F.	4.30			4.30	4.73
0910	Steel tubular, heavy duty shoring, buy									
0920	Frames 5′ high 2′ wide				Ea.	81.50			81.50	90
0925	5′ high 4′ wide					93			93	102
0930	6′ high 2′ wide					93.50			93.50	103
0935	6′ high 4′ wide					109			109	120

01 54 23 – Temporary Scaffolding and Platforms

01 54 23.70 Scaffolding		Crew	Daily Output	Labor-Hours	Unit	2016 Bare Costs Material	Labor	Equipment	Total	Total Incl O&P
0940	Accessories									
0945	Cross braces				Ea.	15.50			15.50	17.05
0950	U-head, 8" x 8"					19.10			19.10	21
0955	J-head, 4" x 8"					13.90			13.90	15.30
0960	Base plate, 8" x 8"					15.50			15.50	17.05
0965	Leveling jack					33.50			33.50	36.50
1000	Steel tubular, regular, buy									
1100	Frames 3' high 5' wide				Ea.	89			89	97.50
1150	5' high 5' wide					104			104	114
1200	6'-4" high 5' wide					147			147	162
1350	7'-6" high 6' wide					158			158	173
1500	Accessories cross braces					18.10			18.10	19.90
1550	Guardrail post					16.40			16.40	18.05
1600	Guardrail 7' section					8.70			8.70	9.60
1650	Screw jacks & plates					21.50			21.50	24
1700	Sidearm brackets					30.50			30.50	33.50
1750	8" casters					31			31	34
1800	Plank 2" x 10" x 16'-0"					59.50			59.50	65.50
1900	Stairway section					286			286	315
1910	Stairway starter bar					32			32	35
1920	Stairway inside handrail					58.50			58.50	64
1930	Stairway outside handrail					86.50			86.50	95
1940	Walk-thru frame guardrail					41.50			41.50	46
2000	Steel tubular, regular, rent/mo.									
2100	Frames 3' high 5' wide				Ea.	5			5	5.50
2150	5' high 5' wide					5			5	5.50
2200	6'-4" high 5' wide					5.35			5.35	5.90
2250	7'-6" high 6' wide					10			10	11
2500	Accessories, cross braces					1			1	1.10
2550	Guardrail post					1			1	1.10
2600	Guardrail 7' section					1			1	1.10
2650	Screw jacks & plates					2			2	2.20
2700	Sidearm brackets					2			2	2.20
2750	8" casters					8			8	8.80
2800	Outrigger for rolling tower					3			3	3.30
2850	Plank 2" x 10" x 16'-0"					10			10	11
2900	Stairway section					35			35	38.50
2940	Walk-thru frame guardrail					2.50			2.50	2.75
3000	Steel tubular, heavy duty shoring, rent/mo.									
3250	5' high 2' & 4' wide				Ea.	8.50			8.50	9.35
3300	6' high 2' & 4' wide					8.50			8.50	9.35
3500	Accessories, cross braces					1			1	1.10
3600	U - head, 8" x 8"					2.50			2.50	2.75
3650	J - head, 4" x 8"					2.50			2.50	2.75
3700	Base plate, 8" x 8"					1			1	1.10
3750	Leveling jack					2.50			2.50	2.75
5700	Planks, 2" x 10" x 16'-0", labor only to erect & remove to 50' H	3 Carp	72	.333			12.60		12.60	21
5800	Over 50' high	4 Carp	80	.400			15.10		15.10	25.50

01 54 Construction Aids

01 54 23 – Temporary Scaffolding and Platforms

01 54 23.75 Scaffolding Specialties		Crew	Daily Output	Labor-Hours	Unit	2016 Bare Costs Material	Labor	Equipment	Total	Total Incl O&P
0010	**SCAFFOLDING SPECIALTIES**									
5000	Motorized work platform, mast climber									
5050	Base unit, 50′ W, less than 100′ tall, rent/mo				Ea.	2,875			2,875	3,175
5100	Less than 200′ tall, rent/mo					3,500			3,500	3,850
5150	Less than 300′ tall, rent/mo					4,375			4,375	4,800
5200	Less than 400′ tall, rent/mo					5,125			5,125	5,625
5250	Set up and demob, per unit, less than 100′ tall	B-68F	16.60	1.446	C.S.F.		56.50	20	76.50	117
5300	Less than 200′ tall		25	.960			37.50	13.35	50.85	77.50
5350	Less than 300′ tall		30	.800			31	11.15	42.15	65
5400	Less than 400′ tall		33	.727			28.50	10.10	38.60	58.50
5500	Mobilization (price includes freight in and out) per unit				Ea.				1,100	1,225

01 54 23.80 Staging Aids		Crew	Daily Output	Labor-Hours	Unit	Material	Labor	Equipment	Total	Total Incl O&P
0010	**STAGING AIDS** and fall protection equipment									
0110	Cost each per day, based on 250 days use				Day	.20			.20	.22
0200	Guard post, buy				Ea.	52			52	57
0210	Cost each per day, based on 250 days use				Day	.21			.21	.23
0300	End guard chains, buy per pair				Pair	38			38	41.50
0310	Cost per set per day, based on 250 days use				Day	.20			.20	.22
1010	Cost each per day, based on 250 days use				"	.04			.04	.04
1100	Wood bracket, buy				Ea.	21			21	23.50
1110	Cost each per day, based on 250 days use				Day	.08			.08	.09
2010	Cost per pair per day, based on 250 days use				"	.49			.49	.53
2100	Steel siderail jack, buy per pair				Pair	97.50			97.50	107
2110	Cost per pair per day, based on 250 days use				Day	.39			.39	.43
3010	Cost each per day, based on 250 days use				"	.24			.24	.26
3100	Aluminum scaffolding plank, 20″ wide x 24′ long, buy				Ea.	790			790	870
3110	Cost each per day, based on 250 days use				Day	3.16			3.16	3.48
4010	Cost each per day, based on 250 days use				"	.70			.70	.77
4100	Rope for safety line, 5/8″ x 100′ nylon, buy				Ea.	56			56	61.50
4110	Cost each per day, based on 250 days use				Day	.22			.22	.25
4200	Permanent U-Bolt roof anchor, buy				Ea.	35.50			35.50	39.50
4300	Temporary (one use) roof ridge anchor, buy				"	32			32	35.50
5000	Installation (setup and removal) of staging aids									
5010	Sidewall staging bracket	2 Carp	64	.250	Ea.		9.45		9.45	15.85
5020	Guard post with 2 wood rails	"	64	.250			9.45		9.45	15.85
5030	End guard chains, set	1 Carp	64	.125			4.73		4.73	7.90
5100	Roof shingling bracket		96	.083			3.15		3.15	5.25
5200	Ladder jack		64	.125			4.73		4.73	7.90
5300	Wood plank, 2″ x 10″ x 16′	2 Carp	80	.200			7.55		7.55	12.65
5310	Aluminum scaffold plank, 20″ x 24′	"	40	.400			15.10		15.10	25.50
5410	Safety rope	1 Carp	40	.200			7.55		7.55	12.65
5420	Permanent U-Bolt roof anchor (install only)	2 Carp	40	.400			15.10		15.10	25.50
5430	Temporary roof ridge anchor (install only)	1 Carp	64	.125			4.73		4.73	7.90

01 54 36 – Equipment Mobilization

01 54 36.50 Mobilization		Crew	Daily Output	Labor-Hours	Unit	Material	Labor	Equipment	Total	Total Incl O&P
0010	**MOBILIZATION** (Use line item again for demobilization) R015436-50									
0015	Up to 25 mi. haul dist. (50 mi. RT for mob/demob crew)									
1200	Small equipment, placed in rear of, or towed by pickup truck	A-3A	4	2	Ea.		75.50	39	114.50	168
1300	Equipment hauled on 3-ton capacity towed trailer	A-3Q	2.67	3			114	67.50	181.50	263
1400	20-ton capacity	B-34U	2	8			288	236	524	735
1500	40-ton capacity	B-34N	2	8			294	375	669	900

01 54 Construction Aids

01 54 36 – Equipment Mobilization

01 54 36.50 Mobilization		Crew	Daily Output	Labor-Hours	Unit	Material	2016 Bare Costs Labor	Equipment	Total	Total Incl O&P
1600	50-ton capacity	B-34V	1	24	Ea.		900	1,050	1,950	2,625
1700	Crane, truck-mounted, up to 75 ton (driver only)	1 Eqhv	4	2	↓		80.50		80.50	133
2500	For each additional 5 miles haul distance, add						10%	10%		
3000	For large pieces of equipment, allow for assembly/knockdown									
3100	For mob/demob of micro-tunneling equip, see Section 33 05 23.19									

01 56 Temporary Barriers and Enclosures

01 56 13 – Temporary Air Barriers

01 56 13.60 Tarpaulins

		Crew	Daily Output	Labor-Hours	Unit	Material	Labor	Equipment	Total	Total Incl O&P
0010	**TARPAULINS**									
0020	Cotton duck, 10 oz. to 13.13 oz. per S.Y., 6′ x 8′				S.F.	.88			.88	.97
0050	30′ x 30′					.64			.64	.70
0200	Reinforced polyethylene 3 mils thick, white					.04			.04	.04
0300	4 mils thick, white, clear or black					.11			.11	.12
0730	Polyester reinforced w/integral fastening system, 11 mils thick				↓	.22			.22	.24

01 56 16 – Temporary Dust Barriers

01 56 16.10 Dust Barriers, Temporary

		Crew	Daily Output	Labor-Hours	Unit	Material	Labor	Equipment	Total	Total Incl O&P
0010	**DUST BARRIERS, TEMPORARY**									
0020	Spring loaded telescoping pole & head, to 12′, erect and dismantle	1 Clab	240	.033	Ea.		.97		.97	1.63
0025	Cost per day (based upon 250 days)				Day	.26			.26	.29
0030	To 21′, erect and dismantle	1 Clab	240	.033	Ea.		.97		.97	1.63
0035	Cost per day (based upon 250 days)				Day	.54			.54	.60
0040	Accessories, caution tape reel, erect and dismantle	1 Clab	480	.017	Ea.		.49		.49	.82
0045	Cost per day (based upon 250 days)				Day	.36			.36	.40
0060	Foam rail and connector, erect and dismantle	1 Clab	240	.033	Ea.		.97		.97	1.63
0065	Cost per day (based upon 250 days)				Day	.12			.12	.13
0070	Caution tape	1 Clab	384	.021	C.L.F.	3.13	.61		3.74	4.46
0080	Zipper, standard duty		60	.133	Ea.	8	3.89		11.89	15.30
0090	Heavy duty		48	.167	″	9.75	4.87		14.62	18.90
0100	Polyethylene sheet, 4 mil		37	.216	Sq.	2.55	6.30		8.85	13.35
0110	6 mil	↓	37	.216	″	3.70	6.30		10	14.60
1000	Dust partition, 6 mil polyethylene, 1″ x 3″ frame	2 Carp	2000	.008	S.F.	.30	.30		.60	.83
1080	2″ x 4″ frame	″	2000	.008	″	.33	.30		.63	.87

01 71 Examination and Preparation

01 71 23 – Field Engineering

01 71 23.13 Construction Layout

		Crew	Daily Output	Labor-Hours	Unit	Material	Labor	Equipment	Total	Total Incl O&P
0010	**CONSTRUCTION LAYOUT**									
1100	Crew for layout of building, trenching or pipe laying, 2 person crew	A-6	1	16	Day		600	43.50	643.50	1,050
1200	3 person crew	A-7	1	24	″		965	43.50	1,008.50	1,650

01 74 Cleaning and Waste Management

01 74 13 – Progress Cleaning

01 74 13.20 Cleaning Up		Crew	Daily Output	Labor-Hours	Unit	Material	2016 Bare Costs Labor	Equipment	Total	Total Incl O&P
0010	**CLEANING UP**									
0020	After job completion, allow, minimum				Job				.30%	.30%
0040	Maximum				"				1%	1%

01 76 Protecting Installed Construction

01 76 13 – Temporary Protection of Installed Construction

01 76 13.20 Temporary Protection		Crew	Daily Output	Labor-Hours	Unit	Material	Labor	Equipment	Total	Total Incl O&P
0010	**TEMPORARY PROTECTION**									
0020	Flooring, 1/8" tempered hardboard, taped seams	2 Carp	1500	.011	S.F.	.42	.40		.82	1.14
0030	Peel away carpet protection	1 Clab	3200	.003	"	.13	.07		.20	.26

Division Notes

		CREW	DAILY OUTPUT	LABOR-HOURS	UNIT	BARE COSTS MAT.	BARE COSTS LABOR	BARE COSTS EQUIP.	BARE COSTS TOTAL	TOTAL INCL O&P

Estimating Tips

02 30 00 Subsurface Investigation

In preparing estimates on structures involving earthwork or foundations, all information concerning soil characteristics should be obtained. Look particularly for hazardous waste, evidence of prior dumping of debris, and previous stream beds.

02 40 00 Demolition and Structure Moving

The costs shown for selective demolition do not include rubbish handling or disposal. These items should be estimated separately using RSMeans data or other sources.

- Historic preservation often requires that the contractor remove materials from the existing structure, rehab them, and replace them. The estimator must be aware of any related measures and precautions that must be taken when doing selective demolition and cutting and patching. Requirements may include special handling and storage, as well as security.
- In addition to Subdivision 02 41 00, you can find selective demolition items in each division. Example: Roofing demolition is in Division 7.
- Absent of any other specific reference, an approximate demolish-in-place cost can be obtained by halving the new-install labor cost. To remove for reuse, allow the entire new-install labor figure.

02 40 00 Building Deconstruction

This section provides costs for the careful dismantling and recycling of most of low-rise building materials.

02 50 00 Containment of Hazardous Waste

This section addresses on-site hazardous waste disposal costs.

02 80 00 Hazardous Material Disposal/ Remediation

This subdivision includes information on hazardous waste handling, asbestos remediation, lead remediation, and mold remediation. See reference R028213-20 and R028319-60 for further guidance in using these unit price lines.

02 90 00 Monitoring Chemical Sampling, Testing Analysis

This section provides costs for on-site sampling and testing hazardous waste.

Reference Numbers

Reference numbers are shown at the beginning of some major classifications. These numbers refer to related items in the Reference Section. The reference information may be an estimating procedure, an alternate pricing method, or technical information.

Note: Not all subdivisions listed here necessarily appear. ■

02 21 Surveys

02 21 13 – Site Surveys

02 21 13.09 Topographical Surveys		Crew	Daily Output	Labor-Hours	Unit	Material	2016 Bare Costs Labor	Equipment	Total	Total Incl O&P
0010	**TOPOGRAPHICAL SURVEYS**									
0020	Topographical surveying, conventional, minimum	A-7	3.30	7.273	Acre	20	293	13.25	326.25	525
0100	Maximum	A-8	.60	53.333	″	60	2,100	73	2,233	3,650

02 21 13.13 Boundary and Survey Markers

Line	Description	Crew	Daily Output	Labor-Hours	Unit	Material	Labor	Equipment	Total	Total Incl O&P
0010	**BOUNDARY AND SURVEY MARKERS**									
0300	Lot location and lines, large quantities, minimum	A-7	2	12	Acre	35	485	22	542	870
0320	Average	″	1.25	19.200		55	775	35	865	1,400
0400	Small quantities, maximum	A-8	1	32		75	1,250	44	1,369	2,225
0600	Monuments, 3′ long	A-7	10	2.400	Ea.	40	96.50	4.37	140.87	210
0800	Property lines, perimeter, cleared land	″	1000	.024	L.F.	.05	.97	.04	1.06	1.72
0900	Wooded land	A-8	875	.037	″	.07	1.44	.05	1.56	2.53

02 32 Geotechnical Investigations

02 32 13 – Subsurface Drilling and Sampling

02 32 13.10 Boring and Exploratory Drilling

Line	Description	Crew	Daily Output	Labor-Hours	Unit	Material	Labor	Equipment	Total	Total Incl O&P
0010	**BORING AND EXPLORATORY DRILLING**									
0020	Borings, initial field stake out & determination of elevations	A-6	1	16	Day		600	43.50	643.50	1,050
0100	Drawings showing boring details				Total		335		335	425
0200	Report and recommendations from P.E.						775		775	970
0300	Mobilization and demobilization	B-55	4	4			125	273	398	505
0350	For over 100 miles, per added mile		450	.036	Mile		1.11	2.43	3.54	4.51
0600	Auger holes in earth, no samples, 2-1/2″ diameter		78.60	.204	L.F.		6.35	13.90	20.25	26
0800	Cased borings in earth, with samples, 2-1/2″ diameter		55.50	.288	″	15	9	19.70	43.70	53
1400	Borings, earth, drill rig and crew with truck mounted auger		1	16	Day		500	1,100	1,600	2,025
1500	For inner city borings add, minimum								10%	10%
1510	Maximum								20%	20%

02 41 Demolition

02 41 13 – Selective Site Demolition

02 41 13.17 Demolish, Remove Pavement and Curb

Line	Description	Crew	Daily Output	Labor-Hours	Unit	Material	Labor	Equipment	Total	Total Incl O&P
0010	**DEMOLISH, REMOVE PAVEMENT AND CURB** R024119-10									
5010	Pavement removal, bituminous roads, up to 3″ thick	B-38	690	.035	S.Y.		1.12	.79	1.91	2.73
5050	4″ to 6″ thick		420	.057			1.83	1.30	3.13	4.48
5100	Bituminous driveways		640	.038			1.20	.85	2.05	2.94
5200	Concrete to 6″ thick, hydraulic hammer, mesh reinforced		255	.094			3.02	2.14	5.16	7.40
5300	Rod reinforced		200	.120			3.85	2.73	6.58	9.40
5600	With hand held air equipment, bituminous, to 6″ thick	B-39	1900	.025	S.F.		.75	.12	.87	1.38
5700	Concrete to 6″ thick, no reinforcing		1600	.030			.89	.14	1.03	1.64
5800	Mesh reinforced		1400	.034			1.01	.16	1.17	1.88
5900	Rod reinforced		765	.063			1.85	.30	2.15	3.43
6000	Curbs, concrete, plain	B-6	360	.067	L.F.		2.14	1.02	3.16	4.68
6100	Reinforced		275	.087			2.80	1.33	4.13	6.10
6200	Granite		360	.067			2.14	1.02	3.16	4.68
6300	Bituminous		528	.045			1.46	.69	2.15	3.19

02 41 13.23 Utility Line Removal

Line	Description	Crew	Daily Output	Labor-Hours	Unit	Material	Labor	Equipment	Total	Total Incl O&P
0010	**UTILITY LINE REMOVAL**									
0015	No hauling, abandon catch basin or manhole	B-6	7	3.429	Ea.		110	52	162	241
0020	Remove existing catch basin or manhole, masonry		4	6			192	91.50	283.50	420
0030	Catch basin or manhole frames and covers, stored		13	1.846			59	28	87	130

02 41 Demolition

02 41 13 – Selective Site Demolition

02 41 13.23 Utility Line Removal		Crew	Daily Output	Labor-Hours	Unit	Material	2016 Bare Costs Labor	Equipment	Total	Total Incl O&P
0040	Remove and reset	B-6	7	3.429	Ea.		110	52	162	241
2900	Pipe removal, sewer/water, no excavation, 12″ diameter		175	.137	L.F.		4.40	2.09	6.49	9.65
2930	15″-18″ diameter	B-12Z	150	.160			5.25	10.55	15.80	20.50
2960	21″-24″ diameter	″	120	.200			6.60	13.20	19.80	25.50

02 41 13.30 Minor Site Demolition		Crew	Daily Output	Labor-Hours	Unit	Material	Labor	Equipment	Total	Total Incl O&P
0010	**MINOR SITE DEMOLITION** R024119-10									
1000	Masonry walls, block, solid	B-5	1800	.022	C.F.		.70	.79	1.49	2.04
1200	Brick, solid		900	.044			1.41	1.58	2.99	4.09
1400	Stone, with mortar		900	.044			1.41	1.58	2.99	4.09
1500	Dry set		1500	.027			.84	.95	1.79	2.46
4000	Sidewalk removal, bituminous, 2″ thick	B-6	350	.069	S.Y.		2.20	1.04	3.24	4.82
4010	2-1/2″ thick		325	.074			2.37	1.12	3.49	5.20
4050	Brick, set in mortar		185	.130			4.16	1.98	6.14	9.10
4100	Concrete, plain, 4″		160	.150			4.81	2.28	7.09	10.50
4110	Plain, 5″		140	.171			5.50	2.61	8.11	12
4120	Plain, 6″		120	.200			6.40	3.05	9.45	14.05
4200	Mesh reinforced, concrete, 4″		150	.160			5.15	2.44	7.59	11.25
4210	5″ thick		131	.183			5.90	2.79	8.69	12.85
4220	6″ thick		112	.214			6.85	3.26	10.11	15.05
4300	Slab on grade removal, plain	B-5	45	.889	C.Y.		28	31.50	59.50	82
4310	Mesh reinforced		33	1.212			38.50	43	81.50	112
4320	Rod reinforced		25	1.600			50.50	57	107.50	147
4400	For congested sites or small quantities, add up to								200%	200%
4450	For disposal on site, add	B-11A	232	.069			2.36	6	8.36	10.55
4500	To 5 miles, add	B-34D	76	.105			3.59	9.75	13.34	16.65

02 41 13.60 Selective Demolition Fencing		Crew	Daily Output	Labor-Hours	Unit	Material	Labor	Equipment	Total	Total Incl O&P
0010	**SELECTIVE DEMOLITION FENCING** R024119-10									
1600	Fencing, barbed wire, 3 strand	2 Clab	430	.037	L.F.		1.09		1.09	1.82
1650	5 strand	″	280	.057			1.67		1.67	2.79
1700	Chain link, posts & fabric, 8′ to 10′ high, remove only	B-6	445	.054			1.73	.82	2.55	3.78

02 41 16 – Structure Demolition

02 41 16.13 Building Demolition		Crew	Daily Output	Labor-Hours	Unit	Material	Labor	Equipment	Total	Total Incl O&P
0010	**BUILDING DEMOLITION** Large urban projects, incl. 20 mi. haul R024119-10									
0500	Small bldgs, or single bldgs, no salvage included, steel	B-3	14800	.003	C.F.		.11	.17	.28	.37
0600	Concrete	″	11300	.004	″		.14	.23	.37	.48
0605	Concrete, plain	B-5	33	1.212	C.Y.		38.50	43	81.50	112
0610	Reinforced		25	1.600			50.50	57	107.50	147
0615	Concrete walls		34	1.176			37	42	79	108
0620	Elevated slabs		26	1.538			48.50	55	103.50	142
0650	Masonry	B-3	14800	.003	C.F.		.11	.17	.28	.37
0700	Wood	″	14800	.003			.11	.17	.28	.37
0750	For buildings with no interior walls, deduct								30%	30%
1000	Demoliton single family house, one story, wood 1600 S.F.	B-3	1	48	Ea.		1,575	2,575	4,150	5,450
1020	3200 S.F.		.50	96			3,150	5,150	8,300	10,900
1200	Demoliton two family house, two story, wood 2400 S.F.		.67	71.964			2,375	3,875	6,250	8,175
1220	4200 S.F.		.38	128			4,200	6,875	11,075	14,600
1300	Demoliton three family house, three story, wood 3200 S.F.		.50	96			3,150	5,150	8,300	10,900
1320	5400 S.F.		.30	160			5,250	8,600	13,850	18,200
5000	For buildings with no interior walls, deduct								30%	30%

02 41 Demolition

02 41 16 – Structure Demolition

02 41 16.17 Building Demolition Footings and Foundations		Crew	Daily Output	Labor-Hours	Unit	Material	Labor	Equipment	Total	Total Incl O&P
						2016 Bare Costs				
0010	BUILDING DEMOLITION FOOTINGS AND FOUNDATIONS R024119-10									
0200	Floors, concrete slab on grade,									
0240	4" thick, plain concrete	B-13L	5000	.003	S.F.		.13	.41	.54	.66
0280	Reinforced, wire mesh		4000	.004			.16	.51	.67	.83
0300	Rods		4500	.004			.14	.46	.60	.74
0400	6" thick, plain concrete		4000	.004			.16	.51	.67	.83
0420	Reinforced, wire mesh		3200	.005			.20	.64	.84	1.04
0440	Rods		3600	.004			.18	.57	.75	.93
1000	Footings, concrete, 1' thick, 2' wide		300	.053	L.F.		2.15	6.85	9	11.05
1080	1'-6" thick, 2' wide		250	.064			2.58	8.20	10.78	13.25
1120	3' wide		200	.080			3.22	10.25	13.47	16.60
1200	Average reinforcing, add								10%	10%
2000	Walls, block, 4" thick	B-13L	8000	.002	S.F.		.08	.26	.34	.41
2040	6" thick		6000	.003			.11	.34	.45	.56
2080	8" thick		4000	.004			.16	.51	.67	.83
2100	12" thick		3000	.005			.21	.68	.89	1.10
2400	Concrete, plain concrete, 6" thick		4000	.004			.16	.51	.67	.83
2420	8" thick		3500	.005			.18	.59	.77	.94
2440	10" thick		3000	.005			.21	.68	.89	1.10
2500	12" thick		2500	.006			.26	.82	1.08	1.33
2600	For average reinforcing, add								10%	10%
4000	For congested sites or small quantities, add up to								200%	200%
4200	Add for disposal, on site	B-11A	232	.069	C.Y.		2.36	6	8.36	10.55
4250	To five miles	B-30	220	.109	"		3.91	10.95	14.86	18.50

02 41 19 – Selective Demolition

02 41 19.13 Selective Building Demolition		Crew	Daily Output	Labor-Hours	Unit	Material	Labor	Equipment	Total	Total Incl O&P
0010	SELECTIVE BUILDING DEMOLITION									
0020	Costs related to selective demolition of specific building components									
0025	are included under Common Work Results (XX 05)									
0030	in the component's appropriate division.									

02 41 19.16 Selective Demolition, Cutout		Crew	Daily Output	Labor-Hours	Unit	Material	Labor	Equipment	Total	Total Incl O&P
0010	SELECTIVE DEMOLITION, CUTOUT R024119-10									
0020	Concrete, elev. slab, light reinforcement, under 6 C.F.	B-9	65	.615	C.F.		18.20	3.52	21.72	34.50
0050	Light reinforcing, over 6 C.F.		75	.533	"		15.80	3.05	18.85	30
0200	Slab on grade to 6" thick, not reinforced, under 8 S.F.		85	.471	S.F.		13.95	2.69	16.64	26.50
0250	8 – 16 S.F.		175	.229	"		6.75	1.31	8.06	12.80
0255	For over 16 S.F. see Line 02 41 16.17 0400									
0600	Walls, not reinforced, under 6 C.F.	B-9	60	.667	C.F.		19.75	3.81	23.56	37
0650	6 – 12 C.F.	"	80	.500	"		14.80	2.86	17.66	28
0655	For over 12 C.F. see Line 02 41 16.17 2500									
1000	Concrete, elevated slab, bar reinforced, under 6 C.F.	B-9	45	.889	C.F.		26.50	5.10	31.60	49.50
1050	Bar reinforced, over 6 C.F.		50	.800	"		23.50	4.58	28.08	44.50
1200	Slab on grade to 6" thick, bar reinforced, under 8 S.F.		75	.533	S.F.		15.80	3.05	18.85	30
1250	8 – 16 S.F.		150	.267	"		7.90	1.53	9.43	14.90
1255	For over 16 S.F. see Line 02 41 16.17 0440									
1400	Walls, bar reinforced, under 6 C.F.	B-9	50	.800	C.F.		23.50	4.58	28.08	44.50
1450	6 – 12 C.F.	"	70	.571	"		16.90	3.27	20.17	32
1455	For over 12 C.F. see Lines 02 41 16.17 2500 and 2600									
2000	Brick, to 4 S.F. opening, not including toothing									
2040	4" thick	B-9	30	1.333	Ea.		39.50	7.65	47.15	74.50
2060	8" thick		18	2.222			66	12.70	78.70	124

02 41 19.16 Selective Demolition, Cutout		Crew	Daily Output	Labor-Hours	Unit	Material	2016 Bare Costs Labor	Equipment	Total	Total Incl O&P
2080	12" thick	B-9	10	4	Ea.		118	23	141	223
2400	Concrete block, to 4 S.F. opening, 2" thick		35	1.143			34	6.55	40.55	63.50
2420	4" thick		30	1.333			39.50	7.65	47.15	74.50
2440	8" thick		27	1.481			44	8.45	52.45	83
2460	12" thick		24	1.667			49.50	9.55	59.05	93
2600	Gypsum block, to 4 S.F. opening, 2" thick		80	.500			14.80	2.86	17.66	28
2620	4" thick		70	.571			16.90	3.27	20.17	32
2640	8" thick		55	.727			21.50	4.16	25.66	40.50
2800	Terra cotta, to 4 S.F. opening, 4" thick		70	.571			16.90	3.27	20.17	32
2840	8" thick		65	.615			18.20	3.52	21.72	34.50
2880	12" thick		50	.800			23.50	4.58	28.08	44.50
3000	Toothing masonry cutouts, brick, soft old mortar	1 Brhe	40	.200	V.L.F.		5.95		5.95	9.90
3100	Hard mortar		30	.267			7.90		7.90	13.20
3200	Block, soft old mortar		70	.114			3.39		3.39	5.65
3400	Hard mortar		50	.160			4.74		4.74	7.90
6000	Walls, interior, not including re-framing,									
6010	openings to 5 S.F.									
6100	Drywall to 5/8" thick	1 Clab	24	.333	Ea.		9.75		9.75	16.30
6200	Paneling to 3/4" thick		20	.400			11.70		11.70	19.55
6300	Plaster, on gypsum lath		20	.400			11.70		11.70	19.55
6340	On wire lath		14	.571			16.70		16.70	28
7000	Wood frame, not including re-framing, openings to 5 S.F.									
7200	Floors, sheathing and flooring to 2" thick	1 Clab	5	1.600	Ea.		46.50		46.50	78
7310	Roofs, sheathing to 1" thick, not including roofing		6	1.333			39		39	65
7410	Walls, sheathing to 1" thick, not including siding		7	1.143			33.50		33.50	56

02 41 19.19 Selective Demolition		Crew	Daily Output	Labor-Hours	Unit	Material	Labor	Equipment	Total	Total Incl O&P
0010	**SELECTIVE DEMOLITION**, Rubbish Handling R024119-10									
0020	The following are to be added to the demolition prices									
0050	The following are components for a complete chute system									
0100	Top chute circular steel, 4' long, 18" diameter R024119-30	B-1C	15	1.600	Ea.	266	48	30	344	405
0102	23" diameter		15	1.600		288	48	30	366	430
0104	27" diameter		15	1.600		310	48	30	388	455
0106	30" diameter		15	1.600		330	48	30	408	480
0108	33" diameter		15	1.600		355	48	30	433	505
0110	36" diameter		15	1.600		375	48	30	453	530
0112	Regular chute, 18" diameter		15	1.600		199	48	30	277	330
0114	23" diameter		15	1.600		222	48	30	300	355
0116	27" diameter		15	1.600		243	48	30	321	380
0118	30" diameter		15	1.600		266	48	30	344	405
0120	33" diameter		15	1.600		288	48	30	366	430
0122	36" diameter		15	1.600		310	48	30	388	455
0124	Control door chute, 18" diameter		15	1.600		375	48	30	453	530
0126	23" diameter		15	1.600		400	48	30	478	555
0128	27" diameter		15	1.600		420	48	30	498	580
0130	30" diameter		15	1.600		445	48	30	523	605
0132	33" diameter		15	1.600		465	48	30	543	625
0134	36" diameter		15	1.600		490	48	30	568	650
0136	Chute liners, 14 ga., 18-30" diameter		15	1.600		223	48	30	301	360
0138	33-36" diameter		15	1.600		280	48	30	358	425
0140	17% thinner chute, 30" diameter		15	1.600		222	48	30	300	355
0142	33% thinner chute, 30" diameter		15	1.600		167	48	30	245	296
0144	Top chute cover	1 Clab	24	.333		153	9.75		162.75	184

02 41 19.19 Selective Demolition		Crew	Daily Output	Labor-Hours	Unit	2016 Bare Costs Material	Labor	Equipment	Total	Total Incl O&P
0146	Door chute cover	1 Clab	24	.333	Ea.	153	9.75		162.75	184
0148	Top chute trough	2 Clab	12	1.333		510	39		549	625
0150	Bolt down frame & counter weights, 250 lb.	B-1	4	6		4,250	179		4,429	4,950
0152	500 lb.		4	6		6,200	179		6,379	7,125
0154	750 lb.		4	6		9,100	179		9,279	10,300
0156	1000 lb.		2.67	8.989		10,200	268		10,468	11,700
0158	1500 lb.		2.67	8.989		13,000	268		13,268	14,800
0160	Chute warning light system, 5 stories	B-1C	4	6		8,975	179	112	9,266	10,300
0162	10 stories	"	2	12		14,300	360	224	14,884	16,500
0164	Dust control device for dumpsters	1 Clab	8	1		102	29		131	161
0166	Install or replace breakaway cord		8	1		27	29		56	78.50
0168	Install or replace warning sign		16	.500		11.45	14.60		26.05	37
0600	Dumpster, weekly rental, 1 dump/week, 6 C.Y. capacity (2 Tons)				Week	415			415	455
0700	10 C.Y. capacity (3 Tons)					480			480	530
0725	20 C.Y. capacity (5 Tons) R024119-20					565			565	625
0800	30 C.Y. capacity (7 Tons)					730			730	800
0840	40 C.Y. capacity (10 Tons)					775			775	850
2000	Load, haul, dump and return, 0 – 50′ haul, hand carried	2 Clab	24	.667	C.Y.		19.45		19.45	32.50
2005	Wheeled		37	.432			12.65		12.65	21
2040	0 – 100′ haul, hand carried		16.50	.970			28.50		28.50	47.50
2045	Wheeled		25	.640			18.70		18.70	31.50
2080	Haul and return, add per each extra 100′ haul, hand carried		35.50	.451			13.15		13.15	22
2085	Wheeled		54	.296			8.65		8.65	14.50
2120	For travel in elevators, up to 10 floors, add		140	.114			3.34		3.34	5.60
2130	0 – 50′ haul, incl. up to 5 riser stair, hand carried		23	.696			20.50		20.50	34
2135	Wheeled		35	.457			13.35		13.35	22.50
2140	6 – 10 riser stairs, hand carried		22	.727			21		21	35.50
2145	Wheeled		34	.471			13.75		13.75	23
2150	11 – 20 riser stairs, hand carried		20	.800			23.50		23.50	39
2155	Wheeled		31	.516			15.05		15.05	25
2160	21 – 40 riser stairs, hand carried		16	1			29		29	49
2165	Wheeled		24	.667			19.45		19.45	32.50
2170	0 – 100′ haul, incl. 5 riser stair, hand carried		15	1.067			31		31	52
2175	Wheeled		23	.696			20.50		20.50	34
2180	6 – 10 riser stair, hand carried		14	1.143			33.50		33.50	56
2185	Wheeled		21	.762			22.50		22.50	37.50
2190	11 – 20 riser stair, hand carried		12	1.333			39		39	65
2195	Wheeled		18	.889			26		26	43.50
2200	21 – 40 riser stair, hand carried		8	2			58.50		58.50	98
2205	Wheeled		12	1.333			39		39	65
2210	Haul and return, add per each extra 100′ haul, hand carried		35.50	.451			13.15		13.15	22
2215	Wheeled		54	.296			8.65		8.65	14.50
2220	For each additional flight of stairs, up to 5 risers, add		550	.029	Flight		.85		.85	1.42
2225	6 – 10 risers, add		275	.058			1.70		1.70	2.85
2230	11 – 20 risers, add		138	.116			3.39		3.39	5.65
2235	21 – 40 risers, add		69	.232			6.75		6.75	11.35
3000	Loading & trucking, including 2 mile haul, chute loaded	B-16	45	.711	C.Y.		22	15.35	37.35	53.50
3040	Hand loading truck, 50′ haul	"	48	.667			20.50	14.40	34.90	50.50
3080	Machine loading truck	B-17	120	.267			8.70	6.50	15.20	21.50
5000	Haul, per mile, up to 8 C.Y. truck	B-34B	1165	.007			.23	.59	.82	1.04
5100	Over 8 C.Y. truck	"	1550	.005			.18	.45	.63	.78

02 41 Demolition

02 41 19 – Selective Demolition

02 41 19.20 Selective Demolition, Dump Charges		Crew	Daily Output	Labor-Hours	Unit	Material	2016 Bare Costs Labor	Equipment	Total	Total Incl O&P
0010	**SELECTIVE DEMOLITION, DUMP CHARGES** R024119-10									
0020	Dump charges, typical urban city, tipping fees only									
0100	Building construction materials				Ton	74			74	81
0200	Trees, brush, lumber					63			63	69.50
0300	Rubbish only					63			63	69.50
0500	Reclamation station, usual charge				↓	74			74	81

02 41 19.21 Selective Demolition, Gutting

		Crew	Daily Output	Labor-Hours	Unit	Material	Labor	Equipment	Total	Total Incl O&P
0010	**SELECTIVE DEMOLITION, GUTTING** R024119-10									
0020	Building interior, including disposal, dumpster fees not included									
0500	Residential building									
0560	Minimum	B-16	400	.080	SF Flr.		2.48	1.73	4.21	6.05
0580	Maximum	"	360	.089	"		2.75	1.92	4.67	6.70
0900	Commercial building									
1000	Minimum	B-16	350	.091	SF Flr.		2.83	1.97	4.80	6.90
1020	Maximum	"	250	.128	"		3.96	2.76	6.72	9.65

02 82 Asbestos Remediation

02 82 13 – Asbestos Abatement

02 82 13.39 Asbestos Remediation Plans and Methods

		Crew	Daily Output	Labor-Hours	Unit	Material	Labor	Equipment	Total	Total Incl O&P
0010	**ASBESTOS REMEDIATION PLANS AND METHODS**									
0100	Building Survey-Commercial Building				Ea.				2,200	2,400
0200	Asbestos Abatement Remediation Plan				"				1,350	1,475

02 82 13.41 Asbestos Abatement Equipment

		Crew	Daily Output	Labor-Hours	Unit	Material	Labor	Equipment	Total	Total Incl O&P
0010	**ASBESTOS ABATEMENT EQUIPMENT** R028213-20									
0011	Equipment and supplies, buy									
0200	Air filtration device, 2000 CFM				Ea.	975			975	1,075
0250	Large volume air sampling pump, minimum					355			355	390
0260	Maximum					340			340	375
0300	Airless sprayer unit, 2 gun					3,275			3,275	3,600
0350	Light stand, 500 watt				↓	51			51	56
0400	Personal respirators									
0410	Negative pressure, 1/2 face, dual operation, min.				Ea.	27			27	29.50
0420	Maximum					30			30	33
0450	P.A.P.R., full face, minimum					127			127	140
0460	Maximum					169			169	185
0470	Supplied air, full face, incl. air line, minimum					172			172	189
0480	Maximum					410			410	450
0500	Personnel sampling pump					230			230	254
1500	Power panel, 20 unit, incl. GFI					615			615	675
1600	Shower unit, including pump and filters					1,300			1,300	1,450
1700	Supplied air system (type C)					3,475			3,475	3,825
1750	Vacuum cleaner, HEPA, 16 gal., stainless steel, wet/dry					440			440	485
1760	55 gallon					570			570	630
1800	Vacuum loader, 9 – 18 ton/hr.					96,000			96,000	106,000
1900	Water atomizer unit, including 55 gal. drum					286			286	315
2000	Worker protection, whole body, foot, head cover & gloves, plastic					9			9	9.90
2500	Respirator, single use					25.50			25.50	28
2550	Cartridge for respirator					5.85			5.85	6.45
2570	Glove bag, 7 mil, 50" x 64"					9.35			9.35	10.30
2580	10 mil, 44" x 60"				↓	5.90			5.90	6.45

02 82 13.41 Asbestos Abatement Equipment

02 82 13.41 Asbestos Abatement Equipment		Crew	Daily Output	Labor-Hours	Unit	2016 Bare Costs Material	Labor	Equipment	Total	Total Incl O&P
3000	HEPA vacuum for work area, minimum				Ea.	310			310	345
3050	Maximum					830			830	915
6000	Disposable polyethylene bags, 6 mil, 3 C.F.					.84			.84	.92
6300	Disposable fiber drums, 3 C.F.					18.55			18.55	20.50
6400	Pressure sensitive caution labels, 3" x 5"					3.54			3.54	3.89
6450	11" x 17"					7.50			7.50	8.25
6500	Negative air machine, 1800 CFM					865			865	950

02 82 13.42 Preparation of Asbestos Containment Area

02 82 13.42 Preparation of Asbestos Containment Area		Crew	Daily Output	Labor-Hours	Unit	Material	Labor	Equipment	Total	Total Incl O&P
0010	**PREPARATION OF ASBESTOS CONTAINMENT AREA**									
0100	Pre-cleaning, HEPA vacuum and wet wipe, flat surfaces	A-9	12000	.005	S.F.	.02	.21		.23	.38
0200	Protect carpeted area, 2 layers 6 mil poly on 3/4" plywood	"	1000	.064		2	2.53		4.53	6.50
0300	Separation barrier, 2" x 4" @ 16", 1/2" plywood ea. side, 8' high	2 Carp	400	.040		3.40	1.51		4.91	6.25
0310	12' high		320	.050		3.40	1.89		5.29	6.90
0320	16' high		200	.080		2.30	3.02		5.32	7.60
0400	Personnel decontam. chamber, 2" x 4" @ 16", 3/4" ply ea. side		280	.057		4.40	2.16		6.56	8.45
0450	Waste decontam. chamber, 2" x 4" studs @ 16", 3/4" ply ea. side		360	.044		4.40	1.68		6.08	7.65
0500	Cover surfaces with polyethylene sheeting									
0501	Including glue and tape									
0550	Floors, each layer, 6 mil	A-9	8000	.008	S.F.	.04	.32		.36	.58
0551	4 mil		9000	.007		.03	.28		.31	.51
0560	Walls, each layer, 6 mil		6000	.011		.04	.42		.46	.76
0561	4 mil		7000	.009		.03	.36		.39	.64
0570	For heights above 12', add						20%			
0575	For heights above 20', add						30%			
0580	For fire retardant poly, add					100%				
0590	For large open areas, deduct					10%	20%			
0600	Seal floor penetrations with foam firestop to 36 sq. in.	2 Carp	200	.080	Ea.	9.50	3.02		12.52	15.55
0610	36 sq. in. to 72 sq. in.		125	.128		19.05	4.84		23.89	29
0615	72 sq. in. to 144 sq. in.		80	.200		38	7.55		45.55	54.50
0620	Wall penetrations, to 36 square inches		180	.089		9.50	3.36		12.86	16.15
0630	36 sq. in. to 72 sq. in.		100	.160		19.05	6.05		25.10	31
0640	72 sq. in. to 144 sq. in.		60	.267		38	10.10		48.10	59
0800	Caulk seams with latex	1 Carp	230	.035	L.F.	.17	1.31		1.48	2.39
0900	Set up neg. air machine, 1-2k CFM/25 M.C.F. volume	1 Asbe	4.30	1.860	Ea.		73.50		73.50	125
0950	Set up and remove portable shower unit	2 Asbe	4	4	"		158		158	268

02 82 13.43 Bulk Asbestos Removal

02 82 13.43 Bulk Asbestos Removal		Crew	Daily Output	Labor-Hours	Unit	Material	Labor	Equipment	Total	Total Incl O&P
0010	**BULK ASBESTOS REMOVAL**									
0020	Includes disposable tools and 2 suits and 1 respirator filter/day/worker									
0100	Beams, W 10 x 19	A-9	235	.272	L.F.	.81	10.75		11.56	19.20
0110	W 12 x 22		210	.305		.91	12.05		12.96	21.50
0120	W 14 x 26		180	.356		1.06	14.05		15.11	25
0130	W 16 x 31		160	.400		1.19	15.80		16.99	28.50
0140	W 18 x 40		140	.457		1.36	18.10		19.46	32
0150	W 24 x 55		110	.582		1.74	23		24.74	41
0160	W 30 x 108		85	.753		2.25	30		32.25	53
0170	W 36 x 150		72	.889		2.65	35		37.65	62.50
0200	Boiler insulation		480	.133	S.F.	.47	5.25		5.72	9.45
0210	With metal lath, add				%				50%	50%
0300	Boiler breeching or flue insulation	A-9	520	.123	S.F.	.37	4.87		5.24	8.65
0310	For active boiler, add				%				100%	100%
0400	Duct or AHU insulation	A-10B	440	.073	S.F.	.22	2.88		3.10	5.15
0500	Duct vibration isolation joints, up to 24 sq. in. duct	A-9	56	1.143	Ea.	3.41	45		48.41	80.50

02 82 Asbestos Remediation

02 82 13 – Asbestos Abatement

02 82 13.43 Bulk Asbestos Removal		Crew	Daily Output	Labor-Hours	Unit	Material	2016 Bare Costs Labor	Equipment	Total	Total Incl O&P
0520	25 sq. in. to 48 sq. in. duct	A-9	48	1.333	Ea.	3.98	53		56.98	94
0530	49 sq. in. to 76 sq. in. duct		40	1.600		4.77	63.50		68.27	112
0600	Pipe insulation, air cell type, up to 4" diameter pipe		900	.071	L.F.	.21	2.81		3.02	5
0610	4" to 8" diameter pipe		800	.080		.24	3.16		3.40	5.60
0620	10" to 12" diameter pipe		700	.091		.27	3.62		3.89	6.45
0630	14" to 16" diameter pipe		550	.116		.35	4.60		4.95	8.20
0650	Over 16" diameter pipe		650	.098	S.F.	.29	3.90		4.19	6.90
0700	With glove bag up to 3" diameter pipe		200	.320	L.F.	9.30	12.65		21.95	32
1000	Pipe fitting insulation up to 4" diameter pipe		320	.200	Ea.	.60	7.90		8.50	14.05
1100	6" to 8" diameter pipe		304	.211		.63	8.35		8.98	14.85
1110	10" to 12" diameter pipe		192	.333		.99	13.20		14.19	23.50
1120	14" to 16" diameter pipe		128	.500		1.49	19.80		21.29	35
1130	Over 16" diameter pipe		176	.364	S.F.	1.08	14.40		15.48	25.50
1200	With glove bag, up to 8" diameter pipe		75	.853	L.F.	6.40	34		40.40	64.50
2000	Scrape foam fireproofing from flat surface		2400	.027	S.F.	.08	1.06		1.14	1.88
2100	Irregular surfaces		1200	.053		.16	2.11		2.27	3.75
3000	Remove cementitious material from flat surface		1800	.036		.11	1.41		1.52	2.51
3100	Irregular surface		1000	.064		.14	2.53		2.67	4.45
4000	Scrape acoustical coating/fireproofing, from ceiling		3200	.020		.06	.79		.85	1.41
5000	Remove VAT and mastic from floor by hand		2400	.027		.08	1.06		1.14	1.88
5100	By machine	A-11	4800	.013		.04	.53	.01	.58	.94
5150	For 2 layers, add				%				50%	50%
6000	Remove contaminated soil from crawl space by hand	A-9	400	.160	C.F.	.48	6.35		6.83	11.25
6100	With large production vacuum loader	A-12	700	.091	"	.27	3.62	1.08	4.97	7.65
7000	Radiator backing, not including radiator removal	A-9	1200	.053	S.F.	.16	2.11		2.27	3.75
8000	Cement-asbestos transite board and cement wall board	2 Asbe	1000	.016		.15	.63		.78	1.23
8100	Transite shingle siding	A-10B	750	.043		.22	1.69		1.91	3.11
8200	Shingle roofing	"	2000	.016		.08	.63		.71	1.17
8250	Built-up, no gravel, non-friable	B-2	1400	.029		.08	.85		.93	1.51
8260	Bituminous flashing	1 Rofc	300	.027		.08	.86		.94	1.67
8300	Asbestos millboard, flat board and VAT contaminated plywood	2 Asbe	1000	.016		.08	.63		.71	1.16
9000	For type B (supplied air) respirator equipment, add				%				10%	10%

02 82 13.44 Demolition In Asbestos Contaminated Area

02 82 13.44		Crew	Daily Output	Labor-Hours	Unit	Material	Labor	Equipment	Total	Total Incl O&P
0010	**DEMOLITION IN ASBESTOS CONTAMINATED AREA**									
0200	Ceiling, including suspension system, plaster and lath	A-9	2100	.030	S.F.	.09	1.21		1.30	2.15
0210	Finished plaster, leaving wire lath		585	.109		.33	4.33		4.66	7.70
0220	Suspended acoustical tile		3500	.018		.05	.72		.77	1.29
0230	Concealed tile grid system		3000	.021		.06	.84		.90	1.50
0240	Metal pan grid system		1500	.043		.13	1.69		1.82	3
0250	Gypsum board		2500	.026		.08	1.01		1.09	1.80
0260	Lighting fixtures up to 2' x 4'		72	.889	Ea.	2.65	35		37.65	62.50
0400	Partitions, non load bearing									
0410	Plaster, lath, and studs	A-9	690	.093	S.F.	.91	3.67		4.58	7.20
0450	Gypsum board and studs	"	1390	.046	"	14	1.02		1.96	3.24
9000	For type B (supplied air) respirator equipment, add				%				10%	10%

02 82 13.45 OSHA Testing

02 82 13.45		Crew	Daily Output	Labor-Hours	Unit	Material	Labor	Equipment	Total	Total Incl O&P
0010	**OSHA TESTING**									
0100	Certified technician, minimum				Day				200	220
0110	Maximum								300	330
0120	Industrial hygienist, minimum								250	250
0130	Maximum								400	440
0200	Asbestos sampling and PCM analysis, NIOSH 7400, minimum	1 Asbe	8	1	Ea.	14	39.50		53.50	82.50

02 82 Asbestos Remediation

02 82 13 – Asbestos Abatement

02 82 13.45 OSHA Testing		Crew	Daily Output	Labor-Hours	Unit	Material	Labor	Equipment	Total	Total Incl O&P
						2016 Bare Costs				
0210	Maximum	1 Asbe	4	2	Ea.	22	79		101	158
1000	Cleaned area samples		8	1		200	39.50		239.50	287
1100	PCM air sample analysis, NIOSH 7400, minimum		8	1		14	39.50		53.50	82.50
1110	Maximum	↓	4	2		2.20	79		81.20	136
1200	TEM air sample analysis, NIOSH 7402, minimum								80	106
1210	Maximum				↓				360	450
02 82 13.46 Decontamination of Asbestos Containment Area										
0010	**DECONTAMINATION OF ASBESTOS CONTAINMENT AREA**									
0100	Spray exposed substrate with surfactant (bridging)									
0200	Flat surfaces	A-9	6000	.011	S.F.	.40	.42		.82	1.16
0250	Irregular surfaces		4000	.016	"	.40	.63		1.03	1.51
0300	Pipes, beams, and columns		2000	.032	L.F.	.60	1.27		1.87	2.81
1000	Spray encapsulate polyethylene sheeting		8000	.008	S.F.	.40	.32		.72	.98
1100	Roll down polyethylene sheeting		8000	.008	"		.32		.32	.54
1500	Bag polyethylene sheeting		400	.160	Ea.	1	6.35		7.35	11.85
2000	Fine clean exposed substrate, with nylon brush		2400	.027	S.F.		1.06		1.06	1.79
2500	Wet wipe substrate		4800	.013			.53		.53	.89
2600	Vacuum surfaces, fine brush	↓	6400	.010	↓		.40		.40	.67
3000	Structural demolition									
3100	Wood stud walls	A-9	2800	.023	S.F.		.90		.90	1.53
3500	Window manifolds, not incl. window replacement		4200	.015			.60		.60	1.02
3600	Plywood carpet protection	↓	2000	.032	↓		1.27		1.27	2.15
4000	Remove custom decontamination facility	A-10A	8	3	Ea.	15	119		134	219
4100	Remove portable decontamination facility	3 Asbe	12	2	"	13.35	79		92.35	149
5000	HEPA vacuum, shampoo carpeting	A-9	4800	.013	S.F.	.10	.53		.63	1
9000	Final cleaning of protected surfaces	A-10A	8000	.003	"		.12		.12	.20
02 82 13.47 Asbestos Waste Pkg., Handling, and Disp.										
0010	**ASBESTOS WASTE PACKAGING, HANDLING, AND DISPOSAL**									
0100	Collect and bag bulk material, 3 C.F. bags, by hand	A-9	400	.160	Ea.	.84	6.35		7.19	11.65
0200	Large production vacuum loader	A-12	880	.073		1	2.88	.86	4.74	6.95
1000	Double bag and decontaminate	A-9	960	.067		.84	2.64		3.48	5.40
2000	Containerize bagged material in drums, per 3 C.F. drum	"	800	.080		18.55	3.16		21.71	26
3000	Cart bags 50′ to dumpster	2 Asbe	400	.040	↓		1.58		1.58	2.68
5000	Disposal charges, not including haul, minimum				C.Y.				61	67
5020	Maximum				"				355	395
9000	For type B (supplied air) respirator equipment, add				%				10%	10%
02 82 13.48 Asbestos Encapsulation With Sealants										
0010	**ASBESTOS ENCAPSULATION WITH SEALANTS**									
0100	Ceilings and walls, minimum	A-9	21000	.003	S.F.	.40	.12		.52	.64
0110	Maximum		10600	.006		.50	.24		.74	.96
0200	Columns and beams, minimum		13300	.005		.40	.19		.59	.76
0210	Maximum		5325	.012	↓	.60	.48		1.08	1.47
0300	Pipes to 12″ diameter including minor repairs, minimum		800	.080	L.F.	.50	3.16		3.66	5.90
0310	Maximum	↓	400	.160	"	1.20	6.35		7.55	12.05

02 83 19 – Lead-Based Paint Remediation

02 83 19.21 Lead Paint Remediation Plans and Methods

	02 83 19.21 Lead Paint Remediation Plans and Methods	Crew	Daily Output	Labor-Hours	Unit	2016 Bare Costs Material	Labor	Equipment	Total	Total Incl O&P
0010	**LEAD PAINT REMEDIATION PLANS AND METHODS**									
0100	Building Survey-Commercial Building				Ea.				2,050	2,250
0200	Lead Abatement Remediation Plan								1,225	1,350
0300	Lead Paint Testing, AAS Analysis								51	56
0400	Lead Paint Testing, X-Ray Fluorescence								51	56

02 83 19.23 Encapsulation of Lead-Based Paint

	02 83 19.23 Encapsulation of Lead-Based Paint	Crew	Daily Output	Labor-Hours	Unit	Material	Labor	Equipment	Total	Total Incl O&P
0010	**ENCAPSULATION OF LEAD-BASED PAINT**									
0020	Interior, brushwork, trim, under 6"	1 Pord	240	.033	L.F.	2.30	1.06		3.36	4.28
0030	6" to 12" wide		180	.044		3	1.42		4.42	5.65
0040	Balustrades		300	.027		2	.85		2.85	3.60
0050	Pipe to 4" diameter		500	.016		1.20	.51		1.71	2.16
0060	To 8" diameter		375	.021		1.50	.68		2.18	2.77
0070	To 12" diameter		250	.032		2.20	1.02		3.22	4.10
0080	To 16" diameter		170	.047		3.30	1.50		4.80	6.10
0090	Cabinets, ornate design		200	.040	S.F.	3	1.28		4.28	5.40
0100	Simple design		250	.032	"	2.30	1.02		3.32	4.21
0110	Doors, 3' x 7', both sides, incl. frame & trim									
0120	Flush	1 Pord	6	1.333	Ea.	28	42.50		70.50	101
0130	French, 10 – 15 lite		3	2.667		6	85		91	147
0140	Panel		4	2		35	64		99	144
0150	Louvered		2.75	2.909		32	93		125	188
0160	Windows, per interior side, per 15 S.F.									
0170	1 to 6 lite	1 Pord	14	.571	Ea.	20	18.25		38.25	52
0180	7 to 10 lite		7.50	1.067		22	34		56	80
0190	12 lite		5.75	1.391		30	44.50		74.50	106
0200	Radiators		8	1		70	32		102	130
0210	Grilles, vents		275	.029	S.F.	2	.93		2.93	3.73
0220	Walls, roller, drywall or plaster		1000	.008		.60	.26		.86	1.08
0230	With spunbonded reinforcing fabric		720	.011		.70	.35		1.05	1.35
0240	Wood		800	.010		.70	.32		1.02	1.30
0250	Ceilings, roller, drywall or plaster		900	.009		.70	.28		.98	1.24
0260	Wood		700	.011		.80	.36		1.16	1.48
0270	Exterior, brushwork, gutters and downspouts		300	.027	L.F.	1.90	.85		2.75	3.49
0280	Columns		400	.020	S.F.	1.40	.64		2.04	2.59
0290	Spray, siding		600	.013	"	1	.43		1.43	1.80
0300	Miscellaneous									
0310	Electrical conduit, brushwork, to 2" diameter	1 Pord	500	.016	L.F.	1.20	.51		1.71	2.16
0320	Brick, block or concrete, spray		500	.016	S.F.	1.20	.51		1.71	2.16
0330	Steel, flat surfaces and tanks to 12"		500	.016		1.20	.51		1.71	2.16
0340	Beams, brushwork		400	.020		1.40	.64		2.04	2.59
0350	Trusses		400	.020		1.40	.64		2.04	2.59

02 83 19.26 Removal of Lead-Based Paint

	02 83 19.26 Removal of Lead-Based Paint	Crew	Daily Output	Labor-Hours	Unit	Material	Labor	Equipment	Total	Total Incl O&P
0010	**REMOVAL OF LEAD-BASED PAINT** R028319-60									
0011	By chemicals, per application									
0050	Baseboard, to 6" wide	1 Pord	64	.125	L.F.	.80	3.99		4.79	7.45
0070	To 12" wide		32	.250	"	1.47	8		9.47	14.75
0200	Balustrades, one side		28	.286	S.F.	1.50	9.10		10.60	16.65
1400	Cabinets, simple design		32	.250		1.37	8		9.37	14.65
1420	Ornate design		25	.320		1.60	10.20		11.80	18.55
1600	Cornice, simple design		60	.133		1.60	4.25		5.85	8.75
1620	Ornate design		20	.400		5.50	12.75		18.25	27
2800	Doors, one side, flush		84	.095		1.73	3.04		4.77	6.90

02 83 Lead Remediation

02 83 19 – Lead-Based Paint Remediation

02 83 19.26 Removal of Lead-Based Paint		Crew	Daily Output	Labor-Hours	Unit	2016 Bare Costs Material	Labor	Equipment	Total	Total Incl O&P
2820	Two panel	1 Pord	80	.100	S.F.	1.37	3.19		4.56	6.75
2840	Four panel		45	.178		1.43	5.65		7.08	10.90
2880	For trim, one side, add		64	.125	L.F.	.77	3.99		4.76	7.40
3000	Fence, picket, one side		30	.267	S.F.	1.37	8.50		9.87	15.50
3200	Grilles, one side, simple design		30	.267		1.37	8.50		9.87	15.50
3220	Ornate design		25	.320		1.47	10.20		11.67	18.40
3240	Handrails		90	.089	L.F.	1.37	2.84		4.21	6.20
4400	Pipes, to 4" diameter		90	.089		1.73	2.84		4.57	6.60
4420	To 8" diameter		50	.160		3.47	5.10		8.57	12.20
4440	To 12" diameter		36	.222		5.20	7.10		12.30	17.40
4460	To 16" diameter		20	.400		6.95	12.75		19.70	28.50
4500	For hangers, add		40	.200	Ea.	2.50	6.40		8.90	13.25
4800	Siding		90	.089	S.F.	1.30	2.84		4.14	6.10
5000	Trusses, open		55	.145	SF Face	2	4.64		6.64	9.85
6200	Windows, one side only, double hung, 1/1 light, 24" x 48" high		4	2	Ea.	24	64		88	132
6220	30" x 60" high		3	2.667		32	85		117	175
6240	36" x 72" high		2.50	3.200		38	102		140	210
6280	40" x 80" high		2	4		47	128		175	262
6400	Colonial window, 6/6 light, 24" x 48" high		2	4		47	128		175	262
6420	30" x 60" high		1.50	5.333		62	170		232	350
6440	36" x 72" high		1	8		93	255		348	520
6480	40" x 80" high		1	8		93	255		348	520
6600	8/8 light, 24" x 48" high		2	4		47	128		175	262
6620	40" x 80" high		1	8		93	255		348	520
6800	12/12 light, 24" x 48" high		1	8		93	255		348	520
6820	40" x 80" high		.75	10.667		125	340		465	700
6840	Window frame & trim items, included in pricing above									
7000	Hand scraping and HEPA vacuum, less than 4 S.F.	1 Pord	8	1	Ea.	.84	32		32.84	53.50
8000	Collect and bag bulk material, 3 C.F. bags, by hand	"	30	.267	"	.84	8.50		9.34	14.90

02 85 Mold Remediation

02 85 16 – Mold Remediation Preparation and Containment

02 85 16.50 Preparation of Mold Containment Area		Crew	Daily Output	Labor-Hours	Unit	Material	Labor	Equipment	Total	Total Incl O&P
0010	**PREPARATION OF MOLD CONTAINMENT AREA**									
0100	Pre-cleaning, HEPA vacuum and wet wipe, flat surfaces	A-9	12000	.005	S.F.	.02	.21		.23	.38
0300	Separation barrier, 2" x 4" @ 16", 1/2" plywood ea. side, 8' high	2 Carp	400	.040		3.40	1.51		4.91	6.25
0310	12' high		320	.050		3.40	1.89		5.29	6.90
0320	16' high		200	.080		2.30	3.02		5.32	7.60
0400	Personnel decontam. chamber, 2" x 4" @ 16", 3/4" ply ea. side		280	.057		4.40	2.16		6.56	8.45
0450	Waste decontam. chamber, 2" x 4" studs @ 16", 3/4" ply each side		360	.044		4.40	1.68		6.08	7.65
0500	Cover surfaces with polyethylene sheeting									
0501	Including glue and tape									
0550	Floors, each layer, 6 mil	A-9	8000	.008	S.F.	.04	.32		.36	.58
0551	4 mil		9000	.007		.03	.28		.31	.51
0560	Walls, each layer, 6 mil		6000	.011		.04	.42		.46	.76
0561	4 mil		7000	.009		.03	.36		.39	.64
0570	For heights above 12', add						20%			
0575	For heights above 20', add						30%			
0580	For fire retardant poly, add					100%				
0590	For large open areas, deduct					10%	20%			
0600	Seal floor penetrations with foam firestop to 36 sq. in.	2 Carp	200	.080	Ea.	9.50	3.02		12.52	15.55

02 85 16 – Mold Remediation Preparation and Containment

02 85 16.50 Preparation of Mold Containment Area		Crew	Daily Output	Labor-Hours	Unit	Material	Labor	Equipment	Total	Total Incl O&P
						2016 Bare Costs				
0610	36 sq. in. to 72 sq. in.	2 Carp	125	.128	Ea.	19.05	4.84		23.89	29
0615	72 sq. in. to 144 sq. in.		80	.200		38	7.55		45.55	54.50
0620	Wall penetrations, to 36 square inches		180	.089		9.50	3.36		12.86	16.15
0630	36 sq. in. to 72 sq. in.		100	.160		19.05	6.05		25.10	31
0640	72 sq. in. to 144 sq. in.	↓	60	.267	↓	38	10.10		48.10	59
0800	Caulk seams with latex caulk	1 Carp	230	.035	L.F.	.17	1.31		1.48	2.39
0900	Set up neg. air machine, 1-2k CFM/25 M.C.F. volume	1 Asbe	4.30	1.860	Ea.		73.50		73.50	125

02 85 33 – Removal and Disposal of Materials with Mold

02 85 33.50 Demolition in Mold Contaminated Area		Crew	Daily Output	Labor-Hours	Unit	Material	Labor	Equipment	Total	Total Incl O&P
0010	**DEMOLITION IN MOLD CONTAMINATED AREA**									
0200	Ceiling, including suspension system, plaster and lath	A-9	2100	.030	S.F.	.09	1.21		1.30	2.15
0210	Finished plaster, leaving wire lath		585	.109		.33	4.33		4.66	7.70
0220	Suspended acoustical tile		3500	.018		.05	.72		.77	1.29
0230	Concealed tile grid system		3000	.021		.06	.84		.90	1.50
0240	Metal pan grid system		1500	.043		.13	1.69		1.82	3
0250	Gypsum board		2500	.026		.08	1.01		1.09	1.80
0255	Plywood		2500	.026	↓	.08	1.01		1.09	1.80
0260	Lighting fixtures up to 2′ x 4′	↓	72	.889	Ea.	2.65	35		37.65	62.50
0400	Partitions, non load bearing									
0410	Plaster, lath, and studs	A-9	690	.093	S.F.	.91	3.67		4.58	7.20
0450	Gypsum board and studs		1390	.046		.14	1.82		1.96	3.24
0465	Carpet & pad		1390	.046	↓	.14	1.82		1.96	3.24
0600	Pipe insulation, air cell type, up to 4″ diameter pipe		900	.071	L.F.	.21	2.81		3.02	5
0610	4″ to 8″ diameter pipe		800	.080		.24	3.16		3.40	5.60
0620	10″ to 12″ diameter pipe		700	.091		.27	3.62		3.89	6.45
0630	14″ to 16″ diameter pipe		550	.116	↓	.35	4.60		4.95	8.20
0650	Over 16″ diameter pipe	↓	650	.098	S.F.	.29	3.90		4.19	6.90
9000	For type B (supplied air) respirator equipment, add				%				10%	10%

Division Notes

		CREW	DAILY OUTPUT	LABOR-HOURS	UNIT	BARE COSTS MAT.	LABOR	EQUIP.	TOTAL	TOTAL INCL O&P

Estimating Tips

General

- Carefully check all the plans and specifications. Concrete often appears on drawings other than structural drawings, including mechanical and electrical drawings for equipment pads. The cost of cutting and patching is often difficult to estimate. See Subdivision 03 81 for Concrete Cutting, Subdivision 02 41 19.16 for Cutout Demolition, Subdivision 03 05 05.10 for Concrete Demolition, and Subdivision 02 41 19.19 for Rubbish Handling (handling, loading and hauling of debris).
- Always obtain concrete prices from suppliers near the job site. A volume discount can often be negotiated, depending upon competition in the area. Remember to add for waste, particularly for slabs and footings on grade.

03 10 00 Concrete Forming and Accessories

- A primary cost for concrete construction is forming. Most jobs today are constructed with prefabricated forms. The selection of the forms best suited for the job and the total square feet of forms required for efficient concrete forming and placing are key elements in estimating concrete construction. Enough forms must be available for erection to make efficient use of the concrete placing equipment and crew.
- Concrete accessories for forming and placing depend upon the systems used. Study the plans and specifications to ensure that all special accessory requirements have been included in the cost estimate, such as anchor bolts, inserts, and hangers.
- Included within costs for forms-in-place are all necessary bracing and shoring.

03 20 00 Concrete Reinforcing

- Ascertain that the reinforcing steel supplier has included all accessories, cutting, bending, and an allowance for lapping, splicing, and waste. A good rule of thumb is 10% for lapping, splicing, and waste. Also, 10% waste should be allowed for welded wire fabric.
- The unit price items in the subdivisions for Reinforcing In Place, Glass Fiber Reinforcing, and Welded Wire Fabric include the labor to install accessories such as beam and slab bolsters, high chairs, and bar ties and tie wire. The material cost for these accessories is not included; they may be obtained from the Accessories Subdivisions.

03 30 00 Cast-In-Place Concrete

- When estimating structural concrete, pay particular attention to requirements for concrete additives, curing methods, and surface treatments. Special consideration for climate, hot or cold, must be included in your estimate. Be sure to include requirements for concrete placing equipment, and concrete finishing.
- For accurate concrete estimating, the estimator must consider each of the following major components individually: forms, reinforcing steel, ready-mix concrete, placement of the concrete, and finishing of the top surface. For faster estimating, Subdivision 03 30 53.40 for Concrete-In-Place can be used; here, various items of concrete work are presented that include the costs of all five major components (unless specifically stated otherwise).

03 40 00 Precast Concrete
03 50 00 Cast Decks and Underlayment

- The cost of hauling precast concrete structural members is often an important factor. For this reason, it is important to get a quote from the nearest supplier. It may become economically feasible to set up precasting beds on the site if the hauling costs are prohibitive.

Reference Numbers

Reference numbers are shown at the beginning of some major classifications. These numbers refer to related items in the Reference Section. The reference information may be an estimating procedure, an alternate pricing method, or technical information.

Note: Not all subdivisions listed here necessarily appear. ■

03 01 Maintenance of Concrete

03 01 30 – Maintenance of Cast-In-Place Concrete

03 01 30.64 Floor Patching		Crew	Daily Output	Labor-Hours	Unit	Material	Labor	Equipment	Total	Total Incl O&P
						2016 Bare Costs				
0010	**FLOOR PATCHING**									
0012	Floor patching, 1/4" thick, small areas, regular	1 Cefi	170	.047	S.F.	3.07	1.68		4.75	6.10
0100	Epoxy	"	100	.080	"	7.50	2.85		10.35	12.85

03 05 Common Work Results for Concrete

03 05 05 – Selective Demolition for Concrete

03 05 05.10 Selective Demolition, Concrete

		Crew	Daily Output	Labor-Hours	Unit	Material	Labor	Equipment	Total	Total Incl O&P
0010	**SELECTIVE DEMOLITION, CONCRETE** R024119-10									
0012	Excludes saw cutting, torch cutting, loading or hauling									
0050	Break into small pieces, reinf. less than 1% of cross-sectional area	B-9	24	1.667	C.Y.		49.50	9.55	59.05	93
0060	Reinforcing 1% to 2% of cross-sectional area		16	2.500			74	14.30	88.30	140
0070	Reinforcing more than 2% of cross-sectional area		8	5			148	28.50	176.50	280
0150	Remove whole pieces, up to 2 tons per piece	E-18	36	1.111	Ea.		44.50	27	71.50	110
0160	2 – 5 tons per piece		30	1.333			53.50	32	85.50	132
0170	5 – 10 tons per piece		24	1.667			67	40.50	107.50	166
0180	10 – 15 tons per piece		18	2.222			89.50	53.50	143	220
0250	Precast unit embedded in masonry, up to 1 C.F.	D-1	16	1			33		33	55
0260	1 – 2 C.F.		12	1.333			44		44	73
0270	2 – 5 C.F.		10	1.600			52.50		52.50	87.50
0280	5 – 10 C.F.		8	2			66		66	110

03 05 13 – Basic Concrete Materials

03 05 13.85 Winter Protection

		Crew	Daily Output	Labor-Hours	Unit	Material	Labor	Equipment	Total	Total Incl O&P
0010	**WINTER PROTECTION**									
0012	For heated ready mix, add				C.Y.	4.45			4.45	4.90
0100	Temporary heat to protect concrete, 24 hours	2 Clab	50	.320	M.S.F.	200	9.35		209.35	235
0200	Temporary shelter for slab on grade, wood frame/polyethylene sheeting									
0201	Build or remove, light framing for short spans	2 Carp	10	1.600	M.S.F.	292	60.50		352.50	420
0210	Large framing for long spans	"	3	5.333	"	380	202		582	760
0710	Electrically, heated pads, 15 watts/S.F., 20 uses				S.F.	.55			.55	.61

03 11 Concrete Forming

03 11 13 – Structural Cast-In-Place Concrete Forming

03 11 13.45 Forms In Place, Footings

		Crew	Daily Output	Labor-Hours	Unit	Material	Labor	Equipment	Total	Total Incl O&P
0010	**FORMS IN PLACE, FOOTINGS** R031113-40									
0020	Continuous wall, plywood, 1 use	C-1	375	.085	SFCA	6.80	2.83		9.63	12.25
0150	4 use	"	485	.066	"	2.21	2.19		4.40	6.10
1500	Keyway, 4 use, tapered wood, 2" x 4" R031113-60	1 Carp	530	.015	L.F.	.21	.57		.78	1.19
1550	2" x 6"	"	500	.016	"	.30	.60		.90	1.34
5000	Spread footings, job-built lumber, 1 use	C-1	305	.105	SFCA	2.15	3.48		5.63	8.20
5150	4 use	"	414	.077	"	.70	2.57		3.27	5.05

03 11 13.65 Forms In Place, Slab On Grade

		Crew	Daily Output	Labor-Hours	Unit	Material	Labor	Equipment	Total	Total Incl O&P
0010	**FORMS IN PLACE, SLAB ON GRADE** R031113-40									
1000	Bulkhead forms w/keyway, wood, 6" high, 1 use R031113-60	C-1	510	.063	L.F.	.97	2.08		3.05	4.56
1400	Bulkhead form for slab, 4-1/2" high, exp metal, incl keyway & stakes G		1200	.027		.91	.89		1.80	2.49
1410	5-1/2" high G		1100	.029		1.15	.97		2.12	2.89
1420	7-1/2" high G		960	.033		1.26	1.11		2.37	3.25
1430	9-1/2" high G		840	.038		1.38	1.26		2.64	3.64
2000	Curb forms, wood, 6" to 12" high, on grade, 1 use		215	.149	SFCA	2.64	4.94		7.58	11.20

03 11 13 – Structural Cast-In-Place Concrete Forming

03 11 13.65 Forms In Place, Slab On Grade		Crew	Daily Output	Labor-Hours	Unit	Material	Labor	Equipment	Total	Total Incl O&P
						2016 Bare Costs				
2150	4 use	C-1	275	.116	SFCA	.86	3.86		4.72	7.45
3000	Edge forms, wood, 4 use, on grade, to 6" high		600	.053	L.F.	.33	1.77		2.10	3.34
3050	7" to 12" high		435	.074	SFCA	.83	2.44		3.27	5
4000	For slab blockouts, to 12" high, 1 use		200	.160	L.F.	.78	5.30		6.08	9.75
4100	Plastic (extruded), to 6" high, multiple use, on grade		800	.040	"	8.20	1.33		9.53	11.25
8760	Void form, corrugated fiberboard, 4" x 12", 4' long G		3000	.011	S.F.	3.33	.35		3.68	4.26
8770	6" x 12", 4' long		3000	.011		4.10	.35		4.45	5.10
8780	1/4" thick hardboard protective cover for void form	2 Carp	1500	.011		.63	.40		1.03	1.37

03 11 13.85 Forms In Place, Walls		Crew	Daily Output	Labor-Hours	Unit	Material	Labor	Equipment	Total	Total Incl O&P
0010	**FORMS IN PLACE, WALLS** R031113-10									
0100	Box out for wall openings, to 16" thick, to 10 S.F.	C-2	24	2	Ea.	25.50	67		92.50	141
0150	Over 10 S.F. (use perimeter) R031113-40	"	280	.171	L.F.	2.20	5.75		7.95	12
0250	Brick shelf, 4" w, add to wall forms, use wall area above shelf									
0260	1 use R031113-60	C-2	240	.200	SFCA	2.37	6.70		9.07	13.80
0350	4 use		300	.160	"	.95	5.35		6.30	10
0500	Bulkhead, wood with keyway, 1 use, 2 piece		265	.181	L.F.	2.07	6.05		8.12	12.45
0600	Bulkhead forms with keyway, 1 piece expanded metal, 8" wall G	C-1	1000	.032		1.26	1.06		2.32	3.17
0610	10" wall G		800	.040		1.38	1.33		2.71	3.75
0620	12" wall G		525	.061		1.66	2.02		3.68	5.20
2000	Wall, job-built plywood, to 8' high, 1 use	C-2	370	.130	SFCA	2.68	4.34		7.02	10.25
2050	2 use		435	.110		1.70	3.69		5.39	8.05
2100	3 use		495	.097		1.24	3.24		4.48	6.80
2150	4 use		505	.095		1	3.18		4.18	6.45
2400	Over 8' to 16' high, 1 use		280	.171		2.96	5.75		8.71	12.85
2450	2 use		345	.139		1.29	4.65		5.94	9.20
2500	3 use		375	.128		.92	4.28		5.20	8.20
2550	4 use		395	.122		.76	4.06		4.82	7.65
7800	Modular prefabricated plywood, based on 20 uses of purchased									
7820	forms, and 4 uses of bracing lumber									
7860	To 8' high	C-2	800	.060	SFCA	.90	2.01		2.91	4.36
8060	Over 8' to 16' high	"	600	.080	"	.95	2.67		3.62	5.55

03 11 19 – Insulating Concrete Forming

03 11 19.10 Insulating Forms, Left In Place		Crew	Daily Output	Labor-Hours	Unit	Material	Labor	Equipment	Total	Total Incl O&P
0010	**INSULATING FORMS, LEFT IN PLACE**									
0020	S.F. is for exterior face, but includes forms for both faces (total R22)									
2000	4" wall, straight block, 16" x 48" (5.33 S.F.) G	2 Carp	90	.178	Ea.	20	6.70		26.70	33.50
2010	90 corner block, exterior 16" x 38" x 22" (6.67 S.F.) G		75	.213		23.50	8.05		31.55	39.50
2020	45 corner block, exterior 16" x 34" x 18" (5.78 S.F.) G		75	.213		23.50	8.05		31.55	39.50
2100	6" wall, straight block, 16" x 48" (5.33 S.F.) G		90	.178		19.70	6.70		26.40	33
2110	90 corner block, exterior 16" x 32" x 24" (6.22 S.F.) G		75	.213		25	8.05		33.05	41
2120	45 corner block, exterior 16" x 26" x 18" (4.89 S.F.) G		75	.213		22.50	8.05		30.55	38
2130	Brick ledge block, 16" x 48" (5.33 S.F.) G		80	.200		25.50	7.55		33.05	40.50
2140	Taper top block, 16" x 48" (5.33 S.F.) G		80	.200		24.50	7.55		32.05	39.50
2200	8" wall, straight block, 16" x 48" (5.33 S.F.) G		90	.178		20.50	6.70		27.20	34
2210	90 corner block, exterior 16" x 34" x 26" (6.67 S.F.) G		75	.213		27	8.05		35.05	43.50
2220	45 corner block, exterior 16" x 28" x 20" (5.33 S.F.) G		75	.213		24	8.05		32.05	39.50
2230	Brick ledge block, 16" x 48" (5.33 S.F.) G		80	.200		26	7.55		33.55	41.50
2240	Taper top block, 16" x 48" (5.33 S.F.) G		80	.200		25	7.55		32.55	40

03 11 Concrete Forming

03 11 23 – Permanent Stair Forming

03 11 23.75 Forms In Place, Stairs		Crew	Daily Output	Labor-Hours	Unit	2016 Bare Costs Material	Labor	Equipment	Total	Total Incl O&P
0010	FORMS IN PLACE, STAIRS R031113-40									
0015	(Slant length x width), 1 use	C-2	165	.291	S.F.	5.80	9.70		15.50	22.50
0150	4 use R031113-60		190	.253		2.04	8.45		10.49	16.40
2000	Stairs, cast on sloping ground (length x width), 1 use		220	.218		2.22	7.30		9.52	14.70
2025	2 use		232	.207		1.22	6.90		8.12	12.95
2050	3 use		244	.197		.89	6.55		7.44	12
2100	4 use		256	.188		.72	6.25		6.97	11.30

03 15 Concrete Accessories

03 15 05 – Concrete Forming Accessories

03 15 05.85 Stair Tread Inserts

		Crew	Daily Output	Labor-Hours	Unit	Material	Labor	Equipment	Total	Total Incl O&P
0010	STAIR TREAD INSERTS									
0105	Cast nosing insert, abrasive surface, pre-drilled, includes screws									
0110	Aluminum, 3″ wide x 3′ long	1 Cefi	32	.250	Ea.	57.50	8.90		66.40	77.50
0120	4′ long		31	.258		73	9.20		82.20	95.50
0130	5′ long		30	.267		90.50	9.50		100	115
0135	Extruded nosing insert, black abrasive strips, continuous anchor									
0140	Aluminum, 3″ wide x 3′ long	1 Cefi	64	.125	Ea.	37	4.45		41.45	48.50
0150	4′ long		60	.133		57.50	4.75		62.25	70.50
0160	5′ long		56	.143		64.50	5.10		69.60	79
0165	Extruded nosing insert, black abrasive strips, pre-drilled, incl. screws									
0170	Aluminum, 3″ wide x 3′ long	1 Cefi	32	.250	Ea.	51	8.90		59.90	70.50
0180	4′ long		31	.258		66	9.20		75.20	88
0190	5′ long		30	.267		77.50	9.50		87	101

03 15 13 – Waterstops

03 15 13.50 Waterstops

		Crew	Daily Output	Labor-Hours	Unit	Material	Labor	Equipment	Total	Total Incl O&P
0010	**WATERSTOPS**, PVC and Rubber									
0020	PVC, ribbed 3/16″ thick, 4″ wide	1 Carp	155	.052	L.F.	1.32	1.95		3.27	4.72
0050	6″ wide		145	.055		2.29	2.09		4.38	6
0500	With center bulb, 6″ wide, 3/16″ thick		135	.059		2.02	2.24		4.26	5.95
0550	3/8″ thick		130	.062		3.79	2.33		6.12	8.05
0600	9″ wide x 3/8″ thick		125	.064		6.15	2.42		8.57	10.80

03 15 16 – Concrete Construction Joints

03 15 16.20 Control Joints, Saw Cut

		Crew	Daily Output	Labor-Hours	Unit	Material	Labor	Equipment	Total	Total Incl O&P
0010	CONTROL JOINTS, SAW CUT									
0100	Sawcut control joints in green concrete									
0120	1″ depth	C-27	2000	.008	L.F.	.04	.28	.08	.40	.59
0140	1-1/2″ depth		1800	.009		.05	.32	.09	.46	.67
0160	2″ depth		1600	.010		.07	.36	.11	.54	.78
0180	Sawcut joint reservoir in cured concrete									
0182	3/8″ wide x 3/4″ deep, with single saw blade	C-27	1000	.016	L.F.	.05	.57	.17	.79	1.17
0184	1/2″ wide x 1″ deep, with double saw blades		900	.018		.10	.63	.19	.92	1.35
0186	3/4″ wide x 1-1/2″ deep, with double saw blades		800	.020		.21	.71	.21	1.13	1.62
0190	Water blast joint to wash away laitance, 2 passes	C-29	2500	.003			.09	.03	.12	.19
0200	Air blast joint to blow out debris and air dry, 2 passes	C-28	2000	.004			.14	.01	.15	.24
0300	For backer rod, see Section 07 91 23.10									
0342	For joint sealant, see Section 07 92 13.20									
0900	For replacement of joint sealant, see Section 07 01 90.81									

03 15 Concrete Accessories

03 15 19 – Cast-In Concrete Anchors

03 15 19.05 Anchor Bolt Accessories			Crew	Daily Output	Labor-Hours	Unit	Material	Labor	Equipment	Total	Total Incl O&P
							2016 Bare Costs				
0010	**ANCHOR BOLT ACCESSORIES**										
0015	For anchor bolts set in fresh concrete, see Section 03 15 19.10										
8150	Anchor bolt sleeve, plastic, 1″ diam. bolts		1 Carp	60	.133	Ea.	12.95	5.05		18	22.50
8500	1-1/2″ diameter			28	.286		19.20	10.80		30	39
8600	2″ diameter			24	.333		19.45	12.60		32.05	42.50
8650	3″ diameter			20	.400		35.50	15.10		50.60	64.50

03 15 19.10 Anchor Bolts

Line	Description		Crew	Daily Output	Labor-Hours	Unit	Material	Labor	Equipment	Total	Total Incl O&P
0010	**ANCHOR BOLTS**										
0015	Made from recycled materials										
0025	Single bolts installed in fresh concrete, no templates										
0030	Hooked w/nut and washer, 1/2″ diameter, 8″ long	G	1 Carp	132	.061	Ea.	1.32	2.29		3.61	5.30
0040	12″ long	G		131	.061		1.47	2.31		3.78	5.50
0050	5/8″ diameter, 8″ long	G		129	.062		3.09	2.34		5.43	7.35
0060	12″ long	G		127	.063		3.81	2.38		6.19	8.20
0070	3/4″ diameter, 8″ long	G		127	.063		3.81	2.38		6.19	8.20
0080	12″ long	G		125	.064		4.76	2.42		7.18	9.30
0090	2-bolt pattern, including job-built 2-hole template, per set										
0100	J-type, incl. hex nut & washer, 1/2″ diameter x 6″ long	G	1 Carp	21	.381	Set	5.55	14.40		19.95	30
0110	12″ long	G		21	.381		6.15	14.40		20.55	31
0120	18″ long	G		21	.381		7	14.40		21.40	31.50
0130	3/4″ diameter x 8″ long	G		20	.400		10.80	15.10		25.90	37.50
0140	12″ long	G		20	.400		12.70	15.10		27.80	39.50
0150	18″ long	G		20	.400		15.55	15.10		30.65	42.50
0160	1″ diameter x 12″ long	G		19	.421		22	15.90		37.90	50.50
0170	18″ long	G		19	.421		25.50	15.90		41.40	54.50
0180	24″ long	G		19	.421		30.50	15.90		46.40	60
0190	36″ long	G		18	.444		40.50	16.80		57.30	72.50
0200	1-1/2″ diameter x 18″ long	G		17	.471		42	17.80		59.80	76
0210	24″ long	G		16	.500		49	18.90		67.90	85.50
0300	L-type, incl. hex nut & washer, 3/4″ diameter x 12″ long	G		20	.400		13.75	15.10		28.85	40.50
0310	18″ long	G		20	.400		16.70	15.10		31.80	44
0320	24″ long	G		20	.400		19.65	15.10		34.75	47
0330	30″ long	G		20	.400		24	15.10		39.10	52
0340	36″ long	G		20	.400		27	15.10		42.10	55.50
0350	1″ diameter x 12″ long	G		19	.421		21	15.90		36.90	49.50
0360	18″ long	G		19	.421		25.50	15.90		41.40	54.50
0370	24″ long	G		19	.421		30.50	15.90		46.40	60
0380	30″ long	G		19	.421		35.50	15.90		51.40	65.50
0390	36″ long	G		18	.444		40	16.80		56.80	72
0400	42″ long	G		18	.444		48	16.80		64.80	81
0410	48″ long	G		18	.444		53.50	16.80		70.30	87
0420	1-1/4″ diameter x 18″ long	G		18	.444		30	16.80		46.80	61
0430	24″ long	G		18	.444		35	16.80		51.80	66.50
0440	30″ long	G		17	.471		40	17.80		57.80	74
0450	36″ long	G		17	.471		45	17.80		62.80	79.50
1000	4-bolt pattern, including job-built 4-hole template, per set										
1100	J-type, incl. hex nut & washer, 1/2″ diameter x 6″ long	G	1 Carp	19	.421	Set	7.90	15.90		23.80	35
1110	12″ long	G		19	.421		9.05	15.90		24.95	36.50
1120	18″ long	G		18	.444		10.85	16.80		27.65	40
1130	3/4″ diameter x 8″ long	G		17	.471		18.40	17.80		36.20	50.50
1140	12″ long	G		17	.471		22	17.80		39.80	54.50
1150	18″ long	G		17	.471		28	17.80		45.80	60.50

03 15 Concrete Accessories

03 15 19 – Cast-In Concrete Anchors

03 15 19.10 Anchor Bolts			Crew	Daily Output	Labor-Hours	Unit	Material	Labor	Equipment	Total	Total Incl O&P
							2016 Bare Costs				
1160	1" diameter x 12" long	G	1 Carp	16	.500	Set	40.50	18.90		59.40	76
1170	18" long	G		15	.533		48	20		68	87
1180	24" long	G		15	.533		58	20		78	97.50
1190	36" long	G		15	.533		78	20		98	120
1200	1-1/2" diameter x 18" long	G		13	.615		80.50	23.50		104	128
1210	24" long	G		12	.667		95	25		120	147
1300	L-type, incl. hex nut & washer, 3/4" diameter x 12" long	G		17	.471		24.50	17.80		42.30	56.50
1310	18" long	G		17	.471		30	17.80		47.80	63
1320	24" long	G		17	.471		36	17.80		53.80	70
1330	30" long	G		16	.500		45	18.90		63.90	81
1340	36" long	G		16	.500		51	18.90		69.90	87.50
1350	1" diameter x 12" long	G		16	.500		38.50	18.90		57.40	74
1360	18" long	G		15	.533		47.50	20		67.50	86
1370	24" long	G		15	.533		58	20		78	97.50
1380	30" long	G		15	.533		67.50	20		87.50	109
1390	36" long	G		15	.533		77	20		97	119
1400	42" long	G		14	.571		93	21.50		114.50	138
1410	48" long	G		14	.571		104	21.50		125.50	150
1420	1-1/4" diameter x 18" long	G		14	.571		57	21.50		78.50	98.50
1430	24" long	G		14	.571		67	21.50		88.50	110
1440	30" long	G		13	.615		77	23.50		100.50	124
1450	36" long	G	↓	13	.615	↓	87	23.50		110.50	135

03 21 Reinforcement Bars

03 21 11 – Plain Steel Reinforcement Bars

03 21 11.60 Reinforcing In Place			Crew	Daily Output	Labor-Hours	Unit	Material	Labor	Equipment	Total	Total Incl O&P
0010	**REINFORCING IN PLACE**, 50-60 ton lots, A615 Grade 60 R032110-10										
0020	Includes labor, but not material cost, to install accessories										
0030	Made from recycled materials										
0500	Footings, #4 to #7	G	4 Rodm	2.10	15.238	Ton	960	605		1,565	2,075
0550	#8 to #18 R032110-20	G		3.60	8.889		960	355		1,315	1,650
0700	Walls, #3 to #7	G		3	10.667		960	425		1,385	1,775
0750	#8 to #18	G	↓	4	8	↓	960	320		1,280	1,575
0900	For other than 50 – 60 ton lots										
1000	Under 10 ton job, #3 to #7, add						25%	10%			
1010	#8 to #18, add						20%	10%			
1050	10 – 50 ton job, #3 to #7, add						10%				
1060	#8 to #18, add						5%				
1100	60 – 100 ton job, #3 to #7, deduct						5%				
1110	#8 to #18, deduct						10%				
1150	Over 100 ton job, #3 to #7, deduct						10%				
1160	#8 to #18, deduct						15%				
2400	Dowels, 2 feet long, deformed, #3	G	2 Rodm	520	.031	Ea.	.40	1.22		1.62	2.51
2410	#4 R032110-40	G		480	.033		.71	1.32		2.03	3.02
2420	#5	G		435	.037		1.10	1.46		2.56	3.68
2430	#6 R032110-50	G	↓	360	.044	↓	1.59	1.77		3.36	4.72
2600	Dowel sleeves for CIP concrete, 2-part system R032110-70										
2610	Sleeve base, plastic, for 5/8" smooth dowel sleeve, fasten to edge form		1 Rodm	200	.040	Ea.	.53	1.59		2.12	3.27
2615	Sleeve, plastic, 12" long, for 5/8" smooth dowel, snap onto base			400	.020		1.33	.80		2.13	2.80
2620	Sleeve base, for 3/4" smooth dowel sleeve R032110-80			175	.046		.53	1.82		2.35	3.65
2625	Sleeve, 12" long, for 3/4" smooth dowel		↓	350	.023	↓	1.61	.91		2.52	3.31

03 21 Reinforcement Bars

03 21 11 – Plain Steel Reinforcement Bars

03 21 11.60 Reinforcing In Place		Crew	Daily Output	Labor-Hours	Unit	Material	2016 Bare Costs Labor	Equipment	Total	Total Incl O&P
2630	Sleeve base, for 1" smooth dowel sleeve	1 Rodm	150	.053	Ea.	.68	2.12		2.80	4.33
2635	Sleeve, 12" long, for 1" smooth dowel		300	.027		1.42	1.06		2.48	3.35
2700	Dowel caps, visual warning only, plastic, #3 to #8	2 Rodm	800	.020		.30	.80		1.10	1.67
2720	#8 to #18		750	.021		.74	.85		1.59	2.25
2750	Impalement protective, plastic, #4 to #9		800	.020		1.66	.80		2.46	3.17

03 21 21 – Composite Reinforcement Bars

03 21 21.11 Glass Fiber-Reinforced Polymer Reinf. Bars		Crew	Daily Output	Labor-Hours	Unit	Material	Labor	Equipment	Total	Total Incl O&P
0010	**GLASS FIBER-REINFORCED POLYMER REINFORCEMENT BARS**									
0020	Includes labor, but not material cost, to install accessories									
0050	#2 bar, .043 lb./L.F.	4 Rodm	9500	.003	L.F.	.40	.13		.53	.67
0100	#3 bar, .092 lb./L.F.		9300	.003		.49	.14		.63	.77
0150	#4 bar, .160 lb./L.F.		9100	.004		.71	.14		.85	1.02
0200	#5 bar, .258 lb./L.F.		8700	.004		1.09	.15		1.24	1.45
0250	#6 bar, .372 lb./L.F.		8300	.004		1.41	.15		1.56	1.81
0300	#7 bar, .497 lb./L.F.		7900	.004		1.84	.16		2	2.29
0350	#8 bar, .620 lb./L.F.		7400	.004		2.38	.17		2.55	2.91
0400	#9 bar, .800 lb./L.F.		6800	.005		3.10	.19		3.29	3.73
0450	#10 bar, 1.08 lb./L.F.		5800	.006		3.75	.22		3.97	4.50
0500	For Bends, add per bend				Ea.	1.48			1.48	1.63

03 22 Fabric and Grid Reinforcing

03 22 11 – Plain Welded Wire Fabric Reinforcing

03 22 11.10 Plain Welded Wire Fabric		Crew	Daily Output	Labor-Hours	Unit	Material	Labor	Equipment	Total	Total Incl O&P
0010	**PLAIN WELDED WIRE FABRIC** ASTM A185 R032205-30									
0020	Includes labor, but not material cost, to install accessories									
0030	Made from recycled materials									
0050	Sheets									
0100	6 x 6 - W1.4 x W1.4 (10 x 10) 21 lb. per C.S.F. G	2 Rodm	35	.457	C.S.F.	13.50	18.15		31.65	45.50
0300	6 x 6 - W2.9 x W2.9 (6 x 6) 42 lb. per C.S.F. G		29	.552		23	22		45	62
0500	4 x 4 - W1.4 x W1.4 (10 x 10) 31 lb. per C.S.F. G		31	.516		20	20.50		40.50	56.50
0750	Rolls									
0900	2 x 2 - #12 galv. for gunite reinforcing G	2 Rodm	6.50	2.462	C.S.F.	62.50	98		160.50	234

03 22 13 – Galvanized Welded Wire Fabric Reinforcing

03 22 13.10 Galvanized Welded Wire Fabric		Crew	Daily Output	Labor-Hours	Unit	Material	Labor	Equipment	Total	Total Incl O&P
0010	**GALVANIZED WELDED WIRE FABRIC**									
0100	Add to plain welded wire pricing for galvanized welded wire				Lb.	.23			.23	.25

03 22 16 – Epoxy-Coated Welded Wire Fabric Reinforcing

03 22 16.10 Epoxy-Coated Welded Wire Fabric		Crew	Daily Output	Labor-Hours	Unit	Material	Labor	Equipment	Total	Total Incl O&P
0010	**EPOXY-COATED WELDED WIRE FABRIC**									
0100	Add to plain welded wire pricing for epoxy-coated welded wire				Lb.	.22			.22	.24

03 23 Stressed Tendon Reinforcing

03 23 05 – Prestressing Tendons

03 23 05.50 Prestressing Steel		Crew	Daily Output	Labor-Hours	Unit	Material	Labor	Equipment	Total	Total Incl O&P
						2016 Bare Costs				
0010	PRESTRESSING STEEL R034136-90									
3000	Slabs on grade, 0.5-inch diam. non-bonded strands, HDPE sheathed,									
3050	attached dead-end anchors, loose stressing-end anchors									
3100	25′ x 30′ slab, strands @ 36″ O.C., placing	2 Rodm	2940	.005	S.F.	.60	.22		.82	1.03
3105	Stressing	C-4A	3750	.004			.17	.01	.18	.30
3110	42″ O.C., placing	2 Rodm	3200	.005		.53	.20		.73	.93
3115	Stressing	C-4A	4040	.004			.16	.01	.17	.28
3120	48″ O.C., placing	2 Rodm	3510	.005		.47	.18		.65	.82
3125	Stressing	C-4A	4390	.004			.14	.01	.15	.25
3150	25′ x 40′ slab, strands @ 36″ O.C., placing	2 Rodm	3370	.005		.58	.19		.77	.96
3155	Stressing	C-4A	4360	.004			.15	.01	.16	.26
3160	42″ O.C., placing	2 Rodm	3760	.004		.50	.17		.67	.84
3165	Stressing	C-4A	4820	.003			.13	.01	.14	.23
3170	48″ O.C., placing	2 Rodm	4090	.004		.45	.16		.61	.75
3175	Stressing	C-4A	5190	.003			.12	.01	.13	.22
3200	30′ x 30′ slab, strands @ 36″ O.C., placing	2 Rodm	3260	.005		.58	.20		.78	.97
3205	Stressing	C-4A	4190	.004			.15	.01	.16	.27
3210	42″ O.C., placing	2 Rodm	3530	.005		.52	.18		.70	.88
3215	Stressing	C-4A	4500	.004			.14	.01	.15	.25
3220	48″ O.C., placing	2 Rodm	3840	.004		.47	.17		.64	.79
3225	Stressing	C-4A	4850	.003			.13	.01	.14	.23
3230	30′ x 40′ slab, strands @ 36″ O.C., placing	2 Rodm	3780	.004		.56	.17		.73	.90
3235	Stressing	C-4A	4920	.003			.13	.01	.14	.23
3240	42″ O.C., placing	2 Rodm	4190	.004		.49	.15		.64	.80
3245	Stressing	C-4A	5410	.003			.12	.01	.13	.21
3250	48″ O.C., placing	2 Rodm	4520	.004		.45	.14		.59	.73
3255	Stressing	C-4A	5790	.003			.11	.01	.12	.20
3260	30′ x 50′ slab, strands @ 36″ O.C., placing	2 Rodm	4300	.004		.53	.15		.68	.84
3265	Stressing	C-4A	5650	.003			.11	.01	.12	.20
3270	42″ O.C., placing	2 Rodm	4720	.003		.47	.13		.60	.75
3275	Stressing	C-4A	6150	.003			.10	.01	.11	.18
3280	48″ O.C., placing	2 Rodm	5240	.003		.42	.12		.54	.66
3285	Stressing	C-4A	6760	.002	▼		.09	.01	.10	.17

03 24 Fibrous Reinforcing

03 24 05 – Reinforcing Fibers

03 24 05.30 Synthetic Fibers		Crew	Daily Output	Labor-Hours	Unit	Material	Labor	Equipment	Total	Total Incl O&P
0010	SYNTHETIC FIBERS									
0100	Synthetic fibers, add to concrete				Lb.	4.40			4.40	4.84
0110	1-1/2 lb. per C.Y.				C.Y.	6.80			6.80	7.50

03 24 05.70 Steel Fibers		Crew	Daily Output	Labor-Hours	Unit	Material	Labor	Equipment	Total	Total Incl O&P
0010	STEEL FIBERS									
0140	ASTM A850, Type V, continuously deformed, 1-1/2″ long x 0.045″ diam.									
0150	Add to price of ready mix concrete G				Lb.	1.13			1.13	1.24
0205	Alternate pricing, dosing at 5 lb. per C.Y., add to price of RMC G				C.Y.	5.65			5.65	6.20
0210	10 lb. per C.Y. G					11.30			11.30	12.45
0215	15 lb. per C.Y. G					16.95			16.95	18.65
0220	20 lb. per C.Y. G					22.50			22.50	25
0225	25 lb. per C.Y. G					28.50			28.50	31
0230	30 lb. per C.Y. G					34			34	37.50
0235	35 lb. per C.Y. G				▼	39.50			39.50	43.50

03 24 Fibrous Reinforcing

03 24 05 – Reinforcing Fibers

03 24 05.70 Steel Fibers		Crew	Daily Output	Labor-Hours	Unit	2016 Bare Costs Material	Labor	Equipment	Total	Total Incl O&P
0240	40 lb. per C.Y. G				C.Y.	45			45	49.50
0250	50 lb. per C.Y. G					56.50			56.50	62
0275	75 lb. per C.Y. G					85			85	93
0300	100 lb. per C.Y. G					113			113	124

03 30 Cast-In-Place Concrete

03 30 53 – Miscellaneous Cast-In-Place Concrete

03 30 53.40 Concrete In Place		Crew	Daily Output	Labor-Hours	Unit	2016 Bare Costs Material	Labor	Equipment	Total	Total Incl O&P
0010	**CONCRETE IN PLACE** R033053-10									
0020	Including forms (4 uses), Grade 60 rebar, concrete (Portland cement R033105-10									
0050	Type I), placement and finishing unless otherwise indicated R033105-20									
0500	Chimney foundations (5000 psi), over 5 C.Y. R033105-50	C-14C	32.22	3.476	C.Y.	155	124	.97	279.97	380
0510	(3500 psi), under 5 C.Y. R033105-70	"	23.71	4.724	"	182	168	1.32	351.32	480
3540	Equipment pad (3000 psi), 3′ x 3′ x 6″ thick	C-14H	45	1.067	Ea.	47.50	39	.70	87.20	118
3550	4′ x 4′ x 6″ thick		30	1.600		70.50	58.50	1.06	130.06	177
3560	5′ x 5′ x 8″ thick		18	2.667		124	98	1.76	223.76	300
3570	6′ x 6′ x 8″ thick		14	3.429		167	126	2.26	295.26	395
3580	8′ x 8′ x 10″ thick		8	6		360	220	3.96	583.96	765
3590	10′ x 10′ x 12″ thick		5	9.600		615	350	6.35	971.35	1,275
3800	Footings (3000 psi), spread under 1 C.Y.	C-14C	28	4	C.Y.	171	142	1.12	314.12	425
3825	1 C.Y. to 5 C.Y.		43	2.605		207	93	.73	300.73	385
3850	Over 5 C.Y. R033105-80		75	1.493		190	53	.42	243.42	298
3900	Footings, strip (3000 psi), 18″ x 9″, unreinforced	C-14L	40	2.400		131	84	.79	215.79	285
3920	18″ x 9″, reinforced R033105-85	C-14C	35	3.200		154	114	.90	268.90	360
3925	20″ x 10″, unreinforced	C-14L	45	2.133		127	74.50	.70	202.20	265
3930	20″ x 10″, reinforced	C-14C	40	2.800		146	99.50	.78	246.28	330
3935	24″ x 12″, unreinforced	C-14L	55	1.745		125	61	.58	186.58	241
3940	24″ x 12″, reinforced	C-14C	48	2.333		144	83	.65	227.65	299
3945	36″ x 12″, unreinforced	C-14L	70	1.371		121	48	.45	169.45	213
3950	36″ x 12″, reinforced	C-14C	60	1.867		138	66.50	.52	205.02	264
4000	Foundation mat (3000 psi), under 10 C.Y.		38.67	2.896		209	103	.81	312.81	405
4050	Over 20 C.Y.		56.40	1.986		184	70.50	.56	255.06	320
4520	Handicap access ramp (4000 psi), railing both sides, 3′ wide	C-14H	14.58	3.292	L.F.	315	121	2.17	438.17	555
4525	5′ wide		12.22	3.928		325	144	2.59	471.59	605
4530	With 6″ curb and rails both sides, 3′ wide		8.55	5.614		325	206	3.71	534.71	710
4535	5′ wide		7.31	6.566		330	241	4.33	575.33	770
4751	Slab on grade (3500 psi), incl. troweled finish, not incl. forms									
4760	or reinforcing, over 10,000 S.F., 4″ thick	C-14F	3425	.021	S.F.	1.43	.71	.01	2.15	2.74
4820	6″ thick	"	3350	.021	"	2.08	.72	.01	2.81	3.49
5000	Slab on grade (3000 psi), incl. broom finish, not incl. forms									
5001	or reinforcing, 4″ thick	C-14G	2873	.019	S.F.	1.39	.65	.01	2.05	2.60
5010	6″ thick		2590	.022		2.18	.72	.01	2.91	3.59
5020	8″ thick		2320	.024		2.84	.80	.01	3.65	4.45
6800	Stairs (3500 psi), not including safety treads, free standing, 3′-6″ wide	C-14H	83	.578	LF Nose	5.70	21	.38	27.08	42
6850	Cast on ground		125	.384	"	4.73	14.10	.25	19.08	29
7000	Stair landings, free standing		200	.240	S.F.	4.61	8.80	.16	13.57	19.90
7050	Cast on ground		475	.101	"	3.61	3.70	.07	7.38	10.25

03 31 Structural Concrete

03 31 13 – Heavyweight Structural Concrete

03 31 13.25 Concrete, Hand Mix		Crew	Daily Output	Labor-Hours	Unit	Material	2016 Bare Costs Labor	Equipment	Total	Total Incl O&P
0010	**CONCRETE, HAND MIX** for small quantities or remote areas									
0050	Includes bulk local aggregate, bulk sand, bagged Portland									
0060	cement (Type I) and water, using gas powered cement mixer									
0125	2500 psi	C-30	135	.059	C.F.	3.50	1.73	1.28	6.51	8.15
0130	3000 psi		135	.059		3.75	1.73	1.28	6.76	8.45
0135	3500 psi		135	.059		3.89	1.73	1.28	6.90	8.60
0140	4000 psi		135	.059		4.06	1.73	1.28	7.07	8.80
0145	4500 psi		135	.059		4.26	1.73	1.28	7.27	9
0150	5000 psi		135	.059		4.53	1.73	1.28	7.54	9.30
0300	Using pre-bagged dry mix and wheelbarrow (80-lb. bag = 0.6 C.F.)									
0340	4000 psi	1 Clab	48	.167	C.F.	6.50	4.87		11.37	15.30

03 31 13.30 Concrete, Volumetric Site-Mixed		Crew	Daily Output	Labor-Hours	Unit	Material	Labor	Equipment	Total	Total Incl O&P
0010	**CONCRETE, VOLUMETRIC SITE-MIXED**									
0015	Mixed on-site in volumetric truck									
0020	Includes local aggregate, sand, Portland cement (Type I) and water									
0025	Excludes all additives and treatments									
0100	3000 psi, 1 C.Y. mixed and discharged				C.Y.	193			193	213
0110	2 C.Y.					147			147	162
0120	3 C.Y.					131			131	144
0130	4 C.Y.					118			118	130
0140	5 C.Y.					110			110	121
0200	For truck holding/waiting time past first 2 on-site hours, add				Hr.	88			88	97
0210	For trip charge beyond first 20 miles, each way, add				Mile	3.50			3.50	3.85
0220	For each additional increase of 500 psi, add				Ea.	4.17			4.17	4.59

03 31 13.35 Heavyweight Concrete, Ready Mix		Crew	Daily Output	Labor-Hours	Unit	Material	Labor	Equipment	Total	Total Incl O&P
0010	**HEAVYWEIGHT CONCRETE, READY MIX**, delivered R033105-10									
0012	Includes local aggregate, sand, Portland cement (Type I) and water									
0015	Excludes all additives and treatments R033105-20									
0020	2000 psi				C.Y.	102			102	113
0100	2500 psi R033105-30					105			105	115
0150	3000 psi					107			107	118
0200	3500 psi R033105-40					110			110	121
0300	4000 psi					113			113	124
0350	4500 psi R033105-50					116			116	127
0400	5000 psi					119			119	131
0411	6000 psi					122			122	134
0412	8000 psi					129			129	141
0413	10,000 psi					135			135	149
0414	12,000 psi					142			142	156
1000	For high early strength (Portland cement Type III), add					10%				
1300	For winter concrete (hot water), add					4.45			4.45	4.90
1410	For mid-range water reducer, add					3.33			3.33	3.66
1420	For high-range water reducer/superplasticizer, add					6			6	6.60
1430	For retarder, add					3.23			3.23	3.55
1440	For non-Chloride accelerator, add					5.70			5.70	6.25
1450	For Chloride accelerator, per 1%, add					3			3	3.30
1460	For fiber reinforcing, synthetic (1 lb./C.Y.), add					6.70			6.70	7.35
1500	For Saturday delivery, add					7.50			7.50	8.25
1510	For truck holding/waiting time past 1st hour per load, add				Hr.	96.50			96.50	106
1520	For short load (less than 4 C.Y.), add per load				Ea.	74.50			74.50	82
2000	For all lightweight aggregate, add				C.Y.	45%				

03 31 13 – Heavyweight Structural Concrete

03 31 13.70 Placing Concrete		Crew	Daily Output	Labor-Hours	Unit	2016 Bare Costs Material	Labor	Equipment	Total	Total Incl O&P
0010	**PLACING CONCRETE**									
0020	Includes labor and equipment to place, level (strike off) and consolidate									
1400	Elevated slabs, less than 6" thick, pumped	C-20	140	.457	C.Y.		14.40	5.65	20.05	30
1450	With crane and bucket (R033105-70)	C-7	95	.758			24	12.75	36.75	54.50
1500	6" to 10" thick, pumped	C-20	160	.400			12.60	4.94	17.54	26.50
1550	With crane and bucket	C-7	110	.655			21	11	32	47
1600	Slabs over 10" thick, pumped	C-20	180	.356			11.20	4.39	15.59	23.50
1650	With crane and bucket	C-7	130	.554			17.70	9.30	27	40
1900	Footings, continuous, shallow, direct chute	C-6	120	.400			12.25	.53	12.78	21
1950	Pumped	C-20	150	.427			13.45	5.25	18.70	28.50
2000	With crane and bucket	C-7	90	.800			25.50	13.45	38.95	57.50
2400	Footings, spread, under 1 C.Y., direct chute	C-6	55	.873			26.50	1.15	27.65	46
2600	Over 5 C.Y., direct chute		120	.400			12.25	.53	12.78	21
2900	Foundation mats, over 20 C.Y., direct chute		350	.137			4.20	.18	4.38	7.20
4300	Slab on grade, up to 6" thick, direct chute		110	.436			13.35	.58	13.93	22.50
4350	Pumped	C-20	130	.492			15.50	6.10	21.60	32.50
4400	With crane and bucket	C-7	110	.655			21	11	32	47
4900	Walls, 8" thick, direct chute	C-6	90	.533			16.30	.70	17	28
4950	Pumped	C-20	100	.640			20	7.90	27.90	42
5000	With crane and bucket	C-7	80	.900			29	15.10	44.10	64.50
5050	12" thick, direct chute	C-6	100	.480			14.70	.63	15.33	25
5100	Pumped	C-20	110	.582			18.35	7.20	25.55	38.50
5200	With crane and bucket	C-7	90	.800			25.50	13.45	38.95	57.50
5600	Wheeled concrete dumping, add to placing costs above									
5610	Walking cart, 50' haul, add	C-18	32	.281	C.Y.		8.25	1.88	10.13	15.90
5620	150' haul, add		24	.375			11.05	2.50	13.55	21.50
5700	250' haul, add		18	.500			14.70	3.34	18.04	28
5800	Riding cart, 50' haul, add	C-19	80	.113			3.31	1.25	4.56	6.90
5810	150' haul, add		60	.150			4.41	1.66	6.07	9.25
5900	250' haul, add		45	.200			5.90	2.22	8.12	12.30
6000	Concrete in-fill for pan-type metal stairs and landings. Manual placement									
6010	includes up to 50' horizontal haul from point of concrete discharge.									
6100	Stair pan treads, 2" deep									
6110	Flights in 1st floor level up/down from discharge point	C-8A	3200	.015	S.F.		.48		.48	.79
6120	2nd floor level		2500	.019			.61		.61	1.01
6130	3rd floor level		2000	.024			.76		.76	1.26
6140	4th floor level		1800	.027			.84		.84	1.40
6200	Intermediate stair landings, pan-type 4" deep									
6210	Flights in 1st floor level up/down from discharge point	C-8A	2000	.024	S.F.		.76		.76	1.26
6220	2nd floor level		1500	.032			1.01		1.01	1.68
6230	3rd floor level		1200	.040			1.27		1.27	2.10
6240	4th floor level		1000	.048			1.52		1.52	2.52

03 35 Concrete Finishing

03 35 13 – High-Tolerance Concrete Floor Finishing

03 35 13.30 Finishing Floors, High Tolerance		Crew	Daily Output	Labor-Hours	Unit	Material (2016 Bare Costs)	Labor (2016 Bare Costs)	Equipment (2016 Bare Costs)	Total (2016 Bare Costs)	Total Incl O&P
0010	**FINISHING FLOORS, HIGH TOLERANCE**									
0012	Finishing of fresh concrete flatwork requires that concrete									
0013	first be placed, struck off & consolidated									
0015	Basic finishing for various unspecified flatwork									
0100	Bull float only	C-10	4000	.006	S.F.		.20		.20	.33
0125	Bull float & manual float		2000	.012			.40		.40	.66
0150	Bull float, manual float, & broom finish, w/edging & joints		1850	.013			.43		.43	.71
0200	Bull float, manual float & manual steel trowel	↓	1265	.019	↓		.63		.63	1.04
0210	For specified Random Access Floors in ACI Classes 1, 2, 3 and 4 to achieve									
0215	Composite Overall Floor Flatness and Levelness values up to FF35/FL25									
0250	Bull float, machine float & machine trowel (walk-behind)	C-10C	1715	.014	S.F.		.47	.03	.50	.80
0300	Power screed, bull float, machine float & trowel (walk-behind)	C-10D	2400	.010			.33	.05	.38	.60
0350	Power screed, bull float, machine float & trowel (ride-on)	C-10E	4000	.006	↓		.20	.06	.26	.40
0352	For specified Random Access Floors in ACI Classes 5, 6, 7 and 8 to achieve									
0354	Composite Overall Floor Flatness and Levelness values up to FF50/FL50									
0356	Add for two-dimensional restraightening after power float	C-10	6000	.004	S.F.		.13		.13	.22
0358	For specified Random or Defined Access Floors in ACI Class 9 to achieve									
0360	Composite Overall Floor Flatness and Levelness values up to FF100/FL100									
0362	Add for two-dimensional restraightening after bull float & power float	C-10	3000	.008	S.F.		.27		.27	.44
0364	For specified Superflat Defined Access Floors in ACI Class 9 to achieve									
0366	Minimum Floor Flatness and Levelness values of FF100/FL100									
0368	Add for 2-dim'l restraightening after bull float, power float, power trowel	C-10	2000	.012	S.F.		.40		.40	.66

03 35 23 – Exposed Aggregate Concrete Finishing

03 35 23.30 Finishing Floors, Exposed Aggregate		Crew	Daily Output	Labor-Hours	Unit	Material	Labor	Equipment	Total	Total Incl O&P
0010	**FINISHING FLOORS, EXPOSED AGGREGATE**									
1600	Exposed local aggregate finish, seeded on fresh concrete, 3 lb. per S.F.	1 Cefi	625	.013	S.F.	.20	.46		.66	.96
1650	4 lb. per S.F.	"	465	.017	"	.33	.61		.94	1.36

03 35 29 – Tooled Concrete Finishing

03 35 29.30 Finishing Floors, Tooled		Crew	Daily Output	Labor-Hours	Unit	Material	Labor	Equipment	Total	Total Incl O&P
0010	**FINISHING FLOORS, TOOLED**									
4400	Stair finish, fresh concrete, float finish	1 Cefi	275	.029	S.F.		1.04		1.04	1.68
4500	Steel trowel finish	"	200	.040	"		1.42		1.42	2.31

03 35 29.60 Finishing Walls		Crew	Daily Output	Labor-Hours	Unit	Material	Labor	Equipment	Total	Total Incl O&P
0010	**FINISHING WALLS**									
0020	Break ties and patch voids	1 Cefi	540	.015	S.F.	.04	.53		.57	.91
0050	Burlap rub with grout	"	450	.018		.04	.63		.67	1.08
0300	Bush hammer, green concrete	B-39	1000	.048			1.42	.23	1.65	2.62
0350	Cured concrete	"	650	.074			2.18	.35	2.53	4.04
0500	Acid etch	1 Cefi	575	.014	↓	.14	.50		.64	.96
0850	Grind form fins flush	1 Clab	700	.011	L.F.		.33		.33	.56

03 35 33 – Stamped Concrete Finishing

03 35 33.50 Slab Texture Stamping		Crew	Daily Output	Labor-Hours	Unit	Material	Labor	Equipment	Total	Total Incl O&P
0010	**SLAB TEXTURE STAMPING**									
0050	Stamping requires that concrete first be placed, struck off, consolidated,									
0060	bull floated and free of bleed water. Decorative stamping tasks include:									
0100	Step 1 - first application of dry shake colored hardener	1 Cefi	6400	.001	S.F.	.44	.04		.48	.56
0110	Step 2 - bull float		6400	.001			.04		.04	.07
0130	Step 3 - second application of dry shake colored hardener	↓	6400	.001		.22	.04		.26	.31
0140	Step 4 - bull float, manual float & steel trowel	3 Cefi	1280	.019			.67		.67	1.08
0150	Step 5 - application of dry shake colored release agent	1 Cefi	6400	.001		.09	.04		.13	.17
0160	Step 6 - place, tamp & remove mats	3 Cefi	2400	.010	↓	.88	.36		1.24	1.55

03 35 Concrete Finishing

03 35 33 – Stamped Concrete Finishing

03 35 33.50 Slab Texture Stamping		Crew	Daily Output	Labor-Hours	Unit	Material	2016 Bare Costs Labor	Equipment	Total	Total Incl O&P
0170	Step 7 - touch up edges, mat joints & simulated grout lines	1 Cefi	1280	.006	S.F.		.22		.22	.36
0300	Alternate stamping estimating method includes all tasks above	4 Cefi	800	.040		1.63	1.42		3.05	4.11
0400	Step 8 - pressure wash @ 3000 psi after 24 hours	1 Cefi	1600	.005			.18		.18	.29
0500	Step 9 - roll 2 coats cure/seal compound when dry	″	800	.010		.63	.36		.99	1.27

03 35 43 – Polished Concrete Finishing

03 35 43.10 Polished Concrete Floors

		Crew	Daily Output	Labor-Hours	Unit	Material	Labor	Equipment	Total	Total Incl O&P
0010	**POLISHED CONCRETE FLOORS** R033543-10									
0015	Processing of cured concrete to include grinding, honing,									
0020	and polishing of interior floors with 22″ segmented diamond									
0025	planetary floor grinder (2 passes in different directions per grit)									
0100	Removal of pre-existing coatings, dry, with carbide discs using									
0105	dry vacuum pick-up system, final hand sweeping									
0110	Glue, adhesive or tar	J-4	1.60	15	M.S.F.	20.50	500	139	659.50	995
0120	Paint, epoxy, 1 coat		3.60	6.667		20.50	223	61.50	305	455
0130	2 coats		1.80	13.333		20.50	445	123	588.50	890
0200	Grinding and edging, wet, including wet vac pick-up and auto									
0205	scrubbing between grit changes									
0210	40-grit diamond/metal matrix	J-4A	1.60	20	M.S.F.	32.50	650	289	971.50	1,425
0220	80-grit diamond/metal matrix		2	16		32.50	520	231	783.50	1,150
0230	120-grit diamond/metal matrix		2.40	13.333		32.50	430	193	655.50	960
0240	200-grit diamond/metal matrix		2.80	11.429		32.50	370	165	567.50	830
0300	Spray on dye or stain (1 coat)	1 Cefi	16	.500		220	17.80		237.80	271
0400	Spray on densifier/hardener (2 coats)	″	8	1		350	35.50		385.50	445
0410	Auto scrubbing after 2nd coat, when dry	J-4B	16	.500			14.60	15.05	29.65	41
0500	Honing and edging, wet, including wet vac pick-up and auto									
0505	scrubbing between grit changes									
0510	100-grit diamond/resin matrix	J-4A	2.80	11.429	M.S.F.	32.50	370	165	567.50	830
0520	200-grit diamond/resin matrix	″	2.80	11.429	″	32.50	370	165	567.50	830
0530	Dry, including dry vacuum pick-up system, final hand sweeping									
0540	400-grit diamond/resin matrix	J-4A	2.80	11.429	M.S.F.	32.50	370	165	567.50	830
0600	Polishing and edging, dry, including dry vac pick-up and hand									
0605	sweeping between grit changes									
0610	800-grit diamond/resin matrix	J-4A	2.80	11.429	M.S.F.	32.50	370	165	567.50	830
0620	1500-grit diamond/resin matrix		2.80	11.429		32.50	370	165	567.50	830
0630	3000-grit diamond/resin matrix		2.80	11.429		32.50	370	165	567.50	830
0700	Auto scrubbing after final polishing step	J-4B	16	.500			14.60	15.05	29.65	41

03 39 Concrete Curing

03 39 13 – Water Concrete Curing

03 39 13.50 Water Curing

		Crew	Daily Output	Labor-Hours	Unit	Material	Labor	Equipment	Total	Total Incl O&P
0010	**WATER CURING**									
0015	With burlap, 4 uses assumed, 7.5 oz.	2 Clab	55	.291	C.S.F.	13.95	8.50		22.45	29.50
0100	10 oz.	″	55	.291	″	25	8.50		33.50	42
0400	Curing blankets, 1″ to 2″ thick, buy				S.F.	.26			.26	.29

03 39 23 – Membrane Concrete Curing

03 39 23.13 Chemical Compound Membrane Concrete Curing

		Crew	Daily Output	Labor-Hours	Unit	Material	Labor	Equipment	Total	Total Incl O&P
0010	**CHEMICAL COMPOUND MEMBRANE CONCRETE CURING**									
0300	Sprayed membrane curing compound	2 Clab	95	.168	C.S.F.	12.70	4.92		17.62	22.50

03 39 Concrete Curing

03 39 23 – Membrane Concrete Curing

03 39 23.23 Sheet Membrane Concrete Curing		Crew	Daily Output	Labor-Hours	Unit	Material	Labor	Equipment	Total	Total Incl O&P
						2016 Bare Costs				
0010	**SHEET MEMBRANE CONCRETE CURING**									
0200	Curing blanket, burlap/poly, 2-ply	2 Clab	70	.229	C.S.F.	29	6.65		35.65	42.50

03 41 Precast Structural Concrete

03 41 13 – Precast Concrete Hollow Core Planks

03 41 13.50 Precast Slab Planks

Line	Description	Crew	Daily Output	Labor-Hours	Unit	Material	Labor	Equipment	Total	Total Incl O&P
0010	**PRECAST SLAB PLANKS** R034105-30									
0020	Prestressed roof/floor members, grouted, solid, 4" thick	C-11	2400	.023	S.F.	6.15	.91	.78	7.84	9.15
0050	6" thick		2800	.020		6.90	.78	.67	8.35	9.65
0100	Hollow, 8" thick		3200	.018		7.70	.68	.58	8.96	10.25
0150	10" thick		3600	.016		8	.60	.52	9.12	10.40
0200	12" thick		4000	.014		8.30	.54	.47	9.31	10.50

03 41 23 – Precast Concrete Stairs

03 41 23.50 Precast Stairs

Line	Description	Crew	Daily Output	Labor-Hours	Unit	Material	Labor	Equipment	Total	Total Incl O&P
0010	**PRECAST STAIRS**									
0020	Precast concrete treads on steel stringers, 3' wide	C-12	75	.640	Riser	135	24	8.80	167.80	197
0300	Front entrance, 5' wide with 48" platform, 2 risers		16	3	Flight	485	111	41	637	760
0350	5 risers		12	4		765	148	55	968	1,150
0500	6' wide, 2 risers		15	3.200		535	119	44	698	835
0550	5 risers		11	4.364		845	162	60	1,067	1,250
0700	7' wide, 2 risers		14	3.429		685	127	47	859	1,025
1200	Basement entrance stairwell, 6 steps, incl. steel bulkhead door	B-51	22	2.182		1,450	66	11.30	1,527.30	1,700
1250	14 steps	"	11	4.364		2,400	132	22.50	2,554.50	2,875

03 48 Precast Concrete Specialties

03 48 43 – Precast Concrete Trim

03 48 43.40 Precast Lintels

Line	Description	Crew	Daily Output	Labor-Hours	Unit	Material	Labor	Equipment	Total	Total Incl O&P
0010	**PRECAST LINTELS**, smooth gray, prestressed, stock units only									
0800	4" wide, 8" high, x 4' long	D-10	28	1.143	Ea.	25	41	17.10	83.10	115
0850	8' long		24	1.333		63	48	19.95	130.95	172
1000	6" wide, 8" high, x 4' long		26	1.231		37	44.50	18.45	99.95	135
1050	10' long		22	1.455		92.50	52.50	22	167	213

03 48 43.90 Precast Window Sills

Line	Description	Crew	Daily Output	Labor-Hours	Unit	Material	Labor	Equipment	Total	Total Incl O&P
0010	**PRECAST WINDOW SILLS**									
0600	Precast concrete, 4" tapers to 3", 9" wide	D-1	70	.229	L.F.	14.65	7.50		22.15	28.50
0650	11" wide	"	60	.267	"	24	8.75		32.75	40.50

03 54 Cast Underlayment

03 54 13 – Gypsum Cement Underlayment

03 54 13.50 Poured Gypsum Underlayment		Crew	Daily Output	Labor-Hours	Unit	Material	2016 Bare Costs Labor	Equipment	Total	Total Incl O&P
0010	**POURED GYPSUM UNDERLAYMENT**									
0400	Underlayment, gypsum based, self-leveling 2500 psi, pumped, 1/2" thick	C-8	24000	.002	S.F.	.36	.08	.03	.47	.56
0500	3/4" thick		20000	.003		.54	.09	.04	.67	.79
0600	1" thick		16000	.004		.72	.11	.05	.88	1.04
1400	Hand placed, 1/2" thick	C-18	450	.020		.36	.59	.13	1.08	1.54
1500	3/4" thick	"	300	.030		.54	.88	.20	1.62	2.30

03 63 Epoxy Grouting

03 63 05 – Grouting of Dowels and Fasteners

03 63 05.10 Epoxy Only		Crew	Daily Output	Labor-Hours	Unit	Material	Labor	Equipment	Total	Total Incl O&P
0010	**EPOXY ONLY**									
1500	Chemical anchoring, epoxy cartridge, excludes layout, drilling, fastener									
1530	For fastener 3/4" diam. x 6" embedment	2 Skwk	72	.222	Ea.	4.84	8.50		13.34	19.60
1535	1" diam. x 8" embedment		66	.242		7.25	9.30		16.55	23.50
1540	1-1/4" diam. x 10" embedment		60	.267		14.50	10.20		24.70	33
1545	1-3/4" diam. x 12" embedment		54	.296		24	11.35		35.35	45.50
1550	14" embedment		48	.333		29	12.75		41.75	53.50
1555	2" diam. x 12" embedment		42	.381		38.50	14.60		53.10	67
1560	18" embedment		32	.500		48.50	19.15		67.65	85

03 82 Concrete Boring

03 82 13 – Concrete Core Drilling

03 82 13.10 Core Drilling		Crew	Daily Output	Labor-Hours	Unit	Material	Labor	Equipment	Total	Total Incl O&P
0010	**CORE DRILLING**									
0015	Includes bit cost, layout and set-up time									
0020	Reinforced concrete slab, up to 6" thick									
0100	1" diameter core	B-89A	17	.941	Ea.	.18	32	6.90	39.08	61.50
0150	For each additional inch of slab thickness in same hole, add		1440	.011		.03	.38	.08	.49	.75
0200	2" diameter core		16.50	.970		.28	32.50	7.10	39.88	63
0250	For each additional inch of slab thickness in same hole, add		1080	.015		.05	.50	.11	.66	1.01
0300	3" diameter core		16	1		.38	34	7.30	41.68	65
0350	For each additional inch of slab thickness in same hole, add		720	.022		.06	.75	.16	.97	1.51
0500	4" diameter core		15	1.067		.50	36	7.80	44.30	69.50
0550	For each additional inch of slab thickness in same hole, add		480	.033		.08	1.12	.24	1.44	2.25
0700	6" diameter core		14	1.143		.76	38.50	8.35	47.61	74.50
0750	For each additional inch of slab thickness in same hole, add		360	.044		.13	1.50	.32	1.95	3.02
0900	8" diameter core		13	1.231		1.04	41.50	9	51.54	80.50
0950	For each additional inch of slab thickness in same hole, add		288	.056		.17	1.88	.41	2.46	3.79
1760	For horizontal holes, add to above						20%	20%		
1770	Prestressed hollow core plank, 8" thick									
1780	1" diameter core	B-89A	17.50	.914	Ea.	.24	31	6.70	37.94	59.50
1790	For each additional inch of plank thickness in same hole, add		3840	.004		.03	.14	.03	.20	.30
1794	2" diameter core		17.25	.928		.38	31.50	6.80	38.68	60.50
1796	For each additional inch of plank thickness in same hole, add		2880	.006		.05	.19	.04	.28	.40
1800	3" diameter core		17	.941		.51	32	6.90	39.41	61.50
1810	For each additional inch of plank thickness in same hole, add		1920	.008		.06	.28	.06	.40	.61
1820	4" diameter core		16.50	.970		.66	32.50	7.10	40.26	63.50
1830	For each additional inch of plank thickness in same hole, add		1280	.013		.08	.42	.09	.59	.90
1840	6" diameter core		15.50	1.032		1.02	35	7.55	43.57	68

03 82 Concrete Boring

03 82 13 – Concrete Core Drilling

03 82 13.10 Core Drilling		Crew	Daily Output	Labor-Hours	Unit	Material	Labor	Equipment	Total	Total Incl O&P
1850	For each additional inch of plank thickness in same hole, add	B-89A	960	.017	Ea.	.13	.56	.12	.81	1.21
1860	8" diameter core		15	1.067		1.39	36	7.80	45.19	70.50
1870	For each additional inch of plank thickness in same hole, add	↓	768	.021	↓	.17	.70	.15	1.02	1.54

(Material, Labor, Equipment and Total are 2016 Bare Costs.)

03 82 16 – Concrete Drilling

03 82 16.10 Concrete Impact Drilling

		Crew	Daily Output	Labor-Hours	Unit	Material	Labor	Equipment	Total	Total Incl O&P
0010	**CONCRETE IMPACT DRILLING**									
0020	Includes bit cost, layout and set-up time, no anchors									
0050	Up to 4" deep in concrete/brick floors/walls									
0100	Holes, 1/4" diameter	1 Carp	75	.107	Ea.	.07	4.03		4.10	6.85
0150	For each additional inch of depth in same hole, add		430	.019		.02	.70		.72	1.20
0200	3/8" diameter		63	.127		.06	4.80		4.86	8.10
0250	For each additional inch of depth in same hole, add		340	.024		.01	.89		.90	1.51
0300	1/2" diameter		50	.160		.06	6.05		6.11	10.20
0350	For each additional inch of depth in same hole, add		250	.032		.02	1.21		1.23	2.05
0400	5/8" diameter		48	.167		.10	6.30		6.40	10.65
0450	For each additional inch of depth in same hole, add		240	.033		.03	1.26		1.29	2.14
0500	3/4" diameter		45	.178		.13	6.70		6.83	11.40
0550	For each additional inch of depth in same hole, add		220	.036		.03	1.37		1.40	2.34
0600	7/8" diameter		43	.186		.18	7.05		7.23	12
0650	For each additional inch of depth in same hole, add		210	.038		.04	1.44		1.48	2.46
0700	1" diameter		40	.200		.16	7.55		7.71	12.80
0750	For each additional inch of depth in same hole, add		190	.042		.04	1.59		1.63	2.71
0800	1-1/4" diameter		38	.211		.27	7.95		8.22	13.65
0850	For each additional inch of depth in same hole, add		180	.044		.07	1.68		1.75	2.88
0900	1-1/2" diameter		35	.229		.41	8.65		9.06	14.90
0950	For each additional inch of depth in same hole, add	↓	165	.048	↓	.10	1.83		1.93	3.18
1000	For ceiling installations, add						40%			

Division 4 Masonry

Estimating Tips

04 05 00 Common Work Results for Masonry

- The terms mortar and grout are often used interchangeably, and incorrectly. Mortar is used to bed masonry units, seal the entry of air and moisture, provide architectural appearance, and allow for size variations in the units. Grout is used primarily in reinforced masonry construction and is used to bond the masonry to the reinforcing steel. Common mortar types are M(2500 psi), S(1800 psi), N(750 psi), and O(350 psi), and conform to ASTM C270. Grout is either fine or coarse and conforms to ASTM C476, and in-place strengths generally exceed 2500 psi. Mortar and grout are different components of masonry construction and are placed by entirely different methods. An estimator should be aware of their unique uses and costs.
- Mortar is included in all assembled masonry line items. The mortar cost, part of the assembled masonry material cost, includes all ingredients, all labor, and all equipment required. Please see reference number R040513-10.
- Waste, specifically the loss/droppings of mortar and the breakage of brick and block, is included in all unit cost lines that include mortar and masonry units in this division. A factor of 25% is added for mortar and 3% for brick and concrete masonry units.
- Scaffolding or staging is not included in any of the Division 4 costs. Refer to Subdivision 01 54 23 for scaffolding and staging costs.

04 20 00 Unit Masonry

- The most common types of unit masonry are brick and concrete masonry. The major classifications of brick are building brick (ASTM C62), facing brick (ASTM C216), glazed brick, fire brick, and pavers. Many varieties of texture and appearance can exist within these classifications, and the estimator would be wise to check local custom and availability within the project area. For repair and remodeling jobs, matching the existing brick may be the most important criteria.
- Brick and concrete block are priced by the piece and then converted into a price per square foot of wall. Openings less than two square feet are generally ignored by the estimator because any savings in units used is offset by the cutting and trimming required.
- It is often difficult and expensive to find and purchase small lots of historic brick. Costs can vary widely. Many design issues affect costs, selection of mortar mix, and repairs or replacement of masonry materials. Cleaning techniques must be reflected in the estimate.
- All masonry walls, whether interior or exterior, require bracing. The cost of bracing walls during construction should be included by the estimator, and this bracing must remain in place until permanent bracing is complete. Permanent bracing of masonry walls is accomplished by masonry itself, in the form of pilasters or abutting wall corners, or by anchoring the walls to the structural frame. Accessories in the form of anchors, anchor slots, and ties are used, but their supply and installation can be by different trades. For instance, anchor slots on spandrel beams and columns are supplied and welded in place by the steel fabricator, but the ties from the slots into the masonry are installed by the bricklayer. Regardless of the installation method, the estimator must be certain that these accessories are accounted for in pricing.

Reference Numbers

Reference numbers are shown at the beginning of some major classifications. These numbers refer to related items in the Reference Section. The reference information may be an estimating procedure, an alternate pricing method, or technical information.

Note: Not all subdivisions listed here necessarily appear. ■

04 01 Maintenance of Masonry

04 01 20 – Maintenance of Unit Masonry

04 01 20.20 Pointing Masonry

	04 01 20.20 Pointing Masonry	Crew	Daily Output	Labor-Hours	Unit	Material	Labor	Equipment	Total	Total Incl O&P
						2016 Bare Costs				
0010	**POINTING MASONRY**									
0300	Cut and repoint brick, hard mortar, running bond	1 Bric	80	.100	S.F.	.58	3.61		4.19	6.65
0320	Common bond		77	.104		.58	3.75		4.33	6.90
0360	Flemish bond		70	.114		.61	4.13		4.74	7.55
0400	English bond		65	.123		.61	4.44		5.05	8.05
0600	Soft old mortar, running bond		100	.080		.58	2.89		3.47	5.45
0620	Common bond		96	.083		.58	3.01		3.59	5.65
0640	Flemish bond		90	.089		.61	3.21		3.82	6
0680	English bond		82	.098		.61	3.52		4.13	6.50
0700	Stonework, hard mortar		140	.057	L.F.	.77	2.06		2.83	4.29
0720	Soft old mortar		160	.050	"	.77	1.81		2.58	3.86
1000	Repoint, mask and grout method, running bond		95	.084	S.F.	.77	3.04		3.81	5.90
1020	Common bond		90	.089		.77	3.21		3.98	6.20
1040	Flemish bond		86	.093		.81	3.36		4.17	6.50
1060	English bond		77	.104		.81	3.75		4.56	7.15
2000	Scrub coat, sand grout on walls, thin mix, brushed		120	.067		3.22	2.41		5.63	7.55
2020	Troweled		98	.082		4.49	2.95		7.44	9.85

04 01 20.30 Pointing CMU

	04 01 20.30 Pointing CMU	Crew	Daily Output	Labor-Hours	Unit	Material	Labor	Equipment	Total	Total Incl O&P
0010	**POINTING CMU**									
0300	Cut and repoint block, hard mortar, running bond	1 Bric	190	.042	S.F.	.24	1.52		1.76	2.80
0310	Stacked bond		200	.040		.24	1.44		1.68	2.67
0600	Soft old mortar, running bond		230	.035		.24	1.26		1.50	2.35
0610	Stacked bond		245	.033		.24	1.18		1.42	2.23

04 01 30 – Unit Masonry Cleaning

04 01 30.60 Brick Washing

	04 01 30.60 Brick Washing	Crew	Daily Output	Labor-Hours	Unit	Material	Labor	Equipment	Total	Total Incl O&P
0010	**BRICK WASHING** R040130-10									
0012	Acid cleanser, smooth brick surface	1 Bric	560	.014	S.F.	.05	.52		.57	.91
0050	Rough brick		400	.020		.06	.72		.78	1.27
0060	Stone, acid wash		600	.013		.08	.48		.56	.88
1000	Muriatic acid, price per gallon in 5 gallon lots				Gal.	9.45			9.45	10.35

04 05 Common Work Results for Masonry

04 05 05 – Selective Demolition for Masonry

04 05 05.10 Selective Demolition

	04 05 05.10 Selective Demolition	Crew	Daily Output	Labor-Hours	Unit	Material	Labor	Equipment	Total	Total Incl O&P
0010	**SELECTIVE DEMOLITION** R024119-10									
0200	Bond beams, 8" block with #4 bar	2 Clab	32	.500	L.F.		14.60		14.60	24.50
0300	Concrete block walls, unreinforced, 2" thick		1200	.013	S.F.		.39		.39	.65
0310	4" thick		1150	.014			.41		.41	.68
0320	6" thick		1100	.015			.42		.42	.71
0330	8" thick		1050	.015			.45		.45	.75
0340	10" thick		1000	.016			.47		.47	.78
0360	12" thick		950	.017			.49		.49	.82
0380	Reinforced alternate courses, 2" thick		1130	.014			.41		.41	.69
0390	4" thick		1080	.015			.43		.43	.72
0400	6" thick		1035	.015			.45		.45	.76
0410	8" thick		990	.016			.47		.47	.79
0420	10" thick		940	.017			.50		.50	.83
0430	12" thick		890	.018			.53		.53	.88
0440	Reinforced alternate courses & vertically 48" OC, 4" thick		900	.018			.52		.52	.87
0450	6" thick		850	.019			.55		.55	.92

04 05 05 – Selective Demolition for Masonry

04 05 05.10 Selective Demolition		Crew	Daily Output	Labor-Hours	Unit	Material	2016 Bare Costs Labor	Equipment	Total	Total Incl O&P
0460	8" thick	2 Clab	800	.020	S.F.		.58		.58	.98
0480	10" thick		750	.021			.62		.62	1.04
0490	12" thick		700	.023			.67		.67	1.12
1000	Chimney, 16" x 16", soft old mortar	1 Clab	55	.145	C.F.		4.25		4.25	7.10
1020	Hard mortar		40	.200			5.85		5.85	9.80
1030	16" x 20", soft old mortar		55	.145			4.25		4.25	7.10
1040	Hard mortar		40	.200			5.85		5.85	9.80
1050	16" x 24", soft old mortar		55	.145			4.25		4.25	7.10
1060	Hard mortar		40	.200			5.85		5.85	9.80
1080	20" x 20", soft old mortar		55	.145			4.25		4.25	7.10
1100	Hard mortar		40	.200			5.85		5.85	9.80
1110	20" x 24", soft old mortar		55	.145			4.25		4.25	7.10
1120	Hard mortar		40	.200			5.85		5.85	9.80
1140	20" x 32", soft old mortar		55	.145			4.25		4.25	7.10
1160	Hard mortar		40	.200			5.85		5.85	9.80
1200	48" x 48", soft old mortar		55	.145			4.25		4.25	7.10
1220	Hard mortar		40	.200			5.85		5.85	9.80
1250	Metal, high temp steel jacket, 24" diameter	E-2	130	.369	V.L.F.		14.90	11.70	26.60	40
1260	60" diameter	"	60	.800			32	25.50	57.50	86
1280	Flue lining, up to 12" x 12"	1 Clab	200	.040			1.17		1.17	1.96
1282	Up to 24" x 24"		150	.053			1.56		1.56	2.61
2000	Columns, 8" x 8", soft old mortar		48	.167			4.87		4.87	8.15
2020	Hard mortar		40	.200			5.85		5.85	9.80
2060	16" x 16", soft old mortar		16	.500			14.60		14.60	24.50
2100	Hard mortar		14	.571			16.70		16.70	28
2140	24" x 24", soft old mortar		8	1			29		29	49
2160	Hard mortar		6	1.333			39		39	65
2200	36" x 36", soft old mortar		4	2			58.50		58.50	98
2220	Hard mortar		3	2.667			78		78	130
2230	Alternate pricing method, soft old mortar		30	.267	C.F.		7.80		7.80	13.05
2240	Hard mortar		23	.348	"		10.15		10.15	17
3000	Copings, precast or masonry, to 8" wide									
3020	Soft old mortar	1 Clab	180	.044	L.F.		1.30		1.30	2.17
3040	Hard mortar	"	160	.050	"		1.46		1.46	2.45
3100	To 12" wide									
3120	Soft old mortar	1 Clab	160	.050	L.F.		1.46		1.46	2.45
3140	Hard mortar	"	140	.057	"		1.67		1.67	2.79
4000	Fireplace, brick, 30" x 24" opening									
4020	Soft old mortar	1 Clab	2	4	Ea.		117		117	196
4040	Hard mortar		1.25	6.400			187		187	315
4100	Stone, soft old mortar		1.50	5.333			156		156	261
4120	Hard mortar		1	8			234		234	390
4150	Up to 48" fireplace, 15' chimney and foundation		.28	28.571			835		835	1,400
4400	Premanufactured, up to 48"	2 Clab	14	1.143			33.50		33.50	56
5000	Veneers, brick, soft old mortar	1 Clab	140	.057	S.F.		1.67		1.67	2.79
5020	Hard mortar		125	.064			1.87		1.87	3.13
5050	Glass block, up to 4" thick		500	.016			.47		.47	.78
5100	Granite and marble, 2" thick		180	.044			1.30		1.30	2.17
5120	4" thick		170	.047			1.37		1.37	2.30
5140	Stone, 4" thick		180	.044			1.30		1.30	2.17
5160	8" thick		175	.046			1.33		1.33	2.24
5400	Alternate pricing method, stone, 4" thick		60	.133	C.F.		3.89		3.89	6.50
5420	8" thick		85	.094			2.75		2.75	4.60

04 05 Common Work Results for Masonry

04 05 05 – Selective Demolition for Masonry

04 05 05.10 Selective Demolition		Crew	Daily Output	Labor-Hours	Unit	2016 Bare Costs Material	Labor	Equipment	Total	Total Incl O&P
5450	Solid masonry	1 Clab	130	.062	C.F.		1.80		1.80	3.01
5460	Stone or precast sills, treads, copings		130	.062			1.80		1.80	3.01
5470	Solid stone or precast		110	.073			2.12		2.12	3.56
5500	Remove and reset steel lintel	1 Bric	40	.200	L.F.		7.20		7.20	12.05
5600	Vent box removal	1 Clab	50	.160	S.F.		4.67		4.67	7.80
5700	Remove block pilaster for fence, 6′ high		2.33	3.433	Ea.		100		100	168
5800	Remove 12″ x 12″ step flashing from mortar joints		240	.033	C.F.		.97		.97	1.63

04 05 13 – Masonry Mortaring

04 05 13.10 Cement		Crew	Daily Output	Labor-Hours	Unit	Material	Labor	Equipment	Total	Total Incl O&P
0010	**CEMENT** R040513-10									
0100	Masonry, 70 lb. bag, T.L. lots				Bag	12			12	13.20
0150	L.T.L. lots					12.75			12.75	14
0200	White, 70 lb. bag, T.L. lots					15.70			15.70	17.30
0250	L.T.L. lots					16.55			16.55	18.20

04 05 16 – Masonry Grouting

04 05 16.30 Grouting		Crew	Daily Output	Labor-Hours	Unit	Material	Labor	Equipment	Total	Total Incl O&P
0010	**GROUTING**									
0011	Bond beams & lintels, 8″ deep, 6″ thick, 0.15 C.F. per L.F.	D-4	1480	.027	L.F.	.69	.83	.09	1.61	2.25
0020	8″ thick, 0.2 C.F. per L.F.		1400	.029		1.11	.88	.10	2.09	2.80
0050	10″ thick, 0.25 C.F. per L.F.		1200	.033		1.16	1.03	.11	2.30	3.11
0060	12″ thick, 0.3 C.F. per L.F.		1040	.038		1.39	1.19	.13	2.71	3.65
0200	Concrete block cores, solid, 4″ thk., by hand, 0.067 C.F./S.F. of wall	D-8	1100	.036	S.F.	.31	1.22		1.53	2.37
0210	6″ thick, pumped, 0.175 C.F. per S.F.	D-4	720	.056		.81	1.71	.19	2.71	3.96
0250	8″ thick, pumped, 0.258 C.F. per S.F.		680	.059		1.19	1.81	.20	3.20	4.56
0300	10″ thick, pumped, 0.340 C.F. per S.F.		660	.061		1.57	1.87	.20	3.64	5.05
0350	12″ thick, pumped, 0.422 C.F. per S.F.		640	.063		1.95	1.93	.21	4.09	5.60

04 05 19 – Masonry Anchorage and Reinforcing

04 05 19.05 Anchor Bolts		Crew	Daily Output	Labor-Hours	Unit	Material	Labor	Equipment	Total	Total Incl O&P
0010	**ANCHOR BOLTS**									
0015	Installed in fresh grout in CMU bond beams or filled cores, no templates									
0020	Hooked, with nut and washer, 1/2″ diam., 8″ long	1 Bric	132	.061	Ea.	1.32	2.19		3.51	5.10
0030	12″ long		131	.061		1.47	2.20		3.67	5.30
0040	5/8″ diameter, 8″ long		129	.062		3.09	2.24		5.33	7.15
0050	12″ long		127	.063		3.81	2.27		6.08	8
0060	3/4″ diameter, 8″ long		127	.063		3.81	2.27		6.08	8
0070	12″ long		125	.064		4.76	2.31		7.07	9.10

04 05 19.16 Masonry Anchors		Crew	Daily Output	Labor-Hours	Unit	Material	Labor	Equipment	Total	Total Incl O&P
0010	**MASONRY ANCHORS**									
0020	For brick veneer, galv., corrugated, 7/8″ x 7″, 22 Ga.	1 Bric	10.50	.762	C	15.15	27.50		42.65	62.50
0100	24 Ga.		10.50	.762		9.90	27.50		37.40	57
0150	16 Ga.		10.50	.762		29	27.50		56.50	77.50
0200	Buck anchors, galv., corrugated, 16 ga., 2″ bend, 8″ x 2″		10.50	.762		56.50	27.50		84	108
0250	8″ x 3″		10.50	.762		58.50	27.50		86	111
0660	Cavity wall, Z-type, galvanized, 6″ long, 1/8″ diam.		10.50	.762		25	27.50		52.50	73.50
0670	3/16″ diameter		10.50	.762		30.50	27.50		58	79.50
0680	1/4″ diameter		10.50	.762		39	27.50		66.50	89
0850	8″ long, 3/16″ diameter		10.50	.762		26.50	27.50		54	75
0855	1/4″ diameter		10.50	.762		48	27.50		75.50	99
1000	Rectangular type, galvanized, 1/4″ diameter, 2″ x 6″		10.50	.762		73	27.50		100.50	127
1050	4″ x 6″		10.50	.762		88	27.50		115.50	143
1100	3/16″ diameter, 2″ x 6″		10.50	.762		45.50	27.50		73	96

04 05 Common Work Results for Masonry

04 05 19 – Masonry Anchorage and Reinforcing

04 05 19.16 Masonry Anchors		Crew	Daily Output	Labor-Hours	Unit	Material	2016 Bare Costs Labor	Equipment	Total	Total Incl O&P
1150	4" x 6"	1 Bric	10.50	.762	C	52	27.50		79.50	104
1500	Rigid partition anchors, plain, 8" long, 1" x 1/8"		10.50	.762		243	27.50		270.50	315
1550	1" x 1/4"		10.50	.762		285	27.50		312.50	360
1580	1-1/2" x 1/8"		10.50	.762		267	27.50		294.50	340
1600	1-1/2" x 1/4"		10.50	.762		335	27.50		362.50	410
1650	2" x 1/8"		10.50	.762		315	27.50		342.50	390
1700	2" x 1/4"		10.50	.762		415	27.50		442.50	500

04 05 19.26 Masonry Reinforcing Bars		Crew	Daily Output	Labor-Hours	Unit	Material	Labor	Equipment	Total	Total Incl O&P
0010	**MASONRY REINFORCING BARS** R040519-50									
0015	Steel bars A615, placed horiz., #3 & #4 bars	1 Bric	450	.018	Lb.	.48	.64		1.12	1.60
0050	Placed vertical, #3 & #4 bars		350	.023		.48	.83		1.31	1.91
0060	#5 & #6 bars		650	.012		.48	.44		.92	1.27
0200	Joint reinforcing, regular truss, to 6" wide, mill std galvanized		30	.267	C.L.F.	22.50	9.65		32.15	41
0250	12" wide		20	.400		26.50	14.45		40.95	53
0400	Cavity truss with drip section, to 6" wide		30	.267		23	9.65		32.65	41.50
0450	12" wide		20	.400		26.50	14.45		40.95	53.50

04 05 23 – Masonry Accessories

04 05 23.13 Masonry Control and Expansion Joints		Crew	Daily Output	Labor-Hours	Unit	Material	Labor	Equipment	Total	Total Incl O&P
0010	**MASONRY CONTROL AND EXPANSION JOINTS**									
0020	Rubber, for double wythe 8" minimum wall (Brick/CMU)	1 Bric	400	.020	L.F.	2.08	.72		2.80	3.49
0025	"T" shaped		320	.025		1.25	.90		2.15	2.89
0030	Cross-shaped for CMU units		280	.029		1.47	1.03		2.50	3.34
0050	PVC, for double wythe 8" minimum wall (Brick/CMU)		400	.020		1.48	.72		2.20	2.83
0120	"T" shaped		320	.025		.78	.90		1.68	2.37
0160	Cross-shaped for CMU units		280	.029		.97	1.03		2	2.79

04 21 Clay Unit Masonry

04 21 13 – Brick Masonry

04 21 13.13 Brick Veneer Masonry		Crew	Daily Output	Labor-Hours	Unit	Material	Labor	Equipment	Total	Total Incl O&P
0010	**BRICK VENEER MASONRY**, T.L. lots, excl. scaff., grout & reinforcing									
0015	Material costs incl. 3% brick and 25% mortar waste									
2000	Standard, sel. common, 4" x 2-2/3" x 8", (6.75/S.F.)	D-8	230	.174	S.F.	4.34	5.85		10.19	14.45
2020	Red, 4" x 2-2/3" x 8", running bond		220	.182		3.93	6.10		10.03	14.45
2050	Full header every 6th course (7.88/S.F.)		185	.216		4.58	7.25		11.83	17.15
2100	English, full header every 2nd course (10.13/S.F.)		140	.286		5.85	9.60		15.45	22.50
2150	Flemish, alternate header every course (9.00/S.F.)		150	.267		5.20	8.95		14.15	20.50
2200	Flemish, alt. header every 6th course (7.13/S.F.)		205	.195		4.15	6.55		10.70	15.45
2250	Full headers throughout (13.50/S.F.)		105	.381		7.80	12.75		20.55	30
2300	Rowlock course (13.50/S.F.)		100	.400		7.80	13.40		21.20	31
2350	Rowlock stretcher (4.50/S.F.)		310	.129		2.64	4.33		6.97	10.10
2400	Soldier course (6.75/S.F.)		200	.200		3.93	6.70		10.63	15.50
2450	Sailor course (4.50/S.F.)		290	.138		2.64	4.62		7.26	10.60
2600	Buff or gray face, running bond (6.75/S.F.)		220	.182		4.15	6.10		10.25	14.70
2700	Glazed face brick, running bond		210	.190		12.20	6.40		18.60	24
2750	Full header every 6th course (7.88/S.F.)		170	.235		14.20	7.90		22.10	29
3000	Jumbo, 6" x 4" x 12" running bond (3.00/S.F.)		435	.092		4.85	3.08		7.93	10.50
3050	Norman, 4" x 2-2/3" x 12" running bond (4.5/S.F.)		320	.125		6.05	4.19		10.24	13.65
3100	Norwegian, 4" x 3-1/5" x 12" (3.75/S.F.)		375	.107		5.35	3.58		8.93	11.85
3150	Economy, 4" x 4" x 8" (4.50/S.F.)		310	.129		4.24	4.33		8.57	11.85
3200	Engineer, 4" x 3-1/5" x 8" (5.63/S.F.)		260	.154		3.76	5.15		8.91	12.75

04 21 Clay Unit Masonry

04 21 13 – Brick Masonry

04 21 13.13 Brick Veneer Masonry		Crew	Daily Output	Labor-Hours	Unit	2016 Bare Costs Material	Labor	Equipment	Total	Total Incl O&P
3250	Roman, 4" x 2" x 12" (6.00/S.F.)	D-8	250	.160	S.F.	7.20	5.35		12.55	16.85
3300	SCR, 6" x 2-2/3" x 12" (4.50/S.F.)		310	.129		6.20	4.33		10.53	14.05
3350	Utility, 4" x 4" x 12" (3.00/S.F.)		360	.111		4.78	3.72		8.50	11.45
3360	For less than truck load lots, add				M	15%				
3400	For cavity wall construction, add						15%			
3450	For stacked bond, add						10%			
3500	For interior veneer construction, add						15%			
3550	For curved walls, add						30%			

04 21 13.14 Thin Brick Veneer

		Crew	Daily Output	Labor-Hours	Unit	Material	Labor	Equipment	Total	Total Incl O&P
0010	**THIN BRICK VENEER**									
0015	Material costs incl. 3% brick and 25% mortar waste									
0020	On & incl. metal panel support sys, modular, 2-2/3" x 5/8" x 8", red	D-7	92	.174	S.F.	9.35	5.40		14.75	18.95
0100	Closure, 4" x 5/8" x 8"		110	.145		9.10	4.50		13.60	17.30
0110	Norman, 2-2/3" x 5/8" x 12"		110	.145		9.15	4.50		13.65	17.35
0120	Utility, 4" x 5/8" x 12"		125	.128		8.85	3.96		12.81	16.15
0130	Emperor, 4" x 3/4" x 16"		175	.091		10	2.83		12.83	15.60
0140	Super emperor, 8" x 3/4" x 16"		195	.082		10.30	2.54		12.84	15.45
0150	For L shaped corners with 4" return, add				L.F.	9.25			9.25	10.20
0200	On masonry/plaster back-up, modular, 2-2/3" x 5/8" x 8", red	D-7	137	.117	S.F.	4.36	3.61		7.97	10.65
0210	Closure, 4" x 5/8" x 8"		165	.097		4.12	3		7.12	9.40
0220	Norman, 2-2/3" x 5/8" x 12"		165	.097		4.16	3		7.16	9.45
0230	Utility, 4" x 5/8" x 12"		185	.086		3.88	2.68		6.56	8.60
0240	Emperor, 4" x 3/4" x 16"		260	.062		5	1.90		6.90	8.60
0250	Super emperor, 8" x 3/4" x 16"		285	.056		5.35	1.74		7.09	8.65
0260	For L shaped corners with 4" return, add				L.F.	9.25			9.25	10.20
0270	For embedment into pre-cast concrete panels, add				S.F.	14.40			14.40	15.85

04 21 13.15 Chimney

		Crew	Daily Output	Labor-Hours	Unit	Material	Labor	Equipment	Total	Total Incl O&P
0010	**CHIMNEY**, excludes foundation, scaffolding, grout and reinforcing									
0100	Brick, 16" x 16", 8" flue	D-1	18.20	.879	V.L.F.	25	29		54	75.50
0150	16" x 20" with one 8" x 12" flue		16	1		39	33		72	98
0200	16" x 24" with two 8" x 8" flues		14	1.143		57	37.50		94.50	126
0250	20" x 20" with one 12" x 12" flue		13.70	1.168		46.50	38.50		85	115
0300	20" x 24" with two 8" x 12" flues		12	1.333		65	44		109	144
0350	20" x 32" with two 12" x 12" flues		10	1.600		81.50	52.50		134	178

04 21 13.18 Columns

		Crew	Daily Output	Labor-Hours	Unit	Material	Labor	Equipment	Total	Total Incl O&P
0010	**COLUMNS**, solid, excludes scaffolding, grout and reinforcing									
0050	Brick, 8" x 8", 9 brick per V.L.F.	D-1	56	.286	V.L.F.	5	9.40		14.40	21
0100	12" x 8", 13.5 brick per V.L.F.		37	.432		7.50	14.20		21.70	32
0200	12" x 12", 20 brick per V.L.F.		25	.640		11.10	21		32.10	47.50
0300	16" x 12", 27 brick per V.L.F.		19	.842		15	27.50		42.50	62.50
0400	16" x 16", 36 brick per V.L.F.		14	1.143		20	37.50		57.50	84.50
0500	20" x 16", 45 brick per V.L.F.		11	1.455		25	48		73	108
0600	20" x 20", 56 brick per V.L.F.		9	1.778		31	58.50		89.50	132
0700	24" x 20", 68 brick per V.L.F.		7	2.286		38	75		113	167
0800	24" x 24", 81 brick per V.L.F.		6	2.667		45	87.50		132.50	196
1000	36" x 36", 182 brick per V.L.F.		3	5.333		101	175		276	405

04 21 13.30 Oversized Brick

		Crew	Daily Output	Labor-Hours	Unit	Material	Labor	Equipment	Total	Total Incl O&P
0010	**OVERSIZED BRICK**, excludes scaffolding, grout and reinforcing									
0100	Veneer, 4" x 2.25" x 16"	D-8	387	.103	S.F.	5.30	3.46		8.76	11.65
0102	8" x 2.25" x 16", multicell		265	.151		16.65	5.05		21.70	27
0105	4" x 2.75" x 16"		412	.097		5.50	3.25		8.75	11.55
0107	8" x 2.75" x 16", multicell		295	.136		16.70	4.55		21.25	26

04 21 Clay Unit Masonry

04 21 13 – Brick Masonry

04 21 13.30 Oversized Brick

	04 21 13.30 Oversized Brick	Crew	Daily Output	Labor-Hours	Unit	2016 Bare Costs Material	Labor	Equipment	Total	Total Incl O&P
0110	4" x 4" x 16"	D-8	460	.087	S.F.	3.61	2.91		6.52	8.85
0120	4" x 8" x 16"		533	.075		4.25	2.52		6.77	8.85
0122	4" x 8" x 16" multicell		327	.122		15.65	4.10		19.75	24
0125	Loadbearing, 6" x 4" x 16", grouted and reinforced		387	.103		10.70	3.46		14.16	17.55
0130	8" x 4" x 16", grouted and reinforced		327	.122		11.75	4.10		15.85	19.80
0132	10" x 4" x 16", grouted and reinforced		327	.122		23.50	4.10		27.60	33
0135	6" x 8" x 16", grouted and reinforced		440	.091		13.90	3.05		16.95	20.50
0140	8" x 8" x 16", grouted and reinforced		400	.100		14.85	3.35		18.20	22
0145	Curtainwall/reinforced veneer, 6" x 4" x 16"		387	.103		15.25	3.46		18.71	22.50
0150	8" x 4" x 16"		327	.122		18.55	4.10		22.65	27.50
0152	10" x 4" x 16"		327	.122		26	4.10		30.10	36
0155	6" x 8" x 16"		440	.091		18.95	3.05		22	26
0160	8" x 8" x 16"	↓	400	.100		26.50	3.35		29.85	34.50
0200	For 1 to 3 slots in face, add					15%				
0210	For 4 to 7 slots in face, add					25%				
0220	For bond beams, add					20%				
0230	For bullnose shapes, add					20%				
0240	For open end knockout, add					10%				
0250	For white or gray color group, add					10%				
0260	For 135 degree corner, add				↓	250%				

04 21 13.35 Common Building Brick

	04 21 13.35 Common Building Brick	Crew	Daily Output	Labor-Hours	Unit	Material	Labor	Equipment	Total	Total Incl O&P
0010	**COMMON BUILDING BRICK**, C62, TL lots, material only R042110-20									
0020	Standard				M	495			495	545
0050	Select				"	520			520	570

04 21 13.45 Face Brick

	04 21 13.45 Face Brick	Crew	Daily Output	Labor-Hours	Unit	Material	Labor	Equipment	Total	Total Incl O&P
0010	**FACE BRICK** Material Only, C216, TL lots R042110-20									
0300	Standard modular, 4" x 2-2/3" x 8"				M	460			460	505
2170	For less than truck load lots, add					15			15	16.50
2180	For buff or gray brick, add				↓	16			16	17.60

04 22 Concrete Unit Masonry

04 22 10 – Concrete Masonry Units

04 22 10.11 Autoclave Aerated Concrete Block

	04 22 10.11 Autoclave Aerated Concrete Block	Crew	Daily Output	Labor-Hours	Unit	Material	Labor	Equipment	Total	Total Incl O&P
0010	**AUTOCLAVE AERATED CONCRETE BLOCK**, excl. scaffolding, grout & reinforcing									
0050	Solid, 4" x 8" x 24", incl. mortar G	D-8	600	.067	S.F.	1.55	2.23		3.78	5.45
0060	6" x 8" x 24" G		600	.067		2.34	2.23		4.57	6.30
0070	8" x 8" x 24" G		575	.070		3.11	2.33		5.44	7.30
0080	10" x 8" x 24" G		575	.070		3.80	2.33		6.13	8.05
0090	12" x 8" x 24" G	↓	550	.073	↓	4.66	2.44		7.10	9.20

04 22 10.14 Concrete Block, Back-Up

	04 22 10.14 Concrete Block, Back-Up	Crew	Daily Output	Labor-Hours	Unit	Material	Labor	Equipment	Total	Total Incl O&P
0010	**CONCRETE BLOCK, BACK-UP**, C90, 2000 psi									
0020	Normal weight, 8" x 16" units, tooled joint 1 side									
0050	Not-reinforced, 2000 psi, 2" thick	D-8	475	.084	S.F.	1.53	2.82		4.35	6.40
0200	4" thick		460	.087		1.84	2.91		4.75	6.90
0300	6" thick		440	.091		2.40	3.05		5.45	7.75
0350	8" thick		400	.100		2.56	3.35		5.91	8.40
0400	10" thick	↓	330	.121		3.03	4.06		7.09	10.15
0450	12" thick	D-9	310	.155		4.17	5.10		9.27	13.10
1000	Reinforced, alternate courses, 4" thick	D-8	450	.089		2	2.98		4.98	7.15
1100	6" thick	↓	430	.093	↓	2.56	3.12		5.68	8

04 22 Concrete Unit Masonry

04 22 10 – Concrete Masonry Units

04 22 10.14 Concrete Block, Back-Up		Crew	Daily Output	Labor-Hours	Unit	2016 Bare Costs Material	Labor	Equipment	Total	Total Incl O&P
1150	8" thick	D-8	395	.101	S.F.	2.76	3.39		6.15	8.70
1200	10" thick	↓	320	.125		3.19	4.19		7.38	10.50
1250	12" thick	D-9	300	.160	↓	4.34	5.25		9.59	13.55

04 22 10.16 Concrete Block, Bond Beam

		Crew	Daily Output	Labor-Hours	Unit	Material	Labor	Equipment	Total	Total Incl O&P
0010	**CONCRETE BLOCK, BOND BEAM**, C90, 2000 psi									
0020	Not including grout or reinforcing									
0125	Regular block, 6" thick	D-8	584	.068	L.F.	2.73	2.30		5.03	6.85
0130	8" high, 8" thick	"	565	.071		2.82	2.37		5.19	7.05
0150	12" thick	D-9	510	.094		3.83	3.09		6.92	9.35
0525	Lightweight, 6" thick	D-8	592	.068	↓	2.85	2.26		5.11	6.90

04 22 10.19 Concrete Block, Insulation Inserts

		Crew	Daily Output	Labor-Hours	Unit	Material	Labor	Equipment	Total	Total Incl O&P
0010	**CONCRETE BLOCK, INSULATION INSERTS**									
0100	Styrofoam, plant installed, add to block prices									
0200	8" x 16" units, 6" thick				S.F.	1.20			1.20	1.32
0250	8" thick					1.35			1.35	1.49
0300	10" thick					1.40			1.40	1.54
0350	12" thick					1.55			1.55	1.71
0500	8" x 8" units, 8" thick					1.20			1.20	1.32
0550	12" thick				↓	1.40			1.40	1.54

04 22 10.23 Concrete Block, Decorative

		Crew	Daily Output	Labor-Hours	Unit	Material	Labor	Equipment	Total	Total Incl O&P
0010	**CONCRETE BLOCK, DECORATIVE**, C90, 2000 psi									
0020	Embossed, simulated brick face									
0100	8" x 16" units, 4" thick	D-8	400	.100	S.F.	2.80	3.35		6.15	8.70
0200	8" thick		340	.118		3.06	3.94		7	9.95
0250	12" thick	↓	300	.133	↓	5.15	4.47		9.62	13.15
0400	Embossed both sides									
0500	8" thick	D-8	300	.133	S.F.	4.13	4.47		8.60	12
0550	12" thick	"	275	.145	"	5.40	4.88		10.28	14.10
1000	Fluted high strength									
1100	8" x 16" x 4" thick, flutes 1 side,	D-8	345	.116	S.F.	3.98	3.89		7.87	10.90
1150	Flutes 2 sides		335	.119		4.57	4		8.57	11.70
1200	8" thick	↓	300	.133		5.35	4.47		9.82	13.35
1250	For special colors, add				↓	.64			.64	.71
1400	Deep grooved, smooth face									
1450	8" x 16" x 4" thick	D-8	345	.116	S.F.	2.56	3.89		6.45	9.30
1500	8" thick	"	300	.133	"	4.05	4.47		8.52	11.90
2000	Formblock, incl. inserts & reinforcing									
2100	8" x 16" x 8" thick	D-8	345	.116	S.F.	3.57	3.89		7.46	10.45
2150	12" thick	"	310	.129	"	4.73	4.33		9.06	12.40
2500	Ground face									
2600	8" x 16" x 4" thick	D-8	345	.116	S.F.	3.88	3.89		7.77	10.75
2650	6" thick		325	.123		4.30	4.13		8.43	11.65
2700	8" thick	↓	300	.133		4.74	4.47		9.21	12.65
2750	12" thick	D-9	265	.181	↓	6.10	5.95		12.05	16.65
2900	For special colors, add, minimum					15%				
2950	For special colors, add, maximum					45%				
4000	Slump block									
4100	4" face height x 16" x 4" thick	D-1	165	.097	S.F.	4.14	3.19		7.33	9.85
4150	6" thick		160	.100		6.25	3.29		9.54	12.35
4200	8" thick		155	.103		6	3.39		9.39	12.25
4250	10" thick		140	.114		11.55	3.76		15.31	19
4300	12" thick	↓	130	.123	↓	13.10	4.05		17.15	21

04 22 10.23 Concrete Block, Decorative		Crew	Daily Output	Labor-Hours	Unit	Material	2016 Bare Costs Labor	Equipment	Total	Total Incl O&P
4400	6" face height x 16" x 6" thick	D-1	155	.103	S.F.	5.80	3.39		9.19	12
4450	8" thick		150	.107		8.75	3.51		12.26	15.50
4500	10" thick		130	.123		13.40	4.05		17.45	21.50
4550	12" thick		120	.133		14.05	4.38		18.43	23
5000	Split rib profile units, 1" deep ribs, 8 ribs									
5100	8" x 16" x 4" thick	D-8	345	.116	S.F.	3.98	3.89		7.87	10.90
5150	6" thick		325	.123		4.50	4.13		8.63	11.85
5200	8" thick		300	.133		5.10	4.47		9.57	13.05
5250	12" thick	D-9	275	.175		5.95	5.75		11.70	16.10
5400	For special deeper colors, 4" thick, add					1.29			1.29	1.42
5450	12" thick, add					1.33			1.33	1.47
5600	For white, 4" thick, add					1.29			1.29	1.42
5650	6" thick, add					1.31			1.31	1.44
5700	8" thick, add					1.34			1.34	1.48
5750	12" thick, add					1.39			1.39	1.53
6000	Split face									
6100	8" x 16" x 4" thick	D-8	350	.114	S.F.	3.59	3.83		7.42	10.35
6150	6" thick		325	.123		4.09	4.13		8.22	11.40
6200	8" thick		300	.133		4.60	4.47		9.07	12.50
6250	12" thick	D-9	270	.178		5.55	5.85		11.40	15.85
6300	For scored, add					.38			.38	.42
6400	For special deeper colors, 4" thick, add					.62			.62	.68
6450	6" thick, add					.75			.75	.82
6500	8" thick, add					.77			.77	.85
6550	12" thick, add					.79			.79	.87
6650	For white, 4" thick, add					1.28			1.28	1.40
6700	6" thick, add					1.29			1.29	1.42
6750	8" thick, add					1.30			1.30	1.43
6800	12" thick, add					1.33			1.33	1.47
7000	Scored ground face, 2 to 5 scores									
7100	8" x 16" x 4" thick	D-8	340	.118	S.F.	7.85	3.94		11.79	15.20
7150	6" thick		310	.129		8.70	4.33		13.03	16.75
7200	8" thick		290	.138		9.80	4.62		14.42	18.50
7250	12" thick	D-9	265	.181		13.15	5.95		19.10	24.50
8000	Hexagonal face profile units, 8" x 16" units									
8100	4" thick, hollow	D-8	340	.118	S.F.	3.66	3.94		7.60	10.65
8200	Solid		340	.118		4.69	3.94		8.63	11.75
8300	6" thick, hollow		310	.129		3.96	4.33		8.29	11.55
8350	8" thick, hollow		290	.138		4.53	4.62		9.15	12.70
8500	For stacked bond, add						26%			
8550	For high rise construction, add per story	D-8	67.80	.590	M.S.F.		19.80		19.80	33
8600	For scored block, add					10%				
8650	For honed or ground face, per face, add				Ea.	1.18			1.18	1.29
8700	For honed or ground end, per end, add				"	1.18			1.18	1.29
8750	For bullnose block, add					10%				
8800	For special color, add					13%				

04 22 10.24 Concrete Block, Exterior		Crew	Daily Output	Labor-Hours	Unit	Material	Labor	Equipment	Total	Total Incl O&P
0010	**CONCRETE BLOCK, EXTERIOR**, C90, 2000 psi									
0020	Reinforced alt courses, tooled joints 2 sides									
0100	Normal weight, 8" x 16" x 6" thick	D-8	395	.101	S.F.	2.35	3.39		5.74	8.25
0200	8" thick		360	.111		3.62	3.72		7.34	10.20
0250	10" thick		290	.138		4.19	4.62		8.81	12.30

04 22 Concrete Unit Masonry

04 22 10 – Concrete Masonry Units

04 22 10.24 Concrete Block, Exterior

		Crew	Daily Output	Labor-Hours	Unit	2016 Bare Costs Material	Labor	Equipment	Total	Total Incl O&P
0300	12″ thick	D-9	250	.192	S.F.	4.89	6.30		11.19	15.95

04 22 10.26 Concrete Block Foundation Wall

		Crew	Daily Output	Labor-Hours	Unit	Material	Labor	Equipment	Total	Total Incl O&P
0010	**CONCRETE BLOCK FOUNDATION WALL**, C90/C145									
0050	Normal-weight, cut joints, horiz joint reinf, no vert reinf.									
0200	Hollow, 8″ x 16″ x 6″ thick	D-8	455	.088	S.F.	3.02	2.95		5.97	8.25
0250	8″ thick		425	.094		3.24	3.15		6.39	8.80
0300	10″ thick		350	.114		3.70	3.83		7.53	10.45
0350	12″ thick	D-9	300	.160		4.88	5.25		10.13	14.10
0500	Solid, 8″ x 16″ block, 6″ thick	D-8	440	.091		3.08	3.05		6.13	8.50
0550	8″ thick	″	415	.096		4.29	3.23		7.52	10.10
0600	12″ thick	D-9	350	.137		6.15	4.51		10.66	14.25

04 22 10.32 Concrete Block, Lintels

		Crew	Daily Output	Labor-Hours	Unit	Material	Labor	Equipment	Total	Total Incl O&P
0010	**CONCRETE BLOCK, LINTELS**, C90, normal weight									
0100	Including grout and horizontal reinforcing									
0200	8″ x 8″ x 8″, 1 #4 bar	D-4	300	.133	L.F.	3.87	4.11	.45	8.43	11.60
0250	2 #4 bars		295	.136		4.09	4.18	.46	8.73	12
0400	8″ x 16″ x 8″, 1 #4 bar		275	.145		3.90	4.49	.49	8.88	12.35
0450	2 #4 bars		270	.148		4.12	4.57	.50	9.19	12.75
1000	12″ x 8″ x 8″, 1 #4 bar		275	.145		5.30	4.49	.49	10.28	13.90
1100	2 #4 bars		270	.148		5.50	4.57	.50	10.57	14.25
1150	2 #5 bars		270	.148		5.75	4.57	.50	10.82	14.55
1200	2 #6 bars		265	.151		6.05	4.66	.51	11.22	14.95
1500	12″ x 16″ x 8″, 1 #4 bar		250	.160		5.95	4.94	.54	11.43	15.40
1600	2 #3 bars		245	.163		6	5.05	.55	11.60	15.60
1650	2 #4 bars		245	.163		6.15	5.05	.55	11.75	15.80
1700	2 #5 bars		240	.167		6.40	5.15	.56	12.11	16.25

04 22 10.33 Lintel Block

		Crew	Daily Output	Labor-Hours	Unit	Material	Labor	Equipment	Total	Total Incl O&P
0010	**LINTEL BLOCK**									
3481	Lintel block 6″ x 8″ x 8″	D-1	300	.053	Ea.	1.32	1.75		3.07	4.37
3501	6″ x 16″ x 8″		275	.058		2.07	1.91		3.98	5.45
3521	8″ x 8″ x 8″		275	.058		1.19	1.91		3.10	4.50
3561	8″ x 16″ x 8″		250	.064		2.09	2.10		4.19	5.80

04 22 10.34 Concrete Block, Partitions

		Crew	Daily Output	Labor-Hours	Unit	Material	Labor	Equipment	Total	Total Incl O&P
0010	**CONCRETE BLOCK, PARTITIONS**, excludes scaffolding									
1000	Lightweight block, tooled joints, 2 sides, hollow									
1100	Not reinforced, 8″ x 16″ x 4″ thick	D-8	440	.091	S.F.	1.85	3.05		4.90	7.15
1150	6″ thick		410	.098		2.63	3.27		5.90	8.35
1200	8″ thick		385	.104		3.20	3.48		6.68	9.30
1250	10″ thick		370	.108		3.88	3.62		7.50	10.30
1300	12″ thick	D-9	350	.137		4.09	4.51		8.60	12
4000	Regular block, tooled joints, 2 sides, hollow									
4100	Not reinforced, 8″ x 16″ x 4″ thick	D-8	430	.093	S.F.	1.74	3.12		4.86	7.10
4150	6″ thick		400	.100		2.30	3.35		5.65	8.15
4200	8″ thick		375	.107		2.47	3.58		6.05	8.65
4250	10″ thick		360	.111		2.93	3.72		6.65	9.45
4300	12″ thick	D-9	340	.141		4.08	4.64		8.72	12.25

04 23 Glass Unit Masonry

04 23 13 – Vertical Glass Unit Masonry

	04 23 13.10 Glass Block	Crew	Daily Output	Labor-Hours	Unit	2016 Bare Costs Material	Labor	Equipment	Total	Total Incl O&P
0010	**GLASS BLOCK**									
0100	Plain, 4" thick, under 1,000 S.F., 6" x 6"	D-8	115	.348	S.F.	27.50	11.65		39.15	50
0150	8" x 8"		160	.250		16	8.40		24.40	31.50
0160	end block		160	.250		60.50	8.40		68.90	80.50
0170	90 deg corner		160	.250		56	8.40		64.40	75.50
0180	45 deg corner		160	.250		46	8.40		54.40	64.50
0200	12" x 12"		175	.229		24	7.65		31.65	39.50
0210	4" x 8"		160	.250		30	8.40		38.40	47
0220	6" x 8"	↓	160	.250	↓	21.50	8.40		29.90	37.50
0700	For solar reflective blocks, add					100%				
1000	Thinline, plain, 3-1/8" thick, under 1,000 S.F., 6" x 6"	D-8	115	.348	S.F.	22.50	11.65		34.15	44
1050	8" x 8"		160	.250		12.55	8.40		20.95	28
1400	For cleaning block after installation (both sides), add	↓	1000	.040	↓	.16	1.34		1.50	2.42

04 24 Adobe Unit Masonry

04 24 16 – Manufactured Adobe Unit Masonry

	04 24 16.06 Adobe Brick		Crew	Daily Output	Labor-Hours	Unit	Material	Labor	Equipment	Total	Total Incl O&P
0010	**ADOBE BRICK**, Semi-stabilized, with cement mortar										
0060	Brick, 10" x 4" x 14", 2.6/S.F.	G	D-8	560	.071	S.F.	4.60	2.39		6.99	9.05
0080	12" x 4" x 16", 2.3/S.F.	G		580	.069		6.90	2.31		9.21	11.45
0100	10" x 4" x 16", 2.3/S.F.	G		590	.068		6.45	2.27		8.72	10.90
0120	8" x 4" x 16", 2.3/S.F.	G		560	.071		4.94	2.39		7.33	9.45
0140	4" x 4" x 16", 2.3/S.F.	G		540	.074		4.82	2.48		7.30	9.45
0160	6" x 4" x 16", 2.3/S.F.	G		540	.074		4.48	2.48		6.96	9.05
0180	4" x 4" x 12", 3.0/S.F.	G		520	.077		5.20	2.58		7.78	10
0200	8" x 4" x 12", 3.0/S.F.	G	↓	520	.077	↓	4.24	2.58		6.82	8.95

04 27 Multiple-Wythe Unit Masonry

04 27 10 – Multiple-Wythe Masonry

		Crew	Daily Output	Labor-Hours	Unit	Material	Labor	Equipment	Total	Total Incl O&P
	04 27 10.10 Cornices									
0010	**CORNICES**									
0110	Face bricks, 12 brick/S.F.	D-1	30	.533	SF Face	5.85	17.55		23.40	35.50
0150	15 brick/S.F.	"	23	.696	"	7	23		30	45.50
	04 27 10.30 Brick Walls									
0010	**BRICK WALLS**, including mortar, excludes scaffolding									
0800	Face brick, 4" thick wall, 6.75 brick/S.F.	D-8	215	.186	S.F.	3.86	6.25		10.11	14.65
0850	Common brick, 4" thick wall, 6.75 brick/S.F.		240	.167		4.12	5.60		9.72	13.85
0900	8" thick, 13.50 bricks per S.F.		135	.296		8.55	9.95		18.50	26
1000	12" thick, 20.25 bricks per S.F.		95	.421		12.05	14.10		26.95	37.50
1050	16" thick, 27.00 bricks per S.F.		75	.533		17.40	17.90		35.30	49
1200	Reinforced, face brick, 4" thick wall, 6.75 brick/S.F.		210	.190		4.02	6.40		10.42	15.05
1220	Common brick, 4" thick wall, 6.75 brick/S.F.		235	.170		4.28	5.70		9.98	14.20
1250	8" thick, 13.50 bricks per S.F.		130	.308		8.85	10.30		19.15	27
1300	12" thick, 20.25 bricks per S.F.		90	.444		13.35	14.90		28.25	39.50
1350	16" thick, 27.00 bricks per S.F.	↓	70	.571	↓	18.05	19.15		37.20	52
	04 27 10.40 Steps									
0010	**STEPS**									
0012	Entry steps, select common brick	D-1	.30	53.333	M	520	1,750		2,270	3,500

04 41 Dry-Placed Stone

04 41 10 – Dry Placed Stone

04 41 10.10 Rough Stone Wall			Crew	Daily Output	Labor-Hours	Unit	2016 Bare Costs Material	Labor	Equipment	Total	Total Incl O&P
0011	**ROUGH STONE WALL**, Dry										
0012	Dry laid (no mortar), under 18″ thick	G	D-1	60	.267	C.F.	12.65	8.75		21.40	28.50
0100	Random fieldstone, under 18″ thick	G	D-12	60	.533		12.65	17.55		30.20	43
0150	Over 18″ thick	G	″	63	.508		15.15	16.70		31.85	44.50
0500	Field stone veneer	G	D-8	120	.333	S.F.	12.60	11.15		23.75	32.50
0600	Rubble stone walls, in mortar bed, up to 18″ thick	G	D-11	75	.320	C.F.	15.25	10.85		26.10	35

04 43 Stone Masonry

04 43 10 – Masonry with Natural and Processed Stone

04 43 10.45 Granite		Crew	Daily Output	Labor-Hours	Unit	Material	Labor	Equipment	Total	Total Incl O&P
0010	**GRANITE**, cut to size									
0050	Veneer, polished face, 3/4″ to 1-1/2″ thick									
0150	Low price, gray, light gray, etc.	D-10	130	.246	S.F.	26.50	8.85	3.69	39.04	48.50
0180	Medium price, pink, brown, etc.		130	.246		29.50	8.85	3.69	42.04	51.50
0220	High price, red, black, etc.		130	.246		42	8.85	3.69	54.54	65.50
0300	1-1/2″ to 2-1/2″ thick, veneer									
0350	Low price, gray, light gray, etc.	D-10	130	.246	S.F.	28.50	8.85	3.69	41.04	50.50
0500	Medium price, pink, brown, etc.		130	.246		33.50	8.85	3.69	46.04	56
0550	High price, red, black, etc.		130	.246		52.50	8.85	3.69	65.04	76.50
0700	2-1/2″ to 4″ thick, veneer									
0750	Low price, gray, light gray, etc.	D-10	110	.291	S.F.	38.50	10.50	4.36	53.36	64.50
0850	Medium price, pink, brown, etc.		110	.291		44	10.50	4.36	58.86	70.50
0950	High price, red, black, etc.		110	.291		63	10.50	4.36	77.86	91.50
1000	For bush hammered finish, deduct					5%				
1050	Coarse rubbed finish, deduct					10%				
1100	Honed finish, deduct					5%				
1150	Thermal finish, deduct					18%				
2500	Steps, copings, etc., finished on more than one surface									
2550	Low price, gray, light gray, etc.	D-10	50	.640	C.F.	91	23	9.60	123.60	149
2575	Medium price, pink, brown, etc.		50	.640		118	23	9.60	150.60	179
2600	High price, red, black, etc.		50	.640		146	23	9.60	178.60	209
2800	Pavers, 4″ x 4″ x 4″ blocks, split face and joints									
2850	Low price, gray, light gray, etc.	D-11	80	.300	S.F.	13.10	10.20		23.30	31.50
2875	Medium price, pinks, browns, etc.		80	.300		21	10.20		31.20	40
2900	High price, red, black, etc.		80	.300		29	10.20		39.20	49
4000	Soffits, 2″ thick, low price, gray, light gray	D-13	35	1.371		37.50	48	13.70	99.20	137
4050	Medium price, pink, brown, etc.		35	1.371		64.50	48	13.70	126.20	166
4100	High price, red, black, etc.		35	1.371		91.50	48	13.70	153.20	195
4200	Low price, gray, light gray, etc.		35	1.371		63	48	13.70	124.70	165
4250	Medium price, pink, brown, etc.		35	1.371		91	48	13.70	152.70	195
4300	High price, red, black, etc.		35	1.371		119	48	13.70	180.70	226
5000	Reclaimed or Antique									
5010	Treads, up to 12″ wide	D-10	100	.320	L.F.	42.50	11.55	4.79	58.84	71.50
5020	Up to 18″ wide		100	.320		38.50	11.55	4.79	54.84	67
5030	Capstone, size varies		50	.640		30.50	23	9.60	63.10	82.50
5040	Posts		30	1.067	V.L.F.	30.50	38.50	16	85	115

04 43 10.55 Limestone		Crew	Daily Output	Labor-Hours	Unit	Material	Labor	Equipment	Total	Total Incl O&P
0010	**LIMESTONE**, cut to size									
0020	Veneer facing panels									
0500	Texture finish, light stick, 4-1/2″ thick, 5′ x 12′	D-4	300	.133	S.F.	20.50	4.11	.45	25.06	30
0750	5″ thick, 5′ x 14′ panels	D-10	275	.116		21.50	4.19	1.74	27.43	33

04 43 Stone Masonry

04 43 10 – Masonry with Natural and Processed Stone

04 43 10.55 Limestone

	04 43 10.55 Limestone	Crew	Daily Output	Labor-Hours	Unit	Material	Labor	Equipment	Total	Total Incl O&P
						2016 Bare Costs				
1000	Sugarcube finish, 2″ thick, 3′ x 5′ panels	D-10	275	.116	S.F.	29	4.19	1.74	34.93	41
1050	3″ thick, 4′ x 9′ panels		275	.116		25.50	4.19	1.74	31.43	37.50
1200	4″ thick, 5′ x 11′ panels		275	.116		31.50	4.19	1.74	37.43	43.50
1400	Sugarcube, textured finish, 4-1/2″ thick, 5′ x 12′		275	.116		32.50	4.19	1.74	38.43	45
1450	5″ thick, 5′ x 14′ panels		275	.116		34	4.19	1.74	39.93	46
2000	Coping, sugarcube finish, top & 2 sides		30	1.067	C.F.	68.50	38.50	16	123	157
2100	Sills, lintels, jambs, trim, stops, sugarcube finish, simple		20	1.600		68.50	57.50	24	150	198
2150	Detailed		20	1.600		68.50	57.50	24	150	198
2300	Steps, extra hard, 14″ wide, 6″ rise		50	.640	L.F.	25	23	9.60	57.60	76.50
3000	Quoins, plain finish, 6″ x 12″ x 12″	D-12	25	1.280	Ea.	41	42		83	116
3050	6″ x 16″ x 24″	″	25	1.280	″	55	42		97	131

04 43 10.60 Marble

	04 43 10.60 Marble	Crew	Daily Output	Labor-Hours	Unit	Material	Labor	Equipment	Total	Total Incl O&P
0011	**MARBLE**, ashlar, split face, 4″ + or - thick, random									
0040	Lengths 1′ to 4′ & heights 2″ to 7-1/2″, average	D-8	175	.229	S.F.	18.30	7.65		25.95	33
1000	Facing, polished finish, cut to size, 3/4″ to 7/8″ thick									
1050	Carrara or equal	D-10	130	.246	S.F.	22.50	8.85	3.69	35.04	44
1100	Arabescato or equal	″	130	.246	″	39.50	8.85	3.69	52.04	62.50
2200	Window sills, 6″ x 3/4″ thick	D-1	85	.188	L.F.	11.10	6.20		17.30	22.50
2500	Flooring, polished tiles, 12″ x 12″ x 3/8″ thick									
2510	Thin set, Giallo Solare or equal	D-11	90	.267	S.F.	17.50	9.05		26.55	34.50
2600	Sky Blue or equal		90	.267		16	9.05		25.05	32.50
2700	Mortar bed, Giallo Solare or equal		65	.369		17.50	12.55		30.05	40.50
2740	Sky Blue or equal		65	.369		16	12.55		28.55	38.50
2780	Travertine, 3/8″ thick, Sierra or equal	D-10	130	.246		9.25	8.85	3.69	21.79	29
2790	Silver or equal	″	130	.246		25.50	8.85	3.69	38.04	47
3500	Thresholds, 3′ long, 7/8″ thick, 4″ to 5″ wide, plain	D-12	24	1.333	Ea.	35.50	44		79.50	112
3550	Beveled		24	1.333	″	71.50	44		115.50	152
3700	Window stools, polished, 7/8″ thick, 5″ wide		85	.376	L.F.	21.50	12.40		33.90	44.50

04 43 10.75 Sandstone or Brownstone

	04 43 10.75 Sandstone or Brownstone	Crew	Daily Output	Labor-Hours	Unit	Material	Labor	Equipment	Total	Total Incl O&P
0011	**SANDSTONE OR BROWNSTONE**									
0100	Sawed face veneer, 2-1/2″ thick, to 2′ x 4′ panels	D-10	130	.246	S.F.	19.35	8.85	3.69	31.89	40.50
0150	4″ thick, to 3′-6″ x 8′ panels		100	.320		19.35	11.55	4.79	35.69	46
0300	Split face, random sizes		100	.320		14.15	11.55	4.79	30.49	40
0350	Cut stone trim (limestone)									
0360	Ribbon stone, 4″ thick, 5′ pieces	D-8	120	.333	Ea.	160	11.15		171.15	195
0370	Cove stone, 4″ thick, 5′ pieces		105	.381		161	12.75		173.75	199
0380	Cornice stone, 10″ to 12″ wide		90	.444		199	14.90		213.90	244
0390	Band stone, 4″ thick, 5′ pieces		145	.276		109	9.25		118.25	135
0410	Window and door trim, 3″ to 4″ wide		160	.250		93.50	8.40		101.90	117
0420	Key stone, 18″ long		60	.667		93.50	22.50		116	141

04 43 10.80 Slate

	04 43 10.80 Slate	Crew	Daily Output	Labor-Hours	Unit	Material	Labor	Equipment	Total	Total Incl O&P
0010	**SLATE**									
3100	Stair landings, 1″ thick, black, clear	D-1	65	.246	S.F.	21	8.10		29.10	37
3200	Ribbon	″	65	.246	″	23.50	8.10		31.60	39
3500	Stair treads, sand finish, 1″ thick x 12″ wide									
3600	3 L.F. to 6 L.F.	D-10	120	.267	L.F.	25	9.60	3.99	38.59	48
3700	Ribbon, sand finish, 1″ thick x 12″ wide									
3750	To 6 L.F.	D-10	120	.267	L.F.	21	9.60	3.99	34.59	44

04 43 10.85 Window Sill

	04 43 10.85 Window Sill	Crew	Daily Output	Labor-Hours	Unit	Material	Labor	Equipment	Total	Total Incl O&P
0010	**WINDOW SILL**									
0020	Bluestone, thermal top, 10″ wide, 1-1/2″ thick	D-1	85	.188	S.F.	10	6.20		16.20	21.50
0050	2″ thick		75	.213	″	10	7		17	22.50

04 43 Stone Masonry

04 43 10 – Masonry with Natural and Processed Stone

04 43 10.85 Window Sill		Crew	Daily Output	Labor-Hours	Unit	2016 Bare Costs Material	Labor	Equipment	Total	Total Incl O&P
0100	Cut stone, 5″ x 8″ plain	D-1	48	.333	L.F.	12.55	10.95		23.50	32
0200	Face brick on edge, brick, 8″ wide		80	.200		5.25	6.60		11.85	16.75
0400	Marble, 9″ wide, 1″ thick		85	.188		8.95	6.20		15.15	20
0900	Slate, colored, unfading, honed, 12″ wide, 1″ thick		85	.188		8.50	6.20		14.70	19.65
0950	2″ thick		70	.229		8.50	7.50		16	22

04 51 Flue Liner Masonry

04 51 10 – Clay Flue Lining

04 51 10.10 Flue Lining		Crew	Daily Output	Labor-Hours	Unit	Material	Labor	Equipment	Total	Total Incl O&P
0010	**FLUE LINING**, including mortar									
0020	Clay, 8″ x 8″	D-1	125	.128	V.L.F.	5.70	4.21		9.91	13.25
0100	8″ x 12″		103	.155		8.45	5.10		13.55	17.80
0200	12″ x 12″		93	.172		10.80	5.65		16.45	21.50
0300	12″ x 18″		84	.190		22	6.25		28.25	34.50
0400	18″ x 18″		75	.213		28	7		35	42.50
0500	20″ x 20″		66	.242		42.50	7.95		50.45	60
0600	24″ x 24″		56	.286		54.50	9.40		63.90	75.50
1000	Round, 18″ diameter		66	.242		38.50	7.95		46.45	56
1100	24″ diameter		47	.340		74.50	11.20		85.70	101

04 57 Masonry Fireplaces

04 57 10 – Brick or Stone Fireplaces

04 57 10.10 Fireplace		Crew	Daily Output	Labor-Hours	Unit	Material	Labor	Equipment	Total	Total Incl O&P
0010	**FIREPLACE**									
0100	Brick fireplace, not incl. foundations or chimneys									
0110	30″ x 29″ opening, incl. chamber, plain brickwork	D-1	.40	40	Ea.	570	1,325		1,895	2,825
0200	Fireplace box only (110 brick)	″	2	8	″	162	263		425	620
0300	For elaborate brickwork and details, add					35%	35%			
0400	For hearth, brick & stone, add	D-1	2	8	Ea.	212	263		475	675
0410	For steel, damper, cleanouts, add		4	4		18	132		150	239
0600	Plain brickwork, incl. metal circulator		.50	32		900	1,050		1,950	2,750
0800	Face brick only, standard size, 8″ x 2-2/3″ x 4″		.30	53.333	M	600	1,750		2,350	3,575
0900	Stone fireplace, fieldstone, add				SF Face	8.50			8.50	9.35
1000	Cut stone, add				″	8			8	8.80

04 72 Cast Stone Masonry

04 72 10 – Cast Stone Masonry Features

04 72 10.10 Coping		Crew	Daily Output	Labor-Hours	Unit	Material	Labor	Equipment	Total	Total Incl O&P
0010	**COPING**, stock units									
0050	Precast concrete, 10″ wide, 4″ tapers to 3-1/2″, 8″ wall	D-1	75	.213	L.F.	17.45	7		24.45	31
0100	12″ wide, 3-1/2″ tapers to 3″, 10″ wall		70	.229		18.80	7.50		26.30	33
0110	14″ wide, 4″ tapers to 3-1/2″, 12″ wall		65	.246		21.50	8.10		29.60	37
0150	16″ wide, 4″ tapers to 3-1/2″, 14″ wall		60	.267		23	8.75		31.75	40
0300	Limestone for 12″ wall, 4″ thick		90	.178		14.70	5.85		20.55	26
0350	6″ thick		80	.200		22	6.60		28.60	35.50
0500	Marble, to 4″ thick, no wash, 9″ wide		90	.178		12.85	5.85		18.70	24
0550	12″ wide		80	.200		18	6.60		24.60	31

04 72 Cast Stone Masonry

04 72 10 – Cast Stone Masonry Features

04 72 10.10 Coping		Crew	Daily Output	Labor-Hours	Unit	Material	Labor	Equipment	Total	Total Incl O&P
						2016 Bare Costs				
0700	Terra cotta, 9" wide	D-1	90	.178	L.F.	6.45	5.85		12.30	16.85
0750	12" wide		80	.200		8.80	6.60		15.40	20.50
0800	Aluminum, for 12" wall		80	.200		8.90	6.60		15.50	21

04 72 20 – Cultured Stone Veneer

04 72 20.10 Cultured Stone Veneer Components		Crew	Daily Output	Labor-Hours	Unit	Material	Labor	Equipment	Total	Total Incl O&P
0010	**CULTURED STONE VENEER COMPONENTS**									
0110	On wood frame and sheathing substrate, random sized cobbles, corner stones	D-8	70	.571	V.L.F.	10.25	19.15		29.40	43.50
0120	Field stones		140	.286	S.F.	7.45	9.60		17.05	24
0130	Random sized flats, corner stones		70	.571	V.L.F.	10.55	19.15		29.70	43.50
0140	Field stones		140	.286	S.F.	8.75	9.60		18.35	25.50
0150	Horizontal lined ledgestones, corner stones		75	.533	V.L.F.	10.25	17.90		28.15	41.50
0160	Field stones		150	.267	S.F.	7.45	8.95		16.40	23
0170	Random shaped flats, corner stones		65	.615	V.L.F.	10.25	20.50		30.75	46
0180	Field stones		150	.267	S.F.	7.45	8.95		16.40	23
0190	Random shaped/textured face, corner stones		65	.615	V.L.F.	10.25	20.50		30.75	46
0200	Field stones		130	.308	S.F.	7.45	10.30		17.75	25.50
0210	Random shaped river rock, corner stones		65	.615	V.L.F.	10.25	20.50		30.75	46
0220	Field stones		130	.308	S.F.	7.45	10.30		17.75	25.50
0240	On concrete or CMU substrate, random sized cobbles, corner stones		70	.571	V.L.F.	9.60	19.15		28.75	42.50
0250	Field stones		140	.286	S.F.	7.10	9.60		16.70	24
0260	Random sized flats, corner stones		70	.571	V.L.F.	9.90	19.15		29.05	43
0270	Field stones		140	.286	S.F.	8.45	9.60		18.05	25.50
0280	Horizontal lined ledgestones, corner stones		75	.533	V.L.F.	9.60	17.90		27.50	40.50
0290	Field stones		150	.267	S.F.	7.10	8.95		16.05	22.50
0300	Random shaped flats, corner stones		70	.571	V.L.F.	9.60	19.15		28.75	42.50
0310	Field stones		140	.286	S.F.	7.10	9.60		16.70	24
0320	Random shaped/textured face, corner stones		65	.615	V.L.F.	9.60	20.50		30.10	45
0330	Field stones		130	.308	S.F.	7.10	10.30		17.40	25
0340	Random shaped river rock, corner stones		65	.615	V.L.F.	9.60	20.50		30.10	45
0350	Field stones		130	.308	S.F.	7.10	10.30		17.40	25
0360	Cultured stone veneer, #15 felt weather resistant barrier	1 Clab	3700	.002	Sq.	5.30	.06		5.36	5.95
0370	Expanded metal lath, diamond, 2.5 lb./S.Y., galvanized	1 Lath	85	.094	S.Y.	2.82	3.47		6.29	8.70
0390	Water table or window sill, 18" long	1 Bric	80	.100	Ea.	9.80	3.61		13.41	16.80

Division Notes

		CREW	DAILY OUTPUT	LABOR-HOURS	UNIT	BARE COSTS MAT.	LABOR	EQUIP.	TOTAL	TOTAL INCL O&P

Division 5 Metals

Estimating Tips

05 05 00 Common Work Results for Metals

- Nuts, bolts, washers, connection angles, and plates can add a significant amount to both the tonnage of a structural steel job and the estimated cost. As a rule of thumb, add 10% to the total weight to account for these accessories.
- Type 2 steel construction, commonly referred to as "simple construction," consists generally of field-bolted connections with lateral bracing supplied by other elements of the building, such as masonry walls or x-bracing. The estimator should be aware, however, that shop connections may be accomplished by welding or bolting. The method may be particular to the fabrication shop and may have an impact on the estimated cost.

05 10 00 Structural Steel

- Steel items can be obtained from two sources: a fabrication shop or a metals service center. Fabrication shops can fabricate items under more controlled conditions than can crews in the field. They are also more efficient and can produce items more economically. Metal service centers serve as a source of long mill shapes to both fabrication shops and contractors.
- Most line items in this structural steel subdivision, and most items in 05 50 00 Metal Fabrications, are indicated as being shop fabricated. The bare material cost for these shop fabricated items is the "Invoice Cost" from the shop and includes the mill base price of steel plus mill extras, transportation to the shop, shop drawings and detailing where warranted, shop fabrication and handling, sandblasting and a shop coat of primer paint, all necessary structural bolts, and delivery to the job site. The bare labor cost and bare equipment cost for these shop fabricated items is for field installation or erection.
- Line items in Subdivision 05 12 23.40 Lightweight Framing, and other items scattered in Division 5, are indicated as being field fabricated. The bare material cost for these field fabricated items is the "Invoice Cost" from the metals service center and includes the mill base price of steel plus mill extras, transportation to the metals service center, material handling, and delivery of long lengths of mill shapes to the job site. Material costs for structural bolts and welding rods should be added to the estimate. The bare labor cost and bare equipment cost for these items is for both field fabrication and field installation or erection, and include time for cutting, welding and drilling in the fabricated metal items. Drilling into concrete and fasteners to fasten field fabricated items to other work are not included and should be added to the estimate.

05 20 00 Steel Joist Framing

- In any given project the total weight of open web steel joists is determined by the loads to be supported and the design. However, economies can be realized in minimizing the amount of labor used to place the joists. This is done by maximizing the joist spacing, and therefore minimizing the number of joists required to be installed on the job. Certain spacings and locations may be required by the design, but in other cases maximizing the spacing and keeping it as uniform as possible will keep the costs down.

05 30 00 Steel Decking

- The takeoff and estimating of metal deck involves more than simply the area of the floor or roof and the type of deck specified or shown on the drawings. Many different sizes and types of openings may exist. Small openings for individual pipes or conduits may be drilled after the floor/roof is installed, but larger openings may require special deck lengths as well as reinforcing or structural support. The estimator should determine who will be supplying this reinforcing. Additionally, some deck terminations are part of the deck package, such as screed angles and pour stops, and others will be part of the steel contract, such as angles attached to structural members and cast-in-place angles and plates. The estimator must ensure that all pieces are accounted for in the complete estimate.

05 50 00 Metal Fabrications

- The most economical steel stairs are those that use common materials, standard details, and most importantly, a uniform and relatively simple method of field assembly. Commonly available A36/A992 channels and plates are very good choices for the main stringers of the stairs, as are angles and tees for the carrier members. Risers and treads are usually made by specialty shops, and it is most economical to use a typical detail in as many places as possible. The stairs should be pre-assembled and shipped directly to the site. The field connections should be simple and straightforward to be accomplished efficiently, and with minimum equipment and labor.

Reference Numbers

Reference numbers are shown at the beginning of some major classifications. These numbers refer to related items in the Reference Section. The reference information may be an estimating procedure, an alternate pricing method, or technical information.

Note: Not all subdivisions listed here necessarily appear. ■

05 01 Maintenance of Metals

05 01 10 – Maintenance of Structural Metal Framing

05 01 10.51 Cleaning of Structural Metal Framing		Crew	Daily Output	Labor-Hours	Unit	2016 Bare Costs Material	Labor	Equipment	Total	Total Incl O&P
0010	**CLEANING OF STRUCTURAL METAL FRAMING**									
6125	Steel surface treatments, PDCA guidelines									
6235	Com'l blast (SSPC-SP6), loose scale, fine pwder rust, 2.0#/S.F. sand	E-11	1200	.027	S.F.	.35	.88	.19	1.42	2.14
6240	Tight mill scale, little/no rust, 3.0#/S.F. sand		1000	.032		.52	1.06	.23	1.81	2.69
6245	Exist coat blistered/pitted, 4.0#/S.F. sand		875	.037		.70	1.21	.27	2.18	3.19

05 05 Common Work Results for Metals

05 05 05 – Selective Demolition for Metals

05 05 05.10 Selective Demolition, Metals

		Crew	Daily Output	Labor-Hours	Unit	Material	Labor	Equipment	Total	Total Incl O&P
0010	**SELECTIVE DEMOLITION, METALS** R024119-10									
0015	Excludes shores, bracing, cutting, loading, hauling, dumping									
0020	Remove nuts only up to 3/4" diameter	1 Sswk	480	.017	Ea.		.67		.67	1.22
0030	7/8" to 1-1/4" diameter		240	.033			1.33		1.33	2.44
0040	1-3/8" to 2" diameter		160	.050			2		2	3.66
0060	Unbolt and remove structural bolts up to 3/4" diameter		240	.033			1.33		1.33	2.44
0070	7/8" to 2" diameter		160	.050			2		2	3.66
0140	Light weight framing members, remove whole or cut up, up to 20 lb.		240	.033			1.33		1.33	2.44
0150	21 – 40 lb.	2 Sswk	210	.076			3.04		3.04	5.60
0160	41 – 80 lb.	3 Sswk	180	.133			5.30		5.30	9.75
0170	81 – 120 lb.	4 Sswk	150	.213			8.50		8.50	15.60
0230	Structural members, remove whole or cut up, up to 500 lb.	E-19	48	.500			19.95	20	39.95	57.50
0240	1/4 – 2 tons	E-18	36	1.111			44.50	27	71.50	110
0250	2 – 5 tons	E-24	30	1.067			42.50	25	67.50	104
0260	5 – 10 tons	E-20	24	2.667			106	49.50	155.50	243
0270	10 – 15 tons	E-2	18	2.667			107	84.50	191.50	287
0340	Fabricated item, remove whole or cut up, up to 20 lb.	1 Sswk	96	.083			3.32		3.32	6.10
0350	21 – 40 lb.	2 Sswk	84	.190			7.60		7.60	13.95
0360	41 – 80 lb.	3 Sswk	72	.333			13.30		13.30	24.50
0370	81 – 120 lb.	4 Sswk	60	.533			21.50		21.50	39
0380	121 – 500 lb.	E-19	48	.500			19.95	20	39.95	57.50
0500	Steel roof decking, uncovered, bare	B-2	5000	.008	S.F.		.24		.24	.40

05 05 13 – Shop-Applied Coatings for Metal

05 05 13.50 Paints and Protective Coatings

		Crew	Daily Output	Labor-Hours	Unit	Material	Labor	Equipment	Total	Total Incl O&P
0010	**PAINTS AND PROTECTIVE COATINGS**									
5900	Galvanizing structural steel in shop, under 1 ton				Ton	550			550	605
5950	1 ton to 20 tons					505			505	555
6000	Over 20 tons					460			460	505

05 05 19 – Post-Installed Concrete Anchors

05 05 19.10 Chemical Anchors

		Crew	Daily Output	Labor-Hours	Unit	Material	Labor	Equipment	Total	Total Incl O&P
0010	**CHEMICAL ANCHORS**									
0020	Includes layout & drilling									
1430	Chemical anchor, w/rod & epoxy cartridge, 3/4" diam. x 9-1/2" long	B-89A	27	.593	Ea.	9.25	20	4.33	33.58	48.50
1435	1" diameter x 11-3/4" long		24	.667		16.95	22.50	4.87	44.32	62
1440	1-1/4" diameter x 14" long		21	.762		34	25.50	5.55	65.05	86.50
1445	1-3/4" diameter x 15" long		20	.800		63.50	27	5.85	96.35	122
1450	18" long		17	.941		76.50	32	6.90	115.40	145
1455	2" diameter x 18" long		16	1		101	34	7.30	142.30	176
1460	24" long		15	1.067		132	36	7.80	175.80	214

05 05 19 – Post-Installed Concrete Anchors

05 05 19.20 Expansion Anchors			Crew	Daily Output	Labor-Hours	Unit	2016 Bare Costs Material	Labor	Equipment	Total	Total Incl O&P
0010	**EXPANSION ANCHORS**										
0100	Anchors for concrete, brick or stone, no layout and drilling										
0200	Expansion shields, zinc, 1/4" diameter, 1-5/16" long, single	G	1 Carp	90	.089	Ea.	.44	3.36		3.80	6.15
0300	1-3/8" long, double	G		85	.094		.55	3.56		4.11	6.55
0400	3/8" diameter, 1-1/2" long, single	G		85	.094		.66	3.56		4.22	6.70
0500	2" long, double	G		80	.100		1.20	3.78		4.98	7.65
0600	1/2" diameter, 2-1/16" long, single	G		80	.100		1.21	3.78		4.99	7.70
0700	2-1/2" long, double	G		75	.107		1.94	4.03		5.97	8.90
0800	5/8" diameter, 2-5/8" long, single	G		75	.107		2.01	4.03		6.04	8.95
0900	2-3/4" long, double	G		70	.114		2.74	4.32		7.06	10.25
1000	3/4" diameter, 2-3/4" long, single	G		70	.114		3.18	4.32		7.50	10.75
1100	3-15/16" long, double	G		65	.123		5.10	4.65		9.75	13.40
2100	Hollow wall anchors for gypsum wall board, plaster or tile										
2500	3/16" diameter, short	G	1 Carp	150	.053	Ea.	.46	2.02		2.48	3.89
3000	Toggle bolts, bright steel, 1/8" diameter, 2" long	G		85	.094		.21	3.56		3.77	6.20
3100	4" long	G		80	.100		.27	3.78		4.05	6.65
3200	3/16" diameter, 3" long	G		80	.100		.28	3.78		4.06	6.65
3300	6" long	G		75	.107		.39	4.03		4.42	7.20
3400	1/4" diameter, 3" long	G		75	.107		.38	4.03		4.41	7.15
3500	6" long	G		70	.114		.55	4.32		4.87	7.85
3600	3/8" diameter, 3" long	G		70	.114		.93	4.32		5.25	8.25
3700	6" long	G		60	.133		1.43	5.05		6.48	10
3800	1/2" diameter, 4" long	G		60	.133		2	5.05		7.05	10.65
3900	6" long	G		50	.160		2.59	6.05		8.64	13
4000	Nailing anchors										
4100	Nylon nailing anchor, 1/4" diameter, 1" long		1 Carp	3.20	2.500	C	22	94.50		116.50	182
4200	1-1/2" long			2.80	2.857		25.50	108		133.50	209
4300	2" long			2.40	3.333		30	126		156	244
4400	Metal nailing anchor, 1/4" diameter, 1" long	G		3.20	2.500		18.60	94.50		113.10	179
4500	1-1/2" long	G		2.80	2.857		24.50	108		132.50	208
4600	2" long	G		2.40	3.333		29	126		155	243
5000	Screw anchors for concrete, masonry,										
5100	stone & tile, no layout or drilling included										
5700	Lag screw shields, 1/4" diameter, short	G	1 Carp	90	.089	Ea.	.37	3.36		3.73	6.05
5800	Long	G		85	.094		.41	3.56		3.97	6.40
5900	3/8" diameter, short	G		85	.094		.71	3.56		4.27	6.75
6000	Long	G		80	.100		.94	3.78		4.72	7.40
6100	1/2" diameter, short	G		80	.100		1.01	3.78		4.79	7.45
6200	Long	G		75	.107		1.31	4.03		5.34	8.20
6300	5/8" diameter, short	G		70	.114		1.57	4.32		5.89	9
6400	Long	G		65	.123		2	4.65		6.65	10
6600	Lead, #6 & #8, 3/4" long	G		260	.031		.19	1.16		1.35	2.16
6700	#10 - #14, 1-1/2" long	G		200	.040		.39	1.51		1.90	2.96
6800	#16 & #18, 1-1/2" long	G		160	.050		.42	1.89		2.31	3.63
6900	Plastic, #6 & #8, 3/4" long			260	.031		.04	1.16		1.20	1.99
7000	#8 & #10, 7/8" long			240	.033		.04	1.26		1.30	2.15
7100	#10 & #12, 1" long			220	.036		.05	1.37		1.42	2.36
7200	#14 & #16, 1-1/2" long			160	.050		.08	1.89		1.97	3.26
8950	Self-drilling concrete screw, hex washer head, 3/16" diam. x 1-3/4" long	G		300	.027		.21	1.01		1.22	1.92
8960	2-1/4" long	G		250	.032		.24	1.21		1.45	2.29
8970	Phillips flat head, 3/16" diam. x 1-3/4" long	G		300	.027		.21	1.01		1.22	1.92
8980	2-1/4" long	G		250	.032		.24	1.21		1.45	2.29

05 05 Common Work Results for Metals

05 05 21 – Fastening Methods for Metal

05 05 21.15 Drilling Steel		Crew	Daily Output	Labor-Hours	Unit	Material	2016 Bare Costs Labor	Equipment	Total	Total Incl O&P
0010	**DRILLING STEEL**									
1910	Drilling & layout for steel, up to 1/4" deep, no anchor									
1920	Holes, 1/4" diameter	1 Sswk	112	.071	Ea.	.10	2.85		2.95	5.35
1925	For each additional 1/4" depth, add		336	.024		.10	.95		1.05	1.85
1930	3/8" diameter		104	.077		.09	3.07		3.16	5.75
1935	For each additional 1/4" depth, add		312	.026		.09	1.02		1.11	1.98
1940	1/2" diameter		96	.083		.11	3.32		3.43	6.25
1945	For each additional 1/4" depth, add		288	.028		.11	1.11		1.22	2.16
1950	5/8" diameter		88	.091		.16	3.63		3.79	6.85
1955	For each additional 1/4" depth, add		264	.030		.16	1.21		1.37	2.40
1960	3/4" diameter		80	.100		.20	3.99		4.19	7.50
1965	For each additional 1/4" depth, add		240	.033		.20	1.33		1.53	2.66
1970	7/8" diameter		72	.111		.26	4.43		4.69	8.45
1975	For each additional 1/4" depth, add		216	.037		.26	1.48		1.74	3
1980	1" diameter		64	.125		.23	4.99		5.22	9.40
1985	For each additional 1/4" depth, add		192	.042		.23	1.66		1.89	3.30
1990	For drilling up, add						40%			

05 05 23 – Metal Fastenings

05 05 23.10 Bolts and Hex Nuts			Crew	Daily Output	Labor-Hours	Unit	Material	Labor	Equipment	Total	Total Incl O&P
0010	**BOLTS & HEX NUTS**, Steel, A307										
0100	1/4" diameter, 1/2" long	G	1 Sswk	140	.057	Ea.	.06	2.28		2.34	4.25
0200	1" long	G		140	.057		.07	2.28		2.35	4.26
0300	2" long	G		130	.062		.10	2.46		2.56	4.61
0400	3" long	G		130	.062		.15	2.46		2.61	4.67
0500	4" long	G		120	.067		.17	2.66		2.83	5.05
0600	3/8" diameter, 1" long	G		130	.062		.14	2.46		2.60	4.66
0700	2" long	G		130	.062		.18	2.46		2.64	4.70
0800	3" long	G		120	.067		.24	2.66		2.90	5.15
0900	4" long	G		120	.067		.30	2.66		2.96	5.20
1000	5" long	G		115	.070		.38	2.78		3.16	5.50
1100	1/2" diameter, 1-1/2" long	G		120	.067		.40	2.66		3.06	5.30
1200	2" long	G		120	.067		.46	2.66		3.12	5.40
1300	4" long	G		115	.070		.75	2.78		3.53	5.95
1400	6" long	G		110	.073		1.05	2.90		3.95	6.45
1500	8" long	G		105	.076		1.38	3.04		4.42	7.10
1600	5/8" diameter, 1-1/2" long	G		120	.067		.85	2.66		3.51	5.80
1700	2" long	G		120	.067		.94	2.66		3.60	5.90
1800	4" long	G		115	.070		1.34	2.78		4.12	6.60
1900	6" long	G		110	.073		1.72	2.90		4.62	7.20
2000	8" long	G		105	.076		2.55	3.04		5.59	8.40
2100	10" long	G		100	.080		3.20	3.19		6.39	9.35
2200	3/4" diameter, 2" long	G		120	.067		1.15	2.66		3.81	6.15
2300	4" long	G		110	.073		1.65	2.90		4.55	7.10
2400	6" long	G		105	.076		2.12	3.04		5.16	7.95
2500	8" long	G		95	.084		3.20	3.36		6.56	9.65
2600	10" long	G		85	.094		4.20	3.76		7.96	11.50
2700	12" long	G		80	.100		4.92	3.99		8.91	12.70
2800	1" diameter, 3" long	G		105	.076		2.69	3.04		5.73	8.55
2900	6" long	G		90	.089		3.94	3.55		7.49	10.85
3000	12" long	G		75	.107		7.10	4.26		11.36	15.60
3100	For galvanized, add						75%				
3200	For stainless, add						350%				

05 05 23 – Metal Fastenings

05 05 23.30 Lag Screws			Crew	Daily Output	Labor-Hours	Unit	2016 Bare Costs Material	Labor	Equipment	Total	Total Incl O&P
0010	LAG SCREWS										
0020	Steel, 1/4" diameter, 2" long	G	1 Carp	200	.040	Ea.	.09	1.51		1.60	2.63
0100	3/8" diameter, 3" long	G		150	.053		.32	2.02		2.34	3.73
0200	1/2" diameter, 3" long	G		130	.062		.68	2.33		3.01	4.65
0300	5/8" diameter, 3" long	G	↓	120	.067	↓	1.22	2.52		3.74	5.55

05 05 23.50 Powder Actuated Tools and Fasteners

			Crew	Daily Output	Labor-Hours	Unit	Material	Labor	Equipment	Total	Total Incl O&P
0010	POWDER ACTUATED TOOLS & FASTENERS										
0020	Stud driver, .22 caliber, single shot					Ea.	149			149	164
0100	.27 caliber, semi automatic, strip					"	450			450	495
0300	Powder load, single shot, .22 cal, power level 2, brown					C	5.50			5.50	6.05
0400	Strip, .27 cal, power level 4, red						7.90			7.90	8.70
0600	Drive pin, .300 x 3/4" long	G	1 Carp	4.80	1.667		4.23	63		67.23	111
0700	.300 x 3" long with washer	G	"	4	2	↓	13	75.50		88.50	141

05 05 23.55 Rivets

			Crew	Daily Output	Labor-Hours	Unit	Material	Labor	Equipment	Total	Total Incl O&P
0010	RIVETS										
0100	Aluminum rivet & mandrel, 1/2" grip length x 1/8" diameter	G	1 Carp	4.80	1.667	C	8.05	63		71.05	115
0200	3/16" diameter	G		4	2		11.90	75.50		87.40	140
0300	Aluminum rivet, steel mandrel, 1/8" diameter	G		4.80	1.667		10.30	63		73.30	117
0400	3/16" diameter	G		4	2		16.50	75.50		92	145
0500	Copper rivet, steel mandrel, 1/8" diameter	G		4.80	1.667		9.50	63		72.50	116
0800	Stainless rivet & mandrel, 1/8" diameter	G		4.80	1.667		25.50	63		88.50	134
0900	3/16" diameter	G		4	2		40.50	75.50		116	172
1000	Stainless rivet, steel mandrel, 1/8" diameter	G		4.80	1.667		15.70	63		78.70	123
1100	3/16" diameter	G		4	2		25.50	75.50		101	155
1200	Steel rivet and mandrel, 1/8" diameter	G		4.80	1.667		7.75	63		70.75	115
1300	3/16" diameter	G	↓	4	2	↓	12.55	75.50		88.05	141
1400	Hand riveting tool, standard					Ea.	76.50			76.50	84
1500	Deluxe						385			385	425
1600	Power riveting tool, standard						500			500	555
1700	Deluxe					↓	1,650			1,650	1,825

05 05 23.70 Structural Blind Bolts

			Crew	Daily Output	Labor-Hours	Unit	Material	Labor	Equipment	Total	Total Incl O&P
0010	STRUCTURAL BLIND BOLTS										
0100	1/4" diameter x 1/4" grip	G	1 Sswk	240	.033	Ea.	1.44	1.33		2.77	4.02
0150	1/2" grip	G		216	.037		1.55	1.48		3.03	4.42
0200	3/8" diameter x 1/2" grip	G		232	.034		2.04	1.38		3.42	4.76
0250	3/4" grip	G		208	.038		2.14	1.53		3.67	5.15
0300	1/2" diameter x 1/2" grip	G		224	.036		4.66	1.42		6.08	7.75
0350	3/4" grip	G		200	.040		6.50	1.60		8.10	10.10
0400	5/8" diameter x 3/4" grip	G		216	.037		9.60	1.48		11.08	13.30
0450	1" grip	G	↓	192	.042	↓	11.05	1.66		12.71	15.20

05 12 Structural Steel Framing

05 12 23 – Structural Steel for Buildings

05 12 23.10 Ceiling Supports			Crew	Daily Output	Labor-Hours	Unit	2016 Bare Costs Material	Labor	Equipment	Total	Total Incl O&P
0010	**CEILING SUPPORTS**										
1000	Entrance door/folding partition supports, shop fabricated	G	E-4	60	.533	L.F.	26.50	21.50	2.45	50.45	71
1100	Linear accelerator door supports	G		14	2.286		121	92.50	10.50	224	315
1200	Lintels or shelf angles, hung, exterior hot dipped galv.	G		267	.120		18.10	4.84	.55	23.49	29.50
1250	Two coats primer paint instead of galv.	G		267	.120		15.65	4.84	.55	21.04	27
1400	Monitor support, ceiling hung, expansion bolted	G		4	8	Ea.	420	325	37	782	1,100
1450	Hung from pre-set inserts	G		6	5.333		450	215	24.50	689.50	915
1600	Motor supports for overhead doors	G		4	8		214	325	37	576	870
1700	Partition support for heavy folding partitions, without pocket	G		24	1.333	L.F.	60.50	54	6.15	120.65	172
1750	Supports at pocket only	G		12	2.667		121	108	12.25	241.25	345
2000	Rolling grilles & fire door supports	G		34	.941		51.50	38	4.33	93.83	132
2100	Spider-leg light supports, expansion bolted to ceiling slab	G		8	4	Ea.	172	162	18.40	352.40	505
2150	Hung from pre-set inserts	G		12	2.667	″	186	108	12.25	306.25	415
2400	Toilet partition support	G		36	.889	L.F.	60.50	36	4.09	100.59	137
2500	X-ray travel gantry support	G		12	2.667	″	207	108	12.25	327.25	440

05 12 23.15 Columns, Lightweight			Crew	Daily Output	Labor-Hours	Unit	Material	Labor	Equipment	Total	Total Incl O&P
0010	**COLUMNS, LIGHTWEIGHT**										
8000	Lally columns, to 8′, 3-1/2″ diameter		2 Carp	24	.667	Ea.	66	25		91	115
8080	4″ diameter		″	20	.800	″	81	30		111	140

05 12 23.17 Columns, Structural			Crew	Daily Output	Labor-Hours	Unit	Material	Labor	Equipment	Total	Total Incl O&P
0010	**COLUMNS, STRUCTURAL**	R051223-10									
0015	Made from recycled materials										
0020	Shop fab'd for 100-ton, 1-2 story project, bolted connections										
0800	Steel, concrete filled, extra strong pipe, 3-1/2″ diameter		E-2	660	.073	L.F.	43.50	2.93	2.30	48.73	56
0830	4″ diameter			780	.062		48.50	2.48	1.95	52.93	60
0890	5″ diameter			1020	.047		58	1.90	1.49	61.39	69
0930	6″ diameter			1200	.040		77	1.61	1.27	79.88	89
0940	8″ diameter			1100	.044		77	1.76	1.38	80.14	89
1100	For galvanizing, add					Lb.	.25			.25	.28
1300	For web ties, angles, etc., add per added lb.		1 Sswk	945	.008		1.33	.34		1.67	2.08
1500	Steel pipe, extra strong, no concrete, 3″ to 5″ diameter	G	E-2	16000	.003		1.33	.12	.10	1.55	1.78
1600	6″ to 12″ diameter	G		14000	.003		1.33	.14	.11	1.58	1.83
5100	Structural tubing, rect., 5″ to 6″ wide, light section	G		8000	.006		1.33	.24	.19	1.76	2.11
5200	Heavy section	G		12000	.004		1.33	.16	.13	1.62	1.89
8090	For projects 75 to 99 tons, add					%	10%				
8092	50 to 74 tons, add						20%				
8094	25 to 49 tons, add						30%	10%			
8096	10 to 24 tons, add						50%	25%			
8098	2 to 9 tons, add						75%	50%			
8099	Less than 2 tons, add						100%	100%			

05 12 23.45 Lintels			Crew	Daily Output	Labor-Hours	Unit	Material	Labor	Equipment	Total	Total Incl O&P
0010	**LINTELS**										
0015	Made from recycled materials										
0020	Plain steel angles, shop fabricated, under 500 lb.	G	1 Bric	550	.015	Lb.	1.02	.53		1.55	2
0100	500 to 1000 lb.	G		640	.013	″	.99	.45		1.44	1.84
2000	Steel angles, 3-1/2″ x 3″, 1/4″ thick, 2′-6″ long	G		47	.170	Ea.	14.30	6.15		20.45	26
2100	4′-6″ long	G		26	.308		26	11.10		37.10	47
2600	4″ x 3-1/2″, 1/4″ thick, 5′-0″ long	G		21	.381		33	13.75		46.75	59
2700	9′-0″ long	G		12	.667		59	24		83	105

05 12 Structural Steel Framing

05 12 23 – Structural Steel for Buildings

05 12 23.65 Plates		Crew	Daily Output	Labor-Hours	Unit	Material	2016 Bare Costs Labor	Equipment	Total	Total Incl O&P
0010	**PLATES**									
0015	Made from recycled materials									
0020	For connections & stiffener plates, shop fabricated									
0050	1/8" thick (5.1 lb./S.F.) G				S.F.	6.75			6.75	7.45
0100	1/4" thick (10.2 lb./S.F.) G					13.50			13.50	14.85
0300	3/8" thick (15.3 lb./S.F.) G					20.50			20.50	22.50
0400	1/2" thick (20.4 lb./S.F.) G					27			27	29.50
0450	3/4" thick (30.6 lb./S.F.) G					40.50			40.50	44.50
0500	1" thick (40.8 lb./S.F.) G					54			54	59.50
2000	Steel plate, warehouse prices, no shop fabrication									
2100	1/4" thick (10.2 lb./S.F.) G				S.F.	6.85			6.85	7.55

05 12 23.77 Structural Steel Projects		Crew	Daily Output	Labor-Hours	Unit	Material	Labor	Equipment	Total	Total Incl O&P
0010	**STRUCTURAL STEEL PROJECTS** R050516-30									
0015	Made from recycled materials									
0020	Shop fab'd for 100-ton, 1-2 story project, bolted connections									
0201	Apartments, nursing homes, etc., 1 to 2 stories R050523-10 G	E-2	6.45	7.442	Ton	2,650	300	236	3,186	3,725
0701	Offices, hospitals, etc., steel bearing, 1 to 2 stories G		6.45	7.442		2,650	300	236	3,186	3,725
3101	Roof trusses, simple connections G		8.13	5.904		3,700	238	187	4,125	4,700
3200	Moment/composite connections G	E-5	8.30	8.675		4,500	350	201	5,051	5,800
5390	For projects 75 to 99 tons, add R051223-20					10%				
5392	50 to 74 tons, add					20%				
5394	25 to 49 tons, add R051223-25					30%	10%			
5396	10 to 24 tons, add					50%	25%			
5398	2 to 9 tons, add R051223-30					75%	50%			
5399	Less than 2 tons, add					100%	100%			

05 12 23.78 Structural Steel Secondary Members		Crew	Daily Output	Labor-Hours	Unit	Material	Labor	Equipment	Total	Total Incl O&P
0010	**STRUCTURAL STEEL SECONDARY MEMBERS**									
0015	Made from recycled materials									
0020	Shop fabricated for 20-ton girt/purlin framing package, materials only									
0100	Girts/purlins, C/Z-shapes, includes clips and bolts									
0110	6" x 2-1/2" x 2-1/2", 16 ga., 3.0 lb./L.F.				L.F.	3.58			3.58	3.94
0115	14 ga., 3.5 lb./L.F.					4.17			4.17	4.59
0120	8" x 2-3/4" x 2-3/4", 16 ga., 3.4 lb./L.F.					4.05			4.05	4.46
0125	14 ga., 4.1 lb./L.F.					4.89			4.89	5.40
0130	12 ga., 5.6 lb./L.F.					6.70			6.70	7.35
0135	10" x 3-1/2" x 3-1/2", 14 ga., 4.7 lb./L.F.					5.60			5.60	6.15
0140	12 ga., 6.7 lb./L.F.					8			8	8.80
0145	12" x 3-1/2" x 3-1/2", 14 ga., 5.3 lb./L.F.					6.30			6.30	6.95
0150	12 ga., 7.4 lb./L.F.					8.80			8.80	9.70
0200	Eave struts, C-shape, includes clips and bolts									
0210	6" x 4" x 3", 16 ga., 3.1 lb./L.F.				L.F.	3.70			3.70	4.07
0215	14 ga., 3.9 lb./L.F.					4.65			4.65	5.10
0220	8" x 4" x 3", 16 ga., 3.5 lb./L.F.					4.17			4.17	4.59
0225	14 ga., 4.4 lb./L.F.					5.25			5.25	5.75
0230	12 ga., 6.2 lb./L.F.					7.40			7.40	8.15
0235	10" x 5" x 3", 14 ga., 5.2 lb./L.F.					6.20			6.20	6.80
0240	12 ga., 7.3 lb./L.F.					8.70			8.70	9.60
0245	12" x 5" x 4", 14 ga., 6.0 lb./L.F.					7.15			7.15	7.85
0250	12 ga., 8.4 lb./L.F.					10			10	11
0300	Rake/base angle, excludes concrete drilling and expansion anchors									
0310	2" x 2", 14 ga., 1.0 lb./L.F.	2 Sswk	640	.025	L.F.	1.19	1		2.19	3.14
0315	3" x 2", 14 ga., 1.3 lb./L.F.		535	.030		1.55	1.19		2.74	3.90

05 12 Structural Steel Framing

05 12 23 – Structural Steel for Buildings

05 12 23.78 Structural Steel Secondary Members

	05 12 23.78 Structural Steel Secondary Members	Crew	Daily Output	Labor-Hours	Unit	Material	2016 Bare Costs Labor	Equipment	Total	Total Incl O&P
0320	3" x 3", 14 ga., 1.6 lb./L.F.	2 Sswk	500	.032	L.F.	1.91	1.28		3.19	4.44
0325	4" x 3", 14 ga., 1.8 lb./L.F.		480	.033		2.15	1.33		3.48	4.80
0600	Installation of secondary members, erection only									
0610	Girts, purlins, eave struts, 16 ga., 6" deep	E-18	100	.400	Ea.		16.10	9.65	25.75	39.50
0615	8" deep		80	.500			20	12.10	32.10	49.50
0620	14 ga., 6" deep		80	.500			20	12.10	32.10	49.50
0625	8" deep		65	.615			24.50	14.90	39.40	61
0630	10" deep		55	.727			29	17.60	46.60	72
0635	12" deep		50	.800			32	19.35	51.35	79.50
0640	12 ga., 8" deep		50	.800			32	19.35	51.35	79.50
0645	10" deep		45	.889			35.50	21.50	57	88
0650	12" deep		40	1			40	24	64	99
0900	For less than 20-ton job lots									
0905	For 15 to 19 tons, add				%	10%				
0910	For 10 to 14 tons, add					25%				
0915	For 5 to 9 tons, add					50%	50%	50%		
0920	For 1 to 4 tons, add					75%	75%	75%		
0925	For less than 1 ton, add					100%	100%	100%		

05 12 23.79 Structural Steel

	05 12 23.79 Structural Steel	Crew	Daily Output	Labor-Hours	Unit	Material	Labor	Equipment	Total	Total Incl O&P
0010	**STRUCTURAL STEEL** R050516-30									
0020	Shop fab'd for 100-ton, 1-2 story project, bolted conn's.									
0050	Beams, W 6 x 9 R050521-20 G	E-2	720	.067	L.F.	14.30	2.69	2.11	19.10	23
0100	W 8 x 10 G		720	.067		15.90	2.69	2.11	20.70	24.50
0200	Columns, W 6 x 15 R051223-10 G		540	.089		26	3.58	2.82	32.40	38
0250	W 8 x 31 G		540	.089		53.50	3.58	2.82	59.90	68
7990	For projects 75 to 99 tons, add R051223-20				All	10%				
7992	50 to 75 tons, add					20%				
7994	25 to 49 tons, add R051223-25					30%	10%			
7996	10 to 24 tons, add					50%	25%			
7998	2 to 9 tons, add R051223-30					75%	50%			
7999	Less than 2 tons, add					100%	100%			

05 15 Wire Rope Assemblies

05 15 16 – Steel Wire Rope Assemblies

05 15 16.70 Temporary Cable Safety Railing

	05 15 16.70 Temporary Cable Safety Railing	Crew	Daily Output	Labor-Hours	Unit	Material	Labor	Equipment	Total	Total Incl O&P
0010	**TEMPORARY CABLE SAFETY RAILING**, Each 100' strand incl.									
0020	2 eyebolts, 1 turnbuckle, 100' cable, 2 thimbles, 6 clips									
0025	Made from recycled materials									
0100	One strand using 1/4" cable & accessories G	2 Sswk	4	4	C.L.F.	194	160		354	505
0200	1/2" cable & accessories G	"	2	8	"	425	320		745	1,050

05 21 Steel Joist Framing

05 21 19 – Open Web Steel Joist Framing

05 21 19.10 Open Web Joists			Crew	Daily Output	Labor-Hours	Unit	Material	2016 Bare Costs Labor	Equipment	Total	Total Incl O&P
0010	**OPEN WEB JOISTS**										
0015	Made from recycled materials										
0050	K series, 40-ton lots, horiz. bridging, spans to 30', shop primer	G	E-7	12	6	Ton	1,650	241	151	2,042	2,425
0130	8K1, 5.1 lb./L.F.	G		1200	.060	L.F.	4.22	2.41	1.51	8.14	10.70
0140	10K1, 5.0 lb./L.F.	G		1200	.060		4.14	2.41	1.51	8.06	10.60
0160	12K3, 5.7 lb./L.F.	G		1500	.048		4.72	1.93	1.21	7.86	10.05
0180	14K3, 6.0 lb./L.F.	G		1500	.048		4.97	1.93	1.21	8.11	10.30
0200	16K3, 6.3 lb./L.F.	G		1800	.040		5.20	1.61	1.01	7.82	9.75
0220	16K6, 8.1 lb./L.F.	G		1800	.040		6.70	1.61	1.01	9.32	11.40
0240	18K5, 7.7 lb./L.F.	G		2000	.036		6.40	1.45	.91	8.76	10.60
0260	18K9, 10.2 lb./L.F.	G		2000	.036		8.45	1.45	.91	10.81	12.90
0440	K series, 30' to 50' spans	G		17	4.235	Ton	1,625	170	107	1,902	2,225
0500	20K5, 8.2 lb./L.F.	G		2000	.036	L.F.	6.65	1.45	.91	9.01	10.95
0520	20K9, 10.8 lb./L.F.	G		2000	.036		8.80	1.45	.91	11.16	13.25
0540	22K5, 8.8 lb./L.F.	G		2000	.036		7.15	1.45	.91	9.51	11.45
0560	22K9, 11.3 lb./L.F.	G		2000	.036		9.20	1.45	.91	11.56	13.70
0580	24K6, 9.7 lb./L.F.	G		2200	.033		7.90	1.31	.82	10.03	11.95
0600	24K10, 13.1 lb./L.F.	G		2200	.033		10.65	1.31	.82	12.78	15
0620	26K6, 10.6 lb./L.F.	G		2200	.033		8.60	1.31	.82	10.73	12.75
0640	26K10, 13.8 lb./L.F.	G		2200	.033		11.20	1.31	.82	13.33	15.65
0660	28K8, 12.7 lb./L.F.	G		2400	.030		10.30	1.20	.76	12.26	14.35
0680	28K12, 17.1 lb./L.F.	G		2400	.030		13.90	1.20	.76	15.86	18.30
0700	30K8, 13.2 lb./L.F.	G		2400	.030		10.75	1.20	.76	12.71	14.80
0720	30K12, 17.6 lb./L.F.	G		2400	.030		14.30	1.20	.76	16.26	18.75
0800	For less than 40-ton job lots										
0802	For 30 to 39 tons, add					%	10%				
0804	20 to 29 tons, add						20%				
0806	10 to 19 tons, add						30%				
0807	5 to 9 tons, add						50%	25%			
0808	1 to 4 tons, add						75%	50%			
0809	Less than 1 ton, add						100%	100%			
6200	For shop prime paint other than mfrs. standard, add						20%				
6400	Individual steel bearing plate, 6" x 6" x 1/4" with J-hook	G	1 Bric	160	.050	Ea.	7.95	1.81		9.76	11.75

05 31 Steel Decking

05 31 13 – Steel Floor Decking

05 31 13.50 Floor Decking			Crew	Daily Output	Labor-Hours	Unit	Material	Labor	Equipment	Total	Total Incl O&P
0010	**FLOOR DECKING**										
0015	Made from recycled materials										
5100	Non-cellular composite decking, galvanized, 1-1/2" deep, 16 ga.	G	E-4	3500	.009	S.F.	3.48	.37	.04	3.89	4.56
5120	18 ga.	G		3650	.009		2.81	.35	.04	3.20	3.78
5140	20 ga.	G		3800	.008		2.24	.34	.04	2.62	3.12
5200	2" deep, 22 ga.	G		3860	.008		1.95	.33	.04	2.32	2.80
5300	20 ga.	G		3600	.009		2.15	.36	.04	2.55	3.08
5400	18 ga.	G		3380	.009		2.75	.38	.04	3.17	3.78
5500	16 ga.	G		3200	.010		3.44	.40	.05	3.89	4.57
5700	3" deep, 22 ga.	G		3200	.010		2.13	.40	.05	2.58	3.13
5800	20 ga.	G		3000	.011		2.37	.43	.05	2.85	3.44
5900	18 ga.	G		2850	.011		2.93	.45	.05	3.43	4.11
6000	16 ga.	G		2700	.012		3.91	.48	.05	4.44	5.25

05 31 Steel Decking

05 31 23 – Steel Roof Decking

05 31 23.50 Roof Decking			Crew	Daily Output	Labor-Hours	Unit	Material	Labor	Equipment	Total	Total Incl O&P
							2016 Bare Costs				
0010	**ROOF DECKING**										
0015	Made from recycled materials										
2100	Open type, 1-1/2" deep, Type B, wide rib, galv., 22 ga., under 50 sq.	G	E-4	4500	.007	S.F.	2.07	.29	.03	2.39	2.85
2600	20 ga., under 50 squares	G		3865	.008		2.42	.33	.04	2.79	3.31
2900	18 ga., under 50 squares	G		3800	.008		3.12	.34	.04	3.50	4.09
3050	16 ga., under 50 squares	G		3700	.009		4.21	.35	.04	4.60	5.30
3200	3" deep, Type N, 22 ga., under 50 squares	G		3600	.009		3.04	.36	.04	3.44	4.05
3300	20 ga., under 50 squares	G		3400	.009		3.28	.38	.04	3.70	4.36
3400	18 ga., under 50 squares	G		3200	.010		4.25	.40	.05	4.70	5.45
3500	16 ga., under 50 squares	G		3000	.011		5.60	.43	.05	6.08	7
3700	4-1/2" deep, Type J, 20 ga., over 50 squares	G		2700	.012		3.68	.48	.05	4.21	4.99
4100	6" deep, Type H, 18 ga., over 50 squares	G		2000	.016		5.80	.65	.07	6.52	7.65
4500	7-1/2" deep, Type H, 18 ga., over 50 squares	G		1690	.019		6.90	.76	.09	7.75	9.10

05 31 33 – Steel Form Decking

05 31 33.50 Form Decking			Crew	Daily Output	Labor-Hours	Unit	Material	Labor	Equipment	Total	Total Incl O&P
0010	**FORM DECKING**										
0015	Made from recycled materials										
6100	Slab form, steel, 28 ga., 9/16" deep, Type UFS, uncoated	G	E-4	4000	.008	S.F.	1.56	.32	.04	1.92	2.35
6200	Galvanized	G		4000	.008		1.38	.32	.04	1.74	2.15
6220	24 ga., 1" deep, Type UF1X, uncoated	G		3900	.008		1.50	.33	.04	1.87	2.30
6240	Galvanized	G		3900	.008		1.76	.33	.04	2.13	2.59
6300	24 ga., 1-5/16" deep, Type UFX, uncoated	G		3800	.008		1.59	.34	.04	1.97	2.41
6400	Galvanized	G		3800	.008		1.87	.34	.04	2.25	2.72
6500	22 ga., 1-5/16" deep, uncoated	G		3700	.009		2.02	.35	.04	2.41	2.90
6600	Galvanized	G		3700	.009		2.06	.35	.04	2.45	2.95
6700	22 ga., 2" deep, uncoated	G		3600	.009		2.62	.36	.04	3.02	3.59
6800	Galvanized	G		3600	.009		2.57	.36	.04	2.97	3.54

05 41 Structural Metal Stud Framing

05 41 13 – Load-Bearing Metal Stud Framing

05 41 13.05 Bracing			Crew	Daily Output	Labor-Hours	Unit	Material	Labor	Equipment	Total	Total Incl O&P
0010	**BRACING**, shear wall X-bracing, per 10′ x 10′ bay, one face										
0015	Made of recycled materials										
0120	Metal strap, 20 ga. x 4" wide	G	2 Carp	18	.889	Ea.	18.90	33.50		52.40	77.50
0130	6" wide	G		18	.889		31.50	33.50		65	91
0160	18 ga. x 4" wide	G		16	1		33	38		71	99.50
0170	6" wide	G		16	1		48.50	38		86.50	117
0410	Continuous strap bracing, per horizontal row on both faces										
0420	Metal strap, 20 ga. x 2" wide, studs 12" O.C.	G	1 Carp	7	1.143	C.L.F.	57	43		100	136
0430	16" O.C.	G		8	1		57	38		95	127
0440	24" O.C.	G		10	.800		57	30		87	114
0450	18 ga. x 2" wide, studs 12" O.C.	G		6	1.333		79	50.50		129.50	172
0460	16" O.C.	G		7	1.143		79	43		122	160
0470	24" O.C.	G		8	1		79	38		117	151

05 41 13.10 Bridging			Crew	Daily Output	Labor-Hours	Unit	Material	Labor	Equipment	Total	Total Incl O&P
0010	**BRIDGING**, solid between studs w/1-1/4" leg track, per stud bay										
0015	Made from recycled materials										
0200	Studs 12" O.C., 18 ga. x 2-1/2" wide	G	1 Carp	125	.064	Ea.	.95	2.42		3.37	5.10
0210	3-5/8" wide	G		120	.067		1.16	2.52		3.68	5.50
0220	4" wide	G		120	.067		1.22	2.52		3.74	5.55

05 41 13.10 Bridging

05 41 13.10 Bridging			Crew	Daily Output	Labor-Hours	Unit	2016 Bare Costs Material	Labor	Equipment	Total	Total Incl O&P
0230	6" wide	G	1 Carp	115	.070	Ea.	1.60	2.63		4.23	6.15
0240	8" wide	G		110	.073		1.97	2.75		4.72	6.75
0300	16 ga. x 2-1/2" wide	G		115	.070		1.21	2.63		3.84	5.75
0310	3-5/8" wide	G		110	.073		1.47	2.75		4.22	6.20
0320	4" wide	G		110	.073		1.57	2.75		4.32	6.35
0330	6" wide	G		105	.076		2.01	2.88		4.89	7.05
0340	8" wide	G		100	.080		2.51	3.02		5.53	7.80
1200	Studs 16" O.C., 18 ga. x 2-1/2" wide	G		125	.064		1.22	2.42		3.64	5.40
1210	3-5/8" wide	G		120	.067		1.49	2.52		4.01	5.85
1220	4" wide	G		120	.067		1.57	2.52		4.09	5.95
1230	6" wide	G		115	.070		2.05	2.63		4.68	6.65
1240	8" wide	G		110	.073		2.52	2.75		5.27	7.40
1300	16 ga. x 2-1/2" wide	G		115	.070		1.55	2.63		4.18	6.10
1310	3-5/8" wide	G		110	.073		1.88	2.75		4.63	6.65
1320	4" wide	G		110	.073		2.01	2.75		4.76	6.80
1330	6" wide	G		105	.076		2.57	2.88		5.45	7.65
1340	8" wide	G		100	.080		3.22	3.02		6.24	8.60
2200	Studs 24" O.C., 18 ga. x 2-1/2" wide	G		125	.064		1.77	2.42		4.19	6
2210	3-5/8" wide	G		120	.067		2.15	2.52		4.67	6.60
2220	4" wide	G		120	.067		2.27	2.52		4.79	6.70
2230	6" wide	G		115	.070		2.96	2.63		5.59	7.65
2240	8" wide	G		110	.073		3.65	2.75		6.40	8.60
2300	16 ga. x 2-1/2" wide	G		115	.070		2.24	2.63		4.87	6.85
2310	3-5/8" wide	G		110	.073		2.72	2.75		5.47	7.60
2320	4" wide	G		110	.073		2.91	2.75		5.66	7.80
2330	6" wide	G		105	.076		3.72	2.88		6.60	8.90
2340	8" wide	G		100	.080		4.65	3.02		7.67	10.15
3000	Continuous bridging, per row										
3100	16 ga. x 1-1/2" channel thru studs 12" O.C.	G	1 Carp	6	1.333	C.L.F.	51.50	50.50		102	142
3110	16" O.C.	G		7	1.143		51.50	43		94.50	130
3120	24" O.C.	G		8.80	.909		51.50	34.50		86	115
4100	2" x 2" angle x 18 ga., studs 12" O.C.	G		7	1.143		80.50	43		123.50	161
4110	16" O.C.	G		9	.889		80.50	33.50		114	145
4120	24" O.C.	G		12	.667		80.50	25		105.50	131
4200	16 ga., studs 12" O.C.	G		5	1.600		101	60.50		161.50	212
4210	16" O.C.	G		7	1.143		101	43		144	184
4220	24" O.C.	G		10	.800		101	30		131	162

05 41 13.25 Framing, Boxed Headers/Beams

05 41 13.25 Framing, Boxed Headers/Beams			Crew	Daily Output	Labor-Hours	Unit	Material	Labor	Equipment	Total	Total Incl O&P
0010	**FRAMING, BOXED HEADERS/BEAMS**										
0015	Made from recycled materials										
0200	Double, 18 ga. x 6" deep	G	2 Carp	220	.073	L.F.	5.50	2.75		8.25	10.65
0210	8" deep	G		210	.076		6.05	2.88		8.93	11.45
0220	10" deep	G		200	.080		7.40	3.02		10.42	13.20
0230	12" deep	G		190	.084		8.05	3.18		11.23	14.20
0300	16 ga. x 8" deep	G		180	.089		7	3.36		10.36	13.35
0310	10" deep	G		170	.094		8.45	3.56		12.01	15.20
0320	12" deep	G		160	.100		9.15	3.78		12.93	16.45
0400	14 ga. x 10" deep	G		140	.114		9.75	4.32		14.07	17.95
0410	12" deep	G		130	.123		10.65	4.65		15.30	19.55
1210	Triple, 18 ga. x 8" deep	G		170	.094		8.75	3.56		12.31	15.60
1220	10" deep	G		165	.097		10.60	3.67		14.27	17.80
1230	12" deep	G		160	.100		11.60	3.78		15.38	19.10

05 41 Structural Metal Stud Framing

05 41 13 – Load-Bearing Metal Stud Framing

05 41 13.25 Framing, Boxed Headers/Beams			Crew	Daily Output	Labor-Hours	Unit	2016 Bare Costs Material	Labor	Equipment	Total	Total Incl O&P
1300	16 ga. x 8" deep	G	2 Carp	145	.110	L.F.	10.15	4.17		14.32	18.15
1310	10" deep	G		140	.114		12.15	4.32		16.47	20.50
1320	12" deep	G		135	.119		13.25	4.48		17.73	22
1400	14 ga. x 10" deep	G		115	.139		13.25	5.25		18.50	23.50
1410	12" deep	G	↓	110	.145	↓	14.65	5.50		20.15	25.50

05 41 13.30 Framing, Stud Walls

			Crew	Daily Output	Labor-Hours	Unit	Material	Labor	Equipment	Total	Total Incl O&P
0010	**FRAMING, STUD WALLS** w/top & bottom track, no openings,										
0020	Headers, beams, bridging or bracing										
0025	Made from recycled materials										
4100	8' high walls, 18 ga. x 2-1/2" wide, studs 12" O.C.	G	2 Carp	54	.296	L.F.	9.10	11.20		20.30	29
4110	16" O.C.	G		77	.208		7.25	7.85		15.10	21
4120	24" O.C.	G		107	.150		5.45	5.65		11.10	15.45
4130	3-5/8" wide, studs 12" O.C.	G		53	.302		10.85	11.40		22.25	31
4140	16" O.C.	G		76	.211		8.70	7.95		16.65	23
4150	24" O.C.	G		105	.152		6.55	5.75		12.30	16.85
4160	4" wide, studs 12" O.C.	G		52	.308		11.30	11.65		22.95	32
4170	16" O.C.	G		74	.216		9.05	8.15		17.20	23.50
4180	24" O.C.	G		103	.155		6.80	5.85		12.65	17.35
4190	6" wide, studs 12" O.C.	G		51	.314		14.40	11.85		26.25	35.50
4200	16" O.C.	G		73	.219		11.55	8.30		19.85	26.50
4210	24" O.C.	G		101	.158		8.70	6		14.70	19.65
4220	8" wide, studs 12" O.C.	G		50	.320		17.45	12.10		29.55	39.50
4230	16" O.C.	G		72	.222		14	8.40		22.40	29.50
4240	24" O.C.	G		100	.160		10.60	6.05		16.65	22
4300	16 ga. x 2-1/2" wide, studs 12" O.C.	G		47	.340		10.85	12.85		23.70	33.50
4310	16" O.C.	G		68	.235		8.60	8.90		17.50	24.50
4320	24" O.C.	G		94	.170		6.35	6.45		12.80	17.70
4330	3-5/8" wide, studs 12" O.C.	G		46	.348		12.95	13.15		26.10	36
4340	16" O.C.	G		66	.242		10.25	9.15		19.40	26.50
4350	24" O.C.	G		92	.174		7.55	6.55		14.10	19.30
4360	4" wide, studs 12" O.C.	G		45	.356		13.55	13.45		27	37.50
4370	16" O.C.	G		65	.246		10.75	9.30		20.05	27.50
4380	24" O.C.	G		90	.178		7.95	6.70		14.65	19.95
4390	6" wide, studs 12" O.C.	G		44	.364		17.05	13.75		30.80	42
4400	16" O.C.	G		64	.250		13.55	9.45		23	31
4410	24" O.C.	G		88	.182		10.05	6.85		16.90	22.50
4420	8" wide, studs 12" O.C.	G		43	.372		21	14.05		35.05	46.50
4430	16" O.C.	G		63	.254		16.60	9.60		26.20	34.50
4440	24" O.C.	G		86	.186		12.30	7.05		19.35	25.50
5100	10' high walls, 18 ga. x 2-1/2" wide, studs 12" O.C.	G		54	.296		10.90	11.20		22.10	31
5110	16" O.C.	G		77	.208		8.65	7.85		16.50	22.50
5120	24" O.C.	G		107	.150		6.35	5.65		12	16.45
5130	3-5/8" wide, studs 12" O.C.	G		53	.302		13	11.40		24.40	33.50
5140	16" O.C.	G		76	.211		10.30	7.95		18.25	24.50
5150	24" O.C.	G		105	.152		7.60	5.75		13.35	18
5160	4" wide, studs 12" O.C.	G		52	.308		13.55	11.65		25.20	34.50
5170	16" O.C.	G		74	.216		10.75	8.15		18.90	25.50
5180	24" O.C.	G		103	.155		7.95	5.85		13.80	18.55
5190	6" wide, studs 12" O.C.	G		51	.314		17.25	11.85		29.10	39
5200	16" O.C.	G		73	.219		13.70	8.30		22	29
5210	24" O.C.	G		101	.158		10.15	6		16.15	21
5220	8" wide, studs 12" O.C.	G	↓	50	.320	↓	21	12.10		33.10	43.50

05 41 Structural Metal Stud Framing

05 41 13 – Load-Bearing Metal Stud Framing

05 41 13.30 Framing, Stud Walls			Crew	Daily Output	Labor-Hours	Unit	2016 Bare Costs Material	Labor	Equipment	Total	Total Incl O&P
5230	16" O.C.	G	2 Carp	72	.222	L.F.	16.55	8.40		24.95	32.50
5240	24" O.C.	G		100	.160		12.30	6.05		18.35	23.50
5300	16 ga. x 2-1/2" wide, studs 12" O.C.	G		47	.340		13.10	12.85		25.95	36
5310	16" O.C.	G		68	.235		10.30	8.90		19.20	26
5320	24" O.C.	G		94	.170		7.45	6.45		13.90	18.95
5330	3-5/8" wide, studs 12" O.C.	G		46	.348		15.60	13.15		28.75	39
5340	16" O.C.	G		66	.242		12.25	9.15		21.40	29
5350	24" O.C.	G		92	.174		8.90	6.55		15.45	21
5360	4" wide, studs 12" O.C.	G		45	.356		16.35	13.45		29.80	40.50
5370	16" O.C.	G		65	.246		12.85	9.30		22.15	29.50
5380	24" O.C.	G		90	.178		9.35	6.70		16.05	21.50
5390	6" wide, studs 12" O.C.	G		44	.364		20.50	13.75		34.25	45.50
5400	16" O.C.	G		64	.250		16.15	9.45		25.60	33.50
5410	24" O.C.	G		88	.182		11.80	6.85		18.65	24.50
5420	8" wide, studs 12" O.C.	G		43	.372		25	14.05		39.05	51
5430	16" O.C.	G		63	.254		19.80	9.60		29.40	38
5440	24" O.C.	G		86	.186		14.45	7.05		21.50	27.50
6190	12' high walls, 18 ga. x 6" wide, studs 12" O.C.	G		41	.390		20	14.75		34.75	46.50
6200	16" O.C.	G		58	.276		15.80	10.45		26.25	35
6210	24" O.C.	G		81	.198		11.55	7.45		19	25
6220	8" wide, studs 12" O.C.	G		40	.400		24.50	15.10		39.60	52
6230	16" O.C.	G		57	.281		19.15	10.60		29.75	39
6240	24" O.C.	G		80	.200		14	7.55		21.55	28
6390	16 ga. x 6" wide, studs 12" O.C.	G		35	.457		24	17.30		41.30	55.50
6400	16" O.C.	G		51	.314		18.80	11.85		30.65	40.50
6410	24" O.C.	G		70	.229		13.55	8.65		22.20	29.50
6420	8" wide, studs 12" O.C.	G		34	.471		29.50	17.80		47.30	62.50
6430	16" O.C.	G		50	.320		23	12.10		35.10	46
6440	24" O.C.	G		69	.232		16.60	8.75		25.35	33
6530	14 ga. x 3-5/8" wide, studs 12" O.C.	G		34	.471		22.50	17.80		40.30	55
6540	16" O.C.	G		48	.333		17.75	12.60		30.35	40.50
6550	24" O.C.	G		65	.246		12.75	9.30		22.05	29.50
6560	4" wide, studs 12" O.C.	G		33	.485		24	18.35		42.35	57
6570	16" O.C.	G		47	.340		18.75	12.85		31.60	42
6580	24" O.C.	G		64	.250		13.50	9.45		22.95	30.50
6730	12 ga. x 3-5/8" wide, studs 12" O.C.	G		31	.516		31.50	19.50		51	67
6740	16" O.C.	G		43	.372		24.50	14.05		38.55	50
6750	24" O.C.	G		59	.271		17.15	10.25		27.40	36
6760	4" wide, studs 12" O.C.	G		30	.533		33.50	20		53.50	71
6770	16" O.C.	G		42	.381		26	14.40		40.40	52.50
6780	24" O.C.	G		58	.276		18.30	10.45		28.75	37.50
7390	16' high walls, 16 ga. x 6" wide, studs 12" O.C.	G		33	.485		31	18.35		49.35	64.50
7400	16" O.C.	G		48	.333		24	12.60		36.60	47.50
7410	24" O.C.	G		67	.239		17.05	9.05		26.10	34
7420	8" wide, studs 12" O.C.	G		32	.500		38	18.90		56.90	73.50
7430	16" O.C.	G		47	.340		29.50	12.85		42.35	54
7440	24" O.C.	G		66	.242		21	9.15		30.15	38.50
7560	14 ga. x 4" wide, studs 12" O.C.	G		31	.516		31	19.50		50.50	66.50
7570	16" O.C.	G		45	.356		24	13.45		37.45	49
7580	24" O.C.	G		61	.262		17	9.90		26.90	35.50
7590	6" wide, studs 12" O.C.	G		30	.533		39	20		59	77
7600	16" O.C.	G		44	.364		30	13.75		43.75	56
7610	24" O.C.	G	2 Carp	60	.267	L.F.	21.50	10.10		31.60	40.50

05 41 Structural Metal Stud Framing

05 41 13 – Load-Bearing Metal Stud Framing

05 41 13.30 Framing, Stud Walls			Crew	Daily Output	Labor-Hours	Unit	Material	Labor	Equipment	Total	Total Incl O&P
							2016 Bare Costs				
7760	12 ga. x 4" wide, studs 12" O.C.	G	2 Carp	29	.552	L.F.	44	21		65	83
7770	16" O.C.	G		40	.400		33.50	15.10		48.60	62.50
7780	24" O.C.	G		55	.291		23.50	11		34.50	44
7790	6" wide, studs 12" O.C.	G		28	.571		55.50	21.50		77	97
7800	16" O.C.	G		39	.410		42.50	15.50		58	72.50
7810	24" O.C.	G		54	.296		29.50	11.20		40.70	51.50
8590	20' high walls, 14 ga. x 6" wide, studs 12" O.C.	G		29	.552		48	21		69	87.50
8600	16" O.C.	G		42	.381		37	14.40		51.40	64.50
8610	24" O.C.	G		57	.281		26	10.60		36.60	46.50
8620	8" wide, studs 12" O.C.	G		28	.571		52	21.50		73.50	93
8630	16" O.C.	G		41	.390		40	14.75		54.75	68.50
8640	24" O.C.	G		56	.286		28.50	10.80		39.30	49
8790	12 ga. x 6" wide, studs 12" O.C.	G		27	.593		68	22.50		90.50	113
8800	16" O.C.	G		37	.432		52	16.35		68.35	85
8810	24" O.C.	G		51	.314		36	11.85		47.85	59.50
8820	8" wide, studs 12" O.C.	G		26	.615		82.50	23.50		106	130
8830	16" O.C.	G		36	.444		63	16.80		79.80	97.50
8840	24" O.C.	G	↓	50	.320	↓	43.50	12.10		55.60	68.50

05 42 Cold-Formed Metal Joist Framing

05 42 13 – Cold-Formed Metal Floor Joist Framing

05 42 13.05 Bracing

			Crew	Daily Output	Labor-Hours	Unit	Material	Labor	Equipment	Total	Total Incl O&P
0010	**BRACING**, continuous, per row, top & bottom										
0015	Made from recycled materials										
0120	Flat strap, 20 ga. x 2" wide, joists at 12" O.C.	G	1 Carp	4.67	1.713	C.L.F.	60	65		125	174
0130	16" O.C.	G		5.33	1.501		57.50	56.50		114	159
0140	24" O.C.	G		6.66	1.201		55.50	45.50		101	137
0150	18 ga. x 2" wide, joists at 12" O.C.	G		4	2		78.50	75.50		154	214
0160	16" O.C.	G		4.67	1.713		77	65		142	193
0170	24" O.C.	G	↓	5.33	1.501	↓	75.50	56.50		132	178

05 42 13.10 Bridging

			Crew	Daily Output	Labor-Hours	Unit	Material	Labor	Equipment	Total	Total Incl O&P
0010	**BRIDGING**, solid between joists w/1-1/4" leg track, per joist bay										
0015	Made from recycled materials										
0230	Joists 12" O.C., 18 ga. track x 6" wide	G	1 Carp	80	.100	Ea.	1.60	3.78		5.38	8.10
0240	8" wide	G		75	.107		1.97	4.03		6	8.90
0250	10" wide	G		70	.114		2.46	4.32		6.78	9.95
0260	12" wide	G		65	.123		2.78	4.65		7.43	10.85
0330	16 ga. track x 6" wide	G		70	.114		2.01	4.32		6.33	9.45
0340	8" wide	G		65	.123		2.51	4.65		7.16	10.55
0350	10" wide	G		60	.133		3.10	5.05		8.15	11.85
0360	12" wide	G		55	.145		3.58	5.50		9.08	13.15
0440	14 ga. track x 8" wide	G		60	.133		3.13	5.05		8.18	11.90
0450	10" wide	G		55	.145		3.89	5.50		9.39	13.50
0460	12" wide	G		50	.160		4.49	6.05		10.54	15.10
0550	12 ga. track x 10" wide	G		45	.178		5.70	6.70		12.40	17.50
0560	12" wide	G		40	.200		5.80	7.55		13.35	19.05
1230	16" O.C., 18 ga. track x 6" wide	G		80	.100		2.05	3.78		5.83	8.60
1240	8" wide	G		75	.107		2.52	4.03		6.55	9.55
1250	10" wide	G		70	.114		3.15	4.32		7.47	10.70
1260	12" wide	G		65	.123		3.56	4.65		8.21	11.70
1330	16 ga. track x 6" wide	G	↓	70	.114	↓	2.57	4.32		6.89	10.10

05 42 13.10 Bridging

05 42 13.10 Bridging			Crew	Daily Output	Labor-Hours	Unit	Material	Labor	Equipment	Total	Total Incl O&P
							2016 Bare Costs				
1340	8″ wide	G	1 Carp	65	.123	Ea.	3.22	4.65		7.87	11.35
1350	10″ wide	G		60	.133		3.98	5.05		9.03	12.80
1360	12″ wide	G		55	.145		4.59	5.50		10.09	14.25
1440	14 ga. track x 8″ wide	G		60	.133		4.01	5.05		9.06	12.85
1450	10″ wide	G		55	.145		4.98	5.50		10.48	14.70
1460	12″ wide	G		50	.160		5.75	6.05		11.80	16.50
1550	12 ga. track x 10″ wide	G		45	.178		7.30	6.70		14	19.25
1560	12″ wide	G		40	.200		7.45	7.55		15	21
2230	24″ O.C., 18 ga. track x 6″ wide	G		80	.100		2.96	3.78		6.74	9.60
2240	8″ wide	G		75	.107		3.65	4.03		7.68	10.75
2250	10″ wide	G		70	.114		4.56	4.32		8.88	12.25
2260	12″ wide	G		65	.123		5.15	4.65		9.80	13.45
2330	16 ga. track x 6″ wide	G		70	.114		3.72	4.32		8.04	11.35
2340	8″ wide	G		65	.123		4.65	4.65		9.30	12.90
2350	10″ wide	G		60	.133		5.75	5.05		10.80	14.80
2360	12″ wide	G		55	.145		6.65	5.50		12.15	16.50
2440	14 ga. track x 8″ wide	G		60	.133		5.80	5.05		10.85	14.85
2450	10″ wide	G		55	.145		7.20	5.50		12.70	17.15
2460	12″ wide	G		50	.160		8.35	6.05		14.40	19.30
2550	12 ga. track x 10″ wide	G		45	.178		10.55	6.70		17.25	23
2560	12″ wide	G		40	.200		10.80	7.55		18.35	24.50

05 42 13.25 Framing, Band Joist

05 42 13.25 Framing, Band Joist			Crew	Daily Output	Labor-Hours	Unit	Material	Labor	Equipment	Total	Total Incl O&P
0010	**FRAMING, BAND JOIST** (track) fastened to bearing wall										
0015	Made from recycled materials										
0220	18 ga. track x 6″ deep	G	2 Carp	1000	.016	L.F.	1.30	.60		1.90	2.44
0230	8″ deep	G		920	.017		1.61	.66		2.27	2.87
0240	10″ deep	G		860	.019		2.01	.70		2.71	3.39
0320	16 ga. track x 6″ deep	G		900	.018		1.64	.67		2.31	2.93
0330	8″ deep	G		840	.019		2.05	.72		2.77	3.46
0340	10″ deep	G		780	.021		2.53	.78		3.31	4.08
0350	12″ deep	G		740	.022		2.92	.82		3.74	4.58
0430	14 ga. track x 8″ deep	G		750	.021		2.55	.81		3.36	4.16
0440	10″ deep	G		720	.022		3.17	.84		4.01	4.90
0450	12″ deep	G		700	.023		3.66	.86		4.52	5.50
0540	12 ga. track x 10″ deep	G		670	.024		4.64	.90		5.54	6.60
0550	12″ deep	G		650	.025		4.75	.93		5.68	6.75

05 42 13.30 Framing, Boxed Headers/Beams

05 42 13.30 Framing, Boxed Headers/Beams			Crew	Daily Output	Labor-Hours	Unit	Material	Labor	Equipment	Total	Total Incl O&P
0010	**FRAMING, BOXED HEADERS/BEAMS**										
0015	Made from recycled materials										
0200	Double, 18 ga. x 6″ deep	G	2 Carp	220	.073	L.F.	5.50	2.75		8.25	10.65
0210	8″ deep	G		210	.076		6.05	2.88		8.93	11.45
0220	10″ deep	G		200	.080		7.40	3.02		10.42	13.20
0230	12″ deep	G		190	.084		8.05	3.18		11.23	14.20
0300	16 ga. x 8″ deep	G		180	.089		7	3.36		10.36	13.35
0310	10″ deep	G		170	.094		8.45	3.56		12.01	15.20
0320	12″ deep	G		160	.100		9.15	3.78		12.93	16.45
0400	14 ga. x 10″ deep	G		140	.114		9.75	4.32		14.07	17.95
0410	12″ deep	G		130	.123		10.65	4.65		15.30	19.55
0500	12 ga. x 10″ deep	G		110	.145		12.85	5.50		18.35	23.50
0510	12″ deep	G		100	.160		14.15	6.05		20.20	26
1210	Triple, 18 ga. x 8″ deep	G		170	.094		8.75	3.56		12.31	15.60
1220	10″ deep	G		165	.097		10.60	3.67		14.27	17.80

05 42 Cold-Formed Metal Joist Framing

05 42 13 – Cold-Formed Metal Floor Joist Framing

	05 42 13.30 Framing, Boxed Headers/Beams		Crew	Daily Output	Labor-Hours	Unit	2016 Bare Costs Material	Labor	Equipment	Total	Total Incl O&P
1230	12″ deep	G	2 Carp	160	.100	L.F.	11.60	3.78		15.38	19.10
1300	16 ga. x 8″ deep	G		145	.110		10.15	4.17		14.32	18.15
1310	10″ deep	G		140	.114		12.15	4.32		16.47	20.50
1320	12″ deep	G		135	.119		13.25	4.48		17.73	22
1400	14 ga. x 10″ deep	G		115	.139		14.10	5.25		19.35	24.50
1410	12″ deep	G		110	.145		15.50	5.50		21	26.50
1500	12 ga. x 10″ deep	G		90	.178		18.75	6.70		25.45	32
1510	12″ deep	G		85	.188		21	7.10		28.10	35

05 42 13.40 Framing, Joists

			Crew	Daily Output	Labor-Hours	Unit	Material	Labor	Equipment	Total	Total Incl O&P
0010	**FRAMING, JOISTS**, no band joists (track), web stiffeners, headers,										
0020	Beams, bridging or bracing										
0025	Made from recycled materials										
0030	Joists (2″ flange) and fasteners, materials only										
0220	18 ga. x 6″ deep	G				L.F.	1.69			1.69	1.86
0230	8″ deep	G					1.98			1.98	2.18
0240	10″ deep	G					2.34			2.34	2.58
0320	16 ga. x 6″ deep	G					2.07			2.07	2.28
0330	8″ deep	G					2.47			2.47	2.71
0340	10″ deep	G					2.89			2.89	3.18
0350	12″ deep	G					3.28			3.28	3.60
0430	14 ga. x 8″ deep	G					3.10			3.10	3.41
0440	10″ deep	G					3.57			3.57	3.93
0450	12″ deep	G					4.06			4.06	4.47
0540	12 ga. x 10″ deep	G					5.20			5.20	5.70
0550	12″ deep	G					5.90			5.90	6.50
1010	Installation of joists to band joists, beams & headers, labor only										
1220	18 ga. x 6″ deep		2 Carp	110	.145	Ea.		5.50		5.50	9.20
1230	8″ deep			90	.178			6.70		6.70	11.25
1240	10″ deep			80	.200			7.55		7.55	12.65
1320	16 ga. x 6″ deep			95	.168			6.35		6.35	10.65
1330	8″ deep			70	.229			8.65		8.65	14.45
1340	10″ deep			60	.267			10.10		10.10	16.90
1350	12″ deep			55	.291			11		11	18.40
1430	14 ga. x 8″ deep			65	.246			9.30		9.30	15.60
1440	10″ deep			45	.356			13.45		13.45	22.50
1450	12″ deep			35	.457			17.30		17.30	29
1540	12 ga. x 10″ deep			40	.400			15.10		15.10	25.50
1550	12″ deep			30	.533			20		20	34

05 42 13.45 Framing, Web Stiffeners

			Crew	Daily Output	Labor-Hours	Unit	Material	Labor	Equipment	Total	Total Incl O&P
0010	**FRAMING, WEB STIFFENERS** at joist bearing, fabricated from										
0020	Stud piece (1-5/8″ flange) to stiffen joist (2″ flange)										
0025	Made from recycled materials										
2120	For 6″ deep joist, with 18 ga. x 2-1/2″ stud	G	1 Carp	120	.067	Ea.	.91	2.52		3.43	5.20
2130	3-5/8″ stud	G		110	.073		1.08	2.75		3.83	5.80
2140	4″ stud	G		105	.076		1.12	2.88		4	6.05
2150	6″ stud	G		100	.080		1.42	3.02		4.44	6.60
2160	8″ stud	G		95	.084		1.71	3.18		4.89	7.25
2220	8″ deep joist, with 2-1/2″ stud	G		120	.067		1.22	2.52		3.74	5.55
2230	3-5/8″ stud	G		110	.073		1.45	2.75		4.20	6.20
2240	4″ stud	G		105	.076		1.50	2.88		4.38	6.45
2250	6″ stud	G		100	.080		1.90	3.02		4.92	7.15
2260	8″ stud	G		95	.084		2.29	3.18		5.47	7.85

05 42 13.45 Framing, Web Stiffeners		Crew	Daily Output	Labor-Hours	Unit	Material	2016 Bare Costs Labor	Equipment	Total	Total Incl O&P
2320	10″ deep joist, with 2-1/2″ stud	G	1 Carp	110	.073	Ea.	1.51	2.75		4.26	6.25
2330	3-5/8″ stud	G		100	.080		1.79	3.02		4.81	7
2340	4″ stud	G		95	.084		1.86	3.18		5.04	7.40
2350	6″ stud	G		90	.089		2.36	3.36		5.72	8.25
2360	8″ stud	G		85	.094		2.84	3.56		6.40	9.05
2420	12″ deep joist, with 2-1/2″ stud	G		110	.073		1.82	2.75		4.57	6.60
2430	3-5/8″ stud	G		100	.080		2.16	3.02		5.18	7.45
2440	4″ stud	G		95	.084		2.24	3.18		5.42	7.80
2450	6″ stud	G		90	.089		2.84	3.36		6.20	8.75
2460	8″ stud	G		85	.094		3.42	3.56		6.98	9.70
3130	For 6″ deep joist, with 16 ga. x 3-5/8″ stud	G		100	.080		1.34	3.02		4.36	6.50
3140	4″ stud	G		95	.084		1.40	3.18		4.58	6.90
3150	6″ stud	G		90	.089		1.75	3.36		5.11	7.60
3160	8″ stud	G		85	.094		2.14	3.56		5.70	8.30
3230	8″ deep joist, with 3-5/8″ stud	G		100	.080		1.80	3.02		4.82	7.05
3240	4″ stud	G		95	.084		1.88	3.18		5.06	7.40
3250	6″ stud	G		90	.089		2.35	3.36		5.71	8.25
3260	8″ stud	G		85	.094		2.87	3.56		6.43	9.10
3330	10″ deep joist, with 3-5/8″ stud	G		85	.094		2.22	3.56		5.78	8.40
3340	4″ stud	G		80	.100		2.32	3.78		6.10	8.90
3350	6″ stud	G		75	.107		2.91	4.03		6.94	9.95
3360	8″ stud	G		70	.114		3.55	4.32		7.87	11.15
3430	12″ deep joist, with 3-5/8″ stud	G		85	.094		2.68	3.56		6.24	8.90
3440	4″ stud	G		80	.100		2.80	3.78		6.58	9.45
3450	6″ stud	G		75	.107		3.50	4.03		7.53	10.60
3460	8″ stud	G		70	.114		4.28	4.32		8.60	11.95
4230	For 8″ deep joist, with 14 ga. x 3-5/8″ stud	G		90	.089		2.22	3.36		5.58	8.10
4240	4″ stud	G		85	.094		2.35	3.56		5.91	8.55
4250	6″ stud	G		80	.100		2.95	3.78		6.73	9.60
4260	8″ stud	G		75	.107		3.15	4.03		7.18	10.20
4330	10″ deep joist, with 3-5/8″ stud	G		75	.107		2.76	4.03		6.79	9.80
4340	4″ stud	G		70	.114		2.91	4.32		7.23	10.45
4350	6″ stud	G		65	.123		3.65	4.65		8.30	11.80
4360	8″ stud	G		60	.133		3.90	5.05		8.95	12.75
4430	12″ deep joist, with 3-5/8″ stud	G		75	.107		3.32	4.03		7.35	10.40
4440	4″ stud	G		70	.114		3.50	4.32		7.82	11.10
4450	6″ stud	G		65	.123		4.40	4.65		9.05	12.65
4460	8″ stud	G		60	.133		4.70	5.05		9.75	13.60
5330	For 10″ deep joist, with 12 ga. x 3-5/8″ stud	G		65	.123		3.97	4.65		8.62	12.15
5340	4″ stud	G		60	.133		4.23	5.05		9.28	13.10
5350	6″ stud	G		55	.145		5.35	5.50		10.85	15.10
5360	8″ stud	G		50	.160		6.45	6.05		12.50	17.25
5430	12″ deep joist, with 3-5/8″ stud	G		65	.123		4.78	4.65		9.43	13.05
5440	4″ stud	G		60	.133		5.10	5.05		10.15	14.05
5450	6″ stud	G		55	.145		6.45	5.50		11.95	16.30
5460	8″ stud	G		50	.160		7.80	6.05		13.85	18.70

05 42 Cold-Formed Metal Joist Framing

05 42 23 – Cold-Formed Metal Roof Joist Framing

05 42 23.05 Framing, Bracing		Crew	Daily Output	Labor-Hours	Unit	Material	Labor	Equipment	Total	Total Incl O&P
						2016 Bare Costs				
0010	FRAMING, BRACING									
0015	Made from recycled materials									
0020	Continuous bracing, per row									
0100	16 ga. x 1-1/2" channel thru rafters/trusses @ 16" O.C. G	1 Carp	4.50	1.778	C.L.F.	51.50	67		118.50	170
0120	24" O.C. G		6	1.333		51.50	50.50		102	142
0300	2" x 2" angle x 18 ga., rafters/trusses @ 16" O.C. G		6	1.333		80.50	50.50		131	173
0320	24" O.C. G		8	1		80.50	38		118.50	152
0400	16 ga., rafters/trusses @ 16" O.C. G		4.50	1.778		101	67		168	224
0420	24" O.C. G	↓	6.50	1.231	↓	101	46.50		147.50	189

05 42 23.10 Framing, Bridging		Crew	Daily Output	Labor-Hours	Unit	Material	Labor	Equipment	Total	Total Incl O&P
0010	FRAMING, BRIDGING									
0015	Made from recycled materials									
0020	Solid, between rafters w/1-1/4" leg track, per rafter bay									
1200	Rafters 16" O.C., 18 ga. x 4" deep G	1 Carp	60	.133	Ea.	1.57	5.05		6.62	10.15
1210	6" deep G		57	.140		2.05	5.30		7.35	11.15
1220	8" deep G		55	.145		2.52	5.50		8.02	12
1230	10" deep G		52	.154		3.15	5.80		8.95	13.20
1240	12" deep G		50	.160		3.56	6.05		9.61	14.05
2200	24" O.C., 18 ga. x 4" deep G		60	.133		2.27	5.05		7.32	10.95
2210	6" deep G		57	.140		2.96	5.30		8.26	12.15
2220	8" deep G		55	.145		3.65	5.50		9.15	13.20
2230	10" deep G		52	.154		4.56	5.80		10.36	14.75
2240	12" deep G	↓	50	.160	↓	5.15	6.05		11.20	15.80

05 42 23.50 Framing, Parapets		Crew	Daily Output	Labor-Hours	Unit	Material	Labor	Equipment	Total	Total Incl O&P
0010	FRAMING, PARAPETS									
0015	Made from recycled materials									
0100	3' high installed on 1st story, 18 ga. x 4" wide studs, 12" O.C. G	2 Carp	100	.160	L.F.	5.70	6.05		11.75	16.40
0110	16" O.C. G		150	.107		4.85	4.03		8.88	12.10
0120	24" O.C. G		200	.080		4.01	3.02		7.03	9.45
0200	6" wide studs, 12" O.C. G		100	.160		7.30	6.05		13.35	18.20
0210	16" O.C. G		150	.107		6.25	4.03		10.28	13.60
0220	24" O.C. G		200	.080		5.15	3.02		8.17	10.75
1100	Installed on 2nd story, 18 ga. x 4" wide studs, 12" O.C. G		95	.168		5.70	6.35		12.05	16.90
1110	16" O.C. G		145	.110		4.85	4.17		9.02	12.35
1120	24" O.C. G		190	.084		4.01	3.18		7.19	9.75
1200	6" wide studs, 12" O.C. G		95	.168		7.30	6.35		13.65	18.70
1210	16" O.C. G		145	.110		6.25	4.17		10.42	13.85
1220	24" O.C. G		190	.084		5.15	3.18		8.33	11.05
2100	Installed on gable, 18 ga. x 4" wide studs, 12" O.C. G		85	.188		5.70	7.10		12.80	18.15
2110	16" O.C. G		130	.123		4.85	4.65		9.50	13.15
2120	24" O.C. G		170	.094		4.01	3.56		7.57	10.35
2200	6" wide studs, 12" O.C. G		85	.188		7.30	7.10		14.40	19.95
2210	16" O.C. G		130	.123		6.25	4.65		10.90	14.65
2220	24" O.C. G	↓	170	.094	↓	5.15	3.56		8.71	11.65

05 42 23.60 Framing, Roof Rafters		Crew	Daily Output	Labor-Hours	Unit	Material	Labor	Equipment	Total	Total Incl O&P
0010	FRAMING, ROOF RAFTERS									
0015	Made from recycled materials									
0100	Boxed ridge beam, double, 18 ga. x 6" deep G	2 Carp	160	.100	L.F.	5.50	3.78		9.28	12.40
0110	8" deep G		150	.107		6.05	4.03		10.08	13.40
0120	10" deep G		140	.114		7.40	4.32		11.72	15.40
0130	12" deep G	↓	130	.123	↓	8.05	4.65		12.70	16.65

05 42 Cold-Formed Metal Joist Framing

05 42 23 – Cold-Formed Metal Roof Joist Framing

05 42 23.60 Framing, Roof Rafters

	Description		Crew	Daily Output	Labor-Hours	Unit	Material	Labor	Equipment	Total	Total Incl O&P
							2016 Bare Costs				
0200	16 ga. x 6" deep	G	2 Carp	150	.107	L.F.	6.20	4.03		10.23	13.60
0210	8" deep	G		140	.114		7	4.32		11.32	14.95
0220	10" deep	G		130	.123		8.45	4.65		13.10	17.05
0230	12" deep	G		120	.133		9.15	5.05		14.20	18.55
1100	Rafters, 2" flange, material only, 18 ga. x 6" deep	G					1.69			1.69	1.86
1110	8" deep	G					1.98			1.98	2.18
1120	10" deep	G					2.34			2.34	2.58
1130	12" deep	G					2.70			2.70	2.97
1200	16 ga. x 6" deep	G					2.07			2.07	2.28
1210	8" deep	G					2.47			2.47	2.71
1220	10" deep	G					2.89			2.89	3.18
1230	12" deep	G					3.28			3.28	3.60
2100	Installation only, ordinary rafter to 4:12 pitch, 18 ga. x 6" deep		2 Carp	35	.457	Ea.		17.30		17.30	29
2110	8" deep			30	.533			20		20	34
2120	10" deep			25	.640			24		24	40.50
2130	12" deep			20	.800			30		30	50.50
2200	16 ga. x 6" deep			30	.533			20		20	34
2210	8" deep			25	.640			24		24	40.50
2220	10" deep			20	.800			30		30	50.50
2230	12" deep			15	1.067			40.50		40.50	67.50
8100	Add to labor, ordinary rafters on steep roofs							25%			
8110	Dormers & complex roofs							50%			
8200	Hip & valley rafters to 4:12 pitch							25%			
8210	Steep roofs							50%			
8220	Dormers & complex roofs							75%			
8300	Hip & valley jack rafters to 4:12 pitch							50%			
8310	Steep roofs							75%			
8320	Dormers & complex roofs							100%			

05 42 23.70 Framing, Soffits and Canopies

	Description		Crew	Daily Output	Labor-Hours	Unit	Material	Labor	Equipment	Total	Total Incl O&P
0010	**FRAMING, SOFFITS & CANOPIES**										
0015	Made from recycled materials										
0130	Continuous ledger track @ wall, studs @ 16" O.C., 18 ga. x 4" wide	G	2 Carp	535	.030	L.F.	1.05	1.13		2.18	3.04
0140	6" wide	G		500	.032		1.36	1.21		2.57	3.53
0150	8" wide	G		465	.034		1.68	1.30		2.98	4.03
0160	10" wide	G		430	.037		2.10	1.41		3.51	4.67
0230	Studs @ 24" O.C., 18 ga. x 4" wide	G		800	.020		1	.76		1.76	2.37
0240	6" wide	G		750	.021		1.30	.81		2.11	2.78
0250	8" wide	G		700	.023		1.61	.86		2.47	3.22
0260	10" wide	G		650	.025		2.01	.93		2.94	3.77
1000	Horizontal soffit and canopy members, material only										
1030	1-5/8" flange studs, 18 ga. x 4" deep	G				L.F.	1.34			1.34	1.48
1040	6" deep	G					1.70			1.70	1.87
1050	8" deep	G					2.05			2.05	2.26
1140	2" flange joists, 18 ga. x 6" deep	G					1.93			1.93	2.13
1150	8" deep	G					2.27			2.27	2.49
1160	10" deep	G					2.68			2.68	2.94
4030	Installation only, 18 ga., 1-5/8" flange x 4" deep		2 Carp	130	.123	Ea.		4.65		4.65	7.80
4040	6" deep			110	.145			5.50		5.50	9.20
4050	8" deep			90	.178			6.70		6.70	11.25
4140	2" flange, 18 ga. x 6" deep			110	.145			5.50		5.50	9.20
4150	8" deep			90	.178			6.70		6.70	11.25
4160	10" deep			80	.200			7.55		7.55	12.65

05 42 Cold-Formed Metal Joist Framing

05 42 23 – Cold-Formed Metal Roof Joist Framing

05 42 23.70 Framing, Soffits and Canopies			Crew	Daily Output	Labor-Hours	Unit	Material	Labor	Equipment	Total	Total Incl O&P
							2016 Bare Costs				
6010	Clips to attach fascia to rafter tails, 2" x 2" x 18 ga. angle	G	1 Carp	120	.067	Ea.	.95	2.52		3.47	5.25
6020	16 ga. angle	G	"	100	.080	↓	1.20	3.02		4.22	6.35

05 44 Cold-Formed Metal Trusses

05 44 13 – Cold-Formed Metal Roof Trusses

05 44 13.60 Framing, Roof Trusses			Crew	Daily Output	Labor-Hours	Unit	Material	Labor	Equipment	Total	Total Incl O&P
0010	**FRAMING, ROOF TRUSSES**										
0015	Made from recycled materials										
0020	Fabrication of trusses on ground, Fink (W) or King Post, to 4:12 pitch										
0120	18 ga. x 4" chords, 16' span	G	2 Carp	12	1.333	Ea.	62.50	50.50		113	154
0130	20' span	G		11	1.455		78.50	55		133.50	178
0140	24' span	G		11	1.455		94	55		149	195
0150	28' span	G		10	1.600		110	60.50		170.50	222
0160	32' span	G		10	1.600		125	60.50		185.50	239
0250	6" chords, 28' span	G		9	1.778		139	67		206	266
0260	32' span	G		9	1.778		159	67		226	288
0270	36' span	G		8	2		179	75.50		254.50	325
0280	40' span	G		8	2		199	75.50		274.50	345
1120	5:12 to 8:12 pitch, 18 ga. x 4" chords, 16' span	G		10	1.600		71.50	60.50		132	180
1130	20' span	G		9	1.778		89.50	67		156.50	212
1140	24' span	G		9	1.778		108	67		175	231
1150	28' span	G		8	2		125	75.50		200.50	265
1160	32' span	G		8	2		143	75.50		218.50	285
1250	6" chords, 28' span	G		7	2.286		159	86.50		245.50	320
1260	32' span	G		7	2.286		182	86.50		268.50	345
1270	36' span	G		6	2.667		204	101		305	395
1280	40' span	G		6	2.667		227	101		328	420
2120	9:12 to 12:12 pitch, 18 ga. x 4" chords, 16' span	G		8	2		89.50	75.50		165	226
2130	20' span	G		7	2.286		112	86.50		198.50	268
2140	24' span	G		7	2.286		134	86.50		220.50	293
2150	28' span	G		6	2.667		157	101		258	340
2160	32' span	G		6	2.667		179	101		280	365
2250	6" chords, 28' span	G		5	3.200		199	121		320	420
2260	32' span	G		5	3.200		227	121		348	455
2270	36' span	G		4	4		256	151		407	535
2280	40' span	G	↓	4	4		284	151		435	565
5120	Erection only of roof trusses, to 4:12 pitch, 16' span		F-6	48	.833			29	13.70	42.70	63.50
5130	20' span			46	.870			30.50	14.30	44.80	66.50
5140	24' span			44	.909			31.50	14.95	46.45	69.50
5150	28' span			42	.952			33	15.70	48.70	73
5160	32' span			40	1			35	16.45	51.45	76
5170	36' span			38	1.053			36.50	17.35	53.85	80.50
5180	40' span			36	1.111			38.50	18.30	56.80	84.50
5220	5:12 to 8:12 pitch, 16' span			42	.952			33	15.70	48.70	73
5230	20' span			40	1			35	16.45	51.45	76
5240	24' span			38	1.053			36.50	17.35	53.85	80.50
5250	28' span			36	1.111			38.50	18.30	56.80	84.50
5260	32' span			34	1.176			41	19.35	60.35	90
5270	36' span			32	1.250			43.50	20.50	64	95
5280	40' span			30	1.333			46.50	22	68.50	102
5320	9:12 to 12:12 pitch, 16' span		↓	36	1.111	↓		38.50	18.30	56.80	84.50

05 44 Cold-Formed Metal Trusses

05 44 13 – Cold-Formed Metal Roof Trusses

05 44 13.60 Framing, Roof Trusses		Crew	Daily Output	Labor-Hours	Unit	Material	2016 Bare Costs Labor	Equipment	Total	Total Incl O&P
5330	20′ span	F-6	34	1.176	Ea.		41	19.35	60.35	90
5340	24′ span		32	1.250			43.50	20.50	64	95
5350	28′ span		30	1.333			46.50	22	68.50	102
5360	32′ span		28	1.429			50	23.50	73.50	109
5370	36′ span		26	1.538			53.50	25.50	79	118
5380	40′ span		24	1.667			58	27.50	85.50	127

05 51 Metal Stairs

05 51 13 – Metal Pan Stairs

05 51 13.50 Pan Stairs		Crew	Daily Output	Labor-Hours	Unit	Material	Labor	Equipment	Total	Total Incl O&P
0010	**PAN STAIRS**, shop fabricated, steel stringers									
0015	Made from recycled materials									
1700	Pre-erected, steel pan tread, 3′-6″ wide, 2 line pipe rail (G)	E-2	87	.552	Riser	550	22	17.45	589.45	665
1800	With flat bar picket rail (G)	″	87	.552	″	615	22	17.45	654.45	740

05 51 19 – Metal Grating Stairs

05 51 19.50 Grating Stairs		Crew	Daily Output	Labor-Hours	Unit	Material	Labor	Equipment	Total	Total Incl O&P
0010	**GRATING STAIRS**, shop fabricated, steel stringers, safety nosing on treads									
0015	Made from recycled materials									
0020	Grating tread and pipe railing, 3′-6″ wide (G)	E-4	35	.914	Riser	330	37	4.21	371.21	435
0100	4′-0″ wide (G)	″	30	1.067	″	430	43	4.91	477.91	560

05 51 23 – Metal Fire Escapes

05 51 23.50 Fire Escape Stairs		Crew	Daily Output	Labor-Hours	Unit	Material	Labor	Equipment	Total	Total Incl O&P
0010	**FIRE ESCAPE STAIRS**, portable									
0100	Portable ladder				Ea.	122			122	135

05 51 33 – Metal Ladders

05 51 33.13 Vertical Metal Ladders		Crew	Daily Output	Labor-Hours	Unit	Material	Labor	Equipment	Total	Total Incl O&P
0010	**VERTICAL METAL LADDERS**, shop fabricated									
0015	Made from recycled materials									
0020	Steel, 20″ wide, bolted to concrete, with cage (G)	E-4	50	.640	V.L.F.	65.50	26	2.94	94.44	123
0100	Without cage (G)		85	.376		36.50	15.20	1.73	53.43	70
0300	Aluminum, bolted to concrete, with cage (G)		50	.640		120	26	2.94	148.94	183
0400	Without cage (G)		85	.376		51	15.20	1.73	67.93	86

05 51 33.23 Alternating Tread Ladders		Crew	Daily Output	Labor-Hours	Unit	Material	Labor	Equipment	Total	Total Incl O&P
0010	**ALTERNATING TREAD LADDERS**, shop fabricated									
0015	Made from recycled materials									
0800	Alternating tread ladders, 68-degree angle of incline									
0810	8 foot vertical rise, steel, 149 lb., standard paint color	B-68G	3	5.333	Ea.	2,300	213	111	2,624	3,025
0820	Non-standard paint color		3	5.333		2,650	213	111	2,974	3,425
0830	Galvanized		3	5.333		2,675	213	111	2,999	3,450
0840	Stainless		3	5.333		3,875	213	111	4,199	4,775
0850	Aluminum, 87 lb.		3	5.333		2,825	213	111	3,149	3,625
1010	10 foot vertical rise, steel, 181 lb., standard paint color		2.75	5.818		2,775	232	121	3,128	3,600
1020	Non-standard paint color		2.75	5.818		3,175	232	121	3,528	4,050
1030	Galvanized		2.75	5.818		3,225	232	121	3,578	4,100
1040	Stainless		2.75	5.818		4,675	232	121	5,028	5,700
1050	Aluminum, 103 lb.		2.75	5.818		3,425	232	121	3,778	4,300
1210	12 foot vertical rise, steel, 245 lb., standard paint color		2.50	6.400		3,250	255	134	3,639	4,200
1220	Non-standard paint color		2.50	6.400		3,700	255	134	4,089	4,700
1230	Galvanized		2.50	6.400		3,800	255	134	4,189	4,800

05 51 Metal Stairs

05 51 33 – Metal Ladders

05 51 33.23 Alternating Tread Ladders		Crew	Daily Output	Labor-Hours	Unit	2016 Bare Costs Material	Labor	Equipment	Total	Total Incl O&P
1240	Stainless	B-68G	2.50	6.400	Ea.	5,450	255	134	5,839	6,625
1250	Aluminum, 103 lb.		2.50	6.400		4,000	255	134	4,389	5,025
1410	14 foot vertical rise, steel, 281 lb., standard paint color		2.25	7.111		3,725	284	148	4,157	4,775
1420	Non-standard paint color		2.25	7.111		4,225	284	148	4,657	5,325
1430	Galvanized		2.25	7.111		4,350	284	148	4,782	5,475
1440	Stainless		2.25	7.111		6,250	284	148	6,682	7,550
1450	Aluminum, 136 lb.		2.25	7.111		4,575	284	148	5,007	5,700
1610	16 foot vertical rise, steel, 317 lb., standard paint color		2	8		4,225	320	167	4,712	5,425
1620	Non-standard paint color		2	8		4,725	320	167	5,212	5,975
1630	Galvanized		2	8		4,925	320	167	5,412	6,175
1640	Stainless		2	8		7,050	320	167	7,537	8,525
1650	Aluminum, 153 lb.	↓	2	8	↓	5,150	320	167	5,637	6,450

05 52 Metal Railings

05 52 13 – Pipe and Tube Railings

05 52 13.50 Railings, Pipe

			Crew	Daily Output	Labor-Hours	Unit	Material	Labor	Equipment	Total	Total Incl O&P
0010	**RAILINGS, PIPE**, shop fab'd, 3'-6" high, posts @ 5' O.C.										
0015	Made from recycled materials										
0020	Aluminum, 2 rail, satin finish, 1-1/4" diameter	G	E-4	160	.200	L.F.	50.50	8.10	.92	59.52	71.50
0030	Clear anodized	G		160	.200		62	8.10	.92	71.02	84
0040	Dark anodized	G		160	.200		69	8.10	.92	78.02	92
0080	1-1/2" diameter, satin finish	G		160	.200		59	8.10	.92	68.02	81
0090	Clear anodized	G		160	.200		66.50	8.10	.92	75.52	89
0100	Dark anodized	G		160	.200		73	8.10	.92	82.02	96.50
0140	Aluminum, 3 rail, 1-1/4" diam., satin finish	G		137	.234		65	9.45	1.07	75.52	90
0150	Clear anodized	G		137	.234		81.50	9.45	1.07	92.02	108
0160	Dark anodized	G		137	.234		90	9.45	1.07	100.52	117
0200	1-1/2" diameter, satin finish	G		137	.234		77.50	9.45	1.07	88.02	104
0210	Clear anodized	G		137	.234		88.50	9.45	1.07	99.02	116
0220	Dark anodized	G		137	.234		96.50	9.45	1.07	107.02	124
0500	Steel, 2 rail, on stairs, primed, 1-1/4" diameter	G		160	.200		25	8.10	.92	34.02	43.50
0520	1-1/2" diameter	G		160	.200		27	8.10	.92	36.02	46
0540	Galvanized, 1-1/4" diameter	G		160	.200		34	8.10	.92	43.02	53.50
0560	1-1/2" diameter	G		160	.200		38.50	8.10	.92	47.52	58
0580	Steel, 3 rail, primed, 1-1/4" diameter	G		137	.234		37.50	9.45	1.07	48.02	59.50
0600	1-1/2" diameter	G		137	.234		39	9.45	1.07	49.52	61.50
0620	Galvanized, 1-1/4" diameter	G		137	.234		52.50	9.45	1.07	63.02	76
0640	1-1/2" diameter	G		137	.234		60.50	9.45	1.07	71.02	85
0700	Stainless steel, 2 rail, 1-1/4" diam. #4 finish	G		137	.234		110	9.45	1.07	120.52	139
0720	High polish	G		137	.234		178	9.45	1.07	188.52	214
0740	Mirror polish	G		137	.234		223	9.45	1.07	233.52	263
0760	Stainless steel, 3 rail, 1-1/2" diam., #4 finish	G		120	.267		166	10.75	1.23	177.98	204
0770	High polish	G		120	.267		275	10.75	1.23	286.98	325
0780	Mirror finish	G		120	.267		335	10.75	1.23	346.98	390
0900	Wall rail, alum. pipe, 1-1/4" diam., satin finish	G		213	.150		24.50	6.05	.69	31.24	38.50
0905	Clear anodized	G		213	.150		30	6.05	.69	36.74	45
0910	Dark anodized	G		213	.150		35.50	6.05	.69	42.24	51
0915	1-1/2" diameter, satin finish	G		213	.150		27	6.05	.69	33.74	42
0920	Clear anodized	G		213	.150		34	6.05	.69	40.74	49
0925	Dark anodized	G		213	.150		42	6.05	.69	48.74	58
0930	Steel pipe, 1-1/4" diameter, primed	G	↓	213	.150	↓	15	6.05	.69	21.74	28.50

05 52 Metal Railings

05 52 13 – Pipe and Tube Railings

05 52 13.50 Railings, Pipe		Crew	Daily Output	Labor-Hours	Unit	Material	2016 Bare Costs Labor	Equipment	Total	Total Incl O&P
0935	Galvanized	G	E-4	213	.150	L.F.	22	6.05	.69	28.74	36
0940	1-1/2" diameter	G		176	.182		15.50	7.35	.84	23.69	31.50
0945	Galvanized	G		213	.150		22	6.05	.69	28.74	36
0955	Stainless steel pipe, 1-1/2" diam., #4 finish	G		107	.299		88	12.10	1.38	101.48	121
0960	High polish	G		107	.299		179	12.10	1.38	192.48	221
0965	Mirror polish	G		107	.299		212	12.10	1.38	225.48	257
2000	2-line pipe rail (1-1/2" T&B) with 1/2" pickets @ 4-1/2" O.C.,										
2005	attached handrail on brackets										
2010	42" high aluminum, satin finish, straight & level	G	E-4	120	.267	L.F.	257	10.75	1.23	268.98	305
2050	42" high steel, primed, straight & level	G	"	120	.267		127	10.75	1.23	138.98	161
4000	For curved and level rails, add						10%	10%			
4100	For sloped rails for stairs, add						30%	30%			

05 55 Metal Stair Treads and Nosings

05 55 13 – Metal Stair Treads

05 55 13.50 Stair Treads			Crew	Daily Output	Labor-Hours	Unit	Material	Labor	Equipment	Total	Total Incl O&P
0010	**STAIR TREADS**, stringers and bolts not included										
3000	Diamond plate treads, steel, 1/8" thick										
3005	Open riser, black enamel										
3010	9" deep x 36" long	G	2 Sswk	48	.333	Ea.	91.50	13.30		104.80	125
3020	42" long	G		48	.333		96.50	13.30		109.80	131
3030	48" long	G		48	.333		101	13.30		114.30	136
3040	11" deep x 36" long	G		44	.364		96.50	14.50		111	133
3050	42" long	G		44	.364		102	14.50		116.50	139
3060	48" long	G		44	.364		108	14.50		122.50	146
3110	Galvanized, 9" deep x 36" long	G		48	.333		144	13.30		157.30	184
3120	42" long	G		48	.333		154	13.30		167.30	194
3130	48" long	G		48	.333		163	13.30		176.30	204
3140	11" deep x 36" long	G		44	.364		149	14.50		163.50	191
3150	42" long	G		44	.364		163	14.50		177.50	206
3160	48" long	G		44	.364		169	14.50		183.50	213
3200	Closed riser, black enamel										
3210	12" deep x 36" long	G	2 Sswk	40	.400	Ea.	110	15.95		125.95	151
3220	42" long	G		40	.400		119	15.95		134.95	161
3230	48" long	G		40	.400		126	15.95		141.95	168
3240	Galvanized, 12" deep x 36" long	G		40	.400		173	15.95		188.95	221
3250	42" long	G		40	.400		189	15.95		204.95	238
3260	48" long	G		40	.400		198	15.95		213.95	248
4000	Bar grating treads										
4005	Steel, 1-1/4" x 3/16" bars, anti-skid nosing, black enamel										
4010	8-5/8" deep x 30" long	G	2 Sswk	48	.333	Ea.	54	13.30		67.30	81
4020	36" long	G		48	.333		63.50	13.30		76.80	94
4030	48" long	G		48	.333		97.50	13.30		110.80	132
4040	10-15/16" deep x 36" long	G		44	.364		70	14.50		84.50	104
4050	48" long	G		44	.364		100	14.50		114.50	137
4060	Galvanized, 8-5/8" deep x 30" long	G		48	.333		62	13.30		75.30	93
4070	36" long	G		48	.333		73.50	13.30		86.80	106
4080	48" long	G		48	.333		108	13.30		121.30	144
4090	10-15/16" deep x 36" long	G		44	.364		86.50	14.50		101	122
4100	48" long	G		44	.364		112	14.50		126.50	150
4200	Aluminum, 1-1/4" x 3/16" bars, serrated, with nosing										

05 55 Metal Stair Treads and Nosings

05 55 13 – Metal Stair Treads

05 55 13.50 Stair Treads			Crew	Daily Output	Labor-Hours	Unit	2016 Bare Costs Material	Labor	Equipment	Total	Total Incl O&P
4210	7-5/8" deep x 18" long	G	2 Sswk	52	.308	Ea.	50	12.30		62.30	77.50
4220	24" long	G	2 Sswk	52	.308	Ea.	59	12.30		71.30	87.50
4230	30" long	G		52	.308		68.50	12.30		80.80	97.50
4240	36" long	G		52	.308		177	12.30		189.30	218
4250	8-13/16" deep x 18" long	G		48	.333		67	13.30		80.30	98.50
4260	24" long	G		48	.333		99.50	13.30		112.80	134
4270	30" long	G		48	.333		115	13.30		128.30	151
4280	36" long	G		48	.333		194	13.30		207.30	239
4290	10" deep x 18" long	G		44	.364		130	14.50		144.50	170
4300	30" long	G		44	.364		173	14.50		187.50	217
4310	36" long	G		44	.364		210	14.50		224.50	258
5000	Channel grating treads										
5005	Steel, 14 ga., 2-1/2" thick, galvanized										
5010	9" deep x 36" long	G	2 Sswk	48	.333	Ea.	103	13.30		116.30	138
5020	48" long	G	"	48	.333	"	139	13.30		152.30	177

05 55 19 – Metal Stair Tread Covers

05 55 19.50 Stair Tread Covers for Renovation		Crew	Daily Output	Labor-Hours	Unit	Material	Labor	Equipment	Total	Total Incl O&P
0010	**STAIR TREAD COVERS FOR RENOVATION**									
0205	Extruded tread cover with nosing, pre-drilled, includes screws									
0210	Aluminum with black abrasive strips, 9" wide x 3' long	1 Carp	24	.333	Ea.	110	12.60		122.60	142
0220	4' long		22	.364		145	13.75		158.75	182
0230	5' long		20	.400		180	15.10		195.10	224
0240	11" wide x 3' long		24	.333		142	12.60		154.60	177
0250	4' long		22	.364		182	13.75		195.75	224
0260	5' long		20	.400		235	15.10		250.10	284
0305	Black abrasive strips with yellow front strips									
0310	Aluminum, 9" wide x 3' long	1 Carp	24	.333	Ea.	118	12.60		130.60	151
0320	4' long		22	.364		157	13.75		170.75	196
0330	5' long		20	.400		200	15.10		215.10	246
0340	11" wide x 3' long		24	.333		152	12.60		164.60	188
0350	4' long		22	.364		196	13.75		209.75	239
0360	5' long		20	.400		253	15.10		268.10	305
0405	Black abrasive strips with photoluminescent front strips									
0410	Aluminum, 9" wide x 3' long	1 Carp	24	.333	Ea.	141	12.60		153.60	176
0420	4' long		22	.364		175	13.75		188.75	216
0430	5' long		20	.400		219	15.10		234.10	267
0440	11" wide x 3' long		24	.333		156	12.60		168.60	192
0450	4' long		22	.364		208	13.75		221.75	252
0460	5' long		20	.400		260	15.10		275.10	310

05 58 Formed Metal Fabrications

05 58 13 – Column Covers

05 58 13.05 Column Covers			Crew	Daily Output	Labor-Hours	Unit	Material	Labor	Equipment	Total	Total Incl O&P
0010	**COLUMN COVERS**										
0015	Made from recycled materials										
0020	Excludes structural steel, light ga. metal framing, misc. metals, sealants										
0100	Round covers, 2 halves with 2 vertical joints for backer rod and sealant										
0110	Up to 12' high, no horizontal joints										
0120	12" diameter, 0.125" aluminum, anodized/painted finish	G	2 Sswk	32	.500	V.L.F.	31	19.95		50.95	70.50
0130	Type 304 stainless steel, 16 gauge, #4 brushed finish	G		32	.500		46	19.95		65.95	87

05 58 Formed Metal Fabrications

05 58 13 – Column Covers

05 58 13.05 Column Covers			Crew	Daily Output	Labor-Hours	Unit	Material	Labor	Equipment	Total	Total Incl O&P
							2016 Bare Costs				
0140	Type 316 stainless steel, 16 gauge, #4 brushed finish	G	2 Sswk	32	.500	V.L.F.	54.50	19.95		74.45	96.50
0150	18" diameter, aluminum	G		32	.500		46.50	19.95		66.45	87.50
0160	Type 304 stainless steel	G		32	.500		69	19.95		88.95	112
0170	Type 316 stainless steel	G		32	.500		82	19.95		101.95	127
0180	24" diameter, aluminum	G		32	.500		62	19.95		81.95	105
0190	Type 304 stainless steel	G		32	.500		91.50	19.95		111.45	138
0200	Type 316 stainless steel	G		32	.500		109	19.95		128.95	157
0210	30" diameter, aluminum	G		30	.533		77.50	21.50		99	124
0220	Type 304 stainless steel	G		30	.533		115	21.50		136.50	165
0230	Type 316 stainless steel	G		30	.533		137	21.50		158.50	190
0240	36" diameter, aluminum	G		30	.533		93	21.50		114.50	141
0250	Type 304 stainless steel	G		30	.533		138	21.50		159.50	190
0260	Type 316 stainless steel	G	↓	30	.533	↓	164	21.50		185.50	220
0400	Up to 24' high, 2 stacked sections with 1 horizontal joint										
0410	18" diameter, aluminum	G	2 Sswk	28	.571	V.L.F.	48.50	23		71.50	95.50
0450	Type 304 stainless steel	G		28	.571		72	23		95	122
0460	Type 316 stainless steel	G		28	.571		86	23		109	137
0470	24" diameter, aluminum	G		28	.571		65	23		88	114
0480	Type 304 stainless steel	G		28	.571		96.50	23		119.50	148
0490	Type 316 stainless steel	G		28	.571		115	23		138	168
0500	30" diameter, aluminum	G		24	.667		81	26.50		107.50	139
0510	Type 304 stainless steel	G		24	.667		120	26.50		146.50	181
0520	Type 316 stainless steel	G		24	.667		144	26.50		170.50	207
0530	36" diameter, aluminum	G		24	.667		97.50	26.50		124	156
0540	Type 304 stainless steel	G		24	.667		144	26.50		170.50	208
0550	Type 316 stainless steel	G	↓	24	.667	↓	172	26.50		198.50	239

05 58 25 – Formed Lamp Posts

05 58 25.40 Lamp Posts

Line	Description		Crew	Daily Output	Labor-Hours	Unit	Material	Labor	Equipment	Total	Total Incl O&P
0010	**LAMP POSTS**										
0020	Aluminum, 7' high, stock units, post only	G	1 Carp	16	.500	Ea.	85	18.90		103.90	125
0100	Mild steel, plain	G	"	16	.500	"	73.50	18.90		92.40	113

05 71 Decorative Metal Stairs

05 71 13 – Fabricated Metal Spiral Stairs

05 71 13.50 Spiral Stairs

Line	Description		Crew	Daily Output	Labor-Hours	Unit	Material	Labor	Equipment	Total	Total Incl O&P
0010	**SPIRAL STAIRS**										
1805	Shop fabricated, custom ordered										
1810	Aluminum, 5'-0" diameter, plain units	G	E-4	45	.711	Riser	530	28.50	3.27	561.77	640
1820	Fancy units	G		45	.711		925	28.50	3.27	956.77	1,075
1900	Cast iron, 4'-0" diameter, plain units	G		45	.711		665	28.50	3.27	696.77	785
1920	Fancy units	G	↓	25	1.280	↓	1,150	51.50	5.90	1,207.40	1,375
3100	Spiral stair kits, 12 stacking risers to fit exact floor height										

05 73 Decorative Metal Railings

05 73 16 – Wire Rope Decorative Metal Railings

05 73 16.10 Cable Railings		Crew	Daily Output	Labor-Hours	Unit	2016 Bare Costs Material	Labor	Equipment	Total	Total Incl O&P
0010	**CABLE RAILINGS**, with 316 stainless steel 1 x 19 cable, 3/16" diameter									
0015	Made from recycled materials									
0100	1-3/4" diameter stainless steel posts x 42" high, cables 4" OC G	2 Sswk	25	.640	L.F.	43	25.50		68.50	94.50

05 73 23 – Ornamental Railings

05 73 23.50 Railings, Ornamental		Crew	Daily Output	Labor-Hours	Unit	Material	Labor	Equipment	Total	Total Incl O&P
0010	**RAILINGS, ORNAMENTAL**, 3'-6" high, posts @ 6' O.C.									
0020	Bronze or stainless, hand forged, plain G	2 Sswk	24	.667	L.F.	260	26.50		286.50	335
0100	Fancy G		18	.889		520	35.50		555.50	635
0200	Aluminum, panelized, plain G		24	.667		12.10	26.50		38.60	62.50
0300	Fancy G		18	.889		26	35.50		61.50	93.50
0400	Wrought iron, hand forged, plain G		24	.667		81.50	26.50		108	139
0500	Fancy G		18	.889		231	35.50		266.50	320
0550	Steel, panelized, plain G		24	.667		18.65	26.50		45.15	69.50
0560	Fancy G		18	.889		28	35.50		63.50	96
0600	Composite metal/wood/glass, plain		18	.889		143	35.50		178.50	222
0700	Fancy	↓	12	1.333	↓	286	53		339	415

Division 6 Wood, Plastics, & Composites

Estimating Tips

06 05 00 Common Work Results for Wood, Plastics, and Composites

- Common to any wood-framed structure are the accessory connector items such as screws, nails, adhesives, hangers, connector plates, straps, angles, and hold-downs. For typical wood-framed buildings, such as residential projects, the aggregate total for these items can be significant, especially in areas where seismic loading is a concern. For floor and wall framing, the material cost is based on 10 to 25 lbs. of accessory connectors per MBF. Hold-downs, hangers, and other connectors should be taken off by the piece.

  Included with material costs are fasteners for a normal installation. RSMeans engineers use manufacturer's recommendations, written specifications, and/or standard construction practice for size and spacing of fasteners. Prices for various fasteners are shown for informational purposes only. Adjustments should be made if unusual fastening conditions exist.

06 10 00 Carpentry

- Lumber is a traded commodity and therefore sensitive to supply and demand in the marketplace. Even in "budgetary" estimating of wood-framed projects, it is advisable to call local suppliers for the latest market pricing.
- Common quantity units for wood-framed projects are "thousand board feet" (MBF). A board foot is a volume of wood, 1" x 1' x 1', or 144 cubic inches. Board-foot quantities are generally calculated using nominal material dimensions—dressed sizes are ignored. Board foot per lineal foot of any stick of lumber can be calculated by dividing the nominal cross-sectional area by 12. As an example, 2,000 lineal feet of 2 x 12 equates to 4 MBF by dividing the nominal area, 2 x 12, by 12, which equals 2, and multiplying by 2,000 to give 4,000 board feet. This simple rule applies to all nominal dimensioned lumber.
- Waste is an issue of concern at the quantity takeoff for any area of construction. Framing lumber is sold in even foot lengths, i.e., 8', 10', 12', 14', 16' and, depending on spans, wall heights, and the grade of lumber, waste is inevitable. A rule of thumb for lumber waste is 5%–10% depending on material quality and the complexity of the framing.
- Wood in various forms and shapes is used in many projects, even where the main structural framing is steel, concrete, or masonry. Plywood as a back-up partition material and 2x boards used as blocking and cant strips around roof edges are two common examples. The estimator should ensure that the costs of all wood materials are included in the final estimate.

06 20 00 Finish Carpentry

- It is necessary to consider the grade of workmanship when estimating labor costs for erecting millwork and interior finish. In practice, there are three grades: premium, custom, and economy. The RSMeans daily output for base and case moldings is in the range of 200 to 250 L.F. per carpenter per day. This is appropriate for most average custom-grade projects. For premium projects, an adjustment to productivity of 25%–50% should be made, depending on the complexity of the job.

Reference Numbers

Reference numbers are shown at the beginning of some major classifications. These numbers refer to related items in the Reference Section. The reference information may be an estimating procedure, an alternate pricing method, or technical information.

Note: Not all subdivisions listed here necessarily appear. ■

06 05 05 – Selective Demolition for Wood, Plastics, and Composites

06 05 05.10 Selective Demolition Wood Framing		Crew	Daily Output	Labor-Hours	Unit	Material	Labor (2016 Bare Costs)	Equipment (2016 Bare Costs)	Total (2016 Bare Costs)	Total Incl O&P
0010	SELECTIVE DEMOLITION WOOD FRAMING R024119-10									
0100	Timber connector, nailed, small	1 Clab	96	.083	Ea.		2.43		2.43	4.07
0110	Medium		60	.133			3.89		3.89	6.50
0120	Large		48	.167			4.87		4.87	8.15
0130	Bolted, small		48	.167			4.87		4.87	8.15
0140	Medium		32	.250			7.30		7.30	12.25
0150	Large		24	.333			9.75		9.75	16.30
2958	Beams, 2" x 6"	2 Clab	1100	.015	L.F.		.42		.42	.71
2960	2" x 8"		825	.019			.57		.57	.95
2965	2" x 10"		665	.024			.70		.70	1.18
2970	2" x 12"		550	.029			.85		.85	1.42
2972	2" x 14"		470	.034			.99		.99	1.66
2975	4" x 8"	B-1	413	.058			1.74		1.74	2.91
2980	4" x 10"		330	.073			2.17		2.17	3.64
2985	4" x 12"		275	.087			2.61		2.61	4.37
3000	6" x 8"		275	.087			2.61		2.61	4.37
3040	6" x 10"		220	.109			3.26		3.26	5.45
3080	6" x 12"		185	.130			3.88		3.88	6.50
3120	8" x 12"		140	.171			5.10		5.10	8.55
3160	10" x 12"		110	.218			6.50		6.50	10.90
3162	Alternate pricing method		1.10	21.818	M.B.F.		650		650	1,100
3170	Blocking, in 16" OC wall framing, 2" x 4"	1 Clab	600	.013	L.F.		.39		.39	.65
3172	2" x 6"		400	.020			.58		.58	.98
3174	In 24" OC wall framing, 2" x 4"		600	.013			.39		.39	.65
3176	2" x 6"		400	.020			.58		.58	.98
3178	Alt method, wood blocking removal from wood framing		.40	20	M.B.F.		585		585	980
3179	Wood blocking removal from steel framing		.36	22.222	"		650		650	1,075
3180	Bracing, let in, 1" x 3", studs 16" OC		1050	.008	L.F.		.22		.22	.37
3181	Studs 24" OC		1080	.007			.22		.22	.36
3182	1" x 4", studs 16" OC		1050	.008			.22		.22	.37
3183	Studs 24" OC		1080	.007			.22		.22	.36
3184	1" x 6", studs 16" OC		1050	.008			.22		.22	.37
3185	Studs 24" OC		1080	.007			.22		.22	.36
3186	2" x 3", studs 16" OC		800	.010			.29		.29	.49
3187	Studs 24" OC		830	.010			.28		.28	.47
3188	2" x 4", studs 16" OC		800	.010			.29		.29	.49
3189	Studs 24" OC		830	.010			.28		.28	.47
3190	2" x 6", studs 16" OC		800	.010			.29		.29	.49
3191	Studs 24" OC		830	.010			.28		.28	.47
3192	2" x 8", studs 16" OC		800	.010			.29		.29	.49
3193	Studs 24" OC		830	.010			.28		.28	.47
3194	"T" shaped metal bracing, studs at 16" OC		1060	.008			.22		.22	.37
3195	Studs at 24" OC		1200	.007			.19		.19	.33
3196	Metal straps, studs at 16" OC		1200	.007			.19		.19	.33
3197	Studs at 24" OC		1240	.006			.19		.19	.32
3200	Columns, round, 8' to 14' tall		40	.200	Ea.		5.85		5.85	9.80
3202	Dimensional lumber sizes	2 Clab	1.10	14.545	M.B.F.		425		425	710
3250	Blocking, between joists	1 Clab	320	.025	Ea.		.73		.73	1.22
3252	Bridging, metal strap, between joists		320	.025	Pr.		.73		.73	1.22
3254	Wood, between joists		320	.025	"		.73		.73	1.22
3260	Door buck, studs, header & access., 8' high 2" x 4" wall, 3' wide		32	.250	Ea.		7.30		7.30	12.25
3261	4' wide		32	.250			7.30		7.30	12.25
3262	5' wide		32	.250			7.30		7.30	12.25

06 05 05 – Selective Demolition for Wood, Plastics, and Composites

06 05 05.10 Selective Demolition Wood Framing		Crew	Daily Output	Labor-Hours	Unit	2016 Bare Costs Material	Labor	Equipment	Total	Total Incl O&P
3263	6′ wide	1 Clab	32	.250	Ea.		7.30		7.30	12.25
3264	8′ wide		30	.267			7.80		7.80	13.05
3265	10′ wide		30	.267			7.80		7.80	13.05
3266	12′ wide		30	.267			7.80		7.80	13.05
3267	2″ x 6″ wall, 3′ wide		32	.250			7.30		7.30	12.25
3268	4′ wide		32	.250			7.30		7.30	12.25
3269	5′ wide		32	.250			7.30		7.30	12.25
3270	6′ wide		32	.250			7.30		7.30	12.25
3271	8′ wide		30	.267			7.80		7.80	13.05
3272	10′ wide		30	.267			7.80		7.80	13.05
3273	12′ wide		30	.267			7.80		7.80	13.05
3274	Window buck, studs, header & access, 8′ high 2″ x 4″ wall, 2′ wide		24	.333			9.75		9.75	16.30
3275	3′ wide		24	.333			9.75		9.75	16.30
3276	4′ wide		24	.333			9.75		9.75	16.30
3277	5′ wide		24	.333			9.75		9.75	16.30
3278	6′ wide		24	.333			9.75		9.75	16.30
3279	7′ wide		24	.333			9.75		9.75	16.30
3280	8′ wide		22	.364			10.60		10.60	17.80
3281	10′ wide		22	.364			10.60		10.60	17.80
3282	12′ wide		22	.364			10.60		10.60	17.80
3283	2″ x 6″ wall, 2′ wide		24	.333			9.75		9.75	16.30
3284	3′ wide		24	.333			9.75		9.75	16.30
3285	4′ wide		24	.333			9.75		9.75	16.30
3286	5′ wide		24	.333			9.75		9.75	16.30
3287	6′ wide		24	.333			9.75		9.75	16.30
3288	7′ wide		24	.333			9.75		9.75	16.30
3289	8′ wide		22	.364			10.60		10.60	17.80
3290	10′ wide		22	.364			10.60		10.60	17.80
3291	12′ wide		22	.364			10.60		10.60	17.80
3360	Deck or porch decking		825	.010	L.F.		.28		.28	.47
3400	Fascia boards, 1″ x 6″		500	.016			.47		.47	.78
3440	1″ x 8″		450	.018			.52		.52	.87
3480	1″ x 10″		400	.020			.58		.58	.98
3490	2″ x 6″		450	.018			.52		.52	.87
3500	2″ x 8″		400	.020			.58		.58	.98
3510	2″ x 10″		350	.023			.67		.67	1.12
3610	Furring, on wood walls or ceiling		4000	.002	S.F.		.06		.06	.10
3620	On masonry or concrete walls or ceiling		1200	.007	"		.19		.19	.33
3800	Headers over openings, 2 @ 2″ x 6″		110	.073	L.F.		2.12		2.12	3.56
3840	2 @ 2″ x 8″		100	.080			2.34		2.34	3.91
3880	2 @ 2″ x 10″		90	.089			2.60		2.60	4.35
3885	Alternate pricing method		.26	30.651	M.B.F.		895		895	1,500
3920	Joists, 1″ x 4″		1250	.006	L.F.		.19		.19	.31
3930	1″ x 6″		1135	.007			.21		.21	.34
3940	1″ x 8″		1000	.008			.23		.23	.39
3950	1″ x 10″		895	.009			.26		.26	.44
3960	1″ x 12″		765	.010			.31		.31	.51
4200	2″ x 4″	2 Clab	1000	.016			.47		.47	.78
4230	2″ x 6″		970	.016			.48		.48	.81
4240	2″ x 8″		940	.017			.50		.50	.83
4250	2″ x 10″		910	.018			.51		.51	.86
4280	2″ x 12″		880	.018			.53		.53	.89
4281	2″ x 14″		850	.019			.55		.55	.92

06 05 05 – Selective Demolition for Wood, Plastics, and Composites

06 05 05.10 Selective Demolition Wood Framing		Crew	Daily Output	Labor-Hours	Unit	Material	2016 Bare Costs Labor	Equipment	Total	Total Incl O&P
4282	Composite joists, 9-1/2"	2 Clab	960	.017	L.F.		.49		.49	.82
4283	11-7/8"		930	.017			.50		.50	.84
4284	14"		897	.018			.52		.52	.87
4285	16"		865	.019			.54		.54	.90
4290	Wood joists, alternate pricing method		1.50	10.667	M.B.F.		310		310	520
4500	Open web joist, 12" deep		500	.032	L.F.		.93		.93	1.56
4505	14" deep		475	.034			.98		.98	1.65
4510	16" deep		450	.036			1.04		1.04	1.74
4520	18" deep		425	.038			1.10		1.10	1.84
4530	24" deep		400	.040			1.17		1.17	1.96
4550	Ledger strips, 1" x 2"	1 Clab	1200	.007			.19		.19	.33
4560	1" x 3"		1200	.007			.19		.19	.33
4570	1" x 4"		1200	.007			.19		.19	.33
4580	2" x 2"		1100	.007			.21		.21	.36
4590	2" x 4"		1000	.008			.23		.23	.39
4600	2" x 6"		1000	.008			.23		.23	.39
4601	2" x 8 or 2" x 10"		800	.010			.29		.29	.49
4602	4" x 6"		600	.013			.39		.39	.65
4604	4" x 8"		450	.018			.52		.52	.87
5400	Posts, 4" x 4"	2 Clab	800	.020			.58		.58	.98
5405	4" x 6"		550	.029			.85		.85	1.42
5410	4" x 8"		440	.036			1.06		1.06	1.78
5425	4" x 10"		390	.041			1.20		1.20	2.01
5430	4" x 12"		350	.046			1.33		1.33	2.24
5440	6" x 6"		400	.040			1.17		1.17	1.96
5445	6" x 8"		350	.046			1.33		1.33	2.24
5450	6" x 10"		320	.050			1.46		1.46	2.45
5455	6" x 12"		290	.055			1.61		1.61	2.70
5480	8" x 8"		300	.053			1.56		1.56	2.61
5500	10" x 10"		240	.067			1.95		1.95	3.26
5660	Tongue and groove floor planks		2	8	M.B.F.		234		234	390
5682	Rafters, ordinary, 16" OC, 2" x 4"		880	.018	S.F.		.53		.53	.89
5683	2" x 6"		840	.019			.56		.56	.93
5684	2" x 8"		820	.020			.57		.57	.95
5685	2" x 10"		820	.020			.57		.57	.95
5686	2" x 12"		810	.020			.58		.58	.97
5687	24" OC, 2" x 4"		1170	.014			.40		.40	.67
5688	2" x 6"		1117	.014			.42		.42	.70
5689	2" x 8"		1091	.015			.43		.43	.72
5690	2" x 10"		1091	.015			.43		.43	.72
5691	2" x 12"		1077	.015			.43		.43	.73
5795	Rafters, ordinary, 2" x 4" (alternate method)		862	.019	L.F.		.54		.54	.91
5800	2" x 6" (alternate method)		850	.019			.55		.55	.92
5840	2" x 8" (alternate method)		837	.019			.56		.56	.94
5855	2" x 10" (alternate method)		825	.019			.57		.57	.95
5865	2" x 12" (alternate method)		812	.020			.58		.58	.96
5870	Sill plate, 2" x 4"	1 Clab	1170	.007			.20		.20	.33
5871	2" x 6"		780	.010			.30		.30	.50
5872	2" x 8"		586	.014			.40		.40	.67
5873	Alternate pricing method		.78	10.256	M.B.F.		299		299	500
5885	Ridge board, 1" x 4"	2 Clab	900	.018	L.F.		.52		.52	.87
5886	1" x 6"		875	.018			.53		.53	.89
5887	1" x 8"		850	.019			.55		.55	.92

06 05 05 – Selective Demolition for Wood, Plastics, and Composites

06 05 05.10 Selective Demolition Wood Framing		Crew	Daily Output	Labor-Hours	Unit	2016 Bare Costs Material	Labor	Equipment	Total	Total Incl O&P
5888	1" x 10"	2 Clab	825	.019	L.F.		.57		.57	.95
5889	1" x 12"		800	.020			.58		.58	.98
5890	2" x 4"		900	.018			.52		.52	.87
5892	2" x 6"		875	.018			.53		.53	.89
5894	2" x 8"		850	.019			.55		.55	.92
5896	2" x 10"		825	.019			.57		.57	.95
5898	2" x 12"		800	.020			.58		.58	.98
6050	Rafter tie, 1" x 4"		1250	.013			.37		.37	.63
6052	1" x 6"		1135	.014			.41		.41	.69
6054	2" x 4"		1000	.016			.47		.47	.78
6056	2" x 6"		970	.016			.48		.48	.81
6070	Sleepers, on concrete, 1" x 2"	1 Clab	4700	.002			.05		.05	.08
6075	1" x 3"		4000	.002			.06		.06	.10
6080	2" x 4"		3000	.003			.08		.08	.13
6085	2" x 6"		2600	.003			.09		.09	.15
6086	Sheathing from roof, 5/16"	2 Clab	1600	.010	S.F.		.29		.29	.49
6088	3/8"		1525	.010			.31		.31	.51
6090	1/2"		1400	.011			.33		.33	.56
6092	5/8"		1300	.012			.36		.36	.60
6094	3/4"		1200	.013			.39		.39	.65
6096	Board sheathing from roof		1400	.011			.33		.33	.56
6100	Sheathing, from walls, 1/4"		1200	.013			.39		.39	.65
6110	5/16"		1175	.014			.40		.40	.67
6120	3/8"		1150	.014			.41		.41	.68
6130	1/2"		1125	.014			.42		.42	.70
6140	5/8"		1100	.015			.42		.42	.71
6150	3/4"		1075	.015			.43		.43	.73
6152	Board sheathing from walls		1500	.011			.31		.31	.52
6158	Subfloor/roof deck, with boards		2200	.007			.21		.21	.36
6159	Subfloor/roof deck, with tongue & groove boards		2000	.008			.23		.23	.39
6160	Plywood, 1/2" thick		768	.021			.61		.61	1.02
6162	5/8" thick		760	.021			.61		.61	1.03
6164	3/4" thick		750	.021			.62		.62	1.04
6165	1-1/8" thick		720	.022			.65		.65	1.09
6166	Underlayment, particle board, 3/8" thick	1 Clab	780	.010			.30		.30	.50
6168	1/2" thick		768	.010			.30		.30	.51
6170	5/8" thick		760	.011			.31		.31	.51
6172	3/4" thick		750	.011			.31		.31	.52
6200	Stairs and stringers, straight run	2 Clab	40	.400	Riser		11.70		11.70	19.55
6240	With platforms, winders or curves	"	26	.615	"		17.95		17.95	30
6300	Components, tread	1 Clab	110	.073	Ea.		2.12		2.12	3.56
6320	Riser		80	.100	"		2.92		2.92	4.89
6390	Stringer, 2" x 10"		260	.031	L.F.		.90		.90	1.50
6400	2" x 12"		260	.031			.90		.90	1.50
6410	3" x 10"		250	.032			.93		.93	1.56
6420	3" x 12"		250	.032			.93		.93	1.56
6590	Wood studs, 2" x 3"	2 Clab	3076	.005			.15		.15	.25
6600	2" x 4"		2000	.008			.23		.23	.39
6640	2" x 6"		1600	.010			.29		.29	.49
6720	Wall framing, including studs, plates and blocking, 2" x 4"	1 Clab	600	.013	S.F.		.39		.39	.65
6740	2" x 6"		480	.017	"		.49		.49	.82
6750	Headers, 2" x 4"		1125	.007	L.F.		.21		.21	.35
6755	2" x 6"		1125	.007			.21		.21	.35

06 05 05 – Selective Demolition for Wood, Plastics, and Composites

06 05 05.10 Selective Demolition Wood Framing		Crew	Daily Output	Labor-Hours	Unit	Material	2016 Bare Costs Labor	Equipment	Total	Total Incl O&P
6760	2" x 8"	1 Clab	1050	.008	L.F.		.22		.22	.37
6765	2" x 10"		1050	.008			.22		.22	.37
6770	2" x 12"		1000	.008			.23		.23	.39
6780	4" x 10"		525	.015			.45		.45	.75
6785	4" x 12"		500	.016			.47		.47	.78
6790	6" x 8"		560	.014			.42		.42	.70
6795	6" x 10"		525	.015			.45		.45	.75
6797	6" x 12"		500	.016			.47		.47	.78
7000	Trusses									
7050	12' span	2 Clab	74	.216	Ea.		6.30		6.30	10.55
7150	24' span	F-3	66	.606			21	10	31	46
7200	26' span		64	.625			21.50	10.30	31.80	47.50
7250	28' span		62	.645			22	10.60	32.60	48.50
7300	30' span		58	.690			23.50	11.35	34.85	52
7350	32' span		56	.714			24.50	11.75	36.25	54
7400	34' span		54	.741			25.50	12.20	37.70	56
7450	36' span		52	.769			26.50	12.65	39.15	58
8000	Soffit, T & G wood	1 Clab	520	.015	S.F.		.45		.45	.75
8010	Hardboard, vinyl or aluminum	"	640	.013			.37		.37	.61
8030	Plywood	2 Carp	315	.051			1.92		1.92	3.22
9500	See Section 02 41 19.19 for rubbish handling									

06 05 05.20 Selective Demolition Millwork and Trim		Crew	Daily Output	Labor-Hours	Unit	Material	2016 Bare Costs Labor	Equipment	Total	Total Incl O&P
0010	SELECTIVE DEMOLITION MILLWORK AND TRIM R024119-10									
1000	Cabinets, wood, base cabinets, per L.F.	2 Clab	80	.200	L.F.		5.85		5.85	9.80
1020	Wall cabinets, per L.F.	"	80	.200	"		5.85		5.85	9.80
1060	Remove and reset, base cabinets	2 Carp	18	.889	Ea.		33.50		33.50	56.50
1070	Wall cabinets		20	.800			30		30	50.50
1072	Oven cabinet, 7' high		11	1.455			55		55	92
1074	Cabinet door, up to 2' high	1 Clab	66	.121			3.54		3.54	5.95
1076	2' - 4' high	"	46	.174			5.10		5.10	8.50
1100	Steel, painted, base cabinets	2 Clab	60	.267	L.F.		7.80		7.80	13.05
1120	Wall cabinets		60	.267	"		7.80		7.80	13.05
1200	Casework, large area		320	.050	S.F.		1.46		1.46	2.45
1220	Selective		200	.080	"		2.34		2.34	3.91
1500	Counter top, straight runs		200	.080	L.F.		2.34		2.34	3.91
1510	L, U or C shapes		120	.133	"		3.89		3.89	6.50
2000	Paneling, 4' x 8' sheets		2000	.008	S.F.		.23		.23	.39
2100	Boards, 1" x 4"		700	.023			.67		.67	1.12
2120	1" x 6"		750	.021			.62		.62	1.04
2140	1" x 8"		800	.020			.58		.58	.98
3000	Trim, baseboard, to 6" wide		1200	.013	L.F.		.39		.39	.65
3040	Greater than 6" and up to 12" wide		1000	.016			.47		.47	.78
3080	Remove and reset, minimum	2 Carp	400	.040			1.51		1.51	2.53
3090	Maximum	"	300	.053			2.02		2.02	3.38
3100	Ceiling trim	2 Clab	1000	.016			.47		.47	.78
3120	Chair rail		1200	.013			.39		.39	.65
3140	Railings with balusters		240	.067			1.95		1.95	3.26
3160	Wainscoting		700	.023	S.F.		.67		.67	1.12
4000	Curtain rod	1 Clab	80	.100	L.F.		2.92		2.92	4.89

06 05 23 – Wood, Plastic, and Composite Fastenings

06 05 23.10 Nails		Crew	Daily Output	Labor-Hours	Unit	2016 Bare Costs Material	Labor	Equipment	Total	Total Incl O&P
0010	**NAILS**, material only, based upon 50# box purchase									
0020	Copper nails, plain				Lb.	10.10			10.10	11.15
0400	Stainless steel, plain					9.05			9.05	10
0500	Box, 3d to 20d, bright					1.40			1.40	1.54
0520	Galvanized					1.70			1.70	1.87
0600	Common, 3d to 60d, plain					1.37			1.37	1.51
0700	Galvanized					2			2	2.20
0800	Aluminum					10.90			10.90	11.95
1000	Annular or spiral thread, 4d to 60d, plain					2.58			2.58	2.84
1200	Galvanized					3.10			3.10	3.41
1400	Drywall nails, plain					1.40			1.40	1.54
1600	Galvanized					2.10			2.10	2.31
1800	Finish nails, 4d to 10d, plain					1.20			1.20	1.32
2000	Galvanized					1.83			1.83	2.01
2100	Aluminum					5.80			5.80	6.40
2300	Flooring nails, hardened steel, 2d to 10d, plain					3.33			3.33	3.66
2400	Galvanized					4			4	4.40
2500	Gypsum lath nails, 1-1/8", 13 ga. flathead, blued					2.05			2.05	2.26
2600	Masonry nails, hardened steel, 3/4" to 3" long, plain					1.70			1.70	1.87
2700	Galvanized					3.25			3.25	3.58
2900	Roofing nails, threaded, galvanized					1.75			1.75	1.93
3100	Aluminum					4.80			4.80	5.30
3300	Compressed lead head, threaded, galvanized					2.45			2.45	2.70
3600	Siding nails, plain shank, galvanized					2			2	2.20
3800	Aluminum					5.95			5.95	6.55
5000	Add to prices above for cement coating					.12			.12	.13
5200	Zinc or tin plating					.15			.15	.17
5500	Vinyl coated sinkers, 8d to 16d					2.40			2.40	2.64

06 05 23.50 Wood Screws		Crew	Daily Output	Labor-Hours	Unit	Material	Labor	Equipment	Total	Total Incl O&P
0010	**WOOD SCREWS**									
0020	#8 x 1" long, steel				C	2.75			2.75	3.03
0100	Brass					12.25			12.25	13.50
0200	#8, 2" long, steel					4.25			4.25	4.68
0300	Brass					22.50			22.50	24.50
0400	#10, 1" long, steel					3.45			3.45	3.80
0500	Brass					15.60			15.60	17.15
0600	#10, 2" long, steel					5.25			5.25	5.80
0700	Brass					27			27	29.50
0800	#10, 3" long, steel					8.90			8.90	9.80
1000	#12, 2" long, steel					7.35			7.35	8.10
1100	Brass					35			35	38
1500	#12, 3" long, steel					11.20			11.20	12.30
2000	#12, 4" long, steel					19.20			19.20	21

06 05 23.60 Timber Connectors		Crew	Daily Output	Labor-Hours	Unit	Material	Labor	Equipment	Total	Total Incl O&P
0010	**TIMBER CONNECTORS**									
0020	Add up cost of each part for total cost of connection									
0100	Connector plates, steel, with bolts, straight	2 Carp	75	.213	Ea.	30.50	8.05		38.55	47
0110	Tee, 7 ga.		50	.320		35	12.10		47.10	59
0120	T-Strap, 14 ga., 12" x 8" x 2"		50	.320		35	12.10		47.10	59
0150	Anchor plates, 7 ga., 9" x 7"		75	.213		30.50	8.05		38.55	47
0200	Bolts, machine, sq. hd. with nut & washer, 1/2" diameter, 4" long	1 Carp	140	.057		.75	2.16		2.91	4.45
0300	7-1/2" long		130	.062		1.37	2.33		3.70	5.40

06 05 23 – Wood, Plastic, and Composite Fastenings

06 05 23.60 Timber Connectors		Crew	Daily Output	Labor-Hours	Unit	2016 Bare Costs Material	Labor	Equipment	Total	Total Incl O&P
0500	3/4" diameter, 7-1/2" long	1 Carp	130	.062	Ea.	3.20	2.33		5.53	7.40
0610	Machine bolts, w/nut, washer, 3/4" diam., 15" L, HD's & beam hangers		95	.084		5.95	3.18		9.13	11.90
0720	Machine bolts, sq. hd. w/nut & wash		150	.053	Lb.	3.44	2.02		5.46	7.15
0800	Drilling bolt holes in timber, 1/2" diameter		450	.018	Inch		.67		.67	1.13
0900	1" diameter		350	.023	"		.86		.86	1.45
1100	Framing anchor, angle, 3" x 3" x 1-1/2", 12 ga		175	.046	Ea.	2.56	1.73		4.29	5.70
1150	Framing anchors, 18 ga., 4-1/2" x 2-3/4"		175	.046		2.56	1.73		4.29	5.70
1160	Framing anchors, 18 ga., 4-1/2" x 3"		175	.046		2.56	1.73		4.29	5.70
1170	Clip anchors plates, 18 ga., 12" x 1-1/8"		175	.046		2.56	1.73		4.29	5.70
1250	Holdowns, 3 ga. base, 10 ga. body		8	1		25	38		63	91
1260	Holdowns, 7 ga. 11-1/16" x 3-1/4"		8	1		25	38		63	91
1270	Holdowns, 7 ga. 14-3/8" x 3-1/8"		8	1		25	38		63	91
1275	Holdowns, 12 ga. 8" x 2-1/2"		8	1		25	38		63	91
1300	Joist and beam hangers, 18 ga. galv., for 2" x 4" joist		175	.046		.72	1.73		2.45	3.68
1400	2" x 6" to 2" x 10" joist		165	.048		1.36	1.83		3.19	4.57
1600	16 ga. galv., 3" x 6" to 3" x 10" joist		160	.050		2.92	1.89		4.81	6.40
1700	3" x 10" to 3" x 14" joist		160	.050		4.88	1.89		6.77	8.50
1800	4" x 6" to 4" x 10" joist		155	.052		3.04	1.95		4.99	6.60
1900	4" x 10" to 4" x 14" joist		155	.052		4.96	1.95		6.91	8.70
2000	Two-2" x 6" to two-2" x 10" joists		150	.053		4.16	2.02		6.18	7.95
2100	Two-2" x 10" to two-2" x 14" joists		150	.053		4.64	2.02		6.66	8.50
2300	3/16" thick, 6" x 8" joist		145	.055		65	2.09		67.09	75
2400	6" x 10" joist		140	.057		67.50	2.16		69.66	78
2500	6" x 12" joist		135	.059		70.50	2.24		72.74	81.50
2700	1/4" thick, 6" x 14" joist		130	.062		73	2.33		75.33	84.50
2900	Plywood clips, extruded aluminum H clip, for 3/4" panels					.22			.22	.24
3000	Galvanized 18 ga. back-up clip					.20			.20	.22
3200	Post framing, 16 ga. galv. for 4" x 4" base, 2 piece	1 Carp	130	.062		16.35	2.33		18.68	22
3300	Cap		130	.062		22.50	2.33		24.83	29
3500	Rafter anchors, 18 ga. galv., 1-1/2" wide, 5-1/4" long		145	.055		.48	2.09		2.57	4.02
3600	10-3/4" long		145	.055		1.44	2.09		3.53	5.05
3800	Shear plates, 2-5/8" diameter		120	.067		2.44	2.52		4.96	6.90
3900	4" diameter		115	.070		5.70	2.63		8.33	10.70
4000	Sill anchors, embedded in concrete or block, 25-1/2" long		115	.070		12.70	2.63		15.33	18.40
4100	Spike grids, 3" x 6"		120	.067		.94	2.52		3.46	5.25
4400	Split rings, 2-1/2" diameter		120	.067		2.02	2.52		4.54	6.45
4500	4" diameter		110	.073		2.98	2.75		5.73	7.90
4550	Tie plate, 20 ga., 7" x 3 1/8"		110	.073		2.98	2.75		5.73	7.90
4560	Tie plate, 20 ga., 5" x 4 1/8"		110	.073		2.98	2.75		5.73	7.90
4575	Twist straps, 18 ga., 12" x 1 1/4"		110	.073		2.98	2.75		5.73	7.90
4580	Twist straps, 18 ga., 16" x 1 1/4"		110	.073		2.98	2.75		5.73	7.90
4600	Strap ties, 20 ga., 2 -1/16" wide, 12 13/16" long		180	.044		.93	1.68		2.61	3.83
4700	Strap ties, 16 ga., 1-3/8" wide, 12" long		180	.044		.93	1.68		2.61	3.83
4800	21-5/8" x 1-1/4"		160	.050		2.94	1.89		4.83	6.40
5000	Toothed rings, 2-5/8" or 4" diameter		90	.089		1.79	3.36		5.15	7.60
5200	Truss plates, nailed, 20 ga., up to 32' span		17	.471	Truss	12.85	17.80		30.65	44
5400	Washers, 2" x 2" x 1/8"				Ea.	.38			.38	.42
5500	3" x 3" x 3/16"				"	1.04			1.04	1.14

06 05 23.80 Metal Bracing		Crew	Daily Output	Labor-Hours	Unit	Material	Labor	Equipment	Total	Total Incl O&P
0010	**METAL BRACING** R051223-20									
0302	Let-in, "T" shaped, 22 ga. galv. steel, studs at 16" O.C.	1 Carp	580	.014	L.F.	.83	.52		1.35	1.78
0402	Studs at 24" O.C.		600	.013		.83	.50		1.33	1.75

06 05 Common Work Results for Wood, Plastics, and Composites

06 05 23 – Wood, Plastic, and Composite Fastenings

06 05 23.80 Metal Bracing		Crew	Daily Output	Labor-Hours	Unit	Material	2016 Bare Costs Labor	Equipment	Total	Total Incl O&P
0502	Steel straps, 16 ga. galv. steel, studs at 16″ O.C.	1 Carp	600	.013	L.F.	1.07	.50		1.57	2.02
0602	Studs at 24″ O.C.		620	.013		1.07	.49		1.56	2

06 11 Wood Framing

06 11 10 – Framing with Dimensional, Engineered or Composite Lumber

06 11 10.01 Forest Stewardship Council Certification		Crew	Daily Output	Labor-Hours	Unit	Material	Labor	Equipment	Total	Total Incl O&P
0010	**FOREST STEWARDSHIP COUNCIL CERTIFICATION**									
0020	For Forest Stewardship Council (FSC) cert dimension lumber, add G					65%				

06 11 10.02 Blocking		Crew	Daily Output	Labor-Hours	Unit	Material	Labor	Equipment	Total	Total Incl O&P
0010	**BLOCKING**									
1790	Bolted to concrete									
1798	Ledger board, 2″ x 4″	1 Carp	180	.044	L.F.	4.69	1.68		6.37	7.95
1800	2″ x 6″		160	.050		4.70	1.89		6.59	8.30
1810	4″ x 6″		140	.057		8.25	2.16		10.41	12.70
1820	4″ x 8″		120	.067		8.30	2.52		10.82	13.35
1950	Miscellaneous, to wood construction									
2000	2″ x 4″	1 Carp	250	.032	L.F.	.40	1.21		1.61	2.47
2005	Pneumatic nailed		305	.026		.40	.99		1.39	2.10
2050	2″ x 6″		222	.036		.61	1.36		1.97	2.95
2055	Pneumatic nailed		271	.030		.62	1.12		1.74	2.55
2100	2″ x 8″		200	.040		.85	1.51		2.36	3.46
2105	Pneumatic nailed		244	.033		.85	1.24		2.09	3.02
2150	2″ x 10″		178	.045		1.34	1.70		3.04	4.31
2155	Pneumatic nailed		217	.037		1.34	1.39		2.73	3.81
2200	2″ x 12″		151	.053		1.70	2		3.70	5.20
2205	Pneumatic nailed		185	.043		1.71	1.63		3.34	4.62
2300	To steel construction									
2320	2″ x 4″	1 Carp	208	.038	L.F.	.40	1.45		1.85	2.87
2340	2″ x 6″		180	.044		.61	1.68		2.29	3.48
2360	2″ x 8″		158	.051		.85	1.91		2.76	4.13
2380	2″ x 10″		136	.059		1.34	2.22		3.56	5.20
2400	2″ x 12″		109	.073		1.70	2.77		4.47	6.50

06 11 10.04 Wood Bracing		Crew	Daily Output	Labor-Hours	Unit	Material	Labor	Equipment	Total	Total Incl O&P
0010	**WOOD BRACING**									
0012	Let-in, with 1″ x 6″ boards, studs @ 16″ O.C.	1 Carp	150	.053	L.F.	.71	2.02		2.73	4.16
0202	Studs @ 24″ O.C.	″	230	.035	″	.71	1.31		2.02	2.98

06 11 10.06 Bridging		Crew	Daily Output	Labor-Hours	Unit	Material	Labor	Equipment	Total	Total Incl O&P
0010	**BRIDGING**									
0012	Wood, for joists 16″ O.C., 1″ x 3″	1 Carp	130	.062	Pr.	.64	2.33		2.97	4.61
0017	Pneumatic nailed		170	.047		.69	1.78		2.47	3.74
0102	2″ x 3″ bridging		130	.062		.66	2.33		2.99	4.63
0107	Pneumatic nailed		170	.047		.66	1.78		2.44	3.71
0302	Steel, galvanized, 18 ga., for 2″ x 10″ joists at 12″ O.C.		130	.062		.94	2.33		3.27	4.93
0352	16″ O.C.		135	.059		.94	2.24		3.18	4.79
0402	24″ O.C.		140	.057		1.94	2.16		4.10	5.75
0602	For 2″ x 14″ joists at 16″ O.C.		130	.062		1.43	2.33		3.76	5.45
0902	Compression type, 16″ O.C., 2″ x 8″ joists		200	.040		.94	1.51		2.45	3.57
1002	2″ x 12″ joists		200	.040		.94	1.51		2.45	3.57

06 11 10 – Framing with Dimensional, Engineered or Composite Lumber

	06 11 10.10 Beam and Girder Framing	Crew	Daily Output	Labor-Hours	Unit	2016 Bare Costs Material	Labor	Equipment	Total	Total Incl O&P
0010	**BEAM AND GIRDER FRAMING**									
1000	Single, 2" x 6"	2 Carp	700	.023	L.F.	.61	.86		1.47	2.12
1005	Pneumatic nailed		812	.020		.62	.74		1.36	1.93
1020	2" x 8"		650	.025		.85	.93		1.78	2.49
1025	Pneumatic nailed		754	.021		.85	.80		1.65	2.28
1040	2" x 10"		600	.027		1.34	1.01		2.35	3.16
1045	Pneumatic nailed		696	.023		1.34	.87		2.21	2.94
1060	2" x 12"		550	.029		1.70	1.10		2.80	3.71
1065	Pneumatic nailed		638	.025		1.71	.95		2.66	3.47
1080	2" x 14"		500	.032		2.15	1.21		3.36	4.39
1085	Pneumatic nailed		580	.028		2.16	1.04		3.20	4.12
1100	3" x 8"		550	.029		2.72	1.10		3.82	4.83
1120	3" x 10"		500	.032		3.43	1.21		4.64	5.80
1140	3" x 12"		450	.036		4.11	1.34		5.45	6.75
1160	3" x 14"		400	.040		4.80	1.51		6.31	7.85
1170	4" x 6"	F-3	1100	.036		2.90	1.25	.60	4.75	5.95
1180	4" x 8"		1000	.040		2.96	1.37	.66	4.99	6.30
1200	4" x 10"		950	.042		4.70	1.45	.69	6.84	8.35
1220	4" x 12"		900	.044		5.20	1.53	.73	7.46	9.05
1240	4" x 14"		850	.047		6.05	1.62	.77	8.44	10.25
1250	6" x 8"		525	.076		7.50	2.62	1.25	11.37	14
1260	6" x 10"		500	.080		6.55	2.75	1.32	10.62	13.25
1290	8" x 12"		300	.133		15.70	4.58	2.19	22.47	27.50
1300	Treated, single, 2" x 4"	2 Carp	700	.023		.53	.86		1.39	2.03
1320	2" x 6"		700	.023		.78	.86		1.64	2.31
1340	2" x 8"		650	.025		1.06	.93		1.99	2.73
1360	2" x 10"		600	.027		1.36	1.01		2.37	3.18
1380	2" x 12"		550	.029		1.91	1.10		3.01	3.94
1400	2" x 14"		500	.032		2.51	1.21		3.72	4.79
1420	3" x 8"		550	.029		3.70	1.10		4.80	5.90
1440	3" x 10"		500	.032		4.22	1.21		5.43	6.65
1460	3" x 12"		450	.036		5.65	1.34		6.99	8.45
1480	3" x 14"		400	.040		6.60	1.51		8.11	9.80
1500	4" x 8"	F-3	1000	.040		3.83	1.37	.66	5.86	7.25
1520	4" x 10"		950	.042		4.79	1.45	.69	6.93	8.45
1540	4" x 12"		900	.044		5.85	1.53	.73	8.11	9.80
1560	4" x 14"		850	.047		7.65	1.62	.77	10.04	11.95
2000	Double, 2" x 6"	2 Carp	625	.026		1.22	.97		2.19	2.96
2005	Pneumatic nailed		725	.022		1.23	.83		2.06	2.75
2020	2" x 8"		575	.028		1.70	1.05		2.75	3.63
2025	Pneumatic nailed		667	.024		1.71	.91		2.62	3.40
2040	2" x 10"		550	.029		2.67	1.10		3.77	4.78
2045	Pneumatic nailed		638	.025		2.69	.95		3.64	4.55
2060	2" x 12"		525	.030		3.40	1.15		4.55	5.65
2065	Pneumatic nailed		610	.026		3.42	.99		4.41	5.40
2080	2" x 14"		475	.034		4.30	1.27		5.57	6.85
2085	Pneumatic nailed		551	.029		4.32	1.10		5.42	6.60
3000	Triple, 2" x 6"		550	.029		1.83	1.10		2.93	3.86
3005	Pneumatic nailed		638	.025		1.85	.95		2.80	3.62
3020	2" x 8"		525	.030		2.54	1.15		3.69	4.73
3025	Pneumatic nailed		609	.026		2.56	.99		3.55	4.48
3040	2" x 10"		500	.032		4.01	1.21		5.22	6.45

06 11 Wood Framing

06 11 10 – Framing with Dimensional, Engineered or Composite Lumber

06 11 10.10 Beam and Girder Framing		Crew	Daily Output	Labor-Hours	Unit	2016 Bare Costs Material	Labor	Equipment	Total	Total Incl O&P
3045	Pneumatic nailed	2 Carp	580	.028	L.F.	4.03	1.04		5.07	6.20
3060	2" x 12"		475	.034		5.10	1.27		6.37	7.75
3065	Pneumatic nailed		551	.029		5.15	1.10		6.25	7.50
3080	2" x 14"		450	.036		7.15	1.34		8.49	10.10
3085	Pneumatic nailed		522	.031		6.50	1.16		7.66	9.05

06 11 10.12 Ceiling Framing

Line	Description	Crew	Daily Output	Labor-Hours	Unit	Material	Labor	Equipment	Total	Total Incl O&P
0010	CEILING FRAMING									
6000	Suspended, 2" x 3"	2 Carp	1000	.016	L.F.	.38	.60		.98	1.42
6050	2" x 4"		900	.018		.40	.67		1.07	1.57
6100	2" x 6"		800	.020		.61	.76		1.37	1.94
6150	2" x 8"		650	.025		.85	.93		1.78	2.49

06 11 10.14 Posts and Columns

Line	Description	Crew	Daily Output	Labor-Hours	Unit	Material	Labor	Equipment	Total	Total Incl O&P
0010	POSTS AND COLUMNS									
0100	4" x 4"	2 Carp	390	.041	L.F.	1.73	1.55		3.28	4.51
0150	4" x 6"		275	.058		2.90	2.20		5.10	6.85
0200	4" x 8"		220	.073		2.96	2.75		5.71	7.85
0250	6" x 6"		215	.074		5.05	2.81		7.86	10.30
0300	6" x 8"		175	.091		7.50	3.46		10.96	14.05
0350	6" x 10"		150	.107		6.55	4.03		10.58	13.95

06 11 10.18 Joist Framing

Line	Description	Crew	Daily Output	Labor-Hours	Unit	Material	Labor	Equipment	Total	Total Incl O&P
0010	JOIST FRAMING									
2000	Joists, 2" x 4"	2 Carp	1250	.013	L.F.	.40	.48		.88	1.25
2005	Pneumatic nailed		1438	.011		.40	.42		.82	1.14
2100	2" x 6"		1250	.013		.61	.48		1.09	1.48
2105	Pneumatic nailed		1438	.011		.62	.42		1.04	1.38
2150	2" x 8"		1100	.015		.85	.55		1.40	1.85
2155	Pneumatic nailed		1265	.013		.85	.48		1.33	1.74
2200	2" x 10"		900	.018		1.34	.67		2.01	2.60
2205	Pneumatic nailed		1035	.015		1.34	.58		1.92	2.46
2250	2" x 12"		875	.018		1.70	.69		2.39	3.03
2255	Pneumatic nailed		1006	.016		1.71	.60		2.31	2.89
2300	2" x 14"		770	.021		2.15	.79		2.94	3.68
2305	Pneumatic nailed		886	.018		2.16	.68		2.84	3.51
2350	3" x 6"		925	.017		1.89	.65		2.54	3.18
2400	3" x 10"		780	.021		3.43	.78		4.21	5.05
2450	3" x 12"		600	.027		4.11	1.01		5.12	6.20
2500	4" x 6"		800	.020		2.90	.76		3.66	4.46
2550	4" x 10"		600	.027		4.70	1.01		5.71	6.85
2600	4" x 12"		450	.036		5.20	1.34		6.54	7.95
2605	Sister joist, 2" x 6"		800	.020		.61	.76		1.37	1.94
2606	Pneumatic nailed		960	.017		.62	.63		1.25	1.74
3000	Composite wood joist 9-1/2" deep		.90	17.778	M.L.F.	1,850	670		2,520	3,175
3010	11-1/2" deep		.88	18.182		2,050	685		2,735	3,400
3020	14" deep		.82	19.512		2,125	740		2,865	3,575
3030	16" deep		.78	20.513		3,575	775		4,350	5,250
4000	Open web joist 12" deep		.88	18.182		3,325	685		4,010	4,800
4002	Per linear foot		880	.018	L.F.	3.32	.69		4.01	4.80
4004	Treated, per linear foot		880	.018	"	4.21	.69		4.90	5.80
4010	14" deep		.82	19.512	M.L.F.	3,625	740		4,365	5,225
4012	Per linear foot		820	.020	L.F.	3.63	.74		4.37	5.25
4014	Treated, per linear foot		820	.020	"	4.66	.74		5.40	6.40
4020	16" deep		.78	20.513	M.L.F.	3,800	775		4,575	5,475

06 11 Wood Framing

06 11 10 – Framing with Dimensional, Engineered or Composite Lumber

06 11 10.18 Joist Framing

		Crew	Daily Output	Labor-Hours	Unit	Material	Labor	Equipment	Total	Total Incl O&P
4022	Per linear foot	2 Carp	780	.021	L.F.	3.79	.78		4.57	5.45
4024	Treated, per linear foot		780	.021	"	4.97	.78		5.75	6.75
4030	18" deep		.74	21.622	M.L.F.	3,850	815		4,665	5,625
4032	Per linear foot		740	.022	L.F.	3.85	.82		4.67	5.60
4034	Treated, per linear foot		740	.022	"	5.20	.82		6.02	7.05
6000	Composite rim joist, 1-1/4" x 9-1/2"		.90	17.778	M.L.F.	1,975	670		2,645	3,300
6010	1-1/4" x 11-1/2"		.88	18.182		2,275	685		2,960	3,650
6020	1-1/4" x 14-1/2"		.82	19.512		2,750	740		3,490	4,250
6030	1-1/4" x 16-1/2"		.78	20.513		3,175	775		3,950	4,775

06 11 10.24 Miscellaneous Framing

		Crew	Daily Output	Labor-Hours	Unit	Material	Labor	Equipment	Total	Total Incl O&P
0010	**MISCELLANEOUS FRAMING**									
2000	Firestops, 2" x 4"	2 Carp	780	.021	L.F.	.40	.78		1.18	1.74
2005	Pneumatic nailed R061110-30		952	.017		.40	.64		1.04	1.50
2100	2" x 6"		600	.027		.61	1.01		1.62	2.36
2105	Pneumatic nailed		732	.022		.62	.83		1.45	2.06
5000	Nailers, treated, wood construction, 2" x 4"		800	.020		.52	.76		1.28	1.85
5005	Pneumatic nailed		960	.017		.53	.63		1.16	1.64
5100	2" x 6"		750	.021		.77	.81		1.58	2.20
5105	Pneumatic nailed		900	.018		.78	.67		1.45	1.99
5120	2" x 8"		700	.023		1.06	.86		1.92	2.61
5125	Pneumatic nailed		840	.019		1.06	.72		1.78	2.38
5200	Steel construction, 2" x 4"		750	.021		.52	.81		1.33	1.93
5220	2" x 6"		700	.023		.77	.86		1.63	2.30
5240	2" x 8"		650	.025		1.06	.93		1.99	2.72
7000	Rough bucks, treated, for doors or windows, 2" x 6"		400	.040		.77	1.51		2.28	3.38
7005	Pneumatic nailed		480	.033		.78	1.26		2.04	2.97
7100	2" x 8"		380	.042		1.06	1.59		2.65	3.83
7105	Pneumatic nailed		456	.035		1.06	1.33		2.39	3.39
8000	Stair stringers, 2" x 10"		130	.123		1.34	4.65		5.99	9.25
8100	2" x 12"		130	.123		1.70	4.65		6.35	9.65
8150	3" x 10"		125	.128		3.43	4.84		8.27	11.85
8200	3" x 12"		125	.128		4.11	4.84		8.95	12.60

06 11 10.26 Partitions

		Crew	Daily Output	Labor-Hours	Unit	Material	Labor	Equipment	Total	Total Incl O&P
0010	**PARTITIONS**									
0020	Single bottom and double top plate, no waste, std. & better lumber									
0180	2" x 4" studs, 8' high, studs 12" O.C.	2 Carp	80	.200	L.F.	4.37	7.55		11.92	17.45
0185	12" O.C., pneumatic nailed		96	.167		4.40	6.30		10.70	15.40
0200	16" O.C.		100	.160		3.58	6.05		9.63	14.10
0205	16" O.C., pneumatic nailed		120	.133		3.60	5.05		8.65	12.40
0300	24" O.C.		125	.128		2.78	4.84		7.62	11.15
0305	24" O.C., pneumatic nailed		150	.107		2.80	4.03		6.83	9.85
0380	10' high, studs 12" O.C.		80	.200		5.15	7.55		12.70	18.35
0385	12" O.C., pneumatic nailed		96	.167		5.20	6.30		11.50	16.25
0400	16" O.C.		100	.160		4.17	6.05		10.22	14.75
0405	16" O.C., pneumatic nailed		120	.133		4.20	5.05		9.25	13.05
0500	24" O.C.		125	.128		3.18	4.84		8.02	11.60
0505	24" O.C., pneumatic nailed		150	.107		3.20	4.03		7.23	10.25
0580	12' high, studs 12" O.C.		65	.246		5.95	9.30		15.25	22
0585	12" O.C., pneumatic nailed		78	.205		6	7.75		13.75	19.60
0600	16" O.C.		80	.200		4.77	7.55		12.32	17.90
0605	16" O.C., pneumatic nailed		96	.167		4.80	6.30		11.10	15.85
0700	24" O.C.		100	.160		3.58	6.05		9.63	14.10

06 11 10.26 Partitions		Crew	Daily Output	Labor-Hours	Unit	2016 Bare Costs Material	Labor	Equipment	Total	Total Incl O&P
0705	24" O.C., pneumatic nailed	2 Carp	120	.133	L.F.	3.60	5.05		8.65	12.40
0780	2" x 6" studs, 8' high, studs 12" O.C.		70	.229		6.70	8.65		15.35	22
0785	12" O.C., pneumatic nailed		84	.190		6.75	7.20		13.95	19.50
0800	16" O.C.		90	.178		5.50	6.70		12.20	17.30
0805	16" O.C., pneumatic nailed		108	.148		5.55	5.60		11.15	15.50
0900	24" O.C.		115	.139		4.28	5.25		9.53	13.50
0905	24" O.C., pneumatic nailed		138	.116		4.31	4.38		8.69	12.10
0980	10' high, studs 12" O.C.		70	.229		7.95	8.65		16.60	23
0985	12" O.C., pneumatic nailed		84	.190		8	7.20		15.20	21
1000	16" O.C.		90	.178		6.40	6.70		13.10	18.30
1005	16" O.C., pneumatic nailed		108	.148		6.45	5.60		12.05	16.50
1100	24" O.C.		115	.139		4.89	5.25		10.14	14.20
1105	24" O.C., pneumatic nailed		138	.116		4.92	4.38		9.30	12.75
1180	12' high, studs 12" O.C.		55	.291		9.15	11		20.15	28.50
1185	12" O.C., pneumatic nailed		66	.242		9.25	9.15		18.40	25.50
1200	16" O.C.		70	.229		7.35	8.65		16	22.50
1205	16" O.C., pneumatic nailed		84	.190		7.40	7.20		14.60	20
1300	24" O.C.		90	.178		5.50	6.70		12.20	17.30
1305	24" O.C., pneumatic nailed		108	.148		5.55	5.60		11.15	15.50
1400	For horizontal blocking, 2" x 4", add		600	.027		.40	1.01		1.41	2.13
1500	2" x 6", add		600	.027		.61	1.01		1.62	2.36
1600	For openings, add		250	.064			2.42		2.42	4.05
1700	Headers for above openings, material only, add				M.B.F.	640			640	705

06 11 10.28 Porch or Deck Framing		Crew	Daily Output	Labor-Hours	Unit	Material	Labor	Equipment	Total	Total Incl O&P
0010	**PORCH OR DECK FRAMING**									
0100	Treated lumber, posts or columns, 4" x 4"	2 Carp	390	.041	L.F.	1.35	1.55		2.90	4.08
0110	4" x 6"		275	.058		2.02	2.20		4.22	5.90
0120	4" x 8"		220	.073		3.87	2.75		6.62	8.85
0130	Girder, single, 4" x 4"		675	.024		1.35	.90		2.25	2.98
0140	4" x 6"		600	.027		2.02	1.01		3.03	3.91
0150	4" x 8"		525	.030		3.87	1.15		5.02	6.20
0160	Double, 2" x 4"		625	.026		1.07	.97		2.04	2.80
0170	2" x 6"		600	.027		1.59	1.01		2.60	3.43
0180	2" x 8"		575	.028		2.17	1.05		3.22	4.14
0190	2" x 10"		550	.029		2.76	1.10		3.86	4.87
0200	2" x 12"		525	.030		3.87	1.15		5.02	6.20
0210	Triple, 2" x 4"		575	.028		1.61	1.05		2.66	3.53
0220	2" x 6"		550	.029		2.38	1.10		3.48	4.46
0230	2" x 8"		525	.030		3.25	1.15		4.40	5.50
0240	2" x 10"		500	.032		4.13	1.21		5.34	6.60
0250	2" x 12"		475	.034		5.80	1.27		7.07	8.55
0260	Ledger, bolted 4' O.C., 2" x 4"		400	.040		.69	1.51		2.20	3.28
0270	2" x 6"		395	.041		.93	1.53		2.46	3.59
0280	2" x 8"		390	.041		1.21	1.55		2.76	3.93
0290	2" x 10"		385	.042		1.50	1.57		3.07	4.28
0300	2" x 12"		380	.042		2.05	1.59		3.64	4.92
0310	Joists, 2" x 4"		1250	.013		.53	.48		1.01	1.40
0320	2" x 6"		1250	.013		.79	.48		1.27	1.68
0330	2" x 8"		1100	.015		1.08	.55		1.63	2.10
0340	2" x 10"		900	.018		1.37	.67		2.04	2.64
0350	2" x 12"		875	.018		1.64	.69		2.33	2.97
0360	Railings and trim, 1" x 4"	1 Carp	300	.027		.46	1.01		1.47	2.20

06 11 Wood Framing

06 11 10 – Framing with Dimensional, Engineered or Composite Lumber

06 11 10.28 Porch or Deck Framing		Crew	Daily Output	Labor-Hours	Unit	2016 Bare Costs Material	Labor	Equipment	Total	Total Incl O&P
0370	2″ x 2″	1 Carp	300	.027	L.F.	.33	1.01		1.34	2.06
0380	2″ x 4″		300	.027		.53	1.01		1.54	2.27
0390	2″ x 6″		300	.027		.78	1.01		1.79	2.55
0400	Decking, 1″ x 4″		275	.029	S.F.	2.16	1.10		3.26	4.22
0410	2″ x 4″		300	.027		1.78	1.01		2.79	3.64
0420	2″ x 6″		320	.025		1.68	.95		2.63	3.43
0430	5/4″ x 6″		320	.025		2.25	.95		3.20	4.06
0440	Balusters, square, 2″ x 2″	2 Carp	660	.024	L.F.	.34	.92		1.26	1.90
0450	Turned, 2″ x 2″		420	.038		.45	1.44		1.89	2.90
0460	Stair stringer, 2″ x 10″		130	.123		1.37	4.65		6.02	9.30
0470	2″ x 12″		130	.123		1.64	4.65		6.29	9.60
0480	Stair treads, 1″ x 4″		140	.114		2.17	4.32		6.49	9.65
0490	2″ x 4″		140	.114		.53	4.32		4.85	7.85
0500	2″ x 6″		160	.100		.80	3.78		4.58	7.25
0510	5/4″ x 6″		160	.100		1.05	3.78		4.83	7.50
0520	Turned handrail post, 4″ x 4″		64	.250	Ea.	31	9.45		40.45	50
0530	Lattice panel, 4′ x 8′, 1/2″		1600	.010	S.F.	.81	.38		1.19	1.52
0535	3/4″		1600	.010	″	1.21	.38		1.59	1.96
0540	Cedar, posts or columns, 4″ x 4″		390	.041	L.F.	3.96	1.55		5.51	6.95
0550	4″ x 6″		275	.058		7.70	2.20		9.90	12.15
0560	4″ x 8″		220	.073		10.45	2.75		13.20	16.05
0800	Decking, 1″ x 4″		550	.029		1.91	1.10		3.01	3.94
0810	2″ x 4″		600	.027		3.86	1.01		4.87	5.95
0820	2″ x 6″		640	.025		6.95	.95		7.90	9.25
0830	5/4″ x 6″		640	.025		4.45	.95		5.40	6.45
0840	Railings and trim, 1″ x 4″		600	.027		1.91	1.01		2.92	3.79
0860	2″ x 4″		600	.027		3.86	1.01		4.87	5.95
0870	2″ x 6″		600	.027		6.95	1.01		7.96	9.35
0920	Stair treads, 1″ x 4″		140	.114		1.91	4.32		6.23	9.35
0930	2″ x 4″		140	.114		3.86	4.32		8.18	11.50
0940	2″ x 6″		160	.100		6.95	3.78		10.73	14
0950	5/4″ x 6″		160	.100		4.45	3.78		8.23	11.25
0980	Redwood, posts or columns, 4″ x 4″		390	.041		6.45	1.55		8	9.70
0990	4″ x 6″		275	.058		12.60	2.20		14.80	17.55
1000	4″ x 8″		220	.073		23.50	2.75		26.25	30.50
1240	Decking, 1″ x 4″	1 Carp	275	.029	S.F.	3.97	1.10		5.07	6.20
1260	2″ x 6″		340	.024		7.50	.89		8.39	9.75
1270	5/4″ x 6″		320	.025		4.83	.95		5.78	6.90
1280	Railings and trim, 1″ x 4″	2 Carp	600	.027	L.F.	1.17	1.01		2.18	2.98
1310	2″ x 6″		600	.027		7.50	1.01		8.51	9.95
1420	Alternative decking, wood/plastic composite, 5/4″ x 6″ G		640	.025		3.14	.95		4.09	5.05
1440	1″ x 4″ square edge fir		550	.029		2.18	1.10		3.28	4.24
1450	1″ x 4″ tongue and groove fir		450	.036		1.46	1.34		2.80	3.85
1460	1″ x 4″ mahogany		550	.029		2.01	1.10		3.11	4.05
1462	5/4″ x 6″ PVC		550	.029		3.71	1.10		4.81	5.90
1465	Framing, porch or deck, alt deck fastening, screws, add	1 Carp	240	.033	S.F.		1.26		1.26	2.11
1470	Accessories, joist hangers, 2″ x 4″		160	.050	Ea.	.72	1.89		2.61	3.96
1480	2″ x 6″ through 2″ x 12″		150	.053		1.36	2.02		3.38	4.88
1530	Post footing, incl excav, backfill, tube form & concrete, 4′ deep, 8″ dia	F-7	12	2.667		12.65	89.50		102.15	164
1540	10″ diameter		11	2.909		18.40	97.50		115.90	184
1550	12″ diameter		10	3.200		24	107		131	207

06 11 10 – Framing with Dimensional, Engineered or Composite Lumber

06 11 10.30 Roof Framing		Crew	Daily Output	Labor-Hours	Unit	Material	2016 Bare Costs Labor	Equipment	Total	Total Incl O&P
0010	ROOF FRAMING									
1900	Rough fascia, 2" x 6"	2 Carp	250	.064	L.F.	.61	2.42		3.03	4.72
2000	2" x 8"		225	.071		.85	2.69		3.54	5.45
2100	2" x 10"		180	.089		1.34	3.36		4.70	7.10
5000	Rafters, to 4 in 12 pitch, 2" x 6", ordinary		1000	.016		.61	.60		1.21	1.68
5020	On steep roofs		800	.020		.61	.76		1.37	1.94
5040	On dormers or complex roofs		590	.027		.61	1.03		1.64	2.39
5060	2" x 8", ordinary		950	.017		.85	.64		1.49	2
5080	On steep roofs		750	.021		.85	.81		1.66	2.28
5100	On dormers or complex roofs		540	.030		.85	1.12		1.97	2.81
5120	2" x 10", ordinary		630	.025		1.34	.96		2.30	3.08
5140	On steep roofs		495	.032		1.34	1.22		2.56	3.52
5160	On dormers or complex roofs		425	.038		1.34	1.42		2.76	3.85
5180	2" x 12", ordinary		575	.028		1.70	1.05		2.75	3.63
5200	On steep roofs		455	.035		1.70	1.33		3.03	4.10
5220	On dormers or complex roofs		395	.041		1.70	1.53		3.23	4.43
5300	Hip and valley rafters, 2" x 6", ordinary		760	.021		.61	.80		1.41	2
5320	On steep roofs		585	.027		.61	1.03		1.64	2.40
5340	On dormers or complex roofs		510	.031		.61	1.19		1.80	2.66
5360	2" x 8", ordinary		720	.022		.85	.84		1.69	2.34
5380	On steep roofs		545	.029		.85	1.11		1.96	2.79
5400	On dormers or complex roofs		470	.034		.85	1.29		2.14	3.00
5420	2" x 10", ordinary		570	.028		1.34	1.06		2.40	3.25
5440	On steep roofs		440	.036		1.34	1.37		2.71	3.77
5460	On dormers or complex roofs		380	.042		1.34	1.59		2.93	4.14
5480	2" x 12", ordinary		525	.030		1.70	1.15		2.85	3.80
5500	On steep roofs		410	.039		1.70	1.48		3.18	4.34
5520	On dormers or complex roofs		355	.045		1.70	1.70		3.40	4.72
5540	Hip and valley jacks, 2" x 6", ordinary		600	.027		.61	1.01		1.62	2.36
5560	On steep roofs		475	.034		.61	1.27		1.88	2.80
5580	On dormers or complex roofs		410	.039		.61	1.48		2.09	3.14
5600	2" x 8", ordinary		490	.033		.85	1.23		2.08	3
5620	On steep roofs		385	.042		.85	1.57		2.42	3.56
5640	On dormers or complex roofs		335	.048		.85	1.81		2.66	3.95
5660	2" x 10", ordinary		450	.036		1.34	1.34		2.68	3.72
5680	On steep roofs		350	.046		1.34	1.73		3.07	4.36
5700	On dormers or complex roofs		305	.052		1.34	1.98		3.32	4.79
5720	2" x 12", ordinary		375	.043		1.70	1.61		3.31	4.57
5740	On steep roofs		295	.054		1.70	2.05		3.75	5.30
5760	On dormers or complex roofs		255	.063		1.70	2.37		4.07	5.85
5780	Rafter tie, 1" x 4", #3		800	.020		.46	.76		1.22	1.78
5790	2" x 4", #3		800	.020		.40	.76		1.16	1.71
5800	Ridge board, #2 or better, 1" x 6"		600	.027		.71	1.01		1.72	2.47
5820	1" x 8"		550	.029		1.18	1.10		2.28	3.14
5840	1" x 10"		500	.032		1.53	1.21		2.74	3.72
5860	2" x 6"		500	.032		.61	1.21		1.82	2.70
5880	2" x 8"		450	.036		.85	1.34		2.19	3.18
5900	2" x 10"		400	.040		1.34	1.51		2.85	4
5920	Roof cants, split, 4" x 4"		650	.025		1.73	.93		2.66	3.47
5940	6" x 6"		600	.027		5.05	1.01		6.06	7.30
5960	Roof curbs, untreated, 2" x 6"		520	.031		.61	1.16		1.77	2.62
5980	2" x 12"	↓	400	.040	↓	1.70	1.51		3.21	4.40

06 11 10 – Framing with Dimensional, Engineered or Composite Lumber

06 11 10.30 Roof Framing		Crew	Daily Output	Labor-Hours	Unit	2016 Bare Costs Material	Labor	Equipment	Total	Total Incl O&P
6000	Sister rafters, 2″ x 6″	2 Carp	800	.020	L.F.	.61	.76		1.37	1.94
6020	2″ x 8″		640	.025		.85	.95		1.80	2.51
6040	2″ x 10″		535	.030		1.34	1.13		2.47	3.36
6060	2″ x 12″		455	.035		1.70	1.33		3.03	4.10

06 11 10.32 Sill and Ledger Framing

Code	Description	Crew	Daily Output	Labor-Hours	Unit	Material	Labor	Equipment	Total	Total Incl O&P
0010	**SILL AND LEDGER FRAMING**									
0020	Extruded polystyrene sill sealer, 5-1/2″ wide	1 Carp	1600	.005	L.F.	.14	.19		.33	.47
2000	Ledgers, nailed, 2″ x 4″	2 Carp	755	.021		.40	.80		1.20	1.78
2050	2″ x 6″		600	.027		.61	1.01		1.62	2.36
2100	Bolted, not including bolts, 3″ x 6″		325	.049		1.88	1.86		3.74	5.20
2150	3″ x 12″		233	.069		4.08	2.60		6.68	8.85
2600	Mud sills, redwood, construction grade, 2″ x 4″		895	.018		2.26	.68		2.94	3.62
2620	2″ x 6″		780	.021		3.40	.78		4.18	5.05
4000	Sills, 2″ x 4″		600	.027		.39	1.01		1.40	2.12
4050	2″ x 6″		550	.029		.60	1.10		1.70	2.50
4080	2″ x 8″		500	.032		.83	1.21		2.04	2.95
4100	2″ x 10″		450	.036		1.32	1.34		2.66	3.70
4120	2″ x 12″		400	.040		1.68	1.51		3.19	4.38
4200	Treated, 2″ x 4″		550	.029		.52	1.10		1.62	2.41
4220	2″ x 6″		500	.032		.76	1.21		1.97	2.87
4240	2″ x 8″		450	.036		1.04	1.34		2.38	3.40
4280	2″ x 12″		350	.046		1.88	1.73		3.61	4.95
4400	4″ x 4″		450	.036		1.31	1.34		2.65	3.69
4420	4″ x 6″		350	.046		1.96	1.73		3.69	5.05
4460	4″ x 8″		300	.053		3.79	2.02		5.81	7.55
4480	4″ x 10″		260	.062		4.73	2.33		7.06	9.10

06 11 10.34 Sleepers

Code	Description	Crew	Daily Output	Labor-Hours	Unit	Material	Labor	Equipment	Total	Total Incl O&P
0010	**SLEEPERS**									
0100	On concrete, treated, 1″ x 2″	2 Carp	2350	.007	L.F.	.27	.26		.53	.73
0150	1″ x 3″		2000	.008		.44	.30		.74	1
0200	2″ x 4″		1500	.011		.56	.40		.96	1.30
0250	2″ x 6″		1300	.012		.86	.47		1.33	1.72

06 11 10.36 Soffit and Canopy Framing

Code	Description	Crew	Daily Output	Labor-Hours	Unit	Material	Labor	Equipment	Total	Total Incl O&P
0010	**SOFFIT AND CANOPY FRAMING**									
1000	Canopy or soffit framing, 1″ x 4″	2 Carp	900	.018	L.F.	.46	.67		1.13	1.64
1020	1″ x 6″		850	.019		.71	.71		1.42	1.97
1040	1″ x 8″		750	.021		1.18	.81		1.99	2.65
1100	2″ x 4″		620	.026		.40	.98		1.38	2.07
1120	2″ x 6″		560	.029		.61	1.08		1.69	2.48
1140	2″ x 8″		500	.032		.85	1.21		2.06	2.96
1200	3″ x 4″		500	.032		1.14	1.21		2.35	3.29
1220	3″ x 6″		400	.040		1.89	1.51		3.40	4.61
1240	3″ x 10″		300	.053		3.43	2.02		5.45	7.15

06 11 10.38 Treated Lumber Framing Material

Code	Description	Crew	Daily Output	Labor-Hours	Unit	Material	Labor	Equipment	Total	Total Incl O&P
0010	**TREATED LUMBER FRAMING MATERIAL**									
0100	2″ x 4″				M.B.F.	775			775	850
0110	2″ x 6″					765			765	840
0120	2″ x 8″					780			780	860
0130	2″ x 10″					795			795	875
0140	2″ x 12″					940			940	1,025
0200	4″ x 4″					980			980	1,075
0210	4″ x 6″					980			980	1,075

06 11 10.38 Treated Lumber Framing Material		Crew	Daily Output	Labor-Hours	Unit	Material	2016 Bare Costs Labor	Equipment	Total	Total Incl O&P
0220	4" x 8"				M.B.F.	1,425			1,425	1,550

06 11 10.40 Wall Framing		Crew	Daily Output	Labor-Hours	Unit	Material	2016 Bare Costs Labor	Equipment	Total	Total Incl O&P
0010	**WALL FRAMING**									
2000	Headers over openings, 2" x 6"	2 Carp	360	.044	L.F.	.61	1.68		2.29	3.48
2005	2" x 6", pneumatic nailed R061110-30		432	.037		.62	1.40		2.02	3.02
2050	2" x 8"		340	.047		.85	1.78		2.63	3.91
2055	2" x 8", pneumatic nailed		408	.039		.85	1.48		2.33	3.42
2100	2" x 10"		320	.050		1.34	1.89		3.23	4.64
2105	2" x 10", pneumatic nailed		384	.042		1.34	1.58		2.92	4.12
2150	2" x 12"		300	.053		1.70	2.02		3.72	5.25
2155	2" x 12", pneumatic nailed		360	.044		1.71	1.68		3.39	4.69
2180	4" x 8"		260	.062		2.96	2.33		5.29	7.15
2185	4" x 8", pneumatic nailed		312	.051		2.97	1.94		4.91	6.50
2190	4" x 10"		240	.067		4.70	2.52		7.22	9.35
2195	4" x 10", pneumatic nailed		288	.056		4.72	2.10		6.82	8.70
2200	4" x 12"		190	.084		5.20	3.18		8.38	11.05
2205	4" x 12", pneumatic nailed		228	.070		5.20	2.65		7.85	10.20
2240	6" x 10"		165	.097		6.55	3.67		10.22	13.35
2245	6" x 10", pneumatic nailed		198	.081		6.60	3.05		9.65	12.35
2250	6" x 12"		140	.114		8.25	4.32		12.57	16.30
2255	6" x 12", pneumatic nailed		168	.095		8.25	3.60		11.85	15.15
5000	Plates, untreated, 2" x 3"		850	.019		.38	.71		1.09	1.60
5005	2" x 3", pneumatic nailed		1020	.016		.38	.59		.97	1.41
5020	2" x 4"		800	.020		.40	.76		1.16	1.71
5025	2" x 4", pneumatic nailed		960	.017		.40	.63		1.03	1.50
5040	2" x 6"		750	.021		.61	.81		1.42	2.02
5045	2" x 6", pneumatic nailed		900	.018		.62	.67		1.29	1.81
5060	Treated, 2" x 3"		850	.019		.45	.71		1.16	1.69
5065	2" x 3", treated, pneumatic nailed		1020	.016		.45	.59		1.04	1.49
5080	2" x 4"		800	.020		.52	.76		1.28	1.85
5085	2" x 4", treated, pneumatic nailed		960	.017		.53	.63		1.16	1.64
5100	2" x 6"		750	.021		.77	.81		1.58	2.20
5102	2 x 8		725	.022		1.06	.83		1.89	2.56
5103	2" x 6", treated, pneumatic nailed		900	.018		.78	.67		1.45	1.99
5104	2 x 10		700	.023		1.35	.86		2.21	2.93
5105	2" x 12"		670	.024		1.90	.90		2.80	3.60
5107	2" x 8", pneumatic nailed		870	.018		1.06	.70		1.76	2.33
5108	2" x 10", pneumatic nailed		840	.019		1.36	.72		2.08	2.70
5109	2" x 12", pneumatic nailed		804	.020		1.91	.75		2.66	3.36
5120	Studs, 8' high wall, 2" x 3"		1200	.013		.38	.50		.88	1.25
5125	2" x 3", pneumatic nailed		1440	.011		.38	.42		.80	1.12
5140	2" x 4"		1100	.015		.40	.55		.95	1.36
5146	2" x 4", pneumatic nailed		1320	.012		.40	.46		.86	1.21
5160	2" x 6"		1000	.016		.61	.60		1.21	1.68
5166	2" x 6", pneumatic nailed		1200	.013		.62	.50		1.12	1.52
5180	3" x 4"		800	.020		1.14	.76		1.90	2.53
5185	3" x 4", pneumatic nailed		960	.017		1.15	.63		1.78	2.32
5200	Installed on second story, 2" x 3"		1170	.014		.38	.52		.90	1.28
5205	2" x 3", pneumatic nailed		1404	.011		.38	.43		.81	1.14
5220	2" x 4"		1015	.016		.40	.60		1	1.44
5225	2" x 4", pneumatic nailed		1218	.013		.40	.50		.90	1.27
5240	2" x 6"		890	.018		.61	.68		1.29	1.81

06 11 10.40 Wall Framing		Crew	Daily Output	Labor-Hours	Unit	2016 Bare Costs Material	Labor	Equipment	Total	Total Incl O&P
5245	2" x 6", pneumatic nailed	2 Carp	1080	.015	L.F.	.62	.56		1.18	1.62
5260	3" x 4"		800	.020		1.14	.76		1.90	2.53
5265	3" x 4", pneumatic nailed		960	.017		1.15	.63		1.78	2.32
5280	Installed on dormer or gable, 2" x 3"		1045	.015		.38	.58		.96	1.38
5285	2" x 3", pneumatic nailed		1254	.013		.38	.48		.86	1.23
5300	2" x 4"		905	.018		.40	.67		1.07	1.56
5305	2" x 4", pneumatic nailed		1086	.015		.40	.56		.96	1.37
5320	2" x 6"		800	.020		.61	.76		1.37	1.94
5325	2" x 6", pneumatic nailed		960	.017		.62	.63		1.25	1.74
5340	3" x 4"		700	.023		1.14	.86		2	2.71
5345	3" x 4", pneumatic nailed		840	.019		1.15	.72		1.87	2.47
5360	6' high wall, 2" x 3"		970	.016		.38	.62		1	1.45
5365	2" x 3", pneumatic nailed		1164	.014		.38	.52		.90	1.29
5380	2" x 4"		850	.019		.40	.71		1.11	1.63
5385	2" x 4", pneumatic nailed		1020	.016		.40	.59		.99	1.43
5400	2" x 6"		740	.022		.61	.82		1.43	2.04
5405	2" x 6", pneumatic nailed		888	.018		.62	.68		1.30	1.82
5420	3" x 4"		600	.027		1.14	1.01		2.15	2.95
5425	3" x 4", pneumatic nailed		720	.022		1.15	.84		1.99	2.67
5440	Installed on second story, 2" x 3"		950	.017		.38	.64		1.02	1.48
5445	2" x 3", pneumatic nailed		1140	.014		.38	.53		.91	1.31
5460	2" x 4"		810	.020		.40	.75		1.15	1.69
5465	2" x 4", pneumatic nailed		972	.016		.40	.62		1.02	1.48
5480	2" x 6"		700	.023		.61	.86		1.47	2.12
5485	2" x 6", pneumatic nailed		840	.019		.62	.72		1.34	1.89
5500	3" x 4"		550	.029		1.14	1.10		2.24	3.10
5505	3" x 4", pneumatic nailed		660	.024		1.15	.92		2.07	2.79
5520	Installed on dormer or gable, 2" x 3"		850	.019		.38	.71		1.09	1.60
5525	2" x 3", pneumatic nailed		1020	.016		.38	.59		.97	1.41
5540	2" x 4"		720	.022		.40	.84		1.24	1.85
5545	2" x 4", pneumatic nailed		864	.019		.40	.70		1.10	1.61
5560	2" x 6"		620	.026		.61	.98		1.59	2.30
5565	2" x 6", pneumatic nailed		744	.022		.62	.81		1.43	2.04
5580	3" x 4"		480	.033		1.14	1.26		2.40	3.37
5585	3" x 4", pneumatic nailed		576	.028		1.15	1.05		2.20	3.02
5600	3' high wall, 2" x 3"		740	.022		.38	.82		1.20	1.78
5605	2" x 3", pneumatic nailed		888	.018		.38	.68		1.06	1.56
5620	2" x 4"		640	.025		.40	.95		1.35	2.02
5625	2" x 4", pneumatic nailed		768	.021		.40	.79		1.19	1.76
5640	2" x 6"		550	.029		.61	1.10		1.71	2.51
5645	2" x 6", pneumatic nailed		660	.024		.62	.92		1.54	2.21
5660	3" x 4"		440	.036		1.14	1.37		2.51	3.56
5665	3" x 4", pneumatic nailed		528	.030		1.15	1.15		2.30	3.18
5680	Installed on second story, 2" x 3"		700	.023		.38	.86		1.24	1.86
5685	2" x 3", pneumatic nailed		840	.019		.38	.72		1.10	1.63
5700	2" x 4"		610	.026		.40	.99		1.39	2.10
5705	2" x 4", pneumatic nailed		732	.022		.40	.83		1.23	1.82
5720	2" x 6"		520	.031		.61	1.16		1.77	2.62
5725	2" x 6", pneumatic nailed		624	.026		.62	.97		1.59	2.30
5740	3" x 4"		430	.037		1.14	1.41		2.55	3.62
5745	3" x 4", pneumatic nailed		516	.031		1.15	1.17		2.32	3.22
5760	Installed on dormer or gable, 2" x 3"		625	.026		.38	.97		1.35	2.03
5765	2" x 3", pneumatic nailed		750	.021		.38	.81		1.19	1.77

06 11 Wood Framing

06 11 10 – Framing with Dimensional, Engineered or Composite Lumber

06 11 10.40 Wall Framing		Crew	Daily Output	Labor-Hours	Unit	Material	2016 Bare Costs Labor	Equipment	Total	Total Incl O&P
5780	2" x 4"	2 Carp	545	.029	L.F.	.40	1.11		1.51	2.30
5785	2" x 4", pneumatic nailed		654	.024		.40	.92		1.32	1.99
5800	2" x 6"		465	.034		.61	1.30		1.91	2.85
5805	2" x 6", pneumatic nailed		558	.029		.62	1.08		1.70	2.49
5820	3" x 4"		380	.042		1.14	1.59		2.73	3.93
5825	3" x 4", pneumatic nailed	↓	456	.035	↓	1.15	1.33		2.48	3.48
8250	For second story & above, add						5%			
8300	For dormer & gable, add						15%			

06 11 10.42 Furring		Crew	Daily Output	Labor-Hours	Unit	Material	Labor	Equipment	Total	Total Incl O&P
0010	**FURRING**									
0012	Wood strips, 1" x 2", on walls, on wood	1 Carp	550	.015	L.F.	.24	.55		.79	1.19
0015	On wood, pneumatic nailed		710	.011		.24	.43		.67	.98
0300	On masonry		495	.016		.26	.61		.87	1.31
0400	On concrete		260	.031		.26	1.16		1.42	2.24
0600	1" x 3", on walls, on wood		550	.015		.39	.55		.94	1.35
0605	On wood, pneumatic nailed		710	.011		.39	.43		.82	1.14
0700	On masonry		495	.016		.42	.61		1.03	1.48
0800	On concrete		260	.031		.42	1.16		1.58	2.41
0850	On ceilings, on wood		350	.023		.39	.86		1.25	1.88
0855	On wood, pneumatic nailed		450	.018		.39	.67		1.06	1.56
0900	On masonry		320	.025		.42	.95		1.37	2.04
0950	On concrete	↓	210	.038	↓	.42	1.44		1.86	2.87

06 11 10.44 Grounds		Crew	Daily Output	Labor-Hours	Unit	Material	Labor	Equipment	Total	Total Incl O&P
0010	**GROUNDS**									
0020	For casework, 1" x 2" wood strips, on wood	1 Carp	330	.024	L.F.	.24	.92		1.16	1.80
0100	On masonry		285	.028		.26	1.06		1.32	2.07
0200	On concrete		250	.032		.26	1.21		1.47	2.32
0400	For plaster, 3/4" deep, on wood		450	.018		.24	.67		.91	1.40
0500	On masonry		225	.036		.26	1.34		1.60	2.54
0600	On concrete		175	.046		.26	1.73		1.99	3.18
0700	On metal lath	↓	200	.040	↓	.26	1.51		1.77	2.82

06 12 Structural Panels

06 12 10 – Structural Insulated Panels

06 12 10.10 OSB Faced Panels		Crew	Daily Output	Labor-Hours	Unit	Material	Labor	Equipment	Total	Total Incl O&P
0010	**OSB FACED PANELS**									
0100	Structural insul. panels, 7/16" OSB both faces, EPS insul, 3-5/8" T G	F-3	2075	.019	S.F.	3.65	.66	.32	4.63	5.50
0110	5-5/8" thick G		1725	.023		4.10	.80	.38	5.28	6.25
0120	7-3/8" thick G		1425	.028		4.45	.96	.46	5.87	7
0130	9-3/8" thick G		1125	.036		4.75	1.22	.59	6.56	7.95
0140	7/16" OSB one face, EPS insul, 3-5/8" thick G		2175	.018		3.75	.63	.30	4.68	5.50
0150	5-5/8" thick G		1825	.022		4.35	.75	.36	5.46	6.45
0160	7-3/8" thick G		1525	.026		4.85	.90	.43	6.18	7.35
0170	9-3/8" thick G		1225	.033		5.35	1.12	.54	7.01	8.35
0190	7/16" OSB - 1/2" GWB faces, EPS insul, 3-5/8" T G		2075	.019		3.35	.66	.32	4.33	5.15
0200	5-5/8" thick G		1725	.023		4	.80	.38	5.18	6.15
0210	7-3/8" thick G		1425	.028		4.50	.96	.46	5.92	7.05
0220	9-3/8" thick G		1125	.036		5.10	1.22	.59	6.91	8.30
0240	7/16" OSB - 1/2" MRGWB faces, EPS insul, 3-5/8" T G		2075	.019		3.45	.66	.32	4.43	5.25
0250	5-5/8" thick G	↓	1725	.023	↓	4.10	.80	.38	5.28	6.25

06 12 Structural Panels

06 12 10 – Structural Insulated Panels

06 12 10.10 OSB Faced Panels		Crew	Daily Output	Labor-Hours	Unit	2016 Bare Costs Material	Labor	Equipment	Total	Total Incl O&P
0260	7-3/8" thick G	F-3	1425	.028	S.F.	4.55	.96	.46	5.97	7.10
0270	9-3/8" thick G		1125	.036		5.20	1.22	.59	7.01	8.40
0300	For 1/2" GWB added to OSB skin, add G					1.35			1.35	1.49
0310	For 1/2" MRGWB added to OSB skin, add G					1.35			1.35	1.49
0320	For one T1-11 skin, add to OSB-OSB G					1.95			1.95	2.15
0330	For one 19/32" CDX skin, add to OSB-OSB G					1.50			1.50	1.65
0500	Structural insulated panel, 7/16" OSB both sides, straw core									
0510	4-3/8" T, walls (w/sill, splines, plates) G	F-6	2400	.017	S.F.	7.55	.58	.27	8.40	9.55
0520	Floors (w/splines) G		2400	.017		7.55	.58	.27	8.40	9.55
0530	Roof (w/splines) G		2400	.017		7.55	.58	.27	8.40	9.55
0550	7-7/8" T, walls (w/sill, splines, plates) G		2400	.017		11.40	.58	.27	12.25	13.80
0560	Floors (w/splines) G		2400	.017		11.40	.58	.27	12.25	13.80
0570	Roof (w/splines) G		2400	.017		11.40	.58	.27	12.25	13.80

06 12 19 – Composite Shearwall Panels

06 12 19.10 Steel and Wood Composite Shearwall Panels		Crew	Daily Output	Labor-Hours	Unit	Material	Labor	Equipment	Total	Total Incl O&P
0010	**STEEL & WOOD COMPOSITE SHEARWALL PANELS**									
0020	Anchor bolts, 36" long (must be placed in wet concrete)	1 Carp	150	.053	Ea.	34	2.02		36.02	41
0030	On concrete, 2 x 4 & 2 x 6 walls, 7' - 10' high, 360 lb. shear, 12" wide	2 Carp	8	2		460	75.50		535.50	630
0040	715 lb. shear, 15" wide		8	2		510	75.50		585.50	685
0050	1860 lb. shear, 18" wide		8	2		525	75.50		600.50	705
0060	2780 lb. shear, 21" wide		8	2		560	75.50		635.50	740
0070	3790 lb. shear, 24" wide		8	2		640	75.50		715.50	825
0080	2 x 6 walls, 11' to 13' high, 1180 lb. shear, 18" wide		6	2.667		645	101		746	880
0090	1555 lb. shear, 21" wide		6	2.667		720	101		821	960
0100	2280 lb. shear, 24" wide		6	2.667		805	101		906	1,050
0110	For installing above on wood floor frame, add									
0120	Coupler nuts, threaded rods, bolts, shear transfer plate kit	1 Carp	16	.500	Ea.	65	18.90		83.90	103
0130	Framing anchors, angle (2 required)	"	96	.083	"	2.56	3.15		5.71	8.05
0140	For blocking see Section 06 11 10.02									
0150	For installing above, first floor to second floor, wood floor frame, add									
0160	Add stack option to first floor wall panel				Ea.	71.50			71.50	78.50
0170	Threaded rods, bolts, shear transfer plate kit	1 Carp	16	.500		75	18.90		93.90	114
0180	Framing anchors, angle (2 required)	"	96	.083		2.56	3.15		5.71	8.05
0190	For blocking see section 06 11 10.02									
0200	For installing stacked panels, balloon framing									
0210	Add stack option to first floor wall panel				Ea.	71.50			71.50	78.50
0220	Threaded rods, bolts kit	1 Carp	16	.500	"	45	18.90		63.90	81

06 13 Heavy Timber Construction

06 13 23 – Heavy Timber Framing

06 13 23.10 Heavy Framing		Crew	Daily Output	Labor-Hours	Unit	Material	Labor	Equipment	Total	Total Incl O&P
0010	**HEAVY FRAMING**									
0020	Beams, single 6" x 10"	2 Carp	1.10	14.545	M.B.F.	1,525	550		2,075	2,600
0100	Single 8" x 16"		1.20	13.333		2,075	505		2,580	3,150
0200	Built from 2" lumber, multiple 2" x 14"		.90	17.778		910	670		1,580	2,125
0210	Built from 3" lumber, multiple 3" x 6"		.70	22.857		1,250	865		2,115	2,825
0220	Multiple 3" x 8"		.80	20		1,350	755		2,105	2,750
0230	Multiple 3" x 10"		.90	17.778		1,350	670		2,020	2,625
0240	Multiple 3" x 12"		1	16		1,350	605		1,955	2,525
0250	Built from 4" lumber, multiple 4" x 6"		.80	20		1,450	755		2,205	2,850

06 13 Heavy Timber Construction

06 13 23 – Heavy Timber Framing

06 13 23.10 Heavy Framing		Crew	Daily Output	Labor-Hours	Unit	2016 Bare Costs Material	Labor	Equipment	Total	Total Incl O&P
0260	Multiple 4" x 8"	2 Carp	.90	17.778	M.B.F.	1,100	670		1,770	2,325
0270	Multiple 4" x 10"		1	16		1,400	605		2,005	2,575
0280	Multiple 4" x 12"		1.10	14.545		1,300	550		1,850	2,350
0290	Columns, structural grade, 1500f, 4" x 4"		.60	26.667		1,250	1,000		2,250	3,075
0300	6" x 6"		.65	24.615		1,425	930		2,355	3,125
0400	8" x 8"		.70	22.857		1,300	865		2,165	2,875
0500	10" x 10"		.75	21.333		1,500	805		2,305	3,000
0600	12" x 12"		.80	20		1,450	755		2,205	2,875
0800	Floor planks, 2" thick, T & G, 2" x 6"		1.05	15.238		1,600	575		2,175	2,725
0900	2" x 10"		1.10	14.545		1,600	550		2,150	2,675
1100	3" thick, 3" x 6"		1.05	15.238		1,600	575		2,175	2,725
1200	3" x 10"		1.10	14.545		1,600	550		2,150	2,675
1400	Girders, structural grade, 12" x 12"		.80	20		1,450	755		2,205	2,875
1500	10" x 16"		1	16		2,500	605		3,105	3,775
2300	Roof purlins, 4" thick, structural grade	↓	1.05	15.238	↓	1,100	575		1,675	2,175

06 15 Wood Decking

06 15 16 – Wood Roof Decking

06 15 16.10 Solid Wood Roof Decking		Crew	Daily Output	Labor-Hours	Unit	Material	Labor	Equipment	Total	Total Incl O&P
0010	**SOLID WOOD ROOF DECKING**									
0350	Cedar planks, 2" thick	2 Carp	350	.046	S.F.	6.45	1.73		8.18	10
0400	3" thick		320	.050		9.70	1.89		11.59	13.80
0500	4" thick		250	.064		12.90	2.42		15.32	18.25
0550	6" thick		200	.080		19.35	3.02		22.37	26.50
0650	Douglas fir, 2" thick		350	.046		2.73	1.73		4.46	5.90
0700	3" thick		320	.050		4.10	1.89		5.99	7.65
0800	4" thick		250	.064		5.45	2.42		7.87	10.05
0850	6" thick		200	.080		8.20	3.02		11.22	14.05
0950	Hemlock, 2" thick		350	.046		2.78	1.73		4.51	5.95
1000	3" thick		320	.050		4.17	1.89		6.06	7.75
1100	4" thick		250	.064		5.55	2.42		7.97	10.15
1150	6" thick		200	.080		8.35	3.02		11.37	14.20
1250	Western white spruce, 2" thick		350	.046		1.77	1.73		3.50	4.84
1300	3" thick		320	.050		2.66	1.89		4.55	6.10
1400	4" thick		250	.064		3.55	2.42		5.97	7.95
1450	6" thick	↓	200	.080	↓	5.30	3.02		8.32	10.90

06 15 23 – Laminated Wood Decking

06 15 23.10 Laminated Roof Deck		Crew	Daily Output	Labor-Hours	Unit	Material	Labor	Equipment	Total	Total Incl O&P
0010	**LAMINATED ROOF DECK**									
0020	Pine or hemlock, 3" thick	2 Carp	425	.038	S.F.	6.25	1.42		7.67	9.30
0100	4" thick		325	.049		8.15	1.86		10.01	12.05
0300	Cedar, 3" thick		425	.038		7.10	1.42		8.52	10.20
0400	4" thick		325	.049		9.50	1.86		11.36	13.55
0600	Fir, 3" thick		425	.038		5.50	1.42		6.92	8.45
0700	4" thick	↓	325	.049	↓	7.50	1.86		9.36	11.35

06 16 Sheathing

06 16 13 – Insulating Sheathing

06 16 13.10 Insulating Sheathing			Crew	Daily Output	Labor-Hours	Unit	2016 Bare Costs Material	Labor	Equipment	Total	Total Incl O&P
0010	**INSULATING SHEATHING**										
0020	Expanded polystyrene, 1#/C.F. density, 3/4" thick R2.89	G	2 Carp	1400	.011	S.F.	.35	.43		.78	1.11
0030	1" thick R3.85	G		1300	.012		.42	.47		.89	1.24
0040	2" thick R7.69	G		1200	.013		.68	.50		1.18	1.59
0050	Extruded polystyrene, 15 PSI compressive strength, 1" thick, R5	G		1300	.012		.76	.47		1.23	1.61
0060	2" thick, R10	G		1200	.013		.94	.50		1.44	1.87
0070	Polyisocyanurate, 2#/C.F. density, 3/4" thick	G		1400	.011		.62	.43		1.05	1.40
0080	1" thick	G		1300	.012		.63	.47		1.10	1.47
0090	1-1/2" thick	G		1250	.013		.80	.48		1.28	1.69
0100	2" thick	G		1200	.013		.98	.50		1.48	1.92

06 16 23 – Subflooring

06 16 23.10 Subfloor			Crew	Daily Output	Labor-Hours	Unit	Material	Labor	Equipment	Total	Total Incl O&P
0010	**SUBFLOOR**	R061636-20									
0011	Plywood, CDX, 1/2" thick		2 Carp	1500	.011	SF Flr.	.68	.40		1.08	1.43
0015	Pneumatic nailed			1860	.009		.68	.33		1.01	1.29
0100	5/8" thick			1350	.012		.76	.45		1.21	1.59
0105	Pneumatic nailed			1674	.010		.76	.36		1.12	1.45
0200	3/4" thick			1250	.013		.91	.48		1.39	1.81
0205	Pneumatic nailed			1550	.010		.91	.39		1.30	1.65
0300	1-1/8" thick, 2-4-1 including underlayment			1050	.015		2.03	.58		2.61	3.19
0440	With boards, 1" x 6", S4S, laid regular			900	.018		1.54	.67		2.21	2.82
0450	1" x 8", laid regular			1000	.016		1.88	.60		2.48	3.08
0460	Laid diagonal			850	.019		1.88	.71		2.59	3.26
0500	1" x 10", laid regular			1100	.015		1.92	.55		2.47	3.03
0600	Laid diagonal			900	.018		1.92	.67		2.59	3.24
1500	OSB, 5/8" thick			1330	.012	S.F.	.54	.45		.99	1.35
1600	3/4" thick			1230	.013	"	.68	.49		1.17	1.57
8990	Subfloor adhesive, 3/8" bead		1 Carp	2300	.003	L.F.	.12	.13		.25	.35

06 16 26 – Underlayment

06 16 26.10 Wood Product Underlayment			Crew	Daily Output	Labor-Hours	Unit	Material	Labor	Equipment	Total	Total Incl O&P
0010	**WOOD PRODUCT UNDERLAYMENT**										
0015	Plywood, underlayment grade, 1/4" thick		2 Carp	1500	.011	S.F.	.89	.40		1.29	1.66
0018	Pneumatic nailed			1860	.009		.89	.33		1.22	1.52
0030	3/8" thick			1500	.011		.99	.40		1.39	1.77
0070	Pneumatic nailed			1860	.009		.99	.33		1.32	1.63
0100	1/2" thick			1450	.011		1.17	.42		1.59	1.99
0105	Pneumatic nailed			1798	.009		1.17	.34		1.51	1.85
0200	5/8" thick			1400	.011		1.29	.43		1.72	2.14
0205	Pneumatic nailed			1736	.009		1.29	.35		1.64	2
0300	3/4" thick			1300	.012		1.48	.47		1.95	2.41
0305	Pneumatic nailed			1612	.010		1.48	.38		1.86	2.26
0500	Particle board, 3/8" thick	G		1500	.011		.41	.40		.81	1.13
0505	Pneumatic nailed	G		1860	.009		.41	.33		.74	.99
0600	1/2" thick	G		1450	.011		.45	.42		.87	1.20
0605	Pneumatic nailed	G		1798	.009		.45	.34		.79	1.06
0800	5/8" thick	G		1400	.011		.54	.43		.97	1.31
0805	Pneumatic nailed	G		1736	.009		.54	.35		.89	1.17
0900	3/4" thick	G		1300	.012		.68	.47		1.15	1.53
0905	Pneumatic nailed	G		1612	.010		.68	.38		1.06	1.38
1100	Hardboard, underlayment grade, 4' x 4', .215" thick	G		1500	.011		.63	.40		1.03	1.37

06 16 Sheathing

06 16 33 – Wood Board Sheathing

06 16 33.10 Board Sheathing		Crew	Daily Output	Labor-Hours	Unit	Material	2016 Bare Costs Labor	Equipment	Total	Total Incl O&P
0009	**BOARD SHEATHING**									
0010	Roof, 1" x 6" boards, laid horizontal	2 Carp	725	.022	S.F.	1.54	.83		2.37	3.09
0020	On steep roof		520	.031		1.54	1.16		2.70	3.64
0040	On dormers, hips, & valleys		480	.033		1.54	1.26		2.80	3.80
0050	Laid diagonal		650	.025		1.54	.93		2.47	3.25
0070	1" x 8" boards, laid horizontal		875	.018		1.88	.69		2.57	3.23
0080	On steep roof		635	.025		1.88	.95		2.83	3.67
0090	On dormers, hips, & valleys		580	.028		1.92	1.04		2.96	3.86
0100	Laid diagonal		725	.022		1.88	.83		2.71	3.47
0110	Skip sheathing, 1" x 4", 7" OC	1 Carp	1200	.007		.58	.25		.83	1.05
0120	1" x 6", 9" OC		1450	.006		.71	.21		.92	1.13
0180	Tongue and groove sheathing/decking, 1" x 6"		1000	.008		1.82	.30		2.12	2.51
0190	2" x 6"		1000	.008		3.81	.30		4.11	4.70
0200	Walls, 1" x 6" boards, laid regular	2 Carp	650	.025		1.54	.93		2.47	3.25
0210	Laid diagonal		585	.027		1.54	1.03		2.57	3.42
0220	1" x 8" boards, laid regular		765	.021		1.88	.79		2.67	3.39
0230	Laid diagonal		650	.025		1.88	.93		2.81	3.63

06 16 36 – Wood Panel Product Sheathing

06 16 36.10 Sheathing		Crew	Daily Output	Labor-Hours	Unit	Material	Labor	Equipment	Total	Total Incl O&P
0010	**SHEATHING** R061636-20									
0012	Plywood on roofs, CDX									
0030	5/16" thick	2 Carp	1600	.010	S.F.	.62	.38		1	1.32
0035	Pneumatic nailed		1952	.008		.62	.31		.93	1.21
0050	3/8" thick		1525	.010		.62	.40		1.02	1.34
0055	Pneumatic nailed		1860	.009		.62	.33		.95	1.22
0100	1/2" thick		1400	.011		.68	.43		1.11	1.47
0105	Pneumatic nailed		1708	.009		.68	.35		1.03	1.34
0200	5/8" thick		1300	.012		.76	.47		1.23	1.62
0205	Pneumatic nailed		1586	.010		.76	.38		1.14	1.48
0300	3/4" thick		1200	.013		.91	.50		1.41	1.84
0305	Pneumatic nailed		1464	.011		.91	.41		1.32	1.69
0500	Plywood on walls, with exterior CDX, 3/8" thick		1200	.013		.62	.50		1.12	1.52
0505	Pneumatic nailed		1488	.011		.62	.41		1.03	1.36
0600	1/2" thick		1125	.014		.68	.54		1.22	1.65
0605	Pneumatic nailed		1395	.011		.68	.43		1.11	1.48
0700	5/8" thick		1050	.015		.76	.58		1.34	1.80
0705	Pneumatic nailed		1302	.012		.76	.46		1.22	1.62
0800	3/4" thick		975	.016		.91	.62		1.53	2.04
0805	Pneumatic nailed		1209	.013		.91	.50		1.41	1.84
0900	For exterior C-C grade plywood, add					15%				
0920	For application to metal studs, joists, rafters, add						20%			
1000	For shear wall construction, add						20%			
1200	For structural 1 exterior plywood, add				S.F.	10%				
3000	Wood fiber, regular, no vapor barrier, 1/2" thick	2 Carp	1200	.013		.64	.50		1.14	1.54
3100	5/8" thick		1200	.013		.76	.50		1.26	1.68
3300	No vapor barrier, in colors, 1/2" thick		1200	.013		.77	.50		1.27	1.69
3400	5/8" thick		1200	.013		.81	.50		1.31	1.73
3600	With vapor barrier one side, white, 1/2" thick		1200	.013		.63	.50		1.13	1.53
3700	Vapor barrier 2 sides, 1/2" thick		1200	.013		.86	.50		1.36	1.79
3800	Asphalt impregnated, 25/32" thick		1200	.013		.33	.50		.83	1.20
3850	Intermediate, 1/2" thick		1200	.013		.25	.50		.75	1.12
4500	Oriented strand board, on roof, 7/16" thick G		1460	.011		.50	.41		.91	1.24

06 16 Sheathing

06 16 36 – Wood Panel Product Sheathing

06 16 36.10 Sheathing		Crew	Daily Output	Labor-Hours	Unit	Material	Labor	Equipment	Total	Total Incl O&P
						2016 Bare Costs				
4505	Pneumatic nailed G	2 Carp	1780	.009	S.F.	.50	.34		.84	1.12
4550	1/2" thick G		1400	.011		.50	.43		.93	1.27
4555	Pneumatic nailed G		1736	.009		.50	.35		.85	1.13
4600	5/8" thick G		1300	.012		.69	.47		1.16	1.54
4605	Pneumatic nailed G		1586	.010		.69	.38		1.07	1.40
4610	On walls, 7/16" thick G		1200	.013		.50	.50		1	1.39
4615	Pneumatic nailed G		1488	.011		.50	.41		.91	1.23
4620	1/2" thick G		1195	.013		.50	.51		1.01	1.40
4625	Pneumatic nailed G		1325	.012		.50	.46		.96	1.31
4630	5/8" thick G		1050	.015		.69	.58		1.27	1.72
4635	Pneumatic nailed G		1302	.012		.69	.46		1.15	1.54
4700	Oriented strand board, factory laminated W.R. barrier, on roof, 1/2" thick G		1400	.011		.78	.43		1.21	1.58
4705	Pneumatic nailed G		1736	.009		.78	.35		1.13	1.44
4720	5/8" thick G		1300	.012		.93	.47		1.40	1.80
4725	Pneumatic nailed G		1586	.010		.93	.38		1.31	1.66
4730	5/8" thick, T&G G		1150	.014		.93	.53		1.46	1.90
4735	Pneumatic nailed, T&G G		1400	.011		.93	.43		1.36	1.74
4740	On walls, 7/16" thick G		1200	.013		.66	.50		1.16	1.57
4745	Pneumatic nailed G		1488	.011		.66	.41		1.07	1.41
4750	1/2" thick G		1195	.013		.78	.51		1.29	1.71
4755	Pneumatic nailed G		1325	.012		.78	.46		1.24	1.62
4800	Joint sealant tape, 3-1/2"		7600	.002	L.F.	.31	.08		.39	.47
4810	Joint sealant tape, 6"		7600	.002	"	.44	.08		.52	.61

06 16 43 – Gypsum Sheathing

06 16 43.10 Gypsum Sheathing		Crew	Daily Output	Labor-Hours	Unit	Material	Labor	Equipment	Total	Total Incl O&P
0010	**GYPSUM SHEATHING**									
0020	Gypsum, weatherproof, 1/2" thick	2 Carp	1125	.014	S.F.	.46	.54		1	1.41
0040	With embedded glass mats	"	1100	.015	"	.70	.55		1.25	1.69

06 17 Shop-Fabricated Structural Wood

06 17 33 – Wood I-Joists

06 17 33.10 Wood and Composite I-Joists		Crew	Daily Output	Labor-Hours	Unit	Material	Labor	Equipment	Total	Total Incl O&P
0010	**WOOD AND COMPOSITE I-JOISTS**									
0100	Plywood webs, incl. bridging & blocking, panels 24" O.C.									
1200	15' to 24' span, 50 psf live load	F-5	2400	.013	SF Flr.	2.09	.44		2.53	3.04
1300	55 psf live load		2250	.014		2.30	.47		2.77	3.33
1400	24' to 30' span, 45 psf live load		2600	.012		2.40	.40		2.80	3.32
1500	55 psf live load		2400	.013		4.05	.44		4.49	5.20

06 17 53 – Shop-Fabricated Wood Trusses

06 17 53.10 Roof Trusses		Crew	Daily Output	Labor-Hours	Unit	Material	Labor	Equipment	Total	Total Incl O&P
0010	**ROOF TRUSSES**									
5000	Common wood, 2" x 4" metal plate connected, 24" O.C., 4/12 slope									
5010	1' overhang, 12' span	F-5	55	.582	Ea.	41	19.15		60.15	77.50
5050	20' span	F-6	62	.645		70.50	22.50	10.60	103.60	127
5100	24' span		60	.667		76.50	23	10.95	110.45	136
5150	26' span		57	.702		80	24.50	11.55	116.05	142
5200	28' span		53	.755		93	26.50	12.40	131.90	160
5240	30' span		51	.784		100	27.50	12.90	140.40	170
5250	32' span		50	.800		110	28	13.15	151.15	182
5280	34' span		48	.833		114	29	13.70	156.70	189

06 17 Shop-Fabricated Structural Wood

06 17 53 – Shop-Fabricated Wood Trusses

06 17 53.10 Roof Trusses		Crew	Daily Output	Labor-Hours	Unit	Material	2016 Bare Costs Labor	Equipment	Total	Total Incl O&P
5350	8/12 pitch, 1′ overhang, 20′ span	F-6	57	.702	Ea.	77.50	24.50	11.55	113.55	139
5400	24′ span		55	.727		97.50	25.50	11.95	134.95	163
5450	26′ span		52	.769		103	27	12.65	142.65	172
5500	28′ span		49	.816		117	28.50	13.45	158.95	191
5550	32′ span		45	.889		133	31	14.65	178.65	214
5600	36′ span		41	.976		164	34	16.05	214.05	255
5650	38′ span		40	1		184	35	16.45	235.45	278
5700	40′ span		40	1		187	35	16.45	238.45	282

06 18 Glued-Laminated Construction

06 18 13 – Glued-Laminated Beams

06 18 13.10 Laminated Beams

		Crew	Daily Output	Labor-Hours	Unit	Material	Labor	Equipment	Total	Total Incl O&P
0010	**LAMINATED BEAMS**									
0050	3-1/2″ x 18″	F-3	480	.083	L.F.	31	2.86	1.37	35.23	40.50
0100	5-1/4″ x 11-7/8″		450	.089		30.50	3.05	1.46	35.01	40
0150	5-1/4″ x 16″		360	.111		41	3.82	1.83	46.65	53.50
0200	5-1/4″ x 18″		290	.138		46	4.74	2.27	53.01	61.50
0250	5-1/4″ x 24″		220	.182		61.50	6.25	2.99	70.74	81
0300	7″ x 11-7/8″		320	.125		43.50	4.29	2.06	49.85	57
0350	7″ x 16″		260	.154		59	5.30	2.53	66.83	76.50
0400	7″ x 18″		210	.190		68	6.55	3.14	77.69	89.50
0500	For premium appearance, add to S.F. prices					5%				
0550	For industrial type, deduct					15%				
0600	For stain and varnish, add					5%				
0650	For 3/4″ laminations, add					25%				

06 18 13.20 Laminated Framing

		Crew	Daily Output	Labor-Hours	Unit	Material	Labor	Equipment	Total	Total Incl O&P
0010	**LAMINATED FRAMING**									
0020	30 lb., short term live load, 15 lb. dead load									
0200	Straight roof beams, 20′ clear span, beams 8′ O.C.	F-3	2560	.016	SF Flr.	2.08	.54	.26	2.88	3.47
0300	Beams 16′ O.C.		3200	.013		1.52	.43	.21	2.16	2.62
0500	40′ clear span, beams 8′ O.C.		3200	.013		3.95	.43	.21	4.59	5.30
0600	Beams 16′ O.C.		3840	.010		3.25	.36	.17	3.78	4.37
0800	60′ clear span, beams 8′ O.C.	F-4	2880	.014		6.75	.48	.39	7.62	8.70
0900	Beams 16′ O.C.	″	3840	.010		5.05	.36	.29	5.70	6.45
1100	Tudor arches, 30′ to 40′ clear span, frames 8′ O.C.	F-3	1680	.024		8.80	.82	.39	10.01	11.50
1200	Frames 16′ O.C.	″	2240	.018		6.90	.61	.29	7.80	8.95
1400	50′ to 60′ clear span, frames 8′ O.C.	F-4	2200	.018		9.50	.62	.51	10.63	12.05
1500	Frames 16′ O.C.		2640	.015		8.10	.52	.43	9.05	10.25
1700	Radial arches, 60′ clear span, frames 8′ O.C.		1920	.021		8.85	.72	.58	10.15	11.60
1800	Frames 16′ O.C.		2880	.014		7	.48	.39	7.87	8.95
2000	100′ clear span, frames 8′ O.C.		1600	.025		9.15	.86	.70	10.71	12.25
2100	Frames 16′ O.C.		2400	.017		8.05	.57	.47	9.09	10.30
2300	120′ clear span, frames 8′ O.C.		1440	.028		12.15	.95	.78	13.88	15.80
2400	Frames 16′ O.C.		1920	.021		11.10	.72	.58	12.40	14.05
2600	Bowstring trusses, 20′ O.C., 40′ clear span	F-3	2400	.017		5.50	.57	.27	6.34	7.30
2700	60′ clear span	F-4	3600	.011		4.95	.38	.31	5.64	6.45
2800	100′ clear span		4000	.010		7	.34	.28	7.62	8.60
2900	120′ clear span		3600	.011		7.45	.38	.31	8.14	9.20
3100	For premium appearance, add to S.F. prices					5%				
3300	For industrial type, deduct					15%				
3500	For stain and varnish, add					5%				

06 18 Glued-Laminated Construction

06 18 13 – Glued-Laminated Beams

06 18 13.20 Laminated Framing		Crew	Daily Output	Labor-Hours	Unit	2016 Bare Costs Material	Labor	Equipment	Total	Total Incl O&P
3900	For 3/4" laminations, add to straight				SF Flr.	25%				
4100	Add to curved					15%				
4300	Alternate pricing method: (use nominal footage of									
4310	components). Straight beams, camber less than 6"	F-3	3.50	11.429	M.B.F.	2,925	395	188	3,508	4,050
4400	Columns, including hardware		2	20		3,125	685	330	4,140	4,950
4600	Curved members, radius over 32'		2.50	16		3,200	550	263	4,013	4,725
4700	Radius 10' to 32'		3	13.333		3,175	460	219	3,854	4,500
4900	For complicated shapes, add maximum					100%				
5100	For pressure treating, add to straight					35%				
5200	Add to curved					45%				
6000	Laminated veneer members, southern pine or western species									
6050	1-3/4" wide x 5-1/2" deep	2 Carp	480	.033	L.F.	3.44	1.26		4.70	5.90
6100	9-1/2" deep		480	.033		4.74	1.26		6	7.30
6150	14" deep		450	.036		7.85	1.34		9.19	10.90
6200	18" deep		450	.036		10.65	1.34		11.99	14
6300	Parallel strand members, southern pine or western species									
6350	1-3/4" wide x 9-1/4" deep	2 Carp	480	.033	L.F.	5.10	1.26		6.36	7.70
6400	11-1/4" deep		450	.036		5.75	1.34		7.09	8.60
6450	14" deep		400	.040		7.85	1.51		9.36	11.20
6500	3-1/2" wide x 9-1/4" deep		480	.033		16.30	1.26		17.56	20
6550	11-1/4" deep		450	.036		20.50	1.34		21.84	25
6600	14" deep		400	.040		24	1.51		25.51	29
6650	7" wide x 9-1/4" deep		450	.036		34.50	1.34		35.84	40.50
6700	11-1/4" deep		420	.038		43.50	1.44		44.94	50
6750	14" deep		400	.040		52	1.51		53.51	59.50

06 22 Millwork

06 22 13 – Standard Pattern Wood Trim

06 22 13.10 Millwork		Crew	Daily Output	Labor-Hours	Unit	Material	Labor	Equipment	Total	Total Incl O&P
0010	**MILLWORK** R061110-30									
0020	Rule of thumb, milled material equals rough lumber cost x 3									
1020	1" x 12", custom birch				L.F.	4.71			4.71	5.20
1040	Cedar					7.40			7.40	8.15
1060	Oak					4.47			4.47	4.92
1080	Redwood					4.71			4.71	5.20
1100	Southern yellow pine					4.02			4.02	4.42
1120	Sugar pine					4.10			4.10	4.51
1140	Teak					31.50			31.50	35
1160	Walnut					7.40			7.40	8.15
1180	White pine					5.20			5.20	5.75

06 22 13.15 Moldings, Base		Crew	Daily Output	Labor-Hours	Unit	Material	Labor	Equipment	Total	Total Incl O&P
0010	**MOLDINGS, BASE**									
5100	Classic profile, 5/8" x 5-1/2", finger jointed and primed	1 Carp	250	.032	L.F.	1.49	1.21		2.70	3.67
5105	Poplar		240	.033		1.57	1.26		2.83	3.83
5110	Red oak		220	.036		2.41	1.37		3.78	4.95
5115	Maple		220	.036		3.64	1.37		5.01	6.30
5120	Cherry		220	.036		4.32	1.37		5.69	7.05
5125	3/4 x 7-1/2", finger jointed and primed		250	.032		1.89	1.21		3.10	4.11
5130	Poplar		240	.033		2.51	1.26		3.77	4.87
5135	Red oak		220	.036		3.50	1.37		4.87	6.15
5140	Maple		220	.036		4.84	1.37		6.21	7.60

06 22 13 – Standard Pattern Wood Trim

06 22 13.15 Moldings, Base		Crew	Daily Output	Labor-Hours	Unit	Material	2016 Bare Costs Labor	Equipment	Total	Total Incl O&P
5145	Cherry	1 Carp	220	.036	L.F.	5.75	1.37		7.12	8.65
5150	Modern profile, 5/8" x 3-1/2", finger jointed and primed		250	.032		.92	1.21		2.13	3.04
5155	Poplar		240	.033		1	1.26		2.26	3.21
5160	Red oak		220	.036		1.65	1.37		3.02	4.11
5165	Maple		220	.036		2.55	1.37		3.92	5.10
5170	Cherry		220	.036		2.56	1.37		3.93	5.10
5175	Ogee profile, 7/16" x 3", finger jointed and primed		250	.032		.63	1.21		1.84	2.72
5180	Poplar		240	.033		.66	1.26		1.92	2.83
5185	Red oak		220	.036		.93	1.37		2.30	3.32
5200	9/16" x 3-1/2", finger jointed and primed		250	.032		.66	1.21		1.87	2.75
5205	Pine		240	.033		1.13	1.26		2.39	3.35
5210	Red oak		220	.036		2.38	1.37		3.75	4.92
5215	9/16" x 4-1/2", red oak		220	.036		3.56	1.37		4.93	6.20
5220	5/8" x 3-1/2", finger jointed and primed		250	.032		.91	1.21		2.12	3.03
5225	Poplar		240	.033		1	1.26		2.26	3.21
5230	Red oak		220	.036		1.65	1.37		3.02	4.11
5235	Maple		220	.036		2.55	1.37		3.92	5.10
5240	Cherry		220	.036		2.75	1.37		4.12	5.30
5245	5/8" x 4", finger jointed and primed		250	.032		1.17	1.21		2.38	3.31
5250	Poplar		240	.033		1.29	1.26		2.55	3.53
5255	Red oak		220	.036		1.85	1.37		3.22	4.33
5260	Maplo		220	.036		2.85	1.37		4.22	5.45
5265	Cherry		220	.036		2.99	1.37		4.36	5.60
5270	Rectangular profile, oak, 3/8" x 1-1/4"		260	.031		1.20	1.16		2.36	3.27
5275	1/2" x 2-1/2"		255	.031		1.94	1.19		3.13	4.12
5280	1/2" x 3-1/2"		250	.032		2.36	1.21		3.57	4.62
5285	1" x 6"		240	.033		4.21	1.26		5.47	6.75
5290	1" x 8"		240	.033		5.80	1.26		7.06	8.50
5295	Pine, 3/8" x 1-3/4"		260	.031		.46	1.16		1.62	2.45
5300	7/16" x 2-1/2"		255	.031		.74	1.19		1.93	2.80
5305	1" x 6"		240	.033		.83	1.26		2.09	3.02
5310	1" x 8"		240	.033		1.06	1.26		2.32	3.27
5315	Shoe, 1/2" x 3/4", primed		260	.031		.50	1.16		1.66	2.50
5320	Pine		240	.033		.30	1.26		1.56	2.44
5325	Poplar		240	.033		.39	1.26		1.65	2.54
5330	Red oak		220	.036		.53	1.37		1.90	2.88
5335	Maple		220	.036		.70	1.37		2.07	3.07
5340	Cherry		220	.036		.80	1.37		2.17	3.18
5345	11/16" x 1-1/2", pine		240	.033		.70	1.26		1.96	2.88
5350	Caps, 11/16" x 1-3/8", pine		240	.033		.55	1.26		1.81	2.71
5355	3/4" x 1-3/4", finger jointed and primed		260	.031		.71	1.16		1.87	2.73
5360	Poplar		240	.033		1.01	1.26		2.27	3.22
5365	Red oak		220	.036		1.21	1.37		2.58	3.63
5370	Maple		220	.036		1.58	1.37		2.95	4.04
5375	Cherry		220	.036		3.20	1.37		4.57	5.80
5380	Combination base & shoe, 9/16" x 3-1/2" & 1/2" x 3/4", pine		125	.064		1.43	2.42		3.85	5.60
5385	Three piece oak, 6" high		80	.100		5.95	3.78		9.73	12.90
5390	Including 3/4" x 1" base shoe		70	.114		6.10	4.32		10.42	14
5395	Flooring cant strip, 3/4" x 3/4", pre-finished pine	▼	260	.031		.51	1.16		1.67	2.51
5400	For pre-finished, stain and clear coat, add					.53			.53	.58
5405	Clear coat only, add				▼	.45			.45	.50

06 22 13.30 Moldings, Casings		Crew	Daily Output	Labor-Hours	Unit	Material	Labor	Equipment	Total	Total Incl O&P
						2016 Bare Costs				
0010	**MOLDINGS, CASINGS**									
0085	Apron, 9/16" x 2-1/2", pine	1 Carp	250	.032	L.F.	1.63	1.21		2.84	3.82
0090	5/8" x 2-1/2", pine		250	.032		1.63	1.21		2.84	3.82
0110	5/8" x 3-1/2", pine		220	.036		1.86	1.37		3.23	4.34
0300	Band, 11/16" x 1-1/8", pine		270	.030		.75	1.12		1.87	2.70
0310	11/16" x 1-1/2", finger jointed and primed		270	.030		.76	1.12		1.88	2.71
0320	Pine		270	.030		.96	1.12		2.08	2.93
0330	11/16" x 1-3/4", finger jointed and primed		270	.030		.95	1.12		2.07	2.92
0350	Pine		270	.030		1.16	1.12		2.28	3.15
0355	Beaded, 3/4" x 3-1/2", finger jointed and primed		220	.036		1.11	1.37		2.48	3.52
0360	Poplar		220	.036		1.11	1.37		2.48	3.52
0365	Red oak		220	.036		1.65	1.37		3.02	4.11
0370	Maple		220	.036		2.55	1.37		3.92	5.10
0375	Cherry		220	.036		3.09	1.37		4.46	5.70
0380	3/4" x 4", finger jointed and primed		220	.036		1.17	1.37		2.54	3.58
0385	Poplar		220	.036		1.45	1.37		2.82	3.89
0390	Red oak		220	.036		2.05	1.37		3.42	4.55
0395	Maple		220	.036		2.65	1.37		4.02	5.20
0400	Cherry		220	.036		3.33	1.37		4.70	5.95
0405	3/4" x 5-1/2", finger jointed and primed		200	.040		1.17	1.51		2.68	3.81
0410	Poplar		200	.040		1.93	1.51		3.44	4.65
0415	Red oak		200	.040		2.84	1.51		4.35	5.65
0420	Maple		200	.040		3.73	1.51		5.24	6.65
0425	Cherry		200	.040		4.31	1.51		5.82	7.25
0430	Classic profile, 3/4" x 2-3/4", finger jointed and primed		250	.032		.80	1.21		2.01	2.91
0435	Poplar		250	.032		1.02	1.21		2.23	3.15
0440	Red oak		250	.032		1.48	1.21		2.69	3.66
0445	Maple		250	.032		2.09	1.21		3.30	4.33
0450	Cherry		250	.032		2.39	1.21		3.60	4.66
0455	Fluted, 3/4" x 3-1/2", poplar		220	.036		1.12	1.37		2.49	3.53
0460	Red oak		220	.036		1.66	1.37		3.03	4.12
0465	Maple		220	.036		2.53	1.37		3.90	5.10
0470	Cherry		220	.036		3.09	1.37		4.46	5.70
0475	3/4" x 4", poplar		220	.036		1.44	1.37		2.81	3.88
0480	Red oak		220	.036		2.03	1.37		3.40	4.53
0485	Maple		220	.036		2.65	1.37		4.02	5.20
0490	Cherry		220	.036		3.33	1.37		4.70	5.95
0495	3/4" x 5-1/2", poplar		200	.040		1.93	1.51		3.44	4.65
0500	Red oak		200	.040		2.86	1.51		4.37	5.65
0505	Maple		200	.040		3.73	1.51		5.24	6.65
0510	Cherry		200	.040		4.32	1.51		5.83	7.30
0515	3/4" x 7-1/2", poplar		190	.042		1.22	1.59		2.81	4.01
0520	Red oak		190	.042		3.92	1.59		5.51	7
0525	Maple		190	.042		5.60	1.59		7.19	8.80
0530	Cherry		190	.042		6.70	1.59		8.29	10
0535	3/4" x 9-1/2", poplar		180	.044		4.13	1.68		5.81	7.35
0540	Red oak		180	.044		6.55	1.68		8.23	10
0545	Maple		180	.044		8.95	1.68		10.63	12.60
0550	Cherry		180	.044		9.75	1.68		11.43	13.50
0555	Modern profile, 9/16" x 2-1/4", poplar		250	.032		.59	1.21		1.80	2.68
0560	Red oak		250	.032		.79	1.21		2	2.90
0565	11/16" x 2-1/2", finger jointed & primed		250	.032		.84	1.21		2.05	2.95

06 22 13 – Standard Pattern Wood Trim

06 22 13.30 Moldings, Casings

	06 22 13.30 Moldings, Casings	Crew	Daily Output	Labor-Hours	Unit	Material	Labor	Equipment	Total	Total Incl O&P
0570	Pine	1 Carp	250	.032	L.F.	1.28	1.21		2.49	3.44
0575	3/4" x 2-1/2", poplar		250	.032		.90	1.21		2.11	3.02
0580	Red oak		250	.032		1.21	1.21		2.42	3.36
0585	Maple		250	.032		1.81	1.21		3.02	4.02
0590	Cherry		250	.032		2.30	1.21		3.51	4.56
0595	Mullion, 5/16" x 2", pine		270	.030		.86	1.12		1.98	2.82
0600	9/16" x 2-1/2", fingerjointed and primed		250	.032		.98	1.21		2.19	3.11
0605	Pine		250	.032		1.28	1.21		2.49	3.44
0610	Red oak		250	.032		2.86	1.21		4.07	5.15
0615	1-1/16" x 3-3/4", red oak		220	.036		6.95	1.37		8.32	9.90
0620	Ogee, 7/16" x 2-1/2", poplar		250	.032		.58	1.21		1.79	2.67
0625	Red oak		250	.032		.79	1.21		2	2.90
0630	9/16" x 2-1/4", finger jointed and primed		250	.032		.51	1.21		1.72	2.59
0635	Poplar		250	.032		.59	1.21		1.80	2.68
0640	Red oak		250	.032		.78	1.21		1.99	2.89
0645	11/16" x 2-1/2", finger jointed and primed		250	.032		.72	1.21		1.93	2.82
0700	Pine		250	.032		1.41	1.21		2.62	3.58
0701	Red oak		250	.032		2.70	1.21		3.91	5
0730	11/16" x 3-1/2", finger jointed and primed		220	.036		1.11	1.37		2.48	3.52
0750	Pine		220	.036		1.72	1.37		3.09	4.19
0755	3/4" x 2-1/2", finger jointed and primed		250	.032		.93	1.21		2.14	3.05
0760	Poplar		250	.032		.89	1.21		2.10	3.01
0765	Red oak		250	.032		1.20	1.21		2.41	3.35
0770	Maple		250	.032		1.80	1.21		3.01	4.01
0775	Cherry		250	.032		2.27	1.21		3.48	4.52
0780	3/4" x 3-1/2", finger jointed and primed		220	.036		.92	1.37		2.29	3.31
0785	Poplar		220	.036		1.12	1.37		2.49	3.53
0790	Red oak		220	.036		1.66	1.37		3.03	4.12
0795	Maple		220	.036		2.56	1.37		3.93	5.10
0800	Cherry		220	.036		3.11	1.37		4.48	5.70
4700	Square profile, 1" x 1", teak		215	.037		2.26	1.41		3.67	4.84
4800	Rectangular profile, 1" x 3", teak		200	.040		6.20	1.51		7.71	9.40

06 22 13.35 Moldings, Ceilings

	06 22 13.35 Moldings, Ceilings	Crew	Daily Output	Labor-Hours	Unit	Material	Labor	Equipment	Total	Total Incl O&P
0010	**MOLDINGS, CEILINGS**									
0600	Bed, 9/16" x 1-3/4", pine	1 Carp	270	.030	L.F.	1.06	1.12		2.18	3.04
0650	9/16" x 2", pine		270	.030		1.14	1.12		2.26	3.13
0710	9/16" x 1-3/4", oak		270	.030		1.86	1.12		2.98	3.92
1200	Cornice, 9/16" x 1-3/4", pine		270	.030		.96	1.12		2.08	2.93
1300	9/16" x 2-1/4", pine		265	.030		1.27	1.14		2.41	3.30
1350	Cove, 1/2" x 2-1/4", poplar		265	.030		1.04	1.14		2.18	3.05
1360	Red oak		265	.030		1.37	1.14		2.51	3.41
1370	Hard maple		265	.030		1.86	1.14		3	3.95
1380	Cherry		265	.030		2.28	1.14		3.42	4.41
2400	9/16" x 1-3/4", pine		270	.030		.97	1.12		2.09	2.94
2401	Oak		270	.030		1.26	1.12		2.38	3.26
2500	11/16" x 2-3/4", pine		265	.030		1.80	1.14		2.94	3.89
2510	Crown, 5/8" x 5/8", poplar		300	.027		.44	1.01		1.45	2.17
2520	Red oak		300	.027		.52	1.01		1.53	2.26
2530	Hard maple		300	.027		.70	1.01		1.71	2.46
2540	Cherry		300	.027		.74	1.01		1.75	2.50
2600	9/16" x 3-5/8", pine		250	.032		2.02	1.21		3.23	4.25
2700	11/16" x 4-1/4", pine		250	.032		2.90	1.21		4.11	5.20

06 22 Millwork

06 22 13 – Standard Pattern Wood Trim

06 22 13.35 Moldings, Ceilings		Crew	Daily Output	Labor-Hours	Unit	Material	Labor	Equipment	Total	Total Incl O&P
						2016 Bare Costs				
2705	Oak	1 Carp	250	.032	L.F.	6.40	1.21		7.61	9.10
2710	3/4" x 1-3/4", poplar		270	.030		.71	1.12		1.83	2.66
2720	Red oak		270	.030		1.05	1.12		2.17	3.03
2730	Hard maple		270	.030		1.39	1.12		2.51	3.41
2740	Cherry		270	.030		1.71	1.12		2.83	3.76
2750	3/4" x 2", poplar		270	.030		.94	1.12		2.06	2.91
2760	Red oak		270	.030		1.26	1.12		2.38	3.26
2770	Hard maple		270	.030		1.68	1.12		2.80	3.73
2780	Cherry		270	.030		1.88	1.12		3	3.95
2790	3/4" x 2-3/4", poplar		265	.030		1.02	1.14		2.16	3.03
2800	Red oak		265	.030		1.57	1.14		2.71	3.63
2810	Hard maple		265	.030		2.37	1.14		3.51	4.51
2820	Cherry		265	.030		2.40	1.14		3.54	4.55
2830	3/4" x 3-1/2", poplar		250	.032		1.30	1.21		2.51	3.46
2840	Red oak		250	.032		1.96	1.21		3.17	4.18
2850	Hard maple		250	.032		2.56	1.21		3.77	4.84
2860	Cherry		250	.032		3	1.21		4.21	5.35
2870	FJP poplar		250	.032		.92	1.21		2.13	3.04
2880	3/4" x 5", poplar		245	.033		1.93	1.23		3.16	4.19
2890	Red oak		245	.033		2.85	1.23		4.08	5.20
2900	Hard maple		245	.033		3.74	1.23		4.97	6.20
2910	Cherry		245	.033		4.32	1.23		5.55	6.80
2920	FJP poplar		245	.033		1.32	1.23		2.55	3.52
2930	3/4" x 6-1/4", poplar		240	.033		2.29	1.26		3.55	4.63
2940	Red oak		240	.033		3.46	1.26		4.72	5.90
2950	Hard maple		240	.033		4.52	1.26		5.78	7.10
2960	Cherry		240	.033		5.35	1.26		6.61	8
2970	7/8" x 8-3/4", poplar		220	.036		3.56	1.37		4.93	6.20
2980	Red oak		220	.036		4.97	1.37		6.34	7.75
2990	Hard maple		220	.036		6.75	1.37		8.12	9.75
3000	Cherry		220	.036		7.60	1.37		8.97	10.70
3010	1" x 7-1/4", poplar		220	.036		3.77	1.37		5.14	6.45
3020	Red oak		220	.036		5.85	1.37		7.22	8.75
3030	Hard maple		220	.036		8.15	1.37		9.52	11.25
3040	Cherry		220	.036		9.50	1.37		10.87	12.75
3050	1-1/16" x 4-1/4", poplar		250	.032		2.29	1.21		3.50	4.54
3060	Red oak		250	.032		2.69	1.21		3.90	4.98
3070	Hard maple		250	.032		3.56	1.21		4.77	5.95
3080	Cherry		250	.032		5.10	1.21		6.31	7.65
3090	Dentil crown, 3/4" x 5", poplar		250	.032		1.93	1.21		3.14	4.15
3100	Red oak		250	.032		2.85	1.21		4.06	5.15
3110	Hard maple		250	.032		3.74	1.21		4.95	6.15
3120	Cherry		250	.032		4.32	1.21		5.53	6.80
3130	Dentil piece for above, 1/2" x 1/2", poplar		300	.027		2.73	1.01		3.74	4.69
3140	Red oak		300	.027		3.14	1.01		4.15	5.15
3150	Hard maple		300	.027		3.57	1.01		4.58	5.60
3160	Cherry		300	.027		3.89	1.01		4.90	5.95

06 22 13.40 Moldings, Exterior		Crew	Daily Output	Labor-Hours	Unit	Material	Labor	Equipment	Total	Total Incl O&P
0010	MOLDINGS, EXTERIOR									
0100	Band board, cedar, rough sawn, 1" x 2"	1 Carp	300	.027	L.F.	.55	1.01		1.56	2.29
0110	1" x 3"		300	.027		.81	1.01		1.82	2.58
0120	1" x 4"		250	.032		1.08	1.21		2.29	3.22

06 22 13.40 Moldings, Exterior		Crew	Daily Output	Labor-Hours	Unit	Material	Labor	Equipment	Total	Total Incl O&P
						2016 Bare Costs				
0130	1" x 6"	1 Carp	250	.032	L.F.	1.63	1.21		2.84	3.82
0140	1" x 8"		225	.036		2.17	1.34		3.51	4.64
0150	1" x 10"		225	.036		2.70	1.34		4.04	5.20
0160	1" x 12"		200	.040		3.24	1.51		4.75	6.10
0240	STK, 1" x 2"		300	.027		.53	1.01		1.54	2.28
0250	1" x 3"		300	.027		.59	1.01		1.60	2.33
0260	1" x 4"		250	.032		.71	1.21		1.92	2.81
0270	1" x 6"		250	.032		1.13	1.21		2.34	3.27
0280	1" x 8"		225	.036		1.63	1.34		2.97	4.04
0290	1" x 10"		225	.036		2.12	1.34		3.46	4.58
0300	1" x 12"		200	.040		3.62	1.51		5.13	6.50
0310	Pine, #2, 1" x 2"		300	.027		.25	1.01		1.26	1.97
0320	1" x 3"		300	.027		.40	1.01		1.41	2.13
0330	1" x 4"		250	.032		.49	1.21		1.70	2.57
0340	1" x 6"		250	.032		.73	1.21		1.94	2.84
0350	1" x 8"		225	.036		1.20	1.34		2.54	3.57
0360	1" x 10"		225	.036		1.57	1.34		2.91	3.97
0370	1" x 12"		200	.040		1.92	1.51		3.43	4.64
0380	D & better, 1" x 2"		300	.027		.43	1.01		1.44	2.17
0390	1" x 3"		300	.027		.66	1.01		1.67	2.42
0400	1" x 4"		250	.032		.89	1.21		2.10	3.01
0410	1" x 6"		250	.032		1.12	1.21		2.33	3.26
0420	1" x 8"		225	.036		1.62	1.34		2.96	4.03
0430	1" x 10"		225	.036		2.03	1.34		3.37	4.48
0440	1" x 12"		200	.040		2.63	1.51		4.14	5.40
0450	Redwood, clear all heart, 1" x 2"		300	.027		.63	1.01		1.64	2.39
0460	1" x 3"		300	.027		.94	1.01		1.95	2.73
0470	1" x 4"		250	.032		1.19	1.21		2.40	3.34
0480	1" x 6"		252	.032		1.80	1.20		3	3.99
0490	1" x 8"		225	.036		2.36	1.34		3.70	4.85
0500	1" x 10"		225	.036		3.84	1.34		5.18	6.45
0510	1" x 12"		200	.040		4.75	1.51		6.26	7.80
0530	Corner board, cedar, rough sawn, 1" x 2"		225	.036		.55	1.34		1.89	2.85
0540	1" x 3"		225	.036		.81	1.34		2.15	3.14
0550	1" x 4"		200	.040		1.08	1.51		2.59	3.72
0560	1" x 6"		200	.040		1.63	1.51		3.14	4.32
0570	1" x 8"		200	.040		2.17	1.51		3.68	4.92
0580	1" x 10"		175	.046		2.70	1.73		4.43	5.85
0590	1" x 12"		175	.046		3.24	1.73		4.97	6.45
0670	STK, 1" x 2"		225	.036		.53	1.34		1.87	2.84
0680	1" x 3"		225	.036		.59	1.34		1.93	2.89
0690	1" x 4"		200	.040		.69	1.51		2.20	3.29
0700	1" x 6"		200	.040		1.13	1.51		2.64	3.77
0710	1" x 8"		200	.040		1.63	1.51		3.14	4.32
0720	1" x 10"		175	.046		2.12	1.73		3.85	5.20
0730	1" x 12"		175	.046		3.62	1.73		5.35	6.85
0740	Pine, #2, 1" x 2"		225	.036		.25	1.34		1.59	2.53
0750	1" x 3"		225	.036		.40	1.34		1.74	2.69
0760	1" x 4"		200	.040		.49	1.51		2	3.07
0770	1" x 6"		200	.040		.73	1.51		2.24	3.34
0780	1" x 8"		200	.040		1.20	1.51		2.71	3.85
0790	1" x 10"		175	.046		1.57	1.73		3.30	4.61
0800	1" x 12"		175	.046		1.92	1.73		3.65	5

06 22 13.40 Moldings, Exterior		Crew	Daily Output	Labor-Hours	Unit	2016 Bare Costs Material	Labor	Equipment	Total	Total Incl O&P
0810	D & better, 1" x 2"	1 Carp	225	.036	L.F.	.43	1.34		1.77	2.73
0820	1" x 3"		225	.036		.66	1.34		2	2.98
0830	1" x 4"		200	.040		.89	1.51		2.40	3.51
0840	1" x 6"		200	.040		1.12	1.51		2.63	3.76
0850	1" x 8"		200	.040		1.62	1.51		3.13	4.31
0860	1" x 10"		175	.046		2.03	1.73		3.76	5.10
0870	1" x 12"		175	.046		2.63	1.73		4.36	5.80
0880	Redwood, clear all heart, 1" x 2"		225	.036		.63	1.34		1.97	2.95
0890	1" x 3"		225	.036		.94	1.34		2.28	3.29
0900	1" x 4"		200	.040		1.19	1.51		2.70	3.84
0910	1" x 6"		200	.040		1.80	1.51		3.31	4.51
0920	1" x 8"		200	.040		2.36	1.51		3.87	5.15
0930	1" x 10"		175	.046		3.84	1.73		5.57	7.10
0940	1" x 12"		175	.046		4.75	1.73		6.48	8.15
0950	Cornice board, cedar, rough sawn, 1" x 2"		330	.024		.55	.92		1.47	2.13
0960	1" x 3"		290	.028		.81	1.04		1.85	2.64
0970	1" x 4"		250	.032		1.08	1.21		2.29	3.22
0980	1" x 6"		250	.032		1.63	1.21		2.84	3.82
0990	1" x 8"		200	.040		2.17	1.51		3.68	4.92
1000	1" x 10"		180	.044		2.70	1.68		4.38	5.80
1010	1" x 12"		180	.044		3.24	1.68		4.92	6.35
1020	STK, 1" x 2"		330	.024		.53	.92		1.45	2.12
1030	1" x 3"		290	.028		.59	1.04		1.63	2.39
1040	1" x 4"		250	.032		.71	1.21		1.92	2.81
1050	1" x 6"		250	.032		1.13	1.21		2.34	3.27
1060	1" x 8"		200	.040		1.63	1.51		3.14	4.32
1070	1" x 10"		180	.044		2.12	1.68		3.80	5.15
1080	1" x 12"		180	.044		3.62	1.68		5.30	6.80
1500	Pine, #2, 1" x 2"		330	.024		.25	.92		1.17	1.81
1510	1" x 3"		290	.028		.27	1.04		1.31	2.05
1600	1" x 4"		250	.032		.49	1.21		1.70	2.57
1700	1" x 6"		250	.032		.73	1.21		1.94	2.84
1800	1" x 8"		200	.040		1.20	1.51		2.71	3.85
1900	1" x 10"		180	.044		1.57	1.68		3.25	4.53
2000	1" x 12"		180	.044		1.92	1.68		3.60	4.92
2020	D & better, 1" x 2"		330	.024		.43	.92		1.35	2.01
2030	1" x 3"		290	.028		.66	1.04		1.70	2.48
2040	1" x 4"		250	.032		.89	1.21		2.10	3.01
2050	1" x 6"		250	.032		1.12	1.21		2.33	3.26
2060	1" x 8"		200	.040		1.62	1.51		3.13	4.31
2070	1" x 10"		180	.044		2.03	1.68		3.71	5.05
2080	1" x 12"		180	.044		2.63	1.68		4.31	5.70
2090	Redwood, clear all heart, 1" x 2"		330	.024		.63	.92		1.55	2.23
2100	1" x 3"		290	.028		.94	1.04		1.98	2.79
2110	1" x 4"		250	.032		1.19	1.21		2.40	3.34
2120	1" x 6"		250	.032		1.80	1.21		3.01	4.01
2130	1" x 8"		200	.040		2.36	1.51		3.87	5.15
2140	1" x 10"		180	.044		3.84	1.68		5.52	7.05
2150	1" x 12"		180	.044		4.75	1.68		6.43	8.05
2160	3 piece, 1" x 2", 1" x 4", 1" x 6", rough sawn cedar		80	.100		3.27	3.78		7.05	9.95
2180	STK cedar		80	.100		2.37	3.78		6.15	8.95
2200	#2 pine		80	.100		1.47	3.78		5.25	7.95
2210	D & better pine		80	.100		2.44	3.78		6.22	9.05

06 22 13.40 Moldings, Exterior		Crew	Daily Output	Labor-Hours	Unit	Material	Labor	Equipment	Total	Total Incl O&P
						2016 Bare Costs				
2220	Clear all heart redwood	1 Carp	80	.100	L.F.	3.62	3.78		7.40	10.35
2230	1" x 8", 1" x 10", 1" x 12", rough sawn cedar		65	.123		8.10	4.65		12.75	16.70
2240	STK cedar		65	.123		7.35	4.65		12	15.90
2300	#2 pine		65	.123		4.66	4.65		9.31	12.95
2320	D & better pine		65	.123		6.25	4.65		10.90	14.70
2330	Clear all heart redwood		65	.123		10.90	4.65		15.55	19.80
2340	Door/window casing, cedar, rough sawn, 1" x 2"		275	.029		.55	1.10		1.65	2.44
2350	1" x 3"		275	.029		.81	1.10		1.91	2.73
2360	1" x 4"		250	.032		1.08	1.21		2.29	3.22
2370	1" x 6"		250	.032		1.63	1.21		2.84	3.82
2380	1" x 8"		230	.035		2.17	1.31		3.48	4.59
2390	1" x 10"		230	.035		2.70	1.31		4.01	5.15
2395	1" x 12"		210	.038		3.24	1.44		4.68	5.95
2410	STK, 1" x 2"		275	.029		.53	1.10		1.63	2.43
2420	1" x 3"		275	.029		.59	1.10		1.69	2.48
2430	1" x 4"		250	.032		.71	1.21		1.92	2.81
2440	1" x 6"		250	.032		1.13	1.21		2.34	3.27
2450	1" x 8"		230	.035		1.63	1.31		2.94	3.99
2460	1" x 10"		230	.035		2.12	1.31		3.43	4.53
2470	1" x 12"		210	.038		3.62	1.44		5.06	6.40
2550	Pine, #2, 1" x 2"		275	.029		.25	1.10		1.35	2.12
2560	1" x 3"		275	.029		.40	1.10		1.50	2.28
2570	1" x 4"		250	.032		.49	1.21		1.70	2.57
2580	1" x 6"		250	.032		.73	1.21		1.94	2.84
2590	1" x 8"		230	.035		1.20	1.31		2.51	3.52
2600	1" x 10"		230	.035		1.57	1.31		2.88	3.92
2610	1" x 12"		210	.038		1.92	1.44		3.36	4.52
2620	Pine, D & better, 1" x 2"		275	.029		.43	1.10		1.53	2.32
2630	1" x 3"		275	.029		.66	1.10		1.76	2.57
2640	1" x 4"		250	.032		.89	1.21		2.10	3.01
2650	1" x 6"		250	.032		1.12	1.21		2.33	3.26
2660	1" x 8"		230	.035		1.62	1.31		2.93	3.98
2670	1" x 10"		230	.035		2.03	1.31		3.34	4.43
2680	1" x 12"		210	.038		2.63	1.44		4.07	5.30
2690	Redwood, clear all heart, 1" x 2"		275	.029		.63	1.10		1.73	2.54
2695	1" x 3"		275	.029		.94	1.10		2.04	2.88
2710	1" x 4"		250	.032		1.19	1.21		2.40	3.34
2715	1" x 6"		250	.032		1.80	1.21		3.01	4.01
2730	1" x 8"		230	.035		2.36	1.31		3.67	4.80
2740	1" x 10"		230	.035		3.84	1.31		5.15	6.40
2750	1" x 12"		210	.038		4.75	1.44		6.19	7.65
3500	Bellyband, pine, 11/16" x 4-1/4"		250	.032		3.23	1.21		4.44	5.60
3610	Brickmold, pine, 1-1/4" x 2"		200	.040		2.43	1.51		3.94	5.20
3620	FJP, 1-1/4" x 2"		200	.040		1.09	1.51		2.60	3.73
5100	Fascia, cedar, rough sawn, 1" x 2"		275	.029		.55	1.10		1.65	2.44
5110	1" x 3"		275	.029		.81	1.10		1.91	2.73
5120	1" x 4"		250	.032		1.08	1.21		2.29	3.22
5200	1" x 6"		250	.032		1.63	1.21		2.84	3.82
5300	1" x 8"		230	.035		2.17	1.31		3.48	4.59
5310	1" x 10"		230	.035		2.70	1.31		4.01	5.15
5320	1" x 12"		210	.038		3.24	1.44		4.68	5.95
5400	2" x 4"		220	.036		1.07	1.37		2.44	3.47
5500	2" x 6"	▼	220	.036	▼	1.60	1.37		2.97	4.06

06 22 13.40 Moldings, Exterior		Crew	Daily Output	Labor-Hours	Unit	2016 Bare Costs Material	Labor	Equipment	Total	Total Incl O&P
5600	2" x 8"	1 Carp	200	.040	L.F.	2.14	1.51		3.65	4.88
5700	2" x 10"		180	.044		2.66	1.68		4.34	5.75
5800	2" x 12"		170	.047		6.35	1.78		8.13	9.95
6120	STK, 1" x 2"		275	.029		.53	1.10		1.63	2.43
6130	1" x 3"		275	.029		.59	1.10		1.69	2.48
6140	1" x 4"		250	.032		.71	1.21		1.92	2.81
6150	1" x 6"		250	.032		1.13	1.21		2.34	3.27
6160	1" x 8"		230	.035		1.63	1.31		2.94	3.99
6170	1" x 10"		230	.035		2.12	1.31		3.43	4.53
6180	1" x 12"		210	.038		3.62	1.44		5.06	6.40
6185	2" x 2"		260	.031		.65	1.16		1.81	2.67
6190	Pine, #2, 1" x 2"		275	.029		.25	1.10		1.35	2.12
6200	1" x 3"		275	.029		.40	1.10		1.50	2.28
6210	1" x 4"		250	.032		.49	1.21		1.70	2.57
6220	1" x 6"		250	.032		.73	1.21		1.94	2.84
6230	1" x 8"		230	.035		1.20	1.31		2.51	3.52
6240	1" x 10"		230	.035		1.57	1.31		2.88	3.92
6250	1" x 12"		210	.038		1.92	1.44		3.36	4.52
6260	D & better, 1" x 2"		275	.029		.43	1.10		1.53	2.32
6270	1" x 3"		275	.029		.66	1.10		1.76	2.57
6280	1" x 4"		250	.032		.89	1.21		2.10	3.01
6290	1" x 6"		250	.032		1.12	1.21		2.33	3.26
6300	1" x 8"		230	.035		1.62	1.31		2.93	3.98
6310	1" x 10"		230	.035		2.03	1.31		3.34	4.43
6312	1" x 12"		210	.038		2.63	1.44		4.07	5.30
6330	Southern yellow, 1-1/4" x 5"		240	.033		2.43	1.26		3.69	4.78
6340	1-1/4" x 6"		240	.033		2.52	1.26		3.78	4.88
6350	1-1/4" x 8"		215	.037		3.43	1.41		4.84	6.15
6360	1-1/4" x 12"		190	.042		5.05	1.59		6.64	8.20
6370	Redwood, clear all heart, 1" x 2"		275	.029		.63	1.10		1.73	2.54
6380	1" x 3"		275	.029		1.19	1.10		2.29	3.15
6390	1" x 4"		250	.032		1.19	1.21		2.40	3.34
6400	1" x 6"		250	.032		1.80	1.21		3.01	4.01
6410	1" x 8"		230	.035		2.36	1.31		3.67	4.80
6420	1" x 10"		230	.035		3.84	1.31		5.15	6.40
6430	1" x 12"		210	.038		4.75	1.44		6.19	7.65
6440	1-1/4" x 5"		240	.033		1.88	1.26		3.14	4.18
6450	1-1/4" x 6"		240	.033		2.25	1.26		3.51	4.58
6460	1-1/4" x 8"		215	.037		3.53	1.41		4.94	6.25
6470	1-1/4" x 12"		190	.042		7.10	1.59		8.69	10.50
6580	Frieze, cedar, rough sawn, 1" x 2"		275	.029		.55	1.10		1.65	2.44
6590	1" x 3"		275	.029		.81	1.10		1.91	2.73
6600	1" x 4"		250	.032		1.08	1.21		2.29	3.22
6610	1" x 6"		250	.032		1.63	1.21		2.84	3.82
6620	1" x 8"		250	.032		2.17	1.21		3.38	4.42
6630	1" x 10"		225	.036		2.70	1.34		4.04	5.20
6640	1" x 12"		200	.040		3.21	1.51		4.72	6.05
6650	STK, 1" x 2"		275	.029		.53	1.10		1.63	2.43
6660	1" x 3"		275	.029		.59	1.10		1.69	2.48
6670	1" x 4"		250	.032		.71	1.21		1.92	2.81
6680	1" x 6"		250	.032		1.13	1.21		2.34	3.27
6690	1" x 8"		250	.032		1.63	1.21		2.84	3.82
6700	1" x 10"		225	.036		2.12	1.34		3.46	4.58

06 22 13.40 Moldings, Exterior		Crew	Daily Output	Labor-Hours	Unit	Material	2016 Bare Costs Labor	Equipment	Total	Total Incl O&P
6710	1" x 12"	1 Carp	200	.040	L.F.	3.62	1.51		5.13	6.50
6790	Pine, #2, 1" x 2"		275	.029		.25	1.10		1.35	2.12
6800	1" x 3"		275	.029		.40	1.10		1.50	2.28
6810	1" x 4"		250	.032		.49	1.21		1.70	2.57
6820	1" x 6"		250	.032		.73	1.21		1.94	2.84
6830	1" x 8"		250	.032		1.20	1.21		2.41	3.35
6840	1" x 10"		225	.036		1.57	1.34		2.91	3.97
6850	1" x 12"		200	.040		1.92	1.51		3.43	4.64
6860	D & better, 1" x 2"		275	.029		.43	1.10		1.53	2.32
6870	1" x 3"		275	.029		.66	1.10		1.76	2.57
6880	1" x 4"		250	.032		.89	1.21		2.10	3.01
6890	1" x 6"		250	.032		1.12	1.21		2.33	3.26
6900	1" x 8"		250	.032		1.62	1.21		2.83	3.81
6910	1" x 10"		225	.036		2.03	1.34		3.37	4.48
6920	1" x 12"		200	.040		2.63	1.51		4.14	5.40
6930	Redwood, clear all heart, 1" x 2"		275	.029		.63	1.10		1.73	2.54
6940	1" x 3"		275	.029		.94	1.10		2.04	2.88
6950	1" x 4"		250	.032		1.19	1.21		2.40	3.34
6960	1" x 6"		250	.032		1.80	1.21		3.01	4.01
6970	1" x 8"		250	.032		2.36	1.21		3.57	4.63
6980	1" x 10"		225	.036		3.84	1.34		5.18	6.45
6990	1" x 12"		200	.040		4.75	1.51		6.26	7.80
7000	Grounds, 1" x 1", cedar, rough sawn		300	.027		.28	1.01		1.29	2
7010	STK		300	.027		.33	1.01		1.34	2.05
7020	Pine, #2		300	.027		.16	1.01		1.17	1.86
7030	D & better		300	.027		.27	1.01		1.28	1.98
7050	Redwood		300	.027		.39	1.01		1.40	2.11
7060	Rake/verge board, cedar, rough sawn, 1" x 2"		225	.036		.55	1.34		1.89	2.85
7070	1" x 3"		225	.036		.81	1.34		2.15	3.14
7080	1" x 4"		200	.040		1.08	1.51		2.59	3.72
7090	1" x 6"		200	.040		1.63	1.51		3.14	4.32
7100	1" x 8"		190	.042		2.17	1.59		3.76	5.05
7110	1" x 10"		190	.042		2.70	1.59		4.29	5.65
7120	1" x 12"		180	.044		3.24	1.68		4.92	6.35
7130	STK, 1" x 2"		225	.036		.53	1.34		1.87	2.84
7140	1" x 3"		225	.036		.59	1.34		1.93	2.89
7150	1" x 4"		200	.040		.71	1.51		2.22	3.31
7160	1" x 6"		200	.040		1.13	1.51		2.64	3.77
7170	1" x 8"		190	.042		1.63	1.59		3.22	4.46
7180	1" x 10"		190	.042		2.12	1.59		3.71	5
7190	1" x 12"		180	.044		3.62	1.68		5.30	6.80
7200	Pine, #2, 1" x 2"		225	.036		.25	1.34		1.59	2.53
7210	1" x 3"		225	.036		.40	1.34		1.74	2.69
7220	1" x 4"		200	.040		.49	1.51		2	3.07
7230	1" x 6"		200	.040		.73	1.51		2.24	3.34
7240	1" x 8"		190	.042		1.20	1.59		2.79	3.99
7250	1" x 10"		190	.042		1.57	1.59		3.16	4.39
7260	1" x 12"		180	.044		1.92	1.68		3.60	4.92
7340	D & better, 1" x 2"		225	.036		.43	1.34		1.77	2.73
7350	1" x 3"		225	.036		.66	1.34		2	2.98
7360	1" x 4"		200	.040		.89	1.51		2.40	3.51
7370	1" x 6"		200	.040		1.12	1.51		2.63	3.76
7380	1" x 8"		190	.042		1.62	1.59		3.21	4.45

06 22 13.40 Moldings, Exterior		Crew	Daily Output	Labor-Hours	Unit	2016 Bare Costs Material	Labor	Equipment	Total	Total Incl O&P
7390	1" x 10"	1 Carp	190	.042	L.F.	2.03	1.59		3.62	4.90
7400	1" x 12"		180	.044		2.63	1.68		4.31	5.70
7410	Redwood, clear all heart, 1" x 2"		225	.036		.63	1.34		1.97	2.95
7420	1" x 3"		225	.036		.94	1.34		2.28	3.29
7430	1" x 4"		200	.040		1.19	1.51		2.70	3.84
7440	1" x 6"		200	.040		1.80	1.51		3.31	4.51
7450	1" x 8"		190	.042		2.36	1.59		3.95	5.25
7460	1" x 10"		190	.042		3.84	1.59		5.43	6.90
7470	1" x 12"		180	.044		4.75	1.68		6.43	8.05
7480	2" x 4"		200	.040		2.29	1.51		3.80	5.05
7490	2" x 6"		182	.044		3.43	1.66		5.09	6.55
7500	2" x 8"		165	.048		4.56	1.83		6.39	8.05
7630	Soffit, cedar, rough sawn, 1" x 2"	2 Carp	440	.036		.55	1.37		1.92	2.90
7640	1" x 3"		440	.036		.81	1.37		2.18	3.19
7650	1" x 4"		420	.038		1.08	1.44		2.52	3.60
7660	1" x 6"		420	.038		1.63	1.44		3.07	4.20
7670	1" x 8"		420	.038		2.17	1.44		3.61	4.80
7680	1" x 10"		400	.040		2.70	1.51		4.21	5.50
7690	1" x 12"		400	.040		3.24	1.51		4.75	6.10
7700	STK, 1" x 2"		440	.036		.53	1.37		1.90	2.89
7710	1" x 3"		440	.036		.59	1.37		1.96	2.94
7720	1" x 4"		420	.038		.71	1.44		2.15	3.19
7730	1" x 6"		420	.038		1.13	1.44		2.57	3.65
7740	1" x 8"		420	.038		1.63	1.44		3.07	4.20
7750	1" x 10"		400	.040		2.12	1.51		3.63	4.86
7760	1" x 12"		400	.040		3.62	1.51		5.13	6.50
7770	Pine, #2, 1" x 2"		440	.036		.25	1.37		1.62	2.58
7780	1" x 3"		440	.036		.40	1.37		1.77	2.74
7790	1" x 4"		420	.038		.49	1.44		1.93	2.95
7800	1" x 6"		420	.038		.73	1.44		2.17	3.22
7810	1" x 8"		420	.038		1.20	1.44		2.64	3.73
7820	1" x 10"		400	.040		1.57	1.51		3.08	4.25
7830	1" x 12"		400	.040		1.92	1.51		3.43	4.64
7840	D & better, 1" x 2"		440	.036		.43	1.37		1.80	2.78
7850	1" x 3"		440	.036		.66	1.37		2.03	3.03
7860	1" x 4"		420	.038		.89	1.44		2.33	3.39
7870	1" x 6"		420	.038		1.12	1.44		2.56	3.64
7880	1" x 8"		420	.038		1.62	1.44		3.06	4.19
7890	1" x 10"		400	.040		2.03	1.51		3.54	4.76
7900	1" x 12"		400	.040		2.63	1.51		4.14	5.40
7910	Redwood, clear all heart, 1" x 2"		440	.036		.63	1.37		2	3
7920	1" x 3"		440	.036		.94	1.37		2.31	3.34
7930	1" x 4"		420	.038		1.19	1.44		2.63	3.72
7940	1" x 6"		420	.038		1.80	1.44		3.24	4.39
7950	1" x 8"		420	.038		2.36	1.44		3.80	5
7960	1" x 10"		400	.040		3.84	1.51		5.35	6.75
7970	1" x 12"		400	.040		4.75	1.51		6.26	7.80
8050	Trim, crown molding, pine, 11/16" x 4-1/4"	1 Carp	250	.032		4.54	1.21		5.75	7
8060	Back band, 11/16" x 1-1/16"		250	.032		.99	1.21		2.20	3.12
8070	Insect screen frame stock, 1-1/16" x 1-3/4"		395	.020		2.39	.77		3.16	3.91
8080	Dentils, 2-1/2" x 2-1/2" x 4", 6" O.C.		30	.267		1.23	10.10		11.33	18.25
8100	Fluted, 5-1/2"		165	.048		5.05	1.83		6.88	8.60
8110	Stucco bead, 1-3/8" x 1-5/8"		250	.032		2.50	1.21		3.71	4.78

06 22 13 – Standard Pattern Wood Trim

06 22 13.45 Moldings, Trim		Crew	Daily Output	Labor-Hours	Unit	Material	2016 Bare Costs Labor	Equipment	Total	Total Incl O&P
0010	**MOLDINGS, TRIM**									
0200	Astragal, stock pine, 11/16" x 1-3/4"	1 Carp	255	.031	L.F.	1.42	1.19		2.61	3.55
0250	1-5/16" x 2-3/16"		240	.033		2.10	1.26		3.36	4.42
0800	Chair rail, stock pine, 5/8" x 2-1/2"		270	.030		1.58	1.12		2.70	3.62
0900	5/8" x 3-1/2"		240	.033		2.40	1.26		3.66	4.75
1000	Closet pole, stock pine, 1-1/8" diameter		200	.040		1.16	1.51		2.67	3.81
1100	Fir, 1-5/8" diameter		200	.040		2.20	1.51		3.71	4.95
1150	Corner, inside, 5/16" x 1"		225	.036		.33	1.34		1.67	2.61
1160	Outside, 1-1/16" x 1-1/16"		240	.033		1.31	1.26		2.57	3.55
1161	1-5/16" x 1-5/16"		240	.033		1.55	1.26		2.81	3.82
3300	Half round, stock pine, 1/4" x 1/2"		270	.030		.24	1.12		1.36	2.14
3350	1/2" x 1"		255	.031		.73	1.19		1.92	2.79
3400	Handrail, fir, single piece, stock, hardware not included									
3450	1-1/2" x 1-3/4"	1 Carp	80	.100	L.F.	2.56	3.78		6.34	9.15
3470	Pine, 1-1/2" x 1-3/4"		80	.100		2.40	3.78		6.18	9
3500	1-1/2" x 2-1/2"		76	.105		2.46	3.98		6.44	9.35
3600	Lattice, stock pine, 1/4" x 1-1/8"		270	.030		.35	1.12		1.47	2.27
3700	1/4" x 1-3/4"		250	.032		.92	1.21		2.13	3.04
3800	Miscellaneous, custom, pine, 1" x 1"		270	.030		.43	1.12		1.55	2.36
3850	1" x 2"		265	.030		.87	1.14		2.01	2.87
3900	1" x 3"		240	.033		1.30	1.26		2.56	3.54
4100	Birch or oak, nominal 1" x 1"		240	.033		.39	1.26		1.65	2.54
4200	Nominal 1" x 3"		215	.037		1.18	1.41		2.59	3.66
4400	Walnut, nominal 1" x 1"		215	.037		.62	1.41		2.03	3.04
4500	Nominal 1" x 3"		200	.040		1.86	1.51		3.37	4.57
4700	Teak, nominal 1" x 1"		215	.037		2.64	1.41		4.05	5.25
4800	Nominal 1" x 3"		200	.040		7.90	1.51		9.41	11.25
4900	Quarter round, stock pine, 1/4" x 1/4"		275	.029		.24	1.10		1.34	2.10
4950	3/4" x 3/4"		255	.031		.50	1.19		1.69	2.54
5600	Wainscot moldings, 1-1/8" x 9/16", 2' high, minimum		76	.105	S.F.	11.65	3.98		15.63	19.45
5700	Maximum		65	.123	"	15.80	4.65		20.45	25

06 22 13.50 Moldings, Window and Door		Crew	Daily Output	Labor-Hours	Unit	Material	Labor	Equipment	Total	Total Incl O&P
0010	**MOLDINGS, WINDOW AND DOOR**									
2800	Door moldings, stock, decorative, 1-1/8" wide, plain	1 Carp	17	.471	Set	47.50	17.80		65.30	82.50
2900	Detailed		17	.471	"	91.50	17.80		109.30	131
2960	Clear pine door jamb, no stops, 11/16" x 4-9/16"		240	.033	L.F.	5.20	1.26		6.46	7.80
3150	Door trim set, 1 head and 2 sides, pine, 2-1/2 wide		12	.667	Opng.	24	25		49	68
3170	3-1/2" wide		11	.727	"	29	27.50		56.50	78
3250	Glass beads, stock pine, 3/8" x 1/2"		275	.029	L.F.	.33	1.10		1.43	2.20
3270	3/8" x 7/8"		270	.030		.42	1.12		1.54	2.34
4850	Parting bead, stock pine, 3/8" x 3/4"		275	.029		.44	1.10		1.54	2.32
4870	1/2" x 3/4"		255	.031		.42	1.19		1.61	2.45
5000	Stool caps, stock pine, 11/16" x 3 1/2"		200	.040		2.07	1.51		3.58	4.81
5100	1-1/16" x 3-1/4"		150	.053		3.24	2.02		5.26	6.95
5300	Threshold, oak, 3' long, inside, 5/8" x 3-5/8"		32	.250	Ea.	6.50	9.45		15.95	23
5400	Outside, 1-1/2" x 7-5/8"		16	.500	"	45	18.90		63.90	81
5900	Window trim sets, including casings, header, stops,									
5910	stool and apron, 2-1/2" wide, FJP	1 Carp	13	.615	Opng.	31.50	23.50		55	73.50
5950	Pine		10	.800		37	30		67	91.50
6000	Oak		6	1.333		64.50	50.50		115	156

06 22 Millwork

06 22 13 – Standard Pattern Wood Trim

06 22 13.60 Moldings, Soffits		Crew	Daily Output	Labor-Hours	Unit	2016 Bare Costs Material	Labor	Equipment	Total	Total Incl O&P
0010	MOLDINGS, SOFFITS									
0200	Soffits, pine, 1" x 4"	2 Carp	420	.038	L.F.	.46	1.44		1.90	2.92
0210	1" x 6"		420	.038		.71	1.44		2.15	3.19
0220	1" x 8"		420	.038		1.17	1.44		2.61	3.70
0230	1" x 10"		400	.040		1.53	1.51		3.04	4.21
0240	1" x 12"		400	.040		1.88	1.51		3.39	4.60
0250	STK cedar, 1" x 4"		420	.038		.68	1.44		2.12	3.16
0260	1" x 6"		420	.038		1.10	1.44		2.54	3.62
0270	1" x 8"		420	.038		1.60	1.44		3.04	4.17
0280	1" x 10"		400	.040		2.08	1.51		3.59	4.82
0290	1" x 12"		400	.040		3.58	1.51		5.09	6.45
1000	Exterior AC plywood, 1/4" thick		400	.040	S.F.	.95	1.51		2.46	3.58
1050	3/8" thick		400	.040		.99	1.51		2.50	3.62
1100	1/2" thick		400	.040		1.17	1.51		2.68	3.82
1150	Polyvinyl chloride, white, solid	1 Carp	230	.035		2.10	1.31		3.41	4.51
1160	Perforated	"	230	.035		2.10	1.31		3.41	4.51
1170	Accessories, "J" channel 5/8"	2 Carp	700	.023	L.F.	.49	.86		1.35	1.99

06 25 Prefinished Paneling

06 25 13 – Prefinished Hardboard Paneling

06 25 13.10 Paneling, Hardboard

			Crew	Daily Output	Labor-Hours	Unit	Material	Labor	Equipment	Total	Total Incl O&P
0010	PANELING, HARDBOARD										
0050	Not incl. furring or trim, hardboard, tempered, 1/8" thick	G	2 Carp	500	.032	S.F.	.42	1.21		1.63	2.49
0100	1/4" thick	G		500	.032		.65	1.21		1.86	2.75
0300	Tempered pegboard, 1/8" thick	G		500	.032		.42	1.21		1.63	2.49
0400	1/4" thick	G		500	.032		.68	1.21		1.89	2.78
0600	Untempered hardboard, natural finish, 1/8" thick	G		500	.032		.43	1.21		1.64	2.50
0700	1/4" thick	G		500	.032		.55	1.21		1.76	2.64
0900	Untempered pegboard, 1/8" thick	G		500	.032		.46	1.21		1.67	2.54
1000	1/4" thick	G		500	.032		.48	1.21		1.69	2.56
1200	Plastic faced hardboard, 1/8" thick	G		500	.032		.66	1.21		1.87	2.76
1300	1/4" thick	G		500	.032		.89	1.21		2.10	3.01
1500	Plastic faced pegboard, 1/8" thick	G		500	.032		.67	1.21		1.88	2.77
1600	1/4" thick	G		500	.032		.85	1.21		2.06	2.97
1800	Wood grained, plain or grooved, 1/8" thick	G		500	.032		.69	1.21		1.90	2.79
1900	1/4" thick	G		425	.038		1.40	1.42		2.82	3.92
2100	Moldings, wood grained MDF			500	.032	L.F.	.41	1.21		1.62	2.48
2200	Pine			425	.038	"	1.40	1.42		2.82	3.92

06 25 16 – Prefinished Plywood Paneling

06 25 16.10 Paneling, Plywood

		Crew	Daily Output	Labor-Hours	Unit	Material	Labor	Equipment	Total	Total Incl O&P
0010	PANELING, PLYWOOD									
2400	Plywood, prefinished, 1/4" thick, 4' x 8' sheets									
2410	with vertical grooves. Birch faced, economy	2 Carp	500	.032	S.F.	1.45	1.21		2.66	3.63
2420	Average		420	.038		1.20	1.44		2.64	3.73
2430	Custom		350	.046		1.25	1.73		2.98	4.27
2600	Mahogany, African		400	.040		2.80	1.51		4.31	5.60
2700	Philippine (Lauan)		500	.032		.65	1.21		1.86	2.75
2900	Oak		500	.032		1.40	1.21		2.61	3.57
3000	Cherry		400	.040		2.10	1.51		3.61	4.84
3200	Rosewood		320	.050		2.95	1.89		4.84	6.40

06 25 Prefinished Paneling

06 25 16 – Prefinished Plywood Paneling

06 25 16.10 Paneling, Plywood		Crew	Daily Output	Labor-Hours	Unit	Material	Labor	Equipment	Total	Total Incl O&P
						2016 Bare Costs				
3400	Teak	2 Carp	400	.040	S.F.	2.95	1.51		4.46	5.80
3600	Chestnut		375	.043		4.85	1.61		6.46	8.05
3800	Pecan		400	.040		2.55	1.51		4.06	5.35
3900	Walnut, average		500	.032		2.40	1.21		3.61	4.67
3950	Custom		400	.040		5.25	1.51		6.76	8.35
4000	Plywood, prefinished, 3/4" thick, stock grades, economy		320	.050		1.60	1.89		3.49	4.93
4100	Average		224	.071		4.70	2.70		7.40	9.65
4300	Architectural grade, custom		224	.071		5.20	2.70		7.90	10.20
4400	Luxury		160	.100		5.20	3.78		8.98	12.05
4600	Plywood, "A" face, birch, VC, 1/2" thick, natural		450	.036		2.05	1.34		3.39	4.51
4700	Select		450	.036		2.15	1.34		3.49	4.62
4900	Veneer core, 3/4" thick, natural		320	.050		2.24	1.89		4.13	5.65
5000	Select		320	.050		2.44	1.89		4.33	5.85
5200	Lumber core, 3/4" thick, natural		320	.050		3.05	1.89		4.94	6.55
5500	Plywood, knotty pine, 1/4" thick, A2 grade		450	.036		1.70	1.34		3.04	4.12
5600	A3 grade		450	.036		2.15	1.34		3.49	4.62
5800	3/4" thick, veneer core, A2 grade		320	.050		2.20	1.89		4.09	5.60
5900	A3 grade		320	.050		2.45	1.89		4.34	5.85
6100	Aromatic cedar, 1/4" thick, plywood		400	.040		2.25	1.51		3.76	5
6200	1/4" thick, particle board	↓	400	.040	↓	1.15	1.51		2.66	3.80

06 25 26 – Panel System

06 25 26.10 Panel Systems		Crew	Daily Output	Labor-Hours	Unit	Material	Labor	Equipment	Total	Total Incl O&P
0010	**PANEL SYSTEMS**									
0100	Raised panel, eng. wood core w/wood veneer, std., paint grade	2 Carp	300	.053	S.F.	11.20	2.02		13.22	15.70
0110	Oak veneer		300	.053		25.50	2.02		27.52	32
0120	Maple veneer		300	.053		32.50	2.02		34.52	39.50
0130	Cherry veneer		300	.053		37	2.02		39.02	44
0300	Class I fire rated, paint grade		300	.053		13	2.02		15.02	17.70
0310	Oak veneer		300	.053		30	2.02		32.02	36.50
0320	Maple veneer		300	.053		40.50	2.02		42.52	48
0330	Cherry veneer		300	.053		49	2.02		51.02	57
0510	Beadboard, 5/8" MDF, standard, primed		300	.053		8.45	2.02		10.47	12.70
0520	Oak veneer, unfinished		300	.053		13.40	2.02		15.42	18.15
0530	Maple veneer, unfinished		300	.053		14.60	2.02		16.62	19.45
0610	Rustic paneling, 5/8" MDF, standard, maple veneer, unfinished	↓	300	.053	↓	18.20	2.02		20.22	23.50

06 26 Board Paneling

06 26 13 – Profile Board Paneling

06 26 13.10 Paneling, Boards		Crew	Daily Output	Labor-Hours	Unit	Material	Labor	Equipment	Total	Total Incl O&P
0010	**PANELING, BOARDS**									
6400	Wood board paneling, 3/4" thick, knotty pine	2 Carp	300	.053	S.F.	1.95	2.02		3.97	5.55
6500	Rough sawn cedar		300	.053		3.20	2.02		5.22	6.90
6700	Redwood, clear, 1" x 4" boards		300	.053		4.96	2.02		6.98	8.85
6900	Aromatic cedar, closet lining, boards	↓	275	.058	↓	2.40	2.20		4.60	6.30

06 43 Wood Stairs and Railings

06 43 13 – Wood Stairs

06 43 13.20 Prefabricated Wood Stairs		Crew	Daily Output	Labor-Hours	Unit	Material	Labor	Equipment	Total	Total Incl O&P
						2016 Bare Costs				
0010	**PREFABRICATED WOOD STAIRS**									
0100	Box stairs, prefabricated, 3′-0″ wide									
0110	Oak treads, up to 14 risers	2 Carp	39	.410	Riser	92.50	15.50		108	128
0600	With pine treads for carpet, up to 14 risers	"	39	.410	"	59.50	15.50		75	91.50
1100	For 4′ wide stairs, add				Flight	25%				
1550	Stairs, prefabricated stair handrail with balusters	1 Carp	30	.267	L.F.	78.50	10.10		88.60	103
1700	Basement stairs, prefabricated, pine treads									
1710	Pine risers, 3′ wide, up to 14 risers	2 Carp	52	.308	Riser	59.50	11.65		71.15	85
4000	Residential, wood, oak treads, prefabricated		1.50	10.667	Flight	1,200	405		1,605	2,000
4200	Built in place		.44	36.364	"	2,175	1,375		3,550	4,675
4400	Spiral, oak, 4′-6″ diameter, unfinished, prefabricated,									
4500	incl. railing, 9′ high	2 Carp	1.50	10.667	Flight	3,425	405		3,830	4,450

06 43 13.40 Wood Stair Parts		Crew	Daily Output	Labor-Hours	Unit	Material	Labor	Equipment	Total	Total Incl O&P
0010	**WOOD STAIR PARTS**									
0020	Pin top balusters, 1-1/4″, oak, 34″	1 Carp	96	.083	Ea.	5.10	3.15		8.25	10.85
0030	38″		96	.083		5.85	3.15		9	11.70
0040	42″		96	.083		6.35	3.15		9.50	12.25
0050	Poplar, 34″		96	.083		3.05	3.15		6.20	8.60
0060	38″		96	.083		3.88	3.15		7.03	9.50
0070	42″		96	.083		8.85	3.15		12	14.95
0080	Maple, 34″		96	.083		4.90	3.15		8.05	10.65
0090	38″		96	.083		5.60	3.15		8.75	11.45
0100	42″		96	.083		6.50	3.15		9.65	12.40
0130	Primed, 34″		96	.083		3	3.15		6.15	8.55
0140	38″		96	.083		3.62	3.15		6.77	9.25
0150	42″		96	.083		4.42	3.15		7.57	10.10
0180	Box top balusters, 1-1/4″, oak, 34″		60	.133		8.95	5.05		14	18.30
0190	38″		60	.133		9.95	5.05		15	19.40
0200	42″		60	.133		10.95	5.05		16	20.50
0210	Poplar, 34″		60	.133		6.25	5.05		11.30	15.35
0220	38″		60	.133		6.95	5.05		12	16.10
0230	42″		60	.133		7.50	5.05		12.55	16.70
0240	Maple, 34″		60	.133		8.25	5.05		13.30	17.55
0250	38″		60	.133		9	5.05		14.05	18.35
0260	42″		60	.133		10	5.05		15.05	19.45
0290	Primed, 34″		60	.133		7	5.05		12.05	16.15
0300	38″		60	.133		8	5.05		13.05	17.25
0310	42″		60	.133		8.35	5.05		13.40	17.65
0340	Square balusters, cut from lineal stock, pine, 1-1/16″ x 1-1/16″		180	.044	L.F.	1.50	1.68		3.18	4.46
0350	1-5/16″ x 1-5/16″		180	.044		2.10	1.68		3.78	5.10
0360	1-5/8″ x 1-5/8″		180	.044		3.30	1.68		4.98	6.45
0370	Turned newel, oak, 3-1/2″ square, 48″ high		8	1	Ea.	92	38		130	165
0380	62″ high		8	1		92	38		130	165
0390	Poplar, 3-1/2″ square, 48″ high		8	1		54	38		92	123
0400	62″ high		8	1		68	38		106	139
0410	Maple, 3-1/2″ square, 48″ high		8	1		72	38		110	143
0420	62″ high		8	1		92	38		130	165
0430	Square newel, oak, 3-1/2″ square, 48″ high		8	1		54	38		92	123
0440	58″ high		8	1		68	38		106	139
0450	Poplar, 3-1/2″ square, 48″ high		8	1		35	38		73	102
0460	58″ high		8	1		42	38		80	110
0470	Maple, 3″ square, 48″ high		8	1		52	38		90	121

06 43 13 – Wood Stairs

06 43 13.40 Wood Stair Parts		Crew	Daily Output	Labor-Hours	Unit	2016 Bare Costs Material	Labor	Equipment	Total	Total Incl O&P
0480	58″ high	1 Carp	8	1	Ea.	64	38		102	134
0490	Railings, oak, economy		96	.083	L.F.	8.65	3.15		11.80	14.75
0500	Average		96	.083		13	3.15		16.15	19.55
0510	Custom		96	.083		16.50	3.15		19.65	23.50
0520	Maple, economy		96	.083		11	3.15		14.15	17.35
0530	Average		96	.083		13.50	3.15		16.65	20
0540	Custom		96	.083		16.95	3.15		20.10	24
0550	Oak, for bending rail, economy		48	.167		23.50	6.30		29.80	36.50
0560	Average		48	.167		26	6.30		32.30	39
0570	Custom		48	.167		29.50	6.30		35.80	42.50
0580	Maple, for bending rail, economy		48	.167		32	6.30		38.30	45.50
0590	Average		48	.167		32	6.30		38.30	45.50
0600	Custom		48	.167		32	6.30		38.30	45.50
0610	Risers, oak, 3/4″ x 8″, 36″ long		80	.100	Ea.	13	3.78		16.78	20.50
0620	42″ long		70	.114		15.15	4.32		19.47	24
0630	48″ long		63	.127		17.30	4.80		22.10	27
0640	54″ long		56	.143		19.50	5.40		24.90	30.50
0650	60″ long		50	.160		21.50	6.05		27.55	34
0660	72″ long		42	.190		26	7.20		33.20	40.50
0670	Poplar, 3/4″ x 8″, 36″ long		80	.100		12.50	3.78		16.28	20
0680	42″ long		71	.113		14.55	4.26		18.81	23
0690	48″ long		63	.127		16.65	4.80		21.45	26.50
0700	54″ long		56	.143		18.70	5.40		24.10	29.50
0710	60″ long		50	.160		21	6.05		27.05	33
0720	72″ long		42	.190		25	7.20		32.20	39.50
0730	Pine, 1″ x 8″, 36″ long		80	.100		3.52	3.78		7.30	10.20
0740	42″ long		70	.114		4.11	4.32		8.43	11.75
0750	48″ long		63	.127		4.69	4.80		9.49	13.20
0760	54″ long		56	.143		5.30	5.40		10.70	14.85
0770	60″ long		50	.160		5.85	6.05		11.90	16.60
0780	72″ long		42	.190		7.05	7.20		14.25	19.80
0790	Treads, oak, no returns, 1-1/32″ x 11-1/2″ x 36″ long		32	.250		27	9.45		36.45	45.50
0800	42″ long		32	.250		31.50	9.45		40.95	50.50
0810	48″ long		32	.250		36	9.45		45.45	55.50
0820	54″ long		32	.250		40.50	9.45		49.95	60.50
0830	60″ long		32	.250		45	9.45		54.45	65.50
0840	72″ long		32	.250		54	9.45		63.45	75.50
0850	Mitred return one end, 1-1/32″ x 11-1/2″ x 36″ long		24	.333		36	12.60		48.60	60.50
0860	42″ long		24	.333		42	12.60		54.60	67
0870	48″ long		24	.333		48	12.60		60.60	74
0880	54″ long		24	.333		54	12.60		66.60	80.50
0890	60″ long		24	.333		60	12.60		72.60	87
0900	72″ long		24	.333		72	12.60		84.60	100
0910	Mitred return two ends, 1-1/32″ x 11-1/2″ x 36″ long		12	.667		46	25		71	92.50
0920	42″ long		12	.667		53.50	25		78.50	101
0930	48″ long		12	.667		61.50	25		86.50	110
0940	54″ long		12	.667		69	25		94	118
0950	60″ long		12	.667		76.50	25		101.50	127
0960	72″ long		12	.667		92	25		117	143
0970	Starting step, oak, 48″, bullnose		8	1		172	38		210	253
0980	Double end bullnose		8	1		254	38		292	345
1030	Skirt board, pine, 1″ x 10″		55	.145	L.F.	1.53	5.50		7.03	10.90
1040	1″ x 12″		52	.154	″	1.88	5.80		7.68	11.80

06 43 Wood Stairs and Railings

06 43 13 – Wood Stairs

06 43 13.40 Wood Stair Parts		Crew	Daily Output	Labor-Hours	Unit	2016 Bare Costs Material	Labor	Equipment	Total	Total Incl O&P
1050	Oak landing tread, 1-1/16" thick	1 Carp	54	.148	S.F.	9	5.60		14.60	19.30
1060	Oak cove molding		96	.083	L.F.	1	3.15		4.15	6.35
1070	Oak stringer molding		96	.083	"	4	3.15		7.15	9.65
1090	Rail bolt, 5/16" x 3-1/2"		48	.167	Ea.	2.75	6.30		9.05	13.60
1100	5/16" x 4-1/2"		48	.167		2.75	6.30		9.05	13.60
1120	Newel post anchor		16	.500		13	18.90		31.90	46
1130	Tapered plug, 1/2"		240	.033		1	1.26		2.26	3.21
1140	1"	↓	240	.033	↓	.99	1.26		2.25	3.20

06 43 16 – Wood Railings

06 43 16.10 Wood Handrails and Railings		Crew	Daily Output	Labor-Hours	Unit	Material	Labor	Equipment	Total	Total Incl O&P
0010	**WOOD HANDRAILS AND RAILINGS**									
0020	Custom design, architectural grade, hardwood, plain	1 Carp	38	.211	L.F.	12.05	7.95		20	26.50
0100	Shaped		30	.267		62.50	10.10		72.60	86
0300	Stock interior railing with spindles 4" O.C., 4' long		40	.200		38.50	7.55		46.05	55
0400	8' long	↓	48	.167	↓	38.50	6.30		44.80	53

06 44 Ornamental Woodwork

06 44 19 – Wood Grilles

06 44 19.10 Grilles		Crew	Daily Output	Labor-Hours	Unit	Material	Labor	Equipment	Total	Total Incl O&P
0010	**GRILLES** and panels, hardwood, sanded									
0020	2' x 4' to 4' x 8', custom designs, unfinished, economy	1 Carp	38	.211	S.F.	62	7.95		69.95	81.50
0050	Average		30	.267		70	10.10		80.10	94
0100	Custom	↓	19	.421	↓	72	15.90		87.90	106

06 44 33 – Wood Mantels

06 44 33.10 Fireplace Mantels		Crew	Daily Output	Labor-Hours	Unit	Material	Labor	Equipment	Total	Total Incl O&P
0010	**FIREPLACE MANTELS**									
0015	6" molding, 6' x 3'-6" opening, plain, paint grade	1 Carp	5	1.600	Opng.	440	60.50		500.50	580
0100	Ornate, oak		5	1.600		590	60.50		650.50	750
0300	Prefabricated pine, colonial type, stock, deluxe		2	4		1,500	151		1,651	1,900
0400	Economy	↓	3	2.667	↓	690	101		791	930

06 44 33.20 Fireplace Mantel Beam		Crew	Daily Output	Labor-Hours	Unit	Material	Labor	Equipment	Total	Total Incl O&P
0010	**FIREPLACE MANTEL BEAM**									
0020	Rough texture wood, 4" x 8"	1 Carp	36	.222	L.F.	8.30	8.40		16.70	23
0100	4" x 10"		35	.229	"	10.80	8.65		19.45	26.50
0300	Laminated hardwood, 2-1/4" x 10-1/2" wide, 6' long		5	1.600	Ea.	125	60.50		185.50	239
0400	8' long		5	1.600	"	150	60.50		210.50	266
0600	Brackets for above, rough sawn		12	.667	Pr.	12	25		37	55
0700	Laminated	↓	12	.667	"	16	25		41	59.50

06 44 39 – Wood Posts and Columns

06 44 39.10 Decorative Beams		Crew	Daily Output	Labor-Hours	Unit	Material	Labor	Equipment	Total	Total Incl O&P
0010	**DECORATIVE BEAMS**									
0020	Rough sawn cedar, non-load bearing, 4" x 4"	2 Carp	180	.089	L.F.	1.25	3.36		4.61	7.05
0100	4" x 6"		170	.094		1.80	3.56		5.36	7.95
0200	4" x 8"		160	.100		2.40	3.78		6.18	9
0300	4" x 10"		150	.107		3.65	4.03		7.68	10.75
0400	4" x 12"		140	.114		4.61	4.32		8.93	12.30
0500	8" x 8"		130	.123		4.80	4.65		9.45	13.10
0600	Plastic beam, "hewn finish", 6" x 2"		240	.067		3.40	2.52		5.92	7.95
0601	6" x 4"	↓	220	.073	↓	3.75	2.75		6.50	8.75

06 44 Ornamental Woodwork

06 44 39 – Wood Posts and Columns

06 44 39.20 Columns		Crew	Daily Output	Labor-Hours	Unit	Material	2016 Bare Costs Labor	Equipment	Total	Total Incl O&P
0010	**COLUMNS**									
0050	Aluminum, round colonial, 6″ diameter	2 Carp	80	.200	V.L.F.	19	7.55		26.55	33.50
0100	8″ diameter		62.25	.257		22	9.70		31.70	40.50
0200	10″ diameter		55	.291		22.50	11		33.50	43.50
0250	Fir, stock units, hollow round, 6″ diameter		80	.200		29.50	7.55		37.05	45
0300	8″ diameter		80	.200		35.50	7.55		43.05	51.50
0350	10″ diameter		70	.229		44.50	8.65		53.15	63.50
0360	12″ diameter		65	.246		54.50	9.30		63.80	75.50
0400	Solid turned, to 8′ high, 3-1/2″ diameter		80	.200		9.60	7.55		17.15	23
0500	4-1/2″ diameter		75	.213		11.90	8.05		19.95	26.50
0600	5-1/2″ diameter		70	.229		16	8.65		24.65	32
0800	Square columns, built-up, 5″ x 5″		65	.246		14.20	9.30		23.50	31
0900	Solid, 3-1/2″ x 3-1/2″		130	.123		9.60	4.65		14.25	18.35
1600	Hemlock, tapered, T & G, 12″ diam., 10′ high		100	.160		39.50	6.05		45.55	53.50
1700	16′ high		65	.246		73	9.30		82.30	95.50
1900	14″ diameter, 10′ high		100	.160		113	6.05		119.05	135
2000	18′ high		65	.246		103	9.30		112.30	129
2200	18″ diameter, 12′ high		65	.246		165	9.30		174.30	198
2300	20′ high		50	.320		118	12.10		130.10	151
2500	20″ diameter, 14′ high		40	.400		180	15.10		195.10	224
2600	20′ high		35	.457		170	17.30		187.30	216
2800	For flat pilasters, deduct					33%				
3000	For splitting into halves, add				Ea.	106			106	117
4000	Rough sawn cedar posts, 4″ x 4″	2 Carp	250	.064	V.L.F.	3.95	2.42		6.37	8.40
4100	4″ x 6″		235	.068		6.80	2.57		9.37	11.80
4200	6″ x 6″		220	.073		9.90	2.75		12.65	15.50
4300	8″ x 8″		200	.080		19.20	3.02		22.22	26

06 48 Wood Frames

06 48 13 – Exterior Wood Door Frames

06 48 13.10 Exterior Wood Door Frames and Accessories		Crew	Daily Output	Labor-Hours	Unit	Material	Labor	Equipment	Total	Total Incl O&P
0010	**EXTERIOR WOOD DOOR FRAMES AND ACCESSORIES**									
0400	Exterior frame, incl. ext. trim, pine, 5/4 x 4-9/16″ deep	2 Carp	375	.043	L.F.	6.60	1.61		8.21	10
0420	5-3/16″ deep		375	.043		7.90	1.61		9.51	11.35
0440	6-9/16″ deep		375	.043		8.80	1.61		10.41	12.35
0600	Oak, 5/4 x 4-9/16″ deep		350	.046		19.75	1.73		21.48	24.50
0620	5-3/16″ deep		350	.046		21.50	1.73		23.23	27
0640	6-9/16″ deep		350	.046		19.50	1.73		21.23	24.50
1000	Sills, 8/4 x 8″ deep, oak, no horns		100	.160		6.65	6.05		12.70	17.45
1020	2″ horns		100	.160		20.50	6.05		26.55	32.50
1040	3″ horns		100	.160		20.50	6.05		26.55	32.50
1100	8/4 x 10″ deep, oak, no horns		90	.178		6.40	6.70		13.10	18.30
1120	2″ horns		90	.178		26.50	6.70		33.20	40.50
1140	3″ horns		90	.178		26.50	6.70		33.20	40.50
2000	Wood frame & trim, ext, colonial, 3′ opng, fluted pilasters, flat head		22	.727	Ea.	505	27.50		532.50	600
2010	Dentil head		21	.762		580	29		609	685
2020	Ram's head		20	.800		695	30		725	810
2100	5′-4″ opening, in-swing, fluted pilasters, flat head		17	.941		440	35.50		475.50	545
2120	Ram's head		15	1.067		1,400	40.50		1,440.50	1,600
2140	Out swing, fluted pilasters, flat head		17	.941		520	35.50		555.50	630

06 48 Wood Frames

06 48 13 – Exterior Wood Door Frames

06 48 13.10 Exterior Wood Door Frames and Accessories		Crew	Daily Output	Labor-Hours	Unit	2016 Bare Costs Material	Labor	Equipment	Total	Total Incl O&P
2160	Ram's head	2 Carp	15	1.067	Ea.	1,475	40.50		1,515.50	1,700
2400	6'-0" opening, in-swing, fluted pilasters, flat head		16	1		520	38		558	635
2420	Ram's head		10	1.600		1,475	60.50		1,535.50	1,725
2460	Out-swing, fluted pilasters, flat head		16	1		520	38		558	635
2480	Ram's head		10	1.600		1,475	60.50		1,535.50	1,725
2600	For two sidelights, flat head, add		30	.533	Opng.	226	20		246	283
2620	Ram's head, add		20	.800	"	840	30		870	975
2700	Custom birch frame, 3'-0" opening		16	1	Ea.	240	38		278	330
2750	6'-0" opening		16	1		360	38		398	460
2900	Exterior, modern, plain trim, 3' opng., in-swing, FJP		26	.615		46.50	23.50		70	90
2920	Fir		24	.667		55	25		80	102
2940	Oak		22	.727		63	27.50		90.50	116

06 48 16 – Interior Wood Door Frames

06 48 16.10 Interior Wood Door Jamb and Frames		Crew	Daily Output	Labor-Hours	Unit	Material	Labor	Equipment	Total	Total Incl O&P
0010	**INTERIOR WOOD DOOR JAMB AND FRAMES**									
3000	Interior frame, pine, 11/16" x 3-5/8" deep	2 Carp	375	.043	L.F.	4.44	1.61		6.05	7.60
3020	4-9/16" deep		375	.043		4.92	1.61		6.53	8.10
3200	Oak, 11/16" x 3-5/8" deep		350	.046		9.85	1.73		11.58	13.70
3220	4-9/16" deep		350	.046		9.95	1.73		11.68	13.85
3240	5-3/16" deep		350	.046		13.90	1.73		15.63	18.15
3400	Walnut, 11/16" x 3-5/8" deep		350	.046		9	1.73		10.73	12.80
3420	4-9/16" deep		350	.046		9.45	1.73		11.18	13.30
3440	5-3/16" deep		350	.046		9.55	1.73		11.28	13.40
3600	Pocket door frame		16	1	Ea.	81	38		119	153
3800	Threshold, oak, 5/8" x 3-5/8" deep		200	.080	L.F.	3.64	3.02		6.66	9.05
3820	4-5/8" deep		190	.084		4.35	3.18		7.53	10.15
3840	5-5/8" deep		180	.089		6.75	3.36		10.11	13.10

06 49 Wood Screens and Exterior Wood Shutters

06 49 19 – Exterior Wood Shutters

06 49 19.10 Shutters, Exterior		Crew	Daily Output	Labor-Hours	Unit	Material	Labor	Equipment	Total	Total Incl O&P
0010	**SHUTTERS, EXTERIOR**									
0012	Aluminum, louvered, 1'-4" wide, 3'-0" long	1 Carp	10	.800	Pr.	200	30		230	271
0200	4'-0" long		10	.800		240	30		270	315
0300	5'-4" long		10	.800		280	30		310	360
0400	6'-8" long		9	.889		355	33.50		388.50	445
1000	Pine, louvered, primed, each 1'-2" wide, 3'-3" long		10	.800		220	30		250	293
1100	4'-7" long		10	.800		270	30		300	350
1250	Each 1'-4" wide, 3'-0" long		10	.800		220	30		250	293
1350	5'-3" long		10	.800		330	30		360	410
1500	Each 1'-6" wide, 3'-3" long		10	.800		244	30		274	320
1600	4'-7" long		10	.800		325	30		355	405
1620	Cedar, louvered, 1'-2" wide, 5'-7" long		10	.800		315	30		345	395
1630	Each 1'-4" wide, 2'-2" long		10	.800		175	30		205	244
1640	3'-0" long		10	.800		217	30		247	290
1650	3'-3" long		10	.800		228	30		258	300
1660	3'-11" long		10	.800		265	30		295	345
1670	4'-3" long		10	.800		275	30		305	355
1680	5'-3" long		10	.800		290	30		320	370
1690	5'-11" long		10	.800		350	30		380	435

06 49 Wood Screens and Exterior Wood Shutters

06 49 19 – Exterior Wood Shutters

06 49 19.10 Shutters, Exterior		Crew	Daily Output	Labor-Hours	Unit	Material	Labor	Equipment	Total	Total Incl O&P
						2016 Bare Costs				
1700	Door blinds, 6′-9″ long, each 1′-3″ wide	1 Carp	9	.889	Pr.	380	33.50		413.50	470
1710	1′-6″ wide		9	.889		435	33.50		468.50	535
2500	Polystyrene, solid raised panel, each 1′-4″ wide, 3′-3″ long		10	.800		72	30		102	130
2600	3′-11″ long		10	.800		93	30		123	153
2700	4′-7″ long		10	.800		105	30		135	167
2800	5′-3″ long		10	.800		120	30		150	183
2900	6′-8″ long		9	.889		152	33.50		185.50	224
4500	Polystyrene, louvered, each 1′-2″ wide, 3′-3″ long		10	.800		36	30		66	90
4600	4′-7″ long		10	.800		46	30		76	101
4750	5′-3″ long		10	.800		59	30		89	116
4850	6′-8″ long		9	.889		68	33.50		101.50	132
6000	Vinyl, louvered, each 1′-2″ x 4′-7″ long		10	.800		64	30		94	121
6200	Each 1′-4″ x 6′-8″ long		9	.889		76	33.50		109.50	140
8000	PVC exterior rolling shutters									
8100	including crank control	1 Carp	8	1	Ea.	540	38		578	655
8500	Insulative - 6′ x 6′8″ stock unit	"	8	1	"	765	38		803	910

06 51 Structural Plastic Shapes and Plates

06 51 13 – Plastic Lumber

06 51 13.12 Structural Plastic Lumber		Crew	Daily Output	Labor-Hours	Unit	Material	Labor	Equipment	Total	Total Incl O&P
0010	**STRUCTURAL PLASTIC LUMBER**									
1320	Plastic lumber, posts or columns, 4″ x 4″	2 Carp	390	.041	L.F.	9.25	1.55		10.80	12.75
1325	4″ x 6″		275	.058		13.25	2.20		15.45	18.30
1330	4″ x 8″		220	.073		19.35	2.75		22.10	26
1340	Girder, single, 4″ x 4″		675	.024		9.25	.90		10.15	11.65
1345	4″ x 6″		600	.027		13.25	1.01		14.26	16.30
1350	4″ x 8″		525	.030		19.35	1.15		20.50	23.50
1352	Double, 2″ x 4″		625	.026		8.55	.97		9.52	11
1354	2″ x 6″		600	.027		12.85	1.01		13.86	15.85
1356	2″ x 8″		575	.028		16.90	1.05		17.95	20.50
1358	2″ x 10″		550	.029		20.50	1.10		21.60	24.50
1360	2″ x 12″		525	.030		25	1.15		26.15	29.50
1362	Triple, 2″ x 4″		575	.028		12.80	1.05		13.85	15.85
1364	2″ x 6″		550	.029		19.30	1.10		20.40	23
1366	2″ x 8″		525	.030		25.50	1.15		26.65	30
1368	2″ x 10″		500	.032		31	1.21		32.21	36
1370	2″ x 12″		475	.034		37.50	1.27		38.77	43
1372	Ledger, bolted 4′ O.C., 2″ x 4″		400	.040		4.42	1.51		5.93	7.40
1374	2″ x 6″		550	.029		6.50	1.10		7.60	9
1376	2″ x 8″		390	.041		8.55	1.55		10.10	12.05
1378	2″ x 10″		385	.042		10.40	1.57		11.97	14.10
1380	2″ x 12″		300	.042		12.55	1.59		14.14	16.50
1382	Joists, 2″ x 4″		1250	.013		4.27	.48		4.75	5.50
1384	2″ x 6″		1250	.013		6.45	.48		6.93	7.85
1386	2″ x 8″		1100	.015		8.45	.55		9	10.20
1388	2″ x 10″		500	.032		10.40	1.21		11.61	13.50
1390	2″ x 12″		875	.018		12.45	.69		13.14	14.85
1392	Railings and trim , 5/4″ x 4″	1 Carp	300	.027		4.75	1.01		5.76	6.95
1394	2″ x 2″		300	.027		2.40	1.01		3.41	4.33
1396	2″ x 4″		300	.027		4.25	1.01		5.26	6.35
1398	2″ x 6″		300	.027		6.40	1.01		7.41	8.75

06 63 Plastic Railings

06 63 10 – Plastic (PVC) Railings

06 63 10.10 Plastic Railings		Crew	Daily Output	Labor-Hours	Unit	2016 Bare Costs Material	Labor	Equipment	Total	Total Incl O&P
0010	**PLASTIC RAILINGS**									
0100	Horizontal PVC handrail with balusters, 3-1/2" wide, 36" high	1 Carp	96	.083	L.F.	25.50	3.15		28.65	33.50
0150	42" high		96	.083		28.50	3.15		31.65	37
0200	Angled PVC handrail with balusters, 3-1/2" wide, 36" high		72	.111		29	4.20		33.20	39
0250	42" high		72	.111		32.50	4.20		36.70	42.50
0300	Post sleeve for 4 x 4 post		96	.083		13.10	3.15		16.25	19.65
0400	Post cap for 4 x 4 post, flat profile		48	.167	Ea.	13.20	6.30		19.50	25
0450	Newel post style profile		48	.167		24	6.30		30.30	37
0500	Raised corbeled profile		48	.167		36.50	6.30		42.80	50.50
0550	Post base trim for 4 x 4 post		96	.083		18.80	3.15		21.95	26

06 65 Plastic Trim

06 65 10 – PVC Trim

06 65 10.10 PVC Trim, Exterior		Crew	Daily Output	Labor-Hours	Unit	Material	Labor	Equipment	Total	Total Incl O&P
0010	**PVC TRIM, EXTERIOR**									
0100	Cornerboards, 5/4" x 6" x 6"	1 Carp	240	.033	L.F.	8	1.26		9.26	10.90
0110	Door/window casing, 1" x 4"		200	.040		1.46	1.51		2.97	4.14
0120	1" x 6"		200	.040		2.20	1.51		3.71	4.95
0130	1" x 8"		195	.041		2.93	1.55		4.48	5.80
0140	1" x 10"		195	.041		3.81	1.55		5.36	6.80
0150	1" x 12"		190	.042		4.36	1.59		5.95	7.45
0160	5/4" x 4"		195	.041		2.19	1.55		3.74	5
0170	5/4" x 6"		195	.041		3.15	1.55		4.70	6.05
0180	5/4" x 8"		190	.042		4.19	1.59		5.78	7.30
0190	5/4" x 10"		190	.042		5.10	1.59		6.69	8.25
0200	5/4" x 12"		185	.043		5.60	1.63		7.23	8.90
0210	Fascia, 1" x 4"		250	.032		1.46	1.21		2.67	3.64
0220	1" x 6"		250	.032		2.20	1.21		3.41	4.45
0230	1" x 8"		225	.036		2.93	1.34		4.27	5.45
0240	1" x 10"		225	.036		3.81	1.34		5.15	6.45
0250	1" x 12"		200	.040		4.36	1.51		5.87	7.35
0260	5/4" x 4"		240	.033		2.19	1.26		3.45	4.52
0270	5/4" x 6"		240	.033		3.15	1.26		4.41	5.60
0280	5/4" x 8"		215	.037		4.19	1.41		5.60	6.95
0290	5/4" x 10"		215	.037		5.10	1.41		6.51	7.95
0300	5/4" x 12"		190	.042		5.60	1.59		7.19	8.80
0310	Frieze, 1" x 4"		250	.032		1.46	1.21		2.67	3.64
0320	1" x 6"		250	.032		2.20	1.21		3.41	4.45
0330	1" x 8"		225	.036		2.93	1.34		4.27	5.45
0340	1" x 10"		225	.036		3.81	1.34		5.15	6.45
0350	1" x 12"		200	.040		4.36	1.51		5.87	7.35
0360	5/4" x 4"		240	.033		2.19	1.26		3.45	4.52
0370	5/4" x 6"		240	.033		3.15	1.26		4.41	5.60
0380	5/4" x 8"		215	.037		4.19	1.41		5.60	6.95
0390	5/4" x 10"		215	.037		5.10	1.41		6.51	7.95
0400	5/4" x 12"		190	.042		5.60	1.59		7.19	8.80
0410	Rake, 1" x 4"		200	.040		1.46	1.51		2.97	4.14
0420	1" x 6"		200	.040		2.20	1.51		3.71	4.95
0430	1" x 8"		190	.042		2.93	1.59		4.52	5.90
0440	1" x 10"		190	.042		3.81	1.59		5.40	6.85
0450	1" x 12"		180	.044		4.36	1.68		6.04	7.60

06 65 Plastic Trim

06 65 10 – PVC Trim

06 65 10.10 PVC Trim, Exterior		Crew	Daily Output	Labor-Hours	Unit	Material	Labor (2016 Bare Costs)	Equipment (2016 Bare Costs)	Total (2016 Bare Costs)	Total Incl O&P
0460	5/4″ x 4″	1 Carp	195	.041	L.F.	2.19	1.55		3.74	5
0470	5/4″ x 6″		195	.041		3.15	1.55		4.70	6.05
0480	5/4″ x 8″		185	.043		4.19	1.63		5.82	7.35
0490	5/4″ x 10″		185	.043		5.10	1.63		6.73	8.35
0500	5/4″ x 12″		175	.046		5.60	1.73		7.33	9.05
0510	Rake trim, 1″ x 4″		225	.036		1.46	1.34		2.80	3.86
0520	1″ x 6″		225	.036		2.20	1.34		3.54	4.67
0560	5/4″ x 4″		220	.036		2.19	1.37		3.56	4.71
0570	5/4″ x 6″		220	.036		3.15	1.37		4.52	5.75
0610	Soffit, 1″ x 4″	2 Carp	420	.038		1.46	1.44		2.90	4.02
0620	1″ x 6″		420	.038		2.20	1.44		3.64	4.83
0630	1″ x 8″		420	.038		2.93	1.44		4.37	5.65
0640	1″ x 10″		400	.040		3.81	1.51		5.32	6.70
0650	1″ x 12″		400	.040		4.36	1.51		5.87	7.35
0660	5/4″ x 4″		410	.039		2.19	1.48		3.67	4.88
0670	5/4″ x 6″		410	.039		3.15	1.48		4.63	5.95
0680	5/4″ x 8″		410	.039		4.19	1.48		5.67	7.10
0690	5/4″ x 10″		390	.041		5.10	1.55		6.65	8.20
0700	5/4″ x 12″		390	.041		5.60	1.55		7.15	8.75

06 80 Composite Fabrications

06 80 10 – Composite Decking

06 80 10.10 Woodgrained Composite Decking		Crew	Daily Output	Labor-Hours	Unit	Material	Labor	Equipment	Total	Total Incl O&P
0010	**WOODGRAINED COMPOSITE DECKING**									
0100	Woodgrained composite decking, 1″ x 6″	2 Carp	640	.025	L.F.	3.65	.95		4.60	5.60
0110	Grooved edge		660	.024		3.79	.92		4.71	5.70
0120	2″ x 6″		640	.025		3.78	.95		4.73	5.75
0130	Encased, 1″ x 6″		640	.025		3.78	.95		4.73	5.75
0140	Grooved edge		660	.024		3.92	.92		4.84	5.85
0150	2″ x 6″		640	.025		5.10	.95		6.05	7.25

06 81 Composite Railings

06 81 10 – Encased Railings

06 81 10.10 Encased Composite Railings		Crew	Daily Output	Labor-Hours	Unit	Material	Labor	Equipment	Total	Total Incl O&P
0010	**ENCASED COMPOSITE RAILINGS**									
0100	Encased composite railing, 6′ long, 36″ high, incl. balusters	1 Carp	16	.500	Ea.	161	18.90		179.90	209
0110	42″ high, incl. balusters		16	.500		181	18.90		199.90	231
0120	8′ long, 36″ high, incl. balusters		12	.667		161	25		186	219
0130	42″ high, incl. balusters		12	.667		181	25		206	241
0140	Accessories, post sleeve, 4″ x 4″, 39″ long		32	.250		23.50	9.45		32.95	42
0150	96″ long		24	.333		89	12.60		101.60	119
0160	6″ x 6″, 39″ long		32	.250		48	9.45		57.45	69
0170	96″ long		24	.333		141	12.60		153.60	176
0180	Accessories, post skirt, 4″ x 4″		96	.083		4.25	3.15		7.40	9.95
0190	6″ x 6″		96	.083		5	3.15		8.15	10.75
0200	Post cap, 4″ x 4″, flat		48	.167		6.30	6.30		12.60	17.50
0210	Pyramid		48	.167		6.30	6.30		12.60	17.50
0220	Post cap, 6″ x 6″, flat		48	.167		10.20	6.30		16.50	22
0230	Pyramid		48	.167		10.20	6.30		16.50	22

Division Notes

		CREW	DAILY OUTPUT	LABOR-HOURS	UNIT	BARE COSTS MAT.	BARE COSTS LABOR	BARE COSTS EQUIP.	BARE COSTS TOTAL	TOTAL INCL O&P

Estimating Tips

07 10 00 Dampproofing and Waterproofing

- Be sure of the job specifications before pricing this subdivision. The difference in cost between waterproofing and dampproofing can be great. Waterproofing will hold back standing water. Dampproofing prevents the transmission of water vapor. Also included in this section are vapor retarding membranes.

07 20 00 Thermal Protection

- Insulation and fireproofing products are measured by area, thickness, volume or R-value. Specifications may give only what the specific R-value should be in a certain situation. The estimator may need to choose the type of insulation to meet that R-value.

07 30 00 Steep Slope Roofing
07 40 00 Roofing and Siding Panels

- Many roofing and siding products are bought and sold by the square. One square is equal to an area that measures 100 square feet.

  This simple change in unit of measure could create a large error if the estimator is not observant. Accessories necessary for a complete installation must be figured into any calculations for both material and labor.

07 50 00 Membrane Roofing
07 60 00 Flashing and Sheet Metal
07 70 00 Roofing and Wall Specialties and Accessories

- The items in these subdivisions compose a roofing system. No one component completes the installation, and all must be estimated. Built-up or single-ply membrane roofing systems are made up of many products and installation trades. Wood blocking at roof perimeters or penetrations, parapet coverings, reglets, roof drains, gutters, downspouts, sheet metal flashing, skylights, smoke vents, and roof hatches all need to be considered along with the roofing material. Several different installation trades will need to work together on the roofing system. Inherent difficulties in the scheduling and coordination of various trades must be accounted for when estimating labor costs.

07 90 00 Joint Protection

- To complete the weather-tight shell, the sealants and caulkings must be estimated. Where different materials meet—at expansion joints, at flashing penetrations, and at hundreds of other locations throughout a construction project—caulking and sealants provide another line of defense against water penetration. Often, an entire system is based on the proper location and placement of caulking or sealants. The detailed drawings that are included as part of a set of architectural plans show typical locations for these materials. When caulking or sealants are shown at typical locations, this means the estimator must include them for all the locations where this detail is applicable. Be careful to keep different types of sealants separate, and remember to consider backer rods and primers if necessary.

Reference Numbers

Reference numbers are shown at the beginning of some major classifications. These numbers refer to related items in the Reference Section. The reference information may be an estimating procedure, an alternate pricing method, or technical information.

Note: Not all subdivisions listed here necessarily appear. ■

07 01 Operation and Maint. of Thermal and Moisture Protection

07 01 50 – Maintenance of Membrane Roofing

07 01 50.10 Roof Coatings		Crew	Daily Output	Labor-Hours	Unit	2016 Bare Costs Material	Labor	Equipment	Total	Total Incl O&P
0010	**ROOF COATINGS**									
0012	Asphalt, brush grade, material only				Gal.	8.95			8.95	9.80
0200	Asphalt base, fibered aluminum coating G					8.25			8.25	9.10
0300	Asphalt primer, 5 gallon					7.50			7.50	8.25
0600	Coal tar pitch, 200 lb. barrels				Ton	1,250			1,250	1,375
0700	Tar roof cement, 5 gal. lots				Gal.	14.15			14.15	15.55
0800	Glass fibered roof & patching cement, 5 gallon				"	8.25			8.25	9.10
0900	Reinforcing glass membrane, 450 S.F./roll				Ea.	59.50			59.50	65.50
1000	Neoprene roof coating, 5 gal., 2 gal./sq.				Gal.	33.50			33.50	36.50
1100	Roof patch & flashing cement, 5 gallon					8.75			8.75	9.60
1200	Roof resaturant, glass fibered, 3 gal./sq.					8.85			8.85	9.75
1600	Reflective roof coating, white, elastomeric, approx. 50 S.F. per gal. G					17.20			17.20	18.90

07 01 90 – Maintenance of Joint Protection

07 01 90.81 Joint Sealant Replacement		Crew	Daily Output	Labor-Hours	Unit	Material	Labor	Equipment	Total	Total Incl O&P
0010	**JOINT SEALANT REPLACEMENT**									
0050	Control joints in concrete floors/slabs									
0100	Option 1 for joints with hard dry sealant									
0110	Step 1: Sawcut to remove 95% of old sealant									
0112	1/4" wide x 1/2" deep, with single saw blade	C-27	4800	.003	L.F.	.02	.12	.04	.18	.25
0114	3/8" wide x 3/4" deep, with single saw blade		4000	.004		.03	.14	.04	.21	.31
0116	1/2" wide x 1" deep, with double saw blades		3600	.004		.06	.16	.05	.27	.37
0118	3/4" wide x 1-1/2" deep, with double saw blades		3200	.005		.12	.18	.05	.35	.48
0120	Step 2: Water blast joint faces and edges	C-29	2500	.003			.09	.03	.12	.19
0130	Step 3: Air blast joint faces and edges	C-28	2000	.004			.14	.01	.15	.24
0140	Step 4: Sand blast joint faces and edges	E-11	2000	.016			.53	.12	.65	1.06
0150	Step 5: Air blast joint faces and edges	C-28	2000	.004			.14	.01	.15	.24
0200	Option 2 for joints with soft pliable sealant									
0210	Step 1: Plow joint with rectangular blade	B-62	2600	.009	L.F.		.30	.07	.37	.56
0220	Step 2: Sawcut to re-face joint faces									
0222	1/4" wide x 1/2" deep, with single saw blade	C-27	2400	.007	L.F.	.02	.24	.07	.33	.49
0224	3/8" wide x 3/4" deep, with single saw blade		2000	.008		.04	.28	.08	.40	.59
0226	1/2" wide x 1" deep, with double saw blades		1800	.009		.08	.32	.09	.49	.69
0228	3/4" wide x 1-1/2" deep, with double saw blades		1600	.010		.16	.36	.11	.63	.87
0230	Step 3: Water blast joint faces and edges	C-29	2500	.003			.09	.03	.12	.19
0240	Step 4: Air blast joint faces and edges	C-28	2000	.004			.14	.01	.15	.24
0250	Step 5: Sand blast joint faces and edges	E-11	2000	.016			.53	.12	.65	1.06
0260	Step 6: Air blast joint faces and edges	C-28	2000	.004			.14	.01	.15	.24
0290	For saw cutting new control joints, see Section 03 15 16.20									
8910	For backer rod, see Section 07 91 23.10									

07 05 Common Work Results for Thermal and Moisture Protection

07 05 05 – Selective Demolition for Thermal and Moisture Protection

07 05 05.10 Selective Demo., Thermal and Moist. Protection		Crew	Daily Output	Labor-Hours	Unit	Material	Labor	Equipment	Total	Total Incl O&P
0010	**SELECTIVE DEMO., THERMAL AND MOISTURE PROTECTION**									
0020	Caulking/sealant, to 1" x 1" joint R024119-10	1 Clab	600	.013	L.F.		.39		.39	.65
0120	Downspouts, including hangers		350	.023	"		.67		.67	1.12
0220	Flashing, sheet metal		290	.028	S.F.		.81		.81	1.35
0420	Gutters, aluminum or wood, edge hung		240	.033	L.F.		.97		.97	1.63
0520	Built-in		100	.080	"		2.34		2.34	3.91
0620	Insulation, air/vapor barrier		3500	.002	S.F.		.07		.07	.11

07 05 05 – Selective Demolition for Thermal and Moisture Protection

07 05 05.10 Selective Demo., Thermal and Moist. Protection		Crew	Daily Output	Labor-Hours	Unit	Material	2016 Bare Costs Labor	Equipment	Total	Total Incl O&P
0670	Batts or blankets	1 Clab	1400	.006	C.F.		.17		.17	.23
0720	Foamed or sprayed in place	2 Clab	1000	.016	B.F.		.47		.47	.78
0770	Loose fitting	1 Clab	3000	.003	C.F.		.08		.08	.13
0870	Rigid board		3450	.002	B.F.		.07		.07	.11
1120	Roll roofing, cold adhesive		12	.667	Sq.		19.45		19.45	32.50
1170	Roof accessories, adjustable metal chimney flashing		9	.889	Ea.		26		26	43.50
1325	Plumbing vent flashing		32	.250	"		7.30		7.30	12.25
1375	Ridge vent strip, aluminum		310	.026	L.F.		.75		.75	1.26
1620	Skylight to 10 S.F.		8	1	Ea.		29		29	49
2120	Roof edge, aluminum soffit and fascia		570	.014	L.F.		.41		.41	.69
2170	Concrete coping, up to 12" wide	2 Clab	160	.100			2.92		2.92	4.89
2220	Drip edge	1 Clab	1000	.008			.23		.23	.39
2270	Gravel stop		950	.008			.25		.25	.41
2370	Sheet metal coping, up to 12" wide		240	.033			.97		.97	1.63
2470	Roof insulation board, over 2" thick	B-2	7800	.005	B.F.		.15		.15	.25
2520	Up to 2" thick	"	3900	.010	S.F.		.30		.30	.51
2620	Roof ventilation, louvered gable vent	1 Clab	16	.500	Ea.		14.60		14.60	24.50
2670	Remove, roof hatch	G-3	15	2.133			76.50		76.50	128
2675	Rafter vents	1 Clab	960	.008			.24		.24	.41
2720	Soffit vent and/or fascia vent		575	.014	L.F.		.41		.41	.68
2775	Soffit vent strip, aluminum, 3" to 4" wide		160	.050			1.46		1.46	2.45
2820	Roofing accessories, shingle moulding, to 1" x 4"		1600	.005			.15		.15	.24
2870	Cant strip	B-2	2000	.020			.59		.59	.99
2920	Concrete block walkway	1 Clab	230	.035			1.02		1.02	1.70
3070	Roofing, felt paper, 15#		70	.114	Sq.		3.34		3.34	5.60
3125	#30 felt		30	.267	"		7.80		7.80	13.05
3170	Asphalt shingles, 1 layer	B-2	3500	.011	S.F.		.34		.34	.57
3180	2 layers		1750	.023	"		.68		.68	1.13
3370	Modified bitumen		26	1.538	Sq.		45.50		45.50	76.50
3420	Built-up, no gravel, 3 ply		25	1.600			47.50		47.50	79.50
3470	4 ply		21	1.905			56.50		56.50	94.50
3620	5 ply		1600	.025	S.F.		.74		.74	1.24
3720	5 ply, with gravel		890	.045			1.33		1.33	2.23
3725	Loose gravel removal		5000	.008			.24		.24	.40
3730	Embedded gravel removal		2000	.020			.59		.59	.99
3870	Fiberglass sheet		1200	.033			.99		.99	1.65
4120	Slate shingles		1900	.021			.62		.62	1.04
4170	Ridge shingles, clay or slate		2000	.020	L.F.		.59		.59	.99
4320	Single ply membrane, attached at seams		52	.769	Sq.		23		23	38
4370	Ballasted		75	.533			15.80		15.80	26.50
4420	Fully adhered		39	1.026			30.50		30.50	51
4550	Roof hatch, 2'-6" x 3'-0"	1 Clab	10	.800	Ea.		23.50		23.50	39
4670	Wood shingles	B-2	2200	.018	S.F.		.54		.54	.90
4820	Sheet metal roofing	"	2150	.019			.55		.55	.92
4970	Siding, horizontal wood clapboards	1 Clab	380	.021			.61		.61	1.03
5025	Exterior insulation finish system	"	120	.067			1.95		1.95	3.26
5070	Tempered hardboard, remove and reset	1 Carp	380	.021			.80		.80	1.33
5120	Tempered hardboard sheet siding	"	375	.021			.81		.81	1.35
5170	Metal, corner strips	1 Clab	850	.009	L.F.		.27		.27	.46
5225	Horizontal strips		444	.018	S.F.		.53		.53	.88
5320	Vertical strips		400	.020			.58		.58	.98
5520	Wood shingles		350	.023			.67		.67	1.12
5620	Stucco siding		360	.022			.65		.65	1.09

07 05 Common Work Results for Thermal and Moisture Protection

07 05 05 – Selective Demolition for Thermal and Moisture Protection

07 05 05.10 Selective Demo., Thermal and Moist. Protection		Crew	Daily Output	Labor-Hours	Unit	2016 Bare Costs Material	Labor	Equipment	Total	Total Incl O&P
5670	Textured plywood	1 Clab	725	.011	S.F.		.32		.32	.54
5720	Vinyl siding		510	.016			.46		.46	.77
5770	Corner strips		900	.009	L.F.		.26		.26	.43
5870	Wood, boards, vertical		400	.020	S.F.		.58		.58	.98
5920	Waterproofing, protection/drain board	2 Clab	3900	.004	B.F.		.12		.12	.20
5970	Over 1/2" thick		1750	.009	S.F.		.27		.27	.45
6020	To 1/2" thick		2000	.008	"		.23		.23	.39

07 11 Dampproofing

07 11 13 – Bituminous Dampproofing

07 11 13.10 Bituminous Asphalt Coating

		Crew	Daily Output	Labor-Hours	Unit	Material	Labor	Equipment	Total	Total Incl O&P
0010	**BITUMINOUS ASPHALT COATING**									
0030	Brushed on, below grade, 1 coat	1 Rofc	665	.012	S.F.	.22	.39		.61	.96
0100	2 coat		500	.016		.45	.51		.96	1.44
0300	Sprayed on, below grade, 1 coat		830	.010		.22	.31		.53	.82
0400	2 coat		500	.016		.44	.51		.95	1.43
0500	Asphalt coating, with fibers				Gal.	8.25			8.25	9.10
0600	Troweled on, asphalt with fibers, 1/16" thick	1 Rofc	500	.016	S.F.	.36	.51		.87	1.34
0700	1/8" thick		400	.020		.63	.64		1.27	1.88
1000	1/2" thick		350	.023		2.06	.73		2.79	3.62

07 11 16 – Cementitious Dampproofing

07 11 16.20 Cementitious Parging

		Crew	Daily Output	Labor-Hours	Unit	Material	Labor	Equipment	Total	Total Incl O&P
0010	**CEMENTITIOUS PARGING**									
0020	Portland cement, 2 coats, 1/2" thick	D-1	250	.064	S.F.	.35	2.10		2.45	3.89
0100	Waterproofed Portland cement, 1/2" thick, 2 coats	"	250	.064	"	3.95	2.10		6.05	7.85

07 19 Water Repellents

07 19 19 – Silicone Water Repellents

07 19 19.10 Silicone Based Water Repellents

		Crew	Daily Output	Labor-Hours	Unit	Material	Labor	Equipment	Total	Total Incl O&P
0010	**SILICONE BASED WATER REPELLENTS**									
0020	Water base liquid, roller applied	2 Rofc	7000	.002	S.F.	.53	.07		.60	.72
0200	Silicone or stearate, sprayed on CMU, 1 coat	1 Rofc	4000	.002		.37	.06		.43	.52
0300	2 coats	"	3000	.003		.73	.09		.82	.97

07 21 Thermal Insulation

07 21 13 – Board Insulation

07 21 13.10 Rigid Insulation

			Crew	Daily Output	Labor-Hours	Unit	Material	Labor	Equipment	Total	Total Incl O&P
0010	**RIGID INSULATION**, for walls										
0020	Fiberboard, 3/4" thick, R2.08	G	1 Carp	1100	.007	S.F.	.42	.27		.69	.92
0025	1" thick, R2.78	G		800	.010		.55	.38		.93	1.24
0030	2" thick, R5.26	G		730	.011		1.03	.41		1.44	1.82
0040	Fiberglass, 1.5#/C.F., unfaced, 1" thick, R4.1	G		1000	.008		.29	.30		.59	.83
0060	1-1/2" thick, R6.2	G		1000	.008		.38	.30		.68	.93
0080	2" thick, R8.3	G		1000	.008		.49	.30		.79	1.05
0120	3" thick, R12.4	G		800	.010		.56	.38		.94	1.25
0370	3#/C.F., unfaced, 1" thick, R4.3	G		1000	.008		.52	.30		.82	1.08
0390	1-1/2" thick, R6.5	G		1000	.008		.78	.30		1.08	1.37

07 21 Thermal Insulation

07 21 13 – Board Insulation

07 21 13.10 Rigid Insulation			Crew	Daily Output	Labor-Hours	Unit	2016 Bare Costs Material	Labor	Equipment	Total	Total Incl O&P
0400	2" thick, R8.7	G	1 Carp	890	.009	S.F.	1.05	.34		1.39	1.73
0420	2-1/2" thick, R10.9	G		800	.010		1.10	.38		1.48	1.84
0440	3" thick, R13	G		800	.010		1.59	.38		1.97	2.38
0520	Foil faced, 1" thick, R4.3	G		1000	.008		.90	.30		1.20	1.50
0540	1-1/2" thick, R6.5	G		1000	.008		1.35	.30		1.65	2
0560	2" thick, R8.7	G		890	.009		1.69	.34		2.03	2.43
0580	2-1/2" thick, R10.9	G		800	.010		1.98	.38		2.36	2.81
0600	3" thick, R13	G		800	.010		2.18	.38		2.56	3.03
1600	Isocyanurate, 4' x 8' sheet, foil faced, both sides										
1610	1/2" thick	G	1 Carp	800	.010	S.F.	.31	.38		.69	.97
1620	5/8" thick	G		800	.010		.33	.38		.71	.99
1630	3/4" thick	G		800	.010		.38	.38		.76	1.05
1640	1" thick	G		800	.010		.51	.38		.89	1.19
1650	1-1/2" thick	G		730	.011		.59	.41		1	1.34
1660	2" thick	G		730	.011		.75	.41		1.16	1.52
1670	3" thick	G		730	.011		1.85	.41		2.26	2.73
1680	4" thick	G		730	.011		2.10	.41		2.51	3
1700	Perlite, 1" thick, R2.77	G		800	.010		.44	.38		.82	1.11
1750	2" thick, R5.55	G		730	.011		.77	.41		1.18	1.54
1900	Extruded polystyrene, 25 PSI compressive strength, 1" thick, R5	G		800	.010		.56	.38		.94	1.25
1940	2" thick R10	G		730	.011		1.13	.41		1.54	1.93
1960	3" thick, R15	G		730	.011		1.58	.41		1.99	2.43
2100	Expanded polystyrene, 1" thick, R3.85	G		800	.010		.26	.38		.64	.92
2120	2" thick, R7.69	G		730	.011		.52	.41		.93	1.26
2140	3" thick, R11.49	G		730	.011		.78	.41		1.19	1.55
2360	Fiberboard, low density, 1/2" thick, R1.39	G		750	.011		.33	.40		.73	1.04
2400	Wood fiber, 1" thick, R3.85	G		1000	.008		.55	.30		.85	1.12
2410	2" thick, R7.7	G		1000	.008		1.03	.30		1.33	1.64
2680	Mineral fiberboard, rigid, 1" thick, R4.2	G		800	.010		.55	.38		.93	1.24
2700	2" thick, R8.4	G		730	.011		1.03	.41		1.44	1.82

07 21 13.13 Foam Board Insulation			Crew	Daily Output	Labor-Hours	Unit	Material	Labor	Equipment	Total	Total Incl O&P
0010	**FOAM BOARD INSULATION**										
0600	Polystyrene, expanded, 1" thick, R4	G	1 Carp	680	.012	S.F.	.26	.44		.70	1.03
0700	2" thick, R8	G	"	675	.012	"	.52	.45		.97	1.32

07 21 16 – Blanket Insulation

07 21 16.10 Blanket Insulation for Floors/Ceilings			Crew	Daily Output	Labor-Hours	Unit	Material	Labor	Equipment	Total	Total Incl O&P
0010	**BLANKET INSULATION FOR FLOORS/CEILINGS**										
0020	Including spring type wire fasteners										
2000	Fiberglass, blankets or batts, paper or foil backing										
2100	3-1/2" thick, R13	G	1 Carp	700	.011	S.F.	.40	.43		.83	1.16
2150	6-1/4" thick, R19	G		600	.013		.51	.50		1.01	1.40
2210	9-1/2" thick, R30	G		500	.016		.73	.60		1.33	1.81
2220	12" thick, R38	G		475	.017		1.05	.64		1.69	2.23
3000	Unfaced, 3-1/2" thick, R13	G		600	.013		.33	.50		.83	1.20
3010	6-1/4" thick, R19	G		500	.016		.39	.60		.99	1.44
3020	9-1/2" thick, R30	G		450	.018		.60	.67		1.27	1.79
3030	12" thick, R38	G		425	.019		.75	.71		1.46	2.02

07 21 16.20 Blanket Insulation for Walls			Crew	Daily Output	Labor-Hours	Unit	Material	Labor	Equipment	Total	Total Incl O&P
0010	**BLANKET INSULATION FOR WALLS**										
0020	Kraft faced fiberglass, 3-1/2" thick, R11, 15" wide	G	1 Carp	1350	.006	S.F.	.31	.22		.53	.72
0030	23" wide	G		1600	.005		.31	.19		.50	.66
0060	R13, 11" wide	G		1150	.007		.33	.26		.59	.80

07 21 Thermal Insulation

07 21 16 – Blanket Insulation

07 21 16.20 Blanket Insulation for Walls			Crew	Daily Output	Labor-Hours	Unit	2016 Bare Costs Material	Labor	Equipment	Total	Total Incl O&P
0110	R15, 11" wide	G	1 Carp	1150	.007	S.F.	.49	.26		.75	.98
0140	6" thick, R19, 11" wide	G		1150	.007		.44	.26		.70	.92
0182	R21, 11" wide	G		1150	.007		.64	.26		.90	1.14
0184	15" wide	G		1350	.006		.64	.22		.86	1.08
0186	23" wide	G		1600	.005		.64	.19		.83	1.02
0188	9" thick, R30, 11" wide	G		985	.008		.73	.31		1.04	1.31
0200	15" wide	G		1150	.007		.73	.26		.99	1.24
0230	12" thick, R38, 11" wide	G		985	.008		1.05	.31		1.36	1.67
0240	15" wide	G		1150	.007		1.05	.26		1.31	1.60
0410	Foil faced fiberglass, 3-1/2" thick, R13, 11" wide	G		1150	.007		.47	.26		.73	.96
0420	15" wide	G		1350	.006		.47	.22		.69	.90
0442	R15, 11" wide	G		1150	.007		.48	.26		.74	.97
0444	15" wide	G		1350	.006		.48	.22		.70	.91
0448	6" thick, R19, 11" wide	G		1150	.007		.62	.26		.88	1.12
0460	15" wide	G		1350	.006		.62	.22		.84	1.06
0482	R21, 11" wide	G		1150	.007		.64	.26		.90	1.14
0484	15" wide	G		1350	.006		.64	.22		.86	1.08
0488	9" thick, R30, 11" wide	G		985	.008		.94	.31		1.25	1.54
0500	15" wide	G		1150	.007		.94	.26		1.20	1.47
0560	12" thick, R38, 11" wide	G		985	.008		1.15	.31		1.46	1.78
0570	15" wide	G		1150	.007		1.15	.26		1.41	1.71
0620	Unfaced fiberglass, 3-1/2" thick, R13, 11" wide	G		1150	.007		.33	.26		.59	.80
0820	15" wide	G		1350	.006		.33	.22		.55	.74
0832	R15, 11" wide	G		1150	.007		.46	.26		.72	.95
0838	6" thick, R19, 11" wide	G		1150	.007		.39	.26		.65	.87
0860	15" wide	G		1150	.007		.39	.26		.65	.87
0882	R21, 11" wide	G		1150	.007		.50	.26		.76	.99
0886	15" wide	G		1350	.006		.50	.22		.72	.93
0890	9" thick, R30, 11" wide	G		985	.008		.60	.31		.91	1.17
0900	15" wide	G		1150	.007		.60	.26		.86	1.10
0930	12" thick, R38, 11" wide	G		985	.008		.75	.31		1.06	1.34
0940	15" wide	G		1000	.008		.75	.30		1.05	1.34
1300	Wall or ceiling insulation, mineral wool batts										
1320	3-1/2" thick, R15	G	1 Carp	1600	.005	S.F.	.67	.19		.86	1.06
1340	5-1/2" thick, R23	G		1600	.005		1.05	.19		1.24	1.48
1380	7-1/4" thick, R30	G		1350	.006		1.39	.22		1.61	1.91
1700	Non-rigid insul, recycled blue cotton fiber, unfaced batts, R13, 16" wide	G		1600	.005		1.01	.19		1.20	1.43
1710	R19, 16" wide	G		1600	.005		1.38	.19		1.57	1.84
1850	Friction fit wire insulation supports, 16" O.C.			960	.008	Ea.	.07	.31		.38	.61

07 21 19 – Foamed In Place Insulation

07 21 19.10 Masonry Foamed In Place Insulation			Crew	Daily Output	Labor-Hours	Unit	Material	Labor	Equipment	Total	Total Incl O&P
0010	**MASONRY FOAMED IN PLACE INSULATION**										
0100	Amino-plast foam, injected into block core, 6" block	G	G-2A	6000	.004	Ea.	.16	.11	.12	.39	.51
0110	8" block	G		5000	.005		.20	.14	.14	.48	.62
0120	10" block	G		4000	.006		.25	.17	.18	.60	.78
0130	12" block	G		3000	.008		.33	.23	.24	.80	1.03
0140	Injected into cavity wall	G		13000	.002	B.F.	.06	.05	.05	.16	.21
0150	Preparation, drill holes into mortar joint every 4 VLF, 5/8" dia		1 Clab	960	.008	Ea.		.24		.24	.41
0160	7/8" dia			680	.012			.34		.34	.58
0170	Patch drilled holes, 5/8" diameter			1800	.004		.04	.13		.17	.26
0180	7/8" diameter			1200	.007		.05	.19		.24	.39

07 21 Thermal Insulation

07 21 23 – Loose-Fill Insulation

07 21 23.10 Poured Loose-Fill Insulation			Crew	Daily Output	Labor-Hours	Unit	Material	2016 Bare Costs Labor	Equipment	Total	Total Incl O&P
0010	**POURED LOOSE-FILL INSULATION**										
0020	Cellulose fiber, R3.8 per inch	G	1 Carp	200	.040	C.F.	.69	1.51		2.20	3.29
0021	4" thick	G		1000	.008	S.F.	.16	.30		.46	.69
0022	6" thick	G		800	.010	"	.28	.38		.66	.94
0080	Fiberglass wool, R4 per inch	G		200	.040	C.F.	.55	1.51		2.06	3.13
0081	4" thick	G		600	.013	S.F.	.19	.50		.69	1.05
0082	6" thick	G		400	.020	"	.26	.76		1.02	1.56
0100	Mineral wool, R3 per inch	G		200	.040	C.F.	.45	1.51		1.96	3.03
0101	4" thick	G		600	.013	S.F.	.15	.50		.65	1
0102	6" thick	G		400	.020	"	.23	.76		.99	1.52
0300	Polystyrene, R4 per inch	G		200	.040	C.F.	1.60	1.51		3.11	4.29
0301	4" thick	G		600	.013	S.F.	.53	.50		1.03	1.42
0302	6" thick	G		400	.020	"	.80	.76		1.56	2.15
0400	Perlite, R2.78 per inch	G		200	.040	C.F.	5.25	1.51		6.76	8.35
0401	4" thick	G		1000	.008	S.F.	1.75	.30		2.05	2.43
0402	6" thick	G	↓	800	.010	"	2.63	.38		3.01	3.52

07 21 23.20 Masonry Loose-Fill Insulation			Crew	Daily Output	Labor-Hours	Unit	Material	Labor	Equipment	Total	Total Incl O&P
0010	**MASONRY LOOSE-FILL INSULATION**, vermiculite or perlite										
0100	In cores of concrete block, 4" thick wall, .115 C.F./S.F.	G	D-1	4800	.003	S.F.	.60	.11		.71	.84
0700	Foamed in place, urethane in 2-5/8" cavity	G	G-2A	1035	.023		1.38	.66	.69	2.73	3.46
0800	For each 1" added thickness, add	G	"	2372	.010	↓	.53	.29	.30	1.12	1.42

07 21 26 – Blown Insulation

07 21 26.10 Blown Insulation			Crew	Daily Output	Labor-Hours	Unit	Material	Labor	Equipment	Total	Total Incl O&P
0010	**BLOWN INSULATION** Ceilings, with open access										
0020	Cellulose, 3-1/2" thick, R13	G	G-4	5000	.005	S.F.	.24	.14	.08	.46	.59
0030	5-3/16" thick, R19	G		3800	.006		.35	.19	.11	.65	.83
0050	6-1/2" thick, R22	G		3000	.008		.45	.24	.14	.83	1.05
0100	8-11/16" thick, R30	G		2600	.009		.61	.28	.16	1.05	1.30
0120	10-7/8" thick, R38	G		1800	.013		.78	.40	.23	1.41	1.77
1000	Fiberglass, 5.5" thick, R11	G		3800	.006		.19	.19	.11	.49	.64
1050	6" thick, R12	G		3000	.008		.26	.24	.14	.64	.84
1100	8.8" thick, R19	G		2200	.011		.33	.33	.19	.85	1.12
1200	10" thick, R22	G		1800	.013		.38	.40	.23	1.01	1.34
1300	11.5" thick, R26	G		1500	.016		.46	.48	.27	1.21	1.60
1350	13" thick, R30	G		1400	.017		.53	.51	.29	1.33	1.76
1450	16" thick, R38	G		1145	.021		.68	.63	.36	1.67	2.18
1500	20" thick, R49	G	↓	920	.026	↓	.89	.78	.45	2.12	2.78

07 21 27 – Reflective Insulation

07 21 27.10 Reflective Insulation Options			Crew	Daily Output	Labor-Hours	Unit	Material	Labor	Equipment	Total	Total Incl O&P
0010	**REFLECTIVE INSULATION OPTIONS**										
0020	Aluminum foil on reinforced scrim	G	1 Carp	19	.421	C.S.F.	14.90	15.90		30.80	43
0100	Reinforced with woven polyolefin	G		19	.421		22	15.90		37.90	50.50
0500	With single bubble air space, R8.8	G		15	.533		26	20		46	62.50
0600	With double bubble air space, R9.8	G	↓	15	.533	↓	31	20		51	68

07 21 29 – Sprayed Insulation

07 21 29.10 Sprayed-On Insulation			Crew	Daily Output	Labor-Hours	Unit	Material	Labor	Equipment	Total	Total Incl O&P
0010	**SPRAYED-ON INSULATION**										
0300	Closed cell, spray polyurethane foam, 2 pounds per cubic foot density										
0310	1" thick	G	G-2A	6000	.004	S.F.	.53	.11	.12	.76	.91
0320	2" thick	G	↓	3000	.008	↓	1.05	.23	.24	1.52	1.83

07 21 Thermal Insulation

07 21 29 – Sprayed Insulation

07 21 29.10 Sprayed-On Insulation			Crew	Daily Output	Labor-Hours	Unit	Material	Labor	Equipment	Total	Total Incl O&P
							2016 Bare Costs				
0330	3" thick	G	G-2A	2000	.012	S.F.	1.58	.34	.36	2.28	2.73
0335	3-1/2" thick	G		1715	.014		1.84	.40	.42	2.66	3.19
0340	4" thick	G		1500	.016		2.10	.46	.48	3.04	3.64
0350	5" thick	G		1200	.020		2.63	.57	.59	3.79	4.56
0355	5-1/2" thick	G		1090	.022		2.89	.63	.65	4.17	5
0360	6" thick	G		1000	.024		3.15	.68	.71	4.54	5.45

07 22 Roof and Deck Insulation

07 22 16 – Roof Board Insulation

07 22 16.10 Roof Deck Insulation			Crew	Daily Output	Labor-Hours	Unit	Material	Labor	Equipment	Total	Total Incl O&P
0010	**ROOF DECK INSULATION**, fastening excluded										
0016	Asphaltic cover board, fiberglass lined, 1/8" thick		1 Rofc	1400	.006	S.F.	.47	.18		.65	.86
0018	1/4" thick			1400	.006		.94	.18		1.12	1.37
0020	Fiberboard low density, 1/2" thick R1.39	G		1300	.006		.33	.20		.53	.72
0030	1" thick R2.78	G		1040	.008		.55	.25		.80	1.07
0080	1-1/2" thick R4.17	G		1040	.008		.84	.25		1.09	1.38
0100	2" thick R5.56	G		1040	.008		1.10	.25		1.35	1.67
0110	Fiberboard high density, 1/2" thick R1.3	G		1300	.006		.30	.20		.50	.69
0120	1" thick R2.5	G		1040	.008		.56	.25		.81	1.08
0130	1-1/2" thick R3.8	G		1040	.008		.86	.25		1.11	1.41
0200	Fiberglass, 3/4" thick R2.78	G		1300	.006		.61	.20		.81	1.03
0400	15/16" thick R3.70	G		1300	.006		.81	.20		1.01	1.25
0460	1-1/16" thick R4.17	G		1300	.006		1.01	.20		1.21	1.47
0600	1-5/16" thick R5.26	G		1300	.006		1.37	.20		1.57	1.87
0650	2-1/16" thick R8.33	G		1040	.008		1.45	.25		1.70	2.06
0700	2-7/16" thick R10	G		1040	.008		1.68	.25		1.93	2.31
0800	Gypsum cover board, fiberglass mat facer, 1/4" thick			1400	.006		.47	.18		.65	.86
0810	1/2" thick			1300	.006		.57	.20		.77	.99
0820	5/8" thick			1200	.007		.61	.21		.82	1.06
0830	Primed fiberglass mat facer, 1/4" thick			1400	.006		.48	.18		.66	.87
0840	1/2" thick			1300	.006		.57	.20		.77	.99
0850	5/8" thick			1200	.007		.60	.21		.81	1.05
1650	Perlite, 1/2" thick R1.32	G		1365	.006		.26	.19		.45	.64
1655	3/4" thick R2.08	G		1040	.008		.37	.25		.62	.87
1660	1" thick R2.78	G		1040	.008		.53	.25		.78	1.04
1670	1-1/2" thick R4.17	G		1040	.008		.78	.25		1.03	1.32
1680	2" thick R5.56	G		910	.009		1.06	.28		1.34	1.69
1685	2-1/2" thick R6.67	G		910	.009		1.40	.28		1.68	2.06
1690	Tapered for drainage	G		1040	.008	B.F.	1.05	.25		1.30	1.62
1700	Polyisocyanurate, 2#/C.F. density, 3/4" thick	G		1950	.004	S.F.	.46	.13		.59	.75
1705	1" thick	G		1820	.004		.47	.14		.61	.78
1715	1-1/2" thick	G		1625	.005		.64	.16		.80	.99
1725	2" thick	G		1430	.006		.82	.18		1	1.23
1735	2-1/2" thick	G		1365	.006		1.10	.19		1.29	1.56
1745	3" thick	G		1300	.006		1.22	.20		1.42	1.70
1755	3-1/2" thick	G		1300	.006		1.90	.20		2.10	2.45
1765	Tapered for drainage	G		1820	.004	B.F.	.59	.14		.73	.91
1900	Extruded Polystyrene										
1910	15 PSI compressive strength, 1" thick, R5	G	1 Rofc	1950	.004	S.F.	.60	.13		.73	.90
1920	2" thick, R10	G		1625	.005		.78	.16		.94	1.15
1930	3" thick, R15	G		1300	.006		1.56	.20		1.76	2.08

07 22 16.10 Roof Deck Insulation			Crew	Daily Output	Labor-Hours	Unit	Material	2016 Bare Costs Labor	Equipment	Total	Total Incl O&P
1932	4" thick, R20	G	1 Rofc	1300	.006	S.F.	2.10	.20		2.30	2.67
1934	Tapered for drainage	G		1950	.004	B.F.	.62	.13		.75	.92
1940	25 PSI compressive strength, 1" thick, R5	G		1950	.004	S.F.	.69	.13		.82	1
1942	2" thick, R10	G		1625	.005		1.31	.16		1.47	1.73
1944	3" thick, R15	G		1300	.006		2	.20		2.20	2.56
1946	4" thick, R20	G		1300	.006		2.76	.20		2.96	3.40
1948	Tapered for drainage	G		1950	.004	B.F.	.61	.13		.74	.91
1950	40 PSI compressive strength, 1" thick, R5	G		1950	.004	S.F.	.53	.13		.66	.82
1952	2" thick, R10	G		1625	.005		1.01	.16		1.17	1.40
1954	3" thick, R15	G		1300	.006		1.46	.20		1.66	1.96
1956	4" thick, R20	G		1300	.006		1.91	.20		2.11	2.46
1958	Tapered for drainage	G		1820	.004	B.F.	.76	.14		.90	1.10
1960	60 PSI compressive strength, 1" thick, R5	G		1885	.004	S.F.	.74	.14		.88	1.06
1962	2" thick, R10	G		1560	.005		1.41	.16		1.57	1.85
1964	3" thick, R15	G		1270	.006		2.29	.20		2.49	2.89
1966	4" thick, R20	G		1235	.006		2.85	.21		3.06	3.51
1968	Tapered for drainage	G		1820	.004	B.F.	.96	.14		1.10	1.32
2010	Expanded polystyrene, 1#/C.F. density, 3/4" thick, R2.89	G		1950	.004	S.F.	.20	.13		.33	.45
2020	1" thick, R3.85	G		1950	.004		.26	.13		.39	.53
2100	2" thick, R7.69	G		1625	.005		.52	.16		.68	.86
2110	3" thick, R11.49	G		1625	.005		.78	.16		.94	1.15
2120	4" thick, R15.38	G		1625	.005		1.04	.16		1.20	1.43
2130	5" thick, R19.23	G		1495	.005		1.30	.17		1.47	1.75
2140	6" thick, R23.26	G		1495	.005		1.56	.17		1.73	2.04
2150	Tapered for drainage	G		1950	.004	B.F.	.53	.13		.66	.82
2400	Composites with 2" EPS										
2410	1" fiberboard	G	1 Rofc	1325	.006	S.F.	1.40	.19		1.59	1.90
2420	7/16" oriented strand board	G		1040	.008		1.13	.25		1.38	1.70
2430	1/2" plywood	G		1040	.008		1.39	.25		1.64	1.99
2440	1" perlite	G		1040	.008		1.14	.25		1.39	1.71
2450	Composites with 1-1/2" polyisocyanurate										
2460	1" fiberboard	G	1 Rofc	1040	.008	S.F.	1.19	.25		1.44	1.77
2470	1" perlite	G		1105	.007		1.06	.23		1.29	1.60
2480	7/16" oriented strand board	G		1040	.008		.93	.25		1.18	1.48
3000	Fastening alternatives, coated screws, 2" long			3744	.002	Ea.	.06	.07		.13	.20
3010	4" long			3120	.003		.11	.08		.19	.27
3020	6" long			2675	.003		.18	.10		.28	.38
3030	8" long			2340	.003		.27	.11		.38	.50
3040	10" long			1872	.004		.45	.14		.59	.75
3050	Pre-drill and drive wedge spike, 2-1/2"			1248	.006		.39	.21		.60	.81
3060	3-1/2"			1101	.007		.52	.23		.75	1
3070	4-1/2"			936	.009		.64	.27		.91	1.21
3075	3" galvanized deck plates			7488	.001		.08	.03		.11	.15
3080	Spot mop asphalt		G-1	295	.190	Sq.	5.55	5.70	1.96	13.21	18.80
3090	Full mop asphalt		"	192	.292		11.10	8.80	3.01	22.91	32
3110	Low-rise polyurethane adhesive, from 5 gallon kit, 12" OC beads		1 Rofc	45	.178		33	5.70		38.70	47
3120	6" OC beads			32	.250		66	8.05		74.05	87.50
3130	4" OC beads			30	.267		99	8.55		107.55	125

07 24 Exterior Insulation and Finish Systems

07 24 13 – Polymer-Based Exterior Insulation and Finish System

07 24 13.10 Exterior Insulation and Finish Systems			Crew	Daily Output	Labor-Hours	Unit	2016 Bare Costs Material	Labor	Equipment	Total	Total Incl O&P
0010	EXTERIOR INSULATION AND FINISH SYSTEMS										
0095	Field applied, 1" EPS insulation	G	J-1	390	.103	S.F.	1.93	3.37	.36	5.66	8
0100	With 1/2" cement board sheathing	G		268	.149		2.71	4.90	.52	8.13	11.55
0105	2" EPS insulation	G		390	.103		2.19	3.37	.36	5.92	8.30
0110	With 1/2" cement board sheathing	G		268	.149		2.97	4.90	.52	8.39	11.85
0115	3" EPS insulation	G		390	.103		2.45	3.37	.36	6.18	8.60
0120	With 1/2" cement board sheathing	G		268	.149		3.23	4.90	.52	8.65	12.15
0125	4" EPS insulation	G		390	.103		2.71	3.37	.36	6.44	8.90
0130	With 1/2" cement board sheathing	G		268	.149		4.28	4.90	.52	9.70	13.30
0140	Premium finish add			1265	.032		.35	1.04	.11	1.50	2.21
0150	Heavy duty reinforcement add			914	.044		.85	1.44	.15	2.44	3.46
0160	2.5#/S.Y. metal lath substrate add		1 Lath	75	.107	S.Y.	2.82	3.94		6.76	9.45
0170	3.4#/S.Y. metal lath substrate add		"	75	.107	"	4.18	3.94		8.12	10.95
0180	Color or texture change		J-1	1265	.032	S.F.	.80	1.04	.11	1.95	2.70
0190	With substrate leveling base coat		1 Plas	530	.015		.80	.53		1.33	1.74
0210	With substrate sealing base coat		1 Pord	1224	.007		.12	.21		.33	.47
0370	V groove shape in panel face					L.F.	.65			.65	.72
0380	U groove shape in panel face					"	.85			.85	.94
0440	For higher than one story, add							25%			

07 25 Weather Barriers

07 25 10 – Weather Barriers or Wraps

07 25 10.10 Weather Barriers		Crew	Daily Output	Labor-Hours	Unit	Material	Labor	Equipment	Total	Total Incl O&P
0010	WEATHER BARRIERS									
0400	Asphalt felt paper, 15#	1 Carp	37	.216	Sq.	5.30	8.15		13.45	19.55
0401	Per square foot	"	3700	.002	S.F.	.05	.08		.13	.20
0450	Housewrap, exterior, spun bonded polypropylene									
0470	Small roll	1 Carp	3800	.002	S.F.	.15	.08		.23	.30
0480	Large roll	"	4000	.002	"	.14	.08		.22	.28
2100	Asphalt felt roof deck vapor barrier, class 1 metal decks	1 Rofc	37	.216	Sq.	21.50	6.95		28.45	37
2200	For all other decks	"	37	.216		16.40	6.95		23.35	31
2800	Asphalt felt, 50% recycled content, 15 lb., 4 sq. per roll	1 Carp	36	.222		5.50	8.40		13.90	20
2810	30 lb., 2 sq. per roll	"	36	.222		10.85	8.40		19.25	26
3000	Building wrap, spunbonded polyethylene	2 Carp	8000	.002	S.F.	.16	.08		.24	.31

07 26 Vapor Retarders

07 26 10 – Above-Grade Vapor Retarders

07 26 10.10 Vapor Retarders			Crew	Daily Output	Labor-Hours	Unit	Material	Labor	Equipment	Total	Total Incl O&P
0010	VAPOR RETARDERS										
0020	Aluminum and kraft laminated, foil 1 side	G	1 Carp	37	.216	Sq.	11.95	8.15		20.10	27
0100	Foil 2 sides	G		37	.216		14	8.15		22.15	29
0600	Polyethylene vapor barrier, standard, 2 mil	G		37	.216		1.56	8.15		9.71	15.40
0700	4 mil	G		37	.216		2.55	8.15		10.70	16.50
0900	6 mil	G		37	.216		3.70	8.15		11.85	17.75
1200	10 mil	G		37	.216		7.75	8.15		15.90	22.50
1800	Reinf. waterproof, 2 mil polyethylene backing, 1 side			37	.216		6.10	8.15		14.25	20.50
1900	2 sides			37	.216		8	8.15		16.15	22.50

07 27 Air Barriers

07 27 13 – Modified Bituminous Sheet Air Barriers

07 27 13.10 Modified Bituminous Sheet Air Barrier		Crew	Daily Output	Labor-Hours	Unit	Material	2016 Bare Costs Labor	Equipment	Total	Total Incl O&P
0010	**MODIFIED BITUMINOUS SHEET AIR BARRIER**									
0100	SBS Modified sheet laminated to polyethylene sheet, 40 mils, 4″ wide	1 Carp	1200	.007	L.F.	.37	.25		.62	.83
0120	6″ wide		1100	.007		.51	.27		.78	1.02
0140	9″ wide		1000	.008		.71	.30		1.01	1.29
0160	12″ wide		900	.009		.91	.34		1.25	1.56
0180	18″ wide	2 Carp	1700	.009	S.F.	.85	.36		1.21	1.53
0200	36″ wide	″	1800	.009		.83	.34		1.17	1.47
0220	Adhesive for above	1 Carp	1400	.006		.35	.22		.57	.74

07 27 26 – Fluid-Applied Membrane Air Barriers

07 27 26.10 Fluid Applied Membrane Air Barrier		Crew	Daily Output	Labor-Hours	Unit	Material	Labor	Equipment	Total	Total Incl O&P
0010	**FLUID APPLIED MEMBRANE AIR BARRIER**									
0100	Spray applied vapor barrier, 25 S.F./gallon	1 Pord	1375	.006	S.F.	.01	.19		.20	.33

07 31 Shingles and Shakes

07 31 13 – Asphalt Shingles

07 31 13.10 Asphalt Roof Shingles		Crew	Daily Output	Labor-Hours	Unit	Material	Labor	Equipment	Total	Total Incl O&P
0010	**ASPHALT ROOF SHINGLES**									
0100	Standard strip shingles									
0150	Inorganic, class A, 25 year	1 Rofc	5.50	1.455	Sq.	81	46.50		127.50	175
0155	Pneumatic nailed		7	1.143		81	36.50		117.50	157
0200	30 year		5	1.600		96	51.50		147.50	201
0205	Pneumatic nailed		6.25	1.280		96	41		137	182
0250	Standard laminated multi-layered shingles									
0300	Class A, 240-260 lb./square	1 Rofc	4.50	1.778	Sq.	109	57		166	225
0305	Pneumatic nailed		5.63	1.422		109	45.50		154.50	204
0350	Class A, 250-270 lb./square		4	2		109	64		173	238
0355	Pneumatic nailed		5	1.600		109	51.50		160.50	215
0400	Premium, laminated multi-layered shingles									
0450	Class A, 260-300 lb./square	1 Rofc	3.50	2.286	Sq.	158	73.50		231.50	310
0455	Pneumatic nailed		4.37	1.831		158	59		217	282
0500	Class A, 300-385 lb./square		3	2.667		241	85.50		326.50	425
0505	Pneumatic nailed		3.75	2.133		241	68.50		309.50	390
0800	#15 felt underlayment		64	.125		5.30	4.01		9.31	13.25
0825	#30 felt underlayment		58	.138		10.30	4.43		14.73	19.50
0850	Self adhering polyethylene and rubberized asphalt underlayment		22	.364		75	11.65		86.65	104
0900	Ridge shingles		330	.024	L.F.	2.20	.78		2.98	3.86
0905	Pneumatic nailed		412.50	.019	″	2.20	.62		2.82	3.57
1000	For steep roofs (7 to 12 pitch or greater), add						50%			

07 31 16 – Metal Shingles

07 31 16.10 Aluminum Shingles		Crew	Daily Output	Labor-Hours	Unit	Material	Labor	Equipment	Total	Total Incl O&P
0010	**ALUMINUM SHINGLES**									
0020	Mill finish, .019 thick	1 Carp	5	1.600	Sq.	221	60.50		281.50	345
0100	.020″ thick	″	5	1.600		247	60.50		307.50	370
0300	For colors, add					21			21	23
0600	Ridge cap, .024″ thick	1 Carp	170	.047	L.F.	3.68	1.78		5.46	7.05
0700	End wall flashing, .024″ thick		170	.047		2.15	1.78		3.93	5.35
0900	Valley section, .024″ thick		170	.047		3.65	1.78		5.43	7
1000	Starter strip, .024″ thick		400	.020		1.73	.76		2.49	3.17
1200	Side wall flashing, .024″ thick		170	.047		2.10	1.78		3.88	5.30
1500	Gable flashing, .024″ thick		400	.020		1.70	.76		2.46	3.14

07 31 Shingles and Shakes

07 31 26 – Slate Shingles

07 31 26.10 Slate Roof Shingles		Crew	Daily Output	Labor-Hours	Unit	Material	Labor	Equipment	Total	Total Incl O&P
						2016 Bare Costs				
0010	SLATE ROOF SHINGLES R073126-20									
0100	Buckingham Virginia black, 3/16" - 1/4" thick G	1 Rots	1.75	4.571	Sq.	540	147		687	865
0200	1/4" thick G		1.75	4.571		540	147		687	865
0900	Pennsylvania black, Bangor, #1 clear G		1.75	4.571		490	147		637	810
1200	Vermont, unfading, green, mottled green G		1.75	4.571		495	147		642	815
1300	Semi-weathering green & gray G		1.75	4.571		355	147		502	660
1400	Purple G		1.75	4.571		435	147		582	745
1500	Black or gray G		1.75	4.571		475	147		622	795
2700	Ridge shingles, slate		200	.040	L.F.	10	1.29		11.29	13.35

07 31 29 – Wood Shingles and Shakes

07 31 29.13 Wood Shingles

07 31 29.13 Wood Shingles		Crew	Daily Output	Labor-Hours	Unit	Material	Labor	Equipment	Total	Total Incl O&P
0010	WOOD SHINGLES R061110-30									
0012	16" No. 1 red cedar shingles, 5" exposure, on roof	1 Carp	2.50	3.200	Sq.	281	121		402	515
0015	Pneumatic nailed		3.25	2.462		281	93		374	465
0200	7-1/2" exposure, on walls		2.05	3.902		187	148		335	455
0205	Pneumatic nailed		2.67	2.996		187	113		300	395
0300	18" No. 1 red cedar perfections, 5-1/2" exposure, on roof		2.75	2.909		248	110		358	455
0305	Pneumatic nailed		3.57	2.241		248	84.50		332.50	415
0500	7-1/2" exposure, on walls		2.25	3.556		182	134		316	425
0505	Pneumatic nailed		2.92	2.740		182	104		286	375
0600	Resquared and rebutted, 5-1/2" exposure, on roof		3	2.667		282	101		383	480
0605	Pneumatic nailed		3.90	2.051		282	77.50		359.50	440
0900	7-1/2" exposure, on walls		2.45	3.265		207	123		330	435
0905	Pneumatic nailed		3.18	2.516		207	95		302	385
1000	Add to above for fire retardant shingles					56			56	61.50
1060	Preformed ridge shingles	1 Carp	400	.020	L.F.	5	.76		5.76	6.75
2000	White cedar shingles, 16" long, extras, 5" exposure, on roof		2.40	3.333	Sq.	193	126		319	425
2005	Pneumatic nailed		3.12	2.564		193	97		290	375
2050	5" exposure on walls		2	4		193	151		344	465
2055	Pneumatic nailed		2.60	3.077		193	116		309	405
2100	7-1/2" exposure, on walls		2	4		138	151		289	405
2105	Pneumatic nailed		2.60	3.077		138	116		254	345
2150	"B" grade, 5" exposure on walls		2	4		165	151		316	435
2155	Pneumatic nailed		2.60	3.077		165	116		281	375
2300	For 15# organic felt underlayment on roof, 1 layer, add		64	.125		5.30	4.73		10.03	13.75
2400	2 layers, add		32	.250		10.60	9.45		20.05	27.50
2600	For steep roofs (7/12 pitch or greater), add to above						50%			
2700	Panelized systems, No.1 cedar shingles on 5/16" CDX plywood									
2800	On walls, 8' strips, 7" or 14" exposure	2 Carp	700	.023	S.F.	6	.86		6.86	8.05
3000	Ridge shakes or shingle wood	1 Carp	280	.029	L.F.	4.50	1.08		5.58	6.75

07 31 29.16 Wood Shakes

07 31 29.16 Wood Shakes		Crew	Daily Output	Labor-Hours	Unit	Material	Labor	Equipment	Total	Total Incl O&P
0010	WOOD SHAKES									
1100	Hand-split red cedar shakes, 1/2" thick x 24" long, 10" exp. on roof	1 Carp	2.50	3.200	Sq.	287	121		408	520
1105	Pneumatic nailed		3.25	2.462		287	93		380	470
1110	3/4" thick x 24" long, 10" exp. on roof		2.25	3.556		287	134		421	540
1115	Pneumatic nailed		2.92	2.740		287	104		391	490
1200	1/2" thick, 18" long, 8-1/2" exp. on roof		2	4		179	151		330	450
1205	Pneumatic nailed		2.60	3.077		179	116		295	390
1210	3/4" thick x 18" long, 8 1/2" exp. on roof		1.80	4.444		179	168		347	480
1215	Pneumatic nailed		2.34	3.419		179	129		308	415
1255	10" exp. on walls		2	4		173	151		324	445

07 31 Shingles and Shakes

07 31 29 – Wood Shingles and Shakes

07 31 29.16 Wood Shakes		Crew	Daily Output	Labor-Hours	Unit	Material	2016 Bare Costs Labor	Equipment	Total	Total Incl O&P
1260	10″ exposure on walls, pneumatic nailed	1 Carp	2.60	3.077	Sq.	173	116		289	385
1700	Add to above for fire retardant shakes, 24″ long					56			56	61.50
1800	18″ long					56			56	61.50
1810	Ridge shakes	1 Carp	350	.023	L.F.	5.50	.86		6.36	7.50

07 32 Roof Tiles

07 32 13 – Clay Roof Tiles

07 32 13.10 Clay Tiles		Crew	Daily Output	Labor-Hours	Unit	Material	Labor	Equipment	Total	Total Incl O&P
0010	**CLAY TILES**, including accessories									
0300	Flat shingle, interlocking, 15″, 166 pcs/sq, fireflashed blend	3 Rots	6	4	Sq.	465	129		594	745
0500	Terra cotta red		6	4		510	129		639	795
0600	Roman pan and top, 18″, 102 pcs/sq, fireflashed blend		5.50	4.364		500	140		640	810
0640	Terra cotta red	1 Rots	2.40	3.333		555	107		662	810
1100	Barrel mission tile, 18″, 166 pcs/sq, fireflashed blend	3 Rots	5.50	4.364		410	140		550	710
1140	Terra cotta red		5.50	4.364		410	140		550	710
1700	Scalloped edge flat shingle, 14″, 145 pcs/sq, fireflashed blend		6	4		1,125	129		1,254	1,475
1800	Terra cotta red		6	4		1,025	129		1,154	1,350
3010	#15 felt underlayment	1 Rotc	64	.125		5.30	4.01		9.31	13.25
3020	#30 felt underlayment		58	.138		10.30	4.43		14.73	19.50
3040	Polyethylene and rubberized asph. underlayment		22	.364		75	11.65		86.65	104

07 32 16 – Concrete Roof Tiles

07 32 16.10 Concrete Tiles		Crew	Daily Output	Labor-Hours	Unit	Material	Labor	Equipment	Total	Total Incl O&P
0010	**CONCRETE TILES**									
0020	Corrugated, 13″ x 16-1/2″, 90 per sq., 950 lb. per sq.									
0050	Earthtone colors, nailed to wood deck	1 Rots	1.35	5.926	Sq.	103	191		294	465
0150	Blues		1.35	5.926		103	191		294	465
0200	Greens		1.35	5.926		103	191		294	465
0250	Premium colors		1.35	5.926		103	191		294	465
0500	Shakes, 13″ x 16-1/2″, 90 per sq., 950 lb. per sq.									
0600	All colors, nailed to wood deck	1 Rots	1.50	5.333	Sq.	124	171		295	450
1500	Accessory pieces, ridge & hip, 10″ x 16-1/2″, 8 lb. each	″	120	.067	Ea.	3.70	2.14		5.84	8
1700	Rake, 6-1/2″ x 16-3/4″, 9 lb. each					3.70			3.70	4.07
1800	Mansard hip, 10″ x 16-1/2″, 9.2 lb. each					3.70			3.70	4.07
1900	Hip starter, 10″ x 16-1/2″, 10.5 lb. each					10.40			10.40	11.45
2000	3 or 4 way apex, 10″ each side, 11.5 lb. each					12			12	13.20

07 32 19 – Metal Roof Tiles

07 32 19.10 Metal Roof Tiles		Crew	Daily Output	Labor-Hours	Unit	Material	Labor	Equipment	Total	Total Incl O&P
0010	**METAL ROOF TILES**									
0020	Accessories included, .032″ thick aluminum, mission tile	1 Carp	2.50	3.200	Sq.	825	121		946	1,125
0200	Spanish tiles	″	3	2.667	″	540	101		641	765

07 41 Roof Panels

07 41 13 – Metal Roof Panels

07 41 13.10 Aluminum Roof Panels		Crew	Daily Output	Labor-Hours	Unit	2016 Bare Costs Material	Labor	Equipment	Total	Total Incl O&P
0010	**ALUMINUM ROOF PANELS**									
0020	Corrugated or ribbed, .0155" thick, natural	G-3	1200	.027	S.F.	1	.95		1.95	2.69
0300	Painted	"	1200	.027	"	1.45	.95		2.40	3.19

07 41 13.20 Steel Roofing Panels			Crew	Daily Output	Labor-Hours	Unit	Material	Labor	Equipment	Total	Total Incl O&P
0010	**STEEL ROOFING PANELS**										
0012	Corrugated or ribbed, on steel framing, 30 ga. galv		G-3	1100	.029	S.F.	1.64	1.04		2.68	3.54
0100	28 ga.			1050	.030		1.69	1.09		2.78	3.68
0300	26 ga.			1000	.032		1.95	1.15		3.10	4.06
0400	24 ga.			950	.034		2.90	1.21		4.11	5.20
0510	Painted, including fasteners, 18 ga.	G		850	.038		3.75	1.35		5.10	6.40
0520	20 ga.	G		875	.037		3.03	1.31		4.34	5.50
0530	22 ga.	G		900	.036		2.70	1.27		3.97	5.10
0600	Colored, 28 ga.			1050	.030		1.72	1.09		2.81	3.71
0700	26 ga.			1000	.032		2.04	1.15		3.19	4.15
0710	Flat profile, 1-3/4" standing seams, 10" wide, standard finish, 26 ga.			1000	.032		3.75	1.15		4.90	6.05
0715	24 ga.			950	.034		4.35	1.21		5.56	6.80
0720	22 ga.			900	.036		5.40	1.27		6.67	8.10
0725	Zinc aluminum alloy finish, 26 ga.			1000	.032		3	1.15		4.15	5.20
0730	24 ga.			950	.034		3.55	1.21		4.76	5.90
0735	22 ga.			900	.036		4.05	1.27		5.32	6.60
0740	12" wide, standard finish, 26 ga.			1000	.032		3.75	1.15		4.90	6.05
0745	24 ga.			950	.034		4.90	1.21		6.11	7.40
0750	Zinc aluminum alloy finish, 26 ga.			1000	.032		4.26	1.15		5.41	6.60
0755	24 ga.			950	.034		3.50	1.21		4.71	5.85
0840	Flat profile, 1" x 3/8" batten, 12" wide, standard finish, 26 ga.			1000	.032		3.30	1.15		4.45	5.55
0845	24 ga.			950	.034		3.90	1.21		5.11	6.30
0850	22 ga.			900	.036		4.65	1.27		5.92	7.25
0855	Zinc aluminum alloy finish, 26 ga.			1000	.032		3.20	1.15		4.35	5.45
0860	24 ga.			950	.034		3.60	1.21		4.81	5.95
0865	22 ga.			900	.036		4.15	1.27		5.42	6.70
0870	16-1/2" wide, standard finish, 24 ga.			950	.034		3.85	1.21		5.06	6.25
0875	22 ga.			900	.036		4.30	1.27		5.57	6.85
0880	Zinc aluminum alloy finish, 24 ga.			950	.034		3.35	1.21		4.56	5.70
0885	22 ga.			900	.036		3.75	1.27		5.02	6.25
0890	Flat profile, 2" x 2" batten, 12" wide, standard finish, 26 ga.			1000	.032		3.85	1.15		5	6.15
0895	24 ga.			950	.034		4.55	1.21		5.76	7
0900	22 ga.			900	.036		5.55	1.27		6.82	8.25
0905	Zinc aluminum alloy finish, 26 ga.			1000	.032		3.55	1.15		4.70	5.80
0910	24 ga.			950	.034		4.05	1.21		5.26	6.45
0915	22 ga.			900	.036		4.70	1.27		5.97	7.30
0920	16-1/2" wide, standard finish, 24 ga.			950	.034		4.20	1.21		5.41	6.65
0925	22 ga.			900	.036		4.90	1.27		6.17	7.55
0930	Zinc aluminum alloy finish, 24 ga.			950	.034		3.80	1.21		5.01	6.20
0935	22 ga.			900	.036		4.30	1.27		5.57	6.85
0950	Box rib roof panels, painted, including fasteners, 18 ga.	G		850	.038		3.60	1.35		4.95	6.20
0960	20 ga.	G		875	.037		3.13	1.31		4.44	5.65
0970	22 ga.	G		900	.036		2.55	1.27		3.82	4.94
1000	4" rib panel, painted, including fasteners, 22 ga.	G		900	.036		3.05	1.27		4.32	5.50
1010	20 ga.	G		875	.037		3.08	1.31		4.39	5.60
1020	18 ga.	G		850	.038		6.15	1.35		7.50	9
1050	On substrate, 2" standing seam panel, painted, 22 ga.	G		900	.036		3.55	1.27		4.82	6.05
1060	24 ga.	G		950	.034		3.30	1.21		4.51	5.65

07 41 Roof Panels

07 41 13 – Metal Roof Panels

07 41 13.20 Steel Roofing Panels		Crew	Daily Output	Labor-Hours	Unit	Material	Labor	Equipment	Total	Total Incl O&P
						2016 Bare Costs				
1070	26 ga. G	G-3	1000	.032	S.F.	2.58	1.15		3.73	4.75
1200	Ridge, galvanized, 10″ wide G		800	.040	L.F.	3.15	1.43		4.58	5.85
1203	14″ wide G	2 Shee	316	.051		3.78	2.14		5.92	7.75
1205	18″ wide G	″	308	.052		4.41	2.20		6.61	8.50
1210	20″ wide G	G-3	750	.043		4.15	1.53		5.68	7.10

07 41 33 – Plastic Roof Panels

07 41 33.10 Fiberglass Panels		Crew	Daily Output	Labor-Hours	Unit	Material	Labor	Equipment	Total	Total Incl O&P
0010	**FIBERGLASS PANELS**									
0012	Corrugated panels, roofing, 8 oz. per S.F.	G-3	1000	.032	S.F.	2.15	1.15		3.30	4.28
0100	12 oz. per S.F.		1000	.032		4.20	1.15		5.35	6.55
0300	Corrugated siding, 6 oz. per S.F.		880	.036		1.90	1.30		3.20	4.26
0400	8 oz. per S.F.		880	.036		2.15	1.30		3.45	4.54
0500	Fire retardant		880	.036		3.80	1.30		5.10	6.35
0600	12 oz. siding, textured		880	.036		3.75	1.30		5.05	6.30
0700	Fire retardant		880	.036		4.35	1.30		5.65	6.95
0900	Flat panels, 6 oz. per S.F., clear or colors		880	.036		2.40	1.30		3.70	4.81
1100	Fire retardant, class A		880	.036		3.45	1.30		4.75	5.95
1300	8 oz. per S.F., clear or colors		880	.036		2.50	1.30		3.80	4.92

07 42 Wall Panels

07 42 13 – Metal Wall Panels

07 42 13.20 Aluminum Siding		Crew	Daily Output	Labor-Hours	Unit	Material	Labor	Equipment	Total	Total Incl O&P
0011	**ALUMINUM SIDING**									
6040	.024 thick smooth white single 8″ wide	2 Carp	515	.031	S.F.	2.73	1.17		3.90	4.97
6060	Double 4″ pattern		515	.031		2.80	1.17		3.97	5.05
6080	Double 5″ pattern		550	.029		2.72	1.10		3.82	4.83
6120	Embossed white, 8″ wide		515	.031		2.72	1.17		3.89	4.96
6140	Double 4″ pattern		515	.031		2.79	1.17		3.96	5.05
6160	Double 5″ pattern		550	.029		2.77	1.10		3.87	4.89
6170	Vertical, embossed white, 12″ wide		590	.027		2.77	1.03		3.80	4.77
6320	.019 thick, insulated, smooth white, 8″ wide		515	.031		2.45	1.17		3.62	4.67
6340	Double 4″ pattern		515	.031		2.42	1.17		3.59	4.63
6360	Double 5″ pattern		550	.029		2.42	1.10		3.52	4.50
6400	Embossed white, 8″ wide		515	.031		2.82	1.17		3.99	5.05
6420	Double 4″ pattern		515	.031		2.85	1.17		4.02	5.10
6440	Double 5″ pattern		550	.029		2.85	1.10		3.95	4.98
6500	Shake finish 10″ wide white		550	.029		3.06	1.10		4.16	5.20
6600	Vertical pattern, 12″ wide, white		590	.027		2.33	1.03		3.36	4.28
6640	For colors add					.15			.15	.17
6700	Accessories, white									
6720	Starter strip 2-1/8″	2 Carp	610	.026	L.F.	.48	.99		1.47	2.19
6740	Sill trim		450	.036		.65	1.34		1.99	2.97
6760	Inside corner		610	.026		1.70	.99		2.69	3.53
6780	Outside corner post		610	.026		3.60	.99		4.59	5.60
6800	Door & window trim		440	.036		.62	1.37		1.99	2.98
6820	For colors add					.14			.14	.15
6900	Soffit & fascia 1′ overhang solid	2 Carp	110	.145		4.45	5.50		9.95	14.10
6920	Vented		110	.145		4.48	5.50		9.98	14.15
6940	2′ overhang solid		100	.160		6.50	6.05		12.55	17.30
6960	Vented		100	.160		6.50	6.05		12.55	17.30

07 42 Wall Panels

07 42 13 – Metal Wall Panels

07 42 13.30 Steel Siding		Crew	Daily Output	Labor-Hours	Unit	2016 Bare Costs Material	Labor	Equipment	Total	Total Incl O&P
0010	STEEL SIDING									
0020	Beveled, vinyl coated, 8″ wide	1 Carp	265	.030	S.F.	1.85	1.14		2.99	3.95
0050	10″ wide	″	275	.029		1.90	1.10		3	3.93
0080	Galv, corrugated or ribbed, on steel frame, 30 ga.	G-3	800	.040		1.20	1.43		2.63	3.71
0100	28 ga.		795	.040		1.30	1.44		2.74	3.84
0300	26 ga.		790	.041		1.80	1.45		3.25	4.40
0400	24 ga.		785	.041		2	1.46		3.46	4.64
0600	22 ga.		770	.042		2.25	1.49		3.74	4.96
0700	Colored, corrugated/ribbed, on steel frame, 10 yr. finish, 28 ga.		800	.040		1.95	1.43		3.38	4.54
0900	26 ga.		795	.040		2.12	1.44		3.56	4.74
1000	24 ga.		790	.041		2.38	1.45		3.83	5.05
1020	20 ga.		785	.041		3.02	1.46		4.48	5.75

07 46 Siding

07 46 23 – Wood Siding

07 46 23.10 Wood Board Siding

Line	Description	Crew	Daily Output	Labor-Hours	Unit	Material	Labor	Equipment	Total	Total Incl O&P
0010	WOOD BOARD SIDING									
2000	Board & batten, cedar, "B" grade, 1″ x 10″	1 Carp	375	.021	S.F.	5.15	.81		5.96	7.05
2200	Redwood, clear, vertical grain, 1″ x 10″		375	.021		5.35	.81		6.16	7.25
2400	White pine, #2 & better, 1″ x 10″		375	.021		3.18	.81		3.99	4.85
2410	Board & batten siding, white pine #2, 1″ x 12″		420	.019		3.18	.72		3.90	4.71
3200	Wood, cedar bevel, A grade, 1/2″ x 6″		295	.027		4.25	1.03		5.28	6.40
3300	1/2″ x 8″		330	.024		6.50	.92		7.42	8.70
3500	3/4″ x 10″, clear grade		375	.021		6.75	.81		7.56	8.80
3600	"B" grade		375	.021		3.79	.81		4.60	5.50
3800	Cedar, rough sawn, 1″ x 4″, A grade, natural		220	.036		6.55	1.37		7.92	9.55
3900	Stained		220	.036		6.70	1.37		8.07	9.65
4100	1″ x 12″, board & batten, #3 & Btr., natural		420	.019		4.48	.72		5.20	6.15
4200	Stained		420	.019		4.81	.72		5.53	6.50
4400	1″ x 8″ channel siding, #3 & Btr., natural		330	.024		4.57	.92		5.49	6.55
4500	Stained		330	.024		4.92	.92		5.84	6.95
4700	Redwood, clear, beveled, vertical grain, 1/2″ x 4″		220	.036		4.15	1.37		5.52	6.85
4750	1/2″ x 6″		295	.027		4.79	1.03		5.82	6.95
4800	1/2″ x 8″		330	.024		5.20	.92		6.12	7.25
5000	3/4″ x 10″		375	.021		4.90	.81		5.71	6.75
5200	Channel siding, 1″ x 10″, B grade		375	.021		4.50	.81		5.31	6.30
5250	Redwood, T&G boards, B grade, 1″ x 4″		220	.036		3.25	1.37		4.62	5.90
5270	1″ x 8″		330	.024		5.20	.92		6.12	7.25
5400	White pine, rough sawn, 1″ x 8″, natural		330	.024		2.19	.92		3.11	3.94
5500	Stained		330	.024		2.19	.92		3.11	3.94
5600	Tongue and groove, 1″ x 8″		330	.024		2.19	.92		3.11	3.94

07 46 29 – Plywood Siding

07 46 29.10 Plywood Siding Options

Line	Description	Crew	Daily Output	Labor-Hours	Unit	Material	Labor	Equipment	Total	Total Incl O&P
0010	PLYWOOD SIDING OPTIONS									
0900	Plywood, medium density overlaid, 3/8″ thick	2 Carp	750	.021	S.F.	1.29	.81		2.10	2.77
1000	1/2″ thick		700	.023		1.50	.86		2.36	3.10
1100	3/4″ thick		650	.025		1.99	.93		2.92	3.75
1600	Texture 1-11, cedar, 5/8″ thick, natural		675	.024		2.70	.90		3.60	4.47
1700	Factory stained		675	.024		2.80	.90		3.70	4.58
1900	Texture 1-11, fir, 5/8″ thick, natural		675	.024		1.46	.90		2.36	3.11

07 46 Siding

07 46 29 – Plywood Siding

07 46 29.10 Plywood Siding Options		Crew	Daily Output	Labor-Hours	Unit	2016 Bare Costs Material	Labor	Equipment	Total	Total Incl O&P
2000	Factory stained	2 Carp	675	.024	S.F.	1.80	.90		2.70	3.48
2050	Texture 1-11, S.Y.P., 5/8" thick, natural		675	.024		1.37	.90		2.27	3.01
2100	Factory stained		675	.024		1.44	.90		2.34	3.08
2200	Rough sawn cedar, 3/8" thick, natural		675	.024		1.25	.90		2.15	2.88
2300	Factory stained		675	.024		1.50	.90		2.40	3.15
2500	Rough sawn fir, 3/8" thick, natural		675	.024		.90	.90		1.80	2.49
2600	Factory stained		675	.024		1.04	.90		1.94	2.64
2800	Redwood, textured siding, 5/8" thick	↓	675	.024	↓	1.92	.90		2.82	3.61

07 46 33 – Plastic Siding

07 46 33.10 Vinyl Siding		Crew	Daily Output	Labor-Hours	Unit	Material	Labor	Equipment	Total	Total Incl O&P
0010	**VINYL SIDING**									
3995	Clapboard profile, woodgrain texture, .048 thick, double 4	2 Carp	495	.032	S.F.	1.06	1.22		2.28	3.22
4000	Double 5		550	.029		1.06	1.10		2.16	3.01
4005	Single 8		495	.032		1.35	1.22		2.57	3.54
4010	Single 10		550	.029		1.62	1.10		2.72	3.63
4015	.044 thick, double 4		495	.032		1.04	1.22		2.26	3.20
4020	Double 5		550	.029		1.04	1.10		2.14	2.99
4025	.042 thick, double 4		495	.032		1.04	1.22		2.26	3.20
4030	Double 5		550	.029		1.04	1.10		2.14	2.99
4035	Cross sawn texture, .040 thick, double 4		495	.032		.72	1.22		1.94	2.85
4040	Double 5		550	.029		.72	1.10		1.82	2.64
4045	Smooth texture, .042 thick, double 4		495	.032		.79	1.22		2.01	2.92
4050	Double 5		550	.029		.79	1.10		1.89	2.71
4055	Single 8		495	.032		.79	1.22		2.01	2.92
4060	Cedar texture, .044 thick, double 4		495	.032		1.12	1.22		2.34	3.29
4065	Double 6		600	.027		1.12	1.01		2.13	2.93
4070	Dutch lap profile, woodgrain texture, .048 thick, double 5		550	.029		1.06	1.10		2.16	3.01
4075	.044 thick, double 4.5		525	.030		1.06	1.15		2.21	3.10
4080	.042 thick, double 4.5		525	.030		.90	1.15		2.05	2.93
4085	.040 thick, double 4.5		525	.030		.72	1.15		1.87	2.73
4100	Shake profile, 10" wide		400	.040		3.61	1.51		5.12	6.50
4105	Vertical pattern, .046 thick, double 5		550	.029		1.56	1.10		2.66	3.56
4110	.044 thick, triple 3		550	.029		1.74	1.10		2.84	3.76
4115	.040 thick, triple 4		550	.029		1.61	1.10		2.71	3.62
4120	.040 thick, triple 2.66		550	.029		1.74	1.10		2.84	3.76
4125	Insulation, fan folded extruded polystyrene, 1/4"		2000	.008		.29	.30		.59	.83
4130	3/8"		2000	.008	↓	.32	.30		.62	.87
4135	Accessories, J channel, 5/8" pocket		700	.023	L.F.	.50	.86		1.36	2
4140	3/4" pocket		695	.023		.53	.87		1.40	2.04
4145	1-1/4" pocket		680	.024		.84	.89		1.73	2.41
4150	Flexible, 3/4" pocket		600	.027		2.33	1.01		3.34	4.25
4155	Under sill finish trim		500	.032		.54	1.21		1.75	2.62
4160	Vinyl starter strip		700	.023		.65	.86		1.51	2.16
4165	Aluminum starter strip		700	.023		.29	.86		1.15	1.77
4170	Window casing, 2-1/2" wide, 3/4" pocket		510	.031		1.68	1.19		2.87	3.84
4175	Outside corner, woodgrain finish, 4" face, 3/4" pocket		700	.023		2.08	.86		2.94	3.74
4180	5/8" pocket		700	.023		2.08	.86		2.94	3.74
4185	Smooth finish, 4" face, 3/4" pocket		700	.023		2.08	.86		2.94	3.74
4190	7/8" pocket		690	.023		1.96	.88		2.84	3.63
4195	1-1/4" pocket		700	.023		1.38	.86		2.24	2.97
4200	Soffit and fascia, 1' overhang, solid		120	.133		4.65	5.05		9.70	13.55
4205	Vented	↓	120	.133	↓	4.65	5.05		9.70	13.55

07 46 Siding

07 46 33 – Plastic Siding

07 46 33.10 Vinyl Siding		Crew	Daily Output	Labor-Hours	Unit	2016 Bare Costs Material	Labor	Equipment	Total	Total Incl O&P
4207	18″ overhang, solid	2 Carp	110	.145	L.F.	5.45	5.50		10.95	15.20
4208	Vented		110	.145		5.45	5.50		10.95	15.20
4210	2′ overhang, solid		100	.160		6.20	6.05		12.25	17
4215	Vented		100	.160		6.20	6.05		12.25	17
4217	3′ overhang, solid		100	.160		7.75	6.05		13.80	18.70
4218	Vented		100	.160		7.75	6.05		13.80	18.70
4220	Colors for siding and soffits, add				S.F.	.14			.14	.15
4225	Colors for accessories and trim, add				L.F.	.30			.30	.33

07 46 33.20 Polypropylene Siding

		Crew	Daily Output	Labor-Hours	Unit	Material	Labor	Equipment	Total	Total Incl O&P
0010	**POLYPROPYLENE SIDING**									
4090	Shingle profile, random grooves, double 7	2 Carp	400	.040	S.F.	3.11	1.51		4.62	5.95
4092	Cornerpost for above	1 Carp	365	.022	L.F.	12.65	.83		13.48	15.30
4095	Triple 5	2 Carp	400	.040	S.F.	3.11	1.51		4.62	5.95
4097	Cornerpost for above	1 Carp	365	.022	L.F.	11.80	.83		12.63	14.40
5000	Staggered butt, double 7″	2 Carp	400	.040	S.F.	3.61	1.51		5.12	6.50
5002	Cornerpost for above	1 Carp	365	.022	L.F.	12.75	.83		13.58	15.45
5010	Half round, double 6-1/4″	2 Carp	360	.044	S.F.	3.61	1.68		5.29	6.80
5020	Shake profile, staggered butt, double 9″	″	510	.031	″	3.61	1.19		4.80	5.95
5022	Cornerpost for above	1 Carp	365	.022	L.F.	9.60	.83		10.43	11.95
5030	Straight butt, double 7″	2 Carp	400	.040	S.F.	3.61	1.51		5.12	6.50
5032	Cornerpost for above	1 Carp	365	.022	L.F.	12.60	.83		13.43	15.30
6000	Accessories, J channel, 5/8″ pocket	2 Carp	700	.023		.50	.86		1.36	2
6010	3/4″ pocket		695	.023		.53	.87		1.40	2.04
6020	1-1/4″ pocket		680	.024		.84	.89		1.73	2.41
6030	Aluminum starter strip		700	.023		.29	.86		1.15	1.77

07 46 46 – Fiber Cement Siding

07 46 46.10 Fiber Cement Siding

		Crew	Daily Output	Labor-Hours	Unit	Material	Labor	Equipment	Total	Total Incl O&P
0010	**FIBER CEMENT SIDING**									
0020	Lap siding, 5/16″ thick, 6″ wide, 4-3/4″ exposure, smooth texture	2 Carp	415	.039	S.F.	1.29	1.46		2.75	3.85
0025	Woodgrain texture		415	.039		1.29	1.46		2.75	3.85
0030	7-1/2″ wide, 6-1/4″ exposure, smooth texture		425	.038		1.59	1.42		3.01	4.13
0035	Woodgrain texture		425	.038		1.59	1.42		3.01	4.13
0040	8″ wide, 6-3/4″ exposure, smooth texture		425	.038		1.19	1.42		2.61	3.69
0045	Rough sawn texture		425	.038		1.19	1.42		2.61	3.69
0050	9-1/2″ wide, 8-1/4″ exposure, smooth texture		440	.036		1.26	1.37		2.63	3.69
0055	Woodgrain texture		440	.036		1.26	1.37		2.63	3.69
0060	12″ wide, 10-3/8″ exposure, smooth texture		455	.035		2.01	1.33		3.34	4.44
0065	Woodgrain texture		455	.035		2.01	1.33		3.34	4.44
0070	Panel siding, 5/16″ thick, smooth texture		750	.021		1.31	.81		2.12	2.79
0075	Stucco texture		750	.021		1.31	.81		2.12	2.79
0080	Grooved woodgrain texture		750	.021		1.31	.81		2.12	2.79
0085	V - grooved woodgrain texture		750	.021		1.31	.81		2.12	2.79
0088	Shingle siding, 48″ x 15-1/4″ panels, 7″ exposure		700	.023		4.12	.86		4.98	6
0090	Wood starter strip		400	.040	L.F.	.41	1.51		1.92	2.98

07 51 Built-Up Bituminous Roofing

07 51 13 – Built-Up Asphalt Roofing

07 51 13.10 Built-Up Roofing Components

	07 51 13.10 Built-Up Roofing Components	Crew	Daily Output	Labor-Hours	Unit	2016 Bare Costs Material	Labor	Equipment	Total	Total Incl O&P
0010	**BUILT-UP ROOFING COMPONENTS**									
0012	Asphalt saturated felt, #30, 2 square per roll	1 Rofc	58	.138	Sq.	10.30	4.43		14.73	19.50
0200	#15, 4 sq. per roll, plain or perforated, not mopped		58	.138		5.30	4.43		9.73	14
0250	Perforated		58	.138		5.30	4.43		9.73	14
0300	Roll roofing, smooth, #65		15	.533		10.10	17.10		27.20	42.50
0500	#90		12	.667		44	21.50		65.50	87.50
0520	Mineralized		12	.667		35	21.50		56.50	78
0540	D.C. (double coverage), 19" selvage edge		10	.800		54.50	25.50		80	108
0580	Adhesive (lap cement)				Gal.	8.85			8.85	9.75

07 51 13.13 Cold-Applied Built-Up Asphalt Roofing

	07 51 13.13 Cold-Applied Built-Up Asphalt Roofing	Crew	Daily Output	Labor-Hours	Unit	Material	Labor	Equipment	Total	Total Incl O&P
0010	**COLD-APPLIED BUILT-UP ASPHALT ROOFING**									
0020	3 ply system, installation only (components listed below)	G-5	50	.800	Sq.		23.50	3.86	27.36	47.50
0100	Spunbond poly. fabric, 1.35 oz./S.Y., 36"W, 10.8 sq./roll				Ea.	138			138	152
0500	Base & finish coat, 3 gal./sq., 5 gal./can				Gal.	7.85			7.85	8.65
0600	Coating, ceramic granules, 1/2 sq./bag				Ea.	23			23	25.50
0700	Aluminum, 2 gal./sq.				Gal.	13.50			13.50	14.85
0800	Emulsion, fibered or non-fibered, 4 gal./sq.				"	6.50			6.50	7.15

07 51 13.20 Built-Up Roofing Systems

	07 51 13.20 Built-Up Roofing Systems	Crew	Daily Output	Labor-Hours	Unit	Material	Labor	Equipment	Total	Total Incl O&P
0010	**BUILT-UP ROOFING SYSTEMS**									
0120	Asphalt flood coat with gravel/slag surfacing, not including									
0140	Insulation, flashing or wood nailers									
0200	Asphalt base sheet, 3 plies #15 asphalt felt, mopped	G-1	22	2.545	Sq.	99.50	76.50	26.50	202.50	279
0350	On nailable decks		21	2.667		103	80.50	27.50	211	292
0500	4 plies #15 asphalt felt, mopped		20	2.800		137	84.50	29	250.50	335
0550	On nailable decks		19	2.947		121	88.50	30.50	240	330
0700	Coated glass base sheet, 2 plies glass (type IV), mopped		22	2.545		110	76.50	26.50	213	291
0850	3 plies glass, mopped		20	2.800		134	84.50	29	247.50	335
0950	On nailable decks		19	2.947		126	88.50	30.50	245	335
1100	4 plies glass fiber felt (type IV), mopped		20	2.800		166	84.50	29	279.50	370
1150	On nailable decks		19	2.947		150	88.50	30.50	269	365
1200	Coated & saturated base sheet, 3 plies #15 asph. felt, mopped		20	2.800		111	84.50	29	224.50	310
1250	On nailable decks		19	2.947		104	88.50	30.50	223	310
1300	4 plies #15 asphalt felt, mopped		22	2.545		130	76.50	26.50	233	315
2000	Asphalt flood coat, smooth surface									
2200	Asphalt base sheet & 3 plies #15 asphalt felt, mopped	G-1	24	2.333	Sq.	106	70	24	200	273
2400	On nailable decks		23	2.435		98	73.50	25	196.50	271
2600	4 plies #15 asphalt felt, mopped		24	2.333		124	70	24	218	293
2700	On nailable decks		23	2.435		116	73.50	25	214.50	291
2900	Coated glass fiber base sheet, mopped, and 2 plies of									
2910	glass fiber felt (type IV)	G-1	25	2.240	Sq.	106	67.50	23	196.50	266
3100	On nailable decks		24	2.333		100	70	24	194	267
3200	3 plies, mopped		23	2.435		129	73.50	25	227.50	305
3300	On nailable decks		22	2.545		121	76.50	26.50	224	305
3800	4 plies glass fiber felt (type IV), mopped		23	2.435		153	73.50	25	251.50	330
3900	On nailable decks		22	2.545		145	76.50	26.50	248	330
4000	Coated & saturated base sheet, 3 plies #15 asph. felt, mopped		24	2.333		107	70	24	201	275
4200	On nailable decks		23	2.435		99	73.50	25	197.50	272
4300	4 plies #15 organic felt, mopped		22	2.545		125	76.50	26.50	228	310
4500	Coal tar pitch with gravel/slag surfacing									
4600	4 plies #15 tarred felt, mopped	G-1	21	2.667	Sq.	200	80.50	27.50	308	400
4800	3 plies glass fiber felt (type IV), mopped	"	19	2.947	"	165	88.50	30.50	284	380
5000	Coated glass fiber base sheet, and 2 plies of									

07 51 Built-Up Bituminous Roofing

07 51 13 – Built-Up Asphalt Roofing

07 51 13.20 Built-Up Roofing Systems		Crew	Daily Output	Labor-Hours	Unit	Material	Labor	Equipment	Total	Total Incl O&P
						2016 Bare Costs				
5010	glass fiber felt, (type IV), mopped	G-1	19	2.947	Sq.	169	88.50	30.50	288	385
5300	On nailable decks		18	3.111		148	93.50	32	273.50	370
5600	4 plies glass fiber felt (type IV), mopped		21	2.667		229	80.50	27.50	337	430
5800	On nailable decks	↓	20	2.800	↓	207	84.50	29	320.50	415

07 51 13.30 Cants		Crew	Daily Output	Labor-Hours	Unit	Material	Labor	Equipment	Total	Total Incl O&P
0010	**CANTS**									
0012	Lumber, treated, 4" x 4" cut diagonally	1 Rofc	325	.025	L.F.	1.80	.79		2.59	3.44
0300	Mineral or fiber, trapezoidal, 1" x 4" x 48"		325	.025		.30	.79		1.09	1.79
0400	1-1/2" x 5-5/8" x 48"	↓	325	.025	↓	.48	.79		1.27	1.99

07 51 13.40 Felts		Crew	Daily Output	Labor-Hours	Unit	Material	Labor	Equipment	Total	Total Incl O&P
0010	**FELTS**									
0012	Glass fibered roofing felt, #15, not mopped	1 Rofc	58	.138	Sq.	10.85	4.43		15.28	20
0300	Base sheet, #80, channel vented		58	.138		42.50	4.43		46.93	55
0400	#70, coated		58	.138		17.80	4.43		22.23	28
0500	Cap, #87, mineral surfaced		58	.138		81	4.43		85.43	97
0600	Flashing membrane, #65		16	.500		10.10	16.05		26.15	40.50
0800	Coal tar fibered, #15, no mopping		58	.138		18	4.43		22.43	28
0900	Asphalt felt, #15, 4 sq. per roll, no mopping		58	.138		5.30	4.43		9.73	14
1100	#30, 2 sq. per roll		58	.138		10.30	4.43		14.73	19.50
1200	Double coated, #33		58	.138		11	4.43		15.43	20.50
1400	#40, base sheet		58	.138		10.95	4.43		15.38	20
1450	Coated and saturated		58	.138		12	4.43		16.43	21.50
1500	Tarred felt, organic, #15, 4 sq. rolls		58	.138		13.40	4.43		17.83	23
1550	#30, 2 sq. roll	↓	58	.138		26.50	4.43		30.93	37
1700	Add for mopping above felts, per ply, asphalt, 24 lb. per sq.	G-1	192	.292		11.10	8.80	3.01	22.91	32
1800	Coal tar mopping, 30 lb. per sq.		186	.301		18.85	9.05	3.11	31.01	40.50
1900	Flood coat, with asphalt, 60 lb. per sq.		60	.933		28	28	9.65	65.65	93
2000	With coal tar, 75 lb. per sq.	↓	56	1	↓	47	30	10.35	87.35	119

07 52 Modified Bituminous Membrane Roofing

07 52 13 – Atactic-Polypropylene-Modified Bituminous Membrane Roofing

07 52 13.10 APP Modified Bituminous Membrane		Crew	Daily Output	Labor-Hours	Unit	Material	Labor	Equipment	Total	Total Incl O&P
0010	**APP MODIFIED BITUMINOUS MEMBRANE** R075213-30									
0020	Base sheet, #15 glass fiber felt, nailed to deck	1 Rofc	58	.138	Sq.	11.85	4.43		16.28	21
0030	Spot mopped to deck	G-1	295	.190		16.40	5.70	1.96	24.06	31
0040	Fully mopped to deck	"	192	.292		22	8.80	3.01	33.81	43.50
0050	#15 organic felt, nailed to deck	1 Rofc	58	.138		6.30	4.43		10.73	15.05
0060	Spot mopped to deck	G-1	295	.190		10.85	5.70	1.96	18.51	24.50
0070	Fully mopped to deck	"	192	.292	↓	16.40	8.80	3.01	28.21	37.50
2100	APP mod., smooth surf. cap sheet, poly. reinf., torched, 160 mils	G-5	2100	.019	S.F.	.74	.56	.09	1.39	1.94
2150	170 mils		2100	.019		.75	.56	.09	1.40	1.96
2200	Granule surface cap sheet, poly. reinf., torched, 180 mils		2000	.020		.93	.59	.10	1.62	2.21
2250	Smooth surface flashing, torched, 160 mils		1260	.032		.74	.93	.15	1.82	2.70
2300	170 mils		1260	.032		.75	.93	.15	1.83	2.72
2350	Granule surface flashing, torched, 180 mils	↓	1260	.032		.93	.93	.15	2.01	2.91
2400	Fibrated aluminum coating	1 Rofc	3800	.002	↓	.08	.07		.15	.21
2450	Seam heat welding	"	205	.039	L.F.	.08	1.25		1.33	2.40

07 52 Modified Bituminous Membrane Roofing

07 52 16 – Styrene-Butadiene-Styrene Modified Bituminous Membrane Roofing

07 52 16.10 SBS Modified Bituminous Membrane		Crew	Daily Output	Labor-Hours	Unit	2016 Bare Costs Material	Labor	Equipment	Total	Total Incl O&P
0010	**SBS MODIFIED BITUMINOUS MEMBRANE**									
0080	Mod bit rfng, SBS mod, gran surf cap sheet, poly reinf									
0650	120 to 149 mils thick	G-1	2000	.028	S.F.	1.27	.84	.29	2.40	3.27
0750	150 to 160 mils	″	2000	.028		1.71	.84	.29	2.84	3.75
1150	For reflective granules, add					.71	.79	.28	1.78	2.54
1600	Smooth surface cap sheet, mopped, 145 mils	G-1	2100	.027		.80	.80	.28	1.88	2.66
1620	Lightweight base sheet, fiberglass reinforced, 35 to 47 mil		2100	.027		.27	.80	.28	1.35	2.08
1625	Heavyweight base/ply sheet, reinforced, 87 to 120 mil thick		2100	.027		.88	.80	.28	1.96	2.75
1650	Granulated walkpad, 180 – 220 mils	1 Rofc	400	.020		1.77	.64		2.41	3.12
1700	Smooth surface flashing, 145 mils	G-1	1260	.044		.80	1.34	.46	2.60	3.85
1800	150 mils		1260	.044		.50	1.34	.46	2.30	3.52
1900	Granular surface flashing, 150 mils		1260	.044		.70	1.34	.46	2.50	3.74
2000	160 mils		1260	.044		.72	1.34	.46	2.52	3.76
2010	Elastomeric asphalt primer	1 Rofc	2600	.003		.16	.10		.26	.36
2015	Roofing asphalt, 30 lb. per square	G-1	19000	.003		.14	.09	.03	.26	.34
2020	Cold process adhesive, 20 – 30 mils thick	1 Rofc	750	.011		.23	.34		.57	.89
2025	Self adhering vapor retarder, 30 to 45 mils thick	G-5	2150	.019		1	.55	.09	1.64	2.21
2050	Seam heat welding	1 Rofc	205	.039	L.F.	.08	1.25		1.33	2.40

07 53 Elastomeric Membrane Roofing

07 53 16 – Chlorosulfonate-Polyethylene Roofing

07 53 16.10 Chlorosulfonated Polyethylene Roofing		Crew	Daily Output	Labor-Hours	Unit	Material	Labor	Equipment	Total	Total Incl O&P
0010	**CHLOROSULFONATED POLYETHYLENE ROOFING**									
0800	Chlorosulfonated polyethylene (CSPE)									
0900	45 mils, heat welded seams, plate attachment	G-5	35	1.143	Sq.	251	33.50	5.50	290	345
1100	Heat welded seams, plate attachment and ballasted		26	1.538		260	45	7.45	312.45	375
1200	60 mils, heat welded seams, plate attachment		35	1.143		325	33.50	5.50	364	430
1300	Heat welded seams, plate attachment and ballasted		26	1.538		335	45	7.45	387.45	460

07 53 23 – Ethylene-Propylene-Diene-Monomer Roofing

07 53 23.20 Ethylene-Propylene-Diene-Monomer Roofing		Crew	Daily Output	Labor-Hours	Unit	Material	Labor	Equipment	Total	Total Incl O&P
0010	**ETHYLENE-PROPYLENE-DIENE-MONOMER ROOFING (EPDM)**									
3500	Ethylene-propylene-diene-monomer (EPDM), 45 mils, 0.28 psf									
3600	Loose-laid & ballasted with stone (10 psf)	G-5	51	.784	Sq.	83.50	23	3.79	110.29	139
3700	Mechanically attached		35	1.143		78.50	33.50	5.50	117.50	154
3800	Fully adhered with adhesive		26	1.538		111	45	7.45	163.45	213
4500	60 mils, 0.40 psf									
4600	Loose-laid & ballasted with stone (10 psf)	G-5	51	.704	Sq.	95	23	3.79	121.79	152
4700	Mechanically attached		35	1.143		89	33.50	5.50	128	166
4800	Fully adhered with adhesive		26	1.538		122	45	7.45	174.45	225
4810	45 mil, .28 psf, membrane only					50			50	55
4820	60 mil, .40 psf, membrane only					59.50			59.50	65.50
4850	Seam tape for membrane, 3″ x 100′ roll				Ea.	43.50			43.50	48
4900	Batten strips, 10′ sections					3.43			3.43	3.77
4910	Cover tape for batten strips, 6″ x 100′ roll					183			183	202
4930	Plate anchors				M	78			78	86
4970	Adhesive for fully adhered systems, 60 S.F./gal.				Gal.	20			20	22

07 53 Elastomeric Membrane Roofing

07 53 29 – Polyisobutylene Roofing

07 53 29.10 Polyisobutylene Roofing		Crew	Daily Output	Labor-Hours	Unit	2016 Bare Costs Material	Labor	Equipment	Total	Total Incl O&P
0010	POLYISOBUTYLENE ROOFING									
7500	Polyisobutylene (PIB), 100 mils, 0.57 psf									
7600	Loose-laid & ballasted with stone/gravel (10 psf)	G-5	51	.784	Sq.	200	23	3.79	226.79	267
7700	Partially adhered with adhesive		35	1.143		240	33.50	5.50	279	330
7800	Hot asphalt attachment		35	1.143		230	33.50	5.50	269	320
7900	Fully adhered with contact cement	↓	26	1.538	↓	250	45	7.45	302.45	365

07 54 Thermoplastic Membrane Roofing

07 54 19 – Polyvinyl-Chloride Roofing

07 54 19.10 Polyvinyl-Chloride Roofing (PVC)		Crew	Daily Output	Labor-Hours	Unit	Material	Labor	Equipment	Total	Total Incl O&P
0010	POLYVINYL-CHLORIDE ROOFING (PVC)									
8200	Heat welded seams									
8700	Reinforced, 48 mils, 0.33 psf									
8750	Loose-laid & ballasted with stone/gravel (12 psf)	G-5	51	.784	Sq.	117	23	3.79	143.79	176
8800	Mechanically attached		35	1.143		112	33.50	5.50	151	191
8850	Fully adhered with adhesive	↓	26	1.538	↓	154	45	7.45	206.45	261
8860	Reinforced, 60 mils, .40 psf									
8870	Loose-laid & ballasted with stone/gravel (12 psf)	G-5	51	.784	Sq.	119	23	3.79	145.79	177
8880	Mechanically attached		35	1.143		113	33.50	5.50	152	192
8890	Fully adhered with adhesive	↓	26	1.538	↓	156	45	7.45	208.45	262

07 54 23 – Thermoplastic-Polyolefin Roofing

07 54 23.10 Thermoplastic Polyolefin Roofing (T.P.O)		Crew	Daily Output	Labor-Hours	Unit	Material	Labor	Equipment	Total	Total Incl O&P
0010	THERMOPLASTIC POLYOLEFIN ROOFING (T.P.O.)									
0100	45 mil, loose laid & ballasted with stone (1/2 ton/sq.)	G-5	51	.784	Sq.	87	23	3.79	113.79	142
0120	Fully adhered		25	1.600		78	47	7.75	132.75	181
0140	Mechanically attached		34	1.176		77.50	34.50	5.70	117.70	155
0160	Self adhered		35	1.143		84	33.50	5.50	123	161
0180	60 mil membrane, heat welded seams, ballasted		50	.800		100	23.50	3.86	127.36	157
0200	Fully adhered		25	1.600		91.50	47	7.75	146.25	195
0220	Mechanically attached		34	1.176		94.50	34.50	5.70	134.70	174
0240	Self adhered	↓	35	1.143	↓	103	33.50	5.50	142	182

07 54 30 – Ketone Ethylene Ester Roofing

07 54 30.10 Ketone Ethylene Ester Roofing		Crew	Daily Output	Labor-Hours	Unit	Material	Labor	Equipment	Total	Total Incl O&P
0010	KETONE ETHYLENE ESTER ROOFING									
0100	Ketone ethylene ester roofing, 50 mil, fully adhered	G-5	26	1.538	Sq.	190	45	7.45	242.45	300
0120	Mechanically attached		35	1.143		123	33.50	5.50	162	203
0140	Ballasted with stone	↓	51	.784		129	23	3.79	155.79	189
0160	50 mil, fleece backed, adhered w/hot asphalt	G-1	26	2.154	↓	152	65	22.50	239.50	310
0180	Accessories, pipe boot	1 Rofc	32	.250	Ea.	24	8.05		32.05	41.50
0200	Pre-formed corners		32	.250	"	8.50	8.05		16.55	24
0220	Ketone clad metal, including up to 4 bends	↓	330	.024	S.F.	3.88	.78		4.66	5.70
0240	Walkway pad	2 Rofc	800	.020	"	4.08	.64		4.72	5.65
0260	Stripping material	1 Rofc	310	.026	L.F.	.94	.83		1.77	2.56

07 57 Coated Foamed Roofing

07 57 13 – Sprayed Polyurethane Foam Roofing

07 57 13.10 Sprayed Polyurethane Foam Roofing (S.P.F.)		Crew	Daily Output	Labor-Hours	Unit	2016 Bare Costs Material	Labor	Equipment	Total	Total Incl O&P
0010	**SPRAYED POLYURETHANE FOAM ROOFING (S.P.F.)**									
0100	Primer for metal substrate (when required)	G-2A	3000	.008	S.F.	.48	.23	.24	.95	1.20
0200	Primer for non-metal substrate (when required)		3000	.008		.19	.23	.24	.66	.88
0300	Closed cell spray, polyurethane foam, 3 lb. per C.F. density, 1", R6.7		15000	.002		.66	.05	.05	.76	.86
0400	2", R13.4		13125	.002		1.32	.05	.05	1.42	1.60
0500	3", R18.6		11485	.002		1.98	.06	.06	2.10	2.36
0550	4", R24.8		10080	.002		2.64	.07	.07	2.78	3.11
0700	Spray-on silicone coating		2500	.010		1.21	.27	.29	1.77	2.14
0800	Warranty 5-20 year manufacturer's								.15	.15
0900	Warranty 20 year, no dollar limit								.20	.20

07 58 Roll Roofing

07 58 10 – Asphalt Roll Roofing

07 58 10.10 Roll Roofing

		Crew	Daily Output	Labor-Hours	Unit	Material	Labor	Equipment	Total	Total Incl O&P
0010	**ROLL ROOFING**									
0100	Asphalt, mineral surface									
0200	1 ply #15 organic felt, 1 ply mineral surfaced									
0300	Selvage roofing, lap 19", nailed & mopped	G-1	27	2.074	Sq.	77	62.50	21.50	161	223
0400	3 plies glass fiber felt (type IV), 1 ply mineral surfaced									
0500	Selvage roofing, lapped 19", mopped	G-1	25	2.240	Sq.	132	67.50	23	222.50	295
0600	Coated glass fiber base sheet, 2 plies of glass fiber									
0700	Felt (type IV), 1 ply mineral surfaced selvage									
0800	Roofing, lapped 19", mopped	G-1	25	2.240	Sq.	139	67.50	23	229.50	305
0900	On nailable decks	"	24	2.333	"	128	70	24	222	297
1000	3 plies glass fiber felt (type III), 1 ply mineral surfaced									
1100	Selvage roofing, lapped 19", mopped	G-1	25	2.240	Sq.	132	67.50	23	222.50	295

07 61 Sheet Metal Roofing

07 61 13 – Standing Seam Sheet Metal Roofing

07 61 13.10 Standing Seam Sheet Metal Roofing, Field Fab.

		Crew	Daily Output	Labor-Hours	Unit	Material	Labor	Equipment	Total	Total Incl O&P
0010	**STANDING SEAM SHEET METAL ROOFING, FIELD FABRICATED**									
0400	Copper standing seam roofing, over 10 squares, 16 oz., 125 lb. per sq.	1 Shee	1.30	6.154	Sq.	985	261		1,246	1,500
0600	18 oz., 140 lb. per sq.	"	1.20	6.667		1,100	282		1,382	1,675
1200	For abnormal conditions or small areas, add					25%	100%			
1300	For lead-coated copper, add					25%				

07 61 16 – Batten Seam Sheet Metal Roofing

07 61 16.10 Batten Seam Sheet Metal Roofing, Field Fabricated

		Crew	Daily Output	Labor-Hours	Unit	Material	Labor	Equipment	Total	Total Incl O&P
0010	**BATTEN SEAM SHEET METAL ROOFING, FIELD FABRICATED**									
0012	Copper batten seam roofing, over 10 sq., 16 oz., 130 lb. per sq.	1 Shee	1.10	7.273	Sq.	1,250	310		1,560	1,900
0020	Lead batten seam roofing, 5 lb. per S.F.		1.20	6.667		1,350	282		1,632	1,975
0100	Zinc/copper alloy batten seam roofing, .020 thick		1.20	6.667		1,275	282		1,557	1,875
0200	Copper roofing, batten seam, over 10 sq., 18 oz., 145 lb. per sq.		1	8		1,400	340		1,740	2,100
0800	Zinc, copper alloy roofing, batten seam, .027" thick		1.15	6.957		1,600	295		1,895	2,275
0900	.032" thick		1.10	7.273		1,850	310		2,160	2,550
1000	.040" thick		1.05	7.619		2,300	325		2,625	3,075

07 61 Sheet Metal Roofing

07 61 19 – Flat Seam Sheet Metal Roofing

07 61 19.10 Flat Seam Sheet Metal Roofing, Field Fabricated		Crew	Daily Output	Labor-Hours	Unit	2016 Bare Costs Material	Labor	Equipment	Total	Total Incl O&P
0010	**FLAT SEAM SHEET METAL ROOFING, FIELD FABRICATED**									
0900	Copper flat seam roofing, over 10 squares, 16 oz., 115 lb./sq.	1 Shee	1.20	6.667	Sq.	915	282		1,197	1,475
0950	18 oz., 130 lb./sq.		1.15	6.957		1,025	295		1,320	1,625
1000	20 oz., 145 lb./sq.		1.10	7.273		1,200	310		1,510	1,850
1008	Zinc flat seam roofing, .020" thick		1.20	6.667		1,100	282		1,382	1,675
1010	.027" thick		1.15	6.957		1,375	295		1,670	2,000
1020	.032" thick		1.12	7.143		1,575	305		1,880	2,225
1030	.040" thick		1.05	7.619		1,975	325		2,300	2,725
1100	Lead flat seam roofing, 5 lb. per S.F.		1.30	6.154		1,150	261		1,411	1,700

07 65 Flexible Flashing

07 65 10 – Sheet Metal Flashing

07 65 10.10 Sheet Metal Flashing and Counter Flashing		Crew	Daily Output	Labor-Hours	Unit	Material	Labor	Equipment	Total	Total Incl O&P
0010	**SHEET METAL FLASHING AND COUNTER FLASHING**									
0011	Including up to 4 bends									
0020	Aluminum, mill finish, .013" thick	1 Rofc	145	.055	S.F.	.83	1.77		2.60	4.18
0030	.016" thick		145	.055		.95	1.77		2.72	4.32
0060	.019" thick		145	.055		1.25	1.77		3.02	4.65
0100	.032" thick		145	.055		1.35	1.77		3.12	4.76
0200	.040" thick		145	.055		2.31	1.77		4.08	5.80
0300	.050" thick		145	.055		2.64	1.77		4.41	6.15
0325	Mill finish 5" x 7" step flashing, .016" thick		1920	.004	Ea.	.15	.13		.28	.42
0350	Mill finish 12" x 12" step flashing, .016" thick		1600	.005	"	.55	.16		.71	.91
0400	Painted finish, add				S.F.	.32			.32	.35
1000	Mastic-coated 2 sides, .005" thick	1 Rofc	330	.024		1.86	.78		2.64	3.49
1100	.016" thick		330	.024		2.05	.78		2.83	3.70
1600	Copper, 16 oz., sheets, under 1000 lb.		115	.070		7.85	2.23		10.08	12.75
1700	Over 4000 lb.		155	.052		7.85	1.66		9.51	11.70
1900	20 oz. sheets, under 1000 lb.		110	.073		10.30	2.33		12.63	15.65
2000	Over 4000 lb.		145	.055		9.80	1.77		11.57	14
2200	24 oz. sheets, under 1000 lb.		105	.076		14.60	2.45		17.05	20.50
2500	32 oz. sheets, under 1000 lb.		100	.080		18.50	2.57		21.07	25
2700	W shape for valleys, 16 oz., 24" wide		100	.080	L.F.	16.20	2.57		18.77	22.50
5800	Lead, 2.5 lb. per S.F., up to 12" wide		135	.059	S.F.	5.40	1.90		7.30	9.45
5900	Over 12" wide		135	.059		3.99	1.90		5.89	7.90
8900	Stainless steel sheets, 32 ga.,		155	.052		3.25	1.66		4.91	6.65
9000	28 ga.		155	.052		4.20	1.66		5.86	7.70
9100	26 ga.		155	.052		4.45	1.66		6.11	7.95
9200	24 ga.		155	.052		4.95	1.66		6.61	8.50
9290	For mechanically keyed flashing, add					40%				
9320	Steel sheets, galvanized, 20 ga.	1 Rofc	130	.062	S.F.	1.25	1.98		3.23	5
9322	22 ga.		135	.059		1.21	1.90		3.11	4.84
9324	24 ga.		140	.057		.92	1.83		2.75	4.39
9326	26 ga.		148	.054		.80	1.74		2.54	4.08
9328	28 ga.		155	.052		.69	1.66		2.35	3.82
9340	30 ga.		160	.050		.58	1.61		2.19	3.60
9400	Terne coated stainless steel, .015" thick, 28 ga		155	.052		8.05	1.66		9.71	11.90
9500	.018" thick, 26 ga		155	.052		9	1.66		10.66	12.95
9600	Zinc and copper alloy (brass), .020" thick		155	.052		10.20	1.66		11.86	14.25
9700	.027" thick		155	.052		12	1.66		13.66	16.25

07 65 Flexible Flashing

07 65 10 – Sheet Metal Flashing

07 65 10.10 Sheet Metal Flashing and Counter Flashing		Crew	Daily Output	Labor-Hours	Unit	Material	Labor	Equipment	Total	Total Incl O&P
						2016 Bare Costs				
9800	.032" thick	1 Rofc	155	.052	S.F.	15.40	1.66		17.06	20
9900	.040" thick		155	.052		20	1.66		21.66	25

07 65 12 – Fabric and Mastic Flashings

07 65 12.10 Fabric and Mastic Flashing and Counter Flashing

		Crew	Daily Output	Labor-Hours	Unit	Material	Labor	Equipment	Total	Total Incl O&P
0010	**FABRIC AND MASTIC FLASHING AND COUNTER FLASHING**									
1300	Asphalt flashing cement, 5 gallon				Gal.	9.65			9.65	10.60
4900	Fabric, asphalt-saturated cotton, specification grade	1 Rofc	35	.229	S.Y.	3.05	7.35		10.40	16.90
5000	Utility grade		35	.229		1.45	7.35		8.80	15.15
5300	Close-mesh fabric, saturated, 17 oz. per S.Y.		35	.229		2.20	7.35		9.55	15.95
5500	Fiberglass, resin-coated		35	.229		1	7.35		8.35	14.65
8500	Shower pan, bituminous membrane, 7 oz.		155	.052	S.F.	1.72	1.66		3.38	4.95

07 65 13 – Laminated Sheet Flashing

07 65 13.10 Laminated Sheet Flashing

		Crew	Daily Output	Labor-Hours	Unit	Material	Labor	Equipment	Total	Total Incl O&P
0010	**LAMINATED SHEET FLASHING**, Including up to 4 bends									
0500	Aluminum, fabric-backed 2 sides, mill finish, .004" thick	1 Rofc	330	.024	S.F.	1.60	.78		2.38	3.20
0700	.005" thick		330	.024		1.85	.78		2.63	3.48
0750	Mastic-backed, self adhesive		460	.017		3.60	.56		4.16	4.99
0800	Mastic-coated 2 sides, .004" thick		330	.024		1.60	.78		2.38	3.20
2800	Copper, paperbacked 1 side, 2 oz.		330	.024		1.84	.78		2.62	3.46
2900	3 oz.		330	.024		2.20	.78		2.98	3.86
3100	Paperbacked 2 sides, 2 oz.		330	.024		2.05	.78		2.83	3.70
3150	3 oz.		330	.024		2.30	.78		3.08	3.97
3200	5 oz.		330	.024		3.35	.78		4.13	5.15
3400	Mastic-backed 2 sides, copper, 2 oz.		330	.024		1.95	.78		2.73	3.59
3500	3 oz.		330	.024		2.50	.78		3.28	4.19
3700	5 oz.		330	.024		3.80	.78		4.58	5.60
3800	Fabric-backed 2 sides, copper, 2 oz.		330	.024		2.10	.78		2.88	3.75
4000	3 oz.		330	.024		2.80	.78		3.58	4.52
4100	5 oz.		330	.024		3.90	.78		4.68	5.75
4300	Copper-clad stainless steel, .015" thick, under 500 lb.		115	.070		6.40	2.23		8.63	11.15
4400	Over 2000 lb.		155	.052		6.70	1.66		8.36	10.40
4600	.018" thick, under 500 lb.		100	.080		7.85	2.57		10.42	13.40
4700	Over 2000 lb.		145	.055		7.50	1.77		9.27	11.50
8550	Shower pan, 3 ply copper and fabric, 3 oz.		155	.052		4	1.66		5.66	7.45
8600	7 oz.		155	.052		4.20	1.66		5.86	7.70
9300	Stainless steel, paperbacked 2 sides, .005" thick		330	.024		3.88	.78		4.66	5.70

07 65 19 – Plastic Sheet Flashing

07 65 19.10 Plastic Sheet Flashing and Counter Flashing

		Crew	Daily Output	Labor-Hours	Unit	Material	Labor	Equipment	Total	Total Incl O&P
0010	**PLASTIC SHEET FLASHING AND COUNTER FLASHING**									
7300	Polyvinyl chloride, black, 10 mil	1 Rofc	285	.028	S.F.	.21	.90		1.11	1.89
7400	20 mil		285	.028		.26	.90		1.16	1.95
7600	30 mil		285	.028		.34	.90		1.24	2.03
7700	60 mil		285	.028		.82	.90		1.72	2.56
7900	Black or white for exposed roofs, 60 mil		285	.028		1.80	.90		2.70	3.64
8060	PVC tape, 5" x 45 mils, for joint covers, 100 L.F./roll				Ea.	172			172	189
8850	Polyvinyl chloride, 30 mil	1 Rofc	160	.050	S.F.	1.45	1.61		3.06	4.56

07 65 Flexible Flashing

07 65 23 – Rubber Sheet Flashing

07 65 23.10 Rubber Sheet Flashing and Counterflashing		Crew	Daily Output	Labor-Hours	Unit	Material	Labor	Equipment	Total	Total Incl O&P
0010	RUBBER SHEET FLASHING AND COUNTERFLASHING									
4750	EPDM Cured	1 Rofc	285	.028	S.F.	1.25	.90		2.15	3.04
4800	Uncured		285	.028	"	1.75	.90		2.65	3.59
4810	EPDM 90 mils, 1" diameter pipe flashing		32	.250	Ea.	19.35	8.05		27.40	36.50
4820	2" diameter		30	.267		19.30	8.55		27.85	37
4830	3" diameter		28	.286		20.50	9.15		29.65	40
4840	4" diameter		24	.333		24	10.70		34.70	46
4850	6" diameter		22	.364		23.50	11.65		35.15	47.50
7150	Neoprene, 60 mil		285	.028	S.F.	2.08	.90		2.98	3.95
7160	Self-curing		285	.028		2.45	.90		3.35	4.36
7170	Uncured		285	.028		2.11	.90		3.01	3.98
8100	Rubber, butyl, 1/32" thick		285	.028		1.96	.90		2.86	3.82
8200	1/16" thick		285	.028		2.65	.90		3.55	4.58
8300	Neoprene, cured, 1/16" thick		285	.028		2.55	.90		3.45	4.47
8400	1/8" thick		285	.028		6.05	.90		6.95	8.30

07 65 26 – Self-Adhering Sheet Flashing

07 65 26.10 Self-Adhering Sheet or Roll Flashing		Crew	Daily Output	Labor-Hours	Unit	Material	Labor	Equipment	Total	Total Incl O&P
0010	SELF-ADHERING SHEET OR ROLL FLASHING									
0020	Self-adhered flashing, 25 mil cross laminated HDPE, 4" wide	1 Rofc	960	.008	L.F.	.20	.27		.47	.71
0040	6" wide		896	.009		.30	.29		.59	.85
0060	9" wide		832	.010		.44	.31		.75	1.06
0080	12" wide		768	.010		.59	.33		.92	1.27

07 71 Roof Specialties

07 71 16 – Manufactured Counterflashing Systems

07 71 16.10 Roof Drain Boot		Crew	Daily Output	Labor-Hours	Unit	Material	Labor	Equipment	Total	Total Incl O&P
0010	ROOF DRAIN BOOT									
0100	Cast iron, 4" diameter	1 Shee	125	.064	L.F.	50	2.71		52.71	60
0300	4" x 3"		125	.064		57.50	2.71		60.21	68
0400	5" x 4"		125	.064		117	2.71		119.71	134

07 71 19 – Manufactured Gravel Stops and Fasciae

07 71 19.10 Gravel Stop		Crew	Daily Output	Labor-Hours	Unit	Material	Labor	Equipment	Total	Total Incl O&P
0010	GRAVEL STOP									
0020	Aluminum, .050" thick, 4" face height, mill finish	1 Shee	145	.055	L.F.	6.40	2.34		8.74	10.95
0080	Duranodic finish		145	.055		7.35	2.34		9.69	12
0100	Painted		145	.055		7.45	2.34		9.79	12.10
1200	Copper, 16 oz., 3" face height		145	.055		24	2.34		26.34	30.50
1300	6" face height		135	.059		33	2.51		35.51	40.50
1350	Galv steel, 24 ga., 4" leg, plain, with continuous cleat, 4" face		145	.055		6.25	2.34		8.59	10.80
1360	6" face height		145	.055		6.50	2.34		8.84	11.05
1500	Polyvinyl chloride, 6" face height		135	.059		5.70	2.51		8.21	10.45
1800	Stainless steel, 24 ga., 6" face height		135	.059		15.30	2.51		17.81	21

07 71 19.30 Fascia		Crew	Daily Output	Labor-Hours	Unit	Material	Labor	Equipment	Total	Total Incl O&P
0010	FASCIA									
0100	Aluminum, reverse board and batten, .032" thick, colored, no furring incl	1 Shee	145	.055	S.F.	6.75	2.34		9.09	11.35
0200	Residential type, aluminum	1 Carp	200	.040	L.F.	1.95	1.51		3.46	4.68
0220	Vinyl	"	200	.040	"	2.10	1.51		3.61	4.84
0300	Steel, galv and enameled, stock, no furring, long panels	1 Shee	145	.055	S.F.	5.25	2.34		7.59	9.70
0600	Short panels	"	115	.070	"	5.85	2.95		8.80	11.35

07 71 23.10 Downspouts

07 71 23.10 Downspouts		Crew	Daily Output	Labor-Hours	Unit	2016 Bare Costs Material	Labor	Equipment	Total	Total Incl O&P
0010	**DOWNSPOUTS**									
0020	Aluminum, embossed, .020″ thick, 2″ x 3″	1 Shee	190	.042	L.F.	1.05	1.78		2.83	4.14
0100	Enameled		190	.042		1.39	1.78		3.17	4.51
0300	.024″ thick, 2′ x 3″		180	.044		2.05	1.88		3.93	5.40
0400	3″ x 4″		140	.057		2.37	2.42		4.79	6.65
0600	Round, corrugated aluminum, 3″ diameter, .020″ thick		190	.042		2	1.78		3.78	5.20
0700	4″ diameter, .025″ thick		140	.057		2.95	2.42		5.37	7.30
0900	Wire strainer, round, 2″ diameter		155	.052	Ea.	1.84	2.19		4.03	5.65
1000	4″ diameter		155	.052		2.40	2.19		4.59	6.30
1200	Rectangular, perforated, 2″ x 3″		145	.055		2.33	2.34		4.67	6.45
1300	3″ x 4″		145	.055		3.33	2.34		5.67	7.55
1500	Copper, round, 16 oz., stock, 2″ diameter		190	.042	L.F.	7.60	1.78		9.38	11.35
1600	3″ diameter		190	.042		8	1.78		9.78	11.80
1800	4″ diameter		145	.055		9.70	2.34		12.04	14.55
1900	5″ diameter		130	.062		15.75	2.61		18.36	21.50
2100	Rectangular, corrugated copper, stock, 2″ x 3″		190	.042		8.75	1.78		10.53	12.60
2200	3″ x 4″		145	.055		10.40	2.34		12.74	15.35
2400	Rectangular, plain copper, stock, 2″ x 3″		190	.042		11.05	1.78		12.83	15.15
2500	3″ x 4″		145	.055		13.95	2.34		16.29	19.25
2700	Wire strainers, rectangular, 2″ x 3″		145	.055	Ea.	17.05	2.34		19.39	22.50
2800	3″ x 4″		145	.055		17.80	2.34		20.14	23.50
3000	Round, 2″ diameter		145	.055		6.40	2.34		8.74	10.95
3100	3″ diameter		145	.055		7.10	2.34		9.44	11.70
3300	4″ diameter		145	.055		12.10	2.34		14.44	17.20
3400	5″ diameter		115	.070		22.50	2.95		25.45	29.50
3600	Lead-coated copper, round, stock, 2″ diameter		190	.042	L.F.	22	1.78		23.78	27.50
3700	3″ diameter		190	.042		22.50	1.78		24.28	27.50
3900	4″ diameter		145	.055		23.50	2.34		25.84	30
4000	5″ diameter, corrugated		130	.062		23	2.61		25.61	29.50
4200	6″ diameter, corrugated		105	.076		30	3.23		33.23	38.50
4300	Rectangular, corrugated, stock, 2″ x 3″		190	.042		15.70	1.78		17.48	20
4500	Plain, stock, 2″ x 3″		190	.042		24	1.78		25.78	29.50
4600	3″ x 4″		145	.055		33	2.34		35.34	40.50
4800	Steel, galvanized, round, corrugated, 2″ or 3″ diameter, 28 ga.		190	.042		2.10	1.78		3.88	5.30
4900	4″ diameter, 28 ga.		145	.055		2.52	2.34		4.86	6.65
5700	Rectangular, corrugated, 28 ga., 2″ x 3″		190	.042		1.94	1.78		3.72	5.10
5800	3″ x 4″		145	.055		1.82	2.34		4.16	5.90
6000	Rectangular, plain, 28 ga., galvanized, 2″ x 3″		190	.042		3.65	1.78		5.43	7
6100	3″ x 4″		145	.055		4.06	2.34		6.40	8.35
6300	Epoxy painted, 24 ga., corrugated, 2″ x 3″		190	.042		2.33	1.78		4.11	5.55
6400	3″ x 4″		145	.055		2.83	2.34		5.17	7
6600	Wire strainers, rectangular, 2″ x 3″		145	.055	Ea.	17.05	2.34		19.39	22.50
6700	3″ x 4″		145	.055		17.80	2.34		20.14	23.50
6900	Round strainers, 2″ or 3″ diameter		145	.055		3.86	2.34		6.20	8.15
7000	4″ diameter		145	.055		5.80	2.34		8.14	10.25
8200	Vinyl, rectangular, 2″ x 3″		210	.038	L.F.	2.08	1.61		3.69	4.98
8300	Round, 2-1/2″		220	.036	″	1.32	1.54		2.86	4.02

07 71 23.20 Downspout Elbows

07 71 23.20 Downspout Elbows		Crew	Daily Output	Labor-Hours	Unit	Material	Labor	Equipment	Total	Total Incl O&P
0010	**DOWNSPOUT ELBOWS**									
0020	Aluminum, embossed, 2″ x 3″, .020″ thick	1 Shee	100	.080	Ea.	.95	3.39		4.34	6.70
0100	Enameled		100	.080		1.75	3.39		5.14	7.60
0200	Embossed, 3″ x 4″, .025″ thick		100	.080		4.51	3.39		7.90	10.60

07 71 Roof Specialties

07 71 23 – Manufactured Gutters and Downspouts

07 71 23.20 Downspout Elbows		Crew	Daily Output	Labor-Hours	Unit	2016 Bare Costs Material	Labor	Equipment	Total	Total Incl O&P
0300	Enameled	1 Shee	100	.080	Ea.	3.75	3.39		7.14	9.80
0400	Embossed, corrugated, 3" diameter, .020" thick		100	.080		2.74	3.39		6.13	8.65
0500	4" diameter, .025" thick		100	.080		5.80	3.39		9.19	12
0600	Copper, 16 oz., 2" diameter		100	.080		9.45	3.39		12.84	16.05
0700	3" diameter		100	.080		9	3.39		12.39	15.55
0800	4" diameter		100	.080		13.90	3.39		17.29	21
1000	Rectangular, 2" x 3" corrugated		100	.080		9.35	3.39		12.74	15.95
1100	3" x 4" corrugated		100	.080		13.25	3.39		16.64	20
1300	Vinyl, 2-1/2" diameter, 45 or 75 degree bend		100	.080		3.88	3.39		7.27	9.90
1400	Tee Y junction		75	.107		12.90	4.52		17.42	21.50

07 71 23.30 Gutters		Crew	Daily Output	Labor-Hours	Unit	Material	Labor	Equipment	Total	Total Incl O&P
0010	**GUTTERS**									
0012	Aluminum, stock units, 5" K type, .027" thick, plain	1 Shee	125	.064	L.F.	2.81	2.71		5.52	7.60
0100	Enameled		125	.064		2.79	2.71		5.50	7.60
0300	5" K type, .032" thick, plain		125	.064		3.47	2.71		6.18	8.35
0400	Enameled		125	.064		3.59	2.71		6.30	8.45
0700	Copper, half round, 16 oz., stock units, 4" wide		125	.064		9.50	2.71		12.21	14.95
0900	5" wide		125	.064		7.25	2.71		9.96	12.45
1000	6" wide		118	.068		11.85	2.87		14.72	17.80
1200	K type, 16 oz., stock, 5" wide		125	.064		8.05	2.71		10.76	13.35
1300	6" wide		125	.064		8.50	2.71		11.21	13.85
1500	Lead coated copper, 16 oz., half round, stock, 4" wide		125	.064		14.70	2.71		17.41	20.50
1600	6" wide		118	.068		22	2.87		24.87	29
1800	K type, stock, 5" wide		125	.064		18.35	2.71		21.06	24.50
1900	6" wide		125	.064		19.65	2.71		22.36	26
2100	Copper clad stainless steel, K type, 5" wide		125	.064		7.50	2.71		10.21	12.75
2200	6" wide		125	.064		9.45	2.71		12.16	14.90
2400	Steel, galv, half round or box, 28 ga., 5" wide, plain		125	.064		2.10	2.71		4.81	6.85
2500	Enameled		125	.064		2.18	2.71		4.89	6.90
2700	26 ga., stock, 5" wide		125	.064		2.40	2.71		5.11	7.15
2800	6" wide		125	.064		2.78	2.71		5.49	7.60
3000	Vinyl, O.G., 4" wide	1 Carp	115	.070		1.30	2.63		3.93	5.85
3100	5" wide		115	.070		1.60	2.63		4.23	6.15
3200	4" half round, stock units		115	.070		1.35	2.63		3.98	5.90
3250	Joint connectors				Ea.	2.95			2.95	3.25
3300	Wood, clear treated cedar, fir or hemlock, 3" x 4"	1 Carp	100	.080	L.F.	10	3.02		13.02	16.05
3400	4" x 5"	"	100	.080	"	16.75	3.02		19.77	23.50
5000	Accessories, end cap, K type, aluminum 5"	1 Shee	625	.013	Ea.	.70	.54		1.24	1.67
5010	6"		625	.013		1.50	.54		2.04	2.55
5020	Copper, 5"		625	.013		3.22	.54		3.76	4.44
5030	6"		625	.013		3.54	.54		4.08	4.79
5040	Lead coated copper, 5"		625	.013		12.50	.54		13.04	14.65
5050	6"		625	.013		13.35	.54		13.89	15.60
5060	Copper clad stainless steel, 5"		625	.013		3	.54		3.54	4.20
5070	6"		625	.013		3.60	.54		4.14	4.86
5080	Galvanized steel, 5"		625	.013		1.28	.54		1.82	2.31
5090	6"		625	.013		2.18	.54		2.72	3.30
5100	Vinyl, 4"	1 Carp	625	.013		13.15	.48		13.63	15.30
5110	5"	"	625	.013		13.15	.48		13.63	15.30
5120	Half round, copper, 4"	1 Shee	625	.013		4.53	.54		5.07	5.90
5130	5"		625	.013		4.53	.54		5.07	5.90
5140	6"		625	.013		7.65	.54		8.19	9.35

07 71 Roof Specialties

07 71 23 – Manufactured Gutters and Downspouts

07 71 23.30 Gutters		Crew	Daily Output	Labor-Hours	Unit	2016 Bare Costs Material	Labor	Equipment	Total	Total Incl O&P
5150	Lead coated copper, 5"	1 Shee	625	.013	Ea.	14.15	.54		14.69	16.45
5160	6"		625	.013		21	.54		21.54	24.50
5170	Copper clad stainless steel, 5"		625	.013		4.50	.54		5.04	5.85
5180	6"		625	.013		4.50	.54		5.04	5.85
5190	Galvanized steel, 5"		625	.013		2.25	.54		2.79	3.38
5200	6"		625	.013		2.80	.54		3.34	3.98
5210	Outlet, aluminum, 2" x 3"		420	.019		.62	.81		1.43	2.03
5220	3" x 4"		420	.019		1.05	.81		1.86	2.51
5230	2-3/8" round		420	.019		.56	.81		1.37	1.97
5240	Copper, 2" x 3"		420	.019		6.50	.81		7.31	8.50
5250	3" x 4"		420	.019		7.85	.81		8.66	10
5260	2-3/8" round		420	.019		4.55	.81		5.36	6.35
5270	Lead coated copper, 2" x 3"		420	.019		25.50	.81		26.31	29.50
5280	3" x 4"		420	.019		28.50	.81		29.31	33
5290	2-3/8" round		420	.019		25.50	.81		26.31	29.50
5300	Copper clad stainless steel, 2" x 3"		420	.019		6.50	.81		7.31	8.50
5310	3" x 4"		420	.019		7.85	.81		8.66	10
5320	2-3/8" round		420	.019		4.55	.81		5.36	6.35
5330	Galvanized steel, 2" x 3"		420	.019		3.18	.81		3.99	4.85
5340	3" x 4"		420	.019		4.90	.81		5.71	6.75
5350	2-3/8" round		420	.019		3.98	.81		4.79	5.75
5360	K type mitres, aluminum		65	.123		3	5.20		8.20	12
5370	Copper		65	.123		17.40	5.20		22.60	28
5380	Lead coated copper		65	.123		54.50	5.20		59.70	68.50
5390	Copper clad stainless steel		65	.123		27.50	5.20		32.70	39
5400	Galvanized steel		65	.123		22.50	5.20		27.70	33.50
5420	Half round mitres, copper		65	.123		63.50	5.20		68.70	78.50
5430	Lead coated copper		65	.123		89.50	5.20		94.70	107
5440	Copper clad stainless steel		65	.123		56	5.20		61.20	70.50
5450	Galvanized steel		65	.123		25.50	5.20		30.70	36.50
5460	Vinyl mitres and outlets		65	.123		9.75	5.20		14.95	19.45
5470	Sealant		940	.009	L.F.	.01	.36		.37	.61
5480	Soldering		96	.083	"	.20	3.53		3.73	6.10

07 71 23.35 Gutter Guard

		Crew	Daily Output	Labor-Hours	Unit	Material	Labor	Equipment	Total	Total Incl O&P
0010	**GUTTER GUARD**									
0020	6" wide strip, aluminum mesh	1 Carp	500	.016	L.F.	2.50	.60		3.10	3.76
0100	Vinyl mesh	"	500	.016	"	2.70	.60		3.30	3.98

07 71 26 – Reglets

07 71 26.10 Reglets and Accessories

		Crew	Daily Output	Labor-Hours	Unit	Material	Labor	Equipment	Total	Total Incl O&P
0010	**REGLETS AND ACCESSORIES**									
0020	Reglet, aluminum, .025" thick, in parapet	1 Carp	225	.036	L.F.	1.98	1.34		3.32	4.43
0300	16 oz. copper		225	.036		6.50	1.34		7.84	9.40
0400	Galvanized steel, 24 ga.		225	.036		1.36	1.34		2.70	3.75
0600	Stainless steel, .020" thick		225	.036		3.80	1.34		5.14	6.45
0900	Counter flashing for above, 12" wide, .032" aluminum	1 Shee	150	.053		2.20	2.26		4.46	6.20
1200	16 oz. copper		150	.053		6.25	2.26		8.51	10.65
1300	Galvanized steel, 26 ga.		150	.053		1.30	2.26		3.56	5.20
1500	Stainless steel, .020" thick		150	.053		6.30	2.26		8.56	10.70

07 71 Roof Specialties

07 71 43 – Drip Edge

	07 71 43.10 Drip Edge, Rake Edge, Ice Belts	Crew	Daily Output	Labor-Hours	Unit	2016 Bare Costs Material	Labor	Equipment	Total	Total Incl O&P
0010	**DRIP EDGE, RAKE EDGE, ICE BELTS**									
0020	Aluminum, .016" thick, 5" wide, mill finish	1 Carp	400	.020	L.F.	.57	.76		1.33	1.90
0100	White finish		400	.020		.63	.76		1.39	1.96
0200	8" wide, mill finish		400	.020		1.45	.76		2.21	2.87
0300	Ice belt, 28" wide, mill finish		100	.080		7.65	3.02		10.67	13.50
0310	Vented, mill finish		400	.020		1.99	.76		2.75	3.46
0320	Painted finish		400	.020		2.24	.76		3	3.73
0400	Galvanized, 5" wide		400	.020		.59	.76		1.35	1.92
0500	8" wide, mill finish		400	.020		.82	.76		1.58	2.17
0510	Rake edge, aluminum, 1-1/2" x 1-1/2"		400	.020		.33	.76		1.09	1.63
0520	3-1/2" x 1-1/2"	↓	400	.020	↓	.45	.76		1.21	1.77

07 72 Roof Accessories

07 72 26 – Ridge Vents

	07 72 26.10 Ridge Vents and Accessories	Crew	Daily Output	Labor-Hours	Unit	Material	Labor	Equipment	Total	Total Incl O&P
0010	**RIDGE VENTS AND ACCESSORIES**									
2300	Ridge vent strip, mill finish	1 Shee	155	.052	L.F.	3.88	2.19		6.07	7.90

07 72 33 – Roof Hatches

	07 72 33.10 Roof Hatch Options	Crew	Daily Output	Labor-Hours	Unit	Material	Labor	Equipment	Total	Total Incl O&P
0010	**ROOF HATCH OPTIONS**									
0500	2'-6" x 3', aluminum curb and cover	G-3	10	3.200	Ea.	980	115		1,095	1,275
0520	Galvanized steel curb and aluminum cover		10	3.200		635	115		750	890
0540	Galvanized steel curb and cover		10	3.200		705	115		820	965
0600	2'-6" x 4'-6", aluminum curb and cover		9	3.556		1,275	127		1,402	1,625
0800	Galvanized steel curb and aluminum cover		9	3.556		885	127		1,012	1,200
0900	Galvanized steel curb and cover		9	3.556		965	127		1,092	1,275
1100	4' x 4' aluminum curb and cover		8	4		1,650	143		1,793	2,075
1120	Galvanized steel curb and aluminum cover		8	4		1,700	143		1,843	2,125
1140	Galvanized steel curb and cover		8	4		1,175	143		1,318	1,550
1200	2'-6" x 8'-0", aluminum curb and cover		6.60	4.848		1,900	173		2,073	2,375
1400	Galvanized steel curb and aluminum cover		6.60	4.848		1,800	173		1,973	2,275
1500	Galvanized steel curb and cover	↓	6.60	4.848		1,450	173		1,623	1,875
1800	For plexiglass panels, 2'-6" x 3'-0", add to above				↓	450			450	495

07 72 53 – Snow Guards

	07 72 53.10 Snow Guard Options	Crew	Daily Output	Labor-Hours	Unit	Material	Labor	Equipment	Total	Total Incl O&P
0010	**SNOW GUARD OPTIONS**									
0100	Slate & asphalt shingle roofs, fastened with nails	1 Rofc	160	.050	Ea.	12.55	1.61		14.16	16.75
0200	Standing seam metal roofs, fastened with set screws		48	.167		16.95	5.35		22.30	28.50
0300	Surface mount for metal roofs, fastened with solder		48	.167	↓	6.05	5.35		11.40	16.50
0400	Double rail pipe type, including pipe	↓	130	.062	L.F.	34	1.98		35.98	41

07 72 73 – Pitch Pockets

	07 72 73.10 Pitch Pockets, Variable Sizes	Crew	Daily Output	Labor-Hours	Unit	Material	Labor	Equipment	Total	Total Incl O&P
0010	**PITCH POCKETS, VARIABLE SIZES**									
0100	Adjustable, 4" to 7", welded corners, 4" deep	1 Rofc	48	.167	Ea.	16.45	5.35		21.80	28
0200	Side extenders, 6"	"	240	.033	"	2.75	1.07		3.82	5

07 72 Roof Accessories

07 72 80 – Vents

07 72 80.30 Vent Options		Crew	Daily Output	Labor-Hours	Unit	Material	Labor (2016 Bare Costs)	Equipment	Total	Total Incl O&P
0010	VENT OPTIONS									
0020	Plastic, for insulated decks, 1 per M.S.F.	1 Rofc	40	.200	Ea.	21	6.40		27.40	35
0100	Heavy duty		20	.400		38	12.85		50.85	65.50
0300	Aluminum		30	.267		23	8.55		31.55	41.50
0800	Polystyrene baffles, 12" wide for 16" O.C. rafter spacing	1 Carp	90	.089		.44	3.36		3.80	6.15
0900	For 24" O.C. rafter spacing	"	110	.073		.78	2.75		3.53	5.45

07 76 Roof Pavers

07 76 16 – Roof Decking Pavers

07 76 16.10 Roof Pavers and Supports		Crew	Daily Output	Labor-Hours	Unit	Material	Labor	Equipment	Total	Total Incl O&P
0010	ROOF PAVERS AND SUPPORTS									
1000	Roof decking pavers, concrete blocks, 2" thick, natural	1 Clab	115	.070	S.F.	3.37	2.03		5.40	7.10
1100	Colors		115	.070	"	3.73	2.03		5.76	7.50
1200	Support pedestal, bottom cap		960	.008	Ea.	3	.24		3.24	3.71
1300	Top cap		960	.008		4.80	.24		5.04	5.70
1400	Leveling shims, 1/16"		1920	.004		1.20	.12		1.32	1.52
1500	1/8"		1920	.004		1.20	.12		1.32	1.52
1600	Buffer pad		960	.008		2.50	.24		2.74	3.16
1700	PVC legs (4" SDR 35)		2880	.003	Inch	.12	.08		.20	.28
2000	Alternate pricing method, system in place		101	.079	S.F.	7	2.31		9.31	11.55

07 81 Applied Fireproofing

07 81 16 – Cementitious Fireproofing

07 81 16.10 Sprayed Cementitious Fireproofing		Crew	Daily Output	Labor-Hours	Unit	Material	Labor	Equipment	Total	Total Incl O&P
0010	SPRAYED CEMENTITIOUS FIREPROOFING									
0050	Not including canvas protection, normal density									
0100	Per 1" thick, on flat plate steel	G-2	3000	.008	S.F.	.53	.25	.05	.83	1.05
0200	Flat decking		2400	.010		.53	.31	.06	.90	1.17
0400	Beams		1500	.016		.53	.50	.09	1.12	1.51
0500	Corrugated or fluted decks		1250	.019		.80	.60	.11	1.51	1.99
0700	Columns, 1-1/8" thick		1100	.022		.60	.68	.12	1.40	1.92
0800	2-3/16" thick		700	.034		1.35	1.07	.19	2.61	3.47
0900	For canvas protection, add		5000	.005		.10	.15	.03	.28	.39
1000	Not including canvas protection, high density									
1100	Per 1" thick, on flat plate steel	G-2	3000	.008	S.F.	2.13	.25	.05	2.43	2.80
1110	On flat decking		2400	.010		2.13	.31	.06	2.50	2.92
1120	On beams		1500	.016		2.13	.50	.09	2.72	3.26
1130	Corrugated or fluted decks		1250	.019		2.13	.60	.11	2.84	3.45
1140	Columns, 1-1/8" thick		1100	.022		2.39	.68	.12	3.19	3.89
1150	2-3/16" thick		1100	.022		4.79	.68	.12	5.59	6.50
1170	For canvas protection, add		5000	.005		.10	.15	.03	.28	.39
1200	Not including canvas protection, retrofitting									
1210	Per 1" thick, on flat plate steel	G-2	1500	.016	S.F.	.46	.50	.09	1.05	1.42
1220	On flat decking		1200	.020		.46	.63	.11	1.20	1.65
1230	On beams		750	.032		.46	1	.18	1.64	2.35
1240	Corrugated or fluted decks		625	.038		.69	1.20	.22	2.11	2.97
1250	Columns, 1-1/8" thick		550	.044		.51	1.37	.25	2.13	3.09
1260	2-3/16" thick		500	.048		1.03	1.50	.27	2.80	3.90
1400	Accessories, preliminary spattered texture coat		4500	.005		.04	.17	.03	.24	.34

07 81 Applied Fireproofing

07 81 16 – Cementitious Fireproofing

07 81 16.10 Sprayed Cementitious Fireproofing		Crew	Daily Output	Labor-Hours	Unit	Material	2016 Bare Costs Labor	Equipment	Total	Total Incl O&P
1410	Bonding agent	1 Plas	1000	.008	S.F.	.10	.28		.38	.57

07 84 Firestopping

07 84 13 – Penetration Firestopping

07 84 13.10 Firestopping

07 84 13.10 Firestopping		Crew	Daily Output	Labor-Hours	Unit	Material	2016 Bare Costs Labor	Equipment	Total	Total Incl O&P
0010	**FIRESTOPPING** R078413-30									
0100	Metallic piping, non insulated									
0110	Through walls, 2" diameter	1 Carp	16	.500	Ea.	3.60	18.90		22.50	35.50
0120	4" diameter		14	.571		6.30	21.50		27.80	43
0130	6" diameter		12	.667		9.25	25		34.25	52
0140	12" diameter		10	.800		18.45	30		48.45	71
0150	Through floors, 2" diameter		32	.250		1.80	9.45		11.25	17.85
0160	4" diameter		28	.286		3.15	10.80		13.95	21.50
0170	6" diameter		24	.333		4.60	12.60		17.20	26
0180	12" diameter		20	.400		9.25	15.10		24.35	35.50
0190	Metallic piping, insulated									
0200	Through walls, 2" diameter	1 Carp	16	.500	Ea.	6.30	18.90		25.20	38.50
0210	4" diameter		14	.571		9.25	21.50		30.75	46
0220	6" diameter		12	.667		12.40	25		37.40	55.50
0230	12" diameter		10	.800		21	30		51	74
0240	Through floors, 2" diameter		32	.250		3.15	9.45		12.60	19.30
0250	4" diameter		28	.286		4.60	10.80		15.40	23
0260	6" diameter		24	.333		6.20	12.60		18.80	28
0270	12" diameter		20	.400		10.60	15.10		25.70	37
0280	Non metallic piping, non insulated									
0290	Through walls, 2" diameter	1 Carp	12	.667	Ea.	45.50	25		70.50	92
0300	4" diameter		10	.800		87	30		117	146
0310	6" diameter		8	1		170	38		208	251
0330	Through floors, 2" diameter		16	.500		30	18.90		48.90	64.50
0340	4" diameter		6	1.333		55.50	50.50		106	146
0350	6" diameter		6	1.333		103	50.50		153.50	198
0370	Ductwork, insulated & non insulated, round									
0380	Through walls, 6" diameter	1 Carp	12	.667	Ea.	12.40	25		37.40	55.50
0390	12" diameter		10	.800		21	30		51	74
0400	18" diameter		8	1		25.50	38		63.50	91.50
0410	Through floors, 6" diameter		16	.500		6.20	18.90		25.10	38.50
0420	12" diameter		14	.571		10.60	21.50		32.10	47.50
0430	18" diameter		12	.667		12.85	25		37.85	56
0440	Ductwork, insulated & non insulated, rectangular									
0450	With stiffener/closure angle, through walls, 6" x 12"	1 Carp	8	1	Ea.	16.20	38		54.20	81.50
0460	12" x 24"		6	1.333		32.50	50.50		83	120
0470	24" x 48"		4	2		65	75.50		140.50	199
0480	With stiffener/closure angle, through floors, 6" x 12"		10	.800		8.10	30		38.10	59.50
0490	12" x 24"		8	1		16.20	38		54.20	81.50
0500	24" x 48"		6	1.333		32.50	50.50		83	120
0510	Multi trade openings									
0520	Through walls, 6" x 12"	1 Carp	2	4	Ea.	45	151		196	305
0530	12" x 24"	"	1	8		147	300		447	665
0540	24" x 48"	2 Carp	1	16		525	605		1,130	1,600
0550	48" x 96"	"	.75	21.333		1,975	805		2,780	3,525
0560	Through floors, 6" x 12"	1 Carp	2	4		49.50	151		200.50	310

07 84 Firestopping

07 84 13 – Penetration Firestopping

07 84 13.10 Firestopping		Crew	Daily Output	Labor-Hours	Unit	Material	2016 Bare Costs Labor	Equipment	Total	Total Incl O&P
0570	12" x 24"	1 Carp	1	8	Ea.	68	300		368	580
0580	24" x 48"	2 Carp	.75	21.333		138	805		943	1,500
0590	48" x 96"	"	.50	32		390	1,200		1,590	2,450
0600	Structural penetrations, through walls									
0610	Steel beams, W8 x 10	1 Carp	8	1	Ea.	68.50	38		106.50	139
0620	W12 x 14		6	1.333		99	50.50		149.50	194
0630	W21 x 44		5	1.600		138	60.50		198.50	253
0640	W36 x 135		3	2.667		219	101		320	410
0650	Bar joists, 18" deep		6	1.333		40.50	50.50		91	129
0660	24" deep		6	1.333		54	50.50		104.50	144
0670	36" deep		5	1.600		81	60.50		141.50	190
0680	48" deep		4	2		108	75.50		183.50	245
0690	Construction joints, floor slab at exterior wall									
0700	Precast, brick, block or drywall exterior									
0710	2" wide joint	1 Carp	125	.064	L.F.	9.50	2.42		11.92	14.50
0720	4" wide joint	"	75	.107	"	14.85	4.03		18.88	23
0730	Metal panel, glass or curtain wall exterior									
0740	2" wide joint	1 Carp	40	.200	L.F.	9.50	7.55		17.05	23
0750	4" wide joint	"	25	.320	"	14.85	12.10		26.95	37
0760	Floor slab to drywall partition									
0770	Flat joint	1 Carp	100	.080	L.F.	13.20	3.02		16.22	19.55
0780	Fluted joint		50	.160		15.05	6.05		21.10	26.50
0790	Etched fluted joint		75	.107		18.50	4.03		22.53	27.50
0800	Floor slab to concrete/masonry partition									
0810	Flat joint	1 Carp	75	.107	L.F.	13.20	4.03		17.23	21.50
0820	Fluted joint	"	50	.160	"	15.05	6.05		21.10	26.50
0830	Concrete/CMU wall joints									
0840	1" wide	1 Carp	100	.080	L.F.	18.85	3.02		21.87	25.50
0850	2" wide		75	.107		23.50	4.03		27.53	32.50
0860	4" wide		50	.160		33.50	6.05		39.55	46.50
0870	Concrete/CMU floor joints									
0880	1" wide	1 Carp	200	.040	L.F.	18.85	1.51		20.36	23
0890	2" wide		150	.053		23.50	2.02		25.52	29
0900	4" wide		100	.080		33.50	3.02		36.52	41.50

07 91 Preformed Joint Seals

07 91 13 – Compression Seals

07 91 13.10 Compression Seals

		Crew	Daily Output	Labor-Hours	Unit	Material	Labor	Equipment	Total	Total Incl O&P
0010	**COMPRESSION SEALS**									
4900	O-ring type cord, 1/4"	1 Bric	472	.017	L.F.	.43	.61		1.04	1.49
4910	1/2"		440	.018		1.20	.66		1.86	2.41
4920	3/4"		424	.019		2.40	.68		3.08	3.78
4930	1"		408	.020		4.08	.71		4.79	5.65
4940	1-1/4"		384	.021		7.40	.75		8.15	9.40
4950	1-1/2"		368	.022		9.20	.78		9.98	11.40
4960	1-3/4"		352	.023		18.55	.82		19.37	22
4970	2"		344	.023		22	.84		22.84	26

07 91 Preformed Joint Seals

07 91 16 – Joint Gaskets

07 91 16.10 Joint Gaskets		Crew	Daily Output	Labor-Hours	Unit	2016 Bare Costs Material	Labor	Equipment	Total	Total Incl O&P
0010	**JOINT GASKETS**									
4400	Joint gaskets, neoprene, closed cell w/adh, 1/8" x 3/8"	1 Bric	240	.033	L.F.	.32	1.20		1.52	2.36
4500	1/4" x 3/4"		215	.037		.62	1.34		1.96	2.92
4700	1/2" x 1"		200	.040		1.45	1.44		2.89	4.01
4800	3/4" x 1-1/2"	↓	165	.048	↓	1.60	1.75		3.35	4.68

07 91 23 – Backer Rods

07 91 23.10 Backer Rods		Crew	Daily Output	Labor-Hours	Unit	Material	Labor	Equipment	Total	Total Incl O&P
0010	**BACKER RODS**									
0030	Backer rod, polyethylene, 1/4" diameter	1 Bric	4.60	1.739	C.L.F.	2.28	63		65.28	108
0050	1/2" diameter		4.60	1.739		4	63		67	109
0070	3/4" diameter		4.60	1.739		6.50	63		69.50	112
0090	1" diameter	↓	4.60	1.739	↓	11.50	63		74.50	118

07 91 26 – Joint Fillers

07 91 26.10 Joint Fillers		Crew	Daily Output	Labor-Hours	Unit	Material	Labor	Equipment	Total	Total Incl O&P
0010	**JOINT FILLERS**									
4360	Butyl rubber filler, 1/4" x 1/4"	1 Bric	290	.028	L.F.	.22	1		1.22	1.91
4365	1/2" x 1/2"		250	.032		.90	1.16		2.06	2.92
4370	1/2" x 3/4"		210	.038		1.35	1.38		2.73	3.77
4375	3/4" x 3/4"		230	.035		2.02	1.26		3.28	4.31
4380	1" x 1"	↓	180	.044	↓	2.70	1.60		4.30	5.65
4390	For coloring, add					12%				
4980	Polyethylene joint backing, 1/4" x 2"	1 Bric	2.08	3.846	C.L.F.	12.90	139		151.90	246
4990	1/4" x 6"		1.28	6.250	"	28.50	226		254.50	405
5600	Silicone, room temp vulcanizing foam seal, 1/4" x 1/2"		1312	.006	L.F.	.46	.22		.68	.87
5610	1/2" x 1/2"		656	.012		.91	.44		1.35	1.73
5620	1/2" x 3/4"		442	.018		1.37	.65		2.02	2.59
5630	3/4" x 3/4"		328	.024		2.05	.88		2.93	3.72
5640	1/8" x 1"		1312	.006		.46	.22		.68	.87
5650	1/8" x 3"		442	.018		1.37	.65		2.02	2.59
5670	1/4" x 3"		295	.027		2.73	.98		3.71	4.63
5680	1/4" x 6"		148	.054		5.45	1.95		7.40	9.25
5690	1/2" x 6"		82	.098		10.90	3.52		14.42	17.85
5700	1/2" x 9"		52.50	.152		16.40	5.50		21.90	27
5710	1/2" x 12"	↓	33	.242	↓	22	8.75		30.75	38.50

07 92 Joint Sealants

07 92 13 – Elastomeric Joint Sealants

07 92 13.20 Caulking and Sealant Options		Crew	Daily Output	Labor-Hours	Unit	Material	Labor	Equipment	Total	Total Incl O&P
0010	**CAULKING AND SEALANT OPTIONS**									
0050	Latex acrylic based, bulk				Gal.	27			27	29.50
0055	Bulk in place 1/4" x 1/4" bead	1 Bric	300	.027	L.F.	.08	.96		1.04	1.70
0060	1/4" x 3/8"		294	.027		.14	.98		1.12	1.79
0065	1/4" x 1/2"		288	.028		.19	1		1.19	1.88
0075	3/8" x 3/8"		284	.028		.21	1.02		1.23	1.93
0080	3/8" x 1/2"		280	.029		.28	1.03		1.31	2.03
0085	3/8" x 5/8"		276	.029		.35	1.05		1.40	2.14
0095	3/8" x 3/4"		272	.029		.42	1.06		1.48	2.23
0100	1/2" x 1/2"		275	.029		.38	1.05		1.43	2.16
0105	1/2" x 5/8"	↓	269	.030	↓	.47	1.07		1.54	2.31

07 92 Joint Sealants

07 92 13 – Elastomeric Joint Sealants

07 92 13.20 Caulking and Sealant Options		Crew	Daily Output	Labor-Hours	Unit	Material	2016 Bare Costs Labor	Equipment	Total	Total Incl O&P
0110	1/2" x 3/4"	1 Bric	263	.030	L.F.	.56	1.10		1.66	2.45
0115	1/2" x 7/8"		256	.031		.66	1.13		1.79	2.60
0120	1/2" x 1"		250	.032		.75	1.16		1.91	2.76
0125	3/4" x 3/4"		244	.033		.85	1.18		2.03	2.90
0130	3/4" x 1"		225	.036		1.13	1.28		2.41	3.38
0135	1" x 1"		200	.040		1.50	1.44		2.94	4.06
0190	Cartridges				Gal.	32.50			32.50	36
0200	11 fl. oz. cartridge				Ea.	2.80			2.80	3.08
0500	1/4" x 1/2"	1 Bric	288	.028	L.F.	.23	1		1.23	1.92
0600	1/2" x 1/2"		275	.029		.46	1.05		1.51	2.25
0800	3/4" x 3/4"		244	.033		1.03	1.18		2.21	3.10
0900	3/4" x 1"		225	.036		1.37	1.28		2.65	3.65
1000	1" x 1"		200	.040		1.72	1.44		3.16	4.30
1400	Butyl based, bulk				Gal.	35			35	38.50
1500	Cartridges				"	38.50			38.50	42
1700	1/4" x 1/2", 154 L.F./gal.	1 Bric	288	.028	L.F.	.23	1		1.23	1.92
1800	1/2" x 1/2", 77 L.F./gal.	"	275	.029	"	.45	1.05		1.50	2.25
2300	Polysulfide compounds, 1 component, bulk				Gal.	82			82	90
2400	Cartridges				"	128			128	141
2600	1 or 2 component, in place, 1/4" x 1/4", 308 L.F./gal.	1 Bric	300	.027	L.F.	.27	.96		1.23	1.90
2700	1/2" x 1/4", 154 L.F./gal.		288	.028		.53	1		1.53	2.26
2900	3/4" x 3/8", 68 L.F./gal.		272	.029		1.21	1.06		2.27	3.10
3000	1" x 1/2", 38 L.F./gal.		250	.032		2.16	1.16		3.32	4.30
3200	Polyurethane, 1 or 2 component				Gal.	53			53	58.50
3300	Cartridges				"	65			65	71.50
3500	Bulk, in place, 1/4" x 1/4"	1 Bric	300	.027	L.F.	.17	.96		1.13	1.80
3655	1/2" x 1/4"		288	.028		.34	1		1.34	2.05
3800	3/4" x 3/8"		272	.029		.78	1.06		1.84	2.63
3900	1" x 1/2"		250	.032		1.38	1.16		2.54	3.45
4100	Silicone rubber, bulk				Gal.	54			54	59.50
4200	Cartridges				"	53			53	58.50

07 92 19 – Acoustical Joint Sealants

07 92 19.10 Acoustical Sealant		Crew	Daily Output	Labor-Hours	Unit	Material	Labor	Equipment	Total	Total Incl O&P
0010	**ACOUSTICAL SEALANT**									
0020	Acoustical sealant, elastomeric, cartridges				Ea.	8.50			8.50	9.35
0025	In place, 1/4" x 1/4"	1 Bric	300	.027	L.F.	.35	.96		1.31	1.99
0030	1/4" x 1/2"		288	.028		.69	1		1.69	2.43
0035	1/2" x 1/2"		275	.029		1.39	1.05		2.44	3.27
0040	1/2" x 3/4"		263	.030		2.08	1.10		3.18	4.12
0045	3/4" x 3/4"		244	.033		3.12	1.18		4.30	5.40
0050	1" x 1"		200	.040		5.55	1.44		6.99	8.50

Division Notes

	CREW	DAILY OUTPUT	LABOR-HOURS	UNIT	BARE COSTS MAT.	LABOR	EQUIP.	TOTAL	TOTAL INCL O&P

Estimating Tips

08 10 00 Doors and Frames

All exterior doors should be addressed for their energy conservation (insulation and seals).

- Most metal doors and frames look alike, but there may be significant differences among them. When estimating these items, be sure to choose the line item that most closely compares to the specification or door schedule requirements regarding:
  - ☐ type of metal
  - ☐ metal gauge
  - ☐ door core material
  - ☐ fire rating
  - ☐ finish
- Wood and plastic doors vary considerably in price. The primary determinant is the veneer material. Lauan, birch, and oak are the most common veneers. Other variables include the following:
  - ☐ hollow or solid core
  - ☐ fire rating
  - ☐ flush or raised panel
  - ☐ finish
- Door pricing includes bore for cylindrical lockset and mortise for hinges.

08 30 00 Specialty Doors and Frames

- There are many varieties of special doors, and they are usually priced per each. Add frames, hardware, or operators required for a complete installation.

08 40 00 Entrances, Storefronts, and Curtain Walls

- Glazed curtain walls consist of the metal tube framing and the glazing material. The cost data in this subdivision is presented for the metal tube framing alone or the composite wall. If your estimate requires a detailed takeoff of the framing, be sure to add the glazing cost and any tints.

08 50 00 Windows

- Most metal windows are delivered preglazed. However, some metal windows are priced without glass. Refer to 08 80 00 Glazing for glass pricing. The grade C indicates commercial grade windows, usually ASTM C-35.
- All wood windows and vinyl are priced preglazed. The glazing is insulating glass. Add the cost of screens and grills if required, and not already included.

08 70 00 Hardware

- Hardware costs add considerably to the cost of a door. The most efficient method to determine the hardware requirements for a project is to review the door and hardware schedule together. One type of door may have different hardware, depending on the door usage.
- Door hinges are priced by the pair, with most doors requiring 1-1/2 pairs per door. The hinge prices do not include installation labor, because it is included in door installation.

Hinges are classified according to the frequency of use, base material, and finish.

08 80 00 Glazing

- Different openings require different types of glass. The most common types are:
  - ☐ float
  - ☐ tempered
  - ☐ insulating
  - ☐ impact-resistant
  - ☐ ballistic-resistant
- Most exterior windows are glazed with insulating glass. Entrance doors and window walls, where the glass is less than 18" from the floor, are generally glazed with tempered glass. Interior windows and some residential windows are glazed with float glass.
- Coastal communities require the use of impact-resistant glass, dependant on wind speed.
- The insulation or 'u' value is a strong consideration, along with solar heat gain, to determine total energy efficiency.

Reference Numbers

Reference numbers are shown at the beginning of some major classifications. These numbers refer to related items in the Reference Section. The reference information may be an estimating procedure, an alternate pricing method, or technical information.

Note: Not all subdivisions listed here necessarily appear. ■

08 01 Operation and Maintenance of Openings

08 01 53 – Operation and Maintenance of Plastic Windows

	08 01 53.81 Solid Vinyl Replacement Windows	Crew	Daily Output	Labor-Hours	Unit	2016 Bare Costs Material	Labor	Equipment	Total	Total Incl O&P
0010	**SOLID VINYL REPLACEMENT WINDOWS** R085313-20									
0020	Double hung, insulated glass, up to 83 united inches G	2 Carp	8	2	Ea.	345	75.50		420.50	505
0040	84 to 93 G		8	2		385	75.50		460.50	550
0060	94 to 101 G		6	2.667		385	101		486	595
0080	102 to 111 G		6	2.667		405	101		506	615
0100	112 to 120 G		6	2.667		440	101		541	655
0120	For each united inch over 120, add G		800	.020	Inch	5.60	.76		6.36	7.40
0140	Casement windows, one operating sash, 42 to 60 united inches G		8	2	Ea.	240	75.50		315.50	390
0160	61 to 70 G		8	2		270	75.50		345.50	425
0180	71 to 80 G		8	2		290	75.50		365.50	445
0200	81 to 96 G		8	2		310	75.50		385.50	465
0220	Two operating sash, 58 to 78 united inches G		8	2		475	75.50		550.50	650
0240	79 to 88 G		8	2		505	75.50		580.50	680
0260	89 to 98 G		8	2		550	75.50		625.50	730
0280	99 to 108 G		6	2.667		575	101		676	805
0300	109 to 121 G		6	2.667		615	101		716	845
0320	Two operating, one fixed sash, 73 to 108 united inches G		8	2		745	75.50		820.50	945
0340	109 to 118 G		8	2		785	75.50		860.50	990
0360	119 to 128 G		6	2.667		805	101		906	1,050
0380	129 to 138 G		6	2.667		870	101		971	1,125
0400	139 to 156 G		6	2.667		945	101		1,046	1,225
0420	Four operating sash, 98 to 118 united inches G		8	2		1,075	75.50		1,150.50	1,300
0440	119 to 128 G		8	2		1,150	75.50		1,225.50	1,375
0460	129 to 138 G		6	2.667		1,200	101		1,301	1,500
0480	139 to 148 G		6	2.667		1,275	101		1,376	1,575
0500	149 to 168 G		6	2.667		1,350	101		1,451	1,650
0520	169 to 178 G		6	2.667		1,450	101		1,551	1,775
0560	Fixed picture window, up to 63 united inches G		8	2		180	75.50		255.50	325
0580	64 to 83 G		8	2		206	75.50		281.50	355
0600	84 to 101 G		8	2		257	75.50		332.50	410
0620	For each united inch over 101, add G		900	.018	Inch	3.10	.67		3.77	4.54
0800	Cellulose fiber insulation, poured into sash balance cavity G	1 Carp	36	.222	C.F.	.69	8.40		9.09	14.80
0820	Silicone caulking at perimeter G	″	800	.010	L.F.	.17	.38		.55	.82
2000	Impact resistant replacement windows									
2005	Laminated glass, 120 mph rating, measure in united inches									
2010	Installation labor does not cover any rework of the window opening									
2020	Double hung, insulated glass, up to 101 united inches	2 Carp	8	2	Ea.	480	75.50		555.50	650
2025	For each united inch over 101, add		80	.200	Inch	3.25	7.55		10.80	16.25
2100	Casement windows, impact resistant, up to 60 united inches		8	2	Ea.	450	75.50		525.50	620
2120	61 to 70		8	2		480	75.50		555.50	650
2130	71 to 80		8	2		500	75.50		575.50	675
2140	81 to 100		8	2		515	75.50		590.50	690
2150	For each united inch over 100, add		80	.200	Inch	4.55	7.55		12.10	17.65
2200	Awning windows, impact resistant, up to 60 united inches		8	2	Ea.	470	75.50		545.50	640
2220	61 to 70		8	2		480	75.50		555.50	650
2230	71 to 80		8	2		500	75.50		575.50	675
2240	For each united inch over 80, add		80	.200	Inch	4.55	7.55		12.10	17.65
2300	Picture windows, impact resistant, up to 63 united inches		8	2	Ea.	345	75.50		420.50	505
2320	63 to 83		8	2		370	75.50		445.50	530
2330	84 to 101		8	2		415	75.50		490.50	580
2340	For each united inch over 101, add		80	.200	Inch	3.25	7.55		10.80	16.25

08 05 05 – Selective Demolition for Openings

08 05 05.10 Selective Demolition Doors		Crew	Daily Output	Labor-Hours	Unit	Material	2016 Bare Costs Labor	Equipment	Total	Total Incl O&P
0010	**SELECTIVE DEMOLITION DOORS** R024119-10									
0200	Doors, exterior, 1-3/4" thick, single, 3' x 7' high	1 Clab	16	.500	Ea.		14.60		14.60	24.50
0202	Doors, exterior, 3' - 6' wide x 7' high		12	.667			19.45		19.45	32.50
0210	3' x 8' high		10	.800			23.50		23.50	39
0215	Double, 3' x 8' high		6	1.333			39		39	65
0220	Double, 6' x 7' high		12	.667			19.45		19.45	32.50
0500	Interior, 1-3/8" thick, single, 3' x 7' high		20	.400			11.70		11.70	19.55
0520	Double, 6' x 7' high		16	.500			14.60		14.60	24.50
0700	Bi-folding, 3' x 6'-8" high		20	.400			11.70		11.70	19.55
0720	6' x 6'-8" high		18	.444			13		13	21.50
0900	Bi-passing, 3' x 6'-8" high		16	.500			14.60		14.60	24.50
0940	6' x 6'-8" high		14	.571			16.70		16.70	28
1500	Remove and reset, hollow core	1 Carp	8	1			38		38	63.50
1520	Solid		6	1.333			50.50		50.50	84.50
2000	Frames, including trim, metal		8	1			38		38	63.50
2200	Wood	2 Carp	32	.500			18.90		18.90	31.50
2201	Alternate pricing method	1 Carp	200	.040	L.F.		1.51		1.51	2.53
3000	Special doors, counter doors	2 Carp	6	2.667	Ea.		101		101	169
3001	Special doors, counter doors	2 Clab	6	2.667			78		78	130
3200	Floor door (trap type), or access type	2 Carp	8	2			75.50		75.50	127
3300	Glass, sliding, including frames		12	1.333			50.50		50.50	84.50
3400	Overhead, commercial, 12' x 12' high		4	4			151		151	253
3500	Residential, 9' x 7' high		8	2			75.50		75.50	127
3540	16' x 7' high		7	2.286			86.50		86.50	145
3600	Remove and reset, small		4	4			151		151	253
3620	Large		2.50	6.400			242		242	405
3660	Remove and reset elec. garage door opener	1 Carp	8	1			38		38	63.50
3902	Cafe/bar swing door	2 Clab	8	2			58.50		58.50	98
4000	Residential lockset, exterior	1 Carp	28	.286			10.80		10.80	18.10
4010	Residential lockset, exterior w/deadbolt		26	.308			11.65		11.65	19.50
4020	Residential lockset, interior		30	.267			10.10		10.10	16.90
4200	Deadbolt lock		32	.250			9.45		9.45	15.85
4224	Pocket door		8	1			38		38	63.50
5590	Remove mail slot	1 Clab	45	.178			5.20		5.20	8.70
5595	Remove peep hole	"	45	.178			5.20		5.20	8.70
5600	Remove door sidelight	1 Carp	6	1.333			50.50		50.50	84.50
5604	Remove 9' x 7' garage door	2 Carp	9.50	1.684			63.50		63.50	107
5624	16' x 7' garage door		8	2			75.50		75.50	127
5626	9' x 7' swing up garage door		7	2.286			86.50		86.50	145
5628	16' x 7' swing up garage door		7	2.286			86.50		86.50	145
5630	Garage door demolition, remove garage door track	1 Carp	6	1.333			50.50		50.50	84.50
5644	Remove overhead door opener	1 Clab	5	1.600			46.50		46.50	78
6334	Remove shower door unit		16	.500			14.60		14.60	24.50
6384	Remove French door unit		8	1			29		29	49
7100	Remove double swing pneumatic doors, openers and sensors	2 Skwk	.50	32	Opng.		1,225		1,225	2,050
7570	Remove shock absorbing door	2 Sswk	1.90	8.421	"		335		335	615

08 05 05.20 Selective Demolition of Windows		Crew	Daily Output	Labor-Hours	Unit	Material	Labor	Equipment	Total	Total Incl O&P
0010	**SELECTIVE DEMOLITION OF WINDOWS** R024119-10									
0200	Aluminum, including trim, to 12 S.F.	1 Clab	16	.500	Ea.		14.60		14.60	24.50
0240	To 25 S.F.		11	.727			21		21	35.50
0280	To 50 S.F.		5	1.600			46.50		46.50	78
0320	Storm windows/screens, to 12 S.F.		27	.296			8.65		8.65	14.50

08 05 Common Work Results for Openings

08 05 05 – Selective Demolition for Openings

08 05 05.20	Selective Demolition of Windows	Crew	Daily Output	Labor-Hours	Unit	Material	2016 Bare Costs Labor	Equipment	Total	Total Incl O&P
0360	To 25 S.F.	1 Clab	21	.381	Ea.		11.10		11.10	18.65
0400	To 50 S.F.		16	.500			14.60		14.60	24.50
0600	Glass, up to 10 S.F. per window		200	.040	S.F.		1.17		1.17	1.96
0620	Over 10 S.F. per window		150	.053	"		1.56		1.56	2.61
2000	Wood, including trim, to 12 S.F.		22	.364	Ea.		10.60		10.60	17.80
2020	To 25 S.F.		18	.444			13		13	21.50
2060	To 50 S.F.		13	.615			17.95		17.95	30
2065	To 180 S.F.		8	1			29		29	49
4300	Remove bay/bow window	2 Carp	6	2.667			101		101	169
4410	Remove skylight, plstc domes, flush/curb mtd	G-3	395	.081	S.F.		2.90		2.90	4.84
5020	Remove and reset window, up to a 2′ x 2′ window	1 Carp	6	1.333	Ea.		50.50		50.50	84.50
5040	Up to a 3′ x 3′ window		4	2			75.50		75.50	127
5080	Up to a 4′ x 5′ window		2	4			151		151	253
6000	Screening only	1 Clab	4000	.002	S.F.		.06		.06	.10

08 11 Metal Doors and Frames

08 11 16 – Aluminum Doors and Frames

08 11 16.10	Entrance Doors	Crew	Daily Output	Labor-Hours	Unit	Material	Labor	Equipment	Total	Total Incl O&P
0010	**ENTRANCE DOORS** and frame, aluminum, narrow stile									
0011	Including standard hardware, clear finish, no glass									
0012	Top and bottom offset pivots, 1/4″ beveled glass stops, threshold									
0013	Dead bolt lock with inside thumb screw, standard push pull									
0020	3′-0″ x 7′-0″ opening	2 Sswk	2	8	Ea.	935	320		1,255	1,600
0025	Anodizing aluminum entr. door & frame, add					104			104	114
0030	3′-6″ x 7′-0″ opening	2 Sswk	2	8		880	320		1,200	1,550
0100	3′-0″ x 10′-0″ opening, 3′ high transom		1.80	8.889		1,300	355		1,655	2,100
0200	3′-6″ x 10′-0″ opening, 3′ high transom		1.80	8.889		1,375	355		1,730	2,150
0280	5′-0″ x 7′-0″ opening		2	8		1,500	320		1,820	2,225
0300	6′-0″ x 7′-0″ opening		1.30	12.308		1,275	490		1,765	2,300
0301	6′-0″ x 7′-0″ opening		1.30	12.308	Pr.	1,275	490		1,765	2,300
0400	6′-0″ x 10′-0″ opening, 3′ high transom		1.10	14.545	"	1,700	580		2,280	2,950
0520	3′-0″ x 7′-0″ opening, wide stile		2	8	Ea.	1,000	320		1,320	1,675
0540	3′-6″ x 7′-0″ opening		2	8		1,250	320		1,570	1,950
0560	5′-0″ x 7′-0″ opening		2	8		1,625	320		1,945	2,350
0580	6′-0″ x 7′-0″ opening		1.30	12.308	Pr.	1,675	490		2,165	2,725
0600	7′-0″ x 7′-0″ opening		1	16	"	1,800	640		2,440	3,150
1200	For non-standard size, add				Leaf	80%				
1250	For installation of non-standard size, add						20%			
1300	Light bronze finish, add				Leaf	36%				
1400	Dark bronze finish, add					25%				
1500	For black finish, add					40%				
1600	Concealed panic device, add					955			955	1,050
1700	Electric striker release, add				Opng.	286			286	315
1800	Floor check, add				Leaf	660			660	730
1900	Concealed closer, add				"	540			540	595

08 11 Metal Doors and Frames

08 11 63 – Metal Screen and Storm Doors and Frames

08 11 63.23 Aluminum Screen and Storm Doors and Frames		Crew	Daily Output	Labor-Hours	Unit	Material	2016 Bare Costs Labor	Equipment	Total	Total Incl O&P
0010	**ALUMINUM SCREEN AND STORM DOORS AND FRAMES**									
0020	Combination storm and screen									
0420	Clear anodic coating, 2′-8″ wide	2 Carp	14	1.143	Ea.	207	43		250	300
0440	3′-0″ wide	″	14	1.143	″	180	43		223	270
0500	For 7′ door height, add					8%				
1020	Mill finish, 2′-8″ wide	2 Carp	14	1.143	Ea.	240	43		283	335
1040	3′-0″ wide	″	14	1.143		263	43		306	360
1100	For 7′-0″ door, add					8%				
1520	White painted, 2′-8″ wide	2 Carp	14	1.143		288	43		331	390
1540	3′-0″ wide		14	1.143		310	43		353	415
1541	Storm door, painted, alum., insul., 6′-8″ x 2′-6″ wide		14	1.143		260	43		303	360
1545	2′-8″ wide		14	1.143		279	43		322	380
1600	For 7′-0″ door, add					8%				
1800	Aluminum screen door, 6′-8″ x 2′-8″ wide	2 Carp	14	1.143		155	43		198	244
1810	3′-0″ wide	″	14	1.143		260	43		303	360

08 12 Metal Frames

08 12 13 – Hollow Metal Frames

08 12 13.13 Standard Hollow Metal Frames		Crew	Daily Output	Labor-Hours	Unit	Material	Labor	Equipment	Total	Total Incl O&P
0010	**STANDARD HOLLOW METAL FRAMES**									
0020	16 ga., up to 5-3/4″ jamb depth									
0025	3′-0″ x 6′-8″ single G	2 Carp	16	1	Ea.	148	38		186	227
0028	3′-6″ wide, single G		16	1		160	38		198	240
0030	4′-0″ wide, single G		16	1		159	38		197	238
0040	6′-0″ wide, double G		14	1.143		204	43		247	297
0045	8′-0″ wide, double G		14	1.143		213	43		256	305
0100	3′-0″ x 7′-0″ single G		16	1		155	38		193	235
0110	3′-6″ wide, single G		16	1		165	38		203	246
0112	4′-0″ wide, single G		16	1		165	38		203	246
0140	6′-0″ wide, double G		14	1.143		194	43		237	287
0145	8′-0″ wide, double G		14	1.143		229	43		272	325
1000	16 ga., up to 4-7/8″ deep, 3′-0″ x 7′-0″ single G		16	1		166	38		204	247
1140	6′-0″ wide, double G		14	1.143		188	43		231	280
1200	16 ga., 8-3/4″ deep, 3′-0″ x 7′-0″ single G		16	1		199	38		237	283
1240	6′-0″ wide, double G		14	1.143		234	43		277	330
2800	14 ga., up to 3-7/8″ deep, 3′-0″ x 7′-0″ single G		16	1		181	38		219	263
2840	6′-0″ wide, double G		14	1.143		217	43		260	310
3000	14 ga., up to 5-3/4″ deep, 3′-0″ x 6′-8″ single G		16	1		155	38		193	235
3002	3′-6″ wide, single G		16	1		204	38		242	289
3005	4′-0″ wide, single G		16	1		210	38		248	295
3600	up to 5-3/4″ jamb depth, 4′-0″ x 7′-0″ single G		15	1.067		185	40.50		225.50	271
3620	6′-0″ wide, double G		12	1.333		235	50.50		285.50	345
3640	8′-0″ wide, double G		12	1.333		246	50.50		296.50	355
3700	8′-0″ high, 4′-0″ wide, single G		15	1.067		235	40.50		275.50	325
3740	8′-0″ wide, double G		12	1.333		290	50.50		340.50	405
4000	6-3/4″ deep, 4′-0″ x 7′-0″ single G		15	1.067		223	40.50		263.50	315
4020	6′-0″ wide, double G		12	1.333		278	50.50		328.50	390
4040	8′-0″, wide double G		12	1.333		287	50.50		337.50	400
4100	8′-0″ high, 4′-0″ wide, single G		15	1.067		276	40.50		316.50	375
4140	8′-0″ wide, double G		12	1.333		320	50.50		370.50	440

08 12 Metal Frames

08 12 13 – Hollow Metal Frames

08 12 13.13 Standard Hollow Metal Frames			Crew	Daily Output	Labor-Hours	Unit	2016 Bare Costs Material	Labor	Equipment	Total	Total Incl O&P
4400	8-3/4" deep, 4'-0" x 7'-0", single	G	2 Carp	15	1.067	Ea.	252	40.50		292.50	345
4440	8'-0" wide, double	G		12	1.333		315	50.50		365.50	430
4500	4'-0" x 8'-0", single	G		15	1.067		283	40.50		323.50	380
4540	8'-0" wide, double	G		12	1.333		350	50.50		400.50	470
4900	For welded frames, add						63			63	69.50
5380	Steel frames, KD, 14 ga., "B" label, to 5-3/4" throat, to 3'-0" x 7'-0"		2 Carp	15	1.067		221	40.50		261.50	310
5400	14 ga., "B" label, up to 5-3/4" deep, 4'-0" x 7'-0" single	G		15	1.067		211	40.50		251.50	300
5440	8'-0" wide, double	G		12	1.333		274	50.50		324.50	385
5800	6-3/4" deep, 7'-0" high, 4'-0" wide, single	G		15	1.067		218	40.50		258.50	310
5840	8'-0" wide, double	G		12	1.333		390	50.50		440.50	510
6200	8-3/4" deep, 4'-0" x 7'-0" single	G		15	1.067		296	40.50		336.50	395
6240	8'-0" wide, double	G		12	1.333		380	50.50		430.50	505
6300	For "A" label use same price as "B" label										
6400	For baked enamel finish, add						30%	15%			
6500	For galvanizing, add						20%				
6600	For hospital stop, add					Ea.	295			295	325
6620	For hospital stop, stainless steel add					"	380			380	420
7900	Transom lite frames, fixed, add		2 Carp	155	.103	S.F.	54	3.90		57.90	66
8000	Movable, add		"	130	.123	"	68	4.65		72.65	83

08 12 13.15 Borrowed Lites		Crew	Daily Output	Labor-Hours	Unit	Material	Labor	Equipment	Total	Total Incl O&P
0010	**BORROWED LITES**									
0100	Hollow metal section 20 ga., 3-1/2" with glass stop	2 Carp	100	.160	L.F.	8.15	6.05		14.20	19.10
0110	3-3/4" with glass stop		100	.160		8.30	6.05		14.35	19.30
0120	4" with glass stop		100	.160		8.45	6.05		14.50	19.45
0130	4-5/8" with glass stop		100	.160		8.75	6.05		14.80	19.80
0140	4-7/8" with glass stop		100	.160		8.90	6.05		14.95	19.95
0150	5" with glass stop		100	.160		9	6.05		15.05	20
0160	5-3/8" with glass stop		100	.160		9.25	6.05		15.30	20.50
0300	Hollow metal section 18 ga., 3-1/2" with glass stop		80	.200		9	7.55		16.55	22.50
0310	3-3/4" with glass stop		80	.200		9.20	7.55		16.75	23
0320	4" with glass stop		80	.200		9.30	7.55		16.85	23
0330	4-5/8" with glass stop		80	.200		9.75	7.55		17.30	23.50
0340	4-7/8" with glass stop		80	.200		10	7.55		17.55	23.50
0350	5" with glass stop		80	.200		10.10	7.55		17.65	24
0360	5-3/8" with glass stop		80	.200		10.35	7.55		17.90	24
0370	6-5/8" with glass stop		80	.200		11.25	7.55		18.80	25
0380	6-7/8" with glass stop		80	.200		11.40	7.55		18.95	25
0390	7-1/4" with glass stop		80	.200		11.90	7.55		19.45	26
0500	Mullion section 18 ga., 3-1/2" with double stop		50	.320		21	12.10		33.10	43.50
0510	3-3/4" with double stop		50	.320		21.50	12.10		33.60	44
0530	4-5/8" with double stop		50	.320		23	12.10		35.10	45.50
0540	4-7/8" with double stop		50	.320		23.50	12.10		35.60	46
0550	5" with double stop		50	.320		23.50	12.10		35.60	46
0560	5-3/8" with double stop		50	.320		24	12.10		36.10	47
0570	6-5/8" with double stop		50	.320		26	12.10		38.10	49.50
0580	6-7/8" with double stop		50	.320		26.50	12.10		38.60	50
0590	7-1/4" with double stop		50	.320		27	12.10		39.10	50.50
1000	Assembled frame 2'-0" x 2'-0" 16 ga., 3-3/4" with glass stop		10	1.600	Ea.	180	60.50		240.50	299
1010	3'-0" x 2'-0"		10	1.600		180	60.50		240.50	299
1020	4'-0" x 2'-0"		10	1.600		180	60.50		240.50	299
1030	5'-0" x 2'-0"		10	1.600		180	60.50		240.50	299
1040	6'-0" x 2'-0"		8	2		270	75.50		345.50	425

08 12 13 – Hollow Metal Frames

08 12 13.15 Borrowed Lites

	08 12 13.15 Borrowed Lites	Crew	Daily Output	Labor-Hours	Unit	Material	Labor (2016 Bare Costs)	Equipment (2016 Bare Costs)	Total (2016 Bare Costs)	Total Incl O&P
1050	7'-0" x 2'-0"	2 Carp	8	2	Ea.	270	75.50		345.50	425
1060	8'-0" x 2'-0"		8	2		270	75.50		345.50	425
1100	2'-0" x 7'-0"		8	2		270	75.50		345.50	425
1110	3'-0" x 7'-0"		8	2		270	75.50		345.50	425
1120	4'-0" x 7'-0"		8	2		270	75.50		345.50	425
1130	5'-0" x 7'-0"		8	2		270	75.50		345.50	425
1140	6'-0" x 7'-0"		6	2.667		295	101		396	495
1150	7'-0" x 7'-0"		6	2.667		295	101		396	495
1160	8'-0" x 7'-0"		6	2.667		295	101		396	495
1200	Assembled frame 2'-0" x 2'-0" 16 ga., 4-7/8" with glass stop		10	1.600		180	60.50		240.50	299
1210	3'-0" x 2'-0"		10	1.600		180	60.50		240.50	299
1220	4'-0" x 2'-0"		8	2		180	75.50		255.50	325
1230	5'-0" x 2'-0"		8	2		180	75.50		255.50	325
1240	6'-0" x 2'-0"		8	2		270	75.50		345.50	425
1250	7'-0" x 2'-0"		8	2		270	75.50		345.50	425
1260	8'-0" x 2'-0"		8	2		270	75.50		345.50	425
1300	2'-0" x 7'-0"		8	2		270	75.50		345.50	425
1310	3'-0" x 7'-0"		8	2		270	75.50		345.50	425
1320	4'-0" x 7'-0"		8	2		270	75.50		345.50	425
1330	5'-0" x 7'-0"		6	2.667		270	101		371	465
1340	6'-0" x 7'-0"		6	2.667		295	101		396	495
1350	7'-0" x 7'-0"		6	2.667		270	101		371	465
1360	8'-0" x 7'-0"		6	2.667		295	101		396	495
1400	Assembled frame 2'-0" x 3'-0" 16 ga., 6-1/8" with glass stop		6	2.667		180	101		281	365
1410	3'-0" x 3'-0"		8	2		180	75.50		255.50	325
1420	4'-0" x 3'-0"		8	2		180	75.50		255.50	325
1430	5'-0" x 3'-0"		6	2.667		180	101		281	365
1440	6'-0" x 3'-0"		6	2.667		270	101		371	465
1450	7'-0" x 3'-0"		6	2.667		270	101		371	465
1460	8'-0" x 3'-0"	↓	10	1.600	↓	270	60.50		330.50	400

08 12 13.18 Transoms and Sidelights

	08 12 13.18 Transoms and Sidelights	Crew	Daily Output	Labor-Hours	Unit	Material	Labor	Equipment	Total	Total Incl O&P
0010	**TRANSOMS & SIDELIGHTS**									
0160	Transom sash, 6-3/4" thick, 3'-0" x 1'-4"	2 Carp	11	1.455	Ea.	181	55		236	291
0180	3'-4" x 1'-4"		11	1.455		181	55		236	291
0200	6'-0" x 1'-4"		6	2.667		210	101		311	400
0220	4-3/4" thick, 3'-0" x 1'-4"		11	1.455		175	55		230	285
0240	3'-4" x 1'-4"		11	1.455		175	55		230	285
0260	6'-0" x 1'-4"	↓	6	2.667	↓	208	101		309	395

08 12 13.25 Channel Metal Frames

	08 12 13.25 Channel Metal Frames		Crew	Daily Output	Labor-Hours	Unit	Material	Labor	Equipment	Total	Total Incl O&P
0010	**CHANNEL METAL FRAMES**										
0020	Steel channels with anchors and bar stops										
0100	6" channel @ 8.2#/L.F., 3' x 7' door, weighs 150#	G	E-4	13	2.462	Ea.	239	99.50	11.30	349.80	455
0200	8" channel @ 11.5#/L.F., 6' x 8' door, weighs 275#	G		9	3.556		435	144	16.35	595.35	760
0300	8' x 12' door, weighs 400#	G		6.50	4.923		635	199	22.50	856.50	1,100
0400	10" channel @ 15.3#/L.F., 10' x 10' door, weighs 500#	G		6	5.333		795	215	24.50	1,034.50	1,300
0500	12' x 12' door, weighs 600#	G		5.50	5.818		955	235	27	1,217	1,500
0600	12" channel @ 20.7#/L.F., 12' x 12' door, weighs 825#	G		4.50	7.111		1,300	287	32.50	1,619.50	2,000
0700	12' x 16' door, weighs 1000#	G	↓	4	8		1,600	325	37	1,962	2,375
0800	For frames without bar stops, light sections, deduct						15%				
0900	Heavy sections, deduct					↓	10%				

08 13 13.13 Standard Hollow Metal Doors			Crew	Daily Output	Labor-Hours	Unit	2016 Bare Costs Material	Labor	Equipment	Total	Total Incl O&P
0010	STANDARD HOLLOW METAL DOORS R081313-20										
0015	Flush, full panel, hollow core										
0017	When noted doors are prepared but do not include glass or louvers										
0020	1-3/8" thick, 20 ga., 2'-0" x 6'-8"	G	2 Carp	20	.800	Ea.	335	30		365	420
0040	2'-8" x 6'-8"	G		18	.889		350	33.50		383.50	440
0060	3'-0" x 6'-8"	G		17	.941		350	35.50		385.50	445
0100	3'-0" x 7'-0"	G		17	.941		360	35.50		395.50	460
0120	For vision lite, add						96			96	106
0140	For narrow lite, add						104			104	114
0320	Half glass, 20 ga., 2'-0" x 6'-8"	G	2 Carp	20	.800		490	30		520	590
0340	2'-8" x 6'-8"	G		18	.889		515	33.50		548.50	620
0360	3'-0" x 6'-8"	G		17	.941		510	35.50		545.50	620
0400	3'-0" x 7'-0"	G		17	.941		630	35.50		665.50	750
0410	1-3/8" thick, 18 ga., 2'-0" x 6'-8"	G		20	.800		400	30		430	490
0420	3'-0" x 6'-8"	G		17	.941		405	35.50		440.50	505
0425	3'-0" x 7'-0"	G		17	.941		425	35.50		460.50	525
0450	For vision lite, add						96			96	106
0452	For narrow lite, add						104			104	114
0460	Half glass, 18 ga., 2'-0" x 6'-8"	G	2 Carp	20	.800		555	30		585	665
0465	2'-8" x 6'-8"	G		18	.889		575	33.50		608.50	690
0470	3'-0" x 6'-8"	G		17	.941		565	35.50		600.50	680
0475	3'-0" x 7'-0"	G		17	.941		575	35.50		610.50	695
0500	Hollow core, 1-3/4" thick, full panel, 20 ga., 2'-8" x 6'-8"	G		18	.889		420	33.50		453.50	515
0520	3'-0" x 6'-8"	G		17	.941		420	35.50		455.50	520
0640	3'-0" x 7'-0"	G		17	.941		440	35.50		475.50	540
0680	4'-0" x 7'-0"	G		15	1.067		630	40.50		670.50	765
0700	4'-0" x 8'-0"	G		13	1.231		735	46.50		781.50	890
1000	18 ga., 2'-8" x 6'-8"	G		17	.941		490	35.50		525.50	600
1020	3'-0" x 6'-8"	G		16	1		475	38		513	590
1120	3'-0" x 7'-0"	G		17	.941		515	35.50		550.50	630
1180	4'-0" x 7'-0"	G		14	1.143		630	43		673	770
1200	4'-0" x 8'-0"	G		17	.941		735	35.50		770.50	870
1212	For vision lite, add						96			96	106
1214	For narrow lite, add						104			104	114
1230	Half glass, 20 ga., 2'-8" x 6'-8"	G	2 Carp	20	.800		575	30		605	685
1240	3'-0" x 6'-8"	G		18	.889		575	33.50		608.50	690
1260	3'-0" x 7'-0"	G		18	.889		595	33.50		628.50	705
1280	Embossed panel, 1-3/4" thick, poly core, 20 ga., 3'-0" x 7'-0"	G		18	.889		415	33.50		448.50	510
1290	Half glass, 1-3/4" thick, poly core, 20 ga., 3'-0" x 7'-0"	G		18	.889		610	33.50		643.50	725
1320	18 ga., 2'-8" x 6'-8"	G		18	.889		640	33.50		673.50	760
1340	3'-0" x 6'-8"	G		17	.941		635	35.50		670.50	755
1360	3'-0" x 7'-0"	G		17	.941		645	35.50		680.50	770
1380	4'-0" x 7'-0"	G		15	1.067		785	40.50		825.50	935
1400	4'-0" x 8'-0"	G		14	1.143		890	43		933	1,050
1500	Flush full panel, 16 ga., steel hollow core										
1520	2'-0" x 6'-8"	G	2 Carp	20	.800	Ea.	555	30		585	660
1530	2'-8" x 6'-8"	G		20	.800		555	30		585	665
1540	3'-0" x 6'-8"	G		20	.800		550	30		580	655
1560	2'-8" x 7'-0"	G		18	.889		575	33.50		608.50	685
1570	3'-0" x 7'-0"	G		18	.889		560	33.50		593.50	670
1580	3'-6" x 7'-0"	G		18	.889		655	33.50		688.50	775
1590	4'-0" x 7'-0"	G		18	.889		725	33.50		758.50	850
1600	2'-8" x 8'-0"	G		18	.889		700	33.50		733.50	825

08 13 13 – Hollow Metal Doors

08 13 13.13 Standard Hollow Metal Doors			Crew	Daily Output	Labor-Hours	Unit	2016 Bare Costs Material	Labor	Equipment	Total	Total Incl O&P
1620	3'-0" x 8'-0"	G	2 Carp	18	.889	Ea.	720	33.50		753.50	845
1630	3'-6" x 8'-0"	G		18	.889		800	33.50		833.50	935
1640	4'-0" x 8'-0"	G		18	.889		850	33.50		883.50	985
1650	1-13/16", 14 ga., 2'-8" x 7'-0"	G		10	1.600		1,125	60.50		1,185.50	1,350
1670	3'-0" x 7'-0"	G		10	1.600		1,075	60.50		1,135.50	1,300
1690	3'-6" x 7'-0"	G		10	1.600		1,200	60.50		1,260.50	1,400
1700	4'-0" x 7'-0"	G		10	1.600		1,250	60.50		1,310.50	1,475
1720	Insulated, 1-3/4" thick, full panel, 18 ga., 3'-0" x 6'-8"	G		15	1.067		475	40.50		515.50	595
1740	2'-8" x 7'-0"	G		16	1		505	38		543	620
1760	3'-0" x 7'-0"	G		15	1.067		490	40.50		530.50	610
1800	4'-0" x 8'-0"	G		13	1.231		740	46.50		786.50	890
1820	Half glass, 18 ga., 3'-0" x 6'-8"	G		16	1		635	38		673	760
1840	2'-8" x 7'-0"	G		17	.941		660	35.50		695.50	785
1860	3'-0" x 7'-0"	G		16	1		690	38		728	820
1900	4'-0" x 8'-0"	G		14	1.143		670	43		713	810
2000	For vision lite, add						96			96	106
2010	For narrow lite, add						104			104	114
8100	For bottom louver, add						285			285	315
8110	For baked enamel finish, add						30%	15%			
8120	For galvanizing, add						20%				
8190	Commercial door, flush steel, fire door see section 08 13 13 15										
8270	For dutch door with shelf, add to standard door					Ea.	300			300	330

08 13 13.15 Metal Fire Doors			Crew	Daily Output	Labor-Hours	Unit	Material	Labor	Equipment	Total	Total Incl O&P
0010	**METAL FIRE DOORS**	R081313-20									
0015	Steel, flush, "B" label, 90 minute										
0020	Full panel, 20 ga., 2'-0" x 6'-8"		2 Carp	20	.800	Ea.	420	30		450	510
0040	2'-8" x 6'-8"			18	.889		435	33.50		468.50	535
0060	3'-0" x 6'-8"			17	.941		435	35.50		470.50	540
0080	3'-0" x 7'-0"			17	.941		455	35.50		490.50	560
0140	18 ga., 3'-0" x 6'-8"			16	1		495	38		533	610
0160	2'-8" x 7'-0"			17	.941		520	35.50		555.50	630
0180	3'-0" x 7'-0"			16	1		505	38		543	620
0200	4'-0" x 7'-0"			15	1.067		650	40.50		690.50	785
0220	For "A" label, 3 hour, 18 ga., use same price as "B" label										
0240	For vision lite, add					Ea.	156			156	172
0300	Full panel, 16 ga., 2'-0" x 6'-8"		2 Carp	20	.800		545	30		575	650
0310	2'-0" x 6'-8"			18	.889		545	33.50		578.50	655
0320	3'-0" x 6'-8"			17	.941		540	35.50		575.50	655
0350	2'-8" x 7'-0"			17	.941		565	35.50		600.50	680
0360	3'-0" x 7'-0"			16	1		545	38		583	665
0370	4'-0" x 7'-0"			15	1.067		705	40.50		745.50	850
0520	Flush, "B" label 90 min., egress core, 20 ga., 2'-0" x 6'-8"			18	.889		655	33.50		688.50	775
0540	2'-8" x 6'-8"			17	.941		665	35.50		700.50	790
0560	3'-0" x 6'-8"			16	1		665	38		703	795
0580	3'-0" x 7'-0"			16	1		685	38		723	815
0640	Flush, "A" label 3 hour, egress core, 18 ga., 3'-0" x 6'-8"			15	1.067		720	40.50		760.50	865
0660	2'-8" x 7'-0"			16	1		750	38		788	885
0680	3'-0" x 7'-0"			15	1.067		740	40.50		780.50	885
0700	4'-0" x 7'-0"			14	1.143		885	43		928	1,050

08 13 Metal Doors

08 13 13 – Hollow Metal Doors

08 13 13.20 Residential Steel Doors

	08 13 13.20 Residential Steel Doors		Crew	Daily Output	Labor-Hours	Unit	2016 Bare Costs Material	Labor	Equipment	Total	Total Incl O&P
0010	**RESIDENTIAL STEEL DOORS**										
2510	Bi-passing closet, incl. hardware, no frame or trim incl.										
2511	Mirrored, metal frame, 4′-0″ x 6′-8″		2 Carp	10	1.600	Opng.	217	60.50		277.50	340
2512	5′-0″ wide			10	1.600		247	60.50		307.50	370
2513	6′-0″ wide			10	1.600		281	60.50		341.50	410
2514	7′-0″ wide			9	1.778		269	67		336	410
2515	8′-0″ wide			9	1.778		440	67		507	600
2611	Mirrored, metal, 4′-0″ x 8′-0″			10	1.600		320	60.50		380.50	455
2612	5′-0″ wide			10	1.600		330	60.50		390.50	465
2613	6′-0″ wide			10	1.600		385	60.50		445.50	525
2614	7′-0″ wide			9	1.778		400	67		467	555
2615	8′-0″ wide			9	1.778		445	67		512	600

08 13 13.25 Doors Hollow Metal

	08 13 13.25 Doors Hollow Metal		Crew	Daily Output	Labor-Hours	Unit	Material	Labor	Equipment	Total	Total Incl O&P
0010	**DOORS HOLLOW METAL**										
0500	Exterior, commercial, flush, 20 ga., 1-3/4″ x 7′-0″ x 2′-6″ wide	G	2 Carp	15	1.067	Ea.	395	40.50		435.50	500
0530	2′-8″ wide	G		15	1.067		435	40.50		475.50	550
0560	3′-0″ wide	G		14	1.143		435	43		478	555
1000	18 ga., 1-3/4″ x 7′-0″ x 2′-6″ wide	G		15	1.067		495	40.50		535.50	615
1030	2′-8″ wide	G		15	1.067		505	40.50		545.50	625
1060	3′-0″ wide	G		14	1.143		490	43		533	615
1500	16 ga., 1-3/4 x 7′-0″ x 2′-6″ wide	G		15	1.067		570	40.50		610.50	695
1530	2′-8″ wide	G		15	1.067		570	40.50		610.50	700
1560	3′-0″ wide	G		14	1.143		565	43		608	695
1590	3′-6″ wide	G		14	1.143		650	43		693	790
2900	Fire door, "A" label, 18 gauge, 1-3/4″ x 2′-6″ x 7′-0″	G		15	1.067		650	40.50		690.50	785
2930	2′-8″ wide	G		15	1.067		670	40.50		710.50	805
2960	3′-0″ wide	G		14	1.143		640	43		683	780
2990	3′-6″ wide	G		14	1.143		735	43		778	885
3100	"B" label, 2′-6″ wide	G		15	1.067		590	40.50		630.50	715
3130	2′-8″ wide	G		15	1.067		590	40.50		630.50	720
3160	3′-0″ wide	G		14	1.143		585	43		628	720

08 13 16 – Aluminum Doors

08 13 16.10 Commercial Aluminum Doors

	08 13 16.10 Commercial Aluminum Doors	Crew	Daily Output	Labor-Hours	Unit	Material	Labor	Equipment	Total	Total Incl O&P
0010	**COMMERCIAL ALUMINUM DOORS**, flush, no glazing									
5000	Flush panel doors, pair of 2′-6″ x 7′-0″	2 Sswk	2	8	Pr.	1,500	320		1,820	2,225
5050	3′-0″ x 7′-0″, single		2.50	6.400	Ea.	840	255		1,095	1,400
5100	Pair of 3′-0″ x 7′-0″		2	8	Pr.	1,600	320		1,920	2,325
5150	3′-6″ x 7′-0″, single		2.50	6.400	Ea.	1,000	255		1,255	1,600

08 14 Wood Doors

08 14 13 – Carved Wood Doors

08 14 13.10 Types of Wood Doors, Carved

	08 14 13.10 Types of Wood Doors, Carved	Crew	Daily Output	Labor-Hours	Unit	Material	Labor	Equipment	Total	Total Incl O&P
0010	**TYPES OF WOOD DOORS, CARVED**									
3000	Solid wood, 1-3/4″ thick stile and rail									
3020	Mahogany, 3′-0″ x 7′-0″, six panel	2 Carp	14	1.143	Ea.	1,275	43		1,318	1,475
3030	With two lites		10	1.600		1,950	60.50		2,010.50	2,250
3040	3′-6″ x 8′-0″, six panel		10	1.600		1,550	60.50		1,610.50	1,800
3050	With two lites		8	2		2,600	75.50		2,675.50	2,975
3100	Pine, 3′-0″ x 7′-0″, six panel		14	1.143		570	43		613	700
3110	With two lites		10	1.600		825	60.50		885.50	1,000

08 14 Wood Doors

08 14 13 – Carved Wood Doors

08 14 13.10 Types of Wood Doors, Carved		Crew	Daily Output	Labor-Hours	Unit	Material	2016 Bare Costs Labor	Equipment	Total	Total Incl O&P
3120	3'-6" x 8'-0", six panel	2 Carp	10	1.600	Ea.	950	60.50		1,010.50	1,150
3130	With two lites		8	2		1,825	75.50		1,900.50	2,150
3200	Red oak, 3'-0" x 7'-0", six panel		14	1.143		1,850	43		1,893	2,100
3210	With two lites		10	1.600		2,450	60.50		2,510.50	2,800
3220	3'-6" x 8'-0", six panel		10	1.600		2,800	60.50		2,860.50	3,175
3230	With two lites		8	2		3,400	75.50		3,475.50	3,875
4000	Hand carved door, mahogany									
4020	3'-0" x 7'-0", simple design	2 Carp	14	1.143	Ea.	1,750	43		1,793	2,000
4030	Intricate design		11	1.455		3,700	55		3,755	4,175
4040	3'-6" x 8'-0", simple design		10	1.600		3,000	60.50		3,060.50	3,400
4050	Intricate design		8	2		3,700	75.50		3,775.50	4,200
4400	For custom finish, add					475			475	525
4600	Side light, mahogany, 7'-0" x 1'-6" wide, 4 lites	2 Carp	18	.889		1,100	33.50		1,133.50	1,250
4610	6 lites		14	1.143		2,625	43		2,668	2,975
4620	8'-0" x 1'-6" wide, 4 lites		14	1.143		1,800	43		1,843	2,050
4630	6 lites		10	1.600		2,100	60.50		2,160.50	2,400
4640	Side light, oak, 7'-0" x 1'-6" wide, 4 lites		18	.889		1,200	33.50		1,233.50	1,375
4650	6 lites		14	1.143		2,100	43		2,143	2,375
4660	8'-0" x 1'-6" wide, 4 lites		14	1.143		1,100	43		1,143	1,275
4670	6 lites		10	1.600		2,100	60.50		2,160.50	2,400

08 14 16 – Flush Wood Doors

08 14 16.09 Smooth Wood Doors

		Crew	Daily Output	Labor-Hours	Unit	Material	Labor	Equipment	Total	Total Incl O&P
0010	**SMOOTH WOOD DOORS**									
0085	3'-0" x 6'-8"	1 Carp	16	.500	Ea.	54.50	18.90		73.40	91.50
0108	3'-0" x 7'-0"	2 Carp	16	1	"	145	38		183	224
0112	Pair of 3'-0" x 7'-0"		9	1.778	Pr.	125	67		192	250
0206	2'-6" x 7'-0"		16	1	Ea.	125	38		163	202
0210	3'-0" x 7'-0"		16	1	"	138	38		176	216
0214	Pair of 3'-0" x 7'-0"		9	1.778	Pr.	266	67		333	405
2205	1-3/4", 3'-0" x 7'-0"		13	1.231	Ea.	123	46.50		169.50	213
4045	3'-0" x 8'-0"	1 Carp	8	1		430	38		468	540
5000	Wood doors, for vision lite, add					96			96	106
5010	Wood doors, for narrow lite, add					104			104	114
5015	Wood doors, for bottom (or top) louver, add					285			285	315

08 14 16.10 Wood Doors Decorator

		Crew	Daily Output	Labor-Hours	Unit	Material	Labor	Equipment	Total	Total Incl O&P
0010	**WOOD DOORS DECORATOR**									
1800	Exterior, flush, solid wood core, birch 1-3/4" x 2'-6" x 7'-0"	2 Carp	15	1.067	Ea.	320	40.50		360.50	420
1820	2'-8" wide		15	1.067		325	40.50		365.50	430
1840	3'-0" wide		14	1.143		335	43		378	445
1900	Oak faced, 1-3/4" x 2'-6" x 7'-0"		15	1.067		320	40.50		360.50	420
1920	2'-8" wide		15	1.067		330	40.50		370.50	435
1940	3'-0" wide		14	1.143		340	43		383	450
2100	Walnut faced, 1-3/4" x 2'-6"x 7'-0"		15	1.067		385	40.50		425.50	495
2120	2'-8" wide		15	1.067		400	40.50		440.50	510
2140	3'-0" wide		14	1.143		430	43		473	550

08 14 33 – Stile and Rail Wood Doors

08 14 33.10 Wood Doors Paneled

		Crew	Daily Output	Labor-Hours	Unit	Material	Labor	Equipment	Total	Total Incl O&P
0010	**WOOD DOORS PANELED**									
0020	Interior, six panel, hollow core, 1-3/8" thick									
0040	Molded hardboard, 2'-0" x 6'-8"	2 Carp	17	.941	Ea.	62	35.50		97.50	128
0060	2'-6" x 6'-8"		17	.941		64	35.50		99.50	130

08 14 Wood Doors

08 14 33 – Stile and Rail Wood Doors

08 14 33.10 Wood Doors Paneled		Crew	Daily Output	Labor-Hours	Unit	Material	2016 Bare Costs Labor	Equipment	Total	Total Incl O&P
0070	2′-8″ x 6′-8″	2 Carp	17	.941	Ea.	67	35.50		102.50	133
0080	3′-0″ x 6′-8″		17	.941		72	35.50		107.50	139
0140	Embossed print, molded hardboard, 2′-0″ x 6′-8″		17	.941		64	35.50		99.50	130
0160	2′-6″ x 6′-8″		17	.941		64	35.50		99.50	130
0180	3′-0″ x 6′-8″		17	.941		72	35.50		107.50	139
0540	Six panel, solid, 1-3/8″ thick, pine, 2′-0″ x 6′-8″		15	1.067		155	40.50		195.50	239
0560	2′-6″ x 6′-8″		14	1.143		170	43		213	260
0580	3′-0″ x 6′-8″		13	1.231		145	46.50		191.50	238
1020	Two panel, bored rail, solid, 1-3/8″ thick, pine, 1′-6″ x 6′-8″		16	1		270	38		308	360
1040	2′-0″ x 6′-8″		15	1.067		355	40.50		395.50	460
1060	2′-6″ x 6′-8″		14	1.143		400	43		443	515
1340	Two panel, solid, 1-3/8″ thick, fir, 2′-0″ x 6′-8″		15	1.067		160	40.50		200.50	244
1360	2′-6″ x 6′-8″		14	1.143		210	43		253	305
1380	3′-0″ x 6′-8″		13	1.231		415	46.50		461.50	535
1740	Five panel, solid, 1-3/8″ thick, fir, 2′-0″ x 6′-8″		15	1.067		280	40.50		320.50	380
1760	2′-6″ x 6′-8″		14	1.143		420	43		463	535
1780	3′-0″ x 6′-8″		13	1.231		420	46.50		466.50	540

08 14 33.20 Wood Doors Residential		Crew	Daily Output	Labor-Hours	Unit	Material	Labor	Equipment	Total	Total Incl O&P
0010	**WOOD DOORS RESIDENTIAL**									
1420	6′-8″ x 3′-0″ wide, fir	2 Carp	16	1	Ea.	470	38		508	580
1800	Lauan, solid core, 1-3/4″ x 7′-0″ x 2′-4″ wide		16	1		129	38		167	206
1810	2′-6″ wide		15	1.067		134	40.50		174.50	215
1820	2′-8″ wide		9	1.778		139	67		206	266
1830	3′-0″ wide		16	1		152	38		190	232
1840	3′-4″ wide		16	1		242	38		280	330
1850	Pair of 3′-0″ wide		15	1.067	Pr.	305	40.50		345.50	405
1930	For dutch door with shelf, add					140%				
2920	Hardboard, primed 7′-0″ x 4′-0″, wide	2 Carp	12	1.333	Ea.	182	50.50		232.50	285
2930	6′-0″ wide		10	1.600		177	60.50		237.50	296
3100	7′-0″ x 3′-0″ wide		12	1.333		287	50.50		337.50	400
3120	6′-0″ wide		10	1.600		420	60.50		480.50	560
3140	For oak or ash, add					125%				
9000	Passage doors, flush, no frame, birch, solid core, 1-3/8″ x 2′-4″ x 7′-0″	2 Carp	16	1	Ea.	118	38		156	194
9020	2′-8″ wide		16	1		125	38		163	202
9040	3′-0″ wide		16	1		135	38		173	213
9060	3′-4″ wide		15	1.067		156	40.50		196.50	240
9080	Pair of 3′-0″ wide		9	1.778	Pr.	267	67		334	405
9100	Lauan, solid core, 1-3/8″ x 7′-0″ x 2′-4″ wide		16	1	Ea.	137	38		175	214
9120	2′-8″ wide		16	1		146	38		184	225
9140	3′-0″ wide		16	1		155	38		193	234
9160	3′-4″ wide		15	1.067		163	40.50		203.50	247
9180	Pair of 3′-0″ wide		9	1.778	Pr.	281	67		348	425
9200	Hardboard, solid core, 1-3/8″ x 7′-0″ x 2′-4″ wide		16	1	Ea.	169	38		207	250
9220	2′-8″ wide		16	1		175	38		213	257
9240	3′-0″ wide		16	1		180	38		218	262
9260	3′-4″ wide		15	1.067		199	40.50		239.50	287

08 14 Wood Doors

08 14 35 – Torrified Doors

08 14 35.10 Torrified Exterior Doors		Crew	Daily Output	Labor-Hours	Unit	Material	2016 Bare Costs Labor	Equipment	Total	Total Incl O&P
0010	**TORRIFIED EXTERIOR DOORS**									
0020	Wood doors made from torrified wood, exterior									
0030	All doors require a finish be applied, all glass is insulated									
0040	All doors require pilot holes for all fasteners									
0100	6 panel, paint grade poplar, 1-3/4″ x 3′-0″x 6′-8″	2 Carp	12	1.333	Ea.	1,075	50.50		1,125.50	1,250
0120	Half glass 3′-0″x 6′-8″	″	12	1.333		1,175	50.50		1,225.50	1,375
0200	Side lite, full glass, 1-3/4″ x 1′-2″ x 6′-8″					905			905	995
0220	Side lite, half glass, 1-3/4″ x 1′-2″ x 6′-8″					905			905	995
0300	Raised face, 2 panel, paint grade poplar, 1-3/4″ x 3′-0″x 7′-0″	2 Carp	12	1.333		1,275	50.50		1,325.50	1,475
0320	Side lite, raised face, half glass, 1-3/4″ x 1′-2″ x 7′-0″					1,050			1,050	1,175
0500	6 panel, Fir, 1-3/4″ x 3′-0″x 6′-8″	2 Carp	12	1.333		1,550	50.50		1,600.50	1,775
0520	Half glass 3′-0″x 6′-8″	″	12	1.333		1,650	50.50		1,700.50	1,900
0600	Side lite, full glass, 1-3/4″ x 1′-2″ x 6′-8″					1,150			1,150	1,275
0620	Side lite, half glass, 1-3/4″ x 1′-2″ x 6′-8″					1,450			1,450	1,600
0700	6 panel, Mahogany, 1-3/4″ x 3′-0″x 6′-8″	2 Carp	12	1.333		1,625	50.50		1,675.50	1,850
0800	Side lite, full glass, 1-3/4″ x 1′-2″ x 6′-8″					1,225			1,225	1,350
0820	Side lite, half glass, 1-3/4″ x 1′-2″ x 6′-8″					1,200			1,200	1,325

08 14 40 – Interior Cafe Doors

08 14 40.10 Cafe Style Doors		Crew	Daily Output	Labor-Hours	Unit	Material	Labor	Equipment	Total	Total Incl O&P
0010	**CAFE STYLE DOORS**									
6520	Interior cafe doors, 2′-6″ opening, stock, panel pine	2 Carp	16	1	Ea.	218	38		256	305
6540	3′-0″ opening	″	16	1	″	240	38		278	330
6550	Louvered pine									
6560	2′-6″ opening	2 Carp	16	1	Ea.	180	38		218	262
8000	3′-0″ opening		16	1		193	38		231	276
8010	2′-6″ opening, hardwood		16	1		264	38		302	355
8020	3′-0″ opening		16	1		281	38		319	375

08 16 Composite Doors

08 16 13 – Fiberglass Doors

08 16 13.10 Entrance Doors, Fibrous Glass			Crew	Daily Output	Labor-Hours	Unit	Material	Labor	Equipment	Total	Total Incl O&P
0010	**ENTRANCE DOORS, FIBROUS GLASS**										
0020	Exterior, fiberglass, door, 2′-8″ wide x 6′-8″ high	G	2 Carp	15	1.067	Ea.	270	40.50		310.50	365
0040	3′-0″ wide x 6′-8″ high	G		15	1.067		270	40.50		310.50	365
0060	3′-0″ wide x 7′-0″ high	G		15	1.067		460	40.50		500.50	575
0080	3′-0″ wide x 6′-8″ high, with two lites	G		15	1.067		315	40.50		355.50	415
0100	3′-0″ wide x 8′-0″ high, with two lites	G		15	1.067		525	40.50		565.50	650
0110	Half glass, 3′-0″ wide x 6′-8″ high	G		15	1.067		435	40.50		475.50	550
0120	3′-0″ wide x 6′-8″ high, low e	G		15	1.067		465	40.50		505.50	580
0130	3′-0″ wide x 8′-0″ high	G		15	1.067		585	40.50		625.50	715
0140	3′-0″ wide x 8′-0″ high, low e	G		15	1.067		655	40.50		695.50	790
0150	Side lights, 1′-0″ wide x 6′-8″ high	G					266			266	293
0160	1′-0″ wide x 6′-8″ high, low e	G					281			281	310
0180	1′-0″ wide x 6′-8″ high, full glass	G					310			310	340
0190	1′-0″ wide x 6′-8″ high, low e	G					340			340	370

08 16 Composite Doors

08 16 14 – French Doors

08 16 14.10 Exterior Doors With Glass Lites		Crew	Daily Output	Labor-Hours	Unit	2016 Bare Costs Material	Labor	Equipment	Total	Total Incl O&P
0010	**EXTERIOR DOORS WITH GLASS LITES**									
0020	French, Fir, 1-3/4", 3'-0" wide x 6'-8" high	2 Carp	12	1.333	Ea.	600	50.50		650.50	745
0025	Double		12	1.333		1,200	50.50		1,250.50	1,400
0030	Maple, 1-3/4", 3'-0" wide x 6'-8" high		12	1.333		675	50.50		725.50	830
0035	Double		12	1.333		1,350	50.50		1,400.50	1,550
0040	Cherry, 1-3/4", 3'-0" wide x 6'-8" high		12	1.333		790	50.50		840.50	950
0045	Double		12	1.333		1,575	50.50		1,625.50	1,800
0100	Mahogany, 1-3/4", 3'-0" wide x 8'-0" high		10	1.600		800	60.50		860.50	980
0105	Double		10	1.600		1,600	60.50		1,660.50	1,850
0110	Fir, 1-3/4", 3'-0" wide x 8'-0" high		10	1.600		1,200	60.50		1,260.50	1,425
0115	Double		10	1.600		2,400	60.50		2,460.50	2,750
0120	Oak, 1-3/4", 3'-0" wide x 8'-0" high		10	1.600		1,825	60.50		1,885.50	2,100
0125	Double	↓	10	1.600	↓	3,650	60.50		3,710.50	4,100

08 17 Integrated Door Opening Assemblies

08 17 13 – Integrated Metal Door Opening Assemblies

08 17 13.10 Hollow Metal Doors and Frames

Line	Description	Crew	Daily Output	Labor-Hours	Unit	Material	Labor	Equipment	Total	Total Incl O&P
0010	**HOLLOW METAL DOORS AND FRAMES**									
0100	Prehung, flush, 18 ga., 1-3/4" x 6'-8" x 2'-8" wide	1 Carp	16	.500	Ea.	665	18.90		683.90	765
0120	3'-0" wide		15	.533		625	20		645	720
0130	3'-6" wide		13	.615		715	23.50		738.50	825
0140	4'-0" wide		10	.800		775	30		805	900
0300	Double, 1-3/4" x 3'-0" x 6'-8"		5	1.600		700	60.50		760.50	870
0350	3'-0" x 8'-0"	↓	4	2	↓	1,550	75.50		1,625.50	1,825

08 17 13.20 Stainless Steel Doors and Frames

Line	Description	Crew	Daily Output	Labor-Hours	Unit	Material	Labor	Equipment	Total	Total Incl O&P
0010	**STAINLESS STEEL DOORS AND FRAMES**									
0020	Stainless Steel(304) prehung 24 g 2'-6" x 6'-8" door w/16 g frame G	2 Carp	6	2.667	Ea.	1,800	101		1,901	2,150
0025	2'-8" x 6'-8" G		6	2.667		1,825	101		1,926	2,200
0030	3'-0" x 6'-8" G		6	2.667		1,825	101		1,926	2,175
0040	3'-0" x 7'-0" G		6	2.667		1,850	101		1,951	2,200
0050	4'-0" x 7'-0" G		5	3.200		2,400	121		2,521	2,850
0100	Stainless Steel(316) prehung 24 g 2'-6" x 6'-8" door w/16 g frame G		6	2.667		2,700	101		2,801	3,150
0110	2'-8" x 6'-8" G		6	2.667		2,725	101		2,826	3,150
0120	3'-0" x 6'-8" G		6	2.667		2,500	101		2,601	2,925
0150	3'-0" x 7'-0" G		6	2.667		2,675	101		2,776	3,125
0160	4'-0" x 7'-0" G		6	2.667		3,300	101		3,401	3,800
0300	Stainless Steel(304) prehung 18 g 2'-6" x 6'-8" door w/16 g frame G		6	2.667		3,075	101		3,176	3,550
0310	2'-8" x 6'-8" G		6	2.667		3,100	101		3,201	3,575
0320	3'-0" x 6'-8" G		6	2.667		3,150	101		3,251	3,650
0350	3'-0" x 7'-0" G		6	2.667		3,200	101		3,301	3,700
0360	4'-0" x 7'-0" G		5	3.200		3,400	121		3,521	3,925
0500	Stainless steel, prehung door, foam core, 14 ga, 3'-0" x 7'-0" G		5	3.200		3,400	121		3,521	3,950
0600	Stainless steel, prehung double door, foam core, 14 ga, 3'-0" x 7'-0" G	↓	4	4	↓	6,700	151		6,851	7,625

08 17 23 – Integrated Wood Door Opening Assemblies

08 17 23.10 Pre-Hung Doors

Line	Description	Crew	Daily Output	Labor-Hours	Unit	Material	Labor	Equipment	Total	Total Incl O&P
0010	**PRE-HUNG DOORS**									
0300	Exterior, wood, comb. storm & screen, 6'-9" x 2'-6" wide	2 Carp	15	1.067	Ea.	296	40.50		336.50	395
0320	2'-8" wide		15	1.067		296	40.50		336.50	395
0340	3'-0" wide	↓	15	1.067		305	40.50		345.50	405
0360	For 7'-0" high door, add				↓	30			30	33

08 17 Integrated Door Opening Assemblies

08 17 23 – Integrated Wood Door Opening Assemblies

08 17 23.10 Pre-Hung Doors		Crew	Daily Output	Labor-Hours	Unit	2016 Bare Costs Material	Labor	Equipment	Total	Total Incl O&P
1600	Entrance door, flush, birch, solid core									
1620	4-5/8" solid jamb, 1-3/4" x 6'-8" x 2'-8" wide	2 Carp	16	1	Ea.	289	38		327	385
1640	3'-0" wide		16	1		375	38		413	480
1642	5-5/8" jamb		16	1		325	38		363	425
1680	For 7'-0" high door, add					25			25	27.50
2000	Entrance door, colonial, 6 panel pine									
2020	4-5/8" solid jamb, 1-3/4" x 6'-8" x 2'-8" wide	2 Carp	16	1	Ea.	640	38		678	770
2040	3'-0" wide	"	16	1		675	38		713	805
2060	For 7'-0" high door, add					54			54	59
2200	For 5-5/8" solid jamb, add					41.50			41.50	46
2230	French style, exterior, 1 lite, 1-3/4" x 3'-0" x 6'-8"	1 Carp	14	.571		590	21.50		611.50	685
2235	9 lites	"	14	.571		660	21.50		681.50	760
2245	15 lites	2 Carp	14	1.143		640	43		683	780
2250	Double, 15 lites, 2'-0" x 6'-8", 4'-0" opening		7	2.286	Pr.	1,200	86.50		1,286.50	1,475
2260	2'-6" x 6'-8", 5'-0" opening		7	2.286		1,325	86.50		1,411.50	1,600
2280	3'-0" x 6'-8", 6'-0" opening		7	2.286		1,350	86.50		1,436.50	1,625
2430	3'-0" x 7'-0", 15 lites		14	1.143	Ea.	950	43		993	1,125
2432	Two 3'-0" x 7'-0"		7	2.286	Pr.	1,975	86.50		2,061.50	2,300
2435	3'-0" x 8'-0"		14	1.143	Ea.	1,000	43		1,043	1,175
2437	Two, 3'-0" x 8'-0"		7	2.286	Pr.	2,075	86.50		2,161.50	2,450
4000	Interior, passage door, 4-5/8" solid jamb									
4400	Lauan, flush, solid core, 1-3/8" x 6'-8" x 2'-6" wide	2 Carp	17	.941	Ea.	186	35.50		221.50	265
4420	2'-8" wide		17	.941		186	35.50		221.50	265
4440	3'-0" wide		16	1		201	38		239	285
4600	Hollow core, 1-3/8" x 6'-8" x 2'-6" wide		17	.941		125	35.50		160.50	198
4620	2'-8" wide		17	.941		125	35.50		160.50	197
4640	3'-0" wide		16	1		140	38		178	218
4700	For 7'-0" high door, add					35.50			35.50	39
5000	Birch, flush, solid core, 1-3/8" x 6'-8" x 2'-6" wide	2 Carp	17	.941		275	35.50		310.50	365
5020	2'-8" wide		17	.941		201	35.50		236.50	281
5040	3'-0" wide		16	1		300	38		338	395
5200	Hollow core, 1-3/8" x 6'-8" x 2'-6" wide		17	.941		221	35.50		256.50	305
5220	2'-8" wide		17	.941		241	35.50		276.50	325
5240	3'-0" wide		16	1		242	38		280	330
5280	For 7'-0" high door, add					30.50			30.50	33.50
5500	Hardboard paneled, 1-3/8" x 6'-8" x 2'-6" wide	2 Carp	17	.941		145	35.50		180.50	219
5520	2'-8" wide		17	.941		153	35.50		188.50	228
5540	3'-0" wide		16	1		150	38		188	229
6000	Pine paneled, 1-3/8" x 6'-8" x 2'-6" wide		17	.941		255	35.50		290.50	340
6020	2'-8" wide		17	.941		274	35.50		309.50	360
7600	Oak, 6 panel, 1-3/4" x 6'-8" x 3'-0"	1 Carp	17	.471		895	17.80		912.80	1,025
8200	Birch, flush, solid core, 1-3/4" x 6'-8" x 2'-4" wide		17	.471		229	17.80		246.80	282
8220	2'-6" wide		17	.471		227	17.80		244.80	280
8240	2'-8" wide		17	.471		221	17.80		238.80	273
8260	3'-0" wide		16	.500		229	18.90		247.90	284
8280	3'-6" wide		15	.533		360	20		380	430
8500	Pocket door frame with lauan, flush, hollow core, 1-3/8" x 3'-0" x 6'-8"		17	.471		223	17.80		240.80	276

08 31 Access Doors and Panels

08 31 13 – Access Doors and Frames

08 31 13.10 Types of Framed Access Doors		Crew	Daily Output	Labor-Hours	Unit	2016 Bare Costs Material	Labor	Equipment	Total	Total Incl O&P
0010	**TYPES OF FRAMED ACCESS DOORS**									
1000	Fire rated door with lock									
1100	Metal, 12" x 12"	1 Carp	10	.800	Ea.	165	30		195	233
1150	18" x 18"		9	.889		220	33.50		253.50	299
1200	24" x 24"		9	.889		325	33.50		358.50	415
1250	24" x 36"		8	1		335	38		373	430
1300	24" x 48"		8	1		430	38		468	540
1350	36" x 36"		7.50	1.067		495	40.50		535.50	615
1400	48" x 48"		7.50	1.067		635	40.50		675.50	765
1600	Stainless steel, 12" x 12"		10	.800		282	30		312	360
1650	18" x 18"		9	.889		410	33.50		443.50	505
1700	24" x 24"		9	.889		500	33.50		533.50	605
1750	24" x 36"		8	1		640	38		678	770
2000	Flush door for finishing									
2100	Metal 8" x 8"	1 Carp	10	.800	Ea.	38	30		68	92.50
2150	12" x 12"	"	10	.800	"	45	30		75	100
3000	Recessed door for acoustic tile									
3100	Metal, 12" x 12"	1 Carp	4.50	1.778	Ea.	82	67		149	203
3150	12" x 24"		4.50	1.778		100	67		167	223
3200	24" x 24"		4	2		130	75.50		205.50	270
3250	24" x 36"		4	2		180	75.50		255.50	325
4000	Recessed door for drywall									
4100	Metal 12" x 12"	1 Carp	6	1.333	Ea.	80	50.50		130.50	173
4150	12" x 24"		5.50	1.455		114	55		169	217
4200	24" x 36"		5	1.600		182	60.50		242.50	300
6000	Standard door									
6100	Metal, 8" x 8"	1 Carp	10	.800	Ea.	45	30		75	100
6150	12" x 12"		10	.800		50	30		80	106
6200	18" x 18"		9	.889		70	33.50		103.50	134
6250	24" x 24"		9	.889		85	33.50		118.50	150
6300	24" x 36"		8	1		125	38		163	202
6350	36" x 36"		8	1		145	38		183	224
6500	Stainless steel, 8" x 8"		10	.800		85	30		115	144
6550	12" x 12"		10	.800		110	30		140	172
6600	18" x 18"		9	.889		205	33.50		238.50	283
6650	24" x 24"		9	.889		265	33.50		298.50	350

08 31 13.20 Bulkhead/Cellar Doors		Crew	Daily Output	Labor-Hours	Unit	Material	Labor	Equipment	Total	Total Incl O&P
0010	**BULKHEAD/CELLAR DOORS**									
0020	Steel, not incl. sides, 44" x 62"	1 Carp	5.50	1.455	Ea.	600	55		655	750
0100	52" x 73"		5.10	1.569		850	59.50		909.50	1,025
0500	With sides and foundation plates, 57" x 45" x 24"		4.70	1.702		910	64.50		974.50	1,100
0600	42" x 49" x 51"		4.30	1.860		925	70.50		995.50	1,150

08 31 13.30 Commercial Floor Doors		Crew	Daily Output	Labor-Hours	Unit	Material	Labor	Equipment	Total	Total Incl O&P
0010	**COMMERCIAL FLOOR DOORS**									
0020	Aluminum tile, steel frame, one leaf, 2' x 2' opng.	2 Sswk	3.50	4.571	Opng.	880	182		1,062	1,300
0021	Aluminum tile, steel frame, one leaf, 2' x 2' opng.	L-4	3.50	4.571		880	151		1,031	1,225
0050	3'-6" x 3'-6" opening	2 Sswk	3.50	4.571		1,600	182		1,782	2,075
0051	3'-6" x 3'-6" opening	L-4	3.50	4.571		1,600	151		1,751	2,000
0500	Double leaf, 4' x 4' opening	2 Sswk	3	5.333		1,750	213		1,963	2,325
0501	Double leaf, 4' x 4' opening	L-4	3	5.333		1,750	177		1,927	2,225
0550	5' x 5' opening	2 Sswk	3	5.333		3,100	213		3,313	3,800
0551	5' x 5' opening	L-4	3	5.333		3,100	177		3,277	3,700

08 31 Access Doors and Panels

08 31 13 – Access Doors and Frames

08 31 13.35 Industrial Floor Doors		Crew	Daily Output	Labor-Hours	Unit	Material	2016 Bare Costs Labor	Equipment	Total	Total Incl O&P
0010	**INDUSTRIAL FLOOR DOORS**									
0021	Steel 300 psf L.L., single leaf, 2′ x 2′, 175#	L-4	6	2.667	Opng.	760	88.50		848.50	985
0051	3′ x 3′ opening, 300#		5.50	2.909		1,075	96.50		1,171.50	1,325
0301	Double leaf, 4′ x 4′ opening, 455#		5	3.200		2,275	106		2,381	2,675
0351	5′ x 5′ opening, 645#		4.50	3.556		3,300	118		3,418	3,825
1001	Aluminum, 300 psf L.L., single leaf, 2′ x 2′, 60#		6	2.667		800	88.50		888.50	1,025
1051	3′ x 3′ opening, 100#		5.50	2.909		1,300	96.50		1,396.50	1,575
1501	Double leaf, 4′ x 4′ opening, 160#		5	3.200		2,100	106		2,206	2,475
1551	5′ x 5′ opening, 235#		4.50	3.556		2,800	118		2,918	3,275
2001	Aluminum, 150 psf L.L., single leaf, 2′ x 2′, 60#		6	2.667		720	88.50		808.50	940
2051	3′ x 3′ opening, 95#		5.50	2.909		1,200	96.50		1,296.50	1,475
2501	Double leaf, 4′ x 4′ opening, 150#		5	3.200		1,450	106		1,556	1,775
2551	5′ x 5′ opening, 230#		4.50	3.556		1,950	118		2,068	2,325

08 32 Sliding Glass Doors

08 32 13 – Sliding Aluminum-Framed Glass Doors

08 32 13.10 Sliding Aluminum Doors		Crew	Daily Output	Labor-Hours	Unit	Material	Labor	Equipment	Total	Total Incl O&P
0010	**SLIDING ALUMINUM DOORS**									
0350	Aluminum, 5/8″ tempered insulated glass, 6′ wide									
0400	Premium	2 Carp	4	4	Ea.	1,625	151		1,776	2,025
0450	Economy		4	4		820	151		971	1,150
0500	8′ wide, premium		3	5.333		1,675	202		1,877	2,175
0550	Economy		3	5.333		1,450	202		1,652	1,925
0600	12′ wide, premium		2.50	6.400		3,000	242		3,242	3,700
0650	Economy		2.50	6.400		1,550	242		1,792	2,100
4000	Aluminum, baked on enamel, temp glass, 6′-8″ x 10′-0″ wide		4	4		1,100	151		1,251	1,450
4020	Insulating glass, 6′-8″ x 6′-0″ wide		4	4		955	151		1,106	1,300
4040	8′-0″ wide		3	5.333		1,125	202		1,327	1,575
4060	10′-0″ wide		2	8		1,375	300		1,675	2,000
4080	Anodized, temp glass, 6′-8″ x 6′-0″ wide		4	4		455	151		606	755
4100	8′-0″ wide		3	5.333		575	202		777	975
4120	10′-0″ wide		2	8		645	300		945	1,225
5000	Aluminum sliding glass door system									
5010	Sliding door 4′ wide opening single side	2 Carp	2	8	Ea.	6,000	300		6,300	7,100
5015	8′ wide opening single side		2	8		9,000	300		9,300	10,400
5020	Telescoping glass door system, 4′ wide opening biparting		2	8		5,000	300		5,300	6,000
5025	8′ wide opening biparting		2	8		6,000	300		6,300	7,100
5030	Folding glass door, 4′ wide opening biparting		2	8		8,000	300		8,300	9,300
5035	8′ wide opening biparting		2	8		10,000	300		10,300	11,500
5040	ICU-CCU sliding telescoping glass door, 4′ x 7′, single side opening		2	8		3,000	300		3,300	3,800
5045	8′ x 7′, single side opening		2	8		4,625	300		4,925	5,600
7000	Electric swing door operator and control, single door w/sensors	1 Carp	4	2		2,800	75.50		2,875.50	3,200
7005	Double door w/sensors		2	4		5,500	151		5,651	6,300
7010	Electric folding door operator and control, single door w/sensors		4	2		6,500	75.50		6,575.50	7,275
7015	Bi-folding door		4	2		7,800	75.50		7,875.50	8,700
7020	Electric swing door operator and control, single door		4	2		3,000	75.50		3,075.50	3,425

08 32 Sliding Glass Doors

08 32 19 – Sliding Wood-Framed Glass Doors

08 32 19.10 Sliding Wood Doors

		Crew	Daily Output	Labor-Hours	Unit	2016 Bare Costs Material	Labor	Equipment	Total	Total Incl O&P
0010	**SLIDING WOOD DOORS**									
0020	Wood, tempered insul. glass, 6′ wide, premium	2 Carp	4	4	Ea.	1,450	151		1,601	1,850
0100	Economy		4	4		1,175	151		1,326	1,525
0150	8′ wide, wood, premium		3	5.333		1,850	202		2,052	2,375
0200	Economy		3	5.333		1,500	202		1,702	2,000
0235	10′ wide, wood, premium		2.50	6.400		2,600	242		2,842	3,250
0240	Economy		2.50	6.400		2,200	242		2,442	2,825
0250	12′ wide, wood, premium		2.50	6.400		3,000	242		3,242	3,700
0300	Economy		2.50	6.400		2,450	242		2,692	3,100

08 32 19.15 Sliding Glass Vinyl-Clad Wood Doors

		Crew	Daily Output	Labor-Hours	Unit	Material	Labor	Equipment	Total	Total Incl O&P
0010	**SLIDING GLASS VINYL-CLAD WOOD DOORS**									
0020	Glass, sliding vinyl clad, insul. glass, 6′-0″ x 6′-8″ G	2 Carp	4	4	Opng.	1,500	151		1,651	1,925
0025	6′-0″ x 6′-10″ high G		4	4		1,650	151		1,801	2,075
0030	6′-0″ x 8′-0″ high G		4	4		1,975	151		2,126	2,425
0050	5′-0″ x 6′-8″ high G		4	4		1,500	151		1,651	1,900
0100	8′-0″ x 6′-10″ high G		4	4		2,025	151		2,176	2,475
0150	8′-0″ x 8′-0″ high G		4	4		2,225	151		2,376	2,700
0500	4 leaf, 9′-0″ x 6′-10″ high G		3	5.333		3,250	202		3,452	3,925
0550	9′-0″ x 8′-0″ high G		3	5.333		3,775	202		3,977	4,500
0600	12′-0″ x 6′-10″ high G		3	5.333		3,875	202		4,077	4,625
0650	12′-0″ x 8′-0″ high G		3	5.333		3,875	202		4,077	4,625

08 33 Coiling Doors and Grilles

08 33 13 – Coiling Counter Doors

08 33 13.10 Counter Doors, Coiling Type

		Crew	Daily Output	Labor-Hours	Unit	Material	Labor	Equipment	Total	Total Incl O&P
0010	**COUNTER DOORS, COILING TYPE**									
0020	Manual, incl. frame and hardware, galv. stl., 4′ roll-up, 6′ long	2 Carp	2	8	Opng.	1,225	300		1,525	1,850
0300	Galvanized steel, UL label		1.80	8.889		1,300	335		1,635	2,000
0600	Stainless steel, 4′ high roll-up, 6′ long		2	8		2,150	300		2,450	2,850
0700	10′ long		1.80	8.889		2,600	335		2,935	3,425
2000	Aluminum, 4′ high, 4′ long		2.20	7.273		1,550	275		1,825	2,150
2020	6′ long		2	8		1,850	300		2,150	2,525
2040	8′ long		1.90	8.421		2,100	320		2,420	2,825
2060	10′ long		1.80	8.889		2,150	335		2,485	2,925
2080	14′ long		1.40	11.429		2,750	430		3,180	3,750
2100	6′ high, 4′ long		2	8		1,900	300		2,200	2,575
2120	6′ long		1.60	10		1,700	380		2,080	2,500
2140	10′ long		1.40	11.429		2,250	430		2,680	3,200

08 33 16 – Coiling Counter Grilles

08 33 16.10 Coiling Grilles

		Crew	Daily Output	Labor-Hours	Unit	Material	Labor	Equipment	Total	Total Incl O&P
0010	**COILING GRILLES**									
2020	Aluminum, manual operated, mill finish	2 Sswk	82	.195	S.F.	29	7.80		36.80	46
2040	Bronze anodized		82	.195	″	45	7.80		52.80	64
2060	Steel, manual operated, 10′ x 10′ high		1	16	Opng.	2,600	640		3,240	4,025
2080	15′ x 8′ high		.80	20	″	2,900	800		3,700	4,675
3000	For safety edge bottom bar, electric, add				L.F.	49			49	54
8000	For motor operation, add	2 Sswk	5	3.200	Opng.	1,250	128		1,378	1,600

08 33 23 – Overhead Coiling Doors

08 33 23.10 Coiling Service Doors		Crew	Daily Output	Labor-Hours	Unit	Material	2016 Bare Costs Labor	Equipment	Total	Total Incl O&P
0010	**COILING SERVICE DOORS** Steel, manual, 20 ga., incl. hardware									
0051	8' x 8' high	L-4	1.60	10	Ea.	1,175	330		1,505	1,850
0101	10' x 10' high	"	1.40	11.429		1,950	380		2,330	2,775
0130	12' x 12' high, standard	2 Sswk	1.20	13.333		2,200	530		2,730	3,400
0160	10' x 20' high, standard	"	.50	32		2,675	1,275		3,950	5,300
0201	20' x 10' high	L-4	1	16		3,175	530		3,705	4,400
0301	12' x 12' high	"	1.20	13.333		1,975	440		2,415	2,925
0303	Doors, rolling service, steel, manual, 20ga., 12' x 12', incl. hardware	2 Sswk	.92	17.391		1,975	695		2,670	3,450
0401	20' x 12' high	L-4	.90	17.778		2,200	590		2,790	3,425
0420	14' x 12' high	2 Sswk	.80	20		3,025	800		3,825	4,825
0430	14' x 13' high	"	.80	20		3,125	800		3,925	4,925
0501	14' x 14' high	L-4	.80	20	↓	3,150	665		3,815	4,600
0540	14' x 20' high	2 Sswk	.56	28.571	Opng.	4,050	1,150		5,200	6,550
0550	18' x 18' high	"	.60	26.667	"	2,575	1,075		3,650	4,775
0601	20' x 16' high	L-4	.60	26.667	Ea.	3,800	885		4,685	5,650
0700	10' x 20' high	2 Sswk	.50	32	"	2,675	1,275		3,950	5,300
0750	16' x 24' high	"	.48	33.333	Opng.	6,425	1,325		7,750	9,525
1001	12' x 12', crank operated, crank on door side	L-4	.80	20	Ea.	1,825	665		2,490	3,125
1101	Crank thru wall	"	.70	22.857		2,100	755		2,855	3,575
1300	For vision panel, add					370			370	405
1600	3' x 7' pass door within rolling steel door, new construction					1,950			1,950	2,150
1701	Existing construction	L-4	2	8		2,075	265		2,340	2,750
2001	Class A fire doors, manual, 20 ga., 8' x 8' high		1.40	11.429		1,625	380		2,005	2,425
2101	10' x 10' high		1.10	14.545		2,225	480		2,705	3,250
2201	20' x 10' high		.80	20		4,550	665		5,215	6,125
2301	12' x 12' high	↓	1	16		3,500	530		4,030	4,750
2304	Overhead door, roll up, fire rated 12' x 14'	2 Sswk	.90	17.778		3,975	710		4,685	5,675
2401	20' x 12' high	L-4	.80	20		4,925	665		5,590	6,550
2501	14' x 14' high		.60	26.667		3,825	885		4,710	5,700
2601	20' x 16' high		.50	32		6,025	1,050		7,075	8,400
2701	10' x 20' high	↓	.40	40	↓	4,750	1,325		6,075	7,450
2730	Manual steel rollup fire doors	2 Sswk	160	.100	S.F.	35.50	3.99		39.49	46.50
2740	For motor operated, add	"	1	16	Ea.	1,225	640		1,865	2,525
3000	For 18 ga. doors, add				S.F.	1.75			1.75	1.93
3300	For enamel finish, add				"	2.10			2.10	2.31
3600	For safety edge bottom bar, pneumatic, add				L.F.	23			23	25.50
3700	Elctric, add					44			44	48.50
4000	For weatherstripping, extruded rubber, jambs, add					15.20			15.20	16.70
4100	Hood, add					8.55			8.55	9.40
4200	Sill, add				↓	5.85			5.85	6.45
4501	Motor operators, to 14' x 14' opening	L-4	5	3.200	Ea.	1,175	106		1,281	1,475
4601	Over 14' x 14', jack shaft type	"	5	3.200		1,100	106		1,206	1,375
4700	For fire door, additional fusible link, add					30			30	33
5100	Radio control operator to 12' x 12' door	2 Sswk	3	5.333		900	213		1,113	1,375
5120	Receiver to 12' x 12' door		13	1.231		198	49		247	310
5140	Transmitter to 12' x 12' door	↓	49	.327	↓	57	13.05		70.05	87

08 34 Special Function Doors

08 34 16 – Hangar Doors

08 34 16.10 Aircraft Hangar Doors		Crew	Daily Output	Labor-Hours	Unit	Material	Labor	Equipment	Total	Total Incl O&P
						2016 Bare Costs				
0010	**AIRCRAFT HANGAR DOORS**									
0020	Bi-fold, ovhd., 20 psf wind load, incl. elec. oper.									
0101	12′ high x 40′	L-4	240	.067	S.F.	16.60	2.21		18.81	22
0201	16′ high x 60′		230	.070		19.90	2.30		22.20	26
0301	20′ high x 80′		220	.073		21.50	2.41		23.91	28

08 34 36 – Darkroom Doors

08 34 36.10 Various Types of Darkroom Doors		Crew	Daily Output	Labor-Hours	Unit	Material	Labor	Equipment	Total	Total Incl O&P
0010	**VARIOUS TYPES OF DARKROOM DOORS**									
0015	Revolving, standard, 2 way, 36″ diameter	2 Carp	3.10	5.161	Opng.	2,900	195		3,095	3,500
0020	41″ diameter		3.10	5.161		3,000	195		3,195	3,650
0050	3 way, 51″ diameter		1.40	11.429		3,825	430		4,255	4,925
2000	Hinged safety, 2 way, 41″ diameter		2.30	6.957		3,750	263		4,013	4,550
2500	3 way, 51″ diameter		1.40	11.429		4,050	430		4,480	5,200
3000	Pop out safety, 2 way, 41″ diameter		3.10	5.161		4,025	195		4,220	4,750
4000	3 way, 51″ diameter		1.40	11.429		4,875	430		5,305	6,100
5000	Wheelchair-type, pop out, 51″ diameter		1.40	11.429		5,500	430		5,930	6,775

08 34 53 – Security Doors and Frames

08 34 53.20 Steel Door		Crew	Daily Output	Labor-Hours	Unit	Material	Labor	Equipment	Total	Total Incl O&P
0010	**STEEL DOOR** flush with ballistic core and welded frame both 14 ga.									
0120	UL 752 Level 3, 1-3/4″, 3′-0″ x 7′-0″	2 Carp	1.50	10.667	Opng.	2,525	405		2,930	3,450
0125	1-3/4″, 3′-6″ x 7′-0″		1.50	10.667		2,750	405		3,155	3,700
0130	1-3/4″, 4′-0″ x 7′-0″		1.20	13.333		2,950	505		3,455	4,100
0150	UL 752 Level 8, 1-3/4″, 3′-0″ x 7′-0″		1.50	10.667		12,800	405		13,205	14,800
0155	1-3/4″, 3′-6″ x 7′-0″		1.50	10.667		13,900	405		14,305	16,000
0160	1-3/4″, 4′-0″ x 7′-0″		1.20	13.333		14,100	505		14,605	16,300
1000	Safe room sliding door and hardware, 1-3/4″, 3′-0″ x 7′-0″ UL 752 Level 3		.50	32		23,500	1,200		24,700	27,900
1050	Safe room swinging door and hardware, 1-3/4″, 3′-0″ x 7′-0″ UL 752 Level 3		.50	32		28,100	1,200		29,300	32,900

08 34 53.30 Wood Ballistic Doors		Crew	Daily Output	Labor-Hours	Unit	Material	Labor	Equipment	Total	Total Incl O&P
0010	**WOOD BALLISTIC DOORS** with frames and hardware									
0050	Wood, 1-3/4″, 3′-0″ x 7′-0″ UL 752 Level 3	2 Carp	1.50	10.667	Opng.	2,500	405		2,905	3,425

08 34 56 – Security Gates

08 34 56.10 Gates		Crew	Daily Output	Labor-Hours	Unit	Material	Labor	Equipment	Total	Total Incl O&P
0010	**GATES**									
0015	Driveway gates include mounting hardware									
0500	Wood, security gate, driveway, dual, 10′ wide	H-4	.80	25	Opng.	3,900	880		4,780	5,775
0505	12′ wide		.80	25		4,300	880		5,180	6,200
0510	15′ wide		.80	25		5,100	880		5,980	7,075
0600	Steel, security gate, driveway, single, 10′ wide		.80	25		2,150	880		3,030	3,825
0605	12′ wide		.80	25		2,225	880		3,105	3,900
0620	Steel, security gate, driveway, dual, 12′ wide		.80	25		2,050	880		2,930	3,750
0625	14′ wide		.80	25		2,200	880		3,080	3,900
0630	16′ wide		.80	25		2,525	880		3,405	4,250
0700	Aluminum, security gate, driveway, dual, 10′ wide		.80	25		3,000	880		3,880	4,775
0705	12′ wide		.80	25		3,200	880		4,080	5,000
0710	16′ wide		.80	25		3,700	880		4,580	5,550
1000	Security gate, driveway, opener 12 VDC				Ea.	800			800	880
1010	Wireless					1,800			1,800	1,975
1020	Security gate, driveway, opener 24 VDC					1,300			1,300	1,425
1030	Wireless					1,450			1,450	1,600
1040	Security gate, driveway, opener 12 VDC, solar panel 10 watt	1 Elec	2	4		300	176		476	620
1050	20 watt	″	2	4		450	176		626	785

08 34 Special Function Doors

08 34 73 – Sound Control Door Assemblies

08 34 73.10 Acoustical Doors		Crew	Daily Output	Labor-Hours	Unit	Material	2016 Bare Costs Labor	Equipment	Total	Total Incl O&P
0010	**ACOUSTICAL DOORS**									
0020	Including framed seals, 3′ x 7′, wood, 40 STC rating	2 Carp	1.50	10.667	Ea.	1,400	405		1,805	2,225
0100	Steel, 41 STC rating		1.50	10.667		3,300	405		3,705	4,300
0200	45 STC rating		1.50	10.667		3,725	405		4,130	4,775
0300	48 STC rating		1.50	10.667		4,350	405		4,755	5,450
0400	52 STC rating		1.50	10.667		5,000	405		5,405	6,175

08 36 Panel Doors

08 36 13 – Sectional Doors

08 36 13.10 Overhead Commercial Doors		Crew	Daily Output	Labor-Hours	Unit	Material	Labor	Equipment	Total	Total Incl O&P
0010	**OVERHEAD COMMERCIAL DOORS**									
1000	Stock, sectional, heavy duty, wood, 1-3/4″ thick, 8′ x 8′ high	2 Carp	2	8	Ea.	1,150	300		1,450	1,750
1100	10′ x 10′ high		1.80	8.889		1,600	335		1,935	2,350
1200	12′ x 12′ high		1.50	10.667		2,225	405		2,630	3,125
1300	Chain hoist, 14′ x 14′ high		1.30	12.308		3,375	465		3,840	4,475
1400	12′ x 16′ high		1	16		3,325	605		3,930	4,700
1500	20′ x 8′ high		1.30	12.270		2,725	465		3,190	3,775
1600	20′ x 16′ high		.65	24.615		5,650	930		6,580	7,750
1800	Center mullion openings, 8′ high		4	4		1,225	151		1,376	1,600
1900	20′ high		2	8		2,050	300		2,350	2,750
2100	For medium duty custom door, deduct					5%	5%			
2150	For medium duty stock doors, deduct					10%	5%			
2300	Fiberglass and aluminum, heavy duty, sectional, 12′ x 12′ high	2 Carp	1.50	10.667	Ea.	3,050	405		3,455	4,025
2450	Chain hoist, 20′ x 20′ high		.50	32		7,225	1,200		8,425	9,975
2600	Steel, 24 ga. sectional, manual, 8′ x 8′ high		2	8		900	300		1,200	1,500
2650	10′ x 10′ high		1.80	8.889		1,225	335		1,560	1,925
2700	12′ x 12′ high		1.50	10.667		1,475	405		1,880	2,300
2800	Chain hoist, 20′ x 14′ high		.70	22.857		3,425	865		4,290	5,225
2850	For 1-1/4″ rigid insulation and 26 ga. galv.									
2860	back panel, add				S.F.	4.75			4.75	5.25
2900	For electric trolley operator, 1/3 H.P., to 12′ x 12′, add	1 Carp	2	4	Ea.	1,050	151		1,201	1,425
2950	Over 12′ x 12′, 1/2 H.P., add		1	8		1,125	300		1,425	1,750
2980	Overhead, for row of clear lites add		1	8		110	300		410	625

08 38 Traffic Doors

08 38 13 – Flexible Strip Doors

08 38 13.10 Flexible Transparent Strip Doors		Crew	Daily Output	Labor-Hours	Unit	Material	Labor	Equipment	Total	Total Incl O&P
0010	**FLEXIBLE TRANSPARENT STRIP DOORS**									
0100	12″ strip width, 2/3 overlap	3 Shee	135	.178	SF Surf	7.60	7.55		15.15	21
0200	Full overlap		115	.209		9.60	8.85		18.45	25.50
0220	8″ strip width, 1/2 overlap		140	.171		6.20	7.25		13.45	18.90
0240	Full overlap		120	.200		7.80	8.45		16.25	23
0300	Add for suspension system, header mount				L.F.	9.15			9.15	10.05
0400	Wall mount				″	9.45			9.45	10.40

08 38 Traffic Doors

08 38 16 – Flexible Traffic Doors

08 38 16.10 Rubber Doors		Crew	Daily Output	Labor-Hours	Unit	2016 Bare Costs Material	Labor	Equipment	Total	Total Incl O&P
0010	**RUBBER DOORS**									
0015	Incl. frame and hardware, 14″ x 16″ vision panel,									
0020	48″ wear panel, pair of 2′-6″ x 7′	2 Sswk	1.60	10	Ea.	2,225	400		2,625	3,175
0040	Pair of 3′ x 7′		1.50	10.667		2,550	425		2,975	3,575
0060	Pair of 4′ x 7′		1.40	11.429		3,050	455		3,505	4,200

08 38 19 – Rigid Traffic Doors

08 38 19.20 Double Acting Swing Doors		Crew	Daily Output	Labor-Hours	Unit	Material	Labor	Equipment	Total	Total Incl O&P
0010	**DOUBLE ACTING SWING DOORS**									
0020	Including frame, closer, hardware and vision panel									
1000	Polymer, 7′-0″ high, 4′-0″ wide	2 Carp	4.20	3.810	Pr.	2,100	144		2,244	2,550
1025	6′-0″ wide		4	4		2,200	151		2,351	2,675
1050	6′-8″ wide		4	4		2,400	151		2,551	2,900
2000	3/4″ thick, stainless steel									
2010	Stainless steel, 7′ high opening, 4′ wide	2 Carp	4	4	Pr.	2,600	151		2,751	3,100
2050	7′ wide	″	3.80	4.211	″	2,800	159		2,959	3,350

08 38 19.30 Shock Absorbing Doors		Crew	Daily Output	Labor-Hours	Unit	Material	Labor	Equipment	Total	Total Incl O&P
0010	**SHOCK ABSORBING DOORS**									
0020	Rigid, no frame, 1-1/2″ thick, 5′ x 7′	2 Sswk	1.90	8.421	Opng.	1,525	335		1,860	2,300
0100	8′ x 8′		1.80	8.889		2,000	355		2,355	2,850
0500	Flexible, no frame, insulated, .16″ thick, economy, 5′ x 7′		2	8		1,750	320		2,070	2,500
0600	Deluxe		1.90	8.421		2,625	335		2,960	3,500
1000	8′ x 8′ opening, economy		2	8		2,750	320		3,070	3,600
1100	Deluxe		1.90	8.421		3,500	335		3,835	4,475

08 41 Entrances and Storefronts

08 41 13 – Aluminum-Framed Entrances and Storefronts

08 41 13.20 Tube Framing		Crew	Daily Output	Labor-Hours	Unit	Material	Labor	Equipment	Total	Total Incl O&P
0010	**TUBE FRAMING**, For window walls and store fronts, aluminum stock									
0050	Plain tube frame, mill finish, 1-3/4″ x 1-3/4″	2 Glaz	103	.155	L.F.	10.10	5.75		15.85	20.50
0150	1-3/4″ x 4″		98	.163		13.55	6		19.55	25
0200	1-3/4″ x 4-1/2″		95	.168		16.30	6.20		22.50	28.50
0250	2″ x 6″		89	.180		24	6.65		30.65	37.50
0350	4″ x 4″		87	.184		27.50	6.80		34.30	41.50
0400	4-1/2″ x 4-1/2″		85	.188		29	6.95		35.95	43
0450	Glass bead		240	.067		3.15	2.46		5.61	7.55
1000	Flush tube frame, mill finish, 1/4″ glass, 1-3/4″ x 4″, open header		80	.200		13.55	7.40		20.95	27
1050	Open sill		82	.195		11.05	7.20		18.25	24
1100	Closed back header		83	.193		19.40	7.10		26.50	33.50
1150	Closed back sill		85	.188		18.55	6.95		25.50	32
1160	Tube fmg., spandrel cover both sides, alum 1″ wide	1 Sswk	85	.094	S.F.	99	3.76		102.76	116
1170	Tube fmg., spandrel cover both sides, alum 2″ wide	″	85	.094	″	39	3.76		42.76	50
1200	Vertical mullion, one piece	2 Glaz	75	.213	L.F.	20.50	7.85		28.35	35.50
1250	Two piece		73	.219		21.50	8.10		29.60	37
1300	90° or 180° vertical corner post		75	.213		33	7.85		40.85	49
1400	1-3/4″ x 4-1/2″, open header		80	.200		16.30	7.40		23.70	30
1450	Open sill		82	.195		13.80	7.20		21	27
1500	Closed back header		83	.193		19.65	7.10		26.75	33.50
1550	Closed back sill		85	.188		19.35	6.95		26.30	33
1600	Vertical mullion, one piece		75	.213		21.50	7.85		29.35	37
1650	Two piece		73	.219		23	8.10		31.10	39

08 41 Entrances and Storefronts

08 41 13 – Aluminum-Framed Entrances and Storefronts

08 41 13.20 Tube Framing		Crew	Daily Output	Labor-Hours	Unit	Material	2016 Bare Costs Labor	Equipment	Total	Total Incl O&P
1700	90° or 180° vertical corner post	2 Glaz	75	.213	L.F.	23.50	7.85		31.35	38.50
2000	Flush tube frame, mil fin.,ins. glass w/thml brk, 2" x 4-1/2", open header		75	.213		16.70	7.85		24.55	31.50
2050	Open sill		77	.208		14.05	7.65		21.70	28
2100	Closed back header		78	.205		15.85	7.55		23.40	30
2150	Closed back sill		80	.200		15.20	7.40		22.60	29
2200	Vertical mullion, one piece		70	.229		17.65	8.45		26.10	33.50
2250	Two piece		68	.235		19.10	8.70		27.80	35.50
2300	90° or 180° vertical corner post		70	.229		18.15	8.45		26.60	34
5000	Flush tube frame, mill fin., thermal brk., 2-1/4" x 4-1/2", open header		74	.216		17.55	8		25.55	32.50
5050	Open sill		75	.213		15.45	7.85		23.30	30
5100	Vertical mullion, one piece		69	.232		18.20	8.55		26.75	34
5150	Two piece		67	.239		22	8.80		30.80	38.50
5200	90° or 180° vertical corner post		69	.232		19.60	8.55		28.15	35.50
6980	Door stop (snap in)		380	.042		3.45	1.55		5	6.40
7000	For joints, 90°, clip type, add				Ea.	26.50			26.50	29
7050	Screw spline joint, add					24.50			24.50	26.50
7100	For joint other than 90°, add					50.50			50.50	56
8000	For bronze anodized aluminum, add					15%				
8050	For stainless steel materials, add					350%				
8100	For monumental grade, add					53%				
8150	For steel stiffener, add	2 Glaz	200	.080	L.F.	11.60	2.95		14.55	17.65
8200	For 2 to 5 stories, add per story				Story		8%			

08 42 Entrances

08 42 26 – All-Glass Entrances

08 42 26.10 Swinging Glass Doors

		Crew	Daily Output	Labor-Hours	Unit	Material	Labor	Equipment	Total	Total Incl O&P
0010	**SWINGING GLASS DOORS**									
0020	Including hardware, 1/2" thick, tempered, 3' x 7' opening	2 Glaz	2	8	Opng.	2,275	295		2,570	3,000
0100	6' x 7' opening	"	1.40	11.429	"	4,450	420		4,870	5,600

08 42 36 – Balanced Door Entrances

08 42 36.10 Balanced Entrance Doors

		Crew	Daily Output	Labor-Hours	Unit	Material	Labor	Equipment	Total	Total Incl O&P
0010	**BALANCED ENTRANCE DOORS**									
0020	Hardware & frame, alum. & glass, 3' x 7', econ.	2 Sswk	.90	17.778	Ea.	6,700	710		7,410	8,675
0150	Premium	"	.70	22.857	"	8,025	910		8,935	10,500

08 43 Storefronts

08 43 13 – Aluminum-Framed Storefronts

08 43 13.10 Aluminum-Framed Entrance Doors and Frames

		Crew	Daily Output	Labor-Hours	Unit	Material	Labor	Equipment	Total	Total Incl O&P
0010	**ALUMINUM-FRAMED ENTRANCE DOORS AND FRAMES**									
0015	Standard hardware and glass stops but no glass									
0020	Entrance door, 3' x 7' opening, clear anodized finish	2 Sswk	7	2.286	Opng.	565	91		656	785
0200	3'-6" x 7'-0", mill finish		7	2.286		660	91		751	890
0220	Bronze finish		7	2.286		680	91		771	910
0240	Black finish		7	2.286		780	91		871	1,025
0500	6' x 7' opening, clear finish		6	2.667		880	106		986	1,150
0520	Bronze finish		6	2.667		955	106		1,061	1,250
0540	Black finish		6	2.667		1,025	106		1,131	1,350
0600	Door frame for above doors 3'-0" x 7'-0", mill finish		6	2.667		415	106		521	650
0620	Bronze finish		6	2.667		480	106		586	725

08 43 Storefronts

08 43 13 – Aluminum-Framed Storefronts

08 43 13.10 Aluminum-Framed Entrance Doors and Frames		Crew	Daily Output	Labor-Hours	Unit	Material	Labor	Equipment	Total	Total Incl O&P
						2016 Bare Costs				
0640	Black finish	2 Sswk	6	2.667	Opng.	525	106		631	770
0700	3'-6" x 7'-0", mill finish		6	2.667		350	106		456	580
0720	Bronze finish		6	2.667		350	106		456	580
0740	Black finish		6	2.667		350	106		456	580
0800	6'-0" x 7'-0", mill finish		6	2.667		350	106		456	580
0820	Bronze finish		6	2.667		355	106		461	585
0840	Black finish		6	2.667		385	106		491	615
8000	For 8' high doors, add				Ea.	200			200	220

08 43 13.20 Storefront Systems		Crew	Daily Output	Labor-Hours	Unit	Material	Labor	Equipment	Total	Total Incl O&P
0010	**STOREFRONT SYSTEMS**, aluminum frame clear 3/8" plate glass									
0020	incl. 3' x 7' door with hardware (400 sq. ft. max. wall)									
0500	Wall height to 12' high, commercial grade	2 Glaz	150	.107	S.F.	23.50	3.94		27.44	32
0600	Institutional grade		130	.123		28.50	4.54		33.04	39
0700	Monumental grade		115	.139		41	5.15		46.15	53.50
1000	6' x 7' door with hardware, commercial grade		135	.119		30.50	4.37		34.87	41
1100	Institutional grade		115	.139		29	5.15		34.15	40.50
1200	Monumental grade		100	.160		55.50	5.90		61.40	71
1500	For bronze anodized finish, add					15%				
1600	For black anodized finish, add					36%				
1700	For stainless steel framing, add to monumental					78%				

08 43 29 – Sliding Storefronts

08 43 29.10 Sliding Panels		Crew	Daily Output	Labor-Hours	Unit	Material	Labor	Equipment	Total	Total Incl O&P
0010	**SLIDING PANELS**									
0020	Mall fronts, aluminum & glass, 15' x 9' high	2 Glaz	1.30	12.308	Opng.	3,550	455		4,005	4,650
0100	24' x 9' high		.70	22.857		5,025	845		5,870	6,925
0200	48' x 9' high, with fixed panels		.90	17.778		9,275	655		9,930	11,300
0500	For bronze finish, add					17%				

08 45 Translucent Wall and Roof Assemblies

08 45 10 – Translucent Roof Assemblies

08 45 10.10 Skyroofs		Crew	Daily Output	Labor-Hours	Unit	Material	Labor	Equipment	Total	Total Incl O&P
0010	**SKYROOFS**									
1200	Skylights, circular, clear, double glazed acrylic									
1230	30" diameter	2 Carp	3	5.333	Ea.	3,000	202		3,202	3,650
1250	60" diameter		3	5.333		4,000	202		4,202	4,750
1290	96" diameter		2	8		5,000	300		5,300	6,000
1330	3'-0" x 12'-0"	G-3	3	10.667		5,000	380		5,380	6,150
1350	4'-0" x 12'-0"		3	10.667		5,500	380		5,880	6,700
1390	5'-0" x 12'-0"		2	16		6,000	570		6,570	7,550
1430	Square, 3' x 3'		3	10.667		6,000	380		6,380	7,250
1440	4' x 4'		3	10.667		7,000	380		7,380	8,350
1450	5' x 5', glass must be field installed		3	10.667		8,000	380		8,380	9,450
1460	6' x 6', glass must be field installed		2	16		10,000	570		10,570	12,000

08 51 Metal Windows

08 51 13 – Aluminum Windows

08 51 13.10 Aluminum Sash		Crew	Daily Output	Labor-Hours	Unit	Material	2016 Bare Costs Labor	Equipment	Total	Total Incl O&P
0010	**ALUMINUM SASH**									
0020	Stock, grade C, glaze & trim not incl., casement	2 Sswk	200	.080	S.F.	39.50	3.19		42.69	49
0050	Double hung		200	.080		39.50	3.19		42.69	49.50
0100	Fixed casement		200	.080		17.35	3.19		20.54	25
0150	Picture window		200	.080		18.50	3.19		21.69	26.50
0200	Projected window		200	.080		35.50	3.19		38.69	45
0250	Single hung		200	.080		16.55	3.19		19.74	24
0300	Sliding		200	.080		21.50	3.19		24.69	29.50
1000	Mullions for above, tubular		240	.067	L.F.	6.15	2.66		8.81	11.70

08 51 13.20 Aluminum Windows

Line	Description	Crew	Daily Output	Labor-Hours	Unit	Material	Labor	Equipment	Total	Total Incl O&P
0010	**ALUMINUM WINDOWS**, incl. frame and glazing, commercial grade									
1000	Stock units, casement, 3′-1″ x 3′-2″ opening	2 Sswk	10	1.600	Ea.	375	64		439	525
1040	Insulating glass	"	10	1.600		515	64		579	680
1050	Add for storms					120			120	132
1600	Projected, with screen, 3′-1″ x 3′-2″ opening	2 Sswk	10	1.600		355	64		419	505
1650	Insulating glass	"	10	1.600		390	64		454	540
1700	Add for storms					117			117	129
2000	4′-5″ x 5′-3″ opening	2 Sswk	8	2		400	80		480	585
2050	Insulating glass	"	8	2		475	80		555	670
2100	Add for storms					126			126	139
2500	Enamel finish windows, 3′-1″ x 3′-2″	2 Sswk	10	1.600		360	64		424	510
2550	Insulating glass		10	1.600		315	64		379	460
2600	4′-5″ x 5′-3″		8	2		405	80		485	590
2700	Insulating glass		8	2		415	80		495	600
3000	Single hung, 2′ x 3′ opening, enameled, standard glazed		10	1.600		208	64		272	345
3100	Insulating glass		10	1.600		252	64		316	395
3300	2′-8″ x 6′-8″ opening, standard glazed		8	2		365	80		445	545
3400	Insulating glass		8	2		475	80		555	665
3700	3′-4″ x 5′-0″ opening, standard glazed		9	1.778		300	71		371	460
3800	Insulating glass		9	1.778		335	71		406	495
4000	Sliding aluminum, 3′ x 2′ opening, standard glazed		10	1.600		217	64		281	355
4100	Insulating glass		10	1.600		232	64		296	370
4300	5′ x 3′ opening, standard glazed		9	1.778		330	71		401	495
4400	Insulating glass		9	1.778		385	71		456	555
4600	8′ x 4′ opening, standard glazed		6	2.667		350	106		456	580
4700	Insulating glass		6	2.667		565	106		671	815
5000	9′ x 5′ opening, standard glazed		4	4		530	160		690	880
5100	Insulating glass		4	4		850	160		1,010	1,225
5500	Sliding, with thermal barrier and screen, 6′ x 4′, 2 track		8	2		725	80		805	940
5700	4 track		8	2		910	80		990	1,150
6000	For above units with bronze finish, add					15%				
6200	For installation in concrete openings, add					8%				

08 51 13.30 Impact Resistant Aluminum Windows

Line	Description	Crew	Daily Output	Labor-Hours	Unit	Material	Labor	Equipment	Total	Total Incl O&P
0010	**IMPACT RESISTANT ALUMINUM WINDOWS**, incl. frame and glazing									
0100	Single hung, impact resistant, 2′-8″ x 5′-0″	2 Carp	9	1.778	Ea.	1,250	67		1,317	1,500
0120	3′-0″ x 5′-0″		9	1.778		1,300	67		1,367	1,550
0130	4′-0″ x 5′-0″		9	1.778		1,300	67		1,367	1,550
0250	Horizontal slider, impact resistant, 5′-5″ x 5′-2″		9	1.778		1,550	67		1,617	1,825

08 51 Metal Windows

08 51 23 – Steel Windows

08 51 23.10 Steel Sash		Crew	Daily Output	Labor-Hours	Unit	Material	Labor	Equipment	Total	Total Incl O&P
						2016 Bare Costs				
0010	**STEEL SASH** Custom units, glazing and trim not included									
0100	Casement, 100% vented	2 Sswk	200	.080	S.F.	67	3.19		70.19	79.50
0200	50% vented		200	.080		55	3.19		58.19	66.50
0300	Fixed		200	.080		29.50	3.19		32.69	38.50
1000	Projected, commercial, 40% vented		200	.080		51.50	3.19		54.69	62.50
1100	Intermediate, 50% vented		200	.080		58.50	3.19		61.69	70.50
1500	Industrial, horizontally pivoted		200	.080		53.50	3.19		56.69	65
1600	Fixed		200	.080		30.50	3.19		33.69	39.50
2000	Industrial security sash, 50% vented		200	.080		58	3.19		61.19	70
2100	Fixed		200	.080		47	3.19		50.19	57.50
2500	Picture window		200	.080		30.50	3.19		33.69	39.50
3000	Double hung		200	.080		60	3.19		63.19	72
5000	Mullions for above, open interior face		240	.067	L.F.	10.55	2.66		13.21	16.50
5100	With interior cover		240	.067	"	17.40	2.66		20.06	24
6000	Double glazing for above, add	2 Glaz	200	.080	S.F.	13.10	2.95		16.05	19.35
6100	Triple glazing for above, add	"	85	.188	"	12.50	6.95		19.45	25.50

08 51 23.20 Steel Windows

		Crew	Daily Output	Labor-Hours	Unit	Material	Labor	Equipment	Total	Total Incl O&P
0010	**STEEL WINDOWS** Stock, including frame, trim and insul. glass									
1000	Custom units, double hung, 2′-8″ x 4′-6″ opening	2 Sswk	12	1.333	Ea.	705	53		758	880
1100	2′-4″ x 3′-9″ opening		12	1.333		585	53		638	740
1500	Commercial projected, 3′-9″ x 5′-5″ opening		10	1.600		1,225	64		1,289	1,475
1600	6′-9″ x 4′-1″ opening		7	2.286		1,625	91		1,716	1,950
2000	Intermediate projected, 2′-9″ x 4′-1″ opening		12	1.333		685	53		738	855
2100	4′-1″ x 5′-5″ opening		10	1.600		1,400	64		1,464	1,675

08 51 23.40 Basement Utility Windows

		Crew	Daily Output	Labor-Hours	Unit	Material	Labor	Equipment	Total	Total Incl O&P
0010	**BASEMENT UTILITY WINDOWS**									
0015	1′-3″ x 2′-8″	1 Carp	16	.500	Ea.	128	18.90		146.90	172
1100	1′-7″ x 2′-8″		16	.500		140	18.90		158.90	186
1200	1′-11″ x 2′-8″		14	.571		143	21.50		164.50	194

08 51 66 – Metal Window Screens

08 51 66.10 Screens

		Crew	Daily Output	Labor-Hours	Unit	Material	Labor	Equipment	Total	Total Incl O&P
0010	**SCREENS**									
0020	For metal sash, aluminum or bronze mesh, flat screen	2 Sswk	1200	.013	S.F.	4.30	.53		4.83	5.70
0500	Wicket screen, inside window	"	1000	.016	"	6.65	.64		7.29	8.45
0600	Residential, aluminum mesh and frame, 2′ x 3′	2 Carp	32	.500	Ea.	16.30	18.90		35.20	49.50
0610	Rescreen		50	.320		13.25	12.10		25.35	35
0620	3′ x 5′		32	.500		55	18.90		73.90	92
0630	Rescreen		45	.356		33.50	13.45		46.95	59.50
0640	4′ x 8′		25	.640		85	24		109	134
0650	Rescreen		40	.400		50	15.10		65.10	80.50
0660	Patio door		25	.640		200	24		224	261
0680	Rescreening		1600	.010	S.F.	2.50	.38		2.88	3.38
0800	Security screen, aluminum frame with stainless steel cloth	2 Sswk	1200	.013		24	.53		24.53	27
0900	Steel grate, painted, on steel frame		1600	.010		13.10	.40		13.50	15.15
1000	Screens for solar louvers		160	.100		24.50	3.99		28.49	34.50

08 52 Wood Windows

08 52 10 – Plain Wood Windows

08 52 10.10 Wood Windows		Crew	Daily Output	Labor-Hours	Unit	Material	2016 Bare Costs Labor	Equipment	Total	Total Incl O&P
0010	**WOOD WINDOWS**, including frame, screens and grilles									
0020	Residential, stock units									
0050	Awning type, double insulated glass, 2′-10″ x 1′-9″ opening	2 Carp	12	1.333	Opng.	230	50.50		280.50	340
0100	2′-10″ x 6′-0″ opening	1 Carp	8	1		555	38		593	675
0200	4′-0″ x 3′-6″ single pane		10	.800		375	30		405	465
0300	6′ x 5′ single pane		8	1	Ea.	510	38		548	625
1000	Casement, 2′-0″ x 3′-4″ high	2 Carp	20	.800		264	30		294	340
1020	2′-0″ x 4′-0″		18	.889		281	33.50		314.50	365
1040	2′-0″ x 5′-0″		17	.941		310	35.50		345.50	400
1060	2′-0″ x 6′-0″		16	1		305	38		343	405
1080	4′-0″ x 3′-4″		15	1.067		590	40.50		630.50	715
1100	4′-0″ x 4′-0″		15	1.067		660	40.50		700.50	800
1120	4′-0″ x 5′-0″		14	1.143		745	43		788	895
1140	4′-0″ x 6′-0″		12	1.333		845	50.50		895.50	1,000
1600	Casement units, 8′ x 5′, with screens, double insulated glass		2.50	6.400	Opng.	1,500	242		1,742	2,050
1700	Low E glass		2.50	6.400		1,625	242		1,867	2,200
2300	Casements, including screens, 2′-0″ x 3′-4″, dbl. insulated glass		11	1.455		275	55		330	395
2400	Low E glass		11	1.455		275	55		330	395
2600	2 lite, 4′-0″ x 4′-0″, double insulated glass		9	1.778		500	67		567	665
2700	Low E glass		9	1.778		515	67		582	680
2900	3 lite, 5′-2″ x 5′-0″, double insulated glass		7	2.286		760	86.50		846.50	980
3000	Low E glass		7	2.286		805	86.50		891.50	1,025
3200	4 lite, 7′-0″ x 5′-0″, double insulated glass		6	2.667		1,075	101		1,176	1,375
3300	Low E glass		6	2.667		1,150	101		1,251	1,450
3500	5 lite, 8′-6″ x 5′-0″, double insulated glass		5	3.200		1,450	121		1,571	1,775
3600	Low E glass		5	3.200		1,450	121		1,571	1,775
3800	For removable wood grilles, diamond pattern, add				Leaf	39			39	43
3900	Rectangular pattern, add				"	39			39	43
4000	Bow, fixed lites, 8′ x 5′, double insulated glass	2 Carp	3	5.333	Opng.	1,450	202		1,652	1,925
4100	Low E glass	"	3	5.333	"	1,925	202		2,127	2,450
4150	6′-0″ x 5′-0″	1 Carp	8	1	Ea.	1,275	38		1,313	1,475
4300	Fixed lites, 9′-9″ x 5′-0″, double insulated glass	2 Carp	2	8	Opng.	995	300		1,295	1,600
4400	Low E glass	"	2	8	"	1,075	300		1,375	1,700
5000	Bow, casement, 8′-1″ x 4′-8″ high	3 Carp	8	3	Ea.	1,600	113		1,713	1,950
5020	9′-6″ x 4′-8″		8	3		1,700	113		1,813	2,075
5040	8′-1″ x 5′-1″		8	3		1,925	113		2,038	2,325
5060	9′-6″ x 5′-1″		6	4		1,925	151		2,076	2,375
5080	8′-1″ x 6′-0″		6	4		1,925	151		2,076	2,375
5100	9′-6″ x 6′-0″		6	4		2,025	151		2,176	2,475
5800	Skylights, hatches, vents, and sky roofs, see Section 08 62 13.00									

08 52 10.20 Awning Window		Crew	Daily Output	Labor-Hours	Unit	Material	Labor	Equipment	Total	Total Incl O&P
0010	**AWNING WINDOW**, Including frame, screens and grilles									
0100	34″ x 22″, insulated glass	1 Carp	10	.800	Ea.	271	30		301	350
0200	Low E glass		10	.800		272	30		302	350
0300	40″ x 28″, insulated glass		9	.889		315	33.50		348.50	400
0400	Low E Glass		9	.889		340	33.50		373.50	430
0500	48″ x 36″, insulated glass		8	1		470	38		508	580
0600	Low E glass		8	1		495	38		533	605
4000	Impact windows, minimum, add					60%				
4010	Impact windows, maximum, add					160%				

08 52 Wood Windows

08 52 10 – Plain Wood Windows

08 52 10.30 Wood Windows			Crew	Daily Output	Labor-Hours	Unit	2016 Bare Costs Material	Labor	Equipment	Total	Total Incl O&P
0010	**WOOD WINDOWS**, double hung										
0020	Including frame, double insulated glass, screens and grilles										
0040	Double hung, 2′-2″ x 3′-4″ high		2 Carp	15	1.067	Ea.	212	40.50		252.50	300
0060	2′-2″ x 4′-4″			14	1.143		231	43		274	325
0080	2′-6″ x 3′-4″			13	1.231		221	46.50		267.50	320
0100	2′-6″ x 4′-0″			12	1.333		230	50.50		280.50	340
0120	2′-6″ x 4′-8″			12	1.333		253	50.50		303.50	365
0140	2′-10″ x 3′-4″			10	1.600		226	60.50		286.50	350
0160	2′-10″ x 4′-0″			10	1.600		248	60.50		308.50	375
0180	3′-7″ x 3′-4″			9	1.778		256	67		323	395
0200	3′-7″ x 5′-4″			9	1.778		286	67		353	430
0220	3′-10″ x 5′-4″			8	2		515	75.50		590.50	690
3800	Triple glazing for above, add						25%				

08 52 10.40 Casement Window			Crew	Daily Output	Labor-Hours	Unit	Material	Labor	Equipment	Total	Total Incl O&P
0010	**CASEMENT WINDOW**, including frame, screen and grilles										
0100	2′-0″ x 3′-0″ H, dbl. insulated glass	G	1 Carp	10	.800	Ea.	273	30		303	350
0150	Low E glass	G		10	.800		259	30		289	335
0200	2′-0″ x 4′-6″ high, double insulated glass	G		9	.889		370	33.50		403.50	460
0250	Low E glass	G		9	.889		370	33.50		403.50	460
0260	Casement 4′-2″ x 4′-2″ double insulated glass	G		11	.727		890	27.50		917.50	1,025
0270	4′-0″ x 4′-0″ Low E glass	G		11	.727		535	27.50		562.50	635
0290	6′-4″ x 5′-7″ Low E glass	G		9	.889		1,125	33.50		1,158.50	1,300
0300	2′-4″ x 6′-0″ high, double insulated glass	G		8	1		440	38		478	550
0350	Low E glass	G		8	1		440	38		478	550
0600	3′-0″ x 5′-0″	G		8	1		665	38		703	795
0700	4′-0″ x 3′-0″	G		8	1		730	38		768	870
0710	4′-0″ x 4′-0″	G		8	1		625	38		663	750
0720	4′-8″ x 4′-0″	G		8	1		690	38		728	825
0730	4′-8″ x 5′-0″	G		6	1.333		790	50.50		840.50	950
0740	4′-8″ x 6′-0″	G		6	1.333		880	50.50		930.50	1,050
0750	6′-0″ x 4′-0″	G		6	1.333		805	50.50		855.50	970
0800	6′-0″ x 5′-0″	G		6	1.333		895	50.50		945.50	1,075
0900	5′-6″ x 5′-6″	G	2 Carp	15	1.067		1,425	40.50		1,465.50	1,650
2000	Bay, casement units, 8′ x 5′, w/screens, dbl. insul. glass			2.50	6.400	Opng.	1,625	242		1,867	2,200
2100	Low E glass			2.50	6.400	″	1,725	242		1,967	2,275
8190	For installation, add per leaf					Ea.		15%			
8200	For multiple leaf units, deduct for stationary sash										
8220	2′ high					Ea.	23.50			23.50	25.50
8240	4′-6″ high						26.50			26.50	29
8260	6′ high						35			35	38.50
8300	Impact windows, minimum, add						60%				
8310	Impact windows, maximum, add						160%				

08 52 10.50 Double Hung			Crew	Daily Output	Labor-Hours	Unit	Material	Labor	Equipment	Total	Total Incl O&P
0010	**DOUBLE HUNG**, Including frame, screens and grilles										
0100	2′-0″ x 3′-0″ high, low E insul. glass	G	1 Carp	10	.800	Ea.	210	30		240	282
0200	3′-0″ x 4′-0″ high, double insulated glass	G		9	.889		279	33.50		312.50	360
0300	4′-0″ x 4′-6″ high, low E insulated glass	G		8	1		320	38		358	420

08 52 10.55 Picture Window			Crew	Daily Output	Labor-Hours	Unit	Material	Labor	Equipment	Total	Total Incl O&P
0010	**PICTURE WINDOW**, Including frame and grilles										
0100	3′-6″ x 4′-0″ high, dbl. insulated glass		2 Carp	12	1.333	Ea.	420	50.50		470.50	550
0150	Low E glass			12	1.333		435	50.50		485.50	560

08 52 Wood Windows

08 52 10 – Plain Wood Windows

08 52 10.55 Picture Window		Crew	Daily Output	Labor-Hours	Unit	2016 Bare Costs Material	Labor	Equipment	Total	Total Incl O&P
0200	4'-0" x 4'-6" high, double insulated glass	2 Carp	11	1.455	Ea.	550	55		605	695
0250	Low E glass		11	1.455		530	55		585	675
0300	5'-0" x 4'-0" high, double insulated glass		11	1.455		580	55		635	730
0350	Low E glass		11	1.455		605	55		660	755
0400	6'-0" x 4'-6" high, double insulated glass		10	1.600		640	60.50		700.50	800
0450	Low E glass		10	1.600		640	60.50		700.50	805

08 52 10.65 Wood Sash		Crew	Daily Output	Labor-Hours	Unit	Material	Labor	Equipment	Total	Total Incl O&P
0010	**WOOD SASH**, Including glazing but not trim									
0050	Custom, 5'-0" x 4'-0", 1" dbl. glazed, 3/16" thick lites	2 Carp	3.20	5	Ea.	230	189		419	570
0100	1/4" thick lites		5	3.200		245	121		366	475
0200	1" thick, triple glazed		5	3.200		415	121		536	660
0300	7'-0" x 4'-6" high, 1" double glazed, 3/16" thick lites		4.30	3.721		420	141		561	695
0400	1/4" thick lites		4.30	3.721		475	141		616	755
0500	1" thick, triple glazed		4.30	3.721		540	141		681	830
0600	8'-6" x 5'-0" high, 1" double glazed, 3/16" thick lites		3.50	4.571		565	173		738	915
0700	1/4" thick lites		3.50	4.571		620	173		793	970
0800	1" thick, triple glazed		3.50	4.571		625	173		798	975
0900	Window frames only, based on perimeter length				L.F.	4.02			4.02	4.42

08 52 10.70 Sliding Windows		Crew	Daily Output	Labor-Hours	Unit	Material	Labor	Equipment	Total	Total Incl O&P
0010	**SLIDING WINDOWS**									
0100	3'-0" x 3'-0" high, double insulated G	1 Carp	10	.800	Ea.	284	30		314	360
0120	Low E glass G		10	.800		310	30		340	390
0200	4'-0" x 3'-6" high, double insulated G		9	.889		355	33.50		388.50	445
0220	Low E glass G		9	.889		360	33.50		393.50	455
0300	6'-0" x 5'-0" high, double insulated G		8	1		490	38		528	600
0320	Low E glass G		8	1		530	38		568	645
6000	Sliding, insulating glass, including screens,									
6100	3'-0" x 3'-0"	2 Carp	6.50	2.462	Ea.	330	93		423	520
6200	4'-0" x 3'-6"		6.30	2.540		335	96		431	530
6300	5'-0" x 4'-0"		6	2.667		415	101		516	630

08 52 13 – Metal-Clad Wood Windows

08 52 13.10 Awning Windows, Metal-Clad		Crew	Daily Output	Labor-Hours	Unit	Material	Labor	Equipment	Total	Total Incl O&P
0010	**AWNING WINDOWS, METAL-CLAD**									
2000	Metal clad, awning deluxe, double insulated glass, 34" x 22"	1 Carp	9	.889	Ea.	249	33.50		282.50	330
2050	36" x 25"		9	.889		276	33.50		309.50	360
2100	40" x 22"		9	.889		292	33.50		325.50	375
2150	40" x 30"		9	.889		340	33.50		373.50	430
2200	48" x 28"		8	1		350	38		388	450
2250	60" x 36"		8	1		375	38		413	475

08 52 13.20 Casement Windows, Metal-Clad		Crew	Daily Output	Labor-Hours	Unit	Material	Labor	Equipment	Total	Total Incl O&P
0010	**CASEMENT WINDOWS, METAL-CLAD**									
0100	Metal clad, deluxe, dbl. insul. glass, 2'-0" x 3'-0" high G	1 Carp	10	.800	Ea.	285	30		315	365
0120	2'-0" x 4'-0" high G		9	.889		320	33.50		353.50	405
0130	2'-0" x 5'-0" high G		8	1		330	38		368	430
0140	2'-0" x 6'-0" high G		8	1		370	38		408	470
0300	Metal clad, casement, bldrs mdl, 6'-0" x 4'-0", dbl. insltd gls, 3 panels	2 Carp	10	1.600		1,200	60.50		1,260.50	1,425
0310	9'-0" x 4'-0" , 4 panels		8	2		1,550	75.50		1,625.50	1,825
0320	10'-0" x 5'-0", 5 panels		7	2.286		2,100	86.50		2,186.50	2,475
0330	12'-0" x 6'-0", 6 panels		6	2.667		2,700	101		2,801	3,125

08 52 Wood Windows

08 52 13 – Metal-Clad Wood Windows

08 52 13.30 Double-Hung Windows, Metal-clad			Crew	Daily Output	Labor-Hours	Unit	2016 Bare Costs Material	Labor	Equipment	Total	Total Incl O&P
0010	**DOUBLE-HUNG WINDOWS, METAL-CLAD**										
0100	Metal clad, deluxe, dbl. insul. glass, 2'-6" x 3'-0" high	G	1 Carp	10	.800	Ea.	280	30		310	360
0120	3'-0" x 3'-6" high	G		10	.800		320	30		350	400
0140	3'-0" x 4'-0" high	G		9	.889		335	33.50		368.50	425
0160	3'-0" x 4'-6" high	G		9	.889		350	33.50		383.50	440
0180	3'-0" x 5'-0" high	G		8	1		380	38		418	480
0200	3'-6" x 6'-0" high	G		8	1		455	38		493	570

08 52 13.35 Picture and Sliding Windows Metal-Clad

			Crew	Daily Output	Labor-Hours	Unit	Material	Labor	Equipment	Total	Total Incl O&P
0010	**PICTURE AND SLIDING WINDOWS METAL-CLAD**										
2000	Metal clad, dlx picture, dbl. insul. glass, 4'-0" x 4'-0" high		2 Carp	12	1.333	Ea.	380	50.50		430.50	505
2100	4'-0" x 6'-0" high			11	1.455		560	55		615	705
2200	5'-0" x 6'-0" high			10	1.600		620	60.50		680.50	780
2300	6'-0" x 6'-0" high			10	1.600		710	60.50		770.50	880
2400	Metal clad, dlx sliding, double insulated glass, 3'-0" x 3'-0" high	G	1 Carp	10	.800		330	30		360	410
2420	4'-0" x 3'-6" high	G		9	.889		400	33.50		433.50	495
2440	5'-0" x 4'-0" high	G		9	.889		480	33.50		513.50	580
2460	6'-0" x 5'-0" high	G		8	1		730	38		768	865

08 52 13.40 Bow and Bay Windows, Metal-Clad

			Crew	Daily Output	Labor-Hours	Unit	Material	Labor	Equipment	Total	Total Incl O&P
0010	**BOW AND BAY WINDOWS, METAL-CLAD**										
0100	Metal clad, deluxe, dbl. insul. glass, 8'-0" x 5'-0" high, 4 panels		2 Carp	10	1.600	Ea.	1,700	60.50		1,760.50	1,975
0120	10'-0" x 5'-0" high, 5 panels			8	2		1,825	75.50		1,900.50	2,125
0140	10'-0" x 6'-0" high, 5 panels			7	2.286		2,150	86.50		2,236.50	2,525
0160	12'-0" x 6'-0" high, 6 panels			6	2.667		2,925	101		3,026	3,400
0400	Double hung, bldrs. model, bay, 8' x 4' high, dbl. insulated glass			10	1.600		1,350	60.50		1,410.50	1,575
0440	Low E glass			10	1.600		1,450	60.50		1,510.50	1,700
0460	9'-0" x 5'-0" high, double insulated glass			6	2.667		1,450	101		1,551	1,775
0480	Low E glass			6	2.667		1,525	101		1,626	1,850
0500	Metal clad, deluxe, dbl. insul. glass, 7'-0" x 4'-0" high			10	1.600		1,300	60.50		1,360.50	1,525
0520	8'-0" x 4'-0" high			8	2		1,350	75.50		1,425.50	1,600
0540	8'-0" x 5'-0" high			7	2.286		1,375	86.50		1,461.50	1,675
0560	9'-0" x 5'-0" high			6	2.667		1,475	101		1,576	1,800

08 52 16 – Plastic-Clad Wood Windows

08 52 16.10 Bow Window

		Crew	Daily Output	Labor-Hours	Unit	Material	Labor	Equipment	Total	Total Incl O&P
0010	**BOW WINDOW** Including frames, screens, and grilles									
0020	End panels operable									
1000	Bow type, casement, wood, bldrs. mdl., 8' x 5' dbl. insltd glass, 4 panel	2 Carp	10	1.600	Ea.	1,500	60.50		1,560.50	1,750
1050	Low E glass		10	1.600		1,325	60.50		1,385.50	1,550
1100	10'-0" x 5'-0" , double insulated glass, 6 panels		6	2.667		1,350	101		1,451	1,675
1200	Low E glass, 6 panels		6	2.667		1,450	101		1,551	1,775
1300	Vinyl clad, bldrs. model, double insulated glass, 6'-0" x 4'-0", 3 panel		10	1.600		1,025	60.50		1,085.50	1,225
1340	9'-0" x 4'-0", 4 panel		8	2		1,350	75.50		1,425.50	1,625
1380	10'-0" x 6'-0", 5 panels		7	2.286		2,250	86.50		2,336.50	2,625
1420	12'-0" x 6'-0", 6 panels		6	2.667		2,925	101		3,026	3,400
2000	Bay window, 8' x 5', dbl. insul glass		10	1.600		1,925	60.50		1,985.50	2,200
2050	Low E glass		10	1.600		2,325	60.50		2,385.50	2,650
2100	12'-0" x 6'-0" , double insulated glass, 6 panels		6	2.667		2,400	101		2,501	2,800
2200	Low E glass		6	2.667		3,175	101		3,276	3,675
2280	6'-0" x 4'-0"		11	1.455		1,250	55		1,305	1,475
2300	Vinyl clad, premium, double insulated glass, 8'-0" x 5'-0"		10	1.600		1,750	60.50		1,810.50	2,025
2340	10'-0" x 5'-0"		8	2		2,300	75.50		2,375.50	2,650
2380	10'-0" x 6'-0"		7	2.286		2,625	86.50		2,711.50	3,025

08 52 Wood Windows

08 52 16 – Plastic-Clad Wood Windows

08 52 16.10 Bow Window

08 52 16.10 Bow Window		Crew	Daily Output	Labor-Hours	Unit	2016 Bare Costs Material	Labor	Equipment	Total	Total Incl O&P
2420	12′-0″ x 6′-0″	2 Carp	6	2.667	Ea.	3,200	101		3,301	3,700
3300	Vinyl clad, premium, double insulated glass, 7′-0″ x 4′-6″		10	1.600		1,400	60.50		1,460.50	1,625
3340	8′-0″ x 4′-6″		8	2		1,425	75.50		1,500.50	1,675
3380	8′-0″ x 5′-0″		7	2.286		1,500	86.50		1,586.50	1,800
3420	9′-0″ x 5′-0″		6	2.667		1,525	101		1,626	1,850

08 52 16.15 Awning Window Vinyl-Clad

Line	Description	Crew	Daily Output	Labor-Hours	Unit	Material	Labor	Equipment	Total	Total Incl O&P
0010	**AWNING WINDOW VINYL-CLAD** Including frames, screens, and grilles									
0240	Vinyl clad, 34″ x 22″	1 Carp	10	.800	Ea.	267	30		297	345
0280	36″ x 28″		9	.889		305	33.50		338.50	390
0300	36″ x 36″		9	.889		345	33.50		378.50	430
0340	40″ x 22″		10	.800		291	30		321	370
0360	48″ x 28″		8	1		370	38		408	470
0380	60″ x 36″		8	1		500	38		538	615

08 52 16.30 Palladian Windows

Line	Description	Crew	Daily Output	Labor-Hours	Unit	Material	Labor	Equipment	Total	Total Incl O&P
0010	**PALLADIAN WINDOWS**									
0020	Vinyl clad, double insulated glass, including frame and grilles									
0040	3′-2″ x 2′-6″ high	2 Carp	11	1.455	Ea.	1,250	55		1,305	1,475
0060	3′-2″ x 4′-10″		11	1.455		1,750	55		1,805	2,000
0080	3′-2″ x 6′-4″		10	1.600		1,700	60.50		1,760.50	1,975
0100	4′-0″ x 4′-0″		10	1.600		1,525	60.50		1,585.50	1,775
0120	4′-0″ x 5′-4″	3 Carp	10	2.400		1,875	90.50		1,965.50	2,200
0140	4′-0″ x 6′-0″		9	2.667		1,950	101		2,051	2,325
0160	4′-0″ x 7′-4″		9	2.667		2,125	101		2,226	2,500
0180	5′-5″ x 4′-10″		9	2.667		2,275	101		2,376	2,675
0200	5′-5″ x 6′-10″		9	2.667		2,575	101		2,676	3,025
0220	5′-5″ x 7′-9″		9	2.667		2,800	101		2,901	3,250
0240	6′-0″ x 7′-11″		8	3		3,475	113		3,588	4,025
0260	8′-0″ x 6′-0″		8	3		3,075	113		3,188	3,575

08 52 16.35 Double-Hung Window

Line	Description		Crew	Daily Output	Labor-Hours	Unit	Material	Labor	Equipment	Total	Total Incl O&P
0010	**DOUBLE-HUNG WINDOW** Including frames, screens, and grilles										
0300	Vinyl clad, premium, double insulated glass, 2′-6″ x 3′-0″	G	1 Carp	10	.800	Ea.	310	30		340	390
0305	2′-6″ x 4′-0″	G		10	.800		360	30		390	445
0400	3′-0″ x 3′-6″	G		10	.800		335	30		365	420
0500	3′-0″ x 4′-0″	G		9	.889		395	33.50		428.50	490
0600	3′-0″ x 4′-6″	G		9	.889		410	33.50		443.50	505
0700	3′-0″ x 5′-0″	G		8	1		445	38		483	550
0790	3′-4″ x 5′-0″	G		8	1		455	38		493	565
0800	3′-6″ x 6′-0″	G		8	1		490	38		528	605
0820	4′-0″ x 5′-0″	G		7	1.143		560	43		603	690
0830	4′-0″ x 6′-0″	G		7	1.143		700	43		743	840

08 52 16.40 Transom Windows

Line	Description	Crew	Daily Output	Labor-Hours	Unit	Material	Labor	Equipment	Total	Total Incl O&P
0010	**TRANSOM WINDOWS**									
0050	Vinyl clad, premium, double insulated glass, 32″ x 8″	1 Carp	16	.500	Ea.	188	18.90		206.90	238
0100	36″ x 8″		16	.500		199	18.90		217.90	251
0110	36″ x 12″		16	.500		212	18.90		230.90	265
0200	44″ x 48″		12	.667		590	25		615	690
1000	Vinyl clad, premium, dbl. insul. glass, 4′-0″ x 4′-0″	2 Carp	12	1.333		520	50.50		570.50	655
1100	4′-0″ x 6′-0″		11	1.455		955	55		1,010	1,150
1200	5′-0″ x 6′-0″		10	1.600		1,050	60.50		1,110.50	1,275
1300	6′-0″ x 6′-0″		10	1.600		1,125	60.50		1,185.50	1,350

08 52 50 – Window Accessories

08 52 50.10 Window Grille or Muntin		Crew	Daily Output	Labor-Hours	Unit	2016 Bare Costs Material	Labor	Equipment	Total	Total Incl O&P
0010	**WINDOW GRILLE OR MUNTIN**, snap in type									
0020	Standard pattern interior grilles									
2000	Wood, awning window, glass size 28″ x 16″ high	1 Carp	30	.267	Ea.	28	10.10		38.10	48
2060	44″ x 24″ high		32	.250		40	9.45		49.45	60
2100	Casement, glass size, 20″ x 36″ high		30	.267		32	10.10		42.10	52
2180	20″ x 56″ high		32	.250		43	9.45		52.45	63.50
2200	Double hung, glass size, 16″ x 24″ high		24	.333	Set	51	12.60		63.60	77.50
2280	32″ x 32″ high		34	.235	″	131	8.90		139.90	159
2500	Picture, glass size, 48″ x 48″ high		30	.267	Ea.	120	10.10		130.10	149
2580	60″ x 68″ high		28	.286	″	183	10.80		193.80	220
2600	Sliding, glass size, 14″ x 36″ high		24	.333	Set	35.50	12.60		48.10	60
2680	36″ x 36″ high		22	.364	″	43.50	13.75		57.25	70.50

08 52 69 – Wood Storm Windows

08 52 69.10 Storm Windows		Crew	Daily Output	Labor-Hours	Unit	Material	Labor	Equipment	Total	Total Incl O&P
0010	**STORM WINDOWS**, aluminum residential									
0300	Basement, mill finish, incl. fiberglass screen									
0320	1′-10″ x 1′-0″ high G	2 Carp	30	.533	Ea.	36	20		56	73.50
0340	2′-9″ x 1′-6″ high G		30	.533		39	20		59	77
0360	3′-4″ x 2′-0″ high G		30	.533		46	20		66	84.50
1600	Double-hung, combination, storm & screen									
1700	Custom, clear anodic coating, 2′-0″ x 3′-5″ high	2 Carp	30	.533	Ea.	97	20		117	141
1720	2′-6″ x 5′-0″ high		28	.571		117	21.50		138.50	165
1740	4′-0″ x 6′-0″ high		25	.640		235	24		259	300
1800	White painted, 2′-0″ x 3′-5″ high		30	.533		112	20		132	157
1820	2′-6″ x 5′-0″ high		28	.571		173	21.50		194.50	226
1840	4′-0″ x 6′-0″ high		25	.640		286	24		310	355
2000	Clear anodic coating, 2′-0″ x 3′-5″ high G		30	.533		97	20		117	141
2020	2′-6″ x 5′-0″ high G		28	.571		121	21.50		142.50	169
2040	4′-0″ x 6′-0″ high G		25	.640		133	24		157	187
2400	White painted, 2′-0″ x 3′-5″ high G		30	.533		92	20		112	135
2420	2′-6″ x 5′-0″ high G		28	.571		97	21.50		118.50	143
2440	4′-0″ x 6′-0″ high G		25	.640		112	24		136	164
2600	Mill finish, 2′-0″ x 3′-5″ high G		30	.533		87	20		107	130
2620	2′-6″ x 5′-0″ high G		28	.571		92	21.50		113.50	137
2640	4′-0″ x 6′-8″ high G		25	.640		112	24		136	164
4000	Picture window, storm, 1 lite, white or bronze finish									
4020	4′-6″ x 4′-6″ high	2 Carp	25	.640	Ea.	133	24		157	187
4040	5′-8″ x 4′-6″ high		20	.800		150	30		180	216
4400	Mill finish, 4′-6″ x 4′-6″ high		25	.640		133	24		157	187
4420	5′-8″ x 4′-6″ high		20	.800		154	30		184	220
4600	3 lite, white or bronze finish									
4620	4′-6″ x 4′-6″ high	2 Carp	25	.640	Ea.	163	24		187	220
4640	5′-8″ x 4′-6″ high		20	.800		173	30		203	241
4800	Mill finish, 4′-6″ x 4′-6″ high		25	.640		154	24		178	210
4820	5′-8″ x 4′-6″ high		20	.800		163	30		193	230
6000	Sliding window, storm, 2 lite, white or bronze finish									
6020	3′-4″ x 2′-7″ high	2 Carp	28	.571	Ea.	135	21.50		156.50	185
6040	4′-4″ x 3′-3″ high		25	.640		150	24		174	206
6060	5′-4″ x 6′-0″ high		20	.800		240	30		270	315
9000	Interior storm window									
9100	Storm window interior glass	1 Glaz	107	.075	S.F.	6.20	2.76		8.96	11.45

08 53 Plastic Windows

08 53 13 – Vinyl Windows

08 53 13.10 Solid Vinyl Windows

	08 53 13.10 Solid Vinyl Windows		Crew	Daily Output	Labor-Hours	Unit	Material	Labor	Equipment	Total	Total Incl O&P
							2016 Bare Costs				
0010	**SOLID VINYL WINDOWS**										
0020	Double hung, including frame and screen, 2′-0″ x 2′-6″		2 Carp	15	1.067	Ea.	201	40.50		241.50	289
0040	2′-0″ x 3′-6″			14	1.143		214	43		257	310
0060	2′-6″ x 4′-6″			13	1.231		253	46.50		299.50	355
0080	3′-0″ x 4′-0″			10	1.600		225	60.50		285.50	350
0100	3′-0″ x 4′-6″			9	1.778		289	67		356	430
0120	3′-6″ x 4′-6″			8	2		310	75.50		385.50	465
0140	3′-6″ x 6′-0″		↓	7	2.286	↓	380	86.50		466.50	565

08 53 13.20 Vinyl Single Hung Windows

	08 53 13.20 Vinyl Single Hung Windows		Crew	Daily Output	Labor-Hours	Unit	Material	Labor	Equipment	Total	Total Incl O&P
0010	**VINYL SINGLE HUNG WINDOWS**, insulated glass										
0020	Grids, low e, J fin, extension jambs										
0130	25″ x 41″	G	2 Carp	20	.800	Ea.	200	30		230	271
0140	25″ x 49″	G		18	.889		210	33.50		243.50	288
0150	25″ x 57″	G		17	.941		220	35.50		255.50	300
0160	25″ x 65″	G		16	1		240	38		278	330
0170	29″ x 41″	G		18	.889		205	33.50		238.50	283
0180	29″ x 53″	G		18	.889		220	33.50		253.50	299
0190	29″ x 57″	G		17	.941		230	35.50		265.50	315
0200	29″ x 65″	G		16	1		240	38		278	330
0210	33″ x 41″	G		20	.800		215	30		245	288
0220	33″ x 53″	G		18	.889		230	33.50		263.50	310
0230	33″ x 57″	G		17	.941		240	35.50		275.50	325
0240	33″ x 65″	G		16	1		250	38		288	340
0250	37″ x 41″	G		20	.800		242	30		272	315
0260	37″ x 53″	G		18	.889		255	33.50		288.50	340
0270	37″ x 57″	G		17	.941		265	35.50		300.50	350
0280	37″ x 65″	G	↓	16	1	↓	290	38		328	385

08 53 13.30 Vinyl Double Hung Windows

	08 53 13.30 Vinyl Double Hung Windows		Crew	Daily Output	Labor-Hours	Unit	Material	Labor	Equipment	Total	Total Incl O&P
0010	**VINYL DOUBLE HUNG WINDOWS**, insulated glass										
0100	Grids, low E, J fin, ext. jambs, 21″ x 53″	G	2 Carp	18	.889	Ea.	215	33.50		248.50	294
0102	21″ x 37″	G		18	.889		223	33.50		256.50	300
0104	21″ x 41″	G		18	.889		232	33.50		265.50	310
0106	21″ x 49″	G		18	.889		247	33.50		280.50	330
0110	21″ x 57″	G		17	.941		263	35.50		298.50	350
0120	21″ x 65″	G		16	1		279	38		317	370
0128	25″ x 37″	G		20	.800		234	30		264	310
0130	25″ x 41″	G		20	.800		242	30		272	315
0140	25″ x 49″	G		18	.889		258	33.50		291.50	340
0145	25″ x 53″	G		18	.889		265	33.50		298.50	350
0150	25″ x 57″	G		17	.941		274	35.50		309.50	360
0160	25″ x 65″	G		16	1		290	38		328	385
0162	25″ x 69″	G		16	1		297	38		335	390
0164	25″ x 77″	G		16	1		325	38		363	420
0168	29″ x 37″	G		18	.889		242	33.50		275.50	325
0170	29″ x 41″	G		18	.889		250	33.50		283.50	330
0172	29″ x 49″	G		18	.889		267	33.50		300.50	350
0180	29″ x 53″	G		18	.889		271	33.50		304.50	355
0190	29″ x 57″	G		17	.941		280	35.50		315.50	370
0200	29″ x 65″	G		16	1		300	38		338	395
0202	29″ x 69″	G		16	1		310	38		348	405
0205	29″ x 77″	G		16	1		335	38		373	430
0208	33″ x 37″	G	↓	20	.800	↓	255	30		285	330

08 53 Plastic Windows

08 53 13 – Vinyl Windows

08 53 13.30 Vinyl Double Hung Windows			Crew	Daily Output	Labor-Hours	Unit	Material	Labor	Equipment	Total	Total Incl O&P
0210	33" x 41"	G	2 Carp	20	.800	Ea.	263	30		293	340
0215	33" x 49"	G		20	.800		280	30		310	360
0220	33" x 53"	G		18	.889		285	33.50		318.50	370
0230	33" x 57"	G		17	.941		297	35.50		332.50	385
0240	33" x 65"	G		16	1		315	38		353	410
0242	33" x 69"	G		16	1		330	38		368	425
0246	33" x 77"	G		16	1		350	38		388	450
0250	37" x 41"	G		20	.800		289	30		319	370
0255	37" x 49"	G		20	.800		282	30		312	360
0260	37" x 53"	G		18	.889		315	33.50		348.50	400
0270	37" x 57"	G		17	.941		325	35.50		360.50	420
0280	37" x 65"	G		16	1		350	38		388	450
0282	37" x 69"	G		16	1		360	38		398	460
0286	37" x 77"	G	↓	16	1		380	38		418	480
0300	Solid vinyl, average quality, double insulated glass, 2'-0" x 3'-0"	G	1 Carp	10	.800		291	30		321	370
0310	3'-0" x 4'-0"	G		9	.889		209	33.50		242.50	287
0330	Premium, double insulated glass, 2'-6" x 3'-0"	G		10	.800		271	30		301	350
0340	3'-0" x 3'-6"	G		9	.889		305	33.50		338.50	390
0350	3'-0" x 4'-0"	G		9	.889		325	33.50		358.50	410
0360	3'-0" x 4'-6"	G		9	.889		335	33.50		368.50	425
0370	3'-0" x 5'-0"	G		8	1		360	38		398	460
0380	3'-6" x 6'-0"	G	↓	8	1	↓	390	38		428	490

08 53 13.40 Vinyl Casement Windows			Crew	Daily Output	Labor-Hours	Unit	Material	Labor	Equipment	Total	Total Incl O&P
0010	**VINYL CASEMENT WINDOWS**, insulated glass										
0015	Grids, low E, J fin, extension jambs, screens										
0100	One lite, 21" x 41"	G	2 Carp	20	.800	Ea.	295	30		325	375
0110	21" x 47"	G		20	.800		320	30		350	405
0120	21" x 53"	G		20	.800		345	30		375	430
0128	24" x 35"	G		19	.842		283	32		315	365
0130	24" x 41"	G		19	.842		310	32		342	395
0140	24" x 47"	G		19	.842		335	32		367	420
0150	24" x 53"	G		19	.842		360	32		392	450
0158	28" x 35"	G		19	.842		300	32		332	385
0160	28" x 41"	G		19	.842		330	32		362	420
0170	28" x 47"	G		19	.842		350	32		382	440
0180	28" x 53"	G		19	.842		385	32		417	480
0184	28" x 59"	G		19	.842		395	32		427	490
0188	Two lites, 33" x 35"	G		18	.889		470	33.50		503.50	570
0190	33" x 41"	G		18	.889		500	33.50		533.50	605
0200	33" x 47"	G		18	.889		540	33.50		573.50	645
0210	33" x 53"	G		18	.889		575	33.50		608.50	685
0212	33" x 59"	G		18	.889		610	33.50		643.50	725
0215	33" x 72"	G		18	.889		635	33.50		668.50	750
0220	41" x 41"	G		18	.889		545	33.50		578.50	655
0230	41" x 47"	G		18	.889		585	33.50		618.50	695
0240	41" x 53"	G		17	.941		620	35.50		655.50	740
0242	41" x 59"	G		17	.941		650	35.50		685.50	775
0246	41" x 72"	G		17	.941		685	35.50		720.50	815
0250	47" x 41"	G		17	.941		555	35.50		590.50	670
0260	47" x 47"	G		17	.941		590	35.50		625.50	710
0270	47" x 53"	G		17	.941		625	35.50		660.50	750
0272	47" x 59"	G	↓	17	.941	↓	680	35.50		715.50	810

08 53 Plastic Windows

08 53 13 – Vinyl Windows

08 53 13.40 Vinyl Casement Windows			Crew	Daily Output	Labor-Hours	Unit	Material	2016 Bare Costs Labor	Equipment	Total	Total Incl O&P
0280	56″ x 41″	G	2 Carp	15	1.067	Ea.	595	40.50		635.50	725
0290	56″ x 47″	G		15	1.067		625	40.50		665.50	760
0300	56″ x 53″	G		15	1.067		680	40.50		720.50	820
0302	56″ x 59″	G		15	1.067		710	40.50		750.50	850
0310	56″ x 72″	G		15	1.067		770	40.50		810.50	920
0340	Solid vinyl, premium, double insulated glass, 2′-0″ x 3′-0″ high	G	1 Carp	10	.800		270	30		300	350
0360	2′-0″ x 4′-0″ high	G		9	.889		299	33.50		332.50	385
0380	2′-0″ x 5′-0″ high	G		8	1		335	38		373	435

08 53 13.50 Vinyl Picture Windows			Crew	Daily Output	Labor-Hours	Unit	Material	Labor	Equipment	Total	Total Incl O&P
0010	**VINYL PICTURE WINDOWS**, insulated glass										
0120	Grids, low e, J fin, ext. jambs, 47″ x 35″		2 Carp	12	1.333	Ea.	300	50.50		350.50	415
0130	47″ x 41″			12	1.333		390	50.50		440.50	515
0140	47″ x 47″			12	1.333		345	50.50		395.50	460
0150	47″ x 53″			11	1.455		370	55		425	495
0160	71″ x 35″			11	1.455		390	55		445	520
0170	71″ x 41″			11	1.455		410	55		465	540
0180	71″ x 47″			11	1.455		440	55		495	575

08 54 Composite Windows

08 54 13 – Fiberglass Windows

08 54 13.10 Fiberglass Single Hung Windows			Crew	Daily Output	Labor-Hours	Unit	Material	Labor	Equipment	Total	Total Incl O&P
0010	**FIBERGLASS SINGLE HUNG WINDOWS**										
0100	Grids, low E, 18″ x 24″	G	2 Carp	18	.889	Ea.	335	33.50		368.50	425
0110	18″ x 40″	G		17	.941		340	35.50		375.50	435
0130	24″ x 40″	G		20	.800		360	30		390	445
0230	36″ x 36″	G		17	.941		370	35.50		405.50	465
0250	36″ x 48″	G		20	.800		405	30		435	495
0260	36″ x 60″	G		18	.889		445	33.50		478.50	545
0280	36″ x 72″	G		16	1		470	38		508	580
0290	48″ x 40″	G		16	1		470	38		508	580

08 54 13.30 Fiberglass Slider Windows			Crew	Daily Output	Labor-Hours	Unit	Material	Labor	Equipment	Total	Total Incl O&P
0010	**FIBERGLASS SLIDER WINDOWS**										
0100	Grids, low E, 36″ x 24″	G	2 Carp	20	.800	Ea.	340	30		370	425
0110	36″ x 36″	G	″	20	.800	″	390	30		420	480

08 54 13.50 Fiberglass Bay Windows			Crew	Daily Output	Labor-Hours	Unit	Material	Labor	Equipment	Total	Total Incl O&P
0010	**FIBERGLASS BAY WINDOWS**										
0150	48″ x 36″	G	2 Carp	11	1.455	Ea.	1,100	55		1,155	1,300

08 61 Roof Windows

08 61 13 – Metal Roof Windows

08 61 13.10 Roof Windows			Crew	Daily Output	Labor-Hours	Unit	Material	Labor	Equipment	Total	Total Incl O&P
0010	**ROOF WINDOWS**, fixed high perf tmpd glazing, metallic framed										
0020	46″ x 21-1/2″, Flashed for shingled roof		1 Carp	8	1	Ea.	281	38		319	375
0100	46″ x 28″			8	1		305	38		343	400
0125	57″ x 44″			6	1.333		375	50.50		425.50	500
0130	72″ x 28″			7	1.143		375	43		418	490
0150	Fixed, laminated tempered glazing, 46″ x 21-1/2″			8	1		465	38		503	575
0175	46″ x 28″			8	1		515	38		553	630

08 61 Roof Windows

08 61 13 – Metal Roof Windows

08 61 13.10 Roof Windows		Crew	Daily Output	Labor-Hours	Unit	Material	Labor (2016 Bare Costs)	Equipment	Total	Total Incl O&P
0200	57″ x 44″	1 Carp	6	1.333	Ea.	485	50.50		535.50	615
0500	Vented flashing set for shingled roof, 46″ x 21-1/2″		7	1.143		465	43		508	585
0525	46″ x 28″		6	1.333		515	50.50		565.50	650
0550	57″ x 44″		5	1.600		645	60.50		705.50	810
0560	72″ x 28″		5	1.600		645	60.50		705.50	810
0575	Flashing set for low pitched roof, 46″ x 21-1/2″		7	1.143		535	43		578	665
0600	46″ x 28″		7	1.143		585	43		628	720
0625	57″ x 44″		5	1.600		730	60.50		790.50	900
0650	Flashing set for curb 46″ x 21-1/2″		7	1.143		620	43		663	755
0675	46″ x 28″		7	1.143		670	43		713	815
0700	57″ x 44″		5	1.600		825	60.50		885.50	1,000

08 61 16 – Wood Roof Windows

08 61 16.16 Roof Windows, Wood Framed		Crew	Daily Output	Labor-Hours	Unit	Material	Labor (2016 Bare Costs)	Equipment	Total	Total Incl O&P
0010	**ROOF WINDOWS, WOOD FRAMED**									
5600	Roof window incl. frame, flashing, double insulated glass & screens,									
5610	complete unit, 22″ x 38″	2 Carp	3	5.333	Ea.	750	202		952	1,175
5650	2′-5″ x 3′-8″		3.20	5		930	189		1,119	1,350
5700	3′-5″ x 4′-9″		3.40	4.706		1,050	178		1,228	1,450

08 62 Unit Skylights

08 62 13 – Domed Unit Skylights

08 62 13.10 Domed Skylights			Crew	Daily Output	Labor-Hours	Unit	Material	Labor (2016 Bare Costs)	Equipment	Total	Total Incl O&P
0010	**DOMED SKYLIGHTS**										
0020	Skylight, fixed dome type, 22″ x 22″	G	G-3	12	2.667	Ea.	216	95.50		311.50	395
0030	22″ x 46″	G		10	3.200		269	115		384	485
0040	30″ x 30″	G		12	2.667		286	95.50		381.50	475
0050	30″ x 46″	G		10	3.200		380	115		495	610
0110	Fixed, double glazed, 22″ x 27″	G		12	2.667		275	95.50		370.50	465
0120	22″ x 46″	G		10	3.200		293	115		408	510
0130	44″ x 46″	G		10	3.200		430	115		545	665
0210	Operable, double glazed, 22″ x 27″	G		12	2.667		405	95.50		500.50	605
0220	22″ x 46″	G		10	3.200		465	115		580	705
0230	44″ x 46″	G		10	3.200		905	115		1,020	1,200

08 62 13.20 Skylights			Crew	Daily Output	Labor-Hours	Unit	Material	Labor (2016 Bare Costs)	Equipment	Total	Total Incl O&P
0010	**SKYLIGHTS**, flush or curb mounted										
2120	Ventilating insulated plexiglass dome with										
2130	curb mounting, 36″ x 36″	G	G-3	12	2.667	Ea.	480	95.50		575.50	690
2150	52″ x 52″	G		12	2.667		670	95.50		765.50	895
2160	28″ x 52″	G		10	3.200		490	115		605	730
2170	36″ x 52″	G		10	3.200		545	115		660	785
2180	For electric opening system, add	G					315			315	345
2210	Operating skylight, with thermopane glass, 24″ x 48″	G	G-3	10	3.200		585	115		700	835
2220	32″ x 48″	G	″	9	3.556		615	127		742	890
2310	Non venting insulated plexiglass dome skylight with										
2320	Flush mount 22″ x 46″	G	G-3	15.23	2.101	Ea.	330	75		405	490
2330	30″ x 30″	G		16	2		305	71.50		376.50	455
2340	46″ x 46″	G		13.91	2.301		565	82.50		647.50	760
2350	Curb mount 22″ x 46″	G		15.23	2.101		365	75		440	525
2360	30″ x 30″	G		16	2		405	71.50		476.50	570
2370	46″ x 46″	G		13.91	2.301		625	82.50		707.50	830

08 62 Unit Skylights

08 62 13 – Domed Unit Skylights

08 62 13.20 Skylights		Crew	Daily Output	Labor-Hours	Unit	2016 Bare Costs Material	Labor	Equipment	Total	Total Incl O&P
2381	Non-insulated flush mount 22" x 46"	G-3	15.23	2.101	Ea.	225	75		300	375
2382	30" x 30"		16	2		202	71.50		273.50	340
2383	46" x 46"		13.91	2.301		380	82.50		462.50	560
2384	Curb mount 22" x 46"		15.23	2.101		189	75		264	335
2385	30" x 30"		16	2		188	71.50		259.50	325
4000	Skylight, solar tube kit, incl dome, flashing, diffuser, 1 pipe, 10" dia. G	1 Carp	2	4		242	151		393	520
4010	14" diam. G		2	4		335	151		486	625
4020	21" diam. G		2	4		420	151		571	715
4030	Accessories for, 1' long x 9" dia pipe G		24	.333		50	12.60		62.60	76
4040	2' long x 9" dia pipe G		24	.333		43	12.60		55.60	68.50
4050	4' long x 9" dia pipe G		20	.400		75	15.10		90.10	108
4060	1' long x 13" dia pipe G		24	.333		70	12.60		82.60	98
4070	2' long x 13" dia pipe G		24	.333		55	12.60		67.60	81.50
4080	4' long x 13" dia pipe G		20	.400		102	15.10		117.10	138
4090	6.5" turret ext for 21" dia pipe G		16	.500		110	18.90		128.90	153
4100	12' long x 21" dia flexible pipe G		12	.667		88	25		113	139
4110	45 degree elbow, 10" G		16	.500		152	18.90		170.90	199
4120	14" G		16	.500		82	18.90		100.90	122
4130	Interior decorative ring, 9" G		20	.400		40	15.10		55.10	69.50
4140	13" G		20	.400		60	15.10		75.10	91.50

08 63 Metal-Framed Skylights

08 63 13 – Domed Metal-Framed Skylights

08 63 13.20 Skylight Rigid Metal-Framed		Crew	Daily Output	Labor-Hours	Unit	Material	Labor	Equipment	Total	Total Incl O&P
0010	**SKYLIGHT RIGID METAL-FRAMED** skylight framing is aluminum									
0050	Fixed acrylic double domes, curb mount, 25-1/2" x 25-1/2"	G-3	10	3.200	Ea.	210	115		325	420
0060	25-1/2" x 33-1/2"		10	3.200		242	115		357	455
0070	25-1/2" x 49-1/2"		6	5.333		279	191		470	625
0080	33-1/2" x 33-1/2"		8	4		305	143		448	575
0090	37-1/2" x 25-1/2"		8	4		250	143		393	515
0100	37-1/2" x 37-1/2"		8	4		245	143		388	510
0110	37-1/2" x 49-1/2"		6	5.333		420	191		611	780
0120	49-1/2" x 33-1/2"		6	5.333		375	191		566	730
0130	49-1/2" x 49-1/2"		6	5.333		470	191		661	835
1000	Fixed tempered glass, curb mount, 17-1/2" x 33-1/2"		10	3.200		170	115		285	380
1020	17-1/2" x 49-1/2"		6	5.333		190	191		381	530
1030	25-1/2" x 25-1/2"		10	3.200		170	115		285	380
1040	25-1/2" x 33-1/2"		10	3.200		200	115		315	410
1050	25-1/2" x 37-1/2"		8	4		210	143		353	470
1060	25-1/2" x 49-1/2"		6	5.333		222	191		413	565
1070	25-1/2" x 73-1/2"		6	5.333		375	191		566	730
1080	33-1/2" x 33 1/2"		8	4		235	143		378	500
2000	Manual vent tempered glass & screen, curb, 25-1/2" x 25-1/2"		10	3.200		410	115		525	640
2020	25-1/2" x 37-1/2"		10	3.200		465	115		580	700
2030	25-1/2" x 49-1/2"		8	4		505	143		648	795
2040	33-1/2" x 33-1/2"		8	4		535	143		678	830
2050	33-1/2" x 49-1/2"		6	5.333		675	191		866	1,075
2060	37-1/2" x 37-1/2"		6	5.333		615	191		806	995
2070	49-1/2" x 49-1/2"		6	5.333		790	191		981	1,200
3000	Electric vent tempered glass, curb mount, 25-1/2" x 25-1/2"		10	3.200		960	115		1,075	1,250
3020	25-1/2" x 37-1/2"		10	3.200		1,050	115		1,165	1,350

08 63 Metal-Framed Skylights

08 63 13 – Domed Metal-Framed Skylights

08 63 13.20 Skylight Rigid Metal-Framed		Crew	Daily Output	Labor-Hours	Unit	2016 Bare Costs Material	Labor	Equipment	Total	Total Incl O&P
3030	25-1/2″ x 49-1/2″	G-3	8	4	Ea.	1,125	143		1,268	1,475
3040	33-1/2″ x 33-1/2″		8	4		1,125	143		1,268	1,500
3050	33-1/2″ x 49-1/2″		6	5.333		1,225	191		1,416	1,650
3060	37-1/2″ x 37-1/2″		6	5.333		1,200	191		1,391	1,625
3070	49-1/2″ x 49-1/2″		6	5.333		1,325	191		1,516	1,775

08 71 Door Hardware

08 71 13 – Automatic Door Operators

08 71 13.10 Automatic Openers Commercial

Line	Description	Crew	Daily Output	Labor-Hours	Unit	Material	Labor	Equipment	Total	Total Incl O&P
0010	**AUTOMATIC OPENERS COMMERCIAL**									
0020	Pneumatic, incl opener, motion sens, control box, tubing, compressor									
0050	For single swing door, per opening	2 Skwk	.80	20	Ea.	4,550	765		5,315	6,275
0100	Pair, per opening		.50	32	Opng.	7,375	1,225		8,600	10,200
1000	For single sliding door, per opening		.60	26.667		4,975	1,025		6,000	7,200
1300	Bi-parting pair		.50	32		7,475	1,225		8,700	10,300
1420	Electronic door opener incl motion sens, 12 V control box, motor									
1450	For single swing door, per opening	2 Skwk	.80	20	Opng.	3,675	765		4,440	5,325
1500	Pair, per opening		.50	32		6,800	1,225		8,025	9,525
1600	For single sliding door, per opening		.60	26.667		4,500	1,025		5,525	6,675
1700	Bi-parting pair		.50	32		5,475	1,225		6,700	8,075
1750	Handicap actuator buttons, 2, including 12 V DC wiring, add	1 Carp	1.50	5.333	Pr.	470	202		672	855
2000	Electric Panic Button for door	2 Skwk	.50	32	Opng.	203	1,225		1,428	2,275

08 71 20 – Hardware

08 71 20.10 Bolts, Flush

Line	Description	Crew	Daily Output	Labor-Hours	Unit	Material	Labor	Equipment	Total	Total Incl O&P
0010	**BOLTS, FLUSH**									
0020	Standard, concealed	1 Carp	7	1.143	Ea.	23.50	43		66.50	98.50
0800	Automatic fire exit	″	5	1.600		286	60.50		346.50	415
1600	Electrified dead bolt	1 Elec	3	2.667		140	117		257	345
2000	Brass, UL rated, with lever extension	1 Carp	40	.200		31.50	7.55		39.05	47.50
2500	Surface bolt, 6″ long, with strike		32	.250		28.50	9.45		37.95	47.50
3000	Barrel, brass, 2″ long		40	.200		8	7.55		15.55	21.50
3020	4″ long		40	.200		13.20	7.55		20.75	27
3060	6″ long		40	.200		26	7.55		33.55	41

08 71 20.15 Hardware

Line	Description	Crew	Daily Output	Labor-Hours	Unit	Material	Labor	Equipment	Total	Total Incl O&P
0010	**HARDWARE**									
0020	Average percentage for hardware, total job cost									
0500	Total hardware for building, average distribution				Job	85%	15%			
1000	Door hardware, apartment, interior	1 Carp	4	2	Door	465	75.50		540.50	635
1300	Average, door hardware, motel/hotel interior, with access card		4	2		600	75.50		675.50	785
1500	Hospital bedroom, average quality		4	2		650	75.50		725.50	840
2000	High quality		3	2.667		750	101		851	995
2100	Pocket door		6	1.333	Ea.	100	50.50		150.50	195
2250	School, single exterior, incl. lever, incl. panic device		3	2.667	Door	1,325	101		1,426	1,625
2500	Single interior, regular use, lever included		3	2.667		645	101		746	880
2550	Average, door hdwe., school, classroom, ANSI F84, lever handle		3	2.667		865	101		966	1,125
2600	Average, door hdwe. set, school, classroom, ANSI F88, incl. lever		3	2.667		925	101		1,026	1,200
2850	Stairway, single interior		3	2.667		605	101		706	835
3100	Double exterior, with panic device		2	4	Pr.	2,450	151		2,601	2,950
6020	Add for fire alarm door holder, electro-magnetic	1 Elec	4	2	Ea.	103	88		191	257

08 71 Door Hardware

08 71 20 – Hardware

08 71 20.20 Door Protectors

	08 71 20.20 Door Protectors	Crew	Daily Output	Labor-Hours	Unit	Material	2016 Bare Costs Labor	Equipment	Total	Total Incl O&P
0010	**DOOR PROTECTORS**									
0020	1-3/4" x 3/4" U channel	2 Carp	80	.200	L.F.	28	7.55		35.55	43.50
0021	1-3/4" x 1-1/4" U channel		80	.200	"	30	7.55		37.55	45.50
1000	Tear drop, spring-stl, 8" high x 19" long		15	1.067	Ea.	130	40.50		170.50	211
1010	8" high x 32" long		15	1.067		240	40.50		280.50	330
1100	Tear drop, stainless stl., 8" high x 19" long		15	1.067		390	40.50		430.50	500
1200	8" high x 32" long		15	1.067		460	40.50		500.50	575

08 71 20.30 Door Closers

	08 71 20.30 Door Closers	Crew	Daily Output	Labor-Hours	Unit	Material	Labor	Equipment	Total	Total Incl O&P
0010	**DOOR CLOSERS** Adjustable backcheck, multiple mounting									
0015	and rack and pinion									
0020	Standard Regular Arm	1 Carp	6	1.333	Ea.	192	50.50		242.50	296
0023	Install new door closer	"	4.61	1.735		192	65.50		257.50	320
0025	Door closer, adjustable backset, 3-way mount					192			192	211
0040	Hold open arm	1 Carp	6	1.333		200	50.50		250.50	305
0100	Fusible link		6.50	1.231		165	46.50		211.50	260
0210	Light duty, regular arm		6	1.333		112	50.50		162.50	208
0220	Parallel arm		6	1.333		136	50.50		186.50	235
0230	Hold open arm		6	1.333		120	50.50		170.50	217
0240	Fusible link arm		6	1.333		152	50.50		202.50	252
0250	Medium duty, regular arm		6	1.333		120	50.50		170.50	217
0500	Surface mount regular arm		6.50	1.231		158	46.50		204.50	252
0550	Fusible link		6.50	1.231		146	46.50		192.50	239
1000	3'-8" wide					192			192	211
1500	Concealed closers, normal use, head, pivot hung, interior	1 Carp	5.50	1.455		360	55		415	490
1510	Exterior		5.50	1.455		420	55		475	550
1520	Overhead concealed, all sizes, regular arm		5.50	1.455		200	55		255	310
1525	Concealed arm		5	1.600		315	60.50		375.50	445
1530	Concealed in door, all sizes, regular arm		5.50	1.455		325	55		380	450
1535	Concealed arm		5	1.600		255	60.50		315.50	380
1550	Floor		2.20	3.636		275	137		412	535
1560	Floor concealed, all sizes, single acting		2.20	3.636		500	137		637	780
1565	Double acting		2.20	3.636		475	137		612	755
1570	Interior, floor, offset pivot, single acting		3.50	2.286		830	86.50		916.50	1,050
1590	Exterior		3.50	2.286		905	86.50		991.50	1,150
1610	Hold open arm		6	1.333		430	50.50		480.50	560
1620	Double acting, standard arm		6	1.333		780	50.50		830.50	945
1630	Hold open arm		6	1.333		785	50.50		835.50	945
1640	Floor, center hung, single acting, bottom arm		6	1.333		420	50.50		470.50	545
1650	Double acting		6	1.333		475	50.50		525.50	610
1660	Offset hung, single acting, bottom arm		6	1.333		565	50.50		615.50	705
2000	Backcheck and adjustable power, hinge face mount									
5000	For cast aluminum cylinder, deduct				Ea.	35			35	38.50
5010	For delayed action surface mounted, add	1 Carp	6	1.333		36	50.50		86.50	124
5040	For delayed action, add					46			46	50.50
5080	For fusible link arm, add					35			35	38.50
5120	For shock absorbing arm, add					50			50	55
5160	For spring power adjustment, add					40			40	44
6000	Closer-holder, hinge face mount, all sizes, exposed arm	1 Carp	6.50	1.231		205	46.50		251.50	305
6500	Electro magnetic closer/holder									
6510	Single point, no detector	1 Carp	4	2	Ea.	520	75.50		595.50	695
6515	Including detector		4	2		685	75.50		760.50	875
6520	Multi-point, no detector		4	2		910	75.50		985.50	1,125

08 71 Door Hardware

08 71 20 – Hardware

08 71 20.30 Door Closers		Crew	Daily Output	Labor-Hours	Unit	2016 Bare Costs Material	Labor	Equipment	Total	Total Incl O&P
6524	Including detector	1 Carp	4	2	Ea.	1,350	75.50		1,425.50	1,625
6550	Electric automatic operators									
6555	Operator	1 Carp	4	2	Ea.	2,025	75.50		2,100.50	2,350
6570	Wall plate actuator		4	2		222	75.50		297.50	370
8000	Surface mounted, standard duty, parallel arm, primed, traditional		6	1.333		207	50.50		257.50	310
8030	Light duty		6	1.333		151	50.50		201.50	251
8042	Extra duty parallel arm		6	1.333		120	50.50		170.50	217
8044	Hold open arm		6	1.333		160	50.50		210.50	261
8046	Positive stop arm		6	1.333		218	50.50		268.50	325
8050	Heavy duty		6	1.333		242	50.50		292.50	350
8052	Heavy duty, regular arm		6	1.333		220	50.50		270.50	325
8054	Top jamb mount		6	1.333		220	50.50		270.50	325
8056	Extra duty parallel arm		6	1.333		220	50.50		270.50	325
8058	Hold open arm		6	1.333		268	50.50		318.50	380
8060	Positive stop arm		6	1.333		244	50.50		294.50	355
8062	Fusible link arm		6	1.333		264	50.50		314.50	375
8080	Universal heavy duty, regular arm		6	1.333		230	50.50		280.50	340
8084	Parallel arm		6	1.333		230	50.50		280.50	340
8088	Extra duty, parallel arm		6	1.333		264	50.50		314.50	375
8090	Hold open arm		6	1.333		272	50.50		322.50	385
8094	Positive stop arm		6	1.333		282	50.50		332.50	395
8100	Standard duty, parallel arm, modern		6	1.333		252	50.50		302.50	360
8150	Heavy duty		6	1.333		282	50.50		332.50	395

08 71 20.36 Panic Devices		Crew	Daily Output	Labor-Hours	Unit	Material	Labor	Equipment	Total	Total Incl O&P
0010	**PANIC DEVICES** R087110-10									
0015	Touch bars various styles									
0040	Single door exit only, rim device, wide stile									
0050	Economy US28	1 Carp	5	1.600	Ea.	231	60.50		291.50	355
0060	Standard duty US28		5	1.600		665	60.50		725.50	835
0065	US26D		5	1.600		705	60.50		765.50	875
0070	US10		5	1.600		725	60.50		785.50	900
0075	US3		5	1.600		795	60.50		855.50	975
0080	Night Latch, Economy US28		5	1.600		315	60.50		375.50	445
0085	Standard duty, Night Latch US28		5	1.600		970	60.50		1,030.50	1,175
0090	US26D		5	1.600		1,075	60.50		1,135.50	1,275
0095	US10		5	1.600		1,075	60.50		1,135.50	1,300
0100	US3		5	1.600		1,175	60.50		1,235.50	1,400
0500	Single door exit only, rim device, narrow stile									
0540	Economy US28	1 Carp	5	1.600	Ea.	465	60.50		525.50	610
0550	Standard duty US28		5	1.600		830	60.50		890.50	1,000
0565	US26D		5	1.600		900	60.50		960.50	1,100
0570	US10		5	1.600		745	60.50		805.50	920
0575	US3		5	1.600		950	60.50		1,010.50	1,150
0580	Economy with night latch US28		5	1.600		320	60.50		380.50	455
0585	Standard duty US28		5	1.600		870	60.50		930.50	1,050
0590	US26D		5	1.600		1,050	60.50		1,110.50	1,250
0595	US10		5	1.600		870	60.50		930.50	1,050
0600	US32D		5	1.600		1,200	60.50		1,260.50	1,425
1040	Single door exit only, surface vertical rod									
1050	Surface Vertical Rod, Economy US28	1 Carp	4	2	Ea.	525	75.50		600.50	705
1060	Surface Vertical Rod, Standard duty US28		4	2		965	75.50		1,040.50	1,175
1065	US26D		4	2		1,100	75.50		1,175.50	1,350

08 71 20.36 Panic Devices

	08 71 20.36 Panic Devices	Crew	Daily Output	Labor-Hours	Unit	Material	Labor	Equipment	Total	Total Incl O&P
1070	US10	1 Carp	4	2	Ea.	1,025	75.50		1,100.50	1,250
1075	US3		4	2		1,200	75.50		1,275.50	1,450
1080	US32D		4	2		1,200	75.50		1,275.50	1,450
1500	Single door exit, rim, narrow stile, economy, surface vertical rod									
1540	Surface Vertical Rod, Narrow, Economy US28	1 Carp	4	2	Ea.	420	75.50		495.50	590
2040	Single door exit only, concealed vertical rod									
2042	These devices will require separate and extra door preparation									
2060	Concealed Rod, standard duty US28	1 Carp	4	2	Ea.	1,000	75.50		1,075.50	1,225
2065	US26D		4	2		1,125	75.50		1,200.50	1,375
2070	US10		4	2		1,275	75.50		1,350.50	1,525
2560	Concealed Rod, narrow, standard duty US28		4	2		990	75.50		1,065.50	1,225
2565	US26D		4	2		1,075	75.50		1,150.50	1,325
2567	US26		4	2		1,225	75.50		1,300.50	1,475
2570	US10		4	2		1,025	75.50		1,100.50	1,250
2575	US3		4	2		1,225	75.50		1,300.50	1,475
3040	Single door exit only, mortise device, wide stile									
3042	These devices must be combined with a mortise lock									
3060	Mortise, standard duty US28	1 Carp	4	2	Ea.	925	75.50		1,000.50	1,150
3065	US26D		4	2		465	75.50		540.50	640
3067	US26		4	2		1,150	75.50		1,225.50	1,400
3070	US10		4	2		1,000	75.50		1,075.50	1,225
3075	US3		4	2		1,025	75.50		1,100.50	1,250
3080	US32D		4	2		1,025	75.50		1,100.50	1,250
3500	Single door exit only, mortise device, narrow stile									
3510	These devices must be combined with a mortise lock									
3540	Mortise, economy US28	1 Carp	4	2	Ea.	385	75.50		460.50	545

08 71 20.40 Lockset

	08 71 20.40 Lockset	Crew	Daily Output	Labor-Hours	Unit	Material	Labor	Equipment	Total	Total Incl O&P
0010	**LOCKSET**, Standard duty									
0020	Non-keyed, passage, w/sect. trim	1 Carp	12	.667	Ea.	70	25		95	119
0100	Privacy		12	.667		75	25		100	125
0400	Keyed, single cylinder function		10	.800		108	30		138	169
0420	Hotel (see also Section 08 71 20.15)		8	1		200	38		238	284
0500	Lever handled, keyed, single cylinder function		10	.800		128	30		158	192
0600	Bedroom, bathroom and inner office doors		10	.800		143	30		173	208
0900	Apartment, office and corridor doors		10	.800		181	30		211	250
1400	Keyed, single cylinder function		10	.800		185	30		215	255

08 71 20.42 Mortise Locksets

	08 71 20.42 Mortise Locksets	Crew	Daily Output	Labor-Hours	Unit	Material	Labor	Equipment	Total	Total Incl O&P
0010	**MORTISE LOCKSETS**, Comm., wrought knobs & full escutcheon trim									
0015	Assumes mortise is cut									
0020	Non-keyed, passage, grade 3	1 Carp	9	.889	Ea.	149	33.50		182.50	221
0030	Grade 1		8	1		405	38		443	510
0040	Privacy set Grade 3		9	.889		169	33.50		202.50	243
0050	Grade 1		8	1		455	38		493	565
0100	Keyed, office/entrance/apartment, Grade 2		8	1		197	38		235	281
0110	Grade 1		7	1.143		515	43		558	645
0120	Single cylinder, typical, Grade 3		8	1		189	38		227	272
0130	Grade 1		7	1.143		500	43		543	625
0200	Hotel, room, Grade 3		7	1.143		190	43		233	282
0210	Grade 1 (see also Section 08 71 20.15)		6	1.333		510	50.50		560.50	645
0300	Double cylinder, Grade 3		8	1		225	38		263	310
0310	Grade 1		7	1.143		515	43		558	645
1000	Wrought knobs and sectional trim, non-keyed, passage, Grade 3		10	.800		130	30		160	194

08 71 20 – Hardware

08 71 20.42 Mortise Locksets		Crew	Daily Output	Labor-Hours	Unit	Material	2016 Bare Costs Labor	Equipment	Total	Total Incl O&P
1010	Grade 1	1 Carp	9	.889	Ea.	405	33.50		438.50	500
1040	Privacy, Grade 3		10	.800		145	30		175	211
1050	Grade 1		9	.889		455	33.50		488.50	555
1100	Keyed, entrance, office/apartment, Grade 3		9	.889		220	33.50		253.50	299
1103	Install lockset		6.92	1.156		220	43.50		263.50	315
1110	Grade 1		8	1		520	38		558	635
1120	Single cylinder, Grade 3		9	.889		225	33.50		258.50	305
1130	Grade 1	↓	8	1	↓	500	38		538	615
2000	Cast knobs and full escutcheon trim									
2010	Non-keyed, passage, Grade 3	1 Carp	9	.889	Ea.	275	33.50		308.50	360
2020	Grade 1		8	1		380	38		418	485
2040	Privacy, Grade 3		9	.889		320	33.50		353.50	405
2050	Grade 1		8	1		440	38		478	550
2120	Keyed, single cylinder, Grade 3		8	1		330	38		368	430
2123	Mortise lock		6.15	1.301		330	49		379	450
2130	Grade 1		7	1.143		525	43		568	655
3000	Cast knob and sectional trim, non-keyed, passage, Grade 3		10	.800		210	30		240	282
3010	Grade 1		10	.800		380	30		410	470
3040	Privacy, Grade 3		10	.800		225	30		255	299
3050	Grade 1		10	.800		440	30		470	535
3100	Keyed, office/entrance/apartment, Grade 3		9	.889		255	33.50		288.50	340
3110	Grade 1		9	.889		580	33.50		613.50	695
3120	Single cylinder, Grade 3		9	.889		260	33.50		293.50	345
3130	Grade 1	↓	9	.889		525	33.50		558.50	635
3190	For re-core cylinder, add				↓	75			75	82.50
5000	Wrought steel case, brass base, knob US26D									
5020	Closet, nonkeyed passage	1 Carp	8	1	Ea.	275	38		313	370
5040	Bath/bedroom, keyed		8	1		305	38		343	400
5060	Entrance, keyed		8	1		365	38		403	465
5080	Classroom, outside keyed		8	1		390	38		428	490
5100	Storeroom, keyed		8	1		385	38		423	485
5120	Front door, keyed		8	1		375	38		413	480
5140	Dormitory/exit, keyed	↓	8	1	↓	375	38		413	480

08 71 20.45 Peepholes

		Crew	Daily Output	Labor-Hours	Unit	Material	Labor	Equipment	Total	Total Incl O&P
0010	**PEEPHOLES**									
2010	Peephole	1 Carp	32	.250	Ea.	15.80	9.45		25.25	33.50
2020	Peephole, wide view	"	32	.250	"	16.50	9.45		25.95	34

08 71 20.50 Door Stops

		Crew	Daily Output	Labor-Hours	Unit	Material	Labor	Equipment	Total	Total Incl O&P
0010	**DOOR STOPS**									
0020	Holder & bumper, floor or wall	1 Carp	32	.250	Ea.	35	9.45		44.45	55
1300	Wall bumper, 4" diameter, with rubber pad, aluminum		32	.250		12.30	9.45		21.75	29.50
1600	Door bumper, floor type, aluminum		32	.250		8.70	9.45		18.15	25.50
1620	Brass		32	.250		10	9.45		19.45	27
1630	Bronze		32	.250		18.35	9.45		27.80	36
1900	Plunger type, door mounted		32	.250		29.50	9.45		38.95	48.50
2500	Holder, floor type, aluminum		32	.250		35	9.45		44.45	54.50
2520	Wall type, aluminum		32	.250		35.50	9.45		44.95	55
2530	Overhead type, bronze		32	.250		105	9.45		114.45	131
2540	Plunger type, aluminum		32	.250		29.50	9.45		38.95	48.50
2560	Brass		32	.250		45	9.45		54.45	65
3000	Electro-magnetic, wall mounted, US3		3	2.667		263	101		364	460
3020	Floor mounted, US3	↓	3	2.667	↓	315	101		416	520

08 71 20.50 Door Stops		Crew	Daily Output	Labor-Hours	Unit	Material	2016 Bare Costs Labor	Equipment	Total	Total Incl O&P
4000	Doorstop, ceiling mounted	1 Carp	3	2.667	Ea.	20	101		121	191
4030	Doorstop, header		3	2.667		45	101		146	219

08 71 20.55 Push-Pull Plates

Code	Description	Crew	Daily Output	Labor-Hours	Unit	Material	Labor	Equipment	Total	Total Incl O&P
0010	**PUSH-PULL PLATES**									
0090	Push plate, 0.050 thick, 3" x 12", aluminum	1 Carp	12	.667	Ea.	6.25	25		31.25	49
0100	4" x 16"		12	.667		12.90	25		37.90	56
0110	6" x 16"		12	.667		8.50	25		33.50	51.50
0120	8" x 16"		12	.667		9.90	25		34.90	53
0200	Push plate, 0.050 thick, 3" x 12", brass		12	.667		14	25		39	57.50
0210	4" x 16"		12	.667		17.50	25		42.50	61.50
0220	6" x 16"		12	.667		27	25		52	71.50
0230	8" x 16"		12	.667		35	25		60	80.50
0250	Push plate, 0.050 thick, 3" x 12", satin brass		12	.667		14.25	25		39.25	57.50
0260	4" x 16"		12	.667		17.70	25		42.70	61.50
0270	6" x 16"		12	.667		27	25		52	71.50
0280	8" x 16"		12	.667		35	25		60	80.50
0490	Push plate, 0.050 thick, 3" x 12", bronze		12	.667		17	25		42	60.50
0500	4" x 16"		12	.667		25.50	25		50.50	70
0510	6" x 16"		12	.667		32	25		57	77
0520	8" x 16"		12	.667		38.50	25		63.50	84.50
0600	Push plate, antimicrobial copper alloy finish, 3.5" x 15"		13	.615		20.50	23.50		44	62
0610	4" x 16"		13	.615		24.50	23.50		48	65.50
0620	6" x 16"		13	.615		27	23.50		50.50	68.50
0630	6" x 20"		13	.615		27	23.50		50.50	68.50
0740	Push plate, 0.050 thick, 3" x 12", stainless steel		12	.667		12	25		37	55
0750	4" x 16"		12	.667		27.50	25		52.50	72.50
0760	6" x 16"		12	.667		18.65	25		43.65	62.50
0780	8" x 16"		12	.667		23.50	25		48.50	68
0790	Push plate, 0.050 thick, 3" x 12", satin stainless steel		12	.667		7	25		32	49.50
0810	4" x 16"		12	.667		8.50	25		33.50	51.50
0820	6" x 16"		12	.667		11.70	25		36.70	55
0830	8" x 16"		12	.667		16	25		41	59.50
0980	Pull plate, 0.050 thick, 3" x 12", aluminum		12	.667		25.50	25		50.50	70
1000	4" x 16"		12	.667		25.50	25		50.50	70
1050	Pull plate, 0.050 thick, 3" x 12", brass		12	.667		39.50	25		64.50	85.50
1060	4" x 16"		12	.667		34.50	25		59.50	80
1080	Pull plate, 0.050 thick, 3" x 12", bronze		12	.667		50	25		75	97
1100	4" x 16"		12	.667		56.50	25		81.50	104
1180	Pull plate, 0.050 thick, 3" x 12", stainless steel		12	.667		45.50	25		70.50	92
1200	4" x 16"		12	.667		59	25		84	107
1250	Pull plate, 0.050 thick, 3" x 12", chrome		12	.667		44.50	25		69.50	91
1270	4" x 16"		12	.667		46.50	25		71.50	93
1500	Pull handle and push bar, aluminum		11	.727		122	27.50		149.50	180
2000	Bronze		10	.800		164	30		194	231

08 71 20.60 Entrance Locks

Code	Description	Crew	Daily Output	Labor-Hours	Unit	Material	Labor	Equipment	Total	Total Incl O&P
0010	**ENTRANCE LOCKS**									
0015	Cylinder, grip handle deadlocking latch	1 Carp	9	.889	Ea.	175	33.50		208.50	250
0020	Deadbolt		8	1		175	38		213	257
0100	Push and pull plate, dead bolt		8	1		225	38		263	310
0900	For handicapped lever, add					150			150	165

08 71 20 – Hardware

08 71 20.65 Thresholds		Crew	Daily Output	Labor-Hours	Unit	2016 Bare Costs Material	Labor	Equipment	Total	Total Incl O&P
0010	**THRESHOLDS**									
0011	Threshold 3' long saddles aluminum	1 Carp	48	.167	L.F.	9.70	6.30		16	21
0100	Aluminum, 8" wide, 1/2" thick		12	.667	Ea.	49	25		74	95.50
0500	Bronze		60	.133	L.F.	43	5.05		48.05	56
0600	Bronze, panic threshold, 5" wide, 1/2" thick		12	.667	Ea.	158	25		183	215
0700	Rubber, 1/2" thick, 5-1/2" wide		20	.400		40	15.10		55.10	69.50
0800	2-3/4" wide		20	.400		45.50	15.10		60.60	75.50
1950	ADA Compliant Thresholds									
2000	Threshold, wood oak 3-1/2" wide x 24" long	1 Carp	12	.667	Ea.	11	25		36	54
2010	3-1/2" wide x 36" long		12	.667		15	25		40	58.50
2020	3-1/2" wide x 48" long		12	.667		20	25		45	64
2030	4-1/2" wide x 24" long		12	.667		13	25		38	56.50
2040	4-1/2" wide x 36" long		12	.667		19	25		44	63
2050	4-1/2" wide x 48" long		12	.667		25	25		50	69.50
2060	6-1/2" wide x 24" long		12	.667		18	25		43	62
2070	6-1/2" wide x 36" long		12	.667		27	25		52	71.50
2080	6-1/2" wide x 48" long		12	.667		36	25		61	81.50
2090	Threshold, wood cherry 3-1/2" wide x 24" long		12	.667		15	25		40	58.50
2100	3-1/2" wide x 36" long		12	.667		22	25		47	66
2110	3-1/2" wide x 48" long		12	.667		29	25		54	74
2120	4-1/2" wide x 24" long		12	.667		19	25		44	63
2130	4-1/2" wide x 36" long		12	.667		28	25		53	73
2140	4-1/2" wide x 48" long		12	.667		37	25		62	82.50
2150	6-1/2" wide x 24" long		12	.667		27	25		52	71.50
2160	6-1/2" wide x 36" long		12	.667		40	25		65	86
2170	6-1/2" wide x 48" long		12	.667		53	25		78	101
2180	Threshold, wood walnut 3-1/2" wide x 24" long		12	.667		18	25		43	62
2190	3-1/2" wide x 36" long		12	.667		27	25		52	71.50
2200	3-1/2" wide x 48" long		12	.667		36	25		61	81.50
2210	4-1/2" wide x 24" long		12	.667		26	25		51	70.50
2220	4-1/2" wide x 36" long		12	.667		38	25		63	84
2230	4-1/2" wide x 48" long		12	.667		50	25		75	97
2240	6-1/2" wide x 24" long		12	.667		43	25		68	89.50
2250	6-1/2" wide x 36" long		12	.667		63	25		88	112
2260	6-1/2" wide x 48" long		12	.667		84	25		109	135
2300	Threshold, aluminum 4" wide x 36" long		12	.667		33	25		58	78.50
2310	4" wide x 48" long		12	.667		41	25		66	87
2320	4" wide x 72" long		12	.667		66	25		91	115
2330	5" wide x 36" long		12	.667		45	25		70	91.50
2340	5" wide x 48" long		12	.667		56	25		81	104
2350	5" wide x 72" long		12	.667		89	25		114	140
2360	6" wide x 36" long		12	.667		53	25		78	101
2370	6" wide x 48" long		12	.667		68	25		93	117
2380	6" wide x 72" long		12	.667		106	25		131	159
2390	7" wide x 36" long		12	.667		69	25		94	118
2400	7" wide x 48" long		12	.667		90	25		115	141
2410	7" wide x 72" long		12	.667		137	25		162	193
2500	Threshold, ramp, aluminum or rubber 24" x 24"		12	.667		190	25		215	251

08 71 20.75 Door Hardware Accessories		Crew	Daily Output	Labor-Hours	Unit	Material	Labor	Equipment	Total	Total Incl O&P
0010	**DOOR HARDWARE ACCESSORIES**									
0050	Door closing coordinator, 36" (for paired openings up to 56")	1 Carp	8	1	Ea.	105	38		143	180
0060	48" (for paired openings up to 84")		8	1		113	38		151	188

08 71 Door Hardware

08 71 20 – Hardware

08 71 20.75 Door Hardware Accessories		Crew	Daily Output	Labor-Hours	Unit	2016 Bare Costs Material	Labor	Equipment	Total	Total Incl O&P
0070	56" (for paired openings up to 96")	1 Carp	8	1	Ea.	124	38		162	200
1000	Knockers, brass, standard		16	.500		49	18.90		67.90	85.50
1100	Deluxe		10	.800		121	30		151	184
4500	Rubber door silencers		540	.015		.45	.56		1.01	1.44

08 71 20.80 Hasps

Code	Description	Crew	Daily Output	Labor-Hours	Unit	Material	Labor	Equipment	Total	Total Incl O&P
0010	**HASPS**, steel assembly									
0015	3"	1 Carp	26	.308	Ea.	4.90	11.65		16.55	25
0020	4-1/2"		13	.615		6.60	23.50		30.10	46.50
0040	6"		12.50	.640		9.35	24		33.35	51

08 71 20.90 Hinges

Code	Description	Crew	Daily Output	Labor-Hours	Unit	Material	Labor	Equipment	Total	Total Incl O&P
0010	**HINGES**									
0012	Full mortise, avg. freq., steel base, USP, 4-1/2" x 4-1/2"				Pr.	37			37	40.50
0100	5" x 5", USP					57.50			57.50	63
0200	6" x 6", USP					119			119	131
0400	Brass base, 4-1/2" x 4-1/2", US10					60			60	66
0500	5" x 5", US10					86			86	94.50
0600	6" x 6", US10					162			162	178
0800	Stainless steel base, 4-1/2" x 4-1/2", US32					76.50			76.50	84
0900	For non removable pin, add (security item)				Ea.	5.25			5.25	5.80
0910	For floating pin, driven tips, add					3.30			3.30	3.63
0930	For hospital type tip on pin, add					13.95			13.95	15.35
0940	For steeple type tip on pin, add					18.95			18.95	21
0950	Full mortise, high frequency, steel base, 3-1/2" x 3-1/2", US26D				Pr.	30.50			30.50	33.50
1000	4-1/2" x 4-1/2", USP					66			66	73
1100	5" x 5", USP					52			52	57.50
1200	6" x 6", USP					138			138	152
1400	Brass base, 3-1/2" x 3-1/2", US4					51			51	56
1430	4-1/2" x 4-1/2", US10					75.50			75.50	83
1500	5" x 5", US10					126			126	138
1600	6" x 6", US10					169			169	186
1800	Stainless steel base, 4-1/2" x 4-1/2", US32					103			103	114
1810	5" x 4-1/2", US32					137			137	151
1930	For hospital type tip on pin, add				Ea.	14.60			14.60	16.05
1950	Full mortise, low frequency, steel base, 3-1/2" x 3-1/2", US26D				Pr.	25			25	27.50
2000	4-1/2" x 4-1/2", USP					23			23	25
2100	5" x 5", USP					48			48	53
2200	6" x 6", USP					93.50			93.50	103
2300	4-1/2" x 4-1/2", US3					17.35			17.35	19.05
2310	5" x 5", US3					42			42	46
2400	Brass bass, 4-1/2" x 4-1/2", US10					54.50			54.50	59.50
2500	5" x 5", US10					78			78	86
2800	Stainless steel base, 4-1/2" x 4-1/2", US32					76			76	84
8000	Install hinge	1 Carp	34	.235			8.90		8.90	14.90

08 71 20.91 Special Hinges

Code	Description	Crew	Daily Output	Labor-Hours	Unit	Material	Labor	Equipment	Total	Total Incl O&P
0010	**SPECIAL HINGES**									
8000	Continuous hinges									
8010	Steel, piano, 2" x 72"	1 Carp	20	.400	Ea.	22	15.10		37.10	49.50
8020	Brass, piano, 1-1/16" x 30"		30	.267		8	10.10		18.10	25.50
8030	Acrylic, piano, 1-3/4" x 12"		40	.200		15	7.55		22.55	29
8040	Aluminum, door, standard duty, 7'		3	2.667		135	101		236	320
8050	Heavy duty, 7'		3	2.667		142	101		243	325
8060	8'		3	2.667		160	101		261	345

08 71 Door Hardware

08 71 20 – Hardware

	08 71 20.91 Special Hinges	Crew	Daily Output	Labor-Hours	Unit	2016 Bare Costs Material	Labor	Equipment	Total	Total Incl O&P
8070	Steel, door, heavy duty, 7′	1 Carp	3	2.667	Ea.	200	101		301	390
8080	8′		3	2.667		250	101		351	445
8090	Stainless steel, door, heavy duty, 7′		3	2.667		260	101		361	455
8100	8′		3	2.667		280	101		381	480
9200	Continuous geared hinge, aluminum, full mortise, standard duty, 83 inch		3	2.667		130	101		231	310
9250	Continuous geared hinge, aluminum, full mortise, heavy duty, 83 inch		3	2.667		215	101		316	405
	08 71 20.92 Mortised Hinges									
0010	**MORTISED HINGES**									
0200	Average frequency, steel plated, ball bearing, 3-1/2″ x 3-1/2″				Pr.	29.50			29.50	32.50
0300	Bronze, ball bearing					34.50			34.50	38
0900	High frequency, steel plated, ball bearing					78			78	86
1100	Bronze, ball bearing					78			78	86
1300	Average frequency, steel plated, ball bearing, 4-1/2″ x 4-1/2″					37			37	40.50
1500	Bronze, ball bearing, to 36″ wide					39			39	42.50
1700	Low frequency, steel, plated, plain bearing					20.50			20.50	22.50
1900	Bronze, plain bearing					28.50			28.50	31
	08 71 20.95 Kick Plates									
0010	**KICK PLATES**									
0020	Stainless steel, .050, 16 ga., 8″ x 28″, US32	1 Carp	15	.533	Ea.	38	20		58	76
0030	8″ x 30″		15	.533		41	20		61	79
0040	8″ x 34″		15	.533		46	20		66	84.50
0050	10″ x 28″		15	.533		76	20		96	118
0060	10″ x 30″		15	.533		82	20		102	124
0070	10″ x 34″		15	.533		92	20		112	135
0080	Mop/Kick, 4″ x 28″		15	.533		34	20		54	71.50
0090	4″ x 30″		15	.533		36	20		56	73.50
0100	4″ x 34″		15	.533		41	20		61	79
0110	6″ x 28″		15	.533		43	20		63	81.50
0120	6″ x 30″		15	.533		47	20		67	85.50
0130	6″ x 34″		15	.533		53	20		73	92.50
0500	Bronze, .050″, 8″ x 28″		15	.533		67.50	20		87.50	108
0510	8″ x 30″		15	.533		65	20		85	106
0520	8″ x 34″		15	.533		73	20		93	115
0530	10″ x 28″		15	.533		75	20		95	117
0540	10″ x 30″		15	.533		80	20		100	122
0550	10″ x 34″		15	.533		91	20		111	134
0560	Mop/Kick, 4″ x 28″		15	.533		33	20		53	70.50
0570	4″ x 30″		15	.533		36	20		56	73.50
0580	4″ x 34″		15	.533		37	20		57	74.50
0590	6″ x 28″		15	.533		46	20		66	84.50
0600	6″ x 30″		15	.533		52	20		72	91
0610	6″ x 34″		15	.533		56	20		76	95.50
1000	Acrylic, .125″, 8″ x 26″		15	.533		28	20		48	65
1010	8″ x 36″		15	.533		38	20		58	76
1020	8″ x 42″		15	.533		45	20		65	83.50
1030	10″ x 26″		15	.533		35	20		55	72.50
1040	10″ x 36″		15	.533		48	20		68	87
1050	10″ x 42″		15	.533		68.50	20		88.50	110
1060	Mop/Kick, 4″ x 26″		15	.533		17	20		37	52.50
1070	4″ x 36″		15	.533		24	20		44	60.50
1080	4″ x 42″		15	.533		27	20		47	63.50
1090	6″ x 26″		15	.533		23	20		43	59.50

08 71 Door Hardware

08 71 20 – Hardware

08 71 20.95 Kick Plates		Crew	Daily Output	Labor-Hours	Unit	Material	Labor (2016 Bare Costs)	Equipment	Total	Total Incl O&P
1100	6″ x 36″	1 Carp	15	.533	Ea.	34	20		54	71.50
1110	6″ x 42″		15	.533		39	20		59	77
1220	Brass, .050″, 8″ x 26″		15	.533		56	20		76	95.50
1230	8″ x 36″		15	.533		75	20		95	117
1240	8″ x 42″		15	.533		86	20		106	129
1250	10″ x 26″		15	.533		71	20		91	112
1260	10″ x 36″		15	.533		91	20		111	134
1270	10″ x 42″		15	.533		105	20		125	150
1320	Mop/Kick, 4″ x 26″		15	.533		28	20		48	65
1330	4″ x 36″		15	.533		39	20		59	77
1340	4″ x 42″		15	.533		44	20		64	82.50
1350	6″ x 26″		15	.533		38	20		58	76
1360	6″ x 36″		15	.533		48	20		68	87
1370	6″ x 42″		15	.533		55	20		75	94.50
1800	Aluminum, .050″, 8″ x 26″		15	.533		35	20		55	72.50
1810	8″ x 36″		15	.533		40	20		60	78
1820	8″ x 42″		15	.533		47	20		67	85.50
1830	10″ x 26″		15	.533		36	20		56	73.50
1840	10″ x 36″		15	.533		50	20		70	89
1850	10″ x 42″		15	.533		59	20		79	99
1860	Mop/Kick, 4″ x 26″		15	.533		15	20		35	50.50
1070	4″ x 36″		15	.533		20	20		40	56
1880	4″ x 42″		15	.533		24	20		44	60.50
1890	6″ x 26″		15	.533		22	20		42	58
1900	6″ x 36″		15	.533		30	20		50	67
1910	6″ x 42″	↓	15	.533	↓	35	20		55	72.50

08 71 21 – Astragals

08 71 21.10 Exterior Mouldings, Astragals		Crew	Daily Output	Labor-Hours	Unit	Material	Labor	Equipment	Total	Total Incl O&P
0010	**EXTERIOR MOULDINGS, ASTRAGALS**									
4170	Astragal for double doors, aluminum	1 Carp	4	2	Opng.	34.50	75.50		110	165
4174	Bronze	″	4	2	″	47	75.50		122.50	179

08 71 25 – Weatherstripping

08 71 25.10 Mechanical Seals, Weatherstripping		Crew	Daily Output	Labor-Hours	Unit	Material	Labor	Equipment	Total	Total Incl O&P
0010	**MECHANICAL SEALS, WEATHERSTRIPPING**									
1000	Doors, wood frame, interlocking, for 3′ x 7′ door, zinc	1 Carp	3	2.667	Opng.	44	101		145	218
1100	Bronze		3	2.667		56	101		157	231
1300	6′ x 7′ opening, zinc		2	4		54	151		205	315
1400	Bronze	↓	2	4	↓	65	151		216	325
1700	Wood frame, spring type, bronze									
1800	3′ x 7′ door	1 Carp	7.60	1.053	Opng.	23.50	40		63.50	92
1900	6′ x 7′ door		7	1.143		31	43		74	107
1920	Felt, 3′ x 7′ door		14	.571		3.90	21.50		25.40	40.50
1930	6′ x 7′ door		13	.615		4.45	23.50		27.95	44
1950	Rubber, 3′ x 7′ door		7.60	1.053	↓	9.15	40		49.15	76.50
1951	Rubber, 3′ x 7′ door		119	.067	L.F.	.54	2.54		3.08	4.85
1960	6′ x 7′ door	↓	7	1.143	Opng.	11.15	43		54.15	85
2200	Metal frame, spring type, bronze									
2300	3′ x 7′ door	1 Carp	3	2.667	Opng.	47.50	101		148.50	221
2400	6′ x 7′ door	″	2.50	3.200	″	52	121		173	260
2500	For stainless steel, spring type, add					133%				
2700	Metal frame, extruded sections, 3′ x 7′ door, aluminum	1 Carp	3	2.667	Opng.	28	101		129	200
2800	Bronze	↓	3	2.667	↓	82	101		183	259

08 71 Door Hardware

08 71 25 – Weatherstripping

08 71 25.10 Mechanical Seals, Weatherstripping		Crew	Daily Output	Labor-Hours	Unit	2016 Bare Costs Material	Labor	Equipment	Total	Total Incl O&P
3100	6′ x 7′ door, aluminum	1 Carp	1.50	5.333	Opng.	35	202		237	380
3200	Bronze		1.50	5.333		137	202		339	490
3500	Threshold weatherstripping									
3650	Door sweep, flush mounted, aluminum	1 Carp	25	.320	Ea.	19	12.10		31.10	41.50
3700	Vinyl		25	.320		18	12.10		30.10	40.50
5000	Garage door bottom weatherstrip, 12′ aluminum, clear		14	.571		25	21.50		46.50	63.50
5010	Bronze		14	.571		90	21.50		111.50	135
5050	Bottom protection, Rubber		14	.571		37	21.50		58.50	76.50
5100	Threshold		14	.571		73	21.50		94.50	117

08 74 Access Control Hardware

08 74 13 – Card Key Access Control Hardware

08 74 13.60 Entrance Card Systems

Code	Description	Crew	Daily Output	Labor-Hours	Unit	Material	Labor	Equipment	Total	Total Incl O&P
0010	**ENTRANCE CARD SYSTEMS**									
0100	Entrance card, barium ferrite				Ea.	4.70			4.70	5.15
0120	Credential					6.50			6.50	7.15
0140	Proximity					8.90			8.90	9.80
0160	Weigand					9.90			9.90	10.90
0500	Entrance card reader, barium ferrite	R-19	4	5		295	221		516	685
0520	Credential		4	5		240	221		461	625
0540	Proximity		4	5		320	221		541	710
0560	Weigand		4	5		425	221		646	830
0650	Scanner, eye retina		4	5		6,000	221		6,221	6,950
0700	Gate opener, cantilever	R-18	3	8.667		2,300	335		2,635	3,075
0750	Switch, tamper	R-19	6	3.333		22.50	147		169.50	266
0900	Accessories, electric door strike/bolt		5	4		77.50	177		254.50	375
0920	Electromagnetic lock		5	4		171	177		348	475
0940	Keypad for card reader		4	5		615	221		836	1,050

08 75 Window Hardware

08 75 10 – Window Handles and Latches

08 75 10.10 Handles and Latches

Code	Description	Crew	Daily Output	Labor-Hours	Unit	Material	Labor	Equipment	Total	Total Incl O&P
0010	**HANDLES AND LATCHES**									
1000	Handles, surface mounted, aluminum	1 Carp	24	.333	Ea.	4.56	12.60		17.16	26
1020	Brass		24	.333		5.85	12.60		18.45	27.50
1040	Chrome		24	.333		7.30	12.60		19.90	29
1500	Recessed, aluminum		12	.667		2.64	25		27.64	45
1520	Brass		12	.667		4.14	25		29.14	46.50
1540	Chrome		12	.667		3.67	25		28.67	46
2000	Latches, aluminum		20	.400		3.48	15.10		18.58	29.50
2020	Brass		20	.400		4.65	15.10		19.75	30.50
2040	Chrome		20	.400		3.39	15.10		18.49	29

08 75 30 – Weatherstripping

08 75 30.10 Mechanical Weather Seals

Code	Description	Crew	Daily Output	Labor-Hours	Unit	Material	Labor	Equipment	Total	Total Incl O&P
0010	**MECHANICAL WEATHER SEALS**, Window, double hung, 3′ x 5′									
0020	Zinc	1 Carp	7.20	1.111	Opng.	20	42		62	92.50
0100	Bronze		7.20	1.111		40	42		82	115
0200	Vinyl V strip		7	1.143		9.50	43		52.50	83
0500	As above but heavy duty, zinc		4.60	1.739		20	65.50		85.50	132

08 75 Window Hardware

08 75 30 – Weatherstripping

08 75 30.10 Mechanical Weather Seals		Crew	Daily Output	Labor-Hours	Unit	Material	Labor	Equipment	Total	Total Incl O&P
						2016 Bare Costs				
0600	Bronze	1 Carp	4.60	1.739	Opng.	70	65.50		135.50	187

08 79 Hardware Accessories

08 79 20 – Door Accessories

08 79 20.10 Door Hardware Accessories

		Crew	Daily Output	Labor-Hours	Unit	Material	Labor	Equipment	Total	Total Incl O&P
0010	DOOR HARDWARE ACCESSORIES									
0140	Door bolt, surface, 4"	1 Carp	32	.250	Ea.	13	9.45		22.45	30
0160	Door latch	"	12	.667	"	8.55	25		33.55	51.50
0200	Sliding closet door									
0220	Track and hanger, single	1 Carp	10	.800	Ea.	65	30		95	122
0240	Double		8	1		80	38		118	152
0260	Door guide, single		48	.167		30	6.30		36.30	43.50
0280	Double		48	.167		40	6.30		46.30	54.50
0600	Deadbolt and lock cover plate, brass or stainless steel		30	.267		28	10.10		38.10	48
0620	Hole cover plate, brass or chrome		35	.229		8	8.65		16.65	23.50
2240	Mortise lockset, passage, lever handle		9	.889		160	33.50		193.50	233
4000	Security chain, standard		18	.444		10	16.80		26.80	39
4100	Deluxe	↓	18	.444	↓	45	16.80		61.80	77.50

08 81 Glass Glazing

08 81 10 – Float Glass

08 81 10.10 Various Types and Thickness of Float Glass

		Crew	Daily Output	Labor-Hours	Unit	Material	Labor	Equipment	Total	Total Incl O&P
0010	VARIOUS TYPES AND THICKNESS OF FLOAT GLASS R088110-10									
0020	3/16" Plain	2 Glaz	130	.123	S.F.	5.10	4.54		9.64	13.15
0200	Tempered, clear		130	.123		7	4.54		11.54	15.25
0300	Tinted		130	.123		8.10	4.54		12.64	16.45
0600	1/4" thick, clear, plain		120	.133		6.40	4.92		11.32	15.20
0700	Tinted		120	.133		9.10	4.92		14.02	18.15
0800	Tempered, clear		120	.133		9	4.92		13.92	18.05
0900	Tinted		120	.133		11.35	4.92		16.27	20.50
1600	3/8" thick, clear, plain		75	.213		10.40	7.85		18.25	24.50
1700	Tinted		75	.213		15.70	7.85		23.55	30.50
1800	Tempered, clear		75	.213		16.75	7.85		24.60	31.50
1900	Tinted		75	.213		19.10	7.85		26.95	34
2200	1/2" thick, clear, plain		55	.291		17.65	10.75		28.40	37.50
2300	Tinted		55	.291		27.50	10.75		38.25	48
2400	Tempered, clear		55	.291		25	10.75		35.75	45.50
2500	Tinted		55	.291		26	10.75		36.75	47
2800	5/8" thick, clear, plain		45	.356		27.50	13.10		40.60	52
2900	Tempered, clear	↓	45	.356		31.50	13.10		44.60	56.50
8900	For low emissivity coating for 3/16" & 1/4" only, add to above				↓	18%				

08 81 17 – Fire Glass

08 81 17.10 Fire Resistant Glass

		Crew	Daily Output	Labor-Hours	Unit	Material	Labor	Equipment	Total	Total Incl O&P
0010	FIRE RESISTANT GLASS									
0020	Fire Glass Minimum	2 Glaz	40	.400	S.F.	37	14.75		51.75	65
0030	Mid Range		40	.400		80	14.75		94.75	113
0050	High End	↓	40	.400	↓	355	14.75		369.75	415

08 81 Glass Glazing

08 81 20 – Vision Panels

08 81 20.10 Full Vision			Crew	Daily Output	Labor-Hours	Unit	Material	Labor	Equipment	Total	Total Incl O&P
							2016 Bare Costs				
0010	**FULL VISION**, window system with 3/4" glass mullions										
0020	Up to 10' high		H-2	130	.185	S.F.	64	6.35		70.35	81
0100	10' to 20' high, minimum			110	.218		68.50	7.50		76	87.50
0150	Average			100	.240		73.50	8.25		81.75	94.50
0200	Maximum			80	.300		82.50	10.30		92.80	108

08 81 25 – Glazing Variables

08 81 25.10 Applications of Glazing			Crew	Daily Output	Labor-Hours	Unit	Material	Labor	Equipment	Total	Total Incl O&P
0010	**APPLICATIONS OF GLAZING**										
0600	For glass replacement, add					S.F.		100%			
0700	For gasket settings, add					L.F.	5.75			5.75	6.35
0900	For sloped glazing, add					S.F.		26%			
2000	Fabrication, polished edges, 1/4" thick					Inch	.55			.55	.61
2100	1/2" thick						1.30			1.30	1.43
2500	Mitered edges, 1/4" thick						1.30			1.30	1.43
2600	1/2" thick						2.15			2.15	2.37

08 81 30 – Insulating Glass

08 81 30.10 Reduce Heat Transfer Glass			Crew	Daily Output	Labor-Hours	Unit	Material	Labor	Equipment	Total	Total Incl O&P
0010	**REDUCE HEAT TRANSFER GLASS**										
0015	2 lites 1/8" float, 1/2" thk under 15 S.F.										
0100	Tinted	G	2 Glaz	95	.168	S.F.	14.25	6.20		20.45	26
0280	Double glazed, 5/8" thk unit, 3/16" float, 15-30 S.F., clear			90	.178		13.70	6.55		20.25	26
0400	1" thk, dbl. glazed, 1/4" float, 30 to 70 S.F., clear	G		75	.213		17.15	7.85		25	32
0500	Tinted	G		75	.213		23.50	7.85		31.35	39
2000	Both lites, light & heat reflective	G		85	.188		31.50	6.95		38.45	46
2500	Heat reflective, film inside, 1" thick unit, clear	G		85	.188		27.50	6.95		34.45	42
2600	Tinted	G		85	.188		28.50	6.95		35.45	43
3000	Film on weatherside, clear, 1/2" thick unit	G		95	.168		19.60	6.20		25.80	32
3100	5/8" thick unit	G		90	.178		20	6.55		26.55	33
3200	1" thick unit	G		85	.188		27	6.95		33.95	41.50
3350	Clear heat reflective film on inside	G	1 Glaz	50	.160		12.90	5.90		18.80	24
3360	Heat reflective film on inside, tinted	G		25	.320		13.70	11.80		25.50	34.50
3370	Heat reflective film on the inside, metalized	G		20	.400		14	14.75		28.75	40
5000	Spectrally selective film, on ext, blocks solar gain/allows 70% of light	G	2 Glaz	95	.168		14.25	6.20		20.45	26

08 81 55 – Window Glass

08 81 55.10 Sheet Glass		Crew	Daily Output	Labor-Hours	Unit	Material	Labor	Equipment	Total	Total Incl O&P
0010	**SHEET GLASS** (window), clear float, stops, putty bed									
0015	1/8" thick, clear float	2 Glaz	480	.033	S.F.	3.71	1.23		4.94	6.10
0500	3/16" thick, clear		480	.033		6	1.23		7.23	8.65
0600	Tinted		480	.033		7.45	1.23		8.68	10.25
0700	Tempered		480	.033		9.20	1.23		10.43	12.15

08 81 65 – Wire Glass

08 81 65.10 Glass Reinforced With Wire		Crew	Daily Output	Labor-Hours	Unit	Material	Labor	Equipment	Total	Total Incl O&P
0010	**GLASS REINFORCED WITH WIRE**									
0012	1/4" thick rough obscure	2 Glaz	135	.119	S.F.	23.50	4.37		27.87	33.50
1000	Polished wire, 1/4" thick, diamond, clear		135	.119		28	4.37		32.37	38
1500	Pinstripe, obscure		135	.119		41	4.37		45.37	52.50

08 83 Mirrors

08 83 13 – Mirrored Glass Glazing

08 83 13.10 Mirrors		Crew	Daily Output	Labor-Hours	Unit	2016 Bare Costs Material	Labor	Equipment	Total	Total Incl O&P
0010	**MIRRORS**, No frames, wall type, 1/4" plate glass, polished edge									
0100	Up to 5 S.F.	2 Glaz	125	.128	S.F.	9.50	4.72		14.22	18.30
0200	Over 5 S.F.		160	.100		9.25	3.69		12.94	16.35
0500	Door type, 1/4" plate glass, up to 12 S.F.		160	.100		8.80	3.69		12.49	15.85
1000	Float glass, up to 10 S.F., 1/8" thick		160	.100		5.90	3.69		9.59	12.65
1100	3/16" thick		150	.107		7.30	3.94		11.24	14.60
1500	12" x 12" wall tiles, square edge, clear		195	.082		2.43	3.03		5.46	7.70
1600	Veined		195	.082		6.20	3.03		9.23	11.90
2010	Bathroom, unframed, laminated	↓	160	.100	↓	14	3.69		17.69	21.50

08 84 Plastic Glazing

08 84 10 – Plexiglass Glazing

08 84 10.10 Plexiglass Acrylic		Crew	Daily Output	Labor-Hours	Unit	Material	Labor	Equipment	Total	Total Incl O&P
0010	**PLEXIGLASS ACRYLIC**, clear, masked,									
0020	1/8" thick, cut sheets	2 Glaz	170	.094	S.F.	12.20	3.47		15.67	19.15
0200	Full sheets		195	.082		5.15	3.03		8.18	10.70
0500	1/4" thick, cut sheets		165	.097		14.30	3.58		17.88	21.50
0600	Full sheets		185	.086		9.20	3.19		12.39	15.40
0900	3/8" thick, cut sheets		155	.103		20	3.81		23.81	28.50
1000	Full sheets		180	.089		15.30	3.28		18.58	22.50
1300	1/2" thick, cut sheets		135	.119		28.50	4.37		32.87	39
1400	Full sheets		150	.107		20	3.94		23.94	28.50
1700	3/4" thick, cut sheets		115	.139		70.50	5.15		75.65	86
1800	Full sheets		130	.123		40.50	4.54		45.04	52
2100	1" thick, cut sheets		105	.152		79	5.60		84.60	96.50
2200	Full sheets		125	.128		49	4.72		53.72	62
3000	Colored, 1/8" thick, cut sheets		170	.094		18.35	3.47		21.82	26
3200	Full sheets		195	.082		11.20	3.03		14.23	17.35
3500	1/4" thick, cut sheets		165	.097		20.50	3.58		24.08	28.50
3600	Full sheets		185	.086		14.30	3.19		17.49	21
4000	Mirrors, untinted, cut sheets, 1/8" thick		185	.086		12.25	3.19		15.44	18.80
4200	1/4" thick	↓	180	.089	↓	16.30	3.28		19.58	23.50

08 84 20 – Polycarbonate

08 84 20.10 Thermoplastic		Crew	Daily Output	Labor-Hours	Unit	Material	Labor	Equipment	Total	Total Incl O&P
0010	**THERMOPLASTIC**, clear, masked, cut sheets									
0020	1/8" thick	2 Glaz	170	.094	S.F.	14.30	3.47		17.77	21.50
0500	3/16" thick		165	.097		17.35	3.58		20.93	25
1000	1/4" thick		155	.103		17.35	3.81		21.16	25.50
1500	3/8" thick	↓	150	.107	↓	27	3.94		30.94	36

08 85 Glazing Accessories

08 85 10 – Miscellaneous Glazing Accessories

08 85 10.10 Glazing Gaskets		Crew	Daily Output	Labor-Hours	Unit	2016 Bare Costs Material	Labor	Equipment	Total	Total Incl O&P
0010	**GLAZING GASKETS**, Neoprene for glass tongued mullion									
0015	1/4" glass	2 Glaz	200	.080	L.F.	.77	2.95		3.72	5.75
0020	3/8"		200	.080		.96	2.95		3.91	5.95
0040	1/2"		200	.080		1.09	2.95		4.04	6.10
0060	3/4"		180	.089		1.59	3.28		4.87	7.20
0080	1"		180	.089		1.29	3.28		4.57	6.85
1000	Glazing compound, wood	1 Glaz	58	.138		1.10	5.10		6.20	9.65
1005	Glazing compound, per window, up to 30 L.F.					1.10			1.10	1.21
1006	Glazing compound, per window, up to 30 L.F.				Ea.	33			33	36.50
1020	Metal	1 Glaz	58	.138	L.F.	1.10	5.10		6.20	9.65

08 85 20 – Structural Glazing Sealants

08 85 20.10 Structural Glazing Adhesives		Crew	Daily Output	Labor-Hours	Unit	Material	Labor	Equipment	Total	Total Incl O&P
0010	**STRUCTURAL GLAZING ADHESIVES**									
0050	Structural glazing adhesive, 1/4" x 1/4" joint seal	1 Glaz	200	.040	L.F.	14.50	1.48		15.98	18.40

08 87 Glazing Surface Films

08 87 13 – Solar Control Films

08 87 13.10 Solar Films On Glass		Crew	Daily Output	Labor-Hours	Unit	Material	Labor	Equipment	Total	Total Incl O&P
0010	**SOLAR FILMS ON GLASS** (glass not included)									
1000	Bronze, 20% VLT	H-2	180	.133	S.F.	.80	4.58		5.38	8.50
1020	50% VLT		180	.133		1.20	4.58		5.78	8.90
1050	Neutral, 20% VLT		180	.133		1.20	4.58		5.78	8.90
1100	Silver, 15% VLT		180	.133		.85	4.58		5.43	8.55
1120	35% VLT		180	.133		3.21	4.58		7.79	11.15
1150	68% VLT		180	.133		3.74	4.58		8.32	11.70
3000	One way mirror with night vision 5% in and 15% out VLT		180	.133		1.60	4.58		6.18	9.35

08 87 16 – Glass Safety Films

08 87 16.10 Safety Films On Glass		Crew	Daily Output	Labor-Hours	Unit	Material	Labor	Equipment	Total	Total Incl O&P
0010	**SAFETY FILMS ON GLASS** (glass not included)									
0015	Safety film helps hold glass together when broken									
0050	Safety film, clear, 2 mil	H-2	180	.133	S.F.	1.24	4.58		5.82	8.95
0100	Safety film, clear, 4 mil		180	.133		1.71	4.58		6.29	9.50
0110	Safety film, tinted, 4 mil		180	.133		1.34	4.58		5.92	9.05
0150	Safety film, black tint, 8 mil		180	.133		2.04	4.58		6.62	9.85

08 87 53 – Security Films On Glass

08 87 53.10 Security Film Adhered On Glass		Crew	Daily Output	Labor-Hours	Unit	Material	Labor	Equipment	Total	Total Incl O&P
0010	**SECURITY FILM ADHERED ON GLASS** (glass not included)									
0020	Security film, clear, 7 mil	1 Glaz	200	.040	S.F.	1.03	1.48		2.51	3.58
0030	8 mil		200	.040		1.86	1.48		3.34	4.50
0040	9 mil		200	.040		2.07	1.48		3.55	4.73
0050	10 mil		200	.040		2.24	1.48		3.72	4.91
0060	12 mil		200	.040		1.96	1.48		3.44	4.61
0075	15 mil		200	.040		2.38	1.48		3.86	5.05
0100	Security film, sealed with structural adhesive, 7 mil		180	.044		5.55	1.64		7.19	8.80
0110	8 mil		180	.044		6.50	1.64		8.14	9.85
0140	14 mil		180	.044		7.10	1.64		8.74	10.50

08 88 52 – Prefabricated Glass Block Windows

08 88 52.10 Prefabricated Glass Block Windows		Crew	Daily Output	Labor-Hours	Unit	2016 Bare Costs Material	Labor	Equipment	Total	Total Incl O&P
0010	**PREFABRICATED GLASS BLOCK WINDOWS**									
0015	Includes frame and silicone seal									
0020	Glass Block Window 16″ x 8″	2 Glaz	6	2.667	Ea.	185	98.50		283.50	365
0025	16″ x 16″		6	2.667		223	98.50		321.50	410
0030	16″ x 24″		6	2.667		265	98.50		363.50	455
0050	16″ x 48″		4	4		380	148		528	665
0100	Glass Block Window 24″ x 8″		6	2.667		211	98.50		309.50	395
0110	24″ x 16″		6	2.667		265	98.50		363.50	455
0120	24″ x 24″		6	2.667		315	98.50		413.50	510
0150	24″ x 48″		6	2.667		470	98.50		568.50	680
0200	Glass Block Window 32″ x 8″		6	2.667		211	98.50		309.50	395
0210	32″ x 16″		6	2.667		300	98.50		398.50	495
0220	32″ x 24″		4	4		365	148		513	645
0250	32″ x 48″		4	4		555	148		703	855
0300	Glass Block Window 40″ x 8″		6	2.667		271	98.50		369.50	460
0310	40″ x 16″		6	2.667		340	98.50		438.50	540
0320	40″ x 24″		6	2.667		410	98.50		508.50	620
0350	40″ x 48″		6	2.667		645	98.50		743.50	875
0400	Glass Block Window 48″ x 8″		6	2.667		300	98.50		398.50	495
0410	48″ x 16″		6	2.667		380	98.50		478.50	580
0420	48″ x 24″		6	2.667		470	98.50		568.50	680
0450	48″ x 48″		6	2.667		730	98.50		828.50	970
0500	Glass Block Window 56″ x 8″		6	2.667		330	98.50		428.50	530
0510	56″ x 16″		4	4		415	148		563	705
0520	56″ x 24″		4	4		520	148		668	815
0550	56″ x 48″		4	4		825	148		973	1,150

08 88 52.20 Prefabricated Glass Block Windows Hurricane Resistant		Crew	Daily Output	Labor-Hours	Unit	Material	Labor	Equipment	Total	Total Incl O&P
0010	**PREFABRICATED GLASS BLOCK WINDOWS HURRICANE RESISTANT**									
0015	Includes frame and silicone seal									
3020	Glass Block Window 16″ x 16″	2 Glaz	6	2.667	Ea.	500	98.50		598.50	715
3030	16″ x 24″		6	2.667		580	98.50		678.50	805
3040	16″ x 32″		4	4		640	148		788	950
3050	16″ x 40″		4	4		690	148		838	1,000
3060	16″ x 48″		4	4		700	148		848	1,025
3070	16″ x 56″		4	4		880	148		1,028	1,225
3080	Glass Block Window 24″ x 24″		6	2.667		640	98.50		738.50	870
3090	24″ x 32″		4	4		700	148		848	1,025
3100	24″ x 40″		4	4		880	148		1,028	1,225
3110	24″ x 48″		4	4		910	148		1,058	1,250
3120	24″ x 56″		4	4		950	148		1,098	1,300
3130	Glass Block Window 32″ x 32″		4	4		800	148		948	1,125
3140	32″ x 40″		4	4		900	148		1,048	1,225
3150	32″ x 48″		4	4		1,050	148		1,198	1,400
3160	32″ x 56″		4	4		1,200	148		1,348	1,575
3200	Glass Block Window 40″ x 40″		4	4		1,025	148		1,173	1,375
3210	40″ x 48″		3	5.333		1,250	197		1,447	1,700
3220	40″ x 56″		3	5.333		1,400	197		1,597	1,875
3260	Glass Block Window 48″ x 48″		3	5.333		1,450	197		1,647	1,925
3270	48″ x 56″		3	5.333		1,625	197		1,822	2,100
3310	Glass Block Window 56″ x 56″		3	5.333		1,775	197		1,972	2,275

08 88 Special Function Glazing

08 88 56 – Ballistics-Resistant Glazing

08 88 56.10 Laminated Glass		Crew	Daily Output	Labor-Hours	Unit	2016 Bare Costs Material	Labor	Equipment	Total	Total Incl O&P
0010	**LAMINATED GLASS**									
0020	Clear float .03" vinyl 1/4"	2 Glaz	90	.178	S.F.	12.50	6.55		19.05	24.50
0100	3/8" thick		78	.205		22.50	7.55		30.05	37
0200	.06" vinyl, 1/2" thick		65	.246		25.50	9.10		34.60	43
1000	5/8" thick		90	.178		29.50	6.55		36.05	43.50
2000	Bullet-resisting, 1-3/16" thick, to 15 S.F.		16	1		113	37		150	187
2100	Over 15 S.F.		16	1		126	37		163	201
2200	2" thick, to 15 S.F.		10	1.600		140	59		199	252
2300	Over 15 S.F.		10	1.600		120	59		179	230
2500	2-1/4" thick, to 15 S.F.		12	1.333		185	49		234	286
2600	Over 15 S.F.		12	1.333		175	49		224	275
2700	Level 2 (.357 magnum), NIJ and UL		12	1.333		83	49		132	173
2750	Level 3A (.44 magnum) NIJ, UL 3		12	1.333		88	49		137	179
2800	Level 4 (AK-47) NIJ, UL 7 & 8		12	1.333		122	49		171	216
2850	Level 5 (M-16) UL		12	1.333		125	49		174	220
2900	Level 3 (7.62 Armor Piercing) NIJ, UL 4 & 5	↓	12	1.333	↓	147	49		196	244

08 91 Louvers

08 91 16 – Operable Wall Louvers

08 91 16.10 Movable Blade Louvers		Crew	Daily Output	Labor-Hours	Unit	Material	Labor	Equipment	Total	Total Incl O&P
0010	**MOVABLE BLADE LOUVERS**									
0100	PVC, commercial grade, 12" x 12"	1 Shee	20	.400	Ea.	82	16.95		98.95	119
0110	16" x 16"		14	.571		96	24		120	147
0120	18" x 18"		14	.571		99	24		123	150
0130	24" x 24"		14	.571		140	24		164	195
0140	30" x 30"		12	.667		168	28		196	232
0150	36" x 36"		12	.667		197	28		225	264
0300	Stainless steel, commercial grade, 12" x 12"		20	.400		219	16.95		235.95	270
0310	16" x 16"		14	.571		265	24		289	335
0320	18" x 18"		14	.571		282	24		306	350
0330	20" x 20"		14	.571		325	24		349	395
0340	24" x 24"		14	.571		385	24		409	465
0350	30" x 30"		12	.667		430	28		458	520
0360	36" x 36"	↓	12	.667	↓	520	28		548	620

08 91 19 – Fixed Louvers

08 91 19.10 Aluminum Louvers		Crew	Daily Output	Labor-Hours	Unit	Material	Labor	Equipment	Total	Total Incl O&P
0010	**ALUMINUM LOUVERS**									
0020	Aluminum with screen, residential, 8" x 8"	1 Carp	38	.211	Ea.	20	7.95		27.95	35.50
0100	12" x 12"		38	.211		16	7.95		23.95	31
0200	12" x 18"		35	.229		20	8.65		28.65	36.50
0250	14" x 24"		30	.267		29	10.10		39.10	49
0300	18" x 24"		27	.296		32	11.20		43.20	54
0500	24" x 30"		24	.333		60	12.60		72.60	87
0700	Triangle, adjustable, small		20	.400		58	15.10		73.10	89.50
0800	Large		15	.533		79.50	20		99.50	121
2100	Midget, aluminum, 3/4" deep, 1" diameter		85	.094		.77	3.56		4.33	6.80
2150	3" diameter		60	.133		2.48	5.05		7.53	11.20
2200	4" diameter		50	.160		4.95	6.05		11	15.60
2250	6" diameter	↓	30	.267		4.10	10.10		14.20	21.50
3000	PVC, commercial grade, 12" x 12"	1 Shee	20	.400	↓	97	16.95		113.95	136

08 91 Louvers

08 91 19 – Fixed Louvers

08 91 19.10 Aluminum Louvers		Crew	Daily Output	Labor-Hours	Unit	Material	Labor	Equipment	Total	Total Incl O&P
						2016 Bare Costs				
3010	12″ x 18″	1 Shee	20	.400	Ea.	109	16.95		125.95	149
3020	12″ x 24″		20	.400		146	16.95		162.95	190
3030	14″ x 24″		20	.400		152	16.95		168.95	196
3100	Aluminum, commercial grade, 12″ x 12″		20	.400		171	16.95		187.95	217
3110	24″ x 24″		16	.500		249	21		270	310
3120	24″ x 36″		16	.500		300	21		321	365
3130	36″ x 24″		16	.500		310	21		331	375
3140	36″ x 36″		14	.571		390	24		414	470
3150	36″ x 48″		10	.800		400	34		434	495
3160	48″ x 36″		10	.800		390	34		424	485
3170	48″ x 48″		10	.800		435	34		469	530
3180	60″ x 48″		8	1		485	42.50		527.50	605
3190	60″ x 60″	↓	8	1	↓	600	42.50		642.50	730
08 91 19.20 Steel Louvers										
0010	**STEEL LOUVERS**									
3300	Galvanized steel, fixed blades, commercial grade, 18″ x 18″	1 Shee	20	.400	Ea.	197	16.95		213.95	246
3310	24″ x 24″		20	.400		232	16.95		248.95	284
3320	24″ x 36″		16	.500		300	21		321	365
3330	36″ x 24″		16	.500		298	21		319	365
3340	36″ x 36″		14	.571		380	24		404	460
3350	36″ x 48″		10	.800		385	34		419	480
3360	48″ x 36″		10	.800		385	34		419	480
3370	48″ x 48″		10	.800		425	34		459	520
3380	60″ x 48″		10	.800		490	34		524	595
3390	60″ x 60″	↓	10	.800	↓	590	34		624	700

08 91 26 – Door Louvers

08 91 26.10 Steel Louvers, 18 Gauge, Fixed Blade		Crew	Daily Output	Labor-Hours	Unit	Material	Labor	Equipment	Total	Total Incl O&P
0010	**STEEL LOUVERS, 18 GAUGE, FIXED BLADE**									
0050	12″ x 12″, with enamel or powder coat	1 Carp	20	.400	Ea.	82	15.10		97.10	116
0055	18″ x 12″		20	.400		87.50	15.10		102.60	122
0060	18″ x 18″		20	.400		106	15.10		121.10	143
0065	24″ x 12″		20	.400		108	15.10		123.10	145
0070	24″ x 18″		20	.400		117	15.10		132.10	155
0075	24″ x 24″		20	.400		144	15.10		159.10	184
0100	12″ x 12″, galvanized		20	.400		74	15.10		89.10	107
0105	18″ x 12″		20	.400		87.50	15.10		102.60	122
0115	24″ x 12″		20	.400		104	15.10		119.10	140
0125	24″ x 24″	↓	20	.400	↓	144	15.10		159.10	184
3000	Fire rated									
3010	12″ x 12″	1 Carp	10	.800	Ea.	201	30		231	272
3050	18″ x 18″		10	.800		205	30		235	276
3100	24″ x 12″		10	.800		237	30		267	310
3120	24″ x 18″		10	.800		294	30		324	375
3130	24″ x 24″	↓	10	.800	↓	287	30		317	365

08 95 Vents

08 95 13 – Soffit Vents

08 95 13.10 Wall Louvers		Crew	Daily Output	Labor-Hours	Unit	Material	Labor	Equipment	Total	Total Incl O&P
						2016 Bare Costs				
0010	**WALL LOUVERS**									
2400	Under eaves vent, aluminum, mill finish, 16″ x 4″	1 Carp	48	.167	Ea.	1.93	6.30		8.23	12.65
2500	16″ x 8″	″	48	.167	″	2.23	6.30		8.53	13

08 95 16 – Wall Vents

08 95 16.10 Louvers		Crew	Daily Output	Labor-Hours	Unit	Material	Labor	Equipment	Total	Total Incl O&P
0010	**LOUVERS**									
0020	Redwood, 2′-0″ diameter, full circle	1 Carp	16	.500	Ea.	190	18.90		208.90	241
0100	Half circle		16	.500		175	18.90		193.90	225
0200	Octagonal		16	.500		142	18.90		160.90	188
0300	Triangular, 5/12 pitch, 5′-0″ at base		16	.500		200	18.90		218.90	252
1000	Rectangular, 1′-4″ x 1′-3″		16	.500		19.50	18.90		38.40	53
1100	Rectangular, 1′-4″ x 1′-8″		16	.500		28	18.90		46.90	62.50
1200	1′-4″ x 2′-2″		15	.533		31	20		51	68.50
1300	1′-9″ x 2′-2″		15	.533		37.50	20		57.50	75.50
1400	2′-3″ x 2′-2″		14	.571		49.50	21.50		71	90.50
1700	2′-4″ x 2′-11″		13	.615		49.50	23.50		73	93.50
2000	Aluminum, 12″ x 16″		25	.320		20	12.10		32.10	42.50
2010	16″ x 20″		25	.320		27.50	12.10		39.60	51
2020	24″ x 30″		25	.320		59	12.10		71.10	85.50
2100	6′ triangle		12	.667		168	25		193	227
3100	Round, 2′-2″ diameter		16	.500		124	18.90		142.90	168
7000	Vinyl gable vent, 8″ x 8″		38	.211		14	7.95		21.95	29
7020	12″ x 12″		38	.211		28	7.95		35.95	44.50
7080	12″ x 18″		35	.229		37	8.65		45.65	55
7200	18″ x 24″		30	.267		45	10.10		55.10	66.50

Division 9 Finishes

Estimating Tips

General

- Room Finish Schedule: A complete set of plans should contain a room finish schedule. If one is not available, it would be well worth the time and effort to obtain one.

09 20 00 Plaster and Gypsum Board

- Lath is estimated by the square yard plus a 5% allowance for waste. Furring, channels, and accessories are measured by the linear foot. An extra foot should be allowed for each accessory miter or stop.
- Plaster is also estimated by the square yard. Deductions for openings vary by preference, from zero deduction to 50% of all openings over 2 feet in width. The estimator should allow one extra square foot for each linear foot of horizontal interior or exterior angle located below the ceiling level. Also, double the areas of small radius work.
- Drywall accessories, studs, track, and acoustical caulking are all measured by the linear foot. Drywall taping is figured by the square foot. Gypsum wallboard is estimated by the square foot. No material deductions should be made for door or window openings under 32 S.F.

09 60 00 Flooring

- Tile and terrazzo areas are taken off on a square foot basis. Trim and base materials are measured by the linear foot. Accent tiles are listed per each. Two basic methods of installation are used. Mud set is approximately 30% more expensive than thin set. In terrazzo work, be sure to include the linear footage of embedded decorative strips, grounds, machine rubbing, and power cleanup.
- Wood flooring is available in strip, parquet, or block configuration. The latter two types are set in adhesives with quantities estimated by the square foot. The laying pattern will influence labor costs and material waste. In addition to the material and labor for laying wood floors, the estimator must make allowances for sanding and finishing these areas, unless the flooring is prefinished.
- Sheet flooring is measured by the square yard. Roll widths vary, so consideration should be given to use the most economical width, as waste must be figured into the total quantity. Consider also the installation methods available, direct glue down or stretched.

09 70 00 Wall Finishes

- Wall coverings are estimated by the square foot. The area to be covered is measured, length by height of wall above baseboards, to calculate the square footage of each wall. This figure is divided by the number of square feet in the single roll which is being used. Deduct, in full, the areas of openings such as doors and windows. Where a pattern match is required allow 25%–30% waste.

09 80 00 Acoustic Treatment

- Acoustical systems fall into several categories. The takeoff of these materials should be by the square foot of area with a 5% allowance for waste. Do not forget about scaffolding, if applicable, when estimating these systems.

09 90 00 Painting and Coating

- A major portion of the work in painting involves surface preparation. Be sure to include cleaning, sanding, filling, and masking costs in the estimate.
- Protection of adjacent surfaces is not included in painting costs. When considering the method of paint application, an important factor is the amount of protection and masking required. These must be estimated separately and may be the determining factor in choosing the method of application.

Reference Numbers

Reference numbers are shown at the beginning of some major classifications. These numbers refer to related items in the Reference Section. The reference information may be an estimating procedure, an alternate pricing method, or technical information.

Note: Not all subdivisions listed here necessarily appear. ■

09 05 05 – Selective Demolition for Finishes

09 05 05.10 Selective Demolition, Ceilings

	09 05 05.10 Selective Demolition, Ceilings	Crew	Daily Output	Labor-Hours	Unit	Material	2016 Bare Costs Labor	Equipment	Total	Total Incl O&P
0010	**SELECTIVE DEMOLITION, CEILINGS** R024119-10									
0200	Ceiling, drywall, furred and nailed or screwed	2 Clab	800	.020	S.F.		.58		.58	.98
1000	Plaster, lime and horse hair, on wood lath, incl. lath		700	.023			.67		.67	1.12
1200	Suspended ceiling, mineral fiber, 2′ x 2′ or 2′ x 4′		1500	.011			.31		.31	.52
1250	On suspension system, incl. system		1200	.013			.39		.39	.65
1500	Tile, wood fiber, 12″ x 12″, glued		900	.018			.52		.52	.87
1540	Stapled		1500	.011			.31		.31	.52
2000	Wood, tongue and groove, 1″ x 4″		1000	.016			.47		.47	.78
2040	1″ x 8″		1100	.015			.42		.42	.71
2400	Plywood or wood fiberboard, 4′ x 8′ sheets		1200	.013			.39		.39	.65

09 05 05.20 Selective Demolition, Flooring

	09 05 05.20 Selective Demolition, Flooring	Crew	Daily Output	Labor-Hours	Unit	Material	Labor	Equipment	Total	Total Incl O&P
0010	**SELECTIVE DEMOLITION, FLOORING**									
0200	Brick with mortar	2 Clab	475	.034	S.F.		.98		.98	1.65
0400	Carpet, bonded, including surface scraping		2000	.008			.23		.23	.39
0480	Tackless		9000	.002			.05		.05	.09
0550	Carpet tile, releasable adhesive		5000	.003			.09		.09	.16
0560	Permanent adhesive		1850	.009			.25		.25	.42
0800	Resilient, sheet goods		1400	.011			.33		.33	.56
0850	Vinyl or rubber cove base	1 Clab	1000	.008	L.F.		.23		.23	.39
0860	Vinyl or rubber cove base, molded corner	″	1000	.008	Ea.		.23		.23	.39
0870	For glued and caulked installation, add to labor						50%			
0900	Vinyl composition tile, 12″ x 12″	2 Clab	1000	.016	S.F.		.47		.47	.78
2000	Tile, ceramic, thin set		675	.024			.69		.69	1.16
2020	Mud set		625	.026			.75		.75	1.25
3000	Wood, block, on end	1 Carp	400	.020			.76		.76	1.27
3200	Parquet		450	.018			.67		.67	1.13
3400	Strip flooring, interior, 2-1/4″ x 25/32″ thick		325	.025			.93		.93	1.56
3500	Exterior, porch flooring, 1″ x 4″		220	.036			1.37		1.37	2.30
3800	Subfloor, tongue and groove, 1″ x 6″		325	.025			.93		.93	1.56
3820	1″ x 8″		430	.019			.70		.70	1.18
3840	1″ x 10″		520	.015			.58		.58	.97
4000	Plywood, nailed		600	.013			.50		.50	.84
4100	Glued and nailed		400	.020			.76		.76	1.27
4200	Hardboard, 1/4″ thick		760	.011			.40		.40	.67

09 05 05.30 Selective Demolition, Walls and Partitions

	09 05 05.30 Selective Demolition, Walls and Partitions	Crew	Daily Output	Labor-Hours	Unit	Material	Labor	Equipment	Total	Total Incl O&P
0010	**SELECTIVE DEMOLITION, WALLS AND PARTITIONS** R024119-10									
0020	Walls, concrete, reinforced	B-39	120	.400	C.F.		11.80	1.90	13.70	22
0025	Plain	″	160	.300	″		8.85	1.43	10.28	16.40
1000	Drywall, nailed or screwed	1 Clab	1000	.008	S.F.		.23		.23	.39
1010	2 layers		400	.020			.58		.58	.98
1500	Fiberboard, nailed		900	.009			.26		.26	.43
1568	Plenum barrier, sheet lead		300	.027			.78		.78	1.30
2200	Metal or wood studs, finish 2 sides, fiberboard	B-1	520	.046			1.38		1.38	2.31
2250	Lath and plaster		260	.092			2.76		2.76	4.62
2300	Plasterboard (drywall)		520	.046			1.38		1.38	2.31
2350	Plywood		450	.053			1.59		1.59	2.67
2800	Paneling, 4′ x 8′ sheets	1 Clab	475	.017			.49		.49	.82
3000	Plaster, lime and horsehair, on wood lath		400	.020			.58		.58	.98
3020	On metal lath		335	.024			.70		.70	1.17
3450	Plaster, interior gypsum, acoustic, or cement		60	.133	S.Y.		3.89		3.89	6.50
3500	Stucco, on masonry		145	.055			1.61		1.61	2.70
3510	Commercial 3-coat		80	.100			2.92		2.92	4.89

09 05 Common Work Results for Finishes

09 05 05 – Selective Demolition for Finishes

09 05 05.30 Selective Demolition, Walls and Partitions		Crew	Daily Output	Labor-Hours	Unit	2016 Bare Costs Material	Labor	Equipment	Total	Total Incl O&P
3520	Interior stucco	1 Clab	25	.320	S.Y.		9.35		9.35	15.65
3760	Tile, ceramic, on walls, thin set		300	.027	S.F.		.78		.78	1.30
3765	Mud set		250	.032	″		.93		.93	1.56

09 22 Supports for Plaster and Gypsum Board

09 22 03 – Fastening Methods for Finishes

09 22 03.20 Drilling Plaster/Drywall		Crew	Daily Output	Labor-Hours	Unit	Material	Labor	Equipment	Total	Total Incl O&P
0010	**DRILLING PLASTER/DRYWALL**									
1100	Drilling & layout for drywall/plaster walls, up to 1″ deep, no anchor									
1200	Holes, 1/4″ diameter	1 Carp	150	.053	Ea.	.01	2.02		2.03	3.39
1300	3/8″ diameter		140	.057		.01	2.16		2.17	3.63
1400	1/2″ diameter		130	.062		.01	2.33		2.34	3.91
1500	3/4″ diameter		120	.067		.02	2.52		2.54	4.24
1600	1″ diameter		110	.073		.02	2.75		2.77	4.62
1700	1-1/4″ diameter		100	.080		.03	3.02		3.05	5.10
1800	1-1/2″ diameter		90	.089		.05	3.36		3.41	5.70
1900	For ceiling installations, add						40%			

09 22 13 – Metal Furring

09 22 13.13 Metal Channel Furring		Crew	Daily Output	Labor-Hours	Unit	Material	Labor	Equipment	Total	Total Incl O&P
0010	**METAL CHANNEL FURRING**									
0030	Beams and columns, 7/8″ channels, galvanized, 12″ O.C.	1 Lath	155	.052	S.F.	.42	1.90		2.32	3.55
0050	16″ O.C.		170	.047		.35	1.74		2.09	3.19
0070	24″ O.C.		185	.043		.23	1.60		1.83	2.83
0100	Ceilings, on steel, 7/8″ channels, galvanized, 12″ O.C.		210	.038		.38	1.41		1.79	2.69
0300	16″ O.C.		290	.028		.35	1.02		1.37	2.03
0400	24″ O.C.		420	.019		.23	.70		.93	1.39
0600	1-5/8″ channels, galvanized, 12″ O.C.		190	.042		.51	1.55		2.06	3.08
0700	16″ O.C.		260	.031		.46	1.14		1.60	2.35
0900	24″ O.C.		390	.021		.31	.76		1.07	1.56
0930	7/8″ channels with sound isolation clips, 12″ O.C.		120	.067		1.76	2.46		4.22	5.90
0940	16″ O.C.		100	.080		1.32	2.95		4.27	6.20
0950	24″ O.C.		165	.048		.88	1.79		2.67	3.86
0960	1-5/8″ channels, galvanized, 12″ O.C.		110	.073		1.89	2.68		4.57	6.40
0970	16″ O.C.		100	.080		1.42	2.95		4.37	6.35
0980	24″ O.C.		155	.052		.94	1.90		2.84	4.12
1000	Walls, 7/8″ channels, galvanized, 12″ O.C.		235	.034		.38	1.26		1.64	2.45
1200	16″ O.C.		265	.030		.35	1.11		1.46	2.18
1300	24″ O.C.		350	.023		.23	.84		1.07	1.61
1500	1-5/8″ channels, galvanized, 12″ O.C.		210	.038		.51	1.41		1.92	2.84
1600	16″ O.C.		240	.033		.46	1.23		1.69	2.50
1800	24″ O.C.		305	.026		.31	.97		1.28	1.90
1920	7/8″ channels with sound isolation clips, 12″ O.C.		125	.064		1.76	2.36		4.12	5.75
1940	16″ O.C.		100	.080		1.32	2.95		4.27	6.20
1950	24″ O.C.		150	.053		.88	1.97		2.85	4.15
1960	1-5/8″ channels, galvanized, 12″ O.C.		115	.070		1.89	2.57		4.46	6.25
1970	16″ O.C.		95	.084		1.42	3.11		4.53	6.55
1980	24″ O.C.		140	.057		.94	2.11		3.05	4.45

09 22 16.13 Non-Structural Metal Stud Framing		Crew	Daily Output	Labor-Hours	Unit	Material	Labor	Equipment	Total	Total Incl O&P
						2016 Bare Costs				
0010	**NON-STRUCTURAL METAL STUD FRAMING**									
1600	Non-load bearing, galv., 8' high, 25 ga. 1-5/8" wide, 16" O.C.	1 Carp	619	.013	S.F.	.26	.49		.75	1.11
1610	24" O.C.		950	.008		.20	.32		.52	.75
1620	2-1/2" wide, 16" O.C.		613	.013		.33	.49		.82	1.19
1630	24" O.C.		938	.009		.24	.32		.56	.81
1640	3-5/8" wide, 16" O.C.		600	.013		.39	.50		.89	1.27
1650	24" O.C.		925	.009		.29	.33		.62	.87
1660	4" wide, 16" O.C.		594	.013		.44	.51		.95	1.33
1670	24" O.C.		925	.009		.33	.33		.66	.91
1680	6" wide, 16" O.C.		588	.014		.52	.51		1.03	1.43
1690	24" O.C.		906	.009		.39	.33		.72	.99
1700	20 ga. studs, 1-5/8" wide, 16" O.C.		494	.016		.33	.61		.94	1.39
1710	24" O.C.		763	.010		.25	.40		.65	.93
1720	2-1/2" wide, 16" O.C.		488	.016		.42	.62		1.04	1.51
1730	24" O.C.		750	.011		.32	.40		.72	1.03
1740	3-5/8" wide, 16" O.C.		481	.017		.49	.63		1.12	1.58
1750	24" O.C.		738	.011		.36	.41		.77	1.09
1760	4" wide, 16" O.C.		475	.017		.58	.64		1.22	1.71
1770	24" O.C.		738	.011		.43	.41		.84	1.17
1780	6" wide, 16" O.C.		469	.017		.67	.64		1.31	1.82
1790	24" O.C.		725	.011		.51	.42		.93	1.26
2000	Non-load bearing, galv., 10' high, 25 ga. 1-5/8" wide, 16" O.C.		495	.016		.25	.61		.86	1.29
2100	24" O.C.		760	.011		.18	.40		.58	.87
2200	2-1/2" wide, 16" O.C.		490	.016		.31	.62		.93	1.37
2250	24" O.C.		750	.011		.23	.40		.63	.93
2300	3-5/8" wide, 16" O.C.		480	.017		.37	.63		1	1.47
2350	24" O.C.		740	.011		.27	.41		.68	.98
2400	4" wide, 16" O.C.		475	.017		.41	.64		1.05	1.52
2450	24" O.C.		740	.011		.30	.41		.71	1.01
2500	6" wide, 16" O.C.		470	.017		.49	.64		1.13	1.62
2550	24" O.C.		725	.011		.36	.42		.78	1.10
2600	20 ga. studs, 1-5/8" wide, 16" O.C.		395	.020		.32	.77		1.09	1.63
2650	24" O.C.		610	.013		.23	.50		.73	1.09
2700	2-1/2" wide, 16" O.C.		390	.021		.40	.78		1.18	1.74
2750	24" O.C.		600	.013		.30	.50		.80	1.17
2800	3-5/8" wide, 16" OC		385	.021		.46	.79		1.25	1.83
2850	24" O.C.		590	.014		.34	.51		.85	1.23
2900	4" wide, 16" O.C.		380	.021		.55	.80		1.35	1.93
2950	24" O.C.		590	.014		.40	.51		.91	1.30
3000	6" wide, 16" O.C.		375	.021		.64	.81		1.45	2.05
3050	24" O.C.		580	.014		.47	.52		.99	1.39
3060	Non-load bearing, galv., 12' high, 25 ga. 1-5/8" wide, 16" O.C.		413	.019		.24	.73		.97	1.49
3070	24" O.C.		633	.013		.17	.48		.65	.99
3080	2-1/2" wide, 16" O.C.		408	.020		.29	.74		1.03	1.56
3090	24" O.C.		625	.013		.22	.48		.70	1.05
3100	3-5/8" wide, 16" O.C.		400	.020		.35	.76		1.11	1.66
3110	24" O.C.		617	.013		.26	.49		.75	1.10
3120	4" wide, 16" O.C.		396	.020		.39	.76		1.15	1.71
3130	24" O.C.		617	.013		.29	.49		.78	1.14
3140	6" wide, 16" O.C.		392	.020		.47	.77		1.24	1.81
3150	24" O.C.		604	.013		.34	.50		.84	1.22
3160	20 ga. studs, 1-5/8" wide, 16" O.C.		329	.024		.30	.92		1.22	1.87

09 22 Supports for Plaster and Gypsum Board

09 22 16 – Non-Structural Metal Framing

09 22 16.13 Non-Structural Metal Stud Framing		Crew	Daily Output	Labor-Hours	Unit	Material	Labor	Equipment	Total	Total Incl O&P
						2016 Bare Costs				
3170	24" O.C.	1 Carp	508	.016	S.F.	.22	.60		.82	1.24
3180	2-1/2" wide, 16" O.C.		325	.025		.38	.93		1.31	1.98
3190	24" O.C.		500	.016		.28	.60		.88	1.32
3200	3-5/8" wide, 16" O.C.		321	.025		.44	.94		1.38	2.06
3210	24" O.C.		492	.016		.32	.61		.93	1.38
3220	4" wide, 16" O.C.		317	.025		.52	.95		1.47	2.18
3230	24" O.C.		492	.016		.38	.61		.99	1.45
3240	6" wide, 16" O.C.		313	.026		.61	.97		1.58	2.29
3250	24" O.C.	↓	483	.017	↓	.44	.63		1.07	1.54
5000	For load bearing studs, see Section 05 41 13.30									

09 22 26 – Suspension Systems

09 22 26.13 Ceiling Suspension Systems		Crew	Daily Output	Labor-Hours	Unit	Material	Labor	Equipment	Total	Total Incl O&P
0010	**CEILING SUSPENSION SYSTEMS** for gypsum board or plaster									
8000	Suspended ceilings, including carriers									
8200	1-1/2" carriers, 24" O.C. with:									
8300	7/8" channels, 16" O.C.	1 Lath	275	.029	S.F.	.57	1.07		1.64	2.36
8320	24" O.C.		310	.026		.45	.95		1.40	2.03
8400	1-5/8" channels, 16" O.C.		205	.039		.68	1.44		2.12	3.08
8420	24" O.C.	↓	250	.032	↓	.53	1.18		1.71	2.49
8600	2" carriers, 24" O.C. with:									
8700	7/8" channels, 16" O.C.	1 Lath	250	.032	S.F.	.61	1.18		1.79	2.58
8720	24" O.C.		285	.028		.49	1.04		1.53	2.21
8800	1-5/8" channels, 16" O.C.		190	.042		.72	1.55		2.27	3.30
8820	24" O.C.	↓	225	.036	↓	.57	1.31		1.88	2.75

09 22 36 – Lath

09 22 36.13 Gypsum Lath		Crew	Daily Output	Labor-Hours	Unit	Material	Labor	Equipment	Total	Total Incl O&P
0010	**GYPSUM LATH** R092000-50									
0020	Plain or perforated, nailed, 3/8" thick	1 Lath	85	.094	S.Y.	3.06	3.47		6.53	8.95
0100	1/2" thick		80	.100		2.43	3.69		6.12	8.60
0300	Clipped to steel studs, 3/8" thick		75	.107		3.06	3.94		7	9.70
0400	1/2" thick		70	.114		2.43	4.22		6.65	9.45
1500	For ceiling installations, add		216	.037			1.37		1.37	2.21
1600	For columns and beams, add	↓	170	.047	↓		1.74		1.74	2.81

09 22 36.23 Metal Lath		Crew	Daily Output	Labor-Hours	Unit	Material	Labor	Equipment	Total	Total Incl O&P
0010	**METAL LATH** R092000-50									
3600	2.5 lb. diamond painted, on wood framing, on walls	1 Lath	85	.094	S.Y.	4.09	3.47		7.56	10.10
3700	On ceilings		75	.107		4.09	3.94		8.03	10.85
4200	3.4 lb. diamond painted, wired to steel framing		75	.107		4.32	3.94		8.26	11.10
4300	On ceilings		60	.133		4.32	4.92		9.24	12.70
5100	Rib lath, painted, wired to steel, on walls, 2.5 lb.		75	.107		3.48	3.94		7.42	10.20
5200	3.4 lb.		70	.114		4.52	4.22		8.74	11.75
5700	Suspended ceiling system, incl. 3.4 lb. diamond lath, painted		15	.533		4.31	19.70		24.01	36.50
5800	Galvanized	↓	15	.533	↓	4.32	19.70		24.02	37

09 22 36.43 Security Mesh		Crew	Daily Output	Labor-Hours	Unit	Material	Labor	Equipment	Total	Total Incl O&P
0010	**SECURITY MESH**, expanded metal, flat, screwed to framing									
0100	On walls, 3/4", 1.76 lb./S.F.	2 Carp	1500	.011	S.F.	1.92	.40		2.32	2.79
0110	1-1/2", 1.14 lb./S.F.		1600	.010		1.47	.38		1.85	2.25
0200	On ceilings, 3/4", 1.76 lb./S.F.		1350	.012		1.92	.45		2.37	2.86
0210	1-1/2", 1.14 lb./S.F.	↓	1450	.011	↓	1.47	.42		1.89	2.32

09 22 Supports for Plaster and Gypsum Board

09 22 36 – Lath

09 22 36.83 Accessories, Plaster		Crew	Daily Output	Labor-Hours	Unit	Material	Labor	Equipment	Total	Total Incl O&P
						2016 Bare Costs				
0010	**ACCESSORIES, PLASTER**									
0020	Casing bead, expanded flange, galvanized	1 Lath	2.70	2.963	C.L.F.	56.50	109		165.50	239
0200	Foundation weep screed, galvanized	"	2.70	2.963		54.50	109		163.50	237
0900	Channels, cold rolled, 16 ga., 3/4" deep, galvanized					38.50			38.50	42.50
1200	1-1/2" deep, 16 ga., galvanized					51.50			51.50	56.50
1620	Corner bead, expanded bullnose, 3/4" radius, #10, galvanized	1 Lath	2.60	3.077		26	114		140	213
1650	#1, galvanized		2.55	3.137		49.50	116		165.50	242
1670	Expanded wing, 2-3/4" wide, #1, galvanized		2.65	3.019		39	111		150	223
1700	Inside corner (corner rite), 3" x 3", painted		2.60	3.077		20.50	114		134.50	207
1750	Strip-ex, 4" wide, painted		2.55	3.137		24	116		140	214
1800	Expansion joint, 3/4" grounds, limited expansion, galv., 1 piece		2.70	2.963		75.50	109		184.50	260
2100	Extreme expansion, galvanized, 2 piece		2.60	3.077		141	114		255	340

09 23 Gypsum Plastering

09 23 13 – Acoustical Gypsum Plastering

09 23 13.10 Perlite or Vermiculite Plaster

Code	Description	Crew	Daily Output	Labor-Hours	Unit	Material	Labor	Equipment	Total	Total Incl O&P
0010	**PERLITE OR VERMICULITE PLASTER** R092000-50									
0020	In 100 lb. bags, under 200 bags				Bag	17.55			17.55	19.35
0300	2 coats, no lath included, on walls	J-1	92	.435	S.Y.	5.85	14.30	1.53	21.68	31.50
0400	On ceilings		79	.506		5.85	16.65	1.78	24.28	35.50
0900	3 coats, no lath included, on walls		74	.541		6.40	17.75	1.90	26.05	38
1000	On ceilings		63	.635		6.40	21	2.23	29.63	43.50
1700	For irregular or curved surfaces, add to above						30%			
1800	For columns and beams, add to above						50%			
1900	For soffits, add to ceiling prices						40%			

09 23 20 – Gypsum Plaster

09 23 20.10 Gypsum Plaster On Walls and Ceilings

Code	Description	Crew	Daily Output	Labor-Hours	Unit	Material	Labor	Equipment	Total	Total Incl O&P
0010	**GYPSUM PLASTER ON WALLS AND CEILINGS** R092000-50									
0020	80# bag, less than 1 ton				Bag	15.95			15.95	17.55
0300	2 coats, no lath included, on walls	J-1	105	.381	S.Y.	3.70	12.50	1.34	17.54	26
0400	On ceilings		92	.435		3.70	14.30	1.53	19.53	29.50
0900	3 coats, no lath included, on walls		87	.460		5.35	15.10	1.61	22.06	32
1000	On ceilings		78	.513		5.35	16.85	1.80	24	35.50
1600	For irregular or curved surfaces, add						30%			
1800	For columns & beams, add						50%			

09 24 Cement Plastering

09 24 23 – Cement Stucco

09 24 23.40 Stucco

Code	Description	Crew	Daily Output	Labor-Hours	Unit	Material	Labor	Equipment	Total	Total Incl O&P
0010	**STUCCO** R092000-50									
0015	3 coats 1" thick, float finish, with mesh, on wood frame	J-2	63	.762	S.Y.	6.40	25.50	2.22	34.12	51
0100	On masonry construction, no mesh incl.	J-1	67	.597		2.55	19.60	2.10	24.25	37
0150	2 coats, 3/4" thick, float finish, no lath incl.	"	110	.364		2.30	11.95	1.28	15.53	23.50
0300	For trowel finish, add	1 Plas	170	.047			1.65		1.65	2.69
0600	For coloring, add	J-1	685	.058		.40	1.92	.20	2.52	3.81
0700	For special texture, add		200	.200		1.41	6.55	.70	8.66	13.05
1000	Exterior stucco, with bonding agent, 3 coats, on walls, no mesh incl.		200	.200		3.65	6.55	.70	10.90	15.55
1200	Ceilings		180	.222		3.65	7.30	.78	11.73	16.85

09 24 Cement Plastering

09 24 23 – Cement Stucco

09 24 23.40 Stucco		Crew	Daily Output	Labor-Hours	Unit	2016 Bare Costs Material	Labor	Equipment	Total	Total Incl O&P
1300	Beams	J-1	80	.500	S.Y.	3.65	16.45	1.76	21.86	33
1500	Columns		100	.400		3.65	13.15	1.40	18.20	27
1600	Mesh, painted, nailed to wood, 1.8 lb.	1 Lath	60	.133		6.50	4.92		11.42	15.10
1800	3.6 lb.		55	.145		3.85	5.35		9.20	12.95
1900	Wired to steel, painted, 1.8 lb.		53	.151		6.50	5.55		12.05	16.15
2100	3.6 lb.		50	.160		3.85	5.90		9.75	13.80

09 25 Other Plastering

09 25 23 – Lime Based Plastering

09 25 23.10 Venetian Plaster

Code	Description	Crew	Daily Output	Labor-Hours	Unit	Material	Labor	Equipment	Total	Total Incl O&P
0010	**VENETIAN PLASTER**									
0100	Walls, 1 coat primer, roller applied	1 Plas	950	.008	S.F.	.16	.29		.45	.66
0200	Plaster, 3 coats, incl. sanding	2 Plas	700	.023		.46	.80		1.26	1.81
0210	For pigment, light colors add per S.F. plaster					.02			.02	.02
0220	For pigment, dark colors add per S.F. plaster					.05			.05	.06
0300	For sealer/wax coat incl. burnishing, add	1 Plas	300	.027		.41	.93		1.34	1.98

09 26 Veneer Plastering

09 26 13 – Gypsum Veneer Plastering

09 26 13.20 Blueboard

Code	Description	Crew	Daily Output	Labor-Hours	Unit	Material	Labor	Equipment	Total	Total Incl O&P
0010	**BLUEBOARD** For use with thin coat									
0100	plaster application see Section 09 26 13.80									
1000	3/8" thick, on walls or ceilings, standard, no finish included	2 Carp	1900	.008	S.F.	.35	.32		.67	.92
1100	With thin coat plaster finish		875	.018		.46	.69		1.15	1.67
1400	On beams, columns, or soffits, standard, no finish included		675	.024		.40	.90		1.30	1.94
1450	With thin coat plaster finish		475	.034		.51	1.27		1.78	2.70
3000	1/2" thick, on walls or ceilings, standard, no finish included		1900	.008		.34	.32		.66	.90
3100	With thin coat plaster finish		875	.018		.45	.69		1.14	1.66
3300	Fire resistant, no finish included		1900	.008		.34	.32		.66	.90
3400	With thin coat plaster finish		875	.018		.45	.69		1.14	1.66
3450	On beams, columns, or soffits, standard, no finish included		675	.024		.39	.90		1.29	1.93
3500	With thin coat plaster finish		475	.034		.50	1.27		1.77	2.68
3700	Fire resistant, no finish included		675	.024		.39	.90		1.29	1.93
3800	With thin coat plaster finish		475	.034		.50	1.27		1.77	2.68
5000	5/8" thick, on walls or ceilings, fire resistant, no finish included		1900	.008		.34	.32		.66	.90
5100	With thin coat plaster finish		875	.018		.45	.69		1.14	1.66
5500	On beams, columns, or soffits, no finish included		675	.024		.39	.90		1.29	1.93
5600	With thin coat plaster finish		475	.034		.50	1.27		1.77	2.68
6000	For high ceilings, over 8' high, add		3060	.005			.20		.20	.33
6500	For over 3 stories high, add per story		6100	.003			.10		.10	.17

09 26 13.80 Thin Coat Plaster

Code	Description	Crew	Daily Output	Labor-Hours	Unit	Material	Labor	Equipment	Total	Total Incl O&P
0010	**THIN COAT PLASTER** R092000-50									
0012	1 coat veneer, not incl. lath	J-1	3600	.011	S.F.	.11	.37	.04	.52	.76
1000	In 50 lb. bags				Bag	15.05			15.05	16.55

09 29 Gypsum Board

09 29 10 – Gypsum Board Panels

09 29 10.30 Gypsum Board		Crew	Daily Output	Labor-Hours	Unit	Material	Labor	Equipment	Total	Total Incl O&P
						2016 Bare Costs				
0010	**GYPSUM BOARD** on walls & ceilings R092910-10									
0100	Nailed or screwed to studs unless otherwise noted									
0110	1/4" thick, on walls or ceilings, standard, no finish included	2 Carp	1330	.012	S.F.	.36	.45		.81	1.16
0115	1/4" thick, on walls or ceilings, flexible, no finish included		1050	.015		.51	.58		1.09	1.52
0117	1/4" thick, on columns or soffits, flexible, no finish included		1050	.015		.51	.58		1.09	1.52
0130	1/4" thick, standard, no finish included, less than 800 S.F.		510	.031		.36	1.19		1.55	2.39
0150	3/8" thick, on walls, standard, no finish included		2000	.008		.35	.30		.65	.90
0200	On ceilings, standard, no finish included		1800	.009		.35	.34		.69	.95
0250	On beams, columns, or soffits, no finish included		675	.024		.35	.90		1.25	1.89
0300	1/2" thick, on walls, standard, no finish included		2000	.008		.33	.30		.63	.87
0350	Taped and finished (level 4 finish)		965	.017		.38	.63		1.01	1.47
0390	With compound skim coat (level 5 finish)		775	.021		.43	.78		1.21	1.78
0400	Fire resistant, no finish included		2000	.008		.36	.30		.66	.91
0450	Taped and finished (level 4 finish)		965	.017		.41	.63		1.04	1.50
0490	With compound skim coat (level 5 finish)		775	.021		.46	.78		1.24	1.82
0500	Water resistant, no finish included		2000	.008		.41	.30		.71	.96
0550	Taped and finished (level 4 finish)		965	.017		.46	.63		1.09	1.55
0590	With compound skim coat (level 5 finish)		775	.021		.51	.78		1.29	1.87
0600	Prefinished, vinyl, clipped to studs		900	.018		.49	.67		1.16	1.67
0700	Mold resistant, no finish included		2000	.008		.43	.30		.73	.98
0710	Taped and finished (level 4 finish)		965	.017		.48	.63		1.11	1.58
0720	With compound skim coat (level 5 finish)		775	.021		.53	.78		1.31	1.89
1000	On ceilings, standard, no finish included		1800	.009		.33	.34		.67	.92
1050	Taped and finished (level 4 finish)		765	.021		.38	.79		1.17	1.74
1090	With compound skim coat (level 5 finish)		610	.026		.43	.99		1.42	2.13
1100	Fire resistant, no finish included		1800	.009		.36	.34		.70	.96
1150	Taped and finished (level 4 finish)		765	.021		.41	.79		1.20	1.77
1195	With compound skim coat (level 5 finish)		610	.026		.46	.99		1.45	2.17
1200	Water resistant, no finish included		1800	.009		.41	.34		.75	1.01
1250	Taped and finished (level 4 finish)		765	.021		.46	.79		1.25	1.82
1290	With compound skim coat (level 5 finish)		610	.026		.51	.99		1.50	2.22
1310	Mold resistant, no finish included		1800	.009		.43	.34		.77	1.03
1320	Taped and finished (level 4 finish)		765	.021		.48	.79		1.27	1.85
1330	With compound skim coat (level 5 finish)		610	.026		.53	.99		1.52	2.24
1350	Sag resistant, no finish included		1600	.010		.35	.38		.73	1.02
1360	Taped and finished (level 4 finish)		765	.021		.40	.79		1.19	1.76
1370	With compound skim coat (level 5 finish)		610	.026		.45	.99		1.44	2.16
1500	On beams, columns, or soffits, standard, no finish included		675	.024		.38	.90		1.28	1.92
1550	Taped and finished (level 4 finish)		540	.030		.38	1.12		1.50	2.30
1590	With compound skim coat (level 5 finish)		475	.034		.43	1.27		1.70	2.60
1600	Fire resistant, no finish included		675	.024		.36	.90		1.26	1.90
1650	Taped and finished (level 4 finish)		540	.030		.41	1.12		1.53	2.33
1690	With compound skim coat (level 5 finish)		475	.034		.46	1.27		1.73	2.64
1700	Water resistant, no finish included		675	.024		.47	.90		1.37	2.02
1750	Taped and finished (level 4 finish)		540	.030		.46	1.12		1.58	2.38
1790	With compound skim coat (level 5 finish)		475	.034		.51	1.27		1.78	2.69
1800	Mold resistant, no finish included		675	.024		.49	.90		1.39	2.04
1810	Taped and finished (level 4 finish)		540	.030		.48	1.12		1.60	2.41
1820	With compound skim coat (level 5 finish)		475	.034		.53	1.27		1.80	2.71
1850	Sag resistant, no finish included		675	.024		.40	.90		1.30	1.94
1860	Taped and finished (level 4 finish)		540	.030		.40	1.12		1.52	2.32
1870	With compound skim coat (level 5 finish)		475	.034		.45	1.27		1.72	2.63
2000	5/8" thick, on walls, standard, no finish included	↓	2000	.008	↓	.34	.30		.64	.88

09 29 10.30 Gypsum Board		Crew	Daily Output	Labor-Hours	Unit	Material	2016 Bare Costs Labor	Equipment	Total	Total Incl O&P
2050	Taped and finished (level 4 finish)	2 Carp	965	.017	S.F.	.39	.63		1.02	1.48
2090	With compound skim coat (level 5 finish)		775	.021		.44	.78		1.22	1.79
2100	Fire resistant, no finish included		2000	.008		.35	.30		.65	.90
2150	Taped and finished (level 4 finish)		965	.017		.40	.63		1.03	1.49
2195	With compound skim coat (level 5 finish)		775	.021		.45	.78		1.23	1.81
2200	Water resistant, no finish included		2000	.008		.43	.30		.73	.98
2250	Taped and finished (level 4 finish)		965	.017		.48	.63		1.11	1.58
2290	With compound skim coat (level 5 finish)		775	.021		.53	.78		1.31	1.89
2300	Prefinished, vinyl, clipped to studs		900	.018		.76	.67		1.43	1.97
2510	Mold resistant, no finish included		2000	.008		.47	.30		.77	1.03
2520	Taped and finished (level 4 finish)		965	.017		.52	.63		1.15	1.62
2530	With compound skim coat (level 5 finish)		775	.021		.57	.78		1.35	1.94
3000	On ceilings, standard, no finish included		1800	.009		.34	.34		.68	.93
3050	Taped and finished (level 4 finish)		765	.021		.39	.79		1.18	1.75
3090	With compound skim coat (level 5 finish)		615	.026		.44	.98		1.42	2.13
3100	Fire resistant, no finish included		1800	.009		.35	.34		.69	.95
3150	Taped and finished (level 4 finish)		765	.021		.40	.79		1.19	1.76
3190	With compound skim coat (level 5 finish)		615	.026		.45	.98		1.43	2.15
3200	Water resistant, no finish included		1800	.009		.43	.34		.77	1.03
3250	Taped and finished (level 4 finish)		765	.021		.48	.79		1.27	1.85
3290	With compound skim coat (level 5 finish)		615	.026		.53	.98		1.51	2.23
3300	Mold resistant, no finish included		1000	.009		.47	.34		.81	1.08
3310	Taped and finished (level 4 finish)		765	.021		.52	.79		1.31	1.89
3320	With compound skim coat (level 5 finish)		615	.026		.57	.98		1.55	2.28
3500	On beams, columns, or soffits, no finish included		675	.024		.39	.90		1.29	1.93
3550	Taped and finished (level 4 finish)		475	.034		.45	1.27		1.72	2.62
3590	With compound skim coat (level 5 finish)		380	.042		.51	1.59		2.10	3.23
3600	Fire resistant, no finish included		675	.024		.40	.90		1.30	1.94
3650	Taped and finished (level 4 finish)		475	.034		.46	1.27		1.73	2.63
3690	With compound skim coat (level 5 finish)		380	.042		.45	1.59		2.04	3.17
3700	Water resistant, no finish included		675	.024		.49	.90		1.39	2.04
3750	Taped and finished (level 4 finish)		475	.034		.53	1.27		1.80	2.71
3790	With compound skim coat (level 5 finish)		380	.042		.55	1.59		2.14	3.27
3800	Mold resistant, no finish included		675	.024		.54	.90		1.44	2.09
3810	Taped and finished (level 4 finish)		475	.034		.57	1.27		1.84	2.76
3820	With compound skim coat (level 5 finish)		380	.042		.59	1.59		2.18	3.32
4000	Fireproofing, beams or columns, 2 layers, 1/2" thick, incl finish		330	.048		.81	1.83		2.64	3.97
4010	Mold resistant		330	.048		.95	1.83		2.78	4.12
4050	5/8" thick		300	.053		.79	2.02		2.81	4.25
4060	Mold resistant		300	.053		1.03	2.02		3.05	4.52
4100	3 layers, 1/2" thick		225	.071		1.22	2.69		3.91	5.85
4110	Mold resistant		225	.071		1.43	2.69		4.12	6.10
4150	5/8" thick		210	.076		1.19	2.88		4.07	6.15
4160	Mold resistant		210	.076		1.55	2.88		4.43	6.55
5200	For work over 8' high, add		3060	.005			.20		.20	.33
5270	For textured spray, add	2 Lath	1600	.010		.04	.37		.41	.64
5300	For distribution cost over 3 stories high, add per story	2 Carp	6100	.003			.10		.10	.17
5350	For finishing inner corners, add		950	.017	L.F.	.10	.64		.74	1.18
5355	For finishing outer corners, add		1250	.013		.23	.48		.71	1.07
5500	For acoustical sealant, add per bead	1 Carp	500	.016		.04	.60		.64	1.06
5550	Sealant, 1 quart tube				Ea.	7.05			7.05	7.80
6000	Gypsum sound dampening panels									
6010	1/2" thick on walls, multi-layer, light weight, no finish included	2 Carp	1500	.011	S.F.	2.02	.40		2.42	2.90

09 29 10 – Gypsum Board Panels

09 29 10.30 Gypsum Board		Crew	Daily Output	Labor-Hours	Unit	Material	Labor	Equipment	Total	Total Incl O&P
						2016 Bare Costs				
6015	Taped and finished (level 4 finish)	2 Carp	725	.022	S.F.	2.07	.83		2.90	3.67
6020	With compound skim coat (level 5 finish)		580	.028		2.12	1.04		3.16	4.08
6025	5/8" thick on walls, for wood studs, no finish included		1500	.011		2.23	.40		2.63	3.13
6030	Taped and finished (level 4 finish)		725	.022		2.28	.83		3.11	3.91
6035	With compound skim coat (level 5 finish)		580	.028		2.33	1.04		3.37	4.31
6040	For metal stud, no finish included		1500	.011		2.18	.40		2.58	3.08
6045	Taped and finished (level 4 finish)		725	.022		2.23	.83		3.06	3.85
6050	With compound skim coat (level 5 finish)		580	.028		2.28	1.04		3.32	4.26
6055	Abuse resist, no finish included		1500	.011		3.95	.40		4.35	5.05
6060	Taped and finished (level 4 finish)		725	.022		4	.83		4.83	5.80
6065	With compound skim coat (level 5 finish)		580	.028		4.05	1.04		5.09	6.20
6070	Shear rated, no finish included		1500	.011		4.53	.40		4.93	5.65
6075	Taped and finished (level 4 finish)		725	.022		4.58	.83		5.41	6.45
6080	With compound skim coat (level 5 finish)		580	.028		4.63	1.04		5.67	6.85
6085	For SCIF applications, no finish included		1500	.011		4.96	.40		5.36	6.15
6090	Taped and finished (level 4 finish)		725	.022		5	.83		5.83	6.90
6095	With compound skim coat (level 5 finish)		580	.028		5.05	1.04		6.09	7.30
6100	1-3/8" thick on walls, THX Certified, no finish included		1500	.011		8.40	.40		8.80	9.95
6105	Taped and finished (level 4 finish)		725	.022		8.45	.83		9.28	10.70
6110	With compound skim coat (level 5 finish)		580	.028		8.50	1.04		9.54	11.10
6115	5/8" thick on walls, score & snap installation, no finish included		2000	.008		1.79	.30		2.09	2.48
6120	Taped and finished (level 4 finish)		965	.017		1.84	.63		2.47	3.07
6125	With compound skim coat (level 5 finish)		775	.021		1.89	.78		2.67	3.39
7020	5/8" thick on ceilings, for wood joists, no finish included		1200	.013		2.23	.50		2.73	3.29
7025	Taped and finished (level 4 finish)		510	.031		2.28	1.19		3.47	4.50
7030	With compound skim coat (level 5 finish)		410	.039		2.33	1.48		3.81	5.05
7035	For metal joists, no finish included		1200	.013		2.18	.50		2.68	3.24
7040	Taped and finished (level 4 finish)		510	.031		2.23	1.19		3.42	4.44
7045	With compound skim coat (level 5 finish)		410	.039		2.28	1.48		3.76	4.98
7050	Abuse resist, no finish included		1200	.013		3.95	.50		4.45	5.20
7055	Taped and finished (level 4 finish)		510	.031		4	1.19		5.19	6.40
7060	With compound skim coat (level 5 finish)		410	.039		4.05	1.48		5.53	6.95
7065	Shear rated, no finish included		1200	.013		4.53	.50		5.03	5.80
7070	Taped and finished (level 4 finish)		510	.031		4.58	1.19		5.77	7.05
7075	With compound skim coat (level 5 finish)		410	.039		4.63	1.48		6.11	7.55
7080	For SCIF applications, no finish included		1200	.013		4.96	.50		5.46	6.30
7085	Taped and finished (level 4 finish)		510	.031		5	1.19		6.19	7.50
7090	With compound skim coat (level 5 finish)		410	.039		5.05	1.48		6.53	8
8010	5/8" thick on ceilings, score & snap installation, no finish included		1600	.010		1.79	.38		2.17	2.60
8015	Taped and finished (level 4 finish)		680	.024		1.84	.89		2.73	3.51
8020	With compound skim coat (level 5 finish)		545	.029		1.89	1.11		3	3.94

09 29 10.50 High Abuse Gypsum Board		Crew	Daily Output	Labor-Hours	Unit	Material	Labor	Equipment	Total	Total Incl O&P
0010	**HIGH ABUSE GYPSUM BOARD**, fiber reinforced, nailed or									
0100	screwed to studs unless otherwise noted									
0110	1/2" thick, on walls, no finish included	2 Carp	1800	.009	S.F.	.72	.34		1.06	1.35
0120	Taped and finished (level 4 finish)		870	.018		.77	.70		1.47	2
0130	With compound skim coat (level 5 finish)		700	.023		.82	.86		1.68	2.35
0150	On ceilings, no finish included		1620	.010		.72	.37		1.09	1.42
0160	Taped and finished (level 4 finish)		690	.023		.77	.88		1.65	2.31
0170	With compound skim coat (level 5 finish)		550	.029		.82	1.10		1.92	2.74
0210	5/8" thick, on walls, no finish included		1800	.009		.82	.34		1.16	1.46
0220	Taped and finished (level 4 finish)		870	.018		.87	.70		1.57	2.11

09 29 Gypsum Board

09 29 10 – Gypsum Board Panels

09 29 10.50 High Abuse Gypsum Board		Crew	Daily Output	Labor-Hours	Unit	2016 Bare Costs Material	Labor	Equipment	Total	Total Incl O&P
0230	With compound skim coat (level 5 finish)	2 Carp	700	.023	S.F.	.92	.86		1.78	2.46
0250	On ceilings, no finish included		1620	.010		.82	.37		1.19	1.53
0260	Taped and finished (level 4 finish)		690	.023		.87	.88		1.75	2.42
0270	With compound skim coat (level 5 finish)		550	.029		.92	1.10		2.02	2.85
0310	5/8" thick, on walls, very high impact, no finish included		1800	.009		1	.34		1.34	1.66
0320	Taped and finished (level 4 finish)		870	.018		1.05	.70		1.75	2.31
0330	With compound skim coat (level 5 finish)		700	.023		1.10	.86		1.96	2.66
0350	On ceilings, no finish included		1620	.010		1	.37		1.37	1.73
0360	Taped and finished (level 4 finish)		690	.023		1.05	.88		1.93	2.62
0370	With compound skim coat (level 5 finish)		550	.029		1.10	1.10		2.20	3.05
0400	High abuse, gypsum core, paper face									
0410	1/2" thick, on walls, no finish included	2 Carp	1800	.009	S.F.	.63	.34		.97	1.25
0420	Taped and finished (level 4 finish)		870	.018		.68	.70		1.38	1.91
0430	With compound skim coat (level 5 finish)		700	.023		.73	.86		1.59	2.25
0450	On ceilings, no finish included		1620	.010		.63	.37		1	1.32
0460	Taped and finished (level 4 finish)		690	.023		.68	.88		1.56	2.22
0470	With compound skim coat (level 5 finish)		550	.029		.73	1.10		1.83	2.64
0510	5/8" thick, on walls, no finish included		1800	.009		.67	.34		1.01	1.30
0520	Taped and finished (level 4 finish)		870	.018		.72	.70		1.42	1.95
0530	With compound skim coat (level 5 finish)		700	.023		.77	.86		1.63	2.30
0550	On ceilings, no finish included		1620	.010		.67	.37		1.04	1.37
0560	Taped and finished (level 4 finish)		690	.023		.72	.88		1.60	2.26
0570	With compound skim coat (level 5 finish)		550	.029		.77	1.10		1.87	2.69
1000	For high ceilings, over 8' high, add		2750	.006			.22		.22	.37
1010	For distribution cost over 3 stories high, add per story		5500	.003			.11		.11	.18

09 29 15 – Gypsum Board Accessories

09 29 15.10 Accessories, Gypsum Board		Crew	Daily Output	Labor-Hours	Unit	Material	Labor	Equipment	Total	Total Incl O&P
0010	**ACCESSORIES, GYPSUM BOARD**									
0020	Casing bead, galvanized steel	1 Carp	2.90	2.759	C.L.F.	24	104		128	202
0100	Vinyl		3	2.667		22.50	101		123.50	194
0300	Corner bead, galvanized steel, 1" x 1"		4	2		15.45	75.50		90.95	144
0400	1-1/4" x 1-1/4"		3.50	2.286		16.60	86.50		103.10	163
0600	Vinyl		4	2		20.50	75.50		96	150
0900	Furring channel, galv. steel, 7/8" deep, standard		2.60	3.077		33.50	116		149.50	232
1000	Resilient		2.55	3.137		25	119		144	227
1100	J trim, galvanized steel, 1/2" wide		3	2.667		22	101		123	193
1120	5/8" wide		2.95	2.712		32	103		135	207
1160	Screws #6 x 1" A				M	10.25			10.25	11.30
1170	#6 x 1-5/8" A				"	15.10			15.10	16.60
1500	Z stud, galvanized steel, 1-1/2" wide	1 Carp	2.60	3.077	C.L.F.	39.50	116		155.50	239

09 30 13 – Ceramic Tiling

09 30 13.10 Ceramic Tile		Crew	Daily Output	Labor-Hours	Unit	Material	Labor	Equipment	Total	Total Incl O&P
						2016 Bare Costs				
0010	**CERAMIC TILE**									
0020	Backsplash, thinset, average grade tiles	1 Tilf	50	.160	S.F.	2.70	5.55		8.25	11.95
0022	Custom grade tiles		50	.160		5.40	5.55		10.95	14.95
0024	Luxury grade tiles		50	.160		10.80	5.55		16.35	21
0026	Economy grade tiles		50	.160		2.46	5.55		8.01	11.70
0050	Base, using 1′ x 4″ high pc. with 1″ x 1″ tiles, mud set	D-7	82	.195	L.F.	5.10	6.05		11.15	15.40
0100	Thin set	"	128	.125		4.84	3.87		8.71	11.55
0300	For 6″ high base, 1″ x 1″ tile face, add					.77			.77	.85
0400	For 2″ x 2″ tile face, add to above					.42			.42	.46
0600	Cove base, 4-1/4″ x 4-1/4″ high, mud set	D-7	91	.176		4.19	5.45		9.64	13.40
0700	Thin set		128	.125		4.07	3.87		7.94	10.75
0900	6″ x 4-1/4″ high, mud set		100	.160		4.37	4.95		9.32	12.80
1000	Thin set		137	.117		4.25	3.61		7.86	10.50
1200	Sanitary cove base, 6″ x 4-1/4″ high, mud set		93	.172		4.67	5.30		9.97	13.75
1300	Thin set		124	.129		4.55	3.99		8.54	11.45
1500	6″ x 6″ high, mud set		84	.190		5.45	5.90		11.35	15.55
1600	Thin set		117	.137		5.30	4.23		9.53	12.70
1800	Bathroom accessories, average (soap dish, tooth brush holder)		82	.195	Ea.	9.55	6.05		15.60	20.50
1900	Bathtub, 5′, rec. 4-1/4″ x 4-1/4″ tile wainscot, adhesive set 6′ high		2.90	5.517		174	171		345	465
2100	7′ high wainscot		2.50	6.400		198	198		396	540
2200	8′ high wainscot		2.20	7.273		211	225		436	595
2400	Bullnose trim, 4-1/4″ x 4-1/4″, mud set		82	.195	L.F.	4.15	6.05		10.20	14.35
2500	Thin set		128	.125		4.07	3.87		7.94	10.70
2700	2″ x 6″ bullnose trim, mud set		84	.190		4.38	5.90		10.28	14.35
2800	Thin set		124	.129		4.32	3.99		8.31	11.20
3000	Floors, natural clay, random or uniform, thin set, color group 1		183	.087	S.F.	4.15	2.71		6.86	8.95
3100	Color group 2		183	.087		5.75	2.71		8.46	10.75
3255	Floors, glazed, thin set, 6″ x 6″, color group 1		300	.053		4.66	1.65		6.31	7.75
3260	8″ x 8″ tile		300	.053		4.66	1.65		6.31	7.75
3270	12″ x 12″ tile		290	.055		6.05	1.71		7.76	9.45
3280	16″ x 16″ tile		280	.057		7.45	1.77		9.22	11
3281	18″ x 18″ tile		270	.059		6.80	1.83		8.63	10.45
3282	20″ x 20″ tile		260	.062		9.90	1.90		11.80	14
3283	24″ x 24″ tile		250	.064		11.25	1.98		13.23	15.55
3285	Border, 6″ x 12″ tile		200	.080		11.50	2.48		13.98	16.65
3290	3″ x 12″ tile		200	.080		40	2.48		42.48	48
3300	Porcelain type, 1 color, color group 2, 1″ x 1″		183	.087		5.20	2.71		7.91	10.10
3310	2″ x 2″ or 2″ x 1″, thin set		190	.084		6.40	2.61		9.01	11.25
3350	For random blend, 2 colors, add					1			1	1.10
3360	4 colors, add					1.50			1.50	1.65
4300	Specialty tile, 4-1/4″ x 4-1/4″ x 1/2″, decorator finish	D-7	183	.087		10.40	2.71		13.11	15.85
4500	Add for epoxy grout, 1/16″ joint, 1″ x 1″ tile		800	.020		.67	.62		1.29	1.74
4600	2″ x 2″ tile		820	.020		.62	.60		1.22	1.66
4610	Add for epoxy grout, 1/8″ joint, 8″ x 8″ x 3/8″ tile, add		900	.018		1.39	.55		1.94	2.42
4800	Pregrouted sheets, walls, 4-1/4″ x 4-1/4″, 6″ x 4-1/4″									
4810	and 8-1/2″ x 4-1/4″, 4 S.F. sheets, silicone grout	D-7	240	.067	S.F.	5.40	2.06		7.46	9.30
5100	Floors, unglazed, 2 S.F. sheets,									
5110	urethane adhesive	D-7	180	.089	S.F.	5.10	2.75		7.85	10.05
5400	Walls, interior, thin set, 4-1/4″ x 4-1/4″ tile		190	.084		2.49	2.61		5.10	6.95
5500	6″ x 4-1/4″ tile		190	.084		3.01	2.61		5.62	7.55
5700	8-1/2″ x 4-1/4″ tile		190	.084		5.25	2.61		7.86	10
5800	6″ x 6″ tile		175	.091		3.46	2.83		6.29	8.40
5810	8″ x 8″ tile		170	.094		4.75	2.91		7.66	9.90

09 30 Tiling

09 30 13 – Ceramic Tiling

09 30 13.10 Ceramic Tile		Crew	Daily Output	Labor-Hours	Unit	2016 Bare Costs Material	Labor	Equipment	Total	Total Incl O&P
5820	12" x 12" tile	D-7	160	.100	S.F.	4.55	3.10		7.65	10
5830	16" x 16" tile		150	.107		5.05	3.30		8.35	10.95
6000	Decorated wall tile, 4-1/4" x 4-1/4", color group 1		270	.059		3.53	1.83		5.36	6.85
6100	Color group 4		180	.089		52	2.75		54.75	61.50
6600	Crystalline glazed, 4-1/4" x 4-1/4", mud set, plain		100	.160		4.52	4.95		9.47	12.95
6700	4-1/4" x 4-1/4", scored tile		100	.160		6	4.95		10.95	14.60
6900	6" x 6" plain		93	.172		6.80	5.30		12.10	16.10
7000	For epoxy grout, 1/16" joints, 4-1/4" tile, add		800	.020		.41	.62		1.03	1.45
7200	For tile set in dry mortar, add		1735	.009			.29		.29	.46
7300	For tile set in Portland cement mortar, add		290	.055		.16	1.71		1.87	2.94
9300	Ceramic tiles, recycled glass, standard colors, 2" x 2" thru 6" x 6" G		190	.084		21.50	2.61		24.11	27.50
9310	6" x 6" G		175	.091		21.50	2.83		24.33	28.50
9320	8" x 8" G		170	.094		23	2.91		25.91	30
9330	12" x 12" G		160	.100		23	3.10		26.10	30.50
9340	Earthtones, 2" x 2" to 4" x 8" G		190	.084		25.50	2.61		28.11	32
9350	6" x 6" G		175	.091		25.50	2.83		28.33	32.50
9360	8" x 8" G		170	.094		26.50	2.91		29.41	33.50
9370	12" x 12" G		160	.100		26.50	3.10		29.60	34
9380	Deep colors, 2" x 2" to 4" x 8" G		190	.084		30	2.61		32.61	37
9390	6" x 6" G		175	.091		30	2.83		32.83	37.50
9400	8" x 8" G		170	.094		31.50	2.91		34.41	39
9410	12" x 12" G		160	.100		31.50	3.10		34.60	39.50

09 30 13.45 Ceramic Tile Accessories		Crew	Daily Output	Labor-Hours	Unit	Material	Labor	Equipment	Total	Total Incl O&P
0010	**CERAMIC TILE ACCESSORIES**									
0100	Spacers, 1/8"				C	1.98			1.98	2.18
1310	Sealer for natural stone tile, installed	1 Tilf	650	.012	S.F.	.05	.43		.48	.75

09 30 16 – Quarry Tiling

09 30 16.10 Quarry Tile		Crew	Daily Output	Labor-Hours	Unit	Material	Labor	Equipment	Total	Total Incl O&P
0010	**QUARRY TILE**									
0100	Base, cove or sanitary, mud set, to 5" high, 1/2" thick	D-7	110	.145	L.F.	5.85	4.50		10.35	13.75
0300	Bullnose trim, red, mud set, 6" x 6" x 1/2" thick		120	.133		4.70	4.13		8.83	11.85
0400	4" x 4" x 1/2" thick		110	.145		4.83	4.50		9.33	12.60
0600	4" x 8" x 1/2" thick, using 8" as edge		130	.123		4.79	3.81		8.60	11.40
0700	Floors, mud set, 1,000 S.F. lots, red, 4" x 4" x 1/2" thick		120	.133	S.F.	8.05	4.13		12.18	15.55
0900	6" x 6" x 1/2" thick		140	.114		7.95	3.54		11.49	14.50
1000	4" x 8" x 1/2" thick		130	.123		5	3.81		8.81	11.65
1300	For waxed coating, add					.76			.76	.84
1500	For non-standard colors, add					.46			.46	.51
1600	For abrasive surface, add					.52			.52	.57
1800	Brown tile, imported, 6" x 6" x 3/4"	D-7	120	.133		7.40	4.13		11.53	14.85
1900	8" x 8" x 1"		110	.145		8.70	4.50		13.20	16.85
2100	For thin set mortar application, deduct		700	.023			.71		.71	1.15
2700	Stair tread, 6" x 6" x 3/4", plain		50	.320		7.40	9.90		17.30	24
2800	Abrasive		47	.340		6.80	10.55		17.35	24.50
3000	Wainscot, 6" x 6" x 1/2", thin set, red		105	.152		5.55	4.72		10.27	13.75
3100	Non-standard colors		105	.152		5.50	4.72		10.22	13.70
3300	Window sill, 6" wide, 3/4" thick		90	.178	L.F.	5.45	5.50		10.95	14.90
3400	Corners		80	.200	Ea.	6.10	6.20		12.30	16.75

09 30 Tiling

09 30 23 – Glass Mosaic Tiling

09 30 23.10 Glass Mosaics		Crew	Daily Output	Labor-Hours	Unit	2016 Bare Costs Material	Labor	Equipment	Total	Total Incl O&P
0010	**GLASS MOSAICS** 3/4" tile on 12" sheets, standard grout									
1020	1" tile on 12" sheets, opalescent finish	D-7	73	.219	S.F.	17.50	6.80		24.30	30.50
1040	1" x 2" tile on 12" sheet, blend		73	.219		19.05	6.80		25.85	32
1060	2" tile on 12" sheet, blend		73	.219		16.50	6.80		23.30	29
1080	5/8" x random tile, linear, on 12" sheet, blend		73	.219		25	6.80		31.80	38.50
1600	Dots on 12" sheet		73	.219		25	6.80		31.80	38.50
1700	For glass mosaic tiles set in dry mortar, add		290	.055		.45	1.71		2.16	3.26
1720	For glass mosaic tile set in Portland cement mortar, add		290	.055		.01	1.71		1.72	2.77
1730	For polyblend sanded tile grout		96.15	.166	Lb.	2.19	5.15		7.34	10.75

09 30 29 – Metal Tiling

09 30 29.10 Metal Tile		Crew	Daily Output	Labor-Hours	Unit	Material	Labor	Equipment	Total	Total Incl O&P
0010	**METAL TILE** 4' x 4' sheet, 24 ga., tile pattern, nailed									
0200	Stainless steel	2 Carp	512	.031	S.F.	28	1.18		29.18	33
0400	Aluminized steel	"	512	.031	"	15.10	1.18		16.28	18.60

09 30 95 – Tile & Stone Setting Materials and Specialties

09 30 95.10 Moisture Resistant, Anti-Fracture Membrane		Crew	Daily Output	Labor-Hours	Unit	Material	Labor	Equipment	Total	Total Incl O&P
0010	**MOISTURE RESISTANT, ANTI-FRACTURE MEMBRANE**									
0200	Elastomeric membrane, 1/16" thick	D-7	275	.058	S.F.	1.07	1.80		2.87	4.09

09 34 Waterproofing-Membrane Tiling

09 34 13 – Waterproofing-Membrane Ceramic Tiling

09 34 13.10 Ceramic Tile Waterproofing Membrane		Crew	Daily Output	Labor-Hours	Unit	Material	Labor	Equipment	Total	Total Incl O&P
0010	**CERAMIC TILE WATERPROOFING MEMBRANE**									
0020	On floors, including thinset									
0030	Fleece laminated polyethylene grid, 1/8" thick	D-7	250	.064	S.F.	2.28	1.98		4.26	5.70
0040	5/16" thick	"	250	.064	"	2.60	1.98		4.58	6.05
0050	On walls, including thinset									
0060	Fleece laminated polyethylene sheet, 8 mil thick	D-7	480	.033	S.F.	2.28	1.03		3.31	4.17
0070	Accessories, including thinset									
0080	Joint and corner sheet, 4 mils thick, 5" wide	1 Tilf	240	.033	L.F.	1.35	1.16		2.51	3.35
0090	7-1/4" wide		180	.044		1.71	1.54		3.25	4.38
0100	10" wide		120	.067		2.08	2.32		4.40	6.05
0110	Pre-formed corners, inside		32	.250	Ea.	6.95	8.70		15.65	21.50
0120	Outside		32	.250		7.65	8.70		16.35	22.50
0130	2" flanged floor drain with 6" stainless steel grate		16	.500		370	17.40		387.40	440
0140	EPS, sloped shower floor		480	.017	S.F.	4.97	.58		5.55	6.40
0150	Curb		32	.250	L.F.	14.05	8.70		22.75	29.50

09 51 Acoustical Ceilings

09 51 23 – Acoustical Tile Ceilings

09 51 23.10 Suspended Acoustic Ceiling Tiles

	09 51 23.10 Suspended Acoustic Ceiling Tiles	Crew	Daily Output	Labor-Hours	Unit	Material	2016 Bare Costs Labor	Equipment	Total	Total Incl O&P
0010	**SUSPENDED ACOUSTIC CEILING TILES**, not including									
0100	suspension system									
0300	Fiberglass boards, film faced, 2′ x 2′ or 2′ x 4′, 5/8″ thick	1 Carp	625	.013	S.F.	1.25	.48		1.73	2.19
0400	3/4″ thick		600	.013		2.63	.50		3.13	3.73
0500	3″ thick, thermal, R11		450	.018		2.62	.67		3.29	4.01
0600	Glass cloth faced fiberglass, 3/4″ thick		500	.016		2.72	.60		3.32	4
0700	1″ thick		485	.016		3.31	.62		3.93	4.68
0820	1-1/2″ thick, nubby face		475	.017		2.60	.64		3.24	3.93
1110	Mineral fiber tile, lay-in, 2′ x 2′ or 2′ x 4′, 5/8″ thick, fine texture		625	.013		.96	.48		1.44	1.87
1115	Rough textured		625	.013		.86	.48		1.34	1.76
1125	3/4″ thick, fine textured		600	.013		1.99	.50		2.49	3.03
1130	Rough textured		600	.013		1.60	.50		2.10	2.60
1135	Fissured		600	.013		1.85	.50		2.35	2.88
1150	Tegular, 5/8″ thick, fine textured		470	.017		1.04	.64		1.68	2.22
1155	Rough textured		470	.017		1.23	.64		1.87	2.43
1165	3/4″ thick, fine textured		450	.018		2.08	.67		2.75	3.42
1170	Rough textured		450	.018		1.41	.67		2.08	2.68
1175	Fissured		450	.018		2.02	.67		2.69	3.35
1185	For plastic film face, add					.79			.79	.87
1190	For fire rating, add					.45			.45	.50
1600	Solid alum. planks, 3-1/4″x12′, open reveal	1 Carp	500	.016		2.35	.60		2.95	3.60
1650	Closed reveal		500	.016		3	.60		3.60	4.31
1700	7-1/4″ x 12′, open reveal		500	.016		4	.60		4.60	5.40
1750	Closed reveal		500	.016		5.10	.60		5.70	6.60
1775	Metal, open cell, 2′ x 2′, 6″ cell		500	.016		8.25	.60		8.85	10.10
1800	8″ cell		500	.016		9.15	.60		9.75	11.05
1825	2′ x 4′, 6″ cell		500	.016		5.35	.60		5.95	6.90
1850	8″ cell		500	.016		5.35	.60		5.95	6.90
1870	Translucent lay-in panels, 2′ x 2′		500	.016		23	.60		23.60	26
1890	2′ x 6′		500	.016		17.20	.60		17.80	19.90
5020	66 – 78% recycled content, 3/4″ thick G		600	.013		2.03	.50		2.53	3.07
5040	Mylar, 42% recycled content, 3/4″ thick G		600	.013		4.77	.50		5.27	6.10
6000	Remove and replace ceiling tiles, min fiber, 2′ x 2′ or 2′ x 4′, 5/8″thk.		335	.024		.96	.90		1.86	2.57

09 51 23.30 Suspended Ceilings, Complete

	09 51 23.30 Suspended Ceilings, Complete	Crew	Daily Output	Labor-Hours	Unit	Material	Labor	Equipment	Total	Total Incl O&P
0010	**SUSPENDED CEILINGS, COMPLETE**, including standard									
0100	suspension system but not incl. 1-1/2″ carrier channels									
0600	Fiberglass ceiling board, 2′ x 4′ x 5/8″, plain faced	1 Carp	500	.016	S.F.	2.02	.60		2.62	3.23
0700	Offices, 2′ x 4′ x 3/4″		380	.021		3.40	.80		4.20	5.05
1800	Tile, Z bar suspension, 5/8″ mineral fiber tile		150	.053		2.32	2.02		4.34	5.95
1900	3/4″ mineral fiber tile		150	.053		2.35	2.02		4.37	5.95

09 51 53 – Direct-Applied Acoustical Ceilings

09 51 53.10 Ceiling Tile

	09 51 53.10 Ceiling Tile	Crew	Daily Output	Labor-Hours	Unit	Material	Labor	Equipment	Total	Total Incl O&P
0010	**CEILING TILE**, stapled or cemented									
0100	12″ x 12″ or 12″ x 24″, not including furring									
0600	Mineral fiber, vinyl coated, 5/8″ thick	1 Carp	300	.027	S.F.	2.23	1.01		3.24	4.14
0700	3/4″ thick		300	.027		2.93	1.01		3.94	4.91
0900	Fire rated, 3/4″ thick, plain faced		300	.027		1.31	1.01		2.32	3.13
1000	Plastic coated face		300	.027		1.93	1.01		2.94	3.81
1200	Aluminum faced, 5/8″ thick, plain		300	.027		1.74	1.01		2.75	3.60
3300	For flameproofing, add					.10			.10	.11
3400	For sculptured 3 dimensional, add					.31			.31	.34
3900	For ceiling primer, add					.13			.13	.14

09 51 Acoustical Ceilings

09 51 53 – Direct-Applied Acoustical Ceilings

09 51 53.10 Ceiling Tile		Crew	Daily Output	Labor-Hours	Unit	2016 Bare Costs Material	Labor	Equipment	Total	Total Incl O&P
4000	For ceiling cement, add				S.F.	.40			.40	.44

09 53 Acoustical Ceiling Suspension Assemblies

09 53 23 – Metal Acoustical Ceiling Suspension Assemblies

09 53 23.30 Ceiling Suspension Systems

		Crew	Daily Output	Labor-Hours	Unit	Material	Labor	Equipment	Total	Total Incl O&P
0010	**CEILING SUSPENSION SYSTEMS** for boards and tile									
0050	Class A suspension system, 15/16" T bar, 2' x 4' grid	1 Carp	800	.010	S.F.	.77	.38		1.15	1.47
0300	2' x 2' grid		650	.012		.99	.47		1.46	1.87
0310	25% recycled steel, 2' x 4' grid G		800	.010		.80	.38		1.18	1.51
0320	2' x 2' grid G	↓	650	.012		1	.47		1.47	1.88
0350	For 9/16" grid, add					.16			.16	.18
0360	For fire rated grid, add					.09			.09	.10
0370	For colored grid, add					.21			.21	.23
0400	Concealed Z bar suspension system, 12" module	1 Carp	520	.015		.86	.58		1.44	1.92
0600	1-1/2" carrier channels, 4' O.C., add		470	.017		.12	.64		.76	1.21
0650	1-1/2" x 3-1/2" channels	↓	470	.017	↓	.32	.64		.96	1.43
0700	Carrier channels for ceilings with									
0900	recessed lighting fixtures, add	1 Carp	460	.017	S.F.	.22	.66		.88	1.34
3000	Seismic ceiling bracing, IBC Site Class D, Occupancy Category II									
3050	For ceilings less than 2500 S.F.									
3060	Seismic clips at attached walls	1 Carp	180	.044	Ea.	1.06	1.68		2.74	3.97
3100	For ceilings greater than 2500 S.F., add									
3120	Seismic clips, joints at cross tees	1 Carp	120	.067	Ea.	3.02	2.52		5.54	7.55
3140	At cross tees and mains, mains field cut	"	60	.133	"	3.02	5.05		8.07	11.75
3200	Compression posts, telescopic, attached to structure above									
3210	To 30" high	1 Carp	26	.308	Ea.	36.50	11.65		48.15	60
3220	30" to 48" high		25.50	.314		40	11.85		51.85	64
3230	48" to 84" high		25	.320		48.50	12.10		60.60	74
3240	84" to 102" high		24.50	.327		55.50	12.35		67.85	81.50
3250	102" to 120" high		24	.333		79.50	12.60		92.10	108
3260	120" to 144" high	↓	24	.333	↓	88	12.60		100.60	118
3300	Stabilizer bars									
3310	12" long	1 Carp	240	.033	Ea.	1.02	1.26		2.28	3.23
3320	24" long		235	.034		.97	1.29		2.26	3.22
3330	36" long		230	.035		.93	1.31		2.24	3.23
3340	48" long	↓	220	.036	↓	.77	1.37		2.14	3.15
3400	Wire support for light fixtures, per L.F. height to structure above									
3410	Less than 10 lb.	1 Carp	400	.020	L.F.	.28	.76		1.04	1.58
3420	10 lb. to 56 lb.		240	.033	"	.57	1.26		1.83	2.73
5000	Wire hangers, #12 wire	↓	300	.027	Ea.	.06	1.01		1.07	1.76

09 54 Specialty Ceilings

09 54 26 – Suspended Wood Ceilings

09 54 26.10 Wood Ceilings		Crew	Daily Output	Labor-Hours	Unit	Material	2016 Bare Costs Labor	Equipment	Total	Total Incl O&P
0010	**WOOD CEILINGS**									
1000	4" - 6" wood slats on heavy duty 15/16" T-bar grid	2 Carp	250	.064	S.F.	25	2.42		27.42	31

09 54 33 – Decorative Panel Ceilings

09 54 33.20 Metal Panel Ceilings		Crew	Daily Output	Labor-Hours	Unit	Material	Labor	Equipment	Total	Total Incl O&P
0010	**METAL PANEL CEILINGS**									
0020	Lay-in or screwed to furring, not including grid									
0100	Tin ceilings, 2' x 2' or 2' x 4', bare steel finish	2 Carp	300	.053	S.F.	2.51	2.02		4.53	6.15
0120	Painted white finish		300	.053	"	3.69	2.02		5.71	7.45
0140	Copper, chrome or brass finish		300	.053	L.F.	6.50	2.02		8.52	10.55
0200	Cornice molding, 2-1/2" to 3-1/2" wide, 4' long, bare steel finish		200	.080	S.F.	2.30	3.02		5.32	7.60
0220	Painted white finish		200	.080		2.84	3.02		5.86	8.15
0240	Copper, chrome or brass finish		200	.080		4.01	3.02		7.03	9.45
0320	5" to 6-1/2" wide, 4' long, bare steel finish		150	.107		3.23	4.03		7.26	10.30
0340	Painted white finish		150	.107		4.09	4.03		8.12	11.25
0360	Copper, chrome or brass finish		150	.107		6.50	4.03		10.53	13.90
0420	Flat molding, 3-1/2" to 5" wide, 4' long, bare steel finish		250	.064		3.71	2.42		6.13	8.15
0440	Painted white finish		250	.064		3.96	2.42		6.38	8.40
0460	Copper, chrome or brass finish		250	.064		7.50	2.42		9.92	12.30

09 61 Flooring Treatment

09 61 19 – Concrete Floor Staining

09 61 19.40 Floors, Interior		Crew	Daily Output	Labor-Hours	Unit	Material	Labor	Equipment	Total	Total Incl O&P
0010	**FLOORS, INTERIOR**									
0300	Acid stain and sealer									
0310	Stain, one coat	1 Pord	650	.012	S.F.	.12	.39		.51	.79
0320	Two coats		570	.014		.25	.45		.70	1.01
0330	Acrylic sealer, one coat		2600	.003		.23	.10		.33	.42
0340	Two coats		1400	.006		.47	.18		.65	.81

09 62 Specialty Flooring

09 62 19 – Laminate Flooring

09 62 19.10 Floating Floor		Crew	Daily Output	Labor-Hours	Unit	Material	Labor	Equipment	Total	Total Incl O&P
0010	**FLOATING FLOOR**									
8300	Floating floor, laminate, wood pattern strip, complete	1 Clab	133	.060	S.F.	4.50	1.76		6.26	7.90
8310	Components, T & G wood composite strips					4.02			4.02	4.42
8320	Film					.17			.17	.18
8330	Foam					.26			.26	.29
8340	Adhesive					.71			.71	.79
8350	Installation kit					.19			.19	.21
8360	Trim, 2" wide x 3' long				L.F.	4.30			4.30	4.73
8370	Reducer moulding				"	5.70			5.70	6.25

09 62 23 – Bamboo Flooring

09 62 23.10 Flooring, Bamboo		Crew	Daily Output	Labor-Hours	Unit	Material	Labor	Equipment	Total	Total Incl O&P
0010	**FLOORING, BAMBOO**									
8600	Flooring, wood, bamboo strips, unfinished, 5/8" x 4" x 3' G	1 Carp	255	.031	S.F.	4.60	1.19		5.79	7.05
8610	5/8" x 4" x 4' G		275	.029		4.77	1.10		5.87	7.10
8620	5/8" x 4" x 6' G		295	.027		5.25	1.03		6.28	7.45
8630	Finished, 5/8" x 4" x 3' G		255	.031		5.05	1.19		6.24	7.55

09 62 Specialty Flooring

09 62 23 – Bamboo Flooring

09 62 23.10 Flooring, Bamboo			Crew	Daily Output	Labor-Hours	Unit	Material	Labor	Equipment	Total	Total Incl O&P
							2016 Bare Costs				
8640	5/8" x 4" x 4'	G	1 Carp	275	.029	S.F.	5.30	1.10		6.40	7.70
8650	5/8" x 4" x 6'	G		295	.027		4.94	1.03		5.97	7.15
8660	Stair treads, unfinished, 1-1/16" x 11-1/2" x 4'	G		18	.444	Ea.	44	16.80		60.80	76.50
8670	Finished, 1-1/16" x 11-1/2" x 4'	G		18	.444		82	16.80		98.80	119
8680	Stair risers, unfinished, 5/8" x 7-1/2" x 4'	G		18	.444		16.30	16.80		33.10	46
8690	Finished, 5/8" x 7-1/2" x 4'	G		18	.444		31	16.80		47.80	62
8700	Stair nosing, unfinished, 6' long	G		16	.500		36	18.90		54.90	71
8710	Finished, 6' long	G		16	.500		42	18.90		60.90	77.50

09 62 29 – Cork Flooring

09 62 29.10 Cork Tile Flooring			Crew	Daily Output	Labor-Hours	Unit	Material	Labor	Equipment	Total	Total Incl O&P
0010	**CORK TILE FLOORING**										
2200	Cork tile, standard finish, 1/8" thick		1 Tilf	315	.025	S.F.	6.95	.88		7.83	9.10
2250	3/16" thick	G		315	.025		5.55	.88		6.43	7.60
2300	5/16" thick	G		315	.025		8.35	.88		9.23	10.65
2350	1/2" thick	G		315	.025		10.95	.88		11.83	13.45
2500	Urethane finish, 1/8" thick	G		315	.025		8.10	.88		8.98	10.35
2550	3/16" thick	G		315	.025		8.25	.88		9.13	10.50
2600	5/16" thick	G		315	.025		8.85	.88		9.73	11.20
2650	1/2" thick	G		315	.025		12.15	.88		13.03	14.80

09 63 Masonry Flooring

09 63 13 – Brick Flooring

09 63 13.10 Miscellaneous Brick Flooring			Crew	Daily Output	Labor-Hours	Unit	Material	Labor	Equipment	Total	Total Incl O&P
0010	**MISCELLANEOUS BRICK FLOORING**										
0020	Acid-proof shales, red, 8" x 3-3/4" x 1-1/4" thick		D-7	.43	37.209	M	705	1,150		1,855	2,650
0050	2-1/4" thick		D-1	.40	40		965	1,325		2,290	3,250
0200	Acid-proof clay brick, 8" x 3-3/4" x 2-1/4" thick	G		.40	40		925	1,325		2,250	3,225
0250	9" x 4-1/2" x 3"	G		95	.168	S.F.	4.23	5.55		9.78	13.90
0260	Cast ceramic, pressed, 4" x 8" x 1/2", unglazed		D-7	100	.160		6.35	4.95		11.30	14.95
0270	Glazed			100	.160		8.45	4.95		13.40	17.30
0280	Hand molded flooring, 4" x 8" x 3/4", unglazed			95	.168		8.35	5.20		13.55	17.65
0290	Glazed			95	.168		10.50	5.20		15.70	20
0300	8" hexagonal, 3/4" thick, unglazed			85	.188		9.20	5.85		15.05	19.55
0310	Glazed			85	.188		16.60	5.85		22.45	27.50
0450	Acid-proof joints, 1/4" wide		D-1	65	.246		1.46	8.10		9.56	15.10
0500	Pavers, 8" x 4", 1" to 1-1/4" thick, red		D-7	95	.168		3.69	5.20		8.89	12.50
0510	Ironspot		"	95	.168		5.20	5.20		10.40	14.20
0540	1-3/8" to 1-3/4" thick, red		D-1	95	.168		3.56	5.55		9.11	13.15
0560	Ironspot			95	.168		5.15	5.55		10.70	14.95
0580	2-1/4" thick, red			90	.178		3.62	5.85		9.47	13.75
0590	Ironspot			90	.178		5.60	5.85		11.45	15.95
0700	Paver, adobe brick, 6" x 12", 1/2" joint	G		42	.381		1.32	12.55		13.87	22.50
0710	Mexican red, 12" x 12"	G	1 Tilf	48	.167		1.72	5.80		7.52	11.30
0720	Saltillo, 12" x 12"	G	"	48	.167		1.49	5.80		7.29	11.05
0800	For sidewalks and patios with pavers, see Section 32 14 16.10										
0870	For epoxy joints, add		D-1	600	.027	S.F.	2.81	.88		3.69	4.55
0880	For Furan underlayment, add		"	600	.027		2.33	.88		3.21	4.02
0890	For waxed surface, steam cleaned, add		A-1H	1000	.008		.20	.23	.08	.51	.70

09 63 Masonry Flooring

09 63 40 – Stone Flooring

09 63 40.10 Marble		Crew	Daily Output	Labor-Hours	Unit	Material	Labor	Equipment	Total	Total Incl O&P
						2016 Bare Costs				
0010	**MARBLE**									
0020	Thin gauge tile, 12" x 6", 3/8", white Carara	D-7	60	.267	S.F.	14.95	8.25		23.20	30
0100	Travertine		60	.267		9.95	8.25		18.20	24.50
0200	12" x 12" x 3/8", thin set, floors		60	.267		10.30	8.25		18.55	24.50
0300	On walls		52	.308		9.85	9.50		19.35	26.50
1000	Marble threshold, 4" wide x 36" long x 5/8" thick, white		60	.267	Ea.	10.30	8.25		18.55	24.50

09 63 40.20 Slate Tile

		Crew	Daily Output	Labor-Hours	Unit	Material	Labor	Equipment	Total	Total Incl O&P
0010	**SLATE TILE**									
0020	Vermont, 6" x 6" x 1/4" thick, thin set	D-7	180	.089	S.F.	7.50	2.75		10.25	12.70

09 64 Wood Flooring

09 64 16 – Wood Block Flooring

09 64 16.10 End Grain Block Flooring

		Crew	Daily Output	Labor-Hours	Unit	Material	Labor	Equipment	Total	Total Incl O&P
0010	**END GRAIN BLOCK FLOORING**									
0020	End grain flooring, coated, 2" thick	1 Carp	295	.027	S.F.	3.57	1.03		4.60	5.65
0400	Natural finish, 1" thick, fir		125	.064		3.69	2.42		6.11	8.10
0600	1-1/2" thick, pine		125	.064		3.62	2.42		6.04	8.05
0700	2" thick, pine		125	.064		4.49	2.42		6.91	9

09 64 23 – Wood Parquet Flooring

09 64 23.10 Wood Parquet

		Crew	Daily Output	Labor-Hours	Unit	Material	Labor	Equipment	Total	Total Incl O&P
0010	**WOOD PARQUET** flooring									
5200	Parquetry, 5/16" thk, no finish, oak, plain pattern	1 Carp	160	.050	S.F.	5.40	1.89		7.29	9.10
5300	Intricate pattern		100	.080		9.65	3.02		12.67	15.70
5500	Teak, plain pattern		160	.050		6.30	1.89		8.19	10.10
5600	Intricate pattern		100	.080		10.70	3.02		13.72	16.80
5650	13/16" thick, select grade oak, plain pattern		160	.050		9.75	1.89		11.64	13.90
5700	Intricate pattern		100	.080		17.25	3.02		20.27	24
5800	Custom parquetry, including finish, plain pattern		100	.080		17.05	3.02		20.07	24
5900	Intricate pattern		50	.160		25	6.05		31.05	37.50
6700	Parquetry, prefinished white oak, 5/16" thick, plain pattern		160	.050		7.90	1.89		9.79	11.85
6800	Intricate pattern		100	.080		8.45	3.02		11.47	14.30
7000	Walnut or teak, parquetry, plain pattern		160	.050		8.40	1.89		10.29	12.40
7100	Intricate pattern		100	.080		13.50	3.02		16.52	19.90
7200	Acrylic wood parquet blocks, 12" x 12" x 5/16",									
7210	Irradiated, set in epoxy	1 Carp	160	.050	S.F.	10.20	1.89		12.09	14.35

09 64 29 – Wood Strip and Plank Flooring

09 64 29.10 Wood

		Crew	Daily Output	Labor-Hours	Unit	Material	Labor	Equipment	Total	Total Incl O&P
0010	**WOOD**									
0020	Fir, vertical grain, 1" x 4", not incl. finish, grade B & better	1 Carp	255	.031	S.F.	2.79	1.19		3.98	5.05
0100	C grade & better		255	.031		2.63	1.19		3.82	4.88
0300	Flat grain, 1" x 4", not incl. finish, B & better		255	.031		3.19	1.19		4.38	5.50
0400	C & better		255	.031		3.07	1.19		4.26	5.35
4000	Maple, strip, 25/32" x 2-1/4", not incl. finish, select		170	.047		4.90	1.78		6.68	8.40
4100	#2 & better		170	.047		4.42	1.78		6.20	7.85
4300	33/32" x 3-1/4", not incl. finish, #1 grade		170	.047		4.56	1.78		6.34	8
4400	#2 & better		170	.047		4.06	1.78		5.84	7.45
4600	Oak, white or red, 25/32" x 2-1/4", not incl. finish									
4700	#1 common	1 Carp	170	.047	S.F.	3.33	1.78		5.11	6.65
4900	Select quartered, 2-1/4" wide		170	.047		4.19	1.78		5.97	7.60

09 64 Wood Flooring

09 64 29 – Wood Strip and Plank Flooring

09 64 29.10 Wood		Crew	Daily Output	Labor-Hours	Unit	2016 Bare Costs Material	Labor	Equipment	Total	Total Incl O&P
5000	Clear	1 Carp	170	.047	S.F.	4.10	1.78		5.88	7.50
6100	Prefinished, white oak, prime grade, 2-1/4" wide		170	.047		4.83	1.78		6.61	8.30
6200	3-1/4" wide		185	.043		5.25	1.63		6.88	8.50
6400	Ranch plank		145	.055		7.15	2.09		9.24	11.35
6500	Hardwood blocks, 9" x 9", 25/32" thick		160	.050		6	1.89		7.89	9.75
7400	Yellow pine, 3/4" x 3-1/8", T & G, C & better, not incl. finish		200	.040		1.53	1.51		3.04	4.21
7500	Refinish wood floor, sand, 2 coats poly, wax, soft wood	1 Clab	400	.020		.90	.58		1.48	1.97
7600	Hard wood		130	.062		1.34	1.80		3.14	4.48
7800	Sanding and finishing, 2 coats polyurethane		295	.027		.90	.79		1.69	2.32
7900	Subfloor and underlayment, see Section 06 16									
8015	Transition molding, 2 1/4" wide, 5' long	1 Carp	19.20	.417	Ea.	10.85	15.75		26.60	38.50

09 65 Resilient Flooring

09 65 10 – Resilient Tile Underlayment

09 65 10.10 Latex Underlayment

Line	Description	Crew	Daily Output	Labor-Hours	Unit	Material	Labor	Equipment	Total	Total Incl O&P
0010	**LATEX UNDERLAYMENT**									
3600	Latex underlayment, 1/8" thk., cementitious for resilient flooring	1 Tilf	160	.050	S.F.	1.22	1.74		2.96	4.15
4000	Liquid, fortified				Gal.	32			32	35

09 65 13 – Resilient Base and Accessories

09 65 13.13 Resilient Base

Line	Description	Crew	Daily Output	Labor-Hours	Unit	Material	Labor	Equipment	Total	Total Incl O&P
0010	**RESILIENT BASE**									
0690	1/8" vinyl base, 2 1/2" H, straight or cove, standard colors	1 Tilf	315	.025	L.F.	.68	.88		1.56	2.18
0700	4" high		315	.025		1.21	.88		2.09	2.76
0710	6" high		315	.025		1.38	.88		2.26	2.95
0720	Corners, 2 1/2" high		315	.025	Ea.	2.14	.88		3.02	3.78
0730	4" high		315	.025		2.35	.88		3.23	4.02
0740	6" high		315	.025		2.65	.88		3.53	4.35
0800	1/8" rubber base, 2 1/2" H, straight or cove, standard colors		315	.025	L.F.	1.11	.88		1.99	2.65
1100	4" high		315	.025		1.14	.88		2.02	2.68
1110	6" high		315	.025		1.91	.88		2.79	3.54
1150	Corners, 2 1/2" high		315	.025	Ea.	2.25	.88		3.13	3.91
1153	4" high		315	.025		2.30	.88		3.18	3.96
1155	6" high		315	.025		2.86	.88		3.74	4.58
1450	For premium color/finish add					50%				
1500	Millwork profile	1 Tilf	315	.025	L.F.	6.15	.88		7.03	8.20

09 65 16 – Resilient Sheet Flooring

09 65 16.10 Rubber and Vinyl Sheet Flooring

Line	Description	Crew	Daily Output	Labor-Hours	Unit	Material	Labor	Equipment	Total	Total Incl O&P
0010	**RUBBER AND VINYL SHEET FLOORING**									
5500	Linoleum, sheet goods G	1 Tilf	360	.022	S.F.	3.59	.77		4.36	5.20
5900	Rubber, sheet goods, 36" wide, 1/8" thick		120	.067		7.10	2.32		9.42	11.55
5950	3/16" thick		100	.080		10.45	2.78		13.23	16
6000	1/4" thick		90	.089		12.30	3.09		15.39	18.55
8000	Vinyl sheet goods, backed, .065" thick, plain pattern/colors		250	.032		3.69	1.11		4.80	5.85
8050	Intricate pattern/colors		200	.040		4.34	1.39		5.73	7
8100	.080" thick, plain pattern/colors		230	.035		4.12	1.21		5.33	6.50
8150	Intricate pattern/colors		200	.040		6.45	1.39		7.84	9.35
8200	.125" thick, plain pattern/colors		230	.035		3.87	1.21		5.08	6.20
8250	Intricate pattern/colors		200	.040		7.50	1.39		8.89	10.50
8700	Adhesive cement, 1 gallon per 200 to 300 S.F.				Gal.	29.50			29.50	32
8800	Asphalt primer, 1 gallon per 300 S.F.					14			14	15.40

09 65 Resilient Flooring

09 65 16 – Resilient Sheet Flooring

09 65 16.10 Rubber and Vinyl Sheet Flooring		Crew	Daily Output	Labor-Hours	Unit	Material	2016 Bare Costs Labor	Equipment	Total	Total Incl O&P
8900	Emulsion, 1 gallon per 140 S.F.				Gal.	18.85			18.85	20.50

09 65 19 – Resilient Tile Flooring

09 65 19.19 Vinyl Composition Tile Flooring

		Crew	Daily Output	Labor-Hours	Unit	Material	Labor	Equipment	Total	Total Incl O&P
0010	**VINYL COMPOSITION TILE FLOORING**									
7000	Vinyl composition tile, 12" x 12", 1/16" thick	1 Tilf	500	.016	S.F.	1.19	.56		1.75	2.21
7050	Embossed		500	.016		2.21	.56		2.77	3.33
7100	Marbleized		500	.016		2.21	.56		2.77	3.33
7150	Solid		500	.016		2.85	.56		3.41	4.04
7200	3/32" thick, embossed		500	.016		1.43	.56		1.99	2.47
7250	Marbleized		500	.016		2.54	.56		3.10	3.69
7300	Solid		500	.016		2.36	.56		2.92	3.50
7350	1/8" thick, marbleized		500	.016		2.44	.56		3	3.58
7400	Solid		500	.016		1.55	.56		2.11	2.61
7450	Conductive	↓	500	.016	↓	6.05	.56		6.61	7.55

09 65 19.23 Vinyl Tile Flooring

		Crew	Daily Output	Labor-Hours	Unit	Material	Labor	Equipment	Total	Total Incl O&P
0010	**VINYL TILE FLOORING**									
7500	Vinyl tile, 12" x 12", 3/32" thick, standard colors/patterns	1 Tilf	500	.016	S.F.	3.71	.56		4.27	4.98
7550	1/8" thick, standard colors/patterns		500	.016		5.20	.56		5.76	6.60
7600	1/8" thick, premium colors/patterns		500	.016		7	.56		7.56	8.60
7650	Solid colors		500	.016		3.25	.56		3.81	4.48
7700	Marbleized or Travertine pattern		500	.016		5.90	.56		6.46	7.40
7750	Florentine pattern		500	.016		6.30	.56		6.86	7.85
7800	Premium colors/patterns	↓	500	.016	↓	6.25	.56		6.81	7.75

09 65 19.33 Rubber Tile Flooring

		Crew	Daily Output	Labor-Hours	Unit	Material	Labor	Equipment	Total	Total Incl O&P
0010	**RUBBER TILE FLOORING**									
6050	Rubber tile, marbleized colors, 12" x 12", 1/8" thick	1 Tilf	400	.020	S.F.	5.75	.70		6.45	7.45
6100	3/16" thick		400	.020		9.05	.70		9.75	11.10
6300	Special tile, plain colors, 1/8" thick		400	.020		7.95	.70		8.65	9.90
6350	3/16" thick	↓	400	.020	↓	9.05	.70		9.75	11.10

09 65 33 – Conductive Resilient Flooring

09 65 33.10 Conductive Rubber and Vinyl Flooring

		Crew	Daily Output	Labor-Hours	Unit	Material	Labor	Equipment	Total	Total Incl O&P
0010	**CONDUCTIVE RUBBER AND VINYL FLOORING**									
1700	Conductive flooring, rubber tile, 1/8" thick	1 Tilf	315	.025	S.F.	6.95	.88		7.83	9.10
1800	Homogeneous vinyl tile, 1/8" thick	"	315	.025	"	6.70	.88		7.58	8.85

09 65 66 – Resilient Athletic Flooring

09 65 66.10 Resilient Athletic Flooring

		Crew	Daily Output	Labor-Hours	Unit	Material	Labor	Equipment	Total	Total Incl O&P
0010	**RESILIENT ATHLETIC FLOORING**									
1000	Recycled rubber rolled goods, for weight rooms, 3/8" thk.	1 Tilf	315	.025	S.F.	2.97	.88		3.85	4.70
2000	Vinyl sheet flooring, 1/4" thk.	"	315	.025	"	4.39	.88		5.27	6.25

09 66 Terrazzo Flooring

09 66 13 – Portland Cement Terrazzo Flooring

09 66 13.10 Portland Cement Terrazzo		Crew	Daily Output	Labor-Hours	Unit	2016 Bare Costs Material	Labor	Equipment	Total	Total Incl O&P
0010	**PORTLAND CEMENT TERRAZZO**, cast-in-place									
0020	Cove base, 6" high, 16 ga. zinc	1 Mstz	20	.400	L.F.	3.30	13.90		17.20	26
0101	Curb, 6" high and 6" wide	J-3	12	1.333		6.05	42.50	26	74.55	104
0300	Divider strip for floors, 14 ga., 1-1/4" deep, zinc	1 Mstz	375	.021		1.54	.74		2.28	2.89
0400	Brass		375	.021		2.61	.74		3.35	4.07
0600	Heavy top strip 1/4" thick, 1-1/4" deep, zinc		300	.027		2.15	.93		3.08	3.87
1200	For thin set floors, 16 ga., 1/2" x 1/2", zinc		350	.023		1.19	.80		1.99	2.60
1500	Floor, bonded to concrete, 1-3/4" thick, gray cement	J-3	75	.213	S.F.	3.35	6.80	4.13	14.28	19.30
1600	White cement, mud set		75	.213		3.73	6.80	4.13	14.66	19.70
1800	Not bonded, 3" total thickness, gray cement		70	.229		4.17	7.30	4.43	15.90	21.50
1900	White cement, mud set		70	.229		4.86	7.30	4.43	16.59	22
4300	Stone chips, onyx gemstone, per 50 lb. bag				Bag	16.80			16.80	18.50

09 66 16 – Terrazzo Floor Tile

09 66 16.10 Tile or Terrazzo Base

		Crew	Daily Output	Labor-Hours	Unit	Material	Labor	Equipment	Total	Total Incl O&P
0010	**TILE OR TERRAZZO BASE**									
0020	Scratch coat only	1 Mstz	150	.053	S.F.	.43	1.86		2.29	3.47
0500	Scratch and brown coat only	"	75	.107	"	.82	3.71		4.53	6.90

09 66 16.13 Portland Cement Terrazzo Floor Tile

		Crew	Daily Output	Labor-Hours	Unit	Material	Labor	Equipment	Total	Total Incl O&P
0010	**PORTLAND CEMENT TERRAZZO FLOOR TILE**									
1200	Floor tiles, non-slip, 1" thick, 12" x 12"	D-1	60	.267	S.F.	20.50	8.75		29.25	37
1300	1-1/4" thick, 12" x 12"		60	.267		20.50	8.75		29.25	37.50
1500	16" x 16"		50	.320		22.50	10.50		33	42
1600	1-1/2" thick, 16" x 16"		45	.356		20.50	11.70		32.20	42

09 66 16.16 Plastic Matrix Terrazzo Floor Tile

		Crew	Daily Output	Labor-Hours	Unit	Material	Labor	Equipment	Total	Total Incl O&P
0010	**PLASTIC MATRIX TERRAZZO FLOOR TILE**									
0100	12" x 12", 3/16" thick, Floor tiles w/marble chips	1 Tilf	500	.016	S.F.	7.30	.56		7.86	8.95
0200	12" x 12", 3/16" thick, Floor tiles w/glass chips		500	.016		7.90	.56		8.46	9.55
0300	12" x 12", 3/16" thick, Floor tiles w/recycled content		500	.016		6.20	.56		6.76	7.75

09 66 16.30 Terrazzo, Precast

		Crew	Daily Output	Labor-Hours	Unit	Material	Labor	Equipment	Total	Total Incl O&P
0010	**TERRAZZO, PRECAST**									
0020	Base, 6" high, straight	1 Mstz	70	.114	L.F.	12.10	3.98		16.08	19.80
0100	Cove		60	.133		13.20	4.64		17.84	22
0300	8" high, straight		60	.133		11.80	4.64		16.44	20.50
0400	Cove		50	.160		17.35	5.55		22.90	28
0600	For white cement, add					.46			.46	.51
0700	For 16 ga. zinc toe strip, add					1.81			1.81	1.99
0900	Curbs, 4" x 4" high	1 Mstz	40	.200		33.50	6.95		40.45	48
1000	8" x 8" high		30	.267		38.50	9.30		47.80	57.50
4800	Wainscot, 12" x 12" x 1" tiles		12	.667	S.F.	7.10	23		30.10	45.50
4900	16" x 16" x 1-1/2" tiles		8	1	"	14.25	35		49.25	72

09 66 23 – Resinous Matrix Terrazzo Flooring

09 66 23.13 Polyacrylate Mod. Cementitious Terrazzo Flr.

		Crew	Daily Output	Labor-Hours	Unit	Material	Labor	Equipment	Total	Total Incl O&P
0010	**POLYACRYLATE MODIFIED CEMENTITIOUS TERRAZZO FLOORING**									
3150	Polyacrylate, 1/4" thick, granite chips	C-6	735	.065	S.F.	3.52	2	.09	5.61	7.30
3170	Recycled porcelain		480	.100		4.54	3.06	.13	7.73	10.25
3200	3/8" thick, granite chips		620	.077		4.61	2.37	.10	7.08	9.10
3220	Recycled porcelain		480	.100		6.45	3.06	.13	9.64	12.35

09 66 23.16 Epoxy-Resin Terrazzo Flooring

		Crew	Daily Output	Labor-Hours	Unit	Material	Labor	Equipment	Total	Total Incl O&P
0010	**EPOXY-RESIN TERRAZZO FLOORING**									
1800	Epoxy terrazzo, 1/4" thick, chemical resistant, granite chips	J-3	200	.080	S.F.	6	2.56	1.55	10.11	12.45
1900	Recycled porcelain		150	.107		9.10	3.41	2.07	14.58	17.80

09 66 Terrazzo Flooring

09 66 23 – Resinous Matrix Terrazzo Flooring

09 66 23.16 Epoxy-Resin Terrazzo Flooring		Crew	Daily Output	Labor-Hours	Unit	2016 Bare Costs Material	Labor	Equipment	Total	Total Incl O&P
2650	Epoxy terrazzo, 3/8" thick, marble chips	J-3	200	.080	S.F.	5.75	2.56	1.55	9.86	12.20
2675	Glass or mother of pearl	↓	200	.080	↓	6.70	2.56	1.55	10.81	13.20

09 66 33 – Conductive Terrazzo Flooring

09 66 33.19 Conductive Plastic-Matrix Terrazzo Flooring		Crew	Daily Output	Labor-Hours	Unit	Material	Labor	Equipment	Total	Total Incl O&P
0010	CONDUCTIVE PLASTIC-MATRIX TERRAZZO FLOORING									
3300	Conductive, 1/4" thick, granite chips	C-6	450	.107	S.F.	7.55	3.26	.14	10.95	13.90
3330	Recycled porcelain		305	.157		10.05	4.82	.21	15.08	19.30
3350	3/8" thick, granite chips		365	.132		9.95	4.02	.17	14.14	17.80
3370	Recycled porcelain		255	.188		12.85	5.75	.25	18.85	24
3450	Granite, conductive, 1/4" thick, 20% chip		695	.069		9.25	2.11	.09	11.45	13.75
3470	50% chip		420	.114		11.90	3.50	.15	15.55	19.05
3500	3/8" thick, 20% chip		695	.069		13.50	2.11	.09	15.70	18.45
3520	50% chip	↓	380	.126	↓	16.30	3.87	.17	20.34	24.50

09 67 Fluid-Applied Flooring

09 67 13 – Elastomeric Liquid Flooring

09 67 13.13 Elastomeric Liquid Flooring		Crew	Daily Output	Labor-Hours	Unit	Material	Labor	Equipment	Total	Total Incl O&P
0010	ELASTOMERIC LIQUID FLOORING									
0020	Cementitious acrylic, 1/4" thick	C-6	520	.092	S.F.	1.66	2.82	.12	4.60	6.65
0200	Methyl methachrylate, 1/4" thick	C-8A	3000	.016	"	6.50	.51		7.01	8

09 67 26 – Quartz Flooring

09 67 26.26 Quartz Flooring		Crew	Daily Output	Labor-Hours	Unit	Material	Labor	Equipment	Total	Total Incl O&P
0010	QUARTZ FLOORING									
0600	Epoxy, with colored quartz chips, broadcast, 3/8" thick	C-6	675	.071	S.F.	2.81	2.18	.09	5.08	6.80
0700	1/2" thick		490	.098		4.05	3	.13	7.18	9.60
0900	Troweled, minimum		560	.086		3.57	2.62	.11	6.30	8.40
1000	Maximum	↓	480	.100	↓	5.40	3.06	.13	8.59	11.15
1200	Heavy duty epoxy topping, 1/4" thick,									
1300	500 to 1,000 S.F.	C-6	420	.114	S.F.	5.60	3.50	.15	9.25	12.10
1500	1,000 to 2,000 S.F.		450	.107		5.10	3.26	.14	8.50	11.20
1600	Over 10,000 S.F.	↓	480	.100	↓	4.70	3.06	.13	7.89	10.40

09 68 Carpeting

09 68 05 – Carpet Accessories

09 68 05.11 Flooring Transition Strip		Crew	Daily Output	Labor-Hours	Unit	Material	Labor	Equipment	Total	Total Incl O&P
0010	FLOORING TRANSITION STRIP									
0107	Clamp down brass divider, 12' strip, vinyl to carpet	1 Tilf	31.25	.256	Ea.	13.95	8.90		22.85	30
0117	Vinyl to hard surface	"	31.25	.256	"	13.95	8.90		22.85	30

09 68 10 – Carpet Pad

09 68 10.10 Commercial Grade Carpet Pad		Crew	Daily Output	Labor-Hours	Unit	Material	Labor	Equipment	Total	Total Incl O&P
0010	COMMERCIAL GRADE CARPET PAD									
9000	Sponge rubber pad, 20 oz./sq. yd.	1 Tilf	150	.053	S.Y.	4.60	1.85		6.45	8.05
9100	40 to 62 oz./sq. yd.		150	.053		8.75	1.85		10.60	12.65
9200	Felt pad, 20 oz./sq. yd.		150	.053		5.60	1.85		7.45	9.20
9300	32 to 56 oz./sq. yd.		150	.053		9.30	1.85		11.15	13.25
9400	Bonded urethane pad, 2.7 density		150	.053		5.95	1.85		7.80	9.55
9500	13.0 Density		150	.053		8	1.85		9.85	11.80
9600	Prime urethane pad, 2.7 density	↓	150	.053	↓	3.25	1.85		5.10	6.60

09 68 Carpeting

09 68 10 – Carpet Pad

09 68 10.10 Commercial Grade Carpet Pad		Crew	Daily Output	Labor-Hours	Unit	Material	2016 Bare Costs Labor	Equipment	Total	Total Incl O&P
9700	13.0 density	1 Tilf	150	.053	S.Y.	4.97	1.85		6.82	8.45

09 68 13 – Tile Carpeting

09 68 13.10 Carpet Tile		Crew	Daily Output	Labor-Hours	Unit	Material	Labor	Equipment	Total	Total Incl O&P
0010	**CARPET TILE**									
0100	Tufted nylon, 18" x 18", hard back, 20 oz.	1 Tilf	80	.100	S.Y.	24	3.48		27.48	32
0110	26 oz.		80	.100		29	3.48		32.48	37.50
0200	Cushion back, 20 oz.		80	.100		28	3.48		31.48	36
0210	26 oz.	↓	80	.100	↓	43	3.48		46.48	52.50

09 68 16 – Sheet Carpeting

09 68 16.10 Sheet Carpet		Crew	Daily Output	Labor-Hours	Unit	Material	Labor	Equipment	Total	Total Incl O&P
0010	**SHEET CARPET**									
0700	Nylon, level loop, 26 oz., light to medium traffic	1 Tilf	75	.107	S.Y.	21.50	3.71		25.21	29.50
0720	28 oz., light to medium traffic		75	.107		29.50	3.71		33.21	38.50
0900	32 oz., medium traffic		75	.107		39.50	3.71		43.21	49.50
1100	40 oz., medium to heavy traffic		75	.107		45	3.71		48.71	55.50
2920	Nylon plush, 30 oz., medium traffic		75	.107		31	3.71		34.71	40
3000	36 oz., medium traffic		75	.107		35	3.71		38.71	44.50
3100	42 oz., medium to heavy traffic		70	.114		45	3.97		48.97	56
3200	46 oz., medium to heavy traffic		70	.114		50.50	3.97		54.47	62
3300	54 oz., heavy traffic		70	.114		55	3.97		58.97	67
4110	Wool, level loop, 40 oz., medium traffic		70	.114		106	3.97		109.97	123
4500	50 oz., medium to heavy traffic		70	.114		100	3.97		103.97	116
4700	Patterned, 32 oz., medium to heavy traffic		70	.114		91	3.97		94.97	106
4900	48 oz., heavy traffic	↓	70	.114	↓	100	3.97		103.97	116
5000	For less than full roll (approx. 1500 S.F.), add					25%				
5100	For small rooms, less than 12' wide, add						25%			
5200	For large open areas (no cuts), deduct						25%			
5600	For bound carpet baseboard, add	1 Tilf	300	.027	L.F.	3	.93		3.93	4.80
5610	For stairs, not incl. price of carpet, add	"	30	.267	Riser		9.25		9.25	15
5620	For borders and patterns, add to labor						18%			
8950	For tackless, stretched installation, add padding from 09 68 10.10 to above									
9850	For brand-named specific fiber, add				S.Y.	25%				

09 68 20 – Athletic Carpet

09 68 20.10 Indoor Athletic Carpet		Crew	Daily Output	Labor-Hours	Unit	Material	Labor	Equipment	Total	Total Incl O&P
0010	**INDOOR ATHLETIC CARPET**									
3700	Polyethylene, in rolls, no base incl., landscape surfaces	1 Tilf	275	.029	S.F.	4.16	1.01		5.17	6.20
3800	Nylon action surface, 1/8" thick		275	.029		3.76	1.01		4.77	5.80
3900	1/4" thick		275	.029		5.45	1.01		6.46	7.60
4000	3/8" thick	↓	275	.029	↓	6.80	1.01		7.81	9.15

09 72 Wall Coverings

09 72 19 – Textile Wall Coverings

09 72 19.10 Textile Wall Covering		Crew	Daily Output	Labor-Hours	Unit	Material	2016 Bare Costs Labor	Equipment	Total	Total Incl O&P
0010	**TEXTILE WALL COVERING**, including sizing; add 10-30% waste @ takeoff									
0020	Silk	1 Pape	640	.013	S.F.	4.23	.40		4.63	5.30
0030	Cotton		640	.013		6.75	.40		7.15	8.10
0040	Linen		640	.013		1.89	.40		2.29	2.74
0050	Blend		640	.013		3.03	.40		3.43	3.99

09 72 20 – Natural Fiber Wall Covering

09 72 20.10 Natural Fiber Wall Covering		Crew	Daily Output	Labor-Hours	Unit	Material	Labor	Equipment	Total	Total Incl O&P
0010	**NATURAL FIBER WALL COVERING**, including sizing; add 10-30% waste @ takeoff									
0015	Bamboo	1 Pape	640	.013	S.F.	2.31	.40		2.71	3.20
0030	Burlap		640	.013		2.07	.40		2.47	2.94
0045	Jute		640	.013		1.32	.40		1.72	2.11
0060	Sisal		640	.013		1.49	.40		1.89	2.30

09 72 23 – Wallpapering

09 72 23.10 Wallpaper		Crew	Daily Output	Labor-Hours	Unit	Material	Labor	Equipment	Total	Total Incl O&P
0010	**WALLPAPER** including sizing; add 10-30 percent waste @ takeoff									
0050	Aluminum foil	1 Pape	275	.029	S.F.	1.04	.93		1.97	2.68
0100	Copper sheets, .025″ thick, vinyl backing		240	.033		5.55	1.07		6.62	7.85
0300	Phenolic backing		240	.033		7.20	1.07		8.27	9.70
0600	Cork tiles, light or dark, 12″ x 12″ x 3/16″		240	.033		4.66	1.07		5.73	6.90
0700	5/16″ thick		235	.034		3.13	1.09		4.22	5.25
0900	1/4″ basketweave		240	.033		3.50	1.07		4.57	5.60
1000	1/2″ natural, non-directional pattern		240	.033		6.55	1.07		7.62	8.95
1100	3/4″ natural, non-directional pattern		240	.033		11.95	1.07		13.02	14.90
1200	Granular surface, 12″ x 36″, 1/2″ thick		385	.021		1.31	.67		1.98	2.54
1300	1″ thick		370	.022		1.68	.69		2.37	2.99
1500	Polyurethane coated, 12″ x 12″ x 3/16″ thick		240	.033		4.08	1.07		5.15	6.25
1600	5/16″ thick		235	.034		6.60	1.09		7.69	9.05
1800	Cork wallpaper, paperbacked, natural		480	.017		2	.54		2.54	3.08
1900	Colors		480	.017		2.87	.54		3.41	4.04
2100	Flexible wood veneer, 1/32″ thick, plain woods		100	.080		2.44	2.57		5.01	6.90
2200	Exotic woods		95	.084		3.69	2.70		6.39	8.50
2400	Gypsum-based, fabric-backed, fire resistant									
2500	for masonry walls, 21 oz./S.Y.	1 Pape	800	.010	S.F.	.77	.32		1.09	1.38
2600	Average		720	.011		1.27	.36		1.63	1.99
2700	Small quantities		640	.013		.70	.40		1.10	1.43
2750	Acrylic, modified, semi-rigid PVC, .028″ thick	2 Carp	330	.048		1.28	1.83		3.11	4.48
2800	.040″ thick	″	320	.050		1.85	1.89		3.74	5.20
3000	Vinyl wall covering, fabric-backed, lightweight, type 1 (12-15 oz./S.Y.)	1 Pape	640	.013		.97	.40		1.37	1.73
3300	Medium weight, type 2 (20-24 oz./S.Y.)		480	.017		.90	.54		1.44	1.87
3400	Heavy weight, type 3 (28 oz./S.Y.)		435	.018		1.37	.59		1.96	2.48
3600	Adhesive, 5 gal. lots (18 S.Y./gal.)				Gal.	14			14	15.40
3700	Wallpaper, average workmanship, solid pattern, low cost paper	1 Pape	640	.013	S.F.	.59	.40		.99	1.31
3900	Basic patterns (matching required), avg. cost paper		535	.015		1.13	.48		1.61	2.03
4000	Paper at $85 per double roll, quality workmanship		435	.018		1.72	.59		2.31	2.86
4100	Linen wall covering, paper backed									
4150	Flame treatment				S.F.	1.08			1.08	1.19
4180	Stain resistance treatment					1.72			1.72	1.89
4200	Grass cloths with lining paper G	1 Pape	400	.020		.96	.64		1.60	2.12
4300	Premium texture/color G	″	350	.023		2.85	.73		3.58	4.35

09 77 Special Wall Surfacing

09 77 30 – Fiberglass Reinforced Panels

09 77 30.10 Fiberglass Reinforced Plastic Panels		Crew	Daily Output	Labor-Hours	Unit	2016 Bare Costs Material	Labor	Equipment	Total	Total Incl O&P
0010	**FIBERGLASS REINFORCED PLASTIC PANELS**, .090" thick									
0020	On walls, adhesive mounted, embossed surface	2 Carp	640	.025	S.F.	1.11	.95		2.06	2.80
0030	Smooth surface		640	.025		1.37	.95		2.32	3.09
0040	Fire rated, embossed surface		640	.025		1.99	.95		2.94	3.77
0050	Nylon rivet mounted, on drywall, embossed surface		480	.033		1.11	1.26		2.37	3.33
0060	Smooth surface		480	.033		1.37	1.26		2.63	3.62
0070	Fire rated, embossed surface		480	.033		1.99	1.26		3.25	4.30
0080	On masonry, embossed surface		320	.050		1.11	1.89		3	4.39
0090	Smooth surface		320	.050		1.37	1.89		3.26	4.68
0100	Fire rated, embossed surface		320	.050		1.99	1.89		3.88	5.35
0110	Nylon rivet and adhesive mounted, on drywall, embossed surface		240	.067		1.26	2.52		3.78	5.60
0120	Smooth surface		240	.067		1.26	2.52		3.78	5.60
0130	Fire rated, embossed surface		240	.067		2.19	2.52		4.71	6.65
0140	On masonry, embossed surface		190	.084		1.26	3.18		4.44	6.75
0150	Smooth surface		190	.084		1.26	3.18		4.44	6.75
0160	Fire rated, embossed surface		190	.084		2.19	3.18		5.37	7.75
0170	For moldings, add	1 Carp	250	.032	L.F.	.26	1.21		1.47	2.32
0180	On ceilings, for lay in grid system, embossed surface		400	.020	S.F.	1.11	.76		1.87	2.49
0190	Smooth surface		400	.020		1.37	.76		2.13	2.78
0200	Fire rated, embossed surface		400	.020		1.99	.76		2.75	3.46

09 77 43 – Panel Systems

09 77 43.20 Slatwall Panels and Accessories		Crew	Daily Output	Labor-Hours	Unit	Material	Labor	Equipment	Total	Total Incl O&P
0010	**SLATWALL PANELS AND ACCESSORIES**									
0100	Slatwall panel, 4' x 8' x 3/4" T, MDF, paint grade	1 Carp	500	.016	S.F.	1.42	.60		2.02	2.57
0110	Melamine finish		500	.016		1.92	.60		2.52	3.12
0120	High pressure plastic laminate finish		500	.016		3.74	.60		4.34	5.10
0125	Wood veneer		500	.016		3.55	.60		4.15	4.92
0130	Aluminum channel inserts, add					2.59			2.59	2.85
0200	Accessories, corner forms, 8' L				L.F.	6			6	6.60
0210	T-connector, 8' L					8.60			8.60	9.45
0220	J-mold, 8' L					1.25			1.25	1.38
0230	Edge cap, 8' L					1.46			1.46	1.61
0240	Finish end cap, 8' L					3.37			3.37	3.71
0300	Display hook, metal, 4" L				Ea.	.43			.43	.47
0310	6" L					.48			.48	.53
0320	8" L					.52			.52	.57
0330	10" L					.59			.59	.65
0340	12" L					.64			.64	.70
0350	Acrylic, 4" L					.93			.93	1.02
0360	6" L					1.07			1.07	1.18
0370	8" L					1.13			1.13	1.24
0380	10" L					1.28			1.28	1.41
0400	Waterfall hanger, metal, 12" - 16"					3.84			3.84	4.22
0410	Acrylic					11.50			11.50	12.65
0500	Shelf bracket, metal, 8"					1.79			1.79	1.97
0510	10"					1.97			1.97	2.17
0520	12"					2.17			2.17	2.39
0530	14"					2.39			2.39	2.63
0540	16"					2.81			2.81	3.09
0550	Acrylic, 8"					3.81			3.81	4.19
0560	10"					4.06			4.06	4.47
0570	12"					4.29			4.29	4.72

09 77 Special Wall Surfacing

09 77 43 – Panel Systems

09 77 43.20 Slatwall Panels and Accessories		Crew	Daily Output	Labor-Hours	Unit	2016 Bare Costs Material	Labor	Equipment	Total	Total Incl O&P
0580	14"				Ea.	4.52			4.52	4.97
0600	Shelf, acrylic, 12" x 16" x 1/4"					20			20	22
0610	12" x 24" x 1/4"				↓	24.50			24.50	27

09 81 Acoustic Insulation

09 81 16 – Acoustic Blanket Insulation

09 81 16.10 Sound Attenuation Blanket

		Crew	Daily Output	Labor-Hours	Unit	Material	Labor	Equipment	Total	Total Incl O&P
0010	**SOUND ATTENUATION BLANKET**									
0020	Blanket, 1" thick	1 Carp	925	.009	S.F.	.25	.33		.58	.83
0500	1-1/2" thick		920	.009		.25	.33		.58	.83
1000	2" thick		915	.009		.36	.33		.69	.95
1500	3" thick	↓	910	.009	↓	.52	.33		.85	1.13
2000	Wall hung, STC 18 – 21, 1" thick, 4' x 20'	2 Carp	22	.727	Ea.	390	27.50		417.50	475
2010	10' x 20'	"	19	.842		975	32		1,007	1,125
2020	Wall hung, STC 27 – 28, 3" thick, 4' x 20'	3 Carp	12	2		560	75.50		635.50	740
2030	10' x 20'	"	9	2.667	↓	1,400	101		1,501	1,700

09 84 Acoustic Room Components

09 84 13 – Fixed Sound-Absorptive Panels

09 84 13.10 Fixed Panels

		Crew	Daily Output	Labor-Hours	Unit	Material	Labor	Equipment	Total	Total Incl O&P
0010	**FIXED PANELS** Perforated steel facing, painted with									
0100	Fiberglass or mineral filler, no backs, 2-1/4" thick, modular									
0200	space units, ceiling or wall hung, white or colored	1 Carp	100	.080	S.F.	8.90	3.02		11.92	14.85
0300	Fiberboard sound deadening panels, 1/2" thick	"	600	.013	"	.33	.50		.83	1.20
0500	Fiberglass panels, 4' x 8' x 1" thick, with									
0600	glass cloth face for walls, cemented	1 Carp	155	.052	S.F.	8.55	1.95		10.50	12.65
0700	1-1/2" thick, dacron covered, inner aluminum frame,									
0710	wall mounted	1 Carp	300	.027	S.F.	8.90	1.01		9.91	11.50

09 91 Painting

09 91 03 – Paint Restoration

09 91 03.20 Sanding

		Crew	Daily Output	Labor-Hours	Unit	Material	Labor	Equipment	Total	Total Incl O&P
0010	**SANDING** and puttying interior trim, compared to									
0100	Painting 1 coat, on quality work				L.F.		100%			
0300	Medium work						50%			
0400	Industrial grade				↓		25%			
0500	Surface protection, placement and removal									
0510	Basic drop cloths	1 Pord	6400	.001	S.F.		.04		.04	.07
0520	Masking with paper		800	.010		.07	.32		.39	.61
0530	Volume cover up (using plastic sheathing or building paper)	↓	16000	.001	↓		.02		.02	.03

09 91 03.30 Exterior Surface Preparation

		Crew	Daily Output	Labor-Hours	Unit	Material	Labor	Equipment	Total	Total Incl O&P
0010	**EXTERIOR SURFACE PREPARATION**									
0015	Doors, per side, not incl. frames or trim									
0020	Scrape & sand									
0030	Wood, flush	1 Pord	616	.013	S.F.		.41		.41	.68
0040	Wood, detail		496	.016			.51		.51	.85
0050	Wood, louvered	↓	280	.029	↓		.91		.91	1.50

09 91 Painting

09 91 03 – Paint Restoration

09 91 03.30 Exterior Surface Preparation		Crew	Daily Output	Labor-Hours	Unit	Material	2016 Bare Costs Labor	Equipment	Total	Total Incl O&P
0060	Wood, overhead	1 Pord	616	.013	S.F.		.41		.41	.68
0070	Wire brush									
0080	Metal, flush	1 Pord	640	.013	S.F.		.40		.40	.66
0090	Metal, detail		520	.015			.49		.49	.81
0100	Metal, louvered		360	.022			.71		.71	1.17
0110	Metal or fibr., overhead		640	.013			.40		.40	.66
0120	Metal, roll up		560	.014			.46		.46	.75
0130	Metal, bulkhead		640	.013			.40		.40	.66
0140	Power wash, based on 2500 lb. operating pressure									
0150	Metal, flush	A-1H	2240	.004	S.F.		.10	.03	.13	.21
0160	Metal, detail		2120	.004			.11	.04	.15	.22
0170	Metal, louvered		2000	.004			.12	.04	.16	.24
0180	Metal or fibr., overhead		2400	.003			.10	.03	.13	.20
0190	Metal, roll up		2400	.003			.10	.03	.13	.20
0200	Metal, bulkhead		2200	.004			.11	.04	.15	.22
0400	Windows, per side, not incl. trim									
0410	Scrape & sand									
0420	Wood, 1-2 lite	1 Pord	320	.025	S.F.		.80		.80	1.31
0430	Wood, 3-6 lite		280	.029			.91		.91	1.50
0440	Wood, 7-10 lite		240	.033			1.06		1.06	1.75
0450	Wood, 12 lite		200	.040			1.28		1.28	2.10
0460	Wood, Bay/Bow		320	.025			.80		.80	1.31
0470	Wire brush									
0480	Metal, 1-2 lite	1 Pord	480	.017	S.F.		.53		.53	.88
0490	Metal, 3-6 lite		400	.020			.64		.64	1.05
0500	Metal, Bay/Bow		480	.017			.53		.53	.88
0510	Power wash, based on 2500 lb. operating pressure									
0520	1-2 lite	A-1H	4400	.002	S.F.		.05	.02	.07	.11
0530	3-6 lite		4320	.002			.05	.02	.07	.11
0540	7-10 lite		4240	.002			.06	.02	.08	.11
0550	12 lite		4160	.002			.06	.02	.08	.11
0560	Bay/Bow		4400	.002			.05	.02	.07	.11
0600	Siding, scrape and sand, light=10-30%, med.=30-70%									
0610	Heavy=70-100% of surface to sand									
0650	Texture 1-11, light	1 Pord	480	.017	S.F.		.53		.53	.88
0660	Med.		440	.018			.58		.58	.96
0670	Heavy		360	.022			.71		.71	1.17
0680	Wood shingles, shakes, light		440	.018			.58		.58	.96
0690	Med.		360	.022			.71		.71	1.17
0700	Heavy		280	.029			.91		.91	1.50
0710	Clapboard, light		520	.015			.49		.49	.81
0720	Med.		480	.017			.53		.53	.88
0730	Heavy		400	.020			.64		.64	1.05
0740	Wire brush									
0750	Aluminum, light	1 Pord	600	.013	S.F.		.43		.43	.70
0760	Med.		520	.015			.49		.49	.81
0770	Heavy		440	.018			.58		.58	.96
0780	Pressure wash, based on 2500 lb. operating pressure									
0790	Stucco	A-1H	3080	.003	S.F.		.08	.03	.11	.16
0800	Aluminum or vinyl		3200	.003			.07	.02	.09	.15
0810	Siding, masonry, brick & block		2400	.003			.10	.03	.13	.20
1300	Miscellaneous, wire brush									
1310	Metal, pedestrian gate	1 Pord	100	.080	S.F.		2.55		2.55	4.20

09 91 03 – Paint Restoration

09 91 03.30 Exterior Surface Preparation		Crew	Daily Output	Labor-Hours	Unit	Material	2016 Bare Costs Labor	Equipment	Total	Total Incl O&P
1320	Aluminum chain link, both sides	1 Pord	250	.032	S.F.		1.02		1.02	1.68
1400	Existing galvanized surface, clean and prime, prep for painting		380	.021		.12	.67		.79	1.25
8000	For chemical washing, see Section 04 01 30									
8020	For sand blasting, see Sections 03 35 29.60 and 05 01 10.51									

09 91 03.40 Interior Surface Preparation		Crew	Daily Output	Labor-Hours	Unit	Material	2016 Bare Costs Labor	Equipment	Total	Total Incl O&P
0010	**INTERIOR SURFACE PREPARATION**									
0020	Doors, per side, not incl. frames or trim									
0030	Scrape & sand									
0040	Wood, flush	1 Pord	616	.013	S.F.		.41		.41	.68
0050	Wood, detail		496	.016			.51		.51	.85
0060	Wood, louvered		280	.029			.91		.91	1.50
0070	Wire brush									
0080	Metal, flush	1 Pord	640	.013	S.F.		.40		.40	.66
0090	Metal, detail		520	.015			.49		.49	.81
0100	Metal, louvered		360	.022			.71		.71	1.17
0110	Hand wash									
0120	Wood, flush	1 Pord	2160	.004	S.F.		.12		.12	.19
0130	Wood, detailed		2000	.004			.13		.13	.21
0140	Wood, louvered		1360	.006			.19		.19	.31
0150	Metal, flush		2160	.004			.12		.12	.19
0160	Metal, detail		2000	.004			.13		.13	.21
0170	Metal, louvered		1360	.006			.19		.19	.31
0400	Windows, per side, not incl. trim									
0410	Scrape & sand									
0420	Wood, 1-2 lite	1 Pord	360	.022	S.F.		.71		.71	1.17
0430	Wood, 3-6 lite		320	.025			.80		.80	1.31
0440	Wood, 7-10 lite		280	.029			.91		.91	1.50
0450	Wood, 12 lite		240	.033			1.06		1.06	1.75
0460	Wood, Bay/Bow		360	.022			.71		.71	1.17
0470	Wire brush									
0480	Metal, 1-2 lite	1 Pord	520	.015	S.F.		.49		.49	.81
0490	Metal, 3-6 lite		440	.018			.58		.58	.96
0500	Metal, Bay/Bow		520	.015			.49		.49	.81
0600	Walls, sanding, light=10-30%, medium - 30-70%,									
0610	heavy=70-100% of surface to sand									
0650	Walls, sand									
0660	Gypsum board or plaster, light	1 Pord	3077	.003	S.F.		.08		.08	.14
0670	Gypsum board or plaster, medium		2160	.004			.12		.12	.19
0680	Gypsum board or plaster, heavy		923	.009			.28		.28	.46
0690	Wood, T&G, light		2400	.003			.11		.11	.18
0700	Wood, T&G, med.		1600	.005			.16		.16	.26
0710	Wood, T&G, heavy		800	.010			.32		.32	.53
0720	Walls, wash									
0730	Gypsum board or plaster	1 Pord	3200	.003	S.F.		.08		.08	.13
0740	Wood, T&G		3200	.003			.08		.08	.13
0750	Masonry, brick & block, smooth		2800	.003			.09		.09	.15
0760	Masonry, brick & block, coarse		2000	.004			.13		.13	.21
8000	For chemical washing, see Section 04 01 30									
8020	For sand blasting, see Sections 03 35 29.60 and 05 01 10.51									

09 91 03 – Paint Restoration

09 91 03.41 Scrape After Fire Damage		Crew	Daily Output	Labor-Hours	Unit	Material	Labor	Equipment	Total	Total Incl O&P
						2016 Bare Costs				
0010	SCRAPE AFTER FIRE DAMAGE									
0050	Boards, 1″ x 4″	1 Pord	336	.024	L.F.		.76		.76	1.25
0060	1″ x 6″		260	.031			.98		.98	1.62
0070	1″ x 8″		207	.039			1.23		1.23	2.03
0080	1″ x 10″		174	.046			1.47		1.47	2.42
0500	Framing, 2″ x 4″		265	.030			.96		.96	1.59
0510	2″ x 6″		221	.036			1.15		1.15	1.90
0520	2″ x 8″		190	.042			1.34		1.34	2.21
0530	2″ x 10″		165	.048			1.55		1.55	2.55
0540	2″ x 12″		144	.056			1.77		1.77	2.92
1000	Heavy framing, 3″ x 4″		226	.035			1.13		1.13	1.86
1010	4″ x 4″		210	.038			1.22		1.22	2
1020	4″ x 6″		191	.042			1.34		1.34	2.20
1030	4″ x 8″		165	.048			1.55		1.55	2.55
1040	4″ x 10″		144	.056			1.77		1.77	2.92
1060	4″ x 12″		131	.061			1.95		1.95	3.21
2900	For sealing, light damage		825	.010	S.F.	.14	.31		.45	.66
2920	Heavy damage		460	.017	″	.31	.55		.86	1.25

09 91 13 – Exterior Painting

09 91 13.30 Fences

		Crew	Daily Output	Labor-Hours	Unit	Material	Labor	Equipment	Total	Total Incl O&P
0010	FENCES R099100-20									
0100	Chain link or wire metal, one side, water base									
0110	Roll & brush, first coat	1 Pord	960	.008	S.F.	.08	.27		.35	.52
0120	Second coat		1280	.006		.07	.20		.27	.41
0130	Spray, first coat		2275	.004		.08	.11		.19	.27
0140	Second coat		2600	.003		.08	.10		.18	.24
0150	Picket, water base									
0160	Roll & brush, first coat	1 Pord	865	.009	S.F.	.08	.30		.38	.58
0170	Second coat		1050	.008		.08	.24		.32	.49
0180	Spray, first coat		2275	.004		.08	.11		.19	.28
0190	Second coat		2600	.003		.08	.10		.18	.25
0200	Stockade, water base									
0210	Roll & brush, first coat	1 Pord	1040	.008	S.F.	.08	.25		.33	.49
0220	Second coat		1200	.007		.08	.21		.29	.44
0230	Spray, first coat		2275	.004		.08	.11		.19	.28
0240	Second coat		2600	.003		.08	.10		.18	.25

09 91 13.42 Miscellaneous, Exterior

		Crew	Daily Output	Labor-Hours	Unit	Material	Labor	Equipment	Total	Total Incl O&P
0010	MISCELLANEOUS, EXTERIOR									
0015	For painting metals, see Section 09 97 13.23									
0100	Railing, ext., decorative wood, incl. cap & baluster									
0110	Newels & spindles @ 12″ O.C.									
0120	Brushwork, stain, sand, seal & varnish									
0130	First coat	1 Pord	90	.089	L.F.	.85	2.84		3.69	5.60
0140	Second coat	″	120	.067	″	.85	2.13		2.98	4.43
0150	Rough sawn wood, 42″ high, 2″ x 2″ verticals, 6″ O.C.									
0160	Brushwork, stain, each coat	1 Pord	90	.089	L.F.	.26	2.84		3.10	4.96
0170	Wrought iron, 1″ rail, 1/2″ sq. verticals									
0180	Brushwork, zinc chromate, 60″ high, bars 6″ O.C.									
0190	Primer	1 Pord	130	.062	L.F.	.88	1.96		2.84	4.20
0200	Finish coat		130	.062		1.13	1.96		3.09	4.47
0210	Additional coat		190	.042		1.32	1.34		2.66	3.66

09 91 13 – Exterior Painting

09 91 13.42 Miscellaneous, Exterior		Crew	Daily Output	Labor-Hours	Unit	Material	2016 Bare Costs Labor	Equipment	Total	Total Incl O&P
0220	Shutters or blinds, single panel, 2′ x 4′, paint all sides									
0230	Brushwork, primer	1 Pord	20	.400	Ea.	.67	12.75		13.42	21.50
0240	Finish coat, exterior latex		20	.400		.62	12.75		13.37	21.50
0250	Primer & 1 coat, exterior latex		13	.615		1.14	19.65		20.79	34
0260	Spray, primer		35	.229		.98	7.30		8.28	13.10
0270	Finish coat, exterior latex		35	.229		1.31	7.30		8.61	13.45
0280	Primer & 1 coat, exterior latex		20	.400		1.06	12.75		13.81	22
0290	For louvered shutters, add				S.F.	10%				
0300	Stair stringers, exterior, metal									
0310	Roll & brush, zinc chromate, to 14″, each coat	1 Pord	320	.025	L.F.	.38	.80		1.18	1.72
0320	Rough sawn wood, 4″ x 12″									
0330	Roll & brush, exterior latex, each coat	1 Pord	215	.037	L.F.	.09	1.19		1.28	2.06
0340	Trellis/lattice, 2″ x 2″ @ 3″ O.C. with 2″ x 8″ supports									
0350	Spray, latex, per side, each coat	1 Pord	475	.017	S.F.	.09	.54		.63	.98
0450	Decking, ext., sealer, alkyd, brushwork, sealer coat		1140	.007		.10	.22		.32	.48
0460	1st coat		1140	.007		.10	.22		.32	.48
0470	2nd coat		1300	.006		.07	.20		.27	.40
0500	Paint, alkyd, brushwork, primer coat		1140	.007		.12	.22		.34	.50
0510	1st coat		1140	.007		.13	.22		.35	.51
0520	2nd coat		1300	.006		.09	.20		.29	.42
0600	Sand paint, alkyd, brushwork, 1 coat		150	.053		.13	1.70		1.83	2.95

09 91 13.60 Siding Exterior		Crew	Daily Output	Labor-Hours	Unit	Material	Labor	Equipment	Total	Total Incl O&P
0010	**SIDING EXTERIOR**, Alkyd (oil base)									
0450	Steel siding, oil base, paint 1 coat, brushwork	2 Pord	2015	.008	S.F.	.10	.25		.35	.53
0500	Spray		4550	.004		.15	.11		.26	.36
0800	Paint 2 coats, brushwork		1300	.012		.20	.39		.59	.87
1000	Spray		2750	.006		.17	.19		.36	.50
1200	Stucco, rough, oil base, paint 2 coats, brushwork		1300	.012		.20	.39		.59	.87
1400	Roller		1625	.010		.21	.31		.52	.75
1600	Spray		2925	.005		.22	.17		.39	.54
1800	Texture 1-11 or clapboard, oil base, primer coat, brushwork		1300	.012		.15	.39		.54	.82
2000	Spray		4550	.004		.15	.11		.26	.36
2100	Paint 1 coat, brushwork		1300	.012		.15	.39		.54	.81
2200	Spray		4550	.004		.15	.11		.26	.35
2400	Paint 2 coats, brushwork		810	.020		.30	.63		.93	1.36
2600	Spray		2600	.006		.33	.20		.53	.68
3000	Stain 1 coat, brushwork		1520	.011		.09	.34		.43	.65
3200	Spray		5320	.003		.10	.10		.20	.27
3400	Stain 2 coats, brushwork		950	.017		.17	.54		.71	1.07
4000	Spray		3050	.005		.19	.17		.36	.49
4200	Wood shingles, oil base primer coat, brushwork		1300	.012		.14	.39		.53	.80
4400	Spray		3900	.004		.13	.13		.26	.36
4600	Paint 1 coat, brushwork		1300	.012		.12	.39		.51	.79
4800	Spray		3900	.004		.15	.13		.28	.39
5000	Paint 2 coats, brushwork		810	.020		.25	.63		.88	1.31
5200	Spray		2275	.007		.23	.22		.45	.63
5800	Stain 1 coat, brushwork		1500	.011		.09	.34		.43	.66
6000	Spray		3900	.004		.09	.13		.22	.32
6500	Stain 2 coats, brushwork		950	.017		.17	.54		.71	1.07
7000	Spray		2660	.006		.24	.19		.43	.58
8000	For latex paint, deduct					10%				
8100	For work over 12′ H, from pipe scaffolding, add						15%			

09 91 Painting

09 91 13 – Exterior Painting

09 91 13.60 Siding Exterior		Crew	Daily Output	Labor-Hours	Unit	Material	Labor	Equipment	Total	Total Incl O&P
						2016 Bare Costs				
8200	For work over 12′ H, from extension ladder, add				S.F.		25%			
8300	For work over 12′ H, from swing staging, add				S.F.		35%			

09 91 13.62 Siding, Misc.

09 91 13.62 Siding, Misc.		Crew	Daily Output	Labor-Hours	Unit	Material	Labor	Equipment	Total	Total Incl O&P
0010	**SIDING, MISC.**, latex paint									
0100	Aluminum siding									
0110	Brushwork, primer	2 Pord	2275	.007	S.F.	.06	.22		.28	.44
0120	Finish coat, exterior latex		2275	.007		.06	.22		.28	.43
0130	Primer & 1 coat exterior latex		1300	.012		.13	.39		.52	.79
0140	Primer & 2 coats exterior latex		975	.016		.19	.52		.71	1.07
0150	Mineral fiber shingles									
0160	Brushwork, primer	2 Pord	1495	.011	S.F.	.15	.34		.49	.73
0170	Finish coat, industrial enamel		1495	.011		.17	.34		.51	.74
0180	Primer & 1 coat enamel		810	.020		.32	.63		.95	1.39
0190	Primer & 2 coats enamel		540	.030		.48	.95		1.43	2.09
0200	Roll, primer		1625	.010		.17	.31		.48	.71
0210	Finish coat, industrial enamel		1625	.010		.18	.31		.49	.72
0220	Primer & 1 coat enamel		975	.016		.35	.52		.87	1.24
0230	Primer & 2 coats enamel		650	.025		.53	.79		1.32	1.87
0240	Spray, primer		3900	.004		.13	.13		.26	.36
0250	Finish coat, industrial enamel		3900	.004		.15	.13		.28	.38
0260	Primer & 1 coat enamel		2275	.007		.28	.22		.50	.68
0270	Primer & 2 coats enamel		1625	.010		.43	.31		.74	.99
0280	Waterproof sealer, first coat		4485	.004		.10	.11		.21	.30
0290	Second coat		5235	.003		.09	.10		.19	.26
0300	Rough wood incl. shingles, shakes or rough sawn siding									
0310	Brushwork, primer	2 Pord	1280	.013	S.F.	.14	.40		.54	.81
0320	Finish coat, exterior latex		1280	.013		.10	.40		.50	.77
0330	Primer & 1 coat exterior latex		960	.017		.24	.53		.77	1.14
0340	Primer & 2 coats exterior latex		700	.023		.34	.73		1.07	1.58
0350	Roll, primer		2925	.005		.18	.17		.35	.49
0360	Finish coat, exterior latex		2925	.005		.12	.17		.29	.43
0370	Primer & 1 coat exterior latex		1790	.009		.30	.29		.59	.81
0380	Primer & 2 coats exterior latex		1300	.012		.43	.39		.82	1.12
0390	Spray, primer		3900	.004		.15	.13		.28	.39
0400	Finish coat, exterior latex		3900	.004		.10	.13		.23	.33
0410	Primer & 1 coat exterior latex		2600	.006		.25	.20		.45	.59
0420	Primer & 2 coats exterior latex		2080	.008		.34	.25		.59	.78
0430	Waterproof sealer, first coat		4485	.004		.18	.11		.29	.38
0440	Second coat		4485	.004		.10	.11		.21	.30
0450	Smooth wood incl. butt, T&G, beveled, drop or B&B siding									
0460	Brushwork, primer	2 Pord	2325	.007	S.F.	.10	.22		.32	.47
0470	Finish coat, exterior latex		1280	.013		.10	.40		.50	.77
0480	Primer & 1 coat exterior latex		800	.020		.20	.64		.84	1.27
0490	Primer & 2 coats exterior latex		630	.025		.30	.81		1.11	1.66
0500	Roll, primer		2275	.007		.11	.22		.33	.49
0510	Finish coat, exterior latex		2275	.007		.11	.22		.33	.49
0520	Primer & 1 coat exterior latex		1300	.012		.22	.39		.61	.89
0530	Primer & 2 coats exterior latex		975	.016		.33	.52		.85	1.22
0540	Spray, primer		4550	.004		.08	.11		.19	.28
0550	Finish coat, exterior latex		4550	.004		.10	.11		.21	.30
0560	Primer & 1 coat exterior latex		2600	.006		.18	.20		.38	.52
0570	Primer & 2 coats exterior latex		1950	.008		.28	.26		.54	.73

09 91 Painting

09 91 13 – Exterior Painting

09 91 13.62 Siding, Misc.		Crew	Daily Output	Labor-Hours	Unit	2016 Bare Costs Material	Labor	Equipment	Total	Total Incl O&P
0580	Waterproof sealer, first coat	2 Pord	5230	.003	S.F.	.10	.10		.20	.27
0590	Second coat		5980	.003		.10	.09		.19	.25
0600	For oil base paint, add					10%				

09 91 13.70 Doors and Windows, Exterior		Crew	Daily Output	Labor-Hours	Unit	Material	Labor	Equipment	Total	Total Incl O&P
0010	**DOORS AND WINDOWS, EXTERIOR** R099100-10									
0100	Door frames & trim, only									
0110	Brushwork, primer	1 Pord	512	.016	L.F.	.06	.50		.56	.89
0120	Finish coat, exterior latex		512	.016		.08	.50		.58	.90
0130	Primer & 1 coat, exterior latex		300	.027		.14	.85		.99	1.55
0135	2 coats, exterior latex, both sides		15	.533	Ea.	6.85	17		23.85	35.50
0140	Primer & 2 coats, exterior latex		265	.030	L.F.	.22	.96		1.18	1.83
0150	Doors, flush, both sides, incl. frame & trim									
0160	Roll & brush, primer	1 Pord	10	.800	Ea.	4.65	25.50		30.15	47
0170	Finish coat, exterior latex		10	.800		5.85	25.50		31.35	48.50
0180	Primer & 1 coat, exterior latex		7	1.143		10.50	36.50		47	71.50
0190	Primer & 2 coats, exterior latex		5	1.600		16.35	51		67.35	102
0200	Brushwork, stain, sealer & 2 coats polyurethane		4	2		30	64		94	138
0210	Doors, French, both sides, 10-15 lite, incl. frame & trim									
0220	Brushwork, primer	1 Pord	6	1.333	Ea.	2.32	42.50		44.82	72.50
0230	Finish coat, exterior latex		6	1.333		2.93	42.50		45.43	73
0240	Primer & 1 coat, exterior latex		3	2.667		5.25	85		90.25	146
0250	Primer & 2 coats, exterior latex		2	4		8	128		136	219
0260	Brushwork, stain, sealer & 2 coats polyurethane		2.50	3.200		10.70	102		112.70	180
0270	Doors, louvered, both sides, incl. frame & trim									
0280	Brushwork, primer	1 Pord	7	1.143	Ea.	4.65	36.50		41.15	65
0290	Finish coat, exterior latex		7	1.143		5.85	36.50		42.35	66.50
0300	Primer & 1 coat, exterior latex		4	2		10.50	64		74.50	117
0310	Primer & 2 coats, exterior latex		3	2.667		16	85		101	158
0320	Brushwork, stain, sealer & 2 coats polyurethane		4.50	1.778		30	56.50		86.50	127
0330	Doors, panel, both sides, incl. frame & trim									
0340	Roll & brush, primer	1 Pord	6	1.333	Ea.	4.65	42.50		47.15	75
0350	Finish coat, exterior latex		6	1.333		5.85	42.50		48.35	76.50
0360	Primer & 1 coat, exterior latex		3	2.667		10.50	85		95.50	152
0370	Primer & 2 coats, exterior latex		2.50	3.200		16	102		118	186
0380	Brushwork, stain, sealer & 2 coats polyurethane		3	2.667		30	85		115	173
0400	Windows, per ext. side, based on 15 S.F.									
0410	1 to 6 lite									
0420	Brushwork, primer	1 Pord	13	.615	Ea.	.92	19.65		20.57	33.50
0430	Finish coat, exterior latex		13	.615		1.15	19.65		20.80	34
0440	Primer & 1 coat, exterior latex		8	1		2.07	32		34.07	55
0450	Primer & 2 coats, exterior latex		6	1.333		3.16	42.50		45.66	73.50
0460	Stain, sealer & 1 coat varnish		7	1.143		4.23	36.50		40.73	64.50
0470	7 to 10 lite									
0480	Brushwork, primer	1 Pord	11	.727	Ea.	.92	23		23.92	39
0490	Finish coat, exterior latex		11	.727		1.15	23		24.15	39.50
0500	Primer & 1 coat, exterior latex		7	1.143		2.07	36.50		38.57	62.50
0510	Primer & 2 coats, exterior latex		5	1.600		3.16	51		54.16	87.50
0520	Stain, sealer & 1 coat varnish		6	1.333		4.23	42.50		46.73	74.50
0530	12 lite									
0540	Brushwork, primer	1 Pord	10	.800	Ea.	.92	25.50		26.42	43
0550	Finish coat, exterior latex		10	.800		1.15	25.50		26.65	43.50
0560	Primer & 1 coat, exterior latex		6	1.333		2.07	42.50		44.57	72.50

09 91 Painting

09 91 13 – Exterior Painting

09 91 13.70 Doors and Windows, Exterior

		Crew	Daily Output	Labor-Hours	Unit	2016 Bare Costs Material	Labor	Equipment	Total	Total Incl O&P
0570	Primer & 2 coats, exterior latex	1 Pord	5	1.600	Ea.	3.16	51		54.16	87.50
0580	Stain, sealer & 1 coat varnish		6	1.333		4.33	42.50		46.83	75
0590	For oil base paint, add					10%				

09 91 13.80 Trim, Exterior

		Crew	Daily Output	Labor-Hours	Unit	Material	Labor	Equipment	Total	Total Incl O&P
0010	**TRIM, EXTERIOR**									
0100	Door frames & trim (see Doors, interior or exterior)									
0110	Fascia, latex paint, one coat coverage									
0120	1" x 4", brushwork	1 Pord	640	.013	L.F.	.02	.40		.42	.69
0130	Roll		1280	.006		.03	.20		.23	.36
0140	Spray		2080	.004		.02	.12		.14	.22
0150	1" x 6" to 1" x 10", brushwork		640	.013		.08	.40		.48	.75
0160	Roll		1230	.007		.09	.21		.30	.43
0170	Spray		2100	.004		.07	.12		.19	.27
0180	1" x 12", brushwork		640	.013		.08	.40		.48	.75
0190	Roll		1050	.008		.09	.24		.33	.49
0200	Spray		2200	.004		.07	.12		.19	.26
0210	Gutters & downspouts, metal, zinc chromate paint									
0220	Brushwork, gutters, 5", first coat	1 Pord	640	.013	L.F.	.40	.40		.80	1.10
0230	Second coat		960	.008		.38	.27		.65	.85
0240	Third coat		1280	.006		.30	.20		.50	.67
0250	Downspouts, 4", first coat		640	.013		.40	.40		.80	1.10
0260	Second coat		960	.008		.38	.27		.65	.85
0270	Third coat		1280	.006		.30	.20		.50	.67
0280	Gutters & downspouts, wood									
0290	Brushwork, gutters, 5", primer	1 Pord	640	.013	L.F.	.06	.40		.46	.73
0300	Finish coat, exterior latex		640	.013		.07	.40		.47	.74
0310	Primer & 1 coat exterior latex		400	.020		.14	.64		.78	1.20
0320	Primer & 2 coats exterior latex		325	.025		.22	.79		1.01	1.53
0330	Downspouts, 4", primer		640	.013		.06	.40		.46	.73
0340	Finish coat, exterior latex		640	.013		.07	.40		.47	.74
0350	Primer & 1 coat exterior latex		400	.020		.14	.64		.78	1.20
0360	Primer & 2 coats exterior latex		325	.025		.11	.79		.90	1.41
0370	Molding, exterior, up to 14" wide									
0380	Brushwork, primer	1 Pord	640	.013	L.F.	.07	.40		.47	.74
0390	Finish coat, exterior latex		640	.013		.08	.40		.48	.75
0400	Primer & 1 coat exterior latex		400	.020		.17	.64		.81	1.23
0410	Primer & 2 coats exterior latex		315	.025		.17	.81		.98	1.51
0420	Stain & fill		1050	.008		.10	.24		.34	.51
0430	Shellac		1850	.004		.14	.14		.28	.39
0440	Varnish		1275	.006		.10	.20		.30	.44

09 91 13.90 Walls, Masonry (CMU), Exterior

		Crew	Daily Output	Labor-Hours	Unit	Material	Labor	Equipment	Total	Total Incl O&P
0010	**WALLS, MASONRY (CMU), EXTERIOR**									
0360	Concrete masonry units (CMU), smooth surface									
0370	Brushwork, latex, first coat	1 Pord	640	.013	S.F.	.07	.40		.47	.74
0380	Second coat		960	.008		.06	.27		.33	.50
0390	Waterproof sealer, first coat		736	.011		.26	.35		.61	.85
0400	Second coat		1104	.007		.26	.23		.49	.66
0410	Roll, latex, paint, first coat		1465	.005		.09	.17		.26	.38
0420	Second coat		1790	.004		.06	.14		.20	.30
0430	Waterproof sealer, first coat		1680	.005		.26	.15		.41	.53
0440	Second coat		2060	.004		.26	.12		.38	.48
0450	Spray, latex, paint, first coat		1950	.004		.07	.13		.20	.29

09 91 Painting

09 91 13 – Exterior Painting

09 91 13.90 Walls, Masonry (CMU), Exterior		Crew	Daily Output	Labor-Hours	Unit	Material	2016 Bare Costs Labor	Equipment	Total	Total Incl O&P
0460	Second coat	1 Pord	2600	.003	S.F.	.05	.10		.15	.22
0470	Waterproof sealer, first coat		2245	.004		.26	.11		.37	.47
0480	Second coat		2990	.003		.26	.09		.35	.42
0490	Concrete masonry unit (CMU), porous									
0500	Brushwork, latex, first coat	1 Pord	640	.013	S.F.	.14	.40		.54	.82
0510	Second coat		960	.008		.07	.27		.34	.52
0520	Waterproof sealer, first coat		736	.011		.26	.35		.61	.85
0530	Second coat		1104	.007		.26	.23		.49	.66
0540	Roll latex, first coat		1465	.005		.11	.17		.28	.41
0550	Second coat		1790	.004		.07	.14		.21	.31
0560	Waterproof sealer, first coat		1680	.005		.26	.15		.41	.53
0570	Second coat		2060	.004		.26	.12		.38	.48
0580	Spray latex, first coat		1950	.004		.08	.13		.21	.31
0590	Second coat		2600	.003		.05	.10		.15	.22
0600	Waterproof sealer, first coat		2245	.004		.26	.11		.37	.47
0610	Second coat		2990	.003		.26	.09		.35	.42

09 91 23 – Interior Painting

09 91 23.20 Cabinets and Casework

		Crew	Daily Output	Labor-Hours	Unit	Material	Labor	Equipment	Total	Total Incl O&P
0010	**CABINETS AND CASEWORK**									
1000	Primer coat, oil base, brushwork	1 Pord	650	.012	S.F.	.06	.39		.45	.72
2000	Paint, oil base, brushwork, 1 coat		650	.012		.11	.39		.50	.77
2500	2 coats		400	.020		.20	.64		.84	1.28
3000	Stain, brushwork, wipe off		650	.012		.09	.39		.48	.75
4000	Shellac, 1 coat, brushwork		650	.012		.12	.39		.51	.78
4500	Varnish, 3 coats, brushwork, sand after 1st coat		325	.025		.26	.79		1.05	1.57
5000	For latex paint, deduct					10%				

09 91 23.33 Doors and Windows, Interior Alkyd (Oil Base)

		Crew	Daily Output	Labor-Hours	Unit	Material	Labor	Equipment	Total	Total Incl O&P
0010	**DOORS AND WINDOWS, INTERIOR ALKYD (OIL BASE)**									
0500	Flush door & frame, 3′ x 7′, oil, primer, brushwork	1 Pord	10	.800	Ea.	3.70	25.50		29.20	46
1000	Paint, 1 coat		10	.800		4.16	25.50		29.66	46.50
1200	2 coats		6	1.333		4.54	42.50		47.04	75
1400	Stain, brushwork, wipe off		18	.444		1.83	14.20		16.03	25.50
1600	Shellac, 1 coat, brushwork		25	.320		2.47	10.20		12.67	19.50
1800	Varnish, 3 coats, brushwork, sand after 1st coat		9	.889		5.40	28.50		33.90	52.50
2000	Panel door & frame, 3′ x 7′, oil, primer, brushwork		6	1.333		2.41	42.50		44.91	72.50
2200	Paint, 1 coat		6	1.333		4.16	42.50		46.66	74.50
2400	2 coats		3	2.667		10.75	85		95.75	152
2600	Stain, brushwork, panel door, 3′ x 7′, not incl. frame		16	.500		1.83	15.95		17.78	28.50
2800	Shellac, 1 coat, brushwork		22	.364		2.47	11.60		14.07	22
3000	Varnish, 3 coats, brushwork, sand after 1st coat		7.50	1.067		5.40	34		39.40	62
3020	French door, incl. 3′ x 7′, 6 lites, frame & trim									
3022	Paint, 1 coat, over existing paint	1 Pord	5	1.600	Ea.	8.30	51		59.30	93
3024	2 coats, over existing paint		5	1.600		16.15	51		67.15	102
3026	Primer & 1 coat		3.50	2.286		13.15	73		86.15	134
3028	Primer & 2 coats		3	2.667		21.50	85		106.50	164
3032	Varnish or polyurethane, 1 coat		5	1.600		8.40	51		59.40	93.50
3034	2 coats, sanding between		3	2.667		16.85	85		101.85	159
4400	Windows, including frame and trim, per side									
4600	Colonial type, 6/6 lites, 2′ x 3′, oil, primer, brushwork	1 Pord	14	.571	Ea.	.38	18.25		18.63	30.50
5800	Paint, 1 coat		14	.571		.66	18.25		18.91	30.50
6000	2 coats		9	.889		1.28	28.50		29.78	48
6200	3′ x 5′ opening, 6/6 lites, primer coat, brushwork		12	.667		.95	21.50		22.45	36

09 91 23 – Interior Painting

09 91 23.33 Doors and Windows, Interior Alkyd (Oil Base)		Crew	Daily Output	Labor-Hours	Unit	2016 Bare Costs Material	Labor	Equipment	Total	Total Incl O&P
6400	Paint, 1 coat	1 Pord	12	.667	Ea.	1.64	21.50		23.14	37
6600	2 coats		7	1.143		3.19	36.50		39.69	63.50
6800	4' x 8' opening, 6/6 lites, primer coat, brushwork		8	1		2.03	32		34.03	54.50
7000	Paint, 1 coat		8	1		3.50	32		35.50	56.50
7200	2 coats		5	1.600		6.80	51		57.80	91.50
8000	Single lite type, 2' x 3', oil base, primer coat, brushwork		33	.242		.38	7.75		8.13	13.15
8200	Paint, 1 coat		33	.242		.66	7.75		8.41	13.45
8400	2 coats		20	.400		1.28	12.75		14.03	22.50
8600	3' x 5' opening, primer coat, brushwork		20	.400		.95	12.75		13.70	22
8800	Paint, 1 coat		20	.400		1.64	12.75		14.39	23
8900	2 coats		13	.615		3.19	19.65		22.84	36
9200	4' x 8' opening, primer coat, brushwork		14	.571		2.03	18.25		20.28	32
9400	Paint, 1 coat		14	.571		3.50	18.25		21.75	34
9600	2 coats	↓	8	1	↓	6.80	32		38.80	60

09 91 23.35 Doors and Windows, Interior Latex		Crew	Daily Output	Labor-Hours	Unit	Material	Labor	Equipment	Total	Total Incl O&P
0010	**DOORS & WINDOWS, INTERIOR LATEX**									
0100	Doors, flush, both sides, incl. frame & trim									
0110	Roll & brush, primer	1 Pord	10	.800	Ea.	4.15	25.50		29.65	46.50
0120	Finish coat, latex		10	.800		5.20	25.50		30.70	48
0130	Primer & 1 coat latex		7	1.143		9.35	36.50		45.85	70.50
0140	Primer & 2 coats latex		5	1.600		14.25	51		65.25	99.50
0160	Spray, both sides, primer		20	.400		4.36	12.75		17.11	26
0170	Finish coat, latex		20	.400		5.45	12.75		18.20	27
0180	Primer & 1 coat latex		11	.727		9.90	23		32.90	49
0190	Primer & 2 coats latex	↓	8	1	↓	15.10	32		47.10	69
0200	Doors, French, both sides, 10-15 lite, incl. frame & trim									
0210	Roll & brush, primer	1 Pord	6	1.333	Ea.	2.07	42.50		44.57	72.50
0220	Finish coat, latex		6	1.333		2.61	42.50		45.11	73
0230	Primer & 1 coat latex		3	2.667		4.68	85		89.68	145
0240	Primer & 2 coats latex	↓	2	4	↓	7.15	128		135.15	218
0260	Doors, louvered, both sides, incl. frame & trim									
0270	Roll & brush, primer	1 Pord	7	1.143	Ea.	4.15	36.50		40.65	64.50
0280	Finish coat, latex		7	1.143		5.20	36.50		41.70	66
0290	Primer & 1 coat, latex		4	2		9.10	64		73.10	115
0300	Primer & 2 coats, latex		3	2.667		14.55	85		99.55	156
0320	Spray, both sides, primer		20	.400		4.36	12.75		17.11	26
0330	Finish coat, latex		20	.400		5.45	12.75		18.20	27
0340	Primer & 1 coat, latex		11	.727		9.90	23		32.90	49
0350	Primer & 2 coats, latex	↓	8	1	↓	15.45	32		47.45	69.50
0360	Doors, panel, both sides, incl. frame & trim									
0370	Roll & brush, primer	1 Pord	6	1.333	Ea.	4.36	42.50		46.86	75
0380	Finish coat, latex		6	1.333		5.20	42.50		47.70	76
0390	Primer & 1 coat, latex		3	2.667		9.35	85		94.35	150
0400	Primer & 2 coats, latex		2.50	3.200		14.55	102		116.55	184
0420	Spray, both sides, primer		10	.800		4.36	25.50		29.86	47
0430	Finish coat, latex		10	.800		5.45	25.50		30.95	48
0440	Primer & 1 coat, latex		5	1.600		9.90	51		60.90	95
0450	Primer & 2 coats, latex	↓	4	2	↓	15.45	64		79.45	122
0460	Windows, per interior side, based on 15 S.F.									
0470	1 to 6 lite									
0480	Brushwork, primer	1 Pord	13	.615	Ea.	.82	19.65		20.47	33.50
0490	Finish coat, enamel	↓	13	.615	↓	1.03	19.65		20.68	33.50

09 91 23 – Interior Painting

09 91 23.35 Doors and Windows, Interior Latex		Crew	Daily Output	Labor-Hours	Unit	Material	Labor	Equipment	Total	Total Incl O&P
						2016 Bare Costs				
0500	Primer & 1 coat enamel	1 Pord	8	1	Ea.	1.85	32		33.85	54.50
0510	Primer & 2 coats enamel		6	1.333		2.88	42.50		45.38	73
0530	7 to 10 lite									
0540	Brushwork, primer	1 Pord	11	.727	Ea.	.82	23		23.82	39
0550	Finish coat, enamel		11	.727		1.03	23		24.03	39
0560	Primer & 1 coat enamel		7	1.143		1.85	36.50		38.35	62
0570	Primer & 2 coats enamel		5	1.600		2.88	51		53.88	87
0590	12 lite									
0600	Brushwork, primer	1 Pord	10	.800	Ea.	.82	25.50		26.32	43
0610	Finish coat, enamel		10	.800		1.03	25.50		26.53	43
0620	Primer & 1 coat enamel		6	1.333		1.85	42.50		44.35	72
0630	Primer & 2 coats enamel		5	1.600		2.88	51		53.88	87
0650	For oil base paint, add					10%				

09 91 23.39 Doors and Windows, Interior Latex, Zero Voc			Crew	Daily Output	Labor-Hours	Unit	Material	Labor	Equipment	Total	Total Incl O&P
0010	**DOORS & WINDOWS, INTERIOR LATEX, ZERO VOC**										
0100	Doors flush, both sides, incl. frame & trim										
0110	Roll & brush, primer	G	1 Pord	10	.800	Ea.	5.10	25.50		30.60	47.50
0120	Finish coat, latex	G		10	.800		5.55	25.50		31.05	48
0130	Primer & 1 coat latex	G		7	1.143		10.65	36.50		47.15	71.50
0140	Primer & 2 coats latex	G		5	1.600		15.90	51		66.90	101
0160	Spray, both sides, primer	G		20	.400		5.40	12.75		18.15	27
0170	Finish coat, latex	G		20	.400		5.80	12.75		18.55	27.50
0180	Primer & 1 coat latex	G		11	.727		11.25	23		34.25	50.50
0190	Primer & 2 coats latex	G		8	1		16.80	32		48.80	71
0200	Doors, French, both sides, 10-15 lite, incl. frame & trim										
0210	Roll & brush, primer	G	1 Pord	6	1.333	Ea.	2.56	42.50		45.06	73
0220	Finish coat, latex	G		6	1.333		2.77	42.50		45.27	73
0230	Primer & 1 coat latex	G		3	2.667		5.35	85		90.35	146
0240	Primer & 2 coats latex	G		2	4		7.95	128		135.95	219
0360	Doors, panel, both sides, incl. frame & trim										
0370	Roll & brush, primer	G	1 Pord	6	1.333	Ea.	5.40	42.50		47.90	76
0380	Finish coat, latex	G		6	1.333		5.55	42.50		48.05	76
0390	Primer & 1 coat, latex	G		3	2.667		10.65	85		95.65	152
0400	Primer & 2 coats, latex	G		2.50	3.200		16.20	102		118.20	186
0420	Spray, both sides, primer	G		10	.800		5.40	25.50		30.90	48
0430	Finish coat, latex	G		10	.800		5.80	25.50		31.30	48.50
0440	Primer & 1 coat, latex	G		5	1.600		11.25	51		62.25	96.50
0450	Primer & 2 coats, latex	G		4	2		17.15	64		81.15	124
0460	Windows, per interior side, based on 15 S.F.										
0470	1 to 6 lite										
0480	Brushwork, primer	G	1 Pord	13	.615	Ea.	1.01	19.65		20.66	33.50
0490	Finish coat, enamel	G		13	.615		1.09	19.65		20.74	33.50
0500	Primer & 1 coat enamel	G		8	1		2.10	32		34.10	55
0510	Primer & 2 coats enamel	G		6	1.333		3.20	42.50		45.70	73.50

09 91 23.40 Floors, Interior		Crew	Daily Output	Labor-Hours	Unit	Material	Labor	Equipment	Total	Total Incl O&P
0010	**FLOORS, INTERIOR**									
0100	Concrete paint, latex									
0110	Brushwork									
0120	1st coat	1 Pord	975	.008	S.F.	.14	.26		.40	.58
0130	2nd coat		1150	.007		.09	.22		.31	.47
0140	3rd coat		1300	.006		.07	.20		.27	.40
0150	Roll									

09 91 Painting

09 91 23 – Interior Painting

09 91 23.40 Floors, Interior		Crew	Daily Output	Labor-Hours	Unit	Material	Labor	Equipment	Total	Total Incl O&P
						2016 Bare Costs				
0160	1st coat	1 Pord	2600	.003	S.F.	.19	.10		.29	.37
0170	2nd coat		3250	.002		.11	.08		.19	.25
0180	3rd coat		3900	.002		.08	.07		.15	.20
0190	Spray									
0200	1st coat	1 Pord	2600	.003	S.F.	.16	.10		.26	.34
0210	2nd coat		3250	.002		.09	.08		.17	.23
0220	3rd coat		3900	.002		.07	.07		.14	.19

09 91 23.52 Miscellaneous, Interior		Crew	Daily Output	Labor-Hours	Unit	Material	Labor	Equipment	Total	Total Incl O&P
0010	**MISCELLANEOUS, INTERIOR**									
2400	Floors, conc./wood, oil base, primer/sealer coat, brushwork	2 Pord	1950	.008	S.F.	.10	.26		.36	.53
2450	Roller		5200	.003		.10	.10		.20	.27
2600	Spray		6000	.003		.10	.09		.19	.25
2650	Paint 1 coat, brushwork		1950	.008		.10	.26		.36	.54
2800	Roller		5200	.003		.10	.10		.20	.27
2850	Spray		6000	.003		.11	.09		.20	.26
3000	Stain, wood floor, brushwork, 1 coat		4550	.004		.09	.11		.20	.29
3200	Roller		5200	.003		.09	.10		.19	.26
3250	Spray		6000	.003		.09	.09		.18	.24
3400	Varnish, wood floor, brushwork		4550	.004		.09	.11		.20	.28
3450	Roller		5200	.003		.09	.10		.19	.26
3600	Spray		6000	.003		.10	.09		.19	.25
3800	Grilles, per side, oil base, primer coat, brushwork	1 Pord	520	.015		.13	.49		.62	.95
3850	Spray		1140	.007		.13	.22		.35	.52
3880	Paint 1 coat, brushwork		520	.015		.22	.49		.71	1.05
3900	Spray		1140	.007		.24	.22		.46	.64
3920	Paint 2 coats, brushwork		325	.025		.43	.79		1.22	1.76
3940	Spray		650	.012		.49	.39		.88	1.19
4500	Louvers, one side, primer, brushwork		524	.015		.10	.49		.59	.90
4520	Paint one coat, brushwork		520	.015		.10	.49		.59	.91
4530	Spray		1140	.007		.11	.22		.33	.49
4540	Paint two coats, brushwork		325	.025		.18	.79		.97	1.49
4550	Spray		650	.012		.20	.39		.59	.87
4560	Paint three coats, brushwork		270	.030		.27	.95		1.22	1.86
4570	Spray		500	.016		.30	.51		.81	1.17
5000	Pipe, 1" - 4" diameter, primer or sealer coat, oil base, brushwork	2 Pord	1250	.013	L.F.	.10	.41		.51	.78
5100	Spray		2165	.007		.09	.24		.33	.49
5200	Paint 1 coat, brushwork		1250	.013		.11	.41		.52	.79
5300	Spray		2165	.007		.10	.24		.34	.50
5350	Paint 2 coats, brushwork		775	.021		.19	.66		.85	1.30
5400	Spray		1240	.013		.22	.41		.63	.92
5450	5" - 8" diameter, primer or sealer coat, brushwork		620	.026		.20	.82		1.02	1.58
5500	Spray		1085	.015		.33	.47		.80	1.14
5550	Paint 1 coat, brushwork		620	.026		.29	.82		1.11	1.68
5600	Spray		1085	.015		.33	.47		.80	1.14
5650	Paint 2 coats, brushwork		385	.042		.39	1.33		1.72	2.61
5700	Spray		620	.026		.43	.82		1.25	1.83
5750	9" - 12" diameter, primer or sealer coat, brushwork		415	.039		.30	1.23		1.53	2.36
5800	Spray		725	.022		.35	.70		1.05	1.55
5850	Paint 1 coat, brushwork		415	.039		.30	1.23		1.53	2.36
6000	Spray		725	.022		.33	.70		1.03	1.52
6200	Paint 2 coats, brushwork		260	.062		.58	1.96		2.54	3.87
6250	Spray		415	.039		.64	1.23		1.87	2.74

09 91 23 – Interior Painting

09 91 23.52 Miscellaneous, Interior		Crew	Daily Output	Labor-Hours	Unit	Material	2016 Bare Costs Labor	Equipment	Total	Total Incl O&P
6300	13" - 16" diameter, primer or sealer coat, brushwork	2 Pord	310	.052	L.F.	.40	1.65		2.05	3.15
6350	Spray		540	.030		.44	.95		1.39	2.05
6400	Paint 1 coat, brushwork		310	.052		.40	1.65		2.05	3.15
6450	Spray		540	.030		.44	.95		1.39	2.05
6500	Paint 2 coats, brushwork		195	.082		.78	2.62		3.40	5.15
6550	Spray		310	.052		.86	1.65		2.51	3.66
6600	Radiators, per side, primer, brushwork	1 Pord	520	.015	S.F.	.10	.49		.59	.91
6620	Paint, one coat		520	.015		.09	.49		.58	.91
6640	Two coats		340	.024		.18	.75		.93	1.44
6660	Three coats		283	.028		.27	.90		1.17	1.79
7000	Trim, wood, incl. puttying, under 6" wide									
7200	Primer coat, oil base, brushwork	1 Pord	650	.012	L.F.	.03	.39		.42	.68
7250	Paint, 1 coat, brushwork		650	.012		.05	.39		.44	.71
7400	2 coats		400	.020		.11	.64		.75	1.17
7450	3 coats		325	.025		.16	.79		.95	1.46
7500	Over 6" wide, primer coat, brushwork		650	.012		.06	.39		.45	.72
7550	Paint, 1 coat, brushwork		650	.012		.11	.39		.50	.77
7600	2 coats		400	.020		.21	.64		.85	1.28
7650	3 coats		325	.025		.32	.79		1.11	1.64
8000	Cornice, simple design, primer coat, oil base, brushwork		650	.012	S.F.	.06	.39		.45	.72
8250	Paint, 1 coat		650	.012		.11	.39		.50	.77
8300	2 coats		400	.020		.21	.64		.85	1.28
8350	Ornate design, primer coat		350	.023		.06	.73		.79	1.27
8400	Paint, 1 coat		350	.023		.11	.73		.84	1.32
8450	2 coats		400	.020		.21	.64		.85	1.28
8600	Balustrades, primer coat, oil base, brushwork		520	.015		.06	.49		.55	.88
8650	Paint, 1 coat		520	.015		.11	.49		.60	.93
8700	2 coats		325	.025		.21	.79		1	1.52
8900	Trusses and wood frames, primer coat, oil base, brushwork		800	.010		.06	.32		.38	.60
8950	Spray		1200	.007		.07	.21		.28	.42
9000	Paint 1 coat, brushwork		750	.011		.11	.34		.45	.68
9200	Spray		1200	.007		.12	.21		.33	.48
9220	Paint 2 coats, brushwork		500	.016		.21	.51		.72	1.07
9240	Spray		600	.013		.24	.43		.67	.96
9260	Stain, brushwork, wipe off		600	.013		.09	.43		.52	.80
9280	Varnish, 3 coats, brushwork		275	.029		.26	.93		1.19	1.81
9350	For latex paint, deduct					10%				

09 91 23.62 Electrostatic Painting		Crew	Daily Output	Labor-Hours	Unit	Material	Labor	Equipment	Total	Total Incl O&P
0010	**ELECTROSTATIC PAINTING**									
0100	In shop									
0200	Flat surfaces (lockers, casework, elevator doors, etc.)									
0300	One coat	1 Pord	200	.040	S.F.	.57	1.28		1.85	2.72
0400	Two coats	"	120	.067	"	.82	2.13		2.95	4.41
0500	Irregular surfaces (furniture, door frames, etc.)									
0600	One coat	1 Pord	150	.053	S.F.	.57	1.70		2.27	3.42
0700	Two coats	"	100	.080	"	.82	2.55		3.37	5.10
0800	On site									
0900	Flat surfaces (lockers, casework, elevator doors, etc.)									
1000	One coat	1 Pord	150	.053	S.F.	.57	1.70		2.27	3.42
1100	Two coats	"	100	.080	"	.82	2.55		3.37	5.10
1200	Irregular surfaces (furniture, door frames, etc.)									
1300	One coat	1 Pord	115	.070	S.F.	.57	2.22		2.79	4.28

09 91 Painting

09 91 23 – Interior Painting

09 91 23.62 Electrostatic Painting		Crew	Daily Output	Labor-Hours	Unit	Material	Labor	Equipment	Total	Total Incl O&P
						2016 Bare Costs				
1400	Two coats	1 Pord	70	.114	S.F.	.82	3.65		4.47	6.90

09 91 23.72 Walls and Ceilings, Interior

		Crew	Daily Output	Labor-Hours	Unit	Material	Labor	Equipment	Total	Total Incl O&P
0010	**WALLS AND CEILINGS, INTERIOR**									
0100	Concrete, drywall or plaster, latex, primer or sealer coat									
0200	Smooth finish, brushwork	1 Pord	1150	.007	S.F.	.06	.22		.28	.44
0240	Roller		1350	.006		.06	.19		.25	.38
0280	Spray		2750	.003		.05	.09		.14	.21
0300	Sand finish, brushwork		975	.008		.06	.26		.32	.50
0340	Roller		1150	.007		.06	.22		.28	.44
0380	Spray		2275	.004		.05	.11		.16	.25
0400	Paint 1 coat, smooth finish, brushwork		1200	.007		.07	.21		.28	.43
0440	Roller		1300	.006		.07	.20		.27	.40
0480	Spray		2275	.004		.06	.11		.17	.25
0500	Sand finish, brushwork		1050	.008		.06	.24		.30	.47
0540	Roller		1600	.005		.07	.16		.23	.34
0580	Spray		2100	.004		.02	.12		.14	.22
0800	Paint 2 coats, smooth finish, brushwork		680	.012		.14	.38		.52	.77
0840	Roller		800	.010		.14	.32		.46	.68
0880	Spray		1625	.005		.13	.16		.29	.40
0900	Sand finish, brushwork		605	.013		.14	.42		.56	.84
0940	Roller		1020	.008		.14	.25		.39	.56
0980	Spray		1700	.005		.13	.15		.28	.39
1200	Paint 3 coats, smooth finish, brushwork		510	.016		.21	.50		.71	1.05
1240	Roller		650	.012		.21	.39		.60	.88
1280	Spray		1625	.005		.19	.16		.35	.47
1300	Sand finish, brushwork		454	.018		.30	.56		.86	1.26
1340	Roller		680	.012		.31	.38		.69	.96
1380	Spray		1133	.007		.27	.23		.50	.67
1600	Glaze coating, 2 coats, spray, clear		1200	.007		.49	.21		.70	.89
1640	Multicolor		1200	.007		.97	.21		1.18	1.42
1660	Painting walls, complete, including surface prep, primer &									
1670	2 coats finish, on drywall or plaster, with roller	1 Pord	325	.025	S.F.	.21	.79		1	1.52
1700	For oil base paint, add					10%				
1800	For ceiling installations, add						25%			
2000	Masonry or concrete block, primer/sealer, latex paint									
2100	Primer, smooth finish, brushwork	1 Pord	1000	.008	S.F.	.11	.26		.37	.54
2110	Roller		1150	.007		.11	.22		.33	.49
2180	Spray		2400	.003		.10	.11		.21	.29
2200	Sand finish, brushwork		850	.009		.11	.30		.41	.61
2210	Roller		975	.008		.11	.26		.37	.55
2280	Spray		2050	.004		.10	.12		.22	.31
2400	Finish coat, smooth finish, brush		1100	.007		.08	.23		.31	.47
2410	Roller		1300	.006		.08	.20		.28	.41
2480	Spray		2400	.003		.07	.11		.18	.26
2500	Sand finish, brushwork		950	.008		.08	.27		.35	.53
2510	Roller		1090	.007		.08	.23		.31	.48
2580	Spray		2040	.004		.07	.13		.20	.29
2800	Primer plus one finish coat, smooth brush		525	.015		.30	.49		.79	1.13
2810	Roller		615	.013		.19	.42		.61	.89
2880	Spray		1200	.007		.17	.21		.38	.53
2900	Sand finish, brushwork		450	.018		.19	.57		.76	1.14
2910	Roller		515	.016		.19	.50		.69	1.03

09 91 Painting

09 91 23 – Interior Painting

09 91 23.72 Walls and Ceilings, Interior		Crew	Daily Output	Labor-Hours	Unit	Material	2016 Bare Costs Labor	Equipment	Total	Total Incl O&P
2980	Spray	1 Pord	1025	.008	S.F.	.17	.25		.42	.59
3200	Primer plus 2 finish coats, smooth, brush		355	.023		.27	.72		.99	1.48
3210	Roller		415	.019		.27	.62		.89	1.31
3280	Spray		800	.010		.24	.32		.56	.79
3300	Sand finish, brushwork		305	.026		.27	.84		1.11	1.68
3310	Roller		350	.023		.27	.73		1	1.50
3380	Spray		675	.012		.24	.38		.62	.88
3600	Glaze coating, 3 coats, spray, clear		900	.009		.70	.28		.98	1.24
3620	Multicolor		900	.009		1.14	.28		1.42	1.72
4000	Block filler, 1 coat, brushwork		425	.019		.12	.60		.72	1.13
4100	Silicone, water repellent, 2 coats, spray		2000	.004		.37	.13		.50	.62
4120	For oil base paint, add					10%				
8200	For work 8′ - 15′ H, add						10%			
8300	For work over 15′ H, add						20%			
8400	For light textured surfaces, add						10%			
8410	Heavy textured, add						25%			

09 91 23.74 Walls and Ceilings, Interior, Zero VOC Latex			Crew	Daily Output	Labor-Hours	Unit	Material	Labor	Equipment	Total	Total Incl O&P
0010	**WALLS AND CEILINGS, INTERIOR, ZERO VOC LATEX**										
0100	Concrete, dry wall or plaster, latex, primer or sealer coat										
0200	Smooth finish, brushwork	G	1 Pord	1150	.007	S.F.	.07	.22		.29	.44
0240	Roller	G		1350	.006		.06	.19		.25	.38
0280	Spray	G		2750	.003		.05	.09		.14	.20
0300	Sand finish, brushwork	G		975	.008		.06	.26		.32	.50
0340	Roller	G		1150	.007		.07	.22		.29	.45
0380	Spray	G		2275	.004		.05	.11		.16	.25
0400	Paint 1 coat, smooth finish, brushwork	G		1200	.007		.08	.21		.29	.44
0440	Roller	G		1300	.006		.08	.20		.28	.41
0480	Spray	G		2275	.004		.07	.11		.18	.26
0500	Sand finish, brushwork	G		1050	.008		.07	.24		.31	.48
0540	Roller	G		1600	.005		.08	.16		.24	.35
0580	Spray	G		2100	.004		.07	.12		.19	.27
0800	Paint 2 coats, smooth finish, brushwork	G		680	.012		.15	.38		.53	.79
0840	Roller	G		800	.010		.16	.32		.48	.70
0880	Spray	G		1625	.005		.14	.16		.30	.41
0900	Sand finish, brushwork	G		605	.013		.15	.42		.57	.85
0940	Roller	G		1020	.008		.16	.25		.41	.58
0960	Spray	G		1700	.005		.14	.15		.29	.40
1200	Paint 3 coats, smooth finish, brushwork	G		510	.016		.22	.50		.72	1.07
1240	Roller	G		650	.012		.24	.39		.63	.91
1280	Spray	G		1625	.005		.20	.16		.36	.48
1800	For ceiling installations, add	G						25%			
8200	For work 8′ - 15′ H, add							10%			
8300	For work over 15′ H, add							20%			

09 91 23.75 Dry Fall Painting		Crew	Daily Output	Labor-Hours	Unit	Material	Labor	Equipment	Total	Total Incl O&P
0010	**DRY FALL PAINTING** R099100-20									
0100	Sprayed on walls, gypsum board or plaster									
0220	One coat	1 Pord	2600	.003	S.F.	.07	.10		.17	.24
0250	Two coats		1560	.005		.15	.16		.31	.43
0280	Concrete or textured plaster, one coat		1560	.005		.07	.16		.23	.35
0310	Two coats		1300	.006		.15	.20		.35	.48
0340	Concrete block, one coat		1560	.005		.07	.16		.23	.35
0370	Two coats		1300	.006		.15	.20		.35	.48

09 91 Painting

09 91 23 – Interior Painting

09 91 23.75 Dry Fall Painting		Crew	Daily Output	Labor-Hours	Unit	2016 Bare Costs Material	Labor	Equipment	Total	Total Incl O&P
0400	Wood, one coat	1 Pord	877	.009	S.F.	.07	.29		.36	.56
0430	Two coats	↓	650	.012	↓	.15	.39		.54	.81
0440	On ceilings, gypsum board or plaster									
0470	One coat	1 Pord	1560	.005	S.F.	.07	.16		.23	.35
0500	Two coats		1300	.006		.15	.20		.35	.48
0530	Concrete or textured plaster, one coat		1560	.005		.07	.16		.23	.35
0560	Two coats		1300	.006		.15	.20		.35	.48
0570	Structural steel, bar joists or metal deck, one coat		1560	.005		.07	.16		.23	.35
0580	Two coats	↓	1040	.008	↓	.15	.25		.40	.56

09 93 Staining and Transparent Finishing

09 93 23 – Interior Staining and Finishing

09 93 23.10 Varnish		Crew	Daily Output	Labor-Hours	Unit	Material	Labor	Equipment	Total	Total Incl O&P
0010	**VARNISH**									
0012	1 coat + sealer, on wood trim, brush, no sanding included	1 Pord	400	.020	S.F.	.07	.64		.71	1.13
0020	1 coat + sealer, on wood trim, brush, no sanding included, no VOC		400	.020		.21	.64		.85	1.28
0100	Hardwood floors, 2 coats, no sanding included, roller	↓	1890	.004	↓	.15	.13		.28	.39

09 96 High-Performance Coatings

09 96 23 – Graffiti-Resistant Coatings

09 96 23.10 Graffiti Resistant Treatments		Crew	Daily Output	Labor-Hours	Unit	Material	Labor	Equipment	Total	Total Incl O&P
0010	**GRAFFITI RESISTANT TREATMENTS**, sprayed on walls									
0100	Non-sacrificial, permanent non-stick coating, clear, on metals	1 Pord	2000	.004	S.F.	2.06	.13		2.19	2.48
0200	Concrete		2000	.004		2.35	.13		2.48	2.79
0300	Concrete block		2000	.004		3.03	.13		3.16	3.55
0400	Brick		2000	.004		3.44	.13		3.57	3.99
0500	Stone		2000	.004		3.44	.13		3.57	3.99
0600	Unpainted wood		2000	.004		3.97	.13		4.10	4.58
2000	Semi-permanent cross linking polymer primer, on metals		2000	.004		.70	.13		.83	.98
2100	Concrete		2000	.004		.84	.13		.97	1.13
2200	Concrete block		2000	.004		1.05	.13		1.18	1.36
2300	Brick		2000	.004		.84	.13		.97	1.13
2400	Stone		2000	.004		.84	.13		.97	1.13
2500	Unpainted wood		2000	.004		1.16	.13		1.29	1.49
3000	Top coat, on metals		2000	.004		.55	.13		.68	.81
3100	Concrete		2000	.004		.62	.13		.75	.90
3200	Concrete block		2000	.004		.87	.13		1	1.17
3300	Brick		2000	.004		.73	.13		.86	1.01
3400	Stone		2000	.004		.73	.13		.86	1.01
3500	Unpainted wood		2000	.004		.87	.13		1	1.17
5000	Sacrificial, water based, on metal		2000	.004		.32	.13		.45	.57
5100	Concrete		2000	.004		.32	.13		.45	.57
5200	Concrete block		2000	.004		.32	.13		.45	.57
5300	Brick		2000	.004		.32	.13		.45	.57
5400	Stone		2000	.004		.32	.13		.45	.57
5500	Unpainted wood	↓	2000	.004	↓	.32	.13		.45	.57
8000	Cleaner for use after treatment									
8100	Towels or wipes, per package of 30				Ea.	.63			.63	.70
8200	Aerosol spray, 24 oz. can				"	18			18	19.80

09 96 High-Performance Coatings

09 96 23 – Graffiti-Resistant Coatings

09 96 23.10 Graffiti Resistant Treatments		Crew	Daily Output	Labor-Hours	Unit	Material	2016 Bare Costs Labor	Equipment	Total	Total Incl O&P
8500	Graffiti removal with chemicals									
8510	Brush on, spray rinse off, on brick, masonry or stone	A-1H	200	.040	S.F.	.32	1.17	.39	1.88	2.74
8520	On smooth concrete	"	400	.020		.26	.58	.19	1.03	1.47
8530	Wipe on, wipe off, on plastic or painted metal	1 Clab	1500	.005		.63	.16		.79	.96
8540	Brush on, wipe off, on painted wood	"	500	.016	↓	.35	.47		.82	1.16

09 96 46 – Intumescent Painting

09 96 46.10 Coatings, Intumescent		Crew	Daily Output	Labor-Hours	Unit	Material	Labor	Equipment	Total	Total Incl O&P
0010	**COATINGS, INTUMESCENT**, spray applied									
0100	On exterior structural steel, 0.25" d.f.t.	1 Pord	475	.017	S.F.	.41	.54		.95	1.33
0150	0.51" d.f.t.		350	.023		.41	.73		1.14	1.65
0200	0.98" d.f.t.		280	.029		.41	.91		1.32	1.95
0300	On interior structural steel, 0.108" d.f.t.		300	.027		.41	.85		1.26	1.85
0350	0.310" d.f.t.		150	.053		.41	1.70		2.11	3.25
0400	0.670" d.f.t.	↓	100	.080	↓	.41	2.55		2.96	4.65

09 96 53 – Elastomeric Coatings

09 96 53.10 Coatings, Elastomeric		Crew	Daily Output	Labor-Hours	Unit	Material	Labor	Equipment	Total	Total Incl O&P
0010	**COATINGS, ELASTOMERIC**									
0020	High build, water proof, one coat system									
0100	Concrete, brush	1 Pord	650	.012	S.F.	.26	.39		.65	.94

09 96 56 – Epoxy Coatings

09 96 56.20 Wall Coatings		Crew	Daily Output	Labor-Hours	Unit	Material	Labor	Equipment	Total	Total Incl O&P
0010	**WALL COATINGS**									
0100	Acrylic glazed coatings, matte	1 Pord	525	.015	S.F.	.31	.49		.80	1.14
0200	Gloss		305	.026		.65	.84		1.49	2.10
0300	Epoxy coatings, solvent based		525	.015		.40	.49		.89	1.24
0400	Water based		170	.047		.27	1.50		1.77	2.77
0600	Exposed aggregate, troweled on, 1/16" to 1/4", solvent based		235	.034		.62	1.09		1.71	2.47
0700	Water based (epoxy or polyacrylate)		130	.062		1.33	1.96		3.29	4.69
0900	1/2" to 5/8" aggregate, solvent based		130	.062		1.20	1.96		3.16	4.55
1000	Water based		80	.100		2.08	3.19		5.27	7.55
1200	1" aggregate size, solvent based		90	.089		2.12	2.84		4.96	7
1300	Water based		55	.145		3.23	4.64		7.87	11.20
1500	Exposed aggregate, sprayed on, 1/8" aggregate, solvent based		295	.027		.57	.87		1.44	2.06
1600	Water based		145	.055		1.05	1.76		2.81	4.06
1800	High build epoxy, 50 mil, solvent based		390	.021		.68	.65		1.33	1.83
1900	Water based		95	.084		1.15	2.69		3.84	5.70
2100	Laminated epoxy with fiberglass, solvent based		295	.027		.73	.87		1.60	2.23
2200	Water based		145	.055		1.32	1.76		3.08	4.35
2400	Sprayed perlite or vermiculite, 1/16" thick, solvent based		2935	.003		.27	.09		.36	.44
2500	Water based		640	.013		.74	.40		1.14	1.47
2700	Vinyl plastic wall coating, solvent based		735	.011		.33	.35		.68	.93
2800	Water based		240	.033		.82	1.06		1.88	2.65
3000	Urethane on smooth surface, 2 coats, solvent based		1135	.007		.31	.22		.53	.71
3100	Water based		665	.012		.59	.38		.97	1.28
3300	3 coat, solvent based		840	.010		.44	.30		.74	.98
3400	Water based		470	.017		.79	.54		1.33	1.76
3600	Ceramic-like glazed coating, cementitious, solvent based		440	.018		.48	.58		1.06	1.49
3700	Water based		345	.023		.81	.74		1.55	2.11
3900	Resin base, solvent based		640	.013		.38	.40		.78	1.08
4000	Water based	↓	330	.024	↓	.58	.77		1.35	1.91

09 97 Special Coatings

09 97 13 – Steel Coatings

09 97 13.23 Exterior Steel Coatings		Crew	Daily Output	Labor-Hours	Unit	Material (2016 Bare Costs)	Labor (2016 Bare Costs)	Equipment (2016 Bare Costs)	Total (2016 Bare Costs)	Total Incl O&P
0010	**EXTERIOR STEEL COATINGS**									
6101	Cold galvanizing, brush in field	1 Pord	1100	.007	S.F.	.24	.23		.47	.64
6510	Paints & protective coatings, sprayed in field									
6520	Alkyds, primer	2 Psst	3600	.004	S.F.	.10	.14		.24	.37
6540	Gloss topcoats		3200	.005		.08	.16		.24	.39
6560	Silicone alkyd		3200	.005		.17	.16		.33	.48
6610	Epoxy, primer		3000	.005		.28	.17		.45	.63
6630	Intermediate or topcoat		2800	.006		.27	.19		.46	.65
6650	Enamel coat		2800	.006		.34	.19		.53	.73
6700	Epoxy ester, primer		2800	.006		.45	.19		.64	.84
6720	Topcoats		2800	.006		.23	.19		.42	.60
6810	Latex primer		3600	.004		.06	.14		.20	.34
6830	Topcoats		3200	.005		.07	.16		.23	.38
6910	Universal primers, one part, phenolic, modified alkyd		2000	.008		.39	.26		.65	.91
6940	Two part, epoxy spray		2000	.008		.36	.26		.62	.87
7000	Zinc rich primers, self cure, spray, inorganic		1800	.009		.88	.29		1.17	1.51
7010	Epoxy, spray, organic	↓	1800	.009	↓	.27	.29		.56	.84
7020	Above one story, spray painting simple structures, add						25%			
7030	Intricate structures, add						50%			

Estimating Tips

General

- The items in this division are usually priced per square foot or each.
- Many items in Division 10 require some type of support system or special anchors that are not usually furnished with the item. The required anchors must be added to the estimate in the appropriate division.
- Some items in Division 10, such as lockers, may require assembly before installation. Verify the amount of assembly required. Assembly can often exceed installation time.

10 20 00 Interior Specialties

- Support angles and blocking are not included in the installation of toilet compartments, shower/dressing compartments, or cubicles. Appropriate line items from Divisions 5 or 6 may need to be added to support the installations.
- Toilet partitions are priced by the stall. A stall consists of a side wall, pilaster, and door with hardware. Toilet tissue holders and grab bars are extra.
- The required acoustical rating of a folding partition can have a significant impact on costs. Verify the sound transmission coefficient rating of the panel priced to the specification requirements.
- Grab bar installation does not include supplemental blocking or backing to support the required load. When grab bars are installed at an existing facility, provisions must be made to attach the grab bars to solid structure.

Reference Numbers

Reference numbers are shown at the beginning of some major classifications. These numbers refer to related items in the Reference Section. The reference information may be an estimating procedure, an alternate pricing method, or technical information.

Note: Not all subdivisions listed here necessarily appear. ■

10 11 Visual Display Units

10 11 13 – Chalkboards

10 11 13.13 Fixed Chalkboards		Crew	Daily Output	Labor-Hours	Unit	2016 Bare Costs Material	Labor	Equipment	Total	Total Incl O&P
0010	**FIXED CHALKBOARDS** Porcelain enamel steel									
3900	Wall hung									
4000	Aluminum frame and chalktrough									
4300	3′ x 5′	2 Carp	15	1.067	Ea.	325	40.50		365.50	425
4600	4′ x 12′	"	13	1.231	"	605	46.50		651.50	750
4700	Wood frame and chalktrough									
4800	3′ x 4′	2 Carp	16	1	Ea.	199	38		237	283
5300	4′ x 8′	"	13	1.231	"	320	46.50		366.50	430
5400	Liquid chalk, white porcelain enamel, wall hung									
5420	Deluxe units, aluminum trim and chalktrough									
5450	4′ x 4′	2 Carp	16	1	Ea.	269	38		307	360
5550	4′ x 12′	"	12	1.333	"	545	50.50		595.50	685
5700	Wood trim and chalktrough									
5900	4′ x 4′	2 Carp	16	1	Ea.	715	38		753	850
6200	4′ x 8′	"	14	1.143	"	970	43		1,013	1,150

10 11 13.43 Portable Chalkboards

		Crew	Daily Output	Labor-Hours	Unit	Material	Labor	Equipment	Total	Total Incl O&P
0010	**PORTABLE CHALKBOARDS**									
0100	Freestanding, reversible									
0120	Economy, wood frame, 4′ x 6′									
0140	Chalkboard both sides				Ea.	600			600	660
0160	Chalkboard one side, cork other side				"	570			570	625
0200	Standard, lightweight satin finished aluminum, 4′ x 6′									
0220	Chalkboard both sides				Ea.	660			660	725
0240	Chalkboard one side, cork other side				"	665			665	730
0300	Deluxe, heavy duty extruded aluminum, 4′ x 6′									
0320	Chalkboard both sides				Ea.	1,075			1,075	1,175
0340	Chalkboard one side, cork other side				"	985			985	1,075

10 11 16 – Markerboards

10 11 16.53 Electronic Markerboards

		Crew	Daily Output	Labor-Hours	Unit	Material	Labor	Equipment	Total	Total Incl O&P
0010	**ELECTRONIC MARKERBOARDS**									
0100	Wall hung or free standing, 3′ x 4′ to 4′ x 6′	2 Carp	8	2	S.F.	87.50	75.50		163	223
0150	5′ x 6′ to 4′ x 8′		8	2	"	64.50	75.50		140	198
0500	Interactive projection module for existing whiteboards	↓	8	2	Ea.	1,300	75.50		1,375.50	1,550

10 11 23 – Tackboards

10 11 23.10 Fixed Tackboards

		Crew	Daily Output	Labor-Hours	Unit	Material	Labor	Equipment	Total	Total Incl O&P
0010	**FIXED TACKBOARDS**									
0020	Cork sheets, unbacked, no frame, 1/4″ thick	2 Carp	290	.055	S.F.	1.38	2.09		3.47	5
2120	Prefabricated, 1/4″ cork, 3′ x 5′ with aluminum frame		16	1	Ea.	135	38		173	212
2140	Wood frame	↓	16	1	"	159	38		197	239
2300	Glass enclosed cabinets, alum., cork panel, hinged doors									
2600	4′ x 7′, 3 door	2 Carp	10	1.600	Ea.	1,775	60.50		1,835.50	2,050

10 13 Directories

10 13 10 – Building Directories

10 13 10.10 Directory Boards		Crew	Daily Output	Labor-Hours	Unit	2016 Bare Costs Material	Labor	Equipment	Total	Total Incl O&P
0010	**DIRECTORY BOARDS**									
0050	Plastic, glass covered, 30″ x 20″	2 Carp	3	5.333	Ea.	198	202		400	555
0100	36″ x 48″		2	8		845	300		1,145	1,425
0900	Outdoor, weatherproof, black plastic, 36″ x 24″		2	8		760	300		1,060	1,350
1000	36″ x 36″		1.50	10.667		880	405		1,285	1,650

10 14 Signage

10 14 19 – Dimensional Letter Signage

10 14 19.10 Plaques		Crew	Daily Output	Labor-Hours	Unit	2016 Bare Costs Material	Labor	Equipment	Total	Total Incl O&P
0012	**PLAQUES**									
3900	Plaques, custom, 20″ x 30″, for up to 450 letters, cast aluminum	2 Carp	4	4	Ea.	1,850	151		2,001	2,275
4000	Cast bronze		4	4		1,775	151		1,926	2,200
4200	30″ x 36″, up to 900 letters, cast aluminum		3	5.333		2,675	202		2,877	3,275
4300	Cast bronze		3	5.333		4,100	202		4,302	4,850
5100	Exit signs, 24 ga. alum., 14″ x 12″ surface mounted	1 Carp	30	.267		50	10.10		60.10	72
5200	10″ x 7″	″	20	.400		27.50	15.10		42.60	55.50
6400	Replacement sign faces, 6″ or 8″	1 Clab	50	.160		66	4.67		70.67	80.50
8000	Internally illuminated, custom									
8100	On pedestal, 84″ x 30″ x 12″	L-7	.50	52	Ea.	12,600	1,675		14,275	16,700

10 14 23 – Panel Signage

10 14 23.13 Engraved Panel Signage		Crew	Daily Output	Labor-Hours	Unit	2016 Bare Costs Material	Labor	Equipment	Total	Total Incl O&P
0010	**ENGRAVED PANEL SIGNAGE**, interior									
1010	Flexible door sign, adhesive back, w/Braille, 5/8″ letters, 4″ x 4″	1 Clab	32	.250	Ea.	33	7.30		40.30	49
1050	6″ x 6″		32	.250		48.50	7.30		55.80	66
1100	8″ x 2″		32	.250		33	7.30		40.30	49
1150	8″ x 4″		32	.250		43.50	7.30		50.80	60
1200	8″ x 8″		32	.250		53	7.30		60.30	71
1250	12″ x 2″		32	.250		36	7.30		43.30	52
1300	12″ x 6″		32	.250		39	7.30		46.30	55.50
1350	12″ x 12″		32	.250		150	7.30		157.30	177
1500	Graphic symbols, 2″ x 2″		32	.250		12	7.30		19.30	25.50
1550	6″ x 6″		32	.250		31	7.30		38.30	46.50
1600	8″ x 8″		32	.250		39	7.30		46.30	55
2010	Corridor, stock acrylic, 2-sided, with mounting bracket, 2″ x 8″	1 Carp	24	.333		24.50	12.60		37.10	48
2020	2″ x 10″		24	.333		35.50	12.60		48.10	60
2050	3″ x 8″		24	.333		28.50	12.60		41.10	52.50
2060	3″ x 10″		24	.333		38	12.60		50.60	62.50
2070	3″ x 12″		24	.333		37	12.60		49.60	61.50
2100	4″ x 8″		24	.333		21	12.60		33.60	44
2110	4″ x 10″		24	.333		38	12.60		50.60	63
2120	4″ x 12″		24	.333		53.50	12.60		66.10	79.50
7000	Wayfinding signage, custom									
7010	Plastic, flexible, 1/8″ thick, incl. mounting									
7020	6″ x 6″	1 Carp	32	.250	Ea.	16.90	9.45		26.35	34.50
7030	6″ x 12″		32	.250		34.50	9.45		43.95	54
7040	6″ x 24″		28	.286		69	10.80		79.80	94
7050	8″ x 8″		32	.250		30.50	9.45		39.95	50
7060	8″ x 16″		28	.286		61.50	10.80		72.30	85.50
7070	8″ x 24″		26	.308		92	11.65		103.65	121
7080	12″ x 12″		28	.286		69	10.80		79.80	94

10 14 Signage

10 14 23 – Panel Signage

10 14 23.13 Engraved Panel Signage		Crew	Daily Output	Labor-Hours	Unit	2016 Bare Costs Material	Labor	Equipment	Total	Total Incl O&P
7090	12″ x 24″	1 Carp	24	.333	Ea.	138	12.60		150.60	173
7100	12″ x 36″		20	.400		207	15.10		222.10	254
7210	Weather resistant, engraved and color filled									
7220	8″ x 8″	1 Carp	32	.250	Ea.	38	9.45		47.45	58
7230	8″ x 16″		28	.286		61	10.80		71.80	85
7240	8″ x 24″		26	.308		93	11.65		104.65	122
7250	12″ x 12″		28	.286		65	10.80		75.80	89.50
7260	12″ x 24″		24	.333		118	12.60		130.60	150
7270	12″ x 36″		20	.400		191	15.10		206.10	236
7280	16″ x 32″	2 Carp	60	.267		212	10.10		222.10	250
7290	16″ x 48″	″	60	.267		350	10.10		360.10	400
7320	For engraved letters, 1/2″ high, add					.56			.56	.62
7330	1″ high, add					.70			.70	.77
7340	2″ high, add					1.26			1.26	1.39
7350	3″ high, add					1.87			1.87	2.06
7360	For engraved graphic symbols, 3″ high, add					13.35			13.35	14.70
7370	7″ high, add					16.65			16.65	18.30
7380	10″ high, add					23			23	25.50

10 17 Telephone Specialties

10 17 16 – Telephone Enclosures

10 17 16.10 Commercial Telephone Enclosures		Crew	Daily Output	Labor-Hours	Unit	Material	Labor	Equipment	Total	Total Incl O&P
0010	**COMMERCIAL TELEPHONE ENCLOSURES**									
0300	Shelf type, wall hung, recessed	2 Carp	5	3.200	Ea.	745	121		866	1,025
0400	Surface mount	″	5	3.200	″	1,625	121		1,746	1,975

10 21 Compartments and Cubicles

10 21 13 – Toilet Compartments

10 21 13.13 Metal Toilet Compartments		Crew	Daily Output	Labor-Hours	Unit	Material	Labor	Equipment	Total	Total Incl O&P
0010	**METAL TOILET COMPARTMENTS**									
0110	Cubicles, ceiling hung									
0200	Powder coated steel	2 Carp	4	4	Ea.	525	151		676	830
0500	Stainless steel	″	4	4		1,100	151		1,251	1,475
0600	For handicap units, incl. 52″ grab bars, add					450			450	495
0900	Floor and ceiling anchored									
1000	Powder coated steel	2 Carp	5	3.200	Ea.	570	121		691	835
1300	Stainless steel	″	5	3.200		1,225	121		1,346	1,550
1400	For handicap units, incl. 52″ grab bars, add					370			370	405
1610	Floor anchored									
1700	Powder coated steel	2 Carp	7	2.286	Ea.	600	86.50		686.50	805
2000	Stainless steel	″	7	2.286		1,425	86.50		1,511.50	1,725
2100	For handicap units, incl. 52″ grab bars, add					335			335	370
2200	For juvenile units, deduct					41.50			41.50	45.50
2450	Floor anchored, headrail braced									
2500	Powder coated steel	2 Carp	6	2.667	Ea.	380	101		481	590
2804	Stainless steel	″	6	2.667		1,025	101		1,126	1,300
2900	For handicap units, incl. 52″ grab bars, add					370			370	410
3000	Wall hung partitions, powder coated steel	2 Carp	7	2.286		635	86.50		721.50	845
3300	Stainless steel	″	7	2.286		1,650	86.50		1,736.50	1,975

10 21 Compartments and Cubicles

10 21 13 – Toilet Compartments

10 21 13.13 Metal Toilet Compartments		Crew	Daily Output	Labor-Hours	Unit	Material	Labor	Equipment	Total	Total Incl O&P
						2016 Bare Costs				
3400	For handicap units, incl. 52" grab bars, add				Ea.	370			370	410
4000	Screens, entrance, floor mounted, 58" high, 48" wide									
4200	Powder coated steel	2 Carp	15	1.067	Ea.	242	40.50		282.50	335
4500	Stainless steel	"	15	1.067	"	910	40.50		950.50	1,075
4650	Urinal screen, 18" wide									
4704	Powder coated steel	2 Carp	6.15	2.602	Ea.	209	98.50		307.50	395
5004	Stainless steel	"	6.15	2.602	"	535	98.50		633.50	755
5100	Floor mounted, head rail braced									
5300	Powder coated steel	2 Carp	8	2	Ea.	230	75.50		305.50	380
5600	Stainless steel	"	8	2	"	580	75.50		655.50	765
5750	Pilaster, flush									
5800	Powder coated steel	2 Carp	10	1.600	Ea.	270	60.50		330.50	400
6100	Stainless steel		10	1.600		610	60.50		670.50	770
6300	Post braced, powder coated steel		10	1.600		170	60.50		230.50	288
6600	Stainless steel		10	1.600		450	60.50		510.50	595
6700	Wall hung, bracket supported									
6800	Powder coated steel	2 Carp	10	1.600	Ea.	170	60.50		230.50	288
7100	Stainless steel		10	1.600		290	60.50		350.50	420
7400	Flange supported, powder coated steel		10	1.600		114	60.50		174.50	226
7700	Stainless steel		10	1.600		315	60.50		375.50	450
7800	Wedge type, powder coated steel		10	1.600		134	60.50		194.50	249
8100	Stainless steel		10	1.600		575	60.50		635.50	735

10 21 13.14 Metal Toilet Compartment Components

		Crew	Daily Output	Labor-Hours	Unit	Material	Labor	Equipment	Total	Total Incl O&P
0010	**METAL TOILET COMPARTMENT COMPONENTS**									
0100	Pilasters									
0110	Overhead braced, powder coated steel, 7" wide x 82" high	2 Carp	22.20	.721	Ea.	75	27		102	128
0120	Stainless steel		22.20	.721		135	27		162	194
0130	Floor braced, powder coated steel, 7" wide x 70" high		23.30	.687		128	26		154	185
0140	Stainless steel		23.30	.687		253	26		279	320
0150	Ceiling hung, powder coated steel, 7" wide x 83" high		13.30	1.203		137	45.50		182.50	227
0160	Stainless steel		13.30	1.203		283	45.50		328.50	385
0170	Wall hung, powder coated steel, 3" wide x 58" high		18.90	.847		133	32		165	200
0180	Stainless steel		18.90	.847		202	32		234	276
0200	Panels									
0210	Powder coated steel, 31" wide x 58" high	2 Carp	18.90	.847	Ea.	138	32		170	205
0220	Stainless steel		18.90	.847		360	32		392	450
0230	Powder coated steel, 53" wide x 58" high		18.90	.847		173	32		205	244
0240	Stainless steel		18.90	.847		470	32		502	570
0250	Powder coated steel, 63" wide x 58" high		18.90	.847		209	32		241	284
0260	Stainless steel		18.90	.847		540	32		572	650
0300	Doors									
0310	Powder coated steel, 24" wide x 58" high	2 Carp	14.10	1.135	Ea.	145	43		188	231
0320	Stainless steel		14.10	1.135		305	43		348	405
0330	Powder coated steel, 26" wide x 58" high		14.10	1.135		147	43		190	234
0340	Stainless steel		14.10	1.135		315	43		358	415
0350	Powder coated steel, 28" wide x 58" high		14.10	1.135		166	43		209	255
0360	Stainless steel		14.10	1.135		350	43		393	455
0370	Powder coated steel, 36" wide x 58" high		14.10	1.135		180	43		223	270
0380	Stainless steel		14.10	1.135		390	43		433	500
0400	Headrails									
0410	For powder coated steel, 62" long	2 Carp	65	.246	Ea.	22	9.30		31.30	40
0420	Stainless steel		65	.246		22	9.30		31.30	40

10 21 13 – Toilet Compartments

10 21 13.14 Metal Toilet Compartment Components		Crew	Daily Output	Labor-Hours	Unit	Material	Labor	Equipment	Total	Total Incl O&P
						2016 Bare Costs				
0430	For powder coated steel, 84" long	2 Carp	50	.320	Ea.	32	12.10		44.10	55.50
0440	Stainless steel		50	.320		31.50	12.10		43.60	55.50
0450	For powder coated steel, 120" long		30	.533		44	20		64	82.50
0460	Stainless steel		30	.533		44	20		64	82.50

10 21 13.16 Plastic-Laminate-Clad Toilet Compartments

Line	Description	Crew	Daily Output	Labor-Hours	Unit	Material	Labor	Equipment	Total	Total Incl O&P
0010	**PLASTIC-LAMINATE-CLAD TOILET COMPARTMENTS**									
0110	Cubicles, ceiling hung									
0300	Plastic laminate on particle board	2 Carp	4	4	Ea.	525	151		676	835
0600	For handicap units, incl. 52" grab bars, add				"	450			450	495
0900	Floor and ceiling anchored									
1100	Plastic laminate on particle board	2 Carp	5	3.200	Ea.	820	121		941	1,100
1400	For handicap units, incl. 52" grab bars, add				"	370			370	405
1610	Floor mounted									
1800	Plastic laminate on particle board	2 Carp	7	2.286	Ea.	535	86.50		621.50	730
2450	Floor mounted, headrail braced									
2600	Plastic laminate on particle board	2 Carp	6	2.667	Ea.	765	101		866	1,025
3400	For handicap units, incl. 52" grab bars, add					370			370	410
4300	Entrance screen, floor mtd., plas. lam., 58" high, 48" wide	2 Carp	15	1.067		630	40.50		670.50	760
4800	Urinal screen, 18" wide, ceiling braced, plastic laminate		8	2		204	75.50		279.50	350
5400	Floor mounted, headrail braced		8	2		204	75.50		279.50	350
5900	Pilaster, flush, plastic laminate		10	1.600		520	60.50		580.50	675
6400	Post braced, plastic laminate		10	1.600		315	60.50		375.50	450
6700	Wall hung, bracket supported									
6900	Plastic laminate on particle board	2 Carp	10	1.600	Ea.	94.50	60.50		155	205
7450	Flange supported									
7500	Plastic laminate on particle board	2 Carp	10	1.600	Ea.	240	60.50		300.50	365

10 21 13.17 Plastic-Lam. Clad Toilet Compart. Components

Line	Description	Crew	Daily Output	Labor-Hours	Unit	Material	Labor	Equipment	Total	Total Incl O&P
0010	**PLASTIC-LAMINATE CLAD TOILET COMPARTMENT COMPONENTS**									
0100	Pilasters									
0110	Overhead braced, 7" wide x 82" high	2 Carp	22.20	.721	Ea.	101	27		128	157
0130	Floor anchored, 7" wide x 70" high		23.30	.687		101	26		127	155
0150	Ceiling hung, 7" wide x 83" high		13.30	1.203		105	45.50		150.50	192
0180	Wall hung, 3" wide x 58" high		18.90	.847		93	32		125	156
0200	Panels									
0210	31" wide x 58" high	2 Carp	18.90	.847	Ea.	150	32		182	219
0230	51" wide x 58" high		18.90	.847		205	32		237	280
0250	63" wide x 58" high		18.90	.847		239	32		271	315
0300	Doors									
0310	24" wide x 58" high	2 Carp	14.10	1.135	Ea.	148	43		191	235
0330	26" wide x 58" high		14.10	1.135		153	43		196	240
0350	28" wide x 58" high		14.10	1.135		158	43		201	246
0370	36" wide x 58" high		14.10	1.135		188	43		231	279
0400	Headrails									
0410	62" long	2 Carp	65	.246	Ea.	22.50	9.30		31.80	40.50
0430	84" long		60	.267		31	10.10		41.10	51
0450	120" long		30	.533		43	20		63	81.50

10 21 13.19 Plastic Toilet Compartments

Line	Description	Crew	Daily Output	Labor-Hours	Unit	Material	Labor	Equipment	Total	Total Incl O&P
0010	**PLASTIC TOILET COMPARTMENTS**									
0110	Cubicles, ceiling hung									
0250	Phenolic	2 Carp	4	4	Ea.	890	151		1,041	1,225
0600	For handicap units, incl. 52" grab bars, add				"	450			450	495
0900	Floor and ceiling anchored									

10 21 Compartments and Cubicles

10 21 13 – Toilet Compartments

10 21 13.19 Plastic Toilet Compartments		Crew	Daily Output	Labor-Hours	Unit	Material	2016 Bare Costs Labor	Equipment	Total	Total Incl O&P
1050	Phenolic	2 Carp	5	3.200	Ea.	840	121		961	1,125
1400	For handicap units, incl. 52″ grab bars, add				″	370			370	405
1610	Floor mounted									
1750	Phenolic	2 Carp	7	2.286	Ea.	750	86.50		836.50	970
2100	For handicap units, incl. 52″ grab bars, add					335			335	370
2200	For juvenile units, deduct					41.50			41.50	45.50
2450	Floor mounted, headrail braced									
2550	Phenolic	2 Carp	6	2.667	Ea.	750	101		851	995

10 21 13.20 Plastic Toilet Compartment Components

Line	Description	Crew	Daily Output	Labor-Hours	Unit	Material	Labor	Equipment	Total	Total Incl O&P
0010	PLASTIC TOILET COMPARTMENT COMPONENTS									
0100	Pilasters									
0110	Overhead braced, polymer plastic, 7″ wide x 82″ high	2 Carp	22.20	.721	Ea.	129	27		156	188
0120	Phenolic		22.20	.721		152	27		179	214
0130	Floor braced, polymer plastic, 7″ wide x 70″ high		23.30	.687		178	26		204	240
0140	Phenolic		23.30	.687		144	26		170	202
0150	Ceiling hung, polymer plastic, 7″ wide x 83″ high		13.30	1.203		171	45.50		216.50	264
0160	Phenolic		13.30	1.203		163	45.50		208.50	255
0180	Wall hung, phenolic, 3″ wide x 58″ high		18.90	.847		98.50	32		130.50	163
0200	Panels									
0203	Polymer plastic, 18″ high x 55″ high	2 Carp	18.90	.847	Ea.	274	32		306	355
0206	Phenolic, 18″ wide x 58″ high		18.90	.847		244	32		276	325
0210	Polymer plastic, 31″ high x 55″ high		18.90	.847		320	32		352	410
0220	Phenolic, 31″ wide x 58″ high		18.90	.847		274	32		306	355
0223	Polymer plastic, 48″ high x 55″ high		18.90	.847		495	32		527	600
0226	Phenolic, 48″ wide x 58″ high		18.90	.847		465	32		497	565
0230	Polymer plastic, 51″ wide x 55″ high		18.90	.847		435	32		467	535
0240	Phenolic, 51″ wide x 58″ high		18.90	.847		415	32		447	510
0250	Polymer plastic, 63″ wide x 55″ high		18.90	.847		590	32		622	700
0260	Phenolic, 63″ wide x 58″ high		18.90	.847		440	32		472	540
0300	Doors									
0310	Polymer plastic, 24″ wide x 55″ high	2 Carp	14.10	1.135	Ea.	217	43		260	310
0320	Phenolic, 24″ wide x 58″ high		14.10	1.135		305	43		348	405
0330	Polymer plastic, 26″ high x 55″ high		14.10	1.135		232	43		275	325
0340	Phenolic, 26″ wide x 58″ high		14.10	1.135		325	43		368	425
0350	Polymer plastic, 28″ wide x 55″ high		14.10	1.135		254	43		297	350
0360	Phenolic, 28″ wide x 58″ high		14.10	1.135		340	43		383	445
0370	Polymer plastic, 36″ wide x 55″ high		14.10	1.135		300	43		343	400
0380	Phenolic, 36″ wide x 58″ high		14.10	1.135		425	43		468	535
0400	Headrails									
0410	For polymer plastic, 62″ long	2 Carp	65	.246	Ea.	23	9.30		32.30	41
0420	Phenolic		65	.246		22.50	9.30		31.80	40.50
0430	For polymer plastic, 84″ long		50	.320		32.50	12.10		44.60	56.50
0440	Phenolic		50	.320		32	12.10		44.10	56
0450	For polymer plastic, 120″ long		30	.533		44.50	20		64.50	83
0460	Phenolic		30	.533		44	20		64	82.50

10 21 13.40 Stone Toilet Compartments

Line	Description	Crew	Daily Output	Labor-Hours	Unit	Material	Labor	Equipment	Total	Total Incl O&P
0010	STONE TOILET COMPARTMENTS									
0100	Cubicles, ceiling hung, marble	2 Marb	2	8	Ea.	1,800	286		2,086	2,450
0600	For handicap units, incl. 52″ grab bars, add					450			450	495
0800	Floor & ceiling anchored, marble	2 Marb	2.50	6.400		1,975	229		2,204	2,550
1400	For handicap units, incl. 52″ grab bars, add					370			370	405
1600	Floor mounted, marble	2 Marb	3	5.333		1,225	191		1,416	1,675

10 21 Compartments and Cubicles

10 21 13 – Toilet Compartments

10 21 13.40 Stone Toilet Compartments		Crew	Daily Output	Labor-Hours	Unit	2016 Bare Costs Material	Labor	Equipment	Total	Total Incl O&P
2400	Floor mounted, headrail braced, marble	2 Marb	3	5.333	Ea.	1,150	191		1,341	1,600
2900	For handicap units, incl. 52" grab bars, add					370			370	410
4100	Entrance screen, floor mounted marble, 58" high, 48" wide	2 Marb	9	1.778		795	63.50		858.50	975
4600	Urinal screen, 18" wide, ceiling braced, marble	D-1	6	2.667		755	87.50		842.50	975
5100	Floor mounted, head rail braced									
5200	Marble	D-1	6	2.667	Ea.	645	87.50		732.50	855
5700	Pilaster, flush, marble		9	1.778		840	58.50		898.50	1,025
6200	Post braced, marble		9	1.778		825	58.50		883.50	1,000

10 22 Partitions

10 22 19 – Demountable Partitions

10 22 19.43 Demountable Composite Partitions		Crew	Daily Output	Labor-Hours	Unit	Material	Labor	Equipment	Total	Total Incl O&P
0010	**DEMOUNTABLE COMPOSITE PARTITIONS**, add for doors									
0100	Do not deduct door openings from total L.F.									
0900	Demountable gypsum system on 2" to 2-1/2"									
1000	Steel studs, 9' high, 3" to 3-3/4" thick									
1200	Vinyl clad gypsum	2 Carp	48	.333	L.F.	60	12.60		72.60	87
1300	Fabric clad gypsum		44	.364		150	13.75		163.75	188
1500	Steel clad gypsum		40	.400		167	15.10		182.10	210
1600	1.75 system, aluminum framing, vinyl clad hardboard,									
1800	Paper honeycomb core panel, 1-3/4" to 2-1/2" thick									
1900	9' high	2 Carp	48	.333	L.F.	101	12.60		113.60	132
2100	7' high		60	.267		90.50	10.10		100.60	116
2200	5' high		80	.200		76.50	7.55		84.05	96.50
2250	Unitized gypsum system									
2300	Unitized panel, 9' high, 2" to 2-1/2" thick									
2350	Vinyl clad gypsum	2 Carp	48	.333	L.F.	130	12.60		142.60	164
2400	Fabric clad gypsum	"	44	.364	"	214	13.75		227.75	258
2500	Unitized mineral fiber system									
2510	Unitized panel, 9' high, 2-1/4" thick, aluminum frame									
2550	Vinyl clad mineral fiber	2 Carp	48	.333	L.F.	129	12.60		141.60	163
2600	Fabric clad mineral fiber	"	44	.364	"	193	13.75		206.75	235
2800	Movable steel walls, modular system									
2900	Unitized panels, 9' high, 48" wide									
3100	Baked enamel, pre-finished	2 Carp	60	.267	L.F.	146	10.10		156.10	178
3200	Fabric clad steel	"	56	.286	"	212	10.80		222.80	251
5500	For acoustical partitions, add, unsealed				S.F.	2.36			2.36	2.60
5550	Sealed				"	11			11	12.10
5700	For doors, see Sections 08 16									
5800	For door hardware, see Section 08 71									
6100	In-plant modular office system, w/prehung hollow core door									
6200	3" thick polystyrene core panels									
6250	12' x 12', 2 wall	2 Clab	3.80	4.211	Ea.	4,125	123		4,248	4,725
6300	4 wall		1.90	8.421		6,275	246		6,521	7,300
6350	16' x 16', 2 wall		3.60	4.444		6,250	130		6,380	7,100
6400	4 wall		1.80	8.889		8,475	260		8,735	9,750

10 22 Partitions

10 22 23 – Portable Partitions, Screens, and Panels

10 22 23.13 Wall Screens		Crew	Daily Output	Labor-Hours	Unit	2016 Bare Costs Material	Labor	Equipment	Total	Total Incl O&P
0010	**WALL SCREENS**, divider panels, free standing, fiber core									
0020	Fabric face straight									
1500	6'-0" high	2 Carp	125	.128	L.F.	90.50	4.84		95.34	108
1600	6'-0" long, 5'-0" high		162	.099		75.50	3.73		79.23	89.50
3200	Economical panels, fabric face, 4'-0" long, 5'-0" high		132	.121		50	4.58		54.58	62.50
3250	6'-0" high		112	.143		55.50	5.40		60.90	70
3300	5'-0" long, 5'-0" high		150	.107		54.50	4.03		58.53	67
3350	6'-0" high		125	.128		50	4.84		54.84	63
3450	Acoustical panels, 60 to 90 NRC, 3'-0" long, 5'-0" high		90	.178		76.50	6.70		83.20	95.50
3550	6'-0" high		75	.213		89.50	8.05		97.55	112
3600	5'-0" long, 5'-0" high		150	.107		61	4.03		65.03	74
3650	6'-0" high		125	.128		69	4.84		73.84	83.50
3700	6'-0" long, 5'-0" high		162	.099		53	3.73		56.73	65
3750	6'-0" high		138	.116		82	4.38		86.38	97.50
3800	Economy acoustical panels, 40 N.R.C., 4'-0" long, 5'-0" high		132	.121		50	4.58		54.58	62.50
3850	6'-0" high		112	.143		55.50	5.40		60.90	70
3900	5'-0" long, 6'-0" high		125	.128		50	4.84		54.84	63
3950	6'-0" long, 5'-0" high		162	.099		47	3.73		50.73	58.50

10 22 33 – Accordion Folding Partitions

10 22 33.10 Partitions, Accordion Folding		Crew	Daily Output	Labor-Hours	Unit	Material	Labor	Equipment	Total	Total Incl O&P
0010	**PARTITIONS, ACCORDION FOLDING**									
0100	Vinyl covered, over 150 S.F., frame not included									
0300	Residential, 1.25 lb. per S.F., 8' maximum height	2 Carp	300	.053	S.F.	25	2.02		27.02	31
0400	Commercial, 1.75 lb. per S.F., 8' maximum height		225	.071		28.50	2.69		31.19	36
0900	Acoustical, 3 lb. per S.F., 17' maximum height		100	.160		31.50	6.05		37.55	45
1200	5 lb. per S.F., 20' maximum height		95	.168		44	6.35		50.35	59
1500	Vinyl clad wood or steel, electric operation, 5.0 psf		160	.100		63	3.78		66.78	75.50
1900	Wood, non-acoustic, birch or mahogany, to 10' high		300	.053		34	2.02		36.02	40.50

10 22 39 – Folding Panel Partitions

10 22 39.10 Partitions, Folding Panel		Crew	Daily Output	Labor-Hours	Unit	Material	Labor	Equipment	Total	Total Incl O&P
0010	**PARTITIONS, FOLDING PANEL**, acoustic, wood									
0100	Vinyl faced, to 18' high, 6 psf, economy trim	2 Carp	60	.267	S.F.	57	10.10		67.10	79.50
0150	Standard trim		45	.356		68	13.45		81.45	97
0200	Premium trim		30	.533		87.50	20		107.50	131
0400	Plastic laminate or hardwood finish, standard trim		60	.267		58.50	10.10		68.60	81.50
0500	Premium trim		30	.533		62.50	20		82.50	103
0600	Wood, low acoustical type, 4.5 psf, to 14' high		50	.320		42.50	12.10		54.60	67.50

10 26 Wall and Door Protection

10 26 13 – Corner Guards

10 26 13.10 Metal Corner Guards		Crew	Daily Output	Labor-Hours	Unit	Material	Labor	Equipment	Total	Total Incl O&P
0010	**METAL CORNER GUARDS**									
1000	Wall protection, stainless steel, 16 ga, 48" x 36" tall, screwed to studs	2 Skwk	500	.032	S.F.	8.25	1.23		9.48	11.15
1050	Wall end guard, stainless steel, 16 ga, 36" tall, screwed to studs	1 Skwk	30	.267	Ea.	25	10.20		35.20	44.50

10 26 16 – Bumper Guards

10 26 16.10 Wallguard		Crew	Daily Output	Labor-Hours	Unit	Material	Labor	Equipment	Total	Total Incl O&P
0010	**WALLGUARD**									
0400	Rub rail, vinyl, adhesive mounted	1 Carp	185	.043	L.F.	8.65	1.63		10.28	12.30
0500	Neoprene, aluminum backing, 1-1/2" x 2"		110	.073		8.80	2.75		11.55	14.30

10 26 Wall and Door Protection

10 26 16 – Bumper Guards

10 26 16.10 Wallguard		Crew	Daily Output	Labor-Hours	Unit	Material	Labor (2016 Bare Costs)	Equipment	Total	Total Incl O&P
1000	Trolley rail, PVC, clipped to wall, 5" high	1 Carp	185	.043	L.F.	8.80	1.63		10.43	12.45
1050	8" high		180	.044		15	1.68		16.68	19.30
1200	Bed bumper, vinyl acrylic, alum. retainer, 21" long		10	.800	Ea.	40.50	30		70.50	95
1300	53" long with aligner		9	.889	"	102	33.50		135.50	169
1400	Bumper, vinyl cover, alum. retain., cush. mnt., 1-1/2" x 2-3/4"		80	.100	L.F.	14.20	3.78		17.98	22
1500	2" x 4-1/4"		80	.100		20.50	3.78		24.28	29.50
1600	Surface mounted, 1-3/4" x 3-5/8"		80	.100		11.95	3.78		15.73	19.50
1700	Bumper rail, stainless steel, flat bar on brackets, 4" x 1/4"	2 Skwk	120	.133		39.50	5.10		44.60	52
1750	Wallguard stainless steel baseboard, 12" tall, adhesive applied	"	260	.062		34.50	2.36		36.86	41.50
2000	Crash rail, vinyl cover, alum. retainer, 1" x 4"	1 Carp	110	.073		11.05	2.75		13.80	16.75
2100	1" x 8"		90	.089		18.15	3.36		21.51	25.50
2150	Vinyl inserts, aluminum plate, 1" x 2-1/2"		110	.073		14.90	2.75		17.65	21
2200	1" x 5"		90	.089		23	3.36		26.36	31
3000	Handrail/bumper, vinyl cover, alum. retainer									
3010	Bracket mounted, flat rail, 5-1/2"	1 Carp	80	.100	L.F.	18.30	3.78		22.08	26.50
3100	6-1/2"		80	.100		23	3.78		26.78	31.50
3200	Bronze bracket, 1-3/4" diam. rail		80	.100		16.50	3.78		20.28	24.50

10 28 Toilet, Bath, and Laundry Accessories

10 28 13 – Toilet Accessories

10 28 13.13 Commercial Toilet Accessories		Crew	Daily Output	Labor-Hours	Unit	Material	Labor (2016 Bare Costs)	Equipment	Total	Total Incl O&P
0010	**COMMERCIAL TOILET ACCESSORIES**									
0200	Curtain rod, stainless steel, 5' long, 1" diameter	1 Carp	13	.615	Ea.	28	23.50		51.50	70
0300	1-1/4" diameter		13	.615		30.50	23.50		54	73
0400	Diaper changing station, horizontal, wall mounted, plastic		10	.800		245	30		275	320
0500	Dispenser units, combined soap & towel dispensers,									
0510	Mirror and shelf, flush mounted	1 Carp	10	.800	Ea.	310	30		340	395
0600	Towel dispenser and waste receptacle,									
0610	18 gallon capacity	1 Carp	10	.800	Ea.	310	30		340	390
0800	Grab bar, straight, 1-1/4" diameter, stainless steel, 18" long		24	.333		30.50	12.60		43.10	54.50
1100	36" long		20	.400		37	15.10		52.10	66
1105	42" long		20	.400		39	15.10		54.10	68.50
1120	Corner, 36" long		20	.400		90	15.10		105.10	125
2010	Tub/shower/toilet, 2-wall, 36" x 24"		12	.667		93.50	25		118.50	145
3000	Mirror, with stainless steel 3/4" square frame, 18" x 24"		20	.400		49.50	15.10		64.60	80
3100	36" x 24"		15	.533		110	20		130	155
3300	72" x 24"		6	1.333		305	50.50		355.50	420
3500	With 5" stainless steel shelf, 18" x 24"		20	.400		198	15.10		213.10	244
3600	36" x 24"		15	.533		243	20		263	300
3800	72" x 24"		6	1.333		259	50.50		309.50	370
4100	Mop holder strip, stainless steel, 5 holders, 48" long		20	.400		78	15.10		93.10	112
4200	Napkin/tampon dispenser, recessed		15	.533		630	20		650	730
4220	Semi-recessed		6.50	1.231		310	46.50		356.50	420
4250	Napkin receptacle, recessed		6.50	1.231		177	46.50		223.50	273
4300	Robe hook, single, regular		96	.083		19.40	3.15		22.55	27
4400	Heavy duty, concealed mounting		56	.143		22.50	5.40		27.90	34
4600	Soap dispenser, chrome, surface mounted, liquid		20	.400		50	15.10		65.10	80.50
5000	Recessed stainless steel, liquid		10	.800		166	30		196	233
5600	Shelf, stainless steel, 5" wide, 18 ga., 24" long		24	.333		85	12.60		97.60	115
6100	Toilet tissue dispenser, surface mounted, SS, single roll		30	.267		18.95	10.10		29.05	38
6200	Double roll		24	.333		24.50	12.60		37.10	48

10 28 Toilet, Bath, and Laundry Accessories

10 28 13 – Toilet Accessories

10 28 13.13 Commercial Toilet Accessories		Crew	Daily Output	Labor-Hours	Unit	Material	2016 Bare Costs Labor	Equipment	Total	Total Incl O&P
6240	Plastic, twin/jumbo dbl. roll	1 Carp	24	.333	Ea.	28.50	12.60		41.10	52
6400	Towel bar, stainless steel, 18″ long		23	.348		41.50	13.15		54.65	67.50
6500	30″ long		21	.381		66	14.40		80.40	97
6700	Towel dispenser, stainless steel, surface mounted		16	.500		47.50	18.90		66.40	83.50
6800	Flush mounted, recessed		10	.800		282	30		312	360
6900	Plastic, touchless, battery operated		16	.500		89	18.90		107.90	129
7400	Tumbler holder, for tumbler only		30	.267		18.75	10.10		28.85	37.50
7410	Tumbler holder, recessed		20	.400		8	15.10		23.10	34.50
7500	Soap, tumbler & toothbrush		30	.267		19.60	10.10		29.70	38.50
7510	Tumbler & toothbrush holder		20	.400		13.50	15.10		28.60	40.50
7700	Wall urn ash receiver, surface mount, 11″ long		12	.667		95	25		120	146
8000	Waste receptacles, stainless steel, with top, 13 gallon		10	.800		315	30		345	395

10 28 16 – Bath Accessories

10 28 16.20 Medicine Cabinets

Line	Description	Crew	Daily Output	Labor-Hours	Unit	Material	Labor	Equipment	Total	Total Incl O&P
0010	**MEDICINE CABINETS**									
0020	With mirror, sst frame, 16″ x 22″, unlighted	1 Carp	14	.571	Ea.	99	21.50		120.50	145
0100	Wood frame		14	.571		140	21.50		161.50	190
0300	Sliding mirror doors, 20″ x 16″ x 4-3/4″, unlighted		7	1.143		133	43		176	219
0400	24″ x 19″ x 8-1/2″, lighted		5	1.600		184	60.50		244.50	305
0600	Triple door, 30″ x 32″, unlighted, plywood body		7	1.143		325	43		368	435
0700	Steel body		7	1.143		375	43		418	485
0900	Oak door, wood body, beveled mirror, single door		7	1.143		199	43		242	292
1000	Double door		6	1.333		415	50.50		465.50	540
1200	Hotel cabinets, stainless, with lower shelf, unlighted		10	.800		200	30		230	270
1300	Lighted		5	1.600		305	60.50		365.50	435

10 28 19 – Tub and Shower Enclosures

10 28 19.10 Partitions, Shower

Line	Description	Crew	Daily Output	Labor-Hours	Unit	Material	Labor	Equipment	Total	Total Incl O&P
0010	**PARTITIONS, SHOWER** floor mounted, no plumbing									
0400	Cabinet, one piece, fiberglass, 32″ x 32″	2 Carp	5	3.200	Ea.	535	121		656	795
0420	36″ x 36″		5	3.200		625	121		746	890
0440	36″ x 48″		5	3.200		1,375	121		1,496	1,700
0460	Acrylic, 32″ x 32″		5	3.200		340	121		461	580
0480	36″ x 36″		5	3.200		1,050	121		1,171	1,350
0500	36″ x 48″		5	3.200		1,400	121		1,521	1,750
0520	Shower door for above, clear plastic, 24″ wide	1 Carp	8	1		188	38		226	271
0540	28″ wide		8	1		225	38		263	310
0560	Tempered glass, 24″ wide		8	1		220	38		258	305
0580	28″ wide		8	1		252	38		290	340
2400	Glass stalls, with doors, no receptors, chrome on brass	2 Shee	3	5.333		1,600	226		1,826	2,150
2700	Anodized aluminum	″	4	4		1,125	169		1,294	1,500
3200	Receptors, precast terrazzo, 32″ x 32″	2 Marb	14	1.143		335	41		376	440
3300	48″ x 34″		9.50	1.684		440	60.50		500.50	585
3500	Plastic, simulated terrazzo receptor, 32″ x 32″		14	1.143		148	41		189	231
3600	32″ x 48″		12	1.333		222	47.50		269.50	325
3800	Precast concrete, colors, 32″ x 32″		14	1.143		201	41		242	289
3900	48″ x 48″		8	2		277	71.50		348.50	425
4100	Shower doors, economy plastic, 24″ wide	1 Shee	9	.889		144	37.50		181.50	221
4200	Tempered glass door, economy		8	1		258	42.50		300.50	355
4400	Folding, tempered glass, aluminum frame		6	1.333		420	56.50		476.50	555
4700	Deluxe, tempered glass, chrome on brass frame, 42″ to 44″		8	1		415	42.50		457.50	525
4800	39″ to 48″ wide		1	8		675	340		1,015	1,300
4850	On anodized aluminum frame, obscure glass		2	4		575	169		744	915

10 28 Toilet, Bath, and Laundry Accessories

10 28 19 – Tub and Shower Enclosures

10 28 19.10 Partitions, Shower		Crew	Daily Output	Labor-Hours	Unit	Material	Labor	Equipment	Total	Total Incl O&P
						2016 Bare Costs				
4900	Clear glass	1 Shee	1	8	Ea.	675	340		1,015	1,300
5100	Shower enclosure, tempered glass, anodized alum. frame									
5120	2 panel & door, corner unit, 32" x 32"	1 Shee	2	4	Ea.	1,100	169		1,269	1,500
5140	Neo-angle corner unit, 16" x 24" x 16"	"	2	4		1,125	169		1,294	1,500
5200	Shower surround, 3 wall, polypropylene, 32" x 32"	1 Carp	4	2		480	75.50		555.50	655
5220	PVC, 32" x 32"		4	2		380	75.50		455.50	545
5240	Fiberglass		4	2		420	75.50		495.50	585
5250	2 wall, polypropylene, 32" x 32"		4	2		315	75.50		390.50	475
5270	PVC		4	2		395	75.50		470.50	560
5290	Fiberglass		4	2		400	75.50		475.50	565
5300	Tub doors, tempered glass & frame, obscure glass	1 Shee	8	1		229	42.50		271.50	325
5400	Clear glass		6	1.333		535	56.50		591.50	685
5600	Chrome plated, brass frame, obscure glass		8	1		305	42.50		347.50	405
5700	Clear glass		6	1.333		745	56.50		801.50	910
5900	Tub/shower enclosure, temp. glass, alum. frame, obscure glass		2	4		410	169		579	735
6200	Clear glass		1.50	5.333		855	226		1,081	1,325
6500	On chrome-plated brass frame, obscure glass		2	4		565	169		734	910
6600	Clear glass		1.50	5.333		1,200	226		1,426	1,700
6800	Tub surround, 3 wall, polypropylene	1 Carp	4	2		256	75.50		331.50	410
6900	PVC		4	2		390	75.50		465.50	555
7000	Fiberglass, obscure glass		4	2		400	75.50		475.50	565
7100	Clear glass		3	2.667		680	101		781	920

10 31 Manufactured Fireplaces

10 31 13 – Manufactured Fireplace Chimneys

10 31 13.10 Fireplace Chimneys		Crew	Daily Output	Labor-Hours	Unit	Material	Labor	Equipment	Total	Total Incl O&P
0010	**FIREPLACE CHIMNEYS**									
0500	Chimney dbl. wall, all stainless, over 8'-6", 7" diam., add to fireplace	1 Carp	33	.242	V.L.F.	84	9.15		93.15	108
0600	10" diameter, add to fireplace		32	.250		112	9.45		121.45	140
0700	12" diameter, add to fireplace		31	.258		167	9.75		176.75	199
0800	14" diameter, add to fireplace		30	.267		227	10.10		237.10	267
1000	Simulated brick chimney top, 4' high, 16" x 16"		10	.800	Ea.	435	30		465	530
1100	24" x 24"		7	1.143	"	555	43		598	685

10 31 23 – Prefabricated Fireplaces

10 31 23.10 Fireplace, Prefabricated		Crew	Daily Output	Labor-Hours	Unit	Material	Labor	Equipment	Total	Total Incl O&P
0010	**FIREPLACE, PREFABRICATED,** free standing or wall hung									
0100	With hood & screen, painted	1 Carp	1.30	6.154	Ea.	1,475	233		1,708	2,025
0150	Average		1	8		1,700	300		2,000	2,375
0200	Stainless steel		.90	8.889		3,050	335		3,385	3,925
1500	Simulated logs, gas fired, 40,000 BTU, 2' long, manual safety pilot		7	1.143	Set	440	43		483	560
1600	Adjustable flame remote pilot		6	1.333		1,200	50.50		1,250.50	1,400
1700	Electric, 1,500 BTU, 1'-6" long, incandescent flame		7	1.143		205	43		248	298
1800	1,500 BTU, LED flame		6	1.333		320	50.50		370.50	435
2000	Fireplace, built-in, 36" hearth, radiant		1.30	6.154	Ea.	680	233		913	1,150
2100	Recirculating, small fan		1	8		900	300		1,200	1,500
2150	Large fan		.90	8.889		2,000	335		2,335	2,775
2200	42" hearth, radiant		1.20	6.667		915	252		1,167	1,425
2300	Recirculating, small fan		.90	8.889		1,200	335		1,535	1,900
2350	Large fan		.80	10		1,350	380		1,730	2,100
2400	48" hearth, radiant		1.10	7.273		2,175	275		2,450	2,825

10 31 Manufactured Fireplaces

10 31 23 – Prefabricated Fireplaces

10 31 23.10 Fireplace, Prefabricated		Crew	Daily Output	Labor-Hours	Unit	Material	2016 Bare Costs Labor	Equipment	Total	Total Incl O&P
2500	Recirculating, small fan	1 Carp	.80	10	Ea.	2,475	380		2,855	3,350
2550	Large fan		.70	11.429		2,475	430		2,905	3,450
3000	See through, including doors		.80	10		2,200	380		2,580	3,025
3200	Corner (2 wall)	↓	1	8	↓	3,425	300		3,725	4,250

10 43 Emergency Aid Specialties

10 43 13 – Defibrillator Cabinets

10 43 13.05 Defibrillator Cabinets		Crew	Daily Output	Labor-Hours	Unit	Material	Labor	Equipment	Total	Total Incl O&P
0010	**DEFIBRILLATOR CABINETS**, not equipped, stainless steel									
0050	Defibrillator cabinet, stainless steel with strobe & alarm 12" x 27"	1 Carp	10	.800	Ea.	385	30		415	475
0100	Automatic External Defibrillator	"	30	.267	"	1,375	10.10		1,385.10	1,550

10 44 Fire Protection Specialties

10 44 13 – Fire Protection Cabinets

10 44 13.53 Fire Equipment Cabinets		Crew	Daily Output	Labor-Hours	Unit	Material	Labor	Equipment	Total	Total Incl O&P
0010	**FIRE EQUIPMENT CABINETS**, not equipped, 20 ga. steel box									
0040	Recessed, D.S. glass in door, box size given									
1000	Portable extinguisher, single, 8" x 12" x 27", alum. door & frame	Q-12	8	2	Ea.	160	77		237	305
1100	Steel door and frame		8	2		122	77		199	261
2700	Fire blanket & extinguisher cab, inc blanket, rec stl., 14" x 40" x 8"		7	2.286		188	87.50		275.50	350
2800	Fire blanket cab, inc blanket, surf mtd, stl, 15"x10"x5", w/pwdr coat fin	↓	8	2	↓	94	77		171	231
3000	Hose rack assy., 1-1/2" valve & 100' hose, 24" x 40" x 5-1/2"									
3100	Aluminum door and frame	Q-12	6	2.667	Ea.	390	102		492	595
3200	Steel door and frame	"	6	2.667	"	258	102		360	455

10 44 16 – Fire Extinguishers

10 44 16.13 Portable Fire Extinguishers		Crew	Daily Output	Labor-Hours	Unit	Material	Labor	Equipment	Total	Total Incl O&P
0010	**PORTABLE FIRE EXTINGUISHERS**									
0140	CO_2, with hose and "H" horn, 10 lb.				Ea.	288			288	315
1000	Dry chemical, pressurized									
1040	Standard type, portable, painted, 2-1/2 lb.				Ea.	38.50			38.50	42.50
1080	10 lb.					82			82	90
1100	20 lb.					135			135	149
1120	30 lb.					435			435	480
2000	ABC all purpose type, portable, 2-1/2 lb.					21			21	23.50
2080	9-1/2 lb.					44			44	48
3500	Halotron 1, 2-1/2 lb.					127			127	140
3600	5 lb.					199			199	219
3700	11 lb.				↓	395			395	435

10 44 16.16 Wheeled Fire Extinguisher Units		Crew	Daily Output	Labor-Hours	Unit	Material	Labor	Equipment	Total	Total Incl O&P
0010	**WHEELED FIRE EXTINGUISHER UNITS**									
0350	CO_2, portable, with swivel horn									
0360	Wheeled type, cart mounted, 50 lb.				Ea.	1,175			1,175	1,275
0400	100 lb.				"	4,275			4,275	4,700
2200	ABC all purpose type									
2300	Wheeled, 45 lb.				Ea.	735			735	810
2360	150 lb.				"	1,850			1,850	2,025

10 51 Lockers

10 51 13 – Metal Lockers

10 51 13.10 Lockers		Crew	Daily Output	Labor-Hours	Unit	2016 Bare Costs Material	Labor	Equipment	Total	Total Incl O&P
0011	**LOCKERS** steel, baked enamel, pre-assembled									
0110	Single tier box locker, 12" x 15" x 72"	1 Shee	20	.400	Ea.	231	16.95		247.95	283
0120	18" x 15" x 72"		20	.400		245	16.95		261.95	298
0130	12" x 18" x 72"		20	.400		235	16.95		251.95	287
0140	18" x 18" x 72"		20	.400		285	16.95		301.95	345
0410	Double tier, 12" x 15" x 36"		30	.267		233	11.30		244.30	275
0420	18" x 15" x 36"		30	.267		235	11.30		246.30	278
0430	12" x 18" x 36"		30	.267		265	11.30		276.30	310
0440	18" x 18" x 36"		30	.267		225	11.30		236.30	267
0500	Two person, 18" x 15" x 72"		20	.400		297	16.95		313.95	355
0510	18" x 18" x 72"		20	.400		330	16.95		346.95	390
0520	Duplex, 15" x 15" x 72"		20	.400		335	16.95		351.95	400
0530	15" x 21" x 72"		20	.400		370	16.95		386.95	440
1100	Wire meshed wardrobe, floor mtd., open front varsity type		7.50	1.067	↓	277	45		322	380
3250	Rack w/24 wire mesh baskets		1.50	5.333	Set	410	226		636	825
3260	30 baskets		1.25	6.400		360	271		631	845
3270	36 baskets		.95	8.421		470	355		825	1,125
3280	42 baskets	↓	.80	10	↓	455	425		880	1,200
3600	For hanger rods, add				Ea.	2			2	2.20
3650	For number plate kit, 100 plates #1 - #100, add	1 Shee	4	2		82	84.50		166.50	231
3700	For locker base, closed front panel		90	.089		7.20	3.76		10.96	14.20
3710	End panel, bolted		36	.222		9.30	9.40		18.70	26
3800	For sloping top, 12" wide		24	.333		32.50	14.10		46.60	59.50
3810	15" wide		24	.333		35.50	14.10		49.60	62.50
3820	18" wide		24	.333		35.50	14.10		49.60	62.50
3850	Sloping top end panel, 12" deep		72	.111		12.20	4.71		16.91	21.50
3860	15" deep		72	.111		12.20	4.71		16.91	21.50
3870	18" deep		72	.111		12.20	4.71		16.91	21.50
3900	For finish end panels, steel, 60" high, 15" deep		12	.667		36	28		64	86.50
3910	72" high, 12" deep		12	.667		29	28		57	79
3920	18" deep	↓	12	.667	↓	43.50	28		71.50	95
5000	For "ready to assemble" lockers,									
5010	Add to labor						75%			
5020	Deduct from material					20%				

10 55 Postal Specialties

10 55 23 – Mail Boxes

10 55 23.10 Commercial Mail Boxes		Crew	Daily Output	Labor-Hours	Unit	Material	Labor	Equipment	Total	Total Incl O&P
0010	**COMMERCIAL MAIL BOXES**									
0020	Horiz., key lock, 5"H x 6"W x 15"D, alum., rear load	1 Carp	34	.235	Ea.	42.50	8.90		51.40	62
0100	Front loading		34	.235		42.50	8.90		51.40	62
0200	Double, 5"H x 12"W x 15"D, rear loading		26	.308		60.50	11.65		72.15	86
0300	Front loading		26	.308		69	11.65		80.65	95
0500	Quadruple, 10"H x 12"W x 15"D, rear loading		20	.400		101	15.10		116.10	137
0600	Front loading		20	.400		92.50	15.10		107.60	128
1600	Vault type, horizontal, for apartments, 4" x 5"		34	.235		26.50	8.90		35.40	44
1700	Alphabetical directories, 120 names	↓	10	.800	↓	125	30		155	189

10 56 Storage Assemblies

10 56 13 – Metal Storage Shelving

10 56 13.10 Shelving		Crew	Daily Output	Labor-Hours	Unit	Material	2016 Bare Costs Labor	Equipment	Total	Total Incl O&P
0010	**SHELVING**									
0020	Metal, industrial, cross-braced, 3′ wide, 12″ deep	1 Sswk	175	.046	SF Shlf	7.20	1.82		9.02	11.25
0100	24″ deep		330	.024		5.20	.97		6.17	7.45
2200	Wide span, 1600 lb. capacity per shelf, 6′ wide, 24″ deep		380	.021		7.65	.84		8.49	9.95
2400	36″ deep		440	.018		6.75	.73		7.48	8.75
3000	Residential, vinyl covered wire, wardrobe, 12″ deep	1 Carp	195	.041	L.F.	14.95	1.55		16.50	19
3100	16″ deep		195	.041		10.70	1.55		12.25	14.35
3200	Standard, 6″ deep		195	.041		4.25	1.55		5.80	7.30
3300	9″ deep		195	.041		5.80	1.55		7.35	8.95
3400	12″ deep		195	.041		7.70	1.55		9.25	11.05
3500	16″ deep		195	.041		18	1.55		19.55	22.50
3600	20″ deep		195	.041		21.50	1.55		23.05	26
3700	Support bracket		80	.100	Ea.	6.40	3.78		10.18	13.40

10 57 Wardrobe and Closet Specialties

10 57 23 – Closet and Utility Shelving

10 57 23.19 Wood Closet and Utility Shelving		Crew	Daily Output	Labor-Hours	Unit	Material	Labor	Equipment	Total	Total Incl O&P
0010	**WOOD CLOSET AND UTILITY SHELVING**									
0020	Pine, clear grade, no edge band, 1″ x 8″	1 Carp	115	.070	L.F.	3.47	2.63		6.10	8.20
0100	1″ x 10″		110	.073		4.32	2.75		7.07	9.35
0200	1″ x 12″		105	.076		5.20	2.88		8.08	10.55
0600	Plywood, 3/4″ thick with lumber edge, 12″ wide		75	.107		1.91	4.03		5.94	8.85
0700	24″ wide		70	.114		3.39	4.32		7.71	11
0900	Bookcase, clear grade pine, shelves 12″ O.C., 8″ deep, per S.F. shelf		70	.114	S.F.	11.30	4.32		15.62	19.65
1000	12″ deep shelves		65	.123	″	16.95	4.65		21.60	26.50
1200	Adjustable closet rod and shelf, 12″ wide, 3′ long		20	.400	Ea.	11.60	15.10		26.70	38.50
1300	8′ long		15	.533	″	21	20		41	57.50
1500	Prefinished shelves with supports, stock, 8″ wide		75	.107	L.F.	4.89	4.03		8.92	12.15
1600	10″ wide		70	.114	″	5.70	4.32		10.02	13.50

10 73 Protective Covers

10 73 13 – Awnings

10 73 13.10 Awnings, Fabric		Crew	Daily Output	Labor-Hours	Unit	Material	Labor	Equipment	Total	Total Incl O&P
0010	**AWNINGS, FABRIC**									
0020	Including acrylic canvas and frame, standard design									
0100	Door and window, slope, 3′ high, 4′ wide	1 Carp	4.50	1.778	Ea.	755	67		822	945
0110	6′ wide		3.50	2.286		970	86.50		1,056.50	1,225
0120	8′ wide		3	2.667		1,200	101		1,301	1,475
0200	Quarter round convex, 4′ wide		3	2.667		1,175	101		1,276	1,475
0210	6′ wide		2.25	3.556		1,525	134		1,659	1,900
0220	8′ wide		1.80	4.444		1,875	168		2,043	2,325
0300	Dome, 4′ wide		7.50	1.067		455	40.50		495.50	570
0310	6′ wide		3.50	2.286		1,025	86.50		1,111.50	1,275
0320	8′ wide		2	4		1,800	151		1,951	2,250
0350	Elongated dome, 4′ wide		1.33	6.015		1,700	227		1,927	2,250
0360	6′ wide		1.11	7.207		2,025	272		2,297	2,675
0370	8′ wide		1	8		2,375	300		2,675	3,125
1000	Entry or walkway, peak, 12′ long, 4′ wide	2 Carp	.90	17.778		5,475	670		6,145	7,150
1010	6′ wide		.60	26.667		8,450	1,000		9,450	11,000

10 73 Protective Covers

10 73 13 – Awnings

10 73 13.10 Awnings, Fabric		Crew	Daily Output	Labor-Hours	Unit	2016 Bare Costs Material	Labor	Equipment	Total	Total Incl O&P
1020	8′ wide	2 Carp	.40	40	Ea.	11,700	1,500		13,200	15,300
1100	Radius with dome end, 4′ wide		1.10	14.545		4,150	550		4,700	5,500
1110	6′ wide		.70	22.857		6,675	865		7,540	8,800
1120	8′ wide		.50	32		9,500	1,200		10,700	12,400
2000	Retractable lateral arm awning, manual									
2010	To 12′ wide, 8′ - 6″ projection	2 Carp	1.70	9.412	Ea.	1,250	355		1,605	1,950
2020	To 14′ wide, 8′ - 6″ projection		1.10	14.545		1,450	550		2,000	2,525
2030	To 19′ wide, 8′ - 6″ projection		.85	18.824		1,950	710		2,660	3,350
2040	To 24′ wide, 8′ - 6″ projection		.67	23.881		2,475	905		3,380	4,225
2050	Motor for above, add	1 Carp	2.67	3		1,050	113		1,163	1,375
3000	Patio/deck canopy with frame									
3010	12′ wide, 12′ projection	2 Carp	2	8	Ea.	1,750	300		2,050	2,425
3020	16′ wide, 14′ projection	"	1.20	13.333		2,725	505		3,230	3,850
9000	For fire retardant canvas, add					7%				
9010	For lettering or graphics, add					35%				
9020	For painted or coated acrylic canvas, deduct					8%				
9030	For translucent or opaque vinyl canvas, add					10%				
9040	For 6 or more units, deduct					20%	15%			

10 73 16 – Canopies

10 73 16.20 Metal Canopies		Crew	Daily Output	Labor-Hours	Unit	Material	Labor	Equipment	Total	Total Incl O&P
0010	**METAL CANOPIES**									
0020	Wall hung, .032″, aluminum, prefinished, 8′ x 10′	K-2	1.30	18.462	Ea.	2,350	705	236	3,291	4,100
0300	8′ x 20′	"	1.10	21.818	"	3,850	835	279	4,964	6,025
2300	Aluminum entrance canopies, flat soffit, .032″									
2500	3′-6″ x 4′-0″, clear anodized	2 Carp	4	4	Ea.	935	151		1,086	1,275

10 74 Manufactured Exterior Specialties

10 74 23 – Cupolas

10 74 23.10 Wood Cupolas		Crew	Daily Output	Labor-Hours	Unit	Material	Labor	Equipment	Total	Total Incl O&P
0010	**WOOD CUPOLAS**									
0020	Stock units, pine, painted, 18″ sq., 28″ high, alum. roof	1 Carp	4.10	1.951	Ea.	184	74		258	325
0100	Copper roof		3.80	2.105		255	79.50		334.50	415
0300	23″ square, 33″ high, aluminum roof		3.70	2.162		385	81.50		466.50	555
0400	Copper roof		3.30	2.424		505	91.50		596.50	715
0600	30″ square, 37″ high, aluminum roof		3.70	2.162		560	81.50		641.50	750
0700	Copper roof		3.30	2.424		670	91.50		761.50	890
0900	Hexagonal, 31″ wide, 46″ high, copper roof		4	2		930	75.50		1,005.50	1,150
1000	36″ wide, 50″ high, copper roof		3.50	2.286		1,525	86.50		1,611.50	1,825
1200	For deluxe stock units, add to above					25%				
1400	For custom built units, add to above					50%	50%			

10 74 46 – Window Wells

10 74 46.10 Area Window Wells		Crew	Daily Output	Labor-Hours	Unit	Material	Labor	Equipment	Total	Total Incl O&P
0010	**AREA WINDOW WELLS**, Galvanized steel									
0020	20 ga., 3′-2″ wide, 1′ deep	1 Sswk	29	.276	Ea.	17.25	11		28.25	39
0100	2′ deep		23	.348		31.50	13.90		45.40	60
0300	16 ga., 3′-2″ wide, 1′ deep		29	.276		23	11		34	45.50
0400	3′ deep		23	.348		48	13.90		61.90	78
0600	Welded grating for above, 15 lb., painted		45	.178		91.50	7.10		98.60	114
0700	Galvanized		45	.178		124	7.10		131.10	150
0900	Translucent plastic cap for above		60	.133		20.50	5.30		25.80	32.50

10 75 Flagpoles

10 75 16 – Ground-Set Flagpoles

10 75 16.10 Flagpoles		Crew	Daily Output	Labor-Hours	Unit	2016 Bare Costs Material	Labor	Equipment	Total	Total Incl O&P
0010	**FLAGPOLES**, ground set									
0050	Not including base or foundation									
0100	Aluminum, tapered, ground set 20′ high	K-1	2	8	Ea.	1,050	284	154	1,488	1,825
0200	25′ high		1.70	9.412		1,175	335	181	1,691	2,025
0300	30′ high		1.50	10.667		1,350	380	205	1,935	2,350
0500	40′ high	↓	1.20	13.333	↓	2,950	475	256	3,681	4,300

Division Notes

	CREW	DAILY OUTPUT	LABOR-HOURS	UNIT	BARE COSTS MAT.	BARE COSTS LABOR	BARE COSTS EQUIP.	BARE COSTS TOTAL	TOTAL INCL O&P

Division 11 Equipment

Estimating Tips

General

- The items in this division are usually priced per square foot or each. Many of these items are purchased by the owner for installation by the contractor. Check the specifications for responsibilities and include time for receiving, storage, installation, and mechanical and electrical hookups in the appropriate divisions.
- Many items in Division 11 require some type of support system that is not usually furnished with the item. Examples of these systems include blocking for the attachment of casework and support angles for ceiling-hung projection screens. The required blocking or supports must be added to the estimate in the appropriate division.
- Some items in Division 11 may require assembly or electrical hookups. Verify the amount of assembly required or the need for a hard electrical connection and add the appropriate costs.

Reference Numbers

Reference numbers are shown at the beginning of some major classifications. These numbers refer to related items in the Reference Section. The reference information may be an estimating procedure, an alternate pricing method, or technical information.

Note: Not all subdivisions listed here necessarily appear. ■

11 13 Loading Dock Equipment

11 13 13 – Loading Dock Bumpers

11 13 13.10 Dock Bumpers		Crew	Daily Output	Labor-Hours	Unit	Material	Labor (2016 Bare Costs)	Equipment	Total	Total Incl O&P
0010	**DOCK BUMPERS** Bolts not included									
0020	2" x 6" to 4" x 8", average	1 Carp	.30	26.667	M.B.F.	1,350	1,000		2,350	3,175

11 30 Residential Equipment

11 30 13 – Residential Appliances

11 30 13.15 Cooking Equipment

		Crew	Daily Output	Labor-Hours	Unit	Material	Labor	Equipment	Total	Total Incl O&P
0010	**COOKING EQUIPMENT**									
0020	Cooking range, 30" free standing, 1 oven, minimum	2 Clab	10	1.600	Ea.	470	46.50		516.50	595
0050	Maximum		4	4		1,900	117		2,017	2,275
0150	2 oven, minimum		10	1.600		1,250	46.50		1,296.50	1,450
0200	Maximum		10	1.600		2,800	46.50		2,846.50	3,150
0350	Built-in, 30" wide, 1 oven, minimum	1 Elec	6	1.333		900	58.50		958.50	1,075
0400	Maximum	2 Carp	2	8		1,575	300		1,875	2,225
0500	2 oven, conventional, minimum		4	4		1,350	151		1,501	1,725
0550	1 conventional, 1 microwave, maximum		2	8		2,525	300		2,825	3,275
0700	Free-standing, 1 oven, 21" wide range, minimum	2 Clab	10	1.600		505	46.50		551.50	635
0750	21" wide, maximum	"	4	4		580	117		697	835
0900	Countertop cooktops, 4 burner, standard, minimum	1 Elec	6	1.333		350	58.50		408.50	480
0950	Maximum		3	2.667		1,700	117		1,817	2,075
1050	As above, but with grill and griddle attachment, minimum		6	1.333		1,525	58.50		1,583.50	1,775
1100	Maximum		3	2.667		3,900	117		4,017	4,500
1250	Microwave oven, minimum		4	2		119	88		207	275
1300	Maximum		2	4		510	176		686	850
5380	Oven, built in, standard		4	2		900	88		988	1,125
5390	Deluxe		2	4		2,525	176		2,701	3,075

11 30 13.16 Refrigeration Equipment

		Crew	Daily Output	Labor-Hours	Unit	Material	Labor	Equipment	Total	Total Incl O&P
0010	**REFRIGERATION EQUIPMENT**									
2000	Deep freeze, 15 to 23 C.F., minimum	2 Clab	10	1.600	Ea.	570	46.50		616.50	705
2050	Maximum		5	3.200		835	93.50		928.50	1,075
2200	30 C.F., minimum		8	2		675	58.50		733.50	840
2250	Maximum		3	5.333		900	156		1,056	1,250
5200	Icemaker, automatic, 20 lb. per day	1 Plum	7	1.143		1,100	50		1,150	1,275
5350	51 lb. per day	"	2	4		1,600	175		1,775	2,050
5450	Refrigerator, no frost, 6 C.F.	2 Clab	15	1.067		229	31		260	305
5500	Refrigerator, no frost, 10 C.F. to 12 C.F., minimum		10	1.600		475	46.50		521.50	600
5600	Maximum		6	2.667		555	78		633	745
5750	14 C.F. to 16 C.F., minimum		9	1.778		580	52		632	720
5800	Maximum		5	3.200		650	93.50		743.50	870
5950	18 C.F. to 20 C.F., minimum		8	2		735	58.50		793.50	910
6000	Maximum		4	4		1,700	117		1,817	2,075
6150	21 C.F. to 29 C.F., minimum		7	2.286		1,150	66.50		1,216.50	1,350
6200	Maximum		3	5.333		3,850	156		4,006	4,475
6790	Energy-star qualified, 18 C.F., minimum G	2 Carp	4	4		650	151		801	970
6795	Maximum G		2	8		1,250	300		1,550	1,875
6797	21.7 C.F., minimum G		4	4		1,125	151		1,276	1,475
6799	Maximum G		4	4		2,075	151		2,226	2,525

11 30 13.17 Kitchen Cleaning Equipment

		Crew	Daily Output	Labor-Hours	Unit	Material	Labor	Equipment	Total	Total Incl O&P
0010	**KITCHEN CLEANING EQUIPMENT**									
2750	Dishwasher, built-in, 2 cycles, minimum	L-1	4	2.500	Ea.	299	110		409	510
2800	Maximum		2	5		450	219		669	855

11 30 Residential Equipment

11 30 13 – Residential Appliances

11 30 13.17 Kitchen Cleaning Equipment		Crew	Daily Output	Labor-Hours	Unit	Material	Labor	Equipment	Total	Total Incl O&P
						2016 Bare Costs				
2950	4 or more cycles, minimum	L-1	4	2.500	Ea.	425	110		535	650
2960	Average		4	2.500		565	110		675	805
3000	Maximum		2	5		1,225	219		1,444	1,675
3100	Energy-star qualified, minimum G		4	2.500		490	110		600	720
3110	Maximum G		2	5		1,650	219		1,869	2,175

11 30 13.18 Waste Disposal Equipment		Crew	Daily Output	Labor-Hours	Unit	Material	Labor	Equipment	Total	Total Incl O&P
0010	WASTE DISPOSAL EQUIPMENT									
1750	Compactor, residential size, 4 to 1 compaction, minimum	1 Carp	5	1.600	Ea.	690	60.50		750.50	860
1800	Maximum	"	3	2.667		1,150	101		1,251	1,450
3300	Garbage disposal, sink type, minimum	L-1	10	1		93	44		137	174
3350	Maximum	"	10	1		243	44		287	340

11 30 13.19 Kitchen Ventilation Equipment		Crew	Daily Output	Labor-Hours	Unit	Material	Labor	Equipment	Total	Total Incl O&P
0010	KITCHEN VENTILATION EQUIPMENT									
4150	Hood for range, 2 speed, vented, 30" wide, minimum	L-3	5	2	Ea.	83	78		161	222
4200	Maximum		3	3.333		855	130		985	1,150
4300	42" wide, minimum		5	2		166	78		244	310
4330	Custom		5	2		1,625	78		1,703	1,925
4350	Maximum		3	3.333		1,975	130		2,105	2,400
4500	For ventless hood, 2 speed, add					20			20	22
4650	For vented 1 speed, deduct from maximum					55			55	60.50

11 30 13.24 Washers		Crew	Daily Output	Labor-Hours	Unit	Material	Labor	Equipment	Total	Total Incl O&P
0010	WASHERS									
6650	Washing machine, automatic, minimum	1 Plum	3	2.667	Ea.	545	117		662	795
6700	Maximum		1	8		1,650	350		2,000	2,400
6750	Energy star qualified, front loading, minimum G		3	2.667		700	117		817	965
6760	Maximum G		1	8		1,600	350		1,950	2,325
6764	Top loading, minimum G		3	2.667		570	117		687	820
6766	Maximum G		3	2.667		1,400	117		1,517	1,750

11 30 13.25 Dryers		Crew	Daily Output	Labor-Hours	Unit	Material	Labor	Equipment	Total	Total Incl O&P
0010	DRYERS									
6770	Electric, front loading, energy-star qualified, minimum G	L-2	3	5.333	Ea.	475	175		650	820
6780	Maximum G	"	2	8		1,600	263		1,863	2,200
7450	Vent kits for dryers	1 Carp	10	.800		40	30		70	94.50

11 30 15 – Miscellaneous Residential Appliances

11 30 15.13 Sump Pumps		Crew	Daily Output	Labor-Hours	Unit	Material	Labor	Equipment	Total	Total Incl O&P
0010	SUMP PUMPS									
6400	Cellar drainer, pedestal, 1/3 H.P., molded PVC base	1 Plum	3	2.667	Ea.	140	117		257	345
6450	Solid brass	"	2	4	"	297	175		472	615

11 30 15.23 Water Heaters		Crew	Daily Output	Labor-Hours	Unit	Material	Labor	Equipment	Total	Total Incl O&P
0010	WATER HEATERS									
6900	Electric, glass lined, 30 gallon, minimum	L-1	5	2	Ea.	815	87.50		902.50	1,050
6950	Maximum		3	3.333		1,125	146		1,271	1,500
7100	80 gallon, minimum		2	5		1,475	219		1,694	1,975
7150	Maximum		1	10		2,050	440		2,490	2,975
7180	Gas, glass lined, 30 gallon, minimum	2 Plum	5	3.200		1,450	140		1,590	1,825
7220	Maximum		3	5.333		2,025	234		2,259	2,600
7260	50 gallon, minimum		2.50	6.400		1,525	280		1,805	2,125
7300	Maximum		1.50	10.667		2,125	465		2,590	3,125
7310	Water heater, see also Section 22 33 30.13									

11 30 Residential Equipment

11 30 15 – Miscellaneous Residential Appliances

11 30 15.43 Air Quality		Crew	Daily Output	Labor-Hours	Unit	Material	2016 Bare Costs Labor	Equipment	Total	Total Incl O&P
0010	**AIR QUALITY**									
2450	Dehumidifier, portable, automatic, 15 pint	1 Elec	4	2	Ea.	200	88		288	365
2550	40 pint		3.75	2.133		214	94		308	390
3550	Heater, electric, built-in, 1250 watt, ceiling type, minimum		4	2		117	88		205	272
3600	Maximum		3	2.667		204	117		321	415
3700	Wall type, minimum		4	2		191	88		279	355
3750	Maximum		3	2.667		195	117		312	405
3900	1500 watt wall type, with blower		4	2		173	88		261	335
3950	3000 watt		3	2.667		390	117		507	620
4850	Humidifier, portable, 8 gallons per day					152			152	167
5000	15 gallons per day					222			222	244

11 30 33 – Retractable Stairs

11 30 33.10 Disappearing Stairway		Crew	Daily Output	Labor-Hours	Unit	Material	Labor	Equipment	Total	Total Incl O&P
0010	**DISAPPEARING STAIRWAY** No trim included									
0020	One piece, yellow pine, 8′-0″ ceiling	2 Carp	4	4	Ea.	250	151		401	530
0030	9′-0″ ceiling		4	4		284	151		435	565
0040	10′-0″ ceiling		3	5.333		254	202		456	620
0050	11′-0″ ceiling		3	5.333		285	202		487	655
0060	12′-0″ ceiling		3	5.333		360	202		562	735
0100	Custom grade, pine, 8′-6″ ceiling, minimum	1 Carp	4	2		195	75.50		270.50	340
0150	Average		3.50	2.286		244	86.50		330.50	415
0200	Maximum		3	2.667		305	101		406	505
0500	Heavy duty, pivoted, from 7′-7″ to 12′-10″ floor to floor		3	2.667		1,025	101		1,126	1,300
0600	16′-0″ ceiling		2	4		1,600	151		1,751	2,000
0800	Economy folding, pine, 8′-6″ ceiling		4	2		176	75.50		251.50	320
0900	9′-6″ ceiling		4	2		196	75.50		271.50	345
1100	Automatic electric, aluminum, floor to floor height, 8′ to 9′	2 Carp	1	16		8,700	605		9,305	10,600

11 32 Unit Kitchens

11 32 13 – Metal Unit Kitchens

11 32 13.10 Commercial Unit Kitchens		Crew	Daily Output	Labor-Hours	Unit	Material	Labor	Equipment	Total	Total Incl O&P
0010	**COMMERCIAL UNIT KITCHENS**									
1500	Combination range, refrigerator and sink, 30″ wide, minimum	L-1	2	5	Ea.	1,100	219		1,319	1,575
1550	Maximum		1	10		1,250	440		1,690	2,100
1570	60″ wide, average		1.40	7.143		1,425	315		1,740	2,100
1590	72″ wide, average		1.20	8.333		1,500	365		1,865	2,250
1640	Combination range, refrigerator, sink, microwave									
1660	Oven and ice maker	L-1	.80	12.500	Ea.	4,550	550		5,100	5,900

11 41 Foodservice Storage Equipment

11 41 13 – Refrigerated Food Storage Cases

11 41 13.20 Refrigerated Food Storage Equipment		Crew	Daily Output	Labor-Hours	Unit	Material	2016 Bare Costs Labor	Equipment	Total	Total Incl O&P
0010	**REFRIGERATED FOOD STORAGE EQUIPMENT**									
4600	Freezer, pre-fab, 8′ x 8′ w/refrigeration	2 Carp	.45	35.556	Ea.	10,600	1,350		11,950	13,900
4620	8′ x 12′		.35	45.714		12,100	1,725		13,825	16,200
4640	8′ x 16′		.25	64		15,000	2,425		17,425	20,600
4660	8′ x 20′		.17	94.118		21,700	3,550		25,250	29,900
4680	Reach-in, 1 compartment	Q-1	4	4		2,725	158		2,883	3,250
4685	Energy star rated G	R-18	7.80	3.333		2,800	129		2,929	3,275
4700	2 compartment	Q-1	3	5.333		4,275	210		4,485	5,050
4705	Energy star rated G	R-18	6.20	4.194		3,075	162		3,237	3,675
4710	3 compartment	Q-1	3	5.333		5,725	210		5,935	6,650
4715	Energy star rated G	R-18	5.60	4.643		4,475	180		4,655	5,225
8320	Refrigerator, reach-in, 1 compartment		7.80	3.333		2,550	129		2,679	3,025
8325	Energy star rated G		7.80	3.333		2,800	129		2,929	3,275
8330	2 compartment		6.20	4.194		3,900	162		4,062	4,575
8335	Energy star rated G		6.20	4.194		3,075	162		3,237	3,675
8340	3 compartment		5.60	4.643		4,950	180		5,130	5,725
8345	Energy star rated G		5.60	4.643		4,475	180		4,655	5,225
8350	Pre-fab, with refrigeration, 8′ x 8′	2 Carp	.45	35.556		6,875	1,350		8,225	9,825
8360	8′ x 12′		.35	45.714		8,525	1,725		10,250	12,300
8370	8′ x 16′		.25	64		12,500	2,425		14,925	17,800
8380	8′ x 20′		.17	94.118		15,800	3,550		19,350	23,400
8390	Pass-thru/roll-in, 1 compartment	R-18	7.80	3.333		4,525	129		4,654	5,175
8400	2 compartment		6.24	4.167		6,350	161		6,511	7,275
8410	3 compartment		5.60	4.643		8,675	180		8,855	9,825
8420	Walk-in, alum, door & floor only, no refrig, 6′ x 6′ x 7′-6″	2 Carp	1.40	11.429		8,725	430		9,155	10,300
8430	10′ x 6′ x 7′-6″		.55	29.091		12,600	1,100		13,700	15,800
8440	12′ x 14′ x 7′-6″		.25	64		17,400	2,425		19,825	23,200
8450	12′ x 20′ x 7′-6″		.17	94.118		18,700	3,550		22,250	26,600
8460	Refrigerated cabinets, mobile					3,925			3,925	4,325
8470	Refrigerator/freezer, reach-in, 1 compartment	R-18	5.60	4.643		5,750	180		5,930	6,625
8480	2 compartment	″	4.80	5.417		7,200	209		7,409	8,275

11 41 33 – Foodservice Shelving

11 41 33.20 Metal Food Storage Shelving		Crew	Daily Output	Labor-Hours	Unit	Material	2016 Bare Costs Labor	Equipment	Total	Total Incl O&P
0010	**METAL FOOD STORAGE SHELVING**									
8600	Stainless steel shelving, louvered 4-tier, 20″ x 3′	1 Clab	6	1.333	Ea.	1,400	39		1,439	1,625
8605	20″ x 4′		6	1.333		1,575	39		1,614	1,825
8610	20″ x 6′		6	1.333		1,675	39		1,714	1,900
8615	24″ x 3′		6	1.333		1,975	39		2,014	2,250
8620	24″ x 4′		6	1.333		2,350	39		2,389	2,675
8625	24″ x 6′		6	1.333		3,275	39		3,314	3,675
8630	Flat 4-tier, 20″ x 3′		6	1.333		1,150	39		1,189	1,350
8635	20″ x 4′		6	1.333		1,400	39		1,439	1,600
8640	20″ x 5′		6	1.333		1,600	39		1,639	1,825
8645	24″ x 3′		6	1.333		1,250	39		1,289	1,450
8650	24″ x 4′		6	1.333		2,225	39		2,264	2,525
8655	24″ x 6′		6	1.333		2,625	39		2,664	2,975
8700	Galvanized shelving, louvered 4-tier, 20″ x 3′		6	1.333		760	39		799	900
8705	20″ x 4′		6	1.333		860	39		899	1,025
8710	20″ x 6′		6	1.333		1,000	39		1,039	1,175
8715	24″ x 3′		6	1.333		695	39		734	830
8720	24″ x 4′		6	1.333		920	39		959	1,075
8725	24″ x 6′		6	1.333		1,325	39		1,364	1,525

11 41 Foodservice Storage Equipment

11 41 33 – Foodservice Shelving

11 41 33.20 Metal Food Storage Shelving		Crew	Daily Output	Labor-Hours	Unit	2016 Bare Costs Material	Labor	Equipment	Total	Total Incl O&P
8730	Flat 4-tier, 20" x 3'	1 Clab	6	1.333	Ea.	700	39		739	835
8735	20" x 4'		6	1.333		620	39		659	745
8740	20" x 6'		6	1.333		910	39		949	1,075
8745	24" x 3'		6	1.333		690	39		729	820
8750	24" x 4'		6	1.333		715	39		754	855
8755	24" x 6'		6	1.333		820	39		859	970
8760	Stainless steel dunnage rack, 24" x 3'		8	1		395	29		424	485
8765	24" x 4'		8	1		430	29		459	525
8770	Galvanized dunnage rack, 24" x 3'		8	1		169	29		198	235
8775	24" x 4'		8	1		187	29		216	254

11 42 Food Preparation Equipment

11 42 10 – Commercial Food Preparation Equipment

11 42 10.10 Choppers, Mixers and Misc. Equipment		Crew	Daily Output	Labor-Hours	Unit	Material	Labor	Equipment	Total	Total Incl O&P
0010	**CHOPPERS, MIXERS AND MISC. EQUIPMENT**									
1700	Choppers, 5 pounds	R-18	7	3.714	Ea.	2,200	144		2,344	2,650
1720	16 pounds		5	5.200		2,325	201		2,526	2,875
1740	35 to 40 pounds		4	6.500		3,550	251		3,801	4,300
1840	Coffee brewer, 5 burners	1 Plum	3	2.667		1,325	117		1,442	1,675
1860	Single, 3 gallon	"	3	2.667		1,800	117		1,917	2,175
3850	40 quarts	L-7	5.40	4.815		6,450	155		6,605	7,350
4040	80 quarts		3.90	6.667		12,200	215		12,415	13,900
4100	Floor type, 20 quarts		15	1.733		3,225	56		3,281	3,650
4120	60 quarts		14	1.857		10,200	60		10,260	11,300
4140	80 quarts		12	2.167		16,200	70		16,270	17,900
4160	140 quarts		8.60	3.023		27,300	97.50		27,397.50	30,200
6700	Peelers, small	R-18	8	3.250		2,000	126		2,126	2,400
6720	Large	"	6	4.333		4,850	168		5,018	5,600
6800	Pulper/extractor, close coupled, 5 HP	1 Plum	1.90	4.211		3,575	184		3,759	4,250
8580	Slicer with table	R-18	9	2.889		4,675	112		4,787	5,325

11 43 Food Delivery Carts and Conveyors

11 43 13 – Food Delivery Carts

11 43 13.10 Mobile Carts, Racks and Trays		Crew	Daily Output	Labor-Hours	Unit	Material	Labor	Equipment	Total	Total Incl O&P
0010	**MOBILE CARTS, RACKS AND TRAYS**									
1650	Cabinet, heated, 1 compartment, reach-in	R-18	5.60	4.643	Ea.	3,725	180		3,905	4,400
1655	Pass-thru roll-in		5.60	4.643		3,850	180		4,030	4,525
1660	2 compartment, reach-in		4.80	5.417		9,050	209		9,259	10,300
1670	Mobile					3,600			3,600	3,975
2000	Hospital food cart, hot and cold service, 20 tray capacity					15,300			15,300	16,900
6850	Mobile rack w/pan slide					1,400			1,400	1,550
9180	Tray and silver dispenser, mobile	1 Clab	16	.500		925	14.60		939.60	1,050

11 44 Food Cooking Equipment

11 44 13 – Commercial Ranges

11 44 13.10 Cooking Equipment		Crew	Daily Output	Labor-Hours	Unit	2016 Bare Costs Material	Labor	Equipment	Total	Total Incl O&P
0010	**COOKING EQUIPMENT**									
5170	Energy star rated, 50 lb. capacity G	R-18	4	6.500	Ea.	4,625	251		4,876	5,475
5175	85 lb. capacity G	"	4	6.500		10,000	251		10,251	11,400
6200	Iced tea brewer	1 Plum	3.44	2.326		755	102		857	1,000
9160	Pop-up, 2 slot				↓	620			620	685

11 46 Food Dispensing Equipment

11 46 16 – Service Line Equipment

11 46 16.10 Commercial Food Dispensing Equipment

		Crew	Daily Output	Labor-Hours	Unit	Material	Labor	Equipment	Total	Total Incl O&P
0010	**COMMERCIAL FOOD DISPENSING EQUIPMENT**									
1050	Butter pat dispenser	1 Clab	13	.615	Ea.	870	17.95		887.95	990
1100	Bread dispenser, counter top		13	.615		890	17.95		907.95	1,000
1900	Cup and glass dispenser, drop in		4	2		1,075	58.50		1,133.50	1,300
1920	Disposable cup, drop in		16	.500		550	14.60		564.60	630
2650	Dish dispenser, drop in, 12"		11	.727		2,000	21		2,021	2,225
2660	Mobile	↓	10	.800		3,050	23.50		3,073.50	3,400
3600	Well, hot food, built-in, rectangular, 12" x 20"	R-30	10	2.600		805	91		896	1,025
3610	Circular, 7 qt.		10	2.600		415	91		506	610
3620	Refrigerated, 2 compartments		10	2.600		3,050	91		3,141	3,525
3630	3 compartments		9	2.889		3,775	101		3,876	4,325
3640	4 compartments	↓	8	3.250		4,350	114		4,464	5,000
5700	Hot chocolate dispenser	1 Plum	4	2		1,175	87.50		1,262.50	1,450
6300	Juice dispenser, concentrate	R-18	4.50	5.778		1,900	223		2,123	2,450
6690	Milk dispenser, bulk, 2 flavor	R-30	8	3.250		1,750	114		1,864	2,125
6695	3 flavor	"	8	3.250	↓	2,375	114		2,489	2,800
8800	Serving counter, straight	1 Carp	40	.200	L.F.	910	7.55		917.55	1,025
8820	Curved section	"	30	.267	"	1,175	10.10		1,185.10	1,325
8825	Solid surface, see Section 12 36 61.16									
9100	Soft serve ice cream machine, medium	R-18	11	2.364	Ea.	9,200	91.50		9,291.50	10,300
9110	Large	"	9	2.889	"	21,800	112		21,912	24,100

11 46 83 – Ice Machines

11 46 83.10 Commercial Ice Equipment

		Crew	Daily Output	Labor-Hours	Unit	Material	Labor	Equipment	Total	Total Incl O&P
0010	**COMMERCIAL ICE EQUIPMENT**									
5800	Ice cube maker, 50 pounds per day	Q-1	6	2.667	Ea.	1,625	105		1,730	1,950
6050	500 pounds per day		4	4		2,950	158		3,108	3,475
6090	1000 pounds per day, with bin		1	16		5,125	630		5,755	6,700
6100	Ice flakers, 300 pounds per day		1.60	10		3,175	395		3,570	4,125
6120	600 pounds per day		.95	16.842		4,475	665		5,140	6,025
6130	1000 pounds per day		.75	21.333		4,575	840		5,415	6,425
6140	2000 pounds per day	↓	.65	24.615		21,600	970		22,570	25,300
6160	Ice storage bin, 500 pound capacity	Q-5	1	16		1,000	645		1,645	2,175
6180	1000 pound	"	.56	28.571	↓	2,750	1,150		3,900	4,925

11 48 Foodservice Cleaning and Disposal Equipment

11 48 13 – Commercial Dishwashers

11 48 13.10 Dishwashers		Crew	Daily Output	Labor-Hours	Unit	Material	2016 Bare Costs Labor	Equipment	Total	Total Incl O&P
0010	**DISHWASHERS**									
2950	Dishwasher hood, canopy type	L-3A	10	1.200	L.F.	1,025	49		1,074	1,200
2960	Pant leg type	"	2.50	4.800	Ea.	8,700	195		8,895	9,900
6750	Pot sink, 3 compartment	1 Plum	7.25	1.103	L.F.	930	48.50		978.50	1,100
6760	Pot washer, low temp wash/rinse		1.60	5	Ea.	5,075	219		5,294	5,925
6770	High pressure wash, high temperature rinse	↓	1.20	6.667		36,600	292		36,892	40,700
9170	Trash compactor, small, up to 125 lb. compacted weight	L-4	4	4		23,600	133		23,733	26,200
9175	Large, up to 175 lb. compacted weight	"	3	5.333	↓	28,100	177		28,277	31,200

11 52 Audio-Visual Equipment

11 52 13 – Projection Screens

11 52 13.10 Projection Screens, Wall or Ceiling Hung		Crew	Daily Output	Labor-Hours	Unit	Material	Labor	Equipment	Total	Total Incl O&P
0010	**PROJECTION SCREENS, WALL OR CEILING HUNG**, matte white									
0100	Manually operated, economy	2 Carp	500	.032	S.F.	6.45	1.21		7.66	9.10
0300	Intermediate		450	.036		7.85	1.34		9.19	10.90
0400	Deluxe	↓	400	.040	↓	9.75	1.51		11.26	13.30

11 81 Facility Maintenance Equipment

11 81 19 – Vacuum Cleaning Systems

11 81 19.10 Vacuum Cleaning		Crew	Daily Output	Labor-Hours	Unit	Material	Labor	Equipment	Total	Total Incl O&P
0010	**VACUUM CLEANING**									
0020	Central, 3 inlet, residential	1 Skwk	.90	8.889	Total	1,100	340		1,440	1,775
0400	5 inlet system, residential		.50	16		1,550	615		2,165	2,725
0600	7 inlet system, commercial		.40	20		1,750	765		2,515	3,200
0800	9 inlet system, residential	↓	.30	26.667		3,875	1,025		4,900	5,975
4010	Rule of thumb: First 1200 S.F., installed				↓				1,425	1,575
4020	For each additional S.F., add				S.F.				.26	.26

Division 12 Furnishings

Estimating Tips

General

- The items in this division are usually priced per square foot or each. Most of these items are purchased by the owner and installed by the contractor. Do not assume the items in Division 12 will be purchased and installed by the contractor. Check the specifications for responsibilities and include receiving, storage, installation, and mechanical and electrical hookups in the appropriate divisions.
- Some items in this division require some type of support system that is not usually furnished with the item. Examples of these systems include blocking for the attachment of casework and heavy drapery rods. The required blocking must be added to the estimate in the appropriate division.

Reference Numbers

Reference numbers are shown at the beginning of some major classifications. These numbers refer to related items in the Reference Section. The reference information may be an estimating procedure, an alternate pricing method, or technical information.

Note: Not all subdivisions listed here necessarily appear. ■

12 21 Window Blinds

12 21 13 – Horizontal Louver Blinds

12 21 13.13 Metal Horizontal Louver Blinds		Crew	Daily Output	Labor-Hours	Unit	2016 Bare Costs Material	Labor	Equipment	Total	Total Incl O&P
0010	**METAL HORIZONTAL LOUVER BLINDS**									
0020	Horizontal, 1″ aluminum slats, solid color, stock	1 Carp	590	.014	S.F.	4.69	.51		5.20	6

12 21 13.33 Vinyl Horizontal Louver Blinds		Crew	Daily Output	Labor-Hours	Unit	Material	Labor	Equipment	Total	Total Incl O&P
0010	**VINYL HORIZONTAL LOUVER BLINDS**									
0100	2″ composite, 48″ wide, 48″ high	1 Carp	30	.267	Ea.	91	10.10		101.10	117
0120	72″ high		29	.276		123	10.45		133.45	152
0140	96″ high		28	.286		167	10.80		177.80	202
0200	60″ wide, 60″ high		27	.296		114	11.20		125.20	145
0220	72″ high		25	.320		131	12.10		143.10	165
0240	96″ high		24	.333		196	12.60		208.60	237
0300	72″ wide, 72″ high		25	.320		181	12.10		193.10	220
0320	96″ high		23	.348		250	13.15		263.15	297
0400	96″ wide, 96″ high		20	.400		325	15.10		340.10	380
1000	2″ faux wood, 48″ wide, 48″ high		30	.267		60	10.10		70.10	83
1020	72″ high		29	.276		78	10.45		88.45	103
1040	96″ high		28	.286		93.50	10.80		104.30	121
1300	72″ wide, 72″ high		25	.320		141	12.10		153.10	177
1320	96″ high		23	.348		212	13.15		225.15	255
1400	96″ wide, 96″ high		20	.400		223	15.10		238.10	271

12 22 Curtains and Drapes

12 22 16 – Drapery Track and Accessories

12 22 16.10 Drapery Hardware		Crew	Daily Output	Labor-Hours	Unit	Material	Labor	Equipment	Total	Total Incl O&P
0010	**DRAPERY HARDWARE**									
0030	Standard traverse, per foot, minimum	1 Carp	59	.136	L.F.	7.40	5.15		12.55	16.70
0100	Maximum		51	.157	″	10.55	5.95		16.50	21.50
0200	Decorative traverse, 28″-48″, minimum		22	.364	Ea.	26.50	13.75		40.25	52
0220	Maximum		21	.381		58.50	14.40		72.90	88
0300	48″-84″, minimum		20	.400		28.50	15.10		43.60	57
0320	Maximum		19	.421		74.50	15.90		90.40	109
0400	66″-120″, minimum		18	.444		39	16.80		55.80	71
0420	Maximum		17	.471		114	17.80		131.80	155
0500	84″-156″, minimum		16	.500		50	18.90		68.90	86.50
0520	Maximum		15	.533		155	20		175	205
0600	130″-240″, minimum		14	.571		73	21.50		94.50	117
0620	Maximum		13	.615		215	23.50		238.50	276
0700	Slide rings, each, minimum					1.08			1.08	1.19
0720	Maximum					2.11			2.11	2.32
4000	Traverse rods, adjustable, 28″ to 48″	1 Carp	22	.364		30	13.75		43.75	56
4020	48″ to 84″		20	.400		38	15.10		53.10	67.50
4040	66″ to 120″		18	.444		46	16.80		62.80	79
4060	84″ to 156″		16	.500		52.50	18.90		71.40	89.50
4080	100″ to 180″		14	.571		61	21.50		82.50	104
4100	228″ to 312″		13	.615		85	23.50		108.50	133
4500	Curtain rod, 28″ to 48″, single		22	.364		6	13.75		19.75	29.50
4510	Double		22	.364		10.20	13.75		23.95	34
4520	48″ to 86″, single		20	.400		10.25	15.10		25.35	37
4530	Double		20	.400		17.10	15.10		32.20	44.50
4540	66″ to 120″, single		18	.444		17.15	16.80		33.95	47
4550	Double		18	.444		27	16.80		43.80	57.50
4600	Valance, pinch pleated fabric, 12″ deep, up to 54″ long, minimum					39.50			39.50	43.50

12 22 Curtains and Drapes

12 22 16 – Drapery Track and Accessories

12 22 16.10 Drapery Hardware		Crew	Daily Output	Labor-Hours	Unit	Material	Labor	Equipment	Total	Total Incl O&P
4610	Maximum				Ea.	98			98	108
4620	Up to 77" long, minimum					69.50			69.50	76.50
4630	Maximum					158			158	174
5000	Stationary rods, first 2'					8.40			8.40	9.25

Labor, Equipment and Material columns are 2016 Bare Costs.

12 22 16.20 Blast Curtains		Crew	Daily Output	Labor-Hours	Unit	Material	Labor	Equipment	Total	Total Incl O&P
0010	**BLAST CURTAINS** per L.F. horizontal opening width, off-white or gray fabric									
0100	Blast curtains, drapery system, complete, including hardware, minimum	1 Carp	10.25	.780	L.F.	208	29.50		237.50	278
0120	Average		10.25	.780		211	29.50		240.50	282
0140	Maximum		10.25	.780		235	29.50		264.50	310

12 23 Interior Shutters

12 23 10 – Wood Interior Shutters

12 23 10.10 Wood Interior Shutters		Crew	Daily Output	Labor-Hours	Unit	Material	Labor	Equipment	Total	Total Incl O&P
0010	**WOOD INTERIOR SHUTTERS**, louvered									
0200	Two panel, 27" wide, 36" high	1 Carp	5	1.600	Set	158	60.50		218.50	275
0300	33" wide, 36" high		5	1.600		204	60.50		264.50	325
0500	47" wide, 36" high		5	1.600		273	60.50		333.50	400
1000	Four panel, 27" wide, 36" high		5	1.600		158	60.50		218.50	275
1100	33" wide, 36" high		5	1.600		202	60.50		262.50	325
1300	47" wide, 36" high		5	1.600		270	60.50		330.50	400
1400	Plantation shutters, 16" x 48"		5	1.600	Ea.	182	60.50		242.50	300
1450	16" x 96"		4	2		299	75.50		374.50	455
1460	36" x 96"		3	2.667		660	101		761	895

12 23 10.13 Wood Panels		Crew	Daily Output	Labor-Hours	Unit	Material	Labor	Equipment	Total	Total Incl O&P
0010	**WOOD PANELS**									
3000	Wood folding panels with movable louvers, 7" x 20" each	1 Carp	17	.471	Pr.	82.50	17.80		100.30	121
3300	8" x 28" each		17	.471		82.50	17.80		100.30	121
3450	9" x 36" each		17	.471		95	17.80		112.80	134
3600	10" x 40" each		17	.471		103	17.80		120.80	144
4000	Fixed louver type, stock units, 8" x 20" each		17	.471		94	17.80		111.80	133
4150	10" x 28" each		17	.471		79.50	17.80		97.30	118
4300	12" x 36" each		17	.471		94	17.80		111.80	133
4450	18" x 40" each		17	.471		134	17.80		151.80	178
5000	Insert panel type, stock, 7" x 20" each		17	.471		21	17.80		38.80	53
5150	8" x 28" each		17	.471		38	17.80		55.80	72
5300	9" x 36" each		17	.471		48.50	17.80		66.30	83
5450	10" x 40" each		17	.471		52	17.80		69.80	87
5600	Raised panel type, stock, 10" x 24" each		17	.471		132	17.80		149.80	175
5650	12" x 26" each		17	.471		132	17.80		149.80	175
5700	14" x 30" each		17	.471		146	17.80		163.80	190
5750	16" x 36" each		17	.471		161	17.80		178.80	207
6000	For custom built pine, add					22%				
6500	For custom built hardwood blinds, add					42%				

12 24 Window Shades

12 24 13 – Roller Window Shades

12 24 13.10 Shades		Crew	Daily Output	Labor-Hours	Unit	2016 Bare Costs Material	Labor	Equipment	Total	Total Incl O&P
0010	**SHADES**									
0020	Basswood, roll-up, stain finish, 3/8″ slats	1 Carp	300	.027	S.F.	17.15	1.01		18.16	20.50
5011	Insulative shades G		125	.064		12.15	2.42		14.57	17.40
6011	Solar screening, fiberglass G		85	.094		6.90	3.56		10.46	13.55
8011	Interior insulative shutter									
8111	Stock unit, 15″ x 60″ G	1 Carp	17	.471	Pr.	12.25	17.80		30.05	43.50

12 32 Manufactured Wood Casework

12 32 16 – Manufactured Plastic-Laminate-Clad Casework

12 32 16.20 Plastic Laminate Casework Doors

0010	**PLASTIC LAMINATE CASEWORK DOORS**
1100	For casework hardware, see Section 12 32 23.35

12 32 23 – Hardwood Casework

12 32 23.10 Manufactured Wood Casework, Stock Units

Code	Description	Crew	Daily Output	Labor-Hours	Unit	Material	Labor	Equipment	Total	Total Incl O&P
0010	**MANUFACTURED WOOD CASEWORK, STOCK UNITS**									
0700	Kitchen base cabinets, hardwood, not incl. counter tops,									
0710	24″ deep, 35″ high, prefinished									
0800	One top drawer, one door below, 12″ wide	2 Carp	24.80	.645	Ea.	271	24.50		295.50	340
0820	15″ wide		24	.667		282	25		307	350
0840	18″ wide		23.30	.687		305	26		331	385
0860	21″ wide		22.70	.705		315	26.50		341.50	395
0880	24″ wide		22.30	.717		370	27		397	455
1000	Four drawers, 12″ wide		24.80	.645		285	24.50		309.50	355
1020	15″ wide		24	.667		289	25		314	360
1040	18″ wide		23.30	.687		320	26		346	395
1060	24″ wide		22.30	.717		350	27		377	430
1200	Two top drawers, two doors below, 27″ wide		22	.727		400	27.50		427.50	485
1220	30″ wide		21.40	.748		440	28.50		468.50	530
1240	33″ wide		20.90	.766		455	29		484	550
1260	36″ wide		20.30	.788		470	30		500	570
1280	42″ wide		19.80	.808		495	30.50		525.50	595
1300	48″ wide		18.90	.847		535	32		567	640
1500	Range or sink base, two doors below, 30″ wide		21.40	.748		365	28.50		393.50	450
1520	33″ wide		20.90	.766		390	29		419	480
1540	36″ wide		20.30	.788		405	30		435	500
1560	42″ wide		19.80	.808		430	30.50		460.50	520
1580	48″ wide		18.90	.847		450	32		482	550
1800	For sink front units, deduct					161			161	177
2000	Corner base cabinets, 36″ wide, standard	2 Carp	18	.889		670	33.50		703.50	795
2100	Lazy Susan with revolving door	″	16.50	.970		855	36.50		891.50	1,000
4000	Kitchen wall cabinets, hardwood, 12″ deep with two doors									
4050	12″ high, 30″ wide	2 Carp	24.80	.645	Ea.	245	24.50		269.50	310
4100	36″ wide		24	.667		291	25		316	360
4400	15″ high, 30″ wide		24	.667		249	25		274	315
4420	33″ wide		23.30	.687		310	26		336	385
4440	36″ wide		22.70	.705		299	26.50		325.50	375
4450	42″ wide		22.70	.705		335	26.50		361.50	415
4700	24″ high, 30″ wide		23.30	.687		335	26		361	410
4720	36″ wide		22.70	.705		365	26.50		391.50	450
4740	42″ wide		22.30	.717		415	27		442	505

12 32 23.10 Manufactured Wood Casework, Stock Units

	12 32 23.10 Manufactured Wood Casework, Stock Units	Crew	Daily Output	Labor-Hours	Unit	2016 Bare Costs Material	Labor	Equipment	Total	Total Incl O&P
5000	30" high, one door, 12" wide	2 Carp	22	.727	Ea.	234	27.50		261.50	305
5020	15" wide		21.40	.748		246	28.50		274.50	320
5040	18" wide		20.90	.766		270	29		299	345
5060	24" wide		20.30	.788		315	30		345	400
5300	Two doors, 27" wide		19.80	.808		350	30.50		380.50	435
5320	30" wide		19.30	.829		370	31.50		401.50	460
5340	36" wide		18.80	.851		420	32		452	515
5360	42" wide		18.50	.865		455	32.50		487.50	555
5380	48" wide		18.40	.870		510	33		543	620
6000	Corner wall, 30" high, 24" wide		18	.889		360	33.50		393.50	450
6050	30" wide		17.20	.930		380	35		415	480
6100	36" wide		16.50	.970		435	36.50		471.50	540
6500	Revolving Lazy Susan		15.20	1.053		490	40		530	605
7000	Broom cabinet, 84" high, 24" deep, 18" wide		10	1.600		670	60.50		730.50	840
7500	Oven cabinets, 84" high, 24" deep, 27" wide		8	2		1,025	75.50		1,100.50	1,250
7750	Valance board trim		396	.040	L.F.	16.25	1.53		17.78	20.50
7780	Toe kick trim	1 Carp	256	.031	"	3.10	1.18		4.28	5.40
7790	Base cabinet corner filler		16	.500	Ea.	43.50	18.90		62.40	79
7800	Cabinet filler, 3" x 24"		20	.400		17.10	15.10		32.20	44.50
7810	3" x 30"		20	.400		21.50	15.10		36.60	49
7820	3" x 42"		18	.444		30	16.80		46.80	61
7830	3" x 80"		16	.500		57	18.90		75.90	94.50
7850	Cabinet panel		50	.160	S.F.	9.30	6.05		15.35	20.50
9000	For deluxe models of all cabinets, add					40%				
9500	For custom built in place, add					25%	10%			
9558	Rule of thumb, kitchen cabinets not including									
9560	appliances & counter top, minimum	2 Carp	30	.533	L.F.	182	20		202	234
9600	Maximum	"	25	.640	"	405	24		429	485

12 32 23.30 Manufactured Wood Casework Vanities

	12 32 23.30 Manufactured Wood Casework Vanities	Crew	Daily Output	Labor-Hours	Unit	Material	Labor	Equipment	Total	Total Incl O&P
0010	**MANUFACTURED WOOD CASEWORK VANITIES**									
8000	Vanity bases, 2 doors, 30" high, 21" deep, 24" wide	2 Carp	20	.800	Ea.	315	30		345	395
8050	30" wide		16	1		375	38		413	480
8100	36" wide		13.33	1.200		365	45.50		410.50	475
8150	48" wide		11.43	1.400		480	53		533	615
9000	For deluxe models of all vanities, add to above					40%				
9500	For custom built in place, add to above					25%	10%			

12 32 23.35 Manufactured Wood Casework Hardware

	12 32 23.35 Manufactured Wood Casework Hardware	Crew	Daily Output	Labor-Hours	Unit	Material	Labor	Equipment	Total	Total Incl O&P
0010	**MANUFACTURED WOOD CASEWORK HARDWARE**									
1000	Catches, minimum	1 Carp	235	.034	Ea.	1.30	1.29		2.59	3.58
1020	Average		119.40	.067		4	2.53		6.53	8.65
1040	Maximum		80	.100		7.75	3.78		11.53	14.90
2000	Door/drawer pulls, handles									
2200	Handles and pulls, projecting, metal, minimum	1 Carp	48	.167	Ea.	5.20	6.30		11.50	16.25
2220	Average		42	.190		7.90	7.20		15.10	21
2240	Maximum		36	.222		10.65	8.40		19.05	26
2300	Wood, minimum		48	.167		5.30	6.30		11.60	16.40
2320	Average		42	.190		7.10	7.20		14.30	19.85
2340	Maximum		36	.222		9.80	8.40		18.20	25
2600	Flush, metal, minimum		48	.167		5.30	6.30		11.60	16.40
2620	Average		42	.190		7.25	7.20		14.45	20
2640	Maximum		36	.222		9.90	8.40		18.30	25
3000	Drawer tracks/glides, minimum		48	.167	Pr.	8.95	6.30		15.25	20.50

12 32 Manufactured Wood Casework

12 32 23 – Hardwood Casework

12 32 23.35 Manufactured Wood Casework Hardware		Crew	Daily Output	Labor-Hours	Unit	2016 Bare Costs Material	Labor	Equipment	Total	Total Incl O&P
3020	Average	1 Carp	32	.250	Pr.	16	9.45		25.45	33.50
3040	Maximum		24	.333		26	12.60		38.60	49.50
4000	Cabinet hinges, minimum		160	.050		3.10	1.89		4.99	6.60
4020	Average		95.24	.084		7.60	3.18		10.78	13.70
4040	Maximum		68	.118		12.40	4.45		16.85	21

12 36 Countertops

12 36 16 – Metal Countertops

12 36 16.10 Stainless Steel Countertops

Code	Description	Crew	Daily Output	Labor-Hours	Unit	Material	Labor	Equipment	Total	Total Incl O&P
0010	**STAINLESS STEEL COUNTERTOPS**									
3200	Stainless steel, custom	1 Carp	24	.333	S.F.	155	12.60		167.60	192

12 36 19 – Wood Countertops

12 36 19.10 Maple Countertops

Code	Description	Crew	Daily Output	Labor-Hours	Unit	Material	Labor	Equipment	Total	Total Incl O&P
0010	**MAPLE COUNTERTOPS**									
2900	Solid, laminated, 1-1/2" thick, no splash	1 Carp	28	.286	L.F.	76.50	10.80		87.30	103
3000	With square splash		28	.286	"	91	10.80		101.80	118
3400	Recessed cutting block with trim, 16" x 20" x 1"		8	1	Ea.	93	38		131	167
3410	Replace cutting block only		8	1	"	70	38		108	141

12 36 23 – Plastic Countertops

12 36 23.13 Plastic-Laminate-Clad Countertops

Code	Description	Crew	Daily Output	Labor-Hours	Unit	Material	Labor	Equipment	Total	Total Incl O&P
0010	**PLASTIC-LAMINATE-CLAD COUNTERTOPS**									
0020	Stock, 24" wide w/backsplash, minimum	1 Carp	30	.267	L.F.	17.25	10.10		27.35	36
0100	Maximum		25	.320		35	12.10		47.10	59
0300	Custom plastic, 7/8" thick, aluminum molding, no splash		30	.267		30.50	10.10		40.60	50.50
0400	Cove splash		30	.267		29.50	10.10		39.60	49.50
0600	1-1/4" thick, no splash		28	.286		36	10.80		46.80	57.50
0700	Square splash		28	.286		43	10.80		53.80	65.50
0900	Square edge, plastic face, 7/8" thick, no splash		30	.267		33.50	10.10		43.60	54
1000	With splash		30	.267		40	10.10		50.10	61
1200	For stainless channel edge, 7/8" thick, add					3.17			3.17	3.49
1300	1-1/4" thick, add					3.78			3.78	4.16
1500	For solid color suede finish, add					4.14			4.14	4.55
1700	For end splash, add				Ea.	18.65			18.65	20.50
1900	For cut outs, standard, add, minimum	1 Carp	32	.250		12.40	9.45		21.85	29.50
2000	Maximum		8	1		6.20	38		44.20	70.50
2010	Cut out in backsplash for elec. wall outlet		38	.211			7.95		7.95	13.35
2020	Cut out for sink		20	.400			15.10		15.10	25.50
2030	Cut out for stove top		18	.444			16.80		16.80	28
2100	Postformed, including backsplash and front edge		30	.267	L.F.	10.35	10.10		20.45	28.50
2110	Mitred, add		12	.667	Ea.		25		25	42
2200	Built-in place, 25" wide, plastic laminate		25	.320	L.F.	42	12.10		54.10	66.50

12 36 33 – Tile Countertops

12 36 33.10 Ceramic Tile Countertops

Code	Description	Crew	Daily Output	Labor-Hours	Unit	Material	Labor	Equipment	Total	Total Incl O&P
0010	**CERAMIC TILE COUNTERTOPS**									
2300	Ceramic tile mosaic	1 Carp	25	.320	L.F.	34	12.10		46.10	58

12 36 Countertops

12 36 40 – Stone Countertops

12 36 40.10 Natural Stone Countertops		Crew	Daily Output	Labor-Hours	Unit	2016 Bare Costs Material	Labor	Equipment	Total	Total Incl O&P
0010	**NATURAL STONE COUNTERTOPS**									
2500	Marble, stock, with splash, 1/2" thick, minimum	1 Bric	17	.471	L.F.	42.50	17		59.50	75
2700	3/4" thick, maximum	"	13	.615	"	107	22		129	154

12 36 61 – Simulated Stone Countertops

12 36 61.16 Solid Surface Countertops		Crew	Daily Output	Labor-Hours	Unit	2016 Bare Costs Material	Labor	Equipment	Total	Total Incl O&P
0010	**SOLID SURFACE COUNTERTOPS**, Acrylic polymer									
0020	Pricing for orders of 100 L.F. or greater									
0100	25" wide, solid colors	2 Carp	28	.571	L.F.	54.50	21.50		76	96
0200	Patterned colors		28	.571		69	21.50		90.50	112
0300	Premium patterned colors		28	.571		86.50	21.50		108	131
0400	With silicone attached 4" backsplash, solid colors		27	.593		60	22.50		82.50	104
0500	Patterned colors		27	.593		76	22.50		98.50	121
0600	Premium patterned colors		27	.593		94.50	22.50		117	142
0700	With hard seam attached 4" backsplash, solid colors		23	.696		60	26.50		86.50	110
0800	Patterned colors		23	.696		76	26.50		102.50	128
0900	Premium patterned colors		23	.696		94.50	26.50		121	148
1000	Pricing for order of 51 – 99 L.F.									
1100	25" wide, solid colors	2 Carp	24	.667	L.F.	63	25		88	111
1200	Patterned colors		24	.667		79.50	25		104.50	130
1300	Premium patterned colors		24	.667		99.50	25		124.50	151
1400	With silicone attached 4" backsplash, solid colors		23	.696		69	26.50		95.50	120
1500	Patterned colors		23	.696		87.50	26.50		114	140
1600	Premium patterned colors		23	.696		109	26.50		135.50	164
1700	With hard seam attached 4" backsplash, solid colors		20	.800		69	30		99	127
1800	Patterned colors		20	.800		87.50	30		117.50	147
1900	Premium patterned colors		20	.800		109	30		139	171
2000	Pricing for order of 1 – 50 L.F.									
2100	25" wide, solid colors	2 Carp	20	.800	L.F.	73.50	30		103.50	132
2200	Patterned colors		20	.800		93.50	30		123.50	154
2300	Premium patterned colors		20	.800		117	30		147	180
2400	With silicone attached 4" backsplash, solid colors		19	.842		81	32		113	143
2500	Patterned colors		19	.842		102	32		134	167
2600	Premium patterned colors		19	.842		128	32		160	194
2700	With hard seam attached 4" backsplash, solid colors		15	1.067		81	40.50		121.50	157
2800	Patterned colors		15	1.067		102	40.50		142.50	181
2900	Premium patterned colors		15	1.067		128	40.50		168.50	208
3000	Sinks, pricing for order of 100 or greater units									
3100	Single bowl, hard seamed, solid colors, 13" x 17"	1 Carp	3	2.667	Ea.	370	101		471	575
3200	10" x 15"		7	1.143		170	43		213	260
3300	Cutouts for sinks		8	1			38		38	63.50
3400	Sinks, pricing for order of 51 – 99 units									
3500	Single bowl, hard seamed, solid colors, 13" x 17"	1 Carp	2.55	3.137	Ea.	425	119		544	665
3600	10" x 15"		6	1.333		196	50.50		246.50	300
3700	Cutouts for sinks		7	1.143			43		43	72.50
3800	Sinks, pricing for order of 1 – 50 units									
3900	Single bowl, hard seamed, solid colors, 13" x 17"	1 Carp	2	4	Ea.	500	151		651	805
4000	10" x 15"		4.55	1.758		230	66.50		296.50	365
4100	Cutouts for sinks		5.25	1.524			57.50		57.50	96.50
4200	Cooktop cutouts, pricing for 100 or greater units		4	2		27	75.50		102.50	157
4300	51 – 99 units		3.40	2.353		31.50	89		120.50	184
4400	1 – 50 units		3	2.667		36.50	101		137.50	210

12 36 Countertops

12 36 61 – Simulated Stone Countertops

12 36 61.17 Solid Surface Vanity Tops		Crew	Daily Output	Labor-Hours	Unit	2016 Bare Costs Material	Labor	Equipment	Total	Total Incl O&P
0010	SOLID SURFACE VANITY TOPS									
0015	Solid surface, center bowl, 17" x 19"	1 Carp	12	.667	Ea.	190	25		215	251
0020	19" x 25"		12	.667		194	25		219	256
0030	19" x 31"		12	.667		227	25		252	292
0040	19" x 37"		12	.667		264	25		289	335
0050	22" x 25"		10	.800		345	30		375	430
0060	22" x 31"		10	.800		405	30		435	495
0070	22" x 37"		10	.800		470	30		500	570
0080	22" x 43"		10	.800		535	30		565	640
0090	22" x 49"		10	.800		595	30		625	705
0110	22" x 55"		8	1		675	38		713	810
0120	22" x 61"		8	1		770	38		808	915
0220	Double bowl, 22" x 61"		8	1		870	38		908	1,025
0230	Double bowl, 22" x 73"	↓	8	1	↓	950	38		988	1,125
0240	For aggregate colors, add					35%				
0250	For faucets and fittings, see Section 22 41 39.10									

12 36 61.19 Quartz Agglomerate Countertops		Crew	Daily Output	Labor-Hours	Unit	Material	Labor	Equipment	Total	Total Incl O&P
0010	QUARTZ AGGLOMERATE COUNTERTOPS									
0100	25" wide, 4" backsplash, color group A, minimum	2 Carp	15	1.067	L.F.	65	40.50		105.50	139
0110	Maximum		15	1.067		91	40.50		131.50	168
0120	Color group B, minimum		15	1.067		67	40.50		107.50	141
0130	Maximum		15	1.067		95	40.50		135.50	173
0140	Color group C, minimum		15	1.067		78	40.50		118.50	154
0150	Maximum		15	1.067		107	40.50		147.50	186
0160	Color group D, minimum		15	1.067		85	40.50		125.50	161
0170	Maximum	↓	15	1.067	↓	115	40.50		155.50	195

12 48 Rugs and Mats

12 48 13 – Entrance Floor Mats and Frames

12 48 13.13 Entrance Floor Mats			Crew	Daily Output	Labor-Hours	Unit	Material	Labor	Equipment	Total	Total Incl O&P
0010	ENTRANCE FLOOR MATS										
2000	Recycled rubber tire tile, 12" x 12" x 3/8" thick	G	1 Clab	125	.064	S.F.	10.05	1.87		11.92	14.20
2510	Natural cocoa fiber, 1/2" thick	G		125	.064		8.30	1.87		10.17	12.30
2520	3/4" thick	G		125	.064		8.15	1.87		10.02	12.10
2530	1" thick	G	↓	125	.064	↓	11.15	1.87		13.02	15.45

12 93 Interior Public Space Furnishings

12 93 23 – Trash and Litter Receptacles

12 93 23.10 Trash Receptacles			Crew	Daily Output	Labor-Hours	Unit	Material	Labor	Equipment	Total	Total Incl O&P
0010	TRASH RECEPTACLES										
0500	Recycled plastic, var colors, round, 32 gal., 28" x 38" H	G	2 Clab	5	3.200	Ea.	660	93.50		753.50	880
0510	32 gal., 31" x 32" H	G	"	5	3.200	"	760	93.50		853.50	990

Estimating Tips

General

- The items and systems in this division are usually estimated, purchased, supplied, and installed as a unit by one or more subcontractors. The estimator must ensure that all parties are operating from the same set of specifications and assumptions, and that all necessary items are estimated and will be provided. Many times the complex items and systems are covered, but the more common ones, such as excavation or a crane, are overlooked for the very reason that everyone assumes nobody could miss them. The estimator should be the central focus and be able to ensure that all systems are complete.
- Another area where problems can develop in this division is at the interface between systems. The estimator must ensure, for instance, that anchor bolts, nuts, and washers are estimated and included for the air-supported structures and pre-engineered buildings to be bolted to their foundations. Utility supply is a common area where essential items or pieces of equipment can be missed or overlooked, because each subcontractor may feel it is another's responsibility. The estimator should also be aware of certain items which may be supplied as part of a package but installed by others, and ensure that the installing contractor's estimate includes the cost of installation. Conversely, the estimator must also ensure that items are not costed by two different subcontractors, resulting in an inflated overall estimate.

13 30 00 Special Structures

- The foundations and floor slab, as well as rough mechanical and electrical, should be estimated, as this work is required for the assembly and erection of the structure. Generally, as noted in the data set, the pre-engineered building comes as a shell. Pricing is based on the size and structural design parameters stated in the reference section. Additional features, such as windows and doors with their related structural framing, must also be included by the estimator. Here again, the estimator must have a clear understanding of the scope of each portion of the work and all the necessary interfaces.

Reference Numbers

Reference numbers are shown at the beginning of some major classifications. These numbers refer to related items in the Reference Section. The reference information may be an estimating procedure, an alternate pricing method, or technical information.

Note: Not all subdivisions listed here necessarily appear. ■

13 11 Swimming Pools

13 11 13 – Below-Grade Swimming Pools

13 11 13.50 Swimming Pools		Crew	Daily Output	Labor-Hours	Unit	2016 Bare Costs Material	Labor	Equipment	Total	Total Incl O&P
0010	**SWIMMING POOLS** Residential in-ground, vinyl lined									
0020	Concrete sides, w/equip, sand bottom	B-52	300	.187	SF Surf	25.50	6	2	33.50	41
0100	Metal or polystyrene sides R131113-20	B-14	410	.117		21.50	3.63	.89	26.02	30.50
0200	Add for vermiculite bottom					1.64			1.64	1.80
0500	Gunite bottom and sides, white plaster finish									
0600	12′ x 30′ pool	B-52	145	.386	SF Surf	48	12.45	4.13	64.58	78
0720	16′ x 32′ pool		155	.361		43	11.65	3.86	58.51	71
0750	20′ x 40′ pool		250	.224		38.50	7.20	2.39	48.09	57
0810	Concrete bottom and sides, tile finish									
0820	12′ x 30′ pool	B-52	80	.700	SF Surf	48.50	22.50	7.50	78.50	99
0830	16′ x 32′ pool		95	.589		40	19	6.30	65.30	83
0840	20′ x 40′ pool		130	.431		32	13.85	4.60	50.45	63
1100	Motel, gunite with plaster finish, incl. medium									
1150	capacity filtration & chlorination	B-52	115	.487	SF Surf	59	15.70	5.20	79.90	97
1200	Municipal, gunite with plaster finish, incl. high									
1250	capacity filtration & chlorination	B-52	100	.560	SF Surf	76	18.05	6	100.05	121
1600	For water heating system, see Section 23 52 28.10									
1700	Filtration and deck equipment only, as % of total				Total				20%	20%
1800	Deck equipment, rule of thumb, 20′ x 40′ pool				SF Pool				1.18	1.30
3000	Painting pools, preparation + 3 coats, 20′ x 40′ pool, epoxy	2 Pord	.33	48.485	Total	1,725	1,550		3,275	4,450
3100	Rubber base paint, 18 gallons	"	.33	48.485		1,250	1,550		2,800	3,925
3500	42′ x 82′ pool, 75 gallons, epoxy paint	3 Pord	.14	171		7,300	5,475		12,775	17,100
3600	Rubber base paint	"	.14	171		5,225	5,475		10,700	14,700

13 11 23 – On-Grade Swimming Pools

13 11 23.50 Swimming Pools		Crew	Daily Output	Labor-Hours	Unit	Material	Labor	Equipment	Total	Total Incl O&P
0010	**SWIMMING POOLS** Residential above ground, steel construction									
0100	Round, 15′ dia.	B-80A	3	8	Ea.	845	234	102	1,181	1,425
0120	18′ dia.		2.50	9.600		915	280	123	1,318	1,600
0140	21′ dia.		2	12		1,050	350	154	1,554	1,925
0160	24′ dia.		1.80	13.333		1,150	390	171	1,711	2,125
0180	27′ dia.		1.50	16		1,350	465	205	2,020	2,475
0200	30′ dia.		1	24		1,500	700	305	2,505	3,175
0220	Oval, 12′ x 24′		2.30	10.435		1,575	305	134	2,014	2,375
0240	15′ x 30′		1.80	13.333		1,750	390	171	2,311	2,775
0260	18′ x 33′		1	24		1,975	700	305	2,980	3,700

13 11 46 – Swimming Pool Accessories

13 11 46.50 Swimming Pool Equipment		Crew	Daily Output	Labor-Hours	Unit	Material	Labor	Equipment	Total	Total Incl O&P
0010	**SWIMMING POOL EQUIPMENT**									
0020	Diving stand, stainless steel, 3 meter	2 Carp	.40	40	Ea.	15,600	1,500		17,100	19,700
0300	1 meter		2.70	5.926		9,500	224		9,724	10,900
0600	Diving boards, 16′ long, aluminum		2.70	5.926		4,075	224		4,299	4,850
0700	Fiberglass		2.70	5.926		3,275	224		3,499	4,000
0800	14′ long, aluminum		2.70	5.926		3,750	224		3,974	4,500
0850	Fiberglass		2.70	5.926		3,250	224		3,474	3,950
1100	Bulkhead, movable, PVC, 8′-2″ wide	2 Clab	8	2		2,375	58.50		2,433.50	2,725
1120	7′-9″ wide		8	2		2,200	58.50		2,258.50	2,525
1140	7′-3″ wide		8	2		1,975	58.50		2,033.50	2,275
1160	6′-9″ wide		8	2		1,975	58.50		2,033.50	2,275
1200	Ladders, heavy duty, stainless steel, 2 tread	2 Carp	7	2.286		855	86.50		941.50	1,100
1500	4 tread	"	6	2.667		1,125	101		1,226	1,425
2100	Lights, underwater, 12 volt, with transformer, 300 watt	1 Elec	1	8		305	350		655	915
2200	110 volt, 500 watt, standard	"	1	8		293	350		643	895

13 11 Swimming Pools

13 11 46 – Swimming Pool Accessories

13 11 46.50 Swimming Pool Equipment		Crew	Daily Output	Labor-Hours	Unit	Material	Labor	Equipment	Total	Total Incl O&P
						2016 Bare Costs				
3000	Pool covers, reinforced vinyl	3 Clab	1800	.013	S.F.	1.19	.39		1.58	1.96
3100	Vinyl, for winter, 400 S.F. max pool surface		3200	.008		.26	.22		.48	.66
3200	With water tubes, 400 S.F. max pool surface	↓	3000	.008	↓	.31	.23		.54	.73
3300	Slides, tubular, fiberglass, aluminum handrails & ladder, 5'-0", straight	2 Carp	1.60	10	Ea.	3,600	380		3,980	4,600
3320	8'-0", curved	"	3	5.333	"	7,225	202		7,427	8,300

13 12 Fountains

13 12 13 – Exterior Fountains

13 12 13.10 Outdoor Fountains		Crew	Daily Output	Labor-Hours	Unit	Material	Labor	Equipment	Total	Total Incl O&P
0010	**OUTDOOR FOUNTAINS**									
0100	Outdoor fountain, 48" high with bowl and figures	2 Clab	2	8	Ea.	570	234		804	1,025
0200	Commercial, concrete or cast stone, 40-60" H, simple		2	8		760	234		994	1,225
0220	Average		2	8		1,300	234		1,534	1,825
0240	Ornate		2	8		2,475	234		2,709	3,125
0260	Metal, 72" high		2	8		965	234		1,199	1,450
0280	90" High		2	8		1,250	234		1,484	1,800
0300	120" High		2	8		5,000	234		5,234	5,900
0320	Resin or fiberglass, 40-60" H, wall type		2	8		475	234		709	910
0340	Waterfall type	↓	2	8	↓	1,200	234		1,434	1,700

13 12 23 – Interior Fountains

13 12 23.10 Indoor Fountains		Crew	Daily Output	Labor-Hours	Unit	Material	Labor	Equipment	Total	Total Incl O&P
0010	**INDOOR FOUNTAINS**									
0100	Commercial, floor type, resin or fiberglass, lighted, cascade type	2 Clab	2	8	Ea.	292	234		526	710
0120	Tiered type		2	8		287	234		521	705
0140	Waterfall type	↓	2	8	↓	330	234		564	755

13 17 Tubs and Pools

13 17 13 – Hot Tubs

13 17 13.10 Redwood Hot Tub System		Crew	Daily Output	Labor-Hours	Unit	Material	Labor	Equipment	Total	Total Incl O&P
0010	**REDWOOD HOT TUB SYSTEM**									
7050	4' diameter x 4' deep	Q-1	1	16	Ea.	3,225	630		3,855	4,600
7150	6' diameter x 4' deep		.80	20		4,950	790		5,740	6,725
7200	8' diameter x 4' deep	↓	.80	20	↓	7,250	790		8,040	9,275

13 17 33 – Whirlpool Tubs

13 17 33.10 Whirlpool Bath		Crew	Daily Output	Labor-Hours	Unit	Material	Labor	Equipment	Total	Total Incl O&P
0010	**WHIRLPOOL BATH**									
6000	Whirlpool, bath with vented overflow, molded fiberglass									
6100	66" x 36" x 24"	Q-1	1	16	Ea.	3,475	630		4,105	4,875
6400	72" x 36" x 21"		1	16		1,800	630		2,430	3,050
6500	60" x 34" x 21"	↓	1	16	↓	1,800	630		2,430	3,025

13 24 Special Activity Rooms

13 24 16 – Saunas

13 24 16.50 Saunas and Heaters		Crew	Daily Output	Labor-Hours	Unit	2016 Bare Costs Material	Labor	Equipment	Total	Total Incl O&P
0010	**SAUNAS AND HEATERS**									
0020	Prefabricated, incl. heater & controls, 7′ high, 6′ x 4′, C/C	L-7	2.20	11.818	Ea.	5,325	380		5,705	6,500
0050	6′ x 4′, C/P		2	13		4,800	420		5,220	5,975
0400	6′ x 5′, C/C		2	13		5,975	420		6,395	7,275
0450	6′ x 5′, C/P		2	13		5,425	420		5,845	6,650
0600	6′ x 6′, C/C		1.80	14.444		6,325	465		6,790	7,750
0650	6′ x 6′, C/P		1.80	14.444		5,775	465		6,240	7,125
0800	6′ x 9′, C/C		1.60	16.250		8,400	525		8,925	10,100
0850	6′ x 9′, C/P		1.60	16.250		7,650	525		8,175	9,275
1000	8′ x 12′, C/C		1.10	23.636		12,800	760		13,560	15,300
1050	8′ x 12′, C/P		1.10	23.636		11,600	760		12,360	14,000
1200	8′ x 8′, C/C		1.40	18.571		9,625	600		10,225	11,600
1250	8′ x 8′, C/P		1.40	18.571		8,875	600		9,475	10,800
1400	8′ x 10′, C/C		1.20	21.667		10,600	700		11,300	12,900
1450	8′ x 10′, C/P		1.20	21.667		9,700	700		10,400	11,900
1600	10′ x 12′, C/C		1	26		12,800	840		13,640	15,500
1650	10′ x 12′, C/P	↓	1	26		11,600	840		12,440	14,100
1700	Door only, cedar, 2′ x 6′, with 1′ x 4′ tempered insulated glass window	2 Carp	3.40	4.706		780	178		958	1,150
1800	Prehung, incl. jambs, pulls & hardware	″	12	1.333		760	50.50		810.50	925
2500	Heaters only (incl. above), wall mounted, to 200 C.F.					730			730	805
2750	To 300 C.F.					945			945	1,050
3000	Floor standing, to 720 C.F., 10,000 watts, w/controls	1 Elec	3	2.667		2,950	117		3,067	3,450
3250	To 1,000 C.F., 16,000 watts	″	3	2.667	↓	3,900	117		4,017	4,475

13 24 26 – Steam Baths

13 24 26.50 Steam Baths and Components		Crew	Daily Output	Labor-Hours	Unit	Material	Labor	Equipment	Total	Total Incl O&P
0010	**STEAM BATHS AND COMPONENTS**									
0020	Heater, timer & head, single, to 140 C.F.	1 Plum	1.20	6.667	Ea.	2,200	292		2,492	2,875
0500	To 300 C.F.	″	1.10	7.273		2,425	320		2,745	3,200
2700	Conversion unit for residential tub, including door				↓	3,550			3,550	3,925

13 34 Fabricated Engineered Structures

13 34 13 – Glazed Structures

13 34 13.13 Greenhouses		Crew	Daily Output	Labor-Hours	Unit	Material	Labor	Equipment	Total	Total Incl O&P
0010	**GREENHOUSES**, Shell only, stock units, not incl. 2′ stub walls,									
0020	foundation, floors, heat or compartments									
0300	Residential type, free standing, 8′-6″ long x 7′-6″ wide	2 Carp	59	.271	SF Flr.	25	10.25		35.25	44.50
0400	10′-6″ wide		85	.188		41	7.10		48.10	57
0600	13′-6″ wide		108	.148		43.50	5.60		49.10	57
0700	17′-0″ wide		160	.100		47	3.78		50.78	58
0900	Lean-to type, 3′-10″ wide		34	.471		43.50	17.80		61.30	77.50
1000	6′-10″ wide	↓	58	.276	↓	29	10.45		39.45	49.50
1100	Wall mounted to existing window, 3′ x 3′	1 Carp	4	2	Ea.	1,600	75.50		1,675.50	1,875
1120	4′ x 5′	″	3	2.667	″	1,950	101		2,051	2,325
1200	Deluxe quality, free standing, 7′-6″ wide	2 Carp	55	.291	SF Flr.	85	11		96	112
1220	10′-6″ wide		81	.198		63.50	7.45		70.95	82.50
1240	13′-6″ wide		104	.154		53	5.80		58.80	68.50
1260	17′-0″ wide		150	.107		49	4.03		53.03	60.50
1400	Lean-to type, 3′-10″ wide		31	.516		105	19.50		124.50	149
1420	6′-10″ wide		55	.291		70.50	11		81.50	96
1440	8′-0″ wide	↓	97	.165	↓	60	6.25		66.25	76.50

13 34 Fabricated Engineered Structures

13 34 13 – Glazed Structures

13 34 13.13 Greenhouses		Crew	Daily Output	Labor-Hours	Unit	Material	2016 Bare Costs Labor	Equipment	Total	Total Incl O&P
1500	Commercial, custom, truss frame, incl. equip., plumbing, elec.,									
1550	benches and controls, under 2,000 S.F.				SF Flr.	13.30			13.30	14.65
1700	Over 5,000 S.F.				"	12.20			12.20	13.40

13 34 13.19 Swimming Pool Enclosures

		Crew	Daily Output	Labor-Hours	Unit	Material	Labor	Equipment	Total	Total Incl O&P
0010	**SWIMMING POOL ENCLOSURES** Translucent, free standing									
0020	not including foundations, heat or light									
0200	Economy	2 Carp	200	.080	SF Hor.	33.50	3.02		36.52	42
0600	Deluxe	"	70	.229	"	93	8.65		101.65	116

13 34 63 – Natural Fiber Construction

13 34 63.50 Straw Bale Construction

		Crew	Daily Output	Labor-Hours	Unit	Material	Labor	Equipment	Total	Total Incl O&P
0010	**STRAW BALE CONSTRUCTION**									
2020	Straw bales in walls w/modified post and beam frame G	2 Carp	320	.050	S.F.	6.55	1.89		8.44	10.35

Division Notes

I	CREW	DAILY OUTPUT	LABOR-HOURS	UNIT	BARE COSTS MAT.	LABOR	EQUIP.	TOTAL	TOTAL INCL O&P

Division 14 Conveying Equipment

Estimating Tips

General

- Many products in Division 14 will require some type of support or blocking for installation not included with the item itself. Examples are supports for conveyors or tube systems, attachment points for lifts, and footings for hoists or cranes. Add these supports in the appropriate division.

14 10 00 Dumbwaiters 14 20 00 Elevators

- Dumbwaiters and elevators are estimated and purchased in a method similar to buying a car. The manufacturer has a base unit with standard features. Added to this base unit price will be whatever options the owner or specifications require. Increased load capacity, additional vertical travel, additional stops, higher speed, and cab finish options are items to be considered. When developing an estimate for dumbwaiters and elevators, remember that some items needed by the installers may have to be included as part of the general contract.

  Examples are:
  - ☐ shaftway
  - ☐ rail support brackets
  - ☐ machine room
  - ☐ electrical supply
  - ☐ sill angles
  - ☐ electrical connections
  - ☐ pits
  - ☐ roof penthouses
  - ☐ pit ladders

  Check the job specifications and drawings before pricing.
- Installation of elevators and handicapped lifts in historic structures can require significant additional costs. The associated structural requirements may involve cutting into and repairing finishes, moldings, flooring, etc. The estimator must account for these special conditions.

14 30 00 Escalators and Moving Walks

- Escalators and moving walks are specialty items installed by specialty contractors. There are numerous options associated with these items. For specific options, contact a manufacturer or contractor. In a method similar to estimating dumbwaiters and elevators, you should verify the extent of general contract work and add items as necessary.

14 40 00 Lifts 14 90 00 Other Conveying Equipment

- Products such as correspondence lifts, chutes, and pneumatic tube systems, as well as other items specified in this subdivision, may require trained installers. The general contractor might not have any choice as to who will perform the installation, or when it will be performed. Long lead times are often required for these products, making early decisions in scheduling necessary.

Reference Numbers

Reference numbers are shown at the beginning of some major classifications. These numbers refer to related items in the Reference Section. The reference information may be an estimating procedure, an alternate pricing method, or technical information.

Note: Not all subdivisions listed here necessarily appear. ■

14 11 Manual Dumbwaiters

14 11 10 – Hand Operated Dumbwaiters

14 11 10.20 Manual Dumbwaiters		Crew	Daily Output	Labor-Hours	Unit	Material	2016 Bare Costs Labor	Equipment	Total	Total Incl O&P
0010	MANUAL DUMBWAITERS									
0020	2 stop, hand powered, up to 75 lb. capacity	2 Elev	.75	21.333	Ea.	3,100	1,275		4,375	5,475
0100	76 lb. capacity and up		.50	32	"	6,900	1,900		8,800	10,700
0300	For each additional stop, add	↓	.75	21.333	Stop	1,125	1,275		2,400	3,325

14 12 Electric Dumbwaiters

14 12 10 – Dumbwaiters

14 12 10.10 Electric Dumbwaiters		Crew	Daily Output	Labor-Hours	Unit	Material	Labor	Equipment	Total	Total Incl O&P
0010	ELECTRIC DUMBWAITERS									
0020	2 stop, up to 75 lb. capacity	2 Elev	.13	123	Ea.	7,625	7,300		14,925	20,300
0100	76 lb. capacity and up		.11	145	"	23,000	8,625		31,625	39,400
0600	For each additional stop, add	↓	.54	29.630	Stop	3,400	1,750		5,150	6,625

14 21 Electric Traction Elevators

14 21 33 – Electric Traction Residential Elevators

14 21 33.20 Residential Elevators		Crew	Daily Output	Labor-Hours	Unit	Material	Labor	Equipment	Total	Total Incl O&P
0010	RESIDENTIAL ELEVATORS									
7000	Residential, cab type, 1 floor, 2 stop, economy model	2 Elev	.20	80	Ea.	11,400	4,750		16,150	20,400
7100	Custom model		.10	160		19,300	9,500		28,800	36,700
7200	2 floor, 3 stop, economy model		.12	133		17,000	7,925		24,925	31,600
7300	Custom model	↓	.06	266	↓	27,700	15,800		43,500	56,500

14 42 Wheelchair Lifts

14 42 13 – Inclined Wheelchair Lifts

14 42 13.10 Inclined Wheelchair Lifts and Stairclimbers		Crew	Daily Output	Labor-Hours	Unit	Material	Labor	Equipment	Total	Total Incl O&P
0010	INCLINED WHEELCHAIR LIFTS AND STAIRCLIMBERS									
7700	Stair climber (chair lift), single seat, minimum	2 Elev	1	16	Ea.	5,325	950		6,275	7,425
7800	Maximum	"	.20	80	"	7,350	4,750		12,100	15,800

Division 21 Fire Suppression

Estimating Tips

Pipe for fire protection and all uses is located in Subdivisions 21 11 13 and 22 11 13.

The labor adjustment factors listed in Subdivision 22 01 02.20 also apply to Division 21.

Many, but not all, areas in the U.S. require backflow protection in the fire system. It is advisable to check local building codes for specific requirements.

For your reference, the following is a list of the most applicable Fire Codes and Standards which may be purchased from the NFPA, 1 Batterymarch Park, Quincy, MA 02169-7471.

- NFPA 1: Uniform Fire Code
- NFPA 10: Portable Fire Extinguishers
- NFPA 11: Low-, Medium-, and High-Expansion Foam
- NFPA 12: Carbon Dioxide Extinguishing Systems (Also companion 12A)
- NFPA 13: Installation of Sprinkler Systems (Also companion 13D, 13E, and 13R)
- NFPA 14: Installation of Standpipe and Hose Systems
- NFPA 15: Water Spray Fixed Systems for Fire Protection
- NFPA 16: Installation of Foam-Water Sprinkler and Foam-Water Spray Systems
- NFPA 17: Dry Chemical Extinguishing Systems (Also companion 17A)
- NFPA 18: Wetting Agents
- NFPA 20: Installation of Stationary Pumps for Fire Protection
- NFPA 22: Water Tanks for Private Fire Protection
- NFPA 24: Installation of Private Fire Service Mains and their Appurtenances
- NFPA 25: Inspection, Testing and Maintenance of Water-Based Fire Protection

Reference Numbers

Reference numbers are shown at the beginning of some major classifications. These numbers refer to related items in the Reference Section. The reference information may be an estimating procedure, an alternate pricing method, or technical information.

Note: Not all subdivisions listed here necessarily appear. ■

21 05 Common Work Results for Fire Suppression

21 05 23 – General-Duty Valves for Water-Based Fire-Suppression Piping

21 05 23.50 General-Duty Valves		Crew	Daily Output	Labor-Hours	Unit	2016 Bare Costs Material	Labor	Equipment	Total	Total Incl O&P
0010	**GENERAL-DUTY VALVES**, for water-based fire suppression									
6200	Valves and components									
6210	Alarm, includes									
6220	retard chamber, trim, gauges, alarm line strainer									
6260	3″ size	Q-12	3	5.333	Ea.	1,725	205		1,930	2,250
6280	4″ size	″	2	8	″	1,800	305		2,105	2,475

21 11 Facility Fire-Suppression Water-Service Piping

21 11 16 – Facility Fire Hydrants

21 11 16.50 Fire Hydrants for Buildings

		Crew	Daily Output	Labor-Hours	Unit	Material	Labor	Equipment	Total	Total Incl O&P
0010	**FIRE HYDRANTS FOR BUILDINGS**									
3750	Hydrants, wall, w/caps, single, flush, polished brass									
3800	2-1/2″ x 2-1/2″	Q-12	5	3.200	Ea.	224	123		347	450
3840	2-1/2″ x 3″	↓	5	3.200	↓	450	123		573	700
3860	3″ x 3″	↓	4.80	3.333	↓	365	128		493	610
3900	For polished chrome, add				↓	20%				

21 11 19 – Fire-Department Connections

21 11 19.50 Connections for the Fire-Department

		Crew	Daily Output	Labor-Hours	Unit	Material	Labor	Equipment	Total	Total Incl O&P
0010	**CONNECTIONS FOR THE FIRE-DEPARTMENT**									
7140	Standpipe connections, wall, w/plugs & chains									
7160	Single, flush, brass, 2-1/2″ x 2-1/2″	Q-12	5	3.200	Ea.	167	123		290	385
7180	2-1/2″ x 3″	″	5	3.200	″	172	123		295	390
7240	For polished chrome, add					15%				
7280	Double, flush, polished brass									
7300	2-1/2″ x 2-1/2″ x 4″	Q-12	5	3.200	Ea.	545	123		668	805
8800	Sidewalk siamese unit, polished brass, two way									
8820	2-1/2″ x 2-1/2″ x 4″	Q-12	2.50	6.400	Ea.	665	246		911	1,125

21 12 Fire-Suppression Standpipes

21 12 13 – Fire-Suppression Hoses and Nozzles

21 12 13.50 Fire Hoses and Nozzles

		Crew	Daily Output	Labor-Hours	Unit	Material	Labor	Equipment	Total	Total Incl O&P
0010	**FIRE HOSES AND NOZZLES**									
0200	Adapters, rough brass, straight hose threads									
1400	Couplings, sngl. & dbl. jacket, pin lug or rocker lug, cast brass									
1410	1-1/2″				Ea.	56			56	61.50
1420	2-1/2″				″	73.50			73.50	81
1500	For polished brass, add					20%				
1520	For polished chrome, add					40%				
2200	Hose, less couplings									
2260	Synthetic jacket, lined, 300 lb. test, 1-1/2″ diameter	Q-12	2600	.006	L.F.	3.40	.24		3.64	4.13
2280	2-1/2″ diameter	↓	2200	.007	↓	5.85	.28		6.13	6.90
2360	High strength, 500 lb. test, 1-1/2″ diameter	↓	2600	.006	↓	3.51	.24		3.75	4.25
2380	2-1/2″ diameter	↓	2200	.007	↓	6.20	.28		6.48	7.30
5600	Nozzles, brass									
5620	Adjustable fog, 3/4″ booster line				Ea.	119			119	131
5630	1″ booster line				↓	131			131	145
5640	1-1/2″ leader line				↓	80			80	88
5660	2-1/2″ direct connection				↓	164			164	180
5680	2-1/2″ playpipe nozzle				↓	220			220	242

21 12 Fire-Suppression Standpipes

21 12 13 – Fire-Suppression Hoses and Nozzles

21 12 13.50 Fire Hoses and Nozzles		Crew	Daily Output	Labor-Hours	Unit	2016 Bare Costs Material	Labor	Equipment	Total	Total Incl O&P
5780	For chrome plated, add					8%				
5850	Electrical fire, adjustable fog, no shock									
5900	1-1/2"				Ea.	400			400	440
5920	2-1/2"					535			535	590
5980	For polished chrome, add					6%				
6200	Heavy duty, comb. adj. fog and str. stream, with handle									
6210	1" booster line				Ea.	385			385	420
6240	1-1/2"					430			430	475
6260	2-1/2", for playpipe					770			770	845
6280	2-1/2" direct connection					535			535	590
6300	2-1/2" playpipe combination					535			535	590

21 12 16 – Fire-Suppression Hose Reels

21 12 16.50 Fire-Suppression Hose Reels

		Crew	Daily Output	Labor-Hours	Unit	Material	Labor	Equipment	Total	Total Incl O&P
0010	**FIRE-SUPPRESSION HOSE REELS**									
2990	Hose reel, swinging, for 1-1/2" polyester neoprene lined hose									
3000	50' long	Q-12	14	1.143	Ea.	137	44		181	224
3020	100' long		14	1.143		193	44		237	285
3060	For 2-1/2" cotton rubber hose, 75' long		14	1.143		223	44		267	320
3100	150' long		14	1.143		261	44		305	360

21 12 19 – Fire-Suppression Hose Racks

21 12 19.50 Fire Hose Racks

		Crew	Daily Output	Labor-Hours	Unit	Material	Labor	Equipment	Total	Total Incl O&P
0010	**FIRE HOSE RACKS**									
2600	Hose rack, swinging, for 1-1/2" diameter hose,									
2620	Enameled steel, 50' & 75' lengths of hose	Q-12	20	.800	Ea.	64	30.50		94.50	122
2640	100' and 125' lengths of hose		20	.800		64.50	30.50		95	122
2680	Chrome plated, 50' and 75' lengths of hose		20	.800		103	30.50		133.50	165
2700	100' and 125' lengths of hose		20	.800		110	30.50		140.50	171

21 12 23 – Fire-Suppression Hose Valves

21 12 23.70 Fire Hose Valves

		Crew	Daily Output	Labor-Hours	Unit	Material	Labor	Equipment	Total	Total Incl O&P
0010	**FIRE HOSE VALVES**									
0020	Angle, combination pressure adjust/restricting, rough brass									
0030	1-1/2"	1 Spri	12	.667	Ea.	89.50	28.50		118	146
0040	2-1/2"	"	7	1.143	"	184	48.50		232.50	283
0042	Nonpressure adjustable/restricting, rough brass									
0044	1-1/2"	1 Spri	12	.667	Ea.	56	28.50		84.50	109
0046	2-1/2"	"	7	1.143	"	94.50	48.50		143	185
0050	For polished brass, add					30%				
0060	For polished chrome, add					40%				

21 13 Fire-Suppression Sprinkler Systems

21 13 13 – Wet-Pipe Sprinkler Systems

21 13 13.50 Wet-Pipe Sprinkler System Components

		Crew	Daily Output	Labor-Hours	Unit	Material	Labor	Equipment	Total	Total Incl O&P
0010	**WET-PIPE SPRINKLER SYSTEM COMPONENTS**									
1100	Alarm, electric pressure switch (circuit closer)	1 Spri	26	.308	Ea.	93	13.10		106.10	124
1140	For explosion proof, max 20 PSI, contacts close or open		26	.308		560	13.10		573.10	635
1220	Water motor, complete with gong		4	2		435	85.50		520.50	620
1900	Flexible sprinkler head connectors									
1910	Braided stainless steel hose with mounting bracket									
1920	1/2" and 3/4" outlet size									

21 13 Fire-Suppression Sprinkler Systems

21 13 13 – Wet-Pipe Sprinkler Systems

21 13 13.50 Wet-Pipe Sprinkler System Components		Crew	Daily Output	Labor-Hours	Unit	2016 Bare Costs Material	Labor	Equipment	Total	Total Incl O&P
1940	40" length	1 Spri	30	.267	Ea.	74	11.35		85.35	100
1960	60" length	"	22	.364	"	85	15.50		100.50	119
1982	May replace hard-pipe armovers									
1984	For wet, pre-action, deluge or dry pipe systems									
2600	Sprinkler heads, not including supply piping									
3700	Standard spray, pendent or upright, brass, 135°F to 286°F									
3730	1/2" NPT, 7/16" orifice	1 Spri	16	.500	Ea.	16	21.50		37.50	53
5600	Concealed, complete with cover plate									
5620	1/2" NPT, 1/2" orifice, 135°F to 212°F	1 Spri	9	.889	Ea.	25.50	38		63.50	91

21 13 16 – Dry-Pipe Sprinkler Systems

21 13 16.50 Dry-Pipe Sprinkler System Components		Crew	Daily Output	Labor-Hours	Unit	Material	Labor	Equipment	Total	Total Incl O&P
0010	**DRY-PIPE SPRINKLER SYSTEM COMPONENTS**									
0600	Accelerator	1 Spri	8	1	Ea.	795	42.50		837.50	945
0800	Air compressor for dry pipe system, automatic, complete									
0820	30 gal. system capacity, 3/4 HP	1 Spri	1.30	6.154	Ea.	810	262		1,072	1,325
2600	Sprinkler heads, not including supply piping									
2640	Dry, pendent, 1/2" orifice, 3/4" or 1" NPT									
2700	15-1/4" to 18" length	1 Spri	14	.571	Ea.	154	24.50		178.50	210

21 21 Carbon-Dioxide Fire-Extinguishing Systems

21 21 16 – Carbon-Dioxide Fire-Extinguishing Equipment

21 21 16.50 CO2 Fire Extinguishing System		Crew	Daily Output	Labor-Hours	Unit	Material	Labor	Equipment	Total	Total Incl O&P
0010	**CO_2 FIRE EXTINGUISHING SYSTEM**									
0100	Control panel, single zone with batteries (2 zones det., 1 suppr.)	1 Elec	1	8	Ea.	1,725	350		2,075	2,475
0150	Multizone (4) with batteries (8 zones det., 4 suppr.)	"	.50	16		3,275	705		3,980	4,750
1000	Dispersion nozzle, CO_2, 3" x 5"	1 Plum	18	.444		67	19.45		86.45	106
2000	Extinguisher, CO_2 system, high pressure, 75 lb. cylinder	Q-1	6	2.667		1,275	105		1,380	1,575
2100	100 lb. cylinder	"	5	3.200		1,300	126		1,426	1,650
3000	Electro/mechanical release	L-1	4	2.500		167	110		277	365
3400	Manual pull station	1 Plum	6	1.333		60.50	58.50		119	163
4000	Pneumatic damper release	"	8	1		223	44		267	320

21 22 Clean-Agent Fire-Extinguishing Systems

21 22 16 – Clean-Agent Fire-Extinguishing Equipment

21 22 16.50 FM200 Fire Extinguishing System		Crew	Daily Output	Labor-Hours	Unit	Material	Labor	Equipment	Total	Total Incl O&P
0010	**FM200 FIRE EXTINGUISHING SYSTEM**									
1100	Dispersion nozzle FM200, 1-1/2"	1 Plum	14	.571	Ea.	67	25		92	115
2400	Extinguisher, FM200 system, filled, with mounting bracket									
2460	26 lb. container	Q-1	8	2	Ea.	2,300	79		2,379	2,650
2480	44 lb. container		7	2.286		3,050	90		3,140	3,525
2500	63 lb. container		6	2.667		3,575	105		3,680	4,100
2520	101 lb. container		5	3.200		4,775	126		4,901	5,450
2540	196 lb. container		4	4		7,775	158		7,933	8,800
6000	FM200 system, simple nozzle layout, with broad dispersion				C.F.	1.76			1.76	1.94
6020	Complex nozzle layout and/or including underfloor dispersion				"	3.50			3.50	3.85

Division 22 Plumbing

Estimating Tips

22 10 00 Plumbing Piping and Pumps

This subdivision is primarily basic pipe and related materials. The pipe may be used by any of the mechanical disciplines, i.e., plumbing, fire protection, heating, and air conditioning.

Note: CPVC plastic piping approved for fire protection is located in 21 11 13.

- The labor adjustment factors listed in Subdivision 22 01 02.20 apply throughout Divisions 21, 22, and 23. CAUTION: the correct percentage may vary for the same items. For example, the percentage add for the basic pipe installation should be based on the maximum height that the craftsman must install for that particular section. If the pipe is to be located 14' above the floor but it is suspended on threaded rod from beams, the bottom flange of which is 18' high (4' rods), then the height is actually 18' and the add is 20%. The pipe coverer, however, does not have to go above the 14', and so the add should be 10%.
- Most pipe is priced first as straight pipe with a joint (coupling, weld, etc.) every 10' and a hanger usually every 10'. There are exceptions with hanger spacing such as for cast iron pipe (5') and plastic pipe (3 per 10'). Following each type of pipe there are several lines listing sizes and the amount to be subtracted to delete couplings and hangers. This is for pipe that is to be buried or supported together on trapeze hangers. The reason that the couplings are deleted is that these runs are usually long, and frequently longer lengths of pipe are used. By deleting the couplings, the estimator is expected to look up and add back the correct reduced number of couplings.
- When preparing an estimate, it may be necessary to approximate the fittings. Fittings usually run between 25% and 50% of the cost of the pipe. The lower percentage is for simpler runs, and the higher number is for complex areas, such as mechanical rooms.
- For historic restoration projects, the systems must be as invisible as possible, and pathways must be sought for pipes, conduit, and ductwork. While installations in accessible spaces (such as basements and attics) are relatively straightforward to estimate, labor costs may be more difficult to determine when delivery systems must be concealed.

22 40 00 Plumbing Fixtures

- Plumbing fixture costs usually require two lines: the fixture itself and its "rough-in, supply, and waste."
- In the Assemblies Section (Plumbing D2010) for the desired fixture, the System Components Group at the center of the page shows the fixture on the first line. The rest of the list (fittings, pipe, tubing, etc.) will total up to what we refer to in the Unit Price section as "Rough-in, supply, waste, and vent." Note that for most fixtures we allow a nominal 5' of tubing to reach from the fixture to a main or riser.
- Remember that gas- and oil-fired units need venting.

Reference Numbers

Reference numbers are shown at the beginning of some major classifications. These numbers refer to related items in the Reference Section. The reference information may be an estimating procedure, an alternate pricing method, or technical information.

Note: Not all subdivisions listed here necessarily appear. ■

22 01 02 – Labor Adjustments

22 01 02.10 Boilers, General		Crew	Daily Output	Labor-Hours	Unit	Material	2016 Bare Costs Labor	Equipment	Total	Total Incl O&P
0010	**BOILERS, GENERAL**, Prices do not include flue piping, elec. wiring,									
0020	gas or oil piping, boiler base, pad, or tankless unless noted									

22 01 02.20 Labor Adjustment Factors		Crew	Daily Output	Labor-Hours	Unit	Material	2016 Bare Costs Labor	Equipment	Total	Total Incl O&P
0010	**LABOR ADJUSTMENT FACTORS** (For Div. 21, 22 and 23) R220102-20									
0100	Labor factors, The below are reasonable suggestions, however									
0110	each project must be evaluated for its own peculiarities, and									
0120	the adjustments be increased or decreased depending on the									
0130	severity of the special conditions.									
1000	Add to labor for elevated installation (Above floor level)									
1080	10′ to 14.5′ high						10%			
1100	15′ to 19.5′ high						20%			
1120	20′ to 24.5′ high						25%			
1140	25′ to 29.5′ high						35%			
1160	30′ to 34.5′ high						40%			
1180	35′ to 39.5′ high						50%			
1200	40′ and higher						55%			
2000	Add to labor for crawl space									
2100	3′ high						40%			
2140	4′ high						30%			
3000	Add to labor for multi-story building									
3010	For new construction (No elevator available)									
3100	Add for floors 3 thru 10						5%			
3110	Add for floors 11 thru 15						10%			
3120	Add for floors 16 thru 20						15%			
3130	Add for floors 21 thru 30						20%			
3140	Add for floors 31 and up						30%			
3170	For existing structure (Elevator available)									
3180	Add for work on floor 3 and above						2%			
4000	Add to labor for working in existing occupied buildings									
4100	Hospital						35%			
4140	Office building						25%			
4180	School						20%			
4220	Factory or warehouse						15%			
4260	Multi dwelling						15%			
5000	Add to labor, miscellaneous									
5100	Cramped shaft						35%			
5140	Congested area						15%			
5180	Excessive heat or cold						30%			
9000	Labor factors, The above are reasonable suggestions, however									
9010	each project should be evaluated for its own peculiarities.									
9100	Other factors to be considered are:									
9140	Movement of material and equipment through finished areas									
9180	Equipment room									
9220	Attic space									
9260	No service road									
9300	Poor unloading/storage area									
9340	Congested site area/heavy traffic									

22 05 Common Work Results for Plumbing

22 05 05 – Selective Demolition for Plumbing

22 05 05.10 Plumbing Demolition

	22 05 05.10 Plumbing Demolition	Crew	Daily Output	Labor-Hours	Unit	Material	2016 Bare Costs Labor	Equipment	Total	Total Incl O&P
0010	**PLUMBING DEMOLITION**									
1020	Fixtures, including 10′ piping									
1101	Bathtubs, cast iron	1 Clab	4	2	Ea.		58.50		58.50	98
1121	Fiberglass		6	1.333			39		39	65
1141	Steel		5	1.600			46.50		46.50	78
1200	Lavatory, wall hung	1 Plum	10	.800			35		35	58
1221	Counter top	1 Clab	16	.500			14.60		14.60	24.50
1301	Sink, single compartment		16	.500			14.60		14.60	24.50
1321	Double		10	.800			23.50		23.50	39
1400	Water closet, floor mounted	1 Plum	8	1			44		44	72.50
1421	Wall mounted	1 Clab	7	1.143			33.50		33.50	56
2001	Piping, metal, to 1-1/2″ diameter		200	.040	L.F.		1.17		1.17	1.96
2051	2″ thru 3-1/2″ diameter		150	.053			1.56		1.56	2.61
2101	4″ thru 6″ diameter		100	.080			2.34		2.34	3.91
2160	Plastic pipe with fittings, up thru 1-1/2″ diameter	1 Plum	250	.032			1.40		1.40	2.31
2162	2″ thru 3″ diameter	″	200	.040			1.75		1.75	2.89
2164	4″ thru 6″ diameter	Q-1	200	.080			3.15		3.15	5.20
2166	8″ thru 14″ diameter		150	.107			4.21		4.21	6.95
2168	16″ diameter		100	.160			6.30		6.30	10.40
3000	Submersible sump pump	1 Plum	24	.333	Ea.		14.60		14.60	24
6000	Remove and reset fixtures, easy access		6	1.333			58.50		58.50	96.50
6100	Difficult access		4	2			87.50		87.50	145

22 05 23 – General-Duty Valves for Plumbing Piping

22 05 23.20 Valves, Bronze

	22 05 23.20 Valves, Bronze	Crew	Daily Output	Labor-Hours	Unit	Material	Labor	Equipment	Total	Total Incl O&P
0010	**VALVES, BRONZE**									
1300	Ball									
1398	Threaded, 150 psi									
1450	1/2″	1 Plum	22	.364	Ea.	14.15	15.95		30.10	42
1460	3/4″		20	.400		23.50	17.50		41	54.50
1470	1″		19	.421		35	18.45		53.45	69
1480	1-1/4″		15	.533		61	23.50		84.50	106
1490	1-1/2″		13	.615		79.50	27		106.50	132
1500	2″		11	.727		96.50	32		128.50	159
1522	Solder the same price as threaded									
1600	Butterfly, 175 psi, full port, solder or threaded ends									
1610	Stainless steel disc and stem									
1620	1/4″	1 Plum	24	.333	Ea.	18.05	14.60		32.65	44
1630	3/8″		24	.333		18.05	14.60		32.65	44
1640	1/2″		22	.364		19.40	15.95		35.35	48
1650	3/4″		20	.400		31	17.50		48.50	63
1660	1″		19	.421		37.50	18.45		55.95	72
1670	1-1/4″		15	.533		60.50	23.50		84	105
1680	1-1/2″		13	.615		78	27		105	131
1690	2″		11	.727		97.50	32		129.50	160
1750	Check, swing, class 150, regrinding disc, threaded									
1850	1/2″	1 Plum	24	.333	Ea.	79.50	14.60		94.10	112
1860	3/4″		20	.400		105	17.50		122.50	145
1870	1″		19	.421		152	18.45		170.45	198
1880	1-1/4″		15	.533		219	23.50		242.50	280
1900	2″		11	.727		375	32		407	470
2850	Gate, N.R.S., soldered, 125 psi									
2920	1/2″	1 Plum	24	.333	Ea.	56.50	14.60		71.10	86

22 05 Common Work Results for Plumbing

22 05 23 – General-Duty Valves for Plumbing Piping

22 05 23.20	Valves, Bronze	Crew	Daily Output	Labor-Hours	Unit	Material (2016 Bare Costs)	Labor (2016 Bare Costs)	Equipment (2016 Bare Costs)	Total (2016 Bare Costs)	Total Incl O&P
2940	3/4″	1 Plum	20	.400	Ea.	66	17.50		83.50	102
2950	1″		19	.421		81.50	18.45		99.95	121
2960	1-1/4″		15	.533		135	23.50		158.50	187
2970	1-1/2″		13	.615		152	27		179	212
2980	2″		11	.727		198	32		230	270
4250	Threaded, class 150									
4340	3/4″	1 Plum	20	.400	Ea.	77	17.50		94.50	114
4350	1″		19	.421		104	18.45		122.45	145
4360	1-1/4″		15	.533		141	23.50		164.50	194
4370	1-1/2″		13	.615		178	27		205	240
5600	Relief, pressure & temperature, self-closing, ASME, threaded									
5640	3/4″	1 Plum	28	.286	Ea.	206	12.50		218.50	248
5650	1″		24	.333		330	14.60		344.60	390
5660	1-1/4″		20	.400		655	17.50		672.50	750
6400	Pressure, water, ASME, threaded									
6440	3/4″	1 Plum	28	.286	Ea.	181	12.50		193.50	220
6450	1″		24	.333		278	14.60		292.60	330
6460	1-1/4″		20	.400		430	17.50		447.50	500
6470	1-1/2″		18	.444		610	19.45		629.45	700
6480	2″		16	.500		880	22		902	1,000
6490	2-1/2″		15	.533		3,450	23.50		3,473.50	3,850
6900	Reducing, water pressure									
6920	300 psi to 25-75 psi, threaded or sweat									
6940	1/2″	1 Plum	24	.333	Ea.	425	14.60		439.60	490
6950	3/4″		20	.400		435	17.50		452.50	505
6960	1″		19	.421		670	18.45		688.45	770
8350	Tempering, water, sweat connections									
8400	1/2″	1 Plum	24	.333	Ea.	102	14.60		116.60	136
8440	3/4″	″	20	.400	″	136	17.50		153.50	179
8800	Water heater water & gas safety shut off									
8810	Protection against a leaking water heater									
8814	Shut off valve	1 Plum	16	.500	Ea.	171	22		193	224
8818	Water heater dam		32	.250		25	10.95		35.95	45.50
8822	Gas control wiring harness		32	.250		21.50	10.95		32.45	42
8830	Whole house flood safety shut off									
8834	Connections									
8838	3/4″ NPT	1 Plum	12	.667	Ea.	960	29		989	1,100
8842	1″ NPT		11	.727		985	32		1,017	1,125
8846	1-1/4″ NPT		10	.800		1,025	35		1,060	1,175

22 05 76 – Facility Drainage Piping Cleanouts

22 05 76.20	Cleanout Tees	Crew	Daily Output	Labor-Hours	Unit	Material	Labor	Equipment	Total	Total Incl O&P
0010	**CLEANOUT TEES**									
0100	Cast iron, B&S, with countersunk plug									
0220	3″ pipe size	1 Plum	3.60	2.222	Ea.	297	97.50		394.50	485
0240	4″ pipe size	″	3.30	2.424	″	370	106		476	580
0500	For round smooth access cover, same price									
4000	Plastic, tees and adapters. Add plugs									
4010	ABS, DWV									
4020	Cleanout tee, 1-1/2″ pipe size	1 Plum	15	.533	Ea.	8.65	23.50		32.15	48

22 07 19 – Plumbing Piping Insulation

22 07 19.10 Piping Insulation			Crew	Daily Output	Labor-Hours	Unit	Material	2016 Bare Costs Labor	Equipment	Total	Total Incl O&P
0010	**PIPING INSULATION**										
0600	Pipe covering (price copper tube one size less than IPS)										
6600	Fiberglass, with all service jacket										
6840	1" wall, 1/2" iron pipe size	G	Q-14	240	.067	L.F.	.83	2.37		3.20	4.93
6860	3/4" iron pipe size	G		230	.070		.90	2.47		3.37	5.20
6870	1" iron pipe size	G		220	.073		.97	2.59		3.56	5.45
6900	2" iron pipe size	G		200	.080		1.22	2.84		4.06	6.15
6910	2-1/2" iron pipe size	G		190	.084		1.39	2.99		4.38	6.65
6920	3" iron pipe size	G		180	.089		1.49	3.16		4.65	7
6940	4" iron pipe size	G		150	.107		1.97	3.79		5.76	8.60
6960	6" iron pipe size	G		120	.133		2.36	4.74		7.10	10.65
7800	For fiberglass with standard canvas jacket, deduct						5%				
7802	For fittings, add 3 L.F. for each fitting										
7804	plus 4 L.F. for each flange of the fitting										
7879	Rubber tubing, flexible closed cell foam										
8300	3/4" wall, 1/4" iron pipe size	G	1 Asbe	90	.089	L.F.	.85	3.51		4.36	6.90
8330	1/2" iron pipe size	G		89	.090		1.10	3.55		4.65	7.20
8340	3/4" iron pipe size	G		89	.090		1.35	3.55		4.90	7.50
8350	1" iron pipe size	G		88	.091		1.54	3.59		5.13	7.80
8360	1-1/4" iron pipe size	G		87	.092		2.05	3.63		5.68	8.40
8380	2" iron pipe size	G		86	.093		2.68	3.67		6.35	9.20
8390	2-1/2" iron pipe size	G		86	.093		3.58	3.67		7.25	10.20
8400	3" iron pipe size	G		85	.094		4.08	3.72		7.80	10.80
8420	4" iron pipe size	G		80	.100		5.10	3.95		9.05	12.35
8440	6" iron pipe size	G		80	.100		7.45	3.95		11.40	14.90
8444	1" wall, 1/2" iron pipe size	G		86	.093		2.05	3.67		5.72	8.50
8445	3/4" iron pipe size	G		84	.095		2.48	3.76		6.24	9.15
8446	1" iron pipe size	G		84	.095		2.89	3.76		6.65	9.60
8447	1-1/4" iron pipe size	G		82	.098		3.23	3.85		7.08	10.10
8448	1-1/2" iron pipe size	G		82	.098		3.76	3.85		7.61	10.70
8449	2" iron pipe size	G		80	.100		4.95	3.95		8.90	12.15
8450	2-1/2" iron pipe size	G		80	.100		6.45	3.95		10.40	13.80
8456	Rubber insulation tape, 1/8" x 2" x 30'	G				Ea.	21			21	23

22 11 Facility Water Distribution

22 11 13 – Facility Water Distribution Piping

22 11 13.23 Pipe/Tube, Copper		Crew	Daily Output	Labor-Hours	Unit	Material	2016 Bare Costs Labor	Equipment	Total	Total Incl O&P
0010	**PIPE/TUBE, COPPER**, Solder joints									
1000	Type K tubing, couplings & clevis hanger assemblies 10' O.C.									
1180	3/4" diameter	1 Plum	74	.108	L.F.	8.15	4.74		12.89	16.75
1200	1" diameter		66	.121		10.60	5.30		15.90	20.50
1240	1-1/2" diameter		50	.160		16.30	7		23.30	29.50
1260	2" diameter		40	.200		24.50	8.75		33.25	41.50
1280	2-1/2" diameter	Q-1	60	.267		36.50	10.50		47	57.50
1300	3" diameter	"	54	.296		49.50	11.70		61.20	74
2000	Type L tubing, couplings & clevis hanger assemblies 10' O.C.									
2140	1/2" diameter	1 Plum	81	.099	L.F.	3.68	4.33		8.01	11.20
2180	3/4" diameter		76	.105		5.25	4.61		9.86	13.35
2200	1" diameter		68	.118		7.40	5.15		12.55	16.65
2220	1-1/4" diameter		58	.138		10.05	6.05		16.10	21
2240	1-1/2" diameter		52	.154		12.60	6.75		19.35	25

22 11 13 – Facility Water Distribution Piping

22 11 13.23 Pipe/Tube, Copper		Crew	Daily Output	Labor-Hours	Unit	2016 Bare Costs Material	Labor	Equipment	Total	Total Incl O&P
2260	2″ diameter	1 Plum	42	.190	L.F.	18.40	8.35		26.75	34
2280	2-1/2″ diameter	Q-1	62	.258		28	10.20		38.20	47.50
2300	3″ diameter		56	.286		37	11.25		48.25	59
2320	3-1/2″ diameter		43	.372		50.50	14.65		65.15	79.50
2340	4″ diameter		39	.410		57.50	16.20		73.70	90
3000	Type M tubing, couplings & clevis hanger assemblies 10′ O.C.									
3140	1/2″ diameter	1 Plum	84	.095	L.F.	3.21	4.17		7.38	10.45
3180	3/4″ diameter		78	.103		4.39	4.49		8.88	12.20
3200	1″ diameter		70	.114		6.60	5		11.60	15.50
3220	1-1/4″ diameter		60	.133		9.15	5.85		15	19.75
3240	1-1/2″ diameter		54	.148		12.10	6.50		18.60	24
3260	2″ diameter		44	.182		18.15	7.95		26.10	33
3280	2-1/2″ diameter	Q-1	64	.250		26.50	9.85		36.35	46
3300	3″ diameter		58	.276		35	10.90		45.90	57
3320	3-1/2″ diameter		45	.356		49	14		63	76.50
3340	4″ diameter		40	.400		57.50	15.75		73.25	89.50
4000	Type DWV tubing, couplings & clevis hanger assemblies 10′ O.C.									
4100	1-1/4″ diameter	1 Plum	60	.133	L.F.	9.20	5.85		15.05	19.75
4120	1-1/2″ diameter		54	.148		11.30	6.50		17.80	23
4140	2″ diameter		44	.182		14.75	7.95		22.70	29.50
4160	3″ diameter	Q-1	58	.276		26.50	10.90		37.40	47
4180	4″ diameter	″	40	.400		42.50	15.75		58.25	72.50

22 11 13.25 Pipe/Tube Fittings, Copper		Crew	Daily Output	Labor-Hours	Unit	Material	Labor	Equipment	Total	Total Incl O&P
0010	**PIPE/TUBE FITTINGS, COPPER**, Wrought unless otherwise noted									
0040	Solder joints, copper x copper									
0070	90° elbow, 1/4″	1 Plum	22	.364	Ea.	3.28	15.95		19.23	30
0100	1/2″		20	.400		1.13	17.50		18.63	30
0120	3/4″		19	.421		2.53	18.45		20.98	33.50
0130	1″		16	.500		6.20	22		28.20	43
0140	1-1/4″		15	.533		9.20	23.50		32.70	48.50
0150	1-1/2″		13	.615		14.35	27		41.35	60.50
0160	2″		11	.727		26	32		58	81.50
0170	2-1/2″	Q-1	13	1.231		47.50	48.50		96	133
0180	3″		11	1.455		67.50	57.50		125	169
0190	3-1/2″		10	1.600		210	63		273	335
0200	4″		9	1.778		148	70		218	279
0450	Tee, 1/4″	1 Plum	14	.571		6.55	25		31.55	48.50
0480	1/2″		13	.615		1.92	27		28.92	46.50
0500	3/4″		12	.667		4.64	29		33.64	53
0510	1″		10	.800		13.95	35		48.95	73.50
0520	1-1/4″		9	.889		19.25	39		58.25	85
0530	1-1/2″		8	1		29	44		73	105
0540	2″		7	1.143		46	50		96	134
0550	2-1/2″	Q-1	8	2		85	79		164	224
0560	3″		7	2.286		122	90		212	283
0580	4″		5	3.200		286	126		412	525
0612	Tee, reducing on the outlet, 1/4″	1 Plum	15	.533		10.75	23.50		34.25	50.50
0613	3/8″		15	.533		10.40	23.50		33.90	50
0614	1/2″		14	.571		9.65	25		34.65	52
0615	5/8″		13	.615		19.40	27		46.40	66
0616	3/4″		12	.667		6.45	29		35.45	55
0617	1″		11	.727		21	32		53	75.50

22 11 13 – Facility Water Distribution Piping

22 11 13.25 Pipe/Tube Fittings, Copper		Crew	Daily Output	Labor-Hours	Unit	2016 Bare Costs Material	Labor	Equipment	Total	Total Incl O&P
0618	1-1/4"	1 Plum	10	.800	Ea.	24	35		59	84.50
0619	1-1/2"		9	.889		25	39		64	91.50
0620	2"		8	1		40	44		84	117
0621	2-1/2"	Q-1	9	1.778		99	70		169	225
0622	3"		8	2		103	79		182	243
0623	4"		6	2.667		220	105		325	415
0624	5"		5	3.200		1,125	126		1,251	1,450
0625	6"	Q-2	7	3.429		1,550	130		1,680	1,925
0626	8"	"	6	4		5,950	152		6,102	6,800
0630	Tee, reducing on the run, 1/4"	1 Plum	15	.533		13.90	23.50		37.40	54
0631	3/8"		15	.533		19.60	23.50		43.10	60
0632	1/2"		14	.571		13.25	25		38.25	56
0633	5/8"		13	.615		19.55	27		46.55	66
0634	3/4"		12	.667		14.30	29		43.30	63.50
0635	1"		11	.727		16.95	32		48.95	71
0636	1-1/4"		10	.800		27.50	35		62.50	88
0637	1-1/2"		9	.889		47	39		86	116
0638	2"		8	1		60	44		104	139
0639	2-1/2"	Q-1	9	1.778		128	70		198	257
0640	3"		8	2		187	79		266	335
0641	4"		6	2.667		380	105		485	595
0642	5"		5	3.200		1,075	126		1,201	1,375
0643	6"	Q-2	7	3.429		1,625	130		1,755	2,025
0644	8"	"	6	4		5,950	152		6,102	6,800
0650	Coupling, 1/4"	1 Plum	24	.333		.77	14.60		15.37	25
0680	1/2"		22	.364		.85	15.95		16.80	27.50
0700	3/4"		21	.381		1.72	16.70		18.42	29.50
0710	1"		18	.444		3.39	19.45		22.84	35.50
0715	1-1/4"		17	.471		5.85	20.50		26.35	40.50
0716	1-1/2"		15	.533		7.85	23.50		31.35	47
0718	2"		13	.615		13.15	27		40.15	59
0721	2-1/2"	Q-1	15	1.067		25.50	42		67.50	97.50
0726	4"	"	7	2.286		73	90		163	230
2000	DWV, solder joints, copper x copper									
2030	90° Elbow, 1-1/4"	1 Plum	13	.615	Ea.	12.05	27		39.05	58
2050	1-1/2"		12	.667		16	29		45	65.50
2070	2"		10	.800		25.50	35		60.50	86
2090	3"	Q-1	10	1.600		62	63		125	172
2100	4"	"	9	1.778		300	70		370	445
2250	Tee, Sanitary, 1-1/4"	1 Plum	9	.889		21	39		60	87
2290	2"	"	7	1.143		36	50		86	122
2310	3"	Q-1	7	2.286		138	90		228	300

22 11 13.44 Pipe, Steel		Crew	Daily Output	Labor-Hours	Unit	Material	Labor	Equipment	Total	Total Incl O&P
0010	**PIPE, STEEL**									
0050	Schedule 40, threaded, with couplings, and clevis hanger									
0060	assemblies sized for covering, 10' O.C.									
0540	Black, 1/4" diameter	1 Plum	66	.121	L.F.	4.96	5.30		10.26	14.20
0560	1/2" diameter		63	.127		2.90	5.55		8.45	12.35
0570	3/4" diameter		61	.131		3.35	5.75		9.10	13.20
0580	1" diameter		53	.151		4.30	6.60		10.90	15.65
0590	1-1/4" diameter	Q-1	89	.180		5.20	7.10		12.30	17.40
0600	1-1/2" diameter		80	.200		5.90	7.90		13.80	19.45

22 11 13.44 Pipe, Steel

22 11 13.44 Pipe, Steel		Crew	Daily Output	Labor-Hours	Unit	2016 Bare Costs Material	Labor	Equipment	Total	Total Incl O&P
0610	2" diameter	Q-1	64	.250	L.F.	7.45	9.85		17.30	24.50
2000	Welded, sch. 40, on yoke & roll hanger assy's, sized for covering, 10' O.C.									
2040	Black, 1" diameter	Q-15	93	.172	L.F.	4.14	6.80	.62	11.56	16.45
2080	2-1/2" diameter		47	.340		10.55	13.40	1.24	25.19	35
2090	3" diameter		43	.372		12.55	14.65	1.35	28.55	39.50
2100	3-1/2" diameter		39	.410		15.10	16.20	1.49	32.79	44.50
2110	4" diameter		37	.432		17.50	17.05	1.57	36.12	49
2120	5" diameter		32	.500		37	19.70	1.82	58.52	75
2130	6" diameter	Q-16	36	.667		46	27.50	1.61	75.11	97.50
2140	8" diameter		29	.828		73.50	34	2	109.50	139
2150	10" diameter		24	1		95.50	41	2.42	138.92	175
2160	12" diameter		19	1.263		112	51.50	3.06	166.56	211
2170	14" diameter (two rod roll type hanger for 14" diam. and up)		15	1.600		105	65.50	3.87	174.37	227
2180	16" diameter (two rod roll type hanger)		13	1.846		158	75.50	4.47	237.97	305
2190	18" diameter (two rod roll type hanger)		11	2.182		150	89	5.30	244.30	320
2220	24" diameter (two rod roll type hanger)		8	3		194	123	7.25	324.25	425

22 11 13.45 Pipe Fittings, Steel, Threaded

22 11 13.45 Pipe Fittings, Steel, Threaded		Crew	Daily Output	Labor-Hours	Unit	Material	Labor	Equipment	Total	Total Incl O&P
0010	**PIPE FITTINGS, STEEL, THREADED**									
0020	Cast Iron									
0040	Standard weight, black									
0060	90° Elbow, straight									
0100	3/4"	1 Plum	14	.571	Ea.	6.70	25		31.70	49
0110	1"	"	13	.615		7.90	27		34.90	53
0130	1-1/2"	Q-1	20	.800		15.55	31.50		47.05	69
0140	2"	"	18	.889		24.50	35		59.50	84.50
0500	Tee, straight									
0530	1/2"	1 Plum	9	.889	Ea.	10	39		49	75
0540	3/4"		9	.889		11.70	39		50.70	77
0550	1"		8	1		10.40	44		54.40	84
0570	1-1/2"	Q-1	13	1.231		24.50	48.50		73	107
0580	2"	"	11	1.455		34.50	57.50		92	132
6000	For galvanized elbows, tees, and couplings, add					20%				

22 11 13.48 Pipe, Fittings and Valves, Steel, Grooved-Joint

22 11 13.48 Pipe, Fittings and Valves, Steel, Grooved-Joint		Crew	Daily Output	Labor-Hours	Unit	Material	Labor	Equipment	Total	Total Incl O&P
0010	**PIPE, FITTINGS AND VALVES, STEEL, GROOVED-JOINT**									
0012	Fittings are ductile iron. Steel fittings noted.									
0020	Pipe includes coupling & clevis type hanger assemblies, 10' O.C.									
1000	Schedule 40, black									
1040	3/4" diameter	1 Plum	71	.113	L.F.	5	4.94		9.94	13.65
1050	1" diameter		63	.127		4.81	5.55		10.36	14.45
1060	1-1/4" diameter		58	.138		5.95	6.05		12	16.50
1070	1-1/2" diameter		51	.157		6.60	6.85		13.45	18.60
1080	2" diameter		40	.200		7.65	8.75		16.40	23
1090	2-1/2" diameter	Q-1	57	.281		11.60	11.05		22.65	31
1100	3" diameter		50	.320		14	12.60		26.60	36.50
1110	4" diameter		45	.356		19.55	14		33.55	44.50
3990	Fittings: coupling material required at joints not incl. in fitting price.									
3994	Add 1 selected coupling, material only, per joint for installed price.									
4000	Elbow, 90° or 45°, painted									
4030	3/4" diameter	1 Plum	50	.160	Ea.	63.50	7		70.50	81.50
4040	1" diameter		50	.160		34	7		41	49
4050	1-1/4" diameter		40	.200		34	8.75		42.75	52
4060	1-1/2" diameter		33	.242		34	10.60		44.60	55

22 11 Facility Water Distribution

22 11 13 – Facility Water Distribution Piping

22 11 13.48 Pipe, Fittings and Valves, Steel, Grooved-Joint		Crew	Daily Output	Labor-Hours	Unit	2016 Bare Costs Material	Labor	Equipment	Total	Total Incl O&P
4070	2″ diameter	1 Plum	25	.320	Ea.	34	14		48	60.50
4080	2-1/2″ diameter	Q-1	40	.400		34	15.75		49.75	63.50
4090	3″ diameter		33	.485		60	19.10		79.10	97.50
4100	4″ diameter		25	.640		65	25		90	113
4250	For galvanized elbows, add					26%				
4690	Tee, painted									
4700	3/4″ diameter	1 Plum	38	.211	Ea.	68.50	9.20		77.70	90
4740	1″ diameter		33	.242		53	10.60		63.60	75.50
4750	1-1/4″ diameter		27	.296		53	13		66	79.50
4760	1-1/2″ diameter		22	.364		53	15.95		68.95	84.50
4770	2″ diameter		17	.471		53	20.50		73.50	92
4780	2-1/2″ diameter	Q-1	27	.593		53	23.50		76.50	96.50
4790	3″ diameter		22	.727		72	28.50		100.50	127
4800	4″ diameter		17	.941		110	37		147	182
4900	For galvanized tees, add					24%				
7790	Valves: coupling material required at joints not incl. in valve price.									
7794	Add 1 selected coupling, material only, per joint for installed price.									

22 11 13.74 Pipe, Plastic

		Crew	Daily Output	Labor-Hours	Unit	Material	Labor	Equipment	Total	Total Incl O&P
0010	**PIPE, PLASTIC**									
1800	PVC, couplings 10′ O.C., clevis hanger assemblies, 3 per 10′									
1820	Schedule 40									
1860	1/2″ diameter	1 Plum	54	.148	L.F.	4.96	6.50		11.46	16.15
1870	3/4″ diameter		51	.157		5.30	6.85		12.15	17.20
1880	1″ diameter		46	.174		5.95	7.60		13.55	19.10
1890	1-1/4″ diameter		42	.190		6.70	8.35		15.05	21
1900	1-1/2″ diameter		36	.222		7	9.75		16.75	24
1910	2″ diameter	Q-1	59	.271		8.25	10.70		18.95	27
1920	2-1/2″ diameter		56	.286		10.65	11.25		21.90	30.50
1930	3″ diameter		53	.302		13	11.90		24.90	34
1940	4″ diameter		48	.333		16.60	13.15		29.75	40
4100	DWV type, schedule 40, couplings 10′ O.C., clevis hanger assy's, 3 per 10′									
4210	ABS, schedule 40, foam core type									
4212	Plain end black									
4214	1-1/2″ diameter	1 Plum	39	.205	L.F.	5.30	9		14.30	20.50
4216	2″ diameter	Q-1	62	.258		5.70	10.20		15.90	23
4218	3″ diameter		56	.286		8.15	11.25		19.40	27.50
4220	4″ diameter		51	.314		10.45	12.35		22.80	32
4222	6″ diameter		42	.381		18.80	15		33.80	45.50
4240	To delete coupling & hangers, subtract									
4244	1-1/2″ diam. to 6″ diam.					43%	48%			
5300	CPVC, socket joint, couplings 10′ O.C., clevis hanger assemblies, 3 per 10′									
5302	Schedule 40									
5304	1/2″ diameter	1 Plum	54	.148	L.F.	6.05	6.50		12.55	17.35
5305	3/4″ diameter		51	.157		6.95	6.85		13.80	19
5306	1″ diameter		46	.174		8.35	7.60		15.95	22
5307	1-1/4″ diameter		42	.190		10	8.35		18.35	25
5308	1-1/2″ diameter		36	.222		11.20	9.75		20.95	28.50
5309	2″ diameter	Q-1	59	.271		13.30	10.70		24	32.50
5360	CPVC, threaded, couplings 10′ O.C., clevis hanger assemblies, 3 per 10′									
5380	Schedule 40									
5460	1/2″ diameter	1 Plum	54	.148	L.F.	6.90	6.50		13.40	18.30
5470	3/4″ diameter		51	.157		8.35	6.85		15.20	20.50

22 11 13.74 Pipe, Plastic		Crew	Daily Output	Labor-Hours	Unit	2016 Bare Costs Material	Labor	Equipment	Total	Total Incl O&P
5480	1" diameter	1 Plum	46	.174	L.F.	9.85	7.60		17.45	23.50
5490	1-1/4" diameter		42	.190		11.15	8.35		19.50	26
5500	1-1/2" diameter		36	.222		12.20	9.75		21.95	29.50
5510	2" diameter	Q-1	59	.271		14.55	10.70		25.25	33.50
7280	PEX, flexible, no couplings or hangers									
7282	Note: For labor costs add 25% to the couplings and fittings labor total.									
7300	Non-barrier type, hot/cold tubing rolls									
7310	1/4" diameter x 100'				L.F.	.49			.49	.54
7350	3/8" diameter x 100'					.55			.55	.61
7360	1/2" diameter x 100'					.61			.61	.67
7370	1/2" diameter x 500'					.61			.61	.67
7380	1/2" diameter x 1000'					.61			.61	.67
7400	3/4" diameter x 100'					1.11			1.11	1.22
7410	3/4" diameter x 500'					1.11			1.11	1.22
7420	3/4" diameter x 1000'					1.11			1.11	1.22
7460	1" diameter x 100'					1.90			1.90	2.09
7470	1" diameter x 300'					1.90			1.90	2.09
7480	1" diameter x 500'					1.90			1.90	2.09
7500	1-1/4" diameter x 100'					3.23			3.23	3.55
7510	1-1/4" diameter x 300'					3.23			3.23	3.55
7540	1-1/2" diameter x 100'					4.40			4.40	4.84
7550	1-1/2" diameter x 300'					4.40			4.40	4.84
7596	Most sizes available in red or blue									
7700	Non-barrier type, hot/cold tubing straight lengths									
7710	1/2" diameter x 20'				L.F.	.61			.61	.67
7750	3/4" diameter x 20'					1.11			1.11	1.22
7760	1" diameter x 20'					1.90			1.90	2.09
7770	1-1/4" diameter x 20'					3.23			3.23	3.55
7780	1-1/2" diameter x 20'					4.40			4.40	4.84
7790	2" diameter					8.60			8.60	9.45
7796	Most sizes available in red or blue									

22 11 13.76 Pipe Fittings, Plastic		Crew	Daily Output	Labor-Hours	Unit	Material	Labor	Equipment	Total	Total Incl O&P
0010	**PIPE FITTINGS, PLASTIC**									
2700	PVC (white), schedule 40, socket joints									
2760	90° elbow, 1/2"	1 Plum	33.30	.240	Ea.	.50	10.50		11	17.90
2770	3/4"		28.60	.280		.57	12.25		12.82	20.50
2780	1"		25	.320		1.01	14		15.01	24
2790	1-1/4"		22.20	.360		1.79	15.80		17.59	28
2800	1-1/2"		20	.400		1.94	17.50		19.44	31
2810	2"	Q-1	36.40	.440		3.03	17.35		20.38	32
2820	2-1/2"		26.70	.599		9.25	23.50		32.75	49
2830	3"		22.90	.699		11.05	27.50		38.55	57.50
2840	4"		18.20	.879		19.75	34.50		54.25	78.50
3180	Tee, 1/2"	1 Plum	22.20	.360		.63	15.80		16.43	26.50
3190	3/4"		19	.421		.73	18.45		19.18	31.50
3200	1"		16.70	.479		1.36	21		22.36	36
3210	1-1/4"		14.80	.541		2.11	23.50		25.61	41.50
3220	1-1/2"		13.30	.602		2.57	26.50		29.07	46.50
3230	2"	Q-1	24.20	.661		3.74	26		29.74	47
3240	2-1/2"		17.80	.899		12.35	35.50		47.85	72
3250	3"		15.20	1.053		16.20	41.50		57.70	86.50
3260	4"		12.10	1.322		29.50	52		81.50	119

22 11 13.76 Pipe Fittings, Plastic		Crew	Daily Output	Labor-Hours	Unit	2016 Bare Costs Material	Labor	Equipment	Total	Total Incl O&P
3380	Coupling, 1/2"	1 Plum	33.30	.240	Ea.	.33	10.50		10.83	17.70
3390	3/4"		28.60	.280		.46	12.25		12.71	20.50
3400	1"		25	.320		.80	14		14.80	24
3410	1-1/4"		22.20	.360		1.11	15.80		16.91	27
3420	1-1/2"		20	.400		1.19	17.50		18.69	30.50
3430	2"	Q-1	36.40	.440		1.80	17.35		19.15	30.50
3440	2-1/2"		26.70	.599		4	23.50		27.50	43.50
3450	3"		22.90	.699		6.25	27.50		33.75	52.50
3460	4"		18.20	.879		9.10	34.50		43.60	67
4500	DWV, ABS, non pressure, socket joints									
4540	1/4 Bend, 1-1/4"	1 Plum	20.20	.396	Ea.	3.17	17.35		20.52	32
4560	1-1/2"	"	18.20	.440		2.36	19.25		21.61	34.50
4570	2"	Q-1	33.10	.483		3.75	19.05		22.80	35.50
4580	3"		20.80	.769		9.45	30.50		39.95	60.50
4590	4"		16.50	.970		19.60	38		57.60	84.50
4600	6"		10.10	1.584		82.50	62.50		145	194
4650	1/8 Bend, same as 1/4 Bend									
4800	Tee, sanitary									
4020	1 1/4"	1 Plum	13.50	.593	Ea.	4.21	26		30.21	47.50
4830	1-1/2"	"	12.10	.661		3.60	29		32.60	52
4840	2"	Q-1	20	.800		5.55	31.50		37.05	58
4850	3"		13.90	1.151		15.25	45.50		60.75	92
4860	4"		11	1.455		27	57.50		84.50	124
4862	Tee, sanitary, reducing, 2" x 1-1/2"		22	.727		4.84	28.50		33.34	53
4864	3" x 2"		15.30	1.046		9.50	41		50.50	78.50
4868	4" x 3"		12.10	1.322		26	52		78	115
4870	Combination Y and 1/8 bend									
4872	1-1/2"	1 Plum	12.10	.661	Ea.	8.60	29		37.60	57.50
4874	2"	Q-1	20	.800		9.65	31.50		41.15	62.50
4876	3"		13.90	1.151		23	45.50		68.50	100
4878	4"		11	1.455		43.50	57.50		101	143
4880	3" x 1-1/2"		15.50	1.032		22	40.50		62.50	91.50
4882	4" x 3"		12.10	1.322		34.50	52		86.50	124
4900	Wye, 1-1/4"	1 Plum	13.50	.593		4.85	26		30.85	48.50
4902	1-1/2"	"	12.10	.661		5.50	29		34.50	54
4904	2"	Q-1	20	.800		7.20	31.50		38.70	60
4906	3"		13.90	1.151		16.95	45.50		62.45	93.50
4908	4"		11	1.455		34.50	57.50		92	133
4910	6"		6.70	2.388		103	94		197	268
4918	3" x 1-1/2"		15.50	1.032		13.60	40.50		54.10	82
4920	4" x 3"		12.10	1.322		27	52		79	116
4922	6" x 4"		6.90	2.319		82	91.50		173.50	242
4930	Double Wye, 1-1/2"	1 Plum	9.10	.879		17.15	38.50		55.65	82.50
4932	2"	Q-1	16.60	.964		20.50	38		58.50	85
4934	3"		10.40	1.538		49	60.50		109.50	154
4936	4"		8.25	1.939		96.50	76.50		173	232
4940	2" x 1-1/2"		16.80	.952		18.90	37.50		56.40	83
4942	3" x 2"		10.60	1.509		34.50	59.50		94	136
4944	4" x 3"		8.45	1.893		76.50	74.50		151	207
4946	6" x 4"		7.25	2.207		111	87		198	266
4950	Reducer bushing, 2" x 1-1/2"		36.40	.440		1.92	17.35		19.27	30.50
4952	3" x 1-1/2"		27.30	.586		8.15	23		31.15	47
4954	4" x 2"		18.20	.879		15.65	34.50		50.15	74

22 11 Facility Water Distribution

22 11 13 – Facility Water Distribution Piping

22 11 13.76 Pipe Fittings, Plastic		Crew	Daily Output	Labor-Hours	Unit	2016 Bare Costs Material	Labor	Equipment	Total	Total Incl O&P
4956	6″ x 4″	Q-1	11.10	1.441	Ea.	43.50	57		100.50	142
4960	Couplings, 1-1/2″	1 Plum	18.20	.440		1.17	19.25		20.42	33.50
4962	2″	Q-1	33.10	.483		1.57	19.05		20.62	33
4963	3″		20.80	.769		4.41	30.50		34.91	55
4964	4″		16.50	.970		7.95	38		45.95	72
4966	6″		10.10	1.584		33	62.50		95.50	140
4970	2″ x 1-1/2″		33.30	.480		3.39	18.95		22.34	35
4972	3″ x 1-1/2″		21	.762		7.65	30		37.65	58
4974	4″ x 3″		16.70	.958		14.95	38		52.95	79
4978	Closet flange, 4″	1 Plum	32	.250		8.15	10.95		19.10	27
4980	4″ x 3″	″	34	.235		9.85	10.30		20.15	28
5500	CPVC, Schedule 80, threaded joints									
5540	90° Elbow, 1/4″	1 Plum	32	.250	Ea.	12.75	10.95		23.70	32
5560	1/2″		30.30	.264		7.40	11.55		18.95	27.50
5570	3/4″		26	.308		11.05	13.50		24.55	34
5580	1″		22.70	.352		15.55	15.45		31	42.50
5590	1-1/4″		20.20	.396		30	17.35		47.35	61.50
5600	1-1/2″		18.20	.440		32.50	19.25		51.75	67.50
5610	2″	Q-1	33.10	.483		43	19.05		62.05	79
5620	2-1/2″		24.20	.661		134	26		160	190
5630	3″		20.80	.769		144	30.50		174.50	208
5730	Coupling, 1/4″	1 Plum	32	.250		16.25	10.95		27.20	36
5732	1/2″		30.30	.264		13.35	11.55		24.90	34
5734	3/4″		26	.308		21	13.50		34.50	45
5736	1″		22.70	.352		24.50	15.45		39.95	52.50
5738	1-1/4″		20.20	.396		26	17.35		43.35	57
5740	1-1/2″		18.20	.440		28	19.25		47.25	62.50
5742	2″	Q-1	33.10	.483		33	19.05		52.05	67.50
5744	2-1/2″		24.20	.661		59	26		85	108
5746	3″		20.80	.769		69	30.50		99.50	126
5900	CPVC, Schedule 80, socket joints									
5904	90° Elbow, 1/4″	1 Plum	32	.250	Ea.	12.15	10.95		23.10	31.50
5906	1/2″		30.30	.264		4.76	11.55		16.31	24.50
5908	3/4″		26	.308		6.10	13.50		19.60	28.50
5910	1″		22.70	.352		9.65	15.45		25.10	36
5912	1-1/4″		20.20	.396		21	17.35		38.35	51.50
5914	1-1/2″		18.20	.440		23.50	19.25		42.75	57.50
5916	2″	Q-1	33.10	.483		28	19.05		47.05	62.50
5918	2-1/2″		24.20	.661		64.50	26		90.50	114
5920	3″		20.80	.769		73	30.50		103.50	131
5930	45° Elbow, 1/4″	1 Plum	32	.250		18.05	10.95		29	38
5932	1/2″		30.30	.264		5.80	11.55		17.35	25.50
5934	3/4″		26	.308		8.35	13.50		21.85	31
5936	1″		22.70	.352		13.35	15.45		28.80	40
5938	1-1/4″		20.20	.396		26.50	17.35		43.85	57.50
5940	1-1/2″		18.20	.440		27	19.25		46.25	61.50
5942	2″	Q-1	33.10	.483		30.50	19.05		49.55	65
5944	2-1/2″		24.20	.661		62	26		88	111
5946	3″		20.80	.769		79.50	30.50		110	138
5990	Coupling, 1/4″	1 Plum	32	.250		12.90	10.95		23.85	32.50
5992	1/2″		30.30	.264		5.05	11.55		16.60	24.50
5994	3/4″		26	.308		7.05	13.50		20.55	30
5996	1″		22.70	.352		9.45	15.45		24.90	36

22 11 13 – Facility Water Distribution Piping

22 11 13.76 Pipe Fittings, Plastic		Crew	Daily Output	Labor-Hours	Unit	2016 Bare Costs Material	Labor	Equipment	Total	Total Incl O&P
5998	1-1/4"	1 Plum	20.20	.396	Ea.	14.15	17.35		31.50	44
6000	1-1/2"	↓	18.20	.440		17.85	19.25		37.10	51.50
6002	2"	Q-1	33.10	.483		20.50	19.05		39.55	54.50
6004	2-1/2"		24.20	.661		46	26		72	93.50
6006	3"	↓	20.80	.769	↓	50	30.50		80.50	105

22 11 19 – Domestic Water Piping Specialties

22 11 19.38 Water Supply Meters		Crew	Daily Output	Labor-Hours	Unit	Material	Labor	Equipment	Total	Total Incl O&P
0010	**WATER SUPPLY METERS**									
2000	Domestic/commercial, bronze									
2020	Threaded									
2080	3/4" diameter, to 30 GPM	1 Plum	14	.571	Ea.	91	25		116	142
2100	1" diameter, to 50 GPM	"	12	.667	"	138	29		167	200
2300	Threaded/flanged									
2340	1-1/2" diameter, to 100 GPM	1 Plum	8	1	Ea.	340	44		384	445
2360	2" diameter, to 160 GPM	"	6	1.333	"	460	58.50		518.50	600
2600	Flanged, compound									
2640	3" diameter, 320 GPM	Q-1	3	5.333	Ea.	2,150	210		2,360	2,725
2660	4" diameter, to 500 GPM	"	1.50	10.667	"	3,450	420		3,870	4,500

22 11 19.42 Backflow Preventers		Crew	Daily Output	Labor-Hours	Unit	Material	Labor	Equipment	Total	Total Incl O&P
0010	**BACKFLOW PREVENTERS**, Includes valves									
0020	and four test cocks, corrosion resistant, automatic operation									
4000	Reduced pressure principle									
4100	Threaded, bronze, valves are ball									
4120	3/4" pipe size	1 Plum	16	.500	Ea.	450	22		472	530
4140	1" pipe size		14	.571		485	25		510	575
4150	1-1/4" pipe size		12	.667		890	29		919	1,025
4160	1-1/2" pipe size		10	.800		975	35		1,010	1,125
4180	2" pipe size	↓	7	1.143	↓	1,100	50		1,150	1,275
5000	Flanged, bronze, valves are OS&Y									
5060	2-1/2" pipe size	Q-1	5	3.200	Ea.	4,950	126		5,076	5,650
5080	3" pipe size		4.50	3.556		5,625	140		5,765	6,425
5100	4" pipe size	↓	3	5.333		7,050	210		7,260	8,100
5120	6" pipe size	Q-2	3	8	↓	10,200	305		10,505	11,800

22 11 19.50 Vacuum Breakers		Crew	Daily Output	Labor-Hours	Unit	Material	Labor	Equipment	Total	Total Incl O&P
0010	**VACUUM BREAKERS**									
0013	See also backflow preventers Section 22 11 19.42									
1000	Anti-siphon continuous pressure type									
1010	Max. 150 PSI - 210°F									
1020	Bronze body									
1030	1/2" size	1 Stpi	24	.333	Ea.	161	15		176	202
1040	3/4" size		20	.400		161	18		179	207
1050	1" size		19	.421		166	18.95		184.95	214
1060	1-1/4" size		15	.533		325	24		349	400
1070	1-1/2" size		13	.615		405	27.50		432.50	490
1080	2" size	↓	11	.727	↓	415	32.50		447.50	510
1200	Max. 125 PSI with atmospheric vent									
1210	Brass, in-line construction									
1220	1/4" size	1 Stpi	24	.333	Ea.	126	15		141	164
1230	3/8" size	"	24	.333		126	15		141	164
1260	For polished chrome finish, add				↓	13%				
2000	Anti-siphon, non-continuous pressure type									
2010	Hot or cold water 125 PSI - 210°F									

22 11 Facility Water Distribution

22 11 19 – Domestic Water Piping Specialties

22 11 19.50 Vacuum Breakers		Crew	Daily Output	Labor-Hours	Unit	Material	Labor	Equipment	Total	Total Incl O&P
						2016 Bare Costs				
2020	Bronze body									
2030	1/4" size	1 Stpi	24	.333	Ea.	69	15		84	101
2040	3/8" size		24	.333		69	15		84	101
2050	1/2" size		24	.333		78	15		93	110
2060	3/4" size		20	.400		93	18		111	132
2070	1" size		19	.421		144	18.95		162.95	189
2080	1-1/4" size		15	.533		252	24		276	315
2090	1-1/2" size		13	.615		296	27.50		323.50	370
2100	2" size		11	.727		460	32.50		492.50	560
2110	2-1/2" size		8	1		1,325	45		1,370	1,525
2120	3" size	↓	6	1.333	↓	1,750	60		1,810	2,025
2150	For polished chrome finish, add					50%				
3000	Air gap fitting									
3020	1/2" NPT size	1 Plum	19	.421	Ea.	54	18.45		72.45	89.50
3030	1" NPT size	"	15	.533		62	23.50		85.50	107
3040	2" NPT size	Q-1	21	.762		121	30		151	183
3050	3" NPT size		14	1.143		239	45		284	340
3060	4" NPT size	↓	10	1.600	↓	239	63		302	365

22 11 19.54 Water Hammer Arresters/Shock Absorbers		Crew	Daily Output	Labor-Hours	Unit	Material	Labor	Equipment	Total	Total Incl O&P
0010	WATER HAMMER ARRESTERS/SHOCK ABSORBERS									
0490	Copper									
0500	3/4" male I.P.S. For 1 to 11 fixtures	1 Plum	12	.667	Ea.	28	29		57	79
0600	1" male I.P.S. For 12 to 32 fixtures		8	1		47.50	44		91.50	125
0700	1-1/4" male I.P.S. For 33 to 60 fixtures	↓	8	1	↓	48	44		92	126

22 13 Facility Sanitary Sewerage

22 13 16 – Sanitary Waste and Vent Piping

22 13 16.20 Pipe, Cast Iron		Crew	Daily Output	Labor-Hours	Unit	Material	Labor	Equipment	Total	Total Incl O&P
0010	**PIPE, CAST IRON**, Soil, on clevis hanger assemblies, 5' O.C.									
0020	Single hub, service wt., lead & oakum joints 10' O.C.									
2120	2" diameter	Q-1	63	.254	L.F.	11.70	10		21.70	29.50
2140	3" diameter		60	.267		15.45	10.50		25.95	34.50
2160	4" diameter	↓	55	.291		19.65	11.45		31.10	40.50
2180	5" diameter	Q-2	76	.316		26.50	12		38.50	49
2200	6" diameter	"	73	.329		32.50	12.50		45	56
2320	For service weight, double hub, add					10%				
2340	For extra heavy, single hub, add					48%	4%			
2360	For extra heavy, double hub, add				↓	71%	4%			
2400	Lead for caulking (1#/diam. in.)	Q-1	160	.100	Lb.	1.04	3.94		4.98	7.65
2420	Oakum for caulking (1/8#/diam. in.)	"	40	.400	"	4.22	15.75		19.97	30.50
4000	No hub, couplings 10' O.C.									
4100	1-1/2" diameter	Q-1	71	.225	L.F.	11.50	8.90		20.40	27.50
4120	2" diameter		67	.239		12.05	9.40		21.45	29
4140	3" diameter		64	.250		15.45	9.85		25.30	33.50
4160	4" diameter	↓	58	.276		19.70	10.90		30.60	39.50
4180	5" diameter	Q-2	83	.289		27	11		38	47.50
4200	6" diameter	"	79	.304	↓	32.50	11.55		44.05	55

22 13 16 – Sanitary Waste and Vent Piping

	22 13 16.30 Pipe Fittings, Cast Iron	Crew	Daily Output	Labor-Hours	Unit	2016 Bare Costs Material	Labor	Equipment	Total	Total Incl O&P
0010	**PIPE FITTINGS, CAST IRON**, Soil									
0040	Hub and spigot, service weight, lead & oakum joints									
0080	1/4 bend, 2″	Q-1	16	1	Ea.	20.50	39.50		60	87.50
0120	3″		14	1.143		27.50	45		72.50	105
0140	4″		13	1.231		43	48.50		91.50	128
0160	5″	Q-2	18	1.333		60	50.50		110.50	150
0180	6″	″	17	1.412		75	53.50		128.50	171
0500	Sanitary tee, 2″	Q-1	10	1.600		28.50	63		91.50	136
0540	3″		9	1.778		46.50	70		116.50	167
0620	4″		8	2		57	79		136	193
0700	5″	Q-2	12	2		113	76		189	250
0800	6″	″	11	2.182		128	83		211	278
5990	No hub									
6000	Cplg. & labor required at joints not incl. in fitting									
6010	price. Add 1 coupling per joint for installed price									
6020	1/4 Bend, 1-1/2″				Ea.	10.55			10.55	11.65
6060	2″					11.55			11.55	12.70
6080	3″					16.10			16.10	17.70
6120	4″					24			24	26
6140	5″					57.50			57.50	63
6160	6″					57.50			57.50	63.50
6184	1/4 Bend, long sweep, 1-1/2″					27			27	29.50
6186	2″					25			25	27.50
6188	3″					30.50			30.50	33.50
6189	4″					48.50			48.50	53.50
6190	5″					92.50			92.50	102
6191	6″					107			107	118
6380	Sanitary tee, tapped, 1-1/2″					21			21	23
6382	2″ x 1-1/2″					18.60			18.60	20.50
6384	2″					19.95			19.95	22
6386	3″ x 2″					29.50			29.50	32.50
6388	3″					51			51	56.50
6390	4″ x 1-1/2″					26.50			26.50	29
6392	4″ x 2″					30			30	33
6393	4″					30			30	33
6394	6″ x 1-1/2″					68.50			68.50	75.50
6396	6″ x 2″					70			70	77
6459	Sanitary tee, 1-1/2″					14.85			14.85	16.30
6460	2″					15.90			15.90	17.50
6470	3″					19.60			19.60	21.50
6472	4″					37			37	41
6474	5″					86.50			86.50	95.50
6476	6″					88.50			88.50	97.50
8000	Coupling, standard (by CISPI Mfrs.)									
8020	1-1/2″	Q-1	48	.333	Ea.	13.95	13.15		27.10	37
8040	2″		44	.364		15.25	14.35		29.60	40.50
8080	3″		38	.421		17.10	16.60		33.70	46.50
8120	4″		33	.485		19.90	19.10		39	53.50
8160	5″	Q-2	44	.545		41	20.50		61.50	79
8180	6″	″	40	.600		46.50	23		69.50	89

22 13 16.60 Traps		Crew	Daily Output	Labor-Hours	Unit	Material	Labor	Equipment	Total	Total Incl O&P
						2016 Bare Costs				
0010	**TRAPS**									
0030	Cast iron, service weight									
0050	Running P trap, without vent									
1100	2″	Q-1	16	1	Ea.	148	39.50		187.50	228
1150	4″	″	13	1.231		148	48.50		196.50	243
1160	6″	Q-2	17	1.412		655	53.50		708.50	810
3000	P trap, B&S, 2″ pipe size	Q-1	16	1		35	39.50		74.50	104
3040	3″ pipe size	″	14	1.143		52.50	45		97.50	132
4700	Copper, drainage, drum trap									
4840	3″ x 6″ swivel, 1-1/2″ pipe size	1 Plum	16	.500	Ea.	160	22		182	212
5100	P trap, standard pattern									
5200	1-1/4″ pipe size	1 Plum	18	.444	Ea.	78.50	19.45		97.95	118
5240	1-1/2″ pipe size		17	.471		72	20.50		92.50	113
5260	2″ pipe size		15	.533		111	23.50		134.50	161
5280	3″ pipe size		11	.727		281	32		313	365
6710	ABS DWV P trap, solvent weld joint									
6720	1-1/2″ pipe size	1 Plum	18	.444	Ea.	7.75	19.45		27.20	40.50
6722	2″ pipe size		17	.471		10.20	20.50		30.70	45
6724	3″ pipe size		15	.533		40	23.50		63.50	82.50
6726	4″ pipe size		14	.571		80	25		105	130
6732	PVC DWV P trap, solvent weld joint									
6733	1-1/2″ pipe size	1 Plum	18	.444	Ea.	6.10	19.45		25.55	38.50
6734	2″ pipe size		17	.471		8.20	20.50		28.70	43
6735	3″ pipe size		15	.533		28	23.50		51.50	69
6736	4″ pipe size		14	.571		61.50	25		86.50	109
6760	PP DWV, dilution trap, 1-1/2″ pipe size		16	.500		238	22		260	297
6770	P trap, 1-1/2″ pipe size		17	.471		59	20.50		79.50	99
6780	2″ pipe size		16	.500		87.50	22		109.50	132
6790	3″ pipe size		14	.571		160	25		185	218
6800	4″ pipe size		13	.615		270	27		297	340
6860	PVC DWV hub x hub, basin trap, 1-1/4″ pipe size		18	.444		41.50	19.45		60.95	77.50
6870	Sink P trap, 1-1/2″ pipe size		18	.444		11.20	19.45		30.65	44.50
6880	Tubular S trap, 1-1/2″ pipe size		17	.471		26.50	20.50		47	63
6890	PVC sch. 40 DWV, drum trap									
6900	1-1/2″ pipe size	1 Plum	16	.500	Ea.	27.50	22		49.50	66
6910	P trap, 1-1/2″ pipe size		18	.444		6.75	19.45		26.20	39.50
6920	2″ pipe size		17	.471		9.15	20.50		29.65	44
6930	3″ pipe size		15	.533		31	23.50		54.50	72.50
6940	4″ pipe size		14	.571		70.50	25		95.50	119
6950	P trap w/clean out, 1-1/2″ pipe size		18	.444		11.20	19.45		30.65	44.50
6960	2″ pipe size		17	.471		18.80	20.50		39.30	54.50

22 13 16.80 Vent Flashing and Caps

		Crew	Daily Output	Labor-Hours	Unit	Material	Labor	Equipment	Total	Total Incl O&P
0010	**VENT FLASHING AND CAPS**									
0120	Vent caps									
0140	Cast iron									
0180	2-1/2″ - 3-5/8″ pipe	1 Plum	21	.381	Ea.	45	16.70		61.70	77
0190	4″ - 4-1/8″ pipe	″	19	.421	″	55	18.45		73.45	91
0900	Vent flashing									
1000	Aluminum with lead ring									
1040	2″ pipe	1 Plum	18	.444	Ea.	4.97	19.45		24.42	37.50
1060	4″ pipe	″	16	.500	″	6.65	22		28.65	43.50
1350	Copper with neoprene ring									

22 13 Facility Sanitary Sewerage

22 13 16 – Sanitary Waste and Vent Piping

22 13 16.80 Vent Flashing and Caps		Crew	Daily Output	Labor-Hours	Unit	Material	Labor	Equipment	Total	Total Incl O&P
						2016 Bare Costs				
1430	1-1/2" pipe	1 Plum	20	.400	Ea.	68	17.50		85.50	104
1440	2" pipe		18	.444		68	19.45		87.45	107
1460	4" pipe		16	.500		82	22		104	126

22 13 19 – Sanitary Waste Piping Specialties

22 13 19.13 Sanitary Drains		Crew	Daily Output	Labor-Hours	Unit	Material	Labor	Equipment	Total	Total Incl O&P
0010	**SANITARY DRAINS**									
2000	Floor, medium duty, C.I., deep flange, 7" diam. top									
2040	2" and 3" pipe size	Q-1	12	1.333	Ea.	232	52.50		284.50	340
2080	For galvanized body, add					96.50			96.50	106
2120	With polished bronze top					315			315	345
2500	Heavy duty, cleanout & trap w/bucket, C.I., 15" top									
2540	2", 3", and 4" pipe size	Q-1	6	2.667	Ea.	6,575	105		6,680	7,400
2560	For galvanized body, add					1,675			1,675	1,850
2580	With polished bronze top					7,300			7,300	8,025

22 13 19.15 Sink Waste Treatment		Crew	Daily Output	Labor-Hours	Unit	Material	Labor	Equipment	Total	Total Incl O&P
0010	**SINK WASTE TREATMENT**, System for commercial kitchens									
0100	includes clock timer & fittings									
0200	System less chemical, wall mounted cabinet	1 Plum	16	.500	Ea.	580	22		602	675
2000	Chemical, 1 gallon, add					34.50			34.50	38
2100	6 gallons, add					157			157	172
2200	15 gallons, add					425			425	470
2300	30 gallons, add					800			800	880
2400	55 gallons, add					1,350			1,350	1,500

22 13 23 – Sanitary Waste Interceptors

22 13 23.10 Interceptors		Crew	Daily Output	Labor-Hours	Unit	Material	Labor	Equipment	Total	Total Incl O&P
0010	**INTERCEPTORS**									
0150	Grease, fabricated steel, 4 GPM, 8 lb. fat capacity	1 Plum	4	2	Ea.	1,150	87.50		1,237.50	1,425
0200	7 GPM, 14 lb. fat capacity		4	2		1,600	87.50		1,687.50	1,925
1040	15 GPM, 30 lb. fat capacity		4	2		2,800	87.50		2,887.50	3,225
1060	20 GPM, 40 lb. fat capacity		3	2.667		3,425	117		3,542	3,950
3000	Hair, cast iron, 1-1/4" and 1-1/2" pipe connection		8	1		435	44		479	550
3100	For chrome-plated cast iron, add					320			320	355
4000	Oil, fabricated steel, 10 GPM, 2" pipe size	1 Plum	4	2		2,575	87.50		2,662.50	2,975
4100	15 GPM, 2" or 3" pipe size		4	2		3,525	87.50		3,612.50	4,050
4120	20 GPM, 2" or 3" pipe size		3	2.667		4,275	117		4,392	4,900
6000	Solids, precious metals recovery, C.I., 1-1/4" to 2" pipe		4	2		840	87.50		927.50	1,075
6100	Dental Lab., large, C.I., 1-1/2" to 2" pipe		3	2.667		2,950	117		3,067	3,425

22 13 29 – Sanitary Sewerage Pumps

22 13 29.14 Sewage Ejector Pumps		Crew	Daily Output	Labor-Hours	Unit	Material	Labor	Equipment	Total	Total Incl O&P
0010	**SEWAGE EJECTOR PUMPS**, With operating and level controls									
0100	Simplex system incl. tank, cover, pump 15' head									
0500	37 gal. PE tank, 12 GPM, 1/2 HP, 2" discharge	Q-1	3.20	5	Ea.	490	197		687	865
0510	3" discharge		3.10	5.161		535	204		739	925
0600	45 gal. coated stl. tank, 12 GPM, 1/2 HP, 2" discharge		3	5.333		880	210		1,090	1,325
0610	3" discharge		2.90	5.517		915	218		1,133	1,350
0700	70 gal. PE tank, 12 GPM, 1/2 HP, 2" discharge		2.60	6.154		950	243		1,193	1,450
0710	3" discharge		2.40	6.667		1,000	263		1,263	1,525
0730	87 GPM, 0.7 HP, 2" discharge		2.50	6.400		1,225	252		1,477	1,775
0740	3" discharge		2.30	6.957		1,300	274		1,574	1,875
0760	134 GPM, 1 HP, 2" discharge		2.20	7.273		1,325	287		1,612	1,950
0770	3" discharge		2	8		1,425	315		1,740	2,100

22 13 Facility Sanitary Sewerage

22 13 29 – Sanitary Sewerage Pumps

22 13 29.14 Sewage Ejector Pumps

	22 13 29.14 Sewage Ejector Pumps	Crew	Daily Output	Labor-Hours	Unit	2016 Bare Costs Material	Labor	Equipment	Total	Total Incl O&P
1040	Duplex system incl. tank, covers, pumps									
1060	110 gal. fiberglass tank, 24 GPM, 1/2 HP, 2" discharge	Q-1	1.60	10	Ea.	1,900	395		2,295	2,750
1080	3" discharge		1.40	11.429		2,000	450		2,450	2,950
1100	174 GPM, .7 HP, 2" discharge		1.50	10.667		2,450	420		2,870	3,400
1120	3" discharge		1.30	12.308		2,550	485		3,035	3,625
1140	268 GPM, 1 HP, 2" discharge		1.20	13.333		2,675	525		3,200	3,825
1160	3" discharge		1	16		2,775	630		3,405	4,100
1260	135 gal. coated stl. tank, 24 GPM, 1/2 HP, 2" discharge	Q-2	1.70	14.118		1,975	535		2,510	3,025
2000	3" discharge		1.60	15		2,075	570		2,645	3,250
2640	174 GPM, .7 HP, 2" discharge		1.60	15		2,575	570		3,145	3,775
2660	3" discharge		1.50	16		2,700	610		3,310	3,975
2700	268 GPM, 1 HP, 2" discharge		1.30	18.462		2,800	700		3,500	4,225
3040	3" discharge		1.10	21.818		2,975	830		3,805	4,650

22 14 Facility Storm Drainage

22 14 26 – Facility Storm Drains

22 14 26.13 Roof Drains

	22 14 26.13 Roof Drains	Crew	Daily Output	Labor-Hours	Unit	Material	Labor	Equipment	Total	Total Incl O&P
0010	**ROOF DRAINS**									
0140	Cornice, C.I., 45° or 90° outlet									
0200	3" and 4" pipe size	Q-1	12	1.333	Ea.	355	52.50		407.50	475
0260	For galvanized body, add					75.50			75.50	83
0280	For polished bronze dome, add					85.50			85.50	94
3860	Roof, flat metal deck, C.I. body, 12" C.I. dome									
3890	3" pipe size	Q-1	14	1.143	Ea.	425	45		470	540

22 14 29 – Sump Pumps

22 14 29.13 Wet-Pit-Mounted, Vertical Sump Pumps

	22 14 29.13 Wet-Pit-Mounted, Vertical Sump Pumps	Crew	Daily Output	Labor-Hours	Unit	Material	Labor	Equipment	Total	Total Incl O&P
0010	**WET-PIT-MOUNTED, VERTICAL SUMP PUMPS**									
0400	Molded PVC base, 21 GPM at 15' head, 1/3 HP	1 Plum	5	1.600	Ea.	140	70		210	269
0800	Iron base, 21 GPM at 15' head, 1/3 HP		5	1.600		168	70		238	300
1200	Solid brass, 21 GPM at 15' head, 1/3 HP		5	1.600		297	70		367	440
2000	Sump pump, single stage									
2010	25 GPM, 1 HP, 1-1/2" discharge	Q-1	1.80	8.889	Ea.	3,825	350		4,175	4,800
2020	75 GPM, 1-1/2 HP, 2" discharge		1.50	10.667		4,050	420		4,470	5,150
2030	100 GPM, 2 HP, 2-1/2" discharge		1.30	12.308		4,125	485		4,610	5,350
2040	150 GPM, 3 HP, 3" discharge		1.10	14.545		4,125	575		4,700	5,500
2050	200 GPM, 3 HP, 3" discharge		1	16		4,375	630		5,005	5,875
2060	300 GPM, 10 HP, 4" discharge	Q-2	1.20	20		4,725	760		5,485	6,450
2070	500 GPM, 15 HP, 5" discharge		1.10	21.818		5,375	830		6,205	7,275
2080	800 GPM, 20 HP, 6" discharge		1	24		6,350	910		7,260	8,475
2090	1000 GPM, 30 HP, 6" discharge		.85	28.235		6,975	1,075		8,050	9,450
2100	1600 GPM, 50 HP, 8" discharge		.72	33.333		10,900	1,275		12,175	14,100
2110	2000 GPM, 60 HP, 8" discharge	Q-3	.85	37.647		11,100	1,475		12,575	14,700
2202	For general purpose float switch, copper coated float, add	Q-1	5	3.200		103	126		229	320

22 31 Domestic Water Softeners

22 31 13 – Residential Domestic Water Softeners

22 31 13.10 Residential Water Softeners		Crew	Daily Output	Labor-Hours	Unit	Material	2016 Bare Costs Labor	Equipment	Total	Total Incl O&P
0010	**RESIDENTIAL WATER SOFTENERS**									
7350	Water softener, automatic, to 30 grains per gallon	2 Plum	5	3.200	Ea.	400	140		540	670
7400	To 100 grains per gallon	"	4	4	"	850	175		1,025	1,225

22 32 Domestic Water Filtration Equipment

22 32 19 – Domestic-Water Off-Floor Cartridge Filters

22 32 19.10 Water Filters

		Crew	Daily Output	Labor-Hours	Unit	Material	Labor	Equipment	Total	Total Incl O&P
0010	**WATER FILTERS**, Purification and treatment.									
1000	Cartridge style, dirt and rust type	1 Plum	12	.667	Ea.	90.50	29		119.50	148
1200	Replacement cartridge		32	.250		18.40	10.95		29.35	38.50
1600	Taste and odor type		12	.667		123	29		152	183
1700	Replacement cartridge		32	.250		42.50	10.95		53.45	64.50
8000	Commercial, fully automatic or push button automatic									
8200	Iron removal, 660 GPH, 1" pipe size	Q-1	1.50	10.667	Ea.	3,100	420		3,520	4,125
8240	1500 GPH, 1-1/4" pipe size	"	1	16	"	5,225	630		5,855	6,800

22 33 Electric Domestic Water Heaters

22 33 30 – Residential, Electric Domestic Water Heaters

22 33 30.13 Residential, Small-Capacity Elec. Water Heaters

		Crew	Daily Output	Labor-Hours	Unit	Material	Labor	Equipment	Total	Total Incl O&P
0010	**RESIDENTIAL, SMALL-CAPACITY ELECTRIC DOMESTIC WATER HEATERS**									
1000	Residential, electric, glass lined tank, 5 yr., 10 gal., single element	1 Plum	2.30	3.478	Ea.	395	152		547	685
1060	30 gallon, double element		2.20	3.636		905	159		1,064	1,250
1100	52 gallon, double element		2	4		1,075	175		1,250	1,500

22 33 33 – Light-Commercial Electric Domestic Water Heaters

22 33 33.10 Commercial Electric Water Heaters

		Crew	Daily Output	Labor-Hours	Unit	Material	Labor	Equipment	Total	Total Incl O&P
0010	**COMMERCIAL ELECTRIC WATER HEATERS**									
4000	Commercial, 100° rise. NOTE: for each size tank, a range of									
4010	heaters between the ones shown are available									
4020	Electric									
4100	5 gal., 3 kW, 12 GPH, 208 volt	1 Plum	2	4	Ea.	4,275	175		4,450	5,025
4120	10 gal., 6 kW, 25 GPH, 208 volt		2	4		4,750	175		4,925	5,525
4130	30 gal., 24 kW, 98 GPH, 208 volt		1.92	4.167		7,800	183		7,983	8,875
4136	40 gal., 36 kW, 148 GPH, 208 volt		1.88	4.255		9,350	186		9,536	10,600
4140	50 gal., 9 kW, 37 GPH, 208 volt		1.80	4.444		6,525	195		6,720	7,500
4160	50 gal., 36 kW, 148 GPH, 208 volt		1.80	4.444		10,100	195		10,295	11,400
4200	80 gal., 36 kW, 148 GPH, 208 volt		1.50	5.333		11,100	234		11,334	12,600
4220	100 gal., 36 kW, 148 GPH, 208 volt		1.20	6.667		11,600	292		11,892	13,200
4240	120 gal., 36 kW, 148 GPH, 208 volt		1.20	6.667		12,100	292		12,392	13,800
4280	150 gal., 120 kW, 490 GPH, 480 volt		1	8		39,900	350		40,250	44,100
4300	200 gal., 15 kW, 61 GPH, 480 volt	Q-1	1.70	9.412		30,700	370		31,070	34,400

22 34 Fuel-Fired Domestic Water Heaters

22 34 30 – Residential Gas Domestic Water Heaters

22 34 30.13 Residential, Atmos, Gas Domestic Wtr Heaters

	22 34 30.13 Residential, Atmos, Gas Domestic Wtr Heaters	Crew	Daily Output	Labor-Hours	Unit	2016 Bare Costs Material	Labor	Equipment	Total	Total Incl O&P
0010	**RESIDENTIAL, ATMOSPHERIC, GAS DOMESTIC WATER HEATERS**									
2000	Gas fired, foam lined tank, 10 yr., vent not incl.									
2040	30 gallon	1 Plum	2	4	Ea.	1,625	175		1,800	2,075
2100	75 gallon	"	1.50	5.333	"	2,425	234		2,659	3,050
3000	Tank leak safety, water & gas shut off see 22 05 23.20 8800									

22 34 36 – Commercial Gas Domestic Water Heaters

22 34 36.13 Commercial, Atmos., Gas Domestic Water Htrs.

	Description	Crew	Daily Output	Labor-Hours	Unit	Material	Labor	Equipment	Total	Total Incl O&P
0010	**COMMERCIAL, ATMOSPHERIC, GAS DOMESTIC WATER HEATERS**									
6000	Gas fired, flush jacket, std. controls, vent not incl.									
6040	75 MBH input, 73 GPH	1 Plum	1.40	5.714	Ea.	3,425	250		3,675	4,200
6060	98 MBH input, 95 GPH		1.40	5.714		6,700	250		6,950	7,800
6080	120 MBH input, 110 GPH		1.20	6.667		6,950	292		7,242	8,100
6120	140 MBH input, 130 GPH		1	8		8,850	350		9,200	10,300
6140	155 MBH input, 150 GPH		.80	10		8,975	440		9,415	10,600
6180	200 MBH input, 192 GPH		.60	13.333		9,175	585		9,760	11,100
6220	260 MBH input, 250 GPH	Q-1	.80	20		10,400	790		11,190	12,700
6240	360 MBH input, 360 GPH		.80	20		12,300	790		13,090	14,800
6260	500 MBH input, 480 GPH		.70	22.857		17,300	900		18,200	20,500
6280	725 MBH input, 690 GPH		.60	26.667		20,300	1,050		21,350	24,000
6900	For low water cutoff, add	1 Plum	8	1		360	44		404	470
6960	For bronze body hot water circulator, add	"	4	2		2,050	87.50		2,137.50	2,400

22 34 46 – Oil-Fired Domestic Water Heaters

22 34 46.10 Residential Oil-Fired Water Heaters

	Description	Crew	Daily Output	Labor-Hours	Unit	Material	Labor	Equipment	Total	Total Incl O&P
0010	**RESIDENTIAL OIL-FIRED WATER HEATERS**									
3000	Oil fired, glass lined tank, 5 yr., vent not included, 30 gallon	1 Plum	2	4	Ea.	1,175	175		1,350	1,600
3040	50 gallon	"	1.80	4.444	"	1,400	195		1,595	1,850

22 34 46.20 Commercial Oil-Fired Water Heaters

	Description	Crew	Daily Output	Labor-Hours	Unit	Material	Labor	Equipment	Total	Total Incl O&P
0010	**COMMERCIAL OIL-FIRED WATER HEATERS**									
8000	Oil fired, glass lined, UL listed, std. controls, vent not incl.									
8060	140 gal., 140 MBH input, 134 GPH	Q-1	2.13	7.512	Ea.	23,400	296		23,696	26,200
8080	140 gal., 199 MBH input, 191 GPH		2	8		24,200	315		24,515	27,100
8120	140 gal., 270 MBH input, 259 GPH		1.20	13.333		30,700	525		31,225	34,700
8140	140 gal., 400 MBH input, 384 GPH		1	16		31,500	630		32,130	35,700
8180	140 gal., 720 MBH input, 691 GPH		.92	17.391		33,600	685		34,285	38,000
8200	221 gal., 300 MBH input, 288 GPH		.88	18.182		44,400	715		45,115	50,000
8220	221 gal., 600 MBH input, 576 GPH		.86	18.605		49,500	735		50,235	55,500
8240	221 gal., 800 MBH input, 768 GPH		.82	19.512		49,800	770		50,570	56,500

22 41 Residential Plumbing Fixtures

22 41 13 – Residential Water Closets, Urinals, and Bidets

22 41 13.13 Water Closets

	Description	Crew	Daily Output	Labor-Hours	Unit	Material	Labor	Equipment	Total	Total Incl O&P
0010	**WATER CLOSETS**									
0150	Tank type, vitreous china, incl. seat, supply pipe w/stop, 1.6 gpf or noted									
0200	Wall hung									
0960	For rough-in, supply, waste, vent and carrier	Q-1	2.73	5.861	Ea.	945	231		1,176	1,425
0999	Floor mounted									
1020	One piece, low profile	Q-1	5.30	3.019	Ea.	735	119		854	1,000
1100	Two piece, close coupled		5.30	3.019		230	119		349	450
1102	Economy		5.30	3.019		125	119		244	335
1110	Two piece, close coupled, dual flush		5.30	3.019		320	119		439	545

22 41 Residential Plumbing Fixtures

22 41 13 – Residential Water Closets, Urinals, and Bidets

22 41 13.13 Water Closets		Crew	Daily Output	Labor-Hours	Unit	Material	2016 Bare Costs Labor	Equipment	Total	Total Incl O&P
1140	Two piece, close coupled, 1.28 gpf, ADA [G]	Q-1	5.30	3.019	Ea.	310	119		429	535
1960	For color, add					30%				
1980	For rough-in, supply, waste and vent	Q-1	3.05	5.246	Ea.	325	207		532	700

22 41 16 – Residential Lavatories and Sinks

22 41 16.13 Lavatories

		Crew	Daily Output	Labor-Hours	Unit	Material	Labor	Equipment	Total	Total Incl O&P
0010	**LAVATORIES**, With trim, white unless noted otherwise									
0500	Vanity top, porcelain enamel on cast iron									
0600	20″ x 18″	Q-1	6.40	2.500	Ea.	315	98.50		413.50	510
0640	33″ x 19″ oval	″	6.40	2.500	″	475	98.50		573.50	685
0860	For color, add					25%				
1000	Cultured marble, 19″ x 17″, single bowl	Q-1	6.40	2.500	Ea.	151	98.50		249.50	330
1120	25″ x 22″, single bowl	″	6.40	2.500	″	190	98.50		288.50	370
1580	For color, same price									
1900	Stainless steel, self-rimming, 25″ x 22″, single bowl, ledge	Q-1	6.40	2.500	Ea.	340	98.50		438.50	540
1960	17″ x 22″, single bowl		6.40	2.500		330	98.50		428.50	530
2600	Steel, enameled, 20″ x 17″, single bowl		5.80	2.759		150	109		259	345
2900	Vitreous china, 20″ x 16″, single bowl		5.40	2.963		236	117		353	455
3200	22″ x 13″, single bowl		5.40	2.963		243	117		360	460
3580	Rough-in, supply, waste and vent for all above lavatories		2.30	6.957		218	274		492	690
4000	Wall hung									
4040	Porcelain enamel on cast iron, 16″ x 14″, single bowl	Q-1	8	2	Ea.	500	79		579	680
4180	20″ x 18″, single bowl	″	8	2	″	265	79		344	420
4580	For color, add					30%				
6000	Vitreous china, 18″ x 15″, single bowl with backsplash	Q-1	7	2.286	Ea.	186	90		276	355
6060	19″ x 17″, single bowl		7	2.286		148	90		238	310
6960	Rough-in, supply, waste and vent for above lavatories		1.66	9.639		435	380		815	1,100
7000	Pedestal type									
7600	Vitreous china, 27″ x 21″, white	Q-1	6.60	2.424	Ea.	680	95.50		775.50	910
7610	27″ x 21″, colored		6.60	2.424		860	95.50		955.50	1,100
7620	27″ x 21″, premium color		6.60	2.424		975	95.50		1,070.50	1,225
7660	26″ x 20″, white		6.60	2.424		670	95.50		765.50	895
7670	26″ x 20″, colored		6.60	2.424		845	95.50		940.50	1,100
7680	26″ x 20″, premium color		6.60	2.424		960	95.50		1,055.50	1,200
7700	24″ x 20″, white		6.60	2.424		395	95.50		490.50	595
7710	24″ x 20″, colored		6.60	2.424		490	95.50		585.50	700
7720	24″ x 20″, premium color		6.60	2.424		550	95.50		645.50	770
7760	21″ x 18″, white		6.60	2.424		277	95.50		372.50	465
7770	21″ x 18″, colored		6.60	2.424		335	95.50		430.50	530
7990	Rough-in, supply, waste and vent for pedestal lavatories		1.66	9.639		435	380		815	1,100

22 41 16.16 Sinks

		Crew	Daily Output	Labor-Hours	Unit	Material	Labor	Equipment	Total	Total Incl O&P
0010	**SINKS**, With faucets and drain									
2000	Kitchen, counter top style, P.E. on C.I., 24″ x 21″ single bowl	Q-1	5.60	2.857	Ea.	297	113		410	510
2100	31″ x 22″ single bowl		5.60	2.857		550	113		663	785
2200	32″ x 21″ double bowl		4.80	3.333		370	131		501	620
3000	Stainless steel, self rimming, 19″ x 18″ single bowl		5.60	2.857		605	113		718	850
3100	25″ x 22″ single bowl		5.60	2.857		675	113		788	925
3300	43″ x 22″ double bowl		4.80	3.333		1,125	131		1,256	1,475
4000	Steel, enameled, with ledge, 24″ x 21″ single bowl		5.60	2.857		525	113		638	765
4100	32″ x 21″ double bowl		4.80	3.333		520	131		651	785
4960	For color sinks except stainless steel, add					10%				
4980	For rough-in, supply, waste and vent, counter top sinks	Q-1	2.14	7.477		247	295		542	755
5000	Kitchen, raised deck, P.E. on C.I.									

22 41 Residential Plumbing Fixtures

22 41 16 – Residential Lavatories and Sinks

22 41 16.16 Sinks		Crew	Daily Output	Labor-Hours	Unit	2016 Bare Costs Material	Labor	Equipment	Total	Total Incl O&P
5100	32″ x 21″, dual level, double bowl	Q-1	2.60	6.154	Ea.	440	243		683	885
5790	For rough-in, supply, waste & vent, sinks	″	1.85	8.649	″	247	340		587	830

22 41 19 – Residential Bathtubs

22 41 19.10 Baths

		Crew	Daily Output	Labor-Hours	Unit	Material	Labor	Equipment	Total	Total Incl O&P
0010	**BATHS**									
0100	Tubs, recessed porcelain enamel on cast iron, with trim									
0180	48″ x 42″	Q-1	4	4	Ea.	2,675	158		2,833	3,175
0300	Mat bottom									
0380	5′ long	Q-1	4.40	3.636	Ea.	1,150	143		1,293	1,475
0480	Above floor drain, 5′ long		4	4		840	158		998	1,175
0560	Corner 48″ x 44″		4.40	3.636		2,675	143		2,818	3,150
2000	Enameled formed steel, 4′-6″ long		5.80	2.759		485	109		594	715
4600	Module tub & showerwall surround, molded fiberglass									
4610	5′ long x 34″ wide x 76″ high	Q-1	4	4	Ea.	825	158		983	1,175
4750	Handicap with 1-1/2″ OD grab bar, antiskid bottom									
4760	60″ x 32-3/4″ x 72″ high	Q-1	4	4	Ea.	825	158		983	1,175
4770	60″ x 30″ x 71″ high with molded seat		3.50	4.571		860	180		1,040	1,250
9600	Rough-in, supply, waste and vent, for all above tubs, add		2.07	7.729		355	305		660	895

22 41 23 – Residential Showers

22 41 23.20 Showers

		Crew	Daily Output	Labor-Hours	Unit	Material	Labor	Equipment	Total	Total Incl O&P
0010	**SHOWERS**									
1500	Stall, with drain only. Add for valve and door/curtain									
1520	32″ square	Q-1	5	3.200	Ea.	1,175	126		1,301	1,475
1530	36″ square		4.80	3.333		2,925	131		3,056	3,425
1540	Terrazzo receptor, 32″ square		5	3.200		1,350	126		1,476	1,675
1560	36″ square		4.80	3.333		1,475	131		1,606	1,850
1580	36″ corner angle		4.80	3.333		1,725	131		1,856	2,125
3000	Fiberglass, one piece, with 3 walls, 32″ x 32″ square		5.50	2.909		335	115		450	560
3100	36″ x 36″ square		5.50	2.909		385	115		500	610
3200	Handicap, 1-1/2″ O.D. grab bars, nonskid floor									
3210	48″ x 34-1/2″ x 72″ corner seat	Q-1	5	3.200	Ea.	695	126		821	975
3250	64″ x 65-3/4″ x 81-1/2″ fold. seat, whlchr.		3.80	4.211		2,275	166		2,441	2,800
4000	Polypropylene, stall only, w/molded-stone floor, 30″ x 30″		2	8		635	315		950	1,225
4200	Rough-in, supply, waste and vent for above showers		2.05	7.805		345	310		655	890

22 41 23.40 Shower System Components

		Crew	Daily Output	Labor-Hours	Unit	Material	Labor	Equipment	Total	Total Incl O&P
0010	**SHOWER SYSTEM COMPONENTS**									
4500	Receptor only									
4510	For tile, 36″ x 36″	1 Plum	4	2	Ea.	365	87.50		452.50	550
4520	Fiberglass receptor only, 32″ x 32″		8	1		99	44		143	182
4530	34″ x 34″		7.80	1.026		119	45		164	205
4540	36″ x 36″		7.60	1.053		115	46		161	203
4600	Rectangular									
4620	32″ x 48″	1 Plum	7.40	1.081	Ea.	137	47.50		184.50	229
4630	34″ x 54″		7.20	1.111		165	48.50		213.50	262
4640	34″ x 60″		7	1.143		174	50		224	275
5000	Built-in, head, arm, 2.5 GPM valve		4	2		81	87.50		168.50	234
5200	Head, arm, by-pass, integral stops, handles		3.60	2.222		260	97.50		357.50	445
5800	Mixing valve, built-in		6	1.333		132	58.50		190.50	242

22 41 Residential Plumbing Fixtures

22 41 36 – Residential Laundry Trays

	22 41 36.10 Laundry Sinks	Crew	Daily Output	Labor-Hours	Unit	Material	2016 Bare Costs Labor	Equipment	Total	Total Incl O&P
0010	**LAUNDRY SINKS**, With trim									
0020	Porcelain enamel on cast iron, black iron frame									
0050	24" x 21", single compartment	Q-1	6	2.667	Ea.	590	105		695	820
3000	Plastic, on wall hanger or legs									
3100	20" x 24", single compartment	Q-1	6.50	2.462	Ea.	169	97		266	345
3200	36" x 23", double compartment		5.50	2.909		199	115		314	410
9600	Rough-in, supply, waste and vent, for all laundry sinks		2.14	7.477		247	295		542	755

22 41 39 – Residential Faucets, Supplies and Trim

	22 41 39.10 Faucets and Fittings	Crew	Daily Output	Labor-Hours	Unit	Material	Labor	Equipment	Total	Total Incl O&P
0010	**FAUCETS AND FITTINGS**									
0150	Bath, faucets, diverter spout combination, sweat	1 Plum	8	1	Ea.	86	44		130	167
0200	For integral stops, IPS unions, add					109			109	120
0420	Bath, press-bal mix valve w/diverter, spout, shower head, arm/flange	1 Plum	8	1		173	44		217	263
0500	Drain, central lift, 1-1/2" IPS male		20	.400		49	17.50		66.50	83
0600	Trip lever, 1-1/2" IPS male		20	.400		55	17.50		72.50	89.50
1000	Kitchen sink faucets, top mount, cast spout		10	.800		77.50	35		112.50	143
1100	For spray, add		24	.333		16.95	14.60		31.55	42.50
1300	Single control lever handle									
1310	With pull out spray									
1320	Polished chrome	1 Plum	10	.800	Ea.	200	35		235	277
2000	Laundry faucets, shelf type, IPS or copper unions		12	.667		61.50	29		90.50	116
2100	Lavatory faucet, centerset, without drain		10	.800		63.50	35		98.50	128
2120	With pop-up drain		6.66	1.201		83	52.50		135.50	178
2210	Porcelain cross handles and pop-up drain									
2220	Polished chrome	1 Plum	6.66	1.201	Ea.	198	52.50		250.50	305
2230	Polished brass	"	6.66	1.201	"	305	52.50		357.50	420
2260	Single lever handle and pop-up drain									
2280	Satin nickel	1 Plum	6.66	1.201	Ea.	283	52.50		335.50	395
2290	Polished chrome		6.66	1.201		198	52.50		250.50	305
2800	Self-closing, center set		10	.800		148	35		183	221
4000	Shower by-pass valve with union		18	.444		57	19.45		76.45	95
4200	Shower thermostatic mixing valve, concealed, with shower head trim kit		8	1		345	44		389	455
4220	Shower pressure balancing mixing valve									
4230	With shower head, arm, flange and diverter tub spout									
4240	Chrome	1 Plum	6.14	1.303	Ea.	375	57		432	510
4250	Satin nickel		6.14	1.303		505	57		562	650
4260	Polished graphite		6.14	1.303		505	57		562	650
5000	Sillcock, compact, brass, IPS or copper to hose		24	.333		10.20	14.60		24.80	35.50

22 42 Commercial Plumbing Fixtures

22 42 13 – Commercial Water Closets, Urinals, and Bidets

	22 42 13.13 Water Closets	Crew	Daily Output	Labor-Hours	Unit	Material	Labor	Equipment	Total	Total Incl O&P
0010	**WATER CLOSETS**									
3000	Bowl only, with flush valve, seat, 1.6 gpf unless noted									
3100	Wall hung	Q-1	5.80	2.759	Ea.	965	109		1,074	1,225
3200	For rough-in, supply, waste and vent, single WC		2.56	6.250		980	246		1,226	1,475
3300	Floor mounted		5.80	2.759		315	109		424	525
3370	For rough-in, supply, waste and vent, single WC		2.84	5.634		365	222		587	765
3390	Floor mounted children's size, 10-3/4" high									
3392	With automatic flush sensor, 1.6 gpf	Q-1	6.20	2.581	Ea.	625	102		727	855

22 42 Commercial Plumbing Fixtures

22 42 13 – Commercial Water Closets, Urinals, and Bidets

22 42 13.13 Water Closets		Crew	Daily Output	Labor-Hours	Unit	Material	Labor	Equipment	Total	Total Incl O&P
						2016 Bare Costs				
3396	With automatic flush sensor, 1.28 gpf	Q-1	6.20	2.581	Ea.	625	102		727	855
3400	For rough-in, supply, waste and vent, single WC		2.84	5.634		365	222		587	765

22 42 13.16 Urinals

		Crew	Daily Output	Labor-Hours	Unit	Material	Labor	Equipment	Total	Total Incl O&P
0010	**URINALS**									
3000	Wall hung, vitreous china, with self-closing valve									
3100	Siphon jet type	Q-1	3	5.333	Ea.	277	210		487	650
3120	Blowout type		3	5.333		465	210		675	855
3140	Water saving .5 gpf G		3	5.333		555	210		765	955
3300	Rough-in, supply, waste & vent		2.83	5.654		590	223		813	1,025
5000	Stall type, vitreous china, includes valve		2.50	6.400		745	252		997	1,225
6980	Rough-in, supply, waste and vent		1.99	8.040		360	315		675	920

22 42 16 – Commercial Lavatories and Sinks

22 42 16.13 Lavatories

		Crew	Daily Output	Labor-Hours	Unit	Material	Labor	Equipment	Total	Total Incl O&P
0010	**LAVATORIES**, With trim, white unless noted otherwise									
0020	Commercial lavatories same as residential. See Section 22 41 16									

22 42 16.40 Service Sinks

		Crew	Daily Output	Labor-Hours	Unit	Material	Labor	Equipment	Total	Total Incl O&P
0010	**SERVICE SINKS**									
6650	Service, floor, corner, P.E. on C.I., 28" x 28"	Q-1	4.40	3.636	Ea.	1,025	143		1,168	1,350
6750	Vinyl coated rim guard, add					61.50			61.50	68
6760	Mop sink, molded stone, 24" x 36"	1 Plum	3.33	2.402		273	105		378	475
6770	Mop sink, molded stone, 24" x 36", w/rim 3 sides	"	3.33	2.402		278	105		383	480
6790	For rough-in, supply, waste & vent, floor service sinks	Q-1	1.64	9.756		715	385		1,100	1,425

22 42 33 – Wash Fountains

22 42 33.20 Commercial Wash Fountains

		Crew	Daily Output	Labor-Hours	Unit	Material	Labor	Equipment	Total	Total Incl O&P
0010	**COMMERCIAL WASH FOUNTAINS**									
1900	Group, foot control									
2000	Precast terrazzo, circular, 36" diam., 5 or 6 persons	Q-2	3	8	Ea.	7,350	305		7,655	8,575
2100	54" diameter for 8 or 10 persons		2.50	9.600		9,150	365		9,515	10,700
2400	Semi-circular, 36" diam. for 3 persons		3	8		6,450	305		6,755	7,575
2500	54" diam. for 4 or 5 persons		2.50	9.600		8,650	365		9,015	10,100
5610	Group, infrared control, barrier free									
5614	Precast terrazzo									
5620	Semi-circular 36" diam. for 3 persons	Q-2	3	8	Ea.	7,325	305		7,630	8,550
5630	46" diam. for 4 persons		2.80	8.571		7,900	325		8,225	9,225
5640	Circular, 54" diam. for 8 persons, button control		2.50	9.600		9,575	365		9,940	11,100

22 42 39 – Commercial Faucets, Supplies, and Trim

22 42 39.10 Faucets and Fittings

		Crew	Daily Output	Labor-Hours	Unit	Material	Labor	Equipment	Total	Total Incl O&P
0010	**FAUCETS AND FITTINGS**									
0910	Dual flush flushometer	1 Plum	12	.667	Ea.	217	29		246	287
0912	Flushometer retrofit kit	"	18	.444	"	17.10	19.45		36.55	51
0920	Urinal									
0930	Exposed, stall	1 Plum	8	1	Ea.	152	44		196	240
0940	Wall (washout)	"	8	1		139	44		183	226
0945	Flush valve, urinal					139			139	153
0950	Pedestal, top spud	1 Plum	8	1		142	44		186	229
0960	Concealed, stall		8	1		159	44		203	248
0970	Wall (washout)		8	1		172	44		216	262
0971	Automatic flush sensor and operator for									
0972	urinals or water closets, standard G	1 Plum	8	1	Ea.	460	44		504	580
0980	High efficiency water saving									

22 42 Commercial Plumbing Fixtures

22 42 39 – Commercial Faucets, Supplies, and Trim

22 42 39.10 Faucets and Fittings		Crew	Daily Output	Labor-Hours	Unit	Material	Labor	Equipment	Total	Total Incl O&P
						2016 Bare Costs				
0984	Water closets, 1.28 gpf	1 Plum	8	1	Ea.	415	44		459	535
0988	Urinals, .5 gpf	″	8	1	″	415	44		459	535
2790	Faucets for lavatories									
2800	Self-closing, center set	1 Plum	10	.800	Ea.	148	35		183	221
2810	Automatic sensor and operator, with faucet head		6.15	1.301		455	57		512	595
3000	Service sink faucet, cast spout, pail hook, hose end		14	.571		76	25		101	125

22 42 39.30 Carriers and Supports

Line	Description	Crew	Daily Output	Labor-Hours	Unit	Material	Labor	Equipment	Total	Total Incl O&P
0010	**CARRIERS AND SUPPORTS**, For plumbing fixtures									
0500	Drinking fountain, wall mounted									
0600	Plate type with studs, top back plate	1 Plum	7	1.143	Ea.	95	50		145	188
0700	Top front and back plate	″	7	1.143	″	116	50		166	211
3000	Lavatory, concealed arm									
3050	Floor mounted, single									
3100	High back fixture	1 Plum	6	1.333	Ea.	570	58.50		628.50	720
3200	Flat slab fixture		6	1.333		455	58.50		513.50	595
3220	Paraplegic		6	1.333		590	58.50		648.50	745
3250	Floor mounted, back to back									
3300	High back fixtures	1 Plum	5	1.600	Ea.	810	70		880	1,000
3400	Flat slab fixtures		5	1.600		1,000	70		1,070	1,225
3430	Paraplegic		5	1.600		860	70		930	1,050
4600	Sink, floor mounted									
4650	Exposed arm system									
4700	Single heavy fixture	1 Plum	5	1.600	Ea.	880	70		950	1,075
5400	Wall mounted, exposed arms, single heavy fixture		5	1.600		450	70		520	610
6000	Urinal, floor mounted, 2″ or 3″ coupling, blowout type		6	1.333		605	58.50		663.50	760
6100	With fixture or hanger bolts, blowout or washout		6	1.333		435	58.50		493.50	570
6300	Wall mounted, plate type system		6	1.333		345	58.50		403.50	475
6980	Water closet, siphon jet									
7000	Horizontal, adjustable, caulk									
7040	Single, 4″ pipe size	1 Plum	5.33	1.501	Ea.	810	65.50		875.50	1,000
7050	4″ pipe size, paraplegic		5.33	1.501		810	65.50		875.50	1,000
7100	Double, 4″ pipe size		5	1.600		1,650	70		1,720	1,950
7110	4″ pipe size, paraplegic		5	1.600		1,650	70		1,720	1,950
8201	Water closet, residential, vert. centerline, floor mount									
8300	4″ copper sweat, 4″ vent	1 Plum	6	1.333	Ea.	715	58.50		773.50	880
9000	Water cooler (electric), floor mounted									
9100	Plate type with bearing plate, single	1 Plum	6	1.333	Ea.	385	58.50		443.50	520

22 43 Healthcare Plumbing Fixtures

22 43 19 – Healthcare Bathtubs

22 43 19.10 Bathtubs

Line	Description	Crew	Daily Output	Labor-Hours	Unit	Material	Labor	Equipment	Total	Total Incl O&P
0010	**BATHTUBS**									
5300	Whirlpool, porcelain enamel on cast iron, 72″ x 36″	Q-1	1	16	Ea.	3,650	630		4,280	5,075

22 47 Drinking Fountains and Water Coolers

22 47 16 – Pressure Water Coolers

22 47 16.10 Electric Water Coolers		Crew	Daily Output	Labor-Hours	Unit	2016 Bare Costs Material	Labor	Equipment	Total	Total Incl O&P
0010	**ELECTRIC WATER COOLERS**									
0100	Wall mounted, non-recessed									
0180	8.2 GPH	Q-1	4	4	Ea.	805	158		963	1,150
0600	8 GPH hot and cold water		4	4		1,100	158		1,258	1,450
1000	Dual height, 8.2 GPH		3.80	4.211		2,150	166		2,316	2,650
2600	Wheelchair type, 8 GPH		4	4		945	158		1,103	1,275
4600	Floor mounted, flush-to-wall									
4680	8.2 GPH	1 Plum	3	2.667	Ea.	815	117		932	1,100
9800	For supply, waste & vent, all coolers	"	2.21	3.620	"	180	159		339	460

22 51 Swimming Pool Plumbing Systems

22 51 19 – Swimming Pool Water Treatment Equipment

22 51 19.50 Swimming Pool Filtration Equipment		Crew	Daily Output	Labor-Hours	Unit	Material	Labor	Equipment	Total	Total Incl O&P
0010	**SWIMMING POOL FILTRATION EQUIPMENT**									
0900	Filter system, sand or diatomite type, incl. pump, 6,000 gal./hr.	2 Plum	1.80	8.889	Total	2,025	390		2,415	2,875
1020	Add for chlorination system, 800 S.F. pool	"	3	5.333	Ea.	185	234		419	590

Division 23

Heating, Ventilating, & Air Conditioning

Estimating Tips

The labor adjustment factors listed in Subdivision 22 01 02.20 also apply to Division 23.

23 10 00 Facility Fuel Systems

- The prices in this subdivision for above- and below-ground storage tanks do not include foundations or hold-down slabs, unless noted. The estimator should refer to Divisions 3 and 31 for foundation system pricing. In addition to the foundations, required tank accessories, such as tank gauges, leak detection devices, and additional manholes and piping, must be added to the tank prices.

23 50 00 Central Heating Equipment

- When estimating the cost of an HVAC system, check to see who is responsible for providing and installing the temperature control system. It is possible to overlook controls, assuming that they would be included in the electrical estimate.
- When looking up a boiler, be careful on specified capacity. Some manufacturers rate their products on output while others use input.
- Include HVAC insulation for pipe, boiler, and duct (wrap and liner).
- Be careful when looking up mechanical items to get the correct pressure rating and connection type (thread, weld, flange).

23 70 00 Central HVAC Equipment

- Combination heating and cooling units are sized by the air conditioning requirements. (See Reference No. R236000-20 for preliminary sizing guide.)
- A ton of air conditioning is nominally 400 CFM.
- Rectangular duct is taken off by the linear foot for each size, but its cost is usually estimated by the pound. Remember that SMACNA standards now base duct on internal pressure.
- Prefabricated duct is estimated and purchased like pipe: straight sections and fittings.
- Note that cranes or other lifting equipment are not included on any lines in Division 23. For example, if a crane is required to lift a heavy piece of pipe into place high above a gym floor, or to put a rooftop unit on the roof of a four-story building, etc., it must be added. Due to the potential for extreme variation—from nothing additional required to a major crane or helicopter—we feel that including a nominal amount for "lifting contingency" would be useless and detract from the accuracy of the estimate. When using equipment rental cost data from RSMeans, do not forget to include the cost of the operator(s).

Reference Numbers

Reference numbers are shown at the beginning of some major classifications. These numbers refer to related items in the Reference Section. The reference information may be an estimating procedure, an alternate pricing method, or technical information.

Note: Not all subdivisions listed here necessarily appear. ■

Note: Trade Service, in part, has been used as a reference source for some of the material prices used in Division 23.

23 05 Common Work Results for HVAC

23 05 02 – HVAC General

23 05 02.10 Air Conditioning, General		Crew	Daily Output	Labor-Hours	Unit	Material	Labor	Equipment	Total	Total Incl O&P
						2016 Bare Costs				
0010	**AIR CONDITIONING, GENERAL** Prices are for standard efficiencies (SEER 13)									
0020	for upgrade to SEER 14 add					10%				

23 05 05 – Selective Demolition for HVAC

23 05 05.10 HVAC Demolition

		Crew	Daily Output	Labor-Hours	Unit	Material	Labor	Equipment	Total	Total Incl O&P
0010	**HVAC DEMOLITION**									
0100	Air conditioner, split unit, 3 ton	Q-5	2	8	Ea.		325		325	535
0150	Package unit, 3 ton	Q-6	3	8	″		310		310	515
0260	Baseboard, hydronic fin tube, 1/2″	Q-5	117	.137	L.F.		5.55		5.55	9.10
0298	Boilers									
0300	Electric, up thru 148 kW	Q-19	2	12	Ea.		500		500	820
0310	150 thru 518 kW	″	1	24			1,000		1,000	1,650
0320	550 thru 2000 kW	Q-21	.40	80			3,400		3,400	5,600
0330	2070 kW and up	″	.30	106			4,525		4,525	7,450
0340	Gas and/or oil, up thru 150 MBH	Q-7	2.20	14.545			590		590	970
0350	160 thru 2000 MBH		.80	40			1,625		1,625	2,675
0360	2100 thru 4500 MBH		.50	64			2,600		2,600	4,275
0370	4600 thru 7000 MBH		.30	106			4,325		4,325	7,125
0390	12,200 thru 25,000 MBH		.12	266			10,800		10,800	17,800
1000	Ductwork, 4″ high, 8″ wide	1 Clab	200	.040	L.F.		1.17		1.17	1.96
1100	6″ high, 8″ wide		165	.048			1.42		1.42	2.37
1200	10″ high, 12″ wide		125	.064			1.87		1.87	3.13
1300	12″-14″ high, 16″-18″ wide		85	.094			2.75		2.75	4.60
1500	30″ high, 36″ wide		56	.143			4.17		4.17	7
2800	Heat pump, package unit, 3 ton	Q-5	2.40	6.667	Ea.		270		270	445
2840	Split unit, 3 ton		2	8			325		325	535
2950	Tank, steel, oil, 275 gal., above ground		10	1.600			64.50		64.50	107
2960	Remove and reset		3	5.333			216		216	355
5090	Remove refrigerant from system	1 Stpi	40	.200	Lb.		9		9	14.80

23 05 23 – General-Duty Valves for HVAC Piping

23 05 23.20 Valves, Bronze/Brass

		Crew	Daily Output	Labor-Hours	Unit	Material	Labor	Equipment	Total	Total Incl O&P
0010	**VALVES, BRONZE/BRASS**									
0020	Brass									
1300	Ball combination valves, shut-off and union									
1310	Solder, with strainer, drain and PT ports									
1320	1/2″	1 Stpi	17	.471	Ea.	75	21		96	118
1330	3/4″		16	.500		78	22.50		100.50	123
1340	1″		14	.571		123	25.50		148.50	178
1350	1-1/4″		12	.667		142	30		172	206
1360	1-1/2″		10	.800		221	36		257	305
1370	2″		8	1		260	45		305	360
1410	Threaded, with strainer, drain and PT ports									
1420	1/2″	1 Stpi	20	.400	Ea.	75	18		93	112
1430	3/4″		18	.444		78	20		98	119
1440	1″		17	.471		123	21		144	170
1450	1-1/4″		12	.667		142	30		172	206
1460	1-1/2″		11	.727		221	32.50		253.50	297
1470	2″		9	.889		260	40		300	350

23 05 Common Work Results for HVAC

23 05 23 – General-Duty Valves for HVAC Piping

23 05 23.30 Valves, Iron Body		Crew	Daily Output	Labor-Hours	Unit	Material	Labor	Equipment	Total	Total Incl O&P
						2016 Bare Costs				
0010	VALVES, IRON BODY									
1650	Gate, 125 lb., N.R.S.									
2150	Flanged									
2240	2-1/2"	Q-1	5	3.200	Ea.	765	126		891	1,050
2260	3"		4.50	3.556		855	140		995	1,175
2280	4"		3	5.333		1,225	210		1,435	1,700
3550	OS&Y, 125 lb., flanged									
3660	3"	Q-1	4.50	3.556	Ea.	555	140		695	840
3680	4"	"	3	5.333		810	210		1,020	1,225
3700	6"	Q-2	3	8		1,300	305		1,605	1,925
5450	Swing check, 125 lb., threaded									
5500	2"	1 Plum	11	.727	Ea.	450	32		482	550
5950	Flanged									
6000	2"	1 Plum	5	1.600	Ea.	440	70		510	600
6040	2-1/2"	Q-1	5	3.200		400	126		526	655
6050	3"		4.50	3.556		430	140		570	705
6060	4"		3	5.333		675	210		885	1,100
6070	6"	Q-2	3	8		1,150	305		1,455	1,775

23 07 HVAC Insulation

23 07 13 – Duct Insulation

23 07 13.10 Duct Thermal Insulation		Crew	Daily Output	Labor-Hours	Unit	Material	Labor	Equipment	Total	Total Incl O&P
0010	DUCT THERMAL INSULATION									
3000	Ductwork									
3020	Blanket type, fiberglass, flexible									
3030	Fire rated for grease and hazardous exhaust ducts									
3060	1-1/2" thick	Q-14	84	.190	S.F.	4.54	6.75		11.29	16.50
3090	Fire rated for plenums									
3100	1/2" x 24" x 25'	Q-14	1.94	8.247	Roll	167	293		460	680
3110	1/2" x 24" x 25'		98	.163	S.F.	3.35	5.80		9.15	13.55
3120	1/2" x 48" x 25'		1.04	15.385	Roll	335	545		880	1,300
3126	1/2" x 48" x 25'		104	.154	S.F.	3.35	5.45		8.80	13
3140	FSK vapor barrier wrap, .75 lb. density									
3160	1" thick G	Q-14	350	.046	S.F.	.18	1.63		1.81	2.96
3170	1-1/2" thick G	"	320	.050	"	.22	1.78		2	3.26
3210	Vinyl jacket, same as FSK									

23 09 Instrumentation and Control for HVAC

23 09 33 – Electric and Electronic Control System for HVAC

23 09 33.10 Electronic Control Systems		Crew	Daily Output	Labor-Hours	Unit	Material	Labor	Equipment	Total	Total Incl O&P
0010	ELECTRONIC CONTROL SYSTEMS									
0020	For electronic costs, add to Section 23 09 43.10				Ea.				15%	15%

23 09 Instrumentation and Control for HVAC

23 09 43 – Pneumatic Control System for HVAC

23 09 43.10 Pneumatic Control Systems		Crew	Daily Output	Labor-Hours	Unit	Material	2016 Bare Costs Labor	Equipment	Total	Total Incl O&P
0010	**PNEUMATIC CONTROL SYSTEMS**									
0011	Including a nominal 50 ft. of tubing. Add control panelboard if req'd.									
0100	Heating and ventilating, split system									
0200	Mixed air control, economizer cycle, panel readout, tubing									
0220	Up to 10 tons [G]	Q-19	.68	35.294	Ea.	4,275	1,475		5,750	7,125
0240	For 10 to 20 tons [G]		.63	37.915		4,575	1,575		6,150	7,625
0260	For over 20 tons [G]		.58	41.096		4,950	1,700		6,650	8,275
0300	Heating coil, hot water, 3 way valve,									
0320	Freezestat, limit control on discharge, readout	Q-5	.69	23.088	Ea.	3,175	935		4,110	5,025
0500	Cooling coil, chilled water, room									
0520	Thermostat, 3 way valve	Q-5	2	8	Ea.	1,400	325		1,725	2,075
0600	Cooling tower, fan cycle, damper control,									
0620	Control system including water readout in/out at panel	Q-19	.67	35.821	Ea.	5,600	1,500		7,100	8,625
1000	Unit ventilator, day/night operation,									
1100	freezestat, ASHRAE, cycle 2	Q-19	.91	26.374	Ea.	3,100	1,100		4,200	5,225
3000	Boiler room combustion air, damper to 5 S.F., controls		1.37	17.582		2,800	730		3,530	4,275
3500	Fan coil, heating and cooling valves, 4 pipe control system		3	8		1,275	335		1,610	1,950
3600	Heat exchanger system controls		.86	27.907		2,725	1,150		3,875	4,900
3900	Multizone control (one per zone), includes thermostat, damper									
3910	motor and reset of discharge temperature	Q-5	.51	31.373	Ea.	2,800	1,275		4,075	5,175
4000	Pneumatic thermostat, including controlling room radiator valve	″	2.43	6.593		845	267		1,112	1,375
4060	Pump control system	Q-19	3	8		1,300	335		1,635	1,975
4500	Air supply for pneumatic control system									
4600	Tank mounted duplex compressor, starter, alternator,									
4620	piping, dryer, PRV station and filter									
4640	3/4 HP	Q-19	.64	37.383	Ea.	10,900	1,550		12,450	14,600
4650	1 HP		.61	39.539		11,900	1,650		13,550	15,800
4680	3 HP		.55	43.956		17,300	1,825		19,125	22,100
4690	5 HP		.42	57.143		30,200	2,375		32,575	37,200
4800	Main air supply, includes 3/8″ copper main and labor	Q-5	1.82	8.791	C.L.F.	350	355		705	970
4810	If poly tubing used, deduct								30%	30%
8600	VAV boxes, incl. thermostat, damper motor, reheat coil & tubing	Q-5	1.46	10.989	Ea.	1,300	445		1,745	2,150
8610	If no reheat coil, deduct				″				204	204

23 09 53 – Pneumatic and Electric Control System for HVAC

23 09 53.10 Control Components		Crew	Daily Output	Labor-Hours	Unit	Material	Labor	Equipment	Total	Total Incl O&P
0010	**CONTROL COMPONENTS**									
5000	Thermostats									
5030	Manual	1 Stpi	8	1	Ea.	53	45		98	133
5040	1 set back, electric, timed [G]	″	8	1	″	38	45		83	116

23 11 23.20 Gas Piping, Flexible (Csst)		Crew	Daily Output	Labor-Hours	Unit	Material	Labor	Equipment	Total	Total Incl O&P
						2016 Bare Costs				
0010	**GAS PIPING, FLEXIBLE (CSST)**									
0100	Tubing with lightning protection									
0110	3/8″	1 Stpi	65	.123	L.F.	2.60	5.55		8.15	11.95
0120	1/2″		62	.129		2.91	5.80		8.71	12.75
0130	3/4″		60	.133		3.78	6		9.78	14.05
0140	1″		55	.145		5.40	6.55		11.95	16.75
0150	1-1/4″		50	.160		6.75	7.20		13.95	19.30
0160	1-1/2″		45	.178		11.85	8		19.85	26
0170	2″		40	.200		17.05	9		26.05	33.50
0200	Tubing for underground/underslab burial									
0210	3/8″	1 Stpi	65	.123	L.F.	3.60	5.55		9.15	13.05
0220	1/2″		62	.129		4.11	5.80		9.91	14.05
0230	3/4″		60	.133		5.20	6		11.20	15.65
0240	1″		55	.145		7	6.55		13.55	18.50
0250	1-1/4″		50	.160		9.40	7.20		16.60	22
0260	1-1/2″		45	.178		17.65	8		25.65	32.50
0270	2″		40	.200		21	9		30	38
3000	Fittings									
3010	Straight									
3100	Tube to NPT									
3110	3/8″	1 Stpi	29	.276	Ea.	10.35	12.40		22.75	32
3120	1/2″		27	.296		11.20	13.30		24.50	34.50
3130	3/4″		25	.320		15.25	14.40		29.65	40.50
3140	1″		23	.348		23.50	15.65		39.15	52
3150	1-1/4″		20	.400		52.50	18		70.50	87
3160	1-1/2″		17	.471		108	21		129	153
3170	2″		15	.533		183	24		207	242
3200	Coupling									
3210	3/8″	1 Stpi	29	.276	Ea.	18.10	12.40		30.50	40.50
3220	1/2″		27	.296		20.50	13.30		33.80	44.50
3230	3/4″		25	.320		28.50	14.40		42.90	54.50
3240	1″		23	.348		44	15.65		59.65	74
3250	1-1/4″		20	.400		97.50	18		115.50	137
3260	1-1/2″		17	.471		204	21		225	259
3270	2″		15	.533		350	24		374	425
3300	Flange fitting									
3310	3/8″	1 Stpi	25	.320	Ea.	15.90	14.40		30.30	41
3320	1/2″		22	.364		15.20	16.35		31.55	43.50
3330	3/4″		19	.421		19.65	18.95		38.60	52.50
3340	1″		16	.500		27	22.50		49.50	67
3350	1-1/4″		12	.667		63	30		93	119
3400	90° Flange valve									
3410	3/8″	1 Stpi	25	.320	Ea.	30.50	14.40		44.90	57
3420	1/2″		22	.364		31	16.35		47.35	61
3430	3/4″		19	.421		38	18.95		56.95	72.50
4000	Tee									
4120	1/2″	1 Stpi	20.50	.390	Ea.	32.50	17.55		50.05	64.50
4130	3/4″		19	.421		43	18.95		61.95	78.50
4140	1″		17.50	.457		77.50	20.50		98	120
5000	Reducing									
5110	Tube to NPT									
5120	3/4″ to 1/2″ NPT	1 Stpi	26	.308	Ea.	16.35	13.85		30.20	41
5130	1″ to 3/4″ NPT	″	24	.333	″	25.50	15		40.50	52.50

23 11 Facility Fuel Piping

23 11 23 – Facility Natural-Gas Piping

23 11 23.20	Gas Piping, Flexible (Csst)	Crew	Daily Output	Labor-Hours	Unit	Material	Labor	Equipment	Total	Total Incl O&P
						2016 Bare Costs				
5200	Reducing tee									
5210	1/2" x 3/8" x 3/8"	1 Stpi	21	.381	Ea.	43	17.10		60.10	75
5220	3/4" x 1/2" x 1/2"		20	.400		40	18		58	73.50
5230	1" x 3/4" x 1/2"		18	.444		66.50	20		86.50	106
5240	1-1/4" x 1-1/4" x 1"		15.60	.513		178	23		201	233
5250	1-1/2" x 1-1/2" x 1-1/4"		13.30	.602		271	27		298	345
5260	2" x 2" x 1-1/2"		11.80	.678		405	30.50		435.50	500
5300	Manifold with four ports and mounting bracket									
5302	Labor to mount manifold does not include making pipe									
5304	connections which are included in fitting labor.									
5310	3/4" x 1/2" x 1/2"(4)	1 Stpi	76	.105	Ea.	35	4.73		39.73	46.50
5330	1-1/4" x 1" x 3/4"(4)		72	.111		50.50	4.99		55.49	64
5350	2" x 1-1/2" x 1"(4)		68	.118		63	5.30		68.30	78
5600	Protective striker plate									
5610	Quarter plate, 3" x 2"	1 Stpi	88	.091	Ea.	.81	4.09		4.90	7.65
5620	Half plate, 3" x 7"		82	.098		1.58	4.39		5.97	9
5630	Full plate, 3" x 12"		78	.103		3.27	4.61		7.88	11.20

23 13 Facility Fuel-Storage Tanks

23 13 13 – Facility Underground Fuel-Oil, Storage Tanks

23 13 13.09 Single-Wall Steel Fuel-Oil Tanks

Line	Description	Crew	Daily Output	Labor-Hours	Unit	Material	Labor	Equipment	Total	Total Incl O&P
0010	**SINGLE-WALL STEEL FUEL-OIL TANKS**									
5000	Tanks, steel ugnd., sti-p3, not incl. hold-down bars									
5500	Excavation, pad, pumps and piping not included									
5510	Single wall, 500 gallon capacity, 7 ga. shell	Q-5	2.70	5.926	Ea.	2,050	240		2,290	2,650
5520	1,000 gallon capacity, 7 ga. shell	"	2.50	6.400		3,825	259		4,084	4,625
5530	2,000 gallon capacity, 1/4" thick shell	Q-7	4.60	6.957		6,200	281		6,481	7,300
5535	2,500 gallon capacity, 7 ga. shell	Q-5	3	5.333		6,850	216		7,066	7,875
5540	5,000 gallon capacity, 1/4" thick shell	Q-7	3.20	10		11,500	405		11,905	13,400
5560	10,000 gallon capacity, 1/4" thick shell		2	16		12,300	645		12,945	14,700
5610	25,000 gallon capacity, 3/8" thick shell		1.30	24.615		37,900	995		38,895	43,400
5630	40,000 gallon capacity, 3/8" thick shell		.90	35.556		42,000	1,450		43,450	48,600
5640	50,000 gallon capacity, 3/8" thick shell		.80	40		46,600	1,625		48,225	54,000

23 13 13.13 Dbl-Wall Steel, Undrgrnd Fuel-Oil, Stor. Tanks

Line	Description	Crew	Daily Output	Labor-Hours	Unit	Material	Labor	Equipment	Total	Total Incl O&P
0010	**DOUBLE-WALL STEEL, UNDERGROUND FUEL-OIL, STORAGE TANKS**									
6200	Steel, underground, 360°, double wall, U.L. listed,									
6210	with sti-P3 corrosion protection,									
6220	(dielectric coating, cathodic protection, electrical									
6230	isolation) 30 year warranty,									
6240	not incl. manholes or hold-downs.									
6260	1,000 gallon capactiy	Q-5	2.25	7.111	Ea.	6,900	288		7,188	8,050
6280	3,000 gallon capacity	Q-7	3.90	8.205		10,900	330		11,230	12,500
6300	5,000 gallon capacity		2.91	10.997		19,900	445		20,345	22,600
6310	6,000 gallon capacity		2.42	13.223		25,600	535		26,135	29,000
6330	10,000 gallon capacity		1.82	17.582		27,700	710		28,410	31,700
6400	For hold-downs 500-2000 gal., add		16	2	Set	183	81		264	335
6410	For hold-downs 3000-6000 gal., add		12	2.667		370	108		478	590
6420	For hold-downs 8000-12,000 gal., add		11	2.909		445	118		563	685

23 13 Facility Fuel-Storage Tanks

23 13 13 – Facility Underground Fuel-Oil, Storage Tanks

23 13 13.23 Glass-Fiber-Reinfcd-Plastic, Fuel-Oil, Storage		Crew	Daily Output	Labor-Hours	Unit	2016 Bare Costs Material	Labor	Equipment	Total	Total Incl O&P
0010	GLASS-FIBER-REINFCD-PLASTIC, UNDERGRND. FUEL-OIL, STORAGE									
0210	Fiberglass, underground, single wall, U.L. listed, not including									
0220	manway or hold-down strap									
0230	1,000 gallon capacity	Q-5	2.46	6.504	Ea.	4,325	263		4,588	5,175
0250	4,000 gallon capacity	Q-7	3.55	9.014		8,675	365		9,040	10,200
0270	8,000 gallon capacity		2.29	13.974		11,800	565		12,365	13,900
0280	10,000 gallon capacity		2	16		13,400	645		14,045	15,900
0500	For manway, fittings and hold-downs, add					20%	15%			
2210	Fiberglass, underground, single wall, U.L. listed, including									
2220	hold-down straps, no manways									
2230	1,000 gallon capacity	Q-5	1.88	8.511	Ea.	4,800	345		5,145	5,850
2250	4,000 gallon capacity	Q-7	2.90	11.034		9,150	445		9,595	10,800
2270	8,000 gallon capacity		1.78	17.978		12,700	725		13,425	15,200
2280	10,000 gallon capacity		1.60	20		14,400	810		15,210	17,100

23 21 Hydronic Piping and Pumps

23 21 20 – Hydronic HVAC Piping Specialties

23 21 20.46 Expansion Tanks

Code	Description	Crew	Daily Output	Labor-Hours	Unit	Material	Labor	Equipment	Total	Total Incl O&P
0010	EXPANSION TANKS									
1507	Underground fuel-oil storage tanks, see Section 23 13 13									
1512	Tank leak detection systems, see Section 28 33 33.50									
2000	Steel, liquid expansion, ASME, painted, 15 gallon capacity	Q-5	17	.941	Ea.	665	38		703	795
2040	30 gallon capacity		12	1.333		740	54		794	905
2120	100 gallon capacity		6	2.667		1,500	108		1,608	1,850
3000	Steel ASME expansion, rubber diaphragm, 19 gal. cap. accept.		12	1.333		2,525	54		2,579	2,875
3020	31 gallon capacity		8	2		2,800	81		2,881	3,225
3080	119 gallon capacity		4	4		4,250	162		4,412	4,950

23 21 20.58 Hydronic Heating Control Valves

Code	Description	Crew	Daily Output	Labor-Hours	Unit	Material	Labor	Equipment	Total	Total Incl O&P
0010	HYDRONIC HEATING CONTROL VALVES									
0050	Hot water, nonelectric, thermostatic									
0100	Radiator supply, 1/2" diameter	1 Stpi	24	.333	Ea.	68.50	15		83.50	99.50
0120	3/4" diameter		20	.400		71.50	18		89.50	108
0140	1" diameter		19	.421		88	18.95		106.95	128
0160	1-1/4" diameter		15	.533		125	24		149	177
0500	For low pressure steam, add					25%				
1000	Manual, radiator supply									
1010	1/2" pipe size, angle union	1 Stpi	24	.333	Ea.	52.50	15		67.50	82.50
1020	3/4" pipe size, angle union		20	.400		66	18		84	103
1030	1" pipe size, angle union		19	.421		85.50	18.95		104.45	125
1140	Balance and stop valve 1/2" size		22	.364		51.50	16.35		67.85	83.50
1150	3/4" size		20	.400		55.50	18		73.50	91
1160	1" size		19	.421		64.50	18.95		83.45	102
1170	1-1/4" size		15	.533		82.50	24		106.50	131
8000	System balancing and shut-off									
8020	Butterfly, quarter turn, calibrated, threaded or solder									
8040	Bronze, -30°F to +350°F, pressure to 175 psi									
8060	1/2" size	1 Stpi	22	.364	Ea.	19.40	16.35		35.75	48.50
8070	3/4" size		20	.400		31	18		49	63.50
8080	1" size		19	.421		37.50	18.95		56.45	72.50
8090	1-1/4" size		15	.533		60.50	24		84.50	106

23 21 Hydronic Piping and Pumps

23 21 20 – Hydronic HVAC Piping Specialties

23 21 20.58 Hydronic Heating Control Valves		Crew	Daily Output	Labor-Hours	Unit	Material	2016 Bare Costs Labor	Equipment	Total	Total Incl O&P
8110	2" size	1 Stpi	11	.727	Ea.	97.50	32.50		130	161

23 21 20.78 Strainers, Y Type, Iron Body		Crew	Daily Output	Labor-Hours	Unit	Material	Labor	Equipment	Total	Total Incl O&P
0010	**STRAINERS, Y TYPE, IRON BODY**									
0100	1/2" pipe size	1 Stpi	20	.400	Ea.	11.05	18		29.05	41.50
0120	3/4" pipe size		18	.444		12.95	20		32.95	47.50
0140	1" pipe size		16	.500		16.80	22.50		39.30	55.50
0150	1-1/4" pipe size		15	.533		24	24		48	66
0180	2" pipe size		8	1		49	45		94	128
1000	Flanged, 125 lb., 1-1/2" pipe size		11	.727		104	32.50		136.50	168
1020	2" pipe size		8	1		109	45		154	194
1030	2-1/2" pipe size	Q-5	5	3.200		112	129		241	335
1040	3" pipe size		4.50	3.556		179	144		323	435
1060	4" pipe size		3	5.333		296	216		512	680

23 21 23 – Hydronic Pumps

23 21 23.13 In-Line Centrifugal Hydronic Pumps		Crew	Daily Output	Labor-Hours	Unit	Material	Labor	Equipment	Total	Total Incl O&P
0010	**IN-LINE CENTRIFUGAL HYDRONIC PUMPS**									
0600	Bronze, sweat connections, 1/40 HP, in line									
0640	3/4" size	Q-1	16	1	Ea.	225	39.50		264.50	315
1000	Flange connection, 3/4" to 1-1/2" size									
1040	1/12 HP	Q-1	6	2.667	Ea.	580	105		685	815
1060	1/8 HP		6	2.667		1,000	105		1,105	1,275
1100	1/3 HP		6	2.667		1,125	105		1,230	1,400
1140	2" size, 1/6 HP		5	3.200		1,425	126		1,551	1,775
2000	Cast iron, flange connection									
2040	3/4" to 1-1/2" size, in line, 1/12 HP	Q-1	6	2.667	Ea.	375	105		480	590
2060	1/8 HP		6	2.667		625	105		730	865
2101	Pumps, circulating, 3/4" to 1-1/2" size, 1/3 HP		6	2.667		815	105		920	1,075
2140	2" size, 1/6 HP		5	3.200		765	126		891	1,050
2180	2-1/2" size, 1/4 HP		5	3.200		990	126		1,116	1,300
2220	3" size, 1/4 HP		4	4		1,000	158		1,158	1,350
2260	1/3 HP		4	4		1,350	158		1,508	1,750
2300	1/2 HP		4	4		1,400	158		1,558	1,800
2340	3/4 HP		4	4		1,625	158		1,783	2,050
2380	1 HP		4	4		2,350	158		2,508	2,825
4000	Close coupled, end suction, bronze impeller									
4040	1-1/2" size, 1-1/2 HP, to 40 GPM	Q-1	3	5.333	Ea.	2,325	210		2,535	2,925
4090	2" size, 2 HP, to 50 GPM		3	5.333		2,775	210		2,985	3,400
4100	2" size, 3 HP, to 90 GPM		2.30	6.957		2,900	274		3,174	3,625
4190	2-1/2" size, 3 HP, to 150 GPM		2	8		3,125	315		3,440	3,950
4300	3" size, 5 HP, to 225 GPM		1.80	8.889		3,575	350		3,925	4,500
4410	3" size, 10 HP, to 350 GPM		1.60	10		5,225	395		5,620	6,400
4420	4" size, 7-1/2 HP, to 350 GPM		1.60	10		5,275	395		5,670	6,450
5000	Base mounted, bronze impeller, coupling guard									
5190	2-1/2" size, 3 HP, to 150 GPM	Q-1	1.80	8.889	Ea.	8,200	350		8,550	9,600
5300	3" size, 5 HP, to 225 GPM		1.60	10		9,425	395		9,820	11,100
5410	4" size, 5 HP, to 350 GPM		1.50	10.667		10,200	420		10,620	12,000

23 21 Hydronic Piping and Pumps

23 21 29 – Automatic Condensate Pump Units

23 21 29.10 Condensate Removal Pump System		Crew	Daily Output	Labor-Hours	Unit	2016 Bare Costs Material	Labor	Equipment	Total	Total Incl O&P
0010	CONDENSATE REMOVAL PUMP SYSTEM									
0020	Pump with 1 gal. ABS tank									
0100	115 V									
0120	1/50 HP, 200 GPH G	1 Stpi	12	.667	Ea.	197	30		227	267
0140	1/18 HP, 270 GPH G		10	.800		210	36		246	291
0160	1/5 HP, 450 GPH G		8	1		470	45		515	590
0200	230 V									
0240	1/18 HP, 270 GPH	1 Stpi	10	.800	Ea.	217	36		253	299
0260	1/5 HP, 450 GPH G	"	8	1	"	520	45		565	645

23 23 Refrigerant Piping

23 23 16 – Refrigerant Piping Specialties

23 23 16.16 Refrigerant Line Sets		Crew	Daily Output	Labor-Hours	Unit	2016 Bare Costs Material	Labor	Equipment	Total	Total Incl O&P
0010	REFRIGERANT LINE SETS									
0100	Copper tube									
0110	1/2" insulation, both tubes									
0120	Combination 1/4" and 1/2" tubes									
0130	10' set	Q-5	42	.381	Ea.	70.50	15.40		85.90	103
0140	20' set		40	.400		110	16.20		126.20	148
0150	30' set		37	.432		148	17.50		165.50	191
0160	40' set		35	.457		193	18.50		211.50	244
0170	50' set		32	.500		229	20		249	286
0180	100' set		22	.727		520	29.50		549.50	625
0300	Combination 1/4" and 3/4" tubes									
0310	10' set	Q-5	40	.400	Ea.	78.50	16.20		94.70	113
0320	20' set		38	.421		139	17.05		156.05	181
0330	30' set		35	.457		208	18.50		226.50	260
0340	40' set		33	.485		272	19.60		291.60	330
0350	50' set		30	.533		345	21.50		366.50	410
0380	100' set		20	.800		780	32.50		812.50	915
0500	Combination 3/8" & 3/4" tubes									
0510	10' set	Q-5	28	.571	Ea.	90.50	23		113.50	138
0520	20' set		36	.444		141	18		159	185
0530	30' set		34	.471		191	19.05		210.05	243
0540	40' set		31	.516		252	21		273	310
0550	50' set		28	.571		296	23		319	365
0580	100' set		18	.889		875	36		911	1,025
0700	Combination 3/8" & 1-1/8" tubes									
0710	10' set	Q-5	36	.444	Ea.	158	18		176	204
0720	20' set		33	.485		265	19.60		284.60	325
0730	30' set		31	.516		355	21		376	425
0740	40' set		28	.571		535	23		558	625
0750	50' set		26	.615		615	25		640	720
0900	Combination 1/2" & 3/4" tubes									
0910	10' set	Q-5	37	.432	Ea.	95.50	17.50		113	134
0920	20' set		35	.457		169	18.50		187.50	217
0930	30' set		33	.485		255	19.60		274.60	315
0940	40' set		30	.533		335	21.50		356.50	405
0950	50' set		27	.593		420	24		444	505
0980	100' set		17	.941		950	38		988	1,125

23 23 Refrigerant Piping

23 23 16 – Refrigerant Piping Specialties

23 23 16.16 Refrigerant Line Sets		Crew	Daily Output	Labor-Hours	Unit	2016 Bare Costs Material	Labor	Equipment	Total	Total Incl O&P
2100	Combination 1/2" & 1-1/8" tubes									
2110	10' set	Q-5	35	.457	Ea.	163	18.50		181.50	210
2120	20' set		31	.516		279	21		300	340
2130	30' set		29	.552		420	22.50		442.50	495
2140	40' set		25	.640		565	26		591	665
2150	50' set		14	1.143		625	46		671	760
2300	For 1" thick insulation add					30%	15%			

23 31 HVAC Ducts and Casings

23 31 13 – Metal Ducts

23 31 13.13 Rectangular Metal Ducts

Line	Description	Crew	Daily Output	Labor-Hours	Unit	Material	Labor	Equipment	Total	Total Incl O&P
0010	**RECTANGULAR METAL DUCTS**									
0020	Fabricated rectangular, includes fittings, joints, supports,									
0021	allowance for flexible connections and field sketches.									
0030	Does not include "as-built dwgs." or insulation.									
0031	NOTE: Fabrication and installation are combined									
0040	as LABOR cost. Approx. 25% fittings assumed.									
0042	Fabrication/Inst. is to commercial quality standards									
0043	(SMACNA or equiv.) for structure, sealing, leak testing, etc.									
0050	Add to labor for elevated installation									
0051	of fabricated ductwork									
0052	10' to 15' high						6%			
0053	15' to 20' high						12%			
0054	20' to 25' high						15%			
0100	Aluminum, alloy 3003-H14, under 100 lb.	Q-10	75	.320	Lb.	3.20	12.65		15.85	24.50
0140	1,000 to 2,000 lb.		120	.200		1.77	7.90		9.67	15.15
0150	2,000 to 5,000 lb.		130	.185		1.77	7.30		9.07	14.15
0500	Galvanized steel, under 200 lb.		235	.102		.57	4.04		4.61	7.40
0520	200 to 500 lb.		245	.098		.56	3.87		4.43	7.05
0540	500 to 1,000 lb.		255	.094		.55	3.72		4.27	6.80
0560	1,000 to 2,000 lb.		265	.091		.54	3.58		4.12	6.55
0570	2,000 to 5,000 lb.		275	.087		.53	3.45		3.98	6.35

23 31 16 – Nonmetal Ducts

23 31 16.13 Fibrous-Glass Ducts

Line	Description	Crew	Daily Output	Labor-Hours	Unit	Material	Labor	Equipment	Total	Total Incl O&P
0010	**FIBROUS-GLASS DUCTS**									
3490	Rigid fiberglass duct board, foil reinf. kraft facing									
3500	Rectangular, 1" thick, alum. faced, (FRK), std. weight	Q-10	350	.069	SF Surf	.79	2.71		3.50	5.40

23 33 Air Duct Accessories

23 33 13 – Dampers

23 33 13.13 Volume-Control Dampers

Line	Description	Crew	Daily Output	Labor-Hours	Unit	Material	Labor	Equipment	Total	Total Incl O&P
0010	**VOLUME-CONTROL DAMPERS**									
5990	Multi-blade dampers, opposed blade, 8" x 6"	1 Shee	24	.333	Ea.	21.50	14.10		35.60	47.50
5994	8" x 8"		22	.364		22.50	15.40		37.90	50
5996	10" x 10"		21	.381		26.50	16.15		42.65	56.50
6000	12" x 12"		21	.381		29.50	16.15		45.65	59.50
6020	12" x 18"		18	.444		39.50	18.80		58.30	75
6030	14" x 10"		20	.400		29	16.95		45.95	60
6031	14" x 14"		17	.471		35.50	19.95		55.45	72.50

23 33 Air Duct Accessories

23 33 13 – Dampers

23 33 13.13 Volume-Control Dampers		Crew	Daily Output	Labor-Hours	Unit	Material	2016 Bare Costs Labor	Equipment	Total	Total Incl O&P
6033	16" x 12"	1 Shee	17	.471	Ea.	35.50	19.95		55.45	72.50
6035	16" x 16"		16	.500		44	21		65	84
6037	18" x 16"		15	.533		48	22.50		70.50	90.50
6038	18" x 18"		15	.533		52	22.50		74.50	94.50
6070	20" x 16"		14	.571		52	24		76	97.50
6072	20" x 20"		13	.615		62.50	26		88.50	113
6074	22" x 18"		14	.571		62.50	24		86.50	110
6076	24" x 16"		11	.727		61	31		92	119
6078	24" x 20"		8	1		72.50	42.50		115	151
6080	24" x 24"		8	1		85	42.50		127.50	164
6110	26" x 26"		6	1.333		95	56.50		151.50	199
6133	30" x 30"	Q-9	6.60	2.424		137	92.50		229.50	305
6135	32" x 32"	"	6.40	2.500		154	95.50		249.50	330
7500	Variable volume modulating motorized damper, incl. elect. mtr.									
7504	8" x 6"	1 Shee	15	.533	Ea.	118	22.50		140.50	168
7506	10" x 6"		14	.571		118	24		142	171
7510	10" x 10"		13	.615		122	26		148	178
7520	12" x 12"		12	.667		127	28		155	186
7522	12" x 16"		11	.727		127	31		158	192
7524	16" x 10"		12	.667		125	28		153	184
7526	16" x 14"		10	.800		130	34		164	200
7528	16" x 18"		9	.889		135	37.50		172.50	211
7542	18" x 18"		8	1		138	42.50		180.50	222
7544	20" x 14"		8	1		141	42.50		183.50	226
7546	20" x 18"		7	1.143		147	48.50		195.50	243
7560	24" x 12"		8	1		143	42.50		185.50	228
7562	24" x 18"		7	1.143		154	48.50		202.50	250
7564	24" x 24"		6	1.333		160	56.50		216.50	270
7568	28" x 10"		7	1.143		143	48.50		191.50	238
7590	30" x 14"		5	1.600		151	68		219	279
7610	30" x 24"		3.80	2.105		264	89		353	440
7700	For thermostat, add		8	1		31	42.50		73.50	105
8000	Multi-blade dampers, parallel blade									
8100	8" x 8"	1 Shee	24	.333	Ea.	82	14.10		96.10	114
8160	18" x 12"	"	18	.444	"	113	18.80		131.80	156

23 33 13.16 Fire Dampers		Crew	Daily Output	Labor-Hours	Unit	Material	Labor	Equipment	Total	Total Incl O&P
0010	**FIRE DAMPERS**									
3000	Fire damper, curtain type, 1-1/2 hr. rated, vertical, 6" x 6"	1 Shee	24	.333	Ea.	24	14.10		38.10	50
3020	8" x 6"		22	.364		24	15.40		39.40	52
3240	16" x 14"		18	.444		44	18.80		62.80	80

23 33 19 – Duct Silencers

23 33 19.10 Duct Silencers		Crew	Daily Output	Labor-Hours	Unit	Material	Labor	Equipment	Total	Total Incl O&P
0010	**DUCT SILENCERS**									
9000	Silencers, noise control for air flow, duct				MCFM	57			57	63

23 33 46 – Flexible Ducts

23 33 46.10 Flexible Air Ducts		Crew	Daily Output	Labor-Hours	Unit	Material	Labor	Equipment	Total	Total Incl O&P
0010	**FLEXIBLE AIR DUCTS**									
1280	Add to labor for elevated installation									
1282	of prefabricated (purchased) ductwork									
1283	10' to 15' high						10%			
1284	15' to 20' high						20%			

23 33 Air Duct Accessories

23 33 46 – Flexible Ducts

23 33 46.10 Flexible Air Ducts			Crew	Daily Output	Labor-Hours	Unit	2016 Bare Costs Material	Labor	Equipment	Total	Total Incl O&P
1285	20′ to 25′ high							25%			
1286	25′ to 30′ high							35%			
1287	30′ to 35′ high							40%			
1288	35′ to 40′ high							50%			
1289	Over 40′ high							55%			
1300	Flexible, coated fiberglass fabric on corr. resist. metal helix										
1400	pressure to 12″ (WG) UL-181										
1500	Noninsulated, 3″ diameter		Q-9	400	.040	L.F.	1.12	1.53		2.65	3.77
1520	4″ diameter			360	.044		1.16	1.69		2.85	4.11
1540	5″ diameter			320	.050		1.30	1.91		3.21	4.61
1560	6″ diameter			280	.057		1.50	2.18		3.68	5.30
1580	7″ diameter			240	.067		1.53	2.54		4.07	5.90
1600	8″ diameter			200	.080		1.91	3.05		4.96	7.20
1620	9″ diameter			180	.089		2.05	3.39		5.44	7.90
1640	10″ diameter			160	.100		2.46	3.81		6.27	9.05
1900	Insulated, 1″ thick, PE jacket, 3″ diameter	G		380	.042		2.60	1.61		4.21	5.55
1920	5″ diameter	G		300	.053		2.60	2.03		4.63	6.25
1940	6″ diameter	G		260	.062		2.94	2.35		5.29	7.15
1960	7″ diameter	G		220	.073		3.20	2.77		5.97	8.15
1980	8″ diameter	G		180	.089		3.49	3.39		6.88	9.50
2000	9″ diameter	G		160	.100		3.81	3.81		7.62	10.55
2020	10″ diameter	G		140	.114		4.25	4.36		8.61	11.95

23 33 53 – Duct Liners

23 33 53.10 Duct Liner Board			Crew	Daily Output	Labor-Hours	Unit	Material	Labor	Equipment	Total	Total Incl O&P
0010	**DUCT LINER BOARD**										
3490	Board type, fiberglass liner, 3 lb. density										
3940	Board type, non-fibrous foam										
3950	Temperature, bacteria and fungi resistant										
3960	1″ thick	G	Q-14	150	.107	S.F.	2.38	3.79		6.17	9.05
3970	1-1/2″ thick	G		130	.123		3.28	4.38		7.66	11
3980	2″ thick	G		120	.133		3.98	4.74		8.72	12.45

23 34 HVAC Fans

23 34 14 – Blower HVAC Fans

23 34 14.10 Blower Type HVAC Fans		Crew	Daily Output	Labor-Hours	Unit	Material	Labor	Equipment	Total	Total Incl O&P
0010	**BLOWER TYPE HVAC FANS**									
2500	Ceiling fan, right angle, extra quiet, 0.10″ S.P.									
2520	95 CFM	Q-20	20	1	Ea.	315	39.50		354.50	410
2540	210 CFM	″	19	1.053	″	370	41.50		411.50	475

23 34 16 – Centrifugal HVAC Fans

23 34 16.10 Centrifugal Type HVAC Fans		Crew	Daily Output	Labor-Hours	Unit	Material	Labor	Equipment	Total	Total Incl O&P
0010	**CENTRIFUGAL TYPE HVAC FANS**									
5000	Utility set, centrifugal, V belt drive, motor									
5020	1/4″ S.P., 1200 CFM, 1/4 HP	Q-20	6	3.333	Ea.	1,750	131		1,881	2,150
5040	1520 CFM, 1/3 HP		5	4		2,300	157		2,457	2,775
5060	1850 CFM, 1/2 HP		4	5		2,275	197		2,472	2,825
5080	2180 CFM, 3/4 HP		3	6.667		2,700	262		2,962	3,375
7000	Roof exhauster, centrifugal, aluminum housing, 12″ galvanized									
7020	curb, bird screen, back draft damper, 1/4″ S.P.									
7100	Direct drive, 320 CFM, 11″ sq. damper	Q-20	7	2.857	Ea.	725	112		837	980

23 34 16 – Centrifugal HVAC Fans

23 34 16.10 Centrifugal Type HVAC Fans		Crew	Daily Output	Labor-Hours	Unit	2016 Bare Costs Material	Labor	Equipment	Total	Total Incl O&P
7120	600 CFM, 11″ sq. damper	Q-20	6	3.333	Ea.	925	131		1,056	1,250
7140	815 CFM, 13″ sq. damper		5	4		925	157		1,082	1,275
7160	1450 CFM, 13″ sq. damper		4.20	4.762		1,450	187		1,637	1,875
7180	2050 CFM, 16″ sq. damper		4	5		1,800	197		1,997	2,300
7200	V-belt drive, 1650 CFM, 12″ sq. damper		6	3.333		1,350	131		1,481	1,725
7220	2750 CFM, 21″ sq. damper		5	4		1,600	157		1,757	2,000
7230	3500 CFM, 21″ sq. damper		4.50	4.444		1,800	175		1,975	2,275
7240	4910 CFM, 23″ sq. damper		4	5		2,200	197		2,397	2,750

23 34 23 – HVAC Power Ventilators

23 34 23.10 HVAC Power Circulators and Ventilators

Line	Description	Crew	Daily Output	Labor-Hours	Unit	Material	Labor	Equipment	Total	Total Incl O&P
0010	**HVAC POWER CIRCULATORS AND VENTILATORS**									
6000	Propeller exhaust, wall shutter									
6020	Direct drive, one speed, .075″ S.P.									
6100	653 CFM, 1/30 HP	Q-20	10	2	Ea.	200	78.50		278.50	350
6140	1323 CFM, 1/15 HP		8	2.500		335	98.50		433.50	530
6160	2444 CFM, 1/4 HP		7	2.857		415	112		527	645
6650	Residential, bath exhaust, grille, back draft damper									
6660	50 CFM	Q-20	24	.833	Ea.	65	33		98	126
6670	110 CFM	″	22	.909	″	100	35.50		135.50	170

23 34 33 – Air Curtains

23 34 33.10 Air Barrier Curtains

Line	Description	Crew	Daily Output	Labor-Hours	Unit	Material	Labor	Equipment	Total	Total Incl O&P
0010	**AIR BARRIER CURTAINS**, Incl. motor starters, transformers,									
0050	and door switches									
2450	Conveyor openings or service windows									
3000	Service window, 5′ high x 25″ wide	2 Shee	5	3.200	Ea.	315	136		451	575
3100	Environmental separation									
3110	Door heights up to 8′, low profile, super quiet									
3120	Unheated, variable speed									
3130	36″ wide	2 Shee	4	4	Ea.	635	169		804	985
3134	42″ wide		3.80	4.211		665	178		843	1,025
3138	48″ wide		3.60	4.444		690	188		878	1,075
3142	60″ wide		3.40	4.706		715	199		914	1,125
3146	72″ wide	Q-3	4.60	6.957		890	274		1,164	1,425
3150	96″ wide		4.40	7.273		1,350	287		1,637	1,975
3154	120″ wide		4.20	7.619		1,400	300		1,700	2,025
3158	144″ wide		4	8		1,775	315		2,090	2,475
3200	Door heights up to 10′									
3210	Unheated									
3230	36″ wide	2 Shee	3.80	4.211	Ea.	675	178		853	1,050
3234	42″ wide		3.60	4.444		695	188		883	1,075
3238	48″ wide		3.40	4.706		715	199		914	1,125
3242	60″ wide		3.20	5		1,075	212		1,287	1,525
3246	72″ wide	Q-3	4.40	7.273		1,125	287		1,412	1,725
3250	96″ wide		4.20	7.619		1,350	300		1,650	1,975
3254	120″ wide		4	8		1,850	315		2,165	2,575
3258	144″ wide		3.80	8.421		2,000	330		2,330	2,750
3300	Door heights up to 12′									
3310	Unheated									
3334	42″ wide	2 Shee	3.40	4.706	Ea.	1,000	199		1,199	1,425
3338	48″ wide		3.20	5		1,025	212		1,237	1,475
3342	60″ wide		3	5.333		1,025	226		1,251	1,500
3346	72″ wide	Q-3	4.20	7.619		1,825	300		2,125	2,525

23 34 HVAC Fans

23 34 33 – Air Curtains

23 34 33.10 Air Barrier Curtains		Crew	Daily Output	Labor-Hours	Unit	Material	Labor (2016 Bare Costs)	Equipment	Total	Total Incl O&P
3350	96″ wide	Q-3	4	8	Ea.	2,000	315		2,315	2,725
3354	120″ wide		3.80	8.421		2,450	330		2,780	3,250
3358	144″ wide		3.60	8.889		2,625	350		2,975	3,450
3400	Door heights up to 16′									
3410	Unheated									
3438	48″ wide	2 Shee	3	5.333	Ea.	1,175	226		1,401	1,650
3442	60″ wide	″	2.80	5.714		1,225	242		1,467	1,750
3446	72″ wide	Q-3	3.80	8.421		2,100	330		2,430	2,850
3450	96″ wide		3.60	8.889		2,375	350		2,725	3,200
3454	120″ wide		3.40	9.412		3,225	370		3,595	4,150
3458	144″ wide		3.20	10		3,350	395		3,745	4,325
3470	Heated, electric									
3474	48″ wide	2 Shee	2.90	5.517	Ea.	1,950	234		2,184	2,525
3478	60″ wide	″	2.70	5.926		1,975	251		2,226	2,600
3482	72″ wide	Q-3	3.70	8.649		3,400	340		3,740	4,300
3486	96″ wide		3.50	9.143		3,600	360		3,960	4,550
3490	120″ wide		3.30	9.697		4,350	380		4,730	5,425
3494	144″ wide		3.10	10.323		4,700	405		5,105	5,825

23 37 Air Outlets and Inlets

23 37 13 – Diffusers, Registers, and Grilles

23 37 13.10 Diffusers		Crew	Daily Output	Labor-Hours	Unit	Material	Labor (2016 Bare Costs)	Equipment	Total	Total Incl O&P
0010	**DIFFUSERS**, Aluminum, opposed blade damper unless noted									
0100	Ceiling, linear, also for sidewall									
0120	2″ wide	1 Shee	32	.250	L.F.	16.80	10.60		27.40	36
0160	4″ wide		26	.308	″	22.50	13.05		35.55	46
0500	Perforated, 24″ x 24″ lay-in panel size, 6″ x 6″		16	.500	Ea.	157	21		178	208
0520	8″ x 8″		15	.533		165	22.50		187.50	219
0530	9″ x 9″		14	.571		167	24		191	225
0590	16″ x 16″		11	.727		195	31		226	267
1000	Rectangular, 1 to 4 way blow, 6″ x 6″		16	.500		50	21		71	90.50
1010	8″ x 8″		15	.533		58	22.50		80.50	101
1014	9″ x 9″		15	.533		66	22.50		88.50	110
1016	10″ x 10″		15	.533		80	22.50		102.50	125
1020	12″ x 6″		15	.533		71.50	22.50		94	116
1040	12″ x 9″		14	.571		75.50	24		99.50	124
1060	12″ x 12″		12	.667		84.50	28		112.50	140
1070	14″ x 6″		13	.615		77.50	26		103.50	129
1074	14″ x 14″		12	.667		128	28		156	187
1150	18″ x 18″		9	.889		138	37.50		175.50	214
1170	24″ x 12″		10	.800		163	34		197	237
1180	24″ x 24″		7	1.143		270	48.50		318.50	375
1500	Round, butterfly damper, steel, diffuser size, 6″ diameter		18	.444		9.40	18.80		28.20	42
1520	8″ diameter		16	.500		10	21		31	46.50
2000	T bar mounting, 24″ x 24″ lay-in frame, 6″ x 6″		16	.500		72	21		93	115
2020	8″ x 8″		14	.571		72.50	24		96.50	120
2040	12″ x 12″		12	.667		88.50	28		116.50	145
2060	16″ x 16″		11	.727		119	31		150	183
2080	18″ x 18″		10	.800		133	34		167	203
6000	For steel diffusers instead of aluminum, deduct					10%				

23 37 Air Outlets and Inlets

23 37 13 – Diffusers, Registers, and Grilles

23 37 13.30 Grilles		Crew	Daily Output	Labor-Hours	Unit	Material	Labor	Equipment	Total	Total Incl O&P
						2016 Bare Costs				
0010	**GRILLES**									
0020	Aluminum, unless noted otherwise									
1000	Air return, steel, 6″ x 6″	1 Shee	26	.308	Ea.	18.70	13.05		31.75	42
1020	10″ x 6″		24	.333		18.70	14.10		32.80	44
1080	16″ x 8″		22	.364		26.50	15.40		41.90	54.50
1100	12″ x 12″		22	.364		26.50	15.40		41.90	54.50
1120	24″ x 12″		18	.444		36	18.80		54.80	71
1180	16″ x 16″	↓	22	.364	↓	34	15.40		49.40	63

23 37 13.60 Registers		Crew	Daily Output	Labor-Hours	Unit	Material	Labor	Equipment	Total	Total Incl O&P
0010	**REGISTERS**									
0980	Air supply									
3000	Baseboard, hand adj. damper, enameled steel									
3012	8″ x 6″	1 Shee	26	.308	Ea.	5.05	13.05		18.10	27
3020	10″ x 6″		24	.333		5.50	14.10		19.60	29.50
3060	12″ x 6″	↓	23	.348	↓	6.75	14.75		21.50	32
4000	Floor, toe operated damper, enameled steel									
4020	4″ x 8″	1 Shee	32	.250	Ea.	9.40	10.60		20	28
4040	4″ x 12″	″	26	.308	″	11	13.05		24.05	33.50
4300	Spiral pipe supply register									
4310	Steel, with air scoop									
4320	4″ x 12″, for 8″ thru 13″ diameter duct	1 Shee	25	.320	Ea.	74.50	13.55		88.05	104
4330	4″ x 18″, for 8″ thru 13″ diameter duct		18	.444		87	18.80		105.80	127
4340	6″ x 12″, for 14″ thru 21″ diameter duct		19	.421		78.50	17.85		96.35	117
4350	6″ x 16″, for 14″ thru 21″ diameter duct		18	.444		89	18.80		107.80	130
4360	6″ x 20″, for 14″ thru 21″ diameter duct		17	.471		98.50	19.95		118.45	142
4370	6″ x 24″, for 14″ thru 21″ diameter duct		16	.500		116	21		137	163
4380	8″ x 16″, for 22″ thru 31″ diameter duct		19	.421		94	17.85		111.85	133
4390	8″ x 18″, for 22″ thru 31″ diameter duct		18	.444		98.50	18.80		117.30	140
4400	8″ x 24″, for 22″ thru 31″ diameter duct	↓	15	.533	↓	122	22.50		144.50	172

23 37 23 – HVAC Gravity Ventilators

23 37 23.10 HVAC Gravity Air Ventilators		Crew	Daily Output	Labor-Hours	Unit	Material	Labor	Equipment	Total	Total Incl O&P
0010	**HVAC GRAVITY AIR VENTILATORS**, Includes base									
4200	Stationary mushroom, aluminum, 16″ orifice diameter	Q-9	10	1.600	Ea.	645	61		706	810
4240	38″ orifice diameter		5	3.200		2,000	122		2,122	2,400
4250	42″ orifice diameter		4.70	3.404		2,650	130		2,780	3,125
4260	50″ orifice diameter	↓	4.44	3.604	↓	3,150	137		3,287	3,700

23 38 Ventilation Hoods

23 38 13 – Commercial-Kitchen Hoods

23 38 13.10 Hood and Ventilation Equipment		Crew	Daily Output	Labor-Hours	Unit	Material	Labor	Equipment	Total	Total Incl O&P
0010	**HOOD AND VENTILATION EQUIPMENT**									
2970	Exhaust hood, sst, gutter on all sides, 4′ x 4′ x 2′	1 Carp	1.80	4.444	Ea.	4,725	168		4,893	5,475
2980	4′ x 4′ x 7′	″	1.60	5	″	7,525	189		7,714	8,600

23 41 Particulate Air Filtration

23 41 13 – Panel Air Filters

23 41 13.10 Panel Type Air Filters		Crew	Daily Output	Labor-Hours	Unit	2016 Bare Costs Material	Labor	Equipment	Total	Total Incl O&P
0010	PANEL TYPE AIR FILTERS									
2950	Mechanical media filtration units									
3000	High efficiency type, with frame, non-supported G				MCFM	35			35	38.50
3100	Supported type G				"	50			50	55
5500	Throwaway glass or paper media type, 12" x 36" x 1"				Ea.	2.39			2.39	2.63

23 41 16 – Renewable-Media Air Filters

23 41 16.10 Disposable Media Air Filters		Crew	Daily Output	Labor-Hours	Unit	Material	Labor	Equipment	Total	Total Incl O&P
0010	DISPOSABLE MEDIA AIR FILTERS									
5000	Renewable disposable roll				C.S.F.	5.25			5.25	5.75

23 41 19 – Washable Air Filters

23 41 19.10 Permanent Air Filters		Crew	Daily Output	Labor-Hours	Unit	Material	Labor	Equipment	Total	Total Incl O&P
0010	PERMANENT AIR FILTERS									
4500	Permanent washable G				MCFM	20			20	22

23 41 23 – Extended Surface Filters

23 41 23.10 Expanded Surface Filters		Crew	Daily Output	Labor-Hours	Unit	Material	Labor	Equipment	Total	Total Incl O&P
0010	EXPANDED SURFACE FILTERS									
4000	Medium efficiency, extended surface G				MCFM	6			6	6.60

23 42 Gas-Phase Air Filtration

23 42 13 – Activated-Carbon Air Filtration

23 42 13.10 Charcoal Type Air Filtration		Crew	Daily Output	Labor-Hours	Unit	Material	Labor	Equipment	Total	Total Incl O&P
0010	CHARCOAL TYPE AIR FILTRATION									
0050	Activated charcoal type, full flow				MCFM	600			600	660
0060	Full flow, impregnated media 12" deep					225			225	248
0070	HEPA filter & frame for field erection					400			400	440
0080	HEPA filter-diffuser, ceiling install.				↓	300			300	330

23 43 Electronic Air Cleaners

23 43 13 – Washable Electronic Air Cleaners

23 43 13.10 Electronic Air Cleaners		Crew	Daily Output	Labor-Hours	Unit	Material	Labor	Equipment	Total	Total Incl O&P
0010	ELECTRONIC AIR CLEANERS									
2000	Electronic air cleaner, duct mounted									
2150	1000 CFM	1 Shee	4	2	Ea.	420	84.50		504.50	605
2200	1200 CFM		3.80	2.105		485	89		574	685
2250	1400 CFM	↓	3.60	2.222	↓	500	94		594	705

23 51 Breechings, Chimneys, and Stacks

23 51 23 – Gas Vents

23 51 23.10 Gas Chimney Vents		Crew	Daily Output	Labor-Hours	Unit	2016 Bare Costs Material	Labor	Equipment	Total	Total Incl O&P
0010	**GAS CHIMNEY VENTS**, Prefab metal, U.L. listed									
0020	Gas, double wall, galvanized steel									
0080	3″ diameter	Q-9	72	.222	V.L.F.	6.60	8.45		15.05	21.50
0100	4″ diameter		68	.235		8.15	8.95		17.10	24
0120	5″ diameter		64	.250		9.25	9.55		18.80	26
0140	6″ diameter		60	.267		11.05	10.15		21.20	29
0160	7″ diameter		56	.286		21	10.90		31.90	41
0180	8″ diameter		52	.308		22.50	11.75		34.25	44.50
0200	10″ diameter		48	.333		43	12.70		55.70	68.50

23 51 26 – All-Fuel Vent Chimneys

23 51 26.30 All-Fuel Vent Chimneys, Double Wall, St. Stl.

		Crew	Daily Output	Labor-Hours	Unit	Material	Labor	Equipment	Total	Total Incl O&P
0010	**ALL-FUEL VENT CHIMNEYS, DOUBLE WALL, STAINLESS STEEL**									
7780	All fuel, pressure tight, double wall, 4″ insulation, U.L. listed, 1400°F.									
7790	304 stainless steel liner, aluminized steel outer jacket									
7800	6″ diameter	Q-9	60	.267	V.L.F.	65	10.15		75.15	88.50
7806	10″ diameter	″	48	.333	″	83.50	12.70		96.20	113
7880	For 316 stainless steel liner, add				L.F.	30%				

23 52 Heating Boilers

23 52 13 – Electric Boilers

23 52 13.10 Electric Boilers, ASME

		Crew	Daily Output	Labor-Hours	Unit	Material	Labor	Equipment	Total	Total Incl O&P
0010	**ELECTRIC BOILERS, ASME**, Standard controls and trim									
1000	Steam, 6 KW, 20.5 MBH	Q-19	1.20	20	Ea.	4,100	835		4,935	5,875
1160	60 KW, 205 MBH		1	24		6,925	1,000		7,925	9,250
2000	Hot water, 7.5 KW, 25.6 MBH		1.30	18.462		4,950	770		5,720	6,725
2040	30 KW, 102 MBH		1.20	20		5,325	835		6,160	7,225
2070	60 KW, 205 MBH		1.20	20		5,525	835		6,360	7,450

23 52 19 – Pulse Combustion Boilers

23 52 19.20 Pulse Type Combustion Boilers

		Crew	Daily Output	Labor-Hours	Unit	Material	Labor	Equipment	Total	Total Incl O&P
0010	**PULSE TYPE COMBUSTION BOILERS**, High efficiency									
7990	Special feature gas fired boilers									
8000	Pulse combustion, standard controls/trim									
8010	Hot water, DOE MBH output (AFUE %)									
8030	71 MBH, 95.2%	Q-5	1.60	10	Ea.	3,450	405		3,855	4,475
8050	94 MBH, 95.3% G		1.40	11.429		3,825	460		4,285	4,950
8080	139 MBH, 95.6% G		1.20	13.333		4,300	540		4,840	5,625
8090	207 MBH, 95.4%		1.16	13.793		4,850	560		5,410	6,275
8120	270 MBH, 96.4%		1.12	14.286		6,725	580		7,305	8,350
8130	365 MBH, 91.7%		1.07	14.953		7,600	605		8,205	9,350

23 52 23 – Cast-Iron Boilers

23 52 23.20 Gas-Fired Boilers

		Crew	Daily Output	Labor-Hours	Unit	Material	Labor	Equipment	Total	Total Incl O&P
0010	**GAS-FIRED BOILERS**, Natural or propane, standard controls, packaged									
1000	Cast iron, with insulated jacket									
2000	Steam, gross output, 81 MBH	Q-7	1.40	22.857	Ea.	2,450	925		3,375	4,225
2080	203 MBH		.90	35.556		3,825	1,450		5,275	6,575
2180	400 MBH		.56	56.838		5,900	2,300		8,200	10,300
2280	1275 MBH		.34	94.118		17,300	3,800		21,100	25,300
2380	3060 MBH		.19	172		29,300	6,950		36,250	43,800
2400	3570 MBH		.18	181		35,800	7,350		43,150	51,500

23 52 Heating Boilers

23 52 23 – Cast-Iron Boilers

23 52 23.20 Gas-Fired Boilers		Crew	Daily Output	Labor-Hours	Unit	2016 Bare Costs Material	Labor	Equipment	Total	Total Incl O&P
3000	Hot water, gross output, 80 MBH	Q-7	1.46	21.918	Ea.	1,975	885		2,860	3,625
3020	100 MBH		1.35	23.704		2,550	960		3,510	4,400
3023	Install medium efficiency propane boiler, 29.3 kW G	Q-5	.68	23.704		2,550	960		3,510	4,400
3025	Install high efficiency gas boiler, 29.3 kW G		.68	23.704		2,550	960		3,510	4,400
3027	Install high efficiency propane boiler, 29.3 kW G		.68	23.704		2,550	960		3,510	4,400
3080	203 MBH	Q-7	1	32		3,625	1,300		4,925	6,125
3180	400 MBH		.64	50		5,575	2,025		7,600	9,475
3280	1,275 MBH		.36	89.888		17,300	3,625		20,925	25,000
3340	2,312 MBH		.22	148		24,400	6,000		30,400	36,800
3400	3,996 MBH		.16	195		35,800	7,900		43,700	52,500
7000	For tankless water heater, add					10%				

23 52 23.40 Oil-Fired Boilers		Crew	Daily Output	Labor-Hours	Unit	Material	Labor	Equipment	Total	Total Incl O&P
0010	**OIL-FIRED BOILERS**, Standard controls, flame retention burner, packaged									
1000	Cast iron, with insulated flush jacket									
2000	Steam, gross output, 109 MBH	Q-7	1.20	26.667	Ea.	2,325	1,075		3,400	4,325
2060	207 MBH		.90	35.556		3,175	1,450		4,625	5,875
2180	1,084 MBH		.38	85.106		10,900	3,450		14,350	17,700
3000	Hot water, same price as steam									

23 52 26 – Steel Boilers

23 52 26.40 Oil-Fired Boilers		Crew	Daily Output	Labor-Hours	Unit	Material	Labor	Equipment	Total	Total Incl O&P
0010	**OIL-FIRED BOILERS**, Standard controls, flame retention burner									
5000	Steel, with insulated flush jacket									
7000	Hot water, gross output, 103 MBH	Q-6	1.60	15	Ea.	1,825	585		2,410	2,975
7020	122 MBH		1.45	16.506		2,000	645		2,645	3,250
7060	168 MBH		1.30	18.405		2,250	715		2,965	3,650
7080	225 MBH		1.22	19.704		3,250	765		4,015	4,850

23 52 26.70 Packaged Water Tube Boilers		Crew	Daily Output	Labor-Hours	Unit	Material	Labor	Equipment	Total	Total Incl O&P
0010	**PACKAGED WATER TUBE BOILERS**									
2000	Packaged water tube, #2 oil, steam or hot water, gross output									
2040	1200 MBH	Q-7	.50	64	Ea.	22,200	2,600		24,800	28,700
2060	1600 MBH		.40	80		35,100	3,225		38,325	43,900
2080	2400 MBH		.30	106		47,200	4,325		51,525	59,000
2100	3200 MBH		.25	128		54,000	5,175		59,175	68,000
2120	4800 MBH		.20	160		70,500	6,475		76,975	88,000
2200	Gas fired									
2204	200 MBH	Q-6	1.50	16	Ea.	6,025	625		6,650	7,650
2208	275 MBH		1.40	17.143		6,350	670		7,020	8,100
2212	360 MBH		1.10	21.818		6,825	850		7,675	8,900
2216	520 MBH		.65	36.923		8,300	1,450		9,750	11,500
2220	600 MBH		.60	40		8,375	1,550		9,925	11,800
2224	720 MBH		.55	43.636		9,975	1,700		11,675	13,800
2228	960 MBH		.48	50		12,300	1,950		14,250	16,800
2232	1220 MBH	Q-7	.50	64		18,500	2,600		21,100	24,600
2236	1440 MBH		.45	71.111		19,000	2,875		21,875	25,700
2240	1680 MBH		.40	80		26,800	3,225		30,025	34,800
2244	1920 MBH		.35	91.429		26,900	3,700		30,600	35,700
2248	2160 MBH		.33	96.970		27,300	3,925		31,225	36,500
2252	2400 MBH		.30	106		30,700	4,325		35,025	40,800

23 52 Heating Boilers

23 52 28 – Swimming Pool Boilers

23 52 28.10 Swimming Pool Heaters		Crew	Daily Output	Labor-Hours	Unit	2016 Bare Costs Material	Labor	Equipment	Total	Total Incl O&P
0010	**SWIMMING POOL HEATERS**, Not including wiring, external									
0020	piping, base or pad									
0160	Gas fired, input, 155 MBH	Q-6	1.50	16	Ea.	2,100	625		2,725	3,350
0220	250 MBH		.70	34.286		2,425	1,325		3,750	4,875
0240	300 MBH		.60	40		2,550	1,550		4,100	5,375
0280	500 MBH		.40	60		9,325	2,325		11,650	14,200
0370	1,260 MBH		.21	114		17,100	4,450		21,550	26,200
0400	1,800 MBH		.14	171		20,300	6,675		26,975	33,300
2000	Electric, 12 KW, 4,800 gallon pool	Q-19	3	8		2,100	335		2,435	2,875
2020	15 KW, 7,200 gallon pool		2.80	8.571		2,150	355		2,505	2,925
2040	24 KW, 9,600 gallon pool		2.40	10		2,475	415		2,890	3,400
2100	57 KW, 24,000 gallon pool		1.20	20		3,650	835		4,485	5,375

23 52 39 – Fire-Tube Boilers

23 52 39.13 Scotch Marine Boilers		Crew	Daily Output	Labor-Hours	Unit	Material	Labor	Equipment	Total	Total Incl O&P
0010	**SCOTCH MARINE BOILERS**									
1000	Packaged fire tube, #2 oil, gross output									
1006	15 PSI steam									
1020	3348 MBH, 100 HP	Q-7	.21	152	Ea.	83,000	6,175		89,175	101,500
1120	To fire #6, add		.83	38.554		16,000	1,550		17,550	20,200
1140	To fire #6, and gas, add		.42	76.190		23,500	3,075		26,575	30,900

23 52 88 – Burners

23 52 88.10 Replacement Type Burners		Crew	Daily Output	Labor-Hours	Unit	Material	Labor	Equipment	Total	Total Incl O&P
0010	**REPLACEMENT TYPE BURNERS**									
0990	Residential, conversion, gas fired, LP or natural									
1000	Gun type, atmospheric input 50 to 225 MBH	Q-1	2.50	6.400	Ea.	915	252		1,167	1,425
1003	Replace burner		1.92	8.320		915	330		1,245	1,550
1020	100 to 400 MBH		2	8		1,525	315		1,840	2,200
1025	Burner, gas, 100 – 400 MBH G					1,525			1,525	1,675
1040	300 to 1000 MBH	Q-1	1.70	9.412		4,600	370		4,970	5,675
1043	Replace burner G	"	1.30	12.298		4,600	485		5,085	5,875
3000	Flame retention oil fired assembly, input									
3020	.50 to 2.25 GPH	Q-1	2.40	6.667	Ea.	266	263		529	730
3040	2.0 to 5.0 GPH		2	8		335	315		650	890
3060	3.0 to 7.0 GPH		1.80	8.889		575	350		925	1,200
3080	6.0 to 12.0 GPH		1.60	10		920	395		1,315	1,650
4100	Replace burner fireye G	1 Stpi	26.67	.300		234	13.50		247.50	280
4500	Replace burner blower bearing G		4	2		64	90		154	219
4510	Replace conveyor bearing G		3.60	2.222		72	100		172	244
4520	Replace fan bearings G		3.40	2.353		183	106		289	375
4525	Fan bearing G					183			183	202
4600	Gas safety, shut off valve, 3/4" threaded	1 Stpi	20	.400		182	18		200	230
4610	1" threaded		19	.421		176	18.95		194.95	224
4620	1-1/4" threaded		15	.533		198	24		222	258
4630	1-1/2" threaded		13	.615		214	27.50		241.50	281
4640	2" threaded		11	.727		240	32.50		272.50	320
4650	2-1/2" threaded	Q-1	15	1.067		274	42		316	370
4660	3" threaded		13	1.231		375	48.50		423.50	495
4670	4" flanged		3	5.333		2,700	210		2,910	3,300
4680	6" flanged	Q-2	3	8		6,000	305		6,305	7,100
4700	Replace burner ignition transformer G	1 Stpi	16	.500		75.50	22.50		98	120
4705	Burner ignition transformer G					75.50			75.50	83

23 52 Heating Boilers

23 52 88 – Burners

23 52 88.10 Replacement Type Burners			Crew	Daily Output	Labor-Hours	Unit	2016 Bare Costs Material	Labor	Equipment	Total	Total Incl O&P
4710	Replace burner ignition electrode	G	1 Stpi	20	.400	Ea.	14.10	18		32.10	45
4715	Burner ignition electrode	G					14.10			14.10	15.50
4720	Replace burner oil pump	G	1 Stpi	11.43	.700		97.50	31.50		129	159
4725	Burner oil pump	G					97.50			97.50	107
4730	Replace burner nozzle	G	1 Stpi	26.66	.300		6.40	13.50		19.90	29
4735	Burner nozzle	G					6.40			6.40	7
4750	Replace burner ignition transformer	G	1 Stpi	14	.571		75.50	25.50		101	126
4755	Burner ignition transformer	G					75.50			75.50	83
4760	Replace burner ignition electrode	G	1 Stpi	20	.400		14.10	18		32.10	45
4765	Burner ignition electrode	G					14.10			14.10	15.50
4770	Replace burner oil pump	G	1 Stpi	10	.800		115	36		151	186
4775	Burner oil pump	G					115			115	126
4780	Replace burner nozzle	G	1 Stpi	24	.333		9	15		24	34.50
4785	Burner nozzle	G					9			9	9.90
4800	Replace burner ignition transformer	G	1 Stpi	12	.667		75.50	30		105.50	133
4805	Burner ignition transformer	G					75.50			75.50	83
4810	Replace burner ignition electrode	G	1 Stpi	18	.444		14.10	20		34.10	48.50
4815	Burner ignition electrode	G					14.10			14.10	15.50
4820	Replace burner oil pump	G	1 Stpi	8	1		97.50	45		142.50	181
4825	Burner oil pump	G					97.50			97.50	107
4830	Replace burner nozzle	G	1 Stpi	20	.400		9	18		27	39.50
4835	Burner nozzle	G					9			9	9.90
4850	Replace burner oil pump	G	Q-5	6.15	2.602		168	105		273	360
4855	Burner oil pump	G					168			168	185
4860	Replace burner nozzle	G	1 Stpi	14	.571		9	25.50		34.50	52.50
4865	Burner nozzle	G					9			9	9.90
4880	Replace spreader	G	4 Stpi	.25	130		7,400	5,850		13,250	17,800
4885	Pneumatic coal spreader	G					7,400			7,400	8,125
4900	Replace pump seals/bearings	G	1 Stpi	5.33	1.500		30.50	67.50		98	145
4905	Pump seals/bearings	G					30.50			30.50	34

23 54 Furnaces

23 54 13 – Electric-Resistance Furnaces

23 54 13.10 Electric Furnaces

		Crew	Daily Output	Labor-Hours	Unit	Material	Labor	Equipment	Total	Total Incl O&P
0010	**ELECTRIC FURNACES**, Hot air, blowers, std. controls									
0011	not including gas, oil or flue piping									

23 54 16 – Fuel-Fired Furnaces

23 54 16.13 Gas-Fired Furnaces

		Crew	Daily Output	Labor-Hours	Unit	Material	Labor	Equipment	Total	Total Incl O&P
0010	**GAS-FIRED FURNACES**									
3000	Gas, AGA certified, upflow, direct drive models									
3040	60 MBH input	Q-9	3.80	4.211	Ea.	595	161		756	925
3100	100 MBH input		3.20	5		690	191		881	1,075
3130	150 MBH input		2.80	5.714		1,375	218		1,593	1,875
3140	200 MBH input		2.60	6.154		2,950	235		3,185	3,650
4000	For starter plenum, add		16	1		91	38		129	164

23 54 16.16 Oil-Fired Furnaces

		Crew	Daily Output	Labor-Hours	Unit	Material	Labor	Equipment	Total	Total Incl O&P
0010	**OIL-FIRED FURNACES**									
6000	Oil, UL listed, atomizing gun type burner									
6020	56 MBH output	Q-9	3.60	4.444	Ea.	1,700	169		1,869	2,150
6040	95 MBH output		3.40	4.706		2,575	179		2,754	3,125

23 54 Furnaces

23 54 16 – Fuel-Fired Furnaces

23 54 16.16 Oil-Fired Furnaces		Crew	Daily Output	Labor-Hours	Unit	2016 Bare Costs Material	Labor	Equipment	Total	Total Incl O&P
6080	151 MBH output	Q-9	3	5.333	Ea.	2,725	203		2,928	3,350
6100	200 MBH input		2.60	6.154		3,000	235		3,235	3,725

23 55 Fuel-Fired Heaters

23 55 13 – Fuel-Fired Duct Heaters

23 55 13.16 Gas-Fired Duct Heaters		Crew	Daily Output	Labor-Hours	Unit	Material	Labor	Equipment	Total	Total Incl O&P
0010	**GAS-FIRED DUCT HEATERS**, Includes burner, controls, stainless steel									
0020	heat exchanger. Gas fired, electric ignition									
0030	Indoor installation									
0080	100 MBH output	Q-5	5	3.200	Ea.	2,950	129		3,079	3,475
0100	120 MBH output		4	4		3,225	162		3,387	3,825
0130	200 MBH output		2.70	5.926		4,100	240		4,340	4,900
0140	240 MBH output		2.30	6.957		4,300	281		4,581	5,200
0160	280 MBH output		2	8		4,700	325		5,025	5,675
0180	320 MBH output		1.60	10		5,025	405		5,430	6,200
0300	For powered venter and adapter, add					535			535	590

23 55 33 – Fuel-Fired Unit Heaters

23 55 33.16 Gas-Fired Unit Heaters		Crew	Daily Output	Labor-Hours	Unit	Material	Labor	Equipment	Total	Total Incl O&P
0010	**GAS-FIRED UNIT HEATERS**, Cabinet, grilles, fan, ctrls., burner, no piping									
0022	thermostat, no piping. For flue see Section 23 51 23.10									
1000	Gas fired, floor mounted									
1100	60 MBH output	Q-5	10	1.600	Ea.	925	64.50		989.50	1,125
1140	100 MBH output		8	2		1,025	81		1,106	1,250
1180	180 MBH output		6	2.667		1,475	108		1,583	1,775
1500	Rooftop mounted, power vent, stainless steel exchanger									
1520	75 MBH input	Q-6	4	6	Ea.	5,775	234		6,009	6,725
1560	125 MBH input		3.30	7.273		6,300	283		6,583	7,400
1600	175 MBH input		2.60	9.231		6,950	360		7,310	8,250
1620	225 MBH input		2.30	10.435		7,925	405		8,330	9,375
1640	300 MBH input		1.90	12.632		9,125	490		9,615	10,900
1660	350 MBH input		1.40	17.143		9,750	670		10,420	11,800
1720	700 MBH input		.80	30		10,800	1,175		11,975	13,800
1760	1200 MBH input		.30	80		11,300	3,125		14,425	17,600
2000	Suspension mounted, propeller fan, 20 MBH output	Q-5	8.50	1.882		1,150	76		1,226	1,375
2100	130 MBH output		5	3.200		2,250	129		2,379	2,700
2240	320 MBH output		2	8		4,200	325		4,525	5,125
2500	For powered venter and adapter, add					440			440	485
5000	Wall furnace, 17.5 MBH output	Q-5	6	2.667		745	108		853	1,000
5020	24 MBH output		5	3.200		745	129		874	1,025
5040	35 MBH output		4	4		825	162		987	1,175

23 56 Solar Energy Heating Equipment

23 56 16 – Packaged Solar Heating Equipment

23 56 16.40 Solar Heating Systems		Crew	Daily Output	Labor-Hours	Unit	2016 Bare Costs Material	Labor	Equipment	Total	Total Incl O&P
0010	SOLAR HEATING SYSTEMS									
0020	System/Package prices, not including connecting									
0030	pipe, insulation, or special heating/plumbing fixtures									
0500	Hot water, standard package, low temperature									
0540	1 collector, circulator, fittings, 65 gal. tank G	Q-1	.50	32	Ea.	3,700	1,250		4,950	6,150
0580	2 collectors, circulator, fittings, 120 gal. tank G	"	.40	40	"	5,075	1,575		6,650	8,200
0701	Solar energy, medium temperature package									
0720	1 collector, circulator, fittings, 80 gal. tank G	Q-1	.50	32	Ea.	5,050	1,250		6,300	7,625
0740	2 collectors, circulator, fittings, 120 gal. tank G		.40	40		6,500	1,575		8,075	9,775
0780	3 collectors, circulator, fittings, 120 gal. tank G		.30	53.333		7,375	2,100		9,475	11,600
0980	For each additional 120 gal. tank, add G					1,750			1,750	1,950

23 62 Packaged Compressor and Condenser Units

23 62 13 – Packaged Air-Cooled Refrigerant Compressor and Condenser Units

23 62 13.10 Packaged Air-Cooled Refrig. Condensing Units		Crew	Daily Output	Labor-Hours	Unit	Material	Labor	Equipment	Total	Total Incl O&P
0010	PACKAGED AIR-COOLED REFRIGERANT CONDENSING UNITS									
0020	Condensing unit									
0030	Air cooled, compressor, standard controls									
0050	1.5 ton	Q-5	2.50	6.400	Ea.	1,275	259		1,534	1,825
0100	2 ton		2.10	7.619		1,300	310		1,610	1,925
0200	2.5 ton		1.70	9.412		1,425	380		1,805	2,200
0300	3 ton		1.30	12.308		1,475	500		1,975	2,450
0400	4 ton		.90	17.778		1,950	720		2,670	3,300
0500	5 ton		.60	26.667		2,325	1,075		3,400	4,350
0560	8.5 ton		.53	30.189		3,450	1,225		4,675	5,825
0600	10 ton		.50	32		3,700	1,300		5,000	6,200

23 63 Refrigerant Condensers

23 63 13 – Air-Cooled Refrigerant Condensers

23 63 13.10 Air-Cooled Refrig. Condensers		Crew	Daily Output	Labor-Hours	Unit	Material	Labor	Equipment	Total	Total Incl O&P
0010	AIR-COOLED REFRIG. CONDENSERS									
0080	Air cooled, belt drive, propeller fan									
0220	45 ton	Q-6	.70	34.286	Ea.	9,975	1,325		11,300	13,200
0240	50 ton		.69	34.985		10,800	1,375		12,175	14,100
0260	54 ton		.64	37.795		11,800	1,475		13,275	15,300

23 64 Packaged Water Chillers

23 64 13 – Absorption Water Chillers

23 64 13.13 Direct-Fired Absorption Water Chillers		Crew	Daily Output	Labor-Hours	Unit	Material	Labor	Equipment	Total	Total Incl O&P
0010	DIRECT-FIRED ABSORPTION WATER CHILLERS									
3000	Gas fired, air cooled									
3270	10 ton	Q-5	.40	40	Ea.	23,100	1,625		24,725	28,100
23 64 13.16 Indirect-Fired Absorption Water Chillers										
0010	INDIRECT-FIRED ABSORPTION WATER CHILLERS									
0020	Steam or hot water, water cooled									
0050	100 ton	Q-7	.13	240	Ea.	125,000	9,700		134,700	153,500
0100	148 ton	"	.12	258	"	224,500	10,400		234,900	264,000

23 64 Packaged Water Chillers

23 64 16 – Centrifugal Water Chillers

23 64 16.10 Centrifugal Type Water Chillers		Crew	Daily Output	Labor-Hours	Unit	2016 Bare Costs Material	Labor	Equipment	Total	Total Incl O&P
0010	**CENTRIFUGAL TYPE WATER CHILLERS**, With standard controls									
0274	Centrifugal, packaged unit, water cooled, not incl. tower									
0275	Chiller, centrifugal									
0278	200 ton	Q-7	.19	172	Ea.	122,000	6,950		128,950	145,500
0279	300 ton		.14	233		149,000	9,450		158,450	179,500
0280	400 ton		.11	283		150,500	11,500		162,000	185,000
0282	450 ton		.11	290		165,000	11,800		176,800	200,500
0290	500 ton		.11	296		182,500	12,000		194,500	221,000
0292	550 ton		.11	304		213,000	12,300		225,300	254,500
0300	600 ton		.10	310		227,500	12,600		240,100	270,500

23 64 19 – Reciprocating Water Chillers

23 64 19.10 Reciprocating Type Water Chillers		Crew	Daily Output	Labor-Hours	Unit	Material	Labor	Equipment	Total	Total Incl O&P
0010	**RECIPROCATING TYPE WATER CHILLERS**, With standard controls									
0494	Water chillers, integral air cooled condenser									
0980	Water cooled, multiple compressor, semi-hermetic, tower not incl.									
1000	15 ton cooling	Q-6	.36	65.934	Ea.	15,300	2,575		17,875	21,000
1040	30 ton cooling	Q-7	.36	89.888		18,300	3,625		21,925	26,100
1060	35 ton cooling		.31	101		20,200	4,125		24,325	29,000
1090	45 ton cooling		.29	111		22,900	4,500		27,400	32,600
1100	50 ton cooling		.28	113		29,700	4,600		34,300	40,300
1130	75 ton cooling		.23	139		45,300	5,625		50,925	59,000
1150	95 ton cooling		.19	164		51,000	6,675		57,675	67,500
1160	100 ton cooling		.18	179		51,500	7,275		58,775	68,500
4000	Packaged chiller, remote air cooled condensers not incl.									
4020	15 ton cooling	Q-7	.30	108	Ea.	15,500	4,375		19,875	24,300
4040	25 ton cooling		.25	125		17,400	5,100		22,500	27,600
4050	35 ton cooling		.24	133		18,400	5,425		23,825	29,100
4066	45 ton cooling		.21	149		20,600	6,050		26,650	32,600
4070	50 ton cooling		.21	153		29,600	6,225		35,825	42,800
4080	60 ton cooling		.20	164		32,000	6,650		38,650	46,100

23 64 23 – Scroll Water Chillers

23 64 23.10 Scroll Water Chillers		Crew	Daily Output	Labor-Hours	Unit	Material	Labor	Equipment	Total	Total Incl O&P
0010	**SCROLL WATER CHILLERS**, With standard controls									
0480	Packaged w/integral air cooled condenser									
0482	10 ton cooling	Q-7	.34	94.118	Ea.	15,200	3,800		19,000	23,000
0490	15 ton cooling		.37	86.486		18,300	3,500		21,800	26,000
0500	20 ton cooling		.34	94.118		21,700	3,800		25,500	30,100
0510	25 ton cooling		.34	94.118		25,500	3,800		29,300	34,400

23 64 26 – Rotary-Screw Water Chillers

23 64 26.10 Rotary-Screw Type Water Chillers		Crew	Daily Output	Labor-Hours	Unit	Material	Labor	Equipment	Total	Total Incl O&P
0010	**ROTARY-SCREW TYPE WATER CHILLERS**, With standard controls									
0110	Screw, liquid chiller, air cooled, insulated evaporator									
0120	130 ton	Q-7	.14	228	Ea.	93,500	9,250		102,750	118,000
0124	160 ton		.13	246		115,000	9,950		124,950	143,000
0128	180 ton		.13	250		129,500	10,100		139,600	159,000
0132	210 ton		.12	258		142,500	10,400		152,900	174,000
0136	270 ton		.12	266		163,000	10,800		173,800	197,500
0140	320 ton		.12	275		204,500	11,200		215,700	243,500
0200	Packaged unit, water cooled, not incl. tower									
0240	200 ton	Q-7	.13	251	Ea.	82,000	10,200		92,200	107,000

23 65 Cooling Towers

23 65 13 – Forced-Draft Cooling Towers

23 65 13.10 Forced-Draft Type Cooling Towers		Crew	Daily Output	Labor-Hours	Unit	2016 Bare Costs Material	Labor	Equipment	Total	Total Incl O&P
0010	**FORCED-DRAFT TYPE COOLING TOWERS**, Packaged units									
0070	Galvanized steel									
0080	Induced draft, crossflow									
0100	Vertical, belt drive, 61 tons	Q-6	90	.267	TonAC	215	10.40		225.40	254
0150	100 ton		100	.240		208	9.35		217.35	243
0200	115 ton		109	.220		180	8.60		188.60	213
1000	For higher capacities, use multiples									
5000	Fiberglass tower on galvanized steel support structure									
5010	Draw thru									
5100	100 ton	Q-6	1.40	17.143	Ea.	13,900	670		14,570	16,400
5120	120 ton		1.20	20		16,200	780		16,980	19,100
5140	140 ton		1	24		17,500	935		18,435	20,800
5160	160 ton		.80	30		19,400	1,175		20,575	23,300
5180	180 ton		.65	36.923		22,100	1,450		23,550	26,700
5200	200 ton		.48	50		25,000	1,950		26,950	30,700
5300	For stainless steel support structure, add					30%				
5360	For higher capacities, use multiples of each size									
6000	Stainless steel									
6010	Induced draft, crossflow, horizontal, belt drive									
6100	57 ton	Q-6	1.50	16	Ea.	27,100	625		27,725	30,800
6120	91 ton		.99	24.242		32,700	945		33,645	37,600
6140	111 ton		.43	55.814		43,900	2,175		46,075	52,000
6160	126 ton		.22	109		45,600	4,250		49,850	57,000
6170	Induced draft, crossflow, vertical, gear drive									
6180	1016 ton	Q-6	.15	160	Ea.	159,000	6,225		165,225	185,000

23 74 Packaged Outdoor HVAC Equipment

23 74 33 – Dedicated Outdoor-Air Units

23 74 33.10 Rooftop Air Conditioners		Crew	Daily Output	Labor-Hours	Unit	Material	Labor	Equipment	Total	Total Incl O&P
0010	**ROOFTOP AIR CONDITIONERS**, Standard controls, curb, economizer									
1000	Single zone, electric cool, gas heat									
1140	5 ton cooling, 112 MBH heating	Q-5	.56	28.521	Ea.	4,425	1,150		5,575	6,750
1150	7.5 ton cooling, 170 MBH heating		.50	32.258		5,875	1,300		7,175	8,625
1156	8.5 ton cooling, 170 MBH heating		.46	34.783		7,100	1,400		8,500	10,200
1160	10 ton cooling, 200 MBH heating	Q-6	.67	35.982		9,100	1,400		10,500	12,300
1180	15 ton cooling, 270 MBH heating	"	.57	42.032		13,100	1,625		14,725	17,100
1220	30 ton cooling, 540 MBH heating	Q-7	.47	68.376		35,600	2,775		38,375	43,800
2000	Multizone, electric cool, gas heat, economizer									
2102	15 ton cooling, 360 MBH heating SEER 14 G	Q-7	.61	52.545	Ea.	100,500	2,125		102,625	114,000
2122	20 ton cooling, 360 MBH heating SEER 14 G		.53	60.038		107,000	2,425		109,425	121,500
2142	25 ton cooling, 450 MBH heating SEER 14 G		.45	71.910		128,000	2,900		130,900	145,500
2162	30 ton cooling, 450 MBH heating SEER 14 G		.41	79.012		146,000	3,200		149,200	166,000
2182	35 ton cooling, 540 MBH heating SEER 14 G		.37	85.562		164,000	3,450		167,450	185,500
2202	40 ton cooling, 540 MBH heating SEER 14 G		.28	113		187,000	4,600		191,600	213,500
2222	70 ton cooling, 1500 MBH heating SEER 14 G		.16	198		252,000	8,050		260,050	290,500
2242	80 ton cooling, 1500 MBH heating SEER 14 G		.14	228		288,000	9,250		297,250	332,000
2262	90 ton cooling, 1500 MBH heating SEER 14 G		.13	256		302,000	10,400		312,400	349,000
2282	105 ton cooling, 1500 MBH heating SEER 14 G		.11	290		332,500	11,800		344,300	385,500

23 81 13 – Packaged Terminal Air-Conditioners

23 81 13.10 Packaged Cabinet Type Air-Conditioners		Crew	Daily Output	Labor-Hours	Unit	Material	2016 Bare Costs Labor	Equipment	Total	Total Incl O&P
0010	**PACKAGED CABINET TYPE AIR-CONDITIONERS**, Cabinet, wall sleeve,									
0100	louver, electric heat, thermostat, manual changeover, 208 V									
0200	6,000 BTUH cooling, 8800 BTU heat	Q-5	6	2.667	Ea.	825	108		933	1,100
0220	9,000 BTUH cooling, 13,900 BTU heat		5	3.200		1,225	129		1,354	1,575
0240	12,000 BTUH cooling, 13,900 BTU heat		4	4		1,350	162		1,512	1,775
0260	15,000 BTUH cooling, 13,900 BTU heat		3	5.333		1,475	216		1,691	1,975
0320	30,000 BTUH cooling, 10 KW heat		1.40	11.429		2,550	460		3,010	3,550
0340	36,000 BTUH cooling, 10 KW heat		1.25	12.800		2,600	520		3,120	3,700
0360	42,000 BTUH cooling, 10 KW heat		1	16		2,975	645		3,620	4,350
0380	48,000 BTUH cooling, 10 KW heat		.90	17.778		3,150	720		3,870	4,650

23 81 19 – Self-Contained Air-Conditioners

23 81 19.10 Window Unit Air Conditioners		Crew	Daily Output	Labor-Hours	Unit	Material	Labor	Equipment	Total	Total Incl O&P
0010	**WINDOW UNIT AIR CONDITIONERS**									
4000	Portable/window, 15 amp, 125 V grounded receptacle required									
4060	5000 BTUH	1 Carp	8	1	Ea.	330	38		368	425
4340	6000 BTUH		8	1		380	38		418	485
4480	8000 BTUH		6	1.333		475	50.50		525.50	610
4500	10,000 BTUH		6	1.333		590	50.50		640.50	735
4520	12,000 BTUH	L-2	8	2		740	66		806	925

23 81 19.20 Self-Contained Single Package		Crew	Daily Output	Labor-Hours	Unit	Material	Labor	Equipment	Total	Total Incl O&P
0010	**SELF-CONTAINED SINGLE PACKAGE**									
0100	Air cooled, for free blow or duct, not incl. remote condenser									
0110	Constant volume									
0220	5 ton cooling	Q-6	1.20	20	Ea.	4,425	780		5,205	6,125
0221	Self-contained, air cooled, 5 ton cooling, labor only		1.20	20			780		780	1,275
0230	7.5 ton cooling		.90	26.667		5,700	1,050		6,750	8,000
0240	10 ton cooling	Q-7	1	32		7,925	1,300		9,225	10,800
0250	15 ton cooling		.95	33.684		10,300	1,375		11,675	13,600
0260	20 ton cooling		.90	35.556		12,800	1,450		14,250	16,500
0270	25 ton cooling		.85	37.647		19,000	1,525		20,525	23,400
1000	Water cooled for free blow or duct, not including tower									
1010	Constant volume									
1160	20 ton cooling	Q-7	.80	40	Ea.	28,600	1,625		30,225	34,100
1170	25 ton cooling		.75	42.667		33,700	1,725		35,425	40,000
1180	30 ton cooling		.70	45.714		37,900	1,850		39,750	44,800
1200	40 ton cooling		.40	80		48,000	3,225		51,225	58,500
1300	For hot water or steam heat coils, add					12%	10%			

23 81 23 – Computer-Room Air-Conditioners

23 81 23.10 Computer Room Units		Crew	Daily Output	Labor-Hours	Unit	Material	Labor	Equipment	Total	Total Incl O&P
0010	**COMPUTER ROOM UNITS**									
1000	Air cooled, includes remote condenser but not									
1020	interconnecting tubing or refrigerant									
1080	3 ton	Q-5	.50	32	Ea.	19,500	1,300		20,800	23,600
1120	5 ton		.45	35.556		20,900	1,450		22,350	25,300
1160	6 ton		.30	53.333		38,500	2,150		40,650	46,000
1200	8 ton		.27	59.259		38,900	2,400		41,300	46,800
1240	10 ton		.25	64		40,700	2,600		43,300	49,100
1260	12 ton		.24	66.667		42,100	2,700		44,800	51,000
1280	15 ton		.22	72.727		44,800	2,950		47,750	54,000
1290	18 ton		.20	80		51,500	3,225		54,725	62,000
1300	20 ton	Q-6	.26	92.308		53,500	3,600		57,100	65,000

23 81 Decentralized Unitary HVAC Equipment

23 81 23 – Computer-Room Air-Conditioners

23 81 23.10 Computer Room Units		Crew	Daily Output	Labor-Hours	Unit	Material	Labor	Equipment	Total	Total Incl O&P
						2016 Bare Costs				
1320	22 ton	Q-6	.24	100	Ea.	54,500	3,900		58,400	66,000
1360	30 ton		.21	114		67,000	4,450		71,450	81,000
2200	Chilled water, for connection to									
2220	existing chiller system of adequate capacity									
2260	5 ton	Q-5	.74	21.622	Ea.	14,900	875		15,775	17,900

23 81 29 – Variable Refrigerant Flow HVAC Systems

23 81 29.10 Heat Pump, Gas Driven

Line	Description	Crew	Daily Output	Labor-Hours	Unit	Material	Labor	Equipment	Total	Total Incl O&P
0010	**HEAT PUMP, GAS DRIVEN**, Variable refrigerant volume (VRV) type									
0020	Not including interconnecting tubing or multi-zone controls									
1000	For indoor fan VRV type AHU see 23 82 19.40									
1010	Multi-zone split									
1020	Outdoor unit									
1100	8 tons cooling, for up to 17 zones	Q-5	1.30	12.308	Ea.	29,100	500		29,600	32,800
1110	Isolation rails		2.60	6.154	Pair	1,025	249		1,274	1,525
1160	15 tons cooling, for up to 33 zones		1	16	Ea.	38,600	645		39,245	43,500
1170	Isolation rails		2.60	6.154	Pair	1,200	249		1,449	1,725
2000	Packaged unit									
2020	Outdoor unit									
2200	11 tons cooling	Q-5	1.30	12.308	Ea.	37,200	500		37,700	41,700
2210	Roof curb adapter	"	2.60	6.154		645	249		894	1,125
2220	Thermostat	1 Stpi	1.30	6.154		385	277		662	880

23 81 43 – Air-Source Unitary Heat Pumps

23 81 43.10 Air-Source Heat Pumps

Line	Description	Crew	Daily Output	Labor-Hours	Unit	Material	Labor	Equipment	Total	Total Incl O&P
0010	**AIR-SOURCE HEAT PUMPS**, Not including interconnecting tubing									
1000	Air to air, split system, not including curbs, pads, fan coil and ductwork									
1012	Outside condensing unit only, for fan coil see Section 23 82 19.10									
1060	5 ton cooling, 27 MBH heat @ 0°F	Q-5	.50	32	Ea.	2,575	1,300		3,875	4,975
1080	7.5 ton cooling, 33 MBH heat @ 0°F	"	.45	35.556		4,625	1,450		6,075	7,450
1100	10 ton cooling, 50 MBH heat @ 0°F	Q-6	.64	37.500		6,100	1,450		7,550	9,100
1120	15 ton cooling, 64 MBH heat @ 0°F		.50	48		8,450	1,875		10,325	12,400
1130	20 ton cooling, 85 MBH heat @ 0°F		.35	68.571		17,000	2,675		19,675	23,100
1140	25 ton cooling, 119 MBH heat @ 0°F		.25	96		20,400	3,750		24,150	28,700
6000	Air to water, single package, excluding storage tank and ductwork									
6010	Includes circulating water pump, air duct connections, digital temperature									
6020	controller with remote tank temp. probe and sensor for storage tank.									
6040	Water heating - air cooling capacity									

23 81 46 – Water-Source Unitary Heat Pumps

23 81 46.10 Water Source Heat Pumps

Line	Description	Crew	Daily Output	Labor-Hours	Unit	Material	Labor	Equipment	Total	Total Incl O&P
0010	**WATER SOURCE HEAT PUMPS**, Not incl. connecting tubing or water source									
2000	Water source to air, single package									
2220	5 ton cooling, 29 MBH heat @ 75°F	Q-5	.90	17.778	Ea.	3,275	720		3,995	4,775
2240	7.5 ton cooling, 35 MBH heat @ 75°F		.60	26.667		6,825	1,075		7,900	9,275
2260	10 ton cooling, 50 MBH heat @ 75°F		.53	30.189		7,700	1,225		8,925	10,500
2280	15 ton cooling, 64 MBH heat @ 75°F	Q-6	.47	51.064		14,000	2,000		16,000	18,700
2300	20 ton cooling, 100 MBH heat @ 75°F		.41	58.537		15,300	2,275		17,575	20,600
2310	25 ton cooling, 100 MBH heat @ 75°F		.32	75		20,800	2,925		23,725	27,700
2320	30 ton cooling, 128 MBH heat @ 75°F		.24	102		22,800	3,975		26,775	31,700
2340	40 ton cooling, 200 MBH heat @ 75°F		.21	117		32,200	4,550		36,750	42,900
2360	50 ton cooling, 200 MBH heat @ 75°F		.15	160		36,400	6,225		42,625	50,500
3960	For supplementary heat coil, add					10%				

23 82 19.10 Fan Coil Air Conditioning

	23 82 19.10 Fan Coil Air Conditioning	Crew	Daily Output	Labor-Hours	Unit	Material	Labor (2016 Bare Costs)	Equipment (2016 Bare Costs)	Total (2016 Bare Costs)	Total Incl O&P
0010	**FAN COIL AIR CONDITIONING**									
0030	Fan coil AC, cabinet mounted, filters and controls									
0100	Chilled water, 1/2 ton cooling	Q-5	8	2	Ea.	460	81		541	640
0120	1 ton cooling		6	2.667		685	108		793	935
0180	3 ton cooling		4	4		1,625	162		1,787	2,050
0262	For hot water coil, add					40%	10%			
0940	Direct expansion, for use w/air cooled condensing unit, 1.5 ton cooling	Q-5	5	3.200	Ea.	515	129		644	785
1000	5 ton cooling	"	3	5.333		1,025	216		1,241	1,475
1500	For hot water coil, add					40%	10%			
1512	For condensing unit add see Section 23 62									

23 82 19.20 Heating and Ventilating Units

	23 82 19.20 Heating and Ventilating Units	Crew	Daily Output	Labor-Hours	Unit	Material	Labor	Equipment	Total	Total Incl O&P
0010	**HEATING AND VENTILATING UNITS**, Classroom units									
0020	Includes filter, heating/cooling coils, standard controls									
0080	750 CFM, 2 tons cooling	Q-6	2	12	Ea.	4,225	465		4,690	5,425
0100	1000 CFM, 2-1/2 tons cooling		1.60	15		4,975	585		5,560	6,450
0120	1250 CFM, 3 tons cooling		1.40	17.143		5,150	670		5,820	6,750
0140	1500 CFM, 4 tons cooling		.80	30		5,500	1,175		6,675	7,975
0500	For electric heat, add					35%				
1000	For no cooling, deduct					25%	10%			

23 82 19.40 Fan Coil Air Conditioning

	23 82 19.40 Fan Coil Air Conditioning	Crew	Daily Output	Labor-Hours	Unit	Material	Labor	Equipment	Total	Total Incl O&P
0010	**FAN COIL AIR CONDITIONING**, Variable refrigerant volume (VRV) type									
0020	Not including interconnecting tubing or multi-zone controls									
0030	For VRV condensing unit see section 23 81 29.10									
0050	Indoor type, ducted									
0100	Vertical concealed									
0130	1 ton cooling G	Q-5	2.97	5.387	Ea.	2,475	218		2,693	3,075
0140	1.5 ton cooling G		2.77	5.776		2,525	234		2,759	3,150
0150	2 ton cooling G		2.70	5.926		2,725	240		2,965	3,400
0160	2.5 ton cooling G		2.60	6.154		2,875	249		3,124	3,550
0170	3 ton cooling G		2.50	6.400		2,925	259		3,184	3,625
0180	3.5 ton cooling G		2.30	6.957		3,075	281		3,356	3,850
0190	4 ton cooling G		2.19	7.306		3,075	296		3,371	3,875
0200	4.5 ton cooling G		2.08	7.692		3,400	310		3,710	4,250
0230	Outside air connection possible									
1100	Ceiling concealed									
1130	.6 ton cooling G	Q-5	2.60	6.154	Ea.	1,625	249		1,874	2,175
1140	.75 ton cooling G		2.60	6.154		1,675	249		1,924	2,250
1150	1 ton cooling G		2.60	6.154		1,775	249		2,024	2,350
1152	Screening door G	1 Stpi	5.20	1.538		98	69		167	222
1160	1.5 ton cooling G	Q-5	2.60	6.154		1,825	249		2,074	2,425
1162	Screening door G	1 Stpi	5.20	1.538		113	69		182	239
1170	2 ton cooling G	Q-5	2.60	6.154		2,150	249		2,399	2,750
1172	Screening door G	1 Stpi	5.20	1.538		98	69		167	222
1180	2.5 ton cooling G	Q-5	2.60	6.154		2,500	249		2,749	3,150
1182	Screening door G	1 Stpi	5.20	1.538		170	69		239	300
1190	3 ton cooling G	Q-5	2.60	6.154		2,650	249		2,899	3,325
1192	Screening door G	1 Stpi	5.20	1.538		170	69		239	300
1200	4 ton cooling G	Q-5	2.30	6.957		2,725	281		3,006	3,450
1202	Screening door G	1 Stpi	5.20	1.538		170	69		239	300
1220	6 ton cooling G	Q-5	2.08	7.692		4,625	310		4,935	5,600
1230	8 ton cooling G	"	1.89	8.466		5,200	340		5,540	6,300
1260	Outside air connection possible									

23 82 Convection Heating and Cooling Units

23 82 19 – Fan Coil Units

23 82 19.40 Fan Coil Air Conditioning		Crew	Daily Output	Labor-Hours	Unit	Material	Labor	Equipment	Total	Total Incl O&P
						2016 Bare Costs				
4050	Indoor type, duct-free									
4100	Ceiling mounted cassette									
4130	.75 ton cooling G	Q-5	3.46	4.624	Ea.	1,875	187		2,062	2,375
4140	1 ton cooling G		2.97	5.387		2,000	218		2,218	2,550
4150	1.5 ton cooling G		2.77	5.776		2,050	234		2,284	2,625
4160	2 ton cooling G		2.60	6.154		2,200	249		2,449	2,825
4170	2.5 ton cooling G		2.50	6.400		2,250	259		2,509	2,900
4180	3 ton cooling G		2.30	6.957		2,425	281		2,706	3,125
4190	4 ton cooling G		2.08	7.692		2,700	310		3,010	3,500
4198	For ceiling cassette decoration panel, add G	1 Stpi	5.20	1.538		294	69		363	440
4230	Outside air connection possible									
4500	Wall mounted									
4530	.6 ton cooling G	Q-5	5.20	3.077	Ea.	995	124		1,119	1,300
4540	.75 ton cooling G		5.20	3.077		1,050	124		1,174	1,350
4550	1 ton cooling G		4.16	3.846		1,200	156		1,356	1,550
4560	1.5 ton cooling G		3.77	4.244		1,300	172		1,472	1,725
4570	2 ton cooling G		3.46	4.624		1,400	187		1,587	1,850
4590	Condensate pump for the above air handlers G	1 Stpi	6.46	1.238		211	55.50		266.50	325
4700	Floor standing unit									
4730	1 ton cooling G	Q-5	2.60	6.154	Ea.	1,675	249		1,924	2,250
4740	1.5 ton cooling G		2.60	6.154		1,900	249		2,149	2,500
4750	2 ton cooling G		2.60	6.154		2,050	249		2,299	2,650
4800	Floor standing unit, concealed									
4830	1 ton cooling G	Q-5	2.60	6.154	Ea.	1,600	249		1,849	2,175
4840	1.5 ton cooling G		2.60	6.154		1,825	249		2,074	2,400
4850	2 ton cooling G		2.60	6.154		1,950	249		2,199	2,550
4880	Outside air connection possible									
8000	Accessories									
8100	Branch divergence pipe fitting									
8110	Capacity under 76 MBH	1 Stpi	2.50	3.200	Ea.	151	144		295	405
8120	76 MBH to 112 MBH		2.50	3.200		178	144		322	435
8130	112 MBH to 234 MBH		2.50	3.200		505	144		649	790
8200	Header pipe fitting									
8210	Max 4 branches									
8220	Capacity under 76 MBH	1 Stpi	1.70	4.706	Ea.	221	212		433	595
8260	Max 8 branches									
8270	76 MBH to 112 MBH	1 Stpi	1.70	4.706	Ea.	450	212		662	845
8280	112 MBH to 234 MBH	"	1.30	6.154	"	720	277		997	1,250

23 82 27 – Infrared Units

23 82 27.10 Infrared Type Heating Units		Crew	Daily Output	Labor-Hours	Unit	Material	Labor	Equipment	Total	Total Incl O&P
0010	**INFRARED TYPE HEATING UNITS**									
0020	Gas fired, unvented, electric ignition, 100% shutoff.									
0030	Piping and wiring not included									
0120	45 MBH	Q-5	5	3.200	Ea.	1,100	129		1,229	1,425
0160	60 MBH		4	4		1,100	162		1,262	1,475
0180	75 MBH		3	5.333		1,100	216		1,316	1,550
0220	105 MBH		2	8		1,100	325		1,425	1,725
1000	Gas fired, vented, electric ignition, tubular									
1020	Piping and wiring not included, 20' to 80' lengths									
1030	Single stage, input, 60 MBH	Q-6	4.50	5.333	Ea.	1,450	208		1,658	1,950
1040	80 MBH		3.90	6.154		1,450	240		1,690	2,000
1050	100 MBH		3.40	7.059		1,450	275		1,725	2,050

23 82 Convection Heating and Cooling Units

23 82 27 – Infrared Units

23 82 27.10 Infrared Type Heating Units		Crew	Daily Output	Labor-Hours	Unit	2016 Bare Costs Material	Labor	Equipment	Total	Total Incl O&P
1060	125 MBH	Q-6	2.90	8.276	Ea.	1,450	320		1,770	2,125
1070	150 MBH		2.70	8.889		1,450	345		1,795	2,175
1080	170 MBH		2.50	9.600		1,450	375		1,825	2,225
1090	200 MBH		2.20	10.909		1,650	425		2,075	2,525
1100	Note: Final pricing may vary due to									
1110	tube length and configuration package selected									
1130	Two stage, input, 60 MBH high, 45 MBH low	Q-6	4.50	5.333	Ea.	1,775	208		1,983	2,300
1140	80 MBH high, 60 MBH low		3.90	6.154		1,775	240		2,015	2,350
1150	100 MBH high, 65 MBH low		3.40	7.059		1,775	275		2,050	2,400
1160	125 MBH high, 95 MBH low		2.90	8.276		1,775	320		2,095	2,475
1170	150 MBH high, 100 MBH low		2.70	8.889		1,775	345		2,120	2,525
1180	170 MBH high, 125 MBH low		2.50	9.600		2,000	375		2,375	2,825
1190	200 MBH high, 150 MBH low		2.20	10.909		2,000	425		2,425	2,900
1220	Note: Final pricing may vary due to									
1230	tube length and configuration package selected									

23 82 29 – Radiators

23 82 29.10 Hydronic Heating		Crew	Daily Output	Labor-Hours	Unit	Material	Labor	Equipment	Total	Total Incl O&P
0010	**HYDRONIC HEATING**, Terminal units, not incl. main supply pipe									
1000	Radiation									
1100	Panel, baseboard, C.I., including supports, no covers	Q-5	46	.348	L.F.	38	14.05		52.05	64.50

23 82 36 – Finned-Tube Radiation Heaters

23 82 36.10 Finned Tube Radiation		Crew	Daily Output	Labor-Hours	Unit	Material	Labor	Equipment	Total	Total Incl O&P
0010	**FINNED TUBE RADIATION**, Terminal units, not incl. main supply pipe									
1150	Fin tube, wall hung, 14" slope top cover, with damper									
1200	1-1/4" copper tube, 4-1/4" alum. fin	Q-5	38	.421	L.F.	43	17.05		60.05	75.50
1255	2" steel tube, 4-1/4" steel fin		32	.500		43	20		63	80.50
1310	Baseboard, pkgd, 1/2" copper tube, alum. fin, 7" high		60	.267		11.20	10.80		22	30
1320	3/4" copper tube, alum. fin, 7" high		58	.276		7.35	11.15		18.50	26.50
1340	1" copper tube, alum. fin, 8-7/8" high		56	.286		18.85	11.55		30.40	39.50
1360	1-1/4" copper tube, alum. fin, 8-7/8" high		54	.296		28	12		40	50.50
1500	Note: fin tube may also require corners, caps, etc.									

23 82 39 – Unit Heaters

23 82 39.16 Propeller Unit Heaters		Crew	Daily Output	Labor-Hours	Unit	Material	Labor	Equipment	Total	Total Incl O&P
0010	**PROPELLER UNIT HEATERS**									
3950	Unit heaters, propeller, 115 V 2 psi steam, 60°F entering air									
4000	Horizontal, 12 MBH	Q-5	12	1.333	Ea.	385	54		439	515
4060	43.9 MBH		8	2		585	81		666	775
4240	286.9 MBH		2	8		1,775	325		2,100	2,475
4260	364 MBH		1.80	8.889		2,200	360		2,560	3,025
4270	404 MBH		1.60	10		2,300	405		2,705	3,200
4300	Vertical diffuser same price									
4310	Vertical flow, 40 MBH	Q-5	11	1.455	Ea.	575	59		634	730
4326	131.0 MBH	"	4	4		900	162		1,062	1,250
4354	420 MBH,(460 V)	Q-6	1.80	13.333		2,300	520		2,820	3,375
4358	500 MBH, (460 V)		1.71	14.035		3,050	545		3,595	4,250
4362	570 MBH, (460 V)		1.40	17.143		4,175	670		4,845	5,700
4366	620 MBH, (460 V)		1.30	18.462		4,175	720		4,895	5,775
4370	960 MBH, (460 V)		1.10	21.818		7,950	850		8,800	10,100

23 83 Radiant Heating Units

23 83 33 – Electric Radiant Heaters

23 83 33.10 Electric Heating		Crew	Daily Output	Labor-Hours	Unit	Material	Labor	Equipment	Total	Total Incl O&P
						2016 Bare Costs				
0010	**ELECTRIC HEATING**, not incl. conduit or feed wiring									
1100	Rule of thumb: Baseboard units, including control	1 Elec	4.40	1.818	kW	105	80		185	247
1300	Baseboard heaters, 2' long, 350 watt		8	1	Ea.	26	44		70	101
1600	4' long, 1000 watt		6.70	1.194		35.50	52.50		88	125
2000	6' long, 1500 watt		5	1.600		51	70.50		121.50	171
2800	10' long, 1875 watt		3.30	2.424		180	107		287	375
5000	Radiant heating ceiling panels, 2' x 4', 500 watt		16	.500		410	22		432	490
5050	750 watt		16	.500		455	22		477	535
5300	Infrared quartz heaters, 120 volts, 1000 watts		6.70	1.194		345	52.50		397.50	465
5350	1500 watt		5	1.600		345	70.50		415.50	495
5400	240 volts, 1500 watt		5	1.600		345	70.50		415.50	495
5450	2000 watt		4	2		345	88		433	525

23 84 Humidity Control Equipment

23 84 13 – Humidifiers

23 84 13.10 Humidifier Units		Crew	Daily Output	Labor-Hours	Unit	Material	Labor	Equipment	Total	Total Incl O&P
0010	**HUMIDIFIER UNITS**									
0520	Steam, room or duct, filter, regulators, auto. controls, 220 V									
0580	33 lb. per hour	Q-5	4	4	Ea.	3,050	162		3,212	3,650
0620	100 lb. per hour	"	3	5.333	"	4,475	216		4,691	5,275

Estimating Tips

26 05 00 Common Work Results for Electrical

- Conduit should be taken off in three main categories—power distribution, branch power, and branch lighting—so the estimator can concentrate on systems and components, therefore making it easier to ensure all items have been accounted for.
- For cost modifications for elevated conduit installation, add the percentages to labor according to the height of installation, and only to the quantities exceeding the different height levels, not to the total conduit quantities.
- Remember that aluminum wiring of equal ampacity is larger in diameter than copper and may require larger conduit.
- If more than three wires at a time are being pulled, deduct percentages from the labor hours of that grouping of wires.
- When taking off grounding systems, identify separately the type and size of wire, and list each unique type of ground connection.
- The estimator should take the weights of materials into consideration when completing a takeoff. Topics to consider include: How will the materials be supported? What methods of support are available? How high will the support structure have to reach? Will the final support structure be able to withstand the total burden? Is the support material included or separate from the fixture, equipment, and material specified?
- Do not overlook the costs for equipment used in the installation. If scaffolding or highlifts are available in the field, contractors may use them in lieu of the proposed ladders and rolling staging.

26 20 00 Low-Voltage Electrical Transmission

- Supports and concrete pads may be shown on drawings for the larger equipment, or the support system may be only a piece of plywood for the back of a panelboard. In either case, it must be included in the costs.

26 40 00 Electrical and Cathodic Protection

- When taking off cathodic protections systems, identify the type and size of cable, and list each unique type of anode connection.

26 50 00 Lighting

- Fixtures should be taken off room by room, using the fixture schedule, specifications, and the ceiling plan. For large concentrations of lighting fixtures in the same area, deduct the percentages from labor hours.

Reference Numbers

Reference numbers are shown at the beginning of some major classifications. These numbers refer to related items in the Reference Section. The reference information may be an estimating procedure, an alternate pricing method, or technical information.

Note: Not all subdivisions listed here necessarily appear. ■

Note: Trade Service, in part, has been used as a reference source for some of the material prices used in Division 26.

26 01 Operation and Maintenance of Electrical Systems

26 01 02 – Labor Adjustment

26 01 02.20 Labor Adjustment Factors		Crew	Daily Output	Labor-Hours	Unit	Material	2016 Bare Costs Labor	Equipment	Total	Total Incl O&P
0010	**LABOR ADJUSTMENT FACTORS** (For Div. 26, 27 and 28)									
0100	Subtract from labor for Economy of Scale for Wire									
0110	4-5 wires						25%			
0120	6-10 wires						30%			
0130	11-15 wires						35%			
0140	over 15 wires						40%			

26 05 Common Work Results for Electrical

26 05 05 – Selective Demolition for Electrical

26 05 05.10 Electrical Demolition		Crew	Daily Output	Labor-Hours	Unit	Material	2016 Bare Costs Labor	Equipment	Total	Total Incl O&P
0010	**ELECTRICAL DEMOLITION**									
0020	Conduit to 15′ high, including fittings & hangers									
0100	Rigid galvanized steel, 1/2″ to 1″ diameter	1 Elec	242	.033	L.F.		1.46		1.46	2.38
0120	1-1/4″ to 2″	"	200	.040	"		1.76		1.76	2.88
0270	Armored cable (BX) avg. 50′ runs									
0280	#14, 2 wire	1 Elec	690	.012	L.F.		.51		.51	.84
0290	#14, 3 wire		571	.014			.62		.62	1.01
0300	#12, 2 wire		605	.013			.58		.58	.95
0310	#12, 3 wire		514	.016			.69		.69	1.12
0320	#10, 2 wire		514	.016			.69		.69	1.12
0330	#10, 3 wire		425	.019			.83		.83	1.36
0340	#8, 3 wire		342	.023			1.03		1.03	1.69
0350	Non metallic sheathed cable (Romex)									
0360	#14, 2 wire	1 Elec	720	.011	L.F.		.49		.49	.80
0370	#14, 3 wire		657	.012			.54		.54	.88
0380	#12, 2 wire		629	.013			.56		.56	.92
0390	#10, 3 wire		450	.018			.78		.78	1.28
0400	Wiremold raceway, including fittings & hangers									
0420	No. 3000	1 Elec	250	.032	L.F.		1.41		1.41	2.31
0440	No. 4000		217	.037			1.62		1.62	2.66
0460	No. 6000		166	.048			2.12		2.12	3.47
0462	Plugmold with receptacle		114	.070			3.09		3.09	5.05
0465	Telephone/power pole		12	.667	Ea.		29.50		29.50	48
0470	Non-metallic, straight section		480	.017	L.F.		.73		.73	1.20
0500	Channels, steel, including fittings & hangers									
0520	3/4″ x 1-1/2″	1 Elec	308	.026	L.F.		1.14		1.14	1.87
0540	1-1/2″ x 1-1/2″		269	.030			1.31		1.31	2.14
0560	1-1/2″ x 1-7/8″		229	.035			1.54		1.54	2.52
1210	Panel boards, incl. removal of all breakers,									
1220	conduit terminations & wire connections									
1720	Junction boxes, 4″ sq. & oct.	1 Elec	80	.100	Ea.		4.41		4.41	7.20
1740	Handy box		107	.075			3.29		3.29	5.40
1760	Switch box		107	.075			3.29		3.29	5.40
1780	Receptacle & switch plates		257	.031			1.37		1.37	2.24
1800	Wire, THW-THWN-THHN, removed from									
1810	in place conduit, to 15′ high									
1830	#14	1 Elec	65	.123	C.L.F.		5.40		5.40	8.85
1840	#12		55	.145			6.40		6.40	10.50
1850	#10		45.50	.176			7.75		7.75	12.65
2000	Interior fluorescent fixtures, incl. supports									
2010	& whips, to 15′ high									

26 05 Common Work Results for Electrical

26 05 05 – Selective Demolition for Electrical

26 05 05.10 Electrical Demolition		Crew	Daily Output	Labor-Hours	Unit	Material	2016 Bare Costs Labor	Equipment	Total	Total Incl O&P
2100	Recessed drop-in 2′ x 2′, 2 lamp	2 Elec	35	.457	Ea.		20		20	33
2140	2′ x 4′, 4 lamp	"	30	.533	"		23.50		23.50	38.50
2180	Surface mount, acrylic lens & hinged frame									
2220	2′ x 2′, 2 lamp	2 Elec	44	.364	Ea.		16		16	26
2260	2′ x 4′, 4 lamp	"	33	.485	"		21.50		21.50	35
2300	Strip fixtures, surface mount									
2320	4′ long, 1 lamp	2 Elec	53	.302	Ea.		13.30		13.30	22
2380	8′ long, 2 lamp	"	40	.400	"		17.60		17.60	29
2600	Exterior fixtures, incandescent, wall mount									
2620	100 Watt	2 Elec	50	.320	Ea.		14.10		14.10	23

26 05 19 – Low-Voltage Electrical Power Conductors and Cables

26 05 19.13 Undercarpet Electrical Power Cables		Crew	Daily Output	Labor-Hours	Unit	Material	Labor	Equipment	Total	Total Incl O&P
0010	**UNDERCARPET ELECTRICAL POWER CABLES**									
0020	Power System									
0100	Cable flat, 3 conductor, #12, w/attached bottom shield	1 Elec	982	.008	L.F.	5.15	.36		5.51	6.25
0200	Shield, top, steel		1768	.005	"	5.50	.20		5.70	6.40
0250	Splice, 3 conductor		48	.167	Ea.	16.75	7.35		24.10	30.50
0300	Top shield		96	.083		1.49	3.67		5.16	7.65
0350	Tap		40	.200		21.50	8.80		30.30	38
0400	Insulating patch, splice, tap, & end		48	.167		53	7.35		60.35	70
0450	Fold		230	.035			1.53		1.53	2.51
0500	Top shield, tap & fold		96	.083		1.49	3.67		5.16	7.65
0700	Transition, block assembly		77	.104		77	4.58		81.58	92.50
0750	Receptacle frame & base		32	.250		42.50	11		53.50	64.50
0800	Cover receptacle		120	.067		3.61	2.94		6.55	8.75
0850	Cover blank		160	.050		4.24	2.20		6.44	8.25
0860	Receptacle, direct connected, single		25	.320		90.50	14.10		104.60	123
0870	Dual		16	.500		149	22		171	199
0880	Combination high & low, tension		21	.381		109	16.80		125.80	148
0900	Box, floor with cover		20	.400		92	17.60		109.60	130
0920	Floor service w/barrier		4	2		259	88		347	430
1000	Wall, surface, with cover		20	.400		60.50	17.60		78.10	95.50
1100	Wall, flush, with cover	↓	20	.400	↓	42.50	17.60		60.10	75.50
2500	Telephone System									
2510	Transition fitting wall box, surface	1 Elec	24	.333	Ea.	61.50	14.70		76.20	91.50
2520	Flush		24	.333		61.50	14.70		76.20	91.50
2530	Flush, for PC board		24	.333		61.50	14.70		76.20	91.50
2540	Floor service box	↓	4	2		238	88		326	405
2550	Cover, surface					15.90			15.90	17.50
2560	Flush					15.90			15.90	17.50
2570	Flush for PC board					15.90			15.90	17.50
2700	Floor fitting w/duplex jack & cover	1 Elec	21	.381		47	16.80		63.80	79
2720	Low profile		53	.151		17.10	6.65		23.75	29.50
2740	Miniature w/duplex jack		53	.151		26	6.65		32.65	39.50
2760	25 pair kit		21	.381		49	16.80		65.80	81.50
2780	Low profile		53	.151		17.50	6.65		24.15	30
2800	Call director kit for 5 cable		19	.421		78.50	18.55		97.05	117
2820	4 pair kit		19	.421		93.50	18.55		112.05	134
2840	3 pair kit		19	.421		99	18.55		117.55	140
2860	Comb. 25 pair & 3 conductor power		21	.381		78.50	16.80		95.30	114
2880	5 conductor power		21	.381		89.50	16.80		106.30	126
2900	PC board, 8 per 3 pair	↓	161	.050	↓	68	2.19		70.19	78.50

26 05 19.13 Undercarpet Electrical Power Cables		Crew	Daily Output	Labor-Hours	Unit	2016 Bare Costs Material	Labor	Equipment	Total	Total Incl O&P
2920	6 per 4 pair	1 Elec	161	.050	Ea.	68	2.19		70.19	78.50
2940	3 pair adapter		161	.050		62	2.19		64.19	71.50
2950	Plug		77	.104		3.02	4.58		7.60	10.80
2960	Couplers		321	.025		8.50	1.10		9.60	11.15
3000	Bottom shield for 25 pair cable		4420	.002	L.F.	.86	.08		.94	1.08
3020	4 pair		4420	.002		.40	.08		.48	.57
3040	Top shield for 25 pair cable		4420	.002		.86	.08		.94	1.08
3100	Cable assembly, double-end, 50′, 25 pair		11.80	.678	Ea.	240	30		270	315
3110	3 pair		23.60	.339		74	14.95		88.95	106
3120	4 pair		23.60	.339		78.50	14.95		93.45	111
3140	Bulk 3 pair		1473	.005	L.F.	1.27	.24		1.51	1.79
3160	4 pair		1473	.005	"	1.57	.24		1.81	2.12
3500	Data System									
3520	Cable 25 conductor w/connection 40′, 75 ohm	1 Elec	14.50	.552	Ea.	71	24.50		95.50	118
3530	Single lead		22	.364		193	16		209	238
3540	Dual lead		22	.364		240	16		256	290
3560	Shields same for 25 conductor as 25 pair telephone									
3570	Single & dual, none required									
3590	BNC coax connectors, Plug	1 Elec	40	.200	Ea.	11.05	8.80		19.85	26.50
3600	TNC coax connectors, Plug	"	40	.200	"	15.45	8.80		24.25	31.50
3700	Cable-bulk									
3710	Single lead	1 Elec	1473	.005	L.F.	2.70	.24		2.94	3.36
3720	Dual lead	"	1473	.005	"	3.91	.24		4.15	4.69
3730	Hand tool crimp				Ea.	445			445	485
3740	Hand tool notch				"	19.30			19.30	21
3750	Boxes & floor fitting same as telephone									

26 05 19.20 Armored Cable		Crew	Daily Output	Labor-Hours	Unit	Material	Labor	Equipment	Total	Total Incl O&P
0010	**ARMORED CABLE**									
0050	600 volt, copper (BX), #14, 2 conductor, solid	1 Elec	2.40	3.333	C.L.F.	45.50	147		192.50	291
0100	3 conductor, solid		2.20	3.636		73	160		233	345
0120	4 conductor, solid		2	4		101	176		277	400
0150	#12, 2 conductor, solid		2.30	3.478		46	153		199	300
0200	3 conductor, solid		2	4		74.50	176		250.50	370
0220	4 conductor, solid		1.80	4.444		106	196		302	435
0250	#10, 2 conductor, solid		2	4		87.50	176		263.50	385
0300	3 conductor, solid		1.60	5		121	220		341	495
0320	4 conductor, solid		1.40	5.714		188	252		440	615
0340	#8, 2 conductor, stranded		1.50	5.333		229	235		464	635
0350	3 conductor, stranded		1.30	6.154		229	271		500	695
0370	4 conductor, stranded		1.10	7.273		325	320		645	885
9010	600 volt, copper (MC) steel clad, #14, 2 wire	R-1A	8.16	1.961		50.50	77.50		128	183
9020	3 wire		7.21	2.219		74.50	88		162.50	226
9030	4 wire		6.69	2.392		103	95		198	269
9040	#12, 2 wire		7.21	2.219		46.50	88		134.50	195
9050	3 wire		6.69	2.392		79	95		174	242
9060	4 wire		6.04	2.649		121	105		226	305
9070	#10, 2 wire		6.04	2.649		108	105		213	291
9080	3 wire		5.65	2.832		142	112		254	340
9090	4 wire		5.23	3.059		222	121		343	445
9100	#8, 2 wire, stranded		3.94	4.061		194	161		355	475
9110	3 wire, stranded		3.40	4.706		272	187		459	605
9120	4 wire, stranded		2.80	5.714		370	227		597	775

26 05 19 – Low-Voltage Electrical Power Conductors and Cables

26 05 19.55 Non-Metallic Sheathed Cable		Crew	Daily Output	Labor-Hours	Unit	2016 Bare Costs Material	Labor	Equipment	Total	Total Incl O&P
0010	**NON-METALLIC SHEATHED CABLE** 600 volt									
0100	Copper with ground wire (Romex)									
0150	#14, 2 conductor	1 Elec	2.70	2.963	C.L.F.	21.50	131		152.50	237
0200	3 conductor		2.40	3.333		30.50	147		177.50	274
0220	4 conductor		2.20	3.636		44	160		204	310
0250	#12, 2 conductor		2.50	3.200		33	141		174	268
0300	3 conductor		2.20	3.636		46.50	160		206.50	315
0320	4 conductor		2	4		68.50	176		244.50	365
0350	#10, 2 conductor		2.20	3.636		50.50	160		210.50	320
0400	3 conductor		1.80	4.444		73.50	196		269.50	400
0420	4 conductor		1.60	5		106	220		326	475
0430	#8, 2 conductor		1.60	5		79.50	220		299.50	445
0450	3 conductor		1.50	5.333		118	235		353	515
0500	#6, 3 conductor	↓	1.40	5.714	↓	192	252		444	620
0550	SE type SER aluminum cable, 3 RHW and									
0600	1 bare neutral, 3 #8 & 1 #8	1 Elec	1.60	5	C.L.F.	69	220		289	435
0650	3 #6 & 1 #6	"	1.40	5.714		78	252		330	495
0700	3 #4 & 1 #6	2 Elec	2.40	6.667		87.50	294		381.50	575
0750	3 #2 & 1 #4		2.20	7.273		129	320		449	665
0800	3 #1/0 & 1 #2		2	8		195	350		545	790
0850	3 #2/0 & 1 #1		1.80	8.889		229	390		619	890
0900	3 #4/0 & 1 #2/0	↓	1.60	10		325	440		765	1,075
1450	UF underground feeder cable, copper with ground, #14, 2 conductor	1 Elec	4	2		24.50	88		112.50	171
1500	#12, 2 conductor		3.50	2.286		37	101		138	206
1550	#10, 2 conductor		3	2.667		58	117		175	256
1600	#14, 3 conductor		3.50	2.286		35.50	101		136.50	204
1650	#12, 3 conductor		3	2.667		53.50	117		170.50	251
1700	#10, 3 conductor		2.50	3.200		83.50	141		224.50	325
2400	SEU service entrance cable, copper 2 conductors, #8 + #8 neutral		1.50	5.333		107	235		342	505
2600	#6 + #8 neutral		1.30	6.154		158	271		429	620
2800	#6 + #6 neutral	↓	1.30	6.154		180	271		451	645
3000	#4 + #6 neutral	2 Elec	2.20	7.273		248	320		568	800
3200	#4 + #4 neutral		2.20	7.273		280	320		600	835
3400	#3 + #5 neutral		2.10	7.619		345	335		680	930
3600	#3 + #3 neutral		2.10	7.619		370	335		705	960
3800	#2 + #4 neutral		2	8		400	350		750	1,025
4000	#1 + #1 neutral		1.90	8.421		445	370		815	1,100
4200	1/0 + 1/0 neutral		1.80	8.889		695	390		1,085	1,400
4400	2/0 + 2/0 neutral		1.70	9.412		875	415		1,290	1,650
4600	3/0 + 3/0 neutral		1.60	10		1,100	440		1,540	1,925
4620	4/0 + 4/0 neutral	↓	1.45	11.034		1,225	485		1,710	2,150
4000	Aluminum 2 conductors, #8 + #8 neutral	1 Elec	1.60	5		64.50	220		284.50	430
5000	#6 + #6 neutral	"	1.40	5.714		65	252		317	480
5100	#4 + #6 neutral	2 Elec	2.50	6.400		77.50	282		359.50	545
5200	#4 + #4 neutral		2.40	6.667		84.50	294		378.50	575
5300	#2 + #4 neutral		2.30	6.957		103	305		408	615
5400	#2 + #2 neutral		2.20	7.273		113	320		433	650
5450	1/0 + #2 neutral		2.10	7.619		163	335		498	730
5500	1/0 + 1/0 neutral		2	8		172	350		522	765
5550	2/0 + #1 neutral		1.90	8.421		184	370		554	810
5600	2/0 + 2/0 neutral		1.80	8.889		197	390		587	855
5800	3/0 + 1/0 neutral	↓	1.70	9.412	↓	214	415		629	915

26 05 19.55 Non-Metallic Sheathed Cable		Crew	Daily Output	Labor-Hours	Unit	2016 Bare Costs Material	Labor	Equipment	Total	Total Incl O&P
6000	3/0 + 3/0 neutral	2 Elec	1.70	9.412	C.L.F.	235	415		650	940
6200	4/0 + 2/0 neutral		1.60	10		254	440		694	1,000
6400	4/0 + 4/0 neutral		1.60	10		276	440		716	1,025
6500	Service entrance cap for copper SEU									
6600	100 amp	1 Elec	12	.667	Ea.	8.10	29.50		37.60	57
6700	150 amp		10	.800		13.80	35		48.80	72.50
6800	200 amp		8	1		21	44		65	95

26 05 19.70 Portable Cord

		Crew	Daily Output	Labor-Hours	Unit	Material	Labor	Equipment	Total	Total Incl O&P
0010	**PORTABLE CORD** 600 volt									
0100	Type SO, #18, 2 conductor	1 Elec	980	.008	L.F.	.57	.36		.93	1.22
0110	3 conductor		980	.008		.77	.36		1.13	1.44
0120	#16, 2 conductor		840	.010		.75	.42		1.17	1.52
0130	3 conductor		840	.010		1.07	.42		1.49	1.87
0140	4 conductor		840	.010		1.25	.42		1.67	2.07
0240	#14, 2 conductor		840	.010		1.23	.42		1.65	2.04
0250	3 conductor		840	.010		1.53	.42		1.95	2.37
0260	4 conductor		840	.010		1.78	.42		2.20	2.65
0280	#12, 2 conductor		840	.010		1.50	.42		1.92	2.34
0290	3 conductor		840	.010		2.24	.42		2.66	3.15
0300	4 conductor		840	.010		2.64	.42		3.06	3.59
0320	#10, 2 conductor		765	.010		2.54	.46		3	3.54
0330	3 conductor		765	.010		3.05	.46		3.51	4.11
0340	4 conductor		765	.010		3.45	.46		3.91	4.55
0360	#8, 2 conductor		555	.014		5.10	.63		5.73	6.65
0370	3 conductor		540	.015		4.47	.65		5.12	6
0380	4 conductor		525	.015		5.25	.67		5.92	6.90
0400	#6, 2 conductor		525	.015		4.19	.67		4.86	5.70
0410	3 conductor		490	.016		6.85	.72		7.57	8.70
0420	4 conductor		415	.019		7.45	.85		8.30	9.60
0440	#4, 2 conductor	2 Elec	830	.019		7.05	.85		7.90	9.15
0450	3 conductor		700	.023		9.45	1.01		10.46	12.05
0460	4 conductor		660	.024		13.60	1.07		14.67	16.75
0480	#2, 2 conductor		450	.036		10.45	1.57		12.02	14.05
0490	3 conductor		350	.046		10.45	2.01		12.46	14.80
0500	4 conductor		280	.057		18.55	2.52		21.07	24.50
2000	See 26 27 26.20 for Wiring Devices Elements									

26 05 19.75 Modular Flexible Wiring System

		Crew	Daily Output	Labor-Hours	Unit	Material	Labor	Equipment	Total	Total Incl O&P
0010	**MODULAR FLEXIBLE WIRING SYSTEM**									
0020	Commercial system grid ceiling									
0100	Conversion Module	1 Elec	32	.250	Ea.	8.20	11		19.20	27
0120	Fixture cable, for fixture to fixture, 3 conductor, 15′ long		40	.200		34	8.80		42.80	51.50
0150	Extender cable, 3 conductor, 15′ long		48	.167		32	7.35		39.35	47
0200	Switch drop, 1 level, 9′ long		32	.250		36.50	11		47.50	58
0220	2 level, 9′ long		32	.250		30	11		41	51.50
0250	Power tee, 9′ long		32	.250		34	11		45	55
1020	Industrial system open ceiling									
1100	Converter, interface between hardwiring and modular wiring	1 Elec	32	.250	Ea.	17.05	11		28.05	37
1120	Fixture cable, for fixture to fixture, 3 conductor, 21′ long		32	.250		109	11		120	138
1125	3 conductor, 25′ long		24	.333		126	14.70		140.70	163
1130	3 conductor, 31′ long		19	.421		112	18.55		130.55	154
1150	Fixture cord drop, 10′ long		40	.200		30	8.80		38.80	47.50

26 05 19 – Low-Voltage Electrical Power Conductors and Cables

26 05 19.90 Wire		Crew	Daily Output	Labor-Hours	Unit	Material	Labor	Equipment	Total	Total Incl O&P
						2016 Bare Costs				
0010	**WIRE**									
0020	600 volt, copper type THW, solid, #14	1 Elec	13	.615	C.L.F.	7	27		34	52
0030	#12		11	.727		10.65	32		42.65	64.50
0040	#10		10	.800		16.70	35		51.70	76
0050	Stranded, #14		13	.615		8.20	27		35.20	53.50
0100	#12		11	.727		12.55	32		44.55	66.50
0120	#10		10	.800		19.80	35		54.80	79.50
0140	#8		8	1		32.50	44		76.50	108
0400	250 kcmil	3 Elec	6	4		495	176		671	835
0420	300 kcmil		5.70	4.211		560	185		745	920
0450	350 kcmil		5.40	4.444		650	196		846	1,025
0480	400 kcmil		5.10	4.706		745	207		952	1,150
0490	500 kcmil		4.80	5		925	220		1,145	1,375
0540	600 volt, aluminum type THHN, stranded, #6	1 Elec	8	1		36.50	44		80.50	112
0560	#4	2 Elec	13	1.231		45.50	54		99.50	139
0580	#2		10.60	1.509		61.50	66.50		128	177
0600	#1		9	1.778		90	78.50		168.50	227
0620	1/0		8	2		108	88		196	263
0640	2/0		7.20	2.222		128	98		226	300
0680	3/0		6.60	2.424		158	107		265	350
0700	4/0		6.20	2.581		176	114		290	380
0720	250 kcmil	3 Elec	8.70	2.759		215	122		337	435
0740	300 kcmil		8.10	2.963		297	131		428	540
0760	350 kcmil		7.50	3.200		300	141		441	560
0780	400 kcmil		6.90	3.478		355	153		508	640
0800	500 kcmil		6	4		390	176		566	720
0850	600 kcmil		5.70	4.211		495	185		680	845
0880	700 kcmil		5.10	4.706		570	207		777	965
0900	750 kcmil		4.80	5		590	220		810	1,000
0910	1000 kcmil		3.78	6.349		875	280		1,155	1,425
0920	600 volt, copper type THWN-THHN, solid, #14	1 Elec	13	.615		7	27		34	52
0940	#12		11	.727		10.65	32		42.65	64.50
0960	#10		10	.800		16.70	35		51.70	76
1000	Stranded, #14		13	.615		7.95	27		34.95	53.50
1010	2 wire, #14, install wire only		655	.012	L.F.	.16	.54		.70	1.06
1020	3 wire, #14, install wire only		440	.018	"	.24	.80		1.04	1.57
1200	#12		11	.727	C.L.F.	11.30	32		43.30	65
1250	#10		10	.800		17.50	35		52.50	77
1300	#8		8	1		30.50	44		74.50	106
1350	#6		6.50	1.231		52	54		106	146
1400	#4	2 Elec	10.60	1.509		74	66.50		140.50	191

26 05 23 – Control-Voltage Electrical Power Cables

26 05 23.20 Special Wires and Fittings		Crew	Daily Output	Labor-Hours	Unit	Material	Labor	Equipment	Total	Total Incl O&P
0010	**SPECIAL WIRES & FITTINGS**									
0100	Fixture TFFN 600 volt 90°C stranded, #18	1 Elec	13	.615	C.L.F.	9.50	27		36.50	55
0150	#16		13	.615		11.25	27		38.25	57
0500	Thermostat, jacket non-plenum, twisted, #18-2 conductor		8	1		15.60	44		59.60	89
0550	#18-3 conductor		7	1.143		20.50	50.50		71	105
0600	#18-4 conductor		6.50	1.231		30.50	54		84.50	122
0650	#18-5 conductor		6	1.333		34	58.50		92.50	134
0700	#18-6 conductor		5.50	1.455		42	64		106	152
0750	#18-7 conductor		5	1.600		43.50	70.50		114	163

26 05 Common Work Results for Electrical

26 05 23 – Control-Voltage Electrical Power Cables

26 05 23.20 Special Wires and Fittings		Crew	Daily Output	Labor-Hours	Unit	2016 Bare Costs Material	Labor	Equipment	Total	Total Incl O&P
0800	#18-8 conductor	1 Elec	4.80	1.667	C.L.F.	55.50	73.50		129	182

26 05 26 – Grounding and Bonding for Electrical Systems

26 05 26.80 Grounding

		Crew	Daily Output	Labor-Hours	Unit	Material	Labor	Equipment	Total	Total Incl O&P
0010	**GROUNDING**									
0030	Rod, copper clad, 8′ long, 1/2″ diameter	1 Elec	5.50	1.455	Ea.	20	64		84	127
0050	3/4″ diameter		5.30	1.509		34	66.50		100.50	147
0080	10′ long, 1/2″ diameter		4.80	1.667		22.50	73.50		96	145
0100	3/4″ diameter		4.40	1.818		39	80		119	174
0130	15′ long, 3/4″ diameter		4	2		51	88		139	200
0260	Wire ground bare armored, #8-1 conductor		2	4	C.L.F.	69.50	176		245.50	365
0270	#6-1 conductor		1.80	4.444		83.50	196		279.50	410
0390	Bare copper wire, stranded, #8		11	.727		25	32		57	80
0400	#6		10	.800		44.50	35		79.50	106
0600	#2	2 Elec	10	1.600		120	70.50		190.50	247
0800	3/0		6.60	2.424		264	107		371	465
1000	4/0		5.70	2.807		335	124		459	570
1200	250 kcmil	3 Elec	7.20	3.333		395	147		542	675
1800	Water pipe ground clamps, heavy duty									
2000	Bronze, 1/2″ to 1″ diameter	1 Elec	8	1	Ea.	24	44		68	98
2100	1-1/4″ to 2″ diameter		8	1		33	44		77	109
2200	2-1/2″ to 3″ diameter		6	1.333		44.50	58.50		103	145
2800	Brazed connections, #6 wire		12	.667		16.35	29.50		45.85	66
3000	#2 wire		10	.800		22	35		57	81.50
3100	3/0 wire		8	1		32.50	44		76.50	108
3200	4/0 wire		7	1.143		37	50.50		87.50	124
3400	250 kcmil wire		5	1.600		43.50	70.50		114	163

26 05 33 – Raceway and Boxes for Electrical Systems

26 05 33.13 Conduit

		Crew	Daily Output	Labor-Hours	Unit	Material	Labor	Equipment	Total	Total Incl O&P
0010	**CONDUIT** To 15′ high, includes 2 terminations, 2 elbows, R260533-22									
0020	11 beam clamps, and 11 couplings per 100 L.F.									
1750	Rigid galvanized steel, 1/2″ diameter	1 Elec	90	.089	L.F.	2.50	3.92		6.42	9.15
1770	3/4″ diameter		80	.100		2.82	4.41		7.23	10.30
1800	1″ diameter		65	.123		3.89	5.40		9.29	13.15
1830	1-1/4″ diameter		60	.133		5.10	5.85		10.95	15.20
1850	1-1/2″ diameter		55	.145		6.05	6.40		12.45	17.15
1870	2″ diameter		45	.178		7.80	7.85		15.65	21.50
2500	Steel, intermediate conduit (IMC), 1/2″ diameter		100	.080		2.20	3.52		5.72	8.15
2530	3/4″ diameter		90	.089		2.75	3.92		6.67	9.45
2550	1″ diameter		70	.114		3.93	5.05		8.98	12.55
2570	1-1/4″ diameter		65	.123		4.94	5.40		10.34	14.30
2600	1-1/2″ diameter		60	.133		6.45	5.85		12.30	16.70
2630	2″ diameter		50	.160		7.90	7.05		14.95	20.50
2650	2-1/2″ diameter		40	.200		14.10	8.80		22.90	30
2670	3″ diameter	2 Elec	60	.267		18.55	11.75		30.30	39.50
2700	3-1/2″ diameter		54	.296		23	13.05		36.05	47
2730	4″ diameter		50	.320		19.40	14.10		33.50	44.50
5000	Electric metallic tubing (EMT), 1/2″ diameter	1 Elec	170	.047		.70	2.07		2.77	4.16
5020	3/4″ diameter		130	.062		1	2.71		3.71	5.55
5040	1″ diameter		115	.070		1.67	3.06		4.73	6.85
5060	1-1/4″ diameter		100	.080		2.69	3.52		6.21	8.70
5080	1-1/2″ diameter		90	.089		3.43	3.92		7.35	10.20
5100	2″ diameter		80	.100		4.24	4.41		8.65	11.85

26 05 33 – Raceway and Boxes for Electrical Systems

26 05 33.13 Conduit		Crew	Daily Output	Labor-Hours	Unit	2016 Bare Costs Material	Labor	Equipment	Total	Total Incl O&P
5120	2-1/2" diameter	1 Elec	60	.133	L.F.	8.70	5.85		14.55	19.15
5140	3" diameter	2 Elec	100	.160		10.20	7.05		17.25	23
5160	3-1/2" diameter		90	.178		12.85	7.85		20.70	27
5180	4" diameter		80	.200		13.85	8.80		22.65	29.50
9100	PVC, schedule 40, 1/2" diameter	1 Elec	190	.042		.86	1.85		2.71	3.98
9110	3/4" diameter		145	.055		.95	2.43		3.38	5
9120	1" diameter		125	.064		2.06	2.82		4.88	6.90
9130	1-1/4" diameter		110	.073		2.53	3.20		5.73	8.05
9140	1-1/2" diameter		100	.080		2.86	3.52		6.38	8.90
9150	2" diameter		90	.089		3.73	3.92		7.65	10.50
9160	2-1/2" diameter		65	.123		4.59	5.40		9.99	13.90
9170	3" diameter	2 Elec	110	.145		5.20	6.40		11.60	16.25
9880	Heat bender, to 6" diameter				Ea.	1,300			1,300	1,450
9900	Add to labor for higher elevated installation									
9910	15' to 20' high, add						10%			
9920	20' to 25' high, add						20%			
9930	25' to 30' high, add						25%			
9940	30' to 35' high, add						30%			
9950	35' to 40' high, add						35%			
9960	Over 40' high, add						40%			

26 05 33.16 Outlet Boxes		Crew	Daily Output	Labor-Hours	Unit	Material	Labor	Equipment	Total	Total Incl O&P
0010	**OUTLET BOXES**									
0020	Pressed steel, octagon, 4"	1 Elec	20	.400	Ea.	2.22	17.60		19.82	31.50
0060	Covers, blank		64	.125		.93	5.50		6.43	10
0100	Extension rings		40	.200		3.68	8.80		12.48	18.45
0150	Square, 4"		20	.400		2.88	17.60		20.48	32
0200	Extension rings		40	.200		3.72	8.80		12.52	18.50
0250	Covers, blank		64	.125		1	5.50		6.50	10.10
0300	Plaster rings		64	.125		2.04	5.50		7.54	11.25
0550	Handy box		27	.296		2.45	13.05		15.50	24
0560	Covers, device		64	.125		.87	5.50		6.37	9.95
0650	Switchbox		27	.296		3.55	13.05		16.60	25.50
0700	Masonry, 1 gang, 2-1/2" deep		27	.296		6.75	13.05		19.80	29
0710	3-1/2" deep		27	.296		6.80	13.05		19.85	29
0750	2 gang, 2-1/2" deep		20	.400		12.75	17.60		30.35	43
0760	3-1/2" deep		20	.400		10.25	17.60		27.85	40.50
1100	Concrete, floor, 1 gang		5.30	1.509		89.50	66.50		156	208
1400	Cast, 1 gang, FS (2" deep), 1/2" hub		12	.667		18.25	29.50		47.75	68
1410	3/4" hub		12	.667		19.30	29.50		48.80	69
1420	FD (2-11/16" deep), 1/2" hub		12	.667		18.40	29.50		47.90	68.50
1430	3/4" hub		12	.667		20	29.50		49.50	70
1450	2 gang, FS, 1/2" hub		10	.800		33	35		68	93.50
1460	3/4" hub		10	.800		35.50	35		70.50	96.50
1470	FD, 1/2" hub		10	.800		39.50	35		74.50	101
1480	3/4" hub		10	.800		39.50	35		74.50	101
1500	3 gang, FS, 3/4" hub		9	.889		56	39		95	126
1510	Switch cover, 1 gang, FS		64	.125		4.83	5.50		10.33	14.30
1520	2 gang		53	.151		7.35	6.65		14	19
1530	Duplex receptacle cover, 1 gang, FS		64	.125		4.93	5.50		10.43	14.40
1540	2 gang, FS		53	.151		7.50	6.65		14.15	19.15
1542	Weatherproof blank cover, 1 gang		64	.125		1.22	5.50		6.72	10.35
1544	2 gang		53	.151		2.52	6.65		9.17	13.65

26 05 33.16 Outlet Boxes

	26 05 33.16 Outlet Boxes	Crew	Daily Output	Labor-Hours	Unit	2016 Bare Costs Material	Labor	Equipment	Total	Total Incl O&P
1550	Weatherproof switch cover, 1 gang	1 Elec	64	.125	Ea.	6.40	5.50		11.90	16.05
1554	2 gang		53	.151		10.60	6.65		17.25	22.50
1600	Weatherproof receptacle cover, 1 gang		64	.125		4.10	5.50		9.60	13.50
1604	2 gang		53	.151		10.60	6.65		17.25	22.50
1620	Weatherproof receptacle cover, tamper resistant, 1 gang		58	.138		13.50	6.10		19.60	25
1624	2 gang		48	.167		27	7.35		34.35	41.50
1750	FSC, 1 gang, 1/2" hub		11	.727		20	32		52	74.50
1760	3/4" hub		11	.727		22.50	32		54.50	77
1770	2 gang, 1/2" hub		9	.889		30.50	39		69.50	97.50
1780	3/4" hub		9	.889		39	39		78	107
1790	FDC, 1 gang, 1/2" hub		11	.727		23	32		55	77.50
1800	3/4" hub		11	.727		25.50	32		57.50	80.50
1810	2 gang, 1/2" hub		9	.889		47.50	39		86.50	116
1820	3/4" hub		9	.889		42	39		81	110
1850	Weatherproof in-use cover, 1 gang		64	.125		24	5.50		29.50	35.50
1870	2 gang	↓	53	.151	↓	27	6.65		33.65	40.50

26 05 33.17 Outlet Boxes, Plastic

	26 05 33.17 Outlet Boxes, Plastic	Crew	Daily Output	Labor-Hours	Unit	Material	Labor	Equipment	Total	Total Incl O&P
0010	**OUTLET BOXES, PLASTIC**									
0050	4" diameter, round with 2 mounting nails	1 Elec	25	.320	Ea.	2.39	14.10		16.49	25.50
0100	Bar hanger mounted		25	.320		5.10	14.10		19.20	28.50
0200	4", square with 2 mounting nails		25	.320		4.63	14.10		18.73	28
0300	Plaster ring		64	.125		1.98	5.50		7.48	11.20
0400	Switch box with 2 mounting nails, 1 gang		30	.267		4.39	11.75		16.14	24
0500	2 gang		25	.320		3.33	14.10		17.43	26.50
0600	3 gang		20	.400		4.19	17.60		21.79	33.50
1400	PVC, FSS, 1 gang, 1/2" hub		14	.571		13.90	25		38.90	56.50
1410	3/4" hub		14	.571		11.35	25		36.35	53.50
1420	FD, 1 gang for variable terminations		14	.571		9.35	25		34.35	51.50
1450	FS, 2 gang for variable terminations		12	.667		11.05	29.50		40.55	60
1480	Weatherproof blank cover, FS, 1 gang		64	.125		4.09	5.50		9.59	13.50
1500	2 gang		53	.151		4.74	6.65		11.39	16.10
1510	Weatherproof switch cover, FS, 1 gang		64	.125		9.85	5.50		15.35	19.80
1520	2 gang		53	.151		16.20	6.65		22.85	28.50
1530	Weatherproof duplex receptacle cover, FS, 1 gang		64	.125		13.55	5.50		19.05	24
1540	2 gang		53	.151		14.40	6.65		21.05	27
1750	FSC, 1 gang, 1/2" hub		13	.615		9.20	27		36.20	54.50
1760	3/4" hub		13	.615		9.90	27		36.90	55.50
1770	FSC, 2 gang, 1/2" hub		11	.727		14.10	32		46.10	68
1780	3/4" hub		11	.727		14.05	32		46.05	68
1790	FDC, 1 gang, 1/2" hub		13	.615		9.15	27		36.15	54.50
1800	3/4" hub		13	.615		10.35	27		37.35	56
1810	Weatherproof, T box w/3 holes		14	.571		9.90	25		34.90	52
1820	4" diameter round w/5 holes		14	.571		8.85	25		33.85	50.50
1850	In-use cover, 1 gang		64	.125		8.75	5.50		14.25	18.65
1870	2 gang	↓	53	.151	↓	11	6.65		17.65	23

26 05 33.18 Pull Boxes

	26 05 33.18 Pull Boxes	Crew	Daily Output	Labor-Hours	Unit	Material	Labor	Equipment	Total	Total Incl O&P
0010	**PULL BOXES**									
0100	Steel, pull box, NEMA 1, type SC, 6" W x 6" H x 4" D	1 Elec	8	1	Ea.	9.65	44		53.65	82.50
0200	8" W x 8" H x 4" D		8	1		13.20	44		57.20	86.50
0300	10" W x 12" H x 6" D		5.30	1.509		22.50	66.50		89	134
0800	12" W x 16" H x 6" D		4.70	1.702		57	75		132	186
1000	20" W x 20" H x 6" D	↓	3.60	2.222	↓	99	98		197	269

26 05 Common Work Results for Electrical

26 05 33 – Raceway and Boxes for Electrical Systems

26 05 33.23 Wireway

26 05 33.23 Wireway		Crew	Daily Output	Labor-Hours	Unit	Material	2016 Bare Costs Labor	Equipment	Total	Total Incl O&P
0010	**WIREWAY** to 15′ high									
0100	NEMA 1, Screw cover w/fittings and supports, 2-1/2″ x 2-1/2″	1 Elec	45	.178	L.F.	10.80	7.85		18.65	24.50
0200	4″ x 4″	″	40	.200		11.65	8.80		20.45	27
0400	6″ x 6″	2 Elec	60	.267		19.60	11.75		31.35	40.50
0600	8″ x 8″	″	40	.400		28.50	17.60		46.10	60.50
4475	NEMA 3R, Screw cover w/fittings and supports, 4″ x 4″	1 Elec	36	.222		14	9.80		23.80	31.50
4480	6″ x 6″	2 Elec	55	.291		18.75	12.80		31.55	41.50
4485	8″ x 8″		36	.444		29	19.60		48.60	64
4490	12″ x 12″		18	.889		46.50	39		85.50	115

26 05 33.25 Conduit Fittings for Rigid Galvanized Steel

26 05 33.25		Crew	Daily Output	Labor-Hours	Unit	Material	Labor	Equipment	Total	Total Incl O&P
0010	**CONDUIT FITTINGS FOR RIGID GALVANIZED STEEL**									
2280	LB, LR or LL fittings & covers, 1/2″ diameter	1 Elec	16	.500	Ea.	8.85	22		30.85	46
2290	3/4″ diameter		13	.615		11	27		38	56.50
2300	1″ diameter		11	.727		16.05	32		48.05	70
2330	1-1/4″ diameter		8	1		25	44		69	99.50
2350	1-1/2″ diameter		6	1.333		33	58.50		91.50	133
2370	2″ diameter		5	1.600		54	70.50		124.50	175
2380	2-1/2″ diameter		4	2		114	88		202	269
2390	3″ diameter		3.50	2.286		146	101		247	325
2400	3-1/2″ diameter		3	2.667		255	117		372	470
2410	4″ diameter		2.50	3.200		278	141		419	535
5280	Service entrance cap, 1/2″ diameter		16	.500		4.99	22		26.99	41.50
5300	3/4″ diameter		13	.615		5.95	27		32.95	51
5320	1″ diameter		10	.800		7.10	35		42.10	65.50
5340	1-1/4″ diameter		8	1		8.95	44		52.95	82
5360	1-1/2″ diameter		6.50	1.231		17.80	54		71.80	108
5380	2″ diameter		5.50	1.455		26.50	64		90.50	135
5400	2-1/2″ diameter		4	2		71.50	88		159.50	223
5420	3″ diameter		3.40	2.353		117	104		221	299
5440	3-1/2″ diameter		3	2.667		167	117		284	375
5460	4″ diameter		2.70	2.963		217	131		348	450

26 05 33.30 Electrical Nonmetallic Tubing (ENT)

26 05 33.30		Crew	Daily Output	Labor-Hours	Unit	Material	Labor	Equipment	Total	Total Incl O&P
0010	**ELECTRICAL NONMETALLIC TUBING (ENT)**									
0050	Flexible, 1/2″ diameter	1 Elec	270	.030	L.F.	.74	1.31		2.05	2.94
0100	3/4″ diameter		230	.035		1.21	1.53		2.74	3.84
0200	1″ diameter		145	.055		2.03	2.43		4.46	6.20
0210	1-1/4″ diameter		125	.064		1.98	2.82		4.80	6.80
0220	1-1/2″ diameter		100	.080		2.93	3.52		6.45	8.95
0230	2″ diameter		75	.107		3.89	4.70		8.59	12
0300	Connectors, to outlet box, 1/2″ diameter		230	.035	Ea.	1	1.53		2.53	3.61
0310	3/4″ diameter		210	.038		1.82	1.68		3.50	4.75
0320	1″ diameter		200	.040		2.55	1.76		4.31	5.70
0400	Couplings, to conduit, 1/2″ diameter		145	.055		.99	2.43		3.42	5.05
0410	3/4″ diameter		130	.062		1.29	2.71		4	5.85
0420	1″ diameter		125	.064		2.28	2.82		5.10	7.10

26 05 36 – Cable Trays for Electrical Systems

26 05 36.10 Cable Tray Ladder Type

26 05 36.10		Crew	Daily Output	Labor-Hours	Unit	Material	Labor	Equipment	Total	Total Incl O&P
0010	**CABLE TRAY LADDER TYPE** w/ftngs. & supports, 4″ dp., to 15′ elev.									
0160	Galvanized steel tray									
0170	4″ rung spacing, 6″ wide	2 Elec	98	.163	L.F.	18.50	7.20		25.70	32.50
0800	6″ rung spacing, 6″ wide		100	.160		16.85	7.05		23.90	30

26 05 Common Work Results for Electrical

26 05 36 – Cable Trays for Electrical Systems

26 05 36.10 Cable Tray Ladder Type		Crew	Daily Output	Labor-Hours	Unit	2016 Bare Costs Material	Labor	Equipment	Total	Total Incl O&P
0850	9″ wide	2 Elec	94	.170	L.F.	18.50	7.50		26	33
0860	12″ wide		88	.182		19.40	8		27.40	34.50
0880	24″ wide		80	.200		25	8.80		33.80	42

26 05 36.40 Cable Tray, Covers and Dividers

		Crew	Daily Output	Labor-Hours	Unit	Material	Labor	Equipment	Total	Total Incl O&P
0010	**CABLE TRAY, COVERS AND DIVIDERS** To 15′ high									
0100	Covers, ventilated galv. steel, straight, 6″ wide tray size	2 Elec	520	.031	L.F.	8.30	1.36		9.66	11.30
0200	9″ wide tray size		460	.035		8.80	1.53		10.33	12.15
0300	12″ wide tray size		400	.040		15.15	1.76		16.91	19.55
0400	18″ wide tray size		300	.053		28.50	2.35		30.85	35.50
0500	24″ wide tray size		220	.073		47	3.20		50.20	57.50
3250	Covers, aluminum, straight, 6″ wide tray size		520	.031		5.95	1.36		7.31	8.75
3270	9″ wide tray size		460	.035		7.15	1.53		8.68	10.40
3290	12″ wide tray size		400	.040		8.55	1.76		10.31	12.35
3310	18″ wide tray size		320	.050		11.35	2.20		13.55	16.05
3330	24″ wide tray size		260	.062		14.15	2.71		16.86	20

26 05 39 – Underfloor Raceways for Electrical Systems

26 05 39.30 Conduit In Concrete Slab

		Crew	Daily Output	Labor-Hours	Unit	Material	Labor	Equipment	Total	Total Incl O&P
0010	**CONDUIT IN CONCRETE SLAB** Including terminations,									
0020	fittings and supports									
3230	PVC, schedule 40, 1/2″ diameter	1 Elec	270	.030	L.F.	.48	1.31		1.79	2.66
3250	3/4″ diameter		230	.035		.56	1.53		2.09	3.12
3270	1″ diameter		200	.040		.76	1.76		2.52	3.72
3300	1-1/4″ diameter		170	.047		1	2.07		3.07	4.48
3330	1-1/2″ diameter		140	.057		1.21	2.52		3.73	5.45
3350	2″ diameter		120	.067		1.52	2.94		4.46	6.45
4350	Rigid galvanized steel, 1/2″ diameter		200	.040		2.35	1.76		4.11	5.45
4400	3/4″ diameter		170	.047		2.48	2.07		4.55	6.10
4450	1″ diameter		130	.062		3.44	2.71		6.15	8.20
4500	1-1/4″ diameter		110	.073		4.78	3.20		7.98	10.50
4600	1-1/2″ diameter		100	.080		5.35	3.52		8.87	11.60
4800	2″ diameter		90	.089		6.60	3.92		10.52	13.65

26 05 39.40 Conduit In Trench

		Crew	Daily Output	Labor-Hours	Unit	Material	Labor	Equipment	Total	Total Incl O&P
0010	**CONDUIT IN TRENCH** Includes terminations and fittings									
0020	Does not include excavation or backfill, see Section 31 23 16									
0200	Rigid galvanized steel, 2″ diameter	1 Elec	150	.053	L.F.	6.25	2.35		8.60	10.75
0400	2-1/2″ diameter	″	100	.080		12.15	3.52		15.67	19.10
0600	3″ diameter	2 Elec	160	.100		14.15	4.41		18.56	23
0800	3-1/2″ diameter	″	140	.114		18.40	5.05		23.45	29

26 05 80 – Wiring Connections

26 05 80.10 Motor Connections

		Crew	Daily Output	Labor-Hours	Unit	Material	Labor	Equipment	Total	Total Incl O&P
0010	**MOTOR CONNECTIONS**									
0020	Flexible conduit and fittings, 115 volt, 1 phase, up to 1 HP motor	1 Elec	8	1	Ea.	5.55	44		49.55	78
0050	2 HP motor		6.50	1.231		11.35	54		65.35	101
0100	3 HP motor		5.50	1.455		9.85	64		73.85	116
0110	230 volt, 3 phase, 3 HP motor		6.78	1.180		6.85	52		58.85	92.50
0112	5 HP motor		5.47	1.463		5.70	64.50		70.20	111
0114	7-1/2 HP motor		4.61	1.735		8.70	76.50		85.20	135
0120	10 HP motor		4.20	1.905		18.90	84		102.90	158
0150	15 HP motor		3.30	2.424		18.90	107		125.90	196
0200	25 HP motor		2.70	2.963		25	131		156	241
1500	460 volt, 5 HP motor, 3 phase		8	1		5.95	44		49.95	78.50

26 05 Common Work Results for Electrical

26 05 80 – Wiring Connections

26 05 80.10 Motor Connections		Crew	Daily Output	Labor-Hours	Unit	Material	2016 Bare Costs Labor	Equipment	Total	Total Incl O&P
1520	10 HP motor	1 Elec	8	1	Ea.	5.95	44		49.95	78.50
1530	25 HP motor		6	1.333		12.15	58.50		70.65	109

26 05 90 – Residential Applications

26 05 90.10 Residential Wiring

		Crew	Daily Output	Labor-Hours	Unit	Material	Labor	Equipment	Total	Total Incl O&P
0010	**RESIDENTIAL WIRING**									
0020	20′ avg. runs and #14/2 wiring incl. unless otherwise noted									
1000	Service & panel, includes 24′ SE-AL cable, service eye, meter,									
1010	Socket, panel board, main bkr., ground rod, 15 or 20 amp									
1020	1-pole circuit breakers, and misc. hardware									
1100	100 amp, with 10 branch breakers	1 Elec	1.19	6.723	Ea.	485	296		781	1,025
1110	With PVC conduit and wire		.92	8.696		525	385		910	1,200
1120	With RGS conduit and wire		.73	10.959		685	485		1,170	1,550
1200	200 amp, with 18 branch breakers	2 Elec	1.80	8.889		1,050	390		1,440	1,800
1220	With PVC conduit and wire		1.46	10.959		1,125	485		1,610	2,050
1230	With RGS conduit and wire		1.24	12.903		1,525	570		2,095	2,600
1800	Lightning surge suppressor	1 Elec	32	.250		50.50	11		61.50	74
2000	Switch devices									
2100	Single pole, 15 amp, Ivory, with a 1-gang box, cover plate,									
2110	Type NM (Romex) cable	1 Elec	17.10	.468	Ea.	18.70	20.50		39.20	54
2120	Type MC (BX) cable		14.30	.559		27.50	24.50		52	70.50
2130	EMT & wire		5.71	1.401		36	61.50		97.50	141
2150	3-way, #14/3, type NM cable		14.55	.550		12.70	24		36.70	53.50
2170	Type MC cable		12.31	.650		25	28.50		53.50	74.50
2180	EMT & wire		5	1.600		30	70.50		100.50	148
2250	S.P., 20 amp, #12/2, type NM cable		13.33	.600		14.75	26.50		41.25	59
2270	Type MC cable		11.43	.700		21	31		52	73.50
2280	EMT & wire		4.85	1.649		33.50	72.50		106	156
2290	S.P. rotary dimmer, 600W, no wiring		17	.471		31.50	20.50		52	68.50
2300	S.P. rotary dimmer, 600W, type NM cable		14.55	.550		35.50	24		59.50	78.50
2320	Type MC cable		12.31	.650		44.50	28.50		73	95.50
2330	EMT & wire		5	1.600		54.50	70.50		125	175
2350	3-way rotary dimmer, type NM cable		13.33	.600		19.85	26.50		46.35	65
2370	Type MC cable		11.43	.700		28.50	31		59.50	82
2380	EMT & wire		4.85	1.649		39	72.50		111.50	162
2400	Interval timer wall switch, 20 amp, 1-30 min., #12/2									
2410	Type NM cable	1 Elec	14.55	.550	Ea.	57	24		81	103
2420	Type MC cable		12.31	.650		62	28.50		90.50	115
2430	EMT & wire		5	1.600		76	70.50		146.50	199
2500	Decorator style									
2510	S.P., 15 amp, type NM cable	1 Elec	17.10	.468	Ea.	22.50	20.50		43	58.50
2520	Type MC cable		14.30	.559		31.50	24.50		56	75
2530	EMT & wire		5.71	1.401		40	61.50		101.50	145
2550	3-way, #14/3, type NM cable		14.55	.550		16.60	24		40.60	58
2570	Type MC cable		12.31	.650		29	28.50		57.50	79
2580	EMT & wire		5	1.600		33.50	70.50		104	152
2650	S.P., 20 amp, #12/2, type NM cable		13.33	.600		18.60	26.50		45.10	63.50
2670	Type MC cable		11.43	.700		25	31		56	78
2680	EMT & wire		4.85	1.649		37.50	72.50		110	160
2700	S.P., slide dimmer, type NM cable		17.10	.468		24.50	20.50		45	60.50
2720	Type MC cable		14.30	.559		33	24.50		57.50	77
2730	EMT & wire		5.71	1.401		43.50	61.50		105	149
2750	S.P., touch dimmer, type NM cable		17.10	.468		41	20.50		61.50	78.50

26 05 90.10 Residential Wiring		Crew	Daily Output	Labor-Hours	Unit	2016 Bare Costs Material	Labor	Equipment	Total	Total Incl O&P
2770	Type MC cable	1 Elec	14.30	.559	Ea.	50	24.50		74.50	95
2780	EMT & wire		5.71	1.401		60	61.50		121.50	167
2800	3-way touch dimmer, type NM cable		13.33	.600		51.50	26.50		78	99.50
2820	Type MC cable		11.43	.700		60	31		91	117
2830	EMT & wire		4.85	1.649		70	72.50		142.50	196
3000	Combination devices									
3100	S.P. switch/15 amp recpt., Ivory, 1-gang box, plate									
3110	Type NM cable	1 Elec	11.43	.700	Ea.	24	31		55	77
3120	Type MC cable		10	.800		33	35		68	93.50
3130	EMT & wire		4.40	1.818		43	80		123	179
3150	S.P. switch/pilot light, type NM cable		11.43	.700		24.50	31		55.50	77.50
3170	Type MC cable		10	.800		33	35		68	94
3180	EMT & wire		4.43	1.806		43	79.50		122.50	178
3190	2-S.P. switches, 2-#14/2, no wiring		14	.571		13.70	25		38.70	56
3200	2-S.P. switches, 2-#14/2, type NM cables		10	.800		26.50	35		61.50	87
3220	Type MC cable		8.89	.900		40.50	39.50		80	110
3230	EMT & wire		4.10	1.951		45.50	86		131.50	191
3250	3-way switch/15 amp recpt., #14/3, type NM cable		10	.800		31.50	35		66.50	92
3270	Type MC cable		8.89	.900		44	39.50		83.50	114
3280	EMT & wire		4.10	1.951		48.50	86		134.50	195
3300	2-3 way switches, 2-#14/3, type NM cables		8.89	.900		40.50	39.50		80	110
3320	Type MC cable		8	1		61.50	44		105.50	140
3330	EMT & wire		4	2		55.50	88		143.50	205
3350	S.P. switch/20 amp recpt., #12/2, type NM cable		10	.800		31	35		66	92
3370	Type MC cable		8.89	.900		36	39.50		75.50	105
3380	EMT & wire		4.10	1.951		50	86		136	196
4000	Receptacle devices									
4010	Duplex outlet, 15 amp recpt., Ivory, 1-gang box, plate									
4015	Type NM cable	1 Elec	14.55	.550	Ea.	10.30	24		34.30	51
4020	Type MC cable		12.31	.650		19	28.50		47.50	68
4030	EMT & wire		5.33	1.501		28	66		94	139
4050	With #12/2, type NM cable		12.31	.650		12.65	28.50		41.15	61
4070	Type MC cable		10.67	.750		19	33		52	75
4080	EMT & wire		4.71	1.699		31.50	75		106.50	157
4100	20 amp recpt., #12/2, type NM cable		12.31	.650		19.30	28.50		47.80	68.50
4120	Type MC cable		10.67	.750		25.50	33		58.50	82
4130	EMT & wire		4.71	1.699		38	75		113	164
4500	Weather-proof cover for above receptacles, add		32	.250		2.94	11		13.94	21
4550	Air conditioner outlet, 20 amp-240 volt recpt.									
4560	30′ of #12/2, 2 pole circuit breaker									
4570	Type NM cable	1 Elec	10	.800	Ea.	62	35		97	126
4580	Type MC cable		9	.889		70	39		109	141
4590	EMT & wire		4	2		81	88		169	233
4600	Decorator style, type NM cable		10	.800		67	35		102	131
4620	Type MC cable		9	.889		74.50	39		113.50	146
4630	EMT & wire		4	2		85.50	88		173.50	238
4650	Dryer outlet, 30 amp-240 volt recpt., 20′ of #10/3									
4660	2 pole circuit breaker									
4670	Type NM cable	1 Elec	6.41	1.248	Ea.	58	55		113	154
4680	Type MC cable		5.71	1.401		64	61.50		125.50	172
4690	EMT & wire		3.48	2.299		75	101		176	249
4700	Range outlet, 50 amp-240 volt recpt., 30′ of #8/3									
4710	Type NM cable	1 Elec	4.21	1.900	Ea.	85	83.50		168.50	231

26 05 90.10 Residential Wiring		Crew	Daily Output	Labor-Hours	Unit	2016 Bare Costs Material	Labor	Equipment	Total	Total Incl O&P
4720	Type MC cable	1 Elec	4	2	Ea.	126	88		214	282
4730	EMT & wire		2.96	2.703		104	119		223	310
4750	Central vacuum outlet, Type NM cable		6.40	1.250		53.50	55		108.50	149
4770	Type MC cable		5.71	1.401		66.50	61.50		128	175
4780	EMT & wire		3.48	2.299		81	101		182	255
4800	30 amp-110 volt locking recpt., #10/2 circ. bkr.									
4810	Type NM cable	1 Elec	6.20	1.290	Ea.	61.50	57		118.50	161
4820	Type MC cable		5.40	1.481		78.50	65.50		144	193
4830	EMT & wire		3.20	2.500		91.50	110		201.50	281
4900	Low voltage outlets									
4910	Telephone recpt., 20′ of 4/C phone wire	1 Elec	26	.308	Ea.	8.40	13.55		21.95	31
4920	TV recpt., 20′ of RG59U coax wire, F type connector	"	16	.500	"	16	22		38	53.50
4950	Door bell chime, transformer, 2 buttons, 60′ of bellwire									
4970	Economy model	1 Elec	11.50	.696	Ea.	50	30.50		80.50	105
4980	Custom model	"	11.50	.696	"	99.50	30.50		130	159
6000	Lighting outlets									
6050	Wire only (for fixture), type NM cable	1 Elec	32	.250	Ea.	6.35	11		17.35	25
6070	Type MC cable		24	.333		12.55	14.70		27.25	38
6080	EMT & wire		10	.800		20.50	35		55.50	80
6100	Box (4″), and wire (for fixture), type NM cable		25	.320		13.90	14.10		28	38.50
6120	Type MC cable		20	.400		20	17.60		37.60	51
6130	EMT & wire		11	.727		28	32		60	83.50
6200	Fixtures (use with lines 6050 or 6100 above)									
6210	Canopy style, economy grade	1 Elec	40	.200	Ea.	27	8.80		35.80	44
6220	Custom grade		40	.200		53.50	8.80		62.30	73.50
6250	Dining room chandelier, economy grade		19	.421		67.50	18.55		86.05	105
6260	Custom grade		19	.421		315	18.55		333.55	375
6310	Kitchen fixture (fluorescent), economy grade		30	.267		71.50	11.75		83.25	98
6320	Custom grade		25	.320		221	14.10		235.10	267
6350	Outdoor, wall mounted, economy grade		30	.267		25.50	11.75		37.25	47
6360	Custom grade		30	.267		120	11.75		131.75	151
6410	Outdoor PAR floodlights, 1 lamp, 150 watt		20	.400		27.50	17.60		45.10	59
6420	2 lamp, 150 watt each		20	.400		45.50	17.60		63.10	79
6425	Motion sensing, 2 lamp, 150 watt each		20	.400		75.50	17.60		93.10	112
6430	For infrared security sensor, add		32	.250		132	11		143	163
6450	Outdoor, quartz-halogen, 300 watt flood		20	.400		33	17.60		50.60	65.50
6600	Recessed downlight, round, pre-wired, 50 or 75 watt trim		30	.267		68	11.75		79.75	93.50
6610	With shower light trim		30	.267		82.50	11.75		94.25	110
6620	With wall washer trim		28	.286		85	12.60		97.60	114
6700	Porcelain lamp holder		40	.200		2.52	8.80		11.32	17.15
6710	With pull switch		40	.200		7.55	8.80		16.35	22.50
6750	Fluorescent strip, 2-20 watt tube, wrap around diffuser, 24″		24	.333		42	14.70		56.70	70.50
6760	1-34 watt tube, 48″		24	.333		100	14.70		114.70	134
6770	2-34 watt tubes, 48″		20	.400		118	17.60		135.60	159
6800	Bathroom heat lamp, 1-250 watt		28	.286		37.50	12.60		50.10	61.50
6810	2-250 watt lamps		28	.286		59.50	12.60		72.10	86
6820	For timer switch, see Section 26 05 90.10 line 2400									
6900	Outdoor post lamp, incl. post, fixture, 35′ of #14/2									
6910	Type NM cable	1 Elec	3.50	2.286	Ea.	285	101		386	480
6920	Photo-eye, add		27	.296		27.50	13.05		40.55	51.50
6950	Clock dial time switch, 24 hr., w/enclosure, type NM cable		11.43	.700		73	31		104	131
6970	Type MC cable		11	.727		81.50	32		113.50	143
6980	EMT & wire		4.85	1.649		90.50	72.50		163	219

26 05 90.10 Residential Wiring		Crew	Daily Output	Labor-Hours	Unit	2016 Bare Costs Material	Labor	Equipment	Total	Total Incl O&P
7000	Alarm systems									
7050	Smoke detectors, box, #14/3, type NM cable	1 Elec	14.55	.550	Ea.	33.50	24		57.50	76.50
7070	Type MC cable		12.31	.650		44.50	28.50		73	96
7080	EMT & wire		5	1.600		49	70.50		119.50	169
7090	For relay output to security system, add					10.20			10.20	11.20
8000	Residential equipment									
8050	Disposal hook-up, incl. switch, outlet box, 3′ of flex									
8060	20 amp-1 pole circ. bkr., and 25′ of #12/2									
8070	Type NM cable	1 Elec	10	.800	Ea.	36.50	35		71.50	97.50
8080	Type MC cable		8	1		43.50	44		87.50	120
8090	EMT & wire		5	1.600		58.50	70.50		129	179
8100	Trash compactor or dishwasher hook-up, incl. outlet box,									
8110	3′ of flex, 15 amp-1 pole circ. bkr., and 25′ of #14/2									
8130	Type MC cable	1 Elec	8	1	Ea.	31	44		75	106
8140	EMT & wire	"	5	1.600	"	43.50	70.50		114	163
8150	Hot water sink dispenser hook-up, use line 8100									
8200	Vent/exhaust fan hook-up, type NM cable	1 Elec	32	.250	Ea.	6.35	11		17.35	25
8220	Type MC cable		24	.333		12.55	14.70		27.25	38
8230	EMT & wire		10	.800		20.50	35		55.50	80
8250	Bathroom vent fan, 50 CFM (use with above hook-up)									
8260	Economy model	1 Elec	15	.533	Ea.	19.65	23.50		43.15	60
8280	Custom model	"	12	.667	"	131	29.50		160.50	192
8300	Bathroom or kitchen vent fan, 110 CFM									
8310	Economy model	1 Elec	15	.533	Ea.	67.50	23.50		91	113
8320	Low noise model	"	15	.533	"	92	23.50		115.50	140
8350	Paddle fan, variable speed (w/o lights)									
8360	Economy model (AC motor)	1 Elec	10	.800	Ea.	109	35		144	178
8362	With light kit		10	.800		150	35		185	223
8370	Custom model (AC motor)		10	.800		262	35		297	345
8372	With light kit		10	.800		300	35		335	395
8390	Remote speed switch for above, add		12	.667		37.50	29.50		67	89
8500	Whole house exhaust fan, ceiling mount, 36″, variable speed									
8510	Remote switch, incl. shutters, 20 amp-1 pole circ. bkr.									
8520	30′ of #12/2, type NM cable	1 Elec	4	2	Ea.	1,325	88		1,413	1,625
8530	Type MC cable		3.50	2.286		1,350	101		1,451	1,650
8540	EMT & wire		3	2.667		1,350	117		1,467	1,700
8600	Whirlpool tub hook-up, incl. timer switch, outlet box									
8610	3′ of flex, 20 amp-1 pole GFI circ. bkr.									
8620	30′ of #12/2, type NM cable	1 Elec	5	1.600	Ea.	157	70.50		227.50	287
8630	Type MC cable		4.20	1.905		160	84		244	315
8640	EMT & wire		3.40	2.353		172	104		276	360
8650	Hot water heater hook-up, incl. 1-2 pole circ. bkr., box;									
8660	3′ of flex, 20′ of #10/2, type NM cable	1 Elec	5	1.600	Ea.	31.50	70.50		102	150
8670	Type MC cable		4.20	1.905		42.50	84		126.50	184
8680	EMT & wire		3.40	2.353		47	104		151	222
9000	Heating/air conditioning									
9050	Furnace/boiler hook-up, incl. firestat, local on-off switch									
9060	Emergency switch, and 40′ of type NM cable	1 Elec	4	2	Ea.	52	88		140	201
9070	Type MC cable		3.50	2.286		65.50	101		166.50	237
9080	EMT & wire		1.50	5.333		83.50	235		318.50	475
9100	Air conditioner hook-up, incl. local 60 amp disc. switch									
9110	3′ sealtite, 40 amp, 2 pole circuit breaker									
9130	40′ of #8/2, type NM cable	1 Elec	3.50	2.286	Ea.	144	101		245	325

26 05 Common Work Results for Electrical

26 05 90 – Residential Applications

26 05 90.10 Residential Wiring		Crew	Daily Output	Labor-Hours	Unit	Material	Labor	Equipment	Total	Total Incl O&P
						2016 Bare Costs				
9140	Type MC cable	1 Elec	3	2.667	Ea.	204	117		321	415
9150	EMT & wire		1.30	6.154		185	271		456	650
9200	Heat pump hook-up, 1-40 & 1-100 amp 2 pole circ. bkr.									
9210	Local disconnect switch, 3′ sealtite									
9220	40′ of #8/2 & 30′ of #3/2									
9230	Type NM cable	1 Elec	1.30	6.154	Ea.	460	271		731	950
9240	Type MC cable		1.08	7.407		510	325		835	1,100
9250	EMT & wire		.94	8.511		520	375		895	1,175
9500	Thermostat hook-up, using low voltage wire									
9520	Heating only, 25′ of #18-3	1 Elec	24	.333	Ea.	8	14.70		22.70	33
9530	Heating/cooling, 25′ of #18-4	″	20	.400	″	10.50	17.60		28.10	40.50

26 09 Instrumentation and Control for Electrical Systems

26 09 13 – Electrical Power Monitoring

26 09 13.20 Voltage Monitor Systems

Line	Description	Crew	Daily Output	Labor-Hours	Unit	Material	Labor	Equipment	Total	Total Incl O&P
0010	**VOLTAGE MONITOR SYSTEMS** (test equipment)									
0100	AC voltage monitor system, 120/240 V, one-channel				Ea.	2,800			2,800	3,075
0110	Modem adapter					350			350	385
0120	Add-on detector only					1,475			1,475	1,600
0150	AC voltage remote monitor sys., 3 channel, 120, 230, or 480 V					5,075			5,075	5,575
0160	With internal modem					5,350			5,350	5,875
0170	Combination temperature and humidity probe					785			785	865
0180	Add-on detector only					3,675			3,675	4,025
0190	With internal modem					4,000			4,000	4,400

26 09 13.30 Smart Metering

Line	Description		Crew	Daily Output	Labor-Hours	Unit	Material	Labor	Equipment	Total	Total Incl O&P
0010	**SMART METERING,** In panel										
0100	Single phase, 120/208 volt, 100 amp	G	1 Elec	8.78	.911	Ea.	380	40		420	485
0120	200 amp	G		8.78	.911		380	40		420	485
0200	277 volt, 100 amp	G		8.78	.911		405	40		445	515
0220	200 amp	G		8.78	.911		405	40		445	515
1100	Three phase, 120/208 volt, 100 amp	G		4.69	1.706		715	75		790	910
1120	200 amp	G		4.69	1.706		720	75		795	920
1130	400 amp	G		4.69	1.706		730	75		805	930
1140	800 amp	G		4.69	1.706		795	75		870	1,000
1150	1600 amp	G		4.69	1.706		795	75		870	1,000
1200	277/480 volt, 100 amp	G		4.69	1.706		735	75		810	930
1220	200 amp	G		4.69	1.706		795	75		870	1,000
1230	400 amp	G		4.69	1.706		795	75		870	1,000
1240	800 amp	G		4.69	1.706		925	75		1,000	1,150
1250	1600 amp	G		4.69	1.706		925	75		1,000	1,150
2000	Data recorder, 8 meters	G		10.97	.729		1,475	32		1,507	1,675
2100	16 meters	G		8.53	.938		2,025	41.50		2,066.50	2,300
3000	Software package, per meter, basic	G					246			246	271
3100	Premium	G					680			680	750

26 09 23 – Lighting Control Devices

26 09 23.10 Energy Saving Lighting Devices

Line	Description		Crew	Daily Output	Labor-Hours	Unit	Material	Labor	Equipment	Total	Total Incl O&P
0010	**ENERGY SAVING LIGHTING DEVICES**										
0100	Occupancy sensors, passive infrared ceiling mounted	G	1 Elec	7	1.143	Ea.	79	50.50		129.50	170
0110	Ultrasonic ceiling mounted	G		7	1.143		89	50.50		139.50	180
0120	Dual technology ceiling mounted	G		6.50	1.231		134	54		188	236

26 09 Instrumentation and Control for Electrical Systems

26 09 23 – Lighting Control Devices

26 09 23.10 Energy Saving Lighting Devices		Crew	Daily Output	Labor-Hours	Unit	Material	Labor	Equipment	Total	Total Incl O&P
						2016 Bare Costs				
0150	Automatic wall switches G	1 Elec	24	.333	Ea.	75	14.70		89.70	107
0160	Daylighting sensor, manual control, ceiling mounted G		7	1.143		129	50.50		179.50	225
0170	Remote and dimming control with remote controller G		6.50	1.231		159	54		213	264
0200	Remote power pack G		10	.800		32.50	35		67.50	93.50
0250	Photoelectric control, S.P.S.T. 120 V G		8	1		22.50	44		66.50	96.50
0300	S.P.S.T. 208 V/277 V G		8	1		28.50	44		72.50	104
0350	D.P.S.T. 120 V G		6	1.333		217	58.50		275.50	335
0400	D.P.S.T. 208 V/277 V G		6	1.333		202	58.50		260.50	320
0450	S.P.D.T. 208 V/277 V G		6	1.333		243	58.50		301.50	365
0460	Daylight level sensor, wall mounted, on/off or dimming G		8	1		135	44		179	221

26 12 Medium-Voltage Transformers

26 12 19 – Pad-Mounted, Liquid-Filled, Medium-Voltage Transformers

26 12 19.20 Transformer, Liquid-Filled		Crew	Daily Output	Labor-Hours	Unit	Material	Labor	Equipment	Total	Total Incl O&P
0010	**TRANSFORMER, LIQUID-FILLED** Pad mounted									
0020	5 kV or 15 kV primary, 277/480 volt secondary, 3 phase									
0050	225 kVA	R-3	.55	36.364	Ea.	13,900	1,575	255	15,730	18,200
0100	300 kVA		.45	44.444		16,500	1,925	310	18,735	21,600
0200	500 kVA		.40	50		20,800	2,175	350	23,325	26,900
0250	750 kVA		.38	52.632		26,900	2,300	370	29,570	33,800

26 22 Low-Voltage Transformers

26 22 13 – Low-Voltage Distribution Transformers

26 22 13.10 Transformer, Dry-Type		Crew	Daily Output	Labor-Hours	Unit	Material	Labor	Equipment	Total	Total Incl O&P
0010	**TRANSFORMER, DRY-TYPE**									
0050	Single phase, 240/480 volt primary, 120/240 volt secondary									
0100	1 kVA	1 Elec	2	4	Ea.	340	176		516	665
0300	2 kVA		1.60	5		540	220		760	955
0500	3 kVA		1.40	5.714		670	252		922	1,150
0700	5 kVA		1.20	6.667		920	294		1,214	1,475
1100	10 kVA	2 Elec	1.60	10		1,450	440		1,890	2,325
1300	15 kVA		1.20	13.333		1,975	585		2,560	3,100
1700	37.5 kVA		.80	20		2,675	880		3,555	4,400
1900	50 kVA		.70	22.857		3,250	1,000		4,250	5,250
2110	100 kVA	R-3	.90	22.222		5,475	965	156	6,596	7,775
2310	Ventilated, 3 kVA	1 Elec	1	8		1,050	350		1,400	1,725
2900	9 kVA	"	.70	11.429		1,175	505		1,680	2,125
3100	15 kVA	2 Elec	1.10	14.545		1,500	640		2,140	2,700
3300	30 kVA		.90	17.778		1,475	785		2,260	2,900
3700	75 kVA		.70	22.857		2,375	1,000		3,375	4,250
3900	112.5 kVA	R-3	.90	22.222		3,925	965	156	5,046	6,050
4100	150 kVA		.85	23.529		4,750	1,025	165	5,940	7,075
4500	300 kVA		.55	36.364		8,750	1,575	255	10,580	12,500
4700	500 kVA		.45	44.444		14,500	1,925	310	16,735	19,500
5380	3 phase, 5 kV primary, 277/480 volt secondary									
5400	High voltage, 112.5 kVA	R-3	.85	23.529	Ea.	17,200	1,025	165	18,390	20,900
5440	500 kVA		.35	57.143		36,500	2,475	400	39,375	44,600
5480	2000 kVA		.25	80		91,000	3,475	560	95,035	106,500
5590	15 kV primary, 277/480 volt secondary									

26 22 Low-Voltage Transformers

26 22 13 – Low-Voltage Distribution Transformers

26 22 13.10 Transformer, Dry-Type		Crew	Daily Output	Labor-Hours	Unit	Material	2016 Bare Costs Labor	Equipment	Total	Total Incl O&P
5600	High voltage, 112.5 kVA	R-3	.85	23.529	Ea.	26,300	1,025	165	27,490	30,800
5640	500 kVA		.35	57.143		50,500	2,475	400	53,375	60,000
5680	2000 kVA		.25	80		96,500	3,475	560	100,535	112,500

26 22 16 – Low-Voltage Buck-Boost Transformers

26 22 16.10 Buck-Boost Transformer		Crew	Daily Output	Labor-Hours	Unit	Material	Labor	Equipment	Total	Total Incl O&P
0010	**BUCK-BOOST TRANSFORMER**									
0100	Single phase, 120/240 V primary, 12/24 V secondary									
0200	0.10 kVA	1 Elec	8	1	Ea.	127	44		171	212
0400	0.25 kVA		5.70	1.404		186	62		248	305
0600	0.50 kVA		4	2		253	88		341	420
0800	0.75 kVA		3.10	2.581		325	114		439	545
1200	1.5 kVA		1.80	4.444		480	196		676	850
1600	3.0 kVA		1.40	5.714		805	252		1,057	1,300
2000	3 phase, 240 V primary, 208/120 V secondary, 15 kVA	2 Elec	2.40	6.667		2,100	294		2,394	2,775
2200	30 kVA		1.60	10		2,350	440		2,790	3,300
2400	45 kVA		1.40	11.429		2,800	505		3,305	3,925
2600	75 kVA		1.20	13.333		3,400	585		3,985	4,700
3000	150 kVA	R-3	1.10	18.182		5,625	790	128	6,543	7,650
3200	225 kVA	"	1	20		7,350	870	140	8,360	9,650

26 24 Switchboards and Panelboards

26 24 16 – Panelboards

26 24 16.10 Load Centers		Crew	Daily Output	Labor-Hours	Unit	Material	Labor	Equipment	Total	Total Incl O&P
0010	**LOAD CENTERS** (residential type)									
0100	3 wire, 120/240 V, 1 phase, including 1 pole plug-in breakers									
0200	100 amp main lugs, indoor, 8 circuits	1 Elec	1.40	5.714	Ea.	184	252		436	610
0300	12 circuits		1.20	6.667		93	294		387	580
0400	Rainproof, 8 circuits		1.40	5.714		220	252		472	650
0500	12 circuits		1.20	6.667		310	294		604	820
0600	200 amp main lugs, indoor, 16 circuits	R-1A	1.80	8.889		281	350		631	885
0700	20 circuits		1.50	10.667		430	425		855	1,150
0800	24 circuits		1.30	12.308		465	490		955	1,300
1200	Rainproof, 16 circuits		1.80	8.889		405	350		755	1,025
1300	20 circuits		1.50	10.667		400	425		825	1,125
1400	24 circuits		1.30	12.308		730	490		1,220	1,600
1500	30 circuits		1.20	13.333		670	530		1,200	1,600
1600	40 circuits		.80	20		880	795		1,675	2,275
1800	400 amp main lugs, indoor, 42 circuits		.72	22.222		1,025	880		1,905	2,600
1900	Rainproof, 42 circuit		.72	22.222		1,325	880		2,205	2,900
2200	Plug in breakers, 20 amp, 1 pole, 4 wire, 120/208 volts									
2210	125 amp main lugs, indoor, 12 circuits	1 Elec	1.20	6.667	Ea.	265	294		559	770
2300	18 circuits		.80	10		455	440		895	1,225
2400	Rainproof, 12 circuits		1.20	6.667		305	294		599	815
2500	18 circuits		.80	10		495	440		935	1,275
2600	200 amp main lugs, indoor, 24 circuits	R-1A	1.30	12.308		505	490		995	1,350
2700	30 circuits		1.20	13.333		575	530		1,105	1,500
2800	36 circuits		1	16		845	635		1,480	1,975
2900	42 circuits		.80	20		765	795		1,560	2,150
3000	Rainproof, 24 circuits		1.30	12.308		560	490		1,050	1,425
3100	30 circuits		1.20	13.333		630	530		1,160	1,550

26 24 16 – Panelboards

26 24 16.10 Load Centers

26 24 16.10 Load Centers		Crew	Daily Output	Labor-Hours	Unit	2016 Bare Costs Material	Labor	Equipment	Total	Total Incl O&P
3200	36 circuits	R-1A	1	16	Ea.	1,050	635		1,685	2,225
3300	42 circuits		.80	20		1,150	795		1,945	2,550

26 24 16.20 Panelboard and Load Center Circuit Breakers

26 24 16.20 Panelboard and Load Center Circuit Breakers		Crew	Daily Output	Labor-Hours	Unit	Material	Labor	Equipment	Total	Total Incl O&P
0010	**PANELBOARD AND LOAD CENTER CIRCUIT BREAKERS**									
0050	Bolt-on, 10,000 amp I.C., 120 volt, 1 pole									
0100	15 to 50 amp	1 Elec	10	.800	Ea.	19.15	35		54.15	78.50
0200	60 amp		8	1		19.20	44		63.20	93
0300	70 amp		8	1		28	44		72	103
0350	240 volt, 2 pole									
0400	15 to 50 amp	1 Elec	8	1	Ea.	82	44		126	163
0500	60 amp		7.50	1.067		56	47		103	139
0600	80 to 100 amp		5	1.600		107	70.50		177.50	233
0700	3 pole, 15 to 60 amp		6.20	1.290		133	57		190	239
0800	70 amp		5	1.600		168	70.50		238.50	300
0900	80 to 100 amp		3.60	2.222		191	98		289	370
1000	22,000 amp I.C., 240 volt, 2 pole, 70 – 225 amp		2.70	2.963		600	131		731	875
1100	3 pole, 70 – 225 amp		2.30	3.478		595	153		748	905
2000	Plug-in panel or load center, 120/240 volt, to 60 amp, 1 pole		12	.667		9.70	29.50		39.20	58.50
2004	Circuit breaker, 120/240 volt, 20 A, 1 pole with NM cable		6.50	1.231		16.35	54		70.35	106
2006	30 A, 1 pole with NM cable		6.50	1.231		16.35	54		70.35	106
2010	2 pole		9	.889		26	39		65	92.50
2014	50 A, 2 pole with NM cable		5.50	1.455		32.50	64		96.50	141
2020	3 pole		7.50	1.067		76	47		123	161
2030	100 amp, 2 pole		6	1.333		102	58.50		160.50	208
2040	3 pole		4.50	1.778		113	78.50		191.50	252
2050	150 to 200 amp, 2 pole		3	2.667		218	117		335	430

26 24 16.30 Panelboards Commercial Applications

26 24 16.30 Panelboards Commercial Applications		Crew	Daily Output	Labor-Hours	Unit	Material	Labor	Equipment	Total	Total Incl O&P
0010	**PANELBOARDS COMMERCIAL APPLICATIONS**									
0050	NQOD, w/20 amp 1 pole bolt-on circuit breakers									
0600	4 wire, 120/208 volts, 100 amp main lugs, 12 circuits	1 Elec	1	8	Ea.	690	350		1,040	1,325
0650	16 circuits		.75	10.667		790	470		1,260	1,650
0700	20 circuits		.65	12.308		915	540		1,455	1,875
0750	24 circuits		.60	13.333		910	585		1,495	1,950
0800	30 circuits		.53	15.094		1,125	665		1,790	2,350
0850	225 amp main lugs, 32 circuits	2 Elec	.90	17.778		1,275	785		2,060	2,675
0900	34 circuits		.84	19.048		1,300	840		2,140	2,800
0950	36 circuits		.80	20		1,325	880		2,205	2,925
1000	42 circuits		.68	23.529		1,475	1,025		2,500	3,325
1200	NEHB, w/20 amp, 1 pole bolt-on circuit breakers									
1250	4 wire, 277/480 volts, 100 amp main lugs, 12 circuits	1 Elec	.88	9.091	Ea.	1,325	400		1,725	2,100
1300	20 circuits	"	.60	13.333		1,975	585		2,560	3,125
1350	225 amp main lugs, 24 circuits	2 Elec	.90	17.778		2,225	785		3,010	3,725
1400	30 circuits		.80	20		2,675	880		3,555	4,400
1450	36 circuits		.72	22.222		3,125	980		4,105	5,025
1500	42 circuits		.60	26.667		3,550	1,175		4,725	5,825
1510	225 amp main lugs, NEMA 7, 12 circuits		.90	17.778		5,950	785		6,735	7,825
1590	24 circuits		.30	53.333		8,000	2,350		10,350	12,700

26 24 Switchboards and Panelboards

26 24 19 – Motor-Control Centers

26 24 19.30 Motor Control Center		Crew	Daily Output	Labor-Hours	Unit	Material	2016 Bare Costs Labor	Equipment	Total	Total Incl O&P
0010	**MOTOR CONTROL CENTER** Consists of starters & structures									
0050	Starters, class 1, type B, comb. MCP, FVNR, with									
0100	control transformer, 10 HP, size 1, 12" high	1 Elec	2.70	2.963	Ea.	1,650	131		1,781	2,025
0200	25 HP, size 2, 18" high	2 Elec	4	4		1,875	176		2,051	2,350
0300	50 HP, size 3, 24" high		2	8		2,900	350		3,250	3,750
0350	75 HP, size 4, 24" high		1.60	10		3,850	440		4,290	4,950
0400	100 HP, size 4, 30" high		1.40	11.429		5,075	505		5,580	6,400
0500	200 HP, size 5, 48" high		1	16		7,575	705		8,280	9,475

26 24 19.40 Motor Starters and Controls		Crew	Daily Output	Labor-Hours	Unit	Material	Labor	Equipment	Total	Total Incl O&P
0010	**MOTOR STARTERS AND CONTROLS**									
0050	Magnetic, FVNR, with enclosure and heaters, 480 volt									
0080	2 HP, size 00	1 Elec	3.50	2.286	Ea.	197	101		298	380
0100	5 HP, size 0		2.30	3.478		315	153		468	595
0200	10 HP, size 1		1.60	5		267	220		487	655
0300	25 HP, size 2	2 Elec	2.20	7.273		500	320		820	1,075
0400	50 HP, size 3	"	1.80	8.889		820	390		1,210	1,550
1400	Combination, with fused switch, 5 HP, size 0	1 Elec	1.80	4.444		590	196		786	970
1600	10 HP, size 1	"	1.30	6.154		630	271		901	1,150
1800	25 HP, size 2	2 Elec	2	8		1,025	350		1,375	1,700

26 27 Low-Voltage Distribution Equipment

26 27 13 – Electricity Metering

26 27 13.10 Meter Centers and Sockets		Crew	Daily Output	Labor-Hours	Unit	Material	Labor	Equipment	Total	Total Incl O&P
0010	**METER CENTERS AND SOCKETS**									
0100	Sockets, single position, 4 terminal, 100 amp	1 Elec	3.20	2.500	Ea.	43	110		153	228
0200	150 amp		2.30	3.478		48	153		201	305
0300	200 amp		1.90	4.211		92	185		277	405
0400	Transformer rated, 20 amp		3.20	2.500		152	110		262	345
0500	Double position, 4 terminal, 100 amp		2.80	2.857		204	126		330	430
0600	150 amp		2.10	3.810		240	168		408	540
0700	200 amp		1.70	4.706		450	207		657	835
1100	Meter centers and sockets, three phase, single pos, 7 terminal, 100 amp		2.80	2.857		164	126		290	385
1200	200 amp		2.10	3.810		295	168		463	600
1400	400 amp		1.70	4.706		720	207		927	1,125
2590	Basic meter device									
2600	1P 3W 120/240 V 4 jaw 125A sockets, 3 meter	2 Elec	1	16	Ea.	340	705		1,045	1,525
2610	4 meter		.90	17.778		405	785		1,190	1,725
2620	5 meter		.80	20		490	880		1,370	1,975
2630	6 meter		.60	26.667		640	1,175		1,815	2,625
2640	7 meter		.56	28.571		1,400	1,250		2,650	3,600
2650	8 meter		.52	30.769		1,525	1,350		2,875	3,925
2660	10 meter		.48	33.333		1,925	1,475		3,400	4,525
2680	Rainproof 1P 3W 120/240 V 4 jaw 125A sockets									
2690	3 meter	2 Elec	1	16	Ea.	640	705		1,345	1,850
2700	4 meter		.90	17.778		770	785		1,555	2,125
2710	6 meter		.60	26.667		1,100	1,175		2,275	3,150
2720	7 meter		.56	28.571		1,400	1,250		2,650	3,600
2730	8 meter		.52	30.769		1,525	1,350		2,875	3,925
2750	1P 3W 120/240 V 4 jaw sockets									
2760	with 125A circuit breaker, 3 meter	2 Elec	1	16	Ea.	1,200	705		1,905	2,475

26 27 Low-Voltage Distribution Equipment

26 27 13 – Electricity Metering

26 27 13.10 Meter Centers and Sockets		Crew	Daily Output	Labor-Hours	Unit	2016 Bare Costs Material	Labor	Equipment	Total	Total Incl O&P
2770	4 meter	2 Elec	.90	17.778	Ea.	1,525	785		2,310	2,950
2780	5 meter		.80	20		1,900	880		2,780	3,525
2790	6 meter		.60	26.667		2,225	1,175		3,400	4,375
2800	7 meter		.56	28.571		2,725	1,250		3,975	5,025
2810	8 meter		.52	30.769		3,025	1,350		4,375	5,550
2820	10 meter		.48	33.333		3,800	1,475		5,275	6,575
3250	1P 3W 120/240 V 4 jaw sockets									
3260	with 200A circuit breaker, 3 meter	2 Elec	1	16	Ea.	1,800	705		2,505	3,125
3270	4 meter		.90	17.778		2,425	785		3,210	3,950
3290	6 meter		.60	26.667		3,600	1,175		4,775	5,875
3300	7 meter		.56	28.571		4,225	1,250		5,475	6,700
3310	8 meter		.56	28.571		4,875	1,250		6,125	7,400

26 27 16 – Electrical Cabinets and Enclosures

26 27 16.10 Cabinets		Crew	Daily Output	Labor-Hours	Unit	Material	Labor	Equipment	Total	Total Incl O&P
0010	**CABINETS**									
7000	Cabinets, current transformer									
7050	Single door, 24" H x 24" W x 10" D	1 Elec	1.60	5	Ea.	129	220		349	500
7100	30" H x 24" W x 10" D		1.30	6.154		140	271		411	600
7150	36" H x 24" W x 10" D		1.10	7.273		186	320		506	730
7200	30" H x 30" W x 10" D		1	8		250	350		600	850
7250	36" H x 30" W x 10" D		.90	8.889		300	390		690	970
7300	36" H x 36" W x 10" D		.80	10		310	440		750	1,050
7500	Double door, 48" H x 36" W x 10" D		.60	13.333		590	585		1,175	1,600
7550	24" H x 24" W x 12" D		1	8		173	350		523	765

26 27 23 – Indoor Service Poles

26 27 23.40 Surface Raceway		Crew	Daily Output	Labor-Hours	Unit	Material	Labor	Equipment	Total	Total Incl O&P
0010	**SURFACE RACEWAY**									
0090	Metal, straight section									
0100	No. 500	1 Elec	100	.080	L.F.	1.04	3.52		4.56	6.90
0110	No. 700		100	.080		1.17	3.52		4.69	7.05
0200	No. 1000		90	.089		1.45	3.92		5.37	8
0400	No. 1500, small pancake		90	.089		2.15	3.92		6.07	8.75
0600	No. 2000, base & cover, blank		90	.089		2.19	3.92		6.11	8.80
0800	No. 3000, base & cover, blank		75	.107		4.18	4.70		8.88	12.30
1000	No. 4000, base & cover, blank		65	.123		6.80	5.40		12.20	16.35
1200	No. 6000, base & cover, blank		50	.160		11.40	7.05		18.45	24
2400	Fittings, elbows, No. 500		40	.200	Ea.	1.90	8.80		10.70	16.50
2800	Elbow cover, No. 2000		40	.200		3.57	8.80		12.37	18.35
2880	Tee, No. 500		42	.190		3.66	8.40		12.06	17.75
2900	No. 2000		27	.296		11.85	13.05		24.90	34.50
3000	Switch box, No. 500		16	.500		11	22		33	48
3400	Telephone outlet, No. 1500		16	.500		14.10	22		36.10	51.50
3600	Junction box, No. 1500		16	.500		9.65	22		31.65	46.50
3800	Plugmold wired sections, No. 2000									
4000	1 circuit, 6 outlets, 3 ft. long	1 Elec	8	1	Ea.	36	44		80	112
4100	2 circuits, 8 outlets, 6 ft. long	"	5.30	1.509	"	59	66.50		125.50	174
9300	Non-metallic, straight section									
9310	7/16" x 7/8", base & cover, blank	1 Elec	160	.050	L.F.	1.72	2.20		3.92	5.50
9320	Base & cover w/adhesive		160	.050		1.42	2.20		3.62	5.15
9340	7/16" x 1-5/16", base & cover, blank		145	.055		1.88	2.43		4.31	6.05
9350	Base & cover w/adhesive		145	.055		2.17	2.43		4.60	6.35
9370	11/16" x 2-1/4", base & cover, blank		130	.062		2.66	2.71		5.37	7.35

26 27 Low-Voltage Distribution Equipment

26 27 23 – Indoor Service Poles

26 27 23.40 Surface Raceway		Crew	Daily Output	Labor-Hours	Unit	2016 Bare Costs Material	Labor	Equipment	Total	Total Incl O&P
9380	Base & cover w/adhesive	1 Elec	130	.062	L.F.	3.01	2.71		5.72	7.75
9385	1-11/16" x 5-1/4", two compartment base & cover w/screws		80	.100		7.55	4.41		11.96	15.50
9400	Fittings, elbows, 7/16" x 7/8"		50	.160	Ea.	1.88	7.05		8.93	13.60
9410	7/16" x 1-5/16"		45	.178		1.95	7.85		9.80	14.95
9420	11/16" x 2-1/4"		40	.200		2.11	8.80		10.91	16.70
9425	1-11/16" x 5-1/4"		28	.286		11.05	12.60		23.65	32.50
9430	Tees, 7/16" x 7/8"		35	.229		2.42	10.05		12.47	19.10
9440	7/16" x 1-5/16"		32	.250		2.49	11		13.49	20.50
9450	11/16" x 2-1/4"		30	.267		2.56	11.75		14.31	22
9455	1-11/16" x 5-1/4"		24	.333		17.75	14.70		32.45	43.50
9460	Cover clip, 7/16" x 7/8"		80	.100		.49	4.41		4.90	7.75
9470	7/16" x 1-5/16"		72	.111		.44	4.89		5.33	8.50
9480	11/16" x 2-1/4"		64	.125		.75	5.50		6.25	9.85
9484	1-11/16" x 5-1/4"		42	.190		2.61	8.40		11.01	16.55
9486	Wire clip, 1-11/16" x 5-1/4"		68	.118		.45	5.20		5.65	9
9490	Blank end, 7/16" x 7/8"		50	.160		.70	7.05		7.75	12.30
9500	7/16" x 1-5/16"		45	.178		.78	7.85		8.63	13.65
9510	11/16" x 2-1/4"		40	.200		1.18	8.80		9.98	15.70
9515	1-11/16" x 5-1/4"		38	.211		5.45	9.25		14.70	21
9520	Round fixture box, 5.5" dia x 1"		25	.320		11.10	14.10		25.20	35
9530	Device box, 1 gang		30	.267		5	11.75		16.75	24.50
9540	2 gang		25	.320		7.30	14.10		21.40	31

26 27 26 – Wiring Devices

26 27 26.10 Low Voltage Switching		Crew	Daily Output	Labor-Hours	Unit	Material	Labor	Equipment	Total	Total Incl O&P
0010	**LOW VOLTAGE SWITCHING**									
3600	Relays, 120 V or 277 V standard	1 Elec	12	.667	Ea.	41.50	29.50		71	94
3800	Flush switch, standard		40	.200		11.50	8.80		20.30	27
4000	Interchangeable		40	.200		15.05	8.80		23.85	31
4100	Surface switch, standard		40	.200		8.15	8.80		16.95	23.50
4200	Transformer 115 V to 25 V		12	.667		130	29.50		159.50	191
4400	Master control, 12 circuit, manual		4	2		126	88		214	283
4500	25 circuit, motorized		4	2		140	88		228	298
4600	Rectifier, silicon		12	.667		45.50	29.50		75	98
4800	Switchplates, 1 gang, 1, 2 or 3 switch, plastic		80	.100		5	4.41		9.41	12.70
5000	Stainless steel		80	.100		11.35	4.41		15.76	19.70
5400	2 gang, 3 switch, stainless steel		53	.151		23	6.65		29.65	36

26 27 26.20 Wiring Devices Elements		Crew	Daily Output	Labor-Hours	Unit	Material	Labor	Equipment	Total	Total Incl O&P
0010	**WIRING DEVICES ELEMENTS**									
0200	Toggle switch, quiet type, single pole, 15 amp	1 Elec	40	.200	Ea.	9.50	8.80		18.30	25
1650	Dimmer switch, 120 volt, incandescent, 600 watt, 1 pole G		16	.500		22.50	22		44.50	61
2460	Receptacle, duplex, 120 volt, grounded, 15 amp		40	.200		1.26	8.80		10.06	15.80
2470	20 amp		27	.296		7.95	13.05		21	30
2490	Dryer, 30 amp		15	.533		4.46	23.50		27.96	43.50
2500	Range, 50 amp		11	.727		12.15	32		44.15	66
2600	Wall plates, stainless steel, 1 gang		80	.100		2.56	4.41		6.97	10
2800	2 gang		53	.151		4.33	6.65		10.98	15.65
3200	Lampholder, keyless		26	.308		16.85	13.55		30.40	40.50
3400	Pullchain with receptacle		22	.364		20.50	16		36.50	48.50

26 28 Low-Voltage Circuit Protective Devices

26 28 16 – Enclosed Switches and Circuit Breakers

26 28 16.10 Circuit Breakers

	26 28 16.10 Circuit Breakers	Crew	Daily Output	Labor-Hours	Unit	2016 Bare Costs Material	Labor	Equipment	Total	Total Incl O&P
0010	**CIRCUIT BREAKERS** (in enclosure)									
0100	Enclosed (NEMA 1), 600 volt, 3 pole, 30 amp	1 Elec	3.20	2.500	Ea.	520	110		630	750
0200	60 amp		2.80	2.857		635	126		761	905
0400	100 amp		2.30	3.478		730	153		883	1,050
0500	200 amp		1.50	5.333		1,525	235		1,760	2,050
0600	225 amp		1.50	5.333		1,675	235		1,910	2,225
0700	400 amp	2 Elec	1.60	10		2,875	440		3,315	3,900

26 28 16.20 Safety Switches

	26 28 16.20 Safety Switches	Crew	Daily Output	Labor-Hours	Unit	Material	Labor	Equipment	Total	Total Incl O&P
0010	**SAFETY SWITCHES**									
0100	General duty 240 volt, 3 pole NEMA 1, fusible, 30 amp	1 Elec	3.20	2.500	Ea.	62.50	110		172.50	249
0200	60 amp		2.30	3.478		106	153		259	365
0300	100 amp		1.90	4.211		181	185		366	505
0400	200 amp		1.30	6.154		385	271		656	870
0500	400 amp	2 Elec	1.80	8.889		980	390		1,370	1,725

26 29 Low-Voltage Controllers

26 29 13 – Enclosed Controllers

26 29 13.10 Contactors, AC

	26 29 13.10 Contactors, AC	Crew	Daily Output	Labor-Hours	Unit	Material	Labor	Equipment	Total	Total Incl O&P
0010	**CONTACTORS, AC** Enclosed (NEMA 1)									
0050	Lighting, 600 volt 3 pole, electrically held									
0100	20 amp	1 Elec	4	2	Ea.	380	88		468	560
0200	30 amp		3.60	2.222		360	98		458	555
0300	60 amp		3	2.667		675	117		792	930
0400	100 amp		2.50	3.200		1,000	141		1,141	1,350
0500	200 amp		1.40	5.714		2,525	252		2,777	3,175
0600	300 amp	2 Elec	1.60	10		6,775	440		7,215	8,175
0800	600 volt 3 pole, mechanically held, 30 amp	1 Elec	3.60	2.222		445	98		543	650
0900	60 amp		3	2.667		880	117		997	1,150
1000	75 amp		2.80	2.857		1,450	126		1,576	1,800
1100	100 amp		2.50	3.200		1,250	141		1,391	1,600
1200	150 amp		2	4		3,475	176		3,651	4,125
1300	200 amp		1.40	5.714		3,425	252		3,677	4,175
1500	Magnetic with auxiliary contact, size 00, 9 amp		4	2		181	88		269	345
1600	Size 0, 18 amp		4	2		216	88		304	380
1700	Size 1, 27 amp		3.60	2.222		245	98		343	430
1800	Size 2, 45 amp		3	2.667		455	117		572	690
1900	Size 3, 90 amp		2.50	3.200		735	141		876	1,050
2000	Size 4, 135 amp		2.30	3.478		1,675	153		1,828	2,100
2100	Size 5, 270 amp	2 Elec	1.80	8.889		3,525	390		3,915	4,550
2310	Size 8, 1215 amp	″	.80	20		21,600	880		22,480	25,300
2500	Magnetic, 240 volt, 1-2 pole, .75 HP motor	1 Elec	4	2		146	88		234	305
2520	2 HP motor		3.60	2.222		194	98		292	375
2540	5 HP motor		2.50	3.200		395	141		536	665
2560	10 HP motor		1.40	5.714		655	252		907	1,125
2600	240 volt or less, 3 pole, .75 HP motor		4	2		146	88		234	305
2620	5 HP motor		3.60	2.222		181	98		279	360
2640	10 HP motor		3.60	2.222		210	98		308	390
2660	15 HP motor		2.50	3.200		420	141		561	690
2700	25 HP motor		2.50	3.200		420	141		561	690
2720	30 HP motor	2 Elec	2.80	5.714		700	252		952	1,175
2740	40 HP motor		2.80	5.714		700	252		952	1,175

26 29 Low-Voltage Controllers

26 29 13 – Enclosed Controllers

26 29 13.10 Contactors, AC		Crew	Daily Output	Labor-Hours	Unit	Material	2016 Bare Costs Labor	Equipment	Total	Total Incl O&P
2760	50 HP motor	2 Elec	1.60	10	Ea.	700	440		1,140	1,500
2800	75 HP motor		1.60	10		1,650	440		2,090	2,525
2820	100 HP motor		1	16		1,650	705		2,355	2,950
2860	150 HP motor		1	16		3,500	705		4,205	5,000
2880	200 HP motor		1	16		3,500	705		4,205	5,000
3000	600 volt, 3 pole, 5 HP motor	1 Elec	4	2		181	88		269	345
3020	10 HP motor		3.60	2.222		210	98		308	390
3040	25 HP motor		3	2.667		420	117		537	650
3100	50 HP motor		2.50	3.200		700	141		841	1,000
3160	100 HP motor	2 Elec	2.80	5.714		1,650	252		1,902	2,200
3220	200 HP motor	"	1.60	10		3,500	440		3,940	4,575

26 29 13.20 Control Stations		Crew	Daily Output	Labor-Hours	Unit	Material	Labor	Equipment	Total	Total Incl O&P
0010	**CONTROL STATIONS**									
0050	NEMA 1, heavy duty, stop/start	1 Elec	8	1	Ea.	131	44		175	216
0100	Stop/start, pilot light		6.20	1.290		178	57		235	289
0200	Hand/off/automatic		6.20	1.290		97	57		154	200

26 32 Packaged Generator Assemblies

26 32 13 – Engine Generators

26 32 13.13 Diesel-Engine-Driven Generator Sets		Crew	Daily Output	Labor-Hours	Unit	Material	Labor	Equipment	Total	Total Incl O&P
0010	**DIESEL-ENGINE-DRIVEN GENERATOR SETS**									
2000	Diesel engine, including battery, charger,									
2010	muffler, & day tank, 30 kW	R-3	.55	36.364	Ea.	11,200	1,575	255	13,030	15,200
2100	50 kW		.42	47.619		19,800	2,075	335	22,210	25,600
2200	75 kW		.35	57.143		22,800	2,475	400	25,675	29,600
2300	100 kW		.31	64.516		27,600	2,800	455	30,855	35,500

26 32 13.16 Gas-Engine-Driven Generator Sets		Crew	Daily Output	Labor-Hours	Unit	Material	Labor	Equipment	Total	Total Incl O&P
0010	**GAS-ENGINE-DRIVEN GENERATOR SETS**									
0020	Gas or gasoline operated, includes battery,									
0050	charger, & muffler									
0200	3 phase 4 wire, 277/480 volt, 7.5 kW	R-3	.83	24.096	Ea.	8,550	1,050	169	9,769	11,300
0300	11.5 kW		.71	28.169		12,100	1,225	198	13,523	15,500
0400	20 kW		.63	31.746		14,300	1,375	223	15,898	18,200
0500	35 kW		.55	36.364		17,000	1,575	255	18,830	21,600

26 33 Battery Equipment

26 33 43 – Battery Chargers

26 33 43.55 Electric Vehicle Charging			Crew	Daily Output	Labor-Hours	Unit	Material	Labor	Equipment	Total	Total Incl O&P
0010	**ELECTRIC VEHICLE CHARGING**										
0020	Level 2, wall mounted										
2200	Heavy duty	G	R-1A	15.36	1.042	Ea.	2,525	41.50		2,566.50	2,850
2210	with RFID	G		12.29	1.302		3,100	51.50		3,151.50	3,500
2300	Free standing, single connector	G		10.24	1.563		2,900	62		2,962	3,275
2310	with RFID	G		8.78	1.822		3,625	72.50		3,697.50	4,100
2320	Double connector	G		7.68	2.083		4,850	82.50		4,932.50	5,450
2330	with RFID	G		6.83	2.343		6,300	93		6,393	7,075

26 33 Battery Equipment

26 33 53 – Static Uninterruptible Power Supply

26 33 53.10 Uninterruptible Power Supply/Conditioner Trans.		Crew	Daily Output	Labor-Hours	Unit	2016 Bare Costs Material	Labor	Equipment	Total	Total Incl O&P
0010	**UNINTERRUPTIBLE POWER SUPPLY/CONDITIONER TRANSFORMERS**									
0100	Volt. regulating, isolating transf., w/invert. & 10 min. battery pack									
0110	Single-phase, 120 V, 0.35 kVA	1 Elec	2.29	3.493	Ea.	1,100	154		1,254	1,475
0120	0.5 kVA		2	4		1,150	176		1,326	1,575
0130	For additional 55 min. battery, add to 0.35 kVA		2.29	3.493		650	154		804	965
0140	Add to 0.5 kVA		1.14	7.018		685	310		995	1,250
0150	Single-phase, 120 V, 0.75 kVA		.80	10		1,475	440		1,915	2,325
0160	1.0 kVA		.80	10		2,100	440		2,540	3,025
0170	1.5 kVA	2 Elec	1.14	14.035		3,600	620		4,220	4,950
0180	2 kVA	"	.89	17.978		3,900	790		4,690	5,600
0190	3 kVA	R-3	.63	31.746		4,725	1,375	223	6,323	7,725
0200	5 kVA		.42	47.619		6,875	2,075	335	9,285	11,300
0210	7.5 kVA		.33	60.606		8,750	2,625	425	11,800	14,400
0220	10 kVA		.28	71.429		11,300	3,100	500	14,900	18,200
0230	15 kVA		.22	90.909		14,400	3,950	640	18,990	23,100
0240	3 phase, 120/208 V input 120/208 V output, 20 kVA, incl 17 min. battery		.21	95.238		26,600	4,150	670	31,420	36,700
0242	30 kVA, incl 11 min. battery		.20	100		29,700	4,350	700	34,750	40,600
0250	40 kVA, incl 15 min. battery		.20	100		39,700	4,350	700	44,750	51,500
0260	480 V input 277/480 V output, 60 kVA, incl 6 min. battery		.19	105		44,600	4,575	740	49,915	57,500
0262	80 kVA, incl 4 min. battery		.19	105		53,000	4,575	740	58,315	67,000
0400	For additional 34 min./15 min. battery, add to 40 kVA		.71	28.011		13,800	1,225	197	15,222	17,400
0600	For complex & special design systems to meet specific									
0610	requirements, obtain quote from vendor									

26 35 Power Filters and Conditioners

26 35 26 – Harmonic Filters

26 35 26.10 Computer Isolation Transformer

Line	Description	Crew	Daily Output	Labor-Hours	Unit	Material	Labor	Equipment	Total	Total Incl O&P
0010	**COMPUTER ISOLATION TRANSFORMER**									
0100	Computer grade									
0110	Single-phase, 120/240 V, 0.5 kVA	1 Elec	4	2	Ea.	395	88		483	580
0120	1.0 kVA		2.67	2.996		560	132		692	830
0130	2.5 kVA		2	4		855	176		1,031	1,225
0140	5 kVA		1.14	7.018		975	310		1,285	1,575

26 35 26.20 Computer Regulator Transformer

Line	Description	Crew	Daily Output	Labor-Hours	Unit	Material	Labor	Equipment	Total	Total Incl O&P
0010	**COMPUTER REGULATOR TRANSFORMER**									
0100	Ferro-resonant, constant voltage, variable transformer									
0110	Single-phase, 240 V, 0.5 kVA	1 Elec	2.67	2.996	Ea.	400	132		532	655
0120	1.0 kVA		2	4		550	176		726	895
0130	2.0 kVA		1	8		940	350		1,290	1,600
0210	Plug-in unit 120 V, 0.14 kVA		8	1		231	44		275	325
0220	0.25 kVA		8	1		270	44		314	370
0230	0.5 kVA		8	1		400	44		444	510
0240	1.0 kVA		5.33	1.501		550	66		616	715
0250	2.0 kVA		4	2		940	88		1,028	1,175

26 35 26.30 Power Conditioner Transformer

Line	Description	Crew	Daily Output	Labor-Hours	Unit	Material	Labor	Equipment	Total	Total Incl O&P
0010	**POWER CONDITIONER TRANSFORMER**									
0100	Electronic solid state, buck-boost, transformer, w/tap switch									
0110	Single-phase, 115 V, 3.0 kVA, + or - 3% accuracy	2 Elec	1.60	10	Ea.	2,150	440		2,590	3,100
0120	208, 220, 230, or 240 V, 5.0 kVA, + or - 1.5% accuracy	3 Elec	1.60	15		2,800	660		3,460	4,150
0130	5.0 kVA, + or - 6% accuracy	2 Elec	1.14	14.035		2,500	620		3,120	3,750

26 35 Power Filters and Conditioners

26 35 26 – Harmonic Filters

26 35 26.30 Power Conditioner Transformer		Crew	Daily Output	Labor-Hours	Unit	2016 Bare Costs Material	Labor	Equipment	Total	Total Incl O&P
0140	7.5 kVA, + or - 1.5% accuracy	3 Elec	1.50	16	Ea.	3,550	705		4,255	5,050
0150	7.5 kVA, + or - 6% accuracy		1.60	15		2,950	660		3,610	4,325
0160	10.0 kVA, + or - 1.5% accuracy		1.33	18.045		4,725	795		5,520	6,500
0170	10.0 kVA, + or - 6% accuracy		1.41	17.021		4,025	750		4,775	5,650

26 35 26.40 Transient Voltage Suppressor Transformer		Crew	Daily Output	Labor-Hours	Unit	Material	Labor	Equipment	Total	Total Incl O&P
0010	**TRANSIENT VOLTAGE SUPPRESSOR TRANSFORMER**									
0110	Single-phase, 120 V, 1.8 kVA	1 Elec	4	2	Ea.	1,175	88		1,263	1,425
0120	3.6 kVA		4	2		1,750	88		1,838	2,075
0130	7.2 kVA		3.20	2.500		2,050	110		2,160	2,425
0150	240 V, 3.6 kVA		4	2		2,575	88		2,663	2,975
0160	7.2 kVA		4	2		3,000	88		3,088	3,450
0170	14.4 kVA		3.20	2.500		3,400	110		3,510	3,925
0210	Plug-in unit, 120 V, 1.8 kVA		8	1		720	44		764	860

26 35 53 – Voltage Regulators

26 35 53.10 Automatic Voltage Regulators		Crew	Daily Output	Labor-Hours	Unit	Material	Labor	Equipment	Total	Total Incl O&P
0010	**AUTOMATIC VOLTAGE REGULATORS**									
0100	Computer grade, solid state, variable transf. volt. regulator									
0110	Single-phase, 120 V, 8.6 kVA	2 Elec	1.33	12.030	Ea.	5,000	530		5,530	6,375
0120	17.3 kVA		1.14	14.035		5,900	620		6,520	7,500
0130	208/240 V, 7.5/8.6 kVA		1.33	12.030		5,000	530		5,530	6,375
0140	13.5/15.6 kVA		1.33	12.030		5,900	530		6,430	7,375
0150	27.0/31.2 kVA		1.14	14.035		7,450	620		8,070	9,200
0210	Two-phase, single control, 208/240 V, 15.0/17.3 kVA		1.14	14.035		5,900	620		6,520	7,500
0220	Individual phase control, 15.0/17.3 kVA		1.14	14.035		5,900	620		6,520	7,500
0230	30.0/34.6 kVA	3 Elec	1.33	18.045		7,450	795		8,245	9,500
0310	Three-phase single control, 208/240 V, 26/30 kVA	2 Elec	1	16		5,900	705		6,605	7,650
0320	380/480 V, 24/30 kVA	"	1	16		5,900	705		6,605	7,650
0330	43/54 kVA	3 Elec	1.33	18.045		10,700	795		11,495	13,100
0340	Individual phase control, 208 V, 26 kVA	"	1.33	18.045		5,900	795		6,695	7,800
0350	52 kVA	R-3	.91	21.978		7,450	955	154	8,559	9,950
0360	340/480 V, 24/30 kVA	2 Elec	1	16		5,900	705		6,605	7,650
0370	43/54 kVA	"	1	16		7,450	705		8,155	9,350
0380	48/60 kVA	3 Elec	1.33	18.045		10,800	795		11,595	13,200
0390	86/108 kVA	R-3	.91	21.978		12,000	955	154	13,109	14,900
0500	Standard grade, solid state, variable transformer volt. regulator									
0510	Single-phase, 115 V, 2.3 kVA	1 Elec	2.29	3.493	Ea.	2,000	154		2,154	2,450
0520	4.2 kVA		2	4		3,200	176		3,376	3,800
0530	6.6 kVA		1.14	7.018		3,925	310		4,235	4,800
0540	13.0 kVA		1.14	7.018		6,750	310		7,060	7,925
0550	16.6 kVA	2 Elec	1.23	13.008		7,975	575		8,550	9,700
0610	230 V, 8.3 kVA		1.33	12.030		6,750	530		7,280	8,300
0620	21.4 kVA		1.23	13.008		7,975	575		8,550	9,700
0630	29.9 kVA		1.23	13.008		7,975	575		8,550	9,700
0710	460 V, 9.2 kVA		1.33	12.030		6,750	530		7,280	8,300
0720	20.7 kVA		1.23	13.008		7,975	575		8,550	9,700
0810	Three-phase, 230 V, 13.1 kVA	3 Elec	1.41	17.021		6,750	750		7,500	8,650
0820	19.1 kVA		1.41	17.021		7,975	750		8,725	10,000
0830	25.1 kVA		1.23	19.512		7,975	860		8,835	10,200
0840	57.8 kVA	R-3	.95	21.053		14,500	915	148	15,563	17,600
0850	74.9 kVA	"	.91	21.978		14,500	955	154	15,609	17,600
0910	460 V, 14.3 kVA	3 Elec	1.41	17.021		6,750	750		7,500	8,650
0920	19.1 kVA		1.41	17.021		7,975	750		8,725	10,000

26 35 Power Filters and Conditioners

26 35 53 – Voltage Regulators

26 35 53.10 Automatic Voltage Regulators		Crew	Daily Output	Labor-Hours	Unit	2016 Bare Costs Material	Labor	Equipment	Total	Total Incl O&P
0930	27.9 kVA	3 Elec	1.23	19.512	Ea.	7,975	860		8,835	10,200
0940	59.8 kVA	R-3	1	20		14,500	870	140	15,510	17,500
0950	79.7 kVA		.95	21.053		16,200	915	148	17,263	19,500
0960	118 kVA		.95	21.053		17,100	915	148	18,163	20,500
1000	Laboratory grade, precision, electronic voltage regulator									
1110	Single-phase, 115 V, 0.5 kVA	1 Elec	2.29	3.493	Ea.	1,225	154		1,379	1,600
1120	1.0 kVA		2	4		1,300	176		1,476	1,750
1130	3.0 kVA		.80	10		1,825	440		2,265	2,725
1140	6.0 kVA	2 Elec	1.46	10.959		3,375	485		3,860	4,525
1150	10.0 kVA	3 Elec	1	24		4,400	1,050		5,450	6,550
1160	15.0 kVA	″	1.50	16		5,025	705		5,730	6,675
1210	230 V, 3.0 kVA	1 Elec	.80	10		2,075	440		2,515	3,000
1220	6.0 kVA	2 Elec	1.46	10.959		3,450	485		3,935	4,600
1230	10.0 kVA	3 Elec	1.71	14.035		4,600	620		5,220	6,050
1240	15.0 kVA	″	1.60	15		5,200	660		5,860	6,775

26 36 Transfer Switches

26 36 23 – Automatic Transfer Switches

26 36 23.10 Automatic Transfer Switch Devices		Crew	Daily Output	Labor-Hours	Unit	Material	Labor	Equipment	Total	Total Incl O&P
0010	**AUTOMATIC TRANSFER SWITCH DEVICES**									
0100	Switches, enclosed 480 volt, 3 pole, 30 amp	1 Elec	2.30	3.478	Ea.	2,775	153		2,928	3,300
0200	60 amp	″	1.90	4.211	″	2,775	185		2,960	3,350

26 41 Facility Lightning Protection

26 41 13 – Lightning Protection for Structures

26 41 13.13 Lightning Protection for Buildings		Crew	Daily Output	Labor-Hours	Unit	Material	Labor	Equipment	Total	Total Incl O&P
0010	**LIGHTNING PROTECTION FOR BUILDINGS**									
0200	Air terminals & base, copper									
0400	3/8″ diameter x 10″ (to 75′ high)	1 Elec	8	1	Ea.	19	44		63	93
0500	1/2″ diameter x 12″ (over 75′ high)		8	1		23	44		67	97.50
0520	1/2″ diameter x 24″		7.30	1.096		32.50	48.50		81	115
0540	1/2″ diameter x 60″		6.70	1.194		66	52.50		118.50	159
1000	Aluminum, 1/2″ diameter x 12″ (to 75′ high)		8	1		12.85	44		56.85	86
1020	1/2″ diameter x 24″		7.30	1.096		14.55	48.50		63.05	95
1040	1/2″ diameter x 60″		6.70	1.194		21.50	52.50		74	110
1100	5/8″ diameter x 12″ (over 75′ high)		8	1		14.15	44		58.15	87.50
2000	Cable, copper, 220 lb. per thousand ft. (to 75′ high)		320	.025	L.F.	3.03	1.10		4.13	5.15
2100	375 lb. per thousand ft. (over 75′ high)		230	.035		5.50	1.53		7.03	8.55
2500	Aluminum, 101 lb. per thousand ft. (to 75′ high)		280	.029		.95	1.26		2.21	3.11
2600	199 lb. per thousand ft. (over 75′ high)		240	.033		1.40	1.47		2.87	3.94
3000	Arrester, 175 volt AC to ground		8	1	Ea.	149	44		193	236
3100	650 volt AC to ground		6.70	1.194	″	149	52.50		201.50	250
4000	Air terminals, copper									
4010	3/8″ x 10″	1 Elec	50	.160	Ea.	6.70	7.05		13.75	18.95
4020	1/2″ x 12″		38	.211		10.65	9.25		19.90	27
4030	5/8″ x 12″		33	.242		16	10.70		26.70	35
4040	1/2″ x 24″		30	.267		20	11.75		31.75	41
4050	1/2″ x 60″		19	.421		53.50	18.55		72.05	89.50
4060	Air terminals, aluminum									

	26 41 13.13 Lightning Protection for Buildings	Crew	Daily Output	Labor-Hours	Unit	2016 Bare Costs Material	Labor	Equipment	Total	Total Incl O&P
4070	3/8" x 10"	1 Elec	50	.160	Ea.	2.79	7.05		9.84	14.60
4080	1/2" x 12"		38	.211		3.04	9.25		12.29	18.50
4090	5/8" x 12"		33	.242		4.32	10.70		15.02	22
4100	1/2" x 24"		30	.267		4.72	11.75		16.47	24.50
4110	1/2" x 60"		19	.421		11.90	18.55		30.45	43.50
4200	Air terminal bases, copper									
4210	Adhesive bases, 1/2"	1 Elec	15	.533	Ea.	14	23.50		37.50	54
4215	Adhesive base, lightning air terminal					14			14	15.40
4220	Bolted bases	1 Elec	9	.889		12.30	39		51.30	77.50
4230	Hinged bases		7	1.143		24.50	50.50		75	110
4240	Side mounted bases		9	.889		13.10	39		52.10	78.50
4250	Offset point support		9	.889		24.50	39		63.50	91
4260	Concealed base assembly		5	1.600		24.50	70.50		95	142
4270	Tee connector base		13	.615		43	27		70	91.50
4280	Tripod base, 36" for 60" air terminal		7	1.143		28.50	50.50		79	114
4290	Intermediate base, 1/2"		15	.533		27.50	23.50		51	68.50
4300	Air terminal bases, aluminum									
4310	Adhesive bases, 1/2"	1 Elec	15	.533	Ea.	6.40	23.50		29.90	45.50
4320	Bolted bases		9	.889		9.80	39		48.80	75
4330	Hinged bases		7	1.143		17.95	50.50		68.45	102
4340	Side mounted bases		9	.889		9.60	39		48.60	74.50
4350	Offset point support		9	.889		18.20	39		57.20	84
4360	Concealed base assembly		5	1.600		14.55	70.50		85.05	131
4370	Tee connector base		13	.615		9.85	27		36.85	55.50
4380	Tripod base, 36" for 60" air terminal		7	1.143		26.50	50.50		77	112
4390	Intermediate base, 1/2"		15	.533		12	23.50		35.50	51.50
4400	Connector cable, copper									
4410	Through wall connector	1 Elec	7	1.143	Ea.	87	50.50		137.50	178
4420	Through roof connector		5	1.600		80.50	70.50		151	204
4430	Beam connector		7	1.143		28.50	50.50		79	114
4440	Double bolt connector		38	.211		11.25	9.25		20.50	27.50
4450	Bar, copper		40	.200		12.40	8.80		21.20	28
4500	Connector cable, aluminum									
4510	Through wall connector	1 Elec	7	1.143	Ea.	30	50.50		80.50	116
4520	Through roof connector		5	1.600		21.50	70.50		92	139
4530	Beam connector		7	1.143		18.35	50.50		68.85	103
4540	Double bolt connector		38	.211		4.77	9.25		14.02	20.50
4600	Bonding plates, copper									
4610	I Beam, 8" square	1 Elec	7	1.143	Ea.	16.05	50.50		66.55	100
4620	Purlin, 8" square		7	1.143		25.50	50.50		76	111
4630	Large heavy duty, 16" square		5	1.600		31	70.50		101.50	149
4640	Aluminum, I Beam, 8" square		7	1.143		16.60	50.50		67.10	101
4650	Purlin, 8" square		7	1.143		16.40	50.50		66.90	101
4660	Large heavy duty, 16" square		5	1.600		17.95	70.50		88.45	135
4700	Cable support, copper, loop fastener		38	.211		1.08	9.25		10.33	16.35
4710	Adhesive cable holder		38	.211		1.20	9.25		10.45	16.45
4720	Aluminum, loop fastener		38	.211		.59	9.25		9.84	15.80
4730	Adhesive cable holder		38	.211		.54	9.25		9.79	15.75
4740	Bonding strap, copper, 3/4" x 9-1/2"		15	.533		9.65	23.50		33.15	49
4750	Aluminum, 3/4" x 9-1/2"		15	.533		4.62	23.50		28.12	43.50
4760	Swivel adapter, copper, 1/2"		73	.110		13.60	4.83		18.43	23
4770	Aluminum, 1/2"		73	.110		7.15	4.83		11.98	15.80

26 51 13 – Interior Lighting Fixtures, Lamps, and Ballasts

26 51 13.50 Interior Lighting Fixtures			Crew	Daily Output	Labor-Hours	Unit	2016 Bare Costs Material	Labor	Equipment	Total	Total Incl O&P
0010	**INTERIOR LIGHTING FIXTURES** Including lamps, mounting										
0030	hardware and connections										
0100	Fluorescent, C.W. lamps, troffer, recess mounted in grid, RS										
0130	Grid ceiling mount										
0200	Acrylic lens, 1'W x 4'L, two 40 watt		1 Elec	5.70	1.404	Ea.	50	62		112	156
0300	2'W x 2'L, two U40 watt			5.70	1.404		54	62		116	160
0504	Three 40 watt w/armored cable	G		3.40	2.353		70.50	104		174.50	248
0600	2'W x 4'L, four 40 watt			4.70	1.702		60.50	75		135.50	190
0604	Four 40 watt framed w/armored cable	G		3.20	2.500		143	110		253	335
0910	Acrylic lens, 1'W x 4'L, two 32 watt T8	G		5.70	1.404		59.50	62		121.50	167
0930	2'W x 2'L, two U32 watt T8	G		5.70	1.404		93	62		155	203
0940	2'W x 4'L, two 32 watt T8	G		5.30	1.509		77	66.50		143.50	194
0950	2'W x 4'L, three 32 watt T8	G		5	1.600		68	70.50		138.50	190
0960	2'W x 4'L, four 32 watt T8	G		4.70	1.702		72.50	75		147.50	203
1000	Surface mounted, RS										
1350	1' W x 4' L, sim. wood grain, two 40 watt	G	1 Elec	6.20	1.290	Ea.	145	57		202	253
2000	2'W x 8'L, eight 40 watt		2 Elec	6.20	2.581	"	171	114		285	375
2100	Strip fixture										
2130	Surface mounted										
2150	2' long, two 40 watt, RS	G	1 Elec	8	1	Ea.	33.50	44		77.50	109
2200	4' long, one 40 watt, RS			8.50	.941		23	41.50		64.50	93.50
2300	4' long, two 40 watt, RS			8	1		41	44		85	118
2304	4' long, 2 lamp & armored cable	G		4.50	1.778		54	78.50		132.50	188
2600	8' long, one 75 watt, SL		2 Elec	13.40	1.194		50.50	52.50		103	142
2700	8' long, two 75 watt, SL		"	12.40	1.290		61	57		118	160
2704	8' long, 2 lamp & armored cable	G	1 Elec	3.65	2.192		73.50	96.50		170	239
2930	4' long, one 32 watt T8, staggered cove mtd.	G	"	8	1		69	44		113	148
3000	Strip, pendent mounted, industrial, white porcelain enamel										
3402	4' long, three 34 watt	G	1 Elec	7	1.143	Ea.	96	50.50		146.50	189
3404	4' long, four 34 watt	G		6.50	1.231		87.50	54		141.50	185
3405	Parabolic louver, 4' long, continuous row	G		5.70	1.404		98.50	62		160.50	209
3414	4' long, 2 lamp w/cover & armored cable	G		3.70	2.162		98.50	95		193.50	264
3444	8' long, 2 lamp w/cover & armored cable	G		3.10	2.581		182	114		296	385
3465	Hospital type, 4' long, two 40 watt, RS	G		4.80	1.667		205	73.50		278.50	345
3535	Downlight, recess mounted	G		8	1		148	44		192	234
3540	Wall wash reflector, recess mounted	G		8	1		105	44		149	188
3550	Direct/indirect, 4' long, stl., pendent mtd.	G		5	1.600		162	70.50		232.50	293
3560	4' long, alum., pendent mtd.	G		5	1.600		335	70.50		405.50	485
3565	Prefabricated cove, 4' long, stl. continuous row	G		5	1.600		203	70.50		273.50	340
3570	4' long, alum. continuous row	G		5	1.600		345	70.50		415.50	495
4220	Metal halide, integral ballast, ceiling, recess mounted										
4230	prismatic glass lens, floating door										
4240	2'W x 2'L, 250 watt		1 Elec	3.20	2.500	Ea.	315	110		425	530
4250	2'W x 2'L, 400 watt		2 Elec	5.80	2.759		365	122		487	600
4260	Surface mounted, 2'W x 2'L, 250 watt		1 Elec	2.70	2.963		345	131		476	595
4270	400 watt		2 Elec	4.80	3.333		405	147		552	685
4280	High bay, aluminum reflector,										
4290	Single unit, 400 watt		2 Elec	4.60	3.478	Ea.	410	153		563	700
4300	Single unit, 1000 watt			4	4		590	176		766	940
4310	Twin unit, 400 watt			3.20	5		820	220		1,040	1,275
4320	Low bay, aluminum reflector, 250W DX lamp		1 Elec	3.20	2.500		360	110		470	580
4450	Incandescent, high hat can, round alzak reflector, prewired										
4470	100 watt		1 Elec	8	1	Ea.	71.50	44		115.50	151

26 51 Interior Lighting

26 51 13 – Interior Lighting Fixtures, Lamps, and Ballasts

26 51 13.50 Interior Lighting Fixtures		Crew	Daily Output	Labor-Hours	Unit	Material	2016 Bare Costs Labor	Equipment	Total	Total Incl O&P
4480	150 watt	1 Elec	8	1	Ea.	104	44		148	186
5200	Ceiling, surface mounted, opal glass drum									
5300	8", one 60 watt lamp	1 Elec	10	.800	Ea.	44.50	35		79.50	107
5400	10", two 60 watt lamps		8	1		50	44		94	127
5500	12", four 60 watt lamps		6.70	1.194		70.50	52.50		123	164
6900	Mirror light, fluorescent, RS, acrylic enclosure, two 40 watt		8	1		128	44		172	212
6910	One 40 watt		8	1		106	44		150	188
6920	One 20 watt		12	.667		81.50	29.50		111	138

26 51 13.55 Interior LED Fixtures

Line	Description		Crew	Daily Output	Labor-Hours	Unit	Material	Labor	Equipment	Total	Total Incl O&P
0010	**INTERIOR LED FIXTURES** Incl. lamps, and mounting hardware										
0100	Downlight, recess mounted, 7.5" diameter, 25 watt	G	1 Elec	8	1	Ea.	335	44		379	435
0120	10" diameter, 36 watt	G		8	1		360	44		404	465
0160	cylinder, 10 watts	G		8	1		102	44		146	184
0180	20 watts	G		8	1		166	44		210	255
1000	Troffer, recess mounted, 2' x 4', 3200 Lumens	G		5.30	1.509		138	66.50		204.50	260
1010	4800 Lumens	G		5	1.600		179	70.50		249.50	310
1020	6400 Lumens	G		4.70	1.702		198	75		273	340
1100	Troffer retrofit lamp, 38 watt	G		21	.381		238	16.80		254.80	289
1110	60 watt	G		20	.400		340	17.60		357.60	405
1120	100 watt	G		18	.444		510	19.60		529.60	590
1200	Troffer, volumetric recess mounted, 2' x 2'	G		5.70	1.404		279	62		341	405
2000	Strip, surface mounted, one light bar 4' long, 3500K	G		8.50	.941		260	41.50		301.50	355
2010	5000K	G		8	1		260	44		304	360
2020	Two light bar 4' long, 5000K	G		7	1.143		410	50.50		460.50	535
3000	Linear, suspended mounted, one light bar 4' long, 37 watt	G		6.70	1.194		154	52.50		206.50	256
3010	One light bar 8' long, 74 watt	G	2 Elec	12.20	1.311		286	58		344	410
3020	Two light bar 4' long, 74 watt	G	1 Elec	5.70	1.404		305	62		367	440
3030	Two light bar 8' long, 148 watt	G	2 Elec	8.80	1.818		355	80		435	520
4000	High bay, surface mounted, round, 150 watts	G		5.41	2.959		480	130		610	745
4010	2 bars,164 watts	G		5.41	2.959		450	130		580	710
4020	3 bars, 246 watts	G		5.01	3.197		580	141		721	870
4030	4 bars, 328 watts	G		4.60	3.478		710	153		863	1,025
4040	5 bars, 410 watts	G	3 Elec	4.20	5.716		830	252		1,082	1,325
4050	6 bars, 492 watts	G		3.80	6.324		935	279		1,214	1,475
4060	7 bars, 574 watts	G		3.39	7.075		1,025	310		1,335	1,625
4070	8 bars, 656 watts	G		2.99	8.029		1,150	355		1,505	1,825
5000	Track, lighthead, 6 watt	G	1 Elec	32	.250		54.50	11		65.50	78
5010	9 watt	G	"	32	.250		61.50	11		72.50	85.50
6000	Garage, surface mounted, 103 watts	G	2 Elec	6.50	2.462		970	108		1,078	1,250
6100	pendent mounted, 80 watts	G		6.50	2.462		695	108		803	940
6200	95 watts	G		6.50	2.462		800	108		908	1,050
6300	125 watts	G		6.50	2.462		865	108		973	1,125

26 52 Emergency Lighting

26 52 13 – Emergency Lighting Equipments

26 52 13.10 Emergency Lighting and Battery Units		Crew	Daily Output	Labor-Hours	Unit	Material	2016 Bare Costs Labor	Equipment	Total	Total Incl O&P
0010	**EMERGENCY LIGHTING AND BATTERY UNITS**									
0300	Emergency light units, battery operated									
0350	Twin sealed beam light, 25 watt, 6 volt each									
0500	Lead battery operated	1 Elec	4	2	Ea.	162	88		250	325
0700	Nickel cadmium battery operated		4	2		475	88		563	665
0780	Additional remote mount, sealed beam, 25 W 6 V		26.70	.300		28	13.20		41.20	52.50
0790	Twin sealed beam light, 25 W 6 V each		26.70	.300		54.50	13.20		67.70	81.50

26 53 Exit Signs

26 53 13 – Exit Lighting

26 53 13.10 Exit Lighting Fixtures

		Crew	Daily Output	Labor-Hours	Unit	Material	Labor	Equipment	Total	Total Incl O&P
0010	**EXIT LIGHTING FIXTURES**									
0080	Exit light ceiling or wall mount, incandescent, single face	1 Elec	8	1	Ea.	49	44		93	126
0100	Double face		6.70	1.194		43	52.50		95.50	134
0120	Explosion proof		3.80	2.105		550	92.50		642.50	755
0150	Fluorescent, single face		8	1		74	44		118	154
0160	Double face		6.70	1.194		75	52.50		127.50	169

26 55 Special Purpose Lighting

26 55 33 – Hazard Warning Lighting

26 55 33.10 Warning Beacons

		Crew	Daily Output	Labor-Hours	Unit	Material	Labor	Equipment	Total	Total Incl O&P
0010	**WARNING BEACONS**									
0015	Surface mount with colored or clear lens									
0100	Rotating beacon									
0110	120V, 40 watt halogen	1 Elec	3.50	2.286	Ea.	101	101		202	276
0120	24V, 20 watt halogen	″	3.50	2.286	″	211	101		312	400
0200	Steady beacon									
0210	120V, 40 watt halogen	1 Elec	3.50	2.286	Ea.	101	101		202	276
0220	24V, 20W		3.50	2.286		101	101		202	276
0230	12V DC, incandescent		3.50	2.286		101	101		202	276
0300	Flashing beacon									
0310	120V, 40 watt halogen	1 Elec	3.50	2.286	Ea.	101	101		202	276
0320	24V, 20 watt halogen		3.50	2.286		101	101		202	276
0410	12V DC with two 6V lantern batteries		7	1.143		101	50.50		151.50	194

26 55 59 – Display Lighting

26 55 59.10 Track Lighting

		Crew	Daily Output	Labor-Hours	Unit	Material	Labor	Equipment	Total	Total Incl O&P
0010	**TRACK LIGHTING**									
0080	Track, 1 circuit, 4′ section	1 Elec	6.70	1.194	Ea.	44.50	52.50		97	135
0100	8′ section	2 Elec	10.60	1.509		69	66.50		135.50	185
0200	12′ section	″	8.80	1.818		104	80		184	245
0300	3 circuits, 4′ section	1 Elec	6.70	1.194		92.50	52.50		145	188
0400	8′ section	2 Elec	10.60	1.509		109	66.50		175.50	229
0500	12′ section	″	8.80	1.818		159	80		239	305
1000	Feed kit, surface mounting	1 Elec	16	.500		13	22		35	50.50
1100	End cover		24	.333		5.55	14.70		20.25	30
1200	Feed kit, stem mounting, 1 circuit		16	.500		42	22		64	82
1300	3 circuit		16	.500		42	22		64	82
2000	Electrical joiner, for continuous runs, 1 circuit		32	.250		24	11		35	44.50
2100	3 circuit		32	.250		59	11		70	83

26 55 Special Purpose Lighting

26 55 59 – Display Lighting

26 55 59.10 Track Lighting		Crew	Daily Output	Labor-Hours	Unit	Material	2016 Bare Costs Labor	Equipment	Total	Total Incl O&P
2200	Fixtures, spotlight, 75W PAR halogen	1 Elec	16	.500	Ea.	55.50	22		77.50	97
2210	50W MR16 halogen		16	.500		171	22		193	224
3000	Wall washer, 250 watt tungsten halogen		16	.500		132	22		154	181
3100	Low voltage, 25/50 watt, 1 circuit		16	.500		133	22		155	182
3120	3 circuit	↓	16	.500	↓	198	22		220	254

26 56 Exterior Lighting

26 56 13 – Lighting Poles and Standards

26 56 13.10 Lighting Poles

		Crew	Daily Output	Labor-Hours	Unit	Material	Labor	Equipment	Total	Total Incl O&P
0010	**LIGHTING POLES**									
6420	Wood pole, 4-1/2" x 5-1/8", 8' high	1 Elec	6	1.333	Ea.	335	58.50		393.50	465
6440	12' high		5.70	1.404		485	62		547	635
6460	20' high	↓	4	2	↓	685	88		773	900

26 56 16 – Parking Lighting

26 56 16.55 Parking LED Lighting

		Crew	Daily Output	Labor-Hours	Unit	Material	Labor	Equipment	Total	Total Incl O&P
0010	**PARKING LED LIGHTING**									
0100	Round pole mounting, 88 lamp watts G	1 Elec	2	4	Ea.	1,025	176		1,201	1,425
0110	Square pole mounting, 223 lamp watts G	"	2	4	"	1,800	176		1,976	2,275

26 56 23 – Area Lighting

26 56 23.10 Exterior Fixtures

		Crew	Daily Output	Labor-Hours	Unit	Material	Labor	Equipment	Total	Total Incl O&P
0010	**EXTERIOR FIXTURES** With lamps									
0400	Quartz, 500 watt	1 Elec	5.30	1.509	Ea.	52	66.50		118.50	167
1100	Wall pack, low pressure sodium, 35 watt		4	2		234	88		322	400
1150	55 watt	↓	4	2	↓	278	88		366	450

26 56 23.55 Exterior LED Fixtures

		Crew	Daily Output	Labor-Hours	Unit	Material	Labor	Equipment	Total	Total Incl O&P
0010	**EXTERIOR LED FIXTURES**									
0100	Wall mounted, indoor/outdoor, 12 watt G	1 Elec	10	.800	Ea.	166	35		201	240
0110	32 watt G		10	.800		405	35		440	505
0120	66 watt G		10	.800		620	35		655	745
0200	outdoor, 110 watt G		10	.800		970	35		1,005	1,125
0210	220 watt G		10	.800		1,525	35		1,560	1,725
0300	modular, type IV, 120 V, 50 lamp watts G		9	.889		1,000	39		1,039	1,175
0310	101 lamp watts G		9	.889		1,125	39		1,164	1,325
0320	126 lamp watts G		9	.889		1,425	39		1,464	1,650
0330	202 lamp watts G		9	.889		1,625	39		1,664	1,850
0340	240 V, 50 lamp watts G		8	1		1,050	44		1,094	1,225
0350	101 lamp watts G		8	1		1,175	44		1,219	1,375
0360	126 lamp watts G		8	1		1,475	44		1,519	1,700
0370	202 lamp watts G		8	1		1,650	44		1,694	1,900
0400	wall pack, glass, 13 lamp watts G		4	2		390	88		478	570
0410	poly w/photocell, 26 lamp watts G		4	2		271	88		359	440
0420	50 lamp watts G		4	2		685	88		773	895
0430	replacement, 40 watts G		4	2		300	88		388	475
0440	60 watts G	↓	4	2	↓	370	88		458	555

26 56 Exterior Lighting

26 56 26 – Landscape Lighting

26 56 26.20 Landscape Fixtures		Crew	Daily Output	Labor-Hours	Unit	2016 Bare Costs Material	Labor	Equipment	Total	Total Incl O&P
0010	**LANDSCAPE FIXTURES**									
7380	Landscape recessed uplight, incl. housing, ballast, transformer									
7390	& reflector, not incl. conduit, wire, trench									
7420	Incandescent, 250 watt	1 Elec	5	1.600	Ea.	600	70.50		670.50	775
7440	Quartz, 250 watt	"	5	1.600	"	570	70.50		640.50	740

26 56 33 – Walkway Lighting

26 56 33.10 Walkway Luminaire		Crew	Daily Output	Labor-Hours	Unit	Material	Labor	Equipment	Total	Total Incl O&P
0010	**WALKWAY LUMINAIRE**									
6500	Bollard light, lamp & ballast, 42" high with polycarbonate lens									
7200	Incandescent, 150 watt	1 Elec	3	2.667	Ea.	610	117		727	860

26 56 36 – Flood Lighting

26 56 36.55 LED Floodlights		Crew	Daily Output	Labor-Hours	Unit	Material	Labor	Equipment	Total	Total Incl O&P
0010	**LED FLOODLIGHTS** with ballast and lamp,									
0020	Pole mounted, pole not included									
0100	11 watt G	1 Elec	4	2	Ea.	410	88		498	600
0110	46 watt G		4	2		1,175	88		1,263	1,450
0120	90 watt G		4	2		2,000	88		2,088	2,350
0130	288 watt G		4	2		1,800	88		1,888	2,125

26 61 Lighting Systems and Accessories

26 61 23 – Lamps Applications

26 61 23.10 Lamps		Crew	Daily Output	Labor-Hours	Unit	Material	Labor	Equipment	Total	Total Incl O&P
0010	**LAMPS**									
0080	Fluorescent, rapid start, cool white, 2' long, 20 watt	1 Elec	1	8	C	360	350		710	970
0100	4' long, 40 watt		.90	8.889		275	390		665	945
1000	Metal halide, mogul base, 175 watt		.30	26.667		2,175	1,175		3,350	4,325
1100	250 watt		.30	26.667		2,450	1,175		3,625	4,625
1350	High pressure sodium, 70 watt		.30	26.667		2,650	1,175		3,825	4,850
1370	150 watt		.30	26.667		2,825	1,175		4,000	5,050

26 61 23.55 LED Lamps		Crew	Daily Output	Labor-Hours	Unit	Material	Labor	Equipment	Total	Total Incl O&P
0010	**LED LAMPS**									
0100	LED lamp, interior, shape A60, equal to 60 watt G	1 Elec	160	.050	Ea.	20	2.20		22.20	25.50
0200	Globe frosted A60, equal to 60 watt G		160	.050		12	2.20		14.20	16.80
0300	Globe earth, equal to 100 watt G		160	.050		29	2.20		31.20	35.50
1100	MR16, 3 watt, replacement of halogen lamp 25 watt G		130	.062		20	2.71		22.71	26.50
1200	6 watt replacement of halogen lamp 45 watt G		130	.062		21	2.71		23.71	28
2100	10 watt, PAR20, equal to 60 watt G		130	.062		28	2.71		30.71	35.50
2200	15 watt, PAR30, equal to 100 watt G		130	.062		73.50	2.71		76.21	85.50

Division 27 Communications

Estimating Tips

27 20 00 Data Communications
27 30 00 Voice Communications
27 40 00 Audio-Video Communications

When estimating material costs for special systems, it is always prudent to obtain manufacturers' quotations for equipment prices and special installation requirements which will affect the total costs.

Reference Numbers

Reference numbers are shown at the beginning of some major classifications. These numbers refer to related items in the Reference Section. The reference information may be an estimating procedure, an alternate pricing method, or technical information.

Note: Not all subdivisions listed here necessarily appear. ■

Note: Trade Service, in part, has been used as a reference source for some of the material prices used in Division 27.

27 13 Communications Backbone Cabling

27 13 23 – Communications Optical Fiber Backbone Cabling

27 13 23.13 Communications Optical Fiber		Crew	Daily Output	Labor-Hours	Unit	2016 Bare Costs Material	Labor	Equipment	Total	Total Incl O&P
0010	**COMMUNICATIONS OPTICAL FIBER**									
0040	Specialized tools & techniques cause installation costs to vary.									
0070	Fiber optic, cable, bulk simplex, single mode	1 Elec	8	1	C.L.F.	22.50	44		66.50	97
0080	Multi mode		8	1		36	44		80	112
0090	4 strand, single mode		7.34	1.090		38	48		86	120
0095	Multi mode		7.34	1.090		50.50	48		98.50	134
0100	12 strand, single mode		6.67	1.199		76.50	53		129.50	171
0105	Multi mode		6.67	1.199		96.50	53		149.50	193
0150	Jumper				Ea.	33			33	36.50
0200	Pigtail					37			37	40.50
0300	Connector	1 Elec	24	.333		27	14.70		41.70	53.50
0350	Finger splice		32	.250		40	11		51	62.50
0400	Transceiver (low cost bi-directional)		8	1		480	44		524	595
0450	Rack housing, 4 rack spaces, 12 panels (144 fibers)		2	4		550	176		726	895
0500	Patch panel, 12 ports		6	1.333		300	58.50		358.50	425

27 15 Communications Horizontal Cabling

27 15 10 – Special Communications Cabling

27 15 10.23 Sound and Video Cables and Fittings

		Crew	Daily Output	Labor-Hours	Unit	Material	Labor	Equipment	Total	Total Incl O&P
0010	**SOUND AND VIDEO CABLES & FITTINGS**									
0900	TV antenna lead-in, 300 ohm, #20-2 conductor	1 Elec	7	1.143	C.L.F.	22.50	50.50		73	107
0950	Coaxial, feeder outlet		7	1.143		19.95	50.50		70.45	105
1000	Coaxial, main riser		6	1.333		21	58.50		79.50	119
1100	Sound, shielded with drain, #22-2 conductor		8	1		13.95	44		57.95	87.50
1150	#22-3 conductor		7.50	1.067		17.55	47		64.55	96.50
1200	#22-4 conductor		6.50	1.231		22	54		76	113
1250	Nonshielded, #22-2 conductor		10	.800		14.25	35		49.25	73
1300	#22-3 conductor		9	.889		21.50	39		60.50	87.50
1350	#22-4 conductor		8	1		28.50	44		72.50	104
1400	Microphone cable		8	1		42	44		86	119

27 15 13 – Communications Copper Horizontal Cabling

27 15 13.13 Communication Cables and Fittings

		Crew	Daily Output	Labor-Hours	Unit	Material	Labor	Equipment	Total	Total Incl O&P
0010	**COMMUNICATION CABLES AND FITTINGS**									
2200	Telephone twisted, PVC insulation, #22-2 conductor	1 Elec	10	.800	C.L.F.	10.40	35		45.40	69
2250	#22-3 conductor		9	.889		13.65	39		52.65	79
2300	#22-4 conductor		8	1		15.85	44		59.85	89.50
2350	#18-2 conductor		9	.889		16.75	39		55.75	82.50

27 41 Audio-Video Systems

27 41 33 – Master Antenna Television Systems

27 41 33.10 T.V. Systems

		Crew	Daily Output	Labor-Hours	Unit	Material	Labor	Equipment	Total	Total Incl O&P
0010	**T.V. SYSTEMS**, not including rough-in wires, cables & conduits									
0100	Master TV antenna system									
0200	VHF reception & distribution, 12 outlets	1 Elec	6	1.333	Outlet	139	58.50		197.50	249
0800	VHF & UHF reception & distribution, 12 outlets		6	1.333	"	218	58.50		276.50	335
5000	Antenna, small		6	1.333	Ea.	48	58.50		106.50	149
5100	Large		4	2	"	202	88		290	365

27 51 19 – Sound Masking Systems

27 51 19.10 Sound System		Crew	Daily Output	Labor-Hours	Unit	Material	Labor	Equipment	Total	Total Incl O&P
						2016 Bare Costs				
0010	**SOUND SYSTEM**, not including rough-in wires, cables & conduits									
2000	Intercom, 30 station capacity, master station	2 Elec	2	8	Ea.	2,300	350		2,650	3,100
2020	10 station capacity	"	4	4		1,250	176		1,426	1,675
2200	Remote station	1 Elec	8	1		169	44		213	258
2400	Intercom outlets		8	1		99.50	44		143.50	181
2600	Handset		4	2		330	88		418	505
2800	Emergency call system, 12 zones, annunciator		1.30	6.154		990	271		1,261	1,550
3000	Bell		5.30	1.509		102	66.50		168.50	222
3200	Light or relay		8	1		51	44		95	129
3400	Transformer		4	2		225	88		313	390
3600	House telephone, talking station		1.60	5		485	220		705	890
3800	Press to talk, release to listen		5.30	1.509		113	66.50		179.50	233
4000	System-on button					67.50			67.50	74
4200	Door release	1 Elec	4	2		121	88		209	277
4400	Combination speaker and microphone		8	1		205	44		249	298
4600	Termination box		3.20	2.500		64.50	110		174.50	251
4800	Amplifier or power supply		5.30	1.509		740	66.50		806.50	925
5000	Vestibule door unit		16	.500	Name	136	22		158	186
5200	Strip cabinet		27	.296	Ea.	258	13.05		271.05	305
5400	Directory		16	.500		121	22		143	169
6000	Master door, button buzzer type, 100 unit	2 Elec	.54	29.630		1,225	1,300		2,525	3,475
6020	200 unit	"	.30	53.333		2,325	2,350		4,675	6,400

Division Notes

	CREW	DAILY OUTPUT	LABOR-HOURS	UNIT	BARE COSTS MAT.	BARE COSTS LABOR	BARE COSTS EQUIP.	BARE COSTS TOTAL	TOTAL INCL O&P

Division 28 Electronic Safety & Security

Estimating Tips

- When estimating material costs for electronic safety and security systems, it is always prudent to obtain manufacturers' quotations for equipment prices and special installation requirements that affect the total cost.
- Fire alarm systems consist of control panels, annunciator panels, battery with rack, charger, and fire alarm actuating and indicating devices. Some fire alarm systems include speakers, telephone lines, door closer controls, and other components. Be careful not to overlook the costs related to installation for these items. Also be aware of costs for integrated automation instrumentation and terminal devices, control equipment, control wiring, and programming.
- Security equipment includes items such as CCTV, access control, and other detection and identification systems to perform alert and alarm functions. Be sure to consider the costs related to installation for this security equipment, such as for integrated automation instrumentation and terminal devices, control equipment, control wiring, and programming.

Reference Numbers

Reference numbers are shown at the beginning of some major classifications. These numbers refer to related items in the Reference Section. The reference information may be an estimating procedure, an alternate pricing method, or technical information.

Note: Not all subdivisions listed here necessarily appear. ■

Note: Trade Service, in part, has been used as a reference source for some of the material prices used in Division 28.

28 16 Intrusion Detection

28 16 16 – Intrusion Detection Systems Infrastructure

28 16 16.50 Intrusion Detection		Crew	Daily Output	Labor-Hours	Unit	Material	Labor	Equipment	Total	Total Incl O&P
0010	**INTRUSION DETECTION**, not including wires & conduits									
0100	Burglar alarm, battery operated, mechanical trigger	1 Elec	4	2	Ea.	278	88		366	450
0200	Electrical trigger		4	2		330	88		418	510
0400	For outside key control, add		8	1		86	44		130	167
0600	For remote signaling circuitry, add		8	1		137	44		181	222
0800	Card reader, flush type, standard		2.70	2.963		790	131		921	1,075
1000	Multi-code		2.70	2.963		1,200	131		1,331	1,550
1200	Door switches, hinge switch		5.30	1.509		62.50	66.50		129	178
1400	Magnetic switch		5.30	1.509		77	66.50		143.50	194
2800	Ultrasonic motion detector, 12 volt		2.30	3.478		210	153		363	480
3000	Infrared photoelectric detector		4	2		144	88		232	305
3200	Passive infrared detector		4	2		234	88		322	400
3420	Switchmats, 30″ x 5′		5.30	1.509		104	66.50		170.50	223
3440	30″ x 25′		4	2		203	88		291	365
3460	Police connect panel		4	2		269	88		357	440
3480	Telephone dialer		5.30	1.509		385	66.50		451.50	535
3500	Alarm bell		4	2		104	88		192	259
3520	Siren	↓	4	2	↓	146	88		234	305

Column group "2016 Bare Costs" spans Material, Labor, Equipment, Total.

28 31 Fire Detection and Alarm

28 31 23 – Fire Detection and Alarm Annunciation Panels and Fire Stations

28 31 23.50 Alarm Panels and Devices		Crew	Daily Output	Labor-Hours	Unit	Material	Labor	Equipment	Total	Total Incl O&P
0010	**ALARM PANELS AND DEVICES**, not including wires & conduits									
5600	Strobe and horn	1 Elec	5.30	1.509	Ea.	152	66.50		218.50	276
5800	Fire alarm horn		6.70	1.194		61	52.50		113.50	153
6600	Drill switch		8	1		370	44		414	480
6800	Master box		2.70	2.963		6,400	131		6,531	7,275
7800	Remote annunciator, 8 zone lamp	↓	1.80	4.444		209	196		405	550
8000	12 zone lamp	2 Elec	2.60	6.154		335	271		606	815
8200	16 zone lamp	″	2.20	7.273	↓	420	320		740	990

28 31 46 – Smoke Detection Sensors

28 31 46.50 Smoke Detectors		Crew	Daily Output	Labor-Hours	Unit	Material	Labor	Equipment	Total	Total Incl O&P
0010	**SMOKE DETECTORS**									
5200	Smoke detector, ceiling type	1 Elec	6.20	1.290	Ea.	110	57		167	214

28 33 Gas Detection and Alarm

28 33 33 – Gas Detection Sensors

28 33 33.50 Tank Leak Detection Systems		Crew	Daily Output	Labor-Hours	Unit	Material	Labor	Equipment	Total	Total Incl O&P
0010	**TANK LEAK DETECTION SYSTEMS** Liquid and vapor									
0100	For hydrocarbons and hazardous liquids/vapors									
0120	Controller, data acquisition, incl. printer, modem, RS232 port									
0140	24 channel, for use with all probes				Ea.	4,050			4,050	4,475
0160	9 channel, for external monitoring				″	875			875	965
0200	Probes									
0210	Well monitoring									
0220	Liquid phase detection				Ea.	460			460	505
0230	Hydrocarbon vapor, fixed position					480			480	530
0240	Hydrocarbon vapor, float mounted					465			465	510
0250	Both liquid and vapor hydrocarbon				↓	500			500	550

28 33 Gas Detection and Alarm

28 33 33 – Gas Detection Sensors

28 33 33.50 Tank Leak Detection Systems		Crew	Daily Output	Labor-Hours	Unit	2016 Bare Costs Material	Labor	Equipment	Total	Total Incl O&P
0300	Secondary containment, liquid phase									
0310	Pipe trench/manway sump				Ea.	795			795	875
0320	Double wall pipe and manual sump					795			795	875
0330	Double wall fiberglass annular space					365			365	400
0340	Double wall steel tank annular space				▼	365			365	400
0500	Accessories									
0510	Modem, non-dedicated phone line				Ea.	265			265	292
0600	Monitoring, internal									
0610	Automatic tank gauge, incl. overfill				Ea.	1,400			1,400	1,550
0620	Product line				"	1,400			1,400	1,550
0700	Monitoring, special									
0710	Cathodic protection				Ea.	730			730	805
0720	Annular space chemical monitor				"	995			995	1,100

Division Notes

	CREW	DAILY OUTPUT	LABOR-HOURS	UNIT	BARE COSTS MAT.	LABOR	EQUIP.	TOTAL	TOTAL INCL O&P

Division 31 Earthwork

Estimating Tips

31 05 00 Common Work Results for Earthwork

- Estimating the actual cost of performing earthwork requires careful consideration of the variables involved. This includes items such as type of soil, whether water will be encountered, dewatering, whether banks need bracing, disposal of excavated earth, and length of haul to fill or spoil sites, etc. If the project has large quantities of cut or fill, consider raising or lowering the site to reduce costs, while paying close attention to the effect on site drainage and utilities.
- If the project has large quantities of fill, creating a borrow pit on the site can significantly lower the costs.
- It is very important to consider what time of year the project is scheduled for completion. Bad weather can create large cost overruns from dewatering, site repair, and lost productivity from cold weather.

Reference Numbers

Reference numbers are shown at the beginning of some major classifications. These numbers refer to related items in the Reference Section. The reference information may be an estimating procedure, an alternate pricing method, or technical information.

Note: Not all subdivisions listed here necessarily appear. ■

31 05 Common Work Results for Earthwork

31 05 13 – Soils for Earthwork

31 05 13.10 Borrow		Crew	Daily Output	Labor-Hours	Unit	Material	2016 Bare Costs Labor	Equipment	Total	Total Incl O&P
0010	**BORROW**									
0020	Spread, 200 H.P. dozer, no compaction, 2 mi. RT haul									
0200	Common borrow	B-15	600	.047	C.Y.	12.40	1.63	4.63	18.66	21.50

31 05 16 – Aggregates for Earthwork

31 05 16.10 Borrow		Crew	Daily Output	Labor-Hours	Unit	Material	Labor	Equipment	Total	Total Incl O&P
0010	**BORROW**									
0020	Spread, with 200 H.P. dozer, no compaction, 2 mi. RT haul									
0100	Bank run gravel	B-15	600	.047	L.C.Y.	20.50	1.63	4.63	26.76	30.50
0300	Crushed stone (1.40 tons per CY), 1-1/2"		600	.047		24	1.63	4.63	30.26	34.50
0320	3/4"		600	.047		24	1.63	4.63	30.26	34.50
0340	1/2"		600	.047		28	1.63	4.63	34.26	39
0360	3/8"		600	.047		29.50	1.63	4.63	35.76	40.50
0400	Sand, washed, concrete		600	.047		37.50	1.63	4.63	43.76	49.50
0500	Dead or bank sand		600	.047		20	1.63	4.63	26.26	30

31 11 Clearing and Grubbing

31 11 10 – Clearing and Grubbing Land

31 11 10.10 Clear and Grub Site		Crew	Daily Output	Labor-Hours	Unit	Material	Labor	Equipment	Total	Total Incl O&P
0010	**CLEAR AND GRUB SITE**									
0020	Cut & chip light trees to 6" diam.	B-7	1	48	Acre		1,500	1,675	3,175	4,350
0150	Grub stumps and remove	B-30	2	12			430	1,200	1,630	2,025
0200	Cut & chip medium, trees to 12" diam.	B-7	.70	68.571			2,150	2,400	4,550	6,200
0250	Grub stumps and remove	B-30	1	24			860	2,400	3,260	4,075
0300	Cut & chip heavy, trees to 24" diam.	B-7	.30	160			5,000	5,575	10,575	14,500
0350	Grub stumps and remove	B-30	.50	48			1,725	4,825	6,550	8,150
0400	If burning is allowed, deduct cut & chip								40%	40%

31 13 Selective Tree and Shrub Removal and Trimming

31 13 13 – Selective Tree and Shrub Removal

31 13 13.10 Selective Clearing		Crew	Daily Output	Labor-Hours	Unit	Material	Labor	Equipment	Total	Total Incl O&P
0010	**SELECTIVE CLEARING**									
0020	Clearing brush with brush saw	A-1C	.25	32	Acre		935	126	1,061	1,725
0100	By hand	1 Clab	.12	66.667			1,950		1,950	3,250
0300	With dozer, ball and chain, light clearing	B-11A	2	8			274	700	974	1,225
0400	Medium clearing	"	1.50	10.667			365	930	1,295	1,625

31 22 Grading

31 22 16 – Fine Grading

31 22 16.10 Finish Grading		Crew	Daily Output	Labor-Hours	Unit	Material	Labor	Equipment	Total	Total Incl O&P
0010	**FINISH GRADING**									
0012	Finish grading area to be paved with grader, small area	B-11L	400	.040	S.Y.		1.37	1.79	3.16	4.24
1050	For small irregular areas	B-32C	2000	.024			.83	1.18	2.01	2.67
1100	Fine grade for slab on grade, machine	B-11L	1040	.015			.53	.69	1.22	1.64
1160	Hand grading	1 Clab	500	.016	S.F.		.47		.47	.78
3300	Finishing grading slopes, gentle	B-11L	8900	.002	S.Y.		.06	.08	.14	.19
3500	Finish grading lagoon bottoms	"	4	4	M.S.F.		137	179	316	425

31 23 16.13 Excavating, Trench

	31 23 16.13 Excavating, Trench	Crew	Daily Output	Labor-Hours	Unit	Material	2016 Bare Costs Labor	Equipment	Total	Total Incl O&P
0010	**EXCAVATING, TRENCH**									
0011	Or continuous footing									
0050	1′ to 4′ deep, 3/8 C.Y. excavator	B-11C	150	.107	B.C.Y.		3.66	2.44	6.10	8.80
0060	1/2 C.Y. excavator	B-11M	200	.080			2.74	1.98	4.72	6.75
0090	4′ to 6′ deep, 1/2 C.Y. excavator	"	200	.080			2.74	1.98	4.72	6.75
0100	5/8 C.Y. excavator	B-12Q	250	.064			2.22	2.36	4.58	6.30
0300	1/2 C.Y. excavator, truck mounted	B-12J	200	.080			2.78	4.41	7.19	9.45
1352	4′ to 6′ deep, 1/2 C.Y. excavator w/trench box	B-13H	188	.085			2.96	5.10	8.06	10.55
1354	5/8 C.Y. excavator	"	235	.068			2.36	4.10	6.46	8.45
1400	By hand with pick and shovel 2′ to 6′ deep, light soil	1 Clab	8	1			29		29	49
1500	Heavy soil	"	4	2			58.50		58.50	98
5050	1′ to 4′ deep, 3/8 C.Y. tractor loader/backhoe	B-11C	162	.099			3.39	2.26	5.65	8.15
5060	1/2 C.Y. excavator	B-11M	216	.074			2.54	1.83	4.37	6.25
5080	4′ to 6′ deep, 1/2 C.Y. excavator	"	216	.074			2.54	1.83	4.37	6.25
5090	5/8 C.Y. excavator	B-12Q	276	.058			2.01	2.14	4.15	5.70
5130	1/2 C.Y. excavator, truck mounted	B-12J	216	.074			2.57	4.08	6.65	8.75
5352	4′ to 6′ deep, 1/2 C.Y. excavator w/trench box	B-13H	205	.078			2.71	4.70	7.41	9.65
5354	5/8 C.Y. excavator	"	257	.062			2.16	3.75	5.91	7.70
6050	1′ to 4′ deep, 3/8 C.Y. excavator	B-11C	165	.097			3.32	2.21	5.53	8
6060	1/2 C.Y. excavator	B-11M	220	.073			2.49	1.80	4.29	6.10
6080	4′ to 6′ deep, 1/2 C.Y. excavator	"	220	.073			2.49	1.80	4.29	6.10
6090	5/8 C.Y. excavator	B-12Q	275	.058			2.02	2.14	4.16	5.70
6130	1/2 C.Y. excavator, truck mounted	B-12J	220	.073			2.53	4.01	6.54	8.60
6352	4′ to 6′ deep, 1/2 C.Y. excavator w/trench box	B-13H	209	.077			2.66	4.61	7.27	9.45
6354	5/8 C.Y. excavator	"	261	.061			2.13	3.69	5.82	7.60
7020	Dense hard clay with no sheeting or dewatering included									
7050	1′ to 4′ deep, 3/8 C.Y. excavator	B-11C	132	.121	B.C.Y.		4.16	2.77	6.93	9.95
7060	1/2 C.Y. excavator	B-11M	176	.091			3.12	2.25	5.37	7.70
7080	4′ to 6′ deep, 1/2 C.Y. excavator	"	176	.091			3.12	2.25	5.37	7.70
7090	5/8 C.Y. excavator	B-12Q	220	.073			2.53	2.68	5.21	7.15
7130	1/2 C.Y. excavator, truck mounted	B-12J	176	.091			3.16	5	8.16	10.75

31 23 16.14 Excavating, Utility Trench

	31 23 16.14 Excavating, Utility Trench	Crew	Daily Output	Labor-Hours	Unit	Material	Labor	Equipment	Total	Total Incl O&P
0010	**EXCAVATING, UTILITY TRENCH**									
0011	Common earth									
0050	Trenching with chain trencher, 12 H.P., operator walking									
0100	4″ wide trench, 12″ deep	B-53	800	.010	L.F.		.29	.09	.38	.58
1000	Backfill by hand including compaction, add									
1050	4″ wide trench, 12″ deep	A-1G	800	.010	L.F.		.29	.07	.36	.57

31 23 16.16 Structural Excavation for Minor Structures

	31 23 16.16 Structural Excavation for Minor Structures	Crew	Daily Output	Labor-Hours	Unit	Material	Labor	Equipment	Total	Total Incl O&P
0010	**STRUCTURAL EXCAVATION FOR MINOR STRUCTURES**									
0015	Hand, pits to 6′ deep, sandy soil	1 Clab	8	1	B.C.Y.		29		29	49
0100	Heavy soil or clay		4	2			58.50		58.50	98
1100	Hand loading trucks from stock pile, sandy soil		12	.667			19.45		19.45	32.50
1300	Heavy soil or clay		8	1			29		29	49
1500	For wet or muck hand excavation, add to above								50%	50%

31 23 16.42 Excavating, Bulk Bank Measure

	31 23 16.42 Excavating, Bulk Bank Measure	Crew	Daily Output	Labor-Hours	Unit	Material	Labor	Equipment	Total	Total Incl O&P
0010	**EXCAVATING, BULK BANK MEASURE** R312316-40									
0011	Common earth piled									
0020	For loading onto trucks, add								15%	15%
0200	Excavator, hydraulic, crawler mtd., 1 C.Y. cap. = 100 C.Y./hr.	B-12A	800	.020	B.C.Y.		.69	1.02	1.71	2.27
0310	Wheel mounted, 1/2 C.Y. cap. = 40 C.Y./hr.	B-12E	320	.050			1.74	1.39	3.13	4.42
1200	Front end loader, track mtd., 1-1/2 C.Y. cap. = 70 C.Y./hr.	B-10N	560	.014			.56	.94	1.50	1.96

31 23 Excavation and Fill

31 23 16 – Excavation

31 23 16.42 Excavating, Bulk Bank Measure		Crew	Daily Output	Labor-Hours	Unit	Material	2016 Bare Costs Labor	Equipment	Total	Total Incl O&P
1500	Wheel mounted, 3/4 C.Y. cap. = 45 C.Y./hr.	B-10R	360	.022	B.C.Y.		.87	.83	1.70	2.36
5000	Excavating, bulk bank measure, sandy clay & loam piled									
5020	For loading onto trucks, add								15%	15%
5100	Excavator, hydraulic, crawler mtd., 1 C.Y. cap. = 120 C.Y./hr.	B-12A	960	.017	B.C.Y.		.58	.85	1.43	1.89
5610	Wheel mounted, 1/2 C.Y. cap. = 44 C.Y./hr.	B-12E	352	.045	″		1.58	1.27	2.85	4.01
8000	For hauling excavated material, see Section 31 23 23.20									

31 23 19 – Dewatering

31 23 19.20 Dewatering Systems

Line	Description	Crew	Daily Output	Labor-Hours	Unit	Material	Labor	Equipment	Total	Total Incl O&P
0010	**DEWATERING SYSTEMS**									
0020	Excavate drainage trench, 2′ wide, 2′ deep	B-11C	90	.178	C.Y.		6.10	4.06	10.16	14.60
0100	2′ wide, 3′ deep, with backhoe loader	″	135	.119			4.06	2.71	6.77	9.75
0200	Excavate sump pits by hand, light soil	1 Clab	7.10	1.127			33		33	55
0300	Heavy soil	″	3.50	2.286			66.50		66.50	112
0500	Pumping 8 hr., attended 2 hrs. per day, including 20 L.F.									
0550	of suction hose & 100 L.F. discharge hose									
0600	2″ diaphragm pump used for 8 hours	B-10H	4	2	Day		78.50	18.90	97.40	151
0650	4″ diaphragm pump used for 8 hours	B-10I	4	2	″		78.50	31.50	110	165
1600	Sump hole construction, incl. excavation and gravel, pit	B-6	1250	.019	C.F.	1.10	.62	.29	2.01	2.56
1700	With 12″ gravel collar, 12″ pipe, corrugated, 16 ga.	″	70	.343	L.F.	21.50	11	5.20	37.70	47.50

31 23 23 – Fill

31 23 23.13 Backfill

Line	Description	Crew	Daily Output	Labor-Hours	Unit	Material	Labor	Equipment	Total	Total Incl O&P
0010	**BACKFILL**									
0015	By hand, no compaction, light soil	1 Clab	14	.571	L.C.Y.		16.70		16.70	28
0100	Heavy soil		11	.727	″		21		21	35.50
0300	Compaction in 6″ layers, hand tamp, add to above		20.60	.388	E.C.Y.		11.35		11.35	19
0400	Roller compaction operator walking, add	B-10A	100	.080			3.15	1.85	5	7.25
0500	Air tamp, add	B-9D	190	.211			6.25	1.35	7.60	11.95
0600	Vibrating plate, add	A-1D	60	.133			3.89	.60	4.49	7.15
0800	Compaction in 12″ layers, hand tamp, add to above	1 Clab	34	.235			6.85		6.85	11.50
1300	Dozer backfilling, bulk, up to 300′ haul, no compaction	B-10B	1200	.007	L.C.Y.		.26	1.16	1.42	1.71
1400	Air tamped, add	B-11B	80	.200	E.C.Y.		6.70	3.66	10.36	15.20

31 23 23.16 Fill By Borrow and Utility Bedding

Line	Description	Crew	Daily Output	Labor-Hours	Unit	Material	Labor	Equipment	Total	Total Incl O&P
0010	**FILL BY BORROW AND UTILITY BEDDING**									
0049	Utility bedding, for pipe & conduit, not incl. compaction									
0050	Crushed or screened bank run gravel	B-6	150	.160	L.C.Y.	23	5.15	2.44	30.59	36.50
0100	Crushed stone 3/4″ to 1/2″		150	.160		24	5.15	2.44	31.59	37.50
0200	Sand, dead or bank		150	.160		20	5.15	2.44	27.59	33
0500	Compacting bedding in trench	A-1D	90	.089	E.C.Y.		2.60	.40	3	4.79
0600	If material source exceeds 2 miles, add for extra mileage.									
0610	See Section 31 23 23.20 for hauling mileage add.									

31 23 23.17 General Fill

Line	Description	Crew	Daily Output	Labor-Hours	Unit	Material	Labor	Equipment	Total	Total Incl O&P
0010	**GENERAL FILL**									
0011	Spread dumped material, no compaction									
0020	By dozer	B-10B	1000	.008	L.C.Y.		.31	1.40	1.71	2.06
0100	By hand	1 Clab	12	.667	″		19.45		19.45	32.50
0500	Gravel fill, compacted, under floor slabs, 4″ deep	B-37	10000	.005	S.F.	.44	.15	.02	.61	.75
0600	6″ deep		8600	.006		.66	.17	.02	.85	1.04
0700	9″ deep		7200	.007		1.10	.21	.02	1.33	1.58
0800	12″ deep		6000	.008		1.54	.25	.03	1.82	2.13
1000	Alternate pricing method, 4″ deep		120	.400	E.C.Y.	33	12.40	1.36	46.76	58.50
1100	6″ deep		160	.300		33	9.30	1.02	43.32	53

31 23 23.17 General Fill	Crew	Daily Output	Labor-Hours	Unit	Material	2016 Bare Costs Labor	Equipment	Total	Total Incl O&P
1200 9" deep	B-37	200	.240	E.C.Y.	33	7.45	.81	41.26	50
1300 12" deep	↓	220	.218	↓	33	6.75	.74	40.49	48.50

31 23 23.20 Hauling	Crew	Daily Output	Labor-Hours	Unit	Material	Labor	Equipment	Total	Total Incl O&P
0010 **HAULING**									
0011 Excavated or borrow, loose cubic yards									
0012 no loading equipment, including hauling, waiting, loading/dumping									
0013 time per cycle (wait, load, travel, unload or dump & return)									
0014 8 C.Y. truck, 15 MPH ave, cycle 0.5 miles, 10 min. wait/Ld./Uld.	B-34A	320	.025	L.C.Y.		.85	1.30	2.15	2.84
0016 cycle 1 mile		272	.029			1	1.53	2.53	3.35
0018 cycle 2 miles		208	.038			1.31	2.01	3.32	4.38
0020 cycle 4 miles		144	.056			1.90	2.90	4.80	6.30
0022 cycle 6 miles		112	.071			2.44	3.73	6.17	8.15
0024 cycle 8 miles		88	.091			3.10	4.74	7.84	10.30
0026 20 MPH ave, cycle 0.5 mile		336	.024			.81	1.24	2.05	2.71
0028 cycle 1 mile		296	.027			.92	1.41	2.33	3.07
0030 cycle 2 miles		240	.033			1.14	1.74	2.88	3.79
0032 cycle 4 miles		176	.045			1.55	2.37	3.92	5.15
0034 cycle 6 miles		136	.059			2.01	3.07	5.08	6.70
0036 cycle 8 miles		112	.071			2.44	3.73	6.17	8.15
0044 25 MPH ave, cycle 4 miles		192	.042			1.42	2.17	3.59	4.74
0046 cycle 6 miles		160	.050			1.71	2.61	4.32	5.70
0048 cycle 8 miles		128	.063			2.13	3.26	5.39	7.10
0050 30 MPH ave, cycle 4 miles		216	.037			1.26	1.93	3.19	4.22
0052 cycle 6 miles		176	.045			1.55	2.37	3.92	5.15
0054 cycle 8 miles		144	.056			1.90	2.90	4.80	6.30
0114 15 MPH ave, cycle 0.5 mile, 15 min. wait/Ld./Uld.		224	.036			1.22	1.86	3.08	4.06
0116 cycle 1 mile		200	.040			1.37	2.09	3.46	4.55
0118 cycle 2 miles		168	.048			1.63	2.48	4.11	5.40
0120 cycle 4 miles		120	.067			2.28	3.48	5.76	7.60
0122 cycle 6 miles		96	.083			2.85	4.35	7.20	9.50
0124 cycle 8 miles		80	.100			3.42	5.20	8.62	11.40
0126 20 MPH ave, cycle 0.5 mile		232	.034			1.18	1.80	2.98	3.92
0128 cycle 1 mile		208	.038			1.31	2.01	3.32	4.38
0130 cycle 2 miles		184	.043			1.48	2.27	3.75	4.95
0132 cycle 4 miles		144	.056			1.90	2.90	4.80	6.30
0134 cycle 6 miles		112	.071			2.44	3.73	6.17	8.15
0136 cycle 8 miles		96	.083			2.85	4.35	7.20	9.50
0144 25 MPH ave, cycle 4 miles		152	.053			1.80	2.75	4.55	6
0146 cycle 6 miles		128	.063			2.13	3.26	5.39	7.10
0148 cycle 8 miles		112	.071			2.44	3.73	6.17	8.15
0150 30 MPH ave, cycle 4 miles		168	.048			1.63	2.48	4.11	5.40
0152 cycle 6 miles		144	.056			1.90	2.90	4.80	6.30
0154 cycle 8 miles		120	.067			2.28	3.48	5.76	7.60
0214 15 MPH ave, cycle 0.5 mile, 20 min wait/Ld./Uld.		176	.045			1.55	2.37	3.92	5.15
0216 cycle 1 mile		160	.050			1.71	2.61	4.32	5.70
0218 cycle 2 miles		136	.059			2.01	3.07	5.08	6.70
0220 cycle 4 miles		104	.077			2.63	4.01	6.64	8.75
0222 cycle 6 miles		88	.091			3.10	4.74	7.84	10.30
0224 cycle 8 miles		72	.111			3.79	5.80	9.59	12.65
0226 20 MPH ave, cycle 0.5 mile		176	.045			1.55	2.37	3.92	5.15
0228 cycle 1 mile		168	.048			1.63	2.48	4.11	5.40
0230 cycle 2 miles	↓	144	.056	↓		1.90	2.90	4.80	6.30

31 23 23 – Fill

31 23 23.20 Hauling		Crew	Daily Output	Labor-Hours	Unit	Material	2016 Bare Costs Labor	Equipment	Total	Total Incl O&P
0232	cycle 4 miles	B-34A	120	.067	L.C.Y.		2.28	3.48	5.76	7.60
0234	cycle 6 miles		96	.083			2.85	4.35	7.20	9.50
0236	cycle 8 miles		88	.091			3.10	4.74	7.84	10.30
0244	25 MPH ave, cycle 4 miles		128	.063			2.13	3.26	5.39	7.10
0246	cycle 6 miles		112	.071			2.44	3.73	6.17	8.15
0248	cycle 8 miles		96	.083			2.85	4.35	7.20	9.50
0250	30 MPH ave, cycle 4 miles		136	.059			2.01	3.07	5.08	6.70
0252	cycle 6 miles		120	.067			2.28	3.48	5.76	7.60
0254	cycle 8 miles		104	.077			2.63	4.01	6.64	8.75
0314	15 MPH ave, cycle 0.5 mile, 25 min wait/Ld./Uld.		144	.056			1.90	2.90	4.80	6.30
0316	cycle 1 mile		128	.063			2.13	3.26	5.39	7.10
0318	cycle 2 miles		112	.071			2.44	3.73	6.17	8.15
0320	cycle 4 miles		96	.083			2.85	4.35	7.20	9.50
0322	cycle 6 miles		80	.100			3.42	5.20	8.62	11.40
0324	cycle 8 miles		64	.125			4.27	6.50	10.77	14.20
0326	20 MPH ave, cycle 0.5 mile		144	.056			1.90	2.90	4.80	6.30
0328	cycle 1 mile		136	.059			2.01	3.07	5.08	6.70
0330	cycle 2 miles		120	.067			2.28	3.48	5.76	7.60
0332	cycle 4 miles		104	.077			2.63	4.01	6.64	8.75
0334	cycle 6 miles		88	.091			3.10	4.74	7.84	10.30
0336	cycle 8 miles		80	.100			3.42	5.20	8.62	11.40
0344	25 MPH ave, cycle 4 miles		112	.071			2.44	3.73	6.17	8.15
0346	cycle 6 miles		96	.083			2.85	4.35	7.20	9.50
0348	cycle 8 miles		88	.091			3.10	4.74	7.84	10.30
0350	30 MPH ave, cycle 4 miles		112	.071			2.44	3.73	6.17	8.15
0352	cycle 6 miles		104	.077			2.63	4.01	6.64	8.75
0354	cycle 8 miles		96	.083			2.85	4.35	7.20	9.50
0414	15 MPH ave, cycle 0.5 mile, 30 min wait/Ld./Uld.		120	.067			2.28	3.48	5.76	7.60
0416	cycle 1 mile		112	.071			2.44	3.73	6.17	8.15
0418	cycle 2 miles		96	.083			2.85	4.35	7.20	9.50
0420	cycle 4 miles		80	.100			3.42	5.20	8.62	11.40
0422	cycle 6 miles		72	.111			3.79	5.80	9.59	12.65
0424	cycle 8 miles		64	.125			4.27	6.50	10.77	14.20
0426	20 MPH ave, cycle 0.5 mile		120	.067			2.28	3.48	5.76	7.60
0428	cycle 1 mile		112	.071			2.44	3.73	6.17	8.15
0430	cycle 2 miles		104	.077			2.63	4.01	6.64	8.75
0432	cycle 4 miles		88	.091			3.10	4.74	7.84	10.30
0434	cycle 6 miles		80	.100			3.42	5.20	8.62	11.40
0436	cycle 8 miles		72	.111			3.79	5.80	9.59	12.65
0444	25 MPH ave, cycle 4 miles		96	.083			2.85	4.35	7.20	9.50
0446	cycle 6 miles		88	.091			3.10	4.74	7.84	10.30
0448	cycle 8 miles		80	.100			3.42	5.20	8.62	11.40
0450	30 MPH ave, cycle 4 miles		96	.083			2.85	4.35	7.20	9.50
0452	cycle 6 miles		88	.091			3.10	4.74	7.84	10.30
0454	cycle 8 miles		80	.100			3.42	5.20	8.62	11.40
0514	15 MPH ave, cycle 0.5 mile, 35 min wait/Ld./Uld.		104	.077			2.63	4.01	6.64	8.75
0516	cycle 1 mile		96	.083			2.85	4.35	7.20	9.50
0518	cycle 2 miles		88	.091			3.10	4.74	7.84	10.30
0520	cycle 4 miles		72	.111			3.79	5.80	9.59	12.65
0522	cycle 6 miles		64	.125			4.27	6.50	10.77	14.20
0524	cycle 8 miles		56	.143			4.88	7.45	12.33	16.25
0526	20 MPH ave, cycle 0.5 mile		104	.077			2.63	4.01	6.64	8.75
0528	cycle 1 mile		96	.083			2.85	4.35	7.20	9.50

31 23 23.20 Hauling		Crew	Daily Output	Labor-Hours	Unit	Material	2016 Bare Costs Labor	Equipment	Total	Total Incl O&P
0530	cycle 2 miles	B-34A	96	.083	L.C.Y.		2.85	4.35	7.20	9.50
0532	cycle 4 miles		80	.100			3.42	5.20	8.62	11.40
0534	cycle 6 miles		72	.111			3.79	5.80	9.59	12.65
0536	cycle 8 miles		64	.125			4.27	6.50	10.77	14.20
0544	25 MPH ave, cycle 4 miles		88	.091			3.10	4.74	7.84	10.30
0546	cycle 6 miles		80	.100			3.42	5.20	8.62	11.40
0548	cycle 8 miles		72	.111			3.79	5.80	9.59	12.65
0550	30 MPH ave, cycle 4 miles		88	.091			3.10	4.74	7.84	10.30
0552	cycle 6 miles		80	.100			3.42	5.20	8.62	11.40
0554	cycle 8 miles		72	.111			3.79	5.80	9.59	12.65
1014	12 C.Y. truck, cycle 0.5 mile, 15 MPH ave, 15 min. wait/Ld./Uld.	B-34B	336	.024			.81	2.06	2.87	3.60
1016	cycle 1 mile		300	.027			.91	2.30	3.21	4.03
1018	cycle 2 miles		252	.032			1.08	2.74	3.82	4.81
1020	cycle 4 miles		180	.044			1.52	3.84	5.36	6.70
1022	cycle 6 miles		144	.056			1.90	4.80	6.70	8.45
1024	cycle 8 miles		120	.067			2.28	5.75	8.03	10.10
1025	cycle 10 miles		96	.083			2.85	7.20	10.05	12.60
1026	20 MPH ave, cycle 0.5 mile		348	.023			.79	1.99	2.78	3.48
1028	cycle 1 mile		312	.026			.88	2.21	3.09	3.88
1030	cycle 2 miles		276	.029			.99	2.50	3.49	4.38
1032	cycle 4 miles		216	.037			1.26	3.20	4.46	5.60
1034	cycle 6 miles		168	.048			1.63	4.11	5.74	7.20
1036	cycle 8 miles		144	.056			1.90	4.80	6.70	8.45
1038	cycle 10 miles		120	.067			2.28	5.75	8.03	10.10
1040	25 MPH ave, cycle 4 miles		228	.035			1.20	3.03	4.23	5.30
1042	cycle 6 miles		192	.042			1.42	3.60	5.02	6.30
1044	cycle 8 miles		168	.048			1.63	4.11	5.74	7.20
1046	cycle 10 miles		144	.056			1.90	4.80	6.70	8.45
1050	30 MPH ave, cycle 4 miles		252	.032			1.08	2.74	3.82	4.81
1052	cycle 6 miles		216	.037			1.26	3.20	4.46	5.60
1054	cycle 8 miles		180	.044			1.52	3.84	5.36	6.70
1056	cycle 10 miles		156	.051			1.75	4.43	6.18	7.75
1060	35 MPH ave, cycle 4 miles		264	.030			1.03	2.62	3.65	4.59
1062	cycle 6 miles		228	.035			1.20	3.03	4.23	5.30
1064	cycle 8 miles		204	.039			1.34	3.39	4.73	5.95
1066	cycle 10 miles		180	.044			1.52	3.84	5.36	6.70
1068	cycle 20 miles		120	.067			2.28	5.75	8.03	10.10
1069	cycle 30 miles		84	.095			3.25	8.20	11.45	14.40
1070	cycle 40 miles		72	.111			3.79	9.60	13.39	16.80
1072	40 MPH ave, cycle 6 miles		240	.033			1.14	2.88	4.02	5.05
1074	cycle 8 miles		216	.037			1.26	3.20	4.46	5.60
1076	cycle 10 miles		192	.042			1.42	3.60	5.02	6.30
1078	cycle 20 miles		120	.067			2.28	5.75	8.03	10.10
1080	cycle 30 miles		96	.083			2.85	7.20	10.05	12.60
1082	cycle 40 miles		72	.111			3.79	9.60	13.39	16.80
1084	cycle 50 miles		60	.133			4.55	11.50	16.05	20
1094	45 MPH ave, cycle 8 miles		216	.037			1.26	3.20	4.46	5.60
1096	cycle 10 miles		204	.039			1.34	3.39	4.73	5.95
1098	cycle 20 miles		132	.061			2.07	5.25	7.32	9.15
1100	cycle 30 miles		108	.074			2.53	6.40	8.93	11.20
1102	cycle 40 miles		84	.095			3.25	8.20	11.45	14.40
1104	cycle 50 miles		72	.111			3.79	9.60	13.39	16.80
1106	50 MPH ave, cycle 10 miles		216	.037			1.26	3.20	4.46	5.60

31 23 23.20 Hauling		Crew	Daily Output	Labor-Hours	Unit	Material	2016 Bare Costs Labor	Equipment	Total	Total Incl O&P
1108	cycle 20 miles	B-34B	144	.056	L.C.Y.		1.90	4.80	6.70	8.45
1110	cycle 30 miles		108	.074			2.53	6.40	8.93	11.20
1112	cycle 40 miles		84	.095			3.25	8.20	11.45	14.40
1114	cycle 50 miles		72	.111			3.79	9.60	13.39	16.80
1214	15 MPH ave, cycle 0.5 mile, 20 min. wait/Ld./Uld.		264	.030			1.03	2.62	3.65	4.59
1216	cycle 1 mile		240	.033			1.14	2.88	4.02	5.05
1218	cycle 2 miles		204	.039			1.34	3.39	4.73	5.95
1220	cycle 4 miles		156	.051			1.75	4.43	6.18	7.75
1222	cycle 6 miles		132	.061			2.07	5.25	7.32	9.15
1224	cycle 8 miles		108	.074			2.53	6.40	8.93	11.20
1225	cycle 10 miles		96	.083			2.85	7.20	10.05	12.60
1226	20 MPH ave, cycle 0.5 mile		264	.030			1.03	2.62	3.65	4.59
1228	cycle 1 mile		252	.032			1.08	2.74	3.82	4.81
1230	cycle 2 miles		216	.037			1.26	3.20	4.46	5.60
1232	cycle 4 miles		180	.044			1.52	3.84	5.36	6.70
1234	cycle 6 miles		144	.056			1.90	4.80	6.70	8.45
1236	cycle 8 miles		132	.061			2.07	5.25	7.32	9.15
1238	cycle 10 miles		108	.074			2.53	6.40	8.93	11.20
1240	25 MPH ave, cycle 4 miles		192	.042			1.42	3.60	5.02	6.30
1242	cycle 6 miles		168	.048			1.63	4.11	5.74	7.20
1244	cycle 8 miles		144	.056			1.90	4.80	6.70	8.45
1246	cycle 10 miles		132	.061			2.07	5.25	7.32	9.15
1250	30 MPH ave, cycle 4 miles		204	.039			1.34	3.39	4.73	5.95
1252	cycle 6 miles		180	.044			1.52	3.84	5.36	6.70
1254	cycle 8 miles		156	.051			1.75	4.43	6.18	7.75
1256	cycle 10 miles		144	.056			1.90	4.80	6.70	8.45
1260	35 MPH ave, cycle 4 miles		216	.037			1.26	3.20	4.46	5.60
1262	cycle 6 miles		192	.042			1.42	3.60	5.02	6.30
1264	cycle 8 miles		168	.048			1.63	4.11	5.74	7.20
1266	cycle 10 miles		156	.051			1.75	4.43	6.18	7.75
1268	cycle 20 miles		108	.074			2.53	6.40	8.93	11.20
1269	cycle 30 miles		72	.111			3.79	9.60	13.39	16.80
1270	cycle 40 miles		60	.133			4.55	11.50	16.05	20
1272	40 MPH ave, cycle 6 miles		192	.042			1.42	3.60	5.02	6.30
1274	cycle 8 miles		180	.044			1.52	3.84	5.36	6.70
1276	cycle 10 miles		156	.051			1.75	4.43	6.18	7.75
1278	cycle 20 miles		108	.074			2.53	6.40	8.93	11.20
1280	cycle 30 miles		84	.095			3.25	8.20	11.45	14.40
1282	cycle 40 miles		72	.111			3.79	9.60	13.39	16.80
1284	cycle 50 miles		60	.133			4.55	11.50	16.05	20
1294	45 MPH ave, cycle 8 miles		180	.044			1.52	3.84	5.36	6.70
1296	cycle 10 miles		168	.048			1.63	4.11	5.74	7.20
1298	cycle 20 miles		120	.067			2.28	5.75	8.03	10.10
1300	cycle 30 miles		96	.083			2.85	7.20	10.05	12.60
1302	cycle 40 miles		72	.111			3.79	9.60	13.39	16.80
1304	cycle 50 miles		60	.133			4.55	11.50	16.05	20
1306	50 MPH ave, cycle 10 miles		180	.044			1.52	3.84	5.36	6.70
1308	cycle 20 miles		132	.061			2.07	5.25	7.32	9.15
1310	cycle 30 miles		96	.083			2.85	7.20	10.05	12.60
1312	cycle 40 miles		84	.095			3.25	8.20	11.45	14.40
1314	cycle 50 miles		72	.111			3.79	9.60	13.39	16.80
1414	15 MPH ave, cycle 0.5 mile, 25 min. wait/Ld./Uld.		204	.039			1.34	3.39	4.73	5.95
1416	cycle 1 mile	B-34B	192	.042	L.C.Y.		1.42	3.60	5.02	6.30

	31 23 23.20 Hauling	Crew	Daily Output	Labor-Hours	Unit	Material	2016 Bare Costs Labor	Equipment	Total	Total Incl O&P
1418	cycle 2 miles	B-34B	168	.048	L.C.Y.		1.63	4.11	5.74	7.20
1420	cycle 4 miles		132	.061			2.07	5.25	7.32	9.15
1422	cycle 6 miles		120	.067			2.28	5.75	8.03	10.10
1424	cycle 8 miles		96	.083			2.85	7.20	10.05	12.60
1425	cycle 10 miles		84	.095			3.25	8.20	11.45	14.40
1426	20 MPH ave, cycle 0.5 mile		216	.037			1.26	3.20	4.46	5.60
1428	cycle 1 mile		204	.039			1.34	3.39	4.73	5.95
1430	cycle 2 miles		180	.044			1.52	3.84	5.36	6.70
1432	cycle 4 miles		156	.051			1.75	4.43	6.18	7.75
1434	cycle 6 miles		132	.061			2.07	5.25	7.32	9.15
1436	cycle 8 miles		120	.067			2.28	5.75	8.03	10.10
1438	cycle 10 miles		96	.083			2.85	7.20	10.05	12.60
1440	25 MPH ave, cycle 4 miles		168	.048			1.63	4.11	5.74	7.20
1442	cycle 6 miles		144	.056			1.90	4.80	6.70	8.45
1444	cycle 8 miles		132	.061			2.07	5.25	7.32	9.15
1446	cycle 10 miles		108	.074			2.53	6.40	8.93	11.20
1450	30 MPH ave, cycle 4 miles		168	.048			1.63	4.11	5.74	7.20
1452	cycle 6 miles		156	.051			1.75	4.43	6.18	7.75
1454	cycle 8 miles		132	.061			2.07	5.25	7.32	9.15
1456	cycle 10 miles		120	.067			2.28	5.75	8.03	10.10
1460	35 MPH ave, cycle 4 miles		180	.044			1.52	3.84	5.36	6.70
1462	cycle 6 miles		156	.051			1.75	4.43	6.18	7.75
1464	cycle 8 miles		144	.056			1.90	4.80	6.70	8.45
1466	cycle 10 miles		132	.061			2.07	5.25	7.32	9.15
1468	cycle 20 miles		96	.083			2.85	7.20	10.05	12.60
1469	cycle 30 miles		72	.111			3.79	9.60	13.39	16.80
1470	cycle 40 miles		60	.133			4.55	11.50	16.05	20
1472	40 MPH ave, cycle 6 miles		168	.048			1.63	4.11	5.74	7.20
1474	cycle 8 miles		156	.051			1.75	4.43	6.18	7.75
1476	cycle 10 miles		144	.056			1.90	4.80	6.70	8.45
1478	cycle 20 miles		96	.083			2.85	7.20	10.05	12.60
1480	cycle 30 miles		84	.095			3.25	8.20	11.45	14.40
1482	cycle 40 miles		60	.133			4.55	11.50	16.05	20
1484	cycle 50 miles		60	.133			4.55	11.50	16.05	20
1494	45 MPH ave, cycle 8 miles		156	.051			1.75	4.43	6.18	7.75
1496	cycle 10 miles		144	.056			1.90	4.80	6.70	8.45
1498	cycle 20 miles		108	.074			2.53	6.40	8.93	11.20
1500	cycle 30 miles		84	.095			3.25	8.20	11.45	14.40
1502	cycle 40 miles		72	.111			3.79	9.60	13.39	16.80
1504	cycle 50 miles		60	.133			4.55	11.50	16.05	20
1506	50 MPH ave, cycle 10 miles		156	.051			1.75	4.43	6.18	7.75
1508	cycle 20 miles		120	.067			2.28	5.75	8.03	10.10
1510	cycle 30 miles		96	.083			2.85	7.20	10.05	12.60
1512	cycle 40 miles		72	.111			3.79	9.60	13.39	16.80
1514	cycle 50 miles		60	.133			4.55	11.50	16.05	20
1614	15 MPH ave, cycle 0.5 mile, 30 min. wait/Ld./Uld.		180	.044			1.52	3.84	5.36	6.70
1616	cycle 1 mile		168	.048			1.63	4.11	5.74	7.20
1618	cycle 2 miles		144	.056			1.90	4.80	6.70	8.45
1620	cycle 4 miles		120	.067			2.28	5.75	8.03	10.10
1622	cycle 6 miles		108	.074			2.53	6.40	8.93	11.20
1624	cycle 8 miles		84	.095			3.25	8.20	11.45	14.40
1625	cycle 10 miles		84	.095			3.25	8.20	11.45	14.40
1626	20 MPH ave, cycle 0.5 mile		180	.044			1.52	3.84	5.36	6.70

31 23 23.20 Hauling		Crew	Daily Output	Labor-Hours	Unit	Material	2016 Bare Costs Labor	Equipment	Total	Total Incl O&P
1628	cycle 1 mile	B-34B	168	.048	L.C.Y.		1.63	4.11	5.74	7.20
1630	cycle 2 miles		156	.051			1.75	4.43	6.18	7.75
1632	cycle 4 miles		132	.061			2.07	5.25	7.32	9.15
1634	cycle 6 miles		120	.067			2.28	5.75	8.03	10.10
1636	cycle 8 miles		108	.074			2.53	6.40	8.93	11.20
1638	cycle 10 miles		96	.083			2.85	7.20	10.05	12.60
1640	25 MPH ave, cycle 4 miles		144	.056			1.90	4.80	6.70	8.45
1642	cycle 6 miles		132	.061			2.07	5.25	7.32	9.15
1644	cycle 8 miles		108	.074			2.53	6.40	8.93	11.20
1646	cycle 10 miles		108	.074			2.53	6.40	8.93	11.20
1650	30 MPH ave, cycle 4 miles		144	.056			1.90	4.80	6.70	8.45
1652	cycle 6 miles		132	.061			2.07	5.25	7.32	9.15
1654	cycle 8 miles		120	.067			2.28	5.75	8.03	10.10
1656	cycle 10 miles		108	.074			2.53	6.40	8.93	11.20
1660	35 MPH ave, cycle 4 miles		156	.051			1.75	4.43	6.18	7.75
1662	cycle 6 miles		144	.056			1.90	4.80	6.70	8.45
1664	cycle 8 miles		132	.061			2.07	5.25	7.32	9.15
1666	cycle 10 miles		120	.067			2.28	5.75	8.03	10.10
1668	cycle 20 miles		84	.095			3.25	8.20	11.45	14.40
1669	cycle 30 miles		72	.111			3.79	9.60	13.39	16.80
1670	cycle 40 miles		60	.133			4.55	11.50	16.05	20
1672	40 MPH ave, cycle 6 miles		144	.056			1.90	4.80	6.70	8.45
1674	cycle 8 miles		132	.061			2.07	5.25	7.32	9.15
1676	cycle 10 miles		120	.067			2.28	5.75	8.03	10.10
1678	cycle 20 miles		96	.083			2.85	7.20	10.05	12.60
1680	cycle 30 miles		72	.111			3.79	9.60	13.39	16.80
1682	cycle 40 miles		60	.133			4.55	11.50	16.05	20
1684	cycle 50 miles		48	.167			5.70	14.40	20.10	25.50
1694	45 MPH ave, cycle 8 miles		144	.056			1.90	4.80	6.70	8.45
1696	cycle 10 miles		132	.061			2.07	5.25	7.32	9.15
1698	cycle 20 miles		96	.083			2.85	7.20	10.05	12.60
1700	cycle 30 miles		84	.095			3.25	8.20	11.45	14.40
1702	cycle 40 miles		60	.133			4.55	11.50	16.05	20
1704	cycle 50 miles		60	.133			4.55	11.50	16.05	20
1706	50 MPH ave, cycle 10 miles		132	.061			2.07	5.25	7.32	9.15
1708	cycle 20 miles		108	.074			2.53	6.40	8.93	11.20
1710	cycle 30 miles		84	.095			3.25	8.20	11.45	14.40
1712	cycle 40 miles		72	.111			3.79	9.60	13.39	16.80
1714	cycle 50 miles		60	.133			4.55	11.50	16.05	20
2000	Hauling, 8 C.Y. truck, small project cost per hour	B-34A	8	1	Hr.		34	52	86	114
2100	12 C.Y. Truck	B-34B	8	1			34	86.50	120.50	152
2150	16.5 C.Y. Truck	B-34C	8	1			34	90.50	124.50	157
2175	18 C.Y. 8 wheel Truck	B-34I	8	1			34	109	143	177
2200	20 C.Y. Truck	B-34D	8	1			34	92.50	126.50	159
2300	Grading at dump, or embankment if required, by dozer	B-10B	1000	.008	L.C.Y.		.31	1.40	1.71	2.06
2310	Spotter at fill or cut, if required	1 Clab	8	1	Hr.		29		29	49
9014	18 C.Y. truck, 8 wheels,15 min. wait/Ld./Uld.,15 MPH, cycle 0.5 mi.	B-34I	504	.016	L.C.Y.		.54	1.72	2.26	2.79
9016	cycle 1 mile		450	.018			.61	1.93	2.54	3.13
9018	cycle 2 miles		378	.021			.72	2.30	3.02	3.72
9020	cycle 4 miles		270	.030			1.01	3.22	4.23	5.20
9022	cycle 6 miles		216	.037			1.26	4.03	5.29	6.50
9024	cycle 8 miles		180	.044			1.52	4.83	6.35	7.80
9025	cycle 10 miles		144	.056			1.90	6.05	7.95	9.80

31 23 23.20 Hauling		Crew	Daily Output	Labor-Hours	Unit	Material	2016 Bare Costs Labor	Equipment	Total	Total Incl O&P
9026	20 MPH ave, cycle 0.5 mile	B-34I	522	.015	L.C.Y.		.52	1.67	2.19	2.69
9028	cycle 1 mile		468	.017			.58	1.86	2.44	3
9030	cycle 2 miles		414	.019			.66	2.10	2.76	3.40
9032	cycle 4 miles		324	.025			.84	2.68	3.52	4.34
9034	cycle 6 miles		252	.032			1.08	3.45	4.53	5.60
9036	cycle 8 miles		216	.037			1.26	4.03	5.29	6.50
9038	cycle 10 miles		180	.044			1.52	4.83	6.35	7.80
9040	25 MPH ave, cycle 4 miles		342	.023			.80	2.54	3.34	4.12
9042	cycle 6 miles		288	.028			.95	3.02	3.97	4.89
9044	cycle 8 miles		252	.032			1.08	3.45	4.53	5.60
9046	cycle 10 miles		216	.037			1.26	4.03	5.29	6.50
9050	30 MPH ave, cycle 4 miles		378	.021			.72	2.30	3.02	3.72
9052	cycle 6 miles		324	.025			.84	2.68	3.52	4.34
9054	cycle 8 miles		270	.030			1.01	3.22	4.23	5.20
9056	cycle 10 miles		234	.034			1.17	3.72	4.89	6
9060	35 MPH ave, cycle 4 miles		396	.020			.69	2.20	2.89	3.55
9062	cycle 6 miles		342	.023			.80	2.54	3.34	4.12
9064	cycle 8 miles		288	.028			.95	3.02	3.97	4.89
9066	cycle 10 miles		270	.030			1.01	3.22	4.23	5.20
9068	cycle 20 miles		162	.049			1.69	5.35	7.04	8.70
9070	cycle 30 miles		126	.063			2.17	6.90	9.07	11.20
9072	cycle 40 miles		90	.089			3.04	9.65	12.69	15.65
9074	40 MPH ave, cycle 6 miles		360	.022			.76	2.41	3.17	3.91
9076	cycle 8 miles		324	.025			.84	2.68	3.52	4.34
9078	cycle 10 miles		288	.028			.95	3.02	3.97	4.89
9080	cycle 20 miles		180	.044			1.52	4.83	6.35	7.80
9082	cycle 30 miles		144	.056			1.90	6.05	7.95	9.80
9084	cycle 40 miles		108	.074			2.53	8.05	10.58	13
9086	cycle 50 miles		90	.089			3.04	9.65	12.69	15.65
9094	45 MPH ave, cycle 8 miles		324	.025			.84	2.68	3.52	4.34
9096	cycle 10 miles		306	.026			.89	2.84	3.73	4.59
9098	cycle 20 miles		198	.040			1.38	4.39	5.77	7.10
9100	cycle 30 miles		144	.056			1.90	6.05	7.95	9.80
9102	cycle 40 miles		126	.063			2.17	6.90	9.07	11.20
9104	cycle 50 miles		108	.074			2.53	8.05	10.58	13
9106	50 MPH ave, cycle 10 miles		324	.025			.84	2.68	3.52	4.34
9108	cycle 20 miles		216	.037			1.26	4.03	5.29	6.50
9110	cycle 30 miles		162	.049			1.69	5.35	7.04	8.70
9112	cycle 40 miles		126	.063			2.17	6.90	9.07	11.20
9114	cycle 50 miles		108	.074			2.53	8.05	10.58	13
9214	20 min. wait/Ld./Uld.,15 MPH, cycle 0.5 mi.		396	.020			.69	2.20	2.89	3.55
9216	cycle 1 mile		360	.022			.76	2.41	3.17	3.91
9218	cycle 2 miles		306	.026			.89	2.84	3.73	4.59
9220	cycle 4 miles		234	.034			1.17	3.72	4.89	6
9222	cycle 6 miles		198	.040			1.38	4.39	5.77	7.10
9224	cycle 8 miles		162	.049			1.69	5.35	7.04	8.70
9225	cycle 10 miles		144	.056			1.90	6.05	7.95	9.80
9226	20 MPH ave, cycle 0.5 mile		396	.020			.69	2.20	2.89	3.55
9228	cycle 1 mile		378	.021			.72	2.30	3.02	3.72
9230	cycle 2 miles		324	.025			.84	2.68	3.52	4.34
9232	cycle 4 miles		270	.030			1.01	3.22	4.23	5.20
9234	cycle 6 miles		216	.037			1.26	4.03	5.29	6.50
9236	cycle 8 miles		198	.040			1.38	4.39	5.77	7.10

31 23 23.20 Hauling		Crew	Daily Output	Labor-Hours	Unit	Material	2016 Bare Costs Labor	Equipment	Total	Total Incl O&P
9238	cycle 10 miles	B-34I	162	.049	L.C.Y.		1.69	5.35	7.04	8.70
9240	25 MPH ave, cycle 4 miles		288	.028			.95	3.02	3.97	4.89
9242	cycle 6 miles		252	.032			1.08	3.45	4.53	5.60
9244	cycle 8 miles		216	.037			1.26	4.03	5.29	6.50
9246	cycle 10 miles		198	.040			1.38	4.39	5.77	7.10
9250	30 MPH ave, cycle 4 miles		306	.026			.89	2.84	3.73	4.59
9252	cycle 6 miles		270	.030			1.01	3.22	4.23	5.20
9254	cycle 8 miles		234	.034			1.17	3.72	4.89	6
9256	cycle 10 miles		216	.037			1.26	4.03	5.29	6.50
9260	35 MPH ave, cycle 4 miles		324	.025			.84	2.68	3.52	4.34
9262	cycle 6 miles		288	.028			.95	3.02	3.97	4.89
9264	cycle 8 miles		252	.032			1.08	3.45	4.53	5.60
9266	cycle 10 miles		234	.034			1.17	3.72	4.89	6
9268	cycle 20 miles		162	.049			1.69	5.35	7.04	8.70
9270	cycle 30 miles		108	.074			2.53	8.05	10.58	13
9272	cycle 40 miles		90	.089			3.04	9.65	12.69	15.65
9274	40 MPH ave, cycle 6 miles		288	.028			.95	3.02	3.97	4.89
9276	cycle 8 miles		270	.030			1.01	3.22	4.23	5.20
9278	cycle 10 miles		234	.034			1.17	3.72	4.89	6
9280	cycle 20 miles		162	.049			1.69	5.35	7.04	8.70
9282	cycle 30 miles		126	.063			2.17	6.90	9.07	11.20
9284	cycle 40 miles		108	.074			2.53	8.05	10.58	13
9286	cycle 50 miles		90	.089			3.04	9.65	12.69	15.65
9294	45 MPH ave, cycle 8 miles		270	.030			1.01	3.22	4.23	5.20
9296	cycle 10 miles		252	.032			1.08	3.45	4.53	5.60
9298	cycle 20 miles		180	.044			1.52	4.83	6.35	7.80
9300	cycle 30 miles		144	.056			1.90	6.05	7.95	9.80
9302	cycle 40 miles		108	.074			2.53	8.05	10.58	13
9304	cycle 50 miles		90	.089			3.04	9.65	12.69	15.65
9306	50 MPH ave, cycle 10 miles		270	.030			1.01	3.22	4.23	5.20
9308	cycle 20 miles		198	.040			1.38	4.39	5.77	7.10
9310	cycle 30 miles		144	.056			1.90	6.05	7.95	9.80
9312	cycle 40 miles		126	.063			2.17	6.90	9.07	11.20
9314	cycle 50 miles		108	.074			2.53	8.05	10.58	13
9414	25 min. wait/Ld./Uld.,15 MPH, cycle 0.5 mi.		306	.026			.89	2.84	3.73	4.59
9416	cycle 1 mile		288	.028			.95	3.02	3.97	4.89
9418	cycle 2 miles		252	.032			1.08	3.45	4.53	5.60
9420	cycle 4 miles		198	.040			1.38	4.39	5.77	7.10
9422	cycle 6 miles		180	.044			1.52	4.83	6.35	7.80
9424	cycle 8 miles		144	.056			1.90	6.05	7.95	9.80
9425	cycle 10 miles		126	.063			2.17	6.90	9.07	11.20
9426	20 MPH ave, cycle 0.5 mile		324	.025			.84	2.68	3.52	4.34
9428	cycle 1 mile		306	.026			.89	2.84	3.73	4.59
9430	cycle 2 miles		270	.030			1.01	3.22	4.23	5.20
9432	cycle 4 miles		234	.034			1.17	3.72	4.89	6
9434	cycle 6 miles		198	.040			1.38	4.39	5.77	7.10
9436	cycle 8 miles		180	.044			1.52	4.83	6.35	7.80
9438	cycle 10 miles		144	.056			1.90	6.05	7.95	9.80
9440	25 MPH ave, cycle 4 miles		252	.032			1.08	3.45	4.53	5.60
9442	cycle 6 miles		216	.037			1.26	4.03	5.29	6.50
9444	cycle 8 miles		198	.040			1.38	4.39	5.77	7.10
9446	cycle 10 miles		180	.044			1.52	4.83	6.35	7.80
9450	30 MPH ave, cycle 4 miles		252	.032			1.08	3.45	4.53	5.60

31 23 23.20 Hauling		Crew	Daily Output	Labor-Hours	Unit	Material	2016 Bare Costs Labor	Equipment	Total	Total Incl O&P
9452	cycle 6 miles	B-34I	234	.034	L.C.Y.		1.17	3.72	4.89	6
9454	cycle 8 miles		198	.040			1.38	4.39	5.77	7.10
9456	cycle 10 miles		180	.044			1.52	4.83	6.35	7.80
9460	35 MPH ave, cycle 4 miles		270	.030			1.01	3.22	4.23	5.20
9462	cycle 6 miles		234	.034			1.17	3.72	4.89	6
9464	cycle 8 miles		216	.037			1.26	4.03	5.29	6.50
9466	cycle 10 miles		198	.040			1.38	4.39	5.77	7.10
9468	cycle 20 miles		144	.056			1.90	6.05	7.95	9.80
9470	cycle 30 miles		108	.074			2.53	8.05	10.58	13
9472	cycle 40 miles		90	.089			3.04	9.65	12.69	15.65
9474	40 MPH ave, cycle 6 miles		252	.032			1.08	3.45	4.53	5.60
9476	cycle 8 miles		234	.034			1.17	3.72	4.89	6
9478	cycle 10 miles		216	.037			1.26	4.03	5.29	6.50
9480	cycle 20 miles		144	.056			1.90	6.05	7.95	9.80
9482	cycle 30 miles		126	.063			2.17	6.90	9.07	11.20
9484	cycle 40 miles		90	.089			3.04	9.65	12.69	15.65
9486	cycle 50 miles		90	.089			3.04	9.65	12.69	15.65
9494	45 MPH ave, cycle 8 miles		234	.034			1.17	3.72	4.89	6
9496	cycle 10 miles		216	.037			1.26	4.03	5.29	6.50
9498	cycle 20 miles		162	.049			1.69	5.35	7.04	8.70
9500	cycle 30 miles		126	.063			2.17	6.90	9.07	11.20
9502	cycle 40 miles		108	.074			2.53	8.05	10.58	13
9504	cycle 50 miles		90	.089			3.04	9.65	12.69	15.65
9506	50 MPH ave, cycle 10 miles		234	.034			1.17	3.72	4.89	6
9508	cycle 20 miles		180	.044			1.52	4.83	6.35	7.80
9510	cycle 30 miles		144	.056			1.90	6.05	7.95	9.80
9512	cycle 40 miles		108	.074			2.53	8.05	10.58	13
9514	cycle 50 miles		90	.089			3.04	9.65	12.69	15.65
9614	30 min. wait/Ld./Uld.,15 MPH, cycle 0.5 mi.		270	.030			1.01	3.22	4.23	5.20
9616	cycle 1 mile		252	.032			1.08	3.45	4.53	5.60
9618	cycle 2 miles		216	.037			1.26	4.03	5.29	6.50
9620	cycle 4 miles		180	.044			1.52	4.83	6.35	7.80
9622	cycle 6 miles		162	.049			1.69	5.35	7.04	8.70
9624	cycle 8 miles		126	.063			2.17	6.90	9.07	11.20
9625	cycle 10 miles		126	.063			2.17	6.90	9.07	11.20
9626	20 MPH ave, cycle 0.5 mile		270	.030			1.01	3.22	4.23	5.20
9628	cycle 1 mile		252	.032			1.08	3.45	4.53	5.60
9630	cycle 2 miles		234	.034			1.17	3.72	4.89	6
9632	cycle 4 miles		198	.040			1.38	4.39	5.77	7.10
9634	cycle 6 miles		180	.044			1.52	4.83	6.35	7.80
9636	cycle 8 miles		162	.049			1.69	5.35	7.04	8.70
9638	cycle 10 miles		144	.056			1.90	6.05	7.95	9.80
9640	25 MPH ave, cycle 4 miles		216	.037			1.26	4.03	5.29	6.50
9642	cycle 6 miles		198	.040			1.38	4.39	5.77	7.10
9644	cycle 8 miles		180	.044			1.52	4.83	6.35	7.80
9646	cycle 10 miles		162	.049			1.69	5.35	7.04	8.70
9650	30 MPH ave, cycle 4 miles		216	.037			1.26	4.03	5.29	6.50
9652	cycle 6 miles		198	.040			1.38	4.39	5.77	7.10
9654	cycle 8 miles		180	.044			1.52	4.83	6.35	7.80
9656	cycle 10 miles		162	.049			1.69	5.35	7.04	8.70
9660	35 MPH ave, cycle 4 miles		234	.034			1.17	3.72	4.89	6
9662	cycle 6 miles		216	.037			1.26	4.03	5.29	6.50
9664	cycle 8 miles		198	.040			1.38	4.39	5.77	7.10

31 23 Excavation and Fill

31 23 23 – Fill

31 23 23.20 Hauling		Crew	Daily Output	Labor-Hours	Unit	Material	2016 Bare Costs Labor	Equipment	Total	Total Incl O&P
9666	cycle 10 miles	B-34I	180	.044	L.C.Y.		1.52	4.83	6.35	7.80
9668	cycle 20 miles		126	.063			2.17	6.90	9.07	11.20
9670	cycle 30 miles		108	.074			2.53	8.05	10.58	13
9672	cycle 40 miles		90	.089			3.04	9.65	12.69	15.65
9674	40 MPH ave, cycle 6 miles		216	.037			1.26	4.03	5.29	6.50
9676	cycle 8 miles		198	.040			1.38	4.39	5.77	7.10
9678	cycle 10 miles		180	.044			1.52	4.83	6.35	7.80
9680	cycle 20 miles		144	.056			1.90	6.05	7.95	9.80
9682	cycle 30 miles		108	.074			2.53	8.05	10.58	13
9684	cycle 40 miles		90	.089			3.04	9.65	12.69	15.65
9686	cycle 50 miles		72	.111			3.79	12.10	15.89	19.55
9694	45 MPH ave, cycle 8 miles		216	.037			1.26	4.03	5.29	6.50
9696	cycle 10 miles		198	.040			1.38	4.39	5.77	7.10
9698	cycle 20 miles		144	.056			1.90	6.05	7.95	9.80
9700	cycle 30 miles		126	.063			2.17	6.90	9.07	11.20
9702	cycle 40 miles		108	.074			2.53	8.05	10.58	13
9704	cycle 50 miles		90	.089			3.04	9.65	12.69	15.65
9706	50 MPH ave, cycle 10 miles		198	.040			1.38	4.39	5.77	7.10
9708	cycle 20 miles		162	.049			1.69	5.35	7.04	8.70
9710	cycle 30 miles		126	.063			2.17	6.90	9.07	11.20
9712	cycle 40 miles		108	.074			2.53	8.05	10.58	13
9714	cycle 50 miles	↓	90	.089	↓		3.04	9.65	12.69	15.65

31 23 23.24 Compaction, Structural			Crew	Daily Output	Labor-Hours	Unit	Material	Labor	Equipment	Total	Total Incl O&P
0010	**COMPACTION, STRUCTURAL**										
0020	Steel wheel tandem roller, 5 tons	R312323-30	B-10E	8	1	Hr.		39.50	20.50	60	87.50
0050	Air tamp, 6" to 8" lifts, common fill		B-9	250	.160	E.C.Y.		4.74	.92	5.66	8.95
0060	Select fill		"	300	.133			3.95	.76	4.71	7.45
0600	Vibratory plate, 8" lifts, common fill		A-1D	200	.040			1.17	.18	1.35	2.16
0700	Select fill		"	216	.037	↓		1.08	.17	1.25	1.99

31 25 Erosion and Sedimentation Controls

31 25 14 – Stabilization Measures for Erosion and Sedimentation Control

31 25 14.16 Rolled Erosion Control Mats and Blankets			Crew	Daily Output	Labor-Hours	Unit	Material	Labor	Equipment	Total	Total Incl O&P
0010	**ROLLED EROSION CONTROL MATS AND BLANKETS**										
0020	Jute mesh, 100 S.Y. per roll, 4' wide, stapled	G	B-80A	2400	.010	S.Y.	.99	.29	.13	1.41	1.72
0100	Plastic netting, stapled, 2" x 1" mesh, 20 mil	G	B-1	2500	.010		.47	.29		.76	1
0120	Revegetation mat, webbed	G	2 Clab	1000	.016		1.57	.47		2.04	2.51
0200	Polypropylene mesh, stapled, 6.5 oz./S.Y.	G	B-1	2500	.010		1.65	.29		1.94	2.30
0300	Tobacco netting, or jute mesh #2, stapled	G	"	2500	.010	↓	.21	.29		.50	.71
1000	Silt fence, install and maintain, remove	G	B-62	1300	.018	L.F.	.25	.59	.13	.97	1.42
1100	Allow 25% per month for maintenance; 6-month max life										

31 31 Soil Treatment

31 31 16 – Termite Control

31 31 16.13 Chemical Termite Control		Crew	Daily Output	Labor-Hours	Unit	2016 Bare Costs Material	Labor	Equipment	Total	Total Incl O&P
0010	**CHEMICAL TERMITE CONTROL**									
0020	Slab and walls, residential	1 Skwk	1200	.007	SF Flr.	.32	.26		.58	.78
0400	Insecticides for termite control, minimum		14.20	.563	Gal.	68.50	21.50		90	112
0500	Maximum	↓	11	.727	"	117	28		145	176

Division Notes

		CREW	DAILY OUTPUT	LABOR-HOURS	UNIT	BARE COSTS MAT.	LABOR	EQUIP.	TOTAL	TOTAL INCL O&P

Division 32 Exterior Improvements

Estimating Tips

32 01 00 Operations and Maintenance of Exterior Improvements

- Recycling of asphalt pavement is becoming very popular and is an alternative to removal and replacement. It can be a good value engineering proposal if removed pavement can be recycled, either at the project site or at another site that is reasonably close to the project site. Sections on repair of flexible and rigid pavement are included.

32 10 00 Bases, Ballasts, and Paving

- When estimating paving, keep in mind the project schedule. Also note that prices for asphalt and concrete are generally higher in the cold seasons. Lines for pavement markings, including tactile warning systems and fence lines, are included.

32 90 00 Planting

- The timing of planting and guarantee specifications often dictate the costs for establishing tree and shrub growth and a stand of grass or ground cover. Establish the work performance schedule to coincide with the local planting season. Maintenance and growth guarantees can add from 20%–100% to the total landscaping cost and can be contractually cumbersome. The cost to replace trees and shrubs can be as high as 5% of the total cost, depending on the planting zone, soil conditions, and time of year.

Reference Numbers

Reference numbers are shown at the beginning of some major classifications. These numbers refer to related items in the Reference Section. The reference information may be an estimating procedure, an alternate pricing method, or technical information.

Note: Not all subdivisions listed here necessarily appear. ■

32 01 Operation and Maintenance of Exterior Improvements

32 01 13 – Flexible Paving Surface Treatment

32 01 13.61 Slurry Seal (Latex Modified)		Crew	Daily Output	Labor-Hours	Unit	2016 Bare Costs Material	Labor	Equipment	Total	Total Incl O&P
0010	SLURRY SEAL (LATEX MODIFIED)									
3780	Rubberized asphalt (latex) seal	B-45	5000	.003	S.Y.	2.89	.10	.06	3.05	3.42

32 01 13.66 Fog Seal		Crew	Daily Output	Labor-Hours	Unit	Material	Labor	Equipment	Total	Total Incl O&P
0010	FOG SEAL									
0012	Sealcoating, 2 coat coal tar pitch emulsion over 10,000 S.Y.	B-45	5000	.003	S.Y.	.93	.10	.06	1.09	1.26
0030	1000 to 10,000 S.Y.	″	3000	.005		.93	.17	.11	1.21	1.42
0100	Under 1000 S.Y.	B-1	1050	.023		.93	.68		1.61	2.16
0300	Petroleum resistant, over 10,000 S.Y.	B-45	5000	.003		1.34	.10	.06	1.50	1.71
0320	1000 to 10,000 S.Y.	″	3000	.005		1.34	.17	.11	1.62	1.87
0400	Under 1000 S.Y.	B-1	1050	.023		1.34	.68		2.02	2.61

32 06 Schedules for Exterior Improvements

32 06 10 – Schedules for Bases, Ballasts, and Paving

32 06 10.10 Sidewalks, Driveways and Patios		Crew	Daily Output	Labor-Hours	Unit	Material	Labor	Equipment	Total	Total Incl O&P
0010	SIDEWALKS, DRIVEWAYS AND PATIOS No base									
0020	Asphaltic concrete, 2″ thick	B-37	720	.067	S.Y.	7.25	2.07	.23	9.55	11.70
0100	2-1/2″ thick	″	660	.073	″	9.20	2.25	.25	11.70	14.15
0300	Concrete, 3000 psi, CIP, 6 x 6 - W1.4 x W1.4 mesh,									
0310	broomed finish, no base, 4″ thick	B-24	600	.040	S.F.	1.79	1.37		3.16	4.23
0350	5″ thick		545	.044		2.38	1.51		3.89	5.10
0400	6″ thick		510	.047		2.78	1.61		4.39	5.75
0450	For bank run gravel base, 4″ thick, add	B-18	2500	.010		.47	.29	.02	.78	1.02
0520	8″ thick, add	″	1600	.015		.96	.45	.03	1.44	1.83
1000	Crushed stone, 1″ thick, white marble	2 Clab	1700	.009		.42	.27		.69	.92
1050	Bluestone	″	1700	.009		.12	.27		.39	.59
1700	Redwood, prefabricated, 4′ x 4′ sections	2 Carp	316	.051		4.82	1.91		6.73	8.50
1750	Redwood planks, 1″ thick, on sleepers	″	240	.067		4.82	2.52		7.34	9.50

32 06 10.20 Steps		Crew	Daily Output	Labor-Hours	Unit	Material	Labor	Equipment	Total	Total Incl O&P
0010	STEPS									
0011	Incl. excav., borrow & concrete base as required									
0100	Brick steps	B-24	35	.686	LF Riser	15.75	23.50		39.25	56.50
0200	Railroad ties	2 Clab	25	.640		3.59	18.70		22.29	35.50
0300	Bluestone treads, 12″ x 2″ or 12″ x 1-1/2″	B-24	30	.800		33.50	27.50		61	82.50
4025	Steel edge strips, incl. stakes, 1/4″ x 5″	B-1	390	.062	L.F.	5	1.84		6.84	8.60
4050	Edging, landscape timber or railroad ties, 6″ x 8″	2 Carp	170	.094	″	2.11	3.56		5.67	8.25

32 11 Base Courses

32 11 23 – Aggregate Base Courses

32 11 23.23 Base Course Drainage Layers		Crew	Daily Output	Labor-Hours	Unit	Material	Labor	Equipment	Total	Total Incl O&P
0010	BASE COURSE DRAINAGE LAYERS									
0011	For roadways and large areas									
0050	Crushed 3/4″ stone base, compacted, 3″ deep	B-36C	5200	.008	S.Y.	2.35	.28	.80	3.43	3.94
0100	6″ deep		5000	.008		4.71	.29	.83	5.83	6.60
0200	9″ deep		4600	.009		7.05	.32	.91	8.28	9.35
0300	12″ deep		4200	.010		9.45	.35	.99	10.79	12
0301	Crushed 1-1/2″ stone base, compacted to 4″ deep	B-36B	6000	.011		4.04	.37	.80	5.21	5.95
0302	6″ deep		5400	.012		6.05	.42	.89	7.36	8.30
0303	8″ deep		4500	.014		8.10	.50	1.07	9.67	10.90
0304	12″ deep		3800	.017		12.15	.59	1.26	14	15.70

32 11 Base Courses

32 11 23 – Aggregate Base Courses

32 11 23.23 Base Course Drainage Layers		Crew	Daily Output	Labor-Hours	Unit	2016 Bare Costs Material	Labor	Equipment	Total	Total Incl O&P
0350	Bank run gravel, spread and compacted									
0370	6″ deep	B-32	6000	.005	S.Y.	3.96	.20	.39	4.55	5.10
0390	9″ deep		4900	.007		5.95	.24	.48	6.67	7.50
0400	12″ deep		4200	.008		7.95	.28	.56	8.79	9.80
6900	For small and irregular areas, add						50%	50%		
7000	Prepare and roll sub-base, small areas to 2500 S.Y.	B-32A	1500	.016	S.Y.		.58	.94	1.52	1.98

32 11 26 – Asphaltic Base Courses

32 11 26.19 Bituminous-Stabilized Base Courses		Crew	Daily Output	Labor-Hours	Unit	Material	Labor	Equipment	Total	Total Incl O&P
0010	**BITUMINOUS-STABILIZED BASE COURSES**									
0020	And large paved areas									
0800	Prime and seal, cut back asphalt	B-45	6000	.003	Gal.	5.30	.08	.05	5.43	6.05
1620	Bituminous stabilized	B-36E	1400	.034	E.C.Y.	80	1.27	3.01	84.28	93.50
1630	Subbase course, bituminous stabilized	″	1400	.034	″	80	1.27	3.01	84.28	93.50
8900	For small and irregular areas, add						50%	50%		

32 12 Flexible Paving

32 12 16 – Asphalt Paving

32 12 16.14 Paving Asphaltic Concrete		Crew	Daily Output	Labor-Hours	Unit	Material	Labor	Equipment	Total	Total Incl O&P
0010	**PAVING ASPHALTIC CONCRETE**									
0020	6″ stone base, 2″ binder course, 1″ topping	B-25C	9000	.005	S.F.	1.83	.18	.27	2.28	2.60
0025	2″ binder course, 2″ topping		9000	.005		2.28	.18	.27	2.73	3.10
0030	3″ binder course, 2″ topping		9000	.005		2.70	.18	.27	3.15	3.56
0035	4″ binder course, 2″ topping		9000	.005		3.11	.18	.27	3.56	4.01
0040	1.5″ binder course, 1″ topping		9000	.005		1.62	.18	.27	2.07	2.37
0042	3″ binder course, 1″ topping		9000	.005		2.24	.18	.27	2.69	3.06
0045	3″ binder course, 3″ topping		9000	.005		3.15	.18	.27	3.60	4.06
0050	4″ binder course, 3″ topping		9000	.005		3.57	.18	.27	4.02	4.51
0055	4″ binder course, 4″ topping		9000	.005		4.02	.18	.27	4.47	5
0300	Binder course, 1-1/2″ thick		35000	.001		.62	.05	.07	.74	.85
0400	2″ thick		25000	.002		.81	.06	.10	.97	1.11
0500	3″ thick		15000	.003		1.25	.11	.16	1.52	1.73
0600	4″ thick		10800	.004		1.63	.15	.23	2.01	2.28
0800	Sand finish course, 3/4″ thick		41000	.001		.33	.04	.06	.43	.49
0900	1″ thick		34000	.001		.40	.05	.07	.52	.60
1000	Fill pot holes, hot mix, 2″ thick	B-16	4200	.008		.85	.24	.16	1.25	1.51
1100	4″ thick		3500	.009		1.25	.28	.20	1.73	2.06
1120	6″ thick		3100	.010		1.67	.32	.22	2.21	2.62
1140	Cold patch, 2″ thick	B-51	3000	.016		.97	.48	.08	1.53	1.97
1160	4″ thick		2700	.018		1.85	.54	.09	2.48	3.03
1180	6″ thick		1900	.025		2.87	.76	.13	3.76	4.57

32 13 Rigid Paving

32 13 13 – Concrete Paving

32 13 13.23 Concrete Paving Surface Treatment		Crew	Daily Output	Labor-Hours	Unit	2016 Bare Costs Material	Labor	Equipment	Total	Total Incl O&P
0010	**CONCRETE PAVING SURFACE TREATMENT**									
0015	Including joints, finishing and curing									
0020	Fixed form, 12′ pass, unreinforced, 6″ thick	B-26	3000	.029	S.Y.	23	.96	1.17	25.13	28.50
0100	8″ thick		2750	.032		32	1.05	1.28	34.33	38
0110	8″ thick, small area		1375	.064		32	2.10	2.55	36.65	41.50
0410	Conc. pavement, w/jt., fnsh.&curing, fix form, 24′ pass,unreinforced, 6″T		6000	.015		22	.48	.59	23.07	25.50
0430	8″ thick		5500	.016		30	.52	.64	31.16	34.50
0520	Welded wire fabric, sheets for rigid paving 2.33 lb./S.Y.	2 Rodm	389	.041		1.22	1.63		2.85	4.10
0530	Reinforcing steel for rigid paving 12 lb./S.Y.		666	.024		6.05	.95		7	8.25
0540	Reinforcing steel for rigid paving 18 lb./S.Y.		444	.036		9.10	1.43		10.53	12.40
0620	Slip form, 12′ pass, unreinforced, 6″ thick	B-26A	5600	.016		22.50	.51	.66	23.67	26
0624	8″ thick		5300	.017		30.50	.54	.69	31.73	35
0640	Slip form, 24′ pass, unreinforced, 6″ thick		11200	.008		22	.26	.33	22.59	25
0644	8″ thick		10600	.008		29.50	.27	.35	30.12	33
0700	Finishing, broom finish small areas	2 Cefi	120	.133			4.75		4.75	7.70

32 14 Unit Paving

32 14 13 – Precast Concrete Unit Paving

32 14 13.18 Precast Concrete Plantable Pavers		Crew	Daily Output	Labor-Hours	Unit	Material	Labor	Equipment	Total	Total Incl O&P
0010	**PRECAST CONCRETE PLANTABLE PAVERS** (50% grass)									
0300	3/4″ crushed stone base for plantable pavers, 6 inch depth	B-62	1000	.024	S.Y.	4.71	.77	.17	5.65	6.65
0400	8 inch depth		900	.027		6.30	.86	.19	7.35	8.60
0500	10 inch depth		800	.030		7.85	.96	.22	9.03	10.50
0600	12 inch depth		700	.034		9.45	1.10	.25	10.80	12.45
0700	Hydro seeding plantable pavers	B-81A	20	.800	M.S.F.	10.65	25	22	57.65	77
0800	Apply fertilizer and seed to plantable pavers	1 Clab	8	1	″	50.50	29		79.50	105

32 14 16 – Brick Unit Paving

32 14 16.10 Brick Paving		Crew	Daily Output	Labor-Hours	Unit	Material	Labor	Equipment	Total	Total Incl O&P
0010	**BRICK PAVING**									
0012	4″ x 8″ x 1-1/2″, without joints (4.5 brick/S.F.)	D-1	110	.145	S.F.	2.24	4.78		7.02	10.45
0100	Grouted, 3/8″ joint (3.9 brick/S.F.)		90	.178		2.02	5.85		7.87	11.95
0200	4″ x 8″ x 2-1/4″, without joints (4.5 bricks/S.F.)		110	.145		2.33	4.78		7.11	10.55
0300	Grouted, 3/8″ joint (3.9 brick/S.F.)		90	.178		2.02	5.85		7.87	11.95
0455	Pervious brick paving, 4″ x 8″ x 3-1/4″, without joints (4.5 bricks/S.F.)		110	.145		3.49	4.78		8.27	11.85
0500	Bedding, asphalt, 3/4″ thick	B-25	5130	.017		.69	.55	.55	1.79	2.27
0540	Course washed sand bed, 1″ thick	B-18	5000	.005		.35	.14	.01	.50	.64
0580	Mortar, 1″ thick	D-1	300	.053		.80	1.75		2.55	3.80
0620	2″ thick		200	.080		1.60	2.63		4.23	6.15
1500	Brick on 1″ thick sand bed laid flat, 4.5 per S.F.		100	.160		2.85	5.25		8.10	11.90
2000	Brick pavers, laid on edge, 7.2 per S.F.		70	.229		3.93	7.50		11.43	16.90

32 14 23 – Asphalt Unit Paving

32 14 23.10 Asphalt Blocks		Crew	Daily Output	Labor-Hours	Unit	Material	Labor	Equipment	Total	Total Incl O&P
0010	**ASPHALT BLOCKS**									
0020	Rectangular, 6″ x 12″ x 1-1/4″, w/bed & neopr. adhesive	D-1	135	.119	S.F.	5.60	3.90		9.50	12.65
0100	3″ thick		130	.123		7.80	4.05		11.85	15.35
0300	Hexagonal tile, 8″ wide, 1-1/4″ thick		135	.119		5.60	3.90		9.50	12.65
0400	2″ thick		130	.123		7.80	4.05		11.85	15.35
0500	Square, 8″ x 8″, 1-1/4″ thick		135	.119		5.60	3.90		9.50	12.65
0600	2″ thick		130	.123		7.80	4.05		11.85	15.35

32 14 Unit Paving

32 14 40 – Stone Paving

32 14 40.10 Stone Pavers		Crew	Daily Output	Labor-Hours	Unit	Material	2016 Bare Costs Labor	Equipment	Total	Total Incl O&P
0010	**STONE PAVERS**									
1100	Flagging, bluestone, irregular, 1″ thick,	D-1	81	.198	S.F.	9	6.50		15.50	21
1150	Snapped random rectangular, 1″ thick		92	.174		13.65	5.70		19.35	24.50
1200	1-1/2″ thick		85	.188		16.40	6.20		22.60	28.50
1250	2″ thick		83	.193		19.10	6.35		25.45	31.50
1300	Slate, natural cleft, irregular, 3/4″ thick		92	.174		9	5.70		14.70	19.45
1350	Random rectangular, gauged, 1/2″ thick		105	.152		19.50	5		24.50	30
1400	Random rectangular, butt joint, gauged, 1/4″ thick		150	.107		21	3.51		24.51	29
1450	For sand rubbed finish, add					9			9	9.90
1550	Granite blocks, 3-1/2″ x 3-1/2″ x 3-1/2″	D-1	92	.174		15	5.70		20.70	26

32 16 Curbs, Gutters, Sidewalks, and Driveways

32 16 13 – Curbs and Gutters

32 16 13.13 Cast-in-Place Concrete Curbs and Gutters

Line	Description	Crew	Daily Output	Labor-Hours	Unit	Material	Labor	Equipment	Total	Total Incl O&P
0010	**CAST-IN-PLACE CONCRETE CURBS AND GUTTERS**									
0290	Forms only, no concrete									
0300	Concrete, wood forms, 6″ x 18″, straight	C-2	500	.096	L.F.	3.13	3.21		6.34	8.85
0400	6″ x 18″, radius	″	200	.240		3.25	8		11.25	17
0404	Concrete, wood forms, 6″ x 18″, straight & concrete	C-2A	500	.096		6.15	3.49		9.64	12.55
0406	6″ x 18″, radius	″	200	.240		6.25	8.70		14.95	21.50

32 16 13.23 Precast Concrete Curbs and Gutters

Line	Description	Crew	Daily Output	Labor-Hours	Unit	Material	Labor	Equipment	Total	Total Incl O&P
0010	**PRECAST CONCRETE CURBS AND GUTTERS**									
0550	Precast, 6″ x 18″, straight	B-29	700	.069	L.F.	9	2.15	1.26	12.41	14.90
0600	6″ x 18″, radius	″	325	.148	″	10	4.63	2.71	17.34	21.50

32 16 13.33 Asphalt Curbs

Line	Description	Crew	Daily Output	Labor-Hours	Unit	Material	Labor	Equipment	Total	Total Incl O&P
0010	**ASPHALT CURBS**									
0012	Curbs, asphaltic, machine formed, 8″ wide, 6″ high, 40 L.F./ton	B-27	1000	.032	L.F.	1.73	.95	.30	2.98	3.82
0150	Asphaltic berm, 12″ W, 3″-6″ H, 35 L.F./ton, before pavement	″	700	.046		.04	1.36	.42	1.82	2.77
0200	12″ W, 1-1/2″ to 4″ H, 60 L.F. per ton, laid with pavement	B-2	1050	.038		.02	1.13		1.15	1.92

32 16 13.43 Stone Curbs

Line	Description	Crew	Daily Output	Labor-Hours	Unit	Material	Labor	Equipment	Total	Total Incl O&P
0010	**STONE CURBS**									
1000	Granite, split face, straight, 5″ x 16″	D-13	275	.175	L.F.	14	6.10	1.74	21.84	27.50
1100	6″ x 18″	″	250	.192		18.40	6.70	1.92	27.02	33.50
1300	Radius curbing, 6″ x 18″, over 10′ radius	B-29	260	.185		22.50	5.80	3.39	31.69	38.50
1400	Corners, 2′ radius	″	80	.600	Ea.	75.50	18.85	11.05	105.40	127
1600	Edging, 4-1/2″ x 12″, straight	D-13	300	.160	L.F.	7	5.60	1.60	14.20	18.75
1800	Curb inlets, (guttermouth) straight	B-29	41	1.171	Ea.	168	36.50	21.50	226	270
2000	Indian granite (Belgian block)									
2100	Jumbo, 10-1/2″ x 7-1/2″ x 4″, grey	D-1	150	.107	L.F.	7.90	3.51		11.41	14.55
2150	Pink		150	.107		8.60	3.51		12.11	15.30
2200	Regular, 9″ x 4-1/2″ x 4-1/2″, grey		160	.100		4.67	3.29		7.96	10.65
2250	Pink		160	.100		6.15	3.29		9.44	12.25
2300	Cubes, 4″ x 4″ x 4″, grey		175	.091		3.63	3.01		6.64	9
2350	Pink		175	.091		3.81	3.01		6.82	9.20
2400	6″ x 6″ x 6″, pink		155	.103		12.60	3.39		15.99	19.50
2500	Alternate pricing method for Indian granite									
2550	Jumbo, 10-1/2″ x 7-1/2″ x 4″ (30 lb.), grey				Ton	450			450	495
2600	Pink					500			500	550
2650	Regular, 9″ x 4-1/2″ x 4-1/2″ (20 lb.), grey					330			330	365
2700	Pink					430			430	475

32 16 Curbs, Gutters, Sidewalks, and Driveways

32 16 13 – Curbs and Gutters

32 16 13.43 Stone Curbs		Crew	Daily Output	Labor-Hours	Unit	2016 Bare Costs Material	Labor	Equipment	Total	Total Incl O&P
2750	Cubes, 4" x 4" x 4" (5 lb.), grey				Ton	440			440	485
2800	Pink					490			490	540
2850	6" x 6" x 6" (25 lb.), pink					490			490	540
2900	For pallets, add					22			22	24

32 17 Paving Specialties

32 17 23 – Pavement Markings

32 17 23.13 Painted Pavement Markings

		Crew	Daily Output	Labor-Hours	Unit	Material	Labor	Equipment	Total	Total Incl O&P
0010	**PAINTED PAVEMENT MARKINGS**									
0203	Parking striping	B-78	11000	.004	L.F.	.47	.11	.06	.64	.76
0620	Arrows or gore lines	"	2300	.017	S.F.	.22	.51	.26	.99	1.39

32 17 23.14 Pavement Parking Markings

		Crew	Daily Output	Labor-Hours	Unit	Material	Labor	Equipment	Total	Total Incl O&P
0010	**PAVEMENT PARKING MARKINGS**									
0790	Layout of pavement marking	A-2	25000	.001	L.F.		.03	.01	.04	.06
0800	Lines on pvmt., parking stall, paint, white, 4" wide	B-78B	400	.045	Stall	4.62	1.36	.95	6.93	8.40
0825	Parking stall, small quantities	2 Pord	80	.200		9.25	6.40		15.65	20.50
0830	Lines on pvmt., parking stall, thermoplastic, white, 4" wide	B-79	300	.133		14	4.05	4.97	23.02	27.50
1000	Street letters and numbers	B-78B	1600	.011	S.F.	.69	.34	.24	1.27	1.59
1100	Pavement marking letter, 6"	2 Pord	400	.040	Ea.	11	1.28		12.28	14.20
1110	12" letter		272	.059		13.90	1.88		15.78	18.35
1120	24" letter		160	.100		30.50	3.19		33.69	39
1130	36" letter		84	.190		40	6.10		46.10	54
1140	42" letter		84	.190		53	6.10		59.10	68.50
1150	72" letter		40	.400		43.50	12.75		56.25	69
1200	Handicap symbol		40	.400		32	12.75		44.75	56.50
1210	Handicap Parking sign 12" x 18" and post	A-2	12	2		129	61	20.50	210.50	267

32 17 26 – Tactile Warning Surfacing

32 17 26.10 Tactile Warning Surfacing

		Crew	Daily Output	Labor-Hours	Unit	Material	Labor	Equipment	Total	Total Incl O&P
0010	**TACTILE WARNING SURFACING**									
0100	Tactile warning tiles S.F.	2 Clab	400	.040	S.F.	17.25	1.17		18.42	21

32 31 Fences and Gates

32 31 13 – Chain Link Fences and Gates

32 31 13.15 Chain Link Fence

		Crew	Daily Output	Labor-Hours	Unit	Material	Labor	Equipment	Total	Total Incl O&P
0010	**CHAIN LINK FENCE**									
0011	11 ga. wire									
0350	Aluminized steel, 9 ga. wire, 3' high	B-1	185	.130	L.F.	7.60	3.88		11.48	14.90
0400	6' high		115	.209	"	10.40	6.25		16.65	22
0450	Add for gate 3' wide, 1-3/8" frame 3' high		12	2	Ea.	127	59.50		186.50	240
0490	6' high		10	2.400		173	71.50		244.50	310
0500	Add for gate 4' wide, 1-3/8" frame 3' high		10	2.400		132	71.50		203.50	265
0540	6' high		8	3		184	89.50		273.50	350
0860	Tennis courts, 11 ga. wire, 2-1/2" post 10' O.C., 1-5/8" top rail									
0900	2-1/2" corner post, 10' high	B-1	95	.253	L.F.	7.70	7.55		15.25	21
0920	12' high		80	.300	"	7.95	8.95		16.90	24
1000	Add for gate 3' wide, 1-5/8" frame 10' high		10	2.400	Ea.	220	71.50		291.50	360
1040	Aluminized, 11 ga. wire 10' high		95	.253	L.F.	9.30	7.55		16.85	23
1100	12' high		80	.300	"	10.20	8.95		19.15	26

32 31 Fences and Gates

32 31 13 – Chain Link Fences and Gates

	32 31 13.15 Chain Link Fence	Crew	Daily Output	Labor-Hours	Unit	2016 Bare Costs Material	Labor	Equipment	Total	Total Incl O&P
1140	Add for gate 3′ wide, 1-5/8″ frame, 10′ high	B-1	10	2.400	Ea.	142	71.50		213.50	276
1250	Vinyl covered 11 ga. wire, 10′ high		95	.253	L.F.	8.05	7.55		15.60	21.50
1300	12′ high		80	.300	"	9.60	8.95		18.55	25.50
1400	Add for gate 3′ wide, 1-3/8″ frame, 10′ high		10	2.400	Ea.	283	71.50		354.50	430

32 31 23 – Plastic Fences and Gates

	32 31 23.10 Fence, Vinyl	Crew	Daily Output	Labor-Hours	Unit	Material	Labor	Equipment	Total	Total Incl O&P
0010	**FENCE, VINYL**									
0020	Picket, 4″ x 4″ posts @ 6′ - 0″ OC, 3′ high	B-1	140	.171	L.F.	22.50	5.10		27.60	33.50
0030	4′ high		130	.185		25	5.50		30.50	37
0040	5′ high		120	.200		27.50	5.95		33.45	40.50
0100	Board (semi-privacy), 5″ x 5″ posts @ 7′ - 6″ OC, 5′ high		130	.185		25.50	5.50		31	37.50
0120	6′ high		125	.192		28.50	5.75		34.25	41
0200	Basketweave, 5″ x 5″ posts @ 7′ - 6″ OC, 5′ high		160	.150		25.50	4.48		29.98	35.50
0220	6′ high		150	.160		28.50	4.78		33.28	39.50
0300	Privacy, 5″ x 5″ posts @ 7′ - 6″ OC, 5′ high		130	.185		25	5.50		30.50	37
0320	6′ high		150	.160		28.50	4.78		33.28	39.50
0350	Gate, 5′ high		9	2.667	Ea.	315	79.50		394.50	480
0360	6′ high		9	2.667		360	79.50		439.50	530
0400	For posts set in concrete, add		25	.960		8.90	28.50		37.40	58
0500	Post and rail fence, 2 rail		150	.160	L.F.	5.95	4.78		10.73	14.55
0510	3 rail		150	.160		7.70	4.78		12.48	16.45
0515	4 rail		150	.160		9.95	4.78		14.73	18.95

32 31 26 – Wire Fences and Gates

	32 31 26.10 Fences, Misc. Metal	Crew	Daily Output	Labor-Hours	Unit	Material	Labor	Equipment	Total	Total Incl O&P
0010	**FENCES, MISC. METAL**									
0012	Chicken wire, posts @ 4′, 1″ mesh, 4′ high	B-80C	410	.059	L.F.	3.37	1.79	.63	5.79	7.35
0100	2″ mesh, 6′ high		350	.069		3.78	2.09	.73	6.60	8.45
0200	Galv. steel, 12 ga., 2″ x 4″ mesh, posts 5′ O.C., 3′ high		300	.080		2.75	2.44	.86	6.05	8.05
0300	5′ high		300	.080		3.35	2.44	.86	6.65	8.70
0400	14 ga., 1″ x 2″ mesh, 3′ high		300	.080		3.39	2.44	.86	6.69	8.75
0500	5′ high		300	.080		4.52	2.44	.86	7.82	9.95
1000	Kennel fencing, 1-1/2″ mesh, 6′ long, 3′-6″ wide, 6′-2″ high	2 Clab	4	4	Ea.	505	117		622	750
1050	12′ long		4	4		710	117		827	975
1200	Top covers, 1-1/2″ mesh, 6′ long		15	1.067		135	31		166	200
1250	12′ long		12	1.333		189	39		228	273

32 31 29 – Wood Fences and Gates

	32 31 29.10 Fence, Wood	Crew	Daily Output	Labor-Hours	Unit	Material	Labor	Equipment	Total	Total Incl O&P
0010	**FENCE, WOOD**									
0011	Basket weave, 3/8″ x 4″ boards, 2″ x 4″									
0020	stringers on spreaders, 4″ x 4″ posts									
0050	No. 1 cedar, 6′ high	B-80C	160	.150	L.F.	26	4.58	1.60	32.18	38
0860	Open rail fence, split rails, 2 rail 3′ high, no. 1 cedar		160	.150		12.80	4.58	1.60	18.98	23.50
0870	No. 2 cedar		160	.150		11.75	4.58	1.60	17.93	22.50
0880	3 rail, 4′ high, no. 1 cedar		150	.160		13.20	4.88	1.71	19.79	24.50
0890	No. 2 cedar		150	.160		9.35	4.88	1.71	15.94	20.50
1240	Stockade fence, no. 1 cedar, 3-1/4″ rails, 6′ high		160	.150		13.15	4.58	1.60	19.33	24
1260	8′ high		155	.155		18.10	4.72	1.66	24.48	29.50
1320	Gate, 3′-6″ wide		8	3	Ea.	89.50	91.50	32	213	286

32 31 Fences and Gates

32 31 29 – Wood Fences and Gates

32 31 29.20 Fence, Wood Rail		Crew	Daily Output	Labor-Hours	Unit	2016 Bare Costs Material	Labor	Equipment	Total	Total Incl O&P
0010	**FENCE, WOOD RAIL**									
0012	Picket, No. 2 cedar, Gothic, 2 rail, 3′ high	B-1	160	.150	L.F.	8	4.48		12.48	16.30
0050	Gate, 3′-6″ wide	B-80C	9	2.667	Ea.	77	81.50	28.50	187	252
0400	3 rail, 4′ high		150	.160	L.F.	9	4.88	1.71	15.59	19.95
0500	Gate, 3′-6″ wide		9	2.667	Ea.	94	81.50	28.50	204	270
1200	Stockade, No. 2 cedar, treated wood rails, 6′ high		160	.150	L.F.	9	4.58	1.60	15.18	19.25
1250	Gate, 3′ wide		9	2.667	Ea.	96	81.50	28.50	206	273
1300	No. 1 cedar, 3-1/4″ cedar rails, 6′ high		160	.150	L.F.	20.50	4.58	1.60	26.68	32
1500	Gate, 3′ wide		9	2.667	Ea.	220	81.50	28.50	330	410
2700	Prefabricated redwood or cedar, 4′ high		160	.150	L.F.	15.75	4.58	1.60	21.93	26.50
2800	6′ high		150	.160		23.50	4.88	1.71	30.09	35.50
3300	Board, shadow box, 1″ x 6″, treated pine, 6′ high		160	.150		12.50	4.58	1.60	18.68	23
3400	No. 1 cedar, 6′ high		150	.160		25	4.88	1.71	31.59	37
3900	Basket weave, No. 1 cedar, 6′ high		160	.150		34	4.58	1.60	40.18	46.50
4200	Gate, 3′-6″ wide	B-1	9	2.667	Ea.	176	79.50		255.50	325

32 32 Retaining Walls

32 32 13 – Cast-in-Place Concrete Retaining Walls

32 32 13.10 Retaining Walls, Cast Concrete		Crew	Daily Output	Labor-Hours	Unit	Material	Labor	Equipment	Total	Total Incl O&P
0010	**RETAINING WALLS, CAST CONCRETE**									
1800	Concrete gravity wall with vertical face including excavation & backfill									
1850	No reinforcing									
1900	6′ high, level embankment	C-17C	36	2.306	L.F.	78.50	89.50	16.95	184.95	255
2000	33° slope embankment	″	32	2.594	″	91	101	19.10	211.10	290
2800	Reinforced concrete cantilever, incl. excavation, backfill & reinf.									
2900	6′ high, 33° slope embankment	C-17C	35	2.371	L.F.	71.50	92	17.45	180.95	252

32 32 26 – Metal Crib Retaining Walls

32 32 26.10 Metal Bin Retaining Walls		Crew	Daily Output	Labor-Hours	Unit	Material	Labor	Equipment	Total	Total Incl O&P
0010	**METAL BIN RETAINING WALLS**									
0011	Aluminized steel bin, excavation									
0020	and backfill not included, 10′ wide									
0100	4′ high, 5.5′ deep	B-13	650	.074	S.F.	26.50	2.32	1.15	29.97	34
0200	8′ high, 5.5′ deep		615	.078		30.50	2.45	1.22	34.17	39
0300	10′ high, 7.7′ deep		580	.083		34	2.60	1.29	37.89	43
0400	12′ high, 7.7′ deep		530	.091		36.50	2.84	1.41	40.75	46.50
0500	16′ high, 7.7′ deep		515	.093		38.50	2.92	1.45	42.87	49

32 32 29 – Timber Retaining Walls

32 32 29.10 Landscape Timber Retaining Walls		Crew	Daily Output	Labor-Hours	Unit	Material	Labor	Equipment	Total	Total Incl O&P
0010	**LANDSCAPE TIMBER RETAINING WALLS**									
0100	Treated timbers, 6″ x 6″	1 Clab	265	.030	L.F.	3	.88		3.88	4.78
0110	6″ x 8″	″	200	.040	″	3.75	1.17		4.92	6.10
0120	Drilling holes in timbers for fastening, 1/2″	1 Carp	450	.018	Inch		.67		.67	1.13
0130	5/8″	″	450	.018	″		.67		.67	1.13
0140	Reinforcing rods for fastening, 1/2″	1 Clab	312	.026	L.F.	.35	.75		1.10	1.64
0150	5/8″	″	312	.026	″	.55	.75		1.30	1.86
0160	Reinforcing fabric	2 Clab	2500	.006	S.Y.	1.95	.19		2.14	2.46
0170	Gravel backfill		28	.571	C.Y.	18.50	16.70		35.20	48.50
0180	Perforated pipe, 4″ diameter with silt sock		1200	.013	L.F.	1.25	.39		1.64	2.03
0190	Galvanized 60d common nails	1 Clab	625	.013	Ea.	.17	.37		.54	.81

32 32 Retaining Walls

32 32 29 – Timber Retaining Walls

	32 32 29.10 Landscape Timber Retaining Walls	Crew	Daily Output	Labor-Hours	Unit	Material	2016 Bare Costs Labor	Equipment	Total	Total Incl O&P
0200	20d common nails	1 Clab	3800	.002	Ea.	.04	.06		.10	.14

32 32 53 – Stone Retaining Walls

32 32 53.10 Retaining Walls, Stone

		Crew	Daily Output	Labor-Hours	Unit	Material	Labor	Equipment	Total	Total Incl O&P
0010	**RETAINING WALLS, STONE**									
0015	Including excavation, concrete footing and									
0020	stone 3′ below grade. Price is exposed face area.									
0200	Decorative random stone, to 6′ high, 1′-6″ thick, dry set	D-1	35	.457	S.F.	58.50	15.05		73.55	89.50
0300	Mortar set		40	.400		60.50	13.15		73.65	88.50
0500	Cut stone, to 6′ high, 1′-6″ thick, dry set		35	.457		60.50	15.05		75.55	92
0600	Mortar set		40	.400		61.50	13.15		74.65	89.50
0800	Random stone, 6′ to 10′ high, 2′ thick, dry set		45	.356		66.50	11.70		78.20	92.50
0900	Mortar set		50	.320		69	10.50		79.50	93
1100	Cut stone, 6′ to 10′ high, 2′ thick, dry set		45	.356		67	11.70		78.70	93
1200	Mortar set		50	.320		69	10.50		79.50	93.50

32 33 Site Furnishings

32 33 33 – Site Manufactured Planters

32 33 33.10 Planters

		Crew	Daily Output	Labor-Hours	Unit	Material	Labor	Equipment	Total	Total Incl O&P
0010	**PLANTERS**									
0012	Concroto, sandblastod, procast, 48″ diamotor, 24″ high	2 Clab	15	1.067	Ea.	585	31		616	695
0300	Fiberglass, circular, 36″ diameter, 24″ high		15	1.067		655	31		686	775
1200	Wood, square, 48″ side, 24″ high		15	1.067		1,400	31		1,431	1,600
1300	Circular, 48″ diameter, 30″ high		10	1.600		1,025	46.50		1,071.50	1,200
1600	Planter/bench, 72″		5	3.200		3,450	93.50		3,543.50	3,925

32 33 43 – Site Seating and Tables

32 33 43.13 Site Seating

		Crew	Daily Output	Labor-Hours	Unit	Material	Labor	Equipment	Total	Total Incl O&P
0010	**SITE SEATING**									
0012	Seating, benches, park, precast conc., w/backs, wood rails, 4′ long	2 Clab	5	3.200	Ea.	605	93.50		698.50	820
0100	8′ long		4	4		1,025	117		1,142	1,325
0500	Steel barstock pedestals w/backs, 2″ x 3″ wood rails, 4′ long		10	1.600		1,150	46.50		1,196.50	1,350
0510	8′ long		7	2.286		1,500	66.50		1,566.50	1,750
0800	Cast iron pedestals, back & arms, wood slats, 4′ long		8	2		395	58.50		453.50	535
0820	8′ long		5	3.200		1,200	93.50		1,293.50	1,475
1700	Steel frame, fir seat, 10′ long		10	1.600		375	46.50		421.50	495

32 84 Planting Irrigation

32 84 23 – Underground Sprinklers

32 84 23.10 Sprinkler Irrigation System

		Crew	Daily Output	Labor-Hours	Unit	Material	Labor	Equipment	Total	Total Incl O&P
0010	**SPRINKLER IRRIGATION SYSTEM**									
0011	For lawns									
0800	Residential system, custom, 1″ supply	B-20	2000	.012	S.F.	.26	.36		.62	.89
0900	1-1/2″ supply	″	1800	.013	″	.49	.40		.89	1.21

32 91 Planting Preparation

32 91 13 – Soil Preparation

32 91 13.16 Mulching		Crew	Daily Output	Labor-Hours	Unit	2016 Bare Costs Material	Labor	Equipment	Total	Total Incl O&P
0010	**MULCHING**									
0100	Aged barks, 3″ deep, hand spread	1 Clab	100	.080	S.Y.	3.60	2.34		5.94	7.85
0150	Skid steer loader	B-63	13.50	2.963	M.S.F.	400	86.50	12.85	499.35	600
0200	Hay, 1″ deep, hand spread	1 Clab	475	.017	S.Y.	.50	.49		.99	1.37
0250	Power mulcher, small	B-64	180	.089	M.S.F.	55.50	2.77	2.22	60.49	68
0350	Large	B-65	530	.030	″	55.50	.94	1.07	57.51	63.50
0400	Humus peat, 1″ deep, hand spread	1 Clab	700	.011	S.Y.	2.60	.33		2.93	3.42
0450	Push spreader	″	2500	.003	″	2.60	.09		2.69	3.02
0550	Tractor spreader	B-66	700	.011	M.S.F.	289	.43	.38	289.81	320
0600	Oat straw, 1″ deep, hand spread	1 Clab	475	.017	S.Y.	.50	.49		.99	1.37
0650	Power mulcher, small	B-64	180	.089	M.S.F.	55.50	2.77	2.22	60.49	68
0700	Large	B-65	530	.030	″	55.50	.94	1.07	57.51	63.50
0750	Add for asphaltic emulsion	B-45	1770	.009	Gal.	5.80	.29	.18	6.27	7.10
0800	Peat moss, 1″ deep, hand spread	1 Clab	900	.009	S.Y.	2.60	.26		2.86	3.29
0850	Push spreader	″	2500	.003	″	2.60	.09		2.69	3.02
0950	Tractor spreader	B-66	700	.011	M.S.F.	289	.43	.38	289.81	320
1000	Polyethylene film, 6 mil	2 Clab	2000	.008	S.Y.	.41	.23		.64	.84
1100	Redwood nuggets, 3″ deep, hand spread	1 Clab	150	.053	″	2.77	1.56		4.33	5.65
1150	Skid steer loader	B-63	13.50	2.963	M.S.F.	310	86.50	12.85	409.35	500
1200	Stone mulch, hand spread, ceramic chips, economy	1 Clab	125	.064	S.Y.	6.85	1.87		8.72	10.70
1250	Deluxe	″	95	.084	″	10.60	2.46		13.06	15.75
1300	Granite chips	B-1	10	2.400	C.Y.	60.50	71.50		132	187
1400	Marble chips		10	2.400		189	71.50		260.50	330
1600	Pea gravel		28	.857		109	25.50		134.50	163
1700	Quartz	↓	10	2.400	↓	190	71.50		261.50	330
1800	Tar paper, 15 lb. felt	1 Clab	800	.010	S.Y.	.48	.29		.77	1.01
1900	Wood chips, 2″ deep, hand spread	″	220	.036	″	1.80	1.06		2.86	3.76
1950	Skid steer loader	B-63	20.30	1.970	M.S.F.	200	57.50	8.55	266.05	325

32 91 13.26 Planting Beds

		Crew	Daily Output	Labor-Hours	Unit	Material	Labor	Equipment	Total	Total Incl O&P
0010	**PLANTING BEDS**									
0100	Backfill planting pit, by hand, on site topsoil	2 Clab	18	.889	C.Y.		26		26	43.50
0200	Prepared planting mix, by hand	″	24	.667			19.45		19.45	32.50
0300	Skid steer loader, on site topsoil	B-62	340	.071			2.26	.51	2.77	4.33
0400	Prepared planting mix	″	410	.059			1.88	.42	2.30	3.59
1000	Excavate planting pit, by hand, sandy soil	2 Clab	16	1			29		29	49
1100	Heavy soil or clay	″	8	2			58.50		58.50	98
1200	1/2 C.Y. backhoe, sandy soil	B-11C	150	.107			3.66	2.44	6.10	8.80
1300	Heavy soil or clay	″	115	.139			4.77	3.18	7.95	11.45
2000	Mix planting soil, incl. loam, manure, peat, by hand	2 Clab	60	.267		42.50	7.80		50.30	60
2100	Skid steer loader	B-62	150	.160	↓	42.50	5.15	1.16	48.81	57
3000	Pile sod, skid steer loader	″	2800	.009	S.Y.		.27	.06	.33	.53
4000	Remove sod, F.E. loader	B-10S	2000	.004			.16	.19	.35	.47
4100	Sod cutter	B-12K	3200	.005	↓		.17	.31	.48	.63

32 91 19 – Landscape Grading

32 91 19.13 Topsoil Placement and Grading

		Crew	Daily Output	Labor-Hours	Unit	Material	Labor	Equipment	Total	Total Incl O&P
0010	**TOPSOIL PLACEMENT AND GRADING**									
0700	Furnish and place, truck dumped, screened, 4″ deep	B-10S	1300	.006	S.Y.	3.38	.24	.29	3.91	4.43
0800	6″ deep	″	820	.010	″	4.32	.38	.46	5.16	5.90
0900	Fine grading and seeding, incl. lime, fertilizer & seed,									
1000	With equipment	B-14	1000	.048	S.Y.	.45	1.49	.37	2.31	3.38

32 92 Turf and Grasses

32 92 19 – Seeding

32 92 19.13 Mechanical Seeding

32 92 19.13 Mechanical Seeding		Crew	Daily Output	Labor-Hours	Unit	Material	2016 Bare Costs Labor	Equipment	Total	Total Incl O&P
0010	**MECHANICAL SEEDING** R329219-50									
0020	Mechanical seeding, 215 lb./acre	B-66	1.50	5.333	Acre	560	202	176	938	1,150
0100	44 lb./M.S.Y.	"	2500	.003	S.Y.	.20	.12	.11	.43	.54
0101	44 lb./M.S.Y.	1 Clab	13950	.001	S.F.	.02	.02		.04	.05
0300	Fine grading and seeding incl. lime, fertilizer & seed,									
0310	with equipment	B-14	1000	.048	S.Y.	.45	1.49	.37	2.31	3.38
0400	Fertilizer hand push spreader, 35 lb. per M.S.F.	1 Clab	200	.040	M.S.F.	9.75	1.17		10.92	12.70
0600	Limestone hand push spreader, 50 lb. per M.S.F.		180	.044		5.40	1.30		6.70	8.05
0800	Grass seed hand push spreader, 4.5 lb. per M.S.F.	↓	180	.044	↓	22	1.30		23.30	26.50
1000	Hydro or air seeding for large areas, incl. seed and fertilizer	B-81	8900	.002	S.Y.	.50	.06	.08	.64	.73
1100	With wood fiber mulch added	"	8900	.002	"	2	.06	.08	2.14	2.38
1300	Seed only, over 100 lb., field seed, minimum				Lb.	1.80			1.80	1.98
1400	Maximum					1.70			1.70	1.87
1500	Lawn seed, minimum					1.40			1.40	1.54
1600	Maximum				↓	2.50			2.50	2.75
1800	Aerial operations, seeding only, field seed	B-58	50	.480	Acre	585	15.40	63	663.40	740
1900	Lawn seed		50	.480		455	15.40	63	533.40	595
2100	Seed and liquid fertilizer, field seed		50	.480		725	15.40	63	803.40	895
2200	Lawn seed	↓	50	.480	↓	595	15.40	63	673.40	750

32 92 23 – Sodding

32 92 23.10 Sodding Systems

		Crew	Daily Output	Labor-Hours	Unit	Material	Labor	Equipment	Total	Total Incl O&P
0010	**SODDING SYSTEMS**									
0020	Sodding, 1" deep, bluegrass sod, on level ground, over 8 M.S.F.	B-63	22	1.818	M.S.F.	244	53	7.85	304.85	365
0200	4 M.S.F.		17	2.353		260	68.50	10.20	338.70	410
0300	1000 S.F.		13.50	2.963		295	86.50	12.85	394.35	485
0500	Sloped ground, over 8 M.S.F.		6	6.667		244	195	29	468	625
0600	4 M.S.F.		5	8		260	234	34.50	528.50	715
0700	1000 S.F.		4	10		295	292	43.50	630.50	865
1000	Bent grass sod, on level ground, over 6 M.S.F.		20	2		253	58.50	8.65	320.15	385
1100	3 M.S.F.		18	2.222		267	65	9.60	341.60	415
1200	Sodding 1000 S.F. or less		14	2.857		290	83.50	12.35	385.85	475
1500	Sloped ground, over 6 M.S.F.		15	2.667		253	78	11.55	342.55	420
1600	3 M.S.F.		13.50	2.963		267	86.50	12.85	366.35	450
1700	1000 S.F.	↓	12	3.333	↓	290	97.50	14.45	401.95	500

32 93 Plants

32 93 13 – Ground Covers

32 93 13.10 Ground Cover Plants

		Crew	Daily Output	Labor-Hours	Unit	Material	Labor	Equipment	Total	Total Incl O&P
0010	**GROUND COVER PLANTS**									
0012	Plants, pachysandra, in prepared beds	B-1	15	1.600	C	83.50	48		131.50	172
0200	Vinca minor, 1 yr., bare root, in prepared beds		12	2	"	71.50	59.50		131	179
0600	Stone chips, in 50 lb. bags, Georgia marble		520	.046	Bag	3.63	1.38		5.01	6.30
0700	Onyx gemstone		260	.092		17	2.76		19.76	23.50
0800	Quartz		260	.092	↓	17	2.76		19.76	23.50
0900	Pea gravel, truckload lots	↓	28	.857	Ton	27.50	25.50		53	73

32 93 Plants

32 93 33 – Shrubs

32 93 33.10 Shrubs and Trees		Crew	Daily Output	Labor-Hours	Unit	2016 Bare Costs Material	Labor	Equipment	Total	Total Incl O&P
0010	SHRUBS AND TREES									
0011	Evergreen, in prepared beds, B & B									
0100	Arborvitae pyramidal, 4′-5′	B-17	30	1.067	Ea.	101	35	26	162	199
0150	Globe, 12″-15″	B-1	96	.250		24.50	7.45		31.95	39.50
0300	Cedar, blue, 8′-10′	B-17	18	1.778		235	58	43.50	336.50	405
0500	Hemlock, Canadian, 2-1/2′-3′	B-1	36	.667		32	19.90		51.90	68.50
0550	Holly, Savannah, 8′ - 10′ H		9.68	2.479		262	74		336	410
0600	Juniper, andorra, 18″-24″		80	.300		37	8.95		45.95	56
0620	Wiltoni, 15″-18″		80	.300		26	8.95		34.95	44
0640	Skyrocket, 4-1/2′-5′	B-17	55	.582		108	18.95	14.25	141.20	166
0660	Blue pfitzer, 2′-2-1/2′	B-1	44	.545		39	16.30		55.30	70
0680	Ketleerie, 2-1/2′-3′		50	.480		53	14.35		67.35	82
0700	Pine, black, 2-1/2′-3′		50	.480		60.50	14.35		74.85	90.50
0720	Mugo, 18″-24″		60	.400		60.50	11.95		72.45	86.50
0740	White, 4′-5′	B-17	75	.427		52	13.90	10.45	76.35	91.50
0800	Spruce, blue, 18″-24″	B-1	60	.400		68	11.95		79.95	95
0840	Norway, 4′-5′	B-17	75	.427		84.50	13.90	10.45	108.85	128
0900	Yew, denisforma, 12″-15″	B-1	60	.400		35.50	11.95		47.45	59.50
1000	Capitata, 18″-24″		30	.800		33.50	24		57.50	76.50
1100	Hicksi, 2′-2-1/2′		30	.800		79.50	24		103.50	127

32 93 33.20 Shrubs

32 93 33.20 Shrubs		Crew	Daily Output	Labor-Hours	Unit	Material	Labor	Equipment	Total	Total Incl O&P
0010	SHRUBS									
0011	Broadleaf Evergreen, planted in prepared beds									
0100	Andromeda, 15″-18″, container	B-1	96	.250	Ea.	33	7.45		40.45	49
0200	Azalea, 15″ - 18″, container		96	.250		30	7.45		37.45	45.50
0300	Barberry, 9″-12″, container		130	.185		17.95	5.50		23.45	29
0400	Boxwood, 15″-18″, B&B		96	.250		43.50	7.45		50.95	60.50
0500	Euonymus, emerald gaiety, 12″ to 15″, container		115	.209		24.50	6.25		30.75	37
0600	Holly, 15″-18″, B & B		96	.250		37	7.45		44.45	53
0900	Mount laurel, 18″ - 24″, B & B		80	.300		72	8.95		80.95	94
1000	Paxistema, 9″ - 12″ high		130	.185		21.50	5.50		27	33.50
1100	Rhododendron, 18″-24″, container		48	.500		38.50	14.95		53.45	67
1200	Rosemary, 1 gal. container		600	.040		18.05	1.19		19.24	22
2000	Deciduous, planted in prepared beds, amelanchier, 2′-3′, B & B		57	.421		121	12.60		133.60	155
2100	Azalea, 15″-18″, B & B		96	.250		30	7.45		37.45	45
2300	Bayberry, 2′-3′, B & B		57	.421		24.50	12.60		37.10	48
2600	Cotoneaster, 15″-18″, B & B		80	.300		27	8.95		35.95	44.50
2800	Dogwood, 3′-4′, B & B	B-17	40	.800		32.50	26	19.55	78.05	101
2900	Euonymus, alatus compacta, 15″ to 18″, container	B-1	80	.300		26.50	8.95		35.45	44.50
3200	Forsythia, 2′-3′, container	″	60	.400		18.85	11.95		30.80	40.50
3300	Hibiscus, 3′-4′, B & B	B-17	75	.427		49	13.90	10.45	73.35	88.50
3400	Honeysuckle, 3′-4′, B & B	B-1	60	.400		27	11.95		38.95	49.50
3500	Hydrangea, 2′-3′, B & B	″	57	.421		30.50	12.60		43.10	54.50
3600	Lilac, 3′-4′, B & B	B-17	40	.800		28.50	26	19.55	74.05	96.50
3900	Privet, bare root, 18″-24″	B-1	80	.300		14.80	8.95		23.75	31.50
4100	Quince, 2′-3′, B & B	″	57	.421		28.50	12.60		41.10	52.50
4200	Russian olive, 3′-4′, B & B	B-17	75	.427		29.50	13.90	10.45	53.85	67
4400	Spirea, 3′-4′, B & B	B-1	70	.343		21	10.25		31.25	40
4500	Viburnum, 3′-4′, B & B	B-17	40	.800		25.50	26	19.55	71.05	93

32 93 Plants

32 93 43 – Trees

32 93 43.20 Trees			Crew	Daily Output	Labor-Hours	Unit	2016 Bare Costs Material	Labor	Equipment	Total	Total Incl O&P
0010	TREES										
0011	Deciduous, in prep. beds, balled & burlapped (B&B)										
0100	Ash, 2" caliper	G	B-17	8	4	Ea.	180	130	98	408	525
0200	Beech, 5'-6'	G		50	.640		181	21	15.65	217.65	252
0300	Birch, 6'-8', 3 stems	G		20	1.600		163	52	39	254	310
0500	Crabapple, 6'-8'	G		20	1.600		134	52	39	225	278
0600	Dogwood, 4'-5'	G		40	.800		127	26	19.55	172.55	205
0700	Eastern redbud, 4'-5'	G		40	.800		145	26	19.55	190.55	224
0800	Elm, 8'-10'	G		20	1.600		250	52	39	341	405
0900	Ginkgo, 6'-7'	G		24	1.333		144	43.50	32.50	220	266
1000	Hawthorn, 8'-10', 1" caliper	G		20	1.600		157	52	39	248	305
1100	Honeylocust, 10'-12', 1-1/2" caliper	G		10	3.200		200	104	78.50	382.50	480
1300	Larch, 8'	G		32	1		123	32.50	24.50	180	217
1400	Linden, 8'-10', 1" caliper	G		20	1.600		139	52	39	230	283
1500	Magnolia, 4'-5'	G		20	1.600		98	52	39	189	238
1600	Maple, red, 8'-10', 1-1/2" caliper	G		10	3.200		196	104	78.50	378.50	475
1700	Mountain ash, 8'-10', 1" caliper	G		16	2		171	65	49	285	350
1800	Oak, 2-1/2"-3" caliper	G		6	5.333		315	174	130	619	780
2100	Planetree, 9'-11', 1-1/4" caliper	G		10	3.200		229	104	78.50	411.50	510
2200	Plum, 6'-8', 1" caliper	G		20	1.600		79.50	52	39	170.50	217
2300	Poplar, 9'-11', 1-1/4" caliper	G		10	3.200		140	104	78.50	322.50	415
2500	Sumac, 2'-3'	G		75	.427		42.50	13.90	10.45	66.85	81.50
2700	Tulip, 5'-6'	G		40	.800		45.50	26	19.55	91.05	115
2800	Willow, 6'-8', 1" caliper	G		20	1.600		93	52	39	184	232

32 94 Planting Accessories

32 94 13 – Landscape Edging

32 94 13.20 Edging

		Crew	Daily Output	Labor-Hours	Unit	Material	Labor	Equipment	Total	Total Incl O&P
0010	EDGING									
0050	Aluminum alloy, including stakes, 1/8" x 4", mill finish	B-1	390	.062	L.F.	4	1.84		5.84	7.50
0051	Black paint		390	.062		4.64	1.84		6.48	8.20
0052	Black anodized		390	.062		5.35	1.84		7.19	9
0100	Brick, set horizontally, 1-1/2 bricks per L.F.	D-1	370	.043		1.29	1.42		2.71	3.78
0150	Set vertically, 3 bricks per L.F.	"	135	.119		4	3.90		7.90	10.90
0200	Corrugated aluminum, roll, 4" wide	1 Carp	650	.012		2.20	.47		2.67	3.20
0250	6" wide	"	550	.015		2.75	.55		3.30	3.95
0600	Railroad ties, 6" x 8"	2 Carp	170	.094		2.11	3.56		5.67	8.25
0650	7" x 9"		136	.118		2.34	4.45		6.79	10.05
0750	Redwood 2" x 4"		330	.048		2.26	1.83		4.09	5.55
0800	Steel edge strips, incl. stakes, 1/4" x 5"	B-1	390	.062		5	1.84		6.84	8.60
0850	3/16" x 4"	"	390	.062		3.95	1.84		5.79	7.45

32 94 50 – Tree Guying

32 94 50.10 Tree Guying Systems

		Crew	Daily Output	Labor-Hours	Unit	Material	Labor	Equipment	Total	Total Incl O&P
0010	TREE GUYING SYSTEMS									
0015	Tree guying including stakes, guy wire and wrap									
0100	Less than 3" caliper, 2 stakes	2 Clab	35	.457	Ea.	13.10	13.35		26.45	37
0200	3" to 4" caliper, 3 stakes	"	21	.762	"	19.55	22.50		42.05	59
1000	Including arrowhead anchor, cable, turnbuckles and wrap									
1100	Less than 3" caliper, 3" anchors	2 Clab	20	.800	Ea.	28	23.50		51.50	70
1200	3" to 6" caliper, 4" anchors		15	1.067		34	31		65	89.50

32 94 Planting Accessories

32 94 50 – Tree Guying

32 94 50.10 Tree Guying Systems		Crew	Daily Output	Labor-Hours	Unit	Material	Labor	Equipment	Total	Total Incl O&P
						2016 Bare Costs				
1300	6″ caliper, 6″ anchors	2 Clab	12	1.333	Ea.	28.50	39		67.50	96
1400	8″ caliper, 8″ anchors	↓	9	1.778	↓	114	52		166	212

32 96 Transplanting

32 96 23 – Plant and Bulb Transplanting

32 96 23.23 Planting

		Crew	Daily Output	Labor-Hours	Unit	Material	Labor	Equipment	Total	Total Incl O&P
0010	**PLANTING**									
0012	Moving shrubs on site, 12″ ball	B-62	28	.857	Ea.		27.50	6.20	33.70	53
0100	24″ ball	″	22	1.091	″		35	7.90	42.90	67

32 96 23.43 Moving Trees

		Crew	Daily Output	Labor-Hours	Unit	Material	Labor	Equipment	Total	Total Incl O&P
0010	**MOVING TREES**, On site									
0300	Moving trees on site, 36″ ball	B-6	3.75	6.400	Ea.		205	97.50	302.50	445
0400	60″ ball	″	1	24	″		770	365	1,135	1,675

Division 33

Utilities

Estimating Tips

33 10 00 Water Utilities
33 30 00 Sanitary Sewerage Utilities
33 40 00 Storm Drainage Utilities

- Never assume that the water, sewer, and drainage lines will go in at the early stages of the project. Consider the site access needs before dividing the site in half with open trenches, loose pipe, and machinery obstructions. Always inspect the site to establish that the site drawings are complete. Check off all existing utilities on your drawings as you locate them. Be especially careful with underground utilities because appurtenances are sometimes buried during regrading or repaving operations. If you find any discrepancies, mark up the site plan for further research. Differing site conditions can be very costly if discovered later in the project.
- See also Section 33 01 00 for restoration of pipe where removal/replacement may be undesirable. Use of new types of piping materials can reduce the overall project cost. Owners/design engineers should consider the installing contractor as a valuable source of current information on utility products and local conditions that could lead to significant cost savings.

Reference Numbers

Reference numbers are shown at the beginning of some major classifications. These numbers refer to related items in the Reference Section. The reference information may be an estimating procedure, an alternate pricing method, or technical information.

Note: Not all subdivisions listed here necessarily appear. ■

Note: Trade Service, in part, has been used as a reference source for some of the material prices used in Division 33.

33 05 Common Work Results for Utilities

33 05 16 – Utility Structures

33 05 16.13 Precast Concrete Utility Boxes		Crew	Daily Output	Labor-Hours	Unit	2016 Bare Costs Material	Labor	Equipment	Total	Total Incl O&P
0010	**PRECAST CONCRETE UTILITY BOXES**, 6″ thick									
0050	5′ x 10′ x 6′ high, I.D.	B-13	2	24	Ea.	1,775	755	375	2,905	3,600
0350	Hand hole, precast concrete, 1-1/2″ thick									
0400	1′-0″ x 2′-0″ x 1′-9″, I.D., light duty	B-1	4	6	Ea.	410	179		589	750
0450	4′-6″ x 3′-2″ x 2′-0″, O.D., heavy duty	B-6	3	8	″	1,475	257	122	1,854	2,175

33 05 23 – Trenchless Utility Installation

33 05 23.19 Microtunneling		Crew	Daily Output	Labor-Hours	Unit	Material	Labor	Equipment	Total	Total Incl O&P
0010	**MICROTUNNELING**									
0011	Not including excavation, backfill, shoring,									
0020	or dewatering, average 50′/day, slurry method									
0100	24″ to 48″ outside diameter, minimum				L.F.				875	965
0110	Adverse conditions, add				%				50%	50%
1000	Rent microtunneling machine, average monthly lease				Month				97,500	107,000
1010	Operating technician				Day				630	705
1100	Mobilization and demobilization, minimum				Job				41,200	45,900
1110	Maximum				″				445,500	490,500

33 11 Water Utility Distribution Piping

33 11 13 – Public Water Utility Distribution Piping

33 11 13.15 Water Supply, Ductile Iron Pipe		Crew	Daily Output	Labor-Hours	Unit	Material	Labor	Equipment	Total	Total Incl O&P
0010	**WATER SUPPLY, DUCTILE IRON PIPE**									
0020	Not including excavation or backfill									
2000	Pipe, class 50 water piping, 18′ lengths									
2020	Mechanical joint, 4″ diameter	B-21A	200	.200	L.F.	30.50	7.20	2.40	40.10	48
3000	Push-on joint, 4″ diameter		400	.100		15.50	3.59	1.20	20.29	24.50
3020	6″ diameter		333.33	.120		18.25	4.31	1.44	24	28.50
8000	Piping, fittings, mechanical joint, AWWA C110									
8006	90° bend, 4″ diameter	B-20A	16	2	Ea.	155	69.50		224.50	286
8020	6″ diameter		12.80	2.500		229	87		316	395
8200	Wye or tee, 4″ diameter		10.67	2.999		375	104		479	590
8220	6″ diameter		8.53	3.751		565	131		696	835
8398	45° bend, 4″ diameter		16	2		255	69.50		324.50	395
8400	6″ diameter		12.80	2.500		199	87		286	365
8405	8″ diameter		10.67	2.999		291	104		395	495
8450	Decreaser, 6″ x 4″ diameter		14.22	2.250		207	78.50		285.50	360
8460	8″ x 6″ diameter		11.64	2.749		310	95.50		405.50	500
8550	Piping, butterfly valves, cast iron									
8560	4″ diameter	B-20	6	4	Ea.	470	119		589	715
8700	Joint restraint, ductile iron mechanical joints									
8710	4″ diameter	B-20A	32	1	Ea.	17.15	35		52.15	77
8720	6″ diameter		25.60	1.250		21.50	43.50		65	96
8730	8″ diameter		21.33	1.500		30	52		82	120
8740	10″ diameter		18.28	1.751		38.50	61		99.50	144
8750	12″ diameter		16.84	1.900		50	66		116	165
8760	14″ diameter		16	2		53	69.50		122.50	175
8770	16″ diameter		11.64	2.749		58	95.50		153.50	223
8780	18″ diameter		11.03	2.901		73	101		174	248
8785	20″ diameter		9.14	3.501		78	122		200	288
8790	24″ diameter		7.53	4.250		104	148		252	360
9600	Steel sleeve with tap, 4″ diameter	B-20	3	8		405	239		644	845

33 11 Water Utility Distribution Piping

33 11 13 – Public Water Utility Distribution Piping

	33 11 13.15 Water Supply, Ductile Iron Pipe	Crew	Daily Output	Labor-Hours	Unit	Material	2016 Bare Costs Labor	Equipment	Total	Total Incl O&P
9620	6" diameter	B-20	2	12	Ea.	470	360		830	1,125

33 11 13.25 Water Supply, Polyvinyl Chloride Pipe

		Crew	Daily Output	Labor-Hours	Unit	Material	Labor	Equipment	Total	Total Incl O&P
0010	**WATER SUPPLY, POLYVINYL CHLORIDE PIPE**									
0020	Not including excavation or backfill, unless specified									
2100	PVC pipe, Class 150, 1-1/2" diameter	Q-1A	750	.013	L.F.	.50	.59		1.09	1.52
2120	2" diameter		686	.015		.73	.64		1.37	1.86
2140	2-1/2" diameter		500	.020		1.25	.88		2.13	2.84
2160	3" diameter	B-20	430	.056		1.40	1.67		3.07	4.33
8700	PVC pipe, joint restraint									
8710	4" diameter	B-20A	32	1	Ea.	45	35		80	108
8720	6" diameter		25.60	1.250		54.50	43.50		98	133
8730	8" diameter		21.33	1.500		81	52		133	176
8740	10" diameter		18.28	1.751		152	61		213	268
8750	12" diameter		16.84	1.900		159	66		225	285
8760	14" diameter		16	2		237	69.50		306.50	375
8770	16" diameter		11.64	2.749		320	95.50		415.50	510
8780	18" diameter		11.03	2.901		395	101		496	605
8785	20" diameter		9.14	3.501		480	122		602	725
8790	24" diameter		7.53	4.250		565	148		713	865

33 12 Water Utility Distribution Equipment

33 12 13 – Water Service Connections

	33 12 13.15 Tapping, Crosses and Sleeves	Crew	Daily Output	Labor-Hours	Unit	Material	Labor	Equipment	Total	Total Incl O&P
0010	**TAPPING, CROSSES AND SLEEVES**									
4000	Drill and tap pressurized main (labor only)									
4100	6" main, 1" to 2" service	Q-1	3	5.333	Ea.		210		210	345
4150	8" main, 1" to 2" service	"	2.75	5.818	"		229		229	380
4500	Tap and insert gate valve									
4600	8" main, 4" branch	B-21	3.20	8.750	Ea.		274	44	318	510
4650	6" branch		2.70	10.370			325	52	377	600
4800	12" main, 6" branch		2.35	11.915			375	59.50	434.50	690

33 21 Water Supply Wells

33 21 13 – Public Water Supply Wells

	33 21 13.10 Wells and Accessories	Crew	Daily Output	Labor-Hours	Unit	Material	Labor	Equipment	Total	Total Incl O&P
0010	**WELLS & ACCESSORIES**									
0011	Domestic									
0100	Drilled, 4" to 6" diameter	B-23	120	.333	L.F.		9.85	24	33.85	42.50
1500	Pumps, installed in wells to 100' deep, 4" submersible									
1520	3/4 H.P.	Q-1	2.66	6.015	Ea.	855	237		1,092	1,325
1600	1 H.P.	"	2.29	6.987		905	275		1,180	1,450
1800	2 H.P.	Q-22	1.33	12.030		1,725	475	495	2,695	3,225

33 31 Sanitary Utility Sewerage Piping

33 31 13 – Public Sanitary Utility Sewerage Piping

33 31 13.15 Sewage Collection, Concrete Pipe		Crew	Daily Output	Labor-Hours	Unit	Material	2016 Bare Costs Labor	Equipment	Total	Total Incl O&P
0010	SEWAGE COLLECTION, CONCRETE PIPE									
0020	See Section 33 41 13.60 for sewage/drainage collection, concrete pipe									

33 31 13.25 Sewage Collection, Polyvinyl Chloride Pipe		Crew	Daily Output	Labor-Hours	Unit	Material	Labor	Equipment	Total	Total Incl O&P
0010	SEWAGE COLLECTION, POLYVINYL CHLORIDE PIPE									
0020	Not including excavation or backfill									
2000	20′ lengths, SDR 35, B&S, 4″ diameter	B-20	375	.064	L.F.	1.49	1.91		3.40	4.84
2040	6″ diameter		350	.069		3.28	2.05		5.33	7.05
2080	13′ lengths , SDR 35, B&S, 8″ diameter		335	.072		6.15	2.14		8.29	10.35
2120	10″ diameter	B-21	330	.085		11.10	2.66	.43	14.19	17.10

33 36 Utility Septic Tanks

33 36 13 – Utility Septic Tank and Effluent Wet Wells

33 36 13.13 Concrete Utility Septic Tank		Crew	Daily Output	Labor-Hours	Unit	Material	Labor	Equipment	Total	Total Incl O&P
0010	CONCRETE UTILITY SEPTIC TANK									
0011	Not including excavation or piping									
0015	Septic tanks, precast, 1,000 gallon	B-21	8	3.500	Ea.	1,025	110	17.55	1,152.55	1,325
0060	1,500 gallon		7	4		1,575	125	20	1,720	1,975
0100	2,000 gallon		5	5.600		2,200	176	28	2,404	2,750
1150	Leaching field chambers, 13′ x 3′-7″ x 1′-4″, standard	B-13	16	3		485	94	47	626	740
1420	Leaching pit, precast concrete, 6′ dia, 3′ deep	B-21	4.70	5.957		885	187	30	1,102	1,325

33 36 13.19 Polyethylene Utility Septic Tank		Crew	Daily Output	Labor-Hours	Unit	Material	Labor	Equipment	Total	Total Incl O&P
0010	POLYETHYLENE UTILITY SEPTIC TANK									
0015	High density polyethylene, 1,000 gallon	B-21	8	3.500	Ea.	1,225	110	17.55	1,352.55	1,550
0020	1,250 gallon		8	3.500		1,400	110	17.55	1,527.55	1,750
0025	1,500 gallon		7	4		1,550	125	20	1,695	1,925

33 36 19 – Utility Septic Tank Effluent Filter

33 36 19.13 Utility Septic Tank Effluent Tube Filter		Crew	Daily Output	Labor-Hours	Unit	Material	Labor	Equipment	Total	Total Incl O&P
0010	UTILITY SEPTIC TANK EFFLUENT TUBE FILTER									
3000	Effluent filter, 4″ diameter	1 Skwk	8	1	Ea.	43.50	38.50		82	112
3020	6″ diameter		7	1.143		254	44		298	355
3030	8″ diameter		7	1.143		225	44		269	320
3040	8″ diameter, very fine		7	1.143		460	44		504	580
3050	10″ diameter, very fine		6	1.333		240	51		291	350
3060	10″ diameter		6	1.333		280	51		331	395
3080	12″ diameter		6	1.333		670	51		721	825
3090	15″ diameter		5	1.600		1,150	61.50		1,211.50	1,350

33 36 33 – Utility Septic Tank Drainage Field

33 36 33.13 Utility Septic Tank Tile Drainage Field		Crew	Daily Output	Labor-Hours	Unit	Material	Labor	Equipment	Total	Total Incl O&P
0010	UTILITY SEPTIC TANK TILE DRAINAGE FIELD									
0015	Distribution box, concrete, 5 outlets	2 Clab	20	.800	Ea.	82	23.50		105.50	129
0020	7 outlets		16	1		82	29		111	139
0025	9 outlets		8	2		500	58.50		558.50	650
0115	Distribution boxes, HDPE, 5 outlets		20	.800		64.50	23.50		88	110
0117	6 outlets		15	1.067		65	31		96	124
0118	7 outlets		15	1.067		65	31		96	124
0120	8 outlets		10	1.600		68.50	46.50		115	154
0240	Distribution boxes, Outlet Flow Leveler	1 Clab	50	.160		2.35	4.67		7.02	10.40

33 36 Utility Septic Tanks

33 36 50 – Drainage Field Systems

33 36 50.10 Drainage Field Excavation and Fill		Crew	Daily Output	Labor-Hours	Unit	Material	2016 Bare Costs Labor	Equipment	Total	Total Incl O&P
0010	**DRAINAGE FIELD EXCAVATION AND FILL**									
2200	Septic tank & drainage field excavation with 3/4 C.Y. backhoe	B-12F	145	.110	C.Y.		3.83	4.51	8.34	11.30
2400	4′ trench for disposal field, 3/4 C.Y. backhoe	"	335	.048	L.F.		1.66	1.95	3.61	4.91
2600	Gravel fill, run of bank	B-6	150	.160	C.Y.	18.50	5.15	2.44	26.09	31.50
2800	Crushed stone, 3/4″	"	150	.160	"	38.50	5.15	2.44	46.09	53

33 41 Storm Utility Drainage Piping

33 41 13 – Public Storm Utility Drainage Piping

33 41 13.40 Piping, Storm Drainage, Corrugated Metal

		Crew	Daily Output	Labor-Hours	Unit	Material	Labor	Equipment	Total	Total Incl O&P
0010	**PIPING, STORM DRAINAGE, CORRUGATED METAL**									
0020	Not including excavation or backfill									
2040	8″ diameter, 16 ga.	B-14	330	.145	L.F.	8.75	4.51	1.11	14.37	18.35

33 41 13.60 Sewage/Drainage Collection, Concrete Pipe

		Crew	Daily Output	Labor-Hours	Unit	Material	Labor	Equipment	Total	Total Incl O&P
0010	**SEWAGE/DRAINAGE COLLECTION, CONCRETE PIPE**									
0020	Not including excavation or backfill									
1020	8″ diameter	B-14	224	.214	L.F.	8	6.65	1.63	16.28	21.50
1030	10″ diameter	"	216	.222	"	8.85	6.90	1.69	17.44	23
3780	Concrete slotted pipe, class 4 mortar joint									
3800	12″ diameter	B-21	168	.167	L.F.	28	5.25	.84	34.09	40.50
3840	18″ diameter	"	152	.184	"	32	5.80	.92	38.72	45.50
3900	Concrete slotted pipe, Class 4 O-ring joint									
3940	12″ diameter	B-21	168	.167	L.F.	28	5.25	.84	34.09	40.50
3960	18″ diameter	"	152	.184	"	32	5.80	.92	38.72	45.50

33 44 Storm Utility Water Drains

33 44 13 – Utility Area Drains

33 44 13.13 Catch Basins

		Crew	Daily Output	Labor-Hours	Unit	Material	Labor	Equipment	Total	Total Incl O&P
0010	**CATCH BASINS**									
0011	Not including footing & excavation									
1600	Frames & grates, C.I., 24″ square, 500 lb.	B-6	7.80	3.077	Ea.	340	98.50	47	485.50	590

33 46 Subdrainage

33 46 16 – Subdrainage Piping

33 46 16.25 Piping, Subdrainage, Corrugated Metal

		Crew	Daily Output	Labor-Hours	Unit	Material	Labor	Equipment	Total	Total Incl O&P
0010	**PIPING, SUBDRAINAGE, CORRUGATED METAL**									
0021	Not including excavation and backfill									
2010	Aluminum, perforated									
2020	6″ diameter, 18 ga.	B-20	380	.063	L.F.	6.50	1.89		8.39	10.30
2200	8″ diameter, 16 ga.	"	370	.065		8.55	1.94		10.49	12.65
2220	10″ diameter, 16 ga.	B-21	360	.078	↓	10.70	2.44	.39	13.53	16.25
3000	Uncoated galvanized, perforated									
3020	6″ diameter, 18 ga.	B-20	380	.063	L.F.	6.05	1.89		7.94	9.80
3200	8″ diameter, 16 ga.	"	370	.065		8.30	1.94		10.24	12.40
3220	10″ diameter, 16 ga.	B-21	360	.078	↓	8.80	2.44	.39	11.63	14.20
4000	Steel, perforated, asphalt coated									
4020	6″ diameter, 18 ga.	B-20	380	.063	L.F.	6.55	1.89		8.44	10.35
4030	8″ diameter, 18 ga.	"	370	.065	↓	8.60	1.94		10.54	12.70

33 46 Subdrainage

33 46 16 – Subdrainage Piping

33 46 16.25 Piping, Subdrainage, Corrugated Metal		Crew	Daily Output	Labor-Hours	Unit	2016 Bare Costs Material	Labor	Equipment	Total	Total Incl O&P
4040	10" diameter, 16 ga.	B-21	360	.078	L.F.	10.25	2.44	.39	13.08	15.80
4050	12" diameter, 16 ga.		285	.098		11.30	3.08	.49	14.87	18.15
4060	18" diameter, 16 ga.		205	.137		17.40	4.28	.68	22.36	27

33 49 Storm Drainage Structures

33 49 13 – Storm Drainage Manholes, Frames, and Covers

33 49 13.10 Storm Drainage Manholes, Frames and Covers

Line	Description	Crew	Daily Output	Labor-Hours	Unit	Material	Labor	Equipment	Total	Total Incl O&P
0010	**STORM DRAINAGE MANHOLES, FRAMES & COVERS**									
0020	Excludes footing, excavation, backfill (See line items for frame & cover)									
0050	Brick, 4' inside diameter, 4' deep	D-1	1	16	Ea.	535	525		1,060	1,475
1110	Precast, 4' I.D., 4' deep	B-22	4.10	7.317	"	740	234	51.50	1,025.50	1,250

33 51 Natural-Gas Distribution

33 51 13 – Natural-Gas Piping

33 51 13.10 Piping, Gas Service and Distribution, P.E.

Line	Description	Crew	Daily Output	Labor-Hours	Unit	Material	Labor	Equipment	Total	Total Incl O&P
0010	**PIPING, GAS SERVICE AND DISTRIBUTION, POLYETHYLENE**									
0020	Not including excavation or backfill									
1000	60 psi coils, compression coupling @ 100', 1/2" diameter, SDR 11	B-20A	608	.053	L.F.	.50	1.83		2.33	3.59
1010	1" diameter, SDR 11		544	.059		1.13	2.05		3.18	4.64
1040	1-1/4" diameter, SDR 11		544	.059		1.62	2.05		3.67	5.20
1100	2" diameter, SDR 11		488	.066		2.25	2.28		4.53	6.25
1160	3" diameter, SDR 11		408	.078		4.70	2.73		7.43	9.70
1500	60 psi 40' joints with coupling, 3" diameter, SDR 11	B-21A	408	.098		9.40	3.52	1.17	14.09	17.45
1540	4" diameter, SDR 11		352	.114		13.50	4.08	1.36	18.94	23
1600	6" diameter, SDR 11		328	.122		33.50	4.38	1.46	39.34	46
1640	8" diameter, SDR 11		272	.147		46.50	5.30	1.76	53.56	61.50

33 71 Electrical Utility Transmission and Distribution

33 71 16 – Electrical Utility Poles

33 71 16.24 Aluminum Lighting Poles

Line	Description	Crew	Daily Output	Labor-Hours	Unit	Material	Labor	Equipment	Total	Total Incl O&P
0010	**ALUMINUM LIGHTING POLES**									
4000	Aluminum, round, tapered seamless shaft, 20'	R-15A	6.85	7.007	Ea.	615	267	45	927	1,175
4020	25'		6	8		1,150	305	51.50	1,506.50	1,800
4040	30'		5.35	8.972		935	340	58	1,333	1,650

33 71 19 – Electrical Underground Ducts and Manholes

33 71 19.17 Electric and Telephone Underground

Line	Description	Crew	Daily Output	Labor-Hours	Unit	Material	Labor	Equipment	Total	Total Incl O&P
0010	**ELECTRIC AND TELEPHONE UNDERGROUND**									
0011	Not including excavation									
0200	backfill and cast in place concrete									
4200	Underground duct, banks ready for concrete fill, min. of 7.5"									
4400	between conduits, center to center									
4601	PVC, type EB, 2 @ 2" diameter	2 Elec	240	.067	L.F.	1.29	2.94		4.23	6.20
4800	4 @ 2" diameter		120	.133		2.58	5.85		8.43	12.45
5600	4 @ 4" diameter		80	.200		5.65	8.80		14.45	20.50
6200	Rigid galvanized steel, 2 @ 2" diameter		180	.089		12.55	3.92		16.47	20
6400	4 @ 2" diameter		90	.178		25	7.85		32.85	40.50
7400	4 @ 4" diameter		34	.471		80	20.50		100.50	122

Estimating Tips

- When estimating costs for the installation of electrical power generation equipment, factors to review include access to the job site, access and setting at the installation site, required connections, uncrating pads, anchors, leveling, final assembly of the components, and temporary protection from physical damage, including from exposure to the environment.
- Be aware of the cost of equipment supports, concrete pads, and vibration isolators, and cross reference to other trades' specifications. Also, review site and structural drawings for items that must be included in the estimates.
- It is important to include items that are not documented in the plans and specifications but must be priced. These items include, but are not limited to, testing, dust protection, roof penetration, core drilling concrete floors and walls, patching, cleanup, and final adjustments. Add a contingency or allowance for utility company fees for power hookups, if needed.
- The project size and scope of electrical power generation equipment will have a significant impact on cost. The intent of RSMeans cost data is to provide a benchmark cost so that owners, engineers, and electrical contractors will have a comfortable number with which to start a project. Additionally, there are many websites available to use for research and to obtain a vendor's quote to finalize costs.

Reference Numbers

Reference numbers are shown at the beginning of some major classifications. These numbers refer to related items in the Reference Section. The reference information may be an estimating procedure, an alternate pricing method, or technical information.

Note: Not all subdivisions listed here necessarily appear. ■

Note: Trade Service, in part, has been used as a reference source for some of the material prices used in Division 48.

48 15 Wind Energy Electrical Power Generation Equipment

48 15 13 – Wind Turbines

48 15 13.50 Wind Turbines and Components			Crew	Daily Output	Labor-Hours	Unit	2016 Bare Costs Material	Labor	Equipment	Total	Total Incl O&P
0010	**WIND TURBINES & COMPONENTS**										
0500	Complete system, grid connected										
1000	20 kW, 31′ dia, incl. labor & material	G				System				49,900	49,900
1010	Enhanced	G								92,000	92,000
1500	10 kW, 23′ dia, incl. labor & material	G								74,000	74,000
2000	2.4 kW, 12′ dia, incl. labor & material	G				↓				18,000	18,000
2900	Component system										
3000	Turbine, 400 watt, 3′ dia	G	1 Elec	3.41	2.346	Ea.	720	103		823	960
3100	600 watt, 3′ dia	G		2.56	3.125		895	138		1,033	1,200
3200	1000 watt, 9′ dia	G	↓	2.05	3.902	↓	3,700	172		3,872	4,350
3400	Mounting hardware										
3500	30′ guyed tower kit	G	2 Clab	5.12	3.125	Ea.	460	91.50		551.50	660
3505	3′ galvanized helical earth screw	G	1 Clab	8	1		53	29		82	107
3510	Attic mount kit	G	1 Rofc	2.56	3.125		166	100		266	370
3520	Roof mount kit	G	1 Clab	3.41	2.346	↓	152	68.50		220.50	282
8900	Equipment										
9100	DC to AC inverter for, 48 V, 4,000 watt	G	1 Elec	2	4	Ea.	2,300	176		2,476	2,825

Reference Section

All the reference information is in one section, making it easy to find what you need to know . . . and easy to use the data set on a daily basis. This section is visually identified by a vertical black bar on the page edges.

In this Reference Section, we've included Equipment Rental Costs, a listing of rental and operating costs; Crew Listings, a full listing of all crews and equipment, and their costs; Historical Cost Indexes for cost comparisons over time; Location Factors for adjusting costs to the region you are in; Reference Tables, where you will find explanations, estimating information and procedures, or technical data; Change Orders, information on pricing changes to contract documents; and an explanation of all the Abbreviations in the data set.

Table of Contents

Equipment Rental Costs

Estimating Tips

- This section contains the average costs to rent and operate hundreds of pieces of construction equipment. This is useful information when estimating the time and material requirements of any particular operation in order to establish a unit or total cost. Bare equipment costs shown on a unit cost line include not only rental, but also operating costs for equipment under normal use.

Rental Costs

- Equipment rental rates are obtained from the following industry sources throughout North America: contractors, suppliers, dealers, manufacturers, and distributors.
- Rental rates vary throughout the country, with larger cities generally having lower rates. Lease plans for new equipment are available for periods in excess of six months, with a percentage of payments applying toward purchase.
- Monthly rental rates vary from 2% to 5% of the purchase price of the equipment depending on the anticipated life of the equipment and its wearing parts.
- Weekly rental rates are about 1/3 the monthly rates, and daily rental rates are about 1/3 the weekly rate.
- Rental rates can also be treated as reimbursement costs for contractor-owned equipment.

Owned equipment costs include depreciation, loan payments, interest, taxes, insurance, storage, and major repairs.

Operating Costs

- The operating costs include parts and labor for routine servicing, such as repair and replacement of pumps, filters and worn lines. Normal operating expendables, such as fuel, lubricants, tires and electricity (where applicable), are also included.
- Extraordinary operating expendables with highly variable wear patterns, such as diamond bits and blades, are excluded. These costs can be found as material costs in the Unit Price section.
- The hourly operating costs listed do not include the operator's wages.

Equipment Cost/Day

- Any power equipment required by a crew is shown in the Crew Listings with a daily cost.
- This daily cost of equipment needed by a crew includes both the rental cost and the operating cost and is based on dividing the weekly rental rate by 5 (number of working days in the week), and then adding the hourly operating cost times 8 (the number of hours in a day). This "Equipment Cost/ Day" is shown in the far right column of the Equipment Rental section.
- If equipment is needed for only one or two days, it is best to develop your own cost by including components for daily rent and hourly operating cost. This is important when the listed Crew for a task does not contain the equipment needed, such as a crane for lifting mechanical heating/cooling equipment up onto a roof.
- If the quantity of work is less than the crew's Daily Output shown for a Unit Price line item that includes a bare unit equipment cost, it is recommended to estimate one day's rental cost and operating cost for equipment shown in the Crew Listing for that line item.

Mobilization/ Demobilization

- The cost to move construction equipment from an equipment yard or rental company to the job site and back again is not included in equipment rental costs listed in the Reference Section, nor in the bare equipment cost of any unit price line item, nor in any equipment costs shown in the Crew listings.
- Mobilization (to the site) and demobilization (from the site) costs can be found in the Unit Price Section.
- If a piece of equipment is already at the job site, it is not appropriate to utilize mobilization/demobilization costs again in an estimate. ■

01 54 33 | Equipment Rental

			UNIT	HOURLY OPER. COST	RENT PER DAY	RENT PER WEEK	RENT PER MONTH	EQUIPMENT COST/DAY	
10	0010	**CONCRETE EQUIPMENT RENTAL** without operators R015433-10							10
	0200	Bucket, concrete lightweight, 1/2 C.Y.	Ea.	.80	23.50	70	210	20.40	
	0300	1 C.Y.		.85	28	84	252	23.60	
	0400	1-1/2 C.Y.		1.10	38.50	115	345	31.80	
	0500	2 C.Y.		1.20	45	135	405	36.60	
	0580	8 C.Y.		6.00	257	770	2,300	202	
	0600	Cart, concrete, self-propelled, operator walking, 10 C.F.		3.25	56.50	170	510	60	
	0700	Operator riding, 18 C.F.		5.35	95	285	855	99.80	
	0800	Conveyer for concrete, portable, gas, 16″ wide, 26′ long		12.60	123	370	1,100	174.80	
	0900	46′ long		13.00	148	445	1,325	193	
	1000	56′ long		13.15	157	470	1,400	199.20	
	1100	Core drill, electric, 2-1/2 H.P., 1″ to 8″ bit diameter		2.15	78.50	235	705	64.20	
	1150	11 H.P., 8″ to 18″ cores		6.00	115	345	1,025	117	
	1200	Finisher, concrete floor, gas, riding trowel, 96″ wide		12.90	145	435	1,300	190.20	
	1300	Gas, walk-behind, 3 blade, 36″ trowel		2.15	20	60	180	29.20	
	1400	4 blade, 48″ trowel		4.25	27.50	83	249	50.60	
	1500	Float, hand-operated (Bull float) 48″ wide		.08	13.35	40	120	8.65	
	1570	Curb builder, 14 H.P., gas, single screw		14.40	248	745	2,225	264.20	
	1590	Double screw		15.10	292	875	2,625	295.80	
	1600	Floor grinder, concrete and terrazzo, electric, 22″ path		2.70	167	500	1,500	121.60	
	1700	Edger, concrete, electric, 7″ path		1.04	51.50	155	465	39.30	
	1750	Vacuum pick-up system for floor grinders, wet/dry		1.49	81.50	245	735	60.90	
	1800	Mixer, powered, mortar and concrete, gas, 6 C.F., 18 H.P.		8.65	118	355	1,075	140.20	
	1900	10 C.F., 25 H.P.		10.85	143	430	1,300	172.80	
	2000	16 C.F.		11.15	165	495	1,475	188.20	
	2100	Concrete, stationary, tilt drum, 2 C.Y.		6.80	232	695	2,075	193.40	
	2120	Pump, concrete, truck mounted 4″ line 80′ boom		24.70	885	2,650	7,950	727.60	
	2140	5″ line, 110′ boom		32.05	1,175	3,510	10,500	958.40	
	2160	Mud jack, 50 C.F. per hr.		7.25	128	385	1,150	135	
	2180	225 C.F. per hr.		9.55	147	440	1,325	164.40	
	2190	Shotcrete pump rig, 12 C.Y./hr.		15.00	223	670	2,000	254	
	2200	35 C.Y./hr.		17.50	240	720	2,150	284	
	2600	Saw, concrete, manual, gas, 18 H.P.		6.80	45	135	405	81.40	
	2650	Self-propelled, gas, 30 H.P.		13.45	102	305	915	168.60	
	2675	V-groove crack chaser, manual, gas, 6 H.P.		2.35	17.35	52	156	29.20	
	2700	Vibrators, concrete, electric, 60 cycle, 2 H.P.		.41	8.65	26	78	8.50	
	2800	3 H.P.		.60	11.65	35	105	11.80	
	2900	Gas engine, 5 H.P.		2.05	16.35	49	147	26.20	
	3000	8 H.P.		2.80	15.35	46	138	31.60	
	3050	Vibrating screed, gas engine, 8 H.P.		2.58	71.50	215	645	63.65	
	3120	Concrete transit mixer, 6 x 4, 250 H.P., 8 C.Y., rear discharge		61.15	570	1,715	5,150	832.20	
	3200	Front discharge		72.05	700	2,100	6,300	996.40	
	3300	6 x 6, 285 H.P., 12 C.Y., rear discharge		71.10	660	1,980	5,950	964.80	
	3400	Front discharge	↓	74.30	710	2,125	6,375	1,019	
20	0010	**EARTHWORK EQUIPMENT RENTAL** without operators R015433-10							20
	0040	Aggregate spreader, push type 8′ to 12′ wide	Ea.	3.05	25.50	76	228	39.60	
	0045	Tailgate type, 8′ wide		2.90	32.50	98	294	42.80	
	0055	Earth auger, truck-mounted, for fence & sign posts, utility poles		19.10	445	1,340	4,025	420.80	
	0060	For borings and monitoring wells		46.50	690	2,070	6,200	786	
	0070	Portable, trailer mounted		3.25	33	99	297	45.80	
	0075	Truck-mounted, for caissons, water wells		98.70	2,950	8,815	26,400	2,553	
	0080	Horizontal boring machine, 12″ to 36″ diameter, 45 H.P.		24.40	195	585	1,750	312.20	
	0090	12″ to 48″ diameter, 65 H.P.		34.30	340	1,015	3,050	477.40	
	0095	Auger, for fence posts, gas engine, hand held		.60	6	18	54	8.40	
	0100	Excavator, diesel hydraulic, crawler mounted, 1/2 C.Y. cap.		24.45	415	1,250	3,750	445.60	
	0120	5/8 C.Y. capacity		32.30	550	1,655	4,975	589.40	
	0140	3/4 C.Y. capacity		35.25	620	1,860	5,575	654	
	0150	1 C.Y. capacity	↓	49.35	695	2,090	6,275	812.80	

01 54 33 | Equipment Rental

			UNIT	HOURLY OPER. COST	RENT PER DAY	RENT PER WEEK	RENT PER MONTH	EQUIPMENT COST/DAY	
20	0200	1-1/2 C.Y. capacity	Ea.	59.40	925	2,770	8,300	1,029	20
	0300	2 C.Y. capacity		66.90	1,075	3,205	9,625	1,176	
	0320	2-1/2 C.Y. capacity		102.00	1,275	3,830	11,500	1,582	
	0325	3-1/2 C.Y. capacity		143.70	2,150	6,460	19,400	2,442	
	0330	4-1/2 C.Y. capacity		172.45	2,650	7,970	23,900	2,974	
	0335	6 C.Y. capacity		218.85	2,925	8,750	26,300	3,501	
	0340	7 C.Y. capacity		221.10	3,050	9,140	27,400	3,597	
	0342	Excavator attachments, bucket thumbs		3.15	245	735	2,200	172.20	
	0345	Grapples		2.70	193	580	1,750	137.60	
	0346	Hydraulic hammer for boom mounting, 4000 ft lb.		12.15	350	1,045	3,125	306.20	
	0347	5000 ft lb.		14.10	425	1,275	3,825	367.80	
	0348	8000 ft lb.		20.90	625	1,870	5,600	541.20	
	0349	12,000 ft lb.		22.90	750	2,245	6,725	632.20	
	0350	Gradall type, truck mounted, 3 ton @ 15′ radius, 5/8 C.Y.		48.75	820	2,460	7,375	882	
	0370	1 C.Y. capacity		53.85	940	2,825	8,475	995.80	
	0400	Backhoe-loader, 40 to 45 H.P., 5/8 C.Y. capacity		14.85	242	725	2,175	263.80	
	0450	45 H.P. to 60 H.P., 3/4 C.Y. capacity		23.30	298	895	2,675	365.40	
	0460	80 H.P., 1-1/4 C.Y. capacity		25.90	315	945	2,825	396.20	
	0470	112 H.P., 1-1/2 C.Y. capacity		40.25	640	1,915	5,750	705	
	0482	Backhoe-loader attachment, compactor, 20,000 lb.		5.70	142	425	1,275	130.60	
	0485	Hydraulic hammer, 750 ft lb.		3.20	95	285	855	82.60	
	0486	Hydraulic hammer, 1200 ft lb.		6.20	217	650	1,950	179.60	
	0500	Brush chipper, gas engine, 6″ cutter head, 35 H.P.		11.10	103	310	930	150.80	
	0550	Diesel engine, 12″ cutter head, 130 H.P.		27.45	282	845	2,525	388.60	
	0600	15″ cutter head, 165 H.P.		33.00	340	1,020	3,050	468	
	0750	Bucket, clamshell, general purpose, 3/8 C.Y.		1.25	38.50	115	345	33	
	0800	1/2 C.Y.		1.35	45	135	405	37.80	
	0850	3/4 C.Y.		1.50	55	165	495	45	
	0900	1 C.Y.		1.55	58.50	175	525	47.40	
	0950	1-1/2 C.Y.		2.50	80	240	720	68	
	1000	2 C.Y.		2.65	90	270	810	75.20	
	1010	Bucket, dragline, medium duty, 1/2 C.Y.		.70	23.50	70	210	19.60	
	1020	3/4 C.Y.		.75	24.50	73	219	20.60	
	1030	1 C.Y.		.75	26	78	234	21.60	
	1040	1-1/2 C.Y.		1.20	40	120	360	33.60	
	1050	2 C.Y.		1.25	45	135	405	37	
	1070	3 C.Y.		1.95	61.50	185	555	52.60	
	1200	Compactor, manually guided 2-drum vibratory smooth roller, 7.5 H.P.		7.35	210	630	1,900	184.80	
	1250	Rammer/tamper, gas, 8″		2.75	46.50	140	420	50	
	1260	15″		3.10	53.50	160	480	56.80	
	1300	Vibratory plate, gas, 18″ plate, 3000 lb. blow		2.70	24.50	73	219	36.20	
	1350	21″ plate, 5000 lb. blow		3.35	33	99	297	46.60	
	1370	Curb builder/extruder, 14 H.P., gas, single screw		14.40	248	745	2,225	264.20	
	1390	Double screw		15.10	292	875	2,625	295.80	
	1500	Disc harrow attachment, for tractor		.45	75	225	675	48.60	
	1810	Feller buncher, shearing & accumulating trees, 100 H.P.		47.95	755	2,270	6,800	837.60	
	1860	Grader, self-propelled, 25,000 lb.		39.40	665	1,990	5,975	713.20	
	1910	30,000 lb.		43.05	615	1,850	5,550	714.40	
	1920	40,000 lb.		64.75	1,125	3,385	10,200	1,195	
	1930	55,000 lb.		84.00	1,775	5,345	16,000	1,741	
	1950	Hammer, pavement breaker, self-propelled, diesel, 1000 to 1250 lb.		29.55	350	1,055	3,175	447.40	
	2000	1300 to 1500 lb.		44.33	705	2,110	6,325	776.65	
	2050	Pile driving hammer, steam or air, 4150 ft lb. @ 225 bpm		10.25	480	1,435	4,300	369	
	2100	8750 ft lb. @ 145 bpm		12.25	665	2,000	6,000	498	
	2150	15,000 ft lb. @ 60 bpm		13.80	800	2,405	7,225	591.40	
	2200	24,450 ft lb. @ 111 bpm		14.85	890	2,670	8,000	652.80	
	2250	Leads, 60′ high for pile driving hammers up to 20,000 ft lb.		3.25	80	240	720	74	
	2300	90′ high for hammers over 20,000 ft lb.	↓	4.95	141	424	1,275	124.40	

01 54 33 | Equipment Rental

			UNIT	HOURLY OPER. COST	RENT PER DAY	RENT PER WEEK	RENT PER MONTH	EQUIPMENT COST/DAY	
20	2350	Diesel type hammer, 22,400 ft lb.	Ea.	18.00	390	1,170	3,500	378	20
	2400	41,300 ft lb.		27.45	505	1,515	4,550	522.60	
	2450	141,000 ft lb.		47.40	880	2,640	7,925	907.20	
	2500	Vib. elec. hammer/extractor, 200 kW diesel generator, 34 H.P.		54.65	645	1,930	5,800	823.20	
	2550	80 H.P.		100.35	945	2,830	8,500	1,369	
	2600	150 H.P.		190.65	1,800	5,435	16,300	2,612	
	2800	Log chipper, up to 22" diameter, 600 H.P.		63.25	640	1,920	5,750	890	
	2850	Logger, for skidding & stacking logs, 150 H.P.		52.35	830	2,490	7,475	916.80	
	2860	Mulcher, diesel powered, trailer mounted		23.95	213	640	1,925	319.60	
	2900	Rake, spring tooth, with tractor		18.08	345	1,042	3,125	353.05	
	3000	Roller, vibratory, tandem, smooth drum, 20 H.P.		8.95	152	455	1,375	162.60	
	3050	35 H.P.		11.05	252	755	2,275	239.40	
	3100	Towed type vibratory compactor, smooth drum, 50 H.P.		25.70	325	980	2,950	401.60	
	3150	Sheepsfoot, 50 H.P.		27.05	350	1,050	3,150	426.40	
	3170	Landfill compactor, 220 H.P.		88.90	1,550	4,640	13,900	1,639	
	3200	Pneumatic tire roller, 80 H.P.		16.35	360	1,085	3,250	347.80	
	3250	120 H.P.		24.55	615	1,840	5,525	564.40	
	3300	Sheepsfoot vibratory roller, 240 H.P.		67.15	1,125	3,355	10,100	1,208	
	3320	340 H.P.		97.60	1,650	4,960	14,900	1,773	
	3350	Smooth drum vibratory roller, 75 H.P.		24.75	615	1,840	5,525	566	
	3400	125 H.P.		32.10	725	2,175	6,525	691.80	
	3410	Rotary mower, brush, 60", with tractor		22.30	320	955	2,875	369.40	
	3420	Rototiller, walk-behind, gas, 5 H.P.		1.76	51.50	155	465	45.10	
	3422	8 H.P.		2.75	80	240	720	70	
	3440	Scrapers, towed type, 7 C.Y. capacity		5.65	113	340	1,025	113.20	
	3450	10 C.Y. capacity		6.50	155	465	1,400	145	
	3500	15 C.Y. capacity		6.95	180	540	1,625	163.60	
	3525	Self-propelled, single engine, 14 C.Y. capacity		130.55	1,875	5,640	16,900	2,172	
	3550	Dual engine, 21 C.Y. capacity		173.20	2,225	6,645	19,900	2,715	
	3600	31 C.Y. capacity		230.80	3,150	9,445	28,300	3,735	
	3640	44 C.Y. capacity		286.85	4,075	12,220	36,700	4,739	
	3650	Elevating type, single engine, 11 C.Y. capacity		71.80	1,000	2,995	8,975	1,173	
	3700	22 C.Y. capacity		138.90	2,400	7,185	21,600	2,548	
	3710	Screening plant 110 H.P. w/5′ x 10′ screen		19.62	375	1,125	3,375	381.95	
	3720	5′ x 16′ screen		24.98	485	1,450	4,350	489.85	
	3850	Shovel, crawler-mounted, front-loading, 7 C.Y. capacity		257.95	3,575	10,740	32,200	4,212	
	3855	12 C.Y. capacity		419.50	4,950	14,860	44,600	6,328	
	3860	Shovel/backhoe bucket, 1/2 C.Y.		2.40	66.50	200	600	59.20	
	3870	3/4 C.Y.		2.45	73.50	220	660	63.60	
	3880	1 C.Y.		2.55	83.50	250	750	70.40	
	3890	1-1/2 C.Y.		2.70	96.50	290	870	79.60	
	3910	3 C.Y.		3.05	130	390	1,175	102.40	
	3950	Stump chipper, 18" deep, 30 H.P.		6.83	198	593	1,775	173.25	
	4110	Dozer, crawler, torque converter, diesel 80 H.P.		28.80	400	1,205	3,625	471.40	
	4150	105 H.P.		35.30	535	1,605	4,825	603.40	
	4200	140 H.P.		50.95	790	2,365	7,100	880.60	
	4260	200 H.P.		76.95	1,300	3,905	11,700	1,397	
	4310	300 H.P.		100.25	1,800	5,425	16,300	1,887	
	4360	410 H.P.		132.10	2,225	6,675	20,000	2,392	
	4370	500 H.P.		170.65	3,275	9,820	29,500	3,329	
	4380	700 H.P.		280.40	5,175	15,520	46,600	5,347	
	4400	Loader, crawler, torque conv., diesel, 1-1/2 C.Y., 80 H.P.		31.25	460	1,375	4,125	525	
	4450	1-1/2 to 1-3/4 C.Y., 95 H.P.		34.40	600	1,795	5,375	634.20	
	4510	1-3/4 to 2-1/4 C.Y., 130 H.P.		53.50	895	2,680	8,050	964	
	4530	2-1/2 to 3-1/4 C.Y., 190 H.P.		65.50	1,125	3,365	10,100	1,197	
	4560	3-1/2 to 5 C.Y., 275 H.P.		87.45	1,450	4,330	13,000	1,566	
	4610	Front end loader, 4WD, articulated frame, diesel, 1 to 1-1/4 C.Y., 70 H.P.		19.75	235	705	2,125	299	
	4620	1-1/2 to 1-3/4 C.Y., 95 H.P.		24.70	300	905	2,725	378.60	

01 54 33 | Equipment Rental

		01 54 33 Equipment Rental	UNIT	HOURLY OPER. COST	RENT PER DAY	RENT PER WEEK	RENT PER MONTH	EQUIPMENT COST/DAY
20	4650	1-3/4 to 2 C.Y., 130 H.P.	Ea.	29.25	365	1,095	3,275	453
	4710	2-1/2 to 3-1/2 C.Y., 145 H.P.		33.25	415	1,250	3,750	516
	4730	3 to 4-1/2 C.Y., 185 H.P.		42.60	535	1,605	4,825	661.80
	4760	5-1/4 to 5-3/4 C.Y., 270 H.P.		67.20	925	2,770	8,300	1,092
	4810	7 to 9 C.Y., 475 H.P.		113.75	1,750	5,270	15,800	1,964
	4870	9 - 11 C.Y., 620 H.P.		160.25	3,000	9,015	27,000	3,085
	4880	Skid steer loader, wheeled, 10 C.F., 30 H.P. gas		10.40	150	450	1,350	173.20
	4890	1 C.Y., 78 H.P., diesel		20.20	257	770	2,300	315.60
	4892	Skid-steer attachment, auger		.47	77.50	233	700	50.35
	4893	Backhoe		.66	110	329	985	71.10
	4894	Broom		.70	117	351	1,050	75.80
	4895	Forks		.22	37	111	335	23.95
	4896	Grapple		.53	88	264	790	57.05
	4897	Concrete hammer		1.00	167	502	1,500	108.40
	4898	Tree spade		.93	156	467	1,400	100.85
	4899	Trencher		.59	98.50	295	885	63.70
	4900	Trencher, chain, boom type, gas, operator walking, 12 H.P.		5.00	46.50	140	420	68
	4910	Operator riding, 40 H.P.		20.05	290	870	2,600	334.40
	5000	Wheel type, diesel, 4′ deep, 12″ wide		85.15	835	2,510	7,525	1,183
	5100	6′ deep, 20″ wide		91.45	1,925	5,795	17,400	1,891
	5150	Chain type, diesel, 5′ deep, 8″ wide		32.25	470	1,405	4,225	539
	5200	Diesel, 8′ deep, 16″ wide		133.50	2,975	8,960	26,900	2,860
	5202	Rock trencher, wheel type, 6″ wide x 18″ deep		33.65	575	1,730	5,200	615.20
	5206	Chain type, 18″ wide x 7′ deep		110.75	2,850	8,525	25,600	2,591
	5210	Tree spade, self-propelled		12.30	267	800	2,400	258.40
	5250	Truck, dump, 2-axle, 12 ton, 8 C.Y. payload, 220 H.P.		34.30	238	715	2,150	417.40
	5300	Three axle dump, 16 ton, 12 C.Y. payload, 400 H.P.		60.85	340	1,020	3,050	690.80
	5310	Four axle dump, 25 ton, 18 C.Y. payload, 450 H.P.		71.55	495	1,485	4,450	869.40
	5350	Dump trailer only, rear dump, 16-1/2 C.Y.		5.35	138	415	1,250	125.80
	5400	20 C.Y.		5.80	157	470	1,400	140.40
	5450	Flatbed, single axle, 1-1/2 ton rating		25.90	68.50	205	615	248.20
	5500	3 ton rating		31.05	98.50	295	885	307.40
	5550	Off highway rear dump, 25 ton capacity		73.60	1,300	3,905	11,700	1,370
	5600	35 ton capacity		82.20	1,450	4,355	13,100	1,529
	5610	50 ton capacity		102.35	1,725	5,210	15,600	1,861
	5620	65 ton capacity		105.35	1,775	5,360	16,100	1,915
	5630	100 ton capacity		150.95	2,900	8,710	26,100	2,950
	6000	Vibratory plow, 25 H.P., walking	↓	8.50	61.50	185	555	105
40	0010	**GENERAL EQUIPMENT RENTAL** without operators R015433-10						
	0150	Aerial lift, scissor type, to 15′ high, 1000 lb. cap., electric	Ea.	3.15	51.50	155	465	56.20
	0160	To 25′ high, 2000 lb. capacity		3.60	68.50	205	615	69.80
	0170	Telescoping boom to 40′ high, 500 lb. capacity, diesel		13.50	335	1,005	3,025	309
	0180	To 45′ high, 500 lb. capacity		14.75	365	1,090	3,275	336
	0190	To 60′ high, 600 lb. capacity		17.30	515	1,550	4,650	448.40
	0195	Air compressor, portable, 6.5 CFM, electric		.79	13	39	117	14.10
	0196	Gasoline		.70	19.65	59	177	17.40
	0200	Towed type, gas engine, 60 CFM		14.10	50	150	450	142.80
	0300	160 CFM		16.40	51.50	155	465	162.20
	0400	Diesel engine, rotary screw, 250 CFM		16.35	112	335	1,000	197.80
	0500	365 CFM		22.05	137	410	1,225	258.40
	0550	450 CFM		27.95	170	510	1,525	325.60
	0600	600 CFM		49.10	235	705	2,125	533.80
	0700	750 CFM	↓	49.25	245	735	2,200	541
	0800	For silenced models, small sizes, add to rent		3%	5%	5%	5%	
	0900	Large sizes, add to rent		5%	7%	7%	7%	
	0930	Air tools, breaker, pavement, 60 lb.	Ea.	.50	10.35	31	93	10.20
	0940	80 lb.		.50	10.35	31	93	10.20
	0950	Drills, hand (jackhammer) 65 lb.	↓	.60	17.65	53	159	15.40

01 54 33 | Equipment Rental

			UNIT	HOURLY OPER. COST	RENT PER DAY	RENT PER WEEK	RENT PER MONTH	EQUIPMENT COST/DAY	
40	0960	Track or wagon, swing boom, 4" drifter	Ea.	62.25	910	2,735	8,200	1,045	40
	0970	5" drifter		73.80	1,100	3,290	9,875	1,248	
	0975	Track mounted quarry drill, 6" diameter drill		126.40	1,675	4,990	15,000	2,009	
	0980	Dust control per drill		1.04	25	75	225	23.30	
	0990	Hammer, chipping, 12 lb.		.55	26	78	234	20	
	1000	Hose, air with couplings, 50′ long, 3/4" diameter		.03	4.67	14	42	3.05	
	1100	1" diameter		.03	5.65	17	51	3.65	
	1200	1-1/2" diameter		.05	8	24	72	5.20	
	1300	2" diameter		.06	10.65	32	96	6.90	
	1400	2-1/2" diameter		.11	19	57	171	12.30	
	1410	3" diameter		.13	21.50	64	192	13.85	
	1450	Drill, steel, 7/8" x 2′		.06	9.35	28	84	6.10	
	1460	7/8" x 6′		.05	8.65	26	78	5.60	
	1520	Moil points		.02	4	12	36	2.55	
	1525	Pneumatic nailer w/accessories		.55	36.50	109	325	26.20	
	1530	Sheeting driver for 60 lb. breaker		.04	6.65	20	60	4.30	
	1540	For 90 lb. breaker		.14	9	27	81	6.50	
	1550	Spade, 25 lb.		.50	7	21	63	8.20	
	1560	Tamper, single, 35 lb.		.50	33.50	100	300	24	
	1570	Triple, 140 lb.		.75	50	150	450	36	
	1580	Wrenches, impact, air powered, up to 3/4" bolt		.40	12.65	38	114	10.80	
	1590	Up to 1-1/4" bolt		.50	23.50	71	213	18.20	
	1600	Barricades, barrels, reflectorized, 1 to 99 barrels		.03	4.60	13.80	41.50	3	
	1610	100 to 200 barrels		.02	3.53	10.60	32	2.30	
	1620	Barrels with flashers, 1 to 99 barrels		.03	5.25	15.80	47.50	3.40	
	1630	100 to 200 barrels		.03	4.20	12.60	38	2.75	
	1640	Barrels with steady burn type C lights		.04	7	21	63	4.50	
	1650	Illuminated board, trailer mounted, with generator		3.50	130	390	1,175	106	
	1670	Portable barricade, stock, with flashers, 1 to 6 units		.03	5.25	15.80	47.50	3.40	
	1680	25 to 50 units		.03	4.90	14.70	44	3.20	
	1685	Butt fusion machine, wheeled, 1.5 HP electric, 2" - 8" diameter pipe		2.63	167	500	1,500	121.05	
	1690	Tracked, 20 HP diesel, 4"-12" diameter pipe		10.39	490	1,465	4,400	376.10	
	1695	83 HP diesel, 8" - 24" diameter pipe		27.35	975	2,930	8,800	804.80	
	1700	Carts, brick, hand powered, 1000 lb. capacity		.41	68.50	205	615	44.30	
	1800	Gas engine, 1500 lb., 7-1/2′ lift		3.61	108	324	970	93.70	
	1822	Dehumidifier, medium, 6 lb./hr., 150 CFM		1.08	67.50	202.50	610	49.15	
	1824	Large, 18 lb./hr., 600 CFM		2.06	129	385.50	1,150	93.60	
	1830	Distributor, asphalt, trailer mounted, 2000 gal., 38 H.P. diesel		10.25	345	1,035	3,100	289	
	1840	3000 gal., 38 H.P. diesel		11.80	375	1,125	3,375	319.40	
	1850	Drill, rotary hammer, electric		1.07	26.50	80	240	24.55	
	1860	Carbide bit, 1-1/2" diameter, add to electric rotary hammer		.03	4.44	13.33	40	2.90	
	1865	Rotary, crawler, 250 H.P.		147.20	2,150	6,480	19,400	2,474	
	1870	Emulsion sprayer, 65 gal., 5 H.P. gas engine		2.79	101	303	910	82.90	
	1880	200 gal., 5 H.P. engine		8.05	168	505	1,525	165.40	
	1900	Floor auto-scrubbing machine, walk-behind, 28" path		5.13	335	1,000	3,000	241.05	
	1930	Floodlight, mercury vapor, or quartz, on tripod, 1000 watt		.44	20.50	62	186	15.90	
	1940	2000 watt		.83	41.50	124	370	31.45	
	1950	Floodlights, trailer mounted with generator, 1 - 300 watt light		3.60	71.50	215	645	71.80	
	1960	2 - 1000 watt lights		4.80	96.50	290	870	96.40	
	2000	4 - 300 watt lights		4.50	91.50	275	825	91	
	2005	Foam spray rig, incl. box trailer, compressor, generator, proportioner		33.75	495	1,485	4,450	567	
	2020	Forklift, straight mast, 12′ lift, 5000 lb., 2 wheel drive, gas		26.25	207	620	1,850	334	
	2040	21′ lift, 5000 lb., 4 wheel drive, diesel		20.45	250	750	2,250	313.60	
	2050	For rough terrain, 42′ lift, 35′ reach, 9000 lb., 110 H.P.		29.40	500	1,495	4,475	534.20	
	2060	For plant, 4 ton capacity, 80 H.P., 2 wheel drive, gas		16.05	95	285	855	185.40	
	2080	10 ton capacity, 120 H.P., 2 wheel drive, diesel		23.20	165	495	1,475	284.60	
	2100	Generator, electric, gas engine, 1.5 kW to 3 kW		3.70	11.65	35	105	36.60	
	2200	5 kW	↓	4.80	15.35	46	138	47.60	

01 54 33 | Equipment Rental

			UNIT	HOURLY OPER. COST	RENT PER DAY	RENT PER WEEK	RENT PER MONTH	EQUIPMENT COST/DAY
40	2300	10 kW	Ea.	9.05	38.50	115	345	95.40
	2400	25 kW		10.25	86.50	260	780	134
	2500	Diesel engine, 20 kW		11.80	70	210	630	136.40
	2600	50 kW		22.50	105	315	945	243
	2700	100 kW		41.50	135	405	1,225	413
	2800	250 kW		81.70	257	770	2,300	807.60
	2850	Hammer, hydraulic, for mounting on boom, to 500 ft lb.		2.50	75	225	675	65
	2860	1000 ft lb.		4.25	125	375	1,125	109
	2900	Heaters, space, oil or electric, 50 MBH		1.59	7.65	23	69	17.30
	3000	100 MBH		2.93	10.65	32	96	29.85
	3100	300 MBH		8.54	38.50	115	345	91.30
	3150	500 MBH		14.00	45	135	405	139
	3200	Hose, water, suction with coupling, 20′ long, 2″ diameter		.02	3	9	27	1.95
	3210	3″ diameter		.03	4.33	13	39	2.85
	3220	4″ diameter		.03	5	15	45	3.25
	3230	6″ diameter		.11	17.65	53	159	11.50
	3240	8″ diameter		.32	53.50	161	485	34.75
	3250	Discharge hose with coupling, 50′ long, 2″ diameter		.01	1.33	4	12	.90
	3260	3″ diameter		.01	2.33	7	21	1.50
	3270	4″ diameter		.02	3.67	11	33	2.35
	3280	6″ diameter		.06	9.35	28	84	6.10
	3290	8″ diameter		.20	33	99	297	21.40
	3295	Insulation blower		.78	6	18	54	9.85
	3300	Ladders, extension type, 16′ to 36′ long		.14	22.50	68	204	14.70
	3400	40′ to 60′ long		.19	31	93	279	20.10
	3405	Lance for cutting concrete		2.28	64	192	575	56.65
	3407	Lawn mower, rotary, 22″, 5 H.P.		1.55	39.50	118	355	36
	3408	48″ self propelled		3.36	109	327	980	92.30
	3410	Level, electronic, automatic, with tripod and leveling rod		.91	60.50	182	545	43.70
	3430	Laser type, for pipe and sewer line and grade		.73	48.50	145	435	34.85
	3440	Rotating beam for interior control		.78	52	156	470	37.45
	3460	Builder's optical transit, with tripod and rod		.09	14.65	44	132	9.50
	3500	Light towers, towable, with diesel generator, 2000 watt		4.50	91.50	275	825	91
	3600	4000 watt		4.80	96.50	290	870	96.40
	3700	Mixer, powered, plaster and mortar, 6 C.F., 7 H.P.		2.70	20	60	180	33.60
	3800	10 C.F., 9 H.P.		2.85	31.50	95	285	41.80
	3850	Nailer, pneumatic		.55	36.50	109	325	26.20
	3900	Paint sprayers complete, 8 CFM		1.05	70	210	630	50.40
	4000	17 CFM		1.98	132	395	1,175	94.85
	4020	Pavers, bituminous, rubber tires, 8′ wide, 50 H.P., diesel		31.90	510	1,535	4,600	562.20
	4030	10′ wide, 150 H.P.		106.55	1,875	5,635	16,900	1,979
	4050	Crawler, 8′ wide, 100 H.P., diesel		90.15	1,900	5,670	17,000	1,855
	4060	10′ wide, 150 H.P.		113.75	2,300	6,890	20,700	2,288
	4070	Concrete paver, 12′ to 24′ wide, 250 H.P.		100.45	1,550	4,685	14,100	1,741
	4080	Placer-spreader-trimmer, 24′ wide, 300 H.P.		149.30	2,675	8,010	24,000	2,796
	4100	Pump, centrifugal gas pump, 1-1/2″ diam., 65 GPM		4.05	51.50	155	465	63.40
	4200	2″ diameter, 130 GPM		5.45	56.50	170	510	77.60
	4300	3″ diameter, 250 GPM		5.75	58.50	175	525	81
	4400	6″ diameter, 1500 GPM		30.35	180	540	1,625	350.80
	4500	Submersible electric pump, 1-1/4″ diameter, 55 GPM		.40	17	51	153	13.40
	4600	1-1/2″ diameter, 83 GPM		.44	19.35	58	174	15.10
	4700	2″ diameter, 120 GPM		1.55	24.50	73	219	27
	4800	3″ diameter, 300 GPM		2.79	43.50	130	390	48.30
	4900	4″ diameter, 560 GPM		14.32	162	485	1,450	211.55
	5000	6″ diameter, 1590 GPM		21.51	217	650	1,950	302.10
	5100	Diaphragm pump, gas, single, 1-1/2″ diameter		1.08	49.50	148	445	38.25
	5200	2″ diameter		4.35	61.50	185	555	71.80
	5300	3″ diameter	↓	4.35	65	195	585	73.80

01 54 33 | Equipment Rental

			UNIT	HOURLY OPER. COST	RENT PER DAY	RENT PER WEEK	RENT PER MONTH	EQUIPMENT COST/DAY	
40	5400	Double, 4" diameter	Ea.	6.50	110	330	990	118	40
	5450	Pressure washer 5 GPM, 3000 psi		4.95	53.50	160	480	71.60	
	5460	7 GPM, 3000 psi		6.45	61.50	185	555	88.60	
	5500	Trash pump, self-priming, gas, 2" diameter		4.65	22	66	198	50.40	
	5600	Diesel, 4" diameter		8.95	91.50	275	825	126.60	
	5650	Diesel, 6" diameter		24.25	153	460	1,375	286	
	5655	Grout Pump		26.25	272	815	2,450	373	
	5700	Salamanders, L.P. gas fired, 100,000 Btu		3.13	13	39	117	32.85	
	5705	50,000 Btu		1.74	10.35	31	93	20.10	
	5720	Sandblaster, portable, open top, 3 C.F. capacity		.60	26.50	80	240	20.80	
	5730	6 C.F. capacity		.95	40	120	360	31.60	
	5740	Accessories for above		.13	22	66	198	14.25	
	5750	Sander, floor		.98	27.50	83	249	24.45	
	5760	Edger		.53	15.65	47	141	13.65	
	5800	Saw, chain, gas engine, 18" long		2.30	22	66	198	31.60	
	5900	Hydraulic powered, 36" long		.75	65	195	585	45	
	5950	60" long		.75	66.50	200	600	46	
	6000	Masonry, table mounted, 14" diameter, 5 H.P.		1.32	56.50	170	510	44.55	
	6050	Portable cut-off, 8 H.P.		2.50	33.50	100	300	40	
	6100	Circular, hand held, electric, 7-1/4" diameter		.23	4.67	14	42	4.65	
	6200	12" diameter		.23	8	24	72	6.65	
	6250	Wall saw, w/hydraulic power, 10 H.P.		9.75	61.50	185	555	115	
	6275	Shot blaster, walk-behind, 20" wide		4.80	295	885	2,650	215.40	
	6280	Sidewalk broom, walk-behind		2.04	65	195	585	55.30	
	6300	Steam cleaner, 100 gallons per hour		3.80	78.50	235	705	77.40	
	6310	200 gallons per hour		5.40	95	285	855	100.20	
	6340	Tar Kettle/Pot, 400 gallons		16.30	76.50	230	690	176.40	
	6350	Torch, cutting, acetylene-oxygen, 150' hose, excludes gases		.30	15	45	135	11.40	
	6360	Hourly operating cost includes tips and gas		21.00				168	
	6410	Toilet, portable chemical		.13	21.50	64	192	13.85	
	6420	Recycle flush type		.15	25.50	77	231	16.60	
	6430	Toilet, fresh water flush, garden hose,		.19	31	93	279	20.10	
	6440	Hoisted, non-flush, for high rise		.15	25	75	225	16.20	
	6465	Tractor, farm with attachment		21.35	305	910	2,725	352.80	
	6480	Trailers, platform, flush deck, 2 axle, 3 ton capacity		1.45	19.65	59	177	23.40	
	6500	25 ton capacity		5.40	117	350	1,050	113.20	
	6600	40 ton capacity		6.95	165	495	1,475	154.60	
	6700	3 axle, 50 ton capacity		7.50	183	550	1,650	170	
	6800	75 ton capacity		9.30	240	720	2,150	218.40	
	6810	Trailer mounted cable reel for high voltage line work		5.57	265	795	2,375	203.55	
	6820	Trailer mounted cable tensioning rig		11.06	525	1,580	4,750	404.50	
	6830	Cable pulling rig		71.33	2,950	8,850	26,600	2,341	
	6850	Portable cable/wire puller, 8000 lb max pulling capacity		3.71	167	502	1,500	130.10	
	6900	Water tank trailer, engine driven discharge, 5000 gallons		6.95	143	430	1,300	141.60	
	6925	10,000 gallons		9.45	198	595	1,775	194.60	
	6950	Water truck, off highway, 6000 gallons		87.60	780	2,335	7,000	1,168	
	7010	Tram car for high voltage line work, powered, 2 conductor		6.66	144	431	1,300	139.50	
	7020	Transit (builder's level) with tripod		.09	14.65	44	132	9.50	
	7030	Trench box, 3000 lb., 6' x 8'		.56	93.50	280	840	60.50	
	7040	7200 lb., 6' x 20'		.75	125	375	1,125	81	
	7050	8000 lb., 8' x 16'		1.08	180	540	1,625	116.65	
	7060	9500 lb., 8' x 20'		1.21	201	603	1,800	130.30	
	7065	11,000 lb., 8' x 24'		1.27	211	633	1,900	136.75	
	7070	12,000 lb., 10' x 20'		1.50	249	748	2,250	161.60	
	7100	Truck, pickup, 3/4 ton, 2 wheel drive		13.85	58.50	175	525	145.80	
	7200	4 wheel drive		14.10	73.50	220	660	156.80	
	7250	Crew carrier, 9 passenger		19.60	86.50	260	780	208.80	
	7290	Flat bed truck, 20,000 lb. GVW	↓	21.60	125	375	1,125	247.80	

01 54 33 | Equipment Rental

		01 54 33 Equipment Rental	UNIT	HOURLY OPER. COST	RENT PER DAY	RENT PER WEEK	RENT PER MONTH	EQUIPMENT COST/DAY
40	7300	Tractor, 4 x 2, 220 H.P.	Ea.	30.10	197	590	1,775	358.80
	7410	330 H.P.		44.65	272	815	2,450	520.20
	7500	6 x 4, 380 H.P.		51.25	315	950	2,850	600
	7600	450 H.P.		62.10	385	1,150	3,450	726.80
	7610	Tractor, with A frame, boom and winch, 225 H.P.		33.00	273	820	2,450	428
	7620	Vacuum truck, hazardous material, 2500 gallons		11.85	292	875	2,625	269.80
	7625	5,000 gallons		13.20	410	1,230	3,700	351.60
	7650	Vacuum, HEPA, 16 gallon, wet/dry		.85	18	54	162	17.60
	7655	55 gallon, wet/dry		.80	27	81	243	22.60
	7660	Water tank, portable		.16	26.50	80	240	17.30
	7690	Sewer/catch basin vacuum, 14 C.Y., 1500 gallons		17.54	615	1,850	5,550	510.30
	7700	Welder, electric, 200 amp		3.88	16.35	49	147	40.85
	7800	300 amp		5.76	20	60	180	58.10
	7900	Gas engine, 200 amp		14.45	24.50	73	219	130.20
	8000	300 amp		16.55	24.50	74	222	147.20
	8100	Wheelbarrow, any size		.08	13	39	117	8.45
	8200	Wrecking ball, 4000 lb.	▼	2.30	68.50	205	615	59.40
50	0010	**HIGHWAY EQUIPMENT RENTAL** without operators R015433-10						
	0050	Asphalt batch plant, portable drum mixer, 100 ton/hr.	Ea.	80.81	1,450	4,380	13,100	1,522
	0060	200 ton/hr.		92.38	1,550	4,655	14,000	1,670
	0070	300 ton/hr.		109.99	1,825	5,465	16,400	1,973
	0100	Backhoe attachment, long stick, up to 185 H.P., 10.5′ long		.36	23.50	71	213	17.10
	0140	Up to 250 H.P., 12′ long		.39	25.50	77	231	18.50
	0180	Over 250 H.P., 15′ long		.54	35.50	107	320	25.70
	0200	Special dipper arm, up to 100 H.P., 32′ long		1.10	73	219	655	52.60
	0240	Over 100 H.P., 33′ long		1.37	91	273	820	65.55
	0280	Catch basin/sewer cleaning truck, 3 ton, 9 C.Y., 1000 gal.		42.90	390	1,170	3,500	577.20
	0300	Concrete batch plant, portable, electric, 200 C.Y./hr.		22.52	515	1,545	4,625	489.15
	0520	Grader/dozer attachment, ripper/scarifier, rear mounted, up to 135 H.P.		2.90	58.50	175	525	58.20
	0540	Up to 180 H.P.		3.85	88.50	265	795	83.80
	0580	Up to 250 H.P.		5.50	142	425	1,275	129
	0700	Pvmt. removal bucket, for hyd. excavator, up to 90 H.P.		1.85	51.50	155	465	45.80
	0740	Up to 200 H.P.		2.05	68.50	205	615	57.40
	0780	Over 200 H.P.		2.20	85	255	765	68.60
	0900	Aggregate spreader, self-propelled, 187 H.P.		52.46	685	2,050	6,150	829.70
	1000	Chemical spreader, 3 C.Y.		3.25	43.50	130	390	52
	1900	Hammermill, traveling, 250 H.P.		70.05	2,125	6,360	19,100	1,832
	2000	Horizontal borer, 3″ diameter, 13 H.P. gas driven		6.40	55	165	495	84.20
	2150	Horizontal directional drill, 20,000 lb. thrust, 78 H.P. diesel		31.00	680	2,045	6,125	657
	2160	30,000 lb. thrust, 115 H.P.		38.60	1,050	3,135	9,400	935.80
	2170	50,000 lb. thrust, 170 H.P.		55.35	1,325	4,005	12,000	1,244
	2190	Mud trailer for HDD, 1500 gallons, 175 H.P., gas		32.25	153	460	1,375	350
	2200	Hydromulcher, diesel, 3000 gallon, for truck mounting		23.30	247	740	2,225	334.40
	2300	Gas, 600 gallon		8.60	98.50	295	885	127.80
	2400	Joint & crack cleaner, walk behind, 25 H.P.		4.00	50	150	450	62
	2500	Filler, trailer mounted, 400 gallons, 20 H.P.		9.45	212	635	1,900	202.60
	3000	Paint striper, self-propelled, 40 gallon, 22 H.P.		7.25	157	470	1,400	152
	3100	120 gallon, 120 H.P.		23.15	400	1,195	3,575	424.20
	3200	Post drivers, 6″ I-Beam frame, for truck mounting		19.00	385	1,155	3,475	383
	3400	Road sweeper, self-propelled, 8′ wide, 90 H.P.		39.40	585	1,755	5,275	666.20
	3450	Road sweeper, vacuum assisted, 4 C.Y., 220 gallons		75.35	635	1,905	5,725	983.80
	4000	Road mixer, self-propelled, 130 H.P.		47.70	785	2,350	7,050	851.60
	4100	310 H.P.		83.25	2,125	6,395	19,200	1,945
	4220	Cold mix paver, incl. pug mill and bitumen tank, 165 H.P.		98.45	2,400	7,190	21,600	2,226
	4240	Pavement brush, towed		3.15	93.50	280	840	81.20
	4250	Paver, asphalt, wheel or crawler, 130 H.P., diesel		98.15	2,375	7,125	21,400	2,210
	4300	Paver, road widener, gas 1′ to 6′, 67 H.P.	▼	49.10	915	2,750	8,250	942.80

01 54 33 | Equipment Rental

			UNIT	HOURLY OPER. COST	RENT PER DAY	RENT PER WEEK	RENT PER MONTH	EQUIPMENT COST/DAY
50	4400	Diesel, 2′ to 14′, 88 H.P.	Ea.	62.75	1,075	3,205	9,625	1,143
	4600	Slipform pavers, curb and gutter, 2 track, 75 H.P.		61.85	1,000	3,035	9,100	1,102
	4700	4 track, 165 H.P.		41.15	750	2,255	6,775	780.20
	4800	Median barrier, 215 H.P.		66.45	1,225	3,670	11,000	1,266
	4901	Trailer, low bed, 75 ton capacity		10.05	238	715	2,150	223.40
	5000	Road planer, walk behind, 10″ cutting width, 10 H.P.		3.70	33.50	100	300	49.60
	5100	Self-propelled, 12″ cutting width, 64 H.P.		10.50	113	340	1,025	152
	5120	Traffic line remover, metal ball blaster, truck mounted, 115 H.P.		48.45	765	2,295	6,875	846.60
	5140	Grinder, truck mounted, 115 H.P.		54.85	830	2,490	7,475	936.80
	5160	Walk-behind, 11 H.P.		4.20	53.50	160	480	65.60
	5200	Pavement profiler, 4′ to 6′ wide, 450 H.P.		252.75	3,400	10,165	30,500	4,055
	5300	8′ to 10′ wide, 750 H.P.		395.65	4,450	13,350	40,100	5,835
	5400	Roadway plate, steel, 1″ x 8′ x 20′		.08	13.65	41	123	8.85
	5600	Stabilizer, self-propelled, 150 H.P.		48.85	625	1,870	5,600	764.80
	5700	310 H.P.		96.75	1,775	5,295	15,900	1,833
	5800	Striper, truck mounted, 120 gallon paint, 460 H.P.		64.90	485	1,455	4,375	810.20
	5900	Thermal paint heating kettle, 115 gallons		8.29	25.50	77	231	81.70
	6000	Tar kettle, 330 gallon, trailer mounted		12.54	58.50	175	525	135.30
	7000	Tunnel locomotive, diesel, 8 to 12 ton		32.15	595	1,780	5,350	613.20
	7005	Electric, 10 ton		26.90	680	2,035	6,100	622.20
	7010	Muck cars, 1/2 C.Y. capacity		2.10	24.50	74	222	31.60
	7020	1 C.Y. capacity		2.30	33	99	297	38.20
	7030	2 C.Y. capacity		2.45	38.50	115	345	42.60
	7040	Side dump, 2 C.Y. capacity		2.65	45	135	405	48.20
	7050	3 C.Y. capacity		3.60	51.50	155	465	59.80
	7060	5 C.Y. capacity		5.15	65	195	585	80.20
	7100	Ventilating blower for tunnel, 7-1/2 H.P.		2.10	51.50	155	465	47.80
	7110	10 H.P.		2.28	53.50	160	480	50.25
	7120	20 H.P.		3.54	69.50	208	625	69.90
	7140	40 H.P.		5.81	98.50	295	885	105.50
	7160	60 H.P.		8.92	153	460	1,375	163.35
	7175	75 H.P.		11.71	210	630	1,900	219.70
	7180	200 H.P.		23.75	305	920	2,750	374
	7800	Windrow loader, elevating	↓	55.85	1,325	4,000	12,000	1,247
60	0010	**LIFTING AND HOISTING EQUIPMENT RENTAL** without operators R015433-10						
	0120	Aerial lift truck, 2 person, to 80′	Ea.	26.40	745	2,240	6,725	659.20
	0140	Boom work platform, 40′ snorkel R015433-15		11.75	285	855	2,575	265
	0150	Crane, flatbed mounted, 3 ton capacity		16.15	198	595	1,775	248.20
	0200	Crane, climbing, 106′ jib, 6000 lb. capacity, 410 fpm R312316-45		39.85	1,675	5,060	15,200	1,331
	0300	101′ jib, 10,250 lb. capacity, 270 fpm		46.60	2,125	6,410	19,200	1,655
	0500	Tower, static, 130′ high, 106′ jib, 6200 lb. capacity at 400 fpm		43.75	1,950	5,840	17,500	1,518
	0600	Crawler mounted, lattice boom, 1/2 C.Y., 15 tons at 12′ radius		36.96	665	2,000	6,000	695.70
	0700	3/4 C.Y., 20 tons at 12′ radius		49.28	835	2,510	7,525	896.25
	0800	1 C.Y., 25 tons at 12′ radius		65.70	1,125	3,340	10,000	1,194
	0900	1-1/2 C.Y., 40 tons at 12′ radius		65.75	1,125	3,385	10,200	1,203
	1000	2 C.Y., 50 tons at 12′ radius		69.75	1,325	3,940	11,800	1,346
	1100	3 C.Y., 75 tons at 12′ radius		74.30	1,525	4,565	13,700	1,507
	1200	100 ton capacity, 60′ boom		84.05	1,650	4,975	14,900	1,667
	1300	165 ton capacity, 60′ boom		106.80	2,050	6,170	18,500	2,088
	1400	200 ton capacity, 70′ boom		129.75	2,575	7,740	23,200	2,586
	1500	350 ton capacity, 80′ boom		181.25	3,850	11,575	34,700	3,765
	1600	Truck mounted, lattice boom, 6 x 4, 20 tons at 10′ radius		35.42	1,150	3,420	10,300	967.35
	1700	25 tons at 10′ radius		38.28	1,250	3,720	11,200	1,050
	1800	8 x 4, 30 tons at 10′ radius		41.49	1,325	3,960	11,900	1,124
	1900	40 tons at 12′ radius		44.28	1,375	4,140	12,400	1,182
	2000	60 tons at 15′ radius		49.79	1,450	4,380	13,100	1,274
	2050	82 tons at 15′ radius		55.71	1,550	4,680	14,000	1,382
	2100	90 tons at 15′ radius	↓	62.48	1,700	5,100	15,300	1,520

01 54 33 | Equipment Rental

			UNIT	HOURLY OPER. COST	RENT PER DAY	RENT PER WEEK	RENT PER MONTH	EQUIPMENT COST/DAY
60	2200	115 tons at 15′ radius	Ea.	70.50	1,900	5,700	17,100	1,704
	2300	150 tons at 18′ radius		83.50	2,000	6,000	18,000	1,868
	2350	165 tons at 18′ radius		82.77	2,125	6,360	19,100	1,934
	2400	Truck mounted, hydraulic, 12 ton capacity		41.55	545	1,630	4,900	658.40
	2500	25 ton capacity		44.10	660	1,980	5,950	748.80
	2550	33 ton capacity		44.65	675	2,030	6,100	763.20
	2560	40 ton capacity		57.95	795	2,390	7,175	941.60
	2600	55 ton capacity		74.95	870	2,615	7,850	1,123
	2700	80 ton capacity		98.50	1,400	4,210	12,600	1,630
	2720	100 ton capacity		92.65	1,475	4,425	13,300	1,626
	2740	120 ton capacity		107.65	1,625	4,845	14,500	1,830
	2760	150 ton capacity		126.15	2,100	6,320	19,000	2,273
	2800	Self-propelled, 4 x 4, with telescoping boom, 5 ton		17.35	237	710	2,125	280.80
	2900	12-1/2 ton capacity		31.65	375	1,130	3,400	479.20
	3000	15 ton capacity		32.35	395	1,190	3,575	496.80
	3050	20 ton capacity		35.10	455	1,370	4,100	554.80
	3100	25 ton capacity		36.70	520	1,565	4,700	606.60
	3150	40 ton capacity		45.10	585	1,755	5,275	711.80
	3200	Derricks, guy, 20 ton capacity, 60′ boom, 75′ mast		23.97	415	1,238	3,725	439.35
	3300	100′ boom, 115′ mast		37.86	710	2,130	6,400	728.90
	3400	Stiffleg, 20 ton capacity, 70′ boom, 37′ mast		26.57	535	1,610	4,825	534.55
	3500	100′ boom, 47′ mast		41.01	860	2,580	7,750	844.10
	3550	Helicopter, small, lift to 1250 lb. maximum, w/pilot		96.80	3,325	10,000	30,000	2,774
	3600	Hoists, chain type, overhead, manual, 3/4 ton		.15	.33	1	3	1.40
	3900	10 ton		.75	6	18	54	9.60
	4000	Hoist and tower, 5000 lb. cap., portable electric, 40′ high		5.02	238	713	2,150	182.75
	4100	For each added 10′ section, add		.11	18.65	56	168	12.10
	4200	Hoist and single tubular tower, 5000 lb. electric, 100′ high		6.80	330	996	3,000	253.60
	4300	For each added 6′-6″ section, add		.19	32.50	97	291	20.90
	4400	Hoist and double tubular tower, 5000 lb., 100′ high		7.30	365	1,097	3,300	277.80
	4500	For each added 6′-6″ section, add		.21	35.50	107	320	23.10
	4550	Hoist and tower, mast type, 6000 lb., 100′ high		7.87	380	1,137	3,400	290.35
	4570	For each added 10′ section, add		.13	22	66	198	14.25
	4600	Hoist and tower, personnel, electric, 2000 lb., 100′ @ 125 fpm		16.61	1,000	3,030	9,100	738.90
	4700	3000 lb., 100′ @ 200 fpm		18.97	1,150	3,430	10,300	837.75
	4800	3000 lb., 150′ @ 300 fpm		21.02	1,275	3,840	11,500	936.15
	4900	4000 lb., 100′ @ 300 fpm		21.78	1,300	3,920	11,800	958.25
	5000	6000 lb., 100′ @ 275 fpm	↓	23.46	1,375	4,110	12,300	1,010
	5100	For added heights up to 500′, add	L.F.	.01	1.67	5	15	1.10
	5200	Jacks, hydraulic, 20 ton	Ea.	.05	2	6	18	1.60
	5500	100 ton		.40	11.65	35	105	10.20
	6100	Jacks, hydraulic, climbing w/50′ jackrods, control console, 30 ton cap.		2.05	137	410	1,225	98.40
	6150	For each added 10′ jackrod section, add		.05	3.33	10	30	2.40
	6300	50 ton capacity		3.30	220	659	1,975	158.20
	6350	For each added 10′ jackrod section, add		.06	4	12	36	2.90
	6500	125 ton capacity		8.60	575	1,720	5,150	412.80
	6550	For each added 10′ jackrod section, add		.59	39	117	350	28.10
	6600	Cable jack, 10 ton capacity with 200′ cable		1.72	114	343	1,025	82.35
	6650	For each added 50′ of cable, add	↓	.21	13.65	41	123	9.90
70	0010	**WELLPOINT EQUIPMENT RENTAL** without operators (R015433-10)						
	0020	Based on 2 months rental						
	0100	Combination jetting & wellpoint pump, 60 H.P. diesel	Ea.	16.29	340	1,016	3,050	333.50
	0200	High pressure gas jet pump, 200 H.P., 300 psi	″	36.68	289	868	2,600	467.05
	0300	Discharge pipe, 8″ diameter	L.F.	.01	.55	1.65	4.95	.40
	0350	12″ diameter		.01	.81	2.43	7.30	.55
	0400	Header pipe, flows up to 150 GPM, 4″ diameter		.01	.50	1.50	4.50	.40
	0500	400 GPM, 6″ diameter	↓	.01	.58	1.75	5.25	.45

01 54 33 | Equipment Rental

			UNIT	HOURLY OPER. COST	RENT PER DAY	RENT PER WEEK	RENT PER MONTH	EQUIPMENT COST/DAY	
70	0600	800 GPM, 8″ diameter	L.F.	.01	.81	2.43	7.30	.55	70
	0700	1500 GPM, 10″ diameter		.01	.85	2.56	7.70	.60	
	0800	2500 GPM, 12″ diameter		.02	1.61	4.83	14.50	1.15	
	0900	4500 GPM, 16″ diameter		.03	2.06	6.19	18.55	1.50	
	0950	For quick coupling aluminum and plastic pipe, add	↓	.03	2.14	6.41	19.25	1.50	
	1100	Wellpoint, 25′ long, with fittings & riser pipe, 1-1/2″ or 2″ diameter	Ea.	.06	4.26	12.79	38.50	3.05	
	1200	Wellpoint pump, diesel powered, 4″ suction, 20 H.P.		7.16	195	585	1,750	174.30	
	1300	6″ suction, 30 H.P.		9.67	242	726	2,175	222.55	
	1400	8″ suction, 40 H.P.		13.09	330	996	3,000	303.90	
	1500	10″ suction, 75 H.P.		19.62	390	1,164	3,500	389.75	
	1600	12″ suction, 100 H.P.		28.25	615	1,850	5,550	596	
	1700	12″ suction, 175 H.P.	↓	41.13	685	2,050	6,150	739.05	
80	0010	**MARINE EQUIPMENT RENTAL** without operators R015433-10							80
	0200	Barge, 400 Ton, 30′ wide x 90′ long	Ea.	17.40	1,075	3,200	9,600	779.20	
	0240	800 Ton, 45′ wide x 90′ long		21.05	1,300	3,880	11,600	944.40	
	2000	Tugboat, diesel, 100 H.P.		37.80	217	650	1,950	432.40	
	2040	250 H.P.		78.95	395	1,185	3,550	868.60	
	2080	380 H.P.		155.50	1,175	3,535	10,600	1,951	
	3000	Small work boat, gas, 16-foot, 50 H.P.		17.35	60	180	540	174.80	
	4000	Large, diesel, 48-foot, 200 H.P.	↓	88.50	1,250	3,735	11,200	1,455	

Crews - Open Shop

Crew A-1	Bare Costs Hr.	Bare Costs Daily	Incl. Subs O&P Hr.	Incl. Subs O&P Daily	Cost Per Labor-Hour Bare Costs	Cost Per Labor-Hour Incl. O&P
1 Building Laborer	$29.20	$233.60	$48.90	$391.20	$29.20	$48.90
1 Concrete Saw, Gas Manual		81.40		89.54	10.18	11.19
8 L.H., Daily Totals		$315.00		$480.74	$39.38	$60.09

Crew A-1A	Hr.	Daily	Hr.	Daily	Bare Costs	Incl. O&P
1 Skilled Worker	$38.30	$306.40	$64.35	$514.80	$38.30	$64.35
1 Shot Blaster, 20"		215.40		236.94	26.93	29.62
8 L.H., Daily Totals		$521.80		$751.74	$65.22	$93.97

Crew A-1B	Hr.	Daily	Hr.	Daily	Bare Costs	Incl. O&P
1 Building Laborer	$29.20	$233.60	$48.90	$391.20	$29.20	$48.90
1 Concrete Saw		168.60		185.46	21.07	23.18
8 L.H., Daily Totals		$402.20		$576.66	$50.27	$72.08

Crew A-1C	Hr.	Daily	Hr.	Daily	Bare Costs	Incl. O&P
1 Building Laborer	$29.20	$233.60	$48.90	$391.20	$29.20	$48.90
1 Chain Saw, Gas, 18"		31.60		34.76	3.95	4.34
8 L.H., Daily Totals		$265.20		$425.96	$33.15	$53.24

Crew A-1D	Hr.	Daily	Hr.	Daily	Bare Costs	Incl. O&P
1 Building Laborer	$29.20	$233.60	$48.90	$391.20	$29.20	$48.90
1 Vibrating Plate, Gas, 18"		36.20		39.82	4.53	4.98
8 L.H., Daily Totals		$269.80		$431.02	$33.73	$53.88

Crew A-1E	Hr.	Daily	Hr.	Daily	Bare Costs	Incl. O&P
1 Building Laborer	$29.20	$233.60	$48.90	$391.20	$29.20	$48.90
1 Vibrating Plate, Gas, 21"		46.60		51.26	5.83	6.41
8 L.H., Daily Totals		$280.20		$442.46	$35.02	$55.31

Crew A-1F	Hr.	Daily	Hr.	Daily	Bare Costs	Incl. O&P
1 Building Laborer	$29.20	$233.60	$48.90	$391.20	$29.20	$48.90
1 Rammer/Tamper, Gas, 8"		50.00		55.00	6.25	6.88
8 L.H., Daily Totals		$283.60		$446.20	$35.45	$55.77

Crew A-1G	Hr.	Daily	Hr.	Daily	Bare Costs	Incl. O&P
1 Building Laborer	$29.20	$233.60	$48.90	$391.20	$29.20	$48.90
1 Rammer/Tamper, Gas, 15"		56.80		62.48	7.10	7.81
8 L.H., Daily Totals		$290.40		$453.68	$36.30	$56.71

Crew A-1H	Hr.	Daily	Hr.	Daily	Bare Costs	Incl. O&P
1 Building Laborer	$29.20	$233.60	$48.90	$391.20	$29.20	$48.90
1 Exterior Steam Cleaner		77.40		85.14	9.68	10.64
8 L.H., Daily Totals		$311.00		$476.34	$38.88	$59.54

Crew A-1J	Hr.	Daily	Hr.	Daily	Bare Costs	Incl. O&P
1 Building Laborer	$29.20	$233.60	$48.90	$391.20	$29.20	$48.90
1 Cultivator, Walk-Behind, 5 H.P.		45.10		49.61	5.64	6.20
8 L.H., Daily Totals		$278.70		$440.81	$34.84	$55.10

Crew A-1K	Hr.	Daily	Hr.	Daily	Bare Costs	Incl. O&P
1 Building Laborer	$29.20	$233.60	$48.90	$391.20	$29.20	$48.90
1 Cultivator, Walk-Behind, 8 H.P.		70.00		77.00	8.75	9.63
8 L.H., Daily Totals		$303.60		$468.20	$37.95	$58.52

Crew A-1M	Hr.	Daily	Hr.	Daily	Bare Costs	Incl. O&P
1 Building Laborer	$29.20	$233.60	$48.90	$391.20	$29.20	$48.90
1 Snow Blower, Walk-Behind		55.30		60.83	6.91	7.60
8 L.H., Daily Totals		$288.90		$452.03	$36.11	$56.50

Crew A-2	Hr.	Daily	Hr.	Daily	Bare Costs	Incl. O&P
2 Laborers	$29.20	$467.20	$48.90	$782.40	$30.50	$50.80
1 Truck Driver (light)	33.10	264.80	54.60	436.80		
1 Flatbed Truck, Gas, 1.5 Ton		248.20		273.02	10.34	11.38
24 L.H., Daily Totals		$980.20		$1492.22	$40.84	$62.18

Crew A-2A	Hr.	Daily	Hr.	Daily	Bare Costs	Incl. O&P
2 Laborers	$29.20	$467.20	$48.90	$782.40	$30.50	$50.80
1 Truck Driver (light)	33.10	264.80	54.60	436.80		
1 Flatbed Truck, Gas, 1.5 Ton		248.20		273.02		
1 Concrete Saw		168.60		185.46	17.37	19.10
24 L.H., Daily Totals		$1148.80		$1677.68	$47.87	$69.90

Crew A-2B	Hr.	Daily	Hr.	Daily	Bare Costs	Incl. O&P
1 Truck Driver (light)	$33.10	$264.80	$54.60	$436.80	$33.10	$54.60
1 Flatbed Truck, Gas, 1.5 Ton		248.20		273.02	31.02	34.13
8 L.H., Daily Totals		$513.00		$709.82	$64.13	$88.73

Crew A-3A	Hr.	Daily	Hr.	Daily	Bare Costs	Incl. O&P
1 Equip. Oper. (light)	$37.85	$302.80	$62.55	$500.40	$37.85	$62.55
1 Pickup Truck, 4x4, 3/4 Ton		156.80		172.48	19.60	21.56
8 L.H., Daily Totals		$459.60		$672.88	$57.45	$84.11

Crew A-3B	Hr.	Daily	Hr.	Daily	Bare Costs	Incl. O&P
1 Equip. Oper. (medium)	$39.35	$314.80	$65.05	$520.40	$36.75	$60.70
1 Truck Driver (heavy)	34.15	273.20	56.35	450.80		
1 Dump Truck, 12 C.Y., 400 H.P.		690.80		759.88		
1 F.E. Loader, W.M.,2.5 C.Y.		516.00		567.60	75.42	82.97
16 L.H., Daily Totals		$1794.80		$2298.68	$112.18	$143.67

Crew A-3C	Hr.	Daily	Hr.	Daily	Bare Costs	Incl. O&P
1 Equip. Oper. (light)	$37.85	$302.80	$62.55	$500.40	$37.85	$62.55
1 Loader, Skid Steer, 78 H.P.		315.60		347.16	39.45	43.40
8 L.H., Daily Totals		$618.40		$847.56	$77.30	$105.94

Crew A-3D	Hr.	Daily	Hr.	Daily	Bare Costs	Incl. O&P
1 Truck Driver (light)	$33.10	$264.80	$54.60	$436.80	$33.10	$54.60
1 Pickup Truck, 4x4, 3/4 Ton		156.80		172.48		
1 Flatbed Trailer, 25 Ton		113.20		124.52	33.75	37.13
8 L.H., Daily Totals		$534.80		$733.80	$66.85	$91.72

Crew A-3E	Hr.	Daily	Hr.	Daily	Bare Costs	Incl. O&P
1 Equip. Oper. (crane)	$40.25	$322.00	$66.55	$532.40	$37.20	$61.45
1 Truck Driver (heavy)	34.15	273.20	56.35	450.80		
1 Pickup Truck, 4x4, 3/4 Ton		156.80		172.48	9.80	10.78
16 L.H., Daily Totals		$752.00		$1155.68	$47.00	$72.23

Crew A-3F	Hr.	Daily	Hr.	Daily	Bare Costs	Incl. O&P
1 Equip. Oper. (crane)	$40.25	$322.00	$66.55	$532.40	$37.20	$61.45
1 Truck Driver (heavy)	34.15	273.20	56.35	450.80		
1 Pickup Truck, 4x4, 3/4 Ton		156.80		172.48		
1 Truck Tractor, 6x4, 380 H.P.		600.00		660.00		
1 Lowbed Trailer, 75 Ton		223.40		245.74	61.26	67.39
16 L.H., Daily Totals		$1575.40		$2061.42	$98.46	$128.84

Crews - Open Shop

Crew No.	Bare Costs		Incl. Subs O&P		Cost Per Labor-Hour	
Crew A-3G	**Hr.**	**Daily**	**Hr.**	**Daily**	**Bare Costs**	**Incl. O&P**
1 Equip. Oper. (crane)	$40.25	$322.00	$66.55	$532.40	$37.20	$61.45
1 Truck Driver (heavy)	34.15	273.20	56.35	450.80		
1 Pickup Truck, 4x4, 3/4 Ton		156.80		172.48		
1 Truck Tractor, 6x4, 450 H.P.		726.80		799.48		
1 Lowbed Trailer, 75 Ton		223.40		245.74	69.19	76.11
16 L.H., Daily Totals		$1702.20		$2200.90	$106.39	$137.56

Crew A-3H	**Hr.**	**Daily**	**Hr.**	**Daily**	**Bare Costs**	**Incl. O&P**
1 Equip. Oper. (crane)	$40.25	$322.00	$66.55	$532.40	$40.25	$66.55
1 Hyd. Crane, 12 Ton (Daily)		877.40		965.14	109.68	120.64
8 L.H., Daily Totals		$1199.40		$1497.54	$149.93	$187.19

Crew A-3I	**Hr.**	**Daily**	**Hr.**	**Daily**	**Bare Costs**	**Incl. O&P**
1 Equip. Oper. (crane)	$40.25	$322.00	$66.55	$532.40	$40.25	$66.55
1 Hyd. Crane, 25 Ton (Daily)		1013.00		1114.30	126.63	139.29
8 L.H., Daily Totals		$1335.00		$1646.70	$166.88	$205.84

Crew A-3J	**Hr.**	**Daily**	**Hr.**	**Daily**	**Bare Costs**	**Incl. O&P**
1 Equip. Oper. (crane)	$40.25	$322.00	$66.55	$532.40	$40.25	$66.55
1 Hyd. Crane, 40 Ton (Daily)		1259.00		1384.90	157.38	173.11
8 L.H., Daily Totals		$1581.00		$1917.30	$197.63	$239.66

Crew A-3K	**Hr.**	**Daily**	**Hr.**	**Daily**	**Bare Costs**	**Incl. O&P**
1 Equip. Oper. (crane)	$40.25	$322.00	$66.55	$532.40	$37.85	$62.58
1 Equip. Oper. (oiler)	35.45	283.60	58.60	468.80		
1 Hyd. Crane, 55 Ton (Daily)		1470.00		1617.00		
1 P/U Truck, 3/4 Ton (Daily)		170.80		187.88	102.55	112.81
16 L.H., Daily Totals		$2246.40		$2806.08	$140.40	$175.38

Crew A-3L	**Hr.**	**Daily**	**Hr.**	**Daily**	**Bare Costs**	**Incl. O&P**
1 Equip. Oper. (crane)	$40.25	$322.00	$66.55	$532.40	$37.85	$62.58
1 Equip. Oper. (oiler)	35.45	283.60	58.60	468.80		
1 Hyd. Crane, 80 Ton (Daily)		2193.00		2412.30		
1 P/U Truck, 3/4 Ton (Daily)		170.80		187.88	147.74	162.51
16 L.H., Daily Totals		$2969.40		$3601.38	$185.59	$225.09

Crew A-3M	**Hr.**	**Daily**	**Hr.**	**Daily**	**Bare Costs**	**Incl. O&P**
1 Equip. Oper. (crane)	$40.25	$322.00	$66.55	$532.40	$37.85	$62.58
1 Equip. Oper. (oiler)	35.45	283.60	58.60	468.80		
1 Hyd. Crane, 100 Ton (Daily)		2216.00		2437.60		
1 P/U Truck, 3/4 Ton (Daily)		170.80		187.88	149.18	164.09
16 L.H., Daily Totals		$2992.40		$3626.68	$187.03	$226.67

Crew A-3N	**Hr.**	**Daily**	**Hr.**	**Daily**	**Bare Costs**	**Incl. O&P**
1 Equip. Oper. (crane)	$40.25	$322.00	$66.55	$532.40	$40.25	$66.55
1 Tower Cane (monthly)		1145.00		1259.50	143.13	157.44
8 L.H., Daily Totals		$1467.00		$1791.90	$183.38	$223.99

Crew A-3P	**Hr.**	**Daily**	**Hr.**	**Daily**	**Bare Costs**	**Incl. O&P**
1 Equip. Oper. (light)	$37.85	$302.80	$62.55	$500.40	$37.85	$62.55
1 A.T. Forklift, 42' lift		534.20		587.62	66.78	73.45
8 L.H., Daily Totals		$837.00		$1088.02	$104.63	$136.00

Crew A-3Q	**Hr.**	**Daily**	**Hr.**	**Daily**	**Bare Costs**	**Incl. O&P**
1 Equip. Oper. (light)	$37.85	$302.80	$62.55	$500.40	$37.85	$62.55
1 Pickup Truck, 4x4, 3/4 Ton		156.80		172.48		
1 Flatbed trailer, 3 Ton		23.40		25.74	22.52	24.78
8 L.H., Daily Totals		$483.00		$698.62	$60.38	$87.33

Crew A-4	**Hr.**	**Daily**	**Hr.**	**Daily**	**Bare Costs**	**Incl. O&P**
2 Carpenters	$37.80	$604.80	$63.30	$1012.80	$35.83	$59.72
1 Painter, Ordinary	31.90	255.20	52.55	420.40		
24 L.H., Daily Totals		$860.00		$1433.20	$35.83	$59.72

Crew A-5	**Hr.**	**Daily**	**Hr.**	**Daily**	**Bare Costs**	**Incl. O&P**
2 Laborers	$29.20	$467.20	$48.90	$782.40	$29.63	$49.53
.25 Truck Driver (light)	33.10	66.20	54.60	109.20		
.25 Flatbed Truck, Gas, 1.5 Ton		62.05		68.25	3.45	3.79
18 L.H., Daily Totals		$595.45		$959.86	$33.08	$53.33

Crew A-6	**Hr.**	**Daily**	**Hr.**	**Daily**	**Bare Costs**	**Incl. O&P**
1 Instrument Man	$38.30	$306.40	$64.35	$514.80	$37.60	$62.83
1 Rodman/Chainman	36.90	295.20	61.30	490.40		
1 Level, Electronic		43.70		48.07	2.73	3.00
16 L.H., Daily Totals		$645.30		$1053.27	$40.33	$65.83

Crew A-7	**Hr.**	**Daily**	**Hr.**	**Daily**	**Bare Costs**	**Incl. O&P**
1 Chief of Party	$45.60	$364.80	$75.70	$605.60	$40.27	$67.12
1 Instrument Man	38.30	306.40	64.35	514.80		
1 Rodman/Chainman	36.90	295.20	61.30	490.40		
1 Level, Electronic		43.70		48.07	1.82	2.00
24 L.H., Daily Totals		$1010.10		$1658.87	$42.09	$69.12

Crew A-8	**Hr.**	**Daily**	**Hr.**	**Daily**	**Bare Costs**	**Incl. O&P**
1 Chief of Party	$45.60	$364.80	$75.70	$605.60	$39.42	$65.66
1 Instrument Man	38.30	306.40	64.35	514.80		
2 Rodmen/Chainmen	36.90	590.40	61.30	980.80		
1 Level, Electronic		43.70		48.07	1.37	1.50
32 L.H., Daily Totals		$1305.30		$2149.27	$40.79	$67.16

Crew A-9	**Hr.**	**Daily**	**Hr.**	**Daily**	**Bare Costs**	**Incl. O&P**
1 Asbestos Foreman	$40.00	$320.00	$67.85	$542.80	$39.56	$67.11
7 Asbestos Workers	39.50	2212.00	67.00	3752.00		
64 L.H., Daily Totals		$2532.00		$4294.80	$39.56	$67.11

Crew A-10A	**Hr.**	**Daily**	**Hr.**	**Daily**	**Bare Costs**	**Incl. O&P**
1 Asbestos Foreman	$40.00	$320.00	$67.85	$542.80	$39.67	$67.28
2 Asbestos Workers	39.50	632.00	67.00	1072.00		
24 L.H., Daily Totals		$952.00		$1614.80	$39.67	$67.28

Crew A-10B	**Hr.**	**Daily**	**Hr.**	**Daily**	**Bare Costs**	**Incl. O&P**
1 Asbestos Foreman	$40.00	$320.00	$67.85	$542.80	$39.63	$67.21
3 Asbestos Workers	39.50	948.00	67.00	1608.00		
32 L.H., Daily Totals		$1268.00		$2150.80	$39.63	$67.21

Crew A-10C	**Hr.**	**Daily**	**Hr.**	**Daily**	**Bare Costs**	**Incl. O&P**
3 Asbestos Workers	$39.50	$948.00	$67.00	$1608.00	$39.50	$67.00
1 Flatbed Truck, Gas, 1.5 Ton		248.20		273.02	10.34	11.38
24 L.H., Daily Totals		$1196.20		$1881.02	$49.84	$78.38

Crew A-10D	**Hr.**	**Daily**	**Hr.**	**Daily**	**Bare Costs**	**Incl. O&P**
2 Asbestos Workers	$39.50	$632.00	$67.00	$1072.00	$38.67	$64.79
1 Equip. Oper. (crane)	40.25	322.00	66.55	532.40		
1 Equip. Oper. (oiler)	35.45	283.60	58.60	468.80		
1 Hydraulic Crane, 33 Ton		763.20		839.52	23.85	26.23
32 L.H., Daily Totals		$2000.80		$2912.72	$62.52	$91.02

Crews - Open Shop

Crew No.	Bare Costs		Incl. Subs O&P		Cost Per Labor-Hour	
Crew A-11	**Hr.**	**Daily**	**Hr.**	**Daily**	**Bare Costs**	**Incl. O&P**
1 Asbestos Foreman	$40.00	$320.00	$67.85	$542.80	$39.56	$67.11
7 Asbestos Workers	39.50	2212.00	67.00	3752.00		
2 Chip. Hammers, 12 Lb., Elec.		40.00		44.00	0.63	0.69
64 L.H., Daily Totals		$2572.00		$4338.80	$40.19	$67.79
Crew A-12	**Hr.**	**Daily**	**Hr.**	**Daily**	**Bare Costs**	**Incl. O&P**
1 Asbestos Foreman	$40.00	$320.00	$67.85	$542.80	$39.56	$67.11
7 Asbestos Workers	39.50	2212.00	67.00	3752.00		
1 Trk-Mtd Vac, 14 CY, 1500 Gal.		510.30		561.33		
1 Flatbed Truck, 20,000 GVW		247.80		272.58	11.85	13.03
64 L.H., Daily Totals		$3290.10		$5128.71	$51.41	$80.14
Crew A-13	**Hr.**	**Daily**	**Hr.**	**Daily**	**Bare Costs**	**Incl. O&P**
1 Equip. Oper. (light)	$37.85	$302.80	$62.55	$500.40	$37.85	$62.55
1 Trk-Mtd Vac, 14 CY, 1500 Gal.		510.30		561.33		
1 Flatbed Truck, 20,000 GVW		247.80		272.58	94.76	104.24
8 L.H., Daily Totals		$1060.90		$1334.31	$132.61	$166.79
Crew B-1	**Hr.**	**Daily**	**Hr.**	**Daily**	**Bare Costs**	**Incl. O&P**
1 Labor Foreman (outside)	$31.20	$249.60	$52.25	$418.00	$29.87	$50.02
2 Laborers	29.20	467.20	48.90	782.40		
24 L.H., Daily Totals		$716.80		$1200.40	$29.87	$50.02
Crew B-1A	**Hr.**	**Daily**	**Hr.**	**Daily**	**Bare Costs**	**Incl. O&P**
1 Labor Foreman (outside)	$31.20	$249.60	$52.25	$418.00	$29.87	$50.02
2 Laborers	29.20	467.20	48.90	782.40		
2 Cutting Torches		22.80		25.08		
2 Sets of Gases		336.00		369.60	14.95	16.45
24 L.H., Daily Totals		$1075.60		$1595.08	$44.82	$66.46
Crew B-1B	**Hr.**	**Daily**	**Hr.**	**Daily**	**Bare Costs**	**Incl. O&P**
1 Labor Foreman (outside)	$31.20	$249.60	$52.25	$418.00	$32.46	$54.15
2 Laborers	29.20	467.20	48.90	782.40		
1 Equip. Oper. (crane)	40.25	322.00	66.55	532.40		
2 Cutting Torches		22.80		25.08		
2 Sets of Gases		336.00		369.60		
1 Hyd. Crane, 12 Ton		658.40		724.24	31.79	34.97
32 L.H., Daily Totals		$2056.00		$2851.72	$64.25	$89.12
Crew B-1C	**Hr.**	**Daily**	**Hr.**	**Daily**	**Bare Costs**	**Incl. O&P**
1 Labor Foreman (outside)	$31.20	$249.60	$52.25	$418.00	$29.87	$50.02
2 Laborers	29.20	467.20	48.90	782.40		
1 Aerial Lift Truck, 60' Boom		448.40		493.24	18.68	20.55
24 L.H., Daily Totals		$1165.20		$1693.64	$48.55	$70.57
Crew B-1D	**Hr.**	**Daily**	**Hr.**	**Daily**	**Bare Costs**	**Incl. O&P**
2 Laborers	$29.20	$467.20	$48.90	$782.40	$29.20	$48.90
1 Small Work Boat, Gas, 50 H.P.		174.80		192.28		
1 Pressure Washer, 7 GPM		88.60		97.46	16.46	18.11
16 L.H., Daily Totals		$730.60		$1072.14	$45.66	$67.01
Crew B-1E	**Hr.**	**Daily**	**Hr.**	**Daily**	**Bare Costs**	**Incl. O&P**
1 Labor Foreman (outside)	$31.20	$249.60	$52.25	$418.00	$29.70	$49.74
3 Laborers	29.20	700.80	48.90	1173.60		
1 Work Boat, Diesel, 200 H.P.		1455.00		1600.50		
2 Pressure Washers, 7 GPM		177.20		194.92	51.01	56.11
32 L.H., Daily Totals		$2582.60		$3387.02	$80.71	$105.84

Crew No.	Bare Costs		Incl. Subs O&P		Cost Per Labor-Hour	
Crew B-1F	**Hr.**	**Daily**	**Hr.**	**Daily**	**Bare Costs**	**Incl. O&P**
2 Skilled Workers	$38.30	$612.80	$64.35	$1029.60	$35.27	$59.20
1 Laborer	29.20	233.60	48.90	391.20		
1 Small Work Boat, Gas, 50 H.P.		174.80		192.28		
1 Pressure Washer, 7 GPM		88.60		97.46	10.98	12.07
24 L.H., Daily Totals		$1109.80		$1710.54	$46.24	$71.27
Crew B-1G	**Hr.**	**Daily**	**Hr.**	**Daily**	**Bare Costs**	**Incl. O&P**
2 Laborers	$29.20	$467.20	$48.90	$782.40	$29.20	$48.90
1 Small Work Boat, Gas, 50 H.P.		174.80		192.28	10.93	12.02
16 L.H., Daily Totals		$642.00		$974.68	$40.13	$60.92
Crew B-1H	**Hr.**	**Daily**	**Hr.**	**Daily**	**Bare Costs**	**Incl. O&P**
2 Skilled Workers	$38.30	$612.80	$64.35	$1029.60	$35.27	$59.20
1 Laborer	29.20	233.60	48.90	391.20		
1 Small Work Boat, Gas, 50 H.P.		174.80		192.28	7.28	8.01
24 L.H., Daily Totals		$1021.20		$1613.08	$42.55	$67.21
Crew B-1J	**Hr.**	**Daily**	**Hr.**	**Daily**	**Bare Costs**	**Incl. O&P**
1 Labor Foreman (inside)	$29.70	$237.60	$49.70	$397.60	$29.45	$49.30
1 Laborer	29.20	233.60	48.90	391.20		
16 L.H., Daily Totals		$471.20		$788.80	$29.45	$49.30
Crew B-1K	**Hr.**	**Daily**	**Hr.**	**Daily**	**Bare Costs**	**Incl. O&P**
1 Carpenter Foreman (inside)	$38.30	$306.40	$64.10	$512.80	$38.05	$63.70
1 Carpenter	37.80	302.40	63.30	506.40		
16 L.H., Daily Totals		$608.80		$1019.20	$38.05	$63.70
Crew B-2	**Hr.**	**Daily**	**Hr.**	**Daily**	**Bare Costs**	**Incl. O&P**
1 Labor Foreman (outside)	$31.20	$249.60	$52.25	$418.00	$29.60	$49.57
4 Laborers	29.20	934.40	48.90	1564.80		
40 L.H., Daily Totals		$1184.00		$1982.80	$29.60	$49.57
Crew B-2A	**Hr.**	**Daily**	**Hr.**	**Daily**	**Bare Costs**	**Incl. O&P**
1 Labor Foreman (outside)	$31.20	$249.60	$52.25	$418.00	$29.87	$50.02
2 Laborers	29.20	467.20	48.90	782.40		
1 Aerial Lift Truck, 60' Boom		448.40		493.24	18.68	20.55
24 L.H., Daily Totals		$1165.20		$1693.64	$48.55	$70.57
Crew B-3	**Hr.**	**Daily**	**Hr.**	**Daily**	**Bare Costs**	**Incl. O&P**
1 Labor Foreman (outside)	$31.20	$249.60	$52.25	$418.00	$32.88	$54.63
2 Laborers	29.20	467.20	48.90	782.40		
1 Equip. Oper. (medium)	39.35	314.80	65.05	520.40		
2 Truck Drivers (heavy)	34.15	546.40	56.35	901.60		
1 Crawler Loader, 3 C.Y.		1197.00		1316.70		
2 Dump Trucks, 12 C.Y., 400 H.P.		1381.60		1519.76	53.72	59.09
48 L.H., Daily Totals		$4156.60		$5458.86	$86.60	$113.73
Crew B-3A	**Hr.**	**Daily**	**Hr.**	**Daily**	**Bare Costs**	**Incl. O&P**
4 Laborers	$29.20	$934.40	$48.90	$1564.80	$31.23	$52.13
1 Equip. Oper. (medium)	39.35	314.80	65.05	520.40		
1 Hyd. Excavator, 1.5 C.Y.		1029.00		1131.90	25.73	28.30
40 L.H., Daily Totals		$2278.20		$3217.10	$56.95	$80.43
Crew B-3B	**Hr.**	**Daily**	**Hr.**	**Daily**	**Bare Costs**	**Incl. O&P**
2 Laborers	$29.20	$467.20	$48.90	$782.40	$32.98	$54.80
1 Equip. Oper. (medium)	39.35	314.80	65.05	520.40		
1 Truck Driver (heavy)	34.15	273.20	56.35	450.80		
1 Backhoe Loader, 80 H.P.		396.20		435.82		
1 Dump Truck, 12 C.Y., 400 H.P.		690.80		759.88	33.97	37.37
32 L.H., Daily Totals		$2142.20		$2949.30	$66.94	$92.17

Crew No.	Bare Costs		Incl. Subs O&P		Cost Per Labor-Hour	
Crew B-3C	**Hr.**	**Daily**	**Hr.**	**Daily**	**Bare Costs**	**Incl. O&P**
3 Laborers	$29.20	$700.80	$48.90	$1173.60	$31.74	$52.94
1 Equip. Oper. (medium)	39.35	314.80	65.05	520.40		
1 Crawler Loader, 4 C.Y.		1566.00		1722.60	48.94	53.83
32 L.H., Daily Totals		$2581.60		$3416.60	$80.67	$106.77
Crew B-4	**Hr.**	**Daily**	**Hr.**	**Daily**	**Bare Costs**	**Incl. O&P**
1 Labor Foreman (outside)	$31.20	$249.60	$52.25	$418.00	$30.36	$50.70
4 Laborers	29.20	934.40	48.90	1564.80		
1 Truck Driver (heavy)	34.15	273.20	56.35	450.80		
1 Truck Tractor, 220 H.P.		358.80		394.68		
1 Flatbed Trailer, 40 Ton		154.60		170.06	10.70	11.77
48 L.H., Daily Totals		$1970.60		$2998.34	$41.05	$62.47
Crew B-5	**Hr.**	**Daily**	**Hr.**	**Daily**	**Bare Costs**	**Incl. O&P**
1 Labor Foreman (outside)	$31.20	$249.60	$52.25	$418.00	$31.63	$52.80
3 Laborers	29.20	700.80	48.90	1173.60		
1 Equip. Oper. (medium)	39.35	314.80	65.05	520.40		
1 Air Compressor, 250 cfm		197.80		217.58		
2 Breakers, Pavement, 60 lb.		20.40		22.44		
2 -50' Air Hoses, 1.5"		10.40		11.44		
1 Crawler Loader, 3 C.Y.		1197.00		1316.70	35.64	39.20
40 L.H., Daily Totals		$2690.80		$3680.16	$67.27	$92.00
Crew B-5A	**Hr.**	**Daily**	**Hr.**	**Daily**	**Bare Costs**	**Incl. O&P**
1 Labor Foreman (outside)	$31.20	$249.60	$52.25	$418.00	$32.60	$54.25
6 Laborers	29.20	1401.60	48.90	2347.20		
2 Equip. Oper. (medium)	39.35	629.60	65.05	1040.80		
1 Equip. Oper. (light)	37.85	302.80	62.55	500.40		
2 Truck Drivers (heavy)	34.15	546.40	56.35	901.60		
1 Air Compressor, 365 cfm		258.40		284.24		
2 Breakers, Pavement, 60 lb.		20.40		22.44		
8 -50' Air Hoses, 1"		29.20		32.12		
2 Dump Trucks, 8 C.Y., 220 H.P.		834.80		918.28	11.90	13.09
96 L.H., Daily Totals		$4272.80		$6465.08	$44.51	$67.34
Crew B-5B	**Hr.**	**Daily**	**Hr.**	**Daily**	**Bare Costs**	**Incl. O&P**
1 Powderman	$38.30	$306.40	$64.35	$514.80	$36.58	$60.58
2 Equip. Opers. (medium)	39.35	629.60	65.05	1040.80		
3 Truck Drivers (heavy)	34.15	819.60	56.35	1352.40		
1 F.E. Loader, W.M., 2.5 C.Y.		516.00		567.60		
3 Dump Trucks, 12 C.Y., 400 H.P.		2072.40		2279.64		
1 Air Compressor, 365 CFM		258.40		284.24	59.31	65.24
48 L.H., Daily Totals		$4602.40		$6039.48	$95.88	$125.82
Crew B-5C	**Hr.**	**Daily**	**Hr.**	**Daily**	**Bare Costs**	**Incl. O&P**
3 Laborers	$29.20	$700.80	$48.90	$1173.60	$33.87	$56.20
1 Equip. Oper. (medium)	39.35	314.80	65.05	520.40		
2 Truck Drivers (heavy)	34.15	546.40	56.35	901.60		
1 Equip. Oper. (crane)	40.25	322.00	66.55	532.40		
1 Equip. Oper. (oiler)	35.45	283.60	58.60	468.80		
2 Dump Trucks, 12 C.Y., 400 H.P.		1381.60		1519.76		
1 Crawler Loader, 4 C.Y.		1566.00		1722.60		
1 S.P. Crane, 4x4, 25 Ton		606.60		667.26	55.53	61.09
64 L.H., Daily Totals		$5721.80		$7506.42	$89.40	$117.29

Crew No.	Bare Costs		Incl. Subs O&P		Cost Per Labor-Hour	
Crew B-5D	**Hr.**	**Daily**	**Hr.**	**Daily**	**Bare Costs**	**Incl. O&P**
1 Labor Foreman (outside)	$31.20	$249.60	$52.25	$418.00	$32.05	$53.39
3 Laborers	29.20	700.80	48.90	1173.60		
1 Equip. Oper. (medium)	39.35	314.80	65.05	520.40		
1 Truck Driver (heavy)	34.15	273.20	56.35	450.80		
1 Air Compressor, 250 cfm		197.80		217.58		
2 Breakers, Pavement, 60 lb.		20.40		22.44		
2 -50' Air Hoses, 1.5"		10.40		11.44		
1 Crawler Loader, 3 C.Y.		1197.00		1316.70		
1 Dump Truck, 12 C.Y., 400 H.P.		690.80		759.88	44.09	48.50
48 L.H., Daily Totals		$3654.80		$4890.84	$76.14	$101.89
Crew B-6	**Hr.**	**Daily**	**Hr.**	**Daily**	**Bare Costs**	**Incl. O&P**
2 Laborers	$29.20	$467.20	$48.90	$782.40	$32.08	$53.45
1 Equip. Oper. (light)	37.85	302.80	62.55	500.40		
1 Backhoe Loader, 48 H.P.		365.40		401.94	15.23	16.75
24 L.H., Daily Totals		$1135.40		$1684.74	$47.31	$70.20
Crew B-6B	**Hr.**	**Daily**	**Hr.**	**Daily**	**Bare Costs**	**Incl. O&P**
2 Labor Foremen (outside)	$31.20	$499.20	$52.25	$836.00	$29.87	$50.02
4 Laborers	29.20	934.40	48.90	1564.80		
1 S.P. Crane, 4x4, 5 Ton		280.80		308.88		
1 Flatbed Truck, Gas, 1.5 Ton		248.20		273.02		
1 Butt Fusion Mach., 4"-12" diam.		376.10		413.71	18.86	20.74
48 L.H., Daily Totals		$2338.70		$3396.41	$48.72	$70.76
Crew B-6C	**Hr.**	**Daily**	**Hr.**	**Daily**	**Bare Costs**	**Incl. O&P**
2 Labor Foremen (outside)	$31.20	$499.20	$52.25	$836.00	$29.87	$50.02
4 Laborers	29.20	934.40	48.90	1564.80		
1 S.P. Crane, 4x4, 12 Ton		479.20		527.12		
1 Flatbed Truck, Gas, 3 Ton		307.40		338.14		
1 Butt Fusion Mach., 8"-24" diam.		804.80		885.28	33.15	36.47
48 L.H., Daily Totals		$3025.00		$4151.34	$63.02	$86.49
Crew B-7	**Hr.**	**Daily**	**Hr.**	**Daily**	**Bare Costs**	**Incl. O&P**
1 Labor Foreman (outside)	$31.20	$249.60	$52.25	$418.00	$31.23	$52.15
4 Laborers	29.20	934.40	48.90	1564.80		
1 Equip. Oper. (medium)	39.35	314.80	65.05	520.40		
1 Brush Chipper, 12", 130 H.P.		388.60		427.46		
1 Crawler Loader, 3 C.Y.		1197.00		1316.70		
2 Chain Saws, Gas, 36" Long		90.00		99.00	34.91	38.40
48 L.H., Daily Totals		$3174.40		$4346.36	$66.13	$90.55
Crew B-7A	**Hr.**	**Daily**	**Hr.**	**Daily**	**Bare Costs**	**Incl. O&P**
2 Laborers	$29.20	$467.20	$48.90	$782.40	$32.08	$53.45
1 Equip. Oper. (light)	37.85	302.80	62.55	500.40		
1 Rake w/Tractor		353.05		388.36		
2 Chain Saws, Gas, 18"		63.20		69.52	17.34	19.08
24 L.H., Daily Totals		$1186.25		$1740.68	$49.43	$72.53
Crew B-7B	**Hr.**	**Daily**	**Hr.**	**Daily**	**Bare Costs**	**Incl. O&P**
1 Labor Foreman (outside)	$31.20	$249.60	$52.25	$418.00	$31.64	$52.75
4 Laborers	29.20	934.40	48.90	1564.80		
1 Equip. Oper. (medium)	39.35	314.80	65.05	520.40		
1 Truck Driver (heavy)	34.15	273.20	56.35	450.80		
1 Brush Chipper, 12", 130 H.P.		388.60		427.46		
1 Crawler Loader, 3 C.Y.		1197.00		1316.70		
2 Chain Saws, Gas, 36" Long		90.00		99.00		
1 Dump Truck, 8 C.Y., 220 H.P.		417.40		459.14	37.38	41.11
56 L.H., Daily Totals		$3865.00		$5256.30	$69.02	$93.86

Crews - Open Shop

Crew No.	Bare Costs		Incl. Subs O&P		Cost Per Labor-Hour	
Crew B-7C	**Hr.**	**Daily**	**Hr.**	**Daily**	**Bare Costs**	**Incl. O&P**
1 Labor Foreman (outside)	$31.20	$249.60	$52.25	$418.00	$31.64	$52.75
4 Laborers	29.20	934.40	48.90	1564.80		
1 Equip. Oper. (medium)	39.35	314.80	65.05	520.40		
1 Truck Driver (heavy)	34.15	273.20	56.35	450.80		
1 Brush Chipper, 12", 130 H.P.		388.60		427.46		
1 Crawler Loader, 3 C.Y.		1197.00		1316.70		
2 Chain Saws, Gas, 36" Long		90.00		99.00		
1 Dump Truck, 12 C.Y., 400 H.P.		690.80		759.88	42.26	46.48
56 L.H., Daily Totals		$4138.40		$5557.04	$73.90	$99.23

Crew B-8	**Hr.**	**Daily**	**Hr.**	**Daily**	**Bare Costs**	**Incl. O&P**
1 Labor Foreman (outside)	$31.20	$249.60	$52.25	$418.00	$33.80	$56.12
2 Laborers	29.20	467.20	48.90	782.40		
2 Equip. Opers. (medium)	39.35	629.60	65.05	1040.80		
2 Truck Drivers (heavy)	34.15	546.40	56.35	901.60		
1 Hyd. Crane, 25 Ton		748.80		823.68		
1 Crawler Loader, 3 C.Y.		1197.00		1316.70		
2 Dump Trucks, 12 C.Y., 400 H.P.		1381.60		1519.76	59.42	65.36
56 L.H., Daily Totals		$5220.20		$6802.94	$93.22	$121.48

Crew B-9	**Hr.**	**Daily**	**Hr.**	**Daily**	**Bare Costs**	**Incl. O&P**
1 Labor Foreman (outside)	$31.20	$249.60	$52.25	$418.00	$29.60	$49.57
4 Laborers	29.20	934.40	48.90	1564.80		
1 Air Compressor, 250 cfm		197.80		217.58		
2 Breakers, Pavement, 60 lb.		20.40		22.44		
2 -50' Air Hoses, 1.5"		10.40		11.44	5.71	6.29
40 L.H., Daily Totals		$1412.60		$2234.26	$35.31	$55.86

Crew B-9A	**Hr.**	**Daily**	**Hr.**	**Daily**	**Bare Costs**	**Incl. O&P**
2 Laborers	$29.20	$467.20	$48.90	$782.40	$30.85	$51.38
1 Truck Driver (heavy)	34.15	273.20	56.35	450.80		
1 Water Tank Trailer, 5000 Gal.		141.60		155.76		
1 Truck Tractor, 220 H.P.		358.80		394.68		
2 -50' Discharge Hoses, 3"		3.00		3.30	20.98	23.07
24 L.H., Daily Totals		$1243.80		$1786.94	$51.83	$74.46

Crew B-9B	**Hr.**	**Daily**	**Hr.**	**Daily**	**Bare Costs**	**Incl. O&P**
2 Laborers	$29.20	$467.20	$48.90	$782.40	$30.85	$51.38
1 Truck Driver (heavy)	34.15	273.20	56.35	450.80		
2 -50' Discharge Hoses, 3"		3.00		3.30		
1 Water Tank Trailer, 5000 Gal.		141.60		155.76		
1 Truck Tractor, 220 H.P.		358.80		394.68		
1 Pressure Washer		71.60		78.76	23.96	26.35
24 L.H., Daily Totals		$1315.40		$1865.70	$54.81	$77.74

Crew B-9D	**Hr.**	**Daily**	**Hr.**	**Daily**	**Bare Costs**	**Incl. O&P**
1 Labor Foreman (outside)	$31.20	$249.60	$52.25	$418.00	$29.60	$49.57
4 Common Laborers	29.20	934.40	48.90	1564.80		
1 Air Compressor, 250 cfm		197.80		217.58		
2 -50' Air Hoses, 1.5"		10.40		11.44		
2 Air Powered Tampers		48.00		52.80	6.41	7.05
40 L.H., Daily Totals		$1440.20		$2264.62	$36.01	$56.62

Crew B-10	**Hr.**	**Daily**	**Hr.**	**Daily**	**Bare Costs**	**Incl. O&P**
1 Equip. Oper. (medium)	39.35	314.80	65.05	520.40	39.35	65.05
8 L.H., Daily Totals		$314.80		$520.40	$39.35	$65.05

Crew B-10A	**Hr.**	**Daily**	**Hr.**	**Daily**	**Bare Costs**	**Incl. O&P**
1 Equip. Oper. (medium)	$39.35	$314.80	$65.05	$520.40	$39.35	$65.05
1 Roller, 2-Drum, W.B., 7.5 H.P.		184.80		203.28	23.10	25.41
8 L.H., Daily Totals		$499.60		$723.68	$62.45	$90.46

Crew No.	Bare Costs		Incl. Subs O&P		Cost Per Labor-Hour	
Crew B-10B	**Hr.**	**Daily**	**Hr.**	**Daily**	**Bare Costs**	**Incl. O&P**
1 Equip. Oper. (medium)	$39.35	$314.80	$65.05	$520.40	$39.35	$65.05
1 Dozer, 200 H.P.		1397.00		1536.70	174.63	192.09
8 L.H., Daily Totals		$1711.80		$2057.10	$213.97	$257.14

Crew B-10C	**Hr.**	**Daily**	**Hr.**	**Daily**	**Bare Costs**	**Incl. O&P**
1 Equip. Oper. (medium)	$39.35	$314.80	$65.05	$520.40	$39.35	$65.05
1 Dozer, 200 H.P.		1397.00		1536.70		
1 Vibratory Roller, Towed, 23 Ton		401.60		441.76	224.82	247.31
8 L.H., Daily Totals		$2113.40		$2498.86	$264.18	$312.36

Crew B-10D	**Hr.**	**Daily**	**Hr.**	**Daily**	**Bare Costs**	**Incl. O&P**
1 Equip. Oper. (medium)	$39.35	$314.80	$65.05	$520.40	$39.35	$65.05
1 Dozer, 200 H.P.		1397.00		1536.70		
1 Sheepsft. Roller, Towed		426.40		469.04	227.93	250.72
8 L.H., Daily Totals		$2138.20		$2526.14	$267.27	$315.77

Crew B-10E	**Hr.**	**Daily**	**Hr.**	**Daily**	**Bare Costs**	**Incl. O&P**
1 Equip. Oper. (medium)	$39.35	$314.80	$65.05	$520.40	$39.35	$65.05
1 Tandem Roller, 5 Ton		162.60		178.86	20.32	22.36
8 L.H., Daily Totals		$477.40		$699.26	$59.67	$87.41

Crew B-10F	**Hr.**	**Daily**	**Hr.**	**Daily**	**Bare Costs**	**Incl. O&P**
1 Equip. Oper. (medium)	$39.35	$314.80	$65.05	$520.40	$39.35	$65.05
1 Tandem Roller, 10 Ton		239.40		263.34	29.93	32.92
8 L.H., Daily Totals		$554.20		$783.74	$69.28	$97.97

Crew B-10G	**Hr.**	**Daily**	**Hr.**	**Daily**	**Bare Costs**	**Incl. O&P**
1 Equip. Oper. (medium)	$39.35	$314.80	$65.05	$520.40	$39.35	$65.05
1 Sheepsfoot Roller, 240 H.P.		1208.00		1328.80	151.00	166.10
8 L.H., Daily Totals		$1522.80		$1849.20	$190.35	$231.15

Crew B-10H	**Hr.**	**Daily**	**Hr.**	**Daily**	**Bare Costs**	**Incl. O&P**
1 Equip. Oper. (medium)	$39.35	$314.80	$65.05	$520.40	$39.35	$65.05
1 Diaphragm Water Pump, 2"		71.80		78.98		
1 -20' Suction Hose, 2"		1.95		2.15		
2 -50' Discharge Hoses, 2"		1.80		1.98	9.44	10.39
8 L.H., Daily Totals		$390.35		$603.51	$48.79	$75.44

Crew B-10I	**Hr.**	**Daily**	**Hr.**	**Daily**	**Bare Costs**	**Incl. O&P**
1 Equip. Oper. (medium)	$39.35	$314.80	$65.05	$520.40	$39.35	$65.05
1 Diaphragm Water Pump, 4"		118.00		129.80		
1 -20' Suction Hose, 4"		3.25		3.58		
2 -50' Discharge Hoses, 4"		4.70		5.17	15.74	17.32
8 L.H., Daily Totals		$440.75		$658.95	$55.09	$82.37

Crew B-10J	**Hr.**	**Daily**	**Hr.**	**Daily**	**Bare Costs**	**Incl. O&P**
1 Equip. Oper. (medium)	$39.35	$314.80	$65.05	$520.40	$39.35	$65.05
1 Centrifugal Water Pump, 3"		81.00		89.10		
1 -20' Suction Hose, 3"		2.85		3.13		
2 -50' Discharge Hoses, 3"		3.00		3.30	10.86	11.94
8 L.H., Daily Totals		$401.65		$615.93	$50.21	$76.99

Crew B-10K	**Hr.**	**Daily**	**Hr.**	**Daily**	**Bare Costs**	**Incl. O&P**
1 Equip. Oper. (medium)	$39.35	$314.80	$65.05	$520.40	$39.35	$65.05
1 Centr. Water Pump, 6"		350.80		385.88		
1 -20' Suction Hose, 6"		11.50		12.65		
2 -50' Discharge Hoses, 6"		12.20		13.42	46.81	51.49
8 L.H., Daily Totals		$689.30		$932.35	$86.16	$116.54

Crew B-10L	Bare Costs Hr.	Bare Costs Daily	Incl. Subs O&P Hr.	Incl. Subs O&P Daily	Cost Per Labor-Hour Bare Costs	Cost Per Labor-Hour Incl. O&P
1 Equip. Oper. (medium)	$39.35	$314.80	$65.05	$520.40	$39.35	$65.05
1 Dozer, 80 H.P.		471.40		518.54	58.92	64.82
8 L.H., Daily Totals		$786.20		$1038.94	$98.28	$129.87

Crew B-10M	Hr.	Daily	Hr.	Daily	Bare Costs	Incl. O&P
1 Equip. Oper. (medium)	$39.35	$314.80	$65.05	$520.40	$39.35	$65.05
1 Dozer, 300 H.P.		1887.00		2075.70	235.88	259.46
8 L.H., Daily Totals		$2201.80		$2596.10	$275.23	$324.51

Crew B-10N	Hr.	Daily	Hr.	Daily	Bare Costs	Incl. O&P
1 Equip. Oper. (medium)	$39.35	$314.80	$65.05	$520.40	$39.35	$65.05
1 F.E. Loader, T.M., 1.5 C.Y		525.00		577.50	65.63	72.19
8 L.H., Daily Totals		$839.80		$1097.90	$104.97	$137.24

Crew B-10O	Hr.	Daily	Hr.	Daily	Bare Costs	Incl. O&P
1 Equip. Oper. (medium)	$39.35	$314.80	$65.05	$520.40	$39.35	$65.05
1 F.E. Loader, T.M., 2.25 C.Y.		964.00		1060.40	120.50	132.55
8 L.H., Daily Totals		$1278.80		$1580.80	$159.85	$197.60

Crew B-10P	Hr.	Daily	Hr.	Daily	Bare Costs	Incl. O&P
1 Equip. Oper. (medium)	$39.35	$314.80	$65.05	$520.40	$39.35	$65.05
1 Crawler Loader, 3 C.Y.		1197.00		1316.70	149.63	164.59
8 L.H., Daily Totals		$1511.80		$1837.10	$188.97	$229.64

Crew B-10Q	Hr.	Daily	Hr.	Daily	Bare Costs	Incl. O&P
1 Equip. Oper. (medium)	$39.35	$314.80	$65.05	$520.40	$39.35	$65.05
1 Crawler Loader, 4 C.Y.		1566.00		1722.60	195.75	215.32
8 L.H., Daily Totals		$1880.80		$2243.00	$235.10	$280.38

Crew B-10R	Hr.	Daily	Hr.	Daily	Bare Costs	Incl. O&P
1 Equip. Oper. (medium)	$39.35	$314.80	$65.05	$520.40	$39.35	$65.05
1 F.E. Loader, W.M., 1 C.Y.		299.00		328.90	37.38	41.11
8 L.H., Daily Totals		$613.80		$849.30	$76.72	$106.16

Crew B-10S	Hr.	Daily	Hr.	Daily	Bare Costs	Incl. O&P
1 Equip. Oper. (medium)	$39.35	$314.80	$65.05	$520.40	$39.35	$65.05
1 F.E. Loader, W.M., 1.5 C.Y.		378.60		416.46	47.33	52.06
8 L.H., Daily Totals		$693.40		$936.86	$86.67	$117.11

Crew B-10T	Hr.	Daily	Hr.	Daily	Bare Costs	Incl. O&P
1 Equip. Oper. (medium)	$39.35	$314.80	$65.05	$520.40	$39.35	$65.05
1 F.E. Loader, W.M., 2.5 C.Y.		516.00		567.60	64.50	70.95
8 L.H., Daily Totals		$830.80		$1088.00	$103.85	$136.00

Crew B-10U	Hr.	Daily	Hr.	Daily	Bare Costs	Incl. O&P
1 Equip. Oper. (medium)	$39.35	$314.80	$65.05	$520.40	$39.35	$65.05
1 F.E. Loader, W.M., 5.5 C.Y.		1092.00		1201.20	136.50	150.15
8 L.H., Daily Totals		$1406.80		$1721.60	$175.85	$215.20

Crew B-10V	Hr.	Daily	Hr.	Daily	Bare Costs	Incl. O&P
1 Equip. Oper. (medium)	$39.35	$314.80	$65.05	$520.40	$39.35	$65.05
1 Dozer, 700 H.P.		5347.00		5881.70	668.38	735.21
8 L.H., Daily Totals		$5661.80		$6402.10	$707.73	$800.26

Crew B-10W	Hr.	Daily	Hr.	Daily	Bare Costs	Incl. O&P
1 Equip. Oper. (medium)	$39.35	$314.80	$65.05	$520.40	$39.35	$65.05
1 Dozer, 105 H.P.		603.40		663.74	75.42	82.97
8 L.H., Daily Totals		$918.20		$1184.14	$114.78	$148.02

Crew B-10X	Hr.	Daily	Hr.	Daily	Bare Costs	Incl. O&P
1 Equip. Oper. (medium)	$39.35	$314.80	$65.05	$520.40	$39.35	$65.05
1 Dozer, 410 H.P.		2392.00		2631.20	299.00	328.90
8 L.H., Daily Totals		$2706.80		$3151.60	$338.35	$393.95

Crew B-10Y	Hr.	Daily	Hr.	Daily	Bare Costs	Incl. O&P
1 Equip. Oper. (medium)	$39.35	$314.80	$65.05	$520.40	$39.35	$65.05
1 Vibr. Roller, Towed, 12 Ton		566.00		622.60	70.75	77.83
8 L.H., Daily Totals		$880.80		$1143.00	$110.10	$142.88

Crew B-11A	Hr.	Daily	Hr.	Daily	Bare Costs	Incl. O&P
1 Equipment Oper. (med.)	$39.35	$314.80	$65.05	$520.40	$34.27	$56.98
1 Laborer	29.20	233.60	48.90	391.20		
1 Dozer, 200 H.P.		1397.00		1536.70	87.31	96.04
16 L.H., Daily Totals		$1945.40		$2448.30	$121.59	$153.02

Crew B-11B	Hr.	Daily	Hr.	Daily	Bare Costs	Incl. O&P
1 Equipment Oper. (light)	$37.85	$302.80	$62.55	$500.40	$33.52	$55.73
1 Laborer	29.20	233.60	48.90	391.20		
1 Air Powered Tamper		24.00		26.40		
1 Air Compressor, 365 cfm		258.40		284.24		
2 -50' Air Hoses, 1.5"		10.40		11.44	18.30	20.13
16 L.H., Daily Totals		$829.20		$1213.68	$51.83	$75.86

Crew B-11C	Hr.	Daily	Hr.	Daily	Bare Costs	Incl. O&P
1 Equipment Oper. (med.)	$39.35	$314.80	$65.05	$520.40	$34.27	$56.98
1 Laborer	29.20	233.60	48.90	391.20		
1 Backhoe Loader, 48 H.P.		365.40		401.94	22.84	25.12
16 L.H., Daily Totals		$913.80		$1313.54	$57.11	$82.10

Crew B-11K	Hr.	Daily	Hr.	Daily	Bare Costs	Incl. O&P
1 Equipment Oper. (med.)	$39.35	$314.80	$65.05	$520.40	$34.27	$56.98
1 Laborer	29.20	233.60	48.90	391.20		
1 Trencher, Chain Type, 8' D		2860.00		3146.00	178.75	196.63
16 L.H., Daily Totals		$3408.40		$4057.60	$213.03	$253.60

Crew B-11L	Hr.	Daily	Hr.	Daily	Bare Costs	Incl. O&P
1 Equipment Oper. (med.)	$39.35	$314.80	$65.05	$520.40	$34.27	$56.98
1 Laborer	29.20	233.60	48.90	391.20		
1 Grader, 30,000 Lbs.		714.40		785.84	44.65	49.12
16 L.H., Daily Totals		$1262.80		$1697.44	$78.92	$106.09

Crew B-11M	Hr.	Daily	Hr.	Daily	Bare Costs	Incl. O&P
1 Equipment Oper. (med.)	$39.35	$314.80	$65.05	$520.40	$34.27	$56.98
1 Laborer	29.20	233.60	48.90	391.20		
1 Backhoe Loader, 80 H.P.		396.20		435.82	24.76	27.24
16 L.H., Daily Totals		$944.60		$1347.42	$59.04	$84.21

Crew B-11W	Hr.	Daily	Hr.	Daily	Bare Costs	Incl. O&P
1 Equipment Operator (med.)	$39.35	$314.80	$65.05	$520.40	$34.17	$56.45
1 Common Laborer	29.20	233.60	48.90	391.20		
10 Truck Drivers (heavy)	34.15	2732.00	56.35	4508.00		
1 Dozer, 200 H.P.		1397.00		1536.70		
1 Vibratory Roller, Towed, 23 Ton		401.60		441.76		
10 Dump Trucks, 8 C.Y., 220 H.P.		4174.00		4591.40	62.21	68.44
96 L.H., Daily Totals		$9253.00		$11989.46	$96.39	$124.89

Crew No.	Bare Costs		Incl. Subs O&P		Cost Per Labor-Hour	

Crew B-11Y	Hr.	Daily	Hr.	Daily	Bare Costs	Incl. O&P
1 Labor Foreman (outside)	$31.20	$249.60	$52.25	$418.00	$32.81	$54.66
5 Common Laborers	29.20	1168.00	48.90	1956.00		
3 Equipment Operators (med.)	39.35	944.40	65.05	1561.20		
1 Dozer, 80 H.P.		471.40		518.54		
2 Rollers, 2-Drum, W.B., 7.5 H.P.		369.60		406.56		
4 Vibrating Plates, Gas, 21"		186.40		205.04	14.27	15.70
72 L.H., Daily Totals		$3389.40		$5065.34	$47.08	$70.35

Crew B-12A	Hr.	Daily	Hr.	Daily	Bare Costs	Incl. O&P
1 Equip. Oper. (crane)	$40.25	$322.00	$66.55	$532.40	$34.73	$57.73
1 Laborer	29.20	233.60	48.90	391.20		
1 Hyd. Excavator, 1 C.Y.		812.80		894.08	50.80	55.88
16 L.H., Daily Totals		$1368.40		$1817.68	$85.53	$113.61

Crew B-12B	Hr.	Daily	Hr.	Daily	Bare Costs	Incl. O&P
1 Equip. Oper. (crane)	$40.25	$322.00	$66.55	$532.40	$34.73	$57.73
1 Laborer	29.20	233.60	48.90	391.20		
1 Hyd. Excavator, 1.5 C.Y.		1029.00		1131.90	64.31	70.74
16 L.H., Daily Totals		$1584.60		$2055.50	$99.04	$128.47

Crew B-12C	Hr.	Daily	Hr.	Daily	Bare Costs	Incl. O&P
1 Equip. Oper. (crane)	$40.25	$322.00	$66.55	$532.40	$34.73	$57.73
1 Laborer	29.20	233.60	48.90	391.20		
1 Hyd. Excavator, 2 C.Y.		1176.00		1293.60	73.50	80.85
16 L.H., Daily Totals		$1731.60		$2217.20	$108.22	$138.57

Crew B-12D	Hr.	Daily	Hr.	Daily	Bare Costs	Incl. O&P
1 Equip. Oper. (crane)	$40.25	$322.00	$66.55	$532.40	$34.73	$57.73
1 Laborer	29.20	233.60	48.90	391.20		
1 Hyd. Excavator, 3.5 C.Y.		2442.00		2686.20	152.63	167.89
16 L.H., Daily Totals		$2997.60		$3609.80	$187.35	$225.61

Crew B-12E	Hr.	Daily	Hr.	Daily	Bare Costs	Incl. O&P
1 Equip. Oper. (crane)	$40.25	$322.00	$66.55	$532.40	$34.73	$57.73
1 Laborer	29.20	233.60	48.90	391.20		
1 Hyd. Excavator, .5 C.Y.		445.60		490.16	27.85	30.64
16 L.H., Daily Totals		$1001.20		$1413.76	$62.58	$88.36

Crew B-12F	Hr.	Daily	Hr.	Daily	Bare Costs	Incl. O&P
1 Equip. Oper. (crane)	$40.25	$322.00	$66.55	$532.40	$34.73	$57.73
1 Laborer	29.20	233.60	48.90	391.20		
1 Hyd. Excavator, .75 C.Y.		654.00		719.40	40.88	44.96
16 L.H., Daily Totals		$1209.60		$1643.00	$75.60	$102.69

Crew B-12G	Hr.	Daily	Hr.	Daily	Bare Costs	Incl. O&P
1 Equip. Oper. (crane)	$40.25	$322.00	$66.55	$532.40	$34.73	$57.73
1 Laborer	29.20	233.60	48.90	391.20		
1 Crawler Crane, 15 Ton		695.70		765.27		
1 Clamshell Bucket, .5 C.Y.		37.80		41.58	45.84	50.43
16 L.H., Daily Totals		$1289.10		$1730.45	$80.57	$108.15

Crew B-12H	Hr.	Daily	Hr.	Daily	Bare Costs	Incl. O&P
1 Equip. Oper. (crane)	$40.25	$322.00	$66.55	$532.40	$34.73	$57.73
1 Laborer	29.20	233.60	48.90	391.20		
1 Crawler Crane, 25 Ton		1194.00		1313.40		
1 Clamshell Bucket, 1 C.Y.		47.40		52.14	77.59	85.35
16 L.H., Daily Totals		$1797.00		$2289.14	$112.31	$143.07

Crew B-12I	Hr.	Daily	Hr.	Daily	Bare Costs	Incl. O&P
1 Equip. Oper. (crane)	$40.25	$322.00	$66.55	$532.40	$34.73	$57.73
1 Laborer	29.20	233.60	48.90	391.20		
1 Crawler Crane, 20 Ton		896.25		985.88		
1 Dragline Bucket, .75 C.Y.		20.60		22.66	57.30	63.03
16 L.H., Daily Totals		$1472.45		$1932.14	$92.03	$120.76

Crew B-12J	Hr.	Daily	Hr.	Daily	Bare Costs	Incl. O&P
1 Equip. Oper. (crane)	$40.25	$322.00	$66.55	$532.40	$34.73	$57.73
1 Laborer	29.20	233.60	48.90	391.20		
1 Gradall, 5/8 C.Y.		882.00		970.20	55.13	60.64
16 L.H., Daily Totals		$1437.60		$1893.80	$89.85	$118.36

Crew B-12K	Hr.	Daily	Hr.	Daily	Bare Costs	Incl. O&P
1 Equip. Oper. (crane)	$40.25	$322.00	$66.55	$532.40	$34.73	$57.73
1 Laborer	29.20	233.60	48.90	391.20		
1 Gradall, 3 Ton, 1 C.Y.		995.80		1095.38	62.24	68.46
16 L.H., Daily Totals		$1551.40		$2018.98	$96.96	$126.19

Crew B-12L	Hr.	Daily	Hr.	Daily	Bare Costs	Incl. O&P
1 Equip. Oper. (crane)	$40.25	$322.00	$66.55	$532.40	$34.73	$57.73
1 Laborer	29.20	233.60	48.90	391.20		
1 Crawler Crane, 15 Ton		695.70		765.27		
1 F.E. Attachment, .5 C.Y.		59.20		65.12	47.18	51.90
16 L.H., Daily Totals		$1310.50		$1753.99	$81.91	$109.62

Crew B-12M	Hr.	Daily	Hr.	Daily	Bare Costs	Incl. O&P
1 Equip. Oper. (crane)	$40.25	$322.00	$66.55	$532.40	$34.73	$57.73
1 Laborer	29.20	233.60	48.90	391.20		
1 Crawler Crane, 20 Ton		896.25		985.88		
1 F.E. Attachment, .75 C.Y.		63.60		69.96	59.99	65.99
16 L.H., Daily Totals		$1515.45		$1979.43	$94.72	$123.71

Crew B-12N	Hr.	Daily	Hr.	Daily	Bare Costs	Incl. O&P
1 Equip. Oper. (crane)	$40.25	$322.00	$66.55	$532.40	$34.73	$57.73
1 Laborer	29.20	233.60	48.90	391.20		
1 Crawler Crane, 25 Ton		1194.00		1313.40		
1 F.E. Attachment, 1 C.Y.		70.40		77.44	79.03	86.93
16 L.H., Daily Totals		$1820.00		$2314.44	$113.75	$144.65

Crew B-12O	Hr.	Daily	Hr.	Daily	Bare Costs	Incl. O&P
1 Equip. Oper. (crane)	$40.25	$322.00	$66.55	$532.40	$34.73	$57.73
1 Laborer	29.20	233.60	48.90	391.20		
1 Crawler Crane, 40 Ton		1203.00		1323.30		
1 F.E. Attachment, 1.5 C.Y.		79.60		87.56	80.16	88.18
16 L.H., Daily Totals		$1838.20		$2334.46	$114.89	$145.90

Crew B-12P	Hr.	Daily	Hr.	Daily	Bare Costs	Incl. O&P
1 Equip. Oper. (crane)	$40.25	$322.00	$66.55	$532.40	$34.73	$57.73
1 Laborer	29.20	233.60	48.90	391.20		
1 Crawler Crane, 40 Ton		1203.00		1323.30		
1 Dragline Bucket, 1.5 C.Y.		33.60		36.96	77.29	85.02
16 L.H., Daily Totals		$1792.20		$2283.86	$112.01	$142.74

Crew B-12Q	Hr.	Daily	Hr.	Daily	Bare Costs	Incl. O&P
1 Equip. Oper. (crane)	$40.25	$322.00	$66.55	$532.40	$34.73	$57.73
1 Laborer	29.20	233.60	48.90	391.20		
1 Hyd. Excavator, 5/8 C.Y.		589.40		648.34	36.84	40.52
16 L.H., Daily Totals		$1145.00		$1571.94	$71.56	$98.25

Crews - Open Shop

Crew B-12S	Bare Costs Hr.	Bare Costs Daily	Incl. Subs O&P Hr.	Incl. Subs O&P Daily	Cost Per Labor-Hour Bare Costs	Cost Per Labor-Hour Incl. O&P
1 Equip. Oper. (crane)	$40.25	$322.00	$66.55	$532.40	$34.73	$57.73
1 Laborer	29.20	233.60	48.90	391.20		
1 Hyd. Excavator, 2.5 C.Y.		1582.00		1740.20	98.88	108.76
16 L.H., Daily Totals		$2137.60		$2663.80	$133.60	$166.49

Crew B-12T	Hr.	Daily	Hr.	Daily	Bare Costs	Incl. O&P
1 Equip. Oper. (crane)	$40.25	$322.00	$66.55	$532.40	$34.73	$57.73
1 Laborer	29.20	233.60	48.90	391.20		
1 Crawler Crane, 75 Ton		1507.00		1657.70		
1 F.E. Attachment, 3 C.Y.		102.40		112.64	100.59	110.65
16 L.H., Daily Totals		$2165.00		$2693.94	$135.31	$168.37

Crew B-12V	Hr.	Daily	Hr.	Daily	Bare Costs	Incl. O&P
1 Equip. Oper. (crane)	$40.25	$322.00	$66.55	$532.40	$34.73	$57.73
1 Laborer	29.20	233.60	48.90	391.20		
1 Crawler Crane, 75 Ton		1507.00		1657.70		
1 Dragline Bucket, 3 C.Y.		52.60		57.86	97.47	107.22
16 L.H., Daily Totals		$2115.20		$2639.16	$132.20	$164.95

Crew B-12Y	Hr.	Daily	Hr.	Daily	Bare Costs	Incl. O&P
1 Equip. Oper. (crane)	$40.25	$322.00	$66.55	$532.40	$32.88	$54.78
2 Laborers	29.20	467.20	48.90	782.40		
1 Hyd. Excavator, 3.5 C.Y.		2442.00		2686.20	101.75	111.93
24 L.H., Daily Totals		$3231.20		$4001.00	$134.63	$166.71

Crew B-12Z	Hr.	Daily	Hr.	Daily	Bare Costs	Incl. O&P
1 Equip. Oper. (crane)	$40.25	$322.00	$66.55	$532.40	$32.88	$54.78
2 Laborers	29.20	467.20	48.90	782.40		
1 Hyd. Excavator, 2.5 C.Y.		1582.00		1740.20	65.92	72.51
24 L.H., Daily Totals		$2371.20		$3055.00	$98.80	$127.29

Crew B-13	Hr.	Daily	Hr.	Daily	Bare Costs	Incl. O&P
1 Labor Foreman (outside)	$31.20	$249.60	$52.25	$418.00	$31.38	$52.40
4 Laborers	29.20	934.40	48.90	1564.80		
1 Equip. Oper. (crane)	40.25	322.00	66.55	532.40		
1 Hyd. Crane, 25 Ton		748.80		823.68	15.60	17.16
48 L.H., Daily Totals		$2254.80		$3338.88	$46.98	$69.56

Crew B-13A	Hr.	Daily	Hr.	Daily	Bare Costs	Incl. O&P
1 Labor Foreman (outside)	$31.20	$249.60	$52.25	$418.00	$33.80	$56.12
2 Laborers	29.20	467.20	48.90	782.40		
2 Equipment Operators (med.)	39.35	629.60	65.05	1040.80		
2 Truck Drivers (heavy)	34.15	546.40	56.35	901.60		
1 Crawler Crane, 75 Ton		1507.00		1657.70		
1 Crawler Loader, 4 C.Y.		1566.00		1722.60		
2 Dump Trucks, 8 C.Y., 220 H.P.		834.80		918.28	69.78	76.76
56 L.H., Daily Totals		$5800.60		$7441.38	$103.58	$132.88

Crew B-13B	Hr.	Daily	Hr.	Daily	Bare Costs	Incl. O&P
1 Labor Foreman (outside)	$31.20	$249.60	$52.25	$418.00	$31.96	$53.29
4 Laborers	29.20	934.40	48.90	1564.80		
1 Equip. Oper. (crane)	40.25	322.00	66.55	532.40		
1 Equip. Oper. (oiler)	35.45	283.60	58.60	468.80		
1 Hyd. Crane, 55 Ton		1123.00		1235.30	20.05	22.06
56 L.H., Daily Totals		$2912.60		$4219.30	$52.01	$75.34

Crew B-13C	Hr.	Daily	Hr.	Daily	Bare Costs	Incl. O&P
1 Labor Foreman (outside)	$31.20	$249.60	$52.25	$418.00	$31.96	$53.29
4 Laborers	29.20	934.40	48.90	1564.80		
1 Equip. Oper. (crane)	40.25	322.00	66.55	532.40		
1 Equip. Oper. (oiler)	35.45	283.60	58.60	468.80		
1 Crawler Crane, 100 Ton		1667.00		1833.70	29.77	32.74
56 L.H., Daily Totals		$3456.60		$4817.70	$61.73	$86.03

Crew B-13D	Hr.	Daily	Hr.	Daily	Bare Costs	Incl. O&P
1 Laborer	$29.20	$233.60	$48.90	$391.20	$34.73	$57.73
1 Equip. Oper. (crane)	40.25	322.00	66.55	532.40		
1 Hyd. Excavator, 1 C.Y.		812.80		894.08		
1 Trench Box		81.00		89.10	55.86	61.45
16 L.H., Daily Totals		$1449.40		$1906.78	$90.59	$119.17

Crew B-13E	Hr.	Daily	Hr.	Daily	Bare Costs	Incl. O&P
1 Laborer	$29.20	$233.60	$48.90	$391.20	$34.73	$57.73
1 Equip. Oper. (crane)	40.25	322.00	66.55	532.40		
1 Hyd. Excavator, 1.5 C.Y.		1029.00		1131.90		
1 Trench Box		81.00		89.10	69.38	76.31
16 L.H., Daily Totals		$1665.60		$2144.60	$104.10	$134.04

Crew B-13F	Hr.	Daily	Hr.	Daily	Bare Costs	Incl. O&P
1 Laborer	$29.20	$233.60	$48.90	$391.20	$34.73	$57.73
1 Equip. Oper. (crane)	40.25	322.00	66.55	532.40		
1 Hyd. Excavator, 3.5 C.Y.		2442.00		2686.20		
1 Trench Box		81.00		89.10	157.69	173.46
16 L.H., Daily Totals		$3078.60		$3698.90	$192.41	$231.18

Crew B-13G	Hr.	Daily	Hr.	Daily	Bare Costs	Incl. O&P
1 Laborer	$29.20	$233.60	$48.90	$391.20	$34.73	$57.73
1 Equip. Oper. (crane)	40.25	322.00	66.55	532.40		
1 Hyd. Excavator, .75 C.Y.		654.00		719.40		
1 Trench Box		81.00		89.10	45.94	50.53
16 L.H., Daily Totals		$1290.60		$1732.10	$80.66	$108.26

Crew B-13H	Hr.	Daily	Hr.	Daily	Bare Costs	Incl. O&P
1 Laborer	$29.20	$233.60	$48.90	$391.20	$34.73	$57.73
1 Equip. Oper. (crane)	40.25	322.00	66.55	532.40		
1 Gradall, 5/8 C.Y.		882.00		970.20		
1 Trench Box		81.00		89.10	60.19	66.21
16 L.H., Daily Totals		$1518.60		$1982.90	$94.91	$123.93

Crew B-13I	Hr.	Daily	Hr.	Daily	Bare Costs	Incl. O&P
1 Laborer	$29.20	$233.60	$48.90	$391.20	$34.73	$57.73
1 Equip. Oper. (crane)	40.25	322.00	66.55	532.40		
1 Gradall, 3 Ton, 1 C.Y.		995.80		1095.38		
1 Trench Box		81.00		89.10	67.30	74.03
16 L.H., Daily Totals		$1632.40		$2108.08	$102.03	$131.76

Crew B-13J	Hr.	Daily	Hr.	Daily	Bare Costs	Incl. O&P
1 Laborer	$29.20	$233.60	$48.90	$391.20	$34.73	$57.73
1 Equip. Oper. (crane)	40.25	322.00	66.55	532.40		
1 Hyd. Excavator, 2.5 C.Y.		1582.00		1740.20		
1 Trench Box		81.00		89.10	103.94	114.33
16 L.H., Daily Totals		$2218.60		$2752.90	$138.66	$172.06

Crew No.	Bare Costs		Incl. Subs O&P		Cost Per Labor-Hour	
Crew B-13K	**Hr.**	**Daily**	**Hr.**	**Daily**	**Bare Costs**	**Incl. O&P**
2 Equip. Opers. (crane)	$40.25	$644.00	$66.55	$1064.80	$40.25	$66.55
1 Hyd. Excavator, .75 C.Y.		654.00		719.40		
1 Hyd. Hammer, 4000 ft-lb		306.20		336.82		
1 Hyd. Excavator, .75 C.Y.		654.00		719.40	100.89	110.98
16 L.H., Daily Totals		$2258.20		$2840.42	$141.14	$177.53

Crew B-13L	Hr.	Daily	Hr.	Daily	Bare Costs	Incl. O&P
2 Equip. Opers. (crane)	$40.25	$644.00	$66.55	$1064.80	$40.25	$66.55
1 Hyd. Excavator, 1.5 C.Y.		1029.00		1131.90		
1 Hyd. Hammer, 5000 ft-lb		367.80		404.58		
1 Hyd. Excavator, .75 C.Y.		654.00		719.40	128.18	140.99
16 L.H., Daily Totals		$2694.80		$3320.68	$168.43	$207.54

Crew B-13M	Hr.	Daily	Hr.	Daily	Bare Costs	Incl. O&P
2 Equip. Opers. (crane)	$40.25	$644.00	$66.55	$1064.80	$40.25	$66.55
1 Hyd. Excavator, 2.5 C.Y.		1582.00		1740.20		
1 Hyd. Hammer, 8000 ft-lb		541.20		595.32		
1 Hyd. Excavator, 1.5 C.Y.		1029.00		1131.90	197.01	216.71
16 L.H., Daily Totals		$3796.20		$4532.22	$237.26	$283.26

Crew B-13N	Hr.	Daily	Hr.	Daily	Bare Costs	Incl. O&P
2 Equip. Opers. (crane)	$40.25	$644.00	$66.55	$1064.80	$40.25	$66.55
1 Hyd. Excavator, 3.5 C.Y.		2442.00		2686.20		
1 Hyd. Hammer, 12,000 ft-lb		632.20		695.42		
1 Hyd. Excavator, 1.5 C.Y.		1029.00		1131.90	256.45	282.10
16 L.H., Daily Totals		$4747.20		$5578.32	$296.70	$348.64

Crew B-14	Hr.	Daily	Hr.	Daily	Bare Costs	Incl. O&P
1 Labor Foreman (outside)	$31.20	$249.60	$52.25	$418.00	$30.98	$51.73
4 Laborers	29.20	934.40	48.90	1564.80		
1 Equip. Oper. (light)	37.85	302.80	62.55	500.40		
1 Backhoe Loader, 48 H.P.		365.40		401.94	7.61	8.37
48 L.H., Daily Totals		$1852.20		$2885.14	$38.59	$60.11

Crew B-14A	Hr.	Daily	Hr.	Daily	Bare Costs	Incl. O&P
1 Equip. Oper. (crane)	$40.25	$322.00	$66.55	$532.40	$36.57	$60.67
.5 Laborer	29.20	116.80	48.90	195.60		
1 Hyd. Excavator, 4.5 C.Y.		2974.00		3271.40	247.83	272.62
12 L.H., Daily Totals		$3412.80		$3999.40	$284.40	$333.28

Crew B-14B	Hr.	Daily	Hr.	Daily	Bare Costs	Incl. O&P
1 Equip. Oper. (crane)	$40.25	$322.00	$66.55	$532.40	$36.57	$60.67
.5 Laborer	29.20	116.80	48.90	195.60		
1 Hyd. Excavator, 6 C.Y.		3501.00		3851.10	291.75	320.93
12 L.H., Daily Totals		$3939.80		$4579.10	$328.32	$381.59

Crew B-14C	Hr.	Daily	Hr.	Daily	Bare Costs	Incl. O&P
1 Equip. Oper. (crane)	$40.25	$322.00	$66.55	$532.40	$36.57	$60.67
.5 Laborer	29.20	116.80	48.90	195.60		
1 Hyd. Excavator, 7 C.Y.		3597.00		3956.70	299.75	329.73
12 L.H., Daily Totals		$4035.80		$4684.70	$336.32	$390.39

Crew B-14F	Hr.	Daily	Hr.	Daily	Bare Costs	Incl. O&P
1 Equip. Oper. (crane)	$40.25	$322.00	$66.55	$532.40	$36.57	$60.67
.5 Laborer	29.20	116.80	48.90	195.60		
1 Hyd. Shovel, 7 C.Y.		4212.00		4633.20	351.00	386.10
12 L.H., Daily Totals		$4650.80		$5361.20	$387.57	$446.77

Crew B-14G	Hr.	Daily	Hr.	Daily	Bare Costs	Incl. O&P
1 Equip. Oper. (crane)	$40.25	$322.00	$66.55	$532.40	$36.57	$60.67
.5 Laborer	29.20	116.80	48.90	195.60		
1 Hyd. Shovel, 12 C.Y.		6328.00		6960.80	527.33	580.07
12 L.H., Daily Totals		$6766.80		$7688.80	$563.90	$640.73

Crew B-14J	Hr.	Daily	Hr.	Daily	Bare Costs	Incl. O&P
1 Equip. Oper. (medium)	$39.35	$314.80	$65.05	$520.40	$35.97	$59.67
.5 Laborer	29.20	116.80	48.90	195.60		
1 F.E. Loader, 8 C.Y.		1964.00		2160.40	163.67	180.03
12 L.H., Daily Totals		$2395.60		$2876.40	$199.63	$239.70

Crew B-14K	Hr.	Daily	Hr.	Daily	Bare Costs	Incl. O&P
1 Equip. Oper. (medium)	$39.35	$314.80	$65.05	$520.40	$35.97	$59.67
.5 Laborer	29.20	116.80	48.90	195.60		
1 F.E. Loader, 10 C.Y.		3085.00		3393.50	257.08	282.79
12 L.H., Daily Totals		$3516.60		$4109.50	$293.05	$342.46

Crew B-15	Hr.	Daily	Hr.	Daily	Bare Costs	Incl. O&P
1 Equipment Oper. (med.)	$39.35	$314.80	$65.05	$520.40	$34.93	$57.77
.5 Laborer	29.20	116.80	48.90	195.60		
2 Truck Drivers (heavy)	34.15	546.40	56.35	901.60		
2 Dump Trucks, 12 C.Y., 400 H.P.		1381.60		1519.76		
1 Dozer, 200 H.P.		1397.00		1536.70	99.24	109.16
28 L.H., Daily Totals		$3756.60		$4674.06	$134.16	$166.93

Crew B-16	Hr.	Daily	Hr.	Daily	Bare Costs	Incl. O&P
1 Labor Foreman (outside)	$31.20	$249.60	$52.25	$418.00	$30.94	$51.60
2 Laborers	29.20	467.20	48.90	782.40		
1 Truck Driver (heavy)	34.15	273.20	56.35	450.80		
1 Dump Truck, 12 C.Y., 400 H.P.		690.80		759.88	21.59	23.75
32 L.H., Daily Totals		$1680.80		$2411.08	$52.52	$75.35

Crew B-17	Hr.	Daily	Hr.	Daily	Bare Costs	Incl. O&P
2 Laborers	$29.20	$467.20	$48.90	$782.40	$32.60	$54.17
1 Equip. Oper. (light)	37.85	302.80	62.55	500.40		
1 Truck Driver (heavy)	34.15	273.20	56.35	450.80		
1 Backhoe Loader, 48 H.P.		365.40		401.94		
1 Dump Truck, 8 C.Y., 220 H.P.		417.40		459.14	24.46	26.91
32 L.H., Daily Totals		$1826.00		$2594.68	$57.06	$81.08

Crew B-17A	Hr.	Daily	Hr.	Daily	Bare Costs	Incl. O&P
2 Labor Foremen (outside)	$31.20	$499.20	$52.25	$836.00	$31.62	$52.99
6 Laborers	29.20	1401.60	48.90	2347.20		
1 Skilled Worker Foreman (out)	40.30	322.40	67.70	541.60		
1 Skilled Worker	38.30	306.40	64.35	514.80		
80 L.H., Daily Totals		$2529.60		$4239.60	$31.62	$52.99

Crew B-17B	Hr.	Daily	Hr.	Daily	Bare Costs	Incl. O&P
2 Laborers	$29.20	$467.20	$48.90	$782.40	$32.60	$54.17
1 Equip. Oper. (light)	37.85	302.80	62.55	500.40		
1 Truck Driver (heavy)	34.15	273.20	56.35	450.80		
1 Backhoe Loader, 48 H.P.		365.40		401.94		
1 Dump Truck, 12 C.Y., 400 H.P.		690.80		759.88	33.01	36.31
32 L.H., Daily Totals		$2099.40		$2895.42	$65.61	$90.48

Crew B-18	Hr.	Daily	Hr.	Daily	Bare Costs	Incl. O&P
1 Labor Foreman (outside)	$31.20	$249.60	$52.25	$418.00	$29.87	$50.02
2 Laborers	29.20	467.20	48.90	782.40		
1 Vibrating Plate, Gas, 21"		46.60		51.26	1.94	2.14
24 L.H., Daily Totals		$763.40		$1251.66	$31.81	$52.15

Crew No.	Bare Costs Hr.	Bare Costs Daily	Incl. Subs O&P Hr.	Incl. Subs O&P Daily	Cost Per Labor-Hour Bare Costs	Cost Per Labor-Hour Incl. O&P
Crew B-19	**Hr.**	**Daily**	**Hr.**	**Daily**	**Bare Costs**	**Incl. O&P**
1 Pile Driver Foreman (outside)	$39.75	$318.00	$68.60	$548.80	$37.17	$63.52
4 Pile Drivers	37.75	1208.00	65.15	2084.80		
1 Equip. Oper. (crane)	40.25	322.00	66.55	532.40		
1 Building Laborer	29.20	233.60	48.90	391.20		
1 Crawler Crane, 40 Ton		1203.00		1323.30		
1 Lead, 90' High		124.40		136.84		
1 Hammer, Diesel, 22k ft-lb		378.00		415.80	30.45	33.50
56 L.H., Daily Totals		$3787.00		$5433.14	$67.63	$97.02
Crew B-19A	**Hr.**	**Daily**	**Hr.**	**Daily**	**Bare Costs**	**Incl. O&P**
1 Pile Driver Foreman (outside)	$39.75	$318.00	$68.60	$548.80	$37.17	$63.52
4 Pile Drivers	37.75	1208.00	65.15	2084.80		
1 Equip. Oper. (crane)	40.25	322.00	66.55	532.40		
1 Common Laborer	29.20	233.60	48.90	391.20		
1 Crawler Crane, 75 Ton		1507.00		1657.70		
1 Lead, 90' High		124.40		136.84		
1 Hammer, Diesel, 41k ft-lb		522.60		574.86	38.46	42.31
56 L.H., Daily Totals		$4235.60		$5926.60	$75.64	$105.83
Crew B-19B	**Hr.**	**Daily**	**Hr.**	**Daily**	**Bare Costs**	**Incl. O&P**
1 Pile Driver Foreman (outside)	$39.75	$318.00	$68.60	$548.80	$37.17	$63.52
4 Pile Drivers	37.75	1208.00	65.15	2084.80		
1 Equip. Oper. (crane)	40.25	322.00	66.55	532.40		
1 Common Laborer	29.20	233.60	48.90	391.20		
1 Crawler Crane, 40 Ton		1203.00		1323.30		
1 Lead, 90' High		124.40		136.84		
1 Hammer, Diesel, 22k ft-lb		378.00		415.80		
1 Barge, 400 Ton		779.20		857.12	44.37	48.80
56 L.H., Daily Totals		$4566.20		$6290.26	$81.54	$112.33
Crew B-19C	**Hr.**	**Daily**	**Hr.**	**Daily**	**Bare Costs**	**Incl. O&P**
1 Pile Driver Foreman (outside)	$39.75	$318.00	$68.60	$548.80	$37.17	$63.52
4 Pile Drivers	37.75	1208.00	65.15	2084.80		
1 Equip. Oper. (crane)	40.25	322.00	66.55	532.40		
1 Common Laborer	29.20	233.60	48.90	391.20		
1 Crawler Crane, 75 Ton		1507.00		1657.70		
1 Lead, 90' High		124.40		136.84		
1 Hammer, Diesel, 41k ft-lb		522.60		574.86		
1 Barge, 400 Ton		779.20		857.12	52.38	57.62
56 L.H., Daily Totals		$5014.80		$6783.72	$89.55	$121.14
Crew B-20	**Hr.**	**Daily**	**Hr.**	**Daily**	**Bare Costs**	**Incl. O&P**
1 Labor Foreman (outside)	$31.20	$249.60	$52.25	$418.00	$29.87	$50.02
2 Laborers	29.20	467.20	48.90	782.40		
24 L.H., Daily Totals		$716.80		$1200.40	$29.87	$50.02
Crew B-20A	**Hr.**	**Daily**	**Hr.**	**Daily**	**Bare Costs**	**Incl. O&P**
1 Labor Foreman (outside)	$31.20	$249.60	$52.25	$418.00	$34.81	$57.80
1 Laborer	29.20	233.60	48.90	391.20		
1 Plumber	43.80	350.40	72.25	578.00		
1 Plumber Apprentice	35.05	280.40	57.80	462.40		
32 L.H., Daily Totals		$1114.00		$1849.60	$34.81	$57.80
Crew B-21	**Hr.**	**Daily**	**Hr.**	**Daily**	**Bare Costs**	**Incl. O&P**
1 Labor Foreman (outside)	$31.20	$249.60	$52.25	$418.00	$31.35	$52.38
2 Laborers	29.20	467.20	48.90	782.40		
.5 Equip. Oper. (crane)	40.25	161.00	66.55	266.20		
.5 S.P. Crane, 4x4, 5 Ton		140.40		154.44	5.01	5.52
28 L.H., Daily Totals		$1018.20		$1621.04	$36.36	$57.89

Crew No.	Bare Costs Hr.	Bare Costs Daily	Incl. Subs O&P Hr.	Incl. Subs O&P Daily	Cost Per Labor-Hour Bare Costs	Cost Per Labor-Hour Incl. O&P
Crew B-21A	**Hr.**	**Daily**	**Hr.**	**Daily**	**Bare Costs**	**Incl. O&P**
1 Labor Foreman (outside)	$31.20	$249.60	$52.25	$418.00	$35.90	$59.55
1 Laborer	29.20	233.60	48.90	391.20		
1 Plumber	43.80	350.40	72.25	578.00		
1 Plumber Apprentice	35.05	280.40	57.80	462.40		
1 Equip. Oper. (crane)	40.25	322.00	66.55	532.40		
1 S.P. Crane, 4x4, 12 Ton		479.20		527.12	11.98	13.18
40 L.H., Daily Totals		$1915.20		$2909.12	$47.88	$72.73
Crew B-21B	**Hr.**	**Daily**	**Hr.**	**Daily**	**Bare Costs**	**Incl. O&P**
1 Labor Foreman (outside)	$31.20	$249.60	$52.25	$418.00	$31.81	$53.10
3 Laborers	29.20	700.80	48.90	1173.60		
1 Equip. Oper. (crane)	40.25	322.00	66.55	532.40		
1 Hyd. Crane, 12 Ton		658.40		724.24	16.46	18.11
40 L.H., Daily Totals		$1930.80		$2848.24	$48.27	$71.21
Crew B-21C	**Hr.**	**Daily**	**Hr.**	**Daily**	**Bare Costs**	**Incl. O&P**
1 Labor Foreman (outside)	$31.20	$249.60	$52.25	$418.00	$31.96	$53.29
4 Laborers	29.20	934.40	48.90	1564.80		
1 Equip. Oper. (crane)	40.25	322.00	66.55	532.40		
1 Equip. Oper. (oiler)	35.45	283.60	58.60	468.80		
2 Cutting Torches		22.80		25.08		
2 Sets of Gases		336.00		369.60		
1 Lattice Boom Crane, 90 Ton		1520.00		1672.00	33.55	36.91
56 L.H., Daily Totals		$3668.40		$5050.68	$65.51	$90.19
Crew B-22	**Hr.**	**Daily**	**Hr.**	**Daily**	**Bare Costs**	**Incl. O&P**
1 Labor Foreman (outside)	$31.20	$249.60	$52.25	$418.00	$31.94	$53.32
2 Laborers	29.20	467.20	48.90	782.40		
.75 Equip. Oper. (crane)	40.25	241.50	66.55	399.30		
.75 S.P. Crane, 4x4, 5 Ton		210.60		231.66	7.02	7.72
30 L.H., Daily Totals		$1168.90		$1831.36	$38.96	$61.05
Crew B-22A	**Hr.**	**Daily**	**Hr.**	**Daily**	**Bare Costs**	**Incl. O&P**
1 Labor Foreman (outside)	$31.20	$249.60	$52.25	$418.00	$33.63	$56.19
1 Skilled Worker	38.30	306.40	64.35	514.80		
2 Laborers	29.20	467.20	48.90	782.40		
1 Equipment Operator, Crane	40.25	322.00	66.55	532.40		
1 S.P. Crane, 4x4, 5 Ton		280.80		308.88		
1 Butt Fusion Mach., 4"-12" diam.		376.10		413.71	16.42	18.06
40 L.H., Daily Totals		$2002.10		$2970.19	$50.05	$74.25
Crew B-22B	**Hr.**	**Daily**	**Hr.**	**Daily**	**Bare Costs**	**Incl. O&P**
1 Labor Foreman (outside)	$31.20	$249.60	$52.25	$418.00	$33.63	$56.19
1 Skilled Worker	38.30	306.40	64.35	514.80		
2 Laborers	29.20	467.20	48.90	782.40		
1 Equip. Oper. (crane)	40.25	322.00	66.55	532.40		
1 S.P. Crane, 4x4, 5 Ton		280.80		308.88		
1 Butt Fusion Mach., 8"-24" diam.		804.80		885.28	27.14	29.85
40 L.H., Daily Totals		$2430.80		$3441.76	$60.77	$86.04
Crew B-22C	**Hr.**	**Daily**	**Hr.**	**Daily**	**Bare Costs**	**Incl. O&P**
1 Skilled Worker	$38.30	$306.40	$64.35	$514.80	$33.75	$56.63
1 Laborer	29.20	233.60	48.90	391.20		
1 Butt Fusion Mach., 2"-8" diam.		121.05		133.16	7.57	8.32
16 L.H., Daily Totals		$661.05		$1039.16	$41.32	$64.95

Crew No.	Bare Costs		Incl. Subs O&P		Cost Per Labor-Hour	
Crew B-23	**Hr.**	**Daily**	**Hr.**	**Daily**	**Bare Costs**	**Incl. O&P**
1 Labor Foreman (outside)	$31.20	$249.60	$52.25	$418.00	$29.60	$49.57
4 Laborers	29.20	934.40	48.90	1564.80		
1 Drill Rig, Truck-Mounted		2553.00		2808.30		
1 Flatbed Truck, Gas, 3 Ton		307.40		338.14	71.51	78.66
40 L.H., Daily Totals		$4044.40		$5129.24	$101.11	$128.23

Crew B-23A	Hr.	Daily	Hr.	Daily	Bare Costs	Incl. O&P
1 Labor Foreman (outside)	$31.20	$249.60	$52.25	$418.00	$33.25	$55.40
1 Laborer	29.20	233.60	48.90	391.20		
1 Equip. Oper. (medium)	39.35	314.80	65.05	520.40		
1 Drill Rig, Truck-Mounted		2553.00		2808.30		
1 Pickup Truck, 3/4 Ton		145.80		160.38	112.45	123.69
24 L.H., Daily Totals		$3496.80		$4298.28	$145.70	$179.10

Crew B-23B	Hr.	Daily	Hr.	Daily	Bare Costs	Incl. O&P
1 Labor Foreman (outside)	$31.20	$249.60	$52.25	$418.00	$33.25	$55.40
1 Laborer	29.20	233.60	48.90	391.20		
1 Equip. Oper. (medium)	39.35	314.80	65.05	520.40		
1 Drill Rig, Truck-Mounted		2553.00		2808.30		
1 Pickup Truck, 3/4 Ton		145.80		160.38		
1 Centr. Water Pump, 6"		350.80		385.88	127.07	139.77
24 L.H., Daily Totals		$3847.60		$4684.16	$160.32	$195.17

Crew B-24	Hr.	Daily	Hr.	Daily	Bare Costs	Incl. O&P
1 Cement Finisher	$35.60	$284.80	$57.80	$462.40	$34.20	$56.67
1 Laborer	29.20	233.60	48.90	391.20		
1 Carpenter	37.80	302.40	63.30	506.40		
24 L.H., Daily Totals		$820.80		$1360.00	$34.20	$56.67

Crew B-25	Hr.	Daily	Hr.	Daily	Bare Costs	Incl. O&P
1 Labor Foreman (outside)	$31.20	$249.60	$52.25	$418.00	$32.15	$53.61
7 Laborers	29.20	1635.20	48.90	2738.40		
3 Equip. Opers. (medium)	39.35	944.40	65.05	1561.20		
1 Asphalt Paver, 130 H.P.		2210.00		2431.00		
1 Tandem Roller, 10 Ton		239.40		263.34		
1 Roller, Pneum. Whl., 12 Ton		347.80		382.58	31.79	34.97
88 L.H., Daily Totals		$5626.40		$7794.52	$63.94	$88.57

Crew B-25B	Hr.	Daily	Hr.	Daily	Bare Costs	Incl. O&P
1 Labor Foreman (outside)	$31.20	$249.60	$52.25	$418.00	$32.75	$54.56
7 Laborers	29.20	1635.20	48.90	2738.40		
4 Equip. Opers. (medium)	39.35	1259.20	65.05	2081.60		
1 Asphalt Paver, 130 H.P.		2210.00		2431.00		
2 Tandem Rollers, 10 Ton		470.00		526.68		
1 Roller, Pneum. Whl., 12 Ton		347.80		382.58	31.63	34.79
96 L.H., Daily Totals		$6180.60		$8578.26	$64.38	$89.36

Crew B-25C	Hr.	Daily	Hr.	Daily	Bare Costs	Incl. O&P
1 Labor Foreman (outside)	$31.20	$249.60	$52.25	$418.00	$32.92	$54.84
3 Laborers	29.20	700.80	48.90	1173.60		
2 Equip. Opers. (medium)	39.35	629.60	65.05	1040.80		
1 Asphalt Paver, 130 H.P.		2210.00		2431.00		
1 Tandem Roller, 10 Ton		239.40		263.34	51.03	56.13
48 L.H., Daily Totals		$4029.40		$5326.74	$83.95	$110.97

Crew B-25D	Hr.	Daily	Hr.	Daily	Bare Costs	Incl. O&P
1 Labor Foreman (outside)	$31.20	$249.60	$52.25	$418.00	$33.07	$55.08
3 Laborers	29.20	700.80	48.90	1173.60		
2.125 Equip. Opers. (medium)	39.35	668.95	65.05	1105.85		
.125 Truck Driver (heavy)	34.15	34.15	56.35	56.35		
.125 Truck Tractor, 6x4, 380 H.P.		75.00		82.50		
.125 Dist. Tanker, 3000 Gallon		39.92		43.92		
1 Asphalt Paver, 130 H.P.		2210.00		2431.00		
1 Tandem Roller, 10 Ton		239.40		263.34	51.29	56.42
50 L.H., Daily Totals		$4217.82		$5574.56	$84.36	$111.49

Crew B-25E	Hr.	Daily	Hr.	Daily	Bare Costs	Incl. O&P
1 Labor Foreman (outside)	$31.20	$249.60	$52.25	$418.00	$33.21	$55.29
3 Laborers	29.20	700.80	48.90	1173.60		
2.250 Equip. Opers. (medium)	39.35	708.30	65.05	1170.90		
.25 Truck Driver (heavy)	34.15	68.30	56.35	112.70		
.25 Truck Tractor, 6x4, 380 H.P.		150.00		165.00		
.25 Dist. Tanker, 3000 Gallon		79.85		87.83		
1 Asphalt Paver, 130 H.P.		2210.00		2431.00		
1 Tandem Roller, 10 Ton		239.40		263.34	51.52	56.68
52 L.H., Daily Totals		$4406.25		$5822.38	$84.74	$111.97

Crew B-26	Hr.	Daily	Hr.	Daily	Bare Costs	Incl. O&P
1 Labor Foreman (outside)	$31.20	$249.60	$52.25	$418.00	$32.77	$54.61
6 Laborers	29.20	1401.60	48.90	2347.20		
2 Equip. Opers. (medium)	39.35	629.60	65.05	1040.80		
1 Rodman (reinf.)	39.75	318.00	67.15	537.20		
1 Cement Finisher	35.60	284.80	57.80	462.40		
1 Grader, 30,000 Lbs.		714.40		785.84		
1 Paving Mach. & Equip.		2796.00		3075.60	39.89	43.88
88 L.H., Daily Totals		$6394.00		$8667.04	$72.66	$98.49

Crew B-26A	Hr.	Daily	Hr.	Daily	Bare Costs	Incl. O&P
1 Labor Foreman (outside)	$31.20	$249.60	$52.25	$418.00	$32.77	$54.61
6 Laborers	29.20	1401.60	48.90	2347.20		
2 Equip. Opers. (medium)	39.35	629.60	65.05	1040.80		
1 Rodman (reinf.)	39.75	318.00	67.15	537.20		
1 Cement Finisher	35.60	284.80	57.80	462.40		
1 Grader, 30,000 Lbs.		714.40		785.84		
1 Paving Mach. & Equip.		2796.00		3075.60		
1 Concrete Saw		168.60		185.46	41.81	45.99
88 L.H., Daily Totals		$6562.60		$8852.50	$74.58	$100.60

Crew B-26B	Hr.	Daily	Hr.	Daily	Bare Costs	Incl. O&P
1 Labor Foreman (outside)	$31.20	$249.60	$52.25	$418.00	$33.32	$55.48
6 Laborers	29.20	1401.60	48.90	2347.20		
3 Equip. Opers. (medium)	39.35	944.40	65.05	1561.20		
1 Rodman (reinf.)	39.75	318.00	67.15	537.20		
1 Cement Finisher	35.60	284.80	57.80	462.40		
1 Grader, 30,000 Lbs.		714.40		785.84		
1 Paving Mach. & Equip.		2796.00		3075.60		
1 Concrete Pump, 110' Boom		958.40		1054.24	46.55	51.20
96 L.H., Daily Totals		$7667.20		$10241.68	$79.87	$106.68

Crew B-26C	Hr.	Daily	Hr.	Daily	Bare Costs	Incl. O&P
1 Labor Foreman (outside)	$31.20	$249.60	$52.25	$418.00	$32.11	$53.56
6 Laborers	29.20	1401.60	48.90	2347.20		
1 Equip. Oper. (medium)	39.35	314.80	65.05	520.40		
1 Rodman (reinf.)	39.75	318.00	67.15	537.20		
1 Cement Finisher	35.60	284.80	57.80	462.40		
1 Paving Mach. & Equip.		2796.00		3075.60		
1 Concrete Saw		168.60		185.46	37.06	40.76
80 L.H., Daily Totals		$5533.40		$7546.26	$69.17	$94.33

Crews - Open Shop

Crew No.	Bare Costs		Incl. Subs O&P		Cost Per Labor-Hour	
Crew B-27	**Hr.**	**Daily**	**Hr.**	**Daily**	**Bare Costs**	**Incl. O&P**
1 Labor Foreman (outside)	$31.20	$249.60	$52.25	$418.00	$29.70	$49.74
3 Laborers	29.20	700.80	48.90	1173.60		
1 Berm Machine		295.80		325.38	9.24	10.17
32 L.H., Daily Totals		$1246.20		$1916.98	$38.94	$59.91
Crew B-28	**Hr.**	**Daily**	**Hr.**	**Daily**	**Bare Costs**	**Incl. O&P**
2 Carpenters	$37.80	$604.80	$63.30	$1012.80	$34.93	$58.50
1 Laborer	29.20	233.60	48.90	391.20		
24 L.H., Daily Totals		$838.40		$1404.00	$34.93	$58.50
Crew B-29	**Hr.**	**Daily**	**Hr.**	**Daily**	**Bare Costs**	**Incl. O&P**
1 Labor Foreman (outside)	$31.20	$249.60	$52.25	$418.00	$31.38	$52.40
4 Laborers	29.20	934.40	48.90	1564.80		
1 Equip. Oper. (crane)	40.25	322.00	66.55	532.40		
1 Gradall, 5/8 C.Y.		882.00		970.20	18.38	20.21
48 L.H., Daily Totals		$2388.00		$3485.40	$49.75	$72.61
Crew B-30	**Hr.**	**Daily**	**Hr.**	**Daily**	**Bare Costs**	**Incl. O&P**
1 Equip. Oper. (medium)	$39.35	$314.80	$65.05	$520.40	$35.88	$59.25
2 Truck Drivers (heavy)	34.15	546.40	56.35	901.60		
1 Hyd. Excavator, 1.5 C.Y.		1029.00		1131.90		
2 Dump Trucks, 12 C.Y., 400 H.P.		1381.60		1519.76	100.44	110.49
24 L.H., Daily Totals		$3271.80		$4073.66	$136.32	$169.74
Crew B-31	**Hr.**	**Daily**	**Hr.**	**Daily**	**Bare Costs**	**Incl. O&P**
1 Labor Foreman (outside)	$31.20	$249.60	$52.25	$418.00	$29.60	$49.57
4 Laborers	29.20	934.40	48.90	1564.80		
1 Air Compressor, 250 cfm		197.80		217.58		
1 Sheeting Driver		6.50		7.15		
2 -50' Air Hoses, 1.5"		10.40		11.44	5.37	5.90
40 L.H., Daily Totals		$1398.70		$2218.97	$34.97	$55.47
Crew B-32	**Hr.**	**Daily**	**Hr.**	**Daily**	**Bare Costs**	**Incl. O&P**
1 Laborer	$29.20	$233.60	$48.90	$391.20	$36.81	$61.01
3 Equip. Opers. (medium)	39.35	944.40	65.05	1561.20		
1 Grader, 30,000 Lbs.		714.40		785.84		
1 Tandem Roller, 10 Ton		239.40		263.34		
1 Dozer, 200 H.P.		1397.00		1536.70	73.46	80.81
32 L.H., Daily Totals		$3528.80		$4538.28	$110.28	$141.82
Crew B-32A	**Hr.**	**Daily**	**Hr.**	**Daily**	**Bare Costs**	**Incl. O&P**
1 Laborer	$29.20	$233.60	$48.90	$391.20	$35.97	$59.67
2 Equip. Opers. (medium)	39.35	629.60	65.05	1040.80		
1 Grader, 30,000 Lbs.		714.40		785.84		
1 Roller, Vibratory, 25 Ton		691.80		760.98	58.59	64.45
24 L.H., Daily Totals		$2269.40		$2978.82	$94.56	$124.12
Crew B-32B	**Hr.**	**Daily**	**Hr.**	**Daily**	**Bare Costs**	**Incl. O&P**
1 Laborer	$29.20	$233.60	$48.90	$391.20	$35.97	$59.67
2 Equip. Opers. (medium)	39.35	629.60	65.05	1040.80		
1 Dozer, 200 H.P.		1397.00		1536.70		
1 Roller, Vibratory, 25 Ton		691.80		760.98	87.03	95.74
24 L.H., Daily Totals		$2952.00		$3729.68	$123.00	$155.40

Crew No.	Bare Costs		Incl. Subs O&P		Cost Per Labor-Hour	
Crew B-32C	**Hr.**	**Daily**	**Hr.**	**Daily**	**Bare Costs**	**Incl. O&P**
1 Labor Foreman (outside)	$31.20	$249.60	$52.25	$418.00	$34.61	$57.53
2 Laborers	29.20	467.20	48.90	782.40		
3 Equip. Opers. (medium)	39.35	944.40	65.05	1561.20		
1 Grader, 30,000 Lbs.		714.40		785.84		
1 Tandem Roller, 10 Ton		239.40		263.34		
1 Dozer, 200 H.P.		1397.00		1536.70	48.98	53.87
48 L.H., Daily Totals		$4012.00		$5347.48	$83.58	$111.41
Crew B-33A	**Hr.**	**Daily**	**Hr.**	**Daily**	**Bare Costs**	**Incl. O&P**
1 Equip. Oper. (medium)	$39.35	$314.80	$65.05	$520.40	$39.35	$65.05
.25 Equip. Oper. (medium)	39.35	78.70	65.05	130.10		
1 Scraper, Towed, 7 C.Y.		113.20		124.52		
1.250 Dozers, 300 H.P.		2358.75		2594.63	247.19	271.91
10 L.H., Daily Totals		$2865.45		$3369.65	$286.55	$336.96
Crew B-33B	**Hr.**	**Daily**	**Hr.**	**Daily**	**Bare Costs**	**Incl. O&P**
1 Equip. Oper. (medium)	$39.35	$314.80	$65.05	$520.40	$39.35	$65.05
.25 Equip. Oper. (medium)	39.35	78.70	65.05	130.10		
1 Scraper, Towed, 10 C.Y.		145.00		159.50		
1.250 Dozers, 300 H.P.		2358.75		2594.63	250.38	275.41
10 L.H., Daily Totals		$2897.25		$3404.63	$289.73	$340.46
Crew B-33C	**Hr.**	**Daily**	**Hr.**	**Daily**	**Bare Costs**	**Incl. O&P**
1 Equip. Oper. (medium)	$39.35	$314.80	$65.05	$520.40	$39.35	$65.05
.25 Equip. Oper. (medium)	39.35	78.70	65.05	130.10		
1 Scraper, Towed, 15 C.Y.		163.60		179.96		
1.250 Dozers, 300 H.P.		2358.75		2594.63	252.24	277.46
10 L.H., Daily Totals		$2915.85		$3425.09	$291.58	$342.51
Crew B-33D	**Hr.**	**Daily**	**Hr.**	**Daily**	**Bare Costs**	**Incl. O&P**
1 Equip. Oper. (medium)	$39.35	$314.80	$65.05	$520.40	$39.35	$65.05
.25 Equip. Oper. (medium)	39.35	78.70	65.05	130.10		
1 S.P. Scraper, 14 C.Y.		2172.00		2389.20		
.25 Dozer, 300 H.P.		471.75		518.92	264.38	290.81
10 L.H., Daily Totals		$3037.25		$3558.63	$303.73	$355.86
Crew B-33E	**Hr.**	**Daily**	**Hr.**	**Daily**	**Bare Costs**	**Incl. O&P**
1 Equip. Oper. (medium)	$39.35	$314.80	$65.05	$520.40	$39.35	$65.05
.25 Equip. Oper. (medium)	39.35	78.70	65.05	130.10		
1 S.P. Scraper, 21 C.Y.		2715.00		2986.50		
.25 Dozer, 300 H.P.		471.75		518.92	318.68	350.54
10 L.H., Daily Totals		$3580.25		$4155.93	$358.02	$415.59
Crew B-33F	**Hr.**	**Daily**	**Hr.**	**Daily**	**Bare Costs**	**Incl. O&P**
1 Equip. Oper. (medium)	$39.35	$314.80	$65.05	$520.40	$39.35	$65.05
.25 Equip. Oper. (medium)	39.35	78.70	65.05	130.10		
1 Elev. Scraper, 11 C.Y.		1173.00		1290.30		
.25 Dozer, 300 H.P.		471.75		518.92	164.47	180.92
10 L.H., Daily Totals		$2038.25		$2459.72	$203.82	$245.97
Crew B-33G	**Hr.**	**Daily**	**Hr.**	**Daily**	**Bare Costs**	**Incl. O&P**
1 Equip. Oper. (medium)	$39.35	$314.80	$65.05	$520.40	$39.35	$65.05
.25 Equip. Oper. (medium)	39.35	78.70	65.05	130.10		
1 Elev. Scraper, 22 C.Y.		2548.00		2802.80		
.25 Dozer, 300 H.P.		471.75		518.92	301.98	332.17
10 L.H., Daily Totals		$3413.25		$3972.22	$341.32	$397.22

Crews - Open Shop

Crew No.	Bare Costs		Incl. Subs O&P		Cost Per Labor-Hour	
Crew B-33K	**Hr.**	**Daily**	**Hr.**	**Daily**	**Bare Costs**	**Incl. O&P**
1 Equipment Operator (med.)	$39.35	$314.80	$65.05	$520.40	$36.45	$60.44
.25 Equipment Operator (med.)	39.35	78.70	65.05	130.10		
.5 Laborer	29.20	116.80	48.90	195.60		
1 S.P. Scraper, 31 C.Y.		3735.00		4108.50		
.25 Dozer, 410 H.P.		598.00		657.80	309.50	340.45
14 L.H., Daily Totals		$4843.30		$5612.40	$345.95	$400.89
Crew B-34A	**Hr.**	**Daily**	**Hr.**	**Daily**	**Bare Costs**	**Incl. O&P**
1 Truck Driver (heavy)	$34.15	$273.20	$56.35	$450.80	$34.15	$56.35
1 Dump Truck, 8 C.Y., 220 H.P.		417.40		459.14	52.17	57.39
8 L.H., Daily Totals		$690.60		$909.94	$86.33	$113.74
Crew B-34B	**Hr.**	**Daily**	**Hr.**	**Daily**	**Bare Costs**	**Incl. O&P**
1 Truck Driver (heavy)	$34.15	$273.20	$56.35	$450.80	$34.15	$56.35
1 Dump Truck, 12 C.Y., 400 H.P.		690.80		759.88	86.35	94.98
8 L.H., Daily Totals		$964.00		$1210.68	$120.50	$151.34
Crew B-34C	**Hr.**	**Daily**	**Hr.**	**Daily**	**Bare Costs**	**Incl. O&P**
1 Truck Driver (heavy)	$34.15	$273.20	$56.35	$450.80	$34.15	$56.35
1 Truck Tractor, 6x4, 380 H.P.		600.00		660.00		
1 Dump Trailer, 16.5 C.Y.		125.80		138.38	90.72	99.80
8 L.H., Daily Totals		$999.00		$1249.18	$124.88	$156.15
Crew B-34D	**Hr.**	**Daily**	**Hr.**	**Daily**	**Bare Costs**	**Incl. O&P**
1 Truck Driver (heavy)	$34.15	$273.20	$56.35	$450.80	$34.15	$56.35
1 Truck Tractor, 6x4, 380 H.P.		600.00		660.00		
1 Dump Trailer, 20 C.Y.		140.40		154.44	92.55	101.81
8 L.H., Daily Totals		$1013.60		$1265.24	$126.70	$158.16
Crew B-34E	**Hr.**	**Daily**	**Hr.**	**Daily**	**Bare Costs**	**Incl. O&P**
1 Truck Driver (heavy)	$34.15	$273.20	$56.35	$450.80	$34.15	$56.35
1 Dump Truck, Off Hwy., 25 Ton		1370.00		1507.00	171.25	188.38
8 L.H., Daily Totals		$1643.20		$1957.80	$205.40	$244.72
Crew B-34F	**Hr.**	**Daily**	**Hr.**	**Daily**	**Bare Costs**	**Incl. O&P**
1 Truck Driver (heavy)	$34.15	$273.20	$56.35	$450.80	$34.15	$56.35
1 Dump Truck, Off Hwy., 35 Ton		1529.00		1681.90	191.13	210.24
8 L.H., Daily Totals		$1802.20		$2132.70	$225.28	$266.59
Crew B-34G	**Hr.**	**Daily**	**Hr.**	**Daily**	**Bare Costs**	**Incl. O&P**
1 Truck Driver (heavy)	$34.15	$273.20	$56.35	$450.80	$34.15	$56.35
1 Dump Truck, Off Hwy., 50 Ton		1861.00		2047.10	232.63	255.89
8 L.H., Daily Totals		$2134.20		$2497.90	$266.77	$312.24
Crew B-34H	**Hr.**	**Daily**	**Hr.**	**Daily**	**Bare Costs**	**Incl. O&P**
1 Truck Driver (heavy)	$34.15	$273.20	$56.35	$450.80	$34.15	$56.35
1 Dump Truck, Off Hwy., 65 Ton		1915.00		2106.50	239.38	263.31
8 L.H., Daily Totals		$2188.20		$2557.30	$273.52	$319.66
Crew B-34I	**Hr.**	**Daily**	**Hr.**	**Daily**	**Bare Costs**	**Incl. O&P**
1 Truck Driver (heavy)	$34.15	$273.20	$56.35	$450.80	$34.15	$56.35
1 Dump Truck, 18 C.Y., 450 H.P.		869.40		956.34	108.68	119.54
8 L.H., Daily Totals		$1142.60		$1407.14	$142.82	$175.89
Crew B-34J	**Hr.**	**Daily**	**Hr.**	**Daily**	**Bare Costs**	**Incl. O&P**
1 Truck Driver (heavy)	$34.15	$273.20	$56.35	$450.80	$34.15	$56.35
1 Dump Truck, Off Hwy., 100 Ton		2950.00		3245.00	368.75	405.63
8 L.H., Daily Totals		$3223.20		$3695.80	$402.90	$461.98

Crew No.	Bare Costs		Incl. Subs O&P		Cost Per Labor-Hour	
Crew B-34K	**Hr.**	**Daily**	**Hr.**	**Daily**	**Bare Costs**	**Incl. O&P**
1 Truck Driver (heavy)	$34.15	$273.20	$56.35	$450.80	$34.15	$56.35
1 Truck Tractor, 6x4, 450 H.P.		726.80		799.48		
1 Lowbed Trailer, 75 Ton		223.40		245.74	118.78	130.65
8 L.H., Daily Totals		$1223.40		$1496.02	$152.93	$187.00
Crew B-34L	**Hr.**	**Daily**	**Hr.**	**Daily**	**Bare Costs**	**Incl. O&P**
1 Equip. Oper. (light)	$37.85	$302.80	$62.55	$500.40	$37.85	$62.55
1 Flatbed Truck, Gas, 1.5 Ton		248.20		273.02	31.02	34.13
8 L.H., Daily Totals		$551.00		$773.42	$68.88	$96.68
Crew B-34M	**Hr.**	**Daily**	**Hr.**	**Daily**	**Bare Costs**	**Incl. O&P**
1 Equip. Oper. (light)	$37.85	$302.80	$62.55	$500.40	$37.85	$62.55
1 Flatbed Truck, Gas, 3 Ton		307.40		338.14	38.42	42.27
8 L.H., Daily Totals		$610.20		$838.54	$76.28	$104.82
Crew B-34N	**Hr.**	**Daily**	**Hr.**	**Daily**	**Bare Costs**	**Incl. O&P**
1 Truck Driver (heavy)	$34.15	$273.20	$56.35	$450.80	$36.75	$60.70
1 Equip. Oper. (medium)	39.35	314.80	65.05	520.40		
1 Truck Tractor, 6x4, 380 H.P.		600.00		660.00		
1 Flatbed Trailer, 40 Ton		154.60		170.06	47.16	51.88
16 L.H., Daily Totals		$1342.60		$1801.26	$83.91	$112.58
Crew B-34P	**Hr.**	**Daily**	**Hr.**	**Daily**	**Bare Costs**	**Incl. O&P**
1 Pipe Fitter	$44.95	$359.60	$74.10	$592.80	$39.13	$64.58
1 Truck Driver (light)	33.10	264.80	54.60	436.80		
1 Equip. Oper. (medium)	39.35	314.80	65.05	520.40		
1 Flatbed Truck, Gas, 3 Ton		307.40		338.14		
1 Backhoe Loader, 48 H.P.		365.40		401.94	28.03	30.84
24 L.H., Daily Totals		$1612.00		$2290.08	$67.17	$95.42
Crew B-34Q	**Hr.**	**Daily**	**Hr.**	**Daily**	**Bare Costs**	**Incl. O&P**
1 Pipe Fitter	$44.95	$359.60	$74.10	$592.80	$39.43	$65.08
1 Truck Driver (light)	33.10	264.80	54.60	436.80		
1 Equip. Oper. (crane)	40.25	322.00	66.55	532.40		
1 Flatbed Trailer, 25 Ton		113.20		124.52		
1 Dump Truck, 8 C.Y., 220 H.P.		417.40		459.14		
1 Hyd. Crane, 25 Ton		748.80		823.68	53.31	58.64
24 L.H., Daily Totals		$2225.80		$2969.34	$92.74	$123.72
Crew B-34R	**Hr.**	**Daily**	**Hr.**	**Daily**	**Bare Costs**	**Incl. O&P**
1 Pipe Fitter	$44.95	$359.60	$74.10	$592.80	$39.43	$65.08
1 Truck Driver (light)	33.10	264.80	54.60	436.80		
1 Equip. Oper. (crane)	40.25	322.00	66.55	532.40		
1 Flatbed Trailer, 25 Ton		113.20		124.52		
1 Dump Truck, 8 C.Y., 220 H.P.		417.40		459.14		
1 Hyd. Crane, 25 Ton		748.80		823.68		
1 Hyd. Excavator, 1 C.Y.		812.80		894.08	87.17	95.89
24 L.H., Daily Totals		$3038.60		$3863.42	$126.61	$160.98
Crew B-34S	**Hr.**	**Daily**	**Hr.**	**Daily**	**Bare Costs**	**Incl. O&P**
2 Pipe Fitters	$44.95	$719.20	$74.10	$1185.60	$41.08	$67.78
1 Truck Driver (heavy)	34.15	273.20	56.35	450.80		
1 Equip. Oper. (crane)	40.25	322.00	66.55	532.40		
1 Flatbed Trailer, 40 Ton		154.60		170.06		
1 Truck Tractor, 6x4, 380 H.P.		600.00		660.00		
1 Hyd. Crane, 80 Ton		1630.00		1793.00		
1 Hyd. Excavator, 2 C.Y.		1176.00		1293.60	111.27	122.40
32 L.H., Daily Totals		$4875.00		$6085.46	$152.34	$190.17

Crew No.	Bare Costs		Incl. Subs O&P		Cost Per Labor-Hour	
Crew B-34T	**Hr.**	**Daily**	**Hr.**	**Daily**	**Bare Costs**	**Incl. O&P**
2 Pipe Fitters	$44.95	$719.20	$74.10	$1185.60	$41.08	$67.78
1 Truck Driver (heavy)	34.15	273.20	56.35	450.80		
1 Equip. Oper. (crane)	40.25	322.00	66.55	532.40		
1 Flatbed Trailer, 40 Ton		154.60		170.06		
1 Truck Tractor, 6x4, 380 H.P.		600.00		660.00		
1 Hyd. Crane, 80 Ton		1630.00		1793.00	74.52	81.97
32 L.H., Daily Totals		$3699.00		$4791.86	$115.59	$149.75
Crew B-34U	**Hr.**	**Daily**	**Hr.**	**Daily**	**Bare Costs**	**Incl. O&P**
1 Truck Driver (heavy)	$34.15	$273.20	$56.35	$450.80	$36.00	$59.45
1 Equip. Oper. (light)	37.85	302.80	62.55	500.40		
1 Truck Tractor, 220 H.P.		358.80		394.68		
1 Flatbed Trailer, 25 Ton		113.20		124.52	29.50	32.45
16 L.H., Daily Totals		$1048.00		$1470.40	$65.50	$91.90
Crew B-34V	**Hr.**	**Daily**	**Hr.**	**Daily**	**Bare Costs**	**Incl. O&P**
1 Truck Driver (heavy)	$34.15	$273.20	$56.35	$450.80	$37.42	$61.82
1 Equip. Oper. (crane)	40.25	322.00	66.55	532.40		
1 Equip. Oper. (light)	37.85	302.80	62.55	500.40		
1 Truck Tractor, 6x4, 450 H.P.		726.80		799.48		
1 Equipment Trailer, 50 Ton		170.00		187.00		
1 Pickup Truck, 4x4, 3/4 Ton		156.80		172.48	43.90	48.29
24 L.H., Daily Totals		$1951.60		$2642.56	$81.32	$110.11
Crew B-34W	**Hr.**	**Daily**	**Hr.**	**Daily**	**Bare Costs**	**Incl. O&P**
5 Truck Drivers (heavy)	$34.15	$1366.00	$56.35	$2254.00	$35.66	$58.96
2 Equip. Opers. (crane)	40.25	644.00	66.55	1064.80		
1 Equip. Oper. (mechanic)	40.45	323.60	66.85	534.80		
1 Laborer	29.20	233.60	48.90	391.20		
4 Trucks, Tractor, 6x4, 380 H.P.		2400.00		2640.00		
2 Equipment Trailers, 50 Ton		340.00		374.00		
2 Flatbed Trailers, 40 Ton		309.20		340.12		
1 Pickup Truck, 4x4, 3/4 Ton		156.80		172.48		
1 S.P. Crane, 4x4, 20 Ton		554.80		610.28	52.23	57.46
72 L.H., Daily Totals		$6328.00		$8381.68	$87.89	$116.41
Crew B-35	**Hr.**	**Daily**	**Hr.**	**Daily**	**Bare Costs**	**Incl. O&P**
1 Labor Foreman (outside)	$31.20	$249.60	$52.25	$418.00	$36.55	$60.86
1 Skilled Worker	38.30	306.40	64.35	514.80		
1 Welder (plumber)	43.80	350.40	72.25	578.00		
1 Laborer	29.20	233.60	48.90	391.20		
1 Equip. Oper. (crane)	40.25	322.00	66.55	532.40		
1 Welder, Electric, 300 amp		58.10		63.91		
1 Hyd. Excavator, .75 C.Y.		654.00		719.40	17.80	19.58
40 L.H., Daily Totals		$2174.10		$3217.71	$54.35	$80.44
Crew B-35A	**Hr.**	**Daily**	**Hr.**	**Daily**	**Bare Costs**	**Incl. O&P**
1 Labor Foreman (outside)	$31.20	$249.60	$52.25	$418.00	$35.34	$58.83
2 Laborers	29.20	467.20	48.90	782.40		
1 Skilled Worker	38.30	306.40	64.35	514.80		
1 Welder (plumber)	43.80	350.40	72.25	578.00		
1 Equip. Oper. (crane)	40.25	322.00	66.55	532.40		
1 Equip. Oper. (oiler)	35.45	283.60	58.60	468.80		
1 Welder, Gas Engine, 300 amp		147.20		161.92		
1 Crawler Crane, 75 Ton		1507.00		1657.70	29.54	32.49
56 L.H., Daily Totals		$3633.40		$5114.02	$64.88	$91.32

Crew No.	Bare Costs		Incl. Subs O&P		Cost Per Labor-Hour	
Crew B-36	**Hr.**	**Daily**	**Hr.**	**Daily**	**Bare Costs**	**Incl. O&P**
1 Labor Foreman (outside)	$31.20	$249.60	$52.25	$418.00	$33.66	$56.03
2 Laborers	29.20	467.20	48.90	782.40		
2 Equip. Opers. (medium)	39.35	629.60	65.05	1040.80		
1 Dozer, 200 H.P.		1397.00		1536.70		
1 Aggregate Spreader		39.60		43.56		
1 Tandem Roller, 10 Ton		239.40		263.34	41.90	46.09
40 L.H., Daily Totals		$3022.40		$4084.80	$75.56	$102.12
Crew B-36A	**Hr.**	**Daily**	**Hr.**	**Daily**	**Bare Costs**	**Incl. O&P**
1 Labor Foreman (outside)	$31.20	$249.60	$52.25	$418.00	$35.29	$58.61
2 Laborers	29.20	467.20	48.90	782.40		
4 Equip. Opers. (medium)	39.35	1259.20	65.05	2081.60		
1 Dozer, 200 H.P.		1397.00		1536.70		
1 Aggregate Spreader		39.60		43.56		
1 Tandem Roller, 10 Ton		239.40		263.34		
1 Roller, Pneum. Whl., 12 Ton		347.80		382.58	36.14	39.75
56 L.H., Daily Totals		$3999.80		$5508.18	$71.42	$98.36
Crew B-36B	**Hr.**	**Daily**	**Hr.**	**Daily**	**Bare Costs**	**Incl. O&P**
1 Labor Foreman (outside)	$31.20	$249.60	$52.25	$418.00	$35.14	$58.33
2 Laborers	29.20	467.20	48.90	782.40		
4 Equip. Opers. (medium)	39.35	1259.20	65.05	2081.60		
1 Truck Driver (heavy)	34.15	273.20	56.35	450.80		
1 Grader, 30,000 Lbs.		714.40		785.84		
1 F.E. Loader, Crl, 1.5 C.Y.		634.20		697.62		
1 Dozer, 300 H.P.		1887.00		2075.70		
1 Roller, Vibratory, 25 Ton		691.80		760.98		
1 Truck Tractor, 6x4, 450 H.P.		726.80		799.48		
1 Water Tank Trailer, 5000 Gal.		141.60		155.76	74.93	82.43
64 L.H., Daily Totals		$7045.00		$9008.18	$110.08	$140.75
Crew B-36C	**Hr.**	**Daily**	**Hr.**	**Daily**	**Bare Costs**	**Incl. O&P**
1 Labor Foreman (outside)	$31.20	$249.60	$52.25	$418.00	$36.68	$60.75
3 Equip. Opers. (medium)	39.35	944.40	65.05	1561.20		
1 Truck Driver (heavy)	34.15	273.20	56.35	450.80		
1 Grader, 30,000 Lbs.		714.40		785.84		
1 Dozer, 300 H.P.		1887.00		2075.70		
1 Roller, Vibratory, 25 Ton		691.80		760.98		
1 Truck Tractor, 6x4, 450 H.P.		726.80		799.48		
1 Water Tank Trailer, 5000 Gal.		141.60		155.76	104.04	114.44
40 L.H., Daily Totals		$5628.80		$7007.76	$140.72	$175.19
Crew B-36E	**Hr.**	**Daily**	**Hr.**	**Daily**	**Bare Costs**	**Incl. O&P**
1 Labor Foreman (outside)	$31.20	$249.60	$52.25	$418.00	$37.13	$61.47
4 Equip. Opers. (medium)	39.35	1259.20	65.05	2081.60		
1 Truck Driver (heavy)	34.15	273.20	56.35	450.80		
1 Grader, 30,000 Lbs.		714.40		785.84		
1 Dozer, 300 H.P.		1887.00		2075.70		
1 Roller, Vibratory, 25 Ton		691.80		760.98		
1 Truck Tractor, 6x4, 380 H.P.		600.00		660.00		
1 Dist. Tanker, 3000 Gallon		319.40		351.34	87.76	96.54
48 L.H., Daily Totals		$5994.60		$7584.26	$124.89	$158.01
Crew B-37	**Hr.**	**Daily**	**Hr.**	**Daily**	**Bare Costs**	**Incl. O&P**
1 Labor Foreman (outside)	$31.20	$249.60	$52.25	$418.00	$30.98	$51.73
4 Laborers	29.20	934.40	48.90	1564.80		
1 Equip. Oper. (light)	37.85	302.80	62.55	500.40		
1 Tandem Roller, 5 Ton		162.60		178.86	3.39	3.73
48 L.H., Daily Totals		$1649.40		$2662.06	$34.36	$55.46

Crew No.	Bare Costs		Incl. Subs O&P		Cost Per Labor-Hour	
Crew B-37A	**Hr.**	**Daily**	**Hr.**	**Daily**	**Bare Costs**	**Incl. O&P**
2 Laborers	$29.20	$467.20	$48.90	$782.40	$30.50	$50.80
1 Truck Driver (light)	33.10	264.80	54.60	436.80		
1 Flatbed Truck, Gas, 1.5 Ton		248.20		273.02		
1 Tar Kettle, T.M.		135.30		148.83	15.98	17.58
24 L.H., Daily Totals		$1115.50		$1641.05	$46.48	$68.38
Crew B-37B	**Hr.**	**Daily**	**Hr.**	**Daily**	**Bare Costs**	**Incl. O&P**
3 Laborers	$29.20	$700.80	$48.90	$1173.60	$30.18	$50.33
1 Truck Driver (light)	33.10	264.80	54.60	436.80		
1 Flatbed Truck, Gas, 1.5 Ton		248.20		273.02		
1 Tar Kettle, T.M.		135.30		148.83	11.98	13.18
32 L.H., Daily Totals		$1349.10		$2032.25	$42.16	$63.51
Crew B-37C	**Hr.**	**Daily**	**Hr.**	**Daily**	**Bare Costs**	**Incl. O&P**
2 Laborers	$29.20	$467.20	$48.90	$782.40	$31.15	$51.75
2 Truck Drivers (light)	33.10	529.60	54.60	873.60		
2 Flatbed Trucks, Gas, 1.5 Ton		496.40		546.04		
1 Tar Kettle, T.M.		135.30		148.83	19.74	21.71
32 L.H., Daily Totals		$1628.50		$2350.87	$50.89	$73.46
Crew B-37D	**Hr.**	**Daily**	**Hr.**	**Daily**	**Bare Costs**	**Incl. O&P**
1 Laborer	$29.20	$233.60	$48.90	$391.20	$31.15	$51.75
1 Truck Driver (light)	33.10	264.80	54.60	436.80		
1 Pickup Truck, 3/4 Ton		145.80		160.38	9.11	10.02
16 L.H., Daily Totals		$644.20		$988.38	$40.26	$61.77
Crew B-37E	**Hr.**	**Daily**	**Hr.**	**Daily**	**Bare Costs**	**Incl. O&P**
3 Laborers	$29.20	$700.80	$48.90	$1173.60	$33.00	$54.79
1 Equip. Oper. (light)	37.85	302.80	62.55	500.40		
1 Equip. Oper. (medium)	39.35	314.80	65.05	520.40		
2 Truck Drivers (light)	33.10	529.60	54.60	873.60		
4 Barrels w/ Flasher		13.60		14.96		
1 Concrete Saw		168.60		185.46		
1 Rotary Hammer Drill		24.55		27.00		
1 Hammer Drill Bit		2.90		3.19		
1 Loader, Skid Steer, 30 H.P.		173.20		190.52		
1 Conc. Hammer Attach.		108.40		119.24		
1 Vibrating Plate, Gas, 18"		36.20		39.82		
2 Flatbed Trucks, Gas, 1.5 Ton		496.40		546.04	18.28	20.11
56 L.H., Daily Totals		$2871.85		$4194.23	$51.28	$74.90
Crew B-37F	**Hr.**	**Daily**	**Hr.**	**Daily**	**Bare Costs**	**Incl. O&P**
3 Laborers	$29.20	$700.80	$48.90	$1173.60	$30.18	$50.33
1 Truck Driver (light)	33.10	264.80	54.60	436.80		
4 Barrels w/ Flasher		13.60		14.96		
1 Concrete Mixer, 10 C.F.		172.80		190.08		
1 Air Compressor, 60 cfm		142.80		157.08		
1 -50' Air Hose, 3/4"		3.05		3.36		
1 Spade (Chipper)		8.20		9.02		
1 Flatbed Truck, Gas, 1.5 Ton		248.20		273.02	18.40	20.23
32 L.H., Daily Totals		$1554.25		$2257.92	$48.57	$70.56
Crew B-37G	**Hr.**	**Daily**	**Hr.**	**Daily**	**Bare Costs**	**Incl. O&P**
1 Labor Foreman (outside)	$31.20	$249.60	$52.25	$418.00	$30.98	$51.73
4 Laborers	29.20	934.40	48.90	1564.80		
1 Equip. Oper. (light)	37.85	302.80	62.55	500.40		
1 Berm Machine		295.80		325.38		
1 Tandem Roller, 5 Ton		162.60		178.86	9.55	10.51
48 L.H., Daily Totals		$1945.20		$2987.44	$40.52	$62.24

Crew No.	Bare Costs		Incl. Subs O&P		Cost Per Labor-Hour	
Crew B-37H	**Hr.**	**Daily**	**Hr.**	**Daily**	**Bare Costs**	**Incl. O&P**
1 Labor Foreman (outside)	$31.20	$249.60	$52.25	$418.00	$30.98	$51.73
4 Laborers	29.20	934.40	48.90	1564.80		
1 Equip. Oper. (light)	37.85	302.80	62.55	500.40		
1 Tandem Roller, 5 Ton		162.60		178.86		
1 Flatbed Truck, Gas, 1.5 Ton		248.20		273.02		
1 Tar Kettle, T.M.		135.30		148.83	11.38	12.51
48 L.H., Daily Totals		$2032.90		$3083.91	$42.35	$64.25
Crew B-37I	**Hr.**	**Daily**	**Hr.**	**Daily**	**Bare Costs**	**Incl. O&P**
3 Laborers	$29.20	$700.80	$48.90	$1173.60	$33.00	$54.79
1 Equip. Oper. (light)	37.85	302.80	62.55	500.40		
1 Equip. Oper. (medium)	39.35	314.80	65.05	520.40		
2 Truck Drivers (light)	33.10	529.60	54.60	873.60		
4 Barrels w/ Flasher		13.60		14.96		
1 Concrete Saw		168.60		185.46		
1 Rotary Hammer Drill		24.55		27.00		
1 Hammer Drill Bit		2.90		3.19		
1 Air Compressor, 60 cfm		142.80		157.08		
1 -50' Air Hose, 3/4"		3.05		3.36		
1 Spade (Chipper)		8.20		9.02		
1 Loader, Skid Steer, 30 H.P.		173.20		190.52		
1 Conc. Hammer Attach.		108.40		119.24		
1 Concrete Mixer, 10 C.F.		172.80		190.08		
1 Vibrating Plate, Gas, 18"		36.20		39.82		
2 Flatbed Trucks, Gas, 1.5 Ton		496.40		546.04	24.12	26.53
56 L.H., Daily Totals		$3198.70		$4553.77	$57.12	$81.32
Crew B-37J	**Hr.**	**Daily**	**Hr.**	**Daily**	**Bare Costs**	**Incl. O&P**
1 Labor Foreman (outside)	$31.20	$249.60	$52.25	$418.00	$30.98	$51.73
4 Laborers	29.20	934.40	48.90	1564.80		
1 Equip. Oper. (light)	37.85	302.80	62.55	500.40		
1 Air Compressor, 60 cfm		142.80		157.08		
1 -50' Air Hose, 3/4"		3.05		3.36		
2 Concrete Mixers, 10 C.F.		345.60		380.16		
2 Flatbed Trucks, Gas, 1.5 Ton		496.40		546.04		
1 Shot Blaster, 20"		215.40		236.94	25.07	27.57
48 L.H., Daily Totals		$2690.05		$3806.78	$56.04	$79.31
Crew B-37K	**Hr.**	**Daily**	**Hr.**	**Daily**	**Bare Costs**	**Incl. O&P**
1 Labor Foreman (outside)	$31.20	$249.60	$52.25	$418.00	$30.98	$51.73
4 Laborers	29.20	934.40	48.90	1564.80		
1 Equip. Oper. (light)	37.85	302.80	62.55	500.40		
1 Air Compressor, 60 cfm		142.80		157.08		
1 -50' Air Hose, 3/4"		3.05		3.36		
2 Flatbed Trucks, Gas, 1.5 Ton		496.40		546.04		
1 Shot Blaster, 20"		215.40		236.94	17.87	19.65
48 L.H., Daily Totals		$2344.45		$3426.61	$48.84	$71.39
Crew B-38	**Hr.**	**Daily**	**Hr.**	**Daily**	**Bare Costs**	**Incl. O&P**
2 Laborers	$29.20	$467.20	$48.90	$782.40	$32.08	$53.45
1 Equip. Oper. (light)	37.85	302.80	62.55	500.40		
1 Backhoe Loader, 48 H.P.		365.40		401.94		
1 Hyd. Hammer, (1200 lb.)		179.60		197.56	22.71	24.98
24 L.H., Daily Totals		$1315.00		$1882.30	$54.79	$78.43
Crew B-39	**Hr.**	**Daily**	**Hr.**	**Daily**	**Bare Costs**	**Incl. O&P**
1 Labor Foreman (outside)	$31.20	$249.60	$52.25	$418.00	$29.53	$49.46
5 Laborers	29.20	1168.00	48.90	1956.00		
1 Air Compressor, 250 cfm		197.80		217.58		
2 Breakers, Pavement, 60 lb.		20.40		22.44		
2 -50' Air Hoses, 1.5"		10.40		11.44	4.76	5.24
48 L.H., Daily Totals		$1646.20		$2625.46	$34.30	$54.70

Crew No.	Bare Costs		Incl. Subs O&P		Cost Per Labor-Hour	
Crew B-40	**Hr.**	**Daily**	**Hr.**	**Daily**	**Bare Costs**	**Incl. O&P**
1 Pile Driver Foreman (outside)	$39.75	$318.00	$68.60	$548.80	$37.17	$63.52
4 Pile Drivers	37.75	1208.00	65.15	2084.80		
1 Building Laborer	29.20	233.60	48.90	391.20		
1 Equip. Oper. (crane)	40.25	322.00	66.55	532.40		
1 Crawler Crane, 40 Ton		1203.00		1323.30		
1 Vibratory Hammer & Gen.		2612.00		2873.20	68.13	74.94
56 L.H., Daily Totals		$5896.60		$7753.70	$105.30	$138.46

Crew B-40B	Hr.	Daily	Hr.	Daily	Bare Costs	Incl. O&P
1 Labor Foreman (outside)	$31.20	$249.60	$52.25	$418.00	$32.42	$54.02
3 Laborers	29.20	700.80	48.90	1173.60		
1 Equip. Oper. (crane)	40.25	322.00	66.55	532.40		
1 Equip. Oper. (oiler)	35.45	283.60	58.60	468.80		
1 Lattice Boom Crane, 40 Ton		1182.00		1300.20	24.63	27.09
48 L.H., Daily Totals		$2738.00		$3893.00	$57.04	$81.10

Crew B-41	Hr.	Daily	Hr.	Daily	Bare Costs	Incl. O&P
1 Labor Foreman (outside)	$31.20	$249.60	$52.25	$418.00	$30.35	$50.75
4 Laborers	29.20	934.40	48.90	1564.80		
.25 Equip. Oper. (crane)	40.25	80.50	66.55	133.10		
.25 Equip. Oper. (oiler)	35.45	70.90	58.60	117.20		
.25 Crawler Crane, 40 Ton		300.75		330.82	6.84	7.52
44 L.H., Daily Totals		$1636.15		$2563.93	$37.19	$58.27

Crew B-42	Hr.	Daily	Hr.	Daily	Bare Costs	Incl. O&P
1 Labor Foreman (outside)	$31.20	$249.60	$52.25	$418.00	$33.15	$55.24
4 Laborers	29.20	934.40	48.90	1564.80		
1 Equip. Oper. (crane)	40.25	322.00	66.55	532.40		
1 Welder	43.80	350.40	72.25	578.00		
1 Hyd. Crane, 25 Ton		748.80		823.68		
1 Welder, Gas Engine, 300 amp		147.20		161.92		
1 Horz. Boring Csg. Mch.		477.40		525.14	24.52	26.98
56 L.H., Daily Totals		$3229.80		$4603.94	$57.67	$82.21

Crew B-43	Hr.	Daily	Hr.	Daily	Bare Costs	Incl. O&P
1 Labor Foreman (outside)	$31.20	$249.60	$52.25	$418.00	$29.60	$49.57
4 Laborers	29.20	934.40	48.90	1564.80		
1 Drill Rig, Truck-Mounted		2553.00		2808.30	63.83	70.21
40 L.H., Daily Totals		$3737.00		$4791.10	$93.42	$119.78

Crew B-44	Hr.	Daily	Hr.	Daily	Bare Costs	Incl. O&P
1 Pile Driver Foreman (outside)	$39.75	$318.00	$68.60	$548.80	$36.17	$61.69
4 Pile Drivers	37.75	1208.00	65.15	2084.80		
1 Equip. Oper. (crane)	40.25	322.00	66.55	532.40		
2 Laborers	29.20	467.20	48.90	782.40		
1 Crawler Crane, 40 Ton		1203.00		1323.30		
1 Lead, 60' High		74.00		81.40		
1 Hammer, Diesel, 15K ft.-lbs.		591.40		650.54	29.19	32.11
64 L.H., Daily Totals		$4183.60		$6003.64	$65.37	$93.81

Crew B-45	Hr.	Daily	Hr.	Daily	Bare Costs	Incl. O&P
1 Building Laborer	$29.20	$233.60	$48.90	$391.20	$31.68	$52.63
1 Truck Driver (heavy)	34.15	273.20	56.35	450.80		
1 Dist. Tanker, 3000 Gallon		319.40		351.34	19.96	21.96
16 L.H., Daily Totals		$826.20		$1193.34	$51.64	$74.58

Crew B-46	Hr.	Daily	Hr.	Daily	Bare Costs	Incl. O&P
1 Pile Driver Foreman (outside)	$39.75	$318.00	$68.60	$548.80	$33.81	$57.60
2 Pile Drivers	37.75	604.00	65.15	1042.40		
3 Laborers	29.20	700.80	48.90	1173.60		
1 Chain Saw, Gas, 36" Long		45.00		49.50	0.94	1.03
48 L.H., Daily Totals		$1667.80		$2814.30	$34.75	$58.63

Crew B-47	Hr.	Daily	Hr.	Daily	Bare Costs	Incl. O&P
1 Blast Foreman (outside)	$31.20	$249.60	$52.25	$418.00	$30.20	$50.58
1 Driller	29.20	233.60	48.90	391.20		
1 Air Track Drill, 4"		1045.00		1149.50		
1 Air Compressor, 600 cfm		533.80		587.18		
2 -50' Air Hoses, 3"		27.70		30.47	100.41	110.45
16 L.H., Daily Totals		$2089.70		$2576.35	$130.61	$161.02

Crew B-47A	Hr.	Daily	Hr.	Daily	Bare Costs	Incl. O&P
1 Drilling Foreman (outside)	$31.20	$249.60	$52.25	$418.00	$35.63	$59.13
1 Equip. Oper. (heavy)	40.25	322.00	66.55	532.40		
1 Equip. Oper. (oiler)	35.45	283.60	58.60	468.80		
1 Air Track Drill, 5"		1248.00		1372.80	52.00	57.20
24 L.H., Daily Totals		$2103.20		$2792.00	$87.63	$116.33

Crew B-47C	Hr.	Daily	Hr.	Daily	Bare Costs	Incl. O&P
1 Laborer	$29.20	$233.60	$48.90	$391.20	$33.52	$55.73
1 Equip. Oper. (light)	37.85	302.80	62.55	500.40		
1 Air Compressor, 750 cfm		541.00		595.10		
2 -50' Air Hoses, 3"		27.70		30.47		
1 Air Track Drill, 4"		1045.00		1149.50	100.86	110.94
16 L.H., Daily Totals		$2150.10		$2666.67	$134.38	$166.67

Crew B-47E	Hr.	Daily	Hr.	Daily	Bare Costs	Incl. O&P
1 Labor Foreman (outside)	$31.20	$249.60	$52.25	$418.00	$29.70	$49.74
3 Laborers	29.20	700.80	48.90	1173.60		
1 Flatbed Truck, Gas, 3 Ton		307.40		338.14	9.61	10.57
32 L.H., Daily Totals		$1257.80		$1929.74	$39.31	$60.30

Crew B-47G	Hr.	Daily	Hr.	Daily	Bare Costs	Incl. O&P
1 Labor Foreman (outside)	$31.20	$249.60	$52.25	$418.00	$29.87	$50.02
2 Laborers	29.20	467.20	48.90	782.40		
1 Air Track Drill, 4"		1045.00		1149.50		
1 Air Compressor, 600 cfm		533.80		587.18		
2 -50' Air Hoses, 3"		27.70		30.47		
1 Gunite Pump Rig		373.00		410.30	82.48	90.73
24 L.H., Daily Totals		$2696.30		$3377.85	$112.35	$140.74

Crew B-47H	Hr.	Daily	Hr.	Daily	Bare Costs	Incl. O&P
1 Skilled Worker Foreman (out)	$40.30	$322.40	$67.70	$541.60	$38.80	$65.19
3 Skilled Workers	38.30	919.20	64.35	1544.40		
1 Flatbed Truck, Gas, 3 Ton		307.40		338.14	9.61	10.57
32 L.H., Daily Totals		$1549.00		$2424.14	$48.41	$75.75

Crew B-48	Hr.	Daily	Hr.	Daily	Bare Costs	Incl. O&P
1 Labor Foreman (outside)	$31.20	$249.60	$52.25	$418.00	$31.38	$52.40
4 Laborers	29.20	934.40	48.90	1564.80		
1 Equip. Oper. (crane)	40.25	322.00	66.55	532.40		
1 Centr. Water Pump, 6"		350.80		385.88		
1 -20' Suction Hose, 6"		11.50		12.65		
1 -50' Discharge Hose, 6"		6.10		6.71		
1 Drill Rig, Truck-Mounted		2553.00		2808.30	60.86	66.95
48 L.H., Daily Totals		$4427.40		$5728.74	$92.24	$119.35

Crew No.	Bare Costs		Incl. Subs O&P		Cost Per Labor-Hour	
Crew B-49	**Hr.**	**Daily**	**Hr.**	**Daily**	**Bare Costs**	**Incl. O&P**
1 Labor Foreman (outside)	$31.20	$249.60	$52.25	$418.00	$32.55	$54.84
5 Laborers	29.20	1168.00	48.90	1956.00		
1 Equip. Oper. (crane)	40.25	322.00	66.55	532.40		
2 Pile Drivers	37.75	604.00	65.15	1042.40		
1 Hyd. Crane, 25 Ton		748.80		823.68		
1 Centr. Water Pump, 6"		350.80		385.88		
1 -20' Suction Hose, 6"		11.50		12.65		
1 -50' Discharge Hose, 6"		6.10		6.71		
1 Drill Rig, Truck-Mounted		2553.00		2808.30	50.98	56.07
72 L.H., Daily Totals		$6013.80		$7986.02	$83.53	$110.92

Crew B-50	Hr.	Daily	Hr.	Daily	Bare Costs	Incl. O&P
1 Pile Driver Foreman (outside)	$39.75	$318.00	$68.60	$548.80	$34.81	$59.27
6 Pile Drivers	37.75	1812.00	65.15	3127.20		
1 Equip. Oper. (crane)	40.25	322.00	66.55	532.40		
5 Laborers	29.20	1168.00	48.90	1956.00		
1 Crawler Crane, 40 Ton		1203.00		1323.30		
1 Lead, 60' High		74.00		81.40		
1 Hammer, Diesel, 15K ft.-lbs.		591.40		650.54		
1 Air Compressor, 600 cfm		533.80		587.18		
2 -50' Air Hoses, 3"		27.70		30.47		
1 Chain Saw, Gas, 36" Long		45.00		49.50	23.80	26.18
104 L.H., Daily Totals		$6094.90		$8886.79	$58.60	$85.45

Crew B-51	Hr.	Daily	Hr.	Daily	Bare Costs	Incl. O&P
1 Labor Foreman (outside)	$31.20	$249.60	$52.25	$418.00	$30.18	$50.41
4 Laborers	29.20	934.40	48.90	1564.80		
1 Truck Driver (light)	33.10	264.80	54.60	436.80		
1 Flatbed Truck, Gas, 1.5 Ton		248.20		273.02	5.17	5.69
48 L.H., Daily Totals		$1697.00		$2692.62	$35.35	$56.10

Crew B-52	Hr.	Daily	Hr.	Daily	Bare Costs	Incl. O&P
1 Labor Foreman (outside)	$31.20	$249.60	$52.25	$418.00	$32.19	$53.89
1 Carpenter	37.80	302.40	63.30	506.40		
4 Laborers	29.20	934.40	48.90	1564.80		
.5 Rodman (reinf.)	39.75	159.00	67.15	268.60		
.5 Equip. Oper. (medium)	39.35	157.40	65.05	260.20		
.5 Crawler Loader, 3 C.Y.		598.50		658.35	10.69	11.76
56 L.H., Daily Totals		$2401.30		$3676.35	$42.88	$65.65

Crew B-53	Hr.	Daily	Hr.	Daily	Bare Costs	Incl. O&P
1 Building Laborer	$29.20	$233.60	$48.90	$391.20	$29.20	$48.90
1 Trencher, Chain, 12 H.P.		68.00		74.80	8.50	9.35
8 L.H., Daily Totals		$301.60		$466.00	$37.70	$58.25

Crew B-54	Hr.	Daily	Hr.	Daily	Bare Costs	Incl. O&P
1 Equip. Oper. (light)	$37.85	$302.80	$62.55	$500.40	$37.85	$62.55
1 Trencher, Chain, 40 H.P.		334.40		367.84	41.80	45.98
8 L.H., Daily Totals		$637.20		$868.24	$79.65	$108.53

Crew B-54A	Hr.	Daily	Hr.	Daily	Bare Costs	Incl. O&P
.17 Labor Foreman (outside)	$31.20	$42.43	$52.25	$71.06	$38.17	$63.19
1 Equipment Operator (med.)	39.35	314.80	65.05	520.40		
1 Wheel Trencher, 67 H.P.		1183.00		1301.30	126.39	139.03
9.36 L.H., Daily Totals		$1540.23		$1892.76	$164.55	$202.22

Crew B-54B	Hr.	Daily	Hr.	Daily	Bare Costs	Incl. O&P
.25 Labor Foreman (outside)	$31.20	$62.40	$52.25	$104.50	$37.72	$62.49
1 Equipment Operator (med.)	39.35	314.80	65.05	520.40		
1 Wheel Trencher, 150 H.P.		1891.00		2080.10	189.10	208.01
10 L.H., Daily Totals		$2268.20		$2705.00	$226.82	$270.50

Crew B-54D	Hr.	Daily	Hr.	Daily	Bare Costs	Incl. O&P
1 Laborer	$29.20	$233.60	$48.90	$391.20	$34.27	$56.98
1 Equipment Operator (med.)	39.35	314.80	65.05	520.40		
1 Rock Trencher, 6" Width		615.20		676.72	38.45	42.30
16 L.H., Daily Totals		$1163.60		$1588.32	$72.72	$99.27

Crew B-54E	Hr.	Daily	Hr.	Daily	Bare Costs	Incl. O&P
1 Laborer	$29.20	$233.60	$48.90	$391.20	$34.27	$56.98
1 Equipment Operator (med.)	39.35	314.80	65.05	520.40		
1 Rock Trencher, 18" Width		2591.00		2850.10	161.94	178.13
16 L.H., Daily Totals		$3139.40		$3761.70	$196.21	$235.11

Crew B-55	Hr.	Daily	Hr.	Daily	Bare Costs	Incl. O&P
1 Laborer	$29.20	$233.60	$48.90	$391.20	$31.15	$51.75
1 Truck Driver (light)	33.10	264.80	54.60	436.80		
1 Truck-Mounted Earth Auger		786.00		864.60		
1 Flatbed Truck, Gas, 3 Ton		307.40		338.14	68.34	75.17
16 L.H., Daily Totals		$1591.80		$2030.74	$99.49	$126.92

Crew B-56	Hr.	Daily	Hr.	Daily	Bare Costs	Incl. O&P
2 Laborers	$29.20	$467.20	$48.90	$782.40	$29.20	$48.90
1 Air Track Drill, 4"		1045.00		1149.50		
1 Air Compressor, 600 cfm		533.80		587.18		
1 -50' Air Hose, 3"		13.85		15.23	99.54	109.49
16 L.H., Daily Totals		$2059.85		$2534.32	$128.74	$158.39

Crew B-57	Hr.	Daily	Hr.	Daily	Bare Costs	Incl. O&P
1 Labor Foreman (outside)	$31.20	$249.60	$52.25	$418.00	$31.81	$53.10
3 Laborers	29.20	700.80	48.90	1173.60		
1 Equip. Oper. (crane)	40.25	322.00	66.55	532.40		
1 Barge, 400 Ton		779.20		857.12		
1 Crawler Crane, 25 Ton		1194.00		1313.40		
1 Clamshell Bucket, 1 C.Y.		47.40		52.14		
1 Centr. Water Pump, 6"		350.80		385.88		
1 -20' Suction Hose, 6"		11.50		12.65		
20 -50' Discharge Hoses, 6"		122.00		134.20	62.62	68.88
40 L.H., Daily Totals		$3777.30		$4879.39	$94.43	$121.98

Crew B-58	Hr.	Daily	Hr.	Daily	Bare Costs	Incl. O&P
2 Laborers	$29.20	$467.20	$48.90	$782.40	$32.08	$53.45
1 Equip. Oper. (light)	37.85	302.80	62.55	500.40		
1 Backhoe Loader, 48 H.P.		365.40		401.94		
1 Small Helicopter, w/ Pilot		2774.00		3051.40	130.81	143.89
24 L.H., Daily Totals		$3909.40		$4736.14	$162.89	$197.34

Crew B-59	Hr.	Daily	Hr.	Daily	Bare Costs	Incl. O&P
1 Truck Driver (heavy)	$34.15	$273.20	$56.35	$450.80	$34.15	$56.35
1 Truck Tractor, 220 H.P.		358.80		394.68		
1 Water Tank Trailer, 5000 Gal.		141.60		155.76	62.55	68.81
8 L.H., Daily Totals		$773.60		$1001.24	$96.70	$125.16

Crew B-60	Hr.	Daily	Hr.	Daily	Bare Costs	Incl. O&P
1 Labor Foreman (outside)	$31.20	$249.60	$52.25	$418.00	$32.82	$54.67
3 Laborers	29.20	700.80	48.90	1173.60		
1 Equip. Oper. (crane)	40.25	322.00	66.55	532.40		
1 Equip. Oper. (light)	37.85	302.80	62.55	500.40		
1 Crawler Crane, 40 Ton		1203.00		1323.30		
1 Lead, 60' High		74.00		81.40		
1 Hammer, Diesel, 15K ft.-lbs.		591.40		650.54		
1 Backhoe Loader, 48 H.P.		365.40		401.94	46.54	51.19
48 L.H., Daily Totals		$3809.00		$5081.58	$79.35	$105.87

Crews - Open Shop

Crew No.	Bare Costs		Incl. Subs O&P		Cost Per Labor-Hour	
Crew B-61	**Hr.**	**Daily**	**Hr.**	**Daily**	**Bare Costs**	**Incl. O&P**
1 Labor Foreman (outside)	$31.20	$249.60	$52.25	$418.00	$29.60	$49.57
4 Laborers	29.20	934.40	48.90	1564.80		
1 Cement Mixer, 2 C.Y.		193.40		212.74		
1 Air Compressor, 160 cfm		162.20		178.42	8.89	9.78
40 L.H., Daily Totals		$1539.60		$2373.96	$38.49	$59.35

Crew B-62	Hr.	Daily	Hr.	Daily	Bare Costs	Incl. O&P
2 Laborers	$29.20	$467.20	$48.90	$782.40	$32.08	$53.45
1 Equip. Oper. (light)	37.85	302.80	62.55	500.40		
1 Loader, Skid Steer, 30 H.P.		173.20		190.52	7.22	7.94
24 L.H., Daily Totals		$943.20		$1473.32	$39.30	$61.39

Crew B-62A	Hr.	Daily	Hr.	Daily	Bare Costs	Incl. O&P
2 Laborers	$29.20	$467.20	$48.90	$782.40	$32.08	$53.45
1 Equip. Oper. (light)	37.85	302.80	62.55	500.40		
1 Loader, Skid Steer, 30 H.P.		173.20		190.52		
1 Trencher Attachment		63.70		70.07	9.87	10.86
24 L.H., Daily Totals		$1006.90		$1543.39	$41.95	$64.31

Crew B-63	Hr.	Daily	Hr.	Daily	Bare Costs	Incl. O&P
5 Laborers	$29.20	$1168.00	$48.90	$1956.00	$29.20	$48.90
1 Loader, Skid Steer, 30 H.P.		173.20		190.52	4.33	4.76
40 L.H., Daily Totals		$1341.20		$2146.52	$33.53	$53.66

Crew B-63B	Hr.	Daily	Hr.	Daily	Bare Costs	Incl. O&P
1 Labor Foreman (inside)	$29.70	$237.60	$49.70	$397.60	$31.49	$52.51
2 Laborers	29.20	467.20	48.90	782.40		
1 Equip. Oper. (light)	37.85	302.80	62.55	500.40		
1 Loader, Skid Steer, 78 H.P.		315.60		347.16	9.86	10.85
32 L.H., Daily Totals		$1323.20		$2027.56	$41.35	$63.36

Crew B-64	Hr.	Daily	Hr.	Daily	Bare Costs	Incl. O&P
1 Laborer	$29.20	$233.60	$48.90	$391.20	$31.15	$51.75
1 Truck Driver (light)	33.10	264.80	54.60	436.80		
1 Power Mulcher (small)		150.80		165.88		
1 Flatbed Truck, Gas, 1.5 Ton		248.20		273.02	24.94	27.43
16 L.H., Daily Totals		$897.40		$1266.90	$56.09	$79.18

Crew B-65	Hr.	Daily	Hr.	Daily	Bare Costs	Incl. O&P
1 Laborer	$29.20	$233.60	$48.90	$391.20	$31.15	$51.75
1 Truck Driver (light)	33.10	264.80	54.60	436.80		
1 Power Mulcher (Large)		319.60		351.56		
1 Flatbed Truck, Gas, 1.5 Ton		248.20		273.02	35.49	39.04
16 L.H., Daily Totals		$1066.20		$1452.58	$66.64	$90.79

Crew B-66	Hr.	Daily	Hr.	Daily	Bare Costs	Incl. O&P
1 Equip. Oper. (light)	$37.85	$302.80	$62.55	$500.40	$37.85	$62.55
1 Loader-Backhoe, 40 H.P.		263.80		290.18	32.98	36.27
8 L.H., Daily Totals		$566.60		$790.58	$70.83	$98.82

Crew B-67	Hr.	Daily	Hr.	Daily	Bare Costs	Incl. O&P
1 Millwright	$39.85	$318.80	$64.00	$512.00	$38.85	$63.27
1 Equip. Oper. (light)	37.85	302.80	62.55	500.40		
1 Forklift, R/T, 4,000 Lb.		313.60		344.96	19.60	21.56
16 L.H., Daily Totals		$935.20		$1357.36	$58.45	$84.83

Crew B-67B	Hr.	Daily	Hr.	Daily	Bare Costs	Incl. O&P
1 Millwright Foreman (inside)	$40.35	$322.80	$64.80	$518.40	$40.10	$64.40
1 Millwright	39.85	318.80	64.00	512.00		
16 L.H., Daily Totals		$641.60		$1030.40	$40.10	$64.40

Crew B-68	Hr.	Daily	Hr.	Daily	Bare Costs	Incl. O&P
2 Millwrights	$39.85	$637.60	$64.00	$1024.00	$39.18	$63.52
1 Equip. Oper. (light)	37.85	302.80	62.55	500.40		
1 Forklift, R/T, 4,000 Lb.		313.60		344.96	13.07	14.37
24 L.H., Daily Totals		$1254.00		$1869.36	$52.25	$77.89

Crew B-68A	Hr.	Daily	Hr.	Daily	Bare Costs	Incl. O&P
1 Millwright Foreman (inside)	$40.35	$322.80	$64.80	$518.40	$40.02	$64.27
2 Millwrights	39.85	637.60	64.00	1024.00		
1 Forklift, 8,000 Lb.		185.40		203.94	7.72	8.50
24 L.H., Daily Totals		$1145.80		$1746.34	$47.74	$72.76

Crew B-68B	Hr.	Daily	Hr.	Daily	Bare Costs	Incl. O&P
1 Millwright Foreman (inside)	$40.35	$322.80	$64.80	$518.40	$42.25	$68.77
2 Millwrights	39.85	637.60	64.00	1024.00		
2 Electricians	44.05	704.80	72.05	1152.80		
2 Plumbers	43.80	700.80	72.25	1156.00		
1 Forklift, 5,000 Lb.		334.00		367.40	5.96	6.56
56 L.H., Daily Totals		$2700.00		$4218.60	$48.21	$75.33

Crew B-68C	Hr.	Daily	Hr.	Daily	Bare Costs	Incl. O&P
1 Millwright Foreman (inside)	$40.35	$322.80	$64.80	$518.40	$42.01	$68.28
1 Millwright	39.85	318.80	64.00	512.00		
1 Electrician	44.05	352.40	72.05	576.40		
1 Plumber	43.80	350.40	72.25	578.00		
1 Forklift, 5,000 Lb.		334.00		367.40	10.44	11.48
32 L.H., Daily Totals		$1678.40		$2552.20	$52.45	$79.76

Crew B-68D	Hr.	Daily	Hr.	Daily	Bare Costs	Incl. O&P
1 Labor Foreman (inside)	$29.70	$237.60	$49.70	$397.60	$32.25	$53.72
1 Laborer	29.20	233.60	48.90	391.20		
1 Equip. Oper. (light)	37.85	302.80	62.55	500.40		
1 Forklift, 5,000 Lb.		334.00		367.40	13.92	15.31
24 L.H., Daily Totals		$1108.00		$1656.60	$46.17	$69.03

Crew B-68E	Hr.	Daily	Hr.	Daily	Bare Costs	Incl. O&P
1 Struc. Steel Foreman (inside)	$40.40	$323.20	$74.10	$592.80	$40.00	$73.38
3 Struc. Steel Workers	39.90	957.60	73.20	1756.80		
1 Welder	39.90	319.20	73.20	585.60		
1 Forklift, 8,000 Lb.		185.40		203.94	4.63	5.10
40 L.H., Daily Totals		$1785.40		$3139.14	$44.63	$78.48

Crew B-68F	Hr.	Daily	Hr.	Daily	Bare Costs	Incl. O&P
1 Skilled Worker Foreman (out)	$40.30	$322.40	$67.70	$541.60	$38.97	$65.47
2 Skilled Workers	38.30	612.80	64.35	1029.60		
1 Forklift, 5,000 Lb.		334.00		367.40	13.92	15.31
24 L.H., Daily Totals		$1269.20		$1938.60	$52.88	$80.78

Crew B-68G	Hr.	Daily	Hr.	Daily	Bare Costs	Incl. O&P
2 Structural Steel Workers	$39.90	$638.40	$73.20	$1171.20	$39.90	$73.20
1 Forklift, 5,000 Lb.		334.00		367.40	20.88	22.96
16 L.H., Daily Totals		$972.40		$1538.60	$60.77	$96.16

Crew B-69	Hr.	Daily	Hr.	Daily	Bare Costs	Incl. O&P
1 Labor Foreman (outside)	$31.20	$249.60	$52.25	$418.00	$32.42	$54.02
3 Laborers	29.20	700.80	48.90	1173.60		
1 Equip. Oper. (crane)	40.25	322.00	66.55	532.40		
1 Equip. Oper. (oiler)	35.45	283.60	58.60	468.80		
1 Hyd. Crane, 80 Ton		1630.00		1793.00	33.96	37.35
48 L.H., Daily Totals		$3186.00		$4385.80	$66.38	$91.37

Crews - Open Shop

Crew No.	Bare Costs		Incl. Subs O&P		Cost Per Labor-Hour	
Crew B-69A	**Hr.**	**Daily**	**Hr.**	**Daily**	**Bare Costs**	**Incl. O&P**
1 Labor Foreman (outside)	$31.20	$249.60	$52.25	$418.00	$32.29	$53.63
3 Laborers	29.20	700.80	48.90	1173.60		
1 Equip. Oper. (medium)	39.35	314.80	65.05	520.40		
1 Concrete Finisher	35.60	284.80	57.80	462.40		
1 Curb/Gutter Paver, 2-Track		1102.00		1212.20	22.96	25.25
48 L.H., Daily Totals		$2652.00		$3786.60	$55.25	$78.89

Crew No.	Bare Costs		Incl. Subs O&P		Cost Per Labor-Hour	
Crew B-69B	**Hr.**	**Daily**	**Hr.**	**Daily**	**Bare Costs**	**Incl. O&P**
1 Labor Foreman (outside)	$31.20	$249.60	$52.25	$418.00	$32.29	$53.63
3 Laborers	29.20	700.80	48.90	1173.60		
1 Equip. Oper. (medium)	39.35	314.80	65.05	520.40		
1 Cement Finisher	35.60	284.80	57.80	462.40		
1 Curb/Gutter Paver, 4-Track		780.20		858.22	16.25	17.88
48 L.H., Daily Totals		$2330.20		$3432.62	$48.55	$71.51

Crew No.	Bare Costs		Incl. Subs O&P		Cost Per Labor-Hour	
Crew B-70	**Hr.**	**Daily**	**Hr.**	**Daily**	**Bare Costs**	**Incl. O&P**
1 Labor Foreman (outside)	$31.20	$249.60	$52.25	$418.00	$33.84	$56.30
3 Laborers	29.20	700.80	48.90	1173.60		
3 Equip. Opers. (medium)	39.35	944.40	65.05	1561.20		
1 Grader, 30,000 Lbs.		714.40		785.84		
1 Ripper, Beam & 1 Shank		83.80		92.18		
1 Road Sweeper, S.P., 8' wide		666.20		732.82		
1 F.E. Loader, W.M., 1.5 C.Y.		378.60		416.46	32.91	36.20
56 L.H., Daily Totals		$3737.80		$5180.10	$66.75	$92.50

Crew No.	Bare Costs		Incl. Subs O&P		Cost Per Labor-Hour	
Crew B-71	**Hr.**	**Daily**	**Hr.**	**Daily**	**Bare Costs**	**Incl. O&P**
1 Labor Foreman (outside)	$31.20	$249.60	$52.25	$418.00	$33.84	$56.30
3 Laborers	29.20	700.80	48.90	1173.60		
3 Equip. Opers. (medium)	39.35	944.40	65.05	1561.20		
1 Pvmt. Profiler, 750 H.P.		5835.00		6418.50		
1 Road Sweeper, S.P., 8' wide		666.20		732.82		
1 F.E. Loader, W.M., 1.5 C.Y.		378.60		416.46	122.85	135.14
56 L.H., Daily Totals		$8774.60		$10720.58	$156.69	$191.44

Crew No.	Bare Costs		Incl. Subs O&P		Cost Per Labor-Hour	
Crew B-72	**Hr.**	**Daily**	**Hr.**	**Daily**	**Bare Costs**	**Incl. O&P**
1 Labor Foreman (outside)	$31.20	$249.60	$52.25	$418.00	$34.52	$57.39
3 Laborers	29.20	700.80	48.90	1173.60		
4 Equip. Opers. (medium)	39.35	1259.20	65.05	2081.60		
1 Pvmt. Profiler, 750 H.P.		5835.00		6418.50		
1 Hammermill, 250 H.P.		1832.00		2015.20		
1 Windrow Loader		1247.00		1371.70		
1 Mix Paver 165 H.P.		2226.00		2448.60		
1 Roller, Pneum. Whl., 12 Ton		347.80		382.58	179.50	197.45
64 L.H., Daily Totals		$13697.40		$16309.78	$214.02	$254.84

Crew No.	Bare Costs		Incl. Subs O&P		Cost Per Labor-Hour	
Crew B-73	**Hr.**	**Daily**	**Hr.**	**Daily**	**Bare Costs**	**Incl. O&P**
1 Labor Foreman (outside)	$31.20	$249.60	$52.25	$418.00	$35.79	$59.41
2 Laborers	29.20	467.20	48.90	782.40		
5 Equip. Opers. (medium)	39.35	1574.00	65.05	2602.00		
1 Road Mixer, 310 H.P.		1945.00		2139.50		
1 Tandem Roller, 10 Ton		239.40		263.34		
1 Hammermill, 250 H.P.		1832.00		2015.20		
1 Grader, 30,000 Lbs.		714.40		785.84		
.5 F.E. Loader, W.M., 1.5 C.Y.		189.30		208.23		
.5 Truck Tractor, 220 H.P.		179.40		197.34		
.5 Water Tank Trailer, 5000 Gal.		70.80		77.88	80.79	88.86
64 L.H., Daily Totals		$7461.10		$9489.73	$116.58	$148.28

Crew No.	Bare Costs		Incl. Subs O&P		Cost Per Labor-Hour	
Crew B-74	**Hr.**	**Daily**	**Hr.**	**Daily**	**Bare Costs**	**Incl. O&P**
1 Labor Foreman (outside)	$31.20	$249.60	$52.25	$418.00	$35.76	$59.26
1 Laborer	29.20	233.60	48.90	391.20		
4 Equip. Opers. (medium)	39.35	1259.20	65.05	2081.60		
2 Truck Drivers (heavy)	34.15	546.40	56.35	901.60		
1 Grader, 30,000 Lbs.		714.40		785.84		
1 Ripper, Beam & 1 Shank		83.80		92.18		
2 Stabilizers, 310 H.P.		3666.00		4032.60		
1 Flatbed Truck, Gas, 3 Ton		307.40		338.14		
1 Chem. Spreader, Towed		52.00		57.20		
1 Roller, Vibratory, 25 Ton		691.80		760.98		
1 Water Tank Trailer, 5000 Gal.		141.60		155.76		
1 Truck Tractor, 220 H.P.		358.80		394.68	94.00	103.40
64 L.H., Daily Totals		$8304.60		$10409.78	$129.76	$162.65

Crew No.	Bare Costs		Incl. Subs O&P		Cost Per Labor-Hour	
Crew B-75	**Hr.**	**Daily**	**Hr.**	**Daily**	**Bare Costs**	**Incl. O&P**
1 Labor Foreman (outside)	$31.20	$249.60	$52.25	$418.00	$35.99	$59.67
1 Laborer	29.20	233.60	48.90	391.20		
4 Equip. Opers. (medium)	39.35	1259.20	65.05	2081.60		
1 Truck Driver (heavy)	34.15	273.20	56.35	450.80		
1 Grader, 30,000 Lbs.		714.40		785.84		
1 Ripper, Beam & 1 Shank		83.80		92.18		
2 Stabilizers, 310 H.P.		3666.00		4032.60		
1 Dist. Tanker, 3000 Gallon		319.40		351.34		
1 Truck Tractor, 6x4, 380 H.P.		600.00		660.00		
1 Roller, Vibratory, 25 Ton		691.80		760.98	108.49	119.34
56 L.H., Daily Totals		$8091.00		$10024.54	$144.48	$179.01

Crew No.	Bare Costs		Incl. Subs O&P		Cost Per Labor-Hour	
Crew B-76	**Hr.**	**Daily**	**Hr.**	**Daily**	**Bare Costs**	**Incl. O&P**
1 Dock Builder Foreman (outside)	$39.75	$318.00	$68.60	$548.80	$38.27	$65.12
5 Dock Builders	37.75	1510.00	65.15	2606.00		
2 Equip. Opers. (crane)	40.25	644.00	66.55	1064.80		
1 Equip. Oper. (oiler)	35.45	283.60	58.60	468.80		
1 Crawler Crane, 50 Ton		1346.00		1480.60		
1 Barge, 400 Ton		779.20		857.12		
1 Hammer, Diesel, 15K ft.-lbs.		591.40		650.54		
1 Lead, 60' High		74.00		81.40		
1 Air Compressor, 600 cfm		533.80		587.18		
2 -50' Air Hoses, 3"		27.70		30.47	46.56	51.21
72 L.H., Daily Totals		$6107.70		$8375.71	$84.83	$116.33

Crew No.	Bare Costs		Incl. Subs O&P		Cost Per Labor-Hour	
Crew B-76A	**Hr.**	**Daily**	**Hr.**	**Daily**	**Bare Costs**	**Incl. O&P**
1 Labor Foreman (outside)	$31.20	$249.60	$52.25	$418.00	$31.61	$52.74
5 Laborers	29.20	1168.00	48.90	1956.00		
1 Equip. Oper. (crane)	40.25	322.00	66.55	532.40		
1 Equip. Oper. (oiler)	35.45	283.60	58.60	468.80		
1 Crawler Crane, 50 Ton		1346.00		1480.60		
1 Barge, 400 Ton		779.20		857.12	33.21	36.53
64 L.H., Daily Totals		$4148.40		$5712.92	$64.82	$89.26

Crew No.	Bare Costs		Incl. Subs O&P		Cost Per Labor-Hour	
Crew B-77	**Hr.**	**Daily**	**Hr.**	**Daily**	**Bare Costs**	**Incl. O&P**
1 Labor Foreman (outside)	$31.20	$249.60	$52.25	$418.00	$30.38	$50.71
3 Laborers	29.20	700.80	48.90	1173.60		
1 Truck Driver (light)	33.10	264.80	54.60	436.80		
1 Crack Cleaner, 25 H.P.		62.00		68.20		
1 Crack Filler, Trailer Mtd.		202.60		222.86		
1 Flatbed Truck, Gas, 3 Ton		307.40		338.14	14.30	15.73
40 L.H., Daily Totals		$1787.20		$2657.60	$44.68	$66.44

Crews - Open Shop

Crew No.	Bare Costs		Incl. Subs O&P		Cost Per Labor-Hour	
Crew B-78	**Hr.**	**Daily**	**Hr.**	**Daily**	**Bare Costs**	**Incl. O&P**
1 Labor Foreman (outside)	$31.20	$249.60	$52.25	$418.00	$29.60	$49.57
4 Laborers	29.20	934.40	48.90	1564.80		
1 Paint Striper, S.P., 40 Gallon		152.00		167.20		
1 Flatbed Truck, Gas, 3 Ton		307.40		338.14		
1 Pickup Truck, 3/4 Ton		145.80		160.38	15.13	16.64
40 L.H., Daily Totals		$1789.20		$2648.52	$44.73	$66.21

Crew B-78B	Hr.	Daily	Hr.	Daily	Bare Costs	Incl. O&P
2 Laborers	$29.20	$467.20	$48.90	$782.40	$30.16	$50.42
.25 Equip. Oper. (light)	37.85	75.70	62.55	125.10		
1 Pickup Truck, 3/4 Ton		145.80		160.38		
1 Line Rem.,11 H.P.,Walk Behind		65.60		72.16		
.25 Road Sweeper, S.P., 8' wide		166.55		183.21	21.00	23.10
18 L.H., Daily Totals		$920.85		$1323.24	$51.16	$73.51

Crew B-78C	Hr.	Daily	Hr.	Daily	Bare Costs	Incl. O&P
1 Labor Foreman (outside)	$31.20	$249.60	$52.25	$418.00	$30.18	$50.41
4 Laborers	29.20	934.40	48.90	1564.80		
1 Truck Driver (light)	33.10	264.80	54.60	436.80		
1 Paint Striper, T.M., 120 Gal.		810.20		891.22		
1 Flatbed Truck, Gas, 3 Ton		307.40		338.14		
1 Pickup Truck, 3/4 Ton		145.80		160.38	26.32	28.95
48 L.H., Daily Totals		$2712.20		$3809.34	$56.50	$79.36

Crew B-78D	Hr.	Daily	Hr.	Daily	Bare Costs	Incl. O&P
2 Labor Foremen (outside)	$31.20	$499.20	$52.25	$836.00	$29.99	$50.14
7 Laborers	29.20	1635.20	48.90	2738.40		
1 Truck Driver (light)	33.10	264.80	54.60	436.80		
1 Paint Striper, T.M., 120 Gal.		810.20		891.22		
1 Flatbed Truck, Gas, 3 Ton		307.40		338.14		
3 Pickup Trucks, 3/4 Ton		437.40		481.14		
1 Air Compressor, 60 cfm		142.80		157.08		
1 -50' Air Hose, 3/4"		3.05		3.36		
1 Breakers, Pavement, 60 lb.		10.20		11.22	21.39	23.53
80 L.H., Daily Totals		$4110.25		$5893.35	$51.38	$73.67

Crew B-78E	Hr.	Daily	Hr.	Daily	Bare Costs	Incl. O&P
2 Labor Foremen (outside)	$31.20	$499.20	$52.25	$836.00	$29.86	$49.93
9 Laborers	29.20	2102.40	48.90	3520.80		
1 Truck Driver (light)	33.10	264.80	54.60	436.80		
1 Paint Striper, T.M., 120 Gal.		810.20		891.22		
1 Flatbed Truck, Gas, 3 Ton		307.40		338.14		
4 Pickup Trucks, 3/4 Ton		583.20		641.52		
2 Air Compressors, 60 cfm		285.60		314.16		
2 -50' Air Hoses, 3/4"		6.10		6.71		
2 Breakers, Pavement, 60 lb.		20.40		22.44	20.97	23.06
96 L.H., Daily Totals		$4879.30		$7007.79	$50.83	$73.00

Crew B-78F	Hr.	Daily	Hr.	Daily	Bare Costs	Incl. O&P
2 Labor Foremen (outside)	$31.20	$499.20	$52.25	$836.00	$29.76	$49.79
11 Laborers	29.20	2569.60	48.90	4303.20		
1 Truck Driver (light)	33.10	264.80	54.60	436.80		
1 Paint Striper, T.M., 120 Gal.		810.20		891.22		
1 Flatbed Truck, Gas, 3 Ton		307.40		338.14		
7 Pickup Trucks, 3/4 Ton		1020.60		1122.66		
3 Air Compressors, 60 cfm		428.40		471.24		
3 -50' Air Hoses, 3/4"		9.15		10.07		
3 Breakers, Pavement, 60 lb.		30.60		33.66	23.27	25.60
112 L.H., Daily Totals		$5939.95		$8442.99	$53.04	$75.38

Crew No.	Bare Costs		Incl. Subs O&P		Cost Per Labor-Hour	
Crew B-79	**Hr.**	**Daily**	**Hr.**	**Daily**	**Bare Costs**	**Incl. O&P**
1 Labor Foreman (outside)	$31.20	$249.60	$52.25	$418.00	$30.38	$50.71
3 Laborers	29.20	700.80	48.90	1173.60		
1 Truck Driver (light)	33.10	264.80	54.60	436.80		
1 Paint Striper, T.M., 120 Gal.		810.20		891.22		
1 Heating Kettle, 115 Gallon		81.70		89.87		
1 Flatbed Truck, Gas, 3 Ton		307.40		338.14		
2 Pickup Trucks, 3/4 Ton		291.60		320.76	37.27	41.00
40 L.H., Daily Totals		$2706.10		$3668.39	$67.65	$91.71

Crew B-79B	Hr.	Daily	Hr.	Daily	Bare Costs	Incl. O&P
1 Laborer	$29.20	$233.60	$48.90	$391.20	$29.20	$48.90
1 Set of Gases		168.00		184.80	21.00	23.10
8 L.H., Daily Totals		$401.60		$576.00	$50.20	$72.00

Crew B-79C	Hr.	Daily	Hr.	Daily	Bare Costs	Incl. O&P
1 Labor Foreman (outside)	$31.20	$249.60	$52.25	$418.00	$30.04	$50.19
5 Laborers	29.20	1168.00	48.90	1956.00		
1 Truck Driver (light)	33.10	264.80	54.60	436.80		
1 Paint Striper, T.M., 120 Gal.		810.20		891.22		
1 Heating Kettle, 115 Gallon		81.70		89.87		
1 Flatbed Truck, Gas, 3 Ton		307.40		338.14		
3 Pickup Trucks, 3/4 Ton		437.40		481.14		
1 Air Compressor, 60 cfm		142.80		157.08		
1 -50' Air Hose, 3/4"		3.05		3.36		
1 Breaker, Pavement, 60 lb.		10.20		11.22	32.01	35.21
56 L.H., Daily Totals		$3475.15		$4782.82	$62.06	$85.41

Crew B-79D	Hr.	Daily	Hr.	Daily	Bare Costs	Incl. O&P
2 Labor Foremen (outside)	$31.20	$499.20	$52.25	$836.00	$30.19	$50.45
5 Laborers	29.20	1168.00	48.90	1956.00		
1 Truck Driver (light)	33.10	264.80	54.60	436.80		
1 Paint Striper, T.M., 120 Gal.		810.20		891.22		
1 Heating Kettle, 115 Gallon		81.70		89.87		
1 Flatbed Truck, Gas, 3 Ton		307.40		338.14		
4 Pickup Trucks, 3/4 Ton		583.20		641.52		
1 Air Compressor, 60 cfm		142.80		157.08		
1 -50' Air Hose, 3/4"		3.05		3.36		
1 Breaker, Pavement, 60 lb.		10.20		11.22	30.29	33.32
64 L.H., Daily Totals		$3870.55		$5361.20	$60.48	$83.77

Crew B-79E	Hr.	Daily	Hr.	Daily	Bare Costs	Incl. O&P
2 Labor Foremen (outside)	$31.20	$499.20	$52.25	$836.00	$29.99	$50.14
7 Laborers	29.20	1635.20	48.90	2738.40		
1 Truck Driver (light)	33.10	264.80	54.60	436.80		
1 Paint Striper, T.M., 120 Gal.		810.20		891.22		
1 Heating Kettle, 115 Gallon		81.70		89.87		
1 Flatbed Truck, Gas, 3 Ton		307.40		338.14		
5 Pickup Trucks, 3/4 Ton		729.00		801.90		
2 Air Compressors, 60 cfm		285.60		314.16		
2 -50' Air Hoses, 3/4"		6.10		6.71		
2 Breakers, Pavement, 60 lb.		20.40		22.44	28.00	30.81
80 L.H., Daily Totals		$4639.60		$6475.64	$57.99	$80.95

Crew B-80	Hr.	Daily	Hr.	Daily	Bare Costs	Incl. O&P
1 Labor Foreman (outside)	$31.20	$249.60	$52.25	$418.00	$29.87	$50.02
2 Laborers	29.20	467.20	48.90	782.40		
1 Flatbed Truck, Gas, 3 Ton		307.40		338.14		
1 Earth Auger, Truck-Mtd.		420.80		462.88	30.34	33.38
24 L.H., Daily Totals		$1445.00		$2001.42	$60.21	$83.39

Crew No.	Bare Costs		Incl. Subs O&P		Cost Per Labor-Hour	
Crew B-80A	**Hr.**	**Daily**	**Hr.**	**Daily**	**Bare Costs**	**Incl. O&P**
3 Laborers	$29.20	$700.80	$48.90	$1173.60	$29.20	$48.90
1 Flatbed Truck, Gas, 3 Ton		307.40		338.14	12.81	14.09
24 L.H., Daily Totals		$1008.20		$1511.74	$42.01	$62.99

Crew B-80B	Hr.	Daily	Hr.	Daily	Bare Costs	Incl. O&P
3 Laborers	$29.20	$700.80	$48.90	$1173.60	$31.36	$52.31
1 Equip. Oper. (light)	37.85	302.80	62.55	500.40		
1 Crane, Flatbed Mounted, 3 Ton		248.20		273.02	7.76	8.53
32 L.H., Daily Totals		$1251.80		$1947.02	$39.12	$60.84

Crew B-80C	Hr.	Daily	Hr.	Daily	Bare Costs	Incl. O&P
2 Laborers	$29.20	$467.20	$48.90	$782.40	$30.50	$50.80
1 Truck Driver (light)	33.10	264.80	54.60	436.80		
1 Flatbed Truck, Gas, 1.5 Ton		248.20		273.02		
1 Manual Fence Post Auger, Gas		8.40		9.24	10.69	11.76
24 L.H., Daily Totals		$988.60		$1501.46	$41.19	$62.56

Crew B-81	Hr.	Daily	Hr.	Daily	Bare Costs	Incl. O&P
1 Laborer	$29.20	$233.60	$48.90	$391.20	$31.68	$52.63
1 Truck Driver (heavy)	34.15	273.20	56.35	450.80		
1 Hydromulcher, T.M., 3000 Gal.		334.40		367.84		
1 Truck Tractor, 220 H.P.		358.80		394.68	43.33	47.66
16 L.H., Daily Totals		$1200.00		$1604.52	$75.00	$100.28

Crew B-81A	Hr.	Daily	Hr.	Daily	Bare Costs	Incl. O&P
1 Laborer	$29.20	$233.60	$48.90	$391.20	$31.15	$51.75
1 Truck Driver (light)	33.10	264.80	54.60	436.80		
1 Hydromulcher, T.M., 600 Gal.		127.80		140.58		
1 Flatbed Truck, Gas, 3 Ton		307.40		338.14	27.20	29.92
16 L.H., Daily Totals		$933.60		$1306.72	$58.35	$81.67

Crew B-82	Hr.	Daily	Hr.	Daily	Bare Costs	Incl. O&P
1 Laborer	$29.20	$233.60	$48.90	$391.20	$33.52	$55.73
1 Equip. Oper. (light)	37.85	302.80	62.55	500.40		
1 Horiz. Borer, 6 H.P.		84.20		92.62	5.26	5.79
16 L.H., Daily Totals		$620.60		$984.22	$38.79	$61.51

Crew B-82A	Hr.	Daily	Hr.	Daily	Bare Costs	Incl. O&P
2 Laborers	$29.20	$467.20	$48.90	$782.40	$33.52	$55.73
2 Equip. Opers. (light)	37.85	605.60	62.55	1000.80		
2 Dump Trucks, 8 C.Y., 220 H.P.		834.80		918.28		
1 Flatbed Trailer, 25 Ton		113.20		124.52		
1 Horiz. Dir. Drill, 20k lb. Thrust		657.00		722.70		
1 Mud Trailer for HDD, 1500 Gal.		350.00		385.00		
1 Pickup Truck, 4x4, 3/4 Ton		156.80		172.48		
1 Flatbed trailer, 3 Ton		23.40		25.74		
1 Loader, Skid Steer, 78 H.P.		315.60		347.16	76.59	84.25
32 L.H., Daily Totals		$3523.60		$4479.08	$110.11	$139.97

Crew B-82B	Hr.	Daily	Hr.	Daily	Bare Costs	Incl. O&P
2 Laborers	$29.20	$467.20	$48.90	$782.40	$33.52	$55.73
2 Equip. Opers. (light)	37.85	605.60	62.55	1000.80		
2 Dump Trucks, 8 C.Y., 220 H.P.		834.80		918.28		
1 Flatbed Trailer, 25 Ton		113.20		124.52		
1 Horiz. Dir. Drill, 30k lb. Thrust		935.80		1029.38		
1 Mud Trailer for HDD, 1500 Gal.		350.00		385.00		
1 Pickup Truck, 4x4, 3/4 Ton		156.80		172.48		
1 Flatbed trailer, 3 Ton		23.40		25.74		
1 Loader, Skid Steer, 78 H.P.		315.60		347.16	85.30	93.83
32 L.H., Daily Totals		$3802.40		$4785.76	$118.83	$149.56

Crew B-82C	Hr.	Daily	Hr.	Daily	Bare Costs	Incl. O&P
2 Laborers	$29.20	$467.20	$48.90	$782.40	$33.52	$55.73
2 Equip. Opers. (light)	37.85	605.60	62.55	1000.80		
2 Dump Trucks, 8 C.Y., 220 H.P.		834.80		918.28		
1 Flatbed Trailer, 25 Ton		113.20		124.52		
1 Horiz. Dir. Drill, 50k lb. Thrust		1244.00		1368.40		
1 Mud Trailer for HDD, 1500 Gal.		350.00		385.00		
1 Pickup Truck, 4x4, 3/4 Ton		156.80		172.48		
1 Flatbed trailer, 3 Ton		23.40		25.74		
1 Loader, Skid Steer, 78 H.P.		315.60		347.16	94.93	104.42
32 L.H., Daily Totals		$4110.60		$5124.78	$128.46	$160.15

Crew B-82D	Hr.	Daily	Hr.	Daily	Bare Costs	Incl. O&P
1 Equip. Oper. (light)	$37.85	$302.80	$62.55	$500.40	$37.85	$62.55
1 Mud Trailer for HDD, 1500 Gal.		350.00		385.00	43.75	48.13
8 L.H., Daily Totals		$652.80		$885.40	$81.60	$110.68

Crew B-83	Hr.	Daily	Hr.	Daily	Bare Costs	Incl. O&P
1 Tugboat Captain	$39.35	$314.80	$65.05	$520.40	$34.27	$56.98
1 Tugboat Hand	29.20	233.60	48.90	391.20		
1 Tugboat, 250 H.P.		868.60		955.46	54.29	59.72
16 L.H., Daily Totals		$1417.00		$1867.06	$88.56	$116.69

Crew B-84	Hr.	Daily	Hr.	Daily	Bare Costs	Incl. O&P
1 Equip. Oper. (medium)	$39.35	$314.80	$65.05	$520.40	$39.35	$65.05
1 Rotary Mower/Tractor		369.40		406.34	46.17	50.79
8 L.H., Daily Totals		$684.20		$926.74	$85.53	$115.84

Crew B-85	Hr.	Daily	Hr.	Daily	Bare Costs	Incl. O&P
3 Laborers	$29.20	$700.80	$48.90	$1173.60	$32.22	$53.62
1 Equip. Oper. (medium)	39.35	314.80	65.05	520.40		
1 Truck Driver (heavy)	34.15	273.20	56.35	450.80		
1 Aerial Lift Truck, 80'		659.20		725.12		
1 Brush Chipper, 12", 130 H.P.		388.60		427.46		
1 Pruning Saw, Rotary		6.65		7.32	26.36	29.00
40 L.H., Daily Totals		$2343.25		$3304.70	$58.58	$82.62

Crew B-86	Hr.	Daily	Hr.	Daily	Bare Costs	Incl. O&P
1 Equip. Oper. (medium)	$39.35	$314.80	$65.05	$520.40	$39.35	$65.05
1 Stump Chipper, S.P.		173.25		190.57	21.66	23.82
8 L.H., Daily Totals		$488.05		$710.98	$61.01	$88.87

Crew B-86A	Hr.	Daily	Hr.	Daily	Bare Costs	Incl. O&P
1 Equip. Oper. (medium)	$39.35	$314.80	$65.05	$520.40	$39.35	$65.05
1 Grader, 30,000 Lbs.		714.40		785.84	89.30	98.23
8 L.H., Daily Totals		$1029.20		$1306.24	$128.65	$163.28

Crew B-86B	Hr.	Daily	Hr.	Daily	Bare Costs	Incl. O&P
1 Equip. Oper. (medium)	$39.35	$314.80	$65.05	$520.40	$39.35	$65.05
1 Dozer, 200 H.P.		1397.00		1536.70	174.63	192.09
8 L.H., Daily Totals		$1711.80		$2057.10	$213.97	$257.14

Crew B-87	Hr.	Daily	Hr.	Daily	Bare Costs	Incl. O&P
1 Laborer	$29.20	$233.60	$48.90	$391.20	$37.32	$61.82
4 Equip. Opers. (medium)	39.35	1259.20	65.05	2081.60		
2 Feller Bunchers, 100 H.P.		1675.20		1842.72		
1 Log Chipper, 22" Tree		890.00		979.00		
1 Dozer, 105 H.P.		603.40		663.74		
1 Chain Saw, Gas, 36" Long		45.00		49.50	80.34	88.37
40 L.H., Daily Totals		$4706.40		$6007.76	$117.66	$150.19

Crews - Open Shop

Crew B-88	Hr.	Daily	Hr.	Daily	Bare Costs	Incl. O&P
1 Laborer	$29.20	$233.60	$48.90	$391.20	$37.90	$62.74
6 Equip. Opers. (medium)	39.35	1888.80	65.05	3122.40		
2 Feller Bunchers, 100 H.P.		1675.20		1842.72		
1 Log Chipper, 22" Tree		890.00		979.00		
2 Log Skidders, 50 H.P.		1833.60		2016.96		
1 Dozer, 105 H.P.		603.40		663.74		
1 Chain Saw, Gas, 36" Long		45.00		49.50	90.13	99.14
56 L.H., Daily Totals		$7169.60		$9065.52	$128.03	$161.88

Crew B-89	Hr.	Daily	Hr.	Daily	Bare Costs	Incl. O&P
1 Skilled Worker	$38.30	$306.40	$64.35	$514.80	$33.75	$56.63
1 Building Laborer	29.20	233.60	48.90	391.20		
1 Flatbed Truck, Gas, 3 Ton		307.40		338.14		
1 Concrete Saw		168.60		185.46		
1 Water Tank, 65 Gal.		17.30		19.03	30.83	33.91
16 L.H., Daily Totals		$1033.30		$1448.63	$64.58	$90.54

Crew B-89A	Hr.	Daily	Hr.	Daily	Bare Costs	Incl. O&P
1 Skilled Worker	$38.30	$306.40	$64.35	$514.80	$33.75	$56.63
1 Laborer	29.20	233.60	48.90	391.20		
1 Core Drill (Large)		117.00		128.70	7.31	8.04
16 L.H., Daily Totals		$657.00		$1034.70	$41.06	$64.67

Crew B-89B	Hr.	Daily	Hr.	Daily	Bare Costs	Incl. O&P
1 Equip. Oper. (light)	$37.85	$302.80	$62.55	$500.40	$35.48	$58.58
1 Truck Driver (light)	33.10	264.80	54.60	436.80		
1 Wall Saw, Hydraulic, 10 H.P.		115.00		126.50		
1 Generator, Diesel, 100 kW		413.00		454.30		
1 Water Tank, 65 Gal.		17.30		19.03		
1 Flatbed Truck, Gas, 3 Ton		307.40		338.14	53.29	58.62
16 L.H., Daily Totals		$1420.30		$1875.17	$88.77	$117.20

Crew B-90	Hr.	Daily	Hr.	Daily	Bare Costs	Incl. O&P
1 Labor Foreman (outside)	$31.20	$249.60	$52.25	$418.00	$32.85	$54.59
3 Laborers	29.20	700.80	48.90	1173.60		
2 Equip. Opers. (light)	37.85	605.60	62.55	1000.80		
2 Truck Drivers (heavy)	34.15	546.40	56.35	901.60		
1 Road Mixer, 310 H.P.		1945.00		2139.50		
1 Dist. Truck, 2000 Gal.		289.00		317.90	34.91	38.40
64 L.H., Daily Totals		$4336.40		$5951.40	$67.76	$92.99

Crew B-90A	Hr.	Daily	Hr.	Daily	Bare Costs	Incl. O&P
1 Labor Foreman (outside)	$31.20	$249.60	$52.25	$418.00	$35.29	$58.61
2 Laborers	29.20	467.20	48.90	782.40		
4 Equip. Opers. (medium)	39.35	1259.20	65.05	2081.60		
2 Graders, 30,000 Lbs.		1428.80		1571.68		
1 Tandem Roller, 10 Ton		239.40		263.34		
1 Roller, Pneum. Whl., 12 Ton		347.80		382.58	36.00	39.60
56 L.H., Daily Totals		$3992.00		$5499.60	$71.29	$98.21

Crew B-90B	Hr.	Daily	Hr.	Daily	Bare Costs	Incl. O&P
1 Labor Foreman (outside)	$31.20	$249.60	$52.25	$418.00	$34.61	$57.53
2 Laborers	29.20	467.20	48.90	782.40		
3 Equip. Opers. (medium)	39.35	944.40	65.05	1561.20		
1 Roller, Pneum. Whl., 12 Ton		347.80		382.58		
1 Road Mixer, 310 H.P.		1945.00		2139.50	47.77	52.54
48 L.H., Daily Totals		$3954.00		$5283.68	$82.38	$110.08

Crew B-90C	Hr.	Daily	Hr.	Daily	Bare Costs	Incl. O&P
1 Labor Foreman (outside)	$31.20	$249.60	$52.25	$418.00	$33.50	$55.64
4 Laborers	29.20	934.40	48.90	1564.80		
3 Equip. Opers. (medium)	39.35	944.40	65.05	1561.20		
3 Truck Drivers (heavy)	34.15	819.60	56.35	1352.40		
3 Road Mixers, 310 H.P.		5835.00		6418.50	66.31	72.94
88 L.H., Daily Totals		$8783.00		$11314.90	$99.81	$128.58

Crew B-90D	Hr.	Daily	Hr.	Daily	Bare Costs	Incl. O&P
1 Labor Foreman (outside)	$31.20	$249.60	$52.25	$418.00	$32.84	$54.60
6 Laborers	29.20	1401.60	48.90	2347.20		
3 Equip. Opers. (medium)	39.35	944.40	65.05	1561.20		
3 Truck Drivers (heavy)	34.15	819.60	56.35	1352.40		
3 Road Mixers, 310 H.P.		5835.00		6418.50	56.11	61.72
104 L.H., Daily Totals		$9250.20		$12097.30	$88.94	$116.32

Crew B-90E	Hr.	Daily	Hr.	Daily	Bare Costs	Incl. O&P
1 Labor Foreman (outside)	$31.20	$249.60	$52.25	$418.00	$33.36	$55.48
4 Laborers	29.20	934.40	48.90	1564.80		
3 Equip. Opers. (medium)	39.35	944.40	65.05	1561.20		
1 Truck Driver (heavy)	34.15	273.20	56.35	450.80		
1 Road Mixers, 310 H.P.		1945.00		2139.50	27.01	29.72
72 L.H., Daily Totals		$4346.60		$6134.30	$60.37	$85.20

Crew B-91	Hr.	Daily	Hr.	Daily	Bare Costs	Incl. O&P
1 Labor Foreman (outside)	$31.20	$249.60	$52.25	$418.00	$35.14	$58.33
2 Laborers	29.20	467.20	48.90	782.40		
4 Equip. Opers. (medium)	39.35	1259.20	65.05	2081.60		
1 Truck Driver (heavy)	34.15	273.20	56.35	450.80		
1 Dist. Tanker, 3000 Gallon		319.40		351.34		
1 Truck Tractor, 6x4, 380 H.P.		600.00		660.00		
1 Aggreg. Spreader, S.P.		829.70		912.67		
1 Roller, Pneum. Whl., 12 Ton		347.80		382.58		
1 Tandem Roller, 10 Ton		239.40		263.34	36.50	40.16
64 L.H., Daily Totals		$4585.50		$6302.73	$71.65	$98.48

Crew B-91B	Hr.	Daily	Hr.	Daily	Bare Costs	Incl. O&P
1 Laborer	$29.20	$233.60	$48.90	$391.20	$34.27	$56.98
1 Equipment Oper. (med.)	39.35	314.80	65.05	520.40		
1 Road Sweeper, Vac. Assist.		983.80		1082.18	61.49	67.64
16 L.H., Daily Totals		$1532.20		$1993.78	$95.76	$124.61

Crew B-91C	Hr.	Daily	Hr.	Daily	Bare Costs	Incl. O&P
1 Laborer	$29.20	$233.60	$48.90	$391.20	$31.15	$51.75
1 Truck Driver (light)	33.10	264.80	54.60	436.80		
1 Catch Basin Cleaning Truck		577.20		634.92	36.08	39.68
16 L.H., Daily Totals		$1075.60		$1462.92	$67.22	$91.43

Crew B-91D	Hr.	Daily	Hr.	Daily	Bare Costs	Incl. O&P
1 Labor Foreman (outside)	$31.20	$249.60	$52.25	$418.00	$34.02	$56.52
5 Laborers	29.20	1168.00	48.90	1956.00		
5 Equip. Opers. (medium)	39.35	1574.00	65.05	2602.00		
2 Truck Drivers (heavy)	34.15	546.40	56.35	901.60		
1 Aggreg. Spreader, S.P.		829.70		912.67		
2 Trucks, Tractor, 6x4, 380 H.P.		1200.00		1320.00		
2 Dist. Tankers, 3000 Gallon		638.80		702.68		
2 Pavement Brushes, Towed		162.40		178.64		
2 Rollers, Pneum. Whl., 12 Ton		695.60		765.16	33.91	37.30
104 L.H., Daily Totals		$7064.50		$9756.75	$67.93	$93.81

Crew No.	Bare Costs		Incl. Subs O&P		Cost Per Labor-Hour	
Crew B-92	**Hr.**	**Daily**	**Hr.**	**Daily**	**Bare Costs**	**Incl. O&P**
1 Labor Foreman (outside)	$31.20	$249.60	$52.25	$418.00	$29.70	$49.74
3 Laborers	29.20	700.80	48.90	1173.60		
1 Crack Cleaner, 25 H.P.		62.00		68.20		
1 Air Compressor, 60 cfm		142.80		157.08		
1 Tar Kettle, T.M.		135.30		148.83		
1 Flatbed Truck, Gas, 3 Ton		307.40		338.14	20.23	22.26
32 L.H., Daily Totals		$1597.90		$2303.85	$49.93	$72.00

Crew B-93	**Hr.**	**Daily**	**Hr.**	**Daily**	**Bare Costs**	**Incl. O&P**
1 Equip. Oper. (medium)	$39.35	$314.80	$65.05	$520.40	$39.35	$65.05
1 Feller Buncher, 100 H.P.		837.60		921.36	104.70	115.17
8 L.H., Daily Totals		$1152.40		$1441.76	$144.05	$180.22

Crew B-94A	**Hr.**	**Daily**	**Hr.**	**Daily**	**Bare Costs**	**Incl. O&P**
1 Laborer	$29.20	$233.60	$48.90	$391.20	$29.20	$48.90
1 Diaphragm Water Pump, 2"		71.80		78.98		
1 -20' Suction Hose, 2"		1.95		2.15		
2 -50' Discharge Hoses, 2"		1.80		1.98	9.44	10.39
8 L.H., Daily Totals		$309.15		$474.31	$38.64	$59.29

Crew B-94B	**Hr.**	**Daily**	**Hr.**	**Daily**	**Bare Costs**	**Incl. O&P**
1 Laborer	$29.20	$233.60	$48.90	$391.20	$29.20	$48.90
1 Diaphragm Water Pump, 4"		118.00		129.80		
1 -20' Suction Hose, 4"		3.25		3.58		
2 -50' Discharge Hoses, 4"		4.70		5.17	15.74	17.32
8 L.H., Daily Totals		$359.55		$529.75	$44.94	$66.22

Crew B-94C	**Hr.**	**Daily**	**Hr.**	**Daily**	**Bare Costs**	**Incl. O&P**
1 Laborer	$29.20	$233.60	$48.90	$391.20	$29.20	$48.90
1 Centrifugal Water Pump, 3"		81.00		89.10		
1 -20' Suction Hose, 3"		2.85		3.13		
2 -50' Discharge Hoses, 3"		3.00		3.30	10.86	11.94
8 L.H., Daily Totals		$320.45		$486.74	$40.06	$60.84

Crew B-94D	**Hr.**	**Daily**	**Hr.**	**Daily**	**Bare Costs**	**Incl. O&P**
1 Laborer	$29.20	$233.60	$48.90	$391.20	$29.20	$48.90
1 Centr. Water Pump, 6"		350.80		385.88		
1 -20' Suction Hose, 6"		11.50		12.65		
2 -50' Discharge Hoses, 6"		12.20		13.42	46.81	51.49
8 L.H., Daily Totals		$608.10		$803.15	$76.01	$100.39

Crew C-1	**Hr.**	**Daily**	**Hr.**	**Daily**	**Bare Costs**	**Incl. O&P**
2 Carpenters	$37.80	$604.80	$63.30	$1012.80	$33.19	$55.67
1 Carpenter Helper	27.95	223.60	47.20	377.60		
1 Laborer	29.20	233.60	48.90	391.20		
32 L.H., Daily Totals		$1062.00		$1781.60	$33.19	$55.67

Crew C-2	**Hr.**	**Daily**	**Hr.**	**Daily**	**Bare Costs**	**Incl. O&P**
1 Carpenter Foreman (outside)	$39.80	$318.40	$66.65	$533.20	$33.42	$56.09
2 Carpenters	37.80	604.80	63.30	1012.80		
2 Carpenter Helpers	27.95	447.20	47.20	755.20		
1 Laborer	29.20	233.60	48.90	391.20		
48 L.H., Daily Totals		$1604.00		$2692.40	$33.42	$56.09

Crew C-2A	**Hr.**	**Daily**	**Hr.**	**Daily**	**Bare Costs**	**Incl. O&P**
1 Carpenter Foreman (outside)	$39.80	$318.40	$66.65	$533.20	$36.33	$60.54
3 Carpenters	37.80	907.20	63.30	1519.20		
1 Cement Finisher	35.60	284.80	57.80	462.40		
1 Laborer	29.20	233.60	48.90	391.20		
48 L.H., Daily Totals		$1744.00		$2906.00	$36.33	$60.54

Crew C-3	**Hr.**	**Daily**	**Hr.**	**Daily**	**Bare Costs**	**Incl. O&P**
1 Rodman Foreman (outside)	$41.75	$334.00	$70.50	$564.00	$35.81	$60.15
3 Rodmen (reinf.)	39.75	954.00	67.15	1611.60		
1 Equip. Oper. (light)	37.85	302.80	62.55	500.40		
3 Laborers	29.20	700.80	48.90	1173.60		
3 Stressing Equipment		30.60		33.66		
.5 Grouting Equipment		82.20		90.42	1.76	1.94
64 L.H., Daily Totals		$2404.40		$3973.68	$37.57	$62.09

Crew C-4	**Hr.**	**Daily**	**Hr.**	**Daily**	**Bare Costs**	**Incl. O&P**
1 Rodman Foreman (outside)	$41.75	$334.00	$70.50	$564.00	$37.61	$63.42
2 Rodmen (reinf.)	39.75	636.00	67.15	1074.40		
1 Building Laborer	29.20	233.60	48.90	391.20		
3 Stressing Equipment		30.60		33.66	0.96	1.05
32 L.H., Daily Totals		$1234.20		$2063.26	$38.57	$64.48

Crew C-4A	**Hr.**	**Daily**	**Hr.**	**Daily**	**Bare Costs**	**Incl. O&P**
2 Rodmen (reinf.)	$39.75	$636.00	$67.15	$1074.40	$39.75	$67.15
4 Stressing Equipment		40.80		44.88	2.55	2.81
16 L.H., Daily Totals		$676.80		$1119.28	$42.30	$69.95

Crew C-5	**Hr.**	**Daily**	**Hr.**	**Daily**	**Bare Costs**	**Incl. O&P**
1 Rodman Foreman (outside)	$41.75	$334.00	$70.50	$564.00	$36.65	$61.52
2 Rodmen (reinf.)	39.75	636.00	67.15	1074.40		
1 Equip. Oper. (crane)	40.25	322.00	66.55	532.40		
2 Building Laborers	29.20	467.20	48.90	782.40		
1 Hyd. Crane, 25 Ton		748.80		823.68	15.60	17.16
48 L.H., Daily Totals		$2508.00		$3776.88	$52.25	$78.69

Crew C-6	**Hr.**	**Daily**	**Hr.**	**Daily**	**Bare Costs**	**Incl. O&P**
1 Labor Foreman (outside)	$31.20	$249.60	$52.25	$418.00	$30.60	$50.94
4 Laborers	29.20	934.40	48.90	1564.80		
1 Cement Finisher	35.60	284.80	57.80	462.40		
2 Gas Engine Vibrators		63.20		69.52	1.32	1.45
48 L.H., Daily Totals		$1532.00		$2514.72	$31.92	$52.39

Crew C-7	**Hr.**	**Daily**	**Hr.**	**Daily**	**Bare Costs**	**Incl. O&P**
1 Labor Foreman (outside)	$31.20	$249.60	$52.25	$418.00	$31.96	$53.13
5 Laborers	29.20	1168.00	48.90	1956.00		
1 Cement Finisher	35.60	284.80	57.80	462.40		
1 Equip. Oper. (medium)	39.35	314.80	65.05	520.40		
1 Equip. Oper. (oiler)	35.45	283.60	58.60	468.80		
2 Gas Engine Vibrators		63.20		69.52		
1 Concrete Bucket, 1 C.Y.		23.60		25.96		
1 Hyd. Crane, 55 Ton		1123.00		1235.30	16.80	18.48
72 L.H., Daily Totals		$3510.60		$5156.38	$48.76	$71.62

Crew C-8	**Hr.**	**Daily**	**Hr.**	**Daily**	**Bare Costs**	**Incl. O&P**
1 Labor Foreman (outside)	$31.20	$249.60	$52.25	$418.00	$32.76	$54.23
3 Laborers	29.20	700.80	48.90	1173.60		
2 Cement Finishers	35.60	569.60	57.80	924.80		
1 Equip. Oper. (medium)	39.35	314.80	65.05	520.40		
1 Concrete Pump (Small)		727.60		800.36	12.99	14.29
56 L.H., Daily Totals		$2562.40		$3837.16	$45.76	$68.52

Crew C-8A	**Hr.**	**Daily**	**Hr.**	**Daily**	**Bare Costs**	**Incl. O&P**
1 Labor Foreman (outside)	$31.20	$249.60	$52.25	$418.00	$31.67	$52.42
3 Laborers	29.20	700.80	48.90	1173.60		
2 Cement Finishers	35.60	569.60	57.80	924.80		
48 L.H., Daily Totals		$1520.00		$2516.40	$31.67	$52.42

Crew C-8B	Bare Costs Hr.	Bare Costs Daily	Incl. Subs O&P Hr.	Incl. Subs O&P Daily	Cost Per Labor-Hour Bare Costs	Cost Per Labor-Hour Incl. O&P
1 Labor Foreman (outside)	$31.20	$249.60	$52.25	$418.00	$31.63	$52.80
3 Laborers	29.20	700.80	48.90	1173.60		
1 Equip. Oper. (medium)	39.35	314.80	65.05	520.40		
1 Vibrating Power Screed		63.65		70.02		
1 Roller, Vibratory, 25 Ton		691.80		760.98		
1 Dozer, 200 H.P.		1397.00		1536.70	53.81	59.19
40 L.H., Daily Totals		$3417.65		$4479.69	$85.44	$111.99

Crew C-8C	Hr.	Daily	Hr.	Daily	Bare Costs	Incl. O&P
1 Labor Foreman (outside)	$31.20	$249.60	$52.25	$418.00	$32.29	$53.63
3 Laborers	29.20	700.80	48.90	1173.60		
1 Cement Finisher	35.60	284.80	57.80	462.40		
1 Equip. Oper. (medium)	39.35	314.80	65.05	520.40		
1 Shotcrete Rig, 12 C.Y./hr		254.00		279.40		
1 Air Compressor, 160 cfm		162.20		178.42		
4 -50' Air Hoses, 1"		14.60		16.06		
4 -50' Air Hoses, 2"		27.60		30.36	9.55	10.51
48 L.H., Daily Totals		$2008.40		$3078.64	$41.84	$64.14

Crew C-8D	Hr.	Daily	Hr.	Daily	Bare Costs	Incl. O&P
1 Labor Foreman (outside)	$31.20	$249.60	$52.25	$418.00	$33.46	$55.38
1 Laborer	29.20	233.60	48.90	391.20		
1 Cement Finisher	35.60	284.80	57.80	462.40		
1 Equipment Oper. (light)	37.85	302.80	62.55	500.40		
1 Air Compressor, 250 cfm		197.80		217.58		
2 -50' Air Hoses, 1"		7.30		8.03	6.41	7.05
32 L.H., Daily Totals		$1275.90		$1997.61	$39.87	$62.43

Crew C-8E	Hr.	Daily	Hr.	Daily	Bare Costs	Incl. O&P
1 Labor Foreman (outside)	$31.20	$249.60	$52.25	$418.00	$32.04	$53.22
3 Laborers	29.20	700.80	48.90	1173.60		
1 Cement Finisher	35.60	284.80	57.80	462.40		
1 Equipment Oper. (light)	37.85	302.80	62.55	500.40		
1 Shotcrete Rig, 35 C.Y./hr		284.00		312.40		
1 Air Compressor, 250 cfm		197.80		217.58		
4 -50' Air Hoses, 1"		14.60		16.06		
4 -50' Air Hoses, 2"		27.60		30.36	10.92	12.01
48 L.H., Daily Totals		$2062.00		$3130.80	$42.96	$65.22

Crew C-10	Hr.	Daily	Hr.	Daily	Bare Costs	Incl. O&P
1 Laborer	$29.20	$233.60	$48.90	$391.20	$33.47	$54.83
2 Cement Finishers	35.60	569.60	57.80	924.80		
24 L.H., Daily Totals		$803.20		$1316.00	$33.47	$54.83

Crew C-10B	Hr.	Daily	Hr.	Daily	Bare Costs	Incl. O&P
3 Laborers	$29.20	$700.80	$48.90	$1173.60	$31.76	$52.46
2 Cement Finishers	35.60	569.60	57.80	924.80		
1 Concrete Mixer, 10 C.F.		172.80		190.08		
2 Trowels, 48" Walk-Behind		101.20		111.32	6.85	7.54
40 L.H., Daily Totals		$1544.40		$2399.80	$38.61	$59.99

Crew C-10C	Hr.	Daily	Hr.	Daily	Bare Costs	Incl. O&P
1 Laborer	$29.20	$233.60	$48.90	$391.20	$33.47	$54.83
2 Cement Finishers	35.60	569.60	57.80	924.80		
1 Trowel, 48" Walk-Behind		50.60		55.66	2.11	2.32
24 L.H., Daily Totals		$853.80		$1371.66	$35.58	$57.15

Crew C-10D	Hr.	Daily	Hr.	Daily	Bare Costs	Incl. O&P
1 Laborer	$29.20	$233.60	$48.90	$391.20	$33.47	$54.83
2 Cement Finishers	35.60	569.60	57.80	924.80		
1 Vibrating Power Screed		63.65		70.02		
1 Trowel, 48" Walk-Behind		50.60		55.66	4.76	5.24
24 L.H., Daily Totals		$917.45		$1441.68	$38.23	$60.07

Crew C-10E	Hr.	Daily	Hr.	Daily	Bare Costs	Incl. O&P
1 Laborer	$29.20	$233.60	$48.90	$391.20	$33.47	$54.83
2 Cement Finishers	35.60	569.60	57.80	924.80		
1 Vibrating Power Screed		63.65		70.02		
1 Cement Trowel, 96" Ride-On		190.20		209.22	10.58	11.63
24 L.H., Daily Totals		$1057.05		$1595.23	$44.04	$66.47

Crew C-10F	Hr.	Daily	Hr.	Daily	Bare Costs	Incl. O&P
1 Laborer	$29.20	$233.60	$48.90	$391.20	$33.47	$54.83
2 Cement Finishers	35.60	569.60	57.80	924.80		
1 Aerial Lift Truck, 60' Boom		448.40		493.24	18.68	20.55
24 L.H., Daily Totals		$1251.60		$1809.24	$52.15	$75.39

Crew C-11	Hr.	Daily	Hr.	Daily	Bare Costs	Incl. O&P
1 Skilled Worker Foreman	$40.30	$322.40	$67.70	$541.60	$38.86	$65.14
5 Skilled Workers	38.30	1532.00	64.35	2574.00		
1 Equip. Oper. (crane)	40.25	322.00	66.55	532.40		
1 Lattice Boom Crane, 150 Ton		1868.00		2054.80	33.36	36.69
56 L.H., Daily Totals		$4044.40		$5702.80	$72.22	$101.84

Crew C-12	Hr.	Daily	Hr.	Daily	Bare Costs	Incl. O&P
1 Carpenter Foreman (outside)	$39.80	$318.40	$66.65	$533.20	$37.11	$62.00
3 Carpenters	37.80	907.20	63.30	1519.20		
1 Laborer	29.20	233.60	48.90	391.20		
1 Equip. Oper. (crane)	40.25	322.00	66.55	532.40		
1 Hyd. Crane, 12 Ton		658.40		724.24	13.72	15.09
48 L.H., Daily Totals		$2439.60		$3700.24	$50.83	$77.09

Crew C-13	Hr.	Daily	Hr.	Daily	Bare Costs	Incl. O&P
2 Struc. Steel Workers	$39.90	$638.40	$73.20	$1171.20	$39.20	$69.90
1 Carpenter	37.80	302.40	63.30	506.40		
1 Welder, Gas Engine, 300 amp		147.20		161.92	6.13	6.75
24 L.H., Daily Totals		$1088.00		$1839.52	$45.33	$76.65

Crew C-14	Hr.	Daily	Hr.	Daily	Bare Costs	Incl. O&P
1 Carpenter Foreman (outside)	$39.80	$318.40	$66.65	$533.20	$33.69	$56.32
3 Carpenters	37.80	907.20	63.30	1519.20		
2 Carpenter Helpers	27.95	447.20	47.20	755.20		
4 Laborers	29.20	934.40	48.90	1564.80		
2 Rodmen (reinf.)	39.75	636.00	67.15	1074.40		
2 Rodman Helpers	27.95	447.20	47.20	755.20		
2 Cement Finishers	35.60	569.60	57.80	924.80		
1 Equip. Oper. (crane)	40.25	322.00	66.55	532.40		
1 Hyd. Crane, 80 Ton		1630.00		1793.00	11.99	13.18
136 L.H., Daily Totals		$6212.00		$9452.20	$45.68	$69.50

Crew No.	Bare Costs		Incl. Subs O&P		Cost Per Labor-Hour	
Crew C-14A	**Hr.**	**Daily**	**Hr.**	**Daily**	**Bare Costs**	**Incl. O&P**
1 Carpenter Foreman (outside)	$39.80	$318.40	$66.65	$533.20	$37.48	$62.75
16 Carpenters	37.80	4838.40	63.30	8102.40		
4 Rodmen (reinf.)	39.75	1272.00	67.15	2148.80		
2 Laborers	29.20	467.20	48.90	782.40		
1 Cement Finisher	35.60	284.80	57.80	462.40		
1 Equip. Oper. (medium)	39.35	314.80	65.05	520.40		
1 Gas Engine Vibrator		31.60		34.76		
1 Concrete Pump (Small)		727.60		800.36	3.80	4.18
200 L.H., Daily Totals		$8254.80		$13384.72	$41.27	$66.92

Crew C-14B	Hr.	Daily	Hr.	Daily	Bare Costs	Incl. O&P
1 Carpenter Foreman (outside)	$39.80	$318.40	$66.65	$533.20	$37.41	$62.56
16 Carpenters	37.80	4838.40	63.30	8102.40		
4 Rodmen (reinf.)	39.75	1272.00	67.15	2148.80		
2 Laborers	29.20	467.20	48.90	782.40		
2 Cement Finishers	35.60	569.60	57.80	924.80		
1 Equip. Oper. (medium)	39.35	314.80	65.05	520.40		
1 Gas Engine Vibrator		31.60		34.76		
1 Concrete Pump (Small)		727.60		800.36	3.65	4.01
208 L.H., Daily Totals		$8539.60		$13847.12	$41.06	$66.57

Crew C-14C	Hr.	Daily	Hr.	Daily	Bare Costs	Incl. O&P
1 Carpenter Foreman (outside)	$39.80	$318.40	$66.65	$533.20	$35.61	$59.58
6 Carpenters	37.80	1814.40	63.30	3038.40		
2 Rodmen (reinf.)	39.75	636.00	67.15	1074.40		
4 Laborers	29.20	934.40	48.90	1564.80		
1 Cement Finisher	35.60	284.80	57.80	462.40		
1 Gas Engine Vibrator		31.60		34.76	0.28	0.31
112 L.H., Daily Totals		$4019.60		$6707.96	$35.89	$59.89

Crew C-14D	Hr.	Daily	Hr.	Daily	Bare Costs	Incl. O&P
1 Carpenter Foreman (outside)	$39.80	$318.40	$66.65	$533.20	$37.32	$62.44
18 Carpenters	37.80	5443.20	63.30	9115.20		
2 Rodmen (reinf.)	39.75	636.00	67.15	1074.40		
2 Laborers	29.20	467.20	48.90	782.40		
1 Cement Finisher	35.60	284.80	57.80	462.40		
1 Equip. Oper. (medium)	39.35	314.80	65.05	520.40		
1 Gas Engine Vibrator		31.60		34.76		
1 Concrete Pump (Small)		727.60		800.36	3.80	4.18
200 L.H., Daily Totals		$8223.60		$13323.12	$41.12	$66.62

Crew C-14E	Hr.	Daily	Hr.	Daily	Bare Costs	Incl. O&P
1 Carpenter Foreman (outside)	$39.80	$318.40	$66.65	$533.20	$36.15	$60.58
2 Carpenters	37.80	604.80	63.30	1012.80		
4 Rodmen (reinf.)	39.75	1272.00	67.15	2148.80		
3 Laborers	29.20	700.80	48.90	1173.60		
1 Cement Finisher	35.60	284.80	57.80	462.40		
1 Gas Engine Vibrator		31.60		34.76	0.36	0.40
88 L.H., Daily Totals		$3212.40		$5365.56	$36.50	$60.97

Crew C-14F	Hr.	Daily	Hr.	Daily	Bare Costs	Incl. O&P
1 Labor Foreman (outside)	$31.20	$249.60	$52.25	$418.00	$33.69	$55.21
2 Laborers	29.20	467.20	48.90	782.40		
6 Cement Finishers	35.60	1708.80	57.80	2774.40		
1 Gas Engine Vibrator		31.60		34.76	0.44	0.48
72 L.H., Daily Totals		$2457.20		$4009.56	$34.13	$55.69

Crew C-14G	Hr.	Daily	Hr.	Daily	Bare Costs	Incl. O&P
1 Labor Foreman (outside)	$31.20	$249.60	$52.25	$418.00	$33.14	$54.46
2 Laborers	29.20	467.20	48.90	782.40		
4 Cement Finishers	35.60	1139.20	57.80	1849.60		
1 Gas Engine Vibrator		31.60		34.76	0.56	0.62
56 L.H., Daily Totals		$1887.60		$3084.76	$33.71	$55.09

Crew C-14H	Hr.	Daily	Hr.	Daily	Bare Costs	Incl. O&P
1 Carpenter Foreman (outside)	$39.80	$318.40	$66.65	$533.20	$36.66	$61.18
2 Carpenters	37.80	604.80	63.30	1012.80		
1 Rodman (reinf.)	39.75	318.00	67.15	537.20		
1 Laborer	29.20	233.60	48.90	391.20		
1 Cement Finisher	35.60	284.80	57.80	462.40		
1 Gas Engine Vibrator		31.60		34.76	0.66	0.72
48 L.H., Daily Totals		$1791.20		$2971.56	$37.32	$61.91

Crew C-14L	Hr.	Daily	Hr.	Daily	Bare Costs	Incl. O&P
1 Carpenter Foreman (outside)	$39.80	$318.40	$66.65	$533.20	$34.92	$58.32
6 Carpenters	37.80	1814.40	63.30	3038.40		
4 Laborers	29.20	934.40	48.90	1564.80		
1 Cement Finisher	35.60	284.80	57.80	462.40		
1 Gas Engine Vibrator		31.60		34.76	0.33	0.36
96 L.H., Daily Totals		$3383.60		$5633.56	$35.25	$58.68

Crew C-14M	Hr.	Daily	Hr.	Daily	Bare Costs	Incl. O&P
1 Carpenter Foreman (outside)	$39.80	$318.40	$66.65	$533.20	$36.06	$60.13
2 Carpenters	37.80	604.80	63.30	1012.80		
1 Rodman (reinf.)	39.75	318.00	67.15	537.20		
2 Laborers	29.20	467.20	48.90	782.40		
1 Cement Finisher	35.60	284.80	57.80	462.40		
1 Equip. Oper. (medium)	39.35	314.80	65.05	520.40		
1 Gas Engine Vibrator		31.60		34.76		
1 Concrete Pump (Small)		727.60		800.36	11.86	13.05
64 L.H., Daily Totals		$3067.20		$4683.52	$47.92	$73.18

Crew C-15	Hr.	Daily	Hr.	Daily	Bare Costs	Incl. O&P
1 Carpenter Foreman (outside)	$39.80	$318.40	$66.65	$533.20	$34.88	$58.08
2 Carpenters	37.80	604.80	63.30	1012.80		
3 Laborers	29.20	700.80	48.90	1173.60		
2 Cement Finishers	35.60	569.60	57.80	924.80		
1 Rodman (reinf.)	39.75	318.00	67.15	537.20		
72 L.H., Daily Totals		$2511.60		$4181.60	$34.88	$58.08

Crew C-16	Hr.	Daily	Hr.	Daily	Bare Costs	Incl. O&P
1 Labor Foreman (outside)	$31.20	$249.60	$52.25	$418.00	$32.76	$54.23
3 Laborers	29.20	700.80	48.90	1173.60		
2 Cement Finishers	35.60	569.60	57.80	924.80		
1 Equip. Oper. (medium)	39.35	314.80	65.05	520.40		
1 Gunite Pump Rig		373.00		410.30		
2 -50' Air Hoses, 3/4"		6.10		6.71		
2 -50' Air Hoses, 2"		13.80		15.18	7.02	7.72
56 L.H., Daily Totals		$2227.70		$3468.99	$39.78	$61.95

Crew C-16A	Hr.	Daily	Hr.	Daily	Bare Costs	Incl. O&P
1 Laborer	$29.20	$233.60	$48.90	$391.20	$34.94	$57.39
2 Cement Finishers	35.60	569.60	57.80	924.80		
1 Equip. Oper. (medium)	39.35	314.80	65.05	520.40		
1 Gunite Pump Rig		373.00		410.30		
2 -50' Air Hoses, 3/4"		6.10		6.71		
2 -50' Air Hoses, 2"		13.80		15.18		
1 Aerial Lift Truck, 60' Boom		448.40		493.24	26.29	28.92
32 L.H., Daily Totals		$1959.30		$2761.83	$61.23	$86.31

Crews - Open Shop

Crew No.	Bare Costs		Incl. Subs O&P		Cost Per Labor-Hour	
Crew C-17	**Hr.**	**Daily**	**Hr.**	**Daily**	**Bare Costs**	**Incl. O&P**
2 Skilled Worker Foremen (out)	$40.30	$644.80	$67.70	$1083.20	$38.70	$65.02
8 Skilled Workers	38.30	2451.20	64.35	4118.40		
80 L.H., Daily Totals		$3096.00		$5201.60	$38.70	$65.02

Crew C-17A	Hr.	Daily	Hr.	Daily	Bare Costs	Incl. O&P
2 Skilled Worker Foremen (out)	$40.30	$644.80	$67.70	$1083.20	$38.72	$65.04
8 Skilled Workers	38.30	2451.20	64.35	4118.40		
.125 Equip. Oper. (crane)	40.25	40.25	66.55	66.55		
.125 Hyd. Crane, 80 Ton		203.75		224.13	2.52	2.77
81 L.H., Daily Totals		$3340.00		$5492.27	$41.23	$67.81

Crew C-17B	Hr.	Daily	Hr.	Daily	Bare Costs	Incl. O&P
2 Skilled Worker Foremen (out)	$40.30	$644.80	$67.70	$1083.20	$38.74	$65.06
8 Skilled Workers	38.30	2451.20	64.35	4118.40		
.25 Equip. Oper. (crane)	40.25	80.50	66.55	133.10		
.25 Hyd. Crane, 80 Ton		407.50		448.25		
.25 Trowel, 48" Walk-Behind		12.65		13.91	5.12	5.64
82 L.H., Daily Totals		$3596.65		$5796.86	$43.86	$70.69

Crew C-17C	Hr.	Daily	Hr.	Daily	Bare Costs	Incl. O&P
2 Skilled Worker Foremen (out)	$40.30	$644.80	$67.70	$1083.20	$38.76	$65.08
8 Skilled Workers	38.30	2451.20	64.35	4118.40		
.375 Equip. Oper. (crane)	40.25	120.75	66.55	199.65		
.375 Hyd. Crane, 80 Ton		611.25		672.38	7.36	8.10
83 L.H., Daily Totals		$3828.00		$6073.63	$46.12	$73.18

Crew C-17D	Hr.	Daily	Hr.	Daily	Bare Costs	Incl. O&P
2 Skilled Worker Foremen (out)	$40.30	$644.80	$67.70	$1083.20	$38.77	$65.09
8 Skilled Workers	38.30	2451.20	64.35	4118.40		
.5 Equip. Oper. (crane)	40.25	161.00	66.55	266.20		
.5 Hyd. Crane, 80 Ton		815.00		896.50	9.70	10.67
84 L.H., Daily Totals		$4072.00		$6364.30	$48.48	$75.77

Crew C-17E	Hr.	Daily	Hr.	Daily	Bare Costs	Incl. O&P
2 Skilled Worker Foremen (out)	$40.30	$644.80	$67.70	$1083.20	$38.70	$65.02
8 Skilled Workers	38.30	2451.20	64.35	4118.40		
1 Hyd. Jack with Rods		98.40		108.24	1.23	1.35
80 L.H., Daily Totals		$3194.40		$5309.84	$39.93	$66.37

Crew C-18	Hr.	Daily	Hr.	Daily	Bare Costs	Incl. O&P
.125 Labor Foreman (outside)	$31.20	$31.20	$52.25	$52.25	$29.42	$49.27
1 Laborer	29.20	233.60	48.90	391.20		
1 Concrete Cart, 10 C.F.		60.00		66.00	6.67	7.33
9 L.H., Daily Totals		$324.80		$509.45	$36.09	$56.61

Crew C-19	Hr.	Daily	Hr.	Daily	Bare Costs	Incl. O&P
.125 Labor Foreman (outside)	$31.20	$31.20	$52.25	$52.25	$29.42	$49.27
1 Laborer	29.20	233.60	48.90	391.20		
1 Concrete Cart, 18 C.F.		99.80		109.78	11.09	12.20
9 L.H., Daily Totals		$364.60		$553.23	$40.51	$61.47

Crew C-20	Hr.	Daily	Hr.	Daily	Bare Costs	Incl. O&P
1 Labor Foreman (outside)	$31.20	$249.60	$52.25	$418.00	$31.52	$52.45
5 Laborers	29.20	1168.00	48.90	1956.00		
1 Cement Finisher	35.60	284.80	57.80	462.40		
1 Equip. Oper. (medium)	39.35	314.80	65.05	520.40		
2 Gas Engine Vibrators		63.20		69.52		
1 Concrete Pump (Small)		727.60		800.36	12.36	13.59
64 L.H., Daily Totals		$2808.00		$4226.68	$43.88	$66.04

Crew No.	Bare Costs		Incl. Subs O&P		Cost Per Labor-Hour	
Crew C-21	**Hr.**	**Daily**	**Hr.**	**Daily**	**Bare Costs**	**Incl. O&P**
1 Labor Foreman (outside)	$31.20	$249.60	$52.25	$418.00	$31.52	$52.45
5 Laborers	29.20	1168.00	48.90	1956.00		
1 Cement Finisher	35.60	284.80	57.80	462.40		
1 Equip. Oper. (medium)	39.35	314.80	65.05	520.40		
2 Gas Engine Vibrators		63.20		69.52		
1 Concrete Conveyer		199.20		219.12	4.10	4.51
64 L.H., Daily Totals		$2279.60		$3645.44	$35.62	$56.96

Crew C-22	Hr.	Daily	Hr.	Daily	Bare Costs	Incl. O&P
1 Rodman Foreman (outside)	$41.75	$334.00	$70.50	$564.00	$40.04	$67.57
4 Rodmen (reinf.)	39.75	1272.00	67.15	2148.80		
.125 Equip. Oper. (crane)	40.25	40.25	66.55	66.55		
.125 Equip. Oper. (oiler)	35.45	35.45	58.60	58.60		
.125 Hyd. Crane, 25 Ton		93.60		102.96	2.23	2.45
42 L.H., Daily Totals		$1775.30		$2940.91	$42.27	$70.02

Crew C-23	Hr.	Daily	Hr.	Daily	Bare Costs	Incl. O&P
2 Skilled Worker Foremen (out)	$40.30	$644.80	$67.70	$1083.20	$38.61	$64.67
6 Skilled Workers	38.30	1838.40	64.35	3088.80		
1 Equip. Oper. (crane)	40.25	322.00	66.55	532.40		
1 Equip. Oper. (oiler)	35.45	283.60	58.60	468.80		
1 Lattice Boom Crane, 90 Ton		1520.00		1672.00	19.00	20.90
80 L.H., Daily Totals		$4608.80		$6845.20	$57.61	$85.56

Crew C-24	Hr.	Daily	Hr.	Daily	Bare Costs	Incl. O&P
2 Skilled Worker Foremen (out)	$40.30	$644.80	$67.70	$1083.20	$38.61	$64.67
6 Skilled Workers	38.30	1838.40	64.35	3088.80		
1 Equip. Oper. (crane)	40.25	322.00	66.55	532.40		
1 Equip. Oper. (oiler)	35.45	283.60	58.60	468.80		
1 Lattice Boom Crane, 150 Ton		1868.00		2054.80	23.35	25.68
80 L.H., Daily Totals		$4956.80		$7228.00	$61.96	$90.35

Crew C-25	Hr.	Daily	Hr.	Daily	Bare Costs	Incl. O&P
2 Rodmen (reinf.)	$39.75	$636.00	$67.15	$1074.40	$31.93	$55.80
2 Rodmen Helpers	24.10	385.60	44.45	711.20		
32 L.H., Daily Totals		$1021.60		$1785.60	$31.93	$55.80

Crew C-27	Hr.	Daily	Hr.	Daily	Bare Costs	Incl. O&P
2 Cement Finishers	$35.60	$569.60	$57.80	$924.80	$35.60	$57.80
1 Concrete Saw		168.60		185.46	10.54	11.59
16 L.H., Daily Totals		$738.20		$1110.26	$46.14	$69.39

Crew C-28	Hr.	Daily	Hr.	Daily	Bare Costs	Incl. O&P
1 Cement Finisher	$35.60	$284.80	$57.80	$462.40	$35.60	$57.80
1 Portable Air Compressor, Gas		17.40		19.14	2.17	2.39
8 L.H., Daily Totals		$302.20		$481.54	$37.77	$60.19

Crew C-29	Hr.	Daily	Hr.	Daily	Bare Costs	Incl. O&P
1 Laborer	$29.20	$233.60	$48.90	$391.20	$29.20	$48.90
1 Pressure Washer		71.60		78.76	8.95	9.85
8 L.H., Daily Totals		$305.20		$469.96	$38.15	$58.74

Crew C-30	Hr.	Daily	Hr.	Daily	Bare Costs	Incl. O&P
1 Laborer	$29.20	$233.60	$48.90	$391.20	$29.20	$48.90
1 Concrete Mixer, 10 C.F.		172.80		190.08	21.60	23.76
8 L.H., Daily Totals		$406.40		$581.28	$50.80	$72.66

Crew No.	Bare Costs		Incl. Subs O&P		Cost Per Labor-Hour	
Crew C-31	**Hr.**	**Daily**	**Hr.**	**Daily**	**Bare Costs**	**Incl. O&P**
1 Cement Finisher	$35.60	$284.80	$57.80	$462.40	$35.60	$57.80
1 Grout Pump		373.00		410.30	46.63	51.29
8 L.H., Daily Totals		$657.80		$872.70	$82.22	$109.09

Crew C-32	Hr.	Daily	Hr.	Daily	Bare Costs	Incl. O&P
1 Cement Finisher	$35.60	$284.80	$57.80	$462.40	$32.40	$53.35
1 Laborer	29.20	233.60	48.90	391.20		
1 Crack Chaser Saw, Gas, 6 H.P.		29.20		32.12		
1 Vacuum Pick-Up System		60.90		66.99	5.63	6.19
16 L.H., Daily Totals		$608.50		$952.71	$38.03	$59.54

Crew D-1	Hr.	Daily	Hr.	Daily	Bare Costs	Incl. O&P
1 Bricklayer	$36.10	$288.80	$60.20	$481.60	$32.88	$54.83
1 Bricklayer Helper	29.65	237.20	49.45	395.60		
16 L.H., Daily Totals		$526.00		$877.20	$32.88	$54.83

Crew D-2	Hr.	Daily	Hr.	Daily	Bare Costs	Incl. O&P
3 Bricklayers	$36.10	$866.40	$60.20	$1444.80	$33.52	$55.90
2 Bricklayer Helpers	29.65	474.40	49.45	791.20		
40 L.H., Daily Totals		$1340.80		$2236.00	$33.52	$55.90

Crew D-3	Hr.	Daily	Hr.	Daily	Bare Costs	Incl. O&P
3 Bricklayers	$36.10	$866.40	$60.20	$1444.80	$33.72	$56.25
2 Bricklayer Helpers	29.65	474.40	49.45	791.20		
.25 Carpenter	37.80	75.60	63.30	126.60		
42 L.H., Daily Totals		$1416.40		$2362.60	$33.72	$56.25

Crew D-4	Hr.	Daily	Hr.	Daily	Bare Costs	Incl. O&P
1 Bricklayer	$36.10	$288.80	$60.20	$481.60	$30.85	$51.49
3 Bricklayer Helpers	29.65	711.60	49.45	1186.80		
1 Building Laborer	29.20	233.60	48.90	391.20		
1 Grout Pump, 50 C.F./hr.		135.00		148.50	3.38	3.71
40 L.H., Daily Totals		$1369.00		$2208.10	$34.23	$55.20

Crew D-5	Hr.	Daily	Hr.	Daily	Bare Costs	Incl. O&P
1 Block Mason Helper	29.65	237.20	49.45	395.60	29.65	49.45
8 L.H., Daily Totals		$237.20		$395.60	$29.65	$49.45

Crew D-6	Hr.	Daily	Hr.	Daily	Bare Costs	Incl. O&P
3 Bricklayers	$36.10	$866.40	$60.20	$1444.80	$32.88	$54.83
3 Bricklayer Helpers	29.65	711.60	49.45	1186.80		
48 L.H., Daily Totals		$1578.00		$2631.60	$32.88	$54.83

Crew D-7	Hr.	Daily	Hr.	Daily	Bare Costs	Incl. O&P
1 Tile Layer	$34.75	$278.00	$56.25	$450.00	$30.95	$50.10
1 Tile Layer Helper	27.15	217.20	43.95	351.60		
16 L.H., Daily Totals		$495.20		$801.60	$30.95	$50.10

Crew D-8	Hr.	Daily	Hr.	Daily	Bare Costs	Incl. O&P
3 Bricklayers	$36.10	$866.40	$60.20	$1444.80	$33.52	$55.90
2 Bricklayer Helpers	29.65	474.40	49.45	791.20		
40 L.H., Daily Totals		$1340.80		$2236.00	$33.52	$55.90

Crew D-9	Hr.	Daily	Hr.	Daily	Bare Costs	Incl. O&P
3 Bricklayers	$36.10	$866.40	$60.20	$1444.80	$32.88	$54.83
3 Bricklayer Helpers	29.65	711.60	49.45	1186.80		
48 L.H., Daily Totals		$1578.00		$2631.60	$32.88	$54.83

Crew D-10	Hr.	Daily	Hr.	Daily	Bare Costs	Incl. O&P
1 Bricklayer Foreman (outside)	$38.10	$304.80	$63.50	$508.00	$36.02	$59.92
1 Bricklayer	36.10	288.80	60.20	481.60		
1 Bricklayer Helper	29.65	237.20	49.45	395.60		
1 Equip. Oper. (crane)	40.25	322.00	66.55	532.40		
1 S.P. Crane, 4x4, 12 Ton		479.20		527.12	14.98	16.47
32 L.H., Daily Totals		$1632.00		$2444.72	$51.00	$76.40

Crew D-11	Hr.	Daily	Hr.	Daily	Bare Costs	Incl. O&P
2 Bricklayers	$36.10	$577.60	$60.20	$963.20	$33.95	$56.62
1 Bricklayer Helper	29.65	237.20	49.45	395.60		
24 L.H., Daily Totals		$814.80		$1358.80	$33.95	$56.62

Crew D-12	Hr.	Daily	Hr.	Daily	Bare Costs	Incl. O&P
2 Bricklayers	$36.10	$577.60	$60.20	$963.20	$32.88	$54.83
2 Bricklayer Helpers	29.65	474.40	49.45	791.20		
32 L.H., Daily Totals		$1052.00		$1754.40	$32.88	$54.83

Crew D-13	Hr.	Daily	Hr.	Daily	Bare Costs	Incl. O&P
1 Bricklayer Foreman (outside)	$38.10	$304.80	$63.50	$508.00	$34.98	$58.23
2 Bricklayers	36.10	577.60	60.20	963.20		
2 Bricklayer Helpers	29.65	474.40	49.45	791.20		
1 Equip. Oper. (crane)	40.25	322.00	66.55	532.40		
1 S.P. Crane, 4x4, 12 Ton		479.20		527.12	9.98	10.98
48 L.H., Daily Totals		$2158.00		$3321.92	$44.96	$69.21

Crew E-1	Hr.	Daily	Hr.	Daily	Bare Costs	Incl. O&P
2 Struc. Steel Workers	$39.90	$638.40	$73.20	$1171.20	$39.90	$73.20
1 Welder, Gas Engine, 300 amp		147.20		161.92	9.20	10.12
16 L.H., Daily Totals		$785.60		$1333.12	$49.10	$83.32

Crew E-2	Hr.	Daily	Hr.	Daily	Bare Costs	Incl. O&P
1 Struc. Steel Foreman (outside)	$41.90	$335.20	$76.85	$614.80	$40.29	$72.70
4 Struc. Steel Workers	39.90	1276.80	73.20	2342.40		
1 Equip. Oper. (crane)	40.25	322.00	66.55	532.40		
1 Lattice Boom Crane, 90 Ton		1520.00		1672.00	31.67	34.83
48 L.H., Daily Totals		$3454.00		$5161.60	$71.96	$107.53

Crew E-3	Hr.	Daily	Hr.	Daily	Bare Costs	Incl. O&P
1 Struc. Steel Foreman (outside)	$41.90	$335.20	$76.85	$614.80	$40.57	$74.42
2 Struc. Steel Workers	39.90	638.40	73.20	1171.20		
1 Welder, Gas Engine, 300 amp		147.20		161.92	6.13	6.75
24 L.H., Daily Totals		$1120.80		$1947.92	$46.70	$81.16

Crew E-3A	Hr.	Daily	Hr.	Daily	Bare Costs	Incl. O&P
1 Struc. Steel Foreman (outside)	$41.90	$335.20	$76.85	$614.80	$40.57	$74.42
2 Struc. Steel Workers	39.90	638.40	73.20	1171.20		
1 Welder, Gas Engine, 300 amp		147.20		161.92		
1 Aerial Lift Truck, 40' Boom		309.00		339.90	19.01	20.91
24 L.H., Daily Totals		$1429.80		$2287.82	$59.58	$95.33

Crew E-4	Hr.	Daily	Hr.	Daily	Bare Costs	Incl. O&P
1 Struc. Steel Foreman (outside)	$41.90	$335.20	$76.85	$614.80	$40.40	$74.11
3 Struc. Steel Workers	39.90	957.60	73.20	1756.80		
1 Welder, Gas Engine, 300 amp		147.20		161.92	4.60	5.06
32 L.H., Daily Totals		$1440.00		$2533.52	$45.00	$79.17

Crew E-5	Bare Costs Hr.	Bare Costs Daily	Incl. Subs O&P Hr.	Incl. Subs O&P Daily	Bare Costs	Incl. O&P
1 Struc. Steel Foreman (outside)	$41.90	$335.20	$76.85	$614.80	$40.16	$72.87
7 Struc. Steel Workers	39.90	2234.40	73.20	4099.20		
1 Equip. Oper. (crane)	40.25	322.00	66.55	532.40		
1 Lattice Boom Crane, 90 Ton		1520.00		1672.00		
1 Welder, Gas Engine, 300 amp		147.20		161.92	23.16	25.47
72 L.H., Daily Totals		$4558.80		$7080.32	$63.32	$98.34

Crew E-6	Hr.	Daily	Hr.	Daily	Bare Costs	Incl. O&P
1 Struc. Steel Foreman (outside)	$41.90	$335.20	$76.85	$614.80	$39.92	$72.29
12 Struc. Steel Workers	39.90	3830.40	73.20	7027.20		
1 Equip. Oper. (crane)	40.25	322.00	66.55	532.40		
1 Equip. Oper. (light)	37.85	302.80	62.55	500.40		
1 Lattice Boom Crane, 90 Ton		1520.00		1672.00		
1 Welder, Gas Engine, 300 amp		147.20		161.92		
1 Air Compressor, 160 cfm		162.20		178.42		
2 Impact Wrenches		36.40		40.04	15.55	17.10
120 L.H., Daily Totals		$6656.20		$10727.18	$55.47	$89.39

Crew E-7	Hr.	Daily	Hr.	Daily	Bare Costs	Incl. O&P
1 Struc. Steel Foreman (outside)	$41.90	$335.20	$76.85	$614.80	$40.16	$72.87
7 Struc. Steel Workers	39.90	2234.40	73.20	4099.20		
1 Equip. Oper. (crane)	40.25	322.00	66.55	532.40		
1 Lattice Boom Crane, 90 Ton		1520.00		1672.00		
2 Welder, Gas Engine, 300 amp		294.40		323.84	25.20	27.72
72 L.H., Daily Totals		$4706.00		$7242.24	$65.36	$100.59

Crew E-8	Hr.	Daily	Hr.	Daily	Bare Costs	Incl. O&P
1 Struc. Steel Foreman (outside)	$41.90	$335.20	$76.85	$614.80	$40.11	$72.93
9 Struc. Steel Workers	39.90	2872.80	73.20	5270.40		
1 Equip. Oper. (crane)	40.25	322.00	66.55	532.40		
1 Lattice Boom Crane, 90 Ton		1520.00		1672.00		
4 Welder, Gas Engine, 300 amp		588.80		647.68	23.96	26.36
88 L.H., Daily Totals		$5638.80		$8737.28	$64.08	$99.29

Crew E-9	Hr.	Daily	Hr.	Daily	Bare Costs	Incl. O&P
2 Struc. Steel Foremen (outside)	$41.90	$670.40	$76.85	$1229.60	$39.89	$71.89
5 Struc. Steel Workers	39.90	1596.00	73.20	2928.00		
1 Welder Foreman (outside)	41.90	335.20	76.85	614.80		
5 Welders	39.90	1596.00	73.20	2928.00		
1 Equip. Oper. (crane)	40.25	322.00	66.55	532.40		
1 Equip. Oper. (oiler)	35.45	283.60	58.60	468.80		
1 Equip. Oper. (light)	37.85	302.80	62.55	500.40		
1 Lattice Boom Crane, 90 Ton		1520.00		1672.00		
5 Welders, Gas Engine, 300 amp		736.00		809.60	17.63	19.39
128 L.H., Daily Totals		$7362.00		$11683.60	$57.52	$91.28

Crew E-10	Hr.	Daily	Hr.	Daily	Bare Costs	Incl. O&P
1 Struc. Steel Foreman (outside)	$41.90	$335.20	$76.85	$614.80	$40.57	$74.42
2 Struc. Steel Workers	39.90	638.40	73.20	1171.20		
1 Welder, Gas Engine, 300 amp		147.20		161.92		
1 Flatbed Truck, Gas, 3 Ton		307.40		338.14	18.94	20.84
24 L.H., Daily Totals		$1428.20		$2286.06	$59.51	$95.25

Crew E-11	Hr.	Daily	Hr.	Daily	Bare Costs	Incl. O&P
2 Painters, Struc. Steel	$32.60	$521.60	$60.55	$968.80	$33.06	$58.14
1 Building Laborer	29.20	233.60	48.90	391.20		
1 Equip. Oper. (light)	37.85	302.80	62.55	500.40		
1 Air Compressor, 250 cfm		197.80		217.58		
1 Sandblaster, Portable, 3 C.F.		20.80		22.88		
1 Set Sand Blasting Accessories		14.25		15.68	7.28	8.00
32 L.H., Daily Totals		$1290.85		$2116.53	$40.34	$66.14

Crew E-11A	Hr.	Daily	Hr.	Daily	Bare Costs	Incl. O&P
2 Painters, Struc. Steel	$32.60	$521.60	$60.55	$968.80	$33.06	$58.14
1 Building Laborer	29.20	233.60	48.90	391.20		
1 Equip. Oper. (light)	37.85	302.80	62.55	500.40		
1 Air Compressor, 250 cfm		197.80		217.58		
1 Sandblaster, Portable, 3 C.F.		20.80		22.88		
1 Set Sand Blasting Accessories		14.25		15.68		
1 Aerial Lift Truck, 60' Boom		448.40		493.24	21.29	23.42
32 L.H., Daily Totals		$1739.25		$2609.78	$54.35	$81.56

Crew E-11B	Hr.	Daily	Hr.	Daily	Bare Costs	Incl. O&P
2 Painters, Struc. Steel	$32.60	$521.60	$60.55	$968.80	$31.47	$56.67
1 Building Laborer	29.20	233.60	48.90	391.20		
2 Paint Sprayers, 8 C.F.M.		100.80		110.88		
1 Aerial Lift Truck, 60' Boom		448.40		493.24	22.88	25.17
24 L.H., Daily Totals		$1304.40		$1964.12	$54.35	$81.84

Crew E-12	Hr.	Daily	Hr.	Daily	Bare Costs	Incl. O&P
1 Welder Foreman (outside)	$41.90	$335.20	$76.85	$614.80	$39.88	$69.70
1 Equip. Oper. (light)	37.85	302.80	62.55	500.40		
1 Welder, Gas Engine, 300 amp		147.20		161.92	9.20	10.12
16 L.H., Daily Totals		$785.20		$1277.12	$49.08	$79.82

Crew E-13	Hr.	Daily	Hr.	Daily	Bare Costs	Incl. O&P
1 Welder Foreman (outside)	$41.90	$335.20	$76.85	$614.80	$40.55	$72.08
.5 Equip. Oper. (light)	37.85	151.40	62.55	250.20		
1 Welder, Gas Engine, 300 amp		147.20		161.92	12.27	13.49
12 L.H., Daily Totals		$633.80		$1026.92	$52.82	$85.58

Crew E-14	Hr.	Daily	Hr.	Daily	Bare Costs	Incl. O&P
1 Struc. Steel Worker	$39.90	$319.20	$73.20	$585.60	$39.90	$73.20
1 Welder, Gas Engine, 300 amp		147.20		161.92	18.40	20.24
8 L.H., Daily Totals		$466.40		$747.52	$58.30	$93.44

Crew E-16	Hr.	Daily	Hr.	Daily	Bare Costs	Incl. O&P
1 Welder Foreman (outside)	$41.90	$335.20	$76.85	$614.80	$40.90	$75.03
1 Welder	39.90	319.20	73.20	585.60		
1 Welder, Gas Engine, 300 amp		147.20		161.92	9.20	10.12
16 L.H., Daily Totals		$801.60		$1362.32	$50.10	$85.14

Crew E-17	Hr.	Daily	Hr.	Daily	Bare Costs	Incl. O&P
1 Struc. Steel Foreman (outside)	$41.90	$335.20	$76.85	$614.80	$40.90	$75.03
1 Structural Steel Worker	39.90	319.20	73.20	585.60		
16 L.H., Daily Totals		$654.40		$1200.40	$40.90	$75.03

Crew E-18	Hr.	Daily	Hr.	Daily	Bare Costs	Incl. O&P
1 Struc. Steel Foreman (outside)	$41.90	$335.20	$76.85	$614.80	$40.19	$72.30
3 Structural Steel Workers	39.90	957.60	73.20	1756.80		
1 Equipment Operator (med.)	39.35	314.80	65.05	520.40		
1 Lattice Boom Crane, 20 Ton		967.35		1064.09	24.18	26.60
40 L.H., Daily Totals		$2574.95		$3956.09	$64.37	$98.90

Crew E-19	Hr.	Daily	Hr.	Daily	Bare Costs	Incl. O&P
1 Struc. Steel Foreman (outside)	$41.90	$335.20	$76.85	$614.80	$39.88	$70.87
1 Structural Steel Worker	39.90	319.20	73.20	585.60		
1 Equip. Oper. (light)	37.85	302.80	62.55	500.40		
1 Lattice Boom Crane, 20 Ton		967.35		1064.09	40.31	44.34
24 L.H., Daily Totals		$1924.55		$2764.89	$80.19	$115.20

Crew No.	Bare Costs		Incl. Subs O&P		Cost Per Labor-Hour	

Crew E-20	Hr.	Daily	Hr.	Daily	Bare Costs	Incl. O&P
1 Struc. Steel Foreman (outside)	$41.90	$335.20	$76.85	$614.80	$39.64	$71.00
5 Structural Steel Workers	39.90	1596.00	73.20	2928.00		
1 Equip. Oper. (crane)	40.25	322.00	66.55	532.40		
1 Equip. Oper. (oiler)	35.45	283.60	58.60	468.80		
1 Lattice Boom Crane, 40 Ton		1182.00		1300.20	18.47	20.32
64 L.H., Daily Totals		$3718.80		$5844.20	$58.11	$91.32

Crew E-22	Hr.	Daily	Hr.	Daily	Bare Costs	Incl. O&P
1 Skilled Worker Foreman (out)	$40.30	$322.40	$67.70	$541.60	$38.97	$65.47
2 Skilled Workers	38.30	612.80	64.35	1029.60		
24 L.H., Daily Totals		$935.20		$1571.20	$38.97	$65.47

Crew E-24	Hr.	Daily	Hr.	Daily	Bare Costs	Incl. O&P
3 Structural Steel Workers	$39.90	$957.60	$73.20	$1756.80	$39.76	$71.16
1 Equipment Operator (med.)	39.35	314.80	65.05	520.40		
1 Hyd. Crane, 25 Ton		748.80		823.68	23.40	25.74
32 L.H., Daily Totals		$2021.20		$3100.88	$63.16	$96.90

Crew E-25	Hr.	Daily	Hr.	Daily	Bare Costs	Incl. O&P
1 Welder	$39.90	$319.20	$73.20	$585.60	$39.90	$73.20
1 Cutting Torch		11.40		12.54	1.43	1.57
8 L.H., Daily Totals		$330.60		$598.14	$41.33	$74.77

Crew E-26	Hr.	Daily	Hr.	Daily	Bare Costs	Incl. O&P
1 Struc. Steel Foreman (outside)	$41.90	$335.20	$76.85	$614.80	$41.05	$74.09
1 Struc. Steel Worker	39.90	319.20	73.20	585.60		
1 Welder	39.90	319.20	73.20	585.60		
.25 Electrician	44.05	88.10	72.05	144.10		
.25 Plumber	43.80	87.60	72.25	144.50		
1 Welder, Gas Engine, 300 amp		147.20		161.92	5.26	5.78
28 L.H., Daily Totals		$1296.50		$2236.52	$46.30	$79.88

Crew F-3	Hr.	Daily	Hr.	Daily	Bare Costs	Incl. O&P
2 Carpenters	$37.80	$604.80	$63.30	$1012.80	$34.35	$57.51
2 Carpenter Helpers	27.95	447.20	47.20	755.20		
1 Equip. Oper. (crane)	40.25	322.00	66.55	532.40		
1 Hyd. Crane, 12 Ton		658.40		724.24	16.46	18.11
40 L.H., Daily Totals		$2032.40		$3024.64	$50.81	$75.62

Crew F-4	Hr.	Daily	Hr.	Daily	Bare Costs	Incl. O&P
2 Carpenters	$37.80	$604.80	$63.30	$1012.80	$34.35	$57.51
2 Carpenter Helpers	27.95	447.20	47.20	755.20		
1 Equip. Oper. (crane)	40.25	322.00	66.55	532.40		
1 Hyd. Crane, 55 Ton		1123.00		1235.30	28.07	30.88
40 L.H., Daily Totals		$2497.00		$3535.70	$62.42	$88.39

Crew F-5	Hr.	Daily	Hr.	Daily	Bare Costs	Incl. O&P
2 Carpenters	$37.80	$604.80	$63.30	$1012.80	$32.88	$55.25
2 Carpenter Helpers	27.95	447.20	47.20	755.20		
32 L.H., Daily Totals		$1052.00		$1768.00	$32.88	$55.25

Crew F-6	Hr.	Daily	Hr.	Daily	Bare Costs	Incl. O&P
2 Carpenters	$37.80	$604.80	$63.30	$1012.80	$34.85	$58.19
2 Building Laborers	29.20	467.20	48.90	782.40		
1 Equip. Oper. (crane)	40.25	322.00	66.55	532.40		
1 Hyd. Crane, 12 Ton		658.40		724.24	16.46	18.11
40 L.H., Daily Totals		$2052.40		$3051.84	$51.31	$76.30

Crew F-7	Hr.	Daily	Hr.	Daily	Bare Costs	Incl. O&P
2 Carpenters	$37.80	$604.80	$63.30	$1012.80	$33.50	$56.10
2 Building Laborers	29.20	467.20	48.90	782.40		
32 L.H., Daily Totals		$1072.00		$1795.20	$33.50	$56.10

Crew G-1	Hr.	Daily	Hr.	Daily	Bare Costs	Incl. O&P
1 Roofer Foreman (outside)	$34.10	$272.80	$62.90	$503.20	$30.10	$55.51
4 Roofers Composition	32.10	1027.20	59.20	1894.40		
2 Roofer Helpers	24.10	385.60	44.45	711.20		
1 Application Equipment		193.00		212.30		
1 Tar Kettle/Pot		176.40		194.04		
1 Crew Truck		208.80		229.68	10.32	11.36
56 L.H., Daily Totals		$2263.80		$3744.82	$40.42	$66.87

Crew G-2	Hr.	Daily	Hr.	Daily	Bare Costs	Incl. O&P
1 Plasterer	$35.00	$280.00	$57.25	$458.00	$31.28	$51.55
1 Plasterer Helper	29.65	237.20	48.50	388.00		
1 Building Laborer	29.20	233.60	48.90	391.20		
1 Grout Pump, 50 C.F./hr.		135.00		148.50	5.63	6.19
24 L.H., Daily Totals		$885.80		$1385.70	$36.91	$57.74

Crew G-2A	Hr.	Daily	Hr.	Daily	Bare Costs	Incl. O&P
1 Roofer Composition	$32.10	$256.80	$59.20	$473.60	$28.47	$50.85
1 Roofer Helper	24.10	192.80	44.45	355.60		
1 Building Laborer	29.20	233.60	48.90	391.20		
1 Foam Spray Rig, Trailer-Mtd.		567.00		623.70		
1 Pickup Truck, 3/4 Ton		145.80		160.38	29.70	32.67
24 L.H., Daily Totals		$1396.00		$2004.48	$58.17	$83.52

Crew G-3	Hr.	Daily	Hr.	Daily	Bare Costs	Incl. O&P
2 Sheet Metal Workers	$42.35	$677.60	$70.65	$1130.40	$35.77	$59.77
2 Building Laborers	29.20	467.20	48.90	782.40		
32 L.H., Daily Totals		$1144.80		$1912.80	$35.77	$59.77

Crew G-4	Hr.	Daily	Hr.	Daily	Bare Costs	Incl. O&P
1 Labor Foreman (outside)	$31.20	$249.60	$52.25	$418.00	$29.87	$50.02
2 Building Laborers	29.20	467.20	48.90	782.40		
1 Flatbed Truck, Gas, 1.5 Ton		248.20		273.02		
1 Air Compressor, 160 cfm		162.20		178.42	17.10	18.81
24 L.H., Daily Totals		$1127.20		$1651.84	$46.97	$68.83

Crew G-5	Hr.	Daily	Hr.	Daily	Bare Costs	Incl. O&P
1 Roofer Foreman (outside)	$34.10	$272.80	$62.90	$503.20	$29.30	$54.04
2 Roofers Composition	32.10	513.60	59.20	947.20		
2 Roofer Helpers	24.10	385.60	44.45	711.20		
1 Application Equipment		193.00		212.30	4.83	5.31
40 L.H., Daily Totals		$1365.00		$2373.90	$34.13	$59.35

Crew G-6A	Hr.	Daily	Hr.	Daily	Bare Costs	Incl. O&P
2 Roofers Composition	$32.10	$513.60	$59.20	$947.20	$32.10	$59.20
1 Small Compressor, Electric		14.10		15.51		
2 Pneumatic Nailers		52.40		57.64	4.16	4.57
16 L.H., Daily Totals		$580.10		$1020.35	$36.26	$63.77

Crew G-7	Hr.	Daily	Hr.	Daily	Bare Costs	Incl. O&P
1 Carpenter	$37.80	$302.40	$63.30	$506.40	$37.80	$63.30
1 Small Compressor, Electric		14.10		15.51		
1 Pneumatic Nailer		26.20		28.82	5.04	5.54
8 L.H., Daily Totals		$342.70		$550.73	$42.84	$68.84

Crews - Open Shop

Crew No.	Bare Costs		Incl. Subs O&P		Cost Per Labor-Hour	

Crew H-1	Hr.	Daily	Hr.	Daily	Bare Costs	Incl. O&P
2 Glaziers	$36.90	$590.40	$61.30	$980.80	$38.40	$67.25
2 Struc. Steel Workers	39.90	638.40	73.20	1171.20		
32 L.H., Daily Totals		$1228.80		$2152.00	$38.40	$67.25

Crew H-2	Hr.	Daily	Hr.	Daily	Bare Costs	Incl. O&P
2 Glaziers	$36.90	$590.40	$61.30	$980.80	$34.33	$57.17
1 Building Laborer	29.20	233.60	48.90	391.20		
24 L.H., Daily Totals		$824.00		$1372.00	$34.33	$57.17

Crew H-3	Hr.	Daily	Hr.	Daily	Bare Costs	Incl. O&P
1 Glazier	$36.90	$295.20	$61.30	$490.40	$32.42	$54.25
1 Helper	27.95	223.60	47.20	377.60		
16 L.H., Daily Totals		$518.80		$868.00	$32.42	$54.25

Crew H-4	Hr.	Daily	Hr.	Daily	Bare Costs	Incl. O&P
1 Carpenter	$37.80	$302.40	$63.30	$506.40	$35.11	$58.61
1 Carpenter Helper	27.95	223.60	47.20	377.60		
.5 Electrician	44.05	176.20	72.05	288.20		
20 L.H., Daily Totals		$702.20		$1172.20	$35.11	$58.61

Crew J-1	Hr.	Daily	Hr.	Daily	Bare Costs	Incl. O&P
3 Plasterers	$35.00	$840.00	$57.25	$1374.00	$32.86	$53.75
2 Plasterer Helpers	29.65	474.40	48.50	776.00		
1 Mixing Machine, 6 C.F.		140.20		154.22	3.50	3.86
40 L.H., Daily Totals		$1454.60		$2304.22	$36.37	$57.61

Crew J-2	Hr.	Daily	Hr.	Daily	Bare Costs	Incl. O&P
3 Plasterers	$35.00	$840.00	$57.25	$1374.00	$33.53	$54.73
2 Plasterer Helpers	29.65	474.40	48.50	776.00		
1 Lather	36.90	295.20	59.65	477.20		
1 Mixing Machine, 6 C.F.		140.20		154.22	2.92	3.21
48 L.H., Daily Totals		$1749.80		$2781.42	$36.45	$57.95

Crew J-3	Hr.	Daily	Hr.	Daily	Bare Costs	Incl. O&P
1 Terrazzo Worker	$34.80	$278.40	$56.30	$450.40	$31.98	$51.73
1 Terrazzo Helper	29.15	233.20	47.15	377.20		
1 Floor Grinder, 22" Path		121.60		133.76		
1 Terrazzo Mixer		188.20		207.02	19.36	21.30
16 L.H., Daily Totals		$821.40		$1168.38	$51.34	$73.02

Crew J-4	Hr.	Daily	Hr.	Daily	Bare Costs	Incl. O&P
2 Cement Finishers	$35.60	$569.60	$57.80	$924.80	$33.47	$54.83
1 Laborer	29.20	233.60	48.90	391.20		
1 Floor Grinder, 22" Path		121.60		133.76		
1 Floor Edger, 7" Path		39.30		43.23		
1 Vacuum Pick-Up System		60.90		66.99	9.24	10.17
24 L.H., Daily Totals		$1025.00		$1559.98	$42.71	$65.00

Crew J-4A	Hr.	Daily	Hr.	Daily	Bare Costs	Incl. O&P
2 Cement Finishers	$35.60	$569.60	$57.80	$924.80	$32.40	$53.35
2 Laborers	29.20	467.20	48.90	782.40		
1 Floor Grinder, 22" Path		121.60		133.76		
1 Floor Edger, 7" Path		39.30		43.23		
1 Vacuum Pick-Up System		60.90		66.99		
1 Floor Auto Scrubber		241.05		265.15	14.46	15.91
32 L.H., Daily Totals		$1499.65		$2216.34	$46.86	$69.26

Crew J-4B	Hr.	Daily	Hr.	Daily	Bare Costs	Incl. O&P
1 Laborer	$29.20	$233.60	$48.90	$391.20	$29.20	$48.90
1 Floor Auto Scrubber		241.05		265.15	30.13	33.14
8 L.H., Daily Totals		$474.65		$656.36	$59.33	$82.04

Crew J-6	Hr.	Daily	Hr.	Daily	Bare Costs	Incl. O&P
2 Painters	$31.90	$510.40	$52.55	$840.80	$32.71	$54.14
1 Building Laborer	29.20	233.60	48.90	391.20		
1 Equip. Oper. (light)	37.85	302.80	62.55	500.40		
1 Air Compressor, 250 cfm		197.80		217.58		
1 Sandblaster, Portable, 3 C.F.		20.80		22.88		
1 Set Sand Blasting Accessories		14.25		15.68	7.28	8.00
32 L.H., Daily Totals		$1279.65		$1988.54	$39.99	$62.14

Crew J-7	Hr.	Daily	Hr.	Daily	Bare Costs	Incl. O&P
2 Painters	$31.90	$510.40	$52.55	$840.80	$31.90	$52.55
1 Floor Belt Sander		24.45		26.90		
1 Floor Sanding Edger		13.65		15.02	2.38	2.62
16 L.H., Daily Totals		$548.50		$882.71	$34.28	$55.17

Crew K-1	Hr.	Daily	Hr.	Daily	Bare Costs	Incl. O&P
1 Carpenter	$37.80	$302.40	$63.30	$506.40	$35.45	$58.95
1 Truck Driver (light)	33.10	264.80	54.60	436.80		
1 Flatbed Truck, Gas, 3 Ton		307.40		338.14	19.21	21.13
16 L.H., Daily Totals		$874.60		$1281.34	$54.66	$80.08

Crew K-2	Hr.	Daily	Hr.	Daily	Bare Costs	Incl. O&P
1 Struc. Steel Foreman (outside)	$41.90	$335.20	$76.85	$614.80	$38.30	$68.22
1 Struc. Steel Worker	39.90	319.20	73.20	585.60		
1 Truck Driver (light)	33.10	264.80	54.60	436.80		
1 Flatbed Truck, Gas, 3 Ton		307.40		338.14	12.81	14.09
24 L.H., Daily Totals		$1226.60		$1975.34	$51.11	$82.31

Crew L-1	Hr.	Daily	Hr.	Daily	Bare Costs	Incl. O&P
.25 Electrician	$44.05	$88.10	$72.05	$144.10	$43.85	$72.21
1 Plumber	43.80	350.40	72.25	578.00		
10 L.H., Daily Totals		$438.50		$722.10	$43.85	$72.21

Crew L-2	Hr.	Daily	Hr.	Daily	Bare Costs	Incl. O&P
1 Carpenter	$37.80	$302.40	$63.30	$506.40	$32.88	$55.25
1 Carpenter Helper	27.95	223.60	47.20	377.60		
16 L.H., Daily Totals		$526.00		$884.00	$32.88	$55.25

Crew L-3	Hr.	Daily	Hr.	Daily	Bare Costs	Incl. O&P
1 Carpenter	$37.80	$302.40	$63.30	$506.40	$39.05	$65.05
.25 Electrician	44.05	88.10	72.05	144.10		
10 L.H., Daily Totals		$390.50		$650.50	$39.05	$65.05

Crew L-3A	Hr.	Daily	Hr.	Daily	Bare Costs	Incl. O&P
1 Carpenter Foreman (outside)	$39.80	$318.40	$66.65	$533.20	$40.65	$67.98
.5 Sheet Metal Worker	42.35	169.40	70.65	282.60		
12 L.H., Daily Totals		$487.80		$815.80	$40.65	$67.98

Crew L-4	Hr.	Daily	Hr.	Daily	Bare Costs	Incl. O&P
1 Skilled Worker	$38.30	$306.40	$64.35	$514.80	$33.13	$55.77
1 Helper	27.95	223.60	47.20	377.60		
16 L.H., Daily Totals		$530.00		$892.40	$33.13	$55.77

Crew No.	Bare Costs		Incl. Subs O&P		Cost Per Labor-Hour	

Crew L-5	Hr.	Daily	Hr.	Daily	Bare Costs	Incl. O&P
1 Struc. Steel Foreman (outside)	$41.90	$335.20	$76.85	$614.80	$40.24	$72.77
5 Struc. Steel Workers	39.90	1596.00	73.20	2928.00		
1 Equip. Oper. (crane)	40.25	322.00	66.55	532.40		
1 Hyd. Crane, 25 Ton		748.80		823.68	13.37	14.71
56 L.H., Daily Totals		$3002.00		$4898.88	$53.61	$87.48

Crew L-5A	Hr.	Daily	Hr.	Daily	Bare Costs	Incl. O&P
1 Struc. Steel Foreman (outside)	$41.90	$335.20	$76.85	$614.80	$40.49	$72.45
2 Structural Steel Workers	39.90	638.40	73.20	1171.20		
1 Equip. Oper. (crane)	40.25	322.00	66.55	532.40		
1 S.P. Crane, 4x4, 25 Ton		606.60		667.26	18.96	20.85
32 L.H., Daily Totals		$1902.20		$2985.66	$59.44	$93.30

Crew L-5B	Hr.	Daily	Hr.	Daily	Bare Costs	Incl. O&P
1 Struc. Steel Foreman (outside)	$41.90	$335.20	$76.85	$614.80	$41.02	$70.11
2 Structural Steel Workers	39.90	638.40	73.20	1171.20		
2 Electricians	44.05	704.80	72.05	1152.80		
2 Steamfitters/Pipefitters	44.95	719.20	74.10	1185.60		
1 Equip. Oper. (crane)	40.25	322.00	66.55	532.40		
1 Common Laborer	29.20	233.60	48.90	391.20		
1 Hyd. Crane, 80 Ton		1630.00		1793.00	22.64	24.90
72 L.H., Daily Totals		$4583.20		$6841.00	$63.66	$95.01

Crew L-6	Hr.	Daily	Hr.	Daily	Bare Costs	Incl. O&P
1 Plumber	$43.80	$350.40	$72.25	$578.00	$43.88	$72.18
.5 Electrician	44.05	176.20	72.05	288.20		
12 L.H., Daily Totals		$526.60		$866.20	$43.88	$72.18

Crew L-7	Hr.	Daily	Hr.	Daily	Bare Costs	Incl. O&P
1 Carpenter	$37.80	$302.40	$63.30	$506.40	$32.22	$54.07
2 Carpenter Helpers	27.95	447.20	47.20	755.20		
.25 Electrician	44.05	88.10	72.05	144.10		
26 L.H., Daily Totals		$837.70		$1405.70	$32.22	$54.07

Crew L-8	Hr.	Daily	Hr.	Daily	Bare Costs	Incl. O&P
1 Carpenter	$37.80	$302.40	$63.30	$506.40	$35.06	$58.65
1 Carpenter Helper	27.95	223.60	47.20	377.60		
.5 Plumber	43.80	175.20	72.25	289.00		
20 L.H., Daily Totals		$701.20		$1173.00	$35.06	$58.65

Crew L-9	Hr.	Daily	Hr.	Daily	Bare Costs	Incl. O&P
1 Skilled Worker Foreman	$40.30	$322.40	$67.70	$541.60	$34.78	$58.33
1 Skilled Worker	38.30	306.40	64.35	514.80		
2 Helpers	27.95	447.20	47.20	755.20		
.5 Electrician	44.05	176.20	72.05	288.20		
36 L.H., Daily Totals		$1252.20		$2099.80	$34.78	$58.33

Crew L-10	Hr.	Daily	Hr.	Daily	Bare Costs	Incl. O&P
1 Struc. Steel Foreman (outside)	$41.90	$335.20	$76.85	$614.80	$40.68	$72.20
1 Structural Steel Worker	39.90	319.20	73.20	585.60		
1 Equip. Oper. (crane)	40.25	322.00	66.55	532.40		
1 Hyd. Crane, 12 Ton		658.40		724.24	27.43	30.18
24 L.H., Daily Totals		$1634.80		$2457.04	$68.12	$102.38

Crew No.	Bare Costs		Incl. Subs O&P		Cost Per Labor-Hour	

Crew L-11	Hr.	Daily	Hr.	Daily	Bare Costs	Incl. O&P
2 Wreckers	$29.20	$467.20	$51.35	$821.60	$34.13	$57.95
1 Equip. Oper. (crane)	40.25	322.00	66.55	532.40		
1 Equip. Oper. (light)	37.85	302.80	62.55	500.40		
1 Hyd. Excavator, 2.5 C.Y.		1582.00		1740.20		
1 Loader, Skid Steer, 78 H.P.		315.60		347.16	59.30	65.23
32 L.H., Daily Totals		$2989.60		$3941.76	$93.42	$123.18

Crew M-1	Hr.	Daily	Hr.	Daily	Bare Costs	Incl. O&P
3 Elevator Constructors	$59.35	$1424.40	$96.85	$2324.40	$56.39	$92.01
1 Elevator Apprentice	47.50	380.00	77.50	620.00		
5 Hand Tools		48.00		52.80	1.50	1.65
32 L.H., Daily Totals		$1852.40		$2997.20	$57.89	$93.66

Crew M-3	Hr.	Daily	Hr.	Daily	Bare Costs	Incl. O&P
1 Electrician Foreman (outside)	$46.05	$368.40	$75.35	$602.80	$45.16	$74.09
1 Common Laborer	29.20	233.60	48.90	391.20		
.25 Equipment Operator (med.)	39.35	78.70	65.05	130.10		
1 Elevator Constructor	59.35	474.80	96.85	774.80		
1 Elevator Apprentice	47.50	380.00	77.50	620.00		
.25 S.P. Crane, 4x4, 20 Ton		138.70		152.57	4.08	4.49
34 L.H., Daily Totals		$1674.20		$2671.47	$49.24	$78.57

Crew M-4	Hr.	Daily	Hr.	Daily	Bare Costs	Incl. O&P
1 Electrician Foreman (outside)	$46.05	$368.40	$75.35	$602.80	$44.67	$73.31
1 Common Laborer	29.20	233.60	48.90	391.20		
.25 Equipment Operator, Crane	40.25	80.50	66.55	133.10		
.25 Equip. Oper. (oiler)	35.45	70.90	58.60	117.20		
1 Elevator Constructor	59.35	474.80	96.85	774.80		
1 Elevator Apprentice	47.50	380.00	77.50	620.00		
.25 S.P. Crane, 4x4, 40 Ton		177.95		195.75	4.94	5.44
36 L.H., Daily Totals		$1786.15		$2834.84	$49.62	$78.75

Crew Q-1	Hr.	Daily	Hr.	Daily	Bare Costs	Incl. O&P
1 Plumber	$43.80	$350.40	$72.25	$578.00	$39.42	$65.03
1 Plumber Apprentice	35.05	280.40	57.80	462.40		
16 L.H., Daily Totals		$630.80		$1040.40	$39.42	$65.03

Crew Q-1A	Hr.	Daily	Hr.	Daily	Bare Costs	Incl. O&P
.25 Plumber Foreman (outside)	$45.80	$91.60	$75.50	$151.00	$44.20	$72.90
1 Plumber	43.80	350.40	72.25	578.00		
10 L.H., Daily Totals		$442.00		$729.00	$44.20	$72.90

Crew Q-1C	Hr.	Daily	Hr.	Daily	Bare Costs	Incl. O&P
1 Plumber	$43.80	$350.40	$72.25	$578.00	$39.40	$65.03
1 Plumber Apprentice	35.05	280.40	57.80	462.40		
1 Equip. Oper. (medium)	39.35	314.80	65.05	520.40		
1 Trencher, Chain Type, 8' D		2860.00		3146.00	119.17	131.08
24 L.H., Daily Totals		$3805.60		$4706.80	$158.57	$196.12

Crew Q-2	Hr.	Daily	Hr.	Daily	Bare Costs	Incl. O&P
1 Plumber	$43.80	$350.40	$72.25	$578.00	$37.97	$62.62
2 Plumber Apprentices	35.05	560.80	57.80	924.80		
24 L.H., Daily Totals		$911.20		$1502.80	$37.97	$62.62

Crew Q-3	Hr.	Daily	Hr.	Daily	Bare Costs	Incl. O&P
2 Plumbers	$43.80	$700.80	$72.25	$1156.00	$39.42	$65.03
2 Plumber Apprentices	35.05	560.80	57.80	924.80		
32 L.H., Daily Totals		$1261.60		$2080.80	$39.42	$65.03

Crew No.	Bare Costs		Incl. Subs O&P		Cost Per Labor-Hour	
Crew Q-4	**Hr.**	**Daily**	**Hr.**	**Daily**	**Bare Costs**	**Incl. O&P**
2 Plumbers	$43.80	$700.80	$72.25	$1156.00	$41.61	$68.64
1 Welder (plumber)	43.80	350.40	72.25	578.00		
1 Plumber Apprentice	35.05	280.40	57.80	462.40		
1 Welder, Electric, 300 amp		58.10		63.91	1.82	2.00
32 L.H., Daily Totals		$1389.70		$2260.31	$43.43	$70.63

Crew Q-5	Hr.	Daily	Hr.	Daily	Bare Costs	Incl. O&P
1 Steamfitter	$44.95	$359.60	$74.10	$592.80	$40.45	$66.70
1 Steamfitter Apprentice	35.95	287.60	59.30	474.40		
16 L.H., Daily Totals		$647.20		$1067.20	$40.45	$66.70

Crew Q-6	Hr.	Daily	Hr.	Daily	Bare Costs	Incl. O&P
1 Steamfitter	$44.95	$359.60	$74.10	$592.80	$38.95	$64.23
2 Steamfitter Apprentices	35.95	575.20	59.30	948.80		
24 L.H., Daily Totals		$934.80		$1541.60	$38.95	$64.23

Crew Q-7	Hr.	Daily	Hr.	Daily	Bare Costs	Incl. O&P
2 Steamfitters	$44.95	$719.20	$74.10	$1185.60	$40.45	$66.70
2 Steamfitter Apprentices	35.95	575.20	59.30	948.80		
32 L.H., Daily Totals		$1294.40		$2134.40	$40.45	$66.70

Crew Q-8	Hr.	Daily	Hr.	Daily	Bare Costs	Incl. O&P
2 Steamfitters	$44.95	$719.20	$74.10	$1185.60	$42.70	$70.40
1 Welder (steamfitter)	44.95	359.60	74.10	592.80		
1 Steamfitter Apprentice	35.95	287.60	59.30	474.40		
1 Welder, Electric, 300 amp		58.10		63.91	1.82	2.00
32 L.H., Daily Totals		$1424.50		$2316.71	$44.52	$72.40

Crew Q-9	Hr.	Daily	Hr.	Daily	Bare Costs	Incl. O&P
1 Sheet Metal Worker	$42.35	$338.80	$70.65	$565.20	$38.13	$63.60
1 Sheet Metal Apprentice	33.90	271.20	56.55	452.40		
16 L.H., Daily Totals		$610.00		$1017.60	$38.13	$63.60

Crew Q-10	Hr.	Daily	Hr.	Daily	Bare Costs	Incl. O&P
2 Sheet Metal Workers	$42.35	$677.60	$70.65	$1130.40	$39.53	$65.95
1 Sheet Metal Apprentice	33.90	271.20	56.55	452.40		
24 L.H., Daily Totals		$948.80		$1582.80	$39.53	$65.95

Crew Q-11	Hr.	Daily	Hr.	Daily	Bare Costs	Incl. O&P
2 Sheet Metal Workers	$42.35	$677.60	$70.65	$1130.40	$38.13	$63.60
2 Sheet Metal Apprentices	33.90	542.40	56.55	904.80		
32 L.H., Daily Totals		$1220.00		$2035.20	$38.13	$63.60

Crew Q-12	Hr.	Daily	Hr.	Daily	Bare Costs	Incl. O&P
1 Sprinkler Installer	$42.65	$341.20	$70.65	$565.20	$38.38	$63.55
1 Sprinkler Apprentice	34.10	272.80	56.45	451.60		
16 L.H., Daily Totals		$614.00		$1016.80	$38.38	$63.55

Crew Q-13	Hr.	Daily	Hr.	Daily	Bare Costs	Incl. O&P
2 Sprinkler Installers	$42.65	$682.40	$70.65	$1130.40	$38.38	$63.55
2 Sprinkler Apprentices	34.10	545.60	56.45	903.20		
32 L.H., Daily Totals		$1228.00		$2033.60	$38.38	$63.55

Crew Q-14	Hr.	Daily	Hr.	Daily	Bare Costs	Incl. O&P
1 Asbestos Worker	$39.50	$316.00	$67.00	$536.00	$35.55	$60.30
1 Asbestos Apprentice	31.60	252.80	53.60	428.80		
16 L.H., Daily Totals		$568.80		$964.80	$35.55	$60.30

Crew No.	Bare Costs		Incl. Subs O&P		Cost Per Labor-Hour	
Crew Q-15	**Hr.**	**Daily**	**Hr.**	**Daily**	**Bare Costs**	**Incl. O&P**
1 Plumber	$43.80	$350.40	$72.25	$578.00	$39.42	$65.03
1 Plumber Apprentice	35.05	280.40	57.80	462.40		
1 Welder, Electric, 300 amp		58.10		63.91	3.63	3.99
16 L.H., Daily Totals		$688.90		$1104.31	$43.06	$69.02

Crew Q-16	Hr.	Daily	Hr.	Daily	Bare Costs	Incl. O&P
2 Plumbers	$43.80	$700.80	$72.25	$1156.00	$40.88	$67.43
1 Plumber Apprentice	35.05	280.40	57.80	462.40		
1 Welder, Electric, 300 amp		58.10		63.91	2.42	2.66
24 L.H., Daily Totals		$1039.30		$1682.31	$43.30	$70.10

Crew Q-17	Hr.	Daily	Hr.	Daily	Bare Costs	Incl. O&P
1 Steamfitter	$44.95	$359.60	$74.10	$592.80	$40.45	$66.70
1 Steamfitter Apprentice	35.95	287.60	59.30	474.40		
1 Welder, Electric, 300 amp		58.10		63.91	3.63	3.99
16 L.H., Daily Totals		$705.30		$1131.11	$44.08	$70.69

Crew Q-17A	Hr.	Daily	Hr.	Daily	Bare Costs	Incl. O&P
1 Steamfitter	$44.95	$359.60	$74.10	$592.80	$40.38	$66.65
1 Steamfitter Apprentice	35.95	287.60	59.30	474.40		
1 Equip. Oper. (crane)	40.25	322.00	66.55	532.40		
1 Hyd. Crane, 12 Ton		658.40		724.24		
1 Welder, Electric, 300 amp		58.10		63.91	29.85	32.84
24 L.H., Daily Totals		$1685.70		$2387.75	$70.24	$99.49

Crew Q-18	Hr.	Daily	Hr.	Daily	Bare Costs	Incl. O&P
2 Steamfitters	$44.95	$719.20	$74.10	$1185.60	$41.95	$69.17
1 Steamfitter Apprentice	35.95	287.60	59.30	474.40		
1 Welder, Electric, 300 amp		58.10		63.91	2.42	2.66
24 L.H., Daily Totals		$1064.90		$1723.91	$44.37	$71.83

Crew Q-19	Hr.	Daily	Hr.	Daily	Bare Costs	Incl. O&P
1 Steamfitter	$44.95	$359.60	$74.10	$592.80	$41.65	$68.48
1 Steamfitter Apprentice	35.95	287.60	59.30	474.40		
1 Electrician	44.05	352.40	72.05	576.40		
24 L.H., Daily Totals		$999.60		$1643.60	$41.65	$68.48

Crew Q-20	Hr.	Daily	Hr.	Daily	Bare Costs	Incl. O&P
1 Sheet Metal Worker	$42.35	$338.80	$70.65	$565.20	$39.31	$65.29
1 Sheet Metal Apprentice	33.90	271.20	56.55	452.40		
.5 Electrician	44.05	176.20	72.05	288.20		
20 L.H., Daily Totals		$786.20		$1305.80	$39.31	$65.29

Crew Q-21	Hr.	Daily	Hr.	Daily	Bare Costs	Incl. O&P
2 Steamfitters	$44.95	$719.20	$74.10	$1185.60	$42.48	$69.89
1 Steamfitter Apprentice	35.95	287.60	59.30	474.40		
1 Electrician	44.05	352.40	72.05	576.40		
32 L.H., Daily Totals		$1359.20		$2236.40	$42.48	$69.89

Crew Q-22	Hr.	Daily	Hr.	Daily	Bare Costs	Incl. O&P
1 Plumber	$43.80	$350.40	$72.25	$578.00	$39.42	$65.03
1 Plumber Apprentice	35.05	280.40	57.80	462.40		
1 Hyd. Crane, 12 Ton		658.40		724.24	41.15	45.27
16 L.H., Daily Totals		$1289.20		$1764.64	$80.58	$110.29

Crews - Open Shop

Crew No.	Bare Costs		Incl. Subs O&P		Cost Per Labor-Hour	
Crew Q-22A	**Hr.**	**Daily**	**Hr.**	**Daily**	**Bare Costs**	**Incl. O&P**
1 Plumber	$43.80	$350.40	$72.25	$578.00	$37.08	$61.38
1 Plumber Apprentice	35.05	280.40	57.80	462.40		
1 Laborer	29.20	233.60	48.90	391.20		
1 Equip. Oper. (crane)	40.25	322.00	66.55	532.40		
1 Hyd. Crane, 12 Ton		658.40		724.24	20.57	22.63
32 L.H., Daily Totals		$1844.80		$2688.24	$57.65	$84.01

Crew Q-23	Hr.	Daily	Hr.	Daily	Bare Costs	Incl. O&P
1 Plumber Foreman (outside)	$45.80	$366.40	$75.50	$604.00	$42.98	$70.93
1 Plumber	43.80	350.40	72.25	578.00		
1 Equip. Oper. (medium)	39.35	314.80	65.05	520.40		
1 Lattice Boom Crane, 20 Ton		967.35		1064.09	40.31	44.34
24 L.H., Daily Totals		$1998.95		$2766.49	$83.29	$115.27

Crew R-1	Hr.	Daily	Hr.	Daily	Bare Costs	Incl. O&P
1 Electrician Foreman	$44.55	$356.40	$72.90	$583.20	$41.20	$67.39
3 Electricians	44.05	1057.20	72.05	1729.20		
2 Electrician Apprentices	35.25	564.00	57.65	922.40		
48 L.H., Daily Totals		$1977.60		$3234.80	$41.20	$67.39

Crew R-1A	Hr.	Daily	Hr.	Daily	Bare Costs	Incl. O&P
1 Electrician	$44.05	$352.40	$72.05	$576.40	$39.65	$64.85
1 Electrician Apprentice	35.25	282.00	57.65	461.20		
16 L.H., Daily Totals		$634.40		$1037.60	$39.65	$64.85

Crew R-1B	Hr.	Daily	Hr.	Daily	Bare Costs	Incl. O&P
1 Electrician	$44.05	$352.40	$72.05	$576.40	$38.18	$62.45
2 Electrician Apprentices	35.25	564.00	57.65	922.40		
24 L.H., Daily Totals		$916.40		$1498.80	$38.18	$62.45

Crew R-1C	Hr.	Daily	Hr.	Daily	Bare Costs	Incl. O&P
2 Electricians	$44.05	$704.80	$72.05	$1152.80	$39.65	$64.85
2 Electrician Apprentices	35.25	564.00	57.65	922.40		
1 Portable cable puller, 8000 lb.		130.10		143.11	4.07	4.47
32 L.H., Daily Totals		$1398.90		$2218.31	$43.72	$69.32

Crew R-2	Hr.	Daily	Hr.	Daily	Bare Costs	Incl. O&P
1 Electrician Foreman	$44.55	$356.40	$72.90	$583.20	$41.06	$67.27
3 Electricians	44.05	1057.20	72.05	1729.20		
2 Electrician Apprentices	35.25	564.00	57.65	922.40		
1 Equip. Oper. (crane)	40.25	322.00	66.55	532.40		
1 S.P. Crane, 4x4, 5 Ton		280.80		308.88	5.01	5.52
56 L.H., Daily Totals		$2580.40		$4076.08	$46.08	$72.79

Crew R-3	Hr.	Daily	Hr.	Daily	Bare Costs	Incl. O&P
1 Electrician Foreman	$44.55	$356.40	$72.90	$583.20	$43.49	$71.29
1 Electrician	44.05	352.40	72.05	576.40		
.5 Equip. Oper. (crane)	40.25	161.00	66.55	266.20		
.5 S.P. Crane, 4x4, 5 Ton		140.40		154.44	7.02	7.72
20 L.H., Daily Totals		$1010.20		$1580.24	$50.51	$79.01

Crew R-4	Hr.	Daily	Hr.	Daily	Bare Costs	Incl. O&P
1 Struc. Steel Foreman (outside)	$41.90	$335.20	$76.85	$614.80	$41.13	$73.70
3 Struc. Steel Workers	39.90	957.60	73.20	1756.80		
1 Electrician	44.05	352.40	72.05	576.40		
1 Welder, Gas Engine, 300 amp		147.20		161.92	3.68	4.05
40 L.H., Daily Totals		$1792.40		$3109.92	$44.81	$77.75

Crew R-5	Hr.	Daily	Hr.	Daily	Bare Costs	Incl. O&P
1 Electrician Foreman	$44.55	$356.40	$72.90	$583.20	$38.24	$63.09
4 Electrician Linemen	44.05	1409.60	72.05	2305.60		
2 Electrician Operators	44.05	704.80	72.05	1152.80		
4 Electrician Groundmen	27.95	894.40	47.20	1510.40		
1 Crew Truck		208.80		229.68		
1 Flatbed Truck, 20,000 GVW		247.80		272.58		
1 Pickup Truck, 3/4 Ton		145.80		160.38		
.2 Hyd. Crane, 55 Ton		224.60		247.06		
.2 Hyd. Crane, 12 Ton		131.68		144.85		
.2 Earth Auger, Truck-Mtd.		84.16		92.58		
1 Tractor w/Winch		428.00		470.80	16.71	18.39
88 L.H., Daily Totals		$4836.04		$7169.92	$54.95	$81.48

Crew R-6	Hr.	Daily	Hr.	Daily	Bare Costs	Incl. O&P
1 Electrician Foreman	$44.55	$356.40	$72.90	$583.20	$38.24	$63.09
4 Electrician Linemen	44.05	1409.60	72.05	2305.60		
2 Electrician Operators	44.05	704.80	72.05	1152.80		
4 Electrician Groundmen	27.95	894.40	47.20	1510.40		
1 Crew Truck		208.80		229.68		
1 Flatbed Truck, 20,000 GVW		247.80		272.58		
1 Pickup Truck, 3/4 Ton		145.80		160.38		
.2 Hyd. Crane, 55 Ton		224.60		247.06		
.2 Hyd. Crane, 12 Ton		131.68		144.85		
.2 Earth Auger, Truck-Mtd.		84.16		92.58		
1 Tractor w/Winch		428.00		470.80		
3 Cable Trailers		610.65		671.72		
.5 Tensioning Rig		202.25		222.47		
.5 Cable Pulling Rig		1170.50		1287.55	39.25	43.18
88 L.H., Daily Totals		$6819.44		$9351.66	$77.49	$106.27

Crew R-7	Hr.	Daily	Hr.	Daily	Bare Costs	Incl. O&P
1 Electrician Foreman	$44.55	$356.40	$72.90	$583.20	$30.72	$51.48
5 Electrician Groundmen	27.95	1118.00	47.20	1888.00		
1 Crew Truck		208.80		229.68	4.35	4.79
48 L.H., Daily Totals		$1683.20		$2700.88	$35.07	$56.27

Crew R-8	Hr.	Daily	Hr.	Daily	Bare Costs	Incl. O&P
1 Electrician Foreman	$44.55	$356.40	$72.90	$583.20	$38.77	$63.91
3 Electrician Linemen	44.05	1057.20	72.05	1729.20		
2 Electrician Groundmen	27.95	447.20	47.20	755.20		
1 Pickup Truck, 3/4 Ton		145.80		160.38		
1 Crew Truck		208.80		229.68	7.39	8.13
48 L.H., Daily Totals		$2215.40		$3457.66	$46.15	$72.03

Crew R-9	Hr.	Daily	Hr.	Daily	Bare Costs	Incl. O&P
1 Electrician Foreman	$44.55	$356.40	$72.90	$583.20	$36.06	$59.73
1 Electrician Lineman	44.05	352.40	72.05	576.40		
2 Electrician Operators	44.05	704.80	72.05	1152.80		
4 Electrician Groundmen	27.95	894.40	47.20	1510.40		
1 Pickup Truck, 3/4 Ton		145.80		160.38		
1 Crew Truck		208.80		229.68	5.54	6.09
64 L.H., Daily Totals		$2662.60		$4212.86	$41.60	$65.83

Crew R-10	Hr.	Daily	Hr.	Daily	Bare Costs	Incl. O&P
1 Electrician Foreman	$44.55	$356.40	$72.90	$583.20	$41.45	$68.05
4 Electrician Linemen	44.05	1409.60	72.05	2305.60		
1 Electrician Groundman	27.95	223.60	47.20	377.60		
1 Crew Truck		208.80		229.68		
3 Tram Cars		418.50		460.35	13.07	14.38
48 L.H., Daily Totals		$2616.90		$3956.43	$54.52	$82.43

Crews - Open Shop

Crew R-11	Bare Costs Hr.	Bare Costs Daily	Incl. Subs O&P Hr.	Incl. Subs O&P Daily	Cost Per Labor-Hour Bare Costs	Cost Per Labor-Hour Incl. O&P
1 Electrician Foreman	$44.55	$356.40	$72.90	$583.20	$41.46	$68.08
4 Electricians	44.05	1409.60	72.05	2305.60		
1 Equip. Oper. (crane)	40.25	322.00	66.55	532.40		
1 Common Laborer	29.20	233.60	48.90	391.20		
1 Crew Truck		208.80		229.68		
1 Hyd. Crane, 12 Ton		658.40		724.24	15.49	17.03
56 L.H., Daily Totals		$3188.80		$4766.32	$56.94	$85.11

Crew R-12	Hr.	Daily	Hr.	Daily	Bare Costs	Incl. O&P
1 Carpenter Foreman (inside)	$38.30	$306.40	$64.10	$512.80	$35.05	$59.20
4 Carpenters	37.80	1209.60	63.30	2025.60		
4 Common Laborers	29.20	934.40	48.90	1564.80		
1 Equip. Oper. (medium)	39.35	314.80	65.05	520.40		
1 Steel Worker	39.90	319.20	73.20	585.60		
1 Dozer, 200 H.P.		1397.00		1536.70		
1 Pickup Truck, 3/4 Ton		145.80		160.38	17.53	19.29
88 L.H., Daily Totals		$4627.20		$6906.28	$52.58	$78.48

Crew R-15	Hr.	Daily	Hr.	Daily	Bare Costs	Incl. O&P
1 Electrician Foreman	$44.55	$356.40	$72.90	$583.20	$43.10	$70.61
4 Electricians	44.05	1409.60	72.05	2305.60		
1 Equipment Oper. (light)	37.85	302.80	62.55	500.40		
1 Aerial Lift Truck, 40' Boom		309.00		339.90	6.44	7.08
48 L.H., Daily Totals		$2377.80		$3729.10	$49.54	$77.69

Crew R-15A	Hr.	Daily	Hr.	Daily	Bare Costs	Incl. O&P
1 Electrician Foreman	$44.55	$356.40	$72.90	$583.20	$38.15	$62.89
2 Electricians	44.05	704.80	72.05	1152.80		
2 Common Laborers	29.20	467.20	48.90	782.40		
1 Equip. Oper. (light)	37.85	302.80	62.55	500.40		
1 Aerial Lift Truck, 40' Boom		309.00		339.90	6.44	7.08
48 L.H., Daily Totals		$2140.20		$3358.70	$44.59	$69.97

Crew R-18	Hr.	Daily	Hr.	Daily	Bare Costs	Incl. O&P
.25 Electrician Foreman	$44.55	$89.10	$72.90	$145.80	$38.67	$63.25
1 Electrician	44.05	352.40	72.05	576.40		
2 Electrician Apprentices	35.25	564.00	57.65	922.40		
26 L.H., Daily Totals		$1005.50		$1644.60	$38.67	$63.25

Crew R-19	Hr.	Daily	Hr.	Daily	Bare Costs	Incl. O&P
.5 Electrician Foreman	$44.55	$178.20	$72.90	$291.60	$44.15	$72.22
2 Electricians	44.05	704.80	72.05	1152.80		
20 L.H., Daily Totals		$883.00		$1444.40	$44.15	$72.22

Crew R-21	Hr.	Daily	Hr.	Daily	Bare Costs	Incl. O&P
1 Electrician Foreman	$44.55	$356.40	$72.90	$583.20	$44.06	$72.09
3 Electricians	44.05	1057.20	72.05	1729.20		
.1 Equip. Oper. (medium)	39.35	31.48	65.05	52.04		
.1 S.P. Crane, 4x4, 25 Ton		60.66		66.73	1.85	2.03
32.8 L.H., Daily Totals		$1505.74		$2431.17	$45.91	$74.12

Crew R-22	Hr.	Daily	Hr.	Daily	Bare Costs	Incl. O&P
.66 Electrician Foreman	$44.55	$235.22	$72.90	$384.91	$40.34	$65.99
2 Electricians	44.05	704.80	72.05	1152.80		
2 Electrician Apprentices	35.25	564.00	57.65	922.40		
37.28 L.H., Daily Totals		$1504.02		$2460.11	$40.34	$65.99

Crew R-30	Bare Costs Hr.	Bare Costs Daily	Incl. Subs O&P Hr.	Incl. Subs O&P Daily	Cost Per Labor-Hour Bare Costs	Cost Per Labor-Hour Incl. O&P
.25 Electrician Foreman (outside)	$46.05	$92.10	$75.35	$150.70	$35.07	$58.06
1 Electrician	44.05	352.40	72.05	576.40		
2 Laborers (Semi-Skilled)	29.20	467.20	48.90	782.40		
26 L.H., Daily Totals		$911.70		$1509.50	$35.07	$58.06

Historical Cost Indexes

The table below lists both the RSMeans® historical cost index based on Jan. 1, 1993 = 100 as well as the computed value of an index based on Jan. 1, 2016 costs. Since the Jan. 1, 2016 figure is estimated, space is left to write in the actual index figures as they become available through the quarterly *RSMeans Construction Cost Indexes.*

To compute the actual index based on Jan. 1, 2016 = 100, divide the historical cost index for a particular year by the actual Jan. 1, 2016 construction cost index. Space has been left to advance the index figures as the year progresses.

Year	Historical Cost Index Jan. 1, 1993 = 100		Current Index Based on Jan. 1, 2016 = 100		Year	Historical Cost Index Jan. 1, 1993 = 100	Current Index Based on Jan. 1, 2016 = 100		Year	Historical Cost Index Jan. 1, 1993 = 100	Current Index Based on Jan. 1, 2016 = 100	
	Est.	Actual	Est.	Actual		Actual	Est.	Actual		Actual	Est.	Actual
Oct 2016*					July 2001	125.1	60.4		July 1983	80.2	38.7	
July 2016*					2000	120.9	58.3		1982	76.1	36.8	
April 2016*					1999	117.6	56.8		1981	70.0	33.8	
Jan 2016*	207.2		100.0	100.0	1998	115.1	55.6		1980	62.9	30.4	
July 2015		206.2			1997	112.8	54.4		1979	57.8	27.9	
2014		204.9	98.9		1996	110.2	53.2		1978	53.5	25.8	
2013		201.2	97.1		1995	107.6	51.9		1977	49.5	23.9	
2012		194.6	93.9		1994	104.4	50.4		1976	46.9	22.6	
2011		191.2	92.3		1993	101.7	49.1		1975	44.8	21.6	
2010		183.5	88.6		1992	99.4	48.0		1974	41.4	20.0	
2009		180.1	86.9		1991	96.8	46.7		1973	37.7	18.2	
2008		180.4	87.1		1990	94.3	45.5		1972	34.8	16.8	
2007		169.4	81.8		1989	92.1	44.5		1971	32.1	15.5	
2006		162.0	78.2		1988	89.9	43.4		1970	28.7	13.9	
2005		151.6	73.2		1987	87.7	42.3		1969	26.9	13.0	
2004		143.7	69.4		1986	84.2	40.7		1968	24.9	12.0	
2003		132.0	63.7		1985	82.6	39.9		1967	23.5	11.3	
2002		128.7	62.1		1984	82.0	39.6		1966	22.7	11.0	

Adjustments to Costs

The "Historical Cost Index" can be used to convert national average building costs at a particular time to the approximate building costs for some other time.

Time Adjustment Using the Historical Cost Indexes:

$$\frac{\text{Index for Year A}}{\text{Index for Year B}} \times \text{Cost in Year B} = \text{Cost in Year A}$$

Example:

Estimate and compare construction costs for different years in the same city.

To estimate the national average construction cost of a building in 1970, knowing that it cost $900,000 in 2016:

INDEX in 1970 = 28.7

INDEX in 2016 = 207.2

$$\frac{\text{INDEX 1970}}{\text{INDEX 2016}} \times \text{Cost 2016} = \text{Cost 1970}$$

$$\frac{28.7}{207.2} \times \$900{,}000 = .139 \times \$900{,}000 = \$124{,}662$$

The construction cost of the building in 1970 was $124,662.

Note: The city cost indexes for Canada can be used to convert U.S. national averages to local costs in Canadian dollars.

Example:

To estimate and compare the cost of a building in Toronto, ON in 2016 with the known cost of $600,000 (US$) in New York, NY in 2016:

INDEX Toronto = 109.9

INDEX New York = 131.1

$$\frac{\text{INDEX Toronto}}{\text{INDEX New York}} \times \text{Cost New York} = \text{Cost Toronto}$$

$$\frac{109.9}{131.1} \times \$600{,}000 = .841 \times \$600{,}000 = \$502{,}975$$

The construction cost of the building in Toronto is $502,975 (CN$).

*Historical Cost Index updates and other resources are provided on the following website:
http://info.thegordiangroup.com/RSMeans.html

Historical Cost Indexes

Year	National	Alabama					Alaska	Arizona		Arkansas		California				
Year	30 City Average	Birming-ham	Hunts-ville	Mobile	Mont-gomery	Tusca-loosa	Anchor-age	Phoenix	Tuscon	Fort Smith	Little Rock	Anaheim	Bakers-field	Fresno	Los Angeles	Oxnard
Jan 2016	207.2E	185.3E	182.6E	185.9E	183.2E	185E	245.3E	181.5E	178.4E	168E	171.5E	217.9E	215.2E	219.4E	219.1E	216.3E
2015	204.0	184.8	182.1	185.4	182.7	184.5	244.7	181.0	177.9	167.6	171.1	217.3	214.6	218.8	218.5	215.7
2014	203.0	180.3	175.8	170.6	163.1	165.9	241.0	180.0	177.0	165.1	168.4	214.2	215.2	217.0	217.3	214.8
2013	196.9	173.3	168.7	165.7	158.7	161.6	235.3	174.1	169.2	161.2	163.2	207.0	205.6	210.3	210.7	207.8
2012	194.0	169.1	162.7	163.2	153.8	154.6	232.5	172.2	166.4	158.7	161.0	204.1	202.9	207.3	207.2	204.7
2011	185.7	163.0	156.3	156.9	147.6	148.3	225.1	163.8	159.5	152.0	154.1	196.1	194.6	199.8	199.2	197.0
2010	181.6	159.6	152.9	153.1	144.4	144.9	218.3	160.7	157.3	150.3	152.5	192.8	190.8	195.0	194.9	192.8
2009	182.5	162.7	157.2	155.6	148.8	149.4	222.9	161.3	156.8	149.2	156.3	194.5	192.3	195.0	196.6	194.5
2008	171.0	150.3	146.9	143.7	138.4	138.8	210.8	152.2	148.4	138.6	145.4	182.7	179.8	183.2	184.7	182.6
2007	165.0	146.9	143.3	140.3	134.8	135.4	206.8	147.7	143.1	134.8	141.7	175.1	173.7	176.7	177.5	176.2
2006	156.2	135.7	133.6	126.7	124.0	122.2	196.4	137.9	134.8	123.2	127.2	166.8	164.9	169.8	167.3	167.4
2005	146.7	127.9	125.4	119.3	116.6	114.6	185.6	128.5	124.1	115.4	119.4	156.5	153.0	157.9	157.1	156.4
2004	132.8	115.9	112.9	107.0	104.9	102.9	167.0	116.5	113.6	103.8	108.6	142.1	139.5	143.8	142.0	140.9
2003	129.7	113.1	110.7	104.8	102.6	100.8	163.5	113.9	110.6	101.8	105.8	139.4	137.2	142.1	139.6	138.7
2002	126.7	110.0	103.4	103.6	101.7	99.7	159.5	113.3	110.2	100.4	102.1	136.5	133.4	136.0	136.4	136.3
2001	122.2	106.0	100.5	100.6	98.3	95.9	152.6	109.0	106.4	97.3	98.7	132.5	129.3	132.8	132.4	132.4
2000	118.9	104.1	98.9	99.1	94.2	94.6	148.3	106.9	104.9	94.4	95.7	129.4	125.2	129.4	129.9	129.7
1999	116.6	101.2	97.4	97.7	92.5	92.8	145.9	105.5	103.3	93.1	94.4	127.9	123.6	126.9	128.7	128.2
1998	113.6	96.2	94.0	94.8	90.3	89.8	143.8	102.1	101.0	90.6	91.4	125.2	120.3	123.9	125.8	124.7
1997	111.5	94.6	92.4	93.3	88.8	88.3	142.0	101.8	100.7	89.3	90.1	124.0	119.1	122.3	124.6	123.5
1995	105.6	87.8	88.0	88.8	84.5	83.5	138.0	96.1	95.5	85.0	85.6	120.1	115.7	117.5	120.9	120.0
1990	93.2	79.4	77.6	82.7	78.4	75.2	125.8	86.4	87.0	77.1	77.9	107.7	102.7	103.3	107.5	107.4
1985	81.8	71.1	70.5	72.7	70.9	67.9	116.0	78.1	77.7	69.6	71.0	95.4	92.2	92.6	94.6	96.6
1980	60.7	55.2	54.1	56.8	56.8	54.0	91.4	63.7	62.4	53.1	55.8	68.7	69.4	68.7	67.4	69.9
1975	43.7	40.0	40.9	41.8	40.1	37.8	57.3	44.5	45.2	39.0	38.7	47.6	46.6	47.7	48.3	47.0
1970	27.8	24.1	25.1	25.8	25.4	24.2	43.0	27.2	28.5	24.3	22.3	31.0	30.8	31.1	29.0	31.0
1965	21.5	19.6	19.3	19.6	19.5	18.7	34.9	21.8	22.0	18.7	18.5	23.9	23.7	24.0	22.7	23.9
1960	19.5	17.9	17.5	17.8	17.7	16.9	31.7	19.9	20.0	17.0	16.8	21.7	21.5	21.8	20.6	21.7
1955	16.3	14.8	14.7	14.9	14.9	14.2	26.6	16.7	16.7	14.3	14.5	18.2	18.1	18.3	17.3	18.2
1950	13.5	12.2	12.1	12.3	12.3	11.7	21.9	13.8	13.8	11.8	11.6	15.1	15.0	15.1	14.3	15.1
1945	8.6	7.8	7.7	7.8	7.8	7.5	14.0	8.8	8.8	7.5	7.4	9.6	9.5	9.6	9.1	9.6
▼ 1940	6.6	6.0	6.0	6.1	6.0	5.8	10.8	6.8	6.8	5.8	5.7	7.4	7.4	7.4	7.0	7.4

Year	National	California							Colorado			Connecticut				
Year	30 City Average	River-side	Sacra-mento	San Diego	San Francisco	Santa Barbara	Stockton	Vallejo	Colorado Springs	Denver	Pueblo	Bridge-Port	Bristol	Hartford	New Britain	New Haven
Jan 2016	207.2E	217.7E	221.6E	213.7E	250.8E	216.3E	223.4E	230.8E	185.6E	187.1E	184E	224.7E	223.9E	224.8E	223.4E	225.6E
2015	204.0	217.1	221.0	213.1	250.2	215.7	222.8	230.2	185.1	186.6	183.5	224.1	223.3	224.2	222.8	225.0
2014	203.0	214.1	221.5	211.4	248.0	214.7	218.4	227.9	187.8	189.1	184.7	223.8	222.8	223.9	222.4	224.1
2013	196.9	207.0	215.0	203.2	241.0	207.9	211.8	222.2	180.7	183.4	180.0	217.8	217.2	218.4	216.8	218.5
2012	194.0	204.1	211.9	198.9	237.6	204.8	208.8	218.9	179.4	182.2	177.9	213.9	213.3	214.4	212.9	214.7
2011	185.7	195.8	203.1	192.3	227.9	196.2	201.3	210.3	171.7	174.1	170.7	205.2	204.2	205.1	203.9	205.8
2010	181.6	192.7	196.7	188.2	223.0	192.9	195.9	204.6	170.5	172.6	168.1	198.8	199.1	199.8	198.8	200.7
2009	182.5	193.1	198.1	191.6	224.9	193.6	194.5	202.7	168.4	172.0	166.7	199.0	197.7	198.7	197.4	199.3
2008	171.0	181.3	186.0	179.9	210.6	181.5	183.2	191.6	158.8	161.9	157.9	185.8	184.8	185.7	184.5	186.1
2007	165.0	174.7	179.1	173.6	201.1	175.2	178.2	185.9	152.6	155.9	152.3	179.6	178.3	179.6	178.0	179.5
2006	156.2	166.0	172.2	164.0	191.2	166.4	171.0	177.9	146.1	149.2	145.0	169.5	168.0	169.5	167.7	169.8
2005	146.7	155.4	161.1	153.8	179.7	155.7	160.0	167.2	138.3	141.1	136.4	160.8	159.4	159.6	159.1	160.8
2004	132.8	141.0	147.2	139.0	163.6	141.4	144.2	148.6	125.6	127.2	123.5	143.6	142.9	142.4	142.7	144.3
2003	129.7	138.7	144.9	136.6	162.1	139.2	141.8	146.4	123.1	123.7	120.7	141.3	140.2	140.3	140.0	141.6
2002	126.7	135.3	138.4	134.2	157.9	135.9	136.9	143.8	118.7	121.3	116.7	133.8	133.6	133.1	133.3	134.8
2001	122.2	131.4	135.5	129.7	151.8	131.5	134.1	140.2	113.3	117.2	113.1	128.6	128.5	128.5	128.3	128.6
2000	118.9	128.0	131.5	127.1	146.9	128.7	130.7	137.1	109.8	111.8	108.8	122.7	122.6	122.9	122.4	122.7
1999	116.6	126.5	129.4	124.9	145.1	127.2	127.9	135.3	107.0	109.1	107.1	121.1	121.3	121.2	121.1	121.3
1998	113.6	123.7	125.9	121.3	141.9	123.7	124.7	132.5	103.3	106.5	103.7	119.1	119.4	120.0	119.7	120.0
1997	111.5	122.6	124.7	120.3	139.2	122.4	123.3	130.6	101.1	104.4	102.0	119.2	119.5	119.9	119.7	120.0
1995	105.6	119.2	119.5	115.4	133.8	119.0	119.0	122.5	96.1	98.9	96.8	116.0	116.5	116.9	116.3	116.5
1990	93.2	107.0	104.9	105.6	121.8	106.4	105.4	111.4	86.9	88.8	88.8	96.3	95.9	96.6	95.9	96.5
1985	81.8	94.8	92.2	94.2	106.2	93.2	93.7	97.0	79.6	81.0	80.4	86.1	86.2	87.2	86.1	86.3
1980	60.7	68.7	71.3	68.1	75.2	71.1	71.2	71.9	60.7	60.9	59.5	61.5	60.7	61.9	60.7	61.4
1975	43.7	47.3	49.1	47.7	49.8	46.1	47.6	46.5	43.1	42.7	42.5	45.2	45.5	46.0	45.2	46.0
1970	27.8	31.0	32.1	30.7	31.6	31.3	31.8	31.8	27.6	26.1	27.3	29.2	28.2	29.6	28.2	29.3
1965	21.5	23.9	24.8	24.0	23.7	24.1	24.5	24.5	21.3	20.9	21.0	22.4	21.7	22.6	21.7	23.2
1960	19.5	21.7	22.5	21.7	21.5	21.9	22.3	22.2	19.3	19.0	19.1	20.0	19.8	20.0	19.8	20.0
1955	16.3	18.2	18.9	18.2	18.0	18.4	18.7	18.6	16.2	15.9	16.0	16.8	16.6	16.8	16.6	16.8
1950	13.5	15.0	15.6	15.0	14.9	15.2	15.4	15.4	13.4	13.2	13.2	13.9	13.7	13.8	13.7	13.9
1945	8.6	9.6	10.0	9.6	9.5	9.7	9.9	9.8	8.5	8.4	8.4	8.8	8.7	8.8	8.7	8.9
▼ 1940	6.6	7.4	7.7	7.4	7.3	7.5	7.6	7.6	6.6	6.5	6.5	6.8	6.8	6.8	6.8	6.8

Historical Cost Indexes

Year	National	Connecticut			Delaware	D.C.	Florida						Georgia			
	30 City Average	Norwalk	Stamford	Waterbury	Wilmington	Washington	Fort Lauderdale	Jacksonville	Miami	Orlando	Tallahassee	Tampa	Albany	Atlanta	Columbus	Macon
Jan 2016	207.2E	230.8E	230.9E	224.4E	212.2E	197.8E	176.6E	174.7E	178.6E	178.8E	172.9E	179.4E	169E	178.8E	171.7E	170.4E
2015	204.0	230.2	230.3	223.8	211.7	197.3	176.1	174.2	178.1	178.3	172.5	178.9	168.6	178.3	171.3	170
2014	203.0	230.3	230.5	223.2	210.2	197.0	176.5	171.0	178.4	176.5	164.4	183.2	166.6	177.5	170.2	168.2
2013	196.9	223.1	224.2	217.3	203.6	191.7	173.2	168.4	175.8	174.8	160.8	180.2	163.2	173.3	166.7	164.9
2012	194.0	219.8	220.0	213.4	201.0	189.6	170.9	165.8	173.7	172.8	158.5	177.8	158.3	170.7	160.6	159.3
2011	185.7	211.0	211.2	204.6	193.3	182.3	164.4	159.4	167.0	166.1	151.7	171.4	151.8	164.0	154.3	152.7
2010	181.6	198.3	203.8	199.4	187.7	179.0	160.3	155.4	163.3	162.5	148.1	167.5	148.3	160.1	150.9	149.4
2009	182.5	198.8	204.3	198.4	188.9	181.4	160.7	152.0	165.2	163.9	145.1	165.1	152.0	164.5	155.6	153.5
2008	171.0	185.2	189.2	185.3	177.5	170.4	149.6	142.5	152.9	153.0	135.3	155.6	140.4	153.2	143.9	142.5
2007	165.0	178.6	183.3	179.1	173.4	163.1	145.6	139.0	148.9	148.1	131.7	152.0	135.8	148.0	140.1	138.1
2006	156.2	169.7	173.7	169.1	158.3	153.0	136.0	126.5	136.7	134.3	118.4	136.6	123.6	139.9	126.3	124.3
2005	146.7	160.9	164.1	160.5	149.6	142.7	126.0	119.2	127.1	126.0	110.7	127.7	115.7	131.7	118.9	116.1
2004	132.8	143.6	146.5	143.2	136.1	126.8	114.8	107.8	115.6	113.2	100.6	116.6	102.8	119.6	102.0	105.0
2003	129.7	140.6	145.1	140.8	133.0	124.6	109.0	105.6	110.5	107.8	98.0	103.5	100.5	115.8	99.1	102.7
2002	126.7	133.5	136.2	133.9	129.4	120.3	107.1	104.6	107.4	106.7	97.3	102.8	99.7	113.0	98.3	101.9
2001	122.2	128.2	131.5	128.8	124.8	115.9	104.3	100.8	104.6	103.8	94.8	100.3	96.4	109.1	95.5	99.0
2000	118.9	121.5	126.4	123.2	117.4	113.8	102.4	99.0	101.8	101.2	93.6	98.9	94.9	106.1	94.1	97.0
1999	116.6	120.0	122.0	121.2	116.6	111.5	101.4	98.0	100.8	100.1	92.5	97.9	93.5	102.9	92.8	95.8
1998	113.6	118.8	120.8	120.1	112.3	109.6	99.4	96.2	99.0	98.4	90.9	96.3	91.4	100.8	90.4	93.3
1997	111.5	118.9	120.9	120.2	110.7	106.4	98.6	95.4	98.2	97.2	89.9	95.5	89.9	98.5	88.7	91.6
1995	105.6	115.5	117.9	117.3	106.1	102.3	94.0	90.9	93.7	93.5	86.5	92.2	85.0	92.0	82.6	86.8
1990	93.2	96.3	98.9	95.1	92.5	90.4	83.9	81.1	84.0	82.9	78.4	85.0	75.0	80.4	74.0	76.9
1985	81.8	85.3	86.8	85.6	81.1	78.8	76.7	73.4	78.3	73.9	70.6	77.3	66.9	70.3	66.1	68.1
1980	60.7	60.7	60.9	62.3	58.5	59.6	55.3	55.8	56.5	56.7	53.5	57.2	51.7	54.0	51.1	51.3
1975	43.7	44.7	45.0	46.3	42.9	43.7	42.1	40.3	43.2	41.5	38.1	41.3	37.5	38.4	36.2	36.5
1970	27.8	28.1	28.2	28.9	27.0	26.3	25.7	22.8	27.0	26.2	24.5	24.2	23.8	25.2	22.8	23.4
1965	21.5	21.6	21.7	22.2	20.9	21.8	19.8	17.4	19.3	20.2	18.9	18.6	18.3	19.8	17.6	18.0
1960	19.5	19.7	19.7	20.2	18.9	19.4	18.0	15.8	17.6	18.3	17.2	16.9	16.7	17.1	16.0	16.4
1955	16.3	16.5	16.5	17.0	15.9	16.3	15.1	13.2	14.7	15.4	14.4	14.1	14.0	14.4	13.4	13.7
1950	13.5	13.6	13.7	14.0	13.1	13.4	12.5	11.0	12.2	12.7	11.9	11.7	11.5	11.9	11.0	11.3
1945	8.6	8.7	8.7	8.9	8.4	8.6	7.9	7.0	7.8	8.1	7.6	7.5	7.4	7.6	7.0	7.2
▼ 1940	6.6	6.7	6.7	6.9	6.4	6.6	6.1	5.4	6.0	6.2	5.8	5.7	5.7	5.8	5.5	5.6

Year	National	Georgia	Hawaii	Idaho		Illinois						Indiana				
	30 City Average	Savannah	Honolulu	Boise	Pocatello	Chicago	Decatur	Joliet	Peoria	Rockford	Springfield	Anderson	Evansville	Fort Wayne	Gary	Indianapolis
Jan 2016	207.2E	170.9E	251.1E	185.9E	185.8E	239.5E	209.5E	239.3E	214.3E	225.9E	211E	186.9E	190.6E	183.5E	213E	190.1E
2015	204.0	170.5	250.5	185.4	185.3	238.9	209.0	238.7	213.7	225.3	210.5	186.4	190.1	183.0	212.4	189.6
2014	203.0	167.2	239.7	184.1	184.4	238.1	206.0	236.8	212.5	224.4	208.2	184.5	188.1	180.8	210.1	188.4
2013	196.9	163.5	231.5	179.6	179.9	231.1	200.8	228.7	207.2	217.2	203.2	177.8	182.5	175.9	203.6	182.7
2012	194.0	158.9	227.9	170.4	171.5	226.2	195.7	222.7	199.7	211.8	197.2	175.2	177.8	171.8	197.9	180.6
2011	185.7	151.7	219.0	162.8	163.6	217.6	187.9	216.8	193.4	205.8	189.4	168.4	169.7	165.0	191.0	173.0
2010	181.6	147.9	214.6	161.7	162.4	210.6	181.2	208.3	186.2	197.6	182.2	162.8	165.9	160.0	183.7	168.1
2009	182.5	152.8	218.6	162.6	163.4	209.0	182.2	205.9	186.2	196.5	184.0	164.9	166.9	163.0	185.2	170.3
2008	171.0	141.1	205.5	153.2	152.8	195.7	168.7	185.5	172.2	180.2	168.5	152.4	155.8	151.0	169.0	159.2
2007	165.0	136.9	200.4	148.0	148.1	186.5	160.5	177.7	166.0	175.0	159.2	149.0	152.6	147.7	165.4	154.1
2006	156.2	124.0	191.9	141.6	141.2	177.8	153.8	168.7	155.2	163.3	152.8	139.8	144.4	139.1	154.5	144.6
2005	146.7	116.9	181.2	133.8	132.5	164.4	145.0	158.3	146.5	151.2	145.4	131.4	136.2	131.0	146.4	137.4
2004	132.8	105.6	161.2	122.0	120.5	149.3	129.3	145.1	133.9	138.1	130.3	120.9	123.5	120.1	132.2	125.0
2003	129.7	103.0	159.4	120.2	118.7	146.2	127.4	143.5	132.4	137.3	128.4	119.7	121.4	118.2	131.4	122.5
2002	126.7	102.0	157.2	118.3	117.1	141.2	123.8	138.5	129.6	132.9	125.3	117.5	119.0	116.4	129.1	120.5
2001	122.2	99.0	150.0	114.3	113.4	135.8	120.1	133.7	124.3	127.8	119.8	113.4	115.6	112.1	123.4	116.4
2000	118.9	97.5	144.8	112.9	112.1	131.2	115.1	124.6	119.0	122.2	116.2	109.8	111.5	108.4	117.8	113.2
1999	116.6	96.0	143.0	110.2	109.6	129.6	113.0	122.6	116.4	120.7	113.8	107.1	109.3	106.8	112.8	110.6
1998	113.6	93.7	140.4	107.4	107.0	125.2	110.1	119.8	113.8	115.5	111.1	105.0	107.2	104.5	111.1	108.0
1997	111.5	92.0	139.8	104.6	104.7	121.3	107.8	117.5	111.5	113.1	108.9	101.9	104.4	101.8	110.3	105.2
1995	105.6	87.4	130.3	99.5	98.2	114.2	98.5	110.5	102.3	103.6	98.1	96.4	97.2	95.0	100.7	100.1
1990	93.2	77.9	104.7	88.2	88.1	98.4	90.9	98.4	93.7	94.0	90.1	84.6	89.3	83.4	88.4	87.1
1985	81.8	68.9	94.7	78.0	78.0	82.4	81.9	83.4	83.7	83.0	81.5	75.2	79.8	75.0	77.8	77.1
1980	60.7	52.2	68.9	60.3	59.5	62.8	62.3	63.4	64.5	61.6	61.1	56.5	59.0	56.7	59.8	57.9
1975	43.7	36.9	44.6	40.8	40.5	45.7	43.1	44.5	44.7	42.7	42.4	39.5	41.7	39.9	41.9	40.6
1970	27.8	21.0	30.4	26.7	26.6	29.1	28.0	28.6	29.0	27.5	27.6	25.4	26.4	25.5	27.1	26.2
1965	21.5	16.4	21.8	20.6	20.5	22.7	21.5	22.1	22.4	21.2	21.3	19.5	20.4	19.7	20.8	20.7
1960	19.5	14.9	19.8	18.7	18.6	20.2	19.6	20.0	20.3	19.2	19.3	17.7	18.7	17.9	18.9	18.4
1955	16.3	12.5	16.6	15.7	15.6	16.9	16.4	16.8	17.0	16.1	16.2	14.9	15.7	15.0	15.9	15.5
1950	13.5	10.3	13.7	13.0	12.9	14.0	13.6	13.9	14.0	13.3	13.4	12.3	12.9	12.4	13.1	12.8
1945	8.6	6.6	8.8	8.3	8.2	8.9	8.6	8.9	9.0	8.5	8.6	7.8	8.3	7.9	8.4	8.1
▼ 1940	6.6	5.1	6.8	6.4	6.3	6.9	6.7	6.8	6.9	6.5	6.6	6.0	6.4	6.1	6.5	6.3

Historical Cost Indexes

Year	National	Indiana			Iowa					Kansas		Kentucky		Louisiana		
	30 City Average	Muncie	South Bend	Terre Haute	Cedar Rapids	Davenport	Des Moines	Sioux City	Waterloo	Topeka	Wichita	Lexington	Louisville	Baton Rouge	Lake Charles	New Orleans
Jan 2016	207.2E	186.5E	189.1E	191.1E	189.3E	197.4E	189.5E	179.6E	178.3E	176.2E	175.6E	184.9E	186.4E	177.1E	175.9E	178.1E
2015	204.0	186.0	188.6	190.6	188.8	196.9	189.0	179.1	177.8	175.7	175.1	184.4	185.9	176.6	175.4	177.6
2014	203.0	185.0	185.5	189.2	188.3	195.9	188.5	178.9	177.4	173.6	173.7	183.7	186.1	171.3	170.9	177.5
2013	196.9	179.3	179.3	182.6	183.4	190.8	181.8	173.2	171.3	168.8	167.3	179.1	182.4	166.3	166.2	173.1
2012	194.0	175.8	174.3	178.9	180.3	184.6	179.2	168.8	169.2	162.3	161.4	174.8	177.5	164.4	163.8	171.7
2011	185.7	169.2	168.1	171.3	173.8	178.1	172.7	162.3	162.8	156.3	155.3	167.8	170.8	156.9	156.5	162.1
2010	181.6	163.1	163.5	167.8	170.4	175.6	170.2	159.0	160.2	152.8	151.3	158.7	166.9	153.7	153.4	160.3
2009	182.5	164.5	167.1	168.0	165.9	172.0	162.1	156.3	147.6	154.5	152.4	160.7	167.5	156.9	154.4	161.9
2008	171.0	153.0	153.1	155.9	157.3	163.4	152.4	147.8	139.2	144.9	143.1	152.0	156.8	144.0	141.5	149.3
2007	165.0	149.7	149.5	152.7	152.4	157.6	148.9	144.0	135.0	141.0	138.7	147.6	150.9	139.4	138.0	145.2
2006	156.2	141.2	141.3	144.0	145.8	151.1	142.9	137.3	128.4	134.7	132.9	129.0	143.0	129.4	129.8	135.8
2005	146.7	131.0	133.8	135.2	135.7	142.1	133.5	129.4	120.0	125.9	125.6	121.5	135.1	120.8	117.7	126.2
2004	132.8	120.0	120.6	123.0	122.7	128.6	121.6	116.2	108.3	112.6	113.4	110.1	120.3	105.6	109.0	115.3
2003	129.7	118.7	119.2	121.7	120.9	126.4	120.2	114.6	105.9	109.4	111.3	107.8	118.7	103.3	106.1	112.4
2002	126.7	116.7	117.1	119.9	116.0	122.1	116.7	111.4	103.7	107.8	109.6	106.2	116.4	102.5	105.0	110.6
2001	122.2	112.8	111.6	115.5	112.5	117.5	113.2	107.7	100.6	104.3	105.2	103.3	112.7	99.4	101.3	104.6
2000	118.9	109.1	106.9	110.8	108.8	112.6	108.9	99.0	98.1	101.0	101.1	101.4	109.3	97.8	99.7	102.1
1999	116.6	105.7	103.7	107.9	104.1	109.2	107.6	96.7	96.4	98.7	99.3	99.7	106.6	96.1	97.6	99.5
1998	113.6	103.6	102.3	106.0	102.5	106.8	103.8	95.1	94.8	97.4	97.4	97.4	101.5	94.7	96.1	97.7
1997	111.5	101.3	101.2	104.6	101.2	105.6	102.5	93.8	93.5	96.3	96.2	96.1	99.9	93.3	96.8	96.2
1995	105.6	95.6	94.6	97.2	95.4	96.6	94.4	88.2	88.9	91.1	91.0	91.6	94.6	89.6	92.7	91.6
1990	93.2	83.9	85.1	89.1	87.1	86.5	86.9	80.4	80.3	83.2	82.6	82.8	82.6	82.0	84.1	84.5
1985	81.8	75.2	76.0	79.2	75.9	77.6	77.1	72.1	72.3	75.0	74.7	75.5	74.5	74.9	78.5	78.2
1980	60.7	56.1	58.3	59.7	62.7	59.6	61.8	59.3	57.7	58.9	58.0	59.3	59.8	59.1	60.0	57.2
1975	43.7	39.2	40.3	41.9	43.1	41.0	43.1	41.0	40.0	42.0	43.1	42.7	42.5	40.4	40.6	41.5
1970	27.8	25.3	26.2	27.0	28.2	26.9	27.6	26.6	25.9	27.0	25.5	26.9	25.9	25.0	26.7	27.2
1965	21.5	19.5	20.2	20.9	21.7	20.8	21.7	20.5	20.0	20.8	19.6	20.7	20.3	19.4	20.6	20.4
1960	19.5	17.8	18.3	18.9	19.7	18.8	19.5	18.6	18.2	18.9	17.8	18.8	18.4	17.6	18.7	18.5
1955	16.3	14.9	15.4	15.9	16.6	15.8	16.3	15.6	15.2	15.8	15.0	15.8	15.4	14.8	15.7	15.6
1950	13.5	12.3	12.7	13.1	13.7	13.1	13.5	12.9	12.6	13.1	12.4	13.0	12.8	12.2	13.0	12.8
1945	8.6	7.8	8.1	8.4	8.7	8.3	8.6	8.2	8.0	8.3	7.9	8.3	8.1	7.8	8.3	8.2
▼ 1940	6.6	6.0	6.3	6.5	6.7	6.4	6.6	6.3	6.2	6.4	6.1	6.4	6.3	6.0	6.4	6.3

Year	National	Louisiana	Maine		Maryland	Massachusetts									Michigan	
	30 City Average	Shreveport	Lewiston	Portland	Baltimore	Boston	Brockton	Fall River	Lawrence	Lowell	New Bedford	Pittsfield	Springfield	Worcester	Ann Arbor	Dearborn
Jan 2016	207.2E	171.3E	194.2E	197.4E	189.7E	241.3E	228.9E	228.8E	235.1E	233.8E	228E	213.4E	215.2E	227.4E	210.3E	212E
2015	204.0	170.9	193.7	196.9	189.2	240.7	228.3	228.2	234.5	233.2	227.4	212.8	214.6	226.8	209.8	211.5
2014	203.0	167.7	192.4	195.4	187.6	239.1	229.1	229.3	234.8	233.9	228.5	213.3	214.8	227.3	206.1	208.4
2013	196.9	159.0	187.9	189.5	183.2	232.9	221.6	221.8	225.3	223.4	221.1	205.9	206.5	219.3	201.2	202.4
2012	194.0	153.2	180.7	182.4	180.8	229.3	217.2	217.8	221.6	219.3	217.0	200.2	201.9	215.2	196.1	199.2
2011	185.7	146.7	167.4	169.1	173.6	219.3	208.2	206.0	211.6	210.1	205.2	191.2	194.5	206.6	186.7	189.4
2010	181.6	145.1	162.1	163.7	168.3	214.3	202.7	200.7	204.9	204.7	200.0	185.9	189.4	201.1	183.3	186.8
2009	182.5	147.6	159.4	161.8	169.1	211.8	199.2	197.6	201.9	201.9	196.9	185.1	187.1	198.2	179.2	186.9
2008	171.0	135.9	149.1	150.9	157.7	198.6	186.4	185.1	188.7	189.1	184.5	171.3	173.7	184.5	169.6	176.4
2007	165.0	132.4	146.4	148.2	152.5	191.8	180.4	179.7	182.0	181.7	179.1	166.8	169.1	180.4	166.8	172.5
2006	156.2	125.2	140.2	139.8	144.9	180.4	171.6	170.4	172.3	173.5	170.4	157.3	159.7	171.6	161.1	165.5
2005	146.7	117.4	131.7	131.4	135.8	169.6	161.3	159.6	162.2	162.3	159.6	147.6	150.7	158.8	147.5	155.8
2004	132.8	105.7	119.8	119.4	121.4	154.1	144.4	143.6	146.4	146.7	143.6	131.9	136.5	143.2	135.9	142.1
2003	129.7	103.9	117.7	117.3	118.1	150.2	139.6	140.6	143.3	143.6	140.5	128.9	133.8	139.0	134.1	139.0
2002	126.7	102.1	117.5	117.1	115.6	145.6	136.0	135.0	136.9	137.3	134.9	124.7	128.3	134.9	131.4	134.3
2001	122.2	98.3	114.6	114.3	111.7	140.9	132.1	131.4	133.1	132.9	131.3	120.3	124.5	130.1	126.8	129.8
2000	118.9	96.2	105.7	105.4	107.7	138.9	129.4	128.4	129.7	130.3	128.3	116.7	121.0	127.3	124.3	125.8
1999	116.6	94.8	105.0	104.7	106.4	136.2	126.7	126.3	127.3	127.4	126.2	115.0	118.8	123.8	117.5	122.7
1998	113.6	92.0	102.8	102.5	104.1	132.8	125.0	124.7	125.0	125.3	124.6	114.2	117.1	122.6	116.0	119.7
1997	111.5	90.1	101.7	101.4	102.2	132.1	124.4	124.7	125.1	125.6	124.7	114.6	117.4	122.9	113.5	117.9
1995	105.6	86.7	96.8	96.5	96.1	128.6	119.6	117.7	120.5	119.6	117.2	110.7	112.7	114.8	106.1	110.5
1990	93.2	80.3	88.5	88.6	85.6	110.9	103.7	102.9	105.2	102.8	102.6	98.7	101.4	103.2	93.2	96.5
1985	81.8	73.6	76.7	77.0	72.7	92.8	88.8	88.7	89.4	88.2	88.6	85.0	85.6	86.7	80.1	82.7
1980	60.7	58.7	57.3	58.5	53.6	64.0	63.7	64.1	63.4	62.7	63.1	61.8	62.0	62.3	62.9	64.0
1975	43.7	40.5	41.9	42.1	39.8	46.6	45.8	45.7	46.2	45.7	46.1	45.5	45.8	46.0	44.1	44.9
1970	27.8	26.4	26.5	25.8	25.1	29.2	29.1	29.2	29.1	28.9	29.0	28.6	28.5	28.7	28.5	28.9
1965	21.5	20.3	20.4	19.4	20.2	23.0	22.5	22.5	22.5	22.2	22.4	22.0	22.4	22.1	22.0	22.3
1960	19.5	18.5	18.6	17.6	17.5	20.5	20.4	20.4	20.4	20.2	20.3	20.0	20.1	20.1	20.0	20.2
1955	16.3	15.5	15.6	14.7	14.7	17.2	17.1	17.1	17.1	16.9	17.1	16.8	16.9	16.8	16.7	16.9
1950	13.5	12.8	12.9	12.2	12.1	14.2	14.1	14.1	14.1	14.0	14.1	13.8	14.0	13.9	13.8	14.0
1945	8.6	8.1	8.2	7.8	7.7	9.1	9.0	9.0	9.0	8.9	9.0	8.9	8.9	8.9	8.8	8.9
▼ 1940	6.6	6.3	6.3	6.0	6.0	7.0	6.9	7.0	7.0	6.9	6.9	6.8	6.9	6.8	6.8	6.9

Year	National 30 City Average	Michigan: Detroit	Michigan: Flint	Michigan: Grand Rapids	Michigan: Kalamazoo	Michigan: Lansing	Michigan: Saginaw	Minnesota: Duluth	Minnesota: Minneapolis	Minnesota: Rochester	Mississippi: Biloxi	Mississippi: Jackson	Missouri: Kansas City	Missouri: St. Joseph	Missouri: St. Louis	Missouri: Springfield
Jan 2016	207.2E	212.3E	199.4E	191.7E	190.9E	200.1E	194.6E	213.5E	222.6E	209.6E	167.7E	174.7E	210.5E	202.8E	210.6E	192.3E
2015	204.0	211.8	198.9	191.2	190.4	199.6	194.1	212.9	222.0	209.1	167.3	174.2	210.0	202.3	210.1	191.8
2014	203.0	208.7	196.0	187.9	188.4	195.7	191.7	211.1	220.7	207.5	164.6	169.8	210.8	199.3	208.6	189.5
2013	196.9	203.1	190.6	182.4	182.4	190.2	186.5	205.1	216.3	201.4	160.9	164.3	204.7	193.1	202.4	181.6
2012	194.0	200.2	186.3	176.4	178.9	185.9	182.8	203.1	214.7	199.7	157.6	160.5	200.8	189.0	198.0	177.7
2011	185.7	190.3	178.5	169.9	171.9	178.0	174.6	195.7	208.1	193.8	151.4	154.3	191.1	178.8	190.6	168.3
2010	181.6	187.5	176.2	160.6	167.3	175.8	171.9	193.1	203.8	188.9	148.0	151.0	186.1	174.2	185.9	164.3
2009	182.5	187.8	176.0	156.7	166.3	173.7	168.6	191.8	203.1	188.0	152.8	156.6	185.6	172.3	187.2	163.7
2008	171.0	177.0	164.4	140.2	155.7	161.6	158.7	177.4	190.6	175.0	141.8	147.1	175.5	164.2	176.2	152.9
2007	165.0	172.9	161.7	137.2	152.6	158.7	156.1	174.2	184.5	171.4	137.7	131.2	169.0	157.8	170.6	145.3
2006	156.2	165.8	153.8	131.1	146.0	152.5	150.6	166.2	173.9	161.9	123.3	117.1	162.0	152.4	159.3	139.4
2005	146.7	156.2	145.1	123.1	134.9	142.0	140.6	157.3	164.6	152.8	116.0	109.9	151.3	143.1	149.4	131.7
2004	132.8	141.9	129.6	112.8	122.5	129.7	127.6	139.1	150.1	137.0	105.3	99.3	135.1	126.9	135.5	116.3
2003	129.7	138.7	127.9	110.9	120.8	127.7	126.0	136.7	146.5	134.5	101.4	96.7	131.9	124.8	133.3	113.9
2002	126.7	134.2	126.9	107.4	119.9	125.3	124.4	131.9	139.5	130.0	101.0	96.3	128.4	122.7	129.6	112.5
2001	122.2	129.4	122.4	104.3	115.2	120.1	119.7	131.4	136.1	124.9	98.7	93.9	121.8	114.7	125.5	106.7
2000	118.9	125.3	119.8	102.7	111.6	117.6	115.9	124.1	131.1	120.1	97.2	92.9	118.2	111.9	122.4	104.3
1999	116.6	122.6	113.7	100.9	106.1	111.5	110.1	120.3	126.5	117.1	95.7	91.8	114.9	106.1	119.8	101.1
1998	113.6	119.5	112.3	99.7	104.9	110.1	108.7	117.7	124.6	115.5	92.1	89.5	108.2	104.2	115.9	98.9
1997	111.5	117.6	110.8	98.9	104.2	108.7	107.5	115.9	121.9	114.1	90.9	88.3	106.4	102.4	113.2	97.3
1995	105.6	110.1	104.1	91.1	96.4	98.3	102.3	100.3	111.9	102.5	84.2	83.5	99.7	96.1	106.3	89.5
1990	93.2	96.0	91.7	84.0	87.4	90.2	88.9	94.3	100.5	95.6	76.4	75.6	89.5	86.8	94.0	82.6
1985	81.8	81.6	80.7	75.8	79.4	80.0	80.5	85.2	87.9	86.5	69.4	68.5	78.2	77.3	80.8	73.2
1980	60.7	62.5	63.1	59.8	61.7	60.3	63.3	63.4	64.1	64.8	54.3	54.3	59.1	60.4	59.7	56.7
1975	43.7	45.8	44.9	41.8	43.7	44.1	44.8	45.2	46.0	45.3	37.9	37.7	42.7	45.4	44.8	41.1
1970	27.8	29.7	28.5	26.7	28.1	27.8	28.7	28.9	29.5	28.9	24.4	21.3	25.3	28.3	28.4	26.0
1965	21.5	22.1	21.9	20.6	21.7	21.5	22.2	22.3	23.4	22.3	18.8	16.4	20.6	21.8	21.8	20.0
1960	19.5	20.1	20.0	18.7	19.7	19.5	20.1	20.3	20.5	20.2	17.1	14.9	19.0	19.8	19.5	18.2
1955	16.3	16.8	16.7	15.7	16.5	16.3	16.9	17.0	17.2	17.0	14.3	12.5	16.0	16.6	16.3	15.2
1950	13.5	13.9	13.8	13.0	13.6	13.5	13.9	14.0	14.2	14.0	11.8	10.3	13.2	13.7	13.5	12.6
1945	8.6	8.9	8.8	8.3	8.7	8.6	8.9	8.9	9.0	8.9	7.6	6.6	8.4	8.8	8.6	8.0
1940	6.6	6.8	6.8	6.4	6.7	6.6	6.9	6.9	7.0	6.9	5.8	5.1	6.5	6.7	6.7	6.2

Year	National 30 City Average	Montana: Billings	Montana: Great Falls	Nebraska: Lincoln	Nebraska: Omaha	Nevada: Las Vegas	Nevada: Reno	New Hampshire: Manchester	New Hampshire: Nashua	New Jersey: Camden	New Jersey: Jersey City	New Jersey: Newark	New Jersey: Paterson	New Jersey: Trenton	NM: Albuquerque	NY: Albany
Jan 2016	207.2E	188.6E	188.9E	184.2E	185.2E	216.1E	195.4E	199.2E	198.3E	225.1E	226.7E	230.7E	229.1E	228.2E	178.8E	208.6E
2015	204.0	188.1	188.4	183.7	184.7	215.5	194.9	198.7	197.8	224.5	226.1	230.1	228.5	227.6	178.3	208.1
2014	203.0	185.4	186.3	181.1	183.7	210.9	195.7	198.8	197.3	224.6	226.6	230.6	227.8	226.6	178.0	205.1
2013	196.9	180.5	181.1	175.5	180.6	206.8	190.4	193.8	192.5	217.7	219.3	223.6	221.2	218.7	173.6	195.3
2012	194.0	178.6	179.5	169.3	177.1	202.6	186.8	189.2	188.3	213.6	215.2	219.2	217.5	213.7	171.0	191.2
2011	185.7	168.8	170.8	162.8	169.6	195.7	179.8	176.8	176.1	205.5	207.0	210.3	209.2	206.4	163.3	180.9
2010	181.6	166.5	168.6	159.3	165.8	193.7	175.6	172.5	172.0	201.2	203.3	206.3	205.3	200.8	162.5	177.8
2009	182.5	165.3	166.0	161.7	165.0	191.5	176.5	174.0	173.5	196.8	200.5	203.6	202.0	198.6	163.0	178.1
2008	171.0	153.1	154.6	152.0	154.8	176.2	167.0	162.5	161.9	184.0	187.3	189.6	188.7	185.6	152.6	166.0
2007	165.0	147.7	147.8	147.7	150.3	166.7	161.4	159.3	158.7	179.3	183.3	185.2	184.6	181.7	146.7	159.4
2006	156.2	140.5	140.8	132.4	140.7	160.0	154.5	146.7	146.4	167.7	171.0	173.6	172.2	169.9	140.1	151.6
2005	146.7	131.9	132.4	124.1	132.8	149.5	145.4	136.8	136.5	159.0	162.4	163.9	163.0	161.5	130.4	142.2
2004	132.8	118.2	117.9	112.1	119.4	137.1	130.5	124.2	123.9	143.4	145.4	147.2	146.6	146.0	118.3	128.6
2003	129.7	115.8	115.8	110.2	117.3	133.8	128.3	122.4	122.2	142.0	144.3	145.7	145.0	143.0	116.4	126.8
2002	126.7	114.6	114.5	108.2	115.0	131.9	126.4	119.9	119.6	135.9	138.5	140.2	140.1	137.6	114.6	122.6
2001	122.2	117.5	117.7	101.3	111.6	127.8	122.5	116.2	116.2	133.4	136.7	136.9	136.8	136.0	111.4	119.2
2000	118.9	113.7	113.9	98.8	107.0	125.8	118.2	111.9	111.9	128.4	130.5	132.6	132.4	130.6	109.0	116.5
1999	116.6	112.1	112.7	96.6	104.8	121.9	114.4	109.6	109.6	125.3	128.8	131.6	129.5	129.6	106.7	114.6
1998	113.6	109.7	109.1	94.9	101.2	118.1	111.4	110.1	110.0	124.3	127.7	128.9	128.5	127.9	103.8	113.0
1997	111.5	107.4	107.6	93.3	99.5	114.6	109.8	108.6	108.6	121.9	125.1	126.4	126.4	125.2	100.8	110.0
1995	105.6	104.7	104.9	85.6	93.4	108.5	105.0	100.9	100.8	107.0	112.2	111.9	112.1	111.2	96.3	103.6
1990	93.2	92.9	93.8	78.8	85.0	96.3	94.5	86.3	86.3	93.0	93.5	93.8	97.2	94.5	84.9	93.2
1985	81.8	83.9	84.3	71.5	77.5	87.6	85.0	78.1	78.1	81.3	83.3	83.5	84.5	82.8	76.7	79.5
1980	60.7	63.9	64.6	58.5	63.5	64.6	63.0	56.6	56.0	58.6	60.6	60.1	60.0	58.9	59.0	59.5
1975	43.7	43.1	43.8	40.9	43.2	42.8	41.9	41.3	40.8	42.3	43.4	44.2	43.9	43.8	40.3	43.9
1970	27.8	28.5	28.9	26.4	26.8	29.4	28.0	26.2	25.6	27.2	27.8	29.0	27.8	27.4	26.4	28.3
1965	21.5	22.0	22.3	20.3	20.6	22.4	21.6	20.6	19.7	20.9	21.4	23.8	21.4	21.3	20.6	22.3
1960	19.5	20.0	20.3	18.5	18.7	20.2	19.6	18.0	17.9	19.0	19.4	19.4	19.4	19.2	18.5	19.3
1955	16.3	16.7	17.0	15.5	15.7	16.9	16.4	15.1	15.0	16.0	16.3	16.3	16.3	16.1	15.6	16.2
1950	13.5	13.9	14.0	12.8	13.0	14.0	13.6	12.5	12.4	13.2	13.5	13.5	13.5	13.3	12.9	13.4
1945	8.6	8.8	9.0	8.1	8.3	8.9	8.7	8.0	7.9	8.4	8.6	8.6	8.6	8.5	8.2	8.5
1940	6.6	6.8	6.9	6.3	6.4	6.9	6.7	6.2	6.1	6.5	6.6	6.6	6.6	6.6	6.3	6.6

Year	National 30 City Average	New York: Binghamton	Buffalo	New York	Rochester	Schenectady	Syracuse	Utica	Yonkers	North Carolina: Charlotte	Durham	Greensboro	Raleigh	Winston-Salem	N. Dakota: Fargo	Ohio: Akron
Jan 2016	207.2E	202.9E	210.1E	268.7E	204E	207.9E	201.4E	198.6E	244E	173.5E	172.7E	170.9E	168E	171E	181.5E	199.6E
2015	204.0	202.4	209.6	268.0	203.5	207.4	200.9	198.1	243.4	173.1	172.3	170.5	167.6	170.6	181.0	199.1
2014	203.0	202.4	207.8	267.7	202.2	204.2	198.8	196.4	245.6	165.5	165.2	163.9	161.6	164.3	177.1	199.2
2013	196.9	195.5	200.7	259.4	194.3	195.5	193.6	188.5	230.0	159.6	158.4	158.7	157.5	159.1	169.6	191.8
2012	194.0	192.9	198.1	256.3	191.2	191.4	189.6	185.2	227.1	156.4	155.6	156.3	155.4	157.2	167.3	187.7
2011	185.7	176.1	188.1	245.6	181.9	181.5	180.8	173.5	218.9	142.6	143.8	142.2	142.8	140.9	160.3	180.8
2010	181.6	173.8	184.0	241.4	179.6	179.0	176.7	170.5	217.1	140.5	141.6	140.0	140.6	138.8	156.2	174.5
2009	182.5	172.9	184.4	239.9	179.6	178.8	176.9	171.3	217.5	145.1	145.8	144.0	145.1	142.7	154.3	175.7
2008	171.0	160.7	173.9	226.8	168.0	167.2	165.9	161.6	202.1	135.8	136.6	134.6	135.3	133.8	145.0	165.4
2007	165.0	155.7	168.6	215.2	163.5	159.8	159.8	155.0	196.5	132.8	133.7	131.7	132.1	131.0	139.8	159.0
2006	156.2	147.5	159.2	204.5	155.2	151.1	150.9	146.4	186.0	125.1	124.6	123.8	124.2	123.0	133.2	152.8
2005	146.7	137.5	149.4	194.0	147.3	142.2	141.9	136.9	176.1	110.5	112.1	112.1	112.1	111.1	125.3	144.4
2004	132.8	123.5	136.1	177.7	132.0	128.4	127.3	124.4	161.8	98.9	100.0	100.1	100.3	99.4	113.0	131.9
2003	129.7	121.8	132.8	173.4	130.3	126.7	124.7	121.7	160.0	96.2	97.5	97.5	97.8	96.8	110.6	129.6
2002	126.7	119.0	128.5	170.1	127.1	123.0	121.8	118.4	154.8	94.8	96.1	96.1	96.3	95.5	106.3	127.2
2001	122.2	116.0	125.2	164.4	123.1	120.1	118.6	115.3	151.4	91.5	92.9	92.9	93.3	92.3	103.1	123.5
2000	118.9	112.4	122.3	159.2	120.0	117.5	115.1	112.4	144.8	90.3	91.6	91.6	91.9	91.0	97.9	117.8
1999	116.6	108.6	120.2	155.9	116.8	114.7	113.7	108.5	140.6	89.3	90.2	90.3	90.5	90.1	96.8	116.0
1998	113.6	109.0	119.1	154.4	117.2	114.2	113.6	108.6	141.4	88.2	89.1	89.2	89.4	89.0	95.2	113.1
1997	111.5	107.2	115.7	150.3	115.2	111.2	110.6	107.0	138.8	86.8	87.7	87.8	87.9	87.6	93.5	110.6
1995	105.6	99.3	110.1	140.7	106.6	104.6	104.0	97.7	129.4	81.8	82.6	82.6	82.7	82.6	88.4	103.4
1990	93.2	87.0	94.1	118.1	94.6	93.9	91.0	85.6	111.4	74.8	75.4	75.5	75.5	75.3	81.3	94.6
1985	81.8	77.5	83.2	94.9	81.6	80.1	81.0	77.0	92.8	66.9	67.6	67.7	67.6	67.5	73.4	86.8
1980	60.7	58.0	60.6	66.0	60.5	60.3	61.6	58.5	65.9	51.1	52.2	52.5	51.7	50.8	57.4	62.3
1975	43.7	42.5	45.1	49.5	44.8	43.7	44.8	42.7	47.1	36.1	37.0	37.0	37.3	36.1	39.3	44.7
1970	27.8	27.0	28.9	32.8	29.2	27.8	28.5	26.9	30.0	20.9	23.7	23.6	23.4	23.0	25.8	28.3
1965	21.5	20.8	22.2	25.5	22.8	21.4	21.9	20.7	23.1	16.0	18.2	18.2	18.0	17.7	19.9	21.8
1960	19.5	18.9	19.9	21.5	19.7	19.5	19.9	18.8	21.0	14.4	16.6	16.6	16.4	16.1	18.1	19.9
1955	16.3	15.8	16.7	18.1	16.5	16.3	16.7	15.8	17.6	12.1	13.9	13.9	13.7	13.5	15.2	16.6
1950	13.5	13.1	13.8	14.9	13.6	13.5	13.8	13.0	14.5	10.0	11.5	11.5	11.4	11.2	12.5	13.7
1945	8.6	8.3	8.8	9.5	8.7	8.6	8.8	8.3	9.3	6.4	7.3	7.3	7.2	7.1	8.0	8.8
▼ 1940	6.6	6.4	6.8	7.4	6.7	6.7	6.8	6.4	7.2	4.9	5.6	5.6	5.6	5.5	6.2	6.8

Year	National 30 City Average	Ohio: Canton	Cincinnati	Cleveland	Columbus	Dayton	Lorain	Springfield	Toledo	Youngstown	Oklahoma: Lawton	Oklahoma City	Tulsa	Oregon: Eugene	Portland	PA: Allentown
Jan 2016	207.2E	191.6E	188.6E	203.9E	192.9E	188.4E	197.2E	188.6E	201.5E	193.8E	175.6E	177.6E	170.4E	203.6E	204.8E	210.3E
2015	204	191.1	188.1	203.4	192.4	187.9	196.7	188.1	201	193.3	175.1	177.1	170	203.1	204.3	209.8
2014	203.0	190.8	188.1	203.1	192.4	185.8	196.7	187.2	199.7	193.8	169.3	170.9	166.7	200.4	201.4	209.0
2013	196.9	183.9	181.9	195.5	187.1	180.7	187.6	181.5	193.5	186.5	162.7	165.5	161.7	194.7	195.6	203.2
2012	194.0	180.2	178.6	191.4	183.5	177.6	184.1	178.0	190.4	183.3	158.1	158.0	151.3	190.1	191.8	199.3
2011	185.7	172.1	171.3	184.0	175.7	168.4	176.8	168.4	183.1	176.5	151.7	151.6	144.8	184.9	186.6	192.2
2010	181.6	167.5	166.5	180.2	170.3	163.4	173.0	164.6	176.5	171.7	149.5	149.4	143.1	181.5	182.9	187.9
2009	182.5	168.4	168.1	181.6	171.2	165.3	174.0	166.0	178.5	172.8	152.5	153.3	146.8	181.3	183.6	188.3
2008	171.0	158.4	159.0	170.6	160.2	157.5	163.9	157.5	168.2	161.5	141.3	140.6	136.2	172.3	174.7	175.7
2007	165.0	153.2	152.4	164.9	155.1	149.4	158.1	150.3	161.6	156.7	136.2	135.6	132.5	168.5	169.5	169.4
2006	156.2	147.0	144.9	156.9	146.9	142.9	151.8	143.4	154.5	150.3	129.5	129.5	125.7	160.5	161.7	160.1
2005	146.7	137.7	136.8	147.9	138.0	134.5	143.3	134.8	145.5	141.1	121.6	121.6	117.8	150.9	152.2	150.2
2004	132.8	125.6	124.3	135.6	126.2	121.1	131.5	122.0	132.8	129.3	109.3	109.4	107.3	136.8	137.8	133.9
2003	129.7	123.7	121.3	132.7	123.7	119.4	126.1	120.0	130.4	126.2	107.5	107.8	105.0	134.2	135.9	130.3
2002	126.7	120.9	119.2	130.0	120.9	117.5	124.3	118.0	128.4	123.6	106.3	106.0	104.1	132.2	133.9	128.1
2001	122.2	117.7	115.9	125.9	117.1	114.0	120.4	114.7	123.8	119.9	102.0	102.1	99.7	129.8	131.2	123.5
2000	118.9	112.7	110.1	121.3	112.5	109.0	115.0	109.2	115.7	114.1	98.7	98.9	97.5	126.3	127.4	119.9
1999	116.6	110.6	107.9	118.7	109.5	107.1	112.0	106.4	113.6	111.9	97.5	97.4	96.2	120.9	124.3	117.8
1998	113.6	108.5	105.0	114.8	106.8	104.5	109.4	104.1	111.2	109.0	94.3	94.5	94.5	120.0	122.2	115.7
1997	111.5	106.4	102.9	112.9	104.7	102.6	107.4	102.2	108.5	107.2	92.9	93.2	93.0	118.1	119.6	113.9
1995	105.6	98.8	97.1	106.4	99.1	94.6	97.1	92.0	100.6	100.1	85.3	88.0	89.0	112.2	114.3	108.3
1990	93.2	92.1	86.7	95.1	88.1	85.9	89.4	83.4	92.9	91.8	77.1	79.9	80.0	98.0	99.4	95.0
1985	81.8	83.7	78.2	86.2	77.7	76.3	80.7	74.3	84.7	82.7	71.2	73.9	73.8	88.7	90.1	81.9
1980	60.7	60.6	59.8	61.0	58.6	56.6	60.8	56.2	64.0	61.5	52.2	55.5	57.2	68.9	68.4	58.7
1975	43.7	44.0	43.7	44.6	42.2	40.5	42.7	40.0	44.9	44.5	38.6	38.6	39.2	44.6	44.9	42.3
1970	27.8	27.8	28.2	29.6	26.7	27.0	27.5	25.4	29.1	29.6	24.1	23.4	24.1	30.4	30.0	27.3
1965	21.5	21.5	20.7	21.1	20.2	20.0	21.1	19.6	21.3	21.0	18.5	17.8	20.6	23.4	22.6	21.0
1960	19.5	19.5	19.3	19.6	18.7	18.1	19.2	17.8	19.4	19.1	16.9	16.1	18.1	21.2	21.1	19.1
1955	16.3	16.3	16.2	16.5	15.7	15.2	16.1	14.9	16.2	16.0	14.1	13.5	15.1	17.8	17.7	16.0
1950	13.5	13.5	13.3	13.6	13.0	12.5	13.3	12.3	13.4	13.2	11.7	11.2	12.5	14.7	14.6	13.2
1945	8.6	8.6	8.5	8.7	8.3	8.0	8.5	7.8	8.6	8.4	7.4	7.1	8.0	9.4	9.3	8.4
▼ 1940	6.6	6.7	6.5	6.7	6.4	6.2	6.5	6.1	6.6	6.5	5.7	5.5	6.1	7.2	7.2	6.5

Year	National 30 City Average	Pennsylvania						RI	South Carolina		South Dakota		Tennessee			
		Erie	Harrisburg	Philadelphia	Pittsburgh	Reading	Scranton	Providence	Charleston	Columbia	Rapid City	Sioux Falls	Chattanooga	Knoxville	Memphis	Nashville
Jan 2016	207.2E	195E	200.8E	235.3E	210E	204.7E	202.2E	220.5E	171.7E	172.3E	168.1E	168.4E	174.8E	171.4E	178E	180.4E
2015	204.0	194.5	200.3	234.7	209.5	204.2	201.7	219.9	171.3	171.9	167.7	168.0	174.3	171.0	177.5	179.9
2014	203.0	191.4	200.3	232.1	207.1	201.4	201.4	220.3	171.0	162.5	168.6	166.8	174.1	170.9	176.5	178.2
2013	196.9	187.2	192.5	223.6	201.4	196.1	196.0	213.4	166.3	157.9	159.6	161.2	167.4	159.3	169.3	172.8
2012	194.0	182.0	187.9	221.8	196.8	190.8	191.5	209.3	155.3	148.1	156.7	158.8	163.8	155.8	166.3	167.5
2011	185.7	176.0	180.9	212.8	187.7	184.3	185.3	196.8	149.4	142.3	149.5	151.3	157.3	149.3	158.4	160.7
2010	181.6	171.0	175.2	209.3	182.3	179.8	180.8	192.8	147.4	140.2	147.3	149.0	152.4	144.7	154.2	156.7
2009	182.5	171.9	176.5	209.5	181.7	180.7	181.3	192.9	150.8	144.3	148.2	151.1	155.7	148.0	157.2	159.8
2008	171.0	161.0	165.7	196.2	169.2	169.3	167.9	178.2	142.1	134.8	138.4	141.3	136.7	133.4	146.0	147.5
2007	165.0	156.0	159.4	188.4	163.2	164.9	163.1	174.4	139.3	131.7	129.5	133.0	133.7	130.6	143.1	144.4
2006	156.2	148.7	150.7	177.8	155.5	155.0	153.9	163.6	122.4	118.7	122.8	126.6	127.0	123.9	136.5	135.5
2005	146.7	140.6	141.5	166.4	146.6	145.5	142.9	155.6	110.2	109.6	115.2	118.5	116.0	115.1	128.6	127.9
2004	132.8	127.0	126.2	148.4	133.1	128.7	129.0	138.8	98.9	98.4	103.9	107.3	105.0	104.5	115.6	115.8
2003	129.7	124.5	123.8	145.5	130.5	126.3	126.0	135.9	96.1	95.7	101.3	104.0	102.8	102.3	111.1	110.9
2002	126.7	122.1	121.1	142.0	127.8	123.4	124.2	131.1	94.8	93.9	100.2	103.2	102.2	101.2	107.6	109.2
2001	122.2	118.9	118.4	136.9	124.1	120.4	121.1	127.8	92.1	91.3	96.8	99.7	98.2	96.5	102.6	104.3
2000	118.9	114.6	114.0	132.1	120.9	115.9	117.7	122.8	89.9	89.1	94.2	98.0	96.9	95.0	100.8	100.8
1999	116.6	113.8	112.8	129.8	119.6	114.8	116.3	121.6	89.0	88.1	92.4	95.8	95.9	93.4	99.7	98.6
1998	113.6	109.8	110.5	126.6	117.1	112.5	113.7	120.3	88.0	87.2	90.7	93.6	94.9	92.4	97.2	96.8
1997	111.5	108.5	108.8	123.3	113.9	110.8	112.3	118.9	86.5	85.6	89.3	92.1	93.6	90.8	96.1	94.9
1995	105.6	99.8	100.6	117.1	106.3	103.6	103.8	111.1	82.7	82.2	84.0	84.7	89.2	86.0	91.2	87.6
1990	93.2	88.5	89.5	98.5	91.2	90.8	91.3	94.1	73.7	74.5	77.2	78.4	79.9	75.1	81.3	77.1
1985	81.8	79.6	77.2	82.2	81.5	77.7	80.0	83.0	65.9	66.9	69.7	71.2	72.5	67.7	74.3	66.7
1980	60.7	59.4	56.9	58.7	61.3	57.7	58.4	59.2	50.6	51.1	54.8	57.1	55.0	51.6	55.9	53.1
1975	43.7	43.4	42.2	44.5	44.5	43.6	41.9	42.7	35.0	36.4	37.7	39.6	39.1	37.3	40.7	37.0
1970	27.8	27.5	25.8	27.5	28.7	27.1	27.0	27.3	22.8	22.9	24.8	25.8	24.9	22.0	23.0	22.8
1965	21.5	21.1	20.1	21.7	22.4	20.9	20.8	21.9	17.6	17.6	19.1	19.9	19.2	17.1	18.3	17.4
1960	19.5	19.1	18.6	19.4	19.7	19.0	18.9	19.1	16.0	16.0	17.3	18.1	17.4	15.5	16.6	15.8
1955	16.3	16.1	15.6	16.3	16.5	15.9	15.9	16.0	13.4	13.4	14.5	15.1	14.6	13.0	13.9	13.3
1950	13.5	13.3	12.9	13.5	13.6	13.2	13.1	13.2	11.1	11.1	12.0	12.5	12.1	10.7	11.5	10.9
1945	8.6	8.5	8.2	8.6	8.7	8.4	8.3	8.4	7.1	7.1	7.7	8.0	7.7	6.8	7.3	7.0
▼ 1940	6.6	6.5	6.3	6.6	6.7	6.4	6.5	6.5	5.4	5.5	5.9	6.2	6.0	5.3	5.6	5.4

Year	National 30 City Average	Texas														Utah
		Abilene	Amarillo	Austin	Beaumont	Corpus Christi	Dallas	El Paso	Fort Worth	Houston	Lubbock	Odessa	San Antonio	Waco	Wichita Falls	Ogden
Jan 2016	207.2E	171.2E	171.1E	172.7E	174.5E	173.2E	174.9E	169.2E	172.9E	176.9E	173.7E	171.9E	172.2E	167.7E	169.4E	177.6E
2015	204.0	170.8	170.7	172.3	174.0	172.8	174.4	168.8	172.5	176.4	173.3	171.5	171.8	167.3	169.0	177.1
2014	203.0	160.9	167.2	167.5	165.2	167.3	172.4	156.2	169.5	176.0	166.9	159.2	169.5	161.7	160.6	174.2
2013	196.9	155.2	161.7	158.7	160.2	157.7	167.4	151.2	161.7	169.4	159.3	151.5	164.3	156.1	155.9	167.7
2012	194.0	152.5	158.6	154.3	158.2	151.8	165.2	149.1	159.4	167.8	156.6	149.1	160.7	153.8	152.9	165.6
2011	185.7	145.9	151.9	147.6	152.4	145.0	157.9	142.5	152.8	160.8	150.0	142.5	152.6	147.3	146.6	158.7
2010	181.6	144.2	150.1	144.7	150.5	142.1	155.1	140.8	149.6	157.3	148.1	140.7	147.9	145.8	145.0	157.5
2009	182.5	143.5	147.3	146.4	149.8	141.7	155.5	141.1	150.1	160.9	144.6	139.1	150.8	146.7	146.1	155.1
2008	171.0	132.4	137.0	137.5	140.3	133.2	144.1	130.7	138.3	149.1	134.7	128.9	141.0	136.1	136.1	144.8
2007	165.0	128.5	132.2	131.6	137.1	128.8	138.6	126.2	134.8	146.0	130.2	125.0	136.4	132.1	132.1	140.0
2006	156.2	121.6	125.7	125.5	130.1	122.0	131.1	120.3	127.6	138.2	123.2	118.3	129.8	125.2	125.3	133.4
2005	146.7	113.8	117.3	117.9	121.3	114.4	123.7	112.5	119.4	129.0	115.5	110.5	121.3	116.1	117.3	126.0
2004	132.8	102.9	106.3	105.7	108.5	102.9	112.0	101.5	108.4	115.9	104.7	99.9	108.4	104.8	105.1	115.2
2003	129.7	100.5	104.4	103.9	106.1	100.4	109.3	99.5	105.5	113.4	102.3	97.6	104.8	102.7	103.0	112.7
2002	126.7	99.9	101.9	102.4	104.5	98.9	107.9	98.7	104.6	111.5	100.5	95.9	103.7	100.5	101.4	110.8
2001	122.2	93.4	98.4	99.8	102.3	96.6	103.8	95.5	100.9	107.8	97.6	93.4	100.5	97.8	97.3	107.7
2000	118.9	93.4	98.1	99.1	101.6	96.9	102.7	92.4	99.9	106.0	97.6	93.4	99.4	97.3	96.9	104.6
1999	116.6	91.8	94.5	96.0	99.7	94.0	101.0	90.7	97.6	104.6	96.0	92.1	98.0	94.8	95.5	103.3
1998	113.6	89.8	92.7	94.2	97.9	91.8	97.9	88.4	94.5	101.3	93.3	90.1	94.8	92.7	92.9	98.5
1997	111.5	88.4	91.3	92.8	96.8	90.3	96.1	87.0	93.3	100.1	91.9	88.8	93.4	91.4	91.5	96.1
1995	105.6	85.2	87.4	89.3	93.7	87.4	91.4	85.2	89.5	95.9	88.4	85.6	88.9	86.4	86.8	92.2
1990	93.2	78.0	80.1	81.3	86.5	79.3	84.5	76.7	82.1	85.4	81.5	78.6	80.7	79.6	80.3	83.4
1985	81.8	71.1	72.5	74.5	79.3	72.3	77.6	69.4	75.1	79.6	74.0	71.2	73.9	71.7	73.3	75.2
1980	60.7	53.4	55.2	54.5	57.6	54.5	57.9	53.1	57.0	59.4	55.6	57.2	55.0	54.9	55.4	62.2
1975	43.7	37.6	39.0	39.0	39.6	38.1	40.7	38.0	40.4	41.2	38.9	37.9	39.0	38.6	38.0	40.0
1970	27.8	24.5	24.9	24.9	25.7	24.5	25.5	23.7	25.9	25.4	25.1	24.6	23.3	24.8	24.5	26.8
1965	21.5	18.9	19.2	19.2	19.9	18.9	19.9	19.0	19.9	20.0	19.4	19.0	18.5	19.2	18.9	20.6
1960	19.5	17.1	17.4	17.4	18.1	17.1	18.2	17.0	18.1	18.2	17.6	17.3	16.8	17.4	17.2	18.8
1955	16.3	14.4	14.6	14.6	15.1	14.4	15.3	14.3	15.2	15.2	14.8	14.5	14.1	14.6	14.4	15.7
1950	13.5	11.9	12.1	12.1	12.5	11.9	12.6	11.8	12.5	12.6	12.2	12.0	11.6	12.1	11.9	13.0
1945	8.6	7.6	7.7	7.7	8.0	7.6	8.0	7.5	8.0	8.0	7.8	7.6	7.4	7.7	7.6	8.3
▼ 1940	6.6	5.9	5.9	5.9	6.1	5.8	6.2	5.8	6.2	6.2	6.0	5.9	5.7	5.9	5.8	6.4

Historical Cost Indexes

Year	National	Utah	Vermont		Virginia					Washington			West Virginia		Wisconsin	
	30 City Average	Salt Lake City	Burlington	Rutland	Alexandria	Newport News	Norfolk	Richmond	Roanoke	Seattle	Spokane	Tacoma	Charleston	Huntington	Green Bay	Kenosha
Jan 2016	207.2E	182.6E	192.1E	189E	193E	176.3E	177.9E	177.7E	175.8E	210.4E	193.6E	207.7E	200.4E	201.8E	201.3E	205.9E
2015	204.0	182.1	191.6	188.5	192.5	175.8	177.4	177.2	175.3	209.9	193.1	207.2	199.9	201.3	200.8	205.4
2014	203.0	177.1	191.3	188.6	190.5	174.4	175.7	176.4	173.5	209.8	190.8	205.2	197.0	199.5	198.9	205.2
2013	196.9	171.1	178.7	176.6	185.0	170.0	171.5	170.7	168.3	203.4	184.3	199.1	187.2	193.2	193.9	201.2
2012	194.0	168.8	172.4	170.5	182.4	168.3	169.7	169.4	166.4	201.9	181.6	194.3	183.4	186.4	190.0	195.5
2011	185.7	161.8	158.8	157.1	174.5	159.8	161.1	159.0	154.2	194.1	175.5	187.6	177.5	180.3	181.4	187.6
2010	181.6	160.9	156.3	154.6	170.9	156.7	158.0	156.6	151.8	191.8	171.5	186.0	171.6	176.1	179.0	185.1
2009	182.5	160.4	158.3	156.7	172.5	160.0	161.9	160.7	155.6	188.5	173.0	186.3	174.6	177.4	175.5	182.5
2008	171.0	149.7	147.2	145.8	161.8	150.8	150.8	150.9	145.9	176.9	162.4	174.9	162.8	165.7	165.6	172.2
2007	165.0	144.6	144.4	143.2	155.0	146.2	146.1	147.3	142.9	171.4	156.7	168.7	158.7	160.1	159.4	164.8
2006	156.2	137.7	131.7	130.7	146.2	133.7	134.6	134.8	129.7	162.9	150.1	160.7	149.4	151.4	152.7	157.1
2005	146.7	129.4	124.6	123.8	136.5	123.5	124.4	125.4	112.2	153.9	141.9	151.5	140.8	141.0	144.4	148.3
2004	132.8	117.8	113.0	112.3	121.5	108.9	110.2	110.9	99.7	138.0	127.6	134.5	124.8	125.6	128.9	132.4
2003	129.7	116.0	110.7	110.1	119.5	104.8	106.2	108.6	97.0	134.9	125.9	133.0	123.3	123.6	127.4	130.9
2002	126.7	113.7	109.0	108.4	115.1	102.9	104.1	106.6	95.2	132.7	123.9	131.4	121.2	120.9	123.0	127.3
2001	122.2	109.1	105.7	105.2	110.8	99.8	100.3	102.9	92.1	127.9	120.3	125.7	114.6	117.5	119.1	123.6
2000	118.9	106.5	98.9	98.3	108.1	96.5	97.6	100.2	90.7	124.6	118.3	122.9	111.5	114.4	114.6	119.1
1999	116.6	104.5	98.2	97.7	106.1	95.6	96.5	98.8	89.8	123.3	116.7	121.6	110.6	113.4	112.1	115.8
1998	113.6	99.5	97.8	97.3	104.1	93.7	93.9	97.0	88.3	119.4	114.3	118.3	106.7	109.0	109.5	112.9
1997	111.5	97.2	96.6	96.3	101.2	91.6	91.7	92.9	86.9	118.1	111.7	117.2	105.3	107.7	105.6	109.1
1995	105.6	93.1	91.1	90.8	96.3	86.0	86.4	87.8	82.8	113.7	107.4	112.8	95.8	97.2	97.6	97.9
1990	93.2	84.3	83.0	82.9	86.1	76.3	76.7	77.6	76.1	100.1	98.5	100.5	86.1	86.8	86.7	87.8
1985	81.8	75.9	74.8	74.9	75.1	68.7	68.8	69.5	67.2	88.3	89.0	91.2	77.7	77.7	76.7	77.4
1980	60.7	57.0	55.3	58.3	57.3	52.5	52.4	54.3	51.3	67.9	66.3	66.7	57.7	58.3	58.6	58.3
1975	43.7	40.1	41.8	43.9	41.7	37.2	36.9	37.1	37.1	44.9	44.4	44.5	41.0	40.0	40.9	40.5
1970	27.8	26.1	25.4	26.8	26.2	23.9	21.5	22.0	23.7	28.8	29.3	29.6	26.1	25.8	26.4	26.5
1965	21.5	20.0	19.8	20.6	20.2	18.4	17.1	17.2	18.3	22.4	22.5	22.8	20.1	19.9	20.3	20.4
1960	19.5	18.4	18.0	18.8	18.4	16.7	15.4	15.6	16.6	20.4	20.8	20.8	18.3	18.1	18.4	18.6
1955	16.3	15.4	15.1	15.7	15.4	14.0	12.9	13.1	13.9	17.1	17.4	17.4	15.4	15.2	15.5	15.6
1950	13.5	12.7	12.4	13.0	12.7	11.6	10.7	10.8	11.5	14.1	14.4	14.4	12.7	12.5	12.8	12.9
1945	8.6	8.1	7.9	8.3	8.1	7.4	6.8	6.9	7.3	9.0	9.2	9.2	8.1	8.0	8.1	8.2
1940	6.6	6.3	6.1	6.4	6.2	5.7	5.3	5.3	5.7	7.0	7.1	7.1	6.2	6.2	6.3	6.3

Year	National	Wisconsin			Wyoming	Canada									
	30 City Average	Madison	Milwaukee	Racine	Cheyenne	Calgary	Edmonton	Hamilton	London	Montreal	Ottawa	Quebec	Toronto	Vancouver	Winnipeg
Jan 2016	207.2E	202.5E	209.9E	205.6E	175.9E	227.2E	226.9E	221.8E	217.6E	221E	221E	221E	226.5E	217.7E	204.6E
2015	204.0	202.0	209.4	205.1	175.4	226.6	226.3	221.2	217.0	220.4	220.4	220.4	225.9	217.1	204.1
2014	203.0	202.4	210.2	205.3	173.6	228.2	229.1	222.1	218.9	219.8	214.8	218.3	227.4	218.8	202.9
2013	196.9	197.8	204.6	200.5	167.6	222.7	223.3	216.0	213.7	214.5	214.1	212.6	220.9	216.2	200.2
2012	194.0	191.2	202.2	194.9	165.0	219.1	219.6	211.1	208.2	209.6	209.7	207.5	216.6	214.0	200.7
2011	185.7	183.4	191.9	187.0	155.3	212.6	212.4	204.2	200.1	202.1	202.0	200.6	209.6	207.1	192.6
2010	181.6	181.3	187.1	184.4	154.4	200.6	200.7	197.7	193.7	193.9	195.6	192.7	200.7	192.7	182.9
2009	182.5	180.0	187.3	182.8	155.2	205.4	206.6	203.0	198.9	198.8	200.1	197.4	205.6	199.6	188.5
2008	171.0	168.4	176.3	173.0	146.5	190.2	191.0	194.3	188.6	186.7	188.0	186.8	194.6	184.9	174.4
2007	165.0	160.7	168.9	164.7	141.3	183.1	184.4	186.8	182.4	181.7	182.4	182.0	187.7	180.6	170.1
2006	156.2	152.8	158.8	157.2	128.0	163.6	164.8	169.0	164.9	158.6	163.7	159.0	170.6	169.3	155.9
2005	146.7	145.2	148.4	147.9	118.9	154.4	155.7	160.0	156.2	149.2	155.1	150.1	162.6	159.4	146.9
2004	132.8	129.7	134.2	132.7	104.7	138.8	139.4	142.6	140.4	134.5	141.0	135.7	146.0	141.7	129.7
2003	129.7	128.4	131.1	131.5	102.6	133.6	134.2	139.1	137.0	130.2	137.4	131.4	142.3	137.0	124.4
2002	126.7	124.5	128.2	126.5	101.7	122.5	122.2	136.3	134.1	127.2	134.9	128.4	139.8	134.6	121.4
2001	122.2	120.6	123.9	123.3	99.0	117.5	117.4	131.5	129.1	124.4	130.2	125.5	134.7	130.2	117.2
2000	118.9	116.6	120.5	118.7	98.1	115.9	115.8	130.0	127.5	122.8	128.8	124.1	133.1	128.4	115.6
1999	116.6	115.9	117.4	115.5	96.9	115.3	115.2	128.1	125.6	120.8	126.8	121.9	131.2	127.1	115.2
1998	113.6	110.8	113.4	112.7	95.4	112.5	112.4	126.3	123.9	119.0	124.7	119.6	128.5	123.8	113.7
1997	111.5	106.1	110.1	109.3	93.2	110.7	110.6	124.4	121.9	114.6	122.8	115.4	125.7	121.9	111.4
1995	105.6	96.5	103.9	97.8	87.6	107.4	107.4	119.9	117.5	110.8	118.2	111.5	121.6	116.2	107.6
1990	93.2	84.3	88.9	87.3	79.1	98.0	97.1	103.8	101.4	99.0	102.7	96.8	104.6	103.2	95.1
1985	81.8	74.3	77.4	77.0	72.3	90.2	89.1	85.8	84.9	82.4	83.9	79.9	86.5	89.8	83.4
1980	60.7	56.8	58.8	58.1	56.9	64.9	63.3	63.7	61.9	59.2	60.9	59.3	60.9	65.0	61.7
1975	43.7	40.7	43.3	40.7	40.6	42.2	41.6	42.9	41.6	39.7	41.5	39.0	42.2	42.4	39.2
1970	27.8	26.5	29.4	26.5	26.0	28.9	28.6	28.5	27.8	25.6	27.6	26.0	25.6	26.0	23.1
1965	21.5	20.6	21.8	20.4	20.0	22.3	22.0	22.0	21.4	18.7	21.2	20.1	19.4	20.5	17.5
1960	19.5	18.1	19.0	18.6	18.2	20.2	20.0	20.0	19.5	17.0	19.3	18.2	17.6	18.6	15.8
1955	16.3	15.2	15.9	15.6	15.2	17.0	16.8	16.7	16.3	14.3	16.2	15.3	14.8	15.5	13.3
1950	13.5	12.5	13.2	12.9	12.6	14.0	13.9	13.8	13.5	11.8	13.4	12.6	12.2	12.8	10.9
1945	8.6	8.0	8.4	8.2	8.0	9.0	8.8	8.8	8.6	7.5	8.5	8.0	7.8	8.2	7.0
1940	6.6	6.2	6.5	6.3	6.2	6.9	6.8	6.8	6.6	5.8	6.6	6.2	6.0	6.3	5.4

Location Factors - Commercial

Costs shown in RSMeans cost data publications are based on national averages for materials and installation. To adjust these costs to a specific location, simply multiply the base cost by the factor and divide by 100 for that city. The data is arranged alphabetically by state and postal zip code numbers. For a city not listed, use the factor for a nearby city with similar economic characteristics.

STATE/ZIP	CITY	MAT.	INST.	TOTAL
ALABAMA				
350-352	Birmingham	99.3	77.3	89.6
354	Tuscaloosa	98.7	78.1	89.6
355	Jasper	99.2	71.8	87.1
356	Decatur	98.7	76.0	88.7
357-358	Huntsville	98.7	75.3	88.4
359	Gadsden	98.8	67.1	84.8
360-361	Montgomery	97.9	77.0	88.7
362	Anniston	97.3	77.5	88.6
363	Dothan	97.7	76.1	88.2
364	Evergreen	97.3	73.6	86.9
365-366	Mobile	98.3	76.6	88.7
367	Selma	97.4	74.9	87.5
368	Phenix City	98.2	77.2	89.0
369	Butler	97.6	76.7	88.4
ALASKA				
995-996	Anchorage	121.7	115.8	119.1
997	Fairbanks	124.2	116.9	120.9
998	Juneau	123.9	115.8	120.3
999	Ketchikan	134.9	115.8	126.5
ARIZONA				
850,853	Phoenix	100.5	73.0	88.4
851,852	Mesa/Tempe	100.1	72.0	87.7
855	Globe	100.9	69.8	87.2
856-857	Tucson	98.9	70.8	86.5
859	Show Low	101.1	71.9	88.3
860	Flagstaff	102.7	71.9	89.2
863	Prescott	100.4	68.4	86.3
864	Kingman	98.9	71.3	86.7
865	Chambers	98.9	73.5	87.7
ARKANSAS				
716	Pine Bluff	99.6	65.2	84.5
717	Camden	97.6	56.0	79.3
718	Texarkana	98.5	58.7	80.9
719	Hot Springs	96.9	56.9	79.3
720-722	Little Rock	98.2	67.4	84.6
723	West Memphis	97.0	65.3	83.0
724	Jonesboro	97.4	62.6	82.1
725	Batesville	95.3	58.3	79.0
726	Harrison	96.6	57.6	79.4
727	Fayetteville	94.1	55.6	77.1
728	Russellville	95.3	59.7	79.7
729	Fort Smith	97.8	62.3	82.2
CALIFORNIA				
900-902	Los Angeles	100.2	117.2	107.7
903-905	Inglewood	96.0	115.0	104.4
906-908	Long Beach	97.5	115.0	105.2
910-912	Pasadena	97.5	114.8	105.1
913-916	Van Nuys	100.4	114.8	106.7
917-918	Alhambra	99.5	114.8	106.2
919-921	San Diego	101.4	109.8	105.1
922	Palm Springs	98.6	112.9	104.9
923-924	San Bernardino	96.4	112.3	103.4
925	Riverside	100.5	113.9	106.4
926-927	Santa Ana	98.3	112.8	104.7
928	Anaheim	100.6	114.0	106.5
930	Oxnard	99.9	113.5	105.9
931	Santa Barbara	99.2	114.3	105.9
932-933	Bakersfield	100.4	111.4	105.2
934	San Luis Obispo	100.2	113.0	105.9
935	Mojave	97.4	111.9	103.8
936-938	Fresno	100.5	116.5	107.5
939	Salinas	101.0	122.3	110.4
940-941	San Francisco	104.1	146.0	122.5
942,956-958	Sacramento	99.6	120.1	108.8
943	Palo Alto	98.1	133.4	113.7
944	San Mateo	100.5	134.3	115.4
945	Vallejo	99.1	128.9	112.2
946	Oakland	102.1	134.5	116.4
947	Berkeley	102.0	135.3	116.7
948	Richmond	101.3	132.2	114.9
949	San Rafael	103.2	135.4	117.4
950	Santa Cruz	105.9	122.5	113.2
CALIFORNIA (CONT'D)				
951	San Jose	103.9	136.9	118.4
952	Stockton	102.1	117.9	109.0
953	Modesto	102.0	116.6	108.4
954	Santa Rosa	102.6	136.3	117.4
955	Eureka	103.9	118.3	110.2
959	Marysville	103.1	115.9	108.8
960	Redding	112.3	117.8	114.7
961	Susanville	112.0	117.2	114.3
COLORADO				
800-802	Denver	101.6	80.4	92.3
803	Boulder	99.1	80.3	90.8
804	Golden	101.2	77.8	90.9
805	Fort Collins	102.4	77.6	91.5
806	Greeley	99.7	74.0	88.4
807	Fort Morgan	99.7	78.0	90.1
808-809	Colorado Springs	101.3	78.7	91.3
810	Pueblo	102.4	74.7	90.2
811	Alamosa	104.3	73.1	90.5
812	Salida	103.9	75.4	91.3
813	Durango	104.7	74.9	91.6
814	Montrose	103.4	76.5	91.6
815	Grand Junction	106.5	75.6	92.9
816	Glenwood Springs	104.5	76.4	92.1
CONNECTICUT				
060	New Britain	98.3	122.1	108.8
061	Hartford	99.6	122.2	109.6
062	Willimantic	98.9	122.3	109.2
063	New London	95.6	122.4	107.4
064	Meriden	97.5	122.2	108.4
065	New Haven	99.8	122.3	109.7
066	Bridgeport	99.5	122.0	109.4
067	Waterbury	99.0	121.9	109.1
068	Norwalk	99.0	129.7	112.5
069	Stamford	99.1	129.7	112.6
D.C.				
200-205	Washington	100.6	95.9	98.5
DELAWARE				
197	Newark	98.9	109.4	103.5
198	Wilmington	98.7	109.4	103.4
199	Dover	99.0	109.4	103.6
FLORIDA				
320,322	Jacksonville	97.3	69.6	85.1
321	Daytona Beach	97.5	75.5	87.8
323	Tallahassee	98.9	65.0	84.0
324	Panama City	98.8	62.0	82.6
325	Pensacola	101.4	71.0	88.0
326,344	Gainesville	99.0	68.8	85.7
327-328,347	Orlando	99.5	72.0	87.4
329	Melbourne	100.1	76.6	89.8
330-332,340	Miami	98.9	72.3	87.2
333	Fort Lauderdale	97.8	71.2	86.1
334,349	West Palm Beach	96.9	69.9	85.0
335-336,346	Tampa	99.3	72.8	87.6
337	St. Petersburg	101.8	70.2	87.9
338	Lakeland	98.8	72.7	87.3
339,341	Fort Myers	98.1	68.8	85.1
342	Sarasota	100.0	73.8	88.5
GEORGIA				
300-303,399	Atlanta	98.1	76.5	88.6
304	Statesboro	98.2	61.0	81.8
305	Gainesville	96.9	63.6	82.2
306	Athens	96.3	65.2	82.6
307	Dalton	98.0	63.0	82.6
308-309	Augusta	96.8	69.9	85.0
310-312	Macon	95.8	66.2	82.7
313-314	Savannah	97.8	67.6	84.5
315	Waycross	97.1	64.4	82.7
316	Valdosta	97.0	62.3	81.7
317,398	Albany	96.9	63.1	82.0
318-319	Columbus	96.8	66.1	83.3

Location Factors - Commercial

STATE/ZIP	CITY	MAT.	INST.	TOTAL
HAWAII				
967	Hilo	120.2	119.7	120.0
968	Honolulu	125.0	119.7	122.7
STATES & POSS.				
969	Guam	143.0	57.6	105.4
IDAHO				
832	Pocatello	102.5	77.2	91.3
833	Twin Falls	103.8	66.3	87.3
834	Idaho Falls	101.1	74.3	89.3
835	Lewiston	109.4	85.0	98.6
836-837	Boise	101.5	77.8	91.1
838	Coeur d'Alene	109.3	82.2	97.4
ILLINOIS				
600-603	North Suburban	99.4	136.8	115.9
604	Joliet	99.4	139.0	116.8
605	South Suburban	99.4	137.0	115.9
606-608	Chicago	99.9	140.4	117.8
609	Kankakee	96.0	131.6	111.7
610-611	Rockford	99.3	123.5	109.9
612	Rock Island	97.4	101.6	99.2
613	La Salle	98.6	123.4	109.5
614	Galesburg	98.4	107.5	102.4
615-616	Peoria	100.3	109.8	104.5
617	Bloomington	97.7	110.0	103.1
618-619	Champaign	101.2	108.4	104.4
620-622	East St. Louis	95.9	108.4	101.4
623	Quincy	98.1	101.9	99.8
624	Effingham	97.3	106.6	101.4
625	Decatur	98.8	105.8	101.9
626-627	Springfield	99.3	106.9	102.6
628	Centralia	94.9	108.3	100.8
629	Carbondale	94.6	106.7	99.9
INDIANA				
460	Anderson	97.1	83.6	91.1
461-462	Indianapolis	99.0	85.2	93.0
463-464	Gary	98.4	110.0	103.5
465-466	South Bend	97.8	84.1	91.8
467-468	Fort Wayne	97.6	78.6	89.3
469	Kokomo	95.3	81.5	89.2
470	Lawrenceburg	93.8	78.2	86.9
471	New Albany	94.9	74.1	85.7
472	Columbus	97.1	81.3	90.1
473	Muncie	97.3	83.4	91.2
474	Bloomington	98.8	81.2	91.0
475	Washington	95.7	82.6	89.9
476-477	Evansville	96.6	86.6	92.2
478	Terre Haute	97.4	86.6	92.6
479	Lafayette	96.8	81.9	90.2
IOWA				
500-503,509	Des Moines	98.4	83.9	92.0
504	Mason City	97.0	64.4	82.6
505	Fort Dodge	97.2	57.6	79.8
506-507	Waterloo	98.4	73.7	87.5
508	Creston	97.5	78.3	89.1
510-511	Sioux City	99.6	72.3	87.6
512	Sibley	98.7	56.0	79.9
513	Spencer	100.5	56.0	80.9
514	Carroll	97.4	71.0	85.8
515	Council Bluffs	100.6	79.0	91.1
516	Shenandoah	97.8	73.5	87.1
520	Dubuque	99.0	80.6	90.9
521	Decorah	98.5	66.6	84.4
522-524	Cedar Rapids	99.9	82.6	92.3
525	Ottumwa	98.3	74.7	87.9
526	Burlington	97.5	76.9	88.4
527-528	Davenport	99.0	93.6	96.6
KANSAS				
660-662	Kansas City	99.6	95.9	98.0
664-666	Topeka	99.8	69.8	86.6
667	Fort Scott	98.3	73.7	87.5
668	Emporia	98.3	70.1	85.9
669	Belleville	100.1	66.5	85.3
670-672	Wichita	101.0	66.1	85.6
673	Independence	101.6	70.2	87.8
674	Salina	101.8	68.9	87.3
675	Hutchinson	97.1	65.2	83.1
676	Hays	101.2	66.9	86.1
677	Colby	101.8	68.0	86.9

STATE/ZIP	CITY	MAT.	INST.	TOTAL
KANSAS (CONT'D)				
678	Dodge City	103.0	66.1	86.8
679	Liberal	101.1	65.6	85.5
KENTUCKY				
400-402	Louisville	95.1	84.3	90.4
403-405	Lexington	95.5	82.2	89.7
406	Frankfort	96.1	80.2	89.1
407-409	Corbin	93.3	71.9	83.9
410	Covington	95.8	91.3	93.8
411-412	Ashland	94.5	99.8	96.8
413-414	Campton	95.6	77.6	87.7
415-416	Pikeville	96.8	89.3	93.5
417-418	Hazard	95.1	78.6	87.8
420	Paducah	93.7	85.5	90.1
421-422	Bowling Green	95.7	84.1	90.6
423	Owensboro	95.8	88.0	92.4
424	Henderson	93.5	86.3	90.3
425-426	Somerset	93.0	81.7	88.0
427	Elizabethtown	92.5	82.2	87.9
LOUISIANA				
700-701	New Orleans	99.2	70.7	86.6
703	Thibodaux	97.6	65.6	83.5
704	Hammond	95.2	61.0	80.1
705	Lafayette	97.0	66.6	83.6
706	Lake Charles	97.2	69.9	85.2
707-708	Baton Rouge	99.5	69.8	86.4
710-711	Shreveport	100.9	60.7	83.2
712	Monroe	99.6	59.3	81.8
713-714	Alexandria	99.7	63.3	83.7
MAINE				
039	Kittery	94.5	87.4	91.4
040-041	Portland	101.6	89.4	96.3
042	Lewiston	98.5	89.4	94.5
043	Augusta	101.2	81.0	92.3
044	Bangor	97.9	88.8	93.9
045	Bath	96.9	83.3	90.9
046	Machias	96.3	86.9	92.2
047	Houlton	96.5	86.9	92.3
048	Rockland	95.6	86.9	91.7
049	Waterville	96.8	81.0	89.8
MARYLAND				
206	Waldorf	97.1	82.6	90.7
207-208	College Park	97.1	83.8	91.2
209	Silver Spring	96.4	84.0	90.9
210-212	Baltimore	100.9	83.4	93.2
214	Annapolis	101.0	82.7	92.9
215	Cumberland	97.3	84.1	91.5
216	Easton	98.9	71.2	86.7
217	Hagerstown	97.7	85.2	92.2
218	Salisbury	99.4	61.8	82.8
219	Elkton	96.3	84.2	91.0
MASSACHUSETTS				
010-011	Springfield	99.1	110.3	104.0
012	Pittsfield	98.5	109.7	103.4
013	Greenfield	96.9	110.0	102.6
014	Fitchburg	95.6	124.5	108.3
015-016	Worcester	99.0	124.5	110.2
017	Framingham	94.9	132.9	111.6
018	Lowell	98.4	132.3	113.3
019	Lawrence	99.2	132.6	113.9
020-022, 024	Boston	100.6	139.1	117.5
023	Brockton	98.8	125.1	110.4
025	Buzzards Bay	93.7	125.9	107.9
026	Hyannis	95.9	125.9	109.1
027	New Bedford	97.9	126.3	110.4
MICHIGAN				
480,483	Royal Oak	95.1	106.1	99.9
481	Ann Arbor	97.0	107.4	101.6
482	Detroit	98.2	108.2	102.6
484-485	Flint	96.7	95.6	96.2
486	Saginaw	96.3	90.8	93.9
487	Bay City	96.4	90.3	93.7
488-489	Lansing	98.7	94.4	96.8
490	Battle Creek	97.5	86.0	92.4
491	Kalamazoo	97.8	83.9	91.7
492	Jackson	96.0	94.2	95.2
493,495	Grand Rapids	98.8	83.9	92.3
494	Muskegon	96.3	83.2	90.5

Location Factors - Commercial

STATE/ZIP	CITY	MAT.	INST.	TOTAL
MICHIGAN (CONT'D)				
496	Traverse City	95.6	80.9	89.1
497	Gaylord	96.7	83.6	90.9
498-499	Iron Mountain	98.7	85.8	93.0
MINNESOTA				
550-551	Saint Paul	100.7	120.9	109.6
553-555	Minneapolis	100.7	119.9	109.1
556-558	Duluth	101.3	110.5	105.3
559	Rochester	100.3	106.3	103.0
560	Mankato	97.5	102.7	99.8
561	Windom	96.2	93.5	95.0
562	Willmar	95.7	98.8	97.1
563	St. Cloud	96.8	114.0	104.4
564	Brainerd	97.3	100.5	98.7
565	Detroit Lakes	99.4	96.1	98.0
566	Bemidji	98.6	99.1	98.8
567	Thief River Falls	98.2	93.3	96.0
MISSISSIPPI				
386	Clarksdale	96.6	55.3	78.4
387	Greenville	100.3	65.1	84.8
388	Tupelo	98.2	55.6	79.4
389	Greenwood	97.9	52.0	77.7
390-392	Jackson	99.8	68.4	86.0
393	Meridian	97.1	67.9	84.2
394	Laurel	98.8	55.4	79.7
395	Biloxi	98.8	60.5	81.9
396	McComb	97.1	51.5	77.0
397	Columbus	98.6	53.4	78.7
MISSOURI				
630-631	St. Louis	100.2	104.3	102.0
633	Bowling Green	99.1	96.0	97.7
634	Hannibal	97.9	92.5	95.6
635	Kirksville	101.5	89.7	96.3
636	Flat River	100.0	95.1	97.9
637	Cape Girardeau	99.6	92.8	96.6
638	Sikeston	98.4	90.5	94.9
639	Poplar Bluff	97.9	89.7	94.3
640-641	Kansas City	100.2	104.8	102.2
644-645	St. Joseph	99.4	96.7	98.2
646	Chillicothe	96.8	98.6	97.6
647	Harrisonville	96.3	101.0	98.4
648	Joplin	98.2	81.0	90.7
650-651	Jefferson City	99.6	98.4	99.1
652	Columbia	100.3	99.7	100.0
653	Sedalia	100.7	93.9	97.7
654-655	Rolla	98.3	96.7	97.6
656-658	Springfield	101.8	83.3	93.7
MONTANA				
590-591	Billings	103.7	78.8	92.7
592	Wolf Point	103.8	74.2	90.8
593	Miles City	101.6	74.0	89.4
594	Great Falls	105.2	76.8	92.7
595	Havre	102.7	73.0	89.6
596	Helena	102.8	73.6	90.0
597	Butte	103.7	77.1	92.0
598	Missoula	101.1	75.7	89.9
599	Kalispell	100.4	75.5	89.5
NEBRASKA				
680-681	Omaha	99.1	79.5	90.5
683-685	Lincoln	99.8	76.9	89.7
686	Columbus	98.1	77.5	89.0
687	Norfolk	99.6	76.0	89.2
688	Grand Island	99.3	76.6	89.3
689	Hastings	99.1	78.9	90.2
690	McCook	99.0	70.5	86.4
691	North Platte	98.9	75.9	88.7
692	Valentine	101.3	71.8	88.3
693	Alliance	101.2	70.5	87.7
NEVADA				
889-891	Las Vegas	103.6	109.4	106.1
893	Ely	102.4	100.6	101.6
894-895	Reno	102.2	86.5	95.3
897	Carson City	101.8	86.7	95.1
898	Elko	101.1	81.4	92.4
NEW HAMPSHIRE				
030	Nashua	99.3	93.6	96.8
031	Manchester	99.9	93.6	97.1

STATE/ZIP	CITY	MAT.	INST.	TOTAL
NEW HAMPSHIRE (CONT'D)				
032-033	Concord	100.0	93.8	97.3
034	Keene	96.2	75.6	87.1
035	Littleton	96.3	77.6	88.0
036	Charleston	95.7	70.3	84.5
037	Claremont	94.9	70.4	84.1
038	Portsmouth	96.5	93.7	95.3
NEW JERSEY				
070-071	Newark	100.3	129.3	113.1
072	Elizabeth	97.8	128.8	111.5
073	Jersey City	96.7	128.9	110.9
074-075	Paterson	98.5	129.4	112.1
076	Hackensack	96.5	129.0	110.8
077	Long Branch	96.0	124.7	108.6
078	Dover	96.7	129.2	111.0
079	Summit	96.7	128.8	110.9
080,083	Vineland	96.4	123.6	108.4
081	Camden	98.4	124.7	110.0
082,084	Atlantic City	97.2	125.3	109.6
085-086	Trenton	100.2	124.9	111.1
087	Point Pleasant	98.2	124.0	109.6
088-089	New Brunswick	98.8	128.7	112.0
NEW MEXICO				
870-872	Albuquerque	100.3	72.1	87.9
873	Gallup	100.7	72.1	88.1
874	Farmington	101.0	72.1	88.2
875	Santa Fe	101.1	72.1	88.3
877	Las Vegas	99.0	72.1	87.1
878	Socorro	98.6	72.1	86.9
879	Truth/Consequences	98.3	68.8	85.3
880	Las Cruces	97.5	68.1	84.6
881	Clovis	99.9	72.0	87.6
882	Roswell	101.4	72.1	88.5
883	Carrizozo	102.1	72.1	88.9
884	Tucumcari	100.6	72.0	88.0
NEW YORK				
100-102	New York	100.7	169.8	131.1
103	Staten Island	96.8	167.5	128.0
104	Bronx	95.2	166.0	126.4
105	Mount Vernon	95.5	143.1	116.5
106	White Plains	95.2	143.1	116.3
107	Yonkers	99.4	146.3	120.1
108	New Rochelle	95.9	143.0	116.7
109	Suffern	95.5	131.9	111.5
110	Queens	100.3	166.2	129.3
111	Long Island City	101.9	166.2	130.2
112	Brooklyn	102.3	167.9	131.2
113	Flushing	102.4	166.2	130.5
114	Jamaica	100.7	166.2	129.6
115,117,118	Hicksville	100.3	152.7	123.4
116	Far Rockaway	102.5	166.2	130.6
119	Riverhead	101.1	150.9	123.0
120-122	Albany	98.3	104.6	101.1
123	Schenectady	98.1	104.5	101.0
124	Kingston	101.8	117.1	108.6
125-126	Poughkeepsie	101.0	135.2	116.1
127	Monticello	100.3	115.3	106.9
128	Glens Falls	93.3	101.1	96.7
129	Plattsburgh	98.4	96.7	97.7
130-132	Syracuse	98.5	98.2	98.3
133-135	Utica	96.5	97.5	97.0
136	Watertown	98.0	96.7	97.4
137-139	Binghamton	98.2	98.4	98.3
140-142	Buffalo	100.3	106.4	103.0
143	Niagara Falls	98.0	107.2	102.1
144-146	Rochester	100.8	100.0	100.5
147	Jamestown	97.0	93.0	95.3
148-149	Elmira	96.9	98.3	97.6
NORTH CAROLINA				
270,272-274	Greensboro	99.2	61.7	82.7
271	Winston-Salem	98.9	62.1	82.7
275-276	Raleigh	97.7	59.7	81.0
277	Durham	100.7	61.7	83.5
278	Rocky Mount	96.8	61.8	81.4
279	Elizabeth City	97.8	54.5	78.7
280	Gastonia	99.8	62.4	83.3
281-282	Charlotte	99.1	63.8	83.6
283	Fayetteville	102.2	60.6	83.9
284	Wilmington	98.5	61.5	82.2
285	Kinston	97.0	63.0	82.0

Location Factors - Commercial

STATE/ZIP	CITY	MAT.	INST.	TOTAL
NORTH CAROLINA (CONT'D)				
286	Hickory	97.3	61.8	81.7
287-288	Asheville	99.0	62.4	82.9
289	Murphy	98.2	51.0	77.4
NORTH DAKOTA				
580-581	Fargo	101.1	75.0	89.6
582	Grand Forks	101.9	65.1	85.7
583	Devils Lake	101.7	64.8	85.5
584	Jamestown	101.9	64.7	85.5
585	Bismarck	102.1	70.0	87.9
586	Dickinson	102.6	65.9	86.4
587	Minot	101.9	76.9	90.9
588	Williston	101.0	66.2	85.7
OHIO				
430-432	Columbus	98.2	87.4	93.5
433	Marion	95.1	84.6	90.5
434-436	Toledo	98.3	97.6	98.0
437-438	Zanesville	95.6	86.3	91.5
439	Steubenville	96.7	93.9	95.5
440	Lorain	98.9	92.2	96.0
441	Cleveland	99.2	99.4	99.3
442-443	Akron	99.9	93.9	97.3
444-445	Youngstown	99.2	88.1	94.4
446-447	Canton	99.3	89.3	94.9
448-449	Mansfield	97.1	86.2	92.3
450	Hamilton	97.8	84.4	91.9
451-452	Cincinnati	98.2	83.7	91.8
453-454	Dayton	97.9	83.9	91.7
455	Springfield	97.8	84.1	91.8
456	Chillicothe	97.2	92.0	94.9
457	Athens	100.2	83.1	92.6
458	Lima	100.5	86.3	94.3
OKLAHOMA				
730-731	Oklahoma City	99.8	71.3	87.3
734	Ardmore	98.6	64.9	83.7
735	Lawton	100.7	67.5	86.1
736	Clinton	99.8	66.2	85.0
737	Enid	100.2	65.2	84.8
738	Woodward	98.6	68.4	85.3
739	Guymon	99.7	69.1	86.2
740-741	Tulsa	98.3	64.5	83.4
743	Miami	95.2	72.1	85.0
744	Muskogee	97.5	62.9	82.3
745	McAlester	94.8	65.3	81.8
746	Ponca City	95.5	67.6	83.2
747	Durant	95.5	65.9	82.5
748	Shawnee	97.0	65.3	83.0
749	Poteau	94.7	66.4	82.2
OREGON				
970-972	Portland	99.7	101.1	100.3
973	Salem	100.9	99.2	100.2
974	Eugene	99.5	99.2	99.3
975	Medford	101.0	96.9	99.2
976	Klamath Falls	101.3	97.1	99.4
977	Bend	100.3	99.2	99.8
978	Pendleton	96.2	101.4	98.5
979	Vale	94.1	89.1	91.9
PENNSYLVANIA				
150-152	Pittsburgh	100.2	104.6	102.1
153	Washington	97.5	103.6	100.2
154	Uniontown	97.7	102.8	99.9
155	Bedford	98.7	95.3	97.2
156	Greensburg	98.7	102.8	100.5
157	Indiana	97.6	101.4	99.2
158	Dubois	99.1	97.5	98.4
159	Johnstown	98.7	98.1	98.5
160	Butler	92.3	103.2	97.1
161	New Castle	92.4	100.3	95.9
162	Kittanning	92.8	103.1	97.3
163	Oil City	92.3	99.7	95.6
164-165	Erie	94.0	97.5	95.5
166	Altoona	94.1	93.5	93.9
167	Bradford	95.6	97.4	96.4
168	State College	95.1	93.0	94.2
169	Wellsboro	96.2	96.6	96.3
170-171	Harrisburg	98.5	96.5	97.6
172	Chambersburg	95.8	91.2	93.8
173-174	York	95.9	96.8	96.3
175-176	Lancaster	94.6	98.3	96.2

STATE/ZIP	CITY	MAT.	INST.	TOTAL
PENNSYLVANIA (CONT'D)				
177	Williamsport	93.2	96.2	94.5
178	Sunbury	95.3	97.3	96.2
179	Pottsville	94.4	99.1	96.5
180	Lehigh Valley	95.7	112.3	103.0
181	Allentown	97.3	110.6	103.2
182	Hazleton	95.2	100.3	97.5
183	Stroudsburg	95.1	105.0	99.4
184-185	Scranton	98.1	99.8	98.8
186-187	Wilkes-Barre	94.9	101.0	97.6
188	Montrose	94.6	98.7	96.4
189	Doylestown	94.8	127.4	109.1
190-191	Philadelphia	99.5	134.7	115.0
193	Westchester	96.3	126.3	109.5
194	Norristown	95.4	131.2	111.2
195-196	Reading	97.1	104.5	100.4
PUERTO RICO				
009	San Juan	122.0	24.5	79.0
RHODE ISLAND				
028	Newport	97.2	116.8	105.8
029	Providence	99.8	116.8	107.3
SOUTH CAROLINA				
290-292	Columbia	98.2	68.3	85.0
293	Spartanburg	98.3	67.3	84.6
294	Charleston	99.8	66.1	84.9
295	Florence	98.0	67.7	84.7
296	Greenville	98.1	66.0	83.9
297	Rock Hill	98.0	56.5	79.7
298	Aiken	98.9	66.8	84.7
299	Beaufort	99.6	52.5	78.8
SOUTH DAKOTA				
570-571	Sioux Falls	99.0	60.5	82.0
572	Watertown	99.9	49.9	77.9
573	Mitchell	98.6	48.1	76.4
574	Aberdeen	101.0	53.7	80.1
575	Pierre	102.2	59.3	83.3
576	Mobridge	99.4	48.7	77.0
577	Rapid City	100.8	59.6	82.7
TENNESSEE				
370-372	Nashville	98.2	74.1	87.6
373-374	Chattanooga	99.9	66.6	85.2
375,380-381	Memphis	98.9	72.5	87.3
376	Johnson City	99.9	56.4	80.7
377-379	Knoxville	96.5	66.8	83.5
382	McKenzie	97.7	59.4	80.8
383	Jackson	99.0	63.5	83.4
384	Columbia	96.2	67.2	83.4
385	Cookeville	97.5	60.0	81.0
TEXAS				
750	McKinney	98.2	66.9	84.4
751	Waxahackie	98.3	65.4	83.8
752-753	Dallas	99.2	67.9	85.4
754	Greenville	98.4	67.0	84.6
755	Texarkana	97.6	63.6	82.6
756	Longview	98.4	64.3	83.4
757	Tyler	98.4	65.2	83.8
758	Palestine	94.8	64.4	81.4
759	Lufkin	95.6	67.7	83.3
760-761	Fort Worth	98.8	65.7	84.2
762	Denton	99.2	66.5	84.8
763	Wichita Falls	96.5	64.6	82.5
764	Eastland	95.6	64.1	81.8
765	Temple	94.1	61.6	79.8
766-767	Waco	95.9	63.7	81.7
768	Brownwood	99.1	60.6	82.1
769	San Angelo	98.7	59.9	81.6
770-772	Houston	100.7	69.8	87.1
773	Huntsville	99.7	66.9	85.3
774	Wharton	100.6	68.7	86.5
775	Galveston	98.7	68.5	85.4
776-777	Beaumont	98.9	68.9	85.7
778	Bryan	96.4	68.4	84.1
779	Victoria	100.7	66.1	85.5
780	Laredo	99.4	63.5	83.6
781-782	San Antonio	100.3	64.1	84.4
783-784	Corpus Christi	102.0	63.2	84.9
785	McAllen	102.9	59.9	83.9
786-787	Austin	100.0	63.4	83.9

Location Factors - Commercial

STATE/ZIP	CITY	MAT.	INST.	TOTAL
TEXAS (CONT'D)				
788	Del Rio	102.5	63.0	85.1
789	Giddings	99.3	63.8	83.7
790-791	Amarillo	99.6	62.3	83.1
792	Childress	99.7	64.0	84.0
793-794	Lubbock	101.5	63.6	84.8
795-796	Abilene	100.0	62.3	83.4
797	Midland	102.1	64.6	85.6
798-799,885	El Paso	98.3	61.6	82.2
UTAH				
840-841	Salt Lake City	103.8	71.2	89.4
842,844	Ogden	99.3	71.2	86.9
843	Logan	101.3	71.2	88.0
845	Price	102.0	65.3	85.8
846-847	Provo	101.8	71.1	88.2
VERMONT				
050	White River Jct.	98.5	77.3	89.2
051	Bellows Falls	96.9	91.7	94.6
052	Bennington	97.3	92.3	95.1
053	Brattleboro	97.7	91.6	95.0
054	Burlington	102.2	82.5	93.5
056	Montpelier	100.6	81.8	92.3
057	Rutland	99.0	82.5	91.7
058	St. Johnsbury	98.7	76.8	89.1
059	Guildhall	97.2	76.8	88.2
VIRGINIA				
220-221	Fairfax	100.7	89.2	95.6
222	Arlington	101.5	89.4	96.2
223	Alexandria	100.6	89.4	95.6
224-225	Fredericksburg	99.4	87.7	94.2
226	Winchester	100.0	85.2	93.5
227	Culpeper	99.9	85.5	93.6
228	Harrisonburg	100.2	66.9	85.5
229	Charlottesville	100.5	67.9	86.1
230-232	Richmond	99.8	68.7	86.1
233-235	Norfolk	100.1	68.3	86.1
236	Newport News	99.8	68.3	85.9
237	Portsmouth	99.3	65.4	84.4
238	Petersburg	99.8	68.7	86.1
239	Farmville	99.0	52.1	78.3
240-241	Roanoke	101.0	66.6	85.8
242	Bristol	99.4	54.8	79.7
243	Pulaski	99.0	57.5	80.7
244	Staunton	99.7	61.7	83.0
245	Lynchburg	99.9	66.2	85.1
246	Grundy	99.2	52.6	78.7
WASHINGTON				
980-981,987	Seattle	102.3	103.8	103.0
982	Everett	102.2	99.0	100.8
983-984	Tacoma	102.6	99.6	101.3
985	Olympia	101.0	99.7	100.4
986	Vancouver	103.4	94.4	99.4
988	Wenatchee	102.8	85.9	95.3
989	Yakima	102.8	94.9	99.3
990-992	Spokane	104.8	82.6	95.0
993	Richland	104.4	90.7	98.4
994	Clarkston	103.8	81.9	94.1
WEST VIRGINIA				
247-248	Bluefield	98.2	93.6	96.2
249	Lewisburg	99.8	92.8	96.7
250-253	Charleston	99.7	97.4	98.7
254	Martinsburg	100.4	87.0	94.5
255-257	Huntington	101.4	97.8	99.8
258-259	Beckley	98.8	95.1	97.2
260	Wheeling	100.8	94.5	98.0
261	Parkersburg	99.6	92.5	96.5
262	Buckhannon	99.6	94.7	97.4
263-264	Clarksburg	100.2	94.7	97.7
265	Morgantown	100.2	94.8	97.8
266	Gassaway	99.5	94.4	97.3
267	Romney	99.4	89.8	95.1
268	Petersburg	99.3	87.7	94.2
WISCONSIN				
530,532	Milwaukee	98.9	106.9	102.4
531	Kenosha	99.2	101.7	100.3
534	Racine	98.7	102.0	100.2
535	Beloit	98.6	99.8	99.1
537	Madison	99.3	99.1	99.2

STATE/ZIP	CITY	MAT.	INST.	TOTAL
WISCONSIN (CONT'D)				
538	Lancaster	96.8	93.7	95.4
539	Portage	95.3	96.4	95.8
540	New Richmond	97.1	93.8	95.6
541-543	Green Bay	101.2	97.5	99.5
544	Wausau	96.5	96.1	96.3
545	Rhinelander	99.8	91.8	96.3
546	La Crosse	97.6	94.8	96.4
547	Eau Claire	99.2	95.5	97.6
548	Superior	96.8	97.7	97.2
549	Oshkosh	97.2	91.1	94.5
WYOMING				
820	Cheyenne	102.3	66.5	86.5
821	Yellowstone Nat'l Park	100.2	71.5	87.6
822	Wheatland	101.4	65.5	85.6
823	Rawlins	103.1	71.2	89.0
824	Worland	101.0	67.4	86.2
825	Riverton	102.1	66.4	86.4
826	Casper	102.5	63.8	85.5
827	Newcastle	100.8	72.4	88.3
828	Sheridan	103.7	66.5	87.3
829-831	Rock Springs	104.9	68.4	88.8
CANADIAN FACTORS (reflect Canadian currency)				
ALBERTA				
	Calgary	123.8	93.4	110.4
	Edmonton	123.9	93.4	110.5
	Fort McMurray	124.5	91.7	110.0
	Lethbridge	119.1	91.2	106.8
	Lloydminster	113.9	87.9	102.5
	Medicine Hat	114.1	87.2	102.2
	Red Deer	114.6	87.2	102.5
BRITISH COLUMBIA				
	Kamloops	115.0	89.3	103.7
	Prince George	116.0	88.6	104.0
	Vancouver	117.7	89.2	105.1
	Victoria	116.6	86.7	103.4
MANITOBA				
	Brandon	126.4	74.8	103.7
	Portage la Prairie	113.9	73.1	95.9
	Winnipeg	122.6	70.8	99.8
NEW BRUNSWICK				
	Bathurst	111.8	66.3	91.8
	Dalhousie	112.6	66.5	92.3
	Fredericton	118.8	71.5	97.9
	Moncton	112.0	70.7	93.8
	Newcastle	111.9	67.0	92.1
	St. John	114.4	77.8	98.3
NEWFOUNDLAND				
	Corner Brook	131.2	66.6	102.7
	St. John's	123.9	83.9	106.3
NORTHWEST TERRITORIES				
	Yellowknife	136.7	82.9	113.0
NOVA SCOTIA				
	Bridgewater	113.4	73.6	95.9
	Dartmouth	127.5	73.6	103.7
	Halifax	118.2	79.9	101.3
	New Glasgow	125.5	73.6	102.6
	Sydney	123.4	73.6	101.4
	Truro	113.0	73.6	95.6
	Yarmouth	125.4	73.6	102.6
ONTARIO				
	Barrie	118.3	91.7	106.6
	Brantford	115.5	95.5	106.7
	Cornwall	115.4	92.0	105.1
	Hamilton	116.9	95.6	107.5
	Kingston	116.5	92.1	105.7
	Kitchener	109.9	91.8	101.9
	London	116.3	92.2	105.7
	North Bay	128.6	89.8	111.5
	Oshawa	112.5	95.3	104.9
	Ottawa	117.6	92.9	106.7
	Owen Sound	118.4	89.9	105.8
	Peterborough	115.5	91.7	105.0
	Sarnia	115.3	96.1	106.9

Location Factors - Commercial

STATE/ZIP	CITY	MAT.	INST.	TOTAL
ONTARIO (CONT'D)				
	Sault Ste. Marie	110.7	91.1	102.1
	St. Catharines	107.8	93.3	101.5
	Sudbury	107.6	92.1	100.8
	Thunder Bay	109.1	92.3	101.7
	Timmins	115.6	89.8	104.2
	Toronto	116.9	100.9	109.9
	Windsor	108.3	92.3	101.3
PRINCE EDWARD ISLAND				
	Charlottetown	120.4	61.9	94.6
	Summerside	128.1	62.0	99.0
QUEBEC				
	Cap-de-la-Madeleine	112.8	86.4	101.1
	Charlesbourg	112.8	86.4	101.1
	Chicoutimi	111.7	92.0	103.0
	Gatineau	112.5	86.1	100.9
	Granby	112.7	86.1	101.0
	Hull	112.6	86.1	100.9
	Joliette	113.0	86.4	101.3
	Laval	112.8	86.1	101.1
	Montreal	117.9	92.4	106.7
	Quebec City	117.9	92.8	106.8
	Rimouski	112.3	92.0	103.4
	Rouyn-Noranda	112.5	86.1	100.9
	Saint-Hyacinthe	112.0	86.1	100.6
	Sherbrooke	112.8	86.1	101.1
	Sorel	113.0	86.4	101.2
	Saint-Jerome	112.5	86.1	100.9
	Trois-Rivieres	126.1	86.4	108.6
SASKATCHEWAN				
	Moose Jaw	111.2	68.5	92.4
	Prince Albert	110.3	66.8	91.1
	Regina	123.0	89.0	108.0
	Saskatoon	112.3	89.0	102.0
YUKON				
	Whitehorse	137.1	67.5	106.4

R011105-05 Tips for Accurate Estimating

1. Use pre-printed or columnar forms for orderly sequence of dimensions and locations and for recording telephone quotations.
2. Use only the front side of each paper or form except for certain pre-printed summary forms.
3. Be consistent in listing dimensions: For example, length x width x height. This helps in rechecking to ensure that, the total length of partitions is appropriate for the building area.
4. Use printed (rather than measured) dimensions where given.
5. Add up multiple printed dimensions for a single entry where possible.
6. Measure all other dimensions carefully.
7. Use each set of dimensions to calculate multiple related quantities.
8. Convert foot and inch measurements to decimal feet when listing. Memorize decimal equivalents to .01 parts of a foot (1/8″ equals approximately .01′).
9. Do not "round off" quantities until the final summary.
10. Mark drawings with different colors as items are taken off.
11. Keep similar items together, different items separate.
12. Identify location and drawing numbers to aid in future checking for completeness.
13. Measure or list everything on the drawings or mentioned in the specifications.
14. It may be necessary to list items not called for to make the job complete.
15. Be alert for: Notes on plans such as N.T.S. (not to scale); changes in scale throughout the drawings; reduced size drawings; discrepancies between the specifications and the drawings.
16. Develop a consistent pattern of performing an estimate. For example:
    a. Start the quantity takeoff at the lower floor and move to the next higher floor.
    b. Proceed from the main section of the building to the wings.
    c. Proceed from south to north or vice versa, clockwise or counterclockwise.
    d. Take off floor plan quantities first, elevations next, then detail drawings.
17. List all gross dimensions that can be either used again for different quantities, or used as a rough check of other quantities for verification (exterior perimeter, gross floor area, individual floor areas, etc.).
18. Utilize design symmetry or repetition (repetitive floors, repetitive wings, symmetrical design around a center line, similar room layouts, etc.). Note: Extreme caution is needed here so as not to omit or duplicate an area.
19. Do not convert units until the final total is obtained. For instance, when estimating concrete work, keep all units to the nearest cubic foot, then summarize and convert to cubic yards.
20. When figuring alternatives, it is best to total all items involved in the basic system, then total all items involved in the alternates. Therefore you work with positive numbers in all cases. When adds and deducts are used, it is often confusing whether to add or subtract a portion of an item; especially on a complicated or involved alternate.

R011105-50 Metric Conversion Factors

Description: This table is primarily for converting customary U.S. units in the left hand column to SI metric units in the right hand column. In addition, conversion factors for some commonly encountered Canadian and non-SI metric units are included.

	If You Know		Multiply By		To Find
Length	Inches	x	25.4[a]	=	Millimeters
	Feet	x	0.3048[a]	=	Meters
	Yards	x	0.9144[a]	=	Meters
	Miles (statute)	x	1.609	=	Kilometers
Area	Square inches	x	645.2	=	Square millimeters
	Square feet	x	0.0929	=	Square meters
	Square yards	x	0.8361	=	Square meters
Volume (Capacity)	Cubic inches	x	16,387	=	Cubic millimeters
	Cubic feet	x	0.02832	=	Cubic meters
	Cubic yards	x	0.7646	=	Cubic meters
	Gallons (U.S. liquids)[b]	x	0.003785	=	Cubic meters[c]
	Gallons (Canadian liquid)[b]	x	0.004546	=	Cubic meters[c]
	Ounces (U.S. liquid)[b]	x	29.57	=	Milliliters[c, d]
	Quarts (U.S. liquid)[b]	x	0.9464	=	Liters[c, d]
	Gallons (U.S. liquid)[b]	x	3.785	=	Liters[c, d]
Force	Kilograms force[d]	x	9.807	=	Newtons
	Pounds force	x	4.448	=	Newtons
	Pounds force	x	0.4536	=	Kilograms force[d]
	Kips	x	4448	=	Newtons
	Kips	x	453.6	=	Kilograms force[d]
Pressure, Stress, Strength (Force per unit area)	Kilograms force per square centimeter[d]	x	0.09807	=	Megapascals
	Pounds force per square inch (psi)	x	0.006895	=	Megapascals
	Kips per square inch	x	6.895	=	Megapascals
	Pounds force per square inch (psi)	x	0.07031	=	Kilograms force per square centimeter[d]
	Pounds force per square foot	x	47.88	=	Pascals
	Pounds force per square foot	x	4.882	=	Kilograms force per square meter[d]
Flow	Cubic feet per minute	x	0.4719	=	Liters per second
	Gallons per minute	x	0.0631	=	Liters per second
	Gallons per hour	x	1.05	=	Milliliters per second
Bending Moment Or Torque	Inch-pounds force	x	0.01152	=	Meter-kilograms force[d]
	Inch-pounds force	x	0.1130	=	Newton-meters
	Foot-pounds force	x	0.1383	=	Meter-kilograms force[d]
	Foot-pounds force	x	1.356	=	Newton-meters
	Meter-kilograms force[d]	x	9.807	=	Newton-meters
Mass	Ounces (avoirdupois)	x	28.35	=	Grams
	Pounds (avoirdupois)	x	0.4536	=	Kilograms
	Tons (metric)	x	1000	=	Kilograms
	Tons, short (2000 pounds)	x	907.2	=	Kilograms
	Tons, short (2000 pounds)	x	0.9072	=	Megagrams[e]
Mass per Unit Volume	Pounds mass per cubic foot	x	16.02	=	Kilograms per cubic meter
	Pounds mass per cubic yard	x	0.5933	=	Kilograms per cubic meter
	Pounds mass per gallon (U.S. liquid)[b]	x	119.8	=	Kilograms per cubic meter
	Pounds mass per gallon (Canadian liquid)[b]	x	99.78	=	Kilograms per cubic meter
Temperature	Degrees Fahrenheit		(F-32)/1.8	=	Degrees Celsius
	Degrees Fahrenheit		(F+459.67)/1.8	=	Degrees Kelvin
	Degrees Celsius		C+273.15	=	Degrees Kelvin

[a]The factor given is exact
[b]One U.S. gallon = 0.8327 Canadian gallon
[c]1 liter = 1000 milliliters = 1000 cubic centimeters
1 cubic decimeter = 0.001 cubic meter
[d]Metric but not SI unit
[e]Called "tonne" in England and "metric ton" in other metric countries

R011105-60 Weights and Measures

Measures of Length
1 Mile = 1760 Yards = 5280 Feet
1 Yard = 3 Feet = 36 inches
1 Foot = 12 Inches
1 Mil = 0.001 Inch
1 Fathom = 2 Yards = 6 Feet
1 Rod = 5.5 Yards = 16.5 Feet
1 Hand = 4 Inches
1 Span = 9 Inches
1 Micro-inch = One Millionth Inch or 0.000001 Inch
1 Micron = One Millionth Meter + 0.00003937 Inch

Surveyor's Measure
1 Mile = 8 Furlongs = 80 Chains
1 Furlong = 10 Chains = 220 Yards
1 Chain = 4 Rods = 22 Yards = 66 Feet = 100 Links
1 Link = 7.92 Inches

Square Measure
1 Square Mile = 640 Acres = 6400 Square Chains
1 Acre = 10 Square Chains = 4840 Square Yards = 43,560 Sq. Ft.
1 Square Chain = 16 Square Rods = 484 Square Yards = 4356 Sq. Ft.
1 Square Rod = 30.25 Square Yards = 272.25 Square Feet = 625 Square Lines
1 Square Yard = 9 Square Feet
1 Square Foot = 144 Square Inches
An Acre equals a Square 208.7 Feet per Side

Cubic Measure
1 Cubic Yard = 27 Cubic Feet
1 Cubic Foot = 1728 Cubic Inches
1 Cord of Wood = 4 x 4 x 8 Feet = 128 Cubic Feet
1 Perch of Masonry = 16½ x 1½ x 1 Foot = 24.75 Cubic Feet

Avoirdupois or Commercial Weight
1 Gross or Long Ton = 2240 Pounds
1 Net or Short Ton = 2000 Pounds
1 Pound = 16 Ounces = 7000 Grains
1 Ounce = 16 Drachms = 437.5 Grains
1 Stone = 14 Pounds

Power
1 British Thermal Unit per Hour = 0.2931 Watts
1 Ton (Refrigeration) = 3.517 Kilowatts
1 Horsepower (Boiler) = 9.81 Kilowatts
1 Horsepower (550 ft-lb/s) = 0.746 Kilowatts

Shipping Measure
For Measuring Internal Capacity of a Vessel:
1 Register Ton = 100 Cubic Feet

For Measurement of Cargo:
Approximately 40 Cubic Feet of Merchandise is considered a Shipping Ton, unless that bulk would weigh more than 2000 Pounds, in which case Freight Charge may be based upon weight.

40 Cubic Feet = 32.143 U.S. Bushels = 31.16 Imp. Bushels

Liquid Measure
1 Imperial Gallon = 1.2009 U.S. Gallon = 277.42 Cu. In.
1 Cubic Foot = 7.48 U.S. Gallons

R011110-10 Architectural Fees

Tabulated below are typical percentage fees by project size, for good professional architectural service. Fees may vary from those listed depending upon degree of design difficulty and economic conditions in any particular area.

Rates can be interpolated horizontally and vertically. Various portions of the same project requiring different rates should be adjusted proportionately. For alterations, add 50% to the fee for the first $500,000 of project cost and add 25% to the fee for project cost over $500,000.

Architectural fees tabulated below include Structural, Mechanical and Electrical Engineering Fees. They do not include the fees for special consultants such as kitchen planning, security, acoustical, interior design, etc.

Civil Engineering fees are included in the Architectural fee for project sites requiring minimal design such as city sites. However, separate Civil Engineering fees must be added when utility connections require design, drainage calculations are needed, stepped foundations are required, or provisions are required to protect adjacent wetlands.

Building Types	Total Project Size in Thousands of Dollars						
	100	250	500	1,000	5,000	10,000	50,000
Factories, garages, warehouses, repetitive housing	9.0%	8.0%	7.0%	6.2%	5.3%	4.9%	4.5%
Apartments, banks, schools, libraries, offices, municipal buildings	12.2	12.3	9.2	8.0	7.0	6.6	6.2
Churches, hospitals, homes, laboratories, museums, research	15.0	13.6	12.7	11.9	9.5	8.8	8.0
Memorials, monumental work, decorative furnishings	—	16.0	14.5	13.1	10.0	9.0	8.3

General Requirements — R0111 Summary of Work

R011110-30 Engineering Fees

Typical **Structural Engineering Fees** based on type of construction and total project size. These fees are included in Architectural Fees.

Type of Construction	Total Project Size (in thousands of dollars)			
	$500	$500-$1,000	$1,000-$5,000	Over $5000
Industrial buildings, factories & warehouses	Technical payroll times 2.0 to 2.5	1.60%	1.25%	1.00%
Hotels, apartments, offices, dormitories, hospitals, public buildings, food stores	↓	2.00%	1.70%	1.20%
Museums, banks, churches and cathedrals	↓	2.00%	1.75%	1.25%
Thin shells, prestressed concrete, earthquake resistive	↓	2.00%	1.75%	1.50%
Parking ramps, auditoriums, stadiums, convention halls, hangars & boiler houses	↓	2.50%	2.00%	1.75%
Special buildings, major alterations, underpinning & future expansion	↓	Add to above 0.5%	Add to above 0.5%	Add to above 0.5%

For complex reinforced concrete or unusually complicated structures, add 20% to 50%.

Typical **Mechanical and Electrical Engineering Fees** are based on the size of the subcontract. The fee structure for both is shown below. These fees are included in Architectural Fees.

Type of Construction	Subcontract Size							
	$25,000	$50,000	$100,000	$225,000	$350,000	$500,000	$750,000	$1,000,000
Simple structures	6.4%	5.7%	4.8%	4.5%	4.4%	4.3%	4.2%	4.1%
Intermediate structures	8.0	7.3	6.5	5.6	5.1	5.0	4.9	4.8
Complex structures	10.1	9.0	9.0	8.0	7.5	7.5	7.0	7.0

For renovations, add 15% to 25% to applicable fee.

General Requirements — R0121 Allowances

R012153-10 Repair and Remodeling

Cost figures are based on new construction utilizing the most cost-effective combination of labor, equipment and material with the work scheduled in proper sequence to allow the various trades to accomplish their work in an efficient manner.

The costs for repair and remodeling work must be modified due to the following factors that may be present in any given repair and remodeling project.

1. Equipment usage curtailment due to the physical limitations of the project, with only hand-operated equipment being used.
2. Increased requirement for shoring and bracing to hold up the building while structural changes are being made and to allow for temporary storage of construction materials on above-grade floors.
3. Material handling becomes more costly due to having to move within the confines of an enclosed building. For multi-story construction, low capacity elevators and stairwells may be the only access to the upper floors.
4. Large amount of cutting and patching and attempting to match the existing construction is required. It is often more economical to remove entire walls rather than create many new door and window openings. This sort of trade-off has to be carefully analyzed.
5. Cost of protection of completed work is increased since the usual sequence of construction usually cannot be accomplished.
6. Economies of scale usually associated with new construction may not be present. If small quantities of components must be custom fabricated due to job requirements, unit costs will naturally increase. Also, if only small work areas are available at a given time, job scheduling between trades becomes difficult and subcontractor quotations may reflect the excessive start-up and shut-down phases of the job.
7. Work may have to be done on other than normal shifts and may have to be done around an existing production facility which has to stay in production during the course of the repair and remodeling.
8. Dust and noise protection of adjoining non-construction areas can involve substantial special protection and alter usual construction methods.
9. Job may be delayed due to unexpected conditions discovered during demolition or removal. These delays ultimately increase construction costs.
10. Piping and ductwork runs may not be as simple as for new construction. Wiring may have to be snaked through walls and floors.
11. Matching "existing construction" may be impossible because materials may no longer be manufactured. Substitutions may be expensive.
12. Weather protection of existing structure requires additional temporary structures to protect building at openings.
13. On small projects, because of local conditions, it may be necessary to pay a tradesman for a minimum of four hours for a task that is completed in one hour.

All of the above areas can contribute to increased costs for a repair and remodeling project. Each of the above factors should be considered in the planning, bidding and construction stage in order to minimize the increased costs associated with repair and remodeling jobs.

R012909-80 Sales Tax by State

State sales tax on materials is tabulated below (5 states have no sales tax). Many states allow local jurisdictions, such as a county or city, to levy additional sales tax.

Some projects may be sales tax exempt, particularly those constructed with public funds.

State	Tax (%)	State	Tax (%)	State	Tax (%)	State	Tax (%)
Alabama	4	Illinois	6.25	Montana	0	Rhode Island	7
Alaska	0	Indiana	7	Nebraska	5.5	South Carolina	6
Arizona	5.6	Iowa	6	Nevada	6.85	South Dakota	4
Arkansas	6	Kansas	6.15	New Hampshire	0	Tennessee	7
California	7.5	Kentucky	6	New Jersey	7	Texas	6.25
Colorado	2.9	Louisiana	4	New Mexico	5.125	Utah	5.95
Connecticut	6.35	Maine	5.5	New York	4	Vermont	6
Delaware	0	Maryland	6	North Carolina	4.75	Virginia	5.3
District of Columbia	5.75	Massachusetts	6.25	North Dakota	5	Washington	6.5
Florida	6	Michigan	6	Ohio	5.75	West Virginia	6
Georgia	4	Minnesota	6.875	Oklahoma	4.5	Wisconsin	5
Hawaii	4	Mississippi	7	Oregon	0	Wyoming	4
Idaho	6	Missouri	4.225	Pennsylvania	6	Average	5.08 %

Sales Tax by Province (Canada)

GST - a value-added tax, which the government imposes on most goods and services provided in or imported into Canada. PST - a retail sales tax, which three of the provinces impose on the prices of most goods and some services. QST - a value-added tax, similar to the federal GST, which Quebec imposes. HST - Five provinces have combined their retail sales taxes with the federal GST into one harmonized tax.

Province	PST (%)	QST (%)	GST(%)	HST(%)
Alberta	0	0	5	0
British Columbia	7	0	5	0
Manitoba	8	0	5	0
New Brunswick	0	0	0	13
Newfoundland	0	0	0	13
Northwest Territories	0	0	5	0
Nova Scotia	0	0	0	15
Ontario	0	0	0	13
Prince Edward Island	0	0	0	14
Quebec	0	9.975	5	0
Saskatchewan	5	0	5	0
Yukon	0	0	5	0

R012909-85 Unemployment Taxes and Social Security Taxes

State unemployment tax rates vary not only from state to state, but also with the experience rating of the contractor. The federal unemployment tax rate is 6.0% of the first $7,000 of wages. This is reduced by a credit of up to 5.4% for timely payment to the state. The minimum federal unemployment tax is 0.6% after all credits.

Social security (FICA) for 2016 is estimated at time of publication to be 7.65% of wages up to $118,500.

General Requirements — R0129 Payment Procedures

R012909-90 Overtime

One way to improve the completion date of a project or eliminate negative float from a schedule is to compress activity duration times. This can be achieved by increasing the crew size or working overtime with the proposed crew.

To determine the costs of working overtime to compress activity duration times, consider the following examples. Below is an overtime efficiency and cost chart based on a five, six, or seven day week with an eight through twelve hour day. Payroll percentage increases for time and one half and double times are shown for the various working days.

Days per Week	Hours per Day	Production Efficiency					Payroll Cost Factors	
		1st Week	2nd Week	3rd Week	4th Week	Average 4 Weeks	@ 1-1/2 Times	@ 2 Times
5	8	100%	100%	100%	100%	100%	1.000	1.000
	9	100	100	95	90	96	1.056	1.111
	10	100	95	90	85	93	1.100	1.200
	11	95	90	75	65	81	1.136	1.273
	12	90	85	70	60	76	1.167	1.333
6	8	100	100	95	90	96	1.083	1.167
	9	100	95	90	85	93	1.130	1.259
	10	95	90	85	80	88	1.167	1.333
	11	95	85	70	65	79	1.197	1.394
	12	90	80	65	60	74	1.222	1.444
7	8	100	95	85	75	89	1.143	1.286
	9	95	90	80	70	84	1.183	1.365
	10	90	85	75	65	79	1.214	1.429
	11	85	80	65	60	73	1.240	1.481
	12	85	75	60	55	69	1.262	1.524

General Requirements — R0131 Project Management & Coordination

R013113-40 Builder's Risk Insurance

Builder's Risk Insurance is insurance on a building during construction. Premiums are paid by the owner or the contractor. Blasting, collapse and underground insurance would raise total insurance costs above those listed. Floater policy for materials delivered to the job runs $.75 to $1.25 per $100 value. Contractor equipment insurance runs $.50 to $1.50 per $100 value. Insurance for miscellaneous tools to $1,500 value runs from $3.00 to $7.50 per $100 value.

Tabulated below are New England Builder's Risk insurance rates in dollars per $100 value for $1,000 deductible. For $25,000 deductible, rates can be reduced 13% to 34%. On contracts over $1,000,000, rates may be lower than those tabulated. Policies are written annually for the total completed value in place. For "all risk" insurance (excluding flood, earthquake and certain other perils) add $.025 to total rates below.

Coverage	Frame Construction (Class 1)		Brick Construction (Class 4)		Fire Resistive (Class 6)	
	Range	Average	Range	Average	Range	Average
Fire Insurance	$.350 to $ .850	$.600	$.158 to $.189	$.174	$.052 to $.080	$.070
Extended Coverage	.115 to .200	.158	.080 to .105	.101	.081 to .105	.100
Vandalism	.012 to .016	.014	.008 to .011	.011	.008 to .011	.010
Total Annual Rate	$.477 to $1.066	$.772	$.246 to $.305	$.286	$.141 to $.196	$.180

R013113-50 General Contractor's Overhead

There are two distinct types of overhead on a construction project: Project Overhead and Main Office Overhead. Project Overhead includes those costs at a construction site not directly associated with the installation of construction materials. Examples of Project Overhead costs include the following:

1. Superintendent
2. Construction office and storage trailers
3. Temporary sanitary facilities
4. Temporary utilities
5. Security fencing
6. Photographs
7. Clean up
8. Performance and payment bonds

The above Project Overhead items are also referred to as General Requirements and therefore are estimated in Division 1. Division 1 is the first division listed in the CSI MasterFormat but it is usually the last division estimated. The sum of the costs in Divisions 1 through 49 is referred to as the sum of the direct costs.

All construction projects also include indirect costs. The primary components of indirect costs are the contractor's Main Office Overhead and profit. The amount of the Main Office Overhead expense varies depending on the following:

1. Owner's compensation
2. Project managers and estimator's wages
3. Clerical support wages
4. Office rent and utilities
5. Corporate legal and accounting costs
6. Advertising
7. Automobile expenses
8. Association dues
9. Travel and entertainment expenses

These costs are usually calculated as a percentage of annual sales volume. This percentage can range from 35% for a small contractor doing less than $500,000 to 5% for a large contractor with sales in excess of $100 million.

R013113-60 Workers' Compensation Insurance Rates by Trade

The table below tabulates the national averages for workers' compensation insurance rates by trade and type of building. The average "Insurance Rate" is multiplied by the "% of Building Cost" for each trade. This produces the "Workers' Compensation" cost by % of total labor cost, to be added for each trade by building type to determine the weighted average workers' compensation rate for the building types analyzed.

Trade	Insurance Rate (% Labor Cost)		% of Building Cost			Workers' Compensation		
	Range	Average	Office Bldgs.	Schools & Apts.	Mfg.	Office Bldgs.	Schools & Apts.	Mfg.
Excavation, Grading, etc.	3.1 % to 21.5%	9.3%	4.8%	4.9%	4.5%	0.45%	0.46%	0.42%
Piles & Foundations	6.4 to 32.8	14.6	7.1	5.2	8.7	1.04	0.76	1.27
Concrete	4.3 to 32.5	12.6	5.0	14.8	3.7	0.63	1.86	0.47
Masonry	4.6 to 36.8	13.7	6.9	7.5	1.9	0.95	1.03	0.26
Structural Steel	6.5 to 95.8	27.5	10.7	3.9	17.6	2.94	1.07	4.84
Miscellaneous & Ornamental Metals	4.1 to 28.9	11.7	2.8	4.0	3.6	0.33	0.47	0.42
Carpentry & Millwork	5.0 to 42.8	14.4	3.7	4.0	0.5	0.53	0.58	0.07
Metal or Composition Siding	6.5 to 75.9	18.7	2.3	0.3	4.3	0.43	0.06	0.80
Roofing	6.5 to 107.1	31.4	2.3	2.6	3.1	0.72	0.82	0.97
Doors & Hardware	4.1 to 42.8	11.4	0.9	1.4	0.4	0.10	0.16	0.05
Sash & Glazing	4.8 to 28.0	13.1	3.5	4.0	1.0	0.46	0.52	0.13
Lath & Plaster	2.8 to 41.3	10.5	3.3	6.9	0.8	0.35	0.72	0.08
Tile, Marble & Floors	2.9 to 22.4	8.8	2.6	3.0	0.5	0.23	0.26	0.04
Acoustical Ceilings	3.1 to 33.8	8.6	2.4	0.2	0.3	0.21	0.02	0.03
Painting	4.1 to 34.3	11.7	1.5	1.6	1.6	0.18	0.19	0.19
Interior Partitions	5.0 to 42.8	14.4	3.9	4.3	4.4	0.56	0.62	0.63
Miscellaneous Items	2.5 to 105.9	12.6	5.2	3.7	9.7	0.65	0.47	1.22
Elevators	1.4 to 13.6	5.2	2.1	1.1	2.2	0.11	0.06	0.11
Sprinklers	2.8 to 25.9	7.6	0.5	—	2.0	0.04	—	0.15
Plumbing	1.9 to 19.5	6.9	4.9	7.2	5.2	0.34	0.50	0.36
Heat., Vent., Air Conditioning	3.8 to 18.4	8.8	13.5	11.0	12.9	1.19	0.97	1.14
Electrical	2.3 to 14.4	5.6	10.1	8.4	11.1	0.57	0.47	0.62
Total	1.4 % to 107.1%	—	100.0%	100.0%	100.0%	13.01%	12.07%	14.27%
		Overall Weighted Average	13.12%					

Workers' Compensation Insurance Rates by States

The table below lists the weighted average Workers' Compensation base rate for each state with a factor comparing this with the national average of 13.0%.

State	Weighted Average	Factor	State	Weighted Average	Factor	State	Weighted Average	Factor
Alabama	19.5%	150	Kentucky	14.3%	110	North Dakota	9.2%	71
Alaska	13.7	105	Louisiana	18.6	143	Ohio	8.2	63
Arizona	12.6	97	Maine	12.2	94	Oklahoma	12.5	96
Arkansas	8.3	64	Maryland	14.1	108	Oregon	11.8	91
California	25.8	198	Massachusetts	11.6	89	Pennsylvania	17.4	134
Colorado	7.2	55	Michigan	16.3	125	Rhode Island	12.1	93
Connecticut	23.7	182	Minnesota	22.6	174	South Carolina	18.9	145
Delaware	11.9	92	Mississippi	13.2	102	South Dakota	14.9	115
District of Columbia	11.0	85	Missouri	14.8	114	Tennessee	12.0	92
Florida	10.7	82	Montana	8.5	65	Texas	9.1	70
Georgia	28.1	216	Nebraska	17.7	136	Utah	8.8	68
Hawaii	8.1	62	Nevada	8.7	67	Vermont	13.1	101
Idaho	10.3	79	New Hampshire	20.0	154	Virginia	9.5	73
Illinois	27.1	208	New Jersey	14.4	111	Washington	10.5	81
Indiana	5.5	42	New Mexico	14.4	111	West Virginia	8.0	62
Iowa	14.6	112	New York	17.2	132	Wisconsin	13.1	101
Kansas	8.4	65	North Carolina	16.7	128	Wyoming	6.1	47
				Weighted Average for U.S. is	13.7% of payroll = 100%			

The weighted average skilled worker rate for 35 trades is 13.0%. For bidding purposes, apply the full value of Workers' Compensation directly to total labor costs, or if labor is 38%, materials 42% and overhead and profit 20% of total cost, carry 38/80 x 13.0% = 6.2% of cost (before overhead and profit) into overhead. Rates vary not only from state to state but also with the experience rating of the contractor.

Rates are the most current available at the time of publication.

General Requirements — R0131 Project Management & Coordination

R013113-80 Performance Bond

This table shows the cost of a Performance Bond for a construction job scheduled to be completed in 12 months. Add 1% of the premium cost per month for jobs requiring more than 12 months to complete. The rates are "standard" rates offered to contractors that the bonding company considers financially sound and capable of doing the work. Preferred rates are offered by some bonding companies based upon financial strength of the contractor. Actual rates vary from contractor to contractor and from bonding company to bonding company. Contractors should prequalify through a bonding agency before submitting a bid on a contract that requires a bond.

Contract Amount	Building Construction Class B Projects	Highways & Bridges: Class A New Construction	Highways & Bridges: Class A-1 Highway Resurfacing
First $ 100,000 bid	$25.00 per M	$15.00 per M	$9.40 per M
Next 400,000 bid	$ 2,500 plus $15.00 per M	$ 1,500 plus $10.00 per M	$ 940 plus $7.20 per M
Next 2,000,000 bid	8,500 plus 10.00 per M	5,500 plus 7.00 per M	3,820 plus 5.00 per M
Next 2,500,000 bid	28,500 plus 7.50 per M	19,500 plus 5.50 per M	15,820 plus 4.50 per M
Next 2,500,000 bid	47,250 plus 7.00 per M	33,250 plus 5.00 per M	28,320 plus 4.50 per M
Over 7,500,000 bid	64,750 plus 6.00 per M	45,750 plus 4.50 per M	39,570 plus 4.00 per M

General Requirements — R0154 Construction Aids

R015423-10 Steel Tubular Scaffolding

On new construction, tubular scaffolding is efficient up to 60′ high or five stories. Above this it is usually better to use a hung scaffolding if construction permits. Swing scaffolding operations may interfere with tenants. In this case, the tubular is more practical at all heights.

In repairing or cleaning the front of an existing building the cost of tubular scaffolding per S.F. of building front increases as the height increases above the first tier. The first tier cost is relatively high due to leveling and alignment.

The minimum efficient crew for erecting and dismantling is three workers. They can set up and remove 18 frame sections per day up to 5 stories high. For 6 to 12 stories high, a crew of four is most efficient. Use two or more on top and two on the bottom for handing up or hoisting. They can also set up and remove 18 frame sections per day. At 7' horizontal spacing, this will run about 800 S.F. per day of erecting and dismantling. Time for placing and removing planks must be added to the above. A crew of three can place and remove 72 planks per day up to 5 stories. For over 5 stories, a crew of four can place and remove 80 planks per day.

The table below shows the number of pieces required to erect tubular steel scaffolding for 1000 S.F. of building frontage. This area is made up of a scaffolding system that is 12 frames (11 bays) long by 2 frames high.

For jobs under twenty-five frames, add 50% to rental cost. Rental rates will be lower for jobs over three months duration. Large quantities for long periods can reduce rental rates by 20%.

Description of Component	Number of Pieces for 1000 S.F. of Building Front	Unit
5′ Wide Standard Frame, 6′-4″ High	24	Ea.
Leveling Jack & Plate	24	
Cross Brace	44	
Side Arm Bracket, 21″	12	
Guardrail Post	12	
Guardrail, 7′ section	22	
Stairway Section	2	
Stairway Starter Bar	1	
Stairway Inside Handrail	2	
Stairway Outside Handrail	2	
Walk-Thru Frame Guardrail	2	↓

Scaffolding is often used as falsework over 15′ high during construction of cast-in-place concrete beams and slabs. Two foot wide scaffolding is generally used for heavy beam construction. The span between frames depends upon the load to be carried with a maximum span of 5′.

Heavy duty shoring frames with a capacity of 10,000#/leg can be spaced up to 10′ O.C. depending upon form support design and loading.

Scaffolding used as horizontal shoring requires less than half the material required with conventional shoring.

On new construction, erection is done by carpenters.

Rolling towers supporting horizontal shores can reduce labor and speed the job. For maintenance work, catwalks with spans up to 70′ can be supported by the rolling towers.

R015423-20 Pump Staging

Pump staging is generally not available for rent. The table below shows the number of pieces required to erect pump staging for 2400 S.F. of building frontage. This area is made up of a pump jack system that is 3 poles (2 bays) wide by 2 poles high.

Item	Number of Pieces for 2400 S.F. of Building Front	Unit
Aluminum pole section, 24′ long	6	Ea.
Aluminum splice joint, 6′ long	3	↓
Aluminum foldable brace	3	↓
Aluminum pump jack	3	↓
Aluminum support for workbench/back safety rail	3	↓
Aluminum scaffold plank/workbench, 14″ wide x 24′ long	4	↓
Safety net, 22′ long	2	↓
Aluminum plank end safety rail	2	↓

The cost in place for this 2400 S.F. will depend on how many uses are realized during the life of the equipment.

R015433-10 Contractor Equipment

Rental Rates shown elsewhere in the data set pertain to late model high quality machines in excellent working condition, rented from equipment dealers. Rental rates from contractors may be substantially lower than the rental rates from equipment dealers depending upon economic conditions; for older, less productive machines, reduce rates by a maximum of 15%. Any overtime must be added to the base rates. For shift work, rates are lower. Usual rule of thumb is 150% of one shift rate for two shifts; 200% for three shifts.

For periods of less than one week, operated equipment is usually more economical to rent than renting bare equipment and hiring an operator.

Costs to move equipment to a job site (mobilization) or from a job site (demobilization) are not included in rental rates, nor in any Equipment costs on any Unit Price line items or crew listings. These costs can be found elsewhere. If a piece of equipment is already at a job site, it is not appropriate to utilize mob/demob costs in an estimate again.

Rental rates vary throughout the country with larger cities generally having lower rates. Lease plans for new equipment are available for periods in excess of six months with a percentage of payments applying toward purchase.

Rental rates can also be treated as reimbursement costs for contractor-owned equipment. Owned equipment costs include depreciation, loan payments, interest, taxes, insurance, storage, and major repairs.

Monthly rental rates vary from 2% to 5% of the cost of the equipment depending on the anticipated life of the equipment and its wearing parts. Weekly rates are about 1/3 the monthly rates and daily rental rates are about 1/3 the weekly rates.

The hourly operating costs for each piece of equipment include costs to the user such as fuel, oil, lubrication, normal expendables for the equipment, and a percentage of the mechanic's wages chargeable to maintenance. The hourly operating costs listed do not include the operator's wages.

The daily cost for equipment used in the standard crews is figured by dividing the weekly rate by five, then adding eight times the hourly operating cost to give the total daily equipment cost, not including the operator. This figure is in the right hand column of the Equipment listings under Equipment Cost/Day.

Pile Driving rates shown for the pile hammer and extractor do not include leads, cranes, boilers or compressors. Vibratory pile driving requires an added field specialist during set-up and pile driving operation for the electric model. The hydraulic model requires a field specialist for set-up only. Up to 125 reuses of sheet piling are possible using vibratory drivers. For normal conditions, crane capacity for hammer type and size is as follows.

Crane Capacity	Hammer Type and Size		
	Air or Steam	Diesel	Vibratory
25 ton	to 8,750 ft.-lb.		70 H.P.
40 ton	15,000 ft.-lb.	to 32,000 ft.-lb.	170 H.P.
60 ton	25,000 ft.-lb.		300 H.P.
100 ton		112,000 ft.-lb.	

Cranes should be specified for the job by size, building and site characteristics, availability, performance characteristics, and duration of time required.

Backhoes & Shovels rent for about the same as equivalent size cranes but maintenance and operating expenses are higher. The crane operator's rate must be adjusted for high boom heights. Average adjustments: for 150′ boom add 2% per hour; over 185′, add 4% per hour; over 210′, add 6% per hour; over 250′, add 8% per hour and over 295′, add 12% per hour.

Tower Cranes of the climbing or static type have jibs from 50′ to 200′ and capacities at maximum reach range from 4,000 to 14,000 pounds. Lifting capacities increase up to maximum load as the hook radius decreases.

Typical rental rates, based on purchase price, are about 2% to 3% per month.

Erection and dismantling run between 500 and 2000 labor hours. Climbing operation takes 10 labor hours per 20′ climb. Crane dead time is about 5 hours per 40′ climb. If crane is bolted to side of the building add cost of ties and extra mast sections. Climbing cranes have from 80′ to 180′ of mast while static cranes have 80′ to 800′ of mast.

Truck Cranes can be converted to tower cranes by using tower attachments. Mast heights over 400′ have been used.

A single 100′ high material **Hoist and Tower** can be erected and dismantled in about 400 labor hours; a double 100′ high hoist and tower in about 600 labor hours. Erection times for additional heights are 3 and 4 labor hours per vertical foot respectively up to 150′, and 4 to 5 labor hours per vertical foot over 150′ high. A 40′ high portable Buck hoist takes about 160 labor hours to erect and dismantle. Additional heights take 2 labor hours per vertical foot to 80′ and 3 labor hours per vertical foot for the next 100′. Most material hoists do not meet local code requirements for carrying personnel.

A 150′ high **Personnel Hoist** requires about 500 to 800 labor hours to erect and dismantle. Budget erection time at 5 labor hours per vertical foot for all trades. Local code requirements or labor scarcity requiring overtime can add up to 50% to any of the above erection costs.

Earthmoving Equipment: The selection of earthmoving equipment depends upon the type and quantity of material, moisture content, haul distance, haul road, time available, and equipment available. Short haul cut and fill operations may require dozers only, while another operation may require excavators, a fleet of trucks, and spreading and compaction equipment. Stockpiled material and granular material are easily excavated with front end loaders. Scrapers are most economically used with hauls between 300′ and 1-1/2 miles if adequate haul roads can be maintained. Shovels are often used for blasted rock and any material where a vertical face of 8′ or more can be excavated. Special conditions may dictate the use of draglines, clamshells, or backhoes. Spreading and compaction equipment must be matched to the soil characteristics, the compaction required and the rate the fill is being supplied.

R015433-15 Heavy Lifting

Hydraulic Climbing Jacks

The use of hydraulic heavy lift systems is an alternative to conventional type crane equipment. The lifting, lowering, pushing, or pulling mechanism is a hydraulic climbing jack moving on a square steel jackrod from 1-5/8″ to 4″ square, or a steel cable. The jackrod or cable can be vertical or horizontal, stationary or movable, depending on the individual application. When the jackrod is stationary, the climbing jack will climb the rod and push or pull the load along with itself. When the climbing jack is stationary, the jackrod is movable with the load attached to the end and the climbing jack will lift or lower the jackrod with the attached load. The heavy lift system is normally operated by a single control lever located at the hydraulic pump.

The system is flexible in that one or more climbing jacks can be applied wherever a load support point is required, and the rate of lift synchronized.

Economic benefits have been demonstrated on projects such as: erection of ground assembled roofs and floors, complete bridge spans, girders and trusses, towers, chimney liners and steel vessels, storage tanks, and heavy machinery. Other uses are raising and lowering offshore work platforms, caissons, tunnel sections and pipelines.

R015436-50 Mobilization

Costs to move rented construction equipment to a job site from an equipment dealer's or contractor's yard (mobilization) or off the job site (demobilization) are not included in the rental or operating rates, nor in the equipment cost on a unit price line or in a crew listing. These costs can be found consolidated in the Mobilization section of the data and elsewhere in particular site work sections. If a piece of equipment is already on the job site, it is not appropriate to include mob/demob costs in a new estimate that requires use of that equipment. The following table identifies approximate sizes of rented construction equipment that would be hauled on a towed trailer. Because this listing is not all-encompassing, the user can infer as to what size trailer might be required for a piece of equipment not listed.

3-ton Trailer	20-ton Trailer	40-ton Trailer	50-ton Trailer
20 H.P. Excavator	110 H.P. Excavator	200 H.P. Excavator	270 H.P. Excavator
50 H.P. Skid Steer	165 H.P. Dozer	300 H.P. Dozer	Small Crawler Crane
35 H.P. Roller	150 H.P. Roller	400 H.P. Scraper	500 H.P. Scraper
40 H.P. Trencher	Backhoe	450 H.P. Art. Dump Truck	500 H.P. Art. Dump Truck

R024119-10 Demolition Defined

Whole Building Demolition - Demolition of the whole building with no concern for any particular building element, component, or material type being demolished. This type of demolition is accomplished with large pieces of construction equipment that break up the structure, load it into trucks and haul it to a disposal site, but disposal or dump fees are not included. Demolition of below-grade foundation elements, such as footings, foundation walls, grade beams, slabs on grade, etc., is not included. Certain mechanical equipment containing flammable liquids or ozone-depleting refrigerants, electric lighting elements, communication equipment components, and other building elements may contain hazardous waste, and must be removed, either selectively or carefully, as hazardous waste before the building can be demolished.

Foundation Demolition - Demolition of below-grade foundation footings, foundation walls, grade beams, and slabs on grade. This type of demolition is accomplished by hand or pneumatic hand tools, and does not include saw cutting, or handling, loading, hauling, or disposal of the debris.

Gutting - Removal of building interior finishes and electrical/mechanical systems down to the load-bearing and sub-floor elements of the rough building frame, with no concern for any particular building element, component, or material type being demolished. This type of demolition is accomplished by hand or pneumatic hand tools, and includes loading into trucks, but not hauling, disposal or dump fees, scaffolding, or shoring. Certain mechanical equipment containing flammable liquids or ozone-depleting refrigerants, electric lighting elements, communication equipment components, and other building elements may contain hazardous waste, and must be removed, either selectively or carefully, as hazardous waste, before the building is gutted.

Selective Demolition - Demolition of a selected building element, component, or finish, with some concern for surrounding or adjacent elements, components, or finishes (see the first Subdivision (s) at the beginning of appropriate Divisions). This type of demolition is accomplished by hand or pneumatic hand tools, and does not include handling, loading, storing, hauling, or disposal of the debris, scaffolding, or shoring. "Gutting" methods may be used in order to save time, but damage that is caused to surrounding or adjacent elements, components, or finishes may have to be repaired at a later time.

Careful Removal - Removal of a piece of service equipment, building element or component, or material type, with great concern for both the removed item and surrounding or adjacent elements, components or finishes. The purpose of careful removal may be to protect the removed item for later re-use, preserve a higher salvage value of the removed item, or replace an item while taking care to protect surrounding or adjacent elements, components, connections, or finishes from cosmetic and/or structural damage. An approximation of the time required to perform this type of removal is 1/3 to 1/2 the time it would take to install a new item of like kind. This type of removal is accomplished by hand or pneumatic hand tools, and does not include loading, hauling, or storing the removed item, scaffolding, shoring, or lifting equipment.

Cutout Demolition - Demolition of a small quantity of floor, wall, roof, or other assembly, with concern for the appearance and structural integrity of the surrounding materials. This type of demolition is accomplished by hand or pneumatic hand tools, and does not include saw cutting, handling, loading, hauling, or disposal of debris, scaffolding, or shoring.

Rubbish Handling - Work activities that involve handling, loading or hauling of debris. Generally, the cost of rubbish handling must be added to the cost of all types of demolition, with the exception of whole building demolition.

Minor Site Demolition - Demolition of site elements outside the footprint of a building. This type of demolition is accomplished by hand or pneumatic hand tools, or with larger pieces of construction equipment, and may include loading a removed item onto a truck (check the Crew for equipment used). It does not include saw cutting, hauling or disposal of debris, and, sometimes, handling or loading.

R024119-20 Dumpsters

Dumpster rental costs on construction sites are presented in two ways.

The cost per week rental includes the delivery of the dumpster; its pulling or emptying once per week, and its final removal. The assumption is made that the dumpster contractor could choose to empty a dumpster by simply bringing in an empty unit and removing the full one. These costs also include the disposal of the materials in the dumpster.

The Alternate Pricing can be used when actual planned conditions are not approximated by the weekly numbers. For example, these lines can be used when a dumpster is needed for 4 weeks and will need to be emptied 2 or 3 times per week. Conversely the Alternate Pricing lines can be used when a dumpster will be rented for several weeks or months but needs to be emptied only a few times over this period.

R024119-30 Rubbish Handling Chutes

To correctly estimate the cost of rubbish handling chute systems, the individual components must be priced separately. First choose the size of the system; a 30-inch diameter chute is quite common, but the sizes range from 18 to 36 inches in diameter. The 30-inch chute comes in a standard weight and two thinner weights. The thinner weight chutes are sometimes chosen for cost savings, but they are more easily damaged.

There are several types of major chute pieces that make up the chute system. The first component to consider is the top chute section (top intake hopper) where the material is dropped into the chute at the highest point. After determining the top chute, the intermediate chute pieces called the regular chute sections are priced. Next, the number of chute control door sections (intermediate intake hoppers) must be determined. In the more complex systems, a chute control door section is provided at each floor level. The last major component to consider is bolt down frames; these are usually provided at every other floor level.

There are a number of accessories to consider for safe operation and control. There are covers for the top chute and the chute door sections. The top chute can have a trough that allows for better loading of the chute. For the safest operation, a chute warning light system can be added that will warn the other chute intake locations not to load while another is being used. There are dust control devices that spray a water mist to keep down the dust as the debris is loaded into a dumpster. There are special breakaway cords that are used to prevent damage to the chute if the Dumpster is removed without disconnecting from the chute. There are chute liners that can be installed to protect the chute structure from physical damage from rough abrasive materials. Warning signs can be posted at each floor level that is provided with a chute control door section.

In summary, a complete rubbish handling chute system will include one top section, several intermediate regular sections, several intermediate control door (intake hopper) sections and bolt down frames at every other floor level starting with the top floor. If so desired, the system can also include covers and a light warning system for a safer operation. The bottom of the chute should always be above the Dumpster and should be tied off with a breakaway cord to the Dumpster.

Existing Conditions — R0282 Asbestos Remediation

R028213-20 Asbestos Removal Process

Asbestos removal is accomplished by a specialty contractor who understands the federal and state regulations regarding the handling and disposal of the material. The process of asbestos removal is divided into many individual steps. An accurate estimate can be calculated only after all the steps have been priced.

The steps are generally as follows:

1. Obtain an asbestos abatement plan from an industrial hygienist.
2. Monitor the air quality in and around the removal area and along the path of travel between the removal area and transport area. This establishes the background contamination.
3. Construct a two part decontamination chamber at entrance to removal area.
4. Install a HEPA filter to create a negative pressure in the removal area.
5. Install wall, floor and ceiling protection as required by the plan, usually 2 layers of fireproof 6 mil polyethylene.
6. Industrial hygienist visually inspects work area to verify compliance with plan.
7. Provide temporary supports for conduit and piping affected by the removal process.
8. Proceed with asbestos removal and bagging process. Monitor air quality as described in Step #2. Discontinue operations when contaminate levels exceed applicable standards.
9. Document the legal disposal of materials in accordance with EPA standards.
10. Thoroughly clean removal area including all ledges, crevices and surfaces.
11. Post abatement inspection by industrial hygienist to verify plan compliance.
12. Provide a certificate from a licensed industrial hygienist attesting that contaminate levels are within acceptable standards before returning area to regular use.

Existing Conditions — R0283 Lead Remediation

R028319-60 Lead Paint Remediation Methods

Lead paint remediation can be accomplished by the following methods.

1. Abrasive blast
2. Chemical stripping
3. Power tool cleaning with vacuum collection system
4. Encapsulation
5. Remove and replace
6. Enclosure

Each of these methods has strengths and weaknesses depending on the specific circumstances of the project. The following is an overview of each method.

1. **Abrasive blasting** is usually accomplished with sand or recyclable metallic blast. Before work can begin, the area must be contained to ensure the blast material with lead does not escape to the atmosphere. The use of vacuum blast greatly reduces the containment requirements. Lead abatement equipment that may be associated with this work includes a negative air machine. In addition, it is necessary to have an industrial hygienist monitor the project on a continual basis. When the work is complete, the spent blast sand with lead must be disposed of as a hazardous material. If metallic shot was used, the lead is separated from the shot and disposed of as hazardous material. Worker protection includes disposable clothing and respiratory protection.
2. **Chemical stripping** requires strong chemicals to be applied to the surface to remove the lead paint. Before the work can begin, the area under/adjacent to the work area must be covered to catch the chemical and removed lead. After the chemical is applied to the painted surface it is usually covered with paper. The chemical is left in place for the specified period, then the paper with lead paint is pulled or scraped off. The process may require several chemical applications. The paper with chemicals and lead paint adhered to it, plus the containment and loose scrapings collected by a HEPA (High Efficiency Particulate Air Filter) vac, must be disposed of as hazardous material. The chemical stripping process usually requires a neutralizing agent and several wash downs after the paint is removed. Worker protection includes a neoprene or other compatible protective clothing and respiratory protection with face shield. An industrial hygienist is required intermittently during the process.
3. **Power tool cleaning** is accomplished using shrouded needle blasting guns. The shrouding with different end configurations is held up against the surface to be cleaned. The area is blasted with hardened needles and the shroud captures the lead with a HEPA vac and deposits it in a holding tank. An industrial hygienist monitors the project. Protective clothing and a respirator are required until air samples prove otherwise. When the work is complete the lead must be disposed of as hazardous material.
4. **Encapsulation** is a method that leaves the well bonded lead paint in place after the peeling paint has been removed. Before the work can begin, the area under/adjacent to the work must be covered to catch the scrapings. The scraped surface is then washed with a detergent and rinsed. The prepared surface is covered with approximately 10 mils of paint. A reinforcing fabric can also be embedded in the paint covering. The scraped paint and containment must be disposed of as hazardous material. Workers must wear protective clothing and respirators.
5. **Removing and replacing** are effective ways to remove lead paint from windows, gypsum walls and concrete masonry surfaces. The painted materials are removed and new materials are installed. Workers should wear a respirator and tyvek suit. The demolished materials must be disposed of as hazardous waste if it fails the TCLP (Toxicity Characteristic Leachate Process) test.
6. **Enclosure** is the process that permanently seals lead painted materials in place. This process has many applications such as covering lead painted drywall with new drywall, covering exterior construction with tyvek paper then residing, or covering lead painted structural members with aluminum or plastic. The seams on all enclosing materials must be securely sealed. An industrial hygienist monitors the project, and protective clothing and a respirator are required until air samples prove otherwise.

All the processes require clearance monitoring and wipe testing as required by the hygienist.

R031113-10 Wall Form Materials

Aluminum Forms

Approximate weight is 3 lbs. per S.F.C.A. Standard widths are available from 4″ to 36″ with 36″ most common. Standard lengths of 2′, 4′, 6′ to 8′ are available. Forms are lightweight and fewer ties are needed with the wider widths. The form face is either smooth or textured.

Metal Framed Plywood Forms

Manufacturers claim over 75 reuses of plywood and over 300 reuses of steel frames. Many specials such as corners, fillers, pilasters, etc. are available. Monthly rental is generally about 15% of purchase price for first month and 9% per month thereafter with 90% of rental applied to purchase for the first month and decreasing percentages thereafter. Aluminum framed forms cost 25% to 30% more than steel framed.

After the first month, extra days may be prorated from the monthly charge. Rental rates do not include ties, accessories, cleaning, loss of hardware or freight in and out. Approximate weight is 5 lbs. per S.F. for steel; 3 lbs. per S.F. for aluminum.

Forms can be rented with option to buy.

Plywood Forms, Job Fabricated

There are two types of plywood used for concrete forms.

1. Exterior plyform which is completely waterproof. This is face oiled to facilitate stripping. Ten reuses can be expected with this type with 25 reuses possible.
2. An overlaid type consists of a resin fiber fused to exterior plyform. No oiling is required except to facilitate cleaning. This is available in both high density (HDO) and medium density overlaid (MDO). Using HDO, 50 reuses can be expected with 200 possible.

Plyform is available in 5/8″ and 3/4″ thickness. High density overlaid is available in 3/8″, 1/2″, 5/8″ and 3/4″ thickness.

5/8″ thick is sufficient for most building forms, while 3/4″ is best on heavy construction.

Plywood Forms, Modular, Prefabricated

There are many plywood forming systems without frames. Most of these are manufactured from 1-1/8″ (HDO) plywood and have some hardware attached. These are used principally for foundation walls 8′ or less high. With care and maintenance, 100 reuses can be attained with decreasing quality of surface finish.

Steel Forms

Approximate weight is 6-1/2 lbs. per S.F.C.A. including accessories. Standard widths are available from 2″ to 24″, with 24″ most common. Standard lengths are from 2′ to 8′, with 4′ the most common. Forms are easily ganged into modular units.

Forms are usually leased for 15% of the purchase price per month prorated daily over 30 days.

Rental may be applied to sale price, and usually rental forms are bought. With careful handling and cleaning 200 to 400 reuses are possible.

Straight wall gang forms up to 12′ x 20′ or 8′ x 30′ can be fabricated. These crane handled forms usually lease for approx. 9% per month.

Individual job analysis is available from the manufacturer at no charge.

R031113-40 Forms for Reinforced Concrete

Design Economy

Avoid many sizes in proportioning beams and columns.

From story to story avoid changing column dimensions. Gain strength by adding steel or using a richer mix. If a change in size of column is necessary, vary one dimension only to minimize form alterations. Keep beams and columns the same width.

From floor to floor in a multi-story building vary beam depth, not width, as that will leave the slab panel form unchanged. It is cheaper to vary the strength of a beam from floor to floor by means of a steel area than by 2″ changes in either width or depth.

Cost Factors

Material includes the cost of lumber, cost of rent for metal pans or forms if used, nails, form ties, form oil, bolts and accessories.

Labor includes the cost of carpenters to make up, erect, remove and repair, plus common labor to clean and move. Having carpenters remove forms minimizes repairs.

Improper alignment and condition of forms will increase finishing cost. When forms are heavily oiled, concrete surfaces must be neutralized before finishing. Special curing compounds will cause spillages to spall off in first frost. Gang forming methods will reduce costs on large projects.

Materials Used

Boards are seldom used unless their architectural finish is required. Generally, steel, fiberglass and plywood are used for contact surfaces. Labor on plywood is 10% less than with boards. The plywood is backed up with 2 x 4′s at 12″ to 32″ O.C. Walers are generally 2 - 2 x 4′s. Column forms are held together with steel yokes or bands. Shoring is with adjustable shoring or scaffolding for high ceilings.

Reuse

Floor and column forms can be reused four or possibly five times without excessive repair. Remember to allow for 10% waste on each reuse.

When modular sized wall forms are made, up to twenty uses can be expected with exterior plyform.

When forms are reused, the cost to erect, strip, clean and move will not be affected. 10% replacement of lumber should be included and about one hour of carpenter time for repairs on each reuse per 100 S.F.

The reuse cost for certain accessory items normally rented on a monthly basis will be lower than the cost for the first use.

After the fifth use, new material required plus time needed for repair prevent the form cost from dropping further; it may go up. Much depends on care in stripping, the number of special bays, changes in beam or column sizes and other factors.

Costs for multiple use of formwork may be developed as follows:

2 Uses	3 Uses	4 Uses
$\frac{\text{(1st Use + Reuse)}}{2}$ = avg. cost/2 uses	$\frac{\text{(1st Use + 2 Reuses)}}{3}$ = avg. cost/3 uses	$\frac{\text{(1st use + 3 Reuses)}}{4}$ = avg. cost/4 uses

R031113-60 Formwork Labor-Hours

Item	Hours Required				Total Hours	Multiple Use		
	Unit	Fabricate	Erect & Strip	Clean & Move	1 Use	2 Use	3 Use	4 Use
Beam and Girder, interior beams, 12″ wide	100 S.F.	6.4	8.3	1.3	16.0	13.3	12.4	12.0
Hung from steel beams		5.8	7.7	1.3	14.8	12.4	11.6	11.2
Beam sides only, 36″ high		5.8	7.2	1.3	14.3	11.9	11.1	10.7
Beam bottoms only, 24″ wide		6.6	13.0	1.3	20.9	18.1	17.2	16.7
Box out for openings		9.9	10.0	1.1	21.0	16.6	15.1	14.3
Buttress forms, to 8′ high		6.0	6.5	1.2	13.7	11.2	10.4	10.0
Centering, steel, 3/4″ rib lath			1.0		1.0			
3/8″ rib lath or slab form			0.9		0.9			
Chamfer strip or keyway	100 L.F.		1.5		1.5	1.5	1.5	1.5
Columns, fiber tube 8″ diameter			20.6		20.6			
12″			21.3		21.3			
16″			22.9		22.9			
20″			23.7		23.7			
24″			24.6		24.6			
30″			25.6		25.6			
Columns, round steel, 12″ diameter			22.0		22.0	22.0	22.0	22.0
16″			25.6		25.6	25.6	25.6	25.6
20″			30.5		30.5	30.5	30.5	30.5
24″			37.7		37.7	37.7	37.7	37.7
Columns, plywood 8″ x 8″	100 S.F.	7.0	11.0	1.2	19.2	16.2	15.2	14.7
12″ x 12″		6.0	10.5	1.2	17.7	15.2	14.4	14.0
16″ x 16″		5.9	10.0	1.2	17.1	14.7	13.8	13.4
24″ x 24″		5.8	9.8	1.2	16.8	14.4	13.6	13.2
Columns, steel framed plywood 8″ x 8″			10.0	1.0	11.0	11.0	11.0	11.0
12″ x 12″			9.3	1.0	10.3	10.3	10.3	10.3
16″ x 16″			8.5	1.0	9.5	9.5	9.5	9.5
24″ x 24″			7.8	1.0	8.8	8.8	8.8	8.8
Drop head forms, plywood		9.0	12.5	1.5	23.0	19.0	17.7	17.0
Coping forms		8.5	15.0	1.5	25.0	21.3	20.0	19.4
Culvert, box			14.5	4.3	18.8	18.8	18.8	18.8
Curb forms, 6″ to 12″ high, on grade		5.0	8.5	1.2	14.7	12.7	12.1	11.7
On elevated slabs		6.0	10.8	1.2	18.0	15.5	14.7	14.3
Edge forms to 6″ high, on grade	100 L.F.	2.0	3.5	0.6	6.1	5.6	5.4	5.3
7″ to 12″ high	100 S.F.	2.5	5.0	1.0	8.5	7.8	7.5	7.4
Equipment foundations		10.0	18.0	2.0	30.0	25.5	24.0	23.3
Flat slabs, including drops		3.5	6.0	1.2	10.7	9.5	9.0	8.8
Hung from steel		3.0	5.5	1.2	9.7	8.7	8.4	8.2
Closed deck for domes		3.0	5.8	1.2	10.0	9.0	8.7	8.5
Open deck for pans		2.2	5.3	1.0	8.5	7.9	7.7	7.6
Footings, continuous, 12″ high		3.5	3.5	1.5	8.5	7.3	6.8	6.6
Spread, 12″ high		4.7	4.2	1.6	10.5	8.7	8.0	7.7
Pile caps, square or rectangular		4.5	5.0	1.5	11.0	9.3	8.7	8.4
Grade beams, 24″ deep		2.5	5.3	1.2	9.0	8.3	8.0	7.9
Lintel or Sill forms		8.0	17.0	2.0	27.0	23.5	22.3	21.8
Spandrel beams, 12″ wide		9.0	11.2	1.3	21.5	17.5	16.2	15.5
Stairs			25.0	4.0	29.0	29.0	29.0	29.0
Trench forms in floor		4.5	14.0	1.5	20.0	18.3	17.7	17.4
Walls, Plywood, at grade, to 8′ high		5.0	6.5	1.5	13.0	11.0	9.7	9.5
8′ to 16′		7.5	8.0	1.5	17.0	13.8	12.7	12.1
16′ to 20′		9.0	10.0	1.5	20.5	16.5	15.2	14.5
Foundation walls, to 8′ high		4.5	6.5	1.0	12.0	10.3	9.7	9.4
8′ to 16′ high		5.5	7.5	1.0	14.0	11.8	11.0	10.6
Retaining wall to 12′ high, battered		6.0	8.5	1.5	16.0	13.5	12.7	12.3
Radial walls to 12′ high, smooth		8.0	9.5	2.0	19.5	16.0	14.8	14.3
2′ chords		7.0	8.0	1.5	16.5	13.5	12.5	12.0
Prefabricated modular, to 8′ high		—	4.3	1.0	5.3	5.3	5.3	5.3
Steel, to 8′ high		—	6.8	1.2	8.0	8.0	8.0	8.0
8′ to 16′ high		—	9.1	1.5	10.6	10.3	10.2	10.2
Steel framed plywood to 8′ high		—	6.8	1.2	8.0	7.5	7.3	7.2
8′ to 16′ high		—	9.3	1.2	10.5	9.5	9.2	9.0

R032110-10 Reinforcing Steel Weights and Measures

Bar Designation No.**	Nominal Weight Lb./Ft.	U.S. Customary Units Nominal Dimensions*			SI Units Nominal Dimensions*			
		Diameter in.	Cross Sectional Area, in.^{2}	Perimeter in.	Nominal Weight kg/m	Diameter mm	Cross Sectional Area, cm^2	Perimeter mm
3	.376	.375	.11	1.178	.560	9.52	.71	29.9
4	.668	.500	.20	1.571	.994	12.70	1.29	39.9
5	1.043	.625	.31	1.963	1.552	15.88	2.00	49.9
6	1.502	.750	.44	2.356	2.235	19.05	2.84	59.8
7	2.044	.875	.60	2.749	3.042	22.22	3.87	69.8
8	2.670	1.000	.79	3.142	3.973	25.40	5.10	79.8
9	3.400	1.128	1.00	3.544	5.059	28.65	6.45	90.0
10	4.303	1.270	1.27	3.990	6.403	32.26	8.19	101.4
11	5.313	1.410	1.56	4.430	7.906	35.81	10.06	112.5
14	7.650	1.693	2.25	5.320	11.384	43.00	14.52	135.1
18	13.600	2.257	4.00	7.090	20.238	57.33	25.81	180.1

* The nominal dimensions of a deformed bar are equivalent to those of a plain round bar having the same weight per foot as the deformed bar.
** Bar numbers are based on the number of eighths of an inch included in the nominal diameter of the bars.

R032110-20 Metric Rebar Specification - ASTM A615-81

Grade 300 (300 MPa* = 43,560 psi; +8.7% vs. Grade 40)				
Grade 400 (400 MPa* = 58,000 psi; –3.4% vs. Grade 60)				
Bar No.	Diameter mm	Area mm^2	Equivalent in.^{2}	Comparison with U.S. Customary Bars
10M	11.3	100	.16	Between #3 & #4
15M	16.0	200	.31	#5 (.31 in.^{2})
20M	19.5	300	.47	#6 (.44 in.^{2})
25M	25.2	500	.78	#8 (.79 in.^{2})
30M	29.9	700	1.09	#9 (1.00 in.^{2})
35M	35.7	1000	1.55	#11 (1.56 in.^{2})
45M	43.7	1500	2.33	#14 (2.25 in.^{2})
55M	56.4	2500	3.88	#18 (4.00 in.^{2})

* MPa = megapascals

R032110-40 Weight of Steel Reinforcing Per Square Foot of Wall (PSF)

Reinforced Weights: The table below suggests the weights per square foot for reinforcing steel in walls. Weights are approximate and will be the same for all grades of steel bars. For bars in two directions, add weights for each size and spacing.

C/C Spacing in Inches	Bar Size #3 Wt. (PSF)	#4 Wt. (PSF)	#5 Wt. (PSF)	#6 Wt. (PSF)	#7 Wt. (PSF)	#8 Wt. (PSF)	#9 Wt. (PSF)	#10 Wt. (PSF)	#11 Wt. (PSF)
2″	2.26	4.01	6.26	9.01	12.27				
3″	1.50	2.67	4.17	6.01	8.18	10.68	13.60	17.21	21.25
4″	1.13	2.01	3.13	4.51	6.13	8.10	10.20	12.91	15.94
5″	.90	1.60	2.50	3.60	4.91	6.41	8.16	10.33	12.75
6″	.752	1.34	2.09	3.00	4.09	5.34	6.80	8.61	10.63
8″	.564	1.00	1.57	2.25	3.07	4.01	5.10	6.46	7.97
10″	.451	.802	1.25	1.80	2.45	3.20	4.08	5.16	6.38
12″	.376	.668	1.04	1.50	2.04	2.67	3.40	4.30	5.31
18″	.251	.445	.695	1.00	1.32	1.78	2.27	2.86	3.54
24″	.188	.334	.522	.751	1.02	1.34	1.70	2.15	2.66
30″	.150	.267	.417	.600	.817	1.07	1.36	1.72	2.13
36″	.125	.223	.348	.501	.681	.890	1.13	1.43	1.77
42″	.107	.191	.298	.429	.584	.753	.97	1.17	1.52
48″	.094	.167	.261	.376	.511	.668	.85	1.08	1.33

R032110-50 Minimum Wall Reinforcement Weight (PSF)

This table lists the approximate minimum wall reinforcement weights per S.F. according to the specification of .12% of gross area for vertical bars and .20% of gross area for horizontal bars.

Location	Wall Thickness	Bar Size	Horizontal Steel Spacing C/C	Sq. In. Req'd per S.F.	Total Wt. per S.F.	Bar Size	Vertical Steel Spacing C/C	Sq. In. Req'd per S.F.	Total Wt. per S.F.	Horizontal & Vertical Steel Total Weight per S.F.
Both Faces	10″	#4	18″	.24	.89#	#3	18″	.14	.50#	1.39#
	12″	#4	16″	.29	1.00	#3	16″	.17	.60	1.60
	14″	#4	14″	.34	1.14	#3	13″	.20	.69	1.84
	16″	#4	12″	.38	1.34	#3	11″	.23	.82	2.16
	18″	#5	17″	.43	1.47	#4	18″	.26	.89	2.36
One Face	6″	#3	9″	.15	.50	#3	18″	.09	.25	.75
	8″	#4	12″	.19	.67	#3	11″	.12	.41	1.08
	10″	#5	15″	.24	.83	#4	16″	.14	.50	1.34

R032110-70 Bend, Place and Tie Reinforcing

Placing and tying by rodmen for footings and slabs run from nine hrs. per ton for heavy bars to fifteen hrs. per ton for light bars. For beams, columns, and walls, production runs from eight hrs. per ton for heavy bars to twenty hrs. per ton for light bars. The overall average for typical reinforced concrete buildings is about fourteen hrs. per ton. These production figures include the time for placing accessories and usual inserts, but not their material cost (allow 15% of the cost of delivered bent rods). Equipment handling is necessary for the larger-sized bars so that installation costs for the very heavy bars will not decrease proportionately.

Installation costs for splicing reinforcing bars include allowance for equipment to hold the bars in place while splicing as well as necessary scaffolding for iron workers.

R032110-80 Shop-Fabricated Reinforcing Steel

The material prices for reinforcing, shown in the unit cost sections of the data set, are for 50 tons or more of shop-fabricated reinforcing steel and include:

1. Mill base price of reinforcing steel
2. Mill grade/size/length extras
3. Mill delivery to the fabrication shop
4. Shop storage and handling
5. Shop drafting/detailing
6. Shop shearing and bending
7. Shop listing
8. Shop delivery to the job site

Both material and installation costs can be considerably higher for small jobs consisting primarily of smaller bars, while material costs may be slightly lower for larger jobs.

R032205-30 Common Stock Styles of Welded Wire Fabric

This table provides some of the basic specifications, sizes, and weights of welded wire fabric used for reinforcing concrete.

	New Designation	Old Designation		Steel Area per Foot				Approximate Weight per 100 S.F.	
	Spacing — Cross Sectional Area (in.) — (Sq. in. 100)	Spacing — Wire Gauge (in.) — (AS & W)		Longitudinal		Transverse			
				in.	cm	in.	cm	lbs	kg
Rolls	6 x 6 — W1.4 x W1.4	6 x 6 — 10 x 10		.028	.071	.028	.071	21	9.53
	6 x 6 — W2.0 x W2.0	6 x 6 — 8 x 8	1	.040	.102	.040	.102	29	13.15
	6 x 6 — W2.9 x W2.9	6 x 6 — 6 x 6		.058	.147	.058	.147	42	19.05
	6 x 6 — W4.0 x W4.0	6 x 6 — 4 x 4		.080	.203	.080	.203	58	26.91
	4 x 4 — W1.4 x W1.4	4 x 4 — 10 x 10		.042	.107	.042	.107	31	14.06
	4 x 4 — W2.0 x W2.0	4 x 4 — 8 x 8	1	.060	.152	.060	.152	43	19.50
	4 x 4 — W2.9 x W2.9	4 x 4 — 6 x 6		.087	.227	.087	.227	62	28.12
	4 x 4 — W4.0 x W4.0	4 x 4 — 4 x 4		.120	.305	.120	.305	85	38.56
Sheets	6 x 6 — W2.9 x W2.9	6 x 6 — 6 x 6		.058	.147	.058	.147	42	19.05
	6 x 6 — W4.0 x W4.0	6 x 6 — 4 x 4		.080	.203	.080	.203	58	26.31
	6 x 6 — W5.5 x W5.5	6 x 6 — 2 x 2	2	.110	.279	.110	.279	80	36.29
	4 x 4 — W1.4 x W1.4	4 x 4 — 4 x 4		.120	.305	.120	.305	85	38.56

NOTES: 1. Exact W—number size for 8 gauge is W2.1
2. Exact W—number size for 2 gauge is W5.4

The above table was compiled with the following excerpts from the WRI Manual of Standard Practices, 7th Edition, Copyright 2006. Reproduced with permission of the Wire Reinforcement Institute, Inc.:

1. Chapter 3, page 7, Table 1 Common Styles of Metric Wire Reinforcement (WWR) With Equivalent US Customary Units
2. Chapter 6, page 19, Table 5 Customary Units
3. Chapter 6, Page 23, Table 7 Customary Units (in.) Welded Plain Wire Reinforcement
4. Chapter 6, Page 25 Table 8 Wire Size Comparison
5. Chapter 9, Page 30, Table 9 Weight of Longitudinal Wires Weight (Mass) Estimating Tables
6. Chapter 9, Page 31, Table 9M Weight of Longitudinal Wires Weight (Mass) Estimating Tables
7. Chapter 9, Page 32, Table 10 Weight of Transverse Wires Based on 62″ lengths of transverse wire (60″ width plus 1″ overhand each side)
8. Chapter 9, Page 33, Table 10M Weight of Transverse Wires

R033053-10 Spread Footings

General: A spread footing is used to convert a concentrated load (from one superstructure column or substructure grade beams) into an allowable area load on supporting soil.

Because of punching action from the column load, a spread footing is usually thicker than strip footings which support wall loads. One or two story commercial or residential buildings should have no less than 1′ thick spread footings. Heavier loads require no less than 2′ thick. Spread footings may be square, rectangular or octagonal in plan.

Spread footings tend to minimize excavation and foundation materials, as well as labor and equipment. Another advantage is that footings and soil conditions can be readily examined. They are the most widely used type of footing, especially in mild climates and for buildings of four stories or under. This is because they are usually more economical than other types, if suitable soil and site conditions exist.

They are used when suitable supporting soil is located within several feet of the surface or line of subsurface excavation. Suitable soil types include sands and gravels, gravels with a small amount of clay or silt, hardpan, chalk, and rock. Pedestals may be used to bring the column base load down to the top of the footing. Alternately, undesirable soil between the underside of the footing and the top of the bearing level can be removed and replaced with lean concrete mix or compacted granular material.

Depth of footing should be below topsoil, uncompacted fill, muck, etc. It must be lower than frost penetration but should be above the water table. It must not be at the ground surface because of potential surface erosion. If the ground slopes, approximately three horizontal feet of edge protection must remain. Differential footing elevations may overlap soil stresses or cause excavation problems if clear spacing between footings is less than the difference in depth.

Other footing types are usually used for the following reasons:

A. Bearing capacity of soil is low.
B. Very large footings are required, at a cost disadvantage.
C. Soil under footing (shallow or deep) is very compressible, with probability of causing excessive or differential settlement.
D. Good bearing soil is deep.
E. Potential for scour action exists.
F. Varying subsoil conditions within building perimeter.

Cost of spread footings for a building is determined by:

1. The soil bearing capacity.
2. Typical bay size.
3. Total load (live plus dead) per S.F. for roof and elevated floor levels.
4. The size and shape of the building.
5. Footing configuration. Does the building utilize outer spread footings or are there continuous perimeter footings only or a combination of spread footings plus continuous footings?

Soil Bearing Capacity in Kips per S.F.

Bearing Material	Typical Allowable Bearing Capacity
Hard sound rock	120 KSF
Medium hard rock	80
Hardpan overlaying rock	24
Compact gravel and boulder-gravel; very compact sandy gravel	20
Soft rock	16
Loose gravel; sandy gravel; compact sand; very compact sand-inorganic silt	12
Hard dry consolidated clay	10
Loose coarse to medium sand; medium compact fine sand	8
Compact sand-clay	6
Loose fine sand; medium compact sand-inorganic silts	4
Firm or stiff clay	3
Loose saturated sand-clay; medium soft clay	2

R033053-60 Maximum Depth of Frost Penetration in Inches

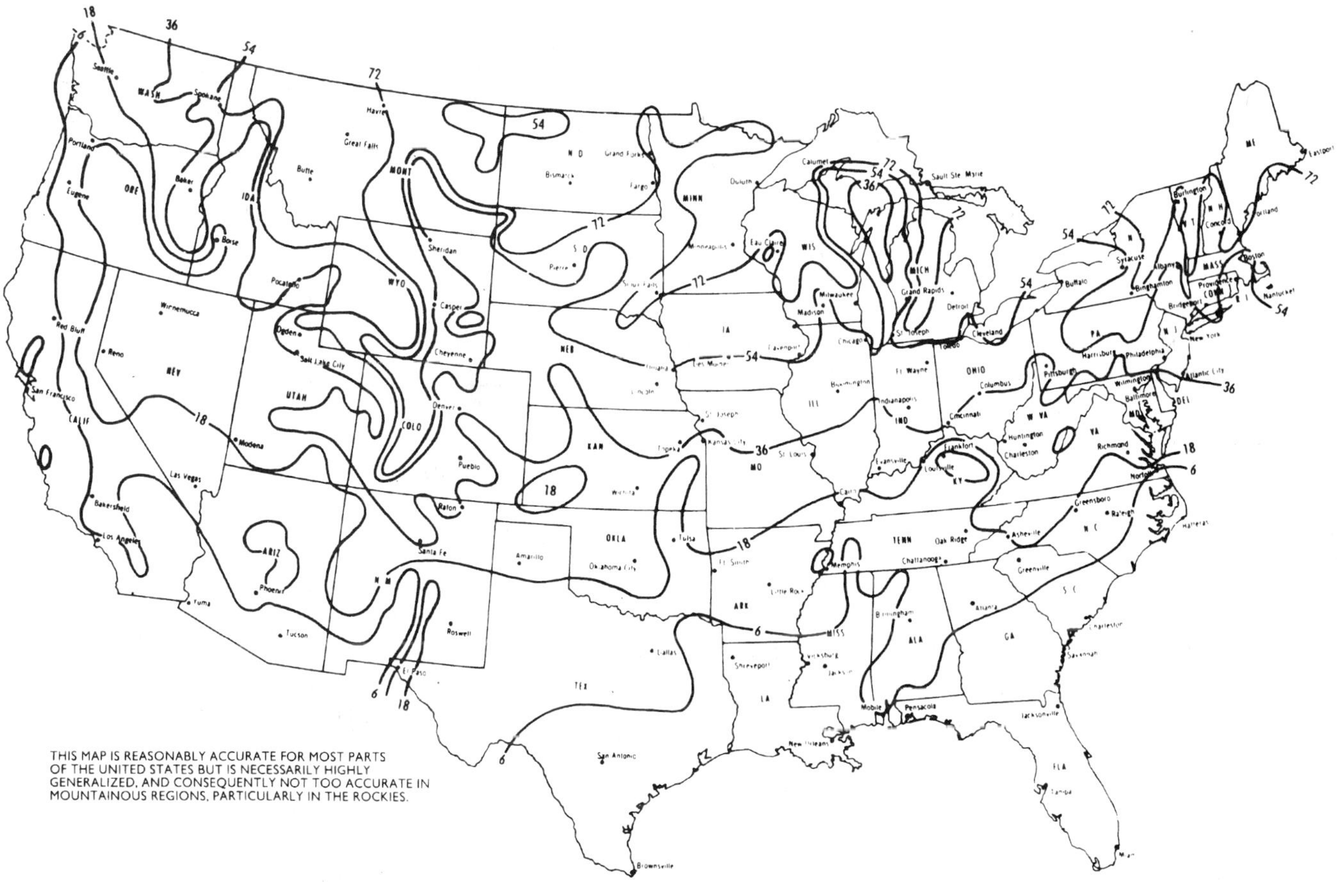

R033105-10 Proportionate Quantities

The tables below show both quantities per S.F. of floor areas as well as form and reinforcing quantities per C.Y. Unusual structural requirements would increase the ratios below. High strength reinforcing would reduce the steel weights. Figures are for 3000 psi concrete and 60,000 psi reinforcing unless specified otherwise.

Type of Construction	Live Load	Span	Per S.F. of Floor Area				Per C.Y. of Concrete		
			Concrete	Forms	Reinf.	Pans	Forms	Reinf.	Pans
Flat Plate	50 psf	15 Ft.	.46 C.F.	1.06 S.F.	1.71lb.		62 S.F.	101 lb.	
		20	.63	1.02	2.40		44	104	
		25	.79	1.02	3.03		35	104	
	100	15	.46	1.04	2.14		61	126	
		20	.71	1.02	2.72		39	104	
		25	.83	1.01	3.47		33	113	
Flat Plate (waffle construction) 20″ domes	50	20	.43	1.00	2.10	.84 S.F.	63	135	53 S.F.
		25	.52	1.00	2.90	.89	52	150	46
		30	.64	1.00	3.70	.87	42	155	37
	100	20	.51	1.00	2.30	.84	53	125	45
		25	.64	1.00	3.20	.83	42	135	35
		30	.76	1.00	4.40	.81	36	160	29
Waffle Construction 30″ domes	50	25	.69	1.06	1.83	.68	42	72	40
		30	.74	1.06	2.39	.69	39	87	39
		35	.86	1.05	2.71	.69	33	85	39
		40	.78	1.00	4.80	.68	35	165	40
Flat Slab (two way with drop panels)	50	20	.62	1.03	2.34		45	102	
		25	.77	1.03	2.99		36	105	
		30	.95	1.03	4.09		29	116	
	100	20	.64	1.03	2.83		43	119	
		25	.79	1.03	3.88		35	133	
		30	.96	1.03	4.66		29	131	
	200	20	.73	1.03	3.03		38	112	
		25	.86	1.03	4.23		32	133	
		30	1.06	1.03	5.30		26	135	
One Way Joists 20″ Pans	50	15	.36	1.04	1.40	.93	78	105	70
		20	.42	1.05	1.80	.94	67	120	60
		25	.47	1.05	2.60	.94	60	150	54
	100	15	.38	1.07	1.90	.93	77	140	66
		20	.44	1.08	2.40	.94	67	150	58
		25	.52	1.07	3.50	.94	55	185	49
One Way Joists 8″ x 16″ filler blocks	50	15	.34	1.06	1.80	.81 Ea.	84	145	64 Ea.
		20	.40	1.08	2.20	.82	73	145	55
		25	.46	1.07	3.20	.83	63	190	49
	100	15	.39	1.07	1.90	.81	74	130	56
		20	.46	1.09	2.80	.82	64	160	48
		25	.53	1.10	3.60	.83	56	190	42
One Way Beam & Slab	50	15	.42	1.30	1.73		84	111	
		20	.51	1.28	2.61		68	138	
		25	.64	1.25	2.78		53	117	
	100	15	.42	1.30	1.90		84	122	
		20	.54	1.35	2.69		68	154	
		25	.69	1.37	3.93		54	145	
	200	15	.44	1.31	2.24		80	137	
		20	.58	1.40	3.30		65	163	
		25	.69	1.42	4.89		53	183	
Two Way Beam & Slab	100	15	.47	1.20	2.26		69	130	
		20	.63	1.29	3.06		55	131	
		25	.83	1.33	3.79		43	123	
	200	15	.49	1.25	2.70		41	149	
		20	.66	1.32	4.04		54	165	
		25	.88	1.32	6.08		41	187	

R033105-10 Proportionate Quantities (cont.)

4000 psi Concrete and 60,000 psi Reinforcing—Form and Reinforcing Quantities per C.Y.

Item	Size	Forms	Reinforcing	Minimum	Maximum
Columns (square tied)	10″ x 10″	130 S.F.C.A.	#5 to #11	220 lbs.	875 lbs.
	12″ x 12″	108	#6 to #14	200	955
	14″ x 14″	92	#7 to #14	190	900
	16″ x 16″	81	#6 to #14	187	1082
	18″ x 18″	72	#6 to #14	170	906
	20″ x 20″	65	#7 to #18	150	1080
	22″ x 22″	59	#8 to #18	153	902
	24″ x 24″	54	#8 to #18	164	884
	26″ x 26″	50	#9 to #18	169	994
	28″ x 28″	46	#9 to #18	147	864
	30″ x 30″	43	#10 to #18	146	983
	32″ x 32″	40	#10 to #18	175	866
	34″ x 34″	38	#10 to #18	157	772
	36″ x 36″	36	#10 to #18	175	852
	38″ x 38″	34	#10 to #18	158	765
	40″ x 40″	32	#10 to #18	143	692

Item	Size	Form	Spiral	Reinforcing	Minimum	Maximum
Columns (spirally reinforced)	12″ diameter	34.5 L.F.	190 lbs.	#4 to #11	165 lbs.	1505 lb.
		34.5	190	#14 & #18	—	1100
	14″	25	170	#4 to #11	150	970
		25	170	#14 & #18	800	1000
	16″	19	160	#4 to #11	160	950
		19	160	#14 & #18	605	1080
	18″	15	150	#4 to #11	160	915
		15	150	#14 & #18	480	1075
	20″	12	130	#4 to #11	155	865
		12	130	#14 & #18	385	1020
	22″	10	125	#4 to #11	165	775
		10	125	#14 & #18	320	995
	24″	9	120	#4 to #11	195	800
		9	120	#14 & #18	290	1150
	26″	7.3	100	#4 to #11	200	729
		7.3	100	#14 & #18	235	1035
	28″	6.3	95	#4 to #11	175	700
		6.3	95	#14 & #18	200	1075
	30″	5.5	90	#4 to #11	180	670
		5.5	90	#14 & #18	175	1015
	32″	4.8	85	#4 to #11	185	615
		4.8	85	#14 & #18	155	955
	34″	4.3	80	#4 to #11	180	600
		4.3	80	#14 & #18	170	855
	36″	3.8	75	#4 to #11	165	570
		3.8	75	#14 & #18	155	865
	40″	3.0	70	#4 to #11	165	500
		3.0	70	#14 & #18	145	765

R033105-10 Proportionate Quantities (cont.)

3000 psi Concrete and 60,000 psi Reinforcing—Form and Reinforcing Quantities per C.Y.

Item	Type	Loading	Height	C.Y./L.F.	Forms/C.Y.	Reinf./C.Y.
Retaining Walls	Cantilever	Level Backfill	4 Ft.	0.2 C.Y.	49 S.F.	35 lbs.
			8	0.5	42	45
			12	0.8	35	70
			16	1.1	32	85
			20	1.6	28	105
		Highway Surcharge	4	0.3	41	35
			8	0.5	36	55
			12	0.8	33	90
			16	1.2	30	120
			20	1.7	27	155
		Railroad Surcharge	4	0.4	28	45
			8	0.8	25	65
			12	1.3	22	90
			16	1.9	20	100
			20	2.6	18	120
	Gravity, with Vertical Face	Level Backfill	4	0.4	37	None
			7	0.6	27	↓
			10	1.2	20	↓
		Sloping Backfill	4	0.3	31	↓
			7	0.8	21	↓
			10	1.6	15	↓

Item	Span	Live Load in Kips per Linear Foot							
		Under 1 Kip		2 to 3 Kips		4 to 5 Kips		6 to 7 Kips	
		Forms	Reinf.	Forms	Reinf.	Forms	Reinf.	Forms	Reinf.
Beams	10 Ft.	—	—	90 S.F.	170 #	85 S.F.	175 #	75 S.F.	185 #
	16	130 S.F.	165 #	85	180	75	180	65	225
	20	110	170	75	185	62	200	51	200
	26	90	170	65	215	62	215	—	—
	30	85	175	60	200	—	—	—	—

Item	Size	Type	Forms per C.Y.	Reinforcing per C.Y.
Spread Footings	Under 1 C.Y.	1,000 psf soil	24 S.F.	44 lbs.
		5,000	24	42
		10,000	24	52
	1 C.Y. to 5 C.Y.	1,000	14	49
		5,000	14	50
		10,000	14	50
	Over 5 C.Y.	1,000	9	54
		5,000	9	52
		10,000	9	56
Pile Caps (30 Ton Concrete Piles)	Under 5 C.Y.	shallow caps	20	65
		medium	20	50
		deep	20	40
	5 C.Y. to 10 C.Y.	shallow	14	55
		medium	15	45
		deep	15	40
	10 C.Y. to 20 C.Y.	shallow	11	60
		medium	11	45
		deep	12	35
	Over 20 C.Y.	shallow	9	60
		medium	9	45
		deep	10	40

R033105-10 Proportionate Quantities (cont.)

3000 psi Concrete and 60,000 psi Reinforcing — Form and Reinforcing Quantities per C.Y.						
Item	**Size**	**Pile Spacing**	**50 T Pile**	**100 T Pile**	**50 T Pile**	**100 T Pile**
Pile Caps (Steel H Piles)	Under 5 C.Y.	24″ O.C.	24 S.F.	24 S.F.	75 lbs.	90 lbs.
		30″	25	25	80	100
		36″	24	24	80	110
	5 C.Y. to 10 C.Y.	24″	15	15	80	110
		30″	15	15	85	110
		36″	15	15	75	90
	Over 10 C.Y.	24″	13	13	85	90
		30″	11	11	85	95
		36″	10	10	85	90

Item		8″ Thick		10″ Thick		12″ Thick		15″ Thick	
	Height	**Forms**	**Reinf.**	**Forms**	**Reinf.**	**Forms**	**Reinf.**	**Forms**	**Reinf.**
Basement Walls	7 Ft.	81 S.F.	44 lbs.	65 S.F.	45 lbs.	54 S.F.	44 lbs.	41 S.F.	43 lbs.
	8	↓	44	↓	45	↓	44	↓	43
	9	↓	46	↓	45	↓	44	↓	43
	10	↓	57	↓	45	↓	44	↓	43
	12	↓	83	↓	50	↓	52	↓	43
	14	↓	116	↓	65	↓	64	↓	51
	16	↓		↓	86	↓	90	↓	65
	18	↓		↓		↓	106	↓	70

R033105-20 Materials for One C.Y. of Concrete

This is an approximate method of figuring quantities of cement, sand and coarse aggregate for a field mix with waste allowance included.

With crushed gravel as coarse aggregate, to determine barrels of cement required, divide 10 by total mix; that is, for 1:2:4 mix, 10 divided by 7 = 1-3/7 barrels. If the coarse aggregate is crushed stone, use 10-1/2 instead of 10 as given for gravel.

To determine tons of sand required, multiply barrels of cement by parts of sand and then by 0.2; that is, for the 1:2:4 mix, as above, 1-3/7 x 2 x .2 = .57 tons.

Tons of crushed gravel are in the same ratio to tons of sand as parts in the mix, or 4/2 x .57 = 1.14 tons.

1 bag cement = 94#
4 bags = 1 barrel

1 C.Y. sand or crushed gravel = 2700#
1 ton sand or crushed gravel = 20 C.F.

1 C.Y. crushed stone = 2575#
1 ton crushed stone = 21 C.F.

Average carload of cement is 692 bags; of sand or gravel is 56 tons.

Do not stack stored cement over 10 bags high.

R033105-30 Metric Equivalents of Cement Content for Concrete Mixes

94 Pound Bags per Cubic Yard	Kilograms per Cubic Meter	94 Pound Bags per Cubic Yard	Kilograms per Cubic Meter
1.0	55.77	7.0	390.4
1.5	83.65	7.5	418.3
2.0	111.5	8.0	446.2
2.5	139.4	8.5	474.0
3.0	167.3	9.0	501.9
3.5	195.2	9.5	529.8
4.0	223.1	10.0	557.7
4.5	251.0	10.5	585.6
5.0	278.8	11.0	613.5
5.5	306.7	11.5	641.3
6.0	334.6	12.0	669.2
6.5	362.5	12.5	697.1

a. If you know the cement content in pounds per cubic yard, multiply by .5933 to obtain kilograms per cubic meter.

b. If you know the cement content in 94 pound bags per cubic yard, multiply by 55.77 to obtain kilograms per cubic meter.

R033105-40 Metric Equivalents of Common Concrete Strengths (to convert other psi values to megapascals, multiply by 0.006895)

U.S. Values psi	SI Values Megapascals	Non-SI Metric Values kgf/cm²*
2000	14	140
2500	17	175
3000	21	210
3500	24	245
4000	28	280
4500	31	315
5000	34	350
6000	41	420
7000	48	490
8000	55	560
9000	62	630
10,000	69	705

* kilograms force per square centimeter

R033105-50 Quantities of Cement, Sand and Stone for One C.Y. of Concrete per Various Mixes

This table can be used to determine the quantities of the ingredients for smaller quantities of site mixed concrete.

	Mix = 1:1:1-3/4			Mix = 1:2:2.25			Mix = 1:2.25:3			Mix = 1:3:4		
Concrete (C.Y.)	Cement (sacks)	Sand (C.Y.)	Stone (C.Y.)	Cement (sacks)	Sand (C.Y.)	Stone (C.Y.)	Cement (sacks)	Sand (C.Y.)	Stone (C.Y.)	Cement (sacks)	Sand (C.Y.)	Stone (C.Y.)
1	10	.37	.63	7.75	.56	.65	6.25	.52	.70	5	.56	.74
2	20	.74	1.26	15.50	1.12	1.30	12.50	1.04	1.40	10	1.12	1.48
3	30	1.11	1.89	23.25	1.68	1.95	18.75	1.56	2.10	15	1.68	2.22
4	40	1.48	2.52	31.00	2.24	2.60	25.00	2.08	2.80	20	2.24	2.96
5	50	1.85	3.15	38.75	2.80	3.25	31.25	2.60	3.50	25	2.80	3.70
6	60	2.22	3.78	46.50	3.36	3.90	37.50	3.12	4.20	30	3.36	4.44
7	70	2.59	4.41	54.25	3.92	4.55	43.75	3.64	4.90	35	3.92	5.18
8	80	2.96	5.04	62.00	4.48	5.20	50.00	4.16	5.60	40	4.48	5.92
9	90	3.33	5.67	69.75	5.04	5.85	56.25	4.68	6.30	45	5.04	6.66
10	100	3.70	6.30	77.50	5.60	6.50	62.50	5.20	7.00	50	5.60	7.40
11	110	4.07	6.93	85.25	6.16	7.15	68.75	5.72	7.70	55	6.16	8.14
12	120	4.44	7.56	93.00	6.72	7.80	75.00	6.24	8.40	60	6.72	8.88
13	130	4.82	8.20	100.76	7.28	8.46	81.26	6.76	9.10	65	7.28	9.62
14	140	5.18	8.82	108.50	7.84	9.10	87.50	7.28	9.80	70	7.84	10.36
15	150	5.56	9.46	116.26	8.40	9.76	93.76	7.80	10.50	75	8.40	11.10
16	160	5.92	10.08	124.00	8.96	10.40	100.00	8.32	11.20	80	8.96	11.84
17	170	6.30	10.72	131.76	9.52	11.06	106.26	8.84	11.90	85	9.52	12.58
18	180	6.66	11.34	139.50	10.08	11.70	112.50	9.36	12.60	90	10.08	13.32
19	190	7.04	11.98	147.26	10.64	12.36	118.76	9.84	13.30	95	10.64	14.06
20	200	7.40	12.60	155.00	11.20	13.00	125.00	10.40	14.00	100	11.20	14.80
21	210	7.77	13.23	162.75	11.76	13.65	131.25	10.92	14.70	105	11.76	15.54
22	220	8.14	13.86	170.05	12.32	14.30	137.50	11.44	15.40	110	12.32	16.28
23	230	8.51	14.49	178.25	12.88	14.95	143.75	11.96	16.10	115	12.88	17.02
24	240	8.88	15.12	186.00	13.44	15.60	150.00	12.48	16.80	120	13.44	17.76
25	250	9.25	15.75	193.75	14.00	16.25	156.25	13.00	17.50	125	14.00	18.50
26	260	9.64	16.40	201.52	14.56	16.92	162.52	13.52	18.20	130	14.56	19.24
27	270	10.00	17.00	209.26	15.12	17.56	168.76	14.04	18.90	135	15.02	20.00
28	280	10.36	17.64	217.00	15.68	18.20	175.00	14.56	19.60	140	15.68	20.72
29	290	10.74	18.28	224.76	16.24	18.86	181.26	15.08	20.30	145	16.24	21.46

R033105-70 Placing Ready-Mixed Concrete

For ground pours allow for 5% waste when figuring quantities.

Prices in the front of the data set assume normal deliveries. If deliveries are made before 8 A.M. or after 5 P.M. or on Saturday afternoons add 30%. Negotiated discounts for large volumes are not included in prices in front of the data set.

For the lower floors without truck access, concrete may be wheeled in rubber-tired buggies, conveyer handled, crane handled or pumped. Pumping is economical if there is top steel. Conveyers are more efficient for thick slabs.

At higher floors the rubber-tired buggies may be hoisted by a hoisting tower and wheeled to the location. Placement by a conveyer is limited to three floors and is best for high-volume pours. Pumped concrete is best when the building has no crane access. Concrete may be pumped directly as high as thirty-six stories using special pumping techniques. Normal maximum height is about fifteen stories. The b

est pumping aggregate is screened and graded bank gravel rather than crushed stone.

Pumping downward is more difficult than pumping upward. The horizontal distance from pump to pour may increase preparation time prior to pour. Placing by cranes, either mobile, climbing or tower types, continues as the most efficient method for high-rise concrete buildings.

R033105-80 Slab on Grade

General: Ground slabs are classified on the basis of use. Thickness is generally controlled by the heaviest concentrated load supported. If load area is greater than 80 sq. in., soil bearing may be important. The base granular fill must be a uniformly compacted material of limited capillarity, such as gravel or crushed rock. Concrete is placed on this surface of the vapor barrier on top of base.

Ground slabs are either single or two course floors. Single course are widely used. Two course floors have a subsequent wear resistant topping.

Reinforcement is provided to maintain tightly closed cracks.

Control joints limit crack locations and provide for differential horizontal movement only. Isolation joints allow both horizontal and vertical differential movement.

Use of Table: Determine appropriate type of slab (A, B, C, or D) by considering type of use or amount of abrasive wear of traffic type.

Determine thickness by maximum allowable wheel load or uniform load, opposite 1st column, thickness. Increase the controlling thickness if details require, and select either plain or reinforced slab thickness and type.

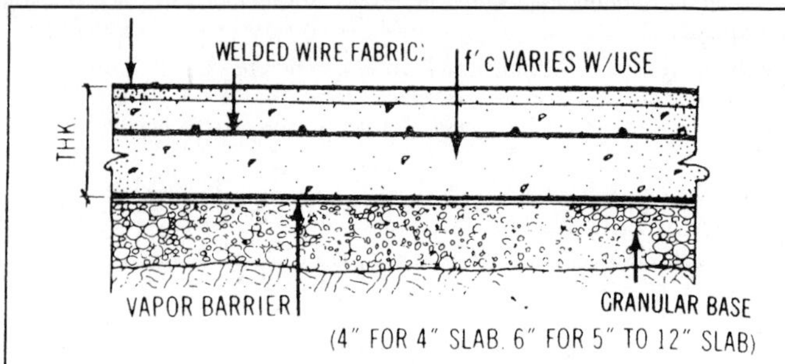

Slab on Grade

Thickness and Loading Assumptions by Type of Use

		A	B	C	D	◄ Slab I.D.
SLAB THICKNESS (IN.)	TYPE	Non	Light	Normal	Heavy	◄ Industrial
		Little	Light	Moderate	Severe	◄ Abrasion
		Foot Only	Pneumatic Wheels	Solid Rubber Wheels	Steel Tires	◄ Type of Traffic
		Load* (K)	Load* (K)	Load* (K)	Load* (K)	Max. Uniform Load to Slab ▼ (PSF)
4″	Reinf. Plain	4K				100
5″	Reinf. Plain	6K	4K			200
6″	Reinf. Plain		8K	6K	6K	500 to 800
7″	Reinf. Plain			9K	8K	1,500
8″	Reinf. Plain				11K	* Max. Wheel Load in Kips (incl. impact)
10″	Reinf. Plain				14K	
12″	Reinf. Plain					
DESIGN ASSUMPTIONS	Concrete, Chuted	f'c = 3.5 KSI	4 KSI	4.5 KSI	Slab @ 3.5 KSI	ASSUMPTIONS BY SLAB TYPE
	Toppings			1″ Integral	1″ Bonded	
	Finish	Steel Trowel	Steel Trowel	Steel Trowel	Screed & Steel Trowel	
	Compacted Granular Base	4″ deep for 4″ slab thickness 6″ deep for 5″ slab thickness & greater				ASSUMPTIONS FOR ALL SLAB TYPES
	Vapor Barrier	6 mil polyethylene				
	Forms & Joints	Allowances included				
	Rein-forcement	WWF as required ≥ 60,000 psi				

R033105-85 Lift Slabs

The cost advantage of the lift slab method is due to placing all concrete, reinforcing steel, inserts and electrical conduit at ground level and in reduction of formwork. Minimum economical project size is about 30,000 S.F. Slabs may be tilted for parking garage ramps.

It is now used in all types of buildings and has gone up to 22 stories high in apartment buildings. The current trend is to use post-tensioned flat plate slabs with spans from 22′ to 35′. Cylindrical void forms are used when deep slabs are required. One pound of prestressing steel is about equal to seven pounds of conventional reinforcing.

To be considered cured for stressing and lifting, a slab must have attained 75% of design strength. Seven days are usually sufficient with four to five days possible if high early strength cement is used. Slabs can be stacked using two coats of a non-bonding agent to insure that slabs do not stick to each other. Lifting is done by companies specializing in this work. Lift rate is 5′ to 15′ per hour with an average of 10′ per hour. Total areas up to 33,000 S.F. have been lifted at one time. 24 to 36 jacking columns are common. Most economical bay sizes are 24′ to 28′ with four to fourteen stories most efficient. Continuous design reduces reinforcing steel cost. Use of post-tensioned slabs allows larger bay sizes.

R033543-10 Polished Concrete Floors

A polished concrete floor has a glossy mirror-like appearance and is created by grinding the concrete floor with finer and finer diamond grits, similar to sanding wood, until the desired level of reflective clarity and sheen are achieved. The technical term for this type of polished concrete is bonded abrasive polished concrete. The basic piece of equipment used in the polishing process is a walk-behind planetary grinder for working large floor areas. This grinder drives diamond-impregnated abrasive discs, which progress from coarse- to fine-grit discs.

The process begins with the use of very coarse diamond segments or discs bonded in a metallic matrix. These segments are coarse enough to allow the removal of pits, blemishes, stains, and light coatings from the floor surface in preparation for final smoothing. The condition of the original concrete surface will dictate the grit coarseness of the initial grinding step which will generally end up being a three- to four-step process using ever finer grits. The purpose of this initial grinding step is to remove surface coatings and blemishes and to cut down into the cream for very fine aggregate exposure, or deeper into the fine aggregate layer just below the cream layer, or even deeper into the coarse aggregate layer. These initial grinding steps will progress up to the 100/120 grit. If wet grinding is done, a waste slurry is produced that must be removed between grit changes and disposed of properly. If dry grinding is done, a high performance vacuum will pick up the dust during grinding and collect it in bags which must be disposed of properly.

The process continues with honing the floor in a series of steps that progresses from 100-grit to 400-grit diamond abrasive discs embedded in a plastic or resin matrix. At some point during, or just prior to, the honing step, one or two coats of stain or dye can be sprayed onto the surface to give color to the concrete, and two coats of densifier/hardener must be applied to the floor surface and allowed to dry. This sprayed-on densifier/hardener will penetrate about 1/8″ into the concrete to make the surface harder, denser and more abrasion-resistant.

The process ends with polishing the floor surface in a series of steps that progresses from resin-impregnated 800-grit (medium polish) to 1500-grit (high polish) to 3000-grit (very high polish), depending on the desired level of reflective clarity and sheen.

The Concrete Polishing Association of America (CPAA) has defined the flooring options available when processing concrete to a desired finish. The first category is aggregate exposure, the grinding of a concrete surface with bonded abrasives, in as many abrasive grits necessary, to achieve one of the following classes:

A. Cream – very little surface cut depth; little aggregate exposure

B. Fine aggregate (salt and pepper) – surface cut depth of 1/16″; fine aggregate exposure with little or no medium aggregate exposure at random locations

C. Medium aggregate – surface cut depth of 1/8″; medium aggregate exposure with little or no large aggregate exposure at random locations

D. Large aggregate – surface cut depth of 1/4″; large aggregate exposure with little or no fine aggregate exposure at random locations

The second CPAA defined category is reflective clarity and sheen, the polishing of a concrete surface with the minimum number of bonded abrasives as indicated to achieve one of the following levels:

1. Ground – flat appearance with none to very slight diffused reflection; none to very low reflective sheen; using a minimum total of 4 grit levels up 100-grit
2. Honed – matte appearance with or without slight diffused reflection; low to medium reflective sheen; using a minimum total of 5 grit levels up to 400-grit
3. Semi-polished – objects being reflected are not quite sharp and crisp but can be easily identified; medium to high reflective sheen; using a minimum total of 6 grit levels up to 800-grit
4. Highly-polished – objects being reflected are sharp and crisp as would be seen in a mirror-like reflection; high to highest reflective sheen; using a minimum total of up to 8 grit levels up to 1500-grit or 3000-grit

The CPAA defines reflective clarity as the degree of sharpness and crispness of the reflection of overhead objects when viewed 5′ above and perpendicular to the floor surface. Reflective sheen is the degree of gloss reflected from a surface when viewed at least 20′ from and at an angle to the floor surface. These terms are relatively subjective. The final outcome depends on the internal makeup and surface condition of the original concrete floor, the experience of the floor polishing crew, and the expectations of the owner. Before the grinding, honing, and polishing work commences on the main floor area, it might be beneficial to do a mock-up panel in the same floor but in an out of the way place to demonstrate the sequence of steps with increasingly fine abrasive grits and to demonstrate the final reflective clarity and reflective sheen. This mock-up panel will be within the area of, and part of, the final work.

R034105-30 Prestressed Precast Concrete Structural Units

Type	Location	Depth	Span in Ft.		Live Load Lb. per S.F.
Double Tee 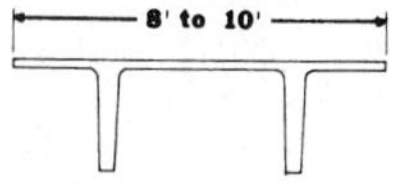	Floor	28″ to 34″	60 to 80		50 to 80
	Roof	12″ to 24″	30 to 50		40
	Wall	Width 8′	Up to 55′ high		Wind
Multiple Tee 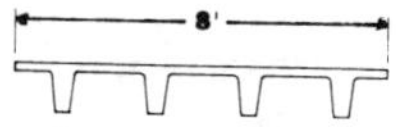	Roof	8″ to 12″	15 to 40		40
	Floor	8″ to 12″	15 to 30		100
Plank 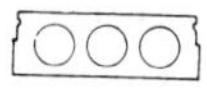 or 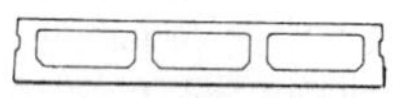	Roof or Floor		Roof	Floor	40 for Roof 100 for Floor
		4″	13	12	
		6″	22	18	
		8″	26	25	
		10″	33	29	
		12″	42	32	
Single Tee 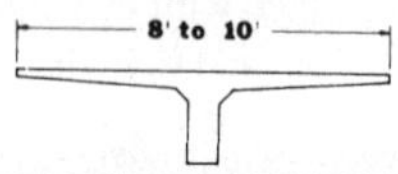	Roof	28″	40		40
		32″	80		
		36″	100		
		48″	120		
AASHO Girder 	Bridges	Type 4	100		Highway
		5	110		
		6	125		
Box Beam 	Bridges	15″ 27″ 33″	40 to 100		Highway

The majority of precast projects today utilize double tees rather than single tees because of speed and ease of installation. As a result casting beds at manufacturing plants are normally formed for double tees. Single tee projects will therefore require an initial set up charge to be spread over the individual single tee costs.

For floors, a 2″ to 3″ topping is field cast over the shapes. For roofs, insulating concrete or rigid insulation is placed over the shapes.

Member lengths up to 40′ are standard haul, 40′ to 60′ require special permits and lengths over 60′ must be escorted. Excessive width and/or length can add up to 100% on hauling costs.

Large heavy members may require two cranes for lifting which would increase erection costs by about 45%. An eight man crew can install 12 to 20 double tees, or 45 to 70 quad tees or planks per day.

Grouting of connections must also be included.

Several system buildings utilizing precast members are available. Heights can go up to 22 stories for apartment buildings. The optimum design ratio is 3 S.F. of surface to 1 S.F. of floor area.

R034136-90 Prestressed Concrete, Post-Tensioned

In post-tensioned concrete the steel tendons are tensioned after the concrete has reached about 3/4 of its ultimate strength. The cableways are grouted after tensioning to provide bond between the steel and concrete. If bond is to be prevented, the tendons are coated with a corrosion-preventative grease and wrapped with waterproofed paper or plastic. Bonded tendons are usually used when ultimate strength (beams & girders) are controlling factors.

High strength concrete is used to fully utilize the steel, thereby reducing the size and weight of the member. A plasticizing agent may be added to reduce water content. Maximum size aggregate ranges from 1/2″ to 1-1/2″ depending on the spacing of the tendons.

The types of steel commonly used are bars and strands. Job conditions determine which is best suited. Bars are best for vertical prestresses since they are easy to support. The trend is for steel manufacturers to supply a finished package, cut to length, which reduces field preparation to a minimum.

Bars vary from 3/4″ to 1-3/8″ diameter. The table below gives time in labor-hours per tendon for placing, tensioning and grouting (if required) a 75′ beam. Tendons used in buildings are not usually grouted; tendons for bridges usually are grouted. For strands the table indicates the labor-hours per pound for typical prestressed units 100′ long. Simple span beams usually require one- end stressing regardless of lengths. Continuous beams are usually stressed from two ends. Long slabs are poured from the center outward and stressed in 75′ increments after the initial 150′ center pour.

Labor Hours per Tendon and per Pound of Prestressed Steel

Length	100′ Beam		75′ Beam		100′ Slab	
Type Steel	**Strand**		**Bars**		**Strand**	
Diameter	**0.5″**		**3/4″**	**1-3/8″**	**0.5″**	**0.6″**
Number	**4**	**12**	**1**	**1**	**1**	**1**
Force in Kips	**100**	**300**	**42**	**143**	**25**	**35**
Preparation & Placing Cables	3.6	7.4	0.9	2.9	0.9	1.1
Stressing Cables	2.0	2.4	0.8	1.6	0.5	0.5
Grouting, if required	2.5	3.0	0.6	1.3		
Total Labor Hours	8.1	12.8	2.3	5.8	1.4	1.6
Prestressing Steel Weights (Lbs.)	215	640	115	380	53	74
Labor-hours per Lb. Bonded	0.038	0.020	0.020	0.015		
Non-bonded					0.026	0.022

Flat slab construction — 4000 psi concrete with span-to-depth ratio between 36 and 44. Two way post-tensioned steel averages 1.0 lb. per S.F. for 24′ to 28′ bays (usually strand) and additional reinforcing steel averages .5 lb. per S.F.

Pan and joist construction — 4000 psi concrete with span-to-depth ratio between 28 to 30. Post-tensioned steel averages .8 lb. per S.F. and reinforcing steel about 1.0 lb. per S.F. Placing and stressing average 40 hours per ton of total material.

Beam construction — 4000 to 5000 psi concrete. Steel weights vary greatly.

Labor cost per pound goes down as the size and length of the tendon increase. The primary economic consideration is the cost per kip for the member.

Post-tensioning becomes feasible for beams and girders over 30′ long; for continuous two-way slabs over 20′ clear; and for transferring upper building loads over longer spans at lower levels. Post-tension suppliers will provide engineering services at no cost to the user. Substantial economies are possible by using post-tensioned lift slabs.

Masonry — R0401 Maintenance of Masonry

R040130-10 Cleaning Face Brick

On smooth brick a person can clean 70 S.F. an hour; on rough brick 50 S.F. per hour. Use one gallon muriatic acid to 20 gallons of water for 1000 S.F. Do not use acid solution until wall is at least seven days old, but a mild soap solution may be used after two days.

Time has been allowed for clean-up in brick prices.

Masonry — R0405 Common Work Results for Masonry

R040513-10 Cement Mortar (material only)

Type N - 1:1:6 mix by volume. Use everywhere above grade except as noted below. - 1:3 mix using conventional masonry cement which saves handling two separate bagged materials.

Type M - 1:1/4:3 mix by volume, or 1 part cement, 1/4 (10% by wt.) lime, 3 parts sand. Use for heavy loads and where earthquakes or hurricanes may occur. Also for reinforced brick, sewers, manholes and everywhere below grade.

Mix Proportions by Volume and Compressive Strength of Mortar

Where Used	Allowable Proportions by Volume					Compressive Strength @ 28 days
	Mortar Type	Portland Cement	Masonry Cement	Hydrated Lime	Masonry Sand	
Plain Masonry		1	1	—	6	
	M	1	—	1/4	3	2500 psi
		1/2	1	—	4	
	S	1	—	1/4 to 1/2	4	1800 psi
		—	1	—	3	
	N	1	—	1/2 to 1-1/4	6	750 psi
		—	1	—	3	
	O	1	—	1-1/4 to 2-1/2	9	350 psi
	K	1	—	2-1/2 to 4	12	75 psi
Reinforced Masonry	PM	1	1	—	6	2500 psi
	PL	1	—	1/4 to 1/2	4	2500 psi

Note: The total aggregate should be between 2.25 to 3 times the sum of the cement and lime used.

The labor cost to mix the mortar is included in the productivity and labor cost of unit price lines in unit cost sections for brickwork, blockwork and stonework.

The material cost of mixed mortar is included in the material cost of those same unit price lines and includes the cost of renting and operating a 10 C.F. mixer at the rate of 200 C.F. per day.

There are two types of mortar color used. One type is the inert additive type with about 100 lbs. per M brick as the typical quantity required. These colors are also available in smaller-batch-sized bags (1 lb. to 15 lb.) which can be placed directly into the mixer without measuring. The other type is premixed and replaces the masonry cement. Dark green color has the highest cost.

R040519-50 Masonry Reinforcing

Horizontal joint reinforcing helps prevent wall cracks where wall movement may occur and in many locations is required by code. Horizontal joint reinforcing is generally not considered to be structural reinforcing and an unreinforced wall may still contain joint reinforcing.

Reinforcing strips come in 10′ and 12′ lengths and in truss and ladder shapes, with and without drips. Field labor runs between 2.7 to 5.3 hours per 1000 L.F. for wall thicknesses up to 12″.

The wire meets ASTM A82 for cold drawn steel wire and the typical size is 9 ga. sides and ties with 3/16″ diameter also available. Typical finish is mill galvanized with zinc coating at .10 oz. per S.F. Class I (.40 oz. per S.F.) and Class III (.80 oz. per S.F.) are also available, as is hot dipped galvanizing at 1.50 oz. per S.F.

R042110-10 Economy in Bricklaying

Have adequate supervision. Be sure bricklayers are always supplied with materials so there is no waiting. Place experienced bricklayers at corners and openings.

Use only screened sand for mortar. Otherwise, labor time will be wasted picking out pebbles. Use seamless metal tubs for mortar as they do not leak or catch the trowel. Locate stack and mortar for easy wheeling.

Have brick delivered for stacking. This makes for faster handling, reduces chipping and breakage, and requires less storage space. Many dealers will deliver select common in 2′ x 3′ x 4′ pallets or face brick packaged. This affords quick handling with a crane or forklift and easy tonging in units of ten, which reduces waste.

Use wider bricks for one wythe wall construction. Keep scaffolding away from the wall to allow mortar to fall clear and not stain the wall.

On large jobs develop specialized crews for each type of masonry unit.

Consider designing for prefabricated panel construction on high rise projects.

Avoid excessive corners or openings. Each opening adds about 50% to the labor cost for area of opening.

Bolting stone panels and using window frames as stops reduce labor costs and speed up erection.

R042110-20 Common and Face Brick

Common building brick manufactured according to ASTM C62 and facing brick manufactured according to ASTM C216 are the two standard bricks available for general building use.

Building brick is made in three grades: SW, where high resistance to damage caused by cyclic freezing is required; MW, where moderate resistance to cyclic freezing is needed; and NW, where little resistance to cyclic freezing is needed. Facing brick is made in only the two grades SW and MW. Additionally, facing brick is available in three types: FBS, for general use; FBX, for general use where a higher degree of precision and lower permissible variation in size than FBS are needed; and FBA, for general use to produce characteristic architectural effects resulting from non-uniformity in size and texture of the units.

In figuring the material cost of brickwork, an allowance of 25% mortar waste and 3% brick breakage was included. If bricks are delivered palletized with 280 to 300 per pallet, or packaged, allow only 1-1/2% for breakage. Packaged or palletized delivery is practical when a job is big enough to have a crane or other equipment available to handle a package of brick. This is so on all industrial work but not always true on small commercial buildings.

The use of buff and gray face is increasing, and there is a continuing trend to the Norman, Roman, Jumbo and SCR brick.

Common red clay brick for backup is not used that often. Concrete block is the most usual backup material with occasional use of sand lime or cement brick. Building brick is commonly used in solid walls for strength and as a fire stop.

Brick panels built on the ground and then crane erected to the upper floors have proven to be economical. This allows the work to be done under cover and without scaffolding.

R042110-50 Brick, Block & Mortar Quantities

Running Bond						For Other Bonds Standard Size Add to S.F. Quantities in Table to Left		
Number of Brick per S.F. of Wall - Single Wythe with 3/8″ Joints				C.F. of Mortar per M Bricks, Waste Included				
Type Brick	Nominal Size (incl. mortar) L H W	Modular Coursing	Number of Brick per S.F.	3/8″ Joint	1/2″ Joint	Bond Type	Description	Factor
Standard	8 x 2-2/3 x 4	3C=8″	6.75	8.1	10.3	Common	full header every fifth course	+20%
Economy	8 x 4 x 4	1C=4″	4.50	9.1	11.6		full header every sixth course	+16.7%
Engineer	8 x 3-1/5 x 4	5C=16″	5.63	8.5	10.8	English	full header every second course	+50%
Fire	9 x 2-1/2 x 4-1/2	2C=5″	6.40	550 # Fireclay	—	Flemish	alternate headers every course	+33.3%
Jumbo	12 x 4 x 6 or 8	1C=4″	3.00	22.5	29.2		every sixth course	+5.6%
Norman	12 x 2-2/3 x 4	3C=8″	4.50	11.2	14.3	Header = W x H exposed		+100%
Norwegian	12 x 3-1/5 x 4	5C=16″	3.75	11.7	14.9	Rowlock = H x W exposed		+100%
Roman	12 x 2 x 4	2C=4″	6.00	10.7	13.7	Rowlock stretcher = L x W exposed		+33.3%
SCR	12 x 2-2/3 x 6	3C=8″	4.50	21.8	28.0	Soldier = H x L exposed		—
Utility	12 x 4 x 4	1C=4″	3.00	12.3	15.7	Sailor = W x L exposed		-33.3%

Concrete Blocks Nominal Size	Approximate Weight per S.F. Standard	Approximate Weight per S.F. Lightweight	Blocks per 100 S.F.	Mortar per M block, waste included: Partitions	Mortar per M block, waste included: Back up
2″ x 8″ x 16″	20 PSF	15 PSF	113	27 C.F.	36 C.F.
4″	30	20		41	51
6″	42	30		56	66
8″	55	38		72	82
10″	70	47		87	97
12″	85	55		102	112

Brick & Mortar Quantities
©Brick Industry Association. 2009 Feb. Technical Notes on Brick Construction 10:
Dimensioning and Estimating Brick Masonry. Reston (VA): BIA. Table 1 Modular Brick Sizes and Table 4 Quantity Estimates for Brick Masonry.

Masonry — R0422 Concrete Unit Masonry

R042210-20 Concrete Block

The material cost of special block such as corner, jamb and head block can be figured at the same price as ordinary block of equal size. Labor on specials is about the same as equal-sized regular block.

Bond beams and 16″ high lintel blocks are more expensive than regular units of equal size. Lintel blocks are 8″ long and either 8″ or 16″ high.

Use of a motorized mortar spreader box will speed construction of continuous walls.

Hollow non-load-bearing units are made according to ASTM C129 and hollow load-bearing units according to ASTM C90.

Metals — R0505 Common Work Results for Metals

R050516-30 Coating Structural Steel

On field-welded jobs, the shop-applied primer coat is necessarily omitted. All painting must be done in the field and usually consists of red oxide rust inhibitive paint or an aluminum paint. The table below shows paint coverage and daily production for field painting.

See Division 09 97 13.23 for hot-dipped galvanizing and for field-applied cold galvanizing and other paints and protective coatings.

See Division 05 01 10.51 for steel surface preparation treatments such as wire brushing, pressure washing and sand blasting.

Type Construction	Surface Area per Ton	Coat	One Gallon Covers		In 8 Hrs. Person Covers		Average per Ton Spray	
			Brush	Spray	Brush	Spray	Gallons	Labor-hours
Light Structural	300 S.F. to 500 S.F.	1st	500 S.F.	455 S.F.	640 S.F.	2000 S.F.	0.9 gals.	1.6 L.H.
		2nd	450	410	800	2400	1.0	1.3
		3rd	450	410	960	3200	1.0	1.0
Medium	150 S.F. to 300 S.F.	All	400	365	1600	3200	0.6	0.6
Heavy Structural	50 S.F. to 150 S.F.	1st	400	365	1920	4000	0.2	0.2
		2nd	400	365	2000	4000	0.2	0.2
		3rd	400	365	2000	4000	0.2	0.2
Weighted Average	225 S.F.	All	400	365	1350	3000	0.6	0.6

R050521-20 Welded Structural Steel

Usual weight reductions with welded design run 10% to 20% compared with bolted or riveted connections. This amounts to about the same total cost compared with bolted structures since field welding is more expensive than bolts. For normal spans of 18′ to 24′ figure 6 to 7 connections per ton.

Trusses — For welded trusses add 4% to weight of main members for connections. Up to 15% less steel can be expected in a welded truss compared to one that is shop bolted. Cost of erection is the same whether shop bolted or welded.

General — Typical electrodes for structural steel welding are E6010, E6011, E60T and E70T. Typical buildings vary between 2# to 8# of weld rod per ton of steel. Buildings utilizing continuous design require about three times as much welding as conventional welded structures. In estimating field erection by welding, it is best to use the average linear feet of weld per ton to arrive at the welding cost per ton. The type, size and position of the weld will have a direct bearing on the cost per linear foot. A typical field welder will deposit 1.8# to 2# of weld rod per hour manually. Using semiautomatic methods can increase production by as much as 50% to 75%.

R050523-10 High Strength Bolts

Common bolts (A307) are usually used in secondary connections (see Division 05 05 23.10).

High strength bolts (A325 and A490) are usually specified for primary connections such as column splices, beam and girder connections to columns, column bracing, connections for supports of operating equipment or of other live loads which produce impact or reversal of stress, and in structures carrying cranes of over 5-ton capacity.

Allow 20 field bolts per ton of steel for a 6 story office building, apartment house or light industrial building. For 6 to 12 stories allow 25 bolts per ton, and above 12 stories, 30 bolts per ton. On power stations, 20 to 25 bolts per ton are needed.

R051223-10 Structural Steel

The bare material prices for structural steel, shown in the unit cost sections of the data set, are for 100 tons of shop-fabricated structural steel and include:

1. Mill base price of structural steel
2. Mill scrap/grade/size/length extras
3. Mill delivery to a metals service center (warehouse)
4. Service center storage and handling
5. Service center delivery to a fabrication shop
6. Shop storage and handling
7. Shop drafting/detailing
8. Shop fabrication
9. Shop coat of primer paint
10. Shop listing
11. Shop delivery to the job site

In unit cost sections of the data set that contain items for field fabrication of steel components, the bare material cost of steel includes:

1. Mill base price of structural steel
2. Mill scrap/grade/size/length extras
3. Mill delivery to a metals service center (warehouse)
4. Service center storage and handling
5. Service center delivery to the job site

R051223-20 Steel Estimating Quantities

One estimate on erection is that a crane can handle 35 to 60 pieces per day. Say the average is 45. With usual sizes of beams, girders, and columns, this would amount to about 20 tons per day. The type of connection greatly affects the speed of erection. Moment connections for continuous design slow down production and increase erection costs.

Short open web bar joists can be set at the rate of 75 to 80 per day, with 50 per day being the average for setting long span joists.

After main members are calculated, add the following for usual allowances: base plates 2% to 3%; column splices 4% to 5%; and miscellaneous details 4% to 5%, for a total of 10% to 13% in addition to main members.

The ratio of column to beam tonnage varies depending on type of steels used, typical spans, story heights and live loads.

It is more economical to keep the column size constant and to vary the strength of the column by using high strength steels. This also saves floor space. Buildings have recently gone as high as ten stories with 8″ high strength columns. For light columns under W8X31 lb. sections, concrete filled steel columns are economical.

High strength steels may be used in columns and beams to save floor space and to meet head room requirements. High strength steels in some sizes sometimes require long lead times.

Round, square and rectangular columns, both plain and concrete filled, are readily available and save floor area, but are higher in cost per pound than rolled columns. For high unbraced columns, tube columns may be less expensive.

Below are average minimum figures for the weights of the structural steel frame for different types of buildings using A36 steel, rolled shapes and simple joints. For economy in domes, rise to span ratio = .13. Open web joist framing systems will reduce weights by 10% to 40%. Composite design can reduce steel weight by up to 25% but additional concrete floor slab thickness may be required. Continuous design can reduce the weights up to 20%. There are many building codes with different live load requirements and different structural requirements, such as hurricane and earthquake loadings, which can alter the figures.

Structural Steel Weights per S.F. of Floor Area

Type of Building	No. of Stories	Avg. Spans	L.L. #/S.F.	Lbs. Per S.F.	Type of Building	No. of Stories	Avg. Spans	L.L. #/S.F.	Lbs. Per S.F.
Steel Frame Mfg.	1	20′x20′	40	8	Apartments	2-8	20′x20′	40	8
		30′x30′		13		9-25			14
		40′x40′		18	Office	to 10	Various	80	10
Parking garage	4	Various	80	8.5		20			18
Domes	1	200′	30	10		30			26
(Schwedler)*		300′		15		over 50			35

R051223-25 Common Structural Steel Specifications

ASTM A992 (formerly A36, then A572 Grade 50) is the all-purpose carbon grade steel widely used in building and bridge construction.

The other high-strength steels listed below may each have certain advantages over ASTM A992 structural carbon steel, depending on the application. They have proven to be economical choices where, due to lighter members, the reduction of dead load and the associated savings in shipping cost can be significant.

ASTM A588 atmospheric weathering, high-strength, low-alloy steels can be used in the bare (uncoated) condition, where exposure to normal atmosphere causes a tightly adherant oxide to form on the surface, protecting the steel from further oxidation. ASTM A242 corrosion-resistant, high-strength, low-alloy steels have enhanced atmospheric corrosion resistance of at least two times that of carbon structural steels with copper, or four times that of carbon structural steels without copper. The reduction or elimination of maintenance resulting from the use of these steels often offsets their higher initial cost.

Steel Type	ASTM Designation	Minimum Yield Stress in KSI	Shapes Available
Carbon	A36	36	All structural shape groups, and plates & bars up through 8″ thick
	A529	50	Structural shape group 1, and plates & bars up through 2″ thick
		50	All structural shape groups
High-Strength Low-Alloy	A913	60	
Quenched & Self-Tempered		65	
		70	
		42	All structural shape groups, and plates & bars up through 6″ thick
High-Strength Low-Alloy	A572	50	All structural shape groups, and plates & bars up through 4″ thick
Columbium-Vanadium		55	Structural shape groups 1 & 2, and plates & bars up through 2″ thick
		60	Structural shape groups 1 & 2, and plates & bars up through 1-1/4″ thick
		65	Structural shape group 1, and plates & bars up through 1-1/4″ thick
High-Strength Low-Alloy	A992	50	All structural shape groups
Columbium-Vanadium			
Weathering		42	Structural shape groups 4 & 5, and plates & bars over 1-1/2″ up through 4″ thick
High-Strength Low-Alloy	A242	46	Structural shape group 3, and plates & bars over 3/4″ up through 1-1/2″ thick
		50	Structural shape groups 1 & 2, and plates & bars up through 3/4″ thick
Weathering		42	Plates & bars over 5″ up through 8″ thick
High-Strength Low-Alloy	A588	46	Plates & bars over 4″ up through 5″ thick
		50	All structural shape groups, and plates & bars up through 4″ thick
Quenched and Tempered	A852	70	Plates & bars up through 4″ thick
Low-Alloy			
Quenched and Tempered	A514	90	Plates & bars over 2-1/2″ up through 6″ thick
Alloy		100	Plates & bars up through 2-1/2″ thick

R051223-30 High Strength Steels

The mill price of high strength steels may be higher than A992 carbon steel, but their proper use can achieve overall savings through total reduced weights. For columns with L/r over 100, A992 steel is best; under 100, high strength steels are economical. For heavy columns, high strength steels are economical when cover plates are eliminated. There is no economy using high strength steels for clip angles or supports or for beams where deflection governs. Thinner members are more economical than thick.

The per ton erection and fabricating costs of the high strength steels will be higher than for A992 since the same number of pieces, but less weight, will be installed.

R053100-10 Decking Descriptions

General - All Deck Products

A steel deck is made by cold forming structural grade sheet steel into a repeating pattern of parallel ribs. The strength and stiffness of the panels are the result of the ribs and the material properties of the steel. Deck lengths can be varied to suit job conditions, but because of shipping considerations, are usually less than 40 feet. Standard deck width varies with the product used but full sheets are usually 12″, 18″, 24″, 30″, or 36″. The deck is typically furnished in a standard width with the ends cut square. Any cutting for width, such as at openings or for angular fit, is done at the job site.

The deck is typically attached to the building frame with arc puddle welds, self-drilling screws, or powder or pneumatically driven pins. Sheet to sheet fastening is done with screws, button punching (crimping), or welds.

Composite Floor Deck

After installation and adequate fastening, a floor deck serves several purposes. It (a) acts as a working platform, (b) stabilizes the frame, (c) serves as a concrete form for the slab, and (d) reinforces the slab to carry the design loads applied during the life of the building. Composite decks are distinguished by the presence of shear connector devices as part of the deck. These devices are designed to mechanically lock the concrete and deck together so that the concrete and the deck work together to carry subsequent floor loads. These shear connector devices can be rolled-in embossments, lugs, holes, or wires welded to the panels. The deck profile can also be used to interlock concrete and steel.

Composite deck finishes are either galvanized (zinc coated) or phosphatized/painted. Galvanized deck has a zinc coating on both the top and bottom surfaces. The phosphatized/painted deck has a bare (phosphatized) top surface that will come into contact with the concrete. This bare top surface can be expected to develop rust before the concrete is placed. The bottom side of the deck has a primer coat of paint.

A composite floor deck is normally installed so the panel ends do not overlap on the supporting beams. Shear lugs or panel profile shapes often prevent a tight metal to metal fit if the panel ends overlap; the air gap caused by overlapping will prevent proper fusion with the structural steel supports when the panel end laps are shear stud welded.

Adequate end bearing of the deck must be obtained as shown on the drawings. If bearing is actually less in the field than shown on the drawings, further investigation is required.

Roof Deck

A roof deck is not designed to act compositely with other materials. A roof deck acts alone in transferring horizontal and vertical loads into the building frame. Roof deck rib openings are usually narrower than floor deck rib openings. This provides adequate support of the rigid thermal insulation board.

A roof deck is typically installed to endlap approximately 2″ over supports. However, it can be butted (or lapped more than 2″) to solve field fit problems. Since designers frequently use the installed deck system as part of the horizontal bracing system (the deck as a diaphragm), any fastening substitution or change should be approved by the designer. Continuous perimeter support of the deck is necessary to limit edge deflection in the finished roof and may be required for diaphragm shear transfer.

Standard roof deck finishes are galvanized or primer painted. The standard factory applied paint for roof decks is a primer paint and is not intended to weather for extended periods of time. Field painting or touching up of abrasions and deterioration of the primer coat or other protective finishes is the responsibility of the contractor.

Cellular Deck

A cellular deck is made by attaching a bottom steel sheet to a roof deck or composite floor deck panel. A cellular deck can be used in the same manner as a floor deck. Electrical, telephone, and data wires are easily run through the chase created between the deck panel and the bottom sheet.

When used as part of the electrical distribution system, the cellular deck must be installed so that the ribs line up and create a smooth cell transition at abutting ends. The joint that occurs at butting cell ends must be taped or otherwise sealed to prevent wet concrete from seeping into the cell. Cell interiors must be free of welding burrs, or other sharp intrusions, to prevent damage to wires.

When used as a roof deck, the bottom flat plate is usually left exposed to view. Care must be maintained during erection to keep good alignment and prevent damage.

A cellular deck is sometimes used with the flat plate on the top side to provide a flat working surface. Installation of the deck for this purpose requires special methods for attachment to the frame because the flat plate, now on the top, can prevent direct access to the deck material that is bearing on the structural steel. It may be advisable to treat the flat top surface to prevent slipping.

A cellular deck is always furnished galvanized or painted over galvanized.

Form Deck

A form deck can be any floor or roof deck product used as a concrete form. Connections to the frame are by the same methods used to anchor floor and roof decks. Welding washers are recommended when welding a deck that is less than 20 gauge thickness.

A form deck is furnished galvanized, prime painted, or uncoated. A galvanized deck must be used for those roof deck systems where a form deck is used to carry a lightweight insulating concrete fill.

Wood, Plastics & Comp. R0611 Wood Framing

R061110-30 Lumber Product Material Prices

The price of forest products fluctuates widely from location to location and from season to season depending upon economic conditions. The bare material prices in the unit cost sections of the data set show the National Average material prices in effect Jan. 1 of this data year. It must be noted that lumber prices in general may change significantly during the year.

Availability of certain items depends upon geographic location and must be checked prior to firm-price bidding.

Wood, Plastics & Comp. R0616 Sheathing

R061636-20 Plywood

There are two types of plywood used in construction: interior, which is moisture-resistant but not waterproofed, and exterior, which is waterproofed.

The grade of the exterior surface of the plywood sheets is designated by the first letter: A, for smooth surface with patches allowed; B, for solid surface with patches and plugs allowed; C, which may be surface plugged or may have knot holes up to 1″ wide; and D, which is used only for interior type plywood and may have knot holes up to 2-1/2″ wide. "Structural Grade" is specifically designed for engineered applications such as box beams. All CC & DD grades have roof and floor spans marked on them.

Underlayment-grade plywood runs from 1/4″ to 1-1/4″ thick. Thicknesses 5/8″ and over have optional tongue and groove joints which eliminate the need for blocking the edges. Underlayment 19/32″ and over may be referred to as Sturd-i-Floor.

The price of plywood can fluctuate widely due to geographic and economic conditions.

Typical uses for various plywood grades are as follows:

AA-AD Interior — cupboards, shelving, paneling, furniture

BB Plyform — concrete form plywood

CDX — wall and roof sheathing

Structural — box beams, girders, stressed skin panels

AA-AC Exterior — fences, signs, siding, soffits, etc.

Underlayment — base for resilient floor coverings

Overlaid HDO — high density for concrete forms & highway signs

Overlaid MDO — medium density for painting, siding, soffits & signs

303 Siding — exterior siding, textured, striated, embossed, etc.

Thermal & Moist. Protec. R0731 Shingles & Shakes

R073126-20 Roof Slate

16″, 18″ and 20″ are standard lengths, and slate usually comes in random widths. For standard 3/16″ thickness use 1-1/2″ copper nails. Allow for 3% breakage.

Thermal & Moist. Protec. R0751 Built-Up Bituminous Roofing

R075113-20 Built-Up Roofing

Asphalt is available in kegs of 100 lbs. each; coal tar pitch in 560 lb. kegs. Prepared roofing felts are available in a wide range of sizes, weights and characteristics. However, the most commonly used are #15 (432 S.F. per roll, 13 lbs. per square) and #30 (216 S.F. per roll, 27 lbs. per square).

Inter-ply bitumen varies from 24 lbs. per sq. (asphalt) to 30 lbs. per sq. (coal tar) per ply, MF4@ 25%. Flood coat bitumen also varies from 60 lbs. per sq. (asphalt) to 75 lbs. per sq. (coal tar), MF4@ 25%. Expendable equipment (mops, brooms, screeds, etc.) runs about 16% of the bitumen cost. For new, inexperienced crews this factor may be much higher.

A rigid insulation board is typically applied in two layers. The first is mechanically attached to nailable decks or spot or solid mopped to non-nailable decks; the second layer is then spot or solid mopped to the first layer. Membrane application follows the insulation, except in protected membrane roofs, where the membrane goes down first and the insulation on top, followed with ballast (stone or concrete pavers). Insulation and related labor costs are NOT included in prices for built-up roofing.

Thermal & Moist. Protec. — R0752 Modified Bituminous Membrane Roofing

R075213-30 Modified Bitumen Roofing

The cost of modified bitumen roofing is highly dependent on the type of installation that is planned. Installation is based on the type of modifier used in the bitumen. The two most popular modifiers are atactic polypropylene (APP) and styrene butadiene styrene (SBS). The modifiers are added to heated bitumen during the manufacturing process to change its characteristics. A polyethylene, polyester or fiberglass reinforcing sheet is then sandwiched between layers of this bitumen. When completed, the result is a pre-assembled, built-up roof that has increased elasticity and weatherability. Some manufacturers include a surfacing material such as ceramic or mineral granules, metal particles or sand.

The preferred method of adhering SBS-modified bitumen roofing to the substrate is with hot-mopped asphalt (much the same as built-up roofing). This installation method requires a tar kettle/pot to heat the asphalt, as well as the labor, tools and equipment necessary to distribute and spread the hot asphalt.

The alternative method for applying APP and SBS modified bitumen is as follows. A skilled installer uses a torch to melt a small pool of bitumen off the membrane. This pool must form across the entire roll for proper adhesion. The installer must unroll the roofing at a pace slow enough to melt the bitumen, but fast enough to prevent damage to the rest of the membrane.

Modified bitumen roofing provides the advantages of both built-up and single-ply roofing. Labor costs are reduced over those of built-up roofing because only a single ply is necessary. The elasticity of single-ply roofing is attained with the reinforcing sheet and polymer modifiers. Modifieds have some self-healing characteristics and because of their multi-layer construction, they offer the reliability and safety of built-up roofing.

Thermal & Moist. Protec. — R0784 Firestopping

R078413-30 Firestopping

Firestopping is the sealing of structural, mechanical, electrical and other penetrations through fire-rated assemblies. The basic components of firestop systems are safing insulation and firestop sealant on both sides of wall penetrations and the top side of floor penetrations.

Pipe penetrations are assumed to be through concrete, grout, or joint compound and can be sleeved or unsleeved. Costs for the penetrations and sleeves are not included. An annular space of 1″ is assumed. Escutchcons are not included.

A metallic pipe is assumed to be copper, aluminum, cast iron or similar metallic material. An insulated metallic pipe is assumed to be covered with a thermal insulating jacket of varying thickness and materials.

A non-metallic pipe is assumed to be PVC, CPVC, FR Polypropylene or similar plastic piping material. Intumescent firestop sealants or wrap strips are included. Collars on both sides of wall penetrations and a sheet metal plate on the underside of floor penetrations are included.

Ductwork is assumed to be sheet metal, stainless steel or similar metallic material. Duct penetrations are assumed to be through concrete, grout or joint compound. Costs for penetrations and sleeves are not included. An annular space of 1/2″ is assumed.

Multi-trade openings include costs for sheet metal forms, firestop mortar, wrap strips, collars and sealants as necessary.

Structural penetrations joints are assumed to be 1/2″ or less. CMU walls are assumed to be within 1-1/2″ of the metal deck. Drywall walls are assumed to be tight to the underside of metal decking.

Metal panel, glass or curtain wall systems include a spandrel area of 5′ filled with mineral wool foil-faced insulation. Fasteners and stiffeners are included.

Openings — R0813 Metal Doors

R081313-20 Steel Door Selection Guide

Standard steel doors are classified into four levels, as recommended by the Steel Door Institute in the chart below. Each of the four levels offers a range of construction models and designs to meet architectural requirements for preference and appearance, including full flush, seamless, and stile & rail. Recommended minimum gauge requirements are also included.

For complete standard steel door construction specifications and available sizes, refer to the Steel Door Institute Technical Data Series, ANSI A250.8-98 (SDI-100), and ANSI A250.4-94 Test Procedure and Acceptance Criteria for Physical Endurance of Steel Door and Hardware Reinforcements.

Level		Model	Construction	For Full Flush or Seamless		
				Min. Gauge	Thickness (in)	Thickness (mm)
I	Standard Duty	1 2	Full Flush Seamless	20	0.032	0.8
II	Heavy Duty	1 2	Full Flush Seamless	18	0.042	1.0
III	Extra Heavy Duty	1 2 3	Full Flush Seamless *Stile & Rail	16	0.053	1.3
IV	Maximum Duty	1 2	Full Flush Seamless	14	0.067	1.6

*Stiles & rails are 16 gauge; flush panels, when specified, are 18 gauge

R085313-20 Replacement Windows

Replacement windows are typically measured per United Inch.

United Inches are calculated by rounding the width and height of the window opening up to the nearest inch, then adding the two figures.

The labor cost for replacement windows includes removal of sash, existing sash balance or weights, parting bead where necessary and installation of new window.

Debris hauling and dump fees are not included.

R087110-10 Hardware Finishes

This table describes hardware finishes used throughout the industry. It also shows the base metal and the respective symbols in the three predominate systems of identification. Many of these are used in pricing descriptions in Division Eight.

US″	BMHA*	CDN^	Base	Description
US P	600	CP	Steel	Primed for Painting
US 1B	601	C1B	Steel	Bright Black Japanned
US 2C	602	C2C	Steel	Zinc Plated
US 2G	603	C2G	Steel	Zinc Plated
US 3	605	C3	Brass	Bright Brass, Clear Coated
US 4	606	C4	Brass	Satin Brass, Clear Coated
US 5	609	C5	Brass	Satin Brass, Blackened, Satin Relieved, Clear Coated
US 7	610	C7	Brass	Satin Brass, Blackened, Bright Relieved, Clear Coated
US 9	611	C9	Bronze	Bright Bronze, Clear Coated
US 10	612	C10	Bronze	Satin Bronze, Clear Coated
US 10A	641	C10A	Steel	Antiqued Bronze, Oiled and Lacquered
US 10B	613	C10B	Bronze	Antiqued Bronze, Oiled
US 11	616	C11	Bronze	Satin Bronze, Blackened, Satin Relieved, Clear Coated
US 14	618	C14	Brass/Bronze	Bright Nickel Plated, Clear Coated
US 15	619	C15	Brass/Bronze	Satin Nickel, Clear Coated
US 15A	620	C15A	Brass/Bronze	Satin Nickel Plated, Blackened, Satin Relieved, Clear Coated
US 17A	621	C17A	Brass/Bronze	Nickel Plated, Blackened, Relieved, Clear Coated
US 19	622	C19	Brass/Bronze	Flat Black Coated
US 20	623	C20	Brass/Bronze	Statuary Bronze, Light
US 20A	624	C20A	Brass/Bronze	Statuary Bronze, Dark
US 26	625	C26	Brass/Bronze	Bright Chromium
US 26D	626	C26D	Brass/Bronze	Satin Chromium
US 20	627	C27	Aluminum	Satin Aluminum Clear
US 28	628	C28	Aluminum	Anodized Dull Aluminum
US 32	629	C32	Stainless Steel	Bright Stainless Steel
US 32D	630	C32D	Stainless Steel	Stainless Steel
US 3	632	C3	Steel	Bright Brass Plated, Clear Coated
US 4	633	C4	Steel	Satin Brass, Clear Coated
US 7	636	C7	Steel	Satin Brass Plated, Blackened, Bright Relieved, Clear Coated
US 9	637	C9	Steel	Bright Bronze Plated, Clear Coated
US 5	638	C5	Steel	Satin Brass Plated, Blackened, Bright Relieved, Clear Coated
US 10	639	C10	Steel	Satin Bronze Plated, Clear Coated
US 10B	640	C10B	Steel	Antique Bronze, Oiled
US 10A	641	C10A	Steel	Antiqued Bronze, Oiled and Lacquered
US 11	643	C11	Steel	Satin Bronze Plated, Blackened, Bright Relieved, Clear Coated
US 14	645	C14	Steel	Bright Nickel Plated, Clear Coated
US 15	646	C15	Steel	Satin Nickel Plated, Clear Coated
US 15A	647	C15A	Steel	Nickel Plated, Blackened, Bright Relieved, Clear Coated
US 17A	648	C17A	Steel	Nickel Plated, Blackened, Relieved, Clear Coated
US 20	649	C20	Steel	Statuary Bronze, Light
US 20A	650	C20A	Steel	Statuary Bronze, Dark
US 26	651	C26	Steel	Bright Chromium Plated
US 26D	652	C26D	Steel	Satin Chromium Plated

* - BMHA Builders Hardware Manufacturing Association
″ - US Equivalent
^ - Canadian Equivalent
Japanning is imitating Asian lacquer work

R088110-10 Glazing Productivity

Some glass sizes are estimated by the "united inch" (height + width). The table below shows the number of lights glazed in an eight-hour period by the crew size indicated, for glass up to 1/4″ thick. Square or nearly square lights are more economical on a S.F. basis. Long slender lights will have a high S.F. installation cost. For insulated glass reduce production by 33%. For 1/2″ float glass reduce production by 50%. Production time for glazing with two glaziers per day averages: 1/4″ float glass 120 S.F.; 1/2″ float glass 55 S.F.; 1/2″ insulated glass 95 S.F.; 3/4″ insulated glass 75 S.F.

Glazing Method	United Inches per Light							
	40″	60″	80″	100″	135″	165″	200″	240″
Number of Men in Crew	1	1	1	1	2	3	3	4
Industrial sash, putty	60	45	24	15	18	—	—	—
With stops, putty bed	50	36	21	12	16	8	4	3
Wood stops, rubber	40	27	15	9	11	6	3	2
Metal stops, rubber	30	24	14	9	9	6	3	2
Structural glass	10	7	4	3	—	—	—	—
Corrugated glass	12	9	7	4	4	4	3	—
Storefronts	16	15	13	11	7	6	4	4
Skylights, putty glass	60	36	21	12	16	—	—	—
Thiokol set	15	15	11	9	9	6	3	2
Vinyl set, snap on	18	18	13	12	12	7	5	4
Maximum area per light	2.8 S.F.	6.3 S.F.	11.1 S.F.	17.4 S.F.	31.6 S.F.	47 S.F.	69 S.F.	100 S.F.

R092000-50 Lath, Plaster and Gypsum Board

Gypsum board lath is available in 3/8″ thick x 16″ wide x 4′ long sheets as a base material for multi-layer plaster applications. It is also available as a base for either multi-layer or veneer plaster applications in 1/2″ and 5/8″ thick–4′ wide x 8′, 10′ or 12′ long sheets. Fasteners are screws or blued ring shank nails for wood framing and screws for metal framing.

Metal lath is available in diamond mesh patterns with flat or self-furring profiles. Paper backing is available for applications where excessive plaster waste needs to be avoided. A slotted mesh ribbed lath should be used in areas where the span between structural supports is greater than normal. Most metal lath comes in 27″ x 96″ sheets. Diamond mesh weighs 1.75, 2.5 or 3.4 pounds per square yard, slotted mesh lath weighs 2.75 or 3.4 pounds per square yard. Metal lath can be nailed, screwed or tied in place.

Many **accessories** are available. Corner beads, flat reinforcing strips, casing beads, control and expansion joints, furring brackets and channels are some examples. Note that accessories are not included in plaster or stucco line items.

Plaster is defined as a material or combination of materials that when mixed with a suitable amount of water, forms a plastic mass or paste. When applied to a surface, the paste adheres to it and subsequently hardens, preserving in a rigid state the form or texture imposed during the period of elasticity.

Gypsum plaster is made from ground calcined gypsum. It is mixed with aggregates and water for use as a base coat plaster.

Vermiculite plaster is a fire-retardant plaster covering used on steel beams, concrete slabs and other heavy construction materials. Vermiculite is a group name for certain clay minerals, hydrous silicates or aluminum, magnesium and iron that have been expanded by heat.

Perlite plaster is a plaster using perlite as an aggregate instead of sand. Perlite is a volcanic glass that has been expanded by heat.

Gauging plaster is a mix of gypsum plaster and lime putty that when applied produces a quick drying finish coat.

Veneer plaster is a one or two component gypsum plaster used as a thin finish coat over special gypsum board.

Keenes cement is a white cementitious material manufactured from gypsum that has been burned at a high temperature and ground to a fine powder. Alum is added to accelerate the set. The resulting plaster is hard and strong and accepts and maintains a high polish, hence it is used as a finishing plaster.

Stucco is a Portland cement based plaster used primarily as an exterior finish.

Plaster is used on both interior and exterior surfaces. Generally it is applied in multiple-coat systems. A three-coat system uses the terms scratch, brown and finish to identify each coat. A two-coat system uses base and finish to describe each coat. Each type of plaster and application system has attributes that are chosen by the designer to best fit the intended use.

Gypsum Plaster Quantities for 100 S.Y.	2 Coat, 5/8″ Thick		3 Coat, 3/4″ Thick		
	Base	Finish	Scratch	Brown	Finish
	1:3 Mix	2:1 Mix	1:2 Mix	1:3 Mix	2:1 MIx
Gypsum plaster	1,300 lb.		1,350 lb.	650 lb.	
Sand	1.75 C.Y.		1.85 C.Y.	1.35 C.Y.	
Finish hydrated lime		340 lb.			340 lb.
Gauging plaster		170 lb.			170 lb.

Vermiculite or Perlite Plaster Quantities for 100 S.Y.	2 Coat, 5/8″ Thick		3 Coat, 3/4″ Thick		
	Base	Finish	Scratch	Brown	Finish
Gypsum plaster	1,250 lb.		1,450 lb.	800 lb.	
Vermiculite or perlite	7.8 bags		8.0 bags	3.3 bags	
Finish hydrated lime		340 lb.			340 lb.
Gauging plaster		170 lb.			170 lb.

Stucco-Three-Coat System Quantities for 100 S.Y.	On Wood Frame	On Masonry
Portland cement	29 bags	21 bags
Sand	2.6 C.Y.	2.0 C.Y.
Hydrated lime	180 lb.	120 lb.

R092910-10 Levels of Gypsum Board Finish

In the past, contract documents often used phrases such as "industry standard" and "workmanlike finish" to specify the expected quality of gypsum board wall and ceiling installations. The vagueness of these descriptions led to unacceptable work and disputes.

In order to resolve this problem, four major trade associations concerned with the manufacture, erection, finish, and decoration of gypsum board wall and ceiling systems developed an industry-wide *Recommended Levels of Gypsum Board Finish.*

The finish of gypsum board walls and ceilings for specific final decoration is dependent on a number of factors. A primary consideration is the location of the surface and the degree of decorative treatment desired. Painted and unpainted surfaces in warehouses and other areas where appearance is normally not critical may simply require the taping of wallboard joints and 'spotting' of fastener heads. Blemish-free, smooth, monolithic surfaces often intended for painted and decorated walls and ceilings in habitated structures, ranging from single-family dwellings through monumental buildings, require additional finishing prior to the application of the final decoration.

Other factors to be considered in determining the level of finish of the gypsum board surface are (1) the type of angle of surface illumination (both natural and artificial lighting), and (2) the paint and method of application or the type and finish of wallcovering specified as the final decoration. Critical lighting conditions, gloss paints, and thin wall coverings require a higher level of gypsum board finish than heavily textured surfaces which are subsequently painted or surfaces which are to be decorated with heavy grade wall coverings.

The following descriptions were developed by the Association of the Wall and Ceiling Industries-International (AWCI), Ceiling & Interior Systems Construction Association (CISCA), Gypsum Association (GA), and Painting and Decorating Contractors of America (PDCA) as a guide.

Level 0: Used in temporary construction or wherever the final decoration has not been determined. Unfinished. No taping, finishing or corner beads are required. Also could be used where non-predecorated panels will be used in demountable-type partitions that are to be painted as a final finish.

Level 1: Frequently used in plenum areas above ceilings, in attics, in areas where the assembly would generally be concealed, or in building service corridors and other areas not normally open to public view. Some degree of sound and smoke control is provided; in some geographic areas, this level is referred to as "fire-taping," although this level of finish does not typically meet fire-resistant assembly requirements. Where a fire resistance rating is required for the gypsum board assembly, details of construction should be in accordance with reports of fire tests of assemblies that have met the requirements of the fire rating acceptable.

All joints and interior angles shall have tape embedded in joint compound. Accessories are optional at specifier discretion in corridors and other areas with pedestrian traffic. Tape and fastener heads need not be covered with joint compound. Surface shall be free of excess joint compound. Tool marks and ridges are acceptable.

Level 2: It may be specified for standard gypsum board surfaces in garages, warehouse storage, or other similar areas where surface appearance is not of primary importance.

All joints and interior angles shall have tape embedded in joint compound and shall be immediately wiped with a joint knife or trowel, leaving a thin coating of joint compound over all joints and interior angles. Fastener heads and accessories shall be covered with a coat of joint compound. Surface shall be free of excess joint compound. Tool marks and ridges are acceptable.

Level 3: Typically used in areas receiving heavy texture (spray or hand applied) finishes before final painting, or where commercial-grade (heavy duty) wall coverings are to be applied as the final decoration. This level of finish should not be used where smooth painted surfaces or where lighter weight wall coverings are specified. The prepared surface shall be coated with a drywall primer prior to the application of final finishes.

All joints and interior angles shall have tape embedded in joint compound and shall be immediately wiped with a joint knife or trowel, leaving a thin coating of joint compound over all joints and interior angles. One additional coat of joint compound shall be applied over all joints and interior angles. Fastener heads and accessories shall be covered with two separate coats of joint compound. All joint compounds shall be smooth and free of tool marks and ridges. The prepared surface shall be covered with a drywall primer prior to the application of the final decoration.

Level 4: This level should be used where residential grade (light duty) wall coverings, flat paints, or light textures are to be applied. The prepared surface shall be coated with a drywall primer prior to the application of final finishes. Release agents for wall coverings are specifically formulated to minimize damage if coverings are subsequently removed.

The weight, texture, and sheen level of the wall covering material selected should be taken into consideration when specifying wall coverings over this level of drywall treatment. Joints and fasteners must be sufficiently concealed if the wall covering material is lightweight, contains limited pattern, has a glossy finish, or has any combination of these features. In critical lighting areas, flat paints applied over light textures tend to reduce joint photographing. Gloss, semi-gloss, and enamel paints are not recommended over this level of finish.

All joints and interior angles shall have tape embedded in joint compound and shall be immediately wiped with a joint knife or trowel, leaving a thin coating of joint compound over all joints and interior angles. In addition, two separate coats of joint compound shall be applied over all flat joints and one separate coat of joint compound applied over interior angles. Fastener heads and accessories shall be covered with three separate coats of joint compound. All joint compounds shall be smooth and free of tool marks and ridges. The prepared surface shall be covered with a drywall primer like Sheetrock® First Coat prior to the application of the final decoration.

Level 5: The highest quality finish is the most effective method to provide a uniform surface and minimize the possibility of joint photographing and of fasteners showing through the final decoration. This level of finish is required where gloss, semi-gloss, or enamel is specified; when flat joints are specified over an untextured surface; or where critical lighting conditions occur. The prepared surface shall be coated with a drywall primer prior to the application of the final decoration.

All joints and interior angles shall have tape embedded in joint compound and be immediately wiped with a joint knife or trowel, leaving a thin coating of joint compound over all joints and interior angles. Two separate coats of joint compound shall be applied over all flat joints and one separate coat of joint compound applied over interior angles. Fastener heads and accessories shall be covered with three separate coats of joint compound.

A thin skim coat of joint compound shall be trowel applied to the entire surface. Excess compound is immediately troweled off, leaving a film or skim coating of compound completely covering the paper. As an alternative to a skim coat, a material manufactured especially for this purpose may be applied such as Sheetrock® Tuff-Hide primer surfacer. The surface must be smooth and free of tool marks and ridges. The prepared surface shall be covered with a drywall primer prior to the application of the final decoration.

Finishes — R0966 Terrazzo Flooring

R096613-10 Terrazzo Floor

The table below lists quantities required for 100 S.F. of 5/8″ terrazzo topping, either bonded or not bonded.

Description	Bonded to Concrete 1-1/8″ Bed, 1:4 Mix	Not Bonded 2-1/8″ Bed and 1/4″ Sand
Portland cement, 94 lb. Bag	6 bags	8 bags
Sand	10 C.F.	20 C.F.
Divider strips, 4′ squares	50 L.F.	50 L.F.
Terrazzo fill, 50 lb. Bag	12 bags	12 bags
15 Lb. tarred felt		1 C.S.F.
Mesh 2 x 2 #14 galvanized		1 C.S.F.
Crew J-3	0.77 days	0.87 days

2′ x 2′ panels require 1.00 L.F. divider strip per S.F.

3′ x 3′ panels require 0.67 L.F. divider strip per S.F.

4′ x 4′ panels require 0.50 L.F. divider strip per S.F.

5′ x 5′ panels require 0.40 L.F. divider strip per S.F.

6′ x 6′ panels require 0.33 L.F. divider strip per S.F.

Finishes — R0972 Wall Coverings

R097223-10 Wall Covering

The table below lists the quantities required for 100 S.F. of wall covering.

Description	Medium-Priced Paper	Expensive Paper
Paper	1.6 dbl. rolls	1.6 dbl. rolls
Wall sizing	0.25 gallon	0.25 gallon
Vinyl wall paste	0.6 gallon	0.6 gallon
Apply sizing	0.3 hour	0.3 hour
Apply paper	1.2 hours	1.5 hours

Most wallpapers now come in double rolls only.
To remove old paper, allow 1.3 hours per 100 S.F.

Finishes — R0991 Painting

R099100-10 Painting Estimating Techniques

Proper estimating methodology is needed to obtain an accurate painting estimate. There is no known reliable shortcut or square foot method. The following steps should be followed:

- List all surfaces to be painted, with an accurate quantity (area) of each. Items having similar surface condition, finish, application method and accessibility may be grouped together.
- List all the tasks required for each surface to be painted, including surface preparation, masking, and protection of adjacent surfaces. Surface preparation may include minor repairs, washing, sanding and puttying.
- Select the proper Means line for each task. Review and consider all adjustments to labor and materials for type of paint and location of work. Apply the height adjustment carefully. For instance, when applying the adjustment for work over 8′ high to a wall that is 12′ high, apply the adjustment only to the area between 8′ and 12′ high, and not to the entire wall.

When applying more than one percent (%) adjustment, apply each to the base cost of the data, rather than applying one percentage adjustment on top of the other.

When estimating the cost of painting walls and ceilings remember to add the brushwork for all cut-ins at inside corners and around windows and doors as a LF measure. One linear foot of cut-in with a brush equals one square foot of painting.

All items for spray painting include the labor for roll-back.

Deduct for openings greater than 100 SF or openings that extend from floor to ceiling and are greater than 5′ wide. Do not deduct small openings.

The cost of brushes, rollers, ladders and spray equipment is considered part of a painting contractor's overhead, and should not be added to the estimate. The cost of rented equipment such as scaffolding and swing staging should be added to the estimate.

Finishes — R0991 Painting

R099100-20 Painting

Item	Coat	One Gallon Covers			In 8 Hours a Laborer Covers			Labor-Hours per 100 S.F.		
		Brush	Roller	Spray	Brush	Roller	Spray	Brush	Roller	Spray
Paint wood siding	prime	250 S.F.	225 S.F.	290 S.F.	1150 S.F.	1300 S.F.	2275 S.F.	.695	.615	.351
	others	270	250	290	1300	1625	2600	.615	.492	.307
Paint exterior trim	prime	400	—	—	650	—	—	1.230	—	—
	1st	475	—	—	800	—	—	1.000	—	—
	2nd	520	—	—	975	—	—	.820	—	—
Paint shingle siding	prime	270	255	300	650	975	1950	1.230	.820	.410
	others	360	340	380	800	1150	2275	1.000	.695	.351
Stain shingle siding	1st	180	170	200	750	1125	2250	1.068	.711	.355
	2nd	270	250	290	900	1325	2600	.888	.603	.307
Paint brick masonry	prime	180	135	160	750	800	1800	1.066	1.000	.444
	1st	270	225	290	815	975	2275	.981	.820	.351
	2nd	340	305	360	815	1150	2925	.981	.695	.273
Paint interior plaster or drywall	prime	400	380	495	1150	2000	3250	.695	.400	.246
	others	450	425	495	1300	2300	4000	.615	.347	.200
Paint interior doors and windows	prime	400	—	—	650	—	—	1.230	—	—
	1st	425	—	—	800	—	—	1.000	—	—
	2nd	450	—	—	975	—	—	.820	—	—

Special Construction — R1311 Swimming Pools

R131113-20 Swimming Pools

Pool prices given per square foot of surface area include pool structure, filter and chlorination equipment, pumps, related piping, ladders/steps, maintenance kit, skimmer and vacuum system. Decks and electrical service to equipment are not included.

Residential in-ground pool construction can be divided into two categories: vinyl lined and gunite. Vinyl lined pool walls are constructed of different materials including wood, concrete, plastic or metal. The bottom is often graded with sand over which the vinyl liner is installed. Vermiculite or soil cement bottoms may be substituted for an added cost.

Gunite pool construction is used both in residential and municipal installations. These structures are steel reinforced for strength and finished with a white cement limestone plaster.

Municipal pools will have a higher cost because plumbing codes require more expensive materials, chlorination equipment and higher filtration rates.

Municipal pools greater than 1,800 S.F. require gutter systems to control waves. This gutter may be formed into the concrete wall. Often a vinyl/stainless steel gutter or gutter/wall system is specified, which will raise the pool cost.

Competition pools usually require tile bottoms and sides with contrasting lane striping, which will also raise the pool cost.

Plumbing — R2201 Operation & Maintenance of Plumbing

R220102-20 Labor Adjustment Factors

Labor Adjustment Factors are provided for Divisions 21, 22, and 23 to assist the mechanical estimator account for the various complexities and special conditions of any particular project. While a single percentage has been entered on each line of Division 22 01 02.20, it should be understood that these are just suggested midpoints of ranges of values commonly used by mechanical estimators. They may be increased or decreased depending on the severity of the special conditions.

The group for "existing occupied buildings" has been the subject of requests for explanation. Actually there are two stages to this group: buildings that are existing and "finished" but unoccupied, and those that also are occupied. Buildings that are "finished" may result in higher labor costs due to the workers having to be more careful not to damage finished walls, ceilings, floors, etc. and may necessitate special protective coverings and barriers. Also corridor bends and doorways may not accommodate long pieces of pipe or larger pieces of equipment. Work above an already hung ceiling can be very time consuming. The addition of occupants may force the work to be done on premium time (nights and/or weekends), eliminate the possible use of some preferred tools such as pneumatic drivers, powder charged drivers, etc. The estimator should evaluate the access to the work area and just how the work is going to be accomplished to arrive at an increase in labor costs over "normal" new construction productivity.

R236000-20 Air Conditioning Requirements

BTU's per hour per S.F. of floor area and S.F. per ton of air conditioning.

Type of Building	BTU/Hr per S.F.	S.F. per Ton	Type of Building	BTU/Hr per S.F.	S.F. per Ton	Type of Building	BTU per S.F.	S.F. per Ton
Apartments, Individual	26	450	Dormitory, Rooms	40	300	Libraries	50	240
Corridors	22	550	Corridors	30	400	Low Rise Office, Exterior	38	320
Auditoriums & Theaters	40	300/18*	Dress Shops	43	280	Interior	33	360
Banks	50	240	Drug Stores	80	150	Medical Centers	28	425
Barber Shops	48	250	Factories	40	300	Motels	28	425
Bars & Taverns	133	90	High Rise Office—Ext. Rms.	46	263	Office (small suite)	43	280
Beauty Parlors	66	180	Interior Rooms	37	325	Post Office, Individual Office	42	285
Bowling Alleys	68	175	Hospitals, Core	43	280	Central Area	46	260
Churches	36	330/20*	Perimeter	46	260	Residences	20	600
Cocktail Lounges	68	175	Hotel, Guest Rooms	44	275	Restaurants	60	200
Computer Rooms	141	85	Corridors	30	400	Schools & Colleges	46	260
Dental Offices	52	230	Public Spaces	55	220	Shoe Stores	55	220
Dept. Stores, Basement	34	350	Industrial Plants, Offices	38	320	Shop'g. Ctrs., Supermarkets	34	350
Main Floor	40	300	General Offices	34	350	Retail Stores	48	250
Upper Floor	30	400	Plant Areas	40	300	Specialty	60	200

*Persons per ton
12,000 BTU = 1 ton of air conditioning

R260533-22 Conductors in Conduit

The table below lists the maximum number of conductors for various sized conduit using THW, TW or THWN insulations.

Copper Wire Size	Conduit Size																										
	½″			¾″			1″			1-¼″			1-½″			2″			2-½″			3″		3-½″		4″	
	TW	THW	THWN	TW	THW	THWN	TW	THW	THWN	TW	THW	THWN	TW	THW	THWN	TW	THW	THWN	TW	THW	THWN	THW	THWN	THW	THWN	THW	THWN
#14	9	6	13	15	10	24	25	16	39	44	29	69	60	40	94	99	65	154	142	93		143		192			
#12	7	4	10	12	8	18	19	13	29	35	24	51	47	32	70	78	53	114	111	76	164	117		157			
#10	5	4	6	9	6	11	15	11	18	26	19	32	36	26	44	60	43	73	85	61	104	95	160	127		163	
#8	2	1	3	4	3	5	7	5	9	12	10	16	17	13	22	28	22	36	40	32	51	49	79	66	106	85	136
#6		1	1		2	4		4	6		7	11		10	15		16	26		23	37	36	57	48	76	62	98
#4		1	1		1	2		3	4		5	7		7	9		12	16		17	22	27	35	36	47	47	60
#3		1	1		1	1		2	3		4	6		6	8		10	13		15	19	23	29	31	39	40	51
#2		1	1		1	1		2	3		4	5		5	7		9	11		13	16	20	25	27	33	34	43
#1					1	1		1	1		3	3		4	5		6	8		9	12	14	18	19	25	25	32
1/0					1	1		1	1		2	3		3	4		5	7		8	10	12	15	16	21	21	27
2/0					1	1		1	1		1	2		3	3		5	6		7	8	10	13	14	17	18	22
3/0					1	1		1	1		1	1		2	3		4	5		6	7	9	11	12	14	15	18
4/0						1		1	1		1	1		1	2		3	4		5	6	7	9	10	12	13	15
250 kcmil								1	1		1	1		1	1		2	3		4	4	6	7	8	10	10	12
300								1	1		1	1		1	1		2	3		3	4	5	6	7	8	9	11
350									1		1	1		1	1		1	2		3	3	4	5	6	7	8	9
400											1	1		1	1		1	1		2	3	4	5	5	6	7	8
500											1	1		1	1		1	1		1	2	3	4	4	5	6	7
600												1		1	1		1	1		1	1	3	3	4	4	5	5
700														1	1		1	1		1	1	2	3	3	4	4	5
750 ▼														1	1		1	1		1	1	2	2	3	3	4	4

R312316-40 Excavating

The selection of equipment used for structural excavation and bulk excavation or for grading is determined by the following factors.

1. Quantity of material
2. Type of material
3. Depth or height of cut
4. Length of haul
5. Condition of haul road
6. Accessibility of site
7. Moisture content and dewatering requirements
8. Availability of excavating and hauling equipment

Some additional costs must be allowed for hand trimming the sides and bottom of concrete pours and other excavation below the general excavation.

When planning excavation and fill, the following should also be considered.

1. Swell factor
2. Compaction factor
3. Moisture content
4. Density requirements

A typical example for scheduling and estimating the cost of excavation of a 15′ deep basement on a dry site when the material must be hauled off the site is outlined below.

Assumptions:

1. Swell factor, 18%
2. No mobilization or demobilization
3. Allowance included for idle time and moving on job
4. No dewatering, sheeting, or bracing
5. No truck spotter or hand trimming

Number of B.C.Y. per truck = 1.5 C.Y. bucket x 8 passes = 12 loose C.Y.

$$= 12 \times \frac{100}{118} = 10.2 \text{ B.C.Y. per truck}$$

Truck Haul Cycle:

Load truck, 8 passes	=	4 minutes
Haul distance, 1 mile	=	9 minutes
Dump time	=	2 minutes
Return, 1 mile	=	7 minutes
Spot under machine	=	1 minute
		23 minute cycle

Fleet Haul Production per day in B.C.Y.

$$4 \text{ trucks} \times \frac{50 \text{ min. hour}}{23 \text{ min. haul cycle}} \times 8 \text{ hrs.} \times 10.2 \text{ B.C.Y.}$$

$$= 4 \times 2.2 \times 8 \times 10.2 = 718 \text{ B.C.Y./day}$$

Add the mobilization and demobilization costs to the total excavation costs. When equipment is rented for more than three days, there is often no mobilization charge by the equipment dealer. On larger jobs outside of urban areas, scrapers can move earth economically provided a dump site or fill area and adequate haul roads are available. Excavation within sheeting bracing or cofferdam bracing is usually done with a clamshell and production is low, since the clamshell may have to be guided by hand between the bracing. When excavating or filling an area enclosed with a wellpoint system, add 10% to 15% to the cost to allow for restricted access. When estimating earth excavation quantities for structures, allow work space outside the building footprint for construction of the foundation and a slope of 1:1 unless sheeting is used.

R312316-45 Excavating Equipment

The table below lists theoretical hourly production in C.Y./hr. bank measure for some typical excavation equipment. Figures assume 50 minute hours, 83% job efficiency, 100% operator efficiency, 90° swing and properly sized hauling units, which must be modified for adverse digging and loading conditions. Actual production costs in the front of the data set average about 50% of the theoretical values listed here.

Equipment	Soil Type	B.C.Y. Weight	% Swell	1 C.Y.	1-1/2 C.Y.	2 C.Y.	2-1/2 C.Y.	3 C.Y.	3-1/2 C.Y.	4 C.Y.
Hydraulic Excavator "Backhoe" 15′ Deep Cut	Moist loam, sandy clay	3400 lb.	40%	85	125	175	220	275	330	380
	Sand and gravel	3100	18	80	120	160	205	260	310	365
	Common earth	2800	30	70	105	150	190	240	280	330
	Clay, hard, dense	3000	33	65	100	130	170	210	255	300
Power Shovel Optimum Cut (Ft.)	Moist loam, sandy clay	3400	40	170	245	295	335	385	435	475
				(6.0)	(7.0)	(7.8)	(8.4)	(8.8)	(9.1)	(9.4)
	Sand and gravel	3100	18	165	225	275	325	375	420	460
				(6.0)	(7.0)	(7.8)	(8.4)	(8.8)	(9.1)	(9.4)
	Common earth	2800	30	145	200	250	295	335	375	425
				(7.8)	(9.2)	(10.2)	(11.2)	(12.1)	(13.0)	(13.8)
	Clay, hard, dense	3000	33	120	175	220	255	300	335	375
				(9.0)	(10.7)	(12.2)	(13.3)	(14.2)	(15.1)	(16.0)
Drag Line Optimum Cut (Ft.)	Moist loam, sandy clay	3400	40	130	180	220	250	290	325	385
				(6.6)	(7.4)	(8.0)	(8.5)	(9.0)	(9.5)	(10.0)
	Sand and gravel	3100	18	130	175	210	245	280	315	375
				(6.6)	(7.4)	(8.0)	(8.5)	(9.0)	(9.5)	(10.0)
	Common earth	2800	30	110	160	190	220	250	280	310
				(8.0)	(9.0)	(9.9)	(10.5)	(11.0)	(11.5)	(12.0)
	Clay, hard, dense	3000	33	90	130	160	190	225	250	280
				(9.3)	(10.7)	(11.8)	(12.3)	(12.8)	(13.3)	(12.0)
				Wheel Loaders				Track Loaders		
				3 C.Y.	4 C.Y.	6 C.Y.	8 C.Y.	2-1/4 C.Y.	3 C.Y.	4 C.Y.
Loading Tractors	Moist loam, sandy clay	3400	40	260	340	510	690	135	180	250
	Sand and gravel	3100	18	245	320	480	650	130	170	235
	Common earth	2800	30	230	300	460	620	120	155	220
	Clay, hard, dense	3000	33	200	270	415	560	110	145	200
	Rock, well-blasted	4000	50	180	245	380	520	100	130	180

R312323-30 Compacting Backfill

Compaction of fill in embankments, around structures, in trenches, and under slabs is important to control settlement. Factors affecting compaction are:

1. Soil gradation
2. Moisture content
3. Equipment used
4. Depth of fill per lift
5. Density required

Production Rate:

$$\frac{1.75' \text{ plate width x 50 F.P.M. x 50 min./hr. x .67' lift}}{27 \text{ C.F. per C.Y.}} = 108.5 \text{ C.Y./hr.}$$

Production Rate for 4 Passes:

$$\frac{108.5 \text{ C.Y.}}{4 \text{ passes}} = 27.125 \text{ C.Y./hr. x 8 hrs. = 217 C.Y./day}$$

Example:

Compact granular fill around a building foundation using a 21″ wide x 24″ vibratory plate in 8″ lifts. Operator moves at 50 F.P.M. working a 50 minute hour to develop 95% Modified Proctor Density with 4 passes.

Exterior Improvements — R3292 Turf & Grasses

R329219-50 Seeding

The type of grass is determined by light, shade and moisture content of soil plus intended use. Fertilizer should be disked 4″ before seeding. For steep slopes disk five tons of mulch and lay two tons of hay or straw on surface per acre after seeding. Surface mulch can be staked, lightly disked or tar emulsion sprayed. Material for mulch can be wood chips, peat moss, partially rotted hay or straw, wood fibers and sprayed emulsions. Hemp seed blankets with fertilizer are also available. For spring seeding, watering is necessary. Late fall seeding may have to be reseeded in the spring. Hydraulic seeding, power mulching, and aerial seeding can be used on large areas.

Utilities — R3311 Water Utility Distribution Piping

R331113-80 Piping Designations

There are several systems currently in use to describe pipe and fittings. The following paragraphs will help to identify and clarify classifications of piping systems used for water distribution.

Piping may be classified by schedule. Piping schedules include 5S, 10S, 10, 20, 30, Standard, 40, 60, Extra Strong, 80, 100, 120, 140, 160 and Double Extra Strong. These schedules are dependent upon the pipe wall thickness. The wall thickness of a particular schedule may vary with pipe size.

Ductile iron pipe for water distribution is classified by Pressure Classes such as Class 150, 200, 250, 300 and 350. These classes are actually the rated water working pressure of the pipe in pounds per square inch (psi). The pipe in these pressure classes is designed to withstand the rated water working pressure plus a surge allowance of 100 psi.

The American Water Works Association (AWWA) provides standards for various types of **plastic pipe.** C-900 is the specification for polyvinyl chloride (PVC) piping used for water distribution in sizes ranging from 4″ through 12″. C-901 is the specification for polyethylene (PE) pressure pipe, tubing and fittings used for water distribution in sizes ranging from 1/2″ through 3″. C-905 is the specification for PVC piping sizes 14″ and greater.

PVC pressure-rated pipe is identified using the standard dimensional ratio (SDR) method. This method is defined by the American Society for Testing and Materials (ASTM) Standard D 2241. This pipe is available in SDR numbers 64, 41, 32.5, 26, 21, 17, and 13.5. A pipe with an SDR of 64 will have the thinnest wall while a pipe with an SDR of 13.5 will have the thickest wall. When the pressure rating (PR) of a pipe is given in psi, it is based on a line supplying water at 73 degrees F.

The National Sanitation Foundation (NSF) seal of approval is applied to products that can be used with potable water. These products have been tested to ANSI/NSF Standard 14.

Valves and strainers are classified by American National Standards Institute (ANSI) Classes. These Classes are 125, 150, 200, 250, 300, 400, 600, 900, 1500 and 2500. Within each class there is an operating pressure range dependent upon temperature. Design parameters should be compared to the appropriate material dependent, pressure-temperature rating chart for accurate valve selection.

Change Orders

Change Order Considerations

A change order is a written document, usually prepared by the design professional, and signed by the owner, the architect/engineer, and the contractor. A change order states the agreement of the parties to: an addition, deletion, or revision in the work; an adjustment in the contract sum, if any; or an adjustment in the contract time, if any. Change orders, or "extras" in the construction process occur after execution of the construction contract and impact architects/ engineers, contractors, and owners.

Change orders that are properly recognized and managed can ensure orderly, professional, and profitable progress for all who are involved in the project. There are many causes for change orders and change order requests. In all cases, change orders or change order requests should be addressed promptly and in a precise and prescribed manner. The following paragraphs include information regarding change order pricing and procedures.

The Causes of Change Orders

Reasons for issuing change orders include:

- Unforeseen field conditions that require a change in the work
- Correction of design discrepancies, errors, or omissions in the contract documents
- Owner-requested changes, either by design criteria, scope of work, or project objectives
- Completion date changes for reasons unrelated to the construction process
- Changes in building code interpretations, or other public authority requirements that require a change in the work
- Changes in availability of existing or new materials and products

Procedures

Properly written contract documents must include the correct change order procedures for all parties—owners, design professionals and contractors—to follow in order to avoid costly delays and litigation.

Being "in the right" is not always a sufficient or acceptable defense. The contract provisions requiring notification and documentation must be adhered to within a defined or reasonable time frame.

The appropriate method of handling change orders is by a written proposal and acceptance by all parties involved. Prior to starting work on a project, all parties should identify their authorized agents who may sign and accept change orders, as well as any limits placed on their authority.

Time may be a critical factor when the need for a change arises. For such cases, the contractor might be directed to proceed on a "time and materials" basis, rather than wait for all paperwork to be processed—a delay that could impede progress. In this situation, the contractor must still follow the prescribed change order procedures including, but not limited to, notification and documentation.

Lack of documentation can be very costly, especially if legal judgments are to be made, and if certain field personnel are no longer available. For time and material change orders, the contractor should keep accurate daily records of all labor and material allocated to the change.

Owners or awarding authorities who do considerable and continual building construction (such as the federal government) realize the inevitability of change orders for numerous reasons, both predictable and unpredictable. As a result, the federal government, the American Institute of Architects (AIA), the Engineers Joint Contract Documents Committee (EJCDC) and other contractor, legal, and technical organizations have developed standards and procedures to be followed by all parties to achieve contract continuance and timely completion, while being financially fair to all concerned.

Pricing Change Orders

When pricing change orders, regardless of their cause, the most significant factor is when the change occurs. The need for a change may be perceived in the field or requested by the architect/engineer *before* any of the actual installation has begun, or may evolve or appear *during* construction when the item of work in question is partially installed. In the latter cases, the original sequence of construction is disrupted, along with all contiguous and supporting systems. Change orders cause the greatest impact when they occur *after* the installation has been completed and must be uncovered, or even replaced. Post-completion changes may be caused by necessary design changes, product failure, or changes in the owner's requirements that are not discovered until the building or the systems begin to function.

Specified procedures of notification and record keeping must be adhered to and enforced regardless of the stage of construction: *before, during,* or *after* installation. Some bidding documents anticipate change orders by requiring that unit prices including overhead and profit percentages—for additional as well as deductible changes—be listed. Generally these unit prices do not fully take into account the ripple effect, or impact on other trades, and should be used for general guidance only.

When pricing change orders, it is important to classify the time frame in which the change occurs. There are two basic time frames for change orders: *pre-installation change orders,* which occur before the start of construction, and *post-installation change orders,* which involve reworking after the original installation. Change orders that occur between these stages may be priced according to the extent of work completed using a combination of techniques developed for pricing *pre-* and *post-installation* changes.

Factors To Consider When Pricing Change Orders

As an estimator begins to prepare a change order, the following questions should be reviewed to determine their impact on the final price.

General

- *Is the change order work* pre-installation *or* post-installation?

  Change order work costs vary according to how much of the installation has been completed. Once workers have the project scoped in their minds, even though they have not started, it can be difficult to refocus. Consequently they may spend more than the normal amount of time understanding the change. Also, modifications to work in place, such as trimming or refitting, usually take more time than was initially estimated. The greater the amount of work in place, the more reluctant workers are to change it. Psychologically they may resent the change and as a result the rework takes longer than normal. Post-installation change order estimates must include demolition of existing work as required to accomplish the change. If the work is performed at a later time, additional obstacles, such as building finishes, may be present which must be protected. Regardless of whether the change occurs pre-installation or post-installation, attempt to isolate the identifiable factors and price them separately. For example, add shipping costs that may be required pre-installation or any demolition required post-installation. Then analyze the potential impact on productivity of psychological and/or learning curve factors and adjust the output rates accordingly. One approach is to break down the typical workday into segments and quantify the impact on each segment.

Change Order Installation Efficiency

The labor-hours expressed (for new construction) are based on average installation time, using an efficiency level. For change order situations, adjustments to this efficiency level should reflect the daily labor-hour allocation for that particular occurrence.

- *Will the change substantially delay the original completion date?*

  A significant change in the project may cause the original completion date to be extended. The extended schedule may subject the contractor to new wage rates dictated by relevant labor contracts. Project supervision and other project overhead must also be extended beyond the original completion date. The schedule extension may also put installation into a new weather season. For example, underground piping scheduled for October installation was delayed until January. As a result, frost penetrated the trench area, thereby changing the degree of difficulty of the task. Changes and delays may have a ripple effect throughout the project. This effect must be analyzed and negotiated with the owner.

- *What is the net effect of a deduct change order?*

  In most cases, change orders resulting in a deduction or credit reflect only bare costs. The contractor may retain the overhead and profit based on the original bid.

Materials

- *Will you have to pay more or less for the new material, required by the change order, than you paid for the original purchase?*

  The same material prices or discounts will usually apply to materials purchased for change orders as new construction. In some instances, however, the contractor may forfeit the advantages of competitive pricing for change orders. Consider the following example:

  A contractor purchased over $20,000 worth of fan coil units for an installation, and obtained the maximum discount. Some time later it was determined the project required an additional matching unit. The contractor has to purchase this unit from the original supplier to ensure a match. The supplier at this time may not discount the unit because of the small quantity, and the fact that he is no longer in a competitive situation. The impact of quantity on purchase can add between 0% and 25% to material prices and/or subcontractor quotes.

- *If materials have been ordered or delivered to the job site, will they be subject to a cancellation charge or restocking fee?*

  Check with the supplier to determine if ordered materials are subject to a cancellation charge. Delivered materials not used as result of a change order may be subject to a restocking fee if returned to the supplier. Common restocking charges run between 20% and 40%. Also, delivery charges to return the goods to the supplier must be added.

Labor

- *How efficient is the existing crew at the actual installation?*

  Is the same crew that performed the initial work going to do the change order? Possibly the change consists of the installation of a unit identical to one already installed; therefore, the change should take less time. Be sure to consider this potential productivity increase and modify the productivity rates accordingly.

- *If the crew size is increased, what impact will that have on supervision requirements?*

  Under most bargaining agreements or management practices, there is a point at which a working foreman is replaced by a nonworking foreman. This replacement increases project overhead by adding a nonproductive worker. If additional workers are added to accelerate the project or to perform changes while maintaining the schedule, be sure to add additional supervision time if warranted. Calculate the hours involved and the additional cost directly if possible.

- *What are the other impacts of increased crew size?*

  The larger the crew, the greater the potential for productivity to decrease. Some of the factors that cause this productivity loss are: overcrowding (producing restrictive conditions in the working space), and possibly a shortage of any special tools and equipment required. Such factors affect not only the crew working on the elements directly involved in the change order, but other crews whose movement may also be hampered. As the crew increases, check its basic composition for changes by the addition or deletion of apprentices or nonworking foreman, and quantify the potential effects of equipment shortages or other logistical factors.

- *As new crews, unfamiliar with the project, are brought onto the site, how long will it take them to become oriented to the project requirements?*

  The orientation time for a new crew to become 100% effective varies with the site and type of project. Orientation is easiest at a new construction site, and most difficult at existing, very restrictive renovation sites. The type of work also affects orientation time. When all elements of the work are exposed, such as concrete or masonry work, orientation is decreased. When the work is concealed or less visible, such as existing electrical systems, orientation takes longer. Usually orientation can be accomplished in one day or less. Costs for added orientation should be itemized and added to the total estimated cost.

- *How much actual production can be gained by working overtime?*

  Short term overtime can be used effectively to accomplish more work in a day. However, as overtime is scheduled to run beyond several weeks, studies have shown marked decreases in output. The following chart shows the effect of long term overtime on worker efficiency. If the anticipated change requires extended overtime to keep the job on schedule, these factors can be used as a guide to predict the impact on time and cost. Add project overhead, particularly supervision, that may also be incurred.

Days per Week	Hours per Day	Production Efficiency					Payroll Cost Factors	
		1st Week	2nd Week	3rd Week	4th Week	Average 4 Weeks	@ 1-1/2 Times	@ 2 Times
5	8	100%	100%	100%	100%	100%	100%	100%
	9	100	100	95	90	96.25	105.6	111.1
	10	100	95	90	85	91.25	110.0	120.0
	11	95	90	75	65	81.25	113.6	127.3
	12	90	85	70	60	76.25	116.7	133.3
6	8	100	100	95	90	96.25	108.3	116.7
	9	100	95	90	85	92.50	113.0	125.9
	10	95	90	85	80	87.50	116.7	133.3
	11	95	85	70	65	78.75	119.7	139.4
	12	90	80	65	60	73.75	122.2	144.4
7	8	100	95	85	75	88.75	114.3	128.6
	9	95	90	80	70	83.75	118.3	136.5
	10	90	85	75	65	78.75	121.4	142.9
	11	85	80	65	60	72.50	124.0	148.1
	12	85	75	60	55	68.75	126.2	152.4

Effects of Overtime

Caution: Under many labor agreements, Sundays and holidays are paid at a higher premium than the normal overtime rate.

The use of long-term overtime is counterproductive on almost any construction job; that is, the longer the period of overtime, the lower the actual production rate. Numerous studies have been conducted, and while they have resulted in slightly different numbers, all reach the same conclusion. The figure above tabulates the effects of overtime work on efficiency.

As illustrated, there can be a difference between the *actual* payroll cost per hour and the *effective* cost per hour for overtime work. This is due to the reduced production efficiency with the increase in weekly hours beyond 40. This difference between actual and effective cost results from overtime work over a prolonged period. Short-term overtime work does not result in as great a reduction in efficiency and, in such cases, effective cost may not vary significantly from the actual payroll cost. As the total hours per week are increased on a regular basis, more time is lost due to fatigue, lowered morale, and an increased accident rate.

As an example, assume a project where workers are working 6 days a week, 10 hours per day. From the figure above (based on productivity studies), the average effective productive hours over a 4-week period are:

$$0.875 \times 60 = 52.5$$

Depending upon the locale and day of week, overtime hours may be paid at time and a half or double time. For time and a half, the overall (average) *actual* payroll cost (including regular and overtime hours) is determined as follows:

$$\frac{40 \text{ reg. hrs.} + (20 \text{ overtime hrs.} \times 1.5)}{60 \text{ hrs.}} = 1.167$$

Based on 60 hours, the payroll cost per hour will be 116.7% of the normal rate at 40 hours per week. However, because the effective production (efficiency) for 60 hours is reduced to the equivalent of 52.5 hours, the effective cost of overtime is calculated as follows:

For time and a half:

$$\frac{40 \text{ reg. hrs.} + (20 \text{ overtime hrs.} \times 1.5)}{52.5 \text{ hrs.}} = 1.33$$

Installed cost will be 133% of the normal rate (for labor).

Thus, when figuring overtime, the actual cost per unit of work will be higher than the apparent overtime payroll dollar increase, due to the reduced productivity of the longer workweek. These efficiency calculations are true only for those cost factors determined by hours worked. Costs that are applied weekly or monthly, such as equipment rentals, will not be similarly affected.

Equipment

- *What equipment is required to complete the change order?*

  Change orders may require extending the rental period of equipment already on the job site, or the addition of special equipment brought in to accomplish the change work. In either case, the additional rental charges and operator labor charges must be added.

Summary

The preceding considerations and others you deem appropriate should be analyzed and applied to a change order estimate. The impact of each should be quantified and listed on the estimate to form an audit trail.

Change orders that are properly identified, documented, and managed help to ensure the orderly, professional and profitable progress of the work. They also minimize potential claims or disputes at the end of the project.

Abbreviations

Abbreviation	Meaning
A	Area Square Feet; Ampere
AAFES	Army and Air Force Exchange Service
ABS	Acrylonitrile Butadiene Stryrene; Asbestos Bonded Steel
A.C., AC	Alternating Current; Air-Conditioning; Asbestos Cement; Plywood Grade A & C
ACI	American Concrete Institute
ACR	Air Conditioning Refrigeration
ADA	Americans with Disabilities Act
AD	Plywood, Grade A & D
Addit.	Additional
Adj.	Adjustable
af	Audio-frequency
AFUE	Annual Fuel Utilization Efficiency
AGA	American Gas Association
Agg.	Aggregate
A.H., Ah	Ampere Hours
A hr.	Ampere-hour
A.H.U., AHU	Air Handling Unit
A.I.A.	American Institute of Architects
AIC	Ampere Interrupting Capacity
Allow.	Allowance
alt., alt	Alternate
Alum.	Aluminum
a.m.	Ante Meridiem
Amp.	Ampere
Anod.	Anodized
ANSI	American National Standards Institute
APA	American Plywood Association
Approx.	Approximate
Apt.	Apartment
Asb.	Asbestos
A.S.B.C.	American Standard Building Code
Asbe.	Asbestos Worker
ASCE	American Society of Civil Engineers
A.S.H.R.A.E.	American Society of Heating, Refrig. & AC Engineers
ASME	American Society of Mechanical Engineers
ASTM	American Society for Testing and Materials
Attchmt.	Attachment
Avg., Ave.	Average
AWG	American Wire Gauge
AWWA	American Water Works Assoc.
Bbl.	Barrel
B&B, BB	Grade B and Better; Balled & Burlapped
B&S	Bell and Spigot
B.&W.	Black and White
b.c.c.	Body-centered Cubic
B.C.Y.	Bank Cubic Yards
BE	Bevel End
B.F.	Board Feet
Bg. cem.	Bag of Cement
BHP	Boiler Horsepower; Brake Horsepower
B.I.	Black Iron
bidir.	bidirectional
Bit., Bitum.	Bituminous
Bit., Conc.	Bituminous Concrete
Bk.	Backed
Bkrs.	Breakers
Bldg., bldg	Building
Blk.	Block
Bm.	Beam
Boil.	Boilermaker
bpm	Blows per Minute
BR	Bedroom
Brg., brng.	Bearing
Brhe.	Bricklayer Helper
Bric.	Bricklayer
Brk., brk	Brick
brkt	Bracket
Brs.	Brass
Brz.	Bronze
Bsn.	Basin
Btr.	Better
BTU	British Thermal Unit
BTUH	BTU per Hour
Bu.	Bushels
BUR	Built-up Roofing
BX	Interlocked Armored Cable
°C	Degree Centigrade
c	Conductivity, Copper Sweat
C	Hundred; Centigrade
C/C	Center to Center, Cedar on Cedar
C-C	Center to Center
Cab	Cabinet
Cair.	Air Tool Laborer
Cal.	Caliper
Calc	Calculated
Cap.	Capacity
Carp.	Carpenter
C.B.	Circuit Breaker
C.C.A.	Chromate Copper Arsenate
C.C.F.	Hundred Cubic Feet
cd	Candela
cd/sf	Candela per Square Foot
CD	Grade of Plywood Face & Back
CDX	Plywood, Grade C & D, exterior glue
Cefi.	Cement Finisher
Cem.	Cement
CF	Hundred Feet
C.F.	Cubic Feet
CFM	Cubic Feet per Minute
CFRP	Carbon Fiber Reinforced Plastic
c.g.	Center of Gravity
CHW	Chilled Water; Commercial Hot Water
C.I., CI	Cast Iron
C.I.P., CIP	Cast in Place
Circ.	Circuit
C.L.	Carload Lot
CL	Chain Link
Clab.	Common Laborer
Clam	Common Maintenance Laborer
C.L.F.	Hundred Linear Feet
CLF	Current Limiting Fuse
CLP	Cross Linked Polyethylene
cm	Centimeter
CMP	Corr. Metal Pipe
CMU	Concrete Masonry Unit
CN	Change Notice
Col.	Column
CO_2	Carbon Dioxide
Comb.	Combination
comm.	Commercial, Communication
Compr.	Compressor
Conc.	Concrete
Cont., cont	Continuous; Continued, Container
Corr.	Corrugated
Cos	Cosine
Cot	Cotangent
Cov.	Cover
C/P	Cedar on Paneling
CPA	Control Point Adjustment
Cplg.	Coupling
CPM	Critical Path Method
CPVC	Chlorinated Polyvinyl Chloride
C.Pr.	Hundred Pair
CRC	Cold Rolled Channel
Creos.	Creosote
Crpt.	Carpet & Linoleum Layer
CRT	Cathode-ray Tube
CS	Carbon Steel, Constant Shear Bar Joist
Csc	Cosecant
C.S.F.	Hundred Square Feet
CSI	Construction Specifications Institute
CT	Current Transformer
CTS	Copper Tube Size
Cu	Copper, Cubic
Cu. Ft.	Cubic Foot
cw	Continuous Wave
C.W.	Cool White; Cold Water
Cwt.	100 Pounds
C.W.X.	Cool White Deluxe
C.Y.	Cubic Yard (27 cubic feet)
C.Y./Hr.	Cubic Yard per Hour
Cyl.	Cylinder
d	Penny (nail size)
D	Deep; Depth; Discharge
Dis., Disch.	Discharge
Db	Decibel
Dbl.	Double
DC	Direct Current
DDC	Direct Digital Control
Demob.	Demobilization
d.f.t.	Dry Film Thickness
d.f.u.	Drainage Fixture Units
D.H.	Double Hung
DHW	Domestic Hot Water
DI	Ductile Iron
Diag.	Diagonal
Diam., Dia	Diameter
Distrib.	Distribution
Div.	Division
Dk.	Deck
D.L.	Dead Load; Diesel
DLH	Deep Long Span Bar Joist
dlx	Deluxe
Do.	Ditto
DOP	Dioctyl Phthalate Penetration Test (Air Filters)
Dp., dp	Depth
D.P.S.T.	Double Pole, Single Throw
Dr.	Drive
DR	Dimension Ratio
Drink.	Drinking
D.S.	Double Strength
D.S.A.	Double Strength A Grade
D.S.B.	Double Strength B Grade
Dty.	Duty
DWV	Drain Waste Vent
DX	Deluxe White, Direct Expansion
dyn	Dyne
e	Eccentricity
E	Equipment Only; East; Emissivity
Ea.	Each
EB	Encased Burial
Econ.	Economy
E.C.Y	Embankment Cubic Yards
EDP	Electronic Data Processing
EIFS	Exterior Insulation Finish System
E.D.R.	Equiv. Direct Radiation
Eq.	Equation
EL	Elevation
Elec.	Electrician; Electrical
Elev.	Elevator; Elevating
EMT	Electrical Metallic Conduit; Thin Wall Conduit
Eng.	Engine, Engineered
EPDM	Ethylene Propylene Diene Monomer
EPS	Expanded Polystyrene
Eqhv.	Equip. Oper., Heavy
Eqlt.	Equip. Oper., Light
Eqmd.	Equip. Oper., Medium
Eqmm.	Equip. Oper., Master Mechanic
Eqol.	Equip. Oper., Oilers
Equip.	Equipment
ERW	Electric Resistance Welded

Abbreviation	Meaning
E.S.	Energy Saver
Est.	Estimated
esu	Electrostatic Units
E.W.	Each Way
EWT	Entering Water Temperature
Excav.	Excavation
excl	Excluding
Exp., exp	Expansion, Exposure
Ext., ext	Exterior; Extension
Extru.	Extrusion
f.	Fiber Stress
F	Fahrenheit; Female; Fill
Fab., fab	Fabricated; Fabric
FBGS	Fiberglass
F.C.	Footcandles
f.c.c.	Face-centered Cubic
f'c.	Compressive Stress in Concrete; Extreme Compressive Stress
F.E.	Front End
FEP	Fluorinated Ethylene Propylene (Teflon)
F.G.	Flat Grain
F.H.A.	Federal Housing Administration
Fig.	Figure
Fin.	Finished
FIPS	Female Iron Pipe Size
Fixt.	Fixture
FJP	Finger jointed and primed
Fl. Oz.	Fluid Ounces
Flr.	Floor
FM	Frequency Modulation; Factory Mutual
Fmg.	Framing
FM/UL	Factory Mutual/Underwriters Labs
Fdn.	Foundation
FNPT	Female National Pipe Thread
Fori.	Foreman, Inside
Foro.	Foreman, Outside
Fount.	Fountain
fpm	Feet per Minute
FPT	Female Pipe Thread
Fr	Frame
F.R.	Fire Rating
FRK	Foil Reinforced Kraft
FSK	Foil/Scrim/Kraft
FRP	Fiberglass Reinforced Plastic
FS	Forged Steel
FSC	Cast Body; Cast Switch Box
Ft., ft	Foot; Feet
Ftng.	Fitting
Ftg.	Footing
Ft lb.	Foot Pound
Furn.	Furniture
FVNR	Full Voltage Non-Reversing
FVR	Full Voltage Reversing
FXM	Female by Male
Fy.	Minimum Yield Stress of Steel
g	Gram
G	Gauss
Ga.	Gauge
Gal., gal.	Gallon
Galv., galv	Galvanized
GC/MS	Gas Chromatograph/Mass Spectrometer
Gen	General
GFI	Ground Fault Interrupter
GFRC	Glass Fiber Reinforced Concrete
Glaz.	Glazier
GPD	Gallons per Day
gpf	Gallon per Flush
GPH	Gallons per Hour
gpm, GPM	Gallons per Minute
GR	Grade
Gran.	Granular
Grnd.	Ground
GVW	Gross Vehicle Weight
GWB	Gypsum Wall Board
H	High Henry
HC	High Capacity
H.D., HD	Heavy Duty; High Density
H.D.O.	High Density Overlaid
HDPE	High Density Polyethylene Plastic
Hdr.	Header
Hdwe.	Hardware
H.I.D., HID	High Intensity Discharge
Help.	Helper Average
HEPA	High Efficiency Particulate Air Filter
Hg	Mercury
HIC	High Interrupting Capacity
HM	Hollow Metal
HMWPE	High Molecular Weight Polyethylene
HO	High Output
Horiz.	Horizontal
H.P., HP	Horsepower; High Pressure
H.P.F.	High Power Factor
Hr.	Hour
Hrs./Day	Hours per Day
HSC	High Short Circuit
Ht.	Height
Htg.	Heating
Htrs.	Heaters
HVAC	Heating, Ventilation & Air-Conditioning
Hvy.	Heavy
HW	Hot Water
Hyd.; Hydr.	Hydraulic
Hz	Hertz (cycles)
I.	Moment of Inertia
IBC	International Building Code
I.C.	Interrupting Capacity
ID	Inside Diameter
I.D.	Inside Dimension; Identification
I.F.	Inside Frosted
I.M.C.	Intermediate Metal Conduit
In.	Inch
Incan.	Incandescent
Incl.	Included; Including
Int.	Interior
Inst.	Installation
Insul., insul	Insulation/Insulated
I.P.	Iron Pipe
I.P.S., IPS	Iron Pipe Size
IPT	Iron Pipe Threaded
I.W.	Indirect Waste
J	Joule
J.I.C.	Joint Industrial Council
K	Thousand; Thousand Pounds; Heavy Wall Copper Tubing, Kelvin
K.A.H.	Thousand Amp. Hours
kcmil	Thousand Circular Mils
KD	Knock Down
K.D.A.T.	Kiln Dried After Treatment
kg	Kilogram
kG	Kilogauss
kgf	Kilogram Force
kHz	Kilohertz
Kip	1000 Pounds
KJ	Kilojoule
K.L.	Effective Length Factor
K.L.F.	Kips per Linear Foot
Km	Kilometer
KO	Knock Out
K.S.F.	Kips per Square Foot
K.S.I.	Kips per Square Inch
kV	Kilovolt
kVA	Kilovolt Ampere
kVAR	Kilovar (Reactance)
KW	Kilowatt
KWh	Kilowatt-hour
L	Labor Only; Length; Long; Medium Wall Copper Tubing
Lab.	Labor
lat	Latitude
Lath.	Lather
Lav.	Lavatory
lb.; #	Pound
L.B., LB	Load Bearing; L Conduit Body
L. & E.	Labor & Equipment
lb./hr.	Pounds per Hour
lb./L.F.	Pounds per Linear Foot
lbf/sq.in.	Pound-force per Square Inch
L.C.L.	Less than Carload Lot
L.C.Y.	Loose Cubic Yard
Ld.	Load
LE	Lead Equivalent
LED	Light Emitting Diode
L.F.	Linear Foot
L.F. Nose	Linear Foot of Stair Nosing
L.F. Rsr	Linear Foot of Stair Riser
Lg.	Long; Length; Large
L & H	Light and Heat
LH	Long Span Bar Joist
L.H.	Labor Hours
L.L., LL	Live Load
L.L.D.	Lamp Lumen Depreciation
lm	Lumen
lm/sf	Lumen per Square Foot
lm/W	Lumen per Watt
LOA	Length Over All
log	Logarithm
L-O-L	Lateralolet
long.	Longitude
L.P., LP	Liquefied Petroleum; Low Pressure
L.P.F.	Low Power Factor
LR	Long Radius
L.S.	Lump Sum
Lt.	Light
Lt. Ga.	Light Gauge
L.T.L.	Less than Truckload Lot
Lt. Wt.	Lightweight
L.V.	Low Voltage
M	Thousand; Material; Male; Light Wall Copper Tubing
M^2CA	Meters Squared Contact Area
m/hr.; M.H.	Man-hour
mA	Milliampere
Mach.	Machine
Mag. Str.	Magnetic Starter
Maint.	Maintenance
Marb.	Marble Setter
Mat; Mat'l.	Material
Max.	Maximum
MBF	Thousand Board Feet
MBH	Thousand BTU's per hr.
MC	Metal Clad Cable
MCC	Motor Control Center
M.C.F.	Thousand Cubic Feet
MCFM	Thousand Cubic Feet per Minute
M.C.M.	Thousand Circular Mils
MCP	Motor Circuit Protector
MD	Medium Duty
MDF	Medium-density fibreboard
M.D.O.	Medium Density Overlaid
Med.	Medium
MF	Thousand Feet
M.F.B.M.	Thousand Feet Board Measure
Mfg.	Manufacturing
Mfrs.	Manufacturers
mg	Milligram
MGD	Million Gallons per Day
MGPH	Million Gallons per Hour
MH, M.H.	Manhole; Metal Halide; Man-Hour
MHz	Megahertz
Mi.	Mile
MI	Malleable Iron; Mineral Insulated
MIPS	Male Iron Pipe Size
mj	Mechanical Joint
m	Meter
mm	Millimeter
Mill.	Millwright
Min., min.	Minimum, Minute

Abbreviations

Abbreviation	Meaning
Misc.	Miscellaneous
ml	Milliliter, Mainline
M.L.F.	Thousand Linear Feet
Mo.	Month
Mobil.	Mobilization
Mog.	Mogul Base
MPH	Miles per Hour
MPT	Male Pipe Thread
MRGWB	Moisture Resistant Gypsum Wallboard
MRT	Mile Round Trip
ms	Millisecond
M.S.F.	Thousand Square Feet
Mstz.	Mosaic & Terrazzo Worker
M.S.Y.	Thousand Square Yards
Mtd., mtd., mtd	Mounted
Mthe.	Mosaic & Terrazzo Helper
Mtng.	Mounting
Mult.	Multi; Multiply
M.V.A.	Million Volt Amperes
M.V.A.R.	Million Volt Amperes Reactance
MV	Megavolt
MW	Megawatt
MXM	Male by Male
MYD	Thousand Yards
N	Natural; North
nA	Nanoampere
NA	Not Available; Not Applicable
N.B.C.	National Building Code
NC	Normally Closed
NEMA	National Electrical Manufacturers Assoc.
NEHB	Bolted Circuit Breaker to 600V.
NFPA	National Fire Protection Association
NLB	Non-Load-Bearing
NM	Non-Metallic Cable
nm	Nanometer
No.	Number
NO	Normally Open
N.O.C.	Not Otherwise Classified
Nose.	Nosing
NPT	National Pipe Thread
NQOD	Combination Plug-on/Bolt on Circuit Breaker to 240V.
N.R.C., NRC	Noise Reduction Coefficient/ Nuclear Regulator Commission
N.R.S.	Non Rising Stem
ns	Nanosecond
nW	Nanowatt
OB	Opposing Blade
OC	On Center
OD	Outside Diameter
O.D.	Outside Dimension
ODS	Overhead Distribution System
O.G.	Ogee
O.H.	Overhead
O&P	Overhead and Profit
Oper.	Operator
Opng.	Opening
Orna.	Ornamental
OSB	Oriented Strand Board
OS&Y	Outside Screw and Yoke
OSHA	Occupational Safety and Health Act
Ovhd.	Overhead
OWG	Oil, Water or Gas
Oz.	Ounce
P.	Pole; Applied Load; Projection
p.	Page
Pape.	Paperhanger
P.A.P.R.	Powered Air Purifying Respirator
PAR	Parabolic Reflector
P.B., PB	Push Button
Pc., Pcs.	Piece, Pieces
P.C.	Portland Cement; Power Connector
P.C.F.	Pounds per Cubic Foot
PCM	Phase Contrast Microscopy
PDCA	Painting and Decorating Contractors of America
P.E., PE	Professional Engineer; Porcelain Enamel; Polyethylene; Plain End
P.E.C.I.	Porcelain Enamel on Cast Iron
Perf.	Perforated
PEX	Cross Linked Polyethylene
Ph.	Phase
P.I.	Pressure Injected
Pile.	Pile Driver
Pkg.	Package
Pl.	Plate
Plah.	Plasterer Helper
Plas.	Plasterer
plf	Pounds Per Linear Foot
Pluh.	Plumber Helper
Plum.	Plumber
Ply.	Plywood
p.m.	Post Meridiem
Pntd.	Painted
Pord.	Painter, Ordinary
pp	Pages
PP, PPL	Polypropylene
P.P.M.	Parts per Million
Pr.	Pair
P.E.S.B.	Pre-engineered Steel Building
Prefab.	Prefabricated
Prefin.	Prefinished
Prop.	Propelled
PSF, psf	Pounds per Square Foot
PSI, psi	Pounds per Square Inch
PSIG	Pounds per Square Inch Gauge
PSP	Plastic Sewer Pipe
Pspr.	Painter, Spray
Psst.	Painter, Structural Steel
P.T.	Potential Transformer
P. & T.	Pressure & Temperature
Ptd.	Painted
Ptns.	Partitions
Pu	Ultimate Load
PVC	Polyvinyl Chloride
Pvmt.	Pavement
PRV	Pressure Relief Valve
Pwr.	Power
Q	Quantity Heat Flow
Qt.	Quart
Quan., Qty.	Quantity
Q.C.	Quick Coupling
r	Radius of Gyration
R	Resistance
R.C.P.	Reinforced Concrete Pipe
Rect.	Rectangle
recpt.	Receptacle
Reg.	Regular
Reinf.	Reinforced
Req'd.	Required
Res.	Resistant
Resi.	Residential
RF	Radio Frequency
RFID	Radio-frequency Identification
Rgh.	Rough
RGS	Rigid Galvanized Steel
RHW	Rubber, Heat & Water Resistant; Residential Hot Water
rms	Root Mean Square
Rnd.	Round
Rodm.	Rodman
Rofc.	Roofer, Composition
Rofp.	Roofer, Precast
Rohe.	Roofer Helpers (Composition)
Rots.	Roofer, Tile & Slate
R.O.W.	Right of Way
RPM	Revolutions per Minute
R.S.	Rapid Start
Rsr	Riser
RT	Round Trip
S.	Suction; Single Entrance; South
SBS	Styrene Butadiere Styrene
SC	Screw Cover
SCFM	Standard Cubic Feet per Minute
Scaf.	Scaffold
Sch., Sched.	Schedule
S.C.R.	Modular Brick
S.D.	Sound Deadening
SDR	Standard Dimension Ratio
S.E.	Surfaced Edge
Sel.	Select
SER, SEU	Service Entrance Cable
S.F.	Square Foot
S.F.C.A.	Square Foot Contact Area
S.F. Flr.	Square Foot of Floor
S.F.G.	Square Foot of Ground
S.F. Hor.	Square Foot Horizontal
SFR	Square Feet of Radiation
S.F. Shlf.	Square Foot of Shelf
S4S	Surface 4 Sides
Shee.	Sheet Metal Worker
Sin.	Sine
Skwk.	Skilled Worker
SL	Saran Lined
S.L.	Slimline
Sldr.	Solder
SLH	Super Long Span Bar Joist
S.N.	Solid Neutral
SO	Stranded with oil resistant inside insulation
S-O-L	Socketolet
sp	Standpipe
S.P.	Static Pressure; Single Pole; Self-Propelled
Spri.	Sprinkler Installer
spwg	Static Pressure Water Gauge
S.P.D.T.	Single Pole, Double Throw
SPF	Spruce Pine Fir; Sprayed Polyurethane Foam
S.P.S.T.	Single Pole, Single Throw
SPT	Standard Pipe Thread
Sq.	Square; 100 Square Feet
Sq. Hd.	Square Head
Sq. In.	Square Inch
S.S.	Single Strength; Stainless Steel
S.S.B.	Single Strength B Grade
sst, ss	Stainless Steel
Sswk.	Structural Steel Worker
Sswl.	Structural Steel Welder
St.; Stl.	Steel
STC	Sound Transmission Coefficient
Std.	Standard
Stg.	Staging
STK	Select Tight Knot
STP	Standard Temperature & Pressure
Stpi.	Steamfitter, Pipefitter
Str.	Strength; Starter; Straight
Strd.	Stranded
Struct.	Structural
Sty.	Story
Subj.	Subject
Subs.	Subcontractors
Surf.	Surface
Sw.	Switch
Swbd.	Switchboard
S.Y.	Square Yard
Syn.	Synthetic
S.Y.P.	Southern Yellow Pine
Sys.	System
t.	Thickness
T	Temperature; Ton
Tan	Tangent
T.C.	Terra Cotta
T & C	Threaded and Coupled
T.D.	Temperature Difference
TDD	Telecommunications Device for the Deaf
T.E.M.	Transmission Electron Microscopy
temp	Temperature, Tempered, Temporary
TFFN	Nylon Jacketed Wire

Abbreviations

Abbreviation	Meaning
TFE	Tetrafluoroethylene (Teflon)
T. & G.	Tongue & Groove; Tar & Gravel
Th., Thk.	Thick
Thn.	Thin
Thrded	Threaded
Tilf.	Tile Layer, Floor
Tilh.	Tile Layer, Helper
THHN	Nylon Jacketed Wire
THW.	Insulated Strand Wire
THWN	Nylon Jacketed Wire
T.L., TL	Truckload
T.M.	Track Mounted
Tot.	Total
T-O-L	Threadolet
tmpd	Tempered
TPO	Thermoplastic Polyolefin
T.S.	Trigger Start
Tr.	Trade
Transf.	Transformer
Trhv.	Truck Driver, Heavy
Trlr	Trailer
Trlt.	Truck Driver, Light
TTY	Teletypewriter
TV	Television
T.W.	Thermoplastic Water Resistant Wire
UCI	Uniform Construction Index
UF	Underground Feeder
UGND	Underground Feeder
UHF	Ultra High Frequency
U.I.	United Inch
U.L., UL	Underwriters Laboratory
Uld.	Unloading
Unfin.	Unfinished
UPS	Uninterruptible Power Supply
URD	Underground Residential Distribution
US	United States
USGBC	U.S. Green Building Council
USP	United States Primed
UTMCD	Uniform Traffic Manual For Control Devices
UTP	Unshielded Twisted Pair
V	Volt
VA	Volt Amperes
VAT	Vinyl Asbestos Tile
V.C.T.	Vinyl Composition Tile
VAV	Variable Air Volume
VC	Veneer Core
VDC	Volts Direct Current
Vent.	Ventilation
Vert.	Vertical
V.F.	Vinyl Faced
V.G.	Vertical Grain
VHF	Very High Frequency
VHO	Very High Output
Vib.	Vibrating
VLF	Vertical Linear Foot
VOC	Volatile Organic Compound
Vol.	Volume
VRP	Vinyl Reinforced Polyester
W	Wire; Watt; Wide; West
w/	With
W.C., WC	Water Column; Water Closet
W.F.	Wide Flange
W.G.	Water Gauge
Wldg.	Welding
W. Mile	Wire Mile
W-O-L	Weldolet
W.R.	Water Resistant
Wrck.	Wrecker
WSFU	Water Supply Fixture Unit
W.S.P.	Water, Steam, Petroleum
WT., Wt.	Weight
WWF	Welded Wire Fabric
XFER	Transfer
XFMR	Transformer
XHD	Extra Heavy Duty
XHHW	Cross-Linked Polyethylene Wire Insulation
XLPE	Cross-Linked Polyethylene Wire Insulation
XLP	Cross-linked Polyethylene
Xport	Transport
Y	Wye
yd	Yard
yr	Year
Δ	Delta
%	Percent
~	Approximately
Ø	Phase; diameter
@	At
#	Pound; Number
<	Less Than
>	Greater Than
Z	Zone

A

B

C

D

E

F

H

M

N

O

P

Q

R

S

X

Y

Z

Notes

Notes

Notes

Notes

Notes

Division Notes

		CREW	DAILY OUTPUT	LABOR-HOURS	UNIT	BARE COSTS MAT.	BARE COSTS LABOR	BARE COSTS EQUIP.	BARE COSTS TOTAL	TOTAL INCL O&P

Division Notes

	CREW	DAILY OUTPUT	LABOR-HOURS	UNIT	BARE COSTS MAT.	LABOR	EQUIP.	TOTAL	TOTAL INCL O&P

Division Notes

		CREW	DAILY OUTPUT	LABOR-HOURS	UNIT	BARE COSTS MAT.	BARE COSTS LABOR	BARE COSTS EQUIP.	BARE COSTS TOTAL	TOTAL INCL O&P

Division Notes

		CREW	DAILY OUTPUT	LABOR-HOURS	UNIT	BARE COSTS MAT.	BARE COSTS LABOR	BARE COSTS EQUIP.	BARE COSTS TOTAL	TOTAL INCL O&P

Division Notes

I	CREW	DAILY OUTPUT	LABOR-HOURS	UNIT	BARE COSTS MAT.	BARE COSTS LABOR	BARE COSTS EQUIP.	BARE COSTS TOTAL	TOTAL INCL O&P

Other RSMeans Products & Services

RSMeans—a tradition of excellence in construction cost information and services since 1942

For more information visit the RSMeans website at www.RSMeans.com

Unit prices according to the latest MasterFormat

Cost Data Selection Guide

The following table provides definitive information on the content of each cost data publication. The number of lines of data provided in each unit price or assemblies division, as well as the number of crews, is listed for each data set. The presence of other elements such as reference tables, square foot models, equipment rental costs, historical cost indexes, and city cost indexes, is also indicated. You can use the table to help select the RSMeans data set that has the quantity and type of information you most need in your work.

Unit Cost Divisions	Building Construction	Mechanical	Electrical	Commercial Renovation	Square Foot	Site Work Landsc.	Green Building	Interior	Concrete Masonry	Open Shop	Heavy Construction	Light Commercial	Facilities Construction	Plumbing	Residential
1	562	383	400	504		498	205	302	445	561	496	245	1040	393	177
2	776	279	85	732		991	206	399	213	775	733	480	1220	286	273
3	1691	340	230	1084		1483	986	354	2037	1691	1693	482	1791	316	389
4	960	21	0	920		725	180	615	1158	928	615	533	1175	0	445
5	1893	158	155	1090		844	1799	1098	720	1893	1037	979	1909	204	746
6	2453	18	18	2111		110	589	1528	281	2449	123	2141	2125	22	2661
7	1594	215	128	1633		580	763	531	523	1591	26	1328	1695	227	1048
8	2111	81	44	2645		257	1136	1757	105	2096	0	2219	2890	0	1534
9	2004	86	45	1828		310	451	2088	391	1944	15	1668	2252	54	1437
10	1049	17	10	645		216	27	864	157	1049	29	539	1141	233	224
11	1096	206	166	551		129	54	933	28	1071	0	231	1116	169	110
12	530	0	2	285		205	133	1533	14	500	0	259	1554	23	217
13	744	149	158	253		367	122	252	78	720	244	108	759	115	103
14	273	36	0	221		0	0	257	0	273	0	12	293	16	6
21	91	0	16	37		0	0	250	0	91	0	73	535	540	220
22	1168	7548	160	1201		1572	1066	848	20	1157	1681	874	7450	9370	718
23	1178	6975	580	929		157	903	778	38	1171	110	879	5213	1922	469
26	1380	458	10150	1033		793	630	1144	55	1372	562	1310	10099	399	619
27	72	0	339	34		13	0	71	0	72	39	52	321	0	22
28	96	58	136	71		0	21	78	0	99	0	40	151	44	25
31	1510	733	610	806		3263	288	7	1217	1455	3277	604	1569	660	613
32	813	49	8	881		4409	353	405	291	784	1826	421	1687	133	468
33	529	1076	538	252		2173	41	0	236	522	2157	128	1697	1283	154
34	107	0	47	4		190	0	0	31	62	214	0	136	0	0
35	18	0	0	0		327	0	0	0	18	442	0	84	0	0
41	60	0	0	33		8	0	22	0	61	31	0	68	14	0
44	75	79	0	0		0	0	0	0	0	0	0	75	75	0
46	23	16	0	0		274	261	0	0	23	264	0	33	33	0
48	12	0	25	0		0	25	0	0	12	17	12	12	0	12
Totals	24868	18981	14050	19783		19894	10239	16114	8038	24440	15631	15617	50090	16531	12690

Assem Div	Building Construction	Mechanical	Elec-trical	Commercial Renovation	Square Foot	Site Work Landscape	Assemblies	Green Building	Interior	Concrete Masonry	Heavy Construction	Light Commercial	Facilities Construction	Plumbing	Asm Div	Residential
A		15	0	188	150	577	598	0	0	536	571	154	24	0	1	378
B		0	0	848	2506	0	5666	56	329	1975	368	2098	174	0	2	211
C		0	0	647	928	0	1309	0	1628	146	0	818	249	0	3	588
D		1067	941	712	1859	72	2538	330	825	0	0	1345	1105	1088	4	851
E		0	0	86	261	0	301	0	5	0	0	258	5	0	5	390
F		0	0	0	114	0	114	0	0	0	0	114	3	0	6	357
G		527	447	318	312	3364	792	0	0	534	1349	205	293	677	7	307
															8	760
															9	80
															10	0
															11	0
															12	0
Totals		1609	1388	2799	6130	4013	11318	386	2787	3191	2288	4992	1853	1765		3922

Reference Section	Building Construction Costs	Mechanical	Electrical	Commercial Renovation	Square Foot	Site Work Landscape	Assem.	Green Building	Interior	Concrete Masonry	Open Shop	Heavy Construction	Light Commercial	Facilities Construction	Plumbing	Resi.
Reference Tables	yes	yes	yes	yes	no	yes	yes	yes	yes	yes	yes	yes	yes	yes	yes	yes
Models					111			25					50			28
Crews	575	575	575	553		575		575	575	575	551	575	551	553	575	551
Equipment Rental Costs	yes	yes	yes	yes		yes		yes	yes	yes	yes	yes	yes	yes	yes	yes
Historical Cost Indexes	yes	yes	yes	yes	yes	yes	yes	yes	yes	yes	yes	yes	yes	yes	yes	no
City Cost Indexes	yes	yes	yes	yes	yes	yes	yes	yes	yes	yes	yes	yes	yes	yes	yes	yes

Unit prices according to the latest MasterFormat

Online · Book · CD · eBook · Online add-on available for additional fee

For more information visit our website at www.RSMeans.com

RSMeans Building Construction Cost Data 2016

Offers you unchallenged unit price reliability in an easy-to-use format. Whether used for verifying complete, finished estimates or for periodic checks, it supplies more cost facts better and faster than any comparable source. More than 24,700 unit prices have been updated for 2016. The City Cost Indexes and Location Factors cover more than 930 areas for indexing to any project location in North America. Order and get *RSMeans Quarterly Update Service* FREE.

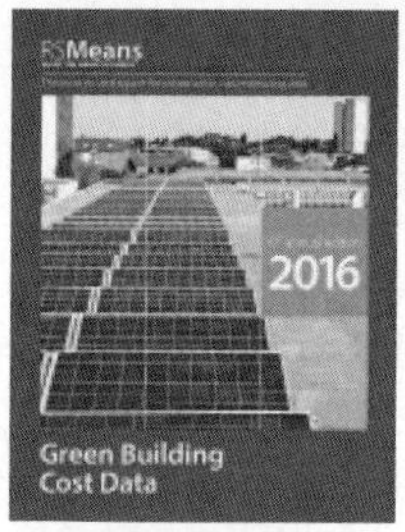

RSMeans Green Building Cost Data 2016

Estimate, plan, and budget the costs of green building for both new commercial construction and renovation work with this sixth edition of *RSMeans Green Building Cost Data*. More than 10,000 unit costs for a wide array of green building products plus assemblies costs. Easily identified cross references to LEED and Green Globes building rating systems criteria.

RSMeans Mechanical Cost Data 2016

Total unit and systems price guidance for mechanical construction—materials, parts, fittings, and complete labor cost information. Includes prices for piping, heating, air conditioning, ventilation, and all related construction.

Plus new 2016 unit costs for:

- Thousands of installed HVAC/controls, sub-assemblies and assemblies
- "On-site" Location Factors for more than 930 cities and towns in the U.S. and Canada
- Crews, labor, and equipment

RSMeans Facilities Construction Cost Data 2016

For the maintenance and construction of commercial, industrial, municipal, and institutional properties. Costs are shown for new and remodeling construction and are broken down into materials, labor, equipment, and overhead and profit. Special emphasis is given to sections on mechanical, electrical, furnishings, site work, building maintenance, finish work, and demolition.

More than 49,800 unit costs, plus assemblies costs and a comprehensive Reference Section are included.

RSMeans Square Foot Costs 2016

Accurate and Easy-to-Use

- Updated price information based on nationwide figures from suppliers, estimators, labor experts, and contractors.
- Green building models
- Realistic graphics, offering true-to-life illustrations of building projects
- Extensive information on using square foot cost data, including sample estimates and alternate pricing methods

RSMeans Commercial Renovation Cost Data 2016

Commercial/Multi-family Residential

Use this valuable tool to estimate commercial and multi-family residential renovation and remodeling.

Includes: Updated costs for hundreds of unique methods, materials, and conditions that only come up in repair and remodeling, PLUS:

- Unit costs for more than 19,600 construction components
- Installed costs for more than 2,700 assemblies
- More than 930 "on-site" localization factors for the U.S. and Canada

RSMeans Electrical Cost Data 2016

Pricing information for every part of electrical cost planning. More than 13,800 unit and systems costs with design tables; clear specifications and drawings; engineering guides; illustrated estimating procedures; complete labor-hour and materials costs for better scheduling and procurement; and the latest electrical products and construction methods.

- A variety of special electrical systems, including cathodic protection
- Costs for maintenance, demolition, HVAC/mechanical, specialties, equipment, and more

RSMeans Electrical Change Order Cost Data 2016

RSMeans Electrical Change Order Cost Data provides you with electrical unit prices exclusively for pricing change orders. Analyze and check your own change order estimates against those prepared when using RSMeans cost data. It also covers productivity analysis and change order cost justifications. With useful information for calculating the effects of change orders and dealing with their administration.

RSMeans Assemblies Cost Data 2016

RSMeans Assemblies Cost Data takes the guesswork out of preliminary or conceptual estimates. Now you don't have to try to calculate the assembled cost by working up individual component costs. We've done all the work for you.

Presents detailed illustrations, descriptions, specifications, and costs for every conceivable building assembly—over 350 types in all—arranged in the easy-to-use UNIFORMAT II system. Each illustrated "assembled" cost includes a complete grouping of materials and associated installation costs, including the installing contractor's overhead and profit.

RSMeans Open Shop Building Construction Cost Data 2016

The latest costs for accurate budgeting and estimating of new commercial and residential construction, renovation work, change orders, and cost engineering.

RSMeans Open Shop "BCCD" will assist you to:

- Develop benchmark prices for change orders.
- Plug gaps in preliminary estimates and budgets.
- Estimate complex projects.
- Substantiate invoices on contracts.
- Price ADA-related renovations.

RSMeans Residential Cost Data 2016

Contains square foot costs for 28 basic home models with the look of today, plus hundreds of custom additions and modifications you can quote right off the page. Includes more than 3,900 costs for 89 residential systems. Complete with blank estimating forms, sample estimates, and step-by-step instructions.

Contains line items for cultured stone and brick, PVC trim, lumber, and TPO roofing.

RSMeans Site Work & Landscape Cost Data 2016

Includes unit and assemblies costs for earthwork, sewerage, piped utilities, site improvements, drainage, paving, trees and shrubs, street openings/repairs, underground tanks, and more. Contains more than 60 types of assemblies costs for accurate conceptual estimates.

Includes:

- Estimating for infrastructure improvements
- Environmentally-oriented construction
- ADA-mandated handicapped access
- Hazardous waste line items

RSMeans Facilities Maintenance & Repair Cost Data 2016

RSMeans Facilities Maintenance & Repair Cost Data gives you a complete system to manage and plan your facility repair and maintenance costs and budget efficiently. Guidelines for auditing a facility and developing an annual maintenance plan. Budgeting is included, along with reference tables on cost and management, and information on frequency and productivity of maintenance operations.

The only nationally recognized source of maintenance and repair costs. Developed in cooperation with the Civil Engineering Research Laboratory (CERL) of the Army Corps of Engineers.

RSMeans Light Commercial Cost Data 2016

Specifically addresses the light commercial market, which is a specialized niche in the construction industry. Aids you, the owner/designer/contractor, in preparing all types of estimates—from budgets to detailed bids. Includes new advances in methods and materials.

The Assemblies section allows you to evaluate alternatives in the early stages of design/planning.

More than 15,500 unit costs ensure that you have the prices you need, when you need them.

RSMeans Concrete & Masonry Cost Data 2016

Provides you with cost facts for virtually all concrete/masonry estimating needs, from complicated formwork to various sizes and face finishes of brick and block—all in great detail. The comprehensive Unit Price Section contains more than 8,000 selected entries. Also contains an Assemblies [Cost] Section, and a detailed Reference Section that supplements the cost data.

RSMeans Labor Rates for the Construction Industry 2016

Complete information for estimating labor costs, making comparisons, and negotiating wage rates by trade for more than 300 U.S. and Canadian cities. With 46 construction trades in each city, and historical wage rates included for comparison. Each city chart lists the county and is alphabetically arranged with handy visual flip tabs for quick reference.

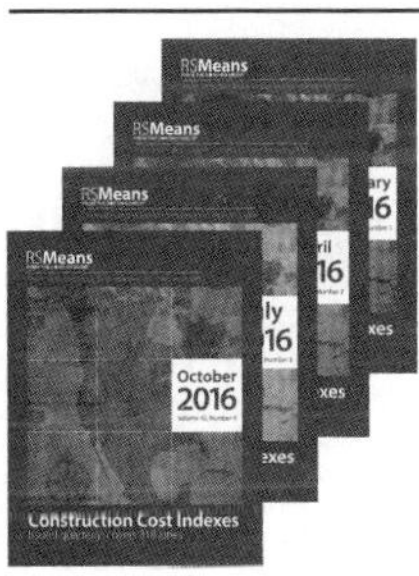

RSMeans Construction Cost Indexes 2016

What materials and labor costs will change unexpectedly this year? By how much?

- Breakdowns for 318 major cities
- National averages for 30 key cities
- Expanded five major city indexes
- Historical construction cost indexes

RSMeans Interior Cost Data 2016

Provides you with prices and guidance needed to make accurate interior work estimates. Contains costs on materials, equipment, hardware, custom installations, furnishings, and labor for new and remodel commercial and industrial interior construction, including updated information on office furnishings, as well as reference information.

RSMeans Heavy Construction Cost Data 2016

A comprehensive guide to heavy construction costs. Includes costs for highly specialized projects such as tunnels, dams, highways, airports, and waterways. Information on labor rates, equipment, and materials costs is included. Features unit price costs, systems costs, and numerous reference tables for costs and design.

RSMeans Plumbing Cost Data 2016

Comprehensive unit prices and assemblies for plumbing, irrigation systems, commercial and residential fire protection, point-of-use water heaters, and the latest approved materials. This publication and its companion, *RSMeans Mechanical Cost Data*, provide full-range cost estimating coverage for all the mechanical trades.

Contains updated costs for potable water, radiant heat systems, and high efficiency fixtures.

RSMeans Online

Competitive Cost Estimates Made Easy

RSMeans Online is a web-based service that provides accurate and up-to-date cost information to help you build competitive estimates or budgets in less time.

Quick, intuitive, easy to use, and automatically updated, RSMeans Online gives you instant access to hundreds of thousands of material, labor, and equipment costs from RSMeans' comprehensive database, delivering the information you need to build budgets and competitive estimates every time.

With RSMeans Online, you can perform quick searches to locate specific costs and adjust costs to reflect prices in your geographic area. Tag and store your favorites for fast access to your frequently used line items and assemblies and clone estimates to save time. System notifications will alert you as updated data becomes available. RSMeans Online is automatically updated throughout the year.

RSMeans Online's visual, interactive estimating features help you create, manage, save and share estimates with ease! It provides increased flexibility with customizable advanced reports. Easily edit custom report templates and import your company logo onto your estimates.

	Standard	Advanced	Professional
Unit Prices	✓	✓	✓
Assemblies	X	✓	✓
Sq Ft Models	X	X	✓
Editable Sq Foot Models	X	X	✓
Editable Assembly Components	X	✓	✓
Custom Cost Data	X	✓	✓
User Defined Components	X	✓	✓
Advanced Reporting & Customization	X	✓	✓
Union Labor Type	✓	✓	✓
Open Shop*	Add on for additional fee	Add on for additional fee	✓
History (3-year lookback)**	Add on for additional fee	Add on for additional fee	Add on for additional fee
Full History (2007)	Add on for additional fee	Add on for additional fee	Add on for additional fee

*Open shop included in Complete Library & Bundles
** 3-year history lookback included with complete library

Estimate with Precision

Find everything you need to develop complete, accurate estimates.

- Verified costs for construction materials
- Equipment rental costs
- Crew sizing, labor hours and labor rates
- Localized costs for U.S. and Canada

Save Time & Increase Efficiency

Make cost estimating and calculating faster and easier than ever with secure, online estimating tools.

- Quickly locate costs in the searchable database
- Create estimates in minutes with RSMeans cost lines
- Tag and store favorites for fast access to frequently used items

Improve Planning & Decision-Making

Back your estimates with complete, accurate and up-to-date cost data for informed business decisions.

- Verify construction costs from third parties
- Check validity of subcontractor proposals
- Evaluate material and assembly alternatives

Increase Profits

Use RSMeans Online to estimate projects quickly and accurately, so you can gain an edge over your competition.

- Create accurate and competitive bids
- Minimize the risk of cost overruns
- Reduce variability
- Gain control over costs

30 Day Free Trial

Register for a free trial at www.RSMeansOnline.com

- Access to unit prices, building assemblies and facilities repair and remodeling items covering every category of construction
- Powerful tools to customize, save and share cost lists, estimates and reports with ease
- Access to square foot models for quick conceptual estimates
- Allows up to 10 users to evaluate the full range of collaborative functionality used items.

2016 RSMeans Seminar Schedule

Note: call for exact dates and details.

Location	Dates	Location	Dates
Seattle, WA	January and August	Bethesda, MD	June
Dallas/Ft. Worth, TX	January	El Segundo, CA	August
Austin, TX	February	Jacksonville, FL	September
Anchorage, AK	March and September	Dallas, TX	September
Las Vegas, NV	March	Philadelphia, PA	October
New Orleans, LA	March	Houston, TX	October
Washington, DC	April and September	Salt Lake City, UT	November
Phoenix, AZ	April	Baltimore, MD	November
Kansas City, MO	April	Orlando, FL	November
Toronto	May	San Diego, CA	December
Denver, CO	May	San Antonio, TX	December
San Francisco, CA	June	Raleigh, NC	December

877-620-6245

Beginning early 2016, RSMeans will provide a suite of online, self-paced training offerings. Check our website www.RSMeans.com for more information.

Professional Development

eLearning Training Sessions

Learn how to use *RSMeans Online*® or the *RSMeans CostWorks*® CD from the convenience of your home or office. Our eLearning training sessions let you join a training conference call and share the instructors' desktops, so you can view the presentation and step-by-step instructions on your own computer screen. The live webinars are held from 9 a.m. to 4 p.m. or 11 a.m. to 6 p.m. eastern standard time, with a one-hour break for lunch.

For these sessions, you must have a computer with high speed Internet access and a compatible Web browser. Learn more at www.RSMeansOnline.com or call for a schedule: 781-422-5115. Webinars are generally held on selected Wednesdays each month.

RSMeans Online	RSMeans CostWorks CD
$299 per person	$299 per person

RSMeans Online Training

Construction estimating is vital to the decision-making process at each state of every project. RSMeansOnline works the way you do. It's systematic, flexible and intuitive. In this one day class you will see how you can estimate any phase of any project faster and better.

Some of what you'll learn:

- Customizing RSMeansOnline
- Making the most of RSMeans "Circle Reference" numbers
- How to integrate your cost data
- Generate reports, exporting estimates to MS Excel, sharing, collaborating and more

Also available: RSMeans Online training webinar

Maintenance & Repair Estimating for Facilities

This *two-day* course teaches attendees how to plan, budget, and estimate the cost of ongoing and preventive maintenance and repair for existing buildings and grounds.

Some of what you'll learn:

- The most financially favorable maintenance, repair, and replacement scheduling and estimating
- Auditing and value engineering facilities
- Preventive planning and facilities upgrading
- Determining both in-house and contract-out service costs
- Annual, asset-protecting M&R plan

Who should attend: facility managers, maintenance supervisors, buildings and grounds superintendents, plant managers, planners, estimators, and others involved in facilities planning and budgeting.

Facilities Construction Estimating

In this *two-day* course, professionals working in facilities management can get help with their daily challenges to establish budgets for all phases of a project.

Some of what you'll learn:

- Determining the full scope of a project
- Identifying the scope of risks and opportunities
- Creative solutions to estimating issues
- Organizing estimates for presentation and discussion
- Special techniques for repair/remodel and maintenance projects
- Negotiating project change orders

Who should attend: facility managers, engineers, contractors, facility tradespeople, planners, and project managers.

Practical Project Management for Construction Professionals

In this *two-day* course, acquire the essential knowledge and develop the skills to effectively and efficiently execute the day-to-day responsibilities of the construction project manager.

Some of what you'll learn:

- General conditions of the construction contract
- Contract modifications: change orders and construction change directives
- Negotiations with subcontractors and vendors
- Effective writing: notification and communications
- Dispute resolution: claims and liens

Who should attend: architects, engineers, owners' representatives, and project managers.

Construction Cost Estimating: Concepts and Practice

This introductory course to improve estimating skills and effectiveness starts with the details of interpreting bid documents and ends with the summary of the estimate and bid submission.

Some of what you'll learn:

- Using the plans and specifications for creating estimates
- The takeoff process—deriving all tasks with correct quantities
- Developing pricing using various sources; how subcontractor pricing fits in
- Summarizing the estimate to arrive at the final number
- Formulas for area and cubic measure, adding waste and adjusting productivity to specific projects
- Evaluating subcontractors' proposals and prices
- Adding insurance and bonds
- Understanding how labor costs are calculated
- Submitting bids and proposals

Who should attend: project managers, architects, engineers, owners' representatives, contractors, and anyone who's responsible for budgeting or estimating construction projects.

Mechanical & Electrical Estimating

This *two-day* course teaches attendees how to prepare more accurate and complete mechanical/electrical estimates, avoid the pitfalls of omission and double-counting, and understand the composition and rationale within the RSMeans mechanical/electrical database.

Some of what you'll learn:

- The unique way mechanical and electrical systems are interrelated
- M&E estimates—conceptual, planning, budgeting, and bidding stages
- Order of magnitude, square foot, assemblies, and unit price estimating
- Comparative cost analysis of equipment and design alternatives

Who should attend: architects, engineers, facilities managers, mechanical and electrical contractors, and others who need a highly reliable method for developing, understanding, and evaluating mechanical and electrical contracts.

Unit Price Estimating

This interactive *two-day* seminar teaches attendees how to interpret project information and process it into final, detailed estimates with the greatest accuracy level.

The most important credential an estimator can take to the job is the ability to visualize construction and estimate accurately.

Some of what you'll learn:

- Interpreting the design in terms of cost
- The most detailed, time-tested methodology for accurate pricing
- Key cost drivers—material, labor, equipment, staging, and subcontracts
- Understanding direct and indirect costs for accurate job cost accounting and change order management

Who should attend: corporate and government estimators and purchasers, architects, engineers, and others who need to produce accurate project estimates.

Conceptual Estimating Using the RSMeans CostWorks CD

This *two-day* class uses the leading industry data and a powerful software package to develop highly accurate conceptual estimates for your construction projects. All attendees must bring a laptop computer loaded with the current year *Square Foot Costs* and the *Assemblies Cost Data* CostWorks titles.

Some of what you'll learn:

- Introduction to conceptual estimating
- Types of conceptual estimates
- Helpful hints
- Order of magnitude estimating
- Square foot estimating
- Assemblies estimating

Who should attend: architects, engineers, contractors, construction estimators, owners' representatives, and anyone looking for an electronic method for performing square foot estimating.

RSMeans CostWorks CD Training

This *one-day* course helps users become more familiar with the functionality of the *RSMeans CostWorks* program. Each menu, icon, screen, and function found in the program is explained in depth. Time is devoted to hands-on estimating exercises.

Some of what you'll learn:

- Searching the database using all navigation methods
- Exporting RSMeans data to your preferred spreadsheet format
- Viewing crews, assembly components, and much more
- Automatically regionalizing the database

This training session requires you to bring a laptop computer to class.

When you register for this course you will receive an outline for your laptop requirements.

Also offering web training for the RSMeans CostWorks CD!

Assessing Scope of Work for Facility Construction Estimating

This *two-day* practical training program addresses the vital importance of understanding the scope of projects in order to produce accurate cost estimates for facility repair and remodeling.

Some of what you'll learn:

- Discussions of site visits, plans/specs, record drawings of facilities, and site-specific lists
- Review of CSI divisions, including means, methods, materials, and the challenges of scoping each topic
- Exercises in scope identification and scope writing for accurate estimating of projects
- Hands-on exercises that require scope, take-off, and pricing

Who should attend: corporate and government estimators, planners, facility managers, and others who need to produce accurate project estimates.

Facilities Estimating Using the RSMeans CostWorks CD

Combines hands-on skill building with best estimating practices and real-life problems. Brings you up-to-date with key concepts and provides tips, pointers, and guidelines to save time and avoid cost oversights and errors.

Some of what you'll learn:

- Estimating process concepts
- Customizing and adapting RSMeans cost data
- Establishing scope of work to account for all known variables
- Budget estimating: when, why, and how
- Site visits: what to look for—what you can't afford to overlook
- How to estimate repair and remodeling variables

This training session requires you to bring a laptop computer to class.

Who should attend: facility managers, architects, engineers, contractors, facility tradespeople, planners, project managers and anyone involved with JOC, SABRE, or IDIQ.

Unit Price Estimating Using the RSMeans CostWorks CD

Step-by-step instructions and practice problems to identify and track key cost drivers—material, labor, equipment, staging, and subcontractors—for each specific task. Learn the most detailed, time-tested methodology for accurately "pricing" these variables, their impact on each other and on total cost.

Some of what you'll learn:

- Unit price cost estimating
- Order of magnitude, square foot, and assemblies estimating
- Quantity takeoff
- Direct and indirect construction costs
- Development of contractors' bill rates
- How to use *RSMeans Building Construction Cost Data*

This training session requires you to bring a laptop computer to class.

Who should attend: architects, engineers, corporate and government estimators, facility managers, and government procurement staff.

Registration Information

Register early and save up to $100!
Register 30 days before the start date of a seminar and save $100 off your total fee. Note: This discount can be applied only once per order. It cannot be applied to team discount registrations or any other special offer.

How to register
Register by phone today! The RSMeans toll-free number for making reservations is 781-422-5115.

Two-day seminar registration fee - $935. One-day RSMeans CostWorks® or RSMeans Online training registration fee - $375.
To register by mail, complete the registration form and return, with your full fee, to: RSMeans Seminars, 1099 Hingham Street, Suite 201, Rockland, MA 02370.

One-day Construction Cost Estimating - $575.

Government pricing
All federal government employees save off the regular seminar price. Other promotional discounts cannot be combined with the government discount.

Team discount program
For over five attendee registrations. Call for pricing: 781-422-5115

Multiple course discounts
When signing up for two or more courses, call for pricing.

Refund policy
Cancellations will be accepted up to ten business days prior to the seminar start. There are no refunds for cancellations received later than ten working days prior to the first day of the seminar. A $150 processing fee will be applied for all cancellations. Written notice of the cancellation is required. Substitutions can be made at any time before the session starts. No-shows are subject to the full seminar fee.

AACE approved courses
Many seminars described and offered here have been approved for 14 hours (1.4 recertification credits) of credit by the AACE International Certification Board toward meeting the continuing education requirements for recertification as a Certified Cost Engineer/Certified Cost Consultant.

AIA Continuing Education
We are registered with the AIA Continuing Education System (AIA/CES) and are committed to developing quality learning activities in accordance with the CES criteria. Many seminars meet the AIA/CES criteria for Quality Level 2. AIA members may receive 14 learning units (LUs) for each two-day RSMeans course.

Daily course schedule
The first day of each seminar session begins at 8:30 a.m. and ends at 4:30 p.m. The second day begins at 8:00 a.m. and ends at 4:00 p.m. Participants are urged to bring a hand-held calculator since many actual problems will be worked out in each session.

Continental breakfast
Your registration includes the cost of a continental breakfast and a morning and afternoon refreshment break. These informal segments allow you to discuss topics of mutual interest with other seminar attendees. (You are free to make your own lunch and dinner arrangements.)

Hotel/transportation arrangements
RSMeans arranges to hold a block of rooms at most host hotels. To take advantage of special group rates when making your reservation, be sure to mention that you are attending the RSMeans seminar. You are, of course, free to stay at the lodging place of your choice. (Hotel reservations and transportation arrangements should be made directly by seminar attendees.)

Important
Class sizes are limited, so please register as soon as possible.

Note: Pricing subject to change.

Registration Form

ADDS-1000

Call 781-422-5115 to register or fax this form to 800-632-6732. Visit our website: www.RSMeans.com

Please register the following people for the RSMeans construction seminars as shown here. We understand that we must make our own hotel reservations if overnight stays are necessary.

❑ Full payment of $_______________ enclosed.

❑ Bill me.

Please print name of registrant(s).

(To appear on certificate of completion)

____

____

____

____

P.O. #: _______________

GOVERNMENT AGENCIES MUST SUPPLY PURCHASE ORDER NUMBER OR TRAINING FORM.

Firm name _______________

Address _______________

City/State/Zip _______________

Telephone no. _______________ Fax no. _______________

E-mail address _______________

Charge registration(s) to: ☐ MasterCard ☐ VISA ☐ American Express

Account no. _______________ Exp. date _______________

Cardholder's signature _______________

Seminar name _______________

Seminar City _______________

Please mail check to: 1099 Hingham Street, Suite 201, Rockland, MA 02370 USA

10 31 Manufactured Fireplaces

10 31 23 – Prefabricated Fireplaces

10 31 23.10 Fireplace, Prefabricated		Crew	Daily Output	Labor-Hours	Unit	Material	2016 Bare Costs Labor	Equipment	Total	Total Incl O&P
2500	Recirculating, small fan	1 Carp	.80	10	Ea.	2,475	380		2,855	3,350
2550	Large fan		.70	11.429		2,475	430		2,905	3,450
3000	See through, including doors		.80	10		2,200	380		2,580	3,025
3200	Corner (2 wall)	↓	1	8	↓	3,425	300		3,725	4,250

10 43 Emergency Aid Specialties

10 43 13 – Defibrillator Cabinets

10 43 13.05 Defibrillator Cabinets		Crew	Daily Output	Labor-Hours	Unit	Material	Labor	Equipment	Total	Total Incl O&P
0010	**DEFIBRILLATOR CABINETS**, not equipped, stainless steel									
0050	Defibrillator cabinet, stainless steel with strobe & alarm 12" x 27"	1 Carp	10	.800	Ea.	385	30		415	475
0100	Automatic External Defibrillator	"	30	.267	"	1,375	10.10		1,385.10	1,550

10 44 Fire Protection Specialties

10 44 13 – Fire Protection Cabinets

10 44 13.53 Fire Equipment Cabinets		Crew	Daily Output	Labor-Hours	Unit	Material	Labor	Equipment	Total	Total Incl O&P
0010	**FIRE EQUIPMENT CABINETS**, not equipped, 20 ga. steel box									
0040	Recessed, D.S. glass in door, box size given									
1000	Portable extinguisher, single, 8" x 12" x 27", alum. door & frame	Q-12	8	2	Ea.	160	77		237	305
1100	Steel door and frame		8	2		122	77		199	261
2700	Fire blanket & extinguisher cab, inc blanket, rec stl., 14" x 40" x 8"		7	2.286		188	87.50		275.50	350
2800	Fire blanket cab, inc blanket, surf mtd, stl, 15"x10"x5", w/pwdr coat fin	↓	8	2	↓	94	77		171	231
3000	Hose rack assy., 1-1/2" valve & 100' hose, 24" x 40" x 5-1/2"									
3100	Aluminum door and frame	Q-12	6	2.667	Ea.	390	102		492	595
3200	Steel door and frame	"	6	2.667	"	258	102		360	455

10 44 16 – Fire Extinguishers

10 44 16.13 Portable Fire Extinguishers		Crew	Daily Output	Labor-Hours	Unit	Material	Labor	Equipment	Total	Total Incl O&P
0010	**PORTABLE FIRE EXTINGUISHERS**									
0140	CO_2, with hose and "H" horn, 10 lb.				Ea.	288			288	315
1000	Dry chemical, pressurized									
1040	Standard type, portable, painted, 2-1/2 lb.				Ea.	38.50			38.50	42.50
1080	10 lb.					82			82	90
1100	20 lb.					135			135	149
1120	30 lb.					435			435	480
2000	ABC all purpose type, portable, 2-1/2 lb.					21			21	23.50
2080	9-1/2 lb.					44			44	48
3500	Halotron 1, 2-1/2 lb.					127			127	140
3600	5 lb.					199			199	219
3700	11 lb.				↓	395			395	435

10 44 16.16 Wheeled Fire Extinguisher Units		Crew	Daily Output	Labor-Hours	Unit	Material	Labor	Equipment	Total	Total Incl O&P
0010	**WHEELED FIRE EXTINGUISHER UNITS**									
0350	CO_2, portable, with swivel horn									
0360	Wheeled type, cart mounted, 50 lb.				Ea.	1,175			1,175	1,275
0400	100 lb.				"	4,275			4,275	4,700
2200	ABC all purpose type									
2300	Wheeled, 45 lb.				Ea.	735			735	810
2360	150 lb.				"	1,850			1,850	2,025

10 51 Lockers

10 51 13 – Metal Lockers

10 51 13.10 Lockers		Crew	Daily Output	Labor-Hours	Unit	Material	2016 Bare Costs Labor	Equipment	Total	Total Incl O&P
0011	**LOCKERS** steel, baked enamel, pre-assembled									
0110	Single tier box locker, 12" x 15" x 72"	1 Shee	20	.400	Ea.	231	16.95		247.95	283
0120	18" x 15" x 72"		20	.400		245	16.95		261.95	298
0130	12" x 18" x 72"		20	.400		235	16.95		251.95	287
0140	18" x 18" x 72"		20	.400		285	16.95		301.95	345
0410	Double tier, 12" x 15" x 36"		30	.267		233	11.30		244.30	275
0420	18" x 15" x 36"		30	.267		235	11.30		246.30	278
0430	12" x 18" x 36"		30	.267		265	11.30		276.30	310
0440	18" x 18" x 36"		30	.267		225	11.30		236.30	267
0500	Two person, 18" x 15" x 72"		20	.400		297	16.95		313.95	355
0510	18" x 18" x 72"		20	.400		330	16.95		346.95	390
0520	Duplex, 15" x 15" x 72"		20	.400		335	16.95		351.95	400
0530	15" x 21" x 72"		20	.400		370	16.95		386.95	440
1100	Wire meshed wardrobe, floor mtd., open front varsity type		7.50	1.067		277	45		322	380
3250	Rack w/24 wire mesh baskets		1.50	5.333	Set	410	226		636	825
3260	30 baskets		1.25	6.400		360	271		631	845
3270	36 baskets		.95	8.421		470	355		825	1,125
3280	42 baskets		.80	10		455	425		880	1,200
3600	For hanger rods, add				Ea.	2			2	2.20
3650	For number plate kit, 100 plates #1 - #100, add	1 Shee	4	2		82	84.50		166.50	231
3700	For locker base, closed front panel		90	.089		7.20	3.76		10.96	14.20
3710	End panel, bolted		36	.222		9.30	9.40		18.70	26
3800	For sloping top, 12" wide		24	.333		32.50	14.10		46.60	59.50
3810	15" wide		24	.333		35.50	14.10		49.60	62.50
3820	18" wide		24	.333		35.50	14.10		49.60	62.50
3850	Sloping top end panel, 12" deep		72	.111		12.20	4.71		16.91	21.50
3860	15" deep		72	.111		12.20	4.71		16.91	21.50
3870	18" deep		72	.111		12.20	4.71		16.91	21.50
3900	For finish end panels, steel, 60" high, 15" deep		12	.667		36	28		64	86.50
3910	72" high, 12" deep		12	.667		29	28		57	79
3920	18" deep		12	.667		43.50	28		71.50	95
5000	For "ready to assemble" lockers,									
5010	Add to labor						75%			
5020	Deduct from material					20%				

10 55 Postal Specialties

10 55 23 – Mail Boxes

10 55 23.10 Commercial Mail Boxes		Crew	Daily Output	Labor-Hours	Unit	Material	Labor	Equipment	Total	Total Incl O&P
0010	**COMMERCIAL MAIL BOXES**									
0020	Horiz., key lock, 5"H x 6"W x 15"D, alum., rear load	1 Carp	34	.235	Ea.	42.50	8.90		51.40	62
0100	Front loading		34	.235		42.50	8.90		51.40	62
0200	Double, 5"H x 12"W x 15"D, rear loading		26	.308		60.50	11.65		72.15	86
0300	Front loading		26	.308		69	11.65		80.65	95
0500	Quadruple, 10"H x 12"W x 15"D, rear loading		20	.400		101	15.10		116.10	137
0600	Front loading		20	.400		92.50	15.10		107.60	128
1600	Vault type, horizontal, for apartments, 4" x 5"		34	.235		26.50	8.90		35.40	44
1700	Alphabetical directories, 120 names		10	.800		125	30		155	189

09 91 Painting

09 91 13 – Exterior Painting

09 91 13.90 Walls, Masonry (CMU), Exterior		Crew	Daily Output	Labor-Hours	Unit	Material	2016 Bare Costs Labor	Equipment	Total	Total Incl O&P
0460	Second coat	1 Pord	2600	.003	S.F.	.05	.10		.15	.22
0470	Waterproof sealer, first coat		2245	.004		.26	.11		.37	.47
0480	Second coat		2990	.003		.26	.09		.35	.42
0490	Concrete masonry unit (CMU), porous									
0500	Brushwork, latex, first coat	1 Pord	640	.013	S.F.	.14	.40		.54	.82
0510	Second coat		960	.008		.07	.27		.34	.52
0520	Waterproof sealer, first coat		736	.011		.26	.35		.61	.85
0530	Second coat		1104	.007		.26	.23		.49	.66
0540	Roll latex, first coat		1465	.005		.11	.17		.28	.41
0550	Second coat		1790	.004		.07	.14		.21	.31
0560	Waterproof sealer, first coat		1680	.005		.26	.15		.41	.53
0570	Second coat		2060	.004		.26	.12		.38	.48
0580	Spray latex, first coat		1950	.004		.08	.13		.21	.31
0590	Second coat		2600	.003		.05	.10		.15	.22
0600	Waterproof sealer, first coat		2245	.004		.26	.11		.37	.47
0610	Second coat		2990	.003		.26	.09		.35	.42

09 91 23 – Interior Painting

09 91 23.20 Cabinets and Casework

Code	Description	Crew	Daily Output	Labor-Hours	Unit	Material	Labor	Equipment	Total	Total Incl O&P
0010	**CABINETS AND CASEWORK**									
1000	Primer coat, oil base, brushwork	1 Pord	650	.012	S.F.	.06	.39		.45	.72
2000	Paint, oil base, brushwork, 1 coat		650	.012		.11	.39		.50	.77
2500	2 coats		400	.020		.20	.64		.84	1.28
3000	Stain, brushwork, wipe off		650	.012		.09	.39		.48	.75
4000	Shellac, 1 coat, brushwork		650	.012		.12	.39		.51	.78
4500	Varnish, 3 coats, brushwork, sand after 1st coat		325	.025		.26	.79		1.05	1.57
5000	For latex paint, deduct					10%				

09 91 23.33 Doors and Windows, Interior Alkyd (Oil Base)

Code	Description	Crew	Daily Output	Labor-Hours	Unit	Material	Labor	Equipment	Total	Total Incl O&P
0010	**DOORS AND WINDOWS, INTERIOR ALKYD (OIL BASE)**									
0500	Flush door & frame, 3′ x 7′, oil, primer, brushwork	1 Pord	10	.800	Ea.	3.70	25.50		29.20	46
1000	Paint, 1 coat		10	.800		4.16	25.50		29.66	46.50
1200	2 coats		6	1.333		4.54	42.50		47.04	75
1400	Stain, brushwork, wipe off		18	.444		1.83	14.20		16.03	25.50
1600	Shellac, 1 coat, brushwork		25	.320		2.47	10.20		12.67	19.50
1800	Varnish, 3 coats, brushwork, sand after 1st coat		9	.889		5.40	28.50		33.90	52.50
2000	Panel door & frame, 3′ x 7′, oil, primer, brushwork		6	1.333		2.41	42.50		44.91	72.50
2200	Paint, 1 coat		6	1.333		4.16	42.50		46.66	74.50
2400	2 coats		3	2.667		10.75	85		95.75	152
2600	Stain, brushwork, panel door, 3′ x 7′, not incl. frame		16	.500		1.83	15.95		17.78	28.50
2800	Shellac, 1 coat, brushwork		22	.364		2.47	11.60		14.07	22
3000	Varnish, 3 coats, brushwork, sand after 1st coat		7.50	1.067		5.40	34		39.40	62
3020	French door, incl. 3′ x 7′, 6 lites, frame & trim									
3022	Paint, 1 coat, over existing paint	1 Pord	5	1.600	Ea.	8.30	51		59.30	93
3024	2 coats, over existing paint		5	1.600		16.15	51		67.15	102
3026	Primer & 1 coat		3.50	2.286		13.15	73		86.15	134
3028	Primer & 2 coats		3	2.667		21.50	85		106.50	164
3032	Varnish or polyurethane, 1 coat		5	1.600		8.40	51		59.40	93.50
3034	2 coats, sanding between		3	2.667		16.85	85		101.85	159
4400	Windows, including frame and trim, per side									
4600	Colonial type, 6/6 lites, 2′ x 3′, oil, primer, brushwork	1 Pord	14	.571	Ea.	.38	18.25		18.63	30.50
5800	Paint, 1 coat		14	.571		.66	18.25		18.91	30.50
6000	2 coats		9	.889		1.28	28.50		29.78	48
6200	3′ x 5′ opening, 6/6 lites, primer coat, brushwork		12	.667		.95	21.50		22.45	36

09 91 23.33 Doors and Windows, Interior Alkyd (Oil Base)		Crew	Daily Output	Labor-Hours	Unit	Material	Labor	Equipment	Total	Total Incl O&P
						2016 Bare Costs				
6400	Paint, 1 coat	1 Pord	12	.667	Ea.	1.64	21.50		23.14	37
6600	2 coats		7	1.143		3.19	36.50		39.69	63.50
6800	4′ x 8′ opening, 6/6 lites, primer coat, brushwork		8	1		2.03	32		34.03	54.50
7000	Paint, 1 coat		8	1		3.50	32		35.50	56.50
7200	2 coats		5	1.600		6.80	51		57.80	91.50
8000	Single lite type, 2′ x 3′, oil base, primer coat, brushwork		33	.242		.38	7.75		8.13	13.15
8200	Paint, 1 coat		33	.242		.66	7.75		8.41	13.45
8400	2 coats		20	.400		1.28	12.75		14.03	22.50
8600	3′ x 5′ opening, primer coat, brushwork		20	.400		.95	12.75		13.70	22
8800	Paint, 1 coat		20	.400		1.64	12.75		14.39	23
8900	2 coats		13	.615		3.19	19.65		22.84	36
9200	4′ x 8′ opening, primer coat, brushwork		14	.571		2.03	18.25		20.28	32
9400	Paint, 1 coat		14	.571		3.50	18.25		21.75	34
9600	2 coats		8	1		6.80	32		38.80	60

09 91 23.35 Doors and Windows, Interior Latex

Line	Description	Crew	Daily Output	Labor-Hours	Unit	Material	Labor	Equipment	Total	Total Incl O&P
0010	**DOORS & WINDOWS, INTERIOR LATEX**									
0100	Doors, flush, both sides, incl. frame & trim									
0110	Roll & brush, primer	1 Pord	10	.800	Ea.	4.15	25.50		29.65	46.50
0120	Finish coat, latex		10	.800		5.20	25.50		30.70	48
0130	Primer & 1 coat latex		7	1.143		9.35	36.50		45.85	70.50
0140	Primer & 2 coats latex		5	1.600		14.25	51		65.25	99.50
0160	Spray, both sides, primer		20	.400		4.36	12.75		17.11	26
0170	Finish coat, latex		20	.400		5.45	12.75		18.20	27
0180	Primer & 1 coat latex		11	.727		9.90	23		32.90	49
0190	Primer & 2 coats latex		8	1		15.10	32		47.10	69
0200	Doors, French, both sides, 10-15 lite, incl. frame & trim									
0210	Roll & brush, primer	1 Pord	6	1.333	Ea.	2.07	42.50		44.57	72.50
0220	Finish coat, latex		6	1.333		2.61	42.50		45.11	73
0230	Primer & 1 coat latex		3	2.667		4.68	85		89.68	145
0240	Primer & 2 coats latex		2	4		7.15	128		135.15	218
0260	Doors, louvered, both sides, incl. frame & trim									
0270	Roll & brush, primer	1 Pord	7	1.143	Ea.	4.15	36.50		40.65	64.50
0280	Finish coat, latex		7	1.143		5.20	36.50		41.70	66
0290	Primer & 1 coat, latex		4	2		9.10	64		73.10	115
0300	Primer & 2 coats, latex		3	2.667		14.55	85		99.55	156
0320	Spray, both sides, primer		20	.400		4.36	12.75		17.11	26
0330	Finish coat, latex		20	.400		5.45	12.75		18.20	27
0340	Primer & 1 coat, latex		11	.727		9.90	23		32.90	49
0350	Primer & 2 coats, latex		8	1		15.45	32		47.45	69.50
0360	Doors, panel, both sides, incl. frame & trim									
0370	Roll & brush, primer	1 Pord	6	1.333	Ea.	4.36	42.50		46.86	75
0380	Finish coat, latex		6	1.333		5.20	42.50		47.70	76
0390	Primer & 1 coat, latex		3	2.667		9.35	85		94.35	150
0400	Primer & 2 coats, latex		2.50	3.200		14.55	102		116.55	184
0420	Spray, both sides, primer		10	.800		4.36	25.50		29.86	47
0430	Finish coat, latex		10	.800		5.45	25.50		30.95	48
0440	Primer & 1 coat, latex		5	1.600		9.90	51		60.90	95
0450	Primer & 2 coats, latex		4	2		15.45	64		79.45	122
0460	Windows, per interior side, based on 15 S.F.									
0470	1 to 6 lite									
0480	Brushwork, primer	1 Pord	13	.615	Ea.	.82	19.65		20.47	33.50
0490	Finish coat, enamel		13	.615		1.03	19.65		20.68	33.50